# Comprehensive Natural Products Chemistry

# Comprehensive Natural Products Chemistry

*Editors-in-Chief*

**Sir Derek Barton†**
*Texas A&M University, USA*

**Koji Nakanishi**
*Columbia University, USA*

*Executive Editor*

**Otto Meth-Cohn**
*University of Sunderland, UK*

**Volume 7**

DNA AND ASPECTS OF MOLECULAR BIOLOGY

*Volume Editor*

**Eric T. Kool**
*University of Rochester, USA*

1999

ELSEVIER

AMSTERDAM – LAUSANNE – NEW YORK – OXFORD – SHANNON – SINGAPORE – TOKYO

Elsevier Science Ltd., The Boulevard, Langford Lane, Kidlington, Oxford,
OX5 1GB, UK

First edition 1999

**Library of Congress Cataloging-in-Publication Data**
Comprehensive natural products chemistry / editors-in-chief, Sir Derek
Barton, Koji Nakanishi ; executive editor, Otto Meth-Cohn. -- 1st ed.
   p. cm.
  Includes index.
  Contents: v. 7. DNA and aspects of molecular biology / volume editor
Eric T. Kool
  1. Natural products. I. Barton, Derek, Sir, 1918-1998. II. Nakanishi,
Koji, 1925- . III. Meth-Cohn, Otto.
QD415.C63 1999
547.7--dc21                                      98-15249


**British Library Cataloguing in Publication Data**
Comprehensive natural products chemistry
  1. Organic compounds
  I. Barton, Sir Derek, 1918-1998 II. Nakanishi Koji III. Meth-Cohn Otto
  572.5


ISBN 0-08-042709-X (set : alk. paper)
ISBN 0-08-043159-3 (Volume 7 : alk. paper)

∞™ The paper used in this publication meets the minimum requirements of the American National Standard for
Information Sciences—Permanence of Paper for Printed Library Materials, ANSI Z39.48–1984.

Typeset by BPC Digital Data Ltd., Glasgow, UK.
Printed and bound in Great Britain by BPC Wheatons Ltd., Exeter, UK.

# Contents

Introduction                                                                                    vii

Preface                                                                                          ix

Sir Derek Barton (an obituary notice)                                                            xi

Contributors to Volume 7                                                                         xiii

Abbreviations                                                                                    xv

Contents of All Volumes                                                                          xix

An Historical Perspective of Natural Products Chemistry                                          xxi

1    Overview                                                                                    1
     E. T. KOOL, *University of Rochester, NY, USA*

2    Thermodynamics and Kinetics of Nucleic Acid Association/Dissociation and Folding            15
     Processes
     G. E. PLUM and K. J. BRESLAUER, *Rutgers University, Piscataway, NJ, USA,*
     and R. W. ROBERTS, *California Institute of Technology, Pasadena, CA, USA*

3    Probing DNA Structure by NMR Spectroscopy                                                   55
     C. DE LOS SANTOS, *State University of New York at Stony Brook, New York, USA*

4    Molecular Probes of DNA Structure                                                           81
     J. T. MILLARD, *Colby College, Waterville, ME, USA*

5    Oligonucleotide Synthesis                                                                   105
     R. P. IYER, *OriGenix, Cambridge, MA, USA*, and S. L. BEAUCAGE, *US Food and
     Drug Administration, Bethesda, MD, USA*

6    Attachment of Reporter and Conjugate Groups to DNA                                          153
     S. L. BEAUCAGE, *US Food and Drug Administration, Bethesda, MD, USA*

7    Use of Nucleoside Analogues to Probe Biochemical Processes                                  251
     L. W. MCLAUGHLIN, M. WILSON, and S. B. HA, *Boston College, Chestnut Hill,
     MA, USA*

8    DNA with Altered Backbones in Antisense Applications                                        285
     Y. S. SANGHVI, *Isis Pharmaceuticals, Carlsbad, CA, USA*

9    DNA with Altered Bases                                                                      313
     G. R. REVANKAR and T. S. RAO, *Gemini Biotech, The Woodlands, TX, USA*

10   Topological Modification of DNA: Circles, Loops, Knots, and Branches                        341
     E. T. KOOL, *University of Rochester, NY, USA*

11   Chemistry of DNA Damage                                                                     371
     M. M. GREENBERG, *Colorado State University, Fort Collins, CO, USA*

12   DNA Intercalators                                                                           427
     W. D. WILSON, *Georgia State University, Atlanta, GA, USA*

13   DNA-binding Peptides                                                                        477
     I. GHOSH, S. YAO, and J. CHMIELEWSKI, *Purdue University, West Lafayette,
     IN, USA*

14   Covalent Modification of DNA by Natural Products                                            491
     K. S. GATES, *University of Missouri–Columbia, Columbia, MO, USA*

15   DNA-damaging Enediyne Compounds                                                             553
     Z. XI and I. H. GOLDBERG, *Harvard Medical School, Boston, MA, USA*

16   DNA Topoisomerase Inhibitors                                                         593
     T. L. MACDONALD, M. A. LABROLI, and J. J. TEPE, *University of Virginia,*
     *Charlottesville, VA, USA*

17   DNA Selection and Amplification                                                      615
     D. SEN, *Simon Fraser University, Burnaby, BC, Canada*

18   Cloning as a Tool for Organic Chemists                                               643
     J. D. PICKERT and B. L. MILLER, *University of Rochester, NY, USA*

Author Index                                                                              675

Subject Index                                                                             707

# Introduction

For many decades, Natural Products Chemistry has been the principal driving force for progress in Organic Chemistry.

In the past, the determination of structure was arduous and difficult. As soon as computing became easy, the application of X-ray crystallography to structural determination quickly surpassed all other methods. Supplemented by the equally remarkable progress made more recently by Nuclear Magnetic Resonance techniques, determination of structure has become a routine exercise. This is even true for enzymes and other molecules of a similar size. Not to be forgotten remains the progress in mass spectrometry which permits another approach to structure and, in particular, to the precise determination of molecular weight.

There have not been such revolutionary changes in the partial or total synthesis of Natural Products. This still requires effort, imagination and time. But remarkable syntheses have been accomplished and great progress has been made in stereoselective synthesis. However, the one hundred percent yield problem is only solved in certain steps in certain industrial processes. Thus there remains a great divide between the reactions carried out in living organisms and those that synthetic chemists attain in the laboratory. Of course Nature edits the accuracy of DNA, RNA, and protein synthesis in a way that does not apply to a multi-step Organic Synthesis.

Organic Synthesis has already a significant component that uses enzymes to carry out specific reactions. This applies particularly to lipases and to oxidation enzymes. We have therefore, given serious attention to enzymatic reactions.

No longer standing in the wings, but already on-stage, are the wonderful tools of Molecular Biology. It is now clear that multi-step syntheses can be carried out in one vessel using multiple cloned enzymes. Thus, Molecular Biology and Organic Synthesis will come together to make economically important Natural Products.

From these preliminary comments it is clear that Natural Products Chemistry continues to evolve in different directions interacting with physical methods, Biochemistry, and Molecular Biology all at the same time.

This new Comprehensive Series has been conceived with the common theme of "How does Nature make all these molecules of life?" The principal idea was to organize the multitude of facts in terms of Biosynthesis rather than structure. The work is not intended to be a comprehensive listing of natural products, nor is it intended that there should be any detail about biological activity. These kinds of information can be found elsewhere.

The work has been planned for eight volumes with one more volume for Indexes. As far as we are aware, a broad treatment of the whole of Natural Products Chemistry has never been attempted before. We trust that our efforts will be useful and informative to all scientific disciplines where Natural Products play a role.

D. H. R. Barton†          K. Nakanishi          O. Meth-Cohn

# Preface

It is surprising indeed that this work is the first attempt to produce a "comprehensive" overview of Natural Products beyond the student text level. However, the awe-inspiring breadth of the topic, which in many respects is still only developing, is such as to make the job daunting to anyone in the field. Fools rush in where angels fear to tread and the particular fool in this case was myself, a lifelong enthusiast and reader of the subject but with no research base whatever in the field!

Having been involved in several of the *Comprehensive* works produced by Pergamon Press, this omission intrigued me and over a period of gestation I put together a rough outline of how such a work could be written and presented it to Pergamon. To my delight they agreed that the project was worthwhile and in short measure Derek Barton was approached and took on the challenge of fleshing out this framework with alacrity. He also brought his long-standing friend and outstanding contributor to the field, Koji Nakanishi, into the team. With Derek's knowledge of the whole field, the subject was broken down into eight volumes and an outstanding team of internationally recognised Volume Editors was appointed.

We used Derek's 80th birthday as a target for finalising the work. Sadly he died just a few months before reaching this milestone. This work therefore is dedicated to the memory of Sir Derek Barton, Natural Products being the area which he loved best of all.

OTTO METH-COHN
Executive Editor

# SIR DEREK BARTON

Sir Derek Barton, who was Distinguished Professor of Chemistry at Texas A&M University and holder of the Dow Chair of Chemical Invention died on March 16, 1998 in College Station, Texas of heart failure. He was 79 years old and had been Chairman of the Executive Board of Editors for Tetrahedron Publications since 1979.

Barton was considered to be one of the greatest organic chemists of the twentieth century whose work continues to have a major influence on contemporary science and will continue to do so for future generations of chemists.

Derek Harold Richard Barton was born on September 8, 1918 in Gravesend, Kent, UK and graduated from Imperial College, London with the degrees of B.Sc. (1940) and Ph.D. (1942). He carried out work on military intelligence during World War II and after a brief period in industry, joined the faculty at Imperial College. It was an early indication of the breadth and depth of his chemical knowledge that his lectureship was in physical chemistry. This research led him into the mechanism of elimination reactions and to the concept of molecular rotation difference to correlate the configurations of steroid isomers. During a sabbatical leave at Harvard in 1949–1950 he published a paper on the "Conformation of the Steroid Nucleus" (*Experientia*, 1950, **6**, 316) which was to bring him the Nobel Prize in Chemistry in 1969, shared with the Norwegian chemist, Odd Hassel. This key paper (only four pages long) altered the way in which chemists thought about the shape and reactivity of molecules, since it showed how the reactivity of functional groups in steroids depends on their axial or equatorial positions in a given conformation. Returning to the UK he held Chairs of Chemistry at Birkbeck College and Glasgow University before returning to Imperial College in 1957, where he developed a remarkable synthesis of the steroid hormone, aldosterone, by a photochemical reaction known as the Barton Reaction (nitrite photolysis). In 1978 he retired from Imperial College and became Director of the Natural Products Institute (CNRS) at Gif-sur-Yvette in France where he studied the invention of new chemical reactions, especially the chemistry of radicals, which opened up a whole new area of organic synthesis involving Gif chemistry. In 1986 he moved to a third career at Texas A&M University as Distinguished Professor of Chemistry and continued to work on novel reactions involving radical chemistry and the oxidation of hydrocarbons, which has become of great industrial importance. In a research career spanning more than five decades, Barton's contributions to organic chemistry included major discoveries which have profoundly altered our way of thinking about chemical structure and reactivity. His chemistry has provided models for the biochemical synthesis of natural products including alkaloids, antibiotics, carbohydrates, and DNA. Most recently his discoveries led to models for enzymes which oxidize hydrocarbons, including methane monooxygenase.

The following are selected highlights from his published work:

The 1950 paper which launched Conformational Analysis was recognized by the Nobel Prize Committee as the key contribution whereby the third dimension was added to chemistry. This work alone transformed our thinking about the connection between stereochemistry and reactivity, and was later adapted from small molecules to macromolecules e.g., DNA, and to inorganic complexes.

Barton's breadth and influence is illustrated in "Biogenetic Aspects of Phenol Oxidation" (*Festschr. Arthur Stoll*, 1957, 117). This theoretical work led to many later experiments on alkaloid biosynthesis and to a set of rules for *ortho-para*-phenolic oxidative coupling which allowed the predication of new natural product systems before they were actually discovered and to the correction of several erroneous structures.

In 1960, his paper on the remarkably short synthesis of the steroid hormone aldosterone (*J. Am. Chem. Soc.*, 1960, **82**, 2641) disclosed the first of many inventions of new reactions—in this case nitrite photolysis—to achieve short, high yielding processes, many of which have been patented and are used worldwide in the pharmaceutical industry.

Moving to 1975, by which time some 500 papers had been published, yet another "Barton reaction" was born—"The Deoxygenation of Secondary Alcohols" (*J. Chem. Soc. Perkin Trans. 1*, 1975, 1574), which has been very widely applied due to its tolerance of quite hostile and complex local environments in carbohydrate and nucleoside chemistry. This reaction is the chemical counterpart to ribonucleotide→deoxyribonucleotide reductase in biochemistry and, until the arrival of the Barton reaction, was virtually impossible to achieve.

In 1985, "Invention of a New Radical Chain Reaction" involved the generation of carbon radicals from carboxylic acids (*Tetrahedron*, 1985, **41**, 3901). The method is of great synthetic utility and has been used many times by others in the burgeoning area of radicals in organic synthesis.

These recent advances in synthetic methodology were remarkable since his chemistry had virtually no precedent in the work of others. The radical methodology was especially timely in light of the significant recent increase in applications for fine chemical syntheses, and Barton gave the organic community an entrée into what will prove to be one of the most important methods of the twenty-first century. He often said how proud he was, at age 71, to receive the ACS Award for Creativity in Organic Synthesis for work published in the preceding five years.

Much of Barton's more recent work is summarized in the articles "The Invention of Chemical Reactions—The Last 5 Years" (*Tetrahedron*, 1992, **48**, 2529) and "Recent Developments in Gif Chemistry" (*Pure Appl. Chem.*, 1997, **69**, 1941).

Working 12 hours a day, Barton's stamina and creativity remained undiminished to the day of his death. The author of more than 1000 papers in chemical journals, Barton also held many successful patents. In addition to the Nobel Prize he received many honors and awards including the Davy, Copley, and Royal medals of the Royal Society of London, and the Roger Adams and Priestley Medals of the American Chemical Society. He held honorary degrees from 34 universities. He was a Fellow of the Royal Societies of London and Edinburgh, Foreign Associate of the National Academy of Sciences (USA), and Foreign Member of the Russian and Chinese Academies of Sciences. He was knighted by Queen Elizabeth in 1972, received the Légion d'Honneur (Chevalier 1972; Officier 1985) from France, and the Order of the Rising Sun from the Emperor of Japan. In his long career, Sir Derek trained over 300 students and postdoctoral fellows, many of whom now hold major positions throughout the world and include some of today's most distinguished organic chemists.

For those of us who were fortunate to know Sir Derek personally there is no doubt that his genius and work ethic were unique. He gave generously of his time to students and colleagues wherever he traveled and engendered such great respect and loyalty in his students and co-workers, that major symposia accompanied his birthdays every five years beginning with the 60th, ending this year with two celebrations just before his 80th birthday.

With the death of Sir Derek Barton, the world of science has lost a major figure, who together with Sir Robert Robinson and Robert B. Woodward, the cofounders of *Tetrahedron*, changed the face of organic chemistry in the twentieth century.

Professor Barton is survived by his wife, Judy, and by a son, William from his first marriage, and three grandchildren.

A. I. SCOTT
*Texas A&M University*

Reprinted from *Tetrahedron*, 1998, **54**, 8847
Photograph courtesy of Library and Information Centre, Royal Society of Chemistry. © The Nobel Foundation

# Contributors to Volume 7

Dr. S. L. Beaucage
Division of Hematologic Products, Center for Biologics Evaluation and Research, US Food and Drug Administration, Bethesda, MD 20892, USA

Professor K. J. Breslauer
Department of Chemistry, Rutgers University, Wright-Rieman Laboratories, Chemistry, 610 Taylor Road, Piscataway, NJ 08854-8087, USA

Professor J. Chmielewski
Department of Chemistry, Purdue University, 1393 Brown Building, West Lafayette, IN 47907, USA

Dr. C. de los Santos
Department of Pharmacology, Basic Science Tower 7-143, SUNY - Stony Brook, Stony Brook, NY 11794-8651, USA

Professor K. S. Gates
Department of Chemistry, 123 Chemistry Building, University of Missouri–Columbia, Columbia, MO 65211, USA

Mr. I. Ghosh
Department of Chemistry, Purdue University, 1393 Brown Building, West Lafayette, IN 47907, USA

Professor I. H. Goldberg
Department of Biological Chemistry and Molecular Pharmacology, Harvard Medical School, 250 Longwood Avenue, Boston, MA 02115, USA

Professor M. M. Greenberg
Department of Chemistry, Colorado State University, Fort Collins, CO 80523-1872, USA

Dr. S. B. Ha
Department of Chemistry, Merkert Chemistry Center, Boston College, Chestnut Hill, MA 02167, USA

Dr. R. P. Iyer
OriGenix, 620 Memorial Drive, Cambridge, MA 02139, USA

Professor E. T. Kool
Department of Chemistry, University of Rochester, Rochester, NY 14627-0216, USA

Mr. M. A. Labroli
Department of Chemistry, University of Virginia, McCormick Road, Charlottesville, VA 22901, USA

Professor T. L. MacDonald
Department of Chemistry, University of Virginia, McCormick Road, Charlottesville, VA 22901, USA

Professor L. W. McLaughlin
Department of Chemistry, Merkert Chemistry Center, Boston College, Chestnut Hill, MA 02167, USA

Professor J. T. Millard
Department of Chemistry, Colby College, 4000 Mayflower Hill, Waterville, ME 04901-8840, USA

Professor B. L. Miller
Department of Chemistry, University of Rochester, Rochester, NY 14627, USA

Dr. J. D. Pickert
Department of Chemistry, University of Rochester, Rochester, NY 14627, USA

Dr. G. E. Plum
Department of Chemistry, Rutgers University, Wright-Rieman Laboratories, Chemistry, 610 Taylor Road, Piscataway, NJ 08854-8087, USA

Dr. T. S. Rao
Gemini Biotech, 3608 Research Forest Drive, The Woodlands, TX 77381, USA

Dr. G. R. Revankar
Gemini Biotech, 3608 Research Forest Drive, The Woodlands, TX 77381, USA

Professor R. W. Roberts
California Institute of Technology, Pasadena, CA 91125, USA

Dr. Y. S. Sanghvi
Isis Pharmaceuticals, 2292 Faraday Avenue, Carlsbad, CA 92008, USA

Professor D. Sen
Institute of Molecular Biology and Biochemistry, Simon Fraser University, Burnaby, BC, Canada, V5A 1S6

Mr. J. J. Tepe
Department of Chemistry, University of Virginia, McCormick Road, Charlottesville, VA, 22901, USA

Dr. M. Wilson
Department of Chemistry, Merkert Chemistry Center, Boston College, Chesnut Hill, MA 02167, USA

Professor W. D. Wilson
Department of Chemistry, Georgia State University, University Plaza, Atlanta, GA 30303, USA

Mr. S. Yao
Department of Chemistry, Purdue University, 1393 Brown Building, West Lafayette, IN 47907, USA

Dr. Z. Xi
Department of Biological Chemistry and Molecular Biology, Harvard Medical School, 250 Longwood Avenue, Boston, MA 02115-5731, USA

# Abbreviations

The most commonly used abbreviations in *Comprehensive Natural Products Chemistry* are listed below. Please note that in some instances these may differ from those used in other branches of chemistry

| | |
|---|---|
| A | adenine |
| ABA | abscisic acid |
| Ac | acetyl |
| ACAC | acetylacetonate |
| ACTH | adrenocorticotropic hormone |
| ADP | adenosine 5'-diphosphate |
| AIBN | 2,2'-azobisisobutyronitrile |
| Ala | alanine |
| AMP | adenosine 5'-monophosphate |
| APS | adenosine 5'-phosphosulfate |
| Ar | aryl |
| Arg | arginine |
| ATP | adenosine 5'-triphosphate |
| | |
| B | nucleoside base (adenine, cylosine, guanine, thymine or uracil) |
| 9-BBN | 9-borabicyclo[3.3.1]nonane |
| BOC | *t*-butoxycarbonyl (or carbo-*t*-butoxy) |
| BSA | *N,O*-bis(trimethylsilyl)acetamide |
| BSTFA | *N,O*-bis(trimethylsilyl)trifluoroacetamide |
| Bu | butyl |
| Bu$^n$ | *n*-butyl |
| Bu$^i$ | isobutyl |
| Bu$^s$ | *s*-butyl |
| Bu$^t$ | *t*-butyl |
| Bz | benzoyl |
| | |
| CAN | ceric ammonium nitrate |
| CD | cyclodextrin |
| CDP | cytidine 5'-diphosphate |
| CMP | cytidine 5'-monophosphate |
| CoA | coenzyme A |
| COD | cyclooctadiene |
| COT | cyclooctatetraene |
| Cp | $\eta^5$-cyclopentadiene |
| Cp* | pentamethylcyclopentadiene |
| 12-Crown-4 | 1,4,7,10-tetraoxacyclododecane |
| 15-Crown-5 | 1,4,7,10,13-pentaoxacyclopentadecane |
| 18-Crown-6 | 1,4,7,10,13,16-hexaoxacyclooctadecane |
| CSA | camphorsulfonic acid |
| CSI | chlorosulfonyl isocyanate |
| CTP | cytidine 5'-triphosphate |
| cyclic AMP | adenosine 3',5'-cyclic monophosphoric acid |
| CySH | cysteine |
| | |
| DABCO | 1,4-diazabicyclo[2.2.2]octane |
| DBA | dibenz[*a,h*]anthracene |
| DBN | 1,5-diazabicyclo[4.3.0]non-5-ene |

DBU            1,8-diazabicyclo[5.4.0]undec-7-ene
DCC            dicyclohexylcarbodiimide
DEAC           diethylaluminum chloride
DEAD           diethyl azodicarboxylate
DET            diethyl tartrate (+ or -)
DHET           dihydroergotoxine
DIBAH          diisobutylaluminum hydride
Diglyme        diethylene glycol dimethyl ether (or bis(2-methoxyethyl)ether)
DiHPhe         2,5-dihydroxyphenylalanine
Dimsyl Na      sodium methylsulfinylmethide
DIOP           2,3-*O*-isopropylidene-2,3-dihydroxy-1,4-bis(diphenylphosphino)butane
dipt           diisopropyl tartrate (+ or -)
DMA            dimethylacetamide
DMAD           dimethyl acetylenedicarboxylate
DMAP           4-dimethylaminopyridine
DME            1,2-dimethoxyethane (glyme)
DMF            dimethylformamide
DMF-DMA        dimethylformamide dimethyl acetal
DMI            1,3-dimethyl-2-imidazalidinone
DMSO           dimethyl sulfoxide
DMTSF          dimethyl(methylthio)sulfonium fluoroborate
DNA            deoxyribonucleic acid
DOCA           deoxycorticosterone acetate

EADC           ethylaluminum dichloride
EDTA           ethylenediaminetetraacetic acid
EEDQ           *N*-ethoxycarbonyl-2-ethoxy-1,2-dihydroquinoline
Et             ethyl
EVK            ethyl vinyl ketone

FAD            flavin adenine dinucleotide
Fl             flavin
FMN            flavin mononucleotide

G              guanine
GABA           4-aminobutyric acid
GDP            guanosine 5'-diphosphate
GLDH           glutamate dehydrogenase
gln            glutamine
Glu            glutamic acid
Gly            glycine
GMP            guanosine 5'-monophosphate
GOD            glucose oxidase
G-6-P          glucose-6-phosphate
GTP            guanosine 5'-triphosphate

Hb             hemoglobin
His            histidine
HMPA           hexamethylphosphoramide
                 (or hexamethylphosphorous triamide)

Ile            isoleucine
INAH           isonicotinic acid hydrazide
IpcBH          isopinocampheylborane
Ipc$_2$BH      diisopinocampheylborane

KAPA           potassium 3-aminopropylamide
K-Slectride    potassium tri-*s*-butylborohydride

| | |
|---|---|
| LAH | lithium aluminum hydride |
| LAP | leucine aminopeptidase |
| LDA | lithium diisopropylamide |
| LDH | lactic dehydrogenase |
| Leu | leucine |
| LICA | lithium isopropylcyclohexylamide |
| L-Selectride | lithium tri-*s*-butylborohydride |
| LTA | lead tetraacetate |
| Lys | lysine |
| | |
| MCPBA | *m*-chloroperoxybenzoic acid |
| Me | methyl |
| MEM | methoxyethoxymethyl |
| MEM-Cl | ß-methoxyethoxymethyl chloride |
| Met | methionine |
| MMA | methyl methacrylate |
| MMC | methyl magnesium carbonate |
| MOM | methoxymethyl |
| Ms | mesyl (or methanesulfonyl) |
| MSA | methanesulfonic acid |
| MsCl | methanesulfonyl chloride |
| MVK | methyl vinyl ketone |
| | |
| NAAD | nicotinic acid adenine dinucleotide |
| NAD | nicotinamide adenine dinucleotide |
| NADH | nicotinamide adenine dinucleotide phosphate, reduced |
| NBS | *N*-bromosuccinimider |
| NMO | *N*-methylmorpholine *N*-oxide monohydrate |
| NMP | *N*-methylpyrrolidone |
| | |
| PCBA | *p*-chlorobenzoic acid |
| PCBC | *p*-chlorobenzyl chloride |
| PCBN | *p*-chlorobenzonitrile |
| PCBTF | *p*-chlorobenzotrifluoride |
| PCC | pyridinium chlorochromate |
| PDC | pyridinium dichromate |
| PG | prostaglandin |
| Ph | phenyl |
| Phe | phenylalanine |
| Phth | phthaloyl |
| PPA | polyphosphoric acid |
| PPE | polyphosphate ester (or ethyl *m*-phosphate) |
| Pr | propyl |
| Pr$^i$ | isopropyl |
| Pro | proline |
| Py | pyridine |
| | |
| RNA | ribonucleic acid |
| Rnase | ribonuclease |
| | |
| Ser | serine |
| Sia$_2$BH | disiamylborane |
| | |
| TAS | tris(diethylamino)sulfonium |
| TBAF | tetra-*n*-butylammonium fluoroborate |
| TBDMS | *t*-butyldimethylsilyl |
| TBDMS-Cl | *t*-butyldimethylsilyl chloride |
| TBDPS | *t*-butyldiphenylsilyl |
| TCNE | tetracyanoethene |

| | |
|---|---|
| TES | triethylsilyl |
| TFA | trifluoracetic acid |
| TFAA | trifluoroacetic anhydride |
| THF | tetrahydrofuran |
| THF | tetrahydrofolic acid |
| THP | tetrahydropyran (or tetrahydropyranyl) |
| Thr | threonine |
| TMEDA | *N,N,N',N'*,tetramethylethylenediamine[1,2-bis(dimethylamino)ethane] |
| TMS | trimethylsilyl |
| TMS-Cl | trimethylsilyl chloride |
| TMS-CN | trimethylsilyl cyanide |
| Tol | toluene |
| TosMIC | tosylmethyl isocyanide |
| TPP | tetraphenylporphyrin |
| Tr | trityl (or triphenylmethyl) |
| Trp | tryptophan |
| Ts | tosyl (or *p*-toluenesulfonyl) |
| TTFA | thallium trifluoroacetate |
| TTN | thallium(III) nitrate |
| Tyr | tyrosine |
| Tyr-OMe | tyrosine methyl ester |
| | |
| U | uridine |
| UDP | uridine 5'-diphosphate |
| UMP | uridine 5'-monophosphate |

# Contents of All Volumes

**Volume 1**   Polyketides and Other Secondary Metabolites Including Fatty Acids and Their Derivatives

**Volume 2**   Isoprenoids Including Carotenoids and Steroids

**Volume 3**   Carbohydrates and Their Derivatives Including Tannins, Cellulose, and Related Lignins

**Volume 4**   Amino Acids, Peptides, Porphyrins, and Alkaloids

**Volume 5**   Enzymes, Enzyme Mechanisms, Proteins, and Aspects of NO Chemistry

**Volume 6**   Prebiotic Chemistry, Molecular Fossils, Nucleosides, and RNA

**Volume 7**   DNA and Aspects of Molecular Biology

**Volume 8**   Miscellaneous Natural Products Including Marine Natural Products, Pheromones, Plant Hormones, and Aspects of Ecology

**Volume 9**   Cumulative Indexes

# An Historical Perspective of Natural Products Chemistry

## KOJI NAKANISHI

*Columbia University, New York, USA*

To give an account of the rich history of natural products chemistry in a short essay is a daunting task. This brief outline begins with a description of ancient folk medicine and continues with an outline of some of the major conceptual and experimental advances that have been made from the early nineteenth century through to about 1960, the start of the modern era of natural products chemistry. Achievements of living chemists are noted only minimally, usually in the context of related topics within the text. More recent developments are reviewed within the individual chapters of the present volumes, written by experts in each field. The subheadings follow, in part, the sequence of topics presented in Volumes 1–8.

### 1. ETHNOBOTANY AND "NATURAL PRODUCTS CHEMISTRY"

Except for minerals and synthetic materials our surroundings consist entirely of organic natural products, either of prebiotic organic origins or from microbial, plant, or animal sources. These materials include polyketides, terpenoids, amino acids, proteins, carbohydrates, lipids, nucleic acid bases, RNA and DNA, etc. Natural products chemistry can be thought of as originating from mankind's curiosity about odor, taste, color, and cures for diseases. Folk interest in treatments for pain, for food-poisoning and other maladies, and in hallucinogens appears to go back to the dawn of humanity

For centuries China has led the world in the use of natural products for healing. One of the earliest health science anthologies in China is the Nei Ching, whose authorship is attributed to the legendary Yellow Emperor (thirtieth century BC), although it is said that the dates were backdated from the third century by compilers. Excavation of a Han Dynasty (206 BC–AD 220) tomb in Hunan Province in 1974 unearthed decayed books, written on silk, bamboo, and wood, which filled a critical gap between the dawn of medicine up to the classic Nei Ching; Book 5 of these excavated documents lists 151 medical materials of plant origin. Generally regarded as the oldest compilation of Chinese herbs is Shen Nung Pen Ts'ao Ching (Catalog of Herbs by Shen Nung), which is believed to have been revised during the Han Dynasty; it lists 365 materials. Numerous revisions and enlargements of Pen Ts'ao were undertaken by physicians in subsequent dynasties, the ultimate being the Pen Ts'ao Kang Mu (General Catalog of Herbs) written by Li Shih-Chen over a period of 27 years during the Ming Dynasty (1573–1620), which records 1898 herbal drugs and 8160 prescriptions. This was circulated in Japan around 1620 and translated, and has made a huge impact on subsequent herbal studies in Japan; however, it has not been translated into English. The number of medicinal herbs used in 1979 in China numbered 5267. One of the most famous of the Chinese folk herbs is the ginseng root *Panax ginseng*, used for health maintenance and treatment of various diseases. The active principles were thought to be the saponins called ginsenosides but this is now doubtful; the effects could well be synergistic between saponins, flavonoids, etc. Another popular folk drug, the extract of the Ginkgo tree, *Ginkgo biloba* L., the only surviving species of the Paleozoic era (250 million years ago) family which became extinct during the last few million years, is mentioned in the Chinese Materia Medica to have an effect in improving memory and sharpening mental alertness. The main constituents responsible for this are now understood to be ginkgolides and flavonoids, but again not much else is known. Clarifying the active constituents and mode of (synergistic) bioactivity of Chinese herbs is a challenging task that has yet to be fully addressed.

The Assyrians left 660 clay tablets describing 1000 medicinal plants used around 1900–400 BC, but the best insight into ancient pharmacy is provided by the two scripts left by the ancient Egyptians, who were masters of human anatomy and surgery because of their extensive mummification practices. The Edwin Smith Surgical Papyrus purchased by Smith in 1862 in Luxor (now in the New York Academy of Sciences collection), is one of the most important medicinal documents of the ancient Nile Valley, and describes the healer's involvement in surgery, prescription, and healing practices using plants, animals, and minerals. The Ebers Papyrus, also purchased by Edwin Smith in 1862, and then acquired by Egyptologist George Ebers in 1872, describes 800 remedies using plants, animals, minerals, and magic. Indian medicine also has a long history, possibly dating back to the second millennium BC. The Indian materia medica consisted mainly of vegetable drugs prepared from plants but also used animals, bones, and minerals such as sulfur, arsenic, lead, copper sulfate, and gold. Ancient Greece inherited much from Egypt, India, and China, and underwent a gradual transition from magic to science. Pythagoras (580–500 BC) influenced the medical thinkers of his time, including Aristotle (384–322 BC), who in turn affected the medical practices of another influential Greek physician Galen (129–216). The Iranian physician Avicenna (980–1037) is noted for his contributions to Aristotelian philosophy and medicine, while the German-Swiss physician and alchemist Paracelsus (1493–1541) was an early champion who established the role of chemistry in medicine.

The rainforests in Central and South America and Africa are known to be particularly abundant in various organisms of interest to our lives because of their rich biodiversity, intense competition, and the necessity for self-defense. However, since folk-treatments are transmitted verbally to the next generation via shamans who naturally have a tendency to keep their plant and animal sources confidential, the recipes tend to get lost, particularly with destruction of rainforests and the encroachment of "civilization." Studies on folk medicine, hallucinogens, and shamanism of the Central and South American Indians conducted by Richard Schultes (Harvard Botanical Museum, emeritus) have led to renewed activity by ethnobotanists, recording the knowledge of shamans, assembling herbaria, and transmitting the record of learning to the village.

Extracts of toxic plants and animals have been used throughout the world for thousands of years for hunting and murder. These include the various arrow poisons used all over the world. *Strychnos* and *Chondrodendron* (containing strychnine, etc.) were used in South America and called "curare," *Strophanthus* (strophantidine, etc.) was used in Africa, the latex of the upas tree *Antiaris toxicaria* (cardiac glycosides) was used in Java, while *Aconitum napellus*, which appears in Greek mythology (aconitine) was used in medieval Europe and Hokkaido (by the Ainus). The Colombian arrow poison is from frogs (batrachotoxins; 200 toxins have been isolated from frogs by B. Witkop and J. Daly at NIH). Extracts of *Hyoscyamus niger* and *Atropa belladonna* contain the toxic tropane alkaloids, for example hyoscyamine, belladonnine, and atropine. The belladonna berry juice (atropine) which dilates the eye pupils was used during the Renaissance by ladies to produce doe-like eyes (belladona means beautiful woman). The Efik people in Calabar, southeastern Nigeria, used extracts of the calabar bean known as esere (physostigmine) for unmasking witches. The ancient Egyptians and Chinese knew of the toxic effect of the puffer fish, fugu, which contains the neurotoxin tetrodotoxin (Y. Hirata, K. Tsuda, R. B. Woodward).

When rye is infected by the fungus *Claviceps purpurea*, the toxin ergotamine and a number of ergot alkaloids are produced. These cause ergotism or the "devil's curse," "St. Anthony's fire," which leads to convulsions, miscarriages, loss of arms and legs, dry gangrene, and death. Epidemics of ergotism occurred in medieval times in villages throughout Europe, killing tens of thousands of people and livestock; Julius Caesar's legions were destroyed by ergotism during a campaign in Gaul, while in AD 994 an estimated 50,000 people died in an epidemic in France. As recently as 1926, a total of 11,000 cases of ergotism were reported in a region close to the Urals. It has been suggested that the witch hysteria that occurred in Salem, Massachusetts, might have been due to a mild outbreak of ergotism. Lysergic acid diethylamide (LSD) was first prepared by A. Hofmann, Sandoz Laboratories, Basel, in 1943 during efforts to improve the physiological effects of the ergot alkaloids when he accidentally inhaled it. "On Friday afternoon, April 16, 1943," he wrote, "I was seized by a sensation of restlessness... ." He went home from the laboratory and "perceived an uninterrupted stream of fantastic dreams ... ." (*Helvetica Chimica Acta*).

Numerous psychedelic plants have been used since ancient times, producing visions, mystical fantasies (cats and tigers also seem to have fantasies?, see nepetalactone below), sensations of flying, glorious feelings in warriors before battle, etc. The ethnobotanists Wasson and Schultes identified "ololiqui," an important Aztec concoction, as the seeds of the morning glory *Rivea corymbosa* and gave the seeds to Hofmann who found that they contained lysergic acid amides similar to but less potent than LSD. Iboga, a powerful hallucinogen from the root of the African shrub *Tabernanthe iboga*, is used by the Bwiti cult in Central Africa who chew the roots to obtain relief from fatigue and hunger; it contains the alkaloid ibogamine. The powerful hallucinogen used for thousands of years by the American Indians, the peyote cactus, contains mescaline and other alkaloids. The Indian hemp plant, *Cannabis sativa*, has been used for making rope since 3000 BC, but when it is used for its pleasure-giving effects it is called cannabis and has been known in central Asia, China, India, and the Near East since ancient times. Marijuana, hashish (named after the Persian founder of the Assassins of the eleventh century, Hasan-e Sabbah), charas, ghanja, bhang, kef, and dagga are names given to various preparations of the hemp plant. The constituent responsible for the mind-altering effect is 1-tetrahydrocannabinol (also referred to as 9-THC) contained in 1%. R. Mechoulam (1930–, Hebrew University) has been the principal worker in the cannabinoids, including structure determination and synthesis of 9-THC (1964 to present); the Israeli police have also made a contribution by providing Mechoulam with a constant supply of marijuana. Opium (morphine) is another ancient drug used for a variety of pain-relievers and it is documented that the Sumerians used poppy as early as 4000 BC; the narcotic effect is present only in seeds before they are fully formed. The irritating secretion of the blister beetles, for example *Mylabris* and the European species *Lytta vesicatoria*, commonly called Spanish fly, was used medically as a topical skin irritant to remove warts but was also a major ingredient in so-called love potions (constituent is cantharidin, stereospecific synthesis in 1951, G. Stork, 1921–; prep. scale high-pressure Diels–Alder synthesis in 1985, W. G. Dauben, 1919–1996).

Plants have been used for centuries for the treatment of heart problems, the most important being the foxgloves *Digitalis purpurea* and *D. lanata* (digitalin, diginin) and *Strophanthus gratus* (ouabain). The bark of cinchona *Cinchona officinalis* (called quina-quina by the Indians) has been used widely among the Indians in the Andes against malaria, which is still one of the major infectious diseases; its most important alkaloid is quinine. The British protected themselves against malaria during the occupation of India through gin and tonic (quinine!). The stimulant coca, used by the Incas around the tenth century, was introduced into Europe by the conquistadors; coca beans are also commonly chewed in West Africa. Wine making was already practiced in the Middle East 6000–8000 years ago; Moors made date wines, the Japanese rice wine, the Vikings honey mead, the Incas maize chicha. It is said that the Babylonians made beer using yeast 5000–6000 years ago. As shown above in parentheses, alkaloids are the major constituents of the herbal plants and extracts used for centuries, but it was not until the early nineteenth century that the active principles were isolated in pure form, for example morphine (1816), strychnine (1817), atropine (1819), quinine (1820), and colchicine (1820). It was a century later that the structures of these compounds were finally elucidated.

## 2. DAWN OF ORGANIC CHEMISTRY, EARLY STRUCTURAL STUDIES, MODERN METHODOLOGY

The term "organic compound" to define compounds made by and isolated from living organisms was coined in 1807 by the Swedish chemist Jons Jacob Berzelius (1779–1848), a founder of today's chemistry, who developed the modern system of symbols and formulas in chemistry, made a remarkably accurate table of atomic weights and analyzed many chemicals. At that time it was considered that organic compounds could not be synthesized from inorganic materials *in vitro*. However, Friedrich Wöhler (1800–1882), a medical doctor from Heidelberg who was starting his chemical career at a technical school in Berlin, attempted in 1828 to make "ammonium cyanate," which had been assigned a wrong structure, by heating the two inorganic salts potassium cyanate and ammonium sulfate; this led to the unexpected isolation of white crystals which were identical to the urea from urine, a typical organic compound. This well-known incident marked the beginning of organic chemistry. With the preparation of acetic acid from inorganic material in 1845 by Hermann Kolbe (1818–1884) at Leipzig, the myth surrounding organic compounds, in which they were associated with some vitalism was brought to an end and organic chemistry became the chemistry of carbon compounds. The same Kolbe was involved in the development of aspirin, one of the earliest and most important success stories in natural products chemistry. Salicylic acid from the leaf of the wintergreen plant had long been used as a pain reliever, especially in treating arthritis and gout. The inexpensive synthesis of salicylic acid from sodium phenolate and carbon dioxide by Kolbe in 1859 led to the industrial production in 1893 by the Bayer Company of acetylsalicylic acid "aspirin," still one of the most popular drugs. Aspirin is less acidic than salicylic acid and therefore causes less irritation in the mouth, throat, and stomach. The remarkable mechanism of the anti-inflammatory effect of aspirin was clarified in 1974 by John Vane (1927–) who showed that it inhibits the biosynthesis of prostaglandins by irreversibly acetylating a serine residue in prostaglandin synthase. Vane shared the 1982 Nobel Prize with Bergström and Samuelsson who determined the structure of prostaglandins (see below).

In the early days, natural products chemistry was focused on isolating the more readily available plant and animal constituents and determining their structures. The course of structure determination in the 1940s was a complex, indirect process, combining evidence from many types of experiments. The first effort was to crystallize the unknown compound or make derivatives such as esters or 2,4-dinitrophenylhydrazones, and to repeat recrystallization until the highest and sharp melting point was reached, since prior to the advent of isolation and purification methods now taken for granted, there was no simple criterion for purity. The only chromatography was through special grade alumina (first used by M. Tswett in 1906, then reintroduced by R. Willstätter). Molecular weight estimation by the Rast method which depended on melting point depression of a sample/camphor mixture, coupled with Pregl elemental microanalysis (see below) gave the molecular formula. Functionalities such as hydroxyl, amino, and carbonyl groups were recognized on the basis of specific derivatization and crystallization, followed by redetermination of molecular formula; the change in molecular composition led to identification of the functionality. Thus, sterically hindered carbonyls, for example the 11-keto group of cortisone, or tertiary hydroxyls, were very difficult to pinpoint, and often had to depend on more searching experiments. Therefore, an entire paper describing the recognition of a single hydroxyl group in a complex natural product would occasionally appear in the literature. An oxygen function suggested from the molecular formula but left unaccounted for would usually be assigned to an ether.

Determination of C-methyl groups depended on Kuhn–Roth oxidation which is performed by drastic oxidation with chromic acid/sulfuric acid, reduction of excess oxidant with hydrazine, neutralization with alkali, addition of phosphoric acid, distillation of the acetic acid originating from the C-methyls, and finally its titration with alkali. However, the results were only approximate, since *gem*-dimethyl groups only yield one equivalent of acetic acid, while primary, secondary, and tertiary methyl groups all give different yields of acetic acid. The skeletal structure of polycyclic compounds were frequently deduced on the basis of dehydrogenation reactions. It is therefore not surprising that the original steroid skeleton put forth by Wieland and Windaus in 1928, which depended a great deal on the production of chrysene upon Pd/C dehydrogenation, had to be revised in 1932 after several discrepancies were found (they received the Nobel prizes in 1927 and 1928 for this "extraordinarily difficult structure determination," see below).

In the following are listed some of the Nobel prizes awarded for the development of methodologies which have contributed critically to the progress in isolation protocols and structure determination. The year in which each prize was awarded is preceded by "Np."

Fritz Pregl, 1869–1930, Graz University, Np 1923. Invention of carbon and hydrogen microanalysis. Improvement of Kuhlmann's microbalance enabled weighing at an accuracy of 1 μg over a 20 g range, and refinement of carbon and hydrogen analytical methods made it possible to perform analysis with 3–4 mg of sample. His microbalance and the monograph *Quantitative Organic Microanalysis* (1916) profoundly influenced subsequent developments in practically all fields of chemistry and medicine.

The Svedberg, 1884–1971, Uppsala, Np 1926. Uppsala was a center for quantitative work on colloids for which the prize was awarded. His extensive study on ultracentrifugation, the first paper of which was published in the year of the award, evolved from a spring visit in 1922 to the University of Wisconsin. The ultracentrifuge together with the electrophoresis technique developed by his student Tiselius, have profoundly influenced subsequent progress in molecular biology and biochemistry.

Arne Tiselius, 1902–1971, Ph.D. Uppsala (T. Svedberg), Uppsala, Np 1948. Assisted by a grant from the Rockefeller Foundation, Tiselius was able to use his early electrophoresis instrument to show four bands in horse blood serum, alpha, beta and gamma globulins in addition to albumin; the first paper published in 1937 brought immediate positive responses.

Archer Martin, 1910–, Ph.D. Cambridge; Medical Research Council, Mill Hill, and Richard Synge, 1914–1994, Ph.D. Cambridge; Rowett Research Institute, Food Research Institute, Np 1952. They developed chromatography using two immiscible phases, gas–liquid, liquid–liquid, and paper chromatography, all of which have profoundly influenced all phases of chemistry.

Frederick Sanger, 1918–, Ph.D. Cambridge (A. Neuberger), Medical Research Council, Cambridge, Np 1958 and 1980. His confrontation with challenging structural problems in proteins and nucleic acids led to the development of two general analytical methods, 1,2,4-fluorodinitrobenzene (DNP) for tagging free amino groups (1945) in connection with insulin sequencing studies, and the dideoxynucleotide method for sequencing DNA (1977) in connection with recombinant DNA. For the latter he received his second Np in chemistry in 1980, which was shared with Paul Berg (1926–, Stanford University) and Walter Gilbert (1932–, Harvard University) for their contributions, respectively, in recombinant DNA and chemical sequencing of DNA. The studies of insulin involved usage of DNP for tagging disulfide bonds as cysteic acid residues (1949), and paper chromatography introduced by Martin and Synge 1944. That it was the first elucidation of any protein structure lowered the barrier for future structure studies of proteins.

Stanford Moore, 1913–1982, Ph.D. Wisconsin (K. P. Link), Rockefeller, Np 1972; and William Stein, 1911–1980, Ph.D. Columbia (E. G. Miller); Rockefeller, Np 1972. Moore and Stein cooperatively developed methods for the rapid quantification of protein hydrolysates by combining partition chromatography, ninhydrin coloration, and drop-counting fraction collector, i.e., the basis for commercial amino acid analyzers, and applied them to analysis of the ribonuclease structure.

Bruce Merrifield, 1921–, Ph.D. UCLA (M. Dunn), Rockefeller, Np 1984. The concept of solid-phase peptide synthesis using porous beads, chromatographic columns, and sequential elongation of peptides and other chains revolutionized the synthesis of biopolymers.

High-performance liquid chromatography (HPLC), introduced around the mid-1960s and now coupled on-line to many analytical instruments, for example UV, FTIR, and MS, is an indispensable daily tool found in all natural products chemistry laboratories.

## 3. STRUCTURES OF ORGANIC COMPOUNDS, NINETEENTH CENTURY

The discoveries made from 1848 to 1874 by Pasteur, Kekulé, van't Hoff, Le Bel, and others led to a revolution in structural organic chemistry. Louis Pasteur (1822–1895) was puzzled about why the potassium salt of tartaric acid (deposited on wine casks during fermentation) was dextrorotatory while the sodium ammonium salt of racemic acid (also deposited on wine casks) was optically inactive although both tartaric acid and "racemic" acid had identical chemical compositions. In 1848, the 25 year old Pasteur examined the racemic acid salt under the microscope and found two kinds of crystals exhibiting a left- and right-hand relation. Upon separation of the left-handed and right-handed crystals, he found that they rotated the plane of polarized light in opposite directions. He had thus performed his famous resolution of a racemic mixture, and had demonstrated the phenomenon of chirality. Pasteur went on to show that the racemic acid formed two kinds of salts with optically active bases such as quinine; this was the first demonstration of diastereomeric resolution. From this work Pasteur concluded that tartaric acid must have an element of asymmetry within the molecule itself. However, a three-dimensional understanding of the enantiomeric pair was only solved 25 years later (see below). Pasteur's own interest shifted to microbiology where he made the crucial discovery of the involvement of "germs" or microorganisms in various processes and proved that yeast induces alcoholic fermentation, while other microorganisms lead to diseases; he thus saved the wine industries of France, originated the process known as "pasteurization," and later developed vaccines for rabies. He was a genius who made many fundamental discoveries in chemistry and in microbiology.

The structures of organic compounds were still totally mysterious. Although Wöhler had synthesized urea, an isomer of ammonium cyanate, in 1828, the structural difference between these isomers was not known. In 1858 August Kekulé (1829–1896; studied with André Dumas and C. A. Wurtz in Paris, taught at Ghent, Heidelberg, and Bonn) published his famous paper in Liebig's *Annalen der Chemie* on the structure of carbon, in which he proposed that carbon atoms could form C–C bonds with hydrogen and other atoms linked to them; his dream on the top deck of a London bus led him to this concept. It was Butlerov who introduced the term "structure theory" in 1861. Further, in 1865 Kekulé conceived the cyclo-hexa-1:3:5-triene structure for benzene ($C_6H_6$) from a dream of a snake biting its own tail. In 1874, two young chemists, van't Hoff (1852–1911, Np 1901) in Utrecht, and Le Bel (1847–1930) in Paris, who had met in 1874 as students of C. A. Wurtz, published the revolutionary three-dimensional (3D) structure of the tetrahedral carbon Cabcd to explain the enantiomeric behavior of Pasteur's salts. The model was welcomed by J. Wislicenus (1835–1902, Zürich, Würzburg, Leipzig) who in 1863 had demonstrated the enantiomeric nature of the two lactic acids found by Scheele in sour milk (1780) and by Berzelius in muscle tissue (1807). This model, however, was criticized by Hermann Kolbe (1818–1884, Leipzig) as an "ingenious but in reality trivial and senseless natural philosophy." After 10 years of heated controversy, the idea of tetrahedral carbon was fully accepted, Kolbe had died and Wislicenus succeeded him in Leipzig.

Emil Fischer (1852–1919, Np 1902) was the next to make a critical contribution to stereochemistry. From the work of van't Hoff and Le Bel he reasoned that glucose should have 16 stereoisomers. Fischer's doctorate work on hydrazines under Baeyer (1835–1917, Np 1905) at Strasbourg had led to studies of osazones which culminated in the brilliant establishment, including configurations, of the Fischer sugar tree starting from D-(+)-glyceraldehyde all the way up to the aldohexoses, allose, altrose, glucose, mannose, gulose, idose, galactose, and talose (from 1884 to 1890). Unfortunately Fischer suffered from the toxic effects of phenylhydrazine for 12 years. The arbitrarily but luckily chosen absolute configuration of D-(+)-glyceraldehyde was shown to be correct sixty years later in 1951 (Johannes-Martin Bijvoet, 1892–1980). Fischer's brilliant correlation of the sugars comprising the Fischer sugar tree was performed using the Kiliani (1855–1945)–Fischer method via cyanohydrin intermediates for elongating sugars. Fischer also made remarkable contributions to the chemistry of amino acids and to nucleic acid bases (see below).

## 4. STRUCTURES OF ORGANIC COMPOUNDS, TWENTIETH CENTURY

The early concept of covalent bonds was provided with a sound theoretical basis by Linus Pauling (1901–1994, Np 1954), one of the greatest intellects of the twentieth century. Pauling's totally interdisciplinary research interests, including proteins and DNA is responsible for our present understanding of molecular structures. His books *Introduction to Quantum Mechanics* (with graduate student E. B. Wilson, 1935) and *The Nature of the Chemical Bond* (1939) have had a profound effect on our understanding of all of chemistry.

The actual 3D shapes of organic molecules which were still unclear in the late 1940s were then brilliantly clarified by Odd Hassel (1897–1981, Oslo University, Np 1969) and Derek Barton (1918–1998, Np 1969). Hassel, an X-ray crystallographer and physical chemist, demonstrated by electron diffraction that cyclohexane adopted the chair form in the gas phase and that it had two kinds of bonds, "standing (axial)" and "reclining (equatorial)" (1943). Because of the German occupation of Norway in 1940, instead of publishing the result in German journals, he published it in a Norwegian journal which was not abstracted in English until 1945. During his 1949 stay at Harvard, Barton attended a seminar by Louis Fieser on steric effects in steroids and showed Fieser that interpretations could be simplified if the shapes ("conformations") of cyclohexane rings were taken into consideration; Barton made these comments because he was familiar with Hassel's study on *cis*- and *trans*-decalins. Following Fieser's suggestion Barton published these ideas in a four-page *Experientia* paper (1950). This led to the joint Nobel prize with Hassel (1969), and established the concept of conformational analysis, which has exerted a profound effect in every field involving organic molecules.

Using conformational analysis, Barton determined the structures of many key terpenoids such as ß-amyrin, cycloartenone, and cycloartenol (Birkbeck College). At Glasgow University (from 1955) he collaborated in a number of cases with Monteath Robertson (1900–1989) and established many challenging structures: limonin, glauconic acid, byssochlamic acid, and nonadrides. Barton was also associated with the Research Institute for Medicine and Chemistry (RIMAC), Cambridge, USA founded by the Schering company, where with J. M. Beaton, he produced 60 g of aldosterone at a time when the world supply of this important hormone was in mg quantities. Aldosterone synthesis ("a good problem") was achieved in 1961 by Beaton ("a good experimentalist") through a nitrite photolysis, which came to be known as the Barton reaction ("a good idea") (quotes from his 1991 autobiography published by the American Chemical Society). From Glasgow, Barton went on to Imperial College, and a year before retirement, in 1977 he moved to France to direct the research at ICSN at Gif-sur-Yvette where he explored the oxidation reaction selectivity for unactivated C–H. After retiring from ICSN he made a further move to Texas A&M University in 1986, and continued his energetic activities, including chairman of the *Tetrahedron* publications. He felt weak during work one evening and died soon after, on March 16, 1998. He was fond of the phrase "gap jumping" by which he meant seeking generalizations between facts that do not seem to be related: "In the conformational analysis story, one had to jump the gap between steroids and chemical physics" (from his autobiography). According to Barton, the three most important qualities for a scientist are "intelligence, motivation, and honesty." His routine at Texas A&M was to wake around 4 a.m., read the literature, go to the office at 7 a.m. and stay there until 7 p.m.; when asked in 1997 whether this was still the routine, his response was that he wanted to wake up earlier because sleep was a waste of time—a remark which characterized this active scientist approaching 80!

Robert B. Woodward (1917–1979, Np 1965), who died prematurely, is regarded by many as the preeminent organic chemist of the twentieth century. He made landmark achievements in spectroscopy, synthesis, structure determination, biogenesis, as well as in theory. His solo papers published in 1941–1942 on empirical rules for estimating the absorption maxima of enones and dienes made the general organic chemical community realize that UV could be used for structural studies, thus launching the beginning of the spectroscopic revolution which soon brought on the applications of IR, NMR, MS, etc. He determined the structures of the following compounds: penicillin in 1945 (through joint UK–USA collaboration, see Hodgkin), strychnine in 1948, patulin in 1949, terramycin, aureomycin, and ferrocene (with G. Wilkinson, Np 1973—shared with E. O. Fischer for sandwich compounds) in 1952, cevine in 1954 (with Barton Np 1966, Jeger and Prelog, Np 1975), magnamycin in 1956, gliotoxin in 1958, oleandomycin in 1960, streptonigrin in 1963, and tetrodotoxin in 1964. He synthesized patulin in 1950, cortisone and cholesterol in 1951, lanosterol, lysergic acid (with Eli Lilly), and strychnine in 1954, reserpine in 1956, chlorophyll in 1960, a tetracycline (with Pfizer) in 1962, cephalosporin in 1965, and vitamin $B_{12}$ in 1972 (with A. Eschenmoser, 1925–, ETH Zürich). He derived biogenetic schemes for steroids in 1953 (with K. Bloch, see below), and for macrolides in 1956, while the Woodward–Hoffmann orbital symmetry rules in 1965 brought order to a large class of seemingly random cyclization reactions.

Another central figure in stereochemistry is Vladimir Prelog (1906–1998, Np 1975), who succeeded Leopold Ruzicka at the ETH Zürich, and continued to build this institution into one of the most active and lively research and discussion centers in the world. The core group of intellectual leaders consisted of P. Plattner (1904–1975), O. Jeger, A. Eschenmoser, J. Dunitz, D. Arigoni, and A. Dreiding (from Zürich University). After completing extensive research on alkaloids, Prelog determined the structures of nonactin, boromycin, ferrioxamins, and rifamycins. His seminal studies in the synthesis and properties of 8–12 membered rings led him into unexplored areas of stereochemisty and chirality. Together with Robert Cahn (1899–1981, London Chemical Society) and Christopher Ingold (1893–1970, University College, London; pioneering mechanistic interpretation of organic reactions), he developed the Cahn Ingold–Prelog (CIP) sequence rules for the unambiguous specification of stereoisomers. Prelog was an excellent story teller, always had jokes to tell, and was respected and loved by all who knew him.

## 4.1 Polyketides and Fatty Acids

Arthur Birch (1915–1995) from Sydney University, Ph.D. with Robert Robinson (Oxford University), then professor at Manchester University and Australian National University, was one of the earliest chemists to perform biosynthetic studies using radiolabels; starting with polyketides he studied the biosynthesis of a variety of natural products such as the $C_6$–$C_3$–$C_6$ backbone of plant phenolics, polyene macrolides, terpenoids, and alkaloids. He is especially known for the Birch reduction of aromatic rings, metal–ammonia reductions leading to 19-norsteroid hormones and other important products (1942–) which were of industrial importance. Feodor Lynen (1911–1979, Np 1964) performed studies on the intermediary metabolism of the living cell that led him to the demonstration of the first step in a chain of reactions resulting in the biosynthesis of sterols and fatty acids.

Prostaglandins, a family of 20-carbon, lipid-derived acids discovered in seminal fluids and accessory genital glands of man and sheep by von Euler (1934), have attracted great interest because of their extremely diverse biological activities. They were isolated and their structures elucidated from 1963 by S. Bergström (1916–, Np 1982) and B. Samuelsson (1934–, Np 1982) at the Karolinska Institute, Stockholm. Many syntheses of the natural prostaglandins and their nonnatural analogues have been published.

Tetsuo Nozoe (1902–1996) who studied at Tohoku University, Sendai, with Riko Majima (1874–1962, see below) went to Taiwan where he stayed until 1948 before returning to Tohoku University. At National Taiwan University he isolated hinokitiol from the essential oil of *taiwanhinoki*. Remembering the resonance concept put forward by Pauling just before World War II, he arrived at the seven-membered nonbenzenoid aromatic structure for hinokitiol in 1941, the first of the troponoids. This highly original work remained unknown to the rest of the world until 1951. In the meantime, during 1945–1948, nonbenzenoid aromatic structures had been assigned to stipitatic acid (isolated by H. Raistrick) by Michael J. S. Dewar (1918–) and to the thujaplicins by Holger Erdtman (1902–1989); the term tropolones was coined by Dewar in 1945. Nozoe continued to work on and discuss troponoids, up to the night before his death, without knowing that he had cancer. He was a remarkably focused and warm scientist, working unremittingly. Erdtman (Royal Institute of Technology, Stockholm) was the central figure in Swedish natural products chemistry who, with his wife Gunhild Aulin Erdtman (dynamic General Secretary of the Swedish Chemistry Society), worked in the area of plant phenolics.

As mentioned in the following and in the concluding sections, classical biosynthetic studies using radioactive isotopes for determining the distribution of isotopes has now largely been replaced by the use of various stable isotopes coupled with NMR and MS. The main effort has now shifted to the identification and cloning of genes, or where possible the gene clusters, involved in the biosynthesis of the natural product. In the case of polyketides (acyclic, cyclic, and aromatic), the focus is on the polyketide synthases.

## 4.2 Isoprenoids, Steroids, and Carotenoids

During his time as an assistant to Kekulé at Bonn, Otto Wallach (1847–1931, Np 1910) had to familiarize himself with the essential oils from plants; many of the components of these oils were compounds for which no structure was known. In 1891 he clarified the relations between 12 different monoterpenes related to pinene. This was summarized together with other terpene chemistry in book form in 1909, and led him to propose the "isoprene rule." These achievements laid the foundation for the future development of terpenoid chemistry and brought order from chaos.

The next period up to around 1950 saw phenomenal advances in natural products chemistry centered on isoprenoids. Many of the best natural products chemists in Europe, including Wieland, Windaus, Karrer, Kuhn, Butenandt, and Ruzicka contributed to this breathtaking pace. Heinrich Wieland (1877–1957) worked on the bile acid structure, which had been studied over a period of 100 years and considered to be one of the most difficult to attack; he received the Nobel Prize in 1927 for these studies. His friend Adolph Windaus (1876–1959) worked on the structure of cholesterol for which he also received the Nobel Prize in 1928. Unfortunately, there were chemical discrepancies in the proposed steroidal skeletal structure, which had a five-membered ring B attached to C-7 and C-9. J. D. Bernal, Mineralogical Museums, Cambridge University, who was examining the X-ray patterns of ergosterol (1932) noted that the dimensions were inconsistent with the Wieland–Windaus formula. A reinterpretation of the production of chrysene from sterols by Pd/C dehydrogenation reported by Diels (see below) in 1927 eventually led Rosenheim and King and Wieland and Dane to deduce the correct structure in 1932. Wieland also worked on the structures of morphine/strychnine alkaloids, phalloidin/amanitin cyclopeptides of toxic mushroom *Amanita phalloides*, and pteridines, the important fluorescent pigments of butterfly wings. Windaus determined the structure of ergosterol and continued structural studies of its irradiation product which exhibited antirachitic activity "vitamin D." The mechanistically complex photochemistry of ergosterol leading to the vitamin D group has been investigated in detail by Egbert Havinga (1927–1988, Leiden University), a leading photochemist and excellent tennis player.

Paul Karrer (1889–1971, Np 1937), established the foundations of carotenoid chemistry through structural determinations of lycopene, carotene, vitamin A, etc. and the synthesis of squalene, carotenoids, and others. George Wald (1906–1997, Np 1967) showed that vitamin A was the key compound in vision during his stay in Karrer's laboratory. Vitamin K (K from "Koagulation"), discovered by Henrik Dam (1895–1976, Polytechnic Institute, Copenhagen, Np 1943) and structurally studied by Edward Doisy (1893–1986, St. Louis University, Np 1943), was also synthesized by Karrer. In addition, Karrer synthesized riboflavin (vitamin $B_2$) and determined the structure and role of nicotinamide adenine dinucleotide phosphate (NADP$^+$) with Otto Warburg. The research on carotenoids and vitamins of Karrer who was at Zürich University overlapped with that of Richard Kuhn (1900–1967, Np 1938) at the ETH Zürich, and the two were frequently rivals. Richard Kuhn, one of the pioneers in using UV-vis spectroscopy for structural studies, introduced the concept of "atropisomerism" in diphenyls, and studied the spectra of a series of diphenyl polyenes. He determined the structures of many natural carotenoids, proved the structure of riboflavin-5-phosphate (flavin-adenine-dinucleotide-5-phosphate) and showed that the combination of NAD-5-phosphate with the carrier protein yielded the yellow oxidation enzyme, thus providing an understanding of the role of a prosthetic group. He also determined the structures of vitamin B complexes, i.e., pyridoxine, *p*-aminobenzoic acid, pantothenic acid. After World War II he went on to structural studies of nitrogen-containing oligosaccharides in human milk that provide immunity for infants, and brain gangliosides. Carotenoid studies in Switzerland were later taken up by Otto Isler (1910–1993), a Ruzicka student at Hoffmann-La Roche, and Conrad Hans Eugster (1921–), a Karrer student at Zürich University.

Adolf Butenandt (1903–1998, Np 1939) initiated and essentially completed isolation and structural studies of the human sex hormones, the insect molting hormone (ecdysone), and the first pheromone, bombykol. With help from industry he was able to obtain large supplies of urine from pregnant women for estrone, sow ovaries for progesterone, and 4,000 gallons of male urine for androsterone (50 mg, crystals). He isolated and determined the structures of two female sex hormones, estrone and progesterone, and the male hormone androsterone all during the period 1934–1939 (!) and was awarded the Nobel prize in 1939. Keen intuition and use of UV data and Pregl's microanalysis all played important roles. He was appointed to a professorship in Danzig at the age of 30. With Peter Karlson he isolated from 500 kg of silkworm larvae 25 mg of α-ecdysone, the prohormone of insect and crustacean molting hormone, and determined its structure as a polyhydroxysteroid (1965); 20-hydroxylation gives the insect and crustacean molting hormone or ß-ecdysone (20-hydroxyecdysteroid). He was also the first to isolate an insect pheromone, bombykol, from female silkworm moths (with E. Hecker). As president of the Max Planck Foundation, he strongly influenced the postwar rebuilding of German science.

The successor to Kuhn, who left ETH Zürich for Heidelberg, was Leopold Ruzicka (1887–1967, Np 1939) who established a close relationship with the Swiss pharmaceutical industry. His synthesis of the 17- and 15-membered macrocyclic ketones, civetone and muscone (the constituents of musk) showed that contrary to Baeyer's prediction, large alicyclic rings could be strainless. He reintroduced and refined the isoprene rule proposed by Wallach (1887) and determined the basic structures of many sesqui-, di-, and triterpenes, as well as the structure of lanosterol, the key intermediate in cholesterol biosynthesis. The "biogenetic isoprene rule" of the ETH group, Albert Eschenmoser, Leopold Ruzicka, Oskar Jeger, and Duilio Arigoni, contributed to a concept of terpenoid cyclization (1955), which was consistent with the mechanistic considerations put forward by Stork as early as 1950. Besides making the ETH group into a center of natural products chemistry, Ruzicka bought many seventeenth century Dutch paintings with royalties accumulated during the war from his Swiss and American patents, and donated them to the Zürich Kunsthaus.

Studies in the isolation, structures, and activities of the antiarthritic hormone, cortisone and related compounds from the adrenal cortex were performed in the mid- to late 1940s during World War II by Edward Kendall (1886–1972, Mayo Clinic, Rochester, Np 1950), Tadeus Reichstein (1897–1996, Basel University, Np 1950), Philip Hench (1896–1965, Mayo Clinic, Rochester, Np 1950), Oskar Wintersteiner (1898–1971, Columbia University, Squibb) and others initiated interest as an adjunct to military medicine as well as to supplement the meager supply from beef adrenal glands by synthesis. Lewis Sarett (1917–, Merck & Co., later president) and co-workers completed the cortisone synthesis in 28 steps, one of the first two totally stereocontrolled syntheses of a natural product; the other was cantharidin (Stork 1951) (see above). The multistep cortisone synthesis was put on the production line by Max Tishler (1906–1989, Merck & Co., later president) who made contributions to the synthesis of a number of drugs, including riboflavin. Besides working on steroid reactions/synthesis and antimalarial agents, Louis F. Fieser (1899–1977) and Mary Fieser (1909–1997) of Harvard University made huge contributions to the chemical community through their outstanding books *Natural Products related to Phenanthrene* (1949), *Steroids* (1959), *Advanced Organic Chemistry* (1961), and *Topics in Organic Chemistry* (1963), as well as their textbooks and an important series of books on Organic Reagents. Carl Djerassi (1923–, Stanford University), a prolific chemist, industrialist, and more recently a novelist, started to work at the Syntex laboratories in Mexico City where he directed the work leading to the first oral contraceptive ("the pill") for women.

Takashi Kubota (1909–, Osaka City University), with Teruo Matsuura (1924–, Kyoto University), determined the structure of the furanoid sesquiterpene, ipomeamarone, from the black rotted portion of spoiled sweet potatoes; this research constitutes the first characterization of a phytoallexin, defense substances produced by plants in response to attack by fungi or physical damage. Damaging a plant and characterizing the defense substances produced may lead to new bioactive compounds. The mechanism of induced biosynthesis of phytoallexins, which is not fully understood, is an interesting biological mechanistic topic that deserves further investigation. Another center of high activity in terpenoids and nucleic acids was headed by Frantisek Sorm (1913–1980, Institute of Organic and Biochemistry, Prague), who determined the structures of many sesquiterpenoids and other natural products; he was not only active scientifically but also was a central figure who helped to guide the careers of many Czech chemists.

The key compound in terpenoid biosynthesis is mevalonic acid (MVA) derived from acetyl-CoA, which was discovered fortuitously in 1957 by the Merck team in Rahway, NJ headed by Karl Folkers (1906–1998). They soon realized and proved that this $C_6$ acid was the precursor of the $C_5$ isoprenoid unit isopentenyl diphosphate (IPP) that ultimately leads to the biosynthesis of cholesterol. In 1952 Konrad Bloch (1912–, Harvard, Np 1964) with R. B. Woodward published a paper suggesting a mechanism of the cyclization of squalene to lanosterol and the subsequent steps to cholesterol, which turned out to be essentially correct. This biosynthetic path from MVA to cholesterol was experimentally clarified in stereochemical detail by John Cornforth (1917–, Np 1975) and George Popják. In 1932, Harold Urey (1893–1981, Np 1934) of Columbia University discovered heavy hydrogen. Urey showed, contrary to common expectation, that isotope separation could be achieved with deuterium in the form of deuterium oxide by fractional electrolysis of water. Urey's separation of the stable isotope deuterium led to the isotopic tracer methodology that revolutionized the protocols for elucidating biosynthetic processes and reaction mechanisms, as exemplified beautifully by the cholesterol studies. Using MVA labeled chirally with isotopes, including chiral methyl, i.e., -CHDT, Cornforth and Popják clarified the key steps in the intricate biosynthetic conversion of mevalonate to cholesterol in stereochemical detail. The chiral methyl group was also prepared independently by Duilio Arigoni (1928–, ETH, Zürich). Cornforth has had great difficulty in hearing and speech since childhood but has been helped expertly by his chemist wife Rita; he is an excellent tennis and chess player, and is renowned for his speed in composing occasional witty limericks.

Although MVA has long been assumed to be the only natural precursor for IPP, a non-MVA pathway in which IPP is formed via the glyceraldehyde phosphate-pyruvate pathway has been discovered (1995–1996) in the ancient bacteriohopanoids by Michel Rohmer, who started working on them with Guy Ourisson (1926–, University of Strasbourg, terpenoid studies, including prebiotic), and by Duilio Arigoni in the ginkgolides, which are present in the ancient *Ginkgo biloba* tree. It is possible that many other terpenoids are biosynthesized via the non-MVA route. In classical biosynthetic experiments, $^{14}$C-labeled acetic acid was incorporated into the microbial or plant product, and location or distribution of the $^{14}$C label was deduced by oxidation or degradation to specific fragments including acetic acid; therefore, it was not possible or extremely difficult to map the distribution of all radioactive carbons. The progress in $^{13}$C NMR made it possible to incorporate $^{13}$C-labeled acetic acid and locate all labeled carbons. This led to the discovery of the nonmevalonate pathway leading to the IPP units. Similarly, NMR and MS have made it possible to use the stable isotopes, e.g., $^{18}$O, $^{2}$H, $^{15}$N, etc., in biosynthetic studies. The current trend of biosynthesis has now shifted to genomic approaches for cloning the genes of various enzyme synthases involved in the biosynthesis.

### 4.3 Carbohydrates and Cellulose

The most important advance in carbohydrate structures following those made by Emil Fischer was the change from acyclic to the current cyclic structure introduced by Walter Haworth (1883–1937). He noticed the presence of α- and ß-anomers, and determined the structures of important disaccharides including cellobiose, maltose, and lactose. He also determined the basic structural aspects of starch, cellulose, inulin, and other polysaccharides, and accomplished the structure determination and synthesis of vitamin C, a sample of which he had received from Albert von Szent-Györgyi (1893–1986, Np 1937). This first synthesis of a vitamin was significant since it showed that a vitamin could be synthesized in the same way as any other organic compound. There was strong belief among leading scientists in the 1910s that cellulose, starch, protein, and rubber were colloidal aggregates of small molecules. However, Hermann Staudinger (1881–1965, Np 1953) who succeeded R. Willstätter and H. Wieland at the ETH Zürich and Freiburg, respectively, showed through viscosity measurements and various molecular weight measurements that macromolecules do exist, and developed the principles of macromolecular chemistry.

In more modern times, Raymond Lemieux (1920–, Universities of Ottawa and Alberta) has been a leader in carbohydrate research. He introduced the concept of *endo-* and *exo-*anomeric effects, accomplished the challenging synthesis of sucrose (1953), pioneered in the use of NMR coupling constants in configuration studies, and most importantly, starting with syntheses of oligosaccharides responsible for human blood group determinants, he prepared antibodies and clarified fundamental aspects of the binding of oligosaccharides by lectins and antibodies. The periodate–potassium permanganate cleavage of double bonds at room temperature (1955) is called the Lemieux reaction.

### 4.4 Amino Acids, Peptides, Porphyrins, and Alkaloids

It is fortunate that we have China's record and practice of herbal medicine over the centuries, which is providing us with an indispensable source of knowledge. China is rapidly catching up in terms of infrastructure and equipment in organic and bioorganic chemistry, and work on isolation, structure determination, and synthesis stemming from these valuable sources has picked up momentum. However, as mentioned above, clarification of the active principles and mode of action of these plant extracts will be quite a challenge since in many cases synergistic action is expected. Wang Yu (1910–1997) who headed the well-equipped Shanghai Institute of Organic Chemistry surprised the world with the total synthesis of bovine insulin performed by his group in 1965; the human insulin was synthesized around the same time by P. G. Katsoyannis, A. Tometsko, and C. Zaut of the Brookhaven National Laboratory (1966).

One of the giants in natural products chemistry during the first half of this century was Robert Robinson (1886–1975, Np 1947) at Oxford University. His synthesis of tropinone, a bicyclic amino ketone related to cocaine, from succindialdehyde, methylamine, and acetone dicarboxylic acid under Mannich reaction conditions was the first biomimetic synthesis (1917). It reduced Willstätter's 1903 13-step synthesis starting with suberone into a single step. This achievement demonstrated Robinson's analytical prowess. He was able to dissect complex molecular structures into simple biosynthetic building blocks, which allowed him to propose the biogenesis of all types of alkaloids and other natural products. His laboratory at Oxford, where he developed the well-known Robinson annulation reaction (1937) in connection with his work on the synthesis of steroids became a world center for natural products study. Robinson was a pioneer in the so-called electronic theory of organic reactions, and introduced the use of curly arrows to show the movements of electrons. His analytical power is exemplified in the structural studies of strychnine and brucine around 1946–1952. Barton clarified the biosynthetic route to the morphine alkaloids, which he saw as an extension of his biomimetic synthesis of usnic acid through a one-electron oxidation; this was later extended to a general phenolate coupling scheme. Morphine total synthesis was brilliantly achieved by Marshall Gates (1915–, University of Rochester) in 1952.

The yield of the Robinson tropinone synthesis was low but Clemens Schöpf (1899–1970) , Ph.D. Munich (Wieland), Universität Darmstadt, improved it to 90% by carrying out the reaction in buffer; he also worked on the stereochemistry of morphine and determined the structure of the steroidal alkaloid salamandarine (1961), the toxin secreted from glands behind the eyes of the salamander.

Roger Adams (1889–1971, University of Illinois), was the central figure in organic chemistry in the USA and is credited with contributing to the rapid development of its chemistry in the late 1930s and 1940s, including training of graduate students for both academe and industry. After earning a Ph.D. in 1912 at Harvard University he did postdoctoral studies with Otto Diels (see below) and Richard Willstätter (see below) in 1913; he once said that around those years in Germany he could cover all *Journal of the American Chemical Society* papers published in a year in a single night. His important work include determination of the structures of tetrahydrocannabinol in marijuana, the toxic gossypol in cottonseed oil, chaulmoogric acid used in treatment of leprosy, and the Senecio alkaloids with Nelson Leonard (1916–, University of Illinois, now at Caltech). He also contributed to many fundamental organic reactions and syntheses. The famous Adams platinum catalyst is not only important for reducing double bonds in industry and in the laboratory, but was central for determining the number of double bonds in a structure. He was also one of the founders of the *Organic Synthesis* (started in 1921) and the *Organic Reactions* series. Nelson Leonard switched interests to bioorganic chemistry and biochemistry, where he has worked with nucleic acid bases and nucleotides, coenzymes, dimensional probes, and fluorescent modifications such as ethenoguanine.

The complicated structures of the medieval plant poisons aconitine (from *Aconitum*) and delphinine (from *Delphinium*) were finally characterized in 1959–1960 by Karel Wiesner (1919–1986, University of New Brunswick), Leo Marion (1899–1979, National Research Council, Ottawa), George Büchi (1921–, mycotoxins, aflatoxin/DNA adduct, synthesis of terpenoids and nitrogen-containing bioactive compounds, photochemistry), and Maria Przybylska (1923–, X-ray).

The complex chlorophyll structure was elucidated by Richard Willstätter (1872–1942, Np 1915). Although he could not join Baeyer's group at Munich because the latter had ceased taking students, a close relation developed between the two. During his chlorophyll studies, Willstätter reintroduced the important technique of column chromatography published in Russian by Michael Tswett (1906). Willstätter further demonstrated that magnesium was an integral part of chlorophyll, clarified the relation between chlorophyll and the blood pigment hemin, and found the wide distribution of carotenoids in tomato, egg yolk, and bovine corpus luteum. Willstätter also synthesized cyclooctatetraene and showed its properties to be wholly unlike benzene but close to those of acyclic polyenes (around 1913). He succeeded Baeyer at Munich in 1915, synthesized the anesthetic cocaine, retired early in protest of anti-Semitism, but remained active until the Hitler era, and in 1938 emigrated to Switzerland.

The hemin structure was determined by another German chemist of the same era, Hans Fischer (1881–1945, Np 1930), who succeeded Windaus at Innsbruck and at Munich. He worked on the structure of hemin from the blood pigment hemoglobin, and completed its synthesis in 1929. He continued Willstätter's structural studies of chlorophyll, and further synthesized bilirubin in 1944. Destruction of his institute at Technische Hochschule München, during World War II led him to take his life in March 1945. The biosynthesis of hemin was elucidated largely by David Shemin (1911–1991).

In the mid 1930s the Department of Biochemistry at Columbia Medical School, which had accepted many refugees from the Third Reich, including Erwin Chargaff, Rudolf Schoenheimer, and others on the faculty, and Konrad Bloch (see above) and David Shemin as graduate students, was a great center of research activity. In 1940, Shemin ingested 66 g of $^{15}$N-labeled glycine over a period of 66 hours in order to determine the half-life of erythrocytes. David Rittenberg's analysis of the heme moiety with his home-made mass spectrometer showed all four pyrrole nitrogens came from glycine. Using $^{14}$C (that had just become available) as a second isotope (see next paragraph), doubly labeled glycine $^{15}NH_2^{14}CH_2COOH$ and other precursors, Shemin showed that glycine and succinic acid condensed to yield δ-aminolevulinate, thus elegantly demonstrating the novel biosynthesis of the porphyrin ring (around 1950). At this time, Bloch was working on the other side of the bench.

Melvin Calvin (1911–1997, Np 1961) at University of California, Berkeley, elucidated the complex photosynthetic pathway in which plants reduce carbon dioxide to carbohydrates. The critical $^{14}CO_2$ had just been made available at Berkeley Lawrence Radiation Laboratory as a result of the pioneering research of Martin Kamen (1913–), while paper chromatography also played crucial roles. Kamen produced $^{14}$C with Sam Ruben (1940), used $^{18}$O to show that oxygen in photosynthesis comes from water and not from carbon dioxide, participated in the *Manhattan* project, testified before the House UnAmerican Activities Committee (1947), won compensatory damages from the US Department of State, and helped build the University of California, La Jolla (1957). The entire structure of the photosynthetic reaction center (>10 000 atoms) from the purple bacterium *Rhodopseudomonas viridis* has been established by X-ray crystallography in the landmark studies performed by Johann Deisenhofer (1943–), Robert Huber (1937–), and Hartmut Michel (1948–) in 1989; this was the first membrane protein structure determined by X-ray, for which they shared the 1988 Nobel prize. The information gained from the full structure of this first membrane protein has been especially rewarding.

The studies on vitamin B$_{12}$, the structure of which was established by crystallographic studies performed by Dorothy Hodgkin (1910–1994, Np 1964), are fascinating. Hodgkin also determined the structure of penicillin (in a joint effort between UK and US scientists during World War II) and insulin. The formidable total synthesis of vitamin B$_{12}$ was completed in 1972 through collaborative efforts between Woodward and Eschenmoser, involving 100 postdoctoral fellows and extending over 10 years. The biosynthesis of fascinating complexity is almost completely solved through studies performed by Alan Battersby (1925–, Cambridge University), Duilio Arigoni, and Ian Scott (1928–, Texas A&M University) and collaborators where advanced NMR techniques and synthesis of labeled precursors is elegantly combined with cloning of enzymes controlling each biosynthetic step. This work provides a beautiful demonstration of the power of the combination of bioorganic chemistry, spectroscopy and molecular biology, a future direction which will become increasingly important for the creation of new "unnatural" natural products.

### 4.5 Enzymes and Proteins

In the early days of natural products chemistry, enzymes and viruses were very poorly understood. Thus, the 1926 paper by James Sumner (1887–1955) at Cornell University on crystalline urease was received with ignorance or skepticism, especially by Willstätter who believed that enzymes were small molecules and not proteins. John Northrop (1891–1987) and co-workers at the Rockefeller Institute went on to crystallize pepsin, trypsin, chymotrypsin, ribonuclease, deoyribonuclease, carboxypeptidase, and other enzymes between 1930 and 1935. Despite this, for many years biochemists did not recognize the significance of these findings, and considered enzymes as being low molecular weight compounds adsorbed onto proteins or colloids. Using Northrop's method for crystalline enzyme preparations, Wendell Stanley (1904–1971) at Princeton obtained tobacco mosaic virus as needles from one ton of tobacco leaves (1935). Sumner, Northrop, and Stanley shared the 1946 Nobel prize in chemistry. All these studies opened a new era for biochemistry.

Meanwhile, Linus Pauling, who in mid-1930 became interested in the magnetic properties of hemoglobin, investigated the configurations of proteins and the effects of hydrogen bonds. In 1949 he showed that sickle cell anemia was due to a mutation of a single amino acid in the hemoglobin molecule, the first correlation of a change in molecular structure with a genetic disease. Starting in 1951 he and colleagues published a series of papers describing the alpha helix structure of proteins; a paper published in the early 1950s with R. B. Corcy on the structure of DNA played an important role in leading Francis Crick and James Watson to the double helix structure (Np 1962).

A further important achievement in the peptide field was that of Vincent Du Vigneaud (1901–1978, Np 1955), Cornell Medical School, who isolated and determined the structure of oxytocin, a posterior pituitary gland hormone, for which a structure involving a disulfide bond was proposed. He synthesized oxytocin in 1953, thereby completing the first synthesis of a natural peptide hormone.

Progress in isolation, purification, crystallization methods, computers, and instrumentation, including cyclotrons, have made X-ray crystallography the major tool in structural. Numerous structures including those of ligand/receptor complexes are being published at an extremely rapid rate. Some of the past major achievements in protein structures are the following. Max Perutz (1914, Np 1962 ) and John Kendrew (1914–1997, Np 1962), both at the Laboratory of Molecular Biology, Cambridge University, determined the structures of hemoglobin and myoglobin, respectively. William Lipscomb (1919–, Np 1976), Harvard University, who has trained many of the world's leaders in protein X-ray crystallography has been involved in the structure determination of many enzymes including carboxypeptidase A (1967); in 1965 he determined the structure of the anticancer bisindole alkaloid, vinblastine. Folding of proteins, an important but still enigmatic phenomenon, is attracting increasing attention. Christian Anfinsen (1916–1995, Np 1972), NIH, one of the pioneers in this area, showed that the amino acid residues in ribonuclease interact in an energetically most favorable manner to produce the unique 3D structure of the protein.

### 4.6 Nucleic Acid Bases, RNA, and DNA

The "Fischer indole synthesis" was first performed in 1886 by Emil Fischer. During the period 1881–1914, he determined the structures of and synthesized uric acid, caffeine, theobromine, xanthine, guanine, hypoxanthine, adenine, guanine, and made theophylline-D-glucoside phosphoric acid, the first synthetic nucleotide. In 1903, he made 5,5-diethylbarbituric acid or Barbital, Dorminal, Veronal, etc. (sedative), and in 1912, phenobarbital or Barbipil, Luminal, Phenobal, etc. (sedative). Many of his syntheses formed the basis of German industrial production of purine bases. In 1912 he showed that tannins are gallates of sugars such as maltose and glucose. Starting in 1899, he synthesized many of the 13 α-amino acids known at that time, including the L- and D-forms, which were separated through fractional crystallization of their salts with optically active bases. He also developed a method for synthesizing fragments of proteins, namely peptides, and made an 18-amino acid peptide. He lost his two sons in World War I, lost his wealth due to postwar inflation, believed he had terminal cancer (a misdiagnosis), and killed himself in July 1919. Fischer was a skilled experimentalist, so that even today, many of the reactions performed by him and his students are so delicately controlled that they are not easy to reproduce. As a result of his suffering by inhaling diethylmercury, and of the poisonous effect of phenylhydrazine, he was one of the first to design fume hoods. He was a superb teacher and was also influential in establishing the Kaiser Wilhelm Institute, which later became the Max Planck Institute. The number and quality of his accomplishments and contributions are hard to believe; he was truly a genius.

Alexander Todd (1907–1997, Np 1957) made critical contributions to the basic chemistry and synthesis of nucleotides. His early experience consisted of an extremely fruitful stay at Oxford in the Robinson group, where he completed the syntheses of many representative anthocyanins, and then at Edinburgh where he worked on the synthesis of vitamin $B_1$. He also prepared the hexacarboxylate of vitamin $B_{12}$ (1954), which was used by D. Hodgkin's group for their X-ray elucidation of this vitamin (1956). M. Wiewiorowski (1918–), Institute for Bioorganic Chemistry, in Poznan, has headed a famous group in nucleic acid chemistry, and his colleagues are now distributed worldwide.

### 4.7 Antibiotics, Pigments, and Marine Natural Products

The concept of one microorganism killing another was introduced by Pasteur who coined the term antibiosis in 1877, but it was much later that this concept was realized in the form of an actual antibiotic. The bacteriologist Alexander Fleming (1881–1955, University of London, Np 1945) noticed that an airborne mold, a *Penicillium* strain, contaminated cultures of *Staphylococci* left on the open bench and formed a transparent circle around its colony due to lysis of *Staphylococci*. He published these results in 1929. The discovery did not attract much interest but the work was continued by Fleming until it was taken up further at Oxford University by pathologist Howard Florey (1898–1968, Np 1945) and biochemist Ernst Chain (1906–1979, Np 1945). The bioactivities of purified "penicillin," the first antibiotic, attracted serious interest in the early 1940s in the midst of World War II. A UK/USA team was formed during the war between academe and industry with Oxford University, Harvard University, ICI, Glaxo, Burroughs Wellcome, Merck, Shell, Squibb, and Pfizer as members. This project resulted in the large scale production of penicillin and determination of its structure (finally by X-ray, D. Hodgkin). John Sheehan (1915–1992) at MIT synthesized 6-aminopenicillanic acid in 1959, which opened the route for the synthesis of a number of analogues. Besides being the first antibiotic to be discovered, penicillin is also the first member of a large number of important antibiotics containing the ß-lactam ring, for example cephalosporins, carbapenems, monobactams, and nocardicins. The strained ß-lactam ring of these antibiotics inactivates the transpeptidase by acylating its serine residue at the active site, thus preventing the enzyme from forming the link between the pentaglycine chain and the D-Ala-D-Ala peptide, the essential link in bacterial cell walls. The overuse of ß-lactam antibiotics, which has given rise to the disturbing appearance of microbial resistant strains, is leading to active research in the design of synthetic ß-lactam analogues to counteract these strains. The complex nature of the important penicillin biosynthesis is being elucidated through efforts combining genetic engineering, expression of biosynthetic genes as well as feeding of synthetic precursors, etc. by Jack Baldwin (1938–, Oxford University), José Luengo (Universidad de León, Spain) and many other groups from industry and academe.

Shortly after the penicillin discovery, Selman Waksman (1888–1973, Rutgers University, Np 1952) discovered streptomycin, the second antibiotic and the first active against the dreaded disease tuberculosis. The discovery and development of new antibiotics continued throughout the world at pharmaceutical companies in Europe, Japan, and the USA from soil and various odd sources: cephalosporin from sewage in Sardinia, cyclosporin from Wisconsin and Norway soil which was carried back to Switzerland, avermectin from the soil near a golf course in Shizuoka Prefecture. People involved in antibiotic discovery used to collect soil samples from various sources during their trips but this has now become severely restricted to protect a country's right to its soil. M. M. Shemyakin (1908–1970, Institute of Chemistry of Natural Products, Moscow) was a grand master of Russian natural products who worked on antibiotics, especially of the tetracycline class; he also worked on cyclic antibiotics composed of alternating sequences of amides and esters and coined the term depsipeptide for these in 1953. He died in 1970 of a sudden heart attack in the midst of the 7th IUPAC Natural Products

Symposium held in Riga, Latvia, which he had organized. The Institute he headed was renamed the Shemyakin Institute.

Indigo, an important vat dye known in ancient Asia, Egypt, Greece, Rome, Britain, and Peru, is probably the oldest known coloring material of plant origin, Indigofera and Isatis. The structure was determined in 1883 and a commercially feasible synthesis was performed in 1883 by Adolf von Baeyer (see above, 1835–1917, Np 1905), who founded the German Chemical Society in 1867 following the precedent of the Chemistry Society of London. In 1872 Baeyer was appointed a professor at Strasbourg where E. Fischer was his student, and in 1875 he succeeded J. Liebig in Munich. Tyrian (or Phoenician) purple, the dibromo derivative of indigo which is obtained from the purple snail Murex bundaris, was used as a royal emblem in connection with religious ceremonies because of its rarity; because of the availability of other cheaper dyes with similar color, it has no commercial value today. K. Venkataraman (1901–1981, University of Bombay then National Chemical Laboratory) who worked with R. Robinson on the synthesis of chromones in his early career, continued to study natural and synthetic coloring matters, including synthetic anthraquinone vat dyes, natural quinonoid pigments, etc. T. R. Seshadri (1900–1975) is another Indian natural products chemist who worked mainly in natural pigments, dyes, drugs, insecticides, and especially in polyphenols. He also studied with Robinson, and with Pregl at Graz, and taught at Delhi University. Seshadri and Venkataraman had a huge impact on Indian chemistry. After a 40 year involvement, Toshio Goto (1929–1990) finally succeeded in solving the mysterious identity of commelinin, the deep-blue flower petal pigment of the Commelina communis isolated by Kozo Hayashi (1958) and protocyanin, isolated from the blue cornflower Centaurea cyanus by E. Bayer (1957). His group elucidated the remarkable structure in its entirety which consisted of six unstable anthocyanins, six flavones and two metals, the molecular weight approaching 10 000; complex stacking and hydrogen bonds were also involved. Thus the pigmentation of petals turned out to be far more complex than the theories put forth by Willstätter (1913) and Robinson (1931). Goto suffered a fatal heart attack while inspecting the first X-ray structure of commelinin; commelinin represents a pinnacle of current natural products isolation and structure determination in terms of subtlety in isolation and complexity of structure.

The study of marine natural products is understandably far behind that of compounds of terrestrial origin due to the difficulty in collection and identification of marine organisms. However, it is an area which has great potentialities for new discoveries from every conceivable source. One pioneer in modern marine chemistry is Paul Scheuer (1915–, University of Hawaii) who started his work with quinones of marine origin and has since characterized a very large number of bioactive compounds from mollusks and other sources. Luigi Minale (1936–1997, Napoli) started a strong group working on marine natural products, concentrating mainly on complex saponins. He was a leading natural products chemist who died prematurely. A. Gonzalez Gonzalez (1917–) who headed the Organic Natural Products Institute at the University of La Laguna, Tenerife, was the first to isolate and study polyhalogenated sesquiterpenoids from marine sources. His group has also carried out extensive studies on terrestrial terpenoids from the Canary Islands and South America. Carotenoids are widely distributed in nature and are of importance as food coloring material and as antioxidants (the detailed mechanisms of which still have to be worked out); new carotenoids continue to be discovered from marine sources, for example by the group of Synnove Liaaen-Jensen, Norwegian Institute of Technology). Yoshimasa Hirata (1915–), who started research at Nagoya University, is a champion in the isolation of nontrivial natural products. He characterized the bioluminescent luciferin from the marine ostracod Cypridina hilgendorfii in 1966 (with his students, Toshio Goto, Yoshito Kishi, and Osamu Shimomura); tetrodotoxin from the fugu fish in 1964 (with Goto and Kishi and co-workers), the structure of which was announced simultaneously by the group of Kyosuke Tsuda (1907–, tetrodotoxin, matrine) and Woodward; and the very complex palytoxin, $C_{129}H_{223}N_3O_{54}$ in 1981–1987 (with Daisuke Uemura and Kishi). Richard E. Moore, University of Hawaii, also announced the structure of palytoxin independently. Jon Clardy (1943–, Cornell University) has determined the X-ray structures of many unique marine natural products, including brevetoxin B (1981), the first of the group of toxins with contiguous _trans_-fused ether rings constituting a stiff ladderlike skeleton. Maitotoxin, $C_{164}H_{256}O_{68}S_2Na_2$, MW 3422, produced by the dinoflagellate _Gambierdiscus toxicus_ is the largest and most toxic of the nonbiopolymeric toxins known; it has 32 alicyclic 6- to 8-membered ethereal rings and acyclic chains. Its isolation (1994) and complete structure determination was accomplished jointly by the groups of Takeshi Yasumoto (Tohoku University), Kazuo Tachibana and Michio Murata (Tokyo University) in 1996. Kishi, Harvard University, also deduced the full structure in 1996.

The well-known excitatory agent for the cat family contained in the volatile oil of catnip, _Nepeta cataria_, is the monoterpene nepetalactone, isolated by S. M. McElvain (1943) and structure determined by Jerrold Meinwald (1954); cats, tigers, and lions start purring and roll on their backs in response to this lactone. Takeo Sakan (1912–1993) investigated the series of monoterpenes neomatatabiols, etc.

from Actinidia, some of which are male lacewing attractants. As little as 1 fg of neomatatabiol attracts lacewings.

The first insect pheromone to be isolated and characterized was bombykol, the sex attractant for the male silkworm, *Bombyx mori* (by Butenandt and co-workers, see above). Numerous pheromones have been isolated, characterized, synthesized, and are playing central roles in insect control and in chemical ecology. The group at Cornell University have long been active in this field: Tom Eisner (1929–, behavior), Jerrold Meinwald (1927–, chemistry), Wendell Roeloff (1939–, electrophysiology, chemistry). Since the available sample is usually minuscule, full structure determination of a pheromone often requires total synthesis; Kenji Mori (1935–, Tokyo University) has been particularly active in this field. Progress in the techniques for handling volatile compounds, including collection, isolation, GC/MS, etc. has started to disclose the extreme complexity of chemical ecology which plays an important role in the lives of all living organisms. In this context, natural products chemistry will be play an increasingly important role in our grasp of the significance of biodiversity.

## 5. SYNTHESIS

Synthesis has been mentioned often in the preceding sections of this essay. In the following, synthetic methods of more general nature are described. The Grignard reaction of Victor Grignard (1871–1935, Np 1912) and then the Diels–Alder reaction by Otto Diels (1876–1954, Np 1950) and Kurt Alder (1902–1956, Np 1950) are extremely versatile reactions. The Diels–Alder reaction can account for the biosynthesis of several natural products with complex structures, and now an enzyme, a Diels–Alderase involved in biosynthesis has been isolated by Akitami Ichihara, Hokkaido University (1997).

The hydroboration reactions of Herbert Brown (1912–, Purdue University, Np 1979) and the Wittig reactions of Georg Wittig (1897–1987, Np 1979) are extremely versatile synthetic reactions. William S. Johnson (1913–1995, University of Wisconsin, Stanford University) developed efficient methods for the cyclization of acyclic polyolefinic compounds for the synthesis of corticoid and other steroids, while Gilbert Stork (1921–, Columbia University) introduced enamine alkylation, regiospecific enolate formation from enones and their kinetic trapping (called "three component coupling" in some cases), and radical cyclization in regio- and stereospecific constructions. Elias J. Corey (1928–, Harvard University, Np 1990) introduced the concept of retrosynthetic analysis and developed many key synthetic reactions and reagents during his synthesis of bioactive compounds, including prostaglandins and gingkolides. A recent development is the ever-expanding supramolecular chemistry stemming from 1967 studies on crown ethers by Charles Pedersen (1904–1989), 1968 studies on cryptates by Jean-Marie Lehn (1939–), and 1973 studies on host–guest chemistry by Donald Cram (1919–); they shared the chemistry Nobel prize in 1987.

## 6. NATURAL PRODUCTS STUDIES IN JAPAN

Since the background of natural products study in Japan is quite different from that in other countries, a brief history is given here. Natural products is one of the strongest areas of chemical research in Japan with probably the world's largest number of chemists pursuing structural studies; these are joined by a healthy number of synthetic and bioorganic chemists. An important Symposium on Natural Products was held in 1957 in Nagoya as a joint event between the faculties of science, pharmacy, and agriculture. This was the beginning of a series of annual symposia held in various cities, which has grown into a three-day event with about 50 talks and numerous papers; practically all achievements in this area are presented at this symposium. Japan adopted the early twentieth century German or European academic system where continuity of research can be assured through a permanent staff in addition to the professor, a system which is suited for natural products research which involves isolation and assay, as well as structure determination, all steps requiring delicate skills and much expertise.

The history of Japanese chemistry is short because the country was closed to the outside world up to 1868. This is when the Tokugawa shogunate which had ruled Japan for 264 years was overthrown and the Meiji era (1868–1912) began. Two of the first Japanese organic chemists sent abroad were Shokei Shibata and Nagayoshi Nagai, who joined the laboratory of A. W. von Hoffmann in Berlin. Upon return to Japan, Shibata (Chinese herbs) started a line of distinguished chemists, Keita and Yuji Shibata (flavones) and Shoji Shibata (1915–, lichens, fungal bisanthraquinonoid pigments, ginsenosides); Nagai returned to Tokyo Science University in 1884, studied ephedrine, and left a big mark in the embryonic era of organic chemistry. Modern natural products chemistry really began when three extraordinary organic chemists returned from Europe in the 1910s and started teaching and research at their respective faculties:

Riko Majima, 1874–1962, C. D. Harries (Kiel University); R. Willstätter (Zürich): Faculty of Science, Tohoku University; studied urushiol, the catecholic mixture of poison ivy irritant.

Yasuhiko Asahina, 1881–1975, R. Willstätter: Faculty of pharmacy, Tokyo University; lichens and Chinese herb.

Umetaro Suzuki, 1874–1943, E. Fischer: Faculty of agriculture, Tokyo University; vitamin $B_1$(thiamine).

Because these three pioneers started research in three different faculties (i.e., science, pharmacy, and agriculture), and because little interfaculty personnel exchange occurred in subsequent years, natural products chemistry in Japan was pursued independently within these three academic domains; the situation has changed now. The three pioneers started lines of first-class successors, but the establishment of a strong infrastructure takes many years, and it was only after the mid-1960s that the general level of science became comparable to that in the rest of the world; the 3rd IUPAC Symposium on the Chemistry of Natural Products, presided over by Munio Kotake (1894–1976, bufotoxins, see below), held in 1964 in Kyoto, was a clear turning point in Japan's role in this area.

Some of the outstanding Japanese chemists not already quoted are the following. Shibasaburo Kitazato (1852–1931), worked with Robert Koch (Np 1905, tuberculosis) and von Behring, antitoxins of diphtheria and tetanus which opened the new field of serology, isolation of microorganism causing dysentery, founder of Kitazato Institute; Chika Kuroda (1884–1968), first female Ph.D., structure of the complex carthamin, important dye in safflower (1930) which was revised in 1979 by Obara *et al.*, although the absolute configuration is still unknown (1998); Munio Kotake (1894–1976), bufotoxins, tryptophan metabolites, nupharidine; Harusada Suginome (1892–1972), aconite alkaloids; Teijiro Yabuta (1888–1977), kojic acid, gibberrelins; Eiji Ochiai (1898–1974), aconite alkaloids; Toshio Hoshino (1899–1979), abrine and other alkaloids; Yusuke Sumiki (1901–1974), gibberrelins; Sankichi Takei (1896–1982), rotenone; Shiro Akabori (1900–1992), peptides, C-terminal hydrazinolysis of amino acid ; Hamao Umezawa (1914–1986), kanamycin, bleomycin, numerous antibiotics; Shojiro Uyeo (1909–1988), lycorine; Tsunematsu Takemoto (1913–1989), inokosterone, kainic acid, domoic acid, quisqualic acid; Tomihide Shimizu (1889–1958), bile acids; Kenichi Takeda (1907–1991), Chinese herbs, sesquiterpenes; Yoshio Ban (1921–1994), alkaloid synthesis; Wataru Nagata (1922–1993), stereocontrolled hydrocyanation.

## 7. CURRENT AND FUTURE TRENDS IN NATURAL PRODUCTS CHEMISTRY

Spectroscopy and X-ray crystallography has totally changed the process of structure determination, which used to generate the excitement of solving a mystery. The first introduction of spectroscopy to the general organic community was Woodward's 1942–1943 empirical rules for estimating the UV maxima of dienes, trienes, and enones, which were extended by Fieser (1959). However, Butenandt had used UV for correctly determining the structures of the sex hormones as early as the early 1930s, while Karrer and Kuhn also used UV very early in their structural studies of the carotenoids. The Beckman DU instruments were an important factor which made UV spectroscopy a common tool for organic chemists and biochemists. With the availability of commercial instruments in 1950, IR spectroscopy became the next physical tool, making the 1950 Colthup IR correlation chart and the 1954 Bellamy monograph indispensable. The IR fingerprint region was analyzed in detail in attempts to gain as much structural information as possible from the molecular stretching and bending vibrations. Introduction of NMR spectroscopy into organic chemistry, first for protons and then for carbons, has totally changed the picture of structure determination, so that now IR is used much less frequently; however, in biopolymer studies, the techniques of difference FTIR and resonance Raman spectroscopy are indispensable.

The dramatic and rapid advancements in mass spectrometry are now drastically changing the protocol of biomacromolecular structural studies performed in biochemistry and molecular biology. Herbert Hauptman (mathematician, 1917–, Medical Foundation, Buffalo, Np 1985) and Jerome Karle (1918–, US Naval Research Laboratory, Washington, DC, Np 1985) developed direct methods for the determination of crystal structures devoid of disproportionately heavy atoms. The direct method together with modern computers revolutionized the X-ray analysis of molecular structures, which has become routine for crystalline compounds, large as well as small. Fred McLafferty (1923–, Cornell University) and Klaus Biemann (1926–, MIT) have made important contributions in the development of organic and bioorganic mass spectrometry. The development of cyclotron-based facilities for crystallographic biology studies has led to further dramatic advances enabling some protein structures to be determined in a single day, while cryoscopic electron micrography developed in 1975 by Richard Henderson and Nigel Unwin has also become a powerful tool for 3D structural determinations of membrane proteins such as bacteriorhodopsin (25 kd) and the nicotinic acetylcholine receptor (270 kd).

Circular dichroism (c.d.), which was used by French scientists Jean B. Biot (1774–1862) and Aimé Cotton during the nineteenth century "deteriorated" into monochromatic measurements at 589 nm after R.W. Bunsen (1811–1899, Heidelberg) introduced the Bunsen burner into the laboratory which readily emitted a 589 nm light characteristic of sodium. The 589 nm $[\alpha]_D$ values, remote from most chromophoric maxima, simply represent the summation of the low-intensity readings of the decreasing end of multiple Cotton effects. It is therefore very difficult or impossible to deduce structural information from $[\alpha]_D$ readings. Chiroptical spectroscopy was reintroduced to organic chemistry in the 1950s by C. Djerassi at Wayne State University (and later at Stanford University) as optical rotatory dispersion (ORD) and by L. Velluz and M. Legrand at Roussel-Uclaf as c.d. Günther Snatzke (1928–1992, Bonn then Ruhr University Bochum) was a major force in developing the theory and application of organic chiroptical spectroscopy. He investigated the chiroptical properties of a wide variety of natural products, including constituents of indigenous plants collected throughout the world, and established semiempirical sector rules for absolute configurational studies. He also established close collaborations with scientists of the former Eastern bloc countries and had a major impact in increasing the interest in c.d. there.

Chiroptical spectroscopy, nevertheless, remains one of the most underutilized physical measurements. Most organic chemists regard c.d. (more popular than ORD because interpretation is usually less ambiguous) simply as a tool for assigning absolute configurations, and since there are only two possibilities in absolute configurations, c.d. is apparently regarded as not as crucial compared to other spectroscopic methods. Moreover, many of the c.d. correlations with absolute configuration are empirical. For such reasons, chiroptical spectroscopy, with its immense potentialities, is grossly underused. However, c.d. curves can now be calculated nonempirically. Moreover, through-space coupling between the electric transition moments of two or more chromophores gives rise to intense Cotton effects split into opposite signs, exciton-coupled c.d.; fluorescence-detected c.d. further enhances the sensitivity by 50- to 100-fold. This leads to a highly versatile nonempirical microscale solution method for determining absolute configurations, etc.

With the rapid advances in spectroscopy and isolation techniques, most structure determinations in natural products chemistry have become quite routine, shifting the trend gradually towards activity-monitored isolation and structural studies of biologically active principles available only in microgram or submicrogram quantities. This in turn has made it possible for organic chemists to direct their attention towards clarifying the mechanistic and structural aspects of the ligand/biopolymeric receptor interactions on a more well-defined molecular structural basis. Until the 1990s, it was inconceivable and impossible to perform such studies.

Why does sugar taste sweet? This is an extremely challenging problem which at present cannot be answered even with major multidisciplinary efforts. Structural characterization of sweet compounds and elucidation of the amino acid sequences in the receptors are only the starting point. We are confronted with a long list of problems such as cloning of the receptors to produce them in sufficient quantities to investigate the physical fit between the active factor (sugar) and receptor by biophysical methods, and the time-resolved change in this physical contact and subsequent activation of G-protein and enzymes. This would then be followed by neurophysiological and ultimately physiological and psychological studies of sensation. How do the hundreds of taste receptors differ in their structures and their physical contact with molecules, and how do we differentiate the various taste sensations? The same applies to vision and to olfactory processes. What are the functions of the numerous glutamate receptor subtypes in our brain? We are at the starting point of a new field which is filled with exciting possibilities.

Familiarity with molecular biology is becoming essential for natural products chemists to plan research directed towards an understanding of natural products biosynthesis, mechanisms of bioactivity triggered by ligand–receptor interactions, etc. Numerous genes encoding enzymes have been cloned and expressed by the cDNA and/or genomic DNA-polymerase chain reaction protocols. This then leads to the possible production of new molecules by gene shuffling and recombinant biosynthetic techniques. Monoclonal catalytic antibodies using haptens possessing a structure similar to a high-energy intermediate of a proposed reaction are also contributing to the elucidation of biochemical mechanisms and the design of efficient syntheses. The technique of photoaffinity labeling, brilliantly invented by Frank Westheimer (1912–, Harvard University), assisted especially by advances in mass spectrometry, will clearly be playing an increasingly important role in studies of ligand–receptor interactions including enzyme–substrate reactions. The combined and sophisticated use of various spectroscopic means, including difference spectroscopy and fast time-resolved spectroscopy, will also become increasingly central in future studies of ligand–receptor studies.

Organic chemists, especially those involved in structural studies have the techniques, imagination, and knowledge to use these approaches. But it is difficult for organic chemists to identify an exciting and worthwhile topic. In contrast, the biochemists, biologists, and medical doctors are daily facing exciting life-related phenomena, frequently without realizing that the phenomena could be understood or at least clarified on a chemical basis. Broad individual expertise and knowledge coupled with multidisciplinary research collaboration thus becomes essential to investigate many of the more important future targets successfully. This approach may be termed "dynamic," as opposed to a "static" approach, exemplified by isolation and structure determination of a single natural product. Fortunately for scientists, nature is extremely complex and hence all the more challenging. Natural products chemistry will be playing an absolutely indispensable role for the future. Conservation of the alarming number of disappearing species, utilization of biodiversity, and understanding of the intricacies of biodiversity are further difficult, but urgent, problems confronting us.

That natural medicines are attracting renewed attention is encouraging from both practical and scientific viewpoints; their efficacy has often been proven over the centuries. However, to understand the mode of action of folk herbs and related products from nature is even more complex than mechanistic clarification of a single bioactive factor. This is because unfractionated or partly fractionated extracts are used, often containing mixtures of materials, and in many cases synergism is most likely playing an important role. Clarification of the active constituents and their modes of action will be difficult. This is nevertheless a worthwhile subject for serious investigations.

*Dedicated to Sir Derek Barton whose amazing insight helped tremendously in the planning of this series, but who passed away just before its completion. It is a pity that he was unable to write this introduction as originally envisaged, since he would have had a masterful overview of the content he wanted, based on his vast experience. I have tried to fulfill his task, but this introduction cannot do justice to his original intention.*

## ACKNOWLEDGMENT

I am grateful to current research group members for letting me take quite a time off in order to undertake this difficult writing assignment with hardly any preparation. I am grateful to Drs. Nina Berova, Reimar Bruening, Jerrold Meinwald, Yoko Naya, and Tetsuo Shiba for their many suggestions.

## 8. BIBLIOGRAPHY

"A 100 Year History of Japanese Chemistry," Chemical Society of Japan, Tokyo Kagaku Dojin, 1978.
K. Bloch, *FASEB J.*, 1996, **10**, 802.
"Britannica Online," 1994–1998.
*Bull. Oriental Healing Arts Inst. USA*, 1980, **5**(7).
L. F. Fieser and M. Fieser, "Advanced Organic Chemistry," Reinhold, New York, 1961.
L. F. Fieser and M. Fieser, "Natural Products Related to Phenanthrene," Reinhold, New York, 1949.
M. Goodman and F. Morehouse, "Organic Molecules in Action," Gordon & Breach, New York, 1973.
L. K. James (ed.), "Nobel Laureates in Chemistry," American Chemical Society and Chemistry Heritage Foundation, 1994.
J. Mann, "Murder, Magic and Medicine," Oxford University Press, New York, 1992.
R. M. Roberts, "Serendipity, Accidental Discoveries in Science," Wiley, New York, 1989.
D. S. Tarbell and T. Tarbell, "The History of Organic Chemistry in the United States, 1875–1955," Folio, Nashville, TN, 1986.

# 7.01
# Overview

ERIC T. KOOL

*University of Rochester, NY, USA*

| 7.01.1 | DNA AS AN ORGANIC NATURAL PRODUCT | 1 |
|---|---|---|
| 7.01.1.1 | *The Structure of DNA* | 1 |
| 7.01.1.2 | *The Functions of DNA* | 2 |
| 7.01.1.3 | *DNA is an Important Synthetic and Design Problem* | 2 |
| 7.01.1.4 | *DNA Chemistry is Important to Medicine* | 4 |
| 7.01.2 | IMPORTANT NATURAL PRODUCTS INTERACT WITH DNA | 5 |
| 7.01.2.1 | *Natural Products that Bind Noncovalently* | 5 |
| 7.01.2.2 | *Natural Products that Damage DNA or Bind Covalently* | 5 |
| 7.01.2.3 | *Natural Products that Interact with DNA:Protein Interfaces* | 5 |
| 7.01.3 | THE RELEVANCE OF MOLECULAR BIOLOGY TO CHEMISTRY | 6 |
| 7.01.3.1 | *Molecular Biology as a Branch of Chemistry* | 6 |
| 7.01.3.2 | *Molecular Biology Offers Important Tools for Handling Large Molecules* | 6 |
| 7.01.3.3 | *What Chemistry Offers to Molecular Biology* | 6 |
| 7.01.4 | WHAT IS COVERED IN THIS VOLUME | 7 |
| 7.01.4.1 | *Physical Properties of DNA* | 7 |
| 7.01.4.2 | *Chemistry of DNA* | 7 |
| 7.01.4.3 | *Molecules that Bind to DNA* | 7 |
| 7.01.4.4 | *Techniques from Molecular Biology* | 8 |
| 7.01.4.5 | *Topics not Covered* | 8 |
| 7.01.5 | THE FUTURE OF DNA CHEMISTRY RESEARCH | 8 |
| 7.01.5.1 | *Natural Products that Interact with DNA* | 9 |
| 7.01.5.2 | *DNA and Analogues as Diagnostic Tools* | 9 |
| 7.01.5.3 | *DNA and Analogues as Therapeutic Agents* | 9 |
| 7.01.5.4 | *DNA as a Molecular Scaffold* | 10 |
| 7.01.5.5 | *DNA as a Ligand* | 10 |
| 7.01.5.6 | *DNA as a Catalyst* | 11 |
| 7.01.5.7 | *DNA as a Computer* | 11 |
| 7.01.6 | REFERENCES | 12 |

## 7.01.1 DNA AS AN ORGANIC NATURAL PRODUCT

### 7.01.1.1 The Structure of DNA

The structure of DNA makes it abundantly clear that this molecule shares many things in common with the smaller compounds that are normally regarded as natural products. DNA contains substructures—sugars and heterocycles—commonly found in many other classes of natural products (Figure 1). Of course, it goes without saying that DNA is in fact a natural product, that is, an organic compound that is biosynthesized in cells for a biological purpose. Despite this, however, chemists sometimes overlook DNA as a natural product.

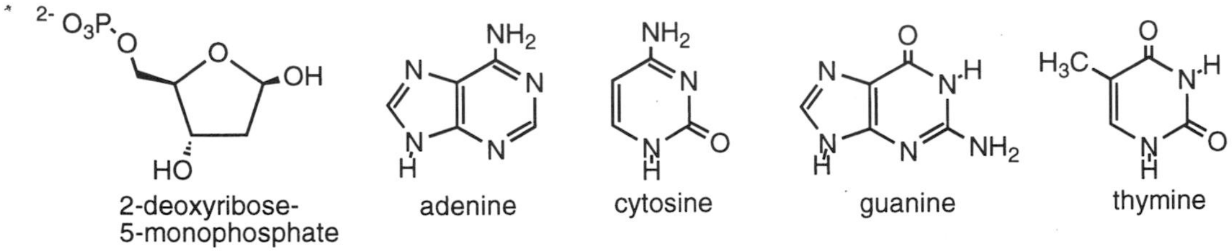

**Figure 1**  The substructures of DNA, a natural product found in all cells.

Why is this the case? It may be that large, charged, polymeric molecules tend to be overlooked in deference to less polar small organic compounds. In part, this is understandable, considering that charged molecules require different methods for synthesis and purification than are commonly used in most synthetically oriented laboratories (which commonly use solution-phase synthesis in nonpolar solvents, and purification by silica column chromatography). In addition, DNA is large enough to make routine structural assignments by NMR spectroscopy a serious undertaking. Moreover, it is usually made and handled on scales much smaller than the multimilligram scales common to most natural products chemists.

Nevertheless, a very strong case can be made for DNA to be considered a natural product very analogous to smaller, less polar molecules. DNA is biosynthesized by a secondary metabolism process, like most natural products. Like such compounds, DNA is biologically active, and DNA interacts with proteins to exert its biological activity. Perhaps the one major difference between DNA and other commonly studied natural products is the fact that DNA is central to the existence, survival, and replication of natural organisms, while many other natural products are not.

### 7.01.1.2  The Functions of DNA

DNA is most widely thought of as the central storehouse of genetic information in each cell, and of course, that is a major function. However, DNA serves several other biological functions as well. It serves also as a replicable template for the copying of cells, as a scaffold for the organization of proteins, and as a polyfunctional switch for control of gene expression. In addition, chemists have used the DNA structure as a starting point for the design of many new molecules with functions other than those listed for naturally occurring DNA. Among these new functions are the ability to act as a ligand for nucleic acids and proteins and smaller molecules, the ability to act as a scaffold for the organization of biological and nonnatural molecules, and the ability to act as sensors and reporters of genetic sequences, and even the ability to act as a computer. Some of these functions are discussed below or later in this volume.

### 7.01.1.3  DNA is an Important Synthetic and Design Problem

Because of the wide applications of DNA in chemistry, biochemistry, biology, medicine, and materials science, there is broad motivation among chemists to develop and apply the chemistry of DNA and related analogues. Early in DNA research, one of the first problems to be tackled was the task of assembling a string of nucleotides in a desired order and length. Although biochemists early on isolated enzymes that could polymerize nucleotides (Figure 2),[1,2] it was difficult or impossible to manipulate the necessary templates in order to utilize these enzymes to make a desired sequence of DNA. Thus, it was not until the development of synthetic methods for constructing specific DNA oligonucleotides on solid supports (Figure 3)[3,4] that it became practical to study defined, varied sequences with purified molecules that were homogeneous in length. Since the 1970s there has been an explosion in the number of chemists working with DNA and in chemistries for manipulating it.

It can be argued that the ability to synthesize DNA oligonucleotides has been as important to the molecular biology revolution as the development of bacterial cloning (see below). Also very important to biochemistry has been the study of DNA damage and damaging agents, and the study of protein–DNA interactions; these were also greatly aided by the ability to synthesize DNA. Thus,

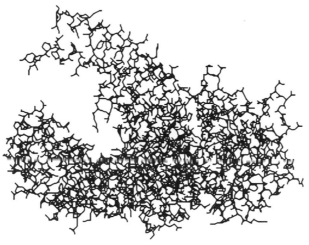

**Figure 2** The structure of the most well-studied and smallest DNA polymerase enzyme, the Klenow fragment of *E. coli* DNA Polymerase I.[1]

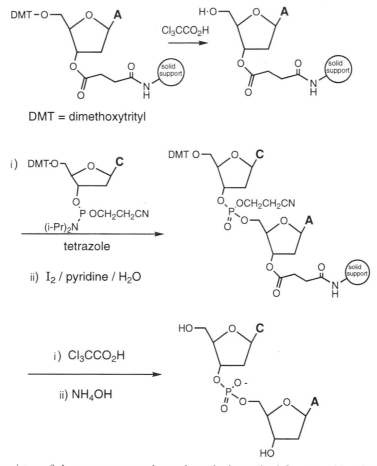

**Figure 3** The chemistry of the most commonly used synthetic method for assembly of DNA strands on a solid support.[4]

the progress made in developing the chemistry of natural DNA has been one of the most important advances in biologically-related fields in the twentieth century.

One of the beautiful aspects of chemistry is that, although chemists respect and admire natural molecules, they are not confined to making and studying only those structures. As a result, the 1980s and 1990s have seen the development of an extremely large number of synthetic modifications to DNA's natural structure (a few examples[5-8] are shown in Figure 4). The wide variety of possible applications of DNA-like molecules has made this framework the subject of many molecular design projects. In addition, the need for introducing many varied structural and functional groups into DNA and DNA-like molecules has brought up the need for the development of new synthetic methods as well. It is also worth pointing out that DNA is not only synthesized on small scales for research purposes, but oligodeoxynucleotides and analogues are also being tested in humans as therapeutic agents.[9,10] This has led to the need for process-level synthetic chemistry aimed at the production of kilogram quantities of specific compounds.[11,12] Overall, then, there are many strong motivations for the continuing development of DNA chemistry.

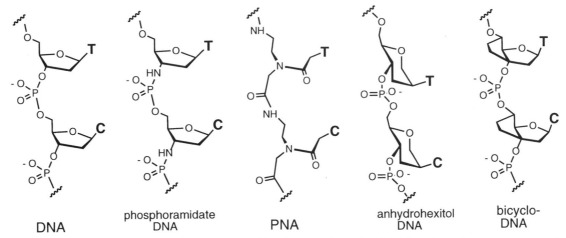

**Figure 4**   Examples of published structural modifications made to the DNA backbone (natural structure is shown on the left).[5-8] All of the analogues shown can bind nucleic acids more tightly than natural DNA can.

### 7.01.1.4   DNA Chemistry is Important to Medicine

As mentioned above, the development of chemistries for manipulating DNAs has been central to the steady growth in biochemistry and molecular biology worldwide. Scientists in these fields have made rapid progress in understanding biological processes such as cellular replication and cell death, gene expression, cell signaling pathways, and DNA damage and repair. These processes (and defects in them) have in many cases been linked directly to pathological conditions in humans. For example, a specific chemical adduct of benzo[a]pyrene to DNA (Figure 5) has been linked to a known, specific point mutation in the p53 gene,[13] which in turn has been suggested as the cause of lung cancer and other cancers.[14,15] The ongoing rapid and large-scale sequencing of DNAs,[16-19] also made possible by the development of DNA chemistry, has also led to the correlation of a growing set of genetic sequences to disease. Many examples of inherited or acquired mutations are now directly linked to human disease states; for example, a number of mutated DNA sequences thought to be responsible for various leukemias are now known.[20,21]

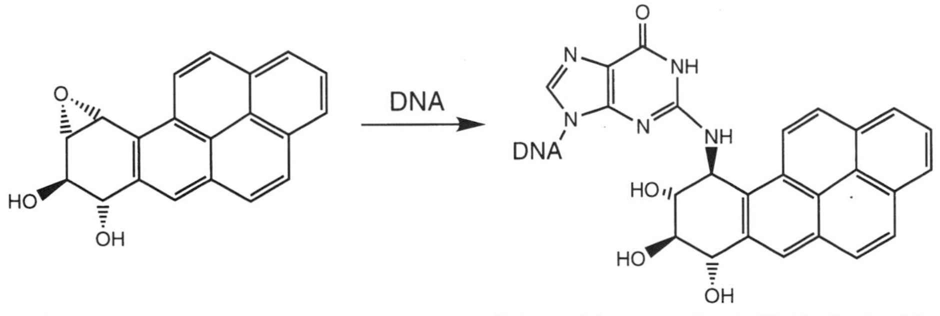

**Figure 5**   The structure of an adduct of benzo[a]pyrene diol epoxide to guanine in DNA. Such adducts in the p53 gene have been directly linked to lung cancer.[13]

In the past one might have said that the chemistry of DNA is relevant only indirectly to medicine, in that it aids research in molecular biology or the understanding of biochemical processes, but this is no longer the case. There are currently a number of clinical trials underway to test the therapeutic potential of synthetic DNA analogues in humans.[9,10] In addition, the US Food and Drug Administration has begun to approve the use of medical diagnostic methods involving nonnaturally modified DNAs for the diagnosis of human disease.

## 7.01.2  IMPORTANT NATURAL PRODUCTS INTERACT WITH DNA

### 7.01.2.1  Natural Products that Bind Noncovalently

There are many natural products that have been shown to interact with DNA by forming noncovalent complexes. Although it may not always be true, this binding may in many cases be responsible for their biological activity. The two major classes of known DNA-binding compounds are the intercalators and the minor groove binders. Note that intercalators are covered in this volume quite comprehensively in Chapter 7.12. While there is no chapter specifically covering minor groove binding, a number of examples of molecules that bind in this fashion are discussed in Chapters 7.13, 7.14, and 7.15.

### 7.01.2.2  Natural Products that Damage DNA or Bind Covalently

Many important natural products take advantage of their DNA binding properties to allow them to react with the DNA in some way. For example, it is common to find that electrophilic molecules react with nucleophilic sites in DNA bases, or that free radicals generated from natural products will react with deoxyribose moiety in DNA. Some of these reaction mechanisms are amazing in their elegance, such as in the enediyne class of natural products (Chapter 7.15). Some such reaction mechanisms result in covalent attachment of a natural product to the DNA (at least temporarily), while others never form a covalent bond with the DNA during their reaction with it. DNA-damaging agents are comprehensively discussed here in Chapter 7.14, and the chemical consequences of this damage are presented in Chapter 7.11.

### 7.01.2.3  Natural Products that Interact with DNA:Protein Interfaces

A relatively new concept in biologically influenced chemistry has been the idea of interfering at the interface between two interacting proteins or between proteins and nucleic acids. One of the most successful classes of natural products in this concept from a medical standpoint has been the topoisomerase inhibitors.[22,23] Topoisomerases are enzymes which alter the topology of DNA by breaking and closing DNA strands,[24–26] and topoisomerase inhibitors can exert biological activity by binding both a topoisomerase and its DNA target. Compounds in this class such as adriamycin (also called doxorubicin) (Figure 6) are useful clinically in the treatment of cancer.[27] Topoisomerase inhibitors are discussed in detail in this volume (see Chapter 7.16).

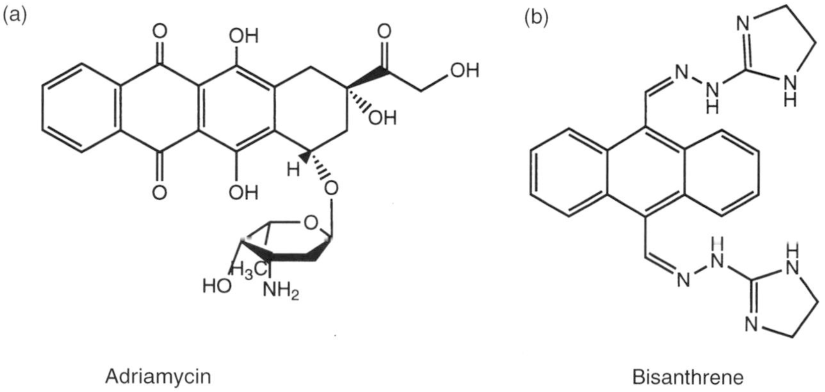

Adriamycin

Bisanthrene

**Figure 6**  The structure of two clinically important topoisomerase inhibitors, (a) adriamycin and (b) bisanthrene. See Chapter 7.16 for details.

### 7.01.3    THE RELEVANCE OF MOLECULAR BIOLOGY TO CHEMISTRY

#### 7.01.3.1    Molecular Biology as a Branch of Chemistry

The burgeoning field of molecular biology has arisen for two reasons: first, there has been a strong desire to understand biological processes on a molecular basis. This requires, in part, identifying and purifying individual proteins and nucleic acids. For example, the observation that certain proteins are made at various times in a cell's replication cycle led to the discovery of operator sequences in DNAs and to the identification of specific proteins such as repressors which can bind them and upregulate or downregulate gene expression.[28,29] This led to the complete chemical characterization of the sequence and structures of many such repressor molecules, and even to the rational design of altered DNA-binding proteins with new amino acid sequences.[30,31] The field called structural biology, which can be considered a subset of molecular biology, strives further to identify the three-dimensional structures of such large purified molecules, usually by X-ray crystallographic methods or by solution-phase NMR spectroscopy.

To a chemist, of course, the ideas of purification, chemical and structural characterization, and molecular design are not new and in fact predate the field called molecular biology. Since the goals and motives are the same, and since the molecules are in both cases organic, it is reasonable to consider molecular biology to be a branch of chemistry as well as a branch of biology. It is also therefore rational to leave the descriptor "chemistry" in the name of the field; hence the use of the term "chemical biology."

#### 7.01.3.2    Molecular Biology Offers Important Tools for Handling Large Molecules

Despite the fact that molecular biology is very closely related to chemistry, it is worth noting that scientists who call themselves molecular biologists have developed a variety of methods and tools for handling large molecules such as proteins and nucleic acids, and some of these methods are particularly useful for work with large charged compounds as opposed to small natural products. Chemists who are interested in working on such large molecules can benefit greatly from using the tools of molecular biology.

For example, there are a number of separation and purification methods that are especially well suited to handling proteins and nucleic acids, and which often do not have direct analogues in small molecule work; among these methods are gel electrophoresis, capillary electrophoresis, isoelectric focusing, affinity chromatography, and size exclusion chromatography. There are also a number of synthesis and preparation methods which are well worth learning if one is working with these large molecules. Among these methods are cloning and overexpression, site-directed mutagenesis, *in vitro* transcription, and amplification via the polymerase chain reaction (PCR). There are important identification methods as well. Among these are direct protein and DNA sequencing, Northern, Southern, and Western blots, and *in situ* hybridization. Of course, a number of analytical methods such as solution-phase synthesis, solid-phase synthesis, NMR spectroscopy, HPLC, and mass spectrometry are useful for working with both small and large molecules.

Finally, there are also a number of combinatorial methods developed for proteins and nucleic acids which are not readily applied to small molecules. Among these methods are *in vitro* selection and evolution of nucleic acids, and phage display selection of peptides and proteins.

#### 7.01.3.3    What Chemistry Offers to Molecular Biology

The above methods are examples of ways that chemistry can benefit from molecular biology. This exchange of information is, however, a two-way street: there are also important ways in which molecular biologists can benefit from the work of chemists. For example, chemists are probably more likely than biologists to develop new analytical methods for molecular purification, identification, and analysis, and yet molecular biologists can and do apply these methods to their own biomolecules of interest. Second, chemists are much more likely to undertake the synthesis of nonnatural molecules, including close analogues of natural biomolecules. Molecular biologists can use such compounds as tools for their biological study, for example, as inhibitors or as probes of molecular interactions and mechanisms.

## 7.01.4 WHAT IS COVERED IN THIS VOLUME

### 7.01.4.1 Physical Properties of DNA

Several chapters in this volume are focused on describing in detail the physical properties of DNA. Chapter 7.02 describes the thermodynamics and kinetics of DNA helix formation. The discussion covers a number of important topics such as the forces which contribute to helix formation, methods for measuring the thermodynamics of DNA helices, and algorithms for predicting the thermodynamics of DNA double helices *a priori*. Also reviewed are the kinetics of DNA helix formation and methods used to measure them.

Chapter 7.03 covers modern methods for the determination of DNA structure by nuclear magnetic resonance spectroscopy. Described are different types of NMR experiments that are helpful in defining structure as well as practical descriptions on just what information can be gained from these experiments.

Because NMR methods are currently limited at times by the size of the molecules under investigation, we felt it prudent also to include a review of other methods for probing structure in solution. Chapter 7.04 describes current chemical probing methods which are widely useful in examination of DNA structure. These methods are particularly useful in larger nucleic acids and in mapping protein-DNA interactions in solution.

### 7.01.4.2 Chemistry of DNA

Several chapters in this volume (Chapters 5–11) are concerned with the varied chemistry of DNA, ranging from its synthetic assembly to its destruction by chemically reactive species. The first two of these, Chapters 7.05 and 7.06, present in-depth coverage of modern methods for the chemical synthesis of DNA oligonucleotides as well as methods for attaching other moieties to such molecules, either during or after oligonucleotide construction.

Chapters 7–10 represent the most design-oriented chapters of the volume. Chapter 7.07 describes nonnatural nucleoside analogues which have been widely useful in probing biochemical mechanisms and interactions. Both the synthesis and the practical utility of these analogues are discussed. Chapters 7.08, 7.09, and 7.10 review the current state of the art in designing analogues of DNA for practical applications. Among the most important of these applications are the use of oligonucleotide analogues as potential therapeutics in the downregulation of specific genes (e.g., antisense oligonucleotides), and the use of oligonucleotide analogues as molecular diagnostic tools for detecting and identifying disease-related genetic sequences. Three different chemical aspects of DNA are discussed separately in these chapters. Chapter 7.08 tackles the summary of a wide body of literature on alterations made to the deoxyribose–phosphate diester backbone of DNA, with a particular eye to the synthesis and nucleic acid binding characteristics of each class of molecule. Likewise, Chapter 7.09 covers the known alterations made to the nucleobases themselves, again with emphasis on how these changes affect binding properties. Finally, Chapter 7.10 covers structural alterations of a higher order than backbone or bases. Described here are topological alterations to the DNA chain (for example, the effects of circular or knotted topologies) and the associated changes in properties, including ability to interact with proteins and nucleic acids.

The final chapter falling under the grouping of the chemistry of DNA is Chapter 7.11, which focuses on the chemistry of DNA damage. This is essentially a compendium of the reactions of DNA in aqueous solution, with a slant toward how this reactivity is relevant to disease states such as cancer. The chapter assesses various mechanisms by which reactive species such as electrophiles and radicals can initiate a cascade of further reactions in the DNA backbone or bases. People interested in the mechanisms of DNA damage will also find Chapters 7.14 and 7.15 useful; here are described specific classes of natural products that bind and react with DNA.

### 7.01.4.3 Molecules that Bind to DNA

The second central grouping of chapters in this volume falls under the heading of molecules that interact with DNA. Chapters 12 through 16 cover this topic in considerable detail. Intercalators are among the broadest classes of DNA-binding molecules; these are reviewed in Chapter 7.12. Another very broad class of DNA-binding compounds in nature are proteins and peptides; although large

DNA-binding proteins are not explicitly discussed here (see below), smaller peptide motifs (including subdomains of active proteins) for binding DNA are covered in Chapter 7.13.

Another medically important class of natural products are those that bind DNA and do damage to the bases or backbone by their reactive nature. Chapter 7.14 surveys the various known classes of reactive compounds and the reactions they promote on DNA substrate; Chapter 7.15 focuses in much more detail on one fascinating class of DNA-reactive compounds, the enediynes.

Finally, also falling under the heading of DNA-binding molecules are those that interact with the interface of proteins and DNA. Chapter 7.16 covers topoisomerase inhibitors, which are clinically important natural products. Their mechanism of action and structure–activity relationships are explored.

### 7.01.4.4   Techniques from Molecular Biology

The last two chapters of this volume are focused on techniques from molecular biology which chemists are finding increasingly useful. Chapter 7.17 covers the rapidly expanding field of DNAs which can act as ligands or as catalysts. These have been discovered using *in vitro* selection methods, which typically involve making randomized libraries of molecules, selecting for subsets which have desirable properties, and amplifying them prior to undergoing another round of selection. Other important and useful methods that are involved in this process are polymerase chain reaction (PCR) amplification of DNAs, and cloning of selected "winner" DNAs to obtain individual sequences.

Finally, Chapter 7.18 covers the application of molecular cloning to chemically oriented problems. Cloning is not only important for obtaining nucleic acid sequences, but it is also highly useful for producing preparative quantities of well-defined proteins or peptides, in natural or modified sequences. The chapter not only covers general cloning strategies which are useful, but also gives specific examples where cloning has been successfully applied to chemical problems.

### 7.01.4.5   Topics not Covered

The topic of the chemistry of DNA is much too broad to be covered in its entirety in one volume of text. Indeed, so many chemists and biochemists have worked in this field since the 1940s that the entire set of volumes from the *Comprehensive Natural Products Chemistry* series could have been filled with reviews on DNA. For that reason it was necessary to narrow the focus of this volume. We have attempted to choose topics most relevant to chemists who are interested in natural products.

Some notable DNA-related topics were not covered in this volume; for example, a comprehensive discussion of DNA-binding proteins, which is a very broad and widely studied topic, was not included. However, the molecular motifs used by proteins in recognizing DNA structure are well covered in Chapter 7.13, and references therein point to studies on whole proteins. Also not discussed in the present volume is the topic of chromatin structure; this is important because DNA in cells is, of course, organized and protected within a multiprotein complex, and changes in this complex are important for gene expression. This topic has been reviewed.[32-35] Higher-order structures of DNAs, such as triple helices,[36-38] tetrastranded structures,[39] and branched junctions,[40] are discussed in parts of several chapters in this volume, and have been reviewed in more detail elsewhere. Telomerase is also an important DNA-related topic of late; this ribonucleoprotein helps protect natural chromosomes by maintaining the ends. Good reviews on telomerase have been published.[41,42] Another important class of compounds that interact with DNA are the minor groove binders. Reviews on minor groove interactions have been written.[43] Finally, the topic of enzymatic DNA repair has received much attention, and more information on this topic is readily available.[44,45] Readers who are interested in one or more of these topics are invited to visit these other sources.

### 7.01.5   THE FUTURE OF DNA CHEMISTRY RESEARCH

The Editor of this volume would like to take this opportunity to speculate on where DNA research is heading over the coming decade. The topics covered are far from comprehensive and are necessarily biased toward my own views. However, this list represents what some chemists believe will be important applications of research on DNA-related topics, and offers a wide array of opportunities for new chemists entering the field to develop useful tools for scientists and therapies for medicine in the future.

### 7.01.5.1 Natural Products that Interact with DNA

While many natural products have already been shown to interact with DNA, it is clear that there exist in nature many more natural products that remain to be identified. Many of these will also be found to bind to DNA as well, and some of them will no doubt be important in a medicinal sense. While many new compounds will be found to bind in already well-defined motifs such as intercalation and minor groove binding, it is highly likely that new molecular motifs for DNA recognition will be discovered through this research, and new modes of DNA binding will be uncovered. An example of this has been found in the interaction of chromomycin with DNA, which was found to form a surprising complex with DNA in the minor groove (Figure 7).[46] Two chromomycin molecules were found to form a co-complex with a magnesium dication. In the future, more such novel DNA-binding modes will undoubtedly be identified. In addition, the imagination of chemists will continue unfettered, and so synthetic molecules inspired by natural products will also be shown to form novel types of complexes with DNA as well.

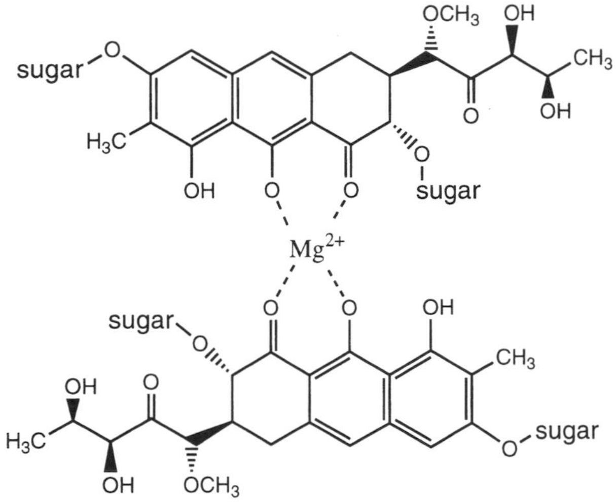

**Figure 7** Chromomycin binds DNA by forming a dimeric complex around a magnesium ion (shown above).[46]

### 7.01.5.2 DNA and Analogues as Diagnostic Tools

The development of DNA-based diagnostic tools is undergoing an explosion, and there will be many more such probes developed. An elegant example of this are the "molecular beacons" of Kramer (Figure 8),[47–49] which are beginning to be widely applied in research. Perhaps the greatest expansion in this field will be in the practical application of many of these new methods to human patients. To date, there are few DNA-based molecular diagnostic methods which are actually approved for use. The completion of the sequencing of the human genome will make available many new genetic targets for identification. Moreover, the future will see the development of rapid, high-throughput diagnostic and sequencing methodologies for individual patients. There is clearly still much chemistry to be done in this area.

### 7.01.5.3 DNA and Analogues as Therapeutic Agents

The concept of the use of small synthetic DNAs as potential therapeutics (antisense and antigene therapy) dates only from the 1970s. After years of basic research in academic labs, the 1980s and 1990s saw the undertaking of commercial development of this idea. Since the mid-1990s there have been some very promising clinical results using such agents in human patients.[9,10] If some of these

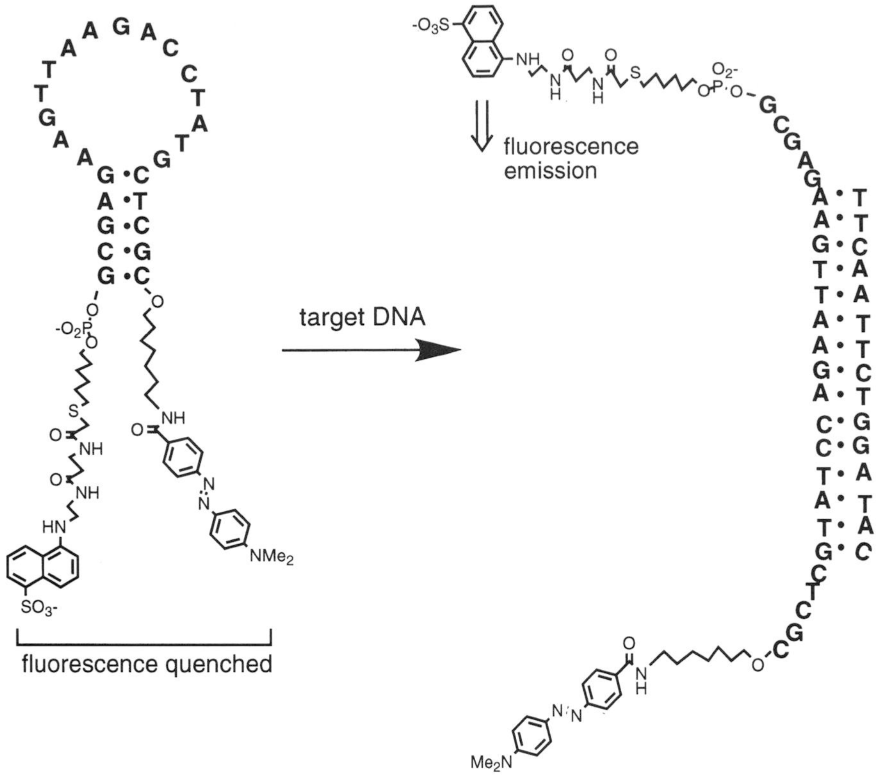

**Figure 8**  Structure of a fluorescent "molecular beacon" for sensing the presence of complementary nucleic acids in solution.[47]

compounds are approved for use, this will drive the development of an even wider number of DNA analogues for application to an increasing number of disease states. Many chemists will therefore remain involved, both in academic and industrial laboratories, in the development of new molecular structures and sequences of DNA analogues. In addition, new genetically targeted and DNA-related molecular strategies will be developed for treating disease. For example, among the newest of these in the literature are a number of strategies which go beyond simple DNA binding to actual gene correction or targeted mutagenesis,[50–52] and these will likely see increasing research over the coming years.

### 7.01.5.4  DNA as a Molecular Scaffold

The regularity and increasing predictability of DNA helical structures has led increasingly to their use as scaffolds in the organization of other molecules or assemblies of molecules. The development of a large number of convenient methods for modification of DNA and conjugation of groups to DNA has made this idea increasingly feasible to carry out. Moreover, the ability of DNA to self-assemble in solution into large, complex, and stable structures makes DNA (as opposed to other possible scaffold-like molecules) very attractive for this purpose. One example of this was the use of DNAs to organize gold nanoparticles;[53] these were use as color-changing sensors of other genetic sequences.[54] Indeed, it seems possible (and even likely) that DNA scaffolds will also be utilized in fields other than the biological or medical ones.[55] For example, materials research can make good use of a regular, predictable scaffold for the organization of molecules on surfaces.

### 7.01.5.5  DNA as a Ligand

The advent of *in vitro* selection methods for single-stranded nucleic acids in the late 1980s[56–58] has led the consensus that DNA (and RNA) may be able to act as ligands for a surprisingly wide variety of molecules. While this may have been initially surprising because of the seeming lack of varied polarity and functionality in DNA, we now know that DNA can fold to form complex shapes which allow it to bind specifically in clefts of large proteins and, in turn, to form small clefts in itself to bind small organic molecules (see Chapter 7.17). Such binding can lead to biological activity (see, for example, the published thrombin inhibitor DNAs (Figure 9)[59]); however, there is, of course, no real reason why this kind of ligand development needs to be limited to the binding of biologically derived molecules. The continuing chemical development of modifications for DNA bases and backbones which can be carried through *in vitro* selection cycles[60] will make possible the tight and specific binding of an even wider variety of molecules in the future.

(a)

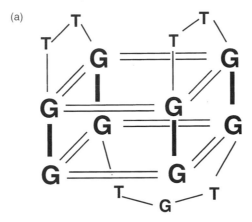

(b)

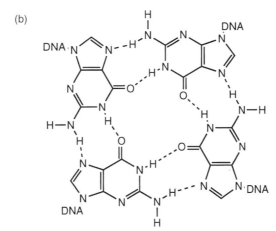

**Figure 9** Structure of a DNA molecule that binds and inhibits human thrombin.[59] (a) Folded structure is shown along with (b) chemical structure of G tetrad portion of structure.

### 7.01.5.6 DNA as a Catalyst

The *in vitro* selection combinatorial strategies described above also make possible the efficient searching for DNAs which fold to form a catalytic site. All that is required to search for catalysts (as opposed to simple ligands) is a means for separating catalytic molecules from inactive ones, and several such strategies are under rapid development. Although there is a lingering perception that natural DNA may be limited as a catalyst because of the lack of suitable functionality (see Chapter 7.17), chemists are developing new functional groups for DNA which may well solve this potential problem.[61,62] It seems possible and even probable, therefore, that synthetic chemists may in the future employ modified DNA molecules as practical catalysts for some enantioselective transformations.

### 7.01.5.7 DNA as a Computer

One of the most surprising applications of DNA research has been the proposal that DNA may be useful in computing of certain types of computationally hard problems.[63] There are now a number of publications describing the successful use of DNAs to solve problems that may have been solved considerably more slowly if a silicon-based computer had tackled them.[63,64] Although it remains to be seen how generally useful this approach will become, it certainly illustrates the broad applicability of DNA, once considered only a natural product.

## ACKNOWLEDGMENT

I thank the National Institutes of Health, the Army Research Office, the New York State Office of Science & Technology, and the Office of Naval Research for support of various aspects of our work involving modified DNAs. I further acknowledge the Dreyfus Foundation for a Teacher-Scholar Award and the Alfred P. Sloan Foundation for a Sloan Fellowship. Finally, I thank all my co-workers, present and past, for their contributions to our work.

## 7.01.6 REFERENCES

1. A. Kornberg, *Science* 1969, **163**, 1410.
2. A. Kornberg, in "DNA Replication," Freeman, San Francisco, CA, 1980, p. 101.
3. R. L. Letsinger and V. Mahadevan, *J. Am. Chem. Soc.*, 1966, **88**, 5319.
4. S. L. Beaucage and M. H. Caruthers, *Tetrahedron Lett.*, 1981, **22**, 1859.
5. S. M. Gryaznov, D. H. Lloyd, J. K. Chen, R. G. Schultz, L. A. DeDionisio, L. Ratmeyer, and W. D. Wilson, *Proc. Natl. Acad. Sci. USA*, 1995, **92**, 5798.
6. P. E. Nielsen, M. Egholm, R. Berg, and O. Buchardt, *Science*, 1991, **254**, 1497.
7. A. Van Aerschot, I. Verheggen, C. Hendrix, and P. Herdewijn, *Angew. Chem., Int. Ed. Engl.*, 1995, **34**, 1338.
8. M. Tarkoy, M. Bolli, B. Schweizer, and C. Leumann, *Helv. Chim. Acta*, 1993, **76**, 481.
9. E. Bayever, P. Iversen, L. Smith, J. Spinolo, and G. Zon, *Antisense Res. Dev.*, 1992, **2**, 109.
10. W. Roush, *Science*, 1997, **276**, 1192.
11. D. Kisner, in "12th International Round Table on Nucleosides, Nucleotides and Their Biological Applications," La Jolla, CA, 1996, OP 46.
12. D. L. Cole, V. T. Ravikumar, A. Krotz, D. C. Capaldi, Z. S. Cheruvallath, A. N. Scozzari, M. Andrade, and E. F. Gritzen, in "12th International Round Table on Nucleosides, Nucleotides and Their Biological Applications," La Jolla, CA, 1996, OP 46.
13. M. F. Denissenko, A. Pao, M-s. Tang, and G. P. Pfeifer, *Science*, 1996, **274**, 430.
14. M. Hollstein, D. Sidransky, B. Vogelstein, and C. C. Harris, *Science*, 1991, **253**, 49.
15. M. S. Greenblatt, W. P. Bennett, M. Hollstein, and C. C. Harris, *Cancer Res.*, 1994, **54**, 4855.
16. S. Beck and P. Sterk, *Curr. Opin. Biotechnol.*, 1998, **9**, 116.
17. C. M. Fraser and R. D. Fleischmann, *Electrophoresis*, 1997, **18**, 1207.
18. K. K. Murray, *J. Mass Spectrom.*, 1996, **31**, 1203.
19. M. J. O'Donnell-Maloney, C. L. Smith, and C. R. Cantor, *Trends Biotechnol.*, 1996, **14**, 401.
20. N. A. Heerema, *Cancer Invest.*, 1998, **16**, 127.
21. M. P. Strout and M. A. Caligiuri, *Curr. Opin. Oncol.*, 1997, **9**, 8.
22. U. Pindur and T. Lemster, *Pharmazie*, 1998, **53**, 79.
23. J. Cummings and J. F. Smyth, *Ann. Oncol.*, 1993, **4**, 533.
24. J. M. Berger, *Curr. Opin. Struct. Biol.*, 1998, **8**, 26.
25. A. D. Bates and A. Maxwell, *Curr. Biol.*, 1997, **7**, R778.
26. J. Tazi, F. Rossi, E. Labourier, I. Gallouzi, C. Brunel, and E. Antoine, *J. Mol. Med.*, 1997, **5**, 786.
27. M. R. Smith, W. A. Peters III, and C. W. Drescher, *Am. J. Obstet. Gynecol.*, 1994, **170**, 1677.
28. M. A. Kercher, P. Lu, and M. Lewis, *Curr. Opin. Struct. Biol.*, 1997, **7**, 76.
29. P. Youderian and D. N. Arvidson, *Gene* 1994, **150**, 1.
30. J. L. Pomerantz, C. O. Pabo, and P. A. Sharp, *Proc. Natl. Acad. Sci. USA*, 1995, **92**, 9752.
31. C. Park, J. L. Campbell, and W. A. Goddard III, *Proc. Natl. Acad. Sci. USA*, 1992, **89**, 9094.
32. A. I. Lamond and W. C. Earnshaw, *Science*, 1998, **280**, 547.
33. K. Luger and T. J. Richmond, *Curr. Opin. Struct. Biol.*, 1998, **8**, 33.
34. M. Grunstein, *Nature*, 1997, **389**, 349.
35. K. Struhl, *Genes Dev.*, 1998, **12**, 599.
36. G. E. Plum, D. S. Pilch, S. F. Singleton, and K. J. Breslauer, *Annu. Rev. Biophys. Biomol. Struct.*, 1995, **24**, 319.
37. J. S. Sun, T. Garestier, and C. Helene, *Curr. Opin. Struct. Biol.*, 1996, **6**, 327.
38. L. J. Maher III, *Bioessays*, 1992, **14**, 807.
39. D. Rhodes and R. Giraldo, *Curr. Opin. Struct. Biol.*, 1995, **5**, 311.
40. N. C. Seeman, *DNA Cell Biol.*, 1991, **10**, 475.
41. C. W. Greider, *Proc. Natl. Acad. Sci. USA*, 1998, **95**, 90.
42. T. R. Cech, T. M. Nakamura, and J. Lingner, Biokhimiya (Moscow), 1997, **62**, 1202.
43. D. E. Wemmer and P. B. Dervan, *Curr. Opin. Struct. Biol.*, 1997, **7**, 355.
44. H. E. Krokan, R. Standal, and G. Slupphaug, *Biochem. J.*, 1997, **325**, 1.
45. S. S. Parikh, C. D. Mol, and J. A. Tainer, *Structure*, 1997, **5**, 1543.
46. X. L. Gao, P. Mirau, and D. J. Patel, *J. Mol. Biol.*, 1992, **223**, 259.
47. S. Tyagi and F. R. Kramer, *Nat. Biotechnol.*, 1996, **14**, 303.
48. G. Leone, H. van Schijndel, B. van Gemen, F. R. Kramer, and C. D. Schoen, *Nucleic Acids Res.*, 1998, **26**, 2150.

49. S. Tyagi, D. P. Bratu, and F. R. Kramer, *Nat. Biotechnol.*, 1998, **16**, 49.
50. E. S. Belousov, I. A. Afonina, M. A. Podyminogin, H. B. Gamper, M. W. Reed, R. M. Wydro, and R. B. Meyer, *Nucleic Acids Res.*, 1997, **25**, 3440.
51. M. Raha, G. Wang, M. M. Seidman, and P. M. Glazer, *Proc. Natl. Acad. Sci. USA*, 1996, **93**, 2941.
52. A. Cole-Strauss, K. Yoon, Y. Xiang, B. C. Byrne, M. C. Rice, J. Gryn, W. K. Holloman, and E. B. Kmiec, *Science*, 1996, **273**, 1386.
53. R. Elghanian, J. J. Storhoff, R. C. Mucic, R. L. Letsinger, and C. A. Mirkin, *Science*, 1997, **277**, 1078.
54. C. A. Mirkin, R. L. Letsinger, R. C. Mucic, and J. J. Storhoff, *Nature*, 1996, **382**, 607.
55. J. H. Chen and N. C. Seeman, *Nature*, 1991, **350**, 631.
56. C. Tuerk and L. Gold, *Science*, 1990, **249**, 505.
57. D. L. Robertson and G. F. Joyce, *Nature*, 1990, **344**, 467.
58. A. D. Ellington and J. W. Szostak, *Nature*, 1990, **346**, 818.
59. L. C. Bock, L. C. Griffin, J. A. Latham, E. H. Vermaas, and J. J. Toole, *Nature*, 1992, **355**, 564.
60. J. A. Latham, R. Johnson, and J. J. Toole, *Nucleic Acids Res.*, 1994, **22**, 2817.
61. J. K. Bashkin, J. K. Gard, and A. S. Modak, *J. Org. Chem.*, 1990, **55**, 5125.
62. G. Wang and D. E. Bergstrom, *Tetrahedron Lett.*, 1993, **34**, 6721.
63. L. M. Adleman, *Science*, 1994, **266**, 1021.
64. F. Guarnieri, M. Fliss, and C. Bancroft, *Science*, 1996, **273**, 220.

# 7.02
# Thermodynamics and Kinetics of Nucleic Acid Association/ Dissociation and Folding Processes

G. ERIC PLUM and KENNETH J. BRESLAUER
*Rutgers University, Piscataway, NJ, USA*

and

RICHARD W. ROBERTS
*California Institute of Technology, Pasadena, CA, USA*

| | | |
|---|---|---:|
| 7.02.1 | NUCLEIC ACID THERMODYNAMICS | 16 |
| | 7.02.1.1 *Importance of Thermodynamic Knowledge of Nucleic Acid Systems* | 16 |
| | 7.02.1.2 *Meaning of Thermodynamic and Extrathermodynamic Parameters* | 16 |
| | 7.02.1.2.1 *Thermodynamic parameters* | 16 |
| | 7.02.1.2.2 *Extrathermodynamic parameters* | 18 |
| | 7.02.1.3 *Equilibrium Constants and Free Energy* | 20 |
| | 7.02.1.4 *Experimental Determination of Thermodynamic Parameters* | 21 |
| | 7.02.1.4.1 *The van't Hoff enthalpy and noncalorimetric methods* | 21 |
| | 7.02.1.4.2 *Differential scanning calorimetry* | 28 |
| | 7.02.1.4.3 *Isothermal titration calorimetry* | 31 |
| | 7.02.1.4.4 *Measurement of volumetric parameters* | 31 |
| | 7.02.1.5 *Interpretation and Application of Thermodynamic Parameters in Context of Various Nucleic Acid Systems* | 32 |
| | 7.02.1.5.1 *Single strands* | 32 |
| | 7.02.1.5.2 *Duplexes* | 33 |
| | 7.02.1.5.3 *Unusual duplex and higher-order structures* | 36 |
| 7.02.2 | NUCLEIC ACID KINETICS | 36 |
| | 7.02.2.1 *Obtaining and Analyzing Kinetic Data on DNA and RNA* | 37 |
| | 7.02.2.1.1 *Relaxation kinetics* | 37 |
| | 7.02.2.1.2 *Mixing kinetics* | 39 |
| | 7.02.2.1.3 *Steady-state kinetics* | 41 |
| | 7.02.2.2 *Interpretation of Kinetic Data* | 41 |
| | 7.02.2.2.1 *Temperature dependence of the rate* | 41 |
| | 7.02.2.2.2 *Reaction order and rate laws* | 41 |
| | 7.02.2.2.3 *Diffusion controlled reactions* | 42 |
| | 7.02.2.3 *Rates of Processes in Nucleic Acids* | 42 |
| | 7.02.2.3.1 *Mononucleotide association in nonaqueous solvents* | 42 |
| | 7.02.2.3.2 *Single-strand stacking and unstacking* | 42 |
| | 7.02.2.3.3 *Formation and dissociation of double helices* | 43 |
| | 7.02.2.3.4 *Formation and denaturation of hairpin helices* | 44 |

7.02.2.3.5    Kinetics of base-pair opening and proton exchange                        45
7.02.2.3.6    Association of substrates with ribozymes                                 45
7.02.2.3.7    Association rates of loop–loop complexes                                 46
7.02.2.3.8    Kinetics of triple helix formation and dissociation                      46
7.02.2.3.9    Higher-order folding kinetics of RNA                                      48
    7.02.2.4    Conclusions                                                             49

7.02.3    REFERENCES                                                                    50

## 7.02.1    NUCLEIC ACID THERMODYNAMICS

### 7.02.1.1    Importance of Thermodynamic Knowledge of Nucleic Acid Systems

Most biological processes involve specific recognition and complex formation between macro-molecules, be they proteins, nucleic acids, or carbohydrates, and other macromolecules or bio-logically active small molecules. A large number of these recognition events involve nucleic acids. Phenomena involving nucleic acids that depend on specific recognition events[1] include the stability of the DNA double helix, its replication and recombination, regulation of gene activity, translation of information encoded in DNA into RNA, and the subsequent transcription of RNA to synthesize proteins. While these recognition events are understood in a descriptive sense, they are not well understood at the level of molecular interactions and the forces which stabilize intermolecular contacts. Knowledge of the thermodynamics of these systems coupled with the growing knowledge of the structures assumed by biological macromolecules should allow for a more profound under-standing of the origins of biological recognition events. With developments in nucleic acid synthetic chemistry,[2,3] spectroscopic[4–8] and calorimetric instrumentation,[9,10] and related methodological advances, the thermodynamic characterization of these recognition events is well under way. The ultimate goal of nucleic acid thermodynamic investigations is the development of the capacity to predict the formation and stability of nucleic acid structures and their complexes with proteins under a wide variety of environmental conditions. In addition to the ability to explain in energetic terms biologically significant recognition and binding events, the thermodynamic characterization of nucleic acid hybridization should result in significant improvement in the systematic design of oligonucleotide (and analog) probes for applications in biotechnology, diagnostics, and therapeutics.

In the first part of this chapter we will focus on the issues associated with nucleic acid hybridization. We define hybridization operationally as an association reaction between two (or more) nucleic acid strands. This review includes brief discussions of the thermodynamic origins of nucleic acid stability and the forces contributing to this stability. In addition, we provide critical evaluation of the methodologies employed to evaluate the thermodynamic origins of nucleic acid stability. We also discuss the assumptions inherent in these methods, as well as reasonable interpretations of the resultant data in light of these assumptions. As part of this discussion, we describe the use of the existing thermodynamics databases to evaluate relative binding affinities for optimizing oligo-nucleotide probe design. We also describe some emerging areas which are not addressed in the existing nucleic acids thermodynamics databases. In addition, we include discussion of some aspects of oligonucleotide design which have significant impact on application of the techniques discussed here.

### 7.02.1.2    Meaning of Thermodynamic and Extrathermodynamic Parameters

In this section, we discuss various thermodynamic and extrathermodynamic parameters, what they represent, and how they can be interpreted in terms of oligonucleotide hybridization processes.

#### 7.02.1.2.1    *Thermodynamic parameters*

*(i)    Free energy change*

The standard free energy change, $\Delta G^\circ$, for a process is the difference in free energy between the state of interest and the standard reference state. In studies of nucleic acid hybridization this is the parameter which quantifies the thermodynamic stability of the nucleic acid complex relative to the single strand reactants. The free energy change associated with nucleic acid complex formation is influenced by, in addition to the properties of the molecules of interest, essentially all environmental variables; these include, but are not limited to, temperature, pressure, and the concentrations of salts and other solution components. Separation of these effects from the intrinsic properties of the molecules which promote complex formation is not a trivial exercise. The origins of nucleic acid complex stability in terms of the molecular forces involved is still controversial.[11,12] Some insight into the origins of free energy changes can be derived from examination of the component enthalpy and entropy terms. Stacking interactions clearly dominate the enthalpy term; however, significant contributions may be derived from hydrogen bonding and solvent interactions. Conformational restrictions, reconfiguration of hydrogen bonds, reorganization of water of hydration, and counterion release probably are dominant contributions to the entropy terms. Because of its simple relation to the equilibrium association (dissociation) constant, $K_{eq}$, the standard free energy change, $\Delta G^\circ$, is accessible experimentally by a variety of methods.

### (ii)  Enthalpy change

The standard enthalpy change, $\Delta H^\circ$, associated with a process represents the temperature sensitivity of $\Delta G^\circ$. Because, under the circumstances normally encountered in nucleic acid hybridization processes, the pressure/volume work is minimal, and because usually for hybridization processes no chemical bonds are formed or broken, $\Delta H^\circ$ reflects primarily changes in the thermal energy of the system. As mentioned above in connection with $\Delta G^\circ$, the primary contribution to $\Delta H^\circ$ for nucleic acid complexation comes from stacking interactions. These stacking interactions in turn are dependent on a variety of forces, including van der Waals contacts, electrostatic interactions, dispersion forces, etc. The dependence of $\Delta H^\circ$ on solution conditions often can be exploited to understand in some detail the origins of complex stability. An example comes from Lohman et al.,[13] who interpret the salt concentration dependence of the enthalpy of protein/nucleic acid complex stability (although the principle can be applied more widely). When no salt concentration dependence of $\Delta H^\circ$ is observed, one can reasonably conclude that the process under investigation involves nonspecific interactions of fully hydrated counterions. In contrast, a salt-dependent $\Delta H^\circ$ value implies specific interactions of counterions with the protein or nucleic acid. As we will see later, the connection between $\Delta G^\circ$ and $K_{eq}$ can be exploited to measure $\Delta H^\circ$ via the van't Hoff equation. Alternatively, direct measurement of the heats of processes by calorimetric methods gives a model-independent measure of $\Delta H^\circ$.

### (iii)  Entropy change

The standard entropy change, $\Delta S^\circ$, is reflective of the conformational restrictions in the duplex relative to the isolated single strands, as well as their differential hydration and differential counterion association. Also, manifest in the entropy term are statistical effects of loops (important in DNA as well as RNA) and of the identity and number of strands involved in the complex. In contrast to the relative ease with which $\Delta H^\circ$ and $\Delta G^\circ$ are determined for a hybridization process, $\Delta S^\circ$ is difficult to obtain accurately.

### (iv)  Heat capacity change

The change in heat capacity at constant pressure, $\Delta C_p^\circ$, associated with a transition quantifies the difference in energy required to heat the reactants and products. As such, it reflects the dependence of $\Delta H^\circ$ and $\Delta S^\circ$ on temperature. Thus, knowledge of $\Delta C_p^\circ$ is necessary for the extrapolation of $\Delta H^\circ$, $\Delta S^\circ$, and $\Delta G^\circ$ to temperatures other than those at which these parameters were measured. Sturtevant[14] has discussed the microscopic origins of heat capacity effects in biological macromolecules, with particular emphasis on molecular vibrations. Other work, primarily from the Record laboratory,[15,16] has related $\Delta C_p^\circ$ to changes in molecular surface area exposed to solvent during a transition. Solution conditions can affect measured values of $\Delta C_p^\circ$ significantly.[17]

*(v)  Volumetric properties*

Recent advances in the sensitivity of instrumentation[18] have made available information on the volumetric properties of nucleic acid complexes which, when coupled with the information derived from the thermal properties, provide access to thermodynamic characterization of these systems with unprecedented completeness. In particular, examination of the adiabatic compressibility, $K_s$, of waters in the hydration shell of a nucleic acid complex can give insight into the quantity and physiochemical quality of hydrating water.[19] In addition, studies of the hydration of counterions, drugs, and proteins in their nucleic acid bound and unbound states can give significant insight into the thermodynamic driving forces of the binding processes.

*(vi)  Standard states and common reference states*

As a practical matter, one cannot measure thermodynamic quantities for nucleic acid hybridization under standard conditions as conventionally defined (1 M concentrations of reactants and products, behaving as if at infinite dilution, at 25 °C and 1 atm). Because these specific conditions are defined arbitrarily, one may choose for the standard reference conditions any reasonable set of conditions. Therefore, it is typical for studies of nucleic acid complexation processes to use internally consistent reference states. While we use the conventional dilute solution standard state notation (°), be advised that here and in virtually all thermodynamic studies of nucleic acids the reference state is defined operationally.

The various parameters which describe a thermodynamic state are represented as differences between the properties of the state of interest and those of a well-defined reference state. While comparisons of data within these studies is straightforward, intercomparisons between studies becomes problematic. Because in applying these data, we are interested only in relative values; i.e., the stability of one DNA duplex vs. another under identical solution conditions, we can avoid this problem by interpreting these data only in terms of differences between the apparent thermodynamic values, $\Delta\Delta G°$, $\Delta\Delta H°$, etc. Deviations from the behavior of the system in the differently defined standard states are assumed to be negated, at least partially, when such differences are computed, thus making intercomparisons between studies more reasonable. This assumption often is inappropriate; therefore, quantitative comparisons between studies with differently defined reference states should be approached with caution.

Figure 1 illustrates some of the issues which arise with respect to standard states in comparison of thermodynamic parameters within a given study and between studies. As described above, investigators usually are most interested in the influence that some small perturbation exerts on the thermodynamic behavior of the nucleic acid complex; therefore, values of $\Delta\Delta X°$, where $X = G$, $H$, $S$, etc., provide appropriate comparisons. The figure is cast in terms of $\Delta G°$; however, the same issues are applicable to changes in any thermodynamic state function. Figure 1(a) represents the changes in free energy associated with transitions between duplex and single stranded states for two duplexes which differ in some way. Because the final single stranded states are thermodynamically equivalent, differences in free energy changes ($\Delta\Delta G°$) reflect thermodynamic differences in the initial duplex states. The common single strand thermodynamic reference state need not be the conventional standard state. Because two processes have a common reference state, the difference between that state and the standard state is identical for both processes. Therefore, when $\Delta\Delta G°$ is calculated the deviation from the standard state is subtracted out. Figure 1(b) represents a case encountered when both the initial and the final states are different. In this case, apparent differences in $\Delta G°$ are not meaningful. This situation might arise when the thermal energy of the system ($kT$) is not sufficient to disrupt fully single strand structures. Alternatively, poor experimental design where solution conditions are not properly matched obviates comparisons of derived thermodynamic parameters.

### 7.02.1.2.2  *Extrathermodynamic parameters*

*(i)  Thermal melting temperature*

In addition to the thermodynamic properties described above, we also consider $T_m$, the thermal melting temperature. The $T_m$, which is defined as the temperature at which one half of a complex is

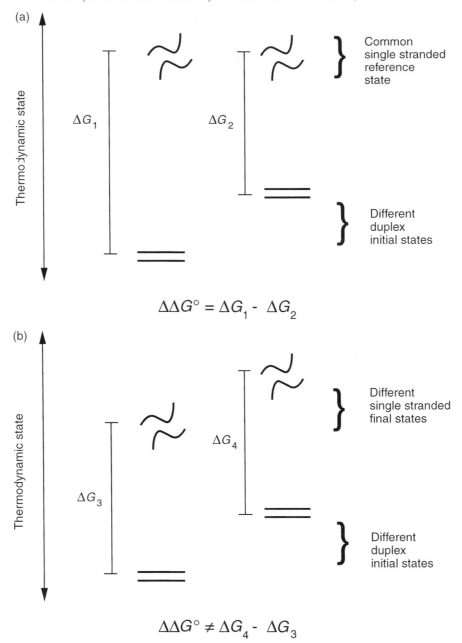

**Figure 1**   (a) Calculation of differences in free energy change associated with two duplex melting transitions based on a common thermodynamic reference state. (b) Failure to attain a common reference state forestalls useful comparison of the free energy changes associated with two duplex melting transitions.

disrupted, is a measure of thermal stability, not thermodynamic stability ($\Delta G^{\circ}$). This is an important distinction because, in the absence of additional data, relative $T_m$ values are not necessarily good predictors of the relative stabilities ($\Delta\Delta G^{\circ}$) at temperatures well removed from $T_m$. If $\Delta H^{\circ}$ for two transitions differs significantly, the temperature dependence of complex stability differs significantly. Direct comparison of $T_m$ does not account for this frequently encountered possibility. The $T_m$ of a nucleic acid complex depends on solution conditions and (usually) concentration, therefore, comparisons must be made with identically prepared solutions.

### (ii) Cooperative unit size

For any biopolymer conformational transition, one can define the cooperative unit as the number of monomers that behave as a single thermodynamic entity. Comparison of $\Delta H^{\circ}$ values obtained by model-independent calorimetric methods with those obtained by model-dependent methods permits evaluation of the size of the cooperative unit. Assessment of the cooperative unit size is important for two reasons. First, meaningful interpretation of noncalorimetric $\Delta H°$ values is dependent on knowledge of the thermodynamic entity to which they refer; that is, the size of the "mole" unit. Second, changes in the cooperative unit size associated with lesions, mismatches, or other anomalous structures provide insight into how such defects alter the ability of the polymer chain to communicate between sites through cooperative effects.

### 7.02.1.3 Equilibrium Constants and Free Energy

For any equilibrium in dilute solution, we can define an equilibrium constant, $K_{eq}$, as the ratio of the product and reactant concentrations

$$K_{eq} = \frac{[\text{products}]}{[\text{reactants}]} \tag{1}$$

To be rigorous we should use activities rather than concentrations; however, for biochemical applications this approximation is rarely the primary source of experimental error. The name equilibrium constant is something of a misnomer; the value of $K_{eq}$ will depend on the experimental conditions. In general, $K_{eq}$ will depend on temperature and pressure. Usually, in aqueous solution the equilibrium is sensitive to temperature over the range conveniently accessible by experiment. To apply pressures sufficient to perturb equilibria in aqueous solutions usually requires very specialized equipment.[20,21]

Note that the above formalism for equilibrium constants is quite general, although some cases are of particular interest to the biophysical community. The reactants and products could be a protein (reactant), its cofactor (reactant) and its complex (product), a protein in its native (reactant) and denatured (product) states, a DNA duplex (reactant) and its constituent single strands (products). Thus, one can assess association constants (binding constants) and order–disorder transition constants via the same formalism.

If one assumes or demonstrates that the process under consideration is reversible, then the equilibrium can be studied by experimental consideration of either the association or the dissociation process; the equilibrium constants describing these processes will be simple reciprocals of each other.

A common error is to assume that accounting for the states of the macromolecules alone defines the equilibrium. It is vitally important to include the small molecules, ions, etc. which participate in the reactions.[22,23] Most macromolecular association reactions involve uptake or release of cations and/or anions of the supporting electrolyte, hydrogen ions, and water. The effects of ion uptake/release, protonation/deprotonation, and hydration/dehydration of surfaces can contribute significantly to the thermodynamics of the processes of interest and thus must be included in one's analysis.

Due to the highly anionic nature of nucleic acids, one of the major contributors to hybridization complex stability is the association of cations. Due to its high degree of negative charge, the concentration of cations in the vicinity of a nucleic acid polymer can be much greater than the bulk solution concentration of cations, to which it is relatively insensitive. This is the so-called counterion condensation effect. Several formalisms have been developed to address the issue of cation binding to nucleic acids.[22,24–29] While they differ significantly in their theoretical underpinnings, all of these formalisms agree in the significant feature that the nonspecific interaction of cations with oligo- and polynucleotides is not governed by the "law of mass action." The array of cations whose binding is reasonably well described by nonspecific binding models includes simple monovalent metal ions, $Na^+$, $K^+$, etc., and multivalent cations such as polyamines, hexaamminecobalt(III), and oligolysines. It is important to note that for oligonucleotides, end effects can significantly lower the amount of cation association.[30,31] Cations also can bind noncovalently to nucleic acids in a site-specific manner. Such cations include some metal ions, a wide variety of drugs, including intercalators and minor groove binders, as well as peptides and proteins. Even $Na^+$ and $K^+$ will bind to specific sites under some circumstances, such as in the formation of guanine tetraplex DNA.[32] While site-specific binding events generally have a significant electrostatic component, other forces tend to dominate the association free energy. It should be noted that anion effects also contribute to the hybridization process; this contribution is indirect and due primarily to the anion influence on the cations.

The formation of canonical Watson–Crick duplexes, under ordinary pH conditions, does not involve a change in the protonation state of the nucleic acids, either in their duplex or single stranded states. Protonation events, however, can be important in equilibria involving damaged nucleotides, triple helical complexes, and a variety of other unusual nucleic acid complexes. Because protonation events lead to changes in the net charge of the oligonucleotide complex, protonation and counterion binding are coupled.[33] This can lead to nucleic acid complex stability which displays complicated dependence on salt and hydrogen ion concentrations.

It is becoming increasingly apparent that macromolecular structure is strongly coupled to structured water at the macromolecule–solvent interface.[34] When the surfaces of two molecules interact in aqueous solution, whether the hydration waters are stripped from the surfaces or the surfaces interact through water-mediated contacts, there is a necessary change in the character and/or extent of hydration of those surfaces. The organization of water molecules around a macromolecule complex can stabilize the complex significantly. In addition, the release of bound water upon complex formation can provide a large favorable entropy change to further promote complex stability.

An equilibrium constant such as defined above can be related to the standard state free energy change of the reaction by the relationship

$$\Delta G^\circ = -RT \ln K_{eq} \tag{2}$$

where $R$ is the gas constant (8.31 J K$^{-1}$ mol$^{-1}$ or 1.98 cal K$^{-1}$ mol$^{-1}$) and $T$ is the temperature in Kelvin units. Because of the logarithmic relation between $K_{eq}$ and $\Delta G^\circ$, small changes in $\Delta G^\circ$ correspond to large changes in $K_{eq}$. Thus, when comparing two hybridization events, small differences in the binding energetics ($\Delta G^\circ$, and its component $\Delta H^\circ$ and $\Delta S^\circ$) can result in very large differences in the amount of hybridized complex which is formed.

### 7.02.1.4 Experimental Determination of Thermodynamic Parameters

In this section, we explore the interrelationships between the various parameters already discussed and how their dependences on environmental variables, with special emphasis on temperature, can be exploited in order to measure them. First, we discuss methods for thermodynamic evaluation of order–disorder transitions by analysis of equilibrium melting profiles. These methods for measuring the thermodynamic parameters of interest are quite general and may be applied to a wide variety of primary data. This wide applicability is tempered by several assumptions which must hold true for the analysis to be valid. Therefore, while convenient, these model-dependent methods are less reliable than direct, model-independent calorimetric methods.

The most information rich methods for studying nucleic acid thermodynamics are based on the direct measurement of heats of association or dissociation by calorimetric techniques.[9,35] In addition to its well-known contributions to the development of thermodynamics and physical organic chemistry, calorimetry has a long history of contribution to the biochemical sciences. Only with the development of exquisitely sensitive instrumentation has calorimetry been applied to nucleic acid hybridization reactions. In later sections, we briefly describe applications to nucleic acid hybridizations of the two most powerful calorimetric methods, differential scanning calorimetry (DSC) and isothermal mixing calorimetry (ITC).

### 7.02.1.4.1 *The van't Hoff enthalpy and noncalorimetric methods*

#### (i) *The van't Hoff enthalpy, $\Delta H^\circ_{vH}$*

As described above, an equilibrium constant can be defined for a variety of systems and it can be related to free energy changes associated with the process of interest. In this section, we will describe how, in the absence of direct calorimetric characterization, one can explore the thermodynamic origins of this free energy change. Recall that the free energy change can be related to the associated enthalpy and entropy changes, at constant temperature,

$$\Delta G^\circ = \Delta H^\circ - T\Delta S^\circ \tag{3}$$

What happens to the equilibrium constant as the temperature is varied? From Equation (2), we can write the following relation:

$$\frac{\partial}{\partial T}\left(\frac{\Delta G^\circ}{T}\right)_{\mathrm{p}} = -R\left(\frac{\partial \ln K_{\mathrm{eq}}}{\partial T}\right)_{\mathrm{p}} \qquad (4)$$

From Equation (3), with the assumption that $\Delta H^\circ$ is independent of temperature (i.e., $\Delta C^\circ_{\mathrm{p}} = 0$), we can write a second relation:

$$\frac{\partial}{\partial T}\left(\frac{\Delta G^\circ}{T}\right)_{\mathrm{p}} = \frac{-\Delta H^\circ}{T^2} \qquad (5)$$

Combining Equations (4) and (5), we get the van't Hoff equation:

$$\Delta H^\circ_{\mathrm{vH}} = RT^2\left(\frac{\partial \ln K_{\mathrm{eq}}}{\partial T}\right)_{\mathrm{p}} \qquad (6\mathrm{a})$$

or

$$\Delta H^\circ_{\mathrm{vH}} = -R\left(\frac{\partial \ln K_{\mathrm{eq}}}{\partial(1/T)}\right)_{\mathrm{p}} \qquad (6\mathrm{b})$$

Thus, a plot of $\ln K_{\mathrm{eq}}$ vs. $1/T$ has a slope of $-\Delta H^\circ_{\mathrm{vH}}/R$. Armed with the ability to determine the equilibrium constant, one can measure $K_{\mathrm{eq}}$ as a function of temperature and, by application of Equation (6a) or (6b), determine the enthalpy change associated with the process. Using this information with Equation (3) provides access to the entropy change associated with the process. When put into practice, care must be exercised to avoid certain statistical deficiencies of this method.[36]

As described above, when determined as a function of temperature, $K_{\mathrm{eq}}$ values can be used in the van't Hoff equation (Equations (6a) and (6b)) to evaluate the enthalpy change associated with a binding reaction. Alternatively, many equilibria (both binding and order/disorder transitions) can be studied by use of some agent to perturb the equilibrium and thus the relative amounts of molecules in the initial and final states. Temperature is a particularly convenient agent of perturbation; the following sections describe how temperature-induced transitions can be exploited to provide thermodynamic data.

*(ii)  Using transition curves to extract thermodynamic data*

Any observable which reflects in a linear fashion the extent of a reaction as a function of temperature may be used to extract van't Hoff thermodynamic parameters. Through the course of a temperature-induced dissociation or order/disorder transition one can define an apparent equilibrium constant by relating the fractional extent of the reaction to the fractional change in the observable. While UV absorbance is most commonly employed, observables which may be utilized include a variety of optical spectroscopies (circular dichroism, fluorescence, infrared, etc.) as well as nonoptical techniques (NMR, etc.). There are several reasons for the widespread use of UV absorbance:[37] the amount of macromolecule required is small, the change in absorbance reflects primarily global transitions of the macromolecule, and the instrumentation required is readily available. In Figure 2 we present a typical UV absorbance monitored melting profile of a DNA oligomer. Note that at temperatures below and above the region of thermal denaturation, the dependence of absorbance on temperature yields linear but not parallel baselines. At the high temperatures above the melting transitions, the final single stranded nucleic acid state is believed to be completely disordered. Thus, two complexes generally are compared relative to a final state in which the constituent single strands of both complexes are assumed to be thermodynamically equivalent.

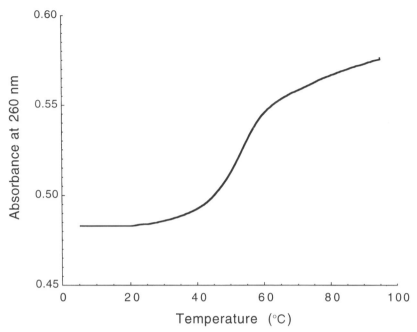

**Figure 2**   Absorbance monitored thermally induced dissociation of the A-G mismatch containing DNA oligonucleotide duplex d(GCGTACGCATGCG)·d(CGCATGAGTACGC). For experimental details, see Plum *et al.*[38]

The analysis of thermally induced transition curves depends on the number of molecules that participate in the process; the so-called molecularity. We will consider two cases: a monomolecular process, in which the initial state is a single molecule as is the final state; and a bimolecular process, in which the initial state is a complex of two molecules and the final state is the two isolated molecules. An example of a monomolecular process is the denaturation of a single oligonucleotide chain containing self-structure, such as a hairpin, for which a single thermally induced transition is observed. An example of a bimolecular process is the disruption of a DNA duplex to form two single strands. Below we will evaluate the dependence of $K_{eq}$ on the extent of reaction and use this dependence to develop an approach for extracting van't Hoff thermodynamic parameters from "melting" transition data. The mono- and bimolecular cases will be developed in parallel. This analysis is readily extended to systems of higher molecularity.[39]

The parameter most easily extracted from thermally induced melting profiles is the thermal stability ($T_m$), which is the temperature at which the process is half complete. While, in addition to $T_m$, melting profiles contain information about the changes in free energy ($\Delta G^\circ$), enthalpy ($\Delta H^\circ$), entropy ($\Delta S^\circ$), and heat capacity ($\Delta C_p^\circ$) associated with a denaturation process, the enthalpy change is obtained most reliably from these data. Therefore, we emphasize enthalpy changes in the following discussion.

### (iii) Extraction of $\Delta H_{vH}^\circ$ from the shape of a transition curve

To begin, we define a parameter $\alpha$, which is the fraction of molecules/complexes in their initial state. In Figure 3 we show the $\alpha$ vs. $T$ curve derived from analysis of the temperature dependent UV absorbance profile shown in Figure 2. The $\alpha$ parameter is computed, at each temperature, by the ratio of two terms—the difference between the extrapolated posttransition baseline and the absorbance signal divided by the difference between the pretransition and posttransition baselines. Note that the equations for the bimolecular case shown below are derived in terms of association constants; this is arbitrary and involves only a change in sign for $\Delta H^\circ$, $\Delta S^\circ$, and $\Delta G^\circ$ relative to the dissociation reaction. Further note that in the reaction models shown below only two thermo-dynamic states are represented, the initial and final states, with no intermediate states being considered. This so-called all-or-none or two-state approximation is not always valid. Intermediate states in which some base pairs (triples, etc.) are not fully formed must be considered when oligonucleotide complexes of moderate length are under investigation. While partially base paired intermediate states can be included in models of these transitions,[40,41] such states are explicitly ignored in most treatments of nucleic acid melting data, including those which follow. Also, note that the bimolecular process described by the following equations is one in which the two strands are different. Due to a statistical effect, self-complementary duplexes require slightly modified equations.[39,40]

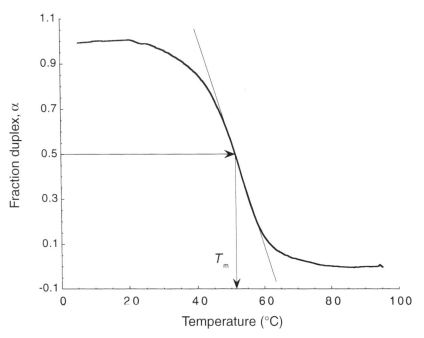

**Figure 3** Temperature dependence of the fraction of duplex, $\alpha$, computed from the data presented in Figure 1. The slope of the line drawn tangent to the melting curve can be used to estimate $\Delta H_{vH}^{\circ}$ by application of Equation (13b).

<table>
<tr><td>*Monomolecular case*</td><td>*Bimolecular case*</td></tr>
</table>

Define reaction

$$A_N \leftrightarrow A_D \qquad (7a)$$

$$A + B \leftrightarrow AB \qquad (7b)$$

Define a total concentration

$$C_t = [A] \qquad (8a)$$

$$C_t = [A] + [B] \qquad (8b)$$

Define equilibrium constant in terms of [reactant] and [product]

$$K_{eq} = \frac{[A_N]}{[A_D]} \qquad (9a)$$

$$K_{eq} = \frac{[AB]}{[A][B]} \qquad (9b)$$

Define equilibrium constant in terms of α, the fraction of reactant in the initial state

$$K_{eq} = \frac{\alpha}{(1-\alpha)} \qquad (10a)$$

$$K_{eq} = \frac{\alpha\left(\frac{C_t}{2}\right)}{\left((1-\alpha)\left(\frac{C_t}{2}\right)\right)^2} = \frac{\alpha}{(1-\alpha)^2\left(\frac{C_t}{2}\right)} \qquad (10b)$$

Substitute into the van't Hoff equation

$$\Delta H_{vH}^\circ = RT^2 \frac{\partial}{\partial T}\left[\ln\left(\frac{\alpha}{(1-\alpha)}\right)\right] \qquad (11a)$$

$$\Delta H_{vH}^\circ = RT^2 \frac{\partial}{\partial T}\left[\ln\left(\frac{\alpha}{(1-\alpha)^2\left(\frac{C_t}{2}\right)}\right)\right] \qquad (11b)$$

Evaluate the above expressions

$$\Delta H_{vH}^\circ = RT^2 \left(\frac{2}{(1-\alpha)}\right)\frac{\partial \alpha}{\partial T} \qquad (12a)$$

$$\Delta H_{vH}^\circ = RT^2 \left(\frac{(1+\alpha)}{\alpha(1-\alpha)}\right)\frac{\partial \alpha}{\partial T} \qquad (12b)$$

At $T_m$, $\alpha = \frac{1}{2}$ by definition

$$\Delta H_{vH}^\circ = 4RT_m^2\left[\frac{\partial \alpha}{\partial T}\right]_{T=T_m} \qquad (13a)$$

$$\Delta H_{vH}^\circ = 6RT_m^2\left[\frac{\partial \alpha}{\partial T}\right]_{T=T_m} \qquad (13b)$$

or

$$\Delta H_{vH}^\circ = -4R\left[\frac{\partial \alpha}{\partial(1/T)}\right]_{T=T_m} \qquad (14a)$$

$$\Delta H_{vH}^\circ = -6R\left[\frac{\partial \alpha}{\partial(1/T)}\right]_{T=T_m} \qquad (14b)$$

Thus, one can determine the van't Hoff enthalpy from a plot of α vs. $T$. To be specific, $\Delta H_{vH}^\circ$ is related in a simple way to the slope at $T_m$. In Figure 3, the slope of the line drawn tangent to the α versus $T$ curve is equal to $[\partial \alpha/\partial T]_{T=T_m}$. This value can be used in Equation (13a) to evaluate $\Delta H_{vH}^\circ$. It is important to note that contrary to common belief, the "sharpness" of the curve tells one nothing about the "cooperativity" or "two-state behavior" of the transition. In the absence of non-model-dependent $\Delta H^\circ$ data from calorimetric measurements, no such conclusion can be justified.

Substituting into the expressions shown above for $K_{eq}$ in terms of α, one can obtain $K_{eq}$ at $T_m$, which we designate $K_{eq}(T_m)$. At $T_m$, for a monomolecular process $K_{eq}(T_m) = 1$; for a bimolecular process involving nonidentical strands $K_{eq}(T_m) = 4/C_t$. Using the integrated form of the van't Hoff equation (Equations (6a) and (6b)) and this knowledge of $K_{eq}$ at $T_m$, one can determine $K_{eq}$ at any temperature.

$$\ln\left[\frac{K_{eq}(T_m)}{K_{eq}(T)}\right] = \frac{\Delta H_{vH}^\circ}{R}\left[\frac{1}{T_m} - \frac{1}{T}\right] \qquad (15)$$

This equation can be evaluated using the appropriate values for $K(T_m)$ shown above (which depend on molecularity and strand identity) to give a relation which allows for estimation of $\Delta G^\circ$ at any temperature.

<table>
<tr><td>*Monomolecular case*</td><td>*Bimolecular case*</td></tr>
</table>

Estimation of $\Delta G^\circ$ from derived parameters

$$\Delta G_{vH}^\circ(T) = \Delta H_{vH}^\circ\left[1 - \frac{T}{T_m}\right] \qquad (16a)$$

$$\Delta G_{vH}^\circ(T) = \Delta H_{vH}^\circ\left[1 - \frac{T}{T_m}\right] + RT\ln(C_t/4) \qquad (16b)$$

Casual examination of Equation (16b) might lead to the conclusion that $\Delta G_{vH}^\circ$ calculated in this manner would depend on concentration; however, because $T_m$ changes in concert with $C_t$, $\Delta G_{vH}^\circ$ should be independent of concentration.

While correct in principle, the method of analysis described above is less reliable than one would like due to difficulties frequently encountered in establishing pre- and posttransition baselines. A more robust method of analysis for $\Delta H_{vH}^\circ$ is based on derivative curves as described below. While the derivation is too lengthy to reproduce here, it is not difficult to derive equations which relate the van't Hoff enthalpy to the shapes of $\partial\alpha/\partial(1/T)$ vs. $T$ curves.[39] The temperature at which $\partial\alpha/\partial(1/T)$ reaches its maximum is designated $T_{max}$. The two temperatures, $T_1$ and $T_2$, which correspond to the temperatures at which $\partial\alpha/\partial(1/T)$ reaches half its maximum value, are defined such that $T_1 < T_{max} < T_2$. An example of a derivative curve is shown in Figure 4. The relevant equations for analysis of such derivative curves are shown below.

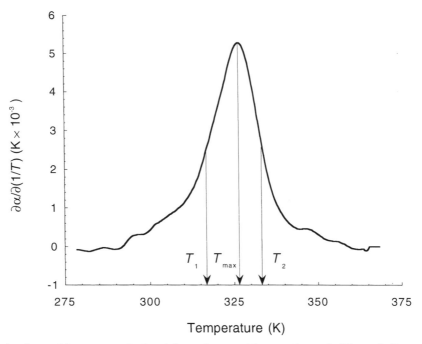

**Figure 4**   Derivative melting curve calculated from the $\alpha$ vs. $T$ curve shown in Figure 3. Temperatures used for estimation of $\Delta H^\circ_{vH}$ by application of Equation (18b) are indicated by arrows.

*Monomolecular case*

$$\Delta H^\circ_{vH} = -R\ln\left[\frac{\alpha_2(1-\alpha_1)}{\alpha_1(1-\alpha_2)}\right]\left[\frac{1}{T_2}-\frac{1}{T_1}\right]^{-1} \qquad (17a)$$

where $\alpha_1 = 0.854$ and $\alpha_2 = 0.146$.

*Bimolecular case*

$$\Delta H^\circ_{vH} = -R\ln\left[\frac{\alpha_2(1-\alpha_1)^2}{\alpha_1(1-\alpha_2)^2}\right]\left[\frac{1}{T_2}-\frac{1}{T_1}\right]^{-1} \qquad (17b)$$

where $\alpha_1 = 0.808$ and $\alpha_2 = 0.106$.

Making the substitutions we find

$$\Delta H^\circ_{vH} = 3.5\,R\left[\frac{1}{T_2}-\frac{1}{T_1}\right]^{-1} \qquad (18a)$$

$$\Delta H^\circ_{vH} = 5.1\,R\left[\frac{1}{T_2}-\frac{1}{T_1}\right]^{-1} \qquad (18b)$$

These equations also are easily generalized to any molecularity.[39]

*(iv) Extraction of $\Delta H^\circ_{vH}$ from the concentration dependence of the transition temperature*

So far we have seen how the definitions of equilibrium constants and the van't Hoff equation can be combined to extract thermodynamic information from single equilibrium transition curves. In this section we will see how the concentration dependence of the transition curve can be exploited to provide thermodynamic information.

In the previous section we saw that for a monomolecular process the value of $K$ does not depend on the concentration of the macromolecule, because $K_{eq} = \alpha/(1-\alpha)$. Therefore, the temperature of the order–disorder transition cannot be shifted by changing $C_t$. This is not the case for bimolecular (and higher molecularity) transitions. For a bimolecular process $K_{eq} = \alpha/((1-\alpha)^2(C_t/2))$.

Combining Equations (2) and (3) we find that

$$-RT \ln K_{eq} = \Delta H^\circ - T\Delta S^\circ \qquad (19)$$

As we have already seen, we can write the equilibrium constant at $T_m$ as

$$K_{eq}(T_m) = 4/C_t \qquad (20)$$

Substituting Equation (20) into Equation (19), we find

$$-RT_m \ln(4/C_t) = \Delta H^\circ - T_m\Delta S^\circ \qquad (21)$$

Dividing by $T_m\Delta H^\circ$ and rearranging leads to

$$\frac{1}{T_m} = \frac{R}{\Delta H^\circ}\ln(C_t) + \frac{\Delta S^\circ - \ln(4)}{\Delta H^\circ} \qquad (22)$$

Given this relation, one readily sees that the slope of a plot of the reciprocal of $T_m$ vs. $\ln(C_t)$ will yield a value for $\Delta H^\circ_{vH}$. In Figure 5, we show such a plot. In principle, in addition to $\Delta H^\circ$, one can also determine $\Delta S^\circ$ from the intercept value (the second term on the right-hand side of Equation (22)). In practice, this is not very reliable because the value of the intercept depends on both $\Delta S^\circ$ and $\Delta H^\circ$. This results in an undesirable statistical coupling between the $\Delta H^\circ$ and $\Delta S^\circ$ values obtained in this fashion.[36] This coupling can lead to a false sense of security in the absolute values of the $\Delta H^\circ$ and $\Delta S^\circ$ data because the coupling will cause these data to yield good estimates of $T_m$. In addition to the coupling, the error in $\Delta H^\circ$ is propagated into $\Delta S^\circ$.

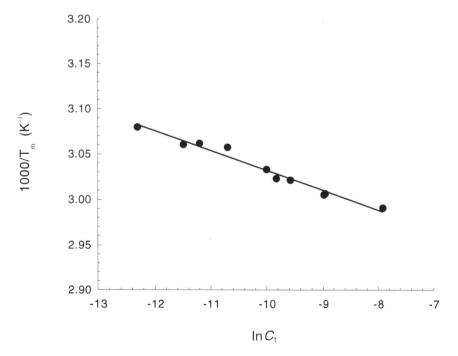

**Figure 5** Oligonucleotide concentration dependence of the d(GCGTACGCATGCG)·d(CGCATGA GTACGC) duplex melting temperature. The slope of the line can be used to estimate $\Delta H^\circ_{vH}$ by application of Equation (22).

The method for extraction of $\Delta H^\circ_{vH}$ just described is often more robust than those methods discussed in the previous section which rely on the shape of the melting transition. Because this method requires only one rather easily obtained point from each transition curve, it can be less susceptible to faulty baseline assignment than methods which depend on the shape of the entire curve. Furthermore, because the logarithm of concentration appears in Equation (22), the $\Delta H^\circ$ estimate does not depend on the accuracy of the extinction coefficient used; the $\Delta S^\circ$ and $\Delta G^\circ$ estimates, however, will be seriously compromised by systematic error in the concentration term. Curvature in such a plot may indicate a significant $\Delta C_p^\circ$ value or systematic error in determination of $T_m$. Also note that this methodology can be generalized for processes of any molecularity greater than 1[39] and that complementary equations in terms of the experimentally more accessible $T_{max}$ rather than $T_m$ can be derived.[38]

Equation (22) can be rearranged to form Equation (23b) which allows for calculation of $T_m$ given values for $\Delta H^\circ$, $\Delta S^\circ$, and the oligonucleotide concentration, $C_t$. Note that Equation (23a), derived from Equation (19) for $K_{eq} = 1$, for the monomolecular process is independent of $C_t$. Again, the bimolecular case shown here applies only to nonidentical strands.

<table>
<tr><td>*Monomolecular case*</td><td>*Bimolecular case*</td></tr>
</table>

Estimation of $T_m$ from derived parameters

$$T_m = \frac{\Delta H^\circ}{\Delta S^\circ} \qquad (23a) \qquad\qquad T_m = \frac{\Delta H^\circ}{\Delta S^\circ + R \ln(C_t/4)} \qquad (23b)$$

Several additional methods are available for evaluating equilibrium melting curves to derive thermodynamic parameters for nucleic acid complexes. These include techniques in which the entire melting curve is analyzed either employing the two-state approximation[42] or allowing for a significant population of intermediate states[40] based on the zipper model.[41] These methods have the apparent advantage of using all of the information available from the melting curves; however, faulty baseline determination can seriously compromise the results. Because the fraction of complex in the initial state at a given temperature can be related in a simple way to $K_{eq}$ (see Equations (10a) and (10b) as examples), direct application of the van't Hoff equation can be used to evaluate $\Delta H^\circ$.[37] This method is particularly sensitive to faulty baseline assignment. Direct determination of $\Delta\Delta G^\circ$ from the dependence of $T_m$ on the nucleic acid complex concentration also has been described.[43]

### 7.02.1.4.2  *Differential scanning calorimetry*

#### (i)  *Measurement of excess heat capacity*

Heat capacity can be measured as a function of temperature using DSC. There are a variety of technical approaches for measuring heat capacity; we will consider here only the approach used for the most sensitive instruments in use for biological systems.[44] These instruments consist of two cells: the sample cell contains the macromolecule solution, the reference cell contains the buffer solution (identical except for the absence of macromolecule). The cells are heated by resistive heaters or Peltier devices. A thermopile, which is sandwiched between the two cells, detects differences in temperature between the sample and reference cells. These temperature differences are in the order of $10^{-6}\,^\circ C$. Each cell is equipped with a feedback heater which attempts to null this temperature difference. The energy employed to equalize the temperature in the two cells is proportional to the heat capacity difference between the cells. At temperatures below the macromolecule's thermally induced transition, the difference in heat capacity (there is always a small difference because of the impossibility of manufacturing exactly matched calorimeter cells) between the two cells is roughly constant; this provides the pretransition baseline. When the temperature increases to the point where the macromolecule undergoes its thermally induced transition, some of the energy introduced by the sample cell heater goes to drive the transition rather than to increasing the temperature of the cell. The resultant temperature difference between the sample cell, which lags in temperature, and the reference cell is nulled by the sample cell feedback heater. The energy introduced by the sample cell feedback heater is proportional to the excess heat capacity of the sample cell at that temperature; this excess heat capacity is due to the thermally induced macromolecular transition. When the transition is complete, the difference in heat capacity between the two cells is again roughly constant; this provides the posttransition baseline.

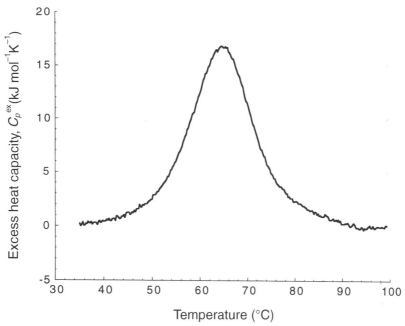

**Figure 6** Differential scanning calorimetry thermogram of the A-G mismatch containing DNA oligonucleotide duplex d(GCGTACGCATGCG)·d(CGCATGAGTACGC). For experimental details, see Plum *et al.*[38]

### (ii) Enthalpy and heat capacity changes

Figure 6 shows a baseline-corrected differential scanning calorimetric excess heat capacity vs. temperature profile. The heat capacity effects observed with the differential scanning calorimeter can be readily related to the enthalpy change associated with the process under investigation.[9,39,44–46] The temperature corresponding to the maximum $C_p$ is $T_{max}$. For a monomolecular process, $T_m = T_{max}$. For processes of higher molecularity, $T_{max} < T_m$. The area under the baseline-corrected curve, where $C_p^{ex}$ indicates the baseline- and concentration-corrected apparent excess molar heat capacity, is the enthalpy change, $\Delta H_{T_{max}}^{\circ}$, of the process at $T_{max}$.

$$\Delta H_{T_{max}}^{\circ} = \int C_p^{ex} \, dT \qquad (24)$$

The difference between the pre- and posttransition baselines for the concentration-corrected curve is the $\Delta C_p^{\circ}$ for the process. This parameter represents the temperature dependence of $\Delta H^{\circ}$. Knowledge of $\Delta C_p^{\circ}$ allows one (assuming that $\Delta C_p^{\circ}$ is independent of temperature, an assumption which usually but not always is justified) to calculate $\Delta H^{\circ}$ at any temperature.

$$\Delta H_T^{\circ} = \Delta H_{T_{max}}^{\circ} - \int_T^{T_{max}} \Delta C_p^{\circ} \, dT \qquad (25)$$

Typically, $\Delta C_p^{\circ}$ is large for protein denaturation and negligible for nucleic acid duplex to single strand transitions. It has been proposed that the $\Delta C_p^{\circ}$ is proportional to the surface area of once-buried hydrophobic groups which becomes exposed to solvent as a result of the observed process.[15] Because both $\Delta H^{\circ}$ and $\Delta C_p^{\circ}$ are dependent on concentration, accurate knowledge of solution concentration is vital.

In principle, $\Delta S^{\circ}$ and $\Delta G^{\circ}$ values can be extracted from the DSC transition profile. Because the parameters so derived are dependent on the same primary data, their values are coupled to that of the enthalpy value. This is similar to the situation encountered in the use of UV melting curves to extract complete thermodynamic profiles of thermally induced transitions. A simple way to minimize the undesirable coupling of the thermodynamic parameters is to use data collected independently by two different techniques. To this end, we have proposed a means of simply combining DSC and UV melting data to calculate $\Delta G^{\circ}$ based on the use of the DSC-derived enthalpy value and the concentration-dependent melting temperature data obtained from UV melting curves.[38,47] In addition to estimates of $\Delta G^\circ$, this method provides a stringent test of the molecularity and/or the two-state assumption for the process under investigation (see below).

Besides the model-independent enthalpy value, van't Hoff enthalpies can be determined directly from the DSC curves in a manner analogous to that described above for noncalorimetric data.[39] A second method by which $\Delta H^\circ_{vH}$ for a monomolecular process may be extracted directly from the DSC curves is embodied in the equation[44]

$$\Delta H^\circ_{vH} = 4RT^2_{max} \frac{C_p^{max}}{\Delta H_{cal}} \tag{26}$$

where $C_p^{max}$ is $C_p$ at $T_{max}$.

### (iii) Comparisons of van't Hoff and calorimetric $\Delta H$ to evaluate cooperativity

Note that the enthalpy change determined by DSC is independent of any model for the process. This stands in contrast to the methods for extracting model-dependent van't Hoff enthalpy changes described above. In fact, comparison between the two enthalpy changes, van't Hoff and calorimetric, is a useful means of assessing the validity of the assumptions explicit and implicit in the van't Hoff expressions. While it is not always possible to discern the origins of discrepancies between the van't Hoff and calorimetric enthalpies,[48] the following discussion highlights some of the most common reasons for such discrepancies.

If the calorimetric enthalpy exhibits a significant temperature dependence (i.e., $\Delta C_p^\circ \neq 0$), one of the assumptions in the van't Hoff analysis as developed above is invalid. It should be emphasized that the van't Hoff equation is not invalidated when $\Delta C_p^\circ \neq 0$, merely the specific application described above must be modified. A plot of $\ln K_{eq}$ vs. $1/T$ would be curved; however, at any temperature on that curve the slope is still the correct value of $-\Delta H^\circ_{vH}/R$ for that temperature.

If the calorimetric enthalpy change is larger than the van't Hoff enthalpy change, then the process has thermodynamically significant intermediate states; that is, the assumption that the transition proceeds in a two-state (all-or-none) fashion is invalid. The mole unit for the calorimetric $\Delta H^\circ$ refers to the measured amount of macromolecule in the solution. The mole unit for the van't Hoff analysis is an implicit mole unit representative of the cooperative melting unit. Thus, the ratio $\Delta H^\circ_{vH}/\Delta H^\circ_{cal}$ is a quantitative measure of the fraction of the macromolecule which operates as a single thermodynamic unit. If the calorimetric $\Delta H^\circ$ is smaller than the van't Hoff $\Delta H^\circ$ ($\Delta H^\circ_{vH}/\Delta H^\circ_{cal} > 1$), then the cooperative unit size is larger than one molecule. Thus, a multimolecular process, such as aggregation, is indicated. A case where $\Delta H^\circ_{vH}/\Delta H^\circ_{cal} > 1$, which does not indicate aggregation, is observed in the duplex to single strand transitions of nucleic acids. Here the calorimetric mole unit is defined in terms of nucleotides (or base pairs) but the thermodynamic melting unit is tens of base pairs. Therefore, the ratio $\Delta H^\circ_{vH}/\Delta H^\circ_{cal}$ gives an estimate of the number of nucleotides (or base pairs) which melt as a single thermodynamic unit. Only in the case where $\Delta H^\circ_{vH} = \Delta H^\circ_{cal}$ can one conclude that the transition proceeds in a two-state manner. This is an important conclusion which can be substantiated only with direct calorimetric determination of $\Delta H^\circ$.

### (iv) Use of DSC to evaluate macromolecular interactions

Generally, methods to assess equilibrium binding constants require significant concentrations of unbound sites and free ligand at each data point. This effectively limits such methods to systems exhibiting moderate binding affinities ($K \sim 10^3$–$10^9$). Many binding reactions which are of interest to biologists are of higher affinity. Brandts and Lin[49] have demonstrated how DSC can be used to expand vastly the range of binding affinities which are accessible experimentally, up to $K \sim 10^{40}$. Application of this method provides access to the thermodynamics ($\Delta G^\circ$, $\Delta H^\circ$, $\Delta S^\circ$, and $\Delta C_p^\circ$) of the binding process as well as the thermodynamics of the order–disorder transitions of the constituent macromolecules.

### 7.02.1.4.3   Isothermal titration calorimetry

Advances in calorimetry have provided a means of studying macromolecule–macromolecule and macromolecule–ligand interactions by direct measurement of reaction heats.[10,50,51] Ultrasensitive titration calorimetry provides a means for complete thermodynamic characterization ($n$, $K_{eq}$, $\Delta G^\circ$, $\Delta H^\circ$, $\Delta S^\circ$) of a system in a single experiment. The heat capacity change of the system, $\Delta C_p^\circ$, may be determined by repeating the experiment at several temperatures. Significantly, the technique is not limited to systems where a convenient optical observable is present, or to optically clear solutions. Also, assumptions, which are inherent in typical van't Hoff analyses (e.g., $\Delta C_p^\circ = 0$) are unnecessary because the heat of reaction is measured directly.

In a titration calorimetric experiment, small aliquots of a ligand solution are added sequentially to a macromolecule solution and the evolution or absorption of heat is monitored. Thus, one measures directly the reaction heat as a function of the ligand to macromolecule ratio. The design and function of the titration calorimeter is very similar to that of the DSC. The primary differences are the apparatus used to inject titrant solution into the sample cell and the isothermal nature of the titration experiment. The cell design and the electronics which provide the feedback and detection are essentially identical.

In many cases, such titration calorimetric data can be analyzed readily using various binding models to obtain a complete thermodynamic profile of a binding event. In principle, any binding model which relates the change in concentration of the ligand–macromolecule complex formed to the change in total ligand concentration, $d[LM]/d[L]_t$, can be used. The heat, $dQ$, absorbed or evolved upon a change in total ligand concentration, $d[L]_t$, due to introduction of titrant into the sample cell is related to the change in ligand–macromolecule complex concentration, the enthalpy change associated with complex formation, $\Delta H^\circ$, and the cell volume, $V$.

$$\frac{dQ}{d[L]_t} = \Delta H^\circ V \frac{d[LM]}{d[L]_t} \tag{27}$$

For virtually any binding model, equations can be developed which relate $d[LM]/d[L]_t$ to the total concentrations of added ligand, $[L]_t$, and macromolecule, $[M]_t$ (both of which are known), the number of ligands which participate in the binding reaction, $n$, and the equilibrium binding constant.[52] Even for simple binding models, these equations are complex and closed form expressions are usually not available; however, the equations are easily dealt with numerically.

Nonlinear least squares fitting of the $Q$ vs. $[L]_t$ curve obtained from the titration calorimetric experiment provides estimates of the binding constant, $K_{eq}$, the enthalpy of binding, $\Delta H^\circ$, and the number of binding sites, $n$. Recalling the relation between the binding constant and Gibbs free energy, $\Delta G^\circ = -RT \ln K_{eq}$, one can determine the free energy change associated with binding. One can then extract the experimentally determined entropy change, $\Delta S^\circ = (\Delta H^\circ - \Delta G^\circ)/T$. Determination of the enthalpy, $\Delta H^\circ$, as a function of temperature provides the constant pressure heat capacity change, $\Delta C_p^\circ$, which can give insight into the relative importance of solvent effects.

### 7.02.1.4.4   Measurement of volumetric parameters

Because most systems of interest respond to temperature changes over a readily accessible range of temperature, thermodynamic studies of nucleic acids generally have relied on either the temperature dependence of some equilibrium property or direct measurement of heat capacity to obtain the thermodynamic parameters which describe the system ($\Delta G^\circ$, $\Delta H^\circ$, $\Delta S^\circ$). An alternative approach, in which pressure is the operational variable, also gives access to the desired thermo-dynamic parameters. Due, however, to the extremely high pressures needed to influence nucleic acid equilibria, this approach has seen only limited application.[20,53] An experimental approach which circumvents this limitation relies on the fact that the speed of sound, $U$, depends in a simple way on the pressure derivative of density, $\rho$.[18,19] Because $U$ is a second pressure derivative of free energy, the relevant thermodynamic data are accessible through sound velocity and density measurements. Development of highly sensitive ultrasonic velocimeters[18] which require sample volumes on the order of 1 mL has led to practical study of nucleic acids by this technique. Densities are measured by a vibrating tube densitometer. Densimetric and ultrasonic velocimetric measurements have several advantages relative to more widely applied methods for studying hydration. The measurements can be performed on dilute solutions and are not complicated by requirements for unusual solution conditions and crystal lattice effects. The observables of these macroscopic methods are highly nonselective and thus sample the broad population of the solvent that is perturbed by the solute. These measurements provide a means of thermodynamic characterization of the solute-perturbed solvent.

Compressibility, like sound velocity, is a second pressure derivative of the Gibbs free energy. The value of compressibility is dependent on intra- and intermolecular interactions. Specifically, measurement of compressibility can lead to insights into interatomic interactions within the solute molecule (the intrinsic compressibility), solute–solute interactions, and solute–solvent interactions (hydration). The apparent molar adiabatic compressibility, $K_s$, can be obtained from measurement of the densities and sound velocities of the solution and solvent (indicated by subscript "o"). First, the apparent molar volume, $V = M/\rho_o - (\rho - \rho_o)/(\rho_o C)$, and the molar sound velocity increment, $[U] = (U - U_o)/(U_o C)$, are determined, where $M$ is the solute molecular weight and $C$ is its molar concentration. The apparent molar adiabatic compressibility is then calculated from the relation, $K_s = (U_o^2 \rho_0)^{-1}(2V - 2[U] - M/\rho_o)$. Apparent molar adiabatic compressibility data can be interpreted in terms of hydration by application of a simple relation, $K_s = K_{sM} + \Delta K_h$, where $K_{sM}$ is the intrinsic adiabatic compressibility of the molecule and $\Delta K_h$ is the compressibility effect of hydration. Further elaboration of this relation leads to $K_s = K_{sM} + \Delta K_h = K_{sM} + n_h (K_{sh} - K_{so})$, where $n_h$ is the number of water molecules in the hydration shell (that is, with altered physicochemical properties relative to bulk water) and $K_{sh}$ and $K_{so}$ are the partial molar adiabatic compressibilities of the water in the hydration shell and the bulk solvent respectively.

### 7.02.1.5 Interpretation and Application of Thermodynamic Parameters in Context of Various Nucleic Acid Systems

In this section we describe briefly a variety of nucleic acid systems which have undergone thermodynamic characterization. This compilation is not exhaustive but rather represents the authors' view of some of the areas in nucleic acid thermodynamics in which progress has been made and in which further attention is needed. Particular emphasis is placed on the potential contributions of single-stranded structure and hydration to the stability of nucleic acid complexes and to the thermodynamic origins of that stability.

#### 7.02.1.5.1 *Single strands*

##### (i) *Single-stranded structures and apparent molecularity*

Single-stranded oligo- and polynucleotides form structures in solution which involve a significant amount of base stacking. This feature comes as no surprise because it has been known for many years that nucleotide monomers in solution will stack, with purines showing a greater propensity than pyrimidines.[54] In addition to the nature of the nucleotide bases, the extent of stacking depends on solution conditions, temperature, etc.[55,56] Due to their low transition enthalpies, structural transitions of single-stranded oligo- and polynucleotides are broad and difficult to treat quantitatively either spectroscopically or calorimetrically.

An additional experimental reality is that the single-strand melting processes are monomolecular. Consequently, the temperature of such transitions cannot be manipulated by nucleotide concentration, traditionally the best way to separate single-strand effects from duplex melting. If the $T_m$ is dependent on oligonucleotide strand concentration, the molecularity of the process must be greater than one. The converse of this statement, however, is not true. Polymer duplex melting is independent of nucleotide concentration, so-called pseudomonomolecular behavior. This is due to the effect of initiation of the duplex, a bimolecular process, being overwhelmed by the propagation of the base pairing of the initiated duplex, a monomolecular process. While the lack of concentration dependence of monomolecular and pseudomonomolecular processes eliminates one parameter available to the experimentalist to manipulate these equilibria, this feature can be used to advantage by decoupling concentration effects from the effects of solution conditions. Several studies of higher-order nucleic acid structures have exploited this property of single-stranded structures by design of oligonucleotides which can fold upon themselves to form stable duplexes, triplexes, and tetraplexes.

Competitive equilibria between the duplex and single-strand hairpin forms of a self-complementary (or nearly so) oligonucleotide can provide a significant impediment to thermodynamic characterization of such nucleic acids.[57] The relative amount of the hairpin and duplex forms is dependent on solution conditions, with the hairpin favored at low salt concentrations and the duplex favored at high salt concentrations. At intermediate salt concentration, complex biphasic melting behavior is observed. The differential salt dependence of the two forms is due to the larger number of negative charges on the duplex. Given these observations, whenever possible self-complementary oligonucleotides should be avoided for thermodynamic characterization. The apparent economy of effort expended in synthesis of a single self-complementary oligonucleotide, as opposed to two nonidentical complementary strands, is more than outweighed by the increased complexity of the thermodynamic characterization. Similar considerations also apply to higher-order nucleic acid complexes, such as triplexes and quadruplexes.

Single-stranded nucleic acid structures also are important in biology. Hairpin loops, bulges, and overhangs are some of the primary secondary structural features of RNA and are involved in a variety of biological processes.[58] Such features are also present in DNA, but to a lesser extent. One might expect the putatively unpaired bases in a loop or overhang region to influence only minimally the stability of the attached base-paired region. Studies reveal the influence on stability of the base composition and sequence of loops[58-60] and overhangs[61-63] can be significant.

### (ii)  Contribution of single-stranded structure to apparent $\Delta C_p^\circ$

As described above, one can plot $\Delta H^\circ$ values derived from ITC experiments as a function of experimental temperature well below $T_m$. A nonzero slope can quite reasonably be interpreted as a change in heat capacity, $\Delta C_p^\circ$.[64-66] The $\Delta C_p^\circ$ determined in this manner may not agree with that derived from DSC data under conditions where $K$ is very small (i.e., at $T_{max}$); yet they may both be correct. Over the range of temperature examined in the ITC experiment, single-stranded structures may contribute significantly to the thermodynamics of the complex formation reaction, whereas, at the elevated temperatures which define the final single-stranded state for the DSC experiment, any single-stranded structure more likely is fully disordered. When comparing the thermodynamics of different duplexes (or any other complex), the observed differences in DSC-derived $\Delta H^\circ$ values generally can be attributed with confidence to differences in the low-temperature fully ordered states, given the reasonable assumption that the final high-temperature states are thermodynamically equivalent. This may not be true for ITC data, in which both the low-temperature single-stranded states and the final complexed states may differ significantly. If given independent knowledge, which may come from DSC, that the heat capacity of the complex is constant over the temperature range of the ITC experiments, one gains access to the otherwise elusive thermodynamics of the single-strand components.

Vesnaver and Breslauer[67] demonstrated, using a combination of DSC and isothermal mixing calorimetry, that by application of Hess' law a thermodynamic cycle could be constructed to evaluate the energetics of the partially ordered single strands and their contribution to duplex stability. They argue that single strands are thermodynamically poised to form duplexes and thus, at least for the case examined, the thermodynamic driving forces toward duplex formation are not as strong as might be concluded from examination of melting transition data alone.

### 7.02.1.5.2  *Duplexes*

### (i)  Prediction of duplex thermodynamics based on base-pair sequence

Of all nucleic acid complexes, DNA and RNA duplexes have been studied most extensively by thermodynamic methods. The resultant database of thermodynamic information has allowed investigators to develop predictive capacity for the energetics and stability of duplexes with Watson–Crick type base pairing.[68-71] This predictive capacity is founded on the observation[72] that, to a good approximation,[73] one can describe the thermodynamic properties of a duplex by summation of nearest-neighbor interactions based on the primary sequence of base pairs. Given this observation, thermodynamic data for oligonucleotide and polymer duplexes, which are selected to represent the distribution of possible nearest neighbors, have been analyzed to extract average contributions (under a given set of solution conditions) of each nearest-neighbor pair to the various thermodynamic parameters, $\Delta H^\circ$, $\Delta S^\circ$, and $\Delta G^\circ$. Alternative parameterizations also have been employed.[74] The primary data generally are derived from melting experiments, monitored by optical and/or calorimetric methods. The analysis of these data is based on the assumed (optical data) or determined (DSC) independence of $\Delta H°$ and $\Delta S°$ from temperature; that is, $\Delta C_p° = 0$.

The practical value of these nearest-neighbor thermodynamic data is that they can be used to predict the stability ($\Delta G°$) and the thermodynamic origins of that stability ($\Delta H°$ and $\Delta S°$) under a given set of conditions. Given these predictions for $\Delta H°$ and $\Delta S°$ and knowledge of the oligo-nucleotide concentration, an estimate for $T_m$ can be obtained from Equation (23b). Use of these data for estimating thermodynamic ($\Delta G°$, $\Delta H°$, $\Delta S°$) and extrathermodynamic ($T_m$) parameters under different solution conditions presents some challenges. Estimates for $T_m$ must be adjusted for differing salt concentrations with the recognition that for oligomeric duplexes of fewer than about 16 base pairs the salt dependence of $T_m$ also is dependent on duplex length.[75,76] Nearest neighbor based $\Delta G°$ predictions, coupled with estimates of free energy contributions of other secondary structure motifs, can be used to predict, based on the primary nucleotide sequence, RNA secondary structure folding patterns.[77,78]

Relatively few thermodynamic data are available for DNA·RNA hybrid duplexes,[75,79–83] despite the unquestionable biological and biotechnological significance of such structures. Comparisons of duplexes with differing ribose versus deoxyribose sugar phosphate backbones but a common base sequence, show that the relative stability of RNA·RNA, DNA·DNA, and DNA·RNA duplexes depends on the base composition (purine/pyrimidine ratio associated with each backbone type) and sequence. A set of nearest-neighbor parameters for DNA·RNA hybrid duplexes has been published.[84]

### (ii)  Effects of damaged, modified, or mispaired bases

Because DNA is the repository of genetic information, defects which result from damage or failures in replication and/or repair fidelity can be costly to the organism. Accumulation of defects in DNA has been implicated in the origins of cancers and a variety of age-related degenerative diseases. Defects in DNA take several forms which include but are not limited to nucleotide base mismatches (non-Watson–Crick base pairs), base insertion or deletion errors due to template misalignment, unmatched bases (bulges, loops), and damaged or missing bases. The structural picture of damaged DNA provided by X-ray crystallography[85] and NMR[86] reveals that the structural effects of damage typically are highly localized, although this depends on sequence context. Yet it has been observed that damage to DNA results in large thermodynamic effects, even in the absence of any apparent lesion-induced structural perturbations.[38] The thermodynamic consequences of damage to DNA have been reviewed.[12,47,87] Typically a reduction in thermal stability is observed. The effects on thermodynamic stability and the origins of those effects are more subtle. Often, large unfavorable enthalpic effects are partially or wholly offset by favorable entropic effects which result in a small net decrease in duplex stability. The origins of this compensating entropic effect are not obvious and probably depend at least in part on damage-induced local reorganization of the solvent as is observed crystallographically.[88] The cross-strand base and the neighboring undamaged base pairs play a significant role in determining the thermodynamic effects of many lesions.

### (iii)  Hydration of duplex DNA

It has long been recognized that the interactions between nucleic acids and their sheath of associated water molecules must play a role in the stability of nucleic acid complexes. However, very little specific knowledge of these interactions was available. Since the early 1990s, advances in X-ray crystallography[89] and NMR[90] have made structural and dynamic information about the interactions of macromolecules and their waters of hydration obtainable experimentally. The thermodynamics of water–nucleic acid interactions are beginning to be addressed by application of densitometry and acoustimetry to gain access to the volumetric properties of nucleic acids and their layer of hydrating water.

A series of DNA polymer duplexes which differ by the GC base pair content were examined by acoustic and densitometric techniques.[91] Water in the vicinity of DNA is perturbed relative to bulk water. It is more dense and exhibits reduced compressibility. The compressibility of the water in the hydration shell of B-form DNA is reduced significantly (20–30%) relative to bulk water. The compressibility of the water in the hydration shell of B-form DNA at 25 °C was found to be linearly related to the density of that water. The density of the hydrating water is estimated to be in the range of 1.17 to $1.26 \, \mathrm{g \, mL}^{-1}$. This observation has been interpreted in terms of electrostatic interactions between the DNA and the hydrating waters.

There has been a long-standing belief that in duplex DNA AT base pairs are more hydrated than GC base pairs. Because the accessible surface area for the AT and GC base pairs in B-form DNA is very similar, the number of water molecules in contact with the surface is also probably the same. Yet the apparent adiabatic compressibility of an all-GC polymer duplex is significantly lower than for all-AT containing polymer duplexes. Thus it is concluded that the waters associated with a GC base pair are more perturbed than those associated with an AT base pair. Therefore, GC is more strongly hydrated than AT. Interestingly, mixed-sequence DNA is less strongly hydrated than either the all-AT or all-GC DNA duplexes. Therefore, hydration-dependent characteristics of DNA cannot be reduced to a weighted sum of contributions from AT and GC base pairs. This assessment of DNA hydration from thermodynamic measurements is in excellent agreement with an analysis of single crystal X-ray diffraction studies of waters associated with DNA[92] and in large part with theoretical work.[93]

The nature of DNA hydration has significant implications for drug[94] and protein binding.[95] A study of the effects of hydration on the binding of the AT-rich minor groove targeting drug netropsin[96] highlights the necessity of quantitative characterization of hydration effects to understand the driving forces for drug binding, and by extension protein binding, to nucleic acid complexes. The changes in volume, $\Delta V$, and adiabatic compressibility, $\Delta K_s$, associated with binding of netropsin to poly(dAdT)·poly(dAdT), poly(dA)·poly(dT), and poly(dA)·poly(dT)$_2$ were measured. Based on an analysis of the macroscopic volumetric data which accounts for ligand-induced changes in solvation, the authors conclude that approximately 22 waters are expelled from the poly(dAdT)·poly(dAdT) hydration shell when netropsin binds. The number of waters released from poly(dA)·poly(dT) and poly(dA)·poly(dT)$_2$ upon netropsin binding are even larger, approximately 40 and 53 respectively. Earlier thermodynamic characterization of the binding of netropsin to poly(dAdT)·poly(dAdT) and poly(dA)·poly(dT) demonstrated the binding of the drug to poly(dA)·poly(dT) is overwhelmingly entropy driven relative to binding to poly(dAdT)·poly(dAdT).[94] Because the hydration properties of poly(dAdT)·poly(dAdT) and poly(dA)·poly(dT) are nearly identical based on the work cited above,[91] the changes in hydration observed upon netropsin binding are interpreted to reflect primarily differences in hydration between the poly(dAdT)·poly(dAdT)-netropsin and the poly(dA)·poly(dT)-netropsin complexes rather than exclusively differences in the unbound polymer duplexes which had been proposed previously. In addition, the data on netropsin-induced water release and the entropy change associated with binding can be combined to estimate that the difference in partial molar entropy of water in the bulk state and in the hydration shell of the two all-AT polymers is about $1.6 \, \mathrm{cal \, K}^{-1} \, \mathrm{mol}^{-1}$. If this value proves general, which it may not, it could be used to estimate the difference in the number of waters released upon drug binding to poly(dAdT)·poly(dAdT) and poly(dA)·poly(dT) simply by measuring the difference in drug binding entropy change, $\Delta\Delta S°$.

### (iv) Probe hybridization

The ability of oligonucleotide or analog probes specifically to recognize and bind to nucleic acids *in vitro* or *in vivo* allows for a variety of applications in biotechnology, diagnostics, and potentially therapeutics. The systematic design of probes with the desired specificity and stability requires thermodynamic information. The goal in probe design is to optimize, and not necessarily to maximize, binding to the target site.

In evaluating candidate probes, it is important to select appropriate criteria for comparison. It is common to evaluate the relative "stability" of hybridization probes based on $T_m$, with the assumption that the complex with the higher $T_m$ will be more stable at some temperature well below $T_m$. Unfortunately, this practice can lead to erroneous conclusions because it fails to account for the possibility of differential temperature dependences of free energies of complex formation. In the absence of accurate $\Delta H°$ data, this procedure may result in incorrect determinations of relative complex stability. Reliance on $T_m$ comparisons is particularly risky when comparing oligonucleotides containing unusual backbones or bases which have not been characterized thermodynamically.

To ensure unique binding sites in a large genome, statistical considerations suggest that oligonucleotides of 12–25 nucleotides be used as antisense or antigene probes. The specificity and stability of a probe–target complex should increase as the number of correct pairing interactions increases. Use of unnecessarily long oligonucleotide probes, however, is undesirable because long probes may potentially form stable complexes even in the absence of complete sequence complementarity. The necessity of using such long oligonucleotides has been challenged by the observation that much shorter molecules (7–8 nucleotides) can specifically target RNA in cells.[97] This is presumably due to the failure of the short oligonucleotides to compete for RNA binding sites involved in secondary and tertiary structures. Shorter oligonucleotides have a significant advantage in ease of synthesis. There may be a disadvantage in that the stability of complexes formed with the short molecules will depend strongly on their concentration, whereas the longer molecules will form complexes which approach the pseudomonomolecular behavior of polymers.

### 7.02.1.5.3  *Unusual duplex and higher-order structures*

A variety of unusual oligonucleotide structures which highlight the remarkable plasticity of nucleic acids have been described. These structures include, but are not limited to, parallel duplexes, duplexes stabilized by $C \cdot C^+$ base pairing interactions, triple helices, circular oligonucleotides (which can form either duplex or triplex structures), stable Holliday junctions, and quadruplexes. The thermodynamic characterization of these structures is in its early stages.

Of the nucleic acid duplex structures which violate the Watson–Crick structural canon, parallel DNA and RNA duplexes have received the most thermodynamic characterization.[40,98] At low pH, protonation of cytosine can lead to duplexes stabilized by $C \cdot C^+$ base pairing interactions.[99] Competitive formation of such duplexes can lead to experimental artifacts under the low pH conditions often employed to induce formation of pyrimidine·purine·pyrimidine triple helices which contain cytosines in the third strand.

Due to the potential uses as antisense or antigene agents and the potential to deliver chemical reactivity in a sequence-specific manner, triple helices of nucleic acids and analogues have received a particularly extensive study. The thermodynamic aspects of triple helix formation have been reviewed.[100–102] Triple helices provide particular experimental challenges to the thermodynamicist due to the possibility of apparent bimolecular or trimolecular melting behavior. By modulation of solution conditions, including counterion valence and concentration and in some cases pH, the apparent molecularity of the transition often can be selected. Construction of a single oligonucleotide chain which can fold upon itself to form a triple helix has provided a means of exploring the salt–pH–temperature "phase diagram" without the complications of oligonucleotide concentration dependence and poorly defined transition molecularities. Roberts and Crothers[103] have reported a set of nearest-neighbor thermodynamic parameters for pyrimidine·purine·pyrimidine triple helices which can be used to predict the stability and energetics of third-strand hybridization for this class of oligonucleotides.

A single oligonucleotide chain can be coupled at the ends to form a circular molecule.[104] Two classes of circular oligonucleotide can be constructed: those with internal base complementarity ("dumbbells") and those which do not form internal base-pairing interactions. The dumbbell molecules are significantly stabilized thermally relative to the corresponding bimolecular Watson–Crick duplex; this is due primarily to the entropic advantage of the reduction of molecularity.[105] Appropriately designed circular oligonucleotides which lack internal complementarity combined with a complementary linear single strand can form triple helices which exhibit a number of desirable properties including enhanced thermal stability and resistance to enzymatic degradation.[104]

Additional higher-order nucleic acid complexes have been examined thermodynamically. Multi-strand DNA complexes which model the Holliday junction, an important intermediate in DNA recombination, have been constructed and characterized thermodynamically.[106–108] DNA quadruplexes, which model the four-stranded complexes found in the telomere regions of eukaryotic chromosomes, also have been characterized thermodynamically. Interestingly, quadruplexes are stabilized by specifically coordinated cations. Thermodynamic aspects of these complexes have been reviewed.[47,109]

## 7.02.2  NUCLEIC ACID KINETICS

Since the mid-1980s, there has been a veritable explosion of interest in the folding of nucleic acids. There are several reasons for this upsurge, including the discovery of catalytic RNA, interest in RNA folding, and the use of oligonucleotides for recognition and therapeutics. The techniques for studying the kinetics of nucleic acid folding are well understood and have been applied with great success to single strands, duplex DNA and RNA, and tertiary folds to provide a basic framework for understanding the mechanism and rates of these processes (see Figure 7). This section begins with an overview of the methods and equations used for obtaining kinetic information on DNA and RNA. We then cover what is known about the basic processes that underlie helix formation and folding of nucleic acids, with an eye toward future areas of interest.

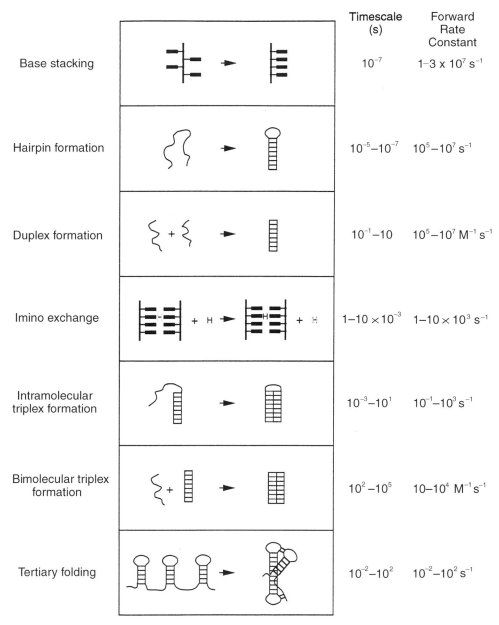

| | Timescale (s) | Forward Rate Constant |
|---|---|---|
| Base stacking | $10^{-7}$ | $1\text{--}3 \times 10^{7}\,\text{s}^{-1}$ |
| Hairpin formation | $10^{-5}\text{--}10^{-7}$ | $10^{5}\text{--}10^{7}\,\text{s}^{-1}$ |
| Duplex formation | $10^{-1}\text{--}10$ | $10^{5}\text{--}10^{7}\,\text{M}^{-1}\,\text{s}^{-1}$ |
| Imino exchange | $1\text{--}10 \times 10^{-3}$ | $1\text{--}10 \times 10^{3}\,\text{s}^{-1}$ |
| Intramolecular triplex formation | $10^{-3}\text{--}10^{1}$ | $10^{-1}\text{--}10^{3}\,\text{s}^{-1}$ |
| Bimolecular triplex formation | $10^{2}\text{--}10^{5}$ | $10\text{--}10^{4}\,\text{M}^{-1}\,\text{s}^{-1}$ |
| Tertiary folding | $10^{-2}\text{--}10^{2}$ | $10^{-2}\text{--}10^{2}\,\text{s}^{-1}$ |

**Figure 7** Kinetics of processes in nucleic acids. Schematic indicating the process, the corresponding forward rate constant, and the timescale on which it would occur at $\sim 1\,\mu\text{M}$ total strand concentration.

## 7.02.2.1 Obtaining and Analyzing Kinetic Data on DNA and RNA

### 7.02.2.1.1 *Relaxation kinetics*

One method that has been very successfully applied to the study of RNA and DNA folding is the analysis of relaxation kinetics. The general principle of this method is somehow to perturb the molecules of interest so the system is no longer at equilibrium, and then observe them as they return to equilibrium. One of the great advantages of this technique is that if the perturbation is small, the relaxation can be represented as a sum of simple exponential terms and relaxation times, even for very complex processes.[110–112] Two of the most popular methods for obtaining relaxation data are temperature-jump kinetics and NMR spectroscopy. These methods give us insight into a broad range of timescales covering processes that are completed in <1 ns to those that occur over many seconds.

### (i) Temperature-jump kinetics

This technique was developed by Eigen in the 1950s for the study of organic reactions (for review, see Turner[113]). In the method, the sample is heated very quickly (1–5 μs), using a capacitor that discharges through the solution. The sample is then observed optically as it returns to equilibrium at the new, higher temperature. Heating can be achieved more quickly ($\sim$ 1 ns) using a pulsed laser at a wavelength which is absorbed by the solvent. The laser heating method has the additional advantage that an ionic strength sufficient to achieve capacitor discharge is not required. The upper bound on the timescale of the experiment is limited to how long the optical signal is stable, usually before it is obscured by convection roll due to contact between the sample (which was heated) and the cell (which often is not).

Temperature-jump kinetics can be applied to a wide variety of kinetic systems. In order to derive the relevant equations for a process, one needs to take advantage of two physical laws: (a) the rate law for the reaction and (b) conservation of mass. The equations relevant for three of the most common reactions in nucleic acids are presented in Table 1: (1) the unimolecular one-step reaction, (2) the bimolecular self-complementary reaction, and (3) the bimolecular non-self-complementary reaction. Treatments of many other reactions exist.[110–112,114]

**Table 1**   Equations for relaxation kinetics.

| Type of reaction | Relaxation time | Determination of rate constants |
|---|---|---|
| **1 Unimolecular one-step reaction**<br><br>$A \underset{k_{-1}}{\overset{k_1}{\rightleftharpoons}} B$ | $\dfrac{1}{\tau} = k_1 + k_{-1}$ | $K_{eq} = k_1/k_{-1}$, determine or know $K_{eq}(T)$, and you can iterate to find $k_1$ and $k_{-1}$. $1/\tau \sim k_1$ well below $t_m$. $1/\tau \sim k_{-1}$ above $t_m$. |
| **2 Bimolecular self-complementary**<br><br>$2A \underset{k_{-1}}{\overset{k_1}{\rightleftharpoons}} B$ | $\dfrac{1}{\tau} = 4k_1[\bar{A}] + k_{-1}$ | Use $K_{eq} = k_1/k_{-1}$. Plot $1/\tau$ vs. $4[\bar{A}]$, slope is $k_1$, intercept is $k_{-1}$. |
| | $\dfrac{1}{\tau^2} = 8k_1 k_{-1} C_t + k_{-1}^{\,2}$ | Plot $1/\tau^2$ vs. $C_t$. Slope $= 8k_1 k_{-1}$, intercept $= k_{-1}^{\,2}$. |
| **3 Bimolecular non-self complementary, [A] = [B][a]**<br><br>$A + B \underset{k_{-1}}{\overset{k_1}{\rightleftharpoons}} C$ | $\dfrac{1}{\tau} = k_1[\bar{A} + \bar{B}] + k_{-1}$ | Use $K_{eq} = k_1/k_{-1}$. Plot $1/\tau$ vs. $[\bar{A} + \bar{B}]$. Slope is $k_1$, intercept is $k_{-1}$. |
| | $\dfrac{1}{\tau^2} = 2k_1 k_{-1} C_t + k_{-1}^{\,2}$ | Plot $1/\tau^2$ vs. $C_t$. Slope $= 2k_1 k_{-1}$, intercept $= k_{-1}^{\,2}$ |

$k_1$ and $k_{-1}$ are the forward and reverse reaction rates. $[\bar{A}]$ and $[\bar{B}]$ are the equilibrium concentrations of A and B after the perturbation. $C_t$ is the total strand concentration.
[a] The equation for $1/\tau$ also holds where $[A]_0 \neq [B]_0$. The equation for $1/\tau^2$ is correct where $[A]_0 = [B]_0$, but also holds well where the ratio of $[A]_0/[B]_0$ is <4.

To derive the rate constants, the relaxation information is usually combined with equilibrium constant data (remembering that $K_{eq} = k_1/k_{-1}$). This combination can be done in several ways. In general, if an estimate is available for the equilibrium constant, it can be used to substitute into the relaxation equations along with the relaxation time ($t$) to obtain values for $k_1$ and $k_{-1}$. These values then are used ($k_1/k_{-1}$) to recalculate a better estimate of $K_{eq}$ and the process is iterated until the values converge. For bimolecular reactions, rate constants can be determined by plotting either $1/t$ or $1/t^2$ vs. the final equilibrium concentration of the reactants after the jump, or vs. the total concentration of strand (reactions 2 and 3 in Table 1). As with unimolecular reactions, the relaxation rate can be combined with the rate constants for the forward and reverse reaction and the equilibrium constant in an iterative procedure. A sample of temperature-jump relaxation kinetic results, taken from data in Pörschke and Eigen,[115] is plotted in Figure 8. The figure shows the rate constants, relaxation times, and equilibrium constants calculated for a nine-base-pair duplex ($A_q \cdot U_q$). The mechanistic information inferred from these data will be discussed later (Section 7.02.2.3.3).

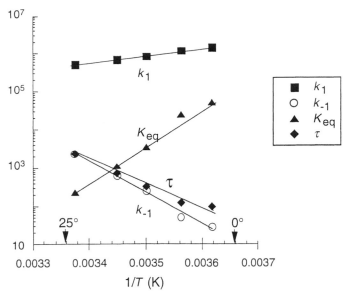

**Figure 8** Comparison of equilibrium and kinetic data from a relaxation study on $rA_q \cdot rU_q$ (data from Pörschke and Eigen[115]). $t$ values calculated for 100 mM total strand concentration. Arrows indicate 0 °C and 25 °C on the *x*-axis. The *y*-axis has units corresponding to the variables plotted; $s^{-1}$ for $k_{-1}$ and $t$, $s^{-1} M^{-1}$ for $k_1$, and $M^{-1}$ for $K_{eq}$.

### (ii) NMR

A wealth of information on the bulk behavior of a macromolecule and the specific atoms and residues can be derived from NMR. Using this technique, a wide variety of kinetic processes have been characterized, including the rates of conformational changes, backbone and base dynamics, and the rates of translation and rotation of the molecule. One of the simplest cases is the kinetics of chemical exchange, where a nucleus explores two or more magnetic environments. The exchange can be due to the atoms physically exchanging (e.g., imino protons exchanging with water), or the molecule's sampling of multiple conformations that place the nuclei in different chemical environments, thereby causing them to exhibit different chemical shifts. Rates of exchange can be determined by analysis of the linewidths, titration of samples with a catalyst to promote chemical exchange, and a variety of experiments where magnetization transfer and relaxation are analyzed. For example, a proton exploring two environments will give two peaks, one peak, or something in-between if the rate of chemical exchange is, respectively, much slower, much faster, or similar to the separation of the two peaks (in Hz). In contrast to other relaxation methods which always include the forward and reverse rates, NMR can measure the lifetime of one state directly (e.g., the forward rate alone). General reviews of dynamics of nucleic acids studied by NMR have been published.[116-119]

### 7.02.2.1.2 Mixing kinetics

A large variety of reactions can be studied by examining the kinetics of the reaction after two or more components are mixed. The mixing may occur mechanically (e.g., stopped flow or Biacore) or manually. Table 2 contains equations for analyzing the data of several of the simplest types of these reactions: (1) the unimolecular one-step reaction; (2) the bimolecular self-complementary reaction; (3) the stoichiometric mixture of a bimolecular non-self-complementary reaction; (4) the 2:1 mixture of type 3; and (5) the *n*:1 mixture of type 3. The expected half lives (Table 2) of each reaction indicate the relative kinetics one would observe in each situation. In particular, comparing reaction types 3 and 4, one can see that for the same reaction, reaction type 4 proceeds to equilibrium much more quickly as it is driven by excess of one reactant. Many additional treatments exist.[114,120–123]

**Table 2**    Equations for forward reactions from mixing kinetics.

| Type of reaction | Time dependence of concentration | Half-life of reaction |
|---|---|---|
| 1  Unimolecular one-step reaction $$[AB] \xrightarrow{k_1} [A]+[B]$$ $$[A] \xrightarrow{k_1} [B]$$ | $[AB] = [AB]_0\, e^{-k_1 t}$ | $t_{1/2} = \dfrac{0.693}{k_1}$ |
| 2  Bimolecular self-complementary $$2[A] \xrightarrow{k_{on}} [B]$$ | $[A] = \dfrac{[A]_0}{1 + 2k_{on}t[A]_0}$ | $t_{1/2} = \dfrac{1}{2k_{on}[A]_0}$ |
| 3  Bimolecular stoichiometric mixture, $[A] = [B]$ $$[A]+[B] \xrightarrow{k_{on}} [AB]$$ | $[A] = \dfrac{1}{k_{on}t + 1/[A]_0}$ | $t_{1/2} = \dfrac{1}{[A]_0 k_{on}}$ |
| 4  Bimolecular non-self-complementary, $[A]/[B] = 2/1$ $$[A]+[B] \xrightarrow{k_{on}} [AB]$$ | $[B] = \dfrac{[B]_0}{2\, e^{[B]_0 k_{on} t} - 1}$ | $t_{1/2} = \dfrac{0.405}{[B]_0 k_{on}}$ |
| 5  Bimolecular non-self-complementary, $[A]/[B] = n$ $$[A]+[B] \xrightarrow{k_{on}} [AB]$$ | $[B] = \dfrac{(n-1)[B]_0}{n\, e^{(n-1)[B]_0 k_{on} t} - 1}$ | $t_{1/2} = \dfrac{\ln(2 - 1/n)}{(n-1)[B]_0 k_{on}}$ |

$[A]_0$ and $[B]_0$ are the concentrations of A and B at time = 0. $n$ is the ratio of $[A]$ over $[B]$. All reactions are given for $aA + bB \rightarrow$ products with no back reaction and $a = b = 1$. Equations can be derived for $a,b > 1$ and other reactions by integrating the corresponding rate laws. Examples including the two-step unimolecular and unimolecular reversible reactions can be found in Levine,[123] Chapter 17. Derivations for other reactions can be found in Frost and Pearson[120] and Moore.[121]

### (i)  Stopped flow

Stopped flow mixing is particularly useful for measuring the kinetics of reactions which proceed too quickly to be followed manually. In this protocol, two solutions are mixed mechanically and one observes the concentration of the components optically as the system comes to equilibrium. The time resolution is limited to the time required for complete mixing of the two solutions, around 1 ms. The upper time bound for such a measurement is the length of time the sample chamber can be considered leak free ($\sim 1000$ s or more). Stopped flow is particularly good at measuring reactions involving the association of two strands when mixed well below their $t_m$, or for measuring dissociations of structures induced by altering the pH, the ionic environment, or by adding denaturant. Stopped flow can also be used to prepare samples for folding studies, such as in pulse/quench mixing experiments which are too quick to be done by hand.[124]

### (ii)  Biacore

A recently developed method for looking at formation and dissociation reactions is the Biacore system (Pharmacia Biosensor).[125–127] This technique examines the interactions of molecules at the surface of a gold sensor chip, by measuring the angle where surface plasmon resonance (SPR) occurs. SPR is a condition where plane polarized light hitting the sensor chip is reflected internally. This is seen as a dip in the reflectance intensity, and occurs only at a specific angle. The angle of SPR is dependent on the refractive index at the surface of the sensor chip. Binding or dissociation of molecules to the surface changes the refractive index (and thus the SPR angle), producing a detectable optical signal. Typically, a ligand is immobilized on the sensor chip, a solution of the analyte is passed over it, and the pseudo-first-order kinetics of the optical effect are monitored.

*(iii)  Manual mixing*

For systems where the reaction proceeds over the course of more than a few seconds, analysis of the kinetics by manual mixing becomes possible. After mixing the reactant's progress can be monitored in many ways, including optically or by either native or denaturing gel electrophoresis. Native gels often allow the separation of different folding isomers of the same molecule.[128,129] Denaturing gels can be used if the molecule is being probed either chemically or enzymatically to demonstrate areas protected by folding of or interaction with other molecules.[130-134] For catalytic molecules, denaturing gels also allow one to follow ribozyme functionally as the reaction proceeds (e.g., by the production of some product once the enzyme has achieved its three-dimensional conformation).[133]

### 7.02.2.1.3  Steady-state kinetics

The equations for steady-state kinetic analysis have been covered extensively elsewhere.[112,114,135-137] The power and necessity for this type of analysis is evident because, even for the Michaelis–Menten mechanism with one intermediate, there is not a convenient analytical solution for the rate equations. This method is based on the assumption that, after the initial phases of the reaction, intermediate species reach a constant concentration, a steady state. Using this approximation, expressions for the rate equations become possible. This type of analysis is particularly useful for enzymatic reactions with one or more intermediates, and therefore is the general method used to examine the chemistry and mechanism of ribozyme reactions.[138,139]

## 7.02.2.2  Interpretation of Kinetic Data

One of the most powerful aspects of kinetic analysis is the ability to probe the mechanism of a reaction. Three of the most important ways of accomplishing this are by determining the temperature dependence, the rate law, and reaction order, and then comparing the reaction with the expected diffusion controlled limit. The ultimate goal is to gain information about the activated complex, the structure whose formation is the rate limiting step (RLS) in both the forward and reverse reactions, and which lies at the energetic saddle point between products and reactants.

### 7.02.2.2.1  Temperature dependence of the rate

The temperature dependence of the reaction rate usually is expressed in terms of the Arrhenius activation energy ($E_a$) or the Eyring transition state enthalpy ($\Delta H^\circ\ddagger$) of the reaction. The derivation of the activation energy is analogous to that of the van't Hoff enthalpy ($\Delta H^\circ$) of a reaction, namely by plotting $\ln(k)$ (instead of $\ln K_{eq}$) vs. $1/T$, with the slope equal to $-E_a/RT$. The $y$-intercept corresponds to the Arrhenius frequency factor, which indicates the frequency of collisions that occur with the correct orientation, and which is difficult to determine accurately. The difference between the $E_a$ for the forward and backward reactions is the $\Delta H^\circ$ of a reaction. The Eyring transition state enthalpy, $\Delta H^\circ\ddagger$, equals $E_a - RT$.

These thermodynamic properties give information about the nature of the activated complex (all the steps which occur prior to the rate-limiting step), which, in turn, gives insight into the mechanism of the reaction. For example, the activation energy of a fundamental reaction step must be positive (i.e., the reaction will go faster if the temperature is increased). A decrease in rate when temperature is increased implies the existence of a complex (multistep) mechanism.

### 7.02.2.2.2  Reaction order and rate laws

The rate law for a reaction gives the rate of the reaction as a function of the concentration of the constituents and a rate constant (e.g., rate $= k[A]^a[B]^b$). Generally, rate laws are determined empirically. The reaction order of a substituent is simply its exponent in the rate law (here $a$ and $b$), and reaction order indicates the number of molecules of that species which are involved prior to the rate-limiting step. As we will see in later sections, knowledge of the rate law and reaction orders can provide detailed information on the mechanism of the reaction. The order of a reaction is determined by plotting log[constituent] vs. log[rate] (or rate constant), with the slope equal to the order. For example, the rate of duplex formation depends on the concentration of both strands, and log-log plots produce a slope of 1 for each. Therefore, the overall reaction is second order, and the reaction order (*a, b*) of each strand is 1. (See Levine[123] for a variety of methods to determine the rate law.)

### 7.02.2.2.3 *Diffusion controlled reactions*

It is useful to compare the observed rate of a reaction under study with the rate one would expect if the reaction were diffusion controlled (i.e., one which occurs every time molecules collide). For uncharged particles, Smoluchowski determined the rate of collision to be

$$k = 4\pi r_0 (D_A + D_B) 10^{-3} N_0 \tag{28}$$

where $D_A$ and $D_B$ are the diffusion coefficients of molecules A and B (usually in the range of $D \sim 10^{-5}$ cm$^2$ s$^{-1}$), $N_0$ is Avogadro's number, $r_0$ is the reaction radius (often just the sum of the individual molecular radii, approximately 20–40 Å). Debye modified this expression for charged molecules by multiplying the right-hand side of Equation (1) by $W/(e^W - 1)$, where $W$ is given by

$$W = \frac{z_A z_B e^2}{\varepsilon r_{AB} k_B T} \tag{29}$$

and is the ratio of the electrostatic interaction energy between two molecules $r_{AB}$ apart and the thermal energy $k_B T$ (the Boltzmann constant and the temperature). $z_A e$ and $z_B e$ are the charges on A and B respectively and $\varepsilon$ is the dielectric constant. For oppositely charged molecules, $W/(e^W - 1)$ increases the diffusion controlled rate by 2–10 times, whereas for like-charged molecules, it decreases the rate by a factor of 0.5–0.01.[123] Thus, normal diffusion-limited rate constants range from $10^8$–$10^{11}$ M$^{-1}$ s$^{-1}$ depending on the size and charge of the species involved.

## 7.02.2.3 Rates of Processes in Nucleic Acids

### 7.02.2.3.1 *Mononucleotide association in nonaqueous solvents*

In aqueous solution, mononucleotides associate too weakly to be measured easily. However, individual base pairs can be studied in nonaqueous media because the strength of hydrogen bonding increases in such solvents. Formation of base pairing changes the dipole moment of the bases. This allows the kinetics of the pairing to be studied by application of a strong electric field and then perturbation of the equilibrium with a smaller oscillating field.[112] Measurement of the kinetics of association by this method yields formation constants of $> 10^9$ M$^{-1}$ s$^{-1}$ (fast enough to be considered diffusion controlled), with little dependence on the bases involved. The dissociation rate constants vary depending of the strength of the base pair, ranging between $10^6$ s$^{-1}$ and $10^8$ s$^{-1}$. Thus, the dissociation rates provide the main source of discrimination between pairs of differing stability, a situation that is common in nucleic acid interactions.

### 7.02.2.3.2 *Single-strand stacking and unstacking*

The rates of stacking and unstacking are too fast to be observed by conventional temperature-jump techniques which involve microsecond heating. In general, one simply sees an absorbance increase in the baseline followed by changes due to slower processes (e.g., hairpin, duplex, or tertiary folding). Two examples where the rates of stacking and unstacking have been measured are laser temperature jump on poly (A) and poly (dA) in low salt[140] and cable discharge temperature jump to study poly (A) and poly (C) in high salt.[141,142] Fitting the data to a two-state model (stacked and unstacked) Dewey and Turner[140] found that the rate constant of stack formation for RNA was slower than that of DNA by a factor of $\sim 4$ ($0.7 \times 10^7$ vs. $2.7 \times 10^7$ s$^{-1}$). The rates of unstacking of each were virtually identical at $3-4 \times 10^6$ s$^{-1}$. At low ionic strength (50 mM) both formation and dissociation have positive activation energies. The activation energy of stacking is small and positive for both the ribo- and deoxyoligomers, 13.4 kJ mol$^{-1}$ and 16.8 kJ mol$^{-1}$ respectively. Unstacking has a larger positive activation energy (63.8 vs. 51.7 kJ mol$^{-1}$) which is likely due to the enthalpy of unstacking the bases. The difference, $E_{a,stack} - E_{a,unstack} \simeq -42$ kJ mol$^{-1}$, is equal to the approximate $\Delta H^\circ$ one observes from melting studies.

These results are qualitatively similar to the stacking rates previously observed by fluorescence quenching experiments on the dimer flavin ethenoadenosine dinucleotide, which found $k_{stack} = 1.7 \times 10^8$ s$^{-1}$ and $k_{unstack} = 1.9 \times 10^7$ s$^{-1}$.[143] Using Equation (1) in Table 1, the relaxation time of the system can be calculated, giving ~5 ns, far too fast for study by conventional temperature jump.

It is interesting to note that the rates for stack formation are much slower than those for backbone rotations measured by phosphorus and proton NMR ($10^8$–$10^{10}$ s$^{-1}$),[144,145] as well as slower than syn–anti conformational changes measured by ultrasonic methods ($10^9$ s$^{-1}$).[146] This feature is likely due to the large entropy of activation ($\Delta S^\circ \ddagger$) of the transition state, and depends little on the ionic conditions of the reaction.[140] Analysis of the observed activation entropy, with either a two-state or an Ising model, predicts that formation of the transition state restricts between three and seven rotational degrees of freedom. Thus, formation of the transition state is analogous to holding several conformations of the molecule fixed while one rotamer crosses one rotational barrier.[146]

### 7.02.2.3.3   *Formation and dissociation of double helices*

#### (i) Formation

Formation of a duplex from two separate strands is a second-order process (first-order in each strand) and occurs more slowly than the diffusion-limited rate.[115,147] Second-order rate constants generally range from $10^5$ to $10^7$ M$^{-1}$ s$^{-1}$ with DNA duplexes forming more quickly than RNA.[148–150] The mechanism of helix formation[115,147] centers around the formation of a stable helical nucleus ($N$) as the rate-limiting step in both association and disproportionation of strands. A schematic that incorporates the basic features of the mechanism is shown in Figure 9. As we shall see, this mechanistic framework also can be applied to the formation of other ordered structures such as triple helices.

The initial investigations of duplex formation confirmed that the overall formation reaction was second order. Analysis of the temperature dependence of the reaction rate showed that GC-rich sequences had small positive activation energies of formation, in line with an Ising lattice model, with the formation of the first base pair being rate limiting, thereby yielding $N = 1$. However, the temperature dependence of AU-rich sequences gave small negative activation energies of formation (between $-21$ and $-63$ kJ mol$^{-1}$). That is, annealing proceeded more quickly as the temperature was reduced. Remembering that the activation energy of a simple elementary step cannot be less than zero, this result indicates that duplex formation is not a simple process, but rather involves at least two steps prior to the rate-limiting step.

How many base pairs are in the stable nucleus? Formation of the first base pair is accompanied by a large positive free energy change (mostly due to an entropic penalty for bringing the two strands together) and little enthalpy change, because no stacking is involved. This step cannot be rate limiting because it would exhibit a positive activation energy. It also is unlikely that assigning formation of the second base pair would produce the observed negative $E_{a,on}$, because the $\Delta H^\circ$ for formation of the first base pair $\sim 0$ (see $\Delta H^\circ$ portion of Figure 9). It is only with formation of the third base pair that the predicted $\Delta H^\circ$ of the nucleus ($N$) is in line with the observed activation energy of formation. Formation of a duplex is thus rate limited by a preequilibrium involving the formation of a short helical nucleus. Once formed, this nucleus rapidly zippers to give a fully bonded duplex.

The mechanism noted above can be treated by assigning separate rate constants for formation of the first ($\kappa_1$) and second ($\kappa_2$) base pairs, and a rate constant for zippering ($s$) to all subsequent base pairs. Formation of the first base pair is the most unfavorable, as it contains the entropic penalty of bringing the two strands together (see $\Delta G^\circ$ portion of Figure 9). Formation of the second base pair is less unfavorable, but still different from formation of all subsequent base pairs, which are downhill in a free energy sense.[112,115,147,151] Craig et al.[147] calculated the rate constant for formation of a single A·U base pair. Given a nucleus of two base pairs, the rate constant was $6$–$16 \times 10^6$ s$^{-1}$; for $N = 3$, the rate constant was $1.5$–$2.4 \times 10^6$ s$^{-1}$. This model explains why helix formation is roughly independent of the length of the oligonucleotides. Formation of a short helical nucleus is

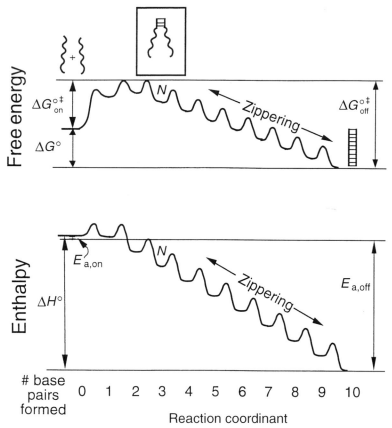

**Figure 9**  Mechanism for duplex formation where the nucleating helix ($N$) contains three base pairs. The $x$-axis is the reaction coordinant for the system indicating the number of base pairs formed. The $y$-axis is the energy of the system, either in free energy (top schematic) or enthalpy (bottom schematic). The enthalpy ($\Delta H^{\circ}$), free energy ($\Delta G^{\circ}$), activation energy ($E_a$), and free energy of forming the nucleating helix ($\Delta G^{\circ\ddagger}$) are indicated.

rate limiting, after which zippering of the helix proceeds spontaneously. Only in the case of long stretches of duplex would the zippering reaction become rate limiting.

Finally, comparison of the expected diffusion limit with the observed rate gives an indication of how many collisions occur prior to formation of the double helix. Comparison between the estimated diffusion limited rate constant ($10^8$–$10^9$ M$^{-1}$ s$^{-1}$) and the observed formation rates ($10^5$–$10^7$ M$^{-1}$ s$^{-1}$) reveals that helix formation occurs once every 10–10 000 collisions.

### (ii) Dissociation

The formalism we have just applied to helix formation can also be applied to helix disproportionation (Figure 9). There, the rate-limiting step is simply fraying of the helix to one base beyond the stable nucleus. As might be expected from the diagram, the $E_a$ of dissociation is large and positive[115,147] and the difference between the $E_{a,on}$ and $E_{a,off}$ equals $\Delta H^{\circ}$. Because the association rate varies only over ~ 100-fold for different complexes (11.8 kJ mol$^{-1}$ at 298 K), it should not be surprising that it is the dissociation rate that provides the main source of discrimination between complexes of differing stability. For example, introduction of single-mismatches into a DNA duplex has little effect on the formation rate, but greatly alters the dissociation rate.[152] Finally, the dissociation rate usually depends more strongly on the counterion concentration, the temperature, and other factors which affect helix stability because it usually involves a much larger stretch of helix than the formation reaction.

### 7.02.2.3.4  *Formation and denaturation of hairpin helices*

Mechanistically, formation of hairpin helices can also be thought of in the same framework as helix formation from individual strands (Figure 9). It is important to realize that the underlying base-pair formation rates for inter- and intramolecular processes are the same, $\sim 10^6$–$10^7 \, \text{s}^{-1}$. Attachment of the strands via the hairpin loop has the effect of raising the local concentration of the strands, increasing the folding rates of these structures and also the stability relative to their parent duplexes. It is only because the intermolecular reaction depends on concentration that the observed rates are much slower. Where one would observe relaxation times for bimolecular duplexes of $\sim 100 \, \text{ms}$, the corresponding times for a hairpin would be in the range of $10^{-3}$–$10^{-5} \, \text{s}$.[43,113,148,153–155] Thus, relaxation kinetics provide a method for determining whether duplex or hairpin is present, and may provide kinetic resolution to transitions that overlap in temperature.[156] As in the duplex case, DNA hairpin formation is somewhat faster than RNA hairpin formation, perhaps due to the higher flexibility of the deoxyribose backbone.[157]

### 7.02.2.3.5  Kinetics of base-pair opening and proton exchange

One of the most general uses of NMR in studying nucleic acids is monitoring the exchangeable imino protons which are involved in base-pairing interactions.[158] The rate of exchange of these protons is measured in two ways, either by magnetization transfer experiments (timescales 2 s to 1 ms) or by physically mixing samples with $D_2O$ and observing the decay of the $^1H$ signal (timescales $> 1$ min).[116] Imino proton exchange is catalyzed by buffer, and becomes faster as the buffer concentration is increased. At low buffer concentration, the exchange lifetimes can be as long as several minutes. The rate of base-pair opening is determined by extrapolating the lifetime to infinite buffer concentration. The rates of exchange depend on the base pair involved, with GU pairs being the fastest, followed by AU/AT pairs (0.5–7 ms) and then GC pairs (4–50 ms) with activation energies of $\sim 80 \, \text{kJ mol}^{-1}$.[116,159–162] The similarity of the activation energies between different bases implies they are related to the chemistry of exchange (the nucleotide acts as the donor and the base acts as the acceptor) rather than the stability of the base pair.[116,161] Some structures have protons that are much more resistant to exchange such as tRNA, Z-DNA, triplexes, or RNA-protein complexes where lifetimes of the order of minutes have been measured.[116,159–161,163]

Several experimenters have used exchange rates in order to assign complex NMR spectra. For tRNA, Crothers *et al.*[156] were able to correlate exchange of groups of imino protons in NMR spectra to relaxation times observed by temperature jump. Leontis and Moore[161] looked at the exchange rates of 5S RNA in the presence and absence of L25 protein to aid in assigning the imino protons and protein binding sites on the helix.

### 7.02.2.3.6  Association of substrates with ribozymes

Ribozymes are RNA molecules capable of acting as catalysts. These RNAs can be roughly segregated into two groups, those that occur naturally (e.g., the group I intron, RNase P, the group II intron, the hammerhead, the hairpin, and hepatitis delta structures) and those that have been discovered via *in vitro* selection experiments (e.g., the class I and II ligases, the kinases, and acylases). Both groups share the feature that often the molecules perform or are selected for a self-modification, but can be redesigned for multiple turnover once the secondary structure is known.

Given that ribozymes are known to be relatively complicated examples of RNA structures, it is perhaps surprising that association of their substrate molecules occurs at about the same rate as simple helix formation in the fastest case and about 100 times more slowly in the slowest (see Table 3). The most surprising example of this is the Group I intron, which is particularly large and contains a significant amount of known tertiary folding.[176,177] There, $k_{on}$ for the matched substrate is 0.3–$7.6 \times 10^6 \, \text{M}^{-1} \, \text{s}^{-1}$,[138,139,164] equal to $k_{cat}/K_m$, and near the oligonucleotide association rate for RNA duplexes. Table 3 gives the $k_{on}$ values for a number of ribozymes with their respective oligonucleotide substrates. Where $k_{on}$ has not been measured, $k_{cat}/K_m$ is given as it has the dimensions of a second-order rate constant and is equal to $k_{on}$ under certain conditions.

It is interesting to note that the substrate association/dissociation rate places an upper bound on the rate of ribozyme reactions involving oligonucleotide substrates. In some sense, ribozyme reactions can never be diffusion controlled because helix formation is not. Taking this into account, ribozymes can be said to represent the perfect enzyme if they modify every oligonucleotide that they bind.[138] However, more substrate binding energy does not always give faster catalysis, as ribozyme reactions are often limited by product release. A good example of this is the Group I intron with a

**Table 3** Association rate constants and catalytic efficiencies of ribozymes.

| Ribozyme | Association rate constant $(k_1)^a$ $(M^{-1} s^{-1})$ | $k_{cat}/K_m^a$ $(M^{-1} s^{-1})$ | Refs. |
|---|---|---|---|
| Group I intron | $0.3–1.5 \times 10^6$ $3.9–7.6 \times 10^6$ | $0.3–1.5 \times 10^6$ | 138, 139, 164 |
| RNase P alone |  | $2.2 \times 10^5$ | 165, 166 |
| RNase P w/ C5 protein |  | $1.4 \times 10^7$ | 165, 166 |
| Hammerhead | $0.2–2.0 \times 10^6$ | $0.2–2.0 \times 10^6$ | 167, 168 |
| Hairpin | $1.6 \times 10^5$ | $1.6 \times 10^5$ | 169, 170 |
| Kinase | $2.0 \times 10^4$ | $1.3 \times 10^3$ | 171 |
| Group II intron |  | $5.5 \times 10^4$ | 172 |
| Hepatitus Delta Virus |  | $8 \times 10^4$ | 173 |
| Class I ligase |  | $1.8 \times 10^5$ | 174 |
| Acylase |  | $6.3 \times 10^4$ | 175 |

$^a$All values are given in $M^{-1} s^{-1}$ for the association rate constants and $k_{cat}/K_m$.

matched and mismatched substrate.[138,139] The presence of the mismatch slightly slows the association rate, but greatly enhances the dissociation rate, thereby giving the enzyme a greatly increased $k_{cat}$ relative to the perfect match. For a quantitative treatment of this effect, see Herschlag.[178]

### 7.02.2.3.7   *Association rates of loop–loop complexes*

In addition to ribozymes, the association rates in other ordered molecules also have been studied. Two examples of this are the association of tRNAs with complementary anticodon loops[179] and the association of oligonucleotides complementary to anticodon loops.[180,181] In both cases, the association rate constants were found to be similar to the oligonucleotide case, with $k_{on}$ of $3–10 \times 10^6$ $M^{-1}$ $s^{-1}$. In the case of the tRNA association between yeast Phe (GmAA) and *E. coli* tRNA Glu (s2UUC), the complex had a much greater stability than expected for a three-base-pair helix. This feature was rationalized in terms of cohelical stacking of the two anticodon stems on top of the base-paired interaction, and the source of the discrimination was found to be the dissociation rate.[179]

### 7.02.2.3.8   *Kinetics of triple helix formation and dissociation*

Triplex formation has been studied by a number of different laboratories, with the bulk of work centering on the $PY \cdot PU \cdot PY$ (two pyrimidine) variety. The results indicate that triple helix formation is mechanistically similar to duplex formation in that it is not diffusion controlled, and that formation of triplex involves a preequilibrium in which formation of a stable helical nucleus ($N$) is the rate-limiting step (just as in Figure 9). The main differences between duplex and triplex formation are (i) that triplex formation (and often dissociation) is $100–10^4$ times slower than for duplex and (ii) the reaction depends on pH for triplexes containing $C^+ \cdot G \cdot C$ triples because the Hoogsteen pairing requires protonation of the third-strand C.

### (i)   Intermolecular $PY \cdot PU \cdot PY$ *triplexes*

The kinetics of intermolecular triplex formation were studied first in $poly(U) + poly(A) + poly(U)$.[182,183] The process was clearly slower than duplex formation, and Fresco and co-workers[183] derived a second-order rate constant for triplex formation of $10^4$ $M^{-1}$ $s^{-1}$. Temperature-jump and stopped-flow studies on oligonucleotides of the form $U_n + A_n$ produced a similar rate, $\sim 100$ times slower than duplex formation.[115]

The field remained little studied until the discovery that triplexes could be used as a means of site-specific recognition of duplex DNA.[184,185] In an effort to understand the kinetics of this process, Dervan and co-workers[186] studied the kinetics of a mixed-sequence oligonucleotide binding to a target duplex using a restriction enzyme assay. They found that triplex formation rates increased with decreasing pH ($\sim$ eightfold per pH unit) and increasing $Mg^{2+}$ concentration. Surprisingly, the rate of triplex formation decreased when the $Na^+$ concentration was increased, perhaps due to competition with the $Mg^{2+}$ and spermine present. The rate of triplex formation at pH 6.8 was found to be $1.8 \times 10^3 \, M^{-1} \, s^{-1}$, slightly slower, but in line with previous studies. In addition, the dissociation rates were found to be very slow, with a half life of $\sim 12$ h ($k_{off} = 1 \times 10^{-5} \, s^{-1}$).

Hélène and co-workers[187] studied the effects of salt and mismatches on triplex kinetics using a hysteresis assay at pH 6.8. The values of $k_{on}$ for the 22-nucleotide third strand were somewhat slower than those measured in other systems, ranging from 13 to 900 $M^{-1} \, s^{-1}$ depending on the $Na^+$ concentration (reaction order = 1.5). This slower rate is almost certainly due to the lack of divalent or multivalent ions, as the formation rate increases to $\sim 10^4 \, M^{-1} \, s^{-1}$ when $Mg^{2+}$ is added. In line with the duplex case, mismatches in the triplex have little effect on the formation rate constant. The major source of discrimination between complexes of differing stability was the dissociation rate constant, which varied over 1000-fold ($10^{-6}$–$10^{-3} \, s^{-1}$) between matched and mismatched triplexes. The temperature dependences of the formation and dissociation rates allowed prediction of a mechanism similar to that in Figure 9 for triplex. One difference is the size of the stable nucleus which contains three to five base triples (based on the large negative value of $E_{a,on} = -92 \, kJ \, mol^{-1}$) instead of only the one to three base pairs needed in the duplex case.

In an effort to understand the differences in stability observed between RNA, DNA, and hybrid triple helices,[81] the kinetics and mechanism of triplexes were studied as a function of backbone (RNA vs. DNA), structure (bulged vs. perfect helix), and sequence (GC-rich vs. AT-rich) using stopped flow.[188,189] This was accomplished by studying the association rates at pH 5.0 and the dissociation rates at pH 7.2. The formation rate constants ($k_{on}$) were second order in strand, 100–1000 times slower than duplex rates, and had small pH dependences (reaction order $[H] \sim 1$) at pH 5.0. At that pH, $k_{on}$ varies $\sim 40$-fold from fastest to slowest, and depends primarily on the nature of the duplex site ($DD > DR > RR > RD$). The relative formation rates parallel the relative stability observed in affinity cleavage experiments in the same series.[190] Introduction of a two-base bulge into the center of the helix to increase accessibility of the major groove[191] actually slows triplex formation two-fold, perhaps indicating that triplexes nucleate preferentially at the ends of the helix where the major groove is more exposed. Finally, at pH 5.0, GC-rich triplexes form $\sim 4$ times more quickly than those rich in AT. However, above pH 5.6, AT-rich triplexes form more quickly due to differences in the pH dependence of $k_{on}$.

As in the case of the mismatched triplex, the dissociation rate constants ($k_{off}$) provide the main source of discrimination between complexes of differing stability, varying over a range $> 13\,000$-fold at pH 7.2.[188,189] It is interesting to note that the $k_{off}$ values mirror but do not fully account for the large differences in equilibrium stability observed in the same triplexes at pH 5.5.[81] The range of $k_{on}/k_{off}$ values predicts a difference in stabilities of $\sim 25 \, kJ \, mol^{-1}$ rather than the 42–50 $kJ \, mol^{-1}$ observed. This difference may be due to the fact that the size of the helical nucleus ($N$) is predicted to be smaller at pH 5.0 (where it is 1–3 triples) than at pH 7.2 (where it is 5–8 triples).[188,189] The increase in $N$ implies $k_{on}$ plays an increasing role in discrimination of complexes at high pH in this system, which would account for the difference in magnitude between the kinetic and equilibrium measurements.

Several other studies of triplex kinetics have been conducted, using techniques as diverse as filter binding,[192] fluorescence energy transfer,[193] the Biacore apparatus,[194] stopped flow,[195] and manual mixing.[196] In general, formation rate constants of $\sim 10^3 \, M^{-1} \, s^{-1}$ are observed with small negative activation energies of formation. One exception is the binding of circular pyrimidine oligonucleotides to a purine target where $k_{on} \sim 10^6 \, M^{-1} \, s^{-1}$, similar to the duplex case.[195] These rates are almost certainly due to the fact that there the RLS is formation of the duplex, after which the triplex zippers together quickly. This is because the expected $t_{1/2}$ of intramolecular triplex formation would be less than or equal to that of the duplex. (For intramolecular triplexes with loops of 4 nt, $k_{on} \sim 500$ at pH 6.0.[188,189] Using the equations in Table 2, $t_{1/2}$ of the formation reaction would be $\sim 1.4$ ms. At the higher pH and $[Mg^{2+}]$,[184] $k_{on}$ should be $> 10 \, s^{-1}$, which would produce a $t_{1/2}$ of $< 70$ ms. Under the conditions of their association experiments, the $t_{1/2}$ of duplex formation would be 70–140 ms (if $k_{on} = 2$–$4 \times 10^6 \, M^{-1} \, s^{-1}$).) This would explain the lack of pH dependence of the formation rate constant, and indicate that the reported rates do not reflect triplex formation, but the $k_{on}$ of the parent duplexes.

*(ii) Intermolecular* PU·PY·PU *triplexes*

Much less work has been done on the kinetics of PU·PY·PU (two-purine) triplexes than their two-pyrimidine counterparts. No mechanistic work is available, but a kinetic study indicates that the formation and dissociation rates are similar to those seen in two-pyrimidine triplexes. Wilson and co-workers[197] found $k_{on} = 2.3 \times 10^4 \, M^{-1} \, s^{-1}$ and $k_{off} = 1.6 \times 10^{-5} \, s^{-1}$, and that triplex formation is greatly aided by the presence of $Mg^{2+}$. This stabilizing effect is thought to manifest itself in the dissociation rate as preformed complexes loaded on gels lacking $Mg^{2+}$ show no triplex, whereas gels containing the cation show triplex formation.

### (iii) Intramolecular PY·PU·PY triplexes

As with the duplex, the fundamental rates of base triple formation are the same in the inter- and intramolecular cases. Similarly, attachment of the third strand via the loop has the effect of both greatly speeding formation and stabilizing the triplex. One difference between triplex and duplex hairpins is that there are two varieties of triplex loops depending on which end of the pyrimidine strand acts as the third strand (Y3 if the 3'-end, Y5 if the 5'-end). Roberts and Crothers[198] investigated both varieties and found that Y3 triplexes form ~ 10 times more quickly than Y5 isomers. However, this kinetic bias can be reversed if the purine strand is extended on the side of the loop. Where Y3 isomers form faster, it is likely due both to positioning of the third strand over the nucleation site, and accessibility of the purine strand in the duplex. Adding the extension to the purine strand effectively blocks the path of the third strand to the major groove in the Y3 isomer, slowing its formation 200-fold. The extension does not obstruct Y5 formation, and thus leaves its formation rate relatively unchanged.

The effect of loop size on the kinetics and stability of intramolecular triplexes also has been examined.[198] There, in contrast to intermolecular duplex and triplex formation, $k_{on}$ plays the major role in distinguishing complexes of differing stability. For Y3 isomers, this is particularly true, as $k_{off}$ is independent of loop size and equal to the value obtained for intermolecular triplex formation. The $k_{off}$ for Y5 isomers shows a slight dependence on loop size, becoming slower and 4–8 kJ mol$^{-1}$ more stable as one moves from 20 to 4 nucleotides. Finally, the $\Delta G^\circ$ loop indicates that triplex loops are more stabilizing than those that close RNA hairpins, providing a higher local concentration of strand (~ 30 mM for 4 nt loops), likely due to the higher preorganization in the triplex.

### 7.02.2.3.9    Higher-order folding kinetics of RNA

In this section we examine what is known about the folding of two well-studied molecules, tRNA and the Group I intron. These processes may involve both standard base pairing and noncanonical interactions. Two general statements can be made about such folding: (i) it usually occurs on a slower timescale than simple helix formation, and (ii) it is often specifically stabilized (relative to duplex) by addition of divalent metal ions, especially $Mg^{2+}$.

### (i) tRNA

The folding of the cloverleaf structure of tRNA is one of the most studied examples of RNA folding. It is perhaps daunting to realize that, even in this relatively small molecule (~ 75 nt), four to five distinct transitions can be observed via temperature jump kinetics.[148,156] Similarly, phase diagrams of tRNA indicate at least four distinct zones depending on the salt concentration and temperature.[199] Perhaps the most difficult part of determining the mechanism of folding and unfolding is assigning the transitions seen in the relaxation kinetics to discrete events in the molecule.

An illustration of how this assignment process was approached can be seen in a series of papers on *E. coli* tRNA$^{fMet}$ from the Crothers laboratory (for a summary see Stein and Crothers[200]). Examination of the molecule indicated that there are four distinct transitions (two slow, low-temperature transitions[201] were later determined to be from a mixture of two tRNA$^{fMet}$ species[156]). To assign the transitions they examined (a) the difference absorbance spectrum of each (to determine the G·C content of helix melted), (b) optical signals from noncloverleaf pairing bases, (c) the predicted stability of the helices, (d) the kinetics of each process,[156,199,201] (e) the correlation between NMR imino proton signals and the observed relaxation times,[156] and (f) comparison of fragment melting to the melting of the entire tRNA.[148]

The low-temperature transition was the slowest of the four and likely corresponds to folding/unfolding of both the tertiary structure and the D helix.[156,201] This process occurs with $k_{on} = 100-$

300 s$^{-1}$, much slower than one would expect for normal hairpin melting.[43] Stein and Crothers[200] demonstrated that this transition is specifically stabilized by both Mg$^{2+}$ and Mn$^{2+}$, although these divalent ions are not required for its formation. In addition, they proposed that formation of the dhU helix acts as a nucleus for formation of the tertiary interactions.

The rest of the transitions occur in the order T$\psi$C, anticodon, and acceptor stem. The relaxation time of each transition is progressively slower, moving from $t = 10^{-5}$ s, $t = 10^{-4}$ s, $t = 10^{-3}$ s, respectively. This can be explained in terms of each hairpin melting from a progressively larger and larger loop, effectively slowing the forward relaxation time.

Finally, there are also much slower rates that can be observed as the molecule isomerizes between forms. A low salt "extended" structure of tRNA$^{Val}$ or tRNA$^{fMet}$ denoted form III can be converted to the native form I structure by a salt jump.[199] This is thought to occur by an isomerization within the tRNA, and occurs with $t \sim 10^{3}$ s.

*(ii)  Group I intron*

The folding and kinetics of the group I ribozyme from *Tetrahymena* have been examined by a large number of laboratories, making it one of the most thoroughly characterized RNA structures available. In particular, the L-21 ScaI intron[176,177,202] has been the subject of particular focus. However, as one may infer from its size ($>400$ nt) and complex secondary and tertiary structure, understanding its folding presents a much more daunting problem than tRNA. In spite of this, a basic framework for the folding of this molecule is beginning to be understood (see Zarrinkar and Williamson[203] and references therein). The kinetics of two types of processes will be discussed here: (a) tertiary interactions involving the P1 helix and (b) formation of the native folded structure from partially denatured versions of the molecule.

As noted in Table 3, association of the substrate with the intron occurs at a rate similar to that seen in oligonucleotide duplexes;[138,139] however, binding is much more stable than expected for that size of duplex.[204] This extra free energy implies a two-step binding process, formation of base pairs followed by some stabilization likely due to tertiary contacts.

A number of studies have been conducted to measure directly the kinetics of binding of a pyrene-labeled substrate to the ribozyme.[164,204–206] The initial work indicates that the binding occurs in two steps, the first corresponding to base pairing of the substrate ($k = 4 \times 10^{6}$ M$^{-1}$ s$^{-1}$) in good agreement with Herschlag and Cech.[138,139] This is followed by docking of the P1 helix onto the ribozyme, which is $\sim 1000$ times slower than intramolecular base pairing, with $k_{on}$ values of 2.5 s$^{-1}$. Combined with $k_{off}$ for this process of 0.02 s$^{-1}$, this rationalizes most of the extra binding energy.

Herschlag[207] also found evidence for two-step binding when looking at J 1/2 mutants. However, the rate for the second step was $\sim 1333$ s$^{-1}$, many times faster than the stopped-flow experiment. Turner and co-workers[206] demonstrated that this difference is likely due to the temperatures of experiments (15 °C vs. 50 °C respectively) as there is a relatively large activation energy of docking ($E_{a,on} = 92$ kJ mol$^{-1}$). This large activation energy implies that docking of the P1 helix does not occur by simple diffusion.[177]

The kinetics of Group I intron folding have been studied by two methods: (a) folding induced by addition of Mg$^{2+}$ at constant temperature,[133,203] and (b) folding brought on by temperature drop.[208] The overall rates of folding one observes differ greatly depending on the method. In the first case, $k_{fold}$ is $\sim 0.010$–$0.012$ s$^{-1}$ for formation of the enzymatically active structure, whereas $k_{fold}$ as measured by kethoxal probing ranges from 1 to $8 \times 10^{-4}$ s$^{-1}$, two orders of magnitude slower. This kinetic difference is likely due to differences in the starting structures in each case.

Using both a catalytic and an RNase H protection assay for folding, Zarrinkar and Williamson[133,203] have been able to propose an overall mechanism for the Mg$^{2+}$-induced pathway of the intron. The first regions to form are the stacked P4-P6 structures, which require Mg$^{2+}$. This process is fast and is followed by the RLS, Mg$^{2+}$-independent folding of the P3-P7 domain. Mutational analysis of this process has illustrated that it is hierarchical, that is P3-P7 formation is dependent on the presence of P4-P6. Furthermore, the triple helix that forms around P4-P6 appears critical for formation of P3-P7, as a single base charge there slows the folding six-fold without changing the rate-limiting step.

### 7.02.2.4  Conclusions

While we have a good understanding of helix formation, we are only beginning to understand the rules that govern higher-order RNA and DNA folding. Knowledge of the kinetic behavior of nucleic acids has many benefits for the understanding of folding and design of functional molecules. An example of this is seen clearly in the Group I intron, where destabilizing the substrate ribozyme interaction actually increases the rate of reaction.[138,139] Kinetic knowledge of newly discovered tertiary interactions (e.g., Cate *et al.*[209,210]) may one day allow us to design RNA molecules with desired folds and predict folding of known RNAs with unknown structures.

### 7.02.3 REFERENCES

1. G. M. Blackburn and M. J. Gait (eds.), "Nucleic Acids in Chemistry and Biology," Oxford University Press, Oxford, 1996.
2. F. Eckstein (ed.), "Oligonucleotides and Analogues: A Practical Approach," IRL Press, Oxford, 1991.
3. M. H. Caruthers, G. Beaton, J. W. Wu, and W. Wiesler, *Methods Enzymol.*, 1992, **211**, 3.
4. J. Feigon, V. Sklenář, E. Wang, D. E. Gilbert, R. F. Macaya, and P. Schultze, *Methods Enzymol.*, 1992, **211**, 235.
5. D. G. Gorenstein, *Methods Enzymol.*, 1992, **211**, 254.
6. E. Taillandier and J. Liquier, *Methods Enzymol.*, 1992, **211**, 307.
7. R. M. Clegg, *Methods Enzymol.*, 1992, **211**, 353.
8. D. M. Gray, R. L. Ratliff, and M. R. Vaughn, *Methods Enzymol.*, 1992, **211**, 389.
9. K. J. Breslauer, E. Friere, and M. Straume, *Methods Enzymol.*, 1992, **211**, 533.
10. H. F. Fisher and N. Singh, *Methods Enzymol.*, 1995, **259**, 194.
11. R. A. Friedman and B. Honig, *Biophys. J.*, 1995, **69**, 1528.
12. D. H. Turner, *Curr. Opin. Struct. Biol.*, 1996, **6**, 299.
13. T. M. Lohman, L. B. Overman, M. E. Ferrari, and A. G. Kozlov, *Biochemistry*, 1996, **35**, 5272.
14. J. M. Sturtevant, *Proc. Natl. Acad. Sci. USA*, 1977, **74**, 2236.
15. J. R. Livingstone, R. S. Spolar, and M. T. Record, Jr., *Biochemistry*, 1991, **30**, 4237.
16. R. S. Spolar and M. T. Record, Jr., *Science*, 1994, **263**, 777.
17. Y. Liu and J. M. Sturtevant, *Biochemistry*, 1996, **35**, 3059.
18. A. P. Sarvazyan, *Annu. Rev. Biophys. Biophys. Chem.*, 1991, **20**, 321.
19. T. V. Chalikian, A. P. Sarvazyan, and K. J. Breslauer, *Biophys. Chem.*, 1994, **51**, 89.
20. C. A. Royer, *Methods Enzymol.*, 1995, **259**, 357.
21. C. R. Robinson and S. G. Sligar, *Methods Enzymol.*, 1995, **259**, 395.
22. M. T. Record, Jr., C. F. Anderson, and T. M. Lohman, *Q. Rev. Biophys.*, 1978, **11**, 103.
23. M. T. Record, Jr., J.-H. Ha, and M. A. Fischer, *Methods Enzymol.*, 1991, **208**, 291.
24. B. Jayaram and D. L. Beveridge, *Annu. Rev. Biophys. Biomol. Struct.*, 1996, **25**, 367.
25. G. S. Manning, *Q. Rev. Biophys.*, 1978, **11**, 179.
26. C. F. Anderson and M. T. Record, Jr., *Annu. Rev. Phys. Chem.*, 1982, **33**, 191.
27. C. F. Anderson and M. T. Record, Jr., *Annu. Rev. Biophys. Biophys. Chem.*, 1990, **19**, 423.
28. I. Rouzina and V. A. Bloomfield, *J. Phys. Chem.*, 1996, **100**, 4292.
29. I. Rouzina and V. A. Bloomfield, *J. Phys. Chem.*, 1996, **100**, 4305.
30. M. C. Olmsted, C. F. Anderson, and M. T. Record, Jr., *Proc. Natl. Acad. Sci. USA*, 1989, **86**, 7766.
31. M. C. Olmsted, C. F. Anderson, and M. T. Record, Jr., *Biopolymers*, 1991, **31**, 1593.
32. J. R. Williamson, *Annu. Rev. Biophys. Biomol. Struct.*, 1994, **23**, 703.
33. M. T. Record, C. P. Woodbury, and T. M. Lohman, *Biopolymers*, 1976, **15**, 893.
34. E. Westhof, "Water and Biological Macromolecules," CRC Press, Boca Raton, FL, 1993.
35. G. E. Plum and K. J. Breslauer, *Curr. Opin. Struct. Biol.*, 1995, **5**, 682.
36. R. R. Krug, W. G. Hunter, and R. A. Grieger, *J. Phys. Chem.*, 1976, **80**, 2335.
37. J. D. Puglisi and I. Tinoco, Jr., *Methods Enzymol.*, 1989, **180**, 304.
38. G. E. Plum, A. P. Grollman, F. Johnson, and K. J. Breslauer, *Biochemistry*, 1995, **34**, 16 148.
39. L. M. Marky and K. J. Breslauer, *Biopolymers*, 1987, **26**, 1601.
40. E. M. Evertsz, K. Rippe, and T. M. Jovin, *Nucleic Acids Res.*, 1994, **22**, 3293.
41. J. Applequist and V. Damle, *J. Chem. Phys.*, 1963, **39**, 2719.
42. M. Petersheim and D. H. Turner, *Biochemistry*, 1983, **22**, 256.
43. J. Gralla and D. M. Crothers, *J. Mol. Biol.*, 1973, **73**, 497.
44. P. L. Privolov and S. A. Potekhin, *Methods Enzymol.*, 1986, **131**, 4.
45. E. Neumann and T. Ackerman, *J. Phys. Chem.*, 1969, **73**, 2170.
46. J. M. Sturtevant, *Annu. Rev. Phys. Chem.*, 1987, **38**, 463.
47. D. S. Pilch, G. E. Plum, and K. J. Breslauer, *Curr. Opin. Struct. Biol.*, 1995, **5**, 334.
48. H. Naghibi, A. Tamura, and J. M. Sturtevant, *Proc. Natl. Acad. Sci. USA*, 1995, **92**, 5597.
49. J. F. Brandts and L.-N. Lin, *Biochemistry*, 1990, **29**, 6927.
50. T. Wiseman, S. Williston, J. F. Brandts, and L.-N. Lin, *Anal. Biochem.*, 1989, **179**, 131.
51. J. E. Ladbury, *Structure*, 1995, **3**, 635.
52. J. Wyman and S. J. Gill, "Binding and Linkage: Functional Chemistry of Biological Macromolecules," University Science Books, Mill Valley, CA, 1990.
53. J. Q. Wu and R. B. Macgregor, Jr., *Biochemistry*, 1993, **32**, 12 531.
54. P. O. P. T'so, in "Basic Principles in Nucleic Acid Chemistry," ed. P. O. P. T'so, Academic Press, New York, 1974, vol. 1, p. 453.
55. D. Pörshke, O. C. Uhlenbeck, and F. H. Martin, *Biopolymers*, 1973, **12**, 1313.
56. D. W. Appleby and N. R. Kallenbach, *Biopolymers*, 1973, **12**, 2093.
57. L. A. Marky, K. S. Blumenfeld, S. Kozlowski, and K. J. Breslauer, *Biopolymers*, 1983, **22**, 1247.

58. G. Varani, *Annu. Rev. Biophys. Biomol. Struct.*, 1995, **24**, 379.
59. M. M. Senior, R. A. Jones, and K. J. Breslauer, *Proc. Natl. Acad. Sci. USA*, 1988, **85**, 6242.
60. S. Wang, M. A. Booher, and E. T. Kool, *Biochemistry*, 1994, **33**, 4639.
61. S. M. Freier, D. Alkema, A. Sinclair, T. Neilson, and D. H. Turner, *Biochemistry*, 1985, **24**, 4533.
62. M. M. Senior, R. A. Jones, and K. J. Breslauer, *Biochemistry*, 1988, **27**, 3879.
63. J. C. Williams, S. C. Case-Green, K. U. Mir, and E. M. Southern, *Nucleic Acids Res.*, 1994, **22**, 1365.
64. M. Lu, Q. Guo, L. A. Marky, N. C. Seeman, and N. R. Kallenbach, *J. Mol. Biol.*, 1992, **223**, 781.
65. J. E. Ladbury, J. M. Sturtevant, and N. B. Leontis, *Biochemistry*, 1994, **33**, 6828.
66. M. Kamiya, H. Torigoe, H. Shindo, and A. Sarai, *J. Am. Chem. Soc.*, 1996, **118**, 4532.
67. G. Vesnaver and K. J. Breslauer, *Proc. Natl. Acad. Sci. USA*, 1991, **88**, 3569.
68. K. J. Breslauer, R. Frank, H. Blöcker, and L. A. Marky, *Proc. Natl. Acad. Sci. USA*, 1986, **83**, 3746.
69. S. M. Freier, R. Kierzek, J. A. Jaeger, N. Sugimoto, M. H. Caruthers, T. Neilson, and D. H. Turner, *Proc. Natl. Acad. Sci. USA*, 1986, **83**, 9373.
70. H. H. Klump, in "Biochemical Thermodynamics," 2nd edn., ed. M. N. Jones, Elsevier, Amsterdam, 1988, p. 100.
71. J. SantaLucia, Jr., H. T. Allawi, and P. A. Seneviratne, *Biochemistry*, 1996, **35**, 3555.
72. H. DeVoe and I. Tinoco, Jr., *J. Mol. Biol.*, 1962, **4**, 500.
73. M. J. Doktycz, M. D. Morris, S. J. Dormady, K. L. Beattie, and K. B. Jacobson, *J. Biol. Chem.*, 1995, **270**, 8439.
74. S. G. Delcourt and R. D. Blake, *J. Biol. Chem.*, 1991, **266**, 15160.
75. J. G. Wetmur, *CRC Crit. Rev. Biochem. Mol. Biol.*, 1991, **26**, 227.
76. K. J. Breslauer, in "Methods in Molecular Biology: Protocols for Oligonucleotide Conjugates," ed. S. Agrawal, Humana Press, NJ, 1994, vol. 26, p. 347.
77. D. H. Turner, N. Sugimoto, and S. M. Freier, *Annu. Rev. Biophys. Biophys. Chem.*, 1988, **17**, 167.
78. M. J. Serra and D. H. Turner, *Methods Enzymol.*, 1995, **259**, 242.
79. F. H. Martin and I. Tinoco, Jr., *Nucleic Acids Res.*, 1980, **8**, 2295.
80. K. B. Hall and L. W. McLaughlin, *Biochemistry*, 1991, **30**, 10606.
81. R. W. Roberts and D. M. Crothers, *Science*, 1992, **258**, 1463.
82. L. Ratmeyer, R. Vinayak, Y. Y. Zhong, G. Zon, and W. D. Wilson, *Biochemistry*, 1994, **33**, 5298.
83. E. A. Lesnik and S. M. Freier, *Biochemistry*, 1995, **34**, 10807.
84. N. Sugimoto, S. Nakano, M. Katoh, A. Matsmura, H. Nakamuta, T. Ohmichi, M. Yoneyama, and M. Sasaki, *Biochemistry*, 1995, **34**, 11211.
85. W. N. Hunter, *Methods Enzymol.*, 1992, **211**, 221.
86. G. V. Fazakerley and Y. Boulard, *Methods Enzymol.*, 1995, **261**, 145.
87. G. E. Plum and K. J. Breslauer, *Ann. N. Y. Acad. Sci.*, 1994, **726**, 45.
88. E. Westhof, *Annu. Rev. Biophys. Biophys. Chem.*, 1988, **17**, 125.
89. H. Berman, *Curr. Opin. Struct. Biol.*, 1994, **4**, 345.
90. M. Kochoyan and J. L. Leroy, *Curr. Opin. Struct. Biol.*, 1995, **5**, 329.
91. T. V. Chalikian, A. P. Sarvazyan, G. E. Plum, and K. J. Breslauer, *Biochemistry*, 1994, **33**, 2394.
92. B. Schneider, D. Cohen, and H. M. Berman, *Biopolymers*, 1992, **32**, 725.
93. A. H. Elcock and J. A. McCammon, *J. Am. Chem. Soc.*, 1995, **117**, 10161.
94. L. M. Marky and K. J. Breslauer, *Proc. Natl. Acad. Sci. USA*, 1987, **84**, 4359.
95. T. Lundbäck and T. Härd, *Proc. Natl. Acad. Sci. USA*, 1996, **93**, 4754.
96. T. V. Chalikian, G. E. Plum, A. P. Sarvazyan, and K. J. Breslauer, *Biochemistry*, 1994, **33**, 8629.
97. R. W. Wagner, M. D. Matteucci, D. Grant, T. Huang, and B. C. Froehler, *Nature Biotechnology*, 1996, **14**, 840.
98. K. Rippe and T. M. Jovin, *Methods Enzymol.*, 1992, **211**, 199.
99. D. S. Pilch and R. H. Shafer, *J. Am. Chem. Soc.*, 1993, **115**, 2565.
100. Y.-K. Cheng and B. M. Pettitt, *Prog. Biophys. Mol. Biol.*, 1992, **58**, 225.
101. G. E. Plum, D. S. Pilch, S. F. Singleton, and K. J. Breslauer, *Annu. Rev. Biophys. Biomol. Struct.*, 1995, **24**, 319.
102. J.-S. Sun, T. Garestier, and C. Hélène, *Curr. Opin. Struct. Biol.*, 1996, **6**, 327.
103. R. W. Roberts and D. M. Crothers, *Proc. Natl. Acad. Sci. USA*, 1996, **93**, 4320.
104. E. T. Kool, *Annu. Rev. Biophys. Biomol. Struct.*, 1996, **25**, 1.
105. D. A. Erie, R. A. Jones, W. K. Olson, and N. K. Sinha, *Biochemistry*, 1989, **28**, 268.
106. L. A. Marky, N. R. Kallenbach, K. A. McDonough, N. C. Seeman, and K. J. Breslauer, *Biopolymers*, 1987, **26**, 1621.
107. D. M. J. Lilley and R. M. Clegg, *Annu. Rev. Biophys. Biomol. Struct.*, 1993, **22**, 299.
108. N. C. Seeman and N. R. Kallenbach, *Annu. Rev. Biophys. Biomol. Struct.*, 1994, **23**, 53.
109. J. R. Williamson, *Curr. Opin. Struct. Biol.*, 1993, **3**, 357.
110. M. Eigen and L. de Maeyer, in "Technique of Organic Chemistry," 2nd edn., eds. S. L. Friess, E. S. Lewis, and A. Weissberger, Interscience, New York, 1963, vol. 8, p. 895.
111. G. G. Hammes and P. R. Schimmel, in "The Enzymes," 3rd edn., ed. P. D. Boyer, Academic Press, New York, 1970, vol. 2, p. 67.
112. C. R. Cantor and P. R. Schimmel, "Biophysical Chemistry Part III: The Behavior of Biological Macromolecules," W. H. Freeman, New York, 1980.
113. D. H. Turner, in "Chemical Kinetics Studies," Invest. Rates Mech. React., Tech. Chem., 4th Pt. 2 edn., New York, 1986, vol. 6, p. 141.
114. G. G. Hammes, "Principles of Chemical Kinetics," Academic Press, New York, 1978.
115. D. Pörschke and M. Eigen, *J. Mol. Biol.*, 1971, **62**, 361.
116. M. Guéron and J.-L. Leroy, *Methods Enzymol.*, 1995, **261**, 383.
117. A. N. Lane, *Methods Enzymol.*, 1995, **261**, 413.
118. G. King, J. W. Harper, and Z. Xi, *Methods Enzymol.*, 1995, **261**, 436.
119. B. H. Robinson and G. P. Drobny, *Methods Enzymol.*, 1995, **261**, 451.
120. A. A. Frost and R. G. Pearson, "Kinetics and Mechanism," 2nd edn., Wiley, New York, 1961.
121. W. J. Moore, "Physical Chemistry," 5th edn., Prentice-Hall, Englewood Cliffs, NJ, 1972.
122. D. Eisenberg and D. M. Crothers, "Physical Chemistry with Applications to the Life Sciences," Benjamin/Cummins, Menlo Park, CA, 1979.

123. I. N. Levine, "Physical Chemistry," 3rd edn., McGraw-Hill, New York, 1988.
124. H. Roder, G. A. Elove, and S. W. Englander, *Nature*, 1988, **335**, 700.
125. U. Jönsson, L. Fägerstam, B. Ivarsson, B. Johnsson, R. Karlsson, K. Lundh, S. Löfås, B. Persson, H. Roos, I. Rönnberg, S. Sjölander, E. Stenberg, R. Ståhlberg, C. Urbaniczky, H. Östlin, and M. Malmqvist, *BioTechniques*, 1991, **11**, 620.
126. L. G. Fagerstam, A. Frostell-Karlsson, R. Karlsson, B. Persson, B. Ronnberg, and I. Ronnberg, *J. Chromatogr.*, 1992, **597**, 397.
127. M. Malmqvist, *Nature*, 1993, **361**, 186.
128. K. A. LeCuyer and D. M. Crothers, *Proc. Natl. Acad. Sci. USA*, 1994, **91**, 3373.
129. H. Weidner and D. M. Crothers, *Nucleic Acids Res.*, 1977, **4**, 3401.
130. D. A. Peattie and W. Gilbert, *Proc. Natl. Acad. Sci. USA*, 1980, **77**, 4679.
131. S. Stern, T. Powers, L. Changchien, and H. F. Noller, *Science*, 1989, **244**, 783.
132. D. Moazed and H. F. Noller, *Proc. Natl. Acad. Sci. USA*, 1991, **88**, 3725.
133. P. P. Zarrinkar and J. R. Williamson, *Science*, 1994, **265**, 918.
134. R. R. Samaha, R. Green, and H. F. Noller, *Nature*, 1995, **377**, 309.
135. W. W. Cleland, in "The Enzymes," ed. P. D. Boyer, Academic Press, New York, 1970, vol. 2, p. 1.
136. W. P. Jencks, "Catalysis in Chemistry and Enzymology," McGraw-Hill, New York, 1969.
137. A. Fersht, "Enzyme Structure and Mechanism," 2nd edn., W. H. Freeman, New York, 1985.
138. D. Herschlag and T. R. Cech, *Biochemistry*, 1990, **29**, 10 159.
139. D. Herschlag and T. R. Cech, *Biochemistry*, 1990, **29**, 10 172.
140. T. G. Dewey and D. H. Turner, *Biochemistry*, 1979, **18**, 5757.
141. D. Pörschke, *Eur. J. Biochem.*, 1973, **39**, 117.
142. D. Pörschke, *Biopolymers*, 1978, **17**, 315.
143. J. R. Barrio, G. L. Tolman, N. J. Leonard, R. D. Spencer, and G. Weber, *Proc. Natl. Acad. Sci. USA*, 1973, **70**, 941.
144. K. Akasaka, *Biopolymers*, 1974, **13**, 2273.
145. P. Davanloo, I. M. Armitage, and D. M. Crothers, *Biopolymers*, 1979, **18**, 663.
146. L. M. Rhodes and P. R. Schimmel, *Biochemistry*, 1971, **10**, 4426.
147. M. E. Craig, D. M. Crothers, and P. Doty, *J. Mol. Biol.*, 1971, **62**, 383.
148. D. Riesner, G. Maass, R. Thiebe, P. Philippsen, and H. G. Zachau, *Eur. J. Biochem.*, 1973, **36**, 76.
149. S. M. Freier, D. D. Albergo, and D. H. Turner, *Biopolymers*, 1983, **22**, 1107.
150. J. W. Nelson and I. J. Tinoco, *Biochemistry*, 1982, **21**, 5289.
151. D. Riesner and R. Romer, in "Physico-chemical Properties of Nucleic Acids," ed. J. Duchesne, Academic Press, New York, 1973, vol. 2, p. 237.
152. N. Tibanyenda, S. De Bruin, C. A. G. Haasnoot, G. A. van der Marel, J. H. van Boom, and C. W. Hilbers, *Eur. J. Biochem.*, 1984, **139**, 19.
153. R. Rigler and W. Wintermeyer, *Ann. Rev. Biophys. Bioeng.*, 1983, **12**, 475.
154. D. Pörschke, *Biophys. Chem.*, 1974, **1**, 381.
155. J. Gralla and D. M. Crothers, *J. Mol. Biol.*, 1973, **78**, 301.
156. D. M. Crothers, P. E. Cole, C. W. Hilbers, and R. G. Shulman, *J. Mol. Biol.*, 1974, **87**, 63.
157. J. H. van Boom, G. A. van der Marel, H. Westerink, C. A. A. van Boeckel, J.-R. Mellema, C. Altona, C. W. Hilbers, C. A. G. Haasnoot, S. H. de Brun, and R. G. Berendsen, *Cold Spring Harbor Symp. Quant. Biol.*, 1982, **47**, 403.
158. D. R. Kearns and R. G. Schulman, *Acc. Chem. Res.*, 1974, **7**, 33.
159. J. L. Leroy, N. Bolo, N. Figueroa, P. Plateau, and M. Guéron, *J. Biomol. Struct. Dyn.*, 1985, **2**, 915.
160. J. L. Leroy, D. Broseta, and M. Guéron, *J. Mol. Biol.*, 1985, **184**, 165.
161. N. B. Leontis and P. B. Moore, *Biochemistry*, 1986, **25**, 5736.
162. M. Guéron, M. Kochoyan, and J. Leroy, *Nature*, 1987, **382**, 89.
163. P. A. Mirau and D. R. Kearns, *Proc. Natl. Acad. Sci. USA*, 1985, **82**, 1594.
164. P. C. Bevilacqua, R. Kierzek, K. A. Johnson, and D. H. Turner, *Science*, 1992, **258**, 1355.
165. C. Reich, G. J. Olsen, B. Pace, and N. R. Pace, *Science*, 1988, **239**, 178.
166. C. Guerrier-Takada, N. Lumelsky, and S. Altman, *Science*, 1989, **246**, 1578.
167. M. J. Fedor and O. C. Uhlenbeck, *Biochemistry*, 1992, **31**, 12 042.
168. K. J. Hertel, D. Herschlag, and O. C. Uhlenbeck, *Biochemistry*, 1994, **33**, 3374.
169. L. A. Hegg and M. J. Fedor, *Biochemistry*, 1995, **34**, 15 813.
170. B. M. Chowrira and J. M. Burke, *Biochemistry*, 1991, **30**, 8518.
171. J. R. Lorsch and J. W. Szostak, *Biochemistry*, 1995, **34**, 15 315.
172. W. J. Michels and A. M. Pyle, *Biochemistry*, 1995, **34**, 2965.
173. M. D. Been, A. T. Perrotta, and S. P. Rosenstein, *Biochemistry*, 1992, **31**, 11 843.
174. E. H. Ekland, J. W. Szostak, and D. P. Bartel, *Science*, 1995, **269**, 364.
175. P. A. Lohse and J. W. Szostak, *Nature*, 1996, **381**, 442.
176. T. R. Cech, D. Herschlag, J. A. Piccirilli, and A. M. Pyle, *J. Biol. Chem.*, 1992, **267**, 17 479.
177. D. H. Turner, Y. Li, M. Fountain, L. Profenno, and P. C. Bevilacqua, in "Nucleic Acids and Molecular Biology," ed. D. J. M. Lilley, Springer-Verlag, Berlin, 1996, vol. 10, p. 19.
178. D. Herschlag, *Proc. Natl. Acad. Sci. USA*, 1991, **88**, 6921.
179. H. Grosjean, D. G. Soll, and D. M. Crothers, *J. Mol. Biol.*, 1976, **103**, 499.
180. K. Yoon, D. H. Turner, and I. J. Tinoco, *J. Mol. Biol.*, 1975, **99**, 507.
181. D. Labuda and D. Pörschke, *Biochemistry*, 1980, **19**, 3799.
182. R. D. Blake and J. R. Fresco, *J. Mol. Biol.*, 1966, **19**, 145.
183. R. D. Blake, L. C. Klotz, and J. R. Fresco, *J. Am. Chem. Soc.*, 1968, **90**, 3556.
184. H. E. Moser and P. B. Dervan, *Science*, 1987, **238**, 645.
185. T. Le Doan, L. Perrouault, D. Praseuth, N. Habhoub, J.-L. Decout, N. T. Thuong, and C. Hélène, *Nucleic Acids Res.*, 1987, **15**, 7749.
186. L. J. I. Maher, P. B. Dervan, and B. J. Wold, *Biochemistry*, 1990, **29**, 8820.
187. M. Rougee, B. Faucon, J. L. Mergny, F. Barcelo, C. Giovannangeli, T. Garestier, and C. Hélène, *Biochemistry*, 1992, **31**, 9269.

188. R. W. Roberts, "Physical Chemistry of Nucleic Acid Triple Helices," Ph.D. Thesis, Yale University, 1993.
189. R. W. Roberts and D. M. Crothers, Manuscript in Preparation, 1996.
190. H. Han and P. B. Dervan, *Proc. Natl. Acad. Sci. USA*, 1993, **90**, 3806.
191. K. M. Weeks and D. M. Crothers, *Science*, 1993, **261**, 1574.
192. H. Shindo, H. Torigoe, and A. Sarai, *Biochemistry*, 1993, **32**, 8963.
193. M. Yang, S. S. Ghosh, and D. P. Millar, *Biochemistry*, 1994, **33**, 15329.
194. P. J. Bates, H. S. Dosanjh, S. Kumar, T. C. Jenkins, C. A. Laughton, and S. Neidle, *Nucleic Acids Res.*, 1995, **23**, 3627.
195. S. Wang, A. E. Friedman, and E. T. Kool, *Biochemistry*, 1995, **34**, 9774.
196. L. E. Xodo, *Eur. J. Biochem.*, 1995, **228**, 918.
197. K. M. Vasquez, T. G. Wensel, M. E. Hogan, and J. H. Wilson, *Biochemistry*, 1995, **34**, 7243.
198. R. W. Roberts and D. M. Crothers, *J. Mol. Biol.*, 1996, **260**, 135.
199. P. E. Cole, S. K. Yang, and D. M. Crothers, *Biochemistry*, 1972, **11**, 4358.
200. A. Stein and D. M. Crothers, *Biochemistry*, 1976, **15**, 160.
201. P. E. Cole and D. M. Crothers, *Biochemistry*, 1972, **11**, 4368.
202. A. J. Zaug, C. A. Grosshans, and T. R. Cech, *Biochemistry*, 1988, **27**, 8924.
203. P. P. Zarrinkar and J. R. Williamson, *Nature Struct. Biol.*, 1996, **3**, 432.
204. P. C. Bevilacqua, Y. Li, and D. H. Turner, *Biochemistry*, 1994, **33**, 11340.
205. P. C. Bevilacqua, K. A. Johnson, and D. H. Turner, *Proc. Natl. Acad. Sci. USA*, 1993, **90**, 8357.
206. Y. Li, P. C. Bevilacqua, D. Mathews, and D. H. Turner, *Biochemistry*, 1995, **34**, 14394.
207. D. Herschlag, *Biochemistry*, 1992, **31**, 1386.
208. A. R. Banerjee and D. H. Turner, *Biochemistry*, 1995, **34**, 6504.
209. J. H. Cate, A. R. Gooding, E. Podell, K. Zhou, B. L. Golden, C. E. Kundrot, T. R. Cech, and J. A. Doudna, *Science*, 1996, **273**, 1678.
210. J. H. Cate, A. R. Gooding, E. Podell, K. Zhou, B. L. Golden, A. A. Szewczak, C. E. Kundrot, T. R. Cech, and J. A. Doudna, *Science*, 1996, **273**, 1696.

# 7.03
# Probing DNA Structure by NMR Spectroscopy

CARLOS DE LOS SANTOS
*State University of New York at Stony Brook, New York, USA*

| | | |
|---|---|---|
| 7.03.1 | INTRODUCTION | 55 |
| 7.03.2 | ELEMENTS OF DNA STRUCTURE | 56 |
| | 7.03.2.1 *Nomenclature* | 56 |
| | 7.03.2.2 *Helical Parameters* | 58 |
| 7.03.3 | ¹H-NMR SPECTRA OF NUCLEIC ACIDS | 59 |
| | 7.03.3.1 *Cross-relaxation Rates and Interproton Distances* | 62 |
| | 7.03.3.2 *Coupling Constants and Torsion Angles* | 64 |
| | 7.03.3.3 *Proton Assignments* | 64 |
| | 7.03.3.3.1 *Nonexchangeable protons* | 64 |
| | 7.03.3.3.2 *Exchangeable protons* | 66 |
| | 7.03.3.4 *Sugar Conformations* | 67 |
| 7.03.4 | ³¹P-NMR OF NUCLEIC ACIDS | 69 |
| | 7.03.4.1 *Backbone Conformation* | 72 |
| 7.03.5 | ¹⁵N- AND ¹³C-NMR OF NUCLEIC ACIDS | 72 |
| | 7.03.5.1 *¹⁵N Spectra* | 72 |
| | 7.03.5.2 *¹³C Spectra* | 73 |
| 7.03.6 | HYDROGEN EXCHANGE AND BASE-PAIR DYNAMICS | 74 |
| | 7.03.6.1 *Exchange Formalism* | 74 |
| 7.03.7 | DNA CONFORMATIONS | 75 |
| | 7.03.7.1 *B-form DNA Structure* | 76 |
| | 7.03.7.2 *A-form DNA Structure* | 76 |
| | 7.03.7.3 *Z-form DNA Structure* | 77 |
| 7.03.8 | REFERENCES | 79 |

## 7.03.1 INTRODUCTION

Since the late 1980s there has been an enormous increase in the number of structures deposited in the Protein Data Bank which, by the end of 1997, amounted to 6491 coordinates released.[1] While X-ray crystallography continues to be the main experimental source of three-dimensional information, about 15% of the structures of macromolecules have been solved in solution using NMR spectroscopy. Notably, the incidence of NMR structures increases to more than 30% in the field of nucleic acids. The relative advantage of NMR spectroscopy in studying RNA and DNA molecules can be attributed to two concurrent factors. On the one hand, advances in the preparation of labeled samples have made it possible to obtain large quantities of isotopically enriched [¹³C] and

[15N]RNA molecules.[2] Although still in a preliminary stage, similar techniques have been used to prepare enriched oligodeoxynucleotide samples.[3] Such progress in labeling techniques opened the door to a large variety of heteronuclear multidimensional NMR experiments, which have solved overlap problems and made possible conformation-independent resonance assignments, especially of RNA molecules.[4] On the other hand, nonglobular DNA and RNA molecules (molecular weight > 2000 Da) are intrinsically "polymorphic" and, in general, it is very difficult to obtain crystals of good diffraction quality. This fact has limited X-ray studies mainly to fiber diffraction data, which yield unit cell and helical parameters, but only limited information about sequence composition effects on local DNA conformation.[5] Finally, it is also important to mention the unique capability of NMR spectroscopy to characterize dynamic processes within a wide timescale ($10^{-9}$–$10^{-3}$ s).[6] A review of the methods used to study dynamic processes in nucleic acids has been published elsewhere.[7]

This chapter describes the use of NMR spectroscopy to characterize the solution structure of oligodeoxynucleotide duplexes. The author starts by defining the nomenclature and structural parameters needed to characterize the conformation of double-stranded DNA molecules. Subsequently, the spectroscopic parameters used to obtain structural information are introduced and their limitations are briefly discussed. Examples of $^1$H, $^{31}$P, $^{15}$N, and $^{13}$C NMR spectra of short oligonucleotide duplexes are used to illustrate the methods employed to assign the NMR signals and the structural information that can be derived from them. Next, the formalism governing water exchange of imino protons and its correlation with base-pair dynamics is presented. Finally, the author describes the solution structure emerging from studies performed on "canonical" Watson–Crick duplex samples, and concludes by discussing the NMR characteristics of A- and Z-form DNA molecules in solution.

NMR methods have long been used to study the structure of RNA and DNA molecules, and many excellent reviews have been published throughout the years. Those interested in learning how to use NMR methods to study RNA molecules are advised to read a review by Varani *et al.*[4] For an exhaustive description of NMR spectroscopy applied to DNA molecules, readers are referred to a series of monographs compiled by James.[8]

## 7.03.2  ELEMENTS OF DNA STRUCTURE

### 7.03.2.1  Nomenclature

Nucleic acids are linear biopolymers composed of a sugar–phosphate backbone containing purine and pyrimidine heterocycles (bases) as side chains. The monomeric units, called nucleotides, are composed of a cyclic sugar (β-D-ribose in RNA or β-D-2′-deoxyribose in DNA), phosphorylated at the O-5′ position and linked to one of four different bases through a β-glycosyl C-1′—N bond. Purine bases, generically denoted "pur" or R, are guanine (Gua or G) and adenine (Ade or A), while pyrimidine bases, denoted "pyr" or Y, are cytosine (Cyt or C), thymine (Thy or T), and uracil (Ura or U, only found on RNA). In the biopolymer, individual nucleotides are linked via a 3′,5′-phosphodiester bond.

Following IUPAC–IUB recommendations,[9] sugar atoms are differentiated from base atoms by primes, and the chain direction goes from C-5′ to C-3′. The backbone atoms are counted following the sequence P→O-5′→C-5′→C-4′→C-3′→O-3′→P and, within the sugar ring, the sequence is C-1′→C-2′→C-3′→C-4′→O-4′→C-1′. As shown in Figure 1, the two hydrogen atoms present at the C-2′ position of the deoxyribose are named H-2′1 (H-2′), and H-2′2 (H-2″), corresponding to the *pro-S* and *pro-R* positions, respectively. In other words, the H-2′ proton and the base are to one side of the plane determined by the sugar ring and H-2″ and H-1′ are to the other. The same convention applies to the hydrogen atoms at the C-5′ position, which are labeled H-5′1 (H-5′) and H-5′2 (H-5″), corresponding to the *pro-S* and *pro-R* positions, respectively. Looking along the O-5′—C-5′ bond, a clockwise rotation gives C-4′, H-5″, H-5′. Figure 1 and Table 1 show the six torsion angles needed to define the conformation of the sugar–phosphate backbone. Each angle describes the rotation along a single bond of the backbone and Greek letters, from α through ζ, denote them. It is a common practice in X-ray crystallography to specify conformational ranges in which the torsion angles lie using the terms *cis*, *trans*, *gauche*⁻ and *gauche*⁺. Their correlation with the Klyne–Prelog notation,[10] recommended by IUPAC–IUB, is shown in Figure 2.

Five endocyclic torsion angles ($v_0$ to $v_4$), defined in Figure 1 and Table 1, can be used to describe the conformation of the sugar. In general, the ribose ring is nonplanar, a fact known as sugar pucker. When four atoms of the sugar ring lie in a plane, this plane is chosen as the reference, and

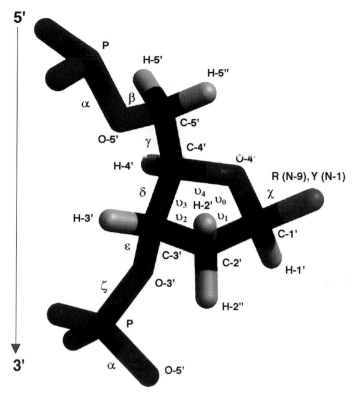

**Figure 1** IUPAC–IUB recommendation for atom nomenclature, chain direction, and definition of sugar and backbone dihedral angles in nucleic acids. Note that $\delta$ and $v_3$ define the same torsion along the C-3′—C-4′ bond.

**Table 1** Definition of torsion angles in nucleic acids.

| Torsion angle | Atoms involved | Torsion angle | Atoms involved |
|---|---|---|---|
| $\alpha$ | $(n-1)$O-3′—P—O-5′—C-5′ | $v_0$ | C-4′—O-4′—C-1′—C-2′ |
| $\beta$ | P—O-5′—C-5′—C-4′ | $v_1$ | O-4′—C-1′—C-2′—C-3′ |
| $\gamma$ | O-5′—C-5′—C-4′—C-3′ | $v_2$ | C-1′—C-2′—C-3′—C-4′ |
| $\delta$ | C-5′—C-4′—C-3′—O-3′ | $v_3$ | C-2′—C-3′—C-4′—O-4′ |
| $\varepsilon$ | C-4′—C-3′—O-3′—P$(n+1)$ | $v_4$ | C-3′—C-4′—O-4′—C-1′ |
| $\zeta$ | C-3′—O-3′—P—O-5′ | | |
| $\chi$ | O-4′—C-1′—N-1—C-2 (pyrimidines) | | |
| | O-4′—C-1′—N-9—C-4 (purines) | | |

the conformation is called envelope (E). If not, the reference plane is that of the three atoms that are closest to the five-atom least-squares plane, and the conformation is called twist (T). These conformations are depicted in Figure 3.

According to convention, the atoms displaced from the reference plane are called *endo* or *exo*, depending on whether they are on the same side as C-5′ or the opposite,[11] and are assigned superscripts and subscripts for the *endo* and *exo* conformations, respectively. Thus, an envelope pucker with the C-3′ atom toward the C-5′ side is written as $^3$E, while a twist conformation with the C-2′ atom on the same side as C-5′ and C-3′ on the opposite is denoted $^2$T$_3$. It is also very convenient to describe sugar puckers using the concept of angle of pseudorotation, which was first introduced to characterize the structure of cyclopentane,[12] and later applied to nucleosides and nucleotides.[13] In this description, endocyclic torsion angles can be reproduced, with an error of less than 0.7°, by a two-parameter equation:

$$v_j = \Phi_m \cos\left[P + 144(j-2)\right] \qquad \text{where } 0 \leqslant j \leqslant 4 \qquad (1)$$

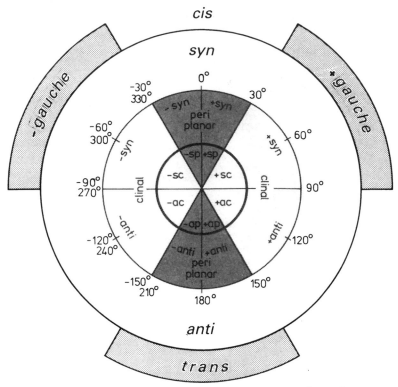

**Figure 2** Pictorial definition of dihedral angle ranges used in X-ray crystallography (*cis, trans, gauche*[+], and *gauche*[−]) and their correlation with the IUPAC–IUB recommended Klyne–Perlog notation (inner circle) (reproduced by permission of Springer-Velag from *Principles of Nucleic Acid Structure*; © Springer-Velag, New York).

Torsion angles are numbered clockwise starting with $v_0$ (C-4′—O-4′—C-1′—C-2′) as previously indicated. The maximum amplitude of pucker, $\Phi_m$, can be calculated from the previous equation by setting $j = 2$. The pseudorotation angle, $P$, takes values from 0° to 360°.

Figure 3 shows the correlation between pseudorotation angle and the envelope and twist conformations. It is a common practice to use the terms *north* and *south* conformations when referring to the C-2′-*exo*/C-3′-*endo* and C-3′-*exo*/C-2′-*endo* ranges, respectively. The orientation of the bases with respect to the sugar is defined by χ, the glycosidic torsion angle (O-4′—C-1′—N-9—C-4 in purines, and O-4′—C-1′—N-9—C-2 in pyrimidines). The preferred conformational range observed in nucleic acids is *anti*, in which the six-membered ring of purines, or the O-2 of pyrimidines, is directed away from the sugar. Another conformational range, frequently observed when studying DNA duplexes containing damaged bases, is *syn*, in which the bulky part of the bases points towards the sugar. Purine and pyrimidine bases can interact through formation of hydrogen bonds, defining what is known as Watson–Crick base-pair alignments, a common, but not unique, motif of secondary structure in DNA. With the bases in *anti* orientation, G pairs with C and A with T in an arrangement that has very similar overall dimensions, as shown in Figure 4.

### 7.03.2.2  Helical Parameters

Since the early determinations of DNA structure originated from fiber diffraction studies, the three-dimensional models were perfect helices. Any sequence-dependent structural perturbation was either averaged out in natural DNA samples or absent in repeating sequences prepared synthetically. Hence it was possible and convenient to identify the long helix axis and reproduce the position of base pairs by describing their vertical separation and relative angle, called rise and twist, respectively. In addition, with the definition of an auxiliary axis within the base pair, it was possible to indicate the inclination and displacement of the bases with respect to the helical axis.[14]

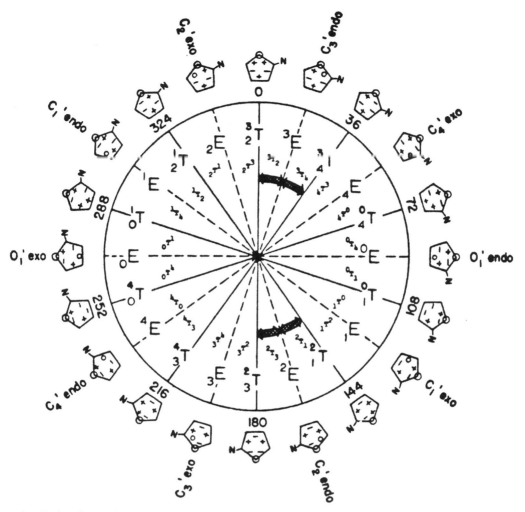

**Figure 3** Cycle of pseudorotation for the ribose ring. The most frequent conformation found in deoxyribonucleic acids is C-2'-endo, on the *south* part of the cycle. Envelope and twist forms alternate every 18° (reproduced by permission of the American Chemical Society from *J. Am. Chem. Soc.*, 1972, **94**, 8206; © American Chemical Society).

However, after the first X-ray structure using a single DNA crystal was reported,[15] it became evident that base sequence composition induces local distortions on the "perfect" helix. Consequently, DNA structure could no longer be depicted with the existing helical parameters, and additional definitions became necessary. The EMBO Workshop on DNA Curvature and Bending, held in Cambridge in 1988, agreed upon the nomenclature and definitions of structural parameters to describe the geometry of nucleic acids.[16] Figure 5 gives a pictorial definition of these parameters.

## 7.03.3 ¹H-NMR SPECTRA OF NUCLEIC ACIDS

Despite the technological advances in multidimensional heteronuclear techniques since the late 1980s,[17] proton NMR spectroscopy remains the central tool for establishing the conformation of macromolecules in solution. While ¹³C- and ¹⁵N-enriched samples have changed the means of assigning proton spectra of proteins and RNA molecules,[4,18] and new pulse sequences have been proposed to determine backbone dihedral angles from homo- and heteronuclear *J* correlation experiments,[19] the set of interproton distances, derived from ¹H spectra, is still the core experimental information implemented during the subsequent structural refinement. When studying DNA structure, the role of proton NMR spectroscopy is even more pronounced, because the strategy for assigning proton signals is still mainly based on homonuclear ¹H spectroscopy.[20]

**Major groove**

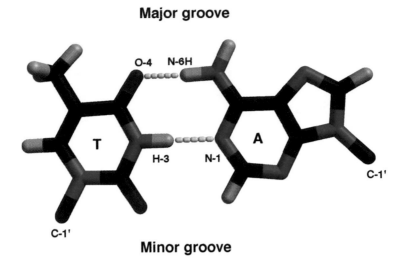

**Minor groove**

**Major groove**

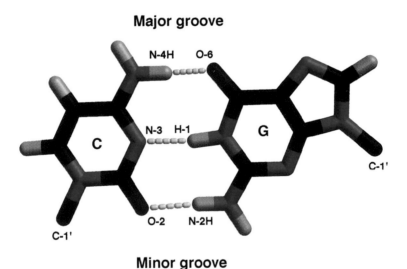

**Minor groove**

**Figure 4**   Watson–Crick base pair alignments in nucleic acids. As the strands run in antiparallel directions, the C-1′ carbons lie on the same edge of the base pair.

Figure 6 shows a typical 1D-proton spectrum of a DNA duplex sample. Generally, the spectra are referenced directly to external TSP (sodium [2,2,3,3-$^2$H$_4$]-3-trimethylsilylpropionate) or, indirectly, using the residual water signal. Proton signals span from about 1 ppm, for thymine methyl protons, to 15 ppm, for thymine imino protons. For practical reasons, it is convenient to divide DNA protons between exchangeable protons, composed of G(N-2,H-2), A(N-6,H-2), and C(N-4,H-2) amino and G(N-1,H) and T(N-3,H) imino protons, and nonexchangeable protons, all the rest. The experimental conditions and/or type of 2D experiment recorded, and also the structural information they provide, are different for each group.

In principle, there are three spectroscopic properties that are correlated with the structure and dynamics of any macromolecule in solution, namely chemical shift, cross-relaxation, and *J* coupling constants. Early studies of DNA structure relied mainly on the use of chemical shift data as the source of structural information,[21,22] an approach that later received severe criticism.[23] The main problem is that the net effect induced by neighboring nuclei on the chemical shift of a signal cannot be correlated with a unique geometrical factor, and the individual contribution of each nucleus cannot be independently deconvoluted without knowledge of the three-dimensional structure.[24] With

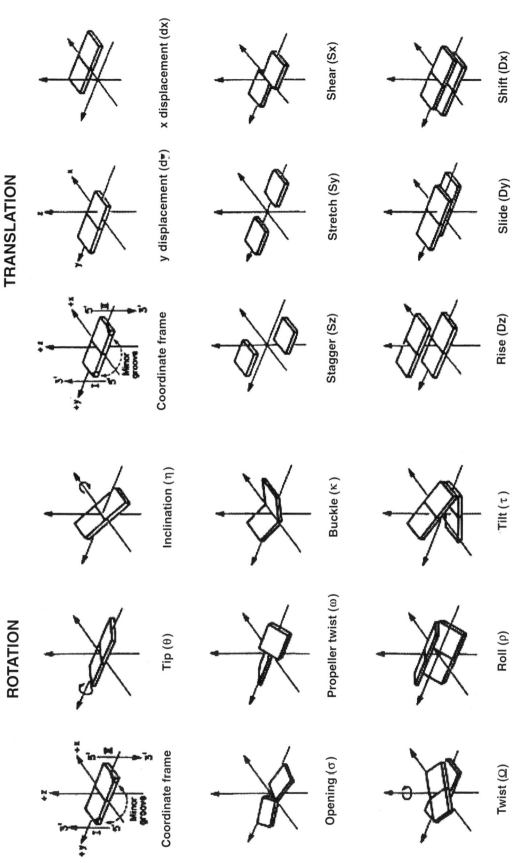

**Figure 5** Pictorial description of helical parameters used to describe nucleic acid structure. The arrows on the coordinate axis indicate the positive direction of translation parameters (reproduced by permission from *J. Mol. Biol.*, 1989, **205**, 787).

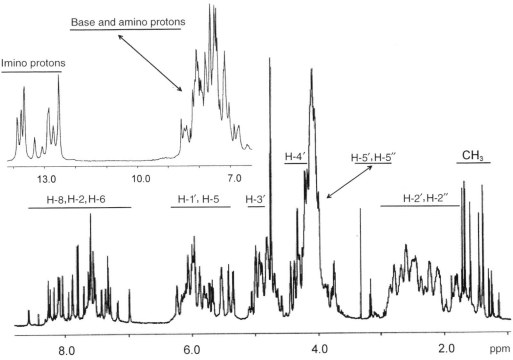

**Figure 6**   $^{1}$H-NMR spectrum of an 11-mer oligodeoxynucleotide duplex recorded in D$_2$O buffer at 30 °C and in 10% D$_2$O buffer at 5 °C (inset).

the development of multidimensional NMR techniques, which make possible the determination of interproton distances and dihedral angles, chemical shifts are no longer used for DNA structure determination. The application of chemical shift data is restricted to a qualitative assessment of local conformations, previously determined using experimental distance and dihedral angle information.

### 7.03.3.1   Cross-relaxation Rates and Interproton Distances

The ability of NMR spectroscopy to aid in the structure determination of the solution structure of nucleic acids and proteins relies on the fact that short interproton distances can be evaluated from NOESY spectra with reasonable accuracy. Furthermore, in the case of DNA molecules, the specific assignment of proton signals is based on NOE spectra, like that shown in Figure 7.

During the mixing period of a 2D NOE experiment, the time course of the longitudinal magnetization is governed by a system of equations:

$$\partial \mathbf{M}/\partial t = -\mathbf{R}\mathbf{M} \tag{2}$$

where $\mathbf{M}$ is the magnetization vector and $\mathbf{R}$ is a matrix that describes the complete relaxation network.[25] Diagonal elements of this matrix are the longitudinal relaxation rates ($\rho_i$), and off-diagonal elements, the cross-relaxation rates ($\sigma_{ij}$). When dipolar interactions are the predominant relaxation mechanism, rates are given by

$$R_{ii} \equiv \rho_i = W_0^{ij} + 3W_1^{ij} + 6W_2^{ij} \tag{3}$$

$$R_{ij} \equiv \sigma_{ij} = 6W_2^{ij} - W_0^{ij} \tag{4}$$

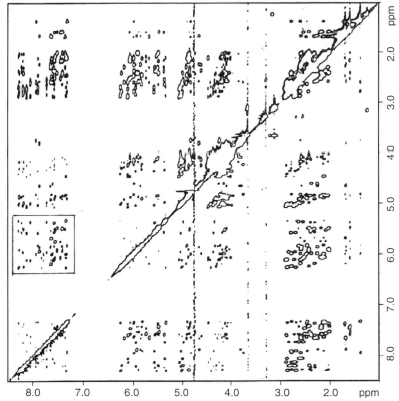

**Figure 7** Contour plot representation of a NOE spectrum (300 ms mixing time) of a d(CTCTAG*G* CTCAC)·d(GTGAGCCTAGAG) 12-mer duplex containing a Pt–d(GG) adduct (at the guanine residues indicated by asterisks) recorded in $D_2O$ buffer at 27°C (after de los Santos and Lippard[43]).

where $W_0^{ij}$, $W_1^{ij}$, and $W_2^{ij}$ represent the zero-, single- and double-quantum probabilities, respectively, which, for a molecule undergoing isotropic random motion, are

$$W_0^{ij} = q\tau_c/r^6 \tag{5}$$

$$W_1^{ij} = (q\tau_c/r^6)\{1/[1+(\omega\tau_c)^2]\} \tag{6}$$

$$W_2^{ij} = (q\tau_c/r^6)\{1/[1+4(\omega\tau_c)^2]\} \tag{7}$$

where $q = 0.1\,\gamma_i^2\,\gamma_j^2\,h^2$ and $\tau_c$ is a correlation time describing the reorientation of the distance vector, $r$, between spins $i$ and $j$. The $1/r^6$ dependence of the relaxation rates (Equations (3)–(7)) explains why NOE interactions diminish rapidly when interproton distances increase. The solution of Equation (2) is

$$\mathbf{M}(t_m) = \mathbf{a}(t_m)\,\mathbf{M}(0) = e^{-\mathbf{R}t_m}\mathbf{M}(0) \tag{8}$$

where $\mathbf{a}$ is a matrix containing the so-called mixing coefficients, which are proportional to the intensity of NOE peaks, and $\mathbf{M}(0)$ is the magnetization at the start of the mixing time, $t_m$. Computation of $\mathbf{a}$ can be achieved by expansion of the exponent into a Taylor series:

$$\text{NOE}_{ij}(t_m) \approx a_{ij}(t_m) \approx 1 - R_{ij}t_m + \tfrac{1}{2}R_{ik}R_{kj}t_m^2 - \tfrac{1}{6}R_{ik}R_{kl}R_{lj}t_m^3 + \cdots + [(-1)^n/n!]R^n t_m^n \tag{9}$$

The first two terms represent the direct interaction between spins $i$ and $j$ and all the others describe the "diffusion" of the magnetization through the entire spin network ("spin-diffusion"). The increase of the $t_m$ exponent with higher terms of the series suggests that spin-diffusion effects are less severe at short mixing times. Truncation of Equation (9) after the second term gives a linear relationship between NOE intensities and relaxation rates. This simplification, called "isolated spin-pair approximation" (ISPA), allows a simple determination of any unknown distance ($r_{ij}$) by comparison with a reference distance ($r_{ref}$) present in the molecule:

$$r_{ij} = r_{ref}(a_{ref}/a_{ij})^{1/6} \qquad (10)$$

where $a_{ref}$ and $a_{ij}$ are the NOE intensities of the reference and unknown cross-peaks, respectively. Early DNA structures were determined using this approach to calculate interproton distances from NOE spectra.[26,27]

However, serious concerns have been raised about the limitations of ISPA in calculating distances in macromolecules.[28–31] In general, a distance will be overestimated if it is shorter than the reference, and underestimated in the opposite case.[30,31] While attenuated at short mixing times, spin diffusion is already present, and can represent a considerable fraction of the cross-peak intensity. In addition, even at short $t_m$, differences in autorelaxation rates are an important source of errors. The ISPA is only applied to short mixing time NOESY data to determine qualitative distances shorter than 3 Å.

An alternative approach to derive distances from NMR data is by back-calculation of NOE intensities using the relaxation matrix. In this method, an initial model is iteratively refined to optimize the agreement between experimental and back-calculated NOE intensities. Once intensity differences have been minimized, a set of interproton distances is obtained. This process can be integrated into the refinement of the final structure,[32,33] or can just be used to generate interproton distances[34] later enforced during molecular dynamics or distance geometry computer simulations. In either case, the advantage of relaxation matrix procedures in obtaining more accurate distances has proven extremely powerful in deriving reliable high-resolution NMR structures of DNA molecules in solution.[35]

### 7.03.3.2 Coupling Constants and Torsion Angles

NMR experiments based on the existence of scalar coupling constants ($J$) among DNA protons provide a complementary source of information useful for studying nucleic acid structure. Detectable coupling constants exist among the sugar protons of any deoxyribonucleotide and, in pyrimidine bases, between the cytidine H-5—H-6 and the thymine H-6—CH$_3$ protons. Several two-dimensional NMR techniques are based on the existence of observable vicinal (three-bond) coupling constants (COSY,[36] double-quantum filtered COSY[37]). In practice, the analysis of $J_{H-1',H-2'}$ and $J_{H-1',H-2''}$ cross-peaks allows the stereospecific assignment of H-2' and H-2'' proton signals, and gives a qualitative picture of sugar conformations. Other 2D experiments, such as relay-COSY[38] and HOHAHA,[39,40] extend the observation of $J$ correlations to the entire proton network. Usually, the analysis of several HOHAHA experiments, recorded with different isotropic mixing times, in combination with COSY and double quantum filtered COSY data, is sufficient to assign all protons within each sugar ring. The identification of H-5' and H-5'' backbone protons is complicated by the severe signal overlap in this region of the NMR spectra and, generally, they are not stereospecifically assigned.

In addition to assisting in the assignment of proton resonances, $J$-correlated spectra can yield direct information about dihedral angles. The magnitude of the vicinal coupling constants is correlated with the torsion angle by a Karplus-type equation, parameterized for deoxyribonucleic acids:

$$^3J = 13.7\cos^2\theta - 0.7\cos\theta + \Sigma\Delta\varepsilon_i\{0.56 - 2.47\cos^2[\xi_i\theta + 16.9|\Delta\varepsilon_i|]\} \qquad (11)$$

where $\theta$ is the torsion angle. The summation runs for all $\alpha$-substituents of the two carbon atoms defining the torsion, $\Delta\varepsilon_i$ is the electronegativity difference between hydrogen and the $i$th $\alpha$-substituents and $\xi_i$ is an orientation factor that can take values of $\pm 1$, exclusively.[41] Generally, determination of coupling constants in DNA samples requires running 2D experiments designed to increase the spectral resolution and diminish partial cancellation of cross-peak components.

### 7.03.3.3 Proton Assignments

#### 7.03.3.3.1 Nonexchangeable protons

The 1D spectrum of a 12-mer DNA duplex, such as that shown in Figure 6, contains more than 260 proton signals, which have to be identified unambiguously before any structural property can be obtained. Even with the sample in $D_2O$ buffer, where exchangeable protons are not observed, more than 200 protons remain to be assigned. This task is readily accomplished by the analysis of two-dimensional NOE experiments, which can be used to assign DNA duplex samples up to 20 base pairs in length.[42] Figure 7 shows a room-temperature NOESY spectrum (300 ms mixing time), recorded in $D_2O$ buffer, of a 12-mer duplex sample containing a single Pt–d(GG) adduct at its center.[43] The diagonal peaks represent the 1D spectrum, and off-diagonal elements are NOE interactions between protons separated by less than 5 Å. Generally, proton assignments will start with the analysis of the so-called "fingerprint" region, shown in Figure 8, which depicts sequential interactions between base and H-1′ sugar protons. In a right-handed helix, a base proton (purine H-8 or pyrimidine H-6) will exhibit NOE peaks to its own and 5′-flanking H-1′ sugar protons.[20,36] The 5′-terminal residues are easily identified, because their base protons will have NOE peaks to just one H-1′ signal. Hence knowing the base composition of the sample, it is possible to "walk" along each strand of the duplex from one end to the other (Figure 8). The strong distance interactions between cytidine H-5–H-6 (2.46 Å separation), readily identified because they are the only proton pair in this region that will originate COSY cross-peaks, serve as alternative starting/control points of the NOE "walk." Additional peaks in this region include NOE connectivities between C(H-5) and the base (purine H-8 or pyrimidine H-6) proton of the 5′-flanking residue. Adenine H-2 residues are within 5 Å of up to four different H-1′ protons, on the same strand, its own and 3′-flanking H-1′ and, in the complementary strand, the H-1′ of its Watson–Crick partner and the 3′-attached residue. Some or all of the above interactions can be seen in a NOESY spectrum (Figure 8). When the glycosidic torsion angle changes to a *syn* conformation, an excellent distance marker is the intraresidue base (H-8/H-6)–H-1′ distance, which becomes <3 Å, making their cross-peak intensity comparable to that of the cytidine H-5–H-6 peak.

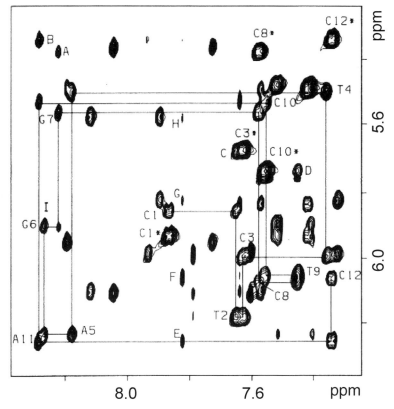

**Figure 8** Expanded region of the NOESY spectrum shown in Figure 7 that depicts distance interactions in the fingerprint region. The NOE "walk" is traced for the modified strand. Only intraresidue connectivities are labeled (see text). C(H-6–H-5) NOE peaks are marked with asterisks. Peaks A–D are sequential NOE peaks involving base and C(H-5) protons and E–G are NOE peaks involving A(H-2) protons. (after de los Santos and Lippard[43]).

Once the base protons have been identified in the fingerprint region, other regions of the spectrum are used to confirm the assignments. A similar directionality in the sequential NOEs is observed between base and H-3′, H-2′, and H-2″ protons. In practice, some of the H-3′ protons resonate very close to (or overlap with) the solvent, and their intensity will be diminished by presaturation of the residual water signal. Hence NOE interactions originating from these protons will be weakened or not observed in the spectrum.

Generally, a change in temperature of $\pm 10\,°C$ moves chemical shifts slightly and allows the detection of the missing peaks. Numerous NOE peaks are commonly observed among the sugar protons of the deoxyribose sugar, and analysis of these interactions yields the stereospecific assignment of H-2′, H-2″ protons. Independent of the sugar conformation, the H-1–H-2″ distance is always shorter, and its NOE peak will be stronger (especially at short mixing times when spin diffusion is less severe), than the H-1′–H-2′ distance. At long mixing times, NOE interactions between sugar protons of flanking residues are also observed. Some of these peaks are the result of distances shorter than 5 Å, but many of them originate from efficient spin diffusion pathways. Table 2 lists nonexchangeable proton distances measured in a B-form DNA duplex.

**Table 2**　Interproton distances in nucleic acids.[a]

| | | | | | *Intraresidue distances* | | | | | | | |
| | *H-6* | *H-8* | *H-2* | *H-5* | *CH₃* | *H-1′* | *H-2′* | *H-2″* | *H-3′* | *H-4′* | *H-5′* | *H-5″* |
|---|---|---|---|---|---|---|---|---|---|---|---|---|
| *H-1′* | 3.7 | 3.9 | 4.5 | 5.4 | 6.3 | | 3.0 | 2.4 | 4.0 | 3.7 | 4.5 | 5.1 |
| *H-2′* | 1.9 | 2.2 | 6.8 | 4.2 | 4.7 | | | 1.8 | 2.4 | 3.9 | 3.8 | 3.9 |
| *H-2″* | 3.4 | 3.7 | 6.4 | 5.5 | 6.3 | | | | 2.7 | 4.1 | 5.0 | 4.9 |
| *H-3′* | 3.9 | 4.2 | 8.3 | 6.4 | 6.8 | | | | | 2.7 | 3.8 | 2.8 |
| *H-4′* | 4.5 | 4.8 | 7.6 | 6.8 | 7.4 | | | | | | 2.6 | 2.3 |
| *H-5′* | 3.3 | 3.5 | 7.5 | 5.1 | 5.5 | | | | | | | 1.8 |
| *H-5″* | 4.1 | 4.4 | 8.9 | 6.2 | 6.5 | | | | | | | |

| | | | | | *Interresidue distances* 5′- | | | | | | | |
| | *H-6* | *H-8* | *H-2* | *H-5* | *CH₃* | *H-1′* | *H-2′* | *H-2″* | *H-3′* | *H-4′* | *H-5′* | *H-5″* |
|---|---|---|---|---|---|---|---|---|---|---|---|---|
| *3′-* | | | | | | | | | | | | |
| *H-6* | 5.1 | 4.8 | 5.4 | 6.5 | 7.2 | 2.8 | 3.9 | 2.4 | 5.0 | 5.9 | 6.9 | 7.1 |
| *H-8* | 4.8 | 5.0 | 5.5 | 6.3 | 6.9 | 2.8 | 3.7 | 2.3 | 5.0 | 5.9 | 6.8 | 7.1 |
| *H-2* | 8.8 | 8.7 | 3.8 | 8.5 | 9.1 | 7.1 | 9 | 8.3 | 10.8 | 10.7 | 10.8 | 11.9 |
| *H-5* | 3.8 | 3.8 | 5.7 | 4.6 | 5.2 | 3.7 | 3.2 | 2.8 | 5.3 | 6.5 | 6.6 | 7.1 |
| *CH₃* | 3.5 | 3.6 | 6.4 | 4.4 | 5.0 | 4.1 | 2.9 | 2.3 | 5.0 | 6.5 | 6.5 | 6.8 |
| *H-1′* | 8.0 | 8.1 | 4.6 | 8.9 | 9.7 | 4.9 | 7.3 | 5.9 | 8.2 | 8.2 | 9.4 | 10.0 |
| *H-2′* | 6.5 | 7.6 | 6.7 | 8.3 | 9.1 | 4.1 | 5.5 | 3.8 | 6.1 | 6.9 | 8.3 | 8.4 |
| *H-2″* | 8.5 | 8.6 | 6.7 | 9.7 | 10.4 | 5.5 | 7.2 | 5.6 | 7.8 | 8.4 | 9.9 | 10.1 |
| *H-3′* | 8.2 | 8.5 | 7.4 | 10.0 | 10.7 | 5.0 | 6.8 | 5.1 | 6.5 | 6.8 | 8.8 | 8.7 |
| *H-4′* | 7.9 | 8.1 | 5.8 | 9.6 | 10.4 | 4.2 | 6.9 | 5.5 | 6.8 | 6.1 | 8.0 | 8.2 |
| *H-5′* | 6.7 | 7.0 | 5.0 | 7.2 | 7.5 | 1.8 | 4.4 | 3.4 | 4.6 | 3.8 | 5.5 | 5.9 |
| *H-5″* | 5.4 | 5.7 | 6.6 | 8.7 | 9.6 | 3.4 | 5.4 | 4.1 | 4.8 | 4.2 | 6.4 | 6.3 |

[a]Distances given in ångstroms (1 Å = 0.1 nm) for a B-form DNA duplex. Methyl distances are the average between the shortest and longest values.

### 7.03.3.3.2　*Exchangeable protons*

It is common practice to collect data and perform the assignment of the exchangeable protons at low temperature. Close to $0\,°C$, the rate of exchange of imino protons with water is greatly reduced, facilitating the observation of NOE interactions. The inset in Figure 6 shows the 1D proton spectrum recorded in a 10% $D_2O$ buffer solution of a duplex sample. Supported by the presence of signals between 12.0 and 15.0 ppm, the sequence-specific assignment of imino and amino protons is based on the assumption that, in solution, base pairs adopt Watson–Crick alignments. An absolute "alignment-independent" assignment of exchangeable protons involves the recording of heteronuclear $^{13}$C-, $^{15}$N-NMR experiments using isotopically enriched samples, which correlate them with the nonexchangeable base and sugar protons. Figure 9(a) shows distance interactions observed between the imino and base/amino proton regions of a NOESY spectrum recorded at $5\,°C$. A·T base pairs are readily identified by the strong NOE interaction between T(N-3H) and A(H-2), which in B-form duplexes are separated by ∼3 Å. It is generally the case that the adenine H-2 protons have been sequence-specifically assigned through their interactions with the sugar H-1′ protons and, therefore, the thymine imino signals can be assigned. Rotation of the CN bond makes the adenine amino proton signals broad, limiting their use for assignment purposes. G·C base pairs are distinguished by the NOE connectivities between G(N-1H) and both the hydrogen-bonded and exposed cytidine amino protons ($\sim 2.5$ Å and $\sim 4.0$ Å, respectively) (Figure 9(a)). In turn, NOE peaks observed in another region of the same spectrum correlate C(N-4H2) and C(H-5) protons within each residue. Thus, the independent assignment of C(H-5) protons, easily obtained from NOESY experiments collected in $D_2O$, identifies all guanine imino protons of the sequence. Owing to their fast exchange with water, signals originating from terminal imino protons are very weak or missing in the spectrum and, generally, are of no use during structure calculations. Usually, sequential NOE interactions detected in the symmetrical imino proton region of the spectrum (Figure 9(b)) confirm the assignments. In theory, these sequential interactions should permit the complete identification of imino protons; however, this has been possible only with short self-complementary duplex sequences where signal overlap is not a serious problem.[27]

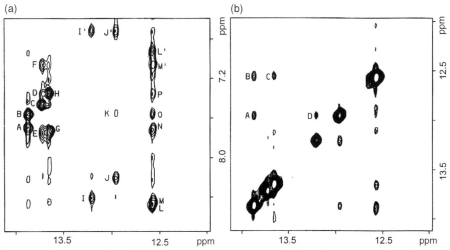

**Figure 9** Expanded regions of a NOESY spectrum of the Pt–d(GG) adduct-containing duplex sample recorded in 10% $D_2O$ buffer at 7 °C. (a) A·T base pairs are identified by distance interactions between T(N-3H) and A(H-2) protons (peaks A–C and G–H) and G·C base pairs by distance interactions between hydrogen-bonded and exposed C(N-4H2) and G(N-1H) protons (peaks I, I′–J, J′ and L, L′–M, M′). (b) NOE cross-peaks between sequential imino protons, indicative of base pair stacking (after de los Santos and Lippard[43]).

Imino protons involved in alternative hydrogen bonding exhibit a wide chemical shift range. In T·T·A triplets, thymine imino protons, which are hydrogen bonded to the purine N-7 atoms (Hoogsteen-type alignment), resonate within the Watson–Crick region.[44] When the hydrogen-bond acceptor is a carbonyl oxygen, the imino proton signal moves upfield and can even appear at < 10 ppm, as was observed in sequences containing mismatched base pairs or DNA lesions.[45,46] In many cases, the identity of the shifted proton signal must be confirmed by additional heteronuclear NMR experiments. Imino protons not involved in hydrogen bonding resonate between 10.5 ppm and 11.5 ppm. In addition, when they are exposed to the solvent, the signals are broad owing to water exchange and, in many instances, only acidification of the sample (which slows the exchange process) allows their observation. In summary, the chemical shift range of imino protons in different structural environments is fairly wide, and their hydrogen-bonding stage must not be assessed based solely on their resonance frequency.

### 7.03.3.4 Sugar Conformations

The conformation of the 2′-deoxyribose sugar ring has a fundamental role in determining the structural properties of DNA molecules. An explicit range of sugar conformations is associated with A-, B-, and Z-form DNA duplexes and, within a determined family (i.e., B-form DNA), local structural variations correlate with the specific sugar pucker. In principle, the determination of $^3J$ coupling constants and sugar dihedral angles is straightforward. Several two-dimensional NMR experiments, such as E.COSY,[47] P.E.COSY,[48] and COSY45, can be implemented to simplify cross-peak patterns and improve coupling constant determination. Subsequent simulation of individual

COSY cross-peaks yields $J$ values and, from a Karplus-type Equation (11), accurate sugar torsion angles.[49]

Figure 10(a) shows the variation of vicinal coupling constants with the pseudorotation angle. Clearly, $J_{H-1',H-2'}$, $J_{H-2'',H-3'}$, and $J_{H-3',H-4'}$ show the largest difference between C-3'-*endo* and C-2'-*endo* conformations and, thus, are the best markers to assess sugar puckers. However, overlap problems and the proximity of H-3' protons and the residual water signal make it difficult to observe clearly $J_{H-2'',H-3'}$ and $J_{H-3',H-4'}$ cross-peaks. A region of a COSY45 spectrum recorded in a Pt–d(GG) adduct-containing duplex sample, depicting $J$ interactions between H-1' and H-2'/H-2'' sugar proton regions, is shown in Figure 11. As commonly observed in DNA duplexes, most of the residues have sugar puckers in the *south* conformation. However, in the Pt–d(GG) adduct-containing duplex, the sugar of the 5'-adducted guanine puckers to the *north*, a fact readily apparent from the absence of a $J_{H-1',H-2'}$ interaction ($J < 2$ Hz), and the different pattern of the $J_{H-1',H-2''}$ cross-peak (Figure 11, peak A).

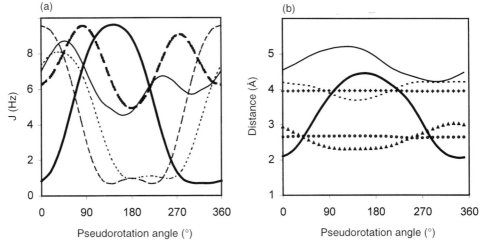

**Figure 10**  (a) Variation of $J$ coupling constants as a function of the pseudorotation angle, as given by Equation (11). Curves are $J_{H-1',H-2'}$ (——); $J_{H-1',H-2''}$ (——); $J_{H-2',H-3'}$ (— —); $J_{H-2'',H-3'}$ (– – –); $J_{H-3',H-4'}$ (· · ·). The amplitude of the pucker was kept constant at 38°. A variation of $\pm 10°$ in the amplitude of pucker changed the $J$ values by as much as 1.5 Hz. For the *cis* couplings, $J_{H-1',H-2''}$ and $J_{H-2',H-3'}$, a Barfield correction with maxima of $-2$ and $-0.5$ Hz, respectively was implemented. (b) Dependence of intraresidue base (purine H-8/pyrimidine H-6) to sugar proton distances as a function of the pseudorotation angle. Values were derived for a purine H-8 proton keeping constant the amplitude of the pucker and the glycosidic torsion angle at 35° and $-105°$, respectively. Distances are H-8–H-1' (+++); H-8–H-2' (▲▲▲) H-8–H-2'' (· · ·); H-8–H-3' (——); H-8–H-4' (——). The curve (●●●) represents the H-8–H-1' distance for a $\chi$ value of 75°, within the *syn* conformation.

An exact description of the 2'-deoxyribose conformation is further complicated by the fact that more than one sugar pucker, in fast exchange on the NMR timescale, is generally present in solution. Hence the observed $J$ value can no longer be associated with a unique conformation, but rather represents a time-averaged parameter:

$$J_{obs} = X_a J_a + X_b J_b + \cdots + X_n J_n \tag{12}$$

where $X_n$ represents the molar fraction of the $n$th population and $J_n$ its coupling constant value. In general, two main populations in the C-3'-*endo* (*north*) and C-2'-*endo* (*south*) ranges are sufficient to reproduce the experimental coupling constant values. Hence a complete description of sugar conformation involves the determination of five independent parameters: *north* and *south* amplitudes and angles of pseudorotation and the molar fraction of one conformer. Methods to fit these parameters simultaneously to a set of experimental coupling constants (or sums of coupling constants) have been developed.[50,51]

In addition to coupling constants, several proton distances can be used to monitor sugar conformations. Figures 10(b) and 12 show the variation of several intraresidue interproton distances with the angle of pseudorotation. The H-1'—H-4' distance varies by more than 1 Å, with a minimum in the O-4'-*endo* conformation, but cannot differentiate between *north* and *south* forms (Figure 12(a)). The H-2''—H-4' distance changes from $<3$ Å in C-3'-*endo* to $\sim 4$ Å in a C-2'-*endo* con-

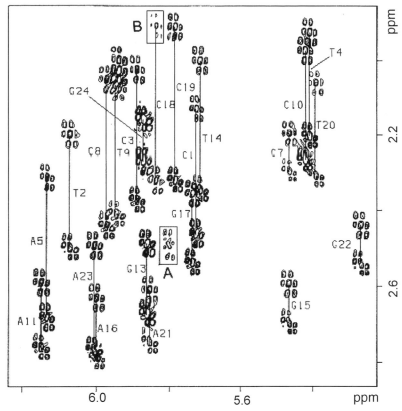

**Figure 11** Expanded region of a COSY45 spectrum of the Pt–d(GG) adduct-containing duplex sample recorded in $D_2O$ buffer at 30 °C, showing $J$ interactions between the H-1′ and H-2′/H-2″ sugar protons. Note the characteristic pattern of a $J_{H-1',H-2''}$ for a sugar in the C-3′-*endo* conformation, shown by peak A (after de los Santos and Lippard[43]).

formation and, thus, is an excellent marker of sugar puckers (Figure 12(b)). Another useful marker is the distance between the base (purine H-8/pyrimidine H-6) and H-3′ protons, which is <3 Å in a C-3′-*endo* and >4 Å in a C-2′-*endo* conformation (Figure 10(b)). Such differences are illustrated in Figure 13, which shows the NOE interactions observed in the base and H-3′ proton regions of the Pt–d(GG) adduct-containing duplex. As expected from its COSY45 spectrum, the H-8–H-3′, distance in the 5′-adducted guanine is short, resulting in a very strong NOE interaction (Figure 13, peak A). Surprisingly, the residue on the 5′-side of the lesion also showed a strong NOE intensity, suggesting an important fraction with a sugar pucker in the *north* region (Figure 13, peak B). In fact, the accurate determination of coupling constant values by simulation of COSY cross-peaks, along with the application of Equation (12), proved that equal molar fractions of C-3′-*endo* and C-2′-*endo* forms exist in solution for this residue.

In summary, a complete description of sugar conformations involves a two-step process. First, a determination of $J$ coupling constants from dedicated 2D NMR experiments and the subsequent computer simulation of the cross-peaks should be performed. Second, the simultaneous fitting of $J$ values according to Equation (12) yields puckers and molar fractions of *north* and *south* conformers. Alternatively, interproton distances and $J$-coupling markers can be used to depict an average conformational range in solution, generally determined by the main conformer. The particular approach to follow will depend on the type of questions to be answered.

## 7.03.4 $^{31}$P-NMR OF NUCLEIC ACIDS

Phosphorus is probably the second most important nucleus for studying the structure and dynamics of oligodeoxynucleotide duplexes in solution by NMR spectroscopy. $^{31}$P has nuclear properties attractive for high-resolution NMR studies: it has spin $^1/_2$, which avoids problems associated with quadrupolar nuclei, and its natural abundance is 100%. Despite having a relative sensitivity

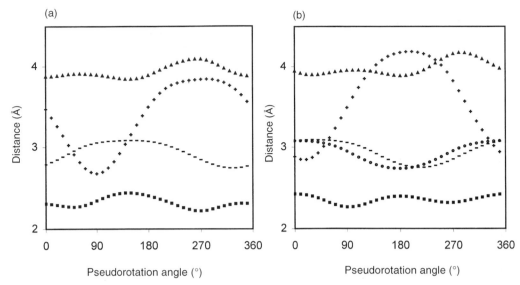

**Figure 12** Plot of interproton distances within the sugar ring as a function of the pseudorotation angle. Values were estimated with a constant sugar pucker amplitude of 35°. Distances are (a), H-1′–H-2′ (---); H-1′–H-2″ (■■■); H-1′–H-3′ (▲▲▲); and H-1′–H-4′ (+++); and (b), H-2′–H-3′ (■■■); H-2″–H-3′ (●●●); H-2′–H-4′ (▲▲▲); H-2″–H-4′ (+++); and H-3–H-4′ (---).

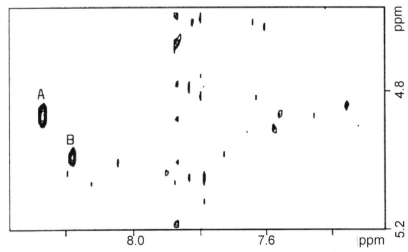

**Figure 13** Expanded region of a NOESY (50 ms mixing time) spectrum of the Pt–d(GG) adduct-containing duplex sample recorded in D₂O buffer at 27 °C, which shows distance interactions between the base and H-3′ proton regions. At this short mixing time, most of the R(H-8)/Y(H-6)–H-3′ NOE peaks are weak, reflecting a C-2′-*endo* sugar conformation, except for residues that have a main population in the C-3′-*endo* form (peak A), and equal populations of C-3′-*endo* and C-2′-*endo* conformers (peak B) (after de los Santos and Lippard[43]).

of 6.6% compared with protons, it is the only other nucleus that can be directly observed in nonenriched oligodeoxynucleotide samples. One- and two-dimensional $^{31}$P spectra can be recorded in about 1 day using samples of less than 1 mmol L$^{-1}$ concentration.

The analysis of $^{31}$P spectra not only completes the spectroscopic characterization of duplex samples, but also provides structural information about the phosphodiester backbone, which is inaccessible from other sources. A Karplus-type equation relates the $J_{P,H-3'}$ and $J_{P,H-5'}$ ($J_{P,H-5''}$) coupling constants to the values of the $\varepsilon$ and $\beta$ dihedral angles, respectively. Furthermore, the theoretical studies have suggested that phosphorus chemical shifts are mainly determined by the conformations of $\alpha$ and $\zeta$ torsion angles. Figure 14(a) shows a proton-decoupled $^{31}$P spectrum of the Pt–d(GG) adduct-containing 12-mer duplex recorded at room temperature. Phosphorus signals are generally referenced to an external 85% solution of orthophosphoric acid or to trimethyl phosphate (3.46 ppm downfield from the orthophosphoric acid signal). Despite the wide $^{31}$P chemical shift range

($\sim$600 ppm), phosphorus signals of DNA duplexes resonate in a narrow region, 3.5–5 ppm upfield from the TMP reference. Provided that the H-3' and H-4' proton resonances have been previously identified from $^1$H-NMR spectra, the assignment of phosphorus signals is readily accomplished by heteronuclear multidimensional techniques. Improved versions of the traditional heteronuclear correlation experiment (HETCOR) (phosphorus detection), which substantially increase sensitivity and resolution in both dimensions, have been reported.[52,53] Alternatively, inverse-detected (proton) HETCOR experiments have proven to be very effective in assigning phosphorus signals of duplex samples of up to 20 base pairs.[54,55] Figure 15 shows a proton-detected [$^1$H–$^{31}$P]-correlation spectrum recorded on the Pt–d(GG) adduct-containing duplex sample. In principle, a phosphorus on the $i$th residue will show cross-peaks to the H-4', H-5', and H-5'' of the $i$th and to the H-3' of the ($i-1$)th residues. However, some coupling constants can be very small, especially those involving the H-5' and H-5'' protons and, in practice, not all possible cross-peaks are observed. In addition, partial overlap of H-3' and H-4' proton signals can hinder the assignment of phosphorus spectra even further. In the early 1990s, a series of two- and three-dimensional [$^1$H–$^{31}$P] cross-polarization experiments (hetero-TOCSY) have been reported.[56] In these experiments, the observation of cross-peaks is extended to the entire proton network, solving overlap problems and facilitating the assignments of phosphorus signals. However, proton-detected HETCOR experiments offer a good compromise of sensitivity and resolution without the need of high spectrometer performance required for hetero-TOCSY experiments.

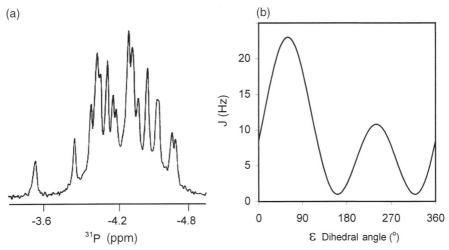

**Figure 14** (a) Proton-decoupled one-dimensional $^{31}$P spectrum of the Pt–d(GG) adduct-containing duplex sample recorded in D$_2$O buffer at 30 °C (after de los Santos and Lippard[43]). (b) Variation of $J_{P,H-3'}$ as a function of the $\varepsilon$ dihedral angle, as given by Equation (13).

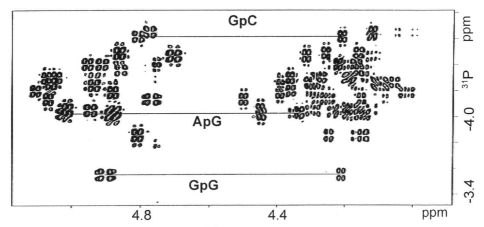

**Figure 15** Contour plot of an inverse-detected $^{31}$P-HETCOR spectrum of the Pt–d(GG) adduct-containing duplex sample recorded in D$_2$O buffer at 30 °C. Identified on the figure are the three phosphorus signals around the lesion (after de los Santos and Lippard[43]).

### 7.03.4.1  Backbone Conformation

In principle, once [31]P signals have been assigned, it is possible to correlate the $J_{P,H-3'}$ coupling constant with the $\varepsilon$ dihedral angle according to a Karplus-type equation:[57]

$$J_{\mathrm{HCOP}} = 15.3\cos^2\theta - 6.1\cos\theta + 1.6 \qquad (13)$$

where $\theta$ is H-3'—C-3'—O-3'—P and $\varepsilon = (-\theta - 120°)$. However, there are some limitations for the direct application of this approach. Figure 14(b) plots the variation of $J_{P,H-3'}$ as a function of the $\varepsilon$ dihedral angle, showing that four different $\varepsilon$ values are possible for each value of $J_{P,H-3'}$. In addition, the predominant $\varepsilon$ conformations observed in crystals,[5] *trans* and *gauche*[-], can have similar $J$ values, making their discrimination impossible. An alternative is to consider only the allowed conformations, which are normally based on crystallographic data. However, the situation could be even more complicated in the case of anomalous DNA molecules or chemically modified residues, where all possible solutions should be considered. Another approach is to use $J_{P,H-3'}$ values to define qualitatively possible conformational ranges of $\varepsilon$ torsion angles, and use them to validate the results obtained during the structural refinement. In principle, a similar Karplus relationship can be applied to $J_{P,H-5'}$ and $J_{P,H-5''}$ to obtain values of the $\beta$ torsion angle. In this case, if both coupling constants are measured simultaneously, a unique $\beta$ value can be determined.

In addition to coupling constants, theoretical studies have suggested that one of the main factors determining [31]P chemical shifts is the conformation of $\alpha$ and $\zeta$ torsion angles.[58–60] In a B$_I$ conformation ($g^-/g^-$; $\zeta/\alpha$), phosphorus resonances are more shielded than in a B$_{II}$ form ($t/g^-$; $\zeta/\alpha$), resulting in chemical shift differences of about 1.6 ppm between them.[61] Furthermore, a correlation between phosphorus chemical shifts and the $J_{P,H-3'}$ coupling constants has been reported in several oligodeoxynucleotide duplexes, where a decrease in coupling constant values was concomitant with a higher field resonance frequency of the phosphorus signal.[58] It was postulated that the relationship observed in DNA crystals between $\zeta$ and $\varepsilon$ torsion angles ($\zeta = -317 - 1.23\varepsilon$)[62,63] is also present in solution. Hence a phosphate in the B$_I$ conformation would have a lower chemical shift value ($\zeta = g^-$) and small $J_{P,H-3'}$ coupling constant ($\varepsilon = t$), whereas in the B$_{II}$ conformation a phosphate would resonate at lower field ($\zeta = t$) and have larger $J_{P,H-3'}$ value ($\varepsilon = g^-$). In practice, the observed chemical shift and $J_{P,H-3'}$ values would result from the weighted average of these two conformers, which are in fast exchange on the NMR timescale.[61]

However, the limitations of the previous approach must be mentioned. Whereas a relationship between [31]P chemical shifts and $J_{P,H-3'}$ was observed in several DNA duplexes, numerous exceptions exist, and the universality (or sequence/experimental conditions) of this correlation remains to be established. Indeed, it has been proven that several other factors can affect the chemical shift of [31]P signals. In particular, differential solvation of DNA phosphates, salt-driven effects, or small variations of the O-5'—P—O-3' bond angle can all induce chemical shift variations of equivalent magnitude.[64–66] Therefore, although [31]P chemical shifts represent the only source of information available to assess the conformation of phosphodiester bonds, the limitations of this method must be kept in mind when discussing the results.

## 7.03.5  [15]N- AND [13]C-NMR OF NUCLEIC ACIDS

### 7.03.5.1  [15]N Spectra

The intrinsically low sensitivity of the [15]N nucleus has been a limiting factor in its use in studying the conformation of oligodeoxynucleotide duplexes in solution. With a natural abundance of only 0.4% and a low gyromagnetic ratio, [15]N is about 10[6]-fold less sensitive than protons at constant field. Indeed, direct observation of nitrogen signals is only possible after the laborious isotopic enrichment of DNA samples.[67–69] However, with the development of two-dimensional inverse-detected multiple quantum correlation techniques, the situation has changed,[70–72] and the observation of protonated nitrogens can be readily accomplished with samples containing millimolar quantities of nonenriched material.[73,74] The emerging studies show that, at low temperature, G(N-1) signals resonate in a region centered around 148 ppm, while T(N-3) nitrogen signals are slightly upfield at 160 ppm relative to liquid ammonia. Similarly, nitrogen signals originating from the amino groups also exhibit a base-specific chemical shift range with G(N-2) centred at 75 ppm, A(N-6) at 82 ppm, and C(N-4) at 98 ppm. As an example of the application of [15]N spectroscopy, the author's group has used the nitrogen chemical shifts to confirm the assignment of shifted imino proton signals. In the expanded region of the $^{15}$N-HMQC spectrum shown in Figure 16, a proton signal at 9.7 ppm could be assigned to a guanine imino proton based on its correlation with a guanine nitrogen at 142 ppm.[46] The alternative, namely a downfield amino proton signal, was absolutely ruled out by the experiment.

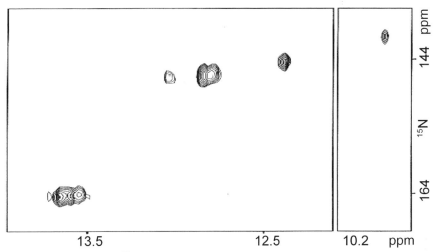

**Figure 16** Expanded region of a $^{15}$N-HMQC spectrum recorded in 10% D$_2$O buffer at 5 °C, of a duplex sample containing a (3,$N^4$-etheno)-dC·dG base pair at its center. The identity of a shifted imino proton signal, at 9.7 ppm, could be confirmed by its correlation with a G(N-1) nitrogen (reproduced with permission of the American Chemical Society from *Biochemistry*, 1997, **36**, 11933; © American Chemical Society).

An enzymatic synthesis of oligodeoxynucleotides that produces cost-effective yields of $^{15}$N, $^{13}$C uniformly enriched samples has been reported.[3] The usefulness of such samples resides in the fact that they will allow a complete assignment of proton signals without making any structural assumption. Additional experimental restraints will greatly improve the accuracy of the NMR-determined structures. In addition, nitrogen chemical shifts are sensitive to electrostatic interactions, and an upfield shift of a few ppm is expected upon hydrogen bonding.[67] Hence the availability of $^{15}$N-enriched samples will greatly assist in the identification of the specific players involved in DNA–protein interactions, as it has in the case of RNA–protein complexes.[75,76]

### 7.03.5.2 $^{13}$C Spectra

While the NMR sensitivity of the $^{13}$C nucleus is slightly higher than that of $^{15}$N (only 10$^4$ times less sensitive than protons at constant field), the routine observation of carbon signals on DNA samples has become possible only after the advent of inverse-detected multiple quantum techniques.[70–72,77] Figure 17 shows a $^{1}$H-detected $^{13}$C-HMQC spectrum recorded in a Pt–d(GG) adduct-containing duplex sample at room temperature. The chemical shift range of protonated carbons is over 140 ppm, from A(C-2) signals around 154 ppm to thymine methyls at 13 ppm, and show defined regions according to carbon type. A similar range was previously observed in unmodified DNA samples.[73,78] A practical application of $^{13}$C-HMQC spectra is to assist in the assignment of proton resonances. It is a frequent situation, when working with samples containing DNA adducts or drug–DNA complexes, that a proton signal can shift by more than 1 ppm from its usual range. In such cases, the possible overlap, or even transposition of proton regions, makes their assignment ambiguous. The chemical shift range of carbons is wider, and signal overlap is generally absent, facilitating the assignment of proton signals based on their $^{1}$H–$^{13}$C interactions.

Similarly to the case with $^{15}$N, fully $^{13}$C-enriched DNA samples will greatly improve the quality of three-dimensional structures determined by NMR spectroscopy. The conformational range of several backbone dihedral angles should be accessible through the determination of $^{3}J_{^{13}C,^{31}P}$ and $^{3}J_{^{13}C,^{1}H}$ coupling constant values. In addition, a large repertoire of half-filtered and double-filtered $^{15}$N, $^{13}$C NMR experiments are available which simplify the proton spectra, and would allow structural studies of large protein–DNA complexes.

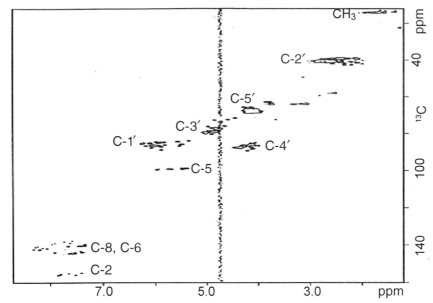

**Figure 17** [13]C-HMQC spectrum of the Pt–d(GG) adduct-containing duplex sample recorded in $D_2O$ buffer at 30°C. Carbon signals show definite chemical shift ranges according to the atom type (after de los Santos and Lippard[43]).

## 7.03.6 HYDROGEN EXCHANGE AND BASE-PAIR DYNAMICS

The formation of base pairs, a ubiquitous feature of nucleic acids, plays a key role in determining their structure and function. Many biological processes, such as DNA replication and transcription, involve the systematic disruption of base pairs from a double-stranded molecule. Several DNA enzymes, which use double-stranded duplexes as their substrate, perform the chemistry over a structural intermediate containing "flipped-out" bases. [1]H-NMR spectroscopy offers the unique possibility of studying the kinetics of base-pair opening at the individual base-pair level, by measuring imino proton exchange rates.

### 7.03.6.1 Exchange Formalism

For a nonpaired nucleotide, the transfer rate of the imino proton from the base to a proton acceptor is given by

$$k_{tr} = k_r[\text{acc}]/[1 + 10^{pK(\text{nu}) - pK(\text{acc})}] \tag{14}$$

where $k_r$ is the diffusion rate of the proton acceptor ($\sim 10^{10}$ L mol$^{-1}$ s$^{-1}$), [acc] is the concentration of a proton acceptor (generally a base catalyst), and $pK$(nu) and $pK$(acc) are the nucleotide (G(N-1), T(N-3)) and catalyst $pK$ values, respectively.[79] When pH < $pK$(nu), the usual conditions of NMR studies, the proton exchange rate, $k_{ex}$, is the same as $k_{tr}$. The situation is different in duplex DNA, when nucleotides are inside the double helix, involved in hydrogen bonding (A·T or C·G). Neither solvent molecules nor a proton acceptor have access to them and, as a result, they cannot exchange. Assuming a two-state model (base pair either closed or open), the exchange proceeds only after transient opening of the base pair, whose rate $k_{op}$ is $1/\tau_0$ ($\tau_0$ is the base-pair lifetime), followed by imino exchange and subsequent closing of the base pair:

$$A \cdot T \leftrightarrow A + TH^{\circ} \leftrightarrow A + TH^{\bullet} \leftrightarrow A \cdot T \tag{15}$$

where H$^{\bullet}$ and H$^{\circ}$ denote two different hydrogen atoms. When $k_{tr} \gg k_{cl}$ (base-pair closing rate), an exchange event will occur every time the base pair opens and, thus, $k_{ex} = k_{op}$. By contrast, if $k_{cl} \gg k_{tr}$, the exchange rate is that of the open state multiplied by the fraction of the time the base pair is in the open state:

$$k_{ex} = k_{op}/[(1+k_{cl}/k_{tr})] \quad \text{or} \quad \tau_{ex} = \tau_0(1+k_{cl}/k_{tr}) \tag{16}$$

where $\tau_{ex} = 1/k_{ex}$ and $k_{tr}$ can be calculated according to Equation (14). Defining $[B]_{eq}$ as the concentration of catalyst for which $k_{cl} = k_{tr}$, one can write

$$\tau_{ex} = \tau_0(1+[B]_{eq}/[B]) \tag{17}$$

Therefore, as shown in Figure 18, a plot of the experimentally determined $\tau_{ex}$ ($k_{ex}^{-1}$) against $1/[B]$ is a straight line, from which $\tau_0$, the inverse of the base-pair opening rate, can be extrapolated from the value of $\tau_{ex}$ at infinite catalyst concentration. Many of the early studies of imino proton exchange were performed under partial catalysis conditions and the reported base-pair opening rates were underestimated.[80–82]

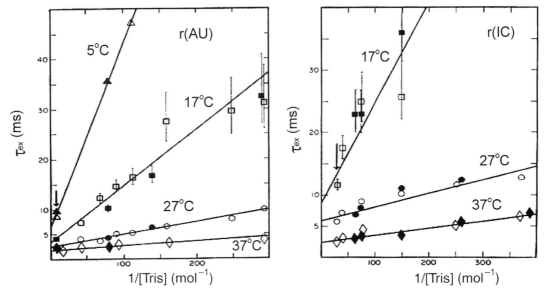

**Figure 18** Variation of exchange time ($\tau_{ex}$) of imino protons as a function of base catalyst concentration for (left) poly(rA)·poly(rU) and (right) poly(rI)·poly(rC). The lifetime of the closed base pair can be extrapolated from $\tau_{ex}$ at infinite catalyst concentration (reproduced by permission of Academic Press from *J. Mol. Biol.*, 1985, **184**, 165; © Academic Press, London).

There are several approaches for measuring the $\tau_{ex}$ of imino protons by NMR. Exchange times longer than 1 min can be measured in real-time experiments similar to those used for tritium exchange.[83] Basically, the sample is lyophilized from an aqueous solution and dissolved in $D_2O$ buffer before NMR data collection. The intensity of the imino proton signals gives the fraction that has exchanged at the time of the experiment.[84] For faster exchange times, spectroscopic methods to measure the magnetization transfer from water protons are needed.[85] Alternatively, methods that measure the differential increment in the linewidth of the imino proton signals as a function of added catalyst can be used. In this case, the contribution of the exchange process to the linewidth of NMR signals is given by

$$\Delta v_{1/2} \text{ (Hz)} = k_{ex}/\pi \tag{18}$$

where $\Delta v_{1/2}$ represents the linewidth difference at half-height between the catalyst-added and control samples. However, extreme caution should be used in analyzing this type of experiment, because the addition of increasing quantities of base catalyst might change the relaxation properties of the sample and increase the linewidth of the imino proton signals for reasons unrelated to the proton exchange rate.[85]

## 7.03.7 DNA CONFORMATIONS

Once the NMR characterization is completed, the *J*-derived values of sugar and backbone dihedral angles and the NOE-calculated interproton distances generate a table of experimental restraints, which are subsequently used to search conformational space for a three-dimensional model compatible with the experimental data set. A variety of computational methods are available to perform such a search, including restrained molecular dynamics,[86–89] a combination of distance geometry and back-calculations,[90,91] and restrained Monte Carlo simulations.[92]

The implementation of molecular dynamics simulations initially restrained by interproton distances and, subsequently, by back-calculation of the NOE intensities generates an ensemble of three-dimensional models, which represent the solution structure of the duplex. Root mean square differences in the atom positions among the models of the ensemble are a measure of the precision of the structural determination, reflecting how well the structure has been defined by the set of experimental restraints and the force field used during the simulation. The accuracy of the models is estimated by their ability to reproduce the experimentally determined dihedral angles and NOE intensities. Since the late 1980s, fundamental advances have been made in all the steps of structure calculation and it is generally accepted that NMR restrained molecular dynamics determine accurate solution structures of short DNA duplexes.[93]

### 7.03.7.1 B-form DNA Structure

B-form DNA is the predominant conformation adopted by unmodified oligodeoxynucleotide duplexes in solution. In contrast to crystallographic studies, a $B_1$ conformation of phosphodiester bonds ($g^-/g^-$; $\zeta/\alpha$) prevails in solution, with little evidence of alternative conformations.[94] Generally, residues have a main population (over 80%) with sugar puckers in the *south* conformation,[95] mainly in the C-2′-*endo*/C-1′-*exo* range, but exceptions to this finding are abundant. Indeed, nucleotides having a main population on the C-3′-*endo* conformation, or similar molar fractions of *north* and *south* conformers, are common, especially when studying chemically modified or anomalous DNA molecules.[43,96] In alternating purine–pyrimidine duplexes a sequence-dependent effect was observed, with angles of pseudorotation of 160–180° for purines and 140–160° for pyrimidines.[97,98]

In general, the availability of several intraresidue distances between the base H-8(H-6) and H-1′, H-2′, H-2″ sugar protons allows the determination of glycosidic angles with a high degree of accuracy. In DNA duplexes in solution, the *anti* range observed in the NMR-refined structures is centered at about −115°, slightly different from the crystallographic mean value of −104°. Rotation along the glycosidic bond to a *syn* range is a conformational adjustment frequently observed in the structure of DNA-containing chemically modified bases or base-pair mismatches.[99–102] As has already been mentioned, this change is easily monitored by the variations of several interproton distances, specifically a shortening of H-8(H-6)–H-1′ and lengthening of H-8(H-6)–H-2′.

Watson–Crick base-pair alignments are clearly established by NMR spectra. However, the assessment of alternate hydrogen bonding generally relies on the existence of additional distance information or results from the structure determination by computer simulations.[46,99,103] Studies of imino proton exchange rates in unmodified DNA duplexes have proven that base pairs open in a noncooperative fashion. Lifetimes are similar for A·T and G·C base pairs, 0.8–7 ms and 7–40 ms, respectively.[85,104] Tracts of four or more consecutive dA residues have increased base pair lifetime (20–120 ms), a fact that was associated with the formation of an alternate structure called B′-DNA, which promotes curvature of the double-stranded helix.[105] In studies of drug–DNA complexes, or unusual DNA structures, base-pair lifetimes of several seconds have been reported.[106,107]

The existence of many sequential distances, mainly between the base H-8(H-6) and sugar protons of the 5′-flanking residue, makes it possible to estimate the base-pair twist fairly accurately by NMR methods. In contrast to the narrow crystallographic range, base-pair twist fluctuates from 24° to 46°, without considering the values for terminal base pairs.

Some sequence-dependent modulation seems to exist, with a broad range observed in TA steps whereas a narrower variation is present in AT and AC steps.[108] ¹H-NMR spectroscopy is less suitable for the evaluation of the other helical parameters because, in general, distances among exchangeable protons are not measured and, as a result, fewer interresidue distances involving base protons are available for the structural refinement. However, it has been shown that not only base-pair twist, but also roll, shift, and slide can be accurately determined from the NMR data.[109] Table 3 gives values of structural parameters measured on unmodified DNA samples.

### 7.03.7.2 A-form DNA Structure

This conformation is readily observed in crystals formed under low levels of hydration, or formed in solution when using alcohol–water mixtures. The main differences with respect to the B-form

**Table 3** Structural parameters of DNA conformations.[a]

| Parameter | A-DNA | B-DNA | Z-DNA CG | Z-DNA GC |
|---|---|---|---|---|
| $X$-Displacement (Å) | −5.4 | −0.7 | −2.4 | −1.4 |
| $Y$-Displacement (Å) | 0.0 | 0.0 | −2.4 | 2.5 |
| Inclination (°) | 19.1 | −5.9 | 9.7 | 3.5 |
| Tip (°) | 6.8 | 0.0 | −177.8 | 177.6 |
| Buckle (°) | 0.0 | 0.0 | 6.1 | −6.1 |
| Propeller (°) | 13.7 | 3.6 | −0.3 | −0.3 |
| Opening (°) | 0.0 | −4.1 | 3.9 | 3.8 |
| Shift (Å) | −4.6 | 0.0 | 1.0 | −1.0 |
| Slide (Å) | 0.0 | 0.0 | 4.9 | −4.9 |
| Rise (Å) | 2.6 | 3.4 | 4.4 | 3.1 |
| Tilt (°) | 0.0 | 0.0 | −6.1 | 6.1 |
| Roll (°) | 0.0 | 0.0 | −4.6 | 4.6 |
| Twist (°) | 32.7 | 36.0 | −1.2 | −58.8 |

| | A-DNA | B-DNA | C | G |
|---|---|---|---|---|
| $\alpha(°)$ | −83.9 | −46.8 | 47.6 | −137.1 |
| $\beta(°)$ | −152.1 | −145.9 | 178.6 | −139.0 |
| $\gamma(°)$ | 45.6 | 36.3 | 55.5 | −169.8 |
| $\delta(°)$ | 84.3 | 156.4 | 138.2 | 99.5 |
| $\varepsilon(°)$ | 179.4 | 155.0 | −93.7 | −104.1 |
| $\zeta(°)$ | −49.1 | −95.1 | 80.9 | −68.6 |
| $\chi(°)$ | −154.3 | −97.9 | −159.1 | 66.5 |
| Pseudorotation (°) | 13.1 | 191.8 | 152.7 | 358.6 |
| Amplitude (°) | 38.9 | 36.3 | 35.2 | 24.3 |

[a] Values taken from fiber diffraction X-ray studies.

structure are a base-pair displacement from the center to the edge of the helix, along with C-3′-*endo* sugar puckers, reduction of base-pair twist and rise, and a small difference in the glycosidic torsion angles.[110] Figure 19 shows side-by-side views of B-, A-, and Z-form DNA structures. Unmodified duplexes in solution show little tendency to adopt an A-form structure, but GC-rich sequences can be induced to change to an A-form conformation, or even to a Z-form structure.[111–113] However, NMR structural studies performed using chemically modified duplexes or DNA triple helixes revealed that some features of A-form DNA structure, especially a change of the sugar puckers to a C-3′-*endo* conformation, can be locally adopted at specific positions of the sequence.[43,96] In addition, NMR spectroscopy and computer simulation studies suggest that the DNA component of DNA–RNA hybrids and chimeric DNA sequences can take some A-form characteristics to adjust to the RNA strand.[114,115] In this case, sugar puckers were in the unusual O-4′-*endo* range (between C-3′-*endo* and C-2′-*endo*), but base-pair inclination and rise were close to A-form values. In summary, although DNA duplexes in solution have not been observed in an A-form conformation, they can assume some A-like characteristics in response to chemical modifications or environmental changes or upon formation of higher-order structures.

### 7.03.7.3 Z-form DNA Structure

Z-form DNA structure is quite different from A- or B-form conformations owing to its unusual left-handed helical twist, a fact initially observed in its circular dichroism spectrum.[116] The first Z-form DNA crystals were obtained working with alternating $(GC)_n$ sequences under high salt conditions. As shown in Table 3, the structural unit is now a dinucleotide (GpC) with remarkable conformational differences for residues within the unit, in addition to the helical parameters for each step (GpC/CpG). Cytosine residues have glycosidic torsion angles in the *anti* range and sugar puckers in the C-2′-*exo*/C-3′-*endo* conformation. Conversely, guanosine residues are *syn* with sugar puckers in the C-2′-*endo*/C-3′-*exo* range.[117,118] A Z-form conformation is also observed with alternating $(AC)_n$ sequences and, in general, is stabilized by high salt concentrations (4 mol L$^{-1}$), divalent

# A-form         B-form         Z-form

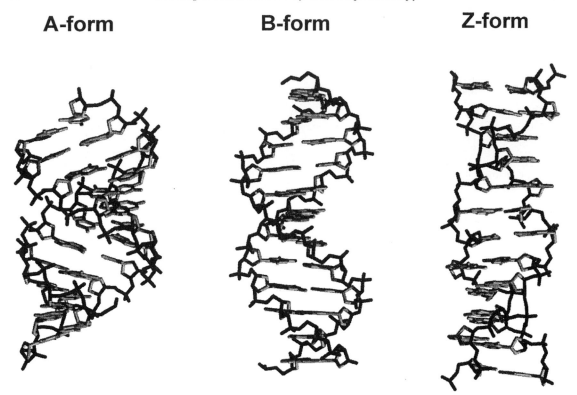

**Figure 19**  Side view of three-dimensional models of A-, B-, and Z-form DNA structures.

ions, alcohol–water mixtures, or the use of modified bases having bulky substituents, such as 8-methyl-dG, 8-bromo-dG, 5-methyl-dC, and 5-bromo-dC.[119] The ¹H-NMR spectra of Z-form DNA have several peculiar characteristics, which have been used as evidence for the presence of this conformation in solution. Owing to shielding by the 3′-neighboring purine base, the H-4′, H-5′, and H-5″ signals of pyrimidine residues are shifted upfield relative to the B-form positions, particularly the H-5′ proton.[120] The glycosidic torsion angle of purine residues in the *syn* conformation is readily monitored by the short G(H-8)–G(H-1′) inter-proton distance. In general, comparison of the spectra recorded on the sample at low and high salt concentrations, or with increasing methanol content, permits the observation of the transition in going from the B- to the Z-form DNA conformation.[121,122] Upon change to a Z-form conformation, the ³¹P spectrum shows a downfield-shifted signal, which was assigned to the (RpY) phosphate, whereas (YpR) shifts slightly upfield. The ¹³C spectrum also has shown downfield-shifted signals, namely the cytosine C-1′ and C-2′ and the guanosine C-1′ carbons.[123] However, relative to the dispersion of ¹³C signals, the magnitudes of these shifts are small and, thus, the value of the C(C-1′,C-2′) and G(C-1′) chemical shifts cannot be used as sufficient evidence for Z-form conformation, as is done with ¹H and ³¹P spectra. In general, when X-ray crystallographic and NMR methods have been used to study the same sample, the Z-form structures have shown few differences between them. The stringent salt and/or solvent conditions needed to induce and observe Z-form DNA result in a rather rigid structure. The fact that base-pair lifetimes are much longer in a Z-form conformation, around 1 s, than in the B-form counterpart, a few milliseconds, points to the same conclusion.[124]

## ACKNOWLEDGMENTS

The author is indebted to Serge Smirnov and David Cullinan for their help during the preparation of the manuscript. The author is supported by NIH Grants CA77094 and CA47995, from the National Cancer Institute.

## 7.03.8 REFERENCES

1. Protein Data Bank, *Q. Newsl.*, 1997, **82**, 1.
2. R. T. Batey, J. L. Battiste, and J. R. Williamson, *Methods Enzymol.*, 1995, **261**, 300.
3. D. P. Zimmer and D. M. Crothers, *Proc. Natl. Acad. Sci. USA*, 1995, **92**, 3091.
4. G. Varani, F. Abul-ela, and F. H.-T. Allain, *Prog. Nucl. Magn. Reson. Spectrosc.*, 1996, **29**, 59.
5. W. Saenger, "Principles of Nucleic Acid Structure," Springer, New York, 1984.
6. K. G. Orrell, V. Sik, and D. Stephenson, *Prog. Nucl. Magn. Reson. Spectrosc.*, 1990, **22**, 141.
7. A. N. Lane, *Prog. Nucl. Magn. Reson. Spectrosc.*, 1993, **25**, 481.
8. T. L. James (ed.), "Nuclear Magnetic Resonance and Nucleic Acids," *Methods Enzymol.*, 1995, **261**.
9. IUPAC–IUB Joint Commission on Biochemical Nomenclature, *Eur. J. Biochem.*, 1983, **131**, 9.
10. W. Klyne and V. Prelog, *Experientia*, 1960, **16**, 521.
11. C. D. Jardetzky, *J. Am. Chem. Soc.*, 1960, **82**, 229.
12. J. E. Kilpatrick, K. S. Pitzer, and R. Spitzer, *J. Am. Chem. Soc.*, 1947, **69**, 2483.
13. C. Altona and M. Sundaralingam, *J. Am. Chem. Soc.*, 1972, **94**, 8205.
14. S. Arnott, *Prog. Biophys. Mol. Biol.*, 1970, **21** 265.
15. R. Wing, H. Drew, T. Takano, C. Broca, K. Itakura, and R. E. Dickerson, *Nature (London)*, 1980, **287**, 6514.
16. R. E. Dickerson, M. Bansal, C. R. Calladine, S. Diekmann, W. N. Hunter, O. Kennard, R. Lavery, H. C. M. Nelson, W. Saenger, Z. Shakked, *et al.*, *J. Mol. Biol.*, 1989, **205**, 787.
17. A. S. Edison, F. Abildgaard, W. M. Westler, E. S. Mooberry, and J. L. Markley, *Methods Enzymol.*, 1994, **239**, 3.
18. G. M. Clore and A. M. Gronenborn, *Methods Enzymol.*, 1994, **239**, 349.
19. A. Bax, G. W. Vuister, S. Grzesiek, F. Delaglio, A. C. Wang, R. Tschudin, and G. Zhu, *Methods Enzymol.*, 1994, **239**, 79.
20. Wuthrich, K., "NMR of Proteins and Nucleic Acids," Wiley, New York, 1986.
21. C. Giessner-Prettre and B. Pullman, *Q. Rev. Biophys*, 1987, **20**, 113.
22. C. K. Mitra, M. H. Sarma, and R. H. Sarma, *J. Am. Chem. Soc.*, 1981, **103**, 6727.
23. D. R. Kearns, *CRC Crit. Rev. Biochem.*, 1984, **15**, 237.
24. R. K. Harris, "Nuclear Magnetic Resonance Spectroscopy: a Physicochemical View", Wiley, New York, 1986.
25. S. Macura and R. R. Ernst, *Mol. Phys.*, 1980, **41**, 95.
26. B. R. Reid, *Q. Rev. Biophys*, 1987, **20**, 1.
27. D. J. Patel, L. Shapiro, and D. Hare, *Q. Rev. Biophys.*, 1987, **20**, 35.
28. D. B. Keepers and T. L. James, *J. Magn. Reson.*, 1984, **57**, 404.
29. G. M. Clore and A. M. Gronenborn, *J. Magn. Reson.*, 1985, **61**, 158.
30. J.-F. Lefèvre, A. N. Lane, and O. Jardetzky, *Biochemistry*, 1987, **26**, 5076.
31. F. J. M. Van de Ven and C. W. Hilbers, *Eur. J. Biochem.*, 1988, **178**, 1.
32. R. Boelens, T. M. G. Koning, G. A. van der Marel, J. H. van Boom, and R. Kaptein, *J. Magn. Reson.*, 1989, **82**, 290.
33. C. B. Post, R. P. Meadows, and D. G. Gorenstein, *J. Am. Chem. Soc.*, 1990, **112**, 6796.
34. B. A. Borgias and T. L. James, *J. Magn. Reson.*, 1990, **87**, 475.
35. T. L. James, *Curr. Opin. Struct. Biol.*, 1991, **1**, 1042.
36. D. Marion and W. Wüthrich, *Biochem. Biophys. Res. Commun.*, 1983, **113**, 967.
37. U. Piantini, O. W. Sørensen, and R. R. Ernst, *J. Am. Chem. Soc.*, 1982, **104**, 6800.
38. G. Eich, G. Bodenhausen, and R. R. Ernst, *J. Am. Chem. Soc.*, 1982, **104**, 3731.
39. L. Braunschweiler and R. R. Ernst, *J. Magn. Reson.*, 1983, **53**, 521.
40. D. Davis and A. Bax, *J. Am. Chem. Soc.*, 1985, **107**, 2821.
41. C. A. G. Hassnoot, T. A. A. M. de Leeuw, and C. Altona, *Tetrahedron*, 1980, **36**, 2783.
42. D. R. Hare, D. E. Wemmer, S. H. Chou, G. Drobny, and B. R. Reid, *J. Mol. Biol.*, 1983, **171**, 319.
43. C. de los Santos and S. Lippard, unpublished results.
44. C. de los Santos, M. Rosen, and D. J. Patel, *Biochemistry*, 1989, **28**, 7282.
45. E. Quignard, G. V. Fazakerley, G. van der Marel, J. H. Boom, and W. Guschlbauer, *Nucleic Acids Res.*, 1987, **15**, 3397.
46. D. Cullinan, F. Johnson, A. P. Grollman, M. Eisenberg, and C. de los Santos, *Biochemistry*, 1997, **36**, 11 933.
47. C. Griessinger, O. W. Sørensen, and R. R. Ernst, *J. Am. Chem. Soc.*, 1986, **107**, 6394.
48. A. Bax and L. Lerner, *J. Magn. Reson.*, 1988, **79**, 429.
49. A. Majumdar and R. V. Hosur, *Prog. Nucl. Magn. Reson. Spectrosc.*, 1992, **24**, 109.
50. L. J. Rinkel and C. Altona, *J. Biomol. Struct. Dyn.*, 1987, **4**, 621.
51. J. van Wijk, B. D. Huckriede, J. H. Ippel, and C. Altona, *Methods Enzymol.*, 1992, **211**, 286.
52. J. M. Fu, S. A. Schroeder, C. R. Jones, R. Santini, and D. G. Gorestein, *J. Magn. Reson.*, 1988, **77**, 577.
53. C. R. Jones, S. A. Schroeder, and D. G. Gorestein, *J. Magn. Reson.*, 1988, **80**, 370.
54. V. Sklenár, H. Miyashiro, G. Zon, H. T. Miles, and A. Bax, *FEBS Lett.*, 1986, **208**, 94.
55. D. Williamson and A. Bax, *J. Magn. Reson.*, 1988, **76**, 174.
56. G. W. Kellogg and B. I. Schweitzer, *J. Biomol. NMR*, 1993, **3**, 577.
57. P. P. Lankhorst, C. A. G. Haasnoot, C. Erkelens, and C. Altona, *J. Biomol. Struct. Dyn.*, 1984, **1**, 1387.
58. F. R. Prado, C. Giessner-Prettre, B. Pullman, and J.-P. Dandley, *J. Am. Chem. Soc.*, 1979, **101**, 1737.
59. D. G. Gorenstein, S. A. Schroeder, J. M. Fu, J. T. Metz, V. R. Roongta, and C. R. Jones, *Biochemistry*, 1988, **27**, 7223.
60. D. G. Gorenstein, *Chem. Rev.*, 1987, **87**, 1047.
61. V. R. Roongta, R. Powers, E. P. Nikonowicz, C. R. Jones, and D. G. Gorenstein, *Biochemistry*, 1990, **29**, 5245.
62. R. E. Dickerson and H. R. Drew, *J. Mol. Biol.*, 1981, **149**, 761.
63. R. E. Dickerson, *J. Mol. Biol.*, 1983, **166**, 416.
64. D. B. Lerner and D. R. Kearns, *J. Am. Chem. Soc.*, 1980, **102**, 7611.
65. A. J. R. Costello, T. Glonek, and J. R. van Wazer, *J. Inorg. Chem.*, 1976, **15**, 972.
66. D. G. Gorenstein, *J. Am. Chem. Soc.*, 1975, **97**, 898.
67. T. L. James, J. L. James, and A. Lapidot, *J. Am. Chem. Soc.*, 1981, **103**, 6748.
68. G. Kupferschmitt, J. Schmidt, T. Schmidt, B. Fera, F. Buck, and H. Ruterjans, *Nucleic Acids Res.*, 1987, **15**, 6225.

69. X. Gao and R. A. Jones, *J. Am. Chem. Soc.*, 1987, **109**, 3169.
70. L. Müller, *J. Am. Chem. Soc.*, 1979, **101**, 4481.
71. G. Bodenhausen and D. J. Ruben, *Chem. Phys. Lett.*, 1980, **69**, 185.
72. A. Bax, M. Ikura, L. E. Kay, D. A. Torchia, and R. Tschudin, *J. Magn. Reson.*, 1990, **86**, 304.
73. J. Ashcroft, D. H. Live, D. J. Patel, and D. Cowburn, *Biopolymers*, 1991, **31**, 45.
74. R. H. Griffey, C. D. Poulter, A. Bax, B. L. Hawkins, Z. Yamaizumi, and S. Nishimura, *Proc. Natl. Acad. Sci. USA*, 1983, **80**, 5895.
75. J. D. Puglisi, R. Tan, B. J. Calnan, A. D. Frankel, and J. R. Williamson, *Science*, 1992, **257**, 72.
76. M. J. Michnicka, J. W. Harper, and G. C. King, *Biochemistry*, 1993, **32**, 395.
77. M. F. Summers, L. G. Marzilli, and A. Bax, *J. Am. Chem. Soc.*, 1986, **108**, 4285.
78. W. Leupin, G. Wagner, W. A. Denny, and K. Wüthrich, *Nucleic Acids Res.*, 1987, **15**, 267.
79. J.-L. Leroy, D. Broseta, and M. Guéron, *J. Mol. Biol.*, 1985, **184**, 165.
80. D. J. Patel, S. Ikuta, S. A. Kozlowski, and K. Itakura, *Proc. Natl. Acad. Sci. USA*, 1983, **80**, 2184.
81. S. Cheung, K. Arndt, and P. Lu, *Proc. Natl. Acad. Sci. USA*, 1983, **81**, 3665.
82. S. H. Chou, D. E. Wemmer, D. R. Hare, and B. R. Reid, *Biochemistry*, 1984, **23**, 2257.
83. H. Teitelbaum and S. W. Eglander, *J. Mol. Biol.*, 1975, **92**, 55.
84. J.-L. Leroy, N. Bolo, N. Figueroa, P. Plateau, and M. Guéron, *J. Biomol. Struct. Dyn.*, 1985, **2**, 915.
85. M. Kochoyan, J.-L. Leroy, and M. Guéron, *J. Mol. Biol.*, 1987, **196**, 599.
86. A. T. Brünger and M. Karplus, *Acc. Chem. Res.*, 1991, **24**, 54.
87. D. A. Pearlman, D. A. Case, J. C. Caldwell, G. L. Seibel, U. C. Singh, P. Weiner, and P. A. Kollman, "AMBER (UCSF), Version 4.0," University of San Francisco, San Francisco, CA, 1990.
88. J. de Vlieg, R. Boelens, R. M. Scheek, R. Kaptein, and W. F. van Gunsteren, *Isr. J. Chem.*, 1986, **27**, 181.
89. A. T. Brünger, "X-PLOR, Version 3.1: a System for X-Ray Crystallography and NMR," Yale University Press, New Haven, CT, 1992.
90. W. Nerdal, D. R. Hare, and B. R. Reid, *Biochemistry*, 1989, **28**, 10 008.
91. R. Nibedita, R. A. Kumar, A. Majumdarm, and R. V. Hosur, *J. Biomol. NMR*, 1992, **2**, 477.
92. N. Ulyanov, U. Schmitz, and T. James, *J. Biomol. NMR*, 1993, **3**, 547.
93. U. Schmitz and T. James, *Methods Enzymol.*, 1995, **261**, 3.
94. C. Karslake, M. V. Botuyan, and D. G. Gorenstein, *Biochemistry*, 1992, **31**, 1849.
95. N. Ulyanov, U. Schmitz, A. Kumar, and T. James, *Biophys. J.*, 1995, **68**, 13.
96. P. Rajagopal and J. Feigon, *Biochemistry*, 1989, **28**, 7859.
97. U. Schmitz, G. Zon, and T. James, *Biochemistry*, 1990, **29**, 2357.
98. J.-W. Cheng, S.-H. Chou, M. Salazar, and B. R. Reid, *J. Mol. Biol.*, 1992, **228**, 118.
99. D. Cullinan, A. Korobka, A. P. Grollman, D. J. Patel, M. Eisenberg, and C. de los Santos, *Biochemistry*, 1996, **35**, 13 319.
100. C. Carbonaux, G. A. van der Marel, J. H. van Boom, W. Guschlbauer, and G. V. Fazakerley, *Biochemistry*, 1991, **30**, 5449.
101. A. N. Lane, T. C. Jenkins, D. J. S. Brown, and T. Brown, *Biochem. J.*, 1991, **279**, 269.
102. X. Gao and D. J. Patel, *J. Am. Chem Soc.*, 1988, **110**, 5178.
103. C. de los Santos, M. Rosen, and D. J. Patel, *Biochemistry*, 1989, **28**, 7282.
104. J.-L. Leroy, M. Kochoyan, T. Huynh-Dinh, and M. Guéron, *J. Mol. Biol.*, 1988, **200**, 223.
105. J.-L. Leroy, E. Charetier, M. Kochoyan, and M. Guéron, *Biochemistry*, 1988, **27**, 8894.
106. J.-L. Leroy, X. Gao, V. Misra, M. Gueron, and D. J. Patel, *Biochemistry*, 1992, **31**, 1407.
107. J.-L. Leroy, M. Guéron, J. L. Mergny, and C. Hélène, *Nucleic Acids Res.*, 1994, **22**, 1600.
108. K. Zakrzewska, *J. Biomol. Struct. Dyn.*, 1992, **9**, 681.
109. A. Lefebvre, S. Fermandjian, and B. Hartmann, *Nucleic Acids Res.*, 1997, **25**, 3855.
110. R. Lavery and H. Sklenar, *J. Biomol. Struct. Dyn.*, 1989, **6**, 655.
111. Y. Wang, G. A. Thomas, and W. L. Peticolas, *J. Biomol. Struct. Dyn.*, 1987, **5**, 249.
112. L. E. Minchenkova, A. K. Schyolkina, B. K. Chernov, and V. I. Ivanov, *J. Biomol. Struct. Dyn.*, 1986, **4**, 463.
113. H. Robinson and A. H. Wang, *Nucleic Acids Res.*, 1996, **24**, 676.
114. O. Y. Fedoroff, M. Salazar, and B. R. Reid, *J. Mol. Biol.*, 1993, **233**, 509.
115. M. Salazar, J. J. Champoux, and B. R. Reid, *Biochemistry*, 1993, **32**, 739.
116. F. M. Pohl and T. M. Jovin, *J. Mol. Biol.*, 1972, **67**, 375.
117. A. H.-J. Wang, G. J. Quigley, F. J. Kolpak, J. L. Crawford, J. H. van Boom, G. van der Marel, and A. Rich, *Nature (London)*, 1979, **282**, 680.
118. H. Drew, T. Takano, S. Tanaka, K. Itakura, and R. E. Dickerson, *Nature (London)*, 1980, **286**, 567.
119. B. H. Johnston, *Methods Enzymol.*, 1992, **211**, 127.
120. C. K. Mitra, M. H. Sarma, and R. H. Sarma, *Biochemistry*, 1981, **20**, 2036.
121. D. J. Patel, S. A. Kozlowski, S. A. Nordhein, and A. Rich, *Proc. Natl. Acad. Sci. USA*, 1982, **79**, 1413.
122. J. Feigon, A. H. Wang, G. van der Marel, J. H. van Boom, and A. Rich, *Nucleic Acids Res.*, 1984, **12**, 1243.
123. V. Sklenar and A. Bax, *J. Am. Chem. Soc.*, 1987, **109**, 2221.
124. M. Kochoyan, J.-L. Leroy, and M. Guéron, *Biochemistry*, 1990, **29**, 4799.

# 7.04
# Molecular Probes of DNA Structure

## JULIE T. MILLARD
### Colby College, Waterville, ME, USA

| | | |
|---|---|---|
| 7.04.1 | INTRODUCTION | 81 |
| 7.04.2 | DNA INTERCALATORS | 83 |
| 7.04.3 | TRANSITION METALS AND METAL COMPLEXES | 85 |
| | 7.04.3.1 *Aqueous Metal Ions* | 86 |
| | 7.04.3.2 *Copper Phenanthroline* | 86 |
| | 7.04.3.3 *Methidiumpropyl-EDTA–Fe^{II}* | 86 |
| | 7.04.3.4 *Chiral Tris(phenanthroline)metal Complexes* | 88 |
| | 7.04.3.5 *Nickel Complexes* | 89 |
| | 7.04.3.6 *Metalloporphyrins* | 89 |
| | 7.04.3.7 *Fe^{II}–EDTA* | 89 |
| 7.04.4 | BASE-SPECIFIC REAGENTS | 91 |
| | 7.04.4.1 *Osmium Tetroxide* | 92 |
| | 7.04.4.2 *Potassium Permanganate* | 92 |
| | 7.04.4.3 *Hydroxylamine and Methoxylamine* | 93 |
| | 7.04.4.4 *Haloacetaldehydes* | 93 |
| | 7.04.4.5 *Diethyl Pyrocarbonate* | 94 |
| | 7.04.4.6 *Dimethyl Sulfate* | 95 |
| | 7.04.4.7 *Other Base-specific Reagents* | 95 |
| 7.04.5 | CROSS-LINKERS | 96 |
| | 7.04.5.1 *Psoralen* | 96 |
| | 7.04.5.2 *Nitrogen Mustards* | 96 |
| 7.04.6 | ANTIBIOTICS | 98 |
| 7.04.7 | CONCLUSIONS | 99 |
| 7.04.8 | REFERENCES | 99 |

## 7.04.1 INTRODUCTION

During early X-ray diffraction studies, Franklin and Gosling observed two distinct structures of DNA: the low humidity, asymmetric A form and the high humidity, symmetric B form.[1] Model building led Watson and Crick to propose an antiparallel, right-handed, double-helical arrangement for DNA with the nucleotide bases on the inside of the structure and the sugar–phosphate backbone on the outside.[2] Specific hydrogen-bonding interactions between the bases (Figure 1) were proposed to account for the fidelity of information transfer from mother-to-daughter DNA strands through semiconservative replication. While the Watson–Crick model was indeed an accurate representation of Franklin's B-DNA, the advent of single-crystal X-ray structural analysis,[3] high-field NMR spectroscopy,[4] and routine chemical synthesis of defined-sequence DNA[5] has led to the characterization of a number of other DNA structures (for reviews, see Dickerson[6,7] and Palecek[8]). Such structures include A-DNA, Z-DNA, cruciforms, bulges, mismatches, bent DNA, kinked DNA, supercoiled DNA, H triplexes, G tetrads, and parallel-stranded DNA, as well as subtle variations of the canonical B-DNA.[9,10] "Unusual" DNA structures (i.e., non-B) are believed to be involved in the recognition of appropriate chromosomal regions during such cellular processes as genetic recombination, transcription of active genes, repair, and replication. For example, formation of cruciforms has been postulated to trigger the initiation of DNA replication.[11] Moreover, the binding of small molecules and proteins can induce structural changes in DNA.[12] These changes in structure may lead to critical changes in function and such devastating cellular disorders as cancer.[13] On the other hand, sequence-specific binding by antitumor drugs can also be beneficial.

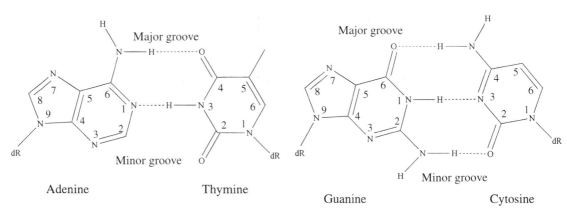

**Figure 1**　Hydrogen-bonding interactions between the bases in DNA.

Eukaryotic DNA structure is further complicated *in vivo* because the majority of nuclear DNA is complexed with histones, forming chromatin, a polymer of repeating nucleosome subunits (for reviews, see van Holde,[14] Wolffe,[15] and Pruss *et al.*[16]). The nucleosome consists of DNA organized in a "core particle" (146 bp of DNA wrapped around a histone octamer) and a "linker" region of variable length depending on the source (20–70 bp of DNA). The incorporation of DNA into the core particle results in some rather dramatic structural changes relative to free B-DNA, including curvature of about 45° per helical turn, sharp bending at several sites, periodic narrowing and widening of the minor groove, underwinding at the center of the core particles, and overwinding elsewhere.

Both natural products, such as antibiotics produced for prokaryotic warfare, and synthetic compounds have been used to probe the structure of DNA *in vivo* and *in vitro*. Such reagents are useful to monitor structural changes induced by DNA-modifying agents and cellular proteins, as well as unusual structures that may serve as biological triggers in the cell. Chemical modification of DNA requires relatively small sample sizes and is quick to perform. Although enzymes can also be used to probe DNA structure, they tend to be more particular to reaction conditions such as temperature, ionic strength, and pH. This chapter will review the current arsenal of chemical reagents available for detecting such non-B structures as A-DNA, Z-DNA, single-stranded DNA, cruciforms, bent DNA, and H triplexes, as well as the more subtle variations that occur in B-DNA. Earlier reviews of chemical probes of free DNA structure include those by Lilley,[17] Wells *et al.*,[18,19] and Nielsen.[20] Probes of chromatin structure have also been reviewed by Hayes[21] and Millard.[22]

A typical experimental strategy for the chemical probing of DNA is a modification of the Maxam–Gilbert sequencing approach (Figure 2).[23] The end-radiolabeled DNA or DNA complex is subjected to treatment with an appropriate reagent, cleavage of the sugar–phosphate backbone at the modification site (either spontaneously or through treatment with piperidine, light, or some other method), and analysis of the resulting fragments through single-base resolution denaturing polyacrylamide–gel electrophoresis (PAGE) with reference Maxam–Gilbert sequencing reactions to elucidate sites of modification. Alternatively, the chemically modified DNA can be used as a template for an RNA or DNA polymerase stop assay, similar to the Sanger method of DNA sequencing (Figure 3).[24] The resulting set of fragments, each of which terminates at a position that defines a site of chemical modification, is analyzed on a sequencing gel. This method has the advantage of amplifying the original signal by incorporating many molecules of radiolabel during the polymerase reaction (for a review, see Htun and Johnston[25]). Both methods require relatively small sample sizes (picomoles to femtomoles[26]), however, and are powerful techniques for analyzing DNA structure.

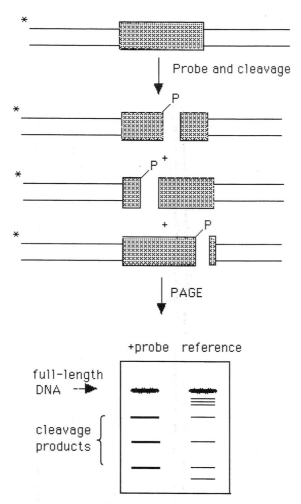

**Figure 2** Schematic of the Maxam–Gilbert-like cleavage assay for the sites targeted by chemical probes of DNA structure. The unusual structure to be probed is denoted by the patterned area in the center of the DNA duplex. End-labeled DNA (radiolabel denoted by an asterisk) is treated with a chemical probe (P) that targets a specific nucleotide (which may appear multiple times) within this structure and then subjected to appropriate conditions for cleavage (often piperidine). The positions of cleavage, and thus the sites targeted by the probe, are determined through denaturing PAGE of the resulting fragments in conjunction with a reference Maxam–Gilbert sequencing reaction.

The crucial decision in using molecular probes of DNA structure is selection of the appropriate reagent(s). An additional criterion for a useful *in vivo* probe is the ability to pass through the plasma membrane. Moreover, possible DNA conformational changes induced by the probe molecules must be considered. Table 1 highlights the probes discussed in this chapter. Both probes that recognize characteristic DNA shapes and probes that target specific bases are available, as discussed below.

## 7.04.2 DNA INTERCALATORS

Historically, the first agents that were studied for their binding to DNA were the aromatic dyes, which could be easily monitored through simple spectroscopic techniques. Intercalators insert themselves between and parallel to the plane of adjacent base pairs, resulting in DNA structural

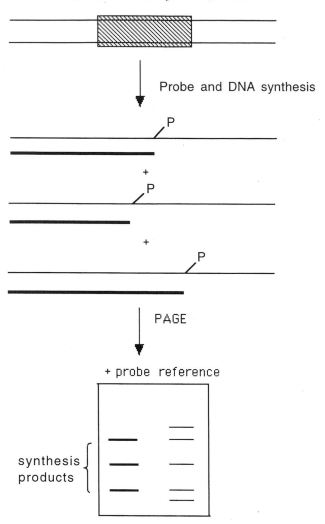

**Figure 3**  Schematic of the Sanger-like primer extension assay for the sites targeted by chemical probes of DNA structure. The unusual structure is denoted by the patterned area in the center of the DNA duplex. The DNA is treated with a chemical probe (P) that targets a specific base (which may appear multiple times) within this structure. A DNA (or RNA) polymerase reaction is then performed with the modified DNA as a template and each of the four bases present (one of which is also present in its radiolabeled form). Synthesis terminates at the sites of chemical modification. The resulting radiolabeled DNA strands are then subjected to denaturing PAGE, where the stop positions are indicative of the sites targeted by the probe.

changes. Intercalating agents have been widely used to probe the DNA structure inside chromatin (for reviews, see Hayes[21] and Millard[22]). In general, intercalators show reduced binding to chromatin relative to free DNA, targeting the linker in preference to the core DNA of nucleosomes because of its greater tolerance for the distortion that accompanies intercalation. Genetically active chromatin shows greater binding than inactive chromatin, attributed to its more "open" conformation.[46] Intercalators can also be used as *in vivo* probes. For example, Hoechst 33258 has been used as a fluorescent probe of DNA structure in *Xenopus laevis* embryos.[47]

Intercalators have also been used to probe the structure of free DNA. Ethidium bromide (EB) (**1**) has been reported to bind preferentially to such distinctive structures as bulges[48] and guanine tetramers as found in telomeres.[49] Moreover, stoichiometric photocleavage and subsequent PAGE allows detection of the sites of EB binding.[50] The resulting binding distribution can provide information about DNA structure. For example, an A tract known to induce bending binds EB poorly, but the ends of the A tract show a high affinity for EB.[50] Photoaffinity labeling, or covalent attachment of photoreactive ethidium analogs to DNA, has also been used to probe DNA structure.[51]

**Table 1** Some of the chemical probes of DNA structure (abbreviations defined in the text).

| Probe | Principal target | Ref.[a] |
|---|---|---|
| EB | Intercalates | 21, 22 |
| $(OP)_2Cu^+$ | Deoxyribose | 27, 28 |
| $MPE-Fe^{II}$ | Deoxyribose | 29 |
| $Ru(phen)_3^{2+}$ | Base | 30, 31 |
| $Co(DIP)_3^+$ | Sugar | 30, 31 |
| NiCR | Guanine | 32, 33 |
| $Fe^{II}-EDTA$ | Deoxyribose | 34 |
| $OsO_4$ | Thymine | 35–37 |
| $KMnO_4$ | Thymine | 38, 39 |
| Hydroxylamine | Cytosine | 40 |
| BAA, CAA | Cytosine, adenine | 41 |
| DEPC | Adenine > guanine | 41 |
| DMS | Guanine | 42 |
| Psoralens | Thymine (5′-TA) | 43, 44 |
| NCS | Deoxyribose | 45 |

[a]References are to papers containing crucial methodology or to reviews. For other references, see text.

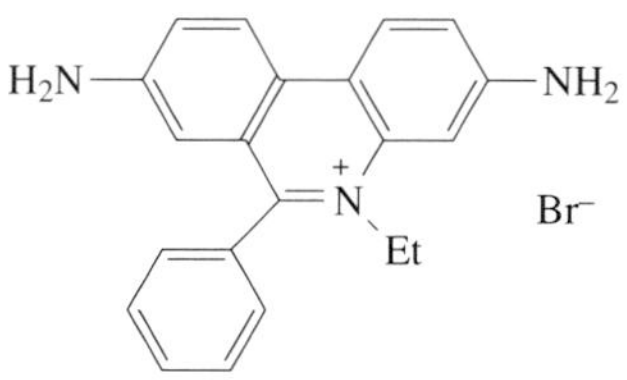

(1)

Other intercalators have also been used to study branched DNA molecules such as might appear during recombination. For example, propidium iodide and the dye Stains-All (4,5:4′,5′-dibenzo-3,3′-diethyl-9-methylthiacarbocyanine bromide) (2) both interact preferentially at the branch point of a four-arm DNA structure.[52,53] These agents are presumed to intercalate into the junction. Another carbocyanine dye, 3,3′-diethyloxadicarbocyanine (DODC), has also been reported to be a powerful spectroscopic probe of hairpin quadruplex structures such as might be found in telomeres.[54] However, the mode of interaction of this agent with DNA remains unclear. Bis-intercalators linked by a rigid linker have also been used to probe the structure of four-way DNA junctions, with changes in accessibility to DNAase I used to analyze structural details.[55]

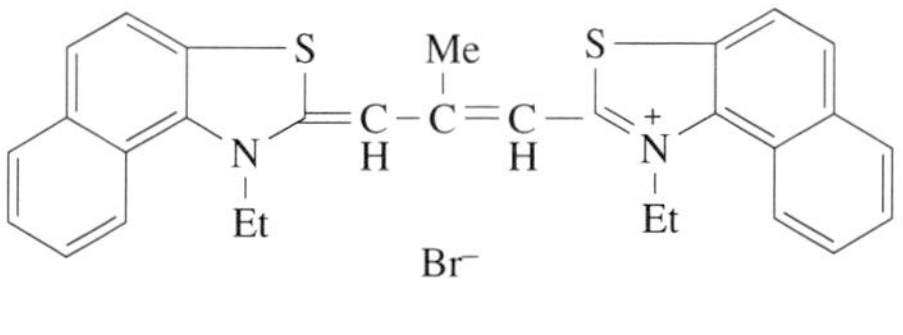

(2)

## 7.04.3 TRANSITION METALS AND METAL COMPLEXES

The use of metals and metal complexes in molecular biology has gained great popularity (for a review, see Tullius[56]). Both simple metal chelates first reported in 1894[57] and elegant complexes rationally designed to bind DNA[58] have proved to be powerful tools for DNA structure analysis. Metal complexes often contain a structural element for DNA recognition in addition to the metal itself, which either provides the capability for DNA cleavage or acts as a spectroscopic probe.

### 7.04.3.1   Aqueous Metal Ions

Aqueous metal ions alone can induce the formation of free radicals, suggesting that metal ion-dependent damage of DNA may be a biologically significant phenomenon. Copper(II) in the presence of ascorbate/$H_2O_2$ or thiols has been reported to mediate the site-specific cleavage of single-stranded DNA.[59,60] Copper ions have been proposed to be tightly and specifically bound to DNA at a stem-and-loop structure. Moreover, ascorbate/Cu[II] cleaves double-stranded DNA under negative torsional stress at specific sites. Although a common targeted conformation could not be identified, the involvement of secondary structure in the copper(II)-induced cleavage is suggested by the dependence on torsional stress, salt, and pH.[61] Additionally, silver(I), platinum(II), and mercury(II) also bind to DNA bases, resulting in changes in absorbance and c.d. spectra and sedimentation rates (for a review, see Marzilli *et al.*[62]). Such changes have been used to probe DNA structure in viruses and chromatin.[63,64] $Hg^{2+}$ can also show site-selective binding to DNA hairpins, forming an intrastrand cross-link between thymine residues.[65]

### 7.04.3.2   Copper Phenanthroline

The first metal complex demonstrated to cleave DNA was the 1,10-phenanthroline–cuprous ion complex (($OP)_2Cu^+$) (**3**), reported by Sigman and co-workers.[66] ($OP)_2Cu^+$ binds to DNA in the minor groove, and in the presence of hydrogen peroxide generates a metal-associated hydroxyl radical-like species, leading to attack at the C-1′ or C-4′ positions of deoxyribose and subsequent backbone scission (for reviews, see Sigman and Chen[27] and Sigman *et al.*[28]). Cleavage is somewhat sequence-dependent, and is believed to reflect minor groove geometry, thereby monitoring local structural variations in B-DNA.[67] This artificial nuclease activity has been used extensively to map DNA conformation and perform footprinting experiments. For example, ($OP)_2Cu^+$ shows a strong preference for gene control regions such as the Pribnow box,[68,69] the junction in a four-arm branched DNA,[70] the linker region of chromatin,[71] and alternating B-DNA.[72] Moreover, Z-DNA is not a substrate for cleavage by ($OP)_2Cu^+$, while A-DNA is cleaved at one-third the rate of B-DNA.[73]

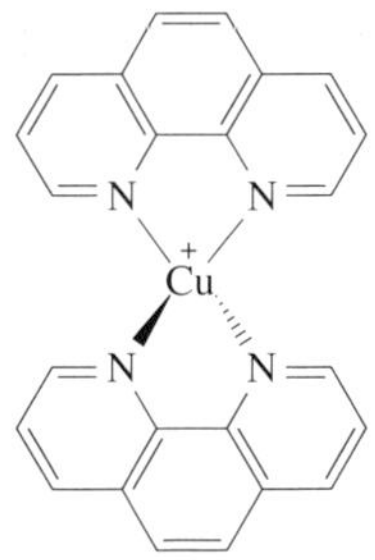

(**3**)

### 7.04.3.3   Methidiumpropyl-EDTA–Fe[II]

An exciting development was made by Dervan and co-workers when they chemically modified the intercalator methidium to produce a rationally designed DNA cleavage agent.[74,75] Methidiumpropyl-EDTA (MPE) (**4**) contains an intercalating ring system covalently bound to the metal chelator EDTA. In the presence of ferrous iron, oxygen, and a reducing agent such as dithiothreitol, the complex induces single-stranded DNA cleavage with little sequence specificity through the production of hydroxyl radical. This reagent has been widely used to footprint DNA because of the small footprint created relative to the often-used enzyme DNase I (Figure 4).[29,76] Other uses of MPE–Fe[II] have been to pinpoint junctions of branched DNA[52,77] and to reveal bulged bases.[78] Preferential cleavage of the linker region of chromatin is also found for MPE, which can be used for chromatin digestion to nucleosomes in lieu of micrococcal nuclease.[79] MPE can also be used for mapping RNA structure.[48]

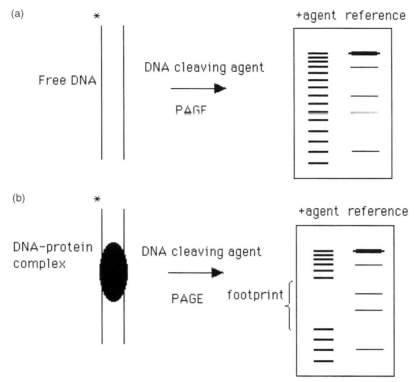

**Figure 4** Schematic of DNA footprinting. End-labeled DNA (radiolabel denoted by an asterisk) is cleaved by a nonsequence-specific reagent, and the resulting fragments are then subjected to denaturing PAGE. Such treatment of free DNA yields a ladder of bands of approximately equal intensity (a). The presence of a protein protects the DNA from cleavage, however, yielding a "footprint" within the cleavage bands (b). The site of this footprint on the DNA can be determined through reference Maxam–Gilbert sequencing reactions.

**(4)**

The $Fe^{II}$–EDTA moiety has likewise been linked to other molecules that show sequence preferences in their binding to DNA, unlike the intercalator methidium. For example, the peptide antibiotics distamycin and netropsin, both of which bind in the minor groove of A/T-rich DNA regions, have been modified to produce "affinity cleavage" reagents.[76] Both $Fe^{II}$ EDTA and copper phenanthroline have also been linked to oligonucleotides capable of binding either a single-stranded target through Watson–Crick hydrogen bonding or a double-stranded target through the formation of a triple helix,[80–82] as well as to DNA-binding proteins.[83]

### 7.04.3.4  Chiral Tris(phenanthroline)metal Complexes

Chiral transition metal complexes have also been designed as DNA structural probes (for reviews, see Pyle and Barton,[30] Chow and Barton,[31] Barton,[58] and Norden *et al.*[84]). These tris(phenanthroline)metal complexes recognize specific characteristic shapes of DNA through noncovalent interactions. Ruthenium-based complexes are sensitive spectroscopic probes of DNA structure. Moreover, photoactivated DNA strand scission can occur with rhodium, ruthenium, and cobalt complexes, allowing easy determination of the sites of binding.

Original reports by Barton and co-workers suggested that the three-bladed propeller structure of tris(1,10-phenanthroline)ruthenium(II) ($Ru(phen)_3^{2+}$)(5) shows enantiomeric discrimination in its binding to B-DNA, with the right-handed $\Delta$ isomer binding more favorably than the left-handed $\Lambda$ isomer.[85] The molecular basis for the preferred binding of the $\Delta$ isomer was attributed to intercalation by both molecules.[86] Intercalation by one of the ligands of the $\Delta$ isomer would result in a favorable disposition of the remaining ligands in the major groove, whereas intercalation by one of the ligands of the $\Lambda$ isomer would result in steric repulsion of its remaining ligands. Both compounds are also postulated to display a second mode of interaction: surface binding to the major groove, favoring the $\Lambda$ isomer.[87] Tris(1,10-phenanthroline)zinc(II) was also reported to show enantiomeric selectivity in binding to DNA.[88] However, the mode of binding of these compounds is somewhat controversial, with other groups reporting electrostatic binding in the major groove rather than intercalation as the predominant mode of interaction with DNA for both enantiomers of $Ru(phen)_3^{2+}$.[89–91]

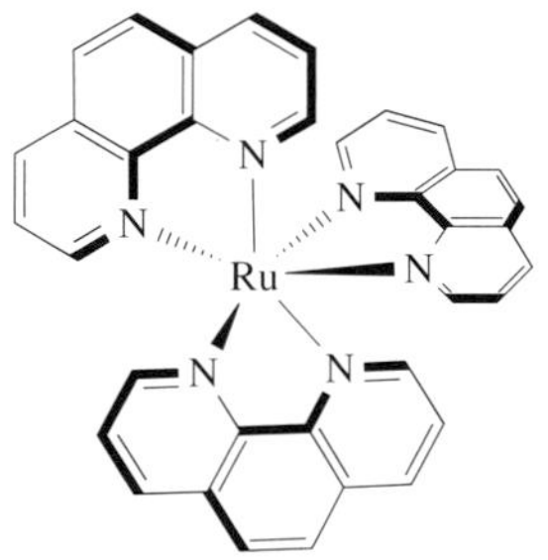

(5)

Other transition metal complexes show different specificities. For example, the left-handed isomer of tris(4,7-diphenylphenanthroline)ruthenium(II) ($\Lambda$-$Ru(DIP)_3^{2+}$) binds avidly to Z-DNA and poorly to B-DNA.[92] Moreover, $Rh(DIP)_3^+$ (6) targets cruciforms,[93] and $\Lambda$-$Co(DIP)_3^{2+}$ upon photoactivation cleaves Z-DNA[94,95] and other unwound structures, such as those found in genetic control regions.[96] Likewise, tris(3,4,7,8-tetramethylphenanthroline)ruthenium(II) ($Ru(TMP)_3^+$), recognizes the shallow minor groove of A-DNA and cleaves A-like structures upon irradiation with visible light.[97,98]

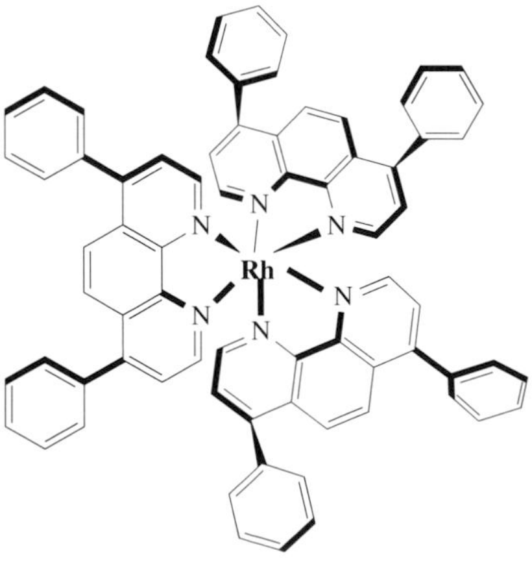

(6)

Transition metal complexes have also been used successfully for photofootprinting,[99] mapping such conformational variabilities of B-DNA as groove widths and propeller twisting,[100,101] monitoring DNA conformational changes induced by protein binding,[102] and probing oxidative damage to DNA.[103] Moreover, bis(phenanthroline) (9,10-phenanthrenequinone diimine)rhodium(III) ($Rh(phen)_2phi^{3+}$) targets triply bonded positions in folded RNA molecules, suggesting its utility in the determination of secondary and tertiary RNA structures.[104] Binuclear ruthenium(II) complexes that show potential as probes for DNA structure have also been synthesized by other laboratories,[105,106] as have $Ru^{II}$–polyaminocarboxylate complexes.[107]

### 7.04.3.5 Nickel Complexes

Complexes of nickel(II) have also been shown to be conformation-specific DNA cleaving agents.[108] Only accessible guanine residues are targeted by a series of square-planar nickel complexes such as NiCR (7), suggesting the utility of such reagents as probes for unusual DNA structures. Indeed, mismatches, bulges, and hairpins show particular reactivity (for reviews, see Burrows and Rokita[32,33]). Moreover, nickel complexes are able to catalyze site-specific oxidative damage to DNA.[109] Such complexes can also be used to probe RNA structure.[110] Analogous cobalt(II) complexes show similar reactivity with unpaired guanine residues in both DNA and RNA.[111]

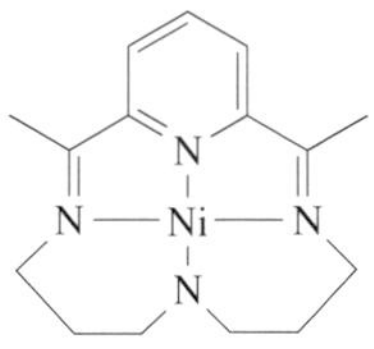

(7)

### 7.04.3.6 Metalloporphyrins

A family of cationic metalloporphyrins has been reported to be useful to probe local DNA structures and conformational changes induced by ligand binding.[112] $Mn^{3+}$, $Fe^{3+}$, and $Co^{3+}$ complexes of *meso* tetrakis($N$-methyl-4-pyridiniumyl)porphine ($H_2T4MPyP$)(8) in the presence of reducing agents produce specific sites of DNA cleavage.[113] Such agents are also useful in probing DNA structure and dynamics (for reviews, see Pasternack and Gibbs[114] and Raner *et al.*[115]).

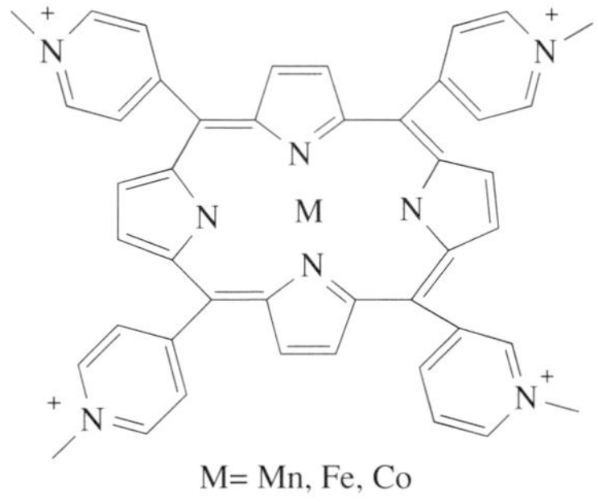

M= Mn, Fe, Co

(8)

### 7.04.3.7 Fe^{II}–EDTA

Unlike the metal complexes described above, EDTA ferrous ($Fe^{II}$–EDTA) is negatively charged and therefore does not bind to DNA. Since the introduction of $Fe^{II}$–EDTA as a DNA cleavage reagent by Tullius and co-workers,[116] it has gained great popularity in the study of DNA structure.[117] $Fe^{II}$–EDTA and hydrogen peroxide in solution generate the hydroxyl radical ($\cdot$OH) through the

Fenton reaction (Equation (1)). Ascorbate is included in the mixture to reduce iron(III) back to iron(II) for further reaction. The production of freely diffusible hydroxyl radicals results in a relatively nonspecific cleavage of duplex DNA through hydrogen abstraction from deoxyribose. The high reactivity and the small size of the hydroxyl radical combine to make it a powerful probe of DNA that allows for a high degree of structural resolution.

$$[Fe(EDTA)]^{2-} + H_2O_2 \longrightarrow [Fe(EDTA)]^- + {}^\bullet OH + OH^- \tag{1}$$

The hydroxyl radical has been widely used for characterizing details of free DNA structure (for reviews, see Tullius[118] and Price and Tullius[119]), to probe the organization of DNA in the nucleosome,[120–122] for footprinting studies,[34] and for mapping the sites of interstrand cross-links (Figures 5 and 6).[123] Other uses of the hydroxyl radical include the characterization of bent DNA,[124,125] the determination of the periodicity of DNA,[116] the characterization of the configuration of a Holliday junction,[119,126] and the analysis of structural distortions induced by protein binding.[127,128]

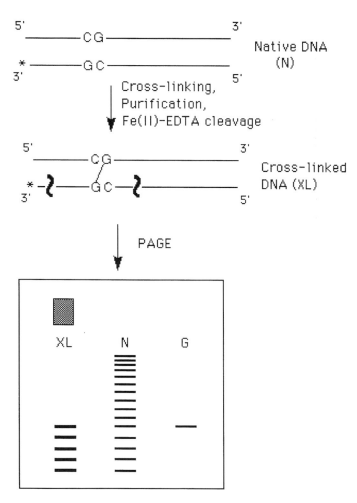

**Figure 5** Schematic of the Fe[II]–EDTA assay for interstrand cross-linking. In this example, there is one potential site for mitomycin C **(9)** cross-linking: 5′-CG, in which the guanine residues on opposite strands of the duplex will be covalently linked. Following incubation with cross-linking agent and gel purification of interstrand cross-linked material (XL), the end-labeled DNA oligomer (radiolabel denoted by an asterisk) is subjected to fragmentation by Fe[II]–EDTA (denoted by curved lines) and denaturing PAGE. Fragmentation of XL yields a ladder of radiolabeled fragments up to the site of cross-linking and then much larger radiolabeled fragments because of the covalent attachment of the nonlabeled strand via the cross-linking agent. The last band in the ladder is diagnostic for the site of cross-linking and can be assigned by a reference Maxam–Gilbert sequencing reaction. Fragmentation of control native DNA (N) yields a ladder of bands of approximately equal intensity.

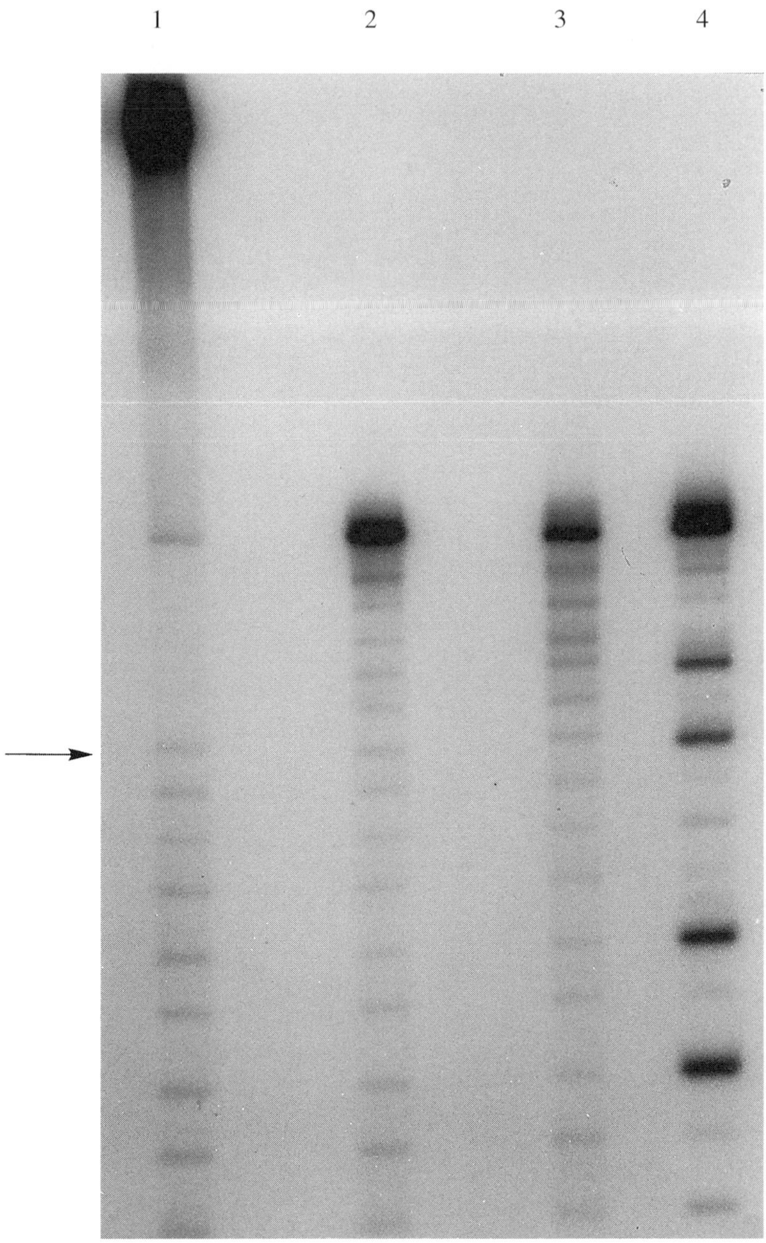

**Figure 6** Denaturing PAGE analysis of the Fe$^{II}$–EDTA cleavage pattern of purified mitomycin C (**9**) interstrand cross-linked (XL) and control native (N) radiolabeled duplex. Lane 1, XL; lane 2, N; lane 3, N; lane 4, Maxam–Gilbert G reaction. The last band in the XL ladder indicates that cross-linking occurred at that G residue in this duplex, as indicated by the arrow.

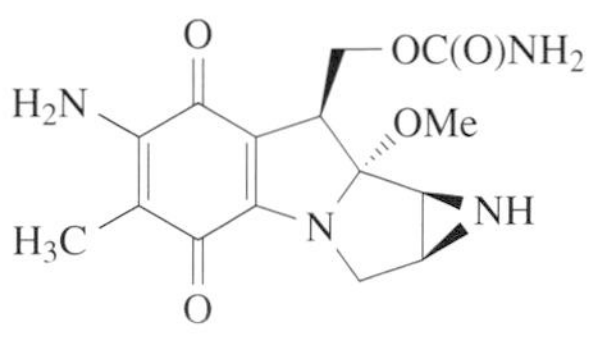

(**9**)

## 7.04.4 BASE-SPECIFIC REAGENTS

The above-described reagents are shape-specific DNA modifiers that are sensitive to helix geometry. Another important type of probe reacts directly with the DNA bases, reading the chemical information presented in the major and minor grooves of DNA. Such reagents can provide information about the local environment of the target bases, as described below.

### 7.04.4.1    Osmium Tetroxide

Osmium tetroxide-2,2'-bipyridine was the first reagent reported to probe DNA structure *in vivo*, providing evidence for the existence of Z-DNA in the cell.[129] Osmium(VIII) tetroxide ($OsO_4$) in the presence of pyridine/bipyridine adds to the 5,6 double bond of thymine to form an osmate ester, which can then be cleaved with piperidine to allow for detection (Equation (2)); for reviews, see Palecek[35-37] In the absence of suitable ligands, $OsO_4$ reacts with thymine to produce chiefly *cis*-thymine glycol (Equation (3)). Because this site is not accessible in B-DNA to the bulky $OsO_4$, this reagent is a sensitive probe for non-B structures such as single-stranded regions. $OsO_4$/pyridine can also be used during sequencing as a thymine-specific reagent for single-stranded DNA.[130,131] Hyperreactivity has been reported toward thymine residues in "open" DNA structures, such as those that are mismatched or bulged,[132-134] at B–Z junctions,[42,135-137] within supercoil-stabilized cruciforms,[138-142] at the Holliday junction,[143] and in other unusual structures (see Palecek[35,36]). For example, $OsO_4$/bipyridine has been used to probe the structure in homopolymeric (hyporeactive) and alternating AT tracts (hyperreactive) with the goal of determining sequence dynamics within promoters[144] as well as to characterize the structural organization of plasmids containing an origin of replication, found to contain both triplexes and Z-DNA.[145] $OsO_4$ hyperreactivity can also indicate an extruded single-stranded loop generated during intramolecular triplex formation[146-152] and has been used to study curvature-inducing DNA sequences[153] as well as structural perturbations induced by incorporation of 6-*O*-methylguanine in DNA.[154] $OsO_4$ probing has also provided evidence for the conformational flexibility of DNA: a single sequence inserted into a plasmid can undergo a transition from B-DNA to either a cruciform or a left-handed helix under different supercoiling conditions.[155] Other complexes of $OsO_4$, such as that with *N,N,N',N'*-tetramethylethylenediamine, have also been shown to be powerful *in situ* probes of DNA structure.[156] Monoclonal antibodies to the $OsO_4$/2,2'-bipyridine adduct have been used to perform immunochemical probing of DNA structure *in situ*.[157] Such studies have provided evidence for regions of open DNA structures in eukaryotic cells.[158]

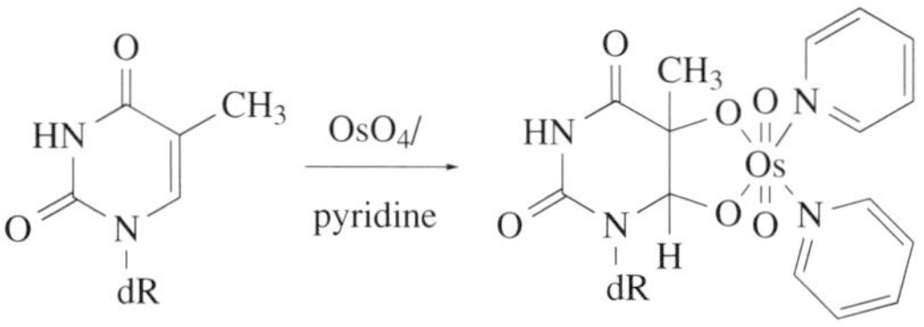

$$(2)$$

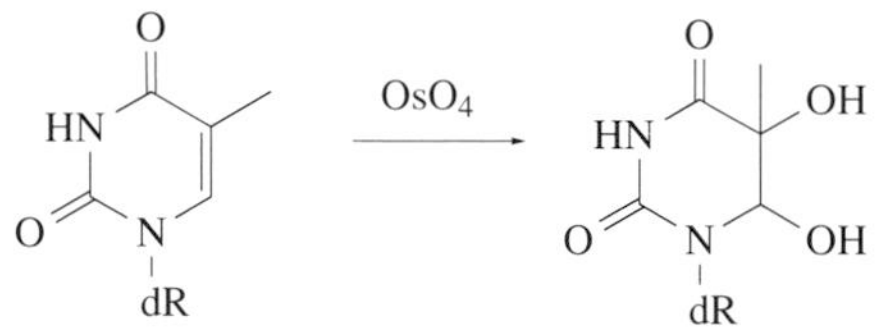

$$(3)$$

### ´7.04.4.2    Potassium Permanganate

Potassium permanganate ($KMnO_4$) also targets the 5,6 double bond of thymine, forming a thymine diol (Equation (4)), although other bases may also be modified.[159] Because base stacking interactions present in B-DNA interfere with accessibility of the principal target site, $KMnO_4$ is a probe for unpaired thymines[160,161] and other altered DNA structures.[38,162,163,164] For example, this reagent was used to demonstrate the formation of single-stranded DNA within the $\mu$ DNA–protein complex during $\mu$ transposition,[165] the existence of left-handed structures adjacent to a genetic control region,[166] and the adoption of an unusual structure by a telomeric structure under conditions of superhelical stress and low pH.[167] $KMnO_4$ hyperreactivity toward purines in one strand of a triple-stranded DNA formed by the RecA protein also allowed elucidation of the structure of a novel three-stranded intermediate formed during homologous recombination.[168] $KMnO_4$ has also been used to quantify Z-DNA *in vivo*[39] and as a T reaction for Maxam–Gilbert sequencing.[169]

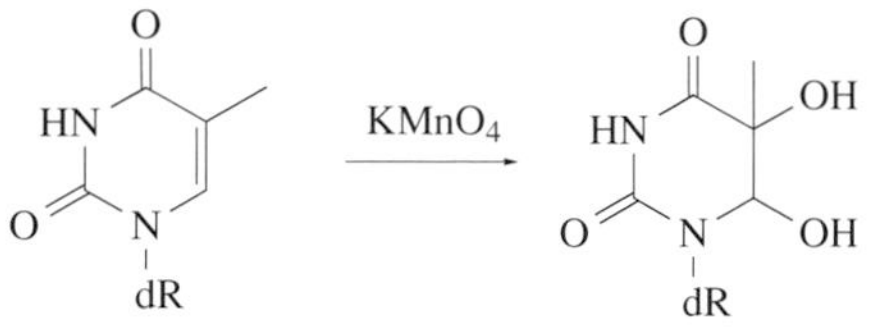

$$\text{(4)}$$

### 7.04.4.3 Hydroxylamine and Methoxylamine

Hydroxylamine and methoxylamine react at the C-4 and C-6 positions of exposed cytosine residues under neutral and acidic pH conditions to form two major products (Scheme 1), particularly in single-stranded regions or non-B-DNA structures.[40] Visualization of sites of modification occurs upon reaction with piperidine. These compounds have been shown to detect B–Z junctions,[42,170] H triplexes,[42,150] and mispaired cytosines[132,134] and has been reported to be useful for Maxam–Gilbert sequencing of denatured DNA.[169] For example, hydroxylamine has been used to monitor the opening of a hairpin tagging the end of an oligodeoxyribonucleotide designed for antisense targeting.[171]

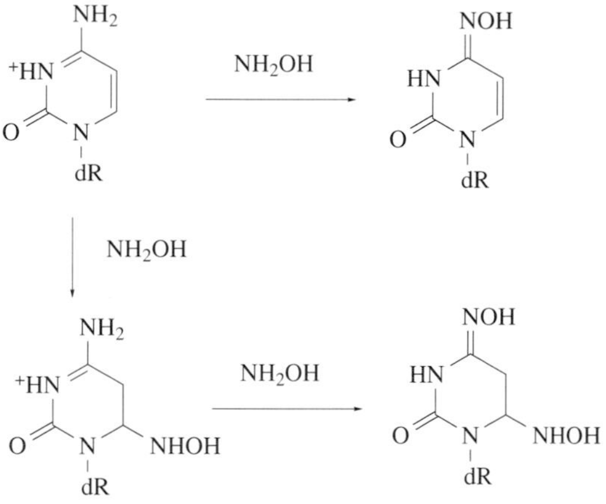

**Scheme 1**

### 7.04.4.4 Haloacetaldehydes

Bromoacetaldehyde (BAA) and chloroacetaldehyde (CAA) both target the N-6 and N-1 positions of adenine or N-3 and N-4 of cytosine to form $1,N^6$-ethenoadenine or $3,N^4$-ethenocytosine derivatives (Scheme 2) only if these bases are not involved in hydrogen bonding interactions during Watson–Crick base pairing.[41] These haloacetaldehydes are therefore specific for single-stranded regions in duplex DNA, adenine and cytosine residues at B–Z junctions in supercoiled DNA,[172–174] triplexes formed by homopurine–homopyrimidine sequences,[175] cruciforms,[174] and other unusual structures.[176] Even a single unpaired or deformed base within many kilobases of DNA can be detected.[177,178] Moreover, haloacetaldehydes have been used to study chromatin structure,[179–181] to analyze the unusual DNA conformations of viral TATA boxes,[182] to probe modifications of DNA structure upon drug binding,[177,183] to provide evidence for H-DNA formation in a supercoiled promoter sequence,[184] to demonstrate drug-stabilized triplex formation *in vivo*,[183] to demonstrate an unusual left-handed structure adopted by the repetitive sequence $d(CA/TG)_n$ distinct from Z-DNA,[185] and to analyze the structure of the hypervariable region upstream of the human insulin gene *in vivo*.[186]

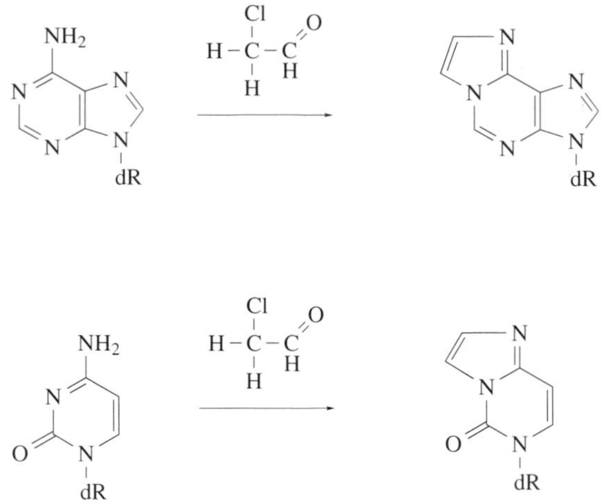

**Scheme 2**

### 7.04.4.5 Diethyl Pyrocarbonate

Diethyl pyrocarbonate (DEPC) (**10**) carboxylates the N-6 and N-7 positions of adenine and, less favorably, N-7 of guanine when these sites are particularly accessible (Scheme 3).[41] Sites of modification can be detected following incubation with piperidine, which results in strand cleavage. DEPC modification occurs preferentially with such unusual DNA structures as Z-DNA, in which the N-7 position of the purines is particularly accessible because of their *syn* conformation,[42,187,188] B–Z junctions,[173] and single-stranded regions of cruciforms.[189,190] DEPC can also pinpoint sequence-dependent variations in B-DNA structure.[38] For example, DEPC hyperreactivity suggests the presence of Z-DNA in the c-Ki-*ras* promoter[184] and in an origin of replication[145] within supercoiled DNA, providing evidence for the role of unusual DNA structures in gene regulation. DEPC has also been used to demonstrate the ability of both a satellite DNA[170] and the myb proto-oncogene[191] to adopt a Z conformation under appropriate topological constraints. Moreover, unwinding of the DNA duplex without disrupting base-pairing interactions also leads to increased DEPC reactivity.[163,192] Lack of DEPC reactivity can indicate Hoogsteen base pairing, such as in a triplex structure.[193] DEPC has also been used to characterize the "alternating" B structure of DNA duplexes containing (GA)$_n$ stretches.[194]

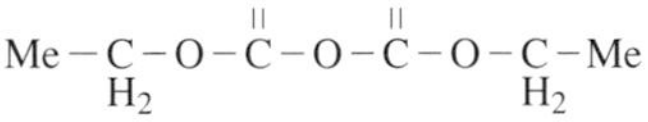

(**10**)

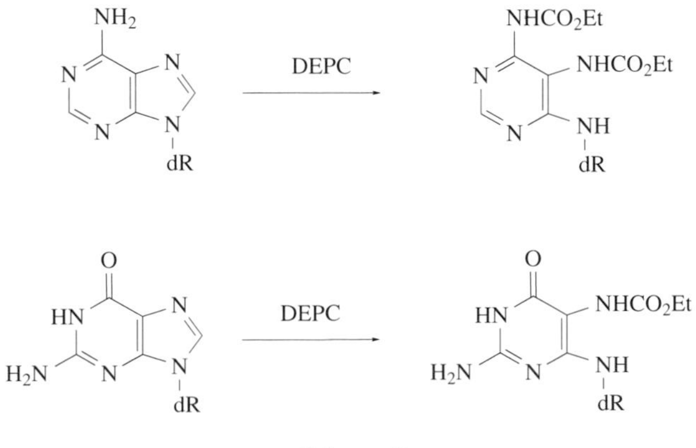

**Scheme 3**

#### 7.04.4.6 Dimethyl Sulfate

Dimethyl sulfate (DMS) (**11**), the Maxam–Gilbert "G reaction" reagent, methylates the N-7 position of guanine. Subsequent incubation with piperidine results in strand cleavage. DMS can also be used for "alkylation interference" studies,[195] in which limited DMS modification of DNA is followed by the separation of molecules that can still form complexes from those that cannot. The sites of modification that prevent complexation are therefore implicated as essential ones for binding. For example, the contacts between the Tet repressor protein and its DNA target were elucidated using DMS methylation interference.[196]

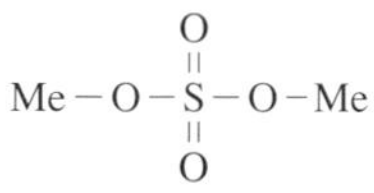

$$Me-O-\overset{\overset{\displaystyle O}{\|}}{\underset{\underset{\displaystyle O}{\|}}{S}}-O-Me$$

**(11)**

DMS is also a useful probe for unusual DNA structures as it reacts preferentially with purines in the *anti* conformation found in Z-DNA.[42,184] Protection from DMS methylation may occur during Hoogsteen base pairing in triplex DNA[147,184,193] and in other unusual structures.[197,198] Lack of protection can also be suggestive in some cases, such as the unusual structures adopted by a telomeric sequence[167] and a purine/pyrimidine mirror repeat found in some promoter regions.[199] DMS can also methylate the N-1 position of adenine and N-3 of cytosine in the absence of Watson–Crick base-pairing interactions. Thus, DMS methylation at these sites can be used as a probe for unpaired bases.[200–204]

#### 7.04.4.7 Other Base-specific Reagents

Glyoxal (**12**) reacts selectively with available guanine residues to form a cyclic adduct between the N-1 position and the exocyclic nitrogen (Equation (5)). Guanines with single-stranded character therefore react preferentially, such as those in cruciforms.[205] Another reagent used to probe the structure of cruciforms is sodium bisulfite, which deaminates unpaired cytosine residues to form deoxyuracil.[206] 2-Hydroperoxytetrahydrofuran (THF-OOH) (**13**) and 2-hydroperoxytetrahydro-pyran (THP-OOH) (**14**) are also sensitive reagents for probing cytosine residues in unusual DNA structures.[207] These reagents can be used as cytosine-specific sequencing reagents, as they react with cytosine residues to yield heat-labile modifications sensitive to cleavage in the presence of heat and piperidine. Like the haloacetaldehydes, THF-OOH and THP-OOH react preferentially with non-Watson–Crick-paired cytosine structures such as in single-stranded DNA, bulges, mismatches, and hairpins.[207,208] Moreover, these compounds also react effectively with 5-methylcytosine.

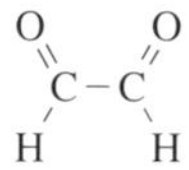

**(12)**

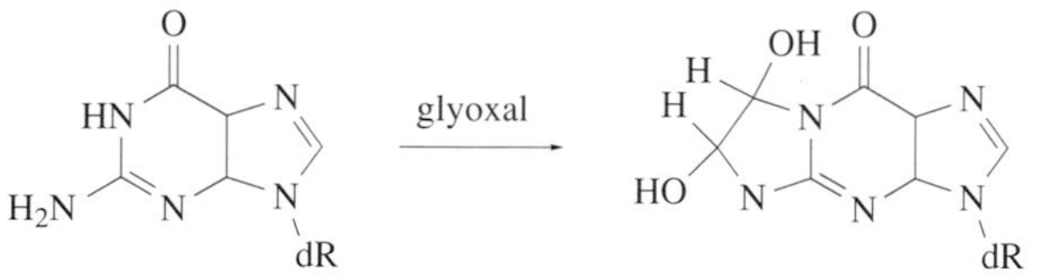

(5)

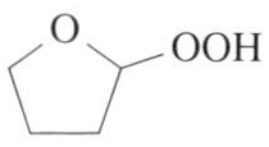

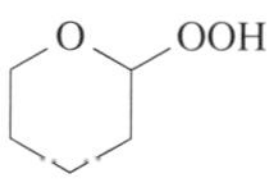

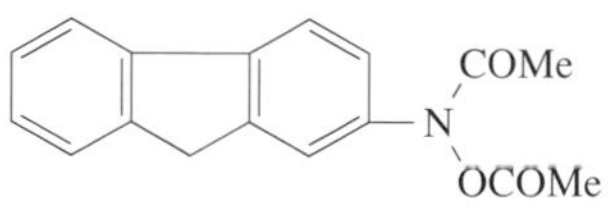

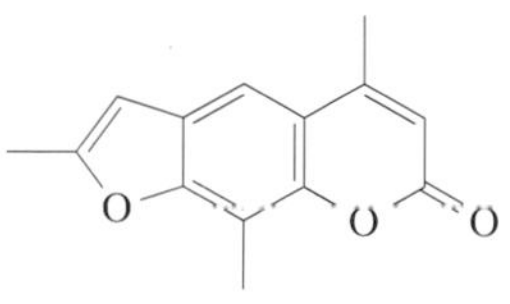

**(13)**      **(14)**      **(15)**      **(16)**

Iodination and bromination of cytosine can also be used to probe unusual cytosine conformations in duplex DNA.[209,210] For example, KBr and $KHSO_5$ generate $Br_2$ *in situ*, which reacts with the 5,6 double bond of cytosine, resulting in a piperidine-sensitive modification product. Single-stranded DNA is the most reactive, indicating that this probe could be used to pinpoint cytosine residues present as bulges, hairpins, or in other accessible regions of duplex DNA.

Derivatives of aminofluorene are profound liver carcinogens, targeting the C-8 position of purines. 2-*N*,*N*-Acetoxyacetylaminofluorene (N-AcO-AAF) (**15**), *N*-hydroxy-2-aminofluorene (N-OH-AF), and structurally related compounds react preferentially with the B form of DNA relative to Z-DNA.[211,212] Hyperreactivity has also been noted with specific guanine residues at the B–Z junction, indicating that these carcinogens can be used to probe for Z-DNA under appropriate topological conditions. 4-Nitroquinoline, another carcinogen that targets the C-8 position of guanine residues, shows a similar B- over Z-DNA preference and hyperreactivity at the B–Z junction, suggesting its potential for use as a probe for DNA polymorphism.[213]

## 7.04.5 CROSS-LINKERS

### 7.04.5.1 Psoralen

Psoralens such as 4,5′-8-trimethylpsoralen (**16**) have been widely used since ancient times to treat a wide variety of skin disorders such as vitiligo and psoriasis (for reviews, see Hearst *et al.*[43] and Song and Tapley[214]). These natural products form a reversible intercalative complex with DNA and can undergo subsequent photoactivation with near-UV light to form first monoadducts and then cross-links through a (2 + 2) cycloaddition with the 5,6 double bond of principally thymine residues (Scheme 4). Interstrand cross-links occur between thymine residues at the duplex site 5′-TA preferentially.[44,123,215] Psoralen photobinding has been used to probe the structure of chromatin *in vivo*[216,217] and also to probe for alternative DNA conformations both *in vitro* and *in vivo*.[218,219] Psoralen reacts exclusively with the linker regions of chromatin, attributed to restricted intercalation into nucleosomal core DNA.[220–222] Non-cross-linked regions of DNA appear as bubbles in electron microscopy, which can therefore be used to map nucleosomes and the organization of bacterial and viral DNA *in vivo*.[223,224] Transcriptionally active DNA is more susceptible to psoralen cross-linking than transcriptionally inactive DNA because of the loss of nucleosome structure, leading to a psoralen-based *in vivo* assay for gene activation.[225]

Psoralen cross-linking has also been used to quantify cruciform formation in DNA,[226] assay for Z-DNA,[227,228] measure the level of supercoiling *in vivo*,[227,229] and probe DNA–protein interactions.[230,231] Moreover, monitoring the ratio of the two major diastereomeric adducts formed upon psoralen monoalkylation of thymine has been used to monitor drug-induced conformational changes of DNA.[232] Psoralen-derivatized antisense/antigene oligonucleoside methylphosphonates have also been shown to be site-specific toward their complementary RNA or DNA, attributed to the secondary and tertiary structures of their targets.[233]

### 7.04.5.2 Nitrogen Mustards

Other cross-linkers have also been used to reveal clues about DNA structure. For example, nitrogen mustards such as mechlorethamine (**17**), the first clinically useful antitumor agent, were originally proposed to alkylate the N-7 position of guanine residues to form interstrand cross-links between adjacent deoxyguanosine residues in duplex DNA at the sequence 5′-GC, which contains the minimal N-7 to N-7 distance in B-DNA.[234] However, both nitrogen mustards and the chemically analogous diepoxyalkanes in fact form cross-links preferentially at the sequence 5′-GNC, providing evidence for the considerable conformational flexibility of DNA in solution.[235–238] Indeed, native PAGE has experimentally confirmed that mustard interstrand cross-linking induces bending of ~15° in its DNA target,[239] implying that capacity for deformation is a requirement for mustard-induced cross-linking. Thus, the constraint induced by the incorporation of DNA into the nucleosome might be expected to negatively impact mustard cross-linking. However, single-base resolution denaturing PAGE analysis of mustard interstrand cross-links formed within defined sequence nucleosomes showed similar sites and efficiencies of cross-linking for free and nucleosomal DNA (Figure 7).[240] Thus, the distortion required to accommodate an interstrand cross-link at the preferred 5′-GNC sequence does not significantly impact cross-linking within the nucleosome, supporting the hypothesis that nucleosomes are in fact dynamic structures that transiently expose DNA.[241]

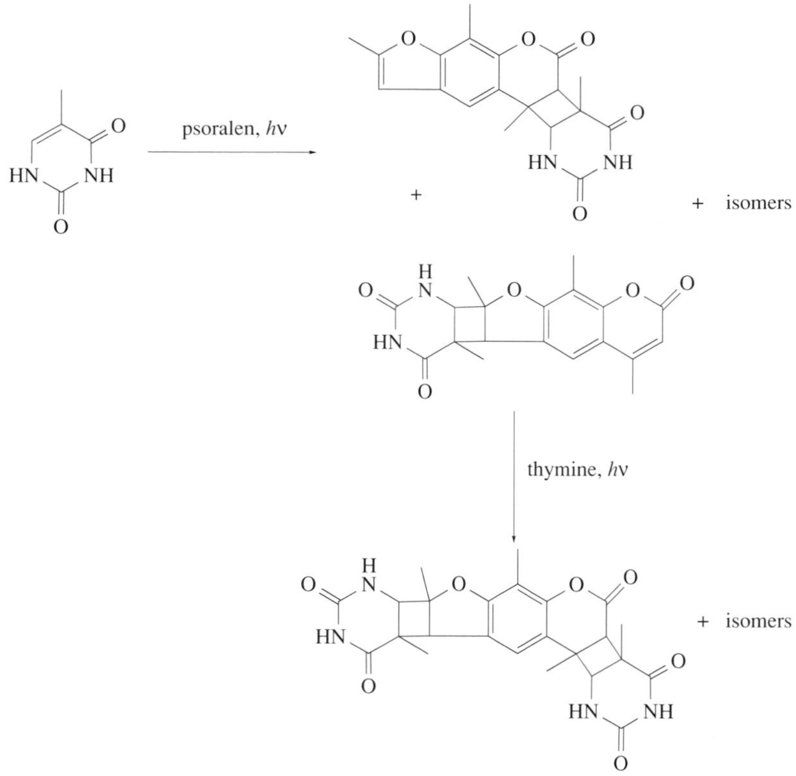

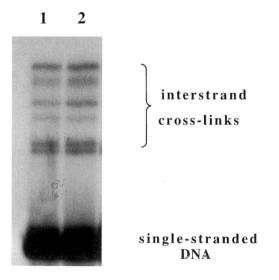

**Scheme 4**

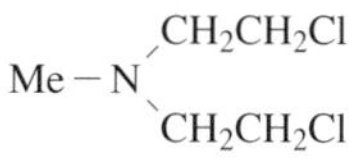

(17)

interstrand
cross-links

single-stranded
DNA

**Figure 7** Denaturing PAGE analysis of nitrogen mustard-treated free and nucleosomal 3′-radiolabeled 154 bp DNA. Lane 1, mechlorethamine-treated free DNA; lane 2, mechlorethamine-treated nucleosomal DNA. Bands of lower mobility than single-stranded DNA correspond to different interstrand cross-linked isomers. The similar positions and intensities of these low-mobility bands suggest that the sites of cross-linking are the same for the free and nucleosomal DNA and that the histone octamer therefore does not alter mustard sequence preferences.

## 7.04.6 ANTIBIOTICS

Neocarzinostatin (NCS) (18), an enediyne antitumor antibiotic, mediates strand breakage of DNA following abstraction of deoxyribose atoms when situated in the minor groove. The naphthoate of NCS may also act as a DNA intercalator. The chromophore of this agent (NCS-Chrom) has been used as a probe for unusual DNA structures such as bulged bases.[78] In the absence of an activating thiol, NCS-Chrom cleaves bulges within duplex and single-stranded DNA site-specifically.[45] The bulged site must contain between two and five nucleotides and be flanked by stabilizing duplexes on both sides.

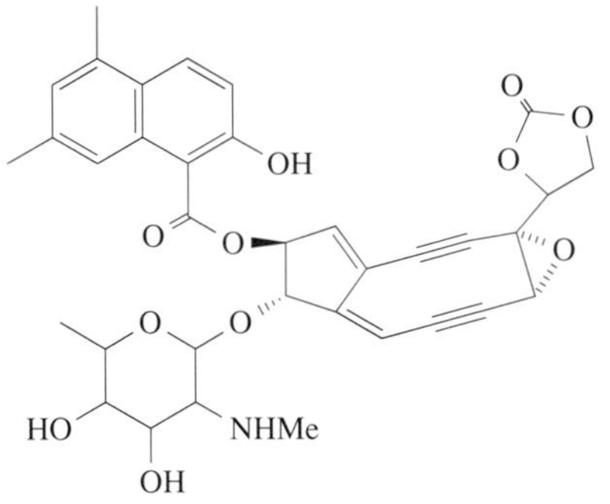

(18)

Dynemicin (19), a hybrid antitumor antibiotic that contains both enediyne and anthracycline cores, mediates strand breakage of DNA which is enhanced in the presence of NADPH or dithiothreitol.[242] Its proposed mechanism involves phenyl diradical formation upon binding in the minor groove or intercalation. Dynemicin has a small degree of sequence specificity but reacts preferentially with such secondary structures as stem and loops,[242] the branch point in DNA junctions,[243] and B–Z junctions.[244]

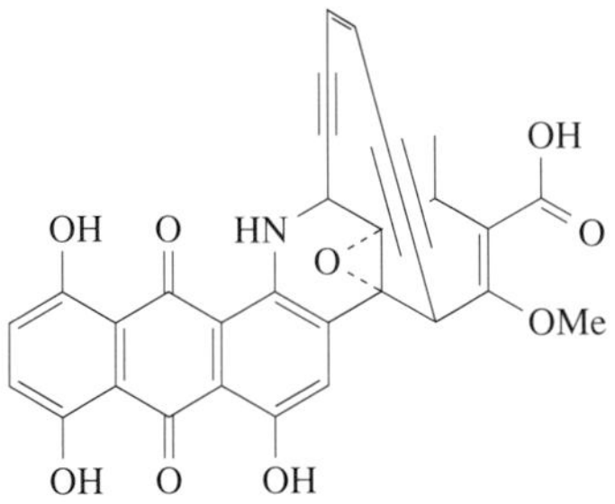

(19)

Bleomycin (BLM) (20) represents a family of glycopeptide antibiotics useful in the treatment of testicular, head, and neck carcinomas.[245] Bleomycin causes DNA fragmentation with predominantly single-stranded breaks occurring preferentially at GpC and GpT sequences. Double-stranded breaks occur less often and are believed to require intercalation of the bithiazole. BLM shows hyperreactivity toward B–Z junctions and Z-DNA,[244] bulged DNA,[78] and other unusual structures.[246] Novel photocleaving agents based on bleomycin that may also show utility for probing DNA structure have also been prepared.[247] Moreover, BLM shows structure-specific cleavage of RNA, suggesting its potential utility as an RNA probe.[245]

(+)-CC-1065 (21) is a minor-groove-binding antibiotic that forms covalent adducts at the N-3 position of adenine in A-T-rich DNA regions, leading to potent antitumor activity.[248] Because binding results in such structural changes as bending, winding, and stiffening of the DNA, (+)-CC-1065 can be used to probe bending induced by DNA-binding proteins.[249,250]

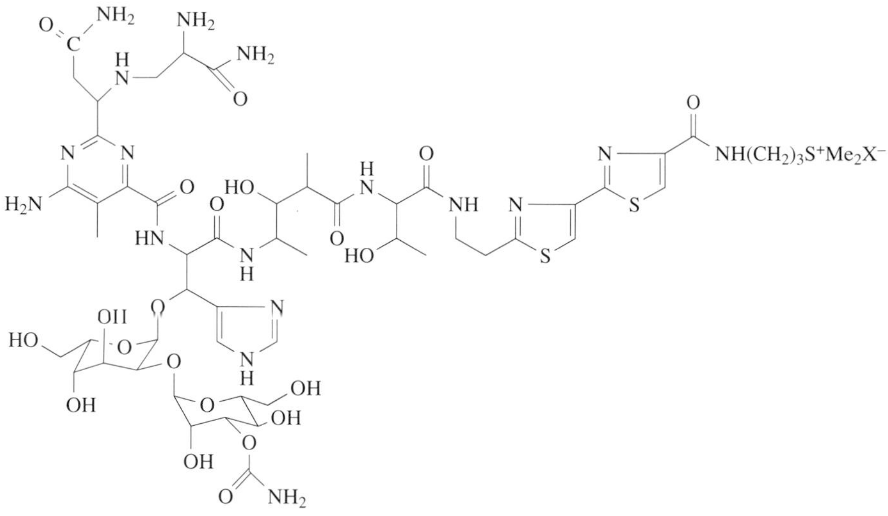

(20)

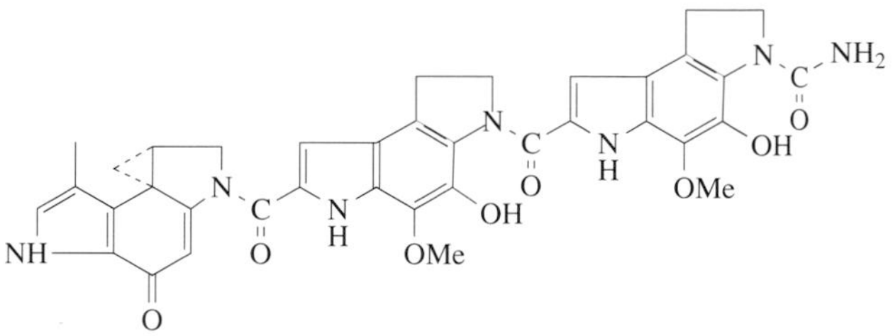

(21)

## 7.04.7 CONCLUSIONS

There are two major classes of chemical probes of DNA structure: those that demonstrate no inherent base preference but are sensitive to DNA geometry, such as groove widths and chirality, and those that are base-specific, detecting bases in distorted regions of the helix. The former bind to DNA through shape-dependent, noncovalent interactions, whereas the latter result in covalent modification of the bases themselves. Numerous examples of each are discussed above and are summarized in Table 1. Enzymes can also be extremely useful to probe DNA structure, and the interested reader is referred to reviews on this topic by Wohlrab[251] and Zacharias[252]. The design of new synthetic probes will continue to be a fertile area of investigation as long as the exact function of unusual DNA structures in the cell continues to be unknown.

## ACKNOWLEDGMENTS

The author would like to thank Susan W. Cole, Jacob E. Conklin, Andrea S. Bassi, Charles R. Langelier, and Elisabeth A. Pimentel for assistance with literature compilation. Erzsebet E. Nagy is also thanked for other contributions.

## 7.04.8 REFERENCES

1. R. E. Franklin and R. G. Gosling, *Nature*, 1953, **171**, 740.
2. J. D. Watson and F. H. C. Crick, *Nature*, 1953, **171**, 737.
3. R. E. Dickerson, H. R. Drew, B. N. Conner, R. M. Wing, A. V. Fratini, and M. L. Kopka, *Science*, 1982, **216**, 475.
4. J. Feigon, V. Sklenar, E. Wang, D. E. Gilbert, R. F. Macaya, and P. Schultze, *Methods Enzymol.*, 1992, **211**, 235.

5.  M. H. Caruthers, A. D. Barone, S. L. Beaucage, D. R. Dodds, E. F. Fisher, L. J. McBride, M. Matteucci, Z. Stabinsky, and J.-Y. Yang, *Methods Enzymol.*, 1987, **154**, 287.
6.  R. E. Dickerson, in "Unusual DNA Structures," eds. R. D. Wells and S. C. Harvey, Springer-Verlag, New, York, 1988, p. 287.
7.  R. E. Dickerson, *Methods Enzymol.*, 1992, **211**, 67.
8.  E. Palecek, *C. R. Biochem. Mol. Biol.*, 1991, **26**, 151.
9.  C. R. Calladine, *J. Mol. Biol.*, 1982, **161**, 343.
10. R. E. Dickerson, *J. Mol. Biol.*, 1983, **166**, 419.
11. C. E. Pearson, H. Zorbas, G. B. Price, and M. Zannis-Hadjopoulos, *J. Cell. Biochem.*, 1996, **63**, 1.
12. K. R. Fox, in "Advances in DNA Sequence Specific Agents," ed. L. H. Hurley, JAI Press, Greenwich, CT, 1992, vol. 1, p. 167.
13. M. J. Waring, *Annu. Rev. Biochem.*, 1981, **50**, 159.
14. K. E. van Holde, "Chromatin," Springer-Verlag, New York, 1989.
15. A. Wolffe, "Chromatin Structure and Function," Academic Press, San Diego, CA, 1992.
16. D. Pruss, J. J. Hayes, and A. P. Wolffe, *BioEssays*, 1995, **17**, 161.
17. D. M. J. Lilley, *Methods Enzymol.*, 1992, **212**, 133.
18. R. D. Wells, S. Amirhaeri, J. A. Blaho, D. A. Collier, J. C. Hanvey, W.-T. Hsieh, A. Jaworski, J. Klysik, J. E. Larson, M. J. McLean, F. Wohlrab, and W. Zacharias, in "Unusual DNA Structures," eds. R. D. Wells and S. C. Harvey, Springer-Verlag, New York, 1988, p. 1.
19. R. D. Wells, D. A. Collier, J. C. Hanvey, M. Shimizu, and F. Wohlrab, *FASEB J.*, 1988, **2**, 2939.
20. P. E. Nielsen, *J. Mol. Recog.*, 1990, **3**, 1.
21. J. J. Hayes, *Chem. Bio.*, 1995, **2**, 127.
22. J. T. Millard, *Biochimie*, 1996, **78**, 803.
23. A. M. Maxam and W. Gilbert, *Methods Enzymol.*, 1980, **65**, 499.
24. F. Sanger, S. Nicklen, and A. R. Coulson, *Proc. Natl. Acad. Sci. USA*, 1977, **74**, 5463.
25. H. Htun and B. H. Johnston, *Methods Enzymol.*, 1992, **212**, 272.
26. J. Sambrook, E. F. Fritsch, and T. Maniatis, "Molecular Cloning," 2nd edn., Cold Spring Harbor Laboratory, Cold Spring Harbor, 1989.
27. D. S. Sigman and C. B. Chen, in "Metal–DNA Chemistry. ACS Symposium Series," ed. T. D. Tullius, American Chemical Society, Washington, DC, 1989, p. 24.
28. D. S. Sigman, R. Landgraf, D. M. Perrin, and L. Pearson, in "Metal Ions in Biological Systems," eds. A. Sigel and H. Sigel, Dekker, New York, 1996, vol. 33, p. 485.
29. P. B. Dervan, *Science*, 1986, **232**, 464.
30. A. M. Pyle and J. K. Barton, *Progr. Inorg. Chem.*, 1990, **38**, 413.
31. C. S. Chow and J. K. Barton, *Methods Enzymol.*, 1992, **212**, 219.
32. C. J. Burrows and S. E. Rokita, *Acc. Chem. Res.*, 1994, **27**, 295.
33. C. J. Burrows and S. E. Rokita, in "Metal Ions in Biological Systems," eds. A. Sigel and H. Sigel, Dekker, New York, 1996, vol. 33, p. 537.
34. W. J. Dixon, J. J. Hayes, J. R. Levin, M. F. Weidner, B. A. Dombroski, and T. D. Tullius, *Methods Enzymol.*, 1991, **208**, 380.
35. E. Palecek, *Methods Enzymol.*, 1992, **212**, 139.
36. E. Palecek, *Methods Enzymol.*, 1992, **212**, 305.
37. E. Palecek, *Nucleic Acids Mol. Biol.*, 1994, **8**, 1.
38. J. G. McCarthy, L. D. Williams, and A. Rich, *Biochemistry*, 1990, **29**, 6071.
39. H. Jiang, W. Zacharias, and S. Amirhaeri, *Nucleic Acids Res.*, 1991, **19**, 6943.
40. B. H. Johnston, *Methods Enzymol.*, 1992, **212**, 180.
41. T. Kohwi-Shigematsu and Y. Kohwi, *Methods Enzymol.*, 1992, **212**, 155.
42. B. H. Johnston and A. Rich, *Cell*, 1985, **42**, 713.
43. J. E. Hearst, S. T. Isaacs, D. Kanne, H. Rapoport, and K. Straub, *Q. Rev. Biophys.*, 1984, **17**, 1.
44. H. Gamper, J. Piette, and J. E. Hearst, *Photochem. Photobiol.*, 1984, **40**, 29.
45. L. S. Kappen and I. H. Goldberg, *Biochemistry*, 1993, **32**, 13 138.
46. V. L. Seligy and P. F. Lurquin, *Nature New Biol.*, 1973, **243**, 20.
47. B. K. Stringer and J. T. Blankemeyer, *Teratogenesis, Carcinog., Mutagen.*, 1995, **15**, 53.
48. J. M. Kean, S. A. White, and D. E. Draper, *Biochemistry*, 1985, **24**, 5062.
49. Q. Guo, M. Lu, L. A. Marky, and N. R. Kallenbach, *Biochemistry*, 1992, **31**, 2451.
50. G. Krishnamurthy, T. Polte, T. Rooney, and M. E. Hogan, *Biochemistry*, 1990, **29**, 981.
51. D. E. Graves, in "Advances in DNA Sequence Specific Agents," eds. L. H. Hurley and J. B. Chaires, JAI Press, Greenwich, CT, 1996, vol. 2, p. 169.
52. Q. Guo, M. Lu, N. C. Seeman, and N. R. Kallenbach, *Biochemistry*, 1990, **29**, 570.
53. M. Lu, Q. Guo, N. C. Seeman, and N. R. Kallenbach, *Biochemistry*, 1990, **29**, 3407.
54. Q. Chen, I. D. Kuntz, and R. H. Shafer, *Proc. Natl. Acad. Sci. USA*, 1996, **93**, 2635.
55. M. L. Carpenter, G. Lowe, and P. R. Cook, *Nucleic Acids Res.*, 1996, **24**, 1594.
56. T. D. Tullius, in "Metal–DNA Chemistry. ACS Symposium Series," ed. T. D. Tullius, American Chemical Society, Washington, DC, 1989, p. 1.
57. H. J. H. Fenton, *J. Chem. Soc.*, 1894, **65**, 899.
58. J. K. Barton, *Science*, 1986, **233**, 727.
59. S. A. Kazakov, T. G. Astashkina, S. V. Mamaev, and V. V. Vlassov, *Nature*, 1988, **335**, 186.
60. D. C. A. John and K. T. Douglas, *Biochem. J.*, 1993, **289**, 463.
61. Y. Wang and B. Van Ness, *Nucleic Acids Res.*, 1989, **17**, 6915.
62. L. G. Marzilli, T. J. Kistenmacher, and G. L. Eichhorn, in "Nucleic Acid–Metal Ion Interactions," ed. T. G. Spiro, Wiley, New York, 1980, p. 179.
63. A. Casadevall and L. A. Day, *Biochemistry*, 1983, **22**, 4831.
64. D. Ding and F. S. Allen, *Biochim. Biophys. Acta*, 1980, **610**, 64.

65. Z. Kuklenyik and L. G. Marzilli, *Inorg. Chem.*, 1996, **35**, 5654.
66. D. S. Sigman, D. R. Graham, V. D'Aurora, and A. M. Stern, *J. Biol. Chem.*, 1979, **254**, 12 269.
67. J. M. Veal and R. L. Rill, *Biochemistry*, 1988, **27**, 1822.
68. A. Spassky, S. Rimsky, H. Buc, and S. Busby, *EMBO J.*, 1988, **7**, 1871.
69. A. Spassky and D. S. Sigman, *Biochemistry*, 1985, **24**, 8050.
70. Q. Guo, M. Lu, and N. R. Kallenbach, *Biopolymers*, 1991, **31**, 359.
71. I. L. Cartwright and S. C. R. Elgin, *Nucleic Acids Res.*, 1982, **10**, 5835.
72. J. W. Suggs and R. W. Wagner, *Nucleic Acids Res.*, 1986, **14**, 3703.
73. L. E. Pope and D. S. Sigman, *Proc. Natl. Acad. Sci. USA*, 1984, **81**, 3.
74. R. P. Hertzberg and P. B. Dervan, *J. Am. Chem. Soc.*, 1982, **104**, 313.
75. R. P. Hertzberg and P. B. Dervan, *Biochemistry*, 1984, **23**, 3934.
76. M. W. Van Dyke and P. B. Dervan, *Nucleic Acids Res.*, 1983, **11**, 5555.
77. Q. Guo, N. C. Seeman, and N. R. Kallenbach, *Biochemistry*, 1989, **28**, 2355.
78. L. D. Williams and I. H. Goldberg, *Biochemistry*, 1988, **27**, 3004.
79. I. L. Cartwright, R. P. Hertzberg, P. B. Dervan, S. C. R. Elgin, *Proc. Natl. Acad. Sci. USA*, 1983, **80**, 3213.
80. G. B. Dreyer and P. B. Dervan, *Proc. Natl. Acad. Sci. USA*, 1985, **82**, 968.
81. C.-H. B. Chen and D. S. Sigman, *Proc. Natl. Acad. Sci. USA*, 1986, **83**, 7147.
82. H. E. Moser and P. B. Dervan, *Science*, 1987, **238**, 645.
83. C.-H. B. Chen and D. S. Sigman, *Science*, 1987, **237**, 1197.
84. B. Norden, P. Lincoln, B. Akerman, and E. Tuite, in "Metal Ions in Biological Systems," eds. A. Sigel and H. Sigel, Dekker, New York, 1996, vol. 33, p. 177.
85. J. K. Barton, A. T. Danishefsky, and J. M. Goldberg, *J. Am. Chem. Soc.*, 1984, **106**, 2172.
86. E. C. Long and J. K. Barton, *Acc. Chem. Res.*, 1990, **23**, 271.
87. J. K. Barton, J. M. Goldberg, C. V. Kumar, and N. J. Turro, *J. Am. Chem. Soc.*, 1986, **108**, 2081.
88. J. K. Barton, J. J. Dannenberg, and A. L. Raphael, *J. Am. Chem. Soc.*, 1982, **104**, 4967.
89. C. Hiort, B. Norden, and A. Rodger, *J. Am. Chem. Soc.*, 1990, **112**, 1971.
90. S. Satyanarayana, J. C. Dabrowiak, and J. B. Chaires, *Biochemistry*, 1992, **31**, 9319.
91. J. E. Coury, J. R. Anderson, L. McFail-Isom, L. D. Williams, and L. A. Bottomley, *J. Am. Chem. Soc.*, 1997, **119**, 3792.
92. J. K. Barton, L. A. Basile, A. Danishefsky, and A. Alexandrescu, *Proc. Natl. Acad. Sci. USA*, 1984, **81**, 1961.
93. M. R. Kirshenbaum, R. Tribolet, and J. K. Barton, *Nucleic Acids Res.*, 1988, **16**, 7943.
94. J. K. Barton and A. L. Raphael, *J. Am. Chem. Soc.*, 1984, **106**, 2466.
95. J. K. Barton and A. L. Raphael, *Proc. Natl. Acad. Sci. USA*, 1985, **82**, 6460.
96. B. C. Muller, A. L. Raphael, and J. K. Barton, *Proc. Natl. Acad. Sci. USA*, 1987, **84**, 1764.
97. H.-Y. Mei and J. K. Barton, *J. Am. Chem. Soc.*, 1986, **108**, 7414.
98. H.-Y. Mei and J. K. Barton, *Proc. Natl. Acad. Sci. USA*, 1988, **85**, 1339.
99. K. Uchida, A. M. Pyle, T. Morii, and J. K. Barton, *Nucleic Acids Res.*, 1989, **17**, 10 259.
100. A. M. Pyle, E. C. Long, and J. K. Barton, *J. Am. Chem. Soc.*, 1989, **111**, 4520.
101. A. M. Pyle, T. Morii, and J. K. Barton, *J. Am. Chem. Soc.*, 1990, **112**, 9432.
102. J. K. Barton and S. R. Paranawithana, *Biochemistry*, 1986, **25**, 2205.
103. E. D. A. Stemp, M. R. Arkin, and J. K. Barton, *J. Am. Chem. Soc.*, 1997, **119**, 2921.
104. C. S. Chow and J. K. Barton, *J. Am. Chem. Soc.*, 1990, **112**, 2839.
105. P. Lincoln and B. Norden, *Chem. Commun.*, 1996, 2145.
106. A. Del Guerzo, A. Kirsch-De Mesmaeker, M. Demeunynck, and J. Lhomme, *J. Phys. Chem. B*, 1997, **101**, 7012.
107. R. E. Shepherd, Y. Chen, S. Zhang, F.-T. Lin, and R. A. Kortes, *Adv. Chem. Ser.*, 1997, **253**, 367.
108. X. Chen, S. E. Rokita, and C. J. Burrows, *J. Am. Chem. Soc.*, 1991, **113**, 5884.
109. S. Routier, J.-L. Bernier, J.-P. Catteau, and C. Bailly, *Bioorg. Med. Chem. Lett.*, 1997, **7**, 63.
110. S. A. Woodson, J. G. Muller, C. J. Burrows, and S. E. Rokita, *Nucleic Acids Res.*, 1993, **21**, 5524.
111. J. G. Muller, P. Zheng, S. E. Rokita, and C. J. Burrows, *J. Am. Chem. Soc.*, 1996, **118**, 2320.
112. B. Ward, A. Skorobogaty, and J. C. Dabrowiak, *Biochemistry*, 1986, **25**, 6875.
113. G. Pratviel, J. Bernadou, and B. Meunier, in "Metal Ions in Biological Systems," eds. A. Sigel and H. Sigel, Dekker, New York, 1996, vol. 33, p. 399.
114. R. F. Pasternack and E. J. Gibbs, in "Metal Ions in Biological Systems," eds. A. Sigel and H. Sigel, Dekker, New York, 1996, vol. 33, p. 367.
115. G. Raner, J. Goodisman, and J. C. Dabrowiak, in "Metal–DNA Chemistry. ACS Symposium Series," ed. T. D. Tullius, American Chemical Society, Washington, DC, 1989, p. 74.
116. T. D. Tullius and B. A. Dombroski, *Science*, 1985, **230**, 679.
117. A. Draganescu and T. Tullius, in "Metal Ions in Biological Systems," eds. A. Sigel and H. Sigel, Dekker, New York, 1996, vol. 33, p. 453.
118. T. D. Tullius, *Trends Biochem. Sci.*, 1987, **12**, 297.
119. M. A. Price and T. D. Tullius, *Methods Enzymol.*, 1992, **212**, 194.
120. J. Bashkin, J. J. Hayes, T. D. Tullius, and A. P. Wolffe, *Biochemistry*, 1993, **32**, 1895.
121. J. J. Hayes, T. D. Tullius, A. P. Wolffe, *Proc. Natl. Acad. Sci. USA*, 1990, **87**, 7405.
122. J. J. Hayes, J. Bashkin, T. D. Tullius, and A. P. Wolffe, *Biochemistry*, 1991, **30**, 8434.
123. M. F. Weidner, J. T. Millard, and P. B. Hopkins, *J. Am. Chem. Soc.*, 1989, **111**, 9270.
124. A. M. Burkhoff and T. D. Tullius, *Cell*, 1987, **48**, 935.
125. A. M. Burkhoff and T. D. Tullius, *Nature*, 1988, **331**, 455.
126. T. D. Tullius, *Free Radic. Res. Commun.*, 1991, **12–13**, 521.
127. M. L. Craig, W.-C. Suh, and M. T. Record, Jr., *Biochemistry*, 1995, **34**, 15 624.
128. D. R. Mernagh and G. G. Kneale, *Nucleic Acids Res.*, 1996, **24**, 4853.
129. E. Palecek, P. Boublikova, and P. Karlovsky, *Gen. Physiol. Biophys.*, 1987, **6**, 593.
130. T. Friedmann and D. M. Brown, *Nucleic Acids Res.*, 1978, **5**, 615.
131. F. Jelen, P. Karlovsky, E. Makaturova, P. Pecinka, and E. Palecek, *Gen. Physiol. Biophys.*, 1991, **10**, 461.

132. A. Bhattacharyya and D. M. J. Lilley, *J. Mol. Biol.*, 1989, **209**, 583.
133. A. Bhattacharyya and D. M. J. Lilley, *Nucleic Acids Res.*, 1989, **17**, 6821.
134. R. G. H. Cotton, N. R. Rodrigues, and R. D. Campbell, *Proc. Natl. Acad. Sci. USA*, 1988, **85**, 4397.
135. K. Nejedly, M. Kwinkowski, G. Galazka, J. Klysik, and E. Palecek, *J. Biomol. Struct. Dyn.*, 1985, **3**, 467.
136. G. Galazka, E. Palecek, R. D. Wells, and J. Klysik, *J. Biol. Chem.*, 1986, **261**, 7093.
137. E. Palecek, E. Rasovska, and P. Boublikova, *Biochem. Biophys. Res. Commun.*, 1988, **150**, 731.
138. J. A. McClellan and D. M. J. Lilley, *J. Mol. Biol.*, 1987, **197**, 707.
139. D. M. J. Lilley and E. Palecek, *EMBO J.*, 1984, **3**, 1187.
140. L. H. Naylor, D. M. J. Lilley, and J. H. van de Sande, *EMBO J.*, 1986, **5**, 2407.
141. J. C. Furlong, K. M. Sullivan, A. I. H. Murchie, G. W. Gough, and D. M. J. Lilley, *Biochemistry*, 1989, **28**, 2009.
142. D. M. J. Lilley, *Chem. Soc. Rev.*, 1989, **18**, 53.
143. D. R. Duckett, A. I. H. Murchie, S. Diekmann, E. von Kitzing, B. Kemper, and D. M. J. Lilley, *Cell*, 1988, **55**, 79.
144. J. K. Sullivan and J. Lebowitz, *Biochemistry*, 1991, **30**, 2664.
145. A. Bianchi, R. D. Wells, N. H. Heintz, and M. S. Caddle, *J. Biol. Chem.*, 1990, **265**, 21 789.
146. M. Vojtiskova and E. Palecek, *J. Biomol. Struct. Dyn.*, 1987, **5**, 283.
147. J. C. Hanvey, M. Shimizu, and R. D. Wells, *Proc. Natl. Acad. Sci. USA*, 1988, **85**, 6292.
148. J. C. Hanvey, J. Klysik, and R. D. Wells, *J. Biol. Chem.*, 1988, **263**, 7386.
149. B. H. Johnston, *Science*, 1988, **241**, 1800.
150. H. Htun and J. E. Dahlberg, *Science*, 1988, **241**, 1791.
151. M. Vojtiskova, S. Mirkin, V. Lyamichev, O. Voloshin, M. Frank-Kamenetskii, and E. Palecek, *FEBS Letts.*, 1988, **234**, 295.
152. K. R. Fox, *Nucleic Acids Res.*, 1990, **18**, 5387.
153. E. Palecek, E. Makaturova-Rasovska, and S. Diekmann, *Gen. Physiol. Biophys.*, 1988, **7**, 379.
154. C.-W. Wong, N.-W. Tan, and B. F. L. Li, *J. Mol. Biol.*, 1992, **228**, 1137.
155. M. J. McLean and R. D. Wells, *J. Biol. Chem.*, 1988, **263**, 7370.
156. P. Boublikova and E. Palecek, *FEBS Lett.*, 1990, **263**, 281.
157. A. E. Kabakov and A. M. Poverenny, *Anal. Biochem.*, 1993, **211**, 224.
158. E. Palecek, M. Robert-Nicoud, and T. M. Jovin, *J. Cell Sci.*, 1993, **104**, 653.
159. S. A. Akman, J. H. Doroshow, and M. Dizdaroglu, *Arch. Biochem. Biophys.*, 1990, **282**, 202.
160. C. Jeppesen and P. E. Nielsen, *Nucleic Acids Res.*, 1989, **17**, 4947.
161. S. Sassa-Dwight and J. D. Gralla, *J. Biol. Chem.*, 1989, **264**, 8074.
162. K. R. Fox and G. W. Grigg, *Nucleic Acids Res.*, 1988, **16**, 2063.
163. C. Jeppesen and P. E. Nielsen, *FEBS Letts.*, 1988, **231**, 172.
164. R. Visse, A. King, G. F. Moolenaar, N. Goosen, and P. van de Putte, *Biochemistry*, 1994, **33**, 9881.
165. Z. Wang, S.-Y. Namgoong, X. Zhang, and R. M. Harshey, *J. Biol. Chem.*, 1996, **271**, 9619.
166. M. P. Kladde, J. D'Cunha, and J. Gorski, *J. Mol. Biol.*, 1993, **229**, 344.
167. V. I. Lyamichev, S. M. Mirkin, O. N. Danilevskaya, O. N. Voloshin, S. V. Balatskaya, V. N. Dobrynin, S. A. Filippov, and M. D. Frank-Kamenetskii, *Nature*, 1989, **339**, 634.
168. S. K. Chiu, B. J. Rao, R. M. Story, C. M. Radding, *Biochemistry*, 1993, **32**, 13 146.
169. C. M. Rubin and C. W. Schmid, *Nucleic Acids Res.*, 1980, **8**, 4613.
170. N. Vogt, N. Rousseau, M. Leng, and B. Malfoy, *J. Biol. Chem.*, 1988, **263**, 11 826.
171. B. Jolles, M. Refregiers, and A. Laigle, *Nucleic Acids Res.*, 1997, **25**, 4608.
172. D. S. Kang and R. D. Wells, *J. Biol. Chem.*, 1985, **260**, 7783.
173. T. Kohwi-Shigematsu, T. Manes, and Y. Kohwi, *Proc. Natl. Acad. Sci. USA*, 1987, **84**, 2223.
174. M. J. McLean, J. E. Larson, F. Wohlrab, and R. D. Wells, *Nucleic Acids Res.*, 1987, **15**, 6917.
175. T. Kohwi-Shigematsu and Y. Kohwi, *Nucleic Acids Res.*, 1991, **19**, 4267.
176. F. Wohlrab, M. J. McLean, and R. D. Wells, *J. Biol. Chem.*, 1987, **262**, 6407.
177. L. Marrot and M. Leng, *Biochemistry*, 1989, **28**, 1454.
178. A. Schwartz, L. Marrot, and M. Leng, *J. Mol. Biol.*, 1989, **207**, 445.
179. T. Kohwi-Shigematsu, R. Gelinas, and H. Weintraub, *Proc. Natl. Acad. Sci. USA*, 1983, **80**, 4389.
180. J. Bode, H.-J. Pucher, and K. Maass, *Eur. J. Biochem.*, 1986, **158**, 393.
181. T. Kohwi-Shigematsu and J. A. Nelson, *Mol. Carcinog.*, 1988, **1**, 20.
182. S. Amirhaeri, F. Wohlrab, E. O. Major, and R. D. Wells, *J. Virol.*, 1988, **62**, 922.
183. G. Duval-Valentin, T. de Bizemont, M. Takasugi, J.-L. Mergny, E. Bisagni, and C. Helene, *J. Mol. Biol.*, 1995, **247**, 847.
184. D. G. Pestov, A. Dayn, E. Y. Siyanova, D. L. George, and S. M. Mirkin, *Nucleic Acids Res.*, 1991, **19**, 6527.
185. M. P. Kladde, Y. Kohwi, T. Kohwi-Shigematsu, and J. Gorski, *Proc. Natl. Acad. Sci. USA*, 1994, **91**, 1898.
186. M. C. U. Hammond-Kosack, M. W. Kilpatrick, and K. Docherty, *J. Mol. Endocrinol.*, 1992, **9**, 221.
187. W. Herr, *Proc. Natl. Acad. Sci. USA*, 1985, **82**, 8009.
188. L. Runkel and A. Nordheim, *J. Mol. Biol.*, 1986, **189**, 487.
189. P. M. Scholten and A. Nordheim, *Nucleic Acids Res.*, 1986, **14**, 3981.
190. J. C. Furlong and D. M. J. Lilley, *Nucleic Acids Res.*, 1986, **14**, 3995.
191. J. Vlach, M. Dvorak, P. Bartunek, V. Pecenka, M. Travnicek, and J. Sponer, *Biochem. Biophys. Res. Commun.*, 1989, **158**, 737.
192. M. J. McLean and M. J. Waring, *J. Mol. Recog.*, 1988, **1**, 138.
193. O. N. Voloshin, S. M. Mirkin, V. I. Lyamichev, B. P. Belotserkovskii, and M. D. Frank-Kamenetskii, *Nature*, 1988, **333**, 475.
194. T. Evans and A. Efstratiadis, *J. Biol. Chem.*, 1986, **261**, 14 771.
195. U. Siebenlist and W. Gilbert, *Proc. Natl. Acad. Sci. USA*, 1980, **77**, 122.
196. C. Heuer and W. Hillen, *J. Mol. Biol.*, 1988, **202**, 407.
197. O. N. Voloshin, A. G. Veselkov, B. P. Belotserkovskii, O. N. Danilevskaya, M. N. Pavlova, V. N. Dobrynin, and M. D. Frank-Kamenetskii, *J. Biomol. Struct. Dyn.*, 1992, **9**, 643.
198. J. E. Mitchell, S. F. Newbury, and J. A. McClellan, *Nucleic Acids Res.*, 1995, **23**, 1876.

199. C. D. McDonald, M. A. Hollingsworth, and L. J. Maher, III, *Gene*, 1994, **150**, 267.
200. A. F. Melnikova, R. Beabealashvilli, and A. D. Mirzabekov, *Eur. J. Biochem.*, 1978, **84**, 301.
201. U. Siebenlist, *Nature*, 1979, **279**, 651.
202. K. Kirkegaard, H. Buc, A. Spassky, and J. C. Wang, *Proc. Natl. Acad. Sci. USA*, 1983, **80**, 2544.
203. M. Buckle and H. Buc, *Biochemistry*, 1989, **28**, 4388.
204. K. Usdin and A. V. Furano, *J. Biol. Chem.*, 1989, **264**, 15 681.
205. D. M. J. Lilley, *IARC Sci. Pub.*, 1986, **70**, 83.
206. G. W. Gough, K. M. Sullivan, and D. M. J. Lilley, *EMBO J.*, 1986, **5**, 191.
207. G. Liang, P. Gannett, and B. Gold, *Nucleic Acids Res.*, 1995, **23**, 713.
208. A. Yu, M. D. Barron, R. M. Romero, M. Christy, B. Gold, J. Dai, D. M. Gray, I. S. Haworth, and M. Mitas, *Biochemistry*, 1997, **36**, 3687.
209. K. F. Jensen, I. F. Nes, and R. D. Wells, *Nucleic Acids Res.*, 1976, **3**, 3143.
210. S. A. Ross and C. J. Burrows, *Nucleic Acids Res.*, 1996, **24**, 5062.
211. P. Rio and M. Leng, *J. Mol. Biol.*, 1986, **191**, 569.
212. L. Marrot, A. Schwartz, E. Hebert, G. Saint-Ruf, and M. Leng, in "Unusual DNA Structures," eds. R. D. Wells and S. C. Harvey, Springer-Verlag, New York, 1988, p. 163.
213. C. Rodolfo, A. Lanza, S. Tornaletti, G. Fronza, and A. M. Pedrini, *Nucleic Acids Res.*, 1994, **22**, 314.
214. P.-S. Song and K. J. Tapley, Jr., *Photochem. Photobiol.*, 1979, **29**, 1177.
215. F. Esposito, R. G. Brankamp, and R. R. Sinden, *J. Biol. Chem.*, 1988, **263**, 11 466.
216. J. O. Carlson, O. Pfenniger, R. R. Sinden, J. M. Lehman, and D. E. Pettijohn, *Nucleic Acids Res.*, 1982, **10**, 2043.
217. S. K. Kondoleon, N. A. Kurkinen, and L. M. Hallick, *Virology*, 1989, **173**, 129.
218. D. W. Ussery, R. W. Hoepfner, and R. R. Sinden, *Methods Enzymol.*, 1992, **212**, 242.
219. R. R. Sinden and D. W. Ussery, *Methods Enzymol.*, 1992, **212**, 319.
220. C. V. Hanson, C.-K. J. Shen, and J. E. Hearst, *Science*, 1976, **193**, 62.
221. T. Cech and M. L. Pardue, *Cell*, 1977, **11**, 631.
222. G. P. Wiesehahn, J. E. Hyde, and J. E. Hearst, *Biochemistry*, 1977, **16**, 925.
223. L. M. Hallick, H. A. Yokota, J. C. Bartholomew, and J. E. Hearst, *J. Virol.*, 1978, **27**, 127.
224. L. M. Hallick, C. V. Hanson, J. O. Cacciapuoti, and J. E. Hearst, *Nucleic Acids Res.*, 1980, **8**, 611.
225. A. Conconi, R. W. Widmer, T. Koller, and J. M. Sogo, *Cell*, 1989, **57**, 753.
226. R. R. Sinden, S. S. Broyles, and D. E. Pettijohn, *Proc. Natl. Acad. Sci. USA*, 1983, **80**, 1797.
227. R. R. Sinden and T. J. Kochel, *Biochemistry*, 1987, **26**, 1343.
228. T. J. Kochel and R. R. Sinden, *J. Mol. Biol.*, 1989, **205**, 91.
229. G. Zheng, T. Kochel, R. W. Hoepfner, S. E. Timmons, and R. R. Sinden, *J. Mol. Biol.*, 1991, **221**, 107.
230. S. S. Sastry and B. M. Ross, *Biochemistry*, 1997, **36**, 3133.
231. S. S. Sastry, B. M. Ross, and A. P'arraga, *J. Biol. Chem.*, 1997, **272**, 3715.
232. J. T. Millard, P. M.-L. Liao, W. J. Bass, and J. W. Suggs, *Bioorg. Chem.*, 1988, **16**, 175.
233. J. M. Kean and P. S. Miller, *Biochemistry*, 1994, **33**, 9178.
234. P. Brookes and P. D. Lawley, *Biochem. J.*, 1961, **80**, 496.
235. J. O. Ojwang, D. A. Grueneberg, and E. L. Loechler, *Cancer Res.*, 1989, **49**, 6529.
236. J. T. Millard, S. Raucher, and P. B. Hopkins, *J. Am. Chem. Soc.*, 1990, **112**, 2459.
237. J. T. Millard and M. W. White, *Biochemistry*, 1993, **32**, 2120.
238. M. J. Yunes, S. E. Charnecki, J. J. Marden, and J. T. Millard, *Chem. Res. Toxicol.*, 1996, **9**, 994.
239. S. M. Rink and P. B. Hopkins, *Biochemistry*, 1995, **34**, 1439.
240. J. T. Millard, R. J. Spencer, and P. B. Hopkins, *Biochemistry*, 1998, **37**, 5211.
241. K. J. Polach and J. Widom, *J. Mol. Biol.*, 1995, **254**, 130.
242. Y. Sugiura, T. Shiraki, M. Konishi, and T. Oki, *Proc. Natl. Acad. Sci. USA*, 1990, **87**, 3831.
243. M. Lu, Q. Guo, and N. R. Kallenbach, *J. Biomol. Struct. Dyn.*, 1991, **9**, 271.
244. Q. Guo, M. Lu, M. Shahrestanifar, R. D. Sheardy, and N. R. Kallenbach, *Biochemistry*, 1991, **30**, 11 735.
245. J.-M. Battigello, M. Cui, and B. J. Carter, in "Metal Ions in Biological Systems," eds. A. Sigel and H. Sigel, Dekker, New York, 1996, vol. 33, p. 593.
246. K. Ueda, S. Kobayashi, and T. Komano, *Nucleic Acids Symp. Ser.*, 1985, **16**, 197.
247. R. Kuroda, H. Satoh, M. Shinomiya, T. Watanabe, and M. Otsuka, *Nucleic Acids Res.*, 1995, **23**, 1524.
248. L. H. Hurley and P. H. Draves, in "Anticancer Drug–DNA Interactions," eds. S. Neidle and M. Waring, CRC Press, Boca Raton, FL, 1993, vol. 1, p. 89.
249. Z.-M. Ding, R. M. Harshey, and L. H. Hurley, *Nucleic Acids Res.*, 1993, **21**, 4281.
250. D. Sun and L. H. Hurley, *Biochemistry*, 1994, **33**, 9578.
251. F. Wohlrab, *Methods Enzymol.*, 1992, **212**, 294.
252. W. Zacharias, *Methods Enzymol.*, 1992, **212**, 336.

# 7.05
# Oligonucleotide Synthesis

RADHAKRISHNAN P. IYER
*OriGenix, Cambridge, MA, USA*

and

SERGE L. BEAUCAGE
*US Food and Drug Administration, Bethesda, MD, USA*

| | | |
|---|---|---:|
| 7.05.1 | INTRODUCTION | 105 |
| | *7.05.1.1 Nucleic Acid Synthesis: A Historical Perspective* | 106 |
| | *7.05.1.1.1 Phosphodiester approach* | 107 |
| | *7.05.1.1.2 Phosphotriester approach* | 107 |
| | *7.05.1.1.3 P(III) approach: the advent of phosphoramidite chemistry* | 108 |
| | *7.05.1.1.4 H-phosphonate chemistry* | 109 |
| 7.05.2 | SOLID SUPPORTS FOR OLIGONUCLEOTIDE SYNTHESIS | 110 |
| | *7.05.2.1 Overview of Developments in Solid Supports Employed in Oligonucleotide Synthesis* | 110 |
| | *7.05.2.2 Solid Supports for the Preparation of Functionalized Oligonucleotides* | 114 |
| 7.05.3 | PROTECTING GROUPS IN OLIGONUCLEOTIDE SYNTHESIS | 117 |
| | *7.05.3.1 General Comments* | 117 |
| | *7.05.3.1.1 Protection of imide/lactam functions of nucleobases* | 117 |
| | *7.05.3.1.2 Protection of the exocyclic amino function of nucleobases* | 119 |
| | *7.05.3.1.3 Protection of purine bases: special considerations* | 121 |
| | *7.05.3.2 Protection of the 5′-Hydroxyl Group* | 126 |
| | *7.05.3.3 Protection of the Phosphate: Formation of the Internucleotidic Linkage by the Phosphoramidite Approach* | 130 |
| 7.05.4 | MECHANISM OF INTERNUCLEOTIDIC COUPLING REACTION | 137 |
| | *7.05.4.1 Phosphoramidite Approach* | 137 |
| | *7.05.4.2 H-phosphonate Approach* | 139 |
| 7.05.5 | SOLID-PHASE OLIGONUCLEOTIDE SYNTHESIS | 142 |
| | *7.05.5.1 Phosphoramidite Approach* | 142 |
| | *7.05.5.2 H-phosphonate Approach* | 144 |
| | *7.05.5.3 Isolation and Purification of the Oligonucleotide* | 144 |
| 7.05.6 | SYNTHESIS OF RNA | 144 |
| 7.05.7 | CONCLUSION | 144 |
| 7.05.8 | REFERENCES | 145 |

## 7.05.1 INTRODUCTION

The chemistry of nucleic acid synthesis goes back to the 1950s in its origin. In this section a brief historical overview of oligonucleotide synthesis is provided.

#### 7.05.1.1 Nucleic Acid Synthesis: A Historical Perspective

In the 1950s, the chemical structure of RNA and DNA began to unfold and culminated in the landmark discovery of the double helical structure of DNA by Watson and Crick.[1] Because the 1950s was an exciting era of "total synthesis," the chemical synthesis of DNA and RNA also became an engaging and passionate mission for chemists. However, the synthesis of a biopolymer like DNA and RNA has both similarities and differences compared with the classical total synthesis of a natural product. In planning a classical total synthesis, a chemist often uses the biosynthetic pathway of the target molecule as a guide and studies the products resulting from chemical degradation of the molecule. Likewise, in retrospect, it can be stated that the understanding of the structure and reactivity of the natural polynucleotides and their enzymatic degradation products provided a retrosynthetic strategy for the synthesis of nucleic acids. The total synthesis of a natural product often involves the combination of stereo-, regio-, and enantiocontrol in C—C bond forming reactions, and interconversion of functional groupings to reach the target. In contrast, the initial challenge in biopolymer synthesis was to achieve controlled polymer assembly through high yielding, rapid coupling of individual monomeric units.

The four major methods of oligonucleotide synthesis, which have evolved since the 1950s, could have been derived by a retrosynthetic approach (Scheme 1). Specifically, the building blocks for nucleic acids would be the nucleosides, which can be obtained from either chemical synthesis or natural sources. Then, "zipping together" the individual nucleosidic units through the formation of internucleotidic phosphate linkages in a rapid and efficient manner would be the heart of nucleic acid synthesis. This challenge appeared at first deceptively simple and, perhaps, dissuaded many exponents of total synthesis from venturing into the field. However, given the plethora of functional groups that decorate the nucleosides, the cross-reactivity of these functional groups, and the demanding requirements for quick and quantitative zipping reactions, the synthesis of nucleic acids turned out to be a formidable challenge.

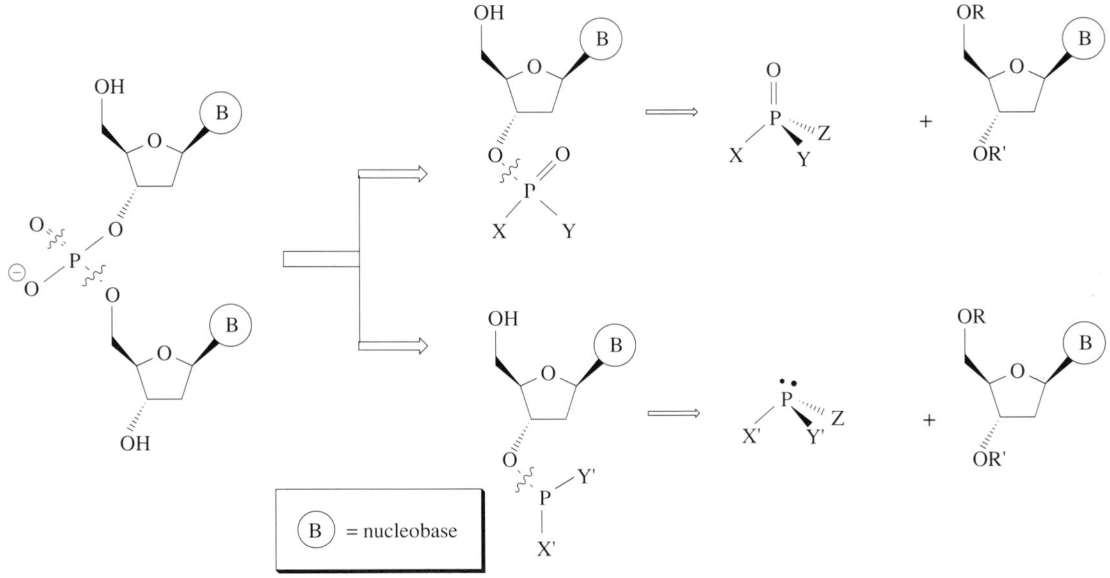

**Scheme 1**

In the 1950s, studies of phosphorylation reactions on nucleotide coenzymes, cofactors, and phospholipids[2] provided the initial impetus to attempt nucleic acid synthesis. Thus, phosphorylating reagents such as diphenylchlorophosphate and dichlorophenylphosphate were introduced (Scheme 2). Early studies by Todd and co-workers[3-6] on the synthesis of adenosine diphosphate and triphosphate revealed the complexity of phosphate linkage. The reactivity of the phosphoryl group had to be precisely controlled to permit the sequential addition of nucleosidic units with minimal "by-products." Seminal contributions were made in this regard by Todd and co-workers,[7,8] whose initial studies laid the foundation for *H*-phosphonate chemistry (see below) and, later, the pioneering work by Khorana and co-workers[9-11] who developed the phosphodiester approach for "gene synthesis."

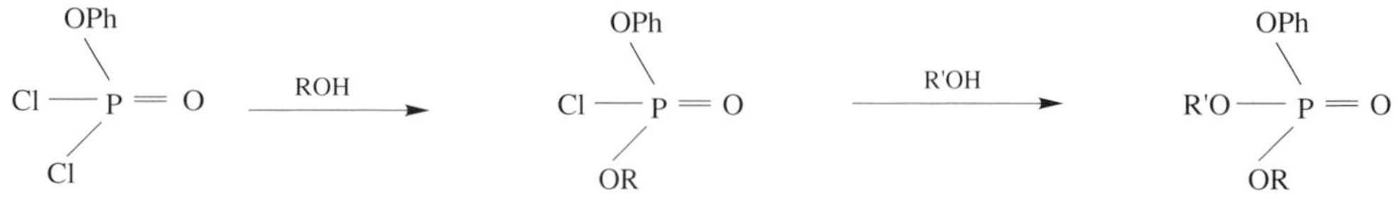

**Scheme 2**

### 7.05.1.1.1 *Phosphodiester approach*

The phosphodiester approach[9–11] employed two reacting entities (Equation (1)): (i) a nucleoside (**1**) having its 5′-hydroxy group protected with the acid-labile triphenylmethyl group and a free 3′-hydroxyl group, and (ii) a nucleoside (**2**) having a 5′-phosphate group and a 3′-acetyl group.

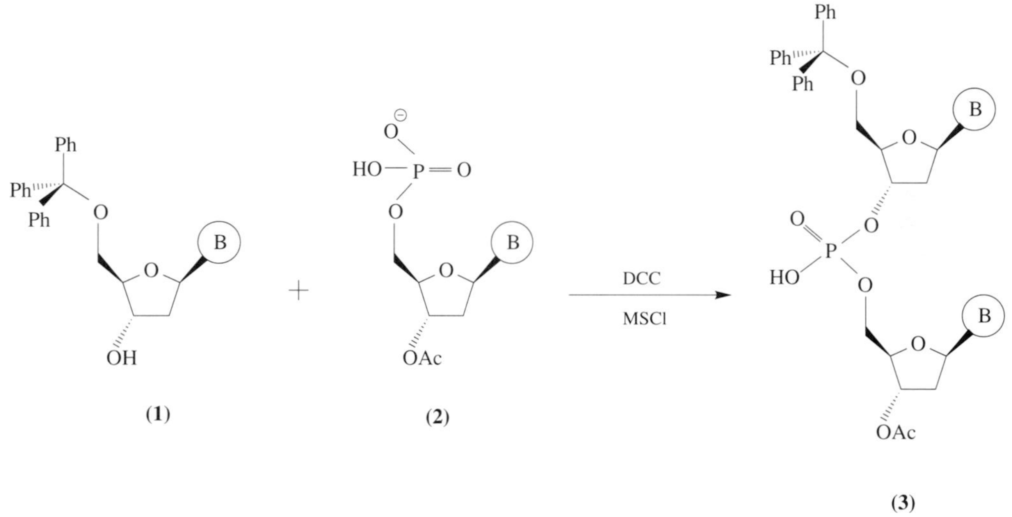

Activation of the nucleoside 5′-phosphate (**2**) in the presence of dicyclohexylcarbodiimide (DCC), mesitylene sulfonyl chloride (MS-Cl), or 2,4,6-triisopropylbenzenesulfonyl chloride, followed by reaction with the nucleoside (**1**), gave the dinucleoside phosphodiester (**3**). Chain extension was accomplished by mild base-induced hydrolysis of the acetyl group in (**3**) followed by the addition of (**2**). This methodology led to the elucidation of the genetic code for which sets of tri- and tetradeoxyribonucleotides were chemically synthesized, and enzymatically converted to oligo-nucleotides for use in *in vitro* protein synthesis systems. Furthermore, this method for oligonucleotide synthesis became an established procedure, and culminated in the landmark synthesis of genes encoding for alanine *t*-RNA from yeast and suppressor *t*-RNA from *Escherichia coli*.[11] From a synthetic perspective, the development of the triphenylmethyl group for the protection of the 5′-hydroxy group, and the robust benzoyl and isobutyryl groups for the protection of the nucleobases have been significant achievements in that these groups could withstand the rigors of multiple synthetic steps. However, the presence of unprotected phosphate and internucleotidic phos-phodiester groups that had to be carried along through the synthesis introduced two problems: (i) formation of oligomers having pyrophosphate linkages and chain-branched products, and (ii) time-consuming separation of the desired products by tedious chromatographic procedures.

### 7.05.1.1.2 *Phosphotriester approach*

The limitations of the phosphodiester approach led Letsinger's[12] and Reese's[13] groups to devise a practical phosphotriester approach (Equation (2)). In their approach, two reacting species were employed: (i) a nucleoside (**4**) having its 3′-hydroxy group derivatized as the β-cyanoethyl phos-phodiester function, and its 5′-hydroxy protected with the *p*-anisyldiphenylmethyl group; and (ii) a nucleoside (**5**) having a free 5′-hydroxy group. The resulting dinucleoside phosphotriester (**6**) could be readily purified by silica gel chromatography, and subsequently employed in chain extensions. This approach to oligonucleotide synthesis was later modified by the introduction of improved activation reagents, such as arenesulfonylimidazolides and arenesulfonyltriazolides, and the use of nucleophilic catalysts like *N*-methylimidazole and pyridine *N*-oxides for rate enhancement of coupling reactions. The phosphotriester approach was exploited in the synthesis of polydeoxyribo- and polyribo-oligonucleotides as well, and has been reviewed by Reese[13] and Ohtsuka *et al.*[14]

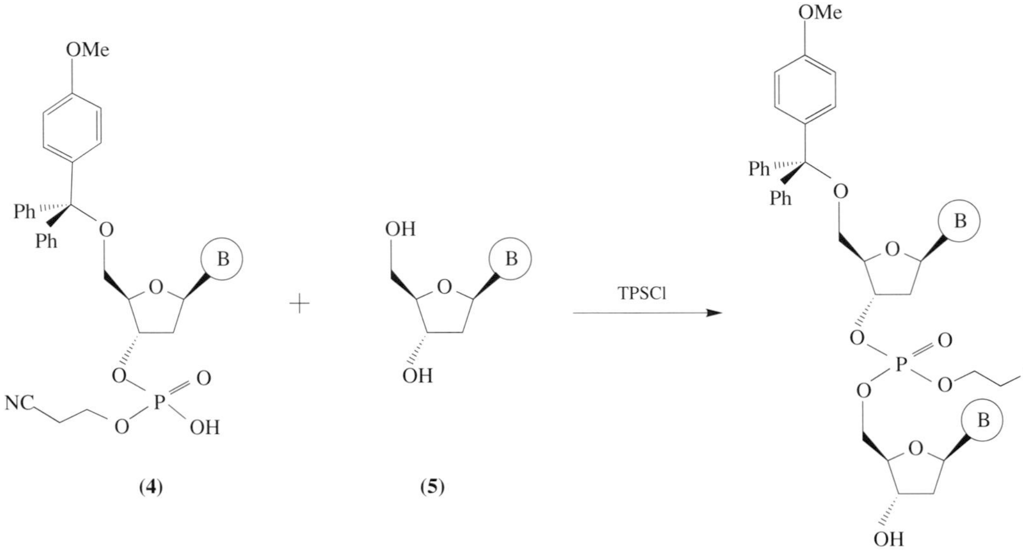

TPSCl = 2,4,6-triisopropylbenzenesulfonyl chloride

The above methodologies were initially developed in solution-phase synthesis. This implicitly meant that, after each step, the products had to be purified by chromatography, which is a tedious process in polymer synthesis. Catlin and Cramer's block condensation strategies[15] provided a partial solution to the problem.

A refreshing approach to polynucleotide synthesis emerged from Letsinger's group. Following Merrifield's ground-breaking work[16] on the solid-phase synthesis of peptides, Letsinger and Mahadevan[17,18] investigated the use of an organic polymer as a solid support for the synthesis of oligonucleotides. Later, Köster introduced inorganic supports, such as silica gel, for the same purpose.[19,20] However, phosphorylation reactions were not efficient with these supports. Additional work by Kössel and Seliger[21] on solid supports would set in motion the application of the phosphotriester approach to automated solid-phase synthesis by Gait,[22,23] Itakura,[24] Köster,[19,20] Narang,[25] and their co-workers. In addition to the production of various primers, probes, and linkers, the above advances in solid-phase oligonucleotide synthesis provided molecular biologists and biochemists with modified oligonucleotides useful in the study of protein–DNA interactions.[25,26]

### 7.05.1.1.3   *P(III) approach: the advent of phosphoramidite chemistry*

It is well known that phosphorus(III) compounds are more reactive than phosphorus(V) compounds. This led Letsinger and Lunsford[27] to investigate the phosphite triester methodology for oligonucleotide synthesis, particularly in the context of rapid-coupling reactions. In their initial approach (Scheme 3), they reacted 5′-*O*-protected thymidine (**7**) with 2,2,2-trichloroethyl phosphorodichloridite at −78 °C to generate the highly reactive nucleoside chlorophosphite (**8**). Upon addition of the 3′-*O*-protected nucleoside (**8a**) to the cold solution, a rapid reaction ensued to give the dinucleoside 2,2,2-trichloroethylphosphite (**9**) within a few minutes. Oxidation with iodine/water gave the dinucleoside phosphotriester (**10**). Attempts were made by Caruthers' group[28,29] to automate this methodology using silica gel as a solid support. However, because the chlorophosphite intermediates were highly reactive, they were difficult to manipulate under normal conditions. Eventually, work by Beaucage and Caruthers[30] led to the development of the nucleoside phosphoramidites (**11**) (Scheme 4) as a new class of stable, but reactive, phosphorus(III) compounds for oligonucleotide synthesis. These phosphoramidites were readily prepared and could be stored as dry powders under an inert atmosphere. Coupling to the nucleoside (**12**) was accomplished by activation of the phosphoramidite by weak acids, such as 1*H*-tetrazole, to give the phosphite triester (**13**). Oxidation gave the phosphotriester (**14**) ready for coupling to the next building block. This pioneering chemistry was improved upon by modifications of the phosphate and nucleobase protecting groups, activation reagents, and improvements in the automated synthesis cycles. Accordingly, oligonucleotide synthesis as it pertains to phosphoramidite chemistry is the major focus of this review.

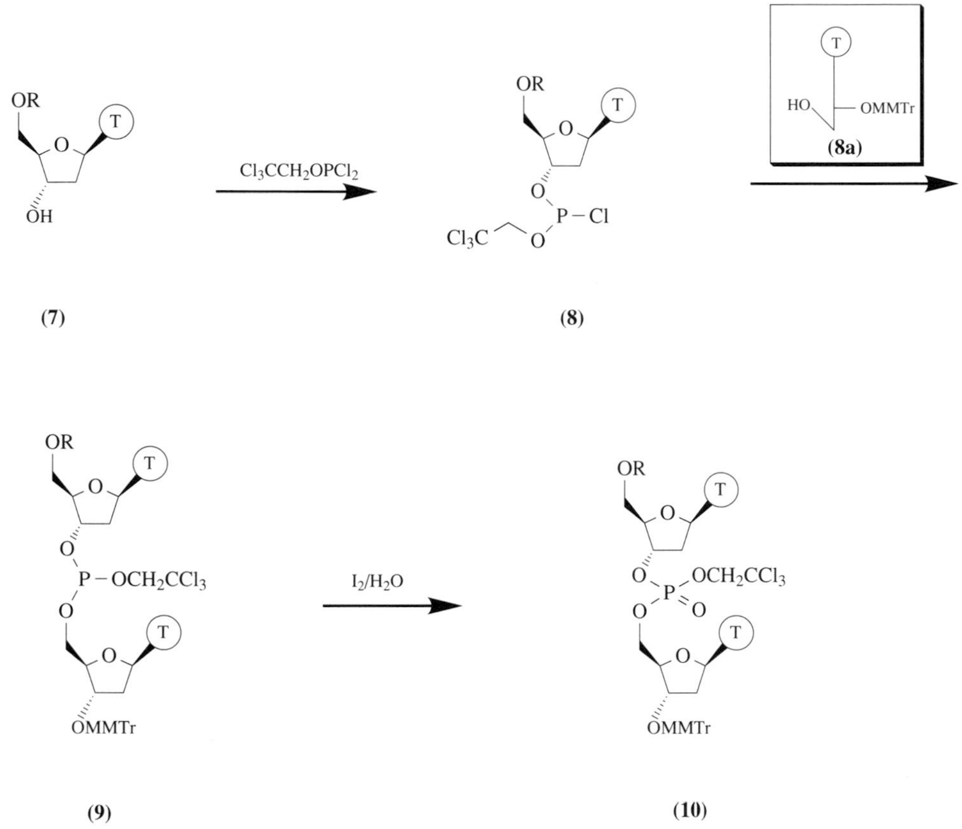

**Scheme 3**

### 7.05.1.1.4 H-*phosphonate chemistry*

While applications for oligonucleotides continued to grow, other synthetic routes for these biomolecules were explored. Two research groups[31,32] improved upon Todd's approach[7,8] to *H*-phosphonate chemistry. In Todd's approach (Scheme 5), the deoxyribonucleoside *H*-phosphonate (**15**) was converted to the reactive chlorophosphate intermediate (**16**). Condensation of a 3′-*O*-protected deoxyribonucleoside with (**16**) produced the dinucleoside phosphotriester (**17**) which under deprotection gave the dinucleotide (**18**). Later improvements primarily involved the use of new coupling reagents and adaptation of the methodology to automated solid-phase oligonucleotide synthesis. A distinctive feature of the *H*-phosphonate approach was that the oxidation of the internucleotidic *H*-phosphonate linkages was performed only once at the end of chain assembly. More importantly, the *H*-phosphonate functions could be converted, for example, to diesters, thioates, amidates, or triesters by simply choosing the appropriate combination of reagents.[33] This provided a number of backbone-modified oligonucleotides for various applications.[34,35]

The following sections provide a concise account of the developments in various aspects of oligonucleotide synthesis and the methods employed in the solid-phase synthesis of these macromolecules.

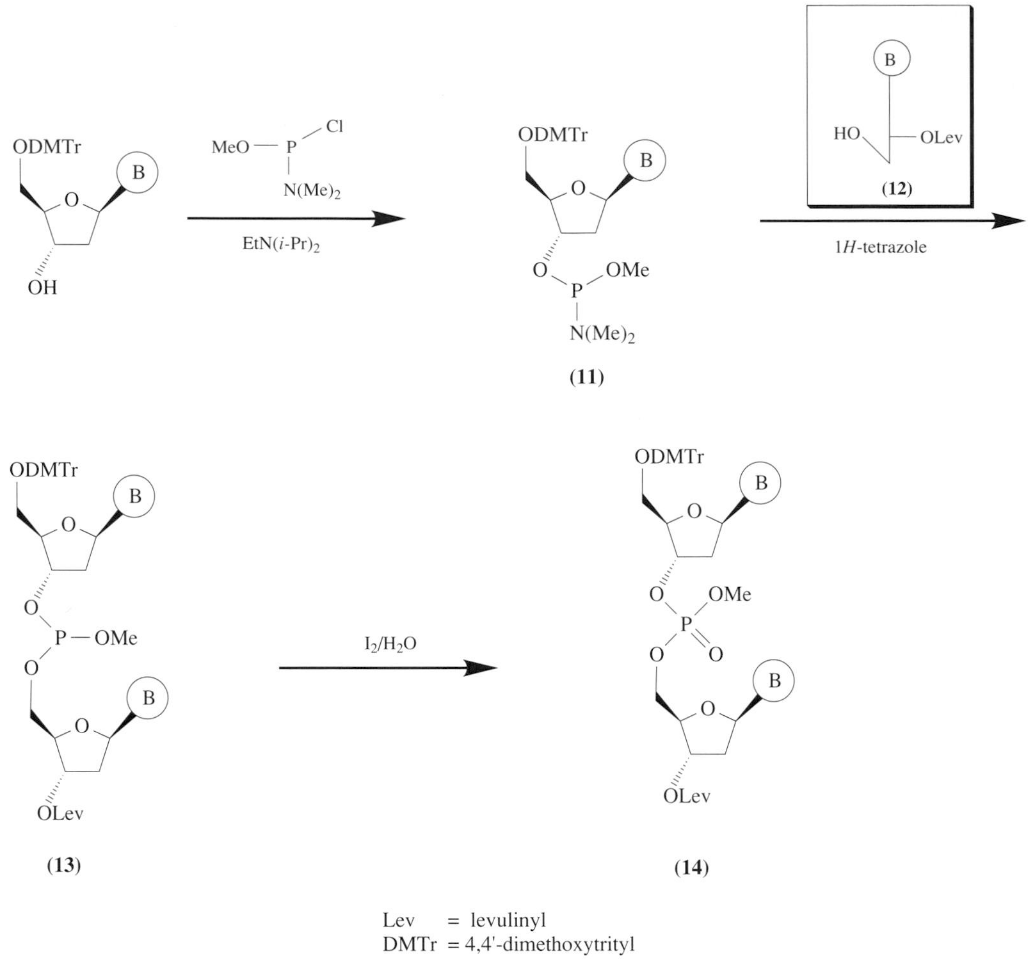

**Scheme 4**

## 7.05.2   SOLID SUPPORTS FOR OLIGONUCLEOTIDE SYNTHESIS

### 7.05.2.1   Overview of Developments in Solid Supports Employed in Oligonucleotide Synthesis

Although oligonucleotide syntheses were performed earlier in solution, the solid-phase approach to these syntheses rapidly became the method of choice even for large-scale (5 to 15 mmol) preparations. Since oligonucleotides are usually assembled by step-wise coupling of individual monomer nucleosides, coupling yields per step must be as close to 100% as possible to obtain practical yields of full-length oligonucleotides. For example, in the synthesis of a 10-mer oligonucleotide, where the step-wise coupling yields are 90%, 97%, and 99%, the overall yields of the full-length product will be 65%, 78%, and 95%, respectively. Thus, given the high step-wise coupling yields and the rapid kinetics (1–2 min), the phosphoramidite, the *H*-phosphonate, and the phosphotriester methods of solid-phase oligonucleotide synthesis became the methods of choice. In this context, the development of suitable solid supports, along with highly efficient methods of coupling reactions, has facilitated the synthesis of oligonucleotides using automated DNA synthesizers (see Section 7.05.1.5).

A support derived from styrene (88%), *p*-vinylbenzoic acid (12%), and *p*-divinylbenzene (0.2%) (known as "popcorn co-polymer") was introduced by Letsinger and Mahadevan in the late 1950s.[17,18] This support was insoluble in water, alkaline solutions, and organic solvents. The leader nucleoside was linked to the carboxy terminus of the polymer through either its 5′- or 3′-hydroxyl group, or

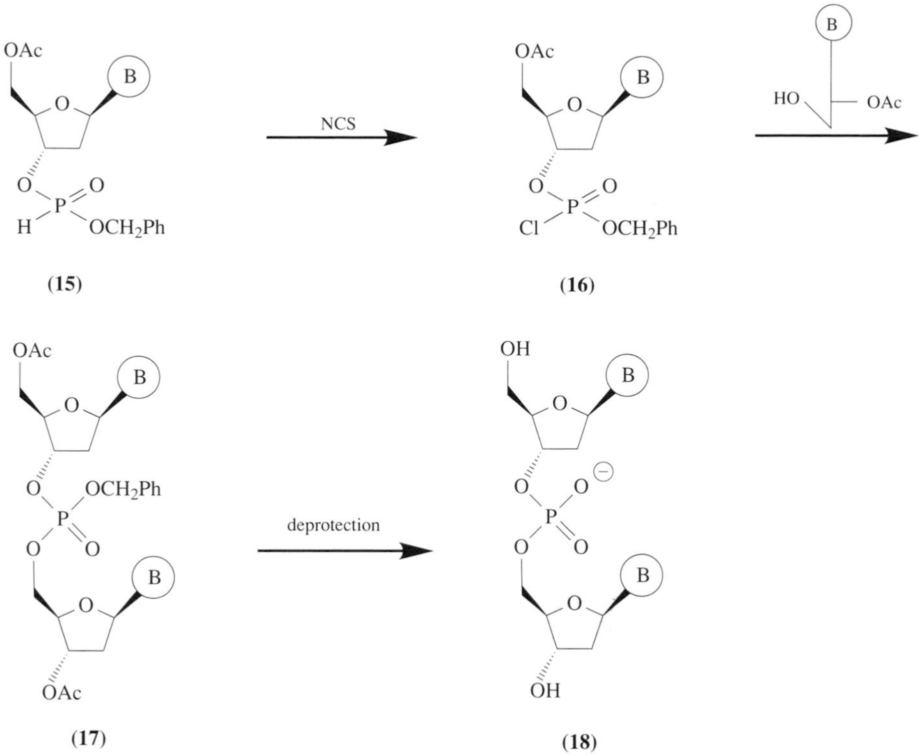

**(15)**      **(16)**

**(17)**      **(18)**

NCS = N-chlorosuccinimide
Ac  = Acetyl

**Scheme 5**

through the exocyclic amino function of a selected nucleobase. Oligonucleotide synthesis was performed according to the phosphodiester or phosphotriester approach. Upon completion of the chain assembly, the oligonucleotide was released from the support **(19)**–**(21)** by base-induced hydrolysis of the anchoring ester or amide function. On the basis of these pioneering studies, several solid supports derived from polystyrene (PS),[17,18] polyamides,[36–38] and cellulose[39,40] were evaluated in the synthesis of oligonucleotides using the phosphotriester approach.

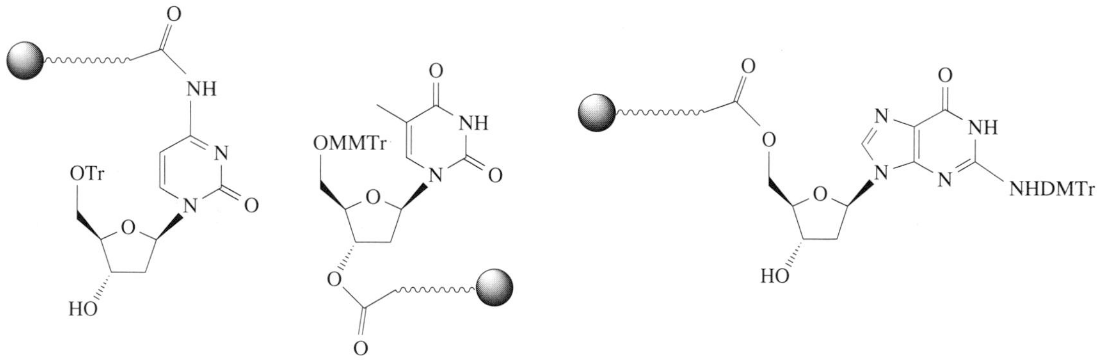

**(19)**      **(20)**      **(21)**

= Poly(styrene-divinylbenzene) copolymer

A major problem with most polymeric supports was the swelling of these polymers in the presence of organic solvents, which impeded the proper diffusion of reagents in and out of the matrix. These considerations led to the development of silica gel-based solid supports,[19,24,41] and composite supports derived from polyamide and silica gel.[42] New techniques for silica gel derivatization in bonded-phase chromatography were also developed.[43–45] These methodologies were adapted for coupling leader nucleosides to silica-based supports (Scheme 6). Thus, treatment of the silica gel support (22) with 3-aminopropyl triethoxysilane gave the functionalized matrix (23), which upon reaction with succinic anhydride gave the carboxylated support (24). The leader nucleoside (25) was linked to the carboxyl function of (24) through its 3′-hydroxyl group by the use of DCC to afford the support-bound nucleoside (26).

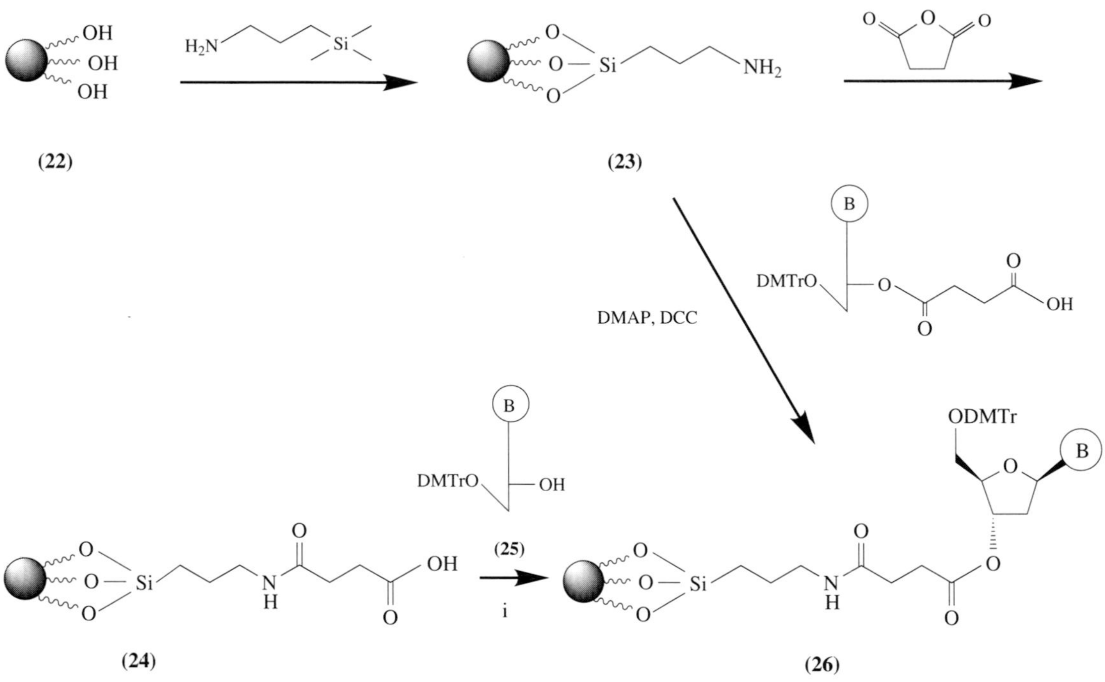

i, DCC/pentachlorophenol/piperidine

**Scheme 6**

Silica gel supports became popular only with the advent of the phosphite approach (phosphorus(III) chemistry) in oligonucleotide synthesis.[28–30,46–51] It was later reported that a "spacer arm" between the matrix and the leader nucleoside in (27) was important for optimum synthetic efficiency and product purity.[52–57] The linker arms were derived from alkylamines, aromatic amino acids and polyethylene glycols (PEGs). Linker arms lacking hydrogen bonding or π interactions and which could adopt extended conformations (as opposed to folded conformations) imparted ideal properties to these supports. However, the fragility of silica gel under the forced flow conditions generated by a DNA synthesizer prompted Adams *et al.* to evaluate long-chain alkylamine controlled-pore glass (CPG) (28) as an alternate support for automated oligonucleotide synthesis.[55,56] While the surface chemistry of CPG was similar to that of silica gel, the CPG could be fashioned into more rigid spherical particles of uniform size that could be etched to produce pore sizes ranging from 75 Å to several thousand Å. It is pertinent to mention that selection of the appropriate pore size of silica supports is important in oligonucleotide synthesis. For example, it was found that during the synthesis of oligonucleotides larger than 100 bases, abrupt chain termination occurred on supports having a pore size of 500 Å. Presumably, steric crowding around pores and channels caused by the growing oligonucleotide chains impaired the diffusion of reagents through the matrix. This problem was circumvented by the use of the supports (29) that had a pore size of 1000 to 2500 Å. Alternatively, nonporous silica gel microbeads[55,56] and rigid nonswelling polystyrene beads[58] have been successfully employed in the synthesis of 150 bases-long oligomers by the phosphoramidite method.

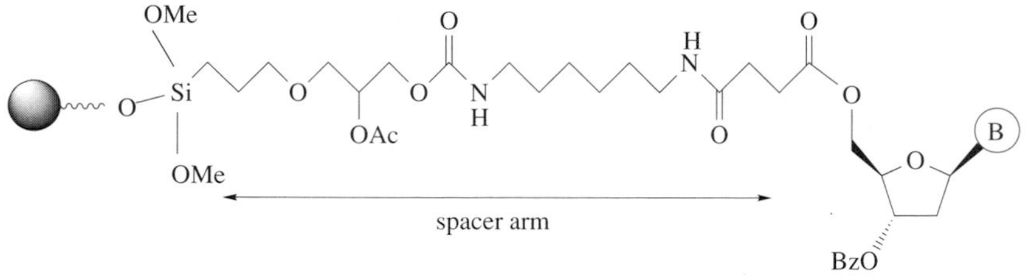

(27)

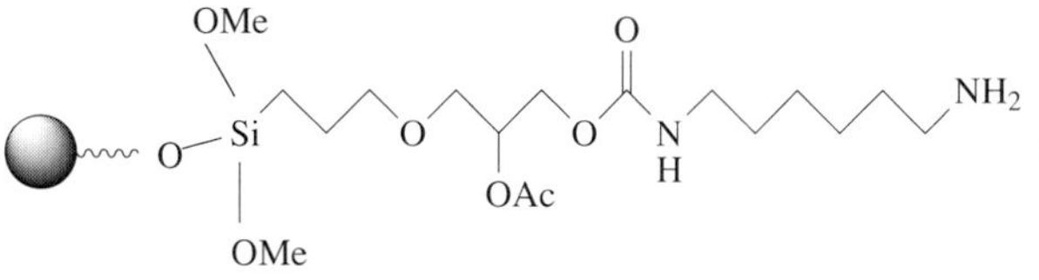

(28)

LCAA-CPG

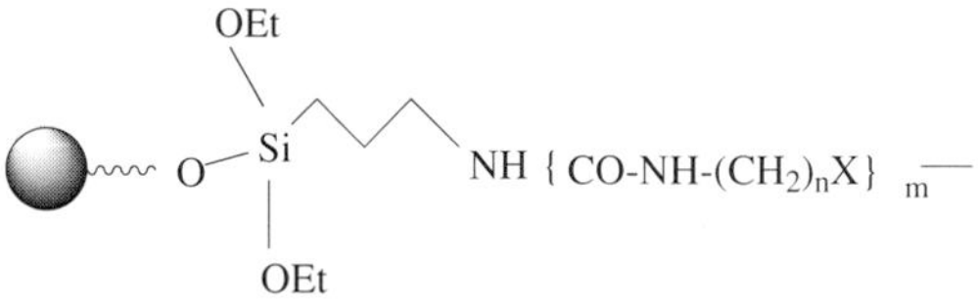

(29)

CPG 500,/Fractosil 500, or 2500
n = 2 to 12
m = 1 to 4
X = O or NH

Despite the popularity of CPG, developmental efforts for new supports have continued. In this context, a highly cross-linked polystyrene support (30) has been reported to provide a rigid non-swelling and nonporous matrix.[58,59] The polymeric support (31) ("tentacle support") has been prepared by grafting PEG chains on the surface of PS beads.[60,61] These PEG-PS beads had uniform size and spherical shape, and allowed "loading" with nucleosides to concentrations as high as 100 to 170 μmol g$^{-1}$. The hydrophilic properties of PEG improved the swelling of the support in those polar solvents employed during automated oligonucleotide synthesis. In the 1990s, Fractogel[62] and various supports have additionally been evaluated in the synthesis of oligonucleotides.[63–67]

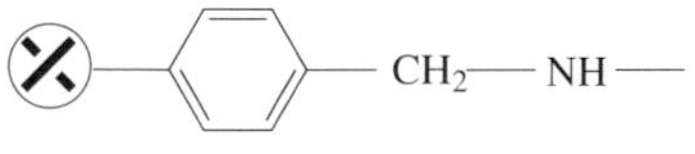

(30)

Highly cross-linked polystyrene

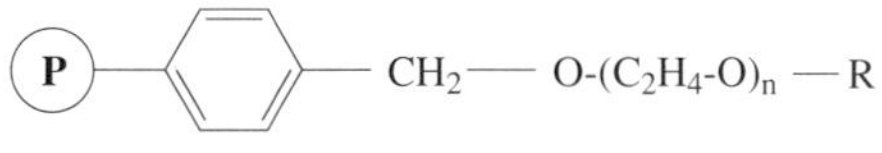

(31)

PEG-polystyrene tentacle support

To obviate the need for the preparation of supports carrying each of the leader nucleosides, "universal" supports (32) and (33) have been proposed.[68–72] In Gough's approach,[68] support-bound oligonucleotides were treated with concentrated NH$_4$OH which unmasked the 2'(3')-OH group. The ensuing intramolecular attack of the 2'-hydroxyl group on the neighboring phosphodiester function resulted in the cleavage of the internucleotidic bridge that linked the oligonucleotides to the *cis*-2',3'-diol system and released the oligonucleotide from the support.

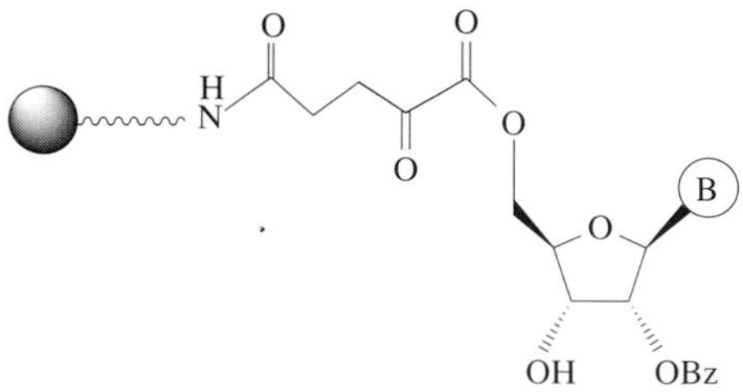

(32)

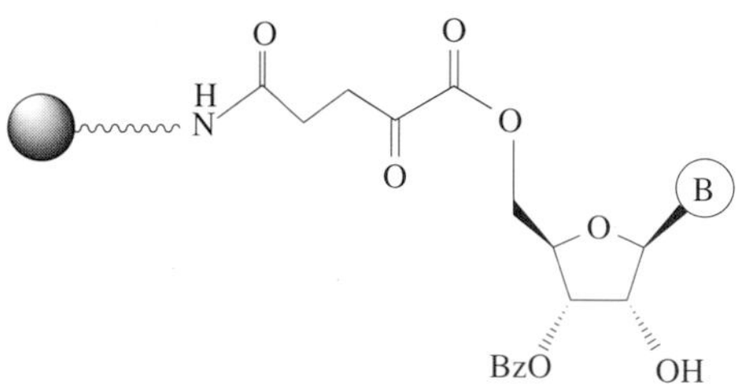

(33)

Replacement of the succinyl linker of support-bound nucleosides by a more stable carbonate or carbamate linker reportedly resulted in a modest increase in oligonucleotide yields by the phosphotriester approach.[73] However, prolonged treatment with aqueous ammonium hydroxide (55 °C, 48 h) was required for the complete release of the oligonucleotides from these supports.

In studies aimed at using the Fmoc group for the 5′-OH protection of nucleosides and 1,8-diazabicyclo[5,4,0]undec-7-ene (DBU) for its deprotection, it was found that the succinyl group linking the leader nucleoside to CPG was unstable to DBU treatments and, hence, unsuitable for oligonucleotide synthesis. Brown *et al.*[74] found that LCAA-CPG derivatized with a succinyl-sarcosyl linker showed less than 5% cleavage upon prolonged treatment with DBU and could be used in oligonucleotide synthesis.

The development of oligonucleotides as potential therapeutic and diagnostic agents have necessitated the synthesis of oligonucleotides with modified backbones. However, certain base-sensitive backbones do not survive the basic conditions (conc. $NH_4OH$) necessary for the cleavage of succinyl-linked oligonucleotides from the support. Alul *et al.*[75] employed an oxalyl linkage for anchoring the leader nucleoside (34) to CPG. Although stepwise coupling yields were slightly lower on (34) than on conventional succinyl-linked supports, rapid cleavage of oligonucleotides from (34) could be achieved under mild conditions using either wet triethylamine or 40% triethylamine in methanol, or 5% $NH_4OH$ in methanol. Thus, the oxalyl-linked CPG supports could be employed for the preparation of modified oligonucleotides having base-sensitive functionalities. The hydroquinone-*O,O*′-diacetic acid (QDA) linker has been reported to permit facile cleavage of oligonucleotides from CPG and highly cross-linked PS supports.[76,77] In conjunction with the QDA linker, automated derivatization of solid supports for oligonucleotide synthesis has also been described.[76,77]

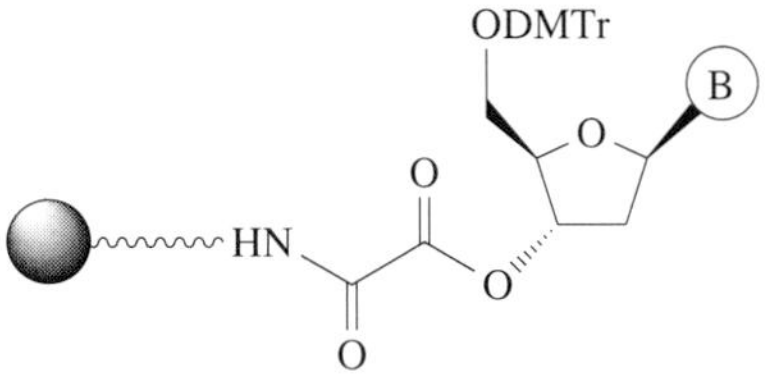

(34)

### 7.05.2.2 Solid Supports for the Preparation of Functionalized Oligonucleotides

The synthesis of oligonucleotides whose terminal 3′-OH group is derivatized by other functionalities, can be accomplished using the appropriately modified anchoring groups (35)–(47) (Figures 1–3). For example, the support (35) or (43) has been used in the synthesis of oligonucleotides with terminal 3′-phosphates.[78–81] A solid support carrying the 2-nitrophenylethyl or a β-thioethanol linker has also been employed for the same purpose.[78,79] Alternatively, oligonucleotides carrying a terminal 3′-phosphate group or a 3′-thioalkyl function were prepared by the use of appropriate disulfide-anchoring groups (36)–(41). In these cases, ammoniacal dithiothreitol released the oligonucleotides bearing the requisite 3′-modification.[78–86]

Various modified anchoring groups (45)–(47) have been employed in the synthesis of 3′-aminoalkylated oligonucleotides. These oligonucleotides could be chemoselectively conjugated to various ligands by taking advantage of the more nucleophilic aliphatic amine group.[35,87–90] Similarly, anchoring groups containing peptides were used in the preparation of oligonucleotide-peptide conjugates.[91–94] Furthermore, anchoring groups containing amino acids (serine, proline) led to the synthesis of oligonucleotides functionalized with terminal 3′-amino or carboxyl groups suitable for conjugation with other ligands.[95–98]

The silyl-linked support (48) has recently been applied to oligonucleotide synthesis.[99] Oligonucleotides were released from the support upon treatment with tetra-*n*-butylammonium fluoride (TBAF).

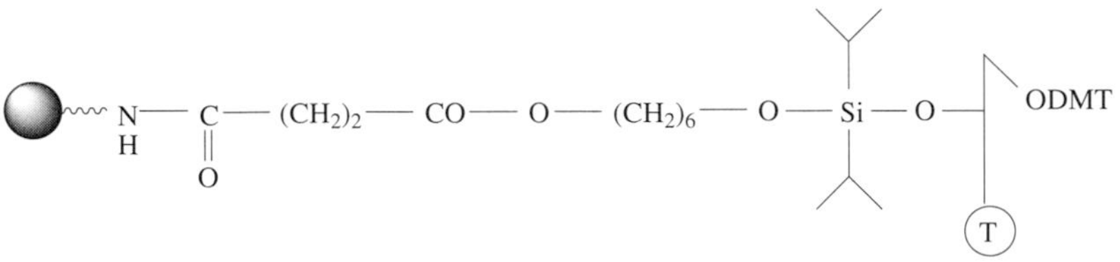

(48)

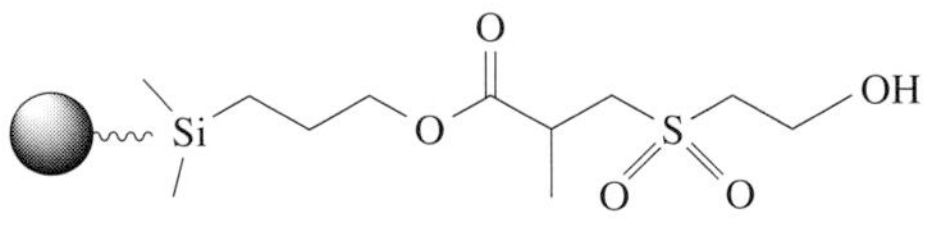

**(35)**

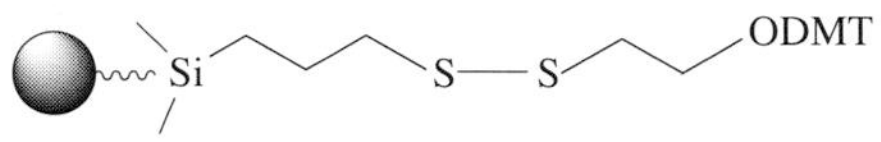

**(36)**

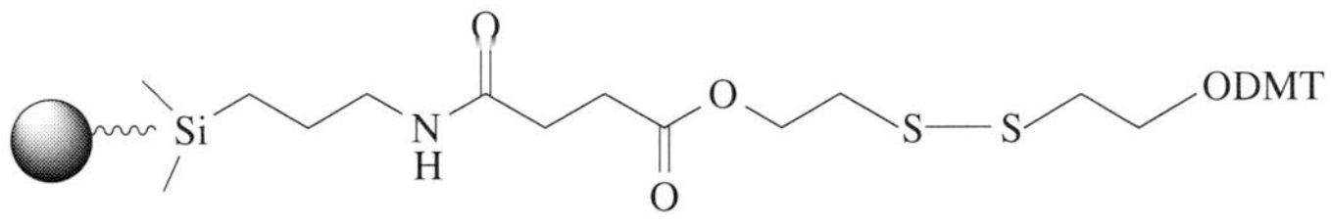

**(37)**

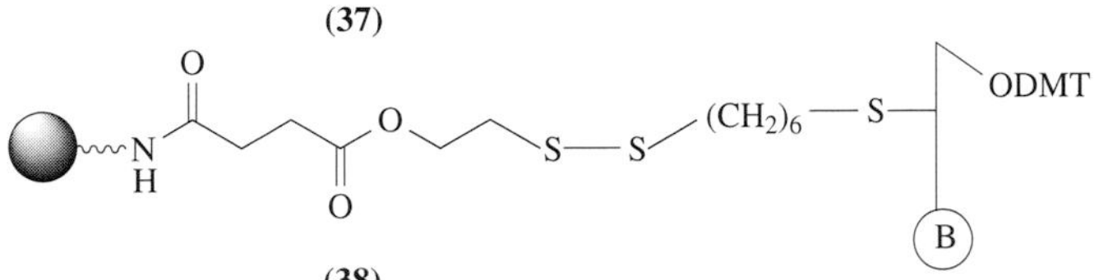

**(38)**

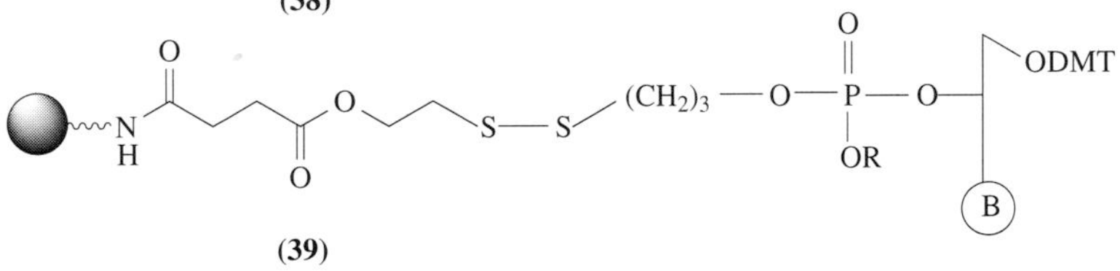

**(39)**

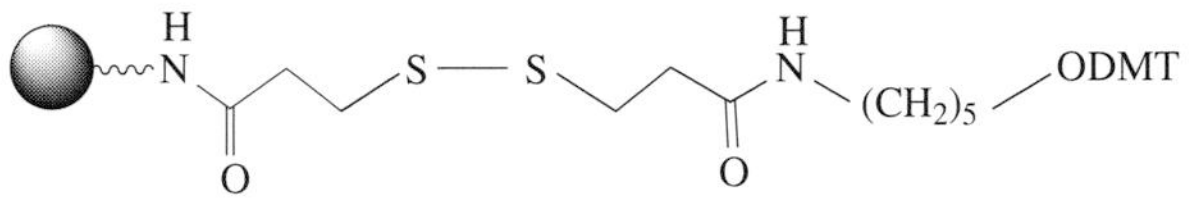

**(40)**

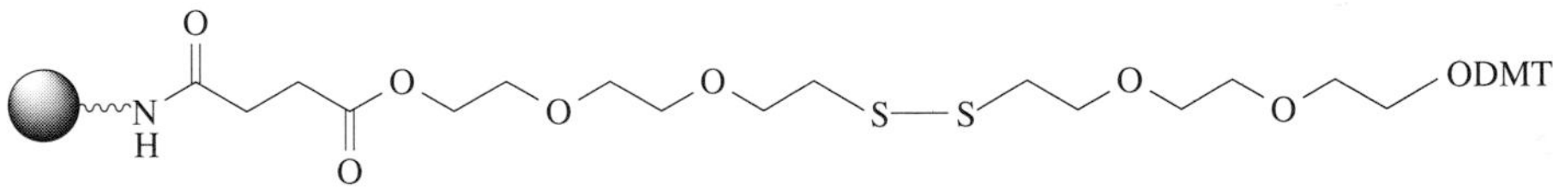

**(41)**

**Figure 1** Linkers with disulfide bridges for the synthesis of 3′-phosphorylated or thiol-terminated oligonucleotides.

Photolabile supports (e.g., (**49**)) have also been described for the synthesis of oligonucleotides and those functionalized with a 3′-alkylamine or a 3′-glycolate terminus.[100–103] Thus, photochemical irradiation of the support-linked oligonucleotides unmasked the 3′-functionalities for subsequent derivatization. Furthermore, the use of a non-nucleosidic support made it possible to tether a specific 3′-function to an oligonucleotide regardless of its sequence (Equation (3)).

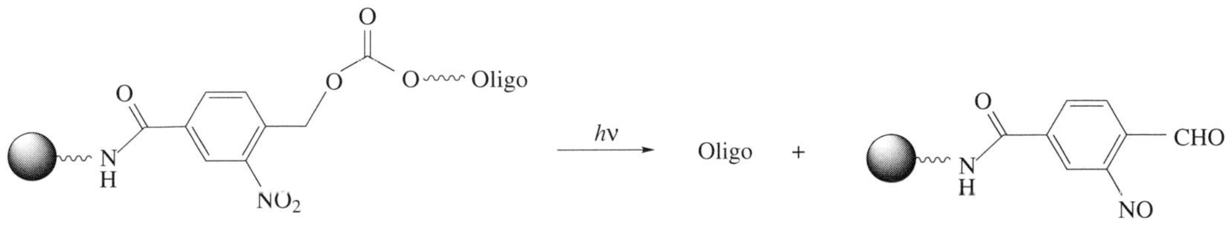

(3)

**(49)**

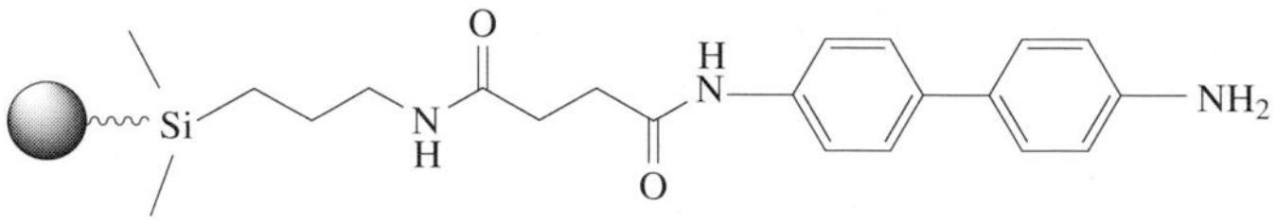

**(42)**

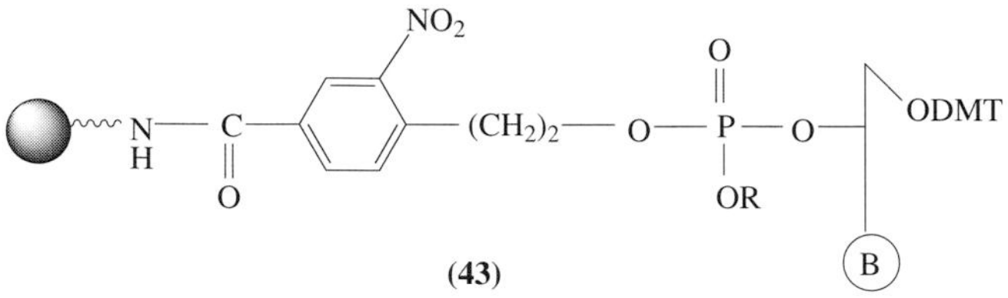

**(43)**

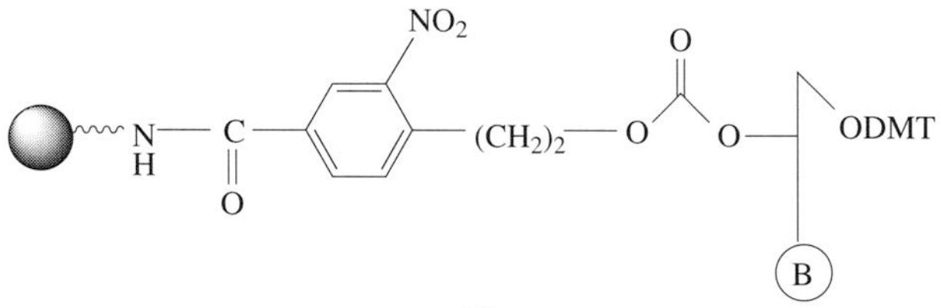

**(44)**

**Figure 2**   Linkers with aromatic tethers.

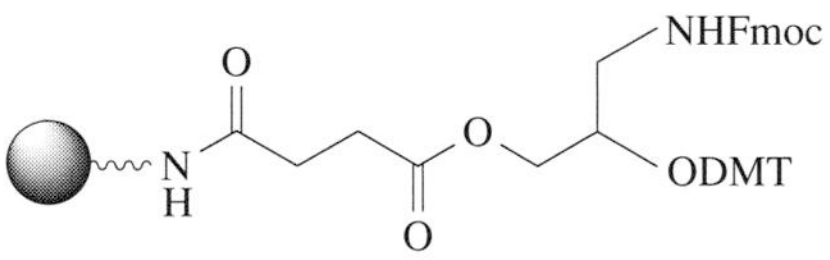

**(45)**

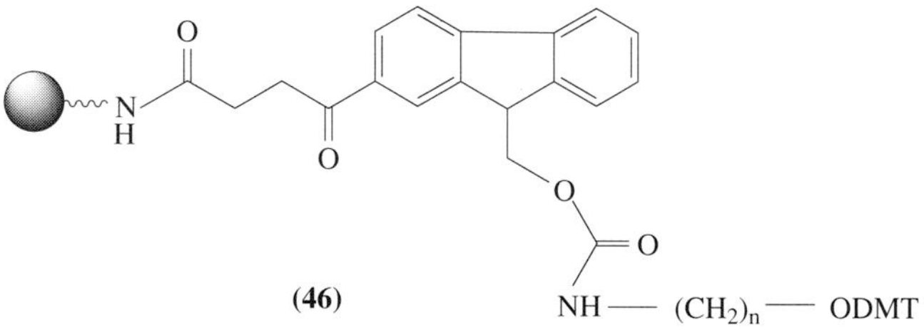

**(46)**

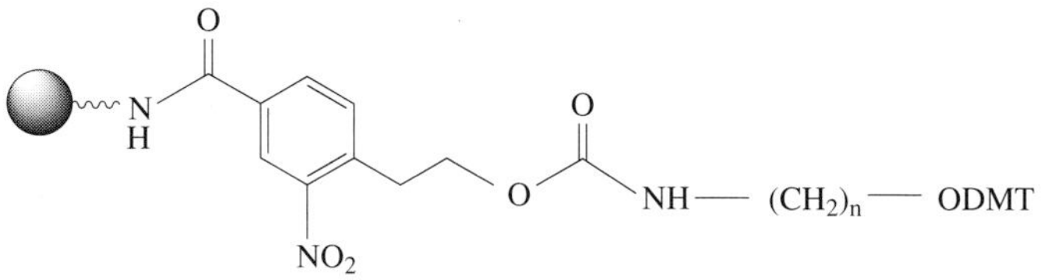

**(47)**

**Figure 3**   Linkers for amino-terminated oligonucleotides.

Anchoring group-containing ligands have also been prepared and have been reviewed.[104] Examples include cholesterol, vitamin E, biotin, fluorescein, rhodamine, and acridine.[104–106] Also, protocols for the functionalization of supports have been delineated.[104]

Finally, the preparation of support-bound oligonucleotides has been achieved by the use of anchoring groups that are resistant to the deprotection conditions employed in oligonucleotide synthesis. Sproat and Brown[73] used 2,6-diisocyanatotoluene to form a urethane bridge between the nucleoside and the support. This linkage was resistant to 28% $NH_4OH$ (16 h, 50 °C) and permitted complete deprotection of oligonucleotides while still being attached to the support (**50**). These support-bound deprotected oligonucleotides can serve as affinity columns and applications in this context have been reported.[107]

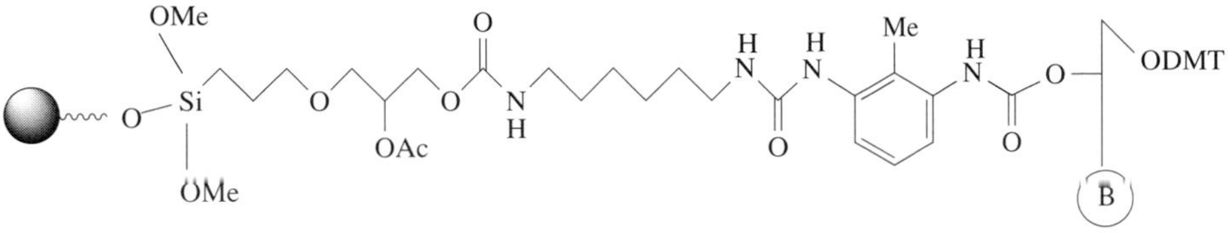

(**50**)

## 7.05.3 PROTECTING GROUPS IN OLIGONUCLEOTIDE SYNTHESIS

### 7.05.3.1 General Comments

Solid-phase synthesis of oligonucleotides is usually accomplished using persistent and transient nucleoside protecting groups. Persistent protecting groups are those employed for the protection of nucleobases (i.e., the aglycone) and internucleotidic phosphodiester functions. These groups are carried throughout the synthesis cycle and are only removed at the end of chain assembly. The transient protecting groups (e.g., the DMTr group for the 5′-hydroxy) are those that are removed just prior to each coupling reaction.

The key step in the formation of internucleotidic phosphodiester linkages is the nucleophilic attack of a hydroxy group on an electrophilic phosphorus(III) or phosphorus(V) atom. In order to achieve chemoselectivity, it was necessary to block all nucleophilic sites on the nucleoside except the intended hydroxy group. This was usually done by the use of persistent protecting groups on the nucleobases. The choice of the appropriate protecting groups is crucial to a successful oligonucleotide synthesis and has been an area of intense activity since the 1950s. While new procedures for adding and removing protecting groups are continuously being developed, the introduction of protecting groups on the nucleobases has created unexpected problems. For example, acylated purine nucleosides are more prone towards depurination, and exposure of the protecting groups to the various reagents used during oligonucleotide synthesis resulted, in some cases, in the formation of irreversible base modifications.[108] The sections that follow provide an overview of the protection strategies that have been employed.

#### 7.05.3.1.1 Protection of imide/lactam functions of nucleobases

Thymine, uracil, and guanine have imide/lactam functionalities. The NH function of thymine and uracil ($pK_a$ 9.8) and guanine ($pK_a$ 9.2) is nucleophilic under basic conditions and reacts with activated phosphates, DCC, MS-Cl, acid chlorides, phosphitylating reagents, and various electrophilic reagents during coupling reactions to give both *N*- and *O*-side-products. For example, it was found that 1-(mesitylenesulfonyl)-3-nitro-1,2,4-triazole (MSNT) reacted with tri-*O*-acetyluridine (**51**) to generate the triazolo derivative (**52**) (Scheme 7).[109,110] Treatment of (**52**) with ammonium hydroxide gave the cytidine derivative (**53**), whereas reaction with the tetramethylguanidinium salt of *syn*-4-nitrobenzaldoxime or TBAF converted (**52**) to uridine.[111] The triazolo derivative (**52**) also reacted with methanol in the presence of DBU to generate the corresponding 4-OMe derivative.[112] Analogous triazolo derivatives were produced upon condensation of guanine nucleosides with MSNT. The O-4 of (**51**) was prone to phosphorylation by phosphorochloridates and, in the presence of pyridine, is converted to the fluorescent pyridinium salt (**54**) (Scheme 8).[113–123]

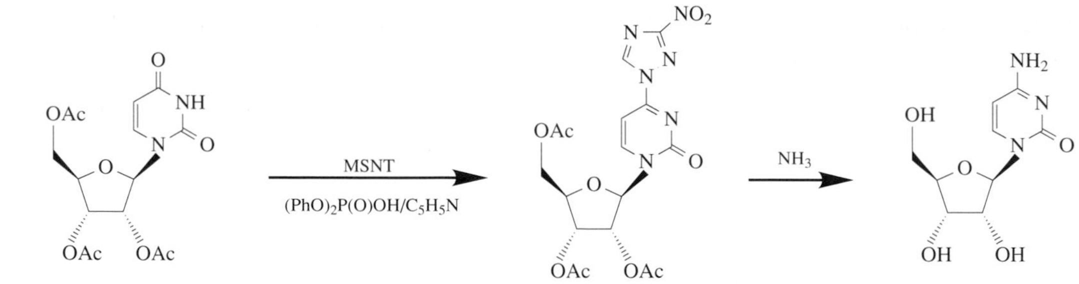

**(51)**     **(52)**     **(53)**

MSNT = 1-(mesitylene-2-sulfonyl)-3-nitro-1,2,4-triazole

**Scheme 7**

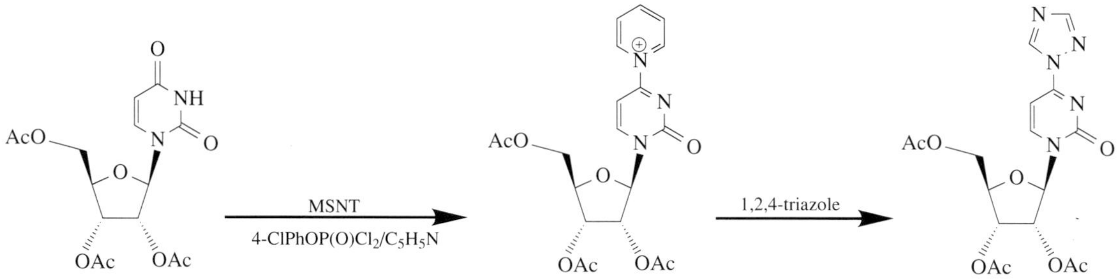

**(51)**     **(54)**     **(55)**

MSNT = 1-(mesitylene-2-sulfonyl)-3-nitro-1,2,4-triazole

**Scheme 8**

The salt (**54**) reacted with triazole to give (**55**). Similarly, condensation of the protected guanosine nucleoside (**56**) with methanesulfonyl chloride produced the crystalline O-6-sulfonated derivative (**57**) (Equation (4)).[124,125]

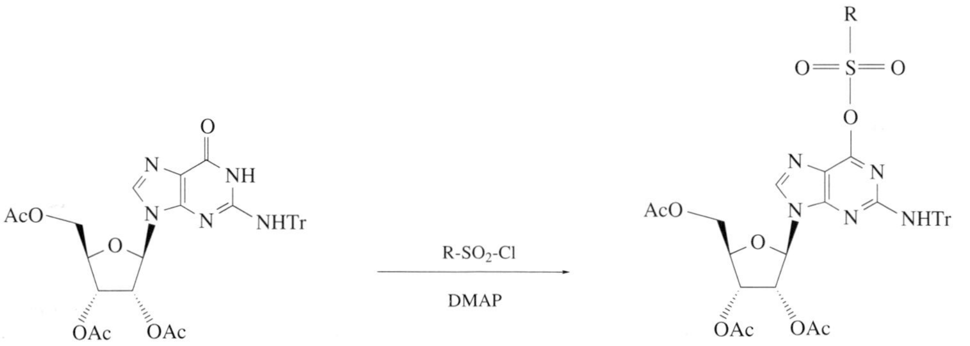

**(56)**     **(57)**     (4)

These observations may have implications in oligonucleotide synthesis using the *H*-phosphonate method, particularly when phosphoryl chlorides are used as activators (see below).

Phosphitylation of guanine at O-6 by methoxyphosphoramidite synthons (phosphorus(III) chemistry) has also been reported.[126–129] The formation of (**58**) ($^{31}$P NMR, $\delta$ 133.95, 133.79 ppm) was noted (Scheme 9) and it has been suggested that during the oxidation of (**58**) with iodine, *O*- to

*N*-phosphoryl migration could occur ((**59**) → (**60**)) and result in extensive depurination during deprotection. O-6-protection could alleviate these problems.

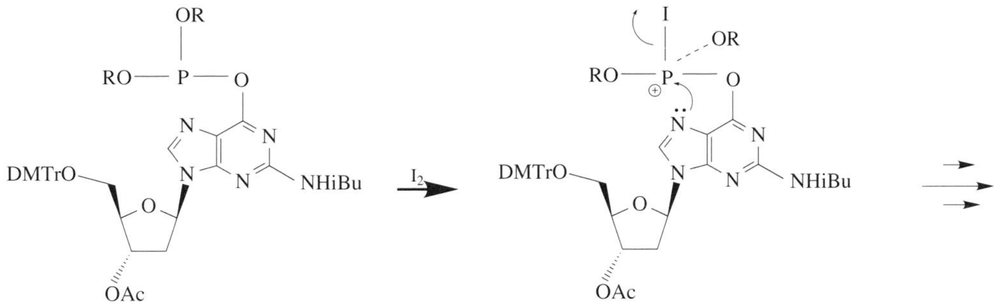

**Scheme 9**

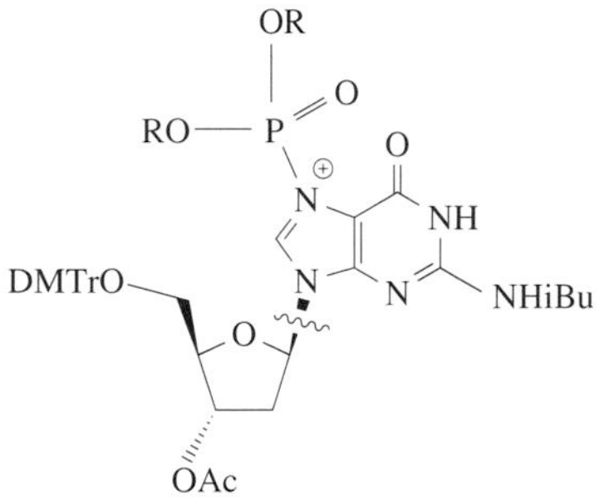

Urdea *et al.*[130] reported that up to 30% of the thymine residues were converted to *N*-3-methyl-thymines during the deprotection of oligonucleoside methylphosphotriesters with concentrated ammonium hydroxide. This problem was overcome by employing thiophenol or 2-thiobenzothiazole for methyl phosphotriester deprotection as recommended by McBride *et al.*[131] and independently by Beaucage and Andrus.[132] These reports have raised concerns and prompted the development of protecting groups for N-3 or O-4 of thymine and uracil, and O-6 of guanine. A number of these protecting groups are listed in Table 1.

It should, however, be pointed out that in modern solid-phase oligonucleotide synthesis the use of automated pulsed delivery of reagents results in short reaction times. Consequently, side-reactions are considerably minimized. Furthermore, during the capping step and/or during the final deprotection step, most nucleobase side-products are destroyed or reverted to the original structure. Thus, imide/lactam protection is not normally employed in solid-phase oligonucleotide synthesis via phosphoramidite chemistry.

### 7.05.3.1.2  *Protection of the exocyclic amino function of nucleobases*

The choice of a suitable protecting group for the exocyclic amino function is governed by several factors: (i) ease of installation, (ii) inertness towards the plethora of reagents employed in oligonucleotide synthesis, and (iii) facile chemoselective removal. Furthermore, the protecting groups should not cause any modifications of the nucleoside during installation, during oligonucleotide synthesis, or during their removal. Many protecting groups such as benzoyl were initially designed as rather robust groups to withstand the harsh reagents employed in phosphodiester[9–11] and -triester chemistry.[12] In the context of oligonucleotide synthesis, the removal of these protecting groups required prolonged heating with 28% $NH_4OH$. However, having withstood scrutiny under a variety of synthesis conditions, the benzoyl groups for adenine and cytosine, and isobutyryl groups for the guanine, have remained immensely popular even with the more modern phosphoramidite and *H*-phosphonate chemistry of oligonucleotide synthesis. Nevertheless, as the applications of oligonucleotides as potential therapeutic and diagnostic agents have continued to grow, new "softer" protecting groups continue to be devised to meet these special requirements.

A number of protecting groups have been devised for the exocyclic amino group (N-4) of cytosine (Table 2). Among nucleobases with exocyclic amino functions, cytosine is the most basic and also the most nucleophilic (pK$_a$ of ca. 4.24) and is usually protected as an acyl group. The acyl protecting groups were installed by peracylation of the nucleoside followed by chemoselective *O*-deacylation.[137,212] This procedure has been largely supplanted by the transient protection approach.[117,213]

Some acyl functions such as benzoyl, α-phenyl cinnamoyl and naphthoyl have been directly incorporated on the nucleobase using the corresponding anhydride.[274,275] The N-4 acetyl[215,216,276] was reportedly too fragile as a protecting group although this group has been reintroduced in the context of "ultrafast" DNA synthesis.[277] The 4-methoxybenzoyl group was used earlier in phosphodiester chemistry. The 2,4-dimethylbenzoyl group was too robust.[215] A number of other novel protecting groups have also been introduced (see Table 2). In some instances, neighboring group participation to accelerate deacylation has been reported.[270–272] However, many of these protecting groups have not been rigorously evaluated, particularly in the context of routine as well as large-scale synthesis. Thus, given the ease of installation and their robustness, the benzoyl group for A and C and the isobutyryl group for G have remained popular protecting groups for routine oligonucleotide synthesis.

The demand for oligodeoxynucleotides, and oligoribonucleotides in diagnostics and therapeutics has revived the search for protecting groups which are less robust in the context of

**Table 1** Groups employed for the protection of the imide/lactam function of guanine, uracil, and thymine.

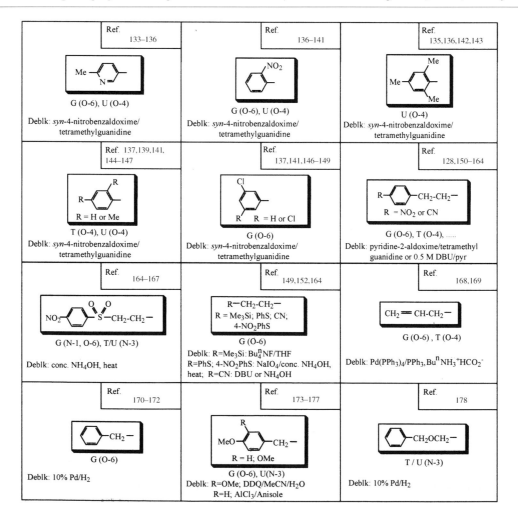

**Table 1** (continued)

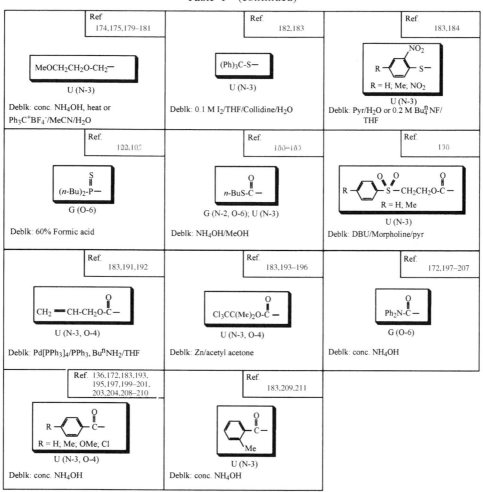

| Ref. 174,175,179–181 | Ref. 182,183 | Ref. 183,184 |
|---|---|---|
| MeOCH₂CH₂O-CH₂— <br> U (N-3) <br> Deblk: conc. NH₄OH, heat or Ph₃C⁺BF₄⁻/MeCN/H₂O | (Ph)₃C-S— <br> U (N-3) <br> Deblk: 0.1 M I₂/THF/Collidine/H₂O | R — S — (NO₂) <br> R = H; Me; NO₂ <br> U (N-3) <br> Deblk: Pyr/H₂O or 0.2 M Buⁿ₄NF/THF |
| Ref. 182,185 | Ref. 182–185 | Ref. 190 |
| (n-Bu)₂-P— <br> G (O-6) <br> Deblk: 60% Formic acid | n-BuS-C— <br> G (N-2, O-6); U (N-3) <br> Deblk: NH₄OH/MeOH | R — S — CH₂CH₂O-C — <br> R = H; Me <br> U (N-3) <br> Deblk: DBU/Morpholine/pyr |
| Ref. 183,191,192 | Ref. 183,193–196 | Ref. 172,197–207 |
| CH₂=CH-CH₂O-C— <br> U (N-3, O-4) <br> Deblk: Pd[PPh₃]₄/PPh₃, BuⁿNH₂/THF | Cl₃CC(Me)₂O-C — <br> U (N-3, O-4) <br> Deblk: Zn/acetyl acetone | Ph₂N-C— <br> G (O-6) <br> Deblk: conc. NH₄OH |
| Ref. 136,172,183,193, 195,197,199–201, 203,204,208–210 | Ref. 183,209,211 | |
| R — C— <br> R = H; Me; OMe; Cl <br> U (N-3, O-4) <br> Deblk: conc. NH₄OH | C— Me <br> U (N-3) <br> Deblk: conc. NH₄OH | |

deprotection, and those which can be chemoselectively removed. Amongst these, phenoxyacetyl (PAC)[216,218,219,257,258] and its derivatives such as *tert*-butylphenoxyacetyl (*t*-PAC)[215,259,260] show accelerated deacylation perhaps because of increased susceptibility of the amide carbonyl group towards nucleophilic attack by inductive withdrawal. The 2-(acetoxymethyl)benzoyl uses the neighboring group effect to promote fast deacylation. The use of an *N*-acetyl protecting group for dC facilitated rapid deprotection using methylamine/ammonia.[277] The allyloxy carbonyl protecting group[168,191,192,231–234] could be chemoselectively removed using Pd(0) and provided the potential for the preparation of support-bound base-deprotected oligonucleotides. The *N*-pent-4-enoyl (PNT) group has been introduced as an acyl protecting group where multiple deprotection strategies could be used for deprotection.[237]

### 7.05.3.1.3 *Protection of purine bases: special considerations*

The development of suitable protecting groups for the purine bases has continued to be quite challenging. The problem is compounded by the facile depurination of purine nucleosides under acidic conditions. Incidentally, the solid-phase synthesis (see below) employs the acid-labile DMTr group for the protection of the 5′-OH group. Prior to each coupling step, the 5′-OH group is uncovered by exposure to a strong acid such as 2% dichloroacetic acid in dichloromethane. Such repeated exposure to acidic solution results in the formation of "depurinated material," and consequent reduced yield of the full-length product. Presumably, in the case of dA, the mechanism of depurination (Scheme 10) involves the initial protonation to reveal N-1 protonated form (**61**),

**Table 2**   Protecting groups for the exocyclic amino function of nucleobases.

| Ref. 214 | Ref. 148,215–219 | Ref. 197,198,200–207 |
|---|---|---|
| $\overset{O}{\underset{\|}{HC-}}$ | $\overset{O}{\underset{\|}{MeC-}}$ | $\overset{O}{\underset{\|}{MeCH_2C-}}$ |
| 7-deaza-6-methyl-G (N-2) | C (N-4) | G (N-2) |
| Deblk: conc. NH₄OH | Deblk: NH₃/EtOH (1:1) | Deblk: conc. NH₄OH/pyr (2:1) |
| Ref. 58,215,217, 220–226 | Ref. 215 | Ref. 227 |
| $\overset{O}{\underset{\|}{(Me)_2CHC-}}$ | $\overset{O}{\underset{\|}{(Me)_3CC-}}$ | $\overset{O}{\underset{\|}{(Me)_2CHCH_2OC-}}$ |
| G (N-2); C (N-4) | C (N-4) | C (N-4) |
| Deblk: conc. NH₄OH, heat | Deblk: 0.2 M NaOH/MeOH (1:1) | Deblk: conc. NH₄OH/pyr (3:1) |
| Ref. 138,196,228–230 | Ref. 168,183, 191,192,231–234 | Ref. 223–225 |
| $\overset{O}{\underset{\|}{Cl_3CC(Me)_2OC-}}$ | $\overset{O}{\underset{\|}{CH_2=CHCH_2OC-}}$ | $\overset{O}{\underset{\|}{MeOCH_2C-}}$ |
| G (N-1, N-2); A (N-6); C (N-4) | G (N-2); A (N-6); C (N-4) | G (N-2) |
| Deblk: LiCo(I) Phthalocyanine | Deblk: Pd[PPh₃]₄/HCO₂H/Et₂NH | Deblk: conc. NH₄OH |
| Ref. 235 | Ref. 236 | Ref. 237,238 |
| $\overset{O}{\underset{\|}{i\text{-PrOCH}_2C-}}$ | $\overset{O\ \ \ \ \ \ O}{\underset{\|\ \ \ \ \ \ \|}{MeCCH_2CH_2C-}}$ | $\overset{O}{\underset{\|}{CH_2=CH(CH_2)_2C-}}$ |
| G-(N-2); A (N-6); C (N-4) | G-(N-2); A (N-6); C (N-4) | G-(N-2); A (N-6); C (N-4) |
| Deblk: conc. NH₄OH | Deblk: 0.5 M NH₂-NH₂.H₂O/ pyr/AcOH (4:1) | Multiple deprot. |

equilibration to the N-7 protonated species (**62**) and (**63**), followed by cleavage of the glycosidic bond to give (**65**) via the formation of the oxonium ion (**64**).[278–281] The depurination occurred only at very low pH (pKₐ −1.48). Interestingly, the presence of the 2′-OH group had a dramatic effect on the depurination. Thus, for example, guanosine and adenosine were more resistant to depurination compared with deoxyadenosine and deoxyguanosine. Unfortunately and unexpectedly, the corresponding *N*-acyl-protected purine nucleosides were more prone towards depurination compared with the unprotected nucleosides. It appears that in the case of the corresponding acyl-protected purine nucleosides, the site of protonation occurs at N-7 making this N-7 protonated species readily available for glycosidic cleavage. For example, the N-6 benzoyl dA is known to depurinate about 10 times more rapidly compared with N-2 isobutyryl dG. Furthermore, depurination occurred to a greater extent during the preparation of longer oligonucleotides because of repetitive exposure of the growing oligonucleotide to acid during detritylation. The rate of depurination and the attendant apurinic side chain breakage was dependent on the position of the purine linkage within the chain. Recent mechanistic studies of the depurination phenomenon suggest that the rate of depurination of purines was several times faster at terminal sites than at internal sites in an oligonucleotide chain.[282]

    To minimize depurination in a growing oligonucleotide chain, alternate protecting groups have been evaluated. The amidine-protected nucleosides (Table 3)[58,161,222,257,283–293] were reportedly ca. 20-fold more resistant to depurination than their benzoyl counterpart.[58,161,222,257,283–291] Presumably, the site of protonation was shifted to the N-1 and N-6 positions in the case of amidine-protected nucleosides instead of N-1 and N-7. The corresponding nucleoside monomers have been employed in the solid-phase synthesis of oligonucleotides. It was reported that during oligonucleotide synthesis,

**Table 2** (continued)

| | | |
|---|---|---|
| Ref. 170,171<br> <br>G (N-2); A (N-6); C (N-4)<br>Deblk: 10% Pd/C/$H_2$ | Ref. 150,151,155, 160,163,239–244<br> <br>G (N-2); A (N-6); C (N-4)<br>Deblk: DBU or DBU/pyr | Ref. 190<br> <br>R =H; Cl; $NO_2$<br>G (N-2); A (N-6); C (N-4)<br>Deblk: Tetramethylguanidine/ morpholine/pyr or DBU/morpholine/ pyr or $NH_4OH$/morpholine/pyr. |
| Ref. 137,139,141,146, 147,215,245–248<br> <br>R = H, C(Me)$_3$<br>G (N-2)<br>Deblk: 5M $NH_3$/ MeOH | Ref. 144,145,149, 209,249–253<br> <br>G (N-2); A (N-6); C (N-4)<br>Deblk: conc. $NH_4OH$/pyr or $Et_3N$/pyr | Ref. 206,254,255<br> <br>G (N-2)<br>Deblk: conc. $NH_4OH$, heat |
| Ref. 256<br> <br>G (N-2); A (N-6); C (N-4)<br>Deblk: conc. $NH_4OH$, heat | Ref. 216,218,219, 223–226, 257,258<br> <br>R = H, Cl<br>G (N-2); A (N-6)<br>Deblk: conc. $NH_4OH$ | Ref. 215,259,260<br> <br>G (N-2)<br>Deblk: 0.2 M NaOH/MeOH (1:1), or conc. $NH_4OH$, or $NH_3$ gas. |
| Ref. 137,146,147,201–205,208, 212, 215,220,221,245,246,261–266<br> <br>R = H, OMe, Cl, $NO_2$, $NMe_2$, $CMe_3$<br>G (N-2); A (N-6); C (N-4)<br>Deblk: conc. $NH_4OH$ | Ref. 206,207,215<br> <br>C (N-4)<br>Deblk: 0.2 M NaOH/MeOH (1:1), or conc. $NH_4OH$/Dioxan (4:1), heat | Ref. 215<br> <br>R = Me, OMe<br>C (N-4)<br>Deblk: 0.2 M NaOH/MeOH (1:1) |
| Ref. 246,262,267<br> <br>A (N-6)<br>Deblk: conc. $NH_4OH$ | Ref. 215<br> <br>G (N-2)<br>Deblk: 0.2 M NaOH/MeOH (1:1) | Ref. 268,269<br> <br>G (N-2); A (N-6); C (N-4)<br>Deblk: conc. $NH_4OH$, heat |

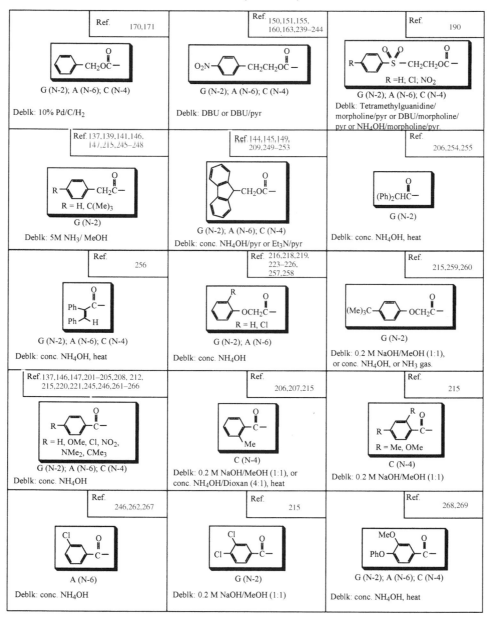

*Continued overleaf*

a dilute solution of iodine had to be employed during the oxidation step to prevent base modification.[293] The imide-type protecting groups, (**66**) and (**67**), as well as the diamide-type protecting groups (**68**) (Figure 4), have been suggested as alternatives.[49,294,295] Although resistant to depurination, these groups were somewhat labile to pyridine-$H_2O$, a reagent combination used during the oxidation step. Certain other protecting groups such as *o*-nitrophenylsulfonyl and tris-benzoyloxytrityl (**69**) groups also appear to be more resistant to depurination.[197,296–298] However, the coupling efficiency of the derived phosphoramidites was reduced compared with other amidites.[299] Deoxyadenosine protected at N-6 with α-phenylcinnamoyl, naphthoyl, 3-methoxy-4-phenoxy-benzoyl, fluorenylmethoxy carbonyl (FMOC), and *t*-PAC (see Table 2) were reportedly less susceptible towards depurination compared with N-6 benzoyl.[108]

Enzyme-mediated site-selective incorporation and removal of a number of protecting groups have been reported.[300] Oligonucleotide synthesis with "no protection option" has also been explored.[301–305]

A trend which has begun to emerge in the context of the "genomics era" is the need for rapid

**Table 2**   (continued)

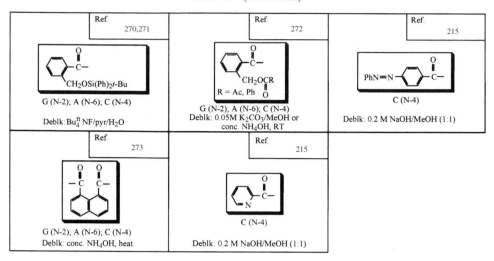

| Ref. 270,271 | Ref. 272 | Ref. 215 |
|---|---|---|
| G (N-2); A (N-6); C (N-4) <br> Deblk: $Bu_4^n NF/pyr/H_2O$ | G (N-2); A (N-6); C (N-4) <br> Deblk: 0.05M $K_2CO_3$/MeOH or <br> conc. $NH_4OH$, RT | C (N-4) <br> Deblk: 0.2 M NaOH/MeOH (1:1) |
| Ref. 273 | Ref. 215 | |
| G (N-2); A (N-6); C (N-4) <br> Deblk: conc. $NH_4OH$, heat | C (N-4) <br> Deblk: 0.2 M NaOH/MeOH (1:1) | |

**(62)**

**(61)**

**(63)**

**(64)**

**(65)**

**Scheme 10**

access to oligodeoxy- and oligoribonucleotides. In turn, this has stimulated the search for protecting groups that are less robust to deprotection. For example, the PAC, *t*-PAC, and formamidine-protected nucleoside phosphoramidites have been used for the rapid synthesis of oligonucleotides. Thus following the oligonucleotide assembly, the PAC groups could be rapidly removed under mildly basic conditions (28% $NH_4OH$, RT),[259] or gaseous amines under pressure.[260] To minimize transacylation during the capping step of oligonucleotide synthesis, *t*-phenoxyacetic anhydride was employed for capping.[259] A PNT moiety has been introduced as a nucleoside protecting group for rapid synthesis of oligonucleotides.[237]

In solid-phase synthesis, additional factors such as the nature of solid-support, the washing cycle, and the composition of the deblock solution, the washing solvent, the synthesis cycle times, etc., seemed to influence further the depurination phenomenon. It was noted that the use of 15% dichloroacetic acid (DCA) in methylene chloride was ideal as a deblocking reagent which induced

**Table 3** Amidine-type protecting groups.

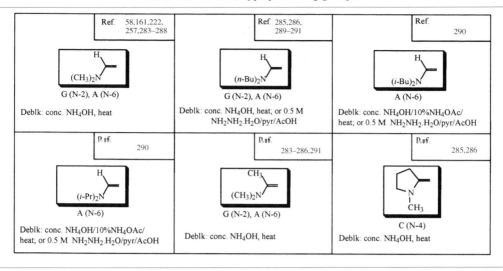

| Ref. 58,161,222, 257,283–288 | Ref. 285,286, 289–291 | Ref. 290 |
|---|---|---|
| G (N-2), A (N-6) | G (N-2), A (N-6) | A (N-6) |
| Deblk: conc. NH₄OH, heat | Deblk: conc. NH₄OH, heat; or 0.5 M NH₂NH₂.H₂O/pyr/AcOH | Deblk: conc. NH₄OH/10%NH₄OAc/ heat; or 0.5 M NH₂NH₂.H₂O/pyr/AcOH |
| Ref. 290 | Ref. 283–286,291 | Ref. 285,286 |
| A (N-6) | G (N-2), A (N-6) | C (N-4) |
| Deblk: conc. NH₄OH/10%NH₄OAc/ heat; or 0.5 M NH₂NH₂.H₂O/pyr/AcOH | Deblk: conc. NH₄OH, heat | Deblk: conc. NH₄OH, heat |

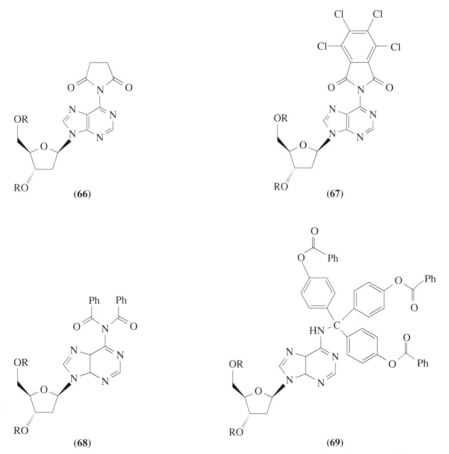

**Figure 4** Some examples of N-protection which reduces depurination of dA.

minimal depurination.[306] It is also pertinent to mention that with the modern-day synthesizers, the synthesis programs are being continuously optimized to minimize depurination. Thus, in spite of the menace of depurination, large-scale routine synthesis of oligonucleotides (25- to 30-mers), has been accomplished using acyl protecting groups.[307]

### 7.05.3.2    Protection of the 5′-Hydroxyl Group

The monomeric synthons used in solid-phase synthesis carry a 5′-protected nucleoside. The 5′-protecting group is removed just prior to the coupling reaction in each synthesis cycle. One of the objectives in designing this protecting group was that once released from the 5′-OH group, it would show up as a "colored species." This way one could have a visual monitor or a more quantitative colorimetric assay of the efficiency of the coupling step that just preceded it. The protecting groups of the family of triphenylmethyl derivatives (trityl),[212] (70)–(73), or of its fused tricyclic counterparts (xanthenyl),[204,208,209,245,246] (74) and (75), have remained most popular because of the bright red-orange to yellow-orange color of their carbo-cations in the presence of acid.

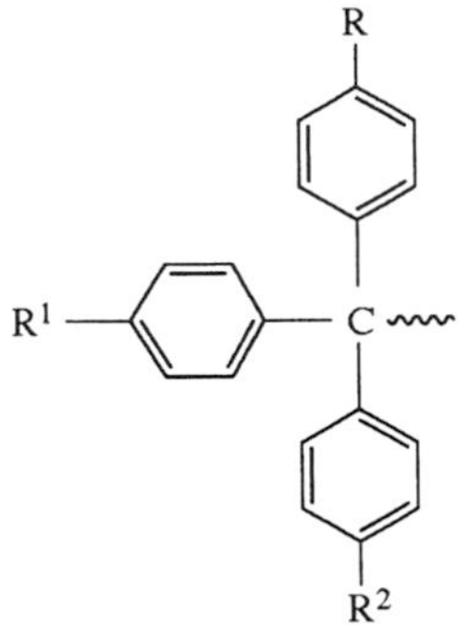

(70)  R, R$^1$, R$^2$ = H
(71)  R,R$^1$ = H; R$^2$ = OMe (MMTr)
(72)  R = H; R$^1$, R$^2$ = OMe (DMTr)
(73)  R, R$^1$, R$^2$ = OMe (TMTr)

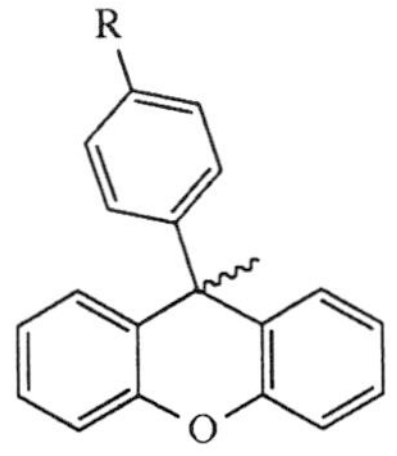

(74)  R = H (Pixyl, Px)
(75)  R = OMe (Moxyl, Mox)

Additionally, their susceptibility towards acid-catalyzed removal could be fine-tuned by the choice of ring substituents, which stabilize the incipient carbo-cation (e.g., electron-releasing methoxy groups). Each *p*-methoxy substituent enhanced the rate of cleavage by a factor of 10.[308] Thus the MMTr and DMTr groups were developed as 5′-hydroxyl protecting groups.[220,308,309] Furthermore, being sterically hindered, these groups could also be selectively introduced at the 5′-hydroxyl group with minimal reaction with the more hindered 3′- or 2′-OH group of the nucleoside. Additionally, in the context of purification of the full-length oligonucleotide, the hydrophobic tether of the 5′-trityl, the 5′-pixyl or the 5′-*S*-pixyl[267] allowed the retention of the "full-length oligonucleotide" on the hydrophobic matrix of the C-18 silica column. These combined attributes have made the DMTr, and to a lesser extent, the 9-phenylxanthen-9-yl (pixyl) or the 9-*p*-methoxyphenylxanthen-9-yl widely used 5′-OH protecting groups.

In addition to the protic acids such as dichloro- and trichloroacetic acids, Lewis acids such as BF$_3$-etherate,[310] zinc bromide,[311,312] AlCl$_3$,[28,29] TiCl$_4$,[28,29] (Et)$_2$AlCl,[313] and (*i*-Bu)$_2$AlCl[313] have been reported to effect the removal of the trityl group.[289] Ceric ammonium nitrate,[314] 1,1,1,3,3,3-hexa-fluoro-2-propanol,[315] and diethyloxomalonate[316] have been developed for the removal of the trityl group.

Several ingenious variations of the trityl protecting groups have also been developed. Thus, Fisher and Caruthers[317] used the DMTr groups for dG, the *p*-anisyl-1-naphthylphenylmethyl (76) for T, the *o*-anisyl-1-naphthylmethyl (77) for dC, and the *p*-tolylmethyl (78) for dA as "color coded" markers—which gave orange, red, yellow, and blue colors, respectively under acidic conditions—and could be used to ascertain the coupling efficiency of individual nucleosides. These were especially useful during the preparation of mixed-sequence probes. In another application, the attachment of an imidazole tether to the trityl group (79) was found to provide rate enhancement of the coupling reaction in the phosphotriester approach.[318,319]

As an alternative to the acid-catalyzed removal of the trityl group, nucleosides protected at the 5′-OH groups with the 4,4′,4″-tris-benzoyloxy trityl (80),[320] the 4,4′,4″-tris-levulinyloxy trityl (81),[321] and the 4,4′,4″-tris-(4,5-dichlorophthalimido) trityl (82)[322,323] have been reported (Figure 5). Trityl groups carrying a single levulinyl ester functionality have been reported for a number of applications.[324,325] Nucleoside hydroxys functionalized as the trityl-levulinyl esters were deprotected using a neutral hydrazine reagent and the liberated tritylphthalazines visualized under acidic con-

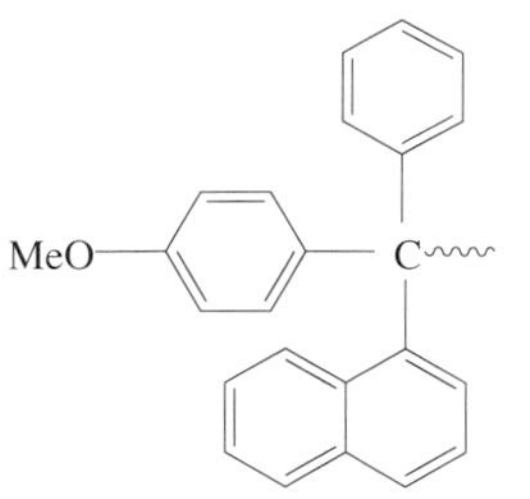

(76) 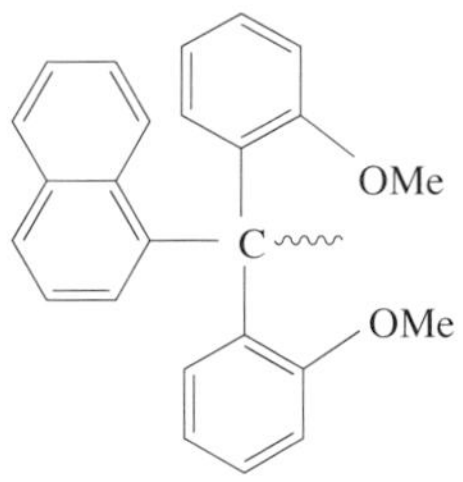          (77) 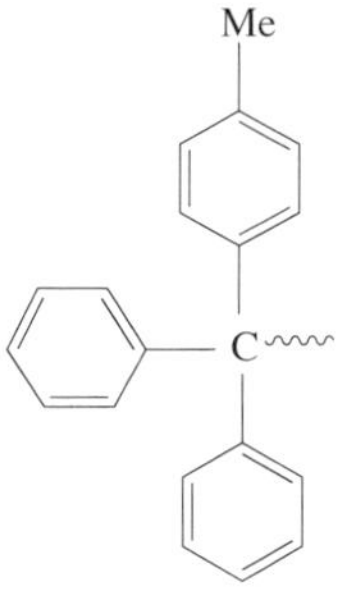          (78)

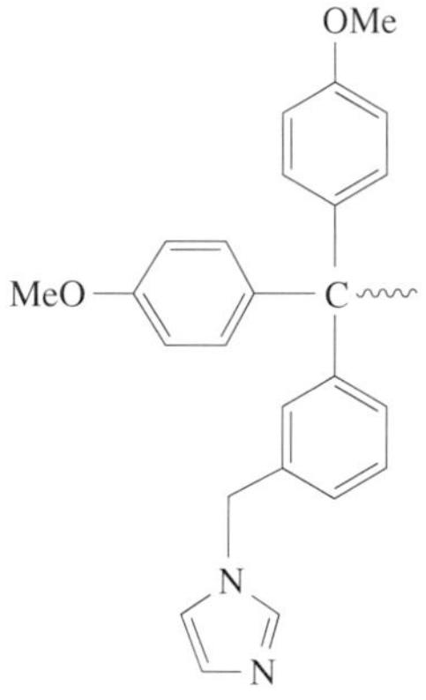

(79)

ditions during oligonucleotide synthesis. Happ and Scalfi-Happ[326] used hybrid FMOC-trityl groups (83) and (84) for the protection of the 5'-OH, the deprotection being effected by DBU-induced β-elimination of the FMOC group accompanied by the cleavage of the trityl group. Seliger *et al.*[327] have investigated *p*-(phenylazo)phenyloxycarbonyl (PAPoc) (85) as a 5'-protecting group (Figure 6). This group could be removed using β-cyanoethanol/(Et)$_3$N/H$_2$O followed by DBU/pyridine/DMAP. Short oligonucleotides have been prepared using this group. The use of the FMOC group for the protection of 5'-OH groups led to the simultaneous development of the sarcosine linker[74] *in lieu* of the succinyl linker (see above) for attachment of the leader nucleoside to the solid support. However, under the conditions used for the removal of the FMOC group, the potential for premature removal of the β-cyanoethyl phosphate protecting group (see below) does exist during solid-phase synthesis although its adverse impact, if any, has not been ascertained. Pfleiderer and co-workers[150,151,328,329] employed the 2,4-dinitrophenylethoxycarbonyl group (86) and the 2-dansylethoxycarbonyl group for 5'-OH protection which can be selectively removed using triethylamine or DBU. Other examples include the 4-(methylthiomethoxy)butyryl (87), 2-(methylthiomethoxymethyl)benzoyl (88), 2-(isopropylthiomethoxymethyl)benzoyl (89), and 2-(2,4-dinitrobenzene sulfenyloxymethyl)benzoyl (90) groups.[137,146,330–333] The presence of the sulfide linkage facilitated the cleavage of these groups by mercury(II) perchlorate or *p*-toluenethiol.

In addition to DMTr and Px groups, trityl groups carrying hydrophobic moieties have been incorporated into nucleosides. Thus, 4-decyloxytrityl, 4-hexadecyloxytrityl, 9-(4-octadecyloxy-phenyl)xanthen-9-yl, and the 1,1-bis(4-methoxyphenyl)-1-pyrenylmethyl groups have been used to protect the 5'-OH group. When incorporated at the 5'-terminus of the oligonucleotide, the hydrophobic group facilitates the retention of the full-length oligonucleotide during purification by reversed-phase HPLC, thus aiding in its separation from the truncated sequences.[334–338] The trityl group carrying the highly hydrophobic polycyclic aromatic hydrocarbon 4-(17-tetrabenzo[*a,c,g,i*] fluorenylmethyl) group (91) was also developed as a hydrophobic handle to aid the purification of oligonucleotides by reversed-phase HPLC.[339,340]

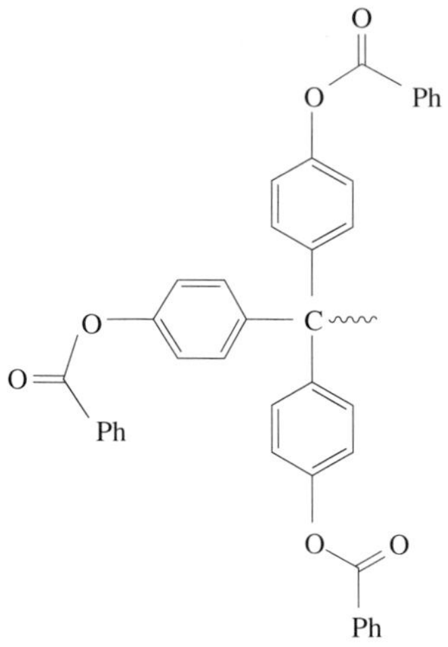

**(80)** 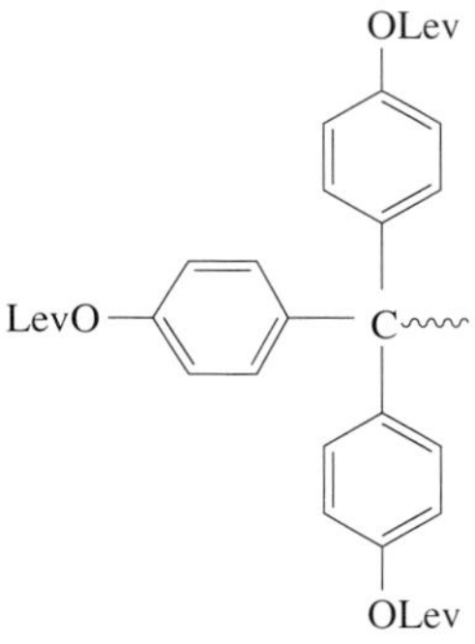 **(81)** 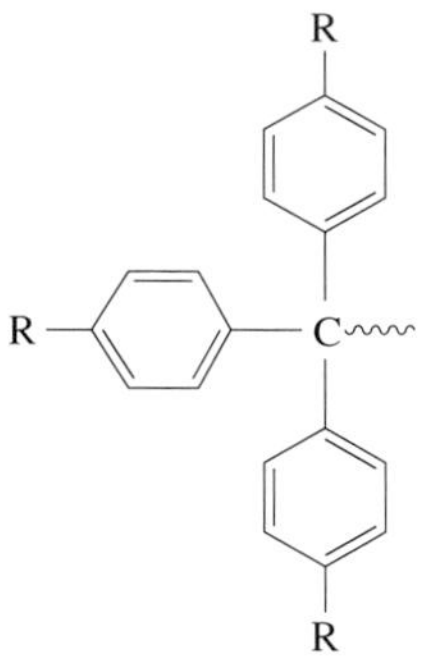 **(82)**

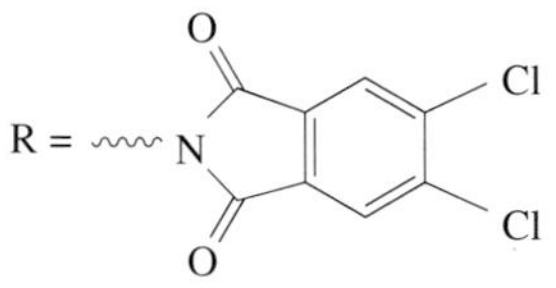

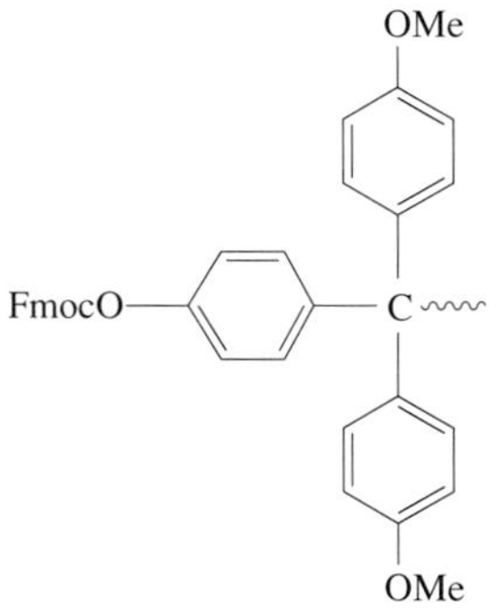

**(83)** 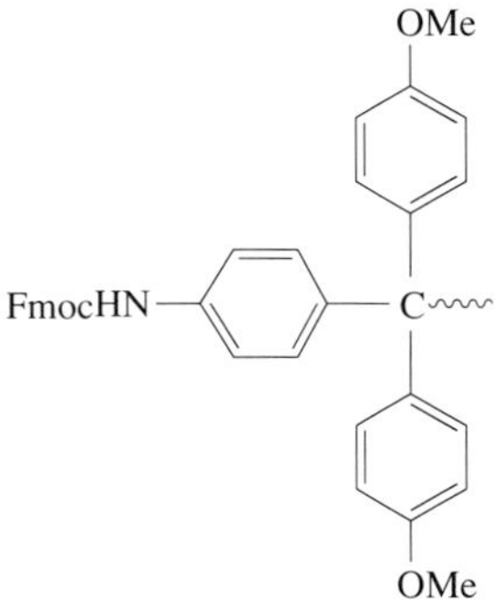 **(84)**

**Figure 5** Examples of base-labile trityl derivatives used for 5′-OH protection.

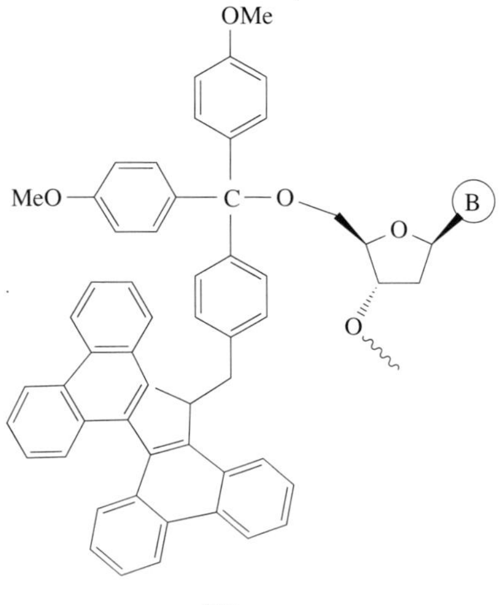

**(91)**

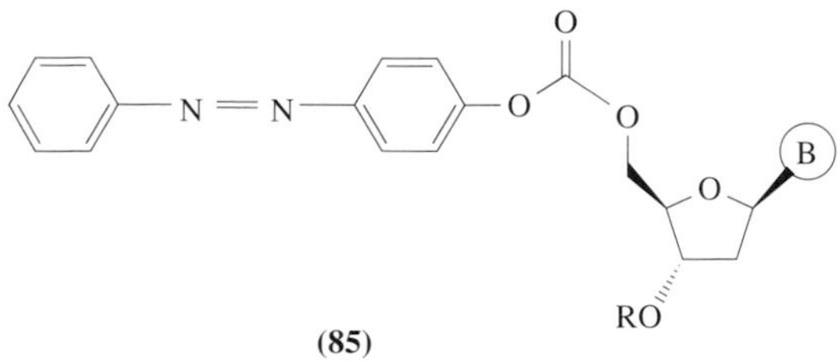

(85)

5'-OH protected by
*p*-(phenyl azo)phenyloxycarbonyl group

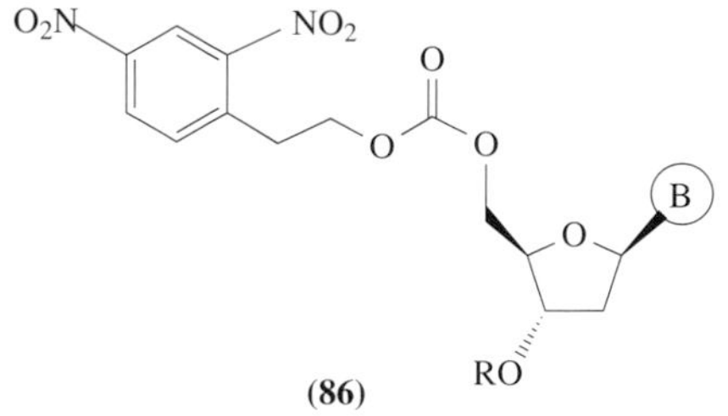

(86)

5'-OH protected by
2,4-dinitrophenyl ethoxycarbonyl group

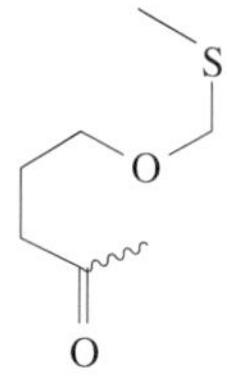

(87)

4-(methylthiomethoxy)-
butyryl

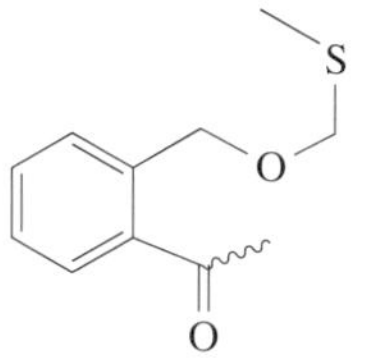

(88)

2-(methylthiomethoxymethyl)-
benzoyl

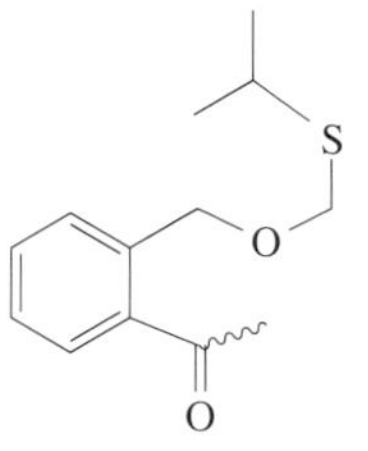

(89)

2-(isopropylthiomethoxymethyl)-
benzoyl

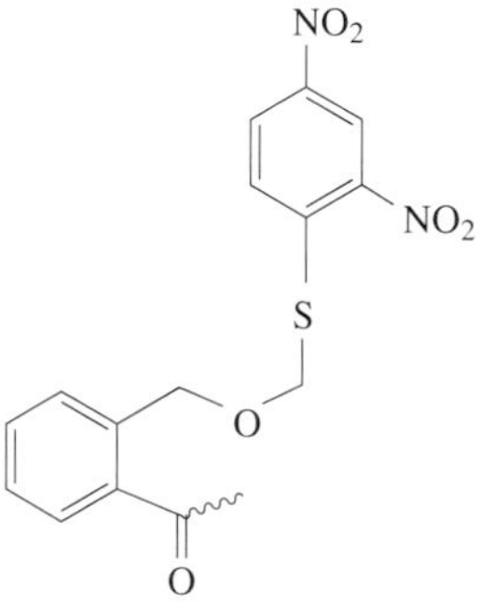

(90)

2-(2,4-dinitrobenzene
sulfenyloxymethyl
benzoyl

**Figure 6** 5'-OH protection by sulfur-containing protecting groups.

The utility of photochemically removable protecting groups at the 5'-position of nucleosides has been demonstrated. Thus, 5'-OH groups of nucleosides derivatized by dimethoxybenzoin (92) were prepared and were found to undergo deprotection by irradiation at $\lambda_{350}$ nm within 1 h (Equation (5)).[341] The derived 3'-phosphoramidites were employed in oligonucleotide synthesis (see below) where the deblocking step in chain elongation consisted of irradiation followed by coupling.

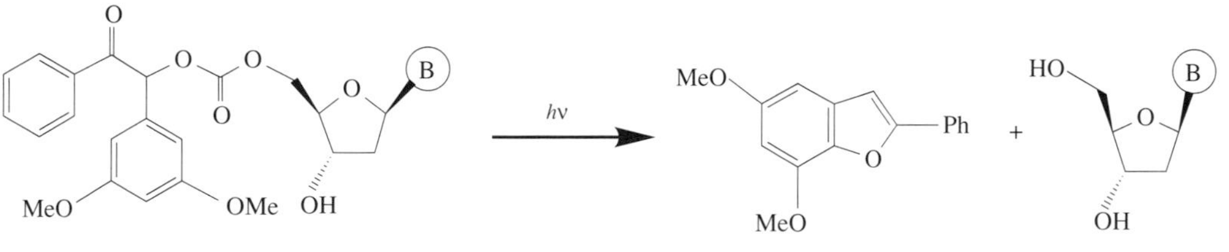

(5)

(92)

These protecting groups were also developed as phosphate protecting groups (see Section 7.05.1.3.3) and used in oligonucleotide synthesis by the phosphotriester approach.[342,343] In a related study,[344] (hydroxystyryl)dimethylsilyl and (hydroxystyryl)diisopropylsilyl were developed as photochemically removable silyl protecting groups of nucleosides (93) (Scheme 11). Deprotection was achieved within 30 min by irradiation at 254 nm and follows the presumed course delineated in Scheme 11.

The trityl group derivatized with a *p*-benzoyl benzoate moiety was used as a "phototrityl group" to cap an oligonucleotide at the 5'-end.[345] Upon irradiation in the presence of bovine serum albumin, the oligonucleotide was covalently linked to the protein. The protein conjugate hybridized to a complementary oligonucleotide.

These innovations apart, for routine oligonucleotide synthesis, the DMTr group has remained the protecting group of choice for the protection of the 5'-hydroxyl group, in the synthesis of oligonucleotides.

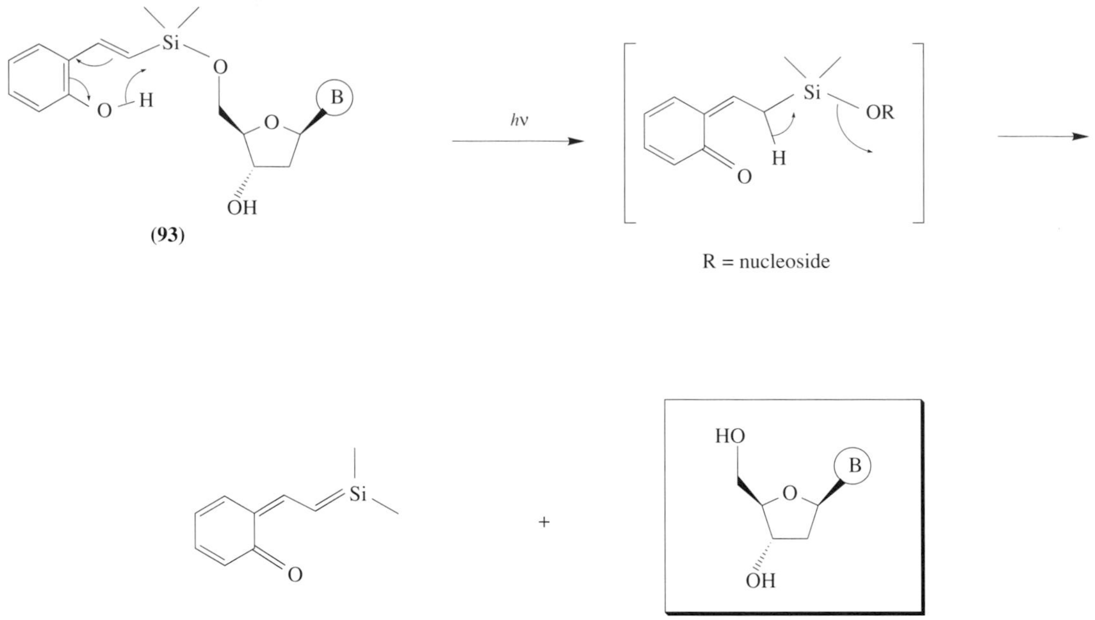

(93)

R = nucleoside

+

**Scheme 11**

### 7.05.3.3  Protection of the Phosphate: Formation of the Internucleotidic Linkage by the Phosphoramidite Approach

The protection of the phosphate function is linked to the coupling chemistry and is therefore discussed together in this section.

As stated in Section 7.05.1.1, Letsinger and Lunsford[27] pioneered the phosphite coupling methodology in the 1970s (Scheme 12). This led to the development of a number of phosphate protecting groups: 2,2,2-trichloroethyl,[346–349] 2,2,2-tribromoethyl,[346,350] β-cyanoethyl,[346,351,352] benzyl,[3,346,353] methyl,[132,346,353,354] p-chlorophenyl,[346,355,356] and p-nitrophenylethyl,[346,357] to name but a few. It was found that the unwanted formation of the 3'-3' product (96) and the 5'-5' product (97) could be minimized by increasing the steric bulk of the chlorophosphite unit, in turn leading to the development of 2,2,2-trichloro-1,1-dimethylethyl phosphorodichloridite (TCDME)[358–360] or 2-cyano-1,1-dimethylethyl phosphorodichloridite[361] as phosphitylation reagents. Following coupling of the suitably protected nucleosides (94), the oxidation of the internucleotidic phosphite linkage was carried out with $I_2$ (Scheme 12) to yield the phosphotriester (95). The 2-cyano-1,1-dimethylethyl phosphate protecting group was easily removed using 0.2 M NaOH in dioxane/methanol and similarly the TCDME group by treatment with tributylphosphine. Subsequently, groups led by Matteucci and Caruthers[311] and Jayaraman and McClaugherty[362] investigated this approach in the solid-phase synthesis of oligonucleotides. The addition of triazole or 1H-tetrazole accelerated the coupling step and also suppressed the formation of the by-products.[28,29] The inherent limitations of this approach were: (i) the sensitivity of the chlorophosphite intermediates to moisture, (ii) the formation of the undesired 3'-3'-coupled by-products, and (iii) the necessity to carry out the reaction at low temperature. Thus, the phosphite methodology clearly needed further refinement before automation could be achieved.

To address these limitations, Beaucage and Caruthers[30] pioneered the development of deoxynucleoside phosphoramidites (98) as a class of relatively stable phosphitylating reagents which were prepared as shown in Equation (6). Known as the phosphoramidite methodology in DNA synthesis, the initial approach (Scheme 13) entailed the reaction of 5'-DMTr-deoxynucleoside phosphoramidite (98) with the 5'-hydroxyl group of a nucleoside (99) using N,N-dimethylaniline hydrochloride as the "activator." Presumably, the reaction proceeded via the chlorophosphite intermediate (100) to give the dinucleoside phosphite (101). Subsequently, 1H-tetrazole was introduced as the activator.[30] The internucleotidic coupling efficiencies ranged from 85 to 100% with

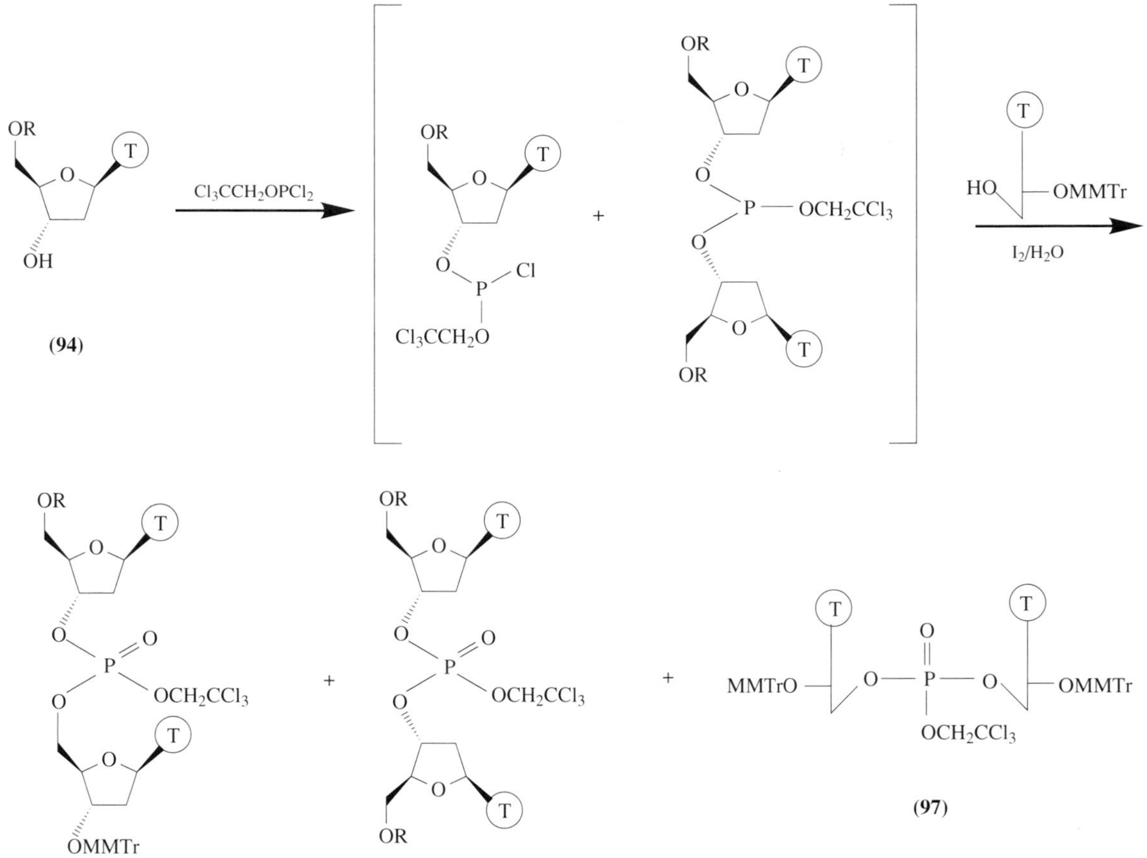

**Scheme 12**

coupling times of less than 5 min at room temperature. The methylphosphate protecting groups were removed by demethylation using thiophenol.[30,353] Later, 2-thiobenzothiazole was developed as a demethylation reagent.[132] This approach was applied to the solid-phase synthesis of oligonucleotides up to 45 bases long.

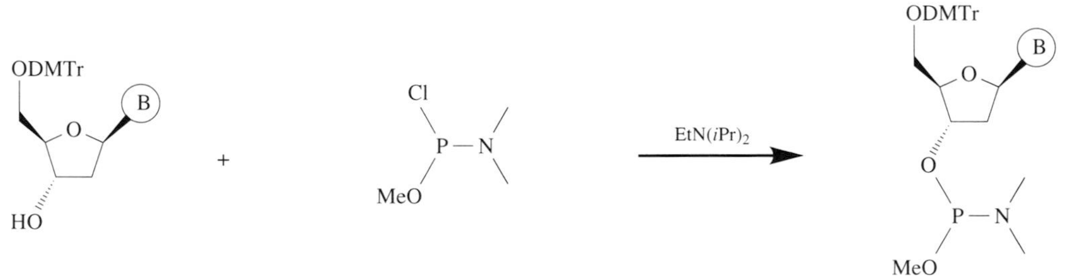

B = Thy, Ade^Bz, Gua^iBu, Cyd^Bz

McBride and Caruthers[363] later investigated a range of substituted phosphoramidites (**102**)–(**105**). The intent was to improve the solution stability of the phosphoramidites. The more hindered *N,N*-diisopropylphosphoramidites (**102**) were stable in solution for extended periods of time and have emerged as the popular class of nucleoside phosphoramidites.

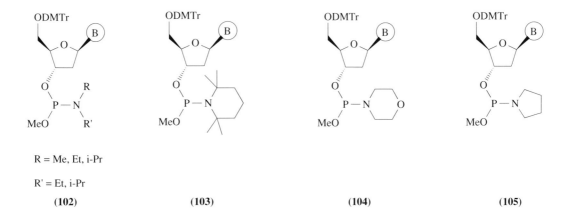

**(98)**                    **(100)**

**(99)**

2,4,6-collidine

**(101)**

**Scheme 13**

R = Me, Et, i-Pr

R' = Et, i-Pr

**(102)**              **(103)**              **(104)**              **(105)**

Fourrey and Varenne[364,365] and Tanaka *et al.*[366] reported alternative methods for the preparation of the phosphoramidite (**108**) from the triazolide (**107**), in turn derived from the bistriazolylphosphite (**106**) or dichlorophosphoramidites (Scheme 14).

Although methoxyphosphoramidites were introduced early on as a stable class of phosphoramidites, the removal of the methyl phosphate protecting groups required the use of the noxious thiophenol[30] or the less noxious 2-thiobenzothiazole.[132] Sinha *et al.*[367,368] introduced β-cyanoethylphosphoramidites (**109a,b**) as an improved class of phosphoramidite reagents. In this case, the cyanoethyl group could be removed by β-elimination under the basic conditions employed

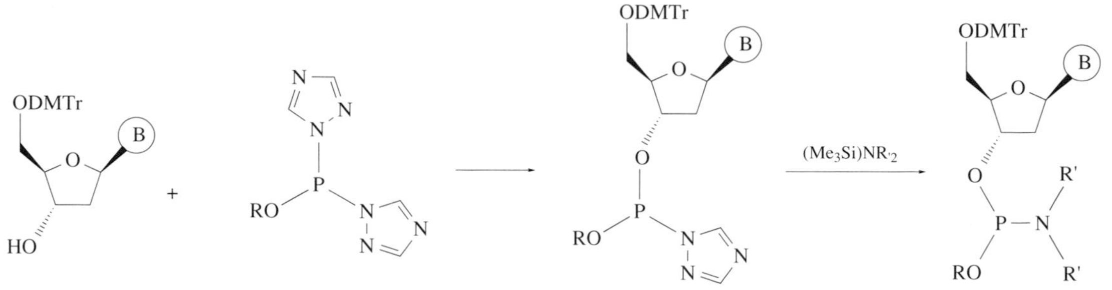

**(106)**                    **(107)**                    **(108)**

R = Me
R = (CH$_2$)$_2$CN
R = (CH$_2$)$_2$SO$_2$Ph
R = *o*-ClPh

**Scheme 14**

in the removal of the nucleobase protecting groups and the cleavage of the oligonucleotide from the support. A range of other nucleoside phosphoramidites incorporating various phosphate protecting groups and different types of *N*-groups have been evaluated[369–383] in internucleotidic coupling reactions and are listed in Table 4. Nucleoside oxazaphospholidines have been introduced as synthons in the synthesis of oligonucleotides[384] and the related chiral oxazaphospholidines were explored in the stereoselective approach towards oligonucleoside phosphorothioates.[385] Oxathiaphospholanes were used by Stec and co-workers as novel reagents for the stereoselective synthesis of phosphorothioates.[386–388]

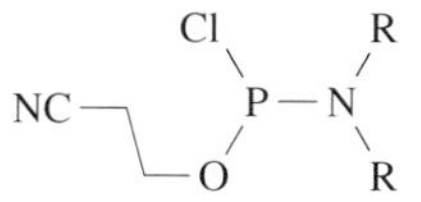

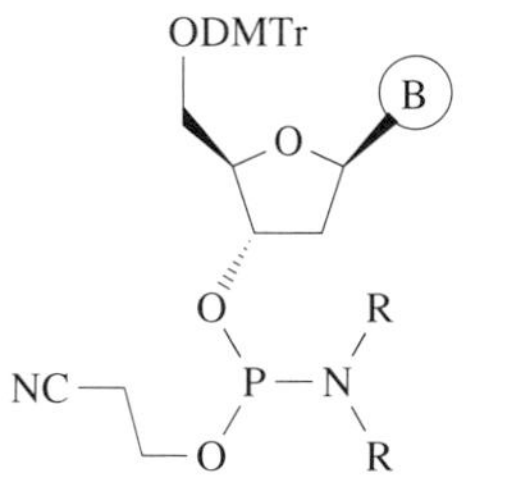

B = Thy, Cyd$^{Bz}$, Ade$^{Bz}$, Gua$^{iBu}$

**(109a)**      R = *i*-Pr
**(109b)**      NR$_2$ = morpholino

In all cases, the monomeric phosphoramidites had to be isolated and purified by chromatography. As an alternative, bis-amidite reagents **(110)** were prepared as stable reagents and were found to undergo stepwise activation by 4,5-dichloroimidazole to form the corresponding imidazole monoamidites **(111)**. The imidazole amidite could be coupled to a nucleoside to form the nucleoside phosphoramidite **(102)** ready for use in DNA synthesis (Scheme 15).[389] Alternatively, for the stepwise activation of phosphoramidites, diisopropyltetrazolides were employed.[390,391] Importantly, the formation of the undesired 3′-3′-phosphites was minimal under these conditions. A variety of such bis-amidite reagents have been reported and are listed in Table 5.

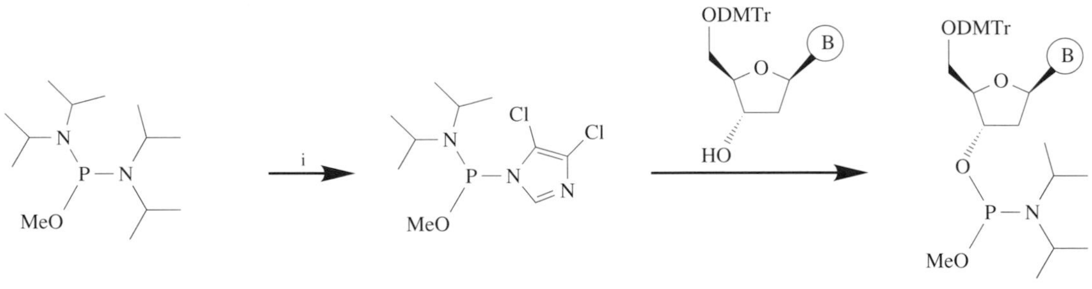

**(110)**               **(111)**               **(102)**

i, 4,5-dichloroimidazole; alternative reagents include tetrazole and diisopropylammonium tetrazolide

**Scheme 15**

**Table 4**    Nucleoside phosphoramidites.

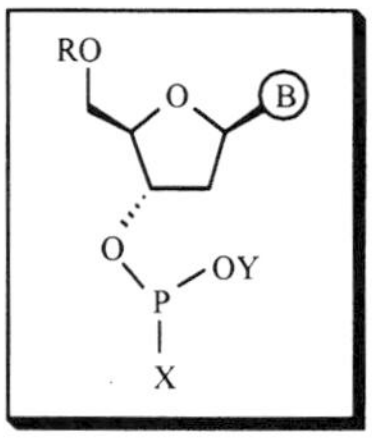

| Ref. 370 | Ref. 370 | Ref. 370 |
|---|---|---|
| $X = -N$ (cyclododecyl ring)    $Y = -Me$ | $X = -N$ (bicyclic ring)    $Y = -Me$ | $X = -N$ (cyclohexyl, isopropyl)    $Y = -Me$ |
| **Ref. 374** | **Ref. 138,375** | **Ref. 376** |
| $X = -N\underset{}{\bigcirc}O$ (morpholino)    $Y = -C(Me)_2CH_2CN$ | $X = -N(Me)_2$    $Y = -C(Me)_2CCl_3$ | $X = -N(CH_2CH_3)_2$    $Y = -C(Me)_2CCl_3$ |
| **Ref. 375** | **Ref. 375** | **Ref. 375** |
| $X = -N(\overset{Me}{CHMe})_2$    $Y = -CMe_2CCl_3$ | $X = -N\underset{}{\bigcirc}O$ (morpholino)    $Y = -C(Me)_2CCl_3$ | $X = -N(Me)_2$    $Y = -CH_2CCl_3$ |
| **Ref. 373** | **Ref. 232,378** | **Ref. 377** |
| $X = -N(\overset{Me}{CHMe})_2$    $Y = -CH_2\text{-}CH_2\text{-}\langle\rangle\text{-}NO_2$ | $X = -N(\overset{Me}{CHMe})_2$    $Y = -CH_2\text{-}CH=CH_2$ | $X = -N(\overset{Me}{CHMe})_2$    $Y = -CH_2CH_2SiMe_3$ |

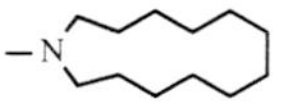

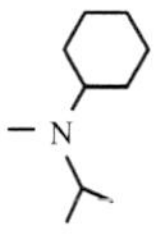

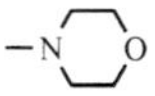

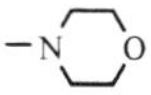

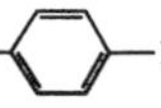

**Table 4** (continued)

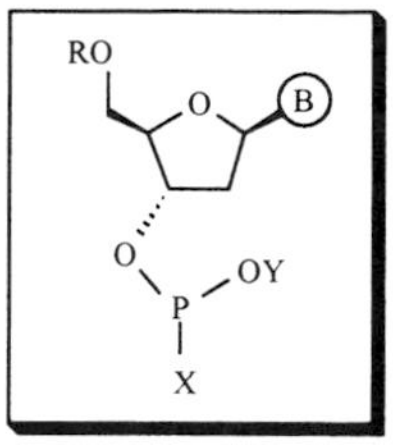

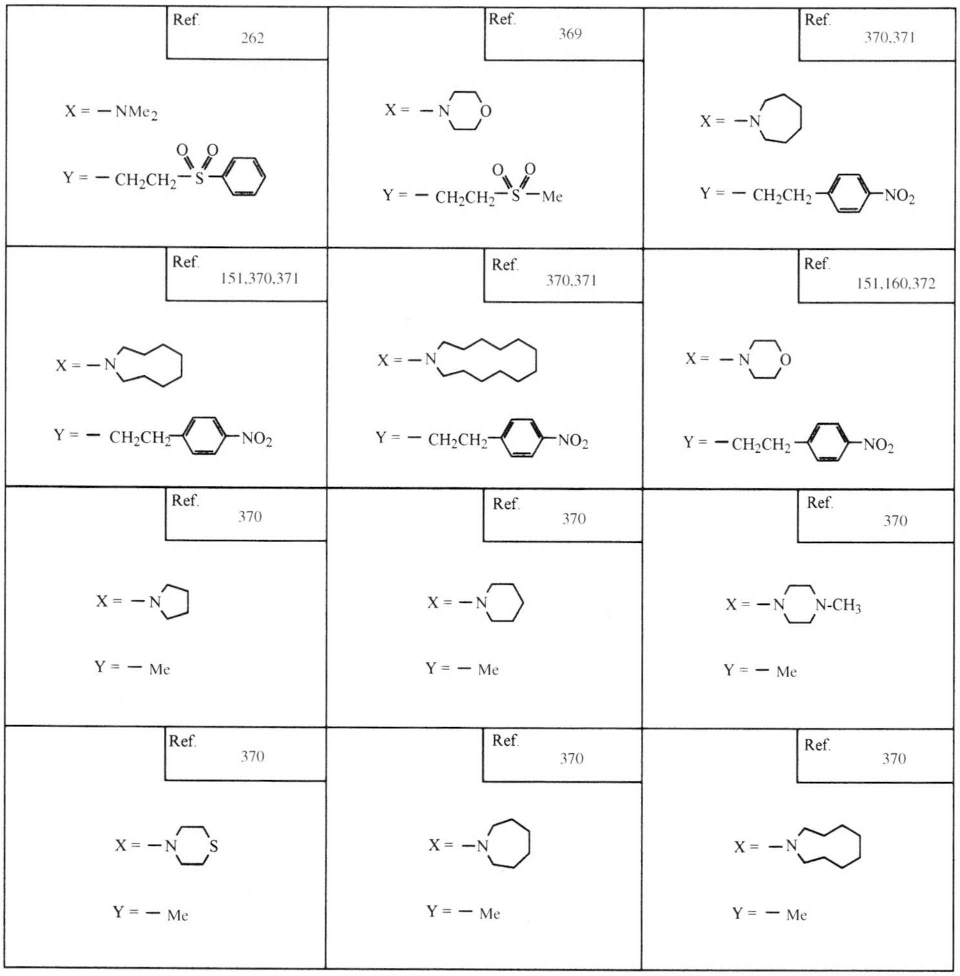

| Ref. 262 | Ref. 369 | Ref. 370,371 |
|---|---|---|
| X = —NMe$_2$ <br><br> Y = —CH$_2$CH$_2$—S(O)(O)—Ph | X = —N⟮morpholine⟯O <br><br> Y = —CH$_2$CH$_2$—S(O)(O)—Me | X = —N⟮azepane⟯ <br><br> Y = —CH$_2$CH$_2$—⟮C$_6$H$_4$⟯—NO$_2$ |
| **Ref. 151,370,371** | **Ref. 370,371** | **Ref. 151,160,372** |
| X = —N⟮ring⟯ <br><br> Y = —CH$_2$CH$_2$—⟮C$_6$H$_4$⟯—NO$_2$ | X = —N⟮ring⟯ <br><br> Y = —CH$_2$CH$_2$—⟮C$_6$H$_4$⟯—NO$_2$ | X = —N⟮morpholine⟯O <br><br> Y = —CH$_2$CH$_2$—⟮C$_6$H$_4$⟯—NO$_2$ |
| **Ref. 370** | **Ref. 370** | **Ref. 370** |
| X = —N⟮pyrrolidine⟯ <br><br> Y = —Me | X = —N⟮piperidine⟯ <br><br> Y = —Me | X = —N⟮piperazine⟯N-CH$_3$ <br><br> Y = —Me |
| **Ref. 370** | **Ref. 370** | **Ref. 370** |
| X = —N⟮thiomorpholine⟯S <br><br> Y = —Me | X = —N⟮azepane⟯ <br><br> Y = —Me | X = —N⟮ring⟯ <br><br> Y = —Me |

*Continued overleaf*

In addition to nucleoside phosphoramidite monomers, the dinucleotide phosphoramidites (**112**) have been prepared and used in solid-phase oligonucleotide synthesis.

Concurrent with the development of *O*-alkyl phosphate protecting groups, a range of *O*-aryl/ *O*-heteroaryl protected phosphoramidites (**113a–j**) have been prepared and employed in internucleotidic coupling reactions (Scheme 16). An added versatility was that the dinucleotide *O*-aryl phosphites could be converted readily to the corresponding *H*-phosphonates. Alternatively, the derived nucleotide (**114**) (also prepared from the bis-amidites) could be converted to the corresponding *H*-phosphonates (**115**).[255] Furthermore, the *H*-phosphonates could be oxidatively converted to an array of phosphomodified analogues, such as amidates, triesters, thioates, etc.[33] Such a mixed phosphoramidite-*H*-phosphonate strategy has been useful for radiolabeling of oligonucleotides.[401]

*O*-Allyl-protected phosphoramidites developed by Hayakawa and co-workers have been employed in oligonucleotide synthesis. The *O*-allyl group could be removed by Pd[PPh$_3$]$_4$.[168,231–234]

**Table 4**   (continued)

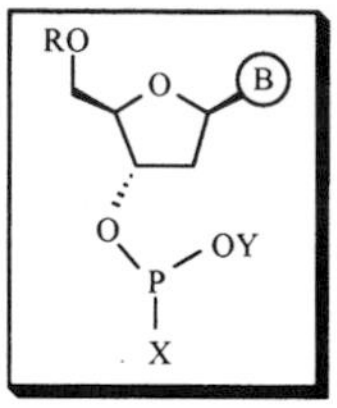

| Ref. 379 | Ref. 380,381 | Ref. 382 |
|---|---|---|
| $X = -N(CHMe)_2$ with Me on N; $Y = $ (cis-alkenyl)—CN | $X = -N(CHMe)_2$ with Me on N; $Y = -CH_2CH_2Si-Me$ with two Ph | $X = -N(CH_2Me)_2$; $Y = -CH_2CH(CF_3)_2$ |

| Ref. 383 | Ref. 384 |
|---|---|
| 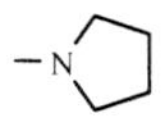 $X = -N$ (pyrrolidine); $OY = -SCH_2CH_2SCOR$ where $R = Ph,\ldots\ldots$ | $-P\langle{}^{OY}_{X} = -P\langle{}^{N(Me)}_{O}$ (oxazolidine) |

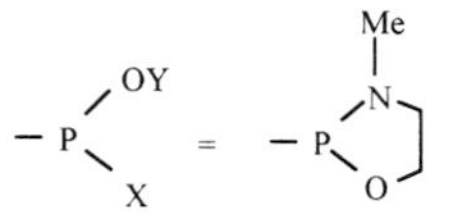

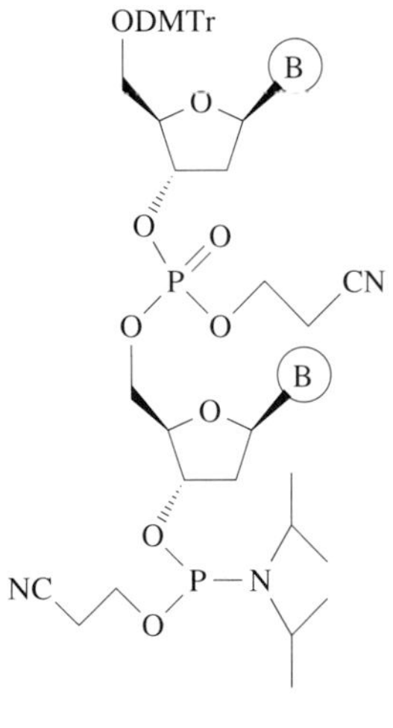

**(112)**

The allyl-protected phosphoramidites in conjunction with allyloxycarbonyl protection of the nucleobases have been employed in oligonucleotide synthesis. Following oligonucleotide synthesis, the protecting groups were removed using palladium complex/triphenylphosphine/formic acid or alternatively under photochemical conditions.[402]

Several research groups have explored the "no-protection option" for oligonucleotide synthesis.[301–305,403,404] Gryaznov and Letsinger elaborated on the no-protection option by employing phosphoramidites, but the synthetic protocol had to be modified to avoid side reactions.[304] Nevertheless, a 20-mer was prepared using this approach.

**Table 5** Bis-phosphoramidites for the preparation of nucleoside phosphoramidites.

| | | |
|---|---|---|
| Ref. 392<br><br>X = −N(CH₂Me)₂<br><br>Y = −CH₂CH₂CN | Ref. 207,392–394<br><br>$\overset{Me}{\underset{|}{}}$<br>X = −N(CHMe)₂<br><br>Y = −CH₂CH₂CN | Ref. 392<br><br>X = −N(CH₂Me)₂<br><br>Y = −CH(Me)CH₂CN |
| Ref. 392<br><br>X = −N(CH₂Me)₂<br><br>Y = −C(Me)₂CH₂CN | Ref. 392<br><br>Me<br>\|<br>X = −N(CHMe)₂<br><br>Y = −C(Me)₂CH₂CN | Ref. 392<br><br>X = −N(CH₂Me)₂<br><br>Y = −CH₂CH₂−S(=O)(=O)−Me |
| Ref. 392<br><br>Me<br>\|<br>X = −N(CHMe)₂<br><br>Y = −CH₂CH₂−S(=O)(=O)−Me | Ref. 395<br><br>X = −N◯O<br><br>Y = −CH₂CH₂−S(=O)(=O)−Me | Ref. 393<br><br>Me<br>\|<br>X = −N(CHMe)₂<br><br>Y = −CH₂CCl₃ |
| Ref. 393<br><br>Me<br>\|<br>X = −N(CHMe)₂<br><br>Y = −C(Me)₂CCl₃ | Ref. 393<br><br>Me<br>\|<br>X = −N(CHMe)₂<br><br>Y = −CH₂CH₂−⬡−NO₂ | Ref. 393,396<br><br>Me<br>\|<br>X = −N(CHMe)₂<br><br>Y = −CH₂CH₂−⬡N |

*Continued overleaf*

## 7.05.4 MECHANISM OF INTERNUCLEOTIDIC COUPLING REACTION

### 7.05.4.1 Phosphoramidite Approach

In the phosphoramidite approach, the mechanism of the coupling reaction effected by 1*H*-tetrazole has been the subject of some scrutiny, and is believed to be acid-catalyzed.[30,405] However, because the coupling reaction was very rapid, the isolation and characterization of the proposed intermediates has not been possible and most accumulated evidence is indirect. In any event, it was possible, by careful experimental design to follow the course of the reaction in the formation of the phosphite by [31]P NMR. For example, Seliger *et al.*[327,406–409] have shown that activation of support-bound phosphoramidites (**116**) with tetrazole generated the putative intermediate (**117**) ([31]P NMR, $\delta$ 126 ppm) which reacted with support-bound nucleoside or alternatively with ethanol to generate the phosphite (**118**) (Scheme 17). Similarly, upon reaction of the phosphoramidite (**102**) ([31]P NMR, $\delta$ 148 ppm) with tetrazole, the phosphorotetrazolide (**117**) ([31]P NMR, $\delta$ 127 ppm) was also observed in solution, and reacted with ethanol to give the corresponding phosphite ([31]P NMR, $\delta$ 138 ppm).

*Oligonucleotide Synthesis*

**Table 5**  (continued)

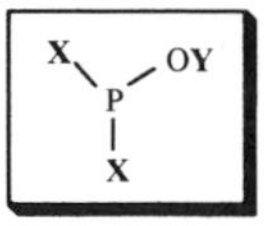

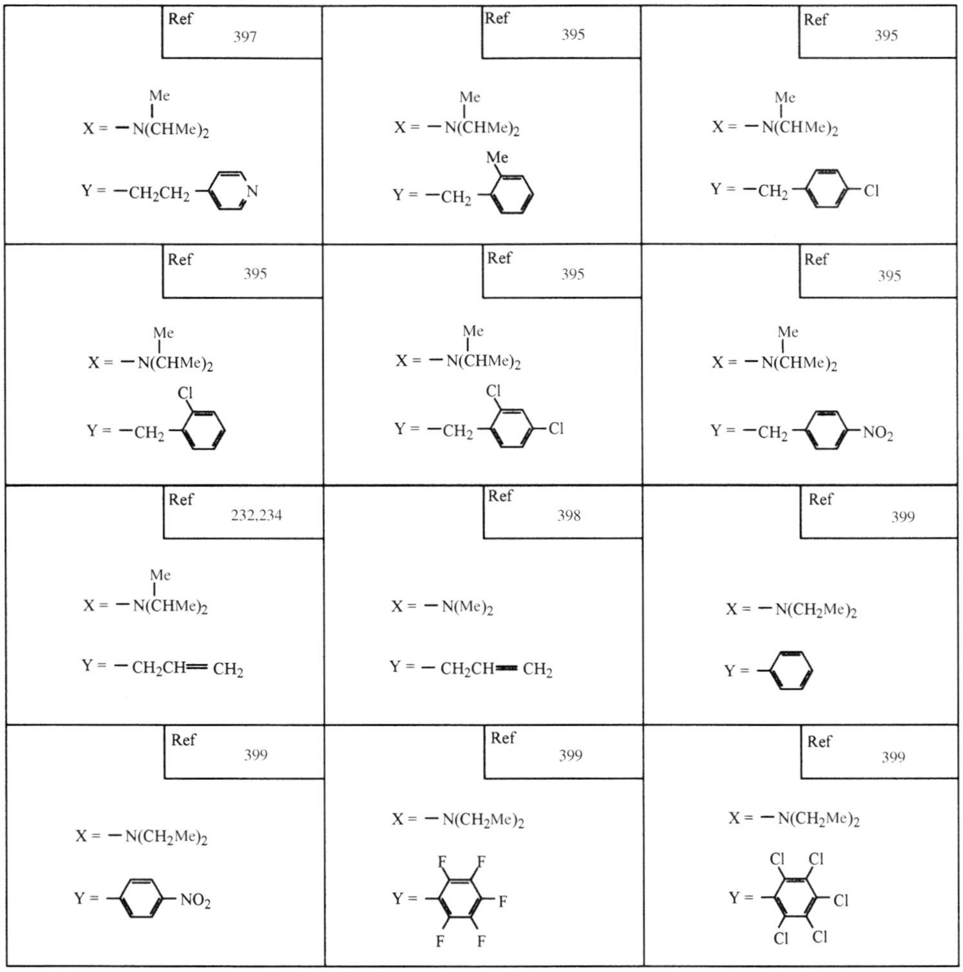

| Ref 397 | Ref 395 | Ref 395 |
|---|---|---|
| X = $-$N(CHMe)$_2$ (Me)<br><br>Y = $-$CH$_2$CH$_2-$(pyridyl) | X = $-$N(CHMe)$_2$ (Me)<br><br>Y = $-$CH$_2-$(o-Me-phenyl) | X = $-$N(CHMe)$_2$ (Me)<br><br>Y = $-$CH$_2-$(p-Cl-phenyl) |

| Ref 395 | Ref 395 | Ref 395 |
|---|---|---|
| X = $-$N(CHMe)$_2$ (Me)<br><br>Y = $-$CH$_2-$(o-Cl-phenyl) | X = $-$N(CHMe)$_2$ (Me)<br><br>Y = $-$CH$_2-$(2,4-Cl-phenyl) | X = $-$N(CHMe)$_2$ (Me)<br><br>Y = $-$CH$_2-$(p-NO$_2$-phenyl) |

| Ref 232,234 | Ref 398 | Ref 399 |
|---|---|---|
| X = $-$N(CHMe)$_2$ (Me)<br><br>Y = $-$CH$_2$CH$=$CH$_2$ | X = $-$N(Me)$_2$<br><br>Y = $-$CH$_2$CH$=$CH$_2$ | X = $-$N(CH$_2$Me)$_2$<br><br>Y = (phenyl) |

| Ref 399 | Ref 399 | Ref 399 |
|---|---|---|
| X = $-$N(CH$_2$Me)$_2$<br><br>Y = (phenyl)$-$NO$_2$ | X = $-$N(CH$_2$Me)$_2$<br><br>Y = (pentafluorophenyl) | X = $-$N(CH$_2$Me)$_2$<br><br>Y = (pentachlorophenyl) |

These observations were consistent with the notion that initial rapid protonation of the phosphoramidite by 1$H$-tetrazole (presumably at the phosphorous center) occurred followed by counter-attack by tetrazole leading to the formation of the intermediate phosphorotetrazolide.[389,409–412] The tetrazolide was then consumed in a second step by reaction with the hydroxyl group of the attacking nucleophile. Apparently, tetrazole was just acidic enough (pK$_a$ 4.8) to protonate the phosphoramidite, but not cause premature detritylation of the 5'-DMTr group. However, protonation by itself was not sufficient to promote coupling. Thus, aliphatic acids such as octanoic acid did not promote the coupling reaction.[413]

The rate of coupling reaction was dependent on the steric and electronic nature of the $O$-and $P$-substituents in the phosphoramidites. Using solution-phase studies, a rough order has been reported by Dahl and co-workers,[414–416] the rate for $N$-substituents being: dimethylamino > diisopropyl > morpholino > anilinomethyl; and for $O$-substituents: methyl > $\beta$-cyanoethyl > 1-methyl-2-cyanoethyl > 1,1-dimethyl-2-cyanoethyl > $o$-chlorophenyl. In addition to 1$H$-tetrazole, nitrophenyl tetrazole, thioethyl tetrazole, $N$-methylanilinium trifluoroacetate, $N$-methylanilinium trichloroacetate, 5-trifluoromethyl-1$H$-tetrazole, 1-hydroxybenzotriazole, $N$-methylimidazole hydrochloride, $N$-methylimidazolium trifluoromethane sulfonate, and benzimidazole trifluoromethane sulfonate have been proposed as activators of the coupling reaction.[108,414–416]

**Table 5** (continued)

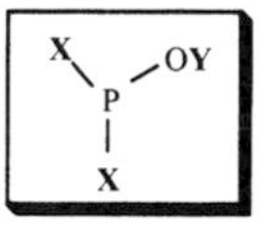

| | | |
|---|---|---|
| Ref 399<br><br>X = −N(CH$_2$Me)$_2$<br><br>Y = (2,4,5-trichlorophenyl) 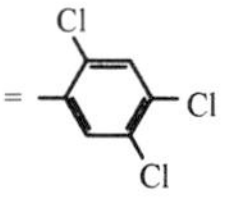 | Ref 399<br><br>X = −N(CH$_2$Me)$_2$<br><br>Y = (2-bromophenyl) 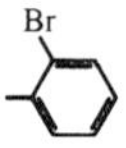 | Ref 399<br><br>X = −N(CH$_2$Me)$_2$<br><br>Y = (4-bromophenyl)  |
| Ref 399<br><br>X = −N(CH$_2$Me)$_2$<br><br>Y = (2-methylphenyl) 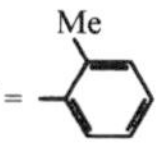 | Ref 399<br><br>X = −N(CH$_2$Me)$_2$<br><br>Y = (2,6-dimethylphenyl) 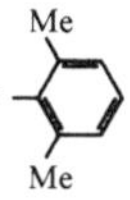 | Ref 399<br><br>X = −N(CH$_2$Me)$_2$<br><br>Y = −CH$_2$(2,4-dinitrophenyl) 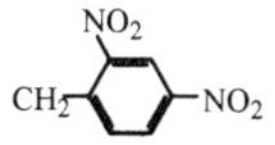 |
| Ref 400<br><br>X = −N(CH$_2$Me)$_2$ (Me)<br><br>Y = −CH(CF$_3$)$_2$ | | |

## 7.05.4.2  *H*-phosphonate Approach

As mentioned earlier, nucleoside *H*-phosphonates were first introduced by Todd and co-workers as important building blocks for the synthesis of polynucleotides.[8] In the earlier work, diphenyl-chlorophosphates were used as the activator for the coupling reaction. The use of hindered acid chlorides such as pivaloyl chloride, and adamantane carbonyl chloride have largely supplanted phenylchlorophosphates as activators.[31–33,417]

The *H*-phosphonate building blocks (**119**) are synthesized using either the tris-triazolylphosphite (**120**), or the 2-chloro-4*H*-1,3,2-benzo-dioxaphosphorin-4-one (**121**) (Scheme 18) and are obtained as their stable triethylammonium or DBU salts. New methods for the synthesis of (**119**) have also been reported.[418–422] For the oligonucleotide synthesis, the *H*-phosphonate monomers are employed as a solution in anhydrous pyridine/acetonitrile (1/1).[33]

The mechanism of the activation reaction has been investigated.[31,33,423–427] The course of the reaction presumably involves the intermediate formation of the mixed phosphonate-carboxylic anhydride (**122**) (Scheme 19). The reaction of this active diester intermediate with the nucleoside 5′-OH group gives the desired internucleotidic *H*-phosphonate linkage. The undesired competing pathways include the formation of the bis-acylphosphite (**123**), or the acylphosphonic-carboxylic anhydride (**124**), and result from "overactivation" of the *H*-phosphonate in the presence of strong bases (Scheme 20). The reaction of these intermediates with the support-bound nucleoside leads to the by-products (**125**) and (**126**). By employing short coupling times, and avoiding nucleophilic catalysts in the coupling reaction, it is possible to achieve optimal coupling efficiency.[33] The *H*-phosphonate chemistry has been useful in the synthesis of backbone-modified oligonucleotide analogues.[33,417,422,428–438] Analogous to the phosphoramidite approach, the no-protection option in oligonucleotide synthesis using the *H*-phosphonate approach has also been explored.[439]

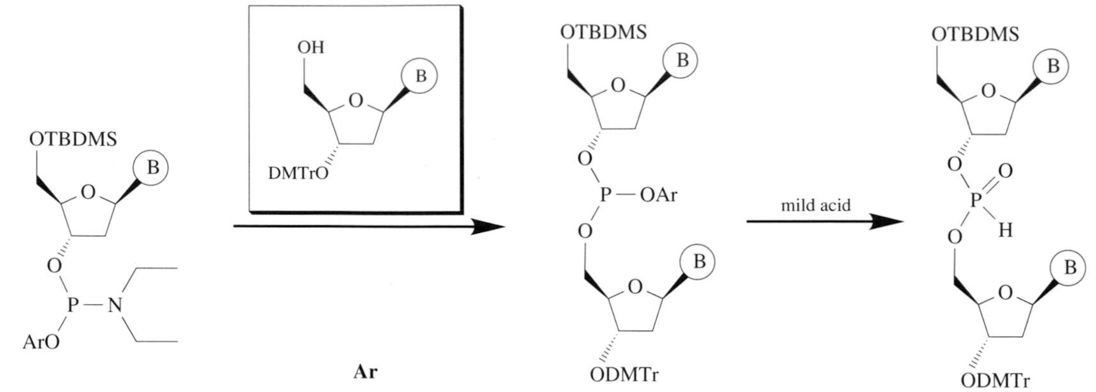

**Ar**

(a) 2,4 -dinitrophenyl
(b) pentafluorophenyl
(c) pentachlorophenyl
(d) 2,4,5-trichlorophenyl
(e) 4-nitrophenyl
(f) 2-bromophenyl
(g) 4-bromophenyl
(h) phenyl
(i) 2-methylphenyl
(j) 2,6-dimethylphenyl

(113 a–j)

(114)

(115)

**Scheme 16**

(116)

(117)

(118)

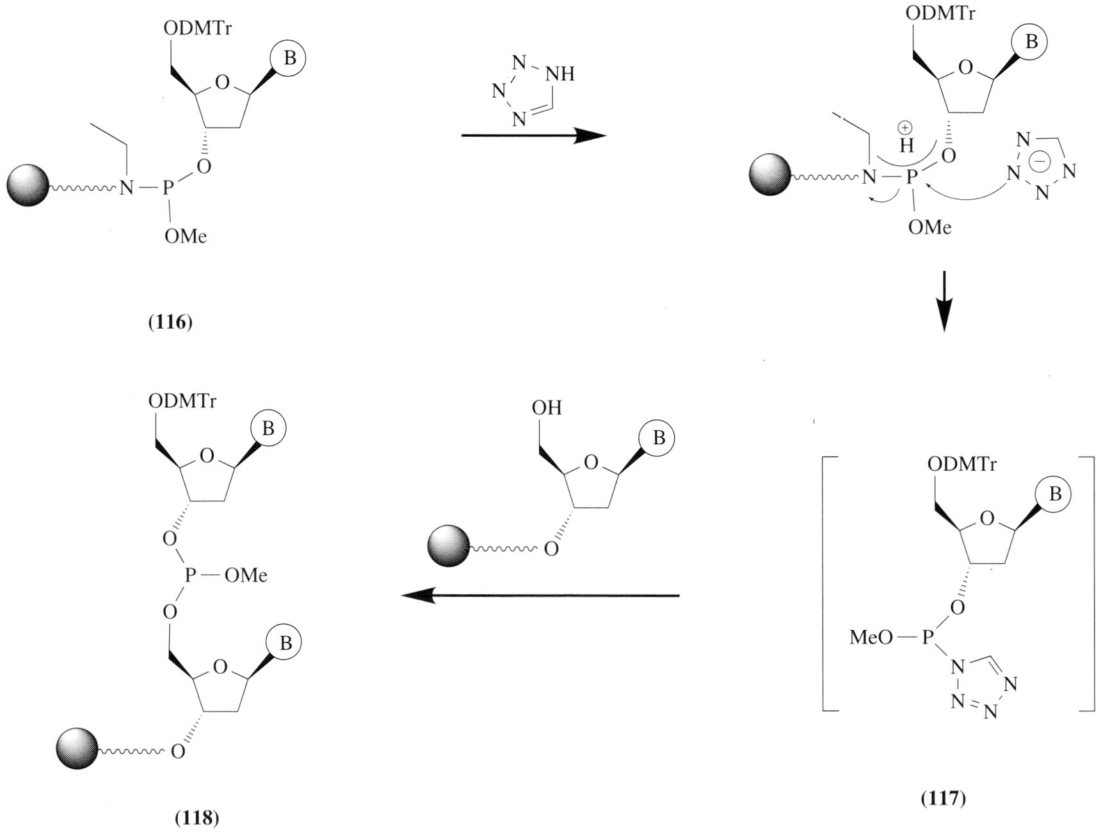

**Scheme 17**

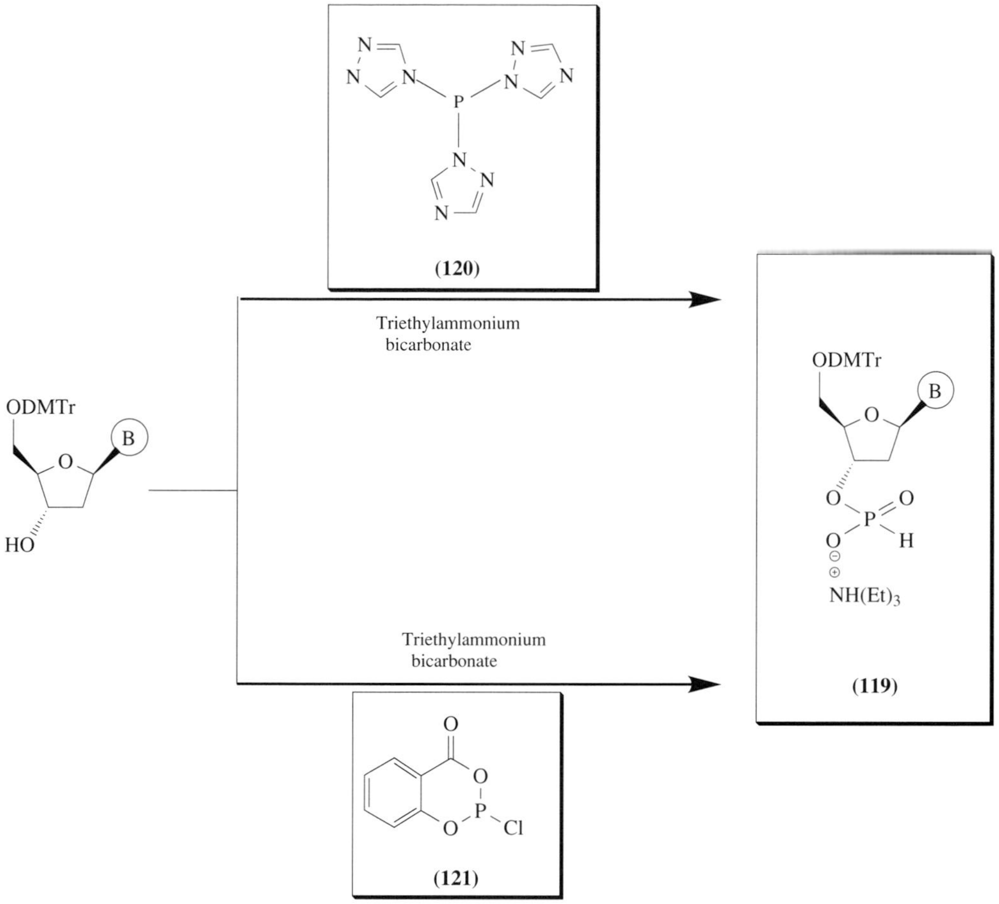

**Scheme 18**

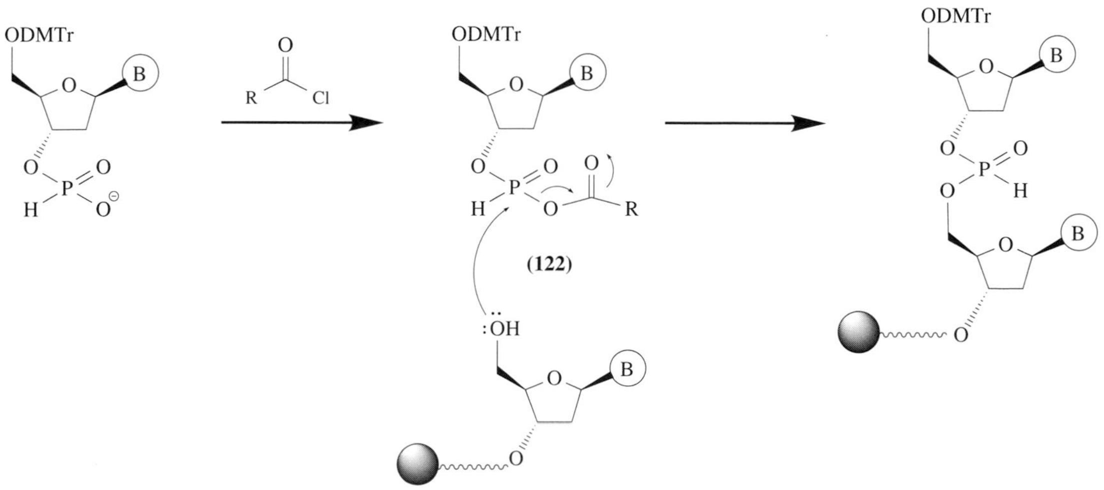

**Scheme 19**

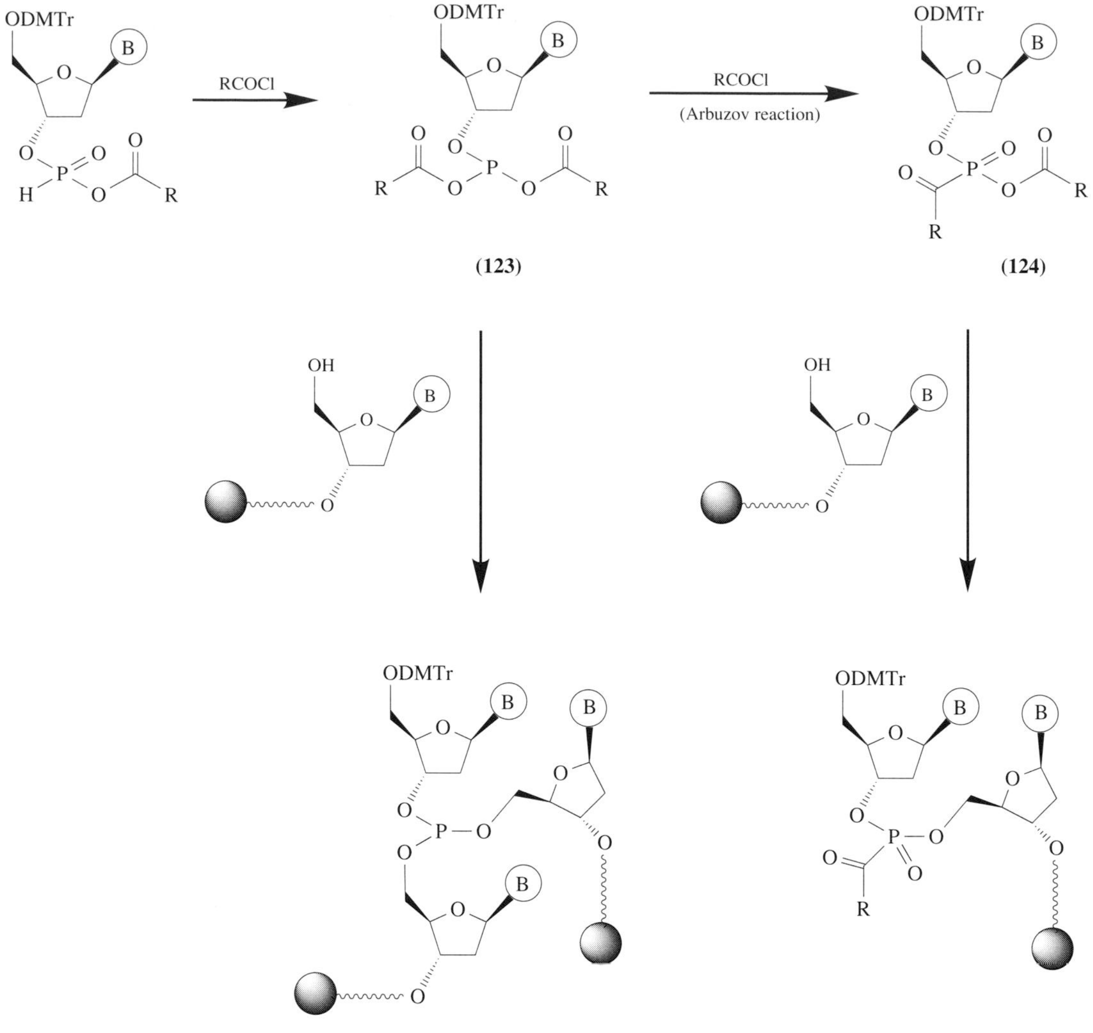

**Scheme 20**

## 7.05.5 SOLID-PHASE OLIGONUCLEOTIDE SYNTHESIS

Oligonucleotides—synthetic DNA—have a number of applications in molecular biology as DNA sequencing primers, probes, adapters, and in gene synthesis.[440] Applications in the area of diagnostics and in forensic testing have also emerged.[441,442] Additionally, the potential therapeutic applications of oligonucleotides[441–445] in the form of "antisense oligonucleotides," ribozymes, and triplex DNA are triggering developments in large-scale synthesis and manufacture.

### 7.05.5.1 Phosphoramidite Approach

Small- as well as large-scale DNA synthesis is carried out on a solid support in the 3′- to 5′-direction mainly using phosphoramidite chemistry in automated DNA synthesizer machines (Scheme 21). The leader 3′-nucleoside is linked to the solid-support generally through a succinyl linker to give the support-bound nucleoside (**127**). The synthesis consists of stepwise coupling of the individual monomeric units (**128**) to (**127**) followed by oxidation of the resulting phosphite (**129**) with iodine to give the phosphotriester (**130a**). In addition to iodine, a host of other reagents have been reported to effect the oxidation of the phosphorus(III) to phosphorus(V).[108] Alternatively, if the oxidation step is replaced by oxidative sulfurization, the corresponding phosphorothioate (**130b**) can be obtained. This solid-phase oxidative sulfurization can be accomplished by elemental sulfur or more conveniently using 3*H*-1,2-benzodithiole-3-one-1,1-dioxide.[446] A plethora of sulfurizing reagents have been reported and evaluated in large-scale synthesis.[447]

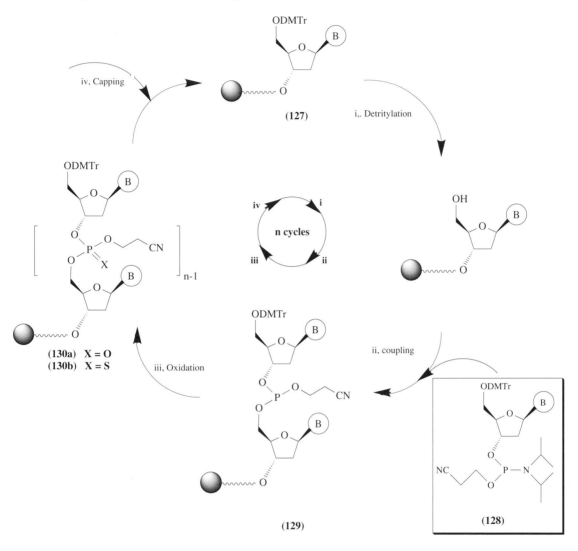

**Scheme 21**

Each synthesis cycle (see Scheme 21) in the synthesis of oligonucleotide consists of the following sequential steps:

   (i) detritylation, i.e., the removal of the DMTr group from the 5′-end of the leader nucleoside with 2% dichloroacetic acid in methylene chloride;

   (ii) washing, with acetonitrile to remove the released trityl cation and to remove the excess reagents;

   (iii) coupling, to form the internucleotidic linkage;

   (iv) washing, with acetonitrile to remove the excess reagent used for coupling;

   (v) oxidation, to convert the phosphite linkage to the phosphotriester linkage (phosphorus(III) to phosphorus(V) conversion);

   (vi) washing, with acetonitrile to remove the excess reagents; and

   (vii) capping, with Ac$_2$O to block the unreacted/free hydroxyl group.

In each step, the reagent is pulse-delivered to the support maintained in a column, and allowed to remain in contact with the support for the appropriate function to take place. The synthesis cycle is performed automatically by the synthesizer using programs recommended by the manufacturer.

In order to ensure complete reaction at each step, excess reagents/solvents are employed in the synthesis cycles in a small scale (0.1 μmol to 1 μmol). Additionally, to make the synthesis cost effective for large scale, considerable efforts have been expended in the optimization of the reagent consumption and cycle times.[448] A procedure for recovering unused phosphoramidites has been reported.[449]

### 7.05.5.2   *H*-phosphonate Approach

This approach has facilitated the preparation of certain analogues not accessible by the phosphoramidite route. In this case, the *H*-phosphonate monomers are employed as a solution in acetonitrile/pyridine (1/1).

In this approach (Scheme 22), the key coupling step consists of the conversion of the monomer *H*-phosphonate (**131**) to a mixed carboxylic-phosphonic anhydride depending on the activation reagent employed.[31–33,417,426] The reaction of this reactive intermediate with the 5′-OH group of the support-bound nucleoside (**132**) gives the coupled *H*-phosphonate (**133**). Capping of the unreacted 5′-hydroxyl group is simultaneously accomplished by the excess acid-chloride present in the activation cycle. Thus, unlike phosphoramidite chemistry, a capping step is optional. However, reports have favored the inclusion of a capping step.[33] At the end of the chain assembly, support-bound oligonucleotide *H*-phosphonate (**133**) is obtained. The oxidative conversion of the *H*-phosphonate linkage to a phosphodiester or a modified backbone is accomplished by the appropriate nucleophilic reagent. A typical synthesis cycle is shown in Scheme 22.

### 7.05.5.3   Isolation and Purification of the Oligonucleotide

Following the synthesis of the oligonucleotide, the base and phosphate protecting groups are removed and the oligonucleotide is released from the support, usually under aqueous basic conditions. The purification of the oligonucleotide can be accomplished using reversed-phase HPLC, ion-exchange chromatography, or polyacrylamide gel electrophoresis using standard protocols.[33,448] Improvements in the purification procedures have been reported.[450–452]

### 7.05.6   SYNTHESIS OF RNA

The presence of a 2′-hydroxy group in ribonucleosides introduces an additional element of complexity in the synthesis of RNA. In RNA, the vicinal 2′- and 3′-hydroxyl groups are configured in a *cis* orientation; consequently, the free 2′-hydroxyl group (or its derivatives) can participate in neighboring group interactions with many 3′-substituents. This results in 3′- to 2′-migration, or alternatively in inducing chain cleavage of an RNA under certain conditions.[453] Because of the notoriety of the 2′-OH group and its derivatives, devising a suitable 2′-protecting group has been a major challenge, and is essentially the crux of RNA synthesis. Along the way, several fascinating aspects of RNA chemistry have also been uncovered.

The protection of the 2′-hydroxy group must meet several criteria: (i) the corresponding 3′-ribonucleoside phosphoramidites or *H*-phosphonates should be amenable to synthesis, (ii) it must remain intact throughout the synthesis, and (iii) the steric bulk of the 2′-protecting group should not impede the internucleotidic 3′- to 5′-coupling reaction. The protection of the 2′-hydroxyl group with 1-[(2-chloro-4-methyl)phenyl]-4′-methoxypiperidin-4-yl group and related derivatives (developed by Reese *et al.*[454]) and 2′-*tert*-butyldimethylsilyl group (developed by Ogilvie and co-workers[455]) have emerged as the most popular. The chemistry leading to the developments of these 2′-hydroxyl protecting groups[108] and the application of the derived phosphoramidites and *H*-phosphonates in the synthesis of RNA are described in *Chemical RNA Synthesis* in Volume 6 of this series and will not be detailed here.

### 7.05.7   CONCLUSION

The development of procedures for the solid-phase synthesis of oligonucleotides and the inevitable automation of the synthesis that accompanied it have led to the ready availability of these

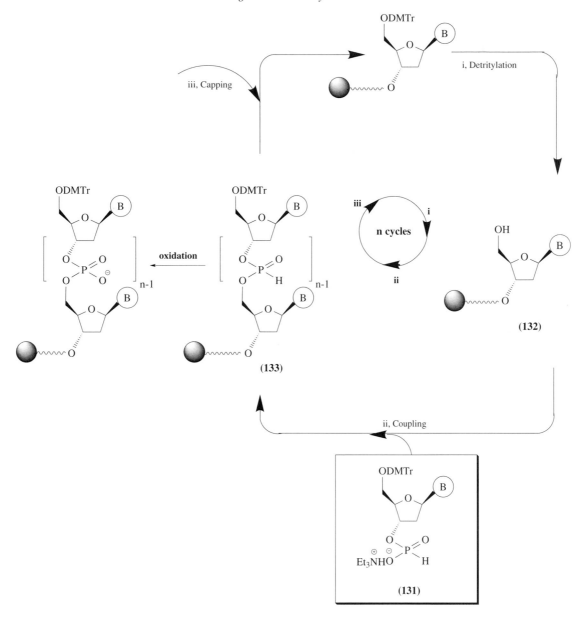

**Scheme 22**

molecules for application in therapeutics and diagnostics, and nurtured tremendous advances in molecular biology. Applications of nucleic acid chemistry continue to grow. Specific applications of the chemistry described in this chapter in the preparation of oligodeoxynucleotide conjugates, modified-oligonucleotide analogues, and oligoribonucleotide conjugates have been covered in reviews.[34,456-458]

## ACKNOWLEDGMENT

This review is dedicated to the memory of the late mathematician Srinivasa Ramanujan.

## 7.05.8 REFERENCES

1. J. D. Watson and F. H. C. Crick, *Nature*, 1953, **171**, 737.
2. E. Baer, *Spec. Publ. Chem. Soc.*, 1957, **n8**, 105.

3. A. R. Todd, *Spec. Publ. Chem. Soc.*, 1957, **n8**, 91.
4. F. R. Atherton, H. T. Openshaw, and A. R. Todd, *J. Chem. Soc.*, 1945, 382.
5. V. M. Clark and A. R. Todd, *J. Chem. Soc.*, 1950, 2023.
6. N. S. Corby, G. W. Kenner, and A. R. Todd, *J. Chem. Soc.*, 1952, 1234.
7. A. M. Michelson and A. R. Todd, *J. Chem. Soc.*, 1955, 2632.
8. R. H. Hall, A. R. Todd, and R. F. Webb, *J. Chem. Soc.*, 1957, 3291.
9. P. T. Gilham and H. G. Khorana, *J. Am. Chem. Soc.*, 1958, **80**, 6212.
10. H. G. Khorana, *Harvey Lect.*, **62**, 79.
11. H. G. Khorana, *Science*, 1979, **203**, 614.
12. R. L. Letsinger and K. K. Ogilvie, *J. Am. Chem. Soc.*, 1969, **91**, 3350.
13. C. B. Reese, *Tetrahedron*, 1978, **34**, 3143.
14. E. Ohtsuka, M. Ikehara, and D. Söll, *Nucleic Acids Res.*, 1982, **10**, 6553.
15. J. C. Catlin and F. Cramer, *J. Org. Chem.*, 1973, **38**, 245.
16. R. B. Merrifield, *Science*, 1965, **150**, 178.
17. R. L. Letsinger and V. Mahadevan, *J. Am. Chem. Soc.*, 1965, **87**, 3526.
18. R. L. Letsinger and V. Mahadevan, *J. Am. Chem. Soc.*, 1966, **88**, 5319.
19. H. Köster, *Tetrahedron Lett.*, 1972, 1527.
20. H. Köster, N. Hoppe, K. H. Kropelin, and K. Kulikowski, *Nucleic Acids Symp. Ser.*, 1980, **n7**, 39.
21. H. Kössel and H. Seliger, *Fortschr. Chem. Org. Naturst.*, 1975, **32**, 297.
22. M. J. Gait, in "Polymer-supported Reactions in Organic Synthesis," eds. P. Hodge and D. C. Sherrington, Wiley, New York, 1980, p. 435.
23. M. J. Gait, S. G. Popov, M. Singh, and R. C. Titmas, *Nucleic Acids Symp. Ser.*, 1980, **n7**, 243.
24. K. Miyoshi and K. Itakura, *Nucleic Acids Symp. Ser.*, 1980, **n7**, 281.
25. R. Wu, C. P. Bahl, and S. A. Narang, *Prog. Nucleic Acids Res. Mol. Biol.*, 1978, **21**, 101.
26. J. W. Engels and E. Uhlmann, *Angew. Chem. Int. Ed. Engl.*, 1989, **28**, 716.
27. R. L. Letsinger and W. B. Lunsford, *J. Am. Chem. Soc.*, 1976, **98**, 3655.
28. M. D. Matteucci and M. H. Caruthers, *J. Am. Chem. Soc.*, 1981, **103**, 2185.
29. M. D. Matteucci and M. H. Caruthers, *Tetrahedron Lett.*, 1980, **21**, 719.
30. S. L. Beaucage and M. H. Caruthers, *Tetrahedron Lett.*, 1981, **22**, 1859.
31. P. J. Garegg, I. Lindh, T. Regberg, J. Stawinski, R. Strömberg, and C. Henrichson, *Tetrahedron Lett.*, 1986, **27**, 4055.
32. B. C. Froehler, P. G. Ng, and M. D. Matteucci, *Nucleic Acids Res.*, 1986, **14**, 5399.
33. B. C. Froehler, in "Protocols for Oligonucleotides and Analogs," ed. S. Agrawal, Humana, Totowa, NJ, 1993, vol. 20, p. 63.
34. S. Agrawal (ed.), "Protocols for Oligonucleotides and Analogs," Humana, Totowa, NJ, vol. 20, 1993.
35. S. Agrawal (ed.), "Protocols for Oligonucleotides and Analogs," Humana, Totowa, NJ, vol. 26, 1994.
36. M. J. Gait and R. C. Sheppard, *Nucleic Acids Res.*, 1977, **4**, 1135.
37. A. F. Markham, M. D. Edge, T. C. Atkinson, A. R. Greene, G. R. Heathcliffe, C. R. Newton, and D. Scanlon, *Nucleic Acids Res.*, 1980, **8**, 5193.
38. F. Bordella, E. Giralt, and E. Pedroso, *Tetrahedron Lett.*, 1990, **31**, 6231.
39. R. Crea and T. Horn, *Nucleic Acids Res.*, 1980, **8**, 2331.
40. R. Frank, W. Heikens, G. Heisterberg-Moutsis, and H. Blöcker, *Nucleic Acids Res.*, 1983, **11**, 4365.
41. R. T. Pon and K. K. Ogilvie, *Tetrahedron Lett.*, 1984, **25**, 713.
42. M. J. Gait, H. W. D. Matthes, M. Singh, B. S. Sproat, and R. C. Titmas, *Nucleic Acids Res.*, 1982, **10**, 6243.
43. R. L. Kaas and J. L. Kardo, *Polym. Eng. Sci.*, 1971, **11**, 11.
44. G. B. Cox, *J. Chromatogr. Sci.*, 1977, **15**, 385.
45. R. E. Majors and M. J. Hopper, *J. Chromatogr. Sci.*, 1974, **12**, 767.
46. G. Alvarado-Urbina, G. M. Sathe, W.-C. Liu, M. F. Gillen, P. D. Duck, R. Bender, and K. K. Ogilvie, *Science*, 1981, **214**, 270.
47. F. Chow, T. Kempe, and G. Palm, *Nucleic Acids Res.*, 1981, **9**, 2807.
48. V. Kohli, A. Balland, M. Wintzerith, R. Sauerwald, A. Staub, and J. P. Lecocq, *Nucleic Acids Res.*, 1982, **10**, 7439.
49. A. Kume, M. Sekine, and T. Hata, *Chem. Lett.*, 1983, 1597.
50. H. Köster, A. Stumpe, and A. Wolter, *Tetrahedron Lett.*, 1983, **24**, 747.
51. H. Köster, J. Biernat, J. McManus, A. Wolter, A. Stumpe, Ch. K. Narang, and N. D. Sinha, *Tetrahedron*, 1984, **40**, 103.
52. J. Katzhendler, S. Cohen, E. Rahamim, I. Ringel, and J. Deutsch, *Tetrahedron*, 1989, **45**, 2777.
53. A. Van Aerschot, P. Herdewijn, and A. Vanderhaeghe, *Nucleosides Nucleotides*, 1988, **7**, 75.
54. L. Arnold, Z. Tocik, E. Bradkova, Z. Hostomsky, V. Paces, and J. Smrt, *Collect. Czech. Chem. Commun.*, 1989, **54**, 523.
55. S. P. Adams, K. S. Kavka, E. J. Wykes, S. B. Holder, and G. R. Galluppi, *J. Am. Chem. Soc.*, 1983, **105**, 661.
56. H. Seliger, U. Kotschi, C. Sharpf, R. Martin, F. Eisenbeiss, J. N. Kinkel, and K. K. Unger, *J. Chromatogr.*, 1989, **476**, 49.
57. R. W. Barnett and H. Erfle, *Nucleic Acids Res.*, 1990, **18**, 3094.
58. H. Vu, C. McCollum, C. Lotys, and A. Andrus, *Nucleic Acids Symp. Ser.*, 1990, **n22**, 63.
59. C. McCollum and A. Andrus, *Tetrahedron Lett.*, 1991, **32**, 4069.
60. E. Bayer, *Angew. Chem. Int. Ed. Engl.*, 1991, **30**, 113.
61. H. Gao, B. L. Gaffney, and R. A. Jones, *Tetrahedron Lett.*, 1991, **32**, 5477.
62. M. P. Reddy, M. A. Michael, F. Farooqui, and N. S. Girgis, *Tetrahedron Lett.*, 1994, **35**, 5771.
63. E. Birch-Hirchfeld, H. Eickhoff, A. Stelzner, K. O. Greulich, Z. Foldes-Papp, H. Seliger, and K.-H. Guhrs, *Collect. Czech. Chem. Commun.*, 1996, **61**, S311.
64. E. Birch-Hirscheld, Z. Foldes-Papp, K.-H. Guhrs, and H. Seliger, *Helv. Chim. Acta*, 1996, **79**, 137.
65. E. Birch-Hirchfeld, Z. Foldes-Papp, K.-H. Guhrs, and H. Seliger, *Nucleic Acids Res.*, 1994, **22**, 1760.
66. G. M. Bonora, *Appl. Biochem. Biotechnol.*, 1995, **54**, 3.
67. G. M. Bonora, A. Baldan, O. Schiavon, P. Ferruti, and F. Veronese, *Tetrahedron Lett.*, 1996, **37**, 4761.

68. G. R. Gough, M. J. Brunden, and P. T. Gilham, *Tetrahedron Lett.*, 1983, **24**, 5321.
69. J. S. DeBear, J. A. Hayes, M. P. Koleck, and G. R. Gough, *Nucleosides Nucleotides*, 1987, **6**, 821.
70. C. S. Larson, C. Rosenbohm, T. J. D. Jorgensen, and J. Wengel, *Nucleosides Nucleotides*, 1997, **16**, 67.
71. X. Zhang and R. A. Jones, *Tetrahedron Lett.*, 1996, **37**, 3789.
72. M. E. Schwartz, R. B. Breaker, G. T. Asteriadis, and G. R. Gough, *Tetrahedron Lett.*, 1995, **36**, 27.
73. B. S. Sproat and D. M. Brown, *Nucleic Acids Res.*, 1985, **13**, 2979.
74. T. Brown, C. E. Pritchard, G. Turner, and S. A. Salisbury, *J. Chem. Soc. Chem. Commun.*, 1989, 891.
75. R. H. Alul, C. N. Singman, G. Zhang, and R. L. Letsinger, *Nucleic Acids Res.*, 1991, **19**, 1527.
76. R. T. Pon and S. Yu, *Tetrahedron Lett.*, 1997, **38**, 3327.
77. R. T. Pon and S. Yu, *Tetrahedron Lett.*, 1997, **38**, 3331.
78. W. T. Markiewicz and T. K. Wyrzykiewicz, *Nucleic Acids Res.*, 1989, **17**, 7149.
79. K. Kamaike, T. Ogawa, and Y. Ishido, *Nucleosides Nucleotides*, 1993, **12**, 1015.
80. F. Fritja, J. Robles, D. Fernandez-Forner, F. Albericio, E. Giralt, and E. Pedroso, *Tetrahedron Lett.*, 1991, **32**, 1511.
81. J. Hovinen, A. Guzaev, A. Azhayev, and H. Lönnberg, *Tetrahedron Lett.*, 1993, **34**, 8169.
82. P. Kumar, N. K. Bose, and K. C. Gupta, *Tetrahedron Lett.*, 1991, **32**, 967.
83. R. Zuckermann, D. Corey, and P. Schultz, *Nucleic Acids Res.*, 1987, **15**, 5305.
84. K. C. Gupta, P. Sharma, S. Sathynarayana, and P. Kumar, *Tetrahedron Lett.*, 1990, **31**, 2471.
85. P. Kumar, D. Bhatia, R. C. Rastogi, and K. C. Gupta, *Bioorg. Med. Chem. Lett.*, 1996, **6**, 683.
86. B. Bonfils and N. T. Thuong, *Tetrahedron Lett.*, 1991, **32**, 3053.
87. U. Asseline and N. T. Thuong, *Tetrahedron Lett.*, 1990, **31**, 81.
88. U. Asseline and N. T. Thuong, *Tetrahedron Lett.*, 1989, **30**, 2521.
89. P. S. Nelson, R. A. Frye, and E. Liu, *Nucleic Acids Res.*, 1989, **17**, 7187.
90. S. M. Gryaznov and R. L. Letsinger, *Tetrahedron Lett.*, 1993, **34**, 1261.
91. J. Haralambidis, L. Duncan, and G. W. Tregear, *Tetrahedron Lett.*, 1987, **28**, 5199.
92. J. Haralambidis, L. Duncan, K. Angus, M. Chai, S. Pownall, and G. Tregear, *Nucleic Acids Symp. Ser.*, 1988, **n20**, 115.
93. J. Haralambidis, L. Duncan, K. Angus, and G. W. Tregear, *Nucleic Acids Res.*, 1990, **18**, 493.
94. J. Haralambidis, K. Angus, S. Pownall, L. Duncan, M. Chai, and G. W. Tregear, *Nucleic Acids Res.*, 1990, **18**, 501.
95. C. D. Juby, C. D. Richardson, and R. Brousseau, *Tetrahedron Lett.*, 1991, **32**, 879.
96. M. W. Reed, A. D. Adams, J. S. Nelson, and R. B. Meyer, Jr., *Bioconjugate Chem.*, 1991, **2**, 217.
97. H. Vu, P. Singh, L. Lewis, J. G. Zendegui, and K. Jayaraman, *Nucleosides Nucleotides*, 1993, **12**, 853.
98. A. Avino, R. G. Garcia, F. Albericio, M. Mann, M. Wilm, G. Neubauer, and R. Eritja, *Bioorg. Med. Chem.*, 1996, **4**, 1649.
99. A. Routledge, M. P. Wallis, K. C. Ross, and W. Fraser, *Bioorg. Med. Chem. Lett.*, 1995, **5**, 2059.
100. H. Venkatesan and M. M. Greenberg, *J. Org. Chem.*, 1996, **91**, 5625.
101. D. L. McMinn and M. M. Greenberg, *Tetrahedron*, 1996, **52**, 3827.
102. M. M. Greenberg, *Tetrahedron*, 1995, **51**, 29.
103. M. M. Greenberg and J. L. Gilmore, *J. Org. Chem.*, 1994, **59**, 746.
104. R. T. Pon, in "Protocols for Oligonucleotides and Analogs," ed. S. Agrawal, Humana, Totowa, NJ, 1994, vol. 26, p. 465.
105. P. Kumar, A. K. Sharma, P. Sharma, B. S. Garg, and K. C. Gupta, *Nucleosides Nucleotides*, 1996, **15**, 879.
106. K. C. Gupta, P. Kumar, D. Bhatia, and A. K. Sharma, *Nucleosides Nucleotides*, 1995, **14**, 829.
107. T. Atkinson, S. Gillam, and M. Smith, *Nucleic Acids Res.*, 1988, **16**, 6232.
108. S. L. Beaucage and R. P. Iyer, *Tetrahedron*, 1992, **48**, 2223.
109. C. B. Reese and A. Ubasawa, *Tetrahedron Lett.*, 1980, **21**, 2265.
110. C. B. Reese and A. Ubasawa, *Nucleic Acids Symp. Ser.*, 1980, **n7**, 5.
111. J. A. J. den Hartog, G. Willie, R. A. Scheublin, and J. H. van Boom, *Biochemistry*, 1982, **21**, 1009.
112. B. F. L. Li, C. B. Reese, and P. F. Swann, *Biochemistry*, 1987, **26**, 1086.
113. T. Huynh-Dinh, B. Langlois d'Estaintot, P. Allard, and J. Igolen, *Tetrahedron Lett.*, 1985, **26**, 431.
114. K. J. Divakar and C. B. Reese, *J. Chem. Soc. Perkin Trans. 1*, 1982, 1171.
115. W. L. Sung, *J. Chem. Soc. Chem. Commun.*, 1981, 1089.
116. W. L. Sung, *Nucleic Acids Res.*, 1981, **9**, 6139.
117. W. L. Sung and S. A. Narang, *Can. J. Chem.*, 1982, **60**, 111.
118. W. L. Sung, *J. Org. Chem.*, 1982, **47**, 3623.
119. R. W. Adamiak, E. Biala, Z. Gdaniec, S. Mielewczyk, and B. Skalski, *Chem. Scr.*, 1986, **26**, 3.
120. R. W. Adamiak, E. Biala, Z. Gdaniec, S. Mielewczyk, and B. Skalski, *Chem. Scr.*, 1986, **26**, 4.
121. C. B. Reese and K. H. Richards, *Tetrahedron Lett.*, 1985, **26**, 2245.
122. H. P. Daskalov, M. Sekine, and T. Hata, *Tetrahedron Lett.*, 1980, **21**, 3899.
123. H. P. Daskalov, M. Sekine, and T. Hata, *Bull. Chem. Soc. Jpn.*, 1981, **54**, 3076.
124. P. K. Bridson, W. Markiewicz, and C. B. Reese, *J. Chem. Soc. Chem. Commun.*, 1977, 447.
125. P. Francois, G. Hamoir, E. Sonveaux, H. Vermeersch, and Y. Ma, *Bull. Soc. Chim. Belg.*, 1985, **94**, 821.
126. R. T. Pon, M. J. Damha, and K. K. Ogilvie, *Tetrahedron Lett.*, 1985, **26**, 2525.
127. R. T. Pon, M. J. Damha, and K. K. Ogilvie, *Nucleic Acids Res.*, 1985, **13**, 6447.
128. J. Nielsen, J. E. Marugg, M. Taagaard, J. H. van Boom, and O. Dahl, *Recl. Trav. Chim. Pays-Bas*, 1986, **105**, 33.
129. J. Nielsen, O. Dahl, G. Remaud, and J. Chattopadhyaya, *Acta Chem. Scand.*, 1987, **B41**, 633.
130. M. S. Urdea, L. Ku, T. Horn, Y. G. Gee, and B. D. Warner, *Nucleic Acids Symp. Ser.*, 1986, **n16**, 257.
131. L. J. McBride, J. S. Eadie, J. W. Efcavitch, and A. Andrus, *Nucleosides Nucleotides*, 1987, **6**, 297.
132. A. Andrus and S. L. Beaucage, *Tetrahedron Lett.*, 1988, **29**, 5479.
133. X.-X. Zhou, C. J. Welch, and J. Chattopadhyaya, *Acta Chem. Scand.*, 1986, **B40**, 806.
134. C. J. Welch, X.-X. Zhou, and J. Chattopadhyaya, *Acta Chem. Scand.*, 1986, **B40**, 817.
135. A. Nyilas, X.-X. Zhou, C. J. Welch, and J. Chattopadhyaya, *Nucleic Acids Symp. Ser.*, 1987, **n18**, 157.
136. X.-X. Zhou and J. Chattopadhyaya, *Tetrahedron*, 1986, **42**, 5149.
137. J. M. Brown, C. Christodoulou, S. S. Jones, A. S. Modak, C. B. Reese, S. Sibanda, and A. Ubasawa, *J. Chem. Soc. Perkin Trans. 1*, 1989, 1735.

138. P. Lemmen, R. Karl, I. Ugi, N. Balgobin, and J. Chattopadhyaya, *Z. Naturforsch.*, 1987, **42C**, 442.
139. S. S. Jones, C. B. Reese, S. Sibanda, and A. Ubasawa, *Tetrahedron Lett.*, 1981, **22**, 4755.
140. X.-X. Zhou, A. Sandström, and J. Chattopadhyaya, *Chem. Scr.*, 1986, **26**, 241.
141. C. B. Reese and P. A. Skone, *J. Chem. Soc. Perkin Trans. 1*, 1984, 1263.
142. A. M. MacMillan and G. L. Verdine, *J. Org. Chem.*, 1990, **55**, 5931.
143. A. M. MacMillan and G. L. Verdine, *Tetrahedron*, 1991, **47**, 2603.
144. C. Scalfi-Happ, E. Happ, and S. Chládek, *Nucleosides Nucleotides*, 1987, **6**, 345.
145. M. D. Hagen, C. Scalfi-Happ, E. Happ, and S. Chládek, *J. Org. Chem.*, 1988, **53**, 5040.
146. J. M. Brown, C. Christodoulou, A. S. Modak, C. B. Reese, and H. T. Serafinowska, *J. Chem. Soc. Perkin Trans. 1*, 1989, 1751.
147. T. S. Rao, C. B. Reese, H. T. Serafinowska, H. Takaku, and G. Zappia, *Tetrahedron Lett.*, 1987, **28**, 4897.
148. E. Happ, C. Scalfi-Happ, G. M. Clore, and A. M. Gronenborn, *Nucleic Acids Symp. Ser.*, 1987, **n18**, 265.
149. M. D. Hagen and S. Chládek, *J. Org. Chem.*, 1989, **54**, 3189.
150. W. Pfleiderer, H. Schirmeister, T. Reiner, M. Pfister, and R. Charubala, in "Biophosphates and Their Analogs—Synthesis, Structure, Metabolism, and Activity," eds. K. S. Bruzik and W. J. Stec, Elsevier, Amsterdam, 1987, p. 133.
151. W. Pfleiderer, M. Schwarz, and H. Schirmeister, *Chem. Scr.*, 1986, **26**, 147.
152. B. L. Gaffney and R. A. Jones, *Tetrahedron Lett.*, 1982, **23**, 2257.
153. T. Trichtinger, R. Charubala, and W. Pfleiderer, *Tetrahedron Lett.*, 1983, **24**, 711.
154. B. S. Schulz and W. Pfleiderer, *Tetrahedron Lett.*, 1983, **24**, 3587.
155. F. Himmelsbach, B. S. Schulz, T. Trichtinger, R. Charubala, and W. Pfleiderer, *Tetrahedron*, 1984, **40**, 59.
156. S. Farkas and W. Pfleiderer, *Nucleic Acids Symp. Ser.*, 1987, **n18**, 153.
157. A. Van Aerschot, M. Mag, P. Herdewijn, and H. Vanderhaeghe, *Nucleosides Nucleotides*, 1989, **8**, 159.
158. A. Van Aerschot, P. Herdewijn, G. Janssen, and H. Vanderhaeghe, *Nucleosides Nucleotides*, 1988, **7**, 519.
159. A. Chollet, E. Ayala, and E. H. Kawashima, *Helv. Chim. Acta*, 1984, **67**, 1356.
160. W. Pfleiderer, F. Himmelsbach, R. Charubala, H. Schirmeister, A. Beiter, B. Schulz, and T. Trichtinger, *Nucleosides Nucleotides*, 1985, **4**, 81.
161. J. Smrt and F. Sorm, *Collect. Czech. Chem. Commun.*, 1967, **32**, 3169.
162. M. Pfister, S. Farkas, R. Charubala, and W. Pfleiderer, *Nucleosides Nucleotides*, 1988, **7**, 595.
163. K.-P. Stengele and W. Pfleiderer, *Nucleic Acids Symp. Ser.*, 1989, **n21**, 101.
164. J. W. Engels and M. Mag, *Nucleosides Nucleotides*, 1982, **6**, 473.
165. C. A. A. Claesen, A. M. A. Pistorius, and G. I. Tesser, *Tetrahedron Lett.*, 1985, **26**, 3859.
166. A. M. A. Pistorius, C. A. A. Claesen, F. J. B. Kremer, E. A. V. Rijk, and G. I. Tesser, *Nucleosides Nucleotides*, 1987, **6**, 389.
167. M. Mag and J. W. Engels, *Nucleic Acids Res.*, 1988, **16**, 3525.
168. Y. Hayakawa, M. Hirose, and R. Noyori, *Nucleic Acids Symp. Ser.*, 1990, **n22**, 1.
169. Y. Hayakawa, M. Hirose, and R. Noyori, *J. Org. Chem.*, 1993, **58**, 5551.
170. B. E. Watkins, J. S. Kiely, and H. Rapoport, *J. Am. Chem. Soc.*, 1982, **104**, 5702.
171. B. E. Watkins and H. Rapoport, *J. Org. Chem.*, 1982, **47**, 4471.
172. H. Tanimura, M. Sekine, and T. Hata, *Tetrahedron Lett.*, 1986, **27**, 4047.
173. H. Takaku, S. Ueda, and Y. Tomita, *Chem. Pharm. Bull.*, 1984, **32**, 2882.
174. H. Takaku, T. Ito, and K. Imai, *Chem. Lett.*, 1986, 1005.
175. H. Takaku, K. Imai, and M. Fujii, *Chem. Scr.*, 1986, **26**, 185.
176. A. Van Aerschot, L. Jie, and P. Herdewijn, *Tetrahedron Lett.*, 1991, **32**, 1905.
177. T. Akiyama, M. Kumegawa, Y. Takesue, H. Nishimoto, and S. Ozaki, *Chem. Lett.*, 1990, 339.
178. M. Krecmerová, H. Hrebabecky, and A. Holy, *Collect. Czech. Chem. Commun.*, 1990, **55**, 2521.
179. T. Ito, S. Ueda, and H. Takaku, *J. Org. Chem.*, 1986, **51**, 931.
180. T. Ito, S. Watanabe, S. Ueda, and H. Takaku, *Nucleic Acids Symp. Ser.*, 1983, **n12**, 51.
181. H. Takaku, S. Ueda, and T. Ito, *Tetrahedron Lett.*, 1983, **24**, 5363.
182. H. Takaku, K. Imai, and M. Nagai, *Chem. Lett.*, 1988, 857.
183. M. Sekine, *J. Org. Chem.*, 1989, **54**, 2321.
184. C. J. Welch, H. Bazin, J. Heikkilä, and J. Chattopadhyaya, *Acta Chem. Scand.*, 1985, **B39**, 203.
185. M. Sekine, J. Matsuzaki, M. Satoh, and T. Hata, *J. Org. Chem.*, 1982, **47**, 571.
186. S. Yamakage, M. Fujii, Y. Horinouchi, and H. Takaku, *Nucleic Acids Symp. Ser.*, 1988, **n19**, 9.
187. M. Fujii, Y. Horinouchi, and H. Takaku, *Chem. Pharm. Bull.*, 1987, **35**, 3066.
188. H. Takaku, K. Imai, and K. Nakayama, *Chem. Lett.*, 1987, 1787.
189. M. Fujii, S. Yamakage, H. Takaku, and T. Hata, *Tetrahedron Lett.*, 1987, **28**, 5713.
190. A. Nyilas, A. Földesi, and J. Chattopadhyaya, *Nucleosides Nucleotides*, 1988, **7**, 787.
191. Y. Hayakawa, H. Kato, M. Uchiyama, H. Kajino, and R. Noyori, *J. Org. Chem.*, 1986, **51**, 2400.
192. Y. Hayakawa, H. Kato, T. Nobori, R. Noyori, and J. Imai, *Nucleic Acids Symp. Ser.*, 1986, **n17**, 97.
193. T. Kamimura, T. Masegi, K. Urakami, S. Honda, M. Sekine, and T. Hata, *Chem. Lett.*, 1983, 1051.
194. T. Kamimura, T. Masegi, and T. Hata, *Chem. Lett.*, 1982, 965.
195. T. Kamimura, T. Masegi, M. Sekine, and T. Hata, *Tetrahedron Lett.*, 1984, **25**, 4241.
196. X.-X. Zhou, I. Ugi, and J. Chattopadhyaya, *Acta Chem. Scand.*, 1985, **B39**, 761.
197. T. Kamimura, M. Tsuchiya, K. Urakami, K. Koura, M. Sekine, K. Shinozaki, K. Miura, and T. Hata, *J. Am. Chem. Soc.*, 1984, **106**, 4552.
198. T. Kamimura, M. Tsuchiya, K. Koura, M. Sekine, and T. Hata, *Tetrahedron Lett.*, 1983, **24**, 2775.
199. K. Kamaike, Y. Hasegawa, and Y. Ishido, *Nucleosides Nucleotides*, 1988, **7**, 37.
200. T. Hata and M. Sekine, in "Natural Products Chemistry 1984," eds. R. I. Zalewski and J. J. Skolik, Elsevier, Amsterdam, 1985, p. 239.
201. M. Sekine and T. Nakanishi, *Chem. Lett.*, 1991, 121.
202. M. Sekine, S. Iimura, and T. Nakanishi, *Tetrahedron Lett.*, 1991, **32**, 395.
203. M. Sekine, J. Heikkila, and T. Hata, *Tetrahedron Lett.*, 1987, **28**, 5691.
204. H. Tanimura, T. Fukazawa, M. Sekine, T. Hata, J. W. Efcavitch, and G. Zon, *Tetrahedron Lett.*, 1988, **29**, 577.

205. M. Sekine, S. Nishiyama, T. Kamimura, Y. Osaki, and T. Hata, *Bull. Chem. Soc. Jpn.*, 1985, **58**, 850.
206. J. E. Marugg, J. Nielsen, O. Dahl, A. Burik, G. A. van der Marel, and J. H. van Boom, *Recl. Trav. Chim. Pays-Bas*, 1987, **106**, 72.
207. J. Nielsen, M. Taagaard, J. E. Marugg, J. H. van Boom, and O. Dahl, *Nucleic Acids Res.*, 1986, **14**, 7391.
208. A. Sandström, M. Kwiatkowski, and J. Chattopadhyaya, *Acta Chem. Scand.*, 1985, **B39**, 273.
209. M. Kwiatkowski, J. Heikkilä, C. J. Welch, and J. Chattopadhyaya, in "Natural Products Chemistry 1984," eds. R. I. Zalewski and J. J. Skolik, Elsevier, Amsterdam, 1985, p. 259.
210. T. Kamimura, T. Masegi, K. Urakami, M. Tsuchiya, K. Koura, J. Matsuzaki, K. Y. Shinozaki, M. Sekine, K. Miura, and T. Hata, *Nucleic Acids Symp. Ser.*, 1983, **n12**, 63.
211. C. J. Welch and J. Chattopadhyaya, *Acta Chem. Scand.*, 1983, **B37**, 147.
212. H. Schaller, G. Weimann, B. Lerch, and H. G. Khorana, *J. Am. Chem. Soc.*, 1963, **85**, 3821.
213. G. S. Ti, B. L. Gaffney, and R. A. Jones, *J. Am. Chem. Soc.*, 1982, **104**, 1316.
214. F. Seela and H. Driller, *Nucleosides Nucleotides*, 1989, **0**, 1.
215. H. Köster, K. Kulikowski, T. Liese, W. Heikens, and V. Kohli, *Tetrahedron*, 1981, **37**, 363.
216. C. Chaix, D. Molko, and R. Téoule, *Tetrahedron Lett.*, 1989, **30**, 71.
217. C. Chaix, A. M. Duplaa, D. Gasparutto, D. Molko, and R. Téoule, *Nucleic Acids Symp. Ser.*, 1989, **n21**, 45.
218. C. Chaix, A. M. Duplaa, D. Molko, and R. Téoule, *Nucleic Acids Res.*, 1989, **17**, 7381.
219. B. Beijer, I. Sulston, B. S. Sproat, P. Rider, A. I. Lamond, and P. Neuner, *Nucleic Acids Res.*, 1990, **18**, 5143.
220. E. L. Brown, R. Belagaje, M. J. Ryan, and H. G. Khorana, in "Methods in Enzymology," ed. R. Wu, Academic Press, New York, 1979, vol. 68, p. 109.
221. H. Büchi and H. G. Khorana, *J. Mol. Biol.*, 1972, **72**, 251.
222. H. Vu, C. McCollum, K. Jacobson, P. Theisen, R. Vinayak, E. Spiess, and A. Andrus, *Tetrahedron Lett.*, 1990, **31**, 7269.
223. J. C. Schulhof, D. Molko, and R. Teoule, *Nucleic Acids Res.*, 1987, **15**, 397.
224. J. C. Schulhof, D. Molko, and R. Teoule, *Tetrahedron Lett.*, 1987, **28**, 51.
225. J. C. Schulhof, D. Molko, and R. Teoule, in "Biophosphates and their Analogs—Synthesis, Structure, Metabolism and Activity," eds. K. S. Bruzik and W. J. Stec, Elsevier, Amsterdam, 1987, p. 143.
226. J. C. Schulhof, D. Molko, and R. Teoule, *Nucleic Acids Res.*, 1988, **16**, 319.
227. R. L. Letsinger and P. S. Miller, *J. Am. Chem. Soc.*, 1969, **91**, 3356.
228. R. G. K. Schneiderwind and I. Ugi, *Z. Naturforsch. B. Anorg. Chem. Org. Chem.*, 1981, **36**, 1173.
229. R. G. K. Schneiderwind and I. Ugi, *Tetrahedron*, 1983, **39**, 2207.
230. H. Eckert, W. Breuer, J. Geller, I. Lagerlund, M. Listl, D. Marquarding, S. Stüber, I. Ugi, S. Zahr, and H. V. Zynchlinski, *Pure Appl. Chem.*, 1979, **51**, 1219.
231. Y. Mitsuhira, S. Tahara, K. Goto, Y. Hayakawa, and R. Noyori, *Nucleic Acids Symp. Ser.*, 1988, **n19**, 25.
232. Y. Hayakawa, S. Wakabayashi, H. Kato, and R. Noyori, *J. Am. Chem. Soc.*, 1990, **112**, 1691.
233. Y. Hayakawa, M. Hirose, and R. Noyori, *Nucleic Acids Symp. Ser.*, 1989, **n21**, 103.
234. Y. Hayakawa, S. Wakabayashi, and R. Noyori, *Nucleic Acids Symp. Ser.*, 1988, **n20**, 75.
235. B. Uznanski, A. Grajkowski, and A. Wilk, *Nucleic Acids Res.*, 1989, **17**, 4863.
236. K. K. Ogilvie, M. J. Nemer, G. H. Hakimelahi, Z. A. Proba, and M. Lucas, *Tetrahedron Lett.*, 1982, **23**, 2615.
237. R. P. Iyer, D. Yu, I. Habus, N.-H. Ho, S. Johnson, T. Devlin, Z. Jiang, W. Zhou, J. Xie, and S. Agrawal, *Tetrahedron*, 1997, **53**, 2731.
238. R. P. Iyer, T. Devlin, I. Habus, D. Yu, S. Johnson, and S. Agrawal, *Tetrahedron Lett.*, 1996, **37**, 1543.
239. M. Pfister and W. Pfleiderer, *Nucleosides Nucleotides*, 1989, **8**, 1001.
240. T. Reiner, and W. Pfleiderer, *Nucleic Acids Symp. Ser.*, 1987, **n18**, 161.
241. M. Pfister and W. Pfleiderer, *Nucleic Acids Symp. Ser.*, 1987, **n18**, 165.
242. R. Charubala and W. Pfleiderer, *Nucleosides Nucleotides*, 1987, **6**, 517.
243. F. Himmelsbach and W. Pfleiderer, *Tetrahedron lett.*, 1983, **24**, 3583.
244. P. Herdewijn, K. Ruf, and W. Pfleiderer, *Helv. Chim. Acta*, 1991, **74**, 7.
245. J. Chattopadhyaya and C. B. Reese, *J. Chem. Soc. Chem. Commun.*, 1978, 639.
246. N. Balgobin, S. Josephson, and J. B. Chattopadhyaya, *Acta Chem. Scand.*, 1981, **B35**, 201.
247. N. Piel, F. Benseler, E. Graeser, and L. W. McLaughlin, *Bioorg. Chem.*, 1985, **13**, 323.
248. F. Benseler and L. W. McLaughlin, *Synthesis*, 1986, 45.
249. J. Heikkilä and J. Chattopadhyaya, *Acta Chem. Scand.*, 1983, **B37**, 263.
250. E. Happ, C. Scalfi-Happ, and S. Chládek, *J. Org. Chem.*, 1987, **52**, 5387.
251. L. H. Koole, H. M. Moody, N. L. H. L. Broeders, P. J. L. M. Quaedflieg, W. H. A. Kuijpers, M. H. P. van Genderen, A. J. J. M. Coenen, S. van der Wal, and H. M. Buck, *J. Org. Chem.*, 1989, **54**, 1657.
252. W. H. A. Kuijpers, J. Huskens, L. H. Koole, and C. A. A. van Boeckel, *Nucleic Acids Res.*, 1990, **18**, 5197.
253. T. R. Webb and M. D. Matteucci, *Nucleic Acids Res.*, 1986, **14**, 7661.
254. J. E. Marugg, M. Tromp, P. Jhurani, C. F. Hoyng, G. A. van der Marel, and J. H. van Boom, *Tetrahedron*, 1984, **40**, 73.
255. J. E. Marugg, A. Burik, M. Tromp, G. A. van der Marel, and J. H. van Boom, *Tetrahedron Lett.*, 1986, **27**, 2271.
256. A. K. Nagaich and K. Misra, *Nucleic Acids Res.*, 1989, **17**, 5125.
257. B. S. Sproat, A. M. Iribarren, R. Guimil Garcia, and B. Beijer, *Nucleic Acids Res.*, 1991, **19**, 733.
258. R. K. Singh and K. Misra, *Indian J. Chem.*, 1988, **27B**, 409.
259. N. D. Sinha, P. Davis, N. Usman, J. Pérez, R. Hodge, J. Kremsky, and R. Casale, *Biochimie*, 1993, **75**, 13.
260. J. H. Boal, A. Wilk, N. Harindranath, E. E. Max, T. Kempe, and S. L. Beaucage, *Nucleic Acids Res.*, 1996, **24**, 3115.
261. J. B. Chattopadhyaya and C. B. Reese, *Nucleic Acids Res.*, 1980, **8**, 2039.
262. N. Balgobin and J. B. Chattopadhyaya, *Acta Chem. Scand.*, 1985, **B39**, 883.
263. H. G. Khorana, A. F. Turner, and J. P. Vizsolyi, *J. Am. Chem. Soc.*, 1961, **83**, 686.
264. J. Matsuzaki, K. Kohno, S. Tahara, M. Sekine, and T. Hata, *Bull. Soc. Chem. Jpn.*, 1987, **60**, 1407.
265. J. Igolen and C. Morin, *J. Org. Chem.*, 1980, **45**, 4802.
266. J. H. van Boom and C. T. J. Wreesmann, in "Oligonucleotides Synthesis: A Practical Approach," ed. M. J. Gait, IRL Press, Oxford, 1984, p. 153.

267. N. Balgobin and J. B. Chattopadhyaya, *Chem. Scr.*, 1982, **19**, 143.
268. R. K. Mishra and K. Misra, *Nucleic Acids Res.*, 1986, **14**, 6197.
269. R. K. Mishra and K. Misra, *Indian J. Chem.*, 1988, **27B**, 817.
270. C. M. Dreef-Tromp, P. Hoogerhout, G. A. van der Marel, and J. H. van Boom, *Tetrahedron Lett.*, 1990, **31**, 427.
271. C. M. Dreef-Tromp, E. M. A. van Dam, H. van den Elst, G. A. van der Marel, and J. H. van Boom, *Nucleic Acids Res.*, 1990, **18**, 6491.
272. W. H. A. Kuijpers, J. Huskens, and C. A. A. van Boeckel, *Tetrahedron Lett.*, 1990, **31**, 6729.
273. A. Dikshit, M. Chaddha, R. K. Singh, and K. Misra, *Can. J. Chem.*, 1988, **66**, 2989.
274. K. A. Watanabe and J. J. Fox, *Angew. Chem. Int. Ed. Engl.*, 1966, **5**, 579.
275. V. Bhat, B. G. Ugarkar, V. A. Sayeed, K. Grimm, N. Kosora, and P. Domenico, *Nucleosides Nucleotides*, 1989, **8**, 179.
276. R. S. Goody and W. T. Walker, *J. Org. Chem.*, 1971, **36**, 727.
277. M. P. Reddy, N. B. Hanna, and F. Farooqui, *Tetrahedron Lett.*, 1994, **35**, 4311.
278. J. A. Zoltewicz, D. F. Clark, T. W. Sharpless, and G. Grahe, *J. Am. Chem. Soc.*, 1970, **92**, 1741.
279. J. A. Zoltewicz and D. F. Clark, *J. Org. Chem.*, 1972, **37**, 1193.
280. M. Oivanen, H. Lönnberg, X.-X. Zhou, and J. Chattopadhyaya, *Tetrahedron*, 1987, **43**, 1133.
281. R. Romero, R. Stein, H. G. Bull, and E. H. Cordes, *J. Am. Chem. Soc.*, 1978, **100**, 7620.
282. T. Suzuki, S. Ohsumi, and K. Makino, *Nucleic Acids Res.*, 1994, **22**, 4997.
283. L. Arnold, Z. Tocík, E. Bradková, Z. Hostomsky, V. Paces, and J. Smrt, *Collect. Czech. Chem. Commun.*, 1989, **54**, 523.
284. L. Arnold, Z. Tocík, E. Bradková, Z. Hostomsky, and J. Smrt, *Nucleic Acids Symp. Ser.*, 1987, **n18**, 181.
285. M. H. Caruthers, L. J. McBride, L. P. Bracco, and J. W. Dubendorff, *Nucleosides Nucleotides*, 1985, **4**, 95.
286. L. J. McBridge, R. Kierzek, S. L. Beaucage, and M. H. Caruthers, *J. Am. Chem. Soc.*, 1986, **108**, 2040.
287. M. D. Hagen and S. Chládek, *J. Org. Chem.*, 1989, **54**, 3189.
288. A. Holy and J. Zemlicka, *Collect. Czech. Chem. Commun.*, 1969, **34**, 2449.
289. B. S. Sproat and M. J. Gait, in "Chemical Synthesis in Molecular Biology," G. B. F. Monographs, eds. H. Blöcker, R. Frank, and H.-J. Fritz, VCH, Weinheim, 1987, vol. 8, p. 13.
290. B. C. Froehler and M. D. Matteucci, *Nucleic Acids Res.*, 1983, **11**, 8031.
291. V. G. Metelev, N. F. Krynetskaya, A. A. Purmal, Z. A. Shabarova, Z. Tocík, L. Arnold, and J. Smrt, *Collect. Czech. Chem. Commun.*, 1990, **55**, 2781.
292. P. Theisen, C. McCollum, and A. Andrus, *Nucleosides Nucleotides*, 1993, **12**, 1033.
293. B. Mullah, A. Andrus, H. Zhao, and R. A. Jones, *Tetrahedron Lett.*, 1995, **36**, 4373.
294. A. Kume, R. Iwase, M. Sekine, and T. Hata, *Nucleic Acids Res.*, 1984, **12**, 8525.
295. A. Kume, M. Sekine, and T. Hata, *Tetrahedron Lett.*, 1982, **23**, 4365.
296. T. Shimidzu and R. L. Letsinger, *J. Org. Chem.*, 1968, **33**, 708.
297. S. Honda, K. Urakami, K. Koura, K. Terada, Y. Sato, K. Kohno, M. Sekine, and T. Hata, *Tetrahedron*, 1984, **40**, 153.
298. T. Shimidzu and R. L. Letsinger, *Bull. Soc. Chem. Jpn.*, 1971, **44**, 1673.
299. M. Sekine, N. Masuda, and T. Hata, *Tetrahedron*, 1985, **41**, 5445.
300. A. K. Prasad and J. Wengel, *Nucleosides Nucleotides*, 1996, **15**, 1347.
301. J.-L. Fourrey and J. Varenne, *Tetrahedron Lett.*, 1985, **26**, 2663.
302. S. A. Narang, K. Itakura, and R. H. Wightman, *Can. J. Chem.*, 1972, **50**, 769.
303. R. W. Adamiak, E. Biala, K. Grzekowiak, P. Kierzek, A. Kraszewski, W. T. Markiewicz, J. Okupniak, J. Stawinski, and M. Wiewiorowski, *Nucleic Acids Res.*, 1978, **5**, 1889.
304. S. M. Gryaznov and R. L. Letsinger, *J. Am. Chem. Soc.*, 1991, **113**, 5876.
305. M. Uchiyama, Y. Aso, and R. Noyori, *J. Org. Chem.*, 1993, **58**, 373.
306. C. H. Paul and A. T. Royappa, *Nucleic Acids Res.*, 1996, **24**, 3048.
307. A. A. Padmapriya, J.-Y. Tang, and S. Agrawal, *Antisense Res. Dev.*, 1994, **4**, 185.
308. M. Smith, D. H. Rammler, I. H. Goldberg, and H. G. Khorana, *J. Am. Chem. Soc.*, 1962, **84**, 430.
309. S. A. Narang, H. M. Hsiung, and R. Brousseau, in "Methods in Enzymology," ed. R. Wu, Academic Press, New York, 1979, p. 90.
310. J. Engels, *Angew. Chem. Int. Ed. Engl.*, 1979, **18**, 148.
311. M. D. Matteucci and M. H. Caruthers, *Tetrahedron Lett.*, 1980, **21**, 3243.
312. V. Kohli, H. Blöcker, and H. Köster, *Tetrahderon Lett.*, 1980, **21**, 2683.
313. H. Köster and N. D. Sinha, *Tetrahedron Lett.*, 1982, **23**, 2641.
314. J. R. Hwu, M. L. Jain, S.-C. Tsay, and G. H. Hakimelahi, *J. Chem. Soc. Chem. Commun.*, 1996, 545.
315. N. J. Leonard and Neelima, *Tetrahedron Lett.*, 1995, **36**, 7833.
316. M. Sekine, *Nucleosides Nucleotides*, 1994, **13**, 1397.
317. E. F. Fisher and M. H. Caruthers, *Nucleic Acids Res.*, 1983, **11**, 1589.
318. M. Sekine, S. Narui, and T. Hata, *Nucleic Acids Symp. Ser.*, 1987, **n18**, 281.
319. M. Sekine, T. Mori, and T. Wade, *Tetrahedron Lett.*, 1993, **34**, 8299.
320. M. Sekine and T. Hata, *J. Org. Chem.*, 1983, **48**, 3011.
321. M. Sekine and T. Hata, *Bull. Chem. Soc. Jpn.*, 1985, **58**, 336.
322. M. Sekine and T. Hata, *J. Am. Chem. Soc.*, 1984, **106**, 5763.
323. M. Sekine and T. Hata, *J. Am. Chem. Soc.*, 1986, **108**, 4581.
324. E. Leikauf, F. Barnekow, and H. Köster, *Tetrahderon*, 1996, **52**, 5557.
325. E. Leikauf, F. Barnekow, and H. Köster, *Tetrahedron*, 1995, **51**, 6913.
326. E. Happ and C. Scalfi-Happ, *Nucleosides Nucleotides*, 1988, **7**, 813.
327. H. Seliger, K. C. Gupta, U. Kotschi, T. Spaney, and D. Zeh, *Chem. Scr.*, 1986, **26**, 561.
328. J. Weiler and W. Pfleiderer, *Nucleosides Nucleotides*, 1995, **14**, 917.
329. F. Bergmann and W. Pfleiderer, *Helv. Chim. Acta*, 1994, **77**, 203.
330. J. M. Brown, C. Christodoulou, C. B. Reese, and G. Sindona, *J. Chem. Soc. Perkin Trans. 1*, 1984, 1785.
331. C. B. Reese, *Nucleosides Nucleotides*, 1985, **4**, 117.
332. C. Christodoulou, S. Agrawal, and M. J. Gait, *Nucleosides Nucleotides*, 1987, **6**, 341.

333. C. Christodoulou, S. Agrawal, and M. J. Gait, in "Biophosphates and Their Analogs—Synthesis, Structure, Metabolism, and Activity," eds. K. S. Bruzik and W. J. Stec, Elsevier, Amsterdam, 1987, p. 99.
334. H. Seliger and G. Schmidt, *J. Chromatogr.*, 1987, **397**, 141.
335. J. L. Fourrey, J. Varenne, C. Blonski, P. Dousset, and D. Shire, *Tetrahderon Lett.*, 1987, **28**, 5157.
336. H.-H. Gortz and H. Seliger, *Angew. Chem. Int. Ed. Engl.*, 1981, **20**, 681.
337. H.-H. Gortz and H. Seliger, *Angew. Chem. Int. Ed. Engl.*, 1981, **20**, 683.
338. M. Kwiatkowski and J. B. Chattopadhyaya, *Acta Chem. Scand.*, 1984, **B38**, 657.
339. R. Ramage and F. O. Wahl, *Tetrahderon Lett.*, 1993, **34**, 7133.
340. S. Takenaka, K. Dohtsa, and M. Takagi, *Anal. Sci.*, 1992, **8**, 3.
341. M. C. Pirrung and J.-C. Bradley, *J. Org. Chem.*, 1995, **60**, 1116.
342. M. C. Pirrung, L. Fallon, D. C. Lever, and S. W. Shuey, *J. Org. Chem.*, 1996, **61**, 2129.
343. M. C. Pirrung and S. W. Shuey, *J. Org. Chem.*, 1994, **59**, 3890.
344. M. C. Pirrung and Y. R. Lee, *J. Org. Chem.*, 1993, **58**, 6961.
345. A. Bidaine, C. Berens, and E. Sonveaux, *Bioorg. Med. Chem. Lett.*, 1996, **6**, 1167.
346. K. K. Ogilvie, N. Y. Theriault, J.-M. Seifert, R. T. Pon, and M. J. Nemer, *Can. J. Chem.*, 1980, **58**, 2686.
347. F. Eckstein and I. Rizk, *Angew. Chem. Int. Ed. Engl.*, 1967, **6**, 695.
348. F. Eckstein and I. Rizk, *Angew. Chem. Int. Ed. Engl.*, 1967, **6**, 949.
349. R. W. Adamiak, E. Biala, K. Grzeskowiak, R. Kierzek, A. Kraszewski, W. T. Markiewicz, J. Stawinski, and M. Wiewiorowski, *Nucleic Acids Res.*, 1977, **4**, 2321.
350. J. H. van Boom, P. M. J. Burgers, R. Crea, G. van der Marel, and G. Willie, *Nucleic Acids Res.*, 1977, **4**, 747.
351. R. L. Letsinger and K. K. Ogilvie, *J. Am. Chem. Soc.*, 1967, **89**, 4801.
352. G. M. Tener, *J. Am. Chem. Soc.*, 1961, **83**, 159.
353. G. W. Daub and E. E. Van Tamelen, *J. Am. Chem. Soc.*, 1977, **99**, 3526.
354. D. J. H. Smith, K. K. Ogilvie, and M. F. Gillen, *Tetrahderon Lett.*, 1980, **21**, 861.
355. K. Itakura, N. Katagiri, C. P. Bahl, R. H. Wightman, and S. A. Narang, *J. Am. Chem. Soc.*, 1975, **97**, 7327.
356. C. B. Reese, R. C. Titmas, and L. Yau, *Tetrahderon Lett.*, 1978, 2727.
357. E. Uhlmann and W. Pfleiderer, *Tetrahedron Lett.*, 1980, **21**, 1181.
358. R. L. Letsinger, E. P. Groody, N. Lander, and T. Tanaka, *Tetrahedron*, 1984, **40**, 137.
359. R. L. Letsinger, E. P. Groody, and T. Tanaka, *J. Am. Chem. Soc.*, 1982, **104**, 6805.
360. R. G. K. Schneiderwind-Stöcklein and I. Ugi, *Z. Naturforsch. B*, 1984, **39**, 968.
361. J. E. Marugg, N. Piel, L. W. McLaughlin, M. Tromp, G. H. Veeneman, G. A. van der Marel, and J. H. van Boom, *Nucleic Acids Res.*, 1984, **12**, 8639.
362. K. Jayaraman and H. McClaugherty, *Tetrahedron Lett.*, 1982, **23**, 5377.
363. L. J. McBride and M. H. Caruthers, *Tetrahedron Lett.*, 1983, **24**, 245.
364. J.-L. Fourrey and J. Varenne, *Tetrahedron Lett.*, 1983, **24**, 1963.
365. J.-L. Fourrey and J. Varenne, *Tetrahedron Lett.*, 1984, **25**, 4511.
366. T. Tanaka, S. Tamatsukuri, and M. Ikehara, *Tetrahedron Lett.*, 1986, **27**, 199.
367. N. D. Sinha, J. Biernat, and H. Köster, *Tetrahedron Lett.*, 1983, **24**, 5843.
368. N. D. Sinha, J. Biernat, J. McManus, and H. Köster, *Nucleic Acids Res.*, 1984, **12**, 4539.
369. C. Claesen, G. I. Tesser, C. E. Dreef, J. E. Marugg, G. A. van der Marel, and J. H. van Boom, *Tetrahedron Lett.*, 1984, **25**, 1307.
370. M. W. Schwarz and W. Pfleiderer, *Tetrahedron Lett.*, 1984, **25**, 5513.
371. M. W. Schwarz and W. Pfleiderer, *Nucleosides Nucleotides*, 1985, **4**, 291.
372. A. H. Beiter and W. Pfleiderer, *Tetrahedron Lett.*, 1984, **25**, 1975.
373. A. M. Avino and R. Eritja, *Nucleosides Nucleotides*, 1994, **13**, 2059.
374. J. E. Marugg, C. E. Dreef, G. A. van der Marel, and J. H. van Boom, *Recl. Trav. Chim. Pays-Bas*, 1984, **103**, 97.
375. G. Hering, R. Stöcklein-Schneiderwind, I. Ugi, T. Pathak, N. Balgobin, and J. Chattopadhyaya, *Nucleosides Nucleotides*, 1985, **4**, 169.
376. I. Ugi, J. Bauer, E. Fontain, J. Götz, G. Hering, P. Jakob, B. Landgraf, R. Karl, P. Lemmen, R. Stöklein-Schneiderwind, R. Schwarz, S. Sluka, N. Balgobin, J. Chattopadhyaya, T. Pathak, and X.-X. Zhou, *Chem. Scr.*, 1986, **26**, 205.
377. T. Wada and M. Sekine, *Tetrahedron Lett.*, 1994, **35**, 757.
378. F. Bergmann, E. Kueng, P. Iaiza, and W. Bannwarth, *Tetrahderon*, 1995, **51**, 6971.
379. V. T. Ravikumar, Z. S. Cheruvallath, and D. L. Cole, *Tetrahedron Lett.*, 1996, **37**, 6643.
380. V. T. Ravikumar, T. K. Wyrzykiewicz, and D. L. Cole, *Tetrahedron*, 1994, **50**, 9255.
381. A. H. Krotz, D. L. Cole, and V. T. Ravikumar, *Tetrahedron Lett.*, 1996, **37**, 1999.
382. S.-G. Kim, K. Eida, and H. Takaku, *Bioorg. Med. Chem. Lett.*, 1995, **5**, 1663.
383. W. T. Wiesler and M. H. Caruthers, *J. Org. Chem.*, 1996, **61**, 4272.
384. R. P. Iyer, D. Yu, T. Devlin, N.-H. Ho, and S. Agrawal, *J. Org. Chem.*, 1995, **60**, 5388.
385. R. P. Iyer, D. Yu, N.-H. Ho, W. Tan, and S. Agrawal, *Tetrahedron: Asymmetry*, 1995, **6**, 1051.
386. W. J. Stec and A. Wilk, *Angew. Chem. Intl. Ed. Engl.*, 1994, **33**, 709.
387. A. Sierzchala, A. Okruszek, and W. J. Stec, *J. Org. Chem.*, 1996, **61**, 6713.
388. A. B. Ouryupin, M. Y. Kadyko, P. V. Petrovskii, E. I. Fedin, A. Okruszek, R. Kinas, and W. J. Stec, *Tetrahedron: Asymmetry*, 1995, **6**, 1813.
389. S. L. Beaucage, *Tetrahedron Lett.*, 1984, **25**, 375.
390. A. D. Barone, J.-Y. Tang, and M. H. Caruthers, *Nucleic Acids Res.*, 1984, **12**, 4051.
391. H.-J. Lee and S.-H. Moon, *Chem. Lett.*, 1984, 1229.
392. J. Nielsen, J. E. Marugg, J. H. van Boom, J. Honnens, M. Taagaard, and O. Dahl, *J. Chem. Res. (S)*, 1986, 26.
393. S. Hamamoto and H. Takaku, *Chem. Lett.*, 1986, 1401.
394. J. Nielsen and O. Dahl, *Nucleic Acids Res.*, 1987, **15**, 3626.
395. M. H. Caruthers, R. Kierzek, and J.-Y. Tang in "Biophosphates and Their Analogs Synthesis, Structure, Metabolism, and Activity," eds. K. S. Bruzik and W. J. Stec, Elsevier, Amsterdam, 1987, p. 3.
396. H. Takaku, T. Watanabe, and S. Hamamoto, *Nucleosides Nucleotides*, 1987, **6**, 293.
397. S. Hamamoto, Y. Shishido, M. Furuta, H. Takaku, M. Kawashima, and M. Takaki, *Nucleosides Nucleotides*, 1989, **8**, 317.

398. Y. Hayakawa, M. Uchiyama, H. Kato, and R. Noyori, *Tetrahderon Lett.*, 1985, **26**, 6505.
399. R. Eritja, V. Smirnov, and M. H. Caruthers, *Tetrahedron*, 1990, **46**, 721.
400. H. Takaku, T. Watanabe, and S. Hamamoto, *Tetrahedron Lett.*, 1988, **29**, 81.
401. S. Agrawal, W. Tan, Z. Jiang, D. Yu, and R. P. Iyer, in "Antisense Oligonucleotides from Technology to Therapy," eds. R. Schlingensiepen, W. Brysch, and K. H. Schlingensiepen, Blackwell Science, Oxford, 1997, p. 59.
402. T. Furuta, H. Torigai, T. Osawa, and M. Iwamura, *Chem. Lett.*, 1993, 1179.
403. K. K. Ogilvie, A. L. Schifman, and C. L. Penney, *Can. J. Chem.*, 1979, **57**, 2230.
404. K. K. Ogilvie, N. Theriault, and K. L. Sadana, *J. Am. Chem. Soc.*, 1977, **99**, 7741.
405. E. E. Nifant'ev and N. L. Ivanov, *Vestn. Mosk. Univ. Ser 2: Khim.*, 1968, **23**, 104 (Engl. Trans. 78).
406. H. Seliger and K. C. Gupta, *Angew Chem. Int. Ed. Engl.*, 1985, **24**, 685.
407. H. Seliger and K. C. Gupta, *Nucleosides Nucleotides*, 1985, **4**, 249.
408. S. Berner, K. Mühlegger, and H. Seliger, *Nucleic Acids Res.*, 1989, **17**, 853.
409. S. Berner, K. Mühlegger, and H. Seliger, *Nucleosides Nucleotides*, 1988, **7**, 763.
410. E. S. Batyeva, V. A. Al'fonsov, G. U. Zamaletdinova, and A. N. Pudovik, *J. Gen. Chem. USSR*, 1976, **46**, 2120.
411. T. A. Van der Knaap and F. Bickelhaupt, *Phosphorus Sulfur*, 1984, **21**, 227.
412. O. Dahl, *Phosphorous Sulfur*, 1983, **18**, 201.
413. W. J. Stec and G. Zon, *Tetrahedron Lett.*, 1984, **25**, 5279.
414. B. H. Dahl, J. Nielsen, and O. Dahl, *Nucleic Acids Res.*, 1987, **15**, 1729.
415. B. H. Dahl, J. Nielsen, and O. Dahl, *Nucleosides Nucleotides*, 1987, **6**, 457.
416. Y. Hayakawa, M. Kataoka, and R. Noyori, *J. Org. Chem.*, 1996, **61**, 7996.
417. A. Andrus and G. Zon, *Nucleic Acids Symp. Ser.*, 1988, **n20**, 121.
418. J. Jankowska, M. Sobkowski, J. Stawinski, and A. Kraszewski, *Tetrahedron Lett.*, 1994, **35**, 3355.
419. Z.-W. Yang, Z.-S. Xu, N.-Z. Shen, and Z.-Q. Fang, *Nucleosides Nucleotides*, 1995, **14**, 167.
420. N. N. Bhongle and J.-Y. Tang, *Tetrahedron Lett.*, 1995, **36**, 6803.
421. T. Szabo, H. Almer, R. Strömberg, and J. Stawinski, *Nucleosides Nucleotides*, 1995, **14**, 715.
422. A. Kers, I. Kers, A. Kraszewski, M. Sobkowski, T. Szabo, M. Thelin, R. Zain, and J. Stawinski, *Nucleosides Nucleotides*, 1996, **15**, 361.
423. B. C. Froehler and M. D. Matteucci, *Nucleosides Nucleotides*, 1987, **6**, 287.
424. P. J. Garegg, T. Regberg, J. Stawinski, and R. Strömberg, *Nucleosides Nucleotides*, 1987, **6**, 283.
425. E. Rozners, R. Renhofa, M. Petrova, J. Popelis, V. Kumkins, and E. Bizdona, *Nucleosides Nucleotides*, 1992, **11**, 1579.
426. J. Stawinski, M. Thelin, and R. Zain, *Tetrahedron Lett.*, 1989, **30**, 2157.
427. V. A. Efimov, A. L. Kalinkina, and O. G. Chakhmakhcheva, *Nucleic Acids Res.*, 1993, **21**, 5337.
428. B. C. Froehler, *Tetrahedron Lett.*, 1986, **27**, 5575.
429. S. Agrawal and J.-Y. Tang, *Tetrahedron Lett.*, 1990, **31**, 7541.
430. R. L. Letsinger, C. N. Singman, G. Histand, and M. Salunkhe, *J. Am. Chem. Soc.*, 1988, **110**, 4470.
431. M. Sobkowski, J. Stawinski, and A. Kraszewski, *Tetrahedron Lett.*, 1995, **36**, 2295.
432. J. Stawinski and R. Zain, *Nucleosides Nucleotides*, 1995, **14**, 711.
433. J. Stawinski and M. Thelin, *J. Org. Chem.*, 1994, **59**, 130.
434. P. H. Seeberger and M. H. Caruthers, *Tetrahedron Lett.*, 1995, **36**, 695.
435. W. K.-D. Brill, *Tetrahderon Lett.*, 1995, **36**, 703.
436. J. Stawinski, R. Strömberg, and R. Zain, *Tetrahedron Lett.*, 1992, **33**, 3185.
437. H. Almer, J. Stawinski, and R. Strömberg, *J. Chem. Soc. Chem. Commun.*, 1994, 1459.
438. J. S. Pudlo, S. Wadhwani, J. F. Milligan, and M. D. Matteucci, *Bioorg. Med. Chem. Lett.*, 1994, **4**, 1025.
439. P.-P. Kung and R. A. Jones, *Tetrahedron Lett.*, 1992, **33**, 5869.
440. R. T. Pon, G. A. Buck, R. L. Niece, M. Robertson, A. J. Smith, and E. Spicer, *BioTechniques*, 1994, **17**, 526.
441. E. Uhlmann and A. Peyman, *Chem. Rev.*, 1990, **90**, 544.
442. S. Agrawal and R. P. Iyer, *Curr. Op. Biotechnol.*, 1995, **6**, 12.
443. S. Agrawal (ed.), "Antisense Therapeutics," Humana, Totowa, NJ, 1996.
444. S. T. Crooke and B. Lebleu (eds.), "Antisense Research and Applications," CRC Press, Boca Raton, FL, 1993.
445. N. T. Thuong and C. Hélène, *Angew Chem. Int. Ed. Engl.*, 1993, **32**, 666.
446. R. P. Iyer, W. Egan, J. B. Regan, and S. L. Beaucage, *J. Am. Chem. Soc.*, 1990, **112**, 1253.
447. T. K. Wyrzykiewicz and V. T. Ravikumar, *Bioorg. Med. Chem. Lett.*, 1994, **12**, 1519; see also, Q. H. Xu, K. Musier-Forsyth, R. P. Hammer, and G. Barany, *Nucleic Acids Res.*, 1996, **24**, 1602.
448. N. D. Sinha, in "Protocols for Oligonucleotides and Analogs," ed. S. Agrawal, Humana, Totowa, NJ, 1993, vol. 20, p. 437.
449. C. L. Scremin, L. Zhou, K. Srinivasachar, and S. L. Beaucage, *J. Org. Chem.*, 1994, **59**, 1963.
450. J. A. Gerstnev, P. Pedroso, J. Morris, and B. J. Bergot, *Nucleic Acids Res.*, 1995, **23**, 2292.
451. A. P. Green, J. Burzynski, N. M. Helveston, G. M. Prior, W. H. Wunner, and J. A. Thompson, *BioTechniques*, 1995, **19**, 836.
452. Z. Foldes-Papp, E. Birch-Hirschfeld, R. Rosch, M. Hartman, A. K. Kleinschmidt, and H. Seliger, *J. Chromatogr. (A)*, 1995, **706**, 405.
453. D. M. Brown, D. I. Magrath, A. H. Nielsen, and A. R. Todd, *Nature*, 1956, **177**, 1124.
454. C. B. Reese, H. T. Serafinowska, and G. Zappia, *Tetrahedron Lett.*, 1986, **27**, 2291.
455. N. Usman, K. K. Ogilvie, M.-Y. Jiang, and R. J. Cedergren, *J. Am. Chem. Soc.*, 1987, **109**, 7845.
456. S. L. Beaucage and R. P. Iyer, *Tetrahedron*, 1993, **49**, 1925.
457. S. L. Beaucage and R. P. Iyer, *Tetrahedron*, 1993, **49**, 6123.
458. S. L. Beaucage and R. P. Iyer, *Tetrahedron*, 1993, **49**, 10441.

# 7.06
# Attachment of Reporter and Conjugate Groups to DNA

## SERGE L. BEAUCAGE
*US Food and Drug Administration, Bethesda, MD, USA*

| | | |
|---|---|---|
| 7.06.1 | INTRODUCTION | 154 |
| 7.06.2 | ATTACHMENT OF REPORTER AND CONJUGATE GROUPS AT EITHER THE 5'- OR 3'-TERMINUS OF DNA OLIGONUCLEOTIDES | 154 |
| 7.06.2.1 | *Functionalization of DNA Oligonucleotides at the 5'-Terminus for Indirect Incorporation of Reporter and Conjugate Groups* | 154 |
| 7.06.2.1.1 | *Aminoalkyl functions* | 154 |
| 7.06.2.1.2 | *Thioalkyl functions* | 167 |
| 7.06.2.1.3 | cis-*Diols and ester functions* | 169 |
| 7.06.2.2 | *Direct Incorporation of Reporter and Conjugate Groups into DNA Oligonucleotides at the 5'-Terminus or at Preselected Sites* | 171 |
| 7.06.2.2.1 | *Biotin derivatives* | 171 |
| 7.06.2.2.2 | *Phosphotyrosine, 2,4-dinitrophenyl, and phenol derivatives* | 174 |
| 7.06.2.2.3 | *Fluorescein and cyanine derivatives* | 177 |
| 7.06.2.2.4 | *Dansyl, pyrene, and acridine derivatives* | 178 |
| 7.06.2.2.5 | *Coumarin, fagaronine, and anthraquinone derivatives* | 182 |
| 7.06.2.2.6 | *Psoralen groups* | 184 |
| 7.06.2.2.7 | *Bipyridine, 1,10-phenanthroline, and bleomycin $A_5$ derivatives* | 185 |
| 7.06.2.2.8 | *Lipophilic groups* | 187 |
| 7.06.2.2.9 | *Polyethylene glycol derivatives* | 190 |
| 7.06.2.3 | *Functionalization of DNA Oligonucleotides at the 3'-Terminus for Either Direct or Indirect Incorporation of Reporter and Conjugate Groups* | 190 |
| 7.06.3 | ATTACHMENT OF REPORTER AND CONJUGATE GROUPS TO THE NUCLEOBASES OF DNA OLIGONUCLEOTIDES | 191 |
| 7.06.3.1 | *Functionalization of Nucleobases for Indirect Incorporation of Reporter and Conjugate Groups into DNA Oligonucleotides* | 191 |
| 7.06.3.1.1 | *Aminoalkylated purines and pyrimidines* | 191 |
| 7.06.3.1.2 | *Thiolated and thioalkylated nucleobases* | 202 |
| 7.06.3.1.3 | *Hydroxylated nucleobases* | 204 |
| 7.06.3.2 | *Modified Nucleobases for Direct Incorporation of Reporter and Conjugate Groups into DNA Oligonucleotides* | 205 |
| 7.06.3.2.1 | *Nucleobases derivatized with fluorescent and nonisotopic markers* | 205 |
| 7.06.3.2.2 | *Nucleobases functionalized with intercalors and DNA/RNA cleavers* | 211 |
| 7.06.3.2.3 | *Copolymerization of pyrrole-containing nucleobases* | 216 |
| 7.06.3.2.4 | *Nucleobases conjugated to cross-linkers* | 216 |
| 7.06.3.2.5 | *Conjugation of nucleobases to nitroxide, heavy atoms, and isotopic markers* | 219 |
| 7.06.3.2.6 | *"Boronated" nucleobases* | 221 |
| 7.06.3.2.7 | *Modification of nucleobases with polycyclic aromatic hydrocarbons* | 222 |
| 7.06.4 | MODIFICATION OF THE CARBOHYDRATE PORTION OF NUCLEOSIDES FOR THE ATTACHMENT OF REPORTER AND CONJUGATE GROUPS TO DNA OLIGONUCLEOTIDES | 226 |
| 7.06.4.1 | *Modification of Deoxyribonucleosides at C-1'* | 226 |
| 7.06.4.2 | *Functionalization of Nucleosides at C-2'* | 228 |
| 7.06.4.3 | *Modification of Deoxyribonucleosides at C-3' and C-4'* | 231 |

7.06.4.4   *Derivatization of Nucleosides at C-5′*                                        233
7.06.4.5   *Modification of Carbonucleosides at C-6′*                                     235

7.06.5   ATTACHMENT OF REPORTER AND CONJUGATE GROUPS TO THE
         INTERNUCLEOTIDIC PHOSPHODIESTER FUNCTIONS OF DNA OLIGONUCLEOTIDES               236

7.06.6   SUMMARY                                                                          241

7.06.7   REFERENCES                                                                       241

## 7.06.1   INTRODUCTION

Since the late 1980s, the functionalization of oligonucleotides with specific labels has rapidly expanded our knowledge of nucleic acid structure and function, and has attracted considerable attention given the potential application of these modified oligonucleotides as diagnostic and therapeutic agents.

Although enzymatic methods have traditionally been applied to the incorporation of labeled nucleoside 5′-triphosphates into oligonucleotides, advances in the chemical synthesis of oligo-nucleotides[1,2] have undoubtedly facilitated the covalent attachment of reporter and conjugate groups to DNA and RNA.[3] Unlike enzymatic methods, the chemical methods for the incorporation of modified nucleosides into oligonucleotides permit a precise control over the type, number, and placement of the reporter and conjugate groups within the sequence of interest. For example, fluorescent markers, intercalators, cross-linkers, and DNA/RNA cleavers are added either to the 5′- or 3′-terminus of oligonucleotides, or inserted anywhere within an oligonucleotidic sequence. A number of these reporter and conjugate groups can be directly incorporated into oligonucleotides mainly via nonnucleosidic and nucleosidic phosphoramidite derivatives. In the case of nucleosidic phosphoramidites, reporter and conjugate groups are attached to the nucleobases, the carbohydrate portion of the nucleosides, or to the phosphorus atom of the phosphoramidites.

However, when the reporter and conjugate groups are not stable in the conditions used during either oligonucleotide synthesis or the subsequent deprotection step, the incorporation of these groups into oligonucleotides is effected according to an indirect approach. Typically, this method involves the insertion of suitably protected aminoalkylated or mercaptoalkylated phosphoramidites in oligonucleotides at predetermined locations. Following deprotection and purification, the func-tionalized oligonucleotides are conjugated to selected electrophilic reporter or potentially reactive groups.

Given the tremendous amount of scientific literature pertaining to the conjugation of various ligands to oligonucleotides, the comprehensiveness of this chapter is not absolute; excellent reports may have inadvertently been omitted. Moreover, peptide– and protein–oligonucleotide conjugates have reluctantly been excluded from this chapter because of the overwhelming number of publi-cations on this topic and obvious space limitations. Nonetheless, this chapter attempts to thoroughly review the scientific literature on the subject matter, and is adequately illustrated to serve as a dependable textual reference.

## 7.06.2   ATTACHMENT OF REPORTER AND CONJUGATE GROUPS AT EITHER THE 5′- OR 3′-TERMINUS OF DNA OLIGONUCLEOTIDES

### 7.06.2.1   Functionalization of DNA Oligonucleotides at the 5′-Terminus for Indirect Incorporation of Reporter and Conjugate Groups

#### 7.06.2.1.1   *Aminoalkyl functions*

Over the years, the phosphoramidites (1)–(12), listed in Table 1, and (13)[31,32] have been developed and used for the aminoalkylation of synthetic oligodeoxyribonucleotides at the 5′-terminus. The condensation of deprotected 5′-aminoalkylated oligonucleotides with the $N$-hydroxysuccinimide[6,22] or $p$-nitrophenyl esters[20] of biotin gave the corresponding 5′-biotinylated oligomers with no detect-able side reaction.[20] Purified 5′-biotinylated oligomers are used as hybridization probes for the detection of specific gene sequences.[4,6] The strong and noncovalent interaction between biotin and avidin or streptavidin generates a high-affinity complex ($K_d \approx 10^{-15}$ M) that is easily detectable by enzymatic methods, and thus offers an alternative to conventional [32]P-radiolabeling. 5′-Biotinylated oligonucleotides have additionally served as primers in polymerase chain reaction (PCR)[6,7,29,33] experiments to facilitate the sequencing of amplified DNA.[34,35]

**Table 1** Phosphoramidite derivatives for the aminoalkylation of oligonucleotides at the 5′-terminus.

$$(Pr^i)_2N \overset{R^1}{\underset{}{\overset{|}{P}}} R^2$$

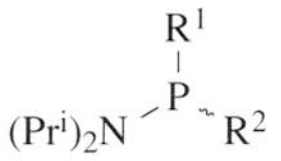

| Compound | $R^1$ | $R^2$ | Ref. |
|---|---|---|---|
| (1) | CF₃C(O)NH(CH₂)₄O | OCH₂CH₂CN | 4,5 |
| (2) | CF₃C(O)NH(CH₂)₆O | OMe | 6–13 |
| (3) | CF₃C(O)NH(CH₂)₆O | OCH₂CH₂CN | 14,15 |
| (4) | FmocNH(CH₂)₂O[a] | OCH₂CH₂CN | 16,17 |
| (5) | FmocNH(CH₂)₆O | OCH₂CH₂CN | 17–19 |
| (6) | MMTrNH(CH₂)₃O[b] | OMe | 20 |
| (7) | MMTrNH(CH₂)₃O | OCH₂CH₂CN | 21 |
| (8) | DMTrNH(CH₂)₅O[c] | OMe | 22 |
| (9) | DMTrNH(CH₂)₆O | OCH₂CH₂CN | 23 |
| (10) | MMTrNH(CH₂)₆O | OCH₂CH₂CN | 14,23–28 |
| (11) | PxNH(CH₂)₆O[d] | OCH₂CH₂CN | 23 |
| (12) | PhtNCH₂CH₂(OCH₂CH₂)₃O[e] | OCH₂CH₂CN | 29 |
| (14) | FmocNH(CH₂)₆O | Me | 30 |

[a] Fmoc, 9-fluorenylmethyloxycarbonyl.  [b] MMTr, (*p*-anisyl)diphenylmethyl.  [c] DMTr, di-(*p*-anisyl)phenylmethyl.
[d] Px, (9-phenyl)xanthen-9-yl.  [e] Pht, phthaloyl.

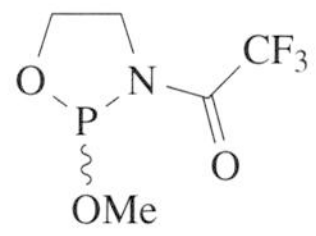

(13)

The usefulness of the phosphoramidite (**15**) has been exemplified by the solid-phase synthesis of an oligonucleotide carrying five consecutive primary amine functions at the 5′-terminus. The repetitive incorporation of the phosphoramidite (**15**) occurred with a coupling efficiency of 95%.[36] Upon removal of the protecting groups by ammonium hydroxide, the aminoalkylated oligonucleotide was biotinylated and then hybridized to complementary M13mp18 DNA immobilized on nitrocellulose filters. Colorimetric detection showed specific hybridization with a sensitivity of 0.5 ng.[36]

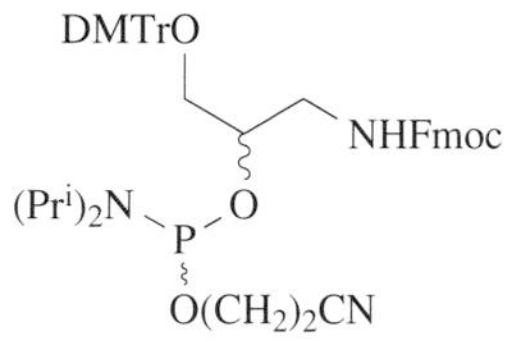

(15)

The phosphoramidite (**16**) (Table 2) has also been used for the automated incorporation of primary amine functions into oligonucleotides at the 5′-terminus. When used in conjunction with (**16**), the phosphoramidite (**17**) led to the insertion of a spacer arm between adjacent primary amine groups to form a comb-like labeling structure.[37] The functionalized oligonucleotides were biotinylated with the *N*-hydroxysuccinimide ester of biotinamidocaproic acid and tested in hybridization assays by the use of complementary sequences covalently linked to polystyrene tubes. Optimal chemiluminescent detection was achieved with a 5′-multifork-like structure carrying eight biotins. The detection limit was ca. 10⁵ molecules.[37]

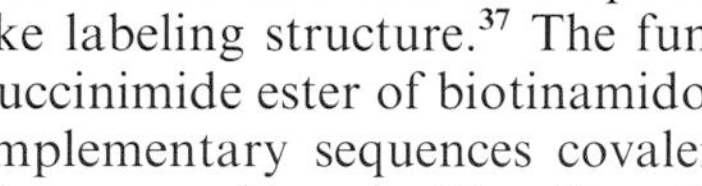

(17)

**Table 2**   Phosphoramidite derivatives for the aminoalkylation of oligonucleotides.

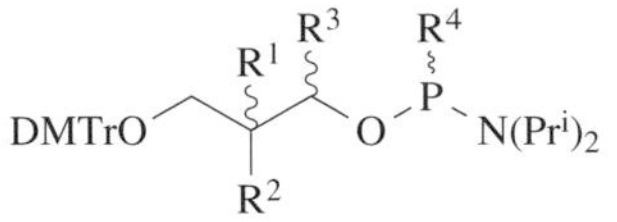

| Compound | $R^1$ | $R^2$ | $R^3$ | $R^4$ | Ref. |
|---|---|---|---|---|---|
| (16) | H | H | $CH_2NHC(O)(CH_2)_5NHC(O)CF_3$ | $OCH_2CH_2CN$ | 37 |
| (18) | $C(O)NH(CH_2)_3CO_2CH_2Ph$ | Me | H | $OCH_2CH_2CN$ | 38 |
| (19) | $NHC(O)(CH_2)_2NHFmoc$ | H | H | $OCH_2CH_2CN$ | 39 |
| (20) | $(CH_2)_4NHFmoc$ | H | H | $OCH_2CH_2CN$ | 40 |
| (21) | NHFmoc | H | Me | Me | 41 |
| (22) | $NHC(O)CH_2NHFmoc$ | H | Me | Me | 41 |
| (23) | $NHC(O)(CH_2)_3NHFmoc$ | H | Me | Me | 41 |
| (24) | $NHC(O)(CH_2)_5NHFmoc$ | H | Me | Me | 41 |
| (25) | $NHC(O)(CH_2)_5NHC(O)CH_2NHFmoc$ | H | Me | Me | 41 |

Multifork-like labeling structures have alternatively been constructed from reiterative additions of the 1,2,6-hexanetriol phosphoramidite derivative (**26**)[42] to the 5′-terminus of oligonucleotides. Primary amine functions were then introduced by phosphitylation of the 5′-polyhydroxylated oligonucleotides with the conventional aminoalkylated phosphoramidite derivative (**5**) (Table 1). Oligonucleotides carrying up to four amino functions at their 5′-end were labeled with fluorescein or digoxigenin according to standard methods.[42]

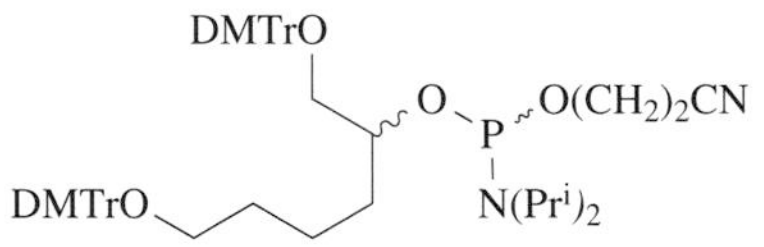

(**26**)

The novel phosphoramidite derivatives (**18**)[38] (Table 2) and (**27**)[43] have been particularly useful for site-selective incorporation of functional groups into oligonucleotides. The benzyl ester function of (**18**) was stable during DNA synthesis but readily reacted with a variety of aminated ligands without releasing the oligonucleotide from the controlled-pore glass (CPG) support.[38] Although the phosphoramidites (**18**) and (**27**) have been extensively applied to the functionalization of oligonucleotides at predetermined sites, these can also be used for single or multiple additions of reporter or conjugate groups to the 5′-terminus.

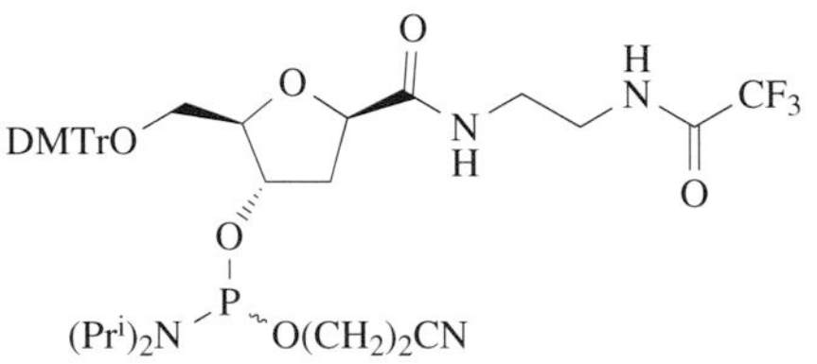

(**27**)

The nonnucleosidic phosphoramidite (**28**) has been prepared from L-threoninol and incorporated into an oligonucleotide at a central site to serve as an artificial abasic nucleoside.[44] Following deprotection, the aminoalkylated oligonucleotide was reacted with the *N*-hydroxysuccinimide ester of *N*-[9-(6-chloro-2-methoxy)acridinyl]-6-aminocaproic acid to generate the conjugate (**29**). It was shown that the modified oligonucleotide (**29**) formed a (1 : 1) duplex with a complementary oligomer. The destabilization of the duplex that resulted from the missing nucleobase was compensated by intercalation of the acridine ring.[44]

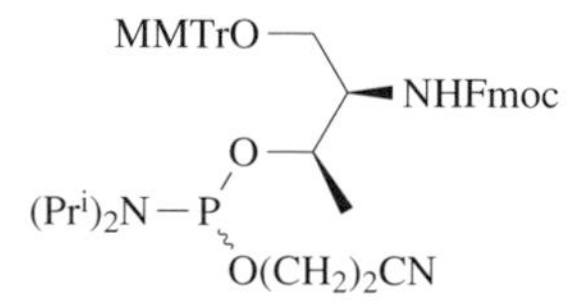

(28)

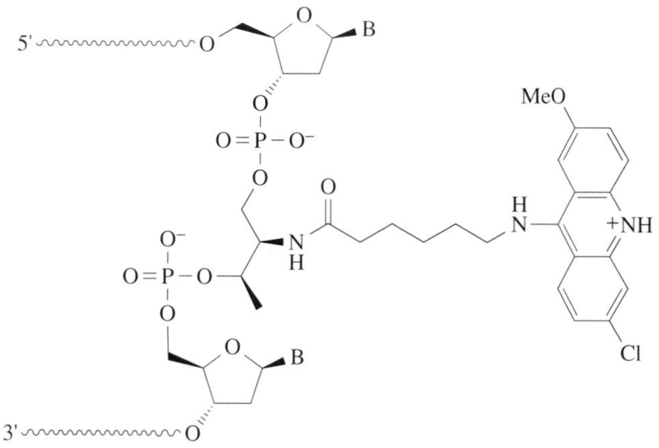

(29)

Insertion of the phosphoramidite (19) (Table 2) at a precise location in oligonucleotides produced, after deprotection and purification, highly reactive aminoalkylated oligonucleotides toward electrophilic reagents such as, for example, fluorescein isothiocyanate.[39] Alternatively, the condensation of these aminoalkylated oligonucleotides with succinic anhydride gave the carboxylated oligonucleotides (30) which could also be functionalized with a variety of ligands.[39]

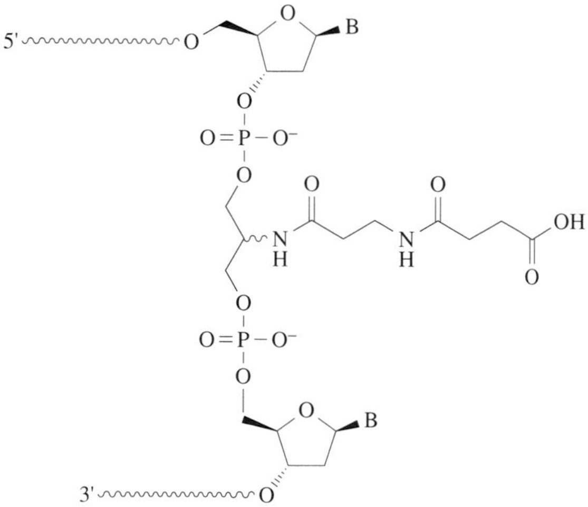

(30)

The recent synthesis of phosphoramidite (31) from achiral precursors allowed the introduction of several amine functions or fluorescent dansyl reporter groups at either the 5'- or the 3'-terminus of synthetic oligonucleotides.[45] The incorporated reporter groups had no or only a minor effect on the hybridization properties of the multilabeled oligonucleotides.

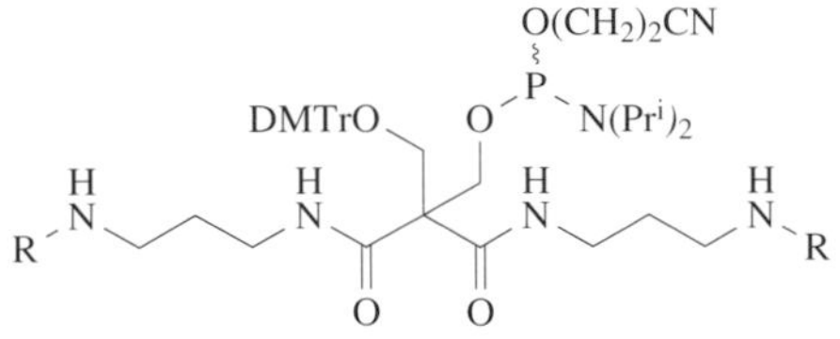

(**31**) R = Fmoc, trifluoroacetyl, or dansyl

Another achiral phosphoramidite for the incorporation of amino functions into oligonucleotides has been prepared from 5-nitroisophthalic acid.[46] Like many reagents of this type, the phosphoramidite (**32**) can be introduced singly or repeatedly at the 5′-end of oligonucleotides. Given the sensitivity of benzyl phosphate esters to standard iodine oxidation, *t*-butyl hydroperoxide was used for the oxidation step of each synthesis cycle. Labeling of the incorporated amino groups with fluorescein isothiocyanate proceeded smoothly.[46]

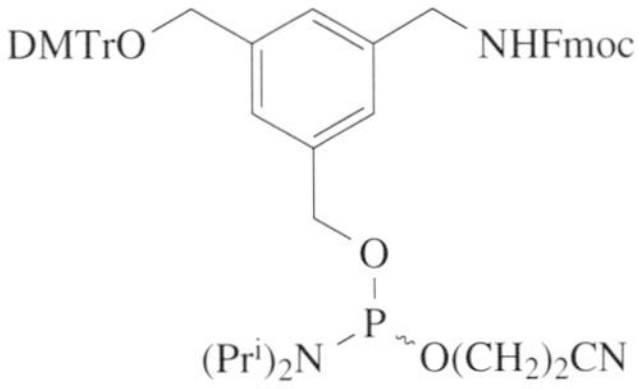

(**32**)

In an attempt to facilitate the detection of PCR-amplified DNA sequences through an improved immobilization of sequence-specific oligonucleotide probes to nylon membranes, Zhang *et al.*[15] described the reaction of a hexaethylene glycol phosphoramidite derivative with the 5′-hydroxy function of an oligonucleotide linked to a solid support. The resulting hexaethylene glycol–oligonucleotide conjugate was aminoalkylated by treatment with the phosphoramidite (**3**) (Table 1) to give, after deprotection, the oligomer (**33**). The purified aminoalkylated oligonucleotide was then efficiently coupled to nylon membranes activated with 1-(3-dimethylaminopropyl)-3-ethyl carbodiimide hydrochloride (EDC).[15]

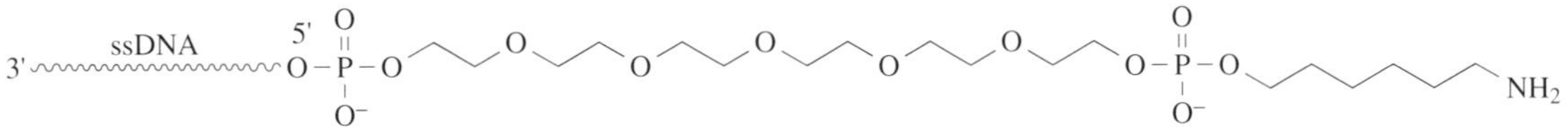

(**33**)

5′-Aminoalkylated oligomers generated from the incorporation of the phosphoramidite (**10**) (Table 1) were, alternatively, activated with cyanuric chloride and then coupled to a nylon support covalently coated with poly(ethyleneimine). This affinity support may permit the capture of specific RNA and DNA targets.[25,26]

The conjugation of 5′-aminoalkylated oligomers to fluorophores has enabled the classification of *Fibrobacter* strains by fluorescence microscopy on the basis of complementarity to partial 16S rRNA sequences.[32] In addition to enhancing the detection of specific DNA sequences by fluorescence amplification,[10] fluorescently labeled oligodeoxyribonucleotides have served in fluorescence energy transfer experiments[9,47] and as primers in automated DNA sequence analysis.[17,31]

In this context, a 5′-aminoalkylated DNA oligomer conjugated to diethylenetriaminepentaacetate and a second DNA oligonucleotide carrying a salicylate group at the 3′-terminus have been designed to form a ternary terbium(III) complex with a target DNA sequence complementary to both DNA conjugates.[48] Upon photoexcitation, the salicylate group of one oligonucleotide transferred the excitation energy to the lanthanide ion of the second oligonucleotide to produce the emission of a terbium(III)-specific long-lived fluorescence.[48] Under these conditions, target DNA was detected at the nanomolar level. This detection sensitivity could potentially be improved by a better choice of fluorescence excitation and emission wavelengths, and by the use of other ligand/lanthanide systems.

The phosphoramidite (**10**) (Table 1) has additionally been used for the attachment of a 5′-aminoalkylated tether to a ribavirin phosphate homopolymer. The aminoalkylated linker allows the potential conjugation of the ribavirin oligomer to a monoclonal antibody.[24]

Similarly, 5′-aminoalkylated oligodeoxyribonucleotides have also been applied to the preparation of oligonucleotide–quinone conjugates.[27,28,49] When hybridized to their complementary sequences, these conjugates are activated by enzymes such as cytochrome c reductase in the presence of NADH, glutathione reductase, or a combination of xanthine oxidase/xanthine, and trigger the formation of cross-linked strands by reductive alkylation.[28,49]

Along these lines, the conjugation of mitomycin C to 5′-aminoalkylated oligonucleotides, that have been prepared from the addition of the phosphoramidite (10) to the 5′-terminus of solid-phase bound DNA oligonucleotides, has been reported.[50] It was shown that the mitomycin C–oligonucleotide conjugates (34) alkylated complementary single-stranded (ss) DNA targets. Specifically, cross-link formation occurred upon reductive activation of the aziridine and carbamate functions of the conjugates with NADPH–cytochrome c reductase/NADPH under anaerobic conditions. The cross-linking efficiency, which was 15%, could be increased to 25% by doubling the length of the aminoalkyl tether.[50] It was also demonstrated that cross-linking occurred through monofunctional alkylation of the 2-amino function of guanines on the complementary DNA targets. These findings suggest that the conjugates (34) may become useful specific inhibitors of cellular or viral gene expression.

(34)

It should be noted that the condensation of 5′-aminoalkylated oligonucleotides with *N*-(iodoacetoxy)succinimide followed by reaction with a pyridoxylpeptide conjugate afforded vitamin $B_6$–peptide–oligonucleotide conjugates.[51] This type of conjugate may utilize cell surface receptors specific for vitamin $B_6$ for cellular uptake through protocytosis,[52] instead of endocytosis, to facilitate therapeutic applications.

The condensation of a 5′-aminoalkylated homopyrimidine oligonucleotide with *N*-(iodoacetoxy)succinimide and then with polyamines such as spermine, *N,N*′-bis(3-aminopropyl)-1-3-propanediamine, and *N,N*′-bis(3-aminopropyl)ethylenediamine, produced oligonucleotides that can form triple helical structures with an appropriate double-stranded (ds) homopurine–homopyrimidine DNA sequence.[53] For example, the spermine-conjugated oligonucleotide generated the most stable triplex structure. However, no triple helix formation was observed when an unmodified third strand was used under similar conditions.[53]

In this context, aminoalkylated foldback triplex-forming oligonucleotides prepared from the incorporation of the phosphoramidite (20) (Table 2) have similarly been evaluated for their relative thermostability and sequence specificity in triplex formation.[40]

An interesting application of 5′-aminoalkylated oligodeoxyribonucleotides is their conjugation to *p*-methoxyphenyl isothiocyanate and their subsequent radioiodination in iodogen-coated polypropylene tubes.[13] This method provides a simple and efficient route to the production of radiolabeled DNA probes useful for hybridization studies.

It should also be pointed out that 5′-aminoalkylated oligodeoxyribonucleotides resulting from incorporation of the phosphoramidite (2) (Table 1) have been conjugated to the photoactivatable cross-linker sulfosuccinimidyl 3-{2-[6-(biotinamido)-2-(*p*-azidobenzamido)hexanamido]ethyl}-dithiopropionate for the quantitation of triple-helix formation.[54] After incubation of an oligonucleotide conjugate with a dsDNA target known to form a purine (purine–pyrimidine) triplex, exposure to light labeled the DNA target with biotin. This technique allows the determination of dissociation constants of triple-helix forming oligonucleotides under a wide variety of conditions via a chemiluminescent assay. In addition, these conjugates readily label associated proteins both in solution and in living cells.[54] This photocross-linking technique may therefore emerge as a screening method of general utility in the study of oligonucleotide–target interactions inside cultured cells.

5′-Aminoalkylated oligodeoxyribonucleotides have additionally been conjugated to the *N*-hydroxysuccinimide ester of 3 *N* (4 carboxybutyl)methylamino-7-dimethylaminophenazathionium chloride, a methylene blue derivative, in order to investigate sequence-specific cleavage of complementary ssDNA under light irradiation.[55] It is also noteworthy to mention that 5′-aminoalkylated oligodeoxyribonucleotides have been conjugated to activated esters of cholic acid,[56,57] to polyamines via a disuccinimidyl suberate linker, and to polyethylene glycol active esters of various averaged molecular weights (550–5000) in an attempt to alter the pharmacokinetic properties of antisense oligonucleotides.[56]

The attachment of lipophilic groups, such as hexadecylaminoacetate and cholesterylaminoacetate, to the 5′-terminal phosphate of oligonucleotides generated conjugates carrying an ester function in the tether structure that can be cleaved by cellular esterases. Loss of the lipophilic anchor inside cells is expected to produce oligonucleotide accumulation. In this context, cholesteryl-conjugated oligonucleotides have been most efficient at inhibiting HIV proliferation in MT-4 cells when targeting the conserved region of the *env* gene in a sequence-specific manner.[58]

The use of (**10**) (Table 1) during the last coupling cycle of automated solid-phase DNA synthesis led to 5′-aminoalkylated oligonucleoside phosphorothioates which were eventually conjugated to the digoxigenin derivative (**35**). The conjugates (**36**) were used to examine intracellular oligonucleotide distribution within Epstein–Barr virus-transformed B cells by immunoelectron microscopy.[59] It was found that oligonucleotides mainly accumulate in lysosomal compartments. Digoxigenin did not interfere with intracellular routing because fluorescein-labeled oligonucleotides showed, by comparison, a similar intracellular distribution.[59]

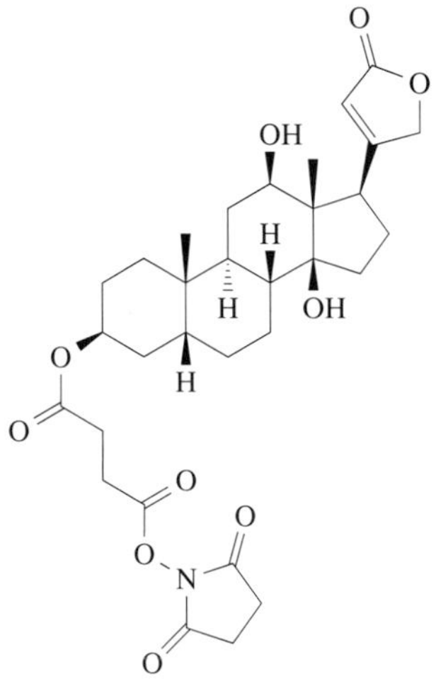

(**35**) 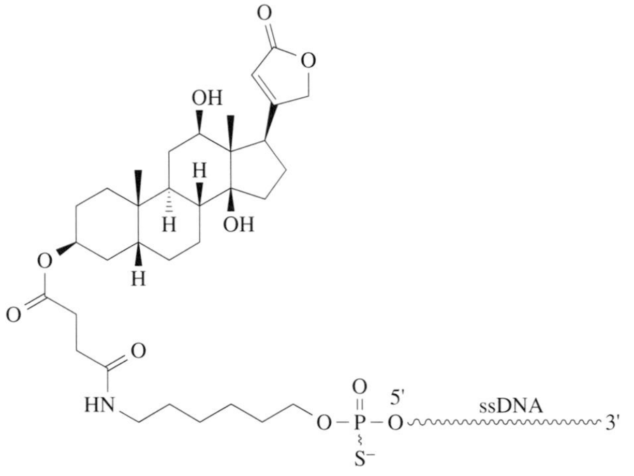　　　　　(**36**)

5′-Aminoalkylated oligonucleotides and their analogues have alternatively been reacted with the Denny–Jaffe reagent to provide the conjugates (**37**).[60] Incubation of the latter conjugates with HL60 cells in culture followed by irradiation with long-wave UV light (366 nm) generated photocrosslinks with associated proteins that were analyzed electrophoretically. Several proteins were labeled but one 75 kDa membrane-associated protein was predominantly labeled. It was also found that intracellular oligonucleotides were mainly associated in vesicles with the same protein to which they bound to on the cell surface. Only a small percentage of non-protein-bound cytosolic oligonucleotides was detected. However, accumulation of oligonucleotides in the nucleus led to the discovery of a unique set of nuclear binding proteins.[60]

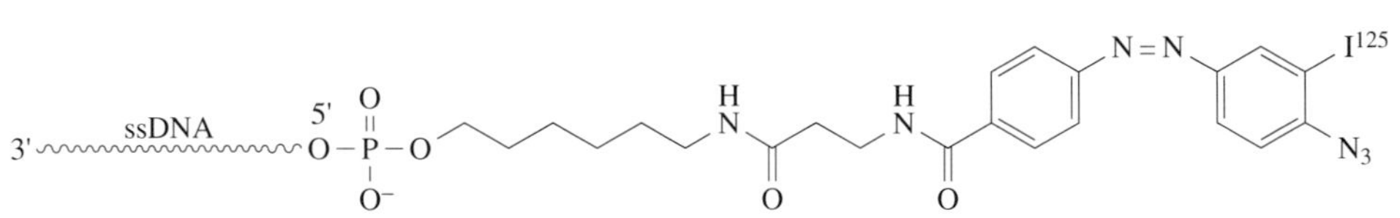

(**37**)

The synthesis of oligodeoxyribonucleoside methylphosphonates from deoxyribonucleoside methylphosphonamidites was reported in the mid-1980s but only since then has the 5′-aminoalkylation of a methylphosphonate oligomer been effected by the methylphosphonamidite (**14**) (Table 1).[30] The deprotected oligomer was biotinylated and, after purification, was immobilized on streptavidin–agarose. Given the stability of the internucleotidic methylphosphonate linkages to nucleases, the biotin–oligonucleoside methylphosphonate conjugate was employed in the capture of maxi U2 RNA/ribonucleoprotein complexes from nuclear extracts of cultured human cells.[30]

The methylphosphonamidite derivatives (21)–(25) (Table 2) were prepared and applied to the incorporation of an aminoalkyl function into oligonucleoside methyl phosphonates, between any two desired nucleobases, by standard phosphonamidite chemistry.[41] In addition, the *N*-hydroxy-succinimide ester of 4′-[(3-carboxypropionamido)methyl]-4,5′,8-trimethylpsoralen (38) has been synthesized for conjugation with aminoalkylated oligonucleotides. Upon hybridization of psoralen–oligonucleotide conjugates with complementary RNA sequences, it was found that the photocross-linking efficiency increased with psoralen–oligomer conjugates prepared from the insertion of the phosphonamidites (21) and (22) (Table 2). These findings indicate that a shorter tether length is more efficient at directing the psoralen entity toward its intercalation site(s). Furthermore, optimal photoaddition occurred when the psoralen moiety was inserted between an adenine that was 5′ to an adjacent thymine in the oligonucleotide sequence.[41] Solid-phase syntheses of oligonucleotides carrying a 5′-aminoalkylphosphonate[61] or a 5′-aminoalkylphosphodiester function[62,63] have also been achieved by the phosphotriester approach.

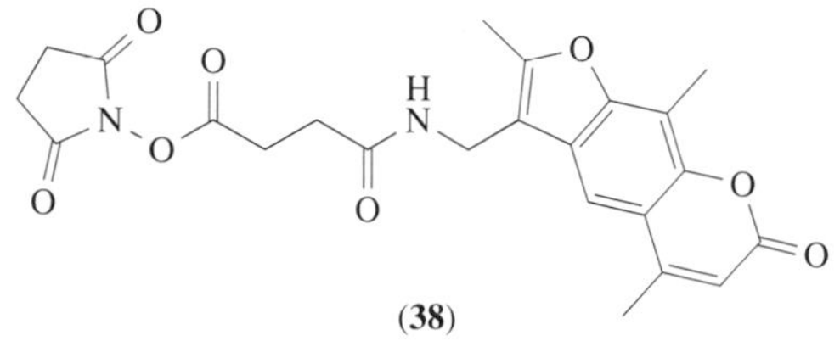

(38)

Conjugation of *N*-(2-hydroxyethyl)phenazinium chloride with a 5′-aminoalkylated oligo-nucleotide gave (39) which led to the formation of a strongly stabilized complex with a comp-lementary DNA sequence.[64] For example, a hybrid composed of a 5′-phenazinium heptanucleotide conjugate and a dodecanucleotide exhibited a thermal denaturation temperature ($T_m$) of 47.4 °C compared with that of 27.2 °C for the unmodified duplex.[64] Similar conjugates may therefore serve as promising candidates in the design of nucleic acid probes.

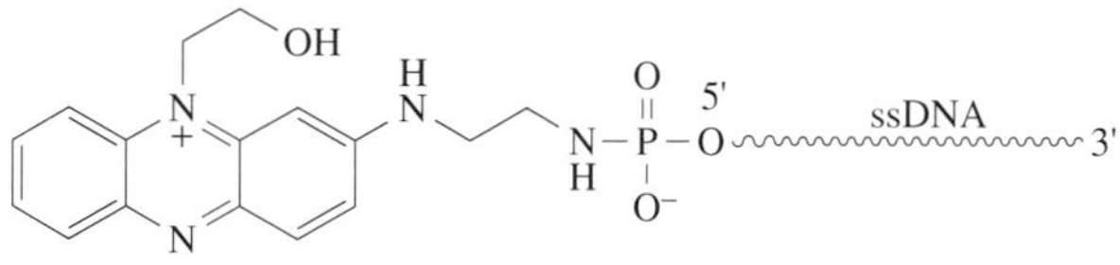

(39)

The reaction of a 5′-aminoalkylated oligodeoxyribonucleotide with the *N*-hydroxysuccinimide ester of *p*-azidotetrafluorobenzoic acid produced the arylazido–oligonucleotide conjugate (40).[65] This conjugate and complementary DNA or RNA targets were photomodified upon irradiation with the light of a high-pressure mercury lamp. Cross-linked product formation was obtained to the extent of 70% with the DNA target. Cross-linking RNA targets with (40) was less efficient possibly because DNA–RNA duplexes are thermodynamically less stable than DNA–DNA duplexes.[65]

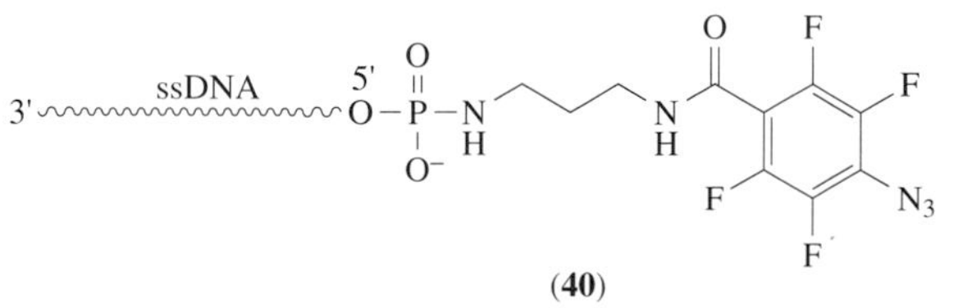

(40)

For similar applications, the conjugation of photoreactive *p*-azidoaniline derivatives to the 5′-terminal phosphate of oligodeoxyribonucleotides has been reported.[66]

Wachter *et al.*[67] further investigated the 5′-aminoalkylation of oligodeoxyribonucleotides by performing the reaction of the 5′-hydroxy function of solid-phase linked oligomers with 1,1′-carbonyldiimidazole and then with hexamethylenediamine. The deprotected 5′-aminoalkylated deoxyribonucleotides (41) were subsequently biotinylated with the *N*-hydroxysuccinimide ester of biotin. Interestingly, the carbamate linkage in (41) survived the basic conditions required for cleavage of the base protecting groups.[67,68] This methodology was also employed by Beck *et al.*[69] toward the development of chemiluminescent detection of DNA, and by others[70] to evaluate cellular uptake and intracellular localization of fluorescently-labeled oligonucleotide–neoglycoprotein conjugates.

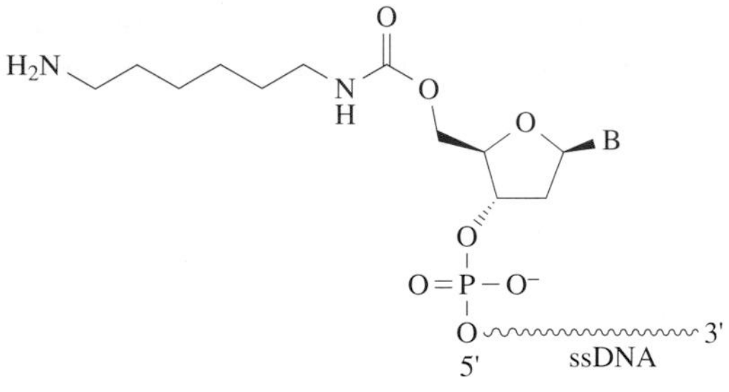

**(41)**

It must also be pointed out that reaction of the 5′-hydroxy function of solid-phase bound oligodeoxyribonucleotides with 1,1′-carbonyldiimidazole followed by diethylenetriamine produced, after release from the support, the aminoalkylated oligonucleotide (**42**).[71,72] Incubation of HPLC-purified (**42**) with a complementary portion of tRNA[phe], at pH 8 and 50 °C, resulted in selective cleavage of the tRNA to the extent of 8 mol.% after 4 h. No tRNA cleavage was observed with (**42**) lacking the polyamine tether or when replaced with a 6-aminohexyl linker.[72] The polyamine entity is therefore necessary for selective RNA cleavage.

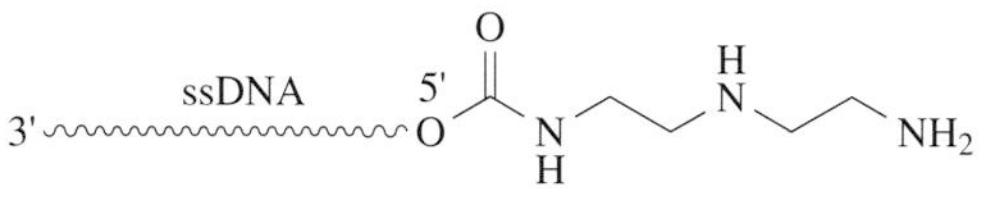

**(42)**

5′-Spin-labeled oligonucleotides have similarly been prepared by condensation of the 5′-hydroxy group of solid-phase bound oligonucleotides with 1,1′-carbonyldiimidazole followed by treatment with either 4-amino-2,2,6,6-tetramethylpiperidine-*N*-oxyl or 4-aminohexylamino-2,2,6,6-tetra-methylpiperidine-*N*-oxyl.[73] These oligonucleotides are useful for monitoring hybrid formation.

5′-Aminoalkylated oligodeoxyribonucleotides prepared according to the method of Wachter *et al.*[67] have been conjugated to activated metallotris(methylpyridinium)porphyrins.[74,75] For example, the ability of a manganese porphyrin linked to triple helix-forming oligonucleotides to cleave dsDNA targets *in vitro* has been studied.[76] It was shown that the nature and the length of the linker between oligonucleotide and the metalloporphyrin cleaving entity were important for cleavage efficiency. When the linker was a long aliphatic chain, such as a dodecamethylenediamine, no efficient cleavage was observed, presumably because of strong hydrophobic interactions. These repulsive interactions were suppressed when spermine containing two protonable sites was used as the linker. In that case, the cleavage yield was as high as 80%.[76]

It must be noted that a manganese porphyrin–oligonucleotide conjugate similar to (**43**) has also been efficient at cleaving a complementary ssDNA oligomer and thus demonstrated its potential for the cleavage of RNA targets *in vitro*.[77]

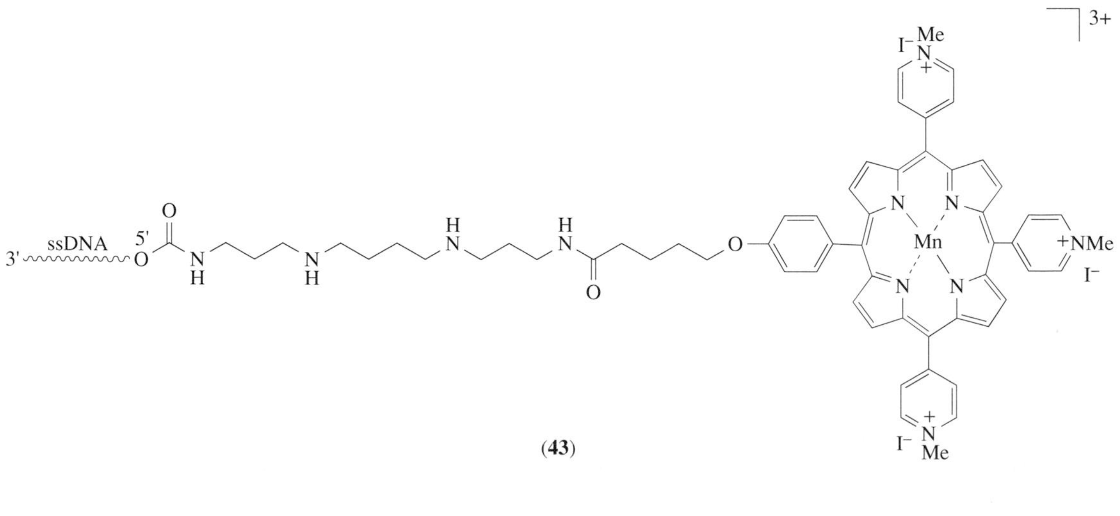

**(43)**

A 5′-aminoalkylated oligonucleotide anchored to a CPG support has been prepared by the use of the phosphoramidite (**7**) (Table 1) which was subsequently conjugated to a water-soluble *meso*-tetra(4-carboxyphenyl)porphine in the presence of EDC and 4-(*N*,*N*-dimethylamino)pyridine (DMAP).[78] Mild deprotection conditions afforded the porphine–oligonucleotide conjugate (**44**) which, upon treatment with iron(II) chloride, was incubated with a complementary DNA sequence. In the presence of dithiothreitol (DTT), the complementary DNA sequence was cleaved with an absolute specificity of 56%.[78] However, the incubation of an RR 1022 rat epithelial cell culture with a 0.6 μM solution of a metal-free porphine–oligonucleotide conjugate produced cytotoxic effects after laser light irradiation at 635 nm.[78]

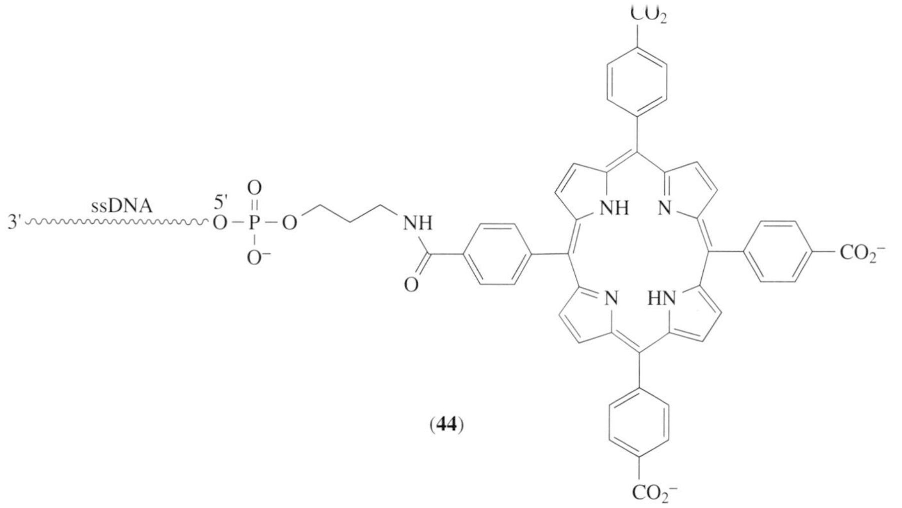

(**44**)

Given the considerable interest in reagents for sequence-specific hydrolysis of RNA, preparation of the europium(III) texaphyrin–oligodeoxyribonucleotide conjugate (**45**) has been achieved via a 5′-aminoalkylated DNA oligonucleotide precursor and the *N*-hydroxysuccinimide ester of a carboxylic acid-functionalized europium(III) texaphyrin complex.[79] Incubation of purified (**45**) with a complementary RNA oligomer resulted in a ca. 30% cleavage of the oligoribonucleotide near the expected location of the europium(III) texaphyrin complex. Control experiments indicated that ambient light, calf thymus DNA, type of buffer used, and the presence or absence of oxygen had no apparent effect on the cleavage efficiency.[79]

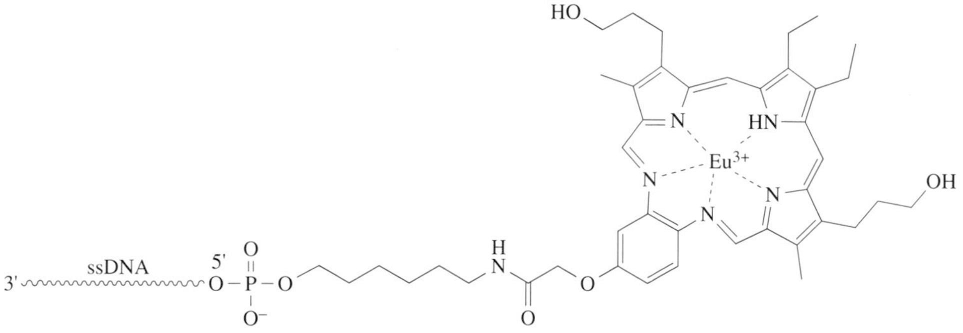

(**45**)

An efficient sequence-specific cleavage of RNA effected by the use of novel europium complexes conjugated to oligonucleotides has additionally been reported.[80,81] Specifically, a 5′-aminoalkylated oligodeoxyribonucleotide was reacted with the *N*-hydroxysuccinimide ester of an 18-membered macrocyclic europium(III) complex to generate the conjugate (**46**). The purified conjugate was mixed with a complementary RNA target, and scission of the RNA strand was observed even when the concentration of (**46**) was as low as 400 nM. After 16 h at 37 °C, (**46**) caused 88% cleavage of the RNA target almost exclusively at the site where the metal complex was expected to be located.[80] Conjugates similar to (**46**) provide a solid basis toward the development of artificial ribonucleases.

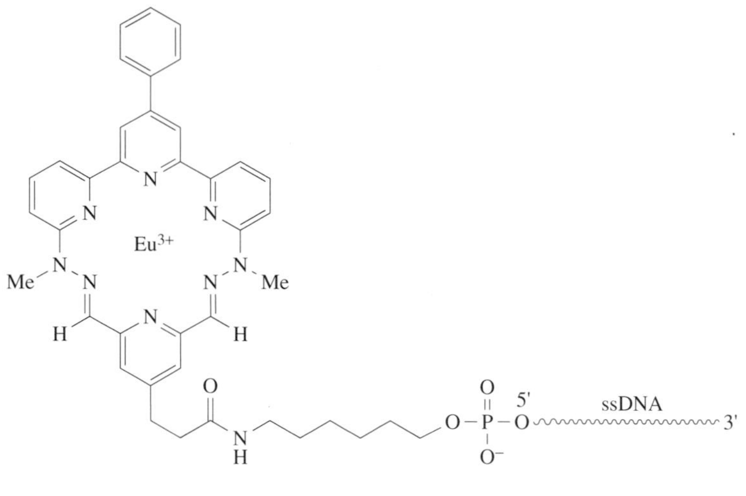

(46)

The conjugation of a lutetium(III) texaphyrin complex to 2′-*O*-methyl oligoribonucleotides has been accomplished in a manner analogous to that described for (**45**).[82] Irradiation of the lutetium(III) texaphyrin–oligonucleotide conjugate and a complementary unmodified DNA oligomer with a dye laser tuned to 732 nm led to extensive cleavage (70–80%) of the DNA strand. The cleavage products were found to comigrate exclusively with bands generated by the Maxam–Gilbert sequencing (G) reaction, and were consistent with a mechanism involving singlet oxygen generation. This is the first example of oligonucleotide-directed photocleavage of DNA with irradiation above 700 nm.[82]

Earlier work has also shown that oligonucleotides covalently linked to methylpyrroporphyrin XXI produced various types of photodamage on complementary DNA targets.[83] Oxidation of guanine nucleobases and cross-linking reactions with the target DNA sequence were recorded. Porphyrins strongly absorb visible light and the excited state produced (singlet or triplet state) can either react with a guanine nucleobase through electron transfer to generate a radical cation or react with oxygen to give singlet oxygen which readily reacts with guanine and, to a lesser extent, with thymine nucleobases.[83] When compared with a methylpyrroporphyrin–oligonucleotide conjugate, a protoporphyrin–oligonucleotide conjugate produced a much higher cross-linking yield with a complementary RNA sequence. The cross-linking sites on the RNA sequence were identified at guanine residues in close proximity to the porphyrin photoactive center in the hybrid. Photocross-linking was less efficient with zinc-porphyrin derivatives than with their corresponding metal-free congeners.[84] It is concluded that site-directed photodamage produced by porphyrins or other sensitizers linked to oligonucleotides may lead to promising systems for the selective inhibition of gene expression.

Similarly, oligonucleotides conjugated to "metallated" methylpyrroporphyrin derivatives,[85] hemin, and deuterohemin entities[86] have been efficient in the selective cleavage of ssDNA targets under defined conditions.

It must also be noted that the reaction of a 5′-aminoalkylated DNA oligonucleotide with the *p*-nitrophenyl ester of *N,N*-bis(ethoxycarbonylmethyl)glycine gave, after saponification of the diethyl ester functions, the conjugate (**47**).[87] Addition of a complementary RNA oligonucleotide to a solution of (**47**) and lanthanide(III) chloride (1:1) produced selective hydrolysis of the RNA oligomer opposite the lanthanide complex at the rate of 17 mol.% after 8 h at 37 °C. Hydrolysis selectivity was lost when the lanthanide(III) ions (lutetium, europium, thulium, or lanthanum) were absent or when the ratio (**47**):lanthanide(III) was 3:1. Dithiothreitol, a well-known promoter of metal-mediated redox reactions, had no measurable effect on the selective hydrolysis of the oligoribonucleotide. Furthermore, hydrolysis proceeded efficiently even under exclusion of molecular oxygen. It is postulated that lanthanides act as Lewis acids and facilitate intramolecular attacks of the 2′-hydroxyl group to effect cleavage.[79]

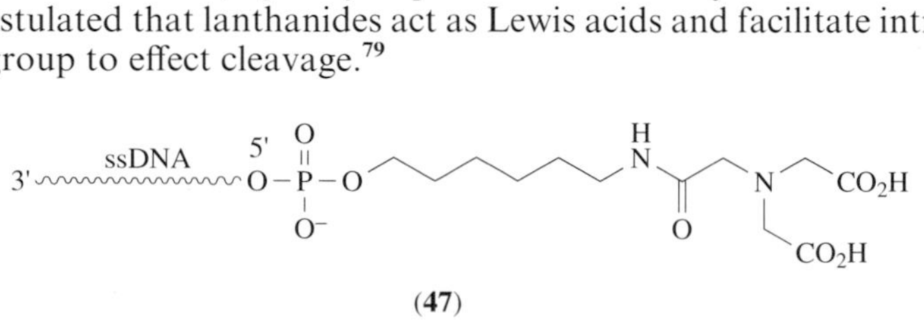

(47)

It should also be pointed out that the condensation of a 5'-aminoalkylated DNA oligomer, still attached to a solid support, with 1,1'-carbonyldiimidazole followed by diethyl iminodiacetate produced, after saponification of the ethyl ester functions and nucleobase deprotection, the conjugate (48).[88] Hybridization of (48) with a complementary unmodified DNA oligomer led, upon incubation with ceric ammonium nitrate at pH 7.5, to efficient cleavage of the strand opposite to (48) at mainly a single site. Minor cleavages at other sites were associated with the dangling motion of the flexible hexamethylene tether in (48).[88] DNA chain scission occurred via hydrolysis of the phosphodiester linkage with no apparent contribution from diffusible radical species. The use of cerium(III) and europium(III) instead of cerium(IV) resulted in much poorer DNA strand cleavages.

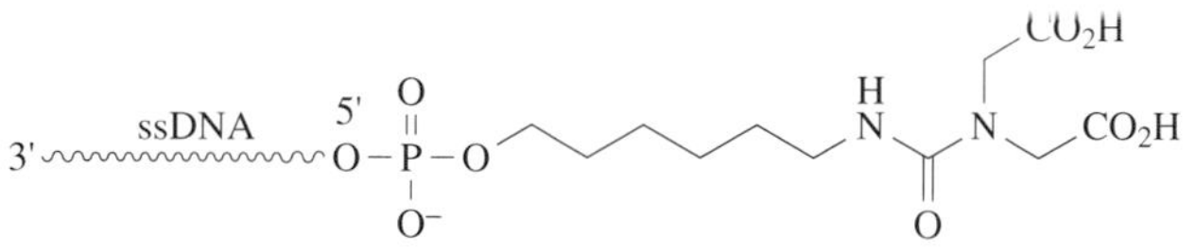

(48)

An oligodeoxyribonucleotide appended to an aminated nonamethylene tether, according to the method of Wachter *et al.*,[67] was reacted with the *N*-hydroxysuccinimide ester of [Rh(phi)$_2$bpy']$^{3+}$ (bpy' is 4-butyric acid 4'-methyl bipyridine and phi is phenanthrenequinone diimine) to give the conjugate (49).[89] Hybridization of the latter conjugate with a complementary DNA strand followed by irradiation of the duplex at low energy (365 nm) led to photoactivated electron transfer from a guanine to the photoexcited tethered rhodium complex. In fact, rhodium-induced photooxidation occurred at a guanine located at least 37 Å away from the site of rhodium intercalation. The site of guanine oxidation was revealed upon treatment of the modified oligomer with piperidine which triggered the release of 8-oxoguanine and strand cleavage to produce 3'- and 5'-phosphate termini on either side of the lesion. These results suggest that oxidative damage to DNA may be induced from a remote site as a result of hole migration through the DNA π-stack.[89]

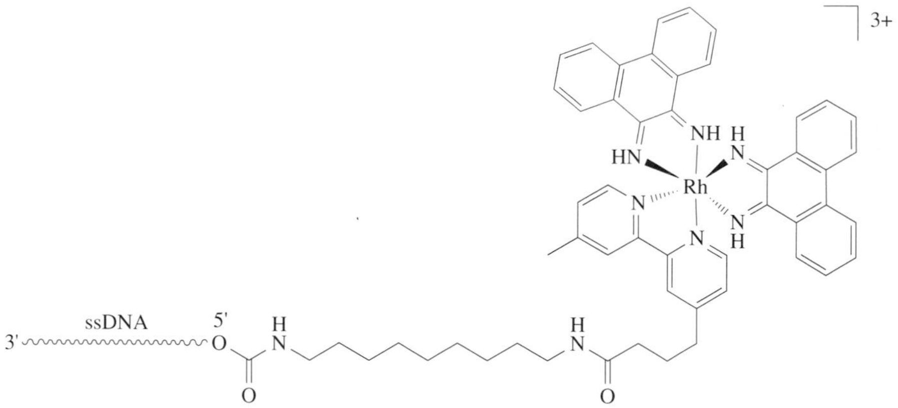

(49)

By virtue of inducible luminescence properties, a number of ruthenium complexes can be successfully applied to the detection of ssDNAs when tethered to the 5'-terminus of complementary DNA oligonucleotides. For example, the coupling of a DNA oligomer functionalized with a hexylamine group at its 5'-terminus to [Ru(phen')$_2$dppz]$^{2+}$ (phen' is 5-(glutaric acid monoamide)-1,10-phenanthroline and dppz is dipyrido[3,2:*a*-2',3':*c*]phenazine) afforded the conjugate (50).[90] Little luminescence was emitted from (50) alone. However, addition of a complementary strand generated an intense luminescence. Intercalation of the phenazine group is responsible for the production of luminescence; without protection of the phenazine from water through stacking, no luminescence would be observed. No luminescence emission was observed when a non-complementary oligonucleotide was added to (50).[90] Furthermore, luminescence emission was not detected when the complementary oligonucleotide was conjugated to the metallointercalator [Rh(phi)$_2$(phen')]$^{3+}$ at the 5'-terminus. Luminescence quenching resulted from rapid photoinduced electron transfer and was observed over a distance greater than 40 Å between the metallointercalators tethered to the 5'-termini of a 15 base pair DNA duplex.[91]

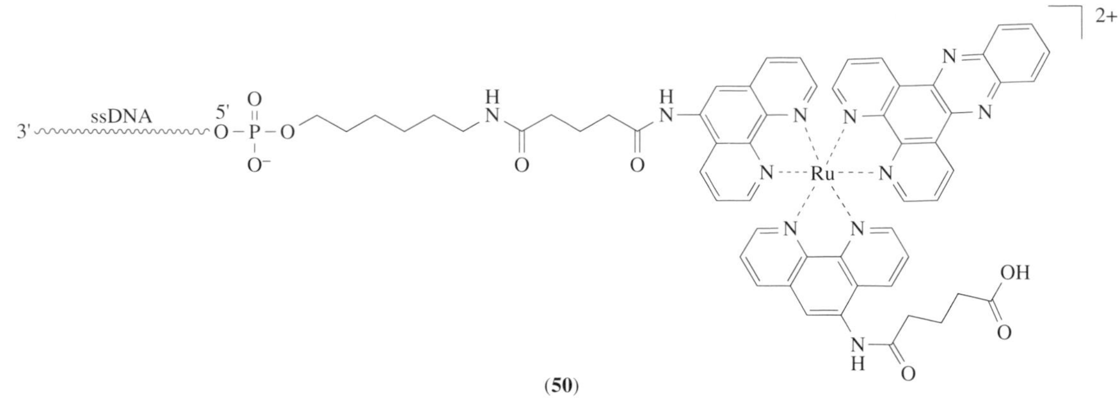

(50)

Thus, ruthenium–oligonucleotide conjugates like (50) can be used to target ssDNA and lead to the development of novel hybridization probes for homogeneous and heterogeneous assays.

5′-Aminoalkylated oligodeoxyribonucleotides that have been prepared by the addition of the phosphoramidite (3) (Table 1) to the 5′-terminus of solid-phase bound DNA oligomers, have been reacted with the *N*-hydroxysuccinimide ester of ferrocenecarboxylic acid to give the ferrocene–oligonucleotide conjugates (51).[92] These modified oligonucleotides are electrochemically active and can be used for the detection of complementary ssDNA and RNA sequences, at femtomolar concentrations, by an HPLC system equipped with an electrochemical detector. The conjugates (51) were found to stabilize triple helix complexes by 2–3 kcal mol$^{-1}$ and should therefore be useful for the detection of dsDNA as well.[92]

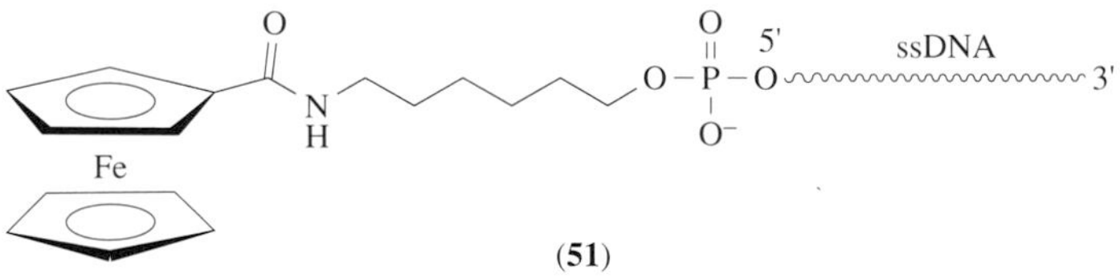

(51)

The synthesis of the deoxyribonucleoside phosphoramidites (52) has been reported to facilitate the incorporation of an aminoalkyl function into oligonucleotides at the 5′-terminus.[68] The (*p*-biphenylyl)isopropyloxycarbonyl group is conveniently removed under the acidic conditions required for the cleavage of a DMTr group. The resulting 5′-aminoalkylated oligomer was biotinylated while the oligonucleotide was either attached to the solid support or in solution after complete deprotection.[68]

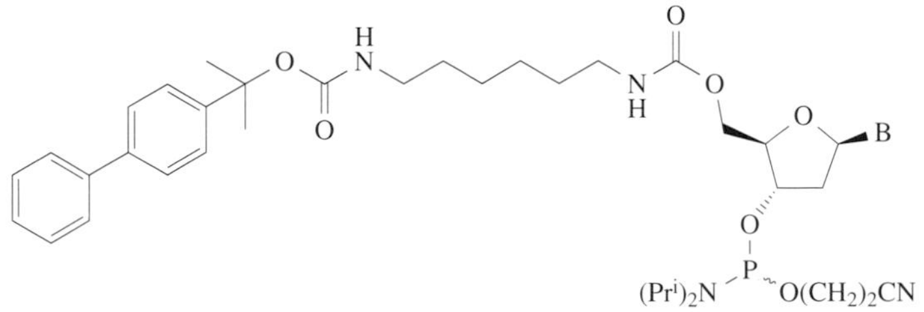

(52)   B = thymin-1-yl or hypoxanthin-9-yl

A versatile linker for the modification of synthetic biomolecules has been described by Gildea *et al.*[93] The linker was incorporated into oligonucleotides via the phosphoramidites (53). The solid-phase synthesis of a pentadecamer was performed, and the oligomer was reacted with 1,6-diamino-hexane followed by the *N*-hydroxysuccinimide ester of biotin. The crude biotinylated oligonucleotide was purified by binding to streptavidin–agarose and washing away nonbiotinylated oligonucleotidic sequences. Treatment of the streptavidin–agarose with aqueous acetic acid efficiently release the captured oligomer.[93] In addition to this specific application, the phosphoramidites (53) can potentially provide an attachment point for a variety of ligands at the 5′-end of oligonucleotides, and offer the unique possibility of recovering the unmodified oligomers under mild acidic conditions.

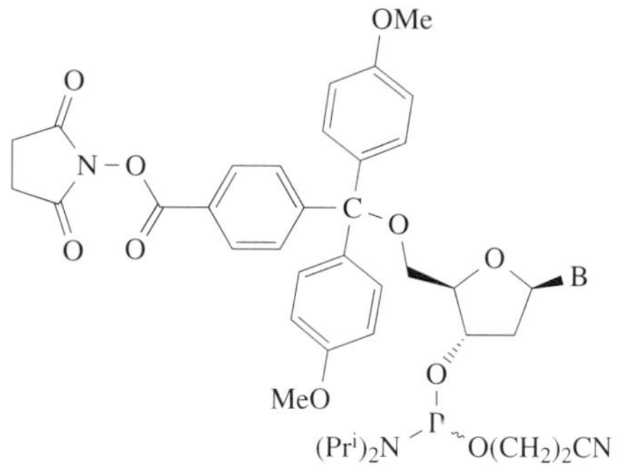

(53) B = thymin-1-yl, $N^4$-benzoylcytosin-1-yl, $N^6$-benzoyladenin-9-yl, or $N^2$-isobutyrylguanin-9-yl

### 7.06.2.1.2 Thioalkyl functions

The thioalkylated phosphoramidites (54)–(61) (Table 3) have been invaluable in the synthesis of 5'-thiolated oligonucleotides which, in turn, can be derivatized with either tetramethylrhodamine iodoacetamide or fluorescein iodoacetamide to give fluorescent M13 primers suitable for automated DNA sequencing.[95,105,106]

**Table 3** Phosphoramidite derivatives for the mercaptoalkylation of oligonucleotides at the 5'-terminus.

$$R^2 - \overset{\overset{\displaystyle R^1}{|}}{P} \sim R^3$$

| Compound | $R^1$ | $R^2$ | $R^3$ | Ref. |
|---|---|---|---|---|
| (54) | $TrS(CH_2)_3O^a$ | 4-morpholino | OMe | 94 |
| (55) | $TrS(CH_2)_6O$ | 4-morpholino | OMe | 94 |
| (56) | $TrS(CH_2)_3O$ | $(Pr^i)_2N$ | $OCH_2CH_2CN$ | 23,95–97 |
| (57) | $TrS(CH_2)_6O$ | $(Pr^i)_2N$ | $OCH_2CH_2CN$ | 14,23,98,99 |
| (58) | $2,4\text{-}NO_2Ph(CH_2)_2S(CH_2)_6O$ | $(Pr^i)_2N$ | $OCH_2CH_2CN$ | 100 |
| (59) | $CH_3C(O)S(CH_2)_6O$ | $(Pr^i)_2N$ | $OCH_2CH_2CN$ | 101 |
| (60) | $PhC(O)S(CH_2)_6O$ | $(Pr^i)_2N$ | $OCH_2CH_2CN$ | 101 |
| (61) | $CH_3C(O)SCH_2CH_2(OCH_2CH_2)_2O$ | $(Pr^i)_2N$ | $OCH_2CH_2CN$ | 102 |
| (62) | $TrS(CH_2)_3O$ | $(Pr^i)_2N$ | OMe | 103 |
| (63) | $TrO(CH_2)_6SS(CH_2)_6O$ | $(Pr^i)_2N$ | $OCH_2CH_2CN$ | 104 |

$^a$ Tr, triphenylmethyl.

In this context, Mori *et al.*[103] described the synthesis of the phosphoramidite (62) (Table 3) and its addition to the 5'-end of synthetic oligonucleotides. The resulting 5'-thioalkylated oligo-nucleotides were condensed with *N*-9-acridinylmaleimide to generate fluorescent oligonucleotides for cellular uptake experiments.[103]

Along these lines, the introduction of the phosphoramidite (58) (Table 3) at the 5'-terminus of oligonucleotides produced, after deprotection, 5'-thioalkylated oligonucleotides that were con-jugated to digoxigenin-3-*O*-succinyl-[2-(*N*-maleimido)]ethylamide.[100] Digoxigenin–oligonucleotide conjugates are easily detected by a conventional antidigoxigenin antibody conjugated to alkaline phosphatase and colorimetric substrates.

Particularly interesting was the reaction of a 5'-thioalkylated oligonucleotide with a bromo-acetylated fullerene derivative.[107] The fullerene–oligonucleotide conjugate was used in the formation of a DNA duplex and a triple helix. Irradiation of the complexes with low-energy light induced the cleavage of both the DNA duplex and the triple helix at well-defined sites (almost exclusively at guanine bases), and was consistent with singlet oxygen mediation. Fullerene–oligonucleotide conjugates are therefore promising photoprobes for experiments that pertain to inhibition of gene transcription and mRNA translation.[107]

The novel disulfide-containing phosphoramidite (63) (Table 3) has also been applied to the derivatization of the 5′-terminus of synthetic oligonucleotides.[104] After deprotection, the disulfide-containing oligonucleotide (64) was converted to the 5′-thioalkylated oligonucleotide (65) by treatment with tributylphosphine. The 5′-thiolated oligonucleotide (65) was then conjugated to a tetraiodoacetylated polyethylene glycol derivative to serve as a model in the treatment of systemic *Lupus Erythematosus.*[104]

(64)

(65)

It is noteworthy that the reaction of 5′-thioalkylated oligodeoxyribonucleotides with tresyl-activated Sepharose 4B or epoxy-activated Sepharose 6B resulted in the covalent attachment of specific oligomers to Sepharose. Addition of complementary unmodified oligodeoxyribonucleotides to these immobilized oligomers generated DNA duplexes as affinity ligands for the isolation of specific DNA binding proteins.[97,108] Alternatively, bromoacetylated Biogel polyacrylamide supports were reacted with 5′-thioalkylated oligonucleotides to produce nucleic acid affinity matrices for the detection of target DNA or RNA sequences by the use of a "sandwich" hybridization format.[109]

It is important to mention that 5′-aminoalkylated oligonucleotides can easily be converted to oligonucleotides carrying a 5′-thiolated function when treated with dithiobis(*N*-succinimidyl) propionate followed by treatment with dithioerythritol.[5] Incidentally, Gaur *et al.*[110] reported the reaction of a 5′-aminoalkylated oligonucleotide with *N*-succinimidyl 3-(2-pyridyldithio)propionate[22] instead of dithiobis(*N*-succinimidyl) propionate. The resulting disulfide-containing oligonucleotide was then reduced to a 5′-thioalkylated oligonucleotide in the presence of DTT.[22,110] The reaction of a 5′-aminoalkylated oligomer with *N*-acetyl-D,L-homocysteine thiolactone[111] or with *N*-succinimidyl-*S*-acetyl thioacetate, followed by hydroxylamine hydrochloride,[112] also efficiently generated the corresponding oligonucleotide carrying a free thiol group. The thiolated oligomer was then derivatized with *N*-(4-dimethylaminobenzene-4′)iodoacetamide or *N*-(3-pyrenyl)maleimide.[111,112]

Condensation of the 5′-phosphate function of a DNA oligomer with cystamine in the presence of EDC additionally produced an aminated disulfide-containing oligonucleotide which, upon reaction with DTT, gave a 5′-thioalkylated oligonucleotide.[113] This modified oligonucleotide was reacted with a 2-pyridyldithiopsoralen derivative to generate a psoralen–oligonucleotide conjugate that was subsequently cross-linked to a specific target site within a large RNA molecule.[113] Particularly noteworthy is the reaction of the 5′-hydroxy function of an oligonucleotide linked to a solid support with 1,1′-carbonyldiimidazole and then with *S*-DMTr–cysteine to afford, after deprotection and purification, an oligomer derivatized with both thiol and carboxylic acid functionalities.[114]

Another nucleophilic sulfur-containing group amenable to facile derivatization is a phosphorothioate monoester. Thuong and Chassignol specifically described the synthesis of oligonucleotides carrying a terminal 5′-thiophosphate monoester function.[115] The procedure entailed the phosphitylation of the 5′-hydroxy function of oligomers anchored to a solid support with bis(2-cyanoethyl)diisopropylphosphoramidite in the presence of 1*H*-tetrazole followed by sulfurization of the phosphite triester intermediate with elemental sulfur. Purified 5′-thiophosphorylated oligonucleotides were reacted with various electrophiles in aqueous solution to effect efficient incorporation of fluorescent and radioactive markers,[115–119] intercalators and chelators,[115–117,120–123] and photoactivatable cross-linkers.[115–117] This method has been particularly useful for the conjugation of proflavin to the 5′-end of α- or β-oligodeoxyribonucleotides in order to direct photosensitized reactions to complementary DNA sequences.[124,125] Similarly, the phosphoramidite (66) enabled the incorporation of a 5'-hexamethylene thiophosphate group into oligonucleotides and subsequent functionalization of the newly thioated oligomers with a phenanthroline–copper chelate.[120]

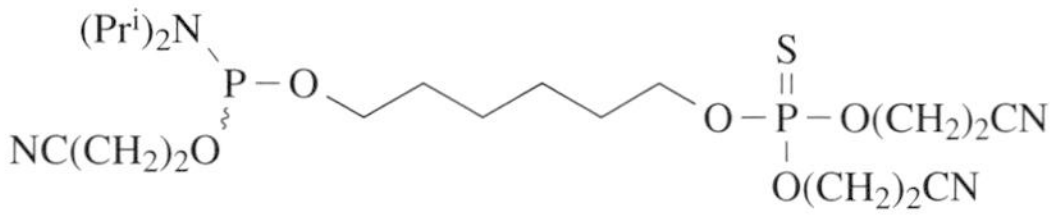

$$\begin{array}{c}(Pr^i)_2N \\ \diagdown \\ P-O \diagup\diagdown\diagup\diagdown\diagup\diagdown O-P-O(CH_2)_2CN \\ \diagup \\ NC(CH_2)_2O \end{array}$$

**(66)**

The nucleophilicity of oligodeoxyribonucleotide-5'-phosphorothioate permitted facile conjugation with the binuclear platinum(II) complex {[*trans*-Pt(NH$_3$)$_2$Cl]$_2$[NH$_2$(CH$_2$)$_n$NH$_2$]}Cl$_2$ ($n = 4$, 5, or 6). These oligonucleotide conjugates cross-linked efficiently and specifically to complementary single-stranded oligonucleotide targets and to polypurine tracts in triple helical DNAs.[126]

Oligodeoxyribonucleotide 5'- or 3'-phosphorothioates have similarly been conjugated to the cyclopropylpyrroloindole portion of the antitumor antibiotic CC-1065 via a bromoacetamidoalkyl tether.[127] That portion of CC-1065 contains an electrophilic cyclopropyl entity that alkylates adjacent adenines at their N-3 position in the minor groove of dsDNA. Thus, hybridization of the 3'-conjugated oligonucleotide (**67**) with a complementary DNA strand triggered rapid and efficient alkylation of adenines near the cyclopropapyrroloindole moiety on the opposite strand.[127] The half-time of the cross-linking reaction was 2 min at 37 °C. Total cross-linking efficiency was almost 50%. Interestingly, the corresponding 5'-conjugated oligonucleotide was inactive under the same conditions. The 3'-conjugated oligonucleotide (**67**) was also inactive when incubated with a non-complementary DNA sequence, and did not efficiently alkylate complementary RNA targets.[127] These properties suggest that conjugates similar to (**67**) may serve as inhibitors of viral ssDNA replication, or as gene-selective inhibitors of transcription initiation, perhaps by binding to an open promoter complex.

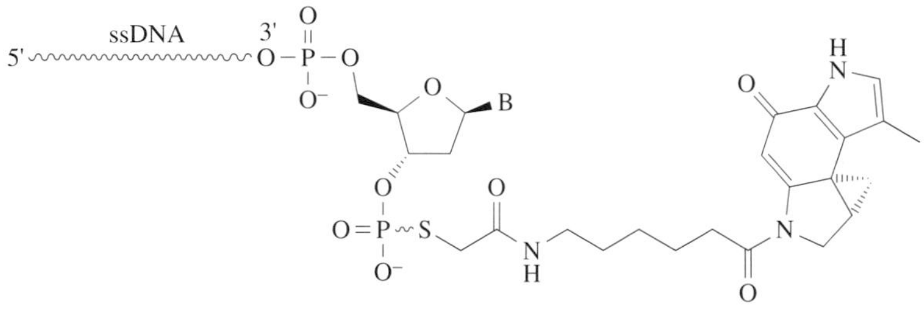

**(67)**

The condensation of a specific 5'-thiophosphorylated oligomer with *p*-azidophenacyl bromide has been performed to demonstrate that the inhibition of SV40 DNA replication in CV-1 cells resulted from the sequence-specific formation of a triple helix between viral DNA and the oligonucleotide conjugated to the cross-linking agent.[128] Psoralen–oligonucleotide conjugates have similarly been prepared from the reaction of 5-(6-iodohexyloxy)psoralen with oligodeoxyribonucleotides carrying a 5'-thiophosphate group.[129,130] The sequence-specific photoinduced cross-linking of the two DNA strands by triple helix-forming psoralen–oligonucleotide conjugates occurred with efficiencies greater than 80%.[129] Psoralen–oligonucleotide conjugates can therefore be designed to recognize and irreversibly modify specific DNA sequences.[129,130] Such modifications are expected to block replication and transcription by preventing opening of the double helix.

### 7.06.2.1.3 cis-*Diols and ester functions*

Alternate routes to the conjugation of nonradioactive labels to the 5'-end of synthetic oligodeoxyribonucleotides have been described. Specifically, the ribonucleoside phosphoramidite (**68**) was employed to generate a (5' → 5')-internucleotidic linkage with an oligomer covalently attached to a solid support.[16] The deprotected oligomer (**69**) was oxidized to (**70**) with sodium metaperiodate.

The reaction of (**70**) with biotin hydrazide, followed by reduction with sodium borohydride, afforded the biotinylated oligonucleotide (**71**). This oligonucleotide conjugate was applied to the detection of M13 DNA immobilized on nitrocellulose filters.[16]

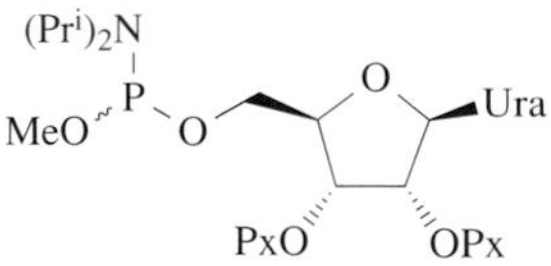

(**68**) 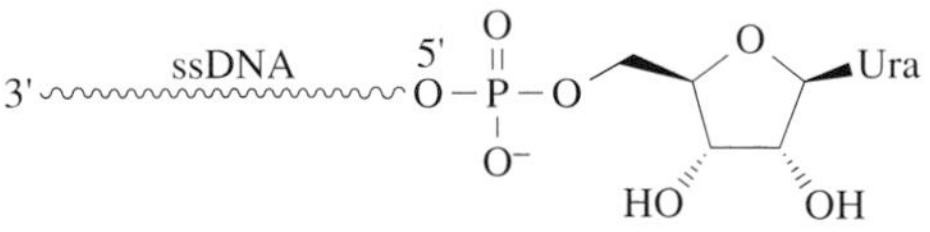          (**69**)

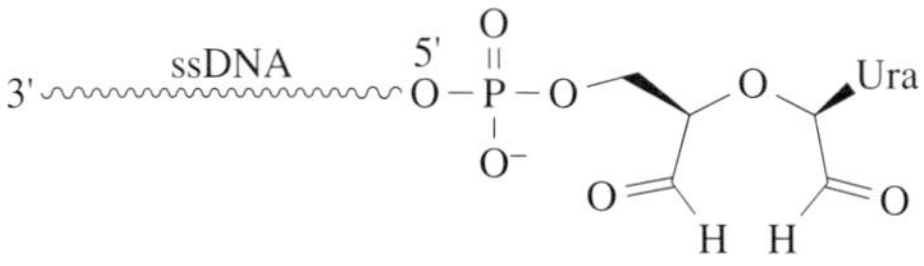

(**70**) 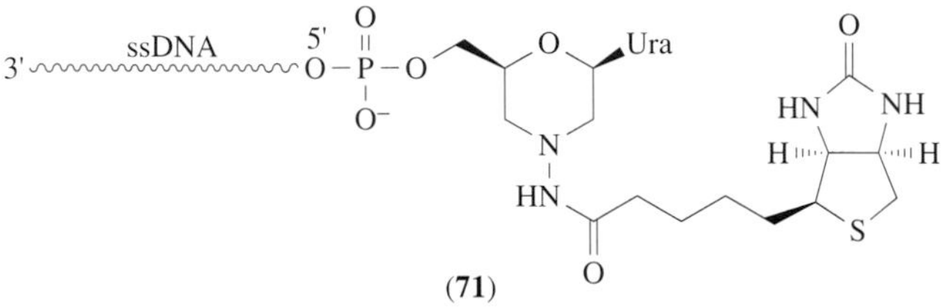          (**71**)

Oligonucleotides analogous to (**70**) were also converted to (**72**) to permit radioiodination of oligomers at the 5'-terminus with iodogen and Na[125]I, or to (**73**) to enable conjugation with maleimide-derivatized antibodies.[131]

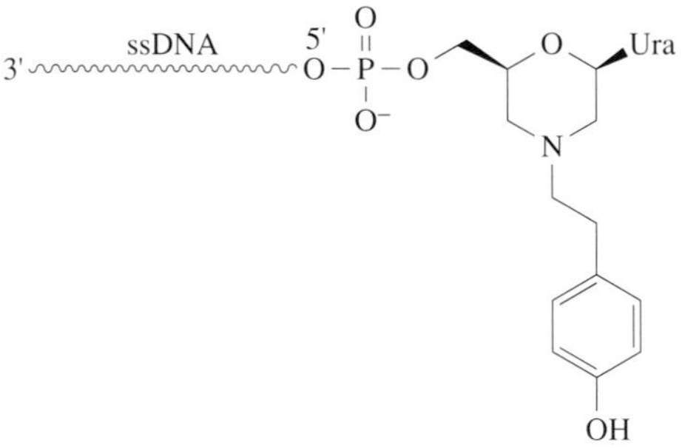

(**72**) 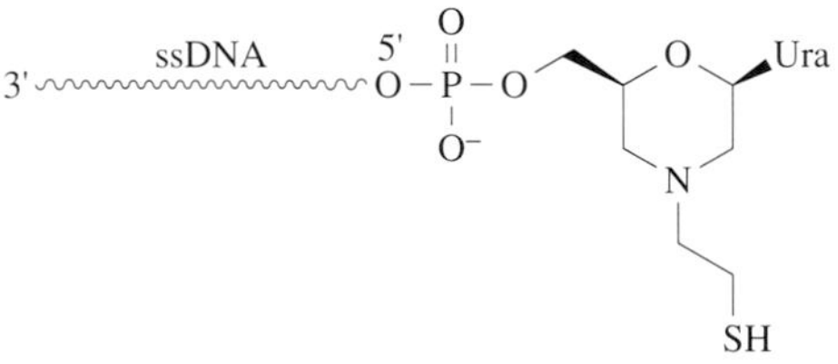          (**73**)

The phosphoramidite (**74**) has been recommended by Kremsky *et al.*[132] for the phosphitylation of the 5'-hydroxy function of an oligonucleotide linked to a solid support. The 5'-carboxyalkylated oligomer (**75**) that resulted from this reaction was reacted with biotin hydrazide in the presence of EDC to generate the biotinylated oligomer (**76**) in ca. 90% yield. Noteworthy procedures for the 5'-biotinylation of synthetic oligonucleotides have also been reported by Kempe *et al.*,[133] Chollet and Kawashima,[134] and Chu and Orgel.[135]

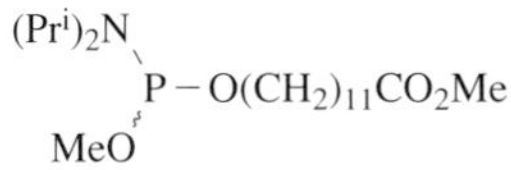

(**74**) 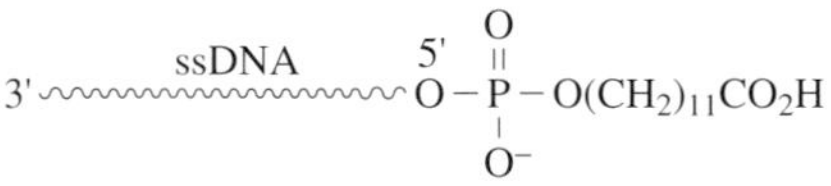          (**75**)

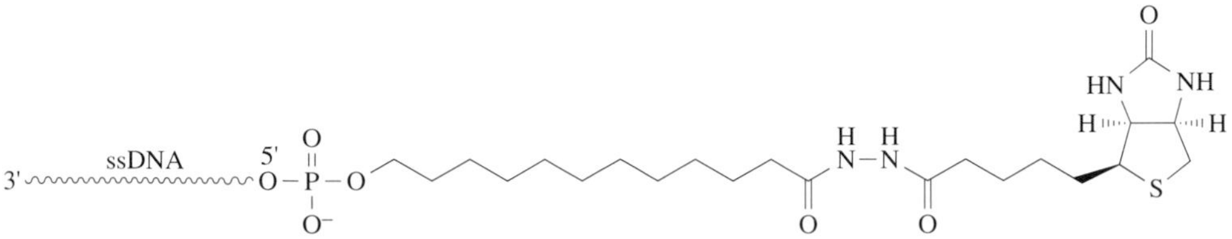

(**76**)

The phosphoramidite derivatives (**77**)–(**79**) have similarly been applied to the functionalization of synthetic oligonucleotides at the 5′-terminus to permit subsequent aminoalkylation of the newly incorporated ester or thioester function with selected amines or diamines.[136–138] The resulting amino-alkylated oligonucleotides can, for example, be conjugated to dabsyl chloride to give the corresponding sulfonamides.[137]

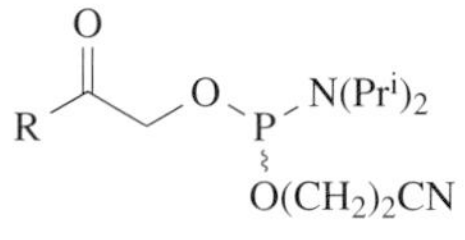

(**77**) R = OCH$_2$CF$_3$
(**78**) R = OCH$_2$CH$_2$Cl
(**79**) R = SCH$_2$Ph

## 7.06.2.2 Direct Incorporation of Reporter and Conjugate Groups into DNA Oligonucleotides at the 5′-Terminus or at Preselected Sites

Similar to the incorporation of aminoalkylated or mercaptoalkylated functions into oligo-nucleotides, one popular approach to the direct incorporation of reporter and conjugate groups into DNA oligonucleotides consists of the conversion of these groups to their corresponding phosphoramidite derivatives and their integration into synthetic oligonucleotides according to standard solid-phase techniques.

### 7.06.2.2.1 Biotin derivatives

Biotinylation of oligonucleotides is easily achieved by the condensation of the phosphoramidite (**80**) with the 5′-hydroxyl group of solid-phase bound oligomers under standard conditions.[139] However, a major limitation in the use of (**80**) is the requirement for mild conditions during oligonucleotide deprotection given the lability of the aromatic amide function. Another approach to the direct incorporation of biotin into oligonucleotides involves the phosphitylation of a biotinol derivative with chloro(methoxy)-*N*-morpholinophosphine to generate the biotinylated phos-phoramidite (**81**).[140] The coupling efficiency of (**81**) with the 5′-hydroxy function of solid-phase linked oligonucleotides can potentially be measured during the removal of the *N*$^1$-DMTr group of the biotin moiety.

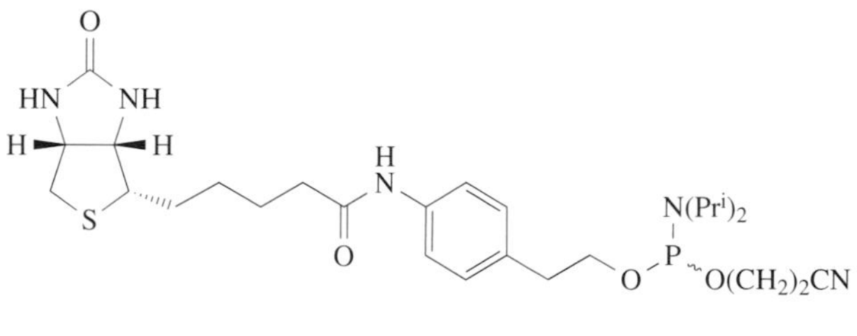

(**80**)

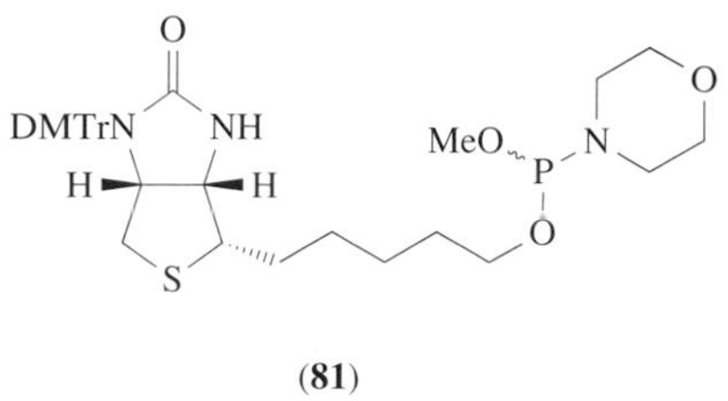

(**81**)

The synthesis of the biotinylated phosphoramidite derivative (**82**, *n* = 6) for automated synthesis of 5′-biotinylated oligonucleotides has been described by Pon.[141] This biotinylated phosphoramidite was incorporated into oligonucleotides with a coupling efficiency of ca. 85–95%. The *N*$^1$-DMTr group provided solubility and purification advantages, whereas the 6-aminohexyl linker facilitated the interaction of biotin with large proteins. The biotinylated oligonucleotides could be heated in concentrated ammonium hydroxide for 24 h at 50 °C without significant loss of biotin.[141] Incidentally, syntheses of the very similar biotin phosphoramidite derivatives (**82**, $n$ = 3, 5) have recently been reported by Kumar *et al.*[142]

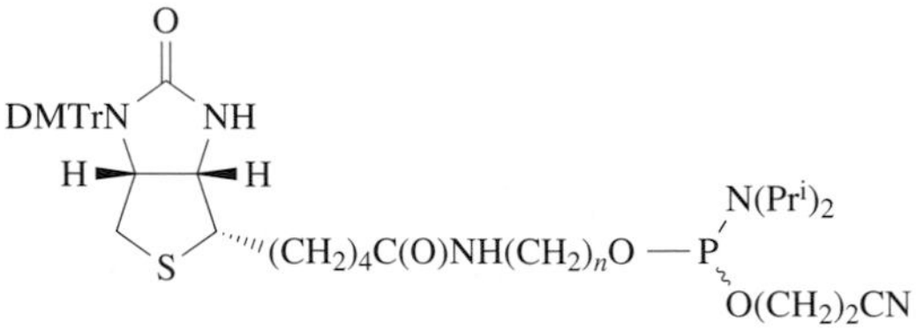

(**82**)   $n$ = 3, 5, or 6

It should be noted that the biotin phosphoramidites (**83**) carrying $N^1$-base labile groups have also been synthesized for the solid-phase biotinylation of oligonucleotides.[143] The $N^1$-base labile protecting groups improved the solubility of (**83**) in acetonitrile for efficient incorporation into oligonucleotides, and were readily cleaved by concentrated ammonium hydroxide under standard deprotection conditions.[143]

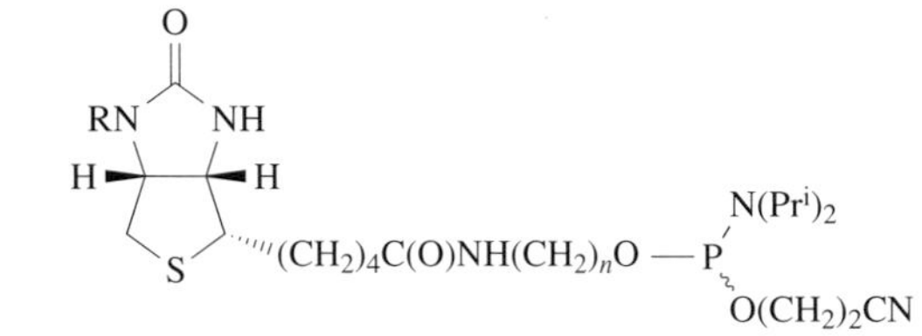

(**83**)

$n$ = 3 or 5
R = benzoyl, *t*-butylbenzoyl, phenoxyacetyl or *trans*-cinnamoyl

Phosphoramidite linkers have been designed for the direct insertion of biotin at multiple sites in synthetic oligonucleotides.[144,145] For example, the biotinyl phosphoramidite (**84**) was prepared in six steps from solketal and was repeatedly added to the 5′-end of a universal M13 sequencing primer with a coupling efficiency of ca. 99%. Oligonucleotides carrying up to eight biotin residues were synthesized. Depending on the assay used, it was shown that an oligomer with eight biotin residues was detected at a concentration 10-fold lower than that of a singly biotinylated oligomer.[144,145]

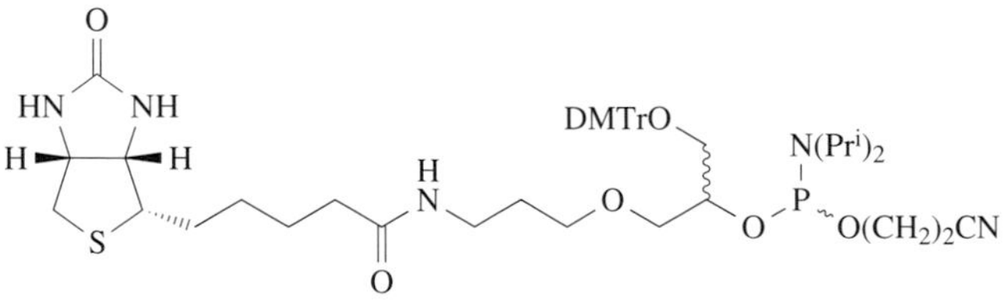

(**84**)

The biotin phosphoramidites (**85**) and (**86**) have additionally been prepared for direct labeling of oligonucleotides during solid-phase synthesis.[146] Biotin can be inserted at either the 5′- or the 3′-terminus, or anywhere within an oligonucleotide sequence.

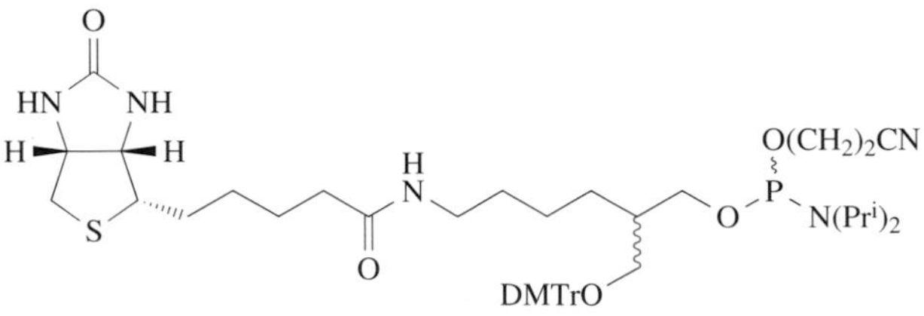

(**85**)

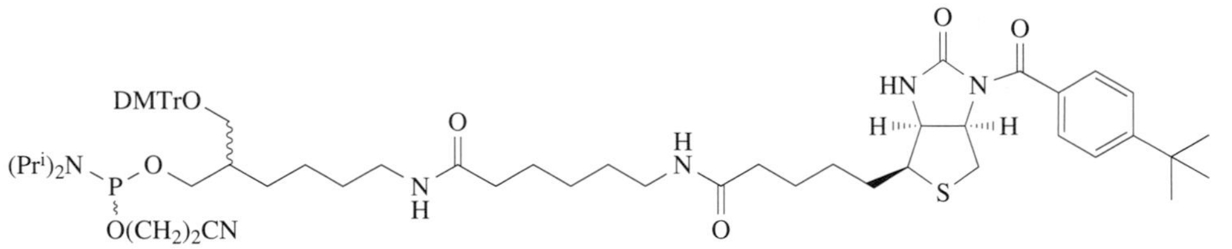

(86)

On the basis of a unique, enantiomerically pure 2'-deoxyribosyl entity, the biotinylated phosphoramidite (87)[43] has recently been used for the direct labeling of oligonucleotides. This phosphoramidite has been designed to preserve anomeric and conformational motifs along with the internucleotide distance of natural oligonucleotides. Like (85) and (86), the phosphoramidite (87) can be incorporated into oligonucleotides at any position with any number of labels.[43]

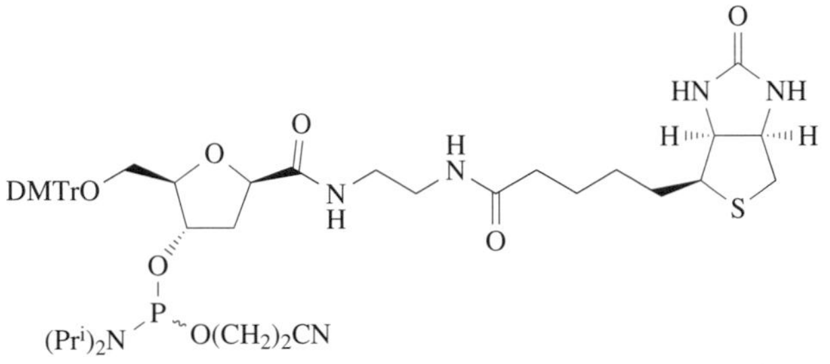

(87)

It must be noted that when the biotin phosphoramidites (84) and (85) are used for labeling oligonucleotides at the 5'-end, the primary nonglycosidic hydroxy function released under acidic conditions can be phosphorylated by $T_4$ polynucleotide kinase.[147] This unexpected enzymatic phosphorylation seems to be stereoselective when (84) is used for 5'-end labeling. However, no stereoselectivity was found when (85) was used for the conjugation reaction.[147]

Solid-phase functionalization of the polythymidylate $dT_{10}$ with the phosphoramidites (88)–(90) also resulted in the addition of a nonnucleosidic primary hydroxy group to the 5'-terminus.[148] This hydroxy function was phosphorylated by $T_4$ polynucleotide kinase in the presence of ATP in 90–100% yield. However, when the phosphoramidites (91) and (92) were used instead of (88)–(90), only S-diastereoisomers were efficiently phosphorylated (60–90% yield) by the kinase.[148]

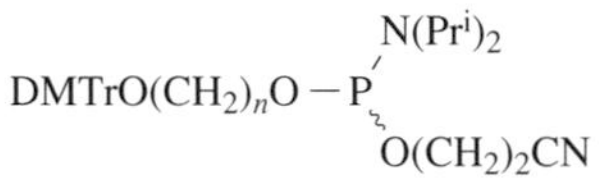

(88) $n = 2$
(89) $n = 3$

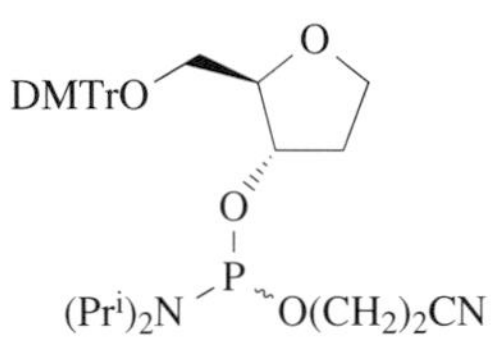

(90)

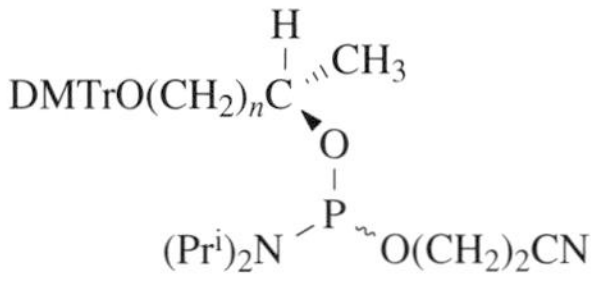

(91) $n = 1$
(92) $n = 2$

The biotin phosphoramidite derivative (93) offers another alternative to the biotinylation of oligonucleotides at the 5'-terminus or at any sites.[149] As expected, the insertion of (93) next to either the 3'- or the 5'-terminus of an oligonucleotide did not significantly affect duplex formation with a complementary sequence. However, insertion of (93) at central locations within an oligomer drastically decreased duplex stability.[149]

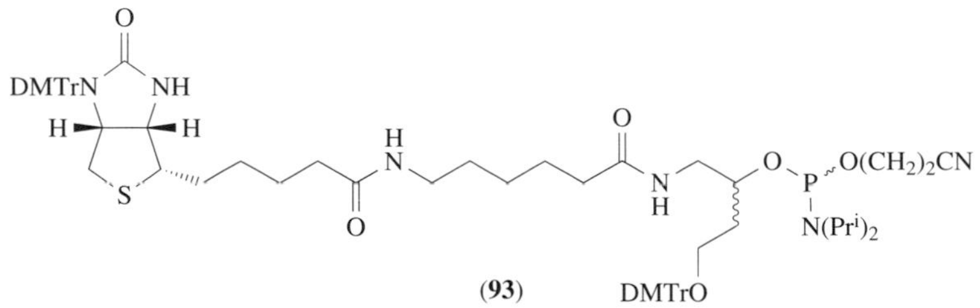

(93)

All attempts to phosphorylate the nonglycosidic hydroxy function generated by the addition of (**93**) with T$_4$ polynucleotide kinase failed.[149] Nonetheless, this difficulty was overcome by the phosphorylative addition of thymidine to this nonglycosidic hydroxy group.[149] For similar applications, the biotin phosphoramidite (**94**) is commercially available.

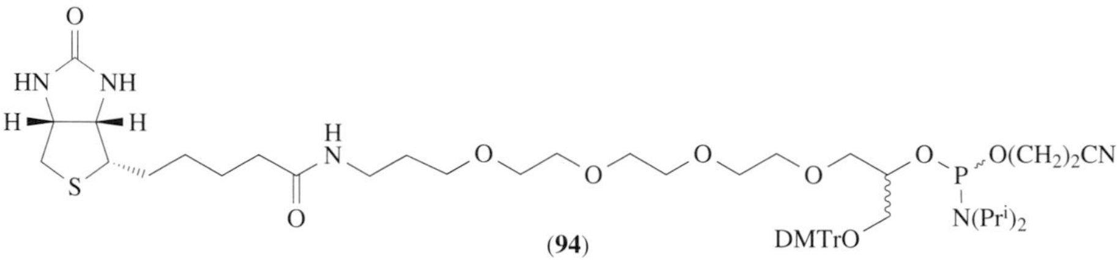

(**94**)

The high stability of the biotin–streptavidin complex can also be exploited for affinity purification and phosphorylation of synthetic oligonucleotides. For example, the nonnucleosidic photocleavable biotin phosphoramidite (**95**) is reacted with the 5′-hydroxy group of a solid-phase bound oligonucleotide according to standard protocols. The biotin group is retained during the cleavage and deprotection steps effected by concentrated ammonium hydroxide. The full length 5′-biotinylated oligonucleotide can then be selectively captured from the crude oligonucleotide mixture by incubation with immobilized streptavidin to form the complex (**96**). Irradiation of (**96**) with 300–350 nm light releases the 5′-phosphorylated and affinity-purified oligonucleotide (**97**) into solution.[150]

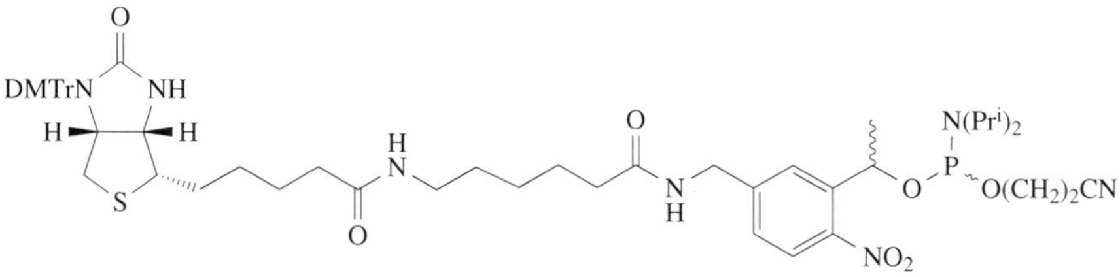

(**95**)

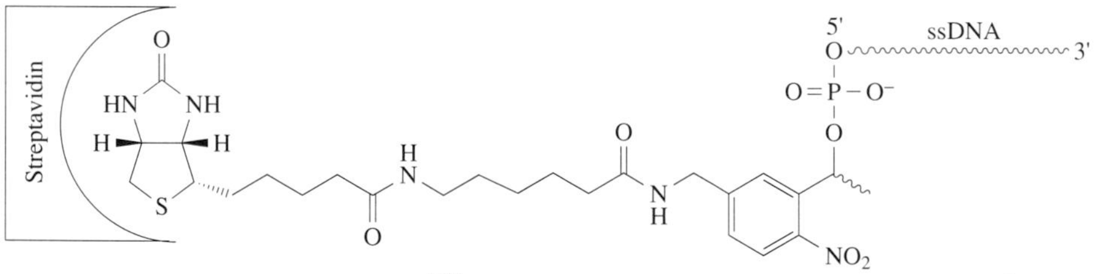

(**96**)

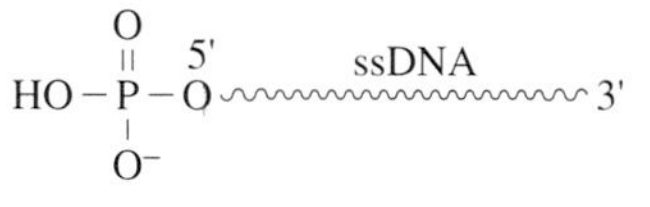

(**97**)

### 7.06.2.2.2 Phosphotyrosine, 2,4-dinitrophenyl, and phenol derivatives

The phosphotyrosinyl phosphoramidite (**98**) has been synthesized in several steps from the *p*-toluenesulfonate salt of L-tyrosine benzyl ester, and reiteratively added to the 5′-terminus of a universal M13 sequencing primer with a coupling efficiency of ca. 96%. Phosphotyrosinyl oligonucleotides were detected by the use of a mouse monoclonal antiphosphotyrosine antibody followed by an alkaline phosphatase secondary antibody system. This method led to the detection of M13 DNA with a signal strength increasing with the number of phosphotyrosinyl residues.[144,145]

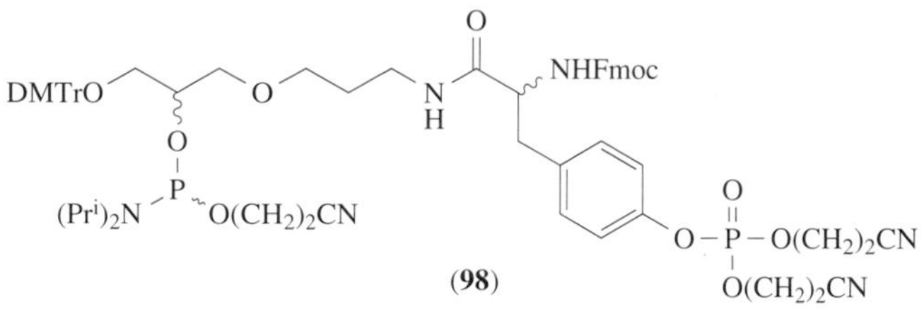

(98)

Will *et al.*[151] reported the synthesis of the phosphoramidites (**99**) and (**100**) for the single or multiple addition of dinitrophenyl (DNP) groups to the 5′-terminus of oligonucleotides. Specifically, the incorporation of (**100**) into oligonucleotides by solid-phase techniques occurred with a coupling efficiency of ca. 97%, and afforded oligonucleotides having up to 10 DNP groups.[151] Given the sensitivity of the DNP labels to concentrated ammonium hydroxide under standard deprotection conditions, FOD-amidites[152] carrying base-protecting groups that can be removed under milder conditions have been used in the synthesis of DNP-labeled oligonucleotides. The facile detection of these oligonucleotides was effected by anti-DNP monoclonal or polyclonal antibodies derivatized with suitable markers.[151] For comparable applications, the DNP phosphoramidite (**101**) is commercially available.

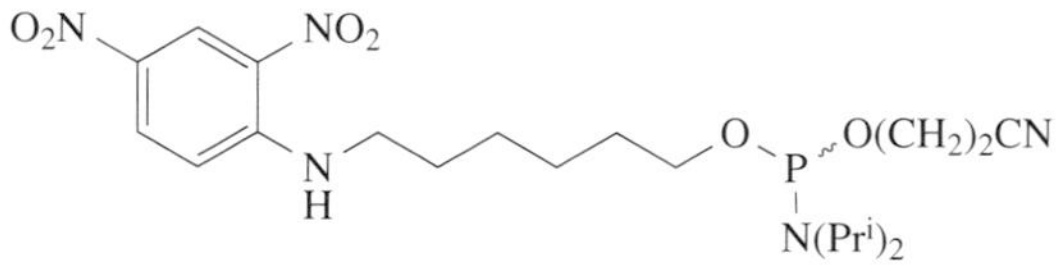

(99)

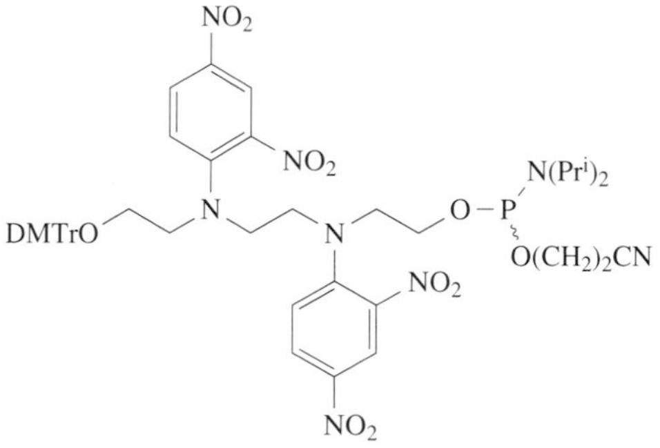

(100)

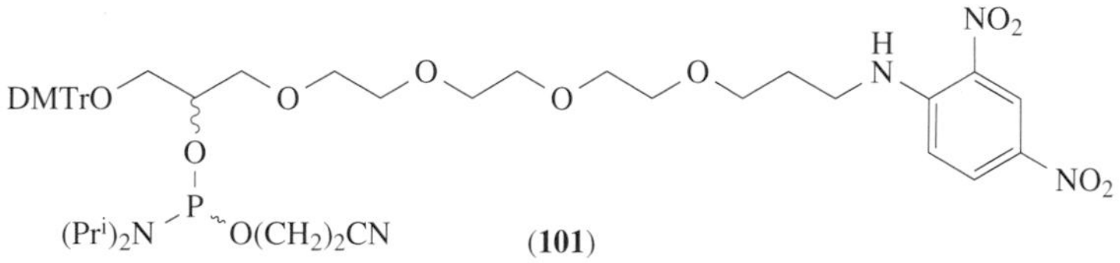

(101)

It is conceivable that the number of DNP groups and the length of the spacer arm linking these groups could affect the detection sensitivity of anti-DNP antibodies.[153] To address this issue properly, the DNP phosphoramidites (**102**)–(**105**) were prepared and each added one, three, or five times to synthetic oligonucleotides. It was demonstrated that the detection sensitivity was about the same for any of the monomers. In fact, three DNP labels linked to an oligonucleotide through a spacer arm of any reasonable length are sufficient to provide optimum detection sensitivity.[153]

In a different context, the DNP phosphoramidite (**106**) has been applied to the synthesis of the dinucleotide (**107**) that is used as a novel fluorogenic substrate for RNases of the pancreatic type.[154] Due to intramolecular quenching, (**107**) exhibited very little fluorescence. However, when the phosphodiester linkage is cleaved 3′- to the uridylyl residue by RNase, a 60-fold increase in fluorescence is observed.[154] The RNase substrate (**107**) has been particularly useful in assessing the specificity constant of various RNases.

A convenient method for the preparation of hapten phosphoramidites has been reported. The method involves the use of a 1,3-diol framework that is constructed from a hapten carboxylic acid precursor.[155] Thus, the hapten phosphoramidites (**108**)–(**110**) are routinely produced on a 20–25 g scale, and are suitable for labeling oligonucleotide probes at preselected sites via standard solid-phase synthesis.[155]

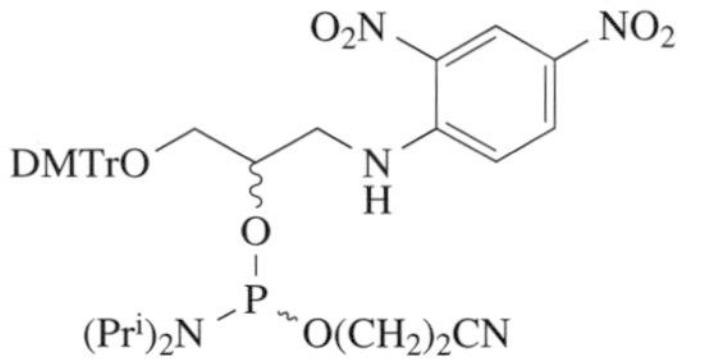

(102)

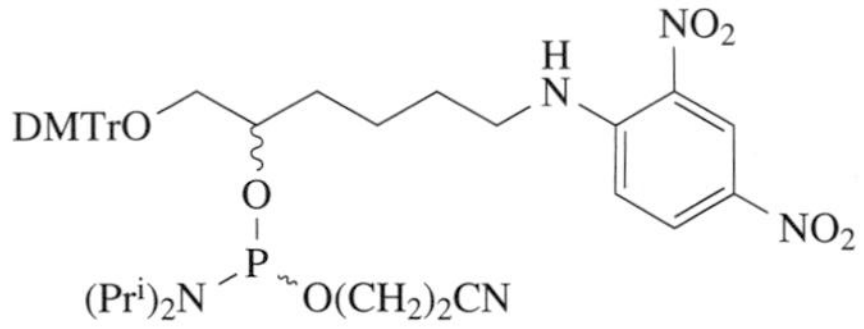

(103)

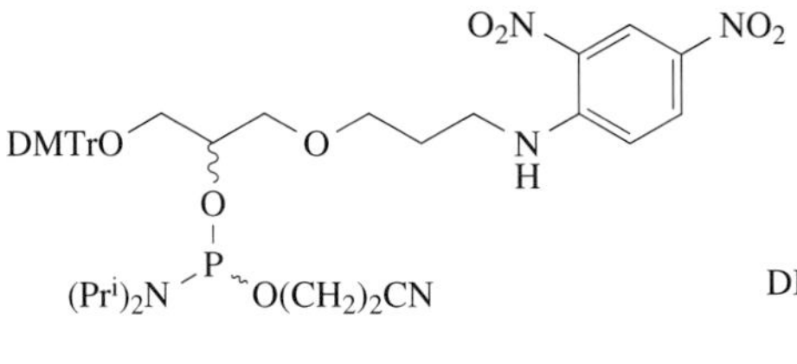

(104)

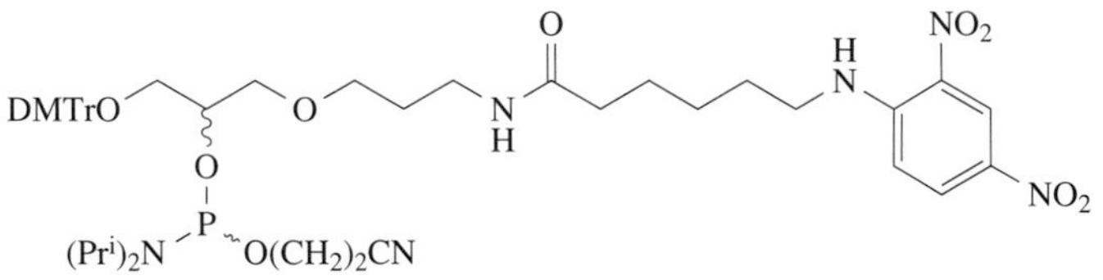

(105)

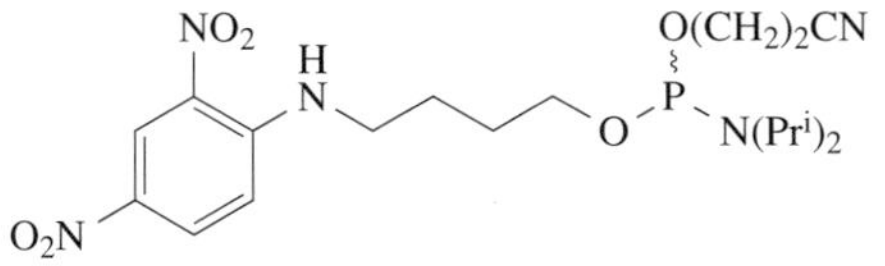

(106)

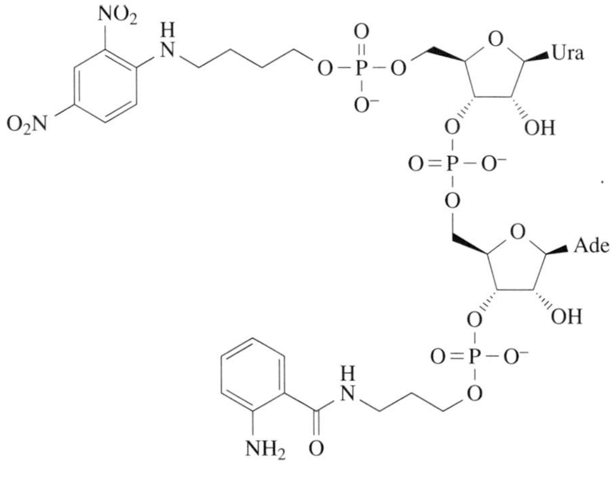

(107)

A common method for the detection of specific nucleic acid sequences entails immobilization of the target nucleic acid on solid supports, such as nylon membranes or nitrocellulose, followed by hybridization with a labeled signal probe. Alternatively, the target nucleic acid is hybridized to a capture oligonucleotide probe.

The novel phosphoramidite (111) has been used for the direct incorporation of phenol groups into oligonucleotides during automated synthesis.[156] Phenol–oligonucleotide conjugates can then be covalently attached to diazotized cellulose to ca 180 pmol cm$^{-2}$ and applied to the screening of different sequences or the detection of point mutations. It should be pointed out that the addition

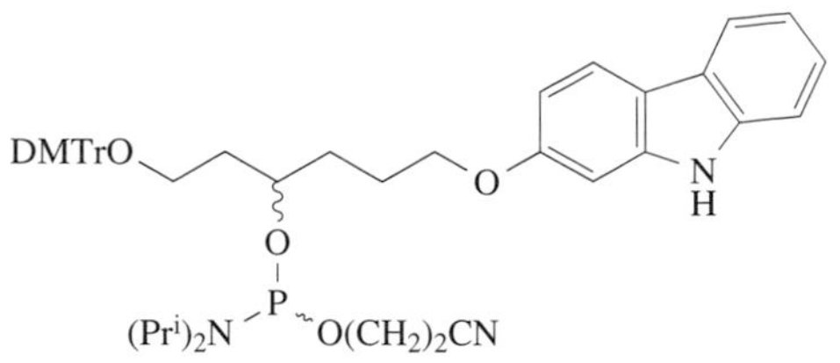

(108)

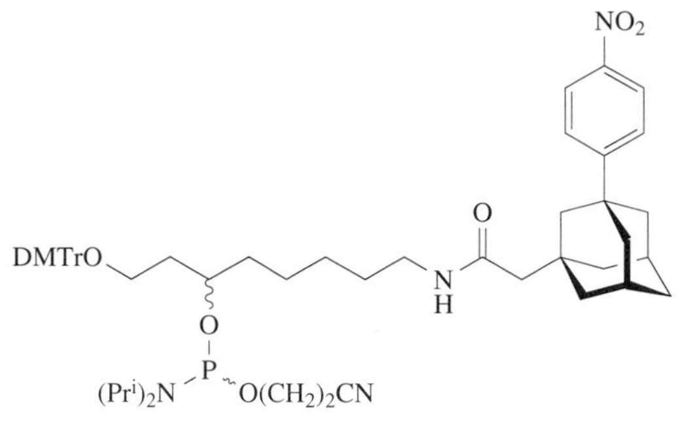

(110)

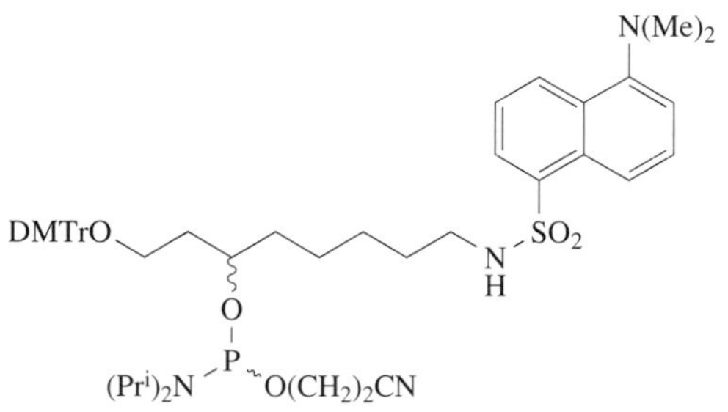

(109)

of (111) to the 5′-terminus of oligonucleotides produced, after deprotection, a nonnucleosidic hydroxy function that can be radiolabeled by phosphorylation with $T_4$-polynucleotide kinase in the presence of [$\gamma$-$^{32}$P]ATP.[156]

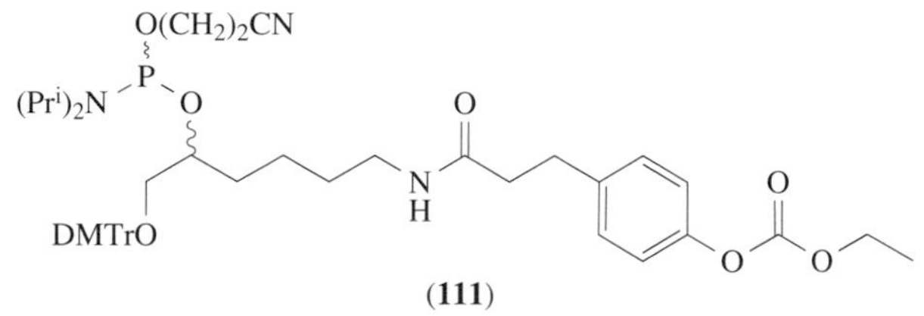

(111)

Coupling of the phenolic phosphoramidite derivative (112) to the 5′-end of an oligonucleotide synthesized on a CPG support afforded the functionalized oligonucleotide (113) after complete deprotection. Radioiodination of (113) to (114) was accomplished with [$^{125}$I]NaI according to the chloramine-T oxidation procedure. The radioactive yield was ca 15% and the specific activity was 2.5 Ci $\mu$mol$^{-1}$. The radiochemical purity was 94% according to thin-layer chromatography.[157] Thus, phenol–oligonucleotide conjugates provide facile access to radioiodinated oligonucleotides.

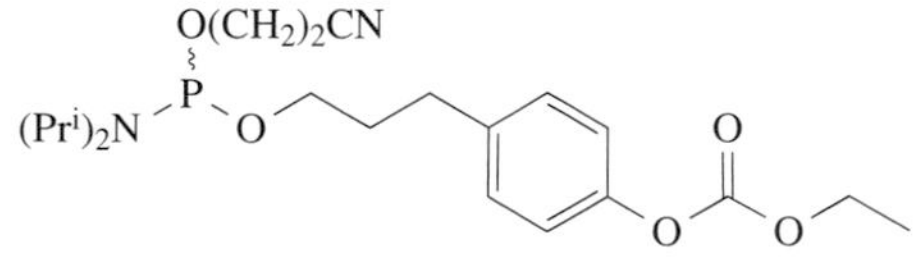

(112)

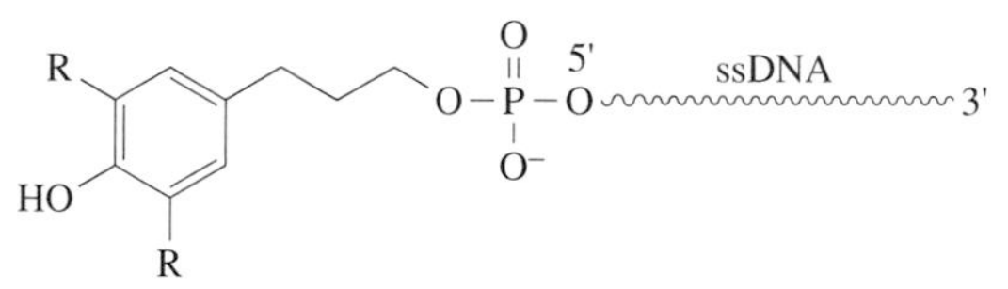

(113)  R = H
(114)  R = $^{125}$I

### 7.06.2.2.3 *Fluorescein and cyanine derivatives*

The preparation of fluorescently labeled oligonucleotides for automated "dideoxy" DNA sequencing has been further simplified by Schubert *et al.*[158] who reported the facile integration of the fluoresceinyl phosphoramidite (115) into various M13mp18 sequencing primers. Coupling yields were equivalent to those recorded for standard deoxyribonucleoside phosphoramidites.[158]

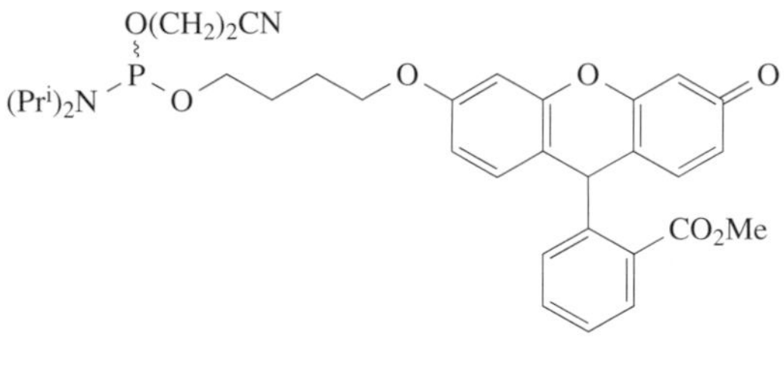

(115)

The laborious synthesis of the 5-carboxyfluoresceinyl phosphoramidite (116) from 5-carboxy-fluorescein and solketal has been described by Theisen *et al.*[159] It is argued that because of the steric hindrance imparted by the DMTr group, the coupling of (116) to the 5'-hydroxy function of a solid-phase linked oligonucleotide required 120 s, and proceeded with an efficiency of only 70%. The fluorophore and its linkage to the oligomer were stable during chain assembly and under the conditions used for complete oligonucleotide deprotection.[159] Like the biotinyl and phosphotyrosinyl phosphoramidites (84) and (98), the fluoresceinyl phosphoramidite (116) can be incorporated at any site into a DNA sequence.

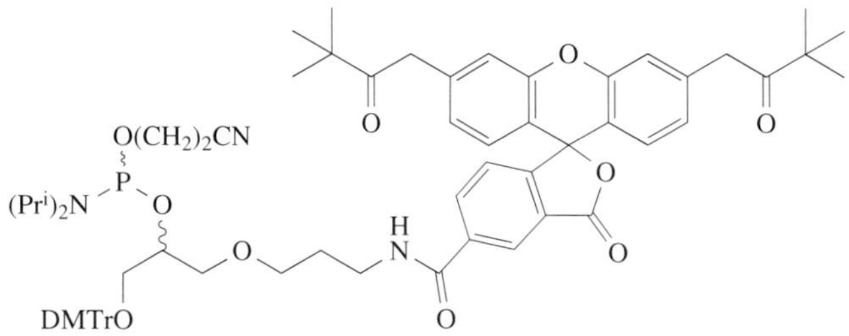

(116)

The fluorescein and cyanine phosphoramidite derivatives (117)–(120) and (121)–(122), respectively, have become commercially available for incorporation into oligonucleotides at the 5'-terminus, and to take advantage of the unique spectral properties of each of these dyes for modern DNA sequencing and genetic analysis applications.[160]

### 7.06.2.2.4  Dansyl, pyrene, and acridine derivatives

An additional strategy to the production of oligonucleotides conjugated to fluorescent groups relates to the dansyl phosphoramidite (123) which has been synthesized from dansyl chloride and (±)-3-amino-1,2-propanediol.[161] Methylation of the sulfonamide provided more stability to the phosphoramidite during its integration into oligonucleotides.

The 5-tris(2-aminoethyl)amino-1-naphthalenesulfonic acid (125) has additionally been applied to the facile fluorescent labeling of oligonucleotides at the 5'-terminus.[162] Typically, activation of the 5'-phosphate function of an oligonucleotide with imidazole hydrochloride and a water-soluble carbodiimide afforded the oligomer phosphoroimidazolide (124). Condensation of (124) with (125) generated the fluorescently labeled oligonucleotide (126) (Equation (1)). Annealing (126) with a complementary DNA oligomer gave a complex that is thermodynamically more stable than the corresponding unlabeled hybrid. Stabilization of the labeled hybrid is presumably due to electrostatic interactions between two positively charged amine functions of the label and the negatively charged phosphodiester backbone of the complementary strand.[162]

In this context, the pyrenyl phosphoramidites (127),[163] (128),[164] and (129)[161] have been designed for the incorporation of pyrene residues into oligonucleotides. It was shown that the presence of one or two pyrene residues at the 5'-terminus did not interfere with hybridization and extension of

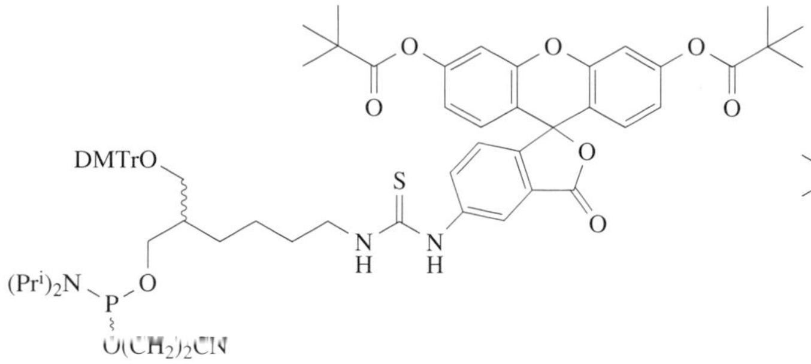

(117)

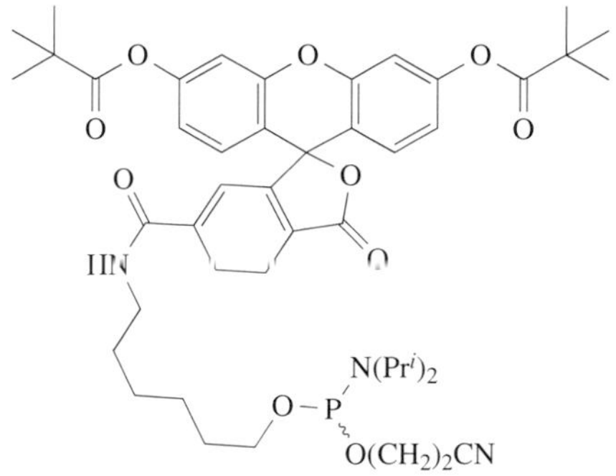

(118)

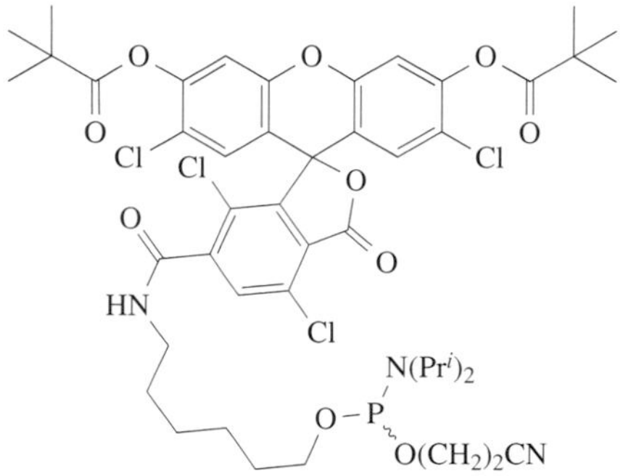

(119)

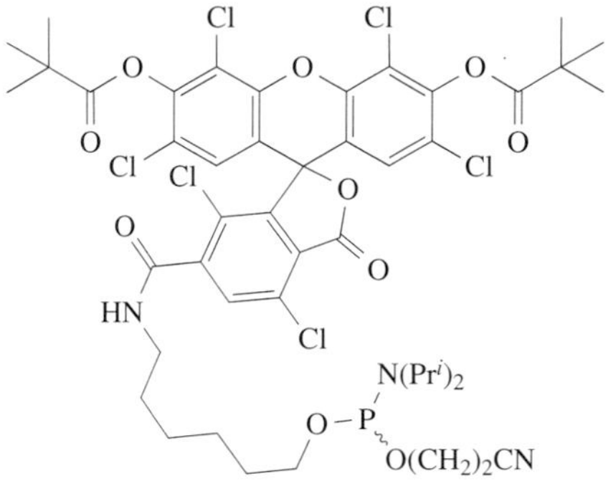

(120)

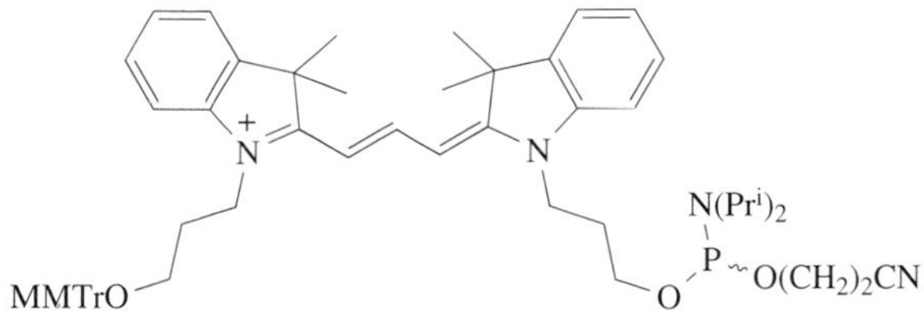

(121)

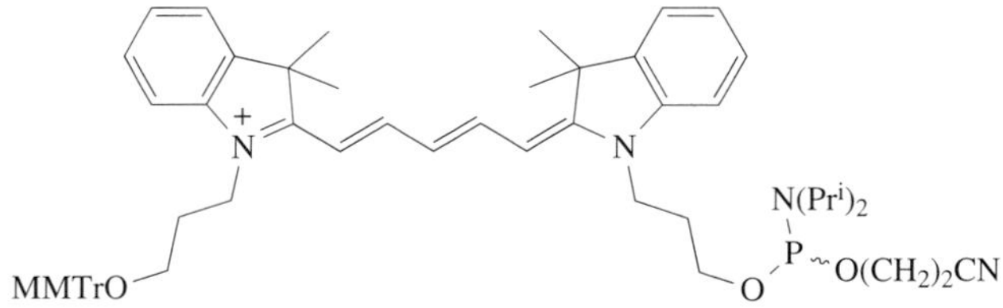

(122)

oligonucleotide primers.[164] It must nonetheless be noted that the attachment of a pyrenyl group to the 5′-terminus of dT$_{15}$ reduced its fluorescence yield to only 9% of that of 4-(1-pyrenyl)butanol. This relative fluorescence yield decreased to 3.7% when the 5′-pyrenylated-dT$_{15}$ was annealed to dA$_{15}$. However, the thermostability of the modified duplex increased ($T_m = 41.9\,°C$, 0.1 M NaCl) relative to that of the unmodified duplex ($T_m = 34.9\,°C$, 0.1 M NaCl). The pyrene-induced stabilization corresponded to a free energy change ($\Delta\Delta G°$) of $-2.6$ kcal mol$^{-1}$, and presumably originated from stacking of the pyrenyl group on the exterior of the ultimate base pair or intercalation between the ultimate and penultimate base pairs.[163]

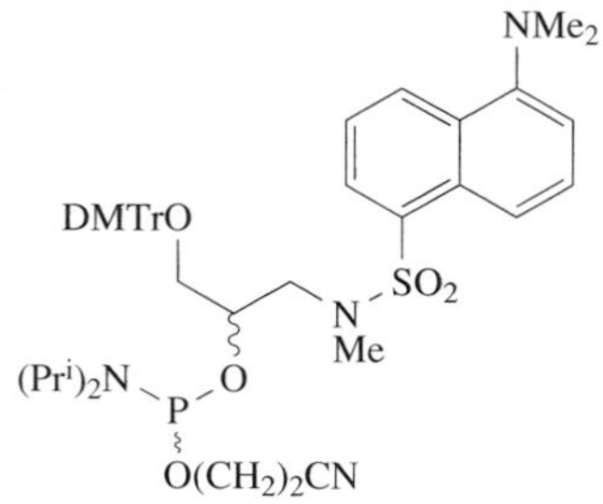

(123)

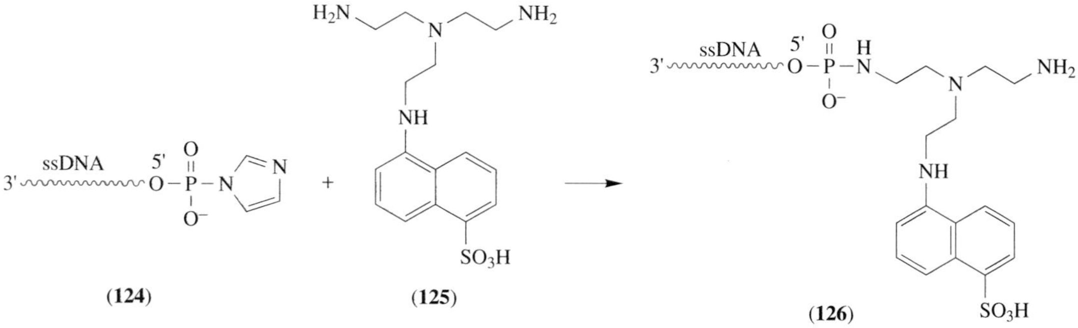

(124)      (125)      (126)      (1)

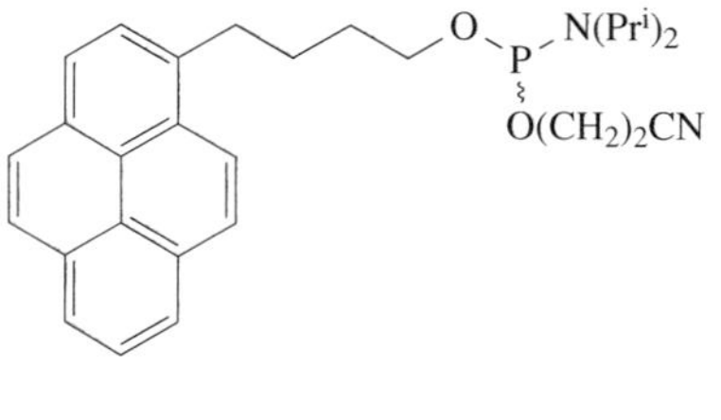

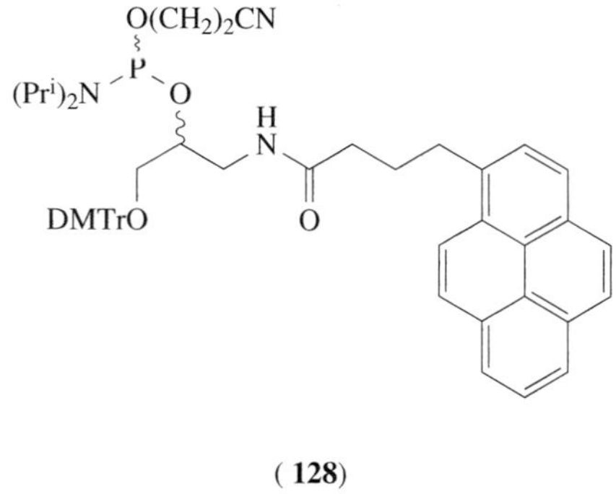

(127)          ( 128)

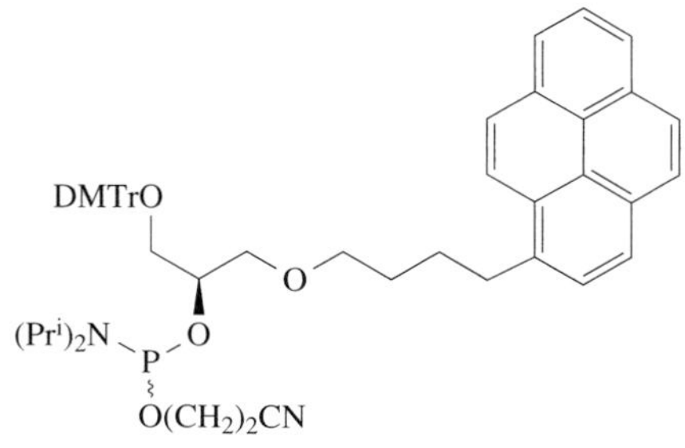

(129)

The derivatization of oligonucleotides with a neutral polycyclic aromatic hydrocarbon such as pyrene is a promising approach toward the development of antisense drugs; pyrene is nonmutagenic and has a high $LD_{50}$ in mice ($250\,mg\,kg^{-1}$). These features are particularly attractive for *in vivo* applications.

The synthesis of the acridinyl phosphoramidites (**130**),[165] (**131**),[166] and (**132**)[166] and their addition to the 5'-terminus of modified and unmodified oligodeoxyribonucleotides have also been documented.[122,167] In addition to providing fluorescence to these oligonucleotides, the acridinyl group stabilized the formation of double and triple helical structures with complementary nucleic acid targets by virtue of its intercalating properties.[118,122,167–171] Fluorescent acridine–oligonucleotide conjugates have been particularly useful in kinetic experiments addressing the cellular uptake of these biomolecules.[166]

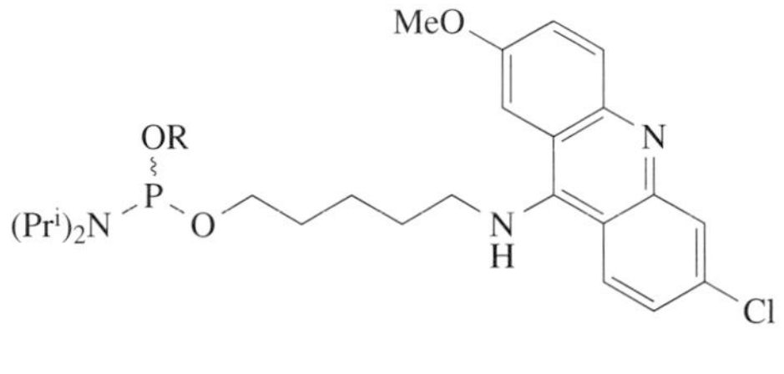

(**130**)   R = CH₂CH₂CN
(**131**)   R = Me

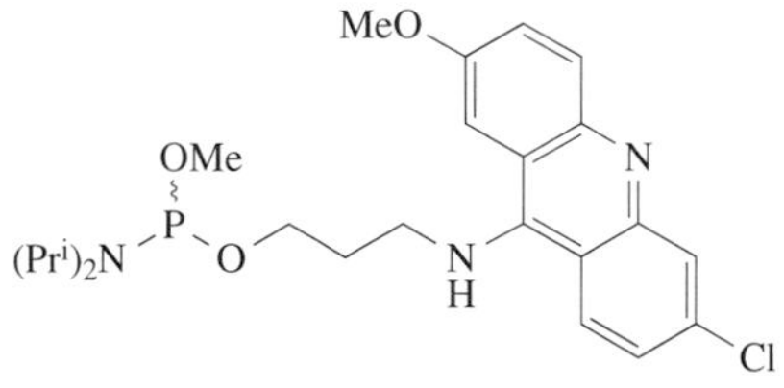

(**132**)

Incorporation of the commercial acridinyl phosphoramidite (**133**) at unique sites into oligonucleotides led to the conjugates (**134**). These conjugates have been shown to strongly stabilize triple helical complexes at base-pair inversions in the target sequences.[172] In fact, fluorescence studies provided evidence that the acridine moiety was intercalated into the triplex structures. The magnitude of triplex stabilization by (**134**) depended on the nature of the nucleobase facing the base-pair inversion site and the position of the acridine relative to this nucleobase.[172]

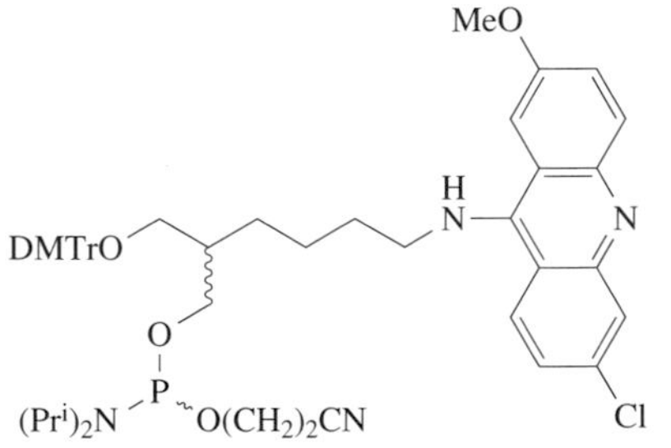

(**133**)

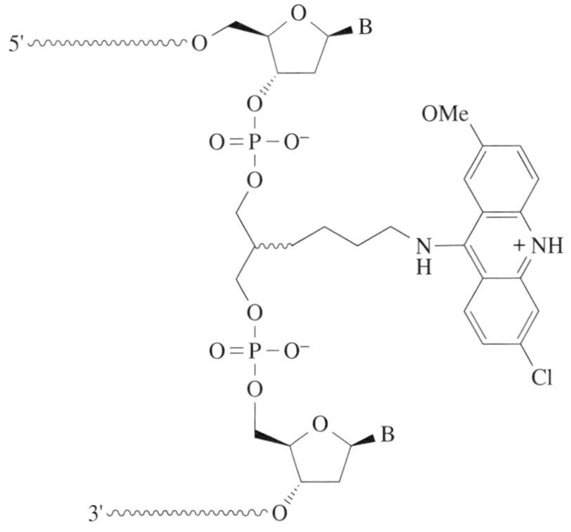

(**134**)

Along these lines, integration of the acridinyl phosphoramidite (**135**) into oligonucleotides prompted the evaluation of the correlation between acridine ring intercalation and the length of the methylene linkers.[173] The coupling efficiency of (**135**) was greater than 96% when a 5 min coupling was used. Deprotection of modified oligonucleotides was performed with concentrated ammonium hydroxide, at ambient temperature, for a short period of time (<2 h) to prevent extensive cleavage of the 9-aminoacridine linkage.

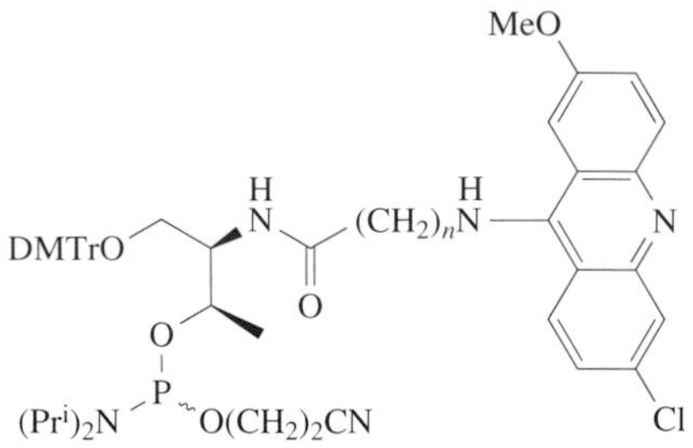

**(135)** $n = 3, 4,$ or 5

All modified oligonucleotides could form dsDNA with complementary oligonucleotides, but only those carrying threoninol residues with the longer methylene chains could favorably permit intercalation of the acridine ring.[173]

### 7.06.2.2.5 Coumarin, fagaronine, and anthraquinone derivatives

In an effort to conjugate oligodeoxyribonucleotides with stabilizing agents, the 5′-phosphomonoester function of DNA oligonucleotides was activated with 2,2′-dipyridyl disulfide and triphenylphosphine in the presence of *N*-methylimidazole, and reacted with aminoalkylated derivatives of coumarinoids.[174] The purified conjugates (136)–(138) (10-mers) were mixed with a complementary DNA strand (20-mer) to evaluate the thermostability of the resulting hybrids. The $T_m$ of each modified complex was significantly higher ($\Delta T_m = 5$–$16\,^{\circ}\mathrm{C}$) than that of the unmodified duplex. In the cases of (136) and (137), the stabilizing effects generated by the coumarin groups were superior to those produced by acridine under similar conditions. However, hybrid stabilization imparted by coumarin derivatives conjugated to larger oligonucleotides (14-mers) was less impressive ($\Delta T_m = 1$–$3\,^{\circ}\mathrm{C}$) but still comparable to that effected by similar acridine–oligonucleotide conjugates.[174]

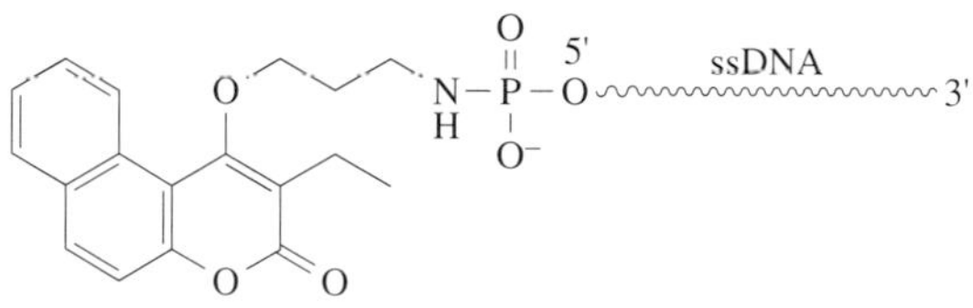

**(136)**

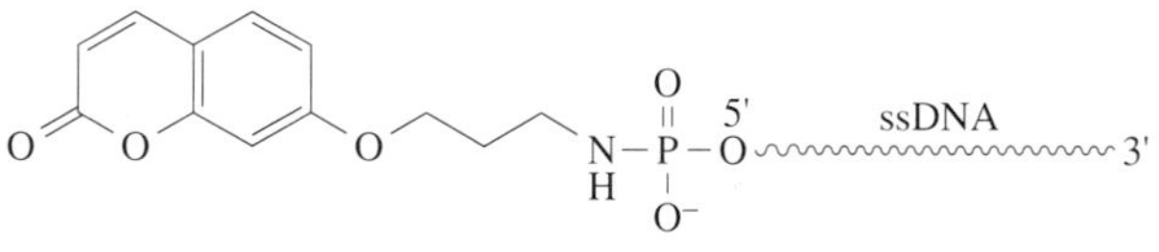

**(137)**

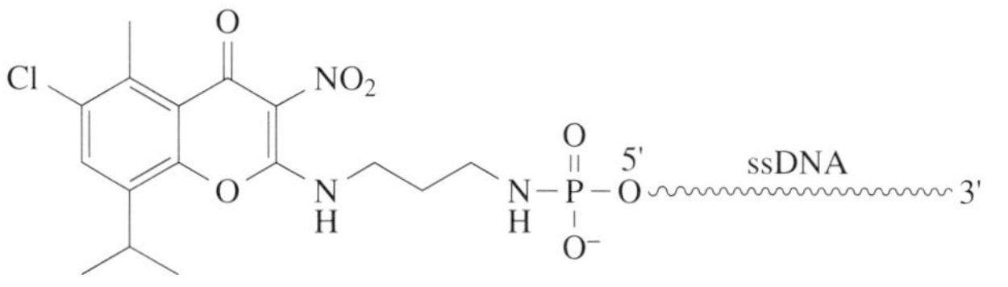

**(138)**

Notably the benzophenanthridine alkaloid fagaronine chloride, isolated from *Fagara zanthoxyloides*, has been found to inhibit the reverse transcriptase activity of several viruses.[175] This discovery provided the impetus to determine whether antisense oligonucleotide–fagaronine conjugates would enhance the inhibition of retroviral reverse transcriptases. Fagaronine chloride was converted to the phosphoramidite (139) to enable phosphitylation of the 5′-hydroxy function of oligonucleotides.[175] Relative to unmodified oligomers, oligonucleotide–fagaronine conjugates hybridized better to complementary RNA sequences, presumably because of the intercalating properties of the fagaronine entity. The conjugates were more potent than unmodified oligomers at inhibiting HIV-reverse transcriptase but were less inhibitory than fagaronine chloride itself.[175]

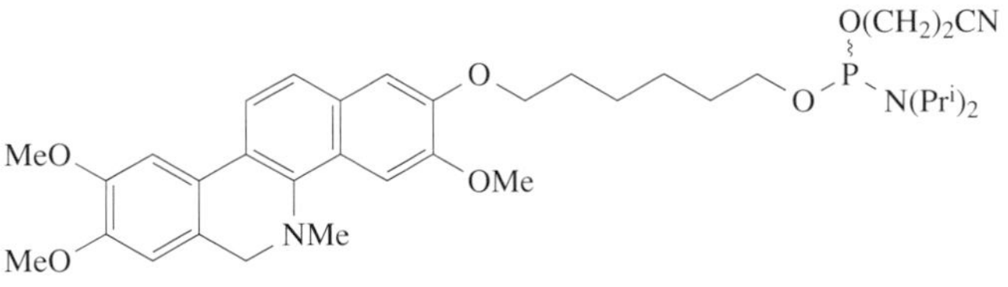

**(139)**

The benzophenanthridine phosphoramidites (**140**)–(**142**) have additionally been synthesized to permit the preparation of oligonucleotides that are conjugated to the benzophenanthridine inter-calator through a variable point of attachment.[176] The effect of the point of attachment on the affinity of oligonucleotide–benzophenanthridine conjugates for target sequences may help define the optimal orientation of the intercalating moiety in a double helix, and thus enhance inhibition of gene expression by similar antisense conjugates.[176]

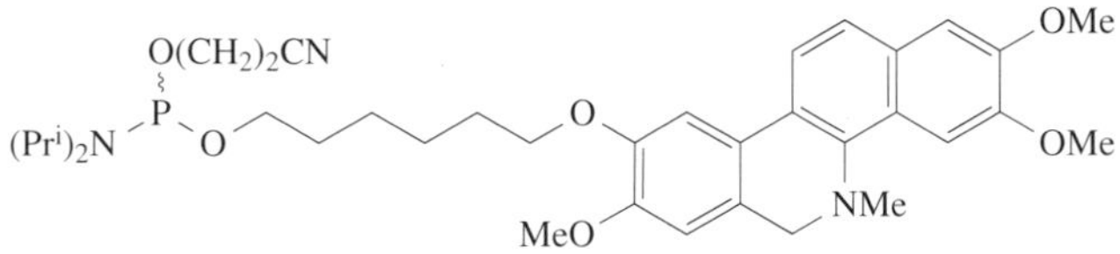

**(140)**

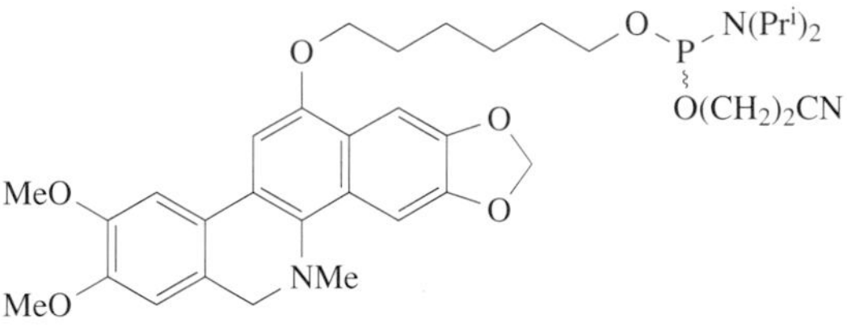

**(141)**

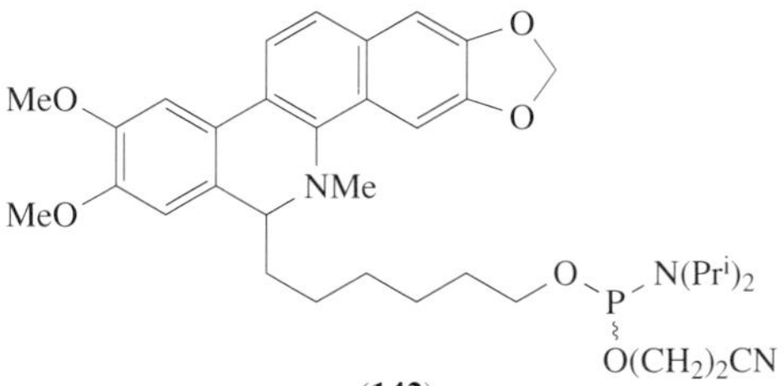

**(142)**

Mori *et al.*[177] developed the phosphoramidite (**143**) for the 5′-derivatization of oligonucleotides with an anthraquinone group. Anthraquinones are potential radical-producing species that do not necessarily require photoactivation or the presence of metal ions. Consequently, anthraquinone–oligonucleotide conjugates may be appropriate for the delivery of radical-producing anticancer drugs to cellular nucleic acid targets.[177]

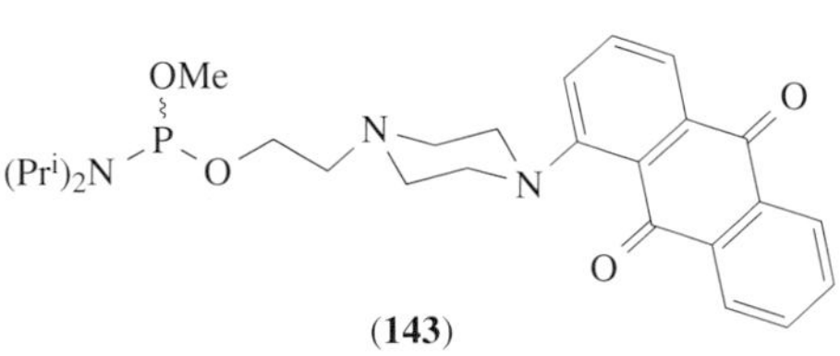

**(143)**

On the basis of computer modeling studies, the anthraquinone phosphoramidite (**144**) was designed for the functionalization of oligonucleotides at the 5′-terminus.[178] Because the amino functions of the anthraquinone group can form, via intercalation, new hydrogen bonds with nucleobases, it is anticipated that these oligonucleotide conjugates will bind more tightly to their complementary nucleic acid targets.[178] Studies on the thermodynamics of duplex binding are still in progress.

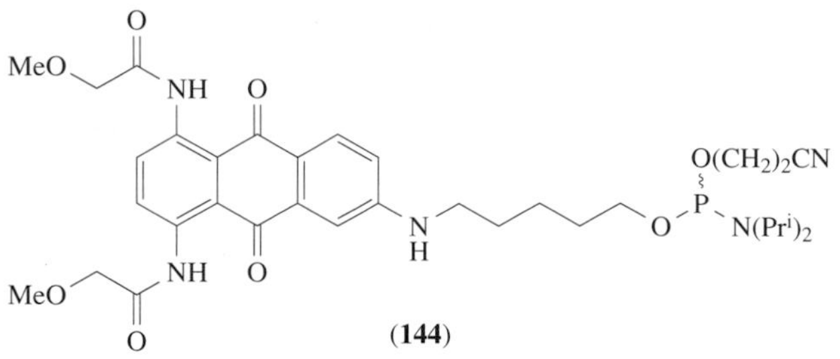

**(144)**

### 7.06.2.2.6   *Psoralen groups*

The direct addition of psoralen derivatives to the 5′-terminus of oligodeoxyribonucleotides has been effected by the phosphoramidite (**145**) via solid-phase methods.[179,180] Psoralen derivatives can intercalate between the base pairs of dsDNA and become covalently attached to thymines on exposure to long-wavelength UV irradiation. Under these conditions, each psoralen–DNA adduct is generated through the formation of a cyclobutane ring. Such adducts have been extensively applied to the elucidation of nucleic acid structures and functions.[129,130,181,182]

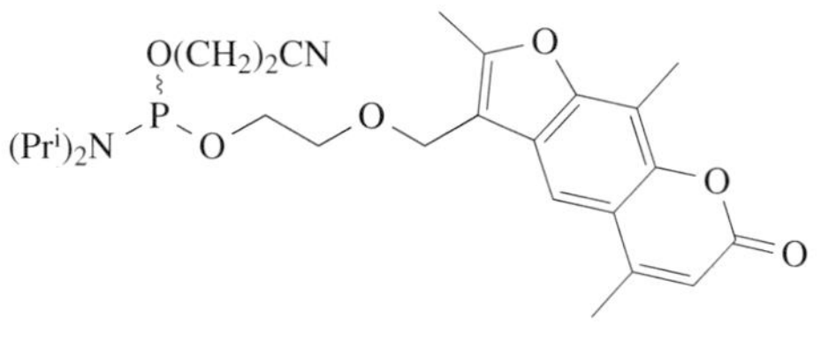

**(145)**

Furthermore, the phosphoramidites (**146**) and (**147**) have recently led to the addition of photo-reactive psoralen derivatives to the 5′-end of triplex-forming oligonucleotides.[183,184] For example, a modified oligonucleotide was employed to target a sequence encoding part of the human *aromatase* gene. Photoadduct formation was found to be sensitive to pH, temperature, cation concentration, and base composition of the third strand. The attached psoralen has been shown to intercalate a triplex–duplex junction mainly in one orientation. Interestingly, oligonucleotides that are func-tionalized with (**146**) formed cross-links faster than those derivatized with (**145**).[183] The phos-phoramidite (**147**) has additionally been used for the introduction of psoralen at the 5′-terminus of oligonucleotides carrying the nucleobase analogue 8-oxo-adenine.[185] These modified oligo-nucleotides can form pyr–pur–pyr type triplexes with dsDNA at pH 7.5. While triplex formation is stabilized by the presence of the psoralen group, irradiation of these triplexes showed the selective formation of a photoadduct with a cytosine of the purine-rich strand of the duplex targets.[185]

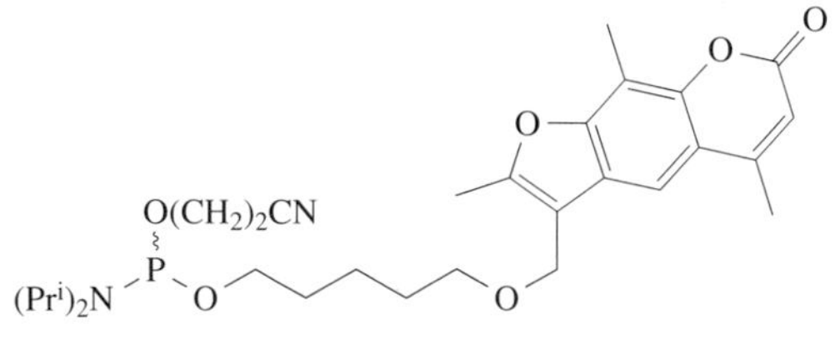

**(146)**

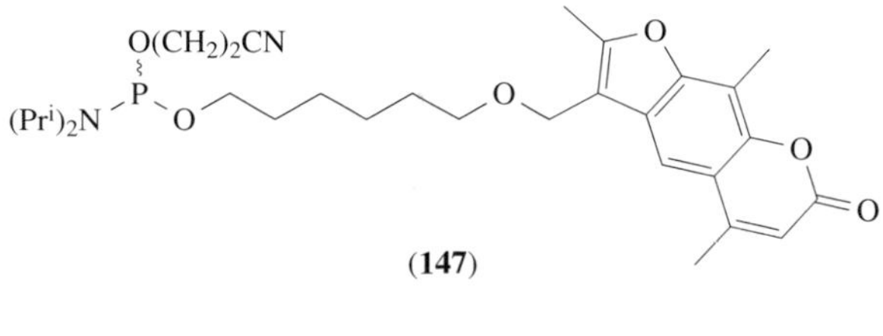

**(147)**

### 7.06.2.2.7  *Bipyridine, 1,10-phenanthroline, and bleomycin $A_5$ derivatives*

The phosphoramidite (**148**) has been prepared by Modak *et al.*[186] and employed in the phosphitylation of 3'-*O*-acetyl-2'-deoxythymidine. After oxidation and purification, the bipyridinyl phosphotriester (**149**) was complexed with copper(II) and incubated with poly $rA_{12-18}$, at 37 °C, for 24 h or 48 h. The RNA substrate was degraded to the extent of 21% within 48 h. Adenosine (2'-3')-cyclic monophosphate was identified as the major hydrolysis product. No degradation of the RNA homopolymer was detected in the absence of copper(II) under similar conditions.[186] The phosphoramidite (**148**) may, therefore, be useful toward the development of sequence-specific RNA cleaving reagents.

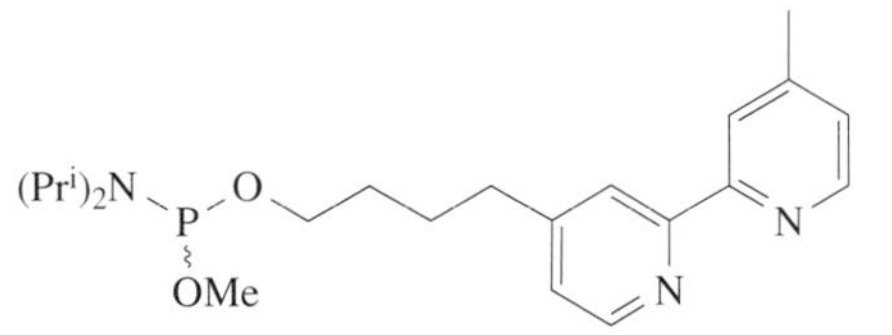

(**148**) 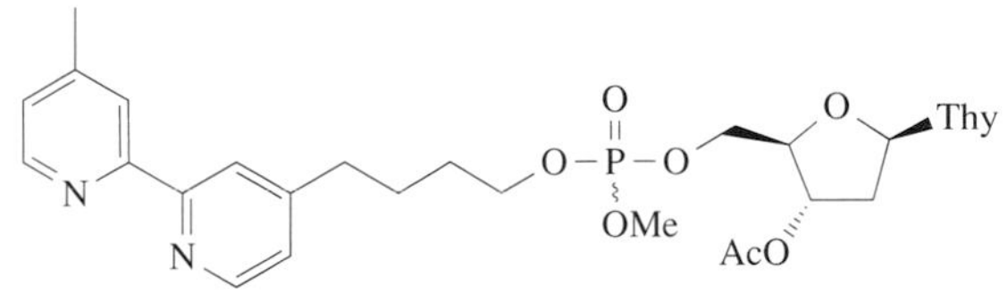                                    (**149**)

The conjugation of 1,10-phenanthroline to the 5'-end of oligonucleotides also mimicked nuclease activity. The synthesis of these oligonucleotide conjugates involved activation of the 5'-terminal phosphate function of an oligonucleotide to the corresponding phosphoroimidazolide (**124**) followed by condensation with 5-glycylamido-1,10-phenanthroline (Equation (2)).[187] After hybridization of the phenanthroline–oligonucleotide conjugate (**150**) with a complementary DNA target, sequence-specific cleavage of the target is observed upon addition of cupric ion and 3-mercaptopropionic acid. This approach provides a method to tailor single-stranded sequences and to analyze DNA or RNA for single-stranded domains.[187]

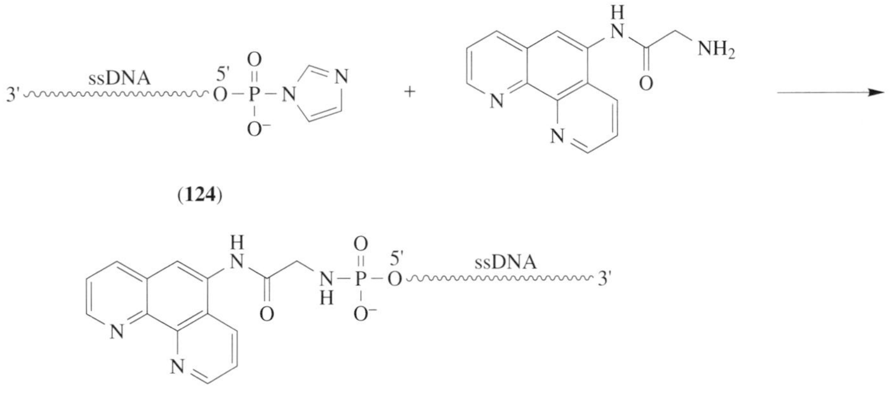

(**124**)

(**150**)                                                                                         (2)

In a different context, a bathophenanthroline–ruthenium(II) complex derivatized as the $\beta$-cyanoethyl*N*,*N*-diisopropyl phosphoramidite derivative (**151**) has been applied to the solid-phase labeling of oligonucleotides.[188] Like europium(III)-labeled oligonucleotides, oligomers labeled with bathophenanthroline–ruthenium(II) complexes can be detected by time-resolved fluorometry.[21]

The phosphoramidites (**152**) and (**153**) have been used to functionalize synthetic oligonucleotides (12-mers) at the 5'-terminus.[189] When hybridized to an ssDNA oligonucleotide (135-mer) in the presence of iron(II), oxygen, and a reducing agent, the modified dodecamers cleaved the complementary DNA chain with a unique and high sequence-specificity.[189] Template cleavage by an oligonucleotide modified with (**152**) occurred mainly at two bases and could result from intercalation of the planar aromatic 2,6-pyridinedicarboxylate–iron(II) complex between the last two base pairs of the double helix. Given that a strong intercalating agent such as ethidium bromide totally inhibited the cleavage, ligand intercalation may therefore be a prerequisite for DNA cleavage.[189]

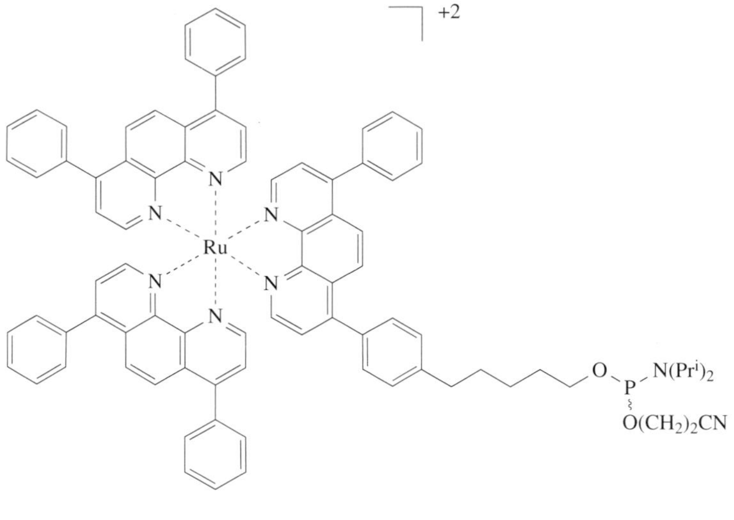

(151)

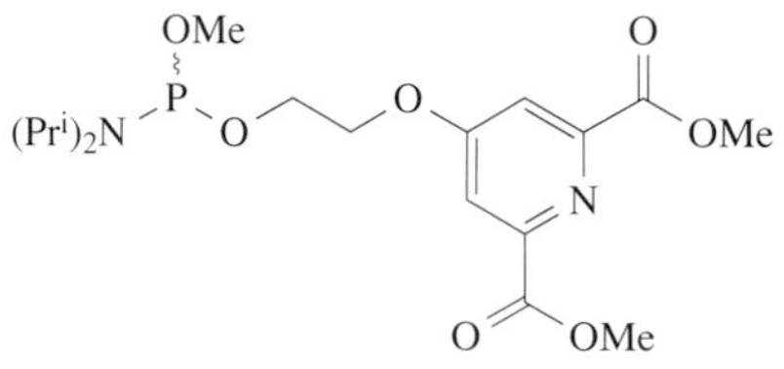

(152)

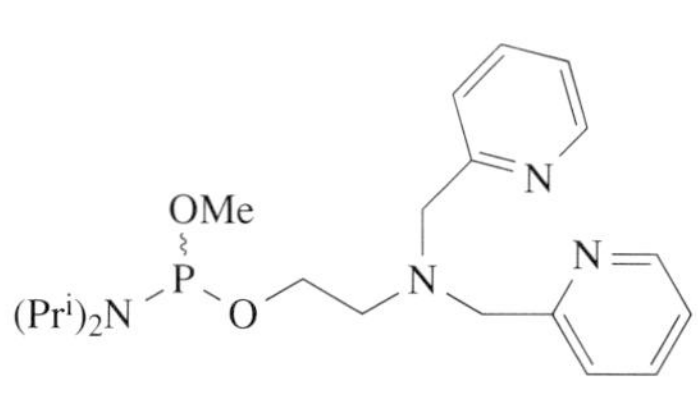

(153)

Bleomycin is a glycopeptide-derived antitumor antibiotic that is believed to disrupt cell proliferation by causing DNA strand scission in a sequence-selective reaction that is dependent on oxygen and a metal ion, such as iron(II).[190,191] While bleomycin preferentially cleaves DNA at 5'-GT and 5'-GC sequences, its conjugation to an oligonucleotide may provide a much higher specificity in the cleavage of complementary DNA targets.

A method for the conjugation of bleomycin $A_5$ to an oligonucleotide has been developed. The synthetic strategy consisted of the activation of the 5'-phosphomonoester function of the oligonucleotide with a mixture of 2,2'-dipyridyl disulfide, triphenylphosphine, and 4-($N,N$-dimethylamino)pyridine 1-oxide, followed by reaction with the copper(II) complex of an aminated bleomycin $A_5$ derivative. The bleomycin–oligonucleotide conjugate (154) was obtained in yields ranging from 60 to 80%.[192]

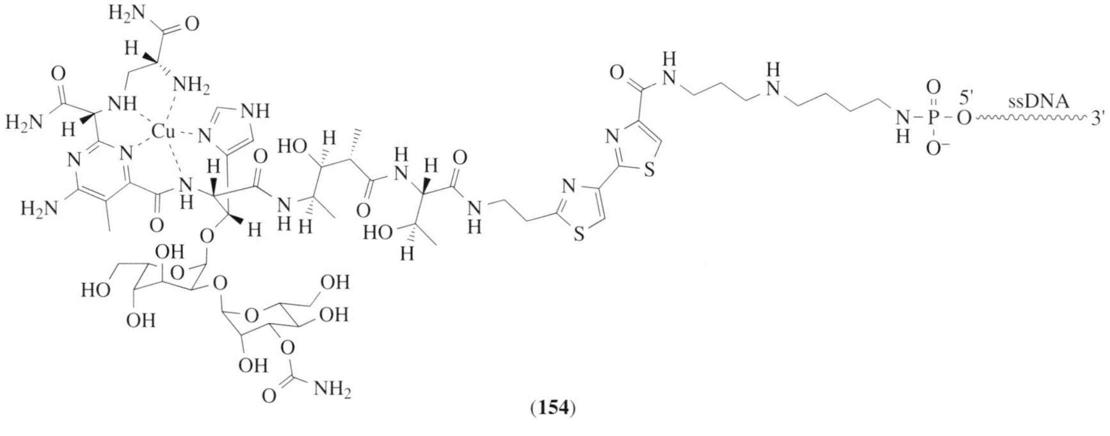

(154)

After removal of the copper(II) ion from (154), the bleomycin $A_5$–oligonucleotide conjugate was mixed with a complementary DNA sequence. In the presence of iron(II) and 2-thioethanol, site-specific cleavage of the DNA target was observed.[192,193] Subsequent studies showed that one bleomycin A$_5$–oligonucleotide conjugate is capable of cleaving three DNA targets.[194] However, the number of DNA targets destroyed by the conjugate was limited by self-degradation of the antibiotic residue.[194] Despite this limitation, bleomycin–oligonucleotide conjugates represent an interesting class of potential therapeutic agents.

### 7.06.2.2.8 *Lipophilic groups*

The use of oligonucleotides in therapeutic applications has been hampered by the low efficiency with which these polyanionic molecules permeate intact cells. It has been rationalized that the attachment of lipophilic groups, known to interact with cell membranes, to oligonucleotides may increase the therapeutic potency of these biopolymers.[195–198] To test this rationale, the cholesteryl phosphoramidite (**155**), along with the adamantylethyl phosphoramidite (**156**) and the polyalkyl phosphoramidites (**157**)–(**159**), has been synthesized and added to the 5′ terminus of an oligonucleoside phosphorothioate that is complementary to a region of the HIV-1 *rev* gene.[195] Preliminary results indicated that duplexes composed of lipophilic oligonucleoside phosphorothioates and complementary unmodified DNA strands are thermodynamically as stable as the corresponding underivatized duplexes.[195] Moreover, the 1,2-di-*O*-hexadecyl-*rac*-glycerol–oligonucleotide conjugate that was prepared via the phosphoramidite (**157**)[195] or the corresponding *H*-phosphonate derivative[199–201] inhibited HIV-1 replication through a nonsequence-specific mechanism.[199] It would appear that lipid-phosphorothioate oligonucleotides may inhibit HIV-1 replication by repressing viral reverse transcriptase activity.[199] In uptake experiments with L929 cells, lipid–oligonucleotide conjugates became 8–10 times more cell-associated than unmodified oligonucleotides.[201]

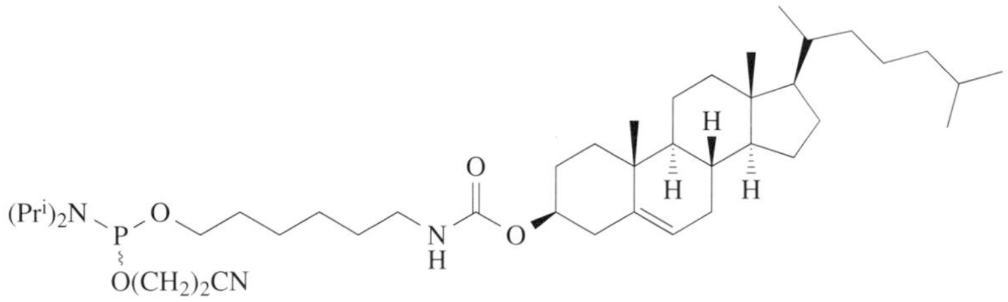

(155)

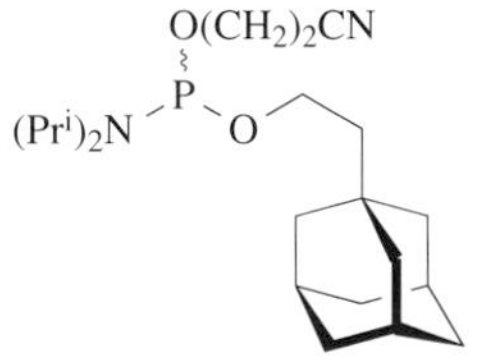

(156)

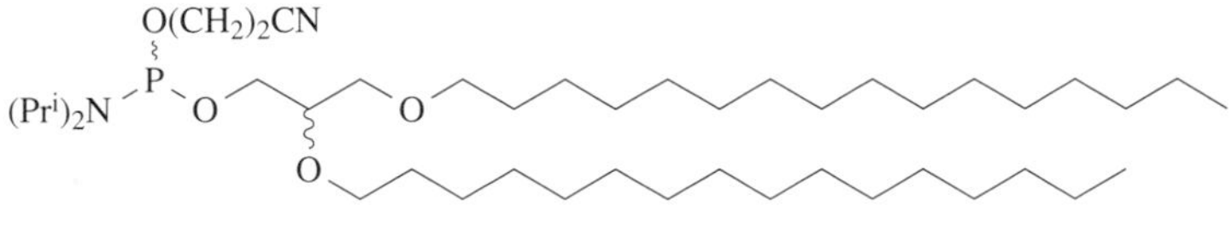

(157)

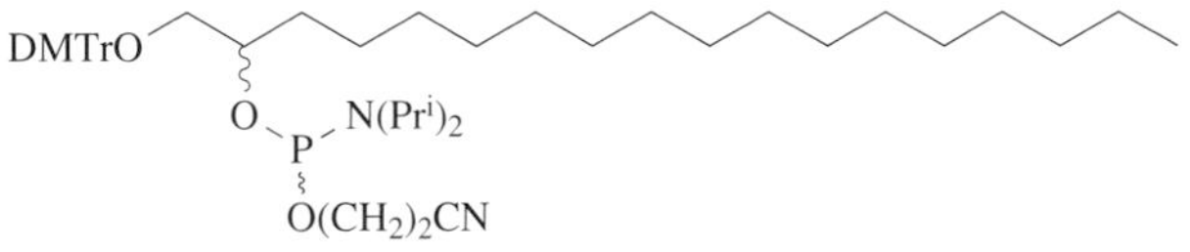

(158)

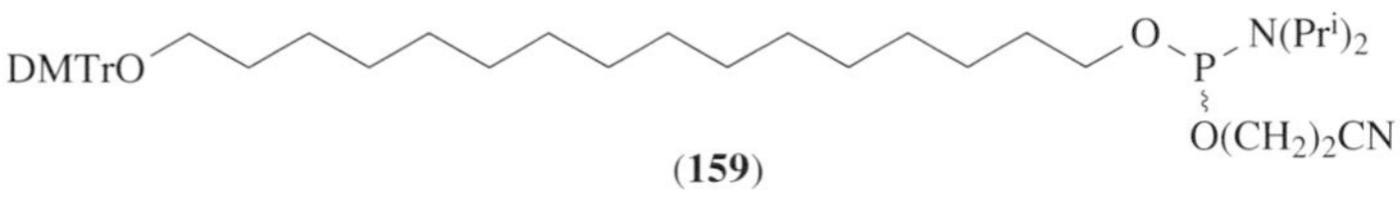

(159)

Along similar lines, Kabanov and co-workers[196,202] reported the synthesis of the phosphoramidite (**160**) to enable the introduction of a hydrophobic group at the 5′-end of oligonucleotides. The targeting of a 5′-undecylated oligonucleotide against a loop-forming site of the influenza A/PR8/34 viral RNA considerably suppressed the development of the virus in permissive MDCK cells relative to that observed with unmodified oligonucleotides under identical conditions.[196,197] The attachment of hydrophobic groups to oligonucleotides has therefore demonstrated its efficacy in improving the biological activity of these biomolecules.

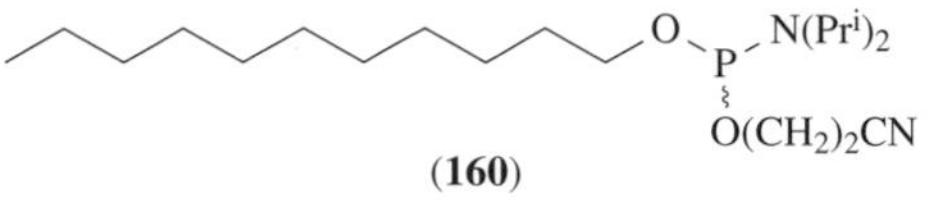

(160)

The cholesteryl phosphoramidite (161) has additionally been synthesized for incorporation into triple helix-forming oligonucleotides.[203] Cellular uptake studies showed that the internal concentration of these cholesterol-modified oligonucleotides exceeded that of the media by a factor of 5–50.[203] It should be noted that lipophilicity can also be imparted to synthetic oligonucleotides via the single or multiple addition of the cholesteryl phosphoramidite (162) or (163)[43] to the 5′-terminus. Incidentally, the phosphoramidite (162) is commercially available.

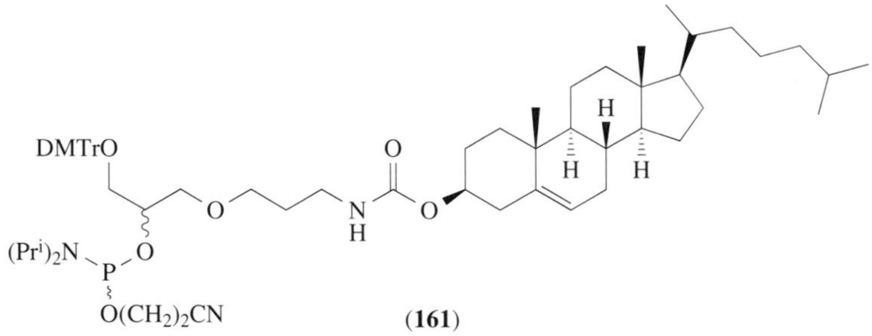

(161)

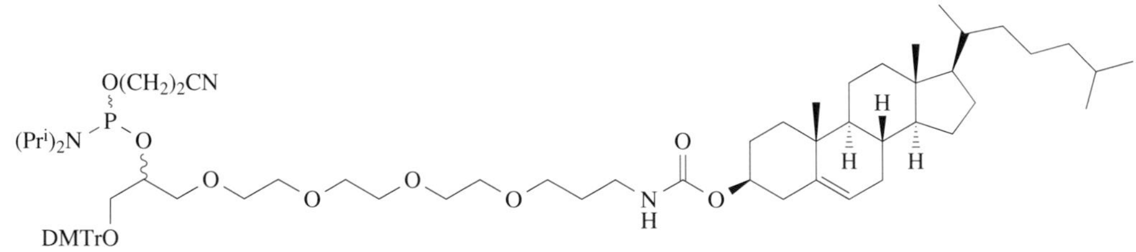

(162)

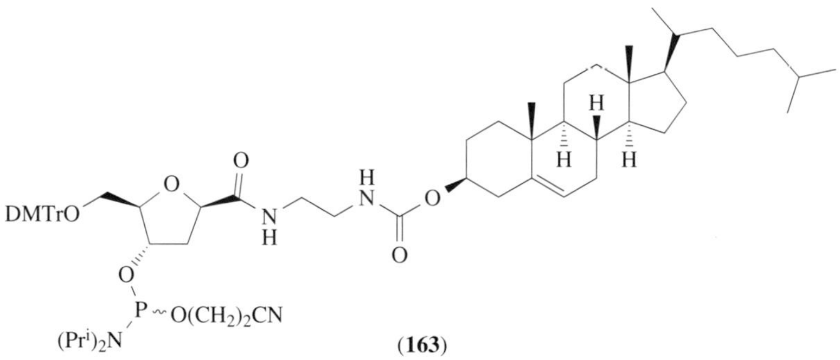

(163)

Another method for the conjugation of cholesterol with the 5′-terminus of oligonucleotides relates to the condensation of the aminated cholesterol derivative (164) with an oligonucleotide 5′-phosphoro-*N*-methylimidazolide.[204] The resulting cholesterol–oligonucleotide conjugate (165) was then incubated with T24 human carcinoma cells. It was found that more than 80% of the conjugate remained on the external surface of the cellular membrane. While internalization of the conjugate proceeded via endocytosis, it remained encapsulated inside stable endosome-like particles. Only about 15% of the cholesterol–oligonucleotide conjugate that penetrated cells reached the nucleus.[204]

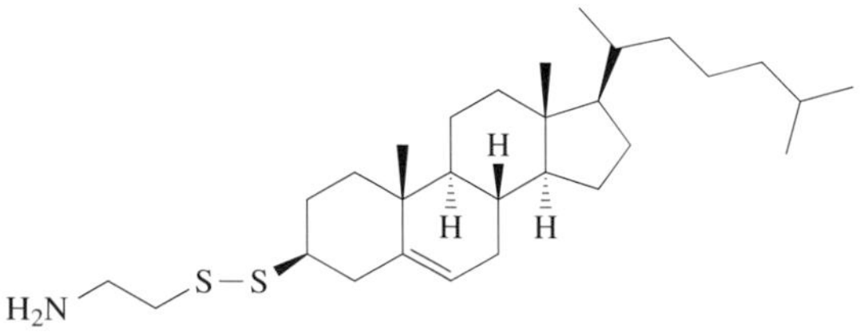

(164)

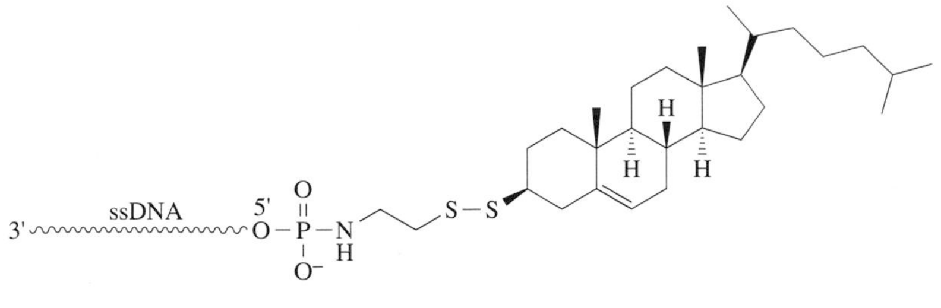

**(165)**

Synthesis of the cholesteryl nucleoside phosphoramidite **(166)** has been performed to provide the antiviral cordycepin core [3′-deoxyadenylyl-(2′-5′)-3′-deoxyadenylyl-(2′-5′)-3′-deoxyadenosine] with improved cell permeability.[205] Comparative biological studies indicated that the 5′-cholesteryl–cordycepin trimer core conjugate was a 1000-fold more potent inhibitor of HIV-1-induced syncytia formation than the unmodified cordycepin trimer core.[205] In this context, it should be mentioned that chemical syntheses of various cordycepin trimers conjugated to vitamin E, $D_2$, and A, through either the 5′- or the 2′-termini, have been reported.[206] The biological properties of these conjugates are currently under investigation.

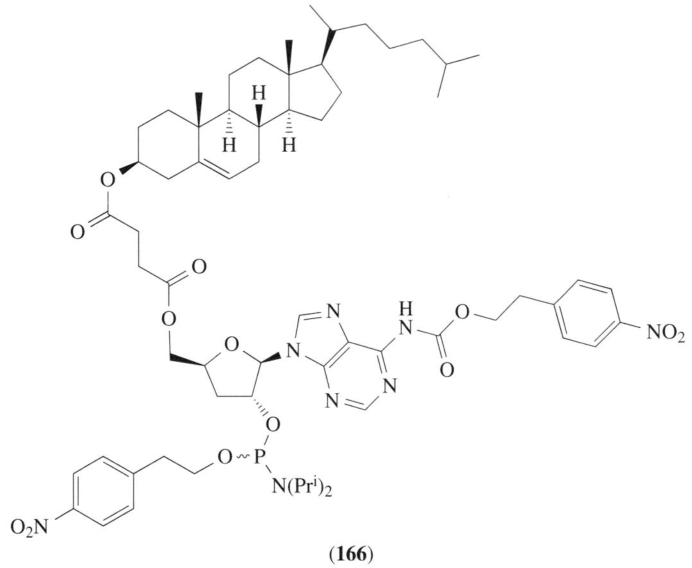

**(166)**

In an additional effort to impart lipophilicity to oligonucleotides, and improve their cellular uptake and distribution, synthesis of the D,L-α-tocopherol (vitamin E) phosphoramidites **(167)** and **(168)** has been described.[198] Like **(160)**, the incorporation of **(167)** and **(168)** into oligonucleotides occurred in yields greater than 95%, according to HPLC analysis of the conjugates.[198] The anti-HIV activity of these conjugates is presently being evaluated in cell cultures.

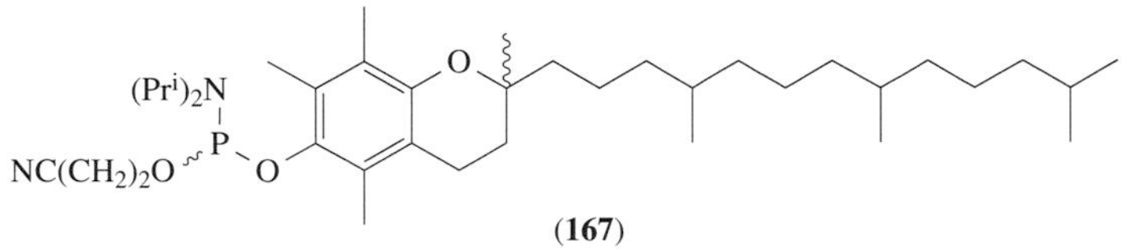

**(167)**

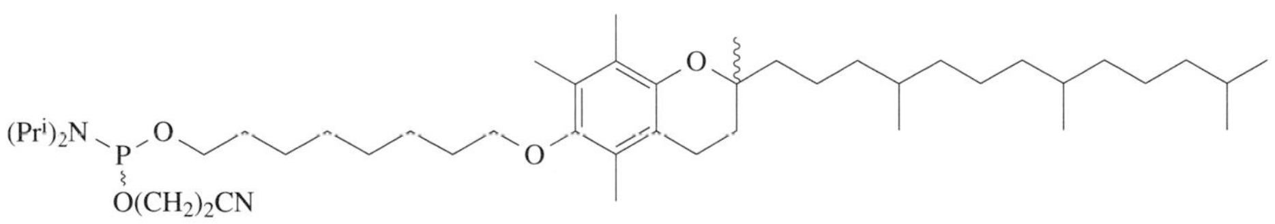

**(168)**

### 7.06.2.2.9 Polyethylene glycol derivatives

Polyethylene glycol (PEG) is known to play an important role in cell fusion and transport processes through cellular membranes.[207] The conjugation of PEG to oligonucleotides may seemingly facilitate their cellular uptake for potential therapeutic applications. Thus, the PEG phosphoramidites (169) and (170) were easily added to the 5'-terminus of oligonucleotides with a coupling efficiency of 70% and 50%, respectively.[208,209] Of particular relevance, oligonucleotides with PEG conjugated to both 3'- and 5'-termini showed more than a 10-fold increase in exonuclease stability.[210] Furthermore, the conjugation of PEG to oligonucleotides had little effect on their affinity for unmodified complementary oligomers. The resulting duplexes can even be extracted in the PEG-rich phase of an aqueous two-phase system composed of PEG and dextran.[211] This unique property of PEG–oligonucleotide conjugates can be applied to the sequence-specific isolation of nucleic acids through hybridization-based affinity partitioning.

$$DMTr(OCH_2CH_2)_nO \underset{\overset{\vphantom{|}}{\underset{O(CH_2)_2CN}{|}}}{\diagdown P \diagup} N(Pr^i)_2$$

(169)   $n = \sim 6\text{–}12$ (PEG 400)
(170)   $n = \sim 16\text{–}32$ (PEG 1000)

In addition to being useful in the synthesis of PEG–oligonucleotide conjugates, PEG phosphoramidites of various sizes have been incorporated into oligonucleotides to effect loop replacements in DNA duplexes and triplexes,[212,213] RNA miniduplexes,[214] and catalytic RNAs.[215,216] It is also noteworthy that PEG has served as a soluble support for the large-scale (100 μmol) production of oligonucleotides.[217]

## 7.06.2.3 Functionalization of DNA Oligonucleotides at the 3'-Terminus for Either Direct or Indirect Incorporation of Reporter and Conjugate Groups

Solid supports have been engineered to enable the derivatization of synthetic oligonucleotides with functional groups, such as masked[70,118,119,218–221] or unmasked 3'-sulfhydryl,[70,218,222–225] 3'-phosphate,[119,218,226–233] 3'-thiophosphate,[47,119,169,226,234–236] 3'-carboxyalkyl,[136,219,220,237–240] 3'-hydroxyalkyl,[202,241–244] 3'-carboxamidoalkyl,[136,219,220] 3'-phosphoglycolate,[245] 3'-phosphoglycaldehyde,[245] 3'-amino,[246] 3'-aminoaralkyl,[46] 3'-(aminoalkyl)imidazole or histamine,[138] 3'-aminoalkyl,[119,136,146,219,220,224–226,240,241,247–260] and 3'-aminoalkylphosphoramidates.[261]

A number of these functional groups can then be further derivatized with various ligands to serve as tools in molecular biology experiments. Solid supports functionalized with acridine,[146,241,262,263] cholesterol,[43,203,241,262,264–267] cholic acid,[266] D,L-α-tocopherol,[198] biotin,[43,146,160,268] fluorescein,[146,158–160] adamantane,[269] dansyl,[45] β-cyclodextrin,[269] multimeric 1,2-dihydro-3H-pyrrolo[3,2-e]indole-7-carboxylates with an N-3 carbamoyl group,[270] tyrosine,[271,272] peptides,[273] polyethylene glycol,[200,208–210] and pyrene[274,275] have alternatively been described for the synthesis of oligonucleotides bearing these ligands at the 3'-terminus.

Oligonucleotides that carry a 3'-terminal uridine such as (171) can easily be functionalized with reporter groups. As shown in Scheme 1, oxidation of (171) with sodium metaperiodate, followed by reductive amination with octamethylene diamine and condensation with acridinyl isothiocyanate, gave the oligonucleotide–acridine conjugate (172).[202] This strategy has been used to conjugate aminoalkylated ellipticine to the 3'-terminus of DNA oligonucleotides in order to study cross-linking and photoinduced cleavage of nucleic acids by these conjugates.[236]

In another experiment, the 3'-terminus of RNA was oxidized to a dialdehyde by sodium metaperiodate and then labeled with fluorescein-5-thiosemicarbazide.[276] The fluorescently-labeled RNA was partially digested with base-specific ribonucleases, and the resulting fragments were separated by polyacrylamide gel electrophoresis. The fluorescent bands were visualized by UV and a partial sequence of yeast 5S rRNA was determined by this method.[276]

Particularly remarkable is the condensation of a 5'-aminoalkylated ribonucleotide analogue with a dialdehyde similar to (70) that was generated from the 3'-terminus of another oligoribonucleotide analogue.[277] After reductive aminoalkylation and purification, the resulting morpholino-linked ribozyme displayed a greater catalytic activity than that of a control ribozyme.[277] Such an approach

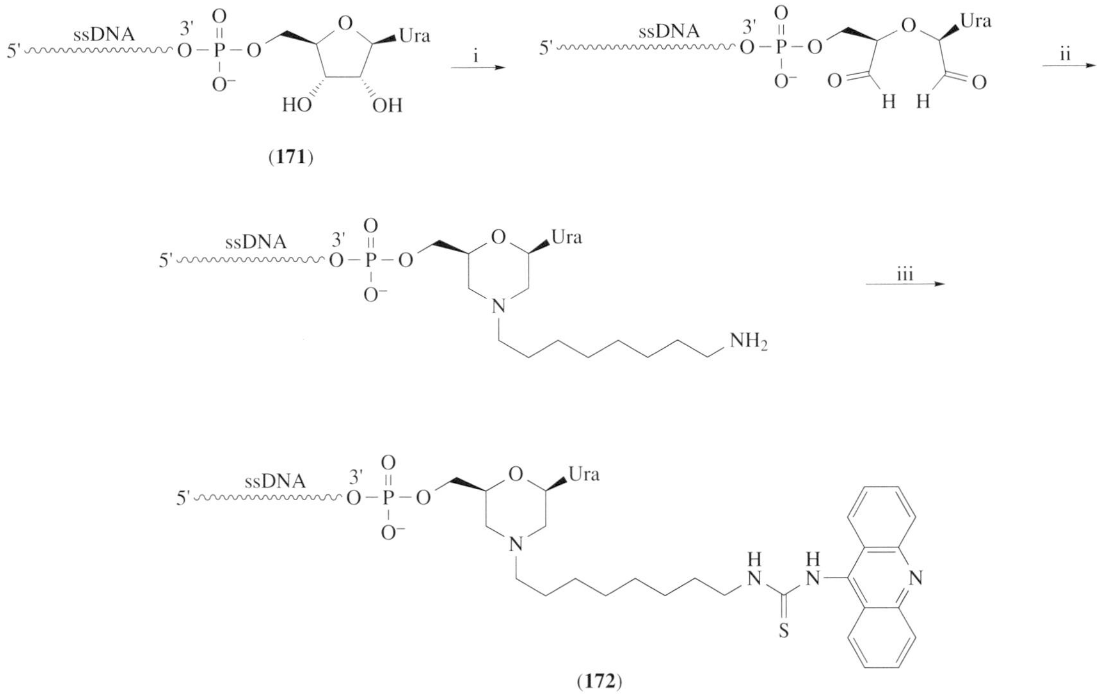

(**171**)

(**172**)

i, NaIO$_4$; ii, NH$_2$(CH$_2$)$_8$NH$_2$/NaBH$_4$; iii, acridine isothiocyanate

**Scheme 1**

to ribozyme synthesis, as opposed to iterative RNA synthesis, facilitates purification of the full-length ribozyme from the halves. It should, however, be understood that the site of conjugation must not interfere with the ribozyme core to ensure retention of full catalytic activity. In addition to the 5′- and 3′-termini, the functionalization of oligonucleotides with reporter and conjugate groups can be accomplished through the nucleobases. This topic will be reviewed in the next section.

## 7.06.3 ATTACHMENT OF REPORTER AND CONJUGATE GROUPS TO THE NUCLEOBASES OF DNA OLIGONUCLEOTIDES

### 7.06.3.1 Functionalization of Nucleobases for Indirect Incorporation of Reporter and Conjugate Groups into DNA Oligonucleotides

#### 7.06.3.1.1 *Aminoalkylated purines and pyrimidines*

The aminoalkylated deoxyribonucleoside phosphoramidites (**173**) and (**174**) originated from either the transamination of 2′-deoxycytidine with 1,3-diaminopropane in the presence of sodium metabisulfite[278,279] or the facile aminoalkylation of $N^4$-$p$-toluenesulfonyl-2′-deoxycytidine derivatives with 1,6-diaminohexane.[280] The phosphoramidites (**174**)[280] and (**175**)[281] were coupled to the 5′-hydroxy function of solid-phase bound DNA oligonucleotides or 2′-$O$-alkyl oligoribonucleotides. After deprotection, these aminoalkylated oligonucleotides were biotinylated and used for the affinity purification of, for example, RNA–protein complexes.[281,282]

Another example of deoxyribonucleoside phosphoramidites dedicated to the preparation of aminoalkylated oligonucleotides is the phosphoramidite (**176**) which has been prepared from $N^6$-phenoxycarbonyl-2′-deoxyadenosine derivatives.[283]

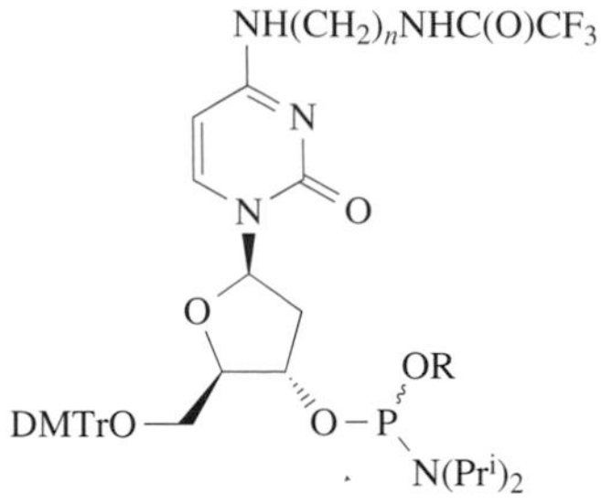

(173)  R = Me; *n* = 3
(174)  R = CH$_2$CH$_2$CN; *n* = 6
(175)  R = CH$_2$CH$_2$CN; *n* = 5

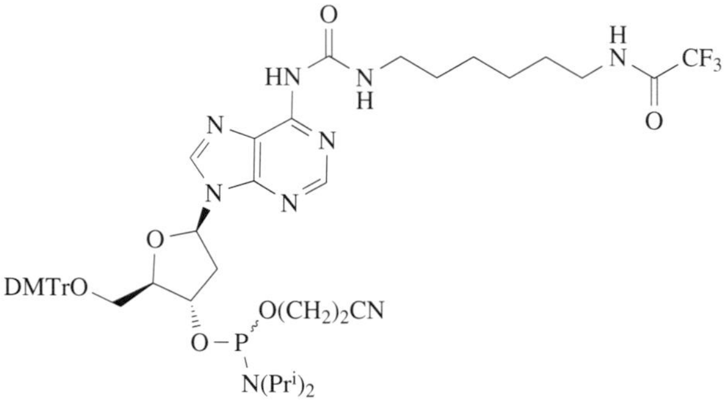

(176)

It should be pointed out that the triazolo function of 4-triazolopyrimidinone nucleosides is notoriously susceptible to nucleophilic displacement by amines.[284,285] Typically, when 4-triazolopyrimidinone nucleosides are incorporated into oligonucleotides at predetermined sites, the triazolyl group of these modified nucleobases can be displaced by 1,10-diaminodecane to generate the corresponding aminoalkylated oligomers.[286] Once biotinylated, the specificity of these oligomers was tested in hybridization assays with plasmid DNA immobilized on nitrocellulose filters. It was observed that the thermostability of the resulting hybrids was not affected by biotin tethered to the 5′-end of the oligomers. Conversely, hybrid stability was strongly impaired by biotin linked to internal sites within an oligonucleotide sequence. The $T_m$ of these hybrids was at least 10 °C lower than that of unmodified duplexes.[286] The 5′-terminus is, therefore, ideal for the incorporation of 4-triazolopyrimidinone nucleosides into oligonucleotides. Urdea *et al.*[287] reported the synthesis of the deoxyribonucleoside phosphoramidites (177) from the corresponding 4-triazolopyrimidinone nucleosides. These derivatives led to the solid-phase synthesis of oligonucleotides carrying site-specific aminoalkyl functions. Reporter groups, such as biotin, fluorescein, and isoluminol, were coupled to the purified aminoalkylated oligomers and their detection limits were assessed.[287,288]

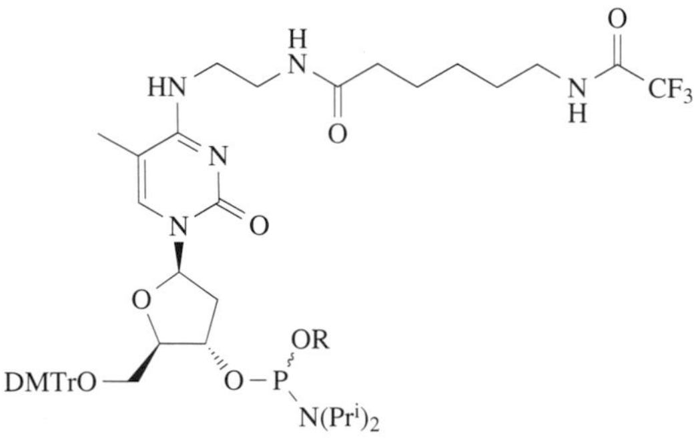

(177)  R = Me or CH$_2$CH$_2$CN

Insertion of the deoxyribonucleoside phosphoramidite (178) at a specific site during oligonucleotide assembly also produced functionally tethered oligodeoxyribonucleotides.[289,290] However, the concomitant use of deoxyribonucleoside phosphoramidites having the phenoxyacetyl group for protection of the exocyclic amino function of adenine and guanine, and the isobutyryl group for that of cytosine, was necessary to ensure rapid oligonucleotide deprotection with concentrated ammonium hydroxide[291] without affecting the $O^4$-arylated uracil residues. Thus, treatment of the modified oligonucleotides with *N*-methylamine, 1,2-diaminoethane, 1,4-diaminobutane, glycine, or bis(aminoethyl)disulfide at 65 °C resulted in functionally tethered oligonucleotides with a conversion efficiency of 89–100%.[289,290] Similarly, oligomers containing $O^4$-alkylthymine, 5-methylcytosine, $N^4$-(dimethylamino)-5-methylcytosine, and 4-thiothymine residues have been generated from oligonucleotides that have initially been modified by incorporation of the triazolo phosphoramidite derivative (**179**).[292]

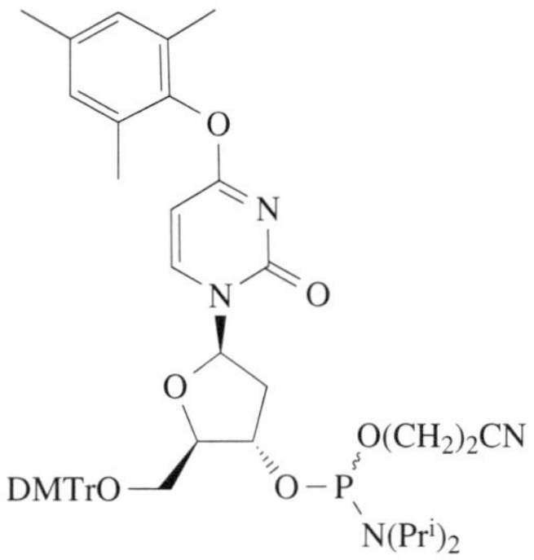

(**178**)

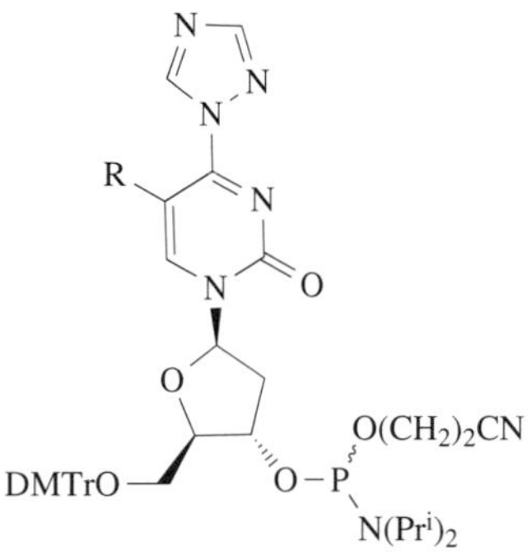

(**179**)   R = H or Me

Incorporation of the modified deoxycytidine phosphoramidite (**180**) into oligodeoxyribonucleotides produced the aminoalkylated oligonucleotide (**181**) after treatment of the partially deprotected oligonucleotide with anhydrous 1,2-bis(2-aminoethoxy)ethane.[293] Purified (**181**) was reacted with fluorescein isothiocyanate to give a fluorescent primer for automated DNA sequencing. The preparation of similarly labeled DNA sequencing primers has also been accomplished by incorporation of the aminoalkylated phosphoramidite (**182**) into oligodeoxyribonucleotides.[294]

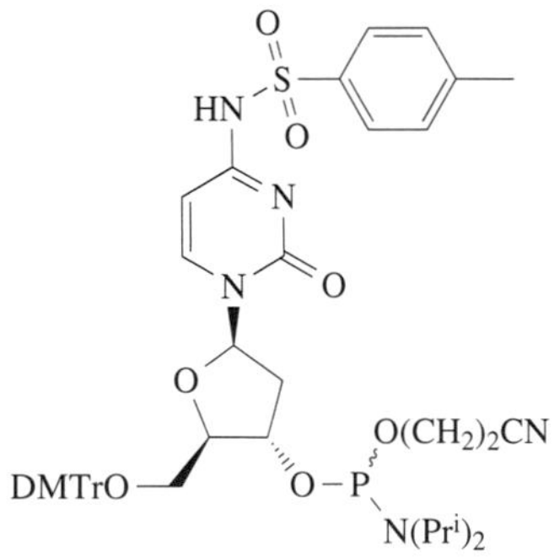

(**180**)

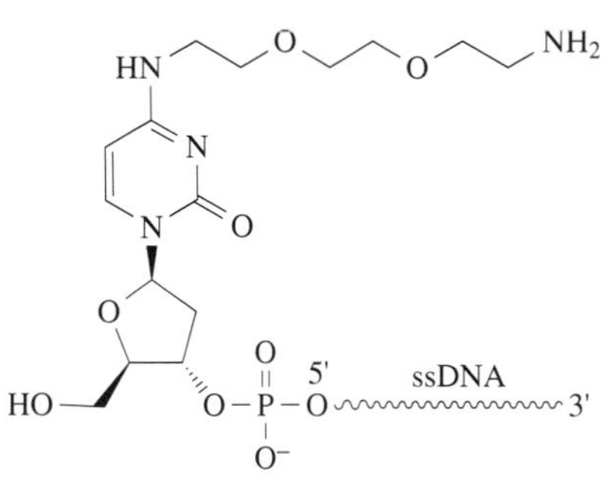

(**181**)

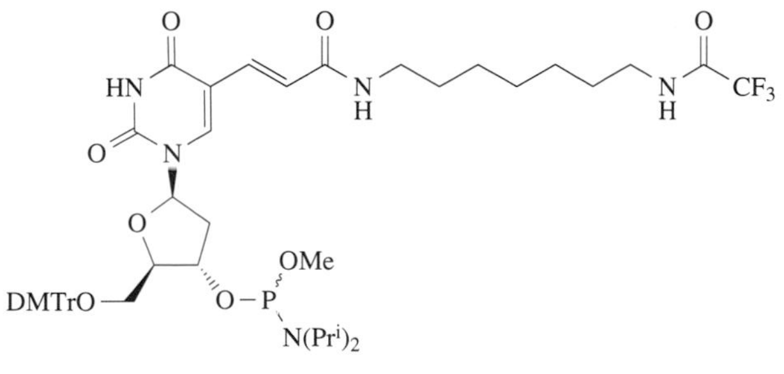

(**182**)

The derivatization of deoxyuridines at C-5 with masked primary aliphatic amino groups has been exploited by Cruickshank and Stockwell[295] and Haralambidis *et al.*[296] The synthetic strategy entailed the reaction of a suitably protected 5-iododeoxyuridine with an *N*-blocked propargylamine in the presence of bis(triphenylphosphine)palladium chloride and cuprous iodide to generate the corresponding alkynyl nucleoside (**183**)[296] or (**185**).[295] Condensation of the deprotected amino function of (**183**) with the *p*-nitrophenyl ester of 6-*t*-butyloxycarbonylamidohexanoic acid gave the alkynyl nucleoside (**184**).[296] The nucleosides (**183**)–(**185**) were converted to their corresponding phosphoramidites (**186**)–(**188**) and incorporated into oligonucleotides by solid-phase synthesis.[295,296] After complete deprotection and purification, the aminoalkylated oligonucleotides were conjugated to fluorescein and mixed with complementary polyadenylated mRNAs.[296] Probes derived from the incorporation of (**186**) did not hybridize as efficiently as the probes derived from the integration of (**187**). Furthermore, probes carrying a single 5′-label hybridized better than the multilabeled probes.[296]

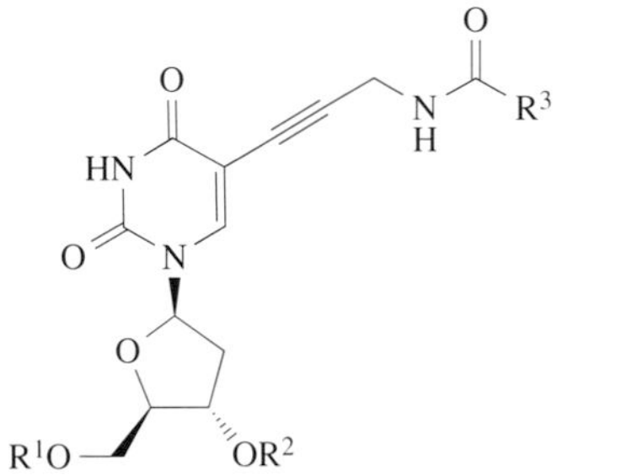

(**183**)  R$^1$ = R$^2$ = toluyl;  R$^3$ = OBu$^t$
(**184**)  R$^1$ = R$^2$ = toluyl;  R$^3$ = (CH$_2$)$_5$NHBoc
(**185**)  R$^1$ = DMTr;  R$^2$ = H;  R$^3$ = CF$_3$

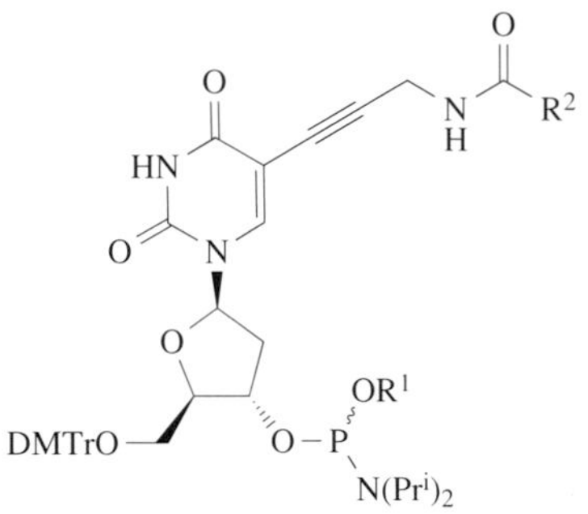

(**186**)  R$^1$ = Me;  R$^2$ = OBu$^t$
(**187**)  R$^1$ = Me;  R$^2$ = (CH$_2$)$_5$NHBoc
(**188**)  R$^1$ = CH$_2$CH$_2$CN;  R$^2$ = CF$_3$

Oligonucleotides with aminoalkylated groups attached to C-5 of uracil residues have also been conjugated to photoactive alkylating or intercalating groups in efforts to investigate DNA–DNA hybrid formation and site-specific modification of nucleic acid targets.[297] In this context and in order to study protein DNA interactions, Gibson and Benkovic[298] described the synthesis of the deoxyribonucleoside phosphoramidite (**189**) from 5-iododeoxyuridine and its incorporation into short oligonucleotides (11-mers). The deprotected undecamers were treated with the cross-linker (*N*-hydroxysuccinimidyl) 5-azido-2-nitrobenzoate to give the conjugates (**191**). Oligomers that carry the photolabile group near the 3′-terminus have served as primers for template-directed DNA synthesis with bacteriophage T4 DNA polymerase, the Klenow fragment of *E. coli* DNA polymerase I, or avian myeloblastosis virus (AMV) reverse transcriptase. Irradiation of the primer elongation mixtures with 302 nm light produced covalent complexes between DNA and the polymerases. Cross-linking with AMV reverse transcriptase was shown to occur predominantly in the "*β*-subunit."[298]

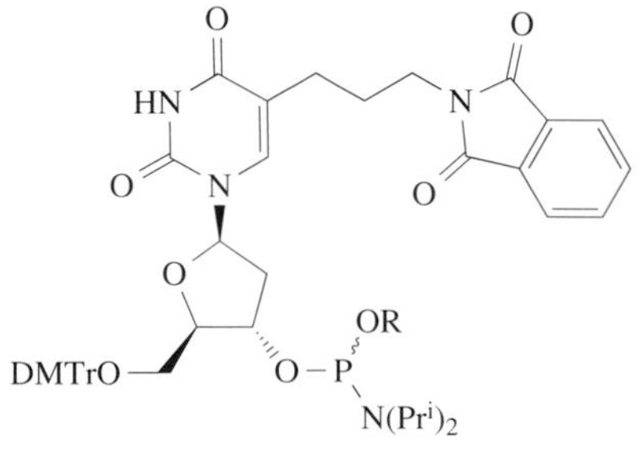

(**189**)  R = Me
(**190**)  R = CH$_2$CH$_2$CN

Along similar lines, incorporation of the deoxyribonucleoside phosphoramidite (**189**) into oligomers has been achieved to generate DNA duplexes biotinylated at specific sites, and to determine the structural requirements for the exonuclease and polymerase activities of prokaryotic and phage DNA polymerases.[299] It was shown that in the presence of avidin the exonucleolytic and polymerase activities of the Klenow fragment of *E. coli* DNA polymerase I required a primer's terminus to be

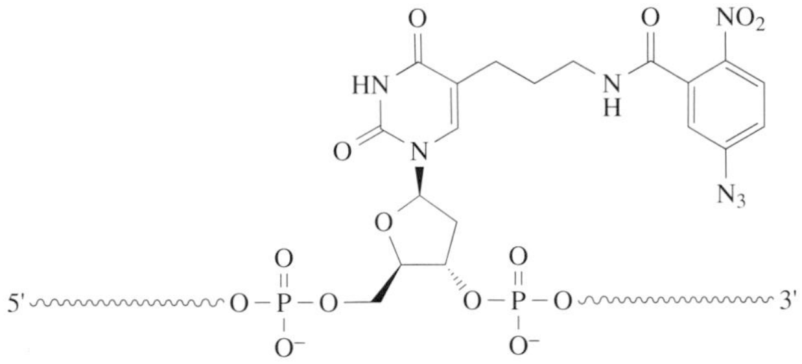

(191)

located at least 15 and 6 base pairs downstream of the biotin–avidin complex, respectively. These results demonstrated that the polymerase and exonuclease sites of the Klenow fragment were physically distinct in solution, and displayed different substrate structural requirements for activity.[299]

The integration of (189) into oligonucleotides allowed their derivatization with mansyl or dansyl chloride.[300,301] Annealing the resulting fluorescent oligomers with complementary oligonucleotides generated duplexes that produced fluorescence emissions, increasing in intensity, upon binding with the above Klenow fragment. Thus, by varying the position of the fluorescent label within the DNA duplexes and observing fluorescence emissions, strong enzyme–DNA contact points were identified.[300,302]

Synthesis of the protected aminoalkylated deoxyribonucleoside phosphoramidite (192) and its incorporation into an oligonucleotide have been outlined by Meyer *et al.*[303] The functionalized oligomer was, after complete deprotection, iodoacetylated by treatment with *N*-hydroxysuccinimidyl iodoacetate to afford (193). Incubation of (193) with its complementary DNA target generated observable interstrand cross-linking within 48 h at 37 °C. Specifically, cross-linking occurred via alkylation of a guanine residue of the complementary strand. Heating the DNA duplex at 51 °C resulted in depurination of the alkylated guanine and scission of the target strand at the depurination site.[303] Moreover, the modified oligonucleotide alkylated selectively the target strand without random alkylation of noncomplementary nucleic acids. This feature is quite attractive for the application of these oligonucleotides as therapeutic agents.

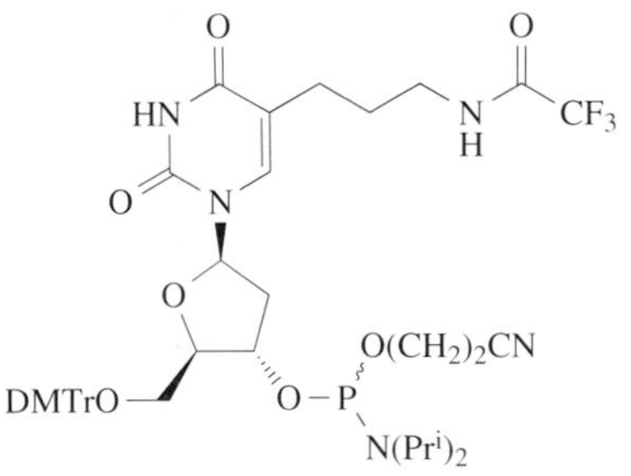

(192)

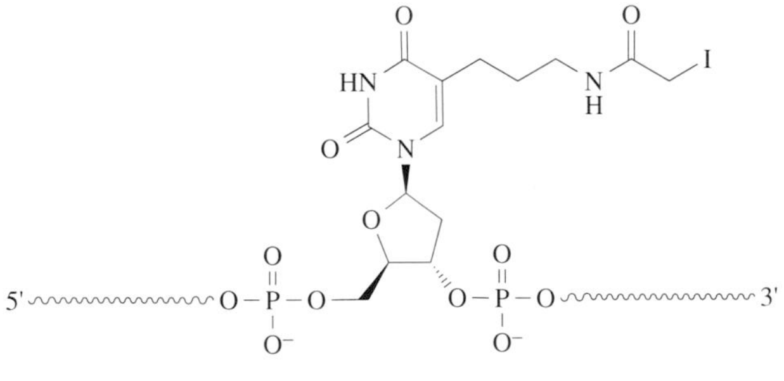

(193)

The deoxyribonucleoside phosphoramidites (190) and (194) have also been integrated into oligonucleotides at a unique site by standard solid-phase synthesis.[304] After removal of the protecting groups, the HPLC-purified aminoalkylated oligonucleotides were reacted with various *N*-hydroxysuccinimidyl haloalkanoates to produce the modified oligonucleotides (195). Hybridization of (195) with complementary oligodeoxyribonucleotides led to the formation of cross-links between the two strands. Ultimately, cleavage of the DNA targets occurred at the cross-link sites via depurination of an alkylated guanine nucleobase.[304] The regiospecificity of alkylation was primarily determined by the length of the electrophilic haloacylamidoalkyl group, whereas the extent of the reaction depended on both the structure of the acylamidoalkyl group and the reactivity of the electrophile. Placement of the electrophile next to the targeted nucleophile, namely, the N-7 of a specific guanine of the target strand, was the most important factor in determining the rate of the reaction. With optimal haloacylamidoalkyl structure and duplex sequence, the most rapid rate of alkylation measured was $t_{1/2} = 1.3$ h at 37 °C.[304]

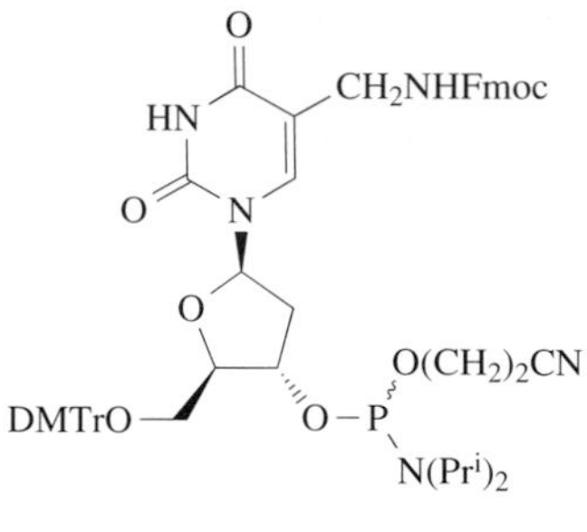

**(194)**

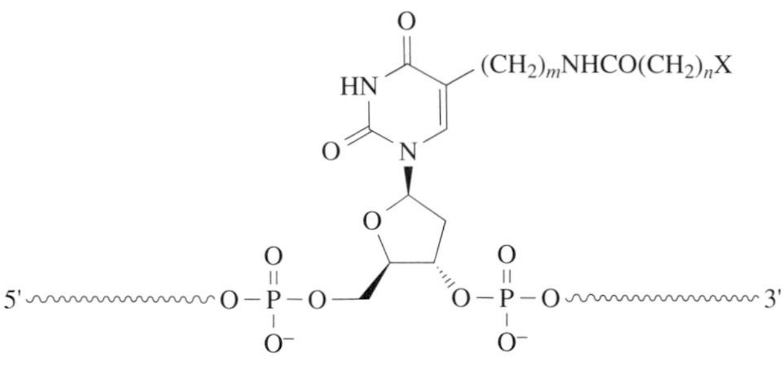

**(195)** $m = 1$ or 3; $n = 1$ or 3; X = Br, Cl, or I

In this context, preparation of the aminoalkylated deoxyribonucleoside phosphoramidite **(196)** from 2-thio-2′-deoxyadenosine has been described by Kido *et al.*[305] The single insertion of **(196)** in oligonucleotides required a coupling time of 300 s. Following deprotection and purification, the aminoalkylated oligomers were iodoacetylated and characterized by enzymatic digestion to the nucleosides.[305] Oligonucleotides functionalized with a 2-(*N*-iodoacetylaminoethyl)thioadenine residue were shown to cross-link and cleave complementary strands at adenines and guanines. It is postulated that alkylation occurred at N-3 of adenine from the minor groove or with reactive guanine groups in the major and minor grooves of the duplex. The efficiency of alkylation was low, probably because of the free rotation of the reactive linker around the sulfur atom.[305] More rigid tethers may provide higher alkylation selectivity.

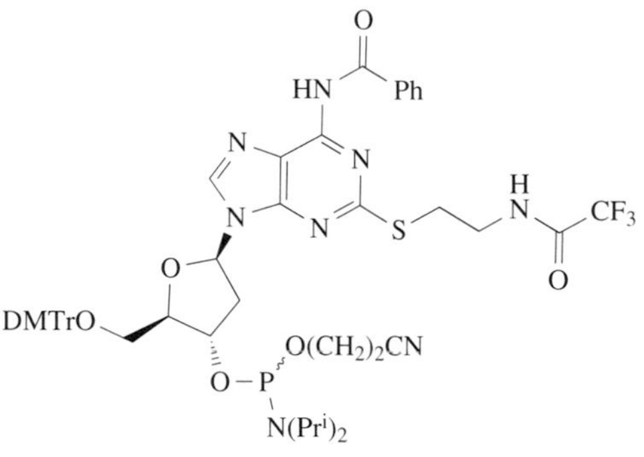

**(196)**

Synthesis of the deoxyribonucleoside phosphoramidite **(197)** and its addition to the 5′-terminus of an oligonucleotide have been reported by Povsic *et al.*[306] The deprotected and modified oligonucleotide was treated with *N*-hydroxysuccinimidyl bromoacetate to give the corresponding *N*-bromoacetylated oligodeoxyribonucleotide. This oligomer hybridized to adjacent inverted purine tracts on double-helical DNA via triple-helix formation, and alkylated a single guanine residue on the Watson–Crick strands. After depurination, double-strand cleavage at a single site within plasmid DNA occurred in yields greater than 85%.[306] This methodology may become a valuable tool for sequence-specific cleavage of large DNA molecules.

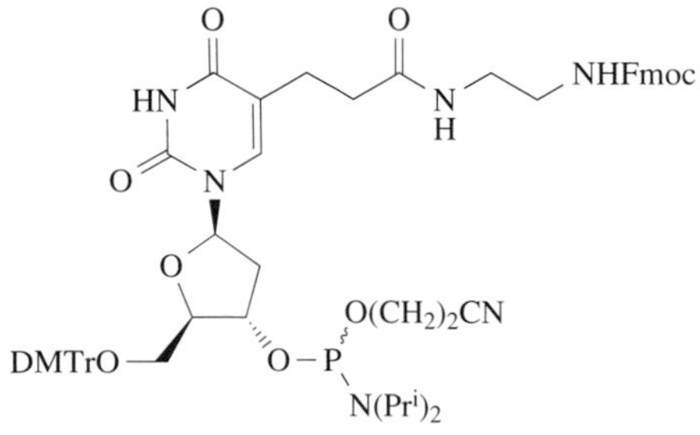

**(197)**

Aminoalkylated oligonucleotides generated from the insertion of the deoxyribonucleoside phosphoramidite (190) at a unique site permitted conjugation with the 2,3,5,6-tetrafluorophenyl ester of chlorambucil to give the modified oligonucleotides (198). These chlorambucil–oligonucleotide conjugates were used as affinity-labeling reagents to study the formation of joint molecules catalyzed by the *E. coli* recombinase *RecA*.[307] Alternatively, conjugation of chlorambucil to 5′-aminoalkylated oligonucleotides produced triplex-forming oligonucleotides that cross-linked to both Watson–Crick strands of a triple-helix structure.[308] Alkylation occurred at the desired guanines flanking the triplex domain. It should therefore be possible to introduce sequence-specific cross-links in the DNA of living cells and, thereby, cause permanent gene inactivation through error-prone repair.[308]

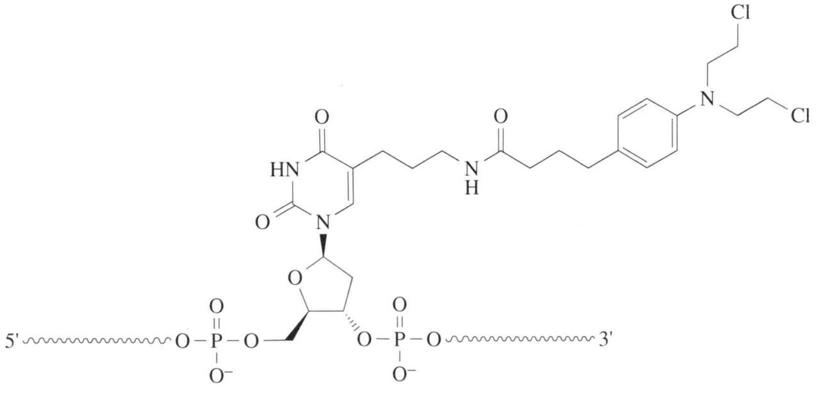

(198)

The incorporation of the phosphoramidite (199) or (200) into oligonucleotides also produced aminoalkylated oligomers suitable for conjugation with the *N*-hydroxysuccinimide ester of *N*-biotinyl-6-aminocaproic acid[309] or digoxigenin-*O*-succinyl-ε-aminocaproic acid.[310,311] Purified biotinylated oligonucleotides were hybridized to target DNA immobilized on microtiter plates, and were detected with a streptavidin–biotinylated horseradish peroxidase complex. Oligonucleotides carrying biotins at or near their termini were more effective probes than those oligomers bearing biotins at internal locations.[309]

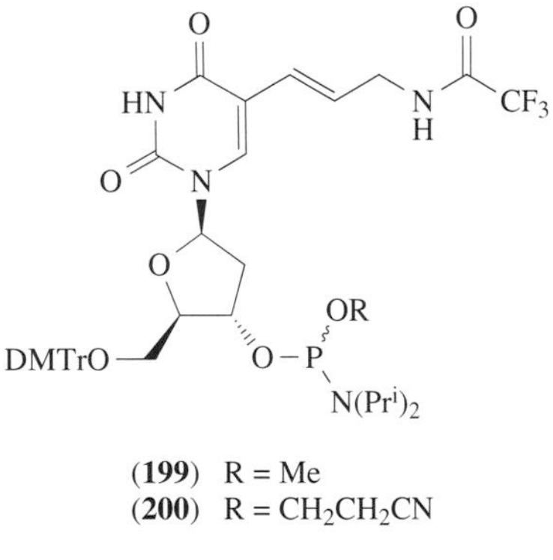

(199)  R = Me
(200)  R = CH₂CH₂CN

The reaction of 5-chloromercury-2′-deoxyuridine with methyl acrylate led to the synthesis of the aminoalkylated deoxyribonucleoside phosphoramidite (201) and its subsequent incorporation into oligonucleotides.[278,279,312] Following deprotection and purification, the oligomers were then derivatized with *N*-succinimidyl 1-pyrenebutyrate, 1-pyrenesulfonyl chloride, fluorescein 5-isothiocyanate, sulfosuccinimidyl 6-(biotinamido)hexanoate, *N*-hydroxysuccinimidyl anthraquinone-2-carboxylate, or the *N*-hydroxysuccinimide ester of 4-carboxy-4′-methyl-2,2′-bipyridine.[278,279,312] Incidentally, bipyridine-labeled oligomers reacted with $Ru(bpy)_2(H_2O)_2^{2+}$ to provide oligonucleotides covalently attached to $Ru(bpy)_3^{2+}$ derivatives.[279] These modified oligonucleotides displayed a normal hybridization behavior, and ruthenium(II) has been known to effect light-induced DNA strand scission.[279]

RNA probes can be easily biotinylated or fluoresceinated by T₇RNA polymerase incorporation of the aminoalkylated adenosine triphosphate (202) followed by treatment of the transcript with

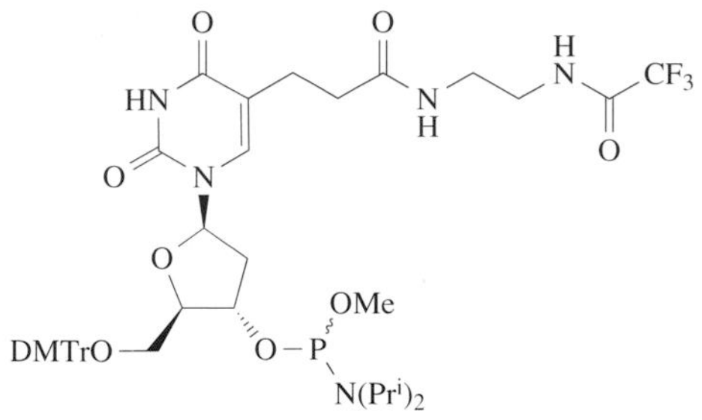

**(201)**

the *N*-hydroxysuccinimide ester of biotin or fluorescein. RNA probes contained, on average, one amine-labeled nucleotide every 12 nucleotides.[313]

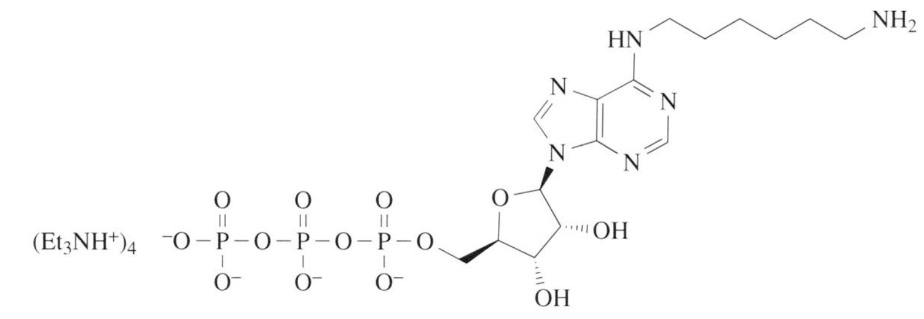

**(202)**

Roduit *et al.*[314] described a different approach for the integration of an amino function into synthetic DNA oligonucleotides. The method involved the reaction of 8-bromodeoxyadenosine with a protected aminoalkylthiolate salt. The modified nucleoside was eventually converted to the phosphoramidite **(203)** and efficiently (96–99%) incorporated into oligonucleotides. Once purified, these aminoalkylated oligonucleotides were biotinylated by treatment with either the *N*-hydroxysuccinimide ester of biotin or caproylamidobiotin in a neutral buffer.[314]

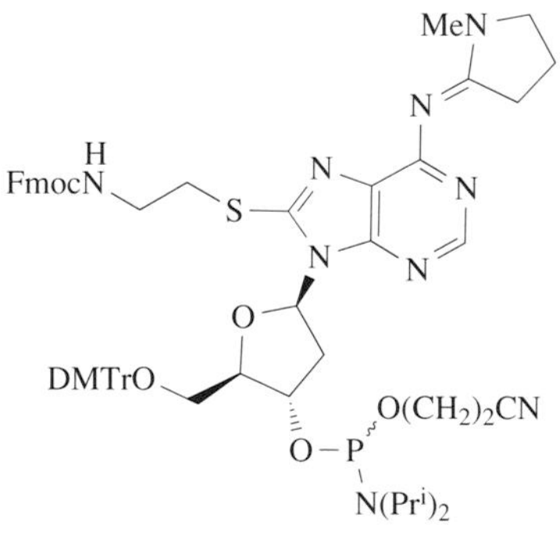

**(203)**

The 3′-termini of ssDNAs can be labeled through the single incorporation of 8-(6-aminohexylamino)adenosine 5′-triphosphate effected by terminal deoxynucleotidyl transferase.[315] Following purification, the 3′-aminoalkylated oligomers were reacted with 1-succinimidyl pyrenebutyrate, fluorescein isothiocyanate, or Texas Red, and used after gel permeation chromatography in the detection of unlabeled DNA targets by competitive hybridization.[315]

Noteworthy is the synthesis of the 5-substituted deoxyuridine phosphoramidite derivative **(204)** and its integration into oligonucleotides which produced, upon reaction with diaminoethane or diaminohexane, oligomers carrying aminoalkylated tethers.[316] These aminoalkylated oligonucleotides were, after purification, reacted with the intercalator *N*-[anthraquinone-2-carbonyloxy]succinimide to generate functionalized oligomers similar to **(205)**. The thermostability of hybrids composed of **(205)** and complementary DNA oligomers is seemingly independent of the attachment site of the anthraquinone group, and higher ($\Delta T_{\mathrm{m}}$ per modification = 6–9 °C) than that of the corresponding unmodified duplexes.[316]

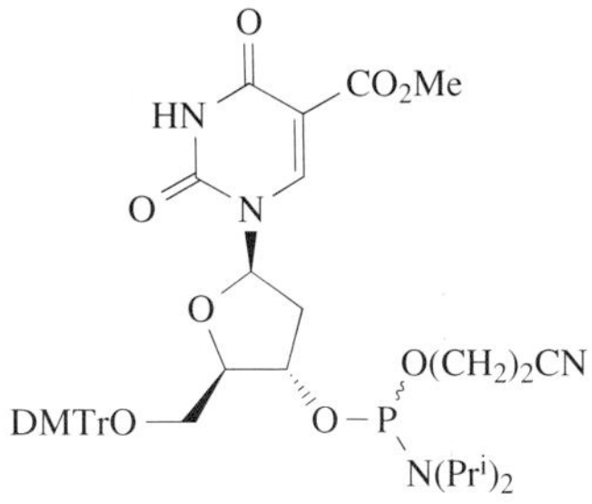

(204)

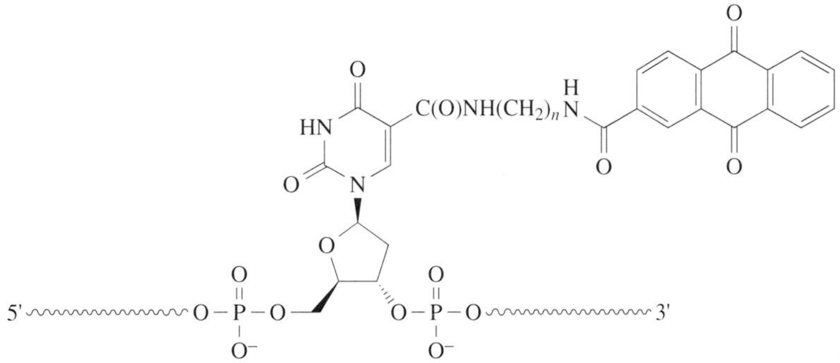

(205)

The incorporation of 6-aminodeoxyuridine into oligodeoxyribonucleotides near the 3′-terminus led to modified oligonucleotides that can be easily biotinylated by commercial reagents.[317,318] Complexation of the biotinylated oligonucleotides with avidin provided complete protection against hydrolysis mediated by serum 3′-exonucleases. This feature may be particularly useful for avidin-mediated drug delivery of oligonucleotides to tissues *in vivo* or to cultured cells *in vitro*.[317–319] Likewise, the biotinylation of oligodeoxyribonucleotides that have been synthesized on commercial [[1-(4,4′-dimethoxytrityl)oxy]-3-[(fluorenylmethoxycarbonyl)amino] propan-2-succinoyl]-long chain alkylamino-CPG imparted, upon complexation with avidin, resistance to serum and cellular 3′-exonucleases. In addition, these oligonucleotide conjugates inactivated mRNA targets via RNase H degradation.[320]

It should be mentioned that the site-specific enzymatic incorporation of unnatural $N^6$-(6-amino-hexyl)isoguanosine into RNA has been reported.[321] Essentially, the incorporation of 2′-deoxy-5-methylisocytidine into a DNA template via the phosphoramidite (206) directed the $T_7$RNA polymerase insertion of the (6-aminohexyl)isoguanosine triphosphate (207) in the transcribed RNA product. Post-transcriptional modification of the newly incorporated amino functions with the N-hydroxysuccinimide ester of biotin or the dianhydride of ethylenediaminetetracetic acid (EDTA) produced site-specifically modified RNA sequences. These sequences should be useful for mapping the structure of folded RNA polymers and RNA–protein complexes by affinity cleavage and affinity labeling.[321]

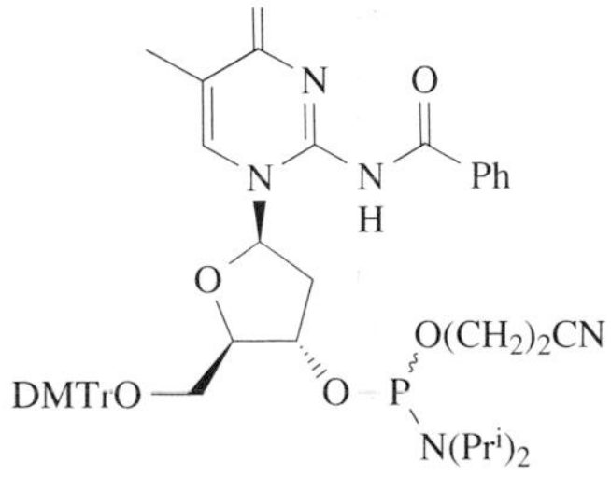

(206)

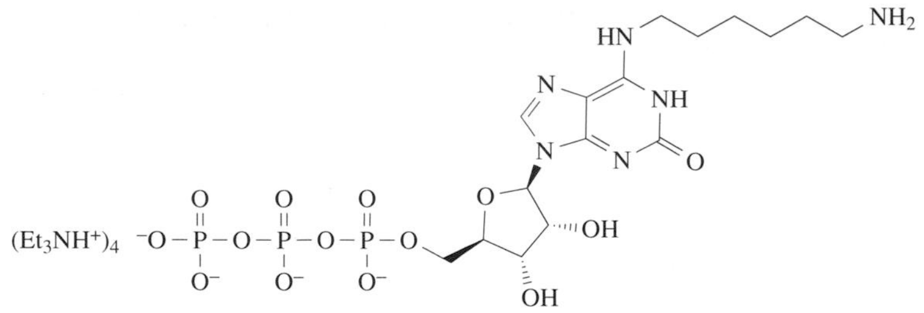

**(207)**

RNAs prepared *in vitro* by the use of 5-(3-aminoallyl)-UTP as the sole source of UTP can readily be conjugated to 1,10-phenanthroline.[322] These RNA conjugates efficiently cleaved complementary ssDNAs in the presence of copper(II) and 3-thiopropionic acid. Sequence-specific double-stranded scission of duplex DNA can also be accomplished by 1,10-phenanthroline-derivatized RNAs within R-loops.[322]

The $N^2$-modified deoxyguanosine phosphoramidite (**208**) has been inserted in a DNA oligomer near the 3'-terminus that binds to the 5'-terminal region of the cytokine-induced ICAM-1 transcript.[323] The deprotected aminoalkylated oligonucleotide was then reacted with the *N*-hydroxy-succinimide ester of α-methyl-*N*-(*S*-benzoyl-2-thioacetyl)glutamic acid to give the modified oligo-nucleotide (**209**) after ligand deprotection. Coordination of copper(II) to the conjugate (**209**) led to the complete and specific removal of the 5'-cap of a complementary 5'-capped RNA oligomer within 120 h at 37 °C.[323] These findings suggest that antisense oligonucleotides equipped with moieties that react with triphosphate linkages or methylated guanosine residues may provide an effective way to incapacitate specific transcripts and thereby prevent expression of the encoded proteins.

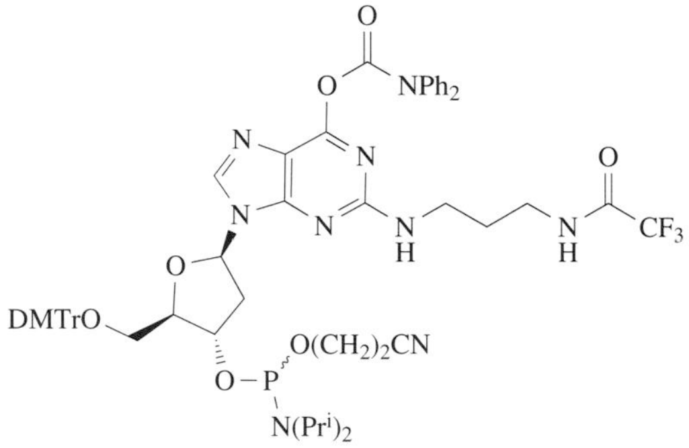

**(208)**

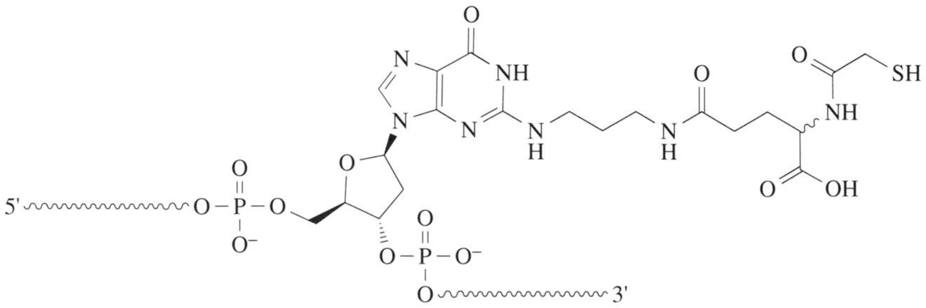

**(209)**

The C-5 modified uridine phosphoramidite (**210**) has been used to incorporate site-specifically a reactive primary amino group into chemically synthesized RNA.[324] Incorporation of the modified phosphoramidite into RNA oligomers occurred with a coupling efficiency greater than 97%. Although the incorporated free amino group(s) can be conjugated to reporter groups, small struc-tural probes, and biological molecules, treatment of the purified aminoalkylated RNA oligomers with an isothiocyanato derivative of EDTA led to site-specific incorporations of a strong chelating agent into these oligonucleotides. The site-specific insertion of Fe–EDTA entities in, for example, the HIV-1 *tar* RNA sequence, can provide valuable structural information about the *Tat–tar* RNA complex by virtue of the proximity-dependent nucleic acid and protein-cleavage properties of Fe–EDTA.[324]

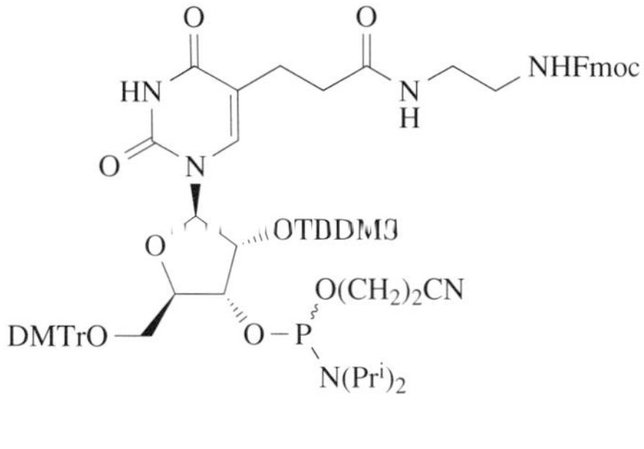

**(210)**

Earlier work reported the integration of the modified ribonucleoside phosphoramidite (**211**) at a specific site into a synthetic RNA fragment of an *E. coli* M1 ribozyme.[325] After purification, the aminoalkylated oligoribonucleotide was treated with ethylenediaminetetraacetic anhydride or diethylenetriaminepentaacetic dianhydride to give (**212**) or (**213**). Both (**212**) and (**213**) promoted the cleavage of a complementary DNA oligonucleotide when incubated with iron(III) or copper(II) and DTT. The cleavage was more efficient with the EDTA derivative and occurred over a seven-base range centered about the complement of the modified nucleobase.[325]

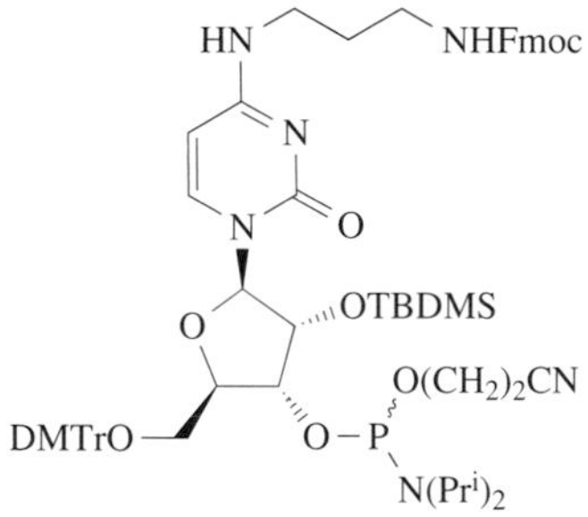

**(211)**

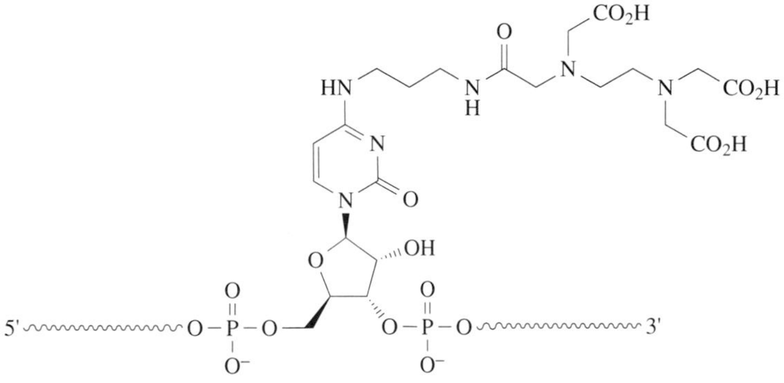

**(212)**

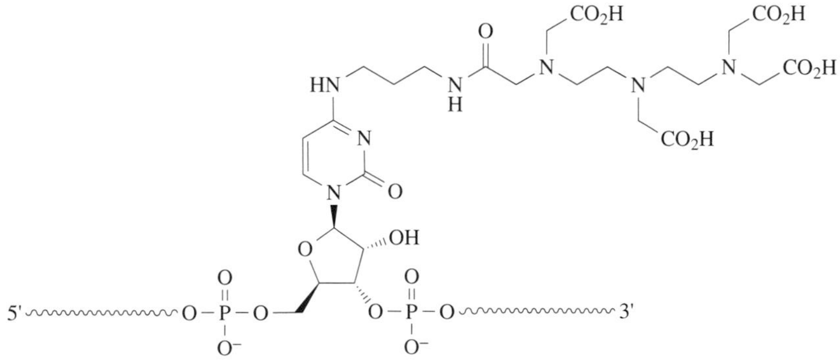

**(213)**

In an effort to prepare europium(III)-labeled DNA probes, the deoxyribonucleoside phosphoramidite (**174**) has repeatedly been coupled, up to 50 times, to the 5′-terminus of a synthetic oligomer (50-mer).[326,327] Following deprotection, the purified oligonucleotide (100-mer) was condensed with the europium(III) chelate (**214**). Hybridization properties of these europium(III)-labeled oligonucleotides were not significantly affected by the labeling process. These probes led to the detection of ca. $10^7$ target molecules in a typical solid-phase hybridization assay,[327] or 200 pg of phage lambda DNA by time-resolved fluorometry.[326]

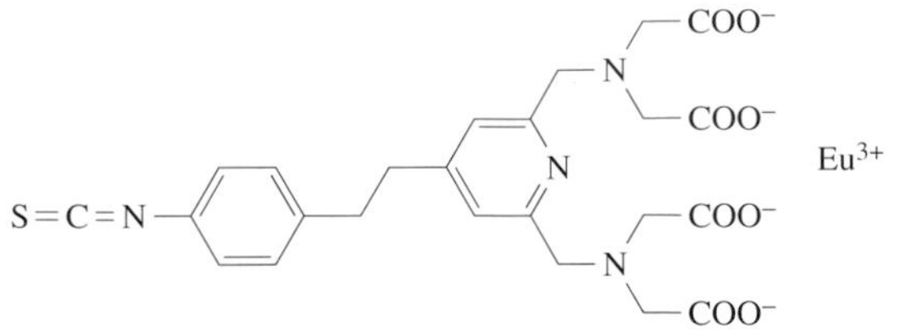

**(214)**

### 7.06.3.1.2  *Thiolated and thioalkylated nucleobases*

4-Thio-2′-deoxyuridine-containing oligonucleotides have allowed tether attachment by post-synthetic *S*-alkylation or mixed disulfide formation.[328,329] Typically, the 4-thioated-2′-deoxyuridine phosphoramidite (**215**) was incorporated into oligonucleotides by solid-phase synthesis. Quantitative *S*-deprotection of solid-phase linked oligonucleotides was effected by treatment with DBU in acetonitrile. The modified oligonucleotides were then completely deprotected and reacted with, for example, *N*-(2-chloroethylthio)phthalimide to afford (**217**) for subsequent derivatization with selected reporter groups.[328]

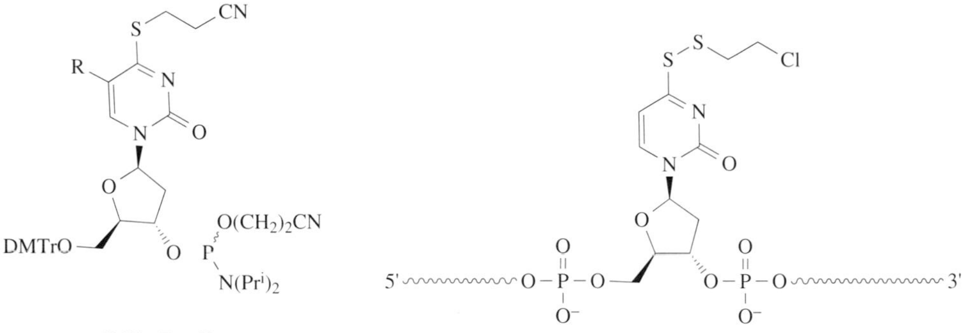

**(215)  R = H**
**(216)  R = Me**

**(217)**

Oligonucleotides containing 4-thio-2′-deoxyuridine residues can be modified chemoselectively with a variety of thiol-specific reagents[329] that would enable the incorporation of diverse functional groups at specific sites. The phosphoramidites (**216**) and (**218**)–(**220**) have analogously been applied to the synthesis of oligonucleotides containing thiolated nucleobases.[330–333]

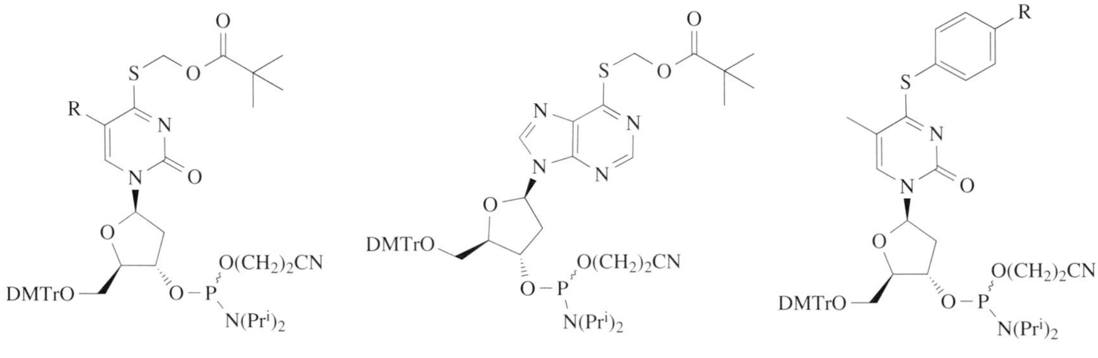

**(218)  R = H or Me**

**(219)**

**(220)  R = H or NO₂**

Synthesis of the 5-thiocyanato-2′-deoxyuridine and uridine phosphoramidites (**221**) and (**222**) has been achieved by Bradley and Hanna[334] for the functionalization of oligonucleotides without altering their Watson–Crick base-pairing abilities. Oligonucleotides that have been modified by the incorporation of (**221**) were deprotected under standard conditions without affecting the 5-thiocyanato groups. However, treatment of these modified oligonucleotides with DTT at 55 °C effected reduction of the thiocyanato functions to 5-thiol groups. Condensation of 5-thiolated nucleobases with either *p*-azidophenacyl bromide or 5-iodoacetamidofluorescein generated functionalized oligonucleotides suitable for the study of protein–nucleic acid interactions.[334] It should nonetheless be understood that various functional groups can be attached to these thiolated oligonucleotides.

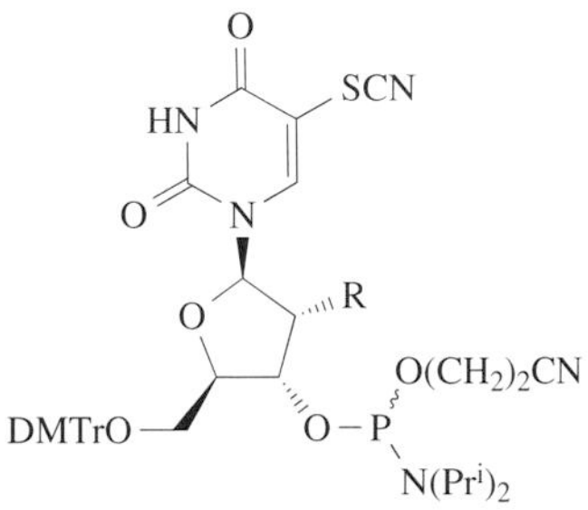

(**221**)  R = H
(**222**)  R = OTBDMS

In a different application, dodecamers containing 4-thiocyanatothymine residues and the recognition site d(GATATC) of the endonuclease *Eco*RV or *Eco*RV methyltransferase were found to cross-link photochemically to either enzyme.[335]

The 5-[*S*-(2,4-dinitrophenyl)thio]-2′-deoxyuridine phosphoramidite (**223**) has been prepared for site-specific incorporation of a reactive thiol group into DNA oligonucleotides.[336] The coupling yield of (**223**) was comparable to that of unmodified phosphoramidites. The dinitrophenyl (DNP)-modified oligonucleotides were deprotected under conditions that required concentrated ammonium hydroxide for an extended period of time (44 h) at ambient temperature. The use of a higher temperature (55 °C) resulted in loss of the DNP group. Further treatment of the modified oligonucleotide (**224**) with *β*-thioethanol produced an oligonucleotide containing a 5-thiodeoxyuridine residue that is amenable to conjugation with 5-(iodoacetamido)fluorescein.[336] The phosphoramidite (**223**) can therefore be considered as a useful synthon for the preparation of site-specifically modified DNA.

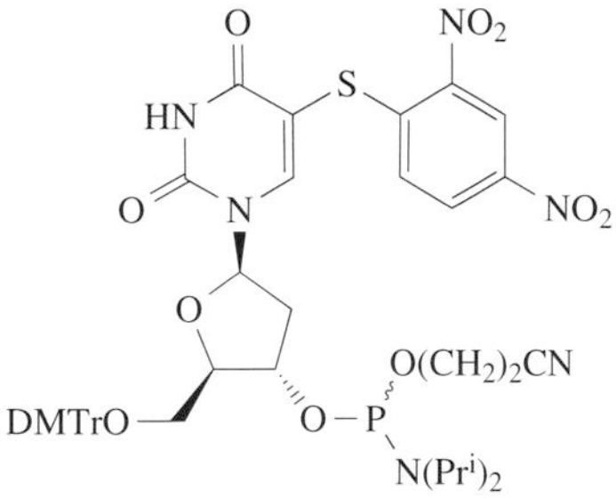

(**223**)

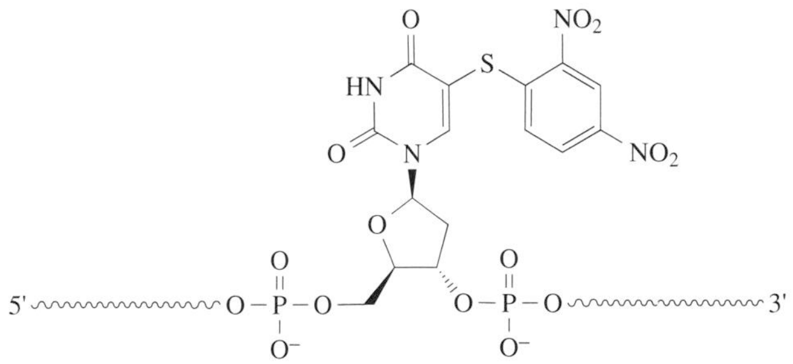

(**224**)

The synthesis of the C-5 thioalkylated deoxyribonucleoside phosphoramidite (**225**) has been accomplished in several steps from 5-iodo-2′-deoxyuridine.[337] The *t*-butyl disulfide protecting group was stable to the reagents used during the incorporation of (**226**) into oligodeoxyribonucleotides. This protecting group was eventually removed by treatment with DTT.[337] The thioalkylated oligonucleotide (**226**) produced a positive Ellman's test, and rapidly reacted with pyridyl-*n*-butyl mixed disulfide to generate the corresponding oligonucleotide–*n*-butyl disulfide conjugate.[337] Thus, mercaptoalkylated oligonucleotides similar to (**226**) may be functionalized further with thiol-specific reporter groups or may be applied to the preparation of DNA–protein or DNA–peptide conjugates. Similar 5-(thioalkyl)uridine phosphoramidites have been prepared and incorporated site-specifically into RNA oligonucleotides.[338]

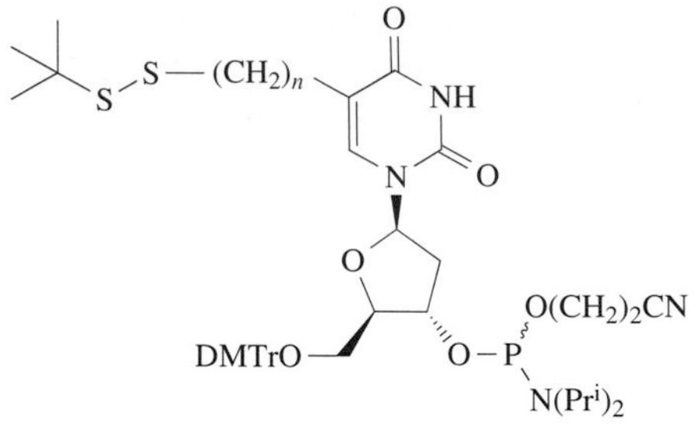

**(225)**  *n* = 1, 2, or 3 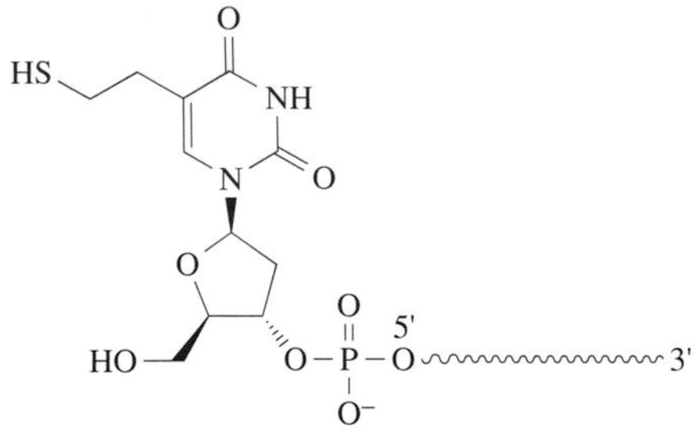                                    **(226)**

### 7.06.3.1.3  Hydroxylated nucleobases

Application of the deoxyribonucleoside phosphoramidite **(227)** to the preparation of polybiotinylated probes has been described by Bazin *et al.*[339] The nonnucleosidic phosphoramidite **(228)**
has also been prepared for the same purpose. Incidentally, the coupling of **(228)** to solid-phase
linked oligonucleotides occurred with an efficiency of 94–97%.[339]

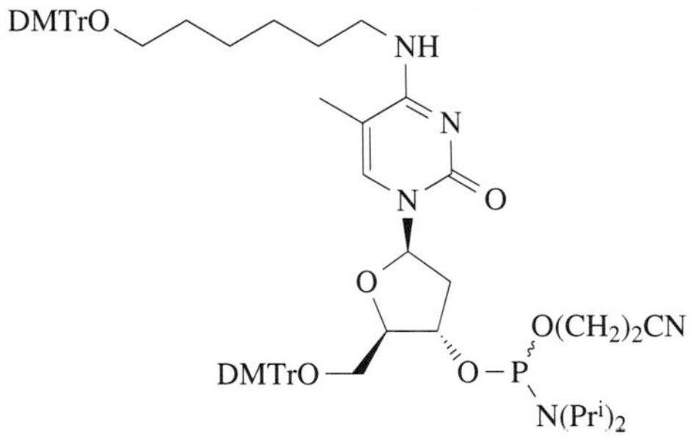

**(227)** 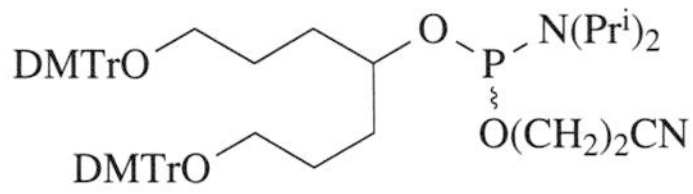                                    **(228)**

The preparation of "branched" DNA molecules, as amplification multimers in bioassays, has
been reported by Horn and Urdea[340] and Chang *et al.*[341] This type of branched DNA molecule is
composed of two distinct domains of oligonucleotide sequences. One domain consists of a target-
specific segment contiguous to an arrangement of branching nucleobases that was generated by
multiple addition of the levulinylated phosphoramidite **(229)** via solid-phase synthesis. Selective
removal of the levulinyl groups unmasks the hydroxy functions necessary for the synthesis of the
second domain of oligonucleotides. Labeled probes are hybridized to this second set of DNA
sequences to provide enhanced detection sensitivity.[340,341] The construction of large branched DNA
amplification multimers ( > 1000 nucleotides) has been achieved by the chemical ligation of single-
stranded linear oligomers to the branches of the original branched DNA molecules.[342]

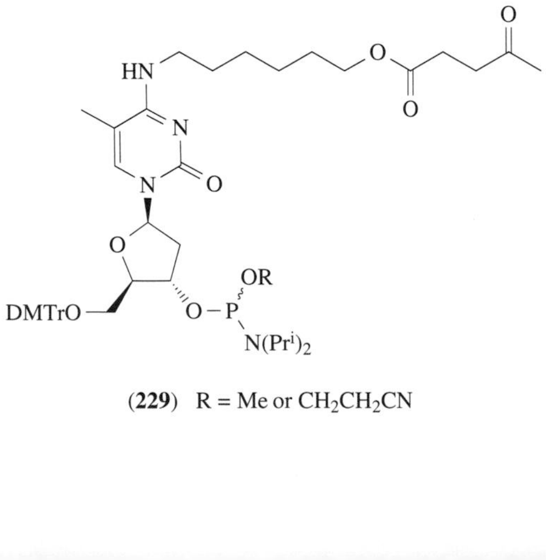

**(229)**   R = Me or CH$_2$CH$_2$CN

The deoxyribonucleoside phosphoramidites (**230**) and (**231**) have been synthesized from 4-thio-thymidine and tyramine, respectively,[157,343] and inserted at specific locations within oligonucleotides. The tyramine residues of the deprotected oligonucleotides were radioiodinated by standard "chloramine T" oxidation. These radiolabeled oligomers have been useful as primers for PCR amplification and for the detection of viral DNA sequences.[343]

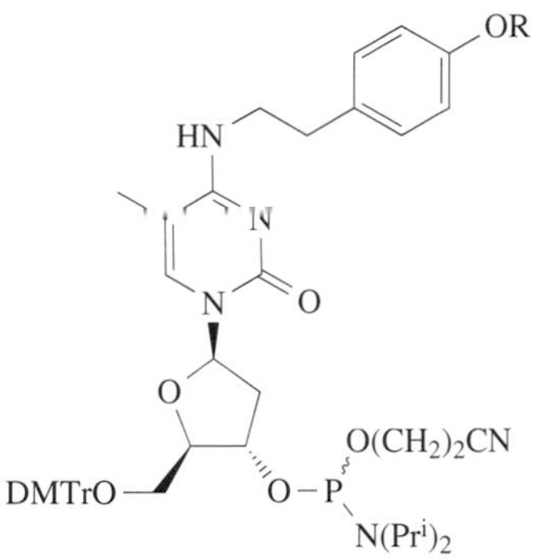

(**230**)  R = C(O)Me
(**231**)  R = CO₂Et

### 7.06.3.2 Modified Nucleobases for Direct Incorporation of Reporter and Conjugate Groups into DNA Oligonucleotides

#### 7.06.3.2.1 *Nucleobases derivatized with fluorescent and nonisotopic markers*

Protected deoxyribonucleoside phosphoramidites functionalized with biotinyl, dinitrophenyl, dansyl, and pyrenyl markers have been developed by Roget *et al.*[344] for labeling synthetic oligo-nucleotides. Specifically, (**232**) was prepared from aminoalkylation of the corresponding 4-thiodeoxyribonucleoside with 1,6-diaminohexane. The reaction of (**232**) with the *N*-hydroxy-succinimide ester of biotin, 1-fluoro-2,4-dinitrobenzene, dansyl chloride, or pyrenesulfonyl chloride, followed by phosphitylation, afforded the labeled phosphoramidites (**233**)–(**236**). These phosphoramidites were incorporated into oligonucleotides with high coupling yields (>97%). Multiple labeling can always be performed on the same oligonucleotide to provide increased detection sensitivity to *in situ* hybridization experiments.[344]

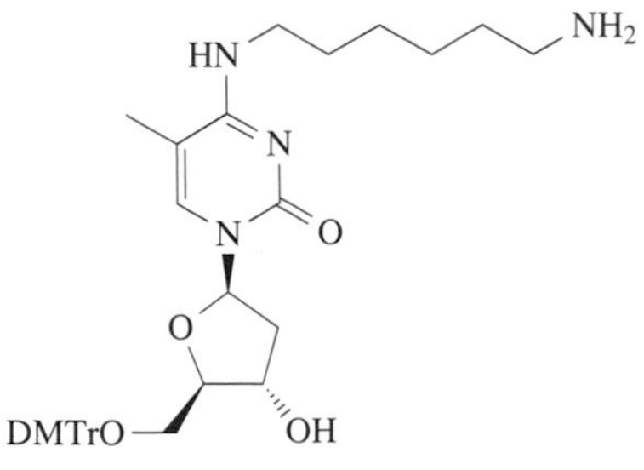

(**232**)

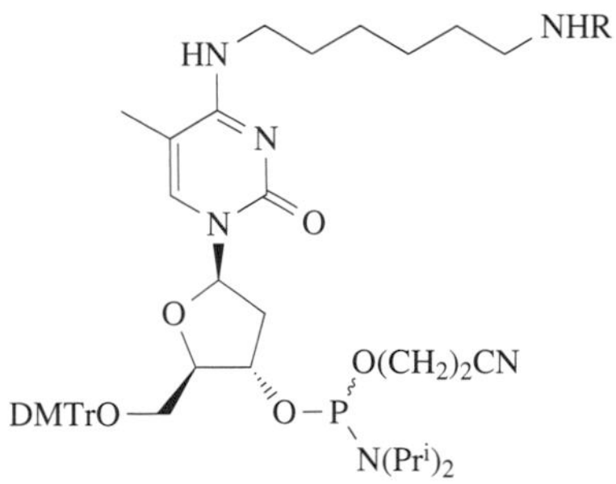

(**233**)  R = biotinyl
(**234**)  R = 2,4-dinitrophenyl
(**235**)  R = dansyl
(**236**)  R = 3-pyrenesulfonyl

This approach has been further refined by the synthesis of the biotinylated deoxyribonucleoside phosphoramidite (**237**).[345] Relative to (**233**), the phosphoramidite (**237**) is readily soluble in aceton-itrile, and carries a larger and more polar spacer arm to facilitate biological interactions. Deprotection of the biotinylated oligonucleotides under standard conditions proceeded without significant loss of biotin. Furthermore, the multiple incorporation of (**237**) produced tetrabiotinylated oligomers that reproducibly gave superior binding and reduced nonspecific binding to streptavidin–agarose when compared with oligonucleotides functionalized with one or two biotin residues.[345]

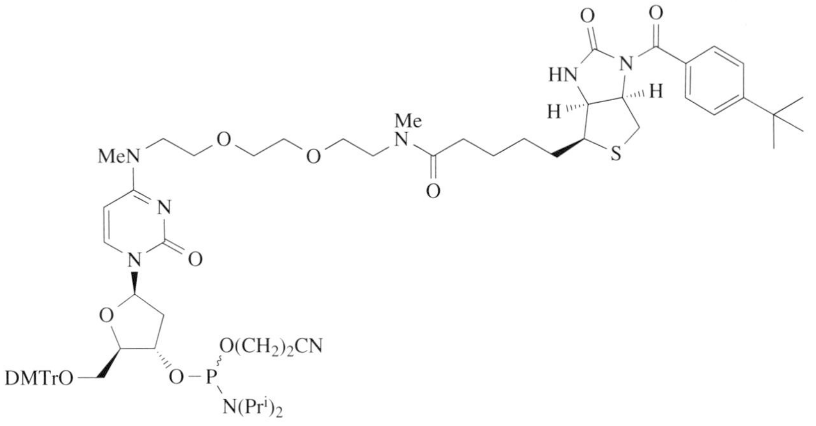

(237)

For similar applications, a number of biotinylated deoxyuridine triphosphate derivatives, such as (238),[346,347] (239),[346] and (240),[346] have been prepared for enzymatic incorporation into DNA. The presence of a disulfide linkage in (239) and (240) provides a means to release DNA macromolecules from biotin–streptavidin complexes upon treatment with DTT.[346]

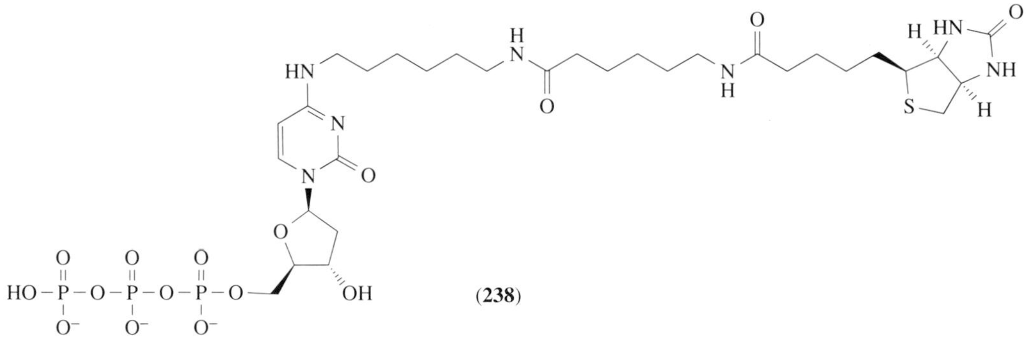

(238)

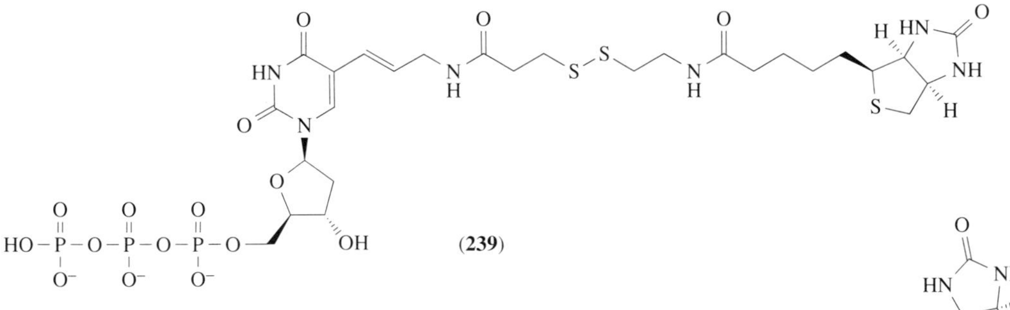

(239)

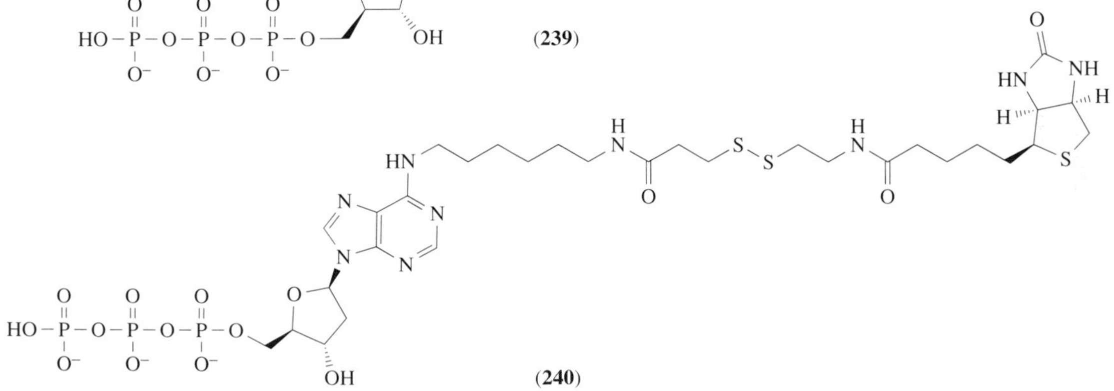

(240)

The novel d(ATP) analogue (241) has also been synthesized for enzymatic insertion of biotin in oligonucleotides by nick translation.[348] The biotinylated oligonucleotides efficiently hybridized to complementary DNA on slot blots, and led to the detection of subpicogram quantities of nucleic acid targets. Unlike the incorporation of commercial N-6 biotinylated d(ATP) into oligonucleotides, the incorporation of (241) had minimal effect on the $T_m$ of a hybrid composed of the biotinylated oligonucleotide and its unmodified complement.[348]

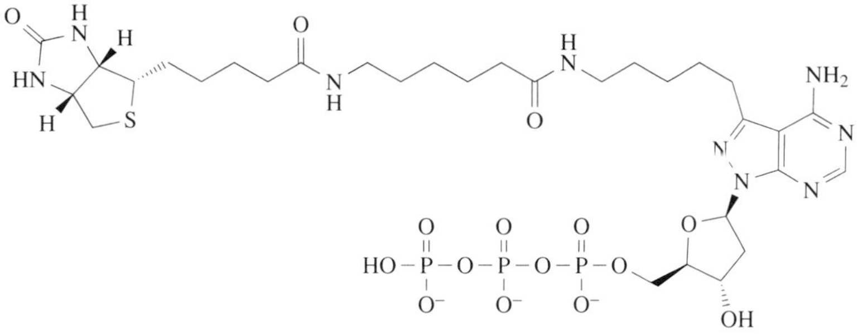

(241)

Similarly, preparation of the labeled deoxycytidine triphosphates (242)–(244) has been accomplished to enable enzymatic insertion of biotin, fluorescein, and photoreactive azide derivatives in oligonucleotides.[349]

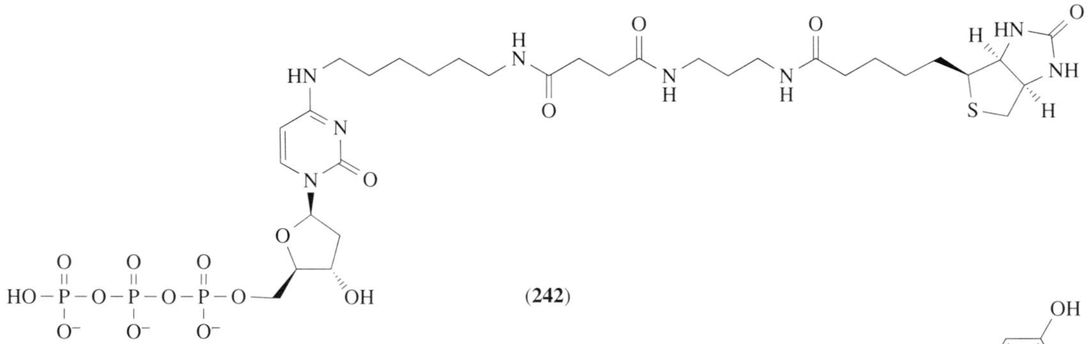

(242)

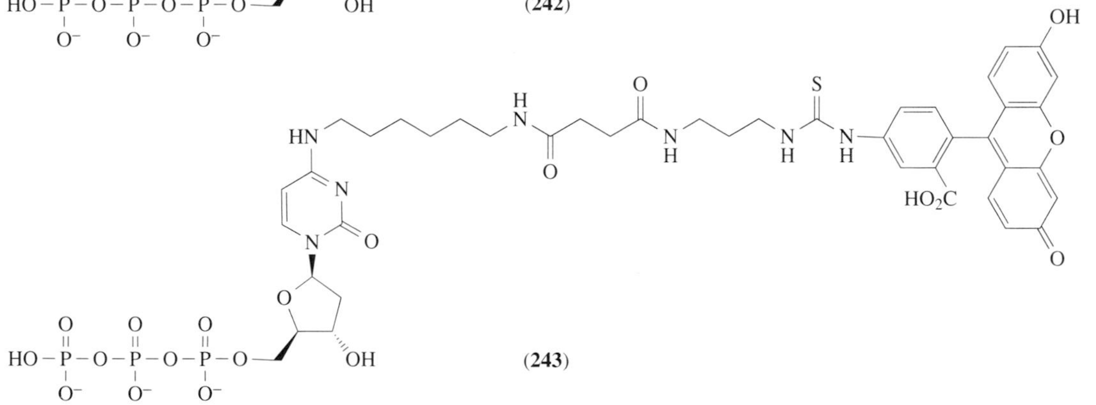

(243)

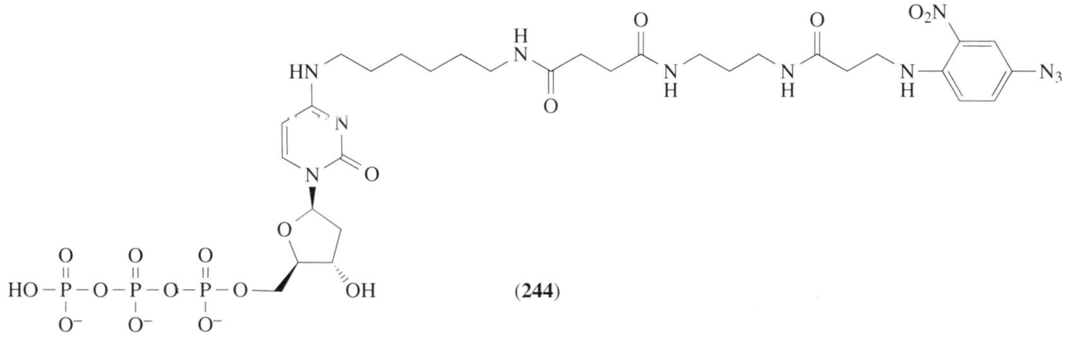

(244)

2,4-Dinitroaniline is an efficient intramolecular fluorescence quencher for fluorescein-labeled oligonucleotides. For example, the fluorescence quantum yield of a hexameric oligothymidylate carrying a fluoresceinated nucleobase at the 3'-terminus, and a 2,4-dinitroanilino function at the 5'-terminus, is about half that of an unquenched labeled oligonucleotide.[350] However, intramolecular fluorescence quenching is affected by hybrid formation with a complementary DNA strand. This trend is even more pronounced when the quencher and the marker are attached to a single nucleobase residue. Thus, incorporation of the modified deoxyguanosine phosphoramidite (245) at a single site into an oligonucleotide (19-mer) produced a fluorescence quantum yield that is 10% that of fluorescein at pH 8. Upon mixing the modified oligonucleotide with an unmodified complementary DNA oligomer, the fluorescence quantum yield of the duplex increased by a factor of about 4.[350]

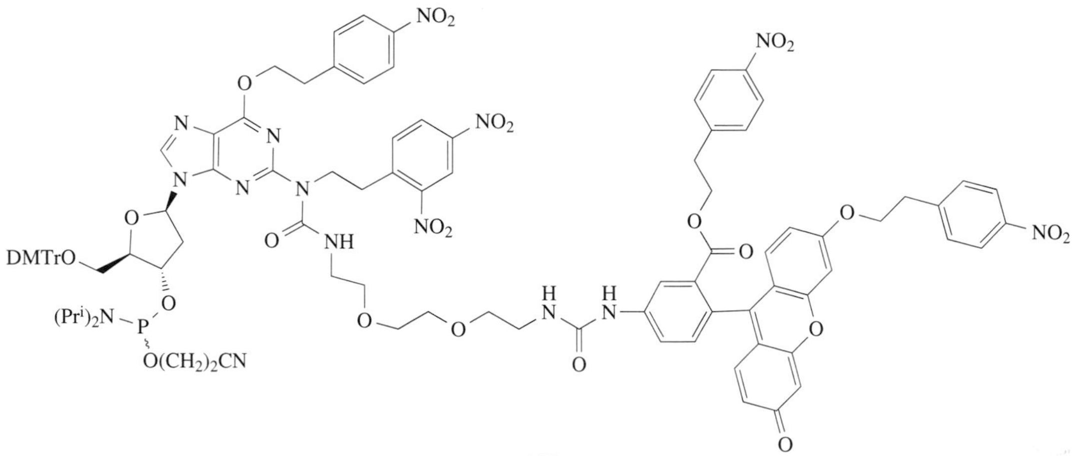

(245)

These findings show that the fluorescence of fluoresceinated DNA probes can be significantly quenched by the presence of other ligands. This feature must therefore be taken into consideration when designing experiments that involve fluoresceinated DNA probes.

Mielewczyk *et al.*[351] reported the synthesis of the deoxyluminarosine phosphoramidite (246) and its application to the preparation of fluorescent oligonucleotides for the nonisotopic detection of specific DNA sequences or the study of nucleic acid stereodynamics. Because of the sensitivity of the luminarosine residue to nucleophilic bases, nucleobase protecting groups that are labile to treatment with concentrated ammonium hydroxide at ambient temperature had to be used. Deoxy-luminarosine-containing oligomers were detected as strong fluorescent bands when analyzed by polyacrylamide gel electrophoresis.[351]

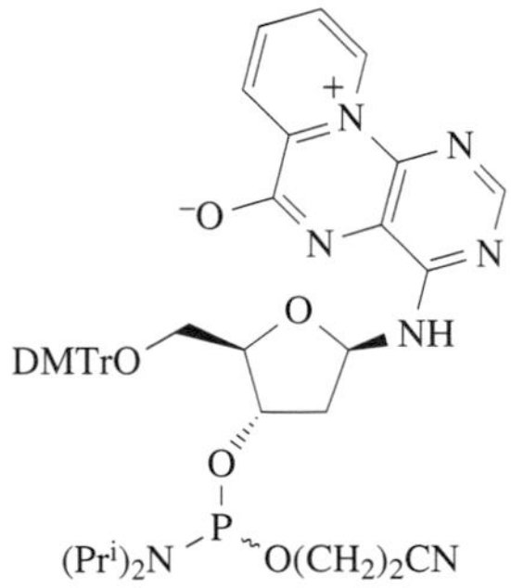

(246)

The highly fluorescent 1,$N^6$-ethenodeoxy- and riboadenosine, and 3,$N^4$-ethenodeoxy- and ribo-cytidine phosphoramidites (247)–(250) have been prepared for incorporation into DNA or RNA oligonucleotides.[352] Given the instability of these etheno derivatives to acidic and basic conditions, 4,4'-dimethoxytrityl and 4,4',4"-trimethoxytrityl groups have been used for 5'-hydroxy protection.

In addition, the phenoxyacetyl group was employed for protection of the exocyclic amino function of adenine and guanine, and isobutyryl for that of cytosine. A short treatment with concentrated ammonium hydroxide (<2 h) for the complete deprotection of these modified oligonucleotides is essential to ensure no degradation of the etheno nucleobases.[352] The characteristic and high fluorescent intensity of etheno nucleobases allowed the detection of ethenoadenine-containing and ethenocytosine-containing oligonucleotides below $10^{-9}$ M and $10^{-7}$ M, respectively.[352] The usefulness of these modified oligonucleotides, as sequencing and amplification primers, has been demonstrated by PCR experiments.

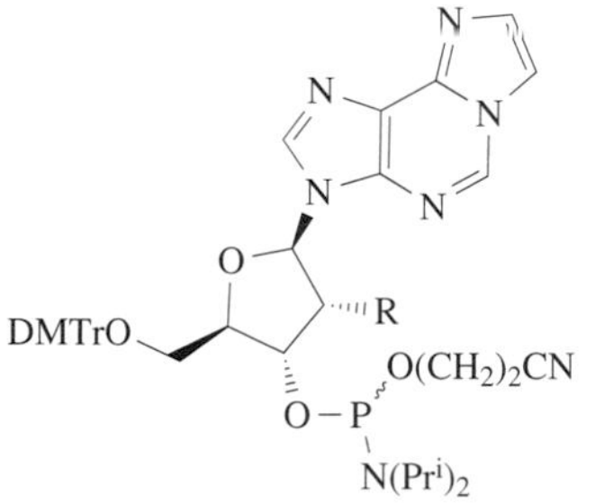

(247) R = H
(248) R = OTBDMS

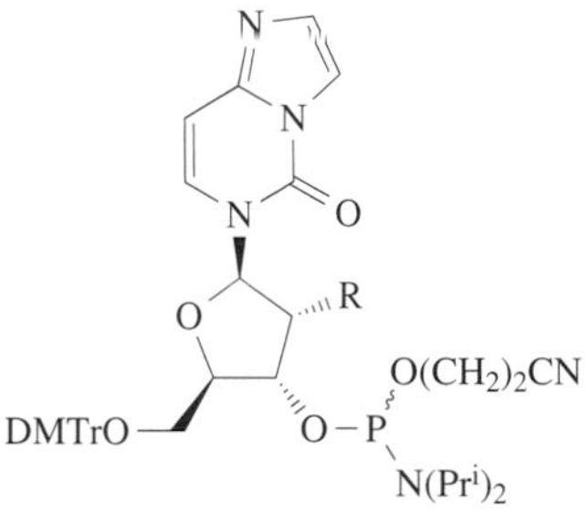

(249) R = H
(250) R = OTBDMS

Fluorescent DNA probes have also been prepared by nick translation and PCR using d(UTP)s such as (251)–(253) that are linked to a fluorescent cyanine label through spacers of different lengths.[353,354] It was demonstrated that the level of incorporation and hybridization fluorescence signal increased when the length of the spacer was increased. Thus, by use of the fluorescent d(UTP) (253) under optimal conditions, it was possible to label up to 28% of the possible incorporation sites on the target DNA by PCR, and 18% by nick translation.[353] These labeled DNA probes may find application in fluorescence *in situ* hybridization which has become an indispensable tool in a variety of areas of research and clinical diagnostics.

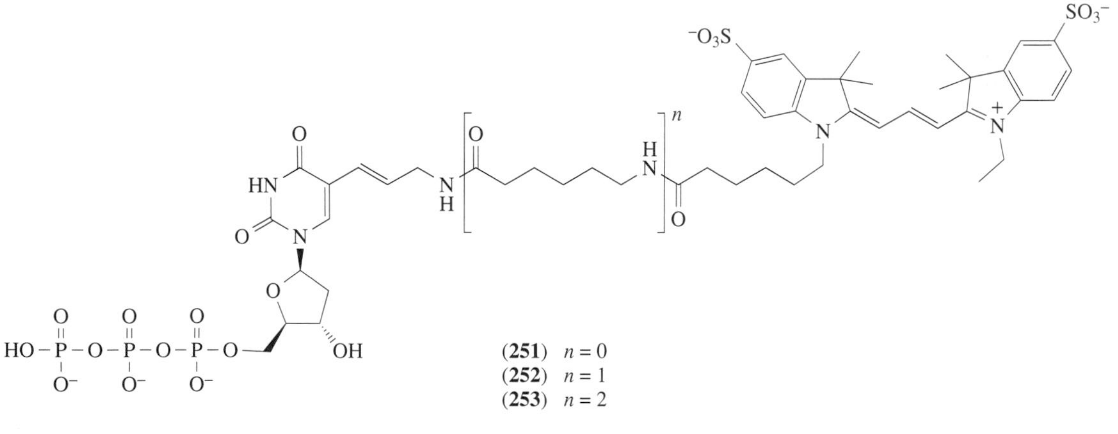

(251) $n = 0$
(252) $n = 1$
(253) $n = 2$

Alternatively, oligonucleotide-containing digoxigenin labels can be detected by an enzyme-linked immunosorbent assay (ELISA) using digoxigenin-specific antibodies linked to, for example, alkaline phosphatase.[310,311] Nonradioactive labeling of oligonucleotides with the digoxigenin-labeled deoxyuridine triphosphate (254)[311] has been accomplished with the Klenow fragment of *E. coli* DNA Polymerase I by primed synthesis or by tailing with terminal deoxynucleotidyl transferase.[355,356] The application of digoxigenin:antidigoxigenin technology, as a bioanalytical indicator system, has been reviewed in detail.[310]

It must be noted that the incorporation of the dansylated deoxyribonucleoside phosphoramidite (255)[357] into oligodeoxyribonucleotides has facilitated the study of exclusive DNA minor-groove events, such as netropsin binding, by monitoring changes in dansyl fluorescence seen from the major groove.[358,359] These findings further underscore the potential application of these fluorescent DNA probes to the study of DNA–drug and DNA–peptide interactions.

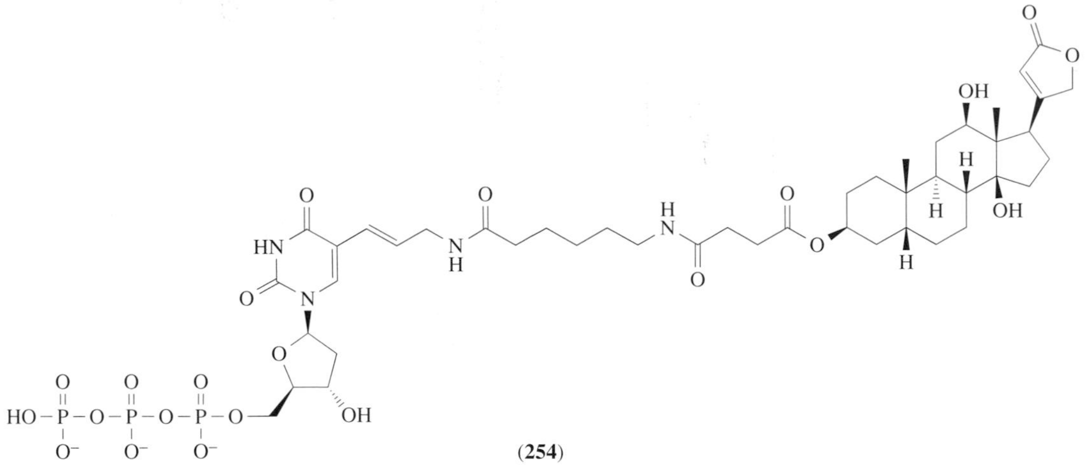

**(254)**

In addition, several mono- and polydansylated d(AT)$_5$ derivatives having the fluorophore attached to the C-8 of adenines, through polymethylene spacers of variable lengths, were synthesized to study their fluorescence properties. It was shown that while fluorescence intensity and emission quantum yields increased with iterative labeling, quenching was minimized through shorter and rigid tethers.[360]

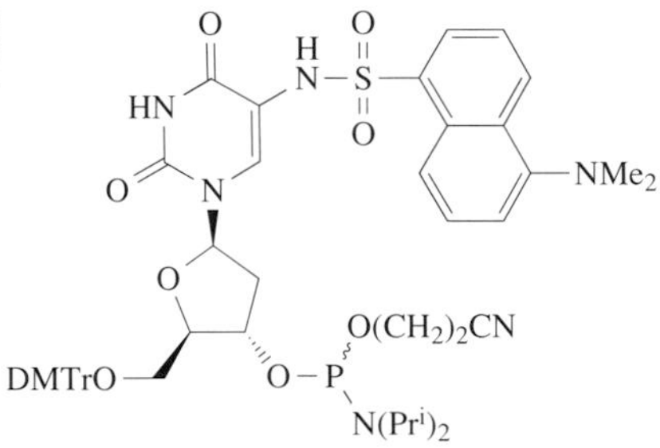

**(255)**

The 2,4-dinitrophenyl (DNP) deoxyribonucleoside phosphoramidite derivatives (256) and (257) have been inserted in oligonucleotides as an alternative to biotin, digoxigenin, or fluorescein for nonradioactive labeling. On the basis of hybridization experiments, it was demonstrated that the incorporation of (257) into an oligomer (39-mer) led to a greater detection sensitivity than that obtained with the insertion of (256).[361] It was further demonstrated that DNP-labeled oligo-nucleotides were at least as effective as digoxigenin-labeled oligonucleotides for the detection of human parvovirus B19. Less than 10 fg of B19 were detected with oligonucleotides having multiple internal and terminal DNPs.[361]

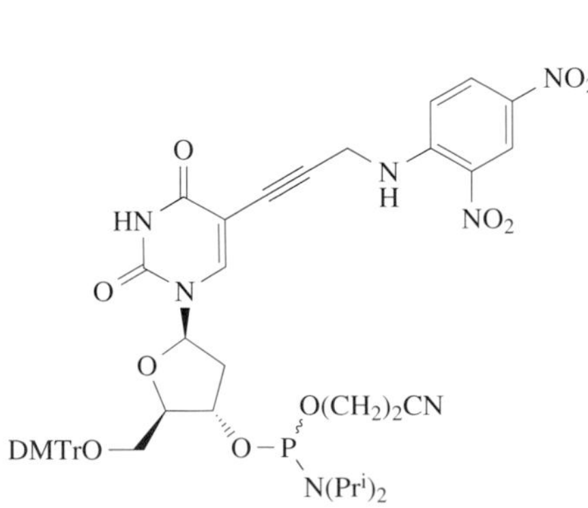

**(256)**

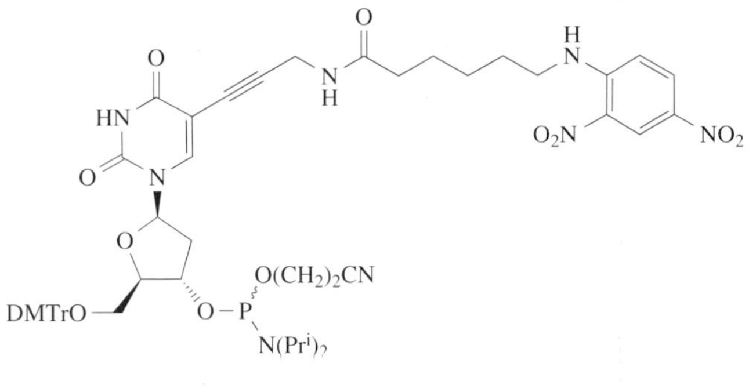

(257)

Of practical importance, the fluorescently-labeled alkynyl dideoxynucleotide triphosphate derivatives (**258**)–(**261**) have been extensively used as chain terminators for rapid DNA sequencing.[362] Typically, AMV reverse transcriptase is employed according to a modified dideoxy-DNA sequencing protocol to produce fluorescently-tagged fragments that are resolved by polyacrylamide gel electrophoresis in one sequencing lane. Given that each of the dideoxynucleotides (**258**)–(**261**) can be distinguished by its unique fluorescence emission, DNA fragments are identified by a fluorescence detection system that allowed base sequence determination to be made by computers.[362]

A convenient route to the preparation of europium(III)-labeled oligonucleotides involves using of the pyrimidine phosphoramidites (**262**) and (**263**) during solid-phase oligonucleotide synthesis.[363] The chelating functions of these modified oligonucleotides formed, after deprotection, strongly fluorescent complexes with europium(III) ions. In fact, oligonucleotides labeled with europium chelates are detected in microtiter wells, at concentrations as low as $10^{-17}$ M, by indirect fluorescence analysis.[363]

### 7.06.3.2.2 *Nucleobases funtionalized with intercalors and DNA/RNA cleavers*

It has been shown that the reaction of 5′-*O*-(4,4′-dimethoxytrityl)-5-bromodeoxyuridine with diaminoalkanes generated the corresponding 5-aminoalkylated deoxyuridine derivatives which were subsequently coupled to 6-chloro-9-phenoxy-2-methoxyacridine.[263] Phosphitylation of the functionalized nucleosides afforded the phosphoramidite (**264**). Insertion of (**264**) in oligonucleotides gave the acridine–oligonucleotide conjugates (**265**) which formed hybrids with their DNA complements. The thermostability of these hybrids depended on the location of the modified nucleobases within the DNA oligomers. Typically, oligonucleotides (17-mers) modified at the 3′-terminus produced a strong stabilization of DNA–DNA duplexes ($\Delta T_\mathrm{m} = 12\,°C$) when compared with unmodified oligonucleotides. However, the insertion of (**264**) at a central site within oligonucleotides led to slight destabilization of DNA–DNA duplexes ($\Delta T_\mathrm{m} = -2\,°C$). The placement of acridine at those internal positions may change the hydration spine inside the groove and detrimentally affect duplex stability.[263]

It is generally accepted that RNA hydrolysis, catalyzed by bovine pancreatic ribonuclease A, requires the participation of the two imidazole groups of histidine residues 12 and 119 in the active site.[364] On this basis, the synthesis of ribonuclease mimics that would combine the RNA hydrolysis activity of imidazole with the ability of oligonucleotides to bind RNA with sequence specificity has been proposed.[365] In this regard, the deoxyribonucleoside phosphoramidites (**266**)[365] and (**267**)[366] have been prepared and incorporated into oligonucleotides at predetermined sites. Interestingly, the 2,4-dinitrophenyl group used for protection of the imidazole group in (**267**) could be readily removed by ammonium hydroxide under conditions similar to those required for the deprotection of synthetic oligonucleotides.[366] The ability of oligonucleotides that have been modified by the insertion of either (**266**) or (**267**) to cleave RNA remains to be determined.

Along these lines, molecular modeling studies indicated that insertion of the modified deoxyguanosine phosphoramidite (**268**) in oligonucleotides would allow the imidazole group to deliver two geometrically distinct nitrogen bases to the vicinity of the 2′-hydroxyl function of a complementary cytosine.[367] Model experiments revealed that the imidazole group of d(T₇G*T₈) did not cleave r(A₁₂₋₁₈) upon hybridization but increased the stability of the complex relative to that obtained with unmodified d(T₇GT₈). These findings support the model in which the imidazole group forms a hydrogen bond with a 2′-hydroxyl group of the opposite RNA strand.[367]

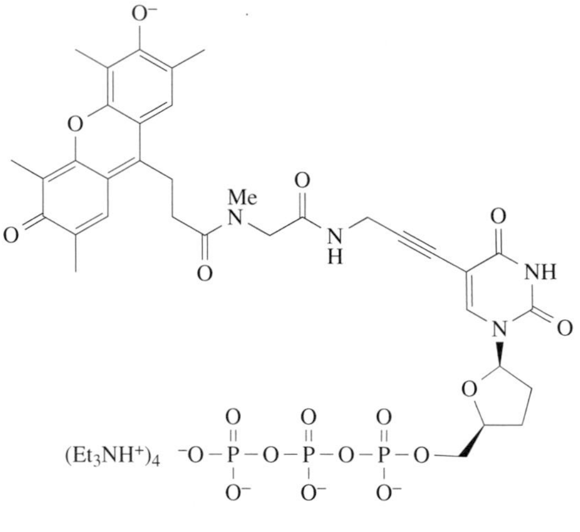

(**258**)

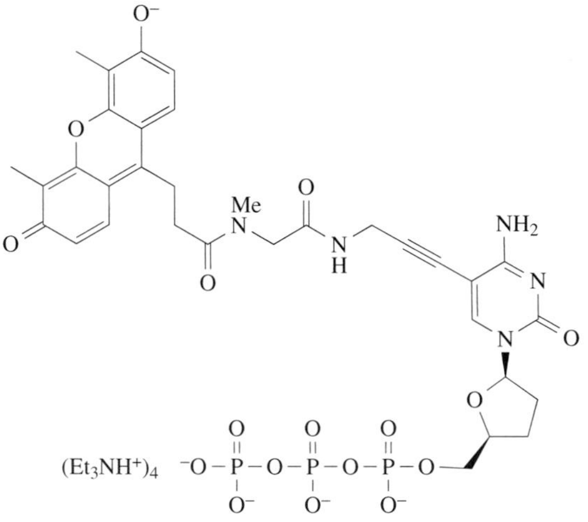

(**259**)

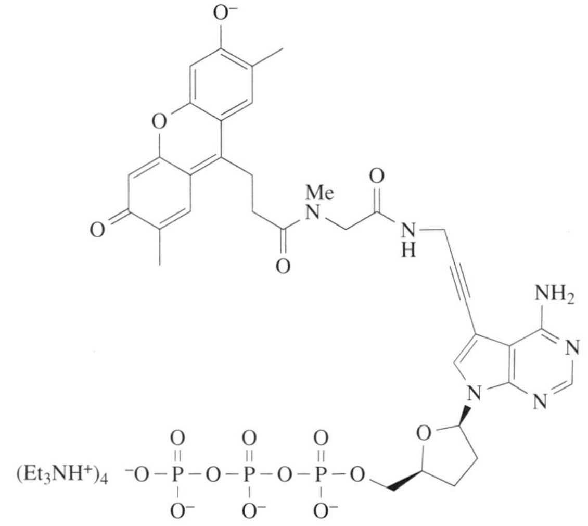

(**260**)

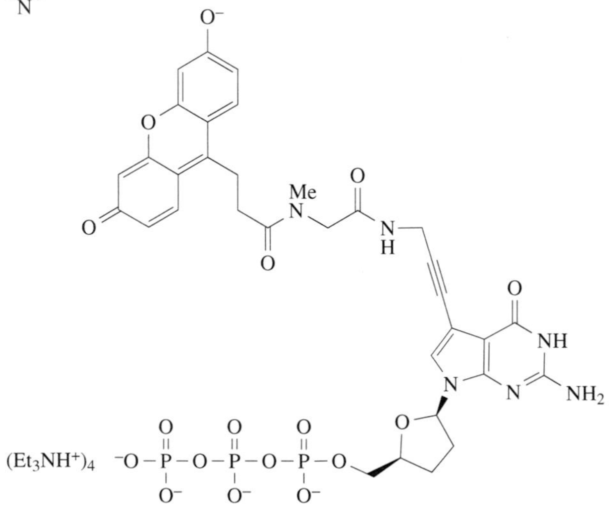

(**261**)

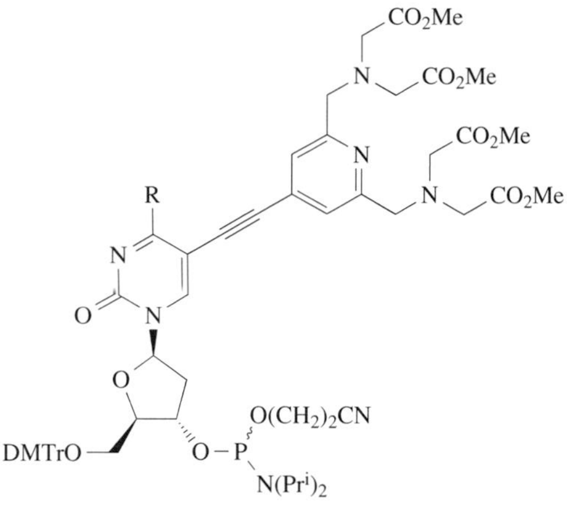

(**262**)  R = OH
(**263**)  R = NHC(O)Ph

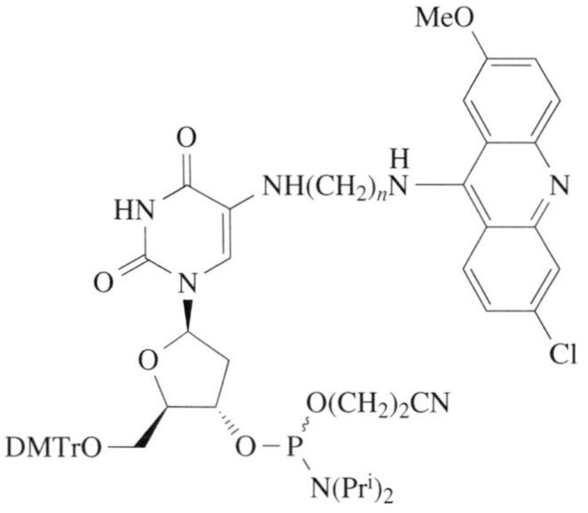

(**264**)  *n* = 4 or 7

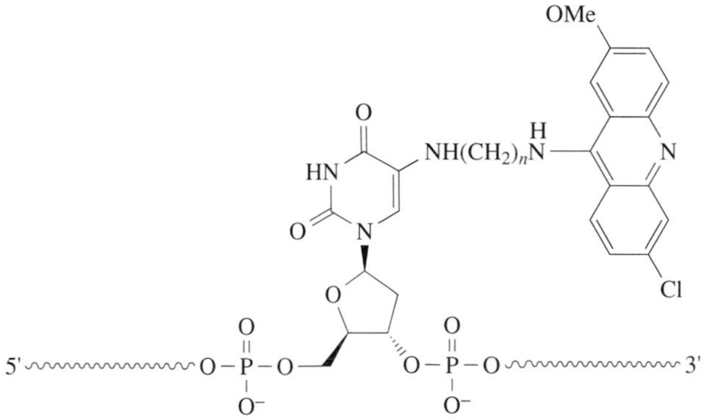

(**265**)  *n* = 4 or 7

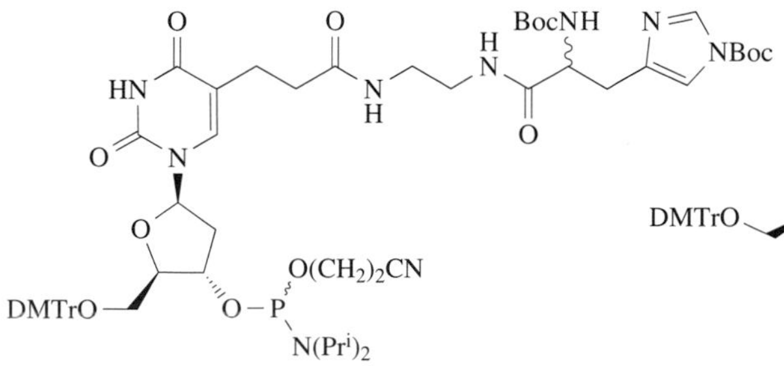

(**266**)

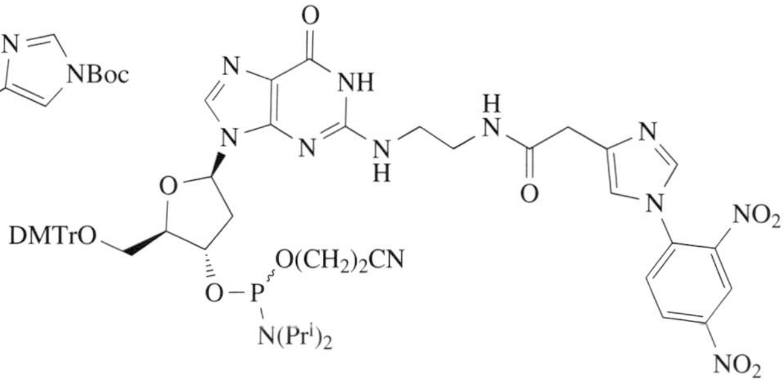

(**267**)

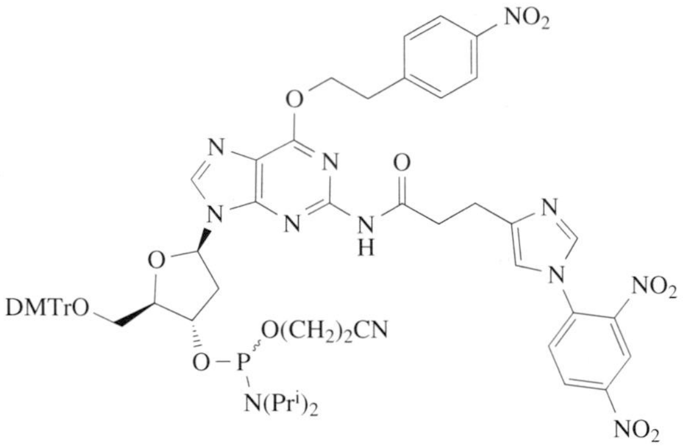

(268)

Oligonucleotides containing $N^2$-imidazolylpropylguanine and $N^2$-imidazolylpropyl-2-amino-adenine residues have been prepared from the multiple insertion of the corresponding deoxyribonucleoside phosphoramidite derivatives (269) and (270). These modified oligonucleotides were hybridized to complementary RNA targets and tested for RNA-cleaving properties.[368] The only RNA-cleaving activity observed was that produced by RNase H. Although the modified DNA oligomers formed hybrids with complementary RNA oligonucleotides that were more stable than the corresponding unmodified DNA–RNA duplexes, the enhancement of binding affinity was much more impressive when these modified oligonucleotides were hybridized to complementary DNA strands.[368]

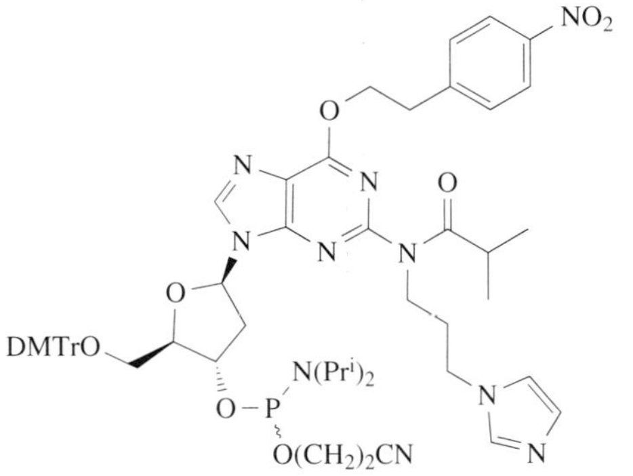

(269) 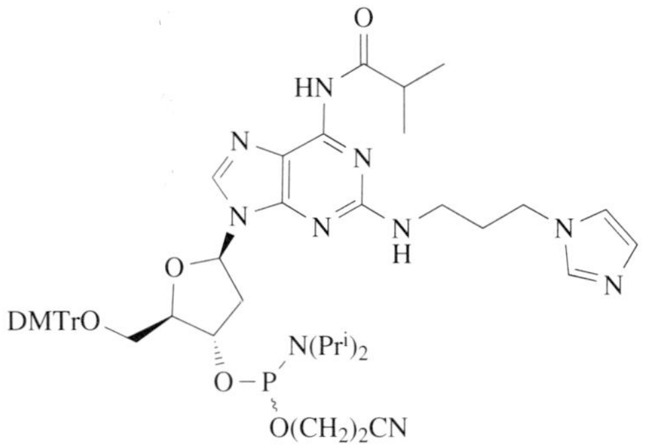                                        (270)

The synthesis of nuclease mimics was further pursued with the synthesis of the bipyridine deoxyribonucleoside phosphoramidite (271). Bipyridine forms stable complexes with metals like copper(II) and zinc(II) that are known to bind and activate phosphodiester functions toward nucleophilic attacks. The tether linking the bipyridine group to the nucleobase in (271) was modeled to position the bipyridine moiety next to a phosphodiester target within an A-type DNA–RNA duplex.[369]

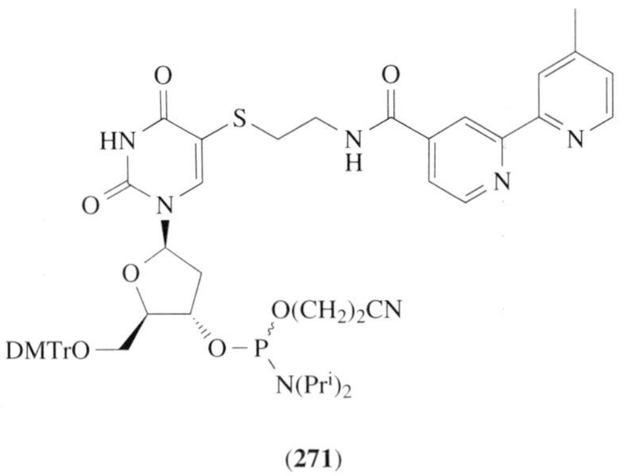

(271)

Incorporation of (271) into DNA oligomers afforded, after deprotection and purification, the conjugate (272).[370] While phenanthroline-complexed copper ion is an effective reagent for the redox-mediated cleavage of both DNA[187] and RNA,[371] essentially no cleavage is observed when (272) is mixed with a complementary DNA oligomer in the presence of copper(II) with or without hydrogen peroxide. Conversely, (272) efficiently cleaved the complementary DNA oligomer (ca. 80%) in the presence of iron(II) or iron(III) and hydrogen peroxide.[370] Bipyridine-conjugated oligonucleotides can therefore provide an opportunity to investigate nucleic acid transformations by positioning a redox-active metal–ligand complex in the major groove of a DNA duplex.

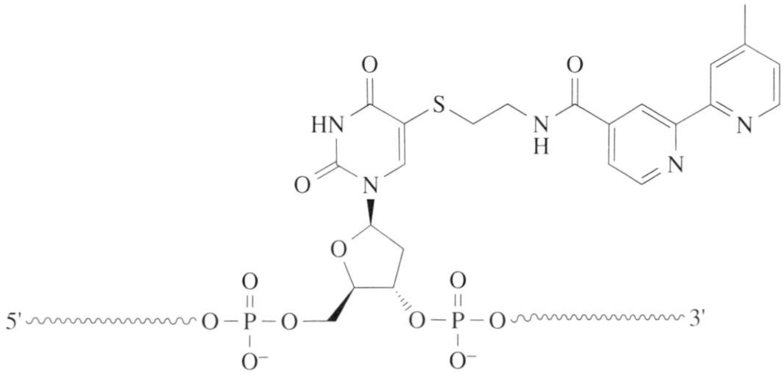

(272)

It should also be pointed out that the terpyridine deoxyribonucleoside phosphoramidite derivative (273) has been synthesized to serve as a functional mimic of ribozymes.[372] The site-specific incorporation of (273) into an oligodeoxyribonucleotide complementary to an RNA target triggered the cleavage of the target, within the duplex region, in the presence of copper(II) ion, whether DTT was present or not. A higher cleavage efficiency (43%) of the RNA target was, however, achieved in the presence of DTT.[372]

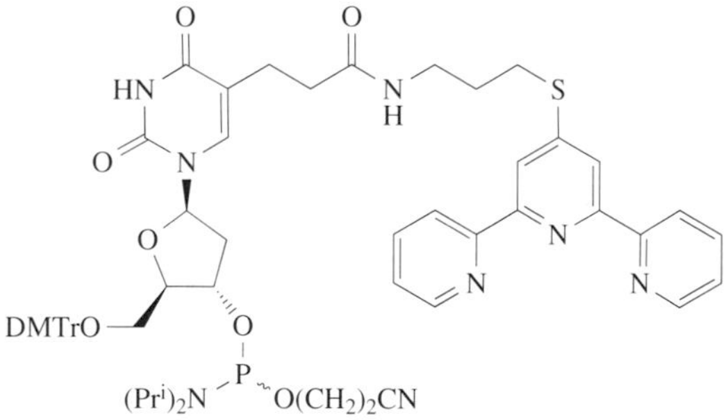

(273)

Incorporation of the $N^2$-modified deoxyguanosine phosphoramidite (274) into oligonucleotides at specific sites also produced phenanthroline conjugates that cleaved complementary DNA oligomers in the presence of ferrous ammonium sulfate and DTT.[373] When (274) is inserted near the middle of the oligonucleotide sequence, a highly specific cleavage of the complementary DNA strand occurred at one or two sites. It would therefore appear that precisely positioned ligands through conformationally restricted linkers could increase cleavage specificity and decrease self-damage.[373]

In their efforts to develop sequence-specific DNA cleaving agents, Dreyer and Dervan[374] reported the synthesis of DNA hybridization probes that carry a thymine residue functionalized with an EDTA moiety. Such a synthesis was achieved by incorporation of the EDTA-functionalized deoxyribonucleoside phosphoramidite (275) into the hybridization probe during solid-phase synthesis.[374,375] In the presence of iron(II), oxygen, and DTT, the probe effected the cleavage of a complementary ssDNA segment.[374] The phosphoramidites (275) and (276) were similarly inserted in triple-helix forming oligodeoxyribonucleotides that mediated the single site-specific cleavage of dsDNA *in vitro*.[376,377]

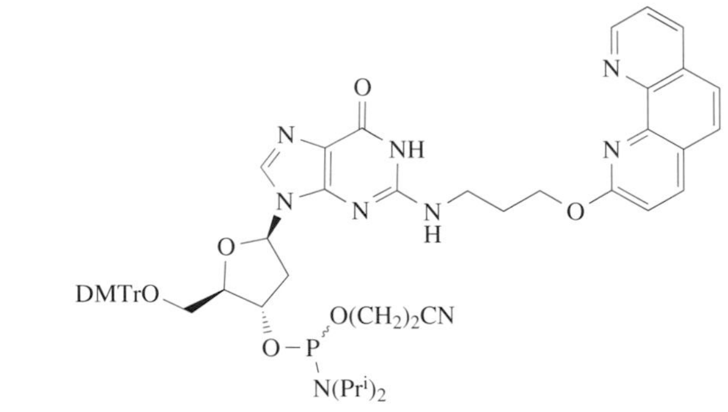

(274)

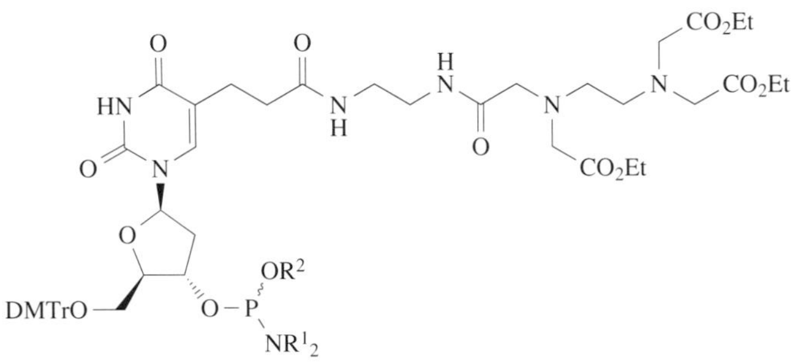

(275)  $R^1 = R^2 = Me$
(276)  $R^1 = Pr^i$;  $R^2 = CH_2CH_2CN$

### 7.06.3.2.3  Copolymerization of pyrrole-containing nucleobases

The pyrrole-containing phosphoramidite (277) was prepared from (232) by treatment with bis-(N-hydroxysuccinimidyl) sebacoate followed by 1-(2-aminoethyl)pyrrole.[378] Phosphitylation of the pyrrole-containing nucleoside afforded (277). Addition of (277) to the 5′-terminus of oligo-nucleotides by solid-phase synthesis generated the pyrrole–oligonucleotide conjugate (278). The copolymerization of (278) with pyrrole was carried out in a classical electrochemical cell to give (279) as a solid film deposited on the surface of the working electrode. Oligonucleotide probes, complementary to those grafted on the matrix, demonstrated that hybridization was feasible; DNA–DNA hybrids were detected by radioactive labeling.[378] Thus, polypyrrole films with covalently linked oligonucleotides may represent attractive candidates for DNA sensors.

### 7.06.3.2.4  Nucleobases conjugated to cross-linkers

Pieles et al.[379] described the attachment of 4,5′,8-trimethylpsoralen at C-8 of deoxyadenosine through a sulfur atom and a five-carbon-atom linker. The modified nucleoside was converted to the phosphoramidite (280) and incorporated into oligonucleotides. Exposure of a hybrid, composed of a psoralen–oligonucleotide conjugate and a complementary unmodified oligomer, to near-UV light (345 nm) resulted in the formation of cross-links between strands to an extent greater than 90%.[379] The potential biological relevance of this *in vitro* cross-linking experiment should provide the impetus to define the parameters required for these events to occur *in vivo*.

The  5-[(4-azidophenacyl)thio]cytidine-5′-triphosphate  (281)  and  5-[(4-azidophenacyl)thio]-uridine-5′-triphosphate  (282)  have  been  synthesized  to  provide  new  photocross-linking reagents.[380,381] These CTP and UTP analogues carry an aryl azide group located at approximately 10 Å from the nucleobase, and can be incorporated into RNA at internal positions by *in vitro* transcription with *E. coli* RNA polymerase. Upon irradiation with long-wavelength UV light, the

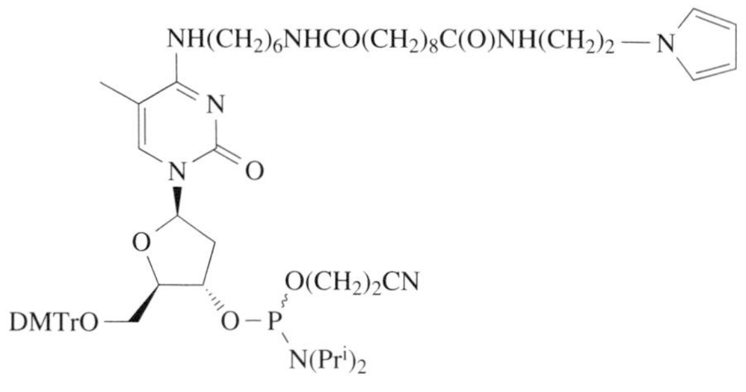

(277)

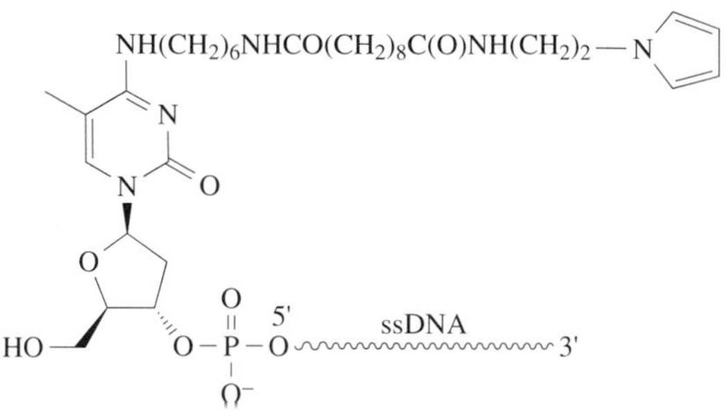

(278)

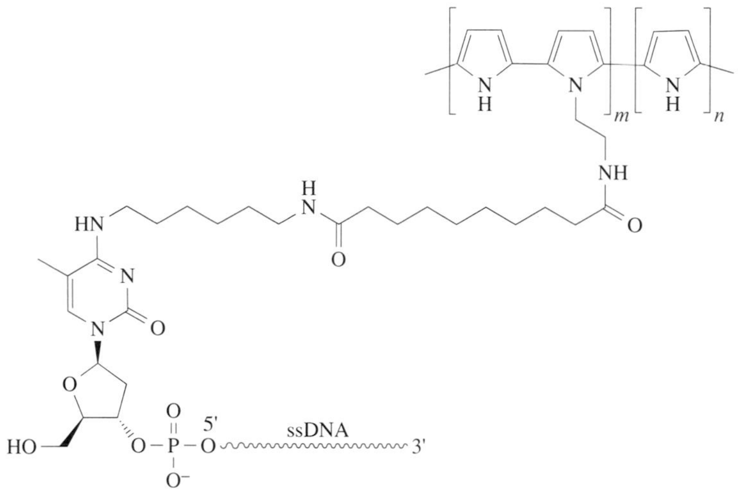

(279) $m \ll n$

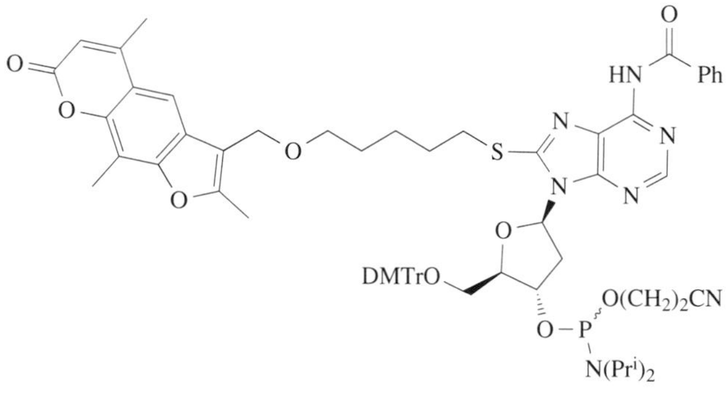

(280)

incorporation of (281) into RNA led to the formation of cross-links with the RNA polymerase. Such cross-linking experiments facilitated analysis of the molecular interactions between the *E. coli* transcription factor *Nus*A and the 3′-end of the nascent RNA in a ternary transcription complex.[381] The above CTP and UTP analogues may therefore be useful for mapping polymerase RNA binding domains and identifying proteins that may be implicated in transcription complexes through interactions with nascent RNAs. It should also be noted that the nucleotide analogue 8-azido-ATP has been used as a substrate toward the identification of contact points between the 3′-terminus of newly transcribed RNAs and *E.coli* RNA polymerase during the early steps of RNA synthesis.[382]

The 2′,3′-dideoxy-*E*-5-[4-(3-trifluoromethyl-3*H*-diazirin-3-yl)styryl]UTP (283) has been employed in photoaffinity labeling experiments that involved human immunodeficiency virus type-1 reverse transcriptase (HIV-1 RT) and poly A:oligo (dT) as the template/primer.[383,384] Specifically, (283)

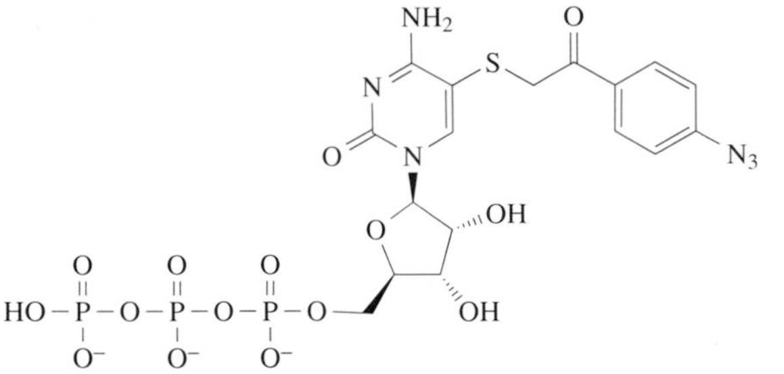

**(281)**

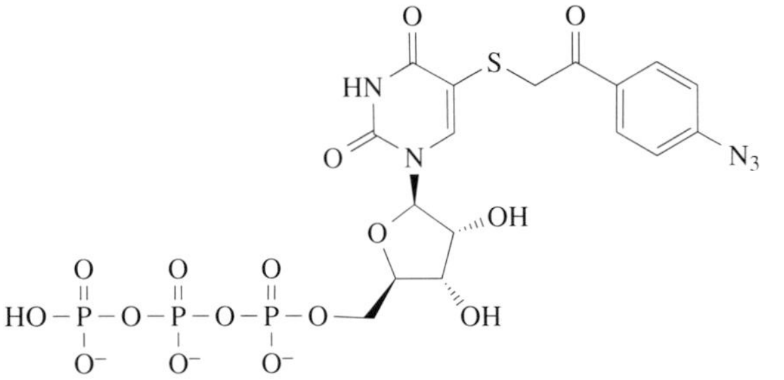

**(282)**

was incorporated into the primer strand at its 3′-terminus via RT activity. When photoirradiated with near-UV light (365 nm), the resulting photolabile primer became covalently linked to the 66 kDa subunit of HIV-1 RT. These results suggest that (283) could serve as a useful tool for characterizing the binding sites of, for example, DNA polymerases and HIV-1 RT to their substrates.[383,384] The photoreactive dCTP analogues (284) and (285) have also been successfully applied to the photoaffinity modification of HIV-1 RT.[385]

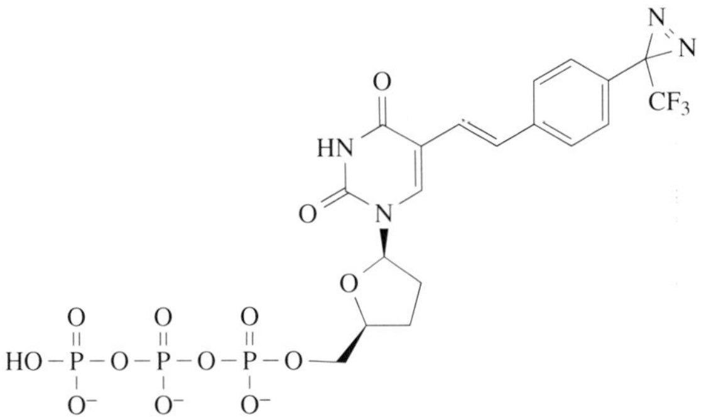

**(283)**

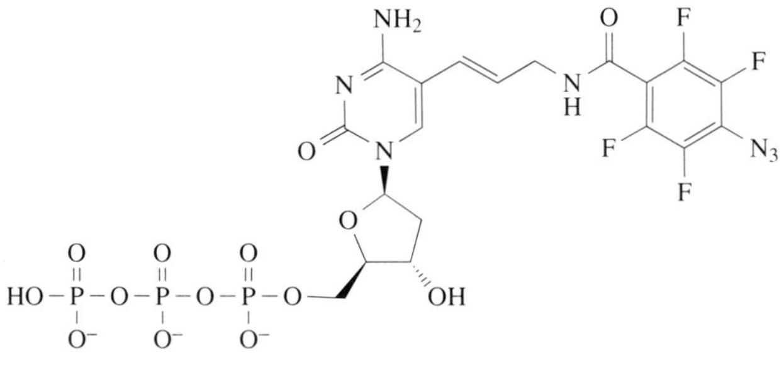

**(284)**

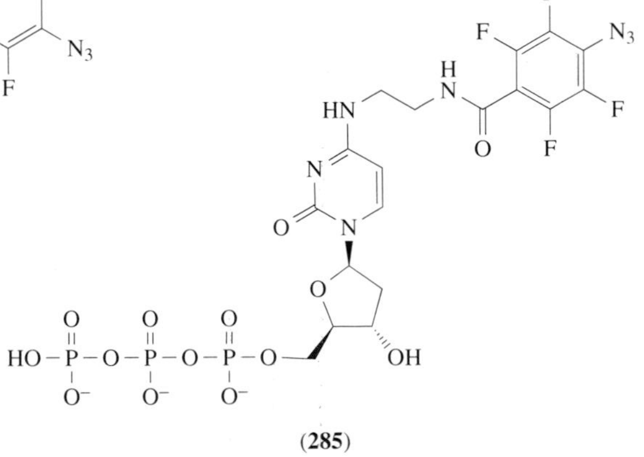

**(285)**

### 7.06.3.2.5   *Conjugation of nucleobases to nitroxide, heavy atoms, and isotopic markers*

The site-specific incorporation of a paramagnetic probe into a self-complementary oligonucleotide has been accomplished by the use of the deoxyribonucleoside phosphoramidites (**286**)[386–388] and (**287**).[389] The bulky and hydrophobic nitroxide moiety that originated from the incorporation of (**286**) did not induce disruption of DNA secondary structures, and was sufficiently constrained to permit the correlation of its motion with that of the carrier DNA. Alternatively, the insertion of (**287**) in the above self-complementary oligonucleotide provided evidence that the length of the alkynyl arm affected the rotation of the nitroxide moiety about the alkyne axis.[389] In fact, when linked through a diacetylene tether, the nitroxide group underwent a rapid uniaxial rotation about the tether together with motion of the DNA that is consistent with global tumbling of the duplex.[390] These probes may be particularly valuable in investigations pertaining to the dynamics of unusual DNA structures and/or sequences of biological importance.

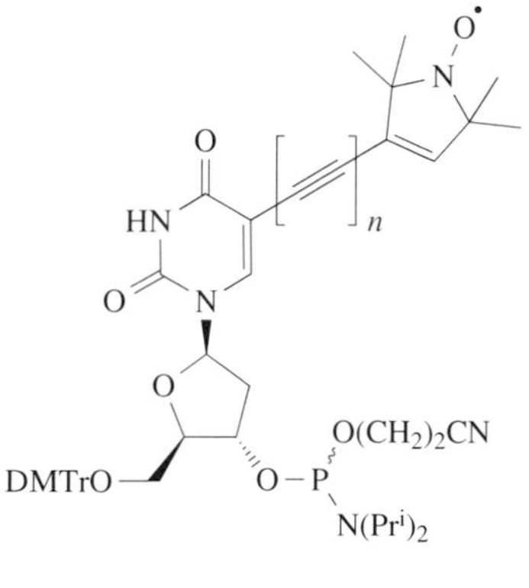

(**286**)   $n = 1$
(**287**)   $n = 2$

Oligonucleotides containing spin-labeled 2′-deoxycytidine(s) and 5-methyl-2′-deoxycytidine(s), as probes for DNA structural motifs, have been prepared by solid-phase mediated insertion of the nucleoside phosphoramidites (**288**) and (**289**) at specific sites.[391] Typically, an oligonucleotide (16-mer) modified by a single insertion of either (**288**) or (**289**), midway within the sequence, formed a relatively unstable complex with a complementary DNA oligomer ($T_m = 49\,°C$) when compared with the corresponding unmodified duplex ($T_m = 63\,°C$). Nonetheless, spin-labeled DNA can be applied to the detection of hybridization processes.[391] For example, at temperatures below 65 °C, relatively broad lines in EPR spectra indicated a strong restriction of label mobility. However, at temperatures above 65 °C, the EPR linewidths and intensities resembled those of the single-stranded oligomer and, consequently, were indicative of strand separation.[391]

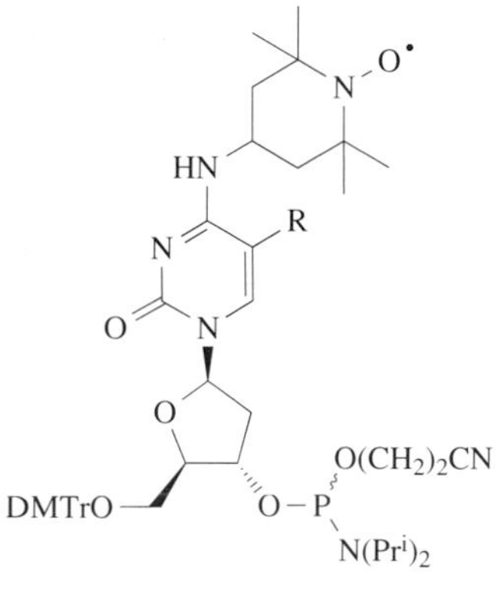

(**288**)   R = H
(**289**)   R = Me

Similar observations were reported earlier, as a consequence of the incorporation of the nitroxide-labeled nucleosides (**290**)[392] and (**291**)[393] into oligothymidylates according to the phosphotriester method.[392,393] Annealing nitroxide-labeled oligothymidylates with oligodeoxyadenylates produced EPR line shape changes consistent with hybridization. These line shape changes primarily originated from a change in local base dynamics.[393]

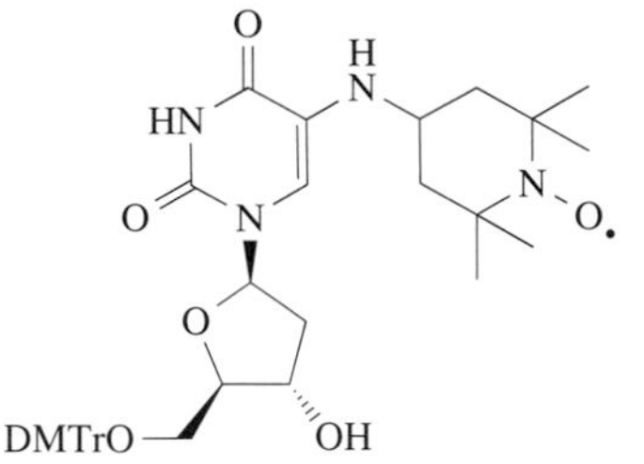

(290) 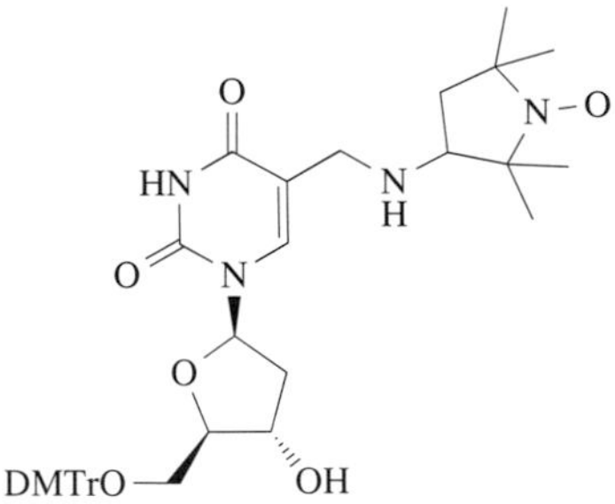          (291)

The spin-labeled deoxycytidine triphosphates (292) and (293) are substrates for *Micrococcus luteus* DNA polymerase and have been enzymatically incorporated into (dG–dC)$_n$. These polymers assumed B- and Z-DNA conformations, according to UV–circular dichroism spectroscopy, under low and high salt conditions, respectively. The EPR line shapes of the labeled copolymers in Z-form are unique and significantly different from those of the B-form. Thus, EPR line shapes of these spin-labeled DNAs are indicative of their local conformations; nucleobase dynamics of Z-DNA are slower than in B-DNA by about a factor of two.[394]

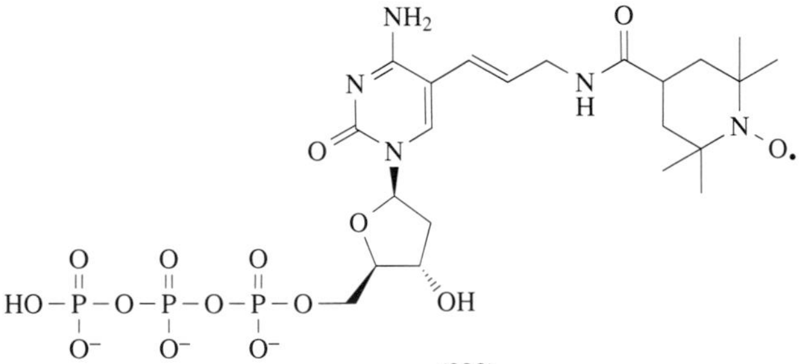

(292)

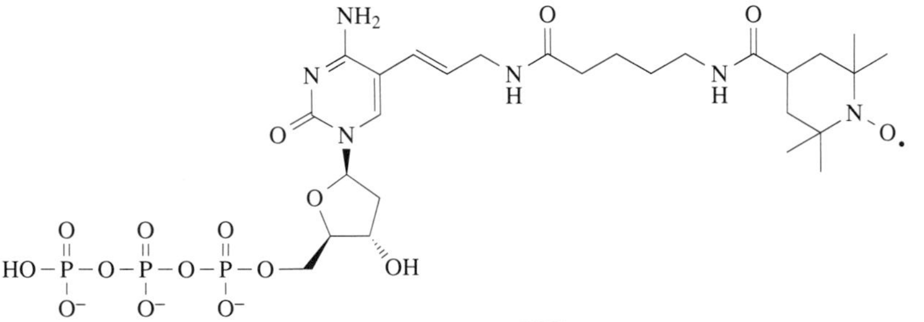

(293)

The study of macromolecular structures can be performed by X-ray scattering methods both in solution and solid state. The presence of a heavy atom label facilitates elucidation of the geometric and dynamic structure of these molecules. In this context and in their efforts to unravel the chemistry of genetic recombination, Sheardy and Seeman[395] described the synthesis of the 5-iododeoxyuridine phosphoramidite (294) and its incorporation into a DNA oligomer corresponding to an analogue of strand 2 of the immobile nucleic acid junction J1.[396,397] The heavy atom-labeled oligonucleotide was deprotected under mild conditions to minimize the loss of iodine.[395]

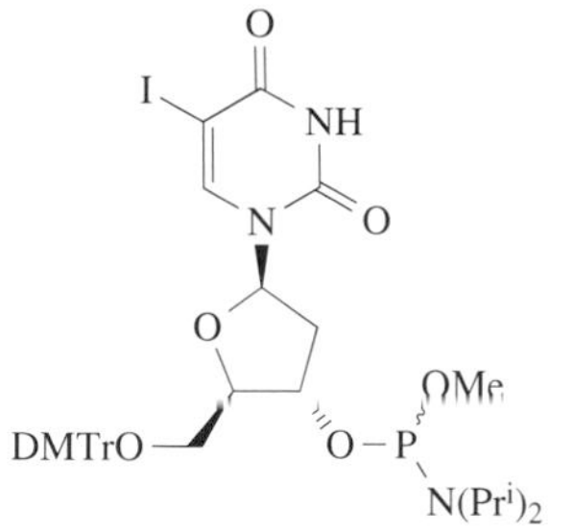

(294)

Iterative incorporation of the trifluorothymidine phosphoramidite (295) into an oligonucleotide, which has been designed to inhibit the expression of gene sequences encoding serine proteases in T-lymphocytes, has been achieved.[398] The modified oligomer d(GAGAT*CT*T*CAT* CT*T*CCCCGG), where T* represents trifluorothymine residues, was detected by [19]F-NMR spectroscopy at a concentration of 10 µM with a signal-to-noise ratio of 10:1.[398] It has been postulated that such limits of detection should be satisfactory for *in vivo* NMR imaging of the cellular uptake and intracellular distribution of the modified oligonucleotide.

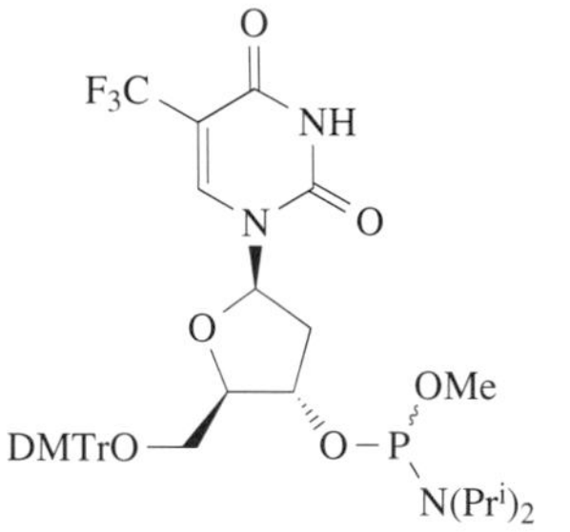

(295)

### 7.06.3.2.6 *"Boronated" nucleobases*

Boron neutron capture therapy (BNCT) may have potential for the treatment of solid tumors for which no therapies exist. BNCT is a binary system that combines two separately nonlethal constituents: a radiosensitizer containing stable boron-10 isotope and nonionizing neutron radiation. When boron-10 is irradiated with low-energy neutrons, a nuclear reaction occurs and produces helium nuclei ($\alpha$-particle), lithium-7 nuclei, and about 100 million times more energy than was put in. The generated radiation destroys malignant cells containing the boron compound and results in a therapeutic effect.

In an effort to obtain nontoxic boron compounds for BNCT, which selectively target proliferating tumor cells for potential incorporation into tumor DNA, the novel 5-substituted pyrimidine nucleosides (296) and (297) have been synthesized.[399] These modified nucleosides were designed to provide better binding to cellular kinases which are essential for the incorporation of these nucleosides into DNA. Additional 5-tethered carborane-containing pyrimidine nucleosides have been prepared and have been the subject of a review.[400] For example, the 5-(*o*-carboran-1-yl)-2′-deoxyuridine phosphoramidite (298) has been incorporated at selected sites into the oligothymidylate $dT_{12}$.[401] The thermostability of duplexes composed of these modified oligonucleotides and either $dA_{12}$ or poly(rA) is affected by the location of the carboranyl nucleotide within the oligothymidylate chain.

Modifications localized to the 5′-terminus of $dT_{12}$ had little effect on the thermostability of the resulting duplexes. However, 3′- or centrally modified $dT_{12}$ generated much less stable complexes with $dA_{12}$ ($\Delta T_m = 8.5\,°C$ or $13.7\,°C$, respectively) than unmodified $dT_{12}$. These findings are in agreement with modeling experiments that revealed steric interactions between the boron cage and the 5′-adjacent nucleobase. Interestingly, modifications targeted to the 3′-terminus of $dT_{12}$ markedly increased resistance to 3′-exonucleases. Furthermore, the presence of one or more carborane clusters in these modified oligonucleotides increased their lipophilicity and made them attractive candidates for BNCT.[401]

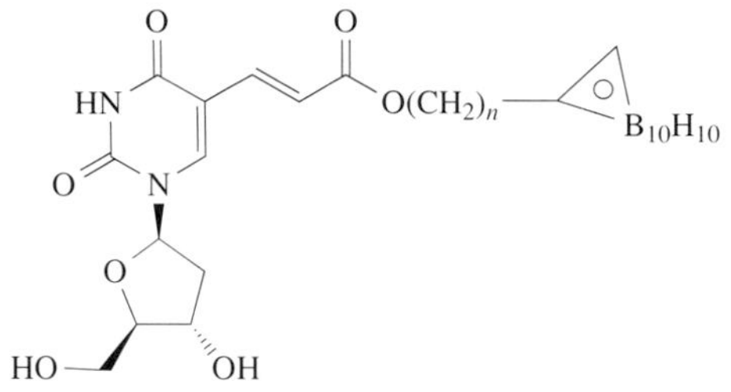

(296) $n$ = 2, 3 or 4

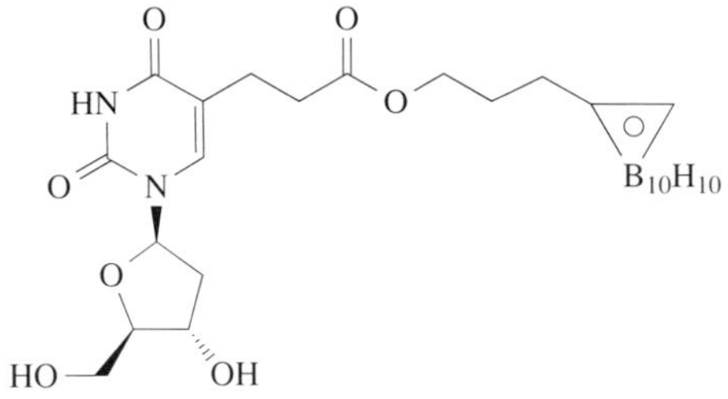

(297)

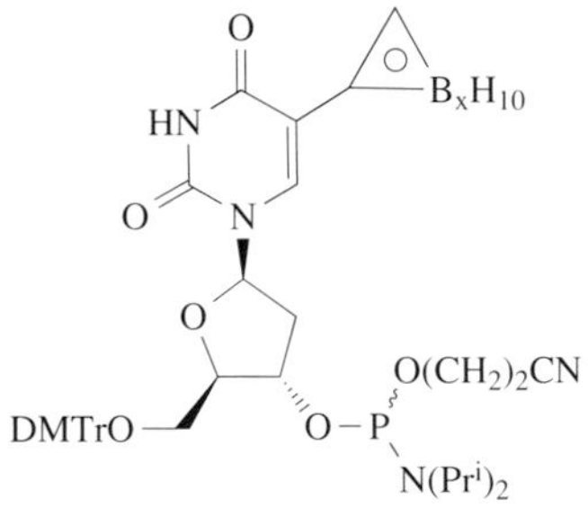

(298) x = 9 or 10

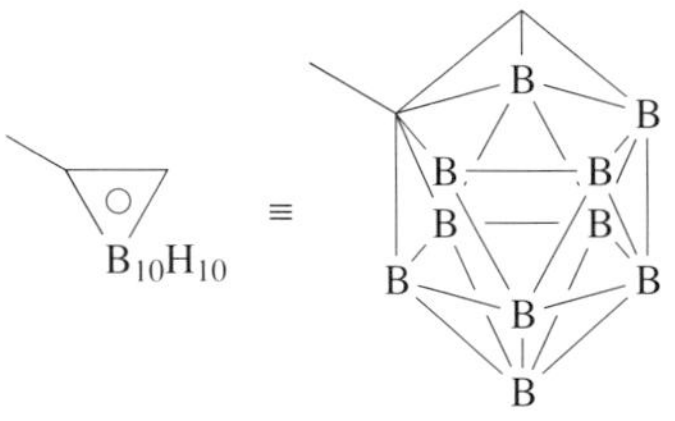

### 7.06.3.2.7 Modification of nucleobases with polycyclic aromatic hydrocarbons

The occurrence of carcinogenic or mutagenic effects associated with polycyclic aromatic hydrocarbons (PAH) has been correlated with the ability of electrophilic metabolites to modify chromosomal DNA and interfere with the normal biochemical events involved in cell division.[402] Several lines of evidence suggest that these electrophiles preferentially react with the exocyclic amino group of guanine.[403] Whether this preferential reaction occurs from optimal positioning of the aromatic electrophile via initial binding within one of the DNA grooves or by transient intercalation of the aromatic ring within the DNA double helix is unclear.[404] In order to provide a better understanding of the interactions of electrophilic PAH with nucleic acids, the synthesis of oligonucleotides carrying well-characterized adducts, at selected sites, has been reported by Casale and McLaughlin.[404] The anthracenyl deoxyguanosine phosphoramidite derivative (299) was prepared from 2′-deoxy-4-desmethylwyosine and incorporated into oligonucleotides by solid-phase synthesis. The coupling efficiency of (299) was modest (65–70%) despite an extended condensation time (ca. 60 min). The circular dichroism spectrum of a DNA duplex (12-mer) containing two anthracen-9-ylmethyl residues (one in each strand) failed to show a well-defined helical structure. Conversely, a duplex (13-mer) carrying only a single polycyclic aromatic residue essentially existed as a normal B-form DNA but exhibited a lower $T_m$ ($50.6\,°C$) than that of an unmodified duplex ($T_m = 56.6\,°C$). The lack of substantial fluorescence quenching and the observed destabilization of the double-stranded 13-mer indicated that the covalent alkylation of DNA by electrophilic aromatic hydrocarbons may not result from the intercalation of the polycyclic aromatic electrophile. Exterior groove binding may be responsible for initial association and/or positioning of the aromatic hydrocarbon prior to alkylation of duplex DNA.[404]

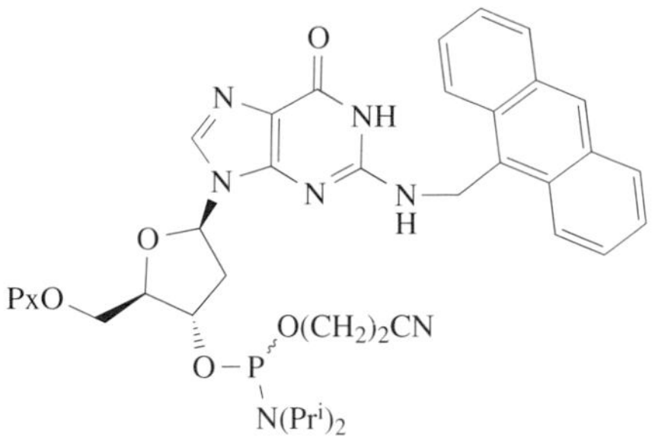

(299)

The anthracenyl deoxyadenosine phosphoramidite (300) has also been inserted in oligonucleotides in an attempt to determine the effect of the bulky PAH group on the conformation and duplex stability of these oligomers.[405] A duplex composed of a pentadecamer carrying one aralkylated residue and a complementary unmodified sequence displayed a lower $T_m$ (46 °C) than that of an unmodified DNA duplex ($T_m = 55$ °C) under the same conditions. The c.d. spectra of unmodified and aralkylated duplexes showed that the conformation of these duplexes was not significantly different from each other in solution.[405]

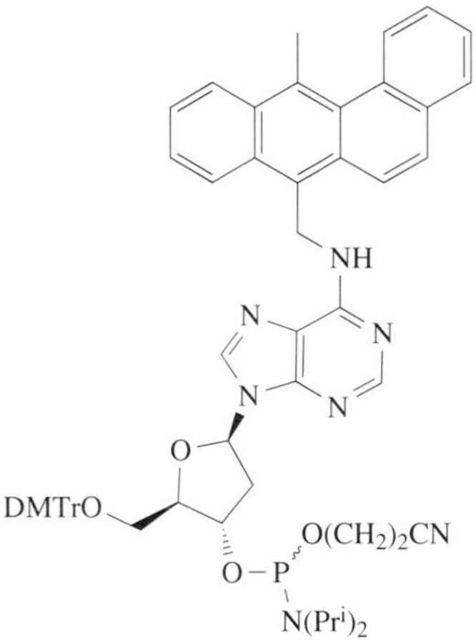

(300)

To further probe the mechanism(s) of cell transformation by PAH, epoxide and diol epoxide adducts of these species to the exocyclic amino function of deoxyadenosine and deoxyguanosine have been synthesized and incorporated into oligonucleotides via the phosphoramidites (301)–(304) and (305), respectively.[406–409] The nucleosidic precursors of these phosphoramidite derivatives were 6-fluoro-9-(2-deoxy-$\beta$-D-erythro-pentofuranosyl)purine[406,407] and 2-fluoro-2'-deoxyinosine.[408] It has been shown that the interaction of the nonanucleotide d(GGTCA*CGAC) (A* represents the altered base derived from (304)) with the complementary strand d(CTCGTGACC) generated a duplex that exhibited a $T_m$ of 23 °C at low ionic strength. Under similar conditions, the $T_m$ of an unmodified DNA duplex was 43 °C.[409] Surprisingly, the thermostability of the modified duplex was unchanged when dG, instead of dT, was opposite the adducted adenine residue. A less stable duplex ($T_m = 14$ °C) was, however, obtained when dA was opposite the modified adenine.[409] It would therefore be interesting to determine whether such preference for specific mismatches in these DNA duplexes is a good predictor of nucleotide misincorporation during replication of the modified DNA.

In an attempt to resolve the role of N-(2'-deoxyguanosin-8-yl)-2-(acetylamino)fluorene in mutagenesis, the corresponding phosphoramidite (306) has been prepared and integrated into several oligonucleotides at selected sites.[410,411]

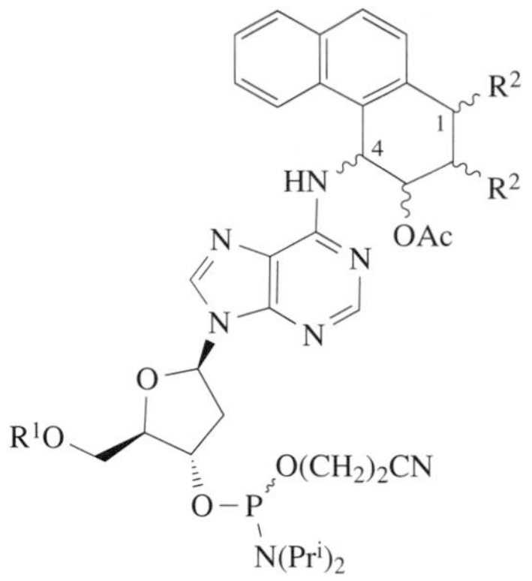

(301)  (3R, 4R);  R¹ = Px,  R² = H
(302)  (3S, 4S);  R¹ = Px,  R² = H
(303)  (3S, 4R);  R¹ = DMTr,  R² = H
(304)  (1R, 2S, 3R, 4S);  R¹ = DMTr,  R² = OAc

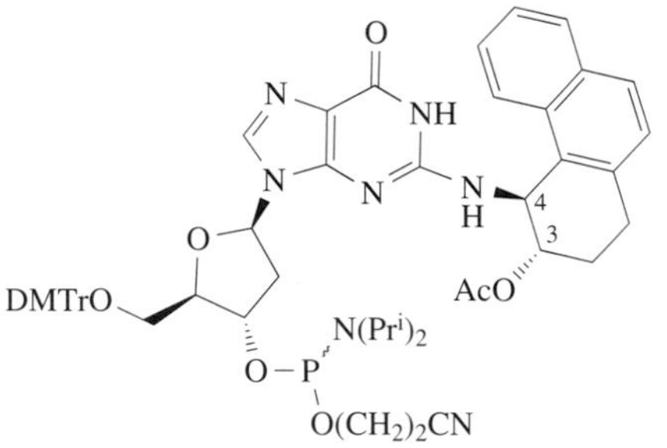

(305)  (3S, 4S)

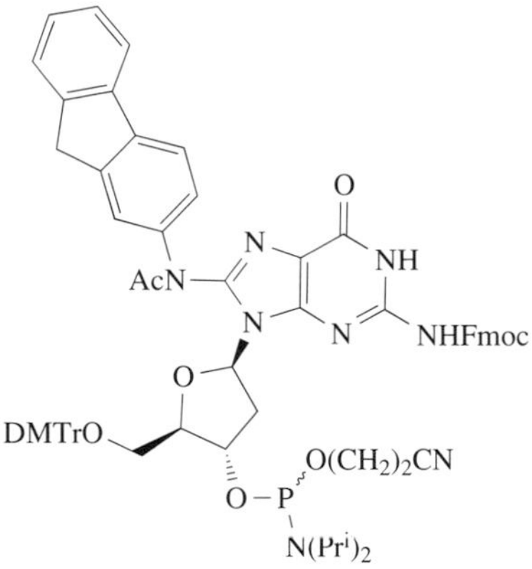

(306)

An oligonucleotide carrying the modified nucleobase at a central location within the sequence formed a 1:1 duplex with a complementary DNA oligomer (12-mer).[411] The stability of the modified hybrid was, however, significantly lower than that of the unmodified duplex ($\Delta T_m = 17\,°C$). Multidimensional NMR studies on these oligonucleotides should establish a correlation between the DNA conformational changes induced by the adducted nucleobase and the biological consequences of these adducted structures.[411]

The pyrenyl phosphoramidite derivatives (307) and (308) have also been prepared and incorporated into oligonucleotides to generate models aimed at investigating the mechanism of PAH carcinogenesis.[412]

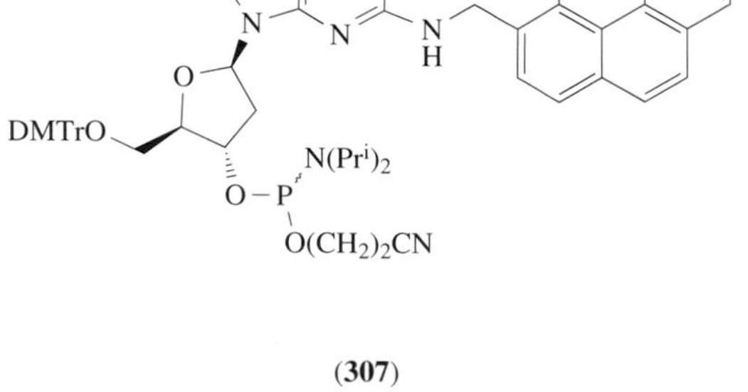

(307)

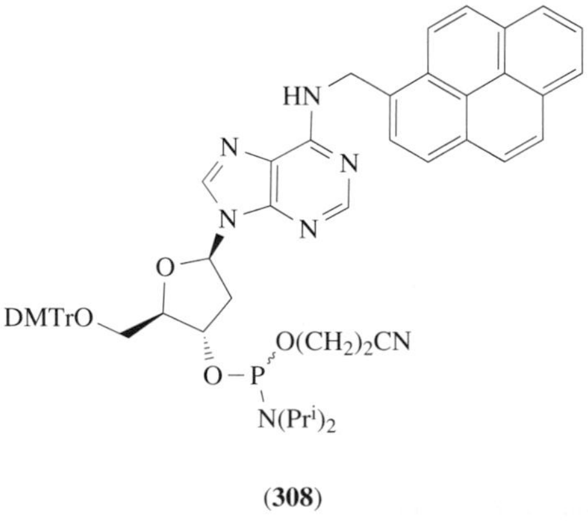

(308)

Like most PAH-phosphoramidite derivatives, the insertion of (**309**) and (**310**) at unique sites in oligonucleotides was accomplished by standard solid-phase synthesis.[413] An extended coupling time (75 s instead of 60 s) and a triple condensation of (**309**) or (**310**) per chain elongation step were, however, required to achieve a coupling efficiency of 98%. Of practical importance, the fluorescence emission spectra (excitation at 250 nm) of the modified oligonucleotides purified by HPLC showed a characteristic band at 389 nm caused by the presence of the phenanthrene chromophore.[413] These modified oligonucleotides can also serve as models in the study of mutagenesis induction by PAH.

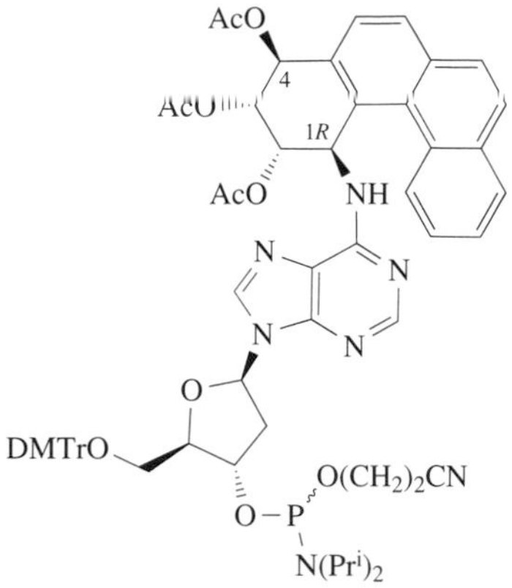

(**309**) 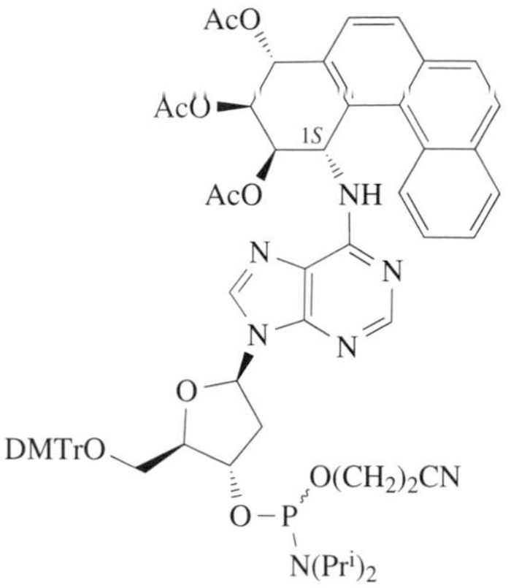                    (**310**)

Benzo[a]pyrene is one PAH that has been extensively studied in an attempt to define the mechanisms of genotoxicity. Once metabolically activated to a 7,8-dihydrodiol 9,10-epoxide derivative, this compound initiates mutagenesis by covalently binding to DNA.[402] Oligodeoxyribonucleotides containing stereospecifically and site-specifically placed lesions are particularly valuable in the study of single-adduct mutagenesis in bacterial and mammalian systems. DNA templates with defined adducts can also be utilized for *in vitro* studies to evaluate the replication capability of various DNA polymerases in the vicinity of the lesions. In this context, Harris and co-workers[414,415] and Kim *et al.*[416] reported the synthesis and incorporation of the halogenated deoxyribonucleoside phosphoramidites (**311**)[414,415] and (**312**)[414–416] into oligonucleotides. Treatment of solid-phase bound oligonucleotides carrying halogenated purine(s) with either D-(−)-phenylglycinol or with amines derived from (±)-*trans*-7,8-dihydroxy-*anti*-9,10-epoxy-8,9,10,11-tetrahydro[a]pyrene and (±)-*trans*-8,9-dihydroxy-*anti*-10,11-epoxy-8,9,10,11-tetrahydro[a]anthracene afforded the corresponding adducted oligomers.[414–416]

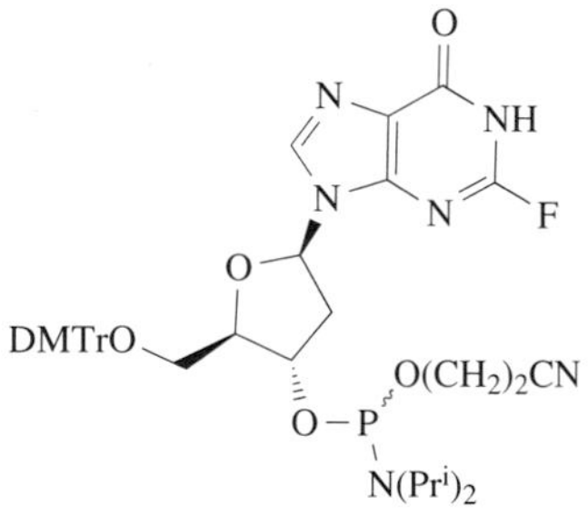

(**311**) 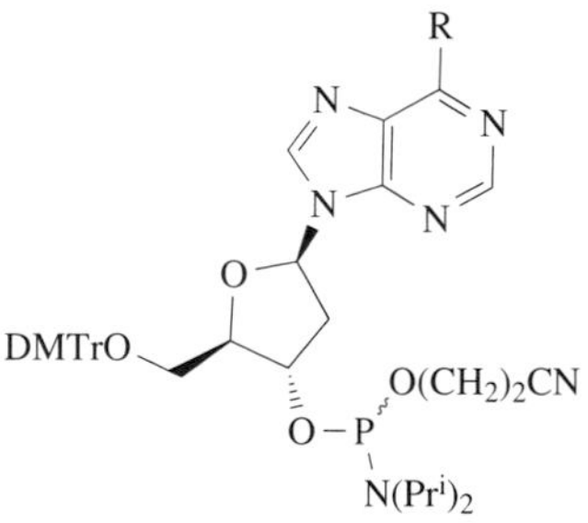                    (**312**)  R = Cl or F

Typically, the diastereoisomeric benzo[a]pyrene-adducted oligonucleotides (**313**) and (**314**) were purified by HPLC and polyacrylamide gel electrophoresis. The purified oligonucleotides were characterized by circular dichroism, capillary gel electrophoresis, and enzyme digestion.[416] This synthetic strategy offers the advantage of recovering unreacted aminotriol from reaction mixtures but suffers from the inherent difficulties of solid-phase reactions that are related to the monitoring of adduction reactions, the lower reactivity of immobilized 6-fluoropurines, and the postsynthesis separation of diastereoisomeric adducts.

An alternative route to the functionalization of oligonucleotides relies on the carbohydrate moiety of nucleosides. The following section will address this issue in detail.

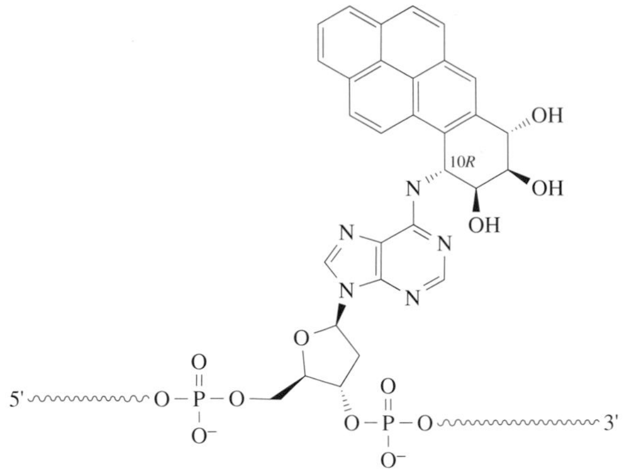

(313)

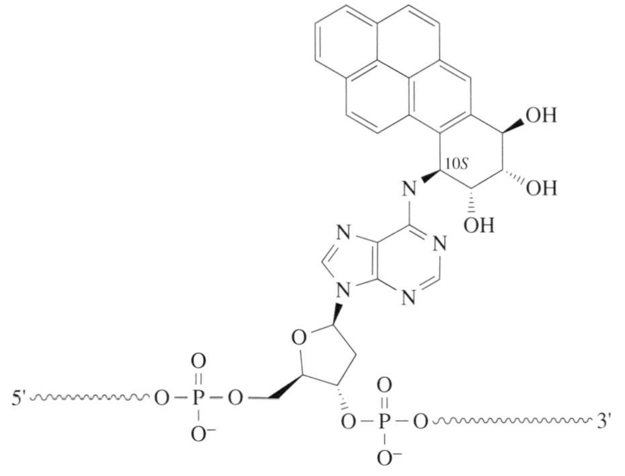

(314)

## 7.06.4　MODIFICATION OF THE CARBOHYDRATE PORTION OF NUCLEOSIDES FOR THE ATTACHMENT OF REPORTER AND CONJUGATE GROUPS TO DNA OLIGONUCLEOTIDES

### 7.06.4.1　Modification of Deoxyribonucleosides at C-1′

The novel 2′-deoxyuridine phosphoramidites (315), that carry aminoalkyl tethers at C-1′ of the sugar, have been synthesized and incorporated into oligodeoxyribonucleotides at selected sites.[417,418] Intercalators, such as anthraquinone and pyrene derivatives, were coupled to the aminoalkyl tethers to produce the conjugates (316). These conjugates and complementary oligoribonucleotides formed hybrids that were substrates for RNase H, and were thermodynamically more stable than the corresponding unmodified duplexes.[417,418] Conversely, hybrids composed of (316) and complementary DNA strands were less stable than the parent unmodified DNA–DNA duplexes.[417] Given that oligonucleotides analogous to (316) are more resistant to nuclease P1 and venom phosphodiesterases than unmodified oligomers, these conjugates may therefore find applications in antisense studies.[417]

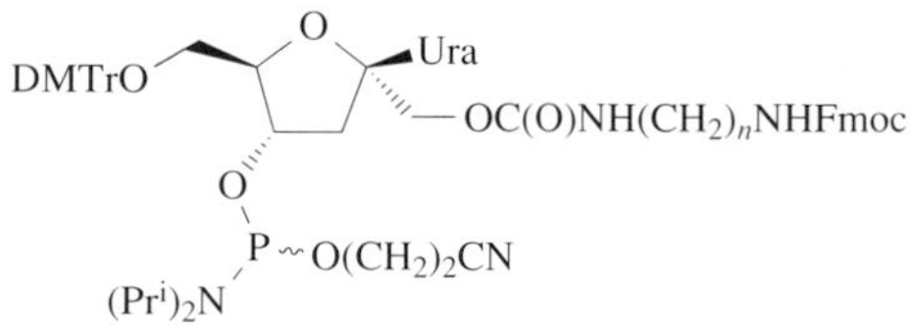

(315)  $n = 4, 6,$ or 8

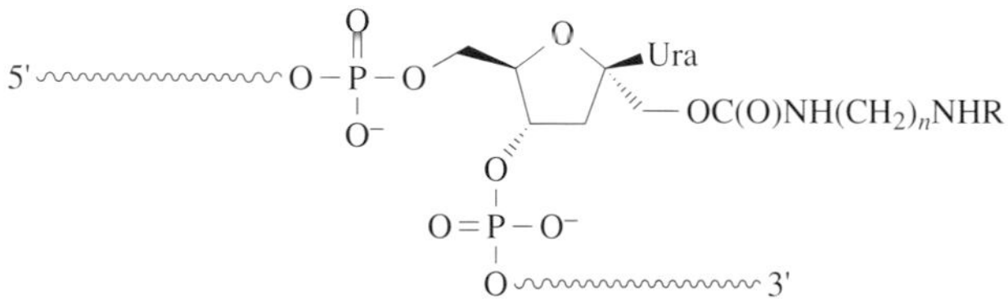

(316)  $n = 4, 6,$ or 8    R =

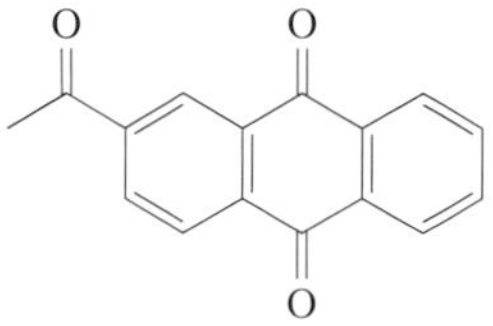

or

(316)  $n = 4, 6,$ or 8    R =

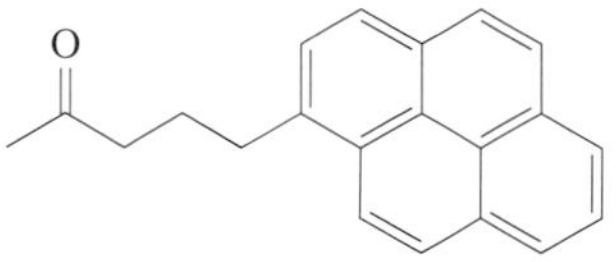

Along these lines, the deoxyribonucleoside phosphoramidite (317) has been prepared from 3′-deoxy-$\beta$-D-psicothymidine, and site-specifically incorporated into oligodeoxyribonucleotides by standard solid-phase synthesis.[419,420] No difference in coupling yields was detected between (317) and commercial unmodified deoxyribonucleoside phosphoramidites.[419] The 1′-*O*-levulinyl protection of the psicothymidine residues was removed by hydrazinolysis. Phosphitylation with the phosphoramidite (3) (Table 1) gave, after iodine oxidation, standard deprotection, and HPLC purification gave aminoalkylated oligonucleotides. These modified oligonucleotides were reacted with fluorescein isothiocyanate to generate the fluorescent conjugates (318).[420] These conjugates, when carrying one or two fluorescein groups, can still hybridize to complementary DNA strands and serve as primers for DNA polymerases.[420]

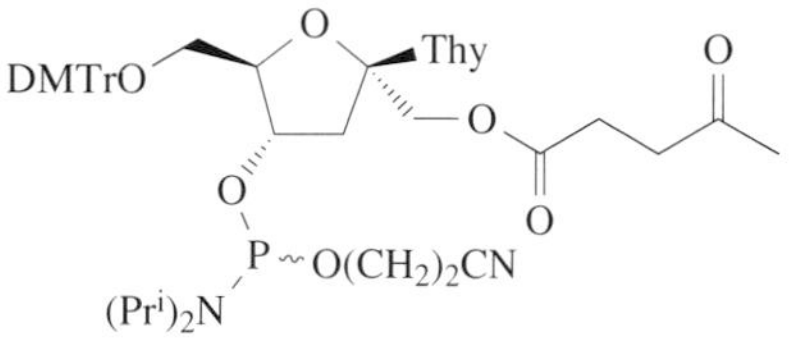

(317)

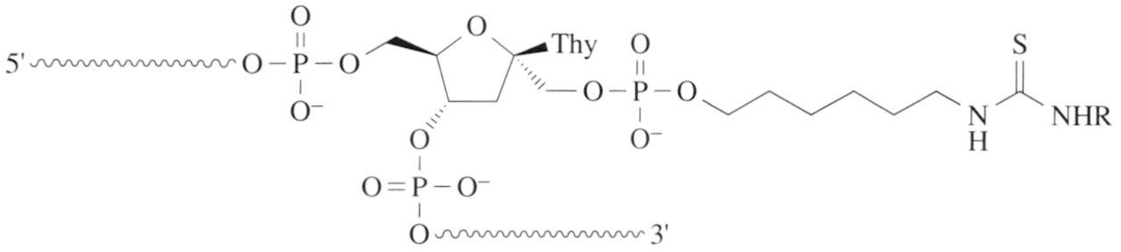

(318)  R = fluorescein

Alternatively, phosphitylation of the above psicothymidine residues with the phosphoramidite (77) led to tethered ester functions that were subsequently reacted with, for example, 1,4-diaminobutane to also give aminoalkylated oligonucleotides. Condensation of these modified oligonucleotides with dabsyl chloride afforded the conjugates (**319**).[137]

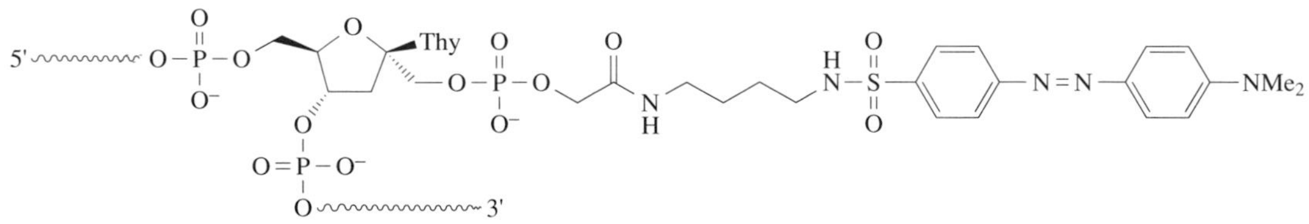

(**319**)

Particularly noteworthy is the C-1′-modified phosphoramidite (**320**) that has been synthesized for potential incorporation into DNA oligonucleotides and designed to deliver hydrolytically active metal complexes across the minor groove of DNA–RNA duplexes.[421]

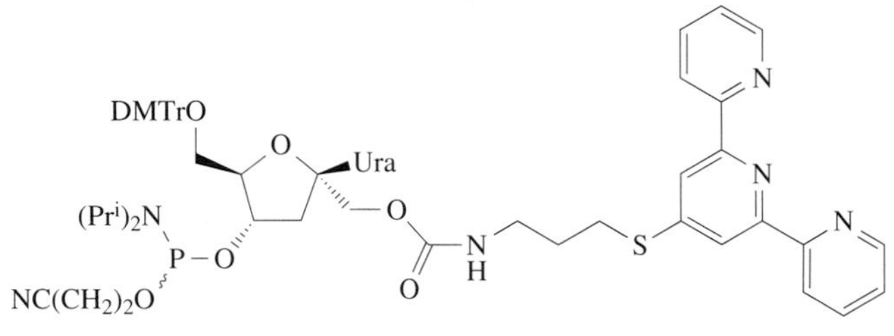

(**320**)

### 7.06.4.2 Functionalization of Nucleosides at C-2′

The anthraquinonylmethyl deoxyribonucleoside phosphoramidites (**321**) and (**322**) have been inserted at a unique site in self-complementary oligodeoxyribonucleotides.[422,423] When compared with identical unmodified DNA–DNA duplexes, self-complementary oligonucleotide–anthraquinone conjugates showed increased thermostability ($\Delta T_m$ = 3.5–8.5°C per modification) as a result of intercalation of the anthraquinone rings into predominantly B-type helices.[422,423] This method for enhancing the affinity of DNA oligonucleotides toward their complementary sequences appears simple and effective.

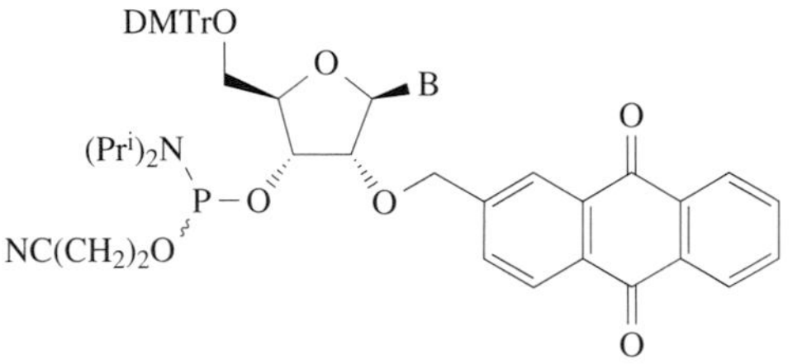

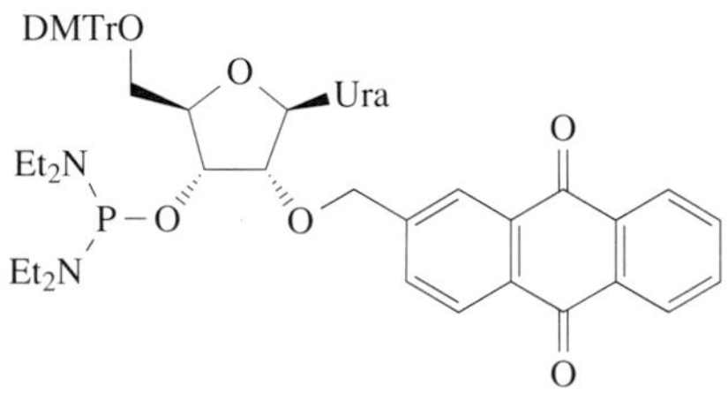

(**321**)  B = $N^6$-benzoyladenin-9-yl          (**322**)

Addition of the 2′-*O*-(1-pyrenylmethyl)uridine phosphordiamidite (**323**) at the 5′-terminus of a nonathymidylate (denoted as U*TTTTTTTT) produced, upon hybridization with poly(rA), a slight increase in the $T_m$ of the resulting duplex relative to that of the corresponding oligothymidylate–poly(rA) duplex ($\Delta T_m$ = 1.6 °C).[424,425] Furthermore, the pyrenylated duplex produced an intense fluorescence that is ca. 20-fold stronger than that generated by U*TTTTTTTT.[424,425] The intense fluorescence of the pyrenylated duplex may be either due to partial intercalation of the pyrene group into the adjacent base pair or to the release from stacking interactions that are fluorescence-quenching in the single-stranded form.[425] This unique feature should lead to increased sensitivity in the detection of sequence-specific gene probes.

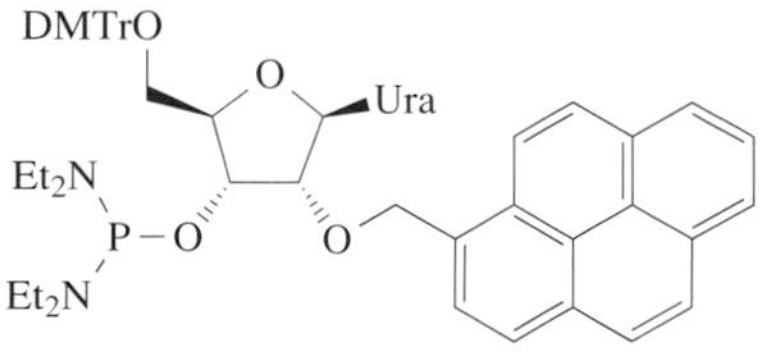

(323)

Alternatively, incorporation of the 2'-*O*-(2-anthracenemethyl)uridine phosphoramidite (324) into DNA oligomers (15-mers) at specific sites was achieved with a coupling efficiency exceeding 98%.[426] The deprotected and purified anthracene-containing oligonucleotides were incubated with either complementary DNA or RNA sequences. Modified DNA–DNA complexes exhibited a greater thermostability than unmodified DNA–DNA duplexes. Furthermore, the stability of these modified hybrids increased with the number of modifications. Conversely, DNA oligomers modified by the incorporation of (324) showed little or no effect on the thermodynamic stability of complexes formed with complementary RNA sequences. Paradoxically, anthracene-modified oligonucleotides showed increased fluorescence when hybridized to complementary RNA oligomers. Even though no increase in fluorescence intensity was observed for the modified DNA–DNA hybrids, anthracene-modified DNA oligomers may be useful as RNA-specific fluorescent probes.[426]

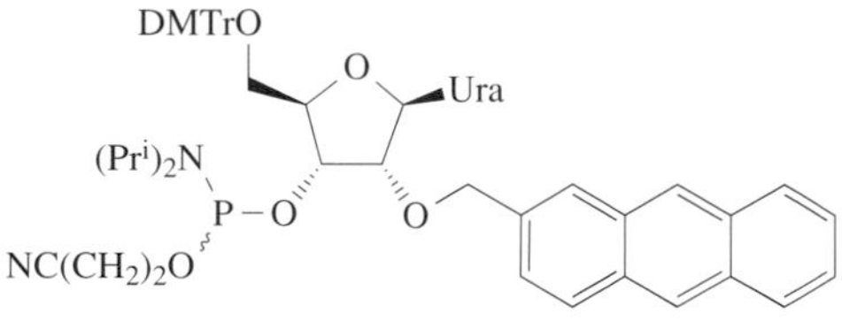

(324)

In a different context, the 2'-*O*-(bipyridinylalkyl)adenosine phosphoramidite (325) has been incorporated into DNA oligonucleotides to serve as ribozyme mimics by delivering metal complexes across the minor groove of DNA–RNA hybrids.[421] The relative hydrolytic abilities of these ribozyme mimics remain, however, to be determined.

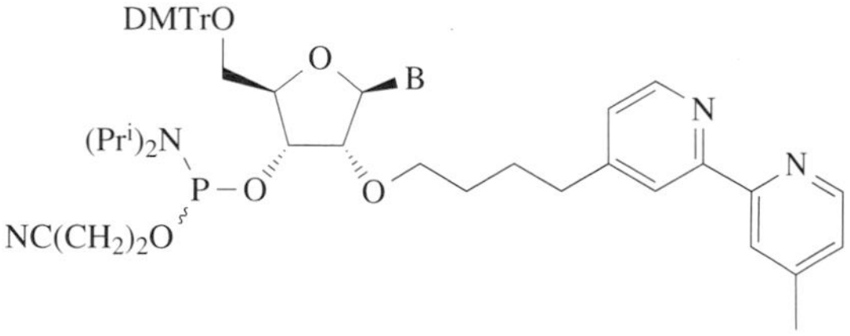

(325)  B = $N^6$-benzoyladenin-9-yl

The synthesis of oligonucleotides bearing methoxyoxalylamido functions has been reported.[427] The 2'-modified nucleoside phosphoramidite (326) was added to the 5'-terminus of an oligo-thymidylate. The coupling efficiency of the phosphoramidite (326) was only 75–80% even when the coupling time was extended to 33 min. The poor coupling yield is attributed to the steric hindrance of the rigid 2'-methoxyoxalylamido group.[427] Treatment of the solid-phase linked oligomer with 1,12-diaminododecane afforded, after elution from the support, the aminoalkylated oligonucleotide (327).[427] This modified oligomer is suitable for conjugation with various reporter groups.

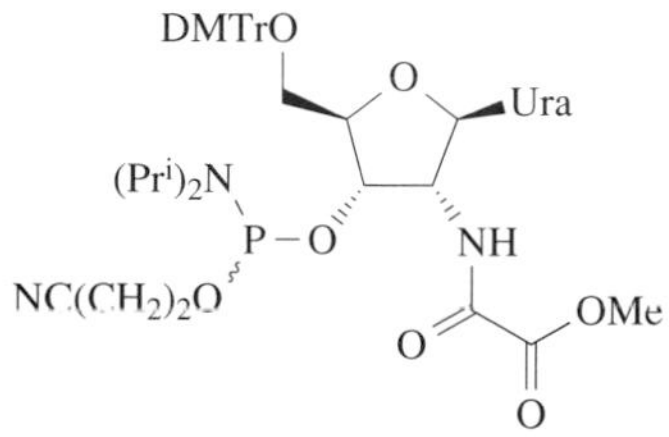

(326)

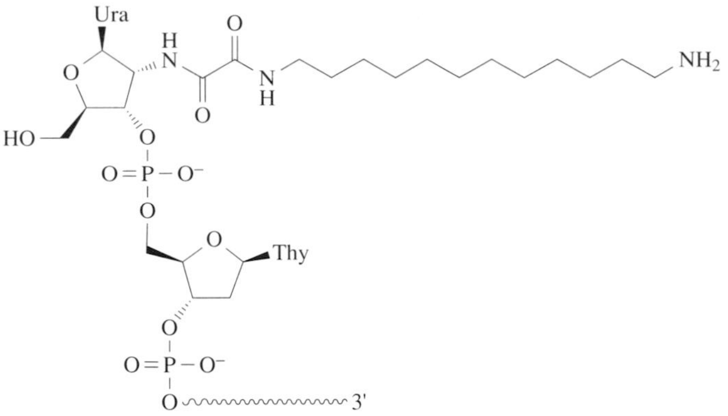

(327)

The functionalization of oligonucleotides with reporter groups via 2'-amido functions has, however, been questioned by Hendrix *et al.*[428] It was demonstrated that oligonucleotides containing 2'-acetamido-2'-deoxyuridine at terminal as well as central sites produced DNA–DNA, DNA–RNA, and RNA–RNA hybrids that were thermodynamically less stable than the corresponding unmodified duplexes. Steric hindrance of the 2'-acetamido group is most probably responsible for the significant destabilization of the modified hybrids.[428] Consequently, the 2'-acetamido function cannot be considered a good linkage for the attachment of reporter groups.

Incorporation of the ribonucleoside phosphoramidite (328) into oligoribonucleotides generated, after deprotection, the 2'-aminated RNA oligomers (329).[429,430] These oligomers can be conjugated to rhodamine upon reaction with rhodamine isothiocyanate[429] or converted to (330) by treatment with (2-isocyanato)ethyl-2-pyridyl disulfide.[430] The disulfide-containing oligoribonucleotide (330) can be further conjugated to other thiols or, after treatment with DTT, to thiol-reactive electrophiles such as monobromobimane.[430]

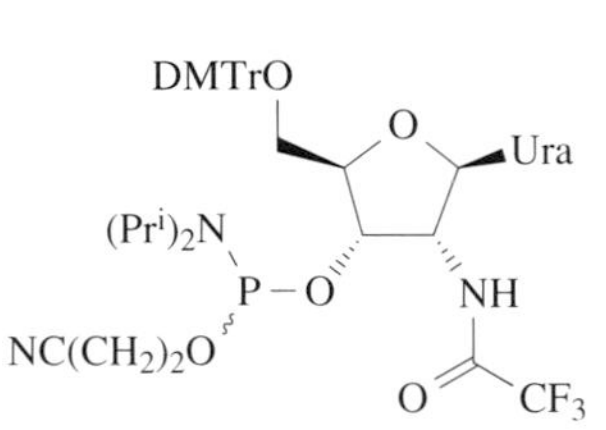

(328)

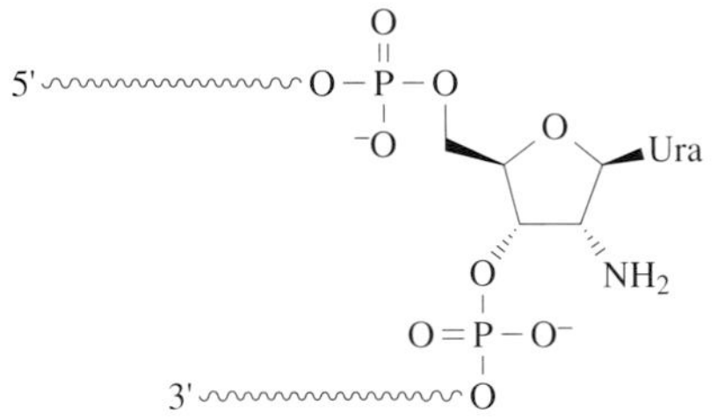

(329)

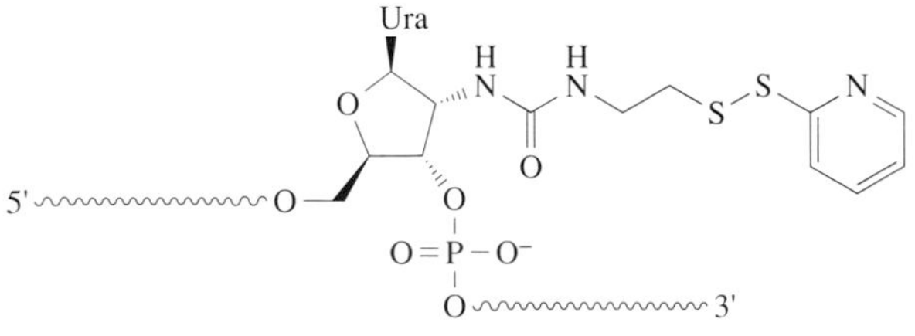

(330)

A versatile method for the introduction of different functional groups into oligonucleotides has been described.[431] Nucleoside phosphoramidite derivatives that carry lipophilic, intercalating, or tertiary amino functions, such as (331), can be incorporated at any desired location into oligonucleotides by standard solid-phase synthesis. Thermal denaturation studies have been performed with these modified oligonucleotides and indicated that the lipophilic *n*-octyl group had little, if any, effect on duplex stability. However, an intercalating substituent, such as 2-aminoanthraquinone, substantially increased duplex stability.[431] Likewise, the incorporation of *N,N*-diethylamino groups into oligonucleotides also increased duplex stability but to a lesser extent than that generated by most intercalators. Finally, the insertion of (331) into oligonucleotides resulted in little or no loss of sequence specificity.[431]

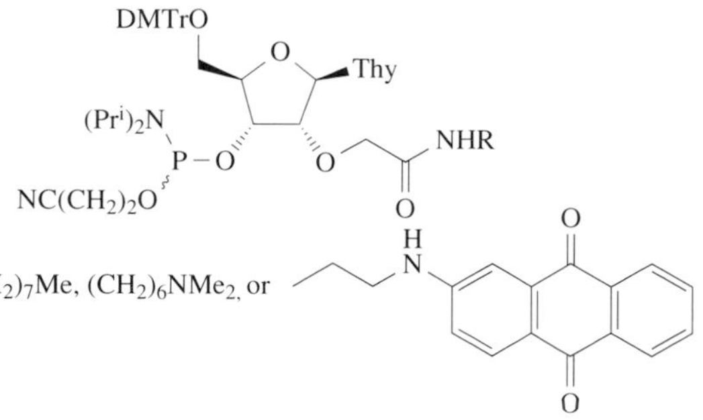

(331)   R = $(CH_2)_7Me$, $(CH_2)_6NMe_2$, or

Manoharan *et al.*[432] outlined the synthesis of the ribonucleoside phosphoramidite (332) from adenosine. The synthetic method consisted of the preferential 2′-*O*-alkylation of the nucleoside with an alkyl halide, followed by nucleobase/5′-*O*-protection, and 3′-*O*-phosphitylation. The insertion of (332) in oligonucleotides (20-mers) proceeded within 10–15 min with a coupling efficiency better than 95%. Standard deprotection produced 2′-*O*-aminoalkylated oligonucleotides which, after purification, were reacted with either fluorescein isothiocyanate or the *N*-hydroxysuccinimide ester of biotin, cholic acid, or digoxigenin-3-*O*-methylcarbonyl-ε-aminocaproic acid. This methodology also led to the functionalization of oligodeoxyribonucleoside phosphorothioates and 2′-*O*-methyl oligoribonucleotides.[432] It must be noted that the length of the 2′-aminoalkylated linker can be selected to facilitate minor groove modifications for potential biological applications.

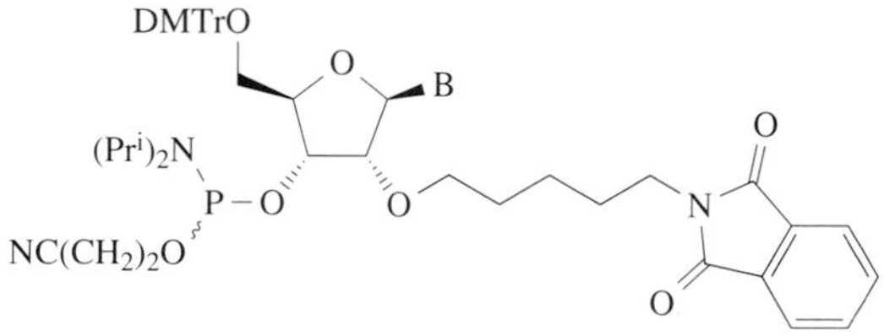

(332)   B = $N^6$-benzoyladenin-9-yl

An alternative approach to labeling oligonucleotides entails incorporation of the 2′-modified nucleoside phosphoramidite (333) at selected sites into oligonucleotides.[433] The *S*-trityl groups were, after oligonucleotide deprotection, cleaved with an aqueous silver nitrate solution under neutral conditions. Treatment with DTT removed silver ions from the thioalkylated oligonucleotides, and allowed conjugation with photoactivatable groups such as phenyl azides. These conjugates should be helpful in mapping interactions with RNA binding proteins.[433]

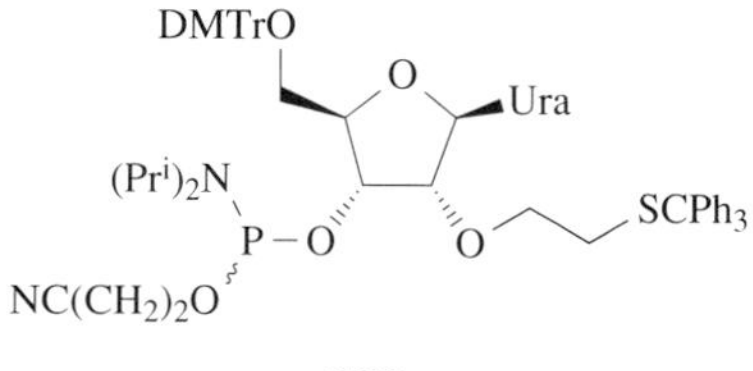

(333)

### 7.06.4.3   Modification of Deoxyribonucleosides at C-3′ and C-4′

The synthesis of 3′-*C*-(hydroxymethyl)thymidine and its conversion to the phosphoramidite (334) have been described.[434] The incorporation of (334) into oligodeoxyribonucleotides was effected in ca. 60% yield during a 12 min coupling reaction. The low coupling yield of (334) is probably due to steric hindrance caused by the bulky silyl protecting group. Hybridization studies that were performed with the hydroxymethylated oligonucleotides (335) revealed that the central insertion of one or two 3′-*C*-(hydroxymethyl)thymidine residues in a heptadecamer did not destabilize hybrid formation with an unmodified DNA complement. However, the insertion of one or two modified nucleosides near the 3′-terminus of the oligonucleotide caused a decrease in $T_m$ of 2 °C per modification.[434,435] The hydroxy functions of (335) can potentially serve as attachment sites for intercalating agents or lipophilic carriers.

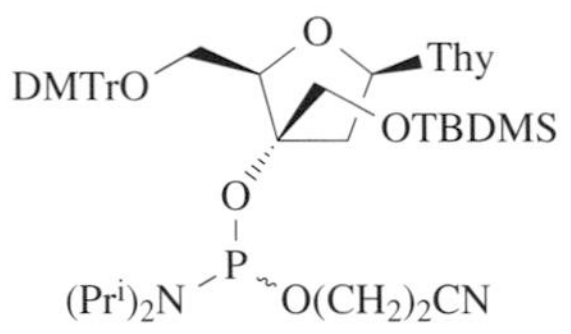

(334) 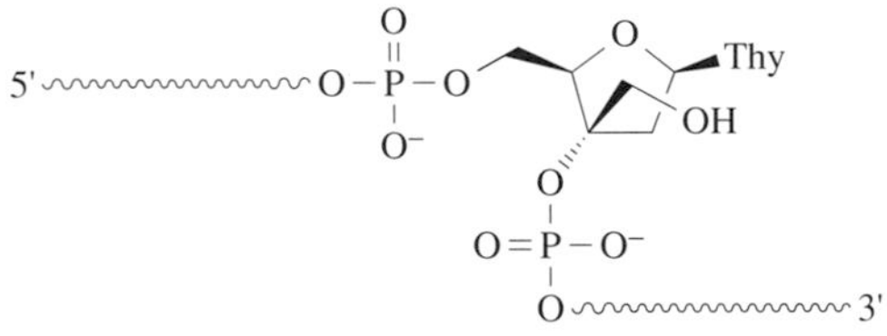                                              (335)

In this context, the preparation of 4′-*C*-(hydroxymethyl)thymidine and its phosphoramidite derivative (336) have been reported by Wengel and co-workers.[436,437] Like (334), the insertion of (336) at selected sites in oligonucleotides occurred with a lower coupling efficiency (ca. 90%) than with unmodified deoxyribonucleoside phosphoramidites, even when the coupling time was extended to 12 min. The modified oligonucleotides (338) exhibited hybridization properties similar to those of (335) with complementary DNA sequences. For example, the insertion of two modified nucleosides near the 3′-terminus of oligomers also caused a decrease in $T_m$ of 2 °C per modification.[436,437] Incidentally, the same decrease in $T_m$ was observed when an oligothymidylate carrying one modified nucleoside was hybridized to a complementary unmodified RNA oligomer under similar conditions.[437] Like the hydroxy functions of (335), those of (338) can also serve as conjugation sites for fluorescent probes or intercalating agents. It should, however, be pointed out that reporter groups attached to the hydroxy functions of (335) will be oriented toward the major grooves of DNA–DNA hybrids, whereas those conjugated to the hydroxy functions of (336) will be facing the minor grooves.[437]

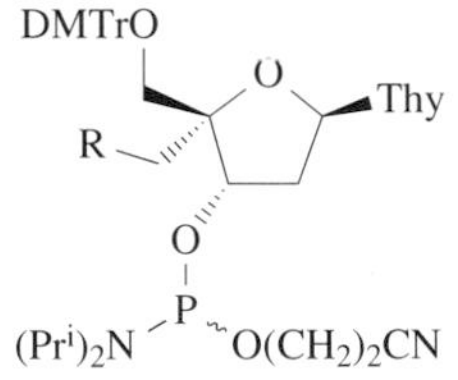

(336)  R = OTBDMS                    (338)  R = OH
(337)  R = NHC(O)CF₃                 (339)  R = NH₂

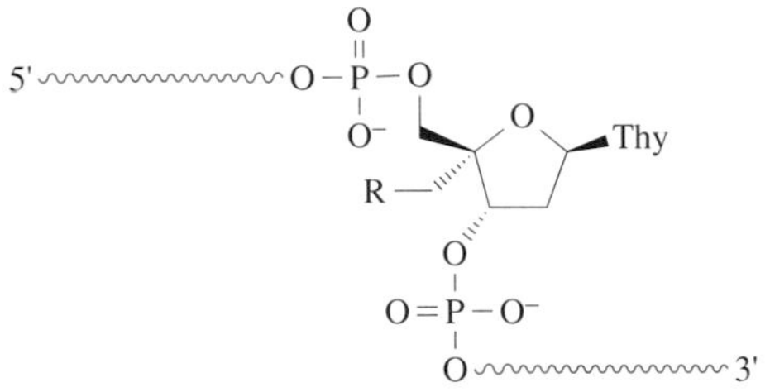

The 4′-*C*-(aminomethyl)thymidine phosphoramidite (337) has also been incorporated site-specifically into oligonucleotides.[438] The coupling efficiency of (337) was 97–99% within a 5 min coupling time. The modified oligonucleotides (339) had similar or better hybridization properties toward DNA and RNA than those of unmodified oligomers. Moreover, the modified oligonucleotides (339) showed better binding affinity for DNA than (338); centrally modified (339) caused a decrease in $T_m$ of less than 1 °C per modification.[438] The amino function(s) of (339), like the hydroxy groups of (335) and (338), can be conjugated to various reporter groups. Of interest, the C-4′ aminoalkylated thymidine phosphoramidite (340) has been synthesized by Maag *et al.*[439] and incorporated up to four times into short oligodeoxyribonucleotides. These modified oligomers have been selected to target a specific sequence in the genome of *Mycobacterium tuberculosis*. Particularly noteworthy is that unlike (336), the coupling efficiency of (340) was not lower than that of unmodified nucleoside phosphoramidites. After deprotection, the aminoalkylated oligonucleotides were treated with the *N*-hydroxysuccinimide ester of biotinyl-ε-caproic acid. The biotinylated probes (341) were tested for their ability to form dsDNA structures with complementary target sequences. It was shown that each additional 4′-modified thymidine decreased the $T_m$ of the corresponding DNA–DNA hybrid by 1.2 °C to 1.7 °C.[439] Furthermore, multilabeled probes provided a stronger detection signal than probes carrying only a single biotin.

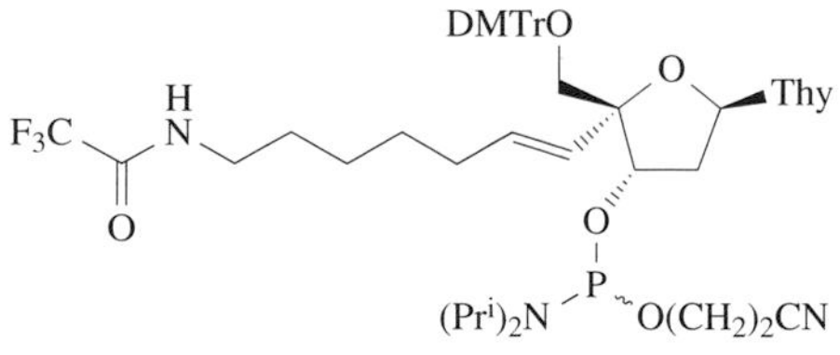

(340)

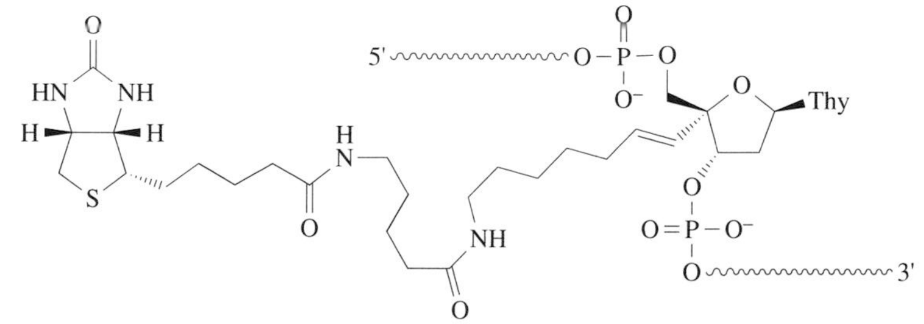

(341)

### 7.06.4.4 Derivatization of Nucleosides at C-5′

The conversion of thymidine to its 5′-(*S*)-epoxy derivative followed by reaction with methanolic ammonia, chemoselective protection steps, and phosphitylation, led to the modified deoxyribonucleoside phosphoramidite (342).[440] Incorporation of (342) into oligodeoxyribonucleotides near the 3′-end produced, after deprotection, aminoalkylated oligonucleotides that formed thermostable hybrids with complementary unmodified DNA and RNA sequences. Each modification decreased the $T_m$ of DNA–DNA and DNA–RNA hybrids by 0.3 °C and 0.4 °C, respectively. In addition, oligonucleotides carrying 5′-*C*-(aminomethyl)thymidine residues near the 3′-terminus were more resistant to snake venom phosphodiesterase than unmodified oligomers by a factor of at least 40.[440] Because these 5′-*C*-aminoalkylated oligonucleotides can potentially be conjugated to reporter groups, they appear quite promising toward the development of oligonucleotide therapeutics and diagnostics.

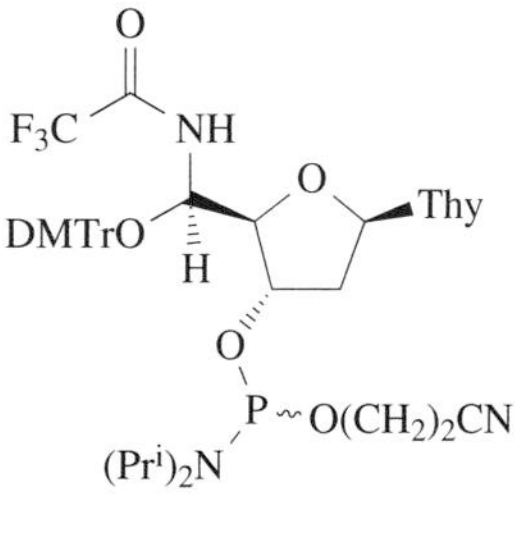

(342)

The synthetic challenge involved with the derivatization of oligonucleotides with nonradioactive markers has led to the development of methods for the generation of aliphatic amino groups at the 5′-terminus of these oligonucleotides. One way to achieve this goal has been proposed by Smith *et al.*[441–443] Their strategy involved the preparation of 5′-amino-5′-dioxythymidine and its conversion to the deoxyribonucleoside phosphoramidites (343)–(345). These phosphoramidites were then added to the 5′-terminus of oligonucleotides by solid-phase synthesis. Once deprotected, the 5′-aminated oligomers were reacted with fluorescein isothiocyanate, tetramethylrhodamine isothiocyanate, 4-fluoro-7-nitrobenzofurazan, or Texas Red to produce the conjugates (346)–(349). These fluorescent oligomers have been extensively used as primers in automated DNA sequence analysis according to the Sanger approach.[441,442,444]

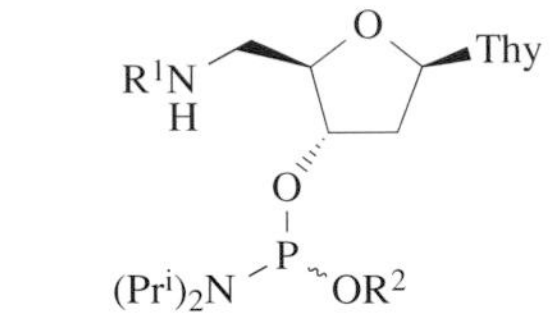

(343) $R^1 = C(O)CF_3$; $R^2 = Me$
(344) $R^1 = Fmoc$; $R^2 = Me$
(345) $R^1 = MMTr$; $R^2 = CH_2CH_2CN$

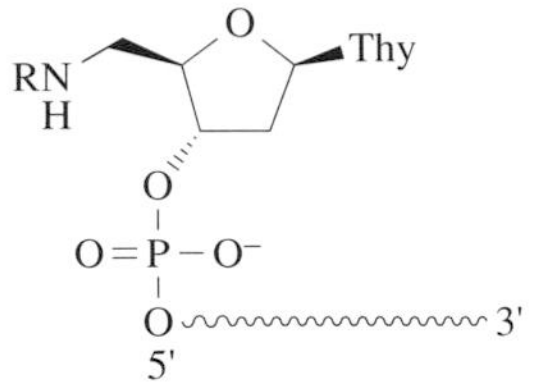

(346) R = fluorescein
(347) R = tetramethylrhodamine
(348) R = 7-nitrobenzofurazan-4-yl
(349) R = Texas Red

Synthetic approaches to the preparation of the 5′-aminated deoxyribonucleoside phosphoramidites (350)–(353) have been developed.[445] The phosphoramidite (351) has specifically been applied to the synthesis of 5′-d(H₂NCCGATATCGG) which, in turn, was reacted with an excess of the pentachlorophenyl ester of a tetrairidium cluster. Co-crystals of the complex formed between the metallated self-complementary decamer and *Eco*RV were obtained. These crystals were suitable for electron microscopy studies and X-ray crystallography.[445] Alternatively, addition of the phosphoramidite (345) to the 5′-terminus of oligonucleotides generated 5′-aminated oligomers that were suitable for condensation with the *N*-hydroxysuccinimide ester of a bathophenanthroline–ruthenium(II) complex. The metallated oligomers that resulted from this reaction were easily detected by time-resolved fluorescence techniques.[21]

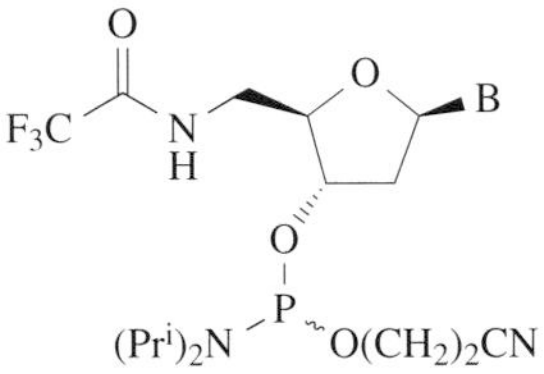

(350) B = thymin-1-yl
(351) B = $N^4$-benzoylcytosin-1-yl
(352) B = $N^6$-pivaloyladenin-9-yl
(353) B = $N^2$-isobutyrylguanin-9-yl

The synthesis of protected 5′-thio-2′,5′-dideoxyribonucleoside-3′-*O*-phosphoramidites (354) and their application to the preparation of 5′-thiolated oligodeoxyribonucleotides have been reported.[446,447] Like 5′-aminated oligomers, 5′-thiolated oligonucleotides, represented by structure (355), can react with a variety of electrophiles, heavy metals, and fluorescent markers.[96]

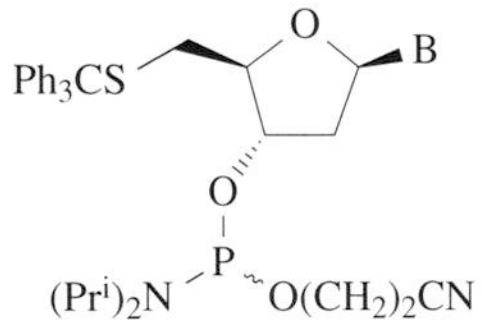

(354) B = thymine: $N^4$-benzoylcytosin-1-yl;
       $N^6$-pivaloyladenin-9-yl, or
       $N^2$-isobutyrylguanin-9-yl

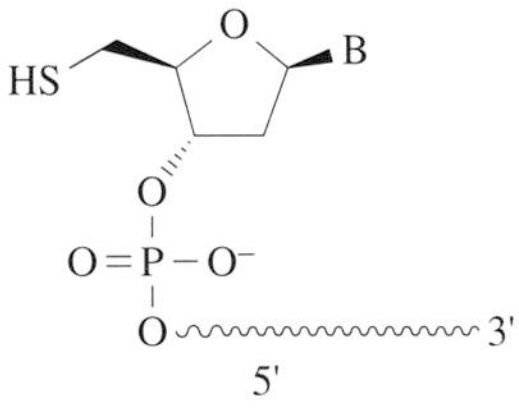

(355) B = thymine, adenine, cytosine,
       or guanine

Addition of the ribonucleoside phosphoramidite (356)[448] to the 5′-end of oligoribonucleotides led to the preparation of various fluorescent conjugates that include, for example, the 5′-pyrenylated RNA oligomers (357). These pyrenylated conjugates were shown to be very sensitive to the environment while causing only minimal perturbations on the thermodynamics of secondary and tertiary structure formation in RNA.[448] Fluorescent oligomers such as (357) are attractive probes for the binding and dynamics of RNA substrates, and should therefore be useful for studying oligomer–oligomer and oligomer–ribozyme interactions. Incidentally, the fluorescence of oligomers similar to (357) increased by a factor of ca. 25 upon binding to the ribozyme isolated from *Tetrahymena thermophilia*.[449]

An alternative route to the addition of a pyrene group to the 5′-terminus of an oligonucleotide involves the use of the (pyrenylmethyl)thymidine phosphordiamidite (358) during the last coupling

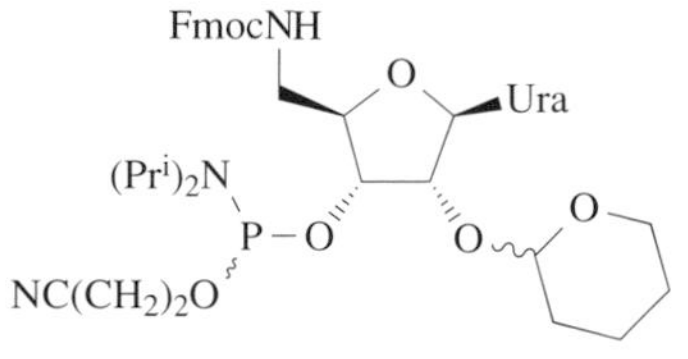

(356)

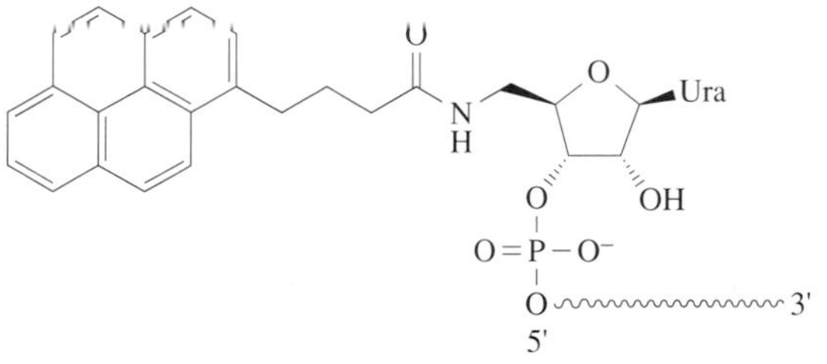

(357)

step of standard solid-phase synthesis.[275] The pyrene–oligonucleotide conjugate (359) was, after purification, hybridized to a complementary DNA oligomer. As in the case of (357), the fluorescence due to pyrene in (359) was not quenched upon hybrid formation but was in fact slightly enhanced.[275] Thus, oligonucleotide conjugates similar to (359) may serve as diagnostic probes.

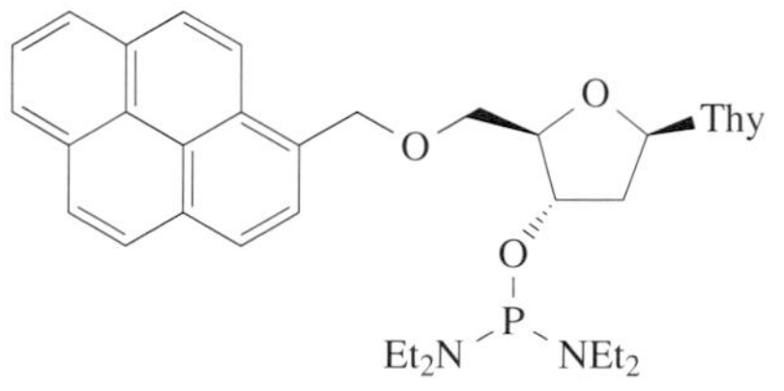

(358)

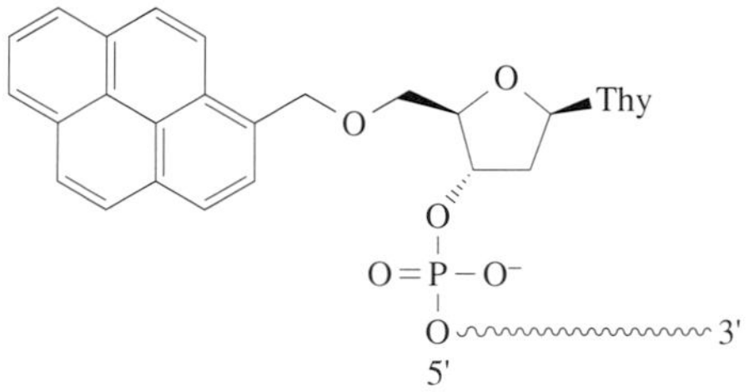

(359)

### 7.06.4.5 Modification of Carbonucleosides at C-6′

The multistep synthesis of the carbocyclic phosphoramidite (360) and its site-specific incorporation into DNA oligomers have been reported.[450] The purified aminoalkylated oligonucleotides (361) were hybridized to complementary RNA strands, and generated hybrids that were less stable ($\Delta T_{\mathrm{m}} = -1.1$ to $-1.9\,°C$ per modification) than the corresponding unmodified DNA–RNA duplexes.[450] Conjugation of (361) to a terpyridine-derived europium complex produced the modified oligonucleotides (362).[81] Interaction of a conjugate similar to (362) with a partially complementary RNA oligomer resulted in cleavage of the RNA (61% within 16 h at $37\,°C$) at or near a bulged site.[81] Since it is conceivable that cleavage within the duplex will result in sufficient destabilization for product release, these findings bring closer the development of genuine artificial ribonucleases capable of catalytic turnover. A similar strategy has been applied to oligonucleotides aminoalkylated through the C-2′ and C-5 positions of selected nucleosides.[81] Other potential targets for the functionalization of oligonucleotides are the internucleotidic linkages. These will be the subject of the next section.

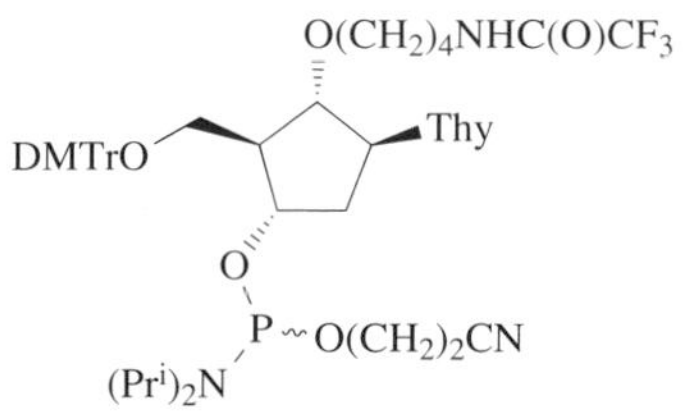

(360)

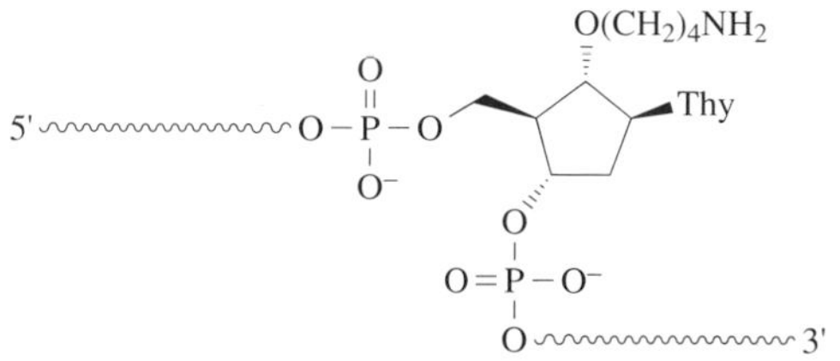

(361)

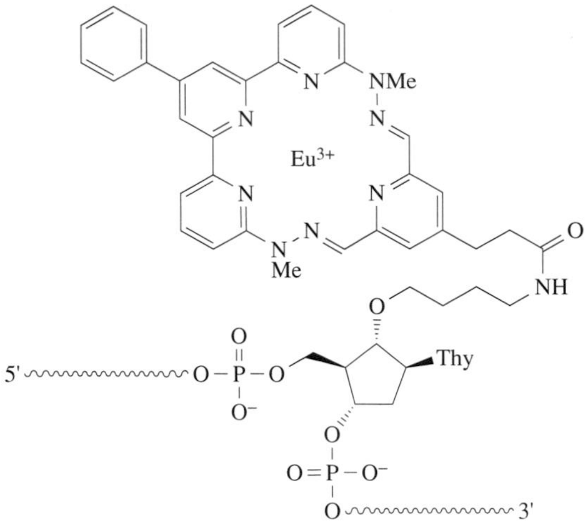

(362)

## 7.06.5 ATTACHMENT OF REPORTER AND CONJUGATE GROUPS TO THE INTERNUCLEOTIDIC PHOSPHODIESTER FUNCTIONS OF DNA OLIGONUCLEOTIDES

The incorporation of intercalators into oligonucleotides at the 3′-terminus has been described by Asseline and Thuong.[226,234] Their method entailed phosphitylation of the solid support (363) with the acridine deoxyribonucleoside phosphoramidite (364) in the presence of 1*H*-tetrazole. Upon completion of oligonucleotide synthesis, the modified oligomers were subsequently released from the support by treatment with DTT under alkaline conditions to give (365). This synthetic strategy also permits the multiple insertion of various ligands within an oligonucleotidic chain.

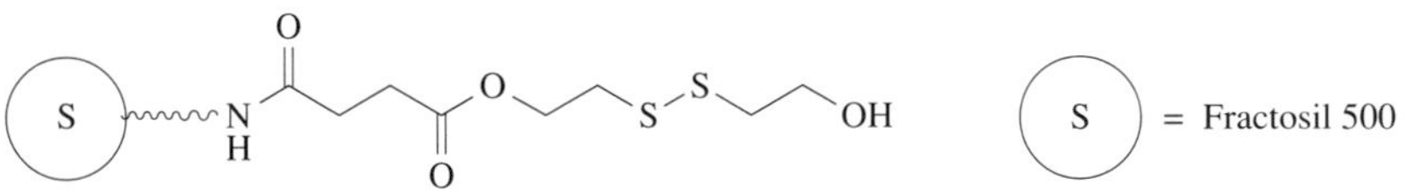

(363)

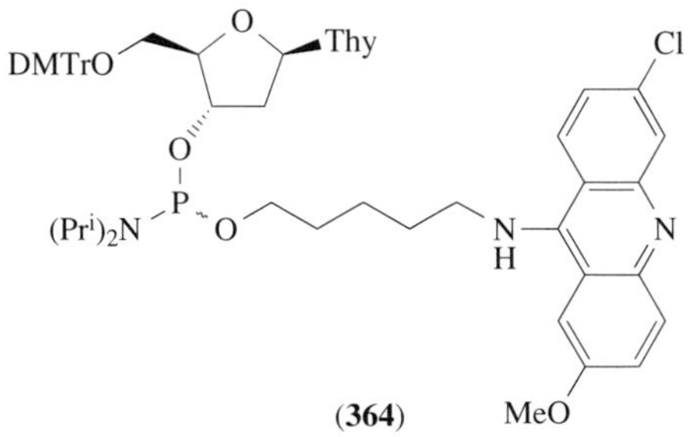

(364)

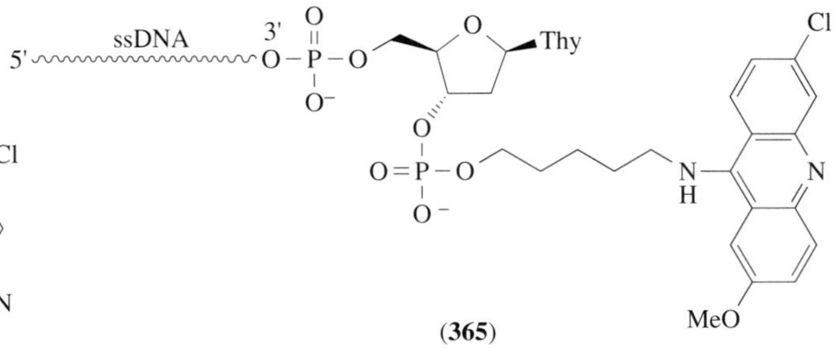

(365)

Seliger *et al.*[451] reported the synthesis of the deoxyribonucleoside phosphoramidite (366) and the dinucleotidic phosphoramidite (367) for the insertion of aminoalkyl functions into oligonucleotides. The monomeric and dimeric phosphoramidites (366) and (367) produced coupling yields of ca. 99% and 87%, respectively. The aminoalkylated oligonucleotides that resulted from these couplings were reacted with the *N*-hydroxysuccinimide ester of biotin.[451] A biotinylated hairpin loop was immobilized on an avidin-coated solid support to demonstrate the potentiality for dsDNA manipulations.[451]

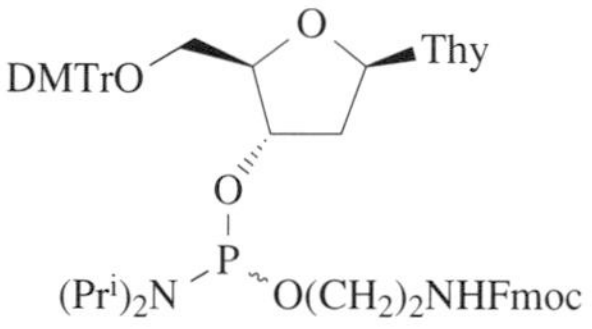

(366) 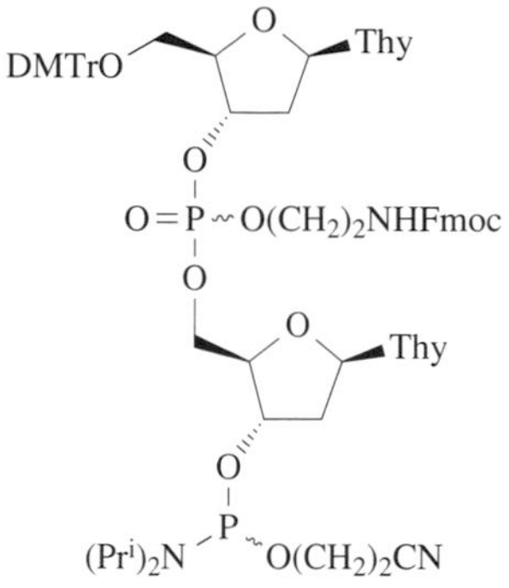          (367)

While an acridine derivative was covalently linked to an internucleotidic phosphodiester function according to the phosphotriester approach for oligonucleotide synthesis in solution,[452] Jäger *et al.*[453] demonstrated that aminoalkylated phosphoramidate oligonucleotides can be prepared by routine solid-phase synthesis using deoxyribonucleoside 3′-*O*-(*N*,*N*-diisopropylamino)methoxyphosphine monomers. This strategy was exemplified by the oxidation of the dinucleoside phosphite triester (368) with iodine and 1,5-diaminopentane to give the aminoalkylated phosphoramidate dimer (369) (Equation (3)). The modified dimer was subsequently conjugated to the intercalator 6-chloro-9-(*p*-chlorophenoxy)-2-methoxyacridine.[453]

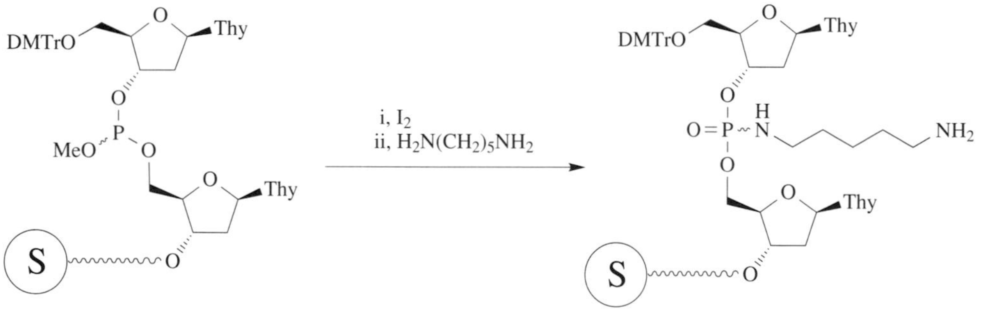

(3)

(368)          (369)

Given the stability of phosphoramidate linkages to the conditions used during standard phosphoramidite solid-phase oligonucleotides synthesis, the combination of deoxyribonucleoside phosphoramidite and *H*-phosphonate monomers has allowed the insertion of cholesteryl,[454,455] tetramethylpiperidine-*N*-oxyl,[456–458] or phenazinyl di-*N*-oxide[459] phosphoramidate links at specific sites in oligonucleotides. The usefulness of this approach has previously been demonstrated by Letsinger *et al.*[460] in a report that delineated the preparation of cationic oligonucleotides. In a similar manner, Agrawal and Tang[461] and others[462,463] introduced either (*N*-1-trifluoroacetyl)hexanediamine arms at selected sites along the oligonucleotidic chains[461] or ethylenediamine and hexamethylenediamine tethers at the 5′-terminus of oligonucleotides.[462,463] After complete deprotection, the aminoalkylated phosphoramidate oligomers were conjugated to ligands such as fluorescein isothiocyanate, rhodamine isothiocyanate, or activated esters of biotin. This method provided multiple labeling capabilities and, thus, increased sensitivity for diagnostic purposes.[461]

It should also be pointed out that the *H*-phosphonate method for oligonucleotide synthesis and the Atherton–Todd reaction have emerged as an efficient approach to the synthesis of oligodeoxyribonucleotides aminoalkylated at specific sites through internucleotidic phosphoramidate linkages.[464] Incidentally, this approach has been successfully applied to the synthesis of viologen-tagged DNA oligonucleotides.[465]

The functionalization of oligonucleotides with two different reporter groups has been described by Agrawal and Zamecnik.[466] Typically, oligonucleotide assembly began with the incorporation of a nucleoside *H*-phosphonate monomer followed by oxidation with carbon tetrachloride and (*N*-1-trifluoroacetyl)hexanediamine to generate a phosphoramidate link. Standard deoxyribonucleoside phosphoramidite monomers were then employed until completion of the synthesis. The last oxidation step was effected with a sulfurizing reagent[467–469] to generate, after oligonucleotide deprotection, a phosphorothioate diester suitable for reaction with monobromobimane. The aminoalkyl function of the purified bimane-labeled oligomer was then reacted with fluorescein isothiocyanate. Such bifunctionalized oligomers may find application in cellular uptake experiments or toward the targeting of specific messenger RNAs.

It should also be noted that the sequence-specific incorporation of a thiol tether into oligodeoxyribonucleotides could also be achieved by oxidation of an internucleotide *H*-phosphonate in the presence of cystamine[470,471] or *N*-triphenylacetylcystamine.[472] After purification of the cystamine-containing oligonucleotide (**370**), a sulfhydryl residue was unmasked by treatment with DTT. The newly generated sulfhydryl function is amenable to modification by a variety of thiol-specific reporter groups, such as the iodoacetamido-PROXYL spin label, and fluorophores like 4-chloro-7-nitrobenz-2-oxa-1,3-diazole, monobromobimane, 1-pyrenemethyl iodoacetate, 6-acryloyl-2-(dimethylamino)naphthalene, and 7-(diethylamino)-3-(4′-maleimidylphenyl)-4-methylcoumarin to give the modified oligonucleotides (**371**).[470,471] Of practical importance, it should be mentioned that the hydrophobic triphenyl moiety of the stereogenic center in (**370**) facilitated the separation of diastereoisomers for sequences shorter than 20 nucleotides.[472]

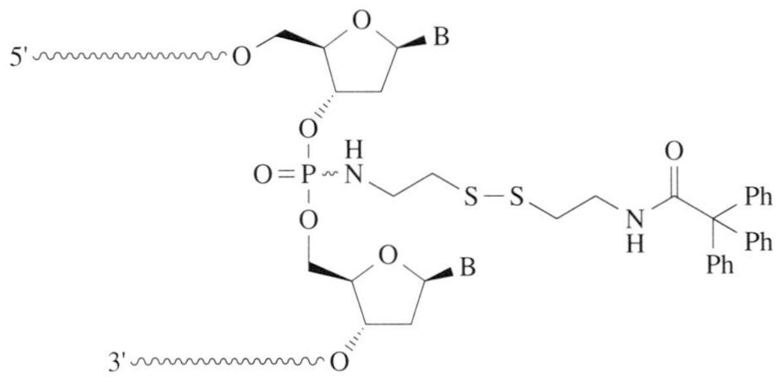

(**370**)

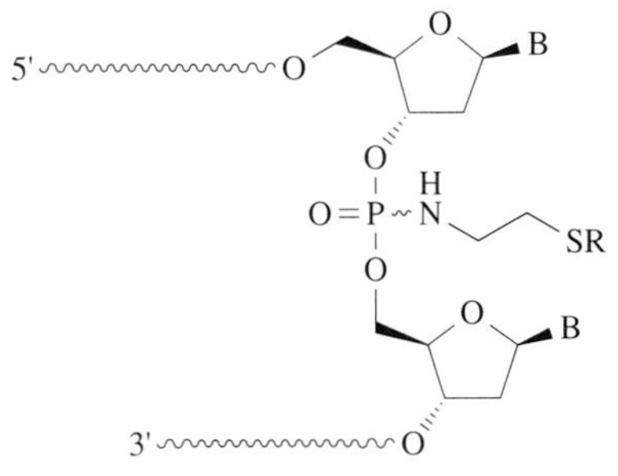

(**371**)   R = reporter group

A small Hoechst-like DNA groove-binding fluorophore carrying a terminal bromoacetamido function has been reacted with each of the two diastereoisomers isolated from a dodecamer analogous to (**370**). In the presence of tris(2-carboxyethyl)phosphine or DTT, the conjugates (**372**) were individually obtained as isomers A and B.[473] One isomer of the DNA–fluorophore conjugates (isomer B) formed a duplex with a complementary ssDNA target that exhibited increased thermostability relative to the corresponding unmodified DNA–DNA duplex and enhanced fluorescence. Specifically, duplex stability was increased by 2–3 °C, while the fluorescence quantum yield was enhanced by a factor of 4. Conversely, isomer A showed reduced helix stability and only slight changes in fluorescence intensity under the same conditions.[473]

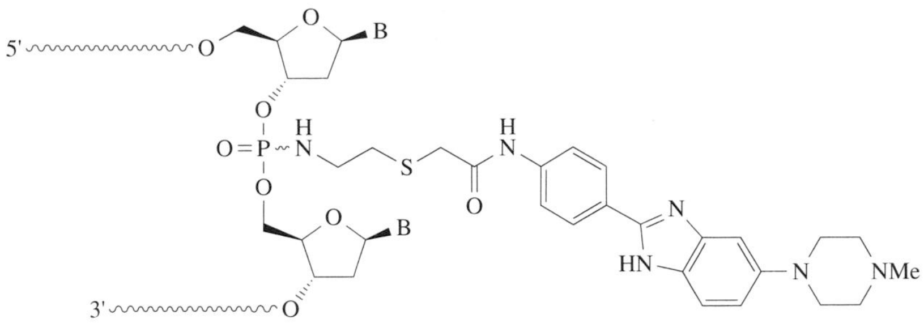

(372)

The lipophilic deoxyribonucleoside phosphoramidites (373)–(375) have been integrated into DNA at selected sites, and sulfurized to generate internucleotidic thiono phosphotriester linkages.[474,475] Oligonucleoside phosphorothioates modified with a single cholesteryl thiono phosphotriester linkage at the 3'-terminus demonstrated potent antihuman cytomegalovirus activity along with enhanced nuclease resistance and cellular association.[474]

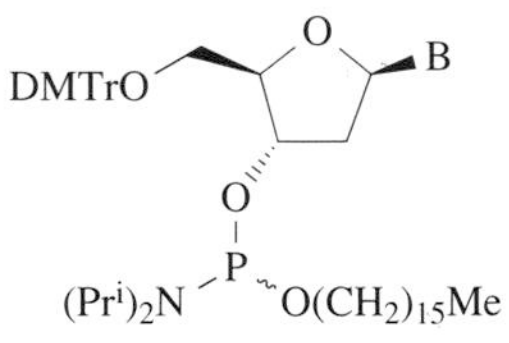

(373)

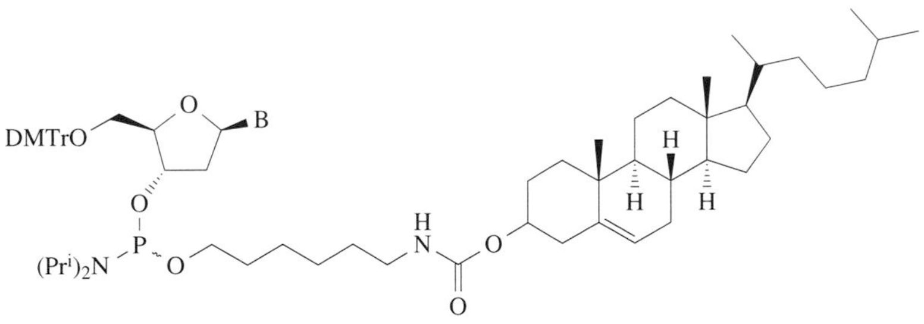

(374)

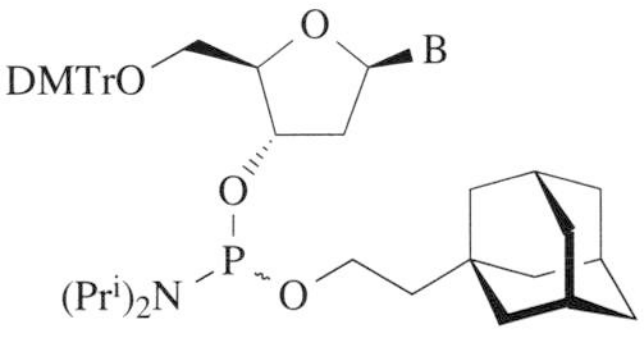

(375)

B = T, A^tBPA, C^tBPA, or G^tBPA; tBPA = *t*-butylphenoxyacetyl

Particularly noteworthy is the creation of a genuine abasic lesion via site-specific incorporation of the nonnucleosidic phosphoramidite (376) into a DNA oligomer. Irradiation of the modified oligomer, followed by reaction with 9-aminoellipticine in the presence of sodium cyanoborohydride, generated the oligonucleotide conjugate (377).[476] This conjugate was prepared in an effort to understand better the structural features of a duplex containing the lesion, and its plausible role in the DNA repair process normally effected by apurinic (AP) endonuclease. It would appear that the accessibility of an abasic lesion to degradation by AP endonuclease is impaired due to the presence of intercalated aminoellipticine in the target DNA.[476]

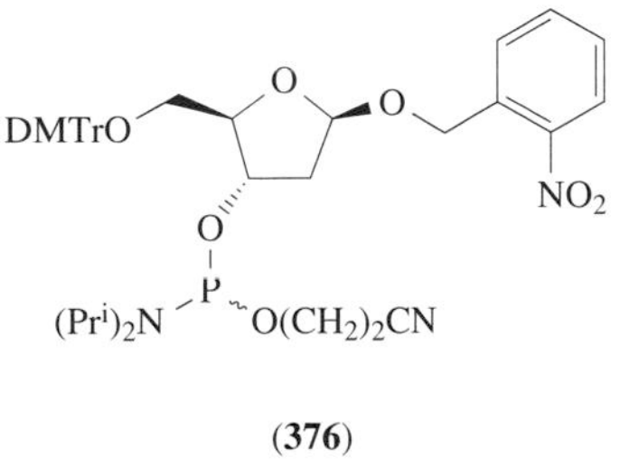

(376)

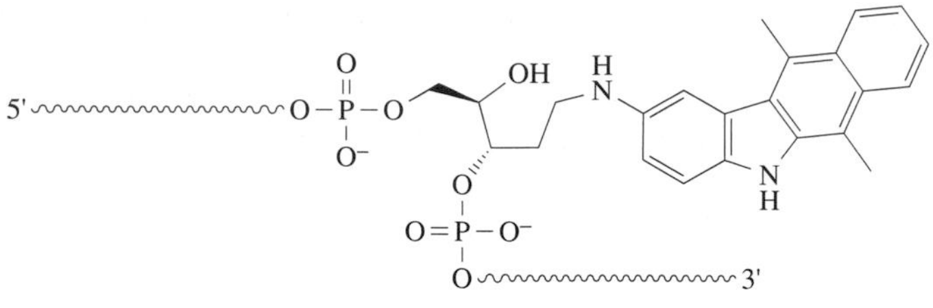

(377)

Numerous reporter and conjugate groups have been introduced at specific sites in oligonucleotides carrying phosphorothioate diester functions.[263,477–480] For example, the reaction of an oligonucleoside phosphorothioate with *N*-dansylaziridine afforded the dansylated phosphorothioate oligomer (378) in high yields.[481] Substituting monobromobimane for *N*-dansylaziridine led to the bimane-labeled phosphorothioate oligomer (379).[482,483] Interestingly, bimane labels have been incorporated at multiple sites into oligonucleoside phosphorothioates while these were still embedded in the polyacrylamide gel matrix after electrophoresis. Oligonucleotides containing 200–400 labels can be detected with the naked eye at low femtomolar concentrations.[482–484]

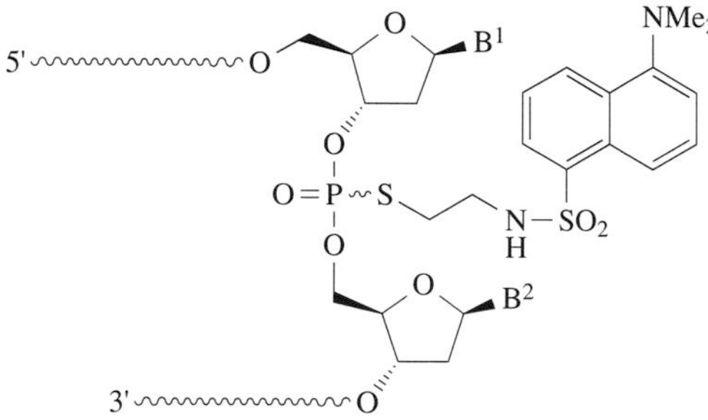

(378)

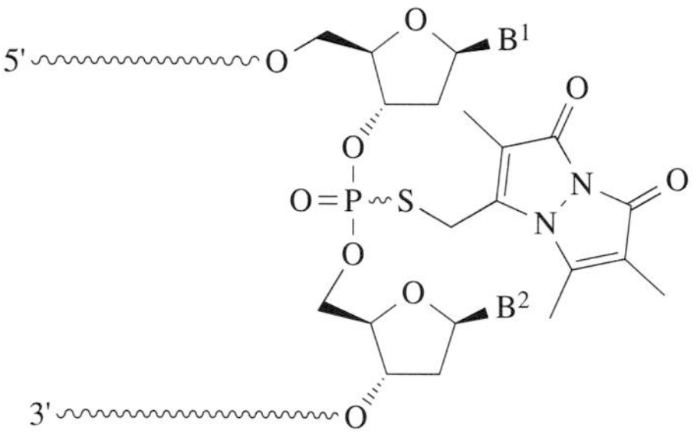

(379)

B$^1$ or B$^2$ = thymine, cytosine, adenine, or guanine

The nonbridging sulfur of internucleotidic phosphorothioate diesters has sufficient thiol character to form adducts with maleimides. In fact, dithymidylyl phosphorothioate reacted with *N*-(1-pyrene) maleimide to generate the corresponding fluorescent conjugate as evidenced by fluorescence emission spectroscopy.[485]

The nucleophilicity of phosphorothioate diesters has also been exploited in the sequence determination of DNA and RNA.[486–489] Specifically, the reaction of oligodeoxyribonucleotides or oligoribonucleotides having randomly distributed phosphorothioate diesters with iodoethanol or 2,3-epoxy-1-propanol generated phosphorothioate triesters that ultimately promoted intramolecular cleavage of oligonucleotidic chains.[486–488] This approach represents an alternative to the current DNA sequencing procedures.

Given that oligonucleotides can potentially target cancer cells by interacting with overexpressed or unique genes found in these cells, the chemical synthesis of the dinucleoside (*o*-carboranyl) methylphosphonate (380) has been described.[400,490]

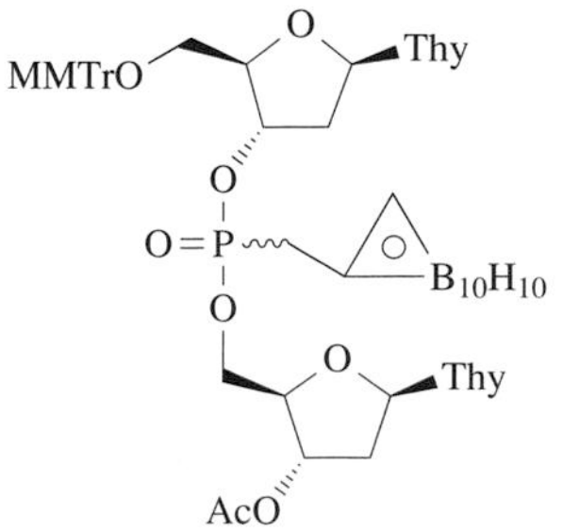

(380)

In consideration of potential BNCT applications, the dinucleotide (380) should serve, after selective deprotection and phosphitylation of its 3′-hydroxy function with 2-cyanoethyl *N*,*N*-diisopropylchlorophosphoramidite, as a building block in solid-phase oligonucleotide synthesis. The lipophilic nature of *o*-carboranyl residues should facilitate the transport of boronated oligonucleotides inside cells to the nucleus. In this context, microdosimetric calculations suggest that there is a five times greater chance for cell killing when boron-10 is primarily confined to the cell nucleus than when it is uniformly distributed throughout the cell.[490]

## 7.06.6 SUMMARY

Advances in the chemical synthesis of modified nucleosides and oligonucleotides have tremendously facilitated the incorporation of reporter and conjugate groups into DNA. Nonnucleosidic and nucleosidic phosphoramidite derivatives have been invaluable for the direct and indirect functionalization of oligonucleotides. These phosphoramidites have allowed fine control of the number, type, and placement of selected ligands within any given oligonucleotidic sequence, and thus stress the versatility of the chemistry.

It is very likely that the attachment of reporter and conjugate groups to DNA and RNA will continue to expand our knowledge of nucleic acid functions and provide the tools necessary for the development of even more sensitive probes and better therapeutic agents.

## 7.06.7 REFERENCES

1. S. L. Beaucage and M. H. Caruthers, in "Bioorganic Chemistry: Nucleic Acids," ed. S. M. Hecht, Oxford University Press, New York, 1996, p. 36.
2. S. L. Beaucage and R. P. Iyer, *Tetrahedron*, 1992, **48**, 2223.
3. S. L. Beaucage and R. P. Iyer, *Tetrahedron*, 1993, **49**, 1925.
4. J. M. Coull, H. L. Weith, and R. Bischoff, *Tetrahedron Lett.*, 1986, **27**, 3991.
5. R. Bischoff, J. M. Coull, and F. E. Regnier, *Anal. Biochem.*, 1987, **164**, 336.
6. M. Bengström, A. Jungell-Nortamo, and A.-C. Syvänen, *Nucleosides Nucleotides*, 1990, **9**, 123.
7. A.-C. Syvänen, M. Bengtström, J. Tenhunen, and H. Söderlund, *Nucleic Acids Res.*, 1988, **16**, 11 327.
8. A. Chollet, *Nucleosides Nucleotides*, 1990, **9**, 957.
9. A. I. H. Murchie, R. M. Clegg, E. von Kitzing, D. R. Duckett, S. Diekmann, and D. M. J. Lilley, *Nature*, 1989, **341**, 763.
10. F. F. Chehab and Y. W. Kan, *Proc. Natl. Acad. Sci. USA*, 1980, **86**, 9178.
11. Y. Shoji, S. Akhtar, A. Periasamy, B. Herman, and R. L. Juliano, *Nucleic Acids Res.*, 1991, **19**, 5543.
12. S. Akhtar, S. Basu, E. Wickstrom, and R. L. Juliano, *Nucleic Acids Res.*, 1991, **19**, 5551.
13. M. K. Dewanjee, A. K. Ghafouripour, R. K. Werner, A. N. Serafini, and G. N. Sfakianakis, *Bioconj. Chem.*, 1991, **2**, 195.
14. N. Sinha and S. Striepeke, in "Oligonucleotides and Analogues: A Practical Approach," ed. F. Eckstein, IRL Press, Oxford, 1991, p. 185.
15. Y. Zhang, M. Y. Coyne, S. G. Will, C. H. Levenson, and E. S. Kawasaki, *Nucleic Acids Res.*, 1991, **19**, 3929.
16. S. Agrawal, C. Christodoulou, and M. Gait, *Nucleic Acids Res.*, 1986, **14**, 6227.
17. R. J. Kaiser, S. L. MacKellar, R. S. Vinayak, J. Z. Sanders, R. A. Saavedra, and L. E. Hood, *Nucleic Acids Res.*, 1989, **17**, 6087.
18. P. C. Emson, H. Arai, S. Agrawal, C. Christodoulou, and M. J. Gait, *Methods Enzymol.*, 1989, **168**, 753.
19. R. A. Cardullo, S. Agrawal, C. Flores, P. C. Zamecnik, and D. E. Wolf, *Proc. Natl. Acad. Sci. USA*, 1988, **85**, 8790.
20. B. A. Connolly, *Nucleic Acids Res.*, 1987, **15**, 3131.
21. W. Bannwarth, D. Schmidt, R. L. Stallard, C. Hornung, R. Knorr, and F. Müller, *Helv. Chim. Acta*, 1988, **71**, 2085.
22. R. K. Gaur, *Nucleosides Nucleotides*, 1991, **10**, 895.
23. N. D. Sinha and R. M. Cook, *Nucleic Acids Res.*, 1988, **16**, 2659.
24. M. I. Dawson, A. N. Jina, S. Torkelson, S. Rhee, S. Moore, D. A. Zarling, and P. D. Hobbs, *Nucleic Acids Res.*, 1990, **18**, 1099.
25. J. Van Ness, S. Kalbfleisch, C. R. Petrie, M. W. Reed, J. C. Tabone, and N. M. J. Vermeulen, *Nucleic Acids Res.*, 1991, **19**, 3345.
26. J. Van Ness and L. Chen, *Nucleic Acids Res.*, 1991, **19**, 5143.
27. M. Chatterjee and S. E. Rokita, *J. Am. Chem. Soc.*, 1990, **112**, 6397.
28. M. Chatterjee and S. E. Rokita, *J. Am. Chem. Soc.*, 1991, **113**, 5116.
29. R. K. Saiki, P. S. Walsh, C. H. Levenson, and H. A. Erlich, *Proc. Natl. Acad. Sci. USA*, 1989, **86**, 6230.
30. S. Agrawal, *Tetrahedron Lett.*, 1989, **30**, 7025.
31. C. Connell, S. Fung, C. Heiner, J. Bridgham, V. Chakerian, E. Heron, B. Jones, S. Menchen, W. Mordan, M. Raff, M. Recknor, L. Smith, J. Springer, S. Woo, and M. Hunkapiller, *Biotechniques*, 1987, **5**, 342.
32. R. I. Amann, L. Krumholz, and D. A. Stahl, *J. Bact.*, 1990, **172**, 762.
33. K. B. Mullis, *Sci. Am.*, 1990, **262**, 56.
34. L. M. Mitchell and C. R. Merril, *Anal. Biochem.*, 1989, **178**, 239.
35. A. J. Cocuzza and R. J. Zagursky, *Nucleosides Nucleotides*, 1991, **10**, 413.

36. P. S. Nelson, R. Sherman-Gold, and R. Leon, *Nucleic Acids Res.*, 1989, **17**, 7179.
37. M. J. De Vos, A. Van Elsen, and A. Bollen, *Nucleosides Nucleotides*, 1994, **13**, 2245.
38. M. Endo, Y. Saga, and M. Komiyama, *Tetrahedron Lett.*, 1994, **35**, 5879.
39. S. I. Antsypovich, T. S. Oretskaya, E. M. Volkov, E. A. Romanova, V. N. Tashlitsky, M. Blumenfeld, and Z. A. Shabarova, *Nucleosides Nucleotides*, 1996, **15**, 923.
40. E. R. Kandimalla, A. N. Manning, G. Venkataraman, V. Sasisekharan, and S. Agrawal, *Nucleic Acids Res.*, 1995, **23**, 4510.
41. M. A. Reynolds, T. A. Beck, R. I. Hogrefe, A. McCaffey, L. J. Arnold, Jr., and M. M. Vaghefi, *Bioconj. Chem.*, 1992, **3**, 366.
42. S. Teigelkamp, S. Ebel, D. W. Will, T. Brown, and J. D. Beggs, *Nucleic Acids Res.*, 1993, **21**, 4651.
43. T. H. Smith, M. A. Kent, S. Muthini, S. J. Boone, and P. S. Nelson, *Nucleosides Nucleotides*, 1996, **15**, 1581.
44. K. Fukui, M. Morimoto, H. Segawa, K. Tanaka, and T. Shimidzu, *Bioconj. Chem.*, 1996, **7**, 349.
45. A. Guzaev, H. Salo, A. Azhayev, and H. Lönnberg, *Bioconj. Chem.*, 1996, **7**, 240.
46. C. Behrens, K. H. Petersen, M. Egholm, J. Nielsen, O. Buchardt, and O. Dahl, *Bioorg. Med. Chem. Lett.*, 1995, **5**, 1785.
47. J.-L. Mergny, A. S. Boutorine, T. Garestier, F. Belloc, M. Rougée, N. V. Bulychev, A. A. Koshkin, J. Bourson, A. V. Lebedev, B. Valeur, N. T. Thuong, and C. Hélène, *Nucleic Acids Res.*, 1994, **22**, 920.
48. A. Oser and G. Valet, *Angew. Chem., Int. Ed. Engl.*, 1990, **29**, 1167.
49. M. Chatterjee and S. E. Rokita, *J. Am. Chem. Soc.*, 1994, **116**, 1690.
50. H. Maruenda and M. Tomasz, *Bioconj. Chem.*, 1996, **7**, 541.
51. T. Zhu and S. Stein, *Bioconj. Chem.*, 1994, **5**, 312.
52. R. G. W. Anderson, B. A. Kamen, K. G. Rothberg, and S. W. Lacey, *Science*, 1992, **255**, 410.
53. C.-H. Tung, K. J. Breslauer, and S. Stein, *Nucleic Acids Res.*, 1993, **21**, 5489.
54. D. A. Geselowitz and R. D. Neumann, *Bioconj. Chem.*, 1995, **6**, 502.
55. F. Schubert, A. Knaf, U. Möller, and D. Cech, *Nucleosides Nucleotides*, 1995, **14**, 1437.
56. M. Manoharan, K. L. Tivel, L. K. Andrade, V. Mohan, T. P. Condon, C. F. Bennett, and P. D. Cook, *Nucleosides Nucleotides*, 1995, **14**, 969.
57. M. Manoharan, L. K. Johnson, C. F. Bennett, T. A. Vickers, D. J. Ecker, L. M. Cowsert, S. M. Freier, and P. D. Cook, *Bioorg. Med. Chem. Lett.*, 1994, **4**, 1053.
58. F. P. Svinarchuk, D. A. Konevetz, O. A. Pliasunova, A. G. Pokrovsky, and V. V. Vlassov, *Biochimie*, 1993, **75**, 49.
59. G. Tarrasón, D. Bellido, R. Eritja, S. Vilaró, and J. Piulats, *Antisen. Res. Dev.*, 1995, **5**, 193.
60. D. A. Geselowitz and L. M. Neckers, *Antisen. Res. Dev.*, 1992, **2**, 17.
61. V. K. Kansal, T. Huynh-Dinh, and J. Igolen, *Tetrahedron Lett.*, 1988, **29**, 5537.
62. T. Tanaka, T. Sakata, K. Fujimoto, and M. Ikehara, *Nucleic Acids Res.*, 1987, **15**, 6209.
63. T. Tanaka, S. Tamatsukuri, and M. Ikehara, *Tetrahedron Lett.*, 1987, **28**, 2611.
64. S. G. Lokhov, M. A. Podyminogin, D. S. Sergeev, V. N. Silnikov, I. V. Kutyavin, G. V. Shishkin, and V. P. Zarytova, *Bioconj. Chem.*, 1992, **3**, 414.
65. A. S. Levina, M. V. Berezovskii, A. G. Venjaminova, M. I. Dobrikov, M. N. Repkova, and V. F. Zarytova, *Biochimie*, 1993, **75**, 25.
66. T. S. Godovikova, V. D. Knorre, G. A. Maksakova, and V. N. Sil'nikov, *Bioconj. Chem.*, 1996, **7**, 343.
67. L. Wachter, J.-A. Jablonski, and K. L. Ramachandran, *Nucleic Acids Res.*, 1986, **14**, 7985.
68. M.-J. De Vos, A. Cravador, J.-P. Lenders, S. Houard, and A. Bollen, *Nucleosides Nucleotides*, 1990, **9**, 259.
69. S. Beck, T. O'Keeffe, J. M. Coull, and H. Köster, *Nucleic Acids Res.*, 1989, **17**, 5115.
70. E. Bonfils, C. Depierreux, P. Midoux, N. T. Thuong, M. Monsigny, and A. C. Roche, *Nucleic Acids Res.*, 1992, **20**, 4621.
71. M. Komiyama and T. Inokawa, *J. Biochem.*, 1994, **116**, 719.
72. M. Komiyama, T. Inokawa, and K. Yoshinari, *J. Chem. Soc., Chem. Commun.*, 1995, 77.
73. A. Murakami, M. Mukae, S. Nagahara, Y. Konishi, H. Ide, and K. Makino, *Free Radical Res. Commun.*, 1993, **19**, S117.
74. B. Mestre, G. Pratviel, and B. Meunier, *Bioconj. Chem.*, 1995, **6**, 466.
75. C. Casas, C. J. Lacey, and B. Meunier, *Bioconj. Chem.*, 1993, **4**, 366.
76. P. Bigey, G. Pratviel, and B. Meunier, *Nucleic Acids Res.*, 1995, **23**, 3894.
77. M. Pitié, C. Casas, C. J. Lacey, G. Pratviel, J. Bernadou, and B. Meunier, *Angew. Chem., Int. Ed. Engl.*, 1993, **32**, 557.
78. J. F. Ramalho Ortigão, A. Rück, K. C. Gupta, R. Rösch, R. Steiner, and H. Seliger, *Biochimie*, 1993, **75**, 29.
79. D. Magda, R. A. Miller, J. L. Sessler, and B. L. Iverson, *J. Am. Chem. Soc.*, 1994, **116**, 7439.
80. J. Hall, D. Hüsken, U. Pieles, H. E. Moser, and R. Häner, *Chem. Biol.*, 1994, **1**, 185.
81. J. Hall, D. Hüsken, and R. Häner, *Nucleic Acids Res.*, 1996, **24**, 3522.
82. D. Magda, M. Wright, R. A. Miller, J. L. Sessler, and P. I. Sansom, *J. Am. Chem. Soc.*, 1995, **117**, 3629.
83. T. Le Doan, D. Praseuth, L. Perrouault, M. Chassignol, N. T. Thuong, and C. Hélène, *Bioconj. Chem.*, 1990, **1**, 108.
84. L. Mastruzzo, A. Woisard, D. D. F. Ma, E. Rizzarelli, A. Favre, and T. Le Doan, *Photochem. Photobiol.*, 1994, **60**, 316.
85. T. Le Doan, L. Perrouault, C. Hélène, M. Chassignol, and N. T. Thuong, *Biochemistry*, 1986, **25**, 6736.
86. E. I. Frolova, E. M. Ivanova, V. F. Zarytova, T. V. Abramova, and V. V. Vlassov, *FEBS Lett.*, 1990, **269**, 101.
87. K. Matsumura, M. Endo, and M. Komiyama, *J. Chem. Soc., Chem. Commun.*, 1994, 2019.
88. M. Komiyama, N. Takeda, T. Shiiba, Y. Takahashi, Y. Matsumoto, and M. Yashiro, *Nucleosides Nucleotides*, 1994, **13**, 1297.
89. D. B. Hall, R. E. Holmlin, and J. K. Barton, *Nature*, 1996, **382**, 731.
90. Y. Jenkins and J. K. Barton, *J. Am. Chem. Soc.*, 1992, **114**, 8736.
91. C. J. Murphy, M. R. Arkin, Y. Jenkins, N. D. Ghatlia, S. H. Bossmann, N. J. Turro, and J. K. Barton, *Science*, 1993, **262**, 1025.
92. T. Ihara, Y. Maruo, S. Takenaka, and M. Takagi, *Nucleic Acids Res.*, 1996, **24**, 4273.
93. B. D. Gildea, J. M. Coull, and H. Köster, *Tetrahedron Lett.*, 1990, **31**, 7095.
94. B. A. Connolly and P. Rider, *Nucleic Acids Res.*, 1985, **13**, 4485.

95. W. Ansorge, A. Rosenthal, B. Sproat, C. Schwager, J. Stegemann, and H. Voss, *Nucleic Acids Res.*, 1988, **16**, 2203.
96. B. S. Sproat, B. Beijer, P. Rider, and P. Neuner, *Nucleosides Nucleotides*, 1988, **7**, 651.
97. R. Blanks and L. W. McLaughlin, *Nucleic Acids Res.*, 1988, **16**, 10 283.
98. R. Eritja, A. Pons, M. Escarceller, E. Giralt, and F. Albericio, *Tetrahedron* 1991, **47**, 4113.
99. N. J. Ede, G. W. Tregear, and J. Haralambidis, *Bioconj. Chem.*, 1994, **5**, 373.
100. B. G. de la Torre, A. M. Aviñó, M. Escarceller, M. Royo, F. Albericio, and R. Eritja, *Nucleosides Nucleotides*, 1993, **12**, 993.
101. P. Kumar, D. Bhatia, R. C. Rastogi, and K. C. Gupta, *Bioorg. Med. Chem. Lett.*, 1996, **6**, 683.
102. W. H. A. Kuijpers and C. A. A. van Boeckel, *Tetrahedron*, 1993, **49**, 10 931.
103. K. Mori, C. Subasinghe, C. A. Stein, and J. S. Cohen, *Nucleosides Nucleotides*, 1989, **8**, 649.
104. D. S. Jones, J. P. Hachmann, S. A. Osgood, M. S. Hayag, P. A. Barstad, G. M. Iverson, and S. M. Coutts, *Bioconj. Chem.*, 1994, **5**, 390.
105. W. Ansorge, B. Sproat, J. Stegemann, C. Schwager, and N. Zenke, *Nucleic Acids Res.*, 1987, **15**, 4593.
106. W. Ansorge, B. S. Sproat, J. Stegemann, and C. Schwager, *J. Biochem. Biophys. Methods*, 1986, **13**, 315.
107. A. S. Boutorine, H. Tokuyama, M. Takasugi, H. Isobe, E. Nakamura, and C. Hélène, *Angew. Chem., Int. Ed. Engl.*, 1994, **33**, 2462.
108. R. Blanks and L. W. McLaughlin, in "Oligonucleotides and Analogues: A Practical Approach," ed. F. Eckstein, IRL Press, Oxford, 1991, p. 241.
109. E. Fahy, G. R. Davis, L. J. DiMichele, and S. S. Ghosh, *Nucleic Acids Res.*, 1993, **21**, 1819.
110. R. K. Gaur, P. Sharma, and K. C. Gupta, *Nucleic Acids Res.*, 1989, **17**, 4404.
111. A. Kumar, S. Advani, H. Dawar, and G. P. Talwar, *Nucleic Acids Res.*, 1991, **19**, 4561.
112. A. Kumar and S. Malhotra, *Nucleosides Nucleotides*, 1992, **11**, 1003.
113. J. Teare and P. Wollenzien, *Nucleic Acids Res.*, 1989, **17**, 3359.
114. A. Kumar and S. Advani, *Nucleosides Nucleotides*, 1992, **11**, 999.
115. N. T. Thuong and M. Chassignol, *Tetrahedron Lett.*, 1987, **28**, 4157.
116. N. T. Thuong and U. Asseline, in "Oligonucleotides and Analogues: A Practical Approach," ed. F. Eckstein, IRL Press, Oxford, 1991, p. 283.
117. C. Cazenave, M. Chevrier, N. T. Thuong, and C. Hélène, *Nucleic Acids Res.*, 1987, **15**, 10 507.
118. E. Bonfils and N. T. Thuong, *Tetrahedron Lett.*, 1991, **32**, 3053.
119. U. Asseline, E. Bonfils, R. Kurfürst, M. Chassignol, V. Roig, and N. T. Thuong, *Tetrahedron*, 1992, **48**, 1233.
120. J.-C. François, T. Saison-Behmoaras, C. Barbier, M. Chassignol, N. T. Thuong, and C. Hélène, *Proc. Natl. Acad. Sci. USA*, 1989, **86**, 9702.
121. J.-C. François, T. Saison-Behmoaras, M. Chassignol, N. T. Thuong, and C. Hélène, *J. Biol. Chem.*, 1989, **264**, 5891.
122. D. A. Collier, J.-L. Mergny, N. T. Thuong, and C. Hélène, *Nucleic Acids Res.*, 1991, **19**, 4219.
123. N. T. Thuong and C. Hélène, *Angew. Chem., Int. Ed. Engl.*, 1993, **32**, 666.
124. D. Praseuth, T. Le Doan, M. Chassignol, J.-L. Decout, N. Habhoub, J. Lhomme, N. T. Thuong, and C. Hélène, *Biochemistry*, 1988, **27**, 3031.
125. C. Hélène and N. T. Thuong, in "Nucleic Acids and Molecular Biology," eds. F. Eckstein and D. M. J. Lilley, Springer, Berlin, 1988, vol. 2, p. 105.
126. E. S. Gruff and L. E. Orgel, *Nucleic Acids Res.*, 1991, **19**, 6849.
127. E. A. Lukhtanov, M. A. Podyminogin, I. V. Kutyavin, R. B. Meyer, Jr., and H. B. Gamper, *Nucleic Acids Res.*, 1996, **24**, 683.
128. F. Birg, D. Praseuth, A. Zerial, N. T. Thuong, U. Asseline, T. Le Doan, and C. Hélène, *Nucleic Acids Res.*, 1990, **18**, 2901.
129. M. Takasugi, A. Guendouz, M. Chassignol, J. L. Decout, J. Lhomme, N. T. Thuong, and C. Hélène, *Proc. Natl. Acad. Sci. USA*, 1991, **88**, 5602.
130. C. Giovannangéli, N. T. Thuong, and C. Hélène, *Nucleic Acids Res.*, 1992, **20**, 4275.
131. W. H. A. Kuijpers, E. S. Bos, F. M. Kaspersen, G. H. Veeneman, and C. A. A. van Boeckel, *Bioconj. Chem.*, 1993, **4**, 94.
132. J. N. Kremsky, J. L. Wooters, J. P. Dougherty, R. E. Meyers, M. Collins, and E. L. Brown, *Nucleic Acids Res.*, 1987, **15**, 2891.
133. T. Kempe, W. I. Sundquist, F. Chow, and S.-L. Hu, *Nucleic Acids Res.*, 1985, **13**, 45.
134. A. Chollet and E. H. Kawashima, *Nucleic Acids Res.*, 1985, **13**, 1529.
135. B. C. F. Chu and L. E. Orgel, *DNA*, 1985, **4**, 327.
136. A. Guzaev, J. Hovinen, A. Azhayev, and H. Lönnberg, *Nucleosides Nucleotides*, 1995, **14**, 833.
137. J. Hovinen, A. Guzaev, A. Azhayev, and H. Lönnberg, *J. Chem. Soc., Perkin Trans. 1*, 1994, 2745.
138. J. Hovinen, A. Guzaev, E. Azhayeva, A. Azhayev, and H. Lönnberg, *J. Org. Chem.*, 1995, **60**, 2205.
139. A. J. Cocuzza, *Tetrahedron Lett.*, 1989, **30**, 6287.
140. A. M. Alves, D. Holland, and M. D. Edge, *Tetrahedron Lett.*, 1989, **30**, 3089.
141. R. T. Pon, *Tetrahedron Lett.*, 1991, **32**, 1715.
142. P. Kumar, D. Bhatia, B. S. Garg, and K. C. Gupta, *Bioorg. Med. Chem. Lett.*, 1994, **4**, 1761.
143. P. Kumar, A. K. Sharma, and K. C. Gupta, *Nucleosides Nucleotides*, 1996, **15**, 1263.
144. K. Misiura, I. Durrant, M. R. Evans, and M. J. Gait, *Nucleic Acids Res.*, 1990, **18**, 4345.
145. K. Misiura, I. Durrant, M. R. Evans, and M. J. Gait, *Nucleosides Nucleotides*, 1991, **10**, 671.
146. P. S. Nelson, M. Kent, and S. Muthini, *Nucleic Acids Res.*, 1992, **20**, 6253.
147. M.-L. Fontanel, H. Bazin, and R. Téoule, *Anal. Biochem.*, 1993, **214**, 338.
148. M.-L. Fontanel, H. Bazin, and R. Téoule, *Nucleic Acids Res.*, 1994, **22**, 2022.
149. P. Neuner, *Bioorg. Med. Chem. Lett.*, 1996, **6**, 147.
150. J. Olejnik, E. Krzymanska-Olejnik, and K. J. Rothschild, *Nucleic Acids Res.*, 1996, **24**, 361.
151. D. W. Will, C. E. Pritchard, and T. Brown, *Carbohydr. Res.*, 1991, **216**, 315.
152. H. Vu, C. McCollum, K. Jacobson, P. Theisen, R. Vinayak, E. Spiess, and A. Andrus, *Tetrahedron Lett.*, 1990, **31**, 7269.
153. J. Grzybowski, D. W. Will, R. E. Randall, C. A. Smith, and T. Brown, *Nucleic Acids Res.*, 1993, **21**, 1705.

154. O. Zelenko, U. Neumann, W. Brill, U. Pieles, H. E. Moser, and J. Hofsteenge, *Nucleic Acids Res.*, 1994, **22**, 2731.
155. J. R. Fino, P. G. Mattingly, and K. A. Ray, *Bioconj. Chem.*, 1996, **7**, 274.
156. M.-L. Fontanel, H. Bazin, and R. Téoule, *Bioconj. Chem.*, 1993, **4**, 380.
157. M.-L. Fontanel, H. Bazin, A. Roget, and R. Téoule, *J. Labelled Compd. Radiopharm.*, 1993, **33**, 717.
158. F. Schubert, K. Ahlert, D. Cech, and A. Rosenthal, *Nucleic Acids Res.*, 1990, **18**, 3427.
159. P. Theisen, C. McCollum, K. Upadhya, K. Jacobson, H. Vu, and A. Andrus, *Tetrahedron Lett.*, 1992, **33**, 5033.
160. *The Glen Report*, 1995, **8**, 1.
161. J. Burmeister, A. Azzawi, and G. von Kiedrowski, *Tetrahedron Lett.*, 1995, **36**, 3667.
162. K. Shinozuka, Y. Seto, H. Kawata, and H. Sawai, *Bioorg. Med. Chem. Lett.*, 1993, **3**, 2883.
163. J. S. Mann, Y. Shibata, and T. Meehan, *Bioconj. Chem.*, 1992, **3**, 554.
164. V. A. Korshun, N. B. Pestov, K. R. Birikh, and Y. A. Berlin, *Bioconj. Chem.*, 1992, **3**, 559.
165. N. T. Thuong and M. Chassignol, *Tetrahedron Lett.*, 1988, **29**, 5905.
166. C. A. Stein, K. Mori, S. L. Loke, C. Subasinghe, K. Shinozuka, J. S. Cohen, and L. M. Neckers, *Gene*, 1988, **72**, 333.
167. D. A. Collier, N. T. Thuong, and C. Hélène, *J. Am. Chem. Soc.*, 1991, **113**, 1457.
168. U. Asseline, N. T. Thuong, and C. Hélène, *C. R. Acad. Sci. Paris, Ser. III*, 1983, **297**, 369.
169. J.-S. Sun, C. Giovannangeli, J.-C. François, R. Kurfurst, T. Montenay-Garestier, U. Asseline, T. Saison-Behmoaras, N. T. Thuong, and C. Hélène, *Proc. Natl. Acad. Sci. USA*, 1991, **88**, 6023.
170. J.-S. Sun, J.-C. François, T. Montenay-Garestier, T. Saison-Behmoaras, V. Roig, N. T. Thuong, and C. Hélène, *Proc. Natl. Acad. Sci. USA*, 1989, **86**, 9198.
171. C. Giovannangeli, T. Montenay-Garestier, M. Rougée, M. Chassignol, N. T. Thuong, and C. Hélène, *J. Am. Chem. Soc.*, 1991, **113**, 7775.
172. B.-W. Zhou, E. Puga, J.-S. Sun, T. Garestier, and C. Hélène, *J. Am. Chem. Soc.*, 1995, **117**, 10425.
173. K. Fukui, K. Iwane, T. Shimidzu, and K. Tanaka, *Tetrahedron Lett.*, 1996, **37**, 4983.
174. A. Balbi, E. Sottofattori, T. Grandi, M. Mazzei, T. V. Abramova, S. G. Lokhov, and A. V. Lebedev, *Tetrahedron*, 1994, **50**, 4009.
175. J.-K. Chen, D. V. Carlson, H. L. Weith, J. A. O'Brien, M. E. Goldman, and M. Cushman, *Tetrahedron Lett.*, 1992, **33**, 2275.
176. J.-K. Chen, H. L. Weith, R. S. Grewal, G. Wang, and M. Cushman, *Bioconj. Chem.*, 1995, **6**, 473.
177. K. Mori, C. Subasinghe, and J. S. Cohen, *FEBS Lett.*, 1989, **249**, 213.
178. L. G. Puskás, J. Czombos, and S. Bottka, *Nucleosides Nucleotides*, 1995, **14**, 967.
179. U. Pieles and U. Englisch, *Nucleic Acids Res.*, 1989, **17**, 285.
180. J. Woo and P. B. Hopkins, *J. Am. Chem. Soc.*, 1991, **113**, 5457.
181. A. J. Courey, S. E. Plon, and J. C. Wang, *Cell*, 1986, **45**, 567.
182. P. E. Nielsen, *Nucleic Acids Res.*, 1987, **15**, 921.
183. P. J. Bates, V. M. Macaulay, M. J. McLean, T. C. Jenkins, A. P. Reszka, C. A. Laughton, and S. Neidle, *Nucleic Acids Res.*, 1995, **23**, 4283.
184. G. Duval-Valentin, N. T. Thuong, and C. Hélène, *Proc. Natl. Acad. Sci. USA*, 1992, **89**, 504.
185. P. S. Miller, G. Bi, S. A. Kipp, V. Fok, and R. K. DeLong, *Nucleic Acids Res.*, 1996, **24**, 730.
186. A. S. Modak, J. K. Gard, M. C. Merriman, K. A. Winkeler, J. K. Bashkin, and M. K. Stern, *J. Am. Chem. Soc.*, 1991, **113**, 283.
187. C.-H. B. Chen and D. C. Sigman, *Proc. Natl. Acad. Sci. USA*, 1986, **83**, 7147.
188. W. Bannwarth and D. Schmidt, *Tetrahedron Lett.*, 1989, **30**, 1513.
189. I. O. Kady and J. T. Groves, *Bioorg. Med. Chem. Lett.*, 1993, **3**, 1367.
190. S. M. Hecht, *Acc. Chem. Res.*, 1986, **19**, 383.
191. J. Stubbe and J. W. Kozarich, *Chem. Rev.*, 1987, **87**, 1107.
192. V. F. Zarytova, D. S. Sergeyev, and T. S. Godovikova, *Bioconj. Chem.*, 1993, **4**, 189.
193. D. S. Sergeyev, T. S. Godovikova, and V. F. Zarytova, *FEBS Lett.*, 1991, **280**, 271.
194. D. S. Sergeyev, T. S. Godovikova, and V. F. Zarytova, *Nucleic Acids Res.*, 1995, **23**, 4400.
195. C. MacKellar, D. Graham, D. W. Will, S. Burgess, and T. Brown, *Nucleic Acids Res.*, 1992, **20**, 3411.
196. A. V. Kabanov, S. V. Vinogradov, A. V. Ovcharenko, A. V. Krivonos, N. S. Melik-Nubarov, V. I. Kiselev, and E. S. Severin, *FEBS Lett.*, 1990, **259**, 327.
197. E. S. Severin, N. S. Melik-Nubarov, A. V. Ovcharenko, S. V. Vinogradov, V. I. Kiselev, and A. V. Kabanov, in "Advances in Enzyme Regulation," ed. G. Weber, Pergamon, Oxford, 1991, vol. 31, p. 417.
198. D. W. Will and T. Brown, *Tetrahedron Lett.*, 1992, **33**, 2729.
199. S.-G. Kim, H. Nakashima, Y. Shoji, T. Inagawa, N. Yamamoto, Y. Kinzuka, K. Takai, and H. Takaku, *Bioorg. Med. Chem.*, 1996, **4**, 603.
200. V. A. Efimov, A. L. Kalinkina, and O. G. Chakhmakhcheva, *Nucleic Acids Res.*, 1993, **21**, 5337.
201. R. G. Shea, J. C. Marsters, and N. Bischofberger, *Nucleic Acids Res.*, 1990, **18**, 3777.
202. S. V. Vinogradov, Y. Suzdaltseva, V. Y. Alakhov, and A. V. Kabanov, *Biochem. Biophys. Res. Commun.*, 1994, **203**, 959.
203. H. Vu, P. Singh, L. Lewis, J. G. Zendegui, and K. Jayaraman, *Nucleosides Nucleotides*, 1993, **12**, 853.
204. A. S. Boutorine and E. V. Kostina, *Biochimie*, 1993, **75**, 35.
205. M. Wasner, E. E. Henderson, R. J. Suhadolnik, and W. Pfleiderer, *Helv. Chim. Acta*, 1994, **77**, 1757.
206. M. Wasner and W. Pfleiderer, *Nucleosides Nucleotides*, 1995, **14**, 1101.
207. D. Hoekstra, L. A. M. Rupert, J. B. F. N. Engberts, S. Nir, H. Hoff, K. Klappe, and S. L. Novick, *Stud. Biophys.*, 1988, **127**, 105.
208. A. Jäschke, J. P. Fürste, D. Cech, and V. A. Erdmann, *Tetrahedron Lett.*, 1993, **34**, 301.
209. A. Jäschke, R. Bald, E. Nordhoff, F. Hillenkamp, D. Cech, V. A. Erdmann, and J. P. Fürste, *Nucleosides Nucleotides*, 1996, **15**, 1519.
210. A. Jäschke, J. P. Fürste, E. Nordhoff, F. Hillenkamp, D. Cech, and V. A. Erdmann, *Nucleic Acids Res.*, 1994, **22**, 4810.
211. A. Jäschke, J. P. Fürste, V. A. Erdmann, and D. Cech, *Nucleic Acids Res.*, 1994, **22**, 1880.
212. M. Durand, K. Chevrie, M. Chassignol, N. T. Thuong, and J. C. Maurizot, *Nucleic Acids Res.*, 1990, **18**, 6353.
213. S. Rumney, IV and E. T. Kool, *J. Am. Chem. Soc.*, 1995, **117**, 5635.

214. M. Y.-X. Ma, L. S. Reid, S. C. Climie, W. C. Lin, R. Kuperman, M. Sumner-Smith, and R. W. Barnett, *Biochemistry*, 1993, **32**, 1751.
215. F. Benseler, D.-J. Fu, J. Ludwig, and L. W. McLaughlin, *J. Am. Chem. Soc.*, 1993, **115**, 8483.
216. J. B. Thomson, T. Tuschl, and F. Eckstein, *Nucleic Acids Res.*, 1993, **21**, 5600.
217. G. M. Bonora, G. Biancotto, M. Maffini, and C. L. Scremin, *Nucleic Acids Res.*, 1993, **21**, 1213.
218. K. C. Gupta, P. Sharma, P. Kumar, and S. Sathyanarayana, *Nucleic Acids Res.*, 1991, **19**, 3019.
219. J. Hovinen, A. Guzaev, A. Azhayev, and H. Lönnberg, *Tetrahedron*, 1994, **50**, 7203.
220. J. Hovinen, A. Guzaev, A. Azhayev, and H. Lönnberg, *Tetrahedron Lett.*, 1993, **34**, 8169.
221. A. Kumar, *Nucleosides Nucleotides*, 1993, **12**, 1047.
222. K. C. Gupta, P. Sharma, S. Sathyanarayana, and P. Kumar, *Tetrahedron Lett.*, 1990, **31**, 2471.
223. A. Kumar, *Nucleosides Nucleotides*, 1993, **12**, 729.
224. A. Kumar, *Nucleosides Nucleotides*, 1994, **13**, 2125.
225. P. Kumar, K. C. Gupta, R. Rosch, and H. Seliger, *Bioorg. Med. Chem. Lett.*, 1996, **6**, 2247.
226. U. Asseline and N. T. Thuong, *Nucleosides Nucleotides*, 1991, **10**, 359.
227. W. T. Markiewicz and T. K. Wyrzykiewicz, *Nucleic Acids Res.*, 1989, **17**, 7149.
228. S. M. Gryaznov and R. L. Letsinger, *Tetrahedron Lett.*, 1992, **33**, 4127.
229. P. Kumar, N. K. Bose, and K. C. Gupta, *Tetrahedron Lett.*, 1991, **32**, 967.
230. M. Gottikh, U. Asseline, and N. T. Thuong, *Tetrahedron Lett.*, 1990, **31**, 6657.
231. R. Eritja, J. Robles, D. Fernandez-Forner, F. Albericio, E. Giralt, and E. Pedroso, *Tetrahedron Lett.*, 1991, **32**, 1511.
232. S. M. Gryaznov and R. L. Letsinger, *Tetrahedron Lett.*, 1992, **33**, 4127.
233. E. Felder, R. Schwyzer, R. Charubala, W. Pfeiderer, and B. Schulz, *Tetrahedron Lett.*, 1984, **25**, 3967.
234. U. Asseline and N. T. Thuong, *Tetrahedron Lett.*, 1989, **30**, 2521.
235. L. Perrouault, U. Asseline, C. Rivalle, N. T. Thuong, E. Bisagni, C. Giovannangeli, T. Le Doan, and C. Hélène, *Nature*, 1990, **344**, 358.
236. T. Le Doan, L. Perrouault, U. Asseline, N. T. Thuong, C. Rivalle, E. Bisagni, and C. Hélène, *Antisen. Res. Dev.*, 1991, **1**, 43.
237. D. Jin Yoo and M. M. Greenberg, *J. Org. Chem.*, 1995, **60**, 3358.
238. M. M. Greenberg, *Tetrahedron*, 1995, **51**, 29.
239. M. M. Greenberg, *Tetrahedron Lett.*, 1993, **34**, 251.
240. J. Hovinen, A. P. Gouzaev, A. V. Azhayev, and H. Lönnberg, *Tetrahedron Lett.*, 1993, **34**, 5163.
241. H. B. Gamper, M. W. Reed, T. Cox, J. S. Virosco, A. D. Adams, A. A. Gall, J. K. Scholler, and R. B. Meyer, Jr., *Nucleic Acids Res.*, 1993, **21**, 145.
242. R. D. Hinrichsen, D. Fraga, and M. W. Reed, *Proc. Natl. Acad. Sci. USA*, 1992, **89**, 8601.
243. T. Saison-Behmoaras, B. Tocqué, I. Rey, M. Chassignol, N. T. Thuong, and C. Hélène, *EMBO J.*, 1991, **10**, 1111.
244. F. Seela and K. Kaiser, *Nucleic Acids Res.*, 1987, **15**, 3113.
245. H. Urata and M. Akagi, *Tetrahedron Lett.*, 1993, **34**, 4015.
246. S. M. Gryaznov and R. L. Letsinger, *Tetrahedron Lett.*, 1993, **34**, 1261.
247. U. Asseline and N. T. Thuong, *Tetrahedron Lett.*, 1990, **31**, 81.
248. J. Haralambidis, L. Duncan, and G. W. Tregear, *Tetrahedron Lett.*, 1987, **28**, 5199.
249. J. Haralambidis, L. Duncan, K. Angus, M. Chai, S. Pownall, and G. Tregear, *Nucleic Acids Res. Symp. Ser. #20*, 1988, 115.
250. P. S. Nelson, R. A. Frye, and E. Liu, *Nucleic Acids Res.*, 1989, **17**, 7187.
251. J. Haralambidis, L. Duncan, K. Angus, and G. W. Tregear, *Nucleic Acids Res.*, 1990, **18**, 493.
252. J. Haralambidis, K. Angus, S. Pownall, L. Duncan, M. Chai, and G. W. Tregear, *Nucleic Acids Res.*, 1990, **18**, 501.
253. C. R. Petrie, M. W. Reed, A. D. Adams, and R. B. Meyer, Jr., *Bioconj. Chem.*, 1992, **3**, 85.
254. J. Thaden and P. S. Miller, *Bioconj. Chem.*, 1993, **4**, 395.
255. D. L. McMinn and M. M. Greenberg, *Tetrahedron*, 1996, **52**, 3827.
256. J. G. Zendegui, K. M. Vasquez, J. H. Tinsley, D. J. Kessler, and M. E. Hogan, *Nucleic Acids Res.*, 1992, **20**, 307.
257. R. Boado and W. M. Partridge, *Bioconj. Chem.*, 1994, **5**, 406.
258. T. Zhu, C.-H. Tung, K. J. Breslauer, W. A. Dickerhof, and S. Stein, *Antisen. Res. Dev.*, 1993, **3**, 349.
259. G. A. Soukup, R. L. Cerny, and L. J. Maher, III, *Bioconj. Chem.*, 1995, **6**, 135.
260. B. Oberhauser and E. Wagner, *Nucleic Acids Res.*, 1992, **20**, 533.
261. S. V. Vinogradov, Y. G. Suzdaltseva, and A. V. Kabanov, *Bioconj. Chem.*, 1996, **7**, 3.
262. M. W. Reed, A. D. Adams, J. S. Nelson, and R. B. Meyer, Jr., *Bioconj. Chem.*, 1991, **2**, 217.
263. U. Asseline, E. Bonfils, D. Dupret, and N. T. Thuong, *Bioconj. Chem.*, 1996, **7**, 369.
264. W. H. Gmeiner, W. Luo, R. T. Pon, and J. W. Lown, *Bioorg. Med. Chem. Lett.*, 1991, **1**, 487.
265. S. M. Gryaznov and D. H. Lloyd, *Nucleic Acids Res.*, 1993, **21**, 5909.
266. T. Y.-K. Chow, C. Juby, and R. Brousseau, *Antisen. Res. Dev.*, 1994, **4**, 81.
267. H. Vu, T. Schmaltz Hill, and K. Jayaraman, *Bioconj. Chem.*, 1994, **5**, 666.
268. G. Tong, J. M. Lawlor, G. W. Tregear, and J. Haralambidis, *J. Org. Chem.*, 1993, **58**, 2223.
269. I. Habus, Q. Zhao, and S. Agrawal, *Bioconj. Chem.*, 1995, **6**, 327.
270. E. A. Lukhtanov, I. V. Kutyavin, H. B. Gamper, and R. B. Meyer, Jr., *Bioconj. Chem.*, 1995, **6**, 418.
271. B. P. Zhao, G. B. Panigrahi, P. D. Sadowski, and J. J. Krepinsky, *Tetrahedron Lett.*, 1996, **37**, 3093.
272. G. Pan, K. Luetke, C. D. Juby, R. Brousseau, and P. Sadowski, *J. Biol. Chem.* 1993, **268**, 3683.
273. S. Soukchareun, G. W. Tregear, and J. Haralambidis, *Bioconj. Chem.*, 1995, **6**, 43.
274. G. Tong, J. M. Lawlor, G. W. Tregear, and J. Haralambidis, *J. Am. Chem. Soc.*, 1995, **117**, 12 151.
275. K. Yamana, K. Nunota, H. Nakano, and O. Sangen, *Tetrahedron Lett.*, 1994, **35**, 2555.
276. T.-P. Wu, K.-C. Ruan, and W.-Y. Liu, *Nucleic Acids Res.*, 1996, **24**, 3472.
277. L. Bellon, C. Workman, J. Scherrer, N. Usman, and F. Wincott, *J. Am. Chem. Soc.*, 1996, **118**, 3771.
278. J. Telser, K. A. Cruickshank, L. E. Morrison, and T. L. Netzel, *J. Am. Chem. Soc.*, 1989, **111**, 6966.
279. J. Telser, K. A. Cruickshank, K. S. Schanze, and T. L. Netzel, *J. Am. Chem. Soc.*, 1989, **111**, 7221.
280. R. Kierzek and W. T. Markiewicz, *Nucleosides Nucleotides*, 1987, **6**, 403.
281. B. S. Sproat, A. I. Lamond, B. Beijer, P. Neuner, and U. Ryder, *Nucleic Acids Res.*, 1989, **17**, 3373.

282. A. M. Iribarren, B. S. Sproat, P. Neuner, I. Sulston, U. Ryder, and A. I. Lamond, *Proc. Natl. Acad. Sci. USA*, 1990, **87**, 7747.
283. E. Krzymanska-Olejnik and R. W. Adamiak, *Nucleosides Nucleotides*, 1991, **10**, 595.
284. C. B. Reese and A. Ubasawa, *Nucleic Acids Res. Symp. Ser. #7*, 1980, 5.
285. W. L. Sung, *Nucleic Acids Res.*, 1981, **9**, 6139.
286. S. Le Brun, N. Duchange, A. Namane, M. M. Zakin, T. Huynh-Dinh, and J. Igolen, *Biochimie*, 1989, **71**, 319.
287. M. S. Urdea, B. D. Warner, J. A. Running, M. Stempien, J. Clyne, and T. Horn, *Nucleic Acids Res.*, 1988, **16**, 4937.
288. M. S. Urdea, J. A. Running, T. Horn, J. Clyne, L. Ku, and B. D. Warner, *Gene*, 1987, **61**, 253.
289. A. M. MacMillan and G. L. Verdine, *Tetrahedron*, 1991, **47**, 2603.
290. A. M. MacMillan and G. L. Verdine, *J. Org. Chem.*, 1990, **55**, 5931.
291. J. C. Schulhof, D. Molko, and R. Teoule, *Tetrahedron Lett.*, 1987, **28**, 51.
292. Y.-Z. Xu, Q. Zheng, and P. F. Swann, *J. Org. Chem.*, 1992, **57**, 3839.
293. W. T. Markiewicz, G. Gröger, R. Rösch, A. Zebrowska, and H. Seliger, *Nucleosides Nucleotides*, 1992, **11**, 1703.
294. J. A. Brumbaugh, L. R. Middendorf, D. L. Grone, and J. L. Ruth, *Proc. Natl. Acad. Sci. USA*, 1988, **85**, 5610.
295. K. A. Cruickshank and D. L. Stockwell, *Tetrahedron Lett.*, 1988, **29**, 5221.
296. J. Haralambidis, M. Chai, and G. W. Tregear, *Nucleic Acids Res.*, 1987, **15**, 4857.
297. A. S. Levina, D. R. Tabatadse, L. M. Khalimskaya, T. A. Prichodko, G. V. Shishkin, L. A. Alexandrova, and V. P. Zatyrova, *Bioconj. Chem.*, 1993, **4**, 319.
298. K. J. Gibson and S. J. Benkovic, *Nucleic Acids Res.*, 1987, **15**, 6455.
299. M. Cowart, K. J. Gibson, D. J. Allen, and S. J. Benkovic, *Biochemistry*, 1989, **28**, 1975.
300. D. J. Allen, P. L. Darke, and S. J. Benkovic, *Biochemistry*, 1989, **28**, 4601.
301. D. J. Allen and S. J. Benkovic, *Biochemistry*, 1989, **28**, 9586.
302. C. R. Guest, R. A. Hochstrasser, C. G. Dupuy, D. J. Allen, S. J. Benkovic, and D. P. Millar, *Biochemistry*, 1991, **30**, 8759.
303. R. B. Meyer, Jr., J. C. Tabone, G. D. Hurst, T. M. Smith, and H. Gamper, *J. Am. Chem. Soc.*, 1989, **111**, 8517.
304. J. C. Tabone, M. R. Stamm, H. B. Gamper, and R. B. Meyer, Jr., *Biochemistry*, 1994, **33**, 375.
305. K. Kido, H. Inoue, and E. Ohtsuka, *Nucleic Acids Res.*, 1992, **20**, 1339.
306. T. J. Povsic, S. A. Strobel, and P. B. Dervan, *J. Am. Chem. Soc.*, 1992, **114**, 5934.
307. M. A. Podyminogin, R. B. Meyer, and H. B. Gamper, *Biochemistry*, 1995, **34**, 13 098.
308. I. V. Kutyavin, H. B. Gamper, A. A. Gall, and R. B. Meyer, Jr., *J. Am. Chem. Soc.*, 1993, **115**, 9303.
309. A. F. Cook, E. Vuocolo, and C. L. Brakel, *Nucleic Acids Res.*, 1988, **16**, 4077.
310. C. Kessler, *Mol. Cell. Probes*, 1991, **5**, 161.
311. K. Mühlegger, E. Huber, H. von Der Eltz, R. Rüger, and C. Kessler, *Biol. Chem. Hoppe-Seyler*, 1990, **371**, 953.
312. J. Telser, K. A. Cruickshank, L. E. Morrison, T. L. Netzel, and C. Chan, *J. Am. Chem. Soc.*, 1989, **111**, 7226.
313. V. Folsom, M. J. Hunkeler, A. Haces, and J. D. Harding, *Anal. Biochem.*, 1989, **182**, 309.
314. J.-P. Roduit, J. Shaw, A. Chollet, and A. Chollet, *Nucleosides Nucleotides*, 1987, **6**, 349.
315. L. E. Morrison, T. C. Halder, and L. M. Stols, *Anal. Biochem.*, 1989, **183**, 231.
316. A. Ono, N. Haginoya, M. Kiyokawa, N. Minakawa, and A. Matsuda, *Bioorg. Med. Chem. Lett.*, 1994, **4**, 361.
317. R. J. Boado and W. M. Partridge, *Bioconj. Chem.*, 1992, **3**, 519.
318. W. M. Partridge and R. J. Boado, *FEBS Lett.*, 1991, **288**, 30.
319. W. M. Partridge, R. J. Boado, and J. L. Buciak, *Drug Delivery*, 1993, **1**, 43.
320. R. J. Boado and W. M. Partridge, *Bioconj. Chem.*, 1994, **5**, 406.
321. Y. Tor and P. B. Dervan, *J. Am. Chem. Soc.*, 1993, **115**, 4461.
322. C.-H. B. Chen, M. B. Gorin, and D. S. Sigman, *Proc. Natl. Acad. Sci. USA*, 1993, **90**, 4206.
323. B. F. Baker, K. Ramasamy, and J. Kiely, *Bioorg. Med. Chem. Lett.*, 1996, **6**, 1647.
324. K. Shah, H. Neenhold, Z. Wang, and T. M. Rana, *Bioconj. Chem.*, 1996, **7**, 283.
325. P. L. Richardson, M. L. Gross, K. J. Light-Wahl, R. D. Smith, and A. Schepartz, *Bioorg. Med. Chem. Lett.*, 1994, **4**, 2133.
326. C. Sund, J. Ylikoski, P. Hurskainen, and M. Kwiatkowski, *Nucleosides Nucleotides*, 1988, **7**, 655.
327. P. Dahlén, L. Liukkonen, M. Kwiatkowski, P. Hurskainen, A. Iitiä, H. Siitari, J. Ylikoski, V.-M. Mukkala, and T. Lövgren, *Bioconj. Chem.*, 1994, **5**, 268, and references therein.
328. R. S. Coleman and J. M. Siedlecki, *J. Am. Chem. Soc.*, 1992, **114**, 9229.
329. R. S. Coleman and E. A. Kesicki, *J. Am. Chem. Soc.*, 1994, **116**, 11 636.
330. T. T. Nikiforov and B. A. Connolly, *Tetrahedron Lett.*, 1992, **33**, 2379.
331. P. Clivio, J.-L. Fourrey, J. Gasche, A. Audic, A. Favre, C. Perrin, and A. Woisard, *Tetrahedron Lett.*, 1992, **33**, 65.
332. A. Woisard, A. Favre, P. Clivio, and J.-L. Fourrey, *J. Am. Chem. Soc.*, 1992, **114**, 10 072.
333. T. T. Nikiforov and B. A. Connolly, *Tetrahedron Lett.*, 1991, **32**, 3851.
334. D. H. Bradley and M. M. Hanna, *Tetrahedron Lett.*, 1992, **33**, 6223.
335. T. T. Nikiforov and B. A. Connolly, *Nucleic Acids Res.*, 1992, **20**, 1209.
336. K. L. Meyer and M. M. Hanna, *Bioconj. Chem.*, 1996, **7**, 401.
337. J. T. Goodwin and G. D. Glick, *Tetrahedron Lett.*, 1993, **34**, 5549.
338. S. Sun, X.-Q. Tang, A. Merchant, P. S. R. Anjaneyulu, and J. A. Piccirilli, *J. Org. Chem.*, 1996, **61**, 5708.
339. H. Bazin, A. Roget, and R. Téoule, *Nucleosides Nucleotides*, 1991, **10**, 363.
340. T. Horn and M. S. Urdea, *Nucleic Acids Res.*, 1989, **17**, 6959.
341. C. Chang, T. Horn, D. Ahle, and M. S. Urdea, *Nucleosides Nucleotides*, 1991, **10**, 389.
342. N. Dolinnaya, S. Gryaznov, D. Ahle, C.-A. Chang, Z. A. Shabarova, M. S. Urdea, and T. Horn, *Bioorg. Med. Chem. Lett.*, 1994, **4**, 1011.
343. S. Sauvaigo, B. Fouqué, A. Roget, T. Livache, H. Bazin, C. Chypre, and R. Téoule, *Nucleic Acids Res.*, 1990, **18**, 3175.
344. A. Roget, H. Bazin, and R. Téoule, *Nucleic Acids Res.*, 1989, **17**, 7643.
345. U. Pieles, B. S. Sproat, and G. M. Lamm, *Nucleic Acids Res.*, 1990, **18**, 4355.
346. L. Klevan and G. Gebeyehu, *Methods Enzymol.*, 1990, **184**, 561.
347. G. Gebeyehu, P. Y. Rao, P. SooChan, D. A. Simms, and L. Klevan, *Nucleic Acids Res.*, 1987, **15**, 4513.
348. C. R. Petrie, A. D. Adams, M. Stamm, J. Van Ness, S. M. Watanabe, and R. B. Meyer, Jr., *Bioconj. Chem.*, 1991, **2**, 441.

349. R. Kierzek and W. T. Markiewicz, *Nucleosides Nucleotides*, 1991, **10**, 1257.
350. T. Maier and W. Pfleiderer, *Nucleosides Nucleotides*, 1995, **14**, 961.
351. S. Mielewczyk, G. Dominiak, Z. Gdaniec, E. Krzymanska-Olejnik, and R. W. Adamiak, *Nucleosides Nucleotides*, 1991, **10**, 263.
352. S. C. Srivastava, S. Kazim Raza, and R. Misra, *Nucleic Acids Res.*, 1994, **22**, 1296.
353. Z. Zhu, J. Chao, H. Yu, and A. S. Waggoner, *Nucleic Acids Res.*, 1994, **22**, 3418.
354. H. Yu, J. Chao, D. Patek, R. Mujumdar, S. Mujumdar, and A. S. Wagonner, *Nucleic Acids Res.*, 1994, **22**, 3226.
355. C. Kessler, H.-J. Höltke, R. Seibl, J. Burg, and K. Mühlegger, *Biol. Chem. Hoppe-Seyler*, 1990, **371**, 917.
356. G. G. Schmitz, T. Walter, R. Seibl, and C. Kessler, *Anal. Biochem.*, 1991, **192**, 222.
357. D. A. Barawkar and K. N. Ganesh, *Bioorg. Med. Chem. Lett.*, 1993, **3**, 347.
358. D. A. Barawkar and K. N. Ganesh, *Biochem. Biophys. Res. Commun.*, 1994, **203**, 53.
359. D. A. Barawkar and K. N. Ganesh, *Nucleic Acids Res.*, 1995, **23**, 159.
360. D. Singh, V. Kumar, and K. N. Ganesh, *Nucleic Acids Res.*, 1990, **18**, 3339.
361. A. Davison, G. Duckworth, M. V. Rao, J. McClean, J. Grzybowski, P. Potier, T. Brown, and H. Cubie, *Nucleosides Nucleotides*, 1995, **14**, 1049.
362. J. M. Prober, G. L. Trainor, R. J. Dam, F. W. Hobbs, C. W. Robertson, R. J. Zagursky, A. J. Cocuzza, M. A. Jensen, and K. Baumeister, *Science*, 1987, **238**, 336.
363. M. Kwiatkowski, M. Samiotaki, U. Lamminmäki, V.-M. Mukkala, and U. Landegren, *Nucleic Acids Res.*, 1994, **22**, 2604.
364. R. Breslow, E. Anslyn, and D.-L. Huang, *Tetrahedron*, 1991, **47**, 2365, and references therein.
365. J. K. Bashkin, J. K. Gard, and A. S. Modak, *J. Org. Chem.*, 1990, **55**, 5125.
366. G. Wang and D. E. Bergstrom, *Tetrahedron Lett.*, 1993, **34**, 6725.
367. N. V. Heeb and S. A. Benner, *Tetrahedron Lett.*, 1994, **35**, 3045.
368. K. S. Ramasamy, M. Zounes, C. Gonzalez, S. M. Freier, E. A. Lesnik, L. L. Cummins, R. H. Griffey, B. P. Monia, and P. D. Cook, *Tetrahedron Lett.*, 1994, **35**, 215.
369. G. Wang and D. E. Bergstrom, *Tetrahedron Lett.*, 1993, **34**, 6721.
370. D. E. Bergstrom and J. Chen, *Bioorg. Med. Chem. Lett.*, 1996, **6**, 2211.
371. C.-H. B. Chen and D. S. Sigman, *J. Am. Chem. Soc.*, 1988, **110**, 6570.
372. J. K. Bashkin, E. I. Frolova, and U. Sampath, *J. Am. Chem. Soc.*, 1994, **116**, 5981.
373. D. E. Bergstrom and N. P. Gerry, *J. Am. Chem. Soc.*, 1994, **116**, 12067.
374. G. B. Dreyer and P. B. Dervan, *Proc. Natl. Acad. Sci. USA*, 1985, **82**, 968.
375. D. A. Horne and P. B. Dervan, *Nucleic Acids Res.*, 1991, **19**, 4963.
376. P. A. Beal and P. B. Dervan, *J. Am. Chem. Soc.*, 1992, **114**, 4976, and references therein.
377. S. F. Singleton and P. B. Dervan, *J. Am. Chem. Soc.*, 1992, **114**, 6957.
378. T. Livache, A. Roget, E. Dejean, C. Barthet, G. Bidan, and R. Téoule, *Nucleic Acids Res.*, 1994, **22**, 2915.
379. U. Pieles, B. S. Sproat, P. Neuner, and F. Cramer, *Nucleic Acids Res.*, 1989, **17**, 8967.
380. M. M. Hanna, Y. Zhang, J. C. Reidling, M. J. Thomas, and J. Jou, *Nucleic Acids Res.*, 1993, **21**, 2073.
381. M. M. Hanna, S. Dissinger, B. D. Williams, and J. E. Colston, *Biochemistry*, 1989, **28**, 5814.
382. C. A. Bowser and M. M. Hanna, *J. Mol. Biol.*, 1991, **220**, 227.
383. T. Yamaguchi and M. Saneyoshi, *Nucleic Acids Res.*, 1996, **24**, 3364.
384. T. Yamaguchi and M. Saneyoshi, *Nucleosides Nucleotides*, 1996, **15**, 607.
385. S. V. Doronin, M. I. Dobrikov, M. Buckle, P. Roux, H. Buc, and O. I. Lavrik, *FEBS Lett.*, 1994, **354**, 200.
386. A. Spaltenstein, B. H. Robinson, and P. B. Hopkins, *J. Am. Chem. Soc.*, 1989, **111**, 2303.
387. A. Spaltenstein, B. H. Robinson, and P. B. Hopkins, *Biochemistry*, 1989, **28**, 9484.
388. A. Spaltenstein, B. H. Robinson, and P. B. Hopkins, *J. Am. Chem. Soc.*, 1988, **110**, 1299.
389. J. J. Kirchner, E. J. Hustedt, B. H. Robinson, and P. B. Hopkins, *Tetrahedron Lett.*, 1990, **31**, 593.
390. E. J. Hustedt, J. J. Kirchner, A. Spaltenstein, P. B. Hopkins, and B. H. Robinson, *Biochemistry*, 1995, **34**, 4369.
391. W. Bannwarth and D. Schmidt, *Bioorg. Med. Chem. Lett.*, 1994, **4**, 977.
392. J.-L. Duh and A. M. Bobst. *Helv. Chim. Acta*, 1991, **74**, 739.
393. O. K. Strobel, D. D. Kryak, E. V. Bobst, and A. M. Bobst, *Bioconj. Chem.*, 1991, **2**, 89.
394. O. K. Strobel, R. S. Keyes, and A. M. Bobst, *Biochemistry*, 1990, **29**, 8522.
395. R. D. Sheardy and N. C. Seeman, *J. Org. Chem.*, 1986, **51**, 4301.
396. N. C. Seeman and N. R. Kallenbach, *Biophys. J.*, 1983, **44**, 201.
397. N. R. Kallenbach, R.-I. Ma, and N. C. Seeman, *Nature*, 1983, **305**, 829.
398. W. H. Gmeiner, R. T. Pon, and J. W. Lown, *J. Org. Chem.*, 1991, **56**, 3602.
399. F.-G. Rong and A. H. Soloway, *Nucleosides Nucleotides*, 1994, **13**, 2021.
400. N. M. Goudgaon, G. Fulcrand El-Kattan, and R. F. Schinazi, *Nucleosides Nucleotides*, 1994, **13**, 849, and references therein.
401. G. Fulcrand El-Kattan, Z. J. Lesnikowski, S. Yao, F. Tanious, W. D. Wilson, and R. F. Schinazi, *J. Am. Chem. Soc.*, 1994, **116**, 7494.
402. A. Dipple, R. C. Moschel, and C. A. H. Bigger, in "Chemical Carcinogens," 2nd edn., ed. C. E. Searle, American Chemical Society, Washington, DC, 1984, vol. 1, ACS Monograph 182, p. 41.
403. M. R. Osborne and N. T. Crosby, in "Benzopyrenes," Cambridge University Press, Cambridge, 1987, p. 147.
404. R. Casale and L. W. McLaughlin, *J. Am. Chem. Soc.*, 1990, **112**, 5264.
405. J. J. Stezowski, G. Joos-Guba, K.-H. Schönwälder, A. Straub, and J. P. Glusker, *J. Biomol. Struct. Dyn.*, 1987, **5**, 615.
406. M. K. Lakshman, J. M. Sayer, and D. M. Jerina, *J. Am. Chem. Soc.*, 1991, **113**, 6589.
407. M. K. Lakshman, J. M. Sayer, and D. M. Jerina, *J. Org. Chem.*, 1992, **57**, 3438.
408. B. Zajc, M. K. Lakshman, J. M. Sayer, and D. M. Jerina, *Tetrahedron Lett.*, 1992, **33**, 3409.
409. M. K. Lakshman, J. M. Sayer, H. Yagi, and D. M. Jerina, *J. Org. Chem.*, 1992, **57**, 4585.
410. Y. Zhou, S. Chládek, and L. J. Romano, *J. Org. Chem.*, 1994, **59**, 556.
411. Y. Zhou and L. J. Romano, *Biochemistry*, 1993, **32**, 14043.
412. H. Lee, M. Hinz, J. J. Stezowski, and R. G. Harvey, *Tetrahedron Lett.*, 1990, **31**, 6773.
413. T. Steinbrecher, A. Becker, J. J. Stezowski, F. Oesch, and A. Seidel, *Tetrahedron Lett.*, 1993, **34**, 1773.

414. C. M. Harris, L. Zhou, E. A. Strand, and T. M. Harris, *J. Am. Chem. Soc.*, 1991, **113**, 4328.
415. S. J. Kim, M. P. Stone, C. M. Harris, and T. M. Harris, *J. Am. Chem. Soc.*, 1992, **114**, 5480.
416. S. J. Kim, K. K. Jajoo, H.-Y. Kim, L. Zhou, P. Horton, C. M. Harris, and T. M. Harris, *Bioorg. Med. Chem.*, 1995, **3**, 811.
417. A. Ono, A. Dan, and A. Matsuda, *Bioconj. Chem.*, 1993, **4**, 499.
418. A. Dan, Y. Yoshimura, A. Ono, and A. Matsuda, *Bioorg. Med. Chem. Lett.*, 1993, **3**, 615.
419. A. Azhayev, A. Gouzaev, J. Hovinen, E. Azhayeva, and H. Lönnberg, *Tetrahedron Lett.*, 1993, **34**, 6435.
420. A. Guzaev, E. Azhayeva, J. Hovinen, A. Azhayev, and H. Lönnberg, *Bioconj. Chem.*, 1994, **5**, 501.
421. J. K. Bashkin, J. Xie, A. T. Daniher, U. Sampath, and J. L.-F. Kao, *J. Org. Chem.*, 1996, **61**, 2314.
422. K. Yamana, Y. Nishijima, T. Ikeda, T. Gokota, H. Ozaki, H. Nakano, O. Sangen, and T. Shimidzu, *Bioconj. Chem.*, 1990, **1**, 319.
423. H. M. Deshmukh, S. P. Joglekar, and A. D. Broom, *Bioconj. Chem.*, 1995, **6**, 578.
424. K. Yamana, Y. Ohashi, K. Nunota, M. Kitamura, H. Nakano, O. Sangen, and T. Shimidzu, *Tetrahedron Lett.*, 1991, **32**, 6347.
425. K. Yamana, T. Gokota, H. Ozaki, H. Nakano, O. Sangen, and T. Shimidzu, *Nucleosides Nucleotides*, 1992, **11**, 383.
426. K. Yamana, R. Aota, and H. Nakano, *Tetrahedron Lett.*, 1995, **36**, 8427.
427. N. N. Polushin, *Tetrahedron Lett.*, 1996, **37**, 3231.
428. C. Hendrix, B. Devreese, J. Rozenski, A. Van Aerschot, A. De Bruyn, J. Van Beeumen, and P. Herdewijn, *Nucleic Acids Res.*, 1995, **23**, 51.
429. H. Aurup, T. Tuschl, F. Benseler, J. Ludwig, and F. Eckstein, *Nucleic Acids Res.*, 1994, **22**, 20.
430. S. T. Sigurdsson and F. Eckstein, *Nucleic Acids Res.*, 1996, **24**, 3129.
431. T. H. Keller and R. Häner, *Nucleic Acids Res.*, 1993, **21**, 4499.
432. M. Manoharan, C. J. Guinosso, and P. D. Cook, *Tetrahedron Lett.*, 1991, **32**, 7171.
433. M. E. Douglas, B. Beijer, and B. Sproat, *Bioorg. Med. Chem. Lett.*, 1994, **4**, 995.
434. P. N. Jørgensen, P. C. Stein, and J. Wengel, *J. Am. Chem. Soc.*, 1994, **116**, 2231.
435. P. N. Jørgensen, M. L. Svendsen, C. Scheuer-Larsen, and J. Wengel, *Tetrahedron*, 1995, **51**, 2155.
436. J. Fensholdt, H. Thrane, and J. Wengel, *Tetrahedron Lett.*, 1995, **36**, 2535.
437. H. Thrane, J. Fensholdt, M. Regner, and J. Wengel, *Tetrahedron*, 1995, **51**, 10 389.
438. G. Wang and W. E. Seifert, *Tetrahedron Lett.*, 1996, **37**, 6515.
439. H. Maag, B. Schmidt, and S. J. Rose, *Tetrahedron Lett.*, 1994, **35**, 6449.
440. G. Wang and P. J. Middleton, *Tetrahedron Lett.*, 1996, **37**, 2739.
441. L. M. Smith, R. J. Kaiser, J. Z. Sanders, and L. E. Hood, *Methods Enzymol.*, 1987, **155**, 260.
442. L. M. Smith, J. Z. Sanders, R. J. Kaiser, P. Hughes, C. Dodd, C. R. Connell, C. Heiner, S. B. H. Kent, and L. E. Hood, *Nature*, 1986, **321**, 674.
443. L. M. Smith, S. Fung, M. W. Hunkapiller, T. J. Hunkapiller, and L. E. Hood, *Nucleic Acids Res.*, 1985, **13**, 2399.
444. F. Sanger, S. Nicklen, and A. R. Coulson, *Proc. Natl. Acad. Sci. USA*, 1977, **74**, 5463.
445. B. S. Sproat, B. Beijer, and P. Rider, *Nucleic Acids Res.*, 1987, **15**, 6181.
446. B. S. Sproat, B. Beijer, P. Rider, and P. Neuner, *Nucleic Acids Res.*, 1987, **15**, 4837.
447. B. S. Sproat, B. Beijer, P. Rider, and P. Neuner, *Nucleic Acid Res. Symp. Ser. #20*, 1988, 117.
448. R. Kierzek, Y. Li, D. H. Turner, and P. C. Bevilacqua, *J. Am. Chem. Soc.*, 1993, **115**, 4985.
449. P. C. Bevilacqua, R. Kierzek, K. A. Johnson, and D. H. Turner, *Science*, 1992, **258**, 1355.
450. K.-H. Altmann, M. O. Bévierre, A. De Mesmaeker, and H. E. Moser, *Bioorg. Med. Chem. Lett.*, 1995, **5**, 431.
451. H. Seliger, B. Krist, and S. Berner, *Nucleosides Nucleotides*, 1991, **10**, 303.
452. U. Asseline and N. T. Thuong, *Nucleosides Nucleotides*, 1988, **7**, 431.
453. A. Jäger, M. J. Levy, and S. M. Hecht, *Biochemistry*, 1988, **27**, 7237.
454. F. Farooqui, P. S. Sarin, D. Sun, and R. L. Letsinger, *Bioconj. Chem.*, 1991, **2**, 422.
455. R. L. Letsinger, G. Zhang, D. K. Sun, T. Ikeuchi, and P. S. Sarin, *Proc. Natl. Acad. Sci. USA*, 1989, **86**, 6553.
456. S. Nagahara, A. Murakami, and K. Makino, *Nucleosides Nucleotides*, 1992, **11**, 889.
457. K. Makino, S. Nagahara, Y. Konishi, M. Mukae, H. Ide, and A. Murakami, *Free Radical Res. Commun.*, 1993, **19**(Suppl.), S109.
458. K. Makino, A. Murakami, S. Nagahara, Y. Nakatsuji, and T. Takeuchi, *Free Radical Res. Commun.*, 1989, **6**, 311.
459. K. Nagai and S. M. Hecht, *J. Biol. Chem.*, 1991, **266**, 23 994.
460. R. L. Letsinger, C. N. Singman, G. Histand, and M. Salunkhe, *J. Am. Chem. Soc.*, 1988, **110**, 4470.
461. S. Agrawal and J.-Y. Tang, *Tetrahedron Lett.*, 1990, **31**, 1543.
462. A. Murakami, M. Nakaura, Y. Nakatsuji, S. Nagahara, Q. Tran-Cong, and K. Makino, *Nucleic Acids Res.*, 1991, **19**, 4097.
463. S. V. Patil and M. M. Salunkhe, *Nucleosides Nucleotides*, 1996, **15**, 1603.
464. S. M. Gryaznov and V. K. Potapov, *Tetrahedron Lett.*, 1991, **32**, 3715.
465. Y. Iso, F. Yoneda, H. Ikeda, K. Tanaka, and K. Fuji, *Tetrahedron Lett.*, 1992, **33**, 503.
466. S. Agrawal and P. C. Zamecnik, *Nucleic Acids Res.*, 1990, **18**, 5419.
467. C. A. Stein, C. Subasinghe, K. Shinozuka, and J. S. Cohen, *Nucleic Acids Res.*, 1988, **16**, 3209.
468. R. P. Iyer, L. R. Phillips, W. Egan, J. B. Regan, and S. L. Beaucage, *J. Org. Chem.*, 1990, **55**, 4693.
469. R. P. Iyer, W. Egan, J. B. Regan, and S. L. Beaucage, *J. Am. Chem. Soc.*, 1990, **112**, 1253.
470. J. A. Fidanza and L. W. McLaughlin, *J. Org. Chem.*, 1992, **57**, 2340.
471. J. A. Fidanza, H. Ozaki, and L. W. McLaughlin, in "Methods in Molecular Biology: Protocols for Oligonucleotide Conjugates," ed. S. Agrawal, Humana Press, Totowa, NJ, 1994, vol. 26, p. 121.
472. M. J. O'Donnell, N. Hebert, and L. W. McLaughlin, *Bioorg. Med. Chem. Lett.*, 1994, **4**, 1001.
473. M. J. O'Donnell, S. B. Rajur, and L. W. McLaughlin, *Bioorg. Med. Chem.*, 1995, **3**, 743.
474. Z. Zhang, J. A. Smith, A. P. Smyth, W. Eisenberg, G. S. Pari, and J. Y. Tang, *Bioorg. Med. Chem. Lett.*, 1996, **6**, 1911.
475. Z. Zhang, J. X. Tang, and J. Y. Tang, *Bioorg. Med. Chem. Lett.*, 1995, **5**, 1735.
476. M. P. Singh, G. C. Hill, D. Péoc'h, B. Rayner, J.-L. Imbach, and J. W. Lown, *Biochemistry*, 1994, **33**, 10 271.
477. R. Cosstick, L. W. McLaughlin, and F. Eckstein, *Nucleic Acids Res.*, 1984, **12**, 1791.
478. N. E. Conway, J. A. Fidanza, M. J. O'Donnell, N. D. Narekian, H. Ozaki, and L. W. McLaughlin, in "Oligonucleotides and Analogues: A Practical Approach," ed. F. Eckstein, IRL Press, Oxford, 1991, p. 211.

479. J. A. Fidanza, H. Ozaki, and L. W. McLaughlin, *J. Am. Chem. Soc.*, 1992, **114**, 5509.
480. H. Ozaki and L. W. McLaughlin, *Nucleic Acids Res.*, 1992, **20**, 5205.
481. J. A. Fidanza and L. W. McLaughlin, *J. Am. Chem. Soc.*, 1989, **111**, 9117.
482. R. R. Hodges, N. E. Conway, and L. W. McLaughlin, *Biochemistry*, 1989, **28**, 261.
483. N. E. Conway, J. Fidanza, and L. W. McLaughlin, *Nucleic Acids Res. Symp. Ser. #21*, 1989, 43.
484. N. E. Conway and L. W. McLaughlin, *Bioconj. Chem.*, 1991, **2**, 452.
485. A. S. Karim, C. S. Johansson, and J. K. Weltman, *Nucleic Acids Res.*, 1995, **23**, 2037.
486. F. Eckstein and G. Gish, *Trends Biochem. Sci.*, 1989, **14**, 97.
487. G. Gish and F. Eckstein, *Science*, 1988, **240**, 1520.
488. K. L. Nakamaye, G. Gish, F. Eckstein, and H.-P. Vosberg, *Nucleic Acids Res.*, 1988, **16**, 9947.
489. P. Richterich, *Nucleic Acids Res.*, 1989, **17**, 2181.
490. Z. J. Lesnikowski and R. F. Schinazi, *J. Org. Chem.*, 1993, **58**, 6531.

# 7.07
# Use of Nucleoside Analogues to Probe Biochemical Processes

LARRY W. McLAUGHLIN, MICHAEL WILSON, and
SEUNG B. HA
*Boston College, Chestnut Hill, MA, USA*

| | | |
|---|---|---:|
| 7.07.1 | INTRODUCTION | 252 |
| | 7.07.1.1 Use of Nucleoside Analogues in Conjunction with Other Techniques to Probe Biochemical Processes | 252 |
| | 7.07.1.1.1 Footprinting | 252 |
| | 7.07.1.1.2 Affinity cleavage | 253 |
| 7.07.2 | DESIGN OF NUCLEOSIDE ANALOGUES | 253 |
| | 7.07.2.1 Alteration of Hydrogen Bonding or Hydrophobic Functionality of the Nucleobases | 253 |
| | 7.07.2.2 Alteration of the Carbohydrate Structure and Functionality | 256 |
| | 7.07.2.3 Alteration of the Phosphate Linkage | 258 |
| 7.07.3 | SYNTHETIC PROCEDURES | 259 |
| | 7.07.3.1 Nucleoside Syntheses | 259 |
| | 7.07.3.1.1 Modification of existing nucleosides | 259 |
| | 7.07.3.1.2 Glycolysis procedures | 260 |
| | 7.07.3.2 Nucleosides with Modifications to the Carbohydrate | 264 |
| | 7.07.3.3 Oligonucleotides with Modified Internucleotide Linkages | 265 |
| 7.07.4 | INTERPRETATION OF BIOCHEMICAL RESULTS | 265 |
| 7.07.5 | PROTEIN–NUCLEIC ACID INTERACTIONS | 266 |
| | 7.07.5.1 Probing Protein–DNA Complexes with Base-modified Nucleoside Analogues | 266 |
| | 7.07.5.1.1 Eco RI restriction endonuclease | 266 |
| | 7.07.5.1.2 The trp repressor | 268 |
| | 7.07.5.1.3 TBP eukaryotic transcription factor | 268 |
| | 7.07.5.2 Probing Protein–Nucleic Acid Complexes with Sugar-modified Nucleoside Analogues | 268 |
| | 7.07.5.3 Probing Protein–Nucleic Acid Complexes with Phosphate-modified Analogues | 268 |
| 7.07.6 | DNA TRIPLE HELICES | 270 |
| | 7.07.6.1 Nucleoside Analogues with Altered Base Residues | 271 |
| | 7.07.6.2 Nucleoside Analogues with Altered Carbohydrate Residues | 273 |
| | 7.07.6.3 Analogues with Altered Phosphate Linkages | 273 |
| 7.07.7 | RNA CATALYSIS | 274 |
| | 7.07.7.1 The Hammerhead, Hairpin, and Tetrahymena Ribozymes | 274 |
| | 7.07.7.2 Use of Base-modified Nucleosides | 276 |

7.07.7.3   Use of Sugar-modified Analogues                                                                278
7.07.7.4   Use of Phosphorothioates                                                                       280

7.07.8   SUMMARY                                                                                          281

7.07.9   REFERENCES                                                                                       281

## 7.07.1   INTRODUCTION

Recognition and catalytic events that involve nucleic acids are affected by a number of factors, both global and local. Global factors include the electrostatic effects that result from the polyanionic nature of the nucleic acid, sequence-dependent conformational effects, and the cooperation exhibited by multicomponent systems. Local effects include the nature and contribution from individual hydrogen bonds, ion–ion interactions, or hydrophobic effects present in the recognition motif or catalytic complex. Nucleoside analogues can be valuable tools to probe a wide variety of local effects in biochemical processes in which "recognition" of individual nucleosides within the nucleic acid polymer is critical to the process under study. In the present chapter we shall consider only those processes which involve the macromolecules DNA and RNA in which the nucleoside analogue typically represents one building block of a relatively large complex. The nucleoside analogue, incorporated into a specific site in the nucleic acid sequence, typically introduces into the biochemical process an incremental alteration in structure or functional group character, and when this alteration is located in a critical portion of the macromolecule one can expect to observe an incremental response by the biochemical process. The use of a series of nucleoside analogues, coupled with an appropriate assay to monitor changes in the biochemical response, permits the localization of critical areas of interaction in large complex systems. The ability to localize the analogue to a particular site in the macromolecule permits a very detailed analysis of the contact sites and measurement of equilibrium dissociation constants, or catalytic parameters, allows for a rigorous determination of energetic values for the identified interactions. The present chapter will focus on the design, preparation, and use of nucleoside analogues placed site specifically in both DNA and RNA sequences to alter recognition or catalytic properties in a defined and targeted manner.

### 7.07.1.1   Use of Nucleoside Analogues in Conjunction with Other Techniques to Probe Biochemical Processes

In describing the use of nucleoside analogues to probe biochemical processes, we do not overlook the variety of other techniques that can also be employed to probe recognition and catalytic processes. In addition to the use of analogue sequences, various footprinting or site-specific cleavage techniques have commonly been used, principally to locate specific protein or ligand binding sites, and even to identify individual functional groups involved in the binding interactions.

#### 7.07.1.1.1   *Footprinting*

Footprinting procedures have typically been applied to sequence-specific protein–nucleic acid complexes,[1] but have also found use in assessing the binding by small ligands to DNA.[2] Footprinting, for example in a DNA–protein complex, can generally be undertaken in two forms, *protection* footprinting and *interference* footprinting. In the former, the protein is bound to the end-labeled target sequence, and this complex is then treated with a reagent that cleaves the DNA in a non-sequence-specific manner. After resolution of the resulting mixture on a high-resolution sequencing gel (and comparison with a series of control sequencing lanes), it becomes apparent that certain areas of the sequence were "protected" from the cleavage reaction and result in an apparent "footprint" along the DNA in the area where the protein was bound. The first of these techniques to be developed was DNase I footprinting.[3] DNase I will cleave double-stranded DNA through binding to the minor groove[4,5] and has been used initially to locate the binding sites of certain regulatory proteins,[6] and subsequently to identify the sequence binding preferences of minor groove binding ligands such as Hoechst 33258.[2] The disadvantage of using as enzyme as the cleavage reagent is twofold: (i) steric interference between the DNA-bound protein and the cleaving enzyme may generate a footprint that is artificially larger than the actual binding site, and (ii) the level of resolution is simply that of protected vs. nonprotected nucleosides. Subsequently, both smaller and site-specific chemical reagents have been employed as the cleavage reagents. Such agents include methidiumpropyl-EDTA/Fe$^{2+}$,[2,7] bis(1,10-phenanthroline)/Cu$^+$,[8] hydroxyl radical,[9] diethyl pyrocarbonate,[10] dimethyl sulfate,[11–13] KMnO$_4$,[14] OsO$_4$,[15] and ethylnitrosourea.[16] In some cases the reagent permits the location of specific contacts in the complex. For example, dimethyl sulfate can be used to footprint the guanine N-7 nitrogens[11–13] and ethylnitrosourea can be used to footprint the internucleotide phosphates.[16]

Interference footprinting also relies upon differential cleavage of the DNA depending upon whether the protein is bound to a specific site,[17,18] but approaches the analysis from a different perspective. In this approach, a series of nucleoside analogues are incorporated into a DNA sequence as a small fraction of the common nucleoside residues using enzymatic polymerization techniques. Ideally, one analogue is present in each sequence. After binding by the protein, the fraction of protein bound and free DNA is resolved using a nondenaturing polyacrylamide gel. The two fractions are then cleaved at the sites of the nucleoside analogues and the resulting mixture is analyzed on a high-resolution sequencing gel. Two complementary patterns are now observed. In the fraction *not bound* by the protein, a strong pattern of bands is apparent on the high-resolution gel indicating those areas of the sequence in which the analogue interfered with protein binding. In the *protein-bound* fraction, a footprint is observed at the binding site, since these sequences interfered with protein binding and were selected out during the initial fractionation.[1] For example, $N^7$-methyl-2′-deoxyguanosine is the reagent of choice to footprint G-residues using the interference footprinting approach.[17]

### 7.07.1.1.2 *Affinity cleavage*

Affinity cleavage techniques have been most commonly applied to the study of ligand–DNA binding processes. Here the concept for cleavage of the DNA is in essence reversed. The binding agent, such as the minor groove ligand distamycin, is prepared such that it contains a DNA cleavage agent, such as Fe$^{2+}$–EDTA.[19] The end-labeled DNA is then treated with the distamycin conjugate, and after binding the Fe$^{2+}$–EDTA moiety is activated with hydrogen peroxide. The hydroxyl radicals generated from the Fenton-like chemistry diffuse a short distance through the aqueous medium before they are destroyed. When the DNA lies within this critical distance, then cleavage results. Analysis of the resulting mixture on a high-resolution DNA sequencing gel (and comparison with a series of control sequencing lanes) results in cleavage bands only at sites near the Fe$^{2+}$–EDTA cleavage agent,[19] which identifies sequences bound by the ligand. Similar affinity cleavage techniques have been employed with ligands that naturally contain alkylating agents such as the CC-1065 derivatives.[20]

Both footprinting and affinity cleavage techniques are very effective in locating a specific binding site within a large DNA or RNA sequence and, in some cases, specific functional group contacts can be identified. The use of sequences containing nucleoside analogues as described in this chapter permits a more detailed analysis of specific functional group contacts present in DNA–protein or DNA–ligand complexes, as well as those of importance for catalytic processes. Rigorous energetic values can also be assigned to identified interactions and such data will assist in characterizing the role of specific contacts in the formation of such complexes.

## 7.07.2 DESIGN OF NUCLEOSIDE ANALOGUES

### 7.07.2.1 Alteration of Hydrogen Bonding or Hydrophobic Functionality of the Nucleobases

In probing the identity and energetics of specific interactions involving the nucleobases of a DNA or RNA sequence, the analogues of most value are those which are best able to mimic the characteristics of the common base residues, but yet lack or alter the properties of a specific functional group potentially involved in a contact that defines affinity within the macromolecular complex. These analogues typically result from the most conservative alterations to the base residues, that of deleting a functional group and replacing it simply with hydrogen, or in other cases a carbon. Such "deletion-modified" analogues include, for example, the 3-deazapurines and the 7-deazapurines in which the N-3 or N-7 nitrogen has been excised and replaced by =CH—. A typical set of analogues that can be used to probe contacts to specific functional groups on the guanine and cytosine bases are illustrated in Figure 1. Special care must be taken in the choice of analogue when the target for the study is double-stranded DNA or RNA. For example, when the target sequence involves a dG–dC base pair (Figure 2(a)), it is unlikely that the internal functional groups, the N-3 nitrogen of dC or the N-1 hydrogen of dG, will be involved in critical contacts (unless the helix is denatured upon complex formation). Moreover, the use of 3-deazacytosine (c³C) in place of C is likely to result in severe destabilization and/or distortion of the DNA helix as the C-3 hydrogen of c³C is opposing the N-1 hydrogen of G (Figure 2(b)). A variety of nucleosides containing altered bases have been prepared by Seela and co-workers.[21-28]

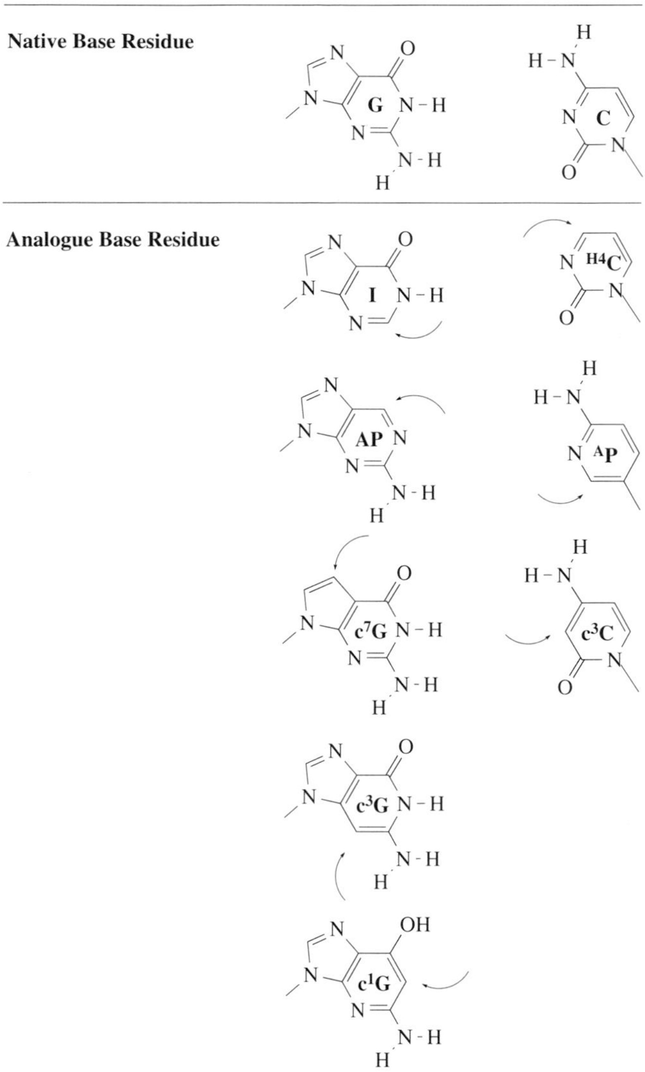

**Figure 1**  Base modified analogues for guanine (G) and cytosine (C). I = hypoxanthene (inosine); AP = 2-aminopurine; c⁷G = 7-deazaguanine; c³G = 3-deazaguanine; c¹G = 1-deazaguanine; ᴴ⁴C = 2-pyrimidinone; ᴬP = 2-aminopyridine; c³C = 3-deazacytosine.

Even in the absence of such clear steric clashes between native and analogue base residues, the "simple" deletion of individual functional groups does not always occur without consequence. For example, tautomeric changes, resulting in the conversion of a hydrogen-bond donor to acceptor (or acceptor to donor), can result at a second site upon the deletion of a single functional group (compare the N-1 nitrogen of 2-aminopurine (AP) and G, and the O-6 oxygen of 1-deazaguanine (c¹G) and G in Figures 2(c) and 2(d)). Such tautomeric effects can be eliminated in some cases by the appropriate choice of analogue. For example, the simplest deletion-modified analogue of dT used to probe the role of the O-2 oxygen in minor-groove complexes would appear to be the 4-pyrimidinone (ᴴ²T illustrated in Figure 3(a)). However, the loss of this carbonyl results in a tautomeric change at the N-3 nitrogen such that it now becomes a hydrogen-bond acceptor. In principle, a base pair might still form between this analogue and dA (Figure 3(a)); it would simply

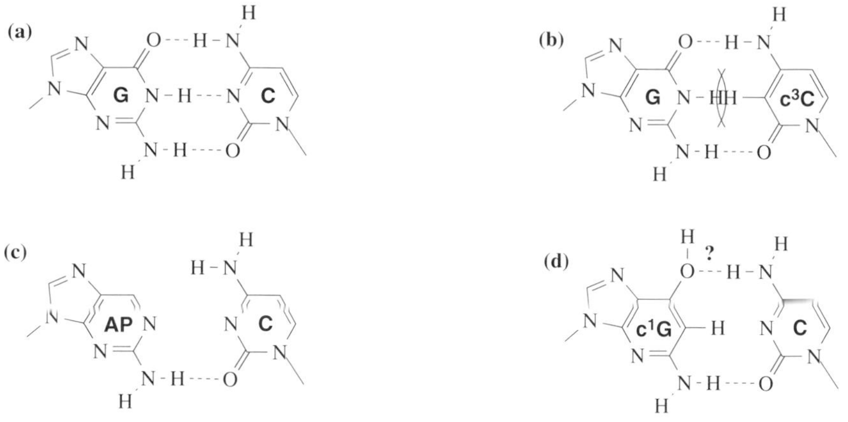

**Figure 2** Analogue G–C base pairs.

lack one of the two interstrand hydrogen bonds. However, in practice DNA duplexes containing this analogue were severely destabilized,[29] presumably owing to the presence of the opposing lone pairs (Figure 3(b)) in the base pair. The problem could be solved in this case by using an analogue prepared as a *C*-nucleoside (Figure 3(c)). Use of the 2-pyridone *C*-nucleoside (2P) maintains the "normal" dA–dT-like interstrand hydrogen bonding,[30,31] yet still permits the deletion of the $O^2$-carbonyl. DNA duplexes containing this dT analogue exhibit $T_m$ values that are virtually indistinguishable from those containing dT.[31]

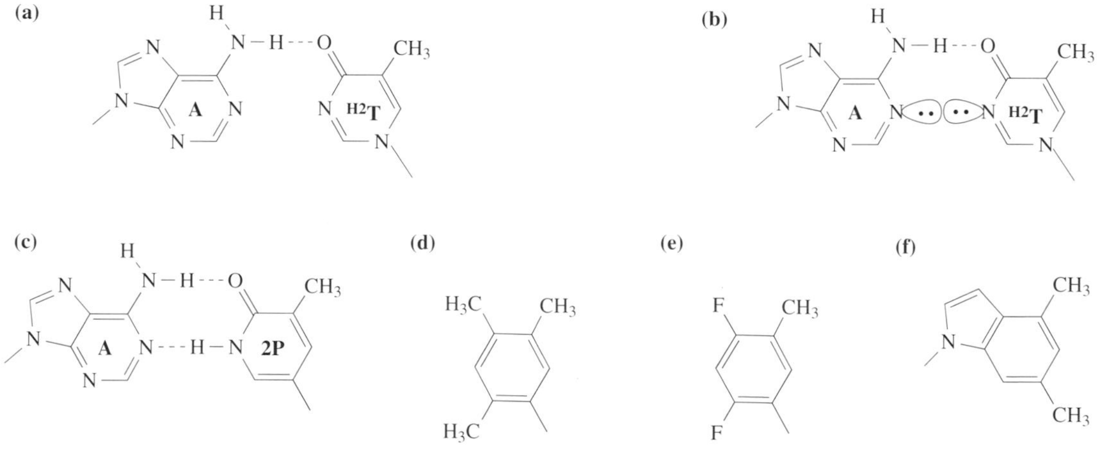

**Figure 3** Analogue A–T pairs (a–c) and T (d, e) and G (f) afunctional isosteres.

A related group of base-modified nucleoside analogues are those which mimic the shape of the base residue, but remove the common hydrogen-bonding acceptor and donor functional groups. These derivatives have typically been prepared as *C*-nucleosides with fluorine atoms or methyl groups to replace the carbonyl and/or exocyclic amine functional groups.[32,33] The derivatives have the potential to probe for enhanced hydrophobic effects, in the absence of hydrogen-bonding interactions, in macromolecular complexes. Modified nucleosides of this type, such as the thymidine and deoxyadenosine isosteres prepared by Kool and co-workers[34,35] (Figure 3(d)–(f)), have been used to investigate the importance of noncovalent interactions in the stabilization of DNA structures (such as hairpin loops). These compounds, unlike their natural counterparts, have little or no hydrogen-bonding capability, but have been found to stabilize significantly double and triple helical DNA structures. This has been attributed to improved stacking of these less polar bases.

Although "deletion-modified" nucleosides can be very valuable for the study of macromolecular complexes containing DNA or RNA, in some cases it is desirable to add functionality to a molecule rather than eliminate it. For example, one of the challenges in the generalization of DNA triplexes has been finding a dC or dm⁵C analogue for the pH-independent formation of C–G–C (or similar) base triplets. In this case it is desirable to change the tautomeric form of the N-3 nitrogen of dC (Figure 4(a)) from a hydrogen-bond acceptor to a hydrogen-bond donor. Three dC pyrimidine-like analogues have been described in which this tautomeric change was effected. In one case the *C*-

nucleoside with the pseudoisocytosine base was employed (Figure 4(b)),[36,37] a related pyrazine *C*-nucleoside also results in a bidentate hydrogen-bond donor (Figure 4(d)),[38] while in a third case a carbonyl was added to the C-6 carbon of m⁵C to control the tautomeric form of the N-3 nitrogen (Figure 4(c)).[39,40] In all cases the analogues result in a change in functionality at the N-3 nitrogen with the presence of an N-3 hydrogen-bond donor.

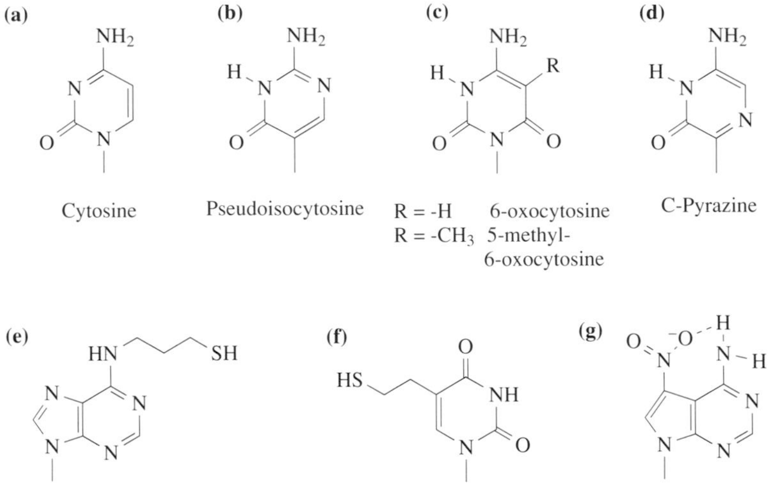

**(a)** Cytosine　**(b)** Pseudoisocytosine　**(c)** R = -H　6-oxocytosine
R = -CH₃　5-methyl-6-oxocytosine　**(d)** C-Pyrazine

**(e)**　**(f)**　**(g)**

**Figure 4**　Cytosine and three neutral analogues containing a formal N-3 hydrogen-bond donor (a–d). A thiol tethered to A (e) or T (f), or a nitro group added to A (g).

The cross-linking of DNA, a useful tool for the study of DNA structure and function, can be achieved by adding a functionality such as an alkanethiol to a specific residue (Figures 4(e) and 4(f)).[41,42] This functionality can form disulfide bonds with other similarly modified bases in adjacent DNA strands, thus forming a cross-linked DNA duplex. It has been found that such interstrand disulfide linkages can significantly stabilize the duplex DNA in what would otherwise be high-energy conformations, and permit the study of the function of unusual DNA structures.

Using base-modified nucleosides, Verdine and co-workers[43] have developed a technique termed template-directed inference (TDI) footprinting for the study of DNA–protein interactions. This method gives information about which DNA bases are contacted by proteins, their groove location, and approximate energies of interaction. For example, a nucleoside analogue containing a nitro group attached to the base moiety (Figure 4(g)) has been used for the TDI footprinting of adenosine.[43] This adenosine analogue shows the same base pairing preference as its native counterpart, and can be selectively degraded by treatment with aqueous piperidine, leading to DNA strand scission. The TDI footprinting method has also been applied to the other major bases.[17,18,44]

## 7.07.2.2　Alteration of the Carbohydrate Structure and Functionality

The carbohydrate portion of the DNA or RNA chain may have a minimal direct effect in terms of "recognition" by sequence-specific binding proteins or in DNA triplex formation, but confirmation effects can impact the relative orientations of base pairs and thus affect interactions with proteins or other sequences of DNA. RNA complexes that are nominally single-stranded may be more directly impacted by carbohydrate-based nucleoside analogues since the 2'-OH represents a functional group that is available for hydrogen bonding interactions or metal chelation.

A number of carbohydrate analogues of nucleosides have been examined in nucleic acid sequences, and those most related to the "deletion-modified" base analogues described above would be the RNA nucleoside analogues containing an altered 2'-OH. For example, the simplest deletion-modified analogue results when the 2'-deoxynucleoside is substituted into the RNA sequence for the native ribo derivative. Although in principle the 2'-deoxynucleoside appears to be a simple deletion modification of the ribo sugar, in fact other effects are present. The ribo nucleoside derivatives prefer the 3'-*endo* conformation[45] (Figure 5(a)) whereas the 2'-deoxy derivatives favor the 2'-*endo* conformation[45] (Figure 5(b)). Therefore, substituting the 2'-deoxynucleoside for the ribo derivative both removes the hydroxyl group from the carbohydrate and additionally alters the conformational preference for the carbohydrate. In the light of the dual effects on functionality and conformation, the 2′-deoxy-2′-fluoro nucleoside derivatives have also been employed as effective analogues.[46,47] The fluoro group provides an electronegative substituent at the 2′-carbon, and the presence of the electronegative group tends to maintain a 3′-*endo* conformation.[45] The minimal hydrogen-bonding capability of fluorine results in a conformationally stable but deletion-modified nucleoside analogue. Use of the 2′-*O*-methyl derivatives[48] (or allyl[49]) also blocks the hydrogen-bond donor capability of the 2′-OH, but the presence of the added methyl (or allyl) group must be accounted for in the interpretation of biological processes in which such derivatives are used. The remaining functionality that can be effectively deleted in the furanose ring is the potential hydrogen-bond acceptor, the O-4′ oxygen. Carbocyclic analogues have been prepared in which the furanose ring is replaced by a cyclic pentane ring,[50] eliminating the furanose ring oxygen.

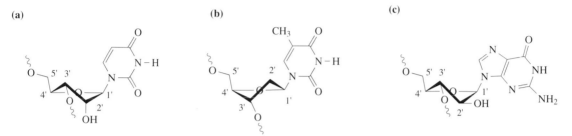

**Figure 5** Sugar conformations: (a) uridine with a C-3′ *endo* conformation: (b) thymidine with a C-2′ *endo* conformation; (c) araguanosine with a C-3′ *endo* conformation.

A second type of nucleoside analogue results when the configuration of the carbohydrate has been altered. This type of analogue is again more prevalent with RNA studies than with DNA. A simple inversion of the C-2′ carbon of a ribonucleoside results in the arabinosyl nucleosides (e.g., araG, Figure 5(c)) and repositions the 2′-hydroxyl without major alterations to the sugar phosphate backbone of the RNA polymer.[51] Other types of configurational isomers are possible, and DNA sequences containing xylo nucleosides[52] and also α-(ribo)nucleosides[53] have both been described.

A third type of nucleoside analogue containing an altered carbohydrate is that which can be considered to be conformationally *less* flexible, or conformationally *more* flexible, than the common ribose derivative. Three types of conformationally less flexible derivatives are illustrated in Figure 6. In one case the C-3′ and C-5′ carbons have been fixed in a second five-membered ring to generate bicylcodeoxynucleosides (Figure 6(a)).[54,55] These derivatives are coupled together using native phosphodiester linkages. A second type of less flexible derivative uses a second five-membered ring that incorporates the 2′-OH and 3′-OH and replaces the phosphodiester linkage with a riboacetal contained within the second five-membered ring (Figure 6(b)).[56] This latter derivative, while also reducing the flexibility of the carbohydrate, must be considered a highly specialized derivative in that the phosphodiester linkage is also replaced by a neutral nonphosphate linkage. The third derivative lacks the C-2′ carbon of the ribose ring, but has a methylene bridge between the C-4′ of the carbohydrate and the C-6 carbon of the pyrimidine base to fix the nucleoside in the *anti* conformation (Figure 6(c)).[57]

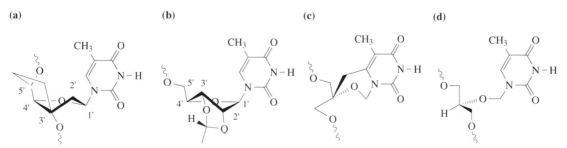

**Figure 6** (a–c) Three thymidine nucleosides with conformationally less flexible carbohydrate analogues and (d) thymidine nucleoside with a conformationally more flexible carbohydrate analogue.

There are a number of conformationally more flexible derivatives with a variety of acyclic carbohydrate linkers. We shall mention here only a single example, one that can adopt a conformation similar to that found in the ribose carbohydrate. In this respect, the simple acyclic linker that results when the C-2′ carbon is excised from the ribose ring produces a highly flexible "carbohydrate" linker (Figure 6(d)) that has been incorporated into nucleic acids for studies of recognition.[58,59]

### 7.07.2.3  Alteration of the Phosphate Linkage

The internucleotide phosphodiesters of DNA and RNA polymers are fully charged at physiological pH values, and provide sites for ionic interactions in complex formation involving proteins, a third strand of DNA, or in the folding of ribozymes. At least two types of analogues can be valuable in probing interactions involving the sugar phosphate backbone of nucleic acids. In both cases the nature of the charged phosphate is incrementally altered. The least perturbing alteration simply involves replacing one of the phosphate oxygens with sulfur.[60] The resulting phosphorothiate can have the sulfur substituted for one of the prochiral nonbridging phosphate oxygens, or it can be placed such that it is in the 3′-bridging[61] or 5′-bridging position.[62,63] The initial studies using phosphorothioates involved placing the sulfur in a nonbridging position.[60] Phosphorothioates of the former type are formed as two diasteromers with *Sp* and *Rp* configurations (Figures 7(a) and (b)). Sulfur is less electronegative than oxygen, and functions as a "softer" ligand than does oxygen—both properties can be expected to alter reactions that involve nucleophilic attack at the phosphorus center, or those involving coordination, or other types of interactions with the nonbridging oxygens. The effects are subtle, such that the biochemical properties of these derivatives are in some cases similar to those of the native derivatives. For example, DNA polymerases will typically use the α-thio-2′-deoxynucleoside triphosphates as substrates for the replication of DNA; however, whereas the polymerization activity remains largely unaltered, the editing function of such polymerases, that is the 3′ → 5′ exonuclease activity, is inhibited.[60] Both the polymerization and the exonuclease activity require a nucleophilic attack at the phosphorus center, in the first case by the incoming 3′-OH of the next dNTP and in the latter case by water.

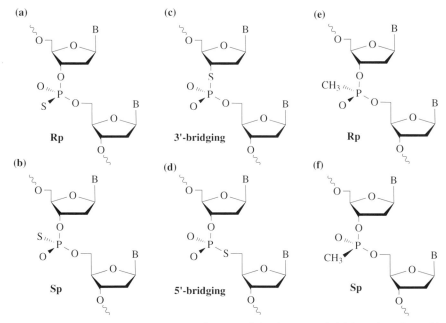

**Figure 7**  Phosphodiester linkage analogues. (a) The *Rp* and (b) the *Sp* nonbridging phosphorothioate diesters; (c) the 3′-bridging phosphorothioate; (d) the 5′-bridging phosphorothioate; (e) the *Rp* and (f) the *Sp* methyl phosphonate linkages.

In many biochemical processes, including the polymerase activity noted above, a metal cofactor is often involved, and one of its roles can be coordination to one of the prochiral nonbridging oxygens. Metal binding sites can be located by preparing the stereochemically pure *Sp* or *Rp* phosphorothioates and examining metal specificity effects. For example, manganese coordinates both oxygen and sulfur ligands about equally, whereas magnesium coordinates oxygen about 30 000 times more effectively than sulfur.[64,65] By using the stereochemically pure *Sp* and *Rp* phosphorothioates, it is possible to probe for metal binding sites in various biochemical processes.[60]

Bridging phosphorothioates result when either the 3'- or 5'-bridging oxygen of the phosphodiester residue is replaced by sulfur (Figures 7(c) and (d)). Both of these derivatives can have significant effects, particularly on biochemical processes, that result in cleavage of one of the phosphodiester bonds. By replacing the leaving group (either the 3'- or 5'-bridging oxygen) by sulfur, for example in a reaction that results in cleavage of the bonds to the sulfur atom,[62,66,67] significant differences in rates can be expected (provided that the bond breakage phenomenon is rate limiting). The $pK_a$ of the alkoxide vs. the thioate leaving group are significantly different (the $pK_a$ of methanol is $\sim 15.5$ whereas that of methanethiol is $\sim 10.5$). These analogue linkages can dramatically alter the rates of strand cleavage on their own,[63,67] but can also be used to probe for metal binding sites in a similar manner to that described above.[66,68,69]

A more dramatic effect on the nature of the phosphodiester linkage results when it is replaced by the methylphosphonate linkage.[70] With this analogue, one of the nonbridging oxygens is replaced by a methyl group resulting in a neutral but chiral linkage (Figures 7(e) and (f)). Such derivatives can be used to probe interactions at specific phosphodiesters because the ionic character of the linkage is completely eliminated.

## 7.07.3 SYNTHETIC PROCEDURES

The synthetic procedures necessary to incorporate a nucleoside analogue into a DNA sequence at a preselected site generally involve, (i) the synthesis of the nucleoside analogue, (ii) its protection as a suitable derivative for DNA synthesis, and (iii) its incorporation into the DNA polymer through assembly, deprotection, and purification protocols. In the ideal case only the first step varies for different analogues; however, in some cases labile analogues may require special procedures for DNA assembly and deprotection. Additionally, analogues that involve alterations to the internucleotide phosphate linkage are often prepared during DNA assembly procedures. Nevertheless, most analogues described in this chapter are first prepared as the nucleoside monomer and we shall concentrate on describing selected procedures for the preparation of such analogue monomers.

### 7.07.3.1 Nucleoside Syntheses

Two general formats can be followed in the preparation of the nucleoside analogue, either (i) an existing nucleoside is suitably modified to alter functionality and generate the desired analogue, or (ii) the analogue is synthesized directly by preparation of the heterocycle and then coupling to the desired carbohydrate moiety. We shall provide examples of each type of approach but our goal here is to illustrate methods and approaches for analogue preparation and not to review the details of such procedures.

#### 7.07.3.1.1 Modification of existing nucleosides

A general procedure can be used for both purine and pyrimidine nucleosides, in essence to convert a carbonyl functionality into an amino group or to remove the carbonyl and leave simply a hydrogen atom. The two procedures are related and will be illustrated for 2'-deoxyguanosine, which can then be converted either into 2,6-diaminopurine-2'-deoxyriboside (C=O → —NH₂) or into 2-amino-purine-2'-deoxyriboside (C=O → —H). The first procedure, described by Gaffney and co-workers,[71,72] involves the sulfonation of a suitably protected derivative of dG. The sulfonylated compound (**1**) reacts with excess trimethylamine to give the trimethylammonium salt (**2**), which with excess liquid ammonia generates (**3**) (Scheme 1). In a related reaction,[73] the 2-aminopurine derivative (**5**) can be prepared via the same sulfonated 2'-deoxyguanosine derivative (**1**). The sulfonated compound (**1**) reacts rapidly with hydrazine in dioxane at ambient temperature to give the 6-hydrazino derivative (**4**). The hydrazino compound (**4**) is oxidized in the presence of silver(I) oxide to generate the protected 2-aminopurine compound (**5**). In a similar manner, dT or dU can be correspondingly sulfonated at the O-4 position and then reacted with either ammonia or hydrazine,[74] the latter followed by oxidation, to generate 2'-deoxy-5-methylcytidine or the related 2-pyrimidinone-2'-deoxyribosides. Other leaving groups can be employed in this type of displacement reaction, for example the 4-triazolyl derivative is a useful intermediate in the preparation of 2'-deoxy-5-methylcytidine from dT.[75]

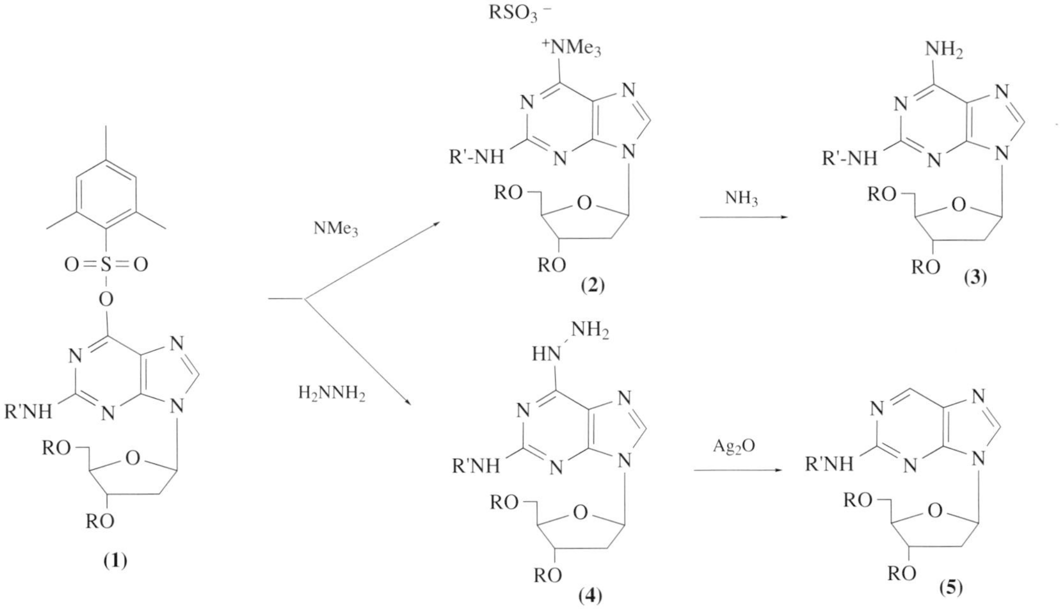

**Scheme 1**

Diazotization of the exocyclic amino group followed by treatment with base can be used to remove an exocyclic amino group and generate a carbonyl at the corresponding site. This reaction has little value for dC or dA, but does provide a valuable route to the preparation of dX (X = xanthosine) from dG.[76]

The 2-thio-,[29] 4-thio-,[77] and 2,4-dithiopyrimidines[78] (or the corresponding 6-thiopurines[79]) can be prepared by reaction of either dT or dU (or dG) with a suitable sulfurizing reagent. It is also possible to remove the sulfur from such compounds (e.g., Raney nickel) to generate, for example, the 4-pyrimidinone ring[29] system in which the O-2 carbonyl is replaced simply by hydrogen.

Finally, in many cases the ribonucleoside is more readily available (or more easily synthesized as a single anomer, see below) than the corresponding 2'-deoxynucleoside. However, the ribonucleoside can now be readily reduced to the 2'-deoxy derivatives by reduction of the 2'-*O*-phenylthiocarbonate derivative with tin hydride.[80]

### 7.07.3.1.2  *Glycolysis procedures*

Glycolysis procedures involve nucleoside preparation by the coupling of the desired heterocycle to the appropriate carbohydrate residue. In most nucleoside syntheses, the heterocycle and the carbohydrate are connected through a C—N bond (*N*-nucleosides), but there is a growing number of important nucleoside analogues in which this bond is a C—C linkage (*C*-nucleosides).[81] A large number of protocols have been developed to accomplish these reactions. For example, the formation of the C—N bond in *N*-nucleosides can be commonly achieved by the Vorbruggen method (modified silyl-Hilbert–Johnson reaction) with acid-catalyzed couplings, high-temperature fusion, and alkylation of the alkali metal (Na or K) salts of the heterocyclic nucleobase.

### (i)  *Acid-catalyzed glycosylation*

A common glycosylation method used to synthesize both pyrimidine[82,83] and purine[84] nucleosides involves the condensation of the silylated heterocycle with a suitably protected carbohydrate by a Lewis acid (Scheme 2). The silylated heterocycle (**7**) is prepared *in situ* simply by refluxing the heterocycle (e.g., (**6**)) with hexamethyldisilazane (and a catalytic amount of ammonium sulfate) or by reaction at ambient temperature with *N,O*-bis(trimethylsilyl)acetamide. Different leaving groups can be present at the C-1′ position of the carbohydrate (**8**), but the 1-chloro-2-deoxy-3,5-di-*O*-toluoyl-α-D-*erythro*-pentofuranose derivative (**11**), where the chloro group is displaced during glycosylation, is commonly used for the synthesis of 2′-deoxynucleosides, since during the chlorination step of the sugar, the α-anomer precipitates out of solution in high purity.[85] A variety of Lewis acids can be employed as catalysts, for example tin(IV) chloride, trimethylsilyl perchlorate, trimethylsilyl trifluoromethanesulfonate and AlCl₃. Ribonucleosides are produced solely as the β-anomer (**10**) by such procedures, while the 2′-deoxynucleosides generally give a mixture of α- and β-anomers (**13** and (**14**)). These results can be explained by a glycosylation reaction that occurs largely through an $S_N1$ mechanism in which initially the chloro leaving group is lost to form an oxonium ion (**9**) or (**12**). In the ribonucleosides, neighboring group stabilization by the 2′-ester permits attack by the heterocycle from only one side of the C-1 carbon (**9**) yielding solely the β-nucleoside (the "*trans*" effect). The absence of a 2′-substituent in the deoxysugar results in attack at either side of C-1 (**12**) generating both α- and β-nucleosides. Glycosylation often occurs preferentially at one of the two possible pyrmidine nitrogens, but regioisomeric mixtures are not uncommon. For example, both m⁵ᵒˣC (**16**) and (**17**) and 6-aminothymidine (**18**) and (**19**), each present as the α- and β-anomers, are generated upon acid-catalyzed glycosylation of 6-aminothymine (Scheme 3).[39,40]

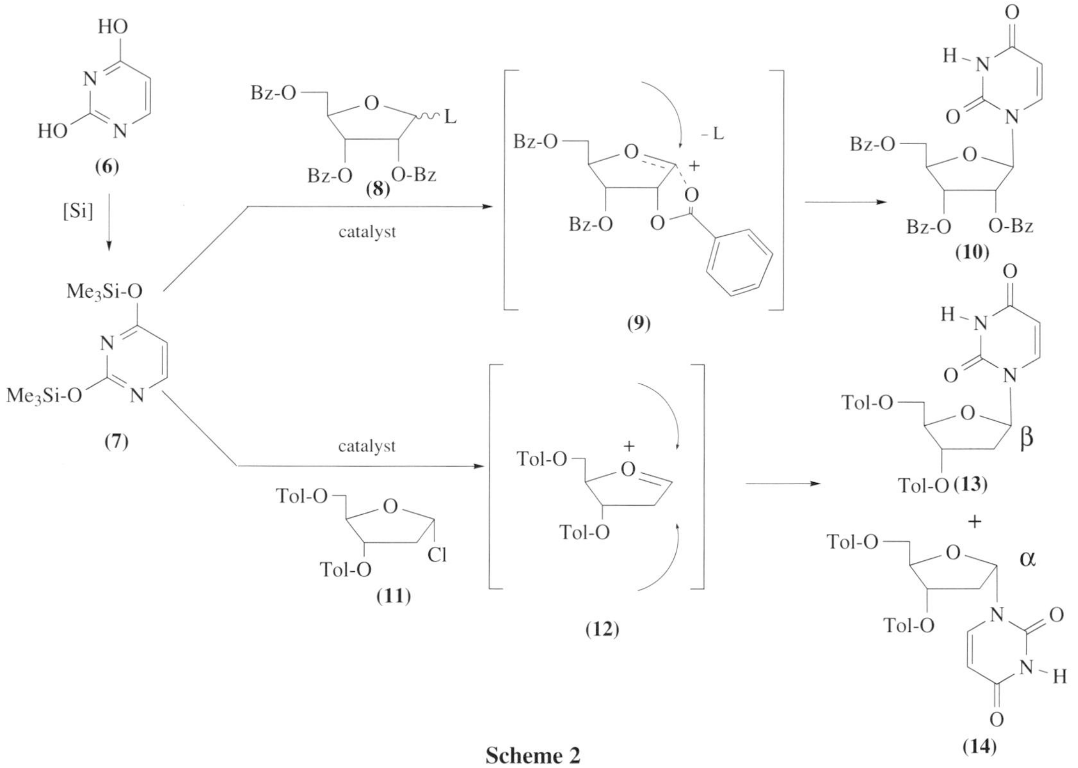

**Scheme 2**

*(ii)  Glycosylation by heavy metal salts of the nucleobases*

   Silver salts of purine bases were first employed in the synthesis of purine nucleosides, but the chloromercuri derivatives are more easily prepared and in subsequent work have been more commonly used. The chloromercuri salt of the nucleobase can either be prepared and isolated or prepared *in situ*, and added to the sugar component containing a suitable leaving group at the C-1 position. As with the acid-catalyzed procedures described above, only the β-anomer of ribonucleoside results, whereas both the α- and β-anomers of 2′-deoxynucleosides are obtained.

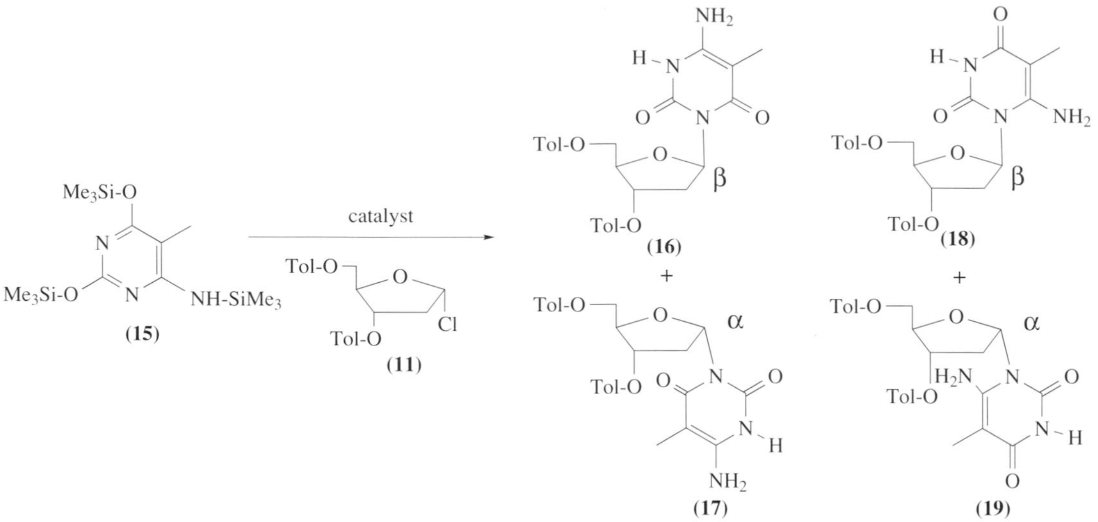

**Scheme 3**

### (iii) Direct glycosylation by alkali metal salts

Preparation of the alkali metal salts of purine bases (commonly with sodium hydride) permits the generation of a strong nitrogen nucleophile on the purine base (Scheme 4). This species can be used for direct $S_N2$ displacement of an appropriate leaving group from the C-1 position of the carbohydrate. When used in conjunction with the $\alpha$-chloro derivative of 2-deoxyribose (**11**), the reaction proceeds with inversion of configuration to produce only the $\beta$-anomer of the 2′-deoxynucleoside (**21**). In some cases regioisomeric products result (see (**24**) and (**25**) in Scheme 4), but such product mixtures can often be more easily separated than the corresponding mixtures of $\alpha$- and $\beta$-anomeric isomers. Owing to the stereochemical specificity of this reaction, it is probably the most effective glycosylation procedure for the preparation of 2′-deoxynucleoside derivatives of the purine heterocycles.

### (iv) Synthesis of C-Nucleosides

Replacing the anomeric nitrogen of a nucleoside with carbon results in a carbon–carbon bond tethering the heterocycle and the carbohydrate portions of the nucleoside. The glycosylation procedures for *C*-nucleosides are less well developed than those for *N*-nucleosides, but a review[81] provides coverage of the various procedures used in the preparation of *C* nucleosides. Here we shall simply highlight a few procedures, in addition to glycosylation, used to prepare *C*-nucleosides for use as analogues in various biochemical studies. In principle there are three common procedures for the generation of *C*-nucleosides: (i) a naturally occurring *C*-nucleoside can be modified to generate the analogue; (ii) the carbon–carbon bond can be introduced very early in the synthesis with the elaboration of this simplified derivative (e.g., a 1-cyano sugar) to the nucleoside; and (iii) the intact heterocycle can be glycosylated with the appropriate sugar derivative.

An example of the first approach is the synthesis of the fully protected phosphoramidite derivative of 2′-*O*-methylpseudoisocytidine from pseudouridine in a six-step synthesis ((**27** → **28**) Scheme 5).[36,37] This approach has the advantage that it by-passes the difficult step of generating the carbon–carbon bond between the heterocycle and the carbohydrate. When the appropriate naturally occurring *C*-nucleoside is available, this approach is likely to be the most efficient, but is limited by the availability of the desired starting materials.

The second approach can also be employed for the synthesis of the same parent *C*-nucleoside. In this case the introduction of the carbon–carbon bond to the carbohydrate occurs at an early point in the synthesis through a relatively simple derivative. For example, the 2,3-*O*-isopropylidene-5-*O*-trityl-D-ribofuranose ((**29**), Scheme 5) was reacted with a Wittig reagent to produce (**30**).[86] This step introduces the requisite carbon–carbon bond, and from this intermediate the six-membered heterocyclic nucleoside is then obtained. This initial derivative was readily cyclized to the ribose

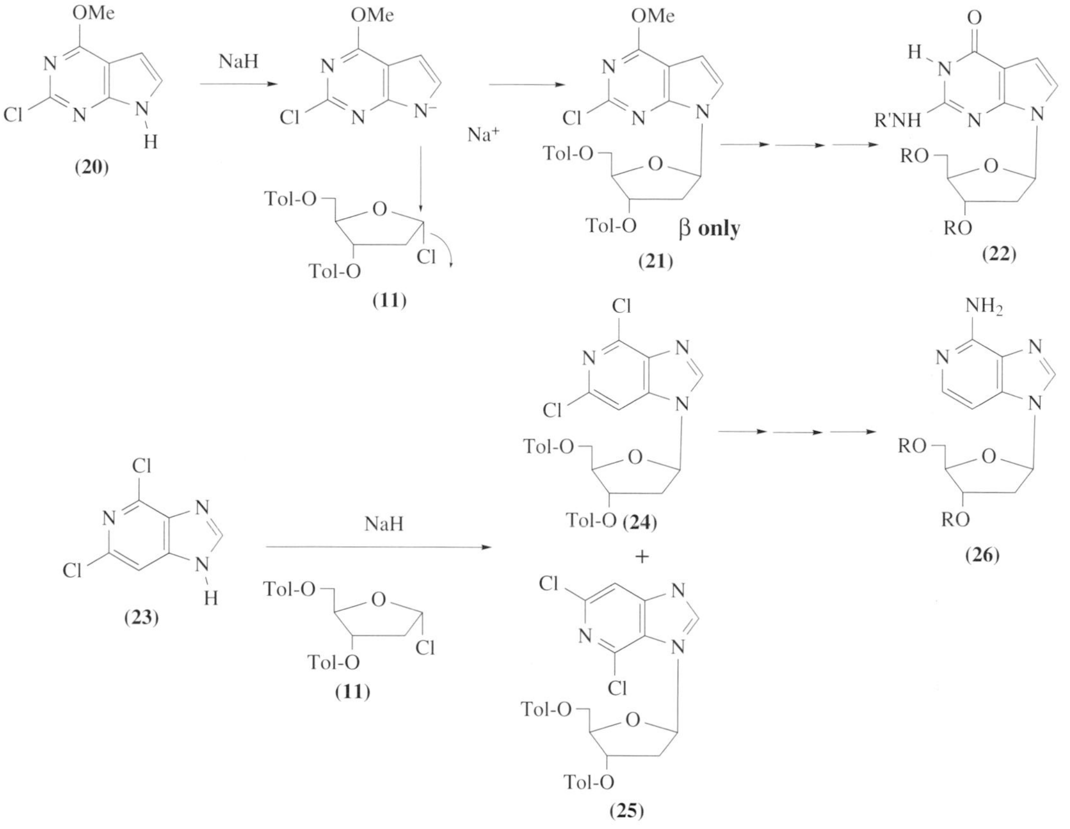

**Scheme 4**

derivative (**31**),[87] which was then converted into the active enolate (**32**).[88] Cyclization of this product with guanidine (**32** → **33**) and acidic deprotection with methanolic HCl generated the *C*-nucleosides (**34**), from which the β-anomer could be selectively crystallized.[89] Similar procedures begin with the carbon–carbon bond formed with a cyano or alkyne at the anomeric position of the sugar followed by further elaboration to form the desired heterocycle and complete the synthesis of the nucleoside analogue.[81]

The third method that will be noted here is the coupling of a preformed heterocycle to the sugar residue by formation of the carbon–carbon bond. This approach most closely parallels that employed for the synthesis of *N*-nucleosides. The simplest approach, that of preparing the carbon anion (lithium or cadmium salt)[90] or a Grignard derivative[32] of the heterocycle and using such intermediates to displace a leaving group on the sugar (e.g., chloro) have not proved to be very successful except in isolated cases. From the other possibilities we shall only note one which we have found particularly useful. Palladium-mediated Heck-type couplings[91–93] can be employed by using the iodo derivative of the heterocycle (e.g., (**36**), Scheme 6) and the 1,2-unsaturated sugar derivative ((**35**), Scheme 6). The glycal is prepared simply by refluxing the desired thymidine derivative in hexamethyldisilazane. Silylation of the thymine heterocycle followed by elimination of the nucleobase generates the desired glycal. The glycal has been most commonly prepared with a bulky group at the 3′-OH (e.g., *t*-butyldimethylsilyl)[91,94] so that the Heck-type coupling proceeds with the heterocycle approaching the sugar derivative from the opposite face, generating solely the β-anomer ((**37**), Scheme 6). Formation of the carbon–carbon bond occurs with yields as high as 90%,[30] but is dependent upon the nature of other factors, for example, the choice of the ancillary palladium ligand. After coupling, and removal of the silyl protecting group, the resulting 3′-keto derivative can be selectively reduced to form the 2′-deoxyribose sugar (e.g., (**38**), Scheme 6). This approach, pioneered by Daves and co-workers,[91–93] has been employed for the synthesis of various derivatives of 2′-deoxypseudouridine. We have described the synthesis of a thymidine analogue ((**38**), Scheme 6), from which the O-2 carbonyl has been deleted, and used this derivative to probe the activity of T7-primase.[31]

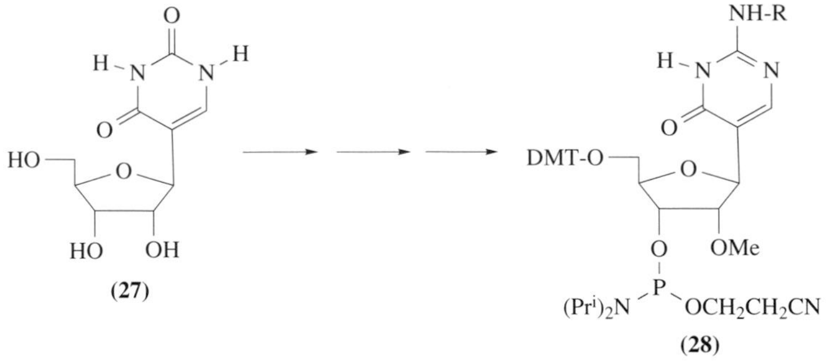

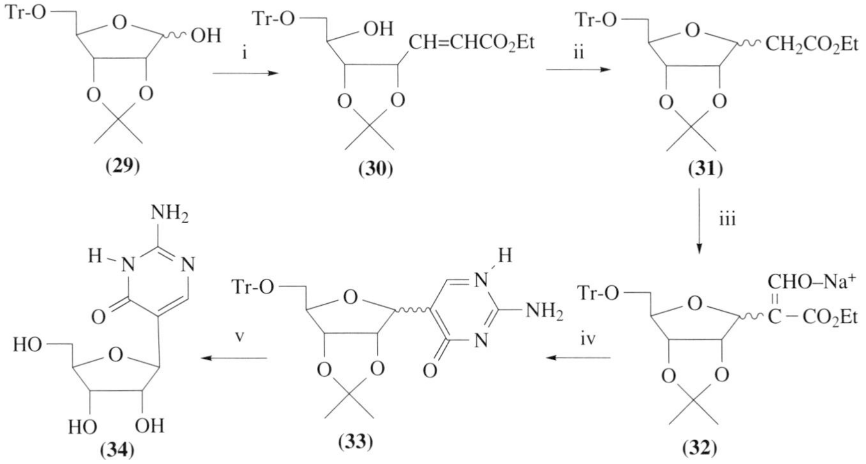

i, Ph₃P=CHCO₂Et/CH₃CN; ii, NaOEt/EtOH; iii, NaH, HCO₂Et;
iv, guanidine/EtOH; v, H⁺, crystallization

**Scheme 5**

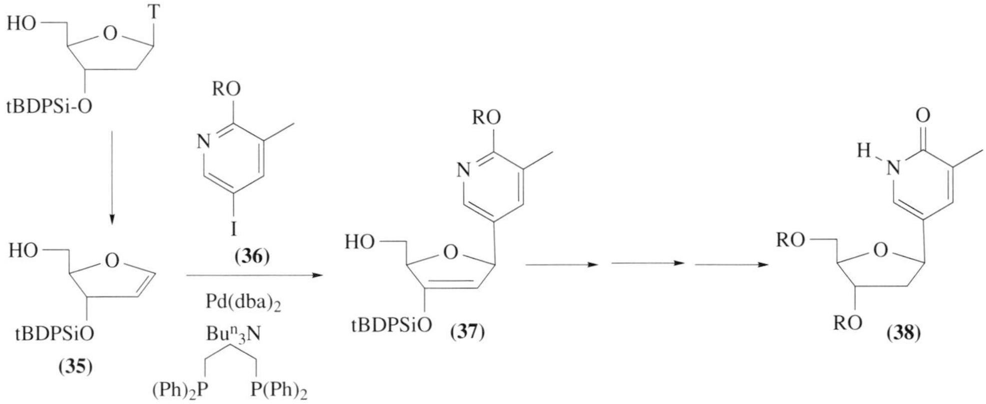

**Scheme 6**

### 7.07.3.2 Nucleosides with Modifications to the Carbohydrate

Nucleoside analogues which contain altered carbohydrates are prepared principally by two methods. The existing nucleoside is transformed directly into the analogue. For example, procedures have been developed to permit the synthesis of 2′-deoxy-2′-fluoro nucleosides, in addition to the corresponding amino derivatives,[95] thiol derivatives,[96] and 2′-*O*-methyl[97] and 2′-*O*-allyl[98] analogues (Figure 8). Alternatively, the carbohydrate can be preformed with the desired modification, and then a coupling reaction, such as those described above, can be employed to attach the heterocycle.

This latter procedure has been employed for the preparation of the arabinosyl[99] and xylo[100] carbohydrate analogues of the ribonucleosides, and also the 4′-deoxy-4′-thio and 4′deoxy derivatives[50] (Figure 8).

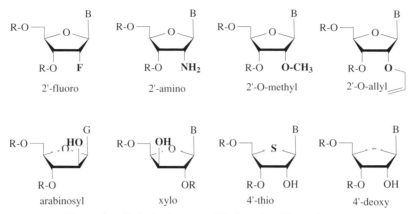

**Figure 8**   Carbohydrate-modified nucleoside analogues.

### 7.07.3.3   Oligonucleotides with Modified Internucleotide Linkages

A wide variety of oligonucleotides containing modified linkages have been described, principally for the generation of sequences useful for antisense studies. Such studies are described in detail in Chapter 7.08. We note here only two types of linkage analogues that have been employed to probe biological processes, the phosphorothioates and the methyl phosphonates (Figure 7). Both types of analogue linkages are introduced either site specifically or universally for each linkage during the assembly of the oligonucleotide. Both the methyl phosphonate and the nonbridging phosphorothioate replace one of the prochiral nonbridging oxygens of the internucleotide phosphate linkage with either a methyl group or sulfur. In both cases, two diastereomers result (Figure 7). When the analogue linkage is present at a single site, it is possible in some cases to resolve the two diastereomeric analogues.[101,102] Otherwise, it is more efficient to prepare a dimer building block, resolve the dimers into the *Rp* and *Sp* diastereomers, and then incorporate each stereochemically pure dimer into the preselected site.[102–104] Methyl phosphonate linkages are prepared using the appropriate methyl phosphonate monomer, and can be introduced into the DNA sequence using standard coupling procedures.[105] Single diastereomeric linkages can be prepared by the synthesis and resolution of the appropriate dimer, its conversion to a phosphoramidite building block and its incorporation into the DNA sequence.[106,107] The nonbridging phosphorothioate is prepared using a common phosphoramidite monomer and oxidation with a sulfur reagent after the coupling step.[102,108]

In addition to the nonbridging phosphorothioates, it is also possible to prepare the phosphorothioate analogues in which the sulfur atom appears in either the 3′- or 5′-bridging position. Both of these analogues require the synthesis of appropriate phosphoramidite building blocks with the sulfur atom present in the desired site.[61–63]

### 7.07.4   INTERPRETATION OF BIOCHEMICAL RESULTS

To understand the impact of the nucleoside analogue on the biochemical process, it is necessary to provide a foundation for the interpretation of the biochemical assays. With catalytic processes one can assess the effects of the analogue by determining kinetic parameters, whereas with simple binding events the easiest approach is simply to compare changes in equilibrium dissociation constants. The simplest interpretation results when the analogue has no effect on the biochemical process. In this case the absence or alteration of the functional group under consideration provides no interaction that affects the catalytic or binding process, neither does it significantly alter DNA structure in a manner that can globally impact the process under study.

If the deletion or alteration of a functional group impacts the biological process, the simplest interpretation is that the functional group is involved in a critical interaction in binding or transition-state stabilization. However, such interpretations are often oversimplified, in that the presence of the nucleoside analogue can also have other indirect effects: it can alter flexibility, hydration, or interstrand hydrogen bonding. Given this caution, nucleoside analogue effects are generally quantified by comparing either kinetic or binding parameters for the native and the analogue sequence.

For catalytic processes the analogue can affect either the binding or the chemical event. Steady-state kinetic analyses can be used to obtain a specificity constant ($k_{cat}/K_m$) which is impacted by either binding effects ($K_m$ is related to but is not the true equilibrium dissociation constant for the enzyme substrate complex) or the mechanistic process ($k_{cat}$).[109] Similarly, single turnover kinetics can be used to obtain a specificity constant defined as $k_1 K_A$, where $k_1$ is the first-order rate constant for the process and $K_A$ is the equilibrium association constant.[109] For a binding process, one can simply determine equilibrium dissociation value, $K_D$. The incremental binding energy for a group Y ($\Delta G_Y$) for the formation of a macromolecular complex can be estimated by comparing the binding parameters ($K_D$) for the native molecule (e.g., A—Y) with those of the analogue (e.g., A—H), where the functional group (Y) has been replaced by hydrogen (H), according to the relationship $\Delta G_Y = RT\ln(K_{A-Y}/K_{A-H})$.[110] For kinetic processes the two relationships $\Delta G_Y = RT\ln[(k_{cat}/K_m)_{A-Y}/(k_{cat}/K_m)_{A-H}]$ gives the apparent energetic contribution for the interaction involving a specific functional group Y using steady-state measurements. Similarly, $\Delta\Delta G^{\ddagger}_Y = RT\ln[(k_1 K_A)_{A-Y}/(k_1 K_A)_{A-H}]$ has been employed[111] to estimate the transition-state energy for single turnover kinetics. This approach permits the assignment of a quantified energetic contribution for the interaction to a selected functional group within a macromolecular complex.

## 7.07.5  PROTEIN–NUCLEIC ACID INTERACTIONS

The analysis of a given protein–nucleic acid complex by single-crystal XRD methods would appear to be the most direct approach to the understanding of macromolecular recognition phenomena between proteins and specific nucleic acid sequences. In spite of the challenges in obtaining high quality crystals of protein–DNA complexes, a number of high-resolution structures have now been obtained.[1] Unfortunately, there are few common recognition elements present in these complexes. The reported structures are characterized by the presence of specific functional group interactions between the protein's amino acid side chains and the nucleobase, carbohydrate, and phosphate residues associated with the nucleic acid,[112] but the nature of the interactions varies enormously.

Although X-ray crystallographic analyses are important in understanding sequence-specific protein–nucleic acid binding, they have also added to its complexity by indicating that numerous modes of binding exist and posing new questions concerning the specificity—and importance—of functional group interactions in such recognition processes. Studies have also indicated that the recognition of DNA sequences by proteins is unlikely to be explained by a simple set of rules. Additionally, it is difficult to assess from crystal structures the relative importance of, or contribution by, a specific interaction to the overall stability of the protein–nucleic acid complex. Although it is easy to assume, simply by its presence, that a specific interaction present in the crystal structure is important for the formation of the high-affinity recognition complex, complementary solution studies using analogue nucleosides have already indicated that the removal of a specific interaction does not necessarily alter the overall specificity or recognition in a given system.[113]

Nucleoside analogue studies can complement crystallographic analyses by examining the dynamics of complex formation (or dissociation) at cognate and noncognate sequences and, in the case of enzymes, the catalytic properties exhibited. In this section we will examine three types of complexes that have been the focus of studies involving nucleoside analogues: (i) a nucleic acid processing enzyme, the Eco RI restriction endonucleases,[114,115] (ii) a prokaryotic regulatory protein complex involving binding to the major groove, the *trp* repressor–operator,[116] and (iii) the eukaryotic transcription factor that binds to the minor groove, the TBP–TATA box complex.[117,118]

### 7.07.5.1  Probing Protein–DNA Complexes with Base-modified Nucleoside Analogues

#### 7.07.5.1.1  *Eco RI restriction endonuclease*

The use of nucleoside analogues to probe the Eco RI and other type II restriction endonuclease has been reviewed.[119] We offer here only a partial analysis of the work performed with the Eco RI restriction endonuclease.

The crystal structure of the Eco RI endonuclease indicates that recognition of the hexameric palindrome d(GAATTC)$_2$ occurs by the contacts to the phosphodiesters and to the edges of the base pairs presented in the major groove.[114,115] The proximity of functional groups suggests that direct amino acid to base residue contacts are made to the exocyclic amino groups of the two dA residues by Glu$_{144}$. Both of the exocyclic amino groups appear to interact with the single glutamic acid residue. Substitution of purine (P) (Figure 9) for adenine at either the "outer" or the "inner" dA–dT base pair results in significantly different effects.[111,120] When present at the outer dA–dT base pair, the $K_m$ for the cleavage reaction is increased more than one order of magnitude, which results in a corresponding reduction in the specificity constant ($k_{cat}/K_m$). By comparison, the loss of the exocyclic amino group from the inner dA–dT base pair, as the result of the substitution of purine for adenine, has almost no effect on the specificity constant. Such results suggest that the interaction involving one of the dA amino groups makes a much larger contribution to recognition of the sequence by the enzyme. The difference in binding of approximately one order of magnitude translates roughly into a 1 kcal (4.184 kJ) difference in binding energy. However, the relative contribution of binding energy by these two amino groups is not the only result that can be obtained with such analogues.

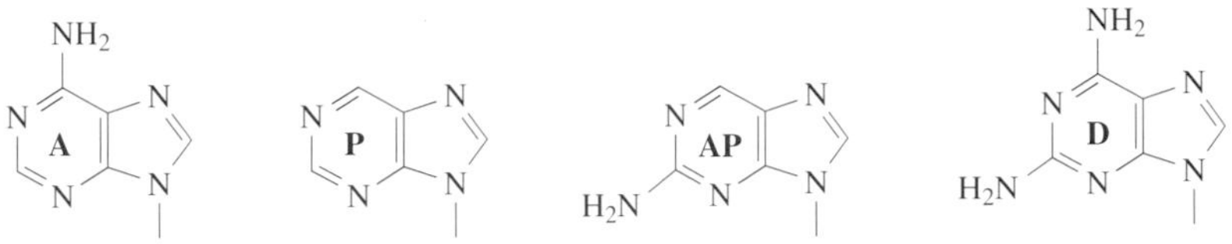

**Figure 9** The base-modified nucleoside analogues purine (P), 2-aminopurine (AP), and 2,6-diaminopurine (D), and the native adenine (A).

The protein interacts with the sequence by complexing in the major groove, so that the introduction of an additional amino group to the dA residues, but in the minor groove, should in principle have relatively little effect upon the process. However, the replacement of the outer dA residue by 2,6-diaminopurine (D) (Figure 9) results in a sequence that cannot be cleaved by the enzyme,[120,121] in spite of the observation that the functional groups present in the major groove are the same as those found in the native sequence. The use of 2-aminopurine (AP) (Figure 9), which lacks the amino group in the major groove, but retains one in the minor groove, and differs from P in that it can form two interstrand hydrogen bonds with the complementary dT residue, is also not cleaved by the enzyme.[121] These results suggest that the amino group of the outer dA nucleoside is a critical contact for the enzyme but, aside from simple hydrogen bonding interactions, the introduction of the exocyclic amino group into the minor groove has a dramatic negative effect upon catalysis. Inspection of the crystal structure reveals that the DNA in the complex is not in the standard B form, but rather it is kinked when it is bound to the enzyme. It is known from other studies, particularly those involving the curvature of DNA as a function of the presence or absence of exocyclic substituents,[122,123] that the presence of a minor groove amino group largely eliminates the ability of the DNA to undergo sequence-dependent curvature. These observations suggest that the analogues have two different effects upon the bimolecular complex. The simple deletion analogue, in which the exocyclic amino group is lost from the DNA, provides information that this amino group represents an important contact by the enzyme. However, the addition of a functional group to the minor groove affects the ability of the protein to kink the DNA and this global structural effect dramatically reduces the activity of the enzyme. Both an important functional group contact and a requisite structural role for the sequence can be identified using the described nucleoside analogues.

In a related portion of this study, the methyl groups of the two thymine residues were removed one at a time using dU as a deletion modified analogue of dT.[111,120,121] Again, the two analogue sequences had different kinetic parameters. The specificity constant ($k_{cat}/K_m$) for d(GAATUC)$_2$ sequence was ~25-fold less than that for the native sequence. By comparison, the d(GAAUTC)$_2$ analogue sequence exhibited near native-like parameters. The enzyme seemed to use a hydrophobic contact to one of the methyl groups, and less so to the other. It "sees" the hydrophobic character of the two thymine residues differently. Substrate activity could be largely restored with the incorporation of 5-bromouracil at either position.[121] In this case, it was initially difficult to correlate these observations with the structure determined by XRD analysis. There appeared to be no electron density in the area of the thymine methyl groups,[114] yet genetic studies[124] had also implicated these methyl groups as critical residues. Subsequently, new crystal forms for the protein–DNA complex[115] now indicate that the protein is capable of making contact with one or both methyl groups.

### 7.07.5.1.2   *The* trp *repressor*

The crystal structure of the *trp* repressor–operator complex represents a simple macromolecular complex in which a prokaryotic regulatory protein binds to the control region of the bacterial operon.[116,125] This protein also interacts with the DNA target by recognizing the structure and functional groups of the major groove, but in contrast to other notable structures, including that of the Eco RI endonuclease described above, there is only a single direct amino acid–nucleobase contact (per monomer) between the protein and the DNA. However, in addition to this direct contact, there were a number of water-mediated contacts present. In three of these contacts (per protein monomer), a water molecule bridged the functional group of the amino acid side chain and an $N^7$-nitrogen present on a specific purine in the operator sequence (see Figure 10). This structure elicited significant controversy[126,127] since the water-mediated contacts suggested that the protein did not "sit" as tightly in the major groove as it should, and the correct sequence dyad for binding was questioned. It seemed plausible that the complex represented a good example of nonspecific, rather than sequence-specific, binding between a protein and duplex DNA. These contacts were probed by using the deletion modified 7-deazapurine analogues.[128] The use of $dc^7A$ and $dc^7G$ permitted the excision of a single N-7 nitrogen from each possible sequence position within the operator binding site.

An assay for the binding affinity of the *trp* repressor for these analogue operators indicated that for sites predicted to have no significant water-mediated or direct contact interaction, binding to the analogue sequence was virtually indistinguishable from the native sequence. Loss of the N-7 nitrogen from $G_{-9}$, the site of a direct Arg–dG interaction (Figure 10), resulted in a 20-fold increase in $K_D$ and a loss of $\sim 1$ kcal (4.184 kJ) of binding energy per analogue. Surprisingly, the only other analogue sequences that resulted in poorer binding than the native sequence were the three sequences in which predicted water-mediated interactions had been disrupted by the 7-deazapurine analogues. The loss of these water-mediated interactions in each case also resulted in the loss of $\sim 1$ kcal (4.184 kJ) of binding energy (Figure 10).[128]

The analogue sequences used in this study provided a very clear picture of a complex that was in question because of the suggested unusual hydrogen bonding interactions. The analogues also permitted the assignment of apparent incremental binding energies for both the direct and water-mediated contacts that assist in defining high-affinity recognition between the *trp* repressor and its operator DNA.

### 7.07.5.1.3   *TBP eukaryotic transcription factor*

A complementary study is under way with the eukaryotic TBP–TATA binding transcriptional factor. This DNA binding protein is unusual in that it appears to bind solely to the minor groove of duplex DNA, it severely distorts the DNA upon binding, and it makes only a few direct hydrogen-bonding contacts to the minor groove.[117,118] The complex is characterized by a large solvent-excluded hydrophobic interfacial surface. Both 3-deazaadenine ($c^3A$) and a thymine analogue, the pyridone *C*-nucleoside (2P) from which the O-2 carbonyl has been excised (Figure 11), are used to probe this interaction.[129] Initial studies indicated that the observed hydrogen-bonding interactions are important for high affinity binding. The nature of the hydrophobic interactions between the minor groove surface and the complementary protein surface is under examination using the three base pairs illustrated in Figure 11.

### 7.07.5.2   Probing Protein–Nucleic Acid Complexes with Sugar-modified Nucleoside Analogues

There have been relatively few studies in which carbohydrate analogues have been placed into DNA or RNA sequences in order to probe the response of protein binding. In most cases, such studies involve the comparison of RNA vs. DNA binding sites to determine whether the helical structure or the presence of the 2′-hydroxyl groups aid or perturb protein binding.[130]

### 7.07.5.3   Probing Protein–Nucleic Acid Complexes with Phosphate-modified Analogues

Two types of analogues have been employed to probe interactions involving the internucleotide phosphodiesters in protein–DNA complexes. The methylphosphonate replaces one of the non-

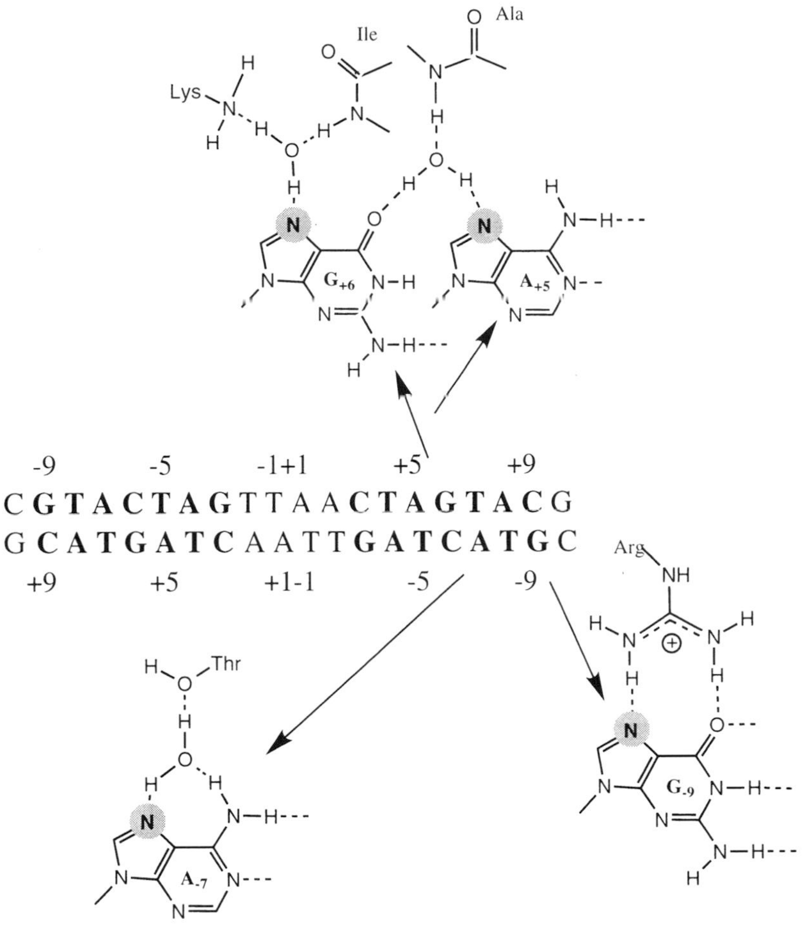

-9 -5 -1+1 +5 +9
CGTACTAGTTAACTAGTACG
GCATGATCAATTGATCATGC
+9 +5 +1-1 -5 -9
Ile
Ala
Lys
Thr
Arg
G+6
A+5
A-7
G-9


**Effects of Site-Specific Base Analogue Substitutions**

| Position | Apparent $K_D$ | $\Delta\Delta G_{22}$(kcal $mol^{-1}$) |
|---|---|---|
| native | 0.024 ± 0.005 μM | |
| -1/+1 | 0.040 ± 0.016 | 0.14 |
| -2/+2 | 0.028 ± 0.005 | - |
| -3/+3 | 0.071 ± 0.012 | 0.32 |
| -4/+4 | 0.024 ± 0.002 | - |
| -5/+5 | 0.65 ± 0.10 | 0.96 |
| -6/+6 | 0.71 ± 0.084 | 0.98 |
| -7/+7 | 0.47 ± 0.068 | 0.87 |
| -8/+8 | 0.027 ± 0.006 | - |
| -9/+9 | 0.56 ± 0.051 | 0.92 |

**Figure 10** The *trp* operator with both the direct and water-mediated contacts to the base residues as predicted from X-ray structural analysis. Below: apparent $K_D$ values determined for various base-modified analogue operators and the energetic contribution from each identified interaction.

bridging oxygens by a methyl group, and results in the loss of the negative charge at that site. It is often the case that an interaction between the protein and the internucleotide phosphodiester occurs stereospecifically to only one of the prochiral oxygens. The methyl phosphonate is a chiral linkage, and for a given interaction, one diastereomer offers a neutral, but hydrogen bond accepting P=O functional group as a partner in the intermolecular interaction, whereas the second diastereomer offers only a hydrophobic methyl group.

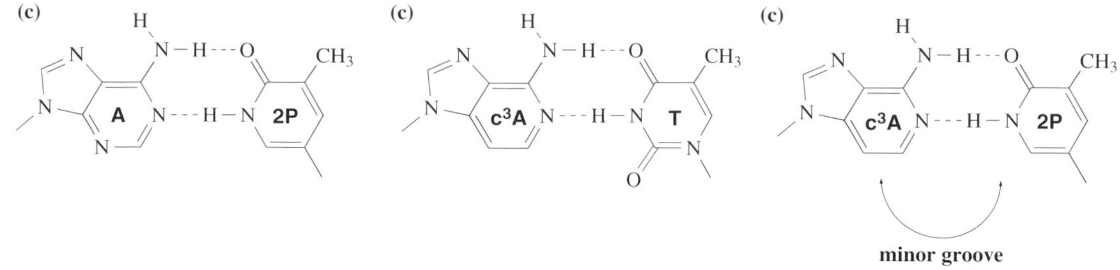

**Figure 11**   Base-modified nucleoside analogues lacking one or both of the minor groove hydrogen-bond acceptors, the N-3 nitrogen of A and the O-2 carbonyl of T.

Phosphorothioates offer a more subtle effect. The prochiral phosphate diester has bond orders that are intermediate between single and double bonds with partial charge on each of the prochiral nonbonded oxygens. By comparison, the phosphorothioate appears to have P=O and P—S$^-$ bonds with the charge localized largely on the sulfur residue.[131] Additionally, the P—S$^-$ bond length is about 0.6 Å longer than that of the internucleotide diester which provides an additional incremental perturbation of the interaction. The chiral nature of the phosphorothioate was first employed to determine which of the two prochiral nonbridging oxygens interacted with the metal cofactor.[102]

The use of stereochemically pure methyl phosphonates and phosphorothioates also offers the possibility for one to probe either of two possible hydrogen-bonding interactions with the two nonbonded phosphate oxygens. This approach has been elegantly demonstrated in the study of the Eco RI restriction endonuclease interaction with its target sequence.[132,133] In this complex, two phosphodiesters appear to be critical for effective binding; the first lies just outside the hexameric recognition site, pGAATTC, and the second is located in the center of the recognition site, GAApTTC. The introduction of either chiral phosphorothioate analogue into the first site results in a decrease in the specificity constant by 14-fold (*Sp* diastereomer) or 45-fold (*Rp* diastereomer). These effects are consistent with the crystal structure analysis in which this phosphate is completely surrounded by protein and each nonbridging oxygen is involved in two hydrogen-bonding interactions with the protein. By comparison, substitution of the *Sp* and *Rp* phosphorothioate diastereomers into the central internucleotide linkage of the recognition site results in two different effects. The *Rp* diastereomer inhibits binding and cleavage whereas the *Sp* diastereomer stimulates binding and cleavage.[132] This result suggests that only one of the nonbridging oxygens is involved in a critical interaction with the protein, and this result is largely borne out by the crystal structure analysis.[114,115]

In a similar context, diastereomeric methyl phosphonates have been used to probe phosphate contacts in the binding by POU eukaryotic transcriptional factor to the sequence 5′-d(TATGCAAATNN)·(NNATTTGCATA) (strand 1·strand 2) using gel shift experiments.[134] Introduction of the methyl phosphonate 5′ to T$^1$ (strand 1) reduced the binding affinity by nearly two orders of magnitude, whereas introduction of the linkage at positions 5′ to A$_6$, A$_7$, A$_8$, or T$_9$ in each case reduced the binding affinity between 5- and 10-fold. By comparison, the analogue linkage present at the other sites in strand 1 had essentially no effect upon protein binding. Similar phosphate mapping studies were also performed in a similar manner on strand 2. Based upon the identified phosphate contacts, the authors were able to support the hypothesis that the alignment of the phage repressor (POU$_S$) and the homeodomains (POU$_{HD}$) were functionally analogous to the phage HTH domains.[134]

### 7.07.6   DNA TRIPLE HELICES

Hydrogen-bonding interactions are important in the stability of nucleic acid complexes, often composed of simpler but fundamental structures of duplexes and/or hairpin loops. Base-modified nucleoside analogues have provided a means to probe the nature of the hydrogen-bonding interactions in these fundamental nucleic acid structures.[135] In higher-order structures such as three-stranded complexes of DNA (triplexes), a third strand of DNA can be viewed as functioning to recognize specific duplex sequences by making hydrogen-bonding contacts to the functional groups on the "edges" of the base pairs in the major groove. A detailed discussion of DNA triplex formation can be found elsewhere in this book. We provide here simply an analysis of how the introduction of specific nucleoside analogues, altered in the nature of the base, carbohydrate, or phosphate linkage, can be used to probe the stability of this type of recognition complex. We shall consider here only one general type of DNA triplex, that of the parallel-stranded pyrimidine–purine–pyrimidine DNA triplex. Until 1996 most of the analogue-based studies on DNA triple helical complexes have been guided using noncrystallographic structural information. However, the crystal structure of small sections of both parallel and antiparallel triple helices have since been obtained, and this information has been extended to create high-resolution crystallographic-based models of more complete triple helices.[136]

Pyrimidine oligodeoxynucleotides can bind to polypurine target sequences in double-stranded DNA by the formation of Hoogsteen hydrogen bonds; at each base pair target, one base from the third strand makes a bidentate interaction with the target purine.[137,138] Hoogsteen hydrogen bonding results in the formation of pyrimidine–purine–pyrimidine base triplets, such as T–A–T and $C^+$–G–C, in which the pyrimidine third strand is oriented *parallel* to the target purine strand.[139,140] The third strand T residue in the T–A–T base triplet provides a complementary hydrogen-bond acceptor and a donor for interaction with the $N^7$-nitrogen and exocyclic amino group of the target adenine. The C residue also provides complementary acceptor and donor functional groups, but the target guanine requires a bidentate hydrogen-bond donor. The C base fulfills this requirement at lowered pH values through protonation of the N-3 nitrogen.[141] However, such protonation means that the stability of the C-containing triplexes is very dependent upon the pH of the solution. Nucleoside analogues with altered base residues have been employed in an attempt to either provide pH-independent hydrogen bonding for the triplets at G–C targets, or otherwise to provide additional stabilization of the three-stranded complexes. We shall briefly describe here the use of selected nucleoside analogues with altered base residues, carbohydrate residues, or phosphate linkages that can either positively or negatively affect the stability of these pyrimidine–purine–pyrimidine DNA triple helices.

### 7.07.6.1 Nucleoside Analogues with Altered Base Residues

Although there have been reports on cytidine analogues designed to enhance the $pK_a$ of the N-3 nitrogen,[142] and thus permit protonation and formation of $C^+$–G–C base triplets at neutral or basic pH values, an effective and easily protonated derivative remains to be found. The addition of a methyl group to the cytosine base ($m^5C$) results in DNA triplexes that can be formed at higher pH values than those formed with C residues.[143,144] Although the methyl group may have a small influence on the $pK_a$ of the N-3 nitrogen, enhanced base-stacking effects (see below) may be the more likely contribution by this derivative to enhanced complex stability. In general, the effects of analogue bases on the $pK_a$ of the N-3 nitrogen appear too minimal, and in some cases the analogue results in other destabilizing interactions. For example, an $N^6$-amino-2′-deoxycytidine analogue was prepared in order to enhance the $pK_a$ of the N-3 nitrogen,[145] but the resulting triplexes were significantly destabilized. Presumably the additional amino group was responsible for unwanted interactions with the phosphate backbone or other functionality that decreased the overall stability.

Base analogues that have been specifically designed to result in pH-independent bidentate recognition of G–C base pairs do not require protonation. A number of such derivatives have been described for this purpose, and here we shall only highlight those which are based upon a pyrimidine or similar six-membered ring structure, in which the analogue can be viewed as being a cytidine analogue with incrementally altered functionality. Three base triplets formed using pyrimidine nucleoside analogues are illustrated in Figure 12. The first pyrimidine analogue capable of functioning as a neutral bidentate hydrogen-bond donor was the *C*-nucleoside containing the pseudo-isocytosine base residue[36,37] (Figure 12(a)). This analogue can be viewed as a subtle analogue of cytosine in which the positions of the N-1 nitrogen and C-5 carbon have been exchanged. This analogue results in the presence of a formal N—H on the N-3 nitrogen (although a tautomeric form with the hydrogen on the N-1 nitrogen is also likely present). The pseudoisocytidine analogue was prepared as its 2′-*O*-methyl derivative and triplexes formed with this derivative were more stable than those containing $m^5C$.[36,37] The enhanced stability is probably the result of two effects, one being the presence of the formal N—H hydrogen-bond donor, in spite of the apparent tautomeric ambiguity at this site. The second is the presence of the 2′-*O*-methylribose, which appears to contribute to helix stability (see below). An alternative cytidine analogue results when a related pyrazine *C*-nucleoside is employed (Figure 12(b)).[38] In this case the analogue can be viewed as resulting from exchanging the N-1 nitrogen and the C-6 carbon. This derivative largely eliminates any significant tautomeric ambiguity at the N-3 nitrogen and is effective in the formation of pH-independent DNA triple helices.

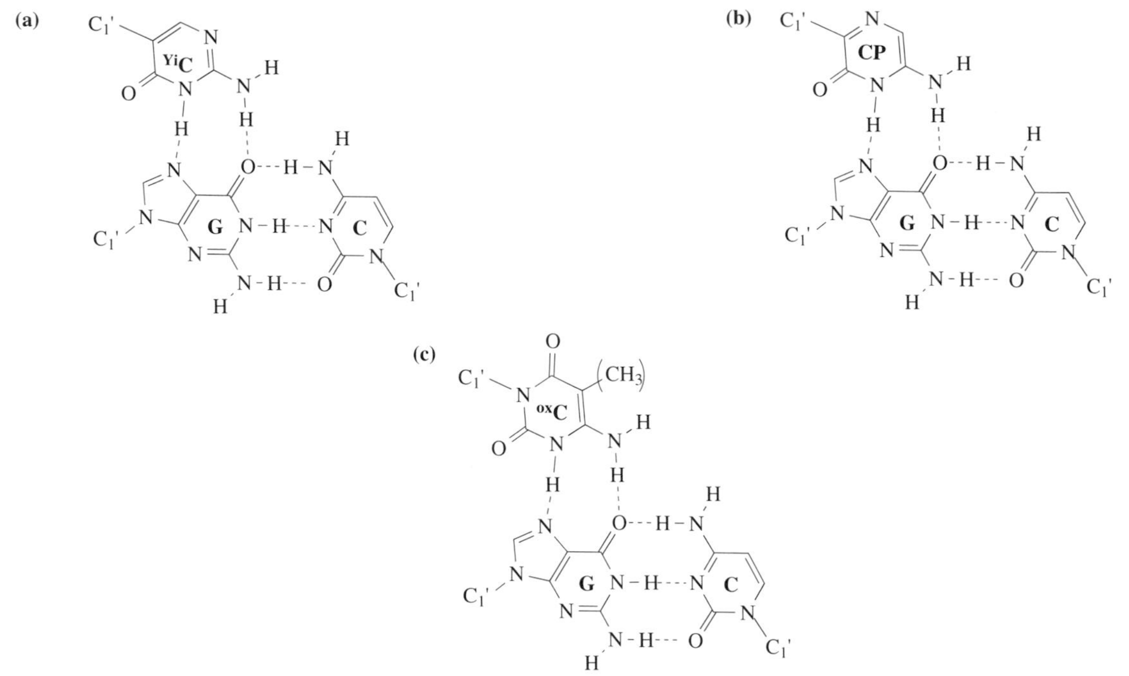

**Figure 12** DNA base triplets for three pyrimidine-like nucleoside analogues. (a) Pseudoisocytosine–G–C; (b) pyrazine *C*-nucleoside–G–C; (c) 6-oxocytosine–G–C.

An alternative cytosine-like analogue results when a carbonyl is introduced at the C-6 position.[39,40,146] Both the methylated[39,40] and nonmethylated[62] analogues have been described (Figure 11(c))—the latter as the 2'-*O*-methyl derivative.[62] Of the possible tautomers for this base residue, $N^3$—H, $O^2$—H or $O^6$—H, the $N^3$—H tautomer is the desired structure for triplex formation, and it appears to be the preferred tautomeric form based upon single-crystal X-ray analysis of the related 4-amino-1-(β-D-ribofuranosyl)2,6-[1*H*,3*H*]pyrimidione.[147] Triplexes containing this analogue form over a wide pH range and are essentially independent of the pH of the solution.[39,40]

In addition to hydrogen-bonding effects, the base residue of the third strand takes part in base-stacking (hydrophobic) interactions. Base analogues have also been prepared to alter base-stacking effects and enhance triplex stability. The observation that complexes containing 5-methyl-2'-deoxycytidine could be formed at higher pH values than the corresponding complexes with 2'-deoxycytidine is probably the result of enhanced hydrophobic base–base interactions present with the 5-methylated derivative.[143] In fact, the use of $m^5C$ with T provides a third strand with methyl groups at each base triplet and this has been suggested to result in a stabilizing hydrophobic spine.[144] Additional analogues have been prepared in which the 5-methyl group has been replaced with propyne to generate the 5-propynylated derivatives.[148] The 5-(1-propynyl)-2'-deoxyuridine analogue results in significant stabilization of DNA triplexes (by roughly 2.5 °C per substitution), whereas the corresponding 5-(1-propynyl)-2'-deoxycytidine results in a loss of stability by about 3 °C per substitution. Enhanced base-stacking effects as the result of the analogue substitution can probably explain the results for the deoxyuridine derivative, particularly since the absence of either propynyl or methyl (dU) also results in significant destabilization of triple helices.[144] In principle, the dC derivative should also experience enhanced base-stacking effects, and this is probably the case, but the propynyl group results in a decrease in the p$K_a$ of the N-3 nitrogen (from 4.35 for $m^5C$ to 3.30 for the propynyl derivative), and loss of the protonated hydrogen bond is more deleterious than the enhanced base-stacking effects.

A number of other nonpyrimidine base analogues have been developed for the recognition of G–C base pairs and these are described more fully elsewhere in this book. We have highlighted here only a selected group in which a small incremental modification of the pyrimidine base is present, and then noted the incremental biological response to the presence of the nucleoside analogue.

### 7.07.6.2 Nucleoside Analogues with Altered Carbohydrate Residues

Although the carbohydrate portion of the nucleoside does not appear to undergo any direct interactions with the target duplex sequence, the conformation of the carbohydrate can affect the position of the third strand base residues and thereby alter the stability of the helix. Nucleoside analogues that alter the functionality of the carbohydrate can incrementally alter the stability of the complex. For example, as noted above, the presence of a 2′-*O*-methyl residue enhances the stability of triplexes relative to corresponding RNA or DNA complexes.[149,150] However, it is unclear whether the substituent directly contributes binding energy (e.g., the hydrophobic character of the methyl group) or simply results in steric and conformational preferences. The carbocyclic analogue of T and m⁵C, those derivatives in which the ring oxygen (O-4′) is replaced by a methylene group (Figure 8),[50,142] were prepared for studies of DNA triplex stability. The carbocyclic derivative of m⁵C resulted in some increased stabilization ($\sim 2$ °C per substitution), and this is explained in part by an apparent increase in the $pK_a$ of the N-3 nitrogen. By comparison the carbocyclic analogue of T results in triplex destabilization (by $\sim 4$ °C per substitution) relative to the native furanose derivative, and this result was explained by the presence of an unfavorable conformation for the cyclopentane ring.

A second group of nucleoside analogues employed in DNA triplex studies are those in which flexibility of the carbohydrate is either increased or decreased. Nucleoside derivatives of T and C have been prepared such that the sugar conformation was restricted to the 2′-*endo*/1′-*exo* range by the introduction of a five-membered ring that included the C-3′ and C-5′ carbons (Figure 6(a)).[55] These two bicyclodeoxynucleoside (bcd) derivatives have different effects upon DNA triplex formation;[54] the presence of bcdT residues results in more stable DNA triplexes, whereas the presence of bcdC residues destabilizes these complexes. Reducing the conformational flexibility within a range that is optimal for triplex formation could enhance the stability of complexes simply by reducing the entropic barrier necessary for the third strand to adopt the desired conformation for targeting the DNA duplex. However, the bcdC analogue results in destabilization of the triple helix, suggesting that reducing the conformational flexibility of the carbohydrate of C may result in a less than optimal conformation for triplex formation. NMR studies have indicated that the sugar conformation for the third strand at C⁺–G–C base triplets differs from that found in T–A–T base triplets.[151] A corresponding more flexible carbohydrate analogue results when the C-2′ carbon is excised from the furanose ring to result in a conformationally flexible acyclic linker (Figure 6(d)).[58] Both nucleoside analogues of T and m⁵ᵒˣC have been prepared containing acyclic linkers (aT and am⁵ᵒˣC). The introduction of aT residues into a third strand results in significant destabilization of the triplex ($\sim -6$ °C per substitution). This result can be explained by an increased entropic barrier (relative to the bcd or furanose residues) necessary for the analogue nucleoside to adopt the requisite conformation. However, a more flexible carbohydrate linker for a protonated C analogue (am⁵ᵒˣC) enhances triplex stability ($\sim +4$ °C per substitution, with two residues present), particularly for targets of contiguous G–C base pairs.[58] Although conformational flexibility should, in principle, result in an unfavorable entropic barrier to complex formation, the resulting base-stacking and/or hydrogen-bonding interactions that result if the flexible linker permits more effective interactions (relative to the bcd or furanose residues) with the target base pair(s) may compensate for the expected unfavorable entropy effects.

A related less flexible analogue results when a five-membered ring is introduced to incorporate the 2′-OH and 3′-OH and replace the phosphodiester linkage with a riboacetal contained within the second five membered ring (Figure 6(b)).[56] These analogues have also been shown to enhance the stability of DNA triplexes. While the less flexible sugar conformation may contribute to this effect, the neutral nonphosphate linkage is also a factor that must be considered.

### 7.07.6.3 Analogues with Altered Phosphate Linkages

There are a wide variety of analogue linkages that have been developed primarily in the search for an effective linkage that will hybridize to target sequences, permit effective passage across the cell membrane, and remain intact in the presence of cellular nucleases. The variety of linkages developed for these purposes are described in detail elsewhere in this book. We note here only two types of analogue linkages, both containing subtle incremental modifications in which one of the nonbridging oxygens of the phosphodiester has been replaced by either a methyl group or a sulfur atom, generating methyl phosphonates and phosphorothioates, respectively.

The use of methyl phosphonate linkages replaces the charged phosphodiester with a neutral linkage and the use of this analogue linkage should reduce charge–charge repulsion effects that lead to the stabilization of the three-stranded complexes. However, the analogue presents an additional complication in that the linkage exists in two stereochemical diastereomers (Figures 7(e) and (f)), and a reliable and effective synthesis of sequences containing all *Sp* or all *Rp* centers remains elusive. Therefore, it has been difficult to separate the potentially stabilizing effects resulting from the presence of a neutral linkage from the potentially destabilizing effects as the result of the presence of a mixture of diastereomers in the third strand. The effects may be even more complex than outlined here. For example, in some studies a third strand containing methyl phosphonate linkages was reported to be unable to form triplexes, or to do so only with dramatically reduced stability.[152] However, a subsequent report suggested that more stable complexes result.[153] It would appear that sequence or substituent effects may have a more dramatic impact upon stability when using methyl phosphonate-linked oligonucleosides.

Similarly, the stability of complexes containing phosphorothioate internucleotide linkages is not well understood. Stability again appears to be dependent upon sequence effects.[154,155] The phosphorothioate linkage is also present in two stereochemical forms, and it is unclear whether one diastereomer provides better triplex stability than the other. A better understanding of the effects of these linkage analogues will result from further developments in stereoselective syntheses.

## 7.07.7 RNA CATALYSIS

Many of the analogue studies involving protein–nucleic acid complexes have been initiated with the availability of significant structural data that could be used to guide the analogue studies. On the other hand, most of the analogue studies performed with ribozymes were performed largely in the absence of structural information, although two crystal structures for the hammerhead ribozyme subsequently became available.[156,157] In many cases, the analogue studies themselves have provided much of the information regarding specific interactions necessary for transition-state stabilization.

Group I ribozymes, first discovered by Cech and co-workers in 1982,[158] are a novel class of metalloenzymes[159,160] but are composed of RNA and function in the absence, or largely in the absence, of associated protein. The mechanism of the cleavage reactions of some ribozymes has been probed using nucleoside analogues, and in this section we discuss how nucleoside analogues can be used to probe the catalytic activity of three such ribozymes: (i) the hammerhead ribozyme, (ii) the hairpin ribozyme, and (iii) the large group I intron, the *Tetrahymena* ribozyme. Although *in vivo* the ribozyme structure arises from the folding of a single RNA molecule, for *in vitro* studies two (or more) strands of RNA can be used to form the ribozyme complex and permit cleavage to occur *in trans*.[161] The fragment which is cleaved becomes the substrate and the intact fragment the ribozyme. In this format, multiple cleavage events can occur and be characterized using kinetic parameters. The ability to use bimolecular complexes *in vitro*, coupled with advances in the automated chemical synthesis of oligonucleotides and modified oligonucleotides, has made the use of nucleoside analogues as RNA probes more convenient.[162]

### 7.07.7.1 The Hammerhead, Hairpin, and Tetrahymena Ribozymes

The satellite RNA of the tobacco ringspot virus undergoes a rolling cycle replication in which the cleavage of multimeric forms of RNA into monomer subunits is catalyzed by a hammerhead ribozyme in the (+) strand and a hairpin in the (−) strand.[163,164] The rates of cleavage by the two ribozymes are comparable under similar conditions, and the reverse reaction, that of ligation, occurs more efficiently with the hairpin ribozyme than with the hammerhead ribozyme.[165–169]

The structure of the hammerhead ribozyme consists of three double-stranded helices (I–III) connected by a nominally single-stranded core region (Figure 13(a)).[160,170,171] Sequence comparisons and extensive mutagenesis have shown that 11 nucleosides in the hammerhead structure (shown in italics), nine of which are in the single-stranded core region, appear to be involved in the catalytic cleavage process.[160,161,171,172] By comparison, the secondary structure of the hairpin ribozyme consists of four double-helical regions (I–IV) and two single-stranded loops (A and B) (Figure 13(b)), with most of the nucleosides essential for cleavage located in the loop regions.[173–177]

The cleavage reaction of the hammerhead ribozyme is similar to that of the hairpin ribozyme in that transesterification occurs via nucleophilic attack of the scissile phosphate by the adjacent 2′-

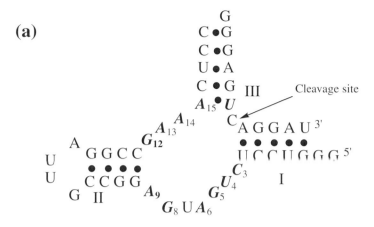

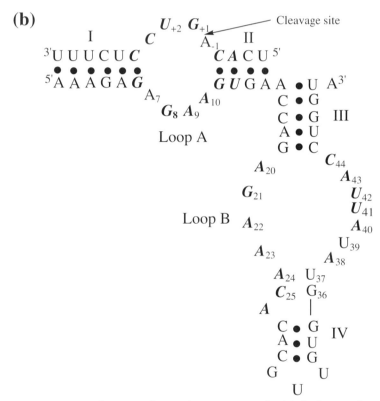

**Figure 13** Consensus sequence and proposed secondary structures for (a) the hammerhead ribozyme and (b) the hairpin ribozyme.

hydroxyl group to produce two fragments, one of which contains a 5′-hydroxyl and the other a 2′,3′-cyclic phosphate (Figure 14(a)).[161] The rate-determining step in the chemical reaction of the cleavage process is thought to be different for these two ribozymes since, for the hammerhead ribozyme, the cleavage rate increases linearly with pH (the rate-determining step is considered to be the abstraction of the 2′-hydroxyl proton),[178,179] whereas for the hairpin ribozyme the rate is relatively unaffected between pH 5.5 and 8.[177]

The third ribozyme discussed in this section, the *Tetrahymena* ribozyme (Figure 14(b)), is derived from the group I intervening sequence of *Tetrahymena thermophila* preribosomal RNA.[158,180] It catalyzes its own excision from the pre-RNA intron and itself retains catalytic activity. The cleavage reaction, which, as for the hairpin ribozyme, is reversible, proceeds via a different mechanism to that for the two ribozymes described above. The cleavage reaction involves nucleophilic attack by

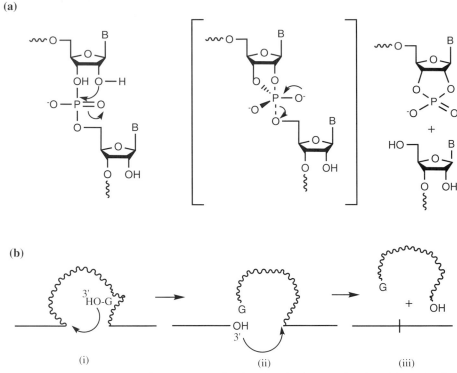

**Figure 14** (a) Proposed mechanism for transesterification for the hammerhead and hairpin ribozymes. (b) Cleavage/ligation process for the *Tetrahymena* ribozyme.

the 3'-hydroxyl group of an external guanosine substrate (or its 5'-phosphorylated form) on the internucleotide phosphate to give a 5'-phosphorylated fragment and a free 3'-hydroxyl fragment (Figure 13(b)). In the first step, the ribozyme binds two substrates, a guanosine (or guanosine triphosphate) and an RNA strand. Cleavage of the RNA occurs by a transesterification reaction in which the guanosine becomes attached to the 5'-end of the 3' cleavage fragment (Figure 14(b)).[181,182]

Although in some cases a ground-state crystal structure analysis has been performed on the ribozyme–substrate complex,[156,157] it remains less clear how the RNA structure can stabilize the transition state in order to effectively permit the described transesterification reactions.

### 7.07.7.2 Use of Base-modified Nucleosides

Nucleosides analogues containing base modifications have been used to probe the role of the base residues of the hammerhead ribozyme structure for catalytic activity. The 7-deazapurine nucleosides (Figures 14(a) and (b)) have been incorporated at single sites in ribozyme–substrate complexes to investigate the importance of specific purine nitrogens for catalytic activity of the hammerhead ribozyme.[183,184] These investigations showed that the loss of N-7 nitrogen of the $A_6$ residue led to a reduction in cleavage efficiency of over 35-fold, suggesting that the N-7 nitrogen of $A_6$ plays a vital role in transition-state stabilization (little effect upon substrate binding was observed), a role which was not apparent from the reported crystal structure of the hammerhead ribozyme.[156,157]

The importance of exocyclic amino groups of individual A or G residues in the hammerhead complex were probed by substituting the base-modified nucleosides, purine and inosine (Figures 15(c) and (d)) at specific single sites in the ribozyme and substrate.[183,184] The reported crystal structures[156,157] of the ground state of the hammerhead complex suggest that the bases $G_8A_9$ and $G_{12}A_{13}$ form a pair of non-Watson–Crick G–A base pairs in which interbase hydrogen-bonding interactions involve the exocyclic amino groups of all four residues. Additionally, the amino group of $G_5$ does not appear to take part in any interaction in the ground-state structure. Kinetic analysis using inosine as an analogue of guanosine identified the amino groups of $G_5$ and $G_{12}$ as being critical in the transition-state stabilization of the complex, with a 25-fold and 50-fold rate reduction observed with the loss of the N-2 amino group from $G_5$ and $G_{12}$, respectively.[183,184] This result highlights how kinetic studies can reveal information about the transition state, which differs from the crystallographic data that provides information about the ground-state structure.

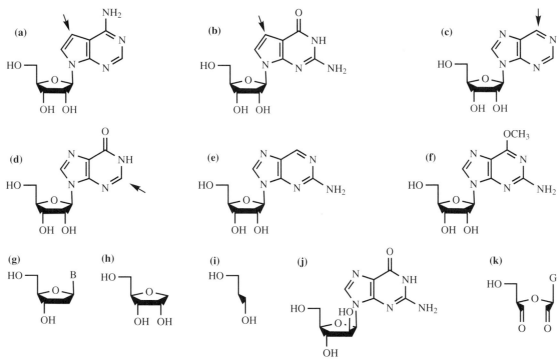

**Figure 15** Base- and carbohydrate-modified nucleoside analogues used to probe ribozyme activity.

In the hairpin ribozyme, the $G_{+1}$ nucleoside has being identified as the only essential nucleotide in the substrate part of loop A.[165,166,174,185] Molecular modeling studies have indicted that the 2-amino group of $G_{+1}$, with the heterocycle in the *syn* conformation, may lie close to both the cleaved P—O bond and the 2'-OH on the adjacent $A_{-1}$ residue.[186] By a methodology similar to that above, the importance of the 2-amino group of $G_{+1}$ has been investigated using substrates in which $G_{+1}$ was replaced by inosine (Figure 15(d)) or 2-aminopurine ribonucleoside (Figure 15(e)).[185] The substrates containing inosine in place of $G_{+1}$ were found to be inactive towards cleavage, whereas the 2-aminopurine-containing substrates retained activity suggesting that the 2-amino group of $G_{+1}$ is a critical element in catalysis.

The 7-deazapurines, lacking the N-7 nitrogen, and purine and inosine, in which the purine exocyclic amino group also has been removed, in addition to the analogue $O^6$-methylguanosine (Figure 15(f)), have all been used to study the functional group requirements of other essential purine residues in the hairpin ribozyme.[187] In contrast to the results found for the hammerhead ribozyme, it was observed that for 11 out of the 12 essential purines in the hairpin ribozyme substitution by a base-modified nucleotide resulted in considerable decrease in the rate of cleavage, largely due to decreases in $k_{cat}$ rather than increases in $K_m$. These rate reductions have been attributed to the loss of single hydrogen bonds to nitrogens in the active complex.

Substitution of the nucleosides $A_{20}$, $G_{21}$, and $U_{37}$–$C_{44}$ of loop B in the hairpin ribozyme by a 2'-deoxynucleoside, an abasic residue, and a propyl linker (Figures 15(g), (h), and (i)) have identified $U_{39}$ as the only residue which could be substituted without significant effect on the cleavage rate, indicating that this residue may act only largely as a spacer and may not play significant role in either folding or catalysis.[188]

In the *Tetrahymena* ribozyme, the interaction between the guanosine substrate and the self-splicing pre-rRNA has been studied using several guanosine analogues, which contained modifications at either the base or sugar moieties (Figure 16).[180] Analogues with modifications to the imidazole ring, such as 7-methylguanosine and 8-azidoguanosine triphosphate, produced no decrease in activity, whereas analogues with modifications to the six-membered ring resulted in a loss of activity. These results suggested that the guanosine base, especially the six-membered ring, is involved in binding the guanosine to a site on the pre-RNA (these modifications would not be expected to effect the reactivity of the 3'-hydroxyl group). The study also indicates that optimal binding is achieved with a hydrogen-bond acceptor at C-6 and a hydrogen bond donor at C-2 in the base moiety. Studies on the kinetics of the splicing reaction mediated by two of the active analogues, inosine and 2-aminopurine ribonucleoside, led the authors to suggest that the free energies of binding of the various nucleoside substrates are the important kinetic parameters and that once the nucleoside is bound to the RNA, its reactivity is independent of the base moiety. Based on these kinetic studies, the authors proposed a model for the nucleoside binding site on the pre-rRNA in which there were four hydrogen bonds between guanosine and the pre-rRNA. The use of base-modified nucleoside analogues within the RNA sequence has permitted the identification of an exocylic amino group of a conserved G–U base pair as being critical to the splicing process.[189]

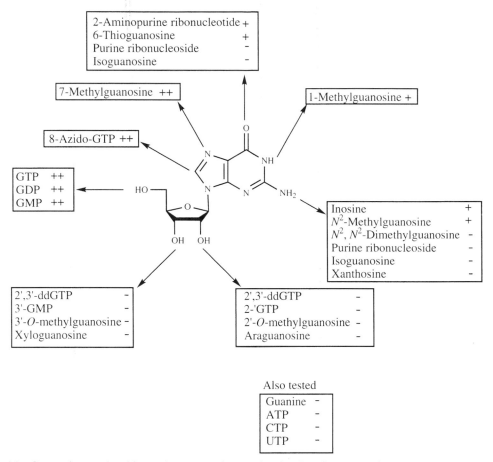

**Figure 16** Guanosine nucleoside analogues used to probe the G cofactor requirements in the *Tetrahymena* ribozyme cleavage/ligation activity. (+ +) Activity near that of native complex; (+) active, but less than the native complex; (−) inactive.

### 7.07.7.3 Use of Sugar-modified Analogues

Nucleosides in which the sugar moiety has been modified have been used to investigate the role of 2′- and 3′-hydroxyl groups in the cleavage process of all three ribozymes. The role of specific 2′-hydroxyl groups in the hammerhead structure has been investigated by using a variety of 2′-modified nucleosides. These analogues included 2′-deoxynucleosides[183,190–193] and nucleosides in which the 2′-hydroxyl group was replaced by a fluorine,[46,47] amine,[194] methoxy,[48] or allyl group.[49] These studies identified the 2′-hydroxyl groups of $G_{5,8}$, $A_9$, $U_{1,16}$, and $C_{17}$ as being critical for efficient cleavage activity, but they offered little with regard to the nature of the optimal positioning of these groups within the catalytic complex. By using the nucleoside arabinosylguanine, in which the stereochemistry of the 2′-hydroxyl group is inverted with respect to guanosine, configurational flexibility of the 2′-hydroxyl groups of $G_5$, $G_8$, and $G_{12}$ could be examined.[51] It was found that the ara$G_5$ complex was essentially inactive, with RNA cleavage by this analogue ribozyme being $\sim 10^5$-fold slower than that by the native complex and, surprisingly, 1000-fold slower than that by either the 2′-deoxy- or 2′-fluoro-substituted ribozymes. In contrast, the araG$_8$ and araG$_{12}$ complexes had first-order cleavage rates only slightly reduced from that of the native ribozyme, and were two orders of magnitude more active than either the 2′-deoxy or 2′-fluoro derivatives. These differences suggest that the 2′-hydroxyl group at G$_8$ and G$_{12}$ can maintain their critical interactions from either the *R* or *S* configuration, unlike the 2′-hydroxyl group of G$_5$.

The 2′-hydroxyl groups in the hairpin ribozyme essential for effective cleavage activity have been probed by substituting 2′-deoxynucleotides and 2′-*O*-methylnucleotides into the ribozyme complex.[186–188,195] In loop A, substitution of A$_{-1}$,[186] A$_{10}$, G$_{11}$, A$_{24}$, and C$_{25}$[165] with 2′-deoxy-or 2′-*O*-methylnucleotides resulted in a significant loss of activity (due to a decrease in $k_{cat}$ rather than an increase in $K_m$), suggesting that these residues bear essential 2′-hydroxyl groups. The activity of the complexes modified at G$_{11}$ and A$_{24}$, however, was found to be restored by increasing the Mg$^{2+}$ concentration, suggesting that the 2′-hydroxyl groups of these residues could be involved in metal ion coordination in the transition state. The residues in loop B that contain essential 2′-hydroxyl groups have been identified as U$_{37}$, A$_{38}$, and U$_{41}$, with a modest reduction of activity being observed when these nucleotides were replaced by 2′-deoxynucleotides.[188]

Information about the three-dimensional structure of the loops in the hairpin ribozyme has also been obtained by using photoaffinity techniques. Cross-linking experiments with substrates containing only 2′-deoxynucleosides with the photolabel deoxy-6-thioinosine substituted for G$_{+1}$ or A$_{-1}$ have been used to show (by irradiation of the complex at 335 nm) that the substrate is linked to the ribozyme at multiple sites.[196] These results have been interpreted as evidence that loop A has a large degree of flexibility.

For the *Tetrahymena* ribozyme, guanosine substrates modified at the 2′-hydroxyl group, namely 2′,3′-dideoxyguanosine triphosphate (2′,3′-ddGTP), 2′-deoxyguanosine triphosphate (2′-dGTP), 2-*O*-methylguanosine, and araguanosine (Figure 16), all proved to be inactive in splicing.[180] This observation could indicate the loss of enhanced reactivity of the 3′-hydroxyl group due to electronic effects of the 2′-hydroxyl group (the p$K_a$ of the 3′-hydroxyl proton of a ribonucleoside is lower than that of the corresponding 2′-deoxynucleoside). Other possible reasons for the loss of self-splicing activity could be that the 2′-hydroxyl group is involved in some form of bonding of the guanosine residue to the RNA or involved in bonding in an intermediate step. If this last explanation is the case then a novel mechanism must be involved because all known mechanisms for other RNA cleavage reactions (such as hairpin and hammerhead ribozymes) give products with unphosphorylated 5′-ends. The orientation of the 2′-hydroxyl group also seems to be important, as indicated from the inactivity of araguanosine.

Self-splicing activity was also lost when 3′-hydroxyl-modified analogues, such as 3′-guanosine monophosphate, 3′-*O*-methylguanosine and xyloguanosine (Figure 15), were used as cofactors. This is not surprising, however, since it is thought that splicing involves nucleophilic attack on phosphate by the 3′-hydroxyl group of guanosine. The inactivity of xyloguanosine also suggests that the orientation of the 3′-hydroxyl group is also crucial in splicing activity.

Competitive inhibitor studies by Bass and Cech[197] further indicate the importance of the 2′- and 3′-hydroxyl groups of guanosine for binding and also that the 2′-hydroxyl group of the guanosine substrate appears necessary for the reaction to occur. This latter suggestion arises from the fact that 2′-deoxyguanosine, which is a competitive inhibitor, has a higher affinity for binding to the RNA than either of the splicing substrates, inosine or 2-aminopurine ribonucleoside.

Because substrates containing the dialdehyde (Figure 15(k)) at the cleavage site retain activity with the *Tetrahymena* ribozyme, Kay and Inoue[198] have proposed that although the ribozyme is designed to recognize a guanosine 2′,3′-*cis*-diol, other hydroxyl groups that are stereochemically equivalent to those of the diol are also capable of participating in cleavage reactions (the authors proposed a mechanism of cleavage whereby the dialdehyde becomes hydrated to give the bis-*gem*-diol).

Although the 2′-hydroxyl group of the guanosine substrate appears essential for ribozyme activity, this is not the case for the oligonucleotide substrate. The *Tetrahymena* ribozyme is capable of catalyzing the cleavage of substrates composed solely of 2′-deoxynucleosides, although cleavage occurs more slowly than for the corresponding RNA substrate.[199] For the DNA substrates, both the binding and the rate of the chemistry are reduced by four orders of magnitude with respect to the RNA analogue, suggesting that one or more 2′-hydroxyl groups of the RNA substrate are involved in binding and catalysis, possibly via hydrogen-bonding interactions or magnesium ion coordination. The maximum rate during steady-state turnover for the DNA substrate, however, is only 10 times slower than that for the RNA substrate, because there is a change in the rate-limiting step when DNA is substituted for RNA.

Tertiary interactions of the 2'-hydroxyl groups in the *Tetrahymena* ribozyme complex have also been studied using 2'-deoxy- and 2'-fluoro-substituted nucleosides.[200,201] These investigations have identified the 2'-hydroxyl groups of residues two and three positions away from the cleavage site as forming major tertiary interactions with residues within the ribozyme core. Further studies have identified a conserved adenine residue ($A_{302}$) in the catalytic core that contributes to the complex stability (by accepting a hydrogen bond from a specific 2'-hydroxyl group in the helix containing the cleavage site).[201]

### 7.07.7.4  Use of Phosphorothioates

The stereochemical course and mechanism of action of ribozyme cleavage have been probed using substrates containing chiral phosphorothioate linkages at the cleavage site (in which a nonbridging oxygen has been replaced by a sulfur). These experiments have shown that an inversion of configuration occurs at the phosphorus center in the cleavage reactions of all three ribozymes.[101,181,202] This change in stereochemistry suggests that attack on the scissile phosphate by the 2'-hydroxyl group, in the case of the hammerhead and hairpin ribozymes, and by the 3'-hydroxyl group of guanosine in the *Tetrahymena* ribozyme, occurs via an in-line $S_N2$ mechanism.

Owing to the different metal ion coordination properties of sulfur and oxygen, phosphorothioate derivatives have also been useful in the investigation of the role of divalent metal ions (usually $Mg^{2+}$) in the RNA catalytic process. Slim and Gait[101] studied the rate of cleavage of phosphorothioate-containing substrates by the hammerhead ribozyme. These experiments showed that substrates containing an *Rp* phosphorothioate linkage were cleaved very slowly in the presence of magnesium ions, whereas substrates containing the *Sp* isomer were cleaved at a rate only slightly lower than that of the unmodified oligonucleotide. These results were interpreted as evidence that the metal cofactor is bound to the pro-*R* nonbridging oxygen in the transition state. The role of the metal ion in the cleavage reaction remains unclear, however, with one possible function being as a Lewis acid in the stabilization of the developing negative charge on the 5'-oxyanion leaving group. This role of the metal ion cofactor has been studied in the laboratory using substrates which contain a 5'-bridging phosphorothioate at the cleavage site.[66,68] These sulfur-containing substrates, which contained all 2'-deoxynucleosides except for a single ribonucleoside adjacent to the phosphorothioate at the cleavage site, were found to be $\sim 10^6$ times more labile than the native RNA linkage. Cleavage studies on these substrates using a variety of metal cations (in the absence of ribozyme) revealed a correlation between the rate of cleavage and the "softness" of the divalent metal, with the rate of cleavage increasing along the series $Mg^{2+}$, $Ca^{2+}$, $Mn^{2+}$, $Co^{2+}$, $Zn^{2+}$, and $Cd^{2+}$. These results are constant with an interaction of the metal with the "soft" mercapto leaving group or the $PO_2$ moiety. In the presence of ribozyme, however, the phosphorothioate-containing substrate was found to be cleaved at a similar rate to that of the native RNA, and showed no preference for a metal cofactor ($Mg^{2+}$ vs. $Mn^{2+}$), which implies that the cleavage of the bond (P—O or P—S) to the 5'-leaving group is not the rate-determining step in the hammerhead ribozyme.

The role of metal ions in the function of the hairpin ribozyme has also been investigated using phosphorothioate derivatives.[203,204] In contrast to the hammerhead ribozyme, the hairpin complex has been found to cleave efficiently RNA containing an *Rp* phosphorothioate, which suggests that the nonbridging oxygen at the cleavage site does not coordinate to $Mg^{2+}$.[186,202] Phosphorothioates have also been used to determine the importance of other pro-*Rp* oxygens in the ribozyme.[186,202] The cleavage activity has been found to remain in ribozymes in which all guanosine, uridine, and cytidine residues have a nonbridging phosphorothioate linkage. When multiple adenosine phosphorothioate linkages were incorporated into the ribozyme, however, a 25-fold loss in activity was observed. Phosphorothioate interference experiments have indicated that this reduction in activity is due mainly to substitutions at $A_7$, $A_9$, and $A_{10}$, which suggests that these pro-*Rp* oxygen atoms may be involved in tertiary contacts within the ribozyme–substrate complex, although induced structural effects cannot be ruled out.[165]

McSwiggen and Cech[181] and others have used substrates containing nonbridging phosphorothioate linkages at the cleavage site to probe the mechanism of the *Tetrahymena* ribozyme. These experiments showed that inversion of configuration of phosphorus occurs during cleavage and also that the presence of a phosphorothioate linkage does not substantially affect the cleavage rate (only a two-fold decrease in reaction rate was observed for the phosphorothioate substrate), which indicates that the chemical step in the cleavage process is not the rate-limiting step (the introduction of a phosphorothioate in a diester should slow the chemical step of a bimolecular nucleophilic displacement reaction by 30–100-fold). The reverse ligation reaction of the *Tetrahymena* ribozyme has also been studied using phosphorothioate-containing oligonucleotides.[204] As expected, the reaction was found to proceed with inversion of configuration at the phosphorus center. It was also noted that replacement of the pro-*R* oxygen of the reactive phosphate by sulfur decreased the rate of reaction by a factor of about 1000, suggesting the possibility of sulfur interference with $Mg^{2+}$ coordination.

The cleavage mechanism was further examined by using the 3′-bridging phosphorothioate analogue linkage in an all-deoxy substrate sequence. With this derivative, the bond that is cleaved during the first transesterification reaction (with the guanosine cofactor) is altered from a P—O to a P—S linkage. The rates of cleavage for the 3′-bridging phosphorothioate in the presence of $Mg^{2+}$ and $Mn^{2+}$ differed 1000 fold, establishing that the metal cofactor participates in the first cleavage reaction by functioning as a Lewis acid and associating with the 3′-leaving group.[205]

The importance of phosphate groups other than that at the cleavage site has been investigated in the *Tetrahymena* ribozyme by monitoring the splicing activity of several RNA samples containing adenosine or uridine-5′-phosphorothioates.[206] The products and unspliced RNA were analyzed for the phosphorothioate content by iodoethanol cleavage, and this study led to the identification of two major and three minor inhibitory phosphorothioate substitutions, all located in the most highly conserved region of the intron.

## 7.07.8 SUMMARY

As can be seen from some of the studies discussed in this chapter, nucleoside analogues can be valuable tools in probing a variety of biological processes. We have tried to highlight different types of analogue studies involving the biological processes of protein–DNA recognition, DNA triplex stability, and RNA ribozyme activity. In each case complexes containing nucleoside analogues have revealed important information about the location, stereochemical consequences, and importance of the interaction or reactions, and helped to identify specific functional groups within the various complexes that are important, or essential, for catalytic activity, binding, and tertiary interactions, in addition to eliminating from consideration other functional groups that appear to take no part in complex stabilization. With the identification of a critical functional group, it is possible in some cases to assign a quantified energetic value for its contribution to the formation of the sequence-specific complex.

## ACKNOWLEDGMENTS

We acknowledge grant support for projects involving the use of nucleoside analogues to probe protein–nucleic acid recognition (NSF-MCB-9707040), DNA triplex formation (NIH-GM53201), and catalytic RNA (NIH-GM47660).

## 7.07.9 REFERENCES

1. C. J. Larson and G. L. Verdine, in "Bioorganic Chemistry: Nucleic Acids," ed. S. M. Hecht, Oxford University Press, New York, 1996, pp. 324–346.
2. K. D. Harshman and P. B. Dervan, *Nucleic Acids Res.*, 1985, **13**, 4825.
3. D. J. Galas and A. Schmitz, *Nucleic Acids Res.*, 1978, **5**, 3157.
4. D. Suck and C. Oefner, *Nature* (*London*), 1986, **321**, 620.
5. D. Suck, A. Lahm, and C. Oefner, *Nature* (*London*), 1988, **332**, 464.
6. L. Hennighausen and H. Lubon, *Methods Enzymol.*, 1987, **152**, 721.
7. M. W. Van Dyke and P. B. Dervan, *Nucleic Acids Res.*, 1983, **11**, 5555.
8. A. Spassky and D. S. Sigman, *Biochemistry*, 1985, **27**, 8050.
9. T. D. Tullius and B. A. Dombrowski, *Science*, 1985, **230**, 679.
10. A. Vincze, R. E. L. Henderson, J. J. McDonald, and N. J. Leonard, *J. Am. Chem. Soc.*, 1973, **95**, 2677.
11. J. Carey, *Biol. Chem.*, 1989, **264**, 1941.
12. R. M. Gronostajski, S. Adhya, K. Nagata, R. A. Guggenheimer, and J. Hurwitz, *Mol. Cell. Biol.*, 1985, **5**, 964.
13. C. Heuer and W. Hillen, *J. Mol. Biol.*, 1988, **202**, 407.
14. C. R. Bayley and A. S. Jones, *Trans. Faraday Soc.*, 1959, **55**, 492.
15. M. Beer, S. Stern, D. Carmalt, and K. H. Hohlhenrich, *Biochemistry*, 1966, **5**, 2283.
16. U. Siebenlist and W. Gilbert, *Proc. Natl. Acad. Sci. USA*, 1980, **77**, 122.
17. K. C. Hayashibara and G. L. Verdine, *J. Am. Chem. Soc.*, 1991, **113**, 5104.
18. K. C. Hayashibara and G. L. Verdine, *Biochemistry*, 1992, **31**, 11265.

19. P. B. Dervan, *Methods Enzymol.*, 1991, **208**, 497.
20. D. L. Boger, S. A. Munk, and H. Zarrinmayeh, *J. Am. Chim. Soc.*, 1991, **113**, 3980.
21. F. Seela, W. Bourgeois, H. Rosemeyer, and T. Wenzel, *Helv. Chim. Acta*, 1996, **79**, 488.
22. F. Seela and P. Leonard, *Helv. Chim. Acta*, 1996, **79**, 477.
23. T. Wenzel and F. Seela, *Helv. Chim. Acta*, 1996, **79**, 169.
24. F. Seela and H. Thomas, *Helv. Chim. Acta*, 1995, **78**, 94.
25. F. Seela and T. Wenzel, *Helv. Chim. Acta*, 1995, **78**, 833.
26. N. Ramzaeva and F. Seela, *Helv. Chim. Acta*, 1995, **78**, 1083.
27. F. Seela, K. Worner, and H. Rosemeyer, *Helv. Chim. Acta*, 1994, **77**, 883.
28. F. Seela and S. Lampe, *Helv. Chim. Acta*, 1994, **77**, 1003.
29. S. B. Rajur and L. W. McLaughlin, *Tetrahedron Lett.*, 1992, **33**, 6081.
30. H. P. Hsieh and L. W. McLaughlin, *J. Org. Chem.*, 1995, **60**, 5356.
31. T. Lan and L. W. McLaughlin, *Nucleic Acids Res.*, in press.
32. B. A. Schweitzer and E. T. Kool, *J. Org. Chem.*, 1994, **59**, 7238.
33. N. C. Chaudhuri and E. T. Kool, *Tetrahedron Lett.*, 1995, **36**, 1795.
34. K. M. Guckian, B. A. Schweitzer, X.-F. Ren, C. J. Sheils, P. L. Paris, D. C. Tahmassebi, and E. T. Kool, *J. Am. Chem. Soc.*, 1996, **118**, 8182.
35. X.-F. Ren, B. A. Schweitzer, C. J. Sheils, and E. T. Kool, *Angew. Chem., Int. Ed. Engl.*, 1996, **35**, 743.
36. A. Ono, P. O. P. Ts'o, and L.-S. Kan, *J. Am. Chem. Soc.*, 1991, **113**, 4032.
37. A. Ono, P. O. P. Ts'o, and L.-S. Kan, *J. Org. Chem.*, 1992, **57**, 3225.
38. U. von Krosigk and S. A. Benner, *J. Am. Chem. Soc.*, 1995, **117**, 5361.
39. G. Xiang, W. Soussou, and L. W. McLaughlin, *J. Am. Chem. Soc.*, 1994, **116**, 11 155.
40. G. Xiang, R. Bogacki, and L. W. McLaughlin, *Nucleic Acids Res.*, 1996, **24**, 1963.
41. A. E. Ferentz and G. L. Verdine, *J. Am. Chem. Soc.*, 1991, **113**, 4000.
42. J. T. Goodwin and G. D. Glick, *Tetrahedron Lett.*, 1993, **34**, 5549.
43. C. Min, T. D. Cushing, and G. L. Verdine, *J. Am. Chem. Soc.*, 1996, **118**, 6116.
44. J. L. Mascarenas, K. C. Hayashibara, and G. L. Verdine, *J. Am. Chem. Soc.*, 1993, **115**, 373.
45. W. Saenger, "Principles of Nucleic Acid Structure", Springer, New York, 1983.
46. D. B. Olsen, F. Benseler, H. Aurup, W. A. Pieken, and F. Eckstein, *Biochemistry*, 1991, **30**, 9735.
47. K. Tyagarajan, J. A. Monforte, and J. E. Hearst, *Biochemistry*, 1991, **30**, 10920.
48. J. Goodchild, *Nucleic Acids Res.*, 1992, **20**, 4607.
49. G. Paolella, B. S. Sproat, and A. I. Lamond, *EMBO J.*, 1992, **11**, 1913.
50. Froehler, B. C. and D. J. Ricca, *J. Am. Chem. Soc.*, 1992, **114**, 8320.
51. D. J. Fu, S. B. Rajur, and L. W. McLaughlin, *Biochemistry*, 1994, **33**, 13903.
52. A. Schoppe, H. J. Hinz, H. Rosemeyer, and F. Seela, *Eur. J. Biochem*, 1996, **239**, 33.
53. F. Depart, G. Tosquellas, G. Rayner, and J. L. Imbach, *Nucleosides Nucleotides*, 1995, **14**, 1015.
54. M. Bolli, and C. Leumann, *Angew. Chem., Int. Ed. Engl.*, 1995, **34**, 694.
55. M. Tarköy and C. Leumann, *Angew, Chem., Int. Ed. Engl.*, 1993, **32**, 1432.
56. R. J. Jones, S. Swaminathan, J. F. Milligan, S. Wadwani, B. C. Froehler, and M. D. Matteucci, *Am. Chem. Soc.*, 1993, **115**, 9816.
57. L.-Y. Hsu, D. S. Wise, L. S. Kucera, J. C. Drach, and L. B. Townsend, *J. Org. Chem.*, 1992, **57**, 3354.
58. G. Xiang and L. W. McLaughlin, *Tetrahedron*, 1998, **54**, 375.
59. K. Schneider and S. A. Benner, *J. Am. Chem. Soc.*, 1990, **112**, 453.
60. F. Eckstein, *Annu. Rev. Biochem.*, 1985, **54**, 367.
61. R. Cosstick and J. A Vyle, *Tetrahedron Lett.*, 1989, **30**, 4693.
62. M. Mag, S. Lüking, and J. W. Engels, *Nucleic Acids Res.*, 1991, **19**, 1437.
63. R. G. Kuimelis and L. W. McLaughlin, *Nucleic Acids Res.*, 1995, **23**, 4753.
64. E. K. Jaffe and M. Cohn, *J. Biol. Chem.*, 1978, **253**, 4823.
65. V. L. Pecoraro, J. D. Hermes and W. W. Clelend, *Biochemistry*, 1984, **23**, 5262.
66. R. G. Kuimelis, L. W. McLaughlin, *J. Am. Chem. Soc.*, 1995, **117**, 11 019.
67. J. S. Vyle, B. A. Connolly, D. Kemp, and R. Cosstick, *Biochemistry*, 1992, **31**, 3012.
68. R. G. Kuimelis and L. W. McLaughlin, *Biochemistry*, 1996, **35**, 5308.
69. D. M. Zhou, N. Usman, F. Wincott, J. Matulicadamic, and K. Taira, *J. Am. Chem. Soc.*, 1996, **118**, 5862.
70. P. S. Miller in "Bioorganic Chemistry: Nucleic Acids," ed S. M. Hecht, Oxford University Press, Oxford, 1996, pp. 347–374.
71. B. L. Gaffney and R. A. Jones, *Tetrahedron Lett.*, 1982, **23**, 2253.
72. B. L. Gaffney, L. A. Marky, and R. A. Jones, *Tetrahedron*, 1984, **40**, 3.
73. L. W. McLaughlin, T. Leong, F. Benseler, and N. Piel, *Nucleic Acids Res.*, 1988, **16**, 5631.
74. B. Gildea and L. W. McLaughlin, *Nucleic Acids Res.*, 1989, **17**, 2261.
75. W. L. Sung, *J. Chem. Soc., Chem. Commun.*, 1981, 1089.
76. R. Eritja, D. M. Horowitz, P. A. Walker, J. P. Ziehler-Martin, M. S. Boosalis, M. F. Goodman, K. Itakura, and B. E. Kaplan, *Nucleic Acids Res.*, 1986, **14**, 8135.
77. E. L., Hancox, B. A. Connolly, and R. T. Walker, *Nucleic Acids Res.*, 1993, **21**, 3485.
78. P. Faeber and K. H. Scheit, *Chem. Ber.*, 1970, **103**, 1307.
79. T. R. Waters and B. A. Connolly, *Nucleosides Nucleotides*, 1992, **11**, 1561.
80. M. J. Robins, J. S. Wilson, and F. Hansske, *J. Am. Chem. Soc.*, 1982, **104**, 4059.
81. K. A. Watanabe, in "Chemistry of Nucleosides and Nucleotides," ed. L. B. Townsend, Plenum Press, New York, 1993, vol. 3, pp. 421–535.
82. U. Niedballa and H. Vorbruggen, *J. Org. Chem.*, 1974, **39**, 3654.
83. U. Niedballa and H. Vorbruggen, *J. Org. Chem.*, 1976, **41**, 2084.
84. H. Vorbruggen, K. Krolikiewicz, and B. C. B. Bennua, *Chem. Ber.*, 1981, **114**, 1234.
85. M. Hoffer, *Chem. Ber.*, 1960, **93**, 2777.
86. R. S. Klein, H. Ohrui, and J. J. Fox, *J. Carbohydr. Nucleoside Nucleotides*, 1974, **1**, 265.

87. J. G. Buchanan, A. R. Edgasr, M. J. Power, and P. D. Theaker, *Carbohydr. Res.*, 1974, **38**, 22.
88. C. K. Chu, K. A. Watanabe, and J. J. Fox, *J. Heterocycl. Chem.*, 1975, **12**, 817.
89. C. K. Chu, I. Wempen, K. A. Watanabe and J. J. Fox., *J. Org. Chem.*, 1976, **41**, 2793.
90. R. Shapiro and R. W. Chambers, *J. Am. Chem. Soc.*, 1961, **83**, 3290.
91. G. D. Daves, Jr., *Acc. Chem. Res.*, 1990, **9**, 653.
92. R. N. Farr, D. I. Kwok, J. C. Y. Cheng, and G. D. Daves, Jr., *J. Org. Chem.*, 1992, **57**, 2093.
93. H. C. Zhang and G. D. Daves, Jr., *J. Org. Chem.*, 1992, **57**, 4690.
94. R. N. Farr and G. D. Daves, Jr., *J. Carbohydr. Chem.*, 1990, **9**, 653.
95. F. Benseler, D. M. Williams, and F. Eckstein, *Nucleosides Nucleotides*, 1992, **11**, 1333.
96. X. Liu and C. B. Reese, *Tetrahedron Lett.*, 1995, **36**, 3413.
97. B. S. Sproat, B. Beijer, and A. M. Irribarren, *Nucleic Acids Res.*, 1990, **18**, 41.
98. B. S. Sproat, A. M. Irribarren, R. G. Garcia, and B. Beijer, *Nucleic Acids Res.*, 1991, **19**, 733.
99. N. B. Hanna, K. Ramasamy, R. K. Robins and G. R. Revankar, *J. Heterocycl. Chem.*, 1988, **25**, 1899.
100. H. Rosemeyer and F. Seela, *Nucleosides Nucleotides*, 1995, **14**, 1041.
101. G. Slim and M. J. Gait, *Nucleic Acids Res.*, 1991, **19**, 1183.
102. B. A. Connolly, B. V. L., Potter, F. Eckstein, A. Pingoud, and L. Grotjahn, *Biochemistry*, 1984, **23**, 3443.
103. J. A. Fidanza, H. Ozaki, and L. W. McLaughlin, *J. Am. Chem. Soc.*, 1992, **114**, 5509.
104. R. Cossitck and F. Eckstein, *Biochemistry*, 1985, **24**, 3630.
105. P. S. Miller, C. D. Cushman and J. T. Levis, "Oligonucleotides and Analogues. A Practical Approach," ed. F. Eckstein, IRL Press, Oxford, 1991, pp. 137–154.
106. M. Schweitzer, W. Samstag, and J. W. Engels, *Nucleosides Nucleotides*, 1995, **14**, 817.
107. S. Smith and L. W. McLaughlin, *Biochemistry*, 1997, **36**, 6046.
108. R. P. Iyer, L. R. Phillips, W. Egan, J. B. Regan, and S. L. Beaucage, *J. Org. Chem.*, 1990, **55**, 4693.
109. A. R. Fersht, "Enzyme Structure and Mechanism," Freeman, New York, 1985.
110 A. R. Fersht, *Trends Biochem. Sci.*, 1987, **12**, 301.
111. D. R. Lesser, M. R. Kurpiewski and L. Jen-Jacobson, *Science*, 1990, **250**, 776.
112. N. C. Seeman, J. M. Rosenberg, and A. Rich, *Proc. Natl. Acad. Sci. USA*, 1976, **73**, 804.
113. J. M. Mazzarelli, S. B. Rajur, P. I. Iadarola, and L. W. McLaughlin, *Biochemistry*, 1992, **31**, 5925.
114. J. A. McClarin, C. A. Frederick, B. C. Wang, P. Green, H. W. Boyer, J. Grabale, and J. M. Rosenberg, *Science*, 1986, **234**, 1526.
115. Y. Kim, J. C. Grable, R. Love, P. J. Green, and J. M. Rosenberg, *Science*, 1990, **249**, 1307.
116. Z. Otwinowski, R. W. Schevitz, R.-G. Zhang, C. L. Lawson, A. Joachimiak, R. Q. Marmorstein, B. F. Luisi, and P. B. Sigler, *Nature (London)*, 1988, **335**, 321.
117. J. L. Kim, D. B. Nikolov, and S. K. Burley, *Nature (London)*, 1993, **365**, 520.
118. Y. Kim, J. H. Geiger, S. Hahn, and P. B. Sigler, *Nature (London)*, 1993, **365**, 512.
119. C. R. Aiken and R. I. Gumport, *Methods Enzymol*, 1991, **208**, 433.
120. L. W. McLaughlin, F. Benseler, E. Graeser, N. Piel, and S. Scholtissek, *Biochemistry*, 1987, **26**, 7238.
121. C. A. Brennan, M. D. Van Cleve, and R. I. Gumport, *J. Biol. Chem.*, 1986, **261**, 7278.
122. S. Diekmann, E. von Kitzing, L. W. McLaughlin, J. Ott, and F. Eckstein, *Proc. Natl. Acad. Sci. USA*, 1987, **84**, 8257.
123. S. Diekmann and L. W. McLaughlin, *J. Mol. Biol.*, 1988, **202**, 823.
124. J. Heitman and P. Model, *Proteins*, 1990, **2**, 185.
125. L. S. Klig, I. P. Crawford, and C. Yanofsky, *Nucleic Acids Res.*, 1987, **15**, 5339.
126. D. Staacke, B. Walter, B. V. Kisters-Wolke, B. Wilcken-Bergmann, and B. Muller-Hill, *EMBO J.*, 1990, **9**, 1963.
127. R. G. Brennan and B. W. Matthews, *J. Biol. Chem.*, 1989, **264**, 267.
128. S. Smith, S. B. Rajur, and L. W. McLaughlin, *Nature Struct. Biol.*, 1994, **1**, 18.
129. T. Lan, L. W. McLaughlin, J. Dennis, and P. Sigler, unpublished results.
130. B. Michel and N. D. Zinder, *Nucleic Acids Res.*, 1989, **17**, 7333.
131. P. A. Frey and R. D Sammons, *Science*, 1985, **228**, 541.
132. D. R. Lesser, A. Grajkowski, M. R. Kurpiewski, M. Koziolkiewicz, W. J. Stec, L. Jen-Jacobson, *J. Biol. Chem.*, 1992, **267**, 24810.
133. M. R. Kurpiewski, M. Koziolkiewicz, A. Wilk, W. J. Stec., and L. Jen-Jacobson, *Biochemistry*, 1996, **35**, 8846.
134. M. C. Botfiled and M. A. Weiss, *Biochemistry*, 1994, **33**, 2349.
135. J. J. Santalucia, R. Kierzek, and D. H. Turner, *Science*, 1992, **256**, 217.
136. D. Vlieghe, L. Van Meervelt, A. Dautant, B. Gallois, G. Precigoux, and O. Kennard, *Science*, 1996, **273**, 1702.
137. N. T. Thuong and C. Helene, *Angew. Chem., Int. Ed. Engl.*, 1993, **32**, 666.
138. H. E. Moser and P. B. Dervan, *Science*, 1987, **238**, 645.
139. S. A. Strobel and P. B. Dervan, *J. Am. Chem. Soc.*, 1989, **111**, 7286.
140. V. I., Syamichev, S. M. Mirkin, M. D. Frank-Kamenetskii, and C. R. Cantor, *Nucleic Acids Res.*, 1988, **16**, 2165.
141. J. S. Lee, D. A Johnson and A. R. Morgan, *Nucleic Acids Res.*, 1979, **6**, 3073.
142. B. C. Froehler and D. J. Ricca, *J. Am. Chem. Soc.*, 1992, **114**, 8320.
143. J. S. Lee, M. L. Woodsworth, L. J. Latimer, and A. R. Morgan, *Nucleic Acids Res.*, 1984, **12**, 6603.
144. T. J. Povsic and P. B. Dervan, *J. Am. Chem. Soc.*, 1989, **111**, 3059.
145. J. S. Pudlo, S. Wadwani, J. F. Milligan, and M. D. Matteuci, *Bioorg. Med. Chem. Lett.*, 1994, **4**, 1025.
146. R. Berressem and J. W. Engels, *Nucleic Acids Res.*, 1995, **23**, 3465.
147. J. Gorski and P. Tollin, *Cryst. Struct. Commun.*, 1982, **11**, 543.
148. B. C. Froehler, S. Wadwani, T. J. Terhorst and S. R. Gerrard, *Tetrahedron Lett.*, 1992, **33**, 5307.
149. C. Escudé, J. S. Sun, M. Rougée, T. Garestier, and C. Hélène, *C. R. Acad. Sci., Ser. III*, 1992, **315**, 521.
150. M. Shimizu, A. Konishi, Y. Shimada, H. Inoue, and E. Ohtsuka, *FEBS Lett.*, 1992, **302**, 155.
151. R. F. Macaya, P. Schultze, and J. Feigon, *J. Am. Chem. Soc.*, 1992, **114**, 781.
152. L. Kiber-Herzog, B. Kell, G. Zon, K. Shinozuka, S. Mizan, and W. D. Wilson, *Nucleic Acids Res.*, 1990, **18**, 3545.
153. L. Kiber-Herzog, G. Zon, G. Whittler, S. Mizan, and W. D. Wilson, *Anticancer Drug. Des.*, 1993, **8**, 65.
154. S. G. Kim, S. Tsukahara, S. Yokoyama, and H. Takaku, *FEBS Lett.*, 1992, **314**, 29.
155. L. J. P. Latimer, K. Hampel, and J. S. Lee, *Nucleic Acids Res.*, 1989, **17**, 1549.

156. H. W. Pley, K. M. Flaherty, and D. B. McKay, *Nature*, 1994, **372**, 68–74.
157. W. G. Scott, J. T. Finch, and A. Klug, *Cell*, 1995, **81**, 991.
158. K. Kruger, P. J. Grabowski, A. J. Zaug, J. Sands, D. E. Gottschling, T. R. Cech, *Cell*, 1982, **31**, 147–157.
159. A. M. Pyle, *Science*, 1993, **261**, 709.
160. R. H. Symons, *Trends Bichem. Sci.*, 1989, **14**, 445.
161. O. C. Uhlenbeck, *Nature (London)*, 1987, **321**, 596.
162. R. G. Kuimelis and L. W. McLaughlin, in "Nucleic Acids and Molecular Biology," eds. F. A. L. D. Eckstein and D. M. J. Lilley, Springer, Berlin, 1996, vol. 10, pp. 197–215.
163. J. M. Buzayan, A. Hampel, and G. Bruening, *Nucleic Acids Res.*, 1986, **14**, 9729.
164. J. M. Buzayan, W. L. Gerlach, and G. Bruening, *Nature (London)*, 1986, **323**, 349.
165. B. M. Chowrira, H. A. Berzal, C. F. Keller, and J. M. Burke, *J. Biol. Chem.*, 1993, **268**, 19458.
166. H. A. Berzal, S. Joseph, and J. M. Burke, *Genes Dev.*, 1992, **6**, 129.
167. P. A. Feldstein and G. Bruening, *Nucleic Acids Res.*, 1993, **21**, 1991.
168. L. A. Hegg and M. J. Fedor, *Biochemistry*, 1995, **34**, 15813.
169. Y. Komatsu, M. Koizumi, A. Sekiguchi, and E. Ohtsuua, *Nucleic Acids Res.*, 1993, **21**, 185.
170. A. C. Forster and R. H. Symons, *Cell*, 1987, **50**, 9.
171. D. E. Ruffner, G. D. Stormo, and O. C. Uhlenbeck, *Biochemistry*, 1990, **29**, 10695.
172. G. Bruening, *Methods Enzymol.*, 1989, **180**, 546.
173. A. Hampel, R. Tritz, M. Hicks, and P. Cruz, *Nucleic Acids Res.*, 1990, **18**, 299.
174. P. Anderson, J. Monforte, R. Tritz, S. Nesbitt, J. Hearst, and A. Hampel, *Nucleic Acids Res.*, 1994, **22**, 1096.
175. M. B. De Young, A. M. Siwkowski, Y. Lian, and A. Hampel, *Biochemistry*, 1995, **34**, 15785.
176. S. Joseph, H. A. Berzal, B. M. Chowrira, S. E. Butcher, and J. M. Burke, *Genes Dev.*, 1993, **7**, 130.
177. A. Hampel and R. Tritz, *Biochemistry*, 1989, **28**, 4929.
178. S. C. Dahm, W. B. Derrick, and O. C. Uhlenbeck, *Biochemistry*, 1993, **32**, 13040.
179. K. J. Hertel and O. C. Uhlenbeck, *Biochemistry*, 1995, **34**, 1744.
180. B. L. Bass and T. R. Cech, *Nature (London)*, 1984, **308**, 820.
181. J. A. McSwiggen and T. R. Cech, *Science*, 1989, **244**, 697.
182. T. R. Cech and B. L. Bass, *Annu. Rev. Biochem.*, 1986, **55**, 599.
183. D. J. Fu and L. W. McLaughlin, *Proc. Natl. Acad. Sci. USA*, 1992, **89**, 3985.
184. D. J. Fu, S. R. Fajur, and L. W. McLaughlin, *Biochemistry*, 1993, **32**, 10629.
185. B. M. Chowrira, H. A. Berzal, and J. M. Burke, *Nature (London)*, 1991, **354**, 320.
186. B. M. Chowrira and J. M. Burke, *Biochemistry*, 1991, **30**, 8518.
187. J. A Grasby, K. Mersmann, M. Singh, and M. J. Gait, *Biochemistry*, 1995, **34**, 4068.
188. S. Schmidt, L. Beigelman, A. Karpeisky, N. Usman, U. S. Sorensen, and M. J. Gait, *Nucleic Acids Res.*, 1996, **24**, 573.
189. S. A. Strobel and T. R. Cech, *Biochemistry*, 1996, **35**, 1201.
190. J. P. Perreault, T. F. Wu, B. Cousineau, K. K. Ogilvie, and R. Cedergren, *Nature (London)*, 1990, **344**, 565.
191. J. H. Yang, J. P. Perreault, D. Labuda, N. Usman, and R. Cedergren, *Biochemistry*, 1990, **29**, 11156.
192. J. P. Perreault, D. Labuda, N. Usman, J. H. Yang, and R. Cedergren, *Biochemistry*, 1991, **30**, 4020.
193. J. H. Yang, N. Usman, P. Chartrand, and R. Cedergren, *Biochemistry*, 1992, **31**, 5005.
194. W. A. Pieken, D. B. Olsen, F. Benseler, H. Audrup, and F. Eckstein, *Science*, 1991, **253**, 314.
195. B. M. Chowrira, H. A. Berzal, C. F. Keller, and J. M. Burke, *J. Biol. Chem.*, 1993, **268**, 19458.
196. D. Vitorino dos Santos, J. L. Fourrey, and A. Favre, *Biochem. Biophys. Res. Commun.*, 1993, **190**, 377.
197. B. L. Bass and T. R. Cech, *Biochemistry*, 1986, **25**, 4473.
198. P. S. Kay and T. Inoue, *Nucleic Acids Res.*, 1987, **15**, 1539.
199. D. Herschlag and T. R. Cech, *Nature (London)*, 1990, 344, 405.
200. A. M. Pyle and T. R. Cech, *Nature (London)*, 1991, **350**, 628.
201. A. M. Pyle, F. L. Murphy, and T. R. Cech, *Nature (London)*, 1992, **358**, 123.
202. H. Van Tol, J. M. Buzayan, P. A. Feldstein, F. Eckstein, and G. Bruening, *Nucleic Acids Res.*, 1990, **18**, 1971.
203. B. M. Chowrira and J. M. Burke, *Nucleic Acids Res.*, 1992, **20**, 2835.
204. J. Rajagopal, J. A. Doudna, and J. W. Szostak, *Science*, 1989, **244**, 692.
205. J. A. Piccirilli, J. S. Vyle, M. H. Caruthers, and T. R Cech, *Nature*, 1993, **361**, 85.
206. R. B. Waring, *Nucleic Acids Res.*, 1989, **17**, 10281.

# 7.08
# DNA with Altered Backbones in Antisense Applications

YOGESH S. SANGHVI

*Isis Pharmaceuticals, Carlsbad, CA, USA*

| 7.08.1 | INTRODUCTION | 285 |
|---|---|---|
| | *7.08.1.1 Principles* | 286 |
| | *7.08.1.2 Scope of the Chapter* | 287 |
| 7.08.2 | WHY ALTER THE DNA BACKBONE? | 287 |
| | *7.08.2.1 Increased Stability Towards Nucleolytic Degradation* | 287 |
| | *7.08.2.2 Enhanced Affinity For Target RNA* | 288 |
| | *7.08.2.3 Improved Cellular Uptake* | 289 |
| | *7.08.2.4 Reduced Cost and Synthetic Feasibility* | 289 |
| 7.08.3 | CLASSIFICATION | 289 |
| | *7.08.3.1 Phosphate-modified Linkages* | 290 |
| | *7.08.3.1.1 Modification of a nonbridging phosphate–oxygen atom* | 291 |
| | *7.08.3.1.2 Modifications of the bridging 3′-oxygen atom* | 295 |
| | *7.08.3.1.3 Modification of the bridging 5′-oxygen atom* | 297 |
| | *7.08.3.1.4 Modifications of the 5′-carbon* | 298 |
| | *7.08.3.2 Nonphosphate or Dephosphono Linkages* | 298 |
| | *7.08.3.2.1 Nitrogen-containing linkages* | 299 |
| | *7.08.3.2.2 Silicon-containing linkages* | 303 |
| | *7.08.3.2.3 Sulfur-containing linkages* | 304 |
| | *7.08.3.2.4 Oxygen-containing linkages* | 304 |
| | *7.08.3.2.5 All-carbon linkages* | 305 |
| | *7.08.3.2.6 Nonphosphate sugar linkage (PNA)* | 306 |
| 7.08.4 | DESIGN CONSIDERATIONS | 306 |
| 7.08.5 | CONCLUSIONS | 307 |
| 7.08.6 | REFERENCES | 308 |

## 7.08.1 INTRODUCTION

Antisense-based oligonucleotide technology is a field of research that has intensified the interest in chemical and biological sciences since the mid-1980s. Evidence of this is the tremendous growth in the number of research publications and review[1-5] articles. In order to obtain a complete picture of the scope and dimension of this technology, readers should consult specialized books[6-10] on this topic. Antisense technology represents a major paradigm shift for pharmaceutical- and biotechnology-based drug discovery, with enormous potential well into the twenty-first century. Advances in the automated synthesis of oligonucleotides (oligos) and commercialization of natural and modified nucleic acid building blocks, allow the generation and screening of unprecedented numbers of novel synthetic oligos. The main emphasis of this chapter will be on advances made

285

towards development of new synthetic methodologies to construct novel oligos containing altered or modified backbone linkages. In addition, this chapter will focus on the success and failure of these altered backbones to serve the intended biological functions. What follows is a discussion and summary of mainly published results together with some unpublished work at the author's institution.

### 7.08.1.1   Principles

Antisense oligos are short ($\sim 20$ mer) synthetic single-stranded molecules, analogues of natural nucleic acids, which inhibit or modulate gene expression. The inhibition of gene expression refers to the binding of an antisense oligo in a sequence-specific manner to preselected RNA targets and therefore blocking the translation to the corresponding protein (Figure 1). The underlying principle for this blockage is governed by simple Watson–Crick hydrogen bonding base-pair rules between the antisense oligo and the cellular mRNA. As the short synthetic oligomer binds to the "sense" mRNA and prevents its use, it is called an "antisense" oligo. These sense–antisense relationships are depicted in Figure 1 for the inhibition of gene expression.

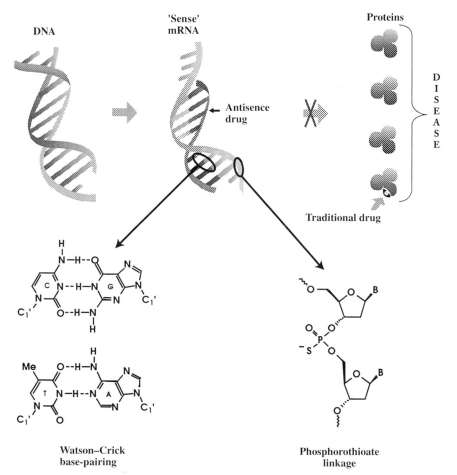

**Figure 1**   Schematic presentation of the antisense concept (top), Watson–Crick base-pairing (bottom left: A–T, G–C), and structure of the phosphorothioate linkage (bottom right) representing the first generation of antisense oligonucleotides.

In principle, this "oligo drug–mRNA" binding recognition motif provides many orders of magnitude higher affinity and specificity than can be achieved using traditional drug-design approaches. Interestingly, both antisense oligos and mRNA targets share common nucleic acid chemistries, such as sugars, bases, and backbones, and since key nucleic acid interactions are well understood, the rational design of antisense oligomers is possible. An antisense oligo can be designed to target any mRNA, potentially creating highly specific drugs for any disease in which genomic sequence information is known. As a result, antisense oligomers have been reported as potential antiviral, antitumor, and anti-inflammatory agents with significant *in vitro* and *in vivo* pharmacological activities.[1]

### 7.08.1.2   Scope of the Chapter

The content of this chapter is restricted to the modifications of the sugar–phosphate moiety that connects the bases in natural nucleic acids. A variety of linkage changes replacing one or both of the nonbridging oxygen atoms, or one or more of the four bridging atoms which connect ($3' \rightarrow 5'$) the two sugar moieties are described in this chapter. This account does not include developments in modifications of unusual or special linkages, such as $2' \rightarrow 5'$ or $3' \rightarrow 3'$ linked oligos, short (3-atoms) or long (5-atoms or more) linkages, α-oligos, and L-oligos.[9] In addition, modifications of sugar and heterocyclic bases are not covered. Chapter 7.09 in this volume describes such base modifications.

Clearly, most of the modifications described herein can be directly or indirectly utilized for other applications, such as construction of synthetic ribozymes, triplexes, circular oligos, and other structured oligos. In order to keep this review short and focused, only antisense-related applications are discussed.

### 7.08.2   WHY ALTER THE DNA BACKBONE?

If natural oligos are considered prototypic, then the available chemically modified antisense oligomers can be regarded as first-, second-, or third-generation antisense molecules, depending upon the complexity of modifications introduced. Early experimentation with unmodified oligos confirmed that use of natural DNA as an antisense molecule had significant limitations. For example, natural DNA or RNA oligos are rapidly degraded under physiological conditions by a variety of cellular nucleases that primarily hydrolyze the phosphodiester internucleosidic linkage. As such, unmodified oligos have only limited utility in advancing antisense research past rudimentary levels. As a consequence, a significant number of modifications of the phosphate backbone have been made.[11] A detailed review of the synthesis of such modified oligomers is beyond the scope of this chapter. However, it is worth highlighting some of the more promising modifications and providing references for some of the well-known modifications. It is believed that for a successful outcome of any backbone modification, the modified oligomer must meet the following criteria, discussed individually.

### 7.08.2.1   Increased Stability Towards Nucleolytic Degradation

A major problem in using oligo-containing natural phosphodiester (PO) linkages in antisense therapeutics, is that these molecules are nuclease sensitive, due to serum and intracellular nucleases that will hydrolyze the PO linkages, resulting in small fragments of nucleotide molecules. Therefore, synthesis of nuclease-resistant antisense oligomers that retain the affinity and specificity for the complementary RNA has been a major focus of antisense research.

As a result, two important first-generation modifications were made to a PO linkage and gave rise to the phosphorothioates[12] (PS) and methylphosphonates[13] (MP). Both modifications successfully increased the nuclease stability of oligomers relative to PO linked oligomers. A detailed account of PS and MP oligomers can be found under individual sections.

It was recognized early on that the modifications of the 3'-hydroxy group of oligo, as well as modifications of the last few 3'-terminal bases or internucleosidic linkages, enhanced their stability against 3'-exonuclease activity.[14] However, use of a "3'-end cap" strategy in antisense oligos may not be adequate for their use *in vivo*, because cellular nucleases also have significant levels of endonuclease activity. On the other hand, efforts have been made to generate a "chimeric antisense molecule" (Figure 2) that displays desirable stability *in vitro* and *in vivo*.[15] Commonly, these oligos contain at least two or more types of modifications. First, a short segment of PS modification (5–10 base pairs) is present for RNase H activity in the middle of the oligo, which is flanked at the 3'- and 5'-ends by segments of another type of modification that provides not only nuclease resistance, but high affinity too. In a chimeric oligo, MP modification was introduced at the 3'- and 5'-ends with the central core of PS modification. Such chimeric oligos have shown favorable half-lives in animals, with potent biological activity as well as reduced binding to plasma proteins.[16]

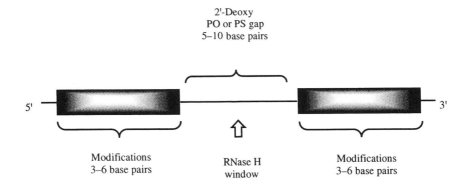

2'-Deoxy
PO or PS gap
5–10 base pairs

5'

3'

Modifications
3–6 base pairs

RNase H
window

Modifications
3–6 base pairs

General requirements:

(i) Nuclease resistant flanking regions
(ii) The 2'-deoxy portion of the molecule has to be sufficiently large
     (5–10 residues in the window)

Potential advantages:

(i)  Increased affinity for targeted mRNA complement
(ii) Increased specificity towards targeted mRNA
(iii) Reduced nonspecific protein interaction

Source: Crooke *et al.*[147]

**Figure 2**   Chimeric oligonucleotide strategy.

In the 1970s, the first efforts to replace completely the PO bridge with nonphosphorus (dephosphono) linkages were made, with an aim towards gaining nuclease resistance. With the advent of antisense research, a tremendous growth in the synthesis of novel dephosphono linkages has occurred. Over 60 types of linkages have been studied since 1994.[17] Comprehensive reviews on this topic have been published.[18–25] Nuclease stability remains a high priority towards the design of novel antisense agents, which becomes clear during the rest of the discussions in this section.

### 7.08.2.2   Enhanced Affinity For Target RNA

Target sites for antisense are discovered using gene walks along the mRNA.[26] This requires synthesis of a large number of oligos. The affinity of an antisense oligo for its RNA complement is traditionally assessed by $T_m$ or melting temperature analysis.[27] The $T_m$ values are highly reproducible under fixed salt and duplex concentration. Therefore, in order to predict the binding affinity of any modified antisense oligo to its RNA target, it is necessary to evaluate hybridization efficiency of such hybrids.

Typically RNA–RNA duplexes show higher $T_m$ compared to DNA–RNA and DNA–DNA duplexes, because of their constrained structure. Based on this fact, chemists working on antisense drug development have pursued several approaches to producing high-affinity molecules that contain the traditional modification around the phosphate atom (via automated procedures), 2'-modifications of the sugar, and complete replacement of the backbone linkage. A more detailed account of how to tune in high affinity ($T_m$) into oligos is discussed under each subheading. Most of the results are described in terms of $T_m$ values increasing or decreasing upon hybridization to the complement RNA. For example, in a closely related group of oligos an increase in $T_m$ of 3–5 °C ($\Delta T_m$) corresponds roughly to a 10-fold enhancement in association constant.

### 7.08.2.3 Improved Cellular Uptake

Unlike traditional drugs, which are of low molecular weight and are uncharged, certain antisense agents are polyanionic and of high molecular weight, which are regarded as presenting a major impediment to traverse through the cellular membrane. Nevertheless, phosphorothioate antisense oligos are taken up reasonably well by most cell types. On the other hand, the transport and distribution of antisense oligos into cellular machinery can be altered by suitable chemical modifications of sugar–base or –backbone. Adsorptive endocytosis and receptor-mediated transport are the two uptake mechanisms by which natural and modified oligos are internalized.[28] Surprisingly, the concentration of free oligos in cytoplasm is low due to their compartmentalization in endocytic vesicles. Nevertheless, oligos do escape the endosomal compartment and produce a biological effect.

Various oligo modification strategies have been utilized to improve cellular uptake and cytoplasmic transport of antisense molecules. These include covalent conjugation of various chemical functionality to an antisense agent to alter its hydrophobicity, charge and amphipathicity.[29] Attachment of specific ligands (such as folic acid and vitamins) to an antisense oligo that binds to certain cellular receptors, may enhance the internalization of such polyanionic molecules. Attempts have been made to increase the membrane permeability of oligos by neutralization of the charge on the PO backbone, such as MP and phosphotriesters. A report indicated that MPs are internalized via an active uptake process and passive diffusion does not play a significant role in their uptake.[30]

Generally, cellular uptake is time- and temperature-dependent. It is also influenced by cell type, cell-culture conditions, media, and sequence (e.g., length).[31] Again, significant differences in subcellular distribution between various types of cells have been noted; therefore, uptake of antisense oligos *in vitro* may not predict their activity *in vivo*.

### 7.08.2.4 Reduced Cost and Synthetic Feasibility

Development of a cost-effective antisense oligo depends upon improving the synthesis and chemistry of its production, increasing the potency and thus reducing the effective dose, and modulating the pharmacokinetic parameters to increase the half-life and concentration at the target site. DNA synthesis and purification technologies can provide oligonucleotides in multigram quantities, sufficient for preclinical and clinical studies.[32] Furthermore, instrumentation for large-scale synthesis is commercially available allowing 100 mmol of 20-mer PS oligos to be prepared in a single run. Pharmacia Biotech AB of Sweden has developed[33] such an instrument, the OligoProcess, shown in Figure 3. One of these mega-synthesizers is currently in use at the Isis GMP manufacturing facilities in Carlsbad, CA, producing ~10 kg of PS oligos to support human clinical trials of five antisense oligos in 1996. Through improved and optimized automated synthesis on OligoProcess and purification conditions, the cost of a 20-mer PS oligo manufactured on a 1 mol (7.5 kg crude/run) scale is estimated to be about US$300/g, making it very attractive for therapeutic use.[34]

Use of certain nonphosphate linkages may have a potential advantage to be more economical and easier to synthesize on a large scale in solution phase rather than the traditional solid-support synthesis of first generation PS oligos.[35] Additionally, use of a nonionic, achiral backbone, which exhibits high affinity for target RNA could result in shortening the length of the oligos without compromising the activity, thus decreasing the production costs.[36] In selecting a completely neutral oligo, water solubility must also be considered. Therefore, an appropriate combination of neutral (nonphosphate) and charged (anionic or even cationic) backbones could be expected to have improved potency and pharmacokinetic properties.

### 7.08.3 CLASSIFICATION

The first use of a synthetic antisense oligonucleotide was reported by Zamecnik and Stephenson in 1978.[37] The apparently simple principle and potential power of the approach led to a rapid proliferation of research. Particularly in the area of backbone modifications, research has been on a fast track due to its usefulness, as discussed in Section 7.08.2. Another obvious reason for the growth of backbone modifications is because it does not directly interfere with the key element of recognition i.e., Watson–Crick base pairing. Broadly stated, all backbone modifications used for antisense research can be classified under two headings: first, phosphorus-containing linkages

**Figure 3**   The OligoProcess instrument installed at Isis Pharmaceuticals.

and modifications thereof, and second, nonphosphate or dephosphono linkages containing other heteroatoms (Figure 4).

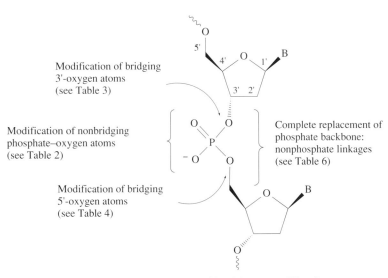

**Figure 4**   The various sites of backbone modifications.

### 7.08.3.1   Phosphate-modified Linkages

Modification of the phosphate backbone is viewed as one of the most fundamental and best-studied modifications in antisense research. The impetus for the PO modification derives from development in two areas. Overall improvements in automated oligo synthesis represents the first, and advanced understanding of their pharmacological and toxicological profile *in vitro* and *in vivo* represents the second. These two aspects taken together allowed the launching of several human clinical trials with first-generation antisense oligos (Table 1).[38] This section is subdivided into four types of modification, depending upon where the modification is placed in relation to the central phosphorus atom.

**Table 1** List of human clinical trials with first-generation antisense oligonucleotides: phosphorothioates (Figure 1).

| Oligonucleotide | Molecular target | Disease indications | Route(s) of administration | Status | Sponsor |
|---|---|---|---|---|---|
| Gem 91 | gag | HIV | intravenous, subcutaneous | II | Hybridon |
| ISIS 2922 | HCMV IE gene product | CMV retenitis | | III | Isis |
| ISIS 2302 | Intercellular adhesion molecule-1 | renal allograft, rheumatoid arthritis, Crohn's disease, ulcerative colitis, psoriasis | intravenous, subcutaneous | II | Isis/Boehringer Ingelheim |
| LR-3001 | C-*myb* | cancer (CML) | intravenous | I/II | Lynx |
| LR-3280 | C-*myc* | restenosis | intracoronary | I | Lynx |
| ISIS 3521/CPG 64128A | protein kinase C-α | cancer (various) | intravenous | I | Isis/Ciba Geigy |
| ISIS 5132/CPG 69846A | C*raf* kinase | cancer (various) | intravenous | I | Isis/Ciba Geigy |
| GPs 0193 | | HIV | intravenous | I | Chugai |
| ? | Ha-*ras* | bladder cancer | bladder instillation | I | Rhone Poulenc Rorer |
| G3139 | BCL-2 | cancer | intravenous | I/IIa | Genta |

### 7.08.3.1.1 *Modification of a nonbridging phosphate–oxygen atom*

The replacement of a nonbridging phosphate–oxygen atom with a heteroatom or a functional group is one of the earliest synthetic manipulations during automated synthesis of oligos, therefore making them easily accessible. It must be realized that any modification of the negatively charged oxygen atom could change the overall charge on the backbone. As a consequence, replacement of this atom with an alkyl group will create a neutral backbone, and replacement with an alkylamine will create a cationic linkage. In principle, three types of molecules can be envisioned by replacement of a nonbridging oxygen atom: (i) negatively charged oligos, (ii) positively charged oligos, and (iii) neutral oligos. The synthesis of these three and related modifications are discussed below, and Table 2 summarizes their properties.

#### (i) Anionic linkages

These oligos are the closest analogues to the natural DNA in terms of structure and charge density. The chemical synthesis of polyanionic oligos is straightforward and has been reviewed extensively.[53] This class of oligos can be further divided into six groups, dependent upon the nature of the substituent, and are discussed below.

*(a) Phosphorothioates (PS).* One of the most versatile and beneficial modifications has been the replacement of one of the nonbridging oxygen atoms with a sulfur atom, thus creating a phosphorothioate group. Although first introduced into DNA enzymatically by Eckstein and co-workers,[54] the phosphoramidite method has greatly facilitated the synthetic aspects of the chemistry.

**Table 2** Properties of the oligos containing modifications of nonbridging phosphate–oxygen atom (charge and stereochemistry are omitted for clarity).

| Name | $X^a$ | $Y^a$ | $T_m^b$ | RNase $H^c$ | $NS^d$ | Chirality | Charge$^e$ | Ref. |
|---|---|---|---|---|---|---|---|---|
| Phosphorothioate | O | S | − | yes | + | yes | -ve | 12 |
| Phosphorodithioate | S | S | − − | yes | + | no | -ve | 39, 40, 41 |
| Boranophosphate | O | $BH_3$ | − | ? | + | yes | -ve | 42 |
| Phosphorofluoridate | O | F | ? | ? | ? | yes | -ve | 43, 44 |
| Phosphoroselenoate | O | Se | ? | ? | ? | yes | -ve | 45 |
| Phosphoramidate | O | NHR | $+^f$ | ? | + | yes | +ve | 46 |
| Aminoalkylphosphonate | O | $CH_2NR$ | $+^f$ | ? | + | yes | +ve | 47 |
| Methylphosphonate | O | Me | $+^f$ | no | + | yes | no | 13 |
| Phosphoramidate (PN) | O | $NH_2$ | − | no | + | yes | no | 48, 49 |
| Phosphoramidimidate | NR | NHR | ? | ? | ? | no | no | 50 |
| Phosphotriester | O | OR | − | ? | + | yes | no | 51, 52 |

$^a$See structure for atom label.  $^b$Melting temperature, − acceptable, − − bad, + good, + + very good, ? not reported.  $^c$Activity of RNase H mediated cleavage.  $^d$Nuclease stability.  $^e$Charge on the backbone.  $^f T_m$ of chirally pure compound.

The chemistry of PS oligos has been reviewed, with particular attention to use of improved reagents, reduction of cost in scale-up, and superior purification methodologies.[55] The current automated synthesis is nonstereospecific, resulting in the creation of a pair of diastereomers so that for $n$ PS linkages, $2^n$ diastereomers will be formed. Stereospecific synthesis of PS oligos has been reported[56] and also reviewed.[57]

Oligos containing PS linkages show considerable reduction in the rate of hydrolysis by a variety of nucleases. Furthermore, nuclease stability of diastereomerically pure PS oligos has been reported.[57] This stability has made them particularly attractive for antisense applications. Generally, $T_m$ of PS oligos are slightly lower than the PO-linked oligos. However, lower affinity of PS oligos can be compensated by increasing the length of the oligo or increasing the GC content. Chirally pure all-$S$p-PS oligos are slightly more stable than those with all-$R$p chirality, which can be utilized when a practical synthesis becomes available. The modest gain in the affinity of chirally pure PS oligos has resulted in development of newer methodologies to prepare them, and has generated some controversy about their clinical applications.[58]

The synthesis of PS oligos via a phosphotriester approach in solution may prove to be a superior method for the synthesis of large quantities of modified oligos and their analogues. Advances reported by scientists from this author's institution and by Reese and co-workers further strengthen this notion of solution phase synthesis[59,60] and proves its utility towards manufacturing of modified antisense agents. High efficiency liquid phase methodology using soluble polyethylene glycol support has been developed by Bonora and Zaramella for the large-scale production of oligos.[61] This method is certainly attractive for the preparation of short oligos and, if combined with other processes, it may become very valuable in making antisense molecules.

Kool and co-workers have reported an exciting method to produce short, sequence-defined oligos using synthetically prepared circular DNA.[62] This novel approach to DNA synthesis may be applicable to the synthesis of antisense molecules, such as PS oligos. A more detailed account of this possibility is discussed in Chapter 7.11 of this volume.

One of the mechanisms by which a PS oligo exerts its antisense effect is due to activation of RNase H, resulting in cleavage of mRNA on formation of hybrid duplexes. Synthesis and study of chimeric oligos is yet another active area of antisense research. Therefore, increasing the hybridization potential of the antisense effector via incorporation of helix-stabilizing modifications (e.g., 2′-O-allyl, neutral dephosphono linkages), in the flanking region have successfully produced molecules that are able to direct RNase H to cleave targeted mRNA. Examples of this approach incorporating alkoxy substitutions with PO linkages in the flanking regions of an oligo have been elegantly demonstrated by Monia et al., by enhanced antitumor activity in vivo.[63] Additionally, such molecules have limited binding to proteins resulting in a favorable profile for sulfur-related toxicity. In addition, methylene (methylimino) linkage and methylphosphonates have been used in the flanking regions as nonionic modifications[36] to lower charge density by incorporating nonionic linkages. It is believed that this will result in reduced binding to proteins.

Studies of PS oligos in animals and humans have shown that they have a very acceptable therapeutic index. Toxicity has only been observed at a very high dose.[64] Moreover, progress in scale-up and cost reduction has been substantial enough to provide confidence that cost of therapy with these first-generation antisense drugs will not be prohibitive. As a result, several PS oligos are in full scale clinical development (see Table 1) and moving along rapidly towards NDA (New Drug Application) filing.

*(b) Phosphorodithioates (PS$_2$).* In an effort to circumvent the problem of chirality with PS oligos, Caruthers and co-workers[39–41] synthesized oligos in which two of the nonbridging oxygens of internucleotidic phosphate linkage were replaced by sulfur, resulting in a phosphorodithioate, which is now devoid of chirality. These isoelectronic analogues of DNA are nuclease resistant and are of interest as inhibitors of viral gene expression. PS$_2$ oligos containing pyrimidines are readily synthesized via a triester solid-phase method, providing PS$_2$ oligo free (<0.5%) of PS contaminations.[65] While PS$_2$ oligos containing mixed bases are not straightforward to make, enough progress has been made by Dahl's and Caruthers' groups independently to obtain such analogues in small amounts, enough for biological screening.

Convincing antisense activity has not been reported with a PS$_2$ oligo, perhaps for the following reasons. First, hybridization $T_m$ of PS$_2$ oligos is lower relative to the PS and PO oligos. It is about 1–2 °C/base destabilizing compared to its PO analogue. Second, because PS$_2$ oligos have a higher sulfur content, they appear to bind proteins tighter than the PS oligos, and to activate Rnase H mediated cleavage with reduced efficiency compared to the PS analogue.[39–41] Finally, synthesis of fully mixed bases in 20-mer sequence has not been accomplished, where all linkages are PS$_2$ and not contaminated with PS linkages. The stability of PS$_2$ oligos is under investigation[66] and improved procedures are needed to guarantee its integrity. These properties taken together argue against PS$_2$ becoming a significant antisense agent.

*(c) Boranophosphates (BP).* The use of boron neutron capture therapy for the treatment of malignant cancer has encouraged Shaw and co-workers[42] to investigate boron incorporation into DNA. They have shown that BP dimers are very stable to nucleases. The BP oligos were synthesized from 2-boranotriphosphates by template-directed primer extension using DNA polymerase *in vitro*. However, chemical synthesis of BP oligos (T$_{10}$) has been reported via automated H-phosphonate chemistry.[67] Like PS oligos, BP oligos are chiral and up until 1997 a stereospecific chemical synthesis had not been reported, except one short communication describing chromatographic separation of diastereomers at a dimer level.[68] The enzymatic preparation of BP oligos provided an *S*p configuration at the chiral center, and exerts no change in $T_m$ compared to DNA. In summary, due to their chirality issue and lack of a good synthetic procedure to provide enough materials for antisense-based gene walk, the therapeutic use of BP oligos may not progress as rapidly as PS oligos.

*(d) Phosphorofluoridates (PF).* Nucleoside phosphorofluoridates were synthesized for the first time by Wittmann[69] in 1963, but it is only in the 1990s that their potential in oligos began to be explored. The attractive features of PF linkages are that they bear a high resemblance to natural PO linkages, that they may provide a new means for a covalent attachment of molecular probes to proteins, or that the fluorine may serve as a reporter group for studying conformational properties of nucleic acids. As indicated, synthesis of PF oligos is still in its infancy and a full-length (~20-mer) oligo has yet to be made. However, Dabkowski and Tworowska[43] and Stawinski and co-workers[44] independently reported a novel entry into PF-linked dimers. The key reaction was an oxidation of H-phosphonate with iodine in the presence of triethylamine trishydrofluoride furnishing a PF dimer in quantitative yield. Similarly, phosphorofluorodithioate was also prepared. In summary, the PF modification is certainly an interesting mimic of natural PO linkage and may find applications in molecular biology, but their application in antisense research is remote at this time. Interestingly, Stec and co-workers pointed out that the PF-linked dimer was unstable under solvolytic conditions and undergoes rapid base-catalyzed hydrolysis.[70]

*(e) Phosphoroselenoates (PSe).* Despite the biological importance of selenium and an indication that metabolism of selenium-containing compounds may involve PSe intermediates, PSe oligos have received relatively little attention. Synthesis of a PSe analogue was accomplished on an automated DNA synthesizer by replacing the sulfurization reagent with potassium selenocyanate, but the yield was low, possibly due to poor solubility of selenone in organic solvent. This problem was solved when Stawinski and Thelin[71] reported the use of 3*H*-1,2-benzothiaselenol-3-one as selenizing reagent. It has been demonstrated that PSe linkage is not nearly as stable as PS linkage, and liberates free selenium upon degradation, which happens to be toxic. Stein and co-workers have described the use of PSe antisense oligos and concluded against their use for the following reasons.[45] First, $T_m$

results were lower than both PS and PO oligos. Second, PSe analogues were found to be non-sequence-specific inhibitors with less activity and much more toxic compared to PS analogues. Therefore, PSe analogues are not suitable for antisense constructs.

### (ii) Cationic linkages

Introduction of cationic groups at various positions throughout the oligo provides a method for producing molecules with widely differing charge densities. Cationic molecules have previously been used to enhance the delivery of a variety of conjugates into cells, such as liposome-mediated transfection protocols for PS oligos.[72] Certain cationic pendant groups have been conjugated to oligos in attempts to improve their cellular uptake or activity *in vivo*.[73] Other methods which utilize base modifications for introduction of a positive charge have been reviewed elsewhere.[74]

*(a) Phosphoramidates and aminoalkylphosphonates.* An interesting way to alter natural anionic PO backbone is to replace it with cationic groups. Letsinger *et al.* first reported[75] the synthesis of a cationic oligo containing alternating PO and basic aminoalkylphosphoramidate linkages, which appeared to be electrically neutral at pH 7. The $T_m$ of these oligos were found to be higher than those of the unmodified oligos at low ionic strength. These studies were carried out with phosphoramidate analogues prepared without stereocontrol at the P—N bonds. In view of this, Letsinger and co-workers reported[46] their study with an oligo containing anionic and stereo-uniform cationic linkages, showing high binding affinity ($\Delta T_m + 2.3\,°C$/modification) for RNA target. The isomeric oligo with opposite configuration binds much more poorly. Similar observations were made by Fathi *et al.* incorporating an aminoethylphosphonate.[47] Their results with alternating cationic/anionic linked oligos indicated that the $R_p$ isomers were capable of forming more stable duplexes with RNA than their natural counterparts. The $S_p$ isomers did not form stable hybrids under similar conditions. Additionally, they indicated that these oligos were stable to S1 nuclease and were taken up by JY cells somewhat better than the control PO oligos. This novel approach certainly has distinct advantages over first-generation PS oligos and may offer a different perspective in terms of its interaction with proteins. However, the vastly different affinities of $R_p$ and $S_p$ isomers emphasizes that a stereospecific synthesis must be developed before their true potential for antisense therapeutics is realized.

A polycationic oligo consisting of PO-linked 12-mer with a cationic tail at the 3′-end and a cholesteryl group at the 5′-end was synthesized and shown to inhibit HSV-1 in Vero cells.[76] Therefore, an appropriate mix of positive and negative charge densities on an antisense oligo may result in some unusual properties, such as higher affinity and improved uptake.

### (iii) Neutral linkages

As indicated earlier, the polyanionic nature of DNA backbone was assumed to be a drawback for its efficient uptake into cells. Therefore, neutral oligos were made in the hope that they would enter the cells via passive diffusion. Neutralization of the charge in natural DNA can be accomplished in several ways. Various approaches towards generating neutral oligos are discussed below, including their antisense-related properties (see Table 2).

*(a) Methylphosphonates (MP).* These nonionic derivatives were first introduced by Miller and Ts'o using modified triester chemistry.[77] However, later methods utilize phosphorus(V) derivatives in which active esters of 3′-methylphosphonates are coupled to 5′-alcohols. It must be realized that most of the older syntheses of MPs resulted in diastereomeric mixtures of oligos. Nevertheless, stereospecific solid-phase syntheses of both $R_p$ and $S_p$ MP oligos are now available.[78]

MP oligos appear to be completely resistant to hydrolysis by endo- and exonucleases in cell-based experiments. The hybrids formed between MP oligo and RNA are not substrate for Rnase H mediated cleavage. However, "chimeric oligos" which contain 4–6 contiguous PO linkages flanked by MP linkages, when bound to RNA, do cleave via RNase H activity (see Figure 2). It has been reported by Lebedev and Wickstrom[57] that all-$R_p$ oligos hybridize much more strongly to a complementary RNA target than do the racemic oligos or all-$S_p$ oligos. As expected, the nonionic nature of MP oligos allows duplex formation in the presence of low or no salt. This salt independence is believed to be due to reduced charge repulsion between the two oligos.

A number of studies[79] have shown that MPs are effective inhibitors of viral protein synthesis in cell culture systems (e.g., VSV, HSV, HIV-1). Although MP oligos have been studied extensively in cell culture systems, their *in vivo* activity has not been followed up in depth. A paper published in 1994 describes inhibition of HSV-1 replication in a mouse model,[80] but animal studies have been few.

The reasons for the lack of intense progress in this area could be summarized as follows. First, it was believed that MPs were taken up by passive diffusion, but later studies indicated nonreceptor endocytosis, possibly absorptive or fluid phase endocytosis.[30,81,82] Release of MP oligos from the endosomal pathway presumably limits the efficacy of these compounds *in vivo*. Second, MPs are not substrate for RNase H, therefore, antisense activity is due to steric block only. Again, lower affinity of racemic MP limits their use *in vivo*. Finally, special procedures for synthesis and deprotection may be required for their production.[81,82] In summary, chirally pure *R*p MP will find its utility in the antisense area as more material becomes available. Scientists from Genta, Inc. have presented interesting results using chirally pure chimeric 2′-*O*-methyl MP/PO oligos, inhibiting BCL2 protein expression in cell culture and in SCID mice.[83]

*(b) Phosphoramidates (PN).* Substitution of an amino or aminoalkyl group for one of the two nonbridging oxygen atoms of the PO linkage creates a nonionic PN internucleosidic linkage. In this regard, Vasseur and co-workers[48] and Iyer *et al.*[49] independently reported a novel nonionic P—NH₂ linkage. It was suggested that the P—NH₂ group could hydrogen bond with water and thus increase their water solubility. The small size of the NH₂ group is believed to create the least steric hindrance around the PN backbone, while hybridizing to complementary RNA. In brief, syntheses of P—NH₂ oligos were accomplished on an automated solid-support DNA synthesizer using both H-phosphonate and phosphoramidite chemistries, in combination with highly base-labile protection strategy. Oligos containing P—NH₂ links formed less stable duplexes with RNA targets ($\Delta T_m$ −2.0 °C/modification) than the corresponding PO oligos. However, these oligos were resistant to cleavage by various nucleases. It was also concluded that this modification does not allow RNase H mediated cleavage of the target RNA. Vasseur and co-workers[84] have described a dramatic change in the affinity ($\Delta T_m$ +1.2 °C/modification) by changing the anomeric configuration to $\alpha$ in P—NH₂ oligos. They did not provide any explanation for these unexpected results in this communication. Syntheses of chimeric PN building blocks have also been reported[85] containing P—NH—OMe and P—NH(CH₂)₅NH acridine functionalities. However, these analogues form less stable duplexes with complementary RNA targets. Given the nature of synthesis and properties of PN oligos, it is unlikely that they will make a useful contribution towards the antisense approach.

*(c) Phosphoramidimates (PN₂).* Fischer and Caruthers have reported the first synthesis of a PN₂ dinucleotide, derived from an aliphatic amine.[50] Their rationale for synthesizing such linkages is in anticipated nuclease resistance, potential for creating a neutral or electropositive backbone by changing substituents, and providing a flexible site for further conjugations. This dimeric PN₂ linked molecule was found to be unstable under acidic and basic conditions, making it very challenging to incorporate into an antisense oligo.

*(d) Phosphotriesters (POR).* These oligos were made as nonionic analogues of DNA even before MP analogues were introduced. Specifically, POR analogues, containing *O*-methyl phosphate or *O*-methyl phosphorothioate, have not been rigorously evaluated for their antisense properties because of their difficult syntheses. The main problem is their base lability during deprotection conditions or cleavage from support. Noyori and co-workers[51] and Iyer *et al.*[52] independently reported the use of alkyloxycarbonyl and *N*-pent-4-enoyl protected nucleosides, respectively. In addition, Tang and co-workers have reported[86] the use of thionotriester containing oligos for antisense utility. These oligos show improved nuclease stability, cellular association, and binding affinity. There was also an implication of these S-triester PS oligos undergoing slow hydrolysis to all-PS oligos, and possibly functioning as prodrugs for corresponding antisense PS oligos. In summary, many synthetic hurdles must be overcome before one can make POR oligos routinely, in sufficient quantity and of good enough quality, for antisense applications.

### 7.08.3.1.2 *Modifications of the bridging 3′-oxygen atom*

The desire for substituting the bridging 3′-oxygen atom of the PO linkage was born out of chirality problems associated with replacement of nonbridging oxygen atoms. Replacement of 3′-oxygen in a PO linkage should result in nuclease stability of the modified backbone and may have potential to alter the sugar conformation in a favorable manner for higher affinity. As a consequence, three major types of modifications have been investigated (Table 3) for their application in antisense research.

**Table 3** Properties of the oligos containing modifications of bridging-3′-oxygen atom.

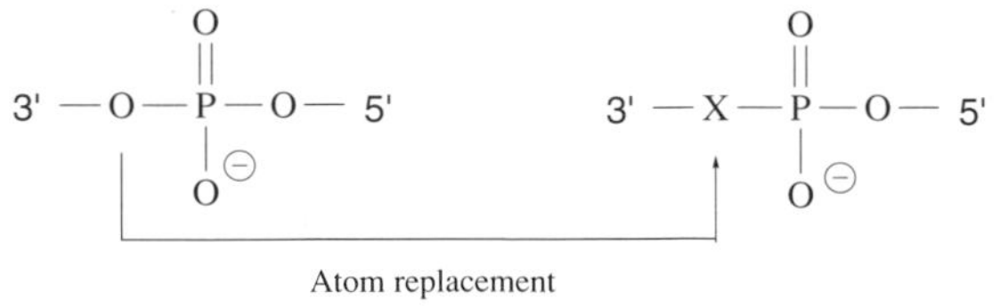

Atom replacement

| Name | $X^a$ | $T_m{}^b$ | RNase $H^c$ | $NS^d$ | Chirality | $Charge^e$ | Ref. |
|---|---|---|---|---|---|---|---|
| Phosphoramidate | NH | $++$ | no | $++$ | no | -ve | 87 |
| Phosphorothioate | S | ? | ? | ? | no | -ve | 88 |
| Phosphinate | $CH_2$ | $-$ | no | $+$ | no | -ve | 89, 90 |

[a] See structure for atom label. [b] Melting temperature, $-$ acceptable, $--$ bad, $+$ good, $++$ very good, ? not reported. [c] Activity of RNase H mediated cleavage. [d] Nuclease stability. [e] Charge on the backbone.

### (i) Phosphoramidates (NP)

Synthesis of NP oligos has been discovered and perfected by Gryaznov and co-workers,[87] utilizing solid-phase methodology and 5′-dimethoxytrityl-protected-3′-amino-2′,3′-deoxyribonucleosides as key monomer units. Oligo chain assembly was based upon a carbon tetrachloride driven oxidative coupling of the appropriately protected 3′-amino-nucleosides with the 5′-H-phosphonate diester group, resulting in the formation of an internucleosidic NP link (**1**). Uniformly modified NP oligos possess some very attractive features for antisense applications,[91] such as negatively charged and achiral phosphorus, good solubility in water, and resistance to nuclease digestion, buttressed by a high affinity for RNA target. Studies have shown that the sugar ring conformation changes from predominantly C-2′-*endo* to C-3′-*endo* when the 3′-*O* is replaced with 3′-*N* functionality, thereby forming a very stable A-type of duplex with complement RNA.[92] Additionally, placement of a 2′-electronegative (F or OMe, X or Y in (**1**)) substituent in combination with the NP backbone enhanced the $T_m$ up to 4 °C per modification.[93]

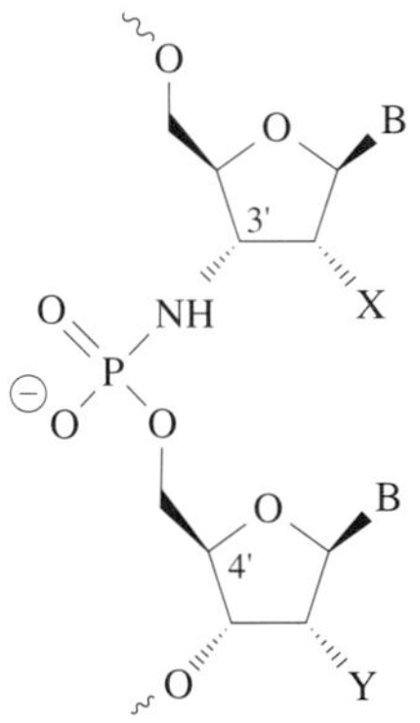

**(1)**
Phosphoramidate

These oligos were used in various *in vitro* and *in vivo* antitumor systems as antisense drugs addressed to different targeted regions of c-myb, c-myc, and bcr-bl mRNAs.[91] In certain examples, up to 90% protein reduction was observed. However, NP oligos do not activate RNase H and observed antisense activity was entirely due to high affinity for the mRNA target. In summary, well-established synthesis protocol and attractive antisense properties, including preorganization, of NP oligos make them very promising candidates for antisense applications.

### (ii) Phosphorothioates (SP)

Analogues where one of the two bridging oxygen atoms in the phosphodiester linkage is replaced by a sulfur atom are particularly attractive for several reasons: (i) they are achiral at phosphorus and thus, no diastereomers are created during their synthesis; (ii) they are electronically and sterically similar to the natural congener; (iii) they provide nuclease resistance to the oligo; and (iv) they are susceptible to specific cleavage under mild conditions. These compounds are synthetically challenging to prepare and only two practical syntheses have been reported. First, Cosstick and co-workers disclosed a procedure to prepare SP dimers from the corresponding thioesters, utilizing a regioselective Michaelis–Arbuzov reaction.[88] The method appears to be compatible with purine and pyrimidine nucleosides, and does not require protection of the hydroxy groups. Second, a complementary procedure has been reported by Reese and co-workers which prepares SP linkage from thioethers using sulfenyl chlorides under acidic conditions.[94] Since a good protocol to prepare SP oligos is available, synthesis of mixed base 20-mer sequences for antisense research should be possible.

### (iii) Phosphinates (CP)

The altered DNA of this class contains a tetravalent carbon moiety in place of the divalent oxygen moiety at the 3′-position of an intermediate linkage. Collingwood and Baxter have reported the synthesis and incorporation of CP-linked dimers into oligos.[89] The synthesis requires base-mediated coupling of a 5′-phosphonous ethyl ester with a 3′-C-aldehydo nucleoside to provide a good yield of dimer. Oligos containing CP-linked dimers were evaluated for their antisense properties. There was an overall destabilizing effect ($\Delta T_m$ $-2.0\,^{\circ}$C/modification) when hybridized to RNA target. However, they were found to be stable to the cleavage by 3′-exonucleases. Abundant patent literature[89,90] is available for the CP-linked oligos claiming the use of these molecules for antisense therapy.

### 7.08.3.1.3 *Modification of the bridging 5′-oxygen atom*

The rationale for studying 5′-O-modified linkage is quite similar to the rationale of the 3′-O-modifications discussed earlier. General properties of these oligos are listed in Table 4.

**Table 4** Properties of the oligos containing modifications of bridging 5′-oxygen atom.

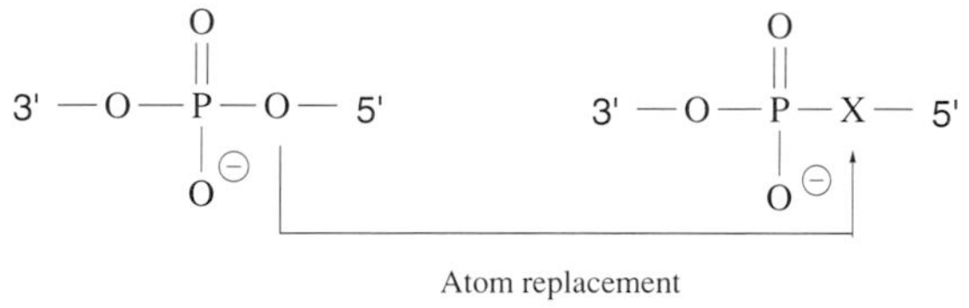

$$3' \mathrm{-O-} \overset{\displaystyle \overset{O}{\|}}{\underset{\displaystyle \underset{O^{\ominus}}{|}}{P}} \mathrm{-O-} 5' \qquad 3' \mathrm{-O-} \overset{\displaystyle \overset{O}{\|}}{\underset{\displaystyle \underset{O^{\ominus}}{|}}{P}} \mathrm{-X-} 5'$$

Atom replacement

| Name | $X^a$ | $T_m{}^b$ | RNase H[c] | NS[d] | Chirality | Charge[e] | Ref. |
|---|---|---|---|---|---|---|---|
| Phosphoramidate | NH | ? | ? | ? | no | -ve | 95 |
| Phosphorothioate | S | ? | ? | ? | no | -ve | 96 |
| Phosphinate | $CH_2$ | – | ? | + | no | -ve | 89 |

[a]See structure for atom label. [b]Melting temperature, − acceptable, − − bad, + good, + + very good, ? not reported. [c]Activity of RNase H mediated cleavage. [d]Nuclease stability. [e]Charge on the backbone.

### (i) Phosphoramidates

It has previously been shown that oligomers bearing nucleoside units linked by 3′-O—P—N-5′ bonds could be synthesized and are stable under neutral and alkaline conditions. Mag and Engels have reported[95] the synthesis of several dimeric phosphoramidates, using the Staudinger reaction followed by Michaelis–Arbuzov type transformation. However, this chemistry has not been applied to the synthesis of uniformly modified oligos; thus, their true potential for antisense research remains uninvestigated.

*(ii) Phosphorothioates*

Synthesis of 3'-O—P—S-5' linked oligos has not been reported (1997). However, Liu and Reese reported[96] synthesis of uridylyl-(3' → 5')-(5'-thiourine), which shows exceptional lability to base treatment, and under neutral conditions as well. Due to the lability of this linkage and lack of viable oligo synthesis, it may not be worthy of consideration towards antisense utility.

*(iii) Phosphinates*

Like many other modifications, replacement of the 5'-oxygen atom with a 5'-methylene group has been superficially studied. Incorporation of a 3'-O—P—CH$_2$-5' linked dimer had a destablizing effect on the duplex formation.[89] In a ribozyme-related study, Matulic-Adamic *et al.* reported[97] synthesis of a 5'-deoxy-5'-difluoromethylphosphonate nucleotide analogue, using a versatile fluoro sugar synthon. This synthetic pathway may be utilized for making an antisense construct.

### 7.08.3.1.4   *Modifications of the 5'-carbon*

Saha *et al.* reported[98] a provocative modification of the DNA backbone, in which a methyl group was introduced at the 5'-carbon of the phosphodiester linkage. They believed that the presence of a 5'-Me group would reduce recognition by nucleases and yet be small enough to conserve hybrid stability. Cellular transport may be facilitated by the lipophilic Me substituent, while adequate water solubility can be retained via the negative charge of the PO linkage. It is worth noting that this linkage contains a chiral carbon atom. This communication reveals that 5'-Me substitution provides extra stability to 3'-exonuclease cleavage and does not adversely affect the $T_m$ ($\Delta T_m$ $-0.2\,°C$/modification). In view of these beneficial results, it may be worthy of consideration to prepare enantiomerically pure backbone and study its antisense potential.

Wang *et al.* reported[99,100] incorporation of chirally pure 5'-(*R*)-*C*-branched (allyl group) and 5'-(*S*)-*C*-branched (methoxy, amino, cyano, and allyl groups) building blocks into oligos. Preliminary results indicate that such modifications have a positive effect toward nuclease stability and comparable affinity for RNA target. The results of incorporation of these modifications in antisense constructs are eagerly awaited. Properties of the oligos containing modifications of the 5'-carbon are shown in Table 5.

**Table 5**   Properties of the oligos containing modifications of 5'-carbon.

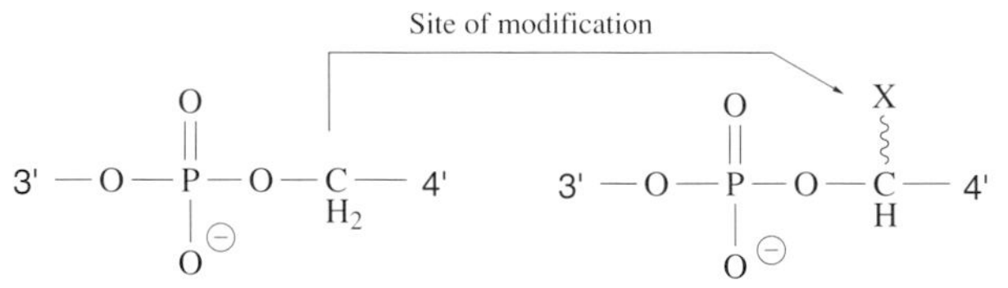

| Name | $X^a$ | $T_m{}^b$ | RNase H$^c$ | NS$^d$ | Chirality | Charge$^e$ | Ref. |
|---|---|---|---|---|---|---|---|
| 5'-Methyl (*R*/*S*) | Me | − | ? | + | yes | -ve | 98 |
| 5'-Methoxymethyl (*S*) | CH$_2$OMe | − | ? | + | yes | -ve | 99, 100 |
| 5'-Aminomethyl (*S*) | CH$_2$NH$_2$ | − | ? | + | yes | -ve | 99, 100 |
| 5'-Cyanomethyl (*S*) | CH$_2$CN | − | ? | + | yes | -ve | 99, 100 |
| 5'-Allyl (*R* or *S*) | CH$_2$CH=CH$_2$ | − | ? | + | yes | -ve | 99, 100 |

$^a$See structure for atom label.   $^b$Melting temperature, − acceptable, − − bad, + good, + + very good, ? not reported.   $^c$Activity of RNase H mediated cleavage.   $^d$Nuclease stability.   $^e$Charge on the backbone.

### 7.08.3.2   Nonphosphate or Dephosphono Linkages

Since the early 1990s, there has been a steady rise in the number of publications and specialized reviews relating to the synthetic methodologies and application of nonionic/achiral linkages that replace the natural PO backbone.[11,17] The data concerning these surrogates of PO linkage, as is discussed in this section, crystallize an understanding of the minimal structural features required not only to mimic a natural PO bond, but also to preorganize these surrogates for the best antisense properties.

It is now well documented that replacement of the PO backbone with a nonphosphate backbone has several distinct advantages in terms of their antisense properties: it automatically confers resistance to cellular nucleases because they are known to cleave only PO linkages; it may provide increased cellular uptake due to reduced charge density compared to polyanionic first-generation PS oligos; it avoids the chirality imposed by the use of PS oligos; and major advantages may be realized in the economics of large-scale solution-phase synthesis.

Chemistries required to produce these linkages are often quite challenging; however, with advances in synthetic/enzymatic methodologies it has become more routine. The most frequently used approach is to prepare a nucleosidic dimer containing the backbone of choice and then incorporate it into an oligo using standard phosphoramidite chemistry. Such procedures allow the synthesis of a polymer linked with alternating PO and novel linkages. Synthesis of fully substituted polymers was not always possible. Laboratories at Isis Pharmaceuticals[36] and several others[101–104] have since taken a major step forward by preparing fully modified polymers devoid of the phosphorus-containing linkage, called 'oligosides.''

This section is limited to developments describing the use of four-atom linkers connecting the two sugar moieties of a nucleosidic dimer in natural configuration. The subsections are created according to the type of heteroatom present in these novel linkers. A summary of the key properties of the nonphosphate backbone-linked oligos is presented in Table 6.

**Table 6** Properties of the oligos containing nonphosphate backbones (Figure 4).

| Name | $3' \rightarrow 5'$ Linkage | $T_m$[a] | RNase H[b] | NS[c] | Chirality | Charge[d] | Ref. |
|---|---|---|---|---|---|---|---|
| Amide 1 | $NRCOCH_2CH_2$ | − | ? | + | no | no | 19 |
| Amide 2 | $CH_2CH_2NHCO$ | − | ? | + | no | no | 19 |
| Amide 3 | $CH_2CONHCH_2$ | + | yes | + + | no | no | 105 |
| Amide 4 | $CH_2NHCOCH_2$ | + | ? | + | no | no | 105 |
| Amide 5 | $CONHCH_2CH_2$ | − | ? | + | no | no | 105 |
| Guanidine 1 | $NHC(-NR)NHCH_2$ | − | ? | + | no | no | 106 |
| Guanidine 2 | $NHC(-NH_2)NHCH_2$ | + | ? | + | no | +ve | 101 |
| Urea | $NHCONHCH_2$ | − | ? | + | no | no | 107 |
| Carbamate | $OCONRCH_2$ | − | ? | + | no | no | 19 |
| Amino 1 | $NHCH_2CH_2CH_2$ | − | ? | + | no | no | 104 |
| Amino 2 | $CH_2CH_2NHCH_2$ | − | ? | + | no | no | 104 |
| Amino 3 | $CH_2NHCH_2CH_2$ | − | ? | + | no | no | 104 |
| Amino 4 | $CH_2N(Me)CH_2CH_2$ | − | ? | + | no | +ve | 108 |
| Amino 5 | $CH_2CH_2N(Me)CH_2$ | − | ? | + | no | +ve | 108 |
| MMI | $CH_2N(Me)OCH_2$ | + +[e] | yes | + + | no | no | 109 |
| Silyl | $OSiR_2OCH_2$ | − | ? | + | no | no | 110 |
| Sulfide | $CH_2CH_2SCH_2$ | − | ? | + | no | no | 111, 112 |
| Sulfone | $CH_2SO_2CH_2CH_2$ | − | ? | + | no | no | 113 |
| Sulfonate | $OSO_2CH_2CH_2$ | − | ? | + | no | no | 114 |
| Sulfonamide | $NHSO_2CH_2CH_2$ | − | ? | + | no | no | 114 |
| Formacetal | $OCH_2OCH_2$ | − | no | + | no | no | 18 |
| Thioformacetal | $SCH_2OCH_2$ | + | no | + | no | no | 18 |
| Ether 1 | $OCH_2CH_2O$ | − | ? | + | no | no | 115 |
| Ether 2 | $OCH_2CH_2CH_2$ | − | ? | + | no | no | 116 |
| Carbon 1 | $CH_2CH_2CH_2CH_2$ | − | ? | + | no | no | 117 |
| Carbon 2 | $CH_2COCH_2CH_2$ | − | ? | + | no | no | 19 |
| Carbon 3 | $CH_2CH=CHCH_2$ | + | ? | + | no | no | 118 |
| Carbon 4 | $CH_2COCH=CH$ | ? | ? | ? | no | no | 119 |

[a]Melting temperature, − acceptable, – bad, + good, + + very good, ? not reported. [b]Activity of RNase H mediated cleavage. [c]Nuclease stability. [d]Charge on the backbone. [e]$T_m$ of chirally pure compound.

### 7.08.3.2.1 Nitrogen-containing linkages

This is one of the most active areas of study for research groups in search of novel surrogates of phosphate backbone. This section is further divided into five major types of nitrogen-containing linkages, depending upon the nature of the functionality produced.

*(i)  Amides*

De Mesmaeker *et al.* designed and synthesized all possible permutations and combinations of an amide functionality within the four-atom space between the two sugar units.[19] Indeed, they have successfully synthesized five flavors of these amide-linked dimers since the early 1990s. An excellent review of the syntheses and their antisense properties has been published.[105] Readers are encouraged to read on this subject to understand the breadth and scope of this large undertaking by Ciba scientists, which clearly lays the foundation for the future of backbone modifications. This hard work has paid off through the emergence of an ideal amide linkage (3′-CH$_2$CONH-5′) that appears to be very attractive for antisense constructs (**2**). The usefulness of this linkage stems from its preorganization of sugar into a high *N*-pucker resulting in an A-type of duplex formation ($\Delta T_{\mathrm{m}}$ +0.1 °C/modification).[120]

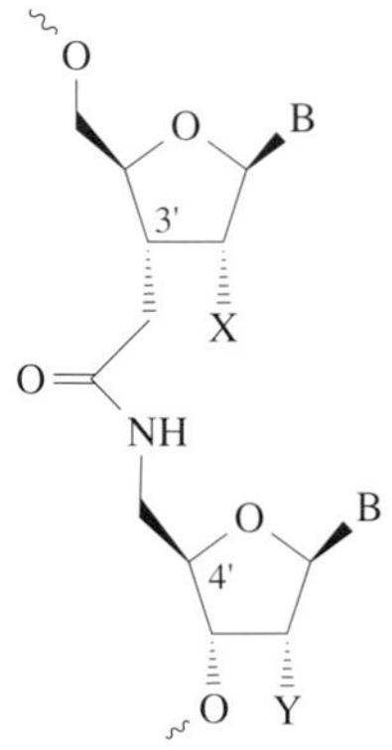

**(2)**

Amide-3

Generally, all amide linkages are resistant to endo- and exonucleases, and are not a substrate for RNase H in a fully alternating motif. RNase H activity can be restored when these modifications are placed in the flanking region of a chimeric oligo, away from the cleavage site on RNA complement. The chimeric oligos containing amide linkages have shown good antisense activity in cell-based experiments.[121] In view of these results, reports of the *in vivo* pharmacokinetic and pharmacodynamic behavior are anxiously awaited.

With the utility of amide backbones established, a novel approach to their synthesis was the focus of attention for Robins *et al.*, described in an elegant manner, via coupling of a 5′-amino nucleoside with 2′,3′-fused-*v*-butyrolactone nucleoside.[122] This procedure describes the synthesis of a ribo analogue of amide-linked dimer, which may be somewhat better than previously described amide modifications, because a 2′-electronegative substituent has been reported[9] to enhance the affinity for RNA target, an asset for antisense applications.

Along similar lines, Wengel and co-workers also made amide-linked dimers containing a piperazine skeleton.[123] Not surprisingly, all modifications were very destabilizing when hybridized to complement RNA. Therefore, special attention should be paid to the fact that too much rigidity may be bad for duplex formation. One needs to achieve a fine balance between rigidity and flexibility.

*(ii)  Guanidines*

In 1993 Herdewijn and co-workers first synthesized thymidine dimers containing a variety of *N*-substituted guanidine linkages as a neutral analogue of PO backbone.[106] However, most of these modifications were quite destabilizing on duplex formation with RNA target, and therefore research was not pursued.

Following the unsuccessful attempts of incorporating *N*-substituted guanidines, a year later Bruice and co-workers looked at the incorporation of a free guanidine moiety into an oligo as a source of positive charge and generated renewed interest in this area of research.[101] They have published a route for the synthesis of a dimeric unit containing a guanidinium [3′-NHC(=NH$_2^+$)NH-5′] linker, and suggested that the methodology is applicable to the preparation of uniformly modified polymer. A pentameric DNA linked via 4-guanidine residues was shown to form a base pair specifically with poly rA with an unprecedented affinity at >100 °C. However, this molecule did not bind to poly G, C, or U tracts. They also proposed a model for this unusual high-affinity interaction, which takes the general conformation of the nucleic acid. This study was unique, and warrants the synthesis of a biologically relevant mixed-base sequence for its evaluation as an antisense construct.

### (iii) Ureas

The logic for synthesizing urea (3'-NRCONR-5') linked dimers was born out of adding more rigidity to the amide (3'-CH$_2$CONH-5') linked dimer and studying the effects on hybridization. Therefore, various *N*-substituted urea analogues were synthesized as dimeric nucleosides and incorporated into oligos.[107,124] As expected, most of the urea-linked oligos were destabilizing when hybridized to RNA complements, compared to amide or PO linkages. Clearly, introduction of extensive conformational restrictions for torsional angles in the backbone has a negative effect on duplex formation. Again, substitution on the 3'-N of the urea linkage was severely destabilizing, whereas substitution on the 5'-N of the urea linkage was well-tolerated in terms of $T_m$. The stability of urea-linked oligos in fetal calf serum was increased by a factor of 15 compared to the unmodified PO-linked oligos. These results taken together indicate that urea-linked oligos may not be a good choice for antisense applications.

### (iv) Carbamates

Various syntheses of 3' → 5' carbamate-bridged (3'-OCONH-5' and 3'-NHCO-*O*-5') dimers and polymers (up to 6-mer) have been reported since the late 1970s. Interestingly, some work was published even before the antisense concept was realized. The chemistry of the carbamate-linked oligo is straightforward and can be performed either in solution or on solid-support. A detailed account of their synthesis is reviewed elsewhere by Sanghvi and Cook.[17] The use of carbamate-linking chemistry has been extended towards the solid-support synthesis of novel oligosides containing morpholino subunits linked by carbamate bridges. Further details and applications of the morpholino type oligosides have been published by Summerton and co-workers.[102]

Carbamate-linked (3'-*O*-CONH-5') polymers with alternating PO linkages were studied for their antisense properties. The overall $T_m$ were destabilized ($\Delta T_m$ −2.0 to −4.0 °C/modification) when hybridized to complement RNA. Isomeric carbamates (3'-NRCOO-5') proved to be worse.[19] No duplex formation was observed with 5'-modifications. The thiocarbamates 3'-SCONH-5') also had a negative effect on the $T_m$ due to longer bond lengths between the 3'C—S atoms compared to 3'C—O and reduced rotational barrier around the 3'-SCO bond.[19] Once again, increasing the overall length of the backbone generally had a negative impact on the affinity for RNA target. Although carbamates provided enhanced nuclease resistance,[103] their poor affinity for RNA targets makes them useless for antisense applications.

### (v) Amines

Researchers have long been interested in utilizing amino-linked oligos in order to control the net charge distribution. It was believed that appropriate substitution of nitrogen atoms could result in considerable change in p$K_a$ values, thus creating a wide range of backbones from neutral to protonated types. All of this eventually would assist in an improved uptake of these molecules. In addition, electrostatic and hydration factors between the neutral or protonated nitrogen atom[46,101] of the amino linkers and the polyanionic backbone of the RNA strand are expected to enhance further the thermal stability of these duplexes.

Several reports have come out of the Sterling group of scientists[104] describing the solution-phase synthesis of amine-bridged dimers and polymers (up to trimers) and studies of their antisense properties. These linkages are all secondary amines: (i) 3'-CH$_2$NHCH$_2$-5', (ii) 3'-CH$_2$CH$_2$NH-5', and (iii) 3'-NHCH$_2$CH$_2$-5', and therefore should be neutral at physiological pH. All of the above modifications, when incorporated into oligos as dimeric nucleosides, destabilized the duplex by −2.0 to −5.0 °C per modification. Further substitution of these nitrogen atoms by a methyl group would make them tertiary amines, thus a cationic linkage instead of neutral. De Mesmaeker *et al.* prepared[108] two such linkages (i) 3'-CH$_2$N(Me)CH$_2$-5' and (ii) 3'-CH$_2$CH$_2$N(Me)-5'. Both of these modifications were equally destabilizing on duplex formation. Unpublished modeling results indicated that these protonated sites were too far away from the negatively charged PO linkage of the complement RNA, therefore charge neutralization could not possibly occur to improve affinity. In addition, presence of a C—C bond within the backbone allows free rotation around the torsional angles, which could lead to a more flexible backbone than necessary. Further restriction of flexibility was achieved with substitution of a methylene group with a carbonyl group (3'-CH$_2$CH$_2$N(Me)-5' to 3'-CH$_2$CON(Me)-5'), which resulted in immediate correction of $T_m$ in a positive manner.[108]

The author's experience with hydroxylamine-linked oligos also follows a similar trail. In the 1990s, substantial progress has been made in utilizing nitrogen-containing backbones, resulting in an extensive structure–activity relationship (SAR) with a broad antisense database. Some of the key linkages studied at the author's institute are: (all 3' → 5') -CH=N-O, -CH$_2$NHO-, -CH$_2$N(Me)O- (MMI), -CH$_2$ON(Me)-, -ON(Me)CH$_2$-, and -CH$_2$N(Me)N(Me)-. As most of the syntheses are published,[17] only unpublished results will be discussed herein.

A number of scientists believe that methylene methylimino (MMI) is one of the most promising backbone modifications for incorporation into antisense oligos (Figure 5(a)).[109] The reasons are as follows. MMI linkages provide a very high degree of nuclease resistance, as hydroxyamino linkages are not a substrate for cellular nucleases. Furthermore, it was found that an MMI linkage confers nuclease stability to the adjacent phosphodiester, which allows the use of dimeric strategy, creating an antisense molecule with alternating MMI and PO linkages. The resulting alternating antisense oligos demonstrated excellent water solubility, good affinity, and specificity for complementary RNA, while reducing the net negative charge of the oligo by 50%. *In vitro* studies indicated that MMI oligos of this type maintain, or in some cases increase, biological activity (e.g., inhibition of PKC-α and H-*ras*) relative to parent PS oligos. Additionally, significant benefits may be realized in the economics of large-scale synthesis in solution or even on solid-support.

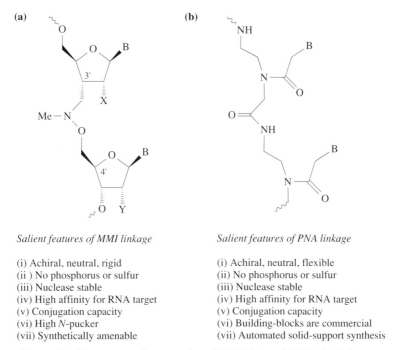

Salient features of MMI linkage

(i) Achiral, neutral, rigid
(ii) No phosphorus or sulfur
(iii) Nuclease stable
(iv) High affinity for RNA target
(v) Conjugation capacity
(vi) High *N*-pucker
(vii) Synthetically amenable

Salient features of PNA linkage

(i) Achiral, neutral, flexible
(ii) No phosphorus or sulfur
(iii) Nuclease stable
(iv) High affinity for RNA target
(v) Conjugation capacity
(vi) Building-blocks are commercial
(vii) Automated solid-support synthesis

**Figure 5** Structure and properties of MMI (a) and PNA (b) linkages.

The conformational flexibility vs. rigidity of MMI linkage, as well as the northern sugar pucker is believed to play a crucial role in the modulation of its binding affinity for RNA target.[36] Extensive NMR and modeling studies have indicated that the 3'-methylene group of the MMI linkage shifted the sugar conformation to a desired 3'-*endo* pucker, via reduced ring *gauche* effects, thus helping the molecule to preorganize into a preferred A-geometry for duplex formation.[125] MMI backbone studies have now been extended to include modifications of the sugar moieties via the introduction of an electronegative 2'-substituent, with an aim to enhance further the *N*-type pucker.[126] It is well-known that thermal stability of hybrids of 2'-OMe or 2'-F RNA with complementary RNA are considerably higher ($\Delta T_m$ +1.0–2.0 °C/modification) than that of the corresponding DNA RNA duplexes.[127] In brief, six dimers with various combinations of 2'-sugar modifications (see Figure

5(a)) were synthesized, their effects on hybridization (Table 7) to RNA were studied, and their % N-pucker was estimated by NMR (Table 8). The novel MMI-sugar modified oligos were found to hybridize to their complement RNA with highly improved affinity and specificity compared to unmodified DNA. The best modification exhibited a $\Delta T_m$ of $\sim 5 \,^\circ C$/modification compared to an antisense PS oligo. This increased stability has been attributed to hydrophobic interactions between substituents in the minor groove, and especially to the higher percentage of the 3'-endo sugar conformation, which resulted in a decreased entropic motion of the sugar while maintaining an optimal preorganized A-type duplex. In essence, an appropriate combination of backbone and sugar modifications allows one to dial in desirable affinity in any antisense construct.[128]

**Table 7** Effect of 2' sugar modification on hybridization of modified backbones (see Figure 5(a) for the structure of MMI dimer).

| | Modifications | | $\Delta T_m$/modification (vs. RNA) | | | |
|---|---|---|---|---|---|---|
| Backbone | 2' of top sugar (x) | 2' of bottom sugar (y) | Oligo seq. A | Oligo seq. B | Oligo seq. C | Average |
| DNA-(PO) | / | / | 0.0 | 0.0 | 0.0 | |
| DNA-(PS) | / | / | −1.12 | | | |
| MMI | H | H | +0.13 | −0.23 | +1.51 | +0.2 |
| MMI | H | F | +1.83 | +1.00 | −0.15 | +1.4 |
| MMI | H | Me | +2.27 | +1.67 | +0.87 | +1.9 |
| MMI | F | F | +3.27 | +2.20 | +1.47 | +2.8 |
| **MMI** | **F** | **Me** | **+3.74** | **+3.01** | **+1.95** | **+3.3** |
| MMI | Me | F | +3.13 | +2.50 | +1.56 | +2.8 |
| **MMI** | **Me** | **Me** | **+3.71** | **+2.78** | **+1.85** | **+3.2** |

Oligo sequence A containing 5 dimeres: GCG T*T T*T T*T T*T T*T GCG.   Oligo sequence B containing 1 dimer: CTC GTA CC T*T TC CGG TCC.   Oligo sequence C containing 2 dimers: CTC GTA C T*T T*T C CGG TCC.

**Table 8** Sugar conformation in MMI dimers (estimated from proton NMR coupling constants).

| | Backbone modification | | | | | | |
|---|---|---|---|---|---|---|---|
| Sugar modification[a] | DNA | | | MMI | | | |
| 2' of top unit (x) | none | none | none | none | F | F | OMe |
| 2' of bottom unit (y) | none | none | OMe | F | F | OMe | OMe |
| % of N-conformation[b] | | | | | | | |
| Top unit | 35 | 68 | 66 | 70 | 96 | 98 | 95 |
| Bottom unit | 28 | 31 | 65 | 92 | 96 | 71 | 76 |

[a]See Figure 5(a) for the structure of MMI dimer.   [b]Fractional popoulation (%) of C-3'-*endo* form calculated by PSEUROT 6.2.

These results encouraged the author and his co-workers to take the MMI linked oligos into animal experiments. The preliminary results indicate that full alternating MMI-PO linked oligos (20-mer) have a distinct biodistribution pattern compared to PS oligos. At 24 h, MMI-PO oligo concentration in kidneys is sixfold less than the PS oligos and also has lower accumulation in liver compared to all-PS oligos.

### 7.08.3.2.2 *Silicon-containing linkages*

The dialkylsilyl internucleoside linkage, first described in 1985 by Ogilvie and Cormier,[129] appeared to be attractive due to its neutral, achiral, size similarity to phosphorus, lipophilic properties, and straightforward synthesis. Saha *et al.*[110] and Maddry *et al.*[130] independently developed convenient synthetic procedures, which allow solution-phase and solid-support synthesis of silicon-linked polymers (up to 10-mer). Although these oligos demonstrated good stability, their binding affinity was poor. In addition, mediocre yields, polarity concerns, and suboptimal binding affinity were the causes of discontinued interest in their use towards antisense research. Nevertheless, use of $^1H$-$^{29}Si$ long-range HMQC NMR spectroscopy may be useful for structural determination of silicon-containing molecules.[131]

### 7.08.3.2.3  Sulfur-containing linkages

There has been an equal amount of excitement in the sulfur-linked backbones (see Table 6), compared to the nitrogen-containing linkages. The key reason is that sulfur is the closest isosteric and isoelectronic replacement for the natural phosphorus atom, offering a variety of options to alter isosteric linkages chemically, thus extending the library of neutral linkers available for development of antisense molecules.

### (i)  Sulfides and sulfones

Some of the early work in this area comes from Just and co-workers describing synthesis and use of $3'$-$CH_2CH_2S$-$5'$ linked dimers. A review article has been published by Just and Kawai entailing their synthetic approaches to obtaining such dimers, and they extended their studies to include the $2'$-sugar modifications, which, as expected, enhanced the binding affinity towards RNA targets.[111,112] Benner and co-workers also reported a full account of the synthesis of dimers containing $3'$-$CH_2SCH_2$-$5'$ linkages at the same time. Their synthetic efforts have been published in detail describing sequences leading to the fully modified octamers in $\sim 1$ mg quantities.[113] These two groups have further oxidized their sulfide-linked molecules to sulfone-linked oligosides. The $3'$-$CH_2CH_2SO_2$-$5'$ linked sulfone described by Just *et al.*[111,112] was found to be base-labile and degraded upon exposure to ammonia at room temperature. However, cleavage problems were circumvented by changing the reaction sequences, i.e., deblock $3'$-$CH_2CH_2S$-$5'$ linked oligo first and then oxidize the last step. The $T_m$ data with these oligos indicated that they bind to RNA poorly ($\Delta T_m$ $-15\,°C$), possibly due to steric interaction of $SO_2$ group and base residues.

It is important to note that Egli and co-workers[132] published the first X-ray crystal structure of a nonionic backbone, which happens to be a $3'$-$CH_2SO_2CH_2$-$5'$ linked ribo dimer placed in the middle of an oligo. These results allay concerns that the atomic substitutions in the sulfone backbone may prevent the formation of Watson–Crick type double helix with a complementary strand. The sulfone octamer prepared by Benner and co-workers displayed a thermal transition in the UV spectrum above $65\,°C$ with a large hyperchromicity, suggesting some secondary structure, possibly a tight hairpin.[113] Interestingly, the $T_m$ of this octamer did not change either in the presence of both complementary RNA and DNA or in their absence. In summary, oligos linked via sulfides and sulfones are difficult to prepare and show very unusual self-association properties which discourages their use in antisense, as clearly stated by Egli and co-workers.[132]

### (ii)  Sulfonates and sulfonamides

Like sulfones, sulfonates ($3'$-$OSO_2CH_2$-$5'$) and sulfonamides ($3'$-$NHSO_2CH_2$-$5'$) linkages are isoelectronic to the natural $3'$-$O$-$PO_2$-$O$-$5'$ linkage. Unlike silicon-linked oligos, sulfonyl-linked oligos are expected to be very polar nonionic congeners. In addition, they possess extra hydrogen bonding capacity which may help towards solubility in water. These oligomers should also share the nuclease stability demonstrated by other nonionic linkages.

Widlanski and co-workers[114] described the synthesis of such linkages (e.g., $3'$-$XSO_2CH_2$-$5'$; $X=O$ or $NH$) in 1994, via standard dimer approach. However, in the case of sulfonate-linked DNA, it was demonstrated that such oligosides can be prepared via solid-support synthesis.[133] Maddry *et al.* also prepared identical sulfonyl-dimers, following a slightly different synthetic strategy.[129] As expected, these oligos were quite resistant to nucleases but hybridized with lower affinity with target DNA. On a side note, Widlanski and co-workers tested oligos containing sulfonamide linkages in place of the sulfonate linkage, expecting to alter the protein–DNA interactions, based on the fact that sulfonamide was a better H-bonding acceptor.[134] Indeed, they were able to modulate protein–DNA binding, which in turn can provide important information about the location, nature, and relative strength of proteins and the specific PO linkage of the DNA to which they bind. Another sulfur-based linkage ($3'$-$OSO_2NH$-$5'$) was reported by Trainor and co-workers in 1992.[135] Not surprisingly, this modification too had a destabilizing effect on duplex formation.

### 7.08.3.2.4  Oxygen-containing linkages

A simple yet provocative replacement of a phosphorus atom with a methylene group created a novel oxygen-containing linkage, commonly known as formacetal ($3'$-$OCH_2O$-$5'$). Conceptually, other ether-linked dimers (3'-OCH$_2$CH$_2$O-4', ethylene glycol; 3'-OCH$_2$CH$_2$CH$_2$-4', 3'-CH$_2$CH$_2$ CH$_2$UC-4', propoxy type) were visualized and finally synthesized. The chemistry and antisense properties (see Table 6) of these analogues are summarized in this section.

### (i) Formacetals

This modification was synthesized and studied by two groups (Matteucci *et al.*[18] and Van Boom *et al.*[136]) simultaneously and independently in 1990. Several publications have appeared on this subject and have been reviewed by Matteucci;[18] therefore, a short summary should suffice herein. The formacetal-linked oligos bind in a sequence-specific manner, but their affinity appears to be somewhat inferior relative to the unmodified oligo. They proposed that this helical instability is due to a strong stereoselective effect, which results in an energy penalty for the *gauche* conformations required for helix formation, pointing out the complexity of phosphate isostere design.[137]

This work took an interesting turn when 3'- and 5'-oxygen atoms of the formacetal linkage were replaced by sulfur, one at a time, thus creating 3'-SCH$_2$O-5' and 3'-OCH$_2$S-5' linked thioformacetals. Again, synthesis of these linkages has been optimized by more than one group (Matteucci[18] and Ducharme and Harrison[138]). The binding affinity of 3'-OCH$_2$S-5' linked oligos was found to be poor and not worth pursuing. On the contrary, the 3'-SCH$_2$O-5' linked oligos had a positive effect on binding to the complement RNA ($\Delta T_m$ +0.8 °C/modification).[18]

It is noteworthy that a subtle change like this one in the backbone could have a dramatic effect on its affinity for RNA target. The differences can be highlighted in terms of longer C—S bond length, change of sugar pucker, and the torsion angles around the anomeric XCH$_2$O center. A 5'-thio derivative produces a bad steric clash with adjacent ribose oxygen, destabilizing the optimal *gauche* conformation of these two heteroatoms. In conclusion, the 3'-thioformacetal linkage appears to be the best among this series of modifications for antisense applications, perhaps even better in combination with 2'-sugar modifications (e.g., OMe, F).

### (ii) Ethers

Teng and Cook reported the synthesis of some unusual ether-linked dimers utilizing Vorbrüggen-type glycosylation reactions.[115] To be specific, they prepared a 3'-OCH$_2$CH$_2$O-4' and 3'-CH$_2$CH$_2$ CH$_2$O-4' linked analogue of thymidine dimers. Incorporation of these dimers into oligos had a destabilizing effect when hybridized to RNA ($\Delta T_m$ −3.0 °C/modification). Detailed NMR and modeling studies of these dimers indicated that the newly formed 4'-oxofuran ring lies in a high S-pucker causing its destabilization during binding to RNA.

Efforts from other researchers[116] include reversal of the 3'-CH$_2$CH$_2$CH$_2$O-4' to 3'-OCH$_2$CH$_2$CH$_2$-4' and its thio analogue (3'-SCH$_2$CH$_2$CH$_2$-4'), which improved $T_m$ compared to its isomeric structural analogue, but $T_m$ remained low compared to unmodified PO DNA. Once again, extra flexibility via freedom of rotation around the $\alpha/\beta$ torsional angles could be the reason for compromised affinity for RNA target. This can be further justified by relatively better $T_m$ with formacetal linkage, which may have reduced freedom of rotation due to the presence of the 5'-*O*-atom, compared to 3'-OCH$_2$CH$_2$CH$_2$-4' linked molecules.

### 7.08.3.2.5 All-carbon linkages

Replacement of the four-atom PO linkage with an all-carbon backbone is synthetically challenging and may turn out to be futile because of too much freedom of rotation and hydrophobic interactions. Nevertheless, first Butterfield and Thomas,[117] and later Lebreton *et al.*[139] have reported the synthesis of all-carbon backbone (3'-CH$_2$CH$_2$CH$_2$-5'). The latter group investigated its affinity for RNA target and found it to be very destabilizing ($\Delta T_m$ −3.2 to −5.9 °C/modification depending upon sequence). Subsequently, the Ciba group reported[19] the $T_m$ studies with 3'-CH$_2$COCH$_2$-5' linked dimers placed in an oligo. This modification had slightly improved affinity ($\Delta T_m$ −2.5 °C/modification) compared to all-carbon linkage, due to reduced rotational freedom around the linkage. Based on this observation, Ciba scientists postulated that an all-carbon linked backbone may be tuned for higher affinity if they introduced a double bond in this linkage, such as 3'-CH$_2$ CH=CHCH$_2$-4', creating *cis* and *trans* isomers. Indeed, they synthesized two thymidine dimers containing *cis* and *trans* linked all-carbon backbones and studied hybridization.[118] As predicted, the *trans*-isomer was able to bind to the complementary RNA ($\Delta T_m$ $-0.6\,°C$/modification) with an improved affinity compared to 3'-CH$_2$CH$_2$CH$_2$-5' linked oligos. On the other hand, the *cis* isomer had a larger destabilizing effect ($\Delta T_m$ $-1.3\,°C$/modification). Lee and Wiemer also reported[119] a synthesis of carbon-bridged ribonucleoside dimer (3'-CH$_2$COCH=CH-4'), which may be worthy of incorporating into oligos. In summary, straight alkane-type linkages are unattractive for antisense utility. Appropriate conformational restriction (e.g., *cis*-isomer) of all-carbon backbone may help, but overall increased hydrophobicity should be a concern in designing such linkages, which invariably results in poor water solubility.

### 7.08.3.2.6  Nonphosphate sugar linkage (PNA)

PNA (peptide nucleic acid) is a DNA analogue in which the phosphate sugar backbone is replaced by a structurally homomorphous pseudopeptide chain, consisting of *N*-(2-aminoethyl)glycine units (Figure 5(b)). Unlike DNA, the PNA backbone carries no charges and has no chiral centers. All four natural DNA bases are retained, and PNA recognizes sequence-complementary nucleic acids specifically, forming Watson–Crick base pairs. The synthesis of PNA has been reviewed[140] and readers are referred to these reviews for a full account.

PNA exhibited exceptional hybridization properties, with $T_m$ well above those of the corresponding natural unmodified duplexes, and stability towards nucleases or proteases. Despite these features, initial efforts to use PNA as an antisense construct have been hampered by poor water solubility, self-aggregation properties, and inability to activate RNase H. Tam and co-workers[141] and van Boom and co-workers[142] have independently reported their syntheses of a PNA–DNA hybrid molecule. They reasoned that natural DNA or PS oligos, covalently linked at both ends (3' and 5') to short segments (4-units) of PNA, would give access to a chimeric type of oligo (PNA–DNA–PNA) discussed earlier in Section 7.08.2.1, a strategy that maintains high affinity and nuclease stability to PNA in flanking regions, and RNase H mediated cleavage via the PO or PS linked central portion of the chimera (see Figure 2). Interestingly, half PNA–half DNA hybrid was able to bind to RNA target with similar affinity to an unmodified DNA. Uhlmann *et al.*[143] have reported the synthesis and properties of DNA–PNA chimeras that are taken up by cells to a similar extent as PS oligos, are largely nuclease resistant, and have excellent binding affinity to complementary RNA. However, the evidence from cell-based experiments indicated that the uptake of pure PNA is very poor; therefore, microinjection techniques have been utilized to demonstrate the antisense efficacy of PNA.

## 7.08.4  DESIGN CONSIDERATIONS

This section summarizes the key elements of novel backbone design considerations for antisense applications, with an emphasis towards pinpointing pitfalls for novices in this area of research. Synthetic backbone linkages have become one of the most attractive areas of antisense-based oligonucleotide research. These isosteres of natural phosphate backbone as discussed in Section 7.08.3 should allow us to begin to understand the key structural elements required for an "ideal" backbone linkage.

Generally, any modification of the phosphate backbone guarantees nuclease resistance to some extent, and complete replacement of the phosphate backbone with a nonphosphate linker assures complete nuclease stability. Therefore, it is almost given that any modification of the backbone would result in stability towards nucleolytic degradation, and should not be a concern anymore towards their design. However, one should avoid ester and peptide linkages because they may be substrate for cellular esterases and peptidases.

The chirality of first-generation PS oligos appears to be of great concern to some researchers and has been the subject of many reviews. Clearly, with a dozen or more PS oligos undergoing clinical trials successfully as mixtures of diastereoisomers, the concern may be only academic. On the other hand, not all of the properties of chirally pure PS oligos are yet known. Thus, the quest for novel routes towards the synthesis of chirally pure PS oligos should continue.[56,57] There has been some excitement in the use of all-$R$p-methylphosphonate due to improved antisense efficacy. Since MP oligos are neutral and may avoid protein binding completely, an effort towards synthesis of chirally pure $R$p dimers or oligomers should be beneficial. Once again, chirality on C-5'-carbon was introduced with some of the interesting backbone modifications. In summary, it is the author's personal bias that chirality in backbone linkers should be avoided.

Solubility of antisense oligo in water is another important feature which should be considered in all design aspects. As has been seen with silyl, sulfone and all-carbon containing backbones, some analogues have poor water solubility. In addition, completely neutral or unchanged antisense molecules have had solubility problems (e.g., methylphosphonates). The general recommendation should be the presence of some charge (positive or negative) on the antisense molecule, enough to dissolve the oligo in water. The experience at the author's institution with the use of neutral oligos (e.g., MMI and PNA), usually 20-mer in length, has shown that at least four charges on the molecule are required for efficient solubility. Particularly for MMI, an alternating PO/MMI motif works very well in all oligos made thus far. Therefore, the message is to incorporate charged species in antisense oligos.

The correct balance of conformational flexibility vs. rigidity of the backbone linkage is one of the most important issues in design considerations. The concept of conformational restriction is a proven method for enhancing the binding affinity of a ligand for its receptor. Therefore, an appropriate preorganization of the backbone linkage can result in enhanced binding affinity and specificity for the target RNA. The basic structure of RNA is well understood; however, those of its secondary and tertiary structural components are not, which plays a major role in recognition during the binding process. From what is known about RNA modified oligo interactions, an RNA mimic (e.g., 2′-F, OMe, O-alkyl, O(CH)$_2$OMe, and others) would serve as a better fit for hybridization, mainly due to its preorganized structure in *N*-type sugar conformation, resulting in A-type duplex formation. Additionally, modifications of the sugar residue (e.g., MMI, amide-3 (**2**), phosphoramidate NP (**1**), 1,5-anhydrohexitol (HNA; (**3**)), and others) can further help this preorganization of the antisense molecule. Van Aerschot *et al.*[144] have reported an elegant example of the preorganization with a 1,5-anhydrohexitol (6′ → 4′) (**3**) linked oligo. These oligos form very stable A-type duplexes with RNA ($\Delta T_m$/HNA unit 1.0–5.0 °C depending on the chain length and the sequence) with good base pair specificity. The enhanced affinity of the HNA for the RNA target can be further explained by increased hydrophobicity in the minor groove of the duplex. Interestingly, in HNA, the base moiety is at the C-2′ axial position instead of the C-1′ position of the pyranose ring. This unusual base–sugar connection causes a distortion of the backbone conformation and gives rise to a favorable helicity for the duplex formation.[24] Similarly, the 2′-fluoro modified phosphoramidate NP oligos (**1**) were found to enhance the $T_m$ ($\Delta T_m$/modification 4.0–5.0 °C) due to their preferred C-3′-*endo* conformation. The phosphoramidate NP oligos–RNA duplex may be further stabilized due to increased hydrogen bonding in the minor groove due to the 3′-amino group.[145] It seems that no single modification can confer all the desirable properties on an oligo; therefore, one must utilize a combination of sugar–backbone modifications. None of the heterocyclic base modifications is discussed in this chapter, but they do have a significant contribution towards design constraints.[146] Ultimately, all second- and third-generation antisense agents are going to be hyper-modified molecules armed with exact amounts of charge, hydration, and preorganization for high affinity and sequence specificity for its complement target RNA.

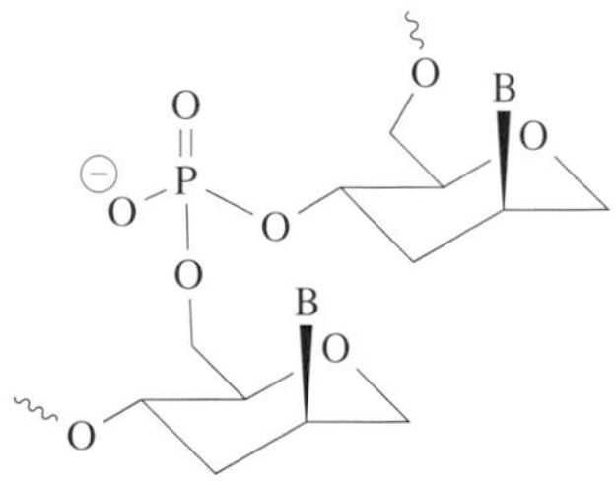

(**3**)

1,5-Anhydrohexitol

## 7.08.5 CONCLUSIONS

Chemical modifications of phosphate backbone at nearly every position have been considered, and numerous potentially interesting nonphosphate backbone linkages have been prepared.

Ironically, there have been more failures than successes reported in this chapter. Now that the concepts and principles of preorganization are better understood, the design of superior linkages should be more amenable and reduce the number of failures. Some of these nonphosphate modifications provide enhanced stability to nucleases, improved affinity ($T_m$), and *in vitro* activity. However, the most rewarding part of this monumental synthetic effort to produce nonphosphate linkages is yet to come, which is their *in vivo* activity based on antisense principles. On the other hand, it is important to note that phosphorothioates have performed exceptionally well as the first-generation antisense agents. As a result, over a dozen PS oligos as potential antisense drugs are undergoing human clinical trials for various diseases. Again, much remains to be learned about PS oligos, but there is no question that novel antisense agents with improved properties have been identified. Although it appears that some of the second- and third-generation modifications have outperformed first-generation PS oligos, it is quite unlikely that any single modification will solve all issues. Therefore, an appropriate combination of base–sugar heterocycle modifications have to be utilized to create a future generation of antisense oligos with improved therapeutic properties. There is every reason to be optimistic that antisense inhibition with altered oligos will be available to the medicinal chemists of the future.

## ACKNOWLEDGMENTS

I would like to thank members of the MMI team (Eric Swayze, Bal Bhat, Didier Peoc'h, and Stu Dimock) for their excellent efforts in making this project successful and allowing me to include some of their unpublished results. Additionally, terrific support from the scientists in the Chemistry and Biology Departments at Isis Pharmaceuticals is much appreciated. The author also wishes to acknowledge Dan Cook and Herb Boswell for their comments and careful reading of the manuscript. Special thanks are also due to Mrs Anna Alessi for typing this manuscript immaculately.

## 7.08.6 REFERENCES

1. S. T. Crooke and C. F. Bennett, *Annu. Rev. Pharmacol. Toxicol.*, 1996, **36**, 107.
2. M. R. Bennett, *J. Drug Dev. Clin. Pract.*, 1995, **7**, 225.
3. S. T. Crooke, *Chem. Ind.*, 1996, 90.
4. S. Agrawal and R. P. Iyer, *Cur. Op. Biotech.*, 1995, **6**, 12.
5. A. De Mesmaeker, R. Häner, P. Martin, and H. Moser, *Acc. Chem. Res.*, 1995, **28**, 366.
6. S. T. Crooke and B. Lebleu (eds.), "Antisense Research and Applications," CRC Press, Boca Raton, FL, 1993.
7. S. Agrawal (ed.), "Methods in Molecular Medicine: Antisense Therapeutics," Humana Press, Totowa, NJ, 1996.
8. F. Eckstein and D. M. J. Lilley (eds.), "Catalytic RNA," Springer-Verlag, New York, 1996.
9. Y. S. Sanghvi and P. D. Cook (eds.), "Carbohydrate Modifications in Antisense Research," ACS Symposium Series 580, ACS, Washington, DC, 1994.
10. V. N. Soyfer and V. N. Potaman (eds.), "Triple Helical Nucleic Acids," Springer-Verlag, New York, 1996.
11. Y. S. Sanghvi and P. D. Cook, in "Nucleosides and Nucleotides as Antitumor and Antiviral Agents," eds. C. K. Chu and D. C. Baker, Plenum, New York, 1993, p. 311.
12. J. S. Cohen, in "Antisense Research and Applications," eds. S. T. Crooke and B. Lebleu, CRC Press, Boca Raton, FL, 1993, p. 205.
13. P. S. Miller, P. O. P. T'so, R. I. Hogrefe, M. A. Reynolds, and L. J. Arnold, Jr., in "Antisense Research and Applications," eds. S. T. Crooke and B. Lebleu, CRC Press, Boca Raton, FL, 1993, p. 189.
14. G. D. Hoke, K. Draper, S. M. Freier, C. Gonzalez, G. B. Driver, M. C. Zounes, and D. J. Ecker, *Nucleic Acids Res.*, 1991, **19**, 5743.
15. B. P. Monia, J. F. Johnston, H. Sasmor, and L. L. Cummins, *J. Biol. Chem.*, 1996, **271**, 14533.
16. K.-H. Altmann, D. Fabbro, N. M. Dean, T. Geiger, B. P. Monia, M. Muller, and P. Nicklin, *Biochem. Soc. Trans.*, 1996, **24**, 630.
17. Y. S. Sanghvi and P. D. Cook, in "Carbohydrate Modifications in Antisense Research," eds. Y. S. Sanghvi and P. D. Cook, ACS Symposium Series 580, ACS, Washington, DC, 1994, p. 1.
18. M. D. Matteucci, in "Perspectives in Drug Discovery and Design," ed. G. L. Trainor, ESCOM, Leiden, 1996, vol. 4, p. 1.
19. A. De Mesmaeker, A. Waldner, V. Fritsch, J. Lebreton, and R. M. Wolf, *Bull. Soc. Chim. Belg.*, 1994, **103**, 705.
20. R. S. Varma, *Synlett*, 1993, 621.
21. E. Uhlman and A. Peyman, in "Protocols for Oligonucleotides and Analogs," Humana Press, Totowa, NJ, 1993, p. 355.
22. R. S. Varma, in "Molecular Biology and Biotechnology: A Comprehensive Desk Reference," ed. R. A. Meyers, VCH Publishers, New York, 1995, p. 617.
23. J. F. Milligan, M. D. Matteucci, and J. C. Martin, *J. Med. Chem.*, 1993, **36**, 1923.
24. P. Herdewijn, *Liebigs Ann.*, 1996, 1337.
25. K.-H. Altmann, N. M. Dean, D. Fabbro, S. M. Freier, T. Geiger, R. Häner, D. Hüsken, P. Martin, B. P. Monia, M. Müller, F. Natt, P. Nicklin, J. Phillips, V. Pieles, H. Sasmor, and H. E. Moser, *Chimia*, 1996, **50**, 168.

26. W. F. Lima, V. B. Driver, M. Fox, R. Hanecak, and T. W. Bruice, *J. Biol. Chem.*, 1997, **272**, 626.
27. S. M. Freier, in "Antisense Research and Applications," eds. S. T. Crooke and B. Lebleu, CRC Press, Boca Raton, FL, 1993, p. 67.
28. J.-P. Leonetti and L. D. Leserman, in "Antisense Research and Applications," eds. S. T. Crooke and B. Lebleu, CRC Press, Boca Raton, FL, 1993, p. 493.
29. M. Manoharan, in "Antisense Research and Applications," eds. S. T. Crooke and B. Lebleu, CRC Press, Boca Raton, FL, 1993, p. 303.
30. J. T. Levis, W. O. Butler, B. Y. Tseng, and P. O. P. Ts'o, *Antisense Res. Develop.*, 1995, **5**, 251.
31. Y. Tojanasakul, *Adv. Drug Delivery Rev.*, 1996, **18**, 115.
32. D. Kisner, in "12th IRT Nucleosides, Nucleotides and their Biological Applications," La Jolla, CA, 1996, OP 46.
33. L. Holmberg, personal communication, Pharmacia Biotech, Inc.
34. D. L. Cole, V. T. Ravikumar, A. Krotz, D. C. Capaldi, Z. S. Cheruvallath, A. N. Scozzari, M. Andrade, and E. F. Gritzen, in "12th IRT Nucleosides, Nucleotides and their Biological Applications," La Jolla, CA, 1996, OP 51.
35. A. M. Alves, K. M. Bowness, D. L. Cole, G. A. Garner, N. Mann, J. R. Parker, J. M. Parsons, B. H. Pope, V. T. Ravikumar, S. Scott, S. Srivasta, P. Wood, and M. J. McLean, in "12th IRT Nucleosides, Nucleotides and their Biological Applications," La Jolla, CA, 1996, OP 53.
36. F. Morvan, Y. S. Sanghvi, M. Perbost, J.-J. Vasseur, and L. Bellon, *J. Am. Chem. Soc.*, 1996, **118**, 255.
37. P. C. Zamecnik and M. L. Stephenson, *Proc. Natl. Acad. Sci. USA*, 1978, **75**, 280.
38. IBC's Fourth Annual International Symposium on Antisense Therapeutics, San Diego, CA, February 6–7, 1997.
39. P. H. Seeberger, E. Yen, and M. H. Caruthers, *J. Am. Chem. Soc.*, 1995, **117**, 1472.
40. L. Cummins, D. Graff, G. Beaton, W. S. Marshall, and M. H. Caruthers, *Biochemistry*, 1996, **35**, 8735.
41. W. T. Wiesler and M. H. Caruthers, *J. Org. Chem.*, 1996, **61**, 4272.
42. H. Li, K. Porter, F. Huang, and B. R. Shaw, *Nucleic Acids Res.*, 1995, **23**, 4495.
43. W. Dabkowski and I. Tworowska, *Chem. Lett.*, 1995, 727.
44. M. Bollmark, R. Zain, and J. Stawinski, *Tetrahedron Lett.*, 1996, **37**, 3537.
45. K. Mori, C. Boiziau, C. Cazename, C. Matsukara, C. Subasinghe, J. S. Cohen, S. Broder, J. J. Coulmé, and C. A. Stein, *Nucleic Acids Res.*, 1989, **17**, 8207.
46. T. Horn, S. Chaturredi, T. N. Balasubramaniam, and R. L. Letsinger, *Tetrahedron Lett.*, 1996, **37**, 743.
47. R. Fathi, Q. Huang, G. Coppola, W. Delancy, R. Teasdale, A. M. Kreig, and A. F. Cook, *Nucleic Acids Res.*, 1994, **22**, 5416.
48. S. Peyrottes, J.-J. Vasseur, J.-L. Imbach, and B. Raynor, *Nucleic Acids Res.*, 1996, **24**, 1841.
49. R. P. Iyer, T. Devlin, I. Habus, D. Yu, S. Johnson, and S. Agrawal, *Tetrahedron Lett.*, 1996, **37**, 1543.
50. R. W. Fischer and M. H. Caruthers, *Tetrahedron Lett.*, 1995, **36**, 6807.
51. Y. Hayakawa, M. Hirose, M. Hayakawa, and R. Noyori, *J. Org. Chem.*, 1996, **60**, 925.
52. R. P. Iyer, D. Yu, N.-H. Ho, T. Devlin, and S. Agrawal, *J. Org. Chem.*, 1995, **60**, 8132.
53. S. L. Beaucage and R. P. Iyer, *Tetrahedron*, 1993, **49**, 6123.
54. F. Eckstein and G. Gish, *Trends Biochem. Sci.*, 1989, 97.
55. G. Zon, in "Methods in Molecular Medicine: Antisense Therapeutics," ed. S. Agrawal, Humana Press, Totowa, NJ, 1996, p. 165.
56. W. J. Stec, A. Grajkowski, A. Kosylanska, B. Karwowski, M. Koziolkiewicz, K. Misiura, A. Okruszek, A. Wilk, P. Guga, and M. Boczkowska, *J. Am. Chem. Soc.*, 1995, **117**, 12 109.
57. A. V. Lebedev and E. Wickstrom, in "Perspectives in Drug Discovery and Design," ed. G. L. Trainor, ESCOM, Leiden, 1996, vol. 4, p. 17.
58. H. Kanehara, T. Wada, M. Mizuguchi, and K. Makino, *Nucleosides Nucleotides*, 1996, **15**, 1169.
59. V. T. Ravikumar and Z. S. Cheruvallath, *Nucleosides Nucleotides*, 1996, **15**, 1149.
60. X. Liu and C. B. Reese, *J. Chem. Soc., Perkins Trans. 1*, 1995, 1685.
61. S. Zaramella and G. M. Bonora, *Nucleosides Nucleotides*, 1995, **14**, 809.
62. D. Liu, S. L. Daubendiek, M. A. Zillman, K. Ryan, and E. T. Kool, *J. Am. Chem. Soc.*, 1996, **118**, 1587.
63. B. P. Monia, J. F. Johnston, T. Geiger, M. Muller, and D. Fabbro, *Nature Med.*, 1996, **2**, 668.
64. S. Agrawal and J. Temsamani, in "Methods in Molecular Medicine: Antisense Therapeutics," ed. S. Agrawal, Humana Press, Totowa, NJ, 1996, p. 247.
65. A. B. Eldrup, J. Felding, J. Kehler, and O. Dahl, *Tetrahedron Lett.*, 1995, **36**, 6127.
66. J. T. Kodra, J. Kehler, and O. Dahl, *Nucleic Acids Res.*, 1995, **23**, 3349.
67. M. H. Caruthers, in "Proceedings of International Congress: Therapeutic Oligonucleotides," Rome, 1996, p. 21.
68. Y.-O. Chen, F.-C. Qu, and Y.-B. Zhang, *Tetrahedron Lett.*, 1995, **36**, 745.
69. R. Wittmann, *Chem. Ber.*, 1963, **96**, 771.
70. K. Misiura, D. Pietrasiak, and W. J. Stec, *J. Chem. Soc., Chem. Commun.*, 1995, 613.
71. J. Stawinski and M. Thelin, *Tetrahedron Lett.*, 1992, **33**, 7255.
72. C. F. Bennett, M. Y. Chiang, H. Chan, and S. Grimm, *J. Liposome Res.*, 1993, **3**, 85.
73. J.-P. Bongartz, A.-M. Aubertin, P. G. Milhand, and B. Lebleu, *Nucleic Acids Res.*, 1994, **22**, 4681.
74. H. Hashimoto, M. G. Nelson, and C. Switzer, *J. Am. Chem. Soc.*, 1993, **115**, 7128.
75. R. L. Letsinger, C. N. Singman, G. Histand, and M. Salunkhe, *J. Am. Chem. Soc.*, 1988, **110**, 4470.
76. S. V. Vinogradov, Y. G. Suzdaltseva, and A. V. Kabanov, *Bioconjugate Chem.*, 1996, **7**, 3.
77. P. S. Miller, in "Bioorganic Chemistry: Nucleic Acids," ed. S. M. Hecht, Oxford University Press, Oxford, 1996, p. 347.
78. C. L. Bec and E. Wickstrom, *J. Org. Chem.*, 1996, **61**, 510.
79. J. M. Kean, S. A. Kipp, P. S. Miller, M. Kulka, and L. Aurelian, *Biochemistry*, 1995, 35.
80. M. Kulka, C. C. Smith, J. Levis, R. Fishelevich, J. C. Hunter, C. Cushman, P. S. Miller, P. O. P. Ts'o, and L. Aurelian, *Antimicrob. Agents Chemother.*, 1994, **38**, 675.
81. B. Y. Tseng and K. D. Brown, *Cancer Gene Therapy*, 1994, **1**, 65.
82. I. Habus, T. Devlin, R. P. Iyer, and S. Agrawal, *Bioorg. Med. Chem. Lett.*, 1996, **6**, 1393.
83. L. J. Arnold, Jr., in "IBC Conference on Antisense Therapeutics," 1996, Coronado, CA.
84. S. Peyrottes, J.-J. Vasseur, J.-L. Imbach, and B. Raynor, *Tetrahedron Lett.*, 1996, **37**, 5869.

85. M. Endo and M. Komiyama, *J. Org. Chem.*, 1996, **61**, 1994.
86. Z. Zhang, J. X. Tang, and J. Y. Tang, *Bioorg. Med. Chem. Lett.*, 1995, **5**, 1735.
87. S. M. Gryaznov, D. H. Lloyd, J.-K. Chen, R. G. Schultz, L. A. DeDionisio, L. Ratmeyer, and W. D. Wilson, *Proc. Natl. Acad. Sci. USA*, 1995, **92**, 5798.
88. A. P. Higson, G. K. Scott, D. J. Earnshaw, A. D. Baxter, R. A. Taylor, and R. Cosstick, *Tetrahedron*, 1996, **52**, 1027.
89. S. P. Collingwood and A. D. Baxter, *Synlett*, 1995, 703.
90. G. Beaton, PCT WO 95/26972, 12 October 1995.
91. S. M. Gryaznov, T. Skorski, C. Cucco, M. Nieborowska-Skorska, C.-Y. Chiu, D. Lloyd, J.-K. Chen, M. Koziolkiewicz, and B. Calabretta, *Nucleic Acids Res.*, 1996, **24**, 1508.
92. D. Ding, S. M. Gryaznov, D. H. Lloyd, S. Chandrasekaran, S. Yao, L. Ratmeyer, Y. Pan, and W. D. Wilson, *Nucleic Acids Res.*, 1996, **24**, 354.
93. S. M. Gryaznov, in "Proceedings of International Congress: Therapeutic Oligonucleotides," Rome, 1996, p. 25.
94. R. Johnson, B. V. Joshi, and C. B. Reese, *J. Chem. Soc., Chem. Commun.*, 1994, 133.
95. M. Mag and J. W. Engels, *Tetrahedron*, 1994, **50**, 10225.
96. X. Liu and C. B. Reese, *Tetrahedron Lett.*, 1995, **36**, 3413.
97. J. Matulic-Adamic, P. Haeberli, and N. Usman, *J. Org. Chem.*, 1995, **60**, 2563.
98. A. K. Saha, T. J. Caulfield, C. Hobbs, D. A. Upson, C. Waychunas, and A. M. Yawman, *J. Org. Chem.*, 1995, **60**, 788.
99. G. Wang, P. J. Middleton, and W. E. Seifert, in "Western Biotech Conference: 31st ACS Western Regional Meeting," 1995, San Diego, CA, Abstract No. 420.
100. G. Wang and W. E. Seifert, *Tetrahedron Lett.*, 1996, **37**, 6515.
101. K. A. Browne, R. O. Dempcy, and T. C. Bruice, *Proc. Natl. Acad. Sci. USA*, 1995, **92**, 7051.
102. J. Summerton and D. D. Weller, *US Pat.* 5185444 (1993).
103. I. Habus, J. Temsamani, and S. Agrawal, *Bioorg. Med. Chem. Lett.*, 1994, **4**, 1065.
104. C. V. C. Prasad, T. J. Caulfield, C. P. Prouty, A. K. Saha, W. C. Schairer, A. Yawman, D. A. Upson, and L. I. Kruse, *Bioorg. Med. Chem. Lett.*, 1995, **5**, 411.
105. A. De Mesmaeker, A. Waldner, J. Labreton, V. Fritsch, and R. M. Wolf, in "Carbohydrate Modifications in Antisense Research," eds. Y. S. Sanghvi and P. D. Cook, ACS Symposium Series 580, ACS, Washington, DC, 1994, p. 24.
106. C. Pannecouque, F. Vandendriessche, J. Rozenski, G. Janssen, R. Busson, A. Van Aerschot, P. Claes, and P. Herdewijn, *Tetrahedron*, 1994, **50**, 7231.
107. A. Waldner, A. De Mesmaeker, J. Lebreton, V. Fritsch, and R. M. Wolf, *Synlett*, 1994, 57.
108. A. De Mesmaeker, A. Waldner, Y. S. Sanghvi, and J. Lebreton, *Bioorg. Med. Chem. Lett.*, 1994, **4**, 395.
109. Y. S. Sanghvi, L. Bellon, F. Morvan, T. Hoshiko, E. Swayze, L. Cummins, S. Freier, N. Dean, B. Monia, R. Griffey, and P. D. Cook, *Nucleosides Nucleotides*, 1995, **16**, 1087.
110. A. K. Saha, M. Sardaro, C. Waychunes, D. Delecki, R. Kutney, P. Cavanaugh, A. Yawman, D. A. Upson, and L. I. Kruse, *J. Org. Chem.*, 1993, **58**, 7827.
111. G. Just and S. H. Kawai, in "Carbohydrate Modifications in Antisense Research," eds. Y. S. Sanghvi and P. D. Cook, ACS Symposium Series 580, ACS, Washington, DC, p. 52.
112. M. J. Damha, B. Meng, D. Wang, C. G. Yannopoulos, and G. Just, *Nucleic Acids Res.*, 1995, **23**, 3967.
113. C. Richert, A. L. Roughton, and S. A. Benner, *J. Am. Chem. Soc.*, 1996, **118**, 4518.
114. E. B. McElroy, R. Bandaru, J. Huang, and T. S. Widlanski, *Bioorg. Med. Chem. Lett.*, 1994, **4**, 1071.
115. K. Teng and P. D. Cook, *J. Org. Chem.*, 1994, **59**, 278.
116. X. Cao and M. D. Matteucci, *Tetrahedron Lett.*, 1994, **35**, 2325.
117. K. Butterfield and E. J. Thomas, *Synlett*, 1993, 411.
118. S. Wendeborn, R. M. Wolf, and A. De Mesmaeker, *Tetrahedron Lett.*, 1995, **36**, 6879.
119. K. Lee and D. F. Wiemer, *J. Org. Chem.*, 1993, **58**, 7808.
120. M. J. J. Blommers, U. Pieles, and A. De Mesmaeker, *Nucleic Acids Res.*, 1994, **22**, 4187.
121. H. E. Moser, in "IBC Conference: Antisense Therapeutics," 1996, Coronado, CA.
122. M. J. Robins, S. Sarker, M. Xie, W. Zhang, and M. A. Peterson, *Tetrahedron Lett.*, 1996, **37**, 3921.
123. G. Viswanadham, G. V. Peterson, and J. Wengel, *Bioorg. Med. Chem. Lett.*, 1996, **6**, 987.
124. K. M. K. Kutterer and G. Just, *Bioorg. Med. Chem. Lett.*, 1994, **4**, 435.
125. V. Mohan, R. H. Griffey, and D. R. Davis, *Tetrahedron*, 1995, **51**, 8655.
126. B. Bhat, E. E. Swayze, P. Wheeler, S. Dimock, M. Perbost, and Y. S. Sanghvi, *J. Org. Chem.*, 1996, **61**, 8186.
127. E. A. Lesnik, C. J. Guinosso, A. M. Kawasaki, H. Sasmor, M. Zounes, L. L. Cummins, D. J. Ecker, P. D. Cook, and S. M. Freier, *Biochemistry*, 1993, **32**, 7832.
128. Y. S. Sanghvi, E. Swayze, D. Peoc'h, B. Bhat, and S. Dimock, *Nucleosides Nucleotides*, 1997, **16**, 907.
129. K. K. Ogilvie and J. F. Cormier, *Tetrahedron Lett.*, 1985, **26**, 4159.
130. J. A. Maddry, R. C. Reynolds, J. A. Montgomery, and J. A. Secrist, III, in "Carbohydrate Modifications in Antisense Research," eds. Y. S. Sanghvi and P. D. Cook, ACS Symposium Series 580, ACS, Washington, DC, p. 40.
131. J. M. Hewitt, W. C. Lenhart, R. N. Moore, A. K. Saha, and A. L. Weis, *Nucleosides Nucleotides*, 1992, **11**, 1661.
132. B. Hyrup, C. Richert, T. Schulte-Herbrüggen, S. A. Benner, and M. Egli, *Nucleic Acids Res.*, 1995, **23**, 2427.
133. J. Huang, E. B. McElroy, and T. S. Widlanski, *J. Org. Chem.*, 1994, **59**, 3520.
134. K. A. Perrin, J. Huang, E. B. McElroy, K. P. Sams, and T. S. Widlanski, *J. Am. Chem. Soc.*, 1994, **116**, 7427.
135. E. M. Huie, M. R. Kirshenbaum, and G. L. Trainor, *J. Org. Chem.*, 1992, **57**, 4569.
136. P. J. L. M. Quaedflieg, C. M. Timmers, G. A. Van der Marel, E. Kuyl-Yeheskiely, and J. H. van Boom, *Synthesis*, 1993, 627.
137. J. M. Veal and F. K. Brown, *J. Am. Chem. Soc.*, 1995, **117**, 1873.
138. Y. Ducharme and K. A. Harrison, *Tetrahedron Lett.*, 1995, **36**, 6643.
139. J. Lebreton, A. De Mesmaeker, and A. Waldner, *Synlett*, 1994, 54.
140. P. E. Nielsen, in "Perspectives in Drug Discovery and Design," eds. G. L. Trainor, ESCOM, Leiden, 1996, vol. 4, p. 76.
141. F. Bergmann, W. Bannworth, and S. Tam, *Tetrahedron Lett.*, 1995, **36**, 6823.
142. A. C. Van der Laan, N. J. Meeuwenoord, E. Kuyl-Yeheskiely, R. S. Oosting, R. Brando, and J. H. van Boom, *Trav. Chim. Pays-Bas*, 1995, **114**, 295.

143. E. Uhlmann, D. W. Will, G. Breipohl, D. Langner, and A. Ryte, *Angew. Chem. Int. Ed. Engl.*, 1996, **35**, 2632.
144. A. Van Aerschot, B. De Bouvere, C. Hendrix, L. Kerremans, H. De Winter, and P. Herdewijn, *Nucleosides Nucleotides*, 1997, **16**, 973.
145. R. G. Schultz and S. M. Gryaznov, *Nucleic Acids Res.*, 1996, **24**, 2966.
146. Y. S. Sanghvi, in "Antisense Research and Applications," eds. S. T. Crooke and B. Lebleu, CRC Press, Boca Raton, FL, 1993, p. 273.
147. S. T. Crooke, K. M. Lemonidis, L. Neilson, R. Griffey, E. Lesnik, and B. Monia, *Biochem. J.*, 1995, **312**, 599.

# 7.09
# DNA with Altered Bases

GANAPATHI R. REVANKAR and T. SUDHAKAR RAO

*Gemini Biotech, The Woodlands, TX, USA*

| | | |
|---|---|---|
| 7.09.1 | INTRODUCTION | 314 |
| 7.09.2 | PYRIMIDINE MODIFICATIONS | 314 |
| | 7.09.2.1   *Thymine Derivatives* | 314 |
| | 7.09.2.1.1   *C-2-Substitution* | 314 |
| | 7.09.2.1.2   *N-3-Substitution* | 315 |
| | 7.09.2.1.3   *C-4-Substitution* | 315 |
| | 7.09.2.1.4   *C-5-Substitution* | 315 |
| | 7.09.2.1.5   *C-6-Substitution* | 319 |
| | 7.09.2.1.6   *6-Azathymine* | 319 |
| | 7.09.2.2   *Cytosine Derivatives* | 320 |
| | 7.09.2.2.1   *C-4-Substitution* | 320 |
| | 7.09.2.2.2   *C-5-Substitution* | 321 |
| | 7.09.2.2.3   *C-6-Substitution* | 322 |
| | 7.09.2.2.4   *6-Azacytosine* | 322 |
| 7.09.3 | PURINE MODIFICATIONS | 323 |
| | 7.09.3.1   *Adenine Derivatives* | 323 |
| | 7.09.3.1.1   *C-2-Substitution* | 323 |
| | 7.09.3.1.2   *C-6-Substitution* | 324 |
| | 7.09.3.1.3   *C-8-Substitution* | 325 |
| | 7.09.3.2   *Hypoxanthines* | 325 |
| | 7.09.3.3   *Xanthines* | 326 |
| | 7.09.3.4   *C-6-Thiopurines* | 327 |
| | 7.09.3.5   *Guanine Derivatives* | 327 |
| | 7.09.3.5.1   *C-2-Amino-substitution* | 327 |
| | 7.09.3.5.2   *C-6-Substitution* | 328 |
| | 7.09.3.5.3   *C-8-Substitution* | 328 |
| | 7.09.3.5.4   *6,8-Disubstitution* | 329 |
| | 7.09.3.5.5   *N-7-Substitution* | 329 |
| 7.09.4 | PURINE ANALOGUES | 329 |
| | 7.09.4.1   *1-Deazapurines* | 329 |
| | 7.09.4.2   *3-Deazapurines* | 329 |
| | 7.09.4.3   *7-Deazapurines* | 330 |
| | 7.09.4.4   *7-Deaza-8-azapurines* | 331 |
| 7.09.5 | C-LINKED NUCLEOSIDES | 332 |
| | 7.09.5.1   *Monocyclic* | 332 |
| | 7.09.5.2   *Bicyclic* | 333 |
| 7.09.6 | MISCELLANEOUS | 333 |
| | 7.09.6.1   *Azoles* | 333 |
| | 7.09.6.2   *Polycyclic Heterocycles* | 334 |

7.09.7  CONCLUDING REMARKS                                                          335

7.09.8  REFERENCES                                                                  336

## 7.09.1  INTRODUCTION

With the advent of antisense and antigene oligonucleotide technology it is becoming increasingly convincing that the activity of any given gene can be selectively turned off or turned on. Such modulation of the expression of genes can result in controlling human diseases of neoplastic, endocrine, immunological, viral or microbial origin. Incorporation of modified DNA bases, in place of the natural bases adenine, cytosine, guanine, or thymine, into oligodeoxynucleotides (ODNs) offers a promising means for generating therapeutic antisense/antigene oligonucleotides with improved biochemical and pharmacological properties. In this chapter, we will focus on the progress made in recent years with ODNs containing altered bases and their hybridization properties.

## 7.09.2  PYRIMIDINE MODIFICATIONS

Principally, there are five sites that are available for chemical modification in the pyrimidine ring system. However, the majority of the modifications in the pyrimidine moiety documented in the literature include C-2, C-4, C-5, and C-6 positions, and only a mention has been made regarding N-3-substitution. In this chapter, we review each modification specifically in connection with synthetic modified DNA bases for altering hybridization properties.

### 7.09.2.1  Thymine Derivatives

#### 7.09.2.1.1  C-2-Substitution

As Watson–Crick base pairing is necessary for complementary hybridization, it may be expected that modification at the C-2 position of the pyrimidines will interfere with the hydrogen bonding, thus affecting the hybridization properties of those oligonucleotides containing the C-2-modified pyrimidines. However, there are a few reports on the modification of the C-2 position of pyrimidine. 2-$O$-Ethylthymine (**1a**)[1] and 2-$O$-methylthymine (**1b**)[2] residues were incorporated into ODNs via phosphoramidite methodology and their hybridization properties were determined by thermal denaturation analysis. The melting temperature value ($T_m$), which is a measure of the binding affinity between the oligonucleotide and its target sequence, shows that 2-$O$-methylthymine can form a much "better" base pair with guanine than adenine, with the $T_m$ values of DNAs containing these base pairs differing by 13.7 °C. The 2-$O$-methyl-T · A pair possibly retains Watson–Crick alignment, but the lack of a proton on N-3 of pyrimidine allows only one hydrogen bond to be formed. Furthermore it is possible that the presence of the 2-$O$-methyl group prevents close approach of the two bases. Both factors may contribute to the low observed $T_m$ value. By contrast, DNA containing 2-$O$-methyl-T · G pair has a high $T_m$ value and this may reflect the formation of a wobble pair, since the formation of a pair with Watson–Crick alignment may be impossible. Thus, neither the 2-$O$-methyl-T · G pair nor 2-$O$-methyl-T · A pair is ideal for DNA polymerase and this may be the reason why the presence of 2-$O$-ethyl-T (**1a**) in the template DNA blocks DNA synthesis. In other studies on C-2-modified pyrimidines, the oxo group was substituted by sulfur to give 2-thiothymine (S$^2$T, (**2a**))[3–8] or 2-thiouracil (**2b**)[6] and these modified pyrimidines were incorporated into ODNs via either phosphoramidite[3–7] or H-phosphonate methodology.[8]

The ODNs containing residues of (**2a**) showed slight deviation in $T_m$ values depending on the length and position of the modification in the oligonucleotide. The palindromic S$^2$T-containing sequences showed increased $T_m$ values relative to control (+4 to 6 °C per substitution).[6] This increase in stability, however, is not a general effect of sulfur substitution and appears to be due to interstrand interactions as opposed to intrastrand interactions. This is revealed by the $T_m$ values for the nonpalindromic S$^2$T containing oligonucleotides, which are slightly lower than the control (−2 to 4 °C). Kutyavin *et al.*[7] have reported the hybridization and strand invasion properties of modified selectively binding complementary (SBC) oligonucleotides containing 2-aminoadenine (A′) and S$^2$T (**2a**), which form stable hybrids with complementary unmodified sequences, yet they do not interact with each other. The natural DNA–DNA hybrid had a $T_m$ of 55 °C, whereas the two alternative SBC–DNA hybrids were found to have approximately 10 °C higher $T_m$ values. The relatively sharp symmetric peaks obtained for the SBC–DNA hybrids indicate that A′·T and A·S²T base pairs did not alter the co-operativity of hybridization.[7] The SBC–SBC hybrid showed a $T_m$ value of nearly 30 °C lower than the DNA–DNA hybrid. The destabilizing effect of an A′·S²T doublet relative to an authentic mismatch was estimated and it was found that the $T_m$ of the control duplex was, on average, depressed by 2.4 °C for each introduction of an A′·S²T doublet versus 6.3 °C for each introduction of a T–T mismatch.

### 7.09.2.1.2   N-3-Substitution

3-*N*-Ethylthymine was incorporated into ODNs utilizing solid-phase phosphoramidite methods. The studies described by Bhanot *et al.*[9] reveal that a single 3-*N*-ethyl-T adduct present in DNA template blocks DNA synthesis by the Klenow fragment of *Escherichia coli* polymerase I (Kf Pol I) either immediately 3′ to the 3-*N*-ethyl-T adduct or after incorporation of a nucleotide opposite to it. The 3-*N*-ethyl-T adduct is formed in both single and double stranded DNA.

### 7.09.2.1.3   C-4-Substitution

Several reports have appeared in the literature[10–18] on the synthesis of ODNs containing 4-*O*-alkyl-T and 4-thio-T. 4-*O*-Methyl-T (**3**) residues were found to pair with both A and G.[18] Melting of the oligomers containing the modification extends over a wider temperature range than that of the respective parents and the melting is biphasic. In addition, the hypochromicity of the sequences was less than that of the unalkylated sequences. The lower and extended melting temperature of oligomers containing (**3**) paired with A or G suggests that the presence of alkylated base destabilizes the duplex, and the lower total hypochromicity suggests that the presence of the methylated base decreases the overall extent of base stacking.

The incorporation of residues of 4-thio-T (**4a**)[4,5,8,16,19–24] and 4-thiouracil (4-thio-U, (**4b**)) into ODNs was described by several groups.[22–26] Owing to the instability of thione functionality during the oligonucleotide synthesis, the thione moiety was protected with different types of protecting groups, e.g., cyanoethyl,[21,25,26] phenyl,[20] 4-nitrophenyl,[19–21] or pivaloyloxymethyl[22,23] groups. Alternatively, these modified ODNs were synthesized by postsynthetic modification[16] of ODNs containing a 4-triazolyl derivative of T. The stability of the duplexes formed by these modified ODNs was evaluated by $T_m$ analysis.[24] The $T_m$ of a self-complementary duplex containing two 4-thio-T·A pairs was only 1 °C lower than the control. However, a non-self-complementary duplex containing only one 4-thio-T·A pair had a $T_m$ of 3 °C lower than the control. It was also found that the presence of either 4-thio-U or 4-thio-T has no significant effect on the melting temperature of duplex structures, when placed opposite to G or A residues.[26]

### 7.09.2.1.4   C-5-Substitution

Several C-5-substituted pyrimidines have been incorporated into ODNs and their hybridization properties have been studied. The modifications at C-5 or C-6 positions of the pyrimidine ring stay in the major groove of the helix and hence should not have any direct effect on the base pairing properties. The simplest modifications that are known are the C-5-halogenated pyrimidines. Substitution of a halogen at C-5 position increases the acidity of the N-3-proton, and thus may increase

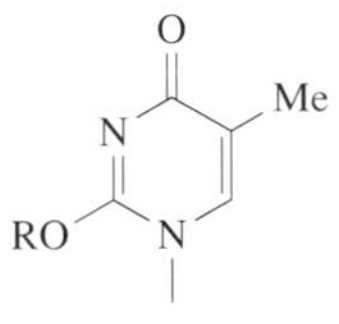

(**1a**) R = Et
(**1b**) R = Me

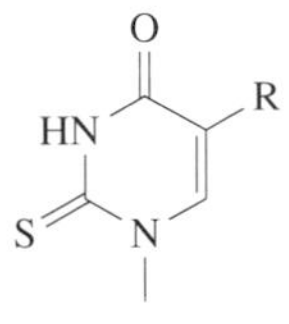

(**2a**) R = Me
(**2b**) R = H

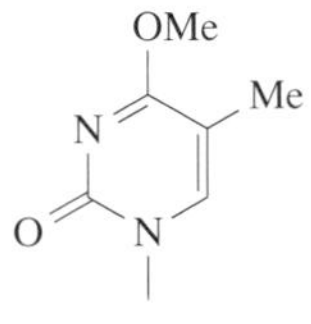

(**3**)

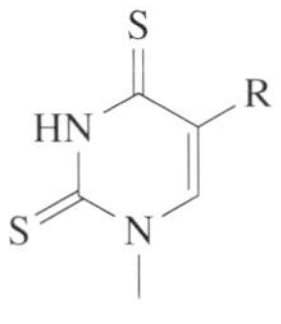

(**4a**) R = Me
(**4b**) R = H

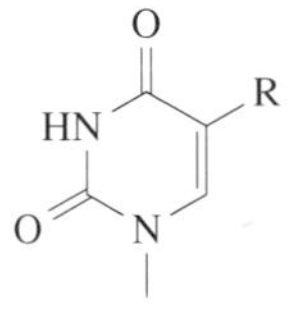

(**5a**) X = F
(**5b**) X = Br
(**5c**) X = I

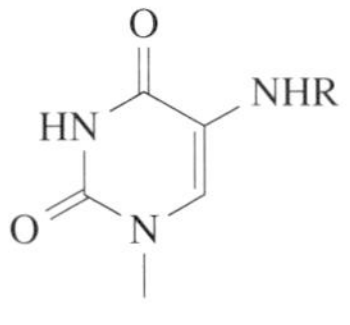

(**6a**) R = H
(**6b**) R = dansyl

the strength of base pairing properties. In addition, base stacking interactions of double-helical polynucleotides appear to be enhanced when halogenated pyrimidines are substituted for the naturally occurring bases. 5-Fluoro (**5a**),[27–32] 5-bromo (**5b**),[33–35] and 5-iodo-U (**5c**)[35,36] derivatives of deoxyuridines have been successfully incorporated into ODNs employing phosphoramidite methodology and their hybridization properties have been studied with respect to duplex and triple helix formation. The duplexes containing 5-fluoro-U ((**5a**), F) were found to be more stable than the corresponding duplexes with thymine. A · F and F · A base pairs were found to be more stable than A · T and T · A base pairs by 1.0 °C and 1.4 °C, respectively. Likewise, G · F and F · G mismatches are 3.1 °C and 3.4 °C more stable than the corresponding G · T and T · G mismatches.[27] 5-Fluoro-U was found to exhibit superior stability and nonselectivity in pairing to A or G compared to the natural pyrimidines C or T. However, [1]H NMR studies[28–30] on two self-complementary DNA dodecamers containing (**5a**) suggested that the A · F base pair was less stable than the A · T base pair. The significant difference between A · F and A · T base pairs is dynamic and not structural. The A · F base pair opens faster than the normal base pairs in the oligonucleotides. This enhancement results from the decreased stacking of the (**5a**) residue and not from the enhanced acidity of N-3-imino proton. In a related study, Sowers *et al.*[29] concluded that a pH-dependent equilibrium exists between wobble and ionized base pairs for F · G mismatch, which is the first observation of an equilibrium between distinct base pairing schemes. According to Sowers *et al.*,[30] this is the first example of a negatively charged base pair in a DNA duplex. In a later study, it has been demonstrated that substitution of (**5b**) in DNA resulted in a very slight destabilization or stabilization depending on the position of the substitution.[31] Substitution of 5-bromo-U (**5b**) or 5-iodo-U (**5c**) for T in DNA has essentially no effect on duplex stability.

The utility of the 5-halogenated pyrimidines in triple helix formation in parallel as well as antiparallel motifs (see Chapter 7.02) were also evaluated. Substitution at position 5 of pyrimidines could alter the hydrophobic driving force, base stacking, and the electronic complementarity of the base pairing for triple strand formation. It has been found that ODNs containing 5-bromo-U substituted for T, in the parallel motif, bind to duplex DNA at the same homopurine target sequence as the parent oligonucleotide, but with greater affinity.[34] The electron-withdrawing bromo substituent increases the acidity of the N-3 proton (better hydrogen donor) and decreases the electron-donating properties of the carbonyl lone pair (poorer hydrogen acceptor) thus increasing the stability of triple stranded structure. The relative stabilities of base triplets are found to be in the order Br-U : AT > T : AT > U : AT. The effects of 5-halogenated pyrimidines on antiparallel triple helix formation was examined in the authors' laboratory.[35] It has been found that 5-fluoro-U has slightly higher affinity for AT base pairs than T, and significantly better than 5-bromo-U and 5-iodo-U. The effects of the C-5-substituent on binding of T analogues to CG base pairs were also studied. In both cases, the effect of the C-5-substituent on relative binding affinities was approximately F > Me > I > Br > H.

Synthesis of ODNs containing 5-amino-U ((**6a**), Ua) using trifluoroacetyl as amino protector has been described.[37] UV melting experiments indicate the formation of duplex with Ua-containing oligonucleotide with slight destabilization ($\Delta T_m = 1.5$ °C) relative to the duplex formed with T at the same position.[38] Similarly, oligonucleotides containing 5-aminodansyl-U (**6b**) were also prepared as fluorescent-labeled compounds and their hybridization properties were studied.[39] The dansylated oligonucleotides exhibited lower $T_m$ values (48 °C) compared to the unmodified dodecamer (60 °C), suggesting a slight destabilizing effect on duplex, locally induced by aminodansyl group in U. The effect of 5-amino-U in the central strand of a DNA triplex in parallel and antiparallel motifs has also been studied. It has been envisaged that the pyrimidine derivative ((**6a**), Ua) would be suitable as the middle base of a triplex triad, since it possesses the electronic requirements for simultaneous recognition of complementary bases of the triad. The availability of one donor and acceptor site in Ua provides orientation specificity for recognition of the Ua · A duplet by a third base in the major groove. The stability of triplexes was measured by UV melting experiments. The parallel triplex having triad A : Ua : A with a modified base in the central strand has higher UV-$T_m$ compared to the control triplex A : T : A in pyrimidine motif; no triplex formation was detected in the corresponding antiparallel mode. In the case of triad G : Ua : A, triplex formation was observed only in the antiparallel orientation, but not in the parallel form. The G : Ua : A antiparallel triplex exhibited a higher stability than G : T : A triplex seen in the parallel motif. Thus, triplexes with modified base Ua exhibited not only higher UV-$T_m$ stability compared to corresponding control T analogues, but also displayed a remarkable orientation selectivity in third-strand recognition.[39] The pyrimidine motif triplexes containing A : Ua : A triad and a A : T : A triad were formed only at pH 5.8 and not detected at pH 7.0. In contrast, the antiparallel purine motif triplex devoid of the base C in the third strand was observed in both pH ranges with higher stability at pH 7.0 ($T_m$ 37 °C) compared to that at 5.8 ($T_m$ 35 °C).

Lown and co-workers[40] described the synthesis and proton NMR studies of ODNs containing trifluorothymine (**7a**). These modified ODNs retain the helix forming properties. Seela *et al.*[41] synthesized dodecanucleotides containing (*E*)-5-(2-bromovinyl)-U (**7b**) and studied the influence of a bulky major groove substituent in duplex stability and found that the modified oligomers exhibited very similar $T_m$ values to the parent compounds. These findings indicate that short DNA fragments can accommodate one bulky bromovinyl residue without changing its secondary structure and as this residue is located in the major groove of the DNA helix, steric interference with other bases is not expected.

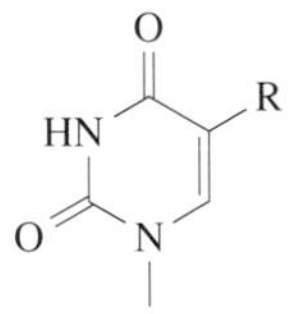

(**7a**) R = CF$_3$
(**7b**) R = CH=CHBr

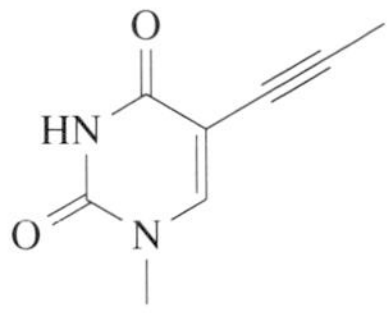

(**8**)

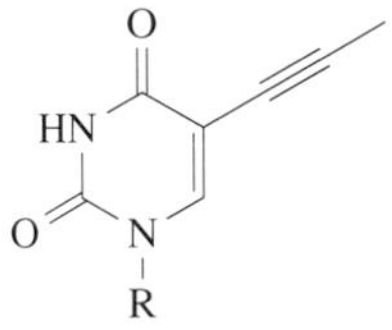

(**9**) R = 2'-*O*-allylribose

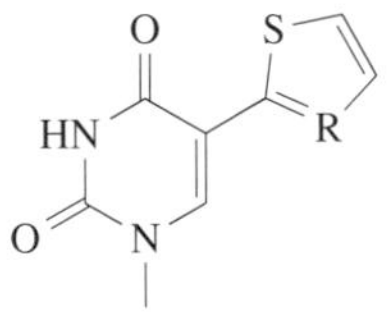

(**10a**) R = N
(**10b**) R = CH

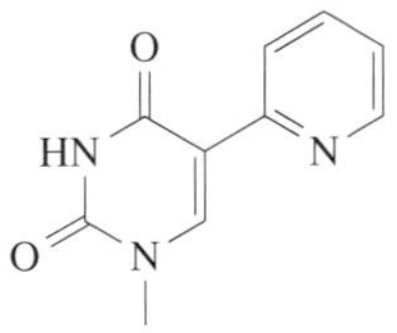

(**11**)

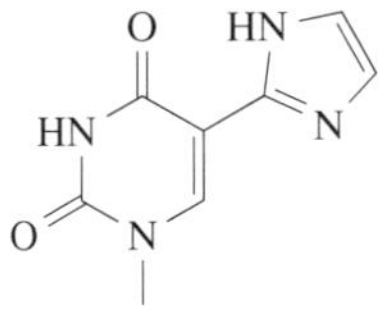

(**12**)

Froehler *et al.*[42] have described the synthesis of ODNs containing C-5-propynylpyrimidines and their hybridization properties with respect to duplex and triple helix formation. The propyne modification is planar with respect to the heterocycle and allows for increased stacking of bases. Additionally the propyne group is more hydrophobic than a methyl group to potentially allow for a further increase in the entropy of binding. C-5-Propynyl-U (**8**) is more hydrophobic than thymine and, therefore, may facilitate passive diffusion of oligonucleotides into cells, making this modification especially useful in therapeutic applications. The stability of DNA/RNA duplex was assessed by $T_m$ analysis and it was found that ODN containing C-5-propynyl-U results in an increase in $T_m$ of 1.7 °C per substitution. The effect of (**8**) on the stability of triple helix was also determined by $T_m$ analysis. ODN containing (**8**) showed an increase in $T_m$ of 2.4 °C per substitution relative to the parent ODN. The oligonucleotides derived from C-5-(1-propynyl)-2'-*O*-allyl-U (**9**) bind with high affinity for the target RNA and increase the $T_m$ by 1.7 °C per substitution relative to the control.[43] In a later report by Gutierrez and Froehler,[44] the stabilities of C-5-alkyne and C-5-thiazole-U (**10a**) ODN–RNA complexes were evaluated by $T_m$ analysis and showed that the propyne and butyne containing ODNs lead to very stable ODN–RNA complexes ($T_m$ = +1.6 °C per substitution and +1.4 °C per substitution, respectively). Increasing the steric bulk on the alkyne substituent resulted in decreased stability. Similar results were observed with the DNA–DNA duplex stabilities with ODN polymers containing increasing alkyl and alkynyl substitutions.[44,45] This is probably due to unfavorable steric interactions of the additional methyl substituents within the canonical double helix and hydrophobic destabilization. It has also been observed by Colocci and Dervan[46] that two short ODNs containing C-5-propynyl-U and C-5-methyl-C bind cooperatively to adjacent sites on double helical DNA by triple helix formation at micromolar concentrations. Possible sources for the cooperative interaction between ODNs containing (**8**) include a structural transition between adjacent sites on DNA and increased base stacking between the C-5-propynyl-Us. The right-handed nature of the triplex allows stacking of the propyne group of the base on the 3'-side of the junction onto the 5'-base across the junction, but not stacking of the propyne group at the 5'-side chain of the junction onto the 3'-adjacent base. The antisense gene inhibition by C-5-propyne pyrimidine-containing ODNs has been described[47,48] and the site and mechanism of action has also been studied.[49]

Froehler and co-workers[50] have also studied the effects of C-5-heteroaryl-U (**10a,b**) on the stability of ODN–RNA duplex. It was found that all $T_m$ transitions of the duplexes containing the C-5-heteroaryl-Us were sharp and the $T_m$ was increased relative to the thymine control. C-5-(Pyridin-2-yl)-U (**11**) and C-5-(thien-2-yl)-U (**10b**) substituted ODNs increased $T_m$ slightly ($T_m = +0.3\,°C$ per substitution and $+0.4\,°C$ per substitution, respectively), and C-5-(imidazol-2-yl)-U (**12**) containing ODN increased the $T_m$ by $+0.7\,°C$ per substitution relative to the thymine control. The largest increase in $T_m$ was observed with the C-5-(thiazol-2-yl)-U containing ODN, which led to a $T_m$ per substitution of $1.7\,°C$ relative to the control and is comparable to the C-5-propynyl-U containing ODN. The authors suggest a probable reason for the increased $T_m$ of these analogues is the increase base stacking interactions.[50] The UV absorbance spectra of these analogues show a large bathochromic shift of the uracil base relative to U. The heteroaryl groups are likely to undergo $\pi$ electron delocalization with the uracil base, suggesting a near planar orientation between the heteroaryl group and uracil base. An increase in the $\pi$ electron delocalization of these pyrimidine rings into the heteroaryl ring would be expected to enhance the base stacking interactions of adjacent base pairs, leading to higher affinity. Additional factors may contribute to the increased affinity of C-5-thiazole-U relative to the other heteroaryl analogues. Pyridine and thiophene both contain one hydrogen *ortho* to the C-5-heteroaryl bond; this may lead to a steric interaction with the C-6-H or C-4-O of uracil, leading to a deviation from planarity. The imidazole moiety of ODN has an exchangeable hydrogen which leads to a more stable duplex but less stable than the thiazole containing ODN. Thiazole containing ODN has no *ortho* hydrogens, therefore the possible steric interactions are eliminated and no restrictions to achieving coplanarity are present. The lower $T_m$ of C-5-imidazole-U containing ODN compared to C-5-thiazole-U containing ODN may be a result of intramolecular hydrogen bonding between the C-4-O and imidazole N—H, leading to a weaker hydrogen bond with the complementary adenine. Substituted C-5-thiazole-U containing ODNs formed stable duplexes with RNA, but in contrast to the alkynes, increased methyl substitution did not automatically result in decreased stability.[44] The C-5-(5-methyl)thiazole-U (**13a**) containing ODN leads to the most stable duplex ($T_m = +2.2\,°C$ per substitution) and C-5-(4-methyl)thiazole-U (**13b**) ODN leads to a relatively less stable duplex ($T_m = +1.4\,°C$ per substitution). The C-5-benzothiazole-U (**14**)[44] containing ODN exhibited the lowest affinity ($T_m = +1.2\,°C$ per substitution) among the thiazole-U substituted ODNs. These results do not support the hypothesis that is proposed above for the highest $T_m$ of thiazole containing ODN, since the C-5-(5-methyl)thiazole-U containing ODN is showing the highest $T_m$ increase in this study.

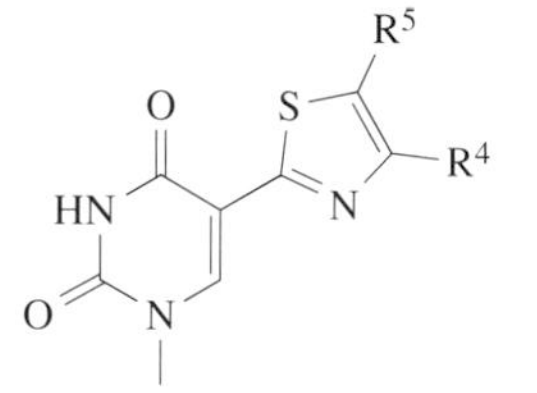

(**13a**) $R^4 = H$, $R^5 = Me$
(**13b**) $R^4 = Me$, $R^5 = H$

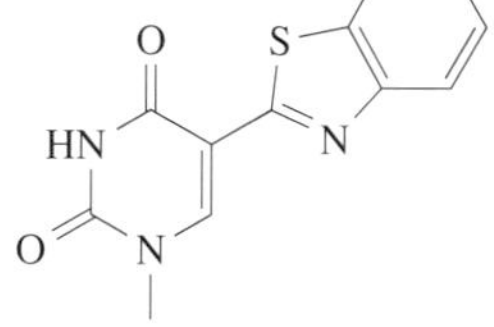

(**14**)

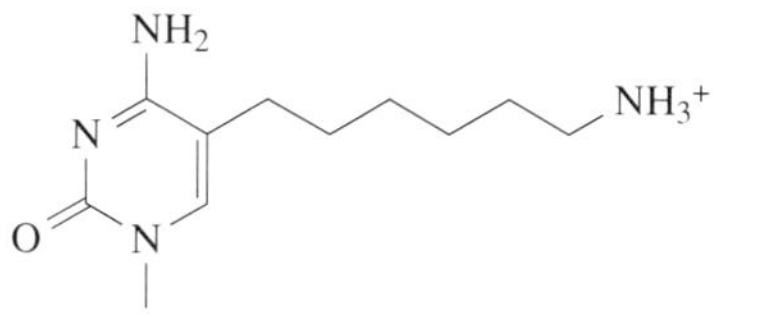

(**15**)

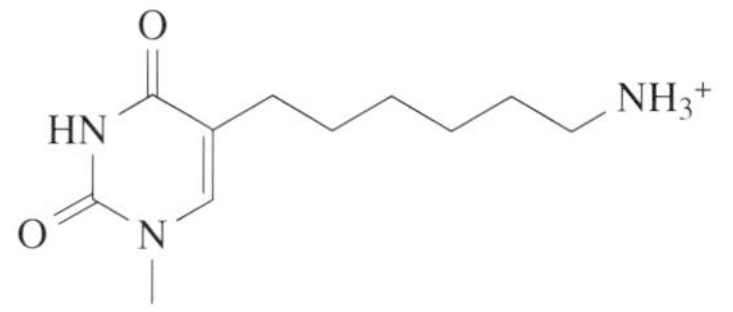

(**16**)

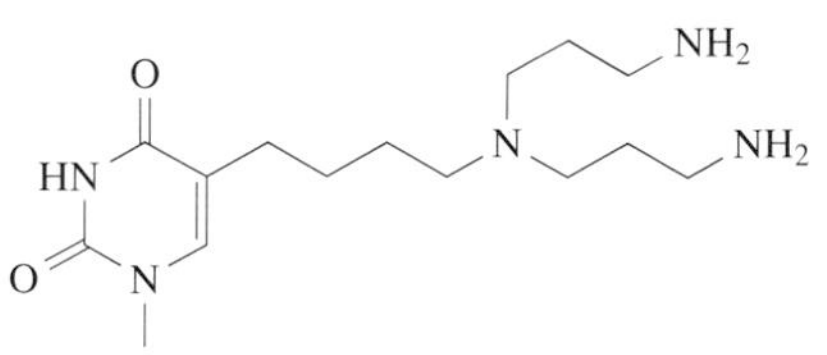

(**17**)

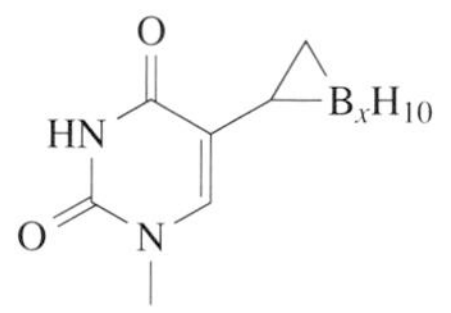

(**18**)

In an effort to make DNA net charge neutral, Switzer and co-workers[51] have prepared the ODNs bearing pyrimidine-$\omega$-aminohexyl substituents. The resultant zwitterionic oligomers were found to bind to natural DNA at low ionic strength as well as or better than does natural DNA with itself, even when all of the nucleotides in a given single strand are rendered zwitterionic. The stabilities of these modified ODN duplexes are relatively insensitive to changes in solution ionic strength as compared to natural DNA. A DNA duplex composed of a fully zwitterionic strand and a natural complementary strand exhibits no change in stability over a 20-fold change in ionic strength. It has also been observed that zwitterionic nucleotides do distinguish matched from mismatched bases in complementary DNA to a degree that is comparable to natural DNA. Interestingly, in all cases a cytidylate zwitterion (**15**)·A mismatch is somewhat more unfavorable ($T_m = 19.4$ and $20.5\,°C$), while uridylate zwitterion (**16**)·G mismatch is somewhat less unfavorable ($T_m = 6.4$ and $4.6\,°C$) than the natural C·A ($T_m = 18.5\,°C$) and natural T·G ($T_m = 7.4\,°C$) mismatches, respectively. The fact that DNA can be made fully zwitterionic by introducing tethered ammonium ions without affecting duplex formation could have important implications for enhancing biological activity of oligonucleotides *in vivo*.

Nara *et al.*[52] have synthesized the heptadecamers containing 5-{4-[*N*,*N*-bis(3-aminopropyl)amino]butyl}-U (**17**) and evaluated their hybridization properties with respect to duplex and triplex formation with the complementary strand and the target duplex by thermal denaturation. $T_m$ of the duplexes were found to be higher than the control duplex, indicating that the modified U has stabilized the duplex formation. When (**17**) was placed at the 5′-end of ODN, it formed a more stable duplex compared to one which has the modified unit in the middle of the ODN. The triplex formation was also stabilized by this modification.

Schinazi and co-workers[53,54] have reported the physicochemical properties of the ODNs containing 5-(*O*-carboran-1-yl)-U (CDU, (**18**)). It has been observed that the thermal stabilities of duplexes formed by these ODNs with natural complementary strand were affected by the location of carboranyluracil within the chain. The melting temperature of the duplexes was highest when the CDU was located at the 5′-end rather than at the 3′-end or in the middle of the ODN. In addition it was found that the presence of one or more CDUs increased their lyphophilicity. These modified ODNs are also found to have increased resistance to exonucleases.

Meyer *et al.*[55] prepared a tetradecadeoxynucleotide containing 5-[3-(iodoacetamido)propyl]-U (**19**) in place of a T and studied the specific cross-linking and cleavage of DNA. Analysis of the cleaved product by gel electrophoresis revealed that reaction occurred predominantly at the guanine base located two base pairs to the 5′-end of the target site. Povsic and Dervan[56] reported the design and synthesis of an ODN equipped with an electrophile (**20**) at the 5′-end that binds to double-helical DNA by triple helix formation and alkylates predominantly at a single guanine base adjacent to the target DNA sequence in high yield. They obtained similar results to those obtained by Meyer *et al.*[55] and suggest that because the oligonucleotide directed triple-helix motif is sufficiently generalizable and specific for the recognition of single sites in genomic DNA, modification of a single base within megabase-sized chromosomes using strictly chemical methods should be possible.

### 7.09.2.1.5  *C-6-Substitution*

Modification of the C-6-position of the thymine with a methyl group (**21a**) destabilized the duplex formed with complementary DNA or RNA.[57–59] Similarly C-6-methyl-U (**21b**) was found to destabilize the duplex. The destabilization of the duplex caused by these modifications may be due to the inability of C-6-methylpyrimidines to adopt the *anti* conformation in the modified ODNs required for appropriate Watson–Crick hydrogen bonding in an A or B-form of duplex.

### 7.09.2.1.6  *6-Azathymine*

6-Azathymine (**22**) was incorporated into several antisense ODNs[57,58] using the conventional phosphoramidite chemistry on an automated synthesizer. The evaluation of the hybridization properties of these modified oligonucleotides showed that incorporation of 6-aza-T lowers the $T_m$ 1 to 2 °C per modification. However, these modified ODNs retain the overall duplex specificity. Incorporation of 6-aza-T at the 3′-end of the oligonucleotides enhanced the nuclease stability.[58]

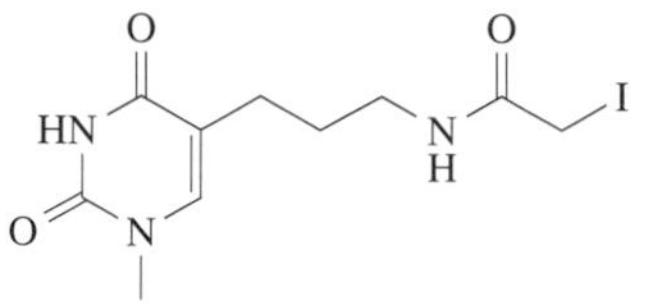

(19) 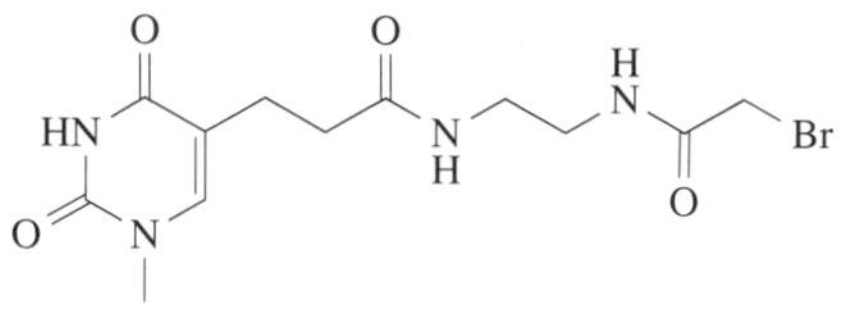                                              (20)

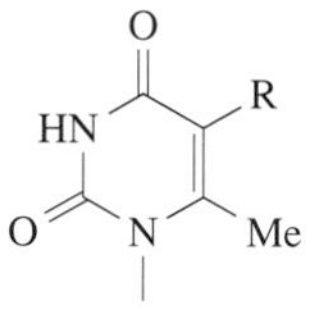

(21a) R = Me
(21b) R = H                                      (22)

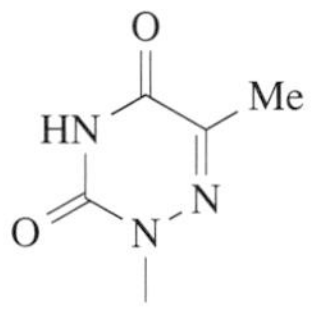

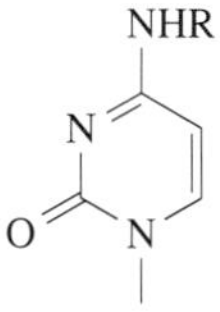

(23)

(a) R = Me; (b) R = (CH$_2$)$_2$NH$_2$; (c) R = (CH$_2$)$_4$NH$_2$; (d) R = (CH$_2$)$_2$OH; (e) R = (CH$_2$)$_2$S–S(CH$_2$)$_2$NH$_2$;
(f) R = CH$_2$CO$_2$H; (g) R = (CH$_2$)$_3$CO$_2$H; (h) R = (CH$_2$)$_3$Me; (i) R = (CH$_2$)$_3$NH$_2$; (j) R = (CH$_2$)$_3$NHAc

### 7.09.2.2 Cytosine Derivatives

#### 7.09.2.2.1 C-4-Substitution

MacMillan and Verdine[60,61] have prepared a number of ODNs containing N-4-alkyl-C and studied the effects of these modifications on duplex stabilities. It was found that N-4-alkyl-C (**23a–e**) weakly destabilize the duplex structure. However, when the alkyl chain contained a carboxyl group (**23f, g**) at the end of the chain, it strongly perturbed the duplex stability, probably due to the charge repulsion between the neighboring carboxylate groups. Miller and Cushman[62] have reported the synthesis of ODNs containing N-4-modified-C. These include N-4-(3-carboxypropyl)-C (**23g**), N-4-(4-aminobutyl)-C (**23c**), and N-4-butyl-C (**23h**). These linker-arm-conjugated oligomers are capable of forming stable duplexes with complementary single-stranded ODNs, as well as stable triplexes with target ODN duplexes. The oligomers which carry a single linker arm at the N-4-position of C were found to form stable duplexes with complementary single-stranded DNA targets. The $T_m$ of these duplexes are only slightly less than that of the duplex formed by unmodified oligomers. The position of the modification within the chain did not appear to affect the stability of the duplex. Although these modified oligomers are capable of forming triplexes with a double-stranded DNA target, the melting temperature of the derivatized third-strand oligomer from the triplex is considerably lower than that of the unmodified third-strand oligomer. This decrease in stability could be due to steric interactions between the side chain and components in the major groove of the triplex. Alternatively, the p$K_a$ of the modified cytosines may be higher than that of parent cytosine as a result of modification. In this case protonation of the modified base would be less likely to occur at pH 7.0 and the modified base would then be capable of forming only a single hydrogen bond to the C-6-O-carbonyl of G in the G·C base pair of the target. Although no triplex formation was observed with modified ODNs containing N-4-(aminobutyl)-C (**23c**), the formation of triplexes with N-4-(3-aminopropyl)-C (**23i**) and N-4-(3-acetamidopropyl)-C (**23j**) modified ODNs has been demonstrated.[63] Examination of the molecular models suggests that the length of the side chain is sufficient to allow either the 3-amino group hydrogen atoms of (**23i**) or the amide hydrogen of (**23j**) to hydrogen bond to the C-6-O-carbonyl of G of the target C·G base pair. It is also noteworthy that a base with flexible side chain such as N-4-(3-acetamidopropyl)-C (**23j**) can also support the triplex formation.

Miller and co-workers[64] have described the effects of N-4-(6-amino-2-pyridinyl)-C (**24**) on triplex formation. This synthon was designed to interact with a single base pair interruption of the purine target tract. This modified nucleobase (**24**) supports stable triple helix formation at pH 7.0 with formation of a (**24**) : $y \cdot z$ triad, where $y \cdot z$ is a base pair in the homopurine tract of the target. Selective interaction was observed when $y \cdot z$ was **C·G**, although A·T and to a lesser extent, T·A and G·C base pairs were also recognized. Removal of the 6-amino group of aminopyridine moiety essentially eliminates triplex formation. These results are consistent with a binding mode in which the 6-amino-2-pyridyl group of (**24**) spans the major groove of the target duplex at the (**24**) : C·G binding site and forms a hydrogen bond with the C-6-O atom of G. An additional stabilizing hydrogen bond could form between the N-4-atom of the imino tautomer of (**24**) and the N-4-amino group of C.

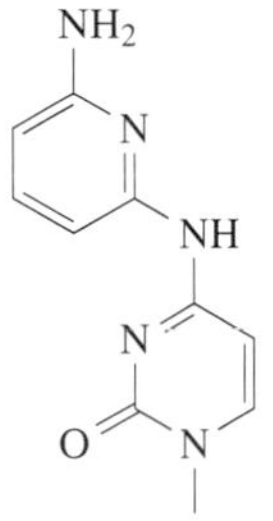

(**24**)

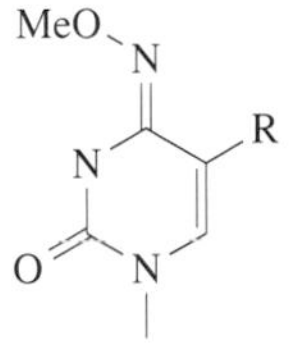

(**25a**) R = H
(**25b**) R = Me

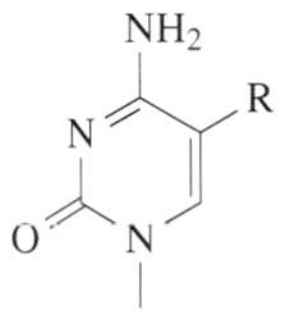

(**26a**) R = Me
(**26b**) R = Br

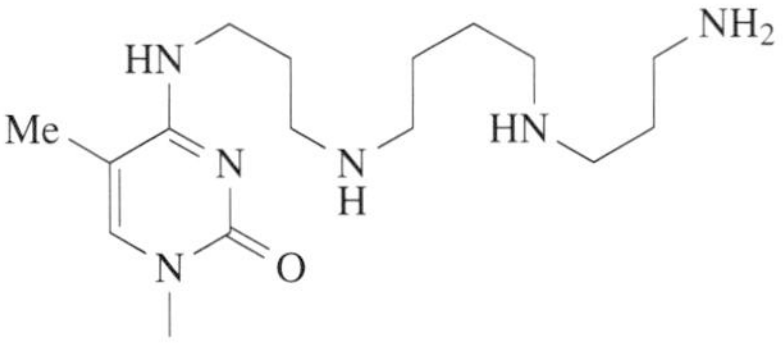

(**27**)

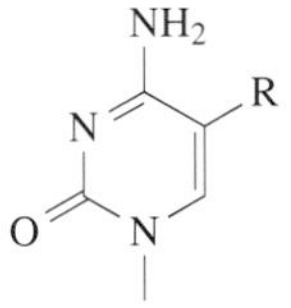

(**28a**) R = C≡C–Me
(**28b**) R = F

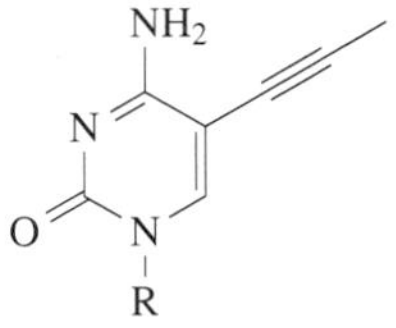

(**29**) R = 2'-*O*-allyl

Brown and co-workers have synthesized the ODNs containing N-4-methoxy-C (M, (**25a**))[65,66] and N-4-methoxy-5-methyl-C (**25b**)[67] and studied the stabilities of the duplexes that are formed by these modifications. It was found that the duplex A·M is more stable than the A·T duplex ($T_m = +4\,°C$). In a comparison of the transition profiles of the duplexes A·M and G·M with the others, strong co-operativity of these two stands out. This is unlikely to be due to stronger hydrogen bonding in these pairs. It is more plausible that the higher $T_m$ values of these duplexes are indicative of increased stacking interactions of the A·M and G·M base pairs, perhaps due to the exocyclic N-4-methoxy group. The difference in thermal stability is small ($\Delta T_m = 2\,°C$) and the free energies of formation of the two duplexes are similar. The duplexes containing two or three modifications demonstrate the essential equivalence of M·G and M·A pairs. Nevertheless duplex stability becomes lower with increasing number of M substitutions. However, inclusion of a C-5-methyl group in the modified base (**25b**) resulted in remarkable destabilization in both the duplexes. It is attributed to the steric inhibition of base pairing, since the methyl group would normally be expected to increase stacking interactions.

### 7.09.2.2.2  C-5-Substitution

In a similar manner as described for C-5-substituted uracils, substitution at C-5-position of cytosine is expected to have considerable effect on the hybridization properties. Substitution of a methyl (**26a**)[68–70] or a bromo (**26b**)[58] group at C-5-position of C had a stabilizing effect ($T_m$ 0.6 to 1.2 °C per modification) compared to C in RNA/DNA duplexes. Similarly, in the case of triplexes the DNA that contained C-5-methyl-C (**26a**) was found to have higher $T_m$ values when compared to C containing DNA.[68,69] In the parallel triplex formation protonation of C is required in order to form a triplex. The p$K_a$ of both C and (**26a**) are 4.5, yet the triplex formed from (**26a**) containing DNA is stable at pH values nearly 2 units higher than the triplex formed from poly(T–C) : poly(G–A). There appears to be no compelling physicochemical explanation for this unexpected observation. However, the discovery of triplex formation at neutral pH is intriguing. If the pyrimidine strands contain significant levels of (26a), then the triplex configuration will be favored under physiological conditions. The ODNs containing C-5-methyl-C-N-4-spermine (X*, (27)) have been shown to form triplexes.[71] Among the spermine-ODNs (sp-ODNs) which differ in the number and sequence-position of modifications, triplexes derived from the third strands containing spermine conjugation towards 5'/3'-termini gave better thermal stability compared to modification in the center. Although the triplex $T_m$ is slightly decreased with the increasing number of modifications, triplex formation was still observable with the trisubstituted sp-ODN at physiological pH and temperature.[71] The stabilities of triplexes from mono sp-ODNs measured with added $MgCl_2$ showed an enhancement of $T_m$ by 6–7 °C. Significantly, triplexes from mono sp-ODNs were as stable as that from ODN with three C-5-methyl-C in the presence of $MgCl_2$. To study the stringency of triplex formation by sp-ODNs triplexes were constituted with duplexes containing all four Watson–Crick base pairs at positions complementary to 5-methyl-C-N-4-spermine of the third strand. A decreased stability was observed in the order X*G : C ~ X*A : T > X*C : G > X*T : A, which is similar to that observed for an unmodified base triad by the affinity cleavage method.[72]

While the C-5-methyl substitution increased the thermal stability of poly(G–C), substitution of methyl group for ethyl group decreased the stability by 9 °C.[73] The substitution of the methyl group at position 5 of C generally improves base stacking whereas the extended ethyl group destabilizes the double helix, presumably through perturbation of its hydration and ion binding sites.

The ODNs containing C-5-propynyl-C (28a) exhibited an increase in $T_m$ of 1.5 °C per substitution in DNA/RNA duplex formation.[42] However it destabilized the parallel triple helical complex, probably due to a decrease in $pK_a$ of the aglycon. Similarly, the ODNs derived from the C-5-(1-propynyl)-2'-O-allyl-C (29) were found to bind with high affinity for the target RNA with an increase of $T_m$ of 2.0 °C per substitution relative to the control.[43] C-5-Fluoro-C (28b) has been incorporated into ODNs using standard phosphoramidite chemistry.[74] These modified ODNs containing (28b) have been shown to inhibit bacterial methyltransferases (MTases).[75–77]

### 7.09.2.2.3   C-6-Substitution

McLaughlin and co-workers[78] synthesized oligonucleotides containing C-5-methyl-6-oxo-C (30a) and evaluated their hybridization properties with respect to parallel triple helix formation in comparison to C and C-5-methyl-C. This modified nucleoside permits the recognition of G · C base pairs by the parallel stranded recognition motif under neutral, mildly acidic, as well as mildly basic pH conditions. Triplex stability was assayed at five different pH values. The $T_m$ values for the C-containing oligonucleotides decreased from 39.0 °C at pH 6.4 to only 14.0 °C at pH 8.5. Replacement of the four C residues by C-5-methyl-C resulted in more stable triplexes, but the pH dependence of triplex formation was still readily apparent. By comparison, the (30a) containing oligomer resulted in largely pH-independent transitions varying by only 2 °C over the range of pH values examined. The $T_m$ values for the C as well as C-5-methyl-C containing triplexes are significantly higher than those observed for the C-5-methyl-6-oxo-C containing triplex at pH 6.4 and 7.0. This observation likely reflects, in part, the differences in hydrogen bonding characteristics. Each C (or C-5-methyl-C) will form one hydrogen bond in the triplex complex that contains a charged H-bonding partner, while the C-5-methyl-6-oxo-C residues form only hydrogen bonds involving uncharged partners. As the pH value of the solution increases to 8.0 and 8.5, the extent of protonation of C (or C-5-methyl-C) decreases, and the C-5-methyl-6-oxo-C containing triplex, with neutral hydrogen bonds, dominates in stability.

Differences in base stacking characteristics are also likely to account for variations in triplex stability. Berressem and Engels[79] have studied the effects of (30b) and its C-5-methyl derivative (30a) on triple helix formation. Their results affirm the formation of triple helices by the modified ODNs containing 6-oxo-C (or C-5-methyl-6-oxo-C) in a pH independent manner.

### 7.09.2.2.4   6-Azacytosine

6-Aza-C (31) has been incorporated into ODNs via phosphoramidite methodology.[57,58] The oligodeoxynucleotides containing 6-aza-C modification exhibited destabilization of the duplex. However, these modified ODNs have enhanced nuclease resistance compared to the unmodified counterpart.

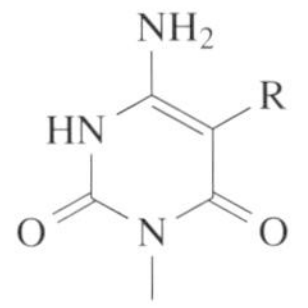

(30a) R = Me
(30b) R = H

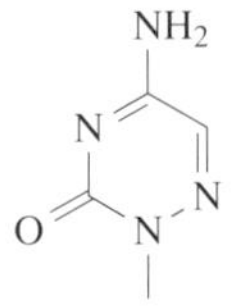

(31)

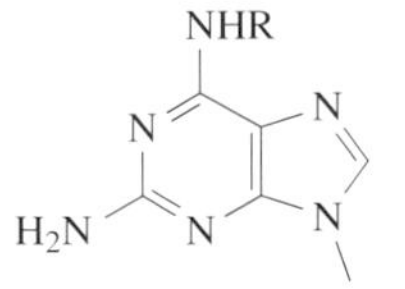

(32a) R = H
(32b) R = OMe

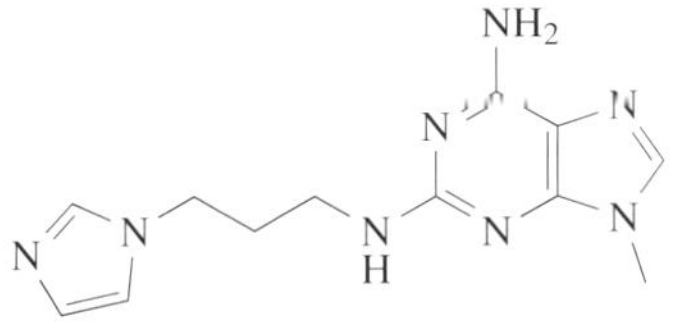

(33)

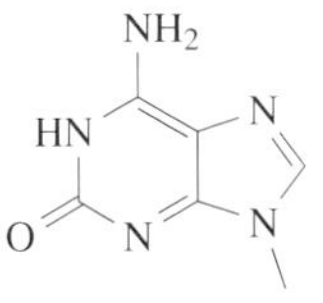

(34)

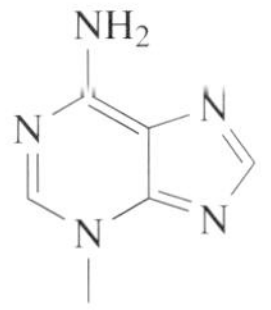

(35)

## 7.09.3 PURINE MODIFICATIONS

The major objective of purine modification in natural ODNs is to improve the efficacy of hybridization to single-stranded nucleic acids. In spite of their synthetic complexities, there is a considerable amount of literature pertaining to the ODNs containing a variety of purine derivatives and analogues. This section is designed to summarize and complement the important developments in the synthesis and pharmacokinetic/pharmacodynamic properties of such modified ODNs.

### 7.09.3.1 Adenine Derivatives

#### 7.09.3.1.1 C-2-Substitution

Oligodeoxynucleotides containing 2-aminoadenine (32a) in place of adenine (A) are expected[80–82] to bind more tightly to their complementary oligonucleotides because of the potential for (32a) to form three hydrogen bonds with T or U bases. Chollet and co-workers[81,83] reported on the synthesis and hybridization properties of ODNs containing (32a) and found that (32a)·T base pairs are intermediate in strength between A·T and G·C base pairs, while Howard and Miles[84] and Lamm *et al.*[85] have shown that the binding enhancement of (32a) over A is greater with RNA than DNA.

The substitution of (32a) for A also resulted in sequence- and target-dependent changes in duplex stability,[86] and facilitated a B → Z transition[80,87,88] or even a B → A transition[89] in DNA. Thermal denaturation studies of short duplexes containing (32a) demonstrated that (32a)·T formed stable base pairs.[90]

Introduction of the N-6-methoxyl function in (32a) (to obtain 2-amino-6-methoxyaminopurine, (32b)) also stabilized the base pair (with T and C) interaction; however, the observed stability was somewhat lower than that with the normal base-pairs A·T and G·C, which has been ascribed *inter alia* to the necessity for the *syn* → *anti* configurational change of the methoxyl group.[91] Appending an imidazolylpropyl group to the C-2-NH$_2$ of (32a) to give (33), within the oligonucleotides enhanced the heteroduplex binding affinity when hybridized to complementary DNA.[92] This enhanced binding affinity may be a result of charge neutralization or the zwitterion effect due to the protonation of the imidazole under assay conditions. Antisense ODNs containing 2-(N-iodoacetylaminoethyl) thioadenine were shown to cross-link and cleave complementary strands at sites of adenine and guanine residues, depending on the torsion angle of the modified base.[93] The alkylating reagent was expected to attack N 3 of adenine from the minor groove, providing the substituted adenine with the *anti*-conformation.

The hydrogen bonding pattern of 2-Oxoadenine (isoG, (34)) contributes considerably to its unique physical and pharmacodynamic properties. For example, poly(isoguanylic acid) forms highly-organized gels in aqueous media,[94,95] as well as in organic solvents of relatively high dielectric constant.[96] Spectroscopic studies showed that the poly(isoG)'s secondary structure (self-associated isoG tetramer) was even stronger than that of polyG helices. A C$_2$-symmetric, hydrogen-bonded tetramer was proposed as the basic unit for the four-stranded poly(isoG) helix.[94] There have been a number of reports from the groups of Benner,[97–100] Dervan,[101] Eckstein,[102,103] Eschenmoser,[104] and Seela[105–107] which describe the incorporation of (34) into ODNs both by chemical and enzymatic procedures, and its pairing with natural, as well as unnatural bases. IsoG forms base pairs with isoC,[108] C-5-methyl-isoC,[109] as well as with T.[97,100] The isoC·isoG base pair has comparable stability to a C–G base pair.[108] However, compared to $(A \cdot T)_6$, the oligomer $(isoG \cdot T)_6$ does not show sigmoidal melting[110] and the $A \cdot U\text{-}isoG \cdot U)_3$ duplex is destabilized in comparison to $(A \cdot U)_6$.[107] Thus, base pairing between isoG and T is comparatively weak. On the other hand, $(isoG \cdot C)_3$ shows a cooperative melting profile ($T_m$ of 32 °C, 1 M NaCl, 100 mM $MgCl_2$), which demonstrates the existence of an isoG·C base pair A in a duplex with parallel chain orientation.[110] The C-5-methyl-isoC·isoG base pair is slightly stronger than C·G base pair (about 2 °C per base pair) and each base can effectively discriminate mismatches.[108,109]

Leonard and co-workers[111] have disclosed the synthesis of a DNA hexamer containing 3-isoA ((35), iA) and found that the oligonucleotide [CG(iA)TCG] forms a well-defined B-type duplex with Hoogsteen-type iA·T base pairs flanked by Watson–Crick pairs. The strength of the iA·T base pair is slightly greater than the normal A·T base pair.

### 7.09.3.1.2   *C-6-Substitution*

Ono and Ueda[69] introduced N-6-methyl-A (36a) in place of A in decamers containing recognition sequences of several endonucleases, employing the solid phase phosphotriester method, albeit in low coupling yields. The results indicate that decamers containing (36a) and N-4-methyl-C (23a) lower the $T_m$ considerably compared with the unmodified decamers. It has been demonstrated[112] that introduction of a single base (36a) in a hexamer also reduces the $T_m$ from 45 °C to 32 °C. Proton NMR studies[113] showed that the normal B-form, right-handed helical structure was retained. It is believed that the N-6-methyl group of (36a) is coplanar with the purine ring, thus its rotation is hindered. Therefore, it is unable to form a Watson–Crick base pair.

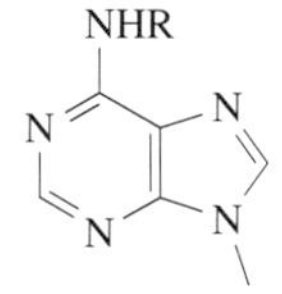

(36a) R = Me
(36b) R = OMe
(36c) R = OH

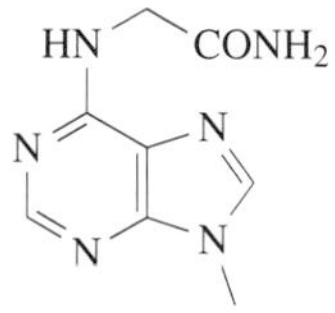

(37)

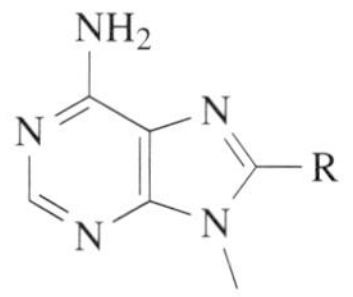

(38a) R = Br
(38b) R = OMe

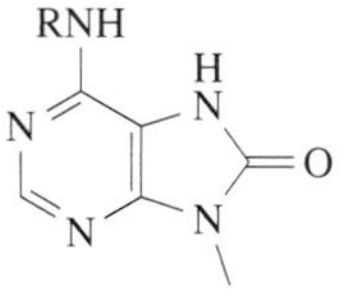

(39a) R = H
(39b) R = Me

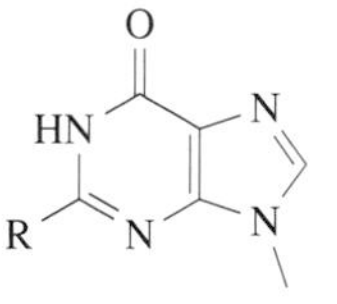

(40a) R = H
(40b) R = F
(40c) R = Spermine

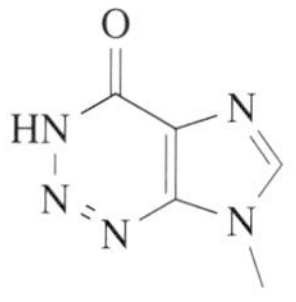

(41)

Ueda and co-workers[114] also synthesized ODNs containing N-6-methoxy-A ((36b), M) using the phosphoramidite methodology. Stabilities of duplexes containing the M·N pairs (N = nucleobases, A, G, T, or C) were studied by thermal denaturation and showed that (36b) formed stable base pairs with A and G, with stabilities similar to those of M·T and M·C pairs. The observed $T_m$ of the duplexes containing the M·N pairs were within the range of 6 °C difference (from 46 °C to 52 °C) which are smaller than those of duplexes containing the mismatches A·N and G·N pairs (11 °C). The duplex stability of M·N base pairs is attributed to the differences in thermal stabilities and stacking abilities of base moiety (36b), as well as its ability to exist in the imino form, where the methoxy group is in the *syn* conformation.[91] Examination of a 15-mer containing one N-6-hydroxy-A ((36c), H) unit[115] revealed that the $T_m$ of the duplex containing the H·T pair ($T_m = 51$ °C) is similar to the $T_m$ of the M·T duplex (50 °C), but lower than the $T_m$ of the A·T duplex (54 °C). This effect may be due to the steric hindrance between the N-7 atom of (36c) and the hydroxyl group. The $T_m$ of the H·C duplex (42 °C) is lower than the $T_m$ of the M·C duplex (46 °C), which may be the result of (**36c**) existing in the amino tautomeric form, which is not suitable for formation of the H · C base pairs.[115] Replacement of adenine base by a N-6-carbamoylmethyladenine residue (**37**) led to a decrease of the $T_m$ values of the oligomers $(A \cdot T)_6$ and ATGAAGCTTCAT.[116] The decrease of the $T_m$ values may be due to the bulkiness of the amino acid substituent which can affect both the association kinetics of the single strands, as well as the helix structure.

### 7.09.3.1.3   *C-8-Substitution*

ODNs containing photoactive congener 8-bromoadenine (**38a**) have been synthesized by Liu and Verdine[117] by employing the solid phase phosphoramidite methodology. The base-labile phenoxyacetyl (PAC) amine-protecting group was used to eliminate possible nucleophilic attack by $NH_3$ to yield 8-amino-A under conditions of elevated temperature or prolonged exposure normally used for the synthesis of ODNs. A 23-mer ODN containing (**38a**) was found not to affect adversely the stability of duplex DNA and function as a protein receptor.[117] Analogues of the $2' \rightarrow 5'$ linked adenylate trimers (p5'A2'p5'A2'p5'A) containing (**38a**), 8-hydroxyadenine (**39a**), and 8-hydroxypropyl-A in the 1st, 2nd, and 3rd nucleotide positions were chemically synthesized and studied by Kanou *et al.*[118,119] for their ability to bind and activate RNase L of mouse L cells. The 8-substituted analogues of $2' \rightarrow 5'$ A were found to be more resistant to degradation by the $(2',5')$phosphodiesterase.

Preparation of various oligomers, replacing A and G bases with (**39a**),[120,121] 8-methoxy-A (**38b**),[121,122] and 8-methoxy-G[121,122] by the H-phosphonate approach, as well as by the phosphoramidite method,[120,123] and the properties of these modified oligomers for recognition by endonucleases at the functional group level have been documented. Octamers with the sequences GGA*ATTCC, GGAA*TTCC, and GG*AATTCC, where A* is (**38a**) or (**38b**) and G* is 8-methoxy-G, exhibited more resistance than natural GGAATTCC to hydrolysis by *Eco*RI. Circular dichroism spectra and $T_m$ values indicate that these modified oligomers are destabilized compared with the unmodified octamer.[122] The detailed analysis of tautomerism, protonation, ionization, and conformation of (**39a**) has been disclosed.[124] Compound (**39a**) exists predominantly in the keto form,[123,124] and the duplex containing (**39a**) conforms to a B-form.[123,125] ODNs containing T and (**39a**) (one and three units) are capable of forming stable triple helices with double-stranded DNA independent of pH.[126-128] The stability of these triplexes is presumed to be due to stacking interactions.[129] $T_m$ and CD studies have also been reported on these triplexes, and it has been shown that 8-oxo-A (**39a**) may be a useful alternative to C or C-5-methyl-C (**26a**) for use in triplex formation at physiological pH.[126,129] Indeed, Matteucci and co-workers[130,131] demonstrated, by DNase I footprinting techniques, stable triple helix formation with an oligonucleotide which contained N-6-methyl-8-oxo-A (**39b**) substituted for C or (**26a**) at physiological pH. Furthermore, compound (**39b**) has been found to be superior to (**26a**) and G in conferring enhanced affinity for duplex DNA under intracellular salt and pH conditions.[131] Thus, ODNs containing (**39b**) have significant promise in the development of therapeutic agents based upon triplex formation *in vivo*.

### 7.09.3.2   Hypoxanthines

ODNs containing hypoxanthine ((**40a**), I) residues have been synthesized both by the phosphotriester[132,133] and phosphoramidite[134,135] methods using a solid support. The thermal stability of ODN duplexes containing (**40a**) residues matched with each of the four major DNA bases.[133] Due to the lack of the C-2-amino group, a I · C base pair is less stable than a G · C base pair.[135] The order of thermodynamic stability at 25 °C for double strand formation is: G · C > A · T > C · G > I · C > I · A > T · A > G · A > I · G > G · G. The observations of Martin *et al.*[136] support the idea that (**40a**) might be an insert or ambiguous base, because it can pair and stack with any of the four major DNA bases at the ambiguous positions, while its matched or mismatched base pairs neither stabilize nor destabilize the DNA duplex.[132,136] A (**40a**) analogue which is more discriminating against mismatches than the normal DNA bases, could be used to increase the selectivity of the probe at positions of unambiguous sequence. The CD spectra of duplexes containing I · C, I · A, G · A, and T · A show an extra transition near 260 nm besides a positive Cotton effect leading to the conclusion that these duplexes have right-handed B-form helices.[133]

2'-Deoxyisoinosine [9-(2-deoxy-*β*-D-*erythro*-pentofuranosyl)purin-2-one, isoI$_d$], which is isomeric with 2'-deoxyinosine has been synthesized[137] and incorporated into ODNs using the phosphoramidite method.[138] One or two isol$_d$ units were introduced into T$_{12}$-mer replacing T either in the middle or at the 3' and 5'-ends. The isol$_d$-containing ODNs were hybridized with a modified A$_{12}$-mer containing A, T, G, C opposite to isol$_d$. The replacement of one isol$_d$ in the center of the duplex reduced the $T_m$ by 15 °C and a decrease of 25 °C was found when two isol$_d$ units were introduced. The isol$_d$ seems to stack in the duplex when T·A base pairs are the neighbors. Incorporation of 2-fluorohypoxanthine moiety (**40b**) into ODNs by the conventional automated solid-phase protocol has been reported.[139,140] The modified ODN containing (**40b**) was further post-synthetically converted to the spermine adduct (**40c**).[140] In low salt concentration, the spermine-containing duplex was found to be 25 °C more stable than the natural one.

An attempt to incorporate 2-azahypoxanthine (**41**) into an oligomer was made,[141] expecting that the acceptor–donor–acceptor configuration of (**41**) would be similar to that of xanthine. However, because (**41**) has a higher p$K_a$ value (6.8), it could stabilize the base pairing. An unexpected but interesting side reaction took place on the deprotection of the oligomer containing protected (**41**) nucleoside, resulting in an AICA riboside, a key intermediate in the *de novo* biosynthesis of purine ribonucleotides.[141]

### 7.09.3.3 Xanthines

Xanthine may be considered as a true universal base, being a purine and large enough to interact with both purines and pyrimidines across the double helix. In addition, it has an acceptor–donor–acceptor hydrogen bonding configuration. The ODNs containing xanthine (**42**) are described by Eritja *et al.*[142] The conventional solid support phosphoramidite procedure was used to prepare the modified oligomers. Nonadecanucleotides containing (**42**) were found to have a low $T_m$ at physiological pH. The observed relative $T_m$ stabilities of base mispairs with (**42**) are: T > G > A ≅ C. The destabilizing effect is surmised to be due to the negative charge present on (**42**) at neutral pH (p$K_a$ 5.3). However, at pH 5.5, xanthine base pairs are more stable.[142]

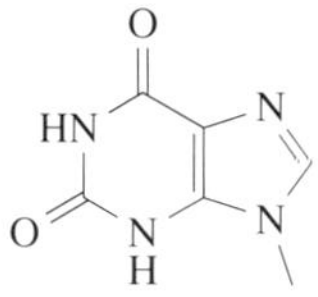

(**42**) 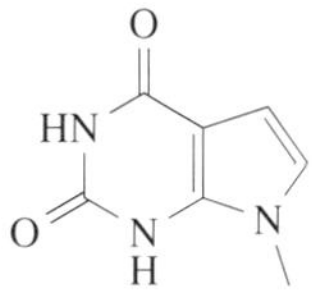          (**43**) 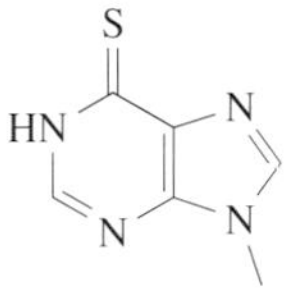          (**44**)

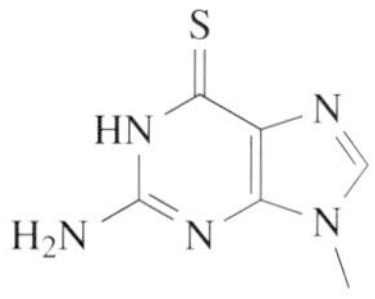

(**45**)

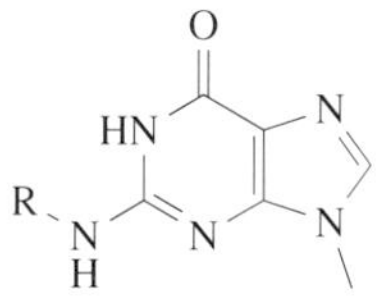

(**46a**) R =

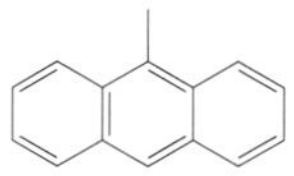

(**46b**) R =

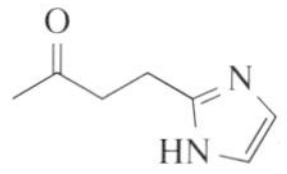

(**46c**) R =

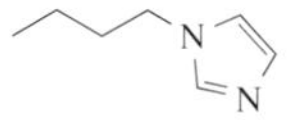

Replacement of the N-7-atom of xanthine with a carbon creates 7-deazaxanthine ($^7$X, (**43**)), which increases the p$K_a$ to 7.3,[143] potentially allowing $^7$X : A : T triple helix formation at physiological pH. Indeed, it was found that ODNs containing (**43**), prepared by standard solid-phase H-phosphonate,[144] as well as phosphoramidite[145] methodology, form triple helices in an antiparallel orientation, with respect to the poly-purine 15-base strand of the target DNA under physiological K$^+$ and Mg$^{2+}$ concentrations and at pH 7.2.[144] It is also found that ODNs containing (**43**) are ineffective in triplex formation above 50 mM K$^+$ concentration.[145] The fact that substitution of $^7$X for T significantly increases the anti-parallel triple helix formation under physiological conditions, may prove useful for the *in vivo* inhibition of gene expression by triple helix formation.[144]

### 7.09.3.4  C-6-Thiopurines

Synthesis of ODNs containing C-6-thiopurine (44) is described by Xu *et al.*[146] A 2,4-dinitrophenyl group was used to protect exocyclic sulfur, which was cleaved from the CPG-support by treatment with concentrated $NH_4OH$ containing mercaptoethanol. Another useful procedure for obtaining dinucleoside phosphates containing (44) is described by Clivio *et al.*[22] who utilized a pivaloyloxymethyl group for thiol protection. The $T_m$ values of oligomers containing (40a), (44) and C-6-thioguanine (45) base paired with either C or T in the complementary strand show that (44) and (45), like (40a) and G, can form better base pairs with C than with T. However, both (44) and (45) form less stable base pairs with C than their oxy analogues (hypoxanthine and guanine), with $T_m$ values of the DNA duplex about 5.2 °C and 6.9 °C lower, respectively, probably because of the larger van der Waals radius of sulfur, relative to oxygen, and its weaker hydrogen bonding capability.[146]

Suitably protected phosphoramidite monomers of C-6-thioguanine (45) for incorporation into ODNs are reported by several groups.[19,135,145,147–154] Protection of the thione function of (45) is of paramount importance in order to prevent oxidative hydrolysis that is expected in sulfur-containing nucleobases. Protection of the thione function with a cyanoethyl group was found to be very practical, since the cyanoethyl group can be readily removed under mild alkaline conditions.[19,148,150,152] Like (40a), C-6-thioguanine containing dodecamer results in the loss of a Watson–Crick hydrogen bonding to its partner base on the complementary strand.[135] The C-6-thioguanine containing ODN has a $T_m$ of 42 °C, 11 °C lower than the 53 °C found for GAC-GATATCGTC. The modified oligomer GAC⁶ˢGATATCGTC exists in B-DNA form and is a poor substrate for the *Eco*RV endonuclease.[152] Site-selective substitution of (45) for G into G-rich triple-helix-forming oligonucleotides (TFO) has been shown to alleviate the guanine quartet-mediated inhibition of triplexes in the presence of physiological levels of $K^+$. This improvement represents a significant step toward creating ODNs that might serve as specific and efficient gene repressors *in vivo*.[145,153,154]

### 7.09.3.5  Guanine Derivatives

#### 7.09.3.5.1  *C-2-Amino-substitution*

Casale and McLaughlin[155] have synthesized site-specifically modified oligodeoxynucleotides containing N-2-(anthracen-9-yl-methyl)guanine ((46a), Gᵃ) and examined the thermal stability of the self-complementary dodecamer (CGCGᵃAATTCGCG)₂ containing two anthracenylmethyl residues at pH 7.0 and 1.0 M sodium chloride solution. No co-operative transition reflecting a well-characterized helix-to-coil transition was obtained. The double-stranded 13-mer containing a single modification with (46a) under the same buffer conditions produced a co-operative helix-to-coil transition with a $T_m$ of 50.6 °C. By comparison, the unmodified 13-mer helix under the same buffer conditions exhibited a $T_m$ of 56.6 °C. From the thermal melting studies it was observed that the introduction of a single base modification into the double-stranded 13-mer not only reduces the $T_m$ but also significantly alters the character of the transition as evidenced by the broadening of the absorbance versus temperature curve.[155] Analysis of both the native and modified 13-mers by CD spectropolarimetry indicated that both sequences adopted an essentially B-form helix. Heeb and Benner[156] prepared ODNs containing G bearing an N-2-(3-imidazolylpropionic acid) ((46b), H), using phosphoramidite methodology. The mixtures of hexadecanucleotide $T_7HT_8$ with the complementary $A_8CA_7$ showed the expected hypochromicity at 260 nm, with the largest effect observed at a 1:1 ratio of the two strands, suggesting that a $T_7HT_8 : A_8CA_7$ pair formed a typical DNA duplex.[156] The $T_m$ (in 100 mM KCl in potassium phosphate buffer, 100 mM, pH 6.0) of the duplex was 37 °C, while the $T_m$ of the duplex between $T_7GT_8$ and $A_8CA_7$ was 43 °C. At 1 M salt concentration, the $T_m$ were 43 °C and 51 °C, respectively. The hybridization of the modified oligonucleotides with the complementary RNA was also assessed.[156] The melting of the duplex between $T_7HT_8$ and ribo$A_{12–18}$ (35 °C) was found to be higher than that between $T_7GT_8$ and ribo$A_{12–18}$. Ramasamy *et al.*[92] have evaluated the effects of N-2-imidazolylpropyl-G containing oligonucleotides on the binding affinity of the modified ODNs when hybridized to complementary DNA or RNA. A 21-mer oligonucleotide containing 1, 3, or 7 modifications (**46c**) was prepared and hybridized to complementary DNA or RNA. Compared to unmodified DNA, the average increase in $T_m$ per modification was 2 °C and 0.3 °C for DNA and RNA, respectively. The relative specificity of hybridization of G or N-2-modified G to C vs. A, G, and U (T) mismatches on an RNA or DNA complement showed that the modified G is more specific to its complementary C. It has also been found that incorporation of three N-2-imidazolylpropyl-G units at the 3′-end of a 15-mer provided an increase in stability to nucleolytic degradation in fetal calf serum compared to the unmodified oligomer.[92]

### 7.09.3.5.2   *C-6-Substitution*

The synthesis of ODNs containing C-6-O-alkyl-G[17,18,24,157,158] has been described and their hybridization properties have been evaluated. These modified ODNs were prepared by either phosphoramidite or phosphotriester methods. C-6-O-Methyl-G (**47a**) was found to pair with T; however, the melting of the oligomers containing (**47a**) was found to extend over a wider temperature range than that of the parent sequences.[18] The melting of these modified ODNs was found to be biphasic. In addition the hypochromicity of the sequences containing C-6-O-methyl-G was less than that of the unalkylated sequences. The lower and extended melting temperature of the modified ODNs suggest that the presence of alkylated base destabilizes the duplex, and the lower total hypochromicity suggests that the presence of the methylated base decreases the overall extent of base stacking.

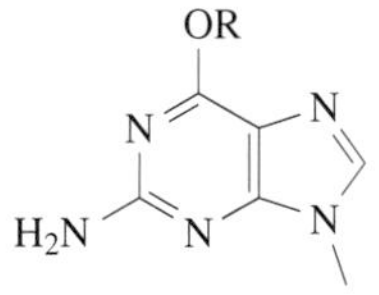

(**47a**) R = Me
(**47b**) R = Bu
(**47c**) R = $C_{16}H_{33}$

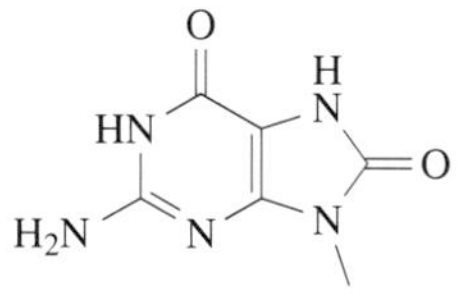

(**48**)

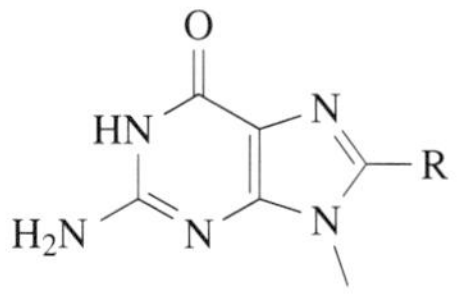

(**49a**) R = $NH_2$
(**49b**) R = OMe

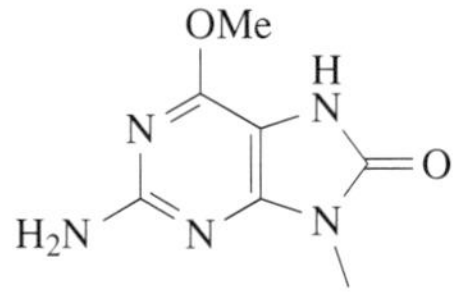

(**50**)

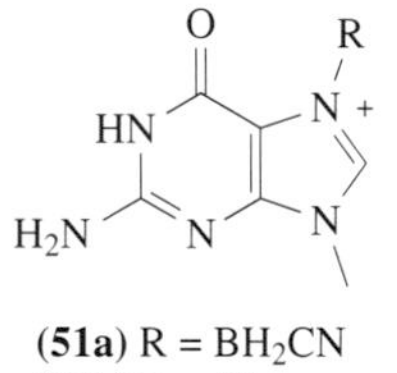

(**51a**) R = $BH_2CN$
(**51b**) R = Me

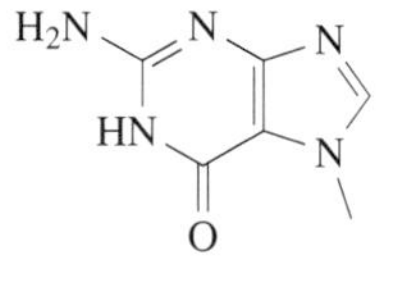

(**52**)

### 7.09.3.5.3   *C-8-Substitution*

Synthesis of oligonucleotides carrying 7-hydro-8-oxo-G (**48**) was carried out by solid-phase phosphoramidite methodology.[159–163] NMR studies[162,163] show that 7-hydro-8-oxo-G exists as two tautomers, or gives rise to two DNA conformations, that are in equilibrium and that interchange slowly. These structural studies demonstrate that 8-oxo-7*H*-G(*syn*) : A(*anti*) forms a stable pair in the interior of the helix. The NMR data also indicate that the 7-hydro-8-oxo-G base takes a 6,8-diketo tautomeric form and is base-paired to C with Watson–Crick type hydrogen bonds in a B-form structure.[162] The thermal stability of the modified DNA was found to be reduced, but the overall structure was found to be the same as that of the unmodified duplex. The destabilization of the duplex by hydroxylation of the C-8-position may be due to steric hindrance between the C-8-substituent and the sugar. 8-Amino-G (**49a**) has been incorporated into oligonucleotides via solid-phase phosphoramidite methodology.[164] Takaku and co-workers[122] have synthesized oligodeoxynucleotides containing C-8-methoxy-G (**49b**) and studied their hybridization properties. The incorporation of (**49b**) into ODN was found to reduce the $T_m$ by 3.3 °C per substitution.

### 7.09.3.5.4   6,8-Disubstitution

Johnson and co-workers[165] have synthesized oligonucleotides containing C-8-oxo-7,8-dihydro-6-O-methyl-G (**50**) using phosphoramidite methodology. The activity of MutY protein toward this modified ODN and a related oligomer containing C-8-methoxy-G has been studied.

### 7.09.3.5.5   N-7-Substitution

Shaw and co-workers[166–168] reported the synthesis of oligonucleotides containing N-7-cyano-borane G (**51a**), enzymatically via triphosphates. UV melting curves were used to determine the $T_m$ values. All UV melting profiles obtained with unmodified and boron-containing hybridized duplexes were found to be monophasic, consistent with a two-state model of DNA melting, and demonstrate reversible melting transitions. The melting curves of boron-containing and normal duplexes were found to be quite comparable, demonstrating that the unmodified and boron-containing duplexes melt with similar degrees of cooperativity. The boron-containing hybrid duplexes have similar $T_m$ values relative to the unmodified duplexes ($-0.7\,^\circ$C per substitution). The synthesis of dinucleotides containing N-7-cyanoborane-G at the 3′, 5′, or both positions of the phosphodiester linkage has been reported using solution phase phosphoramidite methodology[169] and the effect of cyanoborane substitution on the base protons of dinucleotides was evaluated. Ezaz-Nikpay and Verdine[170] have prepared the Dickerson/Drew dodecamer containing N-7-methyl-G (**51b**) via enzymatic pathway and studied the stability of the duplexes by thermal denaturation experiments. The duplex half-melting transition ($T_m$) of the Dickerson/Drew dodecamer was found to be lowered ($-4.5\,^\circ$C per substitution) by the presence of N-7-methyl-G. However, analysis in terms of free energy, enthalpy, and entropy revealed that, at room temperature, duplex stability in 5′-CGCGAATTCGCG is indistinguishable from that in CGCGAATTCGCG. Moreover, the enthalpy of the duplex formation was made significantly more favorable by the presence of N-7-methyl-G, suggesting that the N-7-methyl-G·C base pair is more stable than G·C base pair.[170] The small magnitude of the observed thermodynamic effects suggests that (**51b**) does not significantly perturb duplex DNA structure. The synthesis of the oligonucleotides containing 7-(2-deoxy-$\beta$-D-*erythro*-pentofuranosyl)guanine ((**52**), $^7$NG) has been reported[164,171,172] and its hybridization properties were evaluated with respect to triple helix formation. In the antiparallel triplex mode containing G:GC and T:AT triplets, incorporation of (**52**) into ODNs drastically reduced the binding affinity.[164] However, in the parallel triple helix mode, $^7$NG mimics protonated cytosine and when incorporated into pyrimidine strand, within pyrimidine:purine:pyrimidine triple helix, binds with remarkable specificity to the Watson–Crick guanine·cytosine base pair by triple helix formation.[171,172] The stabilities of triple helical complexes containing $^7$NG were found to decrease in the order $^7$NG:GC >> $^7$NG:CG > $^7$NG:AT, $^7$NG:TA. However, it was found that within the measured pH range (7.0–7.5) the stabilities of $^7$NG containing triple helical structures were strongly dependent on the sequence context and independent from the pH.[172] A third strand oligonucleotide composed of contiguous $^7$NG provides a conveniently accessible solution for targeting contiguous guanosine residues.

## 7.09.4   PURINE ANALOGUES

### 7.09.4.1   1-Deazapurines

Any modification at the N-1 site of purines can be expected to decrease the $T_m$ with complementary nucleobases; therefore, it could have a very destabilizing effect for an antisense ODN. For example, poly 1-deaza-A (**53**), in which N-1 is changed to CH, forms only a Hoogsteen or reverse Hoogsteen, instead of the usual Watson–Crick type 1:1 complex with polyU, which has very low $T_m$ even at high Na$^+$ concentrations.[173,174]

### 7.09.4.2   3-Deazapurines

An improved procedure for the preparation of 2′-deoxy-3-deazaadenosine and its phosphoramidite for the incorporation of 3-deaza-A (**54**) residue into specific sites in oligonucleotides is

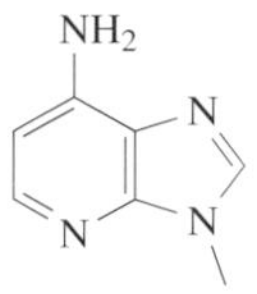

(53) 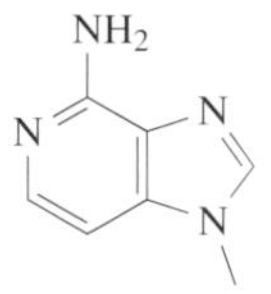                    (54) 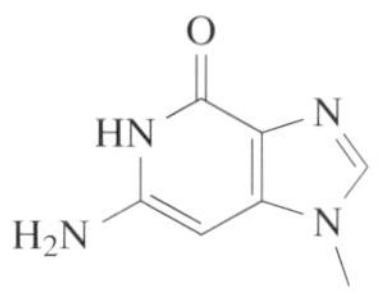                    (55)

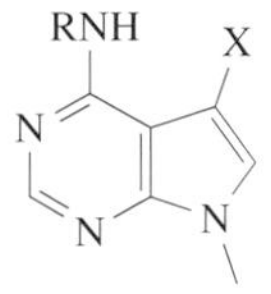

(56a) R = X = H
(56b) R = H; X = halogen or Me
(56c) R = Me; X = H

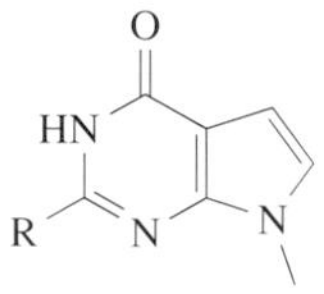

(57a) R = H
(57b) R = NH₂

described by Cosstick *et al.*[175] It is reasoned that the base (54) is able to maintain the Watson–Crick hydrogen bonding pattern but it lacks an essential hydrogen bond acceptor (N-3 of A) from the minor groove. Such a modification is valuable for the study of protein–nucleic acid interactions. The $T_m$ values of dodecamers containing a single base moiety (54) placed in the middle of the sequence show that (54) has no destabilizing effect on the duplex compared with an A in the same position of the sequence.[175] However, the $T_m$ of ODNs containing (54), prepared either by the phosphotriester method[176] or by the phosphonate method,[177] was lower than that of the parent oligomer, probably due to the lower stacking ability of the imidazo[4,5-*c*]pyridine ring system than that of the purine system.[176] Ikehara *et al.*[174] reported high $T_m$ values ($> 80\,°C$) for poly 3-deaza-A : poly-U, which forms a 1:1 or 1:2 complex either in the absence or presence of $Mg^{2+}$. The high $T_m$ is interpreted in terms of the high basicity of 3-deazaadenine ($pK_a$ 6.80, whereas $pK_a$ of A is 3.62).

The first synthesis of an oligonucleotide containing 3-deazaguanine (55) was reported by Acevedo *et al.*[178,179] Phosphoramidite of an imidazole-deoxynucleoside was incorporated into an oligomer, which during the ammonium hydroxide treatment was cyclized to furnish the 3-deazaguanine residue. The ODNs containing single base (55) in the middle of an 18-mer[178] or a dodecamer[135] destabilized the duplex by about 2.9 °C and 6 °C, respectively, compared with natural G at the same site. The lowered $T_m$ was perhaps caused by alterations to base stacking.[135] However, 3-deaza-G maintains the specificity to C vs. A, U, and G mismatches. Seela and Lampe[180] reported an alternative synthesis by employing H-phosphonate chemistry of ODNs containing (55). The self-complementary oligomers were characterized by enzymatic degradation, and the CD spectra support a B-DNA structure.[180]

### 7.09.4.3    7-Deazapurines

Several groups have synthesized a series of phosphoramidite[134,135,149,181–195] and phosphonate[144,177,195] monomers of 2′-deoxy-7-deazapurine nucleosides suitable for incorporation into ODNs. Extensive studies of hybridization properties of these modified ODNs have been reported. Also reported are the studies on nuclease digestion, DNA conformation, and DNA recognition by proteins of ODNs containing 7-deazapurine bases.[196] Replacement of the adenine by the 7-deazaadenine moiety ((56a), c⁷A) results in duplex stabilization of the oligomers $(A \cdot T)_3$ and $(A \cdot T)_6$. This stabilization is demonstrated[184] by an increased $T_m$ and a decreased susceptibility of $(c^7A \cdot T)_3$ and $(c^7A \cdot T)_6$ toward nuclease S1. Decanucleotide duplexes $[(GGCA_6C) \cdot (CGGT_6G)]$ containing various numbers of (56a) in place of A have been prepared.[191] A decreased stability (decrease of $T_m$) of the oligomers with increasing number of (56a) has been observed.[197,198] The difference between A and (56a) is the lack of a proton acceptor site at N-7, as well as an altered π-electron system, which may lead to an altered helix pitch and subsequent destabilization. Also, the major groove of such oligomers is hydrophobized.[199] However, single replacement of one of the A residues by (56a) within the recognition sequence of endonuclease decreased the cleavage activity but retained the duplex specificity.[188–191]

Incorporation of 7-halo (Cl, Br, or I) or 7-methyl-substituted 7-deazaadenine (56b) into ODNs leads to significantly enhanced duplex stabilities compared to the corresponding A or c$^7$A-containing oligomers.[195] The synthesis and properties of ODNs containing C-6-N-methyl-7-deazaadenine (56c) have also been studied by Seela and co-workers.[192] Alternating or palindromic oligonucleotides derived from (A : T)$_6$ or (ATGCAGA*TCTGCA), but containing one (56c) moiety in place of a C-6-N-methyladenine ((36a), A*) were synthesized.[192] Melting temperature experiments showed that the lowering of $T_m$ values was induced by an N-6-methyl group of (56c).

Incorporation of 7-deazahypoxanthine ((57a), c$^7$I) into an alternating GC sequence by the phosphoramidite methodology has been disclosed by Seela and co-workers.[134,187] It is observed that the $T_m$ of the hexamer [GCc$^7$ICGC]$_2$ is lower (23 °C) than that of [GCICGC]$_2$ (27 °C), where I is hypoxanthine (40a). It has been suggested[134] that the low $T_m$ value for the duplex containing (57a) is due to the hydrophobic nature (N-7 → CH) of the modification, which may result in a decreased amount of bound water.

7-Deazaguanine (57b) has been incorporated into several hexamers using phosphoramidite chemistry.[182,185,186,193] Due to the altered π-electron systems of the 7-deazapurine ring system, which affects base stacking and hydrogen bonding pattern, the $T_m$ of the modified duplex is decreased by 10 °C compared to that of the unmodified purine hexamer.[182,186] The hydrogen bonding pattern of (57b) is a consequence of p$K_a$ value. The p$K_a$ of 7-deazaguanine is 10.3, whereas that of guanine is 9.3, indicating (57b) is a less efficient hydrogen donor at N-3 compared with guanine. This would result in a destabilization of the Watson–Crick base-pairing between (57b) and C.[182] However, when (57b) is incorporated into the self-complementary dodecamer GACG*ATATCGTC within the GATATC EcoRV recognition site, it showed no disruption of Watson–Crick hydrogen bonds and virtually identical $T_m$ (52 °C) compared to the complementary strand (53 °C).[135] The substitution of (57b) for guanine substantially decreased the ability of a 15-mer ODN to form triple helices under physiological conditions.[144] The CD-spectra of ODNs containing (57b) in place of G showed that the modified oligomer retained B-conformation.[193]

Hexa- and dodecanucleotides containing C-6-O-methyl-7-deazaguanine (58) have been prepared by solid support synthesis employing methyl as well as cyanoethyl 2′-deoxynucleoside phosphoramidites of (58).[200] Replacement of the G residue within the oligomer CGCGAATTCGCG next to A increased hairpin formation, whereas stable duplexes were formed if one of the outer G moieties was replaced by (58). The preparation of suitably protected phosphoramidite building blocks of 7-deaza-6-thioguanine (59) is reported by Rao et al.[201] Insertion of (59) into G-rich ODNs did not show any enhancement in triple helix formation with the complementary duplex. Thus, in general almost all 7-deazapurine modifications reported so far have destabilizing effects compared with unmodified purines.

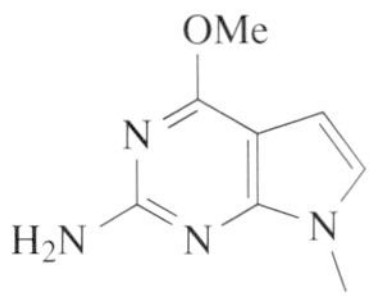

(58) 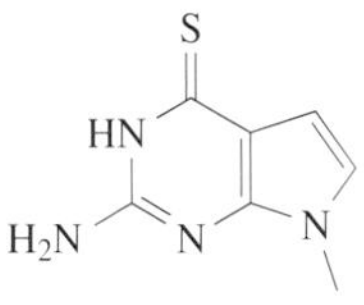  (59) 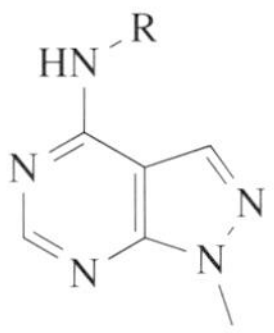  (60a) R = H
(60b) R = Me

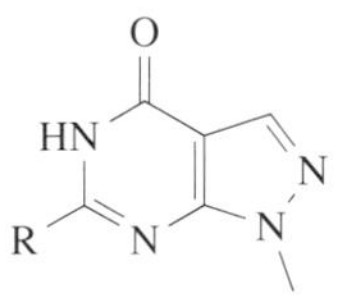

(61a) R = H
(61b) R = NH$_2$  (62a) R = Me
(62b) R = H  (63)

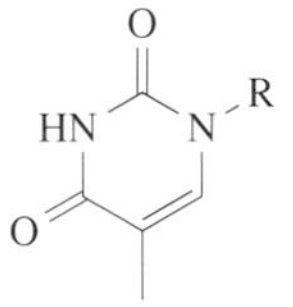

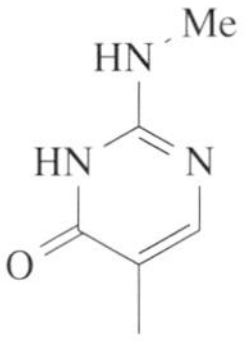

### 7.09.4.4  7-Deaza-8-azapurines

Oligodeoxyribonucleotides with alternating 7-deaza-8-azaadenine (60a) and T residues have been prepared by solid-phase synthesis using phosphoramidite chemistry.[202] Palindromic oligomers derived from CTGGATCCAG containing (60a) in place of A have been generated. Alternating oligomers containing (60a) showed increased $T_m$ values compared to those with A. This phenomenon was attributed to stronger stacking interactions.[202] The different π-electron distribution in (60a) compared to A, which results in altered dipole moments, changes this behavior resulting in a stronger stacking interaction.

The preparation and properties of ODNs containing 7-deaza-8-aza-N-6-methyladenine (60b) have been investigated by Seela *et al.*[192] Melting temperature experiments showed that the lowering of $T_m$ values induced by the N-6-methyl group of (36a) is reversed by the incorporation of (60b). Thermodynamic data of the hexamers GCI*CGC (I* is (61a)), prepared by the phosphoramidite procedure, showed that allopurinol (61a) destabilizes such duplexes (30 °C) less strongly than the parent hypoxanthine (27 °C).[134] This observation demonstrates that allopurinol may be a useful synthon for the preparation of ODNs employed as hybridization probes. Similarly, the hexamers containing 7-deaza-8-azaguanine (61b), prepared by phosphoramidite methodology,[193] showed an increased $T_m$ value (62 °C) compared to the parent hexamer ($T_m$ of 46 °C). The base pair of 7-deaza-8-azaguanine · cytosine which stabilizes the duplex structure exhibits a $\Delta H = -70$ kJ mol$^{-1}$ of stack. It is assumed that the increased stability is due to a better p-electron overlap of the pyrazolo[3,4-*d*]pyrimidine base probably accompanied by an altered helix geometry.[193] It has been shown by CD spectra that the oligomers containing (61b) retain B-conformation.[193]

The octanucleotides derived from GGAATTCC but containing one or two units of (61b) instead of G have been prepared using P$^{III}$ chemistry.[203] These modified ODNs showed increased $T_m$ values (33–37 °C) compared to the unmodified parent oligomer (30 °C). A novel biotinylated adenylate analogue derived from pyrazolo[3,4-*d*]pyrimidine was incorporated into DNA probes and shown to hybridize to complementary targets.[204,205] These modified DNA probes do not cause the loss of hybridization efficiency.[204]

On the basis of the foregoing data, it may be concluded that 7-deaza-8-azapurines, in general, stabilize the duplexes and may be useful in antisense or antigene oligomers.

## 7.09.5   C-LINKED NUCLEOSIDES

### 7.09.5.1   Monocyclic

Although incorporation of 2′-*O*-methylpseudoisocytidine[206,207] or the corresponding pyrazine analogue[208] into ODNs is shown to form stable triple helices at neutral pH, G-rich triplex forming ODNs containing pseudo-T (ψT, (62a)) in place of T did not improve the stability of triple helices in either parallel and antiparallel motifs.[209] In parallel motif, a gradual decrease in the triplex melting transition with increasing number of (62a) residues in the oligomers was noted.[209] Similarly, a large depression of $T_m$ was found for duplexes containing (62a).[210] This is of particular interest because (62a) is isosteric as well as isoelectric to T, but somehow the replacement of a C—N glycosyl bond with a C—C linkage is influencing the conformation of oligonucleotides. When 2′-deoxypseudo-uridine (ψU, (62b)) was incorporated into the second strand of a duplex it gave better stability to the triplex A : ψUC as compared to 2′-deoxyuridine triplex A : UC.[211]

Bhattacharya and Revankar[212] have prepared 2-N-methyl-2′-deoxypseudoisocytidine (63) and incorporated it into an octamer via the solid support phosphoramidite method. No hybridization property of the ODN is disclosed.

It has been shown by Solomon and Hopkins[213] that the pyridone C-nucleoside (64) forms stable base pairs opposite 2-aminopurine (2AP). Circular dichroism spectra of this duplex corresponds to those of B-DNA. The base pairing of (64) to 2AP is preferred over pairing to A, C, G, or T. The thermodynamic stability of (64) · AP is less than the corresponding A · T pair.[213] Insertion of nonpolar isosteres of T ((65a) and (65b)) into synthetic ODNs and their pairing properties have been studied by Schweitzer and Kool.[214] It was found that the hydrophobic, non-hydrogen-bonding base analogues (65a) and (65b) are nonselective in pairing with the four natural nucleobases, but selective for pairing with each other rather than with the natural bases. Compound (65a) selectively pairs with itself and the magnitude of this selectivity is found to be 6.5–9.3 °C in $T_m$ or 1.5–1.8 kcal mol$^{-1}$ in free energy (25 °C). When placed at the end of a duplex, a (65a) · (65a) pair is more stabilizing than a T · A pair, and when situated internally, the affinity of this pair is the same as, or slightly better than, the analogous T · T mismatch pair.[214] This study raises the interesting possibility that hydrogen bonds may not always be required for the formation of stable duplex DNA-like structure.

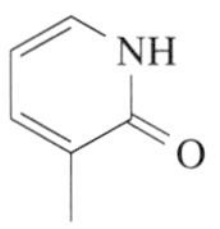

(64) 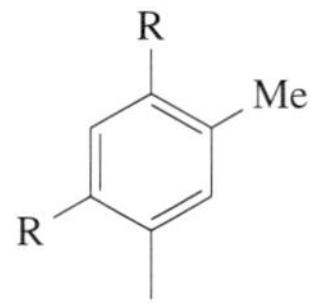     (65a) R = F     (66) 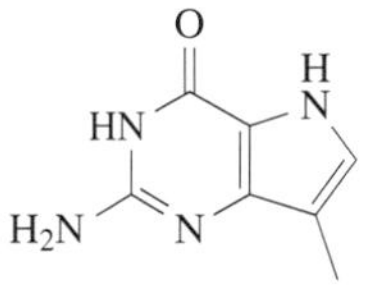     (67) 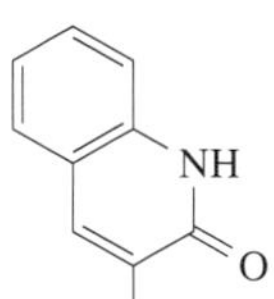     (68)
         (65b) R = Me

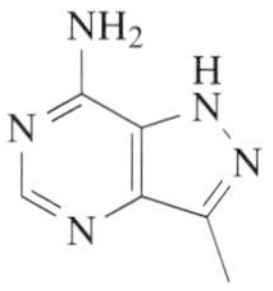

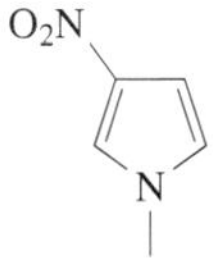

(69) 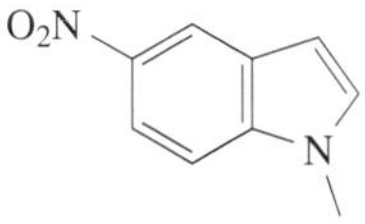     (70) 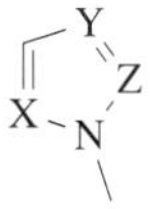     (71a) X = Y = CH; Z = N     (72)
                  (71b) X = Z = CH; Y = N
                  (71c) X = Y = N; Z = CH

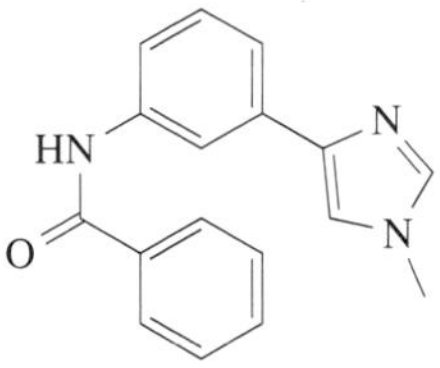

## 7.09.5.2 Bicyclic

2′-Deoxyformycin ((66), F) phosphoramidite was site-specifically incorporated into synthetic DNA via automated solid-phase method.[215] Such modified DNA duplexes containing isolated F·X base pairs (X = A, G, T, or C) are shown to have similar stability to their A·X counterparts. Furthermore A to F substitution into G·A mismatch causes minimal disruption in duplex stability, suggesting that the DNA repair enzyme MutY which recognizes G·A mismatches may also recognize G·F base pairs in DNA.[215] Insertion of three units of (66) into triple helix forming 36-base long ODN (TFO) at CG inversion sites increases the binding affinity tenfold compared to its unmodified counterpart.[216] This significant enhancement of binding affinity resulting from the substitution of (66) into TFOs provides a method to accommodate CG inversion sites within target sites for antiparallel triplex formation.[216] However, ODNs containing 9-deaza-G (67) residues (a 15-mer containing four or 11 units of (67)) reduce triple helix formation but are effective in preventing the G-tetrad formation in G-rich oligonucleotides.[201,217]

C-Linked 2′-deoxynucleoside of quinolin-2-one ((68), Q) was incorporated into synthetic oligonucleotides using automated methodology.[213] Like pyridone C-nucleoside (64), when placed in the middle of the 13-mer synthetic DNA, quinolin-2-one (68) also forms stable base pairs opposite 2-aminopurine (2AP) and this pairing is preferred over pairing to A, G, C, or T. The thermodynamic stability of Q·AP is comparable to that of an A·T base pair.[213]

## 7.09.6 MISCELLANEOUS

### 7.09.6.1 Azoles

2′-Deoxyribonucleoside of 3-nitropyrrole ((69), M) is designed[218] to function as a universal base and is incorporated into ODNs by the conventional phosphoramidite protocol.[219] ODNs containing (69) at several sites were used as primers for sequencing and the polymerase chain reaction. $T_m$ values show that the duplexes formed on hybridization of the sequences 5′-$C_2T_5MT_5G_2$-3′ and 5′-$C_2A_5XA_5G_2$-3′, where X is A, C, G, or T, melted at lower temperature than the corresponding duplexes containing only A·T and C·G base pairs, but showing little variation among different X bases ($T_m$ range 3 °C). The $T_m$ values of sequences containing 5-nitroindole (70) are higher than the corresponding sequences containing 3-nitropyrrole.[220] It has been implicated that the stacking forces are the major contributing factor to the stabilization of duplexes by nitro substituted azoles with an anticipated sacrifice in hydrogen bonding potential.

The ability of certain azole substituted ODNs to promote antiparallel triple helix formation with a duplex target having CG or TA interruptions in the otherwise homopurine sequence is examined by Revankar and co-workers.[221] Pyrazole (71a), imidazole (71b), and 1,2,4-triazole (71c) were inserted into ODNs using phosphoramidite protocol, designed to interact with the nonhomopurine duplex targets. The modified ODNs containing these azoles enhanced the triple helix forming ability considerably and binding to duplex targets containing TA inversion sites is especially noteworthy. The potential for these azoles to maintain some degree of stacking interaction with neighboring base triplets seems likely to contribute to the observed association properties. In addition, the small size of azoles (71a)–(71c) relative to natural bases may reduce or eliminate unfavorable steric hindrance when the modified TFO is bound to the duplex having CG or TA base pairs.[221] The base 4-(3-benzamidophenyl)imidazole ((72), D) binds selectively to TA and CG Watson–Crick base pairs within a pyrimidine–purine–pyrimidine triple helix.[222] From affinity-cleaving analysis, the stabilities of base triplets is shown to decrease in the order: $D \cdot TA \sim D \cdot CG > D \cdot AT > D \cdot GC$. The above studies may provide structural insights into the design of ODNs containing nonnatural azoles with specific recognition properties directed toward a general solution to sequence-specific recognition of double-helical DNA.[222]

### 7.09.6.2  Polycyclic Heterocycles

The desire to enhance stacking interactions/duplex stability has led to a series of pyrimidine modifications that proved very fruitful. An analogue of C, pyrido[2,3-*d*]pyrimidine (73) 2'-deoxy-ribonucleoside, first reported by Inoue *et al.*,[223] has been incorporated into a dodecamer by the phosphotriester method. The thermal stability, CD spectra, and fluorescent properties of the modified oligomer were determined. It has been shown by these authors[223] that (73) forms a stable Watson–Crick base pair with G (by about 8 °C) and a less stable wobble base pair with A within double-helical DNA. The present authors[224] and others[225] have also shown that (73) forms Hoogsteen hydrogen bonding with AT base pair in the parallel as well as antiparallel motif. The presence of multiple hydrogen-bonding groups on (73) and its extended ring system may afford increased stacking interactions with neighboring bases in the third strand of the triplex. In a related binding experiment with ODNs containing a quinazoline-2,4(3*H*)-dione moiety (74),[226] it is shown that the replacement of T with (74) decreases the stability of triple helices in both parallel and antiparallel motifs.

Like the pyridopyrimidine (73), a highly fluorescent guanine analogue 3-methylisoxanthopterin (75) was site-specifically inserted into the double-stranded ODN substrate using the phosphoramidite method.[227] Comparison of the $T_m$ values of (75) containing double-stranded ODN with the $T_m$ of the complementary ODN containing mismatched base pair at equivalent positions show that the modified ODN depresses the $T_m$ to the same degree as a single base pair mismatch.

The synthesis of 6*H*,8*H*-3,4-dihydropyrimido[4,5-*c*]oxazin-7-one (76) and its insertion by established phosphoramidite and H-phosphonate chemistry into ODNs is described by Lin and Brown.[65,228] The $T_m$ values of a range of heptadecamer duplexes containing (76) $\cdot$ A and (76) $\cdot$ G base pairs are compared with the corresponding ones containing N-4-methoxy-C (25a). ODNs containing one (76) residue have lower $T_m$ values by 6–7 °C compared to the complementary strand. Duplex stability becomes lower with increasing number of (76) residues.[65] Proton NMR studies of ODNs carrying (25a) or (76) paired with A and G reveal that the Watson–Crick and wobble base pairs are in slow exchange on the chemical shift timescale.[67]

Since stacking interactions of aromatic heterocycles contribute considerably to the helix stability of DNA and RNA, Matteucci and co-workers[229,230] have synthesized several tricyclic analogues of C and incorporated them into ODNs. The hybridization properties of 15-mer ODNs containing phenoxazine (77a) and phenothiazine (77b) were determined and shown to hybridize specifically with a complementary G. Phenoxazine and phenothiazine in ODNs behave as C analogues with $\Delta T_m$ values of 12 °C and 14.5 °C, respectively, between the G and A targets. Both (77a) and (77b) show higher thermal stability of the resulting duplex for the G target relative to 5-methyl-C (26a).[229] The carbazole analogue (78) base pairs specifically to G.[230]

The successful incorporation of the base labile malondialdehyde-G adduct {pyrimido[1,2-*a*]purin-10(3*H*)-one, (79)} into ODNs has been reported.[231] No structural and functional effects of these modified oligomers are acknowledged. Similarly, the base labile 1,N-6-etheno-A (80) has also been inserted into a 25-base long oligomer using phosphoramidite methodology.[232] A novel DNA base

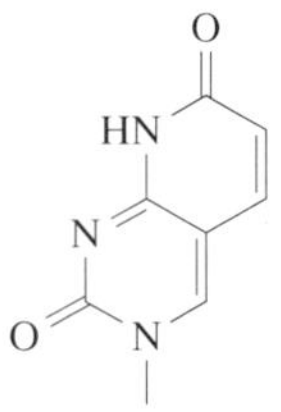

(73) 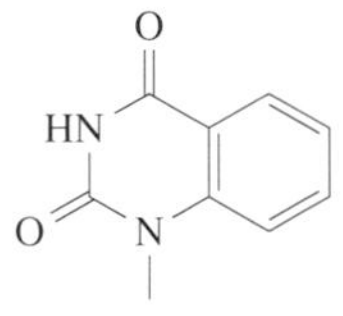　　　　　　　(74) 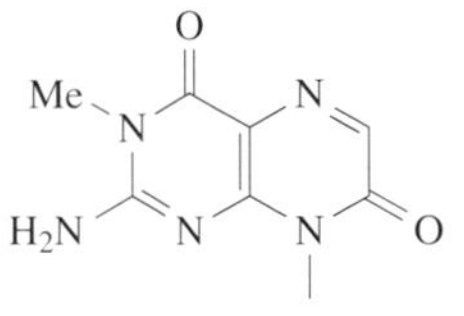　　　　　　　(75)

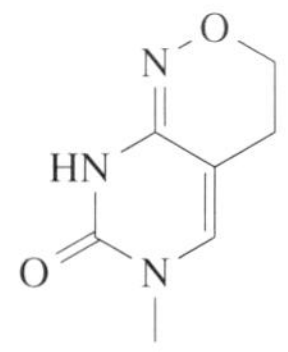

(76) 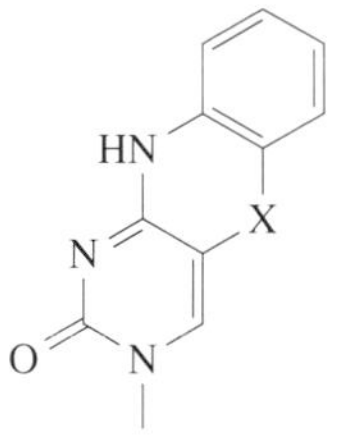　　　　(77a) X = O　　　　(78)
　　　　　　(77b) X = S

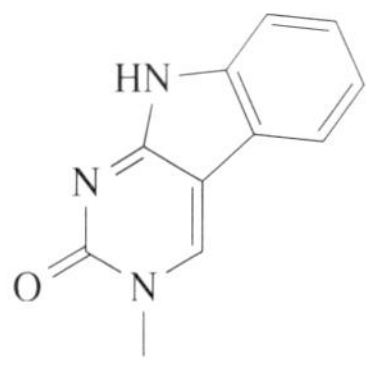

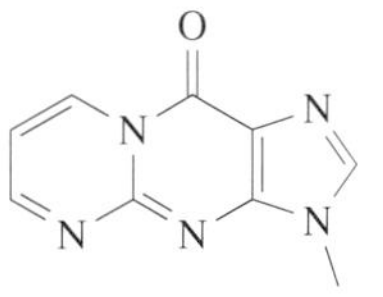

(79) 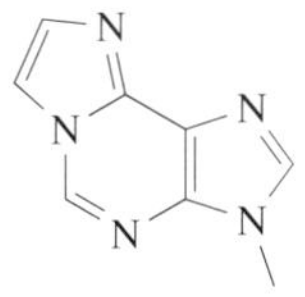　　　　　　　(80) 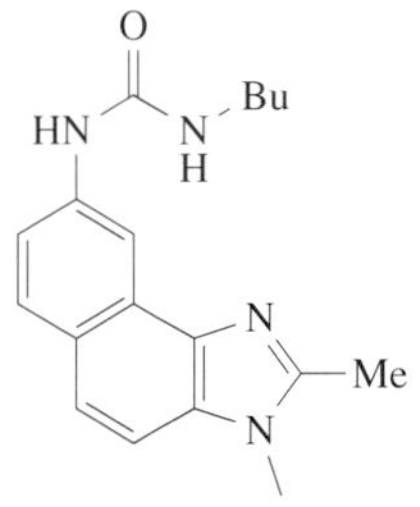　　　　　　　(81)

2-methyl-8-(*N*-butylureido)naphth[1,2-*d*]imidazole (**81**) is shown to bind to a CG base pair in chloroform solution through the formation of simultaneous hydrogen bonds to both nucleobases.[233] This is probably the first reported example of an isolated nonnatural base-triplet in an organic solvent. This type of study allows the quality of the hydrogen bonding interactions to be evaluated without restrictions from base stacking or spatial constraints from preferred backbone conformations.

## 7.09.7　CONCLUDING REMARKS

From the foregoing discussion it is clear that site-specific incorporation of new derivatives and analogues of purine and pyrimidine bases into ODNs can enhance hybridization properties as measured by the $T_m$. Improved methodologies have been developed to prepare modified nucleotide monomers that are compatible with automated DNA synthesizers. Incorporation of such modified nucleobases into ODNs has contributed to a better understanding of the dynamics and thermal stability of DNA duplexes, as well as triplexes in solution and their recognition by proteins.

Further, structure–activity relationship studies will provide impetus to the discovery and development of oligonucleotide based therapeutics. Application of computer-aided molecular modeling techniques will help to design novel heterocycles that recognize base pair sites (i.e., hydrogen bonding and stacking interactions between adjacent bases) with higher affinity, selectivity, and sequence-specificity. Use of base-modified ODNs to generate chimeras that may have stability against nucleases, enhanced hydrophobicity and lipophilicity to facilitate permeability, specificity in target recognition, reduced cytotoxicity, and ability to turn off the gene expression should create a fruitful area of medicinal chemistry of considerable importance over the next few decades.

## ACKNOWLEDGMENTS

We wish to thank Dr Huynh Vu for helpful scientific discussions and Ms Jodie Schrier for expert technical assistance.

## 7.09.8   REFERENCES

1. O. S. Bhanot, P. C. Grevatt, J. M. Donahue, C. N. Gabrielides, and J. J. Solomon, *Nucleic Acids Res.*, 1992, **20**, 587.
2. Y.-Z. Xu and P. F. Swann, *Tetrahedron Lett.*, 1994, **35**, 303.
3. S. B. Rajur and L. W. McLaughlin, *Tetrahedron Lett.*, 1992, **33**, 6081.
4. B. A. Connolly and P. C. Newman, *Nucleic Acids Res.*, 1989, **17**, 4957.
5. P. C. Newman, V. U. Nwosu, D. M. Williams, R. Cosstick, F. Seela, and B. A. Connolly, *Biochemistry*, 1990, **29**, 9891.
6. R. G. Kuimelis and K. P. Nambiar, *Nucleic Acids Res.*, 1994, **22**, 1429.
7. I. V. Kutyavin, R. L. Rhinehart, E. A. Lukhtanov, V. V. Gorn, R. B. Meyer, Jr., and H. B. Gamper, Jr, *Biochemistry*, 1996, **35**, 11 170.
8. T. Ishikawa, F. Yoneda, K. Tanaka, and K. Fuji, *Bioorg. Med. Chem. Lett.*, 1991, **1**, 523.
9. O. S. Bhanot, P. C. Grevatt, J. M. Donahue, C. N. Gabrielides, and J. J. Solomon, *Biochemistry*, 1990, **29**, 10 357.
10. H. Borowy-Borowski and R. W. Chambers, *Biochemistry*, 1989, **28**, 1471.
11. H. C. P. F. Roelen, H. F. Brugghe, H. van den Elst, G. A. van der Marel, and J. H. van Boom, *Recl. Trav. Chim. Pays-Bas*, 1992, **111**, 99.
12. R. Eritja, J. Robles, A. Avino, R. Albericio, and E. Pedroso, *Tetrahedron*, 1992, **20**, 4171.
13. Y.-Z. Xu and P. F. Swann, *Nucleic Acids Res.*, 1990, **18**, 4061.
14. E. Pedroso, D. Fernandez, Y. Palom, and R. Eritja, *Nucleosides Nucleotides*, 1991, **10**, 623.
15. D. Fernandez-Forner, Y. Palom, S. Ikuta, E. Pedroso, and R. Eritja, *Nucleic Acids Res.*, 1990, **18**, 5729.
16. Y.-Z. Xu, Q. Zheng, and P. F. Swann, *J. Org. Chem.*, 1992, **57**, 3839.
17. Y.-Z. Xu and P. F. Swann, *Nucleosides Nucleotides*, 1991, **10**, 315.
18. B. F. L. Li, C. B. Reese, and P. F. Swann, *Biochemistry*, 1987, **26**, 1086.
19. B. A. Connolly, in "Oligonucleotides and Analogs: A Practical Approach," IRL Press, New York, 1991, p. 155.
20. T. T. Nikiforov and B. A. Connolly, *Tetrahedron Lett.*, 1991, **31**, 3851.
21. T. T. Nikiforov and B. A. Connolly, *Tetrahedron Lett.*, 1992, **33**, 2379.
22. P. Clivio, J.-L. Fourrey, J. Gasche, and A. Favre, *Tetrahedron Lett.*, 1992, **33**, 69.
23. P. Clivio, J.-L. Fourrey, J. Gasche, A. Audic, A. Favre, C. Perrin, and A. Woisard, *Tetrahedron Lett.*, 1992, **33**, 65.
24. Y.-Z. Xu, Q. Zheng, and P. F. Swann, *Tetrahedron*, 1992, **48**, 1729.
25. R. S. Coleman and J. M. Siedlecki, *Tetrahedron Let.*, 1991, **32**, 3033.
26. R. S. Coleman and E. A. Kesicki, *J. Am. Chem. Soc.*, 1994, **116**, 11 636.
27. J. F. Habener, C. D. Vo, D. B. Le, G. P. Gryan, L. Ercolani, and A. H.-J. Wang, *Proc. Natl. Acad. Sci. USA*, 1988, **85**, 1735.
28. A. B. Kremer, T. Mikita, and G. P. Beardsley, *Biochemistry*, 1987, **26**, 391.
29. L. C. Sowers, R. Eritja, B. E. Kaplan, M. F. Goodman, and G. V. Fazakerley, *J. Biol. Chem.*, 1987, **262**, 15 436.
30. L. C. Sowers, R. Eritja, B. E. Kaplan, M. F. Goodman, and G. V. Fazakerley, *J. Biol. Chem.*, 1988, **263**, 14 794.
31. P. V. Sahasrabudhe, R. T. Pon, and W. H. Gmeiner, *Nucleic Acids Res.*, 1995, **23**, 3916.
32. T. Hayakawa, A. Ono, and T. Ueda, *Nucleic Acids Res.*, 1988, **16**, 4761.
33. G. V. Fazakerley, L. C. Sowers, R. Eritja, B. E. Kaplan, and M. F. Goodman, *J. Biomol. Struct. Dynam.*, 1987, **5**, 639.
34. T. J. Povsic and P. B. Dervan, *J. Am. Chem. Soc.*, 1989, **111**, 3059.
35. R. H. Durland, T. S. Rao, G. R. Revankar, J. H. Tinsley, M. A. Myrick, D. M. Seth, J. Rayford, P. Singh, and K. Jayaraman, *Nucleic Acids Res.*, 1994, **22**, 3233.
36. R. D. Sheardy and N. C. Seeman, *J. Org. Chem.*, 1986, **51**, 4301.
37. D. A. Barawkar and K. N. Ganesh, *Bioorg. Med. Chem. Lett.*, 1993, **3**, 347.
38. V. S. Rana, D. A. Barawkar, and K. N. Ganesh, *J. Org. Chem.*, 1996, **61**, 3578.
39. D. A. Barawkar and K. N. Ganesh, *Nucleic Acids Res.*, 1995, **23**, 159.
40. W. H. Gmeiner, R. T. Pon, and J. W. Lown, *J. Org. Chem.*, 1991, **56**, 3602.
41. F. Seela, H. Driller, W. Herdering, and E. De Clercq, *Nucleosides Nucleotides*, 1988, **7**, 347.
42. B. C. Froehler, S. Wadwani, T. J. Terhorst, and S. R. Gerrard, *Tetrahedron Lett.*, 1992, **33**, 5307.
43. B. C. Froehler, R. J. Jones, X. Cao, and T. J. Terhorst, *Tetrahedron Lett.*, 1993, **34**, 1003.
44. A. J. Gutierrez and B. C. Froehler, *Tetrahedron Lett.*, 1996, **37**, 3959.
45. J. Sagi, A. Szemzo, K. Ebinger, A. Szabolcs, G. Sagi, E. Ruff, and L. Otvos, *Tetrahedron Lett.*, 1993, **34**, 2191.
46. N. Colocci and P. B. Dervan, *J. Am. Chem. Soc.*, 1994, **116**, 785.
47. R. W. Wagner, M. D. Matteucci, J. G. Lewis, A. J. Gutierrez, C. Moulds, and B. C. Froehler, *Science*, 1993, **260**, 1510.
48. R. W. Wagner, M. D. Matteucci, D. Grant, T. Huang, and B. C. Froehler, *Nature Biotechnol.*, 1996, **14**, 840.
49. C. Moulds, J. G. Lewis, B. C. Froehler, D. Grant, T. Huang, J. F. Milligan, M. D. Matteucci, and R. W. Wagner, *Biochemistry*, 1995, **34**, 5044.
50. A. J. Gutierrez, T. J. Terhorst, M. D. Matteucci, and B. C. Froehler, *J. Am. Chem. Soc.*, 1994, **116**, 5540.
51. H. Hashimoto, M. G. Nelson, and C. Switzer, *J. Am. Chem. Soc.*, 1993, **115**, 7128.
52. H. Nara, A. Ono, and A. Matsuda, *Bioconjugate Chem.*, 1995, **6**, 54.
53. G. F. Kattan, Z. J. Lesnikowski, S. Yao, F. Tanious, W. D. Wilson, and R. F. Schinazi, *J. Am. Chem. Soc.*, 1994, **116**, 7494.
54. Z. J. Lesnikowski, G. Fulcrand, R. M. Lloyd, Jr., A. Juodawlkis, and R. F. Schinazi, *Biochemistry*, 1996, **35**, 5741.
55. R. B. Meyer, Jr., J. C. Tabone, G. D. Hurst, T. M. Smith, and H. Gamper, *J. Am. Chem. Soc.*, 1989, **111**, 8517.
56. T. J. Povsic and P. B. Dervan, *J. Am. Chem. Soc.*, 1990, **112**, 9428.

57. Y. S. Sanghvi, G. D. Hoke, M. C. Zounes, S. M. Freier, J. F. Martin, H. Chan, O. L. Acevedo, D. J. Ecker, C. K. Mirabelli, S. T. Crooke, and P. D. Cook, *Nucleosides Nucleotides*, 1991, **10**, 345.
58. Y. S. Sanghvi, G. D. Hoke, S. M. Freier, M. C. Zounes, C. Gonzalez, L. Cummins, H. Sasmor, and P. D. Cook, *Nucleic Acids Res.*, 1993, **21**, 3197.
59. S. M. Freier, W. F. Lima, Y. S. Sanghvi, T. Vickers, M. Zounes, P. D. Cook, and D. J. Ecker, in "Gene Regulation: Biology of Antisense RNA and DNA," eds. R. P. Erickson and J. G. Izant, Raven Press, New York, 1992, p. 95.
60. A. M. MacMillan and G. L. Verdine, *J. Org. Chem.*, 1990, **55**, 5931.
61. A. M. MacMillan and G. L. Verdine, *Tetrahedron*, 1991, **47**, 2603.
62. P. S. Miller and C. D. Cushman, *Bioconjugate Chem.*, 1992, **3**, 74.
63. C.-Y. Huang, C. D. Cushman, and P. S. Miller, *J. Org. Chem.*, 1993, **58**, 5048.
64. C.-Y. Huang, G. Bi, and P. S. Miller, *Nucleic Acids Res.*, 1996, **24**, 2606.
65. P. K. T. Lin and D. M. Brown, *Nucleic Acids Res.*, 1989, **17**, 10373.
66. N. N. Anand, D. M. Brown, and S. A. Salisbury, *Nucleic Acids Res.*, 1987, **15**, 8167.
67. A. N. R. Nedderman, M. J. Stone, P. K. T. Lin, D. M. Brown, and D. H. Williams, *J. Chem. Soc., Chem. Commun.*, 1991, 1357.
68. J. S. Lee, M. L. Woodworth, L. J. P. Latimer, and A. R. Morgan, *Nucleic Acids Res.*, 1984, **12**, 6603.
69. A. Ono and T. Ueda, *Nucleic Acids Res.*, 1987, **15**, 219.
70. L. E. Xodo, G. Manzini, F. Quadrifoglio, G. A. van der Marel, and J. H. van Boom, *Nucleic Acids Res.*, 1991, **19**, 5625.
71. D. A. Barawkar, K. G. Rajeev, V. A. Kumar, and K. N. Ganesh, *Nucleic Acids Res.*, 1996, **24**, 1229.
72. W. A. Greenberg and P. B. Dervan, *J. Am. Chem. Soc.*, 1995, **117**, 5016.
73. J. Sagi, A. Szemzo, L. Otvos, M. Vorlickova, and J. Kypr, *Int. J. Biol. Macromol.*, 1991, **13**, 329.
74. C. J. Marasco, Jr. and J. R. Sufrin, *J. Org. Chem.*, 1992, **57**, 6363.
75. D. G. Osterman, G. D. DePillis, J. C. Wu, A. Matsuda, and D. V. Santi, *Biochemistry*, 1988, **27**, 5204.
76. L. Chen, A. M. MacMillan, W. Chang, K. Ezaz-Nikpay, W. S. Lane, and G. L. Verdine, *Biochemistry*, 1991, **30**, 11018.
77. S. S. Smith, B. E. Kaplan, L. C. Sowers, and E. M. Newman, *Proc. Natl. Acad. Sci. USA*, 1992, **89**, 4744.
78. G. Xiang, W. Soussou, and L. W. McLaughlin, *J. Am. Chem. Soc.*, 1994, **116**, 11155.
79. R. Berressem and J. W. Engels, *Nucleic Acids Res.*, 1995, **23**, 3465.
80. B. L. Gaffney, L. A. Marky, and R. A. Jones, *Tetrahedron*, 1984, **40**, 3.
81. A. Chollet, A. Chollet-Damerius, and E. H. Kawashima, *Chem. Scr.*, 1986, **26**, 37.
82. C. A. Brennan and R. I. Gumport, *Nucleic Acids Res.*, 1985, **13**, 8665.
83. A. Chollet and E. H. Kawashima, *Nucleic Acids Res.*, 1988, **16**, 305.
84. F. B. Howard and H. T. Miles, *Biochemistry*, 1984, **23**, 6723.
85. G. M. Lamm, B. J. Blencowe, B. S. Sproat, A. M. Iribarren, U. Ryder, and A. I. Lamond, *Nucleic Acids Res.*, 1991, **19**, 3193.
86. S. Gryaznov and R. G. Schultz, *Tetrahedron Lett.*, 1994, **35**, 2489.
87. J. A. Taboury, S. Adam, E. Taillandier, J.-M. Neuman, S. Tran-Dinh, T. Huynh-Dinh, B. Langlois, M. Conti, and J. Igolen, *Nucleic Acids Res.*, 1984, **12**, 6291.
88. M. Coll, A. H. J. Wang, G. A. van der Marel, J. H. van Boom, and A. Rich, *J. Biomol. Struct. Dynam.*, 1986, **4**, 157.
89. B. Borah, J. S. Cohen, F. B. Howard, and H. T. Miles, *Biochemistry*, 1985, **24**, 7456.
90. I. V. Kutyavin, R. L. Rhinehart, E. A. Lukhtanov, V. V. Gorn, R. B. Meyer, Jr., and H. B. Gamper, Jr., *Biochemistry*, 1996, **35**, 11170.
91. D. M. Brown and P. K. T. Lin, *Carbohydr. Res.*, 1991, **216**, 129.
92. K. S. Ramasamy, M. Zounes, C. Gonzalez, S. M. Freier, E. A. Lesnik, L. L. Cummins, R. H. Griffey, B. P. Monia, and P. D. Cook, *Tetrahedron Lett.*, 1994, **35**, 215.
93. K. Kido, H. Inoue, and E. Ohtsuka, *Nucleic Acids Res.*, 1992, **20**, 1339.
94. T. Golas, M. Fikus, Z. Kazimierczuk, and D. Shugar, *Eur. J. Biochem.*, 1976, **65**, 183.
95. J. Sepiol, Z. Kazimierczuk, and D. Shugar, *Z. Naturforsch.*, 1976, 361.
96. J. T. Davis, S. Tirumala, J. R. Jenssen, E. Radler, and D. Fabris, *J. Org. Chem.*, 1995, **60**, 4167.
97. C. Switzer, S. E. Moroney, and S. A. Benner, *J. Am. Chem. Soc.*, 1989, **111**, 8322.
98. J. A. Piccirilli, T. Krauch, S. E. Moroney, and S. A. Benner, *Nature*, 1990, **343**, 33.
99. J. D. Bain, C. Switzer, A. R. Chamberlin, and S. A. Benner, *Nature*, 1992, **356**, 537.
100. C. Y. Switzer, S. E. Moroney, and S. A. Benner, *Biochemistry*, 1993, **32**, 10489.
101. Y. Tor and P. B. Dervan, *J. Am. Chem. Soc.*, 1993, **115**, 4461.
102. T. Tuschl, M. M. P. Ng, W. Pieken, F. Benseler, and F. Eckstein, *Biochemistry*, 1993, **32**, 11658.
103. M. M. P. Ng, F. Benseler, T. Tuschl, and F. Eckstein, *Biochemistry*, 1994, **33**, 12119.
104. A. Eschenmoser, *Pure Appl. Chem.*, 1993, **65**, 1179.
105. Z. Kazimierczuk, R. Mertens, W. Kawczynski, and F. Seela, *Helv. Chim. Acta*, 1991, **74**, 1742.
106. F. Seela, R. Mertens, and Z. Kazimierczuk, *Helv. Chim. Acta*, 1992, **75**, 2298.
107. F. Seela and T. Fröhlich, *Helv. Chim. Acta*, 1994, **77**, 399.
108. C. Roberts, R. Bandaru, and C. Switzer, *Tetrahedron Lett.*, 1995, **36**, 3601.
109. T. Horn, C.-A. Chang, and M. L. Collins, *Tetrahedron Lett.*, 1995, **36**, 2033.
110. F. Seela, B. Gabler, and Z. Kazimierczuk, *Collect. Czech. Chem. Commun.*, 1993, **58**, 170.
111. B. Bhat, Neelima, N. J. Leonard, H. Robinson, and A. H.-J. Wang, *J. Am. Chem. Soc.*, 1996, **118**, 3065.
112. G. V. Fazakerley, R. Te'oule, A. Guy, H. Fritzsche, and W. Guschlbauer, *Biochemistry*, 1985, **24**, 4540.
113. G. V. Fazakerley, A. Guy, R. Te'oule, and W. Guschlbauer, *FEBS Lett.*, 1984, **176**, 449.
114. H. Nishio, A. Ono, A. Matsuda, and T. Ueda, *Nucleic Acids Res.*, 1992, **20**, 777.
115. H. Nishio, A. Ono, A. Matsuda, and T. Ueda, *Chem. Pharm. Bull.*, 1992, **40**, 1355.
116. F. Seela, W. Herdering, and A. Kehne, *Helv. Chim. Acta*, 1987, **70**, 1649.
117. J. Liu and G. L. Verdine, *Tetrahedron Lett.*, 1992, **33**, 4265.
118. M. Kanou, H. Ohomori, H. Takaku, S. Yokoyama, G. Kawai, R. J. Suhadolnik, and R. Sobol, Jr., *Nucleic Acids Res.*, 1990, **18**, 4439.
119. M. Kanou, H. Ohomori, K. Nagai, S. Yokoyama, R. J. Suhadolnik, R. Sobol, Jr., and H. Takaku, *Biochem. Biophys. Res. Commun.*, 1991, **176**, 769.

120. A. Guy, A.-M. Duplaa, P. Harel, and R. Te'oule, *Helv. Chim. Acta*, 1988, **71**, 1566.
121. H. Komatsu, T. Ichikawa, H. Takaku, S. Yokoyama, and G. Kawai, *Nucleic Acids Res.*, 1990, Symp. Ser. No. **22**, 95.
122. H. Komatsu, T. Ichikawa, M. Nakai, and H. Takaku, *Nucleosides Nucleotides*, 1992, **11**, 85.
123. W. Guschlbauer, A.-M. Duplaa, A. Guy, R. Te'oule, and G. V. Fazakerley, *Nucleic Acids Res.*, 1991, **19**, 1753.
124. B. P. Cho and F. E. Evans, *Nucleic Acids Res.*, 1991, **19**, 1041.
125. G. A. Leonard, A. Guy, T. Brown, R. Te'oule, and W. N. Hunter, *Biochemistry*, 1992, **31**, 8415.
126. P. S. Miller, P. Bhan, C. D. Cushman, and T. L. Trapane, *Biochemistry*, 1992, **31**, 6788.
127. P. S. Miller, G. Bi, S. A. Kipp, V. Fok, and R. K. DeLong, *Nucleic Acids Res.*, 1996, **24**, 730.
128. E. C. Davison and K. Johnsson, *Nucleosides Nucleotides*, 1993, **12**, 237.
129. M. C. Jetter and F. W. Hobbs, *Biochemistry*, 1993, **32**, 3249.
130. S. L. Young, S. H. Krawczyk, M. D. Matteucci, and J. J. Toole, *Proc. Natl. Acad. Sci. USA*, 1991, **88**, 10 023.
131. S. H. Krawczyk, J. R. Milligan, S. Wadwani, C. Moulds, B. C. Froehler, and M. D. Matteucci, *Proc. Natl. Acad. Sci. USA*, 1992, **89**, 3761.
132. E. Ohtsuka, S. Matsuki, M. Ikehara, Y. Takahashi, and K. Matsubara, *J. Biol. Chem.*, 1985, **260**, 2605.
133. Y. Kawase, S. Iwai, H. Inoue, K. Miura, and E. Ohtsuka, *Nucleic Acids Res.*, 1986, **14**, 7727.
134. F. Seela and K. Kaiser, *Nucleic Acids Res.*, 1986, **14**, 1825.
135. T. R. Waters and B. A. Connolly, *Biochemistry*, 1994, **33**, 1812.
136. F. H. Martin, M. M. Castro, F. Aboul-ela, and I. Tinoco, Jr., *Nucleic Acids Res.*, 1985, **13**, 8927.
137. F. Seela, Y. Chen, U. Bindig, and Z. Kazimierczuk, *Helv. Chim. Acta*, 1994, **77**, 194.
138. F. Seela and Y. Chen, *Nucleic Acids Res.*, 1995, **23**, 2499.
139. C. M. Harris, L. Zhou, E. A. Strand, and T. M. Harris, *J. Am. Chem. Soc.*, 1991, **113**, 4328.
140. N. Schmid and J.-P. Behr, *Tetrahedron Lett.*, 1995, **36**, 1447.
141. D. Fernandez-Forner, R. Eritja, F. Bardella, C. Ruiz-Perez, X. Solans, E. Giralt, and E. Pedroso, *Tetrahedron*, 1991, **47**, 8917.
142. R. Eritja, D. M. Horowitz, P. A. Walker, J. P. Ziehler-Martin, M. S. Boosalis, M. F. Goodman, K. Itakura, and B. E. Kaplan, *Nucleic Acids Res.*, 1986, **14**, 8135.
143. F. Seela, H. Driller, and U. Liman, *Liebigs Ann. Chem.*, 1985, 312.
144. J. F. Milligan, S. H. Krawczyk, S. Wadwani, and M. D. Matteucci, *Nucleic Acids Res.*, 1993, **21**, 327.
145. W. M. Olivas and L. J. Maher, III, *Nucleic Acids Res.*, 1995, **23**, 1936.
146. Y.-Z. Xu, Q. Zheng, and P. F. Swann, *Tetrahedron Lett.*, 1992, **33**, 5837.
147. H. P. Rappaport, *Nucleic Acids Res.*, 1988, **16**, 7253.
148. M. S. Christopherson and A. D. Broom, *Nucleic Acids Res.*, 1991, **19**, 5719.
149. B. A. Connolly, in "Methods in Enzymology: Chemical Synthesis of DNA," eds. D. M. J. Lilley and J. E. Dahlberg, Academic Press, New York, 1992, vol. 211, p. 36.
150. T. S. Rao, K. Jayaraman, R. H. Durland, and G. R. Revankar, *Tetrahedron Lett.*, 1992, **33**, 7651.
151. T. R. Waters and B. A. Connolly, *Nucleosides Nucleotides*, 1992, **11**, 1561.
152. T. S. Rao, R. H. Durland, D. M. Seth, M. A. Myrick, V. Bodepudi, and G. R. Revankar, *Biochemistry*, 1995, **34**, 765.
153. J. E. Gee, G. R. Revankar, T. S. Rao, and M. E. Hogan, *Biochemistry*, 1995, **34**, 2042.
154. S. Cal, R. G. Nicieza, B. A. Connolly, and J. Sanchez, *Biochemistry*, 1996, **35**, 10 828.
155. R. Casale and L. W. McLaughlin, *J. Am. Chem. Soc.*, 1990, **112**, 5264.
156. N. V. Heeb and S. A. Benner, *Tetrahedron Lett.*, 1994, **35**, 3045.
157. H. Gao, R. Fathi, B. L. Gaffney, B. Goswami, P.-P. Kung, Y. Rhee, R. Jin, and R. A. Jones, *J. Org. Chem.*, 1992, **57**, 6954.
158. M. O. Taktakishvili, A. Tabdjoun, and I. V. Yartseva, *Bioorg. Khim.*, 1990, **16**, 59.
159. V. Bodepudi, C. R. Iden, and F. Johnson, *Nucleosides Nucleotides*, 1991, **10**, 755.
160. J. Tchou, V. Bodepudi, S. Shibutani, I. Antoshechkin, J. Miller, A. P. Grollman, and F. Johnson, *J. Biol. Chem.*, 1994, **269**, 15 318.
161. H. C. P. F. Roelen, C. P. Saris, H. F. Brugghe, H. van den Elst, J. G. Westra, G. A. van der Marel, and J. H. van Boom, *Nucleic Acids Res.*, 1991, **19**, 4631.
162. M. Kouchakdjian, V. Bodepudi, S. Shibutani, M. Eisenberg, F. Johnson, A.P. Grollman, and D. J. Patel, *Biochemistry*, 1991, **30**, 1403.
163. Y. Oda, S. Uesugi, M. Ikehara, S. Nishimura, Y. Kawase, H. Ishikawa, H. Inoue, and E. Ohtsuka, *Nucleic Acids Res.*, 1991, **19**, 1407.
164. T. S. Rao, R. H. Durland, and G. R. Revankar, *J. Heterocycl. Chem.*, 1994, **31**, 935.
165. C. V. Varaprasad, N. Bulychev, A. P. Grollman, and F. Johnson, *Tetrahedron Lett.*, 1996, **37**, 9.
166. K. W. Porter, J. Tomasz, F. Huang, A. Sood, and B. R. Shaw, *Biochemistry*, 1995, **34**, 11 963.
167. J. Tomasz, B. R. Shaw, K. W. Porter, B. F. Spielvogel, and A. Sood, *Angew. Chem. Int. Ed. Engl.*, 1992, **31**, 1373.
168. H. Li, K. Porter, F. Huang, and B. R. Shaw, *Nucleic Acids Res.*, 1995, **23**, 4495.
169. A. Hasan, H. Li, J. Tomasz, and B. R. Shaw, *Nucleic Acids Res.*, 1996, **24**, 2150.
170. K. Ezaz-Nikpay and G. L. Verdine, *J. Am. Chem. Soc.*, 1992, **114**, 6562.
171. J. Hunziker, E. S. Priestley, H. Brunar, and P. B. Dervan, *J. Am. Chem. Soc.*, 1995, **117**, 2661.
172. H. Brunar and P. B. Dervan, *Nucleic Acids Res.*, 1996, **24**, 1987.
173. L. Hagenberg, H. G. Gassen, and H. Matthaei, *Biochem. Biophys. Res. Commun.*, 1973, **50**, 1104.
174. M. Ikehara, T. Fukui, and S. Uesugi, *J. Biochem.*, 1974, **76**, 107.
175. R. Cosstick, X. Li, D. K. Tuli, D. M. Williams, B. A. Connolly, and P. C. Newman, *Nucleic Acids Res.*, 1990, **18**, 4771.
176. A. Ono and T. Ueda, *Nucleic Acids Res.*, 1987, **15**, 3059.
177. F. Seela and T. Grein, *Nucleic Acids Res.*, 1992, **20**, 2297.
178. O. L. Acevedo, R. S. Andrews, R. H. Springer, S. Freier, G. D. Hoke, and P. D. Cook, in "13th International Congress on Heterocyclic Chemistry, Corvallis, OR, August 11–16, 1991," eds. L. Perry, D. Weller, and J. D. White, GE10–11, p..
179. O. L. Acevedo, G. D. Hoke, S. Freier, M. Zounes, C. G. Guinosso, R. H. Springer, and P. D. Cook, in "IUB Conference on Nucleic Acid Therapeutics, Clearwater Beach, FL, January 13–17, 1991," no. 50, p..
180. F. Seela and S. Lampe, *Helv. Chim. Acta*, 1991, **74**, 1790.

181. F. Seela, H.-D. Winkeler, H. Driller, and S. Menkhoff, *Nucleic Acids Res.*, Symp. Ser. **14**, 1984, 245.
182. F. Seela and H. Driller, *Nucleic Acids Res.*, 1985, **13**, 911.
183. F. Seela and A. Kehne, *Tetrahedron*, 1985, **41**, 5387.
184. F. Seela and A. Kehne, *Biochemistry*, 1985, **24**, 7556.
185. F. Seela and H. Driller, *Nucleic Acids Res.*, 1986, **14**, 2319.
186. F. Seela, H. Driller, A. Kehne, and K. Kaiser, *Chim. Scr.*, 1986, **26**, 173.
187. F. Seela and K. Kaiser, *Nucleosides Nucleotides*, 1987, **6**, 447.
188. F. Seela, U. Bindig, H. Driller, W. Herdering, K. Kaiser, A. Kehne, H. Rosemeyer, and H. Steker, *Nucleosides Nucleotides*, 1987, **6**, 11.
189. F. Seela and A. Kehne, *Biochemistry*, 1987, **26**, 2232.
190. A. Rosenthal, H. Billwitz, A. Kehne, and F. Seela, *Nucleic Acids Res.*, 1988, **16**, 1631.
191. F. Seela, H. Berg, and H. Rosemeyer, *Biochemistry*, 1989, **28**, 6193.
192. F. Seela, K. Kaiser, and U. Bindig, *Helv. Chim. Acta*, 1989, **72**, 868.
193. F. Seela and H. Driller, *Nucleic Acids Res.*, 1989, **17**, 901.
194. J. Jiricny, S. Wood, D. Martin, and A. Ubasawa, *Nucleic Acids Res.*, 1986, **14**, 6579.
195. F. Seela and H. Thomas, *Helv. Chim. Acta*, 1995, **78**, 94.
196. F. Seela, Q.-H. Tran-Thi, H. Mentzel, and V. A. Erdmann, *Biochemistry*, 1981, **20**, 2559.
197. J. SantaLucia, Jr., R. Kierzek, and D. H. Turner, *Science*, 1992, **256**, 217.
198. H. A. Heus and A. Pardi, *Science*, 1991, **253**, 191.
199. S. Dickmann and L. W. McLaughlin, *J. Mol. Biol.*, 1988, **202**, 823.
200. F. Seela and H. Driller, *Nucleosides Nucleotides*, 1989, **8**, 1.
201. T. S. Rao, A. F. Lewis, T. S. Hill, and G. R. Revankar, *Nucleosides Nucleotides*, 1995, **14**, 1.
202. F. Seela and K. Kaiser, *Helv. Chim. Acta*, 1988, **71**, 1813.
203. F. Seela and H. Driller, *Helv. Chim. Acta*, 1988, **71**, 1191.
204. C. R. Petrie, A. D. Adams, M. Stamm, J. V. Ness, S. M. Watanabe, and R. B. Meyer, Jr., *Bioconjugate Chem.*, 1991, **2**, 441.
205. R. B. Meyer, Jr., *Methods Mol. Biol.*, 1994, **26**, 73.
206. A. Ono, P. O. P. Ts'o, and L.-S. Kan, *J. Am. Chem. Soc.*, 1991, **113**, 4032.
207. A. Ono, P. O. P. Ts'o, and L.-S. Kan, *J. Org. Chem.*, 1992, **57**, 3225.
208. U. von Krosigk and S. A. Benner, *J. Am. Chem. Soc.*, 1995, **117**, 5361.
209. B. K. Bhattacharya, R. V. Devivar, and G. R. Revankar, *Nucleosides Nucleotides*, 1995, **14**, 1269.
210. I. Rosenberg, J. F. Soler, Z. Tocik, W.-Y. Ren, L. A. Ciszewski, P. Kois, K. W. Pankiewicz, M. Spassova, and K. A. Watanabe, *Nucleosides Nucleotides*, 1993, **12**, 381.
211. T. L. Trapane, M. S. Christopherson, C. D. Roby, P. O. P. Ts'o, and D. Wang, *J. Am. Chem. Soc.*, 1994, **116**, 8412.
212. B. K. Bhattacharya and G. R. Revankar, *Nucleosides Nucleotides*, 1994, **13**, 1721.
213. M. S. Solomon and P. B. Hopkins, *J. Org. Chem.*, 1993, **58**, 2232.
214. B. A. Schweitzer and E. T. Kool, *J. Am. Chem. Soc.*, 1995, **117**, 1863.
215. H. Kuhn, D. P. Smith, and S. S. David, *J. Org. Chem.*, 1995, **60**, 7094.
216. T. S. Rao, M. E. Hogan, and G. R. Revankar, *Nucleosides Nucleotides*, 1994, **13**, 95.
217. T. S. Rao, A. F. Lewis, R. H. Durland, and G. R. Revankar, *Tetrahedron Lett.*, 1993, **34**, 6709.
218. R. Nichols, P. C. Andrews, P. Zhang, and D. E. Bergstrom, *Nature*, 1994, **369**, 492.
219. D. E. Bergstrom, P. Zhang, P. H. Toma, P. C. Andrews, and R. Nichols, *J. Am. Chem. Soc.*, 1995, **117**, 1201.
220. D. Loakes, F. Hill, S. Linde, and D. M. Brown, *Nucleosides Nucleotides*, 1995, **14**, 1001.
221. R. S. Durland, T. S. Rao, V. Bodepudi, D. M. Seth, K. Jayaraman, and G. R. Revankar, *Nucleic Acids Res.*, 1995, **23**, 647.
222. L. C. Griffin, L. L. Kiessling, P. A. Beal, P. Gillespie, and P. B. Dervan, *J. Am. Chem. Soc.*, 1992, **114**, 7976.
223. H. Inoue, A. Imura, and E. Ohtsuka, *Nucleic Acids Res.*, 1985, **13**, 7119.
224. R. H. Durland, T. S. Rao, K. Jayaraman, and G. R. Revankar, *Bioconjugate Chem.*, 1995, **6**, 278.
225. A. B. Staubli and P. B. Dervan, *Nucleic Acids Res.*, 1994, **22**, 2637.
226. B. K. Bhattacharya, M. V. Chari, R. H. Durland, and G. R. Revankar, *Nucleosides Nucleotides*, 1995, **14**, 45.
227. M. E. Hawkins, W. Pfleiderer, A. Mazumder, Y. G. Pommier, and F. M. Balis, *Nucleic Acids Res.*, 1995, **23**, 2872.
228. P. K. T. Lin and D. M. Brown, in "Methods in Molecular Biology: Protocols for Oligonucleotide Conjugates," ed. S. Agrawal, Humana Press, Totowa, NJ, 1994, vol. 26, p. 187.
229. K.-Y. Lin, R. J. Jones, and M. D. Matteucci, *J. Am. Chem. Soc.*, 1995, **117**, 3873.
230. M. D. Matteucci and U. von Krosigk, *Tetrahedron Lett.*, 1996, **37**, 5057.
231. G. R. Reddy and L. J. Marnett, *J. Am. Chem. Soc.*, 1995, **117**, 5007.
232. O. D. Schärer and G. L. Verdine, *J. Am. Chem. Soc.*, 1995, **117**, 10 781.
233. S. C. Zimmerman and P. Schmitt, *J. Am. Chem. Soc.*, 1995, **117**, 10 769.

# 7.10
# Topological Modification of DNA: Circles, Loops, Knots, and Branches

ERIC T. KOOL
*University of Rochester, NY, USA*

| | | |
|---|---|---|
| 7.10.1 | INTRODUCTION | 342 |
| 7.10.1.1 | *The Functions of DNA* | 342 |
| 7.10.1.2 | *Topological Modifications in DNA are Common in Nature* | 343 |
| 7.10.1.3 | *Topological Differences can Greatly Alter Properties* | 343 |
| 7.10.2 | NATURAL TOPOLOGICAL VARIANTS OF NUCLEIC ACIDS | 343 |
| 7.10.2.1 | *Catenated DNAs in Kinetoplastid Mitochondria* | 343 |
| 7.10.2.2 | *Circular, Lariat-shaped, and Pseudoknotted RNAs* | 344 |
| 7.10.3 | TOPOLOGY AND NUCLEIC ACID REPLICATION | 346 |
| 7.10.3.1 | *Problems Associated with Replication of Linear DNAs* | 346 |
| 7.10.3.2 | *Rolling Circles as an Efficient Replication Mechanism* | 346 |
| 7.10.3.3 | *DNA Junctions and Recombination* | 346 |
| 7.10.3.4 | *Topoisomerases and Catenated and Knotted DNAs* | 347 |
| 7.10.4 | NATURAL MOLECULES WHICH RECOGNIZE DNA USING ALTERED TOPOLOGY | 349 |
| 7.10.4.1 | *Topological Variations in Protein–DNA Recognition* | 349 |
| 7.10.4.2 | *Topological Variations in Small Molecule–DNA Recognition* | 351 |
| 7.10.5 | MAN-MADE DNA STRUCTURES HAVING NONLINEAR TOPOLOGY | 352 |
| 7.10.5.1 | *DNA Branches and Junctions* | 352 |
| 7.10.5.2 | *Knotted DNAs* | 353 |
| 7.10.5.3 | *Synthetic Three-dimensional Shapes Built from DNA* | 353 |
| 7.10.5.4 | *Circular DNAs* | 354 |
| 7.10.6 | CIRCULAR PERMUTATION AND DNA BENDING | 354 |
| 7.10.6.1 | *Importance of DNA Bending* | 354 |
| 7.10.6.2 | *Cyclization and Circular Permutation* | 354 |
| 7.10.7 | MOLECULAR RECOGNITION BY TOPOLOGICALLY MODIFIED OLIGONUCLEOTIDES | 355 |
| 7.10.7.1 | *Branched Oligonucleotides: Duplex and Triplex Formation* | 355 |
| 7.10.7.2 | *Circular, Stem-loop and Lariat Oligonucleotides: Triplex Formation* | 356 |
| 7.10.7.3 | *Duplex Formation Involving Loops and Circles* | 356 |
| 7.10.7.4 | *Triplex Formation Between Circles and Single Strands* | 357 |
| 7.10.7.5 | *Multisite Binding by Conformational Switching* | 359 |
| 7.10.7.6 | *Higher Order Structures: Bicyclic Oligonucleotides* | 360 |
| 7.10.7.7 | *Binding Duplex DNA by Threading: Pseudorotaxanes and Catenanes* | 361 |
| 7.10.8 | INTERACTIONS OF PROTEINS WITH TOPOLOGICALLY MODIFIED SYNTHETIC DNA | 361 |
| 7.10.8.1 | *Binding of "Dumbbell" Circular DNAs and RNAs by Proteins* | 362 |

|   |   |
|---|---|
| 7.10.8.2   *Antibody Recognition of a Disulfide-cross-linked DNA Hairpin* | 363 |
| 7.10.8.3   *Polymerase Recognition of Circular DNAs: Rolling Circles.* | 363 |
| 7.10.9   SYNTHESIS OF TOPOLOGICALLY MODIFIED DNA | 363 |
| 7.10.9.1   *Closure of Double-stranded Circles* | 363 |
| 7.10.9.1.1   *Enzymatic methods* | 364 |
| 7.10.9.1.2   *Nonenzymatic (chemical) methods* | 365 |
| 7.10.9.2   *Closure of Single-stranded Circles* | 366 |
| 7.10.9.2.1   *Enzymatic methods* | 366 |
| 7.10.9.2.2   *Nonenzymatic (chemical) methods* | 366 |
| 7.10.9.2.3   *Dumbbell DNA (RNA) cyclizations* | 366 |
| 7.10.9.2.4   *RNA-catalyzed cyclizations* | 367 |
| 7.10.10   CONCLUSIONS AND FUTURE PROSPECTS | 367 |
| 7.10.11   REFERENCES | 367 |

## 7.10.1   INTRODUCTION

### 7.10.1.1   The Functions of DNA

The structure of deoxyribonucleic acid (DNA) is one of the most important and elegant architectures found in nature. Given the multiple functions of DNA in nature, it is difficult to imagine how the structure could have been different—even in the smallest details—and still support life. DNA in the cell serves perhaps primarily as a storehouse of genetic information, and the complementarity of the Watson–Crick base pairing (Figure 1) makes the transfer and copying of this information possible. However, other functions (although interrelated) can also be ascribed to DNA. DNA serves as a recognition element; many proteins as well as small molecules interact directly and indirectly with the various surfaces and functional groups along the helix, and these noncovalent interactions are important in the regulation of gene expression. DNA also serves as a scaffold, aligning and organizing other biomolecules in specific arrangements. Finally, DNA is a catalyst which facilitates the synthesis of copies of itself, crucial property which is of course necessary for life.

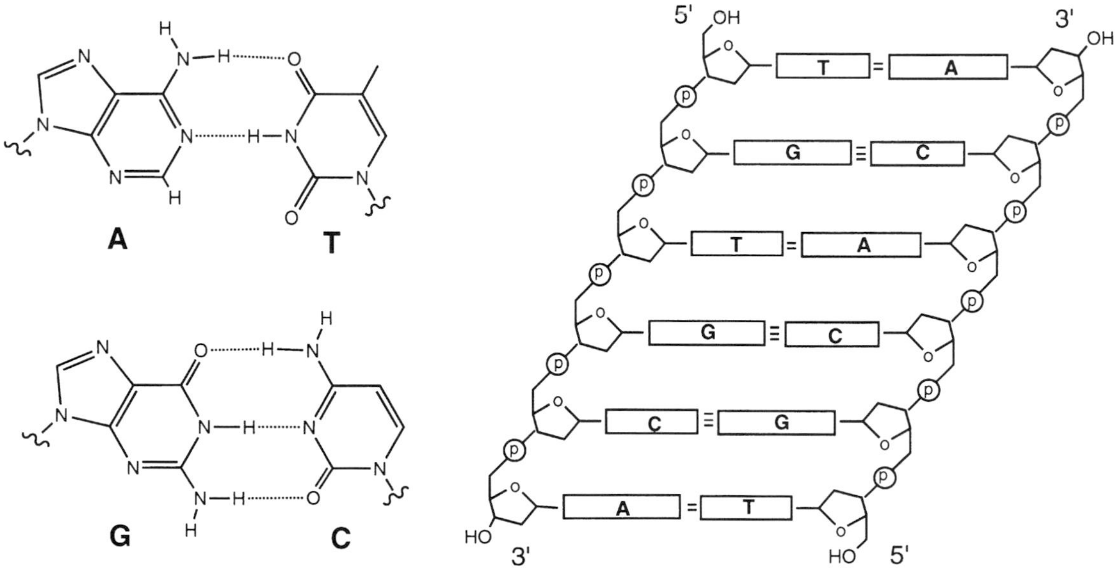

**Figure 1**   The structure of DNA.

Chemists are acutely and increasingly aware that these properties and functions arise directly from the molecular structure of DNA. The structure of organic molecules (and indeed, all molecules) involves the specific arrangement of covalent and noncovalent bonds which connect the atoms; however, chemists have in recent years recognized that a description of what is bonded to what is not always sufficient to describe structure and properties. Two molecules can have the same covalent bonds and still be quite different in structure and properties. One way this can occur is with differences in topology. The structure and dynamics of DNA allow it to be especially susceptible to undergoing topological changes. These topological differences can greatly alter properties and functions, both for DNAs in natural systems and for designed synthetic DNAs in a test tube. Such topological differences are the subject of this chapter.

### 7.10.1.2  Topological Modifications in DNA are Common in Nature

Because DNA occurs in nature as long, flexible polymeric chains, it is common to find topological modifications in natural DNA strands. Indeed, circular DNAs (a simple topological modification) are probably more common than linear ones; the genomes of bacteria and of many viruses are circular,[1] no doubt because this topology gives these organisms benefits in survival. In addition, because there exist enzymes which can break and rejoin DNA chains, further topological changes such as knotting and branching are also commonplace in natural nucleic acids. Because RNAs are more often single strands rather than double helices (the common form of DNA), they fold back on themselves in intricate structures which contain many loops and knots; in addition, RNA can catalyze self-cleavage and self-ligation reactions which can and do alter topology even further. Thus, topological modifications of DNA and RNA are not merely curiosities, but are an integral part of their structure and functions in nature, and understanding such topological issues is important for their possible applications in chemistry, biology, and medicine.

### 7.10.1.3  Topological Differences can Greatly Alter Properties

The topological variations discussed above can greatly affect most of the properties and functions of DNA and RNA. A circular DNA, for example, can store information differently from the way that a linear one can; it also can be recognized by proteins differently from the way a linear variant is; the circular shape necessarily provides a much different scaffold than a linear shape does; and copying a circular DNA is greatly altered relative to copying a linear one. Thus, all these functions are affected by this shape change. The other topological modifications, such as branching, looping, and knotting, can also affect these properties. Specific examples of these effects are discussed below.

## 7.10.2  NATURAL TOPOLOGICAL VARIANTS OF NUCLEIC ACIDS

In this section will be discussed some naturally occurring topological variations seen for DNAs and RNAs. These will serve to demonstrate how properties can be altered by changing topology; some of these properties and principles have been adopted by chemists and biologists in the design of nonnatural nucleic acid structures, and a number of these will be discussed in later sections.

### 7.10.2.1  Catenated DNAs in Kinetoplastid Mitochondria

Two circular DNAs which are covalently interlocked form a catenane (Figure 2).[2] Probably the most well-documented catenanes in nature are found in the mitochondrial DNA of trypanosomes. The trypanosomatid kinetoplast DNA (kDNA) has been found to exist as a massively interlocked network of circular DNAs.[3,4] Two types of DNA are found in this network: larger (sequence-conserved) maxicircles, which are ~20 000 base pairs (bp) in size and are closed circular duplexes; and small minicircles, which are quite variable in sequence and are ~1000 bp closed circular duplexes (Figure 3). The roles of the maxicircles and minicircles are still being uncovered.[4] It is thought that the massively catenated structure exists as an alternative way to link together genetic information in a living organism (thus preserving genes from generation to generation), but still allowing for genetic exchange, which is a vital part of sexual reproduction. The circles are physically linked together by breaking open a chain, passing it through a strand of a closed circle, and reclosing the break to form the catenane; this is accomplished by topoisomerase enzymes (see discussion below).

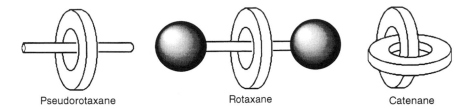

| Pseudorotaxane | Rotaxane | Catenane |

**Figure 2**  The topological structure of pseudorotaxanes, rotaxanes, and catenanes. The first case is a pair of molecules linked only by noncovalent bonds, while the latter two cases are pairs of molecules which, although not directly bonded to each other, can be separated only by breaking covalent bonds.

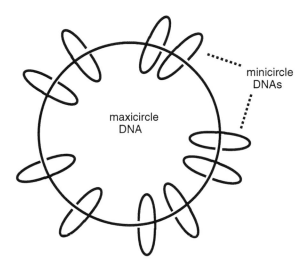

**Figure 3**  Illustration of the topological form of kinetoplastid mitochondrial DNA of trypanosomes. The DNA consists entirely of smaller minicircle duplexes (800 bp to 1300 bp) catenated on maxicircle duplexes ($\sim$ 20 000–40 000 bp). The total array is considerably more complex than shown, because many maxicircles are linked as well.

### 7.10.2.2   Circular, Lariat-shaped, and Pseudoknotted RNAs

Although RNAs are not the primary subject of this volume, it is worth mentioning here that natural topological variations exist not only in DNAs but in RNAs as well. Because of the mainly single-stranded nature of RNA, some topological variants are found which occur only rarely or not at all in DNA. In addition, the 2'-hydroxyl of RNA (which is absent in DNA) allows it to undergo chemistry, and specifically nucleophilic attack, which is not possible for DNA; this also brings about some topological issues.

As is true for DNAs, circular RNAs are not uncommon in nature. A few circular RNAs have been found to arise from natural splicing mechanisms (see below).[5,6] Perhaps the best-known circular RNAs are the viroids and satellite RNAs (also called virusoids).[7–9] This fascinating class of molecules is composed of single-stranded circular RNAs 247 to $\sim$ 1500 nucleotides in size which fold into rod-like structures (Figure 4). Viroids are especially notable because they are the smallest infectious, autonomously replicating species known to man.[9] Viroids are thought mainly to infect plants; however, the hepatitis delta virusoid is infectious to man when co-infected with the hepatitis B virus.[10] The circular topology of viroids gives them two chief advantages: first, the absence of ends allows them to escape some of the degradation which might occur by ribonuclease enzymes which prefer free ends; second, and more importantly, viroids are proposed to replicate by a simple but efficient "rolling circle" pathway (see below), which is only possible because of their topological form.

$_{C}{}^{U}_{U}$ GGGG$^{AAAU}$CUACA GGG CACC $^{CCAAAA}$ACU ACU$^{U}$GCAGG$^{A}$GAG GCCGC$^{U}$UGAG $^{GG}$ AUCC CCGGG $^{GAAA}$ CC$^{U}$CAAGCG$^{AA}$UC$^{U}$GGGA$^{A}$GGG$^{A}$GCG$^{UAC}$CUGGG$^{U}$CG$^{A}$UCG$^{U}$ GC GCG$^{UU}$GGAG$^{GA}$
$_{C}{}^{U}_{U}$ CCCC — GAUGU$_{UU}$CCC$_{U}$GUGG$_{AAAAAAA}$UGA$_{U}$UG$_{U}$UGUCC$_{A}$UUC$_{U}$CGGCG$_{C}$ACUC$_{AACA}$UA GG$_{U}$GGCCC$_{AUCA}$GAGG GUUCGC $_{C}$ AG$_{C}$UCCU CCC CGC $_{C}$ GGCCC GC AGC$_{UU}{}^{C}$G$_{A}$CGC$_{UU}{}^{C}$CUC$_{A}$ G

**Figure 4** Nucleotide sequence of the smallest known replicating agent in nature, the Coconut Cadang Cadang Viroid. It is a circular RNA molecule 246 nucleotides in size which infects coconut palms in Southeast Asia.[9]

Another common topological form for RNAs is the lariat. Biologists and biochemists observed that the final sequence of RNA that codes for protein (mature mRNA) is not the same as the sequence of the RNA just after it is transcribed from the DNA in the nucleus. This pre-mRNA contains extra sequences inserted between the true "coding" sequences of RNA; these extra sequences are called "introns", and the coding sequences are "exons".[11] The mechanism by which introns are spliced out and exons are stitched together has been partially elucidated (Figure 5).[11-13]

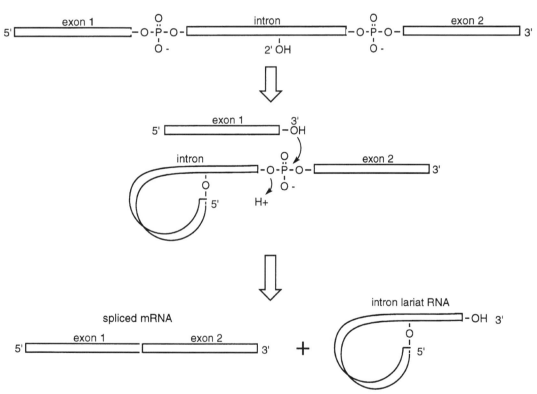

**Figure 5** Scheme showing how introns are spliced out of pre-mRNA to yield mature mRNA. The complex of ribonucleoproteins which carries this out is called the spliceosome. The intron is usually spliced into a lariat form as shown; however, certain sequences can also be spliced into true circular form as well.

The protein/RNA machinery that carries this out is called the spliceosome. It is now known that a branch point at an adenosine in the intron (a 2'-OH group acts as a nucleophile) loops around and cleaves the 5' end of the exon by nucleophilic attack.[12] Further catalytic processing catalyzed by the spliceosome causes the intron to be excised in a lariat-shaped form[13] and the ends of the exons to be covalently joined. It is also worth mentioning here that some introns contain pseudocatalytic sequences with which they splice themselves out; some of these introns can become circular after the splicing is complete.[14]

The last topological variant of RNA to be mentioned here is the pseudoknot.[15] A pseudoknot forms when two loops in separate lengths of RNA base pair with each other. Although a pseudoknot in its most common form is not a true knot and thus is not a distinct topological form, this unusual structure has been recently found to be important to biological functions in nature. Additional information on pseudoknots can be found in much greater detail in the recent literature.[15,16]

### 7.10.3 TOPOLOGY AND NUCLEIC ACID REPLICATION

The topology of a given DNA strand can greatly influence both its ability to be copied, and the products of copying, by a DNA polymerase enzyme. This fact has caused evolution to select for structures which have topologies which can not only be copied efficiently but also with high fidelity. In this section will be discussed the structural requirements for successful DNA replication and the effects of varied topology on replication and associated functions.

#### 7.10.3.1 Problems Associated with Replication of Linear DNAs

For some time it has been recognized that linear double-stranded DNAs can have serious problems both in the efficiency and fidelity of their replication.[1] Since replication is an autocatalytic process, large energetic barriers in the process are undesirable, as they cause reactions to be slowed significantly. The problem with linear duplex DNAs is that the product of copying a strand of DNA is a double-stranded DNA, and yet only a single-stranded DNA can be copied. Thus, energy must be spent in unwinding the product double helix before another copy can be made (Figure 6). Interestingly, the DNA of higher organisms is largely in linear duplex form despite this barrier.[1] In simple replication systems, such as with *in vitro* DNA amplification schemes, however, this product inhibition can be a problem. The widely used polymerase chain reaction (PCR) requires thermal heat–cool cycles and special thermostable enzymes to overcome this problem.[17]

A second and perhaps more serious problem in the replication of linear duplex DNAs is the accurate copying of ends.[1] Life depends on accurate generation-to-generation copying of DNA. DNA polymerases require primers to initiate synthesis, and in a linear DNA there is no guarantee that a primer will bind precisely at the end nucleotide of the template strand. Copying can therefore lead to addition or, more likely, deletion of genetic information at the ends of the DNA strands.

In organisms (such as humans) which have linear duplex DNA genomes, special end structures, such as telomeres,[18] have evolved to overcome some of these kinds of problems. However, another way to overcome the "end problem" is to eliminate the DNA ends altogether. This is probably why bacteria and viruses commonly have their DNA in circular form.

#### 7.10.3.2 Rolling Circles as an Efficient Replication Mechanism

One of the most important advantages of having one's DNA in circular form is the availability of the rolling circle mechanism of replication.[19,20] In this mechanism, a primer can bind anywhere in the circle and the polymerase can begin copying. Multiple copies are made simply by having the enzymes travel around the circular template multiple times, unwinding the duplex it has already synthesized in front of itself as it makes the complementary DNA strand. Thus, at steady state, no net addition of helix is occurring, and no large thermodynamic barrier arises.

Rolling circles also occur with RNA synthesis. The RNA viroids described above probably replicate by this mechanism (Figure 7).[7–9] When a viroid infects a plant it utilizes a plant-derived RNA polymerase to copy itself. The polymerase binds and begins synthesis (RNA polymerases do not need primers), rolling around the circular template multiple times and stringing along a repeating RNA copy of the original circle. To complete the cycle, the repeating viroid RNA actually (in some cases) cleaves itself to monomeric length, and it is then cyclized to circular form to begin another round.

Not only does the rolling circle mechanism of replication serve as an efficient means for copying and amplifying DNA and RNA, but it also overcomes the problem of accuracy in copying ends. As the polymerase travels around the circle, no genetic information is either added or lost, and so generational information transfer is precise.

In the late 1990s, scientists have begun to mimic the rolling circle strategy with smaller synthetic nucleic acids (see below). It is hoped that advantages similar to those in nature will be seen in a test tube as well.

#### 7.10.3.3 DNA Junctions and Recombination

DNA junctions occur at the convergence of three or more strands of DNA. Junctions have topology different from that of linear duplex DNA, and these branched structures are important in

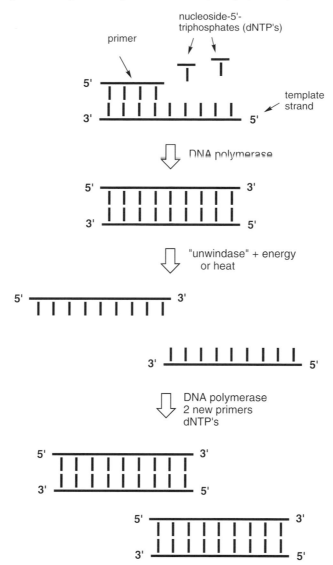

**Figure 6**  Scheme illustrating the DNA polymerase-catalyzed copying of a DNA strand. True replication requires that the original strand be produced again, and so the duplex must somehow be separated (which is thermodynamically unfavorable) in order to complete the replication cycle.

biological processes. Four-way junctions, in particular, have received a good deal of attention[21–24] because they occur during homologous recombination. Two DNA duplexes which have similar (homologous) sequences can undergo stable strand exchange, catalyzed by recombination enzymes such as RecA.[25] The unstable intermediate formed is a four-way junction (Figure 8). This intermediate is unstable because the joint can move along the DNA (and does this as part of the recombination process). Model short DNAs which have four-way junctions have been studied as models of the recombination process;[21–24] careful choice of sequence can prevent movement of the junction, thus rendering the junction structure stable. A good deal of structural study has been carried out with four-way junctions.[23,24]

### 7.10.3.4  Topoisomerases and Catenated and Knotted DNAs

Although topoisomerase enzymes are covered in detail in another chapter (see Chapter 7.16), a brief overview of the structural changes they can effect in DNA is worth mention here. Put simply,

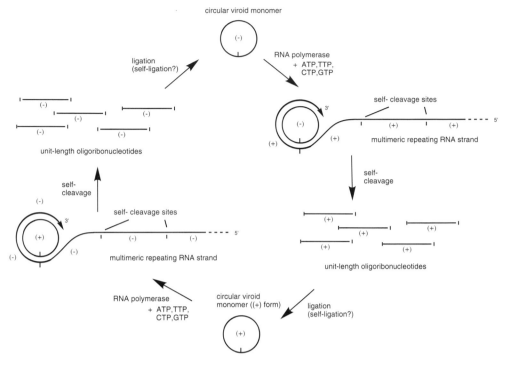

**Figure 7**   Scheme proposed for the rolling replication of circular viroid RNAs. Making more than one copy of the original circle is easily effected by the polymerase travelling around the circle multiple times.

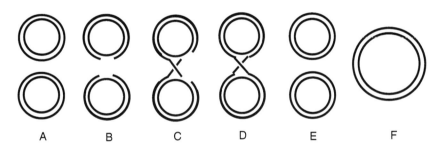

**Figure 8**   Illustration of the mechanism of homologous recombination of two circular duplex DNAs, showing the four-stranded junction intermediate (D). The final products are either two new circular duplexes (E) having some exchanged sequences, or a circular duplex combining both original ones (F).

the job of topoisomerase is to pass one strand of DNA through another (Figure 9). Studies have elucidated the mechanism by which this occurs:[26,27] first, the enzyme binds the strand to be broken and then the strand to be passed through it. It then covalently breaks the first duplex using internal nucleophiles (tyrosine hydroxyls) to break phosphodiester bonds. By undergoing a series of conformational changes, the second strand is passed through the break and the first strand is religated. Because DNA is a flexible structure, this strand passage can lead to topological changes in the DNA.[28–30] If circular duplexes are involved, for example, two circles can be made into a catenated structure (or the reverse can happen). A single circle can also be formed into a knotted structure (Figure 10). Electron microscopy has shown clear photographs of circular plasmid DNAs which are in knotted topologies as a result of topoisomerase action (Figure 11).[28,29] It is therefore clear that topoisomerases are biologically important. The massively catenated kinetoplast DNA of trypanosomatids is, for example, completely dependent on topisomerases for formation of that superstructure.[3]

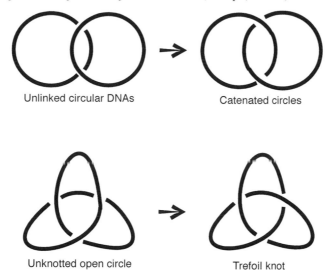

Figure 9  Scheme showing reactions catalyzed by a topoisomerase enzyme. The enzyme can effectively pass one strand of duplex DNA through another by breaking and rejoining one strand. This leads to catenated and knotted structures as shown.

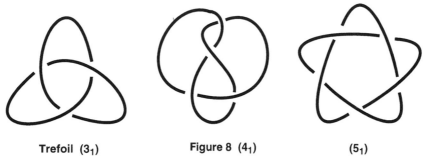

Figure 10  Schematic structures of knots which may be formed in a circular DNA by the action of a topoisomerase.

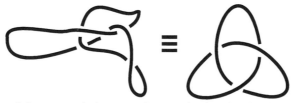

Figure 11  Drawing adapted from actual electron micrograph of a circular duplex DNA having a knotted topology (after Wasserman and Cozzarelli[29]).

## 7.10.4  NATURAL MOLECULES WHICH RECOGNIZE DNA USING ALTERED TOPOLOGY

Perhaps the majority of proteins and small molecules which bind DNA form their complexes simply by diffusing to their target site and forming noncovalent contacts with the DNA major or minor groove surface. However, because of the unique structure of DNA, there are some noteworthy examples where the description is not sufficient to describe the binding event. Indeed, some of these topologically different complexes are crucial to the most important processes in cells.

### 7.10.4.1  Topological Variations in Protein–DNA Recognition

A fascinating story in topology has arisen from the study of DNA polymerase enzymes and the mechanism of DNA replication. The polymerase enzymes whose purpose is to replicate the genome are enzymes with very high processivity; that is, they synthesize very long strands of complementary DNA without dissociating from the template DNA. DNA polymerase III (Pol III), found in *E. coli*, is a good example of such a processive enzyme.[31] This polymerase is a complex machine built from ten different subunit proteins, and until the early 1990s, the purpose of these proteins was not known.

Among the subunits, the core polymerase protein was identified, and measurements of its processivity showed that it had very low processivity relative to its activity in cells.[31] Thus, it seemed likely that the purpose of at least some of the other subunits is to increase the processivity of the enzyme. In 1992 the crystal structure of the $\beta$ subunit (which exists as a $\beta_2$ dimer) was published,[32] and suddenly a very interesting role for it was postulated based on the unusual structure. The $\beta_2$ dimer is a doughnut-shaped complex with a hole in the center large enough for a DNA strand to reside in (Figure 12). Subsequent studies confirmed that the $\beta_2$ dimer binds DNA by being clamped around the duplex; this clamp can easily slide along the DNA for great distances.[33–35]

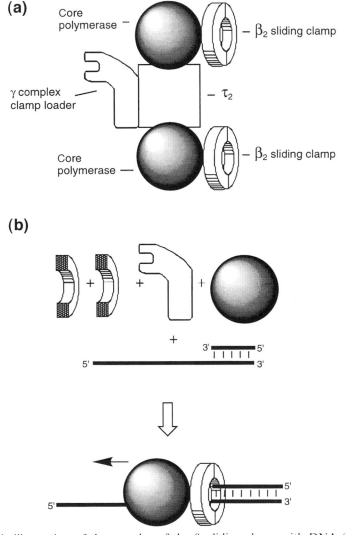

**Figure 12**   Schematic illustration of the complex of the $\beta_2$ sliding clamp with DNA (b), showing how the dimeric protein is clamped around the DNA. This sliding clamp is part of RNA polymerase III holoenzyme complex (a), and helps prevent the polymerase from dissociating from the DNA (thus increasing the processivity, or average strand length synthesized).

It is now known that the sliding clamp in Pol III binds by one face to the rest of the polymerase complex, and is responsible for the high processivity of the polymerase.[33–35] Because it would be extremely difficult to find an end and thread its way onto the DNA, the clamp dimer is instead broken apart and assembled onto the DNA. One of the proteins in the complex serves as a "clamp loader" which puts the clamp on the DNA (Figure 12(a)). More recently it has been shown that other polymerases may also use sliding clamp subunits to increase their processivity.[36]

### 7.10.4.2 Topological Variations in Small Molecule–DNA Recognition

A number of small naturally occurring molecules are also known to take advantage of the structure of the DNA to form topologically unusual complexes. Two examples are nogalamycin[37] and arugomycin,[38] two antibiotics having similar structures (Figure 13). These compounds contain a central core which consists of a flat multiring aromatic system. Such structures have often been observed to intercalate into DNA (see Chapter 7.12) between base pairs, forming stable $\pi$-$\pi$ interactions with the neighboring DNA bases. However, both of these molecules also possess nonplanar carbohydrate groups at both ends of the flat aromatic system. Since the carbohydrates are nonplanar and more bulky than the central aromatic portion, these end groups cannot be intercalated into the DNA.

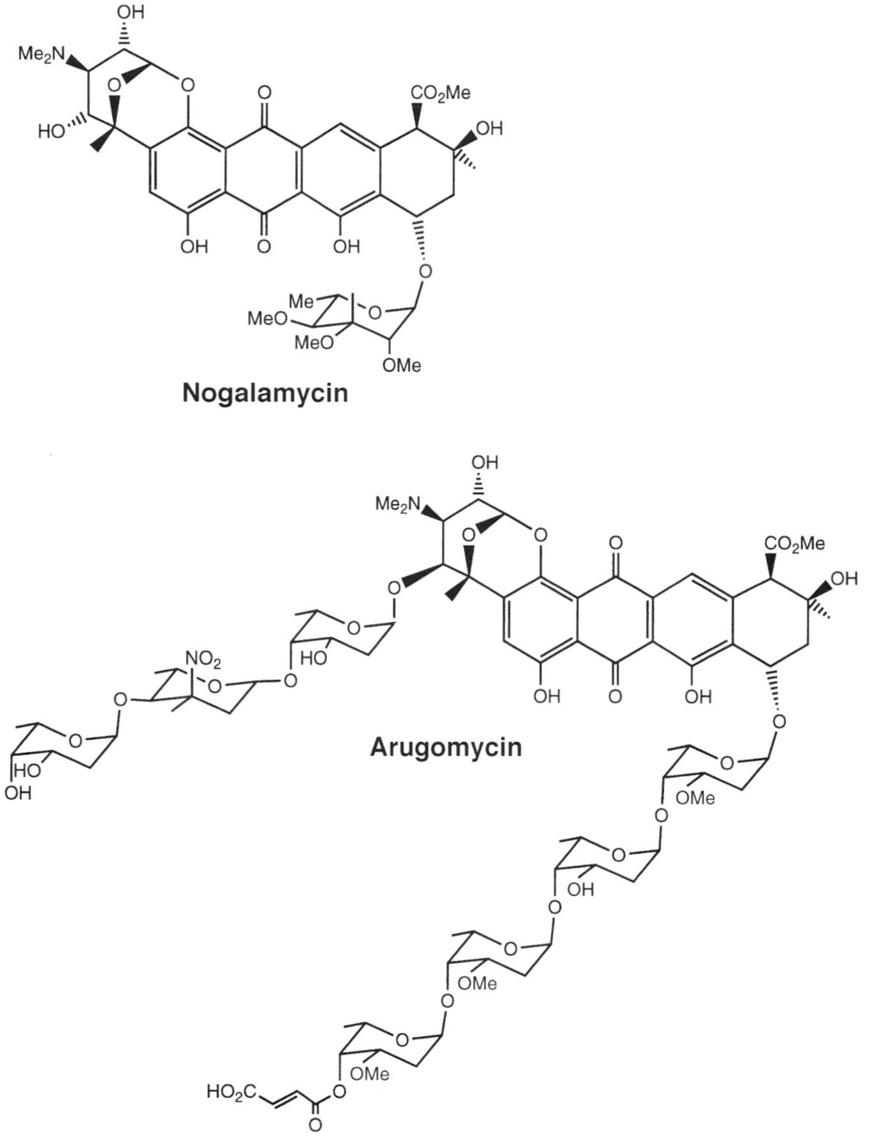

**Nogalamycin**

**Arugomycin**

**Figure 13** Structures of two small DNA intercalating molecules which in order to bind must pass through the two strands of duplex DNA, thus being present both in the major and minor grooves.

For this reason these drugs bind DNA by intercalating the central portion and leaving the bulky end groups out in the grooves of the DNA. An X-ray crystal structure has been published for the nogalamycin complex with DNA,[37] and NMR studies have also been carried out in solution.[39] The arugomycin complex has also been studied by NMR.[38] The most interesting part of these complexes from a topological standpoint is the mechanism by which they form their complexes. Since both carbohydrate groups are too large to fit between base pairs, one of these groups must be physically threaded between the two DNA strands, accompanied by partial separation of the strands. Once the bulky group has passed through, the DNA can reclose and form the intercalated complex. This unusual binding mechanism raises some interesting questions about the rate of transient opening of DNA.[37]

One last topological variation in the binding of DNA by natural products is worth mentioning here, although the topic is covered in detail in Chapter 7.15. There exists a considerable number of peptides (smaller than protein size) whose function is to bind DNA. Since peptides can be quite flexible in solution, the binding affinity for DNA can be lessened by the entropic cost of freezing the bond rotation in the complex. Nature has addressed this problem by restricting the conformational freedom of many of these peptides; perhaps the simplest way this can be done is by cyclizing a given peptide, thus leading to a topological change. A good example of a DNA-binding peptide with restricted conformations is the quinoxaline class of antibiotics.[40] These molecules are cyclic depsi-peptides with additional conformational restriction coming from a cross-link from one side of the macrocycle to the other. Triostin A in this peptide family has been cocrystallized with DNA[41] and it is a beautiful example of a highly specialized structure for which topological changes have a clear purpose.

## 7.10.5    MAN-MADE DNA STRUCTURES HAVING NONLINEAR TOPOLOGY

Just as topological differences in natural DNA structures can have large and biologically impor-tant consequences for biological activity, so too can topological modifications in man-made struc-tures have a large influence on physical and biological properties. Because it often only takes a small alteration of structure to change the topology of a DNA molecule (for example, formation of a single bond which converts a linear molecule to circular form), many laboratories have taken advantage of topology to alter or improve properties.

The goals of researchers in designing and constructing nonnatural DNAs have been quite wide-ranging. For example, branched DNAs have found use in medical diagnostics; knotted DNAs have been useful in the study of DNA topology and topoisomerases; synthetic junctions have been constructed for the study of recombination and in the building of larger architectures; and circular DNAs have been used as ligands for DNA and RNA, as ligands for DNA-binding proteins, and as useful templates for polymerase enzymes.

### 7.10.5.1    DNA Branches and Junctions

As discussed above, branched topologies in DNA are important in genetic recombination. In man-made structures, branches have been constructed both with duplex DNAs and with single-stranded RNAs and DNAs. Single-stranded DNAs (in contrast to double-stranded DNAs) have the ability to hybridize to complementary strands of DNA. Large single-stranded branched DNAs having overall shapes resembling combs and forks have been constructed for medical diagnostic purposes[42] (Figure 14): the presence of multiple copies of a sequence in each molecule allows them the possibility of binding more than one strand of DNA, thus amplifying the signal and increasing the sensitivity of detection.

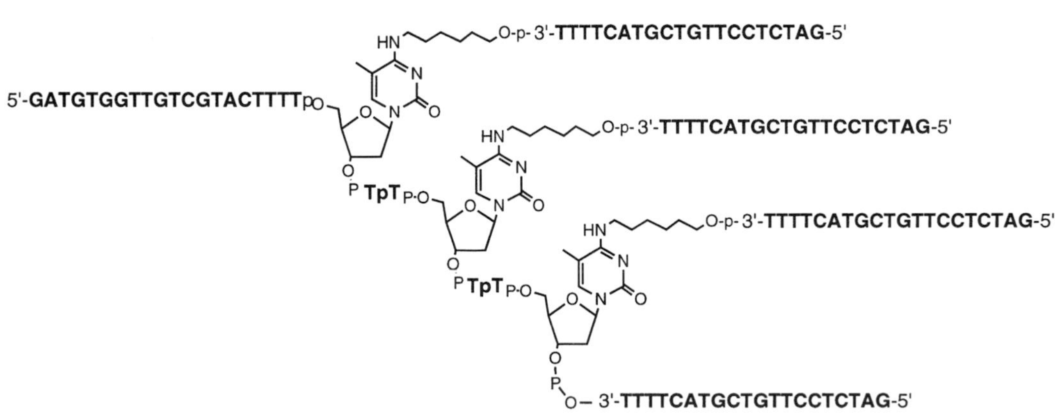

**Figure 14**    Illustration of comb-like branched DNA designed to be used in diagnostic probes of natural nucleic acid sequences.

Branched structures built from duplex DNA have been constructed in a number of laboratories (Figure 15). Four-way junctions (as mentioned above) can be useful in the study of recombination intermediates,[21-24] while other junctions having as many as six arms have been constructed with the aim of using such structures as building blocks in larger synthetic DNA architectures.[43,44]

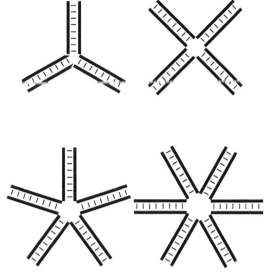

**Figure 15**  Illustrations of several synthetic DNA junctions which have been characterized.

### 7.10.5.2  Knotted DNAs

Several synthetic knotted DNA structures have been constructed from single-stranded circular DNAs. The Seeman laboratory has been the most active group in this area.[45,46] The goal of this research has been the study of DNA topology in general as well as enzymes which interact with topologically modified DNAs. By making use of predicted short helical complementarity, and in some cases altering helical structure by changing solution conditions, the Seeman group has successfully constructed a trefoil knot[45] and a figure-of-eight knot[46] from circular single strands of DNA (Figure 10).

### 7.10.5.3  Synthetic Three-dimensional Shapes Built from DNA

The laboratory of Seeman has also been quite active in the construction of three-dimensional shapes and architectures from duplex DNA (Figure 16).[47,48] The long-term goal of this work is the construction of nanometer-scale molecular scaffolds which may someday be useful in the assembly of molecular materials, molecular computers, and biologically useful structures.[49]

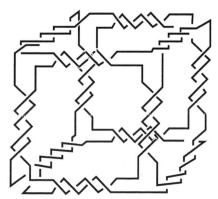

**Figure 16**  Illustration of cube-like structure which was built using duplex DNA as a scaffold.[47] Three-way branched junctions form the corners of the structure.

Two platonic solids built from DNA have been reported by the Seeman group. In 1991 the group reported the synthesis of a DNA structure with the connectivity of a cube (Figure 16).[47] This was an interlocked set of six circular DNAs built from a total of 480 nucleotides. In 1994 the same group reported construction of a more complex solid, a truncated octahedron having 14 sides.[48] To make this larger, more complex structure, the group has developed a solid-phase approach to piecing together subunits of DNA oligonucleotides.[50] A series of enzymatic ligations and restrictions are carried out on the solid phase prior to releasing the DNA structure when it is nearly complete.

### 7.10.5.4   Circular DNAs

The most widely studied topological alteration in synthetic DNAs has been circular oligo-nucleotides. Circular synthetic single-stranded DNAs have been constructed as small as two nucleo-tides[51] and as large as 246 nucleotides.[52] Circular double-stranded DNAs have been constructed as small as a reported 42 base pairs[53] and as large as thousands of nucleotides. Circular single-stranded DNAs have been constructed primarily for their ability to interact with other nucleic acids and proteins, and are discussed elsewhere in this chapter (Section 7.10.8). Circular duplex DNAs have been studied in large part for their relationship to DNA bending, and are also discussed above in relation to their use in the study of topoisomerases.

### 7.10.6   CIRCULAR PERMUTATION AND DNA BENDING

The concept of circular permutation has been a useful one in studying the biochemistry and biophysics of biopolymers.[54–56] Both RNAs[54] and DNAs[55] have been circularly permuted, as have proteins.[56] Circular permutation is done simply by taking a linear chain, cyclizing it, and then re-breaking the chain at another position. This can give information on the importance of chain continuity on structure and folding energetics. The most common use of circular permutation has been in the study of DNA bending, both by specific bent sequences of DNA[57–59] and by proteins which bend DNA when they bind.[60]

### 7.10.6.1   Importance of DNA Bending

Bent DNA structures have biological importance for a number of reasons. It is becoming increasingly recognized that many DNA-binding proteins bend the DNA to which they bind (Figure 17),[60] and that the bending itself causes changes in how other proteins interact. Thus the bending can serve as a switch for turning on and off a specific biological activity. Second, it is also now widely recognized that certain sequences of DNA can be inherently bent in solution.[57] There is evidence that some proteins may bind preferentially to such bent structures.[58]

### 7.10.6.2   Cyclization and Circular Permutation

The method of circular permutation[55] has been widely utilized for analysis of inherent bends in DNA as well as for analysis of bending induced by binding of proteins. The method was first developed (Figure 18) to analyze bending in runs of A in duplex DNA. By this approach it was found that an $A_6$ sequence causes a 17–21° bend toward the minor groove.[57]

The method makes use of the fact that a bent fragment of DNA travels more slowly by gel electrophoresis than an unbent fragment of the same length. If a putative bent sequence is placed within a circular duplex DNA which contains multiple sites for enzymatic cleavage (these can be constructed easily by multiple ligations of smaller fragments), the sequence can then be circularly permuted. The circle is treated with a restriction enzyme such that only one cleavage occurs (on average) per circle. This produces a population of fragments having the putative bent site at different positions relative to the ends. If there is a significant bend, the fragments having the bend near the center will travel considerably more slowly than those with the bend at the end (and the magnitude of the difference depends on the angle of bending). The direction of bending (i.e., towards the major or minor groove) can be determined by phasing the bent site with a known bent sequence; if the bends are in the opposite direction they will tend to cancel out the anomalous gel migration.[57]

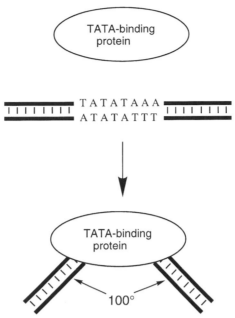

**Figure 17**   Illustration of how the binding of a protein (such as the TATA-binding protein, an important transcription factor) can induce a sharp bend in the DNA.

### 7.10.7   MOLECULAR RECOGNITION BY TOPOLOGICALLY MODIFIED OLIGONUCLEOTIDES

The binding of DNA, RNA, and proteins by oligonucleotides has many potential uses both in the development of molecular diagnostic techniques for detecting and identifying disease in humans and in the potential treatment of important diseases such as cancer and viral infections. Many chemists have undertaken the study of synthetic modifications of DNA structure in order to improve hybridization or pharmacokinetic properties for such applications.[61,62] While the majority of these studies have previously involved alterations to the nucleobase, the sugar, or phosphate backbone of the DNA, an increasing number of researchers have been developing topological modifications of DNA to achieve the same ends. One attractive feature of topological modification (as opposed to structural modification of base or backbone) is that it can be combined with those other structural changes if desired. Topological modification is easily compatible with modified DNA structures, such as the widely studied phosphorothioate DNA modification. Thus, useful improvements in properties of the DNA can be combined to give even greater advantages than topological or structural modifications alone.

In this section a number of important examples of topologically modified synthetic DNA structures will be discussed. These modifications were designed to improve hybridization properties as well as pharmacokinetic properties, and a number of these types of modifications are being studied for potential applications in diagnostics and therapeutics targeted to human disease.

#### 7.10.7.1   Branched Oligonucleotides: Duplex and Triplex Formation

Branched single-stranded oligonucleotides have been constructed by a number of research groups.[42,63–65] Branched and comb-like structures have been synthesized (as described above) for the binding and detection of pathogenic disease-related sequences by duplex formation.[42] A branching group serves well as a method for introducing two similar sequences into DNA, and this has also led to branched structures designed to form intra- and intermolecular triple helical complexes.[63–65] For example, the incorporation of two identical pyrimidine sequences allows the binding of a complementary purine sequence by sandwiching the purine strand between the two binding domains (Figure 19).[63] This leads to the formation of four to five hydrogen bonds to each purine, thus giving strong, cooperative complexes. Branched structures which contain two pyrimidine domains as well as a third purine domain can also form strong intramolecular triplexes which may be useful for structural study (Figure 20).[65]

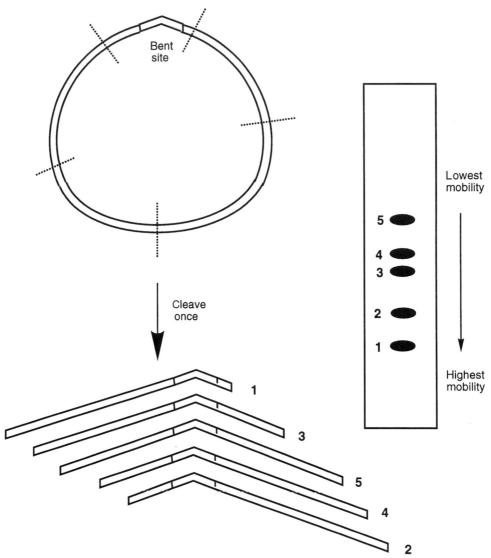

**Figure 18**  Schematic showing how a bend in circularly permuted DNA fragments can give altered mobilities by gel electrophoresis, depending on the position and degree of bending. The closer the bend to the center of the fragment, the slower its mobility; a greater bend angle slows mobility more than a lesser bend.

Branched DNAs can be constructed by inclusion of a branching unit in the strand; this can be a ribose unit, so that the branching group is the 2-hydroxyl,[63] or separate nonnucleoside branching groups (now commercially available) can be utilized as well.[66]

### 7.10.7.2   Circular, Stem-loop and Lariat Oligonucleotides: Triplex Formation

Circular oligonucleotides have been constructed in many laboratories. Some of the studies have been primarily aimed at the development of synthetic procedures for constructing such compounds (see below), while many studies have been aimed at the design of DNAs which can form strong complexes with target strands of DNA or RNA. Particularly successful has been a number of related approaches for binding single strands of DNA or RNA by triplex formation (Figure 21), as described above for branched DNAs.

### 7.10.7.3   Duplex Formation Involving Loops and Circles

Circular or looped oligonucleotides can bind target strands of DNA or RNA by forming either duplex or triplex complexes, depending on sequence and RNA or DNA content.[67,68] Interestingly,

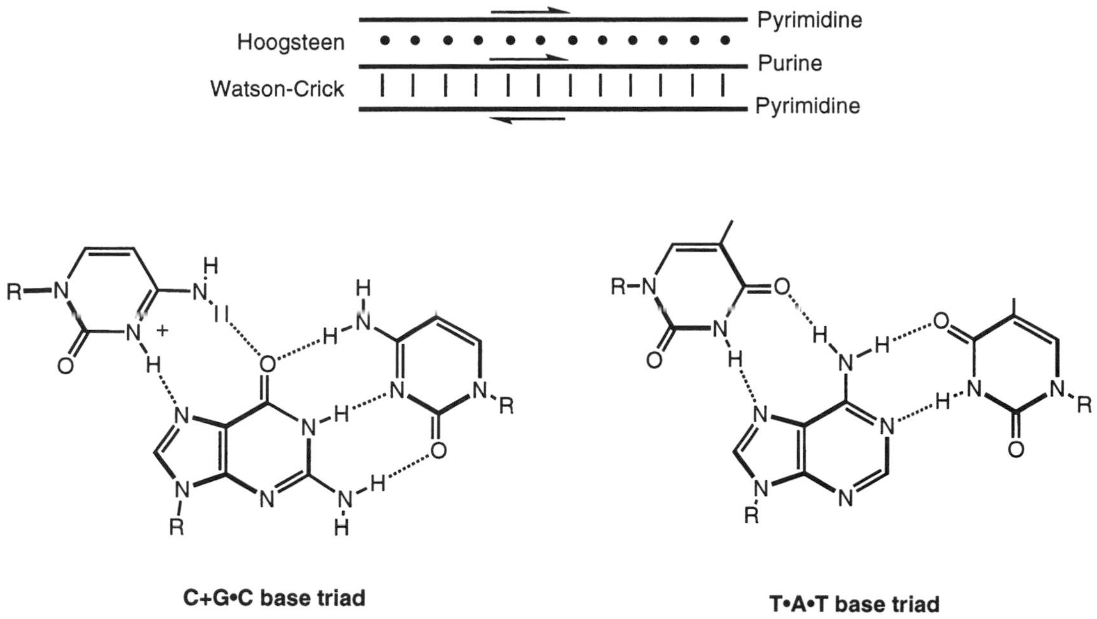

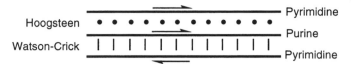

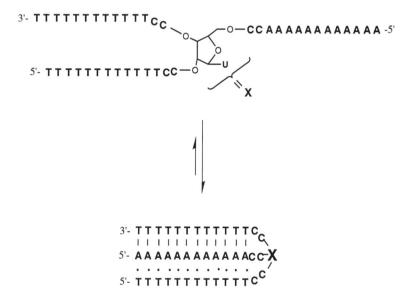

**C+G•C base triad**           **T•A•T base triad**

**Figure 19** Structure of pyrimidine–purine–pyrimidine triple helix.

**Figure 20** A synthetic branched oligonucleotide which folds to form an intramolecular triple helix.[65]

a circular DNA can tightly bind purine-rich targets in DNA by triplex formation, but can form only weaker double stranded complexes with RNA.[68] This RNA/DNA effect was first noticed by Crothers who evaluated composition effects in linear triplexes.[69] Wang and Kool showed that circular oligonucleotides containing some RNA residues can bind tightly by triplex formation.[68,70]

An interesting topological issue is raised by these two binding modes (Figure 22). When a small circle binds a single strand by duplex formation, the strand must pass through the circle (forming a pseudorotaxane) if the binding interaction is ~ 10 bases or longer.[71] However, when triplex formation occurs, no topological link arises no matter how long the binding interaction is.[68]

### 7.10.7.4 Triplex Formation Between Circles and Single Strands

In 1991 two publications described DNA-binding properties of circular oligonucleotides 34 bases in size;[67,72] these were designed to form triplexes with single-stranded target strands (Figures 21 and

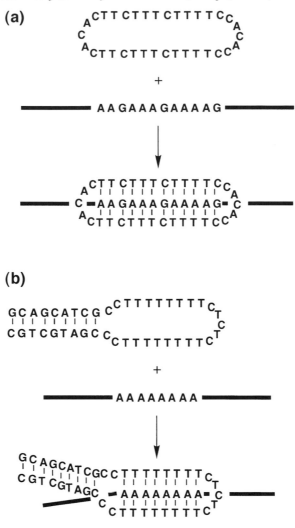

**Figure 21** Sequences of (a) circular and (b) stem-loop oligonucleotides designed to bind single-stranded DNA by triplex formation.

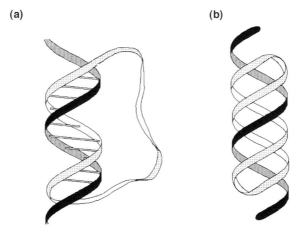

**Figure 22** Illustration of the interaction of a circular oligonucleotide with a single-stranded sequence resulting in either (a) duplex formation or (b) triplex formation. Note that in duplex formation the linear molecule must pass through the circle (forming a pseudorotaxane), while it does not in triplex formation. Which of these complexes is formed depends on the sequence of the circle and on whether the strands are composed of DNA or RNA.[68]

23). These circles were constructed by template-directed nonenzymatic cyclization using the reagents BrCN, imidazole and $Ni^{2+}$. These circular DNAs contain two 12-base stretches of pyrimidines bridged by two 5-base linker sequences. One paper describes the high affinity of complexes formed between such circles and a linear purine complementary strand of DNA;[67] for example, one circle binds six orders of magnitude more tightly than does a standard linear oligonucleotide to the same target sequence. The second paper describes the finding that such circles can have much higher sequence specificity than a linear complement.[72] For example, a circle was shown to display 6–7 kcal mol$^{-1}$ discrimination against a single mismatched base in the target, while a standard linear oligomer shows a smaller 3–4 kcal of discrimination.

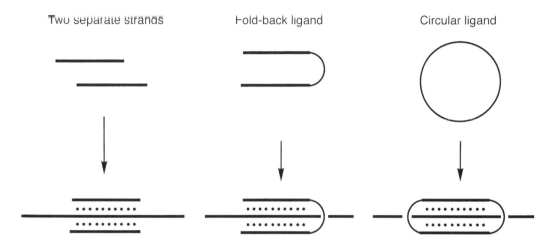

**Figure 23**   Diagram showing strategies for binding a single-stranded DNA by triplex formation. The target strand is bound on two sides in all three approaches; addition of links between the binding domains (one link for fold-back and two for the circle) increases binding affinity by preorganizing the ligand.

Later papers from the same laboratory described further studies on this new binding concept. The studies addressed structural effects in this mode of binding,[73] tested variations in the bridging loops to optimize binding,[74,75] compared the binding of RNA by circular ligands containing RNA or DNA,[68] and measured kinetics of binding and dissociation.[76] In general, such circular oligonucleotides bind complementary purine-rich target sequences with very high affinity and sequence specificity, and can do so at neutral pH and physiological ionic strength.[77]

One manuscript from the same laboratory showed that molecules closed not by covalent bonds, but instead by noncovalent Watson–Crick bonds, can also bind complementary target strands by triplex formation.[78] Stem-loop oligonucleotides (Figure 21) which have a $\sim$6–12 base pair stem and a loop of $\sim$17 mostly pyrimidine nucleotides, were shown to bind purine sequences with affinities nearly as high as those of fully closed circular oligonucleotides.

Another research group has since reported a related strategy in which lariat-shaped oligonucleotides were designed to bind complementary sequences by combined duplex and triplex formation.[79] The loop portion of the lariat can form triplexes with purine-rich sequences, while the loose end of the lariat can hybridize in normal Watson–Crick fashion with an adjacent sequence of DNA.

### 7.10.7.5   Multisite Binding by Conformational Switching

The circular oligonucleotides described in Section 7.10.7.4 were constructed from two pyrimidine-rich binding domains bridged by two 5-base linker domains having (somewhat) arbitrary sequence. The design for triplex formation necessitates that the two binding domains have pseudosequence-symmetry (see Figure 24). Rubin *et al.* realized that the linker sequences could be made useful (beyond the role as simple linkers) if they were made longer and also carried this pseudosymmetry.[80] Such a molecule could be designed to bind two different DNA sequences by switching conformation, alternating the roles of given domains as binders or as linkers (Figure 24). Later an even more complex binding function, using an extension of this strategy, was described.[81] While naturally occurring proteins are known that switch conformation to bind more than one substrate,[82] this was the first case of a naturally designed molecule that could perform such a role in DNA.

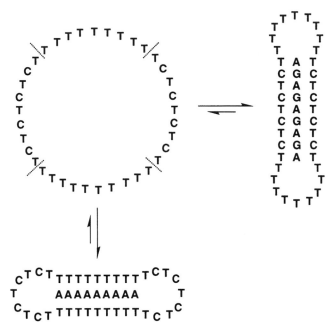

**Figure 24**   Illustration showing how a synthetic circular oligonucleotide can be designed to bind two different sequences of DNA by triplex formation.[80] A conformational switch reverses the roles of binding and bridging domains. A circle which binds six sequences of DNA has also been reported.[81]

### 7.10.7.6   Higher Order Structures: Bicyclic Oligonucleotides

The circular oligonucleotides described in Section 7.10.7.4 were expected to bind more tightly than linear oligonucleotides for a number of reasons. First, they form more hydrogen bonds with the target and form a greater number of base-stacking interactions as well, both of which are enthalpically favorable for complex formation.[77] However, the most important aspect of the design is the fact that the two binding domains are linked together at both ends. It was shown that both these linkages add considerably to the binding affinity.[73] This binding advantage is expected because connection of the binding domains lowers the entropy cost of forming the final complex. Thus, the molecule is preorganized to bind more tightly.

This concept of preorganization of the DNA for triplex formation has now been carried a step further.[83] Additional linking of binding domains (if done with careful design to avoid unfavorable geometry) might be expected further to increase the preorganization of the circular structure and thus further increase binding affinity and sequence specificity. This was accomplished by the synthetic addition of a disulfide bridge across the circle (Figure 25), giving a macrobicyclic structure in some ways analogous to the triostin peptide antibiotics described in Section 7.10.4.2. This was done by synthesis of a thymidine analogue carrying a protected thiol which was incorporated into a circle precursor. Cyclization and disulfide formation were carried out in one pot to yield the final bicyclic molecule.[83]

Measurement of the binding properties for this bicyclic oligonucleotide was then carried out, with comparison being made to the same circular sequence lacking the disulfide bridge. The results showed[83] that the bicyclic molecule binds its target complementary sequence (13 nucleotides long) with affinity 8 kcal mol$^{-1}$ higher than the simple circular molecule at neutral pH, and an impressive 15 kcal mol$^{-1}$ greater than a simple Watson–Crick complement. Moreover, the binding specificity was shown to be increased as well; the bicyclic molecule discriminated against single base mismatches by an equally large 10–12 kcal mol$^{-1}$. This may be the most selective DNA-binding molecule yet characterized.

**(a)**

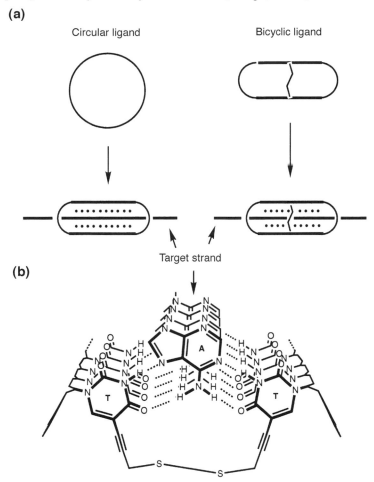

**Figure 25** (a) The strategy of further cross-linking of a circular DNA to increase its binding affinity and selectivity. (b) Cutaway of the triplex structure, showing structure of disulfide cross-link between pyrimidine strands.

### 7.10.7.7 Binding Duplex DNA by Threading: Pseudorotaxanes and Catenanes

Studies in the mid-1990s focused on the question of whether circular DNAs can bind duplex DNA as well as single-stranded targets.[84] These studies have confirmed that circular triplex-forming oligonucleotides do indeed bind double-stranded DNA, albeit with lower affinity than the binding of single-stranded DNAs. However, the most interesting aspect of this complexation was the finding that the preferred complex between a circle and a DNA duplex is one in which the circle threads its way over the end of the DNA and slides down to its binding site to form a triplex (Figure 26).[84] This complex is formally a pseudorotaxane,[85] and is analogous to the sliding clamp $\beta_2$ dimer found in DNA polymerase III (Section 7.10.4.1),[32] except that the DNA circle is covalently closed and must thread its way onto the DNA.

Subsequent to the discovery of this novel DNA-binding mode, Ryan and Kool used it to build a true catenane (Figure 27),[86] by assembling a larger synthetic circular duplex DNA containing one binding site for triplex formation. Prior to closure, the large circle was incubated with the small circle and then ligation was performed. A high yield of the closed catenane was observed.

### 7.10.8 INTERACTIONS OF PROTEINS WITH TOPOLOGICALLY MODIFIED SYNTHETIC DNA

The use of topologically modified DNAs has not been limited to applications of binding to nucleic acids. Since the early 1990s there has been increasing focus on the binding of oligonucleotides to proteins. Binding of synthetic DNA to a protein active site may downregulate the protein's native

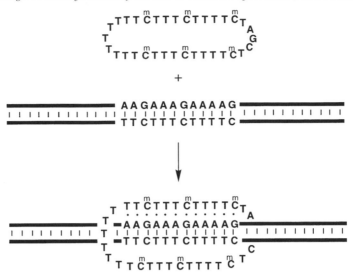

**Figure 26**    Sequences of a target duplex DNA strand and a circular oligonucleotide shown to bind it by threading over the end. This complex is a pseudorotaxane.

**Figure 27**    Molecular model of a DNA catenane which was constructed by allowing a small triplex-forming circle to assemble itself on a linear duplex DNA by threading, and then ligating the duplex DNA closed. The small circle is 34 nucleotides in size; the large duplex circle is 147 base pairs.

activity,[87–90] and so it is possible that one might use topologically modified DNAs as therapeutic agents. In addition, the use of topologically altered DNAs as catalytic templates for DNA and RNA polymerases, a different class of DNA-binding protein, has gained attention.[91–95]

### 7.10.8.1    Binding of "Dumbbell" Circular DNAs and RNAs by Proteins

Dumbbell circular DNAs and RNAs are simply duplex DNAs bridged by nucleotide or even nonnucleotide loops at the ends of the duplex. Dumbbell DNAs were first constructed as stable structures which model unstable hairpin loops that form in DNA;[96–99] however, the focus has since turned to dumbbells which mimic the binding sites of DNA-binding proteins, and thus can serve as decoys for such proteins.[87–90] If this binding were to occur in cells, it might in principle lead to downregulation of gene expression (if the protein is a transcriptional activator) or upregulation (if a repressor). Dumbbell RNAs have also been constructed to resemble secondary structures formed in the HIV virus which interact with specific proteins, and these decoys are meant to downregulate viral replication.[100,101]

### 7.10.8.2 Antibody Recognition of a Disulfide-cross-linked DNA Hairpin

A hairpin structure in DNA often has an alternate conformation (a bulged duplex) available to it, and this ambiguity can lead to difficulty in structural or functional studies.[102] One way to avoid this problem is to seal the end of the hairpin stem with a cross-link. Glick and co-workers have accomplished this by engineering two thiol-containing nucleosides into a hairpin-forming sequence. The resulting hairpin structure is stabilized quite considerably.[103] They have applied this technology to the study of antibody–DNA interactions,[104] which are likely to be important in certain auto-immune diseases such as lupus erythematosus.

### 7.10.8.3 Polymerase Recognition of Circular DNAs: Rolling Circles

It has been found that the "rolling circle" mechanism can be applied to nucleic acid circles even smaller than the smallest (246 nucleotide) viroid described in Sections 7.10.2.2 and 7.10.3.2. It is now known that circles at least as small as 28 nucleotides can be efficient substrates for RNA polymerases even if the circles lack RNA promoter sequences.[93,94] Such a circle is smaller than the polymerase itself. Likewise, circles at least as small as 26 nucleotides can be very good substrates for DNA polymerase enzymes.[91,92] The products of these polymerizations are repeating RNA or DNA strands thousands of nucleotides in length (Figure 28). Studies of the groups of Fire[91] and of Kool[92] have shown that a repeating DNA product strand can be cut into single-unit lengths by use of a restriction endonuclease. Daubendiek and Kool have shown also that repeating RNA products can be cleaved by a ribozyme into unit-length RNA fragments.[94] It is thought that this approach may be useful in the preparation synthesis of short DNAs and RNAs and for generating useful repeating sequences as well.

Finally, Daubendiek and Kool have shown that DNA circles 83 nt in size can be designed to encode the synthesis of catalytic ribozyme sequences.[95] The transcription of these circles with common RNA polymerases produces repeating RNA sequences which cleave themselves into unit-length RNA strands. Moreover, these monomeric RNA strands can perform useful functions such as cleavage of HIV-1 sequences.[95] It seems possible that not only could such circles be used as templates for preparative RNA synthesis in a test tube, but they also might conceivably be used for that purpose inside human cells.

### 7.10.9 SYNTHESIS OF TOPOLOGICALLY MODIFIED DNA

Aside from branched DNAs, the majority of topological modifications reported in the literature involve ligation of DNAs into circular form. This gives access both to simple circular structures as well as to knotted structures and three-dimensional shapes built from DNA. This section will therefore focus on a review of the practical methods developed for cyclization of DNAs, as these methods can lead to a wide variety of structures with a broad spectrum of applications. The construction of both single- and double-stranded circles, and the use of both enzymatic and non-enzymatic methods to perform the closure will be addressed.

### 7.10.9.1 Closure of Double-stranded Circles

The closure of fully double-stranded DNA circles is generally more facile than that of single-stranded circles. The reasons for this are twofold. First, the duplex structure and complementarity is generally taken advantage of to align the ends and thus promote the desired ligation reaction. Second, the duplex structure in the linear precursor for cyclization is usually quite stable and does not have alternate folded structures, whereas a linear precursor for cyclization can often adopt many possible secondary structures which will not successfully cyclize.

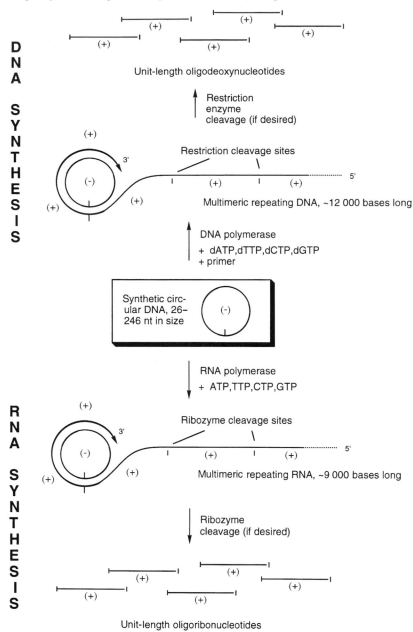

**Figure 28**  Scheme showing how very small single-stranded circular DNAs can behave as catalytic templates for synthesis of DNA (top) or RNA (bottom) oligonucleotides. The polymerase runs around the circle hundreds of times, producing a long repeating DNA[92] or RNA[93] strand, which can then be subsequently cleaved (if desired).

### 7.10.9.1.1  *Enzymatic methods*

By far the most commonly used methods for ligation of double-stranded linear DNAs into closed circular form involve a ligase enzyme, and of these, commercially available T4 DNA ligase is the laboratory standard.[105] This ligase requires a duplex structure and the presence of a phosphate on each 5'-hydroxyl group on each end to be joined. It is common to join both strands of a duplex at once to form a fully closed circular duplex. The enzyme can handle both blunt ends (where both strands of the duplex terminate at the same nucleotide step) and "sticky" ends, where one strand of the duplex overhangs the other by one or more bases (Figure 29).[106] The latter case usually gives more efficient ligation, because sticky ends help the protein organize the reactive ends in the correct conformation for closure.

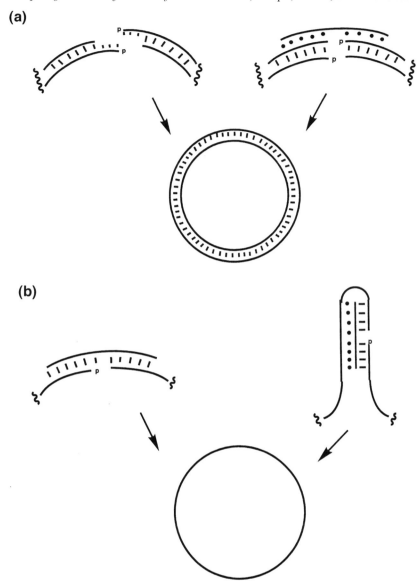

**Figure 29**   Strategies used in the closure of (a) double-stranded circular DNAs and (b) single-stranded circular DNAs.

This method is generally useful for cyclization of plasmid-sized DNAs thousands of base pairs in size, and is a standard part of molecular cloning methods. Smaller circles also can be closed; however, DNA is not infinitely flexible, and circles smaller than ~100 base pairs are extremely difficult if not impossible to close.[55] Indeed, even circles smaller than ~500 bases are slow to close unless there are inherently bent DNA sequences included in the circle precursor.[107,108]

### 7.10.9.1.2   *Nonenzymatic (chemical) methods*

Nonenzymatic methods have been used only rarely for the closure of circular duplex DNAs. In part this is due to the success of the enzymatic ligation when applied to sticky-ended precursors. One method which has shown some success is chemical ligation with *N*-cyanoimidazole and $Zn^{2+}$.[109] A blunt-ended plasmid-length duplex was shown to be closed by use of a triplex-forming splint oligonucleotide. The 30-nucleotide splint was complementary to the two duplex ends when adjacent to each other (Figure 29), and thus triplex formation facilitates the closure reaction.

### 7.10.9.2    Closure of Single-stranded Circles

As mentioned above, the synthesis of single-stranded circles is considerably more problematic than double-stranded circles. One difference from closure of duplex circles is that single-stranded circles can be made considerably smaller because of the added flexibility. However, one problem is that even if a given method for single-strand cyclization is successful in some cases, it is often very difficult to generalize the method because each new sequence may form its own unique secondary structure.

#### 7.10.9.2.1    Enzymatic methods

The simplest enzymatic method for ligation of single-stranded circles is analogous to the method used for duplex DNAs. The difference here is that one uses a separate "splint" oligonucleotide to hybridize to the two reactive ends such that the 5′ phosphate and 3′-hydroxyl are juxtaposed. This can be performed on single-stranded DNAs at least as long as plasmid length, and it can also be successful on circles as small as ∼ 50–60 nucleotides.[91,95] Below that point certain topological questions are raised about whether the splint can bind successfully and whether the enzyme can bind the splint/precircle complex.

#### 7.10.9.2.2    Nonenzymatic (chemical) methods

There are several kinds of chemical method utilized successfully for cyclization of single-stranded circles in the ∼ 26–74 nucleotide size range. One is a water-soluble carbodiimide, which has shown success in the formation of dumbbell DNAs (see next section)[99] but which has not been generally applied to other circle closures. A mixture of cyanogen bromide in a buffer containing $Mg^{2+}$ has also been used in the formation of circles 27–37 nucleotide in size;[110] this reaction is very rapid but success has been dependent on specific kinds of secondary structure in the molecule. A second cyanogen bromide method has been used in the construction of circles 26–74 nucleotides in size.[73,111] In this approach, BrCN is added as a solid in the presence of an imidazole · HCl buffer and $NiCl_2$. However, this approach has shown good success only in the closure of circles which contain a triplex-binding domain; in those cases a purine-rich splint oligonucleotide was added so that the precircle could wrap around it and form a strong triplex (Figure 29).[111] This again places significant sequence requirements on the circle to be made.

Three or four additional approaches to the chemical closure of single-stranded circles are worth mentioning here. First is the random, nontemplated cyclization approach, in which the reactive ends are allowed to diffuse together without assistance of a splint oligomer. This has been successful in the synthesis of very small circular DNAs and RNAs up to ∼ 15–20 bases in size,[112,113] and has been most successful when carried out on a solid support such as polyethylene glycol.[113] The approach gives very low yields for cyclic oligonucleotides larger than this, however. A second approach is the formation of disulfide bonds at the ends of a linear precursor.[114] The reaction can work quite well, but it remains to be seen whether such disulfide bonds are stable in biological media. Another method is the closure of looped DNAs by imine formation.[115] These last two reactions are different from the previous approaches because they produce circular molecules but without continuous DNA chains. The final approach to cyclization is that of Letsinger,[116] who showed that a 5′-tosyl group can survive automated DNA synthesis, and will react quite readily with a phosphorothioate group on the 3′ end of the DNA. This produces a circular DNA having one sulfur replacing one of the phosphodiester oxygens, which would seem to be a relatively small perturbation on the natural structure.

#### 7.10.9.2.3    Dumbbell DNA (RNA) cyclizations

One specific class of circular oligonucleotides which is relatively straightforward to cyclize is the dumbbell structure. Dumbbell structures are relatively easily closed using self-templating effects; that is, placing the gap for the closure near the center of the duplex portion of the molecule. Ligations either by enzymatic or nonenzymatic methods have been successful.[96,99]

### 7.10.9.2.4 *RNA-catalyzed cyclizations*

Because some RNA sequences (for example, hairpin-type ribozymes[117]) are known to be able to catalyze ligations of RNAs, it is possible to construct a catalytic RNA which can ligate two ends of an RNA to closed circular form. It is also possible to design a sequence that will ligate itself to closed form,[118] although this requires a very specific sequence and secondary structure. It is also worth mentioning that splicing reactions can be manipulated to yield circular RNA products.[5,6,14]

### 7.10.10 CONCLUSIONS AND FUTURE PROSPECTS

It is clear that topologically modified DNAs are important in nature, and also that topological modifications will become increasingly useful in the design of new molecules having significant biological activity and application. The design of new DNA-based structures is limited only by the imagination of the researcher and, in the case of single-stranded structures, by the availability of efficient methods for bond formation. As new methods for cyclization of single-stranded DNAs and construction of branched structures become more generally applicable, it seems likely that more topologically modified structures will find practical application in biology and medicine as well as in the design of nanostructured materials and devices.

### ACKNOWLEDGMENT

I thank the National Institutes of Health, the Army Research Office, the New York State Office of Science and Technology, and the Office of Naval Research for support of various aspects of our work involving topologically modified DNAs. I further acknowledge the Dreyfus Foundation for a Teacher-Scholar Award and the Alfred P. Sloan Foundation for a Sloan Fellowship. Finally, I thank all my co-workers, present and past, for their contributions to this work; their names are found in the references.

### 7.10.11 REFERENCES

1. J. D. Watson, N. H. Hopkins, J. W. Roberts, J. A. Steitz, and A. M. Weiner, "Molecular Biology of the Gene," 4th edn., Benjamin/Cummings, Menlo Park, 1987, pp. 193–195.
2. G. Schill, "Catenanes, Rotaxanes, and Knots," Academic Press, New York, 1971.
3. P. Borst, *Trends Genet.*, 1991, **7**, 139.
4. T. A. Shapiro, *Proc. Natl. Acad. Sci. USA*, 1993, **90**, 7809.
5. B. Capel, A. Swain, S. Nicolis, A. Hacker, M. Walter, P. Koopman, P. Goodfellow, and R. Lovell-Badge, *Cell*, 1993, **73**, 1019.
6. B. Bailleul, *Nucl. Acids Res.*, 1996, **24**, 1015.
7. R. H. Symons, *Trends Biochem. Sci.*, 1989, **14**, 445.
8. T. O. Diener, *Trends Microbiol.*, 1993, **1**, 289.
9. D. Hanold and J. W. Randles, *Plant Dis.*, 1991, **75**, 330.
10. M. M. Lai, *Annu. Rev. Biochem.*, 1995, **64**, 259.
11. P. A. Sharp, *Science*, 1987, **235**, 766.
12. H. Domdey, B. Apostol, R. J. Lin, A. Newman, E. Brody, and J. Abelson, *Cell*, 1984, **39**, 611.
13. R. A. Padgett, M. M. Konarska, P. J. Grabowski, S. F. Hardy, and P. A. Sharp, *Science*, 1984, **225**, 898.
14. E. Ford and M. Ares, *Proc. Natl. Acad. Sci. USA*, 1994, **91**, 3117.
15. C. W. Pleij, K. Rietveld, and L. Bosch, *Nucl. Acids Res.*, 1985, **13**, 1717.
16. C. W. Pleij and L. Bosch, *Methods Enzymol.*, 1989, **180**, 289.
17. R. K. Saiki, D. H. Gelfand, S. Stoffel, S. J. Scharf, R. Higuchi, G. T. Horn, K. B. Mullis and H. A. Erlich, *Science*, 1988, **239**, 487.
18. E. H. Blackburn, *Trends Biochem. Sci.*, 1991, **16**, 378.
19. A. D. Branch and H. D. Robertson, *Science*, 1984, **223**, 450.
20. A. Kornberg and T. A. Baker (eds) "DNA Replication," 2nd edn., W. H. Freeman, San Francisco, CA, 1992, p. 113.
21. J. P. Cooper and P. J. Hagerman, *Proc. Natl. Acad. Sci. USA*, 1989, **86**, 7336.
22. N. C. Seeman and N. R. Kalenbach, *Annu. Rev. Biophys. Biomol. Struct.*, 1994, **23**, 53.
23. J. L. Kadrmas, A. J. Ravin, and N. B. Leontis, *Nucl. Acids Res.*, 1995, **23**, 2212.
24. J. A. Pikkemaat, H. van den Elst, J. H. van Boom, and C. Altona, *Biochemistry*, 1994, **33**, 14896.
25. S. C. Kowalczykowski, D. A. Dixon, A. K. Eggleston, S. D. Lauder, and W. M. Rehrauer, *Microbiol. Rev.*, 1994, **58**, 401.
26. A. Luttinger, *Mol. Microbiol.*, 1995, **15**, 601.
27. J. M. Berger, S. J. Gamblin, S. C. Harrison, and J. C. Wang, *Nature*, 1996, **379**, 225.
28. L. F. Liu, L. Perkocha, R. Calendar, and J. C. Wang, *Proc. Natl. Acad. Sci. USA*, 1981, **78**, 5498.

29. S. A. Wasserman and N. R. Cozzarelli, *Science*, 1986, **232**, 951.
30. P. Droge and N. R. Cozzarelli, *Methods Enzymol.*, 1992, **212**, 120.
31. A. Kornberg and T. Baker, "DNA Replications," 2nd edn., W. H. Freeman, New York, 1992, p. 165.
32. X. P. Kong, R. Onrust, M. O'Donnell, and J. Kuriyan, *Cell*, 1992, **69**, 425.
33. P. T. Stukenberg and M. O'Donnell, *J. Biol. Chem.*, 1995, **270**, 13384.
34. J. Kuriyan and M. O'Donnell, *J. Mol. Biol.*, 1993, **234**, 915.
35. Z. Kelman and M. O'Donnell, *Annu. Rev. Biochem.*, 1995, **64**, 171.
36. Z. Kelman and M. O'Donnell, *Nucl. Acids Res.*, 1995, **23**, 3613.
37. L. D. Williams, M. Egli, Q. Gao, P. Bash, G. A. van der Marel, J. H. van Boom, A. Rich, and C. A. Frederick, *Proc. Natl. Acad. Sci. USA*, 1990, **87**, 2225.
38. M. S. Searle, W. Bicknell, L. P. G. Wakelin, and W. A. Denny, *Nucl. Acids Res.*, 1991, **19**, 2897.
39. M. S. Searle, J. G. Hall, W. A. Denny, and L. P. G. Wakelin, *Biochemistry*, 1988, **27**, 4340.
40. K. J. Addess and J. Feigon, *Nucl. Acids Res.*, 1994, **22**, 5484.
41. A. H. Wang, G. Ughetto, G. J. Quigley, T. Hakoshima, G. A. van der Marel, J. H. van Boom, and A. Rich, *Science*, 1984, **225**, 1115.
42. T. Horn and M. S. Urdea, *Nucl. Acids Res.*, 1989, **17**, 6959.
43. Y. L. Wang, J. E. Mueller, B. Kemper, and N. C. Seeman, *Biochemistry*, 1991, **30**, 5667.
44. N. B. Leontis, M. T. Hills, M. Piotto, I. V. Ouporov, A. Malhotra, and D. G. Gorenstein, *Biophys. J.*, 1995, **68**, 251.
45. S. M. Du and N. C. Seeman, *Biopolymers*, 1994, **34**, 31.
46. H. Wang, S. M. Du, and N. C. Seeman, *J. Biomol. Struct. Dyn.*, 1993, **10**, 853.
47. J. Chen and N. C. Seeman, *Nature*, 1991, **350**, 631.
48. Y. Zhang and N. C. Seeman, *J. Am. Chem. Soc.*, 1994, **116**, 1661.
49. B. H. Robinson and N. C. Seeman, *Protein Eng.*, 1987, **1**, 295.
50. Y. Zhang and N. C. Seeman, *J. Am. Chem. Soc.*, 1992, **114**, 2656.
51. M. Egli, R. V. Gessner, L. D. Williams, G. J. Quigley, G. A. van der Marel, J. H. van Boom, A. Rich, and C. A. Frederick, *Proc. Natl. Acad. Sci. USA*, 1990, **87**, 3235.
52. Y. Xu and E. T. Kool, manuscript in preparation.
53. M. Wolters and B. Wittig, *Nucl. Acids Res.*, 1989, **17**, 5163.
54. T. Pan, R. R. Gutell, and O. C. Uhlenbeck, *Science*, 1991, **254**, 1361.
55. S. S. Zinkel and D. M. Crothers, *Nature*, 1987, **328**, 178.
56. K. Luger, U. Hommel, M. Herold, J. Hofsteenge, and K. Kirschner, *Science*, 1989, **243**, 206.
57. K. Zahn and F. R. Blattner, *Science*, 1987, **236**, 416.
58. D. N. Paolella, C. R. Palmer and A. Schepartz, *Science*, 1994, **264**, 1130.
59. J. K. Strauss and L. J. Maher, *Science*, 1994, **266**, 1829.
60. M. H. Werner, A. M. Gronenborn and G. M. Clore, *Science*, 1996, **271**, 778.
61. E. Uhlmann and A. Peyman, *Chem. Rev.*, 1990, **90**, 543.
62. S. L. Beaucage and R. P. Iyer, *Tetrahedron*, 1992, **48**, 2223.
63. R. H. E. Hudson and M. J. Damha, *Nucl. Acids Res. Symp. Ser.*, 1993, **29**, 97.
64. R. H. E. Hudson, A. H. Uddin, and M. J. Damha, *J. Am. Chem. Soc.*, 1995, **117**, 12470.
65. G. Brandenburg, G. V. Petersen, and J. Wengel, *Bioorg. Med. Chem. Lett.*, 1995, **5**, 791.
66. T. Horn, B. D. Warner, J. A. Running, K. Downing, J. Clyne, and M. S. Urdea, *Nucleosides Nucleotides*, 1989, **8**, 875.
67. G. Prakash and E. T. Kool, *J. Chem. Soc., Chem. Commun.*, 1991, 1161.
68. S. Wang and E. T. Kool, *Nucl. Acids Res.*, 1994, **22**, 2326.
69. R. W. Roberts and D. M. Crothers, *Science*, 1992, **258**, 1463.
70. S. Wang and E. T. Kool, *Nucl. Acids Res.*, 1995, **23**, 1157.
71. M. Nilsson, H. Malmgren, M. Samiotaki, M. Kwiatkowski, B. C. Chowdhary, and U. Landegren, *Science*, 1994, **265**, 2085.
72. E. T. Kool, *J. Am. Chem. Soc.*, 1991, **113**, 6265.
73. G. Prakash and E. T. Kool, *J. Am. Chem. Soc.*, 1992, **114**, 3523.
74. S. Wang, M. A. Booher, and E. T. Kool, *Biochemistry*, 1994, **33**, 4639.
75. S. Rumney IV and E. T. Kool, *J. Am. Chem. Soc.*, 1995, **117**, 5635.
76. S. Wang, A. Friedman, and E. T. Kool, *Biochemistry*, 1995, **34**, 9774.
77. D. J. D'Souza and E. T. Kool, *Bioorg. Med. Chem. Lett.*, 1994, **4**, 965.
78. D. J. D'Souza and E. T. Kool, *J. Biomol. Struct. Dyn.*, 1992, **10**, 141.
79. E. Azhayeva, A. Azhayev, A. Guzaev, J. Hovinen, and H. Lonnberg, *Nucl. Acids Res.*, 1995, **23**, 1170.
80. E. Rubin, T. L. McKee, and E. T. Kool, *J. Am. Chem. Soc.*, 1993, **115**, 360.
81. E. Rubin and E. T. Kool, *Angew. Chem.*, 1994, **106**, 1057; *Angew. Chem. Int. Ed.*, 1994, **33**, 1004.
82. H. R. B. Pelham and D. D. Brown, *Proc. Natl. Acad. Sci. USA*, 1980, **77**, 4170.
83. N. C. Chaudhuri and E. T. Kool, *J. Am. Chem. Soc.*, 1995, **117**, 10434.
84. K. Ryan, Ph.D. Thesis, University of Rochester, 1996.
85. D. B. Amabilino and J. F. Stoddardt, *Chem. Rev.*, 1995, **95**, 2725.
86. K. Ryan and E. T. Kool, *Chemistry and Biology*, 1998, **5**, in press.
87. B. C. Chu and L. E. Orgel, *Nucl. Acids Res.*, 1992, **20**, 2497.
88. C. Clusel, E. Ugarte, N. Enjolras, M. Vasseur, and M. Blumenfeld, *Nucl. Acids Res.*, 1993, **21**, 3405.
89. B. C. Chu and L. E. Orgel, *Nucl. Acids Res.*, 1992, **20**, 5857.
90. R. Morishita, G. H. Gibbons, M. Horiuchi, K. E. Elison, M. Nakama, L. Zhang, Y. Kaneda, T. Ogihara, and V. J. Dzau, *Proc. Natl. Acad. Sci. USA*, 1995, **92**, 5855.
91. A. Fire and S.-Q. Xu, *Proc. Natl. Acad. Sci. USA*, 1995, **92**, 4641.
92. D. Liu, S. L. Daubendiek, M. A. Zillmann, K. Ryan, and E. T. Kool, *J. Am. Chem. Soc.*, 1996, **118**, 1587.
93. S. L. Daubendiek, K. Ryan, and E. T. Kool, *J. Am. Chem. Soc.*, 1995, **117**, 7818.
94. S. L. Daubendiek, Ph.D. Thesis, University of Rochester, 1998.
95. S. L. Daubendiek and E. T. Kool, *Nature Biotech.*, 1997, **15**, 273.
96. D. E. Wemmer and A. S. Benight, *Nucl. Acids Res.*, 1985, **13**, 8611.

97. D. A. Erie, R. A Jones, W. K. Olson, N. K. Sinha, and K. J. Breslauer, *Biochemistry*, 1989, **28**, 268.
98. M. J. Koktycz, R. F. Goldstein, T. M. Paner, F. J. Gallo, and A. S. Benight, *Biopolymers*, 1992, **32**, 849.
99. G. W. Ashley and D. M. Kushlam, *Biochemistry*, 1991, **30**, 2927.
100. B. A. Sullenger, H. F. Gallardo, G. E. Ungers, and E. Gilboa, *Cell*, 1990, **63**, 601.
101. M. Y. X Ma, K. McCallum, S. C. Climie, R. Kuperman, W. C. Lin, M. Sumner-Smith, and R. W. Barnett, *Nucl. Acids Res.*, 1993, **21**, 2585.
102. G. D. Glick, S. E. Osborne, D. E. Knitt, and J. P. Marino, *J. Am. Chem. Soc.*, 1992, **114**, 5447.
103. S. Y. Stevens, P. C. Swanson, E. W. Voss, and G. D. Glick, *J. Am. Chem. Soc.*, 1993, **115**, 1585.
104. P. C. Swanson, B. C. Cooper, and G. D. Glick, *J. Immunol.*, 1994, **152**, 2601.
105. V. Sgaramella and S. D. Ehrlich, *Eur. J. Biochem.*, 1978, **86**, 531.
106. J. Sambrook, E. F. Fritsch, and T. Maniatis, "Molecular Cloning," 2nd edn., Cold Spring Harbor Press, Cold Spring Harbor, 1989, p. 1.63.
107. D. Shore, J. Langowski, and R. L. Baldwin, *Proc. Natl. Acad. Sci. USA*, 1981, **78**, 4833.
108. M. A. Livshits and Y. L. Lyubchenko, *Mol. Biol.*, 1994, **28**, 687.
109. K. J. Luebke and P. B. Dervan, *J. Am. Chem. Soc.*, 1989, **111**, 8733.
110. N. G. Dolinnaya, M. Blumenfeld, I. M. Merenkova, T. S. Oretskaya, N. F. Krynetskaya, M. G. Ivanovskaya, M. Vasseur, and Z. A. Shabarova, *Nucl. Acids. Res.*, 1993, **21**, 5403.
111. E. Rubin, S. Rumney IV, and E. T. Kool, *Nucl. Acids Res.*, 1995, **23**, 3547.
112. M. L. Capobianco, A. Carcuro, L. Tondelli, A. Garbesi, and G. M. Bonora, *Nucl. Acids. Res.*, 1990, **18**, 2661.
113. L. DeNapoli, A. Messere, D. Montesarchio, G. Piccialli, and C. Santocroce, *Nucleosides Nucleotides*, 1993, **12**, 21.
114. H. Gao, M. Yang, R. Patel, and A. F. Cook, *Nucl. Acids Res.*, 1995, **23**, 2025.
115. D. Peoc'h, J.-L. Imbach, and B. Rayner, *Nucleosides Nucleotides*, 1995, **14**, 847.
116. M. K. Herrlein, J. S. Nelson, and R. L. Letsinger, *J. Am. Chem. Soc.*, 1995, **117**, 10151.
117. A. Berzal-Herranz, S. Joseph, and J. M. Burke, *Genes Dev.*, 1992, **6**, 129.
118. P. A. Feldstein and G. Bruening, *Nucleic Acids Res.*, 1993, **21**, 1991.

# 7.11
# Chemistry of DNA Damage

MARC M. GREENBERG

*Colorado State University, Fort Collins, CO, USA*

| | | |
|---|---|---|
| 7.11.1 | INTRODUCTION | 372 |
| 7.11.1.1 | *The Prevalence of DNA Damage* | 372 |
| 7.11.1.2 | *General Modes of DNA Damage* | 372 |
| 7.11.2 | GENERATION AND REACTIVITY OF DEOXYRIBOSE-CENTERED REACTIVE INTERMEDIATES | 373 |
| 7.11.2.1 | *Reaction at the C-1′ Position of Nucleotides* | 373 |
| 7.11.2.1.1 | *DNA-damaging agents that react at the C-1′ position* | 373 |
| 7.11.2.1.2 | *Mechanism for the formation of the 2′-deoxyribonolactone (1) intermediate* | 376 |
| 7.11.2.1.3 | *Direct strand break formation vs. 2′-deoxyribonolactone (1) formation* | 380 |
| 7.11.2.2 | *Reaction at the C-2′ Position of Nucleotides* | 380 |
| 7.11.2.2.1 | *Reactivity of C-2′ nucleoside radicals under anaerobic conditions* | 381 |
| 7.11.2.2.2 | *Reactivity of C-2′ nucleoside radicals under aerobic conditions* | 382 |
| 7.11.2.3 | *Reaction at the C-3′ Position of Nucleotides* | 383 |
| 7.11.2.4 | *Reaction at the C-4′ Position of Nucleotides* | 384 |
| 7.11.2.4.1 | *The mechanism of formation of alkaline labile lesions resulting from C-4′ hydrogen atom abstraction* | 387 |
| 7.11.2.4.2 | *The mechanism of formation of direct strand breaks resulting from C-4′ hydrogen atom abstraction* | 387 |
| 7.11.2.4.3 | *Similarities and contrasts between the Criegée and Schulte-Frohlinde/Grob fragmentation pathways for DNA cleavage via C-4′ hydrogen atom abstraction* | 394 |
| 7.11.2.5 | *DNA Damage via Reaction at the C-5′ Position of Nucleotides* | 394 |
| 7.11.2.5.1 | *Agents that react at the C-5′ position* | 394 |
| 7.11.2.5.2 | *The reactivity of C-5′ radicals* | 395 |
| 7.11.3 | GENERATION AND REACTIVITY OF NUCLEOBASE-REACTIVE INTERMEDIATES | 397 |
| 7.11.3.1 | *Hydroxy Radical and Hydrogen Atom Adducts of Pyrimidines* | 398 |
| 7.11.3.1.1 | *Formation of pyrimidine nucleobase radical adducts* | 398 |
| 7.11.3.1.2 | *Hydrogen atom abstraction by pyrimidine nucleobase radical adducts* | 399 |
| 7.11.3.1.3 | *Formation and reactivity of peroxy radicals derived from pyrimidine radical adducts* | 402 |
| 7.11.3.1.4 | *Formation and reactivity of pyrimidine alkene cation radicals* | 405 |
| 7.11.3.1.5 | *The formation and effects of modified pyrimidines on DNA structure and function* | 407 |
| 7.11.3.2 | *Purine-centered Reactive Intermediates* | 408 |
| 7.11.3.2.1 | *Hydroxy radical adducts of purine nucleotides* | 409 |
| 7.11.3.2.2 | *Alkene cation radicals of purines* | 410 |
| 7.11.3.2.3 | *The role of purine nucleobase-centered reactive intermediates in strand scission* | 410 |
| 7.11.3.2.4 | *The formation and effects of modified purines on DNA structure and function* | 410 |
| 7.11.3.3 | *The Role of 2′-Deoxyuridin-5-yl (28) in Strand Damage* | 411 |
| 7.11.4 | DNA DAMAGE SENSITIZATION AND PROTECTION AGENTS | 417 |
| 7.11.4.1 | *The Effects of Metal Ions on DNA Damage* | 417 |
| 7.11.4.2 | *The Effects of Electron Affinic Agents and Nitroxides on DNA Damage* | 417 |
| 7.11.5 | THE FORMATION OF BISTRANDED LESIONS AND DOUBLE STRAND BREAKS IN DNA | 418 |
| 7.11.5.1 | *Double Strand Breaks Produced via Interstrand Hydrogen Atom or Electron Transfer* | 418 |

7.11.5.2   *DNA Damage Amplification*                                                           419
7.11.5.3   *Bistranded Lesions and Double Strand Breaks Produced by Antitumor Antibiotics*        420

7.11.6   REFERENCES                                                                              421

## 7.11.1   INTRODUCTION

### 7.11.1.1   The Prevalence of DNA Damage

DNA damage is induced via a wide spectrum of agents, both natural and made-made. In the context of disease, DNA damage presents a paradox. Depending upon the conditions, DNA damage can be either causative or therapeutic. This dichotomy becomes apparent when one considers the etiology and treatment of cancer, a disease with which DNA damage is typically associated. Nucleobase modification can give rise to mutations. When situated in the region of DNA that codes for an appropriate gene, the mutation can give rise to cancer.[1-4] In contrast, DNA is the target of ionizing radiation and many antitumor antibiotics that are used to treat cancer.

Cancer is not the only disease with which DNA damage is associated. Evidence is accumulating that suggests that there is a relationship between DNA damage produced by reactive oxygen species *in vivo* and aging.[4,5] It appears that the increase in the accumulation of DNA damage as various species age is due to a combination of factors that include increased production of reactive oxygen species and decreased enzymatic repair activity.[6]

The latter factor suggests that DNA damage can be a secondary effect of aging. Evidence is also accumulating that DNA damage may be a secondary result of several diseases. For instance, cystic fibrosis patients exhibit increased susceptibility to DNA damage.[7] Furthermore, additional tissue damage occurs upon reperfusion of ischemic tissue.[8,9] One explanation for damage during reperfusion involves the formation of a burst of reactive oxygen species upon reperfusion. The reactive oxygen species produced during myocardial reperfusion are believed to be formed via metal-catalyzed processes.

In a related vein, it was proposed that the amyloid precursor protein (APP) which is associated with Alzheimer's disease, reduces copper(II) to copper(I).[10-12] As copper(I) is known to be capable of forming reactive oxygen species, it has been postulated that Alzheimer's disease could contribute to copper toxicity by decompartmentalizing copper(I).

These examples serve to illustrate that DNA damage, whether a primary cause, or secondary effect, is associated with a wide range of health-related issues. This chapter is concerned with the varied chemistry of DNA damage. Other reviews of this subject have appeared.[13-15] Hence, efforts will be made to minimize duplication of these other reviews.

### 7.11.1.2   General Modes of DNA Damage

The nucleotides in DNA offer a wide array of reactive sites to oxidizing, reducing, and alkylating agents, despite principally consisting of only four monomeric units. In addition, the phosphate linkage between nucleotides offers a site for hydrolysis. One or more DNA-damaging agents have been characterized that react with the biopolymer by each of these pathways.

Examples abound (mostly in the areas of DNA polymerases and DNA repair) of enzymatic phosphate hydrolysis.[16] In contrast, the number of unnatural molecules that effect DNA phosphate hydrolysis are few in number and are generally much less proficient than their enzymatic counterparts.[17-19] This mode of DNA damage will not be covered in this chapter.

Nonenzymatic DNA damage induced via alkylation of the nucleobases is far more common than phosphodiester hydrolysis. Numerous natural products, derivatives thereof and unnatural products damage DNA via alkylation. Alkylating agents typically do not produce direct strand breaks in DNA, but instead give rise to alkaline labile lesions. Strand scission is induced via elimination of abasic sites that are formed following hydrolysis of the destabilized glycosidic bond of the alkylated nucleotides. The DNA binding of alkylating agents, as well as their subsequent activation and reaction with DNA, is a rich subject that has been reviewed elsewhere.[20-23] The chemistry following hydrolysis of the glycosidic bond is independent of the alkylation event, and will only be discussed here as a component of DNA damage resulting from reaction at the C-1′ position of nuclcotides.

The intent of this chapter is to focus on the chemistry of specific reactive intermediates that are formed within DNA as a result of the interaction between the biopolymer and various damaging agents. While the nature of the agents that generate particular reactive intermediates will be touched upon, emphasis will be placed upon the chemical transformations that occur following the formation of individual intermediates. One goal of this chapter is to compare and contrast the DNA damage resulting from common intermediates produced by different agents. The reactivity of intermediates produced as a result of nucleobase oxidation and reduction, as well as those produced as a result of hydride or hydrogen atom abstraction from the deoxyribose moiety, will be covered. In so doing, DNA damage induced by natural and unnatural products, as well as ionizing radiation and radiomimetic systems, will be discussed. When discussing ionizing radiation, the terms direct effect and indirect effect will be referred to.[13] Damage induced as a result of the indirect effect of ionizing radiation typically involves solvated electrons, hydroxyl radicals, and/or hydrogen atoms derived from the ionization of water. DNA damage that originates with the ionization of the nucleobases will be the focus of the discussion concerning the direct effects of ionizing radiation.

## 7.11.2 GENERATION AND REACTIVITY OF DEOXYRIBOSE-CENTERED REACTIVE INTERMEDIATES

### 7.11.2.1 Reaction at the C-1′ Position of Nucleotides

Consideration of inherent bond strengths of the various carbon–hydrogen bonds in deoxyribose leads to the prediction that the C-1′ position (the anomeric position) should be relatively reactive in DNA. *Ab initio* calculations indicate that the bond dissociation energy of the C-1′ carbon–hydrogen bond is the lowest of all of the carbon–hydrogen bonds in deoxyribose.[24,25] When the accessibility of the C-1′ hydrogen atom, which is located deep in the minor groove of duplex DNA, is included in calculations, its reactivity is predicted to be significantly reduced relative to the C-4′ carbon–hydrogen bond which is of similar strength. In spite of its relatively poor accessibility, the C-1′ position of nucleotides is believed to be directly attacked by a variety of DNA-damaging agents.[26–33] The anomeric position of nucleotides is also believed to be involved in the propagation of DNA damage within a biopolymer.[34,35] This latter issue will be discussed later.

Formation of damaged DNA containing a labile 2′-deoxyribonolactone lesion (**1**) represents formal oxidation of the C-1′ position in the biopolymer.[36] In fact, formation of (**1**), and/or the furanone (**2**) elimination product that is presumably derived from it, are often used as flags for hydrogen atom or hydride abstraction from the C-1′ position of nucleotides (Scheme 1).

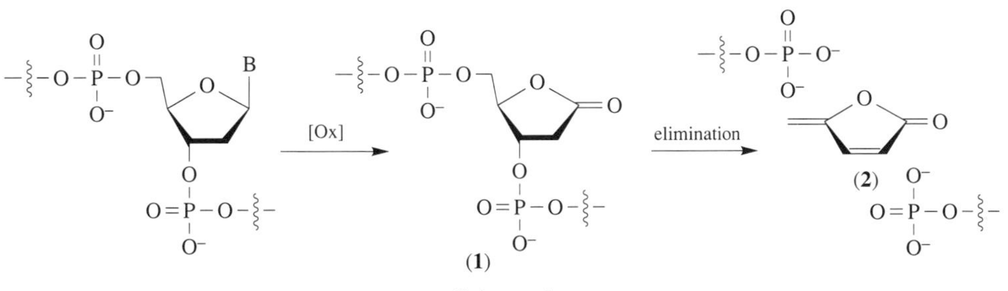

**Scheme 1**

#### 7.11.2.1.1 DNA-damaging agents that react at the C-1′ position

C-1′ hydrogen atom abstraction is believed to comprise a principal reaction pathway for neocarzinostatin (**3**), manganese porphyrins (**4**) and the bisphenanthroline complex of copper (Cu·(OP)$_2$, (**5**)). The former two oxidants give rise to DNA containing a 2′-deoxyribonolactone lesion (**1**) at the site where the initial attack takes place. In contrast, (**5**) gives rise to direct strand breaks which are accompanied by stoichiometric formation of (**2**).[37] The same 2′-deoxyribonolactone (**1**) produced in the DNA damage induced by neocarzinostatin (**3**) and manganese porphyrins (**4**) has been suggested to be an intermediate in Cu(OP)$_2$ (**5**) oxidation of DNA. In keeping with this sequence of transformations, a metastable DNA fragmentation product was observed by gel electrophoresis, which was assigned as the $\alpha,\beta$-unsaturated lactone (**6**).[38] Upon heating, intermediate cleavage product (**6**) was transformed into that containing a 3′-phosphate (Scheme 2). To this reviewer's knowledge, no further characterization of (**6**) has been reported.

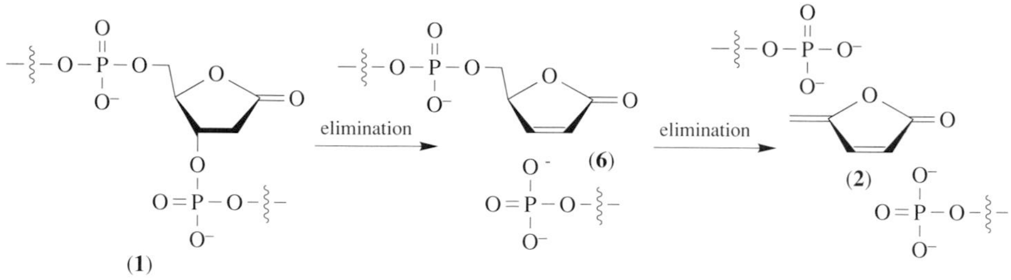

**Scheme 2**

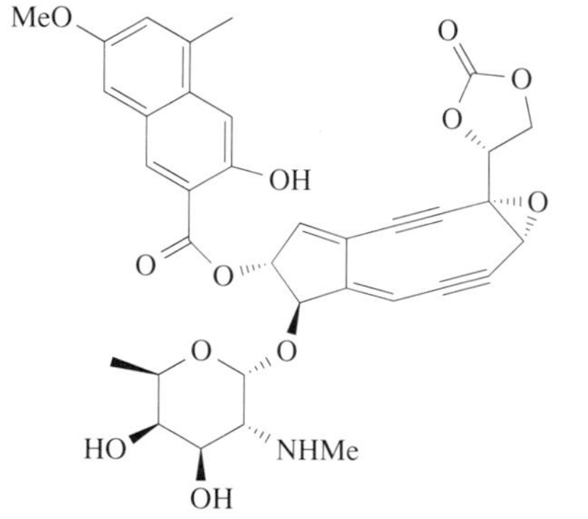

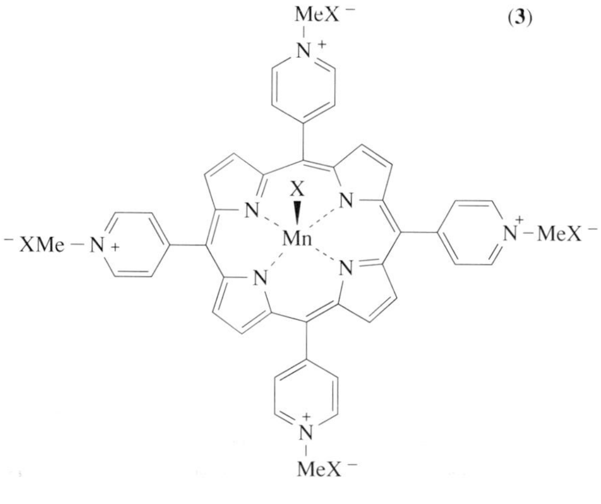

The behavior of a family of oxoruthenium(IV) complexes (e.g., (**7**)–(**9**)), vis-à-vis product formation, is remarkably similar to that observed with Cu(OP)$_2$.[29,39–41] An intermediate fragmentation product, whose gel mobility and chemical behavior was consistent with (**6**), was detected by electrophoresis. The characteristic sugar degradation product, methylene furanone (**2**), was also detected by HPLC. In contrast to neocarzinostatin (**3**), manganese porphyrins (**4**), and Cu(OP)$_2$ (**5**), the oxoruthenium(IV) complexes(**7**)–(**9**) are believed to effect direct two-electron oxidation via hydride abstraction from the anomeric center.[28] This proposal is consistent with mechanistic studies carried out on Ru(tpy)(bpy)O$^{2+}$ (**7**), which was shown to abstract hydrides from organic substrates.[42]

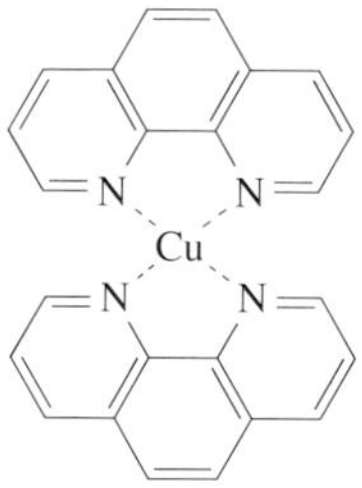

[RuO(tpy)(bpy)]$^{2+}$     [RuO(dppz)(tpy)]$^{2+}$     [RuO(tpy)(η$^2$-tpt)]$^{2+}$

(**7**)            (**8**)             (**9**)

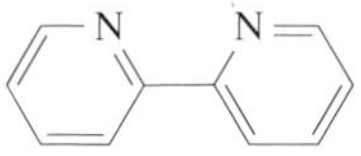

bpy

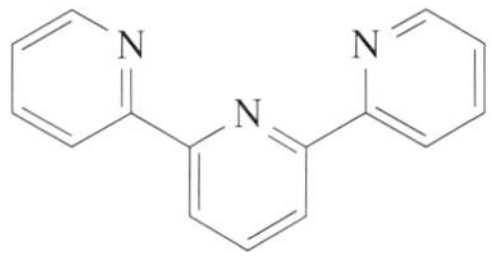

tpy

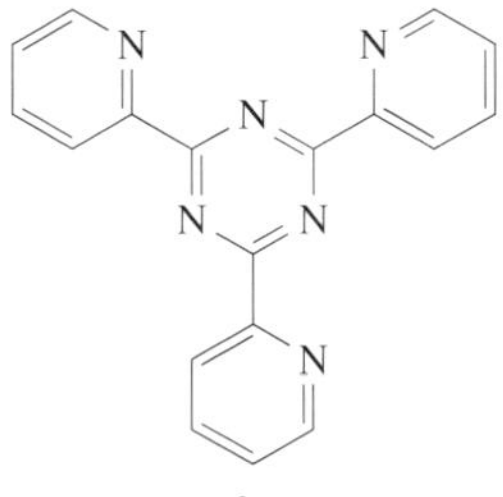

η$^2$-tpt

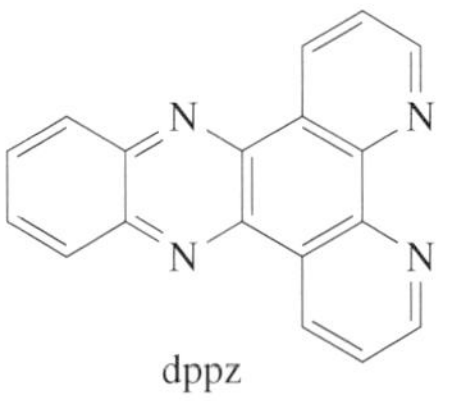

dppz

Observations regarding the photoionization of deoxycytidine suggest that one should exercise caution when proposing a mechanism for DNA damage that involves reaction at the C-1′ position where product analysis is the sole experimental measurement.[43] Irradiation of menadione (2-methyl-1,4-naphthoquinone, (**10**)) results in photoinduced electron transfer from 2′-deoxycytidine (Scheme 3). The formation of small amounts of cytosine (4.5%) and 2′-deoxyribonolactone (**1**) (4.5%) were attributed to deprotonation of the cation radical at C-1′ (**11**). The resulting C-1′-centered radical (**12**) is proposed to give rise to the free base and (**1**). Two caveats must be considered when weighing the relevance of this observation to the chemistry of DNA damage. Stereoelectronic effects controlled via conformational constraints within duplex DNA could affect the relative rate of deprotonation at C-1′ compared to other pathways (e.g., hydration) available to the cation radical. Furthermore, no 2′-deoxyribonolactone (**1**) is observed upon irradiation of menadione in the presence of thymidine. This suggests that formal C-1′ oxidation resulting from nucleobase cation radicals may not be general for all nucleotides in DNA.

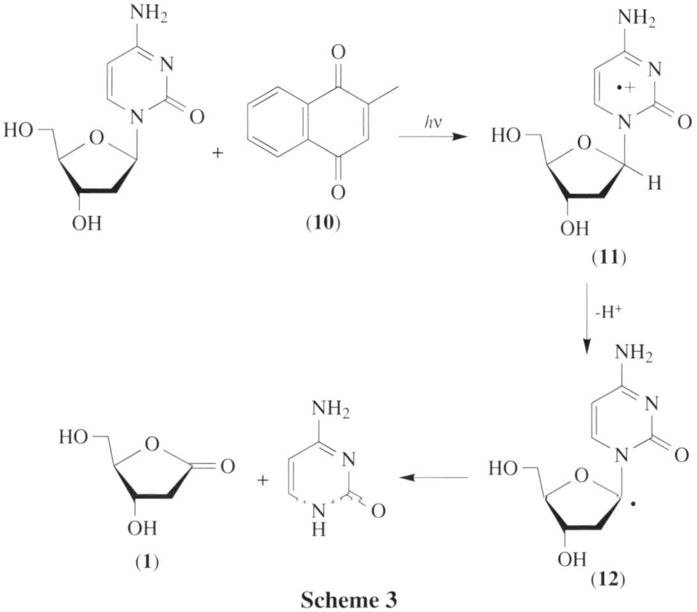

**Scheme 3**

With these caveats in mind, it is reasonable to assume that the C-1′ positions of nucleotides in DNA are targeted by ionizing radiation. In addition to 2′-deoxyribonolactone (**1**) formation, α-nucleoside (e.g., (13)) formation under anaerobic conditions supports C-1′ hydrogen atom abstraction during γ-radiolysis.[44] When present in DNA templates, α-deoxyadenosine (13) has been shown to induce DNA polymerase to misincorporate dATP across from it, classifying (13) as a premutagenic lesion.[45]

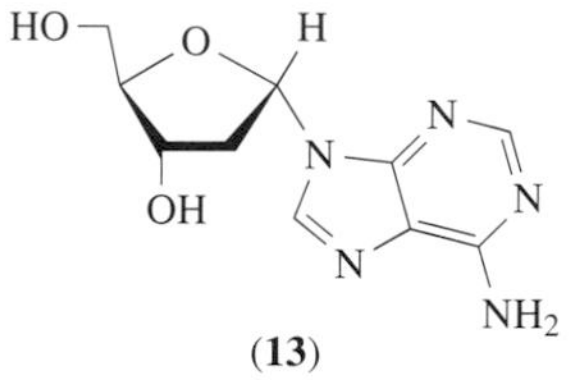

(13)

One question that arises from the formation of the premutagenic α-deoxyadenosine (13) lesion during γ-radiolysis is whether its formation is competitive with trapping by $O_2$, which is present *in vivo*. Trapping of the C-1′ nucleotide radical (12) by $O_2$ gives rise to 2′-deoxyribonolactone (1). (The possible mechanism of this transformation is discussed below.) In principle, one could ascertain whether these processes are competitive by simply measuring the amounts of the respective products produced under aerobic γ-radiolysis conditions in the presence of thiols. However, this is complicated by the possibility that lactone can be formed via other pathways (see above), and that a multitude of other pathways are operative during γ-radiolysis. The latter issue also complicates quantitative product analysis. Such studies are simplified by independent generation of the C-1′ nucleoside radical. Anomeric nucleoside radicals have been generated under radical chain conditions via intramolecular rearrangements ((14–16) and (17–19)), as well as via Norrish Type I, containing photocleavage from Structures (14)–(21) (Scheme 4).[46–48] Photolysis of (20) under aerobic conditions in the presence of varying concentrations of β-thioethanol enabled estimation of the rate of hydrogen atom abstraction by (21) from the thiol to be $3.7 \times 10^6$ $M^{-1}$ $s^{-1}$.[46] Provided that anomeric radicals in DNA are trapped with similar diastereoselectivity as (21), it was concluded that ≈4% of α-deoxynucleotides (e.g., (13) and (22)) would be formed via trapping by thiols present at *in vivo* concentration (5–10 mM) levels. Trapping of (21) by β-thioethanol occurs in a nonstereoselective manner. However, the diastereoselectivity of this process in duplex DNA is unknown, and could be significant. Provided that the anomeric radical (e.g., (21)) is not extrahelical, the DNA duplex may favor hydrogen atom delivery to the α face, selectively forming β-nucleotides. Clearly, the stereoselectivity of this process in duplex DNA is relevant to the importance of α-nucleotides as premutagenic lesions.

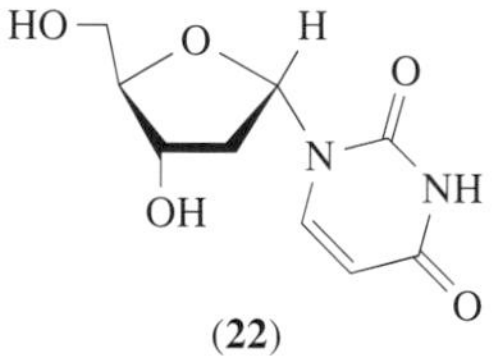

(22)

### 7.11.2.1.2  *Mechanism for the formation of the 2′-deoxyribonolactone (1) intermediate*

Damaged DNA containing the 2′-deoxyribonolactone lesion (1) incorporated within the biopolymers is produced via a variety of oxidative processes. Furthermore, (1) may be a reactive intermediate in other DNA damage processes (Scheme 1). One can envision this lesion arising via a variety of mechanisms.

Direct two-electron oxidation would give rise to the C-1′ carbocation, which upon trapping by $H_2O$ would yield the lactone upon hydrolysis. Indeed, this is the proposed mechanism for strand damage resulting from the family of oxoruthenium(IV) complexes (7)–(9) introduced above (Scheme 5).[28,39,40] Further support that this mechanism accounts for the observed furanone (2) produced by these oxidants could have been provided using $H_2{}^{18}O$ labeling and/or by examining the effect of $O_2$ concentration on product formation. Since the oxoruthenium(IV)-mediated oxidations are typically effected electrochemically, 2′-deoxyribonolactone (1) (and hence furanone, 2) formation should be independent of $O_2$ concentration. Furthermore, furanone formed in $H_2{}^{18}O$ should incorporate oxygen-18, and the level of oxygen-18 incorporation should also be independent of $O_2$ concentration. While these experiments have not yet been reported, such a mechanism is consistent with previously

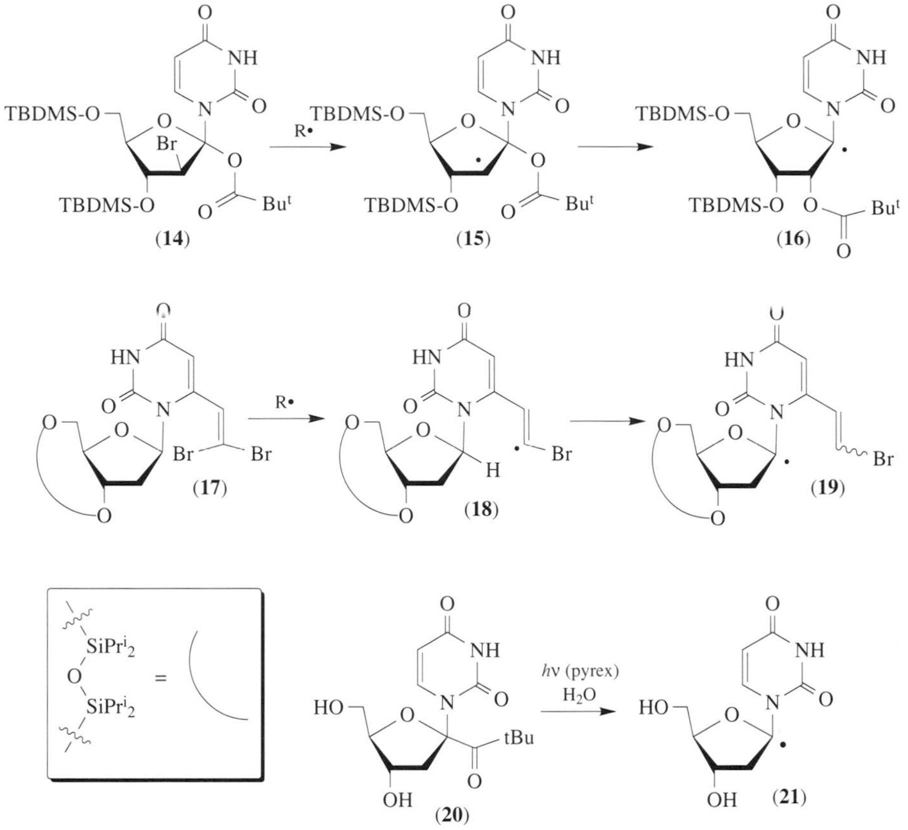

**Scheme 4**

reported studies concerning the mechanism of oxidation of organic substrates of oxoruthenium(IV) complexes.[42]

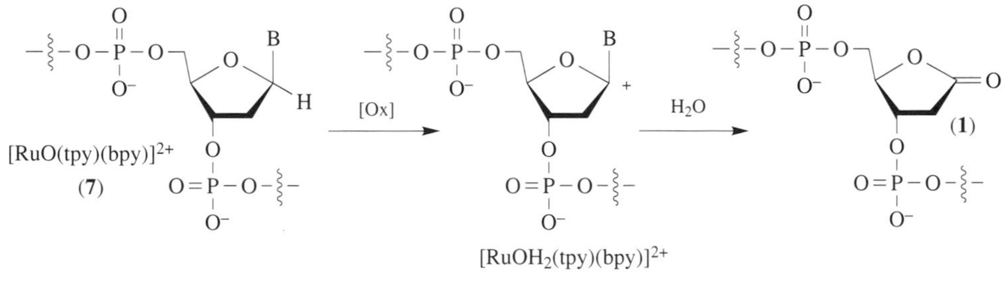

**Scheme 5**

Direct two-electron oxidation by DNA-damaging agents are more the exception than the rule. The majority of DNA damage events involving oxidation at the C-1′ position can be linked to the intermediacy of the C-1′ radical (Scheme 6). Although the oxidation potential ($E_{OX}$) of these radicals are not known, estimates based upon similarly substituted systems suggest that the $E_{OX}$ of C-1′ radicals of nucleotides could be of the order of 0.3 V (Figure 1).[49] This is relatively favorable in comparison to what one might expect for radicals at other positions of the deoxyribose ring of DNA, or the ribose ring of RNA. Hence, one possible route to the 2′-deoxyribonolactone (1) moiety through the C-1′ radical could involve oxidation of the C-1′ radical. Potential oxidants include DNA-damaging agents such as Cu(OP)$_2$ (5), or even free metal ions. The results of such an investigation have not been reported.

Under most circumstances, the question of how 2′-deoxyribonolactone (1) and/or furanone (2) lesions are formed under aerobic conditions is also uncertain. It is tempting to draw a mechanism

| $E_{ox}$ (V) (vs. SCE) | 0.30 | -0.35 | -0.10 | -1.30 |
|---|---|---|---|---|

**Figure 1** Experimental oxidation potentials of radicals analogous to the C-1′ radical of nucleosides.[49]

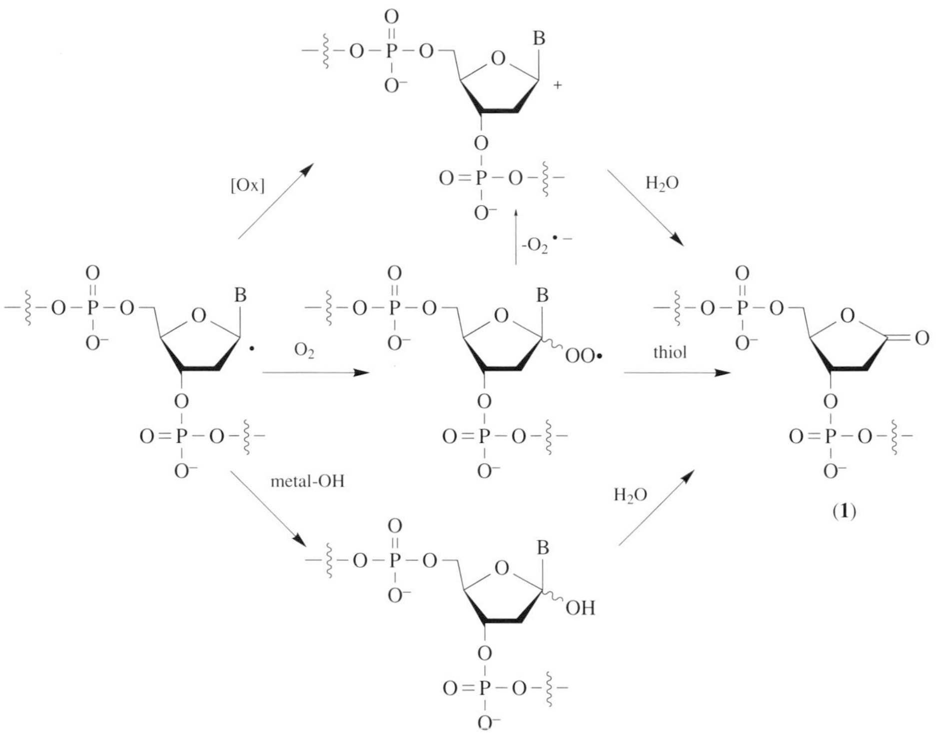

**Scheme 6**

that involves recombination of two peroxy radicals. The intermediate tetroxide would then fragment to two alkoxy radicals en route to 2′-deoxyribonolactone. This mechanism is unlikely to occur in DNA, because it is unlikely that two peroxy radicals would be formed close enough within a strand (or duplex) of DNA to react with each other. It is also unlikely that two peroxy radicals on different strands (or duplexes) of DNA will collide in solution, let alone in a cell. In addition, tertiary peroxy radicals, such as those derived from the C-1′ position of nucleotides, have relatively unfavorable rate constants for dimerization compared to other bimolecular processes that they can participate in (e.g., trapping by thiols).[50,51]

Unimolecular decomposition of an intermediate peroxy radical is a possible alternative pathway that can account for the formation of 2′-deoxyribonolactone (**1**) lesions. Laser flash photolysis studies utilizing conductivity detection estimated the loss of superoxide ($O_2^{\cdot-}$) from (**23**) to be $> 7 \times 10^4$ s$^{-1}$ (Equation (1)).[52] More recent experiments utilizing the peroxy radical derived from the dimethyl acetal of acetaldehyde (**24**) as a model for the C-1′ peroxy radical of a nucleoside indicated that superoxide is cleaved with a rate constant $\sim 6.5 \times 10^4$ s$^{-1}$ at pH 5.0 (Equation (2)).[53] If one assumes that thiols, which are likely hydrogen atom donors *in vivo*, are present at $\leqslant 10$ mM, and that the respective peroxy radicals are trapped by thiols with a rate constant between $\sim 5 \times 10^3$ M$^{-1}$ s$^{-1}$ and $5 \times 10^4$ M$^{-1}$ s$^{-1}$, then unimolecular decomposition of nucleotide peroxy radicals with concomitant $O_2^{\cdot-}$ production will compete with their reduction by thiol.[54,55] Considering the increasing recognition of the biological importance of $O_2^{\cdot-}$, such a process could have significant biological relevance.[56,57] Indeed, examination of the reactivity of (**25**) under aerobic conditions supports the involvement of such a process (Scheme 7).[58] Qualitative evidence for superoxide production was gleaned from the epinephrine (**26**) spectrophotometric assay (Equation (3)). Adrenochrome (**27**) is produced when (**20**) is irradiated in the presence of $O_2$ and (**26**). Quenching of the formation of (**27**) in the presence of superoxide dismutase supported the proposal that superoxide is responsible for the oxidation of (26). When (20) was irradiated in $H_2O$, the ratio of $^{16}O$-(1):$^{18}O$-(1) varied linearly with the concentration of $\beta$-thioethanol.[58] Approximately 4% of (25) eliminates $O_2^{\cdot-}$ in the presence of 5 mM $\beta$-thioethanol. Assuming that the bimolecular trapping rate constant by thiol is between $5 \times 10^3$ $M^{-1}$ $s^{-1}$ and $5 \times 10^4$ $M^{-1}$ $s^{-1}$, the rate of loss of superoxide was estimated to be $> 1.4$ $s^{-1}$. Hence, it is possible that the formation of 2′-deoxyribonolactone lesions is due to the pathway delineated by the model compounds in Equations (1) and (2), as well as in Scheme 6.

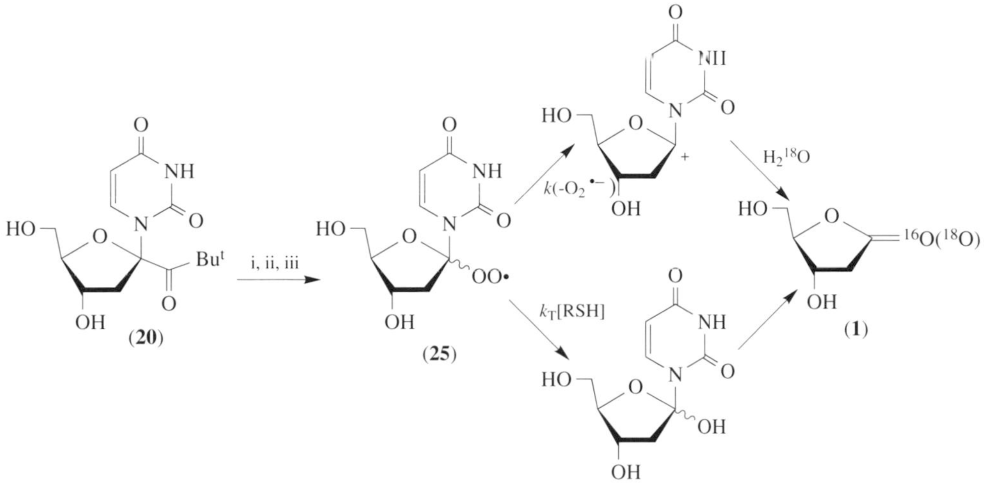

i, *hv*; ii, $O_2$, RSH ; iii, $H_2{}^{18}O$

**Scheme 7**

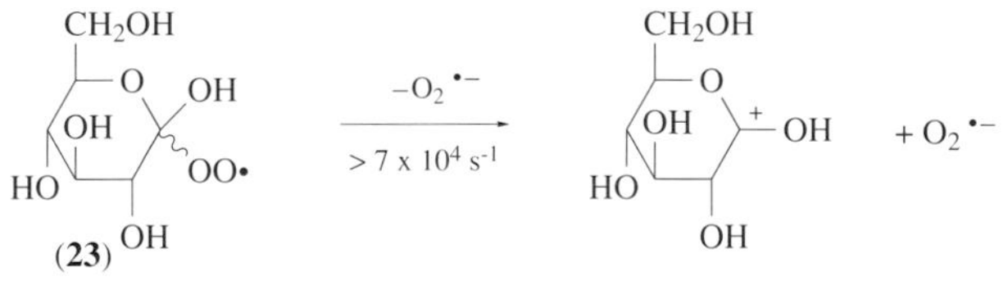

(1)

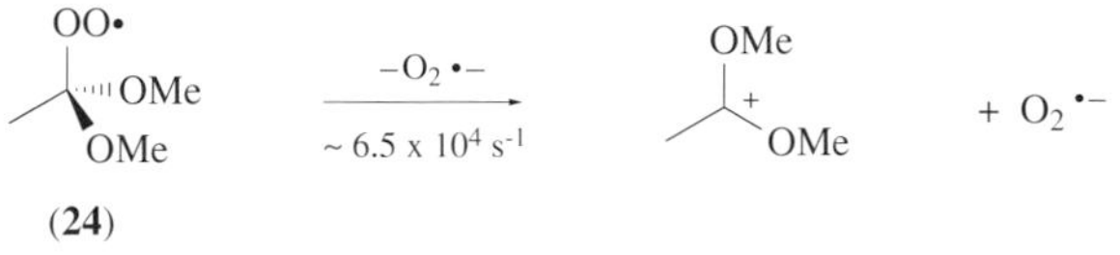

(2)

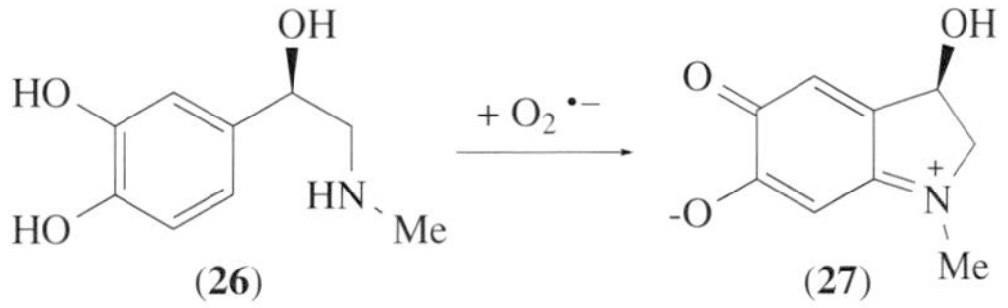

(3)

Metal-mediated DNA damage processes proceeding through the C-1′ radical offer yet another pathway to 2′-deoxyribonolactone lesions (1). Metalloporphyrins (e.g., (4)) are believed to oxidize organic substrates via metal-oxo species.[30–33] Such species are believed to hydroxylate the organic substrates via an oxygen rebound mechanism, in which the oxygen bound to the metal which abstracts the hydrogen atom is the oxygen which becomes associated with the oxygenated substrate. The mechanism by which (4) produces 2′-deoxyribonolactone lesions (1) in DNA was investigated using $^{18}O$-labeling.[32] Analysis of the furanone (2) released upon heating enabled Meunier to distinguish between the three principal possible reaction pathways available to the anomeric radical

(Scheme 6). The incorporation of 50% oxygen-18 in the furanone (**2**) was fully consistent with the oxygen rebound mechanism (Equation (4)).

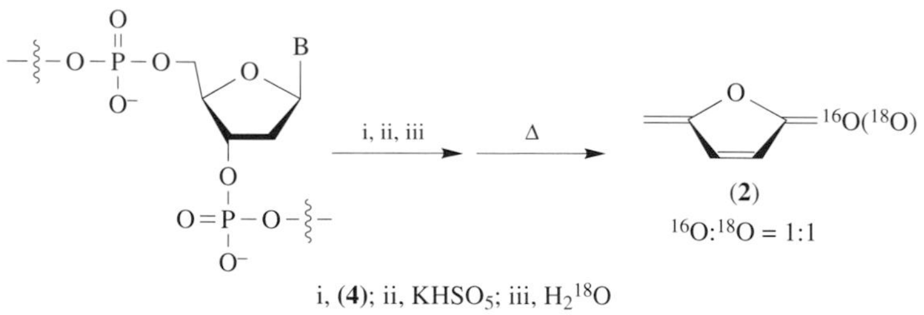

$$\tag{4}$$

i, (**4**); ii, KHSO$_5$; iii, H$_2$$^{18}$O

### 7.11.2.1.3   *Direct strand break formation vs. 2′-deoxyribonolactone (1) formation*

As mentioned above, Cu(OP)$_2$ (**5**) produces direct strand breaks. A variety of oxoruthenium(IV) complexes (**7**)–(**9**) produce direct strand breaks, in addition to alkaline labile lesions. However, in these systems it is uncertain what fraction of the alkaline labile lesions is attributable to base damage vs. hydride abstraction from C-1′. At the other end of the spectrum from Cu(OP)$_2$ (**5**) lie neocarzinostatin (**3**) and compound (**4**), which predominantly produce alkaline labile lesions upon reaction at the C-1′ position. The variety of products produced and pathways available to ionizing radiation prohibit saying with a certainty whether reaction at the C-1′ position of nucleotides produces direct strand breaks, alkaline labile lesions, or both. Ionizing radiation aside, it is unclear why, despite the fact that all of the above DNA-damaging agents react at the C-1′ position, and presumably generate the 2′-deoxyribonolactone lesion (**1**) via one mechanism or another, some agents continue to produce direct strand breaks.

The most common conditions for converting 2′-deoxyribonolactone lesions into strand breaks involve heating at 90 °C for 20–30 min in the presence of piperidine (1 M). However, Thorpe has shown that alkaline labile lesions (presumably (**1**)) are converted into strand breaks by heating at 90 °C in the absence of piperidine.[28,29] Simple amines were shown to convert abasic sites and other lesions into strand breaks under conditions as mild as 37 °C, in 60 min, under buffered conditions, via a general base-catalyzed process.[59]

Elimination of alkaline labile lesions is a common pathway for repair enzymes. In 1995 it was shown that DNA polymerase *β* excises such sites via elimination and not by phosphate hydrolysis.[60] Previously, both endonuclease III from *E. coli* and UV endonuclease V from bacteriophage T4 were shown to induce strand breaks via a general base mechanism that involves a *syn*-elimination of the 3′-phosphate.[61,62] Interestingly, it has long been known that small peptides containing basic amino acids (e.g., Lys–Trp–Lys) induce elimination of abasic DNA lesions.[63,64] In contrast to the aforementioned enzymes, Lys–Trp–Lys effects excision of abasic sites via an *anti*-elimination process.[61]

The observation that small, basic peptides can induce elimination suggests a possible explanation for the disparate types of lesions produced upon reaction of DNA with Cu(OP)$_2$ (**5**) and (to a lesser extent) oxoruthenium(IV) (e.g., (**7**)–(**9**)) complexes in comparison to neocarzinostatin (**3**) and manganese porphyrins (**4**). While no experimental results have been reported that address this issue, one could ask whether the two metal-containing complexes such as (**5**) induce direct strand breaks via base (through their ligands) and/or acid (through their metals) catalysis of elimination from the 2′-deoxyribonolactone lesion (**1**).

### 7.11.2.2   Reaction at the C-2′ Position of Nucleotides

Intuition regarding substituent effects on carbon–hydrogen bond dissociation energy (BDE), and computer calculations on the same lead one to predict that the C-2′ position of nucleotides in DNA should be the least reactive site for hydrogen atom abstraction.[24,25] Consequently, it should come as no surprise that only the most reactive DNA-damaging agents such as hydroxyl and the deoxyuridin-5-yl (**28**) radicals are believed to react at this site. Hydrogen atom abstraction from the C-2′ position of nucleotides by sigma radicals produced from 5-halopyrimidines (**28**) will be addressed later in this chapter (Equation (5)), containing structures (**28**)–(**30**) (see Section 7.11.3.3).

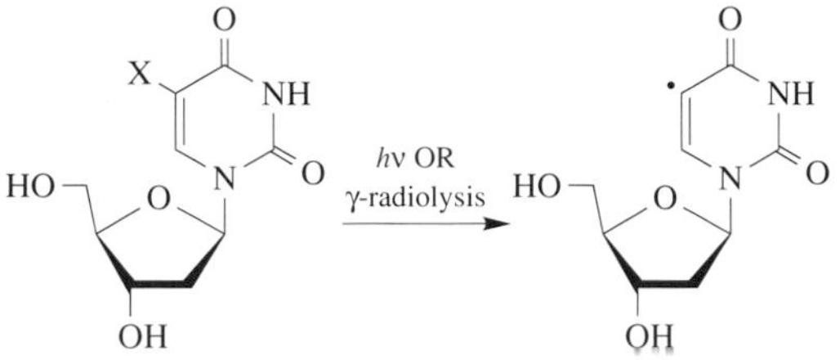

(5)

(**29**) X = Br
(**30**) X = I

(**28**)

### 7.11.2.2.1 *Reactivity of C-2′ nucleoside radicals under anaerobic conditions*

In the absence of $O_2$, a radical centered at the 2′ position of a nucleotide (**31**) in DNA has three likely reaction pathways. Elimination of either the respective nucleobase or 3′-phosphate would yield one of two regioisomeric alkene cation radicals. Based upon the $pK_a$s of the respective leaving groups, cleavage of the phosphate should be significantly faster than that of the nucleobases (Scheme 8). However, such competitive rate data has not been reported, and the tetrahydrofuran oxygen should also influence the rate of nucleobase elimination. The absolute rate constant for elimination of a phosphate monoester $\beta$ to a secondary alkyl radical has not been measured at neutral pH. More stabilized systems (**32**) and (**33**) were shown to eliminate phosphate monoesters and diesters, respectively, with first-order rate constants in the range of $10^3$–$10^4$ s$^{-1}$ (Equations (6) and (7)).[65,66] Phosphate monoester elimination from the 2′ radical in DNA should be significantly slower at neutral pH than either of the above examples, because an $\alpha$-stabilizing substituent on the radical is absent and the phosphate monoester is a poorer leaving group than a phosphate diester.[65–67] While such processes would not be expected to compete with radical trapping by $O_2$, many tumor cells are hypoxic. Hence, even if the rate constant for phosphate elimination is $<10^3$ s$^{-1}$, this process could be involved in the rate-limiting step of DNA cleavage resulting from the indirect effect of ionizing radiation. Pulse radiolysis experiments under anaerobic conditions using conductivity detection reveal a cleavage process that proceeds with a rate constant between 99 and 198 s$^{-1}$ in oligo-deoxynucleotides.[68] While the authors do not propose that the cleavage process involves the C-2′ radical, these experiments do not rule out such a pathway.

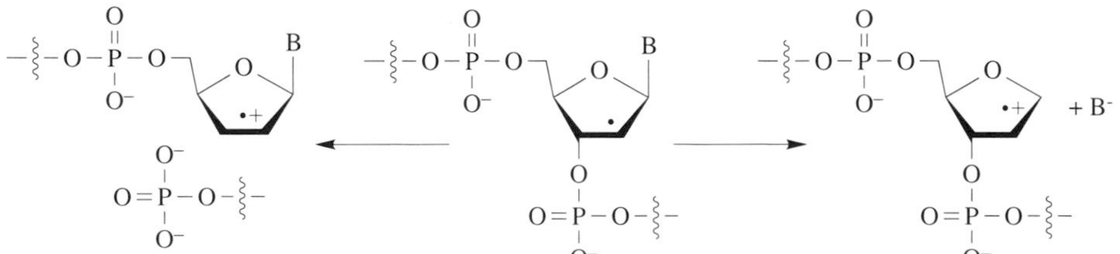

(**31**)

**Scheme 8**

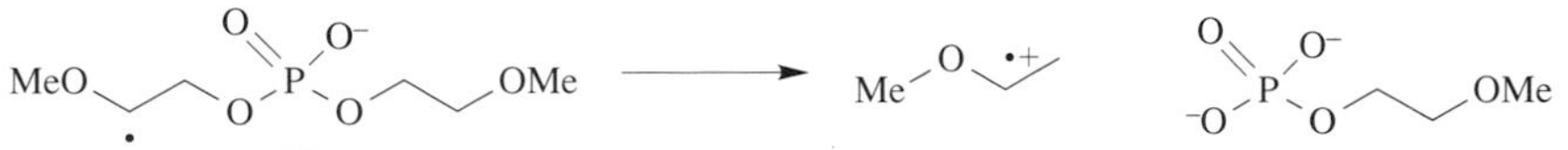

(**32**)

(6)

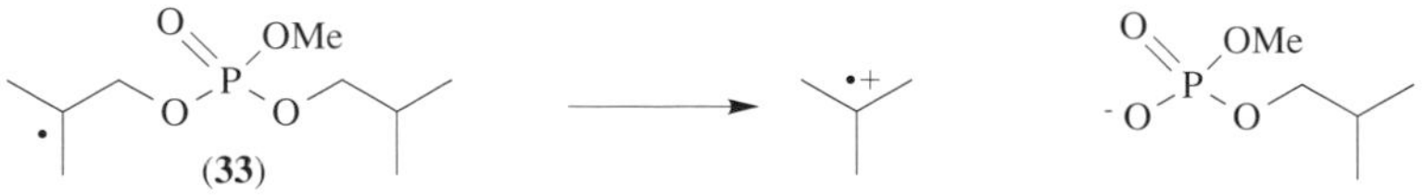

(**33**)

(7)

A third process available to the C-2′ radicals in DNA that has not been characterized within the biopolymer concerns the $\beta$-phosphatoxy rearrangement. Admittedly, such a process would not necessarily result in a strand break. However, the modified nucleic acid structure could have adverse effects on the function of the DNA. Model studies involving phosphate triesters reveal that such rearrangements can occur very rapidly via nondissociative or dissociative mechanisms, depending upon the nature of the aryl phosphate substituents.[69,70]

### 7.11.2.2.2    Reactivity of C-2′ nucleoside radicals under aerobic conditions

2′-Deoxyuridin-2′-yl (**31**, B = uracil) was generated via reaction on 2′-bromo-2′-deoxyuridine (**34**) with solvated electrons that were produced during γ-radiolysis (Scheme 9).[71] Under aerobic conditions, erythrose (**36**) was identified indirectly following NaBD$_4$ reduction through the observation of derivatized erythritol-1-d$_1$ (**37**) by GC–MS. In this study, erythrose (**36**) was suggested to arise via bimolecular decomposition of the nucleoside peroxy radical (**35**) via the respective tetroxide. In light of Giese's recent mechanistic investigations involving the C-4′ nucleoside radicals (see Section 7.11.2.4), and the expected concentrations of peroxy radicals in biopolymers (see Section 7.11.2.1.2), it is unlikely that the tetroxide would form. An alternative explanation for the formation of erythrose (**36**) involves Grob fragmentation of (**38**) (Scheme 10).[72]

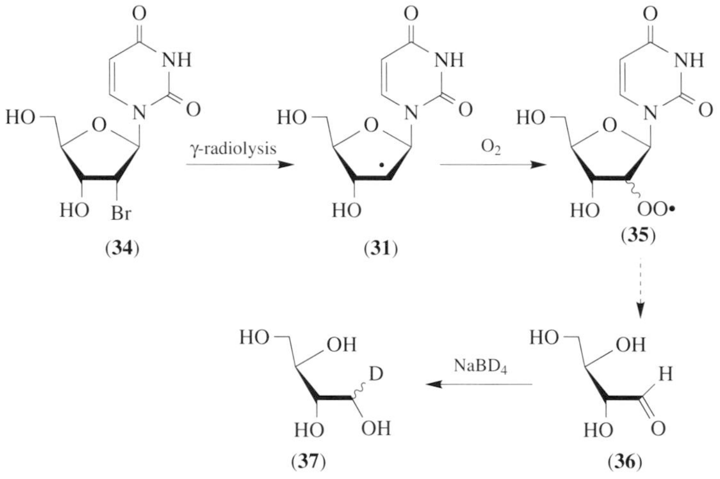

**Scheme 9**

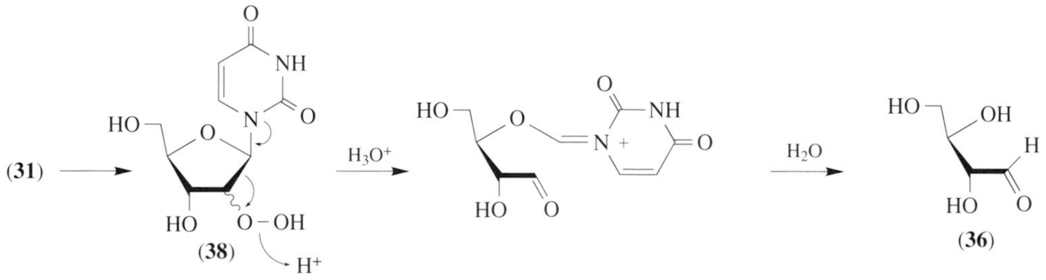

**Scheme 10**

Regardless of the pathway(s) by which erythrose is produced, its formation accounts for only 15% of the expected peroxy radical precursor. A more complete mass balance was obtained upon photolysis of 2′-iodo-2′-deoxyuridine (**39**; Scheme 11, containing Structures (**39**)–(**43**)).[73] Erythrose (**36**) was again only a minor product. The majority of observed products in this study were derived from the C-2′ carbocation (**40**), which is derived from the initially formed radical via single electron transfer within the radical pair (Scheme 11). The carbocation undergoes 1,2-hydride rearrangements ((**41**) and (**42**)) in competition with nucleophilic trapping. The ultimate products derived from the carbocation rearrangements (**1**) and (**43**) are relevant to the mechanisms by which 5-halopyrimidines induce DNA strand breaks, and will be discussed in greater detail later (Section 7.11.3.3.). Although it is unlikely that compound (**40**) would be formed in the manner observed by Saito during DNA damage, this species could be formed via elimination of superoxide from the peroxy radical (**35**), Equation (8). Admittedly, superoxide loss from a secondary alkyl radical is slow. The rate constant for elimination of superoxide from (**35**) would be expected to be $<0.1 \text{ s}^{-1}$.[74] However, one must keep in mind that the lifetimes of peroxy radicals in single- and double-stranded DNA are of the order of a second.[75] Hence, in a biopolymer slow unimolecular processes may have an impact on the chemistry of DNA damage.

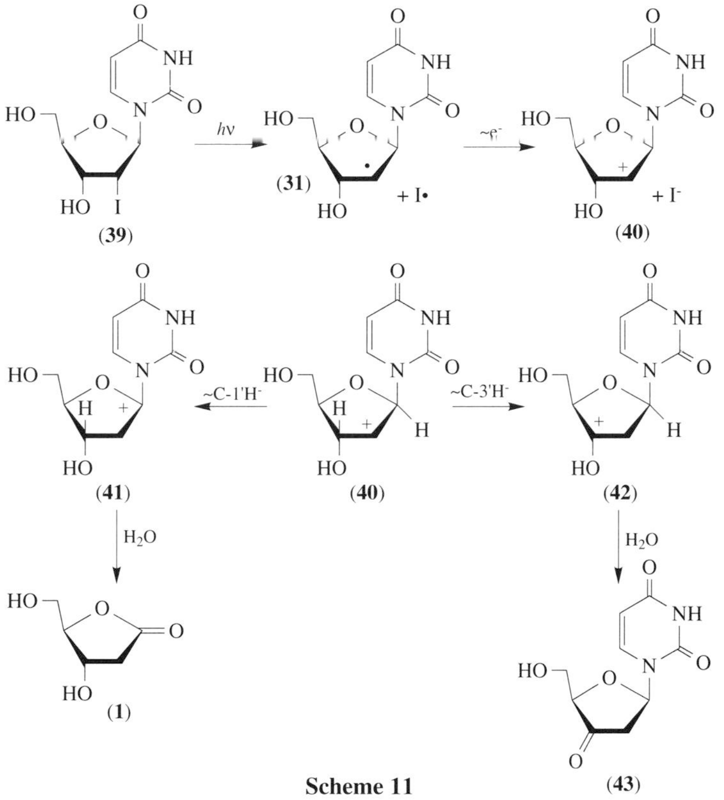

**Scheme 11**

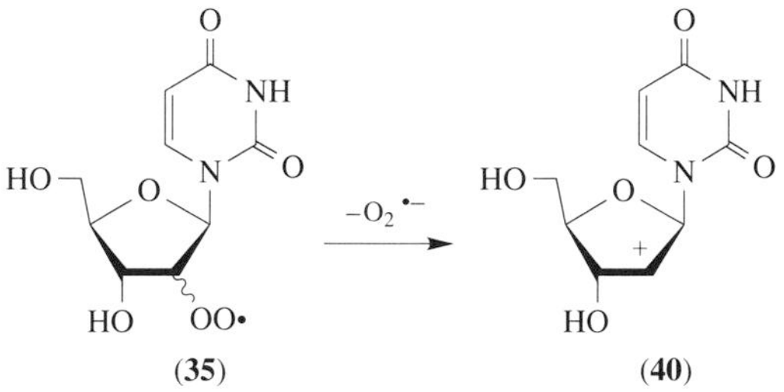

(8)

### 7.11.2.3 Reaction at the C-3′ Position of Nucleotides

The examples of DNA-damaging agents that are believed to react at the C-3′ position are limited to a singular family. This is surprising if one considers that the bond dissociation energy of the C-3′ carbon–hydrogen bond in deoxyribose is expected to be very similar to that at the C-5′ position.[25] Furthermore, when the accessibility of individual hydrogen atoms in B-DNA is considered, the reactivity of the C-3′ hydrogen increases.[24] One can assume that the dearth of examples of DNA-damaging agents that react at the C-3′ position is attributable to the fact that this hydrogen atom lies in the major groove of DNA. The large majority of small molecule agents that oxidize DNA bind in the minor groove of DNA.

The octahedral rhodium complexes examined by the Barton group are prominent exceptions to this rule.[76–78] These molecules cleave DNA upon irradiation with light of wavelength greater than 300 nm. Examination of the cleavage produced on opposite strands in a duplex upon irradiation of

(**44**) yields an asymmetric cleavage pattern that is shifted in the 5′ direction. Previous DNA cleavage experiments enable one to interpret these results to mean that the metal complexes are bound in the major groove of DNA.[79] Free nucleobases, as well as 3′ and 5′-phosphate termini are produced via oxygen-dependent and independent processes (Scheme 12, containing Structures (**44**)–(**50**)). Additional products are also produced via an oxygen-dependent process. Base propenoic acids (**49**) were characterized by reverse-phase HPLC. The structure of the accompanying 3′-phosphoglycaldehyde oligonucleotide termini (**48**) were surmised via the effect of $NaBH_4$ on the gel electrophoretic migration of the 5′-$^{32}$P-labeled fragments. The formation of the products are consistent with a DNA damage process that was initiated by hydrogen atom abstraction from the C-3′ position of (**45**). Phosphoglycaldehyde termini (**48**) and base propenoic acids (**49**) were suggested to result via a Criegée rearrangement via (**46**) and (**47**), analogous to that previously proposed for cleavage induced via C-4′ hydrogen atom abstraction by bleomycin (see Section 7.11.2.4).[80] It was suggested that free nucleobases are derived from an oxygen-independent pathway. However, the labile furanone product (**50**) which may also be formed was not detected.

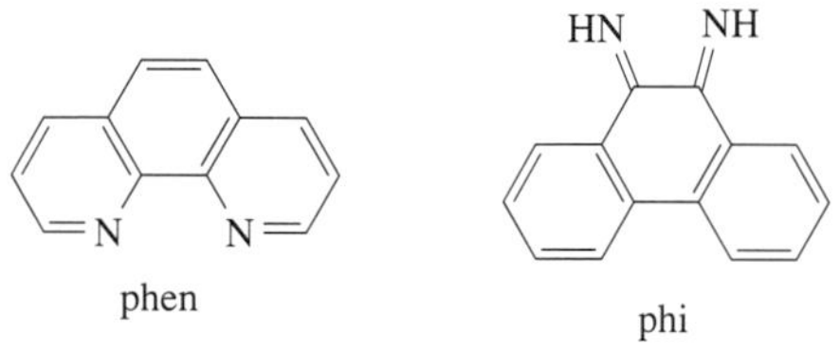

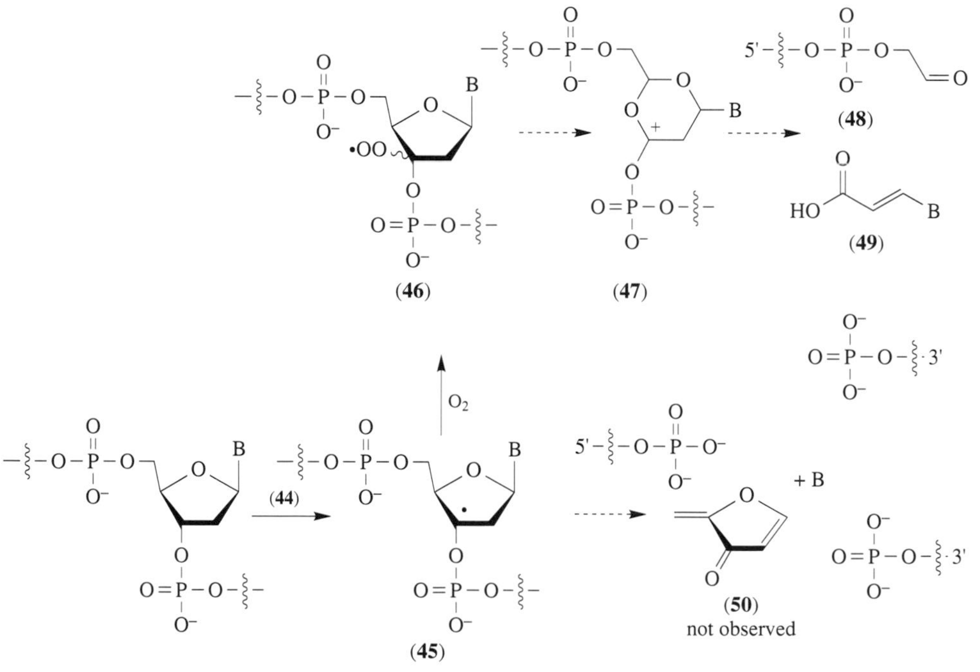

**Scheme 12**

### 7.11.2.4 Reaction at the C-4′ Position of Nucleotides

As expected, based upon calculations concerning the C-4′ carbon–hydrogen BDE in deoxyribose and the accessibility of this hydrogen atom in the minor groove of duplex DNA, the C-4′ hydrogen is involved in a variety of DNA damage processes.[24,25] It is well established that the C-4′ hydrogen atom is the major site of attack by bleomycin (**51**).[80–82] Several other oxidizing agents are believed to react to varying degrees at the C-4′ hydrogen atom. These include neocarzinostatin (**3**), calicheamicin (**52**), esperamicin (**53**), enediyne C1027 (**54**), Cu(OP)$_2$ (**5**), Fe·EDTA derivatives, a dinuclear platinum complex, and of course ionizing radiation.[26,83–91] A variety of tools, such as product analysis, isotopic labeling, and kinetic isotope effects have been employed for determining whether C-4′ hydrogen atom abstraction is involved in a given nucleic acid damage process.

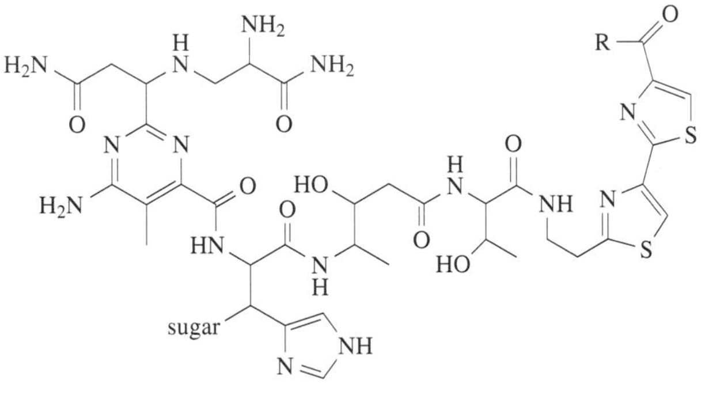

(**51**)

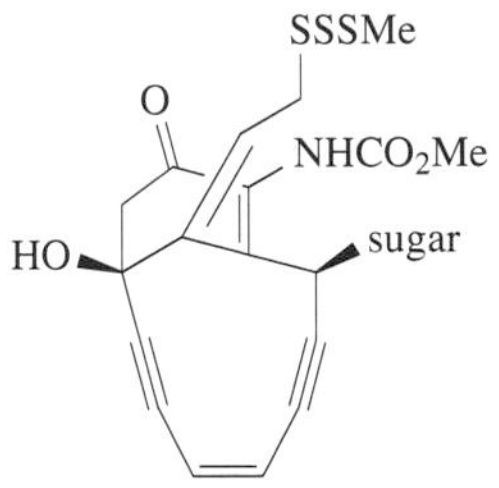

(**52**)

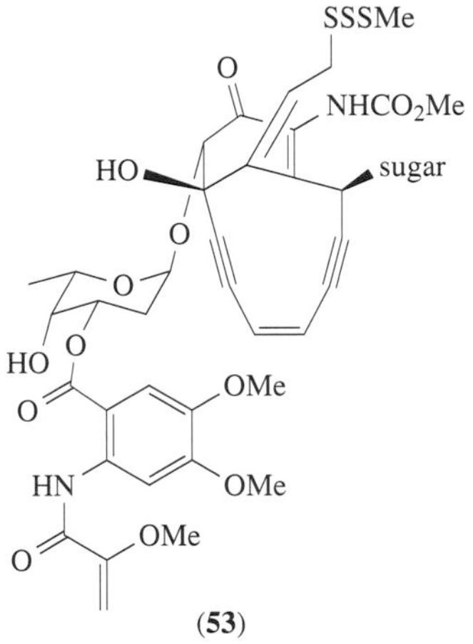

(**53**)

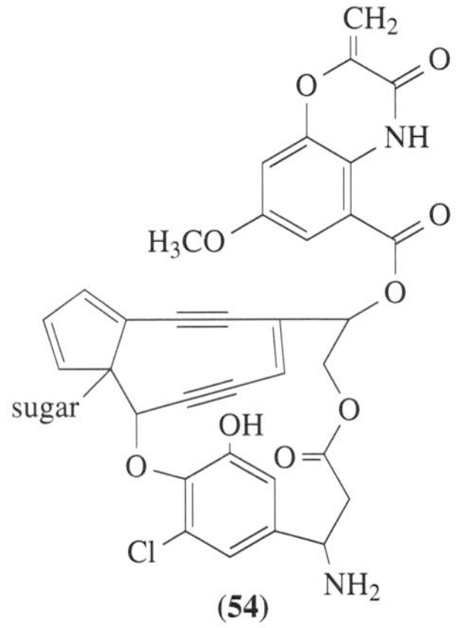

(**54**)

The power of polyacrylamide gel electrophoresis (PAGE) for analyzing nucleic acid damage is most readily apparent by its use in detecting 3′-phosphoglycolate termini (**55**). The observation of DNA fragments containing 3′-phosphoglycolate termini often serve as the signature for C-4′ hydrogen atom abstraction. PAGE analysis of 5′-$^{32}$P-labeled DNA was first employed for establishing the nature of the fragmentation products formed during γ-radiolysis of DNA.[91,92]

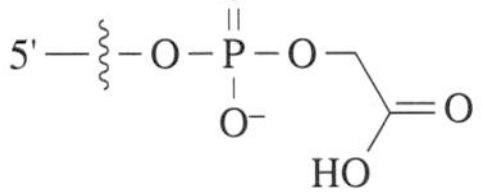

(**55**)

When oxidation is effected by bleomycin, the remaining three carbons of the deoxyribose ring are released as base propenals (**56**).[80] Often, these products are not observed during the oxidation of DNA by other agents which give rise to 3′-phosphoglycolate termini. In many cases, it is not known whether the base propenals are stable to the reaction conditions. Indirect evidence for the possible intermediacy of base propenals during γ-radiolysis is gleaned from the formation of thiobarbituric acid adducts containing long-wavelength chromophores that are characteristic of malondialdehyde, which is a hydrolysis product of (**56**).[13] One caveat which should be noted is that thiobarbituric acid adducts are typically characterized spectrophotometrically. Hence, the assay is not structure specific and other lesions can give rise to positive results.

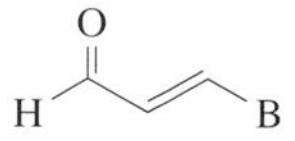

(**56**)

Chemical reactivity can also be used to detect the involvement of the C-4′ position in DNA oxidation. When cleavage can be induced at the 5′-terminal nucleotide of an oligonucleotide, 5′-halo-2′,5′-dideoxynucleotides (**57**) serve as useful radical traps (Scheme 13).[93] A more general method involves trapping (**60**) the alkaline labile lesion (**59**) derived from the C-4′ hemiacetal, which is in turn produced via the respective radical in a number of systems using hydrazine (Scheme 14, containing Structures (**58**)–(**60**)).[86,94,95]

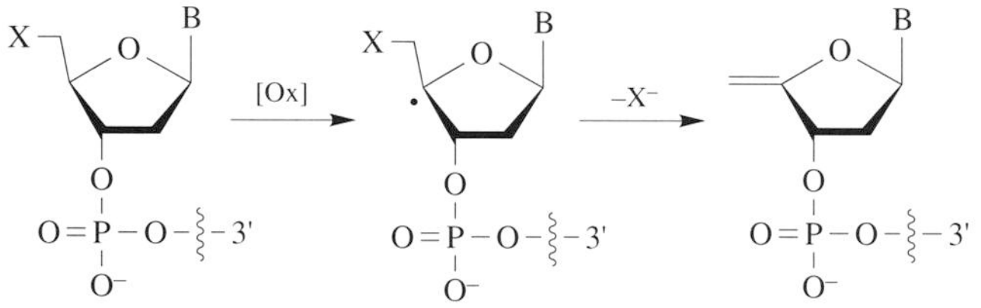

**Scheme 13**

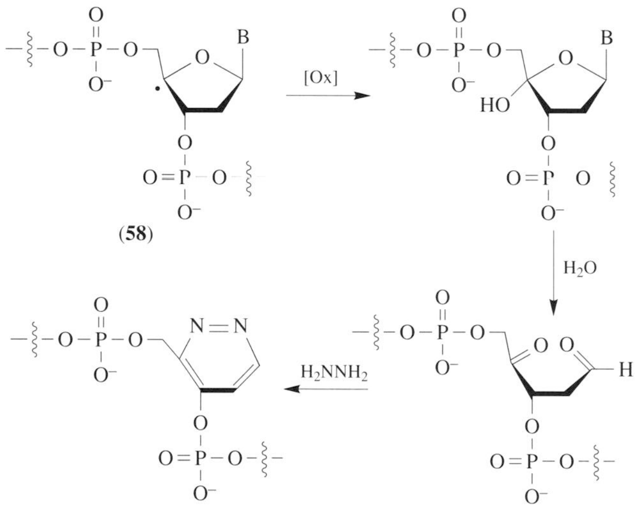

**Scheme 14**

The alkaline labile lesion arising from oxidation at the C-4′ position (**59**) has also been trapped and characterized chemoenzymatically using $NaBH_4$ (**61**) and phosphodiesterases (**62**) (Scheme 15, containing Structures (**58**), (**59**), (**61**), and (**62**)).[96,97]

Finally, the structure of the deoxyribose residue resulting from base-induced cleavage at the alkaline labile site was determined via independent synthesis of an oligonucleotide containing the cyclopentenone product (**63**) resulting from intramolecular aldol condensation of (**59**).[94,98]

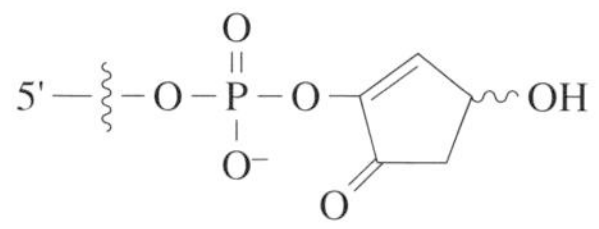

(**63**)

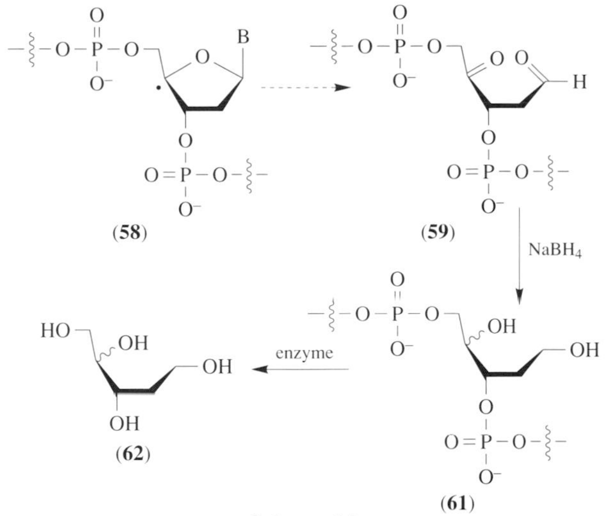

**Scheme 15**

### 7.11.2.4.1 The mechanism of formation of alkaline labile lesions resulting from C-4′ hydrogen atom abstraction

The mechanism of formation of alkaline labile lesions resulting from oxidation at the C-4′ position by bleomycin (**51**) has been investigated. In particular, the sources of the oxygen atoms have been elucidated using $^{18}O$ labeling. Since bleomycin can be activated prior to mixing with DNA, it is possible to distinguish between the $O_2$ required for activation from that involved in chemistry subsequent to hydrogen atom abstraction. By carrying out experiments in either $^{18}O_2$ or $H_2{}^{18}O$, it was proposed that the major pathway for bleomycin-mediated formation of (**59**) involves trapping of a C-4′ carbocation (**64**) by water (Scheme 16).[97] Provided oxygen exchange within the putative iron-oxo intermediate is slow relative to hydrogen atom abstraction, these experiments rule out the possibility of formation of the alkaline labile lesion by a "radical rebound" type mechanism. Such "radical rebound" mechanisms are believed to be involved in DNA damage mediated by (**4**) (see Section 7.11.2.1) and P-450 catalyzed hydroxylations.[32,99] Labeling experiments also detected a minor pathway in which alkaline labile lesion (**59**) is formed via the C-1′ carbocation (**65**).[97] It is suggested that the C-1′ carbocation results from β-cleavage of the C-4′ radical (**58**), followed by oxidation. However, one cannot rule out reversing the order of oxidation and ring opening, which would give rise to (**65**) from (**64**) (Scheme 16).

### 7.11.2.4.2 The mechanism of formation of direct strand breaks resulting from C-4′ hydrogen atom abstraction

A significant amount of work has been reported concerning the transformation of the C-4′ radical produced via bleomycin (**51**) into DNA fragments containing 3′-phosphoglycolate (**55**) termini and base propenals (**56**) under aerobic conditions.[80] The results of these studies, which included extensive use of $^{18}O$-labeling, $^3H$-labeling, and product analysis, demonstrated that direct strand scission results via an oxygen-dependent process that is initiated by C-4′ hydrogen atom abstraction. Tritium (**66**) studies also determined that the 2′-*pro-R* proton is lost during the elimination to introduce the *trans*-alkene component of the base propenal (**56**) (Scheme 17).

A central aspect of the proposal for oxygen-dependent direct strand scission involves cleavage of the C-3′–C-4′ bond via a Criegée rearrangement of a C-4′ hydroperoxy intermediate. Labeling experiments with oxygen-18 were consistent with such a process (Scheme 17).[100,101] These experiments revealed that one oxygen atom in the carboxylate group of the phosphoglycolate is derived from $O_2$, and the other is attributable to the oxygen atom of the deoxyribose ring. The oxygen in the carbonyl group of the base propenal is derived from water. Two mechanisms, both invoking a Criegée rearrangement, had been postulated that were consistent with oxygen-18 labeling studies

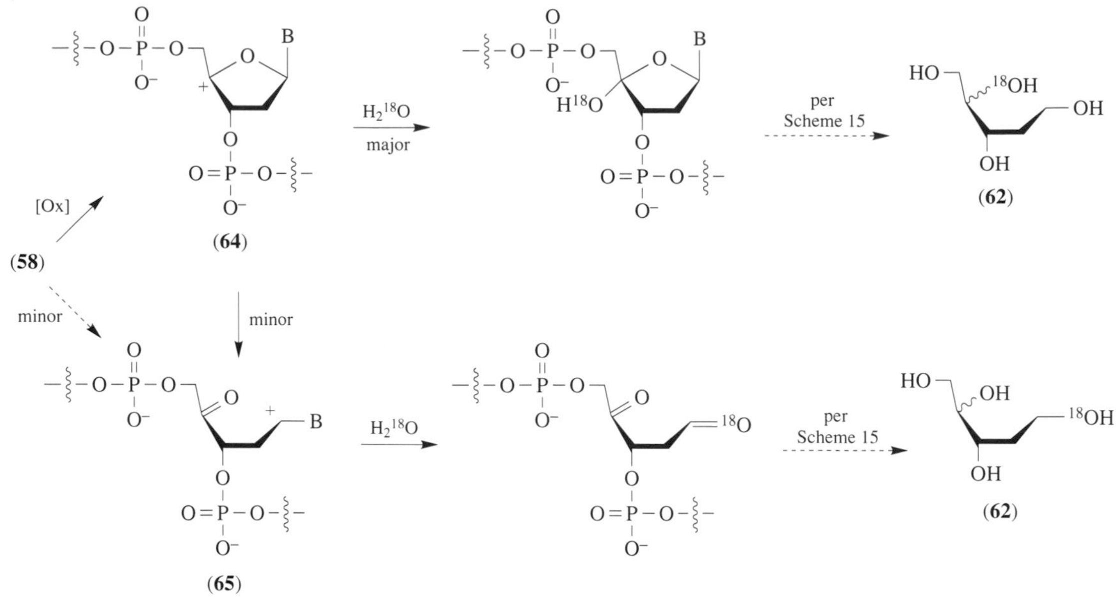

**Scheme 16**

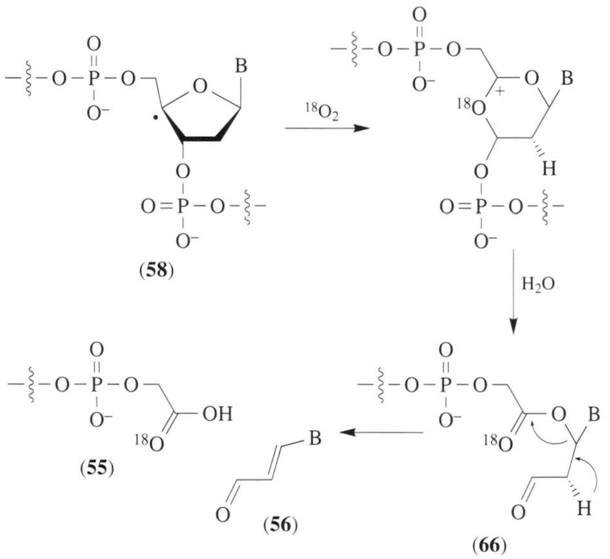

**Scheme 17**

(Schemes 17 and 18). However, only one (Scheme 18) also accommodates the observation that tritium is released from the 2′-*pro-R* deoxyribose position at rates considerably faster than the rate at which base propenals are released, and at rates comparable to that of strand scission.[101,102] This necessitates the intermediacy of (**67**) (Scheme 18) as opposed to (**66**) (Scheme 17).

Substantiation of the Criegée rearrangement has remained the most difficult step to verify in the mechanism for bleomycin-mediated DNA strand scission. Based upon observed rates of strand scission, in order for such a process to be kinetically competent, the Criegée rearrangement must occur at an accelerated rate compared to similar small molecule reactions.[80] One possibility is that the iron-containing bleomycin (**51**) molecule functions as a Lewis acid. Attempts at rationalizing a rate acceleration by invoking stereoelectronic effects were employed in a model system (**68**).[103] However, the observations made using this system were shown (using labeling) to be a result of

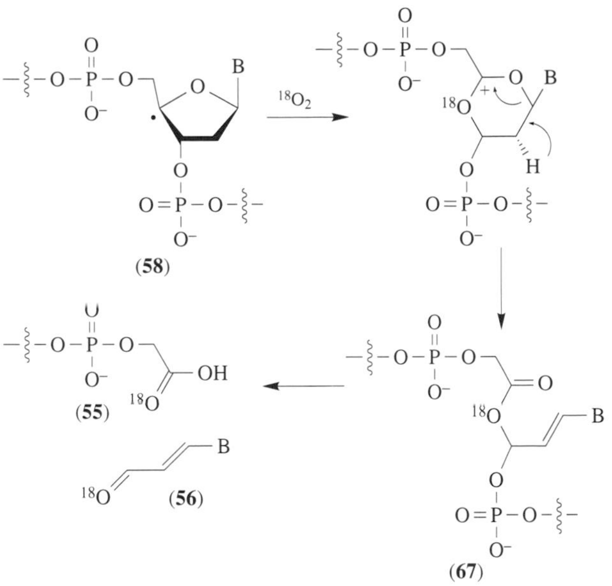

**Scheme 18**

neighboring group participation, and cannot be extrapolated to bleomycin (**51**) oxygen-18-mediated DNA cleavage.[104]

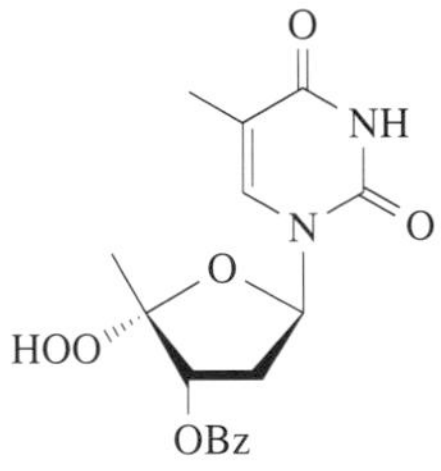

(**68**)

Hydrogen atom abstraction from the C-4′ position of nucleotides in DNA can also give rise to strand scission by a mechanism that is independent of $O_2$. Phosphate elimination from the $\beta$ position of the C-4′ radical in DNA was originally proposed during studies of $\gamma$-radiolysis-induced strand scission (Scheme 19).[105] The alkene cation radicals (**69**) and (**70**) were suggested to ultimately result in a second phosphate elimination, following hydration. This proposal was supported by prior and subsequent model studies.[65–67,106,107] Formation of alkene cation radicals via elimination were found to be strongly dependent upon the substitution pattern of the radical, the nature of the leaving group, and solvent conditions. Rate constants greater than $10^6$ s$^{-1}$ have been detected.[65–67] Such unimolecular processes would compete very effectively with bimolecular trapping by $O_2$, which when present at 0.2 mM would exhibit an effective first-order rate constant of $\sim 4 \times 10^5$ s$^{-1}$.

Giese has clarified and solidified the viability of this pathway in nucleic acid strand scission by independently generating the C-4′ radical under controlled conditions. Initial model studies of DNA utilized thiyl addition to a 4′,5′-unsaturated nucleotide containing various 3′-phosphate diesters as leaving groups (Scheme 20, containing Structures (**71**)–(**74**)).[108] Using an estimated rate constant for trapping of (**71**) by thiophenol ($\sim 10^8$ M$^{-1}$ s$^{-1}$), the rate constant for elimination of the phosphate diester from (**72**) was determined to be $> 10^8$ s$^{-1}$ based upon the formation of (**73**) and (**74**). One should note that the leaving groups in DNA are phosphate monoesters, and their cleavage would not be expected to be as rapid as that observed in the aforementioned model compounds. In contrast, when studying the poorer benzoate leaving group, elimination in (**75**) did not compete with trapping by thiophenol. Subsequent laser flash photolysis experiments on (**76**) and (**77**) using photocurrent detection confirmed the formation of ions.[109] However, rate constants were not reported.

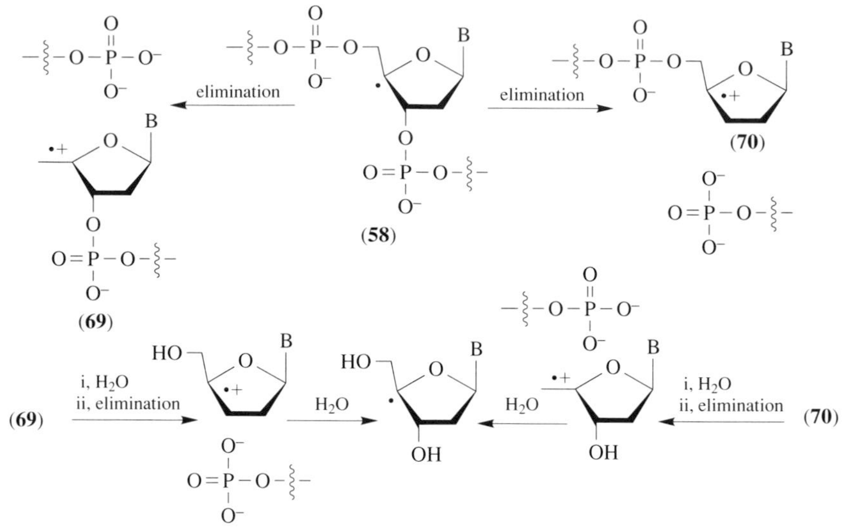

Scheme 19

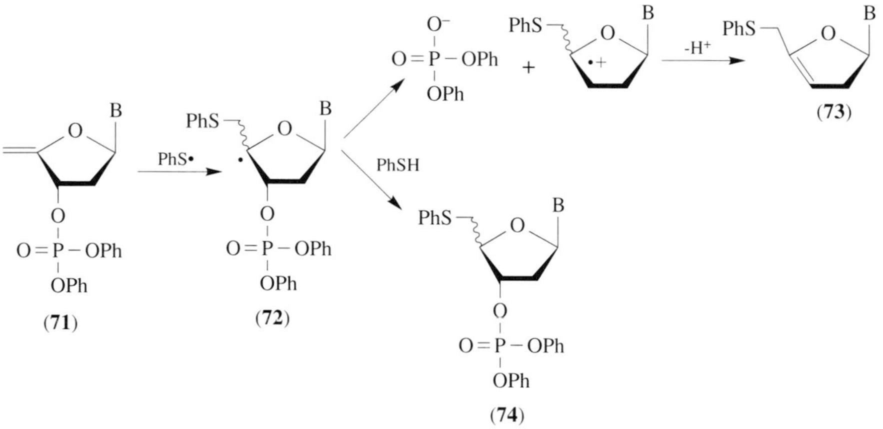

Scheme 20

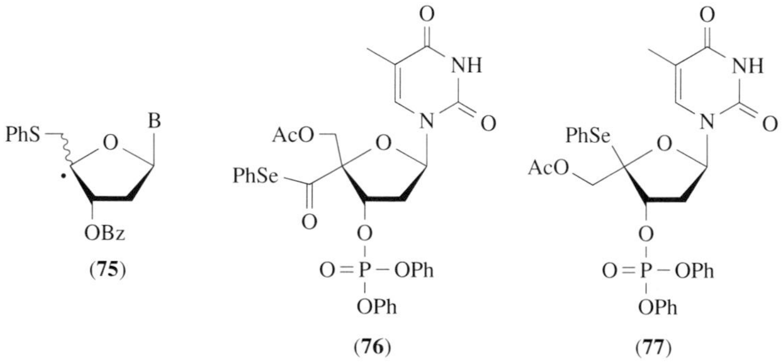

Relative rate constants for phosphate diester elimination from, and hydrogen atom abstraction by, a C-4′ nucleotide radical were obtained using a series of bisphosphates (**78**).[110] Product studies carried out using thiophenol as hydrogen donor supported the intermediacy of cation radicals (Scheme 21, containing Structures (**78**)–(**81**)). The ratio of $k_E/k_H$ from (**79**) was determined by measuring the ratio of (**80**):(**81**). The relationship of the rate constants was found to depend upon the solvent polarity and the acidity of the conjugate acid of the phosphate diester leaving group. The increase in $k_E/k_H$ with increasing solvent polarity probably signifies an even greater increase in $k_E$ than originally assumed, because experiments have shown that hydrogen atom transfer from thiols to alkyl radicals are accelerated in polar protic solvents.[111] Use of the bisphosphates (**78**) also supported the original proposal by Schulte-Frohlinde that the 3′-phosphate should cleave faster than the 5′-phosphate. Preferential formation of (**82**) vs. (**83**) from (**79**) is consistent with studies on simpler systems, and their anticipated relative stability.[107]

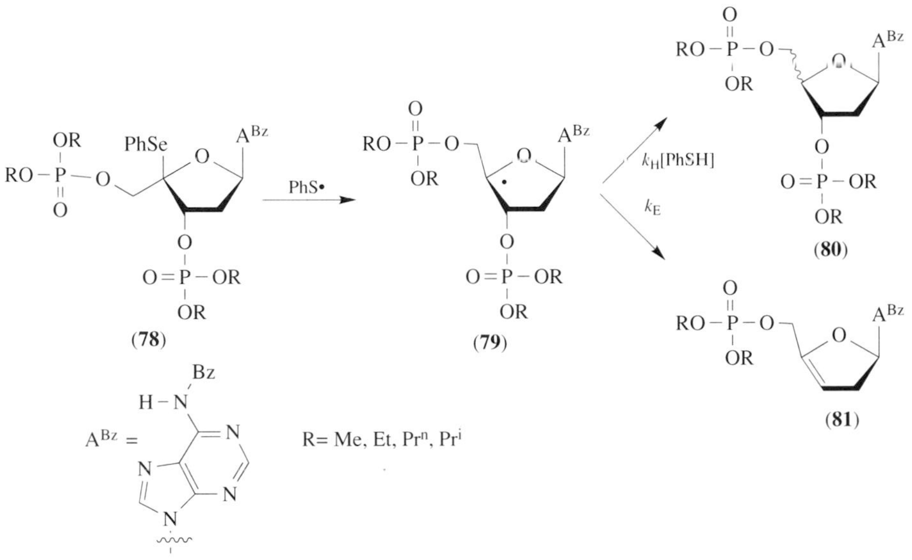

**Scheme 21**

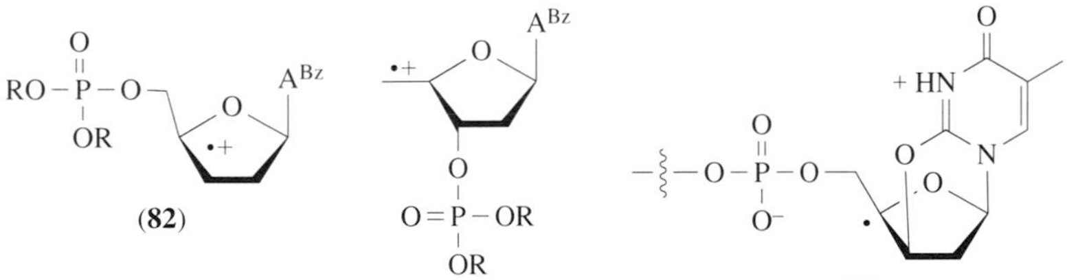

A subtle issue concerning the heterolytic cleavage of phosphate groups is whether the nucleobase (e.g., thymine) provides anchimeric assistance.[112] The observed stereochemistry of the nucleophilic substitution products observed by Giese can be rationalized via either an $S_{RN}2$ reaction or steric influence of the nucleobase on the reactivity of the cation radical formed via an $S_{RN}1$ pathway (**84**). Product mixtures obtained using C-nucleosides ((**85**) and (**86**)) that are incapable of providing anchimeric assistance ($S_{RN}2$) support the $S_{RN}1$ pathway (Scheme 22).[113] The diasteromeric radical precursors yield identical product mixtures upon photolysis, indicating that they proceed through a common intermediate (**87**).

These results have been verified qualitatively by experiments involving single-stranded DNA.[114,115] Irradiation of (**88**) under anaerobic conditions gave rise to (**91**) and (**92**), presumably via (**89**) and (**90**) (Scheme 23). The products were detected by Matrix Assisted Laser Desorption Ionization Time of Flight (MALDI-TOF) mass spectrometry. When the phenyl selenide containing oligonucleotide (**88**) was irradiated in the presence of glutathione, two additional products ((**93**) and (**94**)), attributable to trapping (**89**) and (**90**), respectively, were also observed. In a subsequent study where (**88**) was again used as a substrate, an uncleaved, but modified oligonucleotide (**95**) was detected. Formation of (**95**) was attributed to a radical cage process. One should note that the MALDI-TOF experiment is a qualitative one, and no attempts were made to extrapolate relative rate constants for elimination and hydrogen atom transfer from the data.

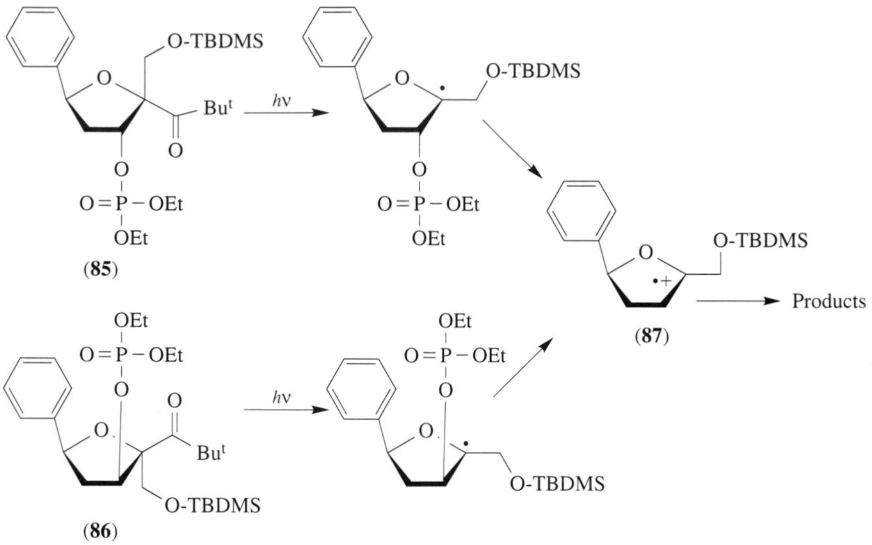

**Scheme 22**

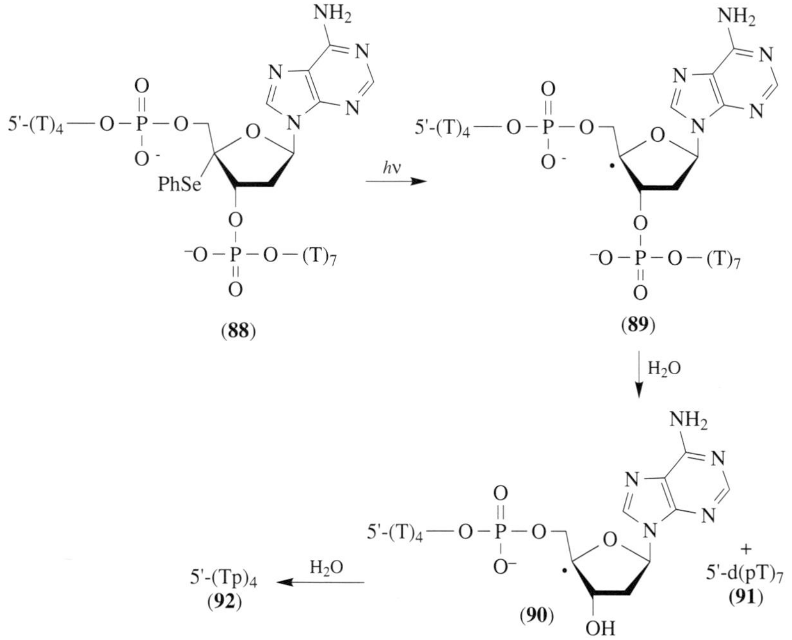

**Scheme 23**

5'-d(T$_4$AT$_7$)        5'-d(T$_4$A)
**(93)**              **(94)**

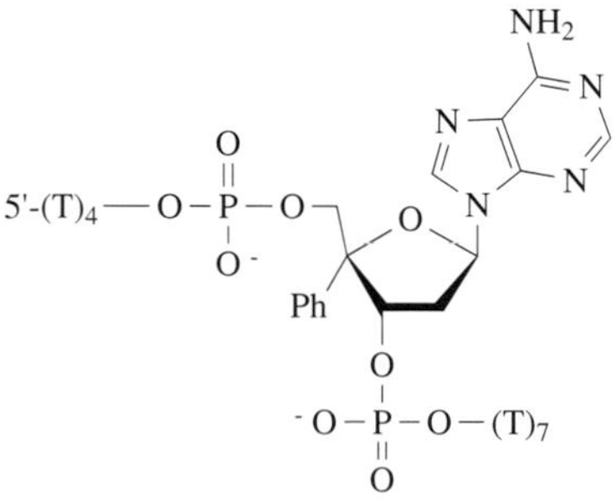

**(95)**

Comparable observations were made utilizing an oligonucleotide containing a *t*-butyl ketone (**96**) as a photolabile precursor for the C-4′ radical in oligonucleotides.[115] The most notable differences observed using the *t*-butyl ketone precursor were the absence of any radical–radical recombination products and the apparently more efficient trapping by glutathione. The latter observation was rationalized in terms of glutathione's greater accessibility to the C-4′ radical when it is generated via Norrish Type I photocleavage, due to the absence of a caged radical pair.[115] No further reports have yet appeared regarding this disparate behavior.

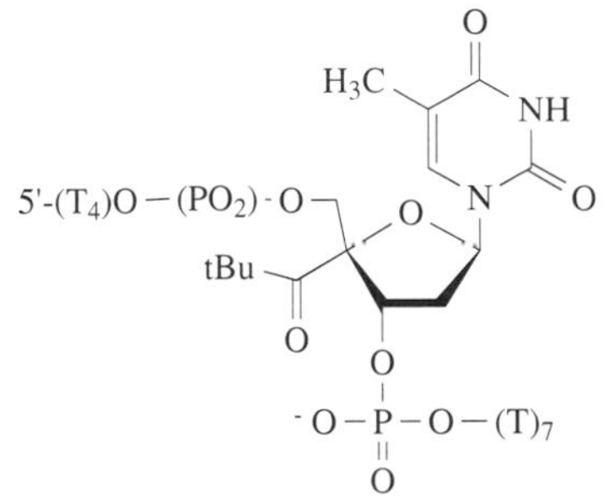

(**96**)

Using these photolabile radical precursors, evocative observations were made regarding the cleavage of DNA via C-4′ radicals under aerobic conditions.[72,115] Photolysis of either (**88**) or (**96**) in the presence of $O_2$ yielded (**97**). Consistent with the results reported when irradiation was carried out in the presence of glutathione, $O_2$ trapped (**98**) from (**96**), but not (**89**) from (**88**). Trapping of (**98**) yielded hydroperoxide (**99**), which was suggested to undergo a Criegée rearrangement en route to (**97**), as proposed previously in studies on bleomycin (**51**).[101,115] In contrast, hydroperoxide (**101**) was believed to be derived from fragmentation of (**89**) (Scheme 24). It was proposed that (**101**) undergoes a Grob fragmentation to yield base propenal (**56**) and (**97**). The plausibility of the Grob fragmentation was supported by examination of the decomposition of model compound (**102**) under mild aqueous conditions.[72] The Grob fragmentation was also consistent with oxygen-18 labeling experiments, in which using MALDI-TOF it was shown that one of the glycolate oxygens was derived from $O_2$, and neither are attributable to $H_2O$.[115] This last result is intriguing. It suggests that nucleophilic attack by $H_2O$ on (**100**) takes place with much higher regioselectivity than does attack by alcohols on model compounds.[110,113]

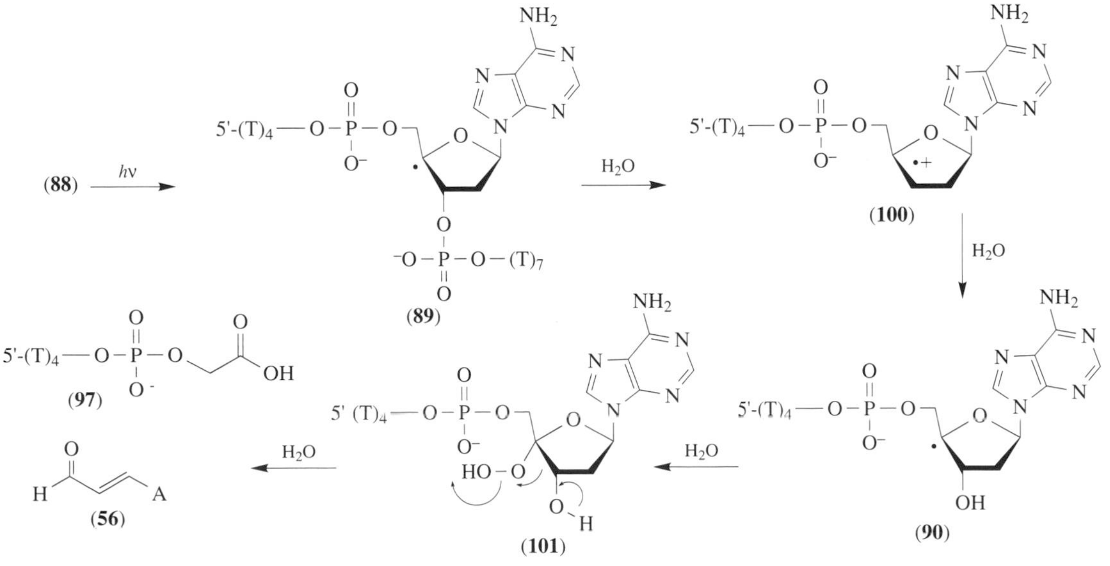

**Scheme 24**

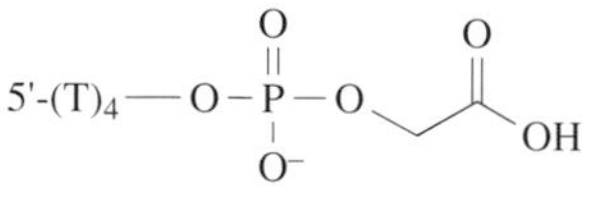

(97)

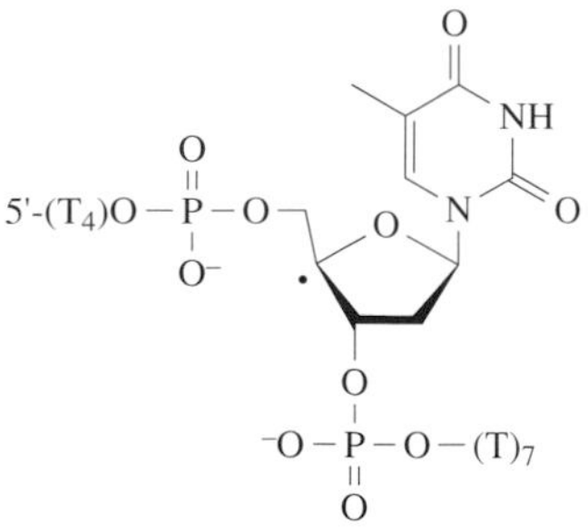

(98)

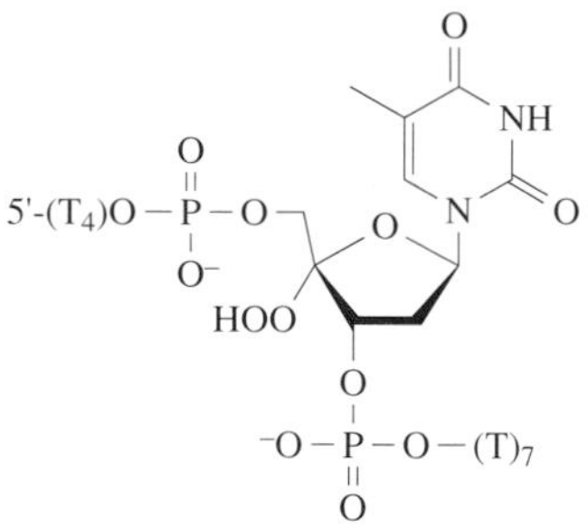

(99)

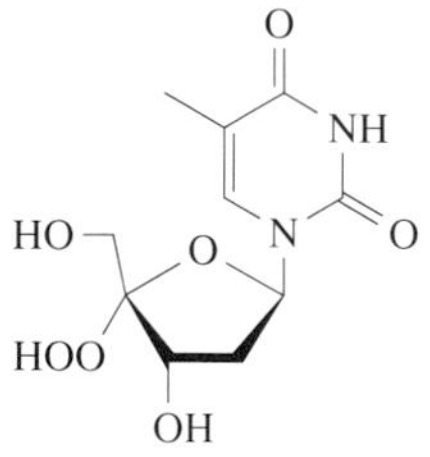

(102)

### 7.11.2.4.3 Similarities and contrasts between the Criegée and Schulte-Frohlinde/Grob fragmentation pathways for DNA cleavage via C-4′ hydrogen atom abstraction

One should be very careful to avoid interpreting the above experiments as indicating that either mechanism is globally correct or incorrect. The conditions under which the studies were conducted were quite different, and some of the observations suggest that cleavage by bleomycin (51) follows a different mechanism than that proposed by Schulte-Frohlinde and documented by Giese. Chief among these differences was the observation that DNA cleavage by bleomycin proceeds concomitantly with release of a proton from the 2′-*pro-R* position of the deoxyribose undergoing oxidation.[80,101,102] Cleavage induced via alkene cation radical formation occurs faster than loss of protons from the C-2′ position. Other differences that one should not lose sight of are:

(i) Direct strand breaks via the Schulte-Frohlinde mechanism are independent of $O_2$. Those produced by bleomycin require $O_2$.

(ii) Bleomycin cleaves double-stranded DNA. Experiments involving independent generation of the C-4′ radical have been carried out on single-stranded DNA. No reports involving duplex DNA have appeared.

Consequently, the presence of bleomycin in the minor groove of DNA upon C-4′ hydrogen atom abstraction may result in a bifurcation of the reactivity of the incipient radical. Under aerobic conditions, this bifurcation could involve the proposed Criegée rearrangement. Such a rearrangement may indeed by catalyzed by the iron in bleomycin.[80] Similarly, it remains to be shown that alkene cation radical formation is the product determining step when the C-4′ radical is generated in duplex DNA. This is particularly pertinent considering that C-4′ hydrogen atom abstraction by neocarzinostatin (3) in duplex DNA produces direct strand breaks and alkaline labile lesions.[116]

### 7.11.2.5 DNA Damage via Reaction at the C-5′ Position of Nucleotides

#### 7.11.2.5.1 Agents that react at the C-5′ position

The enediynes (52)–(54) and neocarzinostatin (3) are the DNA-damaging agents that are most often associated with chemistry at the C-5′ position.[26,84,89,90] The binding and reactivity of these oxidants are discussed in detail in Chapter 7.15. The C-5′ position of DNA is also a site of hydrogen atom abstraction by (4).[30–33] The subsequent chemistry of the C-5′ radical produced appears to be significantly different for these two families of DNA-damaging agents, and is discussed below. Gel electrophoretic analysis and/or product studies indicate that the diplatinum photocleavage agent, $Pt_2(pop)_4^{4-}$, and perhydroxyl radical (HOO·), also abstract hydrogen atoms from the C-5' position of nucleotides in DNA.[83,117]

### 7.11.2.5.2 *The reactivity of C-5' radicals*

The formation of 5'-aldehyde-containing DNA fragments (103) and the liberation of furfural (104) upon alkaline treatment is used as a marker for C-5'-hydrogen atom abstraction, much in the same way that 3'-phosphoglycolate (55) and (97) formation is used to identify lesions that are initiated by reaction at the C-4' position of nucleotides in DNA (Scheme 25).[30,31,89]

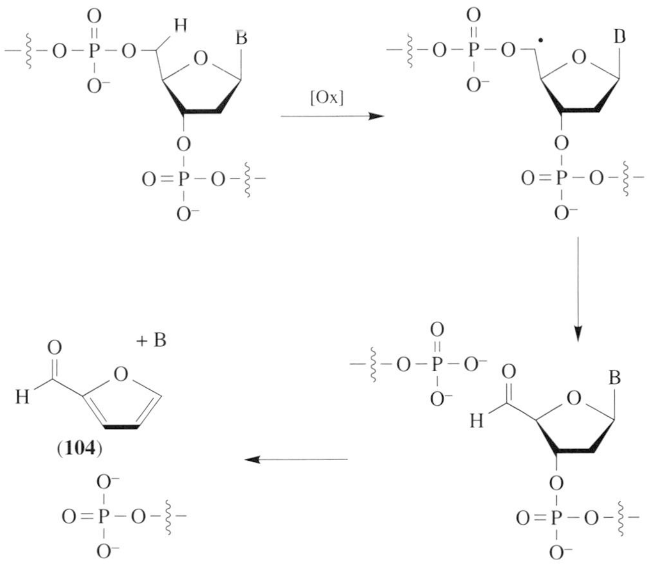

**Scheme 25**

One can readily envision four pathways by which C-5' hydrogen atom abstraction could result in a strand break in which the 3'-terminus contains a phosphate group and the 5'-terminus consists of a nucleoside aldehyde (Scheme 26). Two of these pathways would be independent of oxygen. In principle, oxygen-18 labeling can be employed to distinguish between these pathways.

In the presence of oxone, (4) cleaves DNA via a mechanism that is independent of $O_2$.[32] Data collected concerning C-1' oxidation by (4) imply that the oxygen rebound mechanism could be responsible for strand scission from C-5' oxidation as well. Unfortunately, it was not possible to carry out oxygen-18 labeling studies to differentiate between an oxygen rebound type mechanism and oxidation of the radical to a carbocation involving (4), because the aldehyde carbonyl underwent rapid exchange in water at room temperature. This is in direct contrast to neocarzinostatin (3), in which control experiments showed that the 5'-aldehyde did not undergo significant exchange under those reaction conditions.[118] Another difference between DNA damage induced by (4) and that by neocarzinostatin (3) is that the latter requires $O_2$ for direct strand scission.[116] Oxidation at the C-5' position by (3) gives rise to DNA fragments containing 5'-aldehyde (103) in which the carbonyl oxygen is derived from $O_2$.[118] The source of the carbonyl oxygen was determined via mass spectrometry following reduction of the biopolymer with $NaBH_4$ and enzymatic digestion. It is believed that the peroxy radical is reduced by the thiol which is present to activate the drug. The absence of any oxygen-18 from $H_2^{18}O$ in the aldehyde carbonyl rules out either mechanism in Scheme 26 involving a carbocation intermediate.

The 5'-terminal aldehyde (103) containing DNA fragment is alkaline labile, yielding a 5'-terminal phosphate product and furfural (104). Trapping studies (Scheme 27) support the view that (104) arises via sequential elimination reactions.[33] Intermediate (105) was trapped via reduction with $NaBH_4$, and characterized following hydrogenation to (106). While these studies were carried out using (4) as oxidant, there is no reason to doubt that the same pathway is followed during DNA damage mediated by neocarzinostatin (3).

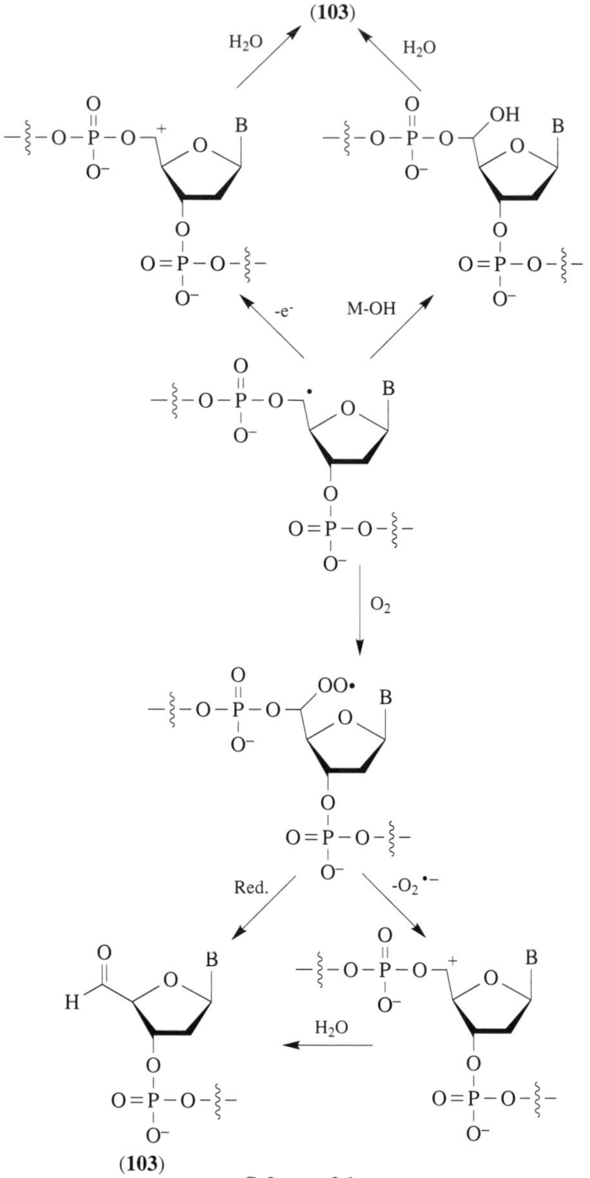

**Scheme 26**

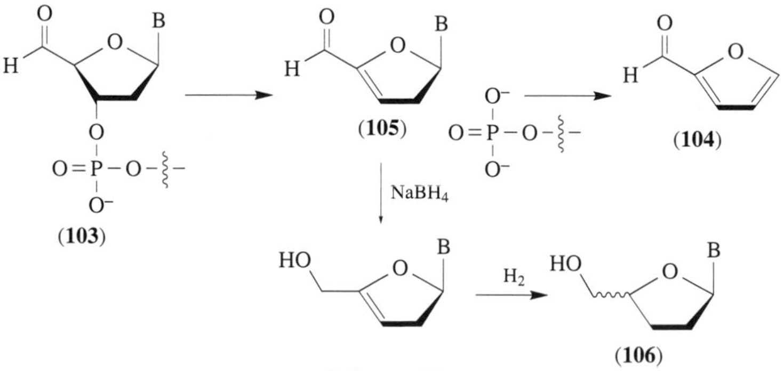

**Scheme 27**

In the case of neocarzinostatin, an additional minor pathway for 5'-mediated damage was detected.[119] Cleavage of a self-complementary hexanucleotide yielded (**107**) and dialdehyde (**108**) as a minor product (Scheme 28). The dialdehyde (**108**) was characterized as the diol following treatment with $NaBH_4$ and alkaline phosphatase digestion. The suggested mechanism for the formation of (**108**) involves a Criegée rearrangement from the C-5' hydroperoxide (**109**). However, in light of Giese's recent work involving the C-4' radical and the likelihood that a Criegée mechanism accounts for base propenal formation under metal-free conditions, it may be worthwhile to consider alternative mechanisms for the formation of (**107**).[115]

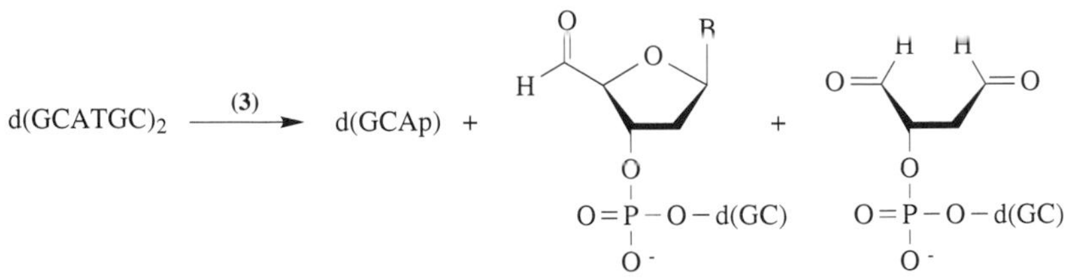

**Scheme 28**

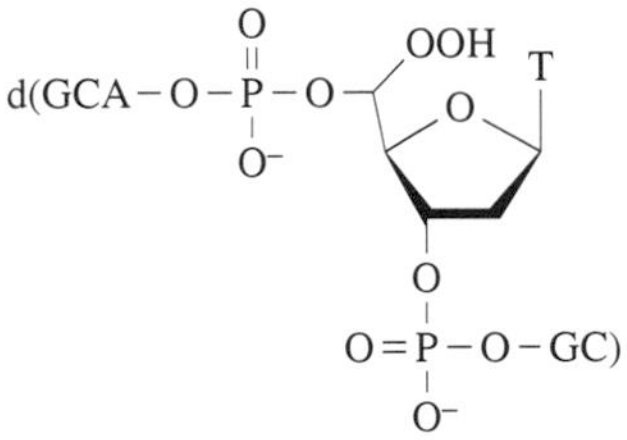

(**109**)

An alternative pathway involving $\beta$-fragmentation following C-5' hydrogen atom abstraction may account for strand breaks induced by neocarzinostatin (**3**) when misonidazole (**110**) is substituted for $O_2$ (Scheme 29).[120] The $\beta$-fragmentation pathway was proposed in order to account for the formation of formic acid under these cleavage conditions. The putative alkoxy radical (**112**) undergoing $\beta$-fragmentation could be derived from the initial misonidazole trapping adduct (**111**). The dialdehyde lesion (**114**) analogous to (**108**) described above could then result from one-electron oxidation (perhaps by another molecule of (**110**)) of the carbon radical formed (**113**) upon $\beta$-fragmentation, followed by hydrolysis. However, the dialdehyde product was not detected at the time that formic acid was observed, and further reports have not appeared.

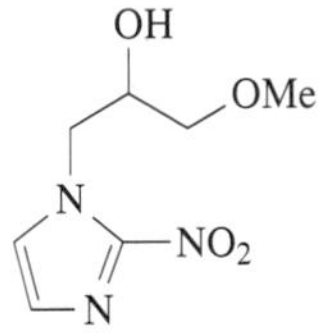

(**110**)

### 7.11.3 GENERATION AND REACTIVITY OF NUCLEOBASE-REACTIVE INTERMEDIATES

Nucleobase-reactive intermediates are most commonly considered when discussing the effects of ionizing radiation on DNA.[13,121,122] In addition, it appears that radiomimetic systems such as Fe·EDTA, Cu(OP)$_2$ (**5**), and other metal chelates also damage DNA via reaction with nucleobases.[123–125] It has even been observed that bleomycin (**51**) induces minor amounts of nucleobase damage.[126] However, the reactive species responsible for this damage is uncertain.

The formation of base-damaged nucleotides in DNA is an important issue. Modified nucleotides can have significant effects on DNA structure and function. The mutagenicity and repair of

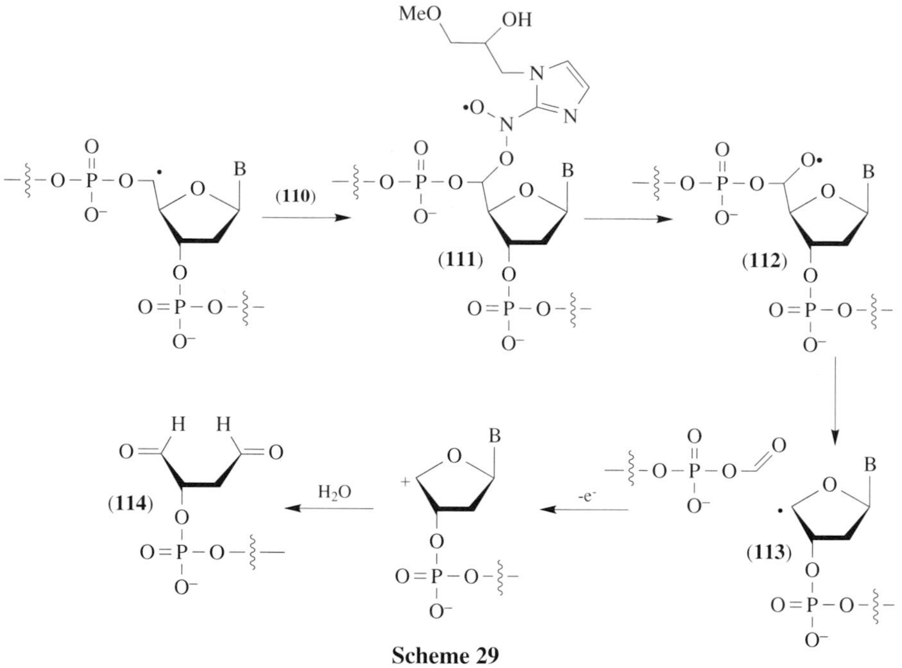

**Scheme 29**

nucleobase-altered nucleotides have been reviewed elsewhere, and will not be discussed in great detail here.[16] The reactivity of nucleobase-centered reactive intermediates which give rise to such lesions are the focus of this chapter.

### 7.11.3.1  Hydroxy Radical and Hydrogen Atom Adducts of Pyrimidines

#### 7.11.3.1.1  *Formation of pyrimidine nucleobase radical adducts*

$\gamma$-Radiolysis of water produces solvated electrons ($e_{aq}^-$), hydrogen atoms (H·), and hydroxyl radicals (HO·). The subsequent reaction of these species with nucleic acids produces damage that is generally referred to as the indirect effect of ionizing radiation. Hydroxy radicals react with nucleosides and nucleotides at or near the diffusion-controlled limit. Hydrogen atoms react with such substrates about an order of magnitude slower than HO·, but at bimolecular rates ($\sim 10^8$ M$^{-1}$ s$^{-1}$) that are still greater than many radical reactions.[13,127]

When HO· and H· react with pyrimidines such as thymidine, the predominant reaction pathway involves addition to the nucleobase double bond.[13] Hydroxy radicals prefer to add to the C-5 carbon ((115) vs. (116)) (Scheme 30, containing Structures (115)–(118)). Hydrogen atoms, being more nucleophilic than HO·, show the opposite regioselectivity ((118) vs. (117)). The HO· and H· adducts are also formed via the direct effects of ionizing radiation. Ionization of pyrimidines yields the alkene cation radicals (119) (Scheme 31). The pyrimidine alkene cation radicals are also produced via UV photolysis in the presence of an appropriate sensitizer such as 2-methyl-1,4-naphthoquinone (menadione, 10) (see Scheme 3).[74,128,129] The addition of SO$_4^-$ and subsequent elimination of SO$_4^{2-}$ and biphotonic ionization are yet other pathways that lead to alkene cation formation.[130–133] The alkene cation radicals are trapped by water. In the case of thymidine, C-6 is the preferential, if not exclusive, site of nucleophilic attack (Scheme 31).[74]

The direct effect of ionizing radiation also ultimately gives rise to the C-6 hydrogen atom adduct (118) (Scheme 32) via (120). During $\gamma$-radiolysis, electrons are believed to migrate over large distances in DNA.[134–137] While deoxyguanosine is often thought of as the electron source, thymidine is considered to be the electron sink.[138,139] The electron adduct of thymidine is irreversibly protonated to yield (118).[134] It has been argued that in cellular DNA, this type of two-step pathway is more prevalent than the formation of nucleobase-reactive intermediates via radical addition to the double bonds.

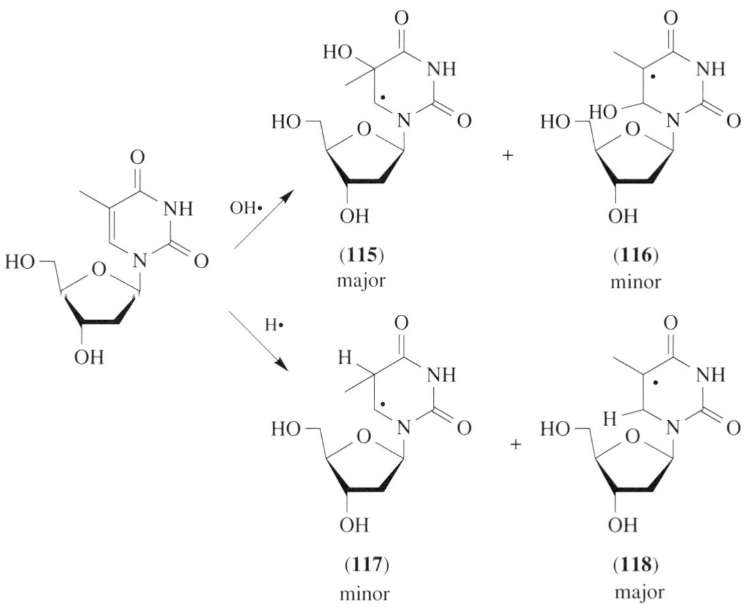

**Scheme 30**

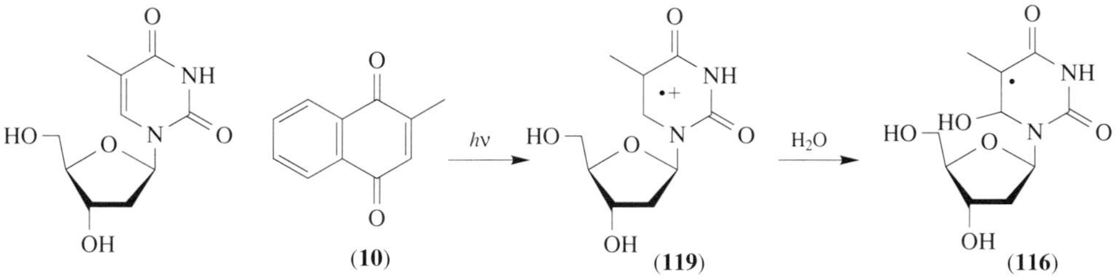

**Scheme 31**

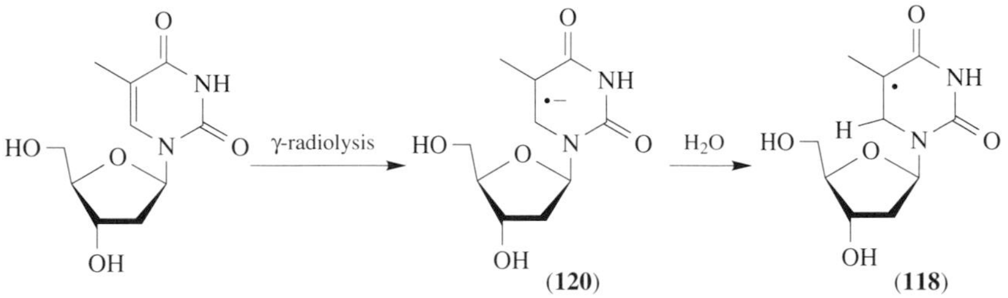

**Scheme 32**

### 7.11.3.1.2 *Hydrogen atom abstraction by pyrimidine nucleobase radical adducts*

A great deal of the research reported regarding pyrimidine-centered reactive intermediates has utilized polythymidylates and polyuridylic acid, as well as the respective monomers, as substrates.[13] The formation of a nucleobase-centered reactive intermediate in itself produces a damaged biopolymer. However, it is the carbohydrate component of the biopolymer that must be oxidized in order to induce a strand break. Consequently, the transposition of spin from the nucleobases to sugar moieties is a question that has received a great deal of attention.

Determination of the efficiency for strand scission of poly(U) induced by HO· under anaerobic conditions (41%) led to the proposal that nucleobase radicals, which account for 80% of the reactions of HO· with the biopolymer, induce strand breaks via hydrogen atom abstraction from the carbohydrate moiety.[140] Similar arguments can be made for H· adducts.[140] That the rate-limiting step in strand scission from the C-5 hydroxyl radical adduct (e.g. (115)) is hydrogen atom abstraction by the nucleobase radical is supported by examination of the effects of dithiothreitol (H· donor) and tetranitromethane (oxidizing agent) on the efficiency of strand scission.[141]

At neutral pH and anoxic conditions, the rate-limiting step in strand scission is believed to be internucleotidyl hydrogen atom abstraction (Scheme 33). Intranucleotidyl hydrogen atom abstraction is disfavored, presumably on kinetic grounds, due to poor stereoelectronic overlap. Electron spin resonance studies indicate that the hydroxyl radical adducts of the monomeric pyrimidines are stable on the millisecond timescale.[130,142] However, this does not preclude their participation in intranucleotidyl hydrogen atom abstraction, because kinetic studies of nucleic acid cleavage (under anaerobic conditions) indicate that the half-life for such a process is greater than 0.1 s.[143–145]

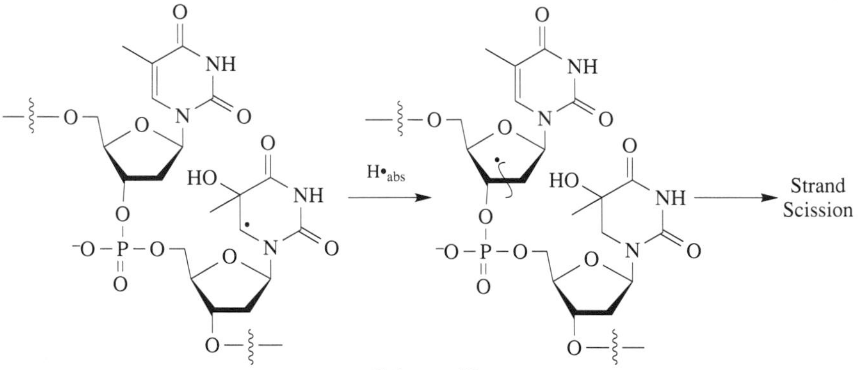

**Scheme 33**

More quantitative analysis of the feasibility that intranucleotidyl hydrogen atom abstraction within the C-5 hydroxyl radical adduct of a pyrimidine nucleoside is kinetically competent to be involved in strand scission was obtained by independently generating (**115**) from (**121**) (Equation (9)).[146,147] Through a combination of experiments utilizing isotopic labeling (**123**) and competitive kinetics, in conjunction with $^2$H NMR and mass spectral analysis of (**122**), it was shown that intramolecular hydrogen atom abstraction within (**115**) was not kinetically competent to be involved in nucleic acid strand scission (Scheme 34, containing Structures (**115**)–(**124**)). Product analyses were carried out on the free nucleobases in order to eliminate possible complications due to multiple reaction pathways available to (**124**). The relevancy of the results obtained using this nucleoside to the issue of intranucleotidyl hydrogen atom abstraction in biopolymers is supported by calculations on 1-amino-2-deoxy-3,5-diphosphateribofuranose.[25] Consideration of bond strengths and sterics suggest that the most likely site of intranucleotidyl hydrogen atom abstraction is the C-3′ position. Computational experiments indicate that this process should be even less thermodynamically favorable in biopolymers, where phosphates are present, than in nucleoside (**115**).[24,25]

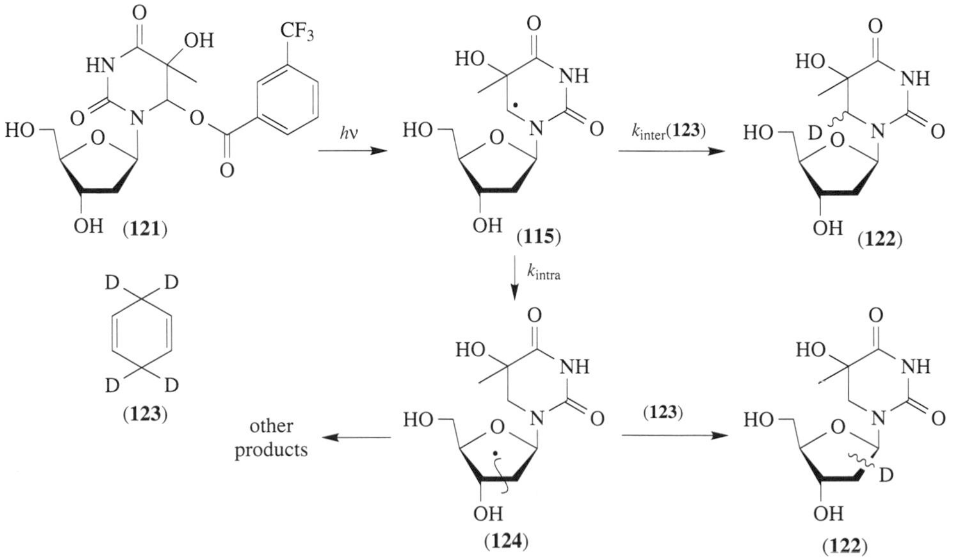

**Scheme 34**

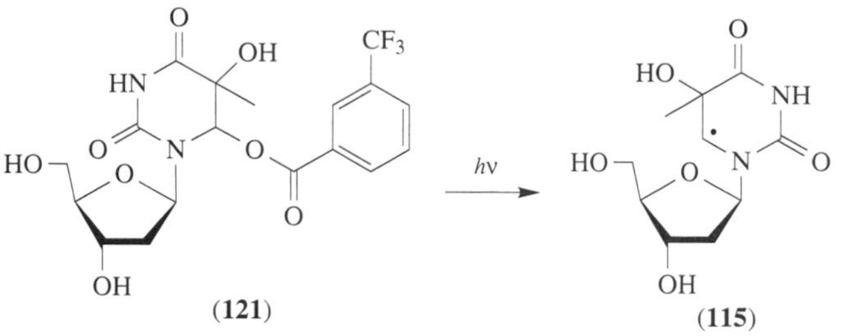

(121)          (115)         (9)

Support for the generation of strand breaks via internucleotidyl hydrogen atom abstraction by pyrimidine C-5 hydroxyl radical adducts (e.g., (115)) was derived from product analysis following γ-radiolysis of thymine, thymidine, thymidine phosphate, and oligomers of thymidine under anoxic conditions.[148] Radiolysis of thymine containing substrates shorter than a dinucleotide revealed that the glycol (125) was the major product and thymidine C-5 hydrate (122) was a minor product. The products were analyzed by mass spectrometry as the free bases following formic acid cleavage of the crude radiolysis mixture. Formation of the glycol (125) is attributed to oxidation of the C-5 hydroxyl radical adduct (115) followed by nucleophilic attack by water. In contrast, the thymidine C-5 hydrate (122) is believed to be derived from hydrogen atom donation or via disproportionation. The contribution of individual pathways cannot be distinguished using the approach employed by these workers. However, observation that the yield of thymidine C-5 hydrate (122) is actually greater when thymine is the substrate than when thymidine phosphate is subjected to radiolysis suggests that intramolecular hydrogen atom abstraction is not responsible for this product. In contrast, when dinucleotide and longer substrates are subjected to anaerobic radiolysis, the yield of (122) increases at the expense of (125) by approximately threefold. These results suggest that adjacent nucleotides provide a hydrogen atom source for (115), and that internucleotidyl hydrogen atom abstraction by (115) competes with oxidation of the radical. Internucleotidyl hydrogen atom transfer by the C-5 hydroxyl radical adduct of thymidine results in damaging two nucleotides in a biopolymer by a single HO·. This result has been referred to as DNA damage amplification. Situations in which DNA damage is amplified via multiple nucleobase modification are discussed later in section 7.11.5.2.

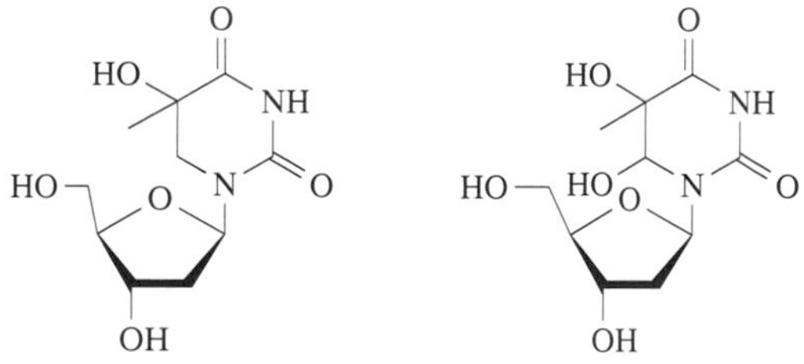

(122)                   (125)

5,6-Dihydrothymidin-5-yl (118) is another nucleobase radical adduct that has been proposed to induce DNA strand breaks via internucleotidyl hydrogen atom abstraction. One could argue that based upon resonance, the C-6 radical adducts of pyrimidines are more stable than the respective C-5 radical adducts. Furthermore, the lower efficiency for strand scission suggests that H· adducts are less reactive than the respective HO· adducts. This latter deduction can be rationalized in terms of the effects of β-substituents on radical reactivity. A comparison of the regioselectivity of chlorination of butanes reveals that electronegative substituents decrease reactivity at the β-position relative to the γ-position (Figure 2).[149] By employing the principle of microscopic reversibility, these data can be interpreted to imply that electronegative substituents destabilize radicals at the β-position. Hence, it is reasonable to assume that (116) should be more reactive than (118). Interpretation of ESR experiments led to the proposal that (118) abstracts the C-1′, C-2′, and/or C-4′ hydrogen atoms of adjacent nucleotides, implying that (116) should as well.[136,150]

This proposal was recently tested in studies on (118). 5,6-Dihydrothymidin-5-yl (118) was independently generated via Norrish Type I photocleavage of (126) (Equation (10)).[151] Chemical trapping with nonexchangeable hydrogen atom donors and $O_2$ in aqueous solvents confirmed that (118) is produced from (126). In order to address the feasibility for hydrogen atom transfer from the carbohydrate components of nucleic acids to (118), propan-2-ol (127) and the dimethylacetal of

|       | α | β | γ |
|-------|----|----|----|
| X —— CH$_2$ — CH$_2$ — CH$_2$ — Me | | | |
| H     | 15 | 35 | 35 |
| Cl    | 10 | 26 | 46 |
| OMe   | 35 | 7  | 43 |

% Reaction

**Figure 2** Substituent effects on hydrogen atom abstraction by Cl·.[149]

glycoaldehyde (**128**) were employed as traps. The acetal and alcohol were used as model compounds of the C-1′ and C-4′ positions of the deoxyribose ring, respectively (Scheme 35). In addition, (**127**) could serve as a model of the C-2′ position of the ribose rings in RNA. Anaerobic photolysis of (**126**) indicated that neither trap, when present at concentrations (5 M) meant to mimic the effective molarity of an adjacent sugar ring in DNA, serves as an effective hydrogen atom donor.[151]

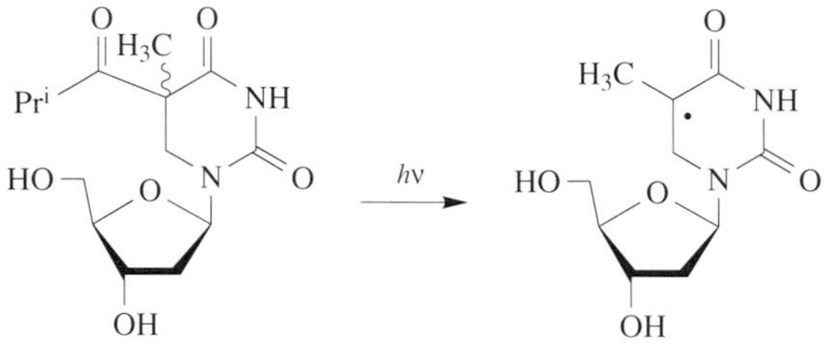

(10)

(126)                                          (118)

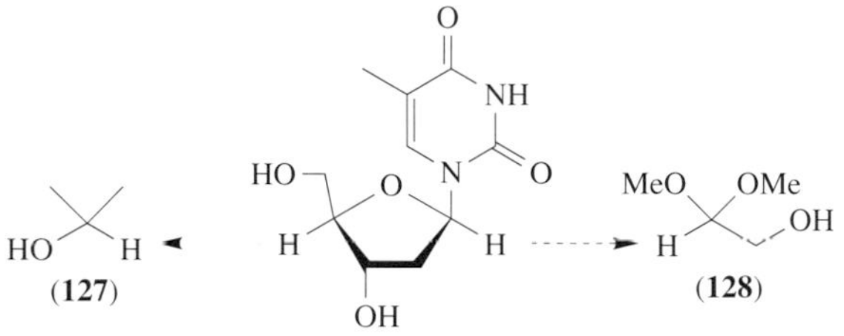

(127)                                          (128)

**Scheme 35**

The inability of (**118**) to effect strand damage in DNA via hydrogen atom abstraction from adjacent nucleotides was investigated further by independently generating the radical from (**126**) within biopolymers.[34,35] Since (**126**) is a dihydropyrimidine, and therefore unstable to concentrated ammonium hydroxide, it was site-specifically incorporated into chemically synthesized DNA using methodology that obviates alkaline deprotection conditions.[152,153] Gel electrophoretic analysis of [32]P-labeled oligonucleotides containing (**126**) that were irradiated under anaerobic conditions showed that no direct strand breaks or alkaline labile lesions were formed.[34,35] These observations strongly suggest that (**118**) is incapable of amplifying DNA damage via internucleotidyl hydrogen atom abstraction.[136,150]

### 7.11.3.1.3 *Formation and reactivity of peroxy radicals derived from pyrimidine radical adducts*

Considering that the rate constant for DNA peroxy radical formation is essentially diffusion-controlled ($\sim 2 \times 10^9$ M$^{-1}$ s$^{-1}$), it is unlikely that hydrogen atom abstraction reactions by the pyrimidine radical adducts will compete with their trapping by O$_2$, whose solubility in aerated water is $\sim 0.2$ mM.[55] The formation of peroxy radicals within γ-irradiated biopolymers was confirmed using ESR, and their involvement in the rate-limiting step of strand scission was corroborated by time-resolved ESR and conductivity measurements.[154,155] These experiments show that the half-life for strand break formation ($< 3$ s) is approximately equal to the half-life for the observed peroxy radicals ($\sim 1.4$ s). While the identity of the specific peroxy radicals that are involved in the rate-limiting step of strand scission is unknown, product studies, and the propensity for radicals derived from water to add to the pyrimidine double bonds, are consistent with the proposal that strand breaks are the result of base peroxy radicals abstracting hydrogen atoms (Scheme 36).[13,154–156] The involvement of pyrimidine base peroxy radicals (e.g., (129)) in strand scission is further supported by time-resolved light scattering and UV transient experiments involving poly(C), poly(U), and poly(A).[157] These experiments also indirectly support the role of pyrimidine base peroxy radicals in strand scission by demonstrating a significant reduction in the fraction of strand breaks attributable to peroxy radicals in poly(A) compared to that observed in polypyrimidines.[68]

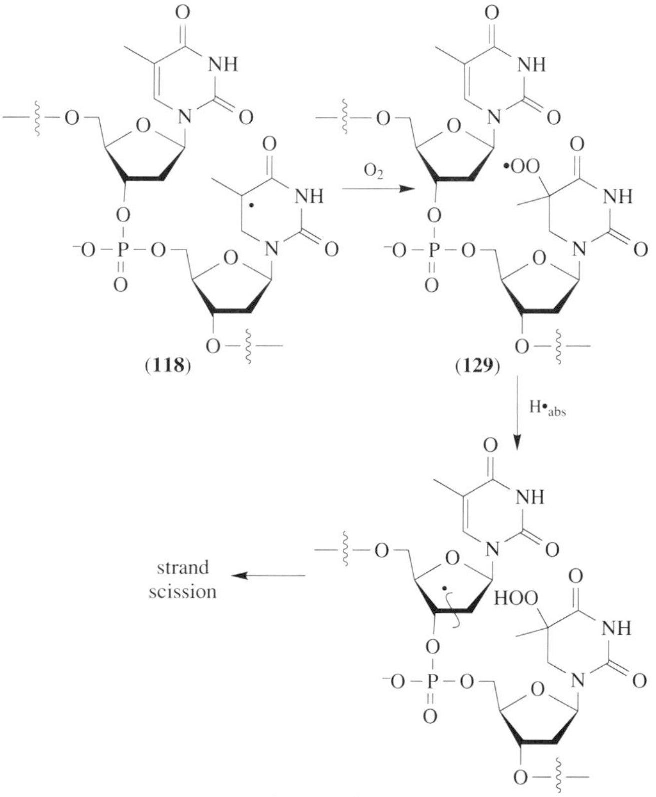

**Scheme 36**

The above experiments strongly implicate pyrimidine base peroxy radicals in nucleic acid strand scission. However, which hydrogen atom(s) is abstracted by the peroxy radicals is unclear. Early work in this area led to the suggestion that the internucleotidyl C-4′ hydrogen atom is abstracted by the C-5 hydroxyl radical adduct of uridine in poly(U).[154,158] This proposal is consistent with recent computational studies which place the C-4′ radical of a substituted deoxyribose only slightly higher in energy than the C-1′ radical.[25] In addition, in an oligonucleotide such as poly(U), C-1′ hydrogen atom abstraction should be disfavored compared to the comparable process in an oligodeoxyribonucleotide, due to the inductive effect of the 2′-hydroxyl group (see Figure 2). However, these predictions have not yet been verified unambiguously.

The most closely related study that has been examined experimentally concerns the reactivity of 5,6-dihydrothymid-5-yl (118) under aerobic conditions.[34,35] When (118) was independently generated under these conditions, direct strand breaks and alkaline labile lesions are produced at the 5′-adjacent nucleotide in a heteropolymer. Kinetic isotope effect experiments implied that cleavage at the 5′-nucleotide adjacent to (129) does not involve C-2′ or C-4′ hydrogen atom abstraction. The lack of involvement of the C-4′ position in strand scission was also reflected by the absence of 3′-phosphoglycolate termini (55).[72,115] However, an approximately-fourfold diminution in direct strand scission was observed upon deuteration of the C-1′ position of the 5′-adjacent nucleotide. The observed kinetic isotope effect suggested that the C-1′ hydrogen atom was abstracted. In poly-thymidylates containing a single molecule of (126), a more diffuse cleavage pattern was observed. Enzymatic end group analysis revealed that the 3′-termini of all fragments produced in both types of biopolymers were phosphates. 3′-End labeling experiments not only revealed that all 5′-termini contain phosphates (by enzymatic end group analysis), but also two products resulted from cleavage at the nucleotide adjacent to the 5′-phosphate of (126).[35] Based upon the observations listed above, and the conversion of the slower migrating product into a DNA fragment one nucleotide shorter that also contains a 5′-phosphate terminus upon alkaline hydrolysis, it was inferred that the two products were (130) and (131) (Scheme 37). The faster moving fragment was believed to contain an intact molecule of (126), and to arise via the effects of singlet oxygen ($^1O_2$) that was produced by quenching the excited state of (126). This scenario was supported by a significant diminution of cleavage upon addition of sodium azide, and independent verification that irradiation of (126) produced $^1O_2$.[35]

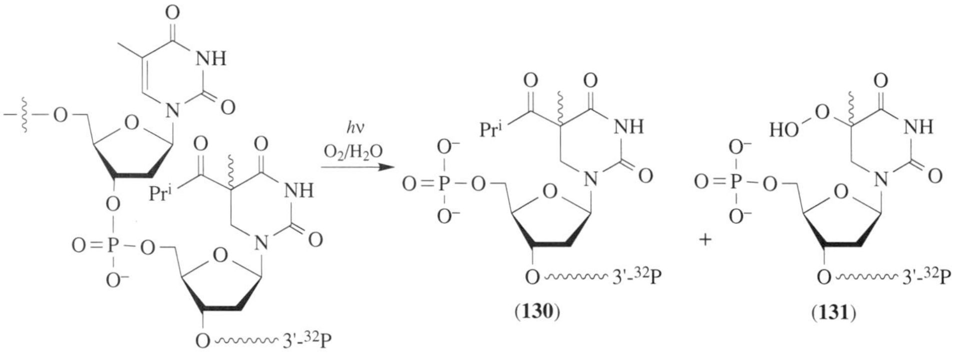

**Scheme 37**

Selective hydrogen atom abstraction from the C-1′ position of the 5′-adjacent nucleotide by the peroxy radical (129) generated from (118) is consistent with what one would expect based upon thermodynamics (Scheme 38). The expected bond dissociation energy of the ROO—H bond that would be formed is 90–92 kcal mol$^{-1}$.[159] Computational studies suggest that the C-1′ and C-4′ carbon–hydrogen bonds of the adjacent nucleotide are the only bonds for which hydrogen atom abstraction by (129) would be thermodynamically downhill.[24,25] The rate constant or efficiency of this process is not yet known.

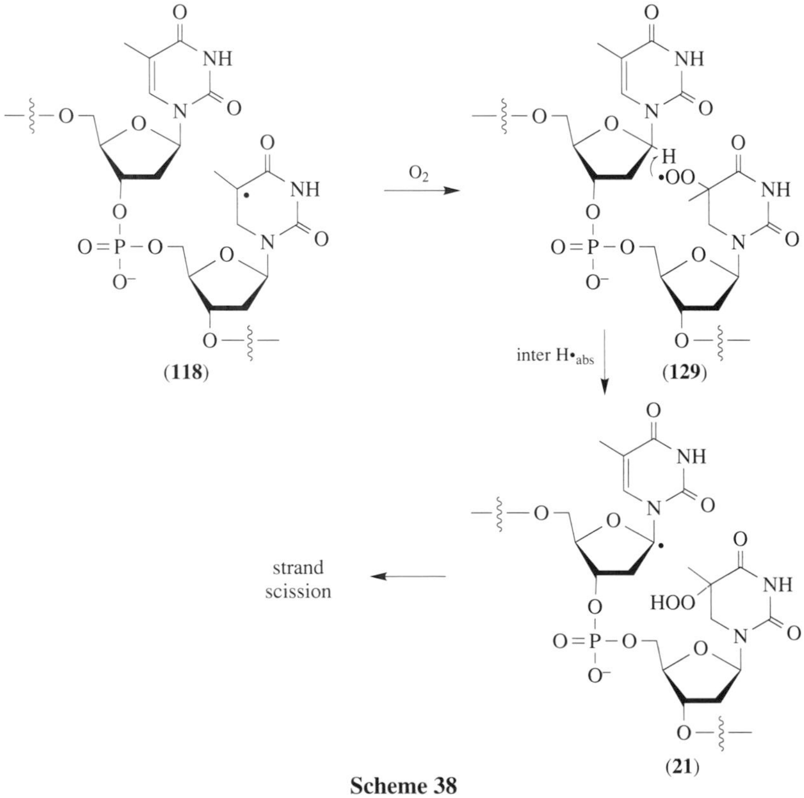

**Scheme 38**

### 7.11.3.1.4  *Formation and reactivity of pyrimidine alkene cation radicals*

As mentioned above, the alkene cation radicals of the pyrimidines are associated with the direct effects of ionizing radiation (Section 7.11.3.1.1). It was recognized some time ago that the direct and indirect effects of ionizing radiation produced a similar spectrum of degradation products. Furthermore, it was also observed that the kinetics and efficiency of strand scission from the two pathways were affected similarly by pH and the hydrogen atom donor, dithiothreitol.[133] Consequently, it was proposed that as the pH is lowered, the increased rate constant for strand scission in poly(U) via the indirect effect is due to an increase in the rate of formation of the alkene cation radical (133).[133,145] This proposal was supported by ESR examination of the reaction products of HO with poly(U), where at pH 7, the C 5 hydroxyl radical adduct (132) analogous to (115) was observed, but (134) was observed at pH ⩽ 4 (Scheme 39).

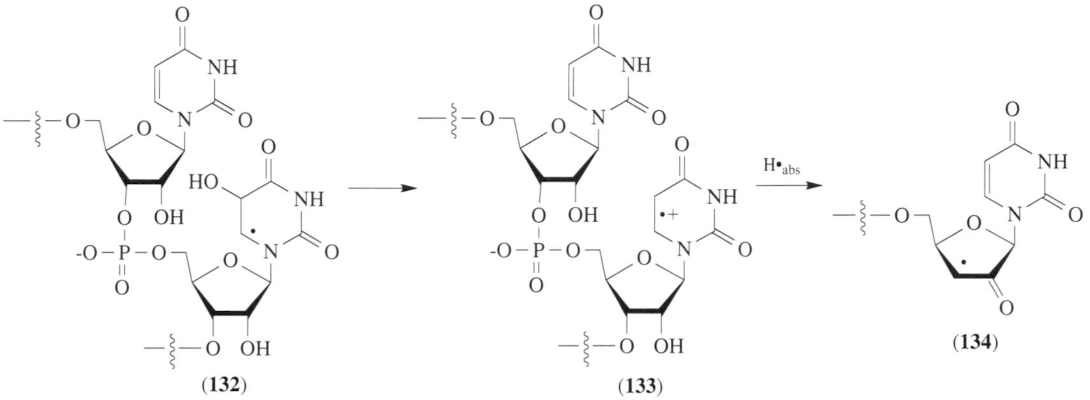

**Scheme 39**

Studies on monomers support the proposal that the alkene cation radicals of ribonucleosides are responsible for the rapid abstraction of hydrogen atoms from the C-2′ and C-3′ positions of the carbohydrate component, giving rise to the radicals (135) and (136) observed in the ESR spectrometer.[130,142,160] In contrast, base-centered radicals are observed by ESR when the alkene cation radical deoxynucleosides are generated via their reaction with $SO_4^{\cdot-}$. The absence of the C-2′ radical in studies on deoxyribonucleosides is rationalizable on the basis of the strength of the respective carbon–hydrogen bond.[24,25] However, the absence of the C-3′ radical in studies on deoxyribonucleosides was ascribed to differences in ring puckering of ribonucleosides and deoxyribonucleosides.[142] Considering the small barriers to the interconversion of ribofuranose conformers, this explanation seems unlikely. As an alternative, it is possible that intramolecular hydrogen bonding between the C-2′ and C-3′ hydroxyl groups of ribonucleosides reduces the bond strengths of the respective carbon–hydrogen bonds. This rationale is supported indirectly by computational experiments.[161] It is also consistent with the absence of sugar radicals when the cation radical of 2′,3′-*O*-isopropylidene uridine (137) is generated in water.[130] More recently, the absence of C-3′ sugar radicals in deoxyribonucleosides has been reconciled with observations made using ribonucleosides by reassigning the spectrum of the C-3′ radical (136) to that of the C-6 hydroxyl radical adduct (138).[160] This work has also resulted in new mechanistic proposals for the formation of (135) which have not yet been further substantiated.

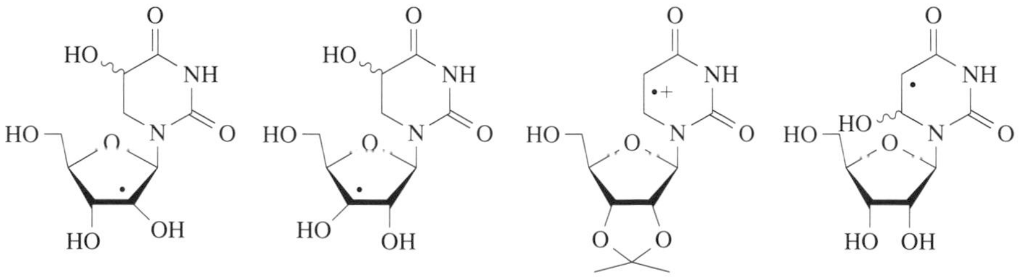

In biopolymers, the pathway to strand scission from the alkene cation radicals of ribonucleosides is believed to fall onto a similar mechanistic manifold as does abstraction of the C-4′ hydrogen (see

Section 7.11.2.4). Following C-2′ hydrogen atom abstraction in polyribonucleotides, phosphate elimination from (139) directly results in strand breaks (Scheme 40).[142,145] Elimination of the nucleobase would yield (140). This process has been postulated in studies on monomers.[160]

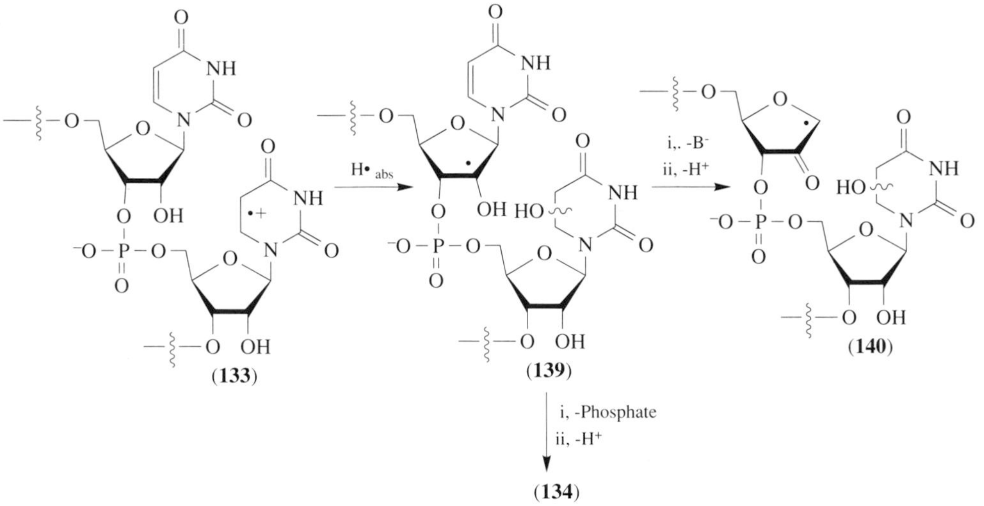

**Scheme 40**

The monomer studies described above suggest that the pyrimidine alkene cation radicals of deoxyribonucleosides are hydrated faster than they undergo intramolecular hydrogen atom abstraction. However, pH studies on DNA suggest that dehydration of hydroxyl radical adducts could be the rate-limiting step in strand scission.[68,133] The viability of this process at neutral pH was investigated by independently generating (115) (Equation (9)).[147] Since the initial hydroxyl group in (115) was not derived from the solvent, it was possible to use $H_2{}^{18}O$ as a label for detecting the intermediacy of (119). Using trapping of (115) by tetradeuterated cyclohexa-1,4,-diene (123) as a kinetic clock, the absence of (141) revealed that dehydration of (115) at neutral pH is not kinetically competent to be involved in strand scission (Scheme 41).

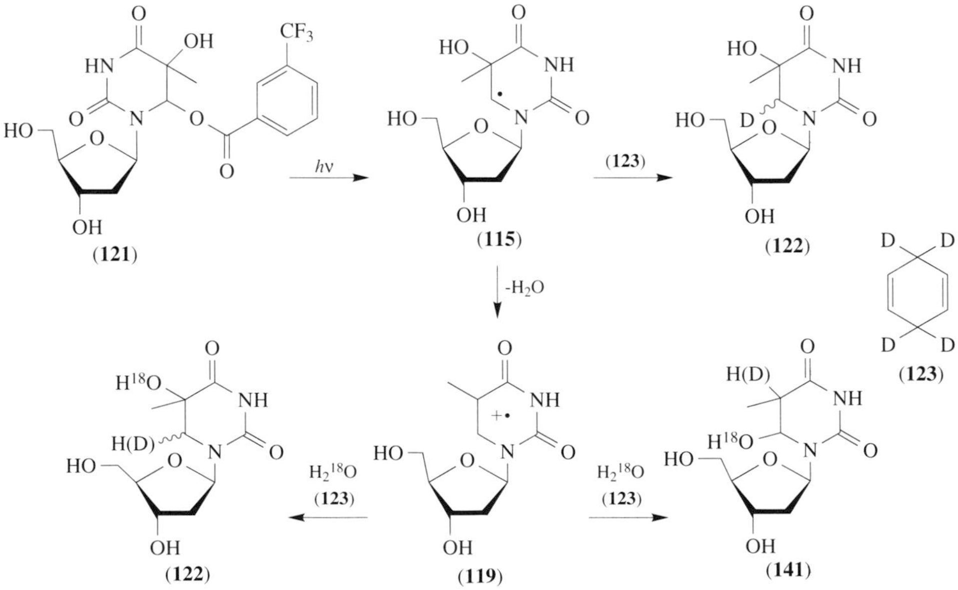

**Scheme 41**

The alkene cation radicals derived from pyrimidines containing a methyl group at the C-5 position (e.g., (119)) have an additional reactive pathway available to them.[162] Cation radical (119) also undergoes deprotonation to yield the allylic radical (142), which is believed to account for as much as 10% of the lesions produced by the indirect effect of ionizing radiation on thymidine (Scheme 42).[74,128,129,142] The involvement (if any) of (142) or the peroxy radical (143) derived from it on strand damage is unknown. Consideration of thermodynamics strongly suggests that (142) would not be proficient at abstracting hydrogen atoms from the carbohydrate component of the nucleic acid. However, (143) should be no less effective than (129) at inducing strand damage via hydrogen atom abstraction.

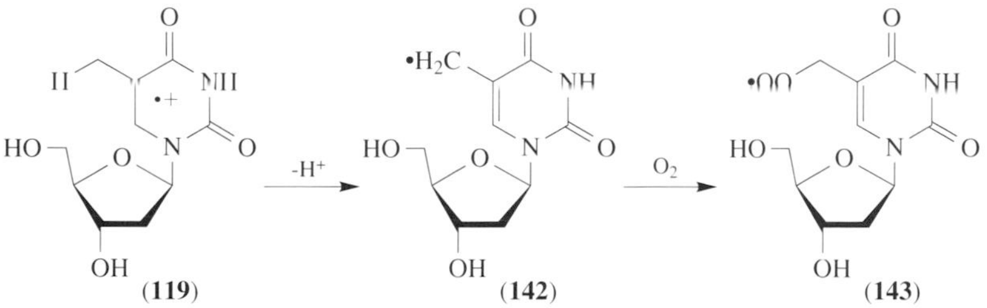

**Scheme 42**

### 7.11.3.1.5 *The formation and effects of modified pyrimidines on DNA structure and function*

The involvement of nucleobase-centered reactive intermediates such as (142) in nucleic acid strand scission may be uncertain. However, the modified nucleotides ultimately produced often exhibit significant biological and structural effects on DNA. For instance, 5-hydroxymethyldeoxyuridine (144) and 5-formyldeoxyuridine (145) are two biologically relevant nucleosides that are derived from (142).[74,128,129] 5-Formyldeoxyuridine (145) is also produced during oxidation of DNA by the nitro-substituted naphthalimide, (146).[163] This modified nucleoside readily undergoes glycosidic cleavage upon heating. 5-Formyldeoxyuridine (145) is believed to result in T → C transversions, and is a substrate for a variety of repair enzymes.[164-166] 5-Hydroxymethyldeoxyuridine (144) is also believed to be a premutagenic lesion in DNA.[167,168] Recently, this molecule was successfully incorporated site-specifically into chemically synthesized oligonucleotides.[169] This should greatly facilitate further examination of the biological and structural effects of (144) on DNA.

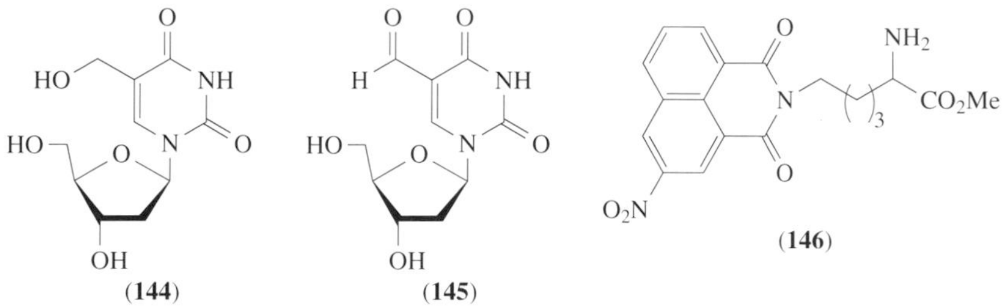

A variety of other modified pyrimidine nucleosides are produced in DNA as a result of oxidative stress. Due to the prevalence of electron and radical addition to the pyrimidine double bonds, dihydropyrimidines comprise a significant fraction of these modified nucleotides. 5,6-Dihydro-thymidine (147), 5,6-dihydro-5-hydroxythymidine (122), and 5,6-dihydro-5,6-dihydroxythymidine (125) are three of the dihydropyrimidine lesions produced from thymidine.[13] Others include the regioisomeric thymidine C-6 hydrate (141) and the hydroperoxides analogous to (125) (e.g., (148)).[170,171] In general, the dihydropyrimidines are substrates for repair enzymes such as endonuclease III.[172] However, the effects of these lesions on polymerase enzyme activity vary significantly. Dihydrothymidine (147) does not appear to inhibit DNA polymerase significantly.[173] In contrast, thymidine glycol (125) was shown to be a significant block of a variety of polymerases *in vitro*.[174]

The disparate effects of (125) and (147) on polymerase enzyme activity were rationalized by computational experiments.[175,176] All of the dihydrothymidines are predicted to maintain the nearly coplanar relationship of hydrogen bonding components, as is the case for the native thymidine. In (147), the methyl substituent in either diastereomer is predicted to adopt a pseudoequatorial position, enabling it to avoid steric interactions with adjacent nucleotides in the oligonucleotide. The resulting prediction is that (147) should present a hydrogen bonding pattern to a polymerase that is remark-

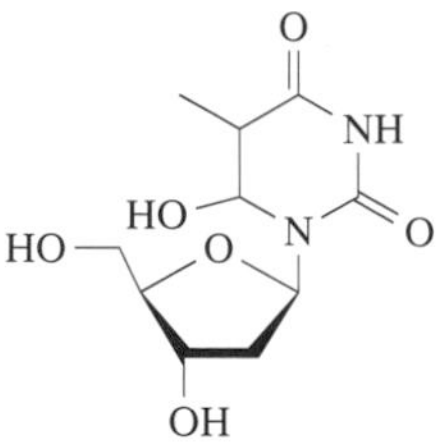

(141) 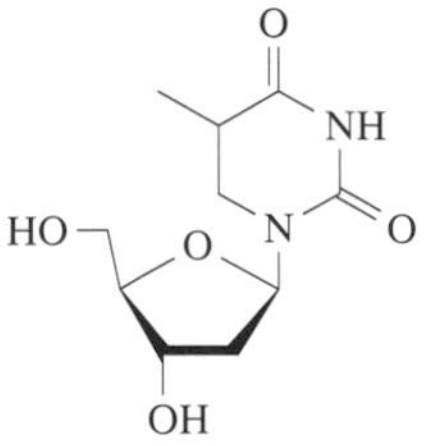    (147) 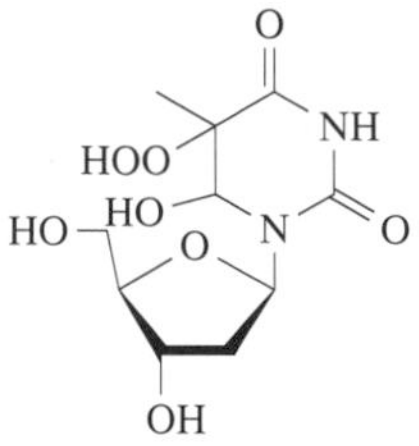    (148)

ably similar to that of thymidine, and should not significantly perturb base stacking within the duplex. The lack of polymerase inhibition by (147) confirms these predictions. The addition of a hydroxyl group at C-5 in (122) and (125) does not affect the nearly coplanar arrangement of the functional groups capable of participating in hydrogen bonds. However, computational experiments predict that the larger methyl group at C-5 will occupy a pseudoaxial position, while the smaller hydroxyl group will be pseudoequatorial (Figure 3). This preferred conformation is rationalized in terms of a favorable dipole–dipole interaction between the carbonyl group at C-4 and the C-5 hydroxyl group in (122) and (125). The predicted outcome of such a spatial arrangement is that, depending upon the configuration at C-5 in (122) and (125), the methyl group will either perturb base stacking with the 5'-adjacent nucleotide (5R) or the 3'-adjacent nucleotide (5S). The inhibition of primer extension past (148) in a synthetic template is also consistent with these predictions. However, since the configuration at C-5 of (125) in the template was not known for certain, and there is an additional hydroxyl group at C-6, these experiments did not unambiguously address the predictions concerning the steric effect of the methyl group at C-5.

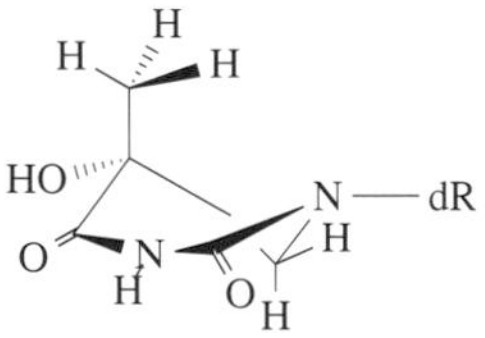

**Figure 3**   Three-dimensional structure of the 5R diastereomer of (122): dR = deoxyribose.[175,176]

This role of the C-5 methyl group was unambiguously addressed by studying the inhibition of Klenow (*exo-*) by the 5R-thymidine C-5 hydrate (149) and isostere analogue (150), which were site-specifically incorporated into chemically synthesized oligonucleotides.[152,177] The predicted coplanar orientation of the imido group was supported by a relatively weak inhibition of translesional synthesis. The most significant inhibition by (149) was observed for primer extension one nucleotide past this lesion. This is consistent with the methyl group at C-5 occupying a pseudoaxial position. Further experimental support for this structural correlation with enzyme inhibition was gleaned from the kinetic analysis of (150) which was essentially identical to that of (149).

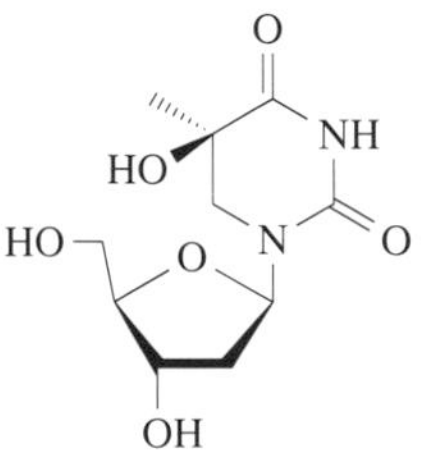

(149) 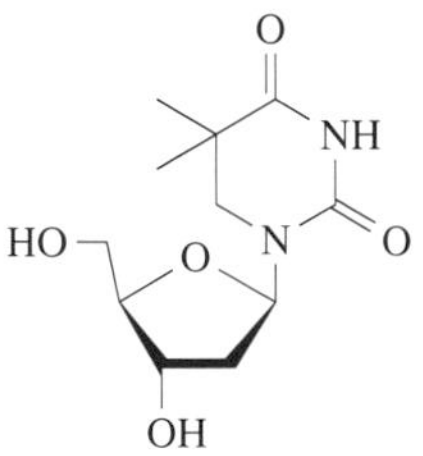    (150)

### 7.11.3.2   Purine-centered Reactive Intermediates

Compared to the chemistry of the pyrimidine nucleotides, significantly less is known about the reactivity of purine nucleobase-centered reactive intermediates.[13] The chemistry of the purines are complicated by the greater variety of reactive sites available for radical attack than in the pyrimidines. The reactivity of the purine radicals are further complicated by the ability of some radicals to act as both reducing and oxidizing agents.[178] One aspect of purine nucleobase chemistry which is less ambiguous than that of the pyrimidines is the commonality of reactive intermediates formed from the direct and indirect effects of ionizing radiation. However, there is some discord regarding the role that purine nucleobase-centered radicals play in nucleic acid strand scission.[56,158,179]

### 7.11.3.2.1  *Hydroxy radical adducts of purine nucleotides*

The adenines and guanines are preferentially attacked by hydroxyl radicals at the C-4 and C-8 positions.[178,180] The C-8 hydroxyl radical adducts (151) and (154) are believed to give rise to the 8-oxopurine and formamidopyrimidine lesions competitively (Scheme 43). The 8-oxopurines (152) and (155) are derived from sequential oxidation deprotonation processes. In contrast, the formamidopyrimidine lesions (153) and (156), which are formally hydration products of the purines, are postulated to result from hydrolysis, followed by reduction and protonation of the ring-opened imidazole. It is not inconceivable that the order of the reduction and hydrolysis processes are reversed.

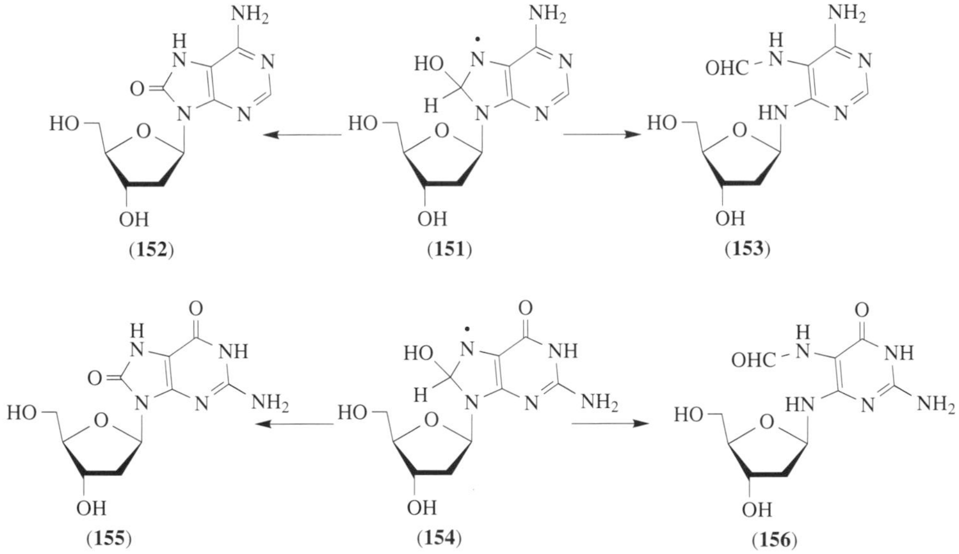

**Scheme 43**

The C-4 hydroxyl radical adducts of deoxyadenosine (157) and deoxyguanosine (158) are believed to give rise to radicals (160) and (162), respectively, via dehydration (Scheme 44). The cation radicals (159) and (161), which are formed via direct ionization, may be intermediates in this transformation.[56,178,179,181]

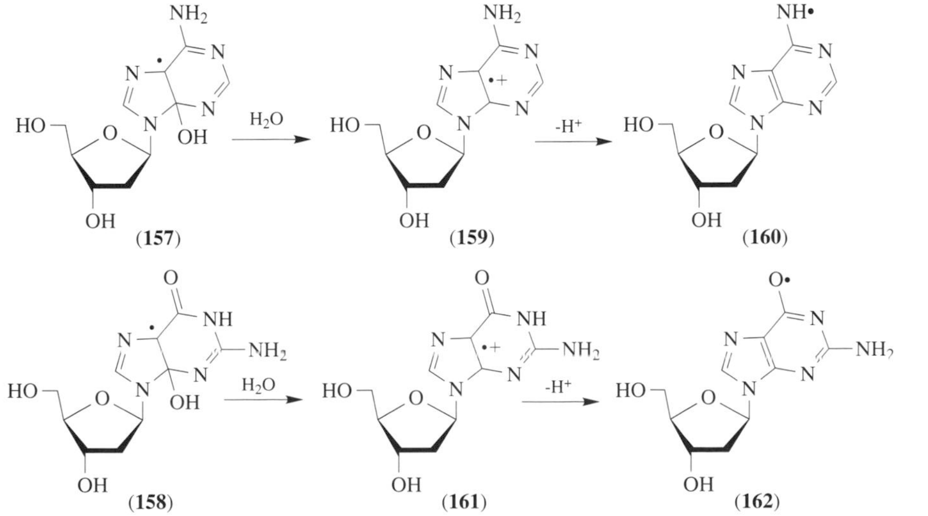

**Scheme 44**

### 7.11.3.2.2    *Alkene cation radicals of purines*

Deoxyguanosine is commonly believed to be the electron source of DNA oxidation due to its relatively favorable oxidation potential.[139,178,182,183] This feature has been taken advantage of in the design of a number of G-selective chemical oxidants.[184–188] It has been shown that the cation radical itself does not give rise to direct strand breaks but instead undergoes hydration to the C-8 hydroxyl radical adduct (**154**).[181] Alternatively, deprotonation of either purine cation radical gives rise to the neutral radicals, which are themselves strong oxidants (Scheme 44).[178,180]

### 7.11.3.2.3    *The role of purine nucleobase-centered reactive intermediates in strand scission*

Studies on poly(A) and poly(dA) homopolymers revealed that hydroxyl radical-induced cleavage of polypurines is significantly less efficient than that of polypyrimidines.[68,189] In addition, cleavage of poly(dA) is more efficient than poly(A). One should note that such relative reactivity is in direct contrast to what is observed for polypyrimidines, where oligoribonucleotides are more susceptible to transposition of spin from the nucleobases to the carbohydrate moiety (see Section 7.11.3.1). The increased reactivity of nucleobase-reactive intermediates in poly(dA) was tentatively ascribed to a conformational effect of unknown origin which facilitates hydrogen atom transfer from the sugar to the base. Neither the hydrogen atom abstracted nor the exact structure of the nucleobase-reactive intermediate involved in hydrogen atom abstraction are identified in these reports.[68,189]

Subsequently, the radicals produced upon deprotonation of the respective purine cation radicals have been discussed as candidates for the species responsible for inducing strand scission via hydrogen atom abstraction from the carbohydrate moiety (Scheme 44).[56,158,178–180] Assuming that (**160**) and (**162**) can be thought of as analogues of aniline and phenol, consideration of thermodynamics suggests that these species should be reluctant to abstract hydrogen atoms from the sugar components of adjacent nucleotides in nucleic acids. The BDE of phenol is only 85 kcal mol$^{-1}$, and nitrogen–hydrogen bonds are typically slightly weaker than oxygen–hydrogen bonds. If the BDEs of these model compounds are reasonably close to the respective nitrogen–hydrogen and oxygen–hydrogen bonds formed during hydrogen atom abstraction by (**160**) and (**162**), then their hydrogen atom abstraction from nucleotides should be thermodynamically uphill. Consistent with this hypothesis is the fact that no experimental data has appeared that directly supports the ability of (**160**) and (**162**) to induce strand breaks via hydrogen atom abstraction. However, experiments have been reported that refute such processes.[158,179] An alternative role for (**160**) and (**162**) in strand scission that takes advantage of their strong oxidizing properties will be discussed in Section 7.11.3.3.

### 7.11.3.2.4    *The formation and effects of modified purines on DNA structure and function*

Direct radical attack or ionization of purines produces the two families of lesions known as the 8-oxopurines ((**152**), (**155**)) and the formamidopyrimidines ((**153**), (**156**)) (Scheme 43). These lesions are also produced via a variety of other forms of oxidative stress.[121,123–126,190,191] The mutagenicity spectrum of the formamidopyrimidine lesions are not yet known. However, the repair of such lesions by the fpg protein (also referred to as the mut M protein and formamidopyrimidine DNA glycosylase) is well documented.[192–195] The fpg protein also excises 8-oxopurine lesions. The biological relevance of at least one of these two families of lesions is reflected by the observation that the expression of the fpg protein in mammalian cells is correlated with the mutagenicity of γ-radiolysis.[196]

There is no ambiguity concerning the mutagenicity of 8-oxodeoxyguanosine (**155**) *in vitro* or *in vivo*.[197,198] This purine lesion gives rise to G → T transversions by instructing DNA polymerase to misinsert deoxyadenosine across from it. The structural and functional effects of (**155**) on DNA are possibly the most well studied of all nucleic acid lesions.[199] The mutagenicity of (**155**) is ascribed to its adoption of a *syn* conformation, which allows it to present a hydrogen bonding pattern to a polymerase enzyme that is remarkably similar to that of thymidine, and enables it to base pair to deoxyadenosine (Equation (11)). In fact, base pairing between (**155**) and deoxyadenosine was recently confirmed by single crystal X-ray diffraction studies.[200] Interestingly, the same *in vivo* study which confirmed the mutagenicity of (**155**) did not find 8-oxodeoxyadenose (**152**) to be mutagenic.[198] This finding was supported by *in vitro* kinetic analysis by DNA polymerase.[201] However, subsequent studies suggest that (**152**) induces misincorporation during *in vitro* DNA synthesis and is mutagenic in at least some cell lines.[202]

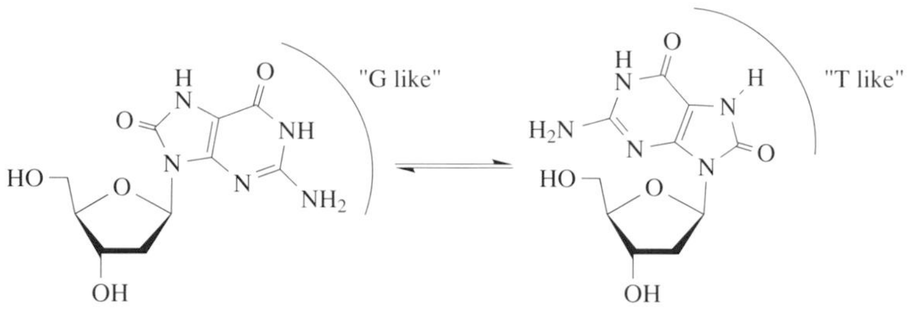

(11)

(155) *anti*                    (155) *syn*

Although the formation of the 8-oxopurine nucleotides and their biological significance have been clearly established, reports suggest that there is more to be learned concerning purine oxidation. In 1994, it was shown that (155) accounts for only a small percentage of the reactivity of deoxyguanosine with hydroxyl radical. Degradation product (163) and its hydrolysis product (164) represent more than 80% of lesions produced via the reaction of hydroxyl radical with deoxyguanosine.[203] Furthermore, under some oxidative stress conditions, 8-oxodeoxyguanosine (155) is subject to further transformation. The facility with which (155) undergoes oxidative modification (e.g., (166)) is underscored by the necessity to deprotect chemically synthesized oligonucleotides with ammonium hydroxide in the presence of the reducing agent dithiothreitol.[204] In particular, in the presence of $^1O_2$, which itself produces 8-oxodeoxyguanosine from deoxyguanosine, (155) then undergoes further transformations.[205–208] Kinetic studies have shown that (155) reacts 100 times more quickly with $^1O_2$ than deoxyguanosine.[205] The dioxetane (165) formed via [2 + 2] cycloaddition of $^1O_2$ with (155) has been identified as a secondary product of deoxyguanosine oxidation, as well as a precursor to a number of other lesions, including (163) and (164) noted above (Scheme 45).[206,208] In addition, it has been suggested that (165) decomposes to (168) via cleavage of the dioxetane ring to an unstable nine-membered ring lesion (167).[206] The cyanuric acid derivative (168) is formed in slightly larger amounts than (163) and (164) from the reaction of $^1O_2$ with 8-oxodeoxyguanosine (155). To date, the effects of these more recently discovered lesions resulting from deoxyguanosine oxidation on nucleic acid structure and function have not yet been determined.

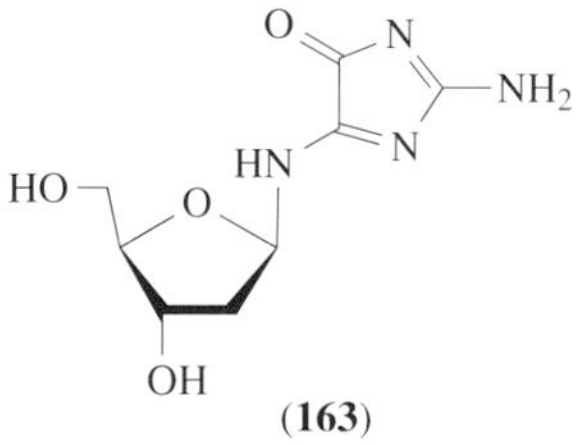

(163) 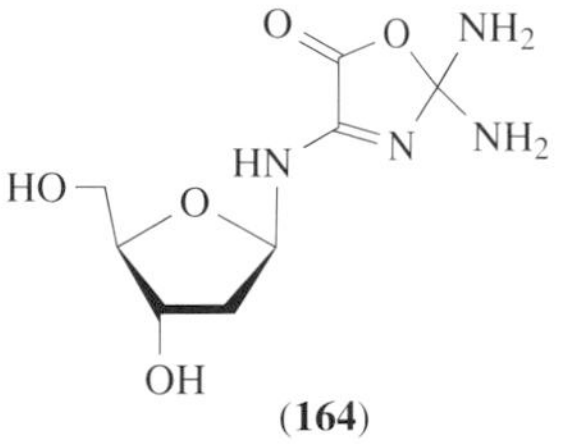                    (164)

Finally, one should note that the effects of $^1O_2$ on deoxyguanosine and 8-oxodeoxyguanosine (155) are well studied, because this nucleoside is oxidized more rapidly than the other common, naturally occurring nucleosides.[209–212] These results show that while $^1O_2$ selectively oxidizes deoxyguanosine, it is not specific for this nucleoside.[210,211] Hence, the possible effects of $^1O_2$ on nucleotides, in particular thymidine, should not be ignored.

### 7.11.3.3 The Role of 2′-Deoxyuridin-5-yl (28) in Strand Damage

5-Bromodeoxyuridine (29) and 5-iododeoxyuridine (30) sensitize nucleic acids to UV radiation and γ-radiolysis-induced damage (Equation (5)).[213–215] The ability of other halogenated nucleotides to act in this capacity have been explored as well, albeit to a lesser extent.[216] The 5-halopyrimidines also form protein–nucleic acid cross-links.[217–219] This aspect of their chemistry will not be examined here. The 5-halopyrimidines are of current clinical interest as radiosensitizers for forming double strand breaks, particularly when used in conjunction with other small molecules.[220–229]

Prior to the routine availability of single nucleotide gel electrophoretic analysis, it was suggested that the uracil-5-yl radical (28) produced from the 5-halopyrimidines induced strand damage via C-2′ hydrogen atom abstraction from the 5′-adjacent nucleotide (Scheme 46).[213] The subsequent reactivity of the C-2′-centered radical was not discussed, and is still not generally well understood (see Section 7.11.2.2). This proposal was based upon the proximity of the C-2′ hydrogen atoms to

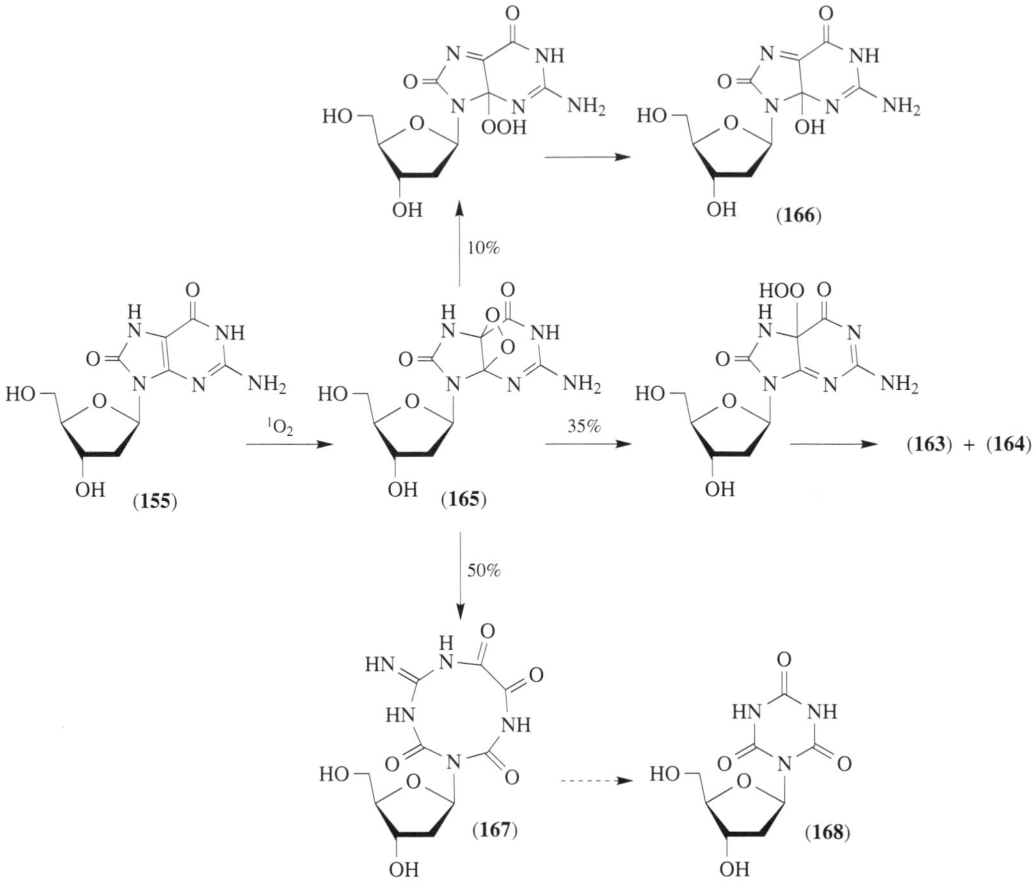

**Scheme 45**

the incipient sigma radical, as suggested by the inspection of models of duplex DNA. Hydrogen atom abstraction at this site by the uracil-5-yl radical (**28**) was also expected to be thermodynamically downhill. However, hydrogen atom abstraction of any deoxyribose hydrogen atom in nucleotides by (**28**) will be downhill.

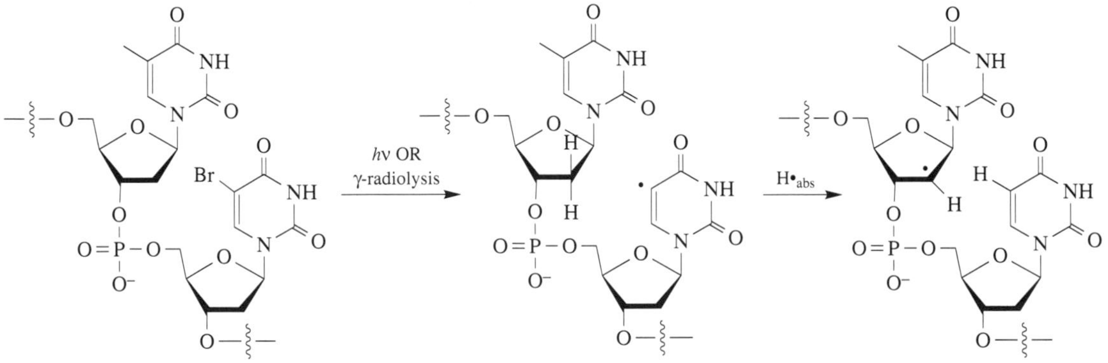

**Scheme 46**

This mechanism remained untested until 1990 when Sugiyama, Saito and co-workers published the first of several papers concerning the strand damage induced upon UV irradiation of oligonucleotides containing 5-halopyrimidine nucleosides. Using hexameric oligonucleotides, in conjunction with product analysis, these workers made several observations which culminated in their putting forth a radically different mechanistic proposal for strand damage involving 5-bromodeoxyuridine.[225,226] Enhancement of UV-sensitized damage was only observed when the nucleotide bonded to the 5′-

phosphate of (28) in duplex DNA was deoxyadenosine. On the basis of this observation, it was proposed that 5-bromodeoxyuridine (29) undergoes a photoinduced electron transfer reaction selectively with a 5′-adjacent deoxyadenosine (Scheme 47). This was somewhat surprising considering the thermodynamically more favorable electron transfer process possible involving deoxyguanosine as electron donor.[139] The radical anion of 5-bromodeoxyuridine was then suggested to yield the same uracil-5-yl radical (28) proposed above, except via loss of bromide ion. However, based upon detection of the deoxyribonolactone alkaline labile lesion (169), it was proposed that the sigma radical abstracts the C-1′ hydrogen and not the C-2′ hydrogen of the 5′-adjacent nucleotide (Scheme 48). This too was a surprising proposal considering that in a static model of duplex DNA, the uracil-5-yl radical lies in the major groove of DNA, and the C-1′ hydrogen of the adjacent nucleotide is deeply imbedded in the minor groove. This mechanistic scheme has been expanded to include hydrogen atom abstraction of both the C-1′ and the C-2′ hydrogen atoms by the sigma radical.[227] This latter study was carried out in the presence of $O_2$, while the initial reports were conducted under $O_2$-limiting conditions. It is unclear what effects $O_2$ had on the hydrogen atom abstraction process, including whether (28) is still the entity responsible for hydrogen atom abstraction. Nonetheless, this modification of the proposal for strand damage by the uracil-5-yl (28) species is more consistent with observations made by the same group during their studies of DNA containing 5-iododeoxyuridine (30).[226,228] Ultraviolet irradiation of hexameric DNA containing 5-iododeoxyuridine (30) produced products resulting from formal oxidation at C-1′ of (169) and C-2′ of (170) of the 5′-adjacent nucleotide (Scheme 49). One significant difference observed during the study of oligonucleotides containing 5-iododeoxyuridine was the absence of the sequence effect exhibited upon photolysis of DNA containing 5-bromodeoxyuridine (29).[225,226] The lack of a sequence preference for photolysis of 5-iododeoxyuridine (30) was ascribed to generation of the uracil-5-yl radical (28) via direct excitation of the halogenated pyrimidine, instead of via a photoinduced electron transfer process, as suggested for (29).

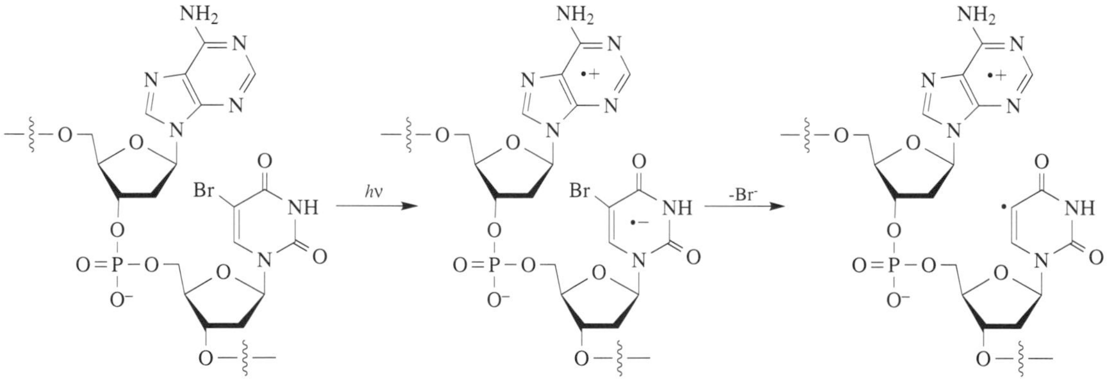

**Scheme 47**

Experiments that addressed the mechanism for direct strand breaks produced upon anaerobic photolysis of duplex DNA containing 5-bromodeoxyuridine gave rise to a mechanism that unifies the discordant proposals described above.[229] Quantitative analysis of duplex DNA substrates containing three helical turns using gel electrophoresis to separate the cleavage products revealed an approximately eightfold preference for direct strand scission at 5′-dABrdU sites relative to 5′-dGBrdU sites. These experiments were consistent with the proposed photoinduced electron transfer process. In contrast, deuterium kinetic isotope effects were more consistent with the original proposals put forth by Hutchinson.[213,214]

Dideuteration of the C-2′ position of the deoxyadenosine on the 5′-side of 5-bromodeoxyuridine (29) resulted in a significant diminution in strand damage. This clearly indicated that hydrogen atom abstraction from the C-2′ position is involved in strand damage. Product analysis of duplex DNA photolyzed under anaerobic conditions yielded an unexpected result.[229] Enzymatic end group analysis of 3′-$^{32}$P-labeled oligonucleotides showed that the 5′-termini of fragments were composed entirely of phosphate groups (171) (Scheme 50). Similar analysis using 5′-$^{32}$P-labeled materials revealed that the 3′-termini were composed of phosphates (173) and a slower migrating product (172), which was converted to 3′-phosphate upon alkaline treatment (Scheme 51). The labile fragmentation product migrated approximately as slowly as a strand scission product containing an additional nucleotide. The respective labile product formed from an otherwise identical

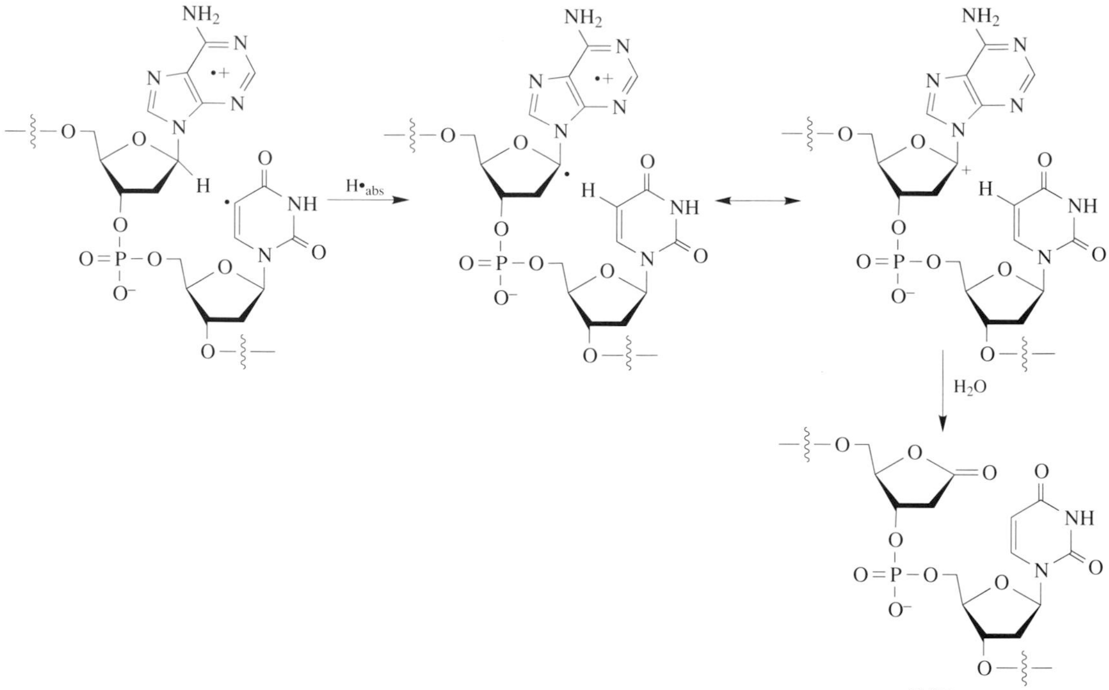

**Scheme 48**

(169)

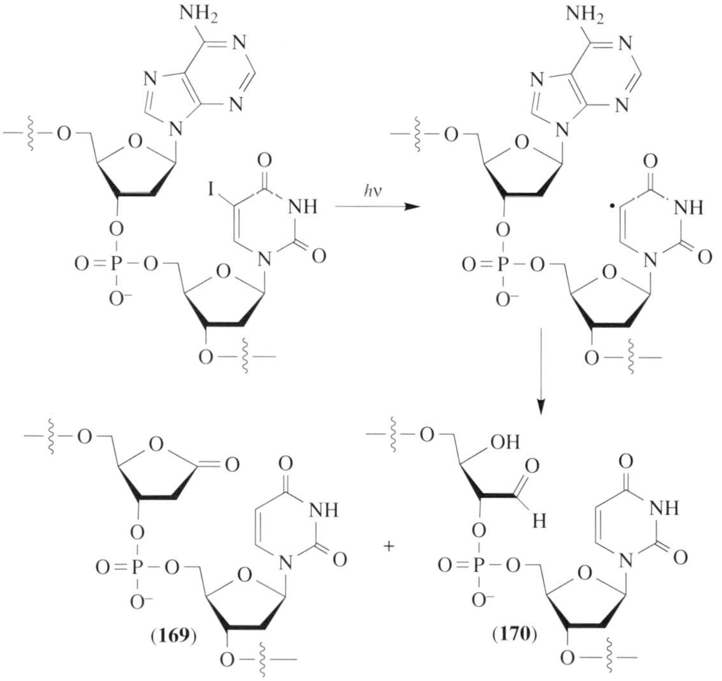

**Scheme 49**

oligonucleotide containing a 5′-dGBrdU sequence migrated slightly faster, demonstrating that the metastable lesions retained the purine nucleobases. Electrospray mass spectral analysis of photolyzed (**174**) did not reveal the presence of a labile fragmentation product containing a purine nucleobase (**175**). However, a product with a mass corresponding to (**176**) was observed. It was postulated that (**176**) was formed during the electrospray ionization process. Based upon these observations, and the results of the kinetic isotope effect experiments discussed below, the slowly migrating, alkaline labile products were assigned as the novel 3′-keto-2′-deoxypurine structures (e.g., (**172**)) resulting from formal oxidation of the C-3′ position of the nucleotide adjacent to the 5′-phosphate of 5-bromodeoxyuridine.

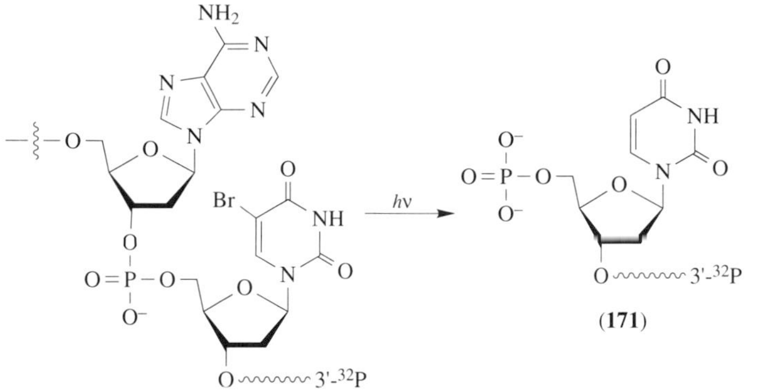

**Scheme 50**

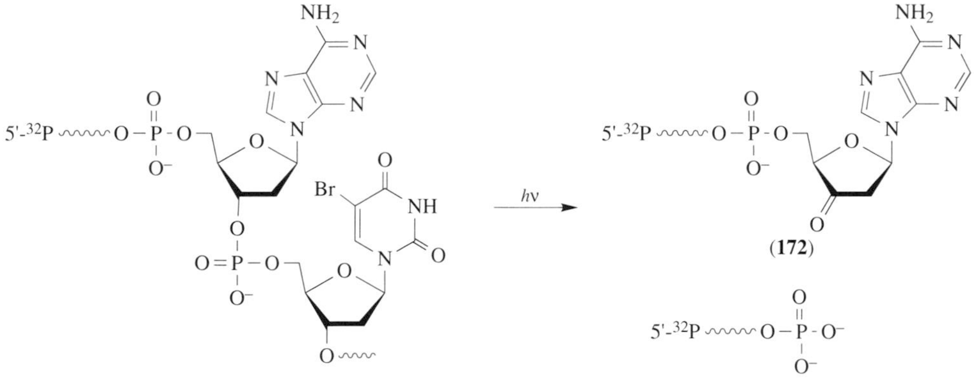

**Scheme 51**                    (173)

5′-d(CGC ATA TGG  CA BrU  GCT)
3′-d(GCG TAT ACC  GT   A   CGA)

(174)

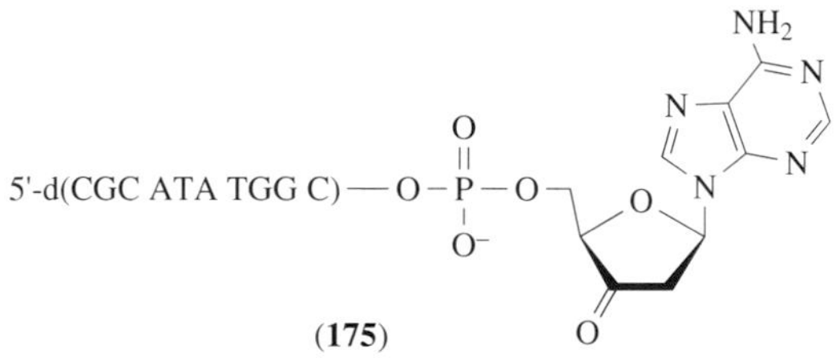

(175) 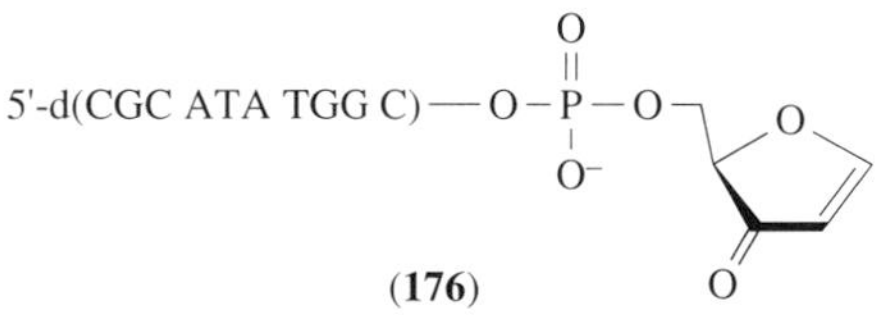                                        (176)

Examination of a static model of duplex DNA suggests that the 3′-hydrogen of the 5′-adjacent nucleotide is in closer proximity to the uracil-5-yl radical than the C-1′ hydrogen atom. However, this is still significantly further than the C-2′ hydrogen atoms. The observed deuterium kinetic isotope effects for direct strand break formation suggest that if C-1′ hydrogen atom abstraction can lead to strand scission (see Section 7.11.2.1), then the uracil-5-yl radical does not abstract a C-1′ hydrogen from the 5′ adjacent nucleotide in duplex DNA.[229] The overall observed kinetic isotope effect upon deuteration of C-1′ was actually slightly inverse. Closer inspection reveals that formation of the 3′-phosphate fragment (173) experiences a slightly normal kinetic isotope effect ($\sim 1.1$), but the kinetic isotope effect for formation of the labile 3′-ketodeoxynucleotide fragment (172) is significantly inverse ($\sim 0.8$). A mechanism that rationalizes these observations involves oxidation of the initially formed C-2′ deoxyribosyl radical (Scheme 52). Oxidation of the C-2′ deoxyribosyl radical by either the deoxyadenosine cation radical or the deprotonated species (160) should be energetically downhill. The reduction potential of the cation radical of deoxyadenosine is $> 1.0$ V.

Based upon the model studies, the oxidation potential of the 2'-deoxyribosyl radical is expected to be ~0.7 V. Competitive 1,2-hydride rearrangements in (**177**) ultimately give rise to the 3'-ketodeoxyadenosine fragment (**172**) and the 3'-phosphate fragment (**173**) via (**169**). (One should note that (**172**) can also give rise to (**173**).) Deuteration at the C-1' position enables the C-3' hydride migration to compete more effectively, increasing the relative amount of the C-3' keto-deoxyadenosine product. This is reflected by the observed inverse kinetic isotope effect for formation of (**172**) and a small normal kinetic isotope effect for formation of (**173**).

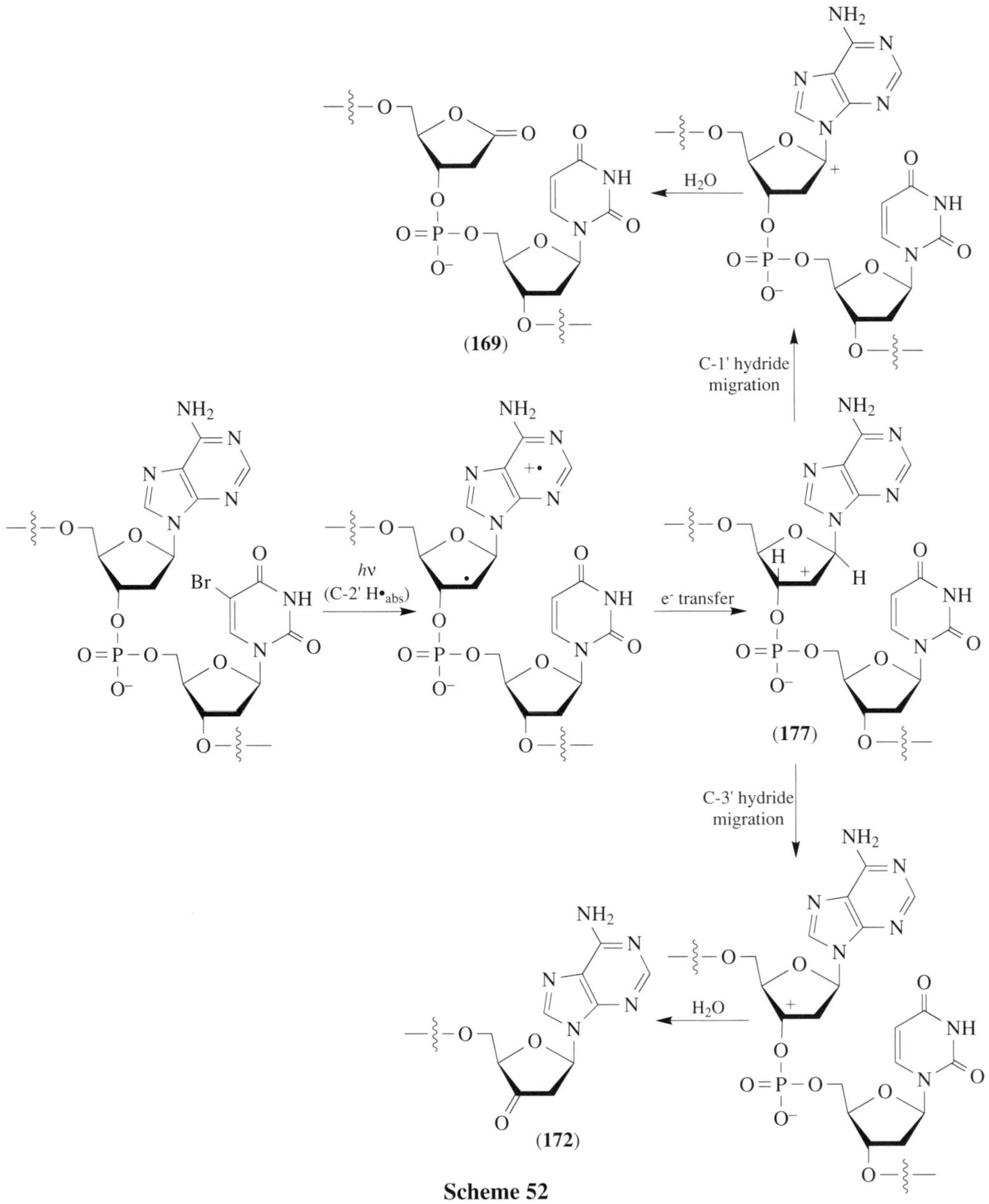

**Scheme 52**

These results suggest that both previously proposed mechanisms are partially correct. Sensitization of DNA containing 5-bromodeoxyuridine to UV irradiation does involve single electron transfer to the halogenated pyrimidine. However, the sigma radical formed following halide loss abstracts the proximal hydrogen atom from the C-2' position of the 5'-adjacent nucleotide. Under anaerobic conditions the deoxyribosyl radical is oxidized to the corresponding carbocation by the adenine moiety, which served as the original electron source. The deoxyribosyl carbocation then yields the C-1' and C-3' formal oxidation products via hydride migrations. The involvement of the C-1' hydrogen atom in strand damage is still possible. Further studies directed at resolving this issue are warranted.

## 7.11.4  DNA DAMAGE SENSITIZATION AND PROTECTION AGENTS

The 5-halopyrimidines are a unique family of molecules that sensitize DNA to damage because they are incorporated within the biopolymer. By and large, protecting and sensitizing agents are exogenous. Thiols are the most common group of exogenous protecting agents. Studies on biologically active DNA suggest that the protecting effect is more involved than simply trapping alkyl radicals, although alkyl radical trapping is certainly a component of the thiol protecting effect.[13] Studies that demonstrate a correlation between the level of intracellular protection and thiol charge support a mechanism in which the thiol reacts with DNA-centered radicals.[230] These experiments suggest that a thiol with overall $+2$ charge (**178**) is more than 20 times as effective as negatively charged glutathione (**179**),

(**178**) 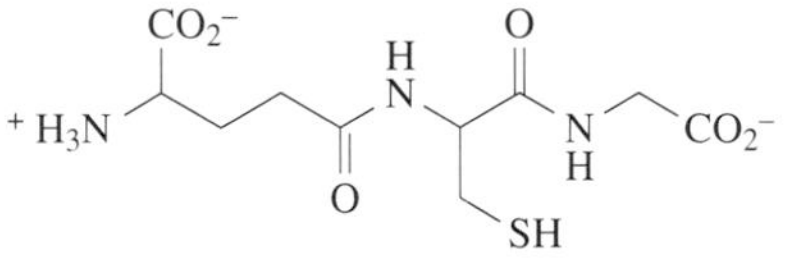                    (**179**)

Regardless of the mechanism by which thiols protect DNA, their role as protecting agents appears to be ambiguous. The same can be said for other families of exogenous agents, including metal ions, nitroxides, and nitroaromatics. Evidence exists for each of these species acting as either sensitizing agents or protecting agents.[231] In some instances, the roles of these agents are further complicated by their acting in tandem with one another.[232]

### 7.11.4.1  The Effects of Metal Ions on DNA Damage

Redox active metal ions are often thought of as sensitizing agents because of their involvement in Haber–Weiss and Fenton chemistries.[13,233] Their effectiveness in generating reactive oxygen species is improved by the coordination of the metal ions to the oppositely charged nucleic acids.[234] Sensitization of nucleic acid damage by metal ions may also be remotely effected in some cases. For example, it has been suggested that the amyloid precursor protein of Alzheimer's disease reduced copper(II) to copper(I), which is capable of reducing $O_2$.[10] However, this conclusion is controversial and is being debated in the literature.[11,12]

Metal ions can influence the path of DNA damage by interacting with the reactive intermediates formed in the biopolymer (Scheme 53). For instance, it is possible that oxidation of the C-5 hydroxyl radical adduct of thymidine (**115**) could divert a strand scission process into one that results in the formation of a repairable DNA lesion (thymidine glycol, **125**).[235] A reduced metal ion, potentially capable of participating in another Fenton reaction, is generated concomitantly. In 1996 evidence was reported which supports such a cycling of redox active metal ions.[236]

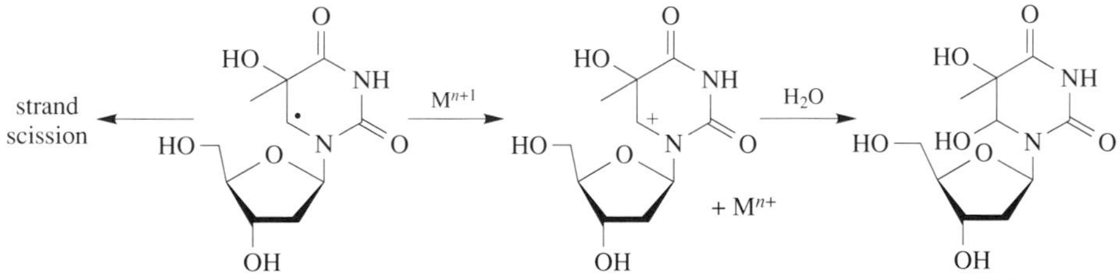

**Scheme 53**

### 7.11.4.2  The Effects of Electron Affinic Agents and Nitroxides on DNA Damage

Oxygen enhances the damage imparted on DNA by $\gamma$-radiolysis.[13] Electron affinic sensitizers were designed to act as $O_2$ surrogates on account of the hypoxic state of some tumor cells.[13] For instance, molecules such as misonidazole (**110**) were designed to react with reducing DNA radicals. Oxidation of DNA radicals via an addition/elimination mechanism was based upon studies with small molecules.[13] However, observation that misonidazole (**110**) did not increase the amount of thymidine glycol (**125**) in $\gamma$-irradiated calf thymus DNA was inconsistent with this proposed mechanism.[237] An alternative mechanism for misonidazole sensitization of DNA cleavage was put forth in studies involving neocarzinostatin (**3**), where this molecule was substituted for $O_2$ (Scheme 29).[238] Product studies were consistent with trapping of the DNA radicals, followed by scission to yield DNA alkoxy radicals and the nitroso analogue of misonidazole (see Chapter 7.15 for a more detailed discussion of neocarzinostatin-mediated DNA damage). Misonidazole has also been suggested to sensitize DNA damage by trapping electrons released during direct ionization, consequently preventing ion recombination.[239]

Finally, $\gamma$-radiolysis experiments have also been reported where misonidazole (**110**) protects DNA from strand damage, presumably via competing with the biopolymer for hydroxyl radicals.[237] A different mechanism is proposed for DNA protection by nitroxides.[240] Instead of scavenging hydroxyl radicals, it was suggested that these molecules oxidize metal ions (e.g., $Fe^{2+}$), preventing Fenton-mediated hydroxyl radical production.

## 7.11.5 THE FORMATION OF BISTRANDED LESIONS AND DOUBLE STRAND BREAKS IN DNA

Double strand breaks occur when opposite strands of duplex DNA are cleaved within $\sim 15$ base pairs of one another. Double strand breaks are more difficult to repair *in vivo* than single strand breaks. As a result, this family of lesions are highly sought after when designing molecules whose mission is to damage DNA. Any molecule which cleaves DNA can in principle be used to effect double strand cleavage by employing a high enough concentration. In practice, one wants to use as low a concentration (dose) of a molecule as possible, since most drugs elicit undesirable side effects. Hence, agents that produce double strand breaks via a single binding/cleaving event are highly desirable.

### 7.11.5.1 Double Strand Breaks Produced via Interstrand Hydrogen Atom or Electron Transfer

The wide array of DNA damage pathways populated by $\gamma$-radiolysis increases the probability of identifying processes that result in a double strand break derived from a single reaction with the biopolymer. Measurement of strand scission by laser light scattering of $\gamma$-irradiated calf thymus DNA revealed a complex linear-quadratic dependence upon dose.[241] At low radiation doses the yield of double strand breaks varied linearly with dose. This observation was independently confirmed using SV40 DNA and electrophoretic analysis.[242] The ratio of single to double strand breaks ($\sim 19:1$) was also much smaller than expected for double strand breaks resulting from two random cleavage events.[241] These observations led to the intriguing proposal that one or more chemical pathways exist in which strand scission initiated via reaction between DNA and a single hydroxyl radical results in a double strand break (Figure 4).[241] Specific chemical transformations were not put forth. However, the key element in double strand break formation is believed to involve transfer of spin from one strand to another via hydrogen atom abstraction by a peroxy radical and/or carbon radical.

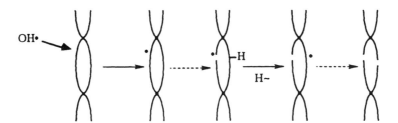

**Figure 4**   Double strand break formation via a single hydroxy radical (OH·).

An alternative explanation for the above observations has been put forth. Ward prefers to ascribe the linear dependence of double strand breaks on dose to a cluster effect in which multiple reactive intermediates are produced in the vicinity of DNA from a single proton.[243–245] The high density of reactive species generated in a localized region of the biopolymer gives rise to multiple lesions, which in turn increase the probability of producing double strand breaks. These two disparate explanations have not yet been reconciled.

Interstrand transfer of DNA damage via a single photon is also believed to be induced in 5-bromodeoxyuridine (**29**) containing DNA through the intermediacy of Hoechst dye 33258 (**180**).[224,246] Neither the chemical mechanism nor products of the double strand breaks formed are known. The dye is believed to serve as an electron shuttle from a nucleotide on the strand opposite that containing (**29**) to the 5-bromodeoxyuridine. This process would result in the formation of a cation radical on the strand opposite that containing (**29**), and a uracil-5-yl (**28**) species on the strand containing 5-bromodeoxyuridine. The involvement of each of these reactive intermediates in strand scission has already been discussed (see Sections 7.11.3.2 and 7.11.3.3). The effect exhibited by (**180**) is also consistent with the binding specificity of this molecule (dAATT sites) and the preference for DNA damage involving photoinduced electron transfer from deoxyadenosine to 5-bromodeoxyuridine.[225–227,229] However, the sequence selectivity for these has not yet been determined.

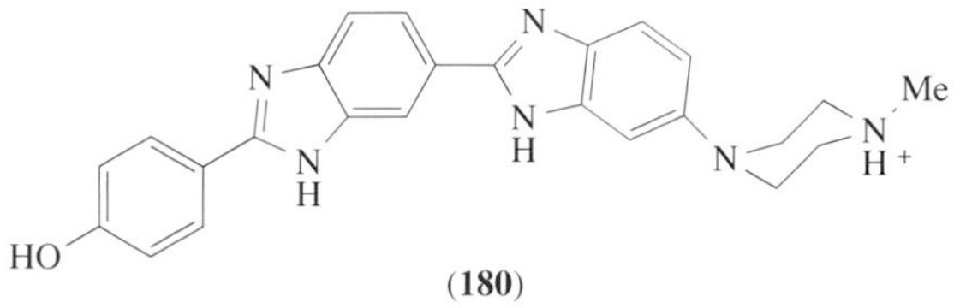

(**180**)

### 7.11.5.2 DNA Damage Amplification

Interstrand hydrogen atom abstraction and the formation of double strand breaks in DNA containing BrdU are examples of DNA damage amplification. An amplification process is defined as an externally initiated (e.g., a photon, an electron, a radical) damage event that results in the modification of two or more sites in DNA. The modifications can be represented by direct strand breaks or alkaline labile lesions. The abstraction of a hydrogen atom from the sugar of the 5'-adjacent nucleotide by the hydroxyl radical adduct of thymidine is an example of a DNA damage amplification process (see Section 7.11.3.1.2). A number of other damage amplification processes that are initiated via electron or radical addition to pyrimidines were discussed in Section 7.11.3.

DNA damage amplification has been studied in detail using very short single-stranded oligonucleotides.[247–249] When dinucleotides, or a tetranucleotide, containing either 5'-dGPy and/or 5'-dPyG sequences, are irradiated under aerobic conditions, double base lesions are formed ((**181**), (**182**)). Thymidine and deoxycytidine are transformed into the formamido fragment. Deoxyguanosine is oxidized to 8-oxodeoxyguanosine (**155**) in each case. The double base lesions are formed linearly as a function of dose, even at very low substrate concentrations. This behavior was interpreted to mean that (**181**) and (**182**) were formed via a single hydroxyl radical. The favored mechanism for these DNA damage amplification processes involves hydroxyl radical addition to the C-6 position of the pyrimidine (**183**) (Scheme 54), despite the fact that this is a minor pathway for the reaction of hydroxyl radical (Scheme 30). The radical adduct is subsequently trapped by $O_2$, and the resulting peroxy radical (**184**) is believed to be reduced by the adjacent deoxyguanosine (**185**). Hydration of the cation radical ultimately yields (**155**). The resulting pyrimidine hydroperoxide

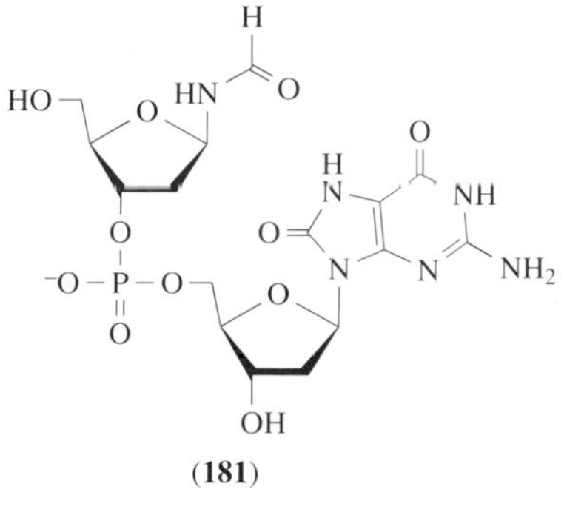

(**181**) 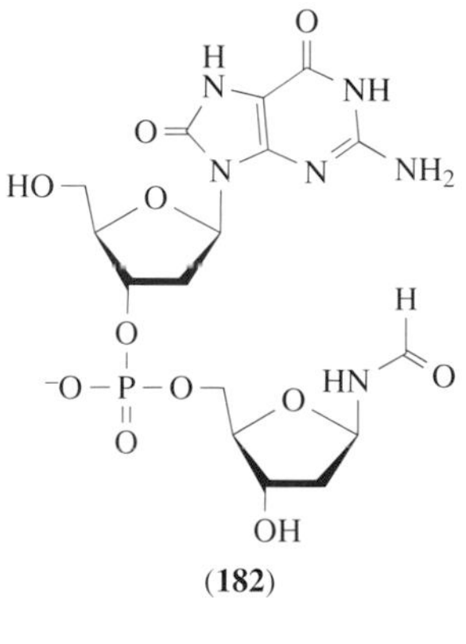                                                  (**182**)

undergoes extensive fragmentation to yield the formamido fragment. Although the mechanism for the formation of these lesions has not been further substantiated, the structure of the double base lesions are a significant departure from the DNA damage amplification processes previously discussed (see Section 7.11.3).

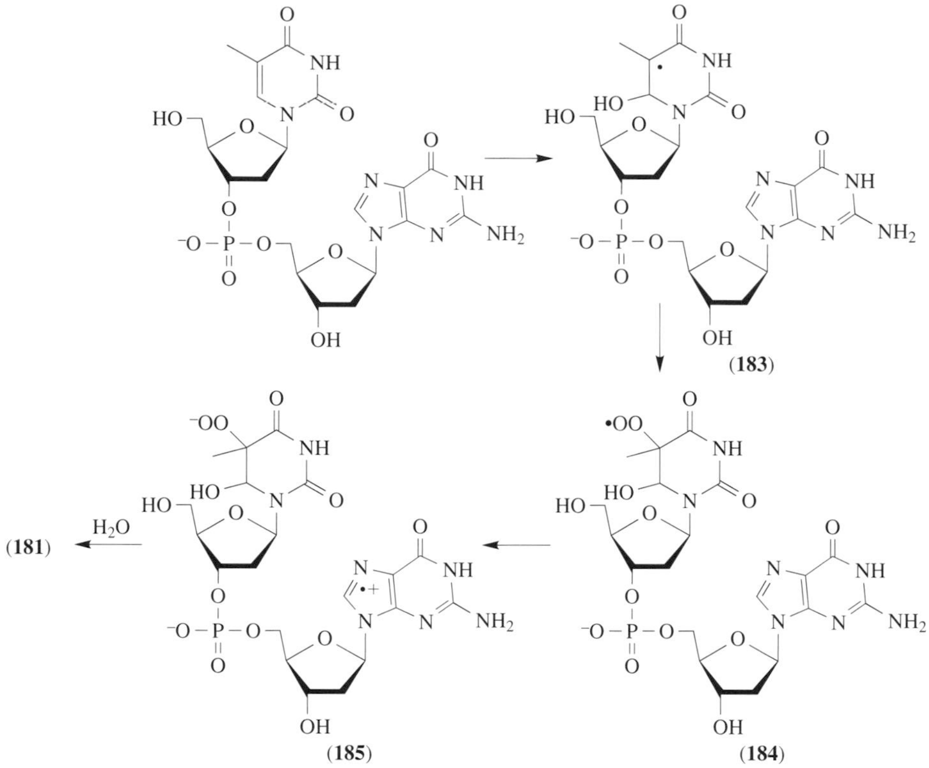

**Scheme 54**

### 7.11.5.3   Bistranded Lesions and Double Strand Breaks Produced by Antitumor Antibiotics

Bleomycin (**51**) is an example of a DNA-damaging agent which has only a single reactive site. In bleomycin (**51**), an iron(oxo) intermediate is believed to initiate strand damage via hydrogen atom abstraction (see Section 7.11.2.4). Despite this limitation, (**51**) produces an inordinately large amount ($\sim$10%) of double strand breaks.[250,251] In addition, the double strand breaks are produced via opposite strand breaks that occur very close to one another. Blunt-ended fragments and single nucleotide 5′-extensions are the predominant double strand cleavage products. These observations led to a proposal for double strand break formation that involved a single bleomycin bound to DNA. It is believed that the bleomycin molecule is reactivated following hydrogen atom abstraction from the primary cleavage site, but prior to its dissociation from duplex DNA. It was suggested that the bleomycin was reactivated via reduction by the metallated drug of an intermediate C-4′ peroxy radical that is produced following the initial hydrogen atom abstraction. Further verification of the central component of this mechanism was obtained in a series of experiments involving hairpin DNA substrates.[252,253] The percentages of double strand breaks obtained with these substrates were even higher ($\sim$5–25%) than observed previously. The involvement of a single bleomycin molecule in double strand break formation was supported by the invariance of the single to double strand break ratio over a 70-fold range in concentration of the drug. An alternative mechanism in which it was proposed that double strand breaks result from two molecules of bleomycin binding simultaneously to duplex DNA was discounted by demonstrating that addition of cobalt bleomycin, which binds to, but does not oxidize DNA, did not affect the ratio of double vs. single strand breaks produced.[252] Kinetic isotope effects and $O_2$ effects also supported a mechanism in which a single bleomycin molecule abstracts the C-4′-hydrogen from each of two different nucleotides via an $O_2$-dependent mechanism.[253]

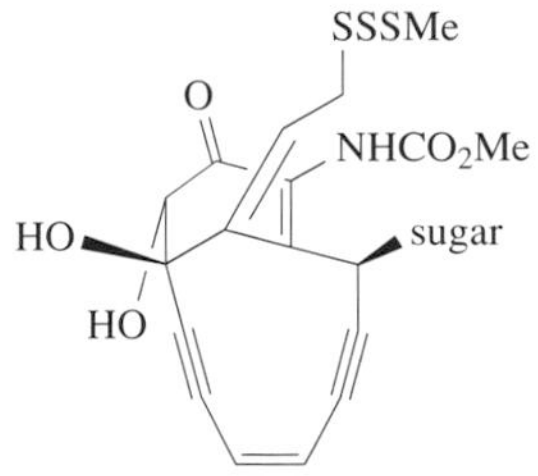

(**186**)

Individual molecules of the activated forms of the enediyne family of DNA-damaging agents contain biradicals. When bound in the minor groove of DNA, the biradicals present two reactive centers capable of abstracting hydrogen atoms from the deoxyribose rings of two different nucleotides. Model studies suggest that the rate of hydrogen atom abstraction by these sigma radicals may be considerably slower than observed for simple sigma radicals.[254] Nonetheless, these species are clearly capable of effecting hydrogen atom abstraction.

Neocarzinostatin (**3**) produces bistranded lesions that consist of a direct strand break and an alkaline labile lesion two nucleotides away on the opposite strand.[255] Direct strand breaks result from C-4' or C-5' hydrogen atom abstraction, whereas alkaline labile lesions are attributed to C-1' hydrogen atom abstraction (see Sections 7.11.2.1, 7.11.2.4, and 7.11.2.5).[116,256] Similarly, C1027 (**54**) induces bistranded lesions that are separated by one or two nucleotides. In this instance direct strand breaks initiated by C-4' hydrogen atom abstraction on one strand are accompanied by either C-1' or C-5' hydrogen atom abstraction on the opposite strand. The latter process results in a double strand break, while the former produces an overall effect similar to that induced by neocarzinostatin (**3**).[84]

The sequence selectivity of (**3**) and C1027 (**54**) are different, and are presumably controlled by the substituents on the reactive core. In the case of the calicheamicin (**52**) and esperamicin (**53**) enediynes, these substituents appear to have a profound effect on the molecule's proclivity towards double strand break formation as well.[89,90,257]

Calicheamicin $\gamma$1 and esperamicins C, D, and E (**186**) are competent at inducing double strand breaks.[89,90,257] Kinetic analysis of DNA cleavage by calicheamicin $\gamma$1 and esperamicin C support a mechanism in which a single molecule is responsible for double strand break formation.[257] Labeling studies revealed that calicheamicin $\gamma$1 effects double strand cleavage via abstraction of the C-5' *S*-hydrogen from one deoxycytidine and the C-4' hydrogen from a second deoxycytidine two nucleotides away in the opposite strand.[90] Esperamicin A (**53**), which differs from esperamicins C, D, and (**186**) by the presence of an anthranilate substituent at C-4 of the molecule's core, is incapable of inducing double strand breaks.[89,257] An explanation for this effect on reactivity has not yet been presented.

## 7.11.6 REFERENCES

1. T. Lindahl, *Nature*, 1993, **362**, 709.
2. P. Modrich, *Science*, 1994, **266**, 1959.
3. P. C. Hanawalt, *Science*, 1994, **266**, 1957.
4. L. J. Marnett and P. C. Burcham, *Chem. Res. Toxicol.*, 1993, **6**, 771.
5. R. L. Rusting, *Sci. Am.*, 1992, 131.
6. R. S. Sohal and R. Weindruch, *Science*, 1996, **273**, 59.
7. R. K. Brown, A. McBurney, J. Lunec, and F. J. Kelley, *Free Radical Biol. Med.*, 1995, **18**, 801.
8. P. J. Simpson, J. C. Fantone, and B. R. Lucchesi, in "Oxygen Radicals and Tissue Injury Symposium [Proceedings]," ed. B. Halliwell, Federation of American Societies for Experimental Biology, Bethesda, MD, 1988, p. 63.
9. B. D. Watson and M. D. Ginsberg, in "Oxygen Radicals and Tissue Injury Symposium [Proceedings]," ed. B. Halliwell, Federation of American Societies for Experimental Biology, Bethesda, MD, 1988, p. 81.
10. G. Multhaup, A. Schlicksupp, L. Hesse, D. Beher, T. Ruppert, C. L. Masters, and K. Beyreuther, *Science*, 1996, **271**, 1406.
11. L. M. Sayre, *Science*, 1996, **274**, 1933.
12. G. Multhaup, *Science*, 1996, **274**, 1934.
13. C. von Sonntag, "The Chemical Basis of Radiation Biology," Taylor & Francis, Philadelphia, PA, 1987.
14. G. Pratviel, J. Bernadou, and B. Meunier, *Angew. Chem., Int. Ed. Engl.*, 1995, **34**, 746.
15. A. P. Breen and J. A. Murphy, *Free Radic. Biol. Med.*, 1995, **18**, 1033.
16. E. C. Friedberg, G. C. Walker, and W. Siede, "DNA Repair and Mutagenesis," ASM Press, Washington DC, 1995.
17. J. Tsang and G. F. Joyce, *Biochemistry*, 1994, **33**, 5966.

18. S. A. Raillard and G. F. Joyce, *Biochemistry*, 1996, **35**, 11 693.
19. L. A. Basile, A. L. Raphael, and J. K. Barton, *J. Am. Chem. Soc.*, 1987, **109**, 7550.
20. L. H. Hurley (ed.), "Advances in DNA Sequence Specific Agents," JAI Press, Greenwich, CT, 1992, vol. 1.
21. L. H. Hurley and J. B. Chaires (eds.), "Advances in DNA Sequence Specific Agents," JAI Press, Greenwich, CT, 1996, vol. 2.
22. S. Neidle and M. Waring (eds.), "Molecular Aspects of Anticancer Drug–DNA Interactions," CRC Press, Boca Raton, FL, 1993, vol. 1.
23. D. L. Boger, *Acc. Chem. Res.*, 1995, **28**, 20.
24. K. Miaskiewicz and R. Osman, *J. Am. Chem. Soc.*, 1994, **116**, 232.
25. A.-O. Colson and M. D. Sevilla, *J. Phys. Chem.*, 1995, **99**, 3867.
26. I. H. Goldberg, *Acc. Chem. Res.*, 1991, **24**, 191.
27. D. S. Sigman, A. Mazumder, and D. M. Perrin, *Chem. Rev.*, 1993, **93**, 2295.
28. G. A. Neyhart, C. C. Cheng, and H. H. Thorp, *J. Am. Chem. Soc.*, 1995, **117**, 1463.
29. C. C. Cheng, J. G. Goll, G. A. Neyhart, T. W. Welch, P. Singh, and H. H. Thorp, *J. Am. Chem. Soc.*, 1995, **117**, 2970.
30. G. Pratviel, M. Pitié, J. Bernadou, and B. Meunier, *Nucleic Acids Res.*, 1991, **19**, 6283.
31. G. Pratviel, M. Pitié, J. Bernadou, and B. Meunier, *Angew. Chem., Int. Ed. Engl.*, 1991, **30**, 702.
32. M. Pitié, J. Bernadou, and B. Meunier, *J. Am. Chem. Soc.*, 1995, **117**, 2935.
33. G. Pratviel, M. Pitié, C. Périgaud, G. Gosselin, J. Bernadou, and B. Meunier, *J. Chem. Soc., Chem. Commun.*, 1993, 149.
34. M. R. Barvian and M. M. Greenberg, *J. Am. Chem. Soc.*, 1995, **117**, 8291.
35. M. M. Greenberg, M. R. Barvian, G. P. Cook, B. K. Goodman, T. J. Matray, C. Tronche, and H. Venkatesan, *J. Am. Chem. Soc.*, 1997, **119**, 1828.
36. L. S. Kappen and I. H. Goldberg, *Biochemistry*, 1989, **28**, 1027.
37. T. E. Goyne and D. S. Sigman, *J. Am. Chem. Soc.*, 1987, **109**, 2846.
38. M. Kuwabara, C. Yoon, T. Goyne, T. Thederahn, and D. S. Sigman, *Biochemistry*, 1986, **25**, 7401.
39. N. Gupta, N. Grover, G. A. Neyhart, W. Liang, P. Singh, and H. H. Thorp, *Angew. Chem., Int. Ed. Engl.*, 1992, **31**, 1048.
40. N. Gupta, N. Grover, G. A. Neyhart, P. Singh, and H. H. Thorp, *Inorg. Chem.*, 1993, **32**, 310.
41. T. W. Welch, G. A. Neyhart, J. G. Goll, S. A. Siftan, and H. H. Thorp, *J. Am. Chem. Soc.*, 1993, **115**, 9311.
42. M. S. Thompson and T. J. Meyer, *J. Am. Chem. Soc.*, 1982, **104**, 4106.
43. C. Decarroz, J. R. Wagner, and J. Cadet, *Free Radic. Res. Commun.*, 1987, **2**, 295.
44. K. B. Lesiak and K. T. Wheeler, *Radiat. Res.*, 1990, **121**, 328.
45. H. Ide, T. Yamaoka, and Y. Kimura, *Biochemistry*, 1994, **33**, 7127.
46. B. K. Goodman and M. M. Greenberg, *J. Org. Chem.*, 1996, **61**, 2.
47. T. Gimisis and C. Chatgilialoglu, *J. Org. Chem.*, 1996, **61**, 1908.
48. T. Gimisis, G. Ialongo, M. Zamboni, and C. Chatgilialoglu, *Tetrahedron Lett.*, 1995, **36**, 6781.
49. D. D. M. Wayner and D. Griller, *Mol. Struct. Energy*, 1989, **11**, 109.
50. C. von Sonntag and H.-P. Schuchmann, *Angew. Chem., Int. Ed. Engl.*, 1990, **30**, 1229.
51. J. E. Bennett, *J. Chem. Soc., Faraday. Trans.*, 1990, **86**, 3247.
52. E. Bothe, D. Schulte-Frohlinde, and C. von Sonntag, *J. Chem. Soc., Perkin Trans 2*, 1978, 416.
53. M. N. Schuchmann, H.-P. Schuchmann, and C. von Sonntag, *J. Am. Chem. Soc.*, 1990, **112**, 403.
54. J. H. B. Chenier, E. Furimsky, and J. A. Howard, *Can. J. Chem.*, 1974, **52**, 3682.
55. P. Neta, *J. Phys. Chem. Ref. Data*, 1990, **19**, 413.
56. C. von Sonntag, *Int. J. Radiat. Biol.*, 1994, **66**, 485.
57. T. Doulci, J. Cadet, and B. N. Ames, *Chem. Res. Toxicol.*, 1996, **9**, 3.
58. K. A. Tallman, C. Tronche, D. J. Yoo, and M. M. Greenberg, *J. Am. Chem. Soc.*, submitted for publication.
59. P. J. McHugh and J. Knowland, *Nucleic Acids Res.*, 1995, **23**, 1664.
60. Y. Matsumoto and K. Kim, *Science*, 1995, **269**, 699.
61. A. Mazumder, J. A. Gerlt, M. J. Absalon, J. Stubbe, R. P. Cunningham, J. Withka, and P. H. Bolton, *Biochemistry*, 1991, **30**, 1119.
62. A. Mazumder, J. A. Gerlt, L. Rabow, M. J. Absalon, J. Stubbe, and P. H. Bolton, *J. Am. Chem. Soc.*, 1989, **111**, 8029.
63. T. Lindahl and A. Andersson, *Biochemistry*, 1972, **11**, 3618.
64. T. Behmoaras, J.-J. Toulmé, and C. Hélène, *Nature*, 1981, **292**, 858.
65. G. Behrens, G. Koltzenburg, A. Ritter, and D. Schulte-Frohlinde, *Int. J. Radiat. Biol.*, 1978, **33**, 163.
66. G. Koltzenburg, G. Behrens, and D. Schulte-Frohlinde, *J. Am. Chem. Soc.*, 1982, **104**, 7311.
67. G. Behrens, G. Koltzenburg, and D. Schulte-Frohlinde, *Z. Naturforsch., Teil C*, 1982, **37**, 1205.
68. M. Adinarayana, E. Bothe, and D. Schulte-Frohlinde, *Int. J. Radiat. Biol.*, 1988, **54**, 723.
69. D. Crich and X.-Y. Jiao, *J. Am. Chem. Soc.*, 1996, **118**, 6666.
70. D. Crich, Q. Yao, and G. F. Filzen, *J. Am. Chem. Soc.*, 1995, **117**, 11 455.
71. A. Hissung, M. Isildar, C. von Sonntag, and H. Witzel, *Int. J. Radiat. Biol.*, 1981, **39**, 185.
72. B. Giese, X. Beyrich-Graf, P. Erdmann, L. Giraud, P. Imwinkelreid, S. N. Muller, and U. Schwitter, *J. Am. Chem. Soc.*, 1995, **117**, 6146.
73. H. Sugiyama, K. Fujimoto, and I. Saito, *J. Am. Chem. Soc.*, 1995, **117**, 2945.
74. J. R. Wagner, J. E. van Lier, and L. J. Johnston, *Photochem. Photobiol.*, 1990, **52**, 333.
75. D. Schulte-Frohlinde, G. Behrens, and A. Onal, *Int. J. Radiat. Biol.*, 1986, **50**, 103.
76. A. M. Pyle, E. C. Long, and J. K. Barton, *J. Am. Chem. Soc.*, 1989, **111**, 4520.
77. A. Sitlani, E. C. Long, A. M. Pyle, and J. K. Barton, *J. Am. Chem. Soc.*, 1992, **114**, 2303.
78. T. P. Shields and J. K. Barton, *Biochemistry*, 1995, **34**, 15 037.
79. P. B. Dervan, *Science*, 1986, **232**, 464.
80. J. Stubbe and J. W. Kozarich, *Chem. Rev.*, 1987, **87**, 1107.
81. J. W. Kozarich, L. Worth, Jr., B. L. Frank, D. F. Christner, D. E. Vanderwall, and J. Stubbe, *Science*, 1989, **245**, 1396.
82. L. Worth, Jr., B. L. Frank, D. F. Christner, M. J. Absalon, J. Stubbe, and J. W. Kozarich, *Biochemistry*, 1993, **32**, 2601.

83. K. M. Breiner, M. A. Daugherty, T. G. Oas, and H. H. Thorp, *J. Am. Chem. Soc.*, 1995, **117**, 11 673.
84. Y. Xu, Z. Xi, Y. Zhen, and I. H. Goldberg, *Biochemistry*, 1995, **34**, 12 451.
85. L. S. Kappen, I. H. Goldberg, B. L. Frank, L. Worth, Jr., D. F. Christner, J. W. Kozarich, and J. Stubbe, *Biochemistry*, 1991, **30**, 2034.
86. I. Saito, H. Kawabata, T. Fujiwara, H. Sugiyama, and T. Matsura, *J. Am. Chem. Soc.*, 1989, **111**, 8302.
87. R. P. Hertzberg and P. B. Dervan, *Biochemistry*, 1984, **23**, 3934.
88. D. S. Sigman, *Acc. Chem. Res.*, 1986, **19**, 180.
89. D. F. Christner, B. L. Frank, J. W. Kozarich, J. Stubbe, J. Golik, T. W. Doyle, I. E. Rosenberg, and B. Krishnan, *J. Am. Chem. Soc.*, 1992, **114**, 8763.
90. J. J. Hangeland, J. J. De Voss, J. A. Heath, and C. A. Townsend, *J. Am. Chem. Soc.*, 1992, **114**, 9200.
91. W. D. Henner, S. M. Grunberg, and W. A. Haseltine, *J. Biol. Chem.*, 1982, **257**, 11 750.
92. W. D. Henner, L. O. Rodriguez, S. M. Hecht, and W. A. Haseltine, *J. Biol. Chem.*, 1983, **258**, 711.
93. H. Sugiyama, K. Ohmori, and I. Saito, *J. Am. Chem. Soc.* 1994 **116** 10 326.
94. H. Sugiyama, C. Xu, N. Murugesan, S. M. Hecht, G. A. van der Marcel, and J. van Bloom, *Biochemistry*, 1988, **27**, 58.
95. H. Sugiyama, H. Kawabata, T. Fujiwara, Y. Dannoue, and I. Saito, *J. Am. Chem. Soc.*, 1990, **112**, 5252.
96. L. E. Rabow, J. Stubbe, and J. W. Kozarich, *J. Am. Chem. Soc.*, 1990, **112**, 3196.
97. L. E. Rabow, G. H. McGall, J. Stubbe, and J. W. Kozarich, *J. Am. Chem. Soc.*, 1990, **112**, 3203.
98. H. Sugiyama, C. Xu, N. Murugesan, and S. M. Hecht, *J. Am. Chem. Soc.*, 1985, **107**, 4104.
99. M. Newcomb, M.-H. Le Tadic-Biadatti, D. L. Chestney, E. S. Roberts, and P. F. Hollenberg, *J. Am. Chem. Soc.*, 1995, **117**, 12 085.
100. G. H. McGall, L. E. Rabow, and J. Stubbe, *J. Am. Chem. Soc.*, 1987, **109**, 2836.
101. G. H. McGall, L. E. Rabow, G. W. Ashley, S. H. Wu, J. W. Kozarich, and J. Stubbe, *J. Am. Chem. Soc.*, 1992, **114**, 4958.
102. R. M. Burger, S. J. Projan, S. B. Horwitz, and J. Peisach, *J. Biol. Chem.*, 1986, **261**, 15 955.
103. I. Saito, T. Morii, and T. Matsuura, *J. Org. Chem.*, 1987, **52**, 1008.
104. G. H. McGall and J. Stubbe, *J. Org. Chem.*, 1991, **56**, 48.
105. M. Dizdaroglu, C. von Sonntag, and D. Schulte-Frohlinde, *J. Am. Chem. Soc.*, 1975, **97**, 2277.
106. S. Steenken, G. Behrens, and D. Schulte-Frohlinde, *Int. J. Radiat. Biol.*, 1975, **25**, 205.
107. B. C. Gilbert, R. O. C. Norman, and P. S. Williams, *J. Chem. Soc., Perkin Trans.* 2, 1981, **514**, 1401.
108. B. Giese, J. Burger, T. W. Kang, C. Kesselheim, and T. Wittmer, *J. Am. Chem. Soc.*, 1992, **114**, 7322.
109. B. Giese, P. Erdmann, L. Girard, T. Göbel, M. Petretta, T. Schäfer, and M. von Raumer, *Tetrahedron Lett.*, 1994, **35**, 2683.
110. B. Giese, X. Beyrich-Graf, J. Burger, C. Kesselheim, M. Senn, and T. Schäfer, *Angew. Chem., Int. Ed. Engl.*, 1993, **32**, 1742.
111. C. Tronche, F. N. Martinez, J. H. Horner, and M. Newcomb, *Tetrahedron Lett.*, 1996, **37**, 5845.
112. H. Zipse, *Angew. Chem., Int. Ed. Engl.*, 1994, **33**, 1985.
113. S. Peukert and B. Giese, *Tetrahedron Lett.*, 1996, **37**, 4365.
114. B. Giese, A. Dussy, C. Elie, P. Erdmann, and U. Schwitter, *Angew. Chem., Int. Ed. Engl.*, 1994, **33**, 1861.
115. B. Giese, X. Beyrich-Graf, P. Erdmann, M. Petretta, and U. Schwitter, *Chem. Biol.*, 1995, **2**, 367.
116. L. S. Kappen, I. H. Goldberg, B. L. Frank, L. Worth, Jr., D. F. Christner, J. W. Kozarich, and J. Stubbe, *Biochemistry*, 1991, **30**, 2034.
117. T. A. Dix, K. M. Hess, M. A. Medina, R. W. Sullivan, S. L. Tilly, and T. L. L. Webb, *Biochemistry*, 1996, **35**, 4578.
118. D.-H. Chin, S. A. Carr, and I. H. Goldberg, *J. Biol. Chem.*, 1994, **259**, 9975.
119. H. Kawabata, H. Takeshita, T. Fujiwara, H. Sugiyama, T. Matsura, and I. Saito, *Tetrahedron Lett.*, 1989, **30**, 4263.
120. D.-H. Chang, L. S. Kappen, and I. H. Goldberg, *Proc. Natl. Acad. Sci. USA*, 1987, **84**, 7070.
121. A. F. Fuciarelli, B. J. Wegher, W. F. Blakely, M. Dizdaroglu, *Int. J. Radiat. Biol.*, 1990, **58**, 397.
122. A. J. S. C. Viera and S. Steenken, *J. Am. Chem. Soc.*, 1990, **112**, 6986.
123. A. Hartwig and R. Schlepegrell, *Carcinogenesis*, 1995, **16**, 3009.
124. O. I. Aruoma, B. Halliwell, E. Gajewski, and M. Dizdaroglu, *J. Biol. Chem.*, 1989, **264**, 20 509.
125. M. Dizdaroglu, O. I. Aruoma, and B. Halliwell, *Biochemistry*, 1990, **29**, 8447.
126. E. Gajewski, O. I. Aruoma, M. Dizdaroglu, and B. Halliwell, *Biochemistry*, 1991, **30**, 2444.
127. G. V. Buxton, C. L. Greenstock, W. P. Helman, and A. B. Ross, *J. Phys. Chem. Ref. Data*, 1988, **17**, 513.
128. C. Decarroz, J. R. Wagner, J. E. van Lier, C. Murali Krishna, P. Riesz, and J. Cadet, *Int. J. Radiat. Biol.*, 1986, **50**, 491.
129. C. Murali Krishna, C. Decarroz, J. R. Wagner, J. Cadet, and P. Riesz, *Photochem. Photobiol.*, 1987, **46**, 175.
130. K. Hildenbrand, G. Behrens, and D. Schulte-Frohlinde, *J. Chem. Soc., Perkin Trans.* 2, 189, 283.
131. R. Rashid, F. Mark, H.-P., Schuchmann, and C. von Sonntag, *Int. J. Radiat. Biol.*, 1991, **59**, 1081.
132. P. Wolf, G. D. D. Jones, L. P. Candeias, and P. O'Neill, *Int. J. Radiat. Biol.*, 1993, **64**, 7.
133. D. Schulte-Frohlinde, J. Opitz, H. Görner, and E. Bothe, *Int. J. Radiat. Biol.*, 1985, **48**, 397.
134. M. C. R. Symons, *J. Chem. Soc., Faraday Trans. I*, 1987, **83**, 1.
135. T. Melvin, S. Botchway, A. W. Parker, and P. O'Neill, *J. Chem. Soc., Chem. Commun.*, 1995, 653.
136. P. M. Cullis, J. D. McClymont, and M. C. R. Symons, *J. Chem. Soc., Faraday Trans.*, 1990, **86**, 591.
137. D. B. Hall, R. E. Holmlin, and J. K. Barton, *Nature*, 1996, **382**, 731.
138. W. Way and M. D. Seville, *Radiat. Res.*, 1994, **138**, 9.
139. A. O. Colson and M. D. Sevilla, *Int. J. Radiat. Biol.*, 1995, **67**, 627.
140. D. G. E. Lemaire, E. Bothe, and D. Schulte-Frohlinde, *Int. J. Radiat. Biol.*, 1984, **45**, 351.
141. D. G. E. Lemaire, E. Bothe, and D. Schulte-Frohlinde, *Int. J. Radiat. Biol.*, 1987, **51**, 319.
142. D. Schulte-Frohlinde and K. Hildenbrand, in "Free Radicals in Synthesis and Biology," ed. F. Minisci, Kluwer, Dordrecht, 1989, p. 335.
143. G. D. D. Jones and P. O'Neill, *Int. J. Radiat. Biol.*, 1991, **59**, 1127.
144. E. Bothe and D. Schulte-Frohlinde, *Z. Naturforsch., Teil C*, 1982, **37**, 1191.
145. K. Hildenbrand and D. Schulte-Frohlinde, *Int. J. Radiat. Biol.*, 1989, **55**, 725.

146. M. R. Barvian and M. M. Greenberg, *Tetrahedron Lett.*, 1992, **33**, 6057.
147. M. R. Barvian, R. M. Barkley, and M. M. Greenberg, *J. Am. Chem. Soc.*, 1995, **117**, 4895.
148. L. R. Karam, M. Dizdaroglu, and M. G. Simic, *Radiat. Res.*, 1988, **116**, 210.
149. G. A. Russell, in "Free Radicals," ed. J. K. Kochi, Wiley, New York, 1973, vol. 1.
150. P. M. Cullis, S. Langman, I. D. Podomore, and M. C. R. Symons, *J. Chem. Soc., Faraday Trans.*, 1990, **86**, 3267.
151. M. R. Barvian and M. M. Greenberg, *J. Org. Chem.*, 1995, **60**, 1916.
152. T. J. Matray and M. M. Greenberg, *J. Am. Chem. Soc.*, 1994, **116**, 6931.
153. H. Venkatesan and M. M. Greenberg, *J. Org. Chem.*, 1996, **61**, 525.
154. D. Schulte-Frohlinde and E. Bothe, *Z. Naturforsch., Teil C*, **39**, 315.
155. E. Bothe, G. Behrens, E. Böhm, B. Sethuram, and D. Schulte-Frohlinde, *Int. J. Radiat. Biol.*, 1986, **49**, 57.
156. D. J. Deeble and C. Von Sonntag, *Int. J. Radiat. Biol.*, 1986, **49**, 927.
157. G. D. D. Jones and P. O'Neill, *Int. J. Radiat. Biol.*, 1990, **57**, 1123.
158. P. J. Prakash Rao, E. Bothe, and D. Schulte-Frohlinde, *Int. J. Radiat. Biol.*, 1992, **61**, 577.
159. S. W. Benson and R. Shaw, in "Organic Peroxides," ed. D. Swern, Interscience, New York, 1971, vol. 1.
160. H. Catterall, M. J. Davies, and B. C. Gilbert, *J. Chem. Soc., Perkin Trans.*, 1992, **2**, 1379.
161. M. L. Steigerwald, W. A. Goddard III, and D. A. Evans, *J. Am. Chem. Soc.*, 1979, **101**, 1994.
162. C. Bienveau, J. R. Wagner, and J. Cadet, *J. Am. Chem. Soc.*, 1996, **118**, 11 406.
163. I. Saito, M. Takayama, and S. Kawanishi, *J. Am. Chem. Soc.*, 1995, **117**, 5590.
164. H. Kasai, A. Iida, Z. Yamaizumi, S. Nishimura, and H. Tanooka, *Mutat. Res.*, 1990, **243**, 249.
165. Q.-M. Zhang, J. Fujimoto, and S. Yonei, *Int. J. Radiat. Biol.*, 1995, **68**, 603.
166. S. Bjelland, L. Eide, R. W. Time, R. Stote, I. Eftedal, G. Voldne, and E. Seeberg, *Biochemistry*, 1995, **34**, 14 758.
167. M. H. Bilimoria and S. V. Gupta, *Mutat. Res.*, 1986, **169**, 123.
168. L. Shirnamé-Moré, T. G. Rossman, W. Troll, G. W. Teebor, and K. Frenkel, *Mutat. Res.*, 1987, **178**, 177.
169. L. C. Sowers and G. P. Beardsley, *J. Org. Chem.*, 1993, **58**, 1664.
170. T. Ganguly and N. J. Duker, *Nucleic Acids Res.*, 1991, **19**, 3319.
171. J. R. Wagner, J. E. van Lier, M. Berger, and J. Cadet, *J. Am. Chem. Soc.*, 1994, **116**, 2235.
172. M. Dizdaroglu, J. Laval, and S. Boiteux, *Biochemistry*, 1993, **32**, 12 105.
173. H. Ide, L. A. Petrullo, Z. Hatahet, and S. S. Wallace, *J. Biol. Chem.*, 1991, **266**, 1469.
174. J. M. Clark and G. P. Beardsley, *Biochemistry*, 1987, **26**, 5398.
175. K. Miaskiewicz, J. Miller, and R. Osman, *Biochim. Biophys. Acta*, 1994, **1218**, 283.
176. K. Miaskiewicz, J. Miller, R. Ornstein, and R. Osman, *Biopolymers*, 1995, **35**, 113.
177. T. J. Matray and M. M. Greenberg, *Biochemistry*, 1997, **36**, 14 071.
178. S. Steenken, *Chem. Rev.*, 1989, **89**, 503.
179. T. Melvin, S. W. Botchway, A. W. Parker, and P. O'Neill, *J. Am. Chem. Soc.*, 1996, **118**, 10031.
180. A. J. S. C. Vieira and S. Steenken, *J. Am. Chem. Soc.*, 1990, **112**, 6986.
181. P. M. Cullis, M. E. Malone, and L. A. Merson-Davies, *J. Am. Chem. Soc.*, 1996, **118**, 2775.
182. M. Hutter and T. Clark, *J. Am. Chem. Soc.*, 1996, **118**, 7574.
183. H. Sugiyama and I. Saito, *J. Am. Chem. Soc.*, 1996, **118**, 7063.
184. J. G. Muller, P. Zheng, S. E. Rokita, and C. J. Burrows, *J. Am. Chem. Soc.*, 1996, **118**, 2320.
185. D. Ly, Y. Kan, B. Armitage, and G. B. Schuster, *J. Am. Chem. Soc.*, 1996, **118**, 8747.
186. D. T. Breslin and G. B. Schuster, *J. Am. Chem. Soc.*, 1996, **118**, 2311.
187. B. Armitage, C. Yu, C. Devadoss, and G. B. Schuster, *J. Am. Chem. Soc.*, 1994, **116**, 9847.
188. I. Saito, M. Takayama, H. Sugiyama, and K. Nakatani, *J. Am. Chem. Soc.*, 1995, **117**, 6406.
189. A. M. Önal, D. G. E. Lemaire, E. Bothe, and D. Schulte-Frolinde, *Int. J. Radiat. Biol.*, 1988, **53**, 787.
190. P. W. Doetsch, T. H. Zastawny, A. M. Martin, and M. Dizdaroglu, *Biochemistry*, 1995, **34**, 737.
191. M. Pflaum, S. Boiteux, and B. Epe, *Carcinogenesis*, 1994, **15**, 297.
192. J. Tchou, V. Bodepudi, S. Shibutani, I. Antoschechkin, J. Miller, A. P. Grollman, and F. Johnson, *J. Biol. Chem.*, 1994, **269**, 15 318.
193. M. Dizdaroglu, T. H. Zastawny, J. R. Carmical, and R. S. Lloyd, *Mutat. Res.*, 1995, **362**, 1.
194. S. Boiteux, T. R. O'Connor, F. Lederer, A. Gouyette, and J. Laval, *J. Biol. Chem.*, 1990, **265**, 3916.
195. J. Tchou and A. P. Grollman, *J. Biol. Chem.*, 1995, **270**, 11 671.
196. F. Laval, *Nucleic Acids Res.*, 1994, **22**, 4943
197. S. Shibutani, M. Takeshita, and A. P. Grollman, *Nature*, 1991, **349**, 431.
198. M. L. Wood, A. Esteve, M. L. Morningstar, G. M. Kuziemko, and J. M. Essigmann, *Nucleic Acids Res.*, 1992, **20**, 6023.
199. A. P. Grollman and K. J. Breslauer, *Biochemistry*, 1995, **34**, 16 148.
200. K. E. McAuley-Hecht, G. A. Leonard, N. J. Gibson, J. B. Thomson, W. P. Watson, W. N. Hunter, and T. Brown, *Biochemistry*, 1994, **33**, 10 266.
201. S. Shibutani, V. Bodepudi, F. Johnson, and A. P. Grollman, *Biochemistry*, 1993, **32**, 4615.
202. H. Kamiya, H. Miura, N. Murata-Kamiya, and H. Ishikawa, *Nucleic Acids Res.*, 1995, **23**, 2893.
203. J. Cadet, M. Berger, G. W. Buchko, P. C. Joshi, S. Raoul, and J.-L. Ravanat, *J. Am. Chem. Soc.*, 1994, **116**, 7403.
204. V. Bodepudi, C. R. Iden, and F. Johnson, *Nucleosides and Nucleotides*, 1991, **10**, 755.
205. C. Sheu and C. S. Foote, *J. Am. Chem. Soc.*, 1995, **117**, 6439.
206. S. Raoul and J. Cadet, *J. Am. Chem. Soc.*, 1996, **118**, 1892.
207. W. Adam, C. R. Saha-Möller, and A. Schönberger, *J. Am. Chem. Soc.*, 1996, **118**, 9233.
208. C. Sheu and C. S. Foote, *J. Am. Chem. Soc.*, 1995, **117**, 474.
209. P. C. C. Lee and M. A. J. Rodgers, *Photochem. Photobiol.*, 1987, **45**, 79.
210. M. I. Simon and H. van Vunakis, *J. Mol. Biol.*, 1962, **4**, 488.
211. M. Rougée and R. V. Benassan, *C. R. Acad. Sci. Ser. II*, 1986, **302**, 1223.
212. T. P. A. Devasagayam, S. Steenken, M. S. W. Obendorf, W. A. Schultz, and H. Sies, *Biochemistry*, 1991, **30**, 6283.
213. F. Hutchinson, *Q. Rev. Biophys.*, 1973, **6**, 201.
214. F. Hutchinson and W. Köhnlein, *Prog. Subcell. Biol.*, 1980, **7**, 1.
215. R. Ogata and W. Gilbert, *Proc. Natl. Acad. Sci. USA*, 1977, **74**, 4973.

216. R. O. Rahn and R. S. Stafford, *Photochem. Photobiol.*, **30**, 449.
217. B. J. Hicke, M. C. Willis, T. H. Koch, and T. R. Cech, *Biochemistry*, 1994, **33**, 3364.
218. M. C. Willis, B. J. Hicke, O. C. Uhlenbeck, T. R. Cech, and T. H. Koch, *Science*, 1993, **262**, 1255.
219. T. M. Dietz, R. J. von Trebra, B. J. Swanson, and T. H. Koch, *J. Am. Chem. Soc.*, 1987, **109**, 1793.
220. R. W. Atcher, A. Russo, W. G. DeGraff, M. Moore, D. J. Grdina, and J. B. Mitchell, *Radiat. Res.*, 1989, **117**, 351.
221. T. S. Lawrence, M. A. Davis, and D. P. Normolle, *Radiat. Res.*, 1995, **144**, 282.
222. E. M. Miller, J. F. Fowler, and T. J. Kinsella, *Radiat. Res.*, 1992, **131**, 81.
223. J. B. Mitchell, A. Russo, T. J. Kinsella, and E. I. Glatstein, *J. Radiat. Oncol. Biol. Phys.*, 1986, **12**, 1513.
224. C. L. Limoli and J. F. Ward, *Int. J. Radiat. Biol.*, 1994, **66**, 717.
225. H. Sugiyama, Y. Tsutsumi, and I. Saito, *J. Am. Chem. Soc.*, 1990, **112**, 6720.
226. I. Saito, *Pure Appl. Chem.*, 1992, **64**, 1305.
227. H. Sugiyama, K. Fujimoto, and I. Saito, *Tetrahedron Lett.*, 1996, **37**, 1805.
228. H. Sugiyama, Y. Tsutsumi, K. Fujimoto, and I. Saito, *J. Am. Chem. Soc.*, 1993, **115**, 4443.
229. G. P. Cook and M. M. Greenberg, *J. Am. Chem. Soc.*, 1996, **118**, 10025.
230. K. M. Prise, N. E. Gillies, A. Whelan, G. L. Newton, R. C. Fahey, and B. D. Michael, *Int. J. Radiat. Biol.*, 1995, **67**, 393.
231. M. Spotheim-Maurizot, F. Garnier, R. Sabattier, and M. Charlier, *Int. J. Radiat. Biol.*, 1992, **62**, 659.
232. M. Basu Roy, P. C. Mandal, and S. N. Bhattacharyya, *Int. J. Radiat. Biol.*, 1996, **69**, 471.
233. G. R. Buettner and B. A. Jurkiewicz, *Radiat. Res.*, 1996, **145**, 532.
234. A. Samuni, M. Chevion, and G. Czapski, *J. Biol. Chem.*, 1981, **256**, 12632.
235. S. Nishimoto, N. Shimbara, M. Kaneto, K. Sakano, and T. Kagiya, *Nucleic Acids Res. Symposium Series No. 16*, 1985, **57**.
236. E. S. Henle, Y. Luo, and S. Linn, *Biochemistry*, 1996, **35**, 12212.
237. G. W. Buchko and M. Weinfeld, *Biochemistry*, 1993, **32**, 2186.
238. P. C. Dedon, Z.-W. Jiang, and I. H. Goldberg, *Biochemistry*, 1992, **31**, 1917.
239. A. T. Al-Kazwini, P. O'Neill, G. E. Admas, and E. M. Fielden, *Radiat. Res.*, 1990, **121**, 149.
240. A. Samuni, D. Godinger, J. Aronovitch, A. Russo, and J. B. Mitchell, *Biochemistry*, 1991, **30**, 555.
241. M. Aslam Siddiqi and E. Bothe, *Radiat. Res.*, 1987, **112**, 449.
242. R. E. Krisch, M. B. Flick, and C. N. Trumbore, *Radiat. Res.*, 1991, **126**, 251.
243. J. F. Ward, *Int. J. Radiat. Biol.*, 1990, **57**, 1141.
244. J. R. Milligan, J. Y.-Y. Ng, C. C. L. Wu, J. A. Aguilera, R. C. Fahey, and J. F. Ward, *Radiat. Res.*, 1995, **143**, 273.
245. D. J. Brenner and J. F. Ward, *Int. J. Radiat. Biol.*, 1992, **61**, 737.
246. C. L. Limoli and J. F. Ward, *Radiat. Res.*, 1993, **134**, 160.
247. H. C. Box, E. E. Budzinski, H. G. Freund, M. S. Evans, H. B. Patrzyc, J. C. Wallace, and A. E. Maccubbin, *Int. J. Radiat. Biol.*, 1993, **64**, 261.
248. H. C. Box, H. G. Freund, E. E. Budzinski, J. C. Wallace, and A. E. Maccubbin, *Radiat. Res.*, 1995, **141**, 91.
249. E. E. Budzinski, J. D. Dawidzik, J. C. Wallace, H. G. Freund, and H. C. Box, *Radiat. Res.*, 1995, **142**, 107.
250. L. F. Povirk, Y.-H. Han, and R. J. Steighner, *Biochemistry*, 1989, **28**, 5808.
251. R. J. Steighner and L. F. Povirk, *Proc. Natl. Acad. Sci. USA*, 1990, **87**, 8350.
252. M. J. Absalon, J. W. Kozarich, and J. Stubbe, *Biochemistry*, 1995, **34**, 2065.
253. M. J. Absalon, W. Wu, J. W. Kozarich, and J. Stubbe, *Biochemistry*, 1995, **334**, 2076.
254. M. J. Schottelius and P. Chen, *J. Am. Chem. Soc.*, 1996, **118**, 4896.
255. S. M. Meschwitz, R. G. Schultz, G. W. Ashley, and I. H. Goldberg, *Biochemistry*, 1992, **31**, 9117.
256. L. S. Krappen, I. H. Goldberg, S. H. Wu, J. Stubbe, L. Worth, Jr., and J. W. Kozarich, *J. Am. Chem. Soc.*, 1990, **112**, 2797.
257. H. Kishikawa, Y.-P. Jiang, J. Goodisman, and J. C. Dabrowiak, *J. Am. Chem. Soc.*, 1991, **113**, 5434.

## W. DAVID WILSON
*Georgia State University, Atlanta, GA, USA*

| | | |
|---|---|---|
| 7.12.1 | INTRODUCTION | 428 |
| 7.12.1.1 | *Discovery of Intercalators: Dyes and Drugs* | 428 |
| 7.12.1.2 | *The Classical Intercalation Model* | 432 |
| 7.12.1.3 | *Initial Tests of the Classical DNA Intercalation Model* | 433 |
| 7.12.1.4 | *Comparison of Intercalation into RNA and DNA Duplexes* | 433 |
| 7.12.1.5 | *Formation of Intercalation Sites* | 434 |
| 7.12.2 | VARIETY OF INTERCALATORS AND INTERCALATION SITES | 435 |
| 7.12.2.1 | *Structural Details for Intercalation Sites* | 435 |
| 7.12.2.2 | *Specificity in Intercalation* | 436 |
| 7.12.3 | MORE COMPLEX INTERCALATORS | 437 |
| 7.12.3.1 | *Threading Intercalators* | 437 |
| 7.12.3.1.1 | *Nogalamycin* | 437 |
| 7.12.3.1.2 | *Synthetic threading intercalators* | 440 |
| 7.12.3.1.3 | *Porphyrins* | 441 |
| 7.12.3.2 | *Bisintercalators and Neighbor Exclusion* | 444 |
| 7.12.3.2.1 | *Ditercalinium* | 444 |
| 7.12.3.2.2 | *TOTO and YOYO* | 448 |
| 7.12.3.3 | *Macrocycles: Threading Bisintercalators* | 449 |
| 7.12.3.4 | *Rigid Bisintercalators* | 449 |
| 7.12.3.5 | *Multi-intercalators* | 451 |
| 7.12.4 | EXTENSION OF THE TYPES OF AROMATIC RING SYSTEM THAT BIND BY INTERCALATION | 452 |
| 7.12.4.1 | *Unfused Aromatic Intercalators* | 452 |
| 7.12.4.2 | *Sequence-dependent Binding Modes* | 456 |
| 7.12.4.3 | *Helix-type Dependent Binding* | 460 |
| 7.12.4.4 | *Self-assembly and Chemically Switched Intercalators* | 461 |
| 7.12.5 | PARTIAL INTERCALATION OF SMALL AROMATIC SYSTEMS | 463 |
| 7.12.5.1 | *Partial Intercalation Model Systems* | 463 |
| 7.12.5.2 | *Partial Intercalation of Protein Side Chains* | 465 |
| 7.12.6 | INTERCALATION INTO MULTISTRANDED HELIXES AND DISTORTED DUPLEXES | 466 |
| 7.12.6.1 | *Triplex Intercalation* | 466 |
| 7.12.6.2 | *Tetraplex Intercalation* | 469 |
| 7.12.6.3 | *Intercalation with Distorted Nucleic Acid Structures* | 469 |
| 7.12.7 | INTERCALATORS THAT REACT WITH DNA | 470 |
| 7.12.7.1 | *Covalent Carcinogens* | 470 |
| 7.12.7.2 | *Covalent Intercalating Drugs* | 470 |

7.12.8  OVERVIEW                                                                    472

7.12.9  REFERENCES                                                                  472

---

## 7.12.1  INTRODUCTION

There are two well-characterized strong binding modes for small molecules with nucleic acids: intercalation with both DNA and RNA and minor-groove binding to DNA.[1-15] Some polycations apparently bind strongly in the major groove of RNA, but this binding mode has not been defined in molecular detail.[16-18] There is also a weaker, amorphous binding of organic cations to the anionic surface of nucleic acids, but this binding mode generally is not significant at salt concentrations of 0.1 M and above with compound concentrations below 100 $\mu$M.[19-21] This chapter focuses specifically on the intercalation binding mode which occurs when adjacent base pairs in a nucleic acid duplex are separated by sufficient space ($\sim 3.4$ Å) to allow an aromatic ring system to be inserted between them. Some common intercalators are shown in Figure 1 and the general shape of an intercalation site can be viewed in Figure 2. There are a number of points in the history of development of our understanding of intercalation of aromatic compounds with nucleic acids where synthetic chemistry, the design and preparation of compounds to answer specific questions about intercalation, has had a major impact, and a number of those points are highlighted in this chapter.

### 7.12.1.1  Discovery of Intercalators: Dyes and Drugs

When we find a colorful wildflower growing through a drab brush pile, we are struck by the beauty and contrast. The ancients noted the same phenomenon and devised methods to dye their clothes to produce a similar effect. Unfortunately, dyeing in that era generally required chopping and grinding a large amount of plant and/or animal material with laborious extraction of the dye or an appropriate precursor. Only the rich could afford to purchase material prepared by such a costly process and brilliantly dyed material remained a trademark of royalty and favored associates until chemists democratized colored materials through the synthesis of low-cost dyes. Chemists of the nineteenth century, particularly in Germany, established this early application of the power of synthetic methods to improve the environment of humans. Through the genius of Paul Erhlich, the synthetic efforts to prepare a wide range of dyes leads directly into the discussion of nucleic acid intercalators in this chapter. Biologists were interested in dyes to label selectively cells and cellular components, and an intense research effort demonstrated the utility of dyes, many of which were intercalators, in the labeling area. Nucleic acids bind a wide variety of dyes, and the fluorescent intercalators quinacrine (Figure 1) and quinacrine mustard are still widely used in cytochemical analysis to help in differentiating chromosomes and preparing cytogenetic maps.[11,25,26] Intercalators such as acridine orange are still used to stain nucleic acids in microscopic analysis and the intercalator ethidium bromide (Figure 1) is widely used to visualize double-stranded nucleic acids on electrophoresis gels.[27-29] A number of other commercial intercalators are useful in the detection of nucleic acids at low levels in a variety of experiments.

Erhlich reasoned that if cells could be differentially stained by dyes, then organisms that cause infectious diseases could be selective stained, or killed at sufficient dose, while the host would be unaffected.[30] He showed that a dye that he appropriately named trypan red could rid mice of the trypanosomal parasite which in humans causes sleeping sickness and which remains a significant health problem. These ideas and observations provided the foundation for the formulation of his famous "magic bullet" hypothesis.[30] Although not the "magic bullets" envisioned by Erhlich, intercalators are very important drugs against diseases from parasitic infections to cancer. Quinacrine (Figure 1) used to be used against malaria, and in his autobiography the entomologist E. O. Wilson notes that when he returned from a long stay in New Guinea in the early 1950s he looked somewhat yellow as a result of taking quinacrine as an antimalarial drug.[31] One wonders if our ancestors tried ingesting such compounds to dye their entire bodies. The results would have certainly been impressive, but given that many dyes are toxic at relatively low dosages, the idea, if it were tried, was probably quickly abandoned.

Intercalators have a number of uses as drugs, particularly against neoplastic diseases, and as probes for nucleic acid structure and dynamics. Anthracycline intercalators (Figure 1) are

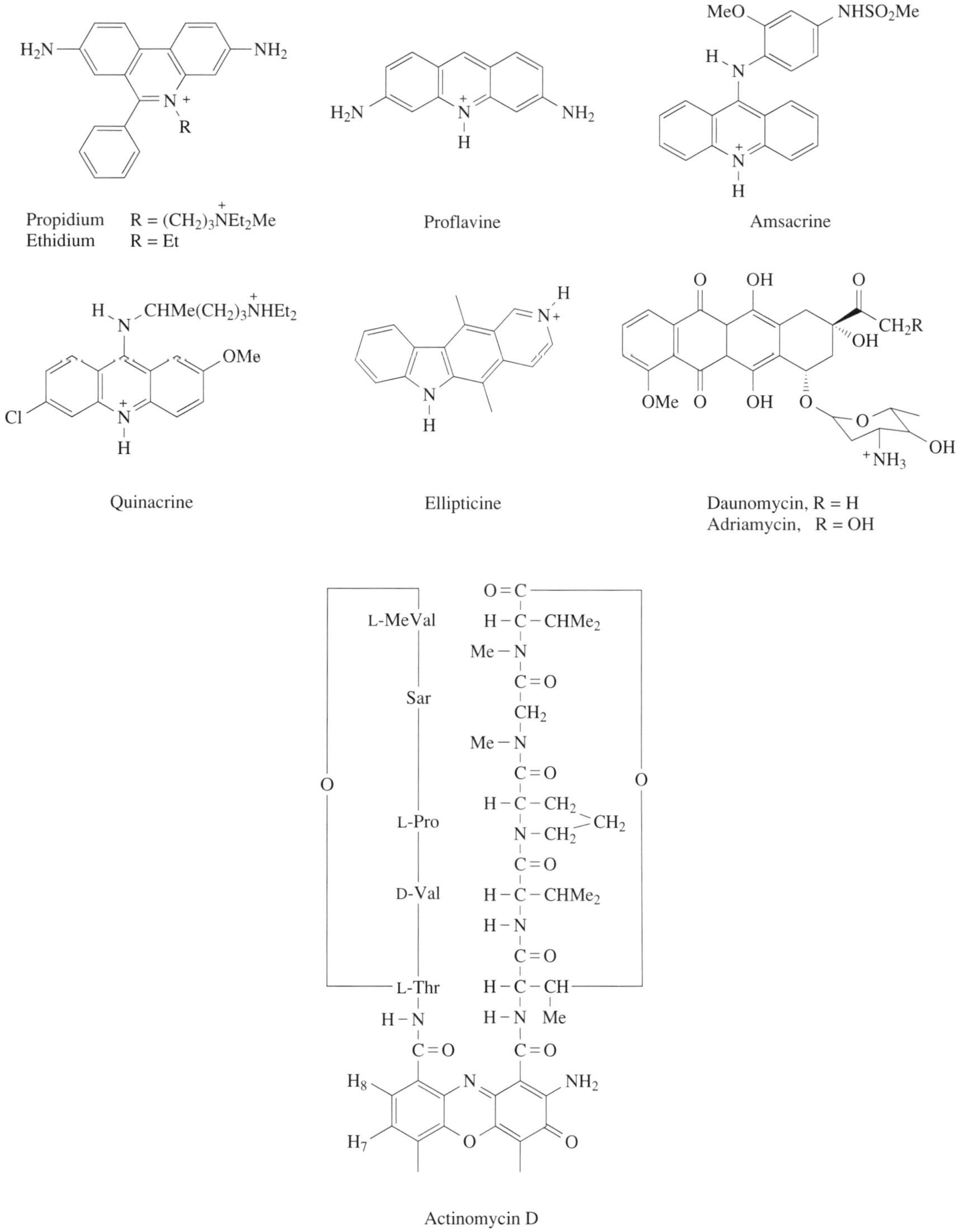

Propidium  R = (CH₂)₃NEt₂Me

Ethidium  R = Et

Proflavine

Amsacrine

Quinacrine

Ellipticine

Daunomycin, R = H
Adriamycin,  R = OH

Actinomycin D

**Figure 1** Structures of some common intercalators that have been extensively studied. The complexity of the structures increases from the top to the bottom of the figure. Ethidium is a common laboratory stain for duplex nucleic acids and is active against the trypanosomal organisms that cause sleeping sickness. Quinacrine was used as an antimalarial drug during World War II and into the 1950s when better drugs were found. The remaining compounds all have significant anticancer activity and have undergone clinical trials. The anthracyclines daunomycin and adriamycin are among the most important anticancer drugs currently available.

particularly important anticancer drugs and are the leading clinical drugs of choice against a number of cancers.[32–34] Actinomycins are complex intercalating antibiotics that were among the first compounds discovered in an intense screen of natural products from soil microorganisms that began in the late 1930s in a search for drugs to combat bacterial infections.[7,30] The actinomycins proved

(a)

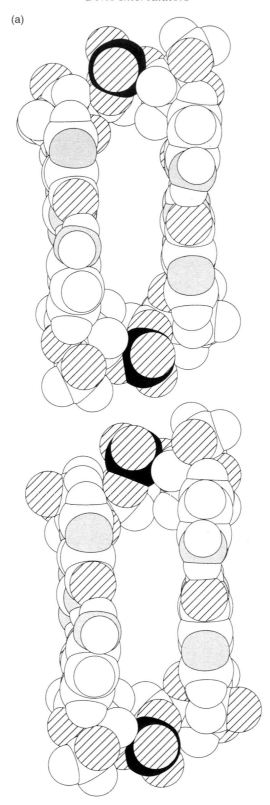

**Figure 2** (a) Space-filling models are shown for CpG intercalation sites from a proflavine–DNA complex[22] (upper) and an ethidium–RNA complex[23] (lower). The intercalators have been deleted from these views to assist in comparison of the intercalation site topologies and the view is from the major groove. The atom shading is as follows: carbon and hydrogen, white; nitrogen, gray; oxygen, diagonal lines; phosphorus, black. (b) The ethidium–RNA complex with intercalated ethidium from the minor groove (lower) and from the major

(b)

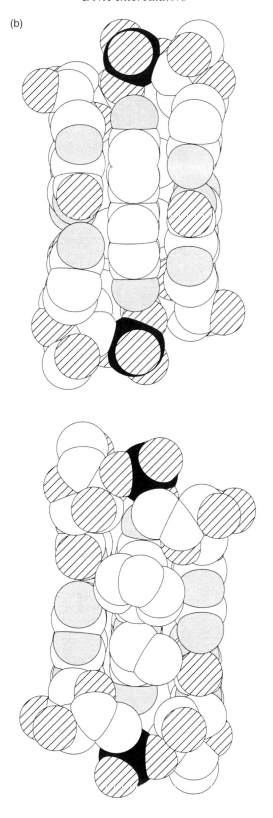

groove (lower). The shading scheme is as in (a), but hydrogens have been deleted to assist in visualization of the intercalating ring system. These models were obtained from the Nucleic Acid Data Base (NDB) at Rutgers University, Department of Chemistry.[24] The NDB has a WWW (world wide web) page that has information on nucleic acids and their complexes and views of a variety of structures. Files are available from the NDB by a variety of methods.

too toxic for use as antibiotics, but actinomycin D was found to be a very effective anticancer drug and is still in use as a drug against certain types of cancer.[30–35] The antibacterial activity of actinomycin and many other intercalators is associated with inhibition of RNA synthesis and/or DNA replication that occur as a result of formation of the intercalation complex.[7]

### 7.12.1.2   The Classical Intercalation Model

The establishment of the double-helical structure of DNA with stacked base pairs provided the initial structural basis for understanding intercalation.[36] The classical intercalation model was formulated and put on a firm experimental basis by Lerman in the early 1960s.[37–40] Quantitative studies of the interactions of planar aromatic cations, such as the acridines in Figure 1, with DNA established two types of binding: (i) a strong binding mode, which we now know to be intercalation, saturated at one compound bound per two base pairs, and (ii) a weaker binding mode with two (or more) compounds bound per base pair at saturation.[1,7] The weaker binding mode is characteristic of nonspecific interactions of planar aromatic cations with polyanions.[19–21] It is best represented as an amorphous stack of the planar cations along the surface of the polyanion. The anionic charges on the nucleic acid neutralize the electrostatic repulsion in the stack of cations, and the self-stacking complex is driven by hydrophobic forces and van der Waals interactions.[7] This type of complex is very dependent on ionic strength, compound concentration, and ring size. It can generally be prevented from forming by working at salt concentrations of 0.1 M and above and at low compound concentrations (below 100 $\mu$M). As a caveat to experimentalists planning studies with intercalators, it should be noted that planar aromatic compounds in water can dimerize, a process enhanced by increased salt concentration, and can bind to solid surfaces of glassware and some plasticware.[41–43] It is generally useful to monitor some absorption spectral signal from solutions of such compounds as a function of time to insure that concentration corrections are made for compound lost as a result of such adsorption, which can be especially significant for samples stored for some time at low temperatures and low concentrations.

Experiments on the strong binding intercalation mode of organic cations with DNA indicated that this interaction accounted for the biological effects of the compounds. In an effort to characterize the intercalation, Lerman[37–40] conducted systematic studies on the DNA complexes of several acridines. He found that the viscosity of a DNA solution was markedly increased by all of the acridine derivatives, whereas simple organic cations and nonplanar organic cations of similar composition caused slight viscosity decreases. From flow dichroism and flow polarized fluorescence, he found that the plane of the bound acridines was essentially parallel to the DNA base pair planes and perpendicular to the axis of the double helix. From the results of these and other experiments, Lerman proposed that an intercalation model best described the acridine–DNA complex. Other structures such as surface attachment of the ligand or insertion of the ligand into the double helix with displacement of a base pair agreed with some but not all of the experimental results. Model building studies indicated that the B-form double helix could be extended, with simple single-bond rotation, to create a space for insertion of a planar aromatic molecule of ~3.4 Å in thickness.[39]

The classical intercalation model as defined for DNA has the following features: (i) the structure and hydrogen bonding of B-form DNA is retained; (ii) two adjacent base pairs are separated to create a planar space for intercalation; (iii) an unwinding of the double helix is required; and (iv) the aromatic system of the intercalator is inserted between and parallel to the base pairs in a way that fills the space created by extension of the double helix as fully as possible. The width of the site created is close to that required to insert an acridine ring or other similar aromatic system with its long axis oriented in the same direction as the long axes of the two adjacent base pairs. In the B-form double helix with ~10 base pairs per turn (360°), each base pair is rotated ~ +36° with respect to the base pair immediately below it.[36] Creation of a space for intercalation results in a reduction of this rotation, the intercalation unwinding angle. Fuller and Waring[44] conducted model building studies on ethidium and concluded that a 12° unwinding angle allowed maximum separation of phosphate groups and the optimum interaction of ethidium with the double helix (hydrogen bonding of ethidium amino groups with the phosphate oxygens of DNA and optimum overlap with base pair aromatic ring systems). Later experimental studies showed that the unwinding angle for ethidium is closer to 26°, leaving the adjacent base pairs at the intercalation site with a twist of 10°.[1,7]

### 7.12.1.3 Initial Tests of the Classical DNA Intercalation Model

The discovery of closed circular superhelical DNA allowed intercalation unwinding angles to be quantitatively determined. As intercalating molecules are added to superhelical DNA, the double helix is unwound and initially the natural right-handed superhelical turns are removed until the DNA has no remaining superhelical structure, at which time it is hydrodynamically equivalent to nicked circular DNA.[7,11] If more of the intercalating drug is added, the double helix continues to unwind with the formation of a left-handed superhelical structure. A maximum in the viscometric titration corresponds to closed circular DNA with no remaining superhelical turns. If both the initial number of superhelical turns and the amount of bound drug required to remove these turns completely are determined, the average unwinding angle of the drug can be calculated.[45] The influence of the various structural features that determine a ligand's unwinding angle is still largely unknown. The 26° value for ethidium is the largest unwinding angle found to date for any intercalating monomer ligand. The classical acridine drugs give smaller unwinding angles, which cluster between 17° and 20°,[46] whereas the anthracycline drugs, such as daunomycin and adriamycin (Figure 1), have smaller unwinding angles, between 10° and 12°, relative to ethidium bromide.[7]

A second key predicted feature of the classical intercalation model is a lengthening of the helix axis on creation of an intercalation site.[7,11,37–40,47,48] This extension causes a change in hydrodynamic properties, including an increase in linear DNA viscosity and a decrease in sedimentation coefficient.[47,48] The flexibility of high-molecular-weight DNA complicates quantitation of the length increase since changes in stiffness can also affect hydrodynamic properties. Sonication of DNA down to a size near the persistence greatly simplifies the interpretation of results from the hydrodynamic experiments.[47] Studies of a number of intercalators (Figure 1) indicated that they caused length increases in sonicated DNA of 70–100% of the calculated value, which assumes a 3.4 Å length increase per bound intercalator.[7,11] Experiments on lengthening and unwinding of the DNA duplex by planar aromatic compounds firmly established the essential validity of the classical intercalation model. Subsequent results on a number of intercalators of different sizes and substituents have indicated that the apparent length increase and unwinding angle on intercalation can vary greatly, and a nonclassical intercalation model to explain these results will be presented below.

### 7.12.1.4 Comparison of Intercalation into RNA and DNA Duplexes

The standard state of DNA in nature is a double-stranded duplex. RNA is single stranded, but in most cases it is folded into extensive hairpin duplexes such that more of the bases are paired in a Watson–Crick fashion than are unpaired.[36,49–52] The duplexes are generally not as regular as with DNA and contain base bulges, internal loops, and base-pair mismatches. As can be seen from Figure 2, it is possible to form very similar intercalation structures in both DNA and RNA. There have been far fewer studies of RNA intercalation than with DNA, for several reasons: (i) a number of antibiotics and anticancer intercalators exert their activity by binding to DNA and there has been a major effort to understand their DNA interactions; (ii) the conformational states of DNA are less amorphous than with RNA and provide simpler model systems for understanding the molecular details of intercalation; and (iii) specific sequence DNA oligomers are easier to prepare and maintain in solution than corresponding RNA oligomers. The difficulty of working with RNA is, however, more than offset by the variety of biological functions that RNAs carry out in organisms from small viruses to humans.[53] The ability to perturb selectively these functions with intercalators (or other ligands) would be of obvious benefit, and the design of such RNA-selective compounds is an important research area for organic chemists.

It is interesting that many intercalating antibiotics have been found that bind to DNA to exert their biological functions, but no similar compounds have been identified that target RNA. Antibiotics such as streptomycin and neomycin act at the ribosome, probably primarily by interactions with ribosomal RNA, but they are nonplanar polycations that do not intercalate.[54,55] Studies with simple intercalators indicate that they can bind to RNA fairly well, as would be expected from the general similarity of DNA and RNA intercalation site structures (Figure 2).[56–59] Most intercalators that have been evaluated bind less well to RNA than to DNA,[56] but this represents a relative small sample of known intercalators and many of the compounds tested (e.g., adriamycin, actinomycin D, nogalamycin) are antibiotics that have structures designed to intercalate with DNA and interfere with DNA metabolic processes. As noted above, no such intercalating antibiotics that are directed to RNA have been found, and the lack of a natural paradigm, as has been available with DNA, greatly complicates the design of RNA-specific intercalators. As with the DNA-directed antibiotics listed above, the specificity of any future RNA-directed intercalators will no doubt have to involve substituents that make extensive contacts in the RNA grooves as part of the intercalation complex. The great variety of RNA structures indicates that it should be possible to design compounds that bind to specific RNAs with high affinity. Structural models for peptides that exhibit high affinity and selectivity for particular RNA structures are available in their RNA complexes[60–63] and these model systems may provide the best initial examples for design of substituents that confer RNA selectivity on intercalators. Thus, although simple intercalators bind fairly well to intercalation sites in both DNA and RNA duplexes, much more is known about the DNA complexes.

### 7.12.1.5  Formation of Intercalation Sites

Before discussing the specifics of intercalation sites and the variety of intercalators, it is worthwhile considering how and at what rate such sites appear in nucleic acids. Thermal motion of the torsional angles of sigma bonds in single-stranded polymers causes them to move rapidly among the set of low-energy states available, and such single-stranded systems generally have a dynamic and amorphous global conformation.[36] Nucleic acids have six backbone torsional angles that take specific values in double-helical or folded states of the polymers such as the B-form of DNA and the A-form duplex of RNA (Table 1). A number of methods, however, indicate that the secondary and tertiary structures of nucleic acids have significant dynamic motions about the global low-energy conformations. Perhaps the most striking observation is that the protons involved in the Watson–Crick hydrogen bonds of base pairs in nucleic acid duplexes exchange with solvent water protons with a half-life in the 10 ms range (the exact value depends on sequence) at 25 °C with 0.1 M salt concentration.[64–66] Concerted thermal motions of the bases and backbone atoms are great enough frequently to cause each base in a long double helix to swing out of the stack and contact water molecules long enough to exchange base NH for water OH protons. Thermal motion can also yield an intercalation site in DNA by concerted changes in backbone torsional angles (see Table 1 and Figure 2), but without the necessity to break base-pair hydrogen bonds. Association rate constants for simple intercalators are considerably slower than for similar compounds that bind tightly in the grooves of a duplex, owing to the need to open a site as part of the mechanism of intercalation.[67–70] At 25 °C and 0.1 M salt concentration, association of simple intercalators with nucleic acid duplexes

**Table 1**  Torsional angles for nucleic acid duplex conformations.[a]

$$-\text{O3}'-\overset{\overset{\text{O}}{|}}{\underset{\underset{\text{O}}{|}}{\text{P}}}\overset{\alpha}{-}\text{O5}'\overset{\beta}{-}\overset{|}{\underset{\underset{\text{O4}'}{|}}{\text{C5}'}}\overset{\gamma}{-}\overset{|}{\underset{\underset{\text{C2}'}{|}}{\text{C4}'}}\overset{\delta}{-}\overset{|}{\text{C3}'}\overset{\varepsilon}{-}\text{O3}'\overset{\zeta}{-}\overset{\overset{\text{O}}{|}}{\underset{\underset{\text{O}}{|}}{\text{P}}}-\text{O5}'-$$

χ: O4'-C1'-N1-C2 pyrimidines
   O4'-C1'-N9-C4 purines

|                | α    | β    | γ  | δ   | ε    | ζ    | χ    |
|----------------|------|------|----|-----|------|------|------|
| B-form (fiber)   | −30  | 136  | 31 | 143 | −141 | −161 | −98  |
| B-form (crystal) | −65  | 167  | 51 | 129 | −157 | −120 | −103 |
| A-form (fiber)   | −52  | 175  | 42 | 79  | −148 | −75  | −157 |
| A-form (crystal) | −73  | 173  | 64 | 78  | −151 | −77  | −165 |
| dCpG·Prof (C1)   | —    | —    | 55 | 84  | −150 | —    | −161 |
| dCpG·Prof (G2)   | −70  | −140 | 46 | 94  | —    | −70  | −109 |
| dCpG·Prof (C3)   | —    | —    | 53 | 79  | −156 | —    | −165 |
| dCpG·Prof (G4)   | −73  | −142 | 72 | 149 | —    | −60  | −75  |
| rCpG·Et (C1)     | —    | —    | 49 | 84  | −138 | —    | −157 |
| rCpG·Et (G2)     | −75  | −136 | 58 | 137 | —    | −71  | −66  |
| rCpG·Et (C3)     | —    | —    | 49 | 82  | −140 | —    | −159 |
| rCpG·Et (G4)     | −61  | −132 | 50 | 135 | —    | −73  | −73  |

[a] For atoms 1–2–3–4 the zero torsional angle for the 2–3 bond is at the fully eclipsed position of atoms 1 and 4, and angles increase for a clockwise rotation of atom 4 with respect to atom 1 when looking down the bond from atom 2 to 3. Negative angle values are for counterclockwise rotation. The B-form and A-form fiber diffraction and crystal angle values are discussed in the literature[4,36] and values in the table are from those references. The dCpG·proflavine and rCpG·ethidium angle values in the table are for the intercalation complexes shown in Figure 2. Each strand in these complexes has only one phosphate and one set of α, β, ε, and ζ angles. Since each strand has two sugar and base residues, there are two γ, δ, and χ angle values.

occurs with a half-life of less than 1 ms,[67] and creation of an intercalation site obviously must occur within that time period. No doubt a range of partially open distorted structures, which do not closely resemble the equilibrium conformations shown in Figure 2, occur as a result of thermal motion and may represent intermediate states to intercalation. On partial insertion of an aromatic ring system, some of these states could proceed to an intercalation conformation while others close down with dissociation of the aromatic molecule and reformation of the native conformation. As one gazes at the beautiful architecture of DNA and RNA structures illustrated in textbooks from basic chemistry to advanced genetics, intercalation sites demonstrate part of the array of dynamic changes that can occur in these structures on the millisecond timescale.

## 7.12.2 VARIETY OF INTERCALATORS AND INTERCALATION SITES

The description and experimental support for the classical intercalation model sparked a tremendous research effort to quantitate its structural features and to determine how the model applies to other planar aromatic drugs, aromatic amino acid side chains of peptides, carcinogens, and similar compounds, and also what modifications in the model could occur in these different complexes. A range of observed variations in intercalation sites and intercalators will be described below.

### 7.12.2.1 Structural Details for Intercalation Sites

NMR and X-ray diffraction studies of intercalation complexes have provided many specific details about intercalation geometry, and have provided insight into the variety of ways in which intercalation sites can be created and into the dynamics of such sites.[2-6] A number of dinucleotides were originally crystallized with relatively simple intercalators such as ethidium and proflavine[4,22,23,71] (Figure 1). Although these small systems are distant from natural DNA and RNA, they are very useful for high-resolution illustration of the varieties of intercalation geometries. In Figure 2(a) two different CpG intercalation sites, one from an RNA–ethidium complex[71] and the other from a DNA–proflavine complex,[22] are shown with the intercalators deleted. In spite of the fact that they are from different nucleic acids with different intercalators, the sites are globally fairly similar and an array of intercalators can be docked into each site with good fit. The sites have different sugars, sugar conformation, and backbone torsional angles (Table 1), but the global topology allows very favorable stacking of planar aromatic ring systems into both types of sites. The two sites in Figure 2 illustrate an extremely important point: intercalation sites with very similar global topology can be created in both DNA and RNA duplex sequences, a finding confirmed by results on longer DNA and RNA sequences. Because the grooves of DNA and RNA are very different,[36] groove-binding compounds generally select either DNA or RNA duplexes, while many intercalators bind fairly well to both DNA and RNA.[56]

An ethidium complex is shown in Figure 2(b) from both the major and minor groove sides. A considerable fraction of the positive charge of ethidium is distributed to the amino substituents on the phenanthridinium ring (Figure 1), and these groups are in close proximity to the phosphate groups on opposite sides of the intercalation site. As can be seen from the views in Figure 2, the phosphate groups are in the plane of the intercalator and they offer the most direct possibility of interaction of the nucleic acid backbone strands with the intercalated ring system. The view into the minor groove illustrates the fit of nonplanar substituents on an intercalating ring into the DNA grooves. Steric clash between the ethyl and phenyl substituents of ethidium causes rotation of the phenyl group to an angle of 70–80° to the phenanthridinium intercalator. As a result of this twist, the phenyl group can contact base pairs on both sides of the intercalation site and it is obvious that appropriate substituents on the phenyl ring could interact favorably with these bases. More complex intercalators have substituents in the grooves that have extensive interactions with bases to give a higher degree of sequence specificity than observed with ethidium or proflavine.[7,8] The origin of the forces stabilizing the intercalation complexes can be understood from these structures and energy calculations.[72,73] The intercalated ring system has van der Waals and electrostatic interactions with the bases and phosphate groups at the intercalation site. The amino nitrogens of ethidium, for example, are located ∼3 Å from the 5′-oxygens of the phosphodiester groups that are on both sides of the intercalation cavity, and this puts them in position to form hydrogen-bond interactions. Ethidium is thus a very well-designed simple intercalator and to improve significantly on its interactions with DNA would not require more complex substituents.[73] The intercalation sites in Figure

2(a) have limited interaction possibilities because of their size and structure and for this reason simple intercalators such as ethidium and proflavine have generally low specificity in binding. In moving to the bottom of Figure 1, the anthracycline drugs and actinomycin D have complex substituents and their interactions in the DNA grooves result in much higher specificity DNA interactions.[1-8]

Compounds with a single aromatic ring (e.g., benzene or pyridine) or compounds with two fused rings (naphthalene or quinoline) do not fill the intercalation space (Figure 2) efficiently. Compounds with three fused rings (e.g., anthracene or acridine) fill the site fairly well while compounds with four aromatic rings (e.g., benzanthracenes and related structures) can efficiently stack with base pairs on both sides of the intercalation site. Further addition of aromatic rings to give compounds with five or more fused ring systems creates compounds that are generally larger than the stacking surface of the adjacent base pairs and should not significantly enhance binding. Steric clash of these larger systems with the DNA backbone can prevent optimum stacking interactions and could actually decrease binding. Dimerization and/or surface adsorption of the larger aromatic compounds can also make them very difficult to study quantitatively in aqueous solutions.

It has proven very difficult to obtain high-resolution structural information on complexes of simple intercalators such as ethidium and proflavine with DNA or RNA strands longer than dinucleotides. The reasons for this difficulty are now fairly clear. The simple systems have low binding specificity and exchange easily among their variety of binding sites. In an oligomer of $n$ base pairs in a DNA or RNA duplex there are $n-1$ potential intercalation sites, and with exchange among the sites, a very complex, heterogeneous mixture is created that is difficult to crystallize or evaluate quantitatively by NMR methods.[74] More complex monointercalators and bisintercalators (two intercalators covalently linked) have been crystallized with longer segments of DNA but not with RNA. Crystal structures of a number of anthracycline drugs (Figure 1) have been determined with self-complementary hexamer duplexes such as d(CGTACG)$_2$ or d(CGATCG)$_2$ that have the intercalator bound at both ends of the duplex at the CpG sites. In addition to their intercalating ring systems, such compounds have complex substituents that interact strongly with the grooves of DNA and reduce the exchange problems that complicate structural studies with simple intercalators.

### 7.12.2.2 Specificity in Intercalation

Compounds that interact with nucleic acids in a sequence-specific fashion do so through formation of specific direct interactions such as hydrogen bonds to groups on bases or by recognition of some unique feature, such as a specific conformation, that is characteristic of a particular nucleic acid sequence.[1-12] Groove-binding compounds recognize specific conformational features of DNA grooves and form hydrogen bonds with groups on bases at the floor of the minor groove.[2,8] Simple intercalators stack between only two bases and intrinsically can only have short-range specificity.[12] In addition, the ability to hydrogen bond with groups on bases requires nonintercalating functional groups in the duplex grooves and more complex intercalator structures.[13-15] Such groups can fit into either the major or minor groove or both grooves for the threading intercalators described below. More complex intercalators such as actinomycin D and the bisintercalating antibiotics triostin and echinomycin obtain specificity through groups that hydrogen bond to bases in the minor groove. Hybrids of intercalators with minor groove binding type substituents have been prepared to increase the specificity of binding of intercalators.[13,14]

Simple intercalators generally vary from slightly GC to slightly AT specific.[11,12] Muller and Crothers[12] have pointed out that because the dipole moment of a GC base pair is greater than that of an AT base pair, highly polarizable compounds exhibit significant GC base-pair specificity. Again, since the intercalating ring itself only contacts two bases, this type of specificity is very short range. For intercalation sites in antiparallel duplexes like B-form DNA or A-form RNA, there are 10 different dinucleotide binding sites available for intercalation[1,7] (the $2^4$ sites are reduced to 10 by symmetry). No simple intercalator is known that binds to only one or two of these 10 possibilities although, as discussed above, highly polarizable intercalators will tend to select sites with more GC base pairs.

The nonspecific binding of intercalators has occasionally been used to advantage to increase the binding free energy of more highly specific compounds such as minor groove binding compounds or nucleotide chains used in antisense or antigene therapeutics.[15] The intercalator can be attached to such compounds with a flexible linker that will allow it to bind in a nonspecific fashion. The complex will then have the same sequence recognition properties but will have enhanced binding affinity.

Most of our knowledge of specificity in intercalator (and other organic compound)–nucleic acid interactions is based on interactions in the DNA minor groove. The major groove of DNA and both grooves of RNA are excellent protein-binding sites and the design of small molecules that can selectively bind to these other grooves with high affinity and specificity is another of the challenges to organic chemists interested in nucleic acid interactions.

## 7.12.3 MORE COMPLEX INTERCALATORS

The emphasis up to this point has been on the creation of intercalation sites and on how such sites are filled by relatively simple aromatic systems. Clearly, it is possible to attach complex substituents to the intercalating ring and such modifications can significantly alter the intercalation complex. In addition, two more intercalating systems could be linked to create multi-intercalators which could also be macrocycles. Some results with more complex intercalators of this type are presented in this section.

### 7.12.3.1 Threading Intercalators

Threading intercalators have substituents on opposite sides of the intercalating ring system such that one substituent must pass between base pairs at the binding site to form the final complex[75–80] (Figure 3). If the substituents are large enough, considerable disruption of the double helix is required to form an intercalation complex. Simple cationic intercalators bind to duplex nucleic acids through a mechanism that involves two basic steps: there is a transient, electrostatic interaction with the anionic charge of the double helix with sliding of the compound along the helix until an intercalation site, opened by thermal motion of the base pairs, is reached.[67] At this point, insertion of the intercalating moiety yields the final complex. The compounds can obviously exit the intercalation site and either dissociate from the nucleic acid or slide to another intercalation site until equilibrium is reached. Exchange among sites continues at equilibrium and can complicate NMR analysis. Threading intercalators with large substituents require larger openings of the duplex with distortion or disruption of base-pair hydrogen bonds in order to form the intercalation complex. Since large distortions of nucleic acid duplexes are much less likely than the creation of a simple intercalation complex, the kinetics of formation and dissociation of threading intercalator complexes are typically significantly slower than those of the formation of complexes with simple intercalators.[67,75–80]

The insertion of a simple aromatic ring system, such as the anthracene of bisantrene (Figure 3), between DNA base pairs to form an intercalation complex requires a separation of adjacent base pairs of 3.4 Å in the thermally induced molecular dynamics of the double helix.[76,77] Addition of a single, simple, cationic side chain at the 9-position of anthracene should not significantly change the complex structure or the mechanism of binding since the chain would be in one of the DNA grooves and the complex would have the long axis of the anthracene ring stacked over the base pairs at the intercalation site for optimum interactions. If side chains are added to both the 9- and 10-positions, on opposite sides of the anthracene ring as in bisantrene, however, one substituent must slide between base pairs during formation of the intercalation complex and this significantly decreases the rates for the binding reaction. The binding affinity could increase or decrease, depending on the interactions of the modified compound with the duplex at equilibrium.

#### 7.12.3.1.1 Nogalamycin

The antibiotic nogalamycin (Figure 3) is a threading intercalator that is very different from bisantrene, and its complex with DNA is characteristic of anthracyclines in general.[81–85] A view of a crystal structure[81] of a complex of two nogalamycins bound at each of the CpG sites in d(CGTACG)$_2$ is shown in Figure 4. Unlike simple intercalators such as ethidium and proflavine, which form complexes with their long axes approximately parallel to the long axes of base pairs at the intercalation site (parallel intercalation), nogalamycin and other anthracyclines, such as

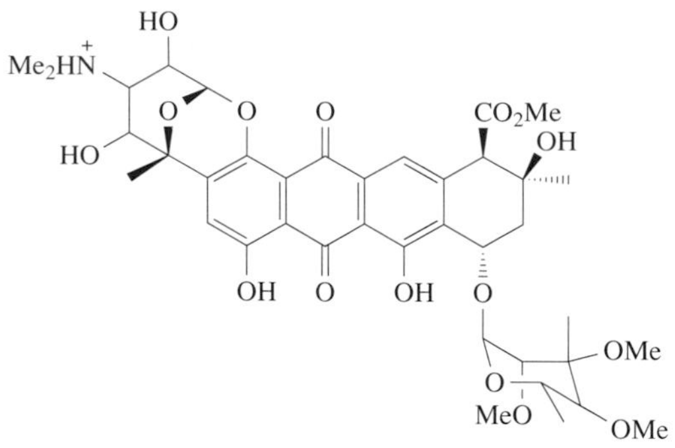

Nogalamycin

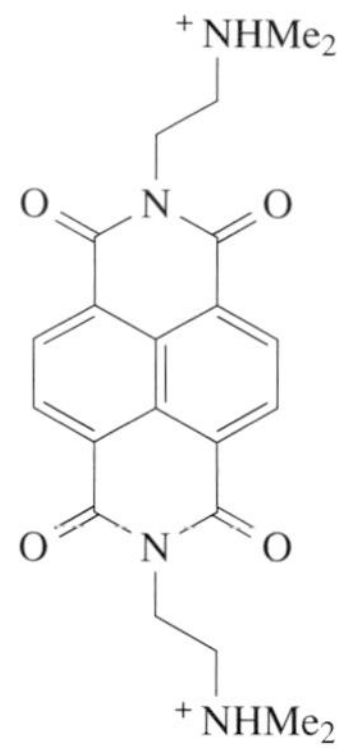

Naphthalene
Diimide

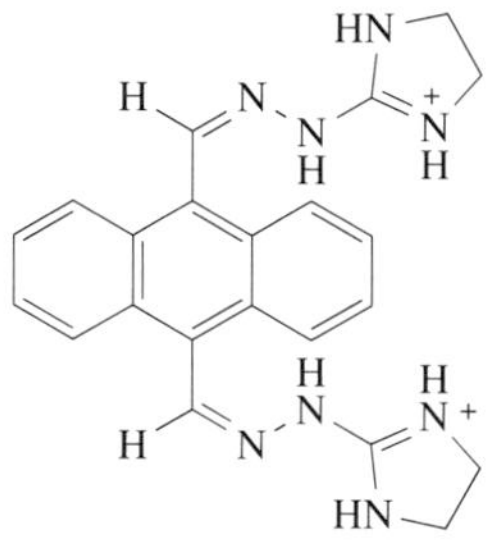

Bisantrene

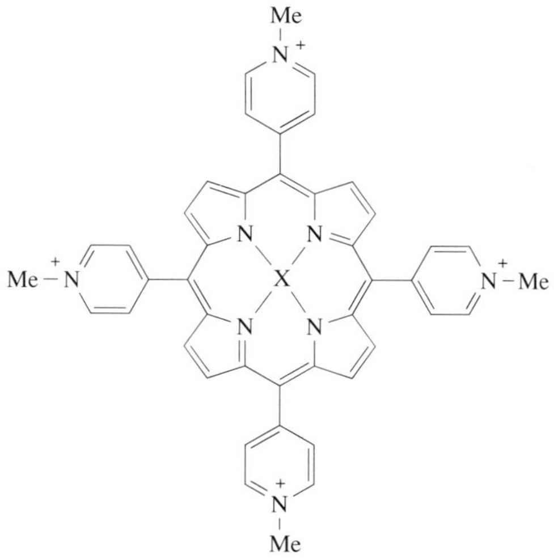

X = 2H, TMPyP(4)
X = Zn$^{II}$, Zn TMPyP (4)
X = Ni$^{II}$, Ni TMPyP (4)
X = Cu$^{II}$, Cu TMPyP (4)

**Figure 3** Structures of some common threading intercalators that have been extensively studied. These compounds have a planar aromatic system that can intercalate and bulky substituents on opposite sides of the ring system. One of the substituents must slide or thread between base pairs in order to form the equilibrium stacked intercalation complex. Bisantrene and the naphthalene diimide are synthetic compounds that have been prepared with a variety of substituents of different size and physical characteristics. As can be seen, the types of compounds that can bind to nucleic acids by threading intercalation are varied. The only requirement is that substituents be fixed at opposite positions on the intercalating ring such that one or more substituents pass through the DNA duplex during the intercalation binding reaction.

daunomycin and adriamycin (Figure 1), bind with their long axes approximately perpendicular to the base-pair long axes (perpendicular intercalation). With nogalamycin this places the cationic amino sugar in the major groove and the uncharged nogalose sugar on the opposite side of the molecule into the minor groove. There is a close fit between the bulky drug and the DNA duplex and the complex is stabilized by numerous van der Waals interactions and hydrogen bonds to the sugars in both grooves in the complex. The base pairs on both sides of the bound nogalamycin are somewhat buckled as a consequence of the orientation of the intercalated ring system. Interestingly, there is little change in DNA winding angle at the binding site, but when the winding angles for all base-pair steps in the complex are determined, an average of 11° unwinding per nogalamycin is obtained, in good agreement with average unwinding results for other anthracycline structures and solution measurements with superhelical DNA.[81]

The intercalation sites for nogalamycin (Figure 4) are topologically similar to those shown in Figure 2 for simple intercalators, but they have a different set of backbone torsional angles and

(a)

(b)

(c)

(d)

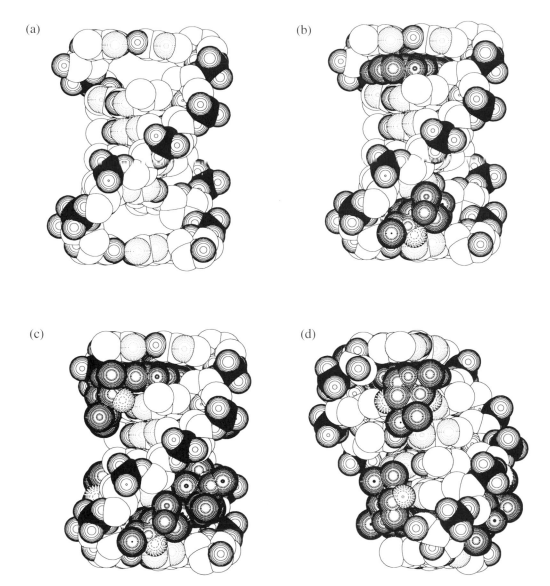

**Figure 4** A space-filling model of the X-ray structure[81] of the nogalamycin complex with d($^{Me}$CGTsA$^{Me}$CG). The DNA contains 5-MeC ($^{Me}$C) and a phosphorothioate linkage(s) at the positions shown. (a) A view of the intercalation site without nogalamycin. The major groove is at the top with the minor groove at the bottom. (b) The same view with the nogalamycin chromophore, and (c) with the complete nogalamycin molecule added. The size of the groups that must pass through the intercalation site can be seen in this view. (d) A view directly into the major groove with the complete nogalamycin molecule. DNA atoms are represented as: carbon – white, nitrogen – dotted concentric circles, oxygen – solid concentric circles, and phosphorous – black. Nogalamycin atoms are shown as concentric circles with radial lines except for the two hydroxyl oxygens in the bicyclo-amino sugar and the carbonyl oxygen of the methyl ester, which have dashed circles. No hydrogens are shown. (Courtesy of M. Egli.)

resulting variations in local base-pair twist, buckle, and other specific conformational parameters relative to the sites in Figure 2. These and other structural studies strongly support the concept that intercalation sites can be created by different sets of duplex torsional angles, and the values of the conformational parameters selected will depend on the local duplex sequence, the compound structure, and the solution conditions so that the complex structure with the global free energy minimum is obtained. Clearly, nucleic acid duplexes have conformational flexibility in accommodating intercalators and a wide variety of aromatic systems can bind by intercalation.

At this point it is worthwhile asking what the essential features of an intercalation site are. It is clear that adjacent base pairs must be separated by ~3.4 Å in order to allow the insertion of an aromatic system. To obtain strong binding, a favorable free energy for complex formation, base-pair hydrogen bonding should not be disrupted, although considerable propeller twist, buckle, and other similar perturbations of base pairs can be accommodated without a great free energy cost. As can be seen from Figure 4, the major and minor grooves are significantly widened at the intercalation site by the separation of base pairs, allowing the insertion of large substituents on the intercalating ring into either the major or minor groove. The design of intercalators with substituents in both the major and minor grooves allows the creation of compounds with much greater recognition ability for nucleic acids and offers an exciting challenge in compound design. Finally, a wide variation in nucleic acid backbone torsional angles can yield acceptable intercalation sites. It should be emphasized, however, that only certain combinations of angles will yield an energetically accessible site, e.g., once five of the angles are set only a limited variation in the final angle is generally possible.

NMR studies with nogalamycin have provided an explanation for the sequence specificity described above. Two-dimensional NMR spectra on the 2:1 complexes formed between noga-lamycin and the oligomer d(GCATGC)[86] and the longer sequence d(AGCATGCT)[87] indicate that intercalation occurs at each of the two CA/TG steps in both oligomers. Although three different orientations of nogalamycin in the 2:1 complex are possible, only a single symmetrical complex is observed. The NMR results support the threading intercalation mode with the bicyclic amino sugar occupying the major groove and the nogalose lying in the minor groove, as observed in the X-ray structure in Figure 4, and indicate that the sugars are directed towards the ends of the duplex. The amino sugar is in close proximity to the floor of the major groove with its two hydroxyl groups positioned to form hydrogen bonds with N-7 and O-6 of G6. Additional hydrogen bonding between the nogalose and the G2 and G6 $NH_2$ groups in the minor groove are predicted from the structure of the complex with d(AGCATGCT).[87] The 2:1 complex formed between nogalamycin and the hexamer d(CGTACG)$_2$ (Figure 4) has been analyzed by NMR spectroscopy and X-ray crystallography.[81,82,88] Both of these studies also show that the nogalose and bicyclic amino sugars are in the minor and major grooves, respectively, but with the nogalose and bicyclic amino sugars directed towards the center not towards the ends of the oligomer. To investigate the molecular basis for these differences, Patel and co-workers[85] conducted NMR studies on duplexes, d($G_1C_2G_3T_4$)·d($A_5C_6G_7C_8$) and d(GCAT)·d(ATGC). The latter oligomer forms complexes similar to those observed with d(AGCATGCT),[87] whereas the d(GCGT) duplex binds nogalamycin in two different complexes in almost equal amounts. In both complexes nogalamycin is intercalated at the CG sequence with the bicyclic and nogalose sugars in the major and minor grooves, respectively, as in previous complexes. In one complex the intercalated ring is stacked on the C6–G7 bases with the sugars directed at the G1·C8 terminus, while in the other complex the ring is stacked with the C2–G3 bases with the sugars pointing to the T4·A5 terminus.[85] All the results are consistent if nogalamycin prefers to bind at the 3′ side of a C and wrap around the GC base pair of the C. Formation of specific hydrogen bonds then establishes the observed specificity and orientation of the drug.

### 7.12.3.1.2 *Synthetic threading intercalators*

Synthetic threading intercalators with substituents of increasing size also provide probes for large dynamic motions of nucleic acid duplexes. Clearly, as the substituents on opposite sides of the intercalator increase in size, larger and larger openings in the double helix would be required to allow insertion of the aromatic ring system.[67–70] After the bulky substituent has passed through the double helix, however, the base pairs on opposite sides of the intercalation site can assume a normal intercalation geometry and can stack on the aromatic system of the intercalator. This threading binding mechanism, which requires significant distortion of the double helix in intermediate states but which produces a fairly standard final intercalation complex, can have energetics of complex formation that are in the range generally expected for intercalation but with rates of both association and dissociation reactions that are markedly reduced. It is thus possible to decouple the qualitative correlation between binding equilibrium constants and the dissociation rate constants frequently observed for simple intercalators.[67]

Several aromatic systems have proven useful for the synthesis of a variety of threading intercalators, such as 9,10-substituted anthracenes and naphthalene diimides (Figure 3). Some of the compounds, such as bisantrene, have shown promising antitumor activity and, as expected, have DNA binding kinetics that are significantly slower than with corresponding monosubstituted derivatives.[76,77] The naphthalene diimide nucleus has proven especially useful for preparation of a series of derivatives with substituents of different size.[67,75,89,90] Kinetics of association of the diimide from Figure 3 are shown in Figure 5 as an example of the characterization of intercalation rates. The dramatic difference between association rate constants for the dicationic threading diimide intercalator and the simple intercalator propidium (Figure 1) is illustrated in Figure 5(c) and emphasizes the effects of threading substituents on intercalator dynamics.

As the size of the diimide substituent increases, the association rate decreases from approximately one-tenth to closer to one-hundredth of that for simple intercalators.[89,90] Interpretation of these results quantitatively in terms of DNA dynamics is complicated by the flexibility of large side chains that have been attached to the diimide and other threading intercalators. Reorientation of the substituent can present barriers of different size to the DNA intercalation site. One might, for example, expect to find a discontinuous decrease in binding kinetics as the substituent size reached the maximum bulk that could slide through a standard intercalation site. Any additional increase in substituent size would require breaking base pairs at the intercalation site to allow threading intercalation to occur. Breaking of the base pairs could be circumvented, however, if a flexible substituent rotated into a conformation that could fit through an intercalation site without breaking base pairs, and this greatly complicates the interpretation of these type of experiments.

The effects of substituent position on threading intercalation and intercalation geometry are illustrated by the anthraquinone derivatives in Figure 6.[78] Kinetic studies indicate that the compounds with single side chains at the 1,8- or 1,4-positions intercalate with both groups in the same groove, whereas the compound with 1,5-substituents binds by threading intercalation. The compounds all have similar equilibrium binding constants in spite of their significantly different dynamics. Designing threading intercalators that are better probes of nucleic acid dynamics and that can provide improved specific recognition through contact in both the major and minor grooves is an exciting challenge for organic chemists conducting research on nucleic acid interactions. The slow off rate of such derivatives (Figure 5) could give them added advantages as enzyme inhibitors in therapeutic use.

### 7.12.3.1.3 *Porphyrins*

The interactions of cationic porphyrins with DNA have been extensively investigated.[91–98] The porphyrin intercalators in Figure 3 have four symmetrically placed pyridyl substituents, and for these compounds to intercalate one or two of the pyridyls would have to slide through the intercalation site. Steric clash prevents these pyridyls from rotating into the porphyrin ring plane and these groups present a significant barrier to intercalation. The compounds can bind both by intercalation, which appears to be limited to GC-rich sequences, and by external interactions. Intercalation at GC sites is limited to porphyrins that do not have a metal ligand or that have a metal without axial ligands that would interfere with stacking in an intercalation complex. The TMPyP(4) compound and its nickel and copper derivatives meet these criteria (Figure 3), but the zinc derivative has axial water ligands in aqueous solution that prevent intercalation. In agreement with the proposed intercalation model for TMPyP(4), the imino proton NMR signals of poly d(G–C)$_2$ in the complex are shifted strongly upfield.[93] The imino proton signals in a normal duplex experience ring-current effects from base pairs below and above the base to which the proton is bonded. In an intercalation complex, the base pair on one side of the imino proton will remain the same, but the base pair on the opposite side will be replaced by an intercalator. If the intercalator has a larger ring current effect at the imino proton than the base pair that was removed, the usual situation, then the proton signal will be shifted upfield. Porphyrins, because of the large size of their aromatic ring systems, have unusually large ring current effects and cause very large shifts of imino protons when they are intercalated.[97] Gorenstein[98] has shown that unwinding of the double-helix backbone by intercalation or other means causes a downfield shift of $^{31}$P NMR signals. The signals for phosphates at the intercalation site typically undergo such downfield shifts as the backbone is unwound and extended to produce an intercalation site. The changes in chemical shifts for imino proton and $^{31}$P NMR signals therefore provide an excellent means for evaluation of the binding mode, and if the signals are assigned in an oligomer duplex, the shifts indicate where the intercalator is bound.[99]

Consistent with intercalation binding, TMPyP(4) complexes with oligomers containing 5′CG3′ intercalation sites have a characteristic $^{31}$P NMR downfield peak; however, complexes with oligomers containing 5′GC3′ or 5′GG3′ sites, but no CG, do not have the downfield signal.[93–95] Addition of TMPyP(4) to oligomers with CpG sites resulted in the appearance of new upfield peaks in the imino proton spectra, again consistent with formation of an intercalation complex, but no such signals were observed in spectra of oligomers without CpG sites.[93] The appearance of new signals at 11.4 ppm and 13.0 ppm and the progressive disappearance of the original G imino proton signals from GC base pairs are easily seen in porphyrin spectra with the oligomer duplex $d(T_1A_2$

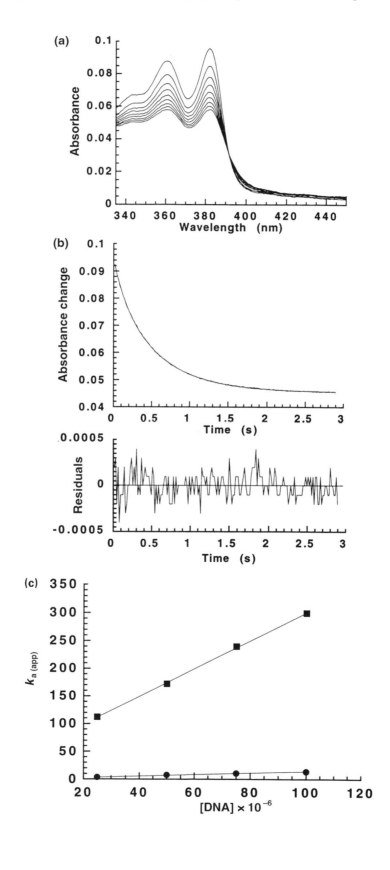

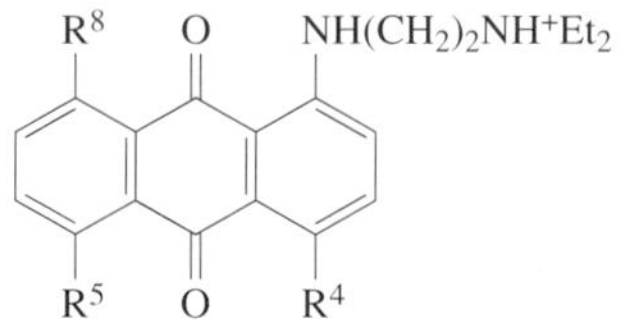

R⁸ O NH(CH₂)₂NH⁺Et₂

R⁵ O R⁴

(**1**) $R^4=R^5=R^8=H$
(**2**) $R^5=R^8=H$; $R^4= -NH(CH_2)_2NH^+Et_2$
(**3**) $R^4=R^8=H$; $R^5= -NH(CH_2)_2NH^+Et_2$
(**4**) $R^4=R^5=H$; $R^8= -NH(CH_2)_2NH^+Et_2$

**Figure 6** Structures of anthraquinone intercalators with the same substituent at different positions on the anthraquinone ring are shown. Compound (**1**) is a control with a single substituent. Compound (**3**) binds by a threading intercalation mode, whereas (**2**) and (**4**) intercalate with both side chains in the same groove.[78]

$T_3A_4T_5G_6C_7GCATATA)_2$. The 11.4 ppm signal was assigned to GC(6) and the 13.0 ppm signal to AT(5). Since [31]P results clearly suggest that the porphyrin is intercalated at the central CG in this oligomer, these large shifts at base pairs removed from the intercalation site at first appear surprising. As noted above, however, porphyrin ring currents[97] are large, and if imino protons lie approximately above the center of the porphyrin ring, as expected for an intercalation complex, the imino protons of GC(7), GC(6), and AT(5) would be located $\sim 3.4$ Å, $\sim 6.8$ Å, and $\sim 10.2$ Å above the porphyrin center. From these distances, upfield shifts of $\sim 4$ ppm, $\sim 1.2$ ppm and $\sim 0.4$ ppm, respectively, would then be predicted for these imino proton signals from the ring current–chemical shift map for tetraphenylporphyrin. The GC(6) and AT(5) shifts of 1.2 ppm and 0.5 ppm, respectively, are in agreement with those predicted. The GC(7) resonance is not observed, but is predicted to occur at $\sim 9$ ppm where TMPyP(4) and base-pair aromatic resonances are found, and signal overlap possibly accounts for the failure to observe that proton.[93] An analysis of observed and predicted shifts in intercalation complexes of TMPyP(4) with other oligomer duplexes is in excellent agreement with intercalation of the porphyrin at CpG sites. As noted above, of the many aromatic systems that have been shown to bind to DNA by intercalation, very few show any significant base-pair site selectivity. Those intercalators which do show selectivity generally have side chains which fit into one of the DNA grooves and hydrogen bond to the edges of the DNA base pairs in the groove. The very high intercalation selectivity of TMPyP(4), which is not capable of forming hydrogen bonds, for 5′CG3′ sites is therefore unique and surprising.

The interactions of Ni[II], Pd[II], and Zn[II] derivatives of TMPyP(4) with some of the same oligomers described above were also investigated by NMR spectroscopy.[94,95] As observed with other biophysical methods, the Ni[II] and Pd[II] porphyrins behave in a very similar manner to TMPyP(4). Addition of ZnTMPyP(4) to the oligomers, however, produces upfield shifts and broadening in [31]P NMR spectra, but no new downfield signals. All biophysical results therefore predict a selective intercalation of planar porphyrin complexes at 5′CG3′ sites and an outside binding mode for these porphyrins at AT sequences. ZnTMPyP(4) binds strongly to AT sequences but binds weakly to GC sequences of any type, and does not bind to DNA by intercalation.

An X-ray structure by Williams and co-workers[96] of a 2:1 complex of the intercalating CuTMPyP(4) (Figure 3) with the duplex d(CGATCG)₂ yielded very surprising results. The two porphyrins are inserted at CG sites with two pyridyls in each groove for each porphyrin, as would

**Figure 5** (a) Absorbance scans for the association reaction of the naphthalene diimide from Figure 3 with poly d(G–C)₂. The scans were collected over a 3 s collection period with the initial scan at the top of the plot. The total concentrations after mixing were $2.5 \times 10^{-5}$ M for polymer (base pairs) and $2.5 \times 10^{-6}$ M for the diimide. (b) A kinetic trace at 382 nm from the data in Figure 5(a) is shown. The plot includes the raw data, a two-exponential fit to the data and a residual plot below. (c) Plots of $k_{a(app)}$ as a function of poly d(G–C)₂ concentrations in base pairs for (●) the naphthalene diimide threading intercalator and (■) the classical intercalator propidium. The slopes of the lines represent the observed second-order association rate constants. Experiments were conducted in 0.1 M MES (2-morpholinoethanesulfonic acid) buffer, pH 6.2, with 0.2 M Na$^{2+}$. Kinetic measurements were conducted on an Olis RSM 1000 stopped-flow spectrometer interfaced with a Gateway 2000 PC computer. The Olis RSM is equipped with a rapid-scanning monochromator which can collect up to 1000 scans per second in each mixing event. (Courtesy of F. Tanious.)

be expected from the discussion above. The copper atom is located near the DNA helix axis and the porphyrin ring system stacks in a fairly standard intercalation geometry on the internal GC base pair. In the terminal position, however, the normal Watson–Crick base pair is disrupted and the cytosine is rotated out of the helical stack. This cytosine interacts with a terminal G from an adjacent duplex in the crystal lattice to form an intermolecular base pair. Hence, each helix has four internal Watson–Crick base pairs and two terminal base pairs that have Gs from the same strand as the internal base pairs and C bases from neighboring helical structures. The porphyrin is intercalated in the standard fashion in the CG site of 5′TCG3′ but the C of 5′CGA3′ is extruded and a different C is inserted into the intercalation complex. The terminal base pair, with bases from two different duplexes, forms Watson–Crick-type hydrogen bonds. Williams and co-workers[96] note that the porphyrin complex is stabilized by very favorable electrostatic interactions, but is destabilized by an unfavorable steric clash of the large porphyrin ring and out-of-plane pyridyl rings with base pairs at the intercalation site. Some of the unfavorable steric interactions can be relieved by extruding the terminal cytosine as seen in this structure. The poor stacking between the porphyrin and base pairs makes this much less costly in energy than would be the case with more classical intercalators. It is not clear how common the intercalated structure with an extruded cytosine is in solution at a nonterminal intercalation site and without an adjacent helix to bind the extruded base. It does seem likely, however, that the steric clash identified in this complex will increase base-pair dynamics at porphyrin intercalation sites and may lead to very favorable binding of porphyrins to some helical structures that have single unpaired bases, base bulges. Porphyrins and other sterically hindered intercalating systems may be designed that are excellent probes for specific distorted conformations in DNA or RNA, and this is discussed in more detail below.

### 7.12.3.2  Bisintercalators and Neighbor Exclusion

Bisintercalators (Figure 7) are another interesting and important variation on simple monointercalating ring systems.[1,7,102] They vary from direct extension of the chemistry of simple intercalators, e.g., 9-aminoacridine derivatives, to complex natural products such as the quinoxaline antibiotics echinomycin and triostin. Perhaps the most important feature of bisintercalators is the dramatic increase in affinity for duplex nucleic acids that occurs owing to the covalent linkage of two monointercalators. In the ideal case the free energy of binding of the bisintercalator would be twice the free energy for binding of its monointercalating units, but steric factors generally prevent both units of a bisintercalator from assuming the optimum geometry of the monointercalators. Many bisintercalators, however, do have exceptionally large equilibrium constants for binding that can be $10^8$–$10^{10}$ $M^{-1}$ or higher with very slow kinetics for complex dissociation.

#### 7.12.3.2.1  Ditercalinium

A crystal structure of the synthetic anticancer drug ditercalinium (Figure 7) complexed with d(CGCG)$_2$ is compared with a structure for the antibiotic triostin complexed with d(GCGT)·d(ACGC) in Figure 8.[103–105] Ditercalinium has a rigid linker that binds in the major groove and forces it to bisintercalate at the two CpG sites and skip the intermediate GpC step in the d(CGCG)$_2$ duplex.[101–103] Triostin forms a similar complex, but the linker is in the minor groove. As illustrated in Figure 9, it has been found in a range of different studies that intercalation at adjacent sites in DNA duplexes is significantly disfavored, probably owing to a combination of steric, electrostatic, and DNA backbone conformational factors, and this effect, as observed with the ditercalinium and triostin complexes, has been termed "neighbor exclusion" in intercalation.[1,7,11,100,106,107] We know from the structures in Figures 5 and 8 that considerable structural

---

**Figure 7**  Structures of a variety of natural and synthetic bisintercalators. A crystal structure of the ditercalinium complex with d(CGCG)$_2$ is shown in Figure 8.[101] The linking chains for all of the compounds, except SN 23349, are long enough to allow the two covalently connected aromatic systems to bind with a vacant site between them, in agreement with the neighbor exclusion principle (Figure 9). Denny and co-workers[100] have connected the acridine rings in SN 23349 by a rigid linker that allows a maximum separation of ~7 Å, which is satisfactory for adjacent bisintercalation, but not for neighbor exclusion binding. The dye TOTO is a very useful stain for the detection of nucleic acids at low levels.[101]

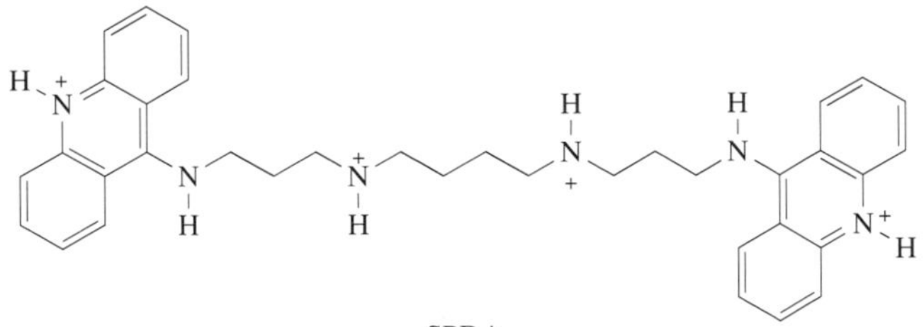

SPDA

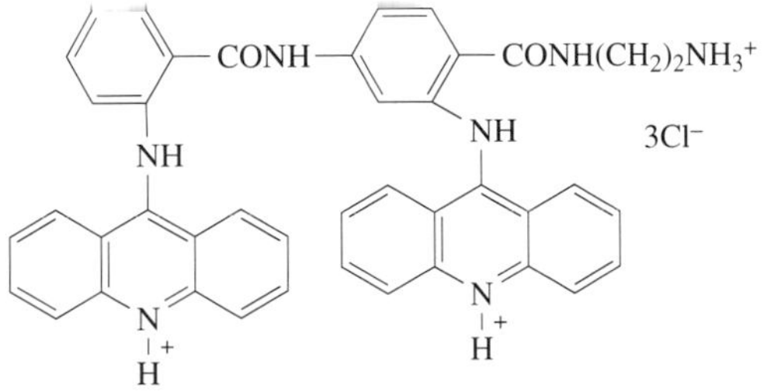

SN 23349

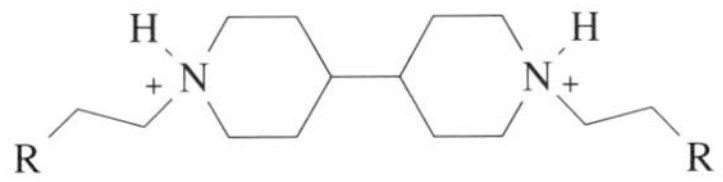

3Cl⁻

Ditercalinium

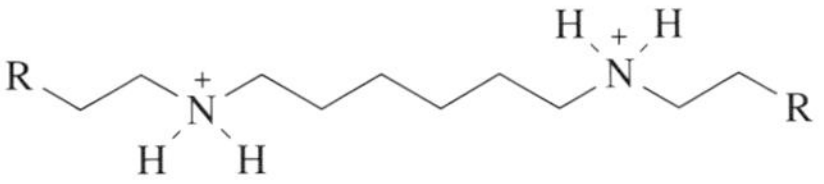

Flexi-Di

R =

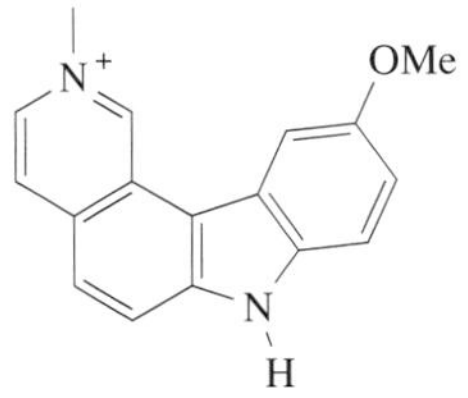

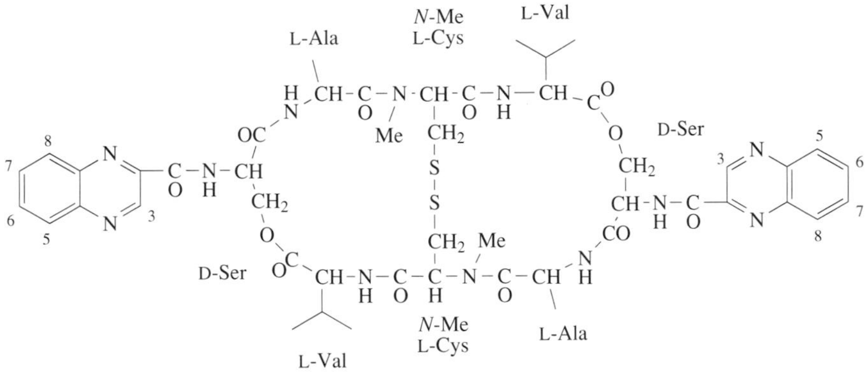

CysMeTANDEM

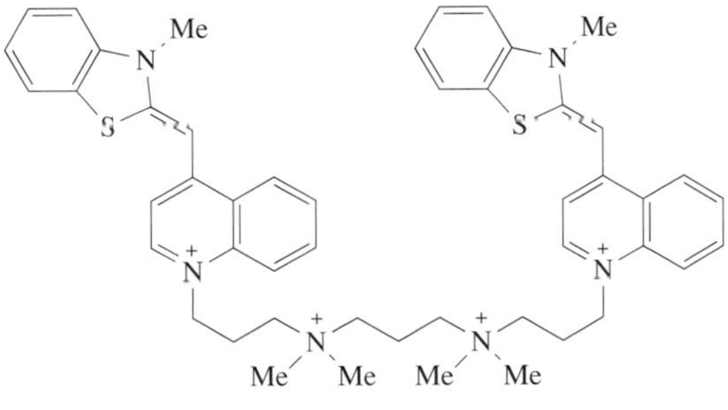

TOTO

perturbation can occur at positions neighboring an intercalation site, and these changes could decrease the free energy for insertion of another intercalator at the adjacent sites (Figure 9). Electrostatic and steric effects can also contribute substantially to the exclusion of intercalation at neighboring sites,[107] and the combined effects lead to anticooperativity in binding.[108,109] As the duplex becomes more crowded with bound intercalators, any single site left between intercalators will also not be able to bind an intercalator (Figure 9), and to make such sites available requires rearrangement of the bound compounds to a less entropically favorable configuration (a less probable configuration).[108,109] This type of site exclusion leads to additional negative cooperativity in binding measurements and the combined effects are observed even for simple intercalators such as ethidium.[106] The structure of ditercalinium places both ring systems in favorable binding sites by skipping the intermediate position. The natural quinoxaline bisintercalators echinomycin and triostin also bisintercalate with neighbor exclusion,[110–114] and ditercalinium with its rigid linker appears to be a minimal mimic of these complex natural products (Figure 8). As noted above, however, ditercalinium intercalates with its linker in the major groove of DNA, whereas the quinoxaline bisintercalators have their linking moiety in the minor groove (Figure 8). As with threading intercalator substituents, insertion of groups into either or both grooves is allowed in bisintercalation.

In the crystal structure of ditercalinium (Figure 8), the duplex is unwound by a total of 36°, the DNA helix axis is kinked by 15°, and the major and minor grooves are significantly wider than in unbound DNA.[103,105] As with the structures of proflavine and ethidium (Figure 2), but unlike the anthracyclines (Figure 4), the long axes of the intercalated ring systems of ditercalinium are oriented roughly parallel to the long axes of base pairs at the intercalation sites. The molecular features that affect the orientation of the intercalated ring causing the observed differences with intercalators of different structure are beginning to be understood. Clearly, all compounds bind to obtain the optimum combination of van der Waals, hydrogen-bonding, and electrostatic interactions. The anthracyclines, however, have bulky substituents on the long axis of the intercalating ring system and to place these substituents in DNA grooves (Figure 5) requires that the intercalator be approximately perpendicular to the base pairs. Ethidium (Figure 2) and ditercalinium (Figure 8) have bulky substituents on their short axes, and placing these in a DNA groove still allows them to bind approximately parallel to the adjacent base pairs and optimize stacking interactions.

An attractive explanation for the 15° kink that occurs in the ditercalinium complex with $d(CGCG)_2$ is that the relatively rigid linking group in ditercalinium is slightly long to allow perfect stacking of the rings of the bisintercalator with the CG intercalation sites and a kink is forced into DNA.[103] To test this idea, Williams and co-workers[115] obtained an X-ray structure of a bisintercalator, Flexi-Di (Figure 7), with a nonrigid, spermine-like linking chain and the same pyridocarbazole intercalating group as with ditercalinium. Surprisingly, the DNA helix in the Flexi-Di complex is bent more than in the ditercalinium structure. Comparison of the two structures suggests that the bending may be caused by pulling the central two base pairs of the complex into the major groove while the external base pairs in the four base-pair intercalation complex are pushed into the minor groove. Hydrogen bonding between the protonated linker amines and the internal guanine bases is the source of the energetics for pulling the internal base pairs into the major groove. Stacking interactions between the pyridocarbazole intercalating groups and the base pairs external to the intercalation site apparently cause these base pairs to move towards the minor groove. The topological consequences of these DNA–intercalator interactions is a bend in the helix axis towards the minor groove. Flexi-Di is able to form more optimized linker–base hydrogen bonds and bends the DNA more than ditercalinium.[115]

---

**Figure 8** Space-filling representations of the crystal structures of the ditercalinium complex with $d(CGCG)_2$ (left) and the triostin complex with $d(GCGT) \cdot d(ACGC)$ (right).[103,105] In both structures the view in (a) is into the major groove and (b) is the same view with the bisintercalators omitted; the view in (c) is into the minor groove and (d) is the same view without the compounds. In DNA the carbon atoms are white, nitrogens are dotted concentric circles, oxygens are lined concentric circles, and phosphorus atoms are dashed. In the bisintercalators solid concentric circles were used for all atoms except for nitrogens which are dashed concentric circles and the sulfur atoms in triostin which are shaded dark (center of c, right). These structures illustrate two dramatically different ways to recognize DNA by bisintercalation with neighbor exclusion (note the unfilled central site). The major groove is filled with the ditercalinium linker (a) while the minor groove is empty (c), but the reverse is observed with the triostin complex. Ditercalinium makes relatively few specific contacts and binds with limited specificity while the triostin peptide linkers make important contacts to the bases in the minor groove to give the observed 5′CG3′ binding specificity of triostin.[100–114] (Courtesy of L. Williams.)

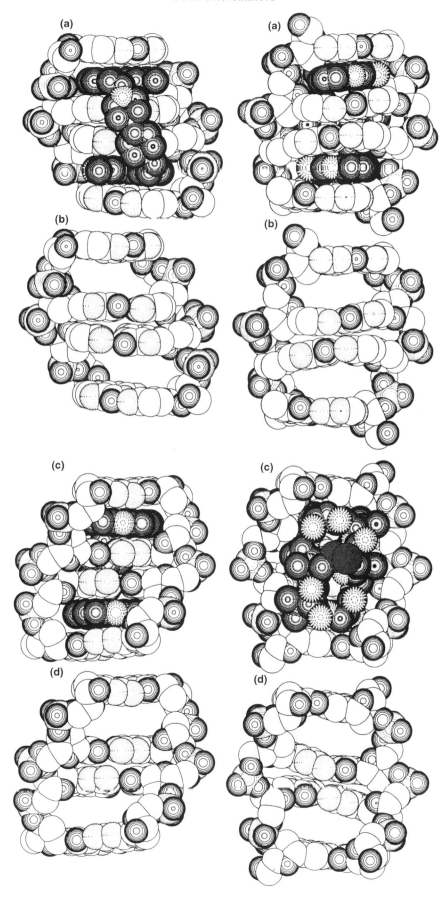

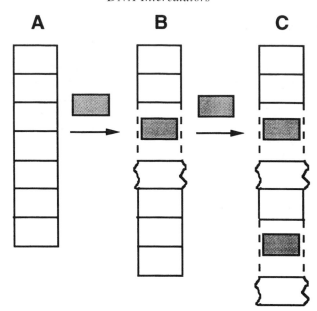

**Figure 9**   Diagrams illustrating the neighbor exclusion principle and its consequences for intercalator–DNA interactions: (a) a free duplex with all sites available for intercalation; (b) the duplex with a single bound intercalator has two sites excluded from binding, i.e., a site with the intercalator bound (dashed lines) and an adjacent site (curved lines) that is blocked by a combination of steric and electrostatic effects; and (c) a duplex that has two bound intercalators but that has five excluded sites. The available site between the two intercalators cannot bind an intercalator (in the configuration shown) since two open sites are required for each bound compound. Rearrangement of the bound compounds is required to make this site available.

### 7.12.3.2.2   *TOTO and YOYO*

The bisintercalator TOTO (Figure 7), which is based on the monomer thiazole orange, and a closely related bisintercalator, YOYO, bind very tightly to nucleic acids and form highly fluorescent complexes that have attracted interest as nonradioactive probes of nucleic acid duplexes.[101] As a result of their strong binding and slow dissociation from nucleic acids, these compounds form complexes that are stable under a variety of gel electrophoresis conditions. The fluorescence enhancement of these compounds on binding to DNA duplexes is over 3000-fold, and this makes them sensitive probes for double-stranded nucleic acids in a variety of experiments. The compounds have an unusual unfused aromatic system with long-range conjugation in a planar conformation, but with the potential for rotation about the bond linking the two aromatic systems (Figure 7).

Analysis of NMR spectra with several oligomers indicated that TOTO binds with significant sequence selectivity at CTAG duplex sites.[116,117] Oligomers without this sequence bind the compound more weakly and chemical exchange among sites causes broadening of NMR signals. Very high-quality NMR spectra have been obtained for TOTO complexed with the self-complementary duplex d(CGCTAGCG)$_2$,[118] which contains the 5′CTAG3′ sequence that is the optimal binding site for TOTO.[116,117] The NMR-derived structure of the TOTO–oligomer complex indicates that the compound bisintercalates at the CT and AG steps of the CTAG sequence with exclusion of binding at the TA central step. The benzothiazole and quinoline rings of TOTO are both stacked between base pairs and provide excellent interactions with the intercalation binding site. In each 5′CT3′·5′AG3′ intercalation site the benzothiazole ring system stacks with the purine bases, whereas the quinoline stacks with the purines. This stacking orientation places the *N*-methyl substituent of the benzothiazole ring in the DNA major groove with the cationic linker in the minor groove. The linker is of the correct length to allow both aromatic systems to intercalate as specified by the neighbor exclusion principle (Figure 9).

The specificity of natural product bisintercalators, such as echinomycin and triostin,[110–114] arises from base interactions with the peptide linkers of the bisintercalator. The synthetic intercalators prepared from acridine rings or ditercalinium have not been shown to exhibit significant sequence-dependent binding. The NMR structural results suggest that the intercalation specificity of TOTO arises from interactions between the methyl groups of thymidine that is stacked on the benzothiazole ring of TOTO, from quinoline–purine stacking, and from the match of the two rings of TOTO to the base-pair propeller twist at the intercalation site.[118] Sites such as 5′CT3′·5′AG3′ provide optimization of all of these interactions with the TOTO aromatic system. Clearly, specificity in binding of both bisintercalators and monointercalators can arise from a variety of concerted interactions.

The TOTO–oligomer complex structure provides a possible explanation for the large fluorescence enhancement on binding of this compound to DNA duplexes.[118] As noted above, the benzothiazole and quinoline ring systems can rotate with respect to each other when the compound is free in solution. When bound to DNA, however, both ring systems are bound tightly between bases at the intercalation site, and rotation of either ring would require significant disruption of the very stable bisintercalation complex. Rotation of the rings about their linking group allows nonradiative decay from the excited state without fluorescence. This rotation is not possible in the DNA complex, and a dramatic fluorescence enhancement is obtained relative to the unbound compound. The sequence specificity of TOTO binding is unique to unfused aromatic systems and offers synthetic chemists interesting new possibilities for the design of monomers and bisintercalators that select specific sequences based on the match of the intercalated system to base-pair propeller twist, stacking, and steric factors at the intercalation site. Unfused aromatic intercalators are discussed in more detail in Section 7.12.4.1.

### 7.12.3.3 Macrocycles: Threading Bisintercalators

An interesting combination of the threading intercalator- and bisintercalator-type systems occurs when two aromatic systems are connected on opposite sides by two linkers to create a macrocycle (Figure 10). Such compounds can only bind to DNA by bisintercalation if extensive base-pair disruption occurs during the binding process to allow one of the linking chains to pass through DNA. If the intercalating moieties bind at adjacent sites, only one base pair has to be disrupted for insertion, but if they bind with neighbor exclusion of sites, then two base pairs would have to be disrupted (Figure 11). If only a single base pair is broken during the binding process, the limiting association rates could be similar to those observed for hydrogen exchange in base pairs and could occur with rates similar to those for other threading intercalators. Zimmerman and co-workers[119,120] have designed and synthesized the macrocycle in Figure 10, and it has been studied along with the related acridine bisintercalator in Figure 7 as a nonmacrocyclic control. If the model in Figure 11 (left) is correct, there should be approximately four base pairs involved in each macrocycle binding site, but spectrophotometric binding results indicate only three base pairs per site consistent with the model in Figure 11 (right). Kinetic studies indicate that dissociation rates of macrocycle complexes are ~10 times slower than those for the related bisintercalator with a single side chain in support of base-pair disruption in macrocycle dissociation.[120] The rate is closer to what would be required for disruption of a single base pair, however, as in Figure 11 (left), and seems too fast to involve two base pairs as in Figure 11 (right). Two-dimensional NMR studies of a macrocycle complex with d(CGCG) are also consistent with the model in Figure 11 (right), but solubility and aggregation problems at NMR concentrations complicated interpretation of the NMR results.[120]

Whichever model in Figure 11 is correct, the inescapable conclusion from studies with the macrocycle in Figure 10 is that bisintercalation, which requires complete base-pair disruption, can occur on accessible timescales. In contrast, a symmetrical acridine macrocycle has been found to bind in the hairpin loop region of a DNA hairpin.[123] It is possible that macrocycles with different aromatic and linker units can be designed to recognize unusual nucleic acid structural features. Complexes of such compounds offer exciting possibilities for the specific recognition of nucleic acid base pairs and groove chemistry in addition to providing important probes of nucleic acid dynamics and distorted DNA structures. The design and preparation of effective macrocycles for such applications are another of the challenges for organic chemists interested in nucleic acid interactions.

### 7.12.3.4 Rigid Bisintercalators

An elegant test of the neighbor exclusion principle was devised by Denny and co-workers[100] with the bisacridine SN 23349 (Figure 7). The compound has a relatively rigid group linking the two acridines that allows them to separate by a maximum of ~7 Å. Neighbor exclusion requires a separation of 10–11 Å and binding at adjacent sites requires a separation of 6–7 Å. The rigid bisacridine must, therefore, either bind with bisintercalation at adjacent sites, e.g., violate the neighbor exclusion principle, or not bind. Physical studies of this compound have been limited

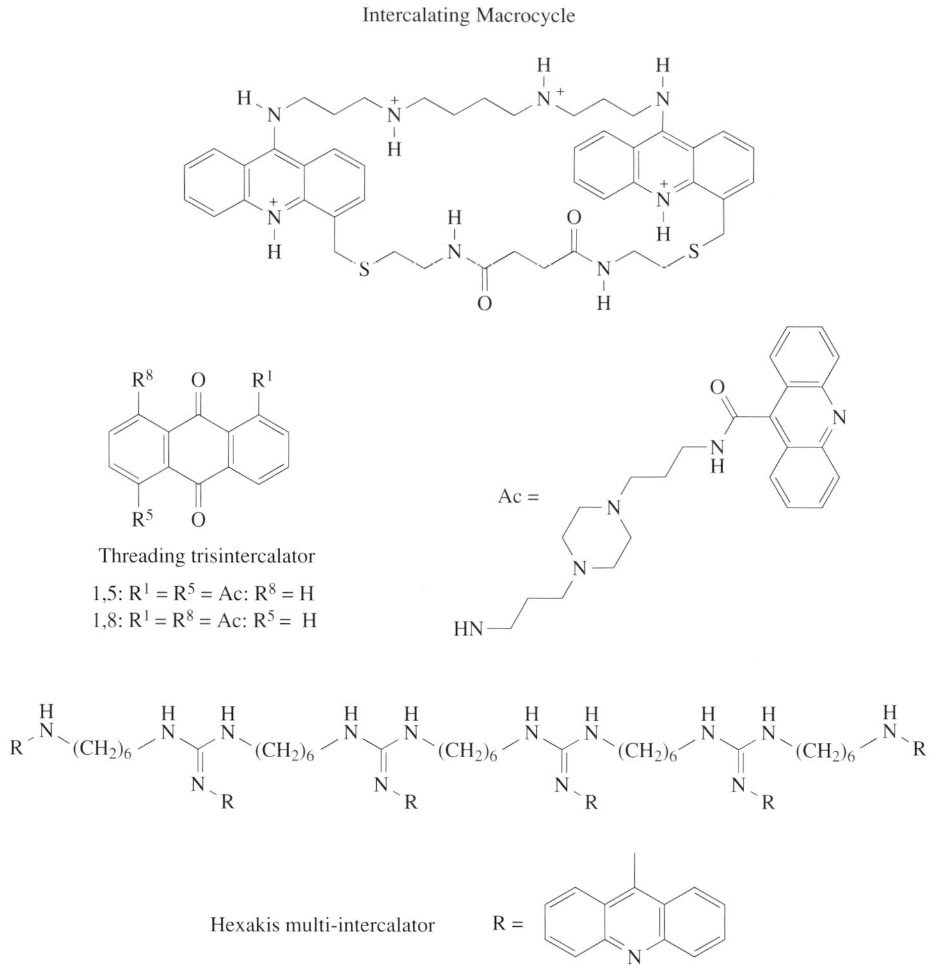

Intercalating Macrocycle

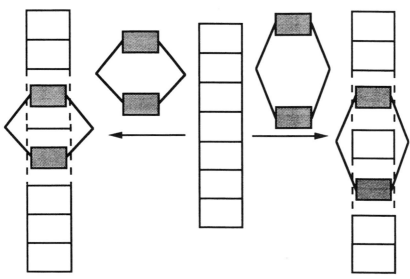

Threading trisintercalator

1,5: $R^1 = R^5 = Ac$: $R^8 = H$
1,8: $R^1 = R^8 = Ac$: $R^5 = H$

Hexakis multi-intercalator   R =

**Figure 10** Structures for a macrocyclic bisintercalator synthesized by Zimmerman and co-workers[119,120] (top), a threading trisintercalator by Takenaka *et al.*[121] (middle), and a hexakisintercalator studied by Norden and co-workers[122] (bottom).

**Figure 11** Diagrams illustrating macrocycle bisintercalation binding modes that require one (left) and two (right) base pairs to be disrupted during the binding reaction. The model assumes sufficient conformational flexibility in the macrocycle such that it could compress (left) or extend (right) to bind by either mechanism.

owing to precipitation of its DNA complexes; however, unwinding experiments indicated that it causes twice the unwinding of DNA as similar monoacridines. It also causes twice the helix extension as similar monoacridines on a molar basis, in agreement with bisintercalation. These results strongly suggest that SN 23349 (Figure 7) binds to DNA with the two acridines at adjacent sites in DNA, in violation of the neighbor exclusion principle.

Chaires *et al.*[124] have designed and synthesized a bisanthracycline based on the structure of complexes of monomeric anthracyclines such as nogalamycin with DNA (Figure 4). The relatively rigid anthracycline monomers were linked with a phenyl system that can place the entire system in the correct orientation to bisintercalate with DNA with optimum interactions. The compound has a very large binding constant ($3 \times 10^{11}$ M$^{-1}$) that results from a very favorable binding enthalpy. An exciting finding from these studies is that the compound is highly active against neoplastic cells, including those which have developed drug resistance against monomeric anthracyclines.[124] This is one of the first studies to use structural and other biophysical results to design a second-generation intercalating drug and demonstrates that a new era in rational drug design is possible. The results for these bisintercalators and the macrocycle in Figure 10 illustrate the power of molecular design and synthetic methods to create compounds that can probe very important aspects of nucleic acid structure and interactions and also induce structural changes in nucleic acids. The activity of the bisanthracycline suggests that bisintercalators are a promising route to the discovery of new anticancer drugs.

### 7.12.3.5 Multi-intercalators

The demonstration of bisintercalation leads to the question of how far the number of covalently linked intercalators can be extended, and how the properties of DNA are modified in multi-intercalator complexes.[99,121,122,125–128] A very thorough analysis of these questions was conducted by Norden and co-workers[122] with a series of synthetic multiacridines linked in a linear array (Figure 10). The reference compound for these studies was the monomer intercalator 9-aminoacridine. Multiacridines with two, three, four, and six acridine units were prepared and their DNA complexes examined with a battery of hydrodynamic, spectroscopic, and affinity methods. The linkers between the acridines (Figure 10) were long enough to allow bisintercalation with an empty intervening site so as to satisfy the neighbor exclusion requirement (Figure 9). The DNA helix extension- and intercalation-induced unwinding of the helix continuously increased on complex formation with mono-, bis-, and trisacridine intercalators but, surprisingly, did not show an additional increase for the multi-intercalators with four and six acridine units. The binding constants increase significantly from the mono- to the bisacridine, but additional affinity increases up to the hexaacridine intercalator are much smaller. The results taken together indicate that both acridines of the bisintercalator and all three of the rings for the trisintercalator are intercalated in the DNA complex. Apparently, however, only three of the acridine units in both the tetrakis- and hexakisacridine compounds can be inserted simultaneously.[122]

Factors that inhibit intercalation in multiaromatic compounds include self-stacking of the ring systems in the unbound compound and local conformational distortions that build up to such an extent that once some (e.g., three) intercalating units are bound, it is very unfavorable to intercalate additional systems in the same region. Binding of the multiacridines is highly cooperative and a binding-induced cooperative DNA conformational change may limit the extent of intercalation at a local level. The difficulty of the synthesis and purification of multi-intercalators and their tendency to undergo extensive self-stacking in aqueous solution has resulted in very limited studies of such compounds.

Takenaka *et al.*[121] have prepared a very interesting series of trisintercalators that are a combination of the multi- and threading intercalator concepts. The compounds (Figure 10) are related to the anthraquinone threading intercalators in Figure 6. As noted above with the anthraquinones, substituents at the 1- and 8-positions are in the same groove, whereas substituents at the 1- and 5-positions are in opposite grooves. Experimental results with the compounds in Figure 10 suggest that trisintercalation occurs with both the 1,8- and 1,5-disubstituted anthraquinones.[121] For the 1,5-disubstituted compounds this requires one of the acridine ring systems to slide through the intercalation binding site of the central anthraquinone, such that in the final structure the acridine above the intercalated anthraquinone is inserted from one groove, while the acridine below the anthraquinone is inserted from the opposite groove.[121] A detailed analysis of these compounds was hampered by low solubility and self-stacking; however, footprinting experiments indicated that the trisintercalators protect approximately 10 base pairs of DNA from enzymatic digestion, as expected if all three ring systems are intercalated.

Threading trisintercalators clamp DNA and offer the potential of very strong binding with sequence recognition for both grooves. Such compounds offer exciting possibilities for organic chemists to vary the affinities, lifetimes, and sequence recognition characteristics in DNA complexes in ways that are very difficult to achieve with other synthetic DNA binding compounds.

## 7.12.4  EXTENSION OF THE TYPES OF AROMATIC RING SYSTEM THAT BIND BY INTERCALATION

We have seen that the array of classical planar aromatic intercalators (Figure 1) has been extended to include compounds with a variety of nonplanar substituents (Figure 3), in addition to bisintercalators (Figure 7) and macrocycles (Figure 10) that have two intercalating ring systems covalently linked. Both of these variations on the intercalation concept maintain the basic planar-fused aromatic intercalating system and, as can be seen from comparison of the structures in Figures 2, 4, and 8, the intercalation sites for all of these compounds have very similar structures. A significant deviation from the classical model occurs with compounds that do not have fused planar aromatic intercalating ring systems, or for partial intercalation of small ring systems such as the aromatic amino acid side chains of proteins. Examples of intercalation of unfused aromatic compounds and of partial intercalation are presented in this section.

### 7.12.4.1  Unfused Aromatic Intercalators

At the same time that the intercalation binding mode was being defined for DNA and RNA, compounds such as netropsin (Figure 12), which have an unfused aromatic system with terminal basic groups, were found to bind strongly in the minor groove of DNA (but not RNA) at AT-rich sequences.[1,4,7,8] These compounds have the torsional flexibility to match the curvature of the groove and the array of N—H donor groups to hydrogen bond with TO2 and AN3 acceptors at the floor of the groove.[8] RNA has the same acceptor groups in the minor groove at all base pairs, but the shape of the groove is significantly wider and shallower than with DNA, and compounds that bind strongly to the DNA minor groove bind much more weakly to the RNA groove.[56] These differences in groove shape are in contrast to intercalation sites in DNA and RNA (Figure 2), which are similar in shape.

Analysis of the structures in Figure 12 raises three important questions for our understanding of the limits of intercalation: (i) do unfused aromatic compounds that do not have the N—H donor hydrogen-bonding groups of netropsin bind to DNA by intercalation instead of groove binding?; (ii) if compounds that have unfused, linked aromatic systems bind to DNAs that are GC rich and do not have sequences of consecutive AT base pairs, do they switch from a groove binding mode to intercalation?; (iii) do such compounds bind to AU or GC sites in RNA by intercalation since the minor groove topology at AU sites in RNA is very different from that at AT sites in DNA? Part of the logic behind these questions arises from our expanded knowledge of the structure of DNA and RNA duplexes. Intercalators stack with base pairs which are themselves unfused aromatic units linked by hydrogen bonds, as illustrated for an A·T/U base pair in Figure 12. Although the Watson–Crick model for DNA has planar base pairs, it is now well established that base pairs can have significant propeller twist that could reduce stacking energetics with planar intercalators.[4,33,129,130] An unfused aromatic system, however, could potentially match the base-pair propeller twist to create a more optimally stacked complex as illustrated in Figure 12. These questions relate directly to DAPI (Figure 12), which binds in the minor groove of DNA in AT sites[131] as with netropsin, but which could switch to an intercalation (or other) binding mode in GC-rich sequences of DNA. The answer to all three of the above questions is "yes," as will be described below, and studies of the unfused aromatic intercalators have led to a nonclassical intercalation model.

An array of unfused aromatic compounds was designed and synthesized by Strekowski and co-workers[132–139] to test the intercalation ability of such compounds and help answer the questions posed above. Three pyrimidine derivatives that are part of this set of compounds are shown in Figure 13. These are other related compounds bind to calf thymus DNA with binding constant ($K$) values of $10^4$–$10^5$ M$^{-1}$. Compound (**2**) has the highest and (**1**) the lowest $K$ of the three pyrimidines

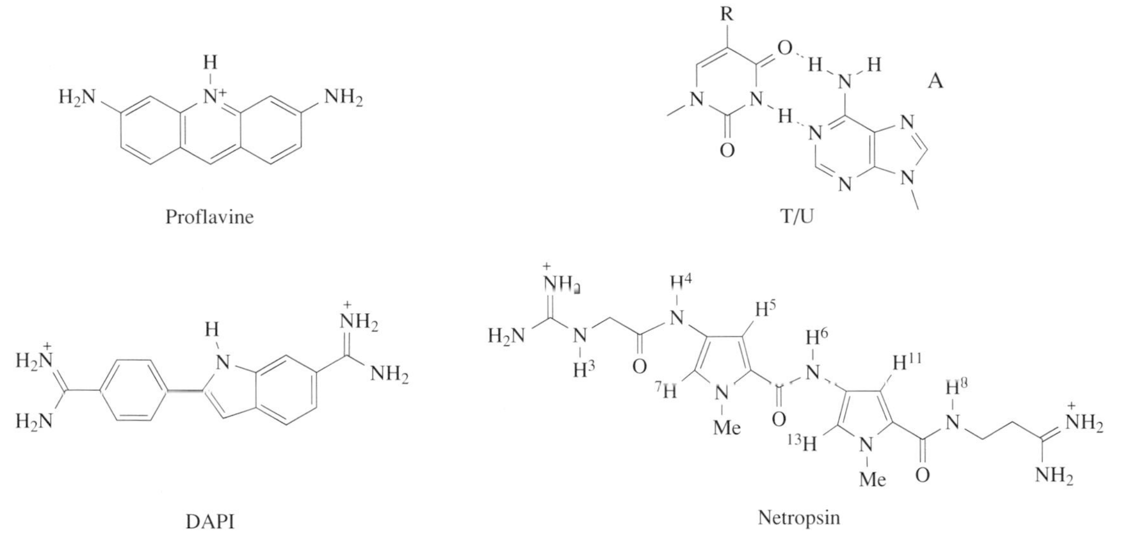

**Figure 12** Structures of the intercalator proflavine and the groove-binding compound netropsin are compared with structures for 4′,6-diamidino-2-phenylindole (DAPI) and an AT (R = Me) or AU (R = H) base pair. The compounds are drawn with the same relative scales so that sizes of the different structures in two dimensions can be directly visualized. Comparison of the planar ring systems of DAPI and proflavine with the base pair indicates that they both are of the appropriate size to form intercalation complexes. Netropsin is too large and its aromatic system has significant twist so that it is unlikely to be able to form a stable intercalation complex. DAPI obviously also shares many structural similarities with netropsin and binds strongly in the minor groove at AT sequences in DNA, as does netropsin.

in Figure 13. The binding of these compounds to DNA induces hypochromism and red shifts in their spectra (illustrated in Figure 14(a) for (**3**)) that are similar to the changes observed for classical intercalators,[7] and these shifts are used to determine the DNA binding constants for the compounds. Additional evidence for the strong binding of these compounds can be seen from the $T_m$ and

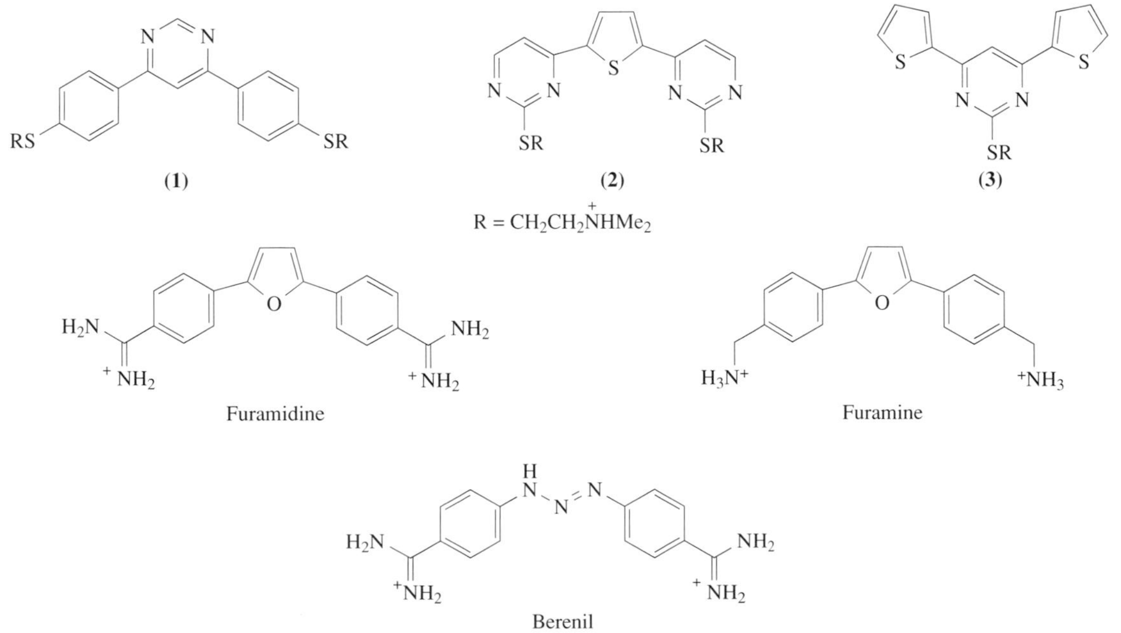

**Figure 13** Structures for a variety of unfused aromatic cations. The pyrimidine derivatives at the top were synthesized by Strekowski and co-workers[132–139] and the diphenylfurans were synthesized by Boykin and co-workers.[140–143] The antitrypanosomal agent berenil, which also has sequence-dependent binding modes,[144,145] is shown at the bottom.

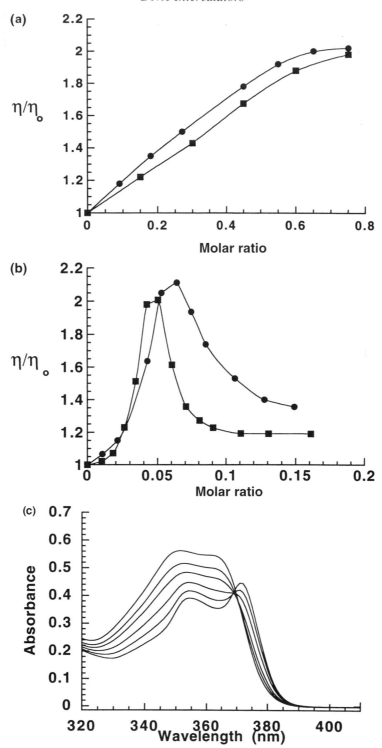

**Figure 14**  (a) Viscosity increases of sonicated calf thymus DNA on formation of complexes of (**3**) from Figure 13 (●) and ethidium bromide (■). The reduced specific viscosity ratio is plotted as a function of molar ratio of compound to DNA base pairs. Experiments were conducted in PIPES buffer at 31 °C at DNA concentrations of $1 \times 10^{-4}$ M base pairs. (b) Viscometric titrations of closed circular supercoiled DNA with (**3**) (●) and ethidium bromide (■). Experiments were conducted in PIPES buffer at 31 °C. (c) Spectral shifts in the visible adsorption region on addition of poly d(A–T)$_2$ to (**3**). The titration was conducted in a 1 cm cuvette at 25 °C in PIPES buffer at a concentration of $3.0 \times 10^{-5}$ M. The polymer concentration ranged from zero (top curve at 350 nm) to $7.6 \times 10^{-5}$ M base pairs (bottom curve at 350 nm). (d) Top: the effect of (**3**) on the melting curve of poly d(A–T)$_2$. The plot shows normalized absorbance vs. temperature. Bottom: the plot shows an approximation of the first derivative of normalized absorbance ($\Delta A/\Delta T$) vs. temperature. Experiments

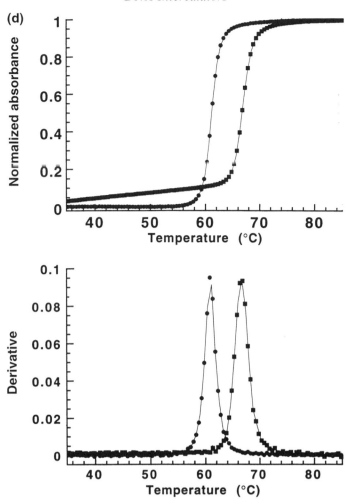

were conducted in PIPES buffer with 0.1 M NaCl. The concentration of the polymer was $5 \times 10^{-5}$ M in base pairs. The molar ratios of drug to polymer base pairs were 0 (●) and 0.6 (■). (e) Proton NMR spectra in the aromatic region of (**2**) from Figure 13 are shown as a function of molar ratio of compound per base pair of added sonicated calf thymus DNA. Experiments were conducted in phosphate buffer at 65 °C. (Courtesy of F. Tanious.)

derivative plots in Figure 14(b). Viscosity and NMR experiments were used to establish the DNA binding mode for the unfused aromatic compounds and the results for (**3**) are shown in Figure 14. The compounds increase the length of linear (sonicated) DNA consistent with an intercalation-induced lengthening of the duplex (Figure 14(c)) and they cause unwinding and rewinding of closed circular DNA (Figure 14(d)), consistent with an intercalation-induced unwinding of the helix at intercalation sites. Particularly strong evidence for intercalation was obtained from NMR experiments with these compounds (Figure 14(e)). Addition of DNA to the compounds caused large upfield shifts in their aromatic but not in their side chain *N*-methyl protons. These results are consistent with stacking of the unfused aromatic systems of the compounds with DNA base pairs in an intercalation complex such that the aromatic ring current effects of the base pairs cause the aromatic proton NMR signals of the compounds to shift upfield. The side chain methyl groups are removed from the strong shielding region of the base pairs and their proton signals experience very small chemical shift changes on binding to DNA. Flow dichroism is negative and similar in magnitude to the dichroism for DNA base pairs for intercalators but can be zero or positive for groove-binding molecules, and dichroism results indicate that the ring systems of (**1**)–(**3**) are stacked parallel to DNA base pairs, as expected for an intercalation complex.

All of the results for pyrimidines (**1**)–(**3**) clearly demonstrate that these compounds bind to DNA by intercalation in both AT and GC sequence regions. Interestingly, for reasons that are not clear, (**2**) binds preferably to GC base pairs, whereas (**3**) binds better to AT sequences (based on com-

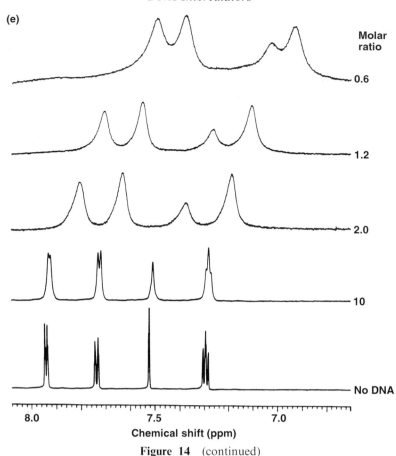

**Figure 14**  (continued)

parative binding studies with poly d(G–C)$_2$ and poly d(A–T)$_2$). These compounds extend the array of possible intercalation structures far beyond the classical fused-ring intercalators of Figure 1 and indicate that the ability to design new intercalators with different properties and biological functions has just begun to be developed. The results indicate that intercalators with significant twist can still bind very strongly to nucleic acid duplexes, and they indicate that the relationship between intercalation and groove-binding modes is much more subtle than previously realized. In summary, the answer to question (i) above is a definite "yes."

### 7.12.4.2  Sequence-dependent Binding Modes

An answer to the second question raised above concerning nonclassical intercalation requires evaluation of the interaction of compounds such as DAPI with DNA sequences that are GC rich and do not have consecutive AT base pairs. As can be seen in Figure 12, DAPI has the torsional freedom and pattern of hydrogen-bonding N—H groups on the inside face of the molecule typically seen with groove-binding molecules. DAPI can, therefore, form hydrogen bonds to thymine O-2 and adenine N-3 atoms in a minor-groove complex, as observed with netropsin and other related molecules.[1,2,6,8] DNAse I footprinting studies indicate that DAPI binds strongly in the DNA minor groove in AT sequences.[146] Dickerson and co-workers[131] have solved the crystal structure of DAPI bound to the self-complementary duplex d(CGCGAATTCGCG)$_2$ and, as with netropsin,[147] DAPI binds in the AATT central region of the oligomer minor groove. It is therefore clear that DAPI forms a selective complex in sequences of three to four AT base pairs which is favorable and is stabilized by the molecular electrostatic potential of the minor groove in AT sequences,[148] by hydrogen bonding, and by van der Waals interactions. As noted above, however, the question with compounds such as DAPI is not whether they bind in the DNA minor groove at AT sequences (it is clear that they do) but how they bind when the sequence has only GC or mixed AT and GC base pairs. Does the binding mode of such compounds change in a sequence-dependent manner?

The answer to this question is also a definite "yes." The most convincing evidence for a dramatic change in binding mode between AT and GC sequences is provided by the summary of NMR proton signal chemical shift changes in Figure 15.[149-151] The aromatic proton NMR signals of DAPI shift by only a small amount and the shifts are a mix of upfield and downfield changes with the poly d(A–T)$_2$ complex, as expected for the formation of a minor-groove complex.[150] On complex formation with poly d(G–C)$_2$, however, all of the aromatic proton signals shift upfield from 0.5 ppm to 0.9 ppm as expected for intercalation.[149-151] The observed variation in shifts is also expected for intercalation since the base-pair ring current effects vary significantly with the position in the complex.[99] Viscosity, kinetics, and dichroism studies also support a minor groove-binding mode for DAPI in AT sequences and intercalation at GC sites.[150] It is clear that compounds such as DAPI bind in a sequence-dependent fashion and a partial explanation for the binding mode changes is apparent from the structural comparisons in Figure 12. DAPI shares conformational features with both the intercalator proflavine and the minor groove-binding agent netropsin. Although it should be able to intercalate at AT base pairs, as proflavine does, it binds in the minor groove, as netropsin does, because the minor groove provides superior interactions and the free energy minimum for complex formation. In GC or mixed AT/GC base-pair regions the interactions in the minor groove are much less favorable, whereas intercalation stacking interactions remain, and the binding mode switches to intercalation. Molecular comparisons (Figure 12) show that netropsin does not have the shape and aromatic surface required for the formation of strong intercalation interactions, and it binds weakly in GC sequences.

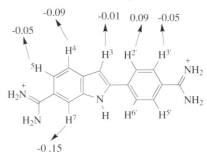

**Figure 15** Proton NMR chemical shift changes that occur when DAPI complexes with AT sites (poly d(A–T)$_2$) or with GC sites (poly d(G–C)$_2$). The small mixed shifts with AT sites are characteristic of groove binding and the significant upfield shifts with GC sites are characteristic of intercalation.[99]

A very interesting and important question at this point is how the DAPI binding affinities compare at the AT and GC sites. DAPI binding to AT and GC DNA polymers was evaluated at several salt concentrations and the binding constant for AT was found to be roughly 1000 times larger than the GC binding constant.[150] This difference is not due to a low GC binding constant for DAPI, since it is in the same range as those for intercalators such as ethidium and quinacrine, but the AT binding constant for DAPI is much higher than the binding constant for most common intercalators. In the formation of an intercalation complex, the free energy that is expended to open the intercalation site and unstack the base pairs must be subtracted from the free energy of complex formation. No such dramatic conformational change and free energy cost is required for binding in the minor groove. Other differences in favor of the AT vs. the GC binding modes may be in the match of DAPI-NH donor groups and hydrogen-bond acceptors on the floor of the minor groove in AT sequences and the extensive contacts of DAPI with the walls of the minor groove. Although the formation of hydrogen bonds involves competition with water, the concerted formation of the hydrogen bonds in the AT complex should provide a favorable term for the binding free energy. Ionic interactions appear similar in the AT and GC complexes and are probably not a major distinguishing factor.

An elegant test of the structural features required for minor-groove interactions vs. intercalation is provided by a series of diphenylfuran derivatives (Figure 13) synthesized by Boykin and co-workers.[140-143] As can be seen by the compound comparisons, the furans have structural features common to the unfused intercalators (**1**)–(**3**) (Figure 13), to DAPI, which binds to DNA in a sequence-dependent manner, and to netropsin (Figure 12), which binds in the minor groove in AT sequences. The diamidine, furamidine, has been crystallized with the DNA duplex d(CGCGAATTCGCG)$_2$ by Neidle and co-workers and is bound in the minor groove of the central

AATT site as with netropsin and DAPI.[142,143] A range of other physical studies also support a minor groove-binding mode for the amidine derivative with AT sequences in DNA.[140,141] Clearly, the shape of the compound matches the DNA minor groove and the hydrogen-bonding amidine groups are placed in order to form good interactions with the minor groove in AT sequences as with netropsin and DAPI. With sequences that do not contain consecutive AT base pairs, however, the compound behaves very much like DAPI and binds by intercalation, another example of a compound with sequence-dependent binding modes.[140,141] In complexes at GC sequences, the amine, furamine, also has all of the characteristics expected of an intercalation binding mode.[140] In particular, its aromatic proton signals shift significantly upfield on titration with GC-rich DNA. The key question then is how this compound interacts with AT sequences. Titration of the furan amine with the AT polymer poly d(A–T)$_2$ results in significant upfield shifts of the compound aromatic proton signals, very different from the results obtained in AT sequences with DAPI (Figure 15) and furamidine, and fully consistent with intercalation of furamine in AT sequences.

Molecular modeling comparisons of furamine and furamidine DNA complexes provide an explanation for the different results in AT sequences, and give some insight into the compound structural features that are responsible for the selection of groove or intercalation binding modes.[140] The planar amidine groups of furamidine are twisted out of the phenyl planes by over 30° in the equilibrium structure of the compound, but can be rotated close to the phenyl plane without a large energy cost. In the X-ray structure of furamidine bound in the minor groove of d(CGCGAATTCGCG)$_2$ of the AATT site, the amidines are at angles of ~10° with respect to the phenyl groups to which they are attached and both amidines form hydrogen bonds to TO2 groups at the floor of the minor groove.[142,143] The amine functions of furamine are rotated closer to 90° with respect to the attached phenyl and the energy cost to rotate them close to the phenyl plane is significantly higher than that with the amidines. This larger twist of the amines in furamine results in a loss of hydrogen bonds to the floor of the minor groove and may decrease some of the normal stacking interactions observed for unfused aromatic systems with the walls of the DNA minor groove. It is possible, however, for the aromatic system of furamine to stack with the base pairs in an intercalation complex at AT sequences and extend the nonplanar groups out into the duplex grooves as observed for the nonplanar groups in other intercalation complexes (Figures 2, 4, and 8). The free energy minimum for the amine derivative therefore switches from a minor-groove complex to an intercalation binding mode in AT sequences. This results in a significant reduction in the binding equilibrium constant for furamine in AT regions, relative to furamidine, much as observed with the differences for furamidine and DAPI binding to AT and GC sequences.[140,150] For the reasons described above, the minor-groove complexes of appropriately structured compounds have significantly better interactions with DNA than can be obtained in intercalation complexes and this results in binding constants that are 100–1000 times greater for minor-groove binding than for intercalation.

Looking back on the pyrimidines (**1**)–(**3**) in Figure 13 from the perspective of the furan results, it is much easier to understand why the pyrimidines are all intercalators as with furamine. Although the pyrimidines have the unfused aromatic systems and terminal cationic groups characteristic of groove-binding compounds such as netropsin and DAPI, they do not have the topological array of hydrogen-bond donors and shape required to select a groove-binding mode over intercalation. Intercalation provides the minimum free energy for complex formation with such compounds and yields binding constants for the unfused aromatic dications that range from ~10$^4$ to ~10$^5$ M$^{-1}$ (at 0.1 M NaCl). Groove-binding constants can be 100–1000 times greater than these values for closely related compounds, but the groove-binding constant undergoes a precipitous drop when the array of hydrogen-bonding groups and interactions with the walls of the minor groove are perturbed. With the pyrimidines (**1**)–(**3**) and the furan, furamine in Figure 13, the groove-binding constant in AT sequences drops below the intercalation value and the binding mode switches from groove binding to intercalation.

Norden and co-workers,[152] based on results from linear dichroism studies, have suggested that furamidine binds in the major groove in GC sequences of DNA rather than by intercalation. The dichroism results indicate that the long axis of the diphenylfuran aromatic system is perpendicular to the DNA helix, as expected for intercalation, but that the short molecular axis is tilted significantly away from a perpendicular orientation with respect to the global DNA helix axis. These results are puzzling given the other evidence in favor of intercalation, and a major groove-binding mode seems particularly unlikely given the shape of the compounds and the dramatic upfield NMR shifts of the furamidine aromatic proton signals on titration with poly d(G–C)$_2$. A possible explanation for all of these results is that the unfused aromatic system of furamidine causes a slight bend or kink in the local DNA helix axis when it intercalates in GC sequences. It is known that bends of this type, if they are not in-phase, leave the global DNA axis essentially unmodified, but at the local level bending can be substantial.

A model that explains all of the results involves intercalation of furamidine, but with a nonclassical intercalation structure that has a substantial bend or kink in the DNA helix axis at the intercalation site. Such bending on formation of intercalation complexes is a well-known phenomenon, and is particularly common for insertion of small aromatic systems between base pairs. Bending of DNA induced by sequences such as $dA_n \cdot dT_n$ tracts has been extensively characterized, and it is clear that DNA can accommodate a large number of substantial, but unphased, local bends without significant changes in physical characteristics such as electrophoretic mobility.[153,154] Dichroism methods report on the global helix axis when they look at DNA base pairs, but report on the local compound orientation, including the *local* helix axis, when the compound wavelength is monitored. The dichroism results[152] are extremely valuable, however, in developing models for how the compounds interact with DNA at nonclassical intercalation sites. Such models must be capable of explaining all spectroscopic and hydrodynamic experimental results, and they must be consistent with the molecular structures of the compounds involved.

To provide additional information to help answer the question of how DAPI binds in GC regions of DNA, Bailly and co-workers[144] conducted a particularly informative set of experiments in which they used dichroism methods in a competition format. The electric dichroism results indicate that classical intercalator proflavine (Figure 1) and a threading acridine intercalator both inhibit the binding of DAPI in GC sequences of DNA. The intercalators have a very small effect on DAPI bound in the minor groove in AT sequences. The authors note that it is unlikely that the intercalators would significantly perturb DAPI complexes in the major groove, but they would directly compete if DAPI binds to GC regions by intercalation, and that is the observed result. DAPI also affects the dichroism results for the two acridine intercalators, and all of the results together strongly support an intercalation binding mode for DAPI and related compounds in GC sequences of DNA. Bailly and co-workers[144] also investigated the diamidine antitrypanosomal drug berenil (Figure 13), which is structurally related to DAPI (Figure 12) and furamidine (Figure 13). As might be expected from its structure, berenil was also able to intercalate in GC regions of DNA. Breslauer and co-workers[145] conducted a detailed calorimetric, spectroscopic, and hydrodynamic investigation of the berenil complex with synthetic DNAs with a variety of sequences and found that, as with DAPI and furamidine, berenil exhibits sequence-dependent binding with a strong minor groove binding mode in AT but with intercalation in GC sequences.

Binding, NMR, and dichroism studies of DAPI oligomer and polymer complexes provide very valuable additional information to define the switch point between groove-binding and intercalation complex formation.[145,155-157] The binding results with oligomers that have a central AATT sequence indicate very tight binding of DAPI. A change in the sequence to replace the first A with 2-aminopurine, P, to give PATT, places the 2-amino group in the minor groove and causes a factor of $>10$ decrease in the affinity of DAPI for the oligomer.[155] For the sequence of APTT, however, the binding constant decreases by a factor of $>100$. These findings are supported by NMR studies of the DAPI 1:1 complex with the oligomer $d(GCGATCGC)_2$.[157] The system is in fast to intermediate exchange and the DAPI affinity is much less than for oligomers with a central AATT sequence. All of these results suggest that DAPI can form minor-groove complexes with sequences that have only two consecutive AT base pairs, but the complexes are much weaker than complexes in sequences with three or more consecutive AT base pairs, and DAPI has a very fast exchange rate in such complexes. The minor-groove complex with the $d(GCGATCGC)_2$ duplex apparently has a slightly more favorable binding free energy than for intercalation, but the affinity is much weaker than observed in longer AT tracts. Footprinting studies of DAPI do not indicate any selective binding to sites with only two consecutive AT base pairs, and this agrees with the observed affinity decreases and increased dynamics of DAPI complexes with two AT base-pair sites. Studies at two or more DAPI per duplex may show a shift to intercalation, since even if the intercalation free energy per bound DAPI were lower than that for groove binding, the total free energy could become favorable for intercalation since intercalation would allow a higher binding density and a more favorable total interaction free energy at higher ratios. It is also possible that DAPI could bind in the minor groove in the center of the oligomer and, at higher ratios, fill intercalation sites in the terminal GC sequences of the duplex.

Additional insight into the switch point between groove binding and intercalation has been provided by spectroscopic analysis of DAPI complexes with oligomers that contain 0–7 AT base pairs with flanking GC sequences.[156] The absorption spectral properties and the DAPI fluorescence quantum yield change dramatically as the number of AT base pairs is increased from two to three. The c.d. spectrum of an oligomer duplex with two AT base pairs is between the spectra for duplexes with none or one AT and duplexes with three, four, and seven AT base pairs. It is clear from these studies that three contiguous AT base pairs are necessary for strong binding of DAPI in the minor groove of AT DNA sequences, in agreement with results described above. Binding of DAPI to DNA oligomer duplexes with none or one AT base pair clearly does not involve minor-groove interactions. Apparently, at two consecutive AT base pairs the minor groove and intercalation binding modes are close in free energy, and the mode that is dominant will depend on the solution conditions and the sequences flanking the AT base pairs. Studies with additional sequences, conditions, and unfused aromatic cation structures will provide additional details for understanding the groove–intercalation binding mode switch point.

### 7.12.4.3  Helix-type Dependent Binding

Two other observations with the furan derivatives and with DAPI also strongly support an intercalation binding mode in sequences that do not contain consecutive AT base pairs. It is now well established that groove-binding compounds are very sensitive to the shape and characteristics of the groove-binding site. As noted above, for example, netropsin binds very strongly in the minor groove in AT sequences, but in GC sequences where the minor groove is wider, the compound binds very weakly (netropsin does not have the continuous aromatic surface and shape to bind to GC sites significantly by intercalation). The grooves in the A-form duplex of RNA are also different in shape from those in DNA.[36] The minor groove in AU base-pair sequences of RNA, for example, has equivalent hydrogen-bond acceptors as on the floor of the minor groove in AT sequences of DNA, but the groove in RNA is much wider and shallower than in DNA. In the same manner, the major groove of RNA is deeper and narrower than the major groove in DNA. Because of these changes, the netropsin binding constant in all sequences of RNA is very low.[56]

The major groove in DNA in either AT or GC sequences also does not provide a good binding site for netropsin and structurally related compounds. The major groove, as with the minor groove in RNA, is too wide to provide good contacts with netropsin and the energetics of complex formation are not favorable.[36] The snug fit provided by the minor groove in AT sequences of DNA is not provided by the minor groove in RNA, even in AU sequences, or in the major groove of either RNA or DNA. Furamidine and DAPI, however, exhibit significant binding to RNA duplexes, much as they do to GC DNA sequences.[158–162] These compounds (Figure 12) both have the contiguous planar aromatic surface and shape that are required to give significant stacking interactions in an intercalation complex. These observations provide the answer to the third question raised above regarding groove vs. intercalation in RNA: unfused aromatic cations such as DAPI and furamidine bind to all sequences in RNA by intercalation, not through a groove complex.

To summarize these observations, ethidium, which is a classical intercalator (Figure 1), binds strongly to both DNA and RNA sites. Given the similarity of intercalation sites in DNA and RNA (Figure 2), this is the expected result. Netropsin, a classical minor-groove binding compound in AT sequences of DNA, binds very weakly to RNA (and to GC sequences in DNA), as expected owing to the very different groove geometry of RNA and DNA. Furamidine and DAPI both bind better than netropsin to GC DNA sequences and to RNA and, based on their structures, which favor intercalation binding away from the AT minor groove in DNA, and the physical results obtained for their DNA GC and RNA complexes, the binding model that best accounts for all of these observations is intercalation.[158–162]

A clear prediction of the intercalation model for unfused aromatic compounds is that the ability of the diphenylfuran aromatic system to intercalate should depend strongly on any modifications of the amidine group. It would seem possible to modify the compound such that the interactions in the DNA minor groove were maintained while RNA intercalation was reduced and visa versa. The results shown in Table 2 provide support for that idea and demonstrate that the RNA binding of the furan derivatives can be significantly decreased while the DNA binding changes by a negligible amount. The strong binding of the compounds in Table 2 to DNA indicates that the three different cationic substituents can fit into the minor groove in AT sequences and allow the N—H donor groups to form hydrogen bonds to groups at the floor of the groove. The intercalation affinity with the RNA duplex varies considerably with the substituent changes, however. The $T_m$ values can be used to calculate approximate binding constants for comparative purposes[163] and these $K$ values range from $0.3 \times 10^5$ M$^{-1}$ for (**5**), the tetrahydropyrimidine, to $3.1 \times 10^5$ M$^{-1}$ for furimidazoline binding to RNA,[161] an order of magnitude difference between the compounds as a result of the change from the five- to the six-membered ring system.

**Table 2** Comparison of DNA and RNA $\Delta T_m$ values for diphenylfuran derivatives.[a]

| Derivative | $\Delta T_m$ | |
| --- | --- | --- |
| | *PolyA·U* | *PolydA·dT* |
| 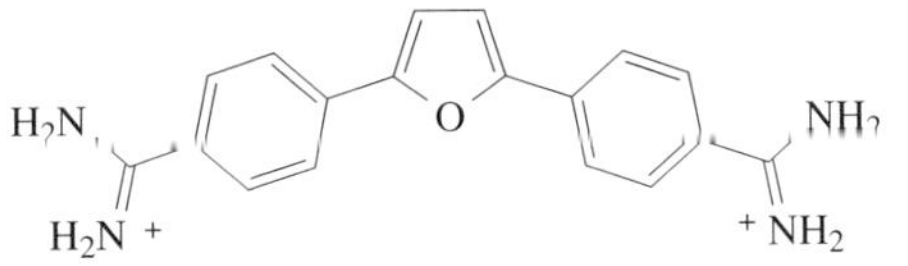 Furamidine | 5.7 | 24.6 |
| 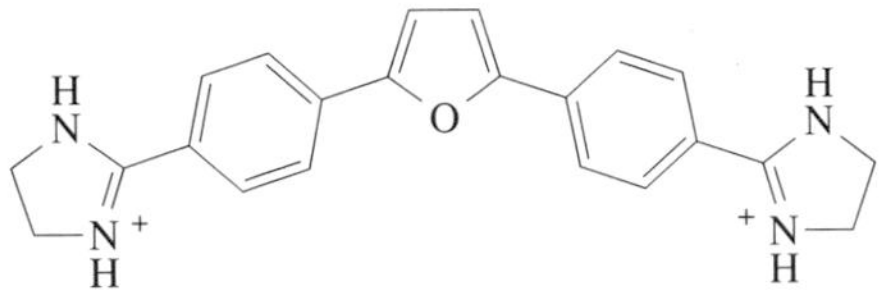 Furimidazoline | 14.4 | 26.1 |
| 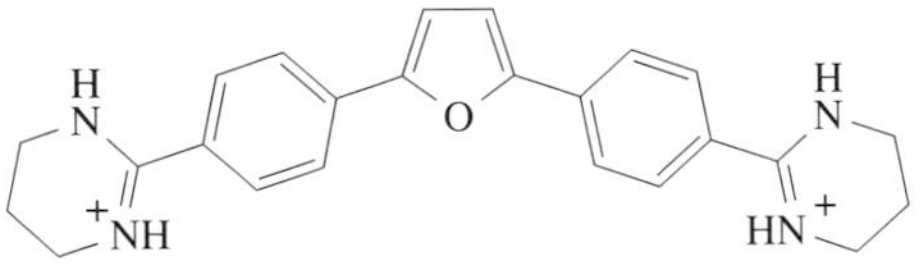 (5) Tetrahydropyrimidine | 2.5 | 26.5 |

[a] $T_m$ values were determined in 0.01 M MES buffer, pH 6.2, with 0.001 M EDTA and 0.1 M NaCl.

Molecular modeling studies provide an explanation for the observed RNA affinity differences for the furan derivatives. The compounds were docked into an intercalation site, constructed in the center of the sequence $A_8 \cdot U_8$ as a model for polyA·polyU, and the complexes were then energy minimized.[160] The two cationic groups of these compounds were placed in the major groove, since an unacceptable steric clash was found with the substituents in the minor groove, and the diphenylfuran aromatic system was stacked into the intercalation site. Analysis of the models of the dications with the intercalation site reveals some clear reasons for the strong binding of fur-imidazoline to RNA. The N—C—N bond angle in the five-membered imidazoline ring is smaller than those in the other three compounds, and this leads to a planar conformation for the imidazoline ring and the phenyl group. The planar imidazoline groups are able to slide more deeply into the intercalation site than the smaller amidine groups which have a 30–35° twist with respect to the phenyl ring. As the cationic group is enlarged to a six-membered ring, the twist with respect to the phenyl ring is again ~30° and the out-of-plane groups of the larger ring cause an additional steric clash with the bases at the intercalation site when the optimum stacking geometry for the diphenylfuran aromatic system is modeled. Hence, the modeling energies and experimental $\Delta T_m$ values both predict the order of stability of the furan–RNA complexes to be furimidazoline > furamidine > (5).[160]

### 7.12.4.4 Self-assembly and Chemically Switched Intercalators

The intercalators discussed up to this point interact with DNA without the need for other compounds to assist in complex formation. A DNA binding site binds a single compound, even in the case of complex compounds such as multi-intercalators. Another class of intercalators could be envisioned that require the interaction of other compounds before the final intercalation complex can form. A molecule X could interact with a preintercalator Y to convert Y into the appropriate structure to bind to DNA by intercalation. X and Y could be identical so that self-association provides the appropriate intercalation structure or they could be different such that the interaction of X with Y converts the entire system to an intercalation conformation. Examples of both of these types of systems are given below.

Synthetic quinobenzoxazines (Figure 16) based on antibacterial quinolones such as norfloxacin (Figure 16) have shown excellent antineoplastic activity, apparently through their ability to inhibit topoisomerase II.[164] The carboxyl group is ionized in these compounds and the most basic amino group is protonated at neutral pH to give a zwitterion in both the quinobenzoxazine and quinolone compounds. The quinolones have excellent antibacterial but little antineoplastic activity.[164] They appear to function by inhibiting the bacterial gyrase enzyme, but they do not form stable complexes with duplex DNA. The quinolones have been proposed to target single-stranded regions of DNA produced in the gyrase mechanism of supercoiling duplex DNA.[164–166]

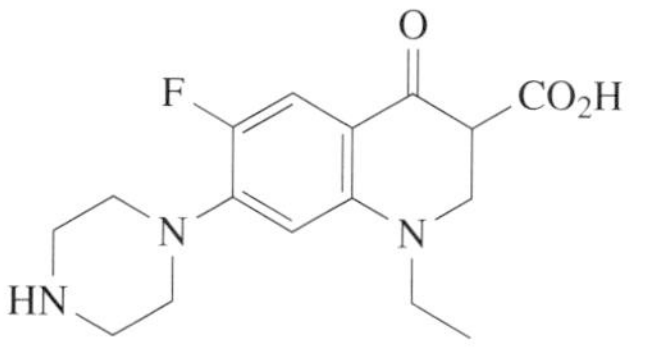

Norfloxacin

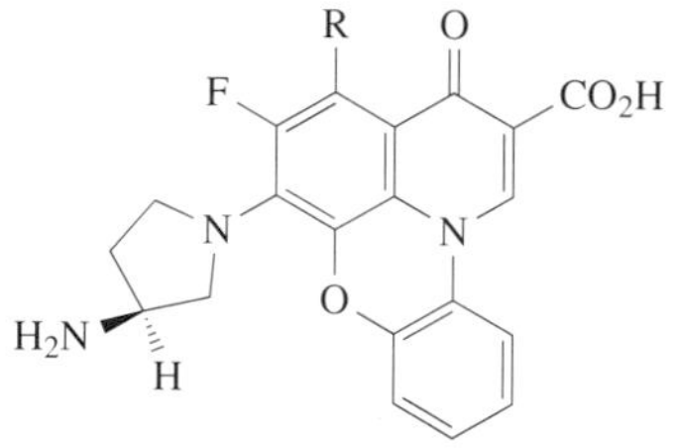

Quinobenzoxazines
A-62176, R = H
A-85226, R = NH$_2$

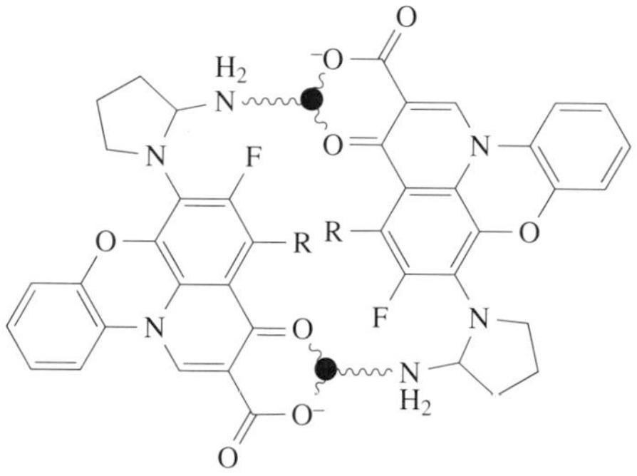

2:2 Mg$^{2+}$-quinobenzoxazine complex

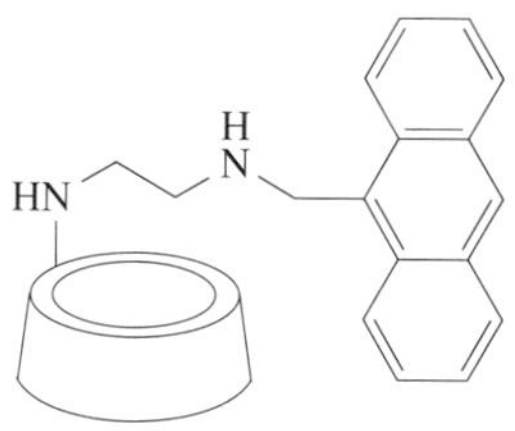

Anthryl(alkylamino)cyclodextrin

**Figure 16**   The quinolone norfloxacin and the quinobenzoxazines A-62176 and A-85226 are shown at the top of the figure. A two-dimensional diagram of the potential interactions in a 2:2 Mg$^{2+}$–quinobenzoxazine complex is shown in the center with the Mg$^{2+}$ ions illustrated as black circles.[165] A chemically switchable intercalator covalently linked to a cyclodextrin[168] is shown at the bottom right.

The quinobenzoxazines (Figure 16) have a larger stacking surface than the quinolones and they look more like the classical intercalators in Figure 1. As the zwitterion, however, they bind only very weakly to duplex DNA. Kerwin and co-workers[165,166] have found, however, that in the presence of Mg$^{2+}$ and some other divalent metal cations, these compounds form strong intercalation complexes with DNA, whereas the quinolones do not form strong intercalation complexes under any conditions. On addition of Mg$^{2+}$ the quinobenzoxazines appear to behave like many of the other intercalators discussed above: they unwind supercoiled DNA, lengthen the DNA helix, and cause upfield shifts of DNA imino protein NMR signals and downfield shifts of DNA $^{31}$P NMR signals. The compounds appear to form an Mg$^{2+}$ complex with the stoichiometry of one magnesium per quinobenzoxazine molecule.

The studies described so far are not particularly unusual and the complexes of quinobenzoxazines would not appear to be different from other intercalators except for several additional observations.[165,166] The stoichiometry of binding, one quinobenzoxazine per base pair, would apparently violate the neighbor exclusion principle (Figure 9). An alternative explanation is that the compounds form a 2:2 dimer with Mg$^{2+}$ and only one of the units of the dimer is bound by intercalation, whereas the other extends out into a DNA groove.[165] The observation of cooperative binding of the quinobenzoxazines to DNA supports this argument. As noted above, the quinolones do not intercalate significantly with duplex DNA; however, they can enhance the intercalation of quinobenzoxazines. Again, this is consistent with a dimer model, in this case a heterodimer, with the quinobenzoxazine intercalated and the quinolone in the groove.

It was found in the experimental work with these compounds that they bind well to GC but not AT sequences. When a quinobenzoxazine intercalates at GC sites, however, it enhances intercalation at neighboring AT sites. Kerwin and co-workers[165,166] used this finding and other observations to propose that the final step in the self-assembly process is a 4:4 quinobenzoxazine–$Mg^{2+}$ complex with two intercalated and two stacked, groove-bound quinobenzoxazines. Stacking of the groove-bound aromatic systems supplies the final free energy of stabilization of the self-assembled bisintercalator. It is the two groove-bound units that can be replaced by nonintercalating compounds such as the quinolones.

The authors note that the self-assembly of the quinobenzoxazines follows the principle of assembly of cytotoxic agents proposed by Rideout.[167] The self-assembled bisintercalator is a new paradigm for nucleic acid interactions and offers exciting possibilities for drug development.[165] They propose that the intercalated units of the complex are for DNA recognition and that the external units are primarily for interaction with topoisomerase II as part of the anticancer action of the compounds. This model implies that the two different units, intercalated and external, have different functions and should be optimized separately to enhance the biological effects. Obviously, a range of self-assembled systems could be created to enhance DNA affinity, binding selectivity, and enzyme inhibition.[165,166] With the quinobenzoxazine paradigm, the only limit is the imagination and molecular design skill of chemists.

A second, very interesting type of system that requires molecular interactions, external to the nucleic acid, to create an intercalator involves addition of an activator to switch a nonintercalator chemically to an intercalating compound. Schneider and co-workers[168] have designed an anthryl(alkylamino)cyclodextrin derivative as an example of this type of system (Figure 16). In aqueous solution the anthracene ring system is inserted into the cyclodextrin cavity and, on addition of DNA, intercalation of the anthracene does not occur. Addition of 1-adamantanol to the anthracene–cyclodextrin, however, causes the anthracene to be displaced. The adamantanol does not bind to DNA, but its addition to the cyclodextrin displaces the anthracene ring system, which then intercalates with DNA, a chemically switched intercalator. The authors note that such systems may find use in drug delivery and in specific nucleic acid reactions.[168] The toxicity of many anticancer intercalators limits their use, and it would seem that the anthracene–cyclodextrin paradigm offers potential ideas for decreasing the nonspecific toxicity. Selective cleavage of macrocycle–intercalator adducts in target cells would be a particularly effective way to deliver the compounds.

## 7.12.5 PARTIAL INTERCALATION OF SMALL AROMATIC SYSTEMS

As was described earlier, intercalation of fused-aromatic compounds is optimized at around four aromatic rings. Compounds with three aromatic rings can still fill the intercalation cavity (Figure 2) fairly well and bind strongly to nucleic acids. Compounds with only two fused rings do not fill the intercalation site and cannot easily balance the energy cost of creating an intercalation site.[169] Such compounds intercalate with low affinity and can give mixed intercalation and external binding. Compounds with single aromatic rings such as phenyl or pyridyl substituents cannot form classical intercalation complexes since the free energy of stacking the small ring system into the site is less than the free energy required to unstack base pairs and open the site. Two questions naturally occur at this point: can small aromatic systems insert between base pairs of nucleic acids through some partial intercalation mechanism and, if so, how does such a partial intercalation complex affect the duplex structure? These questions have particular importance in the context of protein–nucleic acid recognition since the small side chains of aromatic amino acids could partially insert between base pairs in a nucleic acid–protein complex, and such insertion could have profound effects on the overall structure and recognition specificity in the complex (Figure 17).

### 7.12.5.1 Partial Intercalation Model Systems

To answer the above two questions and particularly to begin to understand how important partial intercalation could be to protein–DNA recognition, Gabbay and co-workers[170,171] synthesized a range of model cations with small aromatic systems and studied their interaction with DNA. These were early studies, before the advent of automated nucleic acid synthesis and high-field NMR capabilities, but they provided important insight into how small aromatic systems can interact with

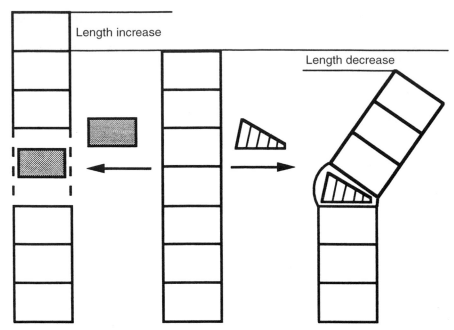

**Figure 17** Diagrams of the effects of classical and partial intercalators on the length of the double helix in complexes with each. The unbound duplex is in the center, the classical complex is on the left, and the partial intercalation complex is on the right. The length increase for the classical complex and the effective length decrease for the partial intercalation complex are illustrated. Proteins that partially insert aromatic amino acid side chains into the duplex when they bind cause bending as illustrated on the right.

base pairs and provided one of the early demonstrations of how the ability to synthesize a specific set of model compounds could answer important questions in nucleic acid chemistry.

Perhaps the most striking early example of partial intercalation was obtained in comparison of the interaction of the simple diastereomeric dipeptide amides L-Lys–L-PheA and L-Lys–D-PheA with DNA.[170] Aromatic proton NMR signals of the phenyl rings were monitored during titration of the peptides with salmon sperm DNA. With the L-Lys–D-PheA peptide, little change in chemical shift and little signal broadening were observed, as expected for a system that does not intercalate. With L-Lys–L-PheA, however, the aromatic signals initially overlap but on addition of DNA the signals for the different aromatic protons shift upfield to different extents so that they are partially resolved, and the signals undergo significant broadening. The difference between the two diastereomers cannot be explained by binding differences since the two peptides have similar DNA binding constants under the conditions of the experiment. A model in which the lysine amino groups provide most of the binding free energy through interactions with the phosphate groups could position the L-PheA group to point into the helix and interact with base pairs, whereas the D-PheA group would point away from the base pairs with the lysine in the same orientation.[170]

The model is supported by a number of additional observations. The variation in upfield shifts for the aromatic proton signals of the L-PheA residue agrees with a partial intercalation model. The *para*- and *meta*-phenyl protons would be inserted more deeply between base pairs and would experience larger upfield chemical shift changes than the *ortho*-protons, which could not insert as far between base pairs. This agrees with the experimental observations. Addition of amino acid residues between L-Lys and L-Phe to create a longer peptide strongly decreases the upfield NMR shifts of the Phe aromatic protons, in agreement with the model in which lysine–phosphate interactions place the Phe ring system in a position to insert between base pairs.[171] The answer to the first question above of whether small ring systems partially intercalate is a definite "yes," and the second question of how the partial insertion affects DNA remains. Addition of dications, including $Mg^{2+}$ and simple aliphatic diamines, to DNA solutions at low salt concentration causes slight reductions in viscosity, caused in part by reduction in DNA stiffness. L-Lys–D-PheA causes decreases in viscosity that are very similar to those of simple amines, in agreement with the NMR results which indicate that the D-PheA residue does not significantly interact with base pairs.[170] With L-Lys–L-PheA, however, much larger viscosity decreases are obtained, in agreement with a model in which partial insertion of the L-PheA residue causes a local kinking or bending of the DNA that results in a net decrease in length of the helix even with opening of the base pairs to allow Phe insertion (Figure 17). Linear dichroism results also show much larger decreases on addition of L-Lys–L-PheA to DNA than for the D-PheA peptide or for simple diamines, again in agreement with a model in which binding of the L-PheA peptide causes DNA bending. All of the results on similar model systems clearly show that small aromatic rings can be placed in the region of base pairs by other functions that provide the free energy for binding, and in such cases the small rings can bind to nucleic acids by partial insertion. Such partial intercalation causes significant perturbations in the local nucleic acid structure and typically causes bending of the helix.

### 7.12.5.2 Partial Intercalation of Protein Side Chains

The biological role of partial intercalation of amino acid side chains in peptides and proteins was not clear until studies into the mechanisms for control of gene expression revealed that in some cases proteins at distant sites on DNA were required to interact as part of transcription control mechanisms.[172] Such interactions obviously require some bending of the DNA helix and it is important to determine what mechanisms are used to control the DNA conformational changes that lead to bending. Nucleic acid double-helical structures are relatively stiff over a period of several hundred base pairs and to bring sites into contact in such a period requires a specific mechanism to bend the helix. The local distortion caused by insertion of the L-PheA in L-Lys–L-PheA,[170] as illustrated in Figure 17, is clearly an attractive mechanism to induce such bending into the DNA helix.

Structural studies of the TATA box binding protein (TBP) with its DNA recognition site provided details for one model of how bending could occur.[173] TBP is required for all three eukaryotic RNA polymerases to initiate transcription. The protein has a curved, eight-stranded antiparallel β-sheet that provides a saddle-like structure for DNA minor-groove recognition. DNA binding to the protein is an induced-fit process in which the DNA is bent in the complex to match the protein molecular saddle. The 5′ end of DNA enters the complex at the underside of the protein saddle. At this point two Phe side chains are inserted into a T–A step in the DNA sequence to initiate the bending and overall conformational transition of the DNA. The DNA is smoothly bent along the underside of the saddle and at the exit point of the DNA from the protein a second kink is induced by partial intercalation of two more Phe residues into the DNA helix. The DNA quickly returns to the B-form helical structure after leaving the protein complex. All base pairs retain Watson–Crick base pairing and numerous contacts in the DNA minor groove provide the free energy for protein binding and DNA bending.[173] Additional transcription control proteins assemble in the TBP–DNA complex to control gene expression at each promoter.

Structural studies on DNA complexes with other proteins involved in eukaryotic transcription control have revealed additional examples of partial insertion of amino acid side chains, including some hydrophobic side chains that are nonaromatic.[172] In all cases substantial bending of DNA occurs as a result of the partial intercalation binding of transcription control proteins and all such cases studied thus far involve insertion of the side chains from the minor groove. The proteins involved have different topology and there does not appear to be a standard protein conformation or set of side chains involved in partial intercalation. Generally parts of several side chains are involved in the partial intercalation and distortion of the DNA sugar–phosphate backbone. As with L-Lys–L-PheA described above, the partial insertion and bending of DNA are generally processes that require some source of free energy to occur. In L-Lys–L-PheA lysine–phosphate interactions provided the driving energetics, whereas in transcription control proteins an array of protein–DNA interactions provide the binding free energy.[172]

The human ETS1 oncoprotein is a helix–turn–helix protein with DNA major groove recognition.[174] The protein also has a helical segment that contacts the minor groove in the same region when the helix–turn–helix motif is in the major groove. The minor groove contacts are primarily hydrophobic and are characterized by intercalation of a tryptophan side chain that binds more similarly to the classical intercalating ring systems in Figure 1 than is typical for protein side chains.[174] The tryptophan side chain inserts between GC base pairs in the DNA recognition sequence and is associated with a significant kink in the DNA. The aromatic amino acid stacks on the base pair on one side of the insertion site, but hydrogen bonding in the base pair on the other side is completely disrupted, although both the G and C bases remain in the helix stack.[174]

The result of partial intercalation of peptide or protein amino acid side chains is generally to introduce a significant local bend in the DNA helix (Figure 17).[170,172] The biological function of the bend generally seems to be to bring together distant protein-binding sites on a DNA sequence for assembly of a complex that, for example, controls gene transcription. The complexes examined have different structures and DNA recognition mechanisms. The array of partial intercalation complexes are at least as varied as those described above for intercalating organic cations.

Protein side chain insertion into a nucleic acid helix can also occur from the major groove. In complexes whose structures have been defined in some detail, the interaction generally occurs in an enzyme complex in which a base has been "flipped out" of the helical stack to allow modification of the base, such as methylation, to occur.[175] Insertion of the amino acid side chain into DNA in this case is primarily to fill the space left by the extruded base and stabilize the open complex. The base remaining in the helix is typically stabilized by hydrogen bonding to the protein amino acid groups. This process is therefore different to that in the usual intercalation complex where the base pairs remain stacked within the helix.

The small peptide and related organic cations synthesized by Gabbay and co-workers[170,171] are the first examples of a systematic set of compounds designed partially to intercalate and distort nucleic acid double helixes. It is clear that much more could be done in this area by combining a partially intercalating group into a more classical binding molecular unit with sufficient interaction free energy to drive the partial intercalation reaction. Both minor and major groove units for partial intercalation could be prepared, and the possibilities are again only limited by imagination and synthetic accessibility. The discovery of cationic cyclophanes[176] that induce base-pair opening in RNA duplexes offers the exciting possibility of combining partial intercalation with base flipping into a macrocycle cavity so that specific chemistry, including mutation repair, could be conducted at desired nucleic acid sequences.[176] In such a complex the flipped-out base would be bound into the macrocycle cavity while the space it left in the double helix would be filled by a specific partial intercalator.

### 7.12.6 INTERCALATION INTO MULTISTRANDED HELIXES AND DISTORTED DUPLEXES

So far, the analysis of intercalation in this chapter has focused on the variety of intercalators that can interact with DNA and RNA duplexes. We shall now shift to analysis of the interaction of intercalators with nucleic acid structures that have more than two strands (triplexes and tetraplexes) or duplexes with distortions such as base bulges. Nucleic acid triplexes and tetraplexes have attracted considerable attention because of their possible applications in biotechnology.[177] Unlike duplexes where the minor groove is a well-defined target for certain small molecules, the grooves of triplexes and tetraplexes have not proven to be particularly strong binding sites for known compounds. Intercalators, on the other hand, have been shown to bind fairly well to these structures and, in some cases, to stabilize them significantly against thermal denaturation. Base bulges in nucleic acid duplexes have also been found to be particularly good sites for intercalator interactions, and results for intercalation into both multistranded helixes and distorted duplexes are presented in this section.

#### 7.12.6.1 Triplex Intercalation

The classical triple-helix structure of nucleic acids with a third strand of pyrimidines forming hydrogen bonds with the purine strand in a purine·pyrimidine duplex sequence (Figure 18, the pyrimidine motif) was discovered shortly after the Watson–Crick model for the double helix,[181,182] and subsequently additional triple-base interactions have been identified.[177] A purine motif with a third strand containing G and T or G and A bases has since been discovered. The interest in triplexes has been heightened by the fact that synthetic oligomers can bind specifically at a sequence in a long duplex DNA molecule through formation of triple-base interactions,[177] and can be used either chemically to cleave DNA[183] or to capture specific sequences through triplex formation.[184,185]

The therapeutic potential of triplex interactions, the antigene strategy, has also generated considerable excitement. An obvious advantage of the formation of a triplex to block transcription of a target gene over devising inhibitors for the products from the gene (antisense strands for mRNA or enzyme inhibitors) is that the single DNA gene generates numerous mRNA and protein copies that must be inhibited to achieve the same effect as transcription inhibition.[177] A recognized limiting feature of antigene therapy is the relatively low stability of many triplexes.[177,186,187] Design of agents, such as intercalators, that bind specifically to the triplex structure and significantly enhance the

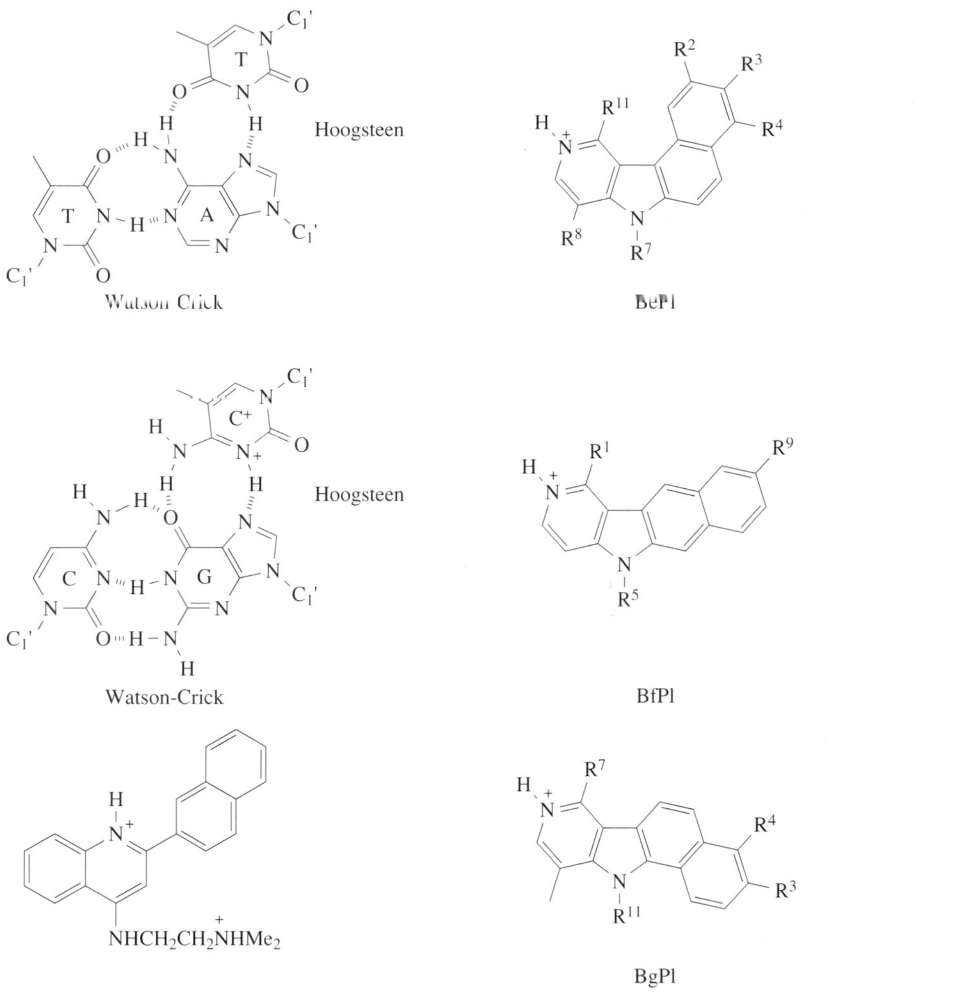

**Figure 18** TAT and CGC base triplets and four triplex-specific ligands. The naphthylquinoline was synthesized by Strekowski, Wilson, and co-workers[178,179] and the indole derivatives were synthesized by Helene and co-workers.[180]

stability of the triplex relative to the corresponding duplex and single-stranded oligomer is an exciting possibility for overcoming these stability problems.

Compared with the extensive database on the interaction of organic cations with duplex nucleic acids, relatively few studies have been conducted on triplex complexes. Groove-binding agents such as netropsin generally destabilize triplex relative to duplex structures.[188–190] For this reason, the drive to find compounds that stabilize triplexes has focused on intercalators. The prototype intercalator ethidium (Figures 1 and 2) stabilizes $T \cdot A \cdot T$-type triplexes,[190,191] but can destabilize pyrimidine-motif triplexes that contain $C^+ \cdot G \cdot C$ triple-base interactions.[177] A study by Wilson *et al.*[190] of the stabilization of pyrimidine triplexes by several classical intercalators indicated significant stabilization of triplexes containing only $T \cdot A \cdot T$ triplets by quinacrine (Figure 1) and coralyne, but relative weak stabilization by anthracycline intercalators (Figure 1). The coralyne stabilization is in agreement with earlier results of Lee *et al.*[192] Addition of $C + \cdot G \cdot C$ triplets to the DNA triplex reduced the stabilization enhancement for all of the intercalators tested. A linear dichroism and c.d. study of 9-aminoacridine and a bisacridine supported an intercalation model for binding of these compounds to triplexes as with duplex DNA.[193] Helene and co-workers[187] have discovered several heterocyclic intercalators that bind with high specificity to some triplex sequences and significantly stabilize the triple helix relative to the corresponding duplex.

Wilson and co-workers[178,179] have developed compounds for the specific recognition of triplex structures, and three essential features were included in the compound design considerations: (i) the compounds should be cations to complement the high negative charge density of the triple helix; (ii) the compounds should have an aromatic surface area which optimally stacks with the three bases in the triplex; and (iii) the compounds should have an unfused aromatic system with torsional flexibility since the three bases in a triplet interaction have to contact points of torsional freedom (propeller twist). A naphthylquinoline triplex intercalator is shown in Figure 18 for comparison with the base triplet structures. This compound gives a large increase in the $T_m$ of triplex DNA ($> 35\,°C$) but a much smaller increase in duplex DNA stabilization ($< 6\,°C$). Decreasing the size of the 2-aryl substituent decreased the triplex stabilization but some larger aromatic substituents at the 2-position of the quinoline gave greater stabilization.[178] Compounds with larger substituents that did not give enhanced triplex stabilization were generally highly twisted so that extended stacking with base triplets would be disrupted in a nucleic acid complex. Replacement of the 4-amino substituent by other groups generally resulted in marked decreases in stabilization of the triplexes. Some bulky groups added to the alkylamine cationic substituent caused a significant increase in triplex stabilization.[194]

The thermal melting results indicating significant stabilization of DNA triplexes have been confirmed by DNAse I footprinting experiments. Cassidy *et al*.[195] used the 2-phenylquinoline in Figure 18 and found that the concentration of the short oligonucleotide sequences $dT_5C_5$ and $dC_5T_5$ required to produce a footprint with their duplex recognition sequences in a long DNA duplex was reduced by at least 100-fold by the triplex-binding compound. Smaller stabilization was obtained for triplexes with the purine motif. Cassidy *et al*.[195] have also investigated the effects of covalently linking an acridine intercalator to a third-strand oligomer on triplex stability. The covalent acridine stabilizes triplex formation, but the stabilizing effect of the naphthylquinoline is significantly less when added with the acridine-linked strand than for the unmodified oligomer, indicating that the naphthylquinoline and the covalent acridine effects are not additive.

Fox *et al*.[196] conducted molecular modeling studies on the binding of anthraquinone derivatives, similar to those in Figure 6, with DNA triplexes. The modeling results predict that placement of the third strand of the triplex in the DNA major groove should significantly inhibit binding of 1,4-substituted anthraquinones to the triple-helical relative to duplex nucleic acid states. With 2,6-disubstituted diamidoanthraquinones, however, there is an extended planar area that could effectively stack with bases in triplex DNA with the anthraquinone side chains in opposite grooves of the triplex. The planar area length is $> 11\,\text{Å}$, which is excessive for duplexes but could be advantageous in the more extended intercalation sites in triplexes. Triplex footprinting studies strongly support the modeling results. The 2,6-substituted compounds enhance triplex footprints at low concentration while even relatively high concentrations of the 1,4-derivatives ($> 30\,\mu m$) do not enhance triplex footprints.[196]

Helene and co-workers[177,180,197–200] have synthesized a series of benzopyridoindole derivatives (Figure 18) for a very thorough and informative structure-binding correlation analysis for pyrimidine motif DNA triplexes differing in fraction of $T \cdot A \cdot T$ and $C^+ \cdot G \cdot C$ base triplets. Derivatives in the BePI and BgPI series (Figure 18) increase the triplex $T_m$ (for a third strand sequence 5′TTCTTCTTTTTTCT3′) by $28\,°C$, whereas no derivative on the BfPI series increases the $T_m$ by as much as $5\,°C$. In the BePI series the best derivative ($\Delta T_m = 28\,°C$ for the triplex) has an OMe at $R^3$, an Me at $R^8$, and an $NH_2$ at $R^{11}$, whereas the worst derivative ($\Delta T_m = 3\,°C$) has an H at $R^3$, an Me at $R^8$, and an alkylamine at $R^{11}$. In contrast, replacing an $NH_2$ ($\Delta T_m = 17\,°C$) at $R^7$ in the BgPI series with an alkylamine ($\Delta T_m = 28\,°C$) dramatically enhances the triplex stability (see Figure 18 for substituent positions).

Helene and co-workers[180] have conducted molecular modeling studies to explain these substituent differences between the BePI and BgPI series. Defining the duplex pyrimidine strand as Watson, the purine strand as Crick, and the third strand as Hoogsteen (Figure 18), modeling results indicate that the alkylamine side chain of BePIs is in the Watson–Hoogsteen groove. The $C^+$ of $C^+ \cdot G \cdot C$ triplets is close to that groove and this proximity explains the unfavorable effects of a cationic substituent on BePIs. The alkylamine substituent of BgPIs, however, is proposed to lie in the more negative Watson–Crick groove, and this explains the favorable effects of cationic substituents in this series and the difference between BePI and BgPI alkylamine derivatives. The poor binding of BfPI derivatives relative to BePI and BgPI compounds results from poor stacking of the more linear BfPI ring system (Figure 18) with base triplets at a triplex intercalation site.

All of the results with triplex intercalators indicate that derivatives which dramatically stabilize $T \cdot A \cdot T$ triplex regions are now available. None of the present compounds, however, are particularly good at stabilizing $C^+ \cdot G \cdot C$ triplets of the pyrimidine motif. Some of the naphthylquinolines and benzopyridoindole compounds stabilize an antiparallel (opposite orientation to the classical pyrimidine motif) purine motif with G and T bases in the third strand that can recognize $G \cdot C$ base pairs. Because of their biotechnology applications, the design of compounds to provide a wider selection of agents to stabilize both the pyrimidine and the purine triplex motifs is certain to continue to be an active area of research in nucleic acid intercalation chemistry.

### 7.12.6.2 Tetraplex Intercalation

The most common DNA tetraplex consists of a stacked system of base tetrads of guanine that are selectively stabilized by $K^+$.[201] Tetraplexes can be made from four single strands, two hairpin strands, or a single folded strand, and this provides a wide array of tetraplexes whose conformation is sequence and condition dependent.[201] The 3′ ends of eukaryotic chromosomes, the telomeres, consist of short sequences of G separated by other bases that are species specific. The G strand is not synthesized in replication, but is added to the chromosome ends by the enzyme telomerase,[202] and the added sequence can fold into a single-stranded G-tetraplex. Telomeres prevent shortening of the coding regions of chromosomes from the 3′ end during replication, and may have several other cellular functions. It has been found that telomerase is present in much higher concentration in cancer cells than in normal cells, and developing inhibitors to this enzyme is an attractive route to the development of new anticancer drugs.[203] One potential way to do this is to develop agents that selectively complex the tetraplex involved in the telomerase mechanism, and this had led to an interest in agents that can intercalate with tetraplexes.

Kallenbach and co-workers[204] have found that ethidium (Figures 1 and 2) causes a large thermal stabilization of a $dT_4G_4$ four-stranded tetraplex. Spectroscopic and footprinting analysis suggest that ethidium binds to the tetraplex by intercalation. An intercalation mechanism is also supported by the observation that ethidium does not bind to fold-back tetraplexes that have short G-stacks, but does bind when the G-stack is lengthened to approximately four tetramers to allow the creation of an intercalation site.

Tetraplexes may also be of importance in nucleic acid associations in processes such as recombination and in folding of some nucleic acid aptamers.[205] The biological importance of telomeres and tetraplexes in general insures that the analysis of their interactions with intercalators and other compounds will continue to be an active area of research.[206] The design of compounds to specifically recognize triplexes and tetraplexes, perhaps by a combination of specific intercalators such as those in Figure 18 with groove-binding units that recognize the unique grooves of multistranded nucleic acids, is one of the challenges for organic chemists in the nucleic acid area.

### 7.12.6.3 Intercalation with Distorted Nucleic Acid Structures

There is obviously a much larger possible variety of distorted nucleic acid structures than uniform duplexes. The nonuniform structures include duplexes with helical junctions, base bulges, loops, and other similar conformations and are much more common in RNA than in DNA.[49–53,207] An interesting feature of many of the distortions in nucleic acids is that they cause enhancement of intercalator binding. A possible reason for the enhancement could be that intercalators extend and unwind nucleic acid duplexes, and any distortion that weakens the duplex, so that an intercalation site can be generated more easily, would enhance intercalation interactions. Relatively few studies have been conducted of intercalation into perturbed duplexes, but the binding of intercalators to duplexes with base bulges has been investigated in several studies.[59,208–213]

Intercalators such as ethidium (Figures 1 and 2) bind more strongly in the region of a bulged base in a nucleic acid duplex.[208–213] Similar distortions such as B–Z helical junctions[214] and branched structures in DNA duplexes[215] also enhance ethidium-binding energetics. White and Draper[59] found that ethidium binding at CG intercalation sites was enhanced when the CG site was near a bulged base in RNA. Ethidium and a zwiterionic derivative have been found to bind better to an RNA duplex model system with a bulged A base from the HIV-1 virus TAR RNA sequence than to a corresponding sequence RNA without the bulged base.[211,212]

Molecular mechanics calculations have been carried out in an attempt to elucidate the molecular basis for the enhanced ethidium binding to RNAs with bulged bases.[212] In an RNA with a bulged A base, the most energetically favorable complex was found to have ethidium bound to an intercalation site at the 3′ side of the bulged A with the bulge base stacked into the duplex. Binding of ethidium at the 5′ side of the bulge or intercalation into the unperturbed duplex gave complexes that were energetically less favorable than the 3′ intercalation structure. Conformational changes in the duplex as a result of 3′ intercalation were found to relieve backbone strain caused by the bulged base. The presence of the bulged base therefore reduces the energetics for the creation of some intercalation sites in the local duplex region and enhances intercalation. Our understanding of the molecular basis of enhanced interactions of intercalators with perturbed duplexes is at a very primitive level and, given the biological importance of many of the perturbations, this is an attractive research area for those interested in nucleic acid interactions.

### 7.12.7    INTERCALATORS THAT REACT WITH DNA

The final class of intercalators to be discussed in this chapter contains compounds that form covalent bonds with bases in nucleic acids, particularly DNA. The DNA duplex is the target for carcinogenic polycyclic aromatic hydrocarbons (PAHs), natural toxins, aromatic amines, and related compounds that are present in a variety of commonly encountered environments and materials. At the opposite end of the spectrum are a variety of antibiotics and synthetic analogues that intercalate with DNA prior to reacting with a neighboring base in the DNA double helix. Many of these latter compounds have excellent anticancer activity and a number are in clinical trials.

#### 7.12.7.1    Covalent Carcinogens

A most unpleasant aspect of intercalation comes from compounds that insert into DNA by intercalation as a step to forming covalent adducts that can lead to mutations and carcinogenesis. Adducts of the chemical carcinogen 2-aminofluorene at the C-8 position of guanine have been studied by two-dimensional NMR methods[216] and the procedures for structure determination for such adducts are described by De Los Santos (Chapter 7.03). The fluorene adduct has two conformations of similar energy, one with the reacted guanine remaining base paired to cytosine while the fluorene is in the major groove, and the other with the fluorene inserted and the adduct–G·C base pair disrupted.

The PAHs are a common by-product of fossil fuel combustion reactions and are the cause of the high incidence of cancer in chimney sweeps.[217] The compounds are metabolically activated to diol epoxide derivatives (Figure 19) that react with DNA after intercalation of the aromatic system. If unrepaired, these adducts can lead to a complex set of mutagenic and carcinogenic effects and the nature of the effects and their severity depend on the isomer involved in the reaction. The primary products of the reaction of the PAHs with DNA are at the purine bases, with the 2-$NH_2$ group of G being the preferred site of reaction of the benzopyrenes. A high-resolution NMR and computational study of the benzo[*a*]pyrene (B[*a*]P) adducts at the G amino group indicated a variety of products.[218–220] With the ($+$)- and ($-$)-*trans-anti* stereoisomers the aromatic group of the G-adduct was positioned in the minor groove, whereas with the ($+$)-*cis-anti* stereoisomers the aromatic moiety was intercalated into the helix with accompanying disruption of the reacted G·C base pair. A similar set of structural studies with benzo[*c*]phenanthrene adducts with the amino group of A indicate that the aromatic hydrocarbon can intercalate without disruption of the A-adduct·T base pair. Conformational changes in the base pair and hydrocarbon allow intercalation to occur with the modified duplex.

The aflatoxins (Figure 19) are natural products from *Aspergillus* that are also potent mutagens and carcinogens.[221–223] These compounds are commonly found as contaminants on stored corn, peanuts, and related food supplies. As with the PAHs, the formation of an aflatoxin epoxide (Figure 19) in a process that requires cytochrome P450 is required for adduct formation between the aflatoxins and DNA. Intercalation is, as with the PAHs, the next step in the mechanism of epoxide–DNA reaction. The intercalation reaction places the epoxide in the major groove near the primary reaction site at the guanine N-7 and adduct formation occurs predominantly at that position (Figure 19). High-resolution NMR and calculation studies indicate that the aflatoxin aromatic group is intercalated into DNA after adduct formation.[221–223] If unrepaired, this lesion can lead to a complex series of mutagenic and carcinogenic cellular events. It is clearly advantageous to minimize the amount of intercalators of this type in addition to PAHs in our environment.

#### 7.12.7.2    Covalent Intercalating Drugs

There are a number of intercalating antibiotics and synthetic compounds that form covalent cross-links to DNA subsequent to intercalation complex formation. Some of these compounds are excellent candidates as anticancer drugs and provide an interesting contrast to the intercalating carcinogens described above. Linking an intercalator with a mustard creates a relatively simple synthetic compound that can bind covalently to DNA.[217] Quinacrine mustard is such a compound and the intercalator enhances the reactivity of the mustard with DNA by significantly increasing the mustard concentration in the local environment of the double helix. Mustards have a low

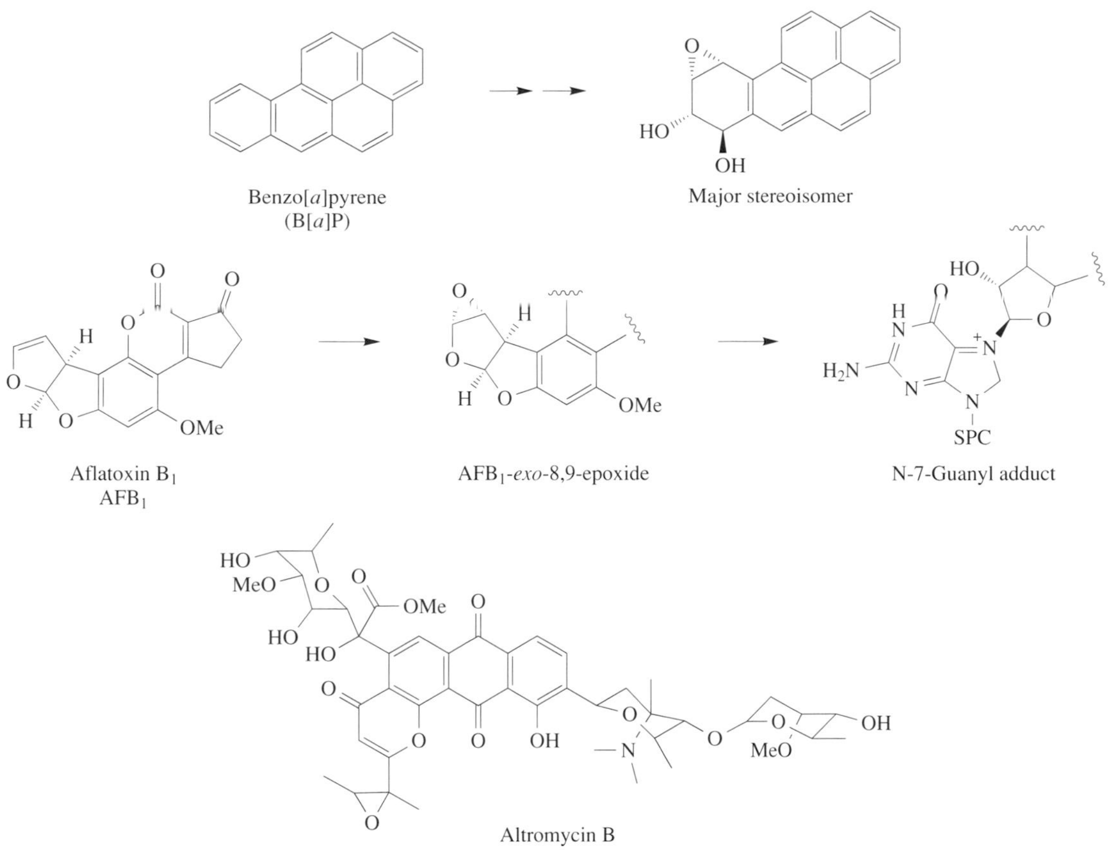

**Figure 19** Benzo[a]pyrene polycyclic aromatic hydrocarbons and a P450-derived epoxide are shown at the top. Aflatoxin B$_1$, its epoxide, and its N-7 guanyl adduct are shown in the center. The pluramycin altromycin B is shown at the bottom. The three compounds bind and alkylate DNA in very different ways, but all do so through an epoxide group, as can be seen from their structures.

selectivity in their reaction spectrum and the intercalator therefore enhances the mustard selectivity for DNA.

Hurley and co-workers[2,224–226] have conducted thorough and very broad studies of a number of antibiotics and synthetic analogues that intercalate and react covalently with DNA. They have conducted detailed NMR structural studies on altromycin B (Figure 19), a member of the pluramycin family of antitumor antibiotics, in complexes with DNA oligomers. Altromycin B has an epoxide group that leads to covalent reaction with DNA, whereas the related compound kidamycin, which does not have an epoxide, intercalates but does not form a covalent complex. Results from a number of studies indicate that alkylation occurs through electrophilic attack of the epoxide on N-7 of guanine. Treatment of the adduct with piperidine results in DNA strand breakage at the alkylation site. Two-dimensional NMR studies indicate that altromycin B intercalates by a threading mechanism (perpendicular intercalation) and positions the epoxide in the major groove in position to alkylate N-7 of guanine.[225] The drug remains intercalated at the 5' side of the modified guanine after reaction and has its disaccharide unit in the minor groove with the monosaccharide substituent in the major groove. The compounds do not appear to have significant sequence specificity in their noncovalent intercalation interactions, and intercalation appears to function to steer the relatively unreactive epoxide into appropriate orientation and proximity to N-7 of guanine.

The position and type of sugar substituents are different for different members of the pluramycin family of antibiotics and their rates of covalent reaction with N-7 of guanine are sequence dependent. The structure of the altromycin B adduct assisted Hurley and co-workers[226] in analysis of the saccharide–DNA groove interactions and in developing an explanation of the molecular basis for the sequence-dependent alkylation rates of the threaded complex. Altromycin B, for example, preferentially alkylates G in 5'AG3' sequences, and this is explained by minor groove interactions of the drug amino sugar with the T in the AT base pair at the 5' side of the modified G. The NH of the sugar amino group is in an optimum position to form a hydrogen bond with the T carbonyl oxygen in the minor groove, while the epoxide is placed in position to alkylate N-7 of G in the major groove. Changes in the sugar substituents affect these interactions and change the reaction selectivity of altromycin B analogues. The structural model also predicts observed differences in reactivity as the length and nature of the epoxide-containing substituent are changed in the pluromycin derivatives.[226] The excellent anticancer activity of these covalent intercalators and the Hurley recognition rules suggest that the design of covalent intercalators that recognize specific DNA sites for alkylation is a promising area for synthetic chemists.

### 7.12.8  OVERVIEW

The first aim of this chapter was to present a thorough view of the state of our understanding of the interactions of intercalators with nucleic acids. The classical model, best illustrated in Figure 2, has been extended to include threading intercalators (Figure 3), macrocyclic and multi-intercalators (Figure 10), intercalators with unfused aromatic ring systems (Figure 13), partial intercalators that distort nucleic acids (Figure 17), and intercalators that react covalently with DNA (Figure 19). A few studies have shown that intercalators can interact strongly with DNA triplexes (Figure 18) and tetraplexes. The difference between the intercalation and groove-binding modes is subtle for some compounds (Figure 12) which bind in the DNA minor groove in AT sequences, but intercalate in GC-rich sequences in DNA and in all sequences in RNA.

The second aim was to illustrate areas in intercalation research where new compounds could significantly enhance our understanding of nucleic acid interactions. Areas such as partial intercalation and binding of intercalators to perturbed duplexes, which are so common with RNA structures, are examples of areas where new intercalators could prove valuable as model systems to understand DNA–protein interactions or as useful new drugs against RNA viruses. Self-assembled and chemically switched intercalators are additional areas where the design of new compounds could yield intercalators with valuable biological properties.

Intercalators have shown promising anticancer activity beginning with the discovery of the actinomycins in the 1940s. The anthracycline intercalators are among the leading currently available anticancer drugs and the design of new bisintercalators based on the structures of their nucleic acid complexes offers exciting new ideas for rational drug design. Some covalent intercalators also have very promising activity against neoplastic cells, and the design of new agents to react specifically with DNA sequences should yield intercalators with improved biological functions. A strong base of understanding, starting with the classical intercalation model of Lerman in the 1960s, can be used as a foundation and guide for future research on nucleic acid interactions in general and on intercalators in particular.

### ACKNOWLEDGMENTS

The research from my laboratory that is described in this chapter was funded by the NIH and much of the equipment used was funded by the Georgia Research Alliance. Several individuals were particularly helpful in the writing of this chapter, and Carol Wilson, Farial Tanious, and Maria Fernandez deserve special thanks for their help and suggestions. My students and postdoctoral associates over the years have carried out the research from my laboratory that is included in this chapter. They have been a uniformly hardworking, bright, and enthusiastic group that have made our laboratory an enjoyable place to study nucleic acids. Our research has benefited greatly from collaborations with several organic chemists, starting with David Boykin and continuing with Lucjan Strekowski, Dabney Dixon, Rick Tidwell, and Jörg Schneider. The compounds from these groups and the research interactions have made enormous differences in our ability to address specific questions about nucleic acid interactions. Figures 5 and 14 were prepared and generously provided by Farial Tanious. Figure 4 was kindly provided by Martin Egli. Figure 8 and numerous helpful suggestions were provided by Loren Williams. I thank Brad Chaires for prepublication information on bisanthracycline anticancer drugs.

### 7.12.9  REFERENCES

1. W. D. Wilson, in "Nucleic Acids in Chemistry and Biology," eds. G. M. Blackburn and M. J. Gait, Oxford University Press, New York, 1996, p. 329.

2. J. A. Mountzouris and L. H. Hurley, in "Bioorganic Chemistry Nucleic Acid," ed. S. M. Hecht, Oxford University Press, New York, 1996, p. 288.
3. T. R. Krugh, *Curr. Opin. Struct. Biol.*, 1994, **4**, 351.
4. S. Neidle, in "DNA Structure and Recognition," eds. D. Rickwood and D. Male, Oxford University Press, New York, 1994.
5. A. Wang, *Curr. Opin. Struct. Biol.*, 1992, **2**, 361.
6. L. Hurley (ed.), "Advances in DNA Sequence Specific Agents," JAI, Greenwich, CT, 1992, vol. 1.
7. M. J. Waring, in "The Molecular Basis of Antibiotic Action," eds. E. F. Gale, E. Cundiffe, P. E. Reynolds, M. H. Richmond, and M. J. Waring, 2nd edn., Wiley, New York, 1981, p. 287.
8. B. H. Geierstanger and D. E. Wemmer, *Annu. Rev. Biophys. Biomol. Struct.*, 1995, **24**, 463.
9. S. Neidle, *CRC Crit. Rev. Biochem.*, 1984, **17**, 73.
10. H. M. Berman and P. R. Young, *Annu. Rev. Biophys. Bioeng.*, 1981, **10**, 87.
11. W. D. Wilson and R. L. Jones, in "Advances in Pharmacology and Chemotherapy," Academic Press, New York, 1981, vol. 18, p. 177.
12. W. Muller and D. M. Crothers, *Eur. J. Biochem.*, 1975, **54**, 267.
13. C. Bailly, C. Michaux, P. Colson, C. Houssier, J.-S. Sun, T. Garestier, C. Helene, J.-P. Henichart, C. Rivalle, E. Bisagni, and M. J. Waring, *Biochemistry*, 1994, **33**, 15 348.
14. A. W. McConnaughie and T. C. Jenkins, *J. Med. Chem.*, 1995, **38**, 3488.
15. D. A. Collier, J.-L. Mergny, N. T. Thuong, and C. Helene, *Nucleic Acids Res.*, 1991, **19**, 4219.
16. M. L. Zapp, S. Stern, and M. R. Green, *Cell*, 1993, **74**, 969.
17. A. W. McConnaughie, J. Spychala, M. Zhao, D. Boykin, and W. D. Wilson, *J. Med. Chem.*, 1994, **37**, 1063.
18. S. Jain, G. Zon, and M. Sundaralingam, *Biochemistry*, 1989, **28**, 2360.
19. M. T. Record, C. F. Anderson, and T. M. Lohman, *Q. Rev. Biophys.*, 1978, **11**, 103.
20. G. S. Manning, *Q. Rev. Biophys.*, 1978, **11**, 179.
21. K. A. Sharp and B. Honig, *Curr. Opin. Struct. Biol.*, 1995, **5**, 323.
22. H. S. Shieh, H. M. Berman, M. Dabrow, and S. Neidle, *Nucleic Acids Res.*, 1980, **8**, 85.
23. S. Neidle, in "Landolt–Bornstein, Nucleic Acids," ed. W. Saenger, Springer, Berlin, 1989, vol. VII/1b, p. 247.
24. H. M. Berman, W. K. Olson, D. L. Beveridge, J. Westbrook, A. Gelbin, T. Demeny, S.-H. Hsieh, A. R. Srinivasan, and B. Schneider, *Biophys. J.*, 1992, **63**, 751.
25. A. T. Sumner, *Methods Mol. Biol.*, 1994, **29**, 83.
26. D. G. Harnden, *Bioassays*, 1996, **18**, 163.
27. J. Sambrook, E. F. Fritsch, and T. Maniatis, in "Molecular Cloning: A Laboratory Manual," ed. C. Nolan, Cold Spring Harbor Laboratory Press, Plainview, NY, 1989, 333.
28. S. Freudenberg, K. I. Fasold, S. R. Muller, D. Siedenberg, G. Kretzmer, K. Schugerl, and M. Gluseppin, *J. Biotechnol.*, 1996, **46**, 256.
29. "BioProbes 23," Molecular Probes, Eugene, OR, 1996.
30. M. Wainwright, "Miracle Cure: The Story of Penicillin and Golden Age of Antibiotics," Blackwell, Oxford, 1991.
31. E. O. Wilson, "Naturalist," Island Press, Washington, DC, 1994.
32. F. Arcamone, "Doxorubicin," Academic Press, New York, 1981.
33. J. W. Lown (ed.) "Anthracycline and Anthracenedione Based Anticancer Agents," Elsevier, Amsterdam, 1988.
34. J. B. Chaires, in "Anthracycline Antibiotics: New Analogues, Methods of Delivery, and Mechanism of Action," ed. W. Priebe, American Chemical Society, Washington, DC, 1995, p. 156.
35. K. C. Agrawal, in "CRC Handbook of Chemotherapeutic Agents," ed. M. Verderame, CRC, Boca Raton, FL, 1986, vol. II, p. 229.
36. W. Saenger, "Principles of Nucleic Acid Structure," Springer, New York, 1984.
37. L. S. Lerman, *J. Mol. Biol.*, 1961, **3**, 18.
38. L. S. Lerman, *Proc. Natl. Acad. Sci. USA*, 1963, **49**, 94.
39. L. S. Lerman, *J. Cell Comp. Physiol.*, 1964, **64**, 1.
40. L. S. Lerman, *J. Mol. Biol.*, 1964, **10**, 367.
41. A. N. Gough, R. L. Jones, and W. D. Wilson, *J. Med. Chem.*, 1979, **22**, 1551.
42. T. Kurucsev and U. P. Strauss, *J. Phys. Chem.*, 1970, **74**, 3081.
43. M. E. Lamm and D. M. Neville, *J. Phys. Chem.*, 1965, **69**, 3872.
44. W. Fuller and M. J. Waring, *Ber. Bunsenges. Phys. Chem.*, 1964, **68**, 805.
45. B. M. J. Revet, M. Schmir, and J. Vinograd, *Nature New Biol.*, 1971, **229**, 10.
46. R. L. Jones, A. L. Lanier, R. A. Keel, and W. D. Wilson, *Nucleic Acids Res.*, 1980, **8**, 1613.
47. G. Cohen and H. Eisenberg, *Biopolymers*, 1969, **8**, 45.
48. J. M. Saucier, B. Festy, and J. B. Le Pecq, *Biochimie*, 1971, **53**, 973.
49. M. Chastain and J. Tinoco, Jr., *Prog. Nucleic Acid Res. Mol. Biol.*, 1991, **41**, 131.
50. D. E. Draper, *Acc. Chem. Res.*, 1992, **25**, 201.
51. A. M. Pyle and J. B. Green, *Curr. Opin. Struct. Biol.*, 1995, **5**, 303.
52. M. J. Serra and D. H. Turner, *Methods Enzymol.*, 1995, **259**, 242.
53. R. F. Gesteland and J. F. Atkins, "The RNA World," Cold Spring Harbor Laboratory Press, Plainview, NY, 1993.
54. K. L. Rinnehart and L. S. Shield, in "Aminocyclitol Antibiotics," eds. K. L. Rinnehart and T. Suami, American Chemical Society, Washington, DC, 1980, p. 1.
55. D. Moazed and H. F. Noller, *Nature (London)*, 1987, **327**, 389.
56. W. D. Wilson, L. Ratmeyer, M. Zhao, L. Strekowski, and D. Boykin, *Biochemistry*, 1993, **32**, 4098.
57. J. W. Nelson and J. Tinoco, Jr., *Biopolymers*, 1984, **23**, 213.
58. J. L. Bresloff and D. M. Crothers, *Biochemistry*, 1981, **20**, 3547.
59. S. A. White and D. E. Draper, *Biochemistry*, 1989, **28**, 1892.
60. J. D. Puglisi, L. Chen, S. Blanchard, and A. D. Frankel, *Science*, 1995, **270**, 1200.
61. X. Ye, R. A. Kumar, and D. J. Patel, *Chem. Biol.*, 1995, **2**, 827.
62. J. L. Battiste, R. Tan, A. D. Frankel, and J. R. Williamson, *J. Biomol. NMR*, 1995, **6**, 375.
63. K. Nagai and I. W. Mattaj (eds.), "RNA–Protein Interactions," Oxford-IRL Press, Oxford, 1994.

64. S. W. Englander and N. R. Kallenbach, *Q. Rev. Biophys.*, 1984, **16**, 521.
65. M. Gueron, E. Charretier, J. Hagerhorst, M. Kochoyan, J. L. Leroy, and A. Moraillon, in "Structure and Methods," eds. R. H. Sarma and M. H. Sarma, Adenine Press, New York, 1990, vol. 3, p. 113.
66. E. Folta-Stogniew and I. Russu, *Biochemistry*, 1996, **35**, 8439.
67. W. D. Wilson and F. A. Tanious, in "Molecular Aspects of Anticancer Drug–DNA Interactions," eds. S. Neidle and M. J. Waring, Macmillan, London, 1994, vol. 2, p. 243.
68. W. D. Wilson, C. R. Krishnamoorthy, Y.-H. Wang, and J. C. Smith, *Biopolymers*, 1985, **24**, 1941.
69. C. R. Krishnamoorthy, S.-F. Yen, J. C. Smith, J. W. Lown, and W. D. Wilson, *Biochemistry*, 1986, **25**, 5933.
70. K. R. Fox and M. J. Waring, *Eur. J. Biochem.*, 1984, **84**, 579.
71. S. C. Jain and H. M. Sobell, *J. Biomol. Struct. Dyn.*, 1984, **1**, 1179.
72. S. Neidle and T. C. Jenkins, *Methods Enzymol.*, 1991, **203**, 433.
73. J. M. Veal and W. D. Wilson, *J. Biomol. Struct. Dyn.*, 1991, **8**, 1119.
74. W. D. Wilson and R. L. Jones, *Biopolymers*, 1986, **25**, 1997.
75. F. A. Tanious, S.-F. Yen, and W. D. Wilson, *Biochemistry*, 1991, **30**, 1813.
76. W. A. Denny and L. P. G. Wakelin, *Anti-Cancer Drug Des.*, 1990, **5**, 189.
77. J. Elliot, W. D. Wilson, R. Shea, J. A. Hartley, K. Reszka, and J. W. Lown, *Anti-Cancer Drug Des.*, 1989, **3**, 271.
78. F. A. Tanious, T. C. Jenkins, S. Neidle, and W. D. Wilson, *Biochemistry*, 1992, **31**, 11632.
79. A. H.-J. Wang, Y.-C. Liaw, H. Robinson, and Y.-G. Gao, in "Molecular Basis of Specificity in Nucleic Acid–Drug Interactions," eds. B. Pullman and J. Jortner, Kluwer, Dordrecht, 1990, vol. 23, p. 1.
80. C. Bourdouxhe-Housiaux, P. Colson, C. Houssier, M. J. Waring, and C. Bailly, *Biochemistry*, 1996, **35**, 4251.
81. M. Egli, L. D. Williams, C. A. Frederick, and A. Rich, *Biochemistry*, 1991, **30**, 1364.
82. Y.-G. Gao, Y.-C. Liaw, H. Robinson, and A. H.-J. Wang, *Biochemistry*, 1990, **29**, 10307.
83. K. R. Fox and M. J. Waring, *Biochem. Biophys. Acta*, 1984, **802**, 162.
84. K. R. Fox, C. Brassett, and M. J. Waring, *Biochem. Biophys. Acta*, 1985, **802**, 383.
85. L. P. A. van Houte, C. J. van Garderen, and D. J. Patel, *Biochemistry*, 1993, **32**, 1667.
86. M. S. Searle, J. G. Hall, W. A. Denny, and L. P. G. Wakelin, *Biochemistry*, 1988, **27**, 4340.
87. X. Zhang and D. S. Patel, *Biochemistry*, 1990, **29**, 9451.
88. H. Robinson, Y.-C. Liaw, G. A. van der Marel, J. H. van Boom, and A. H.-J. Wang, *Nucleic Acids Res.*, 1990, **18**, 4851.
89. S.-F. Yen, E. J. Gabbay, and W. D. Wilson, *Biochemistry*, 1982, **21**, 2070.
90. S. Takenaka, V. Kiselyova, D. W. Dixon, D. Santiago, S.-F. Yen, F. A. Tanious, D. W. Boykin, and W. D. Wilson, in preparation.
91. R. J. Fiel, B. G. Jenkins, and J. L. Alderfer, in "Molecular Basis of Specificity in Nucleic Acid–Drug Interactions," eds. B. Pullman and J. Jortner, Kluwer, Dordrecht, 1990, vol. 23, p. 385.
92. R. F. Pasternack, E. J. Gibbs, and J. J. Villafranca, *Biochemistry*, 1983, **22**, 2406.
93. L. G. Marzilli, D. L. Banville, G. Zon, and W. D. Wilson, *J. Am. Chem. Soc.*, 1986, **108**, 4188.
94. D. L. Banville, L. G. Marzilli, J. A. Strickland, and W. D. Wilson, *Biopolymers*, 1986, **25**, 1837.
95. J. A. Strickland, L. G. Marzilli, and W. D. Wilson, *Biopolymers*, 1990, **29**, 1307.
96. L. A. Lipscomb, F. X. Zhou, S. R. Presnell, R. J. Woo, M. E. Peek, R. R. Plaskon, and L. D. Williams, *Biochemistry*, 1996, **35**, 2818.
97. R. J. Abraham, G. R. Bedford, D. McNeillie, and B. Wright, *Org. Magn. Reson.*, 1980, **14**, 418.
98. D. G. Gorenstein, *Chem. Rev.*, 1994, **94**, 1315.
99. W. D. Wilson, in "Encyclopedia of NMR," eds. D. M. Grant and R. K. Harris, Wiley, New York, 1995, p. 1758.
100. G. J. Atwell, G. M. Stewart, W. Leupin, and W. A. Denny, *J. Am. Chem. Soc.*, 1985, **107**, 4335.
101. R. S. Rye, S. Yue, M. A. Quesada, R. P. Haugland, R. A. Mathies, and A. N. Glazer, *Methods Enzymol.*, 1996, **217**, 414.
102. L. P. G. Wakelin, *Med. Res. Rev.*, 1986, **6**, 275.
103. Q. Gao, L. D. Williams, M. Egli, D. Rabinovich, S.-l. Chen, G. J. Quigley, and A. Rich, *Proc. Natl. Acad. Sci. USA*, 1991, **88**, 2422.
104. A. Delbarre, M. Delepierre, M. Langlois D'Esttaintot, J. Lgolen, and B. P. Roques, *Biopolymers*, 1987, **26**, 1001.
105. L. D. Williams and Q. Gao, *Biochemistry*, 1992, **31**, 4315.
106. D. M. Crothers, *Biopolymers*, 1968, **6**, 575.
107. R. A. G. Friedman and G. S. Manning, *Biopolymers*, 1984, **23**, 2671.
108. J. D. McGhee and P. A. von Hippel, *J. Mol. Biol.*, 1979, **86**, 469.
109. C. R. Cantor and R. R. Schimmel, "Biophysical Chemistry," Freeman, San Francisco, 1980.
110. M. J. Waring, in "Molecular Aspects of Anticancer Drug–DNA Interactions," eds. S. Neidle and M. J. Waring, Macmillan Press, London, 1993, vol. 1, p. 213.
111. C. Bailly, D. Gentle, F. Hamy, M. Purcell, and M. J. Waring, *Biochem. J.*, 1994, **300**, 165.
112. X. Gao and D. J. Patel, *Q. Rev. Biophys.*, 1989, **22**, 93.
113. K. J. Addess, J. S. Sinsheimer, and J. Feigon, *Biochemistry*, 1993, **32**, 2498.
114. M. C. Flecher and K. R. Fox, *Biochemistry*, 1996, **35**, 1064.
115. M. E. Peek, L. A. Lipscomb, J. A. Bertrand, Q. Gao, B. P. Roques, C. Garbay-Jaureguiberry, and L. D. Williams, *Biochemistry*, 1994, **33**, 3794.
116. J. P. Jacobsen, J. B. Pedersen, L. F. Hansen, and D. E. Wemmer, *Nucleic Acids Res.*, 1995, **23**, 753.
117. L. F. Hansen, L. J. Jensen, and J. P. Jacobsen, *Nucleic Acids Res.*, 1996, **24**, 859.
118. H. P. Spielmann, D. E. Wemmer, and J. P. Jacobsen, *Biochemistry*, 1995, **34**, 8542.
119. S. C. Zimmerman, C. R. Lamberson, M. Cory, and T. A. Fairley, *J. Am. Chem. Soc.*, 1989, **111**, 6805.
120. J. M. Veal, Y. Li, S. C. Zimmerman, C. R. Lamberson, M. Cory, G. Zon, and W. D. Wilson, *Biochemistry*, 1990, **29**, 10918.
121. S. Takenaka, S. Nishira, K. Tahara, H. Kondo, and M. Takagi, *Supramol. Chem.*, 1993, **2**, 41.
122. M. Wirth, O. Buchardt, T. Koch, P. E. Nielsen, and B. Norden, *J. Am. Chem. Soc.*, 1988, **110**, 932.
123. A. Slama-Schwok, M.-P. Teulade-Flchou, J.-P. Vigneron, E. Tailandier, and J.-M. Lehn, *J. Am. Chem. Soc.*, 1995, **117**, 6822.

124. F. Leng, J. B. Chaires, and W. Priebe, *Biochemistry*, 1998, **37**, 1743.
125. G. J. Atwell, W. Leupin, S. J. Twigden, and W. A. Denny, *J. Am. Chem. Soc.*, 1983, **105**, 2913.
126. G. J. Atwell, B. C. Baguley, D. Wilmanska, and W. A. Denny, *J. Med. Chem.*, 1986, **29**, 69.
127. P. Laugaa, J. Markovits, A. Delbarre, J.-B. Le Pecq, and B. P. Roques, *Biochemistry*, 1985, **24**, 5567.
128. P. Laugaa, M. Delepierre, B. Dupraz, J. Igolen, and B. P. Roques, *J. Biomol. Struct. Dyn.*, 1988, **6**, 421.
129. K. Yanagi, G. G. Prive, and R. E. Dickerson, *J. Mol. Biol.*, 1991, **217**, 201.
130. R. E. Dickerson, *Sci. Am.*, 1983, **249**, 94.
131. T. A. Larsen, D. S. Goodsell, D. Cascio, K. Grzeskowiak, and R. E. Dickerson, *J. Biomol. Struct. Dyn.*, 1989, **7**, 477.
132. L. Strekowski, S. Chandrasekaran, Y. H. Wang, and W. D. Wilson, *J. Med. Chem.*, 1986, **29**, 1311.
133. L. Strekowski, J. L. Mokrosz, F. A. Tanious, R. A. Watson, D. Harden, M. Mokrosz, W. D. Edwards, and W. D. Wilson, *J. Med. Chem.*, 1988, **31**, 1231.
134. W. D. Wilson, L. Strekowski, F. A. Tanious, R. Watson, J. L. Mokrosz, A. Strekowska, G. Webster, and S. Neidle, *J. Am. Chem. Soc.*, 1988, **110**, 8292.
135. W. D. Wilson, F. A. Tanious, R. A. Watson, H. J. Barton, A. Strekowska, D. B. Harden, and L. Strekowski, *Biochemistry*, 1989, **28**, 1984.
136. W. D. Wilson, H. J. Barton, F. A. Tanious, S. B. Kong, and L. Strekowski, *Biophys. Chem.*, 1990, **35**, 227.
137. L. Strekowski and W. D. Wilson, in "Synergism and Antagonism in Chemotherapy," eds. T.-C. Chou and D. Rideout, Academic Press, New York, 1991, chap. 12.
138. L. Strekowski, W. D. Wilson, J. Mokrosz, M. Mokrosz, D. Harden, F. A. Tanious, R. Wydra, and S. Crow, *J. Med. Chem.*, 1991, **34**, 580.
139. L. Strekowski, J. Mokrosz, W. D. Wilson, M. Mokrosz, and A. Strekowski, *Biochemistry*, 1992, **31**, 10802.
140. W. D. Wilson, F. A. Tannious, H. Buczak, M. K. Venkatramanan, B. P. Das, and D. W. Boykin, in "Molecular Basis of Specificity in Nucleic Acid–Drug Interactions," eds. B. Pullman and J. Jortner, Kluwer, Dordrecht, 1990, vol. 23, p. 331.
141. W. D. Wilson, F. A. Tannious, H. Buczak, L. Ratmeyer, M. K. Venkatramanan, A. Kumar, D. W. Boykin, and B. R. Munson, in "Structure and Function: Nucleic Acids," eds. R. H. Sarma and M. H. Sarma, Adenine Press, New York, 1992, vol. 1, p. 83.
142. D. W. Boykin, A. Kumar, J. Spychala, M. Zhao, R. L. Lombardy, W. D. Wilson, C. C. Dykstra, S. K. Jones, J. E. Hall, R. R. Tidwell, C. Laughton, C. M. Nunn, and S. Neidle, *J. Med. Chem.*, 1995, **38**, 912.
143. C. A. Laughton, F. A. Tanious, C. M. Nunn, D. W. Boykin, W. D. Wilson, and S. Neidle, *Biochemistry*, 1996, **35**, 5611.
144. P. Colson, C. Houssier, and C. Bailly, *J. Biomol. Struct. Dyn.*, 1995, **13**, 351.
145. D. S. Pilch, M. A. Kirolos, X. Liu, G. E. Plum, and K. J. Breslauer, *Biochemistry*, 1995, **34**, 9962.
146. J. Portogal and M. J. Waring, *Biochem. Biophys. Acta*, 1988, **949**, 158.
147. M. L. Kopka, P. Pjura, C. Yoon, D. Goodsell, and R. E. Dickerson, in "Structure and Motion: Membranes, Nucleic Acids and Proteins," eds. E. Clementi, G. Corongiu, M. H. Sarma, and R. Sarma, Adenine Press, New York, 1985, p. 461.
148. A. Pullman and B. Pullman, *Q. Rev. Biophys.*, 1981, **14**, 289.
149. W. D. Wilson, F. Tanious, H. Barton, R. Jones, L. Strekowski, and D. Boykin, *J. Am. Chem. Soc.*, 1989, **111**, 5008.
150. W. D. Wilson, F. A. Tanious, H. J. Barton, R. J. Jones, K. Fox, R. L. Wydra, and L. Strekowski, *Biochemistry*, 1990, **29**, 8452.
151. W. D. Wilson, F. Tanious, H. J. Barton, R. L. Wydra, R. L. Jones, D. Boykin, and L. Strekowski, *Anti-Cancer Drug Des.*, 1990, **5**, 31.
152. K. Jansen, P. Lincoln, and B. Norden, *Biochemistry*, 1993, **32**, 6605.
153. D. M. Crothers and J. Drak, *Methods Enzymol.*, 1992, **212**, 46.
154. P. J. Hagerman, *Biochemistry*, 1985, **24**, 7033.
155. F. G. Loontiens, L. W. McLaughlin, S. Diekmann, and R. M. Clegg, *Biochemistry*, 1991, **30**, 182.
156. K. Jansen, B. Norden, and M. Kubista, *J. Am. Chem. Soc.*, 1993, **115**, 10 527.
157. C. Bailly, P. Colson, C. Houssier, and F. Hamy, *Nucleic Acids Res.*, 1996, **24**, 1460.
158. W. D. Wilson, M. Zhao, S. E. Patterson, R. L. Wydra, and L. Strekowski, *Med. Chem. Res.*, 1992, **2**, 102.
159. F. A. Tanious, J. M. Veal, H. Buczak, L. S. Ratmeyer, and W. D. Wilson, *Biochemistry*, 1992, **31**, 3103.
160. M. Zhao, L. S. Ratmeyer, R. Peloquin, S. Yao, A. Kumar, J. Spychala, D. W. Boykin, and W. D. Wilson, *Bioorg. Med. Chem.*, 1995, **3**, 785.
161. W. D. Wilson, L. Ratmeyer, M. Zhao, D. Ding, A. McConnaughie, A. Kumar, and D. W. Boykin, *J. Mol. Recognit.*, 1996, **9**, 187.
162. L. Ratmeyer, M. Zapp, M. Green, R. Vinayak, A. Kumar, D. Boykin, and W. D. Wilson, *Biochemistry*, 1996, **35**, 13 689.
163. D. M. Crothers, *Biopolymers*, 1971, **10**, 2147.
164. P. A. Permana, R. M. Snapka, L. L. Shen, D. T. W. Chu, J. J. Clement, and J. J. Plattner, *Biochemistry*, 1994, **33**, 11 333.
165. J.-Y. Fan, D. Sun, H. Yu, S. M. Kerwin, and L. H. Hurley, *J. Med. Chem.*, 1995, **38**, 408.
166. H. Yu, L. H. Hurley, and S. M. Kerwin, *J. Am. Chem. Soc.*, 1996, **118**, 7040.
167. D. Rideout, *Science*, 1988, **233**, 561.
168. T. Ikeda, K. Yoshida, and H.-J. Schneider, *J. Am. Chem. Soc.*, 1995, **117**, 1453.
169. R. L. Jones, A. C. Lanier, R. A. Keel, and W. D. Wilson, *Nucleic Acids Res.*, 1980, **8**, 1613.
170. P. Adawadkar, W. D. Wilson, W. Brey, and E. J. Gabbay, *J. Am. Chem. Soc.*, 1975, **97**, 1959.
171. E. J. Gabbay, P. Adawadkar, and W. D. Wilson, *Biochemistry*, 1976, **15**, 146.
172. M. H. Wemer, A. M. Gronenborn, and G. M. Clore, *Science*, 1996, **271**, 778.
173. S. K. Burley, *Curr. Opin. Struct. Biol.*, 1996, **6**, 69.
174. M. H. Wemer, G. M. Clore, C. L. Fisher, R. J. Fisher, L. Trinh, J. Shiloach, and A. M. Gronenborn, *Cell*, 1995, **83**, 761.
175. R. J. Roberts, *Cell*, 1995, **82**, 9.
176. M. Fernandez-Saiz, H.-J. Schneider, J. Sartorius, and W. D. Wilson, *J. Am. Chem. Soc.*, 1996, **118**, 4739.

177. N. T. Thuong and C. Helene, *Angew. Chem., Int. Ed. Engl.*, 1993, **32**, 666.
178. W. D. Wilson, F. A. Tanious, S. Mizan, S. Yao, A. S. Kiselyov, G. Zon, and L. Strekowski, *Biochemistry*, 1993, **32**, 10614.
179. L. Strekowski, Y. Gulevich, T. C. Baranowski, A. N. Parker, A. S. Kiselyov, S.-Y. Lin, F. A. Tanious, and W. D. Wilson, *J. Med. Chem.*, 1996, **39**, 3980.
180. C. Escude, C. H. Nguyen, J.-L. Mergny, J.-S. Sun, E. Blsagni, T. Garestier, and C. Helene, *J. Am. Chem. Soc.*, 1995, **117**, 10212.
181. G. D. Felsenfeld, D. R. Davies, and A. Rich, *J. Am. Chem. Soc.*, 1957, **79**, 2023.
182. G. D. Felsenfeld and H. T. Miles, *Annu. Rev. Biochem.*, 1967, **36**, 407.
183. S. A. Strobel and P. B. Dervan, *Methods Enzymol.*, 1992, **216**, 309.
184. T. Ito, C. L. Smith, and C. R. Cantor, *Nucleic Acids Res.*, 1992, **20**, 3524.
185. T. Takabatake, K. Asada, Y. Uchimura, M. Ohdate, and N. Kusukawa, *Nucleic Acids Res.*, 1992, **20**, 5853.
186. L. J. Maher, III, P. B. Dervan, and B. Wold, *Biochemistry*, 1992, **31**, 70.
187. J. L. Mergny, G. Daval-Valentin, C. H. Nguyen, L. Perrouault, B. Faucon, M. Rougee, T. Montenay-Garestier, E. Bisagni, and C. Helene, *Science*, 1992, **256**, 1681.
188. D. S. Pilch and K. J. Breslauer, *Proc. Natl. Acad. Sci. USA*, 1994, **91**, 9332.
189. M. Durand, N. T. Thuong, and J. C. Maurizot, *J. Biol. Chem.*, 1992, **267**, 24394.
190. W. D. Wilson, S. Mizan, F. A. Tanious, and S. Yao, *J. Mol. Recognit.*, 1994, **7**, 89.
191. P. V. Scaria and R. H. Shafer, *J. Biol. Chem.*, 1991, **266**, 5417.
192. J. S. Lee, L. J. B. Latimer, and K. J. Hampel, *Biochemistry*, 1993, **32**, 5591.
193. H.-K. Kim, J.-M. Kim, S. K. Kim, A. Rodger, and B. Norden, *Biochemistry*, 1996, **35**, 1187.
194. F. A. Tanious, L. Strekowski, and W. D. Wilson, in preparation.
195. S. A. Cassidy, L. Strekowski, W. D. Wilson, and K. R. Fox, *Biochemistry*, 1994, **33**, 15338.
196. K. R. Fox, P. Polucci, T. C. Jenkins, and S. Neidle, *Proc. Natl. Acad. Sci. USA*, 1995, **92**, 7887.
197. D. S. Pilch, M.-T. Martin, C. H. Nguyen, J.-S. Sun, E. Blsagni, T. Garestier, and C. Helene, *J. Am. Chem. Soc.*, 1993, **115**, 9942.
198. C. Marchand, C. Bailly, C. H. Nguyen, E. Blsagni, T. Garestier, C. Helene, and M. J. Waring, *Biochemistry*, 1996, **35**, 5022.
199. C. Marchand, J.-S. Sun, C. H. Nguyen, E. Blsagni, T. Garestier, and C. Helene, *Biochemistry*, 1996, **35**, 5735.
200. G. Duval-Valentin, T. de Bizemont, M. Takasugi, J.-L. Mergny, E. Blsagni, and C. Helene, *J. Mol. Biol.*, 1995, **247**, 847.
201. J. R. Williamson, *Annu. Rev. Biophys. Biomol. Struct.*, 1994, **23**, 703.
202. E. H. Blackburn, *Nature* (*London*), 1991, **350**, 509.
203. N. W. Kim, M. K. Piatyszek, K. R. Prowse, C. B. Harley, M. D. West, P. L. C. Ho, G. M. Coviello, W. E. Wright, S. L. Weinrich, and J. W. Shay, *Science*, 1994, **266**, 2011.
204. Q. Guo, M. Lu, L. A. Marky, and N. R. Kallenbach, *Biochemistry*, 1992, **31**, 2451.
205. C. T. Lauhon and J. W. Szostak, *J. Am. Chem. Soc.*, 1995, **117**, 1246.
206. Q. Chen, J. D. Kuntz, and R. H. Shafer, *Proc. Natl. Acad. Sci. USA*, 1996, **93**, 2635.
207. I. Tinoco, Jr., *J. Phys. Chem.*, 1996, **31**, 13311.
208. J. W. Nelson and I. Tinoco, Jr., *Biochemistry*, 1985, **24**, 6416.
209. L. D. Williams and I. H. Goldberg, *Biochemistry*, 1988, **27**, 3004.
210. S. A. Woodson and D. M. Crothers, *Biochemistry*, 1988, **27**, 8904.
211. L. S. Ratmeyer, R. Vinayak, G. Zon, and W. D. Wilson, *J. Med. Chem.*, 1992, **33**, 966.
212. S. Yao and W. D. Wilson, *J. Biomol. Struct. Dyn.*, 1992, **10**, 367.
213. D. Suh, R. D. Sheardy, and J. B. Chaires, *Biochemistry*, 1991, **30**, 8722.
214. M. Lu, Q. Guo, and N. R. Kallenbach, *Crit. Rev. Biochem. Mol. Biol.*, 1992, **27**, 157.
215. L. I. Hernandez, M. Zhong, S. H. Courtney, L. A. Marky, and N. R. Kallenbach, *Biochemistry*, 1994, **33**, 13140.
216. S. F. Ohandley, D. G. Sanford, R. Xu, C. C. Lester, B. E. Hingerty, S. Broyde, and T. R. Krugh, *Biochemistry*, 1993, **32**, 2481.
217. G. M. Blackburn, in "Nucleic Acids in Chemistry and Biology," eds. G. M. Blackburn and M. J. Gait, Oxford University Press, New York, 1996, chap. 7.
218. M. Cosman, A. Laryea, R. Fiala, B. E. Hingerty, S. Amin, N. E. Geacintov, S. Broyde, and D. J. Patel, *Biochemistry*, 1995, **34**, 1295.
219. M. Cosman, R. Fiala, B. E. Hingerty, S. Amin, N. E. Geacintov, S. Broyde, and D. J. Patel, *Biochemistry*, 1994, **33**, 11507.
220. M. Cosman, R. Fiala, B. E. Hingerty, S. Amin, N. E. Geacintov, S. Broyde, and D. J. Patel, *Biochemistry*, 1994, **33**, 11518.
221. M. P. Stone, S. Gopalakrishnan, K. D. Raney, V. M. Raney, S. Byrd, and T. M. Harris, in "Molecular Basis of Specificity in Nucleic Acid–Drug Interactions," eds. B. Pullman and J. Jortner, Kluwer, Dordrecht, 1990, vol. 23, p. 451.
222. S. Gopalakrishnan, T. M. Harris, and M. P. Stone, *Biochemistry*, 1990, **29**, 10438.
223. S. S. Johnston and M. P. Stone, *Biochemistry*, 1995, **34**, 14037.
224. D. Sun, M. Hansen, J. J. Clement, and L. H. Hurley, *Biochemistry*, 1993, **32**, 8068.
225. M. Hansen and L. H. Hurley, *J. Am. Chem. Soc.*, 1995, **117**, 2421.
226. D. Sun, M. Hansen, and L. H. Hurley, *J. Am. Chem. Soc.*, 1995, **117**, 2430.

# 7.13
# DNA-binding Peptides

## INDRANEEL GHOSH, SHAO YAO, and JEAN CHMIELEWSKI
*Purdue University, West Lafayette, IN, USA*

| 7.13.1 | INTRODUCTION | | 477 |
|---|---|---|---|
| 7.13.2 | DNA-BINDING PEPTIDES BASED ON PROTEIN MOTIFS | | 477 |
| | 7.13.2.1 | α-Helices in the Major Groove | 477 |
| | | 7.13.2.1.1   Helix–turn–helix peptides | 478 |
| | | 7.13.2.1.2   Basic-helix–loop–helix peptides | 480 |
| | | 7.13.2.1.3   Basic-leucine zipper peptides | 482 |
| | | 7.13.2.1.4   Peptides based on the zinc finger motif | 484 |
| | 7.13.2.2 | β-Sheet Peptides in the Major Groove | 486 |
| 7.13.3 | PEPTIDE MINOR GROOVE BINDERS | | 486 |
| 7.13.4 | PEPTIDE INTERCALATORS | | 487 |
| 7.13.5 | CONCLUSIONS | | 488 |
| 7.13.6 | REFERENCES | | 488 |

## 7.13.1   INTRODUCTION

Watson and Crick, upon deducing the structure of DNA, were quick to point out that "there is room between polynucleotide chains for a polypeptide to wind around the same helical axis."[1] Since that time many structures have been solved for protein–DNA complexes. This information has been critical in understanding the nature of protein–DNA interactions and in advancing the design of peptides which make specific interactions with DNA, through either major or minor groove contacts. The focus of this chapter, therefore, is to enumerate the many peptides of approximately 60 amino acid residues or less with the ability to make specific interactions with double-stranded, B-form DNA.

## 7.13.2   DNA-BINDING PEPTIDES BASED ON PROTEIN MOTIFS

### 7.13.2.1   α-Helices in the Major Groove

Numerous DNA-binding proteins rely on interactions between α-helical portions of the protein with the DNA major groove. A variety of structural motifs have been identified within α-helical DNA-binding proteins, including the helix–turn–helix (HTH), basic-helix–loop–helix (bHLH), basic leucine-zipper (bZip), and zinc finger motifs.[2–6] The peptides of these motifs are either small full-length proteins of less than approximately 60 amino acids, or are composed of truncated forms of the DNA-binding protein which contain either the entire binding motif or smaller portions of the motif containing residues responsible for sequence-specific interactions with DNA.

### 7.13.2.1.1  Helix–turn–helix peptides

The HTH motif has been found in a wide range of prokaryotic and eukaryotic transcription factors. The motif is composed of approximately 21 residues which fold into two helices, known as helix-2 and helix-3, which dock onto one another at a 120° angle with an intervening three- or four-residue turn (Figure 1).[7–9] The recognition helix, helix-3, is an integral part of the folded protein structure, but the surface-exposed portions of this helix make contact with the edges of the base pairs in the major groove along with neighboring phosphodiester moieties.

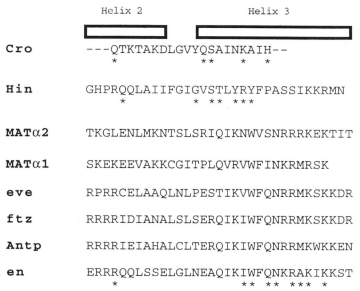

```
                     Helix 2                Helix 3

     Cro         ---QTKTAKDLGVYQSAINKAIH--
                    *           **   *   *

     Hin         GHPRQQLAIIFGIGVSTLYRYFPASSIKKRMN
                     *        * ** ***

    MATα2        TKGLENLMKNTSLSRIQIKNWVSNRRRKEKTIT

    MATα1        SKEKEEVAKKCGITPLQVRVWFINKRMRSK

     eve         RPRRCELAAQLNLPESTIKVWFQNRRMKSKKDR

     ftz         RRRRIDIANALSLSERQIKIWFQNRRMKSKKDR

    Antp         RRRRIEIAHALCLTERQIKIWFQNRRMKWKKEN

     en          ERRRQQLSSELGLNEAQIKIWFQNKRAKIKKST
                    *               ** ** *** *
```

**Figure 1**  Sequence alignment for HTH peptides. Residues with stars interact with DNA.

The HTH motif has been the subject of many review articles.[2–9] Our intent, therefore, is to focus on a few examples of small or truncated HTH proteins, and, where possible, to delineate the point at which DNA binding is lost upon truncation.

Hin recombinase, for example, is composed of a DNA-binding domain which contains an HTH motif, and a recombination domain which results in DNA inversion.[10] Synthesis of a 52 amino acid residue peptide based on the Hin DNA-binding domain, which is comprised of the HTH motif with an additional amino-terminal turn–helix (residues 139–190), resulted in a peptide of sequence that specifically bound to an oligonucleotide fragment containing the *hix*L cross-over site.[11] The dissociation constant obtained for 52mer *hix*L half-site binding was approximately 2 μM compared with the 40 nM dissociation constant obtained with the full-length Hin protein. A cocrystal structure of Hin recombinase with DNA confirmed the presence of the HTH motif and elucidated the interactions between helix-3 and the major groove of DNA (Figure 2).[12] Similar results were obtained with the related protein resolvase recombinase; a carboxy-terminal proteolysis fragment containing the HTH motif of resolvase-bound DNA with a dissociation constant of 0.5–2.0 μM compared with < 0.2 nM for the intact protein.[13] The 52mer of Hin also inhibited Hin inversion by binding to the *hix*L site, although this inhibition could be overcome by increasing the Hin concentration.[11] Further truncation of Hin recombinase to a 31 amino acid peptide composed only of the HTH sequence (residues 160–190) resulted in a complete loss of sequence-specific DNA binding and no inhibition of Hin inversion.[11]

The 66 amino acid cro protein from λ phage contains an HTH motif and recognizes the 17 base pair λ phage O$_R$3 sequence.[14] λ cro binds DNA as a noncovalent dimer with the recognition helices bound into two adjacent DNA half sites. The full-length λ cro, which contains, in addition to the HTH motif, three β-strands and one helix, has a dissociation constant of < 5 nm for the O$_R$3 operator site.[15] Dimerization can be inhibited by the addition of five residues to the C-terminus of λ cro; these additional residues were designed to bind the C-terminal β-strand of λ cro and produce a folded monomeric version.[15] The monomeric cro derived from this procedure was found to fold similarly to the wild-type dimeric protein as determined by circular dichroism and NMR, but showed severely reduced affinity for the O$_R$3 site (> 19 μM). Truncation of λ cro to a peptide containing only the helix–turn–helix motif (residues 16–35)[16] or to a peptide containing just the recognition helix (residues 26–39)[17] resulted in a loss of sequence-specific DNA binding. The peptide

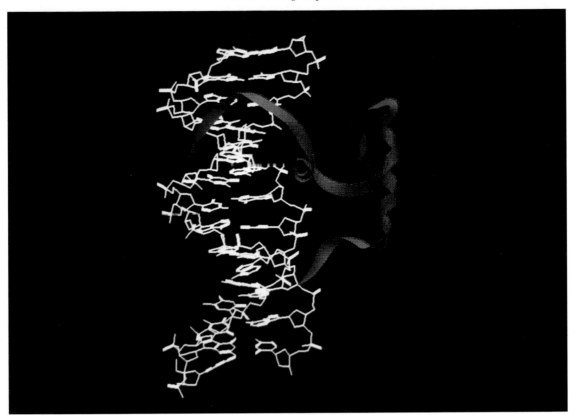

**Figure 2** Cocrystal structure of Hin recombinase and DNA.[12]

containing the HTH motif was not a stable folding unit, as determined by circular dichroism, which showed a helical content of 12%.[16] Attempts to stabilize the HTH structure in this peptide by incorporating a disulfide linkage at the interface of helix-2 and helix-3 between residues 20 and 30 led to only a small increase in the helical content (20%) and no enhanced DNA binding.[16]

The homeodomain is a 60 amino acid region of proteins which are expressed by the homeotic genes. These genes specify the body plan and regulate development in higher organisms, and the homeodomain, which contains an HTH motif, represents the DNA binding domain of the larger homeodomain proteins.[18,19] Peptides corresponding to the full homeodomain of a number of proteins, such as antennapedia, engrailed, even-skipped, ultrabithorax, *fushi tarazu*, and MATa1/MATα2, have been prepared and analyzed for sequence-specific DNA binding. A 68 amino acid homeodomain peptide of the antennapedia protein (residues 297–363), for example, was expressed and was found to bind as a monomer to an oligonucleotide containing an ATTA site with a dissociation constant of 1.6 nM.[20] The solution structure of the 68mer confirmed that antennapedia contained an HTH motif in addition to two other helices.[21] A synthetic 60 amino acid homeodomain peptide had similar activity,[22] whereas truncation of the homeodomain to a peptide containing only the HTH motif (residues 28–55 based on homeodomain) resulted in the loss of sequence-specific binding to DNA.[17]

A 61 residue peptide corresponding to the engrailed homeodomain was prepared and found to bind to an oligonucleotide containing a TAAT subsite with a dissociation constant of 1 nM.[23] A cocrystal structure of the engrailed peptide with DNA confirmed the presence of an HTH motif bound into the major groove of DNA at the TAAT site with an N-terminal tail binding into the minor groove of the DNA and an additional two helices providing stabilizing interaction with the recognition helix (Figure 3).[23] A similar structure of the 60 residue homeodomain peptide of the monomeric even-skipped (Eve) protein was also solved with an oligonucleotide containing an ATTA core sequence.[24] Slightly longer peptide sequences containing the homeodomains of the ultrabithorax (72mer)[25] and *fushi tarazu* (73mer)[26] proteins were also found to bind to oligonucleotides containing TAAT sequences with dissociation constants of 0.1 nM and 0.7 nM, respectively.

While most homeodomain peptides have been found to bind DNA as a monomer, an interesting case of heterodimeric homeodomain DNA binding has been discovered. The homeodomain peptide

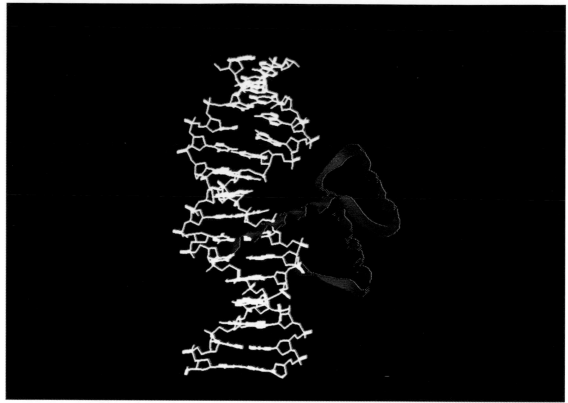

**Figure 3** Cocrystal structure of the engrailed homeodomain and DNA.[23]

(74mer) of the MATα2 protein binds to DNA as a monomer.[27] A similar homeodomain peptide (57mer) derived from the MATa1 protein shows no detectable DNA binding.[28] In the diploid a/α cell type, however, these proteins form a heterodimer which binds to sites upstream of the haploid-specific genes (hsg).[29–31] The truncated homeodomain peptides still maintain sequence-specific binding to the hsg operator with a 10-fold reduction in affinity compared with the full-length proteins.[32] A cocrystal structure of the homeodomain peptides of MATα2 and MATa1 with an oligonucleotide containing α2 and a1 binding sites clearly illustrates the heterodimeric interactions between the two peptides.[32] A C-terminal tail of MATα2 (residues 59–74) binds in a helical conformation between helices 1 and 2 of MATa1, with helix 3 of both peptides binding into the major groove of DNA, and the N-terminal arms binding into the minor groove, as has been observed with other homeodomain proteins.

The results obtained with peptides containing the HTH motif point to a few conclusions concerning how far these peptides may be truncated before losing activity. In all cases where sequence-specific DNA binding was maintained, there was at least one other helix to provide stabilization to the HTH motif, as with Hin recombinase, for example. Truncation to peptides containing only the HTH motif resulted in the loss of a well-defined conformation and also loss of specific DNA affinity. It should be possible to truncate the homeodomain proteins such that they only contain three helices and still maintain binding, as has been observed with Hin, but these experiments have not been carried out to date. Also methods to stabilize the conformation of monomeric HTH sequences would have the potential to produce smaller DNA binding peptides based on this motif.

### 7.13.2.1.2 *Basic-helix–loop–helix peptides*

The bHLH motif is another highly conserved region in a number of transcription factors, which is composed of approximately 60 amino acids.[2–6,33,34] This motif is composed of a dimerization interface known as the helix–loop–helix and a DNA-binding region composed of a number of basic residues. In the case of the bHLH motif, peptides which encompass just this region are able to bind

DNA sequence specifically. A few examples of peptides that have been prepared, which encompass the bHLH motif, are IEB E47, myoD, and USF (Figure 4).

| BASIC REGION | AMPHIPHILIC HELIX I | LOOP | AMPHIPHILIC HELIX II |

E47       RERRMA**NNARER**V**R**VRD**I**N**E**AFRELGRMCQ--MHLKSDKAQT**K**LLILQQAVQVILGLEQQ

MyoD   MELKRKTTNAD**RR**KAA**TM**R**E**R**RR**LSKV**N**EAFETLKRCTS----SNPNQ**R**LP**K**VEILRNAIRYIEGLQALLRD

USF       DEKRRA**QHN**EV**ERRRRD**KINNWIVQLSKIIPDCSME**ST**KSGQS**K**GGILSKACDYIQELRQSNHR

**Figure 4**   Sequence alignment for bHLH peptides. Residues in bold interact with DNA.

IEB E47 is a bHLH protein which plays an important role in activating expression of the immunoglobulin light chain gene by binding to the $\kappa$E2 enhancer site.[35] The E47 bHLH peptide, composed of residues 336–394 of the full-length protein, also bound to an oligonucleotide containing the $\kappa$E2 sequence (CAGGTG) with half-maximal binding occurring at approximately 5 $\mu$M.[36] A cocrystal structure of an E47 homeodomain peptide (residues 335–392) with an oligonucleotide containing a CAGGTG sequence has been solved and confirms the proposed bHLH fold.[37] The two amphiphilic helices of E47 form the dimerization interface, which is composed of a parallel, four-helix bundle, and the basic region forms a helical extension of helix 2 that binds into the major groove of the DNA (Figure 5).

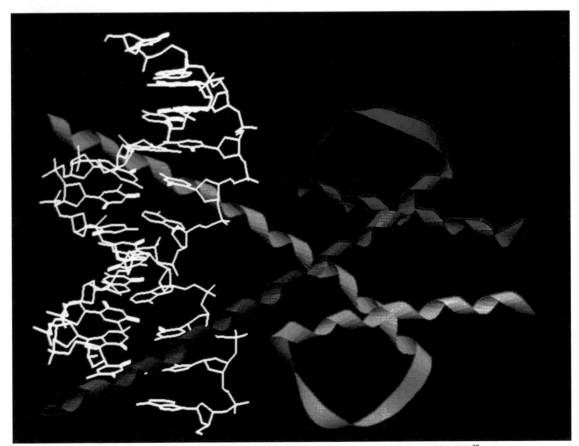

**Figure 5**   Cocrystal structure of the IEB E47 bHLH peptide and DNA.[37]

A peptide composed of the bHLH region of MyoD (residues 102–166) was bacterially expressed, and was also found to bind to an oligonucleotide containing a CAGGTG site and a CACGTG site.[38] By incorporating specific cysteine mutations into the MyoD peptide and performing cross-linking experiments, a parallel orientation of the helices at the dimerization interface was proposed, as was observed with E47. Experiments have been performed to enumerate smaller peptides of MyoD with DNA binding capabilities. By cross-linking and geometrically constraining two peptides corresponding to the basic region of MyoD with a C2 symmetric diol template, it was determined that agents containing the (*R*,*R*) and (*S*,*S*) stereochemistry formed a specific complex with an oligonucleotide containing the MyoD binding site.[39] A monomeric uncross-linked MyoD peptide, on the other hand, showed no affinity for the same DNA sequence.[39]

The USF protein is similar to other bHLH proteins in that it contains a bHLH motif, but in addition it contains a leucine zipper domain (bHLHZ) to provide additional stabilization to the homodimer. Both the bHLH (residues 197–260) and bHLHZ (residues 197–310) portions of USF have been prepared and their DNA affinity evaluated.[40] The bHLHZ peptide binds to an oligo-nucleotide containing a CACGTG sequence with a dissociation constant (1.3 nM) that was indis-tinguishable from the full-length protein. The bHLH peptide did not form an electrophoretically stable complex with the same DNA, but circular dichroism spectroscopy suggests that the bHLH peptide binds to DNA in a sequence-specific manner which is indistinguishable from the full-length protein. A cocrystal structure was also solved for the complex of the bHLH peptide of USF and DNA, and the overall protein fold and interaction with DNA is very similar to that observed with the bHLH peptide of E47.[40]

These experiments serve to illustrate that the bHLH peptides are a fully folded and functional protein motif. The experiments with MyoD also demonstrate that the basic region, as long as it is dimerized in the appropriate conformation, can function as an autonomous DNA binding sequence.

### 7.13.2.1.3 *Basic-leucine zipper peptides*

The bZip DNA-binding motif also relies on a dimerization region, termed the leucine zipper, to mediate the interactions of a highly basic domain with DNA.[2–6,41] The core bZip motif is composed of approximately 60 amino acid residues (Figure 6(a)), and studies of the bZip domains of a variety of proteins such as GCN4, C/EBP, and Jun have shown that these peptides maintain sequence-specific DNA binding.[42–44] A cocrystal structure of the bZip regions of Fos/Jun with an oligo-nucleotide containing an AP-1 site has confirmed the coiled-coil nature of the dimerization interface and has demonstrated that the basic region binds to the major groove of DNA in a helical, scissor-grip binding mode (Figure 6(b)).[45]

New bZip peptides with altered DNA binding specificities have also been obtained by using an *in vitro* selection method.[46] A library of $3.2 \times 10^{-6}$ mutant bZip C/EBP peptides was prepared by randomizing five DNA-binding residues in the basic region, and peptides of the library were selected for binding to either mutant or wild-type DNA sequences. Mutant peptides were found which bind to the corresponding mutant DNA sequences with an affinity similar to that of the wild type bZip C/EBP peptide for the wild type DNA. A "minimalistic" approach has also been applied to the design of a peptide based on the GCN4 bZip domain.[47] Residues of the basic region which were predicted to be nonessential for DNA binding were replaced with Ala residues, and a *de novo* designed heptad repeat replaced a portion of the leucine zipper. The bZip peptide obtained had only 43% sequence homology to GCN4, but bound specifically to an oligonucleotide containing the TRE site.

Truncation of the bZip motif down to peptides containing solely the basic region have generally shown the importance of the dimerization domain for DNA binding. A number of methods have been developed to cross-link the basic regions into a functional "dimer." Disulfide cross-linking at the C-terminus of a basic region peptide of GCN4, for instance, resulted in a peptide which binds to an oligonucleotide containing the GCN4 recognition element (ATGACT) with a dissociation constant of approximately 10 nM at 4 °C, but specific DNA binding was found to be temperature dependent.[48] Further truncation of the basic region down to a dimer of a peptide containing as few as 20 residues of GCN4 resulted in DNA binding with a specificity similar to the intact protein.[49] Metal binding has also been reported as a means of dimerizing the GCN4 basic region.[50] Incor-porating a terpyridyl-moiety at the C-terminus of the GCN4 basic region followed by dimerization of the peptide with addition of $Fe^{2+}$ produced a species which bound to an oligonucleotide containing a CRE binding site with a dissociation constant of 0.13 nM at 4 °C. Ueno *et al.* have employed the noncovalent interactions between $\beta$-cyclodextrin and adamantane as a means to dimerize the basic region of GCN4.[51] Two peptides were prepared in which one peptide contained $\beta$-cyclodextrin at the C-terminus and the other contained an adamantyl unit. DNA binding to an oligonucleotide containing an ATGACT site was only observed when a 1:1 mixture of the two peptides were in solution, with approximately 50% of the DNA in the complexed form at 50 nM at 4 °C.

(a) **GCN4**              ALKRARNTEAARRSRARKLQRMKQL

     **C/EBP**           RVRRERNNIAVRKSRDKAKQRNVET

     **v-Jun**           ERKRMRNRIAASKSRKRKLERIARL

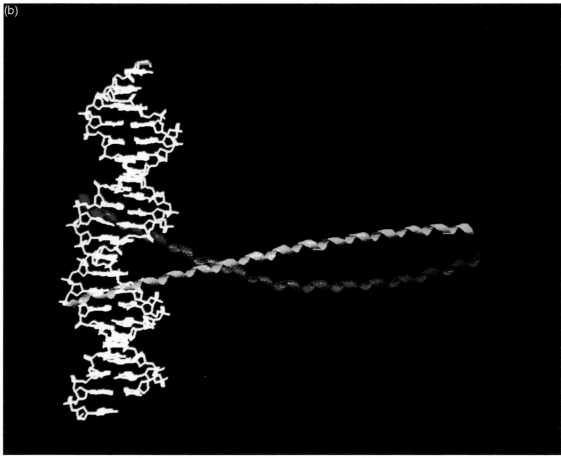

**Figure 6** (a) Basic region sequence alignment for bZip peptides, and (b) cocrystal structure of the Fos/Jun heterodimer and DNA.[45]

In an interesting set of experiments disulfide-linked peptides corresponding to the basic region of Jun have been prepared in which the cross-linking was performed in three different ways: N- to N-termini, C- to C-termini, and N- to C-termini.[52,53] The three cross-linked peptides each bound to oligonucleotides having the appropriate half-site orientation with a dissociation constant of approximately 4 nM at 4 °C. Extension of this work to a peptide containing three cross-linked Jun basic regions, which were disulfide-linked in the C- to C- to N-termini orientation, also provided specific DNA binding to an oligonucleotide which contained three half-binding sites with a dissociation constant of approximately 5 nM at 4 °C.[54] In an analogous set of experiments the basic regions of GCN4 and C/EBP were covalently cross-linked via C-terminal Lys residues to form either the homodimeric peptides or the heterodimeric peptide.[55] The binding affinity and specificity of the homodimeric peptides mirrored the results obtained with the bZip peptides of GCN4 and C/EBP, whereas the heterodimeric peptide bound specifically to an oligonucleotide containing both the GCN4 and C/EBP half sites.

In a recent study, Goddard *et al.* have demonstrated that a monomeric basic peptide from Jun displays affinity for the AP1 site.[56] Taylor *et al.* have prepared peptides corresponding to a single GCN4 basic region with carboxamide cross-links to stabilize the helical conformation and have observed specific DNA binding.[57] Although the binding obtained for these monomeric peptides is weaker than that observed for dimeric peptides, these experiments open up the possibility of designing peptides with increased DNA affinity based on monomeric peptide sequences.

### 7.13.2.1.4  *Peptides based on the zinc finger motif*

The zinc finger motif appears to be one of the most widely used domains of DNA binding proteins.[2-6,58,59] To date, four cocrystal structures between zinc finger proteins and DNA have been solved, and numerous two-dimensional NMR structures have been solved for individual zinc fingers. Zinc finger proteins are composed of three general classes. The first class of zinc finger motifs contains approximately 25 residues, including two Cys and two His residues, which fold into a compact unit composed of a helix packed against a $\beta$-hairpin. The second class is composed of about 30 residues and contains four Cys residues. Together two motifs of this class form a single structural unit with one dimerization helix and one recognition helix interacting in the major groove. A third smaller family of zinc finger motifs contains two zinc ions and six Cys residues. As in the second class there is a dimerization and a recognition helix.

In this chapter the focus will be on determining the minimal zinc finger sequence involved in sequence-specific DNA binding. Peptides are included which contain from one to three zinc finger units, although the triple fingers are longer than our arbitrary size limit of 60 amino acid residue peptides.

### (i)  *Single-zinc-finger peptides*

It was initially believed that peptides, especially of the first class, containing a single zinc finger, were not able to bind sequence specifically to DNA. Recently two examples of single zinc fingers with high-affinity, specific DNA binding have been reported. In one case a peptide from the *Drosophila* transcription factor GAGA (first class) containing residues 310–372 was prepared and bound specifically to an oligonucleotide containing the core consensus sequence GAGAGA.[60] The zinc finger motif in this peptide, however, is flanked on both termini by highly basic regions. Removal of 19 amino acids on the C-terminus had no effect on DNA binding, whereas removal of 27 residues from the N-terminus resulted in complete loss of sequence-specific DNA binding. More structural information will be essential to determine if GAGA interacts with DNA in the same fashion as other zinc-finger proteins of the first class. In a second example a 59 amino acid peptide containing the C-terminal zinc finger with its adjacent basic region derived from the erythroid transcription factor GATA-1 (second class) was synthesized and was found to bind sequence-specifically to an oligonucleotide containing a single GATA motif with an affinity that was only an order of magnitude less than the full-length GATA-1 protein.[61] Removal of six residues from the C-terminal basic region resulted in complete loss of DNA binding. In both of these examples, although the single zinc finger is necessary for DNA binding, the finger alone is not sufficient for high-affinity DNA recognition; an adjacent basic region is also essential. The examples where the zinc finger alone has been used for DNA binding have shown nonspecific affinity for DNA.

### (ii)  *Double-zinc-finger peptides*

Peptide sequences containing one pair of zinc finger motifs have been described with the ability to make sequence-specific interactions with DNA. Two-finger motifs from proteins of the first class have been prepared, such as the *Drosophila melanogaster* regulatory protein Tramtrack, a human enhancer binding protein, MBP-1, and the transcription factor SW15, which contain 66, 57, and 70 amino acids, respectively. The Tramtrack double-zinc-finger peptide binds in a sequence-specific manner to an oligonucleotide containing a natural target site with a dissociation constant of approximately 400 nM.[62] A cocrystal structure has been solved for Tramtrack with DNA, and each of the two zinc fingers forms an independent DNA-binding domain with, at the N-terminus, helix residues binding into the major groove of the DNA (Figure 7).[63] Similarly, the MBP-1 peptide was found to interact with an oligonucleotide comprising a portion of the major histocompatibility complex enhancer sequence, with a dissociation constant of 140 nM.[64] The SW15 peptide, which contains two zinc fingers, maintained specific DNA binding, although much higher peptide concentrations were needed for binding compared to a three-zinc-finger peptide from SW15.[65]

Two-finger peptides from the second and third classes of zinc-finger proteins have also been studied. The 71-residue peptide fragment of the glucocorticoid receptor,[66] the double-zinc-finger peptide from the estrogen receptor,[67] and a 54-residue peptide from the yeast transcriptional activator GAL4[68] (third class) bind specifically to DNA. Interestingly, however, each double-zinc-

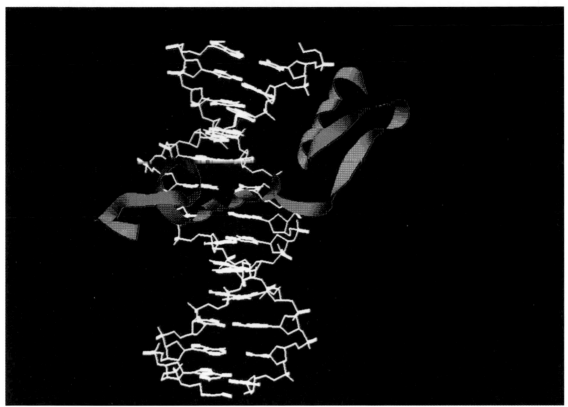

**Figure 7** Cocrystal structure of the Tramtrack zinc finger peptide and DNA.[63]

finger motif of the second and third classes utilizes only one helix when interacting with the major groove of DNA, and dimerization accounts for higher affinity binding.

With a knowledge of the specific orientation of peptide–DNA binding interactions, it should be possible to design fused peptides with unique DNA binding specificities. A novel DNA-binding peptide was designed in this way by covalently linking a 57-residue, two-finger peptide of Zif268 to the 61-residue homeodomain of the Oct-1 protein.[69] The two-peptide fusion protein bound optimally ($K_d$ of 0.8 nM) to an oligonucleotide containing adjacent homeodomain (TAATTA) and zinc finger (TGGGCG) subsites.

### (iii) Triple-zinc-finger peptides

A number of peptides containing three zinc-finger units have been shown to bind specifically to DNA. The main reason for including triple-zinc-finger peptides in this chapter is to point out the design work of Berg *et al.* in which individual zinc fingers from the first class of the motif were mixed and matched to obtain a desired DNA binding specificity (Figure 8).[70] The peptide was designed to contain the consensus sequence of CP-1, but the residues involved in DNA recognition were modified in each of the three zinc fingers. The first finger of the peptide was based on a mutant of the human transcription factor Spl and contained residues Gln13, Asp16, and Arg19. The second finger was based on the transcription factors Zif268 and Spl, and contained residues Arg14, Glu16, and Arg19. The third finger was based on mutants of Spl and Krox-20, and contained residues Arg13, His16, and Arg19. The designed peptide bound sequence specifically to an oligonucleotide containing the predicted binding site 5′-GGG GCG GCT 3′ with a dissociation constant of approximately 2–3 nM. This approach relies on the similarity of the DNA binding structure in the first class of zinc finger proteins, but has great potential for specifically recognizing any length and sequence of DNA at will.

In another set of experiments a library of sequences was used to determine if a code for zinc finger–DNA interactions could be developed.[71] A phage display library was prepared which contained $2.6 \times 10^6$ sequences based on three fingers of Zif268. This library was evaluated for binding to operator sequences in which the middle DNA triplet was altered. The results obtained from this

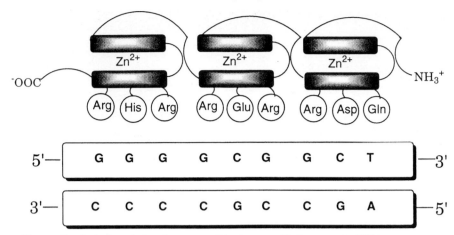

**Figure 8** A designed zinc finger peptide shown with its cognate DNA sequence.

study highlight the fact that there are only three positions involved in the recognition of DNA, and that only a limited set of amino acids are used in these positions. In a similar fashion randomized DNA sequences were used to identify potentially coded zinc finger sequences.[72] The results of these studies were applied to the recognition of a specific oncogenic DNA site by a randomized zinc finger peptide using phage display.[73] A peptide sequence was obtained which bound specifically to the oncogenic site with a dissociation constant of 620 nM.

### 7.13.2.2 β-Sheet Peptides in the Major Groove

A number of protein–DNA cocrystal structures have been solved, for proteins such as TBP, and the Arc and Met repressors, in which the portion of the protein responsible for contacting the DNA exists to a large extent in a β-sheet conformation.[74,75] The Arc repressor is composed of 53 amino acid residues and contains the essential folding unit of a ribbon–helix–helix structure (Figure 9).[76] Dimerization of Arc brings together residues 8–14 from each monomer to form a two-stranded antiparallel β-sheet which inserts into the major groove of an operator half-site. By covalently cross-linking two Arc monomers with a peptide linker, the affinity for the half-site operator was increased from 185 pM for the wild type Arc to 1.7 pM for the cross-linked protein.[77]

To date, there are no examples in the literature of cross-linked β-strand peptides to delineate the smallest folding unit for sequence-specific DNA binding. Surovaya *et al.* have designed constrained peptides which are composed of a β-strand–turn–β-strand motif either with or without disulfide cross-linking.[78] In this case, however, potential DNA-specific residues were incorporated at the ends of the β-strands and in the turn, and this region of the peptide is believed to interact with the minor groove, as demonstrated by distamycin competition binding experiments. Similarly, a novel zinc ribbon motif has been found in the eukaryotic transcription elongation factor TFIIS.[79] This structure is composed of a three-stranded β-sheet with a zinc binding site, and a 55 amino acid residue peptide corresponding to the zinc ribbon motif has a preferred affinity for oligopyrimidine single strands. Whether it is the β-sheet portion of the zinc ribbon or the turns which interact with DNA is yet to be established.

### 7.13.3 PEPTIDE MINOR GROOVE BINDERS

There exist two major classes of peptides which interact specifically within the minor groove of DNA. One class is based upon the minor groove DNA-binding natural products distamycin and netropsin (Figure 10).[80–82] These crescent-shaped molecules consist of repeating units of pyrroles linked by amide bonds, and bind in the minor groove of DNA at sites containing four or five successive A,T base pairs. Many modifications to the natural structures have been made. The other class of minor groove-binding peptides are those derived from larger proteins which contain repeating units of proline and positively charged residues, termed generally the "A,T-hook."[83] These peptides are proposed to have a crescent shape similar to distamycin and netropsin, and interact in

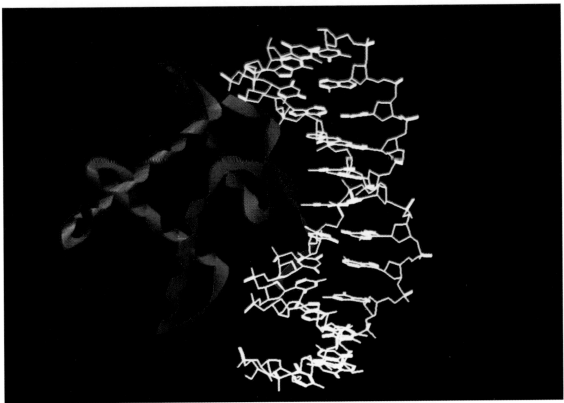

**Figure 9** Cocrystal structure of the Arc peptide and DNA.[76]

the minor groove with the backbone amides, forming hydrogen bonds with the DNA bases in the minor groove.

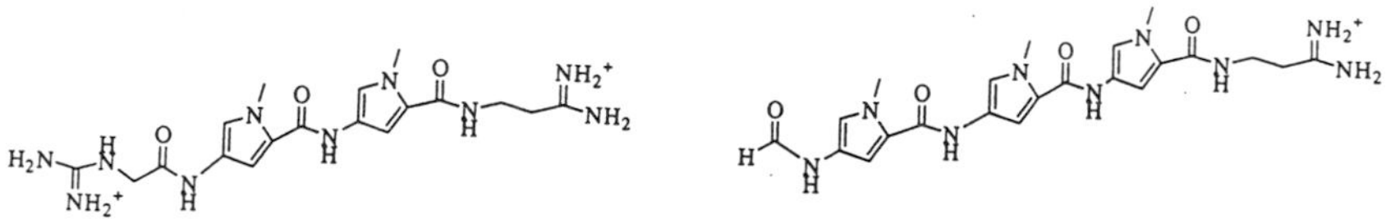

Netropsin                                   Distamycin

**Figure 10** Minor groove DNA-binding natural products.

A,T hook peptides derived from the yeast protein DAT1 and the nonhistone chromosomal protein HMG-I/Y, for example, have been shown to interact specifically in the minor groove of DNA. A 35-residue peptide corresponding to amino acids 2–36 of DAT1 bound in the minor groove at positions containing A-T tracts with a dissociation constant of 0.4 nM.[84] Within this peptide the sequence GRKPG is repeated three times, and the Arg residues were shown to be essential for high-affinity DNA binding. In a similar fashion an 11 amino acid peptide derived from HMG-I/Y with the sequence TPKRPRGRPKK was shown to have the same binding characteristics of the intact protein, and two-dimensional NMR experiments provided evidence that the RGR segment of the peptide is in contact with the minor groove.[85] A number of other DNA-binding proteins contain regions with sequence similarity to the DAT1 and HMG–I/Y peptides, but further experimentation is needed to determine if these sequences are sufficient for high-affinity, sequence-specific DNA binding.

## 7.13.4 PEPTIDE INTERCALATORS

A number of naturally occurring and designed peptides which contain two planar aromatic moieties have been found to interact with DNA by intercalation. The quinoxaline family of anti-tumor antibiotics (Figure 11), for instance, are a class of cyclic octadepsipeptides which contain two quinoxaline moieties attached to the peptide chain. Echinomycin and triostin A are two of the better known drugs of this family which have been shown to bisintercalate into DNA with the quinoxaline chromophores preferentially binding at CpG steps in the minor groove of a double helix.[86–88] A functionally related class of cyclic decadepsipeptides including luzopeptins A-E,[89–91] BBM-928A,[92] quinaldopeptin,[93] and sandramycin,[94] which contain two pendant quinoxaline moieties, have also been shown to interact with DNA by bifunctional intercalation.

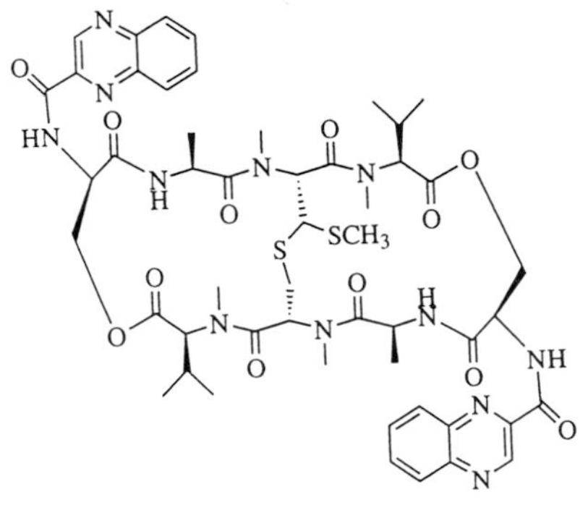

Echinomycin 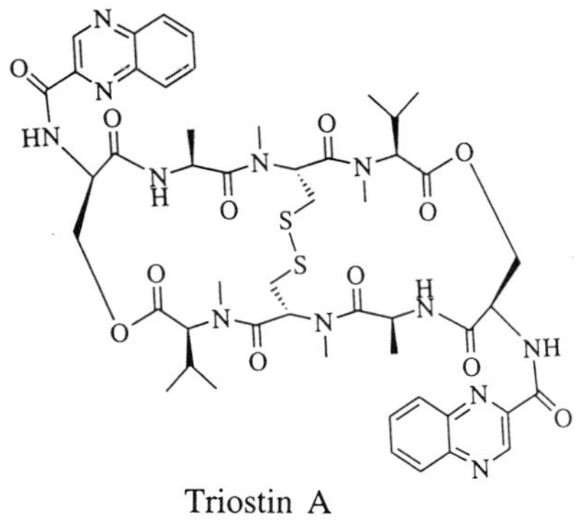          Triostin A

**Figure 11** Peptide intercalators of the quinoxaline family.

Naturally occurring sequences within proteins have been shown to interact with DNA by binding of specific tyrosine residues. A tandem repeat of the sequence SPTSPSY, for instance, has been found in the largest subunit of RNA polymerase II with 26 units in yeast and 52 units in mammals. Synthetic peptides containing two repeating units were found to bind DNA by intercalation of tyrosine residues.[95]

Synthetic linear peptides containing two aromatic moieties have been designed to interact with DNA via intercalation.[96] A synthetic bis(acridine) containing a peptide of the sequence YKKG was found to bind to DNA by intercalation of both chromophores with a 140-fold enhancement of affinity, as compared to 9-aminoacridine, whereas introduction of two $p$-$NO_2$-Phe residues into a peptide with the sequence $KF_{NO_2}AF_{NO_2}$ also provided a peptide with bisintercalation properties.

## 7.13.5 CONCLUSIONS

Most of the examples of DNA-binding peptides presented in this chapter are simply truncated regions of proteins whose sequences correspond exactly to the DNA-binding portions of the proteins. As more structural information of protein–DNA interactions has become available, however, more modifications to DNA-binding motifs have been made. These efforts have led to peptides with unique DNA-binding specificities with the potential for therapeutic application. It is hoped that more examples of *de novo* designed peptides are now within reach.

## 7.13.6 REFERENCES

1. J. D. Watson and F. H. C. Crick, *Nature*, 1953, **171**, 964.
2. P. F. Johnson and S. L. McKnight, *Annu. Rev. Biochem.*, 1989, **58**, 799.
3. T. A. Steitz, *Q. Rev. Biophy.*, 1990, **23**, 205.
4. C. A. Pabo and R. T. Sauer, *Annu. Rev. Biochem.*, 1992, **61**, 1053.
5. T. Ellenberger, *Curr. Opin. Struct. Biol.*, 1994, **4**, 12.
6. S. K. Burley, *Curr. Opin. Struct. Biol.*, 1994, **4**, 3.
7. H. C. M. Nelson, *Curr. Opin. Struct. Biol.*, 1995, **6**, 180.
8. S. C. Harrison and A. K. Aggarwal, *Annu. Rev. Biochem.*, 1990, **59**, 933.
9. R. G. Brennan, *Curr. Opin. Struct. Biol.*, 1991, **1**, 80.
10. R. T. Sauer, R. R. Yocum, R. F. Doolittle, M. Lewis, and C. O. Pabo, *Nature*, 1982, **298**, 447.
11. M. F. Bruist, S. J. Horvath, L. E. Hood, T. A. Steitz, and M. I. Simon, *Science*, 1987, **235**, 777.
12. J. A. Feng, R. C. Johnson, and R. E. Dickerson, *Science*, 1994, **263**, 348.
13. S. S. Abdel-Meguid, N. D. F. Grindley, N. S. Templeton, and T. A. Steitz, *Proc. Natl. Acad. Sci. USA*, 1984, **81**, 2001.
14. R. G. Brennan, S. L. Roderick, Y. Takeda, and B. W. Matthews, *Proc. Natl. Acad. Sci. USA*, 1990, **87**, 8165.
15. M. C. Mossing and R. T. Sauer, *Science*, 1990, **250**, 1712.
16. P. Bishop and J. Chmielewski, unpublished results.
17. R. Mayer, G. Lancelot, and C. Hélène, *FEBS Lett.*, 1983, **153**, 339.

18. W. J. Gehring, Y. Q. Qian, M. Billeter, K. Furukubo-Tokunaga, A. F. Schier, D. Resendex-Perez, M. Affolter, G. Otting, and K. Wüthrich, *Cell*, 1994, **78**, 211.
19. W. J. Gehring, M. Affolter, and T. Bürglin, *Annu. Rev. Biochem.*, 1994, **63**, 487.
20. M. Affolter, A. Percival-Smith, M. Müller, W. Leupin, and W. J. Gehring, *Proc. Natl. Acad. Sci. USA*, 1990, **87**, 4093.
21. G. Otting, Y. Q. Qian, M. Billeter, M. Müller, M. Affolter, W. J. Gehring, and K. Würthrich, *EMBO J.*, 1990, **9**, 3085.
22. H. Mihara and E. T. Kaiser, *Science*, 1988, **242**, 925.
23. C. R. Kissinger, B. Liu, E. Martin-Bianco, T. B. Kornberg, and C. O. Pabo, *Cell*, 1990, **63**, 579.
24. J. A. Hirsch and A. K. Aggarwal, *EMBO J.*, 1995, **14**, 6280.
25. S. C. Ekker, K. E. Young, D. P. von Kessler, and P. A. Beachy, *EMBO J.*, 1991, **10**, 1179.
26. A. Percival-Smith, M. Müller, M. Affolter, and W. J. Gehring, *EMBO J.*, 1990, **9**, 3967.
27. C. Wolberger, A. K. Vershon, B. Liu, A. D. Johnson, and C. O. Pabo, *Cell*, 1991, **67**, 517.
28. C. Goutte and A. D. Johnson, *J. Mol. Biol.*, 1993, **233**, 359.
29. C. Goutte and A. D. Johnson, *Cell* 1988 **52**, 875
30. A. M. Dranginis, *Nature*, 1990, **347**, 682.
31. C. Goutte and A. D. Johnson, *EMBO J.*, 1994, **13**, 1434.
32. T. Li, M. R. Stark, A. D. Johnson, and C. Wolberger, *Science*, 1995, **270**, 262.
33. C. Murre, G. Bain, M. A. van Kijk, I. Engel, B. A. Furnari, M. E. Massari, J. R. Matthews, M. W. Quong, R. R. Rivera, and M. H. Stuiver, *Biochim. Biophys. Acta*, 1994, **1218**, 129.
34. S. E. V. Phillips, *Structure*, 1994, **2**, 1.
35. M. Lenardo, J. W. Pierce, and D. Baltimore, *Science*, 1987, **236**, 1573.
36. P. Bishop, C. Jones, I. Ghosh, and J. Chmielewski, *Int. J. Peptide Protein Res.*, 1995, **46**, 149.
37. T. Ellenberger, D. Fass, M. Arnaud, and S. C. Harrison, *Genes Dev.*, 1994, **8**, 970.
38. S. J. Anthony-Cahill, P. A. Benfield, R. Fairman, Z. R. Wasserman, S. L. Brenner, W. F. Stafford, III, C. Altenbach, W. L. Hubbell, and W. F. DeGrado, *Science*, 1992, **255**, 979.
39. T. Morii, M. Simomura, S. Morimoto, and I. Saito, *J. Am. Chem. Soc.*, 1993, **115**, 1150.
40. A. R. Ferré-D'Amaré, P. P. Pgnonec, R. G. Roeder, and S. K. Burley, *EMBO J.*, 1994, **13**, 180.
41. T. Alber, *Curr. Opin. Genet. Dev.*, 1992, **2**, 205.
42. I. A. Hope and K. Struhl, *Cell*, 1986, **46**, 885.
43. J. A. Nye and B. J. Graves, *Proc. Natl. Acad. Sci. USA*, 1990, **87**, 3993.
44. R. Turner and R. Tjian, *Science*, 1989, **243**, 1689.
45. J. N. M. Glover and S. C. Harrison, *Nature*, 1995, **373**, 257.
46. T. Sera and P. G. Schultz, *Proc. Natl. Acad. Sci. USA*, 1996, **93**, 2920.
47. K. T. O'Neal, R. H. Hoess, and W. F. DeGrado, *Science*, 1990, **249**, 774.
48. R. V. Talanian, C. J. McKnight, and P. S. Kim, *Science*, 1990, **249**, 769.
49. R. V. Talanian, C. J. McKnight, R. Rutkowski, and P. S. Kim, *Biochemistry*, 1992, **31**, 6871.
50. B. Cuenoud and A. Schepartz, *Science*, 1993, **259**, 510.
51. M. Ueno, A. Murakami, K. Makino, and T. Morii, *J. Am. Chem. Soc.*, 1993, **115**, 12 575.
52. C. Park, J. L. Campbell, and W. A. Goddard, III, *Proc. Natl. Acad. Sci. USA*, 1992, **89**, 9094.
53. C. Park, J. L. Campbell, and W. A. Goddard, III, *Proc. Natl. Acad. Sci. USA*, 1993, **90**, 4892.
54. C. Park, J. L. Campbell, and W. A. Goddard, III, *J. Am. Chem. Soc.*, 1995, **117**, 6287.
55. M. Pellegrini and R. H. Ebright, *J. Am. Chem. Soc.*, 1996, **118**, 5831.
56. C. Park, J. L. Campbell, and W. A. Goddard, *J. Am. Chem. Soc.*, 1996, **118**, 4235.
57. B. Y. Wu, B. L. Gaffney, R. A. Jones, and J. W. Taylor, in "Peptides: Chemistry, Structure and Biology," eds. P. T. P. Kaumaya and R. S. Hodges, Mayflower Scientific Ltd., 1996, p. 265.
58. J. M. Berg, *Acc. Chem. Res.*, 1995, **28**, 14.
59. A. Klug and J. W. R. Schwabe, *FASEB J.*, 1995, **9**, 597.
60. P. V. Pedone, R. Ghirlando, G. M. Clore, A. M. Gronenborn, G. Felsenfeld, and J. G. Omichinski, *Proc. Natl. Acad. Sci. USA*, 1996, **93**, 2822.
61. J. G. Omichinski, C. Trainor, T. Evans, A. M. Gronenborn, G. M. Clore, and G. Felsenfeld, *Proc. Natl. Acad. Sci. USA*, 1993, **90**, 1676.
62. L. Fairall, S. D. Harrison, A. A. Travers, and D. Rhodes, *J. Mol. Biol.*, 1992, **226**, 349.
63. L. Fairall, J. W. R. Schwabe, L. Chapman, J. T. Finch, and D. Rhodes, *Nature*, 1993, **366**, 483.
64. K. Sakaguchi, E. Appella, J. G. Omichinski, G. M. Clore, and A. M. Gronenborn, *J. Biol. Chem.*, 1991, **266**, 7306.
65. D. Neuhaus, Y. Nakaseko, K. Nagai, and A. Klug, *FEBS Lett.*, 1990, **262**, 179.
66. T. Härd, E. Kellenbach, R. Boelens, B. A. Maler, K. Dahlman, L. P. Freedman, J. Carlstedt-Duke, K. R. Yamamoto, J.-Ä. Gustafsson, and R. Kaptein, *Science*, 1990, **249**, 157.
67. J. W. R. Schwabe, L. Chapman, J. T. Finch, and D. Rhodes, *Cell*, 1993, **75**, 567.
68. R. Marmorstein, M. Carey, M. Ptashne, and S. C. Harrison, *Nature*, 1992, **356**, 408.
69. J. L. Pomerantz, P. A. Sharp, and C. O. Pabo, *Science*, 1995, **267**, 93.
70. J. R. Desjarlais and J. M. Berg, *Proc. Natl. Acad. Sci. USA*, 1993, **90**, 2256.
71. Y. Choo and A. Klug, *Proc. Natl. Acad. Sci. USA*, 1994, **91**, 11 163.
72. Y. Choo and A. Klug, *Proc. Natl. Acad. Sci. USA*, 1994, **91**, 11 168.
73. Y. Choo, I. Sánchez-García, and A. Klug, *Nature*, 1994, **372**, 642.
74. S. E. Phillips, *Annu. Rev. Biophys. Biomol. Struct.*, 1994, **23**, 671.
75. B. E. Rauman, B. M. Brown, and R. T. Sauer, *Curr. Opin. Struct. Biol.*, 1994, **4**, 36.
76. B. E. Rauman, M. A. Rould, C. O. Pabo, and R. T. Sauer, *Nature*, 1994, **374**, 754.
77. C. R. Robinson and R. T. Sauer, *Biochemistry*, 1996, **35**, 109.
78. A. N. Surovaya, S. L. Gokhovskii, R. V. Brusov, Y. P. Lysov, A. L. Zhuze, and G. V. Gurskii, *Mol. Biol.*, 1995, **28**, 859.
79. X. Qian, C. J. Jeon, H. S. Yoon, K. Agarwal, and M. A. Weiss, *Nature*, 1993, **365**, 277.
80. P. G. Schultz and P. B. Dervan, *J. Biomol. Struct. Dyn.*, 1984, **1**, 1133.
81. P. B. Dervan, *Science*, 1986, **232**, 464.
82. C. Zimmer and U. Wähnert, *Prog. Biophys. Mol. Biol.*, 1986, **47**, 31.

83. R. Reeves and M. S. Nissen, *J. Biol. Chem.*, 1990, **265**, 8573.
84. B. J. Reardon, R. S. Winters, D. Gordon, and E. Winter, *Proc. Natl. Acad. Sci. USA*, 1993, **90**, 11 327.
85. B. H. Geierstanger, B. F. Volkman, W. Kremer, and D. E. Wemmer, *Biochemistry*, 1994, **33**, 5347.
86. A. H.-J. Wang, G. Ughetto, G. J. Quigley, and A. Rich, *J. Biomol. Struct. Dyn.*, 1986, **4**, 319.
87. A. H.-J. Wang, G. Ughetto, G. J. Quigley, T. Hakoshima, G. A. van der Marel, J. H. van Boom, and A. Rich, *Science*, 1984, **225**, 1115.
88. G. J. Quigley, G. Ughetto, G. A. van der Marel, J. H. van Boom, A. H.-J. Wang, and A. Rich, *Science*, 1986, **232**, 1255.
89. H. Ohkuma, F. Sakai, Y. Nishiyama, M. Ohbayashi, H. Imanishi, M. Konishi, T. Miyaki, H. Kosiyama, and H. Kawaguchi, *J. Antibiot.*, 1980, **33**, 1087.
90. M. Konishi, H. Ohkuma, F. Sakai, T. Tsuno, H. Koshiyama, T. Naito, and H. Kawaguchi, *J. Am. Chem. Soc.*, 1981, **103**, 1241.
91. E. Arnold and J. Clardy, *J. Am. Chem. Soc.*, 1981, **103**, 1243.
92. C.-H. Huang, S. Mong, and S. T. Crooke, *Biochemistry*, 1980, **19**, 5537.
93. S. Toda, K. Sugawara, Y. Nishiyama, M. Ohbayashi, N. Ohkusa, H. Yamammoto, K. Konishi, and T. Oki, *J. Antibiot.*, 1990, **43**, 796.
94. D. L. Boger and J.-H. Chen, *J. Am. Chem. Soc.*, 1993, **115**, 11 624.
95. M. Suzuki, *Nature*, 1990, **344**, 562.
96. C. Robledo-Luiggi, W. D. Wilson, E. Pares, M. Vera, C. S. Martinez, and D. Santiago, *Biopolymers*, 1991, **31**, 907.

# 7.14

# Covalent Modification of DNA by Natural Products

## KENT S. GATES
*University of Missouri–Columbia, MO, USA*

7.14.1  INTRODUCTION                                                                            492
    *7.14.1.1   Why Are DNA-damaging Natural Products of Interest?*                         492
    *7.14.1.2   Classification of DNA-damaging Natural Products*                            492
    *7.14.1.3   General Mechanisms of Covalent DNA Modification*                          493

7.14.2  DNA-DAMAGING NATURAL PRODUCTS                                                           495
    *7.14.2.1   Imines*                                                                     495
        *7.14.2.1.1   Pyrrolo[1,4]benzodiazepines*                               495
        *7.14.2.1.2   Saframycins, renieramycins, and safracins*                 497
        *7.14.2.1.3   Oxazolidine-containing natural products*                   499
        *7.14.2.1.4   Barminomycin I*                                            500
        *7.14.2.1.5   Other antibiotics that alkylate DNA via imine formation*   501
    *7.14.2.2   Carbonyl-containing Natural Products*                                       502
        *7.14.2.2.1   Aldehydes*                                                 502
        *7.14.2.2.2   Isochrysohermidine*                                        503
        *7.14.2.2.3   Lactones*                                                  503
    *7.14.2.3   Cyclopropanes*                                                              504
        *7.14.2.3.1   CC-1065 and the duocarmycins*                              504
        *7.14.2.3.2   Myrocin C*                                                 506
        *7.14.2.3.3   Ptaquiloside and the illudins*                             507
    *7.14.2.4   Epoxides*                                                                   508
        *7.14.2.4.1   Pluramycins and pluramycinones*                            508
        *7.14.2.4.2   Kapurimycin A3 and the clecarmycins*                       510
        *7.14.2.4.3   Alkoxyl radicals from vinyl epoxides*                      510
        *7.14.2.4.4   Psorospermin*                                              511
        *7.14.2.4.5   Metabolically activated mycotoxins: the aflatoxins*        511
        *7.14.2.4.6   Other epoxides*                                            513
    *7.14.2.5   Aziridines*                                                                 513
        *7.14.2.5.1   Azinomycin B/carzinophilin*                                513
        *7.14.2.5.2   Azicemicins*                                               514
        *7.14.2.5.3   Other aziridines*                                          514
    *7.14.2.6   Pyrrole-derived Cross-linking Agents*                                       514
        *7.14.2.6.1   The mitomycins*                                            514
        *7.14.2.6.2   FR66979 and FR900482*                                      518
        *7.14.2.6.3   Oxidatively activated pyrrolizidine alkaloids*             518
    *7.14.2.7   Alkenylbenzenes*                                                            519
    *7.14.2.8   N-Nitroso Compounds*                                                        520
        *7.14.2.8.1   Streptozocin*                                              520
        *7.14.2.8.2   Nitrosamines*                                              521
    *7.14.2.9   Nitroaromatics*                                                             521
        *7.14.2.9.1   Azomycin*                                                  521
        *7.14.2.9.2   Aristolochic acid*                                         522
    *7.14.2.10   1,2-Dithiolan-3-one 1-Oxides*                                              522
        *7.14.2.10.1   Leinamycin*                                               522
        *7.14.2.10.2   1,2-Dithiole-3-thiones*                                   526

7.14.2.10.3   *Polysulfides*                                                                 527
7.14.2.11   *Quinones*                                                                       527
    7.14.2.11.1   *Redox cycling of quinones and quinoid compounds*                            528
    7.14.2.11.2   *DNA alkylation by quinone methides*                                         529
    7.14.2.11.3   *Formaldehyde-mediated covalent attachment of anthracyclines to DNA*         530
    7.14.2.11.4   *DNA alkylation by quinones*                                                 531
    7.14.2.11.5   *Other DNA damage by quinones*                                               532
7.14.2.12   *Heterocyclic N-Oxides*                                                           532
    7.14.2.12.1   *Phenazine N-oxides*                                                         532
    7.14.2.12.2   *Quinoxaline N-oxides*                                                       533
7.14.2.13   *Resorcinols*                                                                     535
7.14.2.14   *Bleomycin and Other Metal-binding Antibiotics*                                   536
7.14.2.15   *Various Agents That Reduce Molecular Oxygen to Superoxide*                       540
7.14.2.16   *Diazo and Diazonium Compounds*                                                   541
    7.14.2.16.1   *Kinamycin*                                                                  541
    7.14.2.16.2   *Diazoketones*                                                               542
    7.14.2.16.3   *Benzenediazonium ions*                                                      543
7.14.2.17   *Enediynes*                                                                       544
7.14.2.18   *Photochemically Activated Agents*                                                544
    7.14.2.18.1   *Light-dependent DNA damage not involving covalent adducts*                  544
    7.14.2.18.2   *Formation of covalent photoadducts*                                         545
7.14.2.19   *Restriction and Methylation Enzymes*                                             546

7.14.3   REFERENCES                                                                           546

## 7.14.1   INTRODUCTION

### 7.14.1.1   Why Are DNA-damaging Natural Products of Interest?

This chapter is devoted to the consideration of secondary metabolite natural products that form covalent adducts with DNA and/or engage in reactions that lead to the rupture of covalent bonds in DNA. Agents that cleave DNA via topoisomerase-mediated mechanisms are considered separately, in Chapter 7.16 of this volume.

The utility, if any, of secondary metabolites to the organisms that produce them remains a matter of debate;[1,2] however, it is commonly believed that natural products which display potent biological activity are the result of natural selection and that biosynthesis of these agents confers a selective advantage upon the producing organism.[2] DNA-damaging natural products frequently possess potent cytotoxic, cytostatic, or mutagenic properties and, in nature, may serve as either offensive or defensive weapons in the struggle for survival. Regardless of the reason for their existence, natural products constitute a vast library of organic compounds that can serve as a useful resource.

A practical reason for the longstanding interest in DNA-damaging natural products is the fact that the cytotoxic or cytostatic effects of these agents sometimes endow them with useful medicinal properties, especially as potential anticancer therapeutics.[3–6] Several DNA-damaging natural products are currently in use for the treatment of various cancers and others have served as lead compounds in the development of therapeutic agents.[3–6] Another significant impetus for the study of natural DNA-damaging agents lies in the fact that the chemical reactions and molecular recognition processes employed by these compounds are sometimes unprecedented and strikingly efficient. Thus, natural products provide elegant examples of chemical processes that can effectively, and often selectively, modify DNA in a manner that has significant biological consequences. Finally, natural products with extremely potent biological activities sometimes reveal unforeseen biological pathways and these compounds can become useful tools for elucidating the details of complex life processes. For example, studies with CC-1065[7] and aflatoxin[8] have led to insights regarding mechanisms of cell death and carcinogenesis, respectively.

### 7.14.1.2   Classification of DNA-damaging Natural Products

There are a bewilderingly large number of structurally diverse natural products that damage DNA; however, it is possible to divide this large number of DNA-damaging natural products into a relatively small number of categories if one considers the chemical reactions by which these compounds modify DNA. Although it is not possible to be all-inclusive, this chapter attempts a comprehensive survey of natural products known to covalently modify DNA. The natural products discussed in this chapter are divided into categories based upon the *functional groups* and *chemical reactions* by which these agents damage DNA.

The range of naturally occurring functional groups that have sufficient stability to survive in the cellular environment, yet sufficient reactivity to modify DNA covalently, is quite broad. The chemistry by which these functional groups react with DNA is diverse; some natural products contain functionality that is inherently reactive with DNA, while others require *in vivo* chemical or enzymatic activation; some natural products react with DNA stoichiometrically and others produce DNA damage catalytically. It is not uncommon to find multiple DNA-reactive functional groups in a single natural product or multiple mechanisms of DNA damage mediated by a single functional group.

The emphasis on covalent reaction chemistry in this chapter is not meant to diminish the profound importance that functionality not directly involved in covalent reactions with DNA often has in DNA binding, cell uptake and transport, or triggering of DNA reactions.

### 7.14.1.3 General Mechanisms of Covalent DNA Modification

The majority of all reactions involving covalent modification of DNA can be placed in one of two categories: (i) reaction of electrophiles with nucleophilic sites on DNA, or (ii) reaction of radicals with DNA. A very brief overview of these reactions is presented here. The chemical reactivity of DNA has been discussed in detail in Chapter 7.11 of this volume.

Electrophiles can react at a variety of nucleophilic sites in DNA, although, depending on their chemical structure, electrophiles generally show selectivity for certain nucleophilic sites in DNA.[9-14] Common sites for modification of duplex DNA by electrophilic natural products include N-7 of guanine, N-3 of adenine, N-7 of adenine, the exocyclic $N^2$ amino group of guanine, and N-3 of guanine (Figure 1).[9,11,13,14]

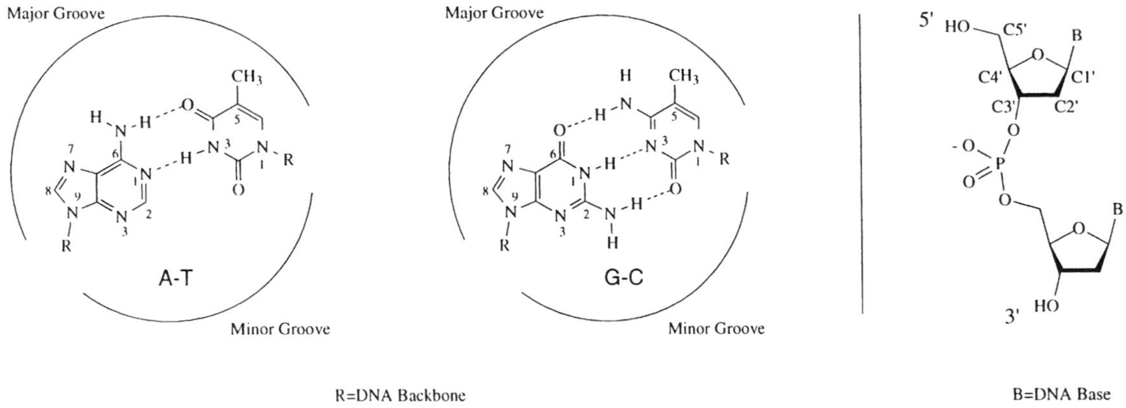

**Figure 1** Watson–Crick base pairs and the sugar–phosphate backbone of DNA.

Electrophilic modification at sites such as N-7 or N-3 of purine residues results in labilization of the glycosidic bond,[14-16] ultimately leading to the formation of an abasic site (**1**) that, under neutral conditions, slowly hydrolyzes to yield a DNA strand break (Scheme 1).[17,18] Abasic sites are rapidly converted to strand breaks under alkaline conditions.[15,16,19] Reaction of electrophilic species with the exocyclic nitrogens or carbonyl oxygens of the DNA bases, or with phosphate oxygens on the DNA backbone, generally affords relatively stable adducts that do not lead to spontaneous strand scission under conditions of moderate pH.[20-22]

Many radical species are capable of reacting with DNA at a variety of positions. Abstraction of hydrogen atoms from the deoxyribose backbone of DNA by radicals represents an important mode of DNA damage that has been the subject of intense investigation.[23-25] Abstraction of hydrogen atoms from the deoxyribose backbone almost universally leads to cleavage of the deoxyribose backbone of DNA through complex reaction cascades (for example, Scheme 2).[24,25] In addition to their reactions with the sugar-phosphate backbone, many radicals react extensively with the DNA

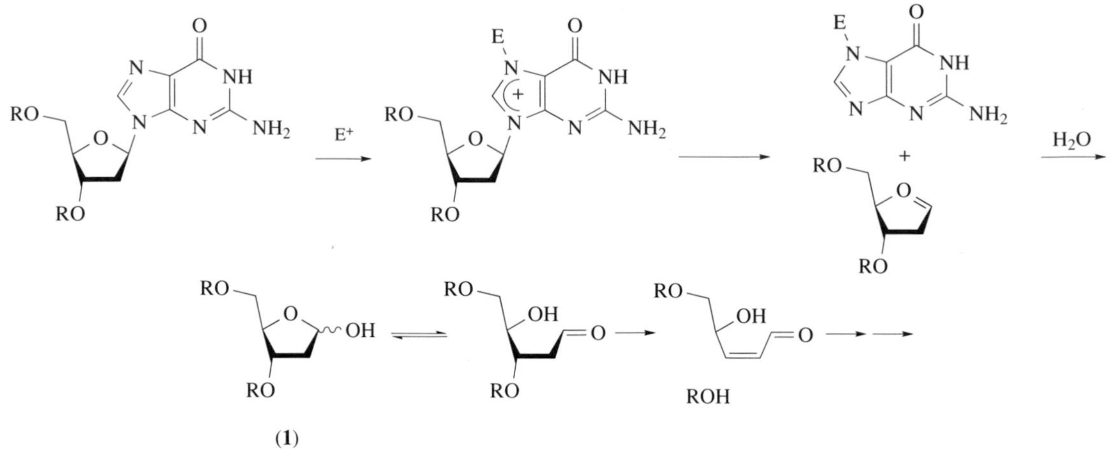

**Scheme 1**

bases.[26–29] These reactions do not necessarily lead to cleavage of the deoxyribose backbone; however, such modification of the structure of DNA bases is biologically important. In the absence of sequence-specific binding by a radical or radical precursor, cleavage of duplex DNA by radicals generally occurs with little sequence or base specificity.[30] This stands in contrast to DNA cleavage resulting from alkylating agents, which commonly shows marked selectivity for one (or more) of the DNA bases. DNA damage by various oxygen-centered,[23–25,31] carbon-centered,[26,32–34] and hydrogen radicals[25] has been reported.

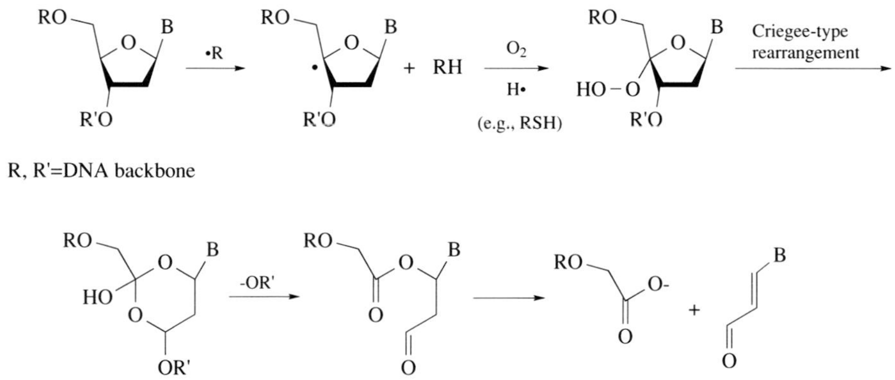

**Scheme 2**

Although the majority of chemical reactions between natural products and DNA involve radicals or electrophiles, species such as carbenes,[35] nitrenes,[35] singlet oxygen,[35–39] strong nucleophiles,[15,40] and photoexcited molecules[35–37] also can damage DNA. Some of these reactions are relevant to certain DNA-damaging natural products and, as such, will be discussed in the appropriate sections below.

This chapter is focused primarily on the chemical reactions of natural products with DNA and, in many cases, does not attempt to address the biological relevance of these reactions. Often, the biological relevance of *in vitro* studies is difficult to assess accurately, as these experiments are not intended to examine issues such as the repair of damaged DNA, cellular uptake of xenobiotics, and xenobiotic metabolism. Similarly, it must be remembered that demonstration of *in vitro* DNA damage by a biologically active natural product by no means indicates that DNA damage is solely responsible for the compound's biological activity. In general, however, damage to cellular DNA has profound biological consequences.[6,7,27,41–48]

## 7.14.2 DNA-DAMAGING NATURAL PRODUCTS

### 7.14.2.1 Imines

#### 7.14.2.1.1 *Pyrrolo[1,4]benzodiazepines*

A number of natural products containing a common pyrrolo[1,4]benzodiazepine (PBD) core have been isolated from various strains of *Streptomyces*.[49] These antibiotics show promising anticancer activity in a variety of assays.[49,50] The cytotoxic properties of the PBDs stem from the formation of covalent adducts with DNA that inhibit its processing by various enzymes.[51,52]

Structure–function analysis of PBDs shows that an imine, carbinolamine, or methanolamine functionality at N-10–C-11 is required for the formation of DNA adducts and for biological activity ((2), (3), Figure 2).[49,53] The imine, carbinolamine, and methanolamine forms of PBDs exist in equilibria, the concentration of each species depending upon conditions (Scheme 3).[49] The imine form of PBDs (4) is generally thought to be the electrophilic species involved in adduct formation with DNA (Scheme 3), although other species such as the iminium ion (5) or ring-opened amino-aldehyde (6) are possible intermediates (Scheme 4). Significant amounts of species other than the imine have not been observed in NMR experiments and, in chemical model reactions utilizing thiophenol as a nucleophile, the rate of reaction of various PBDs was found to correlate with the rates of imine formation.[49,54] The S-configuration found at C-11a in all naturally occurring PBDs confers a slight right-handed twist down the long axis of the molecule which is complementary to that of the minor groove of B-DNA. For certain PBDs, it has been shown that the *R* isomer does not bind DNA and is not biologically active.[55]

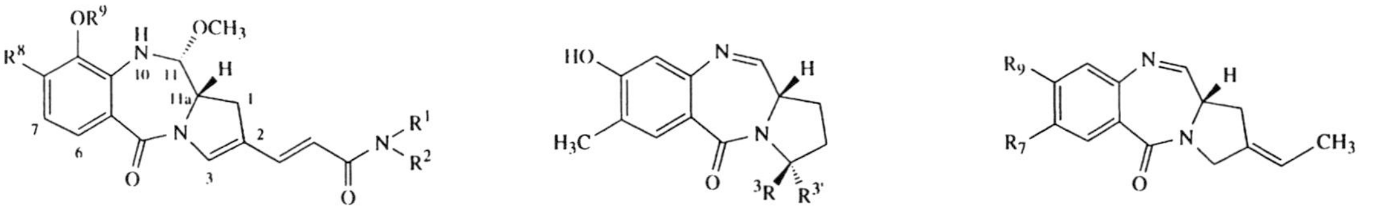

(2) Anthramycin: $R^8$=CH$_3$; $R^9$=R$^1$=R$^2$=H;
Mazethramycin: $R^8$=R$^1$=CH$_3$; $R^9$=R$^2$=H
Porothramycin: $R^8$=H; $R^9$=R$^1$=R$^2$=CH$_3$

Neothramycin A: R$^3$=H; R$^{3'}$=OH
Neothramycin B: R$^3$=OH; R$^{3'}$=H
DC-81: R$^3$=R$^{3'}$=H

Tomaymycin: R$_7$=CH$_3$O; R$_8$=HO
Prothracarcin: R$_7$=R$_8$=H

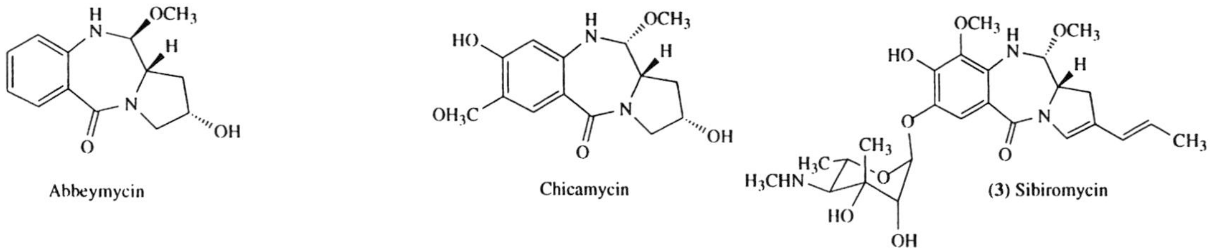

Abbeymycin

Chicamycin

(3) Sibiromycin

**Figure 2** Pyrrolo[1,4]benzodiazepine antibiotics.

Early experiments with the PBDs suggested that covalent attachment involves formation of an aminal linkage (7) with the exocyclic $N^2$ of guanine in the minor groove of duplex DNA (see Scheme 3).[56,57] This conclusion was based upon several experimental observations: (i) no adducts are formed in DNA that does not contain GC base pairs; (ii) the adducts do not lead to strand cleavage upon alkaline work-up, as N7 and N3 attachments would; (iii) adduct formation proceeds normally on T4 DNA, which contains glycosylated 5-hydroxylmethylcytosine residues that place a steric blockade in the major groove of DNA; (iv) 8-$^3$H-guanine-containing DNA shows no loss of tritium upon adduct formation, indicating that attachment is not at C-8 of guanine; and (v) polydI-dC containing DNA does not form adducts. Deoxyinosine (dI, **8**) is structurally analogous to deoxyguanosine, but lacks the exocyclic $N^2$-amino group. Covalent binding of the PBDs to DNA was measured by observing the irreversible association of radiolabeled antibiotics with the double helix.[58] Although isolation of PBD-base adducts (7) has not been possible due to their instability, high field NMR,[59] fluorescence

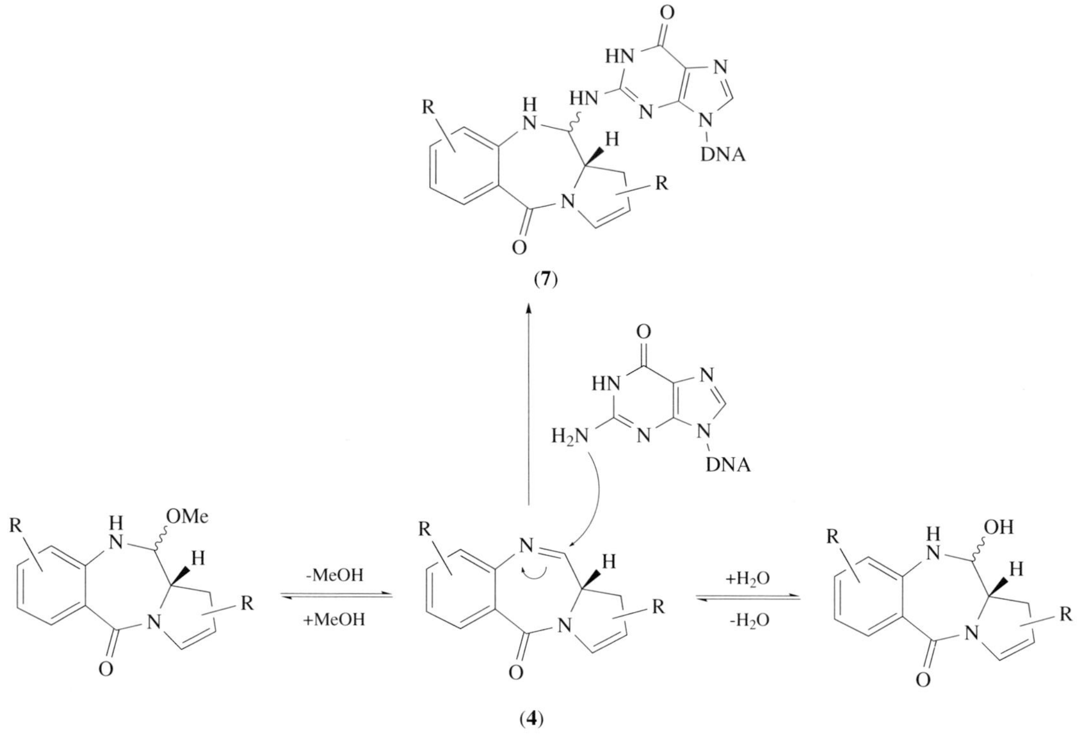

**Scheme 3**

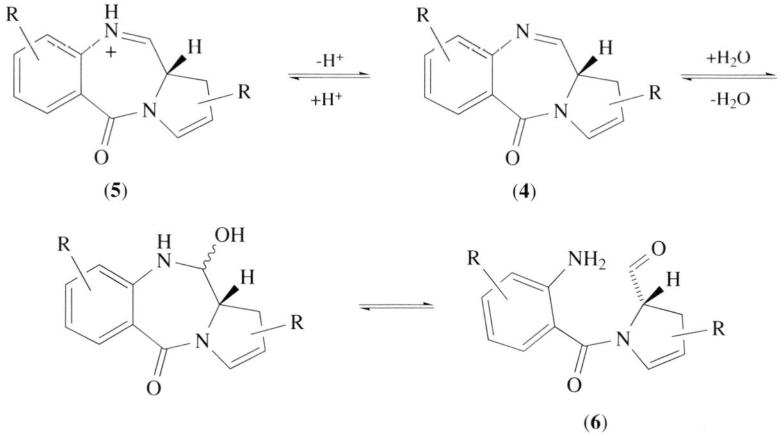

**Scheme 4**

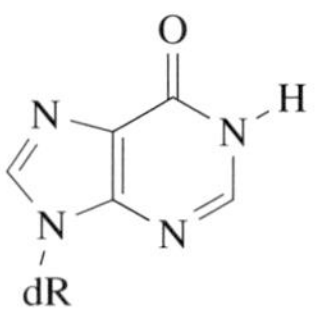

spectroscopy,[59] and, more recently, X-ray crystallography[60] have confirmed that adduct formation involves formation of an aminal linkage between N[2] of guanine and C-10 of the PBDs.

Formation of aminal adducts between PBDs and DNA is acid catalyzed at physiological pH.[61,62] The adducts decompose below pH 5, with the natural product released from the DNA unchanged.[61,62] PBDs do not form adducts with single-stranded DNA or RNA.[61,62] Furthermore, it appears that PBD–DNA adducts, sequestered in the minor groove of duplex DNA, are protected from hydrolysis, as denaturation of PBD-modified DNA results in reversal of adduct formation.[62]

Despite their small size, PBDs display marked sequence selectivity in their reactions with DNA.[49] In general, 3′-PuGPu sequences are favored sites of reaction for PBDs; however, sequence specificities differ somewhat for each PBD. Flanking sequences beyond the two base pairs adjacent to the adducted guanine affect the efficiency of DNA alkylation by PBDs. NMR spectroscopy indicates that the aminal adduct formed by anthramycin (2) has the S-configuration at C-11,[63] although the stereochemistry of PBD adducts, in general, depends upon the sequence context of the alkylated guanine and may differ for each agent.[59] Molecular modeling has been used to rationalize the sequence preferences and adduct stereochemistries for a number of PBDs.[64]

Hurley and co-workers have suggested that the sequence specificity of the PBDs arises from the fact that the natural twist of these compounds does not exactly match that of the minor groove; thus, more flexible sequences, such as the favored 5′-PuGPu sequences, that can easily deform to accommodate the adduct are preferred reaction sites.[52] In agreement with this hypothesis, it has been found that PBD adduct formation in duplex DNA induces a moderate DNA bend of about 6–15°. Also consistent with the notion that DNA flexibility plays a role in PBD sequence specificity is the fact that the amount of bending induced by several PBDs in various sequence contexts correlates with the rate of adduct formation, with sequences that adopt a more bent structure affording higher reaction rates.[65] An alternate hypothesis for the sequence specificity of the PBDs has been put forward suggesting that the 5′-PuGPu binding preference stems from the fact that, in general, purine–purine steps have smaller than average twist angles, thus producing a minor groove shape that closely matches the inherent twist of the PBDs, especially anthramycin (2).[60]

Cardiotoxicity observed for anthramycin (2) and sibiromycin (3) may be due to the formation of *o*-quinoneimines that are capable of producing oxygen radicals through redox-cycling chemistry (see Section 7.14.2.11.1).[66] It is possible that redox cycling of anthramycin and sibiromycin–DNA adducts could cause localized oxidative DNA damage.

Understanding of the covalent and noncovalent chemistry involved in adduct formation by the PBDs has allowed the successful design of novel DNA-cross-linking agents consisting of two covalently tethered PBD units. These cross-linking agents are more cytotoxic than the corresponding monomeric natural products.[67]

### 7.14.2.1.2 Saframycins, renieramycins, and safracins

Some of the saframycin antibiotics (9, 11) form electrophilic imine species that alkylate DNA.[68–71] The early experiments implicating covalent attachment of saframycins to DNA involved measuring the irreversible association of radiolabeled antibiotics with DNA.[71] The structural similarities between the saframycins and the renieramycins (10) and safracins (11) suggests that the chemical reactivity of these agents with DNA will be similar;[50] however, the saframycins are the only members of this group whose DNA-damaging chemistry has been studied in detail. The saframycins and safracins are isolated from *Streptomyces* and the renieramycins are products of marine sponges.[50]

Lown and co-workers described three types of interactions between saframycins and DNA:[69] (i) reversible noncovalent binding; (ii) reversible, acid-catalyzed formation of minor groove adducts with guanine; and (iii) the predominant binding mode involving reversible formation of minor groove adducts with guanine that is promoted by reducing agents, such as dithiothreitol.

Significantly, only saframycins that bear a leaving group (i.e., cyanide or hydroxide) at C-21 are capable of alkylating DNA.[69] Adduct formation for these analogues is significantly enhanced by reduction of the quinone moiety.[68] On the basis of these observations, it was proposed that, for analogues such as saframycin A (9a), reduction of the A-ring quinone to a hydroquinone facilitates ejection of the C-21 leaving group, with corresponding formation of a DNA-alkylating iminium ion (12, Scheme 5).[68]

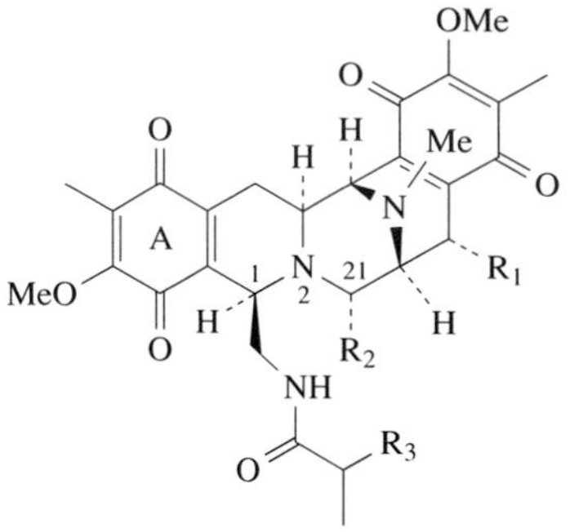

(9) Saframycins
(a) A: R₁=H, R₂=CN, R₃=O
(b) B: R₁=R₂=H, R₃=O
(c) C: R₁=OMe, R₂=H, R₃=O
(d) G: R₁=OH, R₂=CN, R₃=OH
(e) H: R₁=H, R₂=CN, R₃=OH, CH₂COMe
(f) S: R₁=H, R₂=OH, R₃=O

(10) Renieramycins
(a) A: R₁=OH, R₂=H, R₃=CHMe
(b) B: R₁=OC₂H₅, R₂=H, R₃=CHMe
(c) C: R₁=H, R₂=O, R₃=CHMe
(d) D: R₁=OC₂H₅, R₂=O, R₃=CHMe
(e) E: R₁=H, R₂=OH, R₃=CHMe
(f) F: R₁=OC₂H₅, R₂=OH, R₃=CHMe

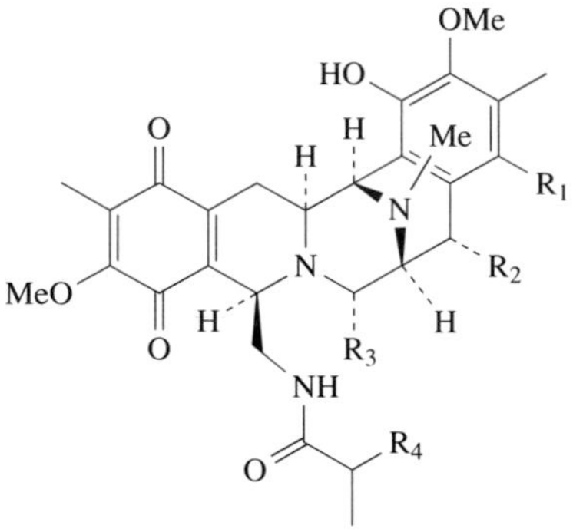

(11) Saframycins
(a) D: R₁=OH, R₂=O, R₃=H, R₄=O
(b) F: R₁=OH, R₂=O, R₃=CN, R₄=O
(c) Mx1: R₁=OH, R₂=OMe, R₃=OH, R₄=NH₂

Safracins
(d) A: R₁= R₂= R₃=H, R₄=NH₂
(e) B: R₁= R₂= H, R₃=CN, R₄=NH₂

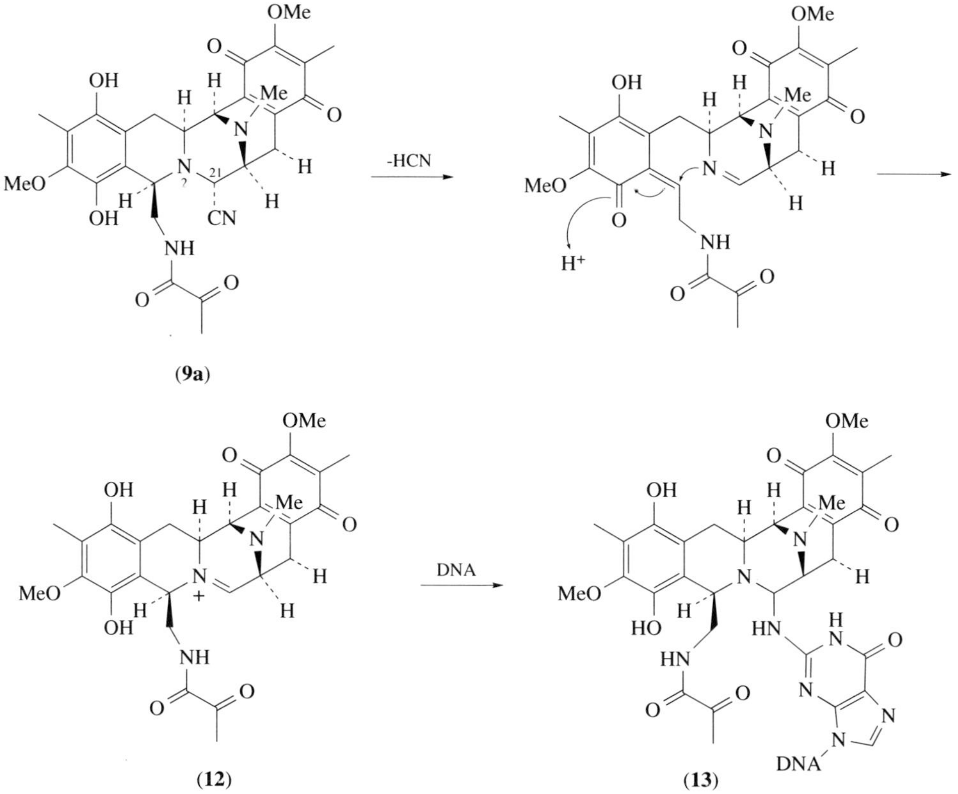

**Scheme 5**

Although saframycin–DNA adducts such as (13) have not been rigorously characterized, experiments strongly suggest that reaction of the saframycin-derived iminium ion (12) with N² of guanine yields aminal attachments similar to those observed for pyrrolo[1,4]benzodiazepines (see Section 7.14.2.1.1).[68] A requirement of GC base pairs is observed for the formation of the slowly reversible covalent adducts and minor groove adduction is suggested by the fact that the DNA alkylation proceeds normally with major groove-glycosylated T4 DNA. Activated saframycin can be trapped by added cyanide ions, suggesting that the reductively activated form is not exclusively associated with DNA.

The quinone moiety of the saframycins, in the presence of molecular oxygen and reducing agents, undergoes redox cycling (see Section 7.14.2.11.1) to produce the DNA-cleaving agent, hydroxyl radical.[68] Such redox properties may explain the cytotoxicity of saframycin derivatives that do not form covalent attachments with DNA. It is not clear whether saframycin–DNA adducts can undergo redox cycling and, thus, efficiently damage DNA via localized production of oxygen radicals.

Molecular modeling has been employed to predict the detailed binding mode of saframycin A (**9a**) in the minor groove of DNA.[77] Footprinting and exonuclease stop assays with various saf-ramycins indicate that covalent attachment is favored at 5'-GGG sites.[70]

### 7.14.2.1.3 Oxazolidine-containing natural products

The family of natural products including naphthyridinomycin (**14a**), the cyanocyclines (**14b**, **14d**), the bioxalomycins (**15**), quinocarcin (**16**), and tetrazomine, isolated from various strains of *Streptomyces*,[50] contain oxazolidine rings that may serve as DNA-damaging iminium ion precursors. Several members of this group of compounds display promising anticancer and antimicrobial properties.[50]

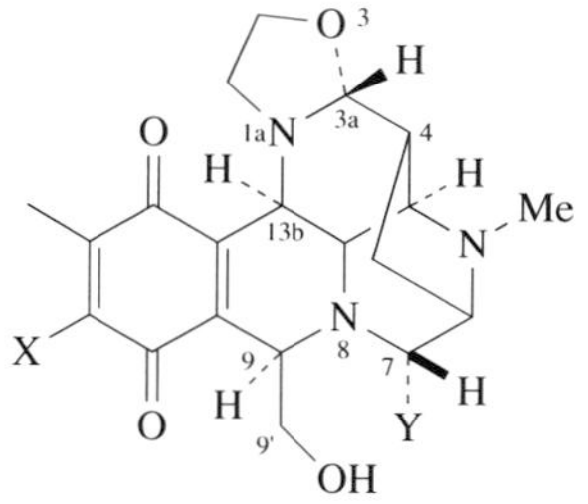

(**14**)

(**a**) Naphthyridinomycin: X=MeO, Y=OH
(**b**) Cyanocycline A:      X=MeO, Y=CN
(**c**) SF-1739HP:           X=OH, Y=OH
(**d**) Cyanocycline F:      X=OH, Y=CN

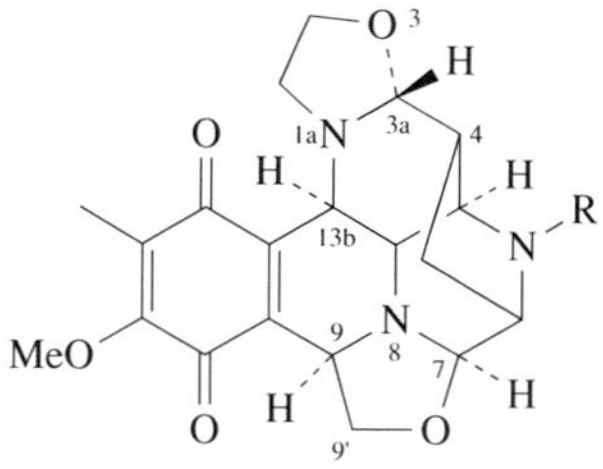

(**15**)

(**a**) Bioxalomycin β1: R=H
(**b**) Bioxalomycin β2: R=Me

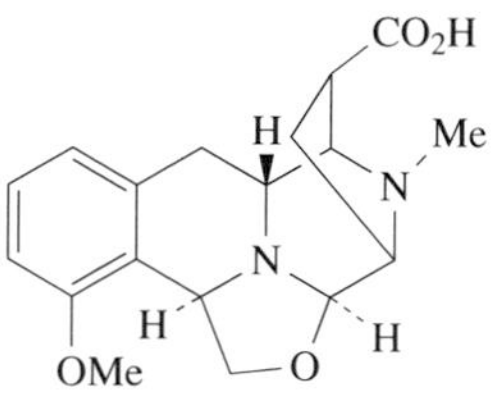

(**16**)

Quinocarcin

Naphthyridinomycin (**14a**) forms covalent adducts with DNA.[73] Experiments similar to those described above for the pyrrolo[1,4]benzodiazepines and the saframycins suggest that these adducts involve attachment through $N^2$ of guanine in the minor groove. Formation of DNA adducts is facilitated under reducing conditions (e.g., 1 mM dithiothreitol).[73] Thus, it was suggested[73] that reduction of the quinone in (**14a**) stimulates the formation of iminium ion (**17**) by facilitating elimination of the leaving group at C-7 (Scheme 6). This mechanism is analogous to that which was later suggested to explain alkylation of DNA by saframycin A (see Scheme 5).[68] The same researchers also suggested an alternative possibility in which reduction of the quinone moiety of (**14a**) promotes noncovalent DNA association, followed by an $S_{N}1_{CA}$ reaction to yield the proposed aminal adduct with guanine. Molecular modeling studies later indicated that both proposed mechanisms are reasonable.[74]

In addition to possible iminium ion formation at C-7 (**17**), the oxazolidine heterocycle in (**14a**) and the cyanocyclines may serve as a precursor for iminium ion formation at C-3a (**18**).[74] Oxazolidine ring-opening to produce the reactive iminium ion at C-3a (see Scheme 6, upper pathway) would be facilitated by reduction, in a manner analogous to imine formation at C-7. Interestingly, treatment of cyanocycline A (**14b**) with acid in the absence of reducing agents results in oxazolidine ring opening with concomitant iminium ion formation at C-3a.[75] Whether DNA attachment by (**14a**) occurs at C-3a, C-7, or both sites remains unknown.

The possibility of successive formation of *two* electrophilic imines from (**14a**) and the cyanocyclines suggest that DNA cross-links are formally possible; however, molecular modeling indicates that such cross-links are unlikely, because the initial covalent attachment at $N^2$ of guanine positions the second electrophilic site outside the minor groove and away from nucleophilic sites on the double helix.[74] Protein–DNA cross-links are a possibility with these agents.

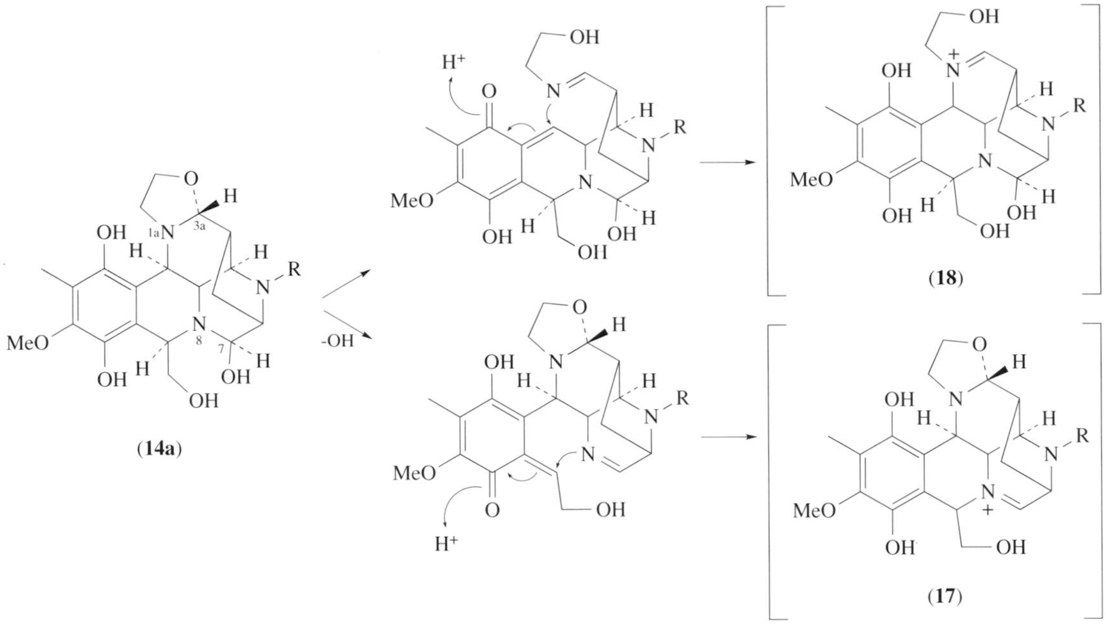

**Scheme 6**

Several natural products, very similar in structure to (**14a**) and the cyanocyclines, except containing two intact oxazolidine rings, were isolated (from a different strain of *Streptomyces* than that which originally produced (**14a**)) and were dubbed the bioxalomycins (**15**).[76] In the course of characterizing the bioxalomycins, researchers at Lederle attempted to isolate (**14a**) from the original producing strain. Interestingly, they were unable to isolate (**14a**) from these cultures, but they did isolate bioxalomycin β2 (**15b**) from the original (**14a**)-producing strain.[76] Thus, although the published spectroscopic data for (**14a**) is clearly distinct from that of the bioxalomycins and was supported by an X-ray crystal structure, it is possible that (**15b**) may be the naturally produced form of (**14a**). It is possible that (**14a**) may have been transiently produced under certain fermentation conditions or may be an artifact of the original isolation process.

The NMR spectra of bioxalomycin α2 (**15a**) indicates that the compound exists in equilibrium with the C-3a ring-opened iminium ion form. Relative peak intensities indicate that 15% of the compound exists in the ring-opened form.[76] These NMR studies suggest that covalent attachment of DNA at C-3a of the bioxalomycins is a strong possibility.

Two additional oxazolidine ring-containing antitumor antibiotics, quinocarcin (**16**)[77] and tetrazomine[78] have the potential to alkylate DNA via iminium ion intermediates. Possible alkylation of DNA by quinocarcin has been examined using molecular modeling,[79] but has not been experimentally verified. Interestingly, a second mechanism for DNA damage, distinct from DNA alkylation, has been observed for quinocarcin and tetrazomine.[80,81] These agents generate DNA-cleaving oxygen radicals through reduction of molecular oxygen to superoxide radical. This chemistry is described further in Section 7.14.2.15.

### 7.14.2.1.4 Barminomycin I

Barminomycin I (**19**), sometimes referred to as SN-07, is a member of the anthracycline family of antibiotics, but is unique among the anthracyclines in that the anthraquinone core is substituted with an unusual carbinolamine-containing sugar moiety. Experiments similar to those described above for the pyrrolo[1,4]benzodiazepines, suggest that (**19**) forms covalent adducts with the $N^2$-amino group of guanine, preferentially at 5′-GC sites.[82] This reaction is mechanistically related to the formaldehyde-mediated cross-linking of anthracyclines to DNA (see Section 7.14.2.11.3).

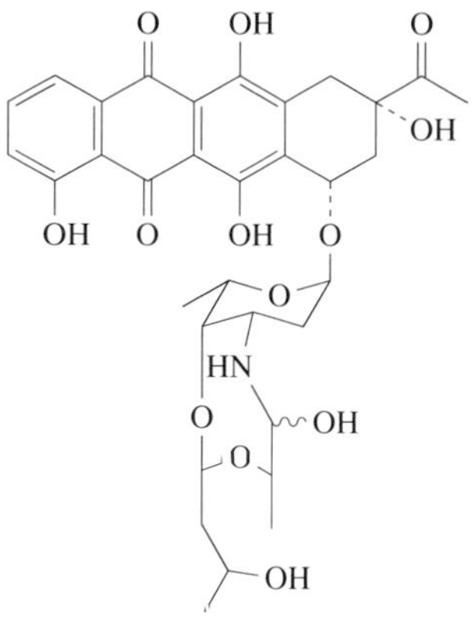

(**19**) Barminomycin I

### 7.14.2.1.5 Other antibiotics that alkylate DNA via imine formation

Alkylation of duplex DNA at $N^2$ of guanines by the novel carbinolamine-containing antibiotic ecteinascidin 743 (**20**), isolated from a Caribbean tunicate, has been reported.[83]

A number of pyrroloiminoquinones such as discorhabdin A (**21**), makaluvamine F (**22**), and epinardin D (**23**), isolated from marine sponges, have the potential to form DNA-reactive imines.[84,85] Imine formation resulting from elimination of the methoxy substituents or opening of the sulfur-containing rings of these natural products would be facilitated by reduction of the iminoquinone moiety. Inhibition of topoisomerase II by these agents has been studied;[85] however, the possibility of covalent adduct formation has not been explored.

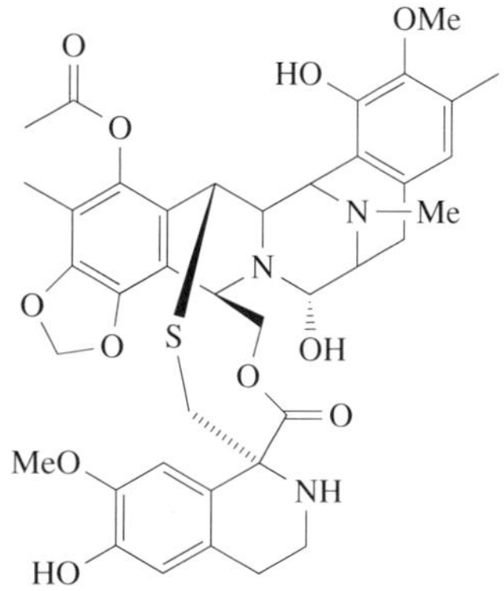

(**20**) Ecteinascidin 743

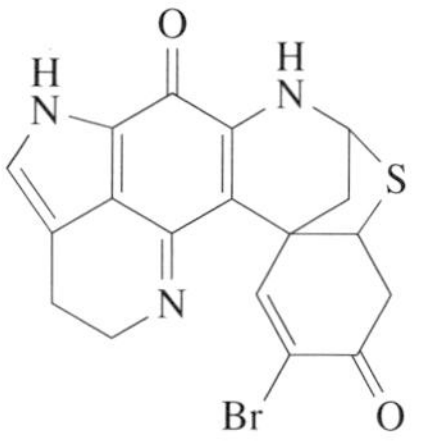

(**21**) Discorhabdin A

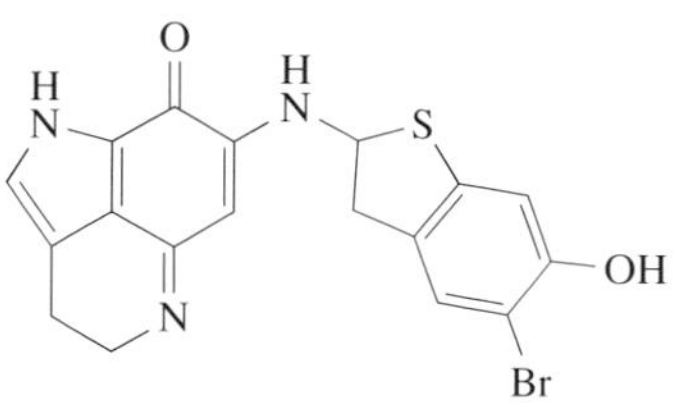

(**22**) Makaluvamine F

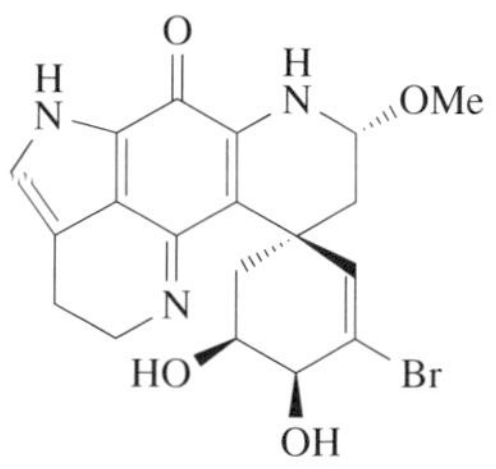

(**23**) Epinardin D

### 7.14.2.2 Carbonyl-containing Natural Products

#### 7.14.2.2.1 *Aldehydes*

A number of aldehyde-containing compounds occur naturally in plants and other organisms.[86–89] Some of these aldehydes are not formally secondary metabolites; rather, they are products of natural oxidative decomposition of lipids. Some naturally occurring aldehydes can react with nucleophilic sites in DNA; for example, α,β-unsaturated aldehydes, such as 2-hexenal, have been isolated from various plant sources, are mutagenic, and form covalent adducts with DNA.[90]

DNA adducts of simple unsaturated aldehydes, such as acrolein and crotonaldehyde (**24**) have been well characterized[91] and may be viewed as models for naturally occurring unsaturated aldehydes. Guanine is the most reactive base toward α,β-unsaturated aldehydes, although adducts with all four of the DNA bases have been isolated (Figure 3). Many of the products obtained from the reaction of DNA bases with these compounds are the apparent result of Michael addition, followed by attack on the aldehyde carbonyl to yield cyclic adducts (Scheme 7).[91]

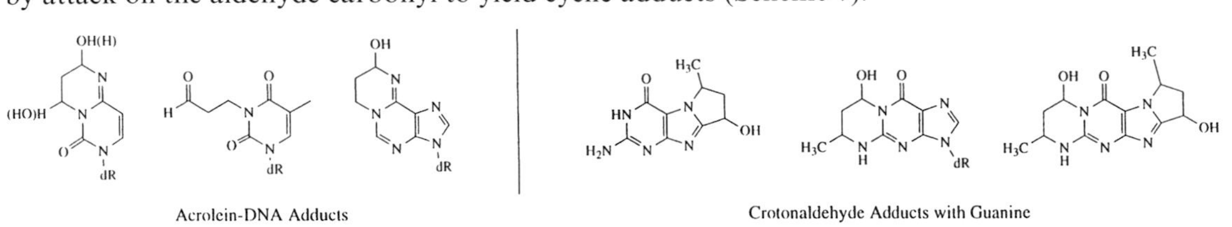

Acrolein-DNA Adducts | Crotonaldehyde Adducts with Guanine

**Figure 3** DNA adducts with α,β-unsaturated aldehydes.

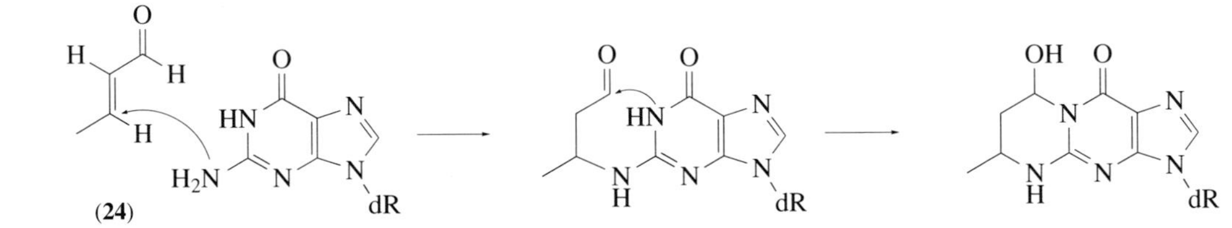

**Scheme 7**

Cyclic adducts attached at N-7 and C-8 of guanine undergo spontaneous depurination to give an abasic site on the DNA. Other adducts, such as those spanning N-1 and N-2 of guanine, are not labile and, when formed *in vivo*, may interfere with enzymatic processing of DNA.

Malondialdehyde is a naturally occurring carcinogen that is formed by endogenous lipid peroxidation.[92] Marnett and co-workers have extensively characterized the adducts of malondialdehyde with DNA bases, examined the mutagenicity of these adducts, and confirmed the existence of malondialdehyde–DNA adducts in human cells.[92]

Formaldehyde is ubiquitous in the environment and, although it exists primarily (~98%) as the hydrated form in aqueous solution, it reacts readily with DNA to form a variety of DNA monoadducts as well as cross-links (**25**).[93,94] Hopkins and co-workers found that 5′-d(AT) sequences are preferred targets for formaldehyde cross-linking, possibly because cross-link formation at these sites requires minimal distortion of the DNA duplex.[94] Formaldehyde also reacts readily with certain anthracycline–DNA complexes to yield drug–DNA cross-links (see Section 7.14.2.11.3).

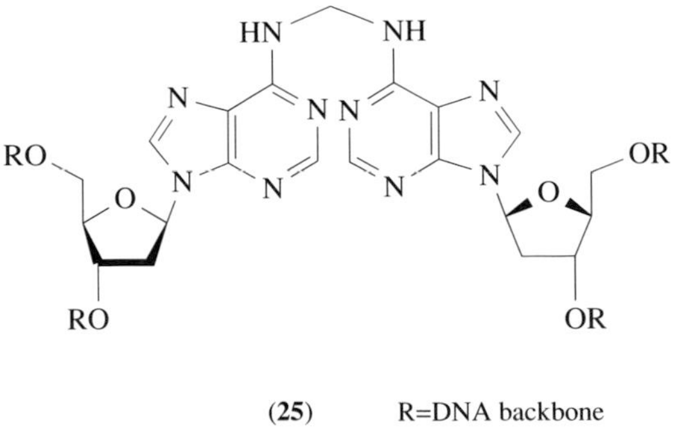

(**25**)    R=DNA backbone

The mutagenic properties of the sesquiterpene lactone hymenovin (**26**) have been ascribed to its bis-hemiacetal functional group.[95] Although adducts of this natural product with DNA have not been reported, hymenovin may be capable of cross-linking DNA, its bis-hemiacetal moiety serving as a masked dialdehyde. Similarly, DNA damage reported for the fungal natural product patulin (**27**)[96] may be due to the formation of a reactive aldehyde from the hemiacetal functional group.[97]

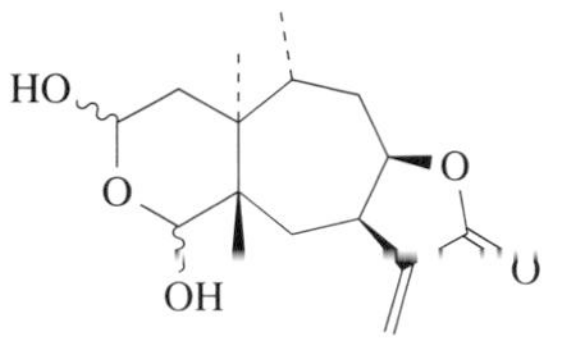

(**26**) Hymenovin 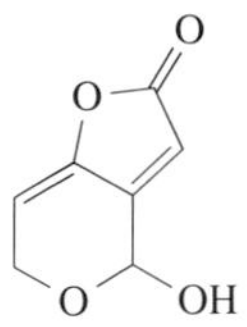                    (**27**) Patulin

### 7.14.2.2.2 *Isochrysohermidine*

Isochrysohermidine (**28**) derives from an oxidative rearrangement of the secondary metabolite hermidin found in certain plant species.[98] The symmetrical dimer (**29**) has been characterized by X-ray crystallography and occurs in the plant as a mixture of the D,L-racemate and the meso compound. The presence of two carbinolamide functional groups suggested to Boger and Baldino that this natural product might cross-link DNA. It was shown that racemic and meso isochrysohermidine, at millimolar concentrations, do, in fact, cross-link DNA.[98] The slow cross-linking reaction is moderately affected by the pH of the reaction mixture, with cross-linking facilitated slightly by either acidic or basic conditions. The pH profile of this reaction suggests that DNA alkylation does not involve acyliminium ion formation, a process that would be acid catalyzed. Rather, it is proposed[98] that acidic or basic conditions catalyze a ring-opening reaction that yields an electrophilic 1,2-dicarbonyl moiety (**29**) on each half of the symmetrical isochrysohermidine molecule (Scheme 8).

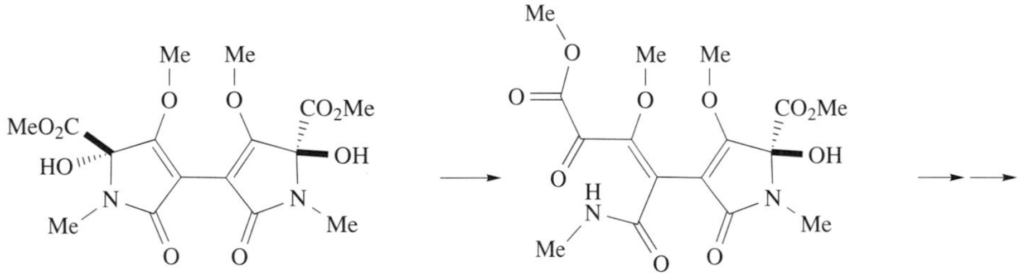

(**28**)                    (**29**)

Electrophilic Dicarbonyl

**Scheme 8**

The vicinal dicarbonyl 2,3-butanedione (diacetyl) is produced by several microbial species and displays mutagenic[88] and antibacterial activity.[99] Bis-tricarbonyl species have been successfully employed as designed DNA-cross-linking agents.[100]

### 7.14.2.2.3 *Lactones*

There are reports that α,β-unsaturated lactones may react with DNA.[97,101] It is noteworthy, however, that studies with the natural α,β-unsaturated lactone helenalin show that, while thiol nucleophiles readily engage in Michael reactions with the α,β-unsaturated lactone helenalin, this natural product does not alkylate purine bases.[102]

### 7.14.2.3    Cyclopropanes

#### *7.14.2.3.1    CC-1065 and the duocarmycins*

Appropriately substituted cyclopropanes are potent electrophiles.[103] A number of cyclopropane-containing antibiotics that derive biological activity through DNA alkylation have been identified.[104] CC-1065 (**30**), isolated in 1974 from *Streptomyces*, is extremely cytotoxic and shows activity against some cancer cell lines at picomolar concentrations.[62,105–107]

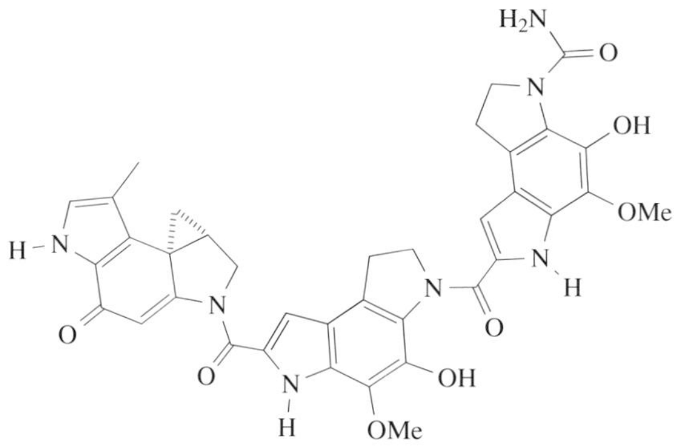

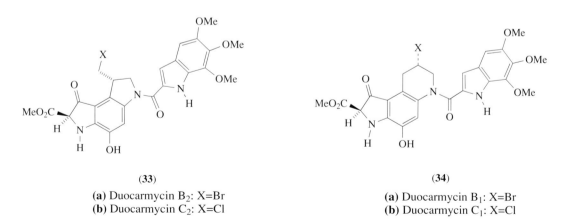

(**30**) CC-1065

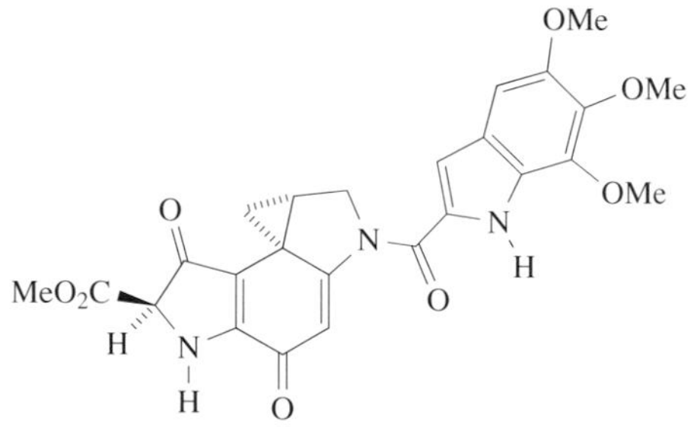

(**31**) Duocarmycin A

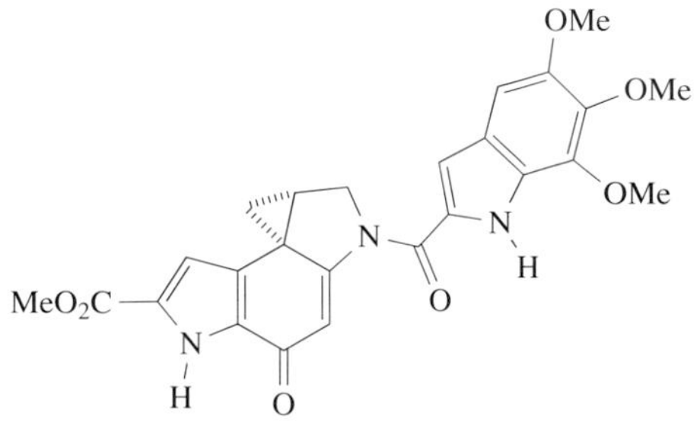

(**32**) Duocarmycin SA

(**33**)

(**a**) Duocarmycin B$_2$: X=Br
(**b**) Duocarmycin C$_2$: X=Cl

(**34**)

(**a**) Duocarmycin B$_1$: X=Br
(**b**) Duocarmycin C$_1$: X=Cl

The cyclopropane ring of (**30**) is electrophilic. Attack of nucleophiles on the strained ring is favored by aromatization of the indole quinone system that results from the reaction (Scheme 9). Early experiments showed that (**30**) forms stable covalent attachments with DNA,[106,107] but that, under thermal treatment, the covalent adducts decompose to yield strand breaks.[108] Isolation and characterization of the adduct (**35**) released by thermal treatment of DNA alkylated by (**30**) clearly demonstrated that covalent modification involves nucleophilic attack of the N-3 nitrogen of adenine on the cyclopropane ring of the natural product (see Scheme 9).[108] DNA alkylation occurs almost exclusively at adenine; however, small amounts of an adduct resulting from the attack of N-3 of guanine have been identified.[109,110] Alkylation of DNA by (30) is essentially irreversible in neutral aqueous solution;[105] however, DNA alkylation by some structurally related cyclopropane-containing antibiotics is reversible.[105,111]

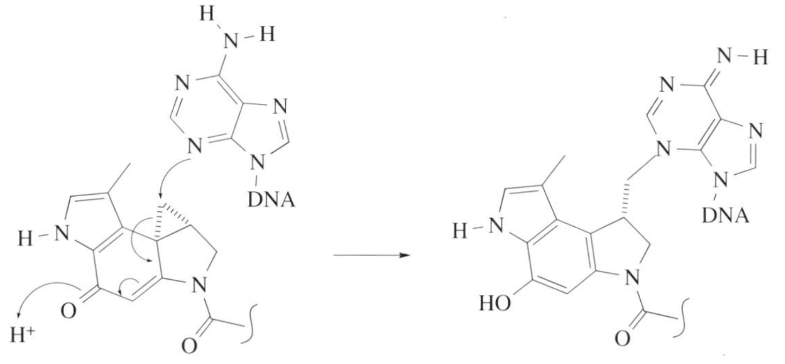

**Scheme 9**

The linked pyrroloindole moieties of (30) lend a right-handed curvature to this antibiotic, reminiscent of the well-known minor groove-binding agents distamycin and netropsin. In accord with its structural similarity to netropsin, early experiments demonstrated that (30) binds in the minor groove of DNA at AT-rich sequences.[106,107] It is shown that netropsin inhibits the binding of (30) to DNA, that major groove glycosylation present in T4 DNA does not inhibit binding, and that the antibiotic shows markedly better binding to homopolymeric AT DNA vs. GC DNA. Subsequent 2D-NMR experiments have conclusively confirmed minor groove binding by this antibiotic.[112]

A number of antibiotics with structures similar to (30) have been reported.[50,111] Duocarmycins A (31) and SA (32) possess a cyclopropane-containing core very similar to (30). Duocarmycins $C_1$, $B_1$ (34), $C_2$, and $B_2$ (33) display properties identical to duocarmycin A and are, in fact, prodrugs that convert to duocarmycin A in aqueous solution. Prior to the discovery of (33b) and (34b) the same compounds had been isolated independently from a different strain of *Streptomyces* than the duocarmycins and named pyrindamycin B and A, respectively.[50,111] In general, the duocarmycins display DNA binding, DNA alkylating, and potent biological properties that are similar to those of (30).

The duocarmycins and (30) display marked sequence specificity in their alkylation of DNA and the molecular origin of this selectivity has been the subject of intense study.[105,111] Preferred sites of alkylation for (30) include sites such as 5'-PuNTT*A* and 5'-AAAA*A*.[105] Both (30) and the duocarmycin antibiotics preferentially bind to AT-rich sites, although binding of these antibiotics to DNA does not necessarily lead to efficient alkylation unless the binding site contains an appropriately positioned nucleophile (i.e., N-3 of adenine). For example, the sequence 5'-GAATT is a strong noncovalent binding site for (30),[113] but is not alkylated with the same efficiency as the comparable binding site 5'-AGTTA.[114]

Interestingly, Hurley *et al.* found that simple CC-1065 analogues, such as (36), alkylate duplex DNA with sequence selectivity nearly identical to the intact natural product.[115,116] This result suggests that sequence selectivity of DNA alkylation by (30) may arise in the chemical bonding steps, possibly stemming from sequence-dependent properties of the DNA target site, such as conformational flexibility or the presence of DNA functional groups at the binding site that catalyze the alkylation reaction.[105,117] Other examples of sequence-dependent effects on the reactivity of DNA are known.[118] Consistent with the notion of DNA-catalysis in the alkylation reaction, NMR studies have identified water molecules coordinated to an AT base pair and a phosphate residue of the target site that might catalyze the alkylation reaction.[105] DNA affinity provided by the pyrroloindole "tail" of (30), while apparently not required for sequence specificity, clearly improves the efficiency of the alkylation reaction, and relatively high concentrations of (36) are required to obtain alkylation levels similar to those achieved by the natural product under a given set of conditions.[115,116]

In results that appear in stark contrast to those described above for (30), Boger's group has reported that simple duocarmycin analogues, such as (37) display markedly reduced sequence

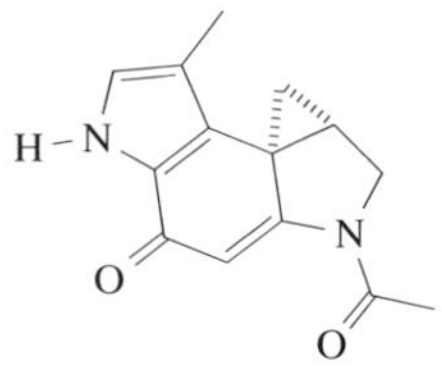

(36)

selectivity relative to intact duocarmycins.[111,119] In addition, analogues of duocarmycin bearing noncyclopropane electrophiles alkylate DNA with sequence selectivity identical to the natural product. These and other results have led Boger and co-workers to suggest that the sequence-selective alkylation of DNA by the duocarmycins derives primarily from noncovalent interactions between DNA and the natural products.[111] The precise roles of noncovalent binding and covalent bonding steps in the high sequence specificity displayed by the duocarmycins and (30) remain a topic of debate and the subject of ongoing experimentation.

Boger's group has proposed that a DNA binding-induced conformational change causes ground state destabilization of (30) and the duocarmycins thereby activating the agents for DNA alkylation.[120] They suggest that this activation is the result of a DNA binding-induced twist in the N-2 amide that links the cyclopropane-containing moiety to the pyrroloindole group on the "right-hand" portion of the molecule.

Analogues of the duocarmycins and (30) have been prepared in which the reactivity of the cyclopropane moiety is altered. Examination of a wide range of natural and synthetic analogues had led Boger's group to the finding that duocarmycin analogues whose cyclopropane groups are more solvolytically stable are more efficient DNA alkylators and are more potent cytotoxins (Figure 4).[111,121] Increased cytotoxicity of the more stable analogues presumably reflects, at least in part, a decrease in the nonproductive hydrolytic destruction of the compounds or a decrease in the nonproductive alkylation of biomolecules. The partitioning of compounds between DNA reaction and hydrolysis has been quantitatively examined in some cases.[109]

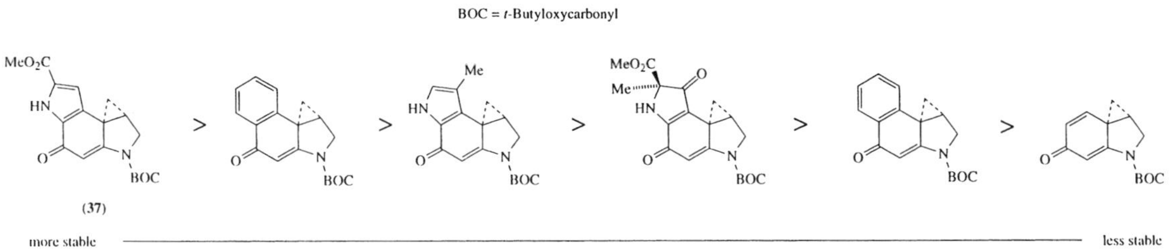

BOC = *t*-Butyloxycarbonyl

more stable ————————————————————————————————————— less stable

**Figure 4**   Relative stability of some duocarmycin and CC-1065 analogues.

The duocarmycins and (30) are exceedingly potent cytotoxins. In the case of (30), as few as three covalent lesions per $10^6$ base pairs of DNA can lead to cell death.[122] The extreme toxicity of these agents even at low levels of DNA damage suggests that adducts of (30) with DNA are particularly lethal. In efforts to elucidate the molecular origin of the biological activity of (30), Hurley's group has extensively characterized the effect that DNA lesions involving (30) have on the processing of DNA by polymerases, helicases, ligases, and other enzymes.[105] In addition, they have characterized DNA bending, winding, and flexibility of DNA adducted by (30).[105] Hurley and co-workers have suggested that the potent toxicity of (30) may derive in part from its ability to act as a surrogate transcriptional activator protein.[7] DNA bending induced by formation of adducts of (30) with DNA may mimic the bending of DNA by transcriptional activators such as Sp1. Thus, formation of adducts with (30) at some sites in genomic DNA may irreversibly activate gene transcription, thereby leading to excessive and unregulated production of messenger RNA and proteins.

### 7.14.2.3.2   *Myrocin C*

Myrocin C (38), isolated from a soil fungus, is a cyclopropane-containing antibiotic that displays antitumor properties.[123] Chu-Moyer and Danishefsky suggested a mechanism by which (38) might alkylate biomolecules.[124] In the proposed mechanism, allylic displacement of the C-9 hydroxyl activates the cyclopropane moiety by bringing this strained ring into conjugation with the C-7 carbonyl group. It was further speculated that the electrophilicity of the cyclopropane moiety could be increased through ring-chain tautomerism to produce (39). Nucleophilic cyclopropane ring opening of (39) would result in favorable aromatization of the B-ring of (38) (Scheme 10).

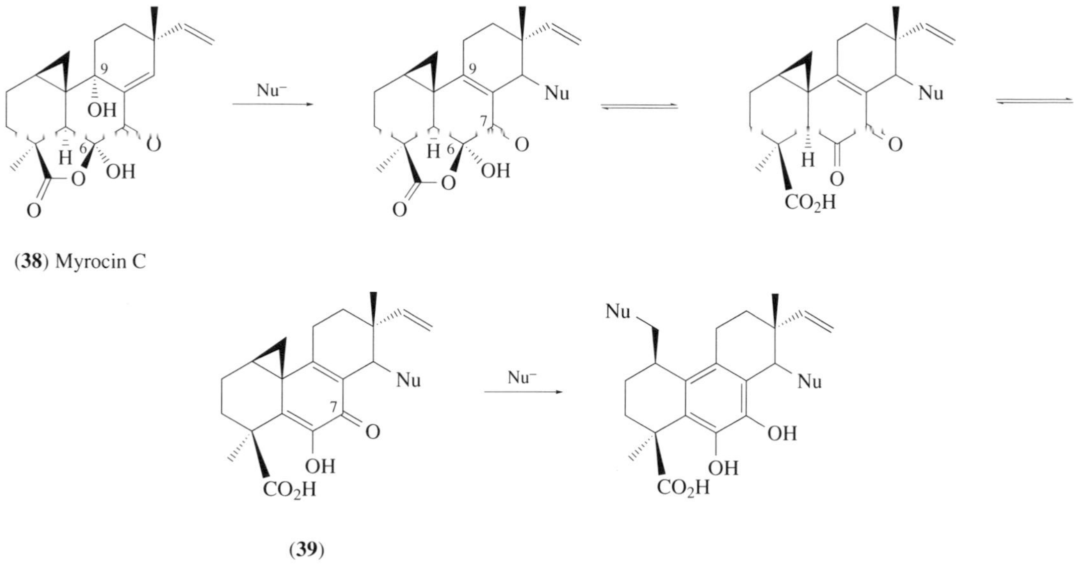

(38) Myrocin C

(39)

**Scheme 10**

The viability of this proposal was supported by chemical model reactions. Utilizing thiophenol as a nucleophile, the expected bis-thiophenol adduct of (38) was obtained.[124] If DNA is, in fact, a biological target for (38), the model reactions suggest that this antibiotic could either serve as a cross-linking agent or as a thiol-activated alkylating agent. As yet, there have been no reports regarding reactions of DNA or other biomolecules with (38).

### 7.14.2.3.3 *Ptaquiloside and the illudins*

Investigations seeking the chemical agents responsible for the carcinogenic properties of bracken fern uncovered the potent carcinogen ptaquiloside (40), containing a spiro-cyclopropane system.[125] Under mildly basic conditions, ptaquiloside eliminates D-glucose to yield the dienone (41).[125] Compound (41) may represent an activated form of ptaquiloside because nucleophilic attack on the cyclopropane ring of this aglycone is favored by the formation of an aromatic ring (Scheme 11), similar to the reactions of CC-1065, the duocarmycins, and myrocin C with nucleophiles.

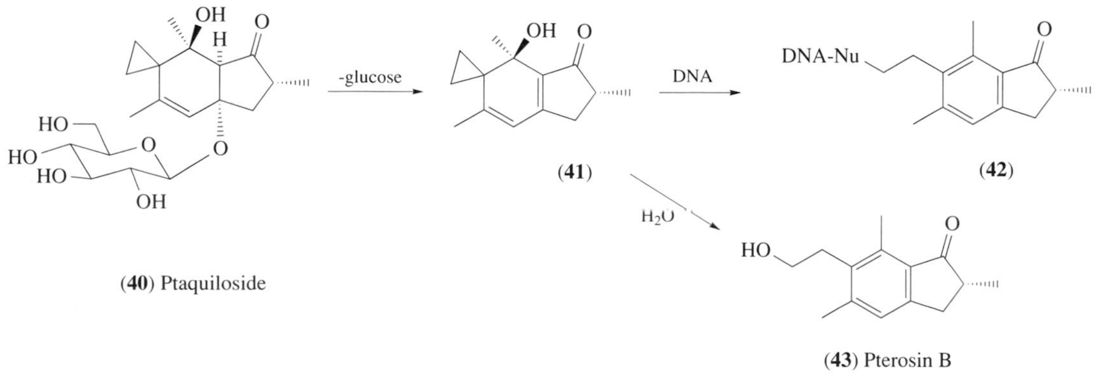

(40) Ptaquiloside

(41)

(42)

(43) Pterosin B

**Scheme 11**

The dienone (**41**) efficiently cleaves DNA at micromolar concentrations, while similar concentrations of (**40**) produce only weak cleavage in a plasmid-nicking assay.[126] Reactions of (**41**) with the oligomer 5′-dGTAC, followed by thermal workup, yield adducts (**42**) attached to guanine N-7 (1.2% yield) and adenine N-3 (0.5% yield). Pterosin B (**43**) is observed as a hydrolysis product.[125,126] Studies with the adducted oligomer show that, in Tris–borate buffer (pH 7.5), the adenine adduct depurinates at a markedly faster rate ($t_{1/2} = 3.2$ h) than the guanine N-7 adduct ($t_{1/2} = 31$ h). On this basis, it is suggested that the relatively stable N-7-guanine adducts may be responsible for the carcinogenic properties of ptaquiloside, because the apurinic sites resulting from the N-3-adenine adducts would be repaired easily *in vivo*.[126]

Although the dienone (**41**) derived from ptaquiloside is a small molecule, with little potential for noncovalent DNA binding, it is reported to alkylate DNA with some sequence selectivity. Flanking sequences affect the efficiency of adenine alkylation, with 5′-dAA*A*T being the most favored alkylation site. The authors of this study note that sequence-dependent differences in the covalent reactivity of DNA may be responsible for the observed sequence selectivity.[126]

The illudins (**44**), isolated from poison mushrooms, are cyclopropane-containing natural products similar in structure to ptaquiloside.[127,128] The illudins are extremely toxic and are carcinogenic. Reaction of these compounds with DNA has been demonstrated[129] and model reactions indicate that, at low pH, these compounds can serve as alkylating agents.[130]

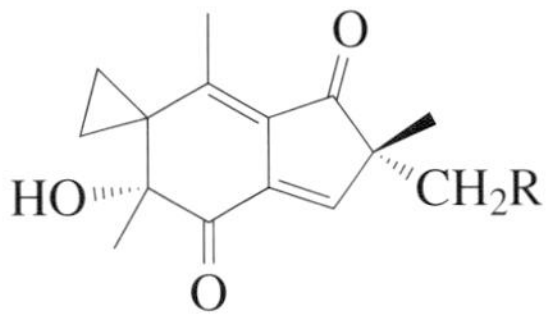

(**44**) Illudin S: R=OH
      Illudin M: R=H

### 7.14.2.4  Epoxides

#### 7.14.2.4.1  *Pluramycins and pluramycinones*

The pluramycins (e.g., (**45**) and (**46**)) were first isolated from *Streptomyces* in the 1950s and found to possess anticancer and antimicrobial activities.[131] Over the years, a large number of antibiotics from this family have been characterized.[132] Similar to simple epoxides,[133] the pluramycins form covalent attachments with DNA that, upon alkaline workup, yield strand breakage at guanines.[134,135] Isolation and characterization of the adduct (**48**) released by thermal workup of altromycin B-treated DNA showed conclusively that covalent attachment involves attack of the N-7 position of guanine on the epoxide of the natural product.[136] The chemistry of DNA attachment for the pluramycins is analogous to that for simple epoxides; however, unlike simple epoxides, the pluramycins alkylate DNA very efficiently and sequence specifically.

Work by Hurley's group has revealed that the planar anthrapyrone-4,7,12-trione moiety of the pluramycins intercalates into DNA on the 5′-side of guanines that become adducted.[132] Similar to some anthraquinone antibiotics, the pluramycins intercalate by a "threading mechanism" with the anthrapyrone system perpendicular to the long axis of the DNA base pair. NMR experiments indicate that the carbohydrate residues of the pluramycins contribute to sequence recognition through hydrogen bonding and hydrophobic interactions with DNA. The preferred sequences for DNA alkylation are 5′-AG for the altromycins and 5′-PyG for the "classical" pluramycins that do not contain sugar residues at the 5-position.[132]

The molecular basis for the sequence-specific alkylation of DNA by the pluramycins has been studied in detail.[132] Neopluramycin, a nonalkylating derivative of pluramycin A, does not show strong DNA binding that can be detected by Dnase I assays. This suggests that sequence-selective alkylation of DNA by pluramycin A does not simply reflect strong sequence-specific binding at the preferred alkylation sites. Rather, it appears that, at favored pluramycin alkylation sites, functional

Pluramycins

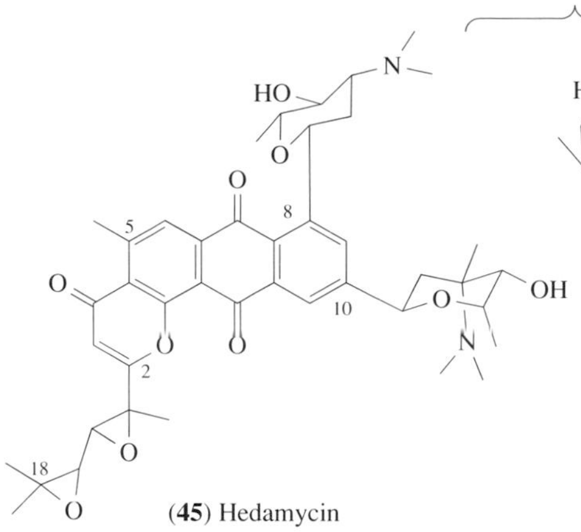

(**45**) Hedamycin
(A "Classical" Pluramycin)

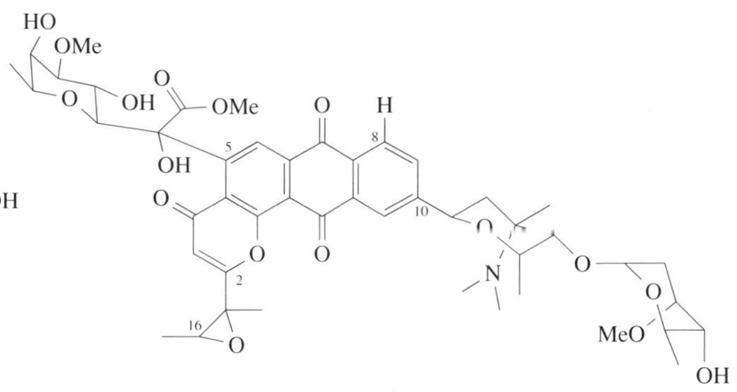

(**46**) Altromycin B

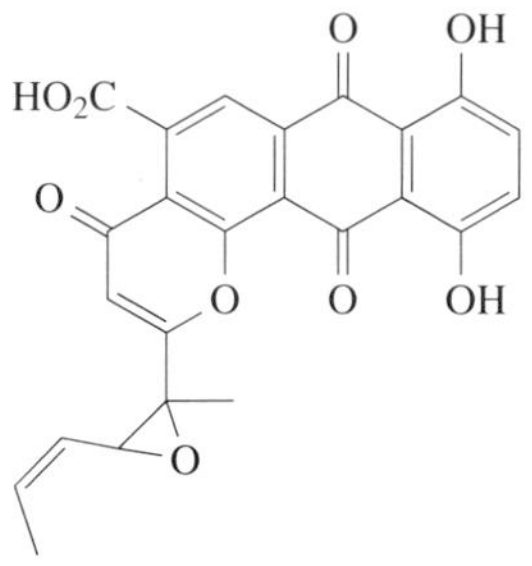

(**47**) Sapurimycin
(A Pluramycinone)

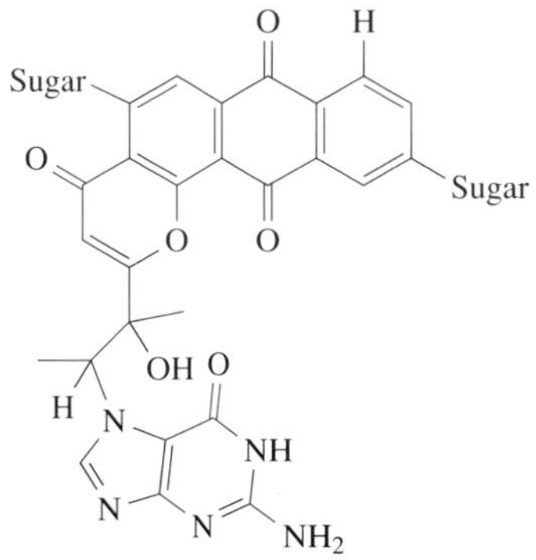

(**48**) Altromycin B–guanine adduct

groups of the DNA bases interact with the natural product to "steer" the electrophilic epoxide into the appropriate position for reaction with N-7 of the target guanine. Less efficient alkylation of guanines at nonpreferred alkylation sites may be the result, not necessarily of weaker binding at these sites, but of inappropriate positioning of the epoxide for reaction with the N-7 position of guanine. The exact interactions involved in binding of the pluramycins and pluramycinones to their preferred target sites differ somewhat for individual members of this class. Flanking sequences outside the intercalation site also play a role in sequence specificity.

Nucleophilic attack of DNA occurs at the C-18 position of hedamycin[137,138] (**45**) and at C-16 of altromycin B (**46**).[136] The higher reactivity of (**45**) and the other diepoxide-containing pluramycins may be explained by the fact that the longer "reach" of the epoxide in these compounds allows favorable alignment of the DNA nucleophile for the stereoelectronically controlled epoxide ring-opening reaction, with minimal distortion of the double helix.[132]

Studies examining the effects of DNA-binding proteins on modification of DNA by the pluramycins indicate that these agents may serve as useful probes for protein-induced changes in DNA conformation.[139] For example, DNA alkylation by the pluramycins is enhanced at sites downstream from TATA binding protein–DNA complexes.

In the presence of reducing agents and molecular oxygen, the pluramycins and pluramycinones may be capable of producing oxygen radicals via redox cycling reactions involving their quinone moieties (see Section 7.14.2.11.1); although, to date, DNA damage by this pathway has not been reported for this class of natural products.

In theory, protein–DNA cross-linking and interstrand DNA cross-linking[133] could be mediated by diepoxides such as hedamycin.

### 7.14.2.4.2  *Kapurimycin A3 and the clecarmycins*

The epoxide-containing antibiotic kapurimycin A3 (**49**) is similar in structure to the pluramycinones and has been found to alkylate guanines in duplex DNA.[140] A covalent adduct of this antibiotic with guanine has been isolated by thermal treatment of (**49**)-treated DNA. Characterization of the product shows attachment of DNA to C-16 of the epoxide through N-7 of guanine.[141] Analysis of the DNA end products resulting from alkylation of a self-complementary oligonucleotide by (**49**) is consistent with the products known to arise from DNA cleavage resulting from alkylation at N-7 of guanine.[141] The sequence specificity of DNA alkylation by (**49**) has not been examined.

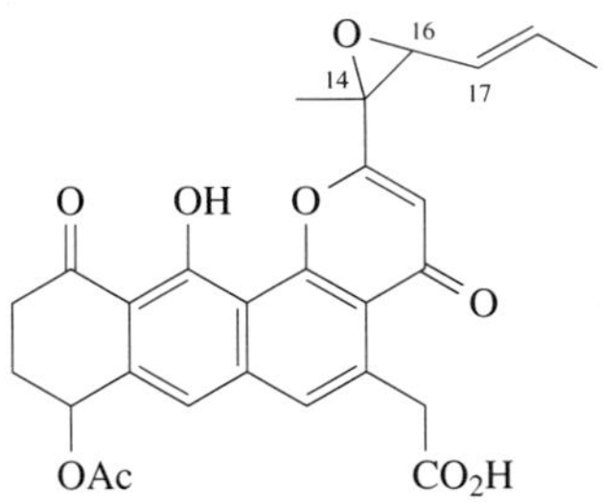

(**49**) Kapurimycin A3

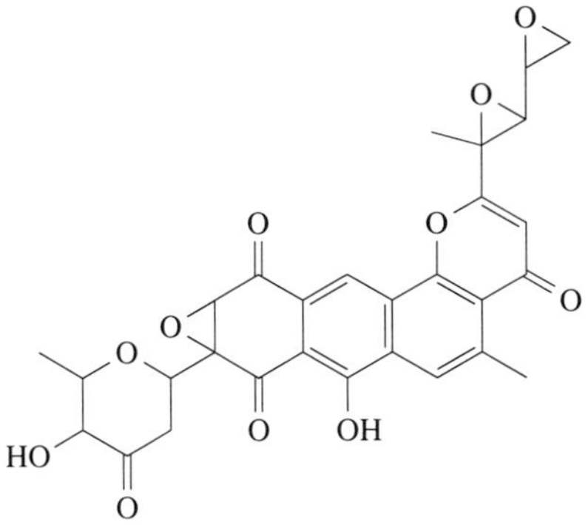

(**50**) Clecarmycin A1

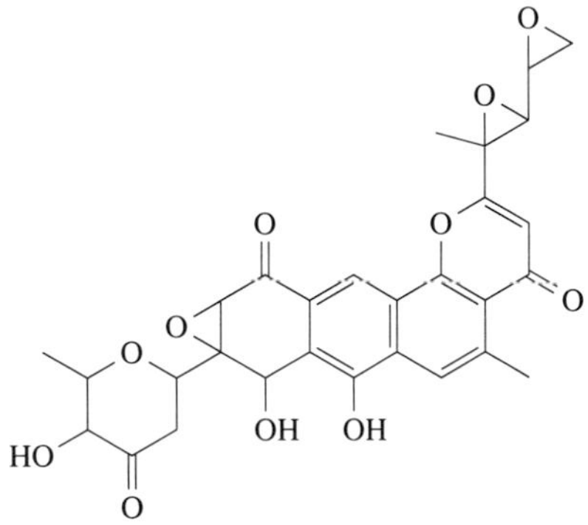

(**51**) Clecarmycin C

The reported clecarmycins (**50**) and (**51**) are similar in structure to (**49**), but bear epoxides on both ends of their aromatic core.[142] If the clecarmycins bind to DNA via threading mechanisms similar to the pluramycins, it can be expected that an epoxide in the diepoxide fragment will alkylate guanine in the major groove of DNA. In addition, it can be imagined that, with some distortion of the double helix, the additional epoxide (on the end *opposite* the diepoxide moiety) of the clecarmycins could alkylate nucleophilic positions in the minor groove of DNA, thus leading to inter- or intrastrand cross-links.

### 7.14.2.4.3  *Alkoxyl radicals from vinyl epoxides*

Studies with synthetic vinyl epoxides[143] (**52**) suggest a second mode of DNA damage, in addition to alkylation of DNA, that may be possible for vinyl epoxide-containing natural products such as (**49**), (**47**), pluramycin A, and rubiflavin A. Thiyl radicals react with the double bond of simple vinyl epoxides, leading to homolytic opening of the epoxide ring (Scheme 12). The resulting alkoxy radical

(**53**) is capable of abstracting hydrogen atoms from a deoxyribose mimic in model reactions.[143] The relevance of this model chemistry to actual DNA cleavage was shown by the finding that a vinyl epoxide covalently linked to a phenanthrolinium intercalator (**54**) mediates DNA cleavage in the presence of thiyl radicals generated by a horseradish peroxidase/hydrogen peroxide/glutathione system.[31] This radical chemistry of vinyl epoxides could be biologically relevant, as thiyl radicals resulting from thiol oxidation may be present in cells.

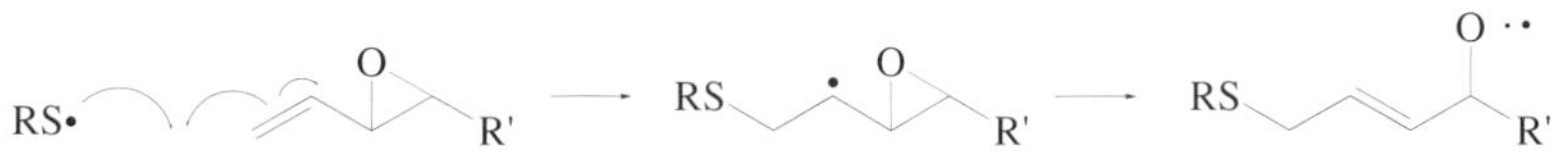

(**52**)                    (**53**)

**Scheme 12**

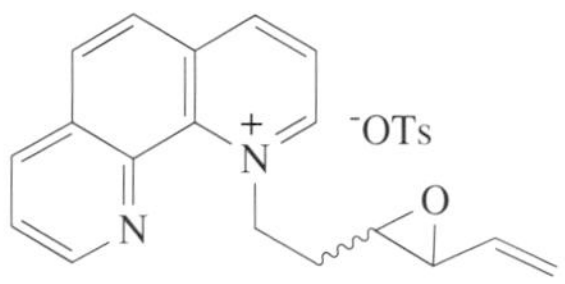

(**54**)

### 7.14.2.4.4 Psorospermin

The epoxide-containing phytotoxin psorospermin (**55**) displays potent anticancer properties and, similar to the other epoxides discussed in this section, has been found to alkylate N-7 of guanine.[144] Experiments utilizing two-dimensional proton NMR show that, similar to the pluramycins, (**55**) intercalates to the 5′-side of guanine residues that become alkylated. Unlike the pluramycins, however, in the (**55**)–DNA adduct, the natural product is intercalated with its long axis parallel to the long axis of the DNA base pairs.[144] Attack of the N-7 position of guanine occurs at C-4′ of (**55**). The preferred alkylation site of (**55**) is 5′-G*G*, a site that is one of the least favored sites for the pluramycins.[144]

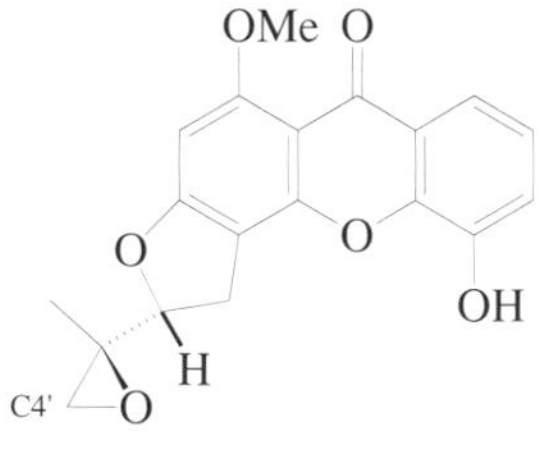

(**55**) Psorospermin

### 7.14.2.4.5 Metabolically activated mycotoxins: the aflatoxins

The aflatoxins, sterigmatocystin, the versicolorins, and the austocystins (**56**)–(**59**) are a group of highly carcinogenic, fungal secondary metabolites that contain a common dihydrofurofuran core.[44,145] These mycotoxins are not inherently reactive with DNA, but are converted to reactive, highly carcinogenic epoxides by cytochrome P450 oxygenase systems *in vivo* (Scheme 13). The DNA reactions of a number of the dihydrofurofuran mycotoxins have been studied, but aflatoxin B$_1$ (**56**) is perhaps the best characterized member of this class.

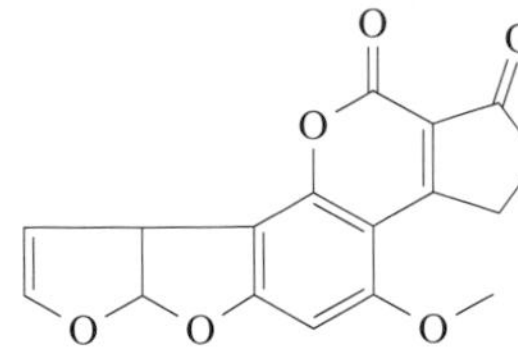

(56) Aflatoxin B$_1$

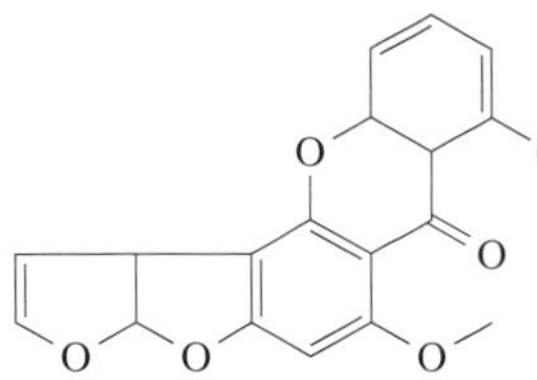

(57) Sterigmatocystin

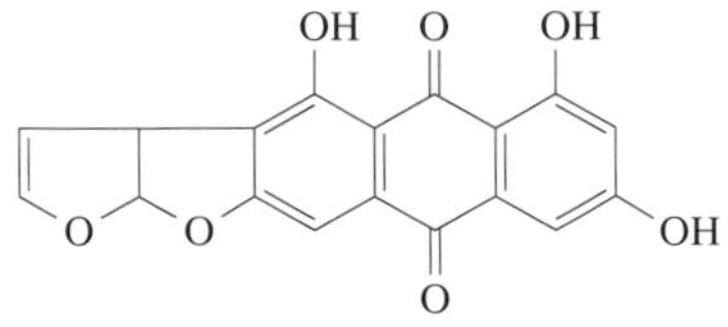

(58) Versicolorin A

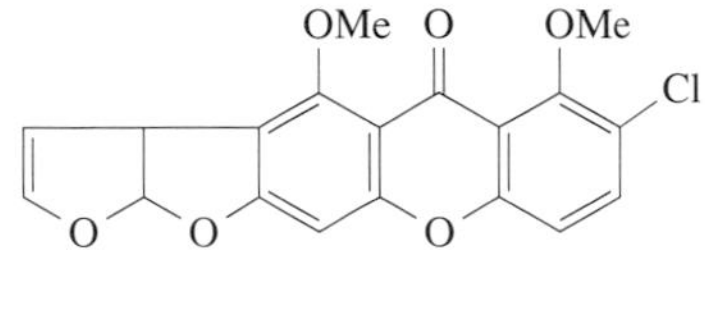

(59) Austocystin A

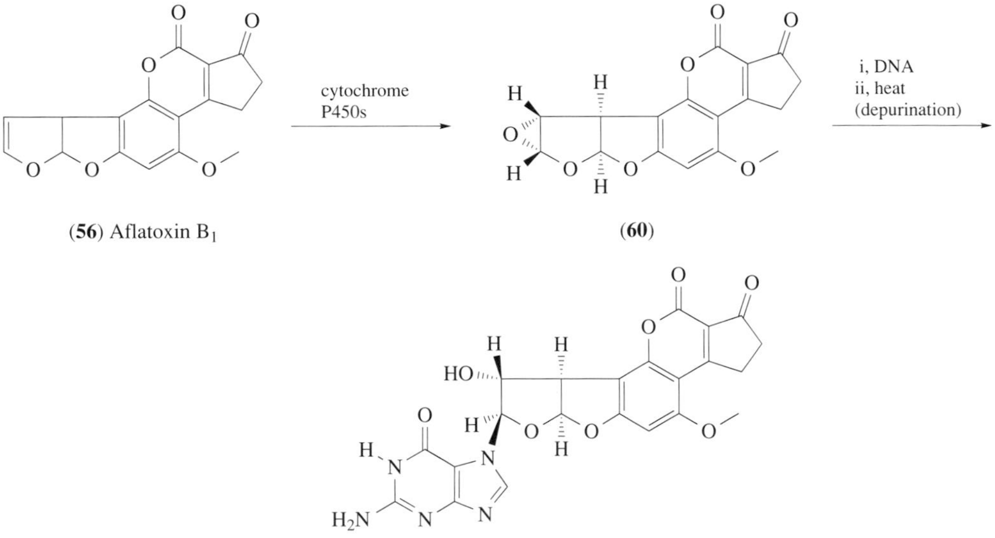

(56) Aflatoxin B$_1$                    (60)

(61)

**Scheme 13**

Enzymatic oxygenation of (56) yields both the exo and endo isomers.[145] Both epoxides are highly reactive; methanolysis of the exo isomer yields a mixture of *cis* and *trans* products, suggesting an S$_N$1 character to this reaction, while the endo isomer produces only the *trans* methanolysis product. The exo epoxide (60) reacts readily with DNA, while the endo epoxide is less reactive. Accordingly, the exo isomer is highly mutagenic and the endo isomer is not.[145]

The major covalent adduct (61) resulting from treatment of DNA with activated aflatoxins has been well characterized and clearly shows that aflatoxin epoxides alkylate guanine at the N-7 position.[146,147] In addition, minor adducts resulting from reaction of activated aflatoxin B$_1$ with N-7 of adenine[148] and with cytosine[149] have been identified.

Due to their instability, the epoxides of (56) have not been isolated from biological systems; however, aflatoxin epoxides may be generated *in situ* by oxidation of (56) with agents such as 3-chloroperbenzoic acid[150] or by conversion of (56) to the dichloride using chlorine gas, followed by the addition of this product to aqueous solution.[145] The exo epoxide has been unambiguously synthesized by oxidation of (56) with dimethyldioxirane in acetone.[151] The half-life of the exo epoxide in water is ∼ 1 s; nonetheless, addition of this material to aqueous, DNA-containing mixtures produces high yields of covalent adducts.[152,153]

Proton NMR experiments using oligomeric DNA duplexes indicate that (56) intercalates into duplex DNA,[154] and that intercalation of the exo epoxide to the 5'-side of a guanine residue would position the epoxide ideally for reaction with N-7 of this guanine.[152] Consistent with this intercalation–alkylation picture, modification of single-stranded DNA, A-form DNA–RNA duplexes, and Z-form DNA duplexes by (56) is inefficient.[155] Investigations of the sequence-specificity of DNA modification by (56) indicate that 5'-GGG sequences are modified, on average, 20 times more efficiently than the least-favored sequence 5'-TGA.[156] It is known that guanine residues that are flanked by guanines possess enhanced nucleophilicity; however, the precise origin of the sequence specificity of the aflatoxins remains a subject of investigation. The reasons for the potent carcinogenicity of the aflatoxins have been reviewed.[8,157]

### 7.14.2.4.6 Other epoxides

A number of environmental carcinogens are metabolically converted to DNA-reactive epoxides in a manner similar to the aflatoxins.[158,159] Azinomycin, which contains a DNA-reactive epoxide, is discussed in Section 7.14.2.5.1. Neocarzinostatin contains an epoxide that is involved in the triggering of enediyne cycloaromatization; however, the epoxide of neocarzinostatin is not thought to react directly with DNA.

### 7.14.2.5 Aziridines

Aziridine and aziridinium ion species have long been known as the active DNA-alkylating species generated by nitrogen mustards.[10,118] Although aziridine-containing natural products are not common, several antibiotics bearing this functional group have been reported.

### 7.14.2.5.1 Azinomycin B/carzinophilin

Carzinophilin was isolated from *Streptomyces* in 1954,[160] although the structure of this natural product remained elusive for many years. The structure of carzinophilin ultimately became clear when it was realized that the natural product was identical in structure to an antibiotic isolated from a different strain of *Streptomyces* in 1986, azinomycin B (62).[161]

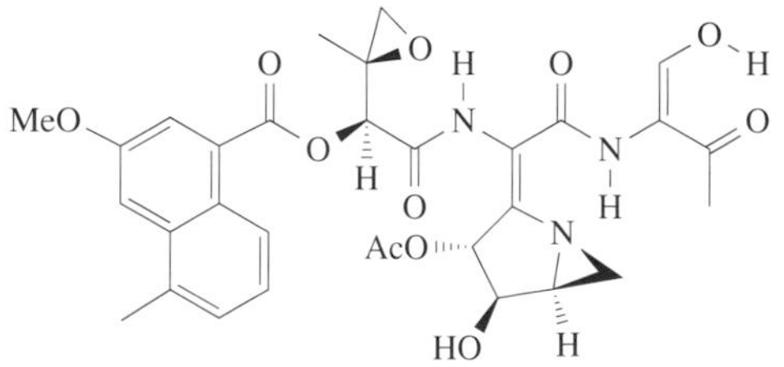

(62) Carzinophilin/Azinomycin B

Compound (62) contains two potentially electrophilic functional groups, an epoxide and an aziridine. Thus, it is not surprising that early experiments showed anomalously facile renaturation of DNA treated with (62), thereby suggesting that this agent forms interstrand DNA cross-links.[162] Work involving electrophoretic isolation of cross-linked oligonucleotides, followed by piperidine treatment and sequencing gel analysis of the resulting cleavage sites on each strand, conclusively demonstrated that reaction of (62) with DNA yields base-labile cleavage sites and that DNA cross-links form at 5'-GNT and 5'-GNC sequences (Figure 5).[161] Cross-linking does not occur at target sequences in oligonucleotide duplexes when the guanine residues normally involved in the cross-link are replaced with 7-deazaguanine (63); however, cross-link formation still occurs at sites where guanine is replaced with inosine (8).[161] Thus, cross-link formation apparently involves covalent attachment of (62) to DNA through the N-7-positions of adenines and guanines in the major groove of DNA. Compound (62) alkylates guanines in single-stranded DNA, but does not react with adenines in single-stranded DNA.[161] This suggests that G–A cross-links may result from initial attachment of the natural product to guanine followed by reaction with adenine on the opposite strand, or that noncovalent binding of (62) to DNA is required to facilitate reaction at adenines.

5'-GNT-3'      5'-GNC-3'
3'-CNA-5'      5'-CNG-3'

**Figure 5**   Sequences cross-linked by (62).

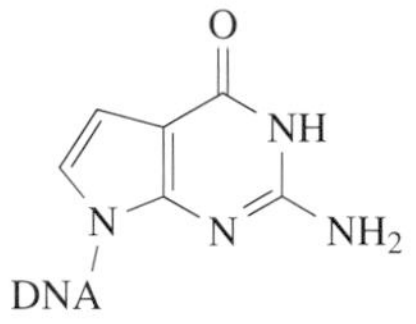

(63) 7-Deazaguanine

### 7.14.2.5.2   *Azicemicins*

An intriguing pair of aziridine-containing antimicrobial agents, the azicemicins (64), have been described.[163] Although there have been no reports regarding the interaction of these agents with DNA, the fact that they contain potentially electrophilic aziridine moieties, along with their structural similarity to the epoxide-containing, DNA-alkylating agents kapurimycin A3 (49) and the pluramycins (e.g., (47)), suggest that these agents may be capable of DNA alkylation.

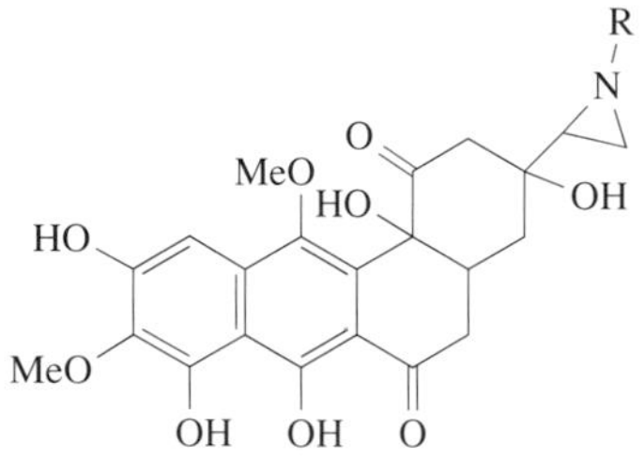

(64)

Azicemicin A: R=Me
Azicemicin B: R=H

### 7.14.2.5.3   *Other aziridines*

The mitomycins FR900482 and FR66979 are DNA-cross-linking agents that contain aziridine moieties. These compounds are discussed in Sections 7.14.2.6.1 and 7.14.2.6.2.

Research indicates that alkylation of phosphate residues in DNA by certain synthetic aziridines may yield substituted aminoethylphosphate triester lesions that lead to hydrolysis of the phosphodiester backbone of DNA.[164]

## 7.14.2.6   Pyrrole-derived Cross-linking Agents

### 7.14.2.6.1   *The mitomycins*

The first mitomycins (Figure 6) were isolated in the 1950s.[165] Mitomycin C (65a) has proven clinically useful in the treatment of certain cancers.[166,167] Early experiments implicated DNA as the biological target of these drugs[167] and, based upon unusually facile renaturation of mitomycin-treated DNA, it was suggested that this agent covalently cross-links opposing strands of double-helical DNA.[168] Pioneering experiments of Iyer and Szybalski showed that cross-link formation was dependent upon the presence of either chemical or enzymatic reducing agents.[169] On this basis, these researchers proposed that reduction of the quinone moiety of (65a) leads to elimination of methanol, thereby increasing the electrophilicity of the aziridine functional group in the natural product.[169] Moore later refined this mechanistic proposal, suggesting that alkylation of DNA by reduced (65a) involves $S_N1$-type reactions with the extended quinone methide (68) formed by aziridine ring opening and with the $\alpha,\beta$-unsaturated iminium ion (70) that results from expulsion of the carbamate group from the monoadduct (69) (Scheme 14).[170] The early proposal of Iyer, Szybalski, and Moore has served as a starting point for many elegant experiments and, over the years, has proven largely correct.[171-175]

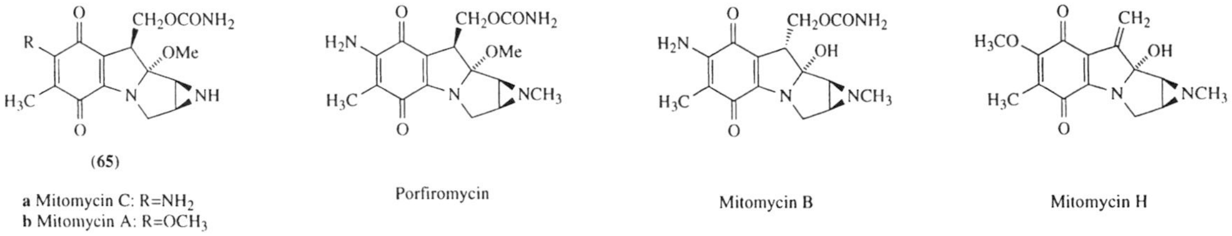

**Figure 6** Mitomycin antibiotics.

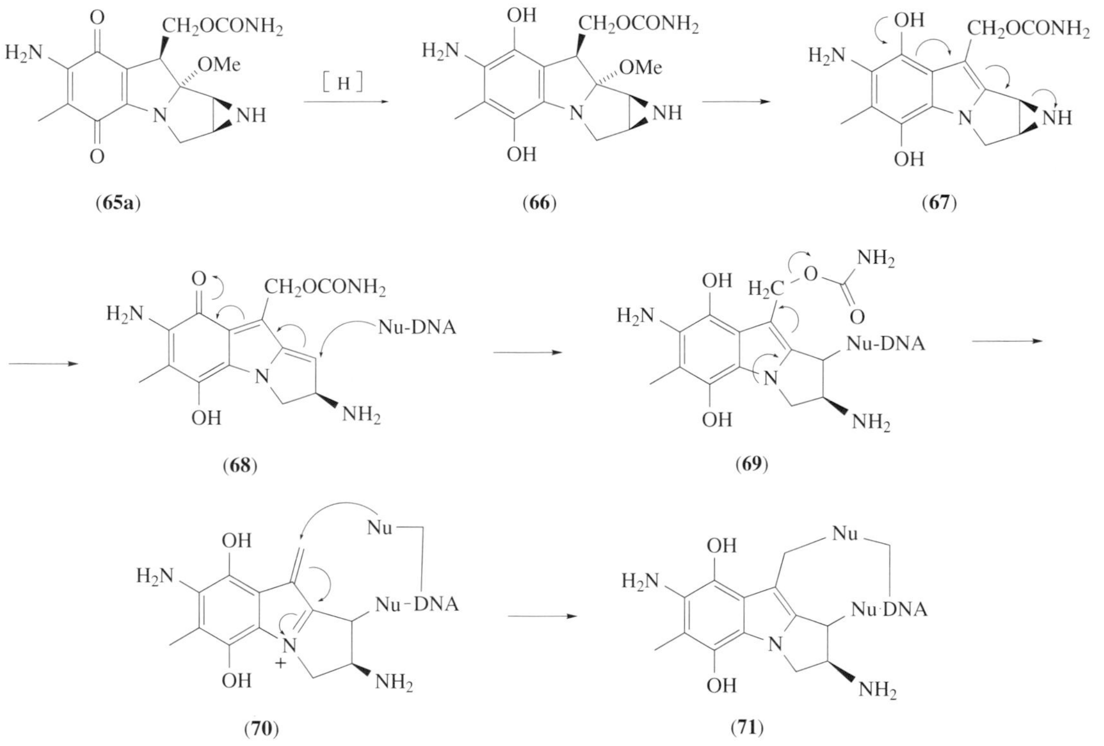

**Scheme 14**

Mitomycin C (65a) is remarkably stable prior to reduction of its quinone functional group. The electron-withdrawing nature of the quinone group diminishes the basicity of the aziridine nitrogen, such that it is essentially unprotonated under physiological conditions ($pK_a$ of the aziridine $R_2NH^+ = 3.2$).[167] In addition, the electron-poor nature of the quinone group stabilizes the hemiaminal group of (65a) by drawing electron density from the pyrrolo nitrogen and preventing the elimination of methanol. Reduction of the quinone of (65a) to the hydroquinone (or semiquinone) form (66) increases the electron density on the pyrrolo nitrogen, thereby facilitating elimination of methanol from the drug, presumably by an E1-type reaction (Scheme 15) to yield the mitosene form of the drug (67). Formation of the resulting C-9—C-9a double bond in (67) increases the electrophilicity of the aziridine ring by bringing this group into conjugation with the hydroquinone ring. Model reactions, as well as characterization of DNA adducts, indicate that the first nucleophilic attack occurs at C-1″ of (68).[171-174] If the conditions are such that the quinone ring remains in the reduced form, the first nucleophilic attack is followed by expulsion of the C-10″ carbamate group to yield an electrophilic iminium ion (70) that can react with a second nucleophile.[176] If the quinone of (69) is oxidized subsequent to the first nucleophilic attack at C-1″, the pyrrolo nitrogen again becomes vinylogously amidic (conjugated with the quinone carbonyl), is unable to facilitate expulsion of the C-10-carbamate group, and the DNA monoadduct (69) is obtained.

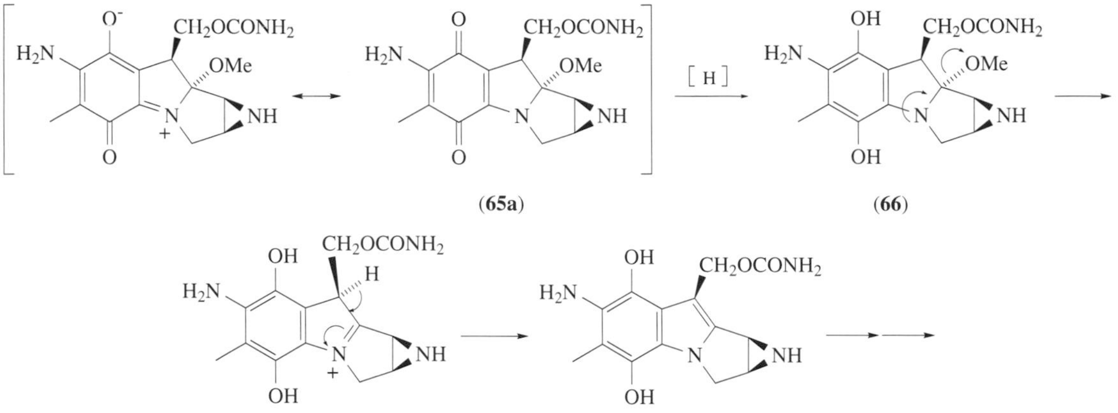

**Scheme 15**

The quinone moiety of (65a) can be reduced chemically by agents such as sodium borohydride and sodium dithionite[167] or a number of enzymes,[166] including DT-diaphorase, NADPH-cytochrome C reductase, and xanthine oxidase.[177] The requirement for reductive activation of (65a) probably plays a key role in its potent antitumor properties. The oxygen sensitivity of the activation step renders this agent selectively toxic to the oxygen-poor (hypoxic) cells found in solid tumors.[166,178] Hypoxic conditions not only favor activation, but further favor formation of DNA cross-links over monoadducts. DNA cross-links are thought to be particularly cytotoxic lesions.[175] It appears likely that either one or two electron reduction is sufficient to activate (65a) for DNA alkylation.[167]

Reductive activation of (65a) in the presence of DNA leads exclusively to covalent attachments at $N^2$ of guanine.[179] The exact nature of the adducts formed depends upon the activation conditions. Monofunctional alkylation of guanines occurs when the second alkylation step is inhibited by oxidation of the activated hydroquinone form of the drug back to the quinone form. The oxidizing agent can be either molecular oxygen or excess, unreduced (quinone form) (65a).[167] Treatment of duplex DNA with (65a) under bifunctional alkylation conditions (for example, anaerobic sodium dithionite) leads to intrastrand[180] and interstrand[182] cross-link formation between exocyclic nitrogens of guanines that are proximal in the DNA double helix. Mono- and bifunctional adducts formed both *in vitro* and *in vivo* have been isolated and completely characterized. In addition to interstrand and intrastrand DNA cross-linking, (65a) can mediate protein–DNA cross-linking.[181]

Not surprisingly, under highly acidic conditions (pH ⩽ 4.5), (65a) can alkylate DNA, without reductive activation. Interestingly, under these conditions alkylation occurs primarily at N-7 of guanine, rather than at $N^2$ of guanine.[182] The shift in the site of adduct formation is rationalized by the fact that, under acidic conditions, protonation of the aziridine leads to formation of a "hard" alkylating species that preferentially reacts with N-7 of guanine.[182]

Noncovalent association of (65a) with DNA prior to covalent reaction is weak and is not highly sequence selective.[167,175] The possible sequences that can be cross-linked by activated mitomycin C are dictated to a large extent by the 3.36 Å distance between the electrophilic C-1″ and C-10″ sites in the activated form of the drug. Due to the fixed distance between the electrophilic centers of activated (65a), and the preference for reaction at $N^2$ of guanines, only two possible sequences for interstrand cross-link formation can reasonably be considered: 5′-GC and 5′-CG (Figure 7). It has been unequivocally demonstrated that the preferred sequence for cross-linking is 5′-CG.[183] In normal B-form DNA, the distance between the guanine nitrogens in the preferred 5′-CG sequence is 3.62 Å vs. 4.1 Å in the less favored 5′-GC site.[167] Thus, the cross-link that induces minimal distortion of the double helix is favored; however, computations show that the two possible cross-linked structures are approximately equal in energy, suggesting that the preference for 5′-CG may be kinetic, rather than thermodynamic.[184] Molecular mechanics calculations on the possible monoadducts suggest that the 5'-GC preference could arise from the existence of unfavorable steric interactions encountered in the first steps of the reaction leading to formation of the 5'-GC cross-link.[183]

5'-CG-3'          5'-GC-3'
3'-GC-5'          3'-CG-5'

**Figure 7** Potential cross-linking sites for (**65a**).

It appears that (**65a**) preferentially alkylates guanines in DNA sequences where cross-link formation is possible. For example, examination of various two-base combinations embedded in a particular sequence showed that the favored sites of monoadduct formation by (**65a**) are as follows: 5'-CG > 5'-GG > 5'-AG ≈ 5'-TG.[167,185] Several lines of evidence indicate that hydrogen bonding of the C-10-carbamate group to the exocyclic $N^2$ amino group of the guanine that subsequently becomes adducted in the preferred 5'-CG sequence during cross-link formation plays a role in "steering" monoadduct formation to sites that can form cross-links (Figure 8).[175] This preference is maintained in a mitomycin C analogue where an appropriate hydrogen bond acceptor (a hydroxyl group) is present at this position.[186] The 5'-CG preference for monoadduct formation is eliminated when inosine (**8**) replaces guanine in the target sequence.[175] Conversely, a preferred monoalkylation site presumably could be created by substitution of 2,6-diaminopurine (**72**) for the adenine opposite the thymine in a 5'-TG sequence. NMR studies on a monoadduct of (**65a**) with a DNA duplex provide further evidence for hydrogen bonding between the C-10 substituent and the adjacent guanine $N^2$ in a 5'-CG sequence.[187]

Only one intrastrand cross-link is possible for (**65a**), with attachment at two adjacent guanines in a 5'-GpG sequence. Intrastrand cross-link formation induces an ∼ 15° bend into duplex DNA.[188]

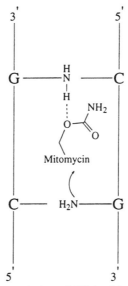

**Figure 8** Possible noncovalent association of (**65a**) with DNA prior to adduct formation.

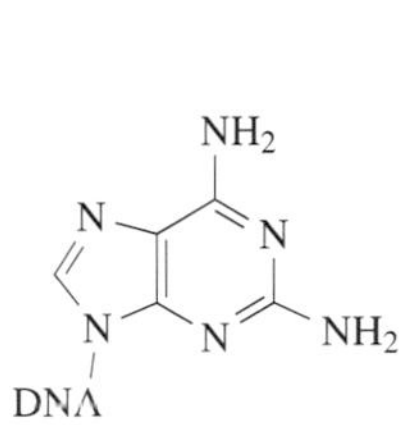

(**72**)

In addition to acting as a redox-sensitive trigger for the alkylation of DNA, the quinone functional group of (**65a**), in the presence of a reducing agent and molecular oxygen, undergoes redox-cycling reactions that lead to oxidative DNA damage (see 7.14.2.11.1).[189,190] Thus, depending upon the conditions, (**65a**) may damage DNA via two distinct mechanisms. Covalent attachment of (**65a**) to DNA, followed by redox cycling, could result in localized regions of oxidative DNA damage.

### 7.14.2.6.2    *FR66979 and FR900482*

Hopkins and co-workers showed that, similar to mitomycin C, FR66979 (**73a**) and FR900482 (**73b**), when treated with a reducing agent such as sodium dithionite, cross-link DNA preferentially at 5′-CG sequences.[191,192] Substitution of deoxyinosine (**8**) for guanine at cross-linking sites eliminates cross-link formation, suggesting that the cross-links, similar to those obtained from (**65a**), involve attachments to DNA at N[2] of guanine.[191] Guanine–guanine cross-links have been isolated by nuclease digestion of cross-linked DNA, followed by reverse-phase HPLC purification. Complete characterization of the isolated cross-links supports the structure (**75**).[192] This structure is consistent with reduction-dependent formation of a mitosene-type reactive intermediate (**74**) from (**73**), as originally proposed by Fukuyama *et al.*[193] (Scheme 16), and is not consistent with other possible mechanisms that have been suggested[194,195] for the alkylation of DNA by (**73a**) and (**73b**).

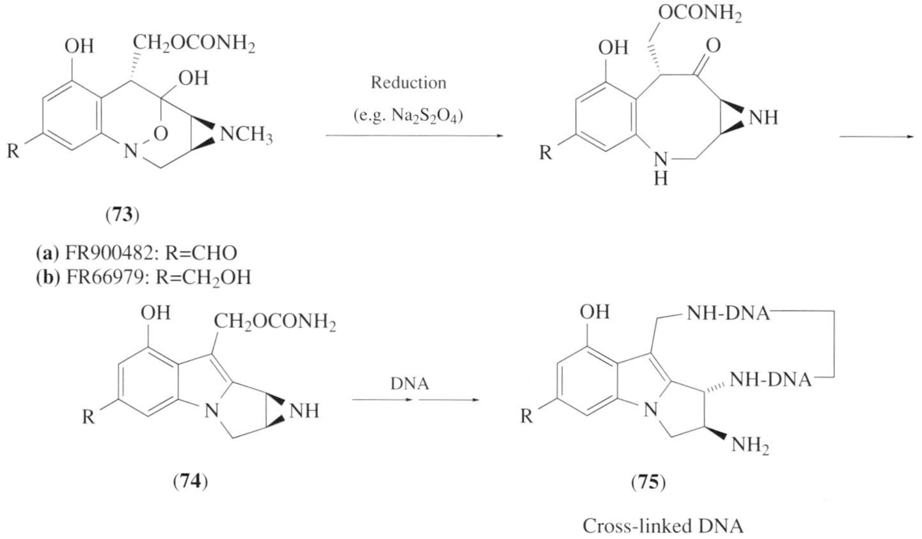

(73)

**(a)** FR900482: R=CHO
**(b)** FR66979: R=CH₂OH

(74)                                      (75)

Cross-linked DNA

**Scheme 16**

### 7.14.2.6.3    *Oxidatively activated pyrrolizidine alkaloids*

In their early work with (**65a**), Iyer and Szybalski[169] noted the structural similarities between this agent and naturally occurring pyrrolizidine alkaloids, such as retrorsine (**76**) and monocrotaline (**77**). These common plant-derived compounds are cytotoxic and carcinogenic.[196,197] Pyrrolizidine alkaloids are relatively unreactive; but *in vivo* oxidative metabolism by cytochrome P450 enzymes affords electrophilic pyrrole analogues (**78**) (Scheme 17).[196,197] It has been demonstrated that, upon oxidative activation, pyrrolizidine alkaloids, such as (**76**) and (**77**), cross-link duplex DNA.[198,199] Monoadducts of these natural products linked to DNA through N[2] of guanine have been isolated.[200] To date (1997), no cross-links derived from natural pyrrolizidine alkaloids have been isolated; however, cross-links involving simple synthetic derivatives of (**78**) have been isolated and characterized.

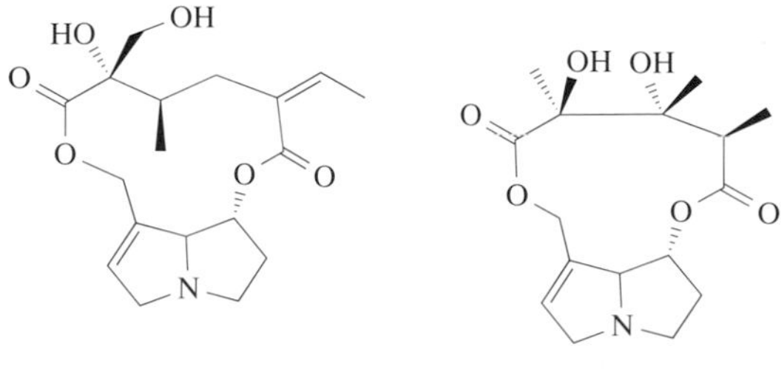

(**76**) Retrorsine                    (**77**) Monocrotaline

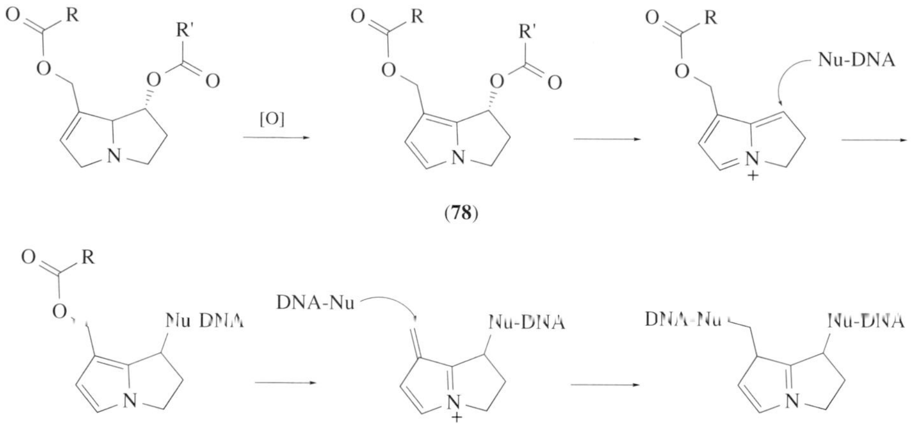

**Scheme 17**

Hopkin's group has shown that dehydroretrorsine and dehydromonocrotaline display a 10:1 preference for 5'-CG sites over 5'-GC sites.[201,202] Although they are not as site selective as mitomycin, work with these pyrrolizidine alkaloids suggests that the common 5'-CG target site may result, in part, simply from the fact that the distance between the electrophilic carbons involved in DNA modification is essentially the same for all pyrrole-derived bifunctional electrophiles (Figure 9).

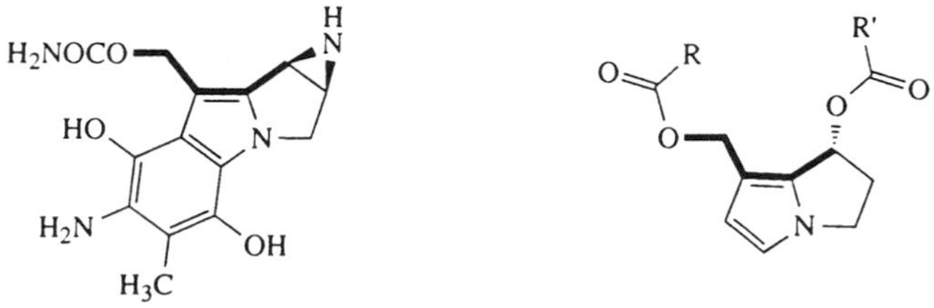

**(65a)**

**Figure 9** Common core structure for pyrrolizidine cross-linking agents.

### 7.14.2.7 Alkenylbenzenes

A number of substituted alkenylbenzenes occur naturally in plants, including some plant species that are used as herbs and spices.[203] For example, safrole (**79**) is a major component of oil of sassafras and estragole (**80**) is found in the oils of tarragon and sweet basil. Both (**79**) and (**80**) are carcinogenic to rats and mice and have been found to modify DNA covalently.[203,204]

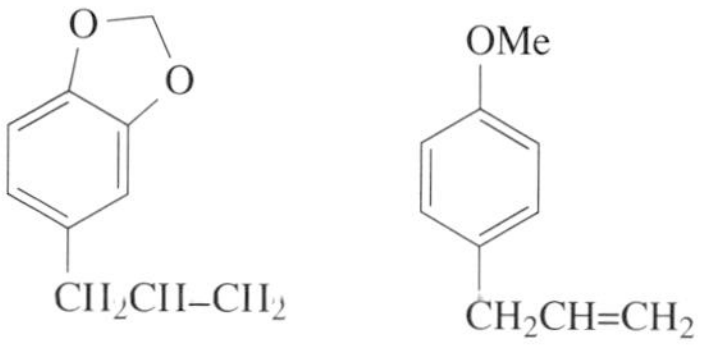

(**79**) Safrole          (**80**) Estragole

Investigations into the carcinogenic action of (**79**) and (**80**) led to the finding that the 1'-hydroxylated metabolites (**81**) of these compounds are more potent carcinogens than the parent alkenylbenzenes. *In vivo*, these allylic alcohols are enzymatically converted to sulfate derivatives (**82**) that are the ultimate carcinogens derived from (**79**) and (**80**).[205] DNA alkylation apparently results from an $S_N1'$ reaction in which departure of the sulfate leaving group yields a resonance-stabilized carbo-cation (83) that reacts with nucleophilic sites on DNA (Scheme 18).[203] Consistent with an $S_N1'$ mechanism, nucleophilic attack occurs at both the 1' and 3' positions of the alkenylbenzene nucleus.

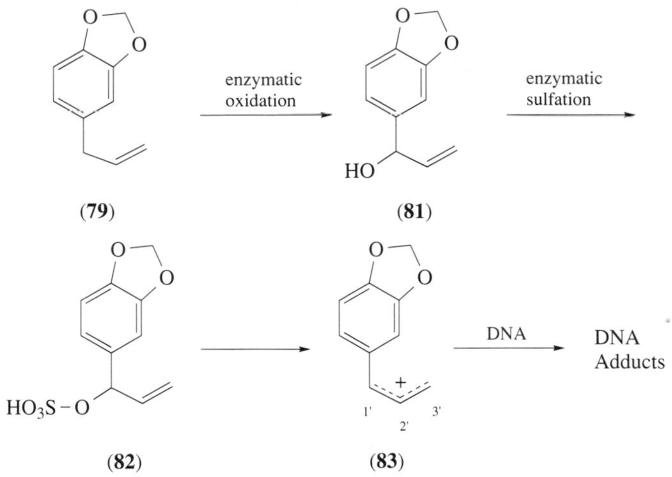

**Scheme 18**

Acetylated derivatives of 1'-hydroxysafrole and 1'-hydroxyestragole serve as useful, easily prepared models for the carcinogenic forms of these molecules.[206] Reaction of these 1'-acetoxy model compounds with DNA *in vitro* yields a number of covalent DNA adducts. For example, adducts of (79) attached to C-8, $N^2$, and N-7 of guanine and $N^6$ of adenine have been reported. Many of the same DNA adducts have been isolated from *in vivo* experiments utilizing the 1'-hydroxyalkenylbenzenes or the parent alkenylbenzenes.[203]

### 7.14.2.8    *N*-Nitroso Compounds

#### 7.14.2.8.1    *Streptozocin*

Streptozocin (84) is an *N*-methyl-*N*-nitrosourea derivative of 2-deoxyglucose that possesses cytotoxic and anticancer properties.[207] Similar to the well-known methylating agent *N*-methyl-*N*-nitrosourea (MNU), (84) decomposes in aqueous solution with release of dinitrogen, presumably via methyldiazoic acid (85) (Scheme 19).[50,208] Treatment of DNA by (84) yields methylation at sites including N-7 and C-6 of guanine and N-1, N-3, and N-7 of adenine.[208] In *in vitro* reactions, the products resulting from alkylation of DNA by (84) are identical to those obtained with MNU; however, in biological experiments (84) is less mutagenic than MNU.[50,208]

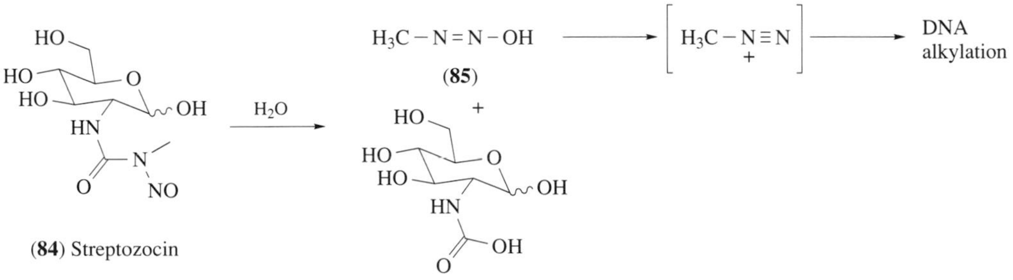

**Scheme 19**

### 7.14.2.8.2  Nitrosamines

Nitrosamines are ubiquitous carcinogens that are formed during fermentation of various foods and tobacco and by the *in vivo* reaction of nitrite preservatives with amines under the acidic conditions found in the gastrointestinal tract.[209] For example, *N*-methyl-*N*-nitrosamines are well-known carcinogens and DNA-methylating agents. DNA alkylation by this class of compounds involves metabolic oxidative activation, followed by decomposition to yield alkylating species (Scheme 20).[209,210] The natural product dephostatin (**86**), isolated from *Streptomyces*, is a tyrosine phosphatase inhibitor that contains the *N*-methyl-*N*-nitrosamino functional group.[211] Inhibition of tyrosine kinases by (**86**) has been proposed to result from nitroso group-transfer to a cysteine thiol of these enzymes.[212] If (**86**) can serve as a substrate for appropriate oxygenases, alkylation of DNA and other biomolecules by this natural product is a possibility. In addition, the semiquinone form of (**86**) may decompose with loss of nitric acid which, in the presence of molecular oxygen, can cause DNA damage.

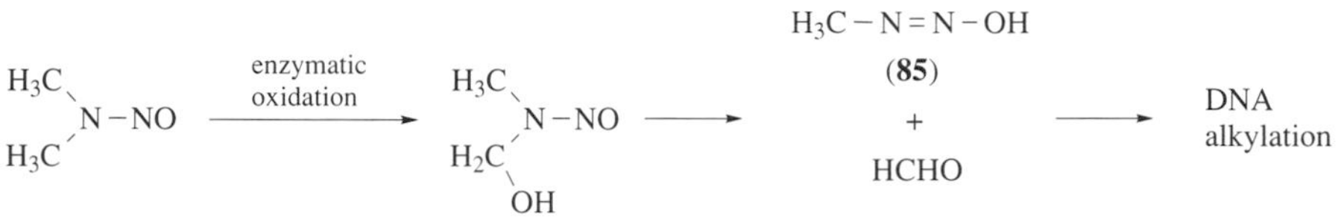

**Scheme 20**

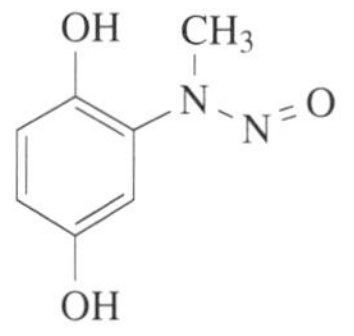

(**86**) Dephostatin

### 7.14.2.9  Nitroaromatics

#### 7.14.2.9.1  Azomycin

Azomycin (**87**) is an antimicrobial agent first isolated in 1953 and structurally characterized in 1955.[213] Studies with (**87**), as well as with substituted 2-nitroimidazoles, strongly suggest that (**87**) may be capable of covalently modifying DNA by several different pathways.

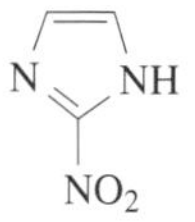

(**87**) Azomycin

Four-electron enzymatic reduction of 2-nitroimidazoles converts the nitro group to a hydroxyamine (**88**).[214] These hydroxyamine derivatives of imidazoles are electrophilic and undergo hydrolysis to produce the known DNA-damaging agent glyoxal (**89**) (Scheme 21) (see Section 7.14.2.2.1).[214] Reductive activation of 2-nitroimidazoles, including (**87**), in the presence of DNA results in the formation of glyoxal adducts with guanine.[215,216] It is possible that electrophilic 2-hydroxyaminoimidazoles may directly alkylate DNA, although this has not been demonstrated.

2-Nitroimidazoles also can cause DNA damage by enzyme-driven redox cycling of the nitro functional group.[217,218] In this process, single-electron enzymatic reduction of the nitro group yields a radical anion that can be oxidized by molecular oxygen to generate superoxide radicals and regenerate the starting nitro compound (Scheme 22). Although enzyme systems protect cells from its deleterious effects, superoxide radical can lead to DNA damage through a cascade of reactions involving dismutation to hydrogen peroxide followed by generation of hydroxyl radical via the Fenton reaction (see Section 7.14.2.15).

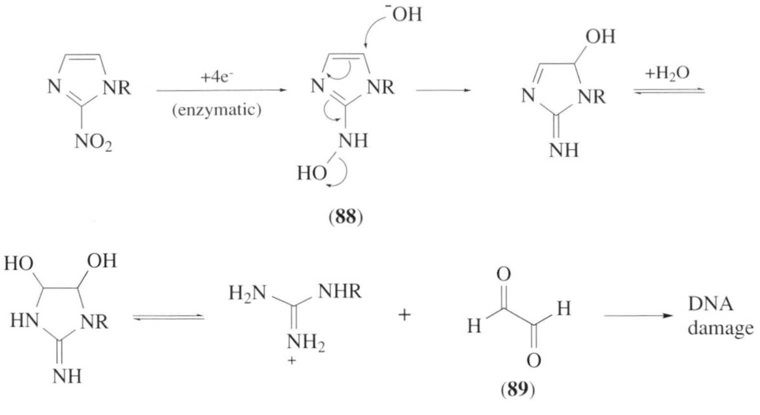

**Scheme 21**

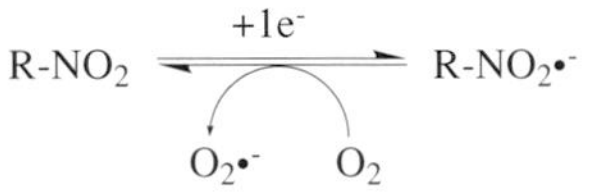

**Scheme 22**

### 7.14.2.9.2 Aristolochic acid

Aristolochic acids (e.g., (**90**)) are carcinogens present in the leaves and roots of *Aristolochia* plant species that have been used medicinally for hundreds of years.[219] Enzymatic reductive activation of aristolochic acids in the presence of DNA leads to the formation of covalent adducts, such as (**93**) and (**94**), attached through N-6 of adenine and $N^2$ of guanine (Scheme 23).[197] DNA adducts have been prepared *in vitro* using a xanthine oxidase enzymatic reduction system[220] and also have been identified by *in vivo* experiments.[221]

Analysis of the DNA adducts formed by aristolochic acid led to the proposal that adduct formation proceeds through reduction of the compound's nitro group to a hydroxylamine (**91**), followed by formation of a cyclic hydroxamide (**92**). It has been suggested that the ultimate alkylating species is a nitrenium ion generated by spontaneous loss of hydroxide from the hydroxamide,[197] although direct reaction of DNA with the hydroxamide is also reasonable. *N*-Chloroaristolactam has been used as a model for the presumed hydroxamide species and affords the same DNA adducts obtained by reductive activation of the natural product.[220] Cyclic hydroxamide formation appears to be crucial for the generation of a DNA-reactive electrophile in these systems; 1,8-nitronaphthoic acid, similar to the aristolochic acids, is a reductively activated mutagen, but the 2,3-substituted analogue is not.[222] Reduction-dependent DNA adduction by aristolochic acid is similar, in some regards, to that by the well-known carcinogens 4-nitroquinoline *N*-oxide and 2-aminofluorene.[223,224]

### 7.14.2.10 1,2-Dithiolan-3-one 1-Oxides

#### 7.14.2.10.1 Leinamycin

Leinamycin (**95**) was isolated from a strain of *Streptomyces* in 1989[225] and its structure elucidated by NMR spectroscopy, chemical analysis, X-ray crystallography,[226] and total chemical synthesis.[227] Compound (**95**) contains an unusual 1,2-dithiolan-3-one 1-oxide heterocycle (**96**) that has not been found in any other natural product. The potent antitumor and antibacterial activity of this sulfur-containing natural product appears to stem from DNA damage.[228]

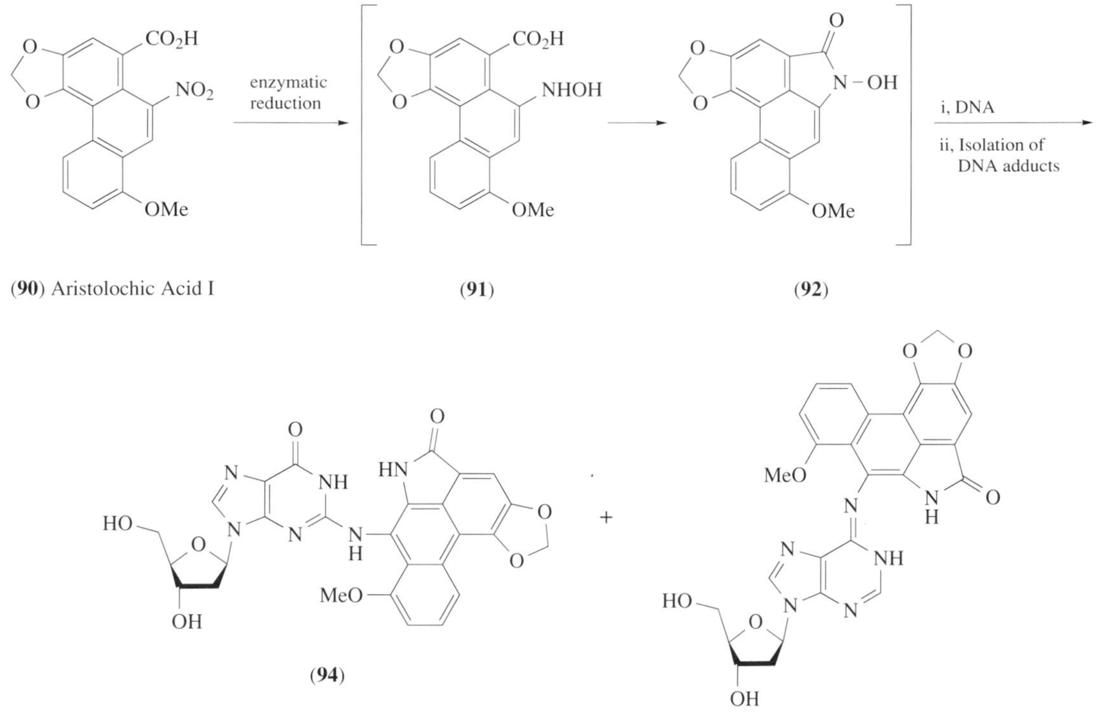

(**90**) Aristolochic Acid I      (**91**)      (**92**)

enzymatic reduction

i, DNA

ii, Isolation of DNA adducts

(**94**)

+

(**93**)

**Scheme 23**

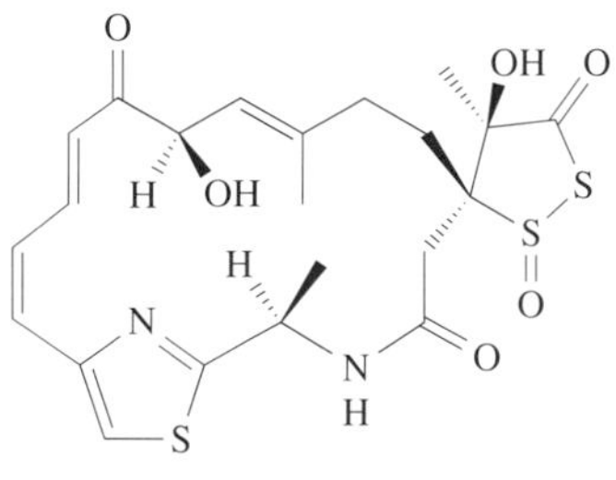

(**95**) Leinamycin 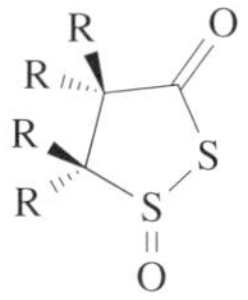      (**96**) 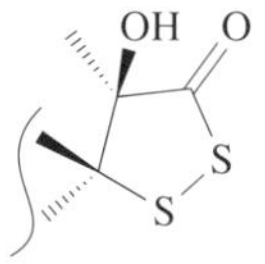      (**97**) *S*-Deoxyleinamycin

Experiments have revealed that (**95**) cleaves DNA *in vitro* and that DNA cleavage requires the presence of thiols in the reaction mixture.[228] Other reducing agents such as NADH and ascorbate do not trigger DNA cleavage by (**95**). It was also shown that *S*-deoxyleinamycin (**97**), prepared by catalytic hydrogenation of the natural product, displays greatly diminished biological activity and does not efficiently cleave DNA *in vitro*.[228] Thus, the unusual 1,2-dithiolan-3-one 1-oxide heterocycle was implicated in the cleavage of DNA by (**95**) and the data suggested that nucleophilic attack of thiols on this sulfur heterocycle triggers subsequent DNA chemistry.

When (**95**) was discovered, little was known about the reactivity of the 1,2-dithiolan-3-one 1-oxide heterocycle. Therefore, chemical model studies involving the reactions of simple 1,2-dithiolan-3-one 1-oxides (**98**) and (**99**) with thiols were able to provide insights regarding the chemistry underlying the thiol-triggered cleavage of DNA by (**95**).[229] The dithiocarboxylic acid (**100**) and polysulfides (**101**) were identified as major products of the reaction of thiols with 1,2-dithiolan-3-one 1-oxides (Scheme 24).[229] It was suggested[229] that these products result from an initial attack of the thiol on the central, sulfenyl sulfur of the heterocycle, followed by cyclization of the resulting sulfenic acid to yield an electrophilic oxathiolanone (**102**) and a hydrodisulfide (**103**). The observed final products of the reaction derive from the attack of excess thiol on the oxathiolanone intermediate and from decomposition of the unstable hydrodisulfide species. Importantly, it was shown[229] that 3H-1,2-benzodithiol-3-one 1-oxide (**98**) and the alicyclic dithiolanone oxide (**99**) react similarly with thiols, indicating that, in these reactions, (98) serves as a reasonable model for the oxygenated sulfur heterocycle of (95). It was noted that the electrophilic oxathiolanone or unstable hydrodisulfide intermediates might be involved in thiol-triggered DNA damage by (95).[229]

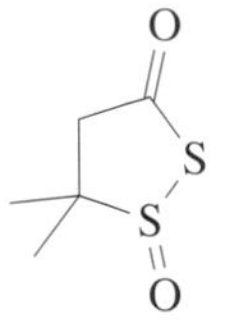

(99)

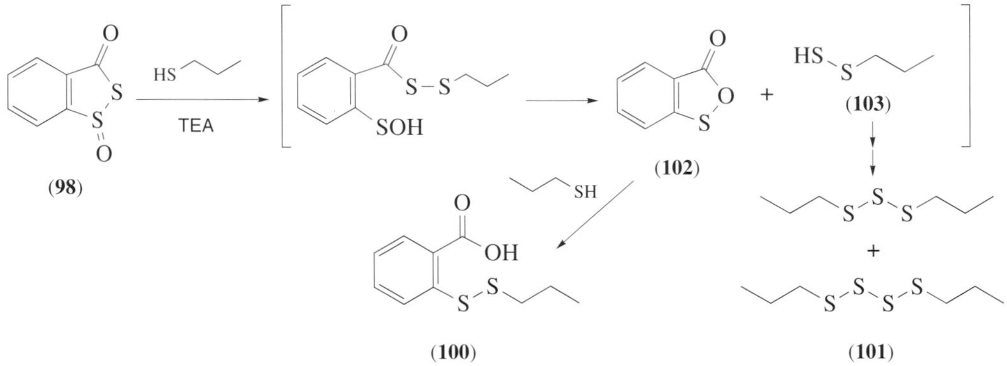

**Scheme 24**

Subsequent studies revealed that the reaction of (95) with DNA in the presence of thiols results in the formation of covalent DNA adducts.[230] Thermal workup of DNA treated with (95) in the presence of a thiol results in release of the adduct. Characterization of the adduct (106) shows that the natural product has undergone a deep-seated rearrangement, while forming a covalent bond with N-7 of a guanine residue (Scheme 25).[230] It was proposed that, analogous to the reaction of thiols with simple 1,2-dithiolan-3-one 1-oxides,[229] the reaction of a thiol with (95) leads to the formation of an electrophilic oxathiolanone. The electrophilic oxathiolanone intermediate of the natural product can be trapped in an intramolecular reaction with the C-6—C-7 alkene of the 18-membered macrocycle resulting in generation of an episulfonium alkylating species (105).[230] This alkene-trapping reaction is precedented in the literature[231] and is analogous to the well-known reaction of alkenes with sulfenyl chlorides. In chemistry similar to that observed for simple sulfur mustards,[232] the thiol-generated episulfonium ion intermediate of (95) alkylates DNA at N-7 of guanine. The sequence specificity of DNA alkylation by (95) has not been reported and it remains unknown whether the 18-membered macrocycle of the natural product serves primarily as a scaffold to position the C-6—C-7 alkene properly for trapping of the thiol-generated oxathiolanone intermediate or whether this macrocycle is also involved in noncovalent association with duplex DNA.

Similar to the natural product, simple analogues of the 1,2-dithiolan-3-one 1-oxide heterocycle in (95), (98), (99), and (108) are thiol-dependent DNA-cleaving agents.[233] Obviously, these leinamycin mimics lack alkene substituents that could be involved in the formation of episulfonium species analogous to (105). Accordingly, the mechanism of DNA cleavage by these 1,2-dithiolan-3-one 1-oxides is clearly distinct from that of the natural product and does not involve DNA alkylation. DNA cleavage by (98), (99), and (108) is inhibited by radical scavengers, the hydrogen peroxide-destroying enzyme catalase, and chelators of adventitious trace metals.[233] These and other results indicate that simple 1,2-dithiolan-3-one 1-oxides, in concert with thiols, mediate the conversion of molecular oxygen to hydrogen peroxide which causes DNA cleavage via a trace metal-dependent Fenton reaction.

These simple 1,2-dithiolan-3-one 1-oxides react rapidly and completely with thiols in aqueous solution and the resulting products mediate thiol-dependent conversion of molecular oxygen to DNA-cleaving oxygen radicals. Examination of each product in the mixture resulting from the reaction of thiols with (98) revealed that polysulfides are potent thiol-dependent DNA-cleaving agents.[234] Mechanistic studies of polysulfides derived from 2-mercaptoethanol and other thiols

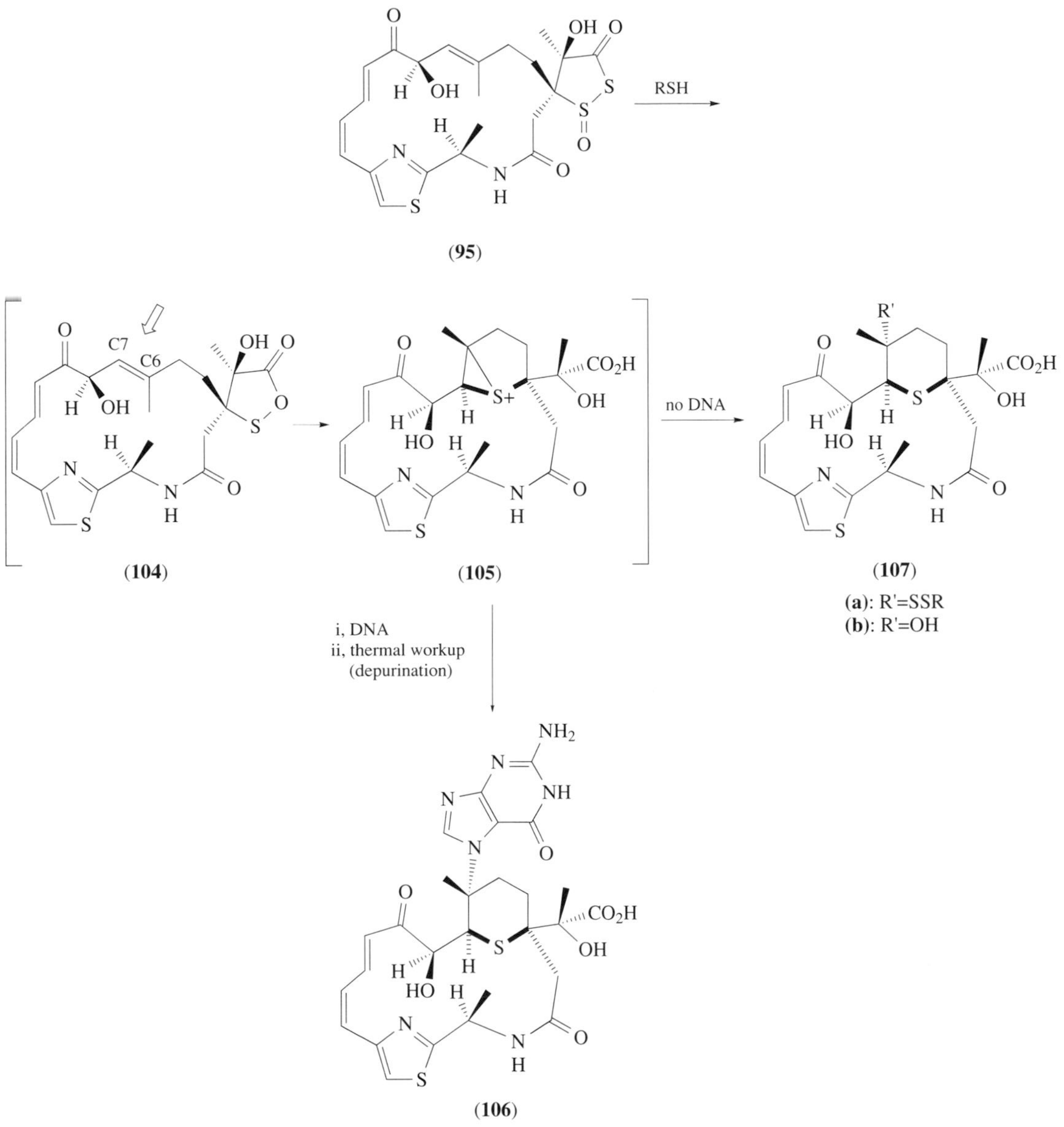

**(95)**

**(104)**     **(105)**     **(107)**

**(a)**: R'=SSR
**(b)**: R'=OH

i, DNA
ii, thermal workup
(depurination)

**(106)**

**Scheme 25**

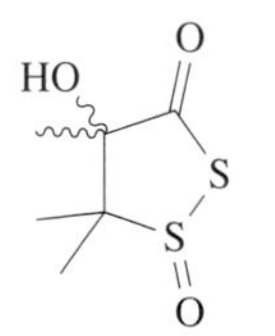

**(108)**

indicated that these compounds are thiol-dependent DNA cleaving agents that mediate the conversion of molecular oxygen to DNA-cleaving oxygen radicals via a trace metal-dependent Fenton reaction.[234] The detailed mechanism by which polysulfides mediate thiol-dependent generation of oxygen radicals remains under investigation, but may involve facile reduction of molecular oxygen to superoxide by thiol-generated hydropolysulfide anions (e.g., **109**) (Scheme 26).[234] It is likely that a number of processes not explicitly shown in Scheme 26, including dimerization of species such as **(109)** and reactions involving hydrogen sulfide and thiol, can lead to the production of superoxide radicals *without* net destruction of polysulfides.

Consistent with Scheme 26, at low concentrations of trisulfide and thiol (1–100 μM), where the ratio of dissolved molecular oxygen (∼200 μM) to hydrodisulfide (**109**) is relatively high, the polysulfide appears to serve as a catalyst for thiol oxidation (and presumably corresponding superoxide production), with one equivalent of trisulfide leading to the production of approximately nine equivalents of disulfide. The hydrogen sulfide by-product derived from the oxygen-independent reaction of thiols with polysulfides was identified by bubbling the reaction headspace through a lead acetate solution, followed by characterization of the resulting lead sulfide precipitate. Superoxide radicals generated by polysulfide/thiol mixtures lead to DNA damage through a cascade of reactions involving dismutation to hydrogen peroxide followed by generation of hydroxyl radicals via the Fenton reaction (see Section 7.14.2.15).

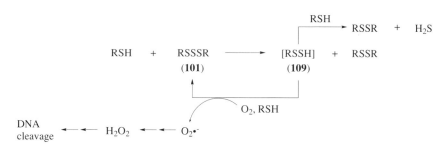

**Scheme 26**

The reaction of thiols with (**95**), similar to the reaction of thiols with (**98**), results in the formation of polysulfides. Consistent with the generation of polysulfides in the reaction of thiols with (**95**), it has been demonstrated that the natural product (**95**), in the presence of physiologically relevant concentrations of thiol, mediates oxidative damage in addition to the DNA alkylation described above.[234] The notion that polysulfides formed by reaction of thiols with (**95**) are derived from a hydrosulfide species (e.g., **109**) is provided by the fact that, when treated with thiol in the absence of DNA, the disulfide adduct (**107a**), resulting from attack of a hydrodisulfide species on the episulfonium ion (**105**), is obtained in 21% yield, along with the expected hydrolysis product (**107b**) (41% yield).[230] As it is known that hydrodisulfides decompose to polysulfides in aqueous solution, the thiol-dependent release of hydrodisulfide from (**95**) explains the formation of DNA-cleaving polysulfides that results from treatment of the natural product with thiol. Thus, (**95**) is capable of damaging DNA by two distinct mechanisms involving either alkylation or the production of polysulfides that generate oxygen radicals in the presence of excess thiol. The relative importance of the two possible pathways will depend upon the conditions under which the thiol-triggering reaction occurs.

1,2-Dithiolan-3-one 1-oxides, such as (**98**), react with anilines and amine compounds under mild conditions, suggesting covalent adducts could result from direct reaction of these sulfur heterocycles with nucleophilic nitrogens in DNA.[235]

Not surprisingly, leinamycin and simple 1,2-dithiolan-3-one 1-oxides are capable of inactivating cysteine-dependent enzymes.[236] The inactivation of cysteine-dependent enzymes clearly can have important biological consequences.

### 7.14.2.10.2   *1,2-Dithiole-3-thiones*

It has been suggested that 1,2-dithiole-3-thiones occur naturally in plants such as broccoli.[237] 1,2-Dithiole-3-thiones, similar to 1,2-dithiolan-3-one 1-oxides, mediate thiol-dependent conversion of molecular oxygen to oxygen radicals and are capable of cleaving DNA in the presence of thiols.[238] Research suggests[238] that production of oxygen radicals might play a role in the induction of cancer-preventive phase II metabolic enzymes by these compounds,[239] perhaps through reaction of the oxidizing radicals with protein transcription factors. The chemical mechanism of oxygen-radical generation by this class of compounds remains under investigation, but appears similar, in many regards, to that of the structurally related dithiolanone oxides (e.g., (**98**) and (**99**)) discussed above.

### 7.14.2.10.3  *Polysulfides*

The finding that polysulfides are thiol-dependent DNA-damaging agents may be relevant to the biological activity of several polysulfide-containing natural product antibiotics, including varacin (**110a**),[240] lissoclinotoxin A (**110b**),[241] the leptosins (**111**),[242] and the sirodesmins (**112**).[243]

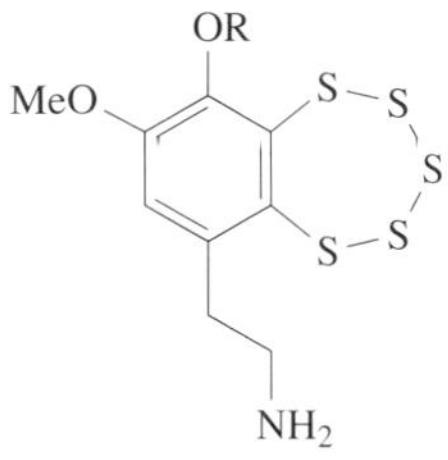

**(110)**

(**a**) Varacin: R=CH$_3$
(**b**) Lissoclinotoxin A: R=H

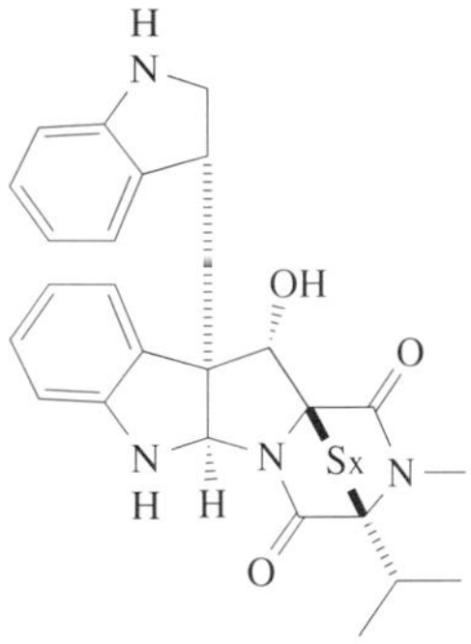

**(111)**

(**a**) Leptosin E: X=3
(**b**) Leptosin F: X=4

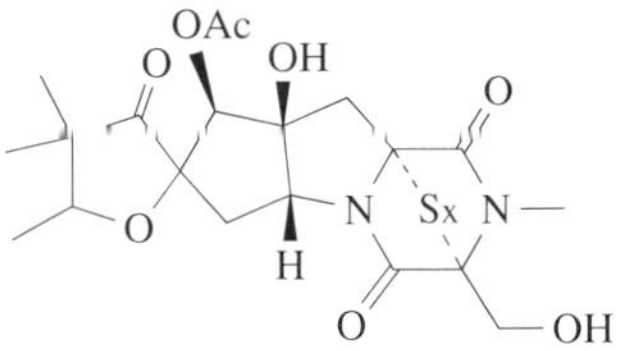

**(112)**

Sirodesmin C: X=3
Sirodesmin B: X=4

Studies show that the varacin analogue, 7-methylbenzopentathiapin, is a potent thiol-dependent DNA-cleaving agent.[242] The reaction of thiols with this pentathiapin results in the formation of a complex mixture of polysulfides. Mechanistic experiments show that DNA cleavage in this system, identical to the polysulfides described above, involves the reduction of molecular oxygen to hydrogen peroxide (via superoxide), followed by a trace metal-dependent Fenton reaction.[244] Interestingly, biological experiments have suggested that DNA damage might play a role in the antitumor properties of the pentathiapin antibiotic varacin.[240]

In addition to their ability to produce oxygen radicals, polysulfides may derive biological activity through reactions with thiol groups on proteins[236] or through chemical reactions that lead to depletion of cellular thiols (Scheme 26).

### 7.14.2.11  Quinones

A large number of quinone-containing natural products have been isolated and characterized.[245,246] Anthraquinone-based natural products, such as daunomycin (daunorubicin) (**113a**) and adriamycin (doxorubicin) (**113b**), have proven useful in the treatment of cancer and, therefore, have been the subject of much investigation.[247,248] In many natural products, including the anthraquinones, the quinone functional group occurs as part of a planar aromatic system that is involved in intercalative binding to DNA.[249] Noncovalent intercalation of these compounds into DNA often has important biological consequences.[247,248] In addition, there are a number of mechanisms by which quinone-containing compounds can covalently modify DNA.

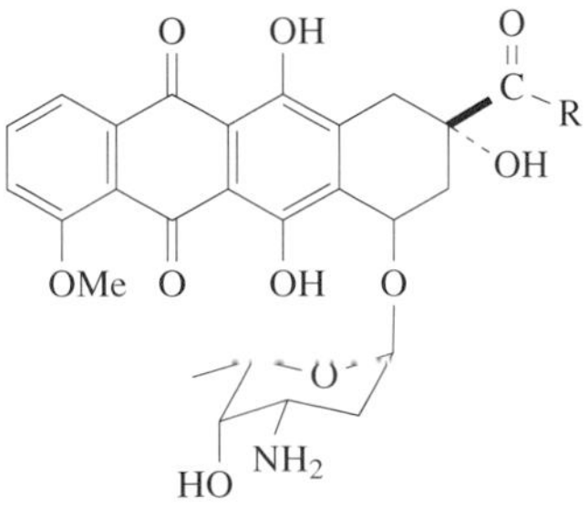

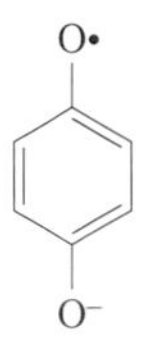

**(113)**

(**a**) Daunomycin: R=Me
(**b**) Adriamycin: R=CH$_2$OH

**(114)**

### 7.14.2.11.1  *Redox cycling of quinones and quinoid compounds*

In the presence of molecular oxygen and appropriate reducing agents, quinones can mediate the transfer of electrons from the reducing agent to molecular oxygen, resulting in the production of superoxide radical anion.[250–253] In this process, reduction of the quinone yields a hydroquinone or semiquinone radical (114), either of which can be oxidized by molecular oxygen to produce superoxide and the quinone starting material (Scheme 27). Importantly, the quinone is not destroyed in this reaction cycle. Superoxide radical leads to DNA damage through a cascade of reactions involving dismutation to hydrogen peroxide followed by generation of hydroxyl radical via the Fenton reaction (see Section 7.14.2.15).

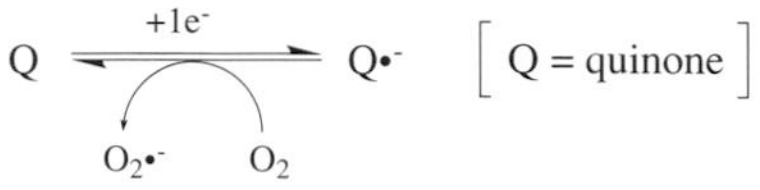

$$ Q \; \underset{O_2^{\bullet-} \qquad O_2}{\overset{+1e^-}{\rightleftharpoons}} \; Q^{\bullet-} \qquad \left[ Q = \text{quinone} \right] $$

**Scheme 27**

Quinones facilitate the reduction of molecular oxygen by a variety of reducing agents. For example, the rate of molecular oxygen reduction by the enzyme NADPH : ferredoxin reductase is increased markedly in the presence of quinones.[254] Similarly, quinones catalyze the reduction of molecular oxygen by thiols.[255]

Quinones can be reduced either chemically or enzymatically.[253] For example, thiols[256] and NADH can serve as chemical reducing agents in redox cycles involving quinones. A variety of enzymes, including NADPH : cytochrome P-450 reductase, xanthine oxidase, NADPH : ferredoxin reductase, and NADPH : quinone acceptor oxidoreductase, have been found to reduce quinones.[250] It remains a matter of debate as to whether intercalated quinones can undergo efficient redox cycling.[256–259] For the anthraquinones, nonintercalative DNA-binding modes may exist, in which the quinone moiety is capable of engaging in redox chemistry.[259]

Quinone antibiotics, such as streptonigrin (115) and adriamycin (113b), chelate transition metal ions that may serve to increase the efficiency of oxygen-radical generation.[259,260] These bound transition metals may serve as intramolecular catalysts for various steps leading to the production of hydroxyl radical, thereby leading to efficient, highly localized generation of oxidizing species.

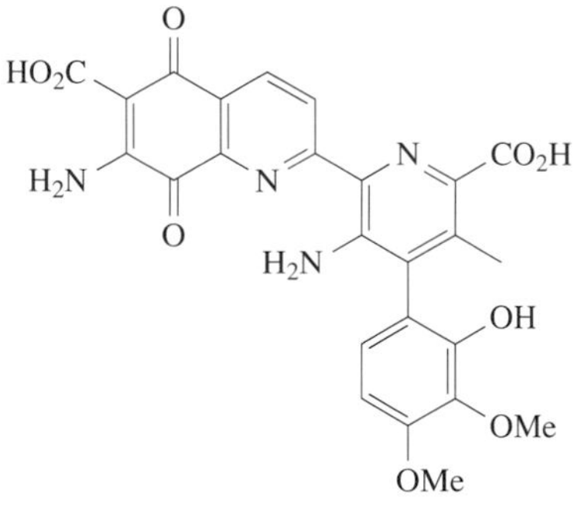

(115) Streptonigrin

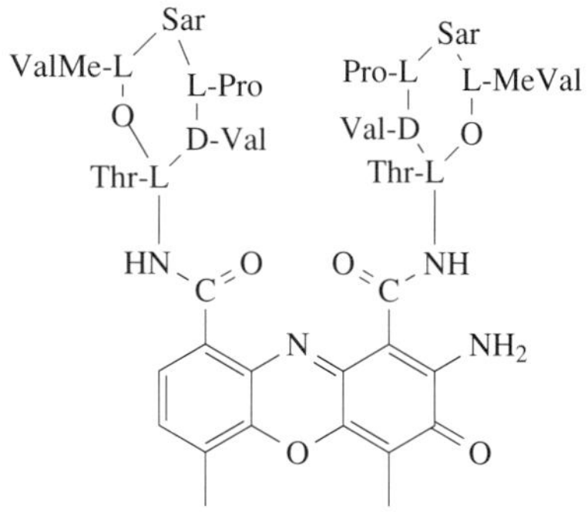

(116) Actinomycin D

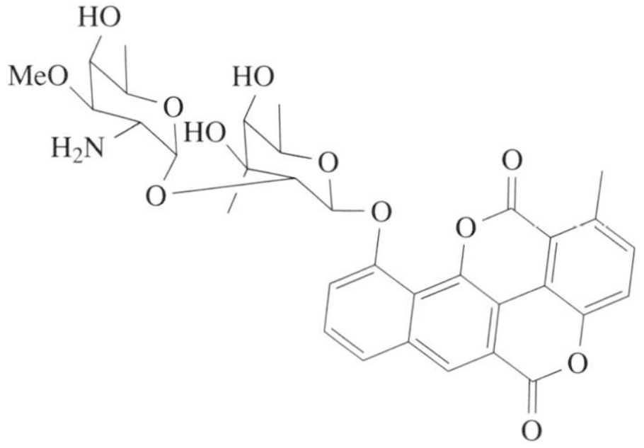

(117) Elsamicin

Redox-dependent DNA damage has been suggested or demonstrated for a variety of quinone and quinoid natural products, including daunomycin (113a),[261] adriamycin (113b),[258] streptonigrin (115),[262] saframycins A (9a) and C (9c),[68] mitomycin C (65a),[189,190] actinomycin D (116),[263,264] elsamycin (117),[265] and many others.[66,85,266-269] Although for many quinone-containing natural products redox chemistry has not been explicitly examined, in many cases toxic or mutagenic properties may result from the production of oxygen radicals.

### 7.14.2.11.2 *DNA alkylation by quinone methides*

Under anaerobic conditions, reduced anthracycline antibiotics eliminate the glycosyl or methoxyl substituents at C-7 to yield reactive quinone methides (118) (Scheme 28).[251] It remains unclear whether elimination occurs from the semiquinone or hydroquinone state. Quinone methides are electrophilic and have been shown to react with nucleophilic sites in DNA bases.[270] Irreversible tautomerization to the 7-deoxyaglycone (119) often competes with the reaction of the quinone methide and nucleophiles.[271] The biological relevance of DNA alkylation by quinone methide species remains unclear; however, there is evidence that anthracyclines do, in fact, form covalent adducts with DNA *in vivo*.[272,273] Koch has noted that the importance of such an attachment could be profound, as it tethers the quinone moiety, a catalyst for the generation of oxygen radicals, in close proximity to DNA.[270]

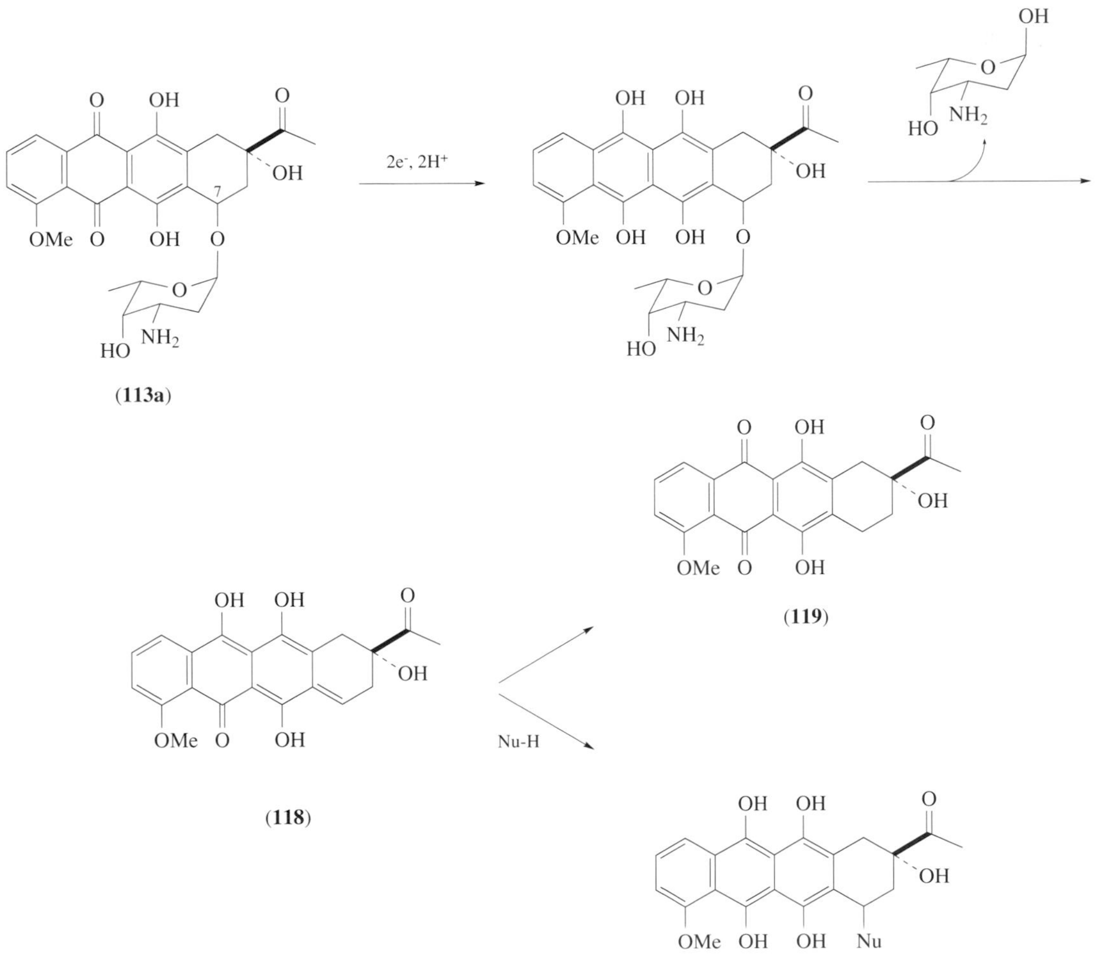

**Scheme 28**

In chemical model reactions, quinone methides have been trapped by nucleophiles including glutathione, imidazole, sulfite, and the exocyclic $N^2$-amino of guanine.[270,271] Quinone methides of menogaril (120) and daunomycin have been observed by UV–vis spectroscopy[274] and the quinone methide of 11-deoxydaunomycin has been observed by NMR.[275] Interestingly, quinone methides are also nucleophilic in character and can react with electrophiles such as protons and aldehydes.[254]

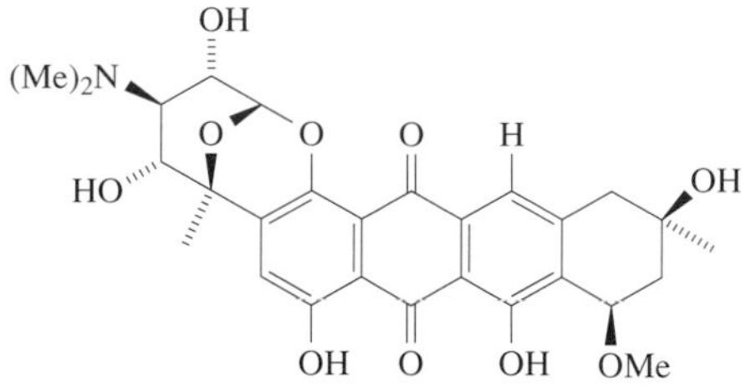

(**120**) Menogaril

Simple, synthetic compounds that yield DNA-damaging quinone methides upon reductive activation have been designed and characterized.[276-278]

### 7.14.2.11.3   *Formaldehyde-mediated covalent attachment of anthracyclines to DNA*

The presence of formaldehyde as an impurity in a crystallization solvent led to the serendipitous discovery that this aldehyde efficiently mediates covalent attachment of a daunomycin analogue to DNA.[279] Subsequent experiments, including a 1.5 Å resolution crystal structure, demonstrated that the natural product daunomycin (**113a**) forms similar formaldehyde-mediated covalent links (**121**) with DNA (Scheme 29).[280] DNA–drug cross-link formation is apparently facilitated by the fact that, in the noncovalent daunomycin–DNA intercalation complex, the N-3'-amino group of the daunosamine glycoside and the exocyclic $N^2$ of guanine in the minor groove are appropriately positioned for bridging by formaldehyde; thus, the resulting methylene bridge between the two nucleophilic sites does not lead to significant distortion of the daunomycin–DNA complex.[280] Formation of the covalent attachment is absolutely dependent upon the N-3' on the anthracycline antibiotic and $N^2$ of guanine on DNA, as shown by the fact that antibiotic analogues or modified DNA substrates lacking these functional groups do not become covalently associated in the presence of formaldehyde.[281]

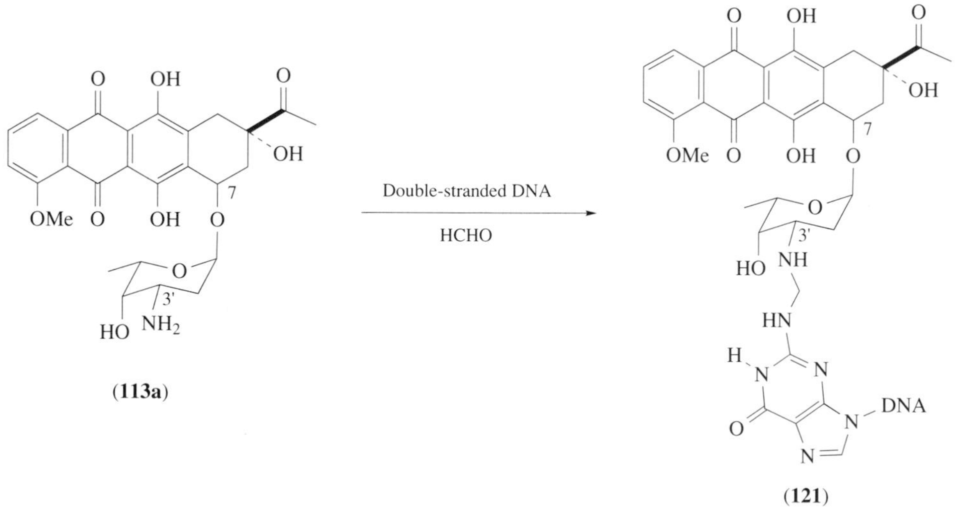

**Scheme 29**

Under conditions that allow redox cycling of the quinone moiety, oxidative decomposition of adriamycin (**113b**) leads to generation of formaldehyde that can mediate subsequent covalent attachment of an intact molecule of (**113b**) to DNA (Scheme 30).[282] Oxidative degradation of (**113b**) yields the by-product (**122**), and it is proposed that the C-9 substituent is converted to formaldehyde. The formaldehyde by-product of this oxidation was detected in the reaction mixture. Electrospray mass spectroscopic analysis of the adriamycin–DNA adducts obtained by incubating adriamycin with dithiothreitol in phosphate buffer is consistent with the adduct structure (**121**) (see Scheme 29). Daunomycin (**113a**) does not undergo a similar oxidative self-degradation to produce formaldehyde.[282]

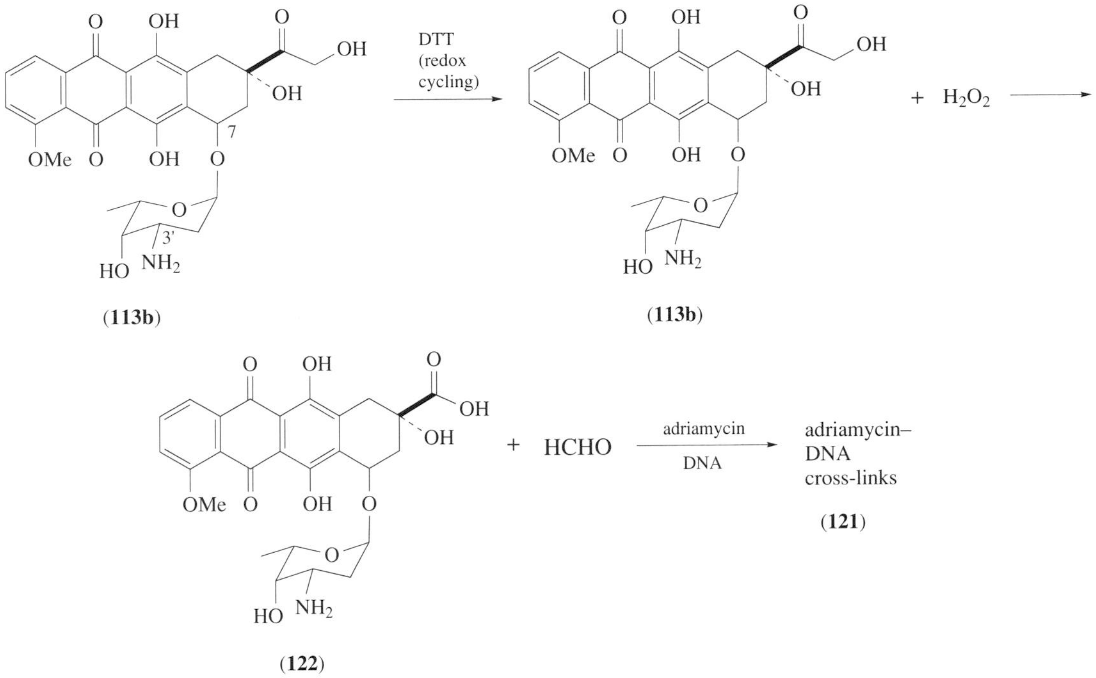

**Scheme 30**

Interestingly, incubation of either daunomycin or adriamycin with DNA and a reducing agent, such as dithiothreitol, in tris(hydroxymethyl)aminomethane (Tris) buffer leads to efficient formation of the covalent adduct (**121**). Thus, it appears that oxygen radical-mediated degradation of Tris buffer produces formaldehyde.[282] In addition, it was observed that the anthracycline–thiol system can degrade the biological polyamine, spermine, to formaldehyde.[282] This finding, coupled with the observation that formaldehyde-mediated anthracycline–DNA cross-linking occurs efficiently at low formaldehyde concentrations, suggests that adduct (**121**) could form under biological conditions.

The formaldehyde-mediated covalent attachment of anthracyclines to DNA is mechanistically similar to the proposed reaction of barminomycin I with DNA (see Section 7.14.2.1.4).

### 7.14.2.11.4 DNA alkylation by quinones

Quinones are electrophilic and, in the case of simple quinones such as benzoquinone, covalent adducts with DNA, such as (**123**) and (**124**), have been isolated and characterized.[283] Similar to $\alpha,\beta$-unsaturated aldehydes (see Section 7.14.2.2.1), simple quinones yield cyclic adducts that are the apparent result of Michael addition and subsequent intramolecular attack of the DNA base on the quinone carbonyl, followed by dehydration.

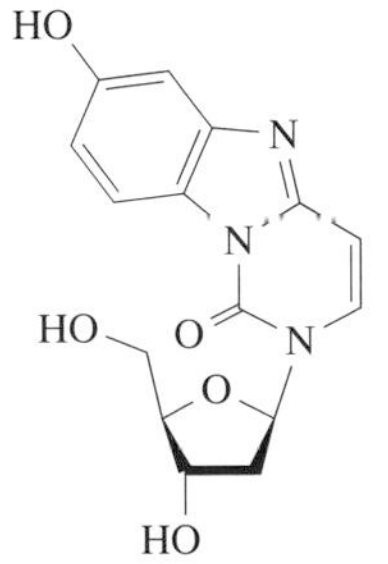

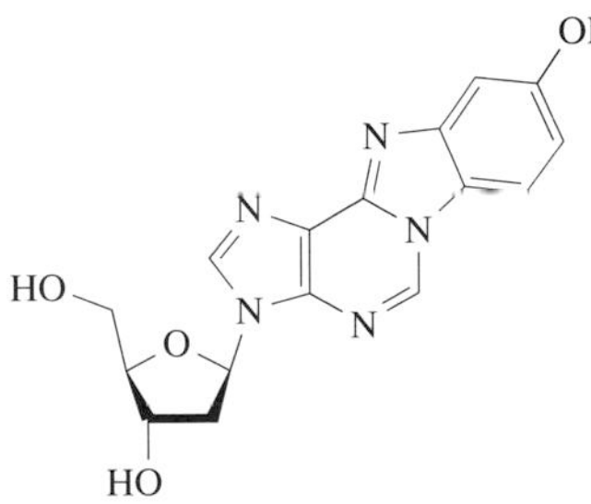

Naturally occurring estrogens, such as (125), undergo oxidative metabolism to catechol estrogens (126).[284] Oxidation of these catechols yields electrophilic *o*-quinone estrogens (127). These *o*-quinone estrogens form a variety of DNA adducts, such as (128), *in vitro* (Scheme 31).[284] In addition, covalent adducts of the quinone estrogen (127) attached at C-8 of adenine have been obtained by *in vitro* reaction of the *o*-semiquinone radical with adenine.[285]

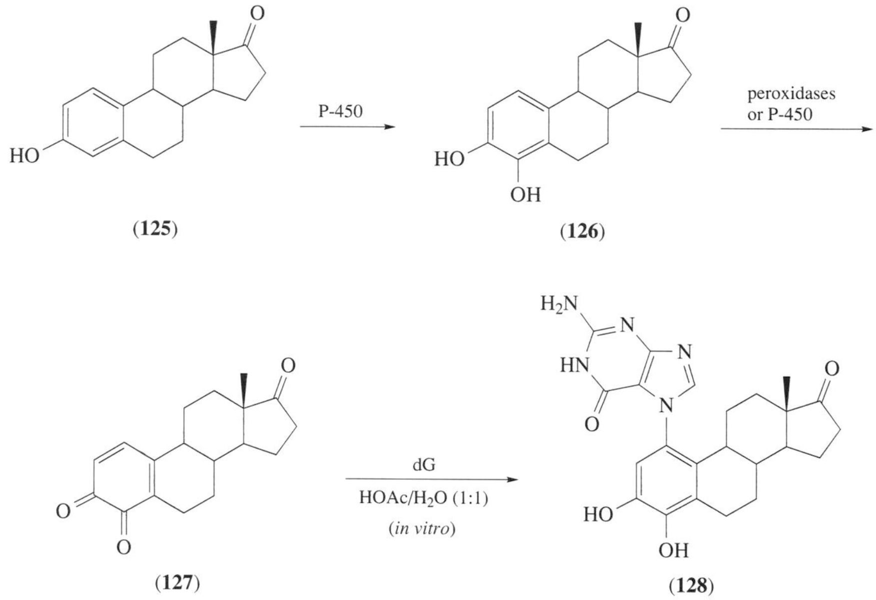

**Scheme 31**

Electrophilic iminoquinones derived from oxidative metabolism of $N^2$-methyl-9-hydroxy-ellipticinium also form covalent adducts with DNA.[286,287]

### 7.14.2.11.5  *Other DNA damage by quinones*

Photochemical DNA damage by anthraquinones has been reported and is discussed in Section 7.14.2.18.

### 7.14.2.12  Heterocyclic *N*-Oxides

Several naturally occurring heterocyclic *N*-oxides have been isolated and found to possess cytotoxic or mutagenic properties.[288] Although no detailed studies of covalent DNA modification by naturally occurring *N*-oxides have been reported, investigations of structurally related synthetic *N*-oxides provide insight into possible mechanisms of DNA damage by the natural products.

### 7.14.2.12.1  *Phenazine* N-*oxides*

Two naturally occurring phenazine *N*-oxides, iodonin (129)[289] and myxin (130),[290] have been reported. *In vitro* experiments demonstrated that the broad spectrum antimicrobial agent[289] (129) inhibits RNA synthesis from a DNA template, probably due to intercalative binding with duplex DNA.[291] In addition, when *E. coli* cells were exposed to (129), single-strand DNA breaks were observed.[292] The DNA-cleavage chemistry of (129) may be analogous to that of a synthetic phenazinedi-*N*-oxide reported by Hecht and co-workers.[293,294]

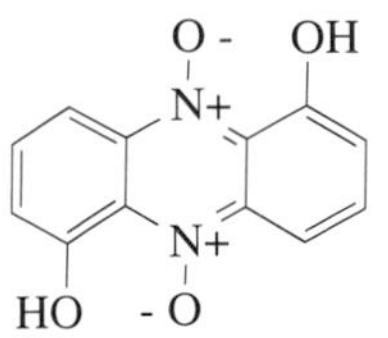

**(129)** Iodinin 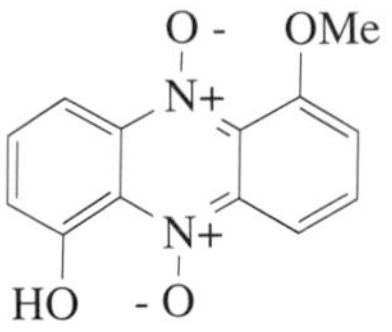          **(130)** Myxin

The phenazine *N*-oxide (**131**) studied by Hecht's group mediates oxidative cleavage of DNA in conjunction with reducing agents such as dithiothreitol and NADPH (Scheme 32).[293] Under aerobic conditions, DNA cleavage appears to involve a redox cycle in which the one electron-reduced phenazinedi-*N*-oxide (**132**) is reoxidized by molecular oxygen to product superoxide radical.[293] Superoxide radical decomposes to yield the DNA-cleaving agent hydroxyl radical (see Section 7.14.2.15). Compound (**131**) also cleaves DNA effectively under anaerobic conditions, in the presence of reducing agents. Under anaerobic conditions, (**131**) undergoes deoxygenation and it is proposed that the one-electron reduced intermediate (**132**) fragments with the release of hydroxyl radical.[293] Deoxygenated products (**133**) and (**134**) were detected only under anaerobic conditions.

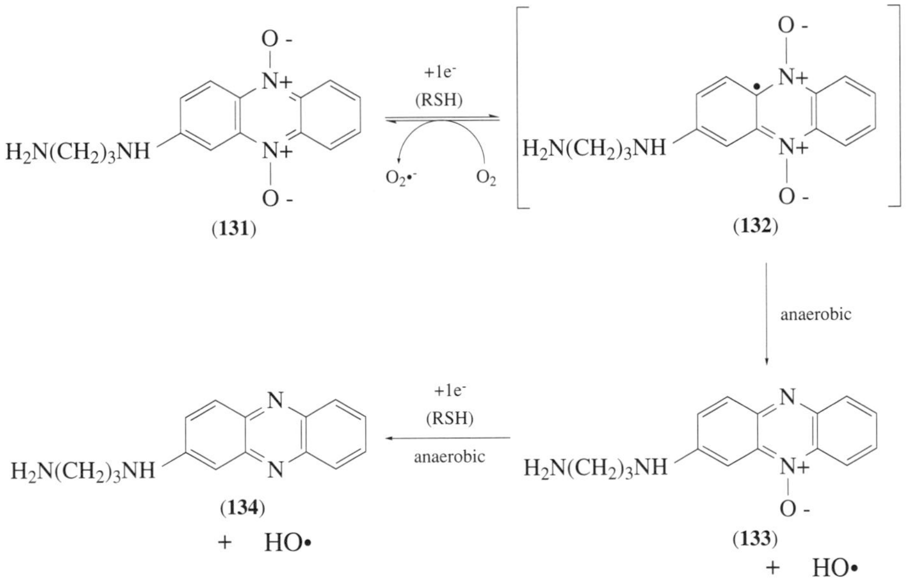

**Scheme 32**

Investigation of phenazinedi-*N*-oxide–oligonucleotide conjugates revealed an interesting DNA-damage pathway distinct from the redox-activated chemistry described above. Hybridization of phenazinedi-*N*-oxide-modified oligonucleotides to complementary deoxyoligonucleotides resulted in the production of heat and alkaline-labile lesions at guanine residues on the complementary strand.[294] All evidence is consistent with DNA cleavage resulting from a mechanism involving alkylation at N-7 of guanine, although alkylation at N-3 of guanine is also a possibility (Figure 10).

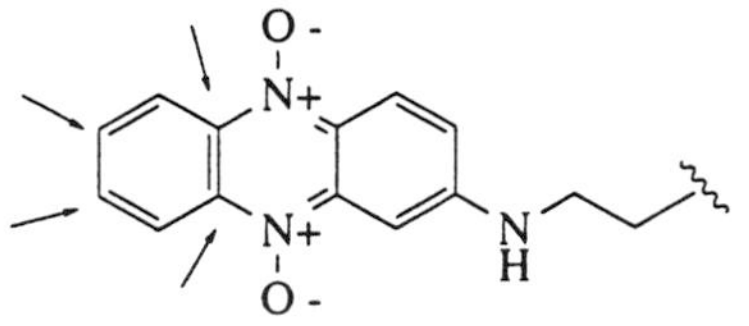

**Figure 10** Potential sites for nucleophilic attack on phenazinedi-*N*-oxides.

### 7.14.2.12.2 *Quinoxaline* N-oxides

2-Carboxyquinoxalinedi-*N*-oxide (**135**) is a fungal metabolite with mutagenic and antibacterial properties.[295] Results obtained with structurally related, synthetic quinoxaline and triazine *N*-oxides suggest that this *N*-oxide-containing natural product may be capable of damaging DNA.

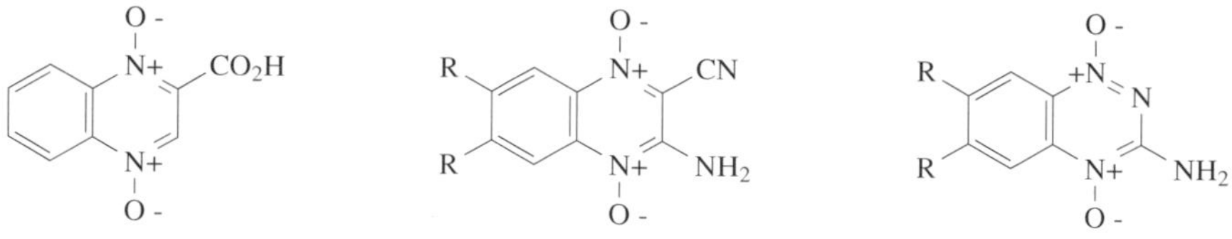

(135)　　　　　(136)　　　　　(137)

In the 1970s, biological experiments with the antibacterial and mutagenic agent quinoxalinedi-*N*-oxide suggested that this compound may cause degradation of DNA *in vivo* and further showed that this *N*-oxide is selectively cytotoxic to oxygen-poor (hypoxic) cells.[296,297] Substituted quinoxaline[298] and triazinedi-*N*-oxides,[299] such as (136) and (137) have been extensively investigated as hypoxia-selective, redox-activated cytotoxins for the possible treatment of solid tumors. *In vitro* cleavage of DNA by quinoxaline-*N*-oxides in a well-defined system has not been reported (1997); however, the chemistry of DNA-cleavage by triazinedi-*N*-oxides has been investigated in some detail.[299,300]

The compound 3-amino-1,2,4-benzotriazine 1,4-dioxide (138) shows selective toxicity toward the hypoxic cells in tumors and has reached phase II and III clinical trials as an antitumor agent.[299] Early experiments suggested that this compound derives its biological activity from the cleavage of cellular DNA and that strand scission may result from a species (139) generated by enzymatic one-electron reduction of the heterocycle.[299] This theory was supported by several observations. In the absence of reducing systems, (138) alone does not damage DNA. In mammalian cells under anaerobic conditions, (138) is ultimately reduced to (140), which is not highly cytotoxic, and the rates of reduction parallel cytotoxicity in several different cell lines.[299] A radical species resulting from the incubation of (138) with rat liver microsomes was observed by ESR;[299] however, no relation between this radical species and DNA cleavage was established. The specific toxicity of (138) toward hypoxic cells may result from the fact that the "activated" radical form of the drug (139) is destroyed by reaction with molecular oxygen.[299] This back-oxidation reaction would produce superoxide, whose *in vivo* cytotoxicity is mitigated by cellular enzymes such as superoxide dismutase and glutathione peroxidase.

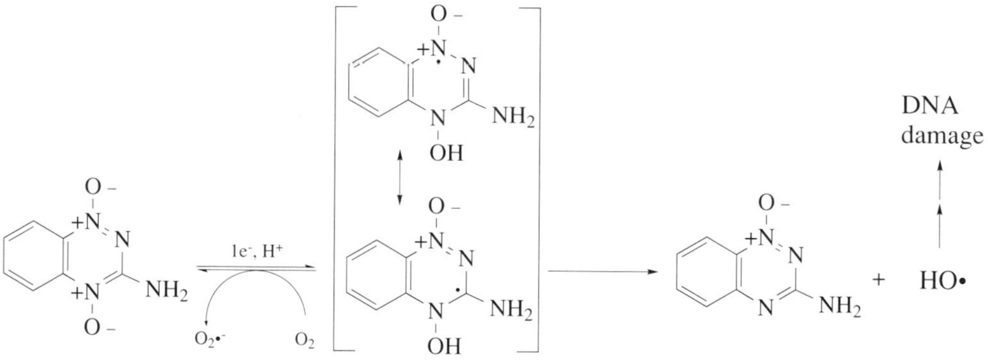

(138)　　　　　(139)　　　　　(140)

**Scheme 33**

Chemical experiments utilizing a xanthine–xanthine oxidase enzyme system as a one-electron reductant show that (138) efficiently cleaves DNA under hypoxic conditions.[300,301] The two-electron-reduced metabolite (140) does not cleave DNA either alone or in the presence of the xanthine oxidase reduction system.[300] Redox-activated DNA cleavage by (138) is effectively inhibited by oxygen-radical scavengers, and strand scission occurs with little or no sequence specificity.[300] In addition, (138), in concert with the xanthine–xanthine oxidase enzyme system, converts dimethyl sulfoxide to methanesulfinic acid, a reaction considered characteristic of hydroxyl radical.[300] Taken together these experiments suggest that one-electron reduction of (138) leads to a fragmentation reaction that produces hydroxyl radical and the deoxygenated metabolite (140). Thus, hydroxyl radical (rather than the radical form of the heterocycle (139)) may be the major DNA-cleaving species generated by one-electron reduction of triazinedi-*N*-oxides, under anaerobic conditions (see Scheme 33). This mechanism is analogous to that suggested by Hecht and co-workers to explain redox-activated DNA cleavage by phenazine *N*-oxides, such as **(131)**.[293] It has been noted[300] that the formation of the high-energy hydroxyl radical from these heterocyclic *N*-oxides may be thermodynamically driven by re-aromatization of the heterocyclic ring systems and by the entropically favorable nature of the fragmentation reaction.

Triazene and quinoxaline di-*N*-oxides also produce photoactivated DNA damage.[302]

### 7.14.2.13 Resorcinols

A variety of 1,3-dihydroxybenzene compounds (resorcinols) have been isolated from plant sources.[303] These natural products display a wide array of biological activities. Hecht's group reported the isolation and characterization of a group of 5-alk(en)yl resorcinols, **(141)**–**(144)**, that efficiently mediate DNA cleavage under aerobic conditions, in the presence of copper(II).[303–306] These researchers noted that the DNA-cleaving ability of the freshly purified resorcinols increased upon incubation with copper(II) in basic, aerobic solution.[304] It was realized that the conditions which enhance the DNA-cleaving ability of these resorcinols are similar to reaction conditions known to cause hydroxylation of phenolic compounds. Thus, it was proposed that, in basic solution containing copper(II) and molecular oxygen, the resorcinol ring system undergoes hydroxylation to yield 1,2,4-trihydroxybenzenes that are the active DNA-cleaving agents.[304] The corresponding trihydroxybenzene compounds were prepared and, consistent with the proposed mechanism, were found to be 50–100 times better DNA-cleaving agents than the analogous resorcinols.[306]

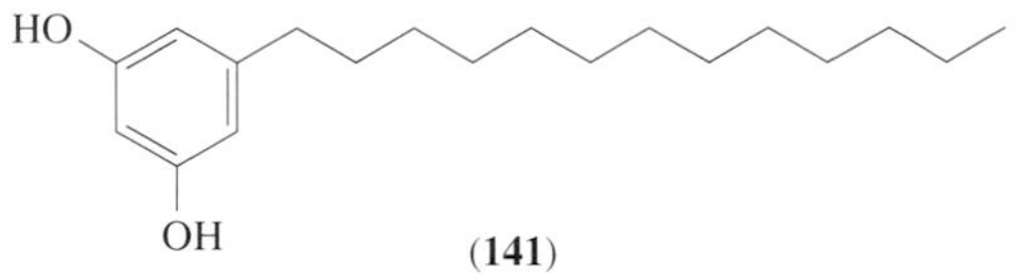

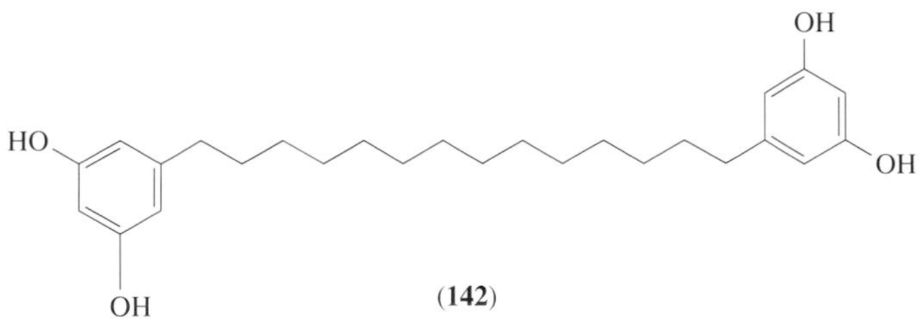

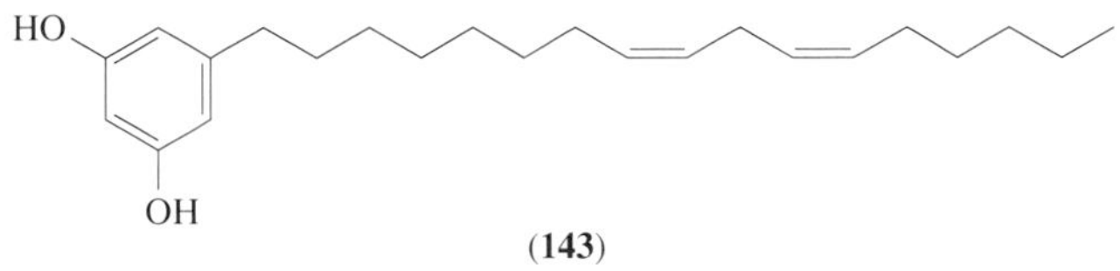

The proposed *in situ* conversion of resorcinols to trihydroxybenzenes requires basic conditions, copper ion, and molecular oxygen. Studies with the trihydroxybenzene compounds demonstrated that the subsequent DNA-cleavage process depends on copper ion and molecular oxygen, but not basic conditions.[306] Experiments designed to elucidate the mechanism of DNA cleavage by the trihydroxybenzenes showed that neocuproine, a specific ligand for copper(I), inhibits cleavage and that oxidation of these trihydroxybenzenes to their quinone form by 2,3-dichloro-5,6-dicyano-1,4-benzoquinone prior to the addition of DNA also prevents strand scission. In addition, the oxygen radical scavenger DMSO and the hydrogen peroxide-destroying enzyme catalase inhibit DNA cleavage by this system.[306]

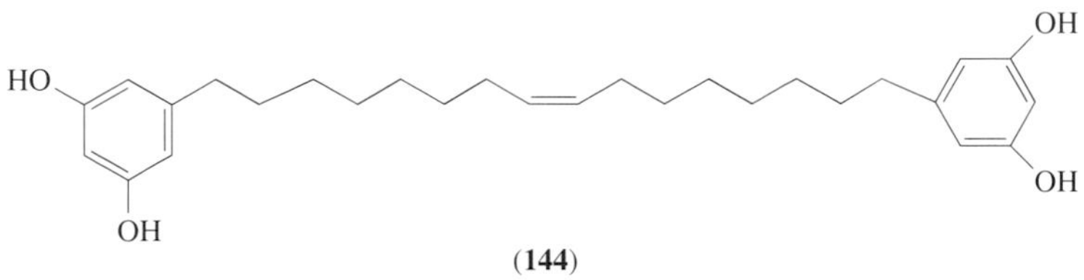

Together, these experiments suggest that DNA cleavage by 5-alkyl resorcinols (**141**)–(**144**), in the presence of copper(II), base, and molecular oxygen, proceeds by initial conversion to the corresponding trihydroxybenzene (**145**), followed by oxidation of the trihydroxybenzene by molecular oxygen to yield the corresponding quinone (**146**) and superoxide radical (Scheme 34).[306] Superoxide radical leads to DNA damage through a cascade of reactions involving dismutation to hydrogen peroxide followed by generation of hydroxyl radical via the metal-dependent Fenton reaction (see Section 7.14.2.15). End-product analysis of resorcinol-cleaved DNA restriction fragments and the detection of DNA free bases as products of these cleavage reactions shows that DNA cleavage by resorcinols is, in this regard, mechanistically identical to that by iron–MPE methidiumpropyl-EDTA, consistent with the notion that hydroxyl radical is the ultimate cleaving agent generated by the resorcinol–copper–oxygen system.[306] It is possible that similar chemistry is involved in copper-dependent DNA damage that has been reported for the natural products tannic acid[307] and urushiol.[308]

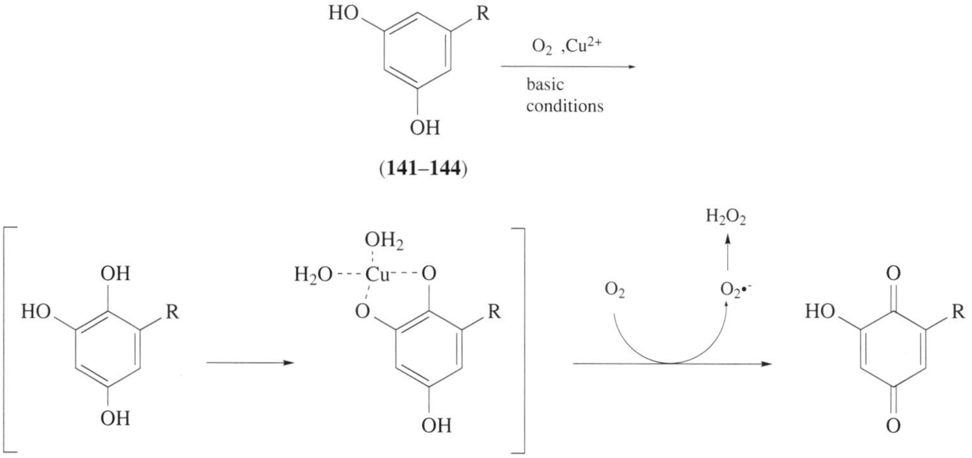

**Scheme 34**

In the cleavage of DNA by 5-alkylresorcinols, copper ion may play multiple roles, converting the resorcinol system to a trihydroxybenzene, possibly catalyzing the oxidation of the trihydroxybenzene to quinone and also possibly catalyzing the decomposition of hydrogen peroxide to hydroxyl radical. This is similar in some regards to the role that has been proposed for transition metal ions chelated by quinone antibiotics such as streptonigrin (**115**) and adriamycin (**113b**) (see Section 7.14.2.11.1).

An interesting aspect of the work with alkylresorcinols, such as (**141**)–(**144**), is the observation that the efficiency of DNA cleavage in this group of compounds is proportional to the length of the alk(en)yl substituent, with analogues bearing longer alk(en)yl side chains producing more efficient DNA cleavage.[306] These resorcinol natural products provide a unique demonstration that simple alk(en)yl substituents can provide significant DNA affinity, presumably through hydrophobic association with the grooves of duplex DNA.

### 7.14.2.14    Bleomycin and Other Metal-binding Antibiotics

The bleomycins (**147**) (BLM) are clinically useful antitumor agents that are known to damage DNA.[309] Bleomycin A$_2$ (**147a**) is a major component of the clinically used BLM mixtures and is probably the most thoroughly studied of the BLMs.[309–311] A subset of the functional groups in the BLMs converge to create a strong metal binding site (Figure 11) that, in the presence of molecular oxygen, reductants (such as thiols or ascorbic acid), and iron catalyzes single and, to a lesser extent, double-stranded cleavage of DNA.[309–311] Although double-stranded cleavage is a relatively infrequent event for the BLMs, these lesions are not easily repaired *in vivo* and are thus probably more cytotoxic than single-strand cleavage.

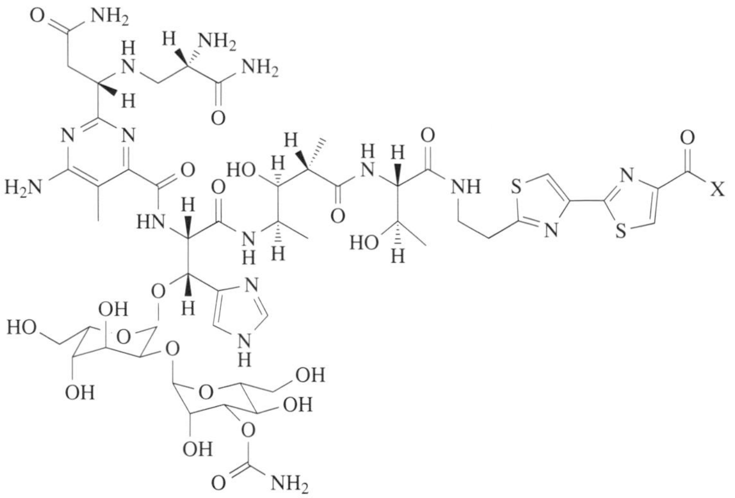

(**147a**)  Bleomycin A$_2$: X=

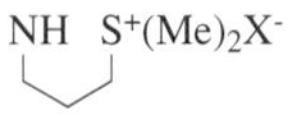

Bleomycin B$_2$: X=

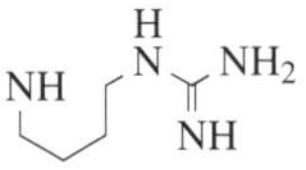

Bleomycin A$_5$: X=

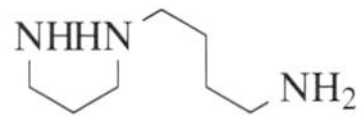

Bleomycin B$_4$: X=

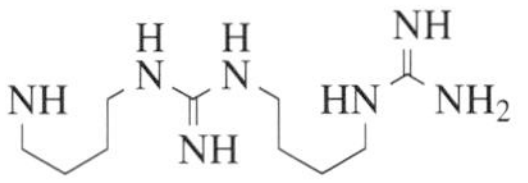

Bleomycin B$_6$: X=

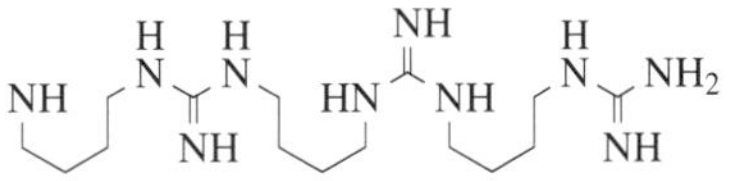

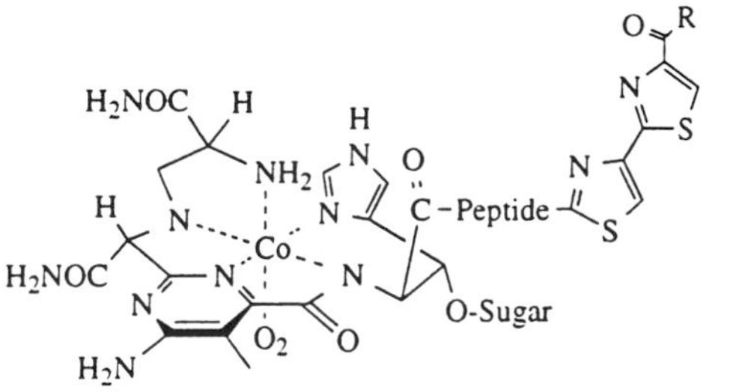

**Figure 11**  Metal binding site of bleomycin.

The exact nature of the reactive species involved in the cleavage of DNA by iron–BLM remains a subject of ongoing investigation; however, research over the last several decades has led to a fairly detailed understanding of the events leading to strand scission (Scheme 35).[310,312] The iron(II)–BLM chelate readily forms a complex (**148**) with molecular oxygen. One-electron reduction of this complex yields an iron(III)–BLM hydroperoxide (**149**) (activated iron–BLM) that either directly cleaves DNA or undergoes heterolytic cleavage of the O—O bond to yield a reactive iron oxene species (**150**) that cleaves DNA. In many regards, the metal binding domain of iron–BLM is analogous to the iron-heme found at the active site of cytochrome P-450. Similar to cytochrome P-450, activated iron–BLM can mediate the oxidation and oxygenation of many organic substrates.[310] Under various conditions, complexes of BLM with iron, cobalt, copper, manganese, vanadium, and nickel can cause DNA cleavage.[310]

The exact structure of the iron(III) intermediates in the DNA-cleavage reaction of BLM

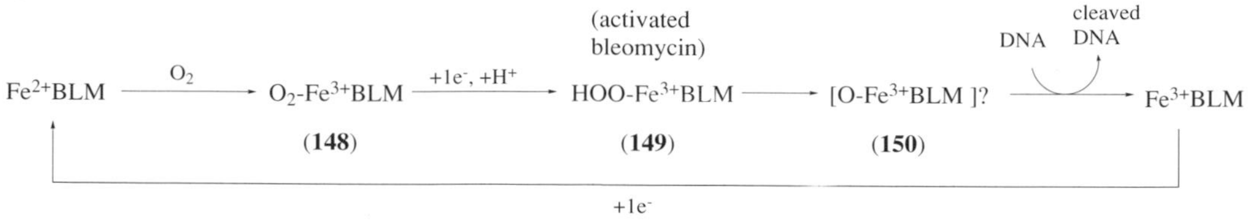
(activated
bleomycin)
cleaved
DNA DNA
Fe2+BLM
O2
O2-Fe3+BLM
+1e-, +H+
HOO-Fe3+BLM
[O-Fe3+BLM ]?
Fe3+BLM
(148)
(149)
(150)
+1e-

**Scheme 35**

remains unclear due to experimental difficulties in analyzing this system; however, zinc(II), iron(II), and cobalt(III) complexes of BLM have been characterized.[310,312] The arrangement of ligands in cobalt(III)–BLM has been deduced from 2D-NMR data (see Figure 11).

Extensive studies involving product analysis and examination of isotope effects have revealed that DNA cleavage by iron–BLM involves exclusive abstraction of the hydrogen from C-4′ of deoxyribose.[310,312,313] The resulting C-4′ deoxyribose radical (**151**) partitions between two distinct strand scission pathways (Scheme 36). The relative predominance of each pathway depends upon the reaction conditions (e.g., oxygen concentrations) and the surrounding environment (sequence context) of the radical.

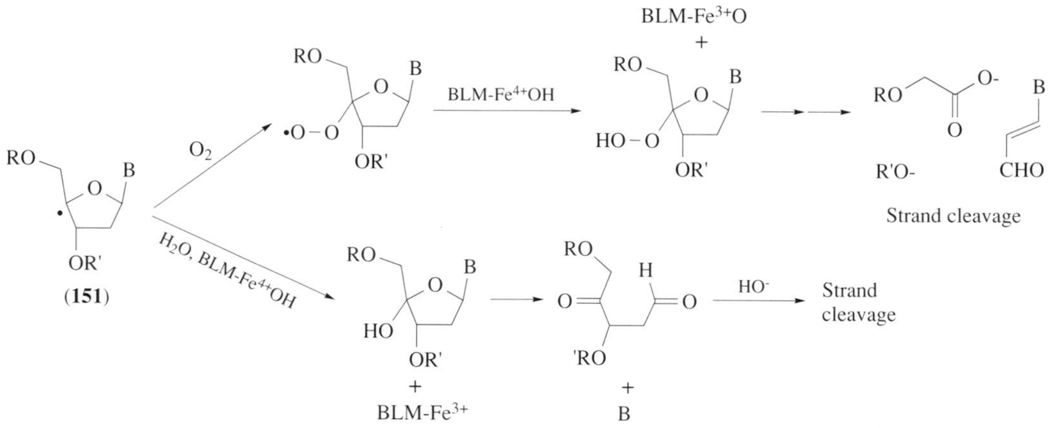
BLM-Fe3+O
BLM-Fe4+OH
O2
H2O, BLM-Fe4+OH
(151)
Strand cleavage
HO-
Strand
cleavage
BLM-Fe3+

**Scheme 36**

BLM binds to DNA with an association constant of $\sim 1 \times 10^5\ M^{-1}$.[314,315] Investigations involving competition binding experiments with agents such as distamycin, suggested that BLM binds in the minor groove of DNA.[316] In addition, BLM binds efficiently to T4 DNA, which is glycosylated in the major groove.[317] Exclusive abstraction of the C-4′ hydrogen of DNA, which protrudes into the minor groove of duplex DNA, is also consistent with minor groove binding by BLM.

BLM cleaves DNA primarily at 5′-dGPy sites, with cleavage occurring between the guanine and adjacent pyrimidine base.[310,312] Experiments with synthetic bithiazoles show that the bithiazole "tail" of BLM does not bind to DNA with any sequence preference; however, this fragment of BLM does provide significant DNA affinity.[318] Certain features of the binding of BLM to DNA are characteristic of intercalation, including unwinding of supercoiled DNA and elongation of the double helix.[315,319] Thus, for many years it has been suspected that the bithiazole moiety of BLM might associate with DNA via intercalative-type binding.

A number of experiments suggested that the metal-binding domain of BLM contains the determinants for sequence-specific DNA cleavage. For example, deglyco-BLM cleaves DNA with specificity essentially identical to the natural product, indicating that the sugar residues are not involved in sequence-specific interactions.[310] Furthermore, investigations of synthetic deglyco-BLM derivatives revealed that changes in the linker between the bithiazole moiety and the metal binding domain have no effect on the sequence specificity of the cleavage reaction.[310] This data implicates the metal-binding domain in sequence-specific DNA binding, because if the bithiazole "tail" controlled specificity, it would be expected that, as linker length between the tail and the cleaving domain increased, sites of DNA cleavage would be displaced relative to the binding site, a phenomenon that is not observed.

Evidence that sequence specificity involves interaction of BLM with the $N^2$-amino group of the guanine adjacent to the DNA cleavage site was provided by the observation that replacement of guanine with inosine (**8**) drastically diminishes DNA cleavage at these sites.[320] Conversely, substitution of diaminopurine (**72**) for adenine converts weak 5′-APy cleavage sites into strong BLM cleavage sites.[321]

A published 2D-NMR structure of cobalt(III)-BLM complexed to a 5′-dCCAGGGCCTGG duplex provides a detailed view of a BLM–DNA complex that is consistent with much of the existing experimental data.[312,322] This structure shows the bithiazole moiety intercalated on the 3′-side of the 5′-GC target site. This NMR structure further suggests that the observed 5′-dGPy sequence preference for BLM cleavage derives from hydrogen-bonding interactions between the pyrimidine functional group of BLM's metal binding domain and the guanine residue at the 5′-GC cleavage site. Thus, a unique, minor groove base triplet (**152**) is formed by hydrogen bonding of N-3 and the amino group of BLM's pyrimidine heterocycle with the exocyclic $N^2$-amino group and N-3 of guanine in the G–C base pair at the target site. The NMR structure places the hydroperoxide proton of cobalt(III)-BLM 2.5 Å from the C-4′ hydrogen that is abstracted in the cleavage reaction of iron–BLM.

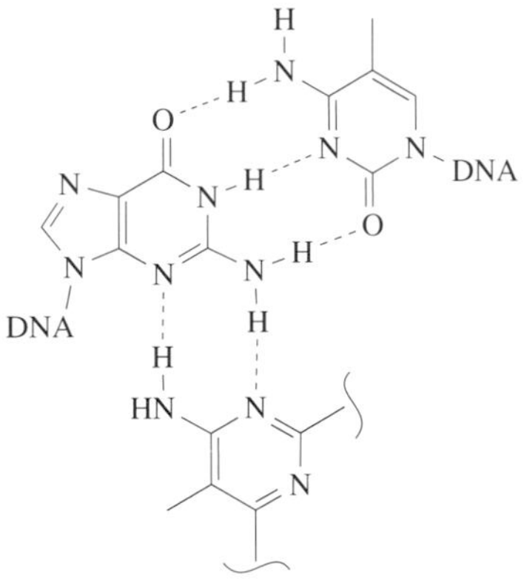

(**152**)

The NMR structure reported by Wu *et al.*[320] nicely accounts for double-stranded cleavage, supporting a previous proposal[311] that cleavage of both strands at certain target sites involves dissociation of the metal-binding domain of BLM from the DNA, while the bithiazole remains bound. A 180° rotation of the metal-binding domain, followed by reassociation with the DNA, would afford a BLM–DNA complex that is able to effect the second cleavage event. Consistent with this picture, palindromic sequences that provide the preferred 5′-dGPy target on each strand are hot spots for double-strand cleavage (Figure 12).[323]

5′-GT̲AC-3′
3′-CAT̲G-5′

**Figure 12** Preferred sequence for double-strand DNA cleavage by bleomycin.

Proton NMR investigations of Zn–BLM complexes have yielded structures distinct from those obtained with the cobalt(III)–BLM complexes, perhaps indicating that multiple binding modes are possible for BLM.[324]

Over the last several years the Hecht group has extensively investigated BLM-mediated cleavage of biologically relevant substrates other than double-stranded DNA, including RNA and DNA–RNA hybrids.[325] Their findings clearly indicate that BLM is able to cleave these substrates, thereby suggesting that damage to nucleic acid targets other than double-stranded DNA could contribute to BLM's biological activity. In addition, Hecht's group has shown that metal-free BLM $A_2$ mediates sequence-specific hydrolysis of RNA.[326]

A number of metal-binding antibiotics are structurally analogous to BLM, including tallysomycin, zorbamycin, the zorbonamycins, victomycin, the platomycins, antibiotic YA-56, cleomycin, and phleomycin.[309] These antibiotics are presumed to possess properties similar to BLM; however, most have not been thoroughly investigated.

### 7.14.2.15  Various Agents That Reduce Molecular Oxygen to Superoxide

A wide variety of naturally occurring organic compounds are capable of reducing molecular oxygen to superoxide radical.[327] Some of these systems, such as quinones (see Section 7.14.2.11.1), are catalysts that mediate the transfer of electrons from reducing agents to molecular oxygen and others stoichiometrically reduce molecular oxygen while being oxidized in the process.

Superoxide exists primarily as the radical anion (153) under physiological conditions.[328,329] Only 0.6% exists as the protonated hydroperoxyl radical (154) ($pK_a = 4.8$) at pH 7 (Equation (1)). Superoxide radical anion (153) is not believed to react directly with DNA; however its conjugate acid, hydroperoxyl radical (154) has been shown to produce sequence-specific DNA damage.[328]

$$HO_2\bullet \; \rightleftharpoons \; O_2\bullet^- \; + \; H^+ \qquad\qquad (pK_a = 4.8) \qquad\qquad (1)$$
$$\textbf{(154)} \qquad\quad \textbf{(153)}$$

Disproportionation (dismutation) of superoxide is of primary importance in the context of DNA damage by this reduced form of molecular oxygen (Equation (2)).[253,328–330] The disproportionation reaction involving a molecule of superoxide radical (153) with a molecule of hydroperoxyl radical (154) is rapid and, although the concentration of (154) is low, this reaction predominates over the dismutation reaction involving two molecules of (153), which is relatively slow (Equation (3)). Hydrogen peroxide generated by the disproportionation of superoxide decomposes via a trace metal-catalyzed Fenton reaction to yield a potent DNA-cleaving agent that either *is* hydroxyl radical or behaves very much like hydroxyl radical (Equation (4)).[331–334] When substoichiometric quantities of transition metal ions are present, efficient generation of hydroxyl radical from hydrogen peroxide requires the presence of a reducing agent that can recycle metal ions back to their active, reduced form. Superoxide radical can serve as such a reducing agent (Equation (5)).[253,335] Interaction of hydrogen peroxide with transition metals also leads to the formation of singlet oxygen, which is capable of reacting with DNA (see Section 7.14.2.18).[38]

$$H^+ \; + \; HO_2\bullet \; + \; O_2\bullet^- \; \longrightarrow \; H_2O_2 \; + \; O_2 \qquad\qquad (7 \times 10^9 \, M^{-1} \, s^{-1}) \qquad\qquad (2)$$

$$2H^+ \; + \; O_2\bullet^- \; + \; O_2\bullet^- \; \longrightarrow \; H_2O_2 \; + \; O_2 \qquad\qquad\qquad\qquad (3)$$

$$H_2O_2 \; + \; M^{n+} \; \longrightarrow \; HO\bullet \; + \; HO^- \; + \; M^{(n+1)+} \qquad (M = Fe: 7.6 \times 10^1 \, M^{-1} \, s^{-1}) \qquad (4)$$

$$M^{(n+1)+} \; + \; O_2\bullet^- \; \longrightarrow \; M^{n+} \; + \; O_2 \qquad\qquad (M = Fe: 3.1 \times 10^5 \, M^{-1} \, s^{-1}) \qquad (5)$$

Several mechanistic tests are commonly used to implicate DNA cleavage by the cascade of reactions shown in Equations (1)–(5), involving superoxide radical, hydrogen peroxide, trace metals, and hydroxyl radical.[330] Addition of the hydrogen peroxide-destroying enzyme catalase to such a DNA-cleavage system results in a significant decrease in strand scission. Chelators of adventitious trace metal ions, such as desferrioxamine B (desferal) or diethylenetriaminepentaacetic acid, will decrease DNA cleavage by sequestering metal ions in a nonredox-active form, thereby inhibiting the Fenton reaction (Equation (5)). The enzyme superoxide dismutase, which catalyzes the disproportionation of superoxide radical (Equation (3)) is also used as a mechanistic tool. In cases where the enzyme superoxide dismutase diminishes DNA cleavage, it is presumably due to the fact that, in these systems, superoxide radical is necessary as a reducing agent to return trace metals to the active form needed in the Fenton reaction (Equation (5)). In systems where other reducing agents, such as thiols or ascorbate, that can serve this function are present, addition of superoxide dismutase may have no effect on, or may even increase, DNA-cleave efficiency. Agents such as ethanol, methanol, mannitol, and dimethyl sulfoxide are commonly employed as "radical scavenging" agents and are sometimes considered diagnostic for the presence of hydroxyl radicals in a DNA-cleaving reaction. While inhibition of DNA cleavage by these agents is consistent with the involvement of hydroxyl radical, caution is necessary in interpreting the results obtained upon addition of "hydroxyl radical scavengers", as they also can react with other radical and nonradical species.[330]

DNA damage involving reduction of molecular oxygen to superoxide has been reported for a large number of organic compounds including thiols,[252,336,337] hydroquinone and hydroquinoid compounds (see Section 7.14.2.11.1), polysulfide/thiol mixtures (see Section 7.14.2.10.1), 1,2-dithiole-3-thiones/thiol mixtures (see Section 7.14.2.10.2), resorcinol/copper mixtures (see Section

7.14.2.13), enzyme/nitroaromatic systems (see Section 7.14.2.9.1), reductant/*N*-oxide systems (see Section 7.14.2.12), ascorbate,[338] 4-hydroxy-5-methyl-3-(2*H*)-furanones,[339] and hydrazines and hydroxylamines.[340] In addition, enzymes such as xanthine oxidase can cause oxidative DNA damage through the reduction of molecular oxygen to superoxide and hydrogen peroxide.[341] There is evidence that neocarzinostatin, under basic conditions, either with or without added thiol, generates superoxide radical, although it appears that this chemistry is of minor importance relative to thiol-dependent, biradical-mediated cleavage of DNA by this natural product.[342]

An interesting example of superoxide production by natural molecules is provided by the oxazolidine-containing antibiotics quinocarcin (16) and tetrazomine.[80,81] Tetrazomine and (16) are expected to be DNA-alkylating agents (see Section 7.14.2.1.3); however, in addition to the expected alkylation chemistry, it has been found that these compounds mediate oxidative DNA damage. Identification of the products resulting from the incubation of (16) and tetrazomine in aqueous solution, along with the study of synthetic analogues, led to the proposal that these agents generate superoxide radical during the course of disproportionation reactions (Scheme 37).

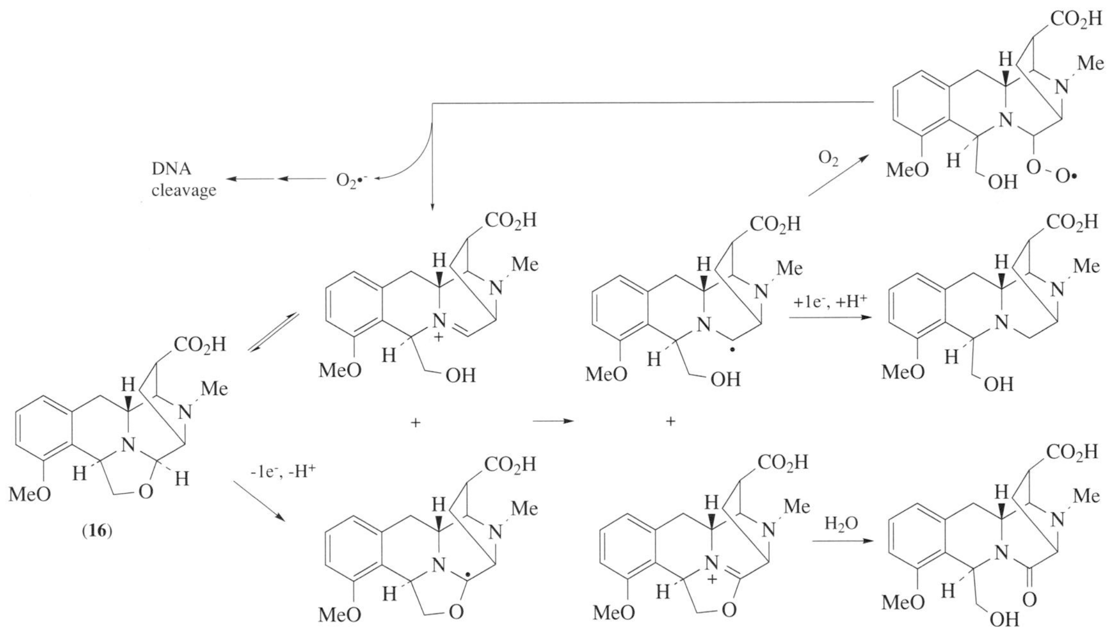

**Scheme 37**

The cytotoxicity of superoxide and hydrogen peroxide is mitigated by the cellular enzymes superoxide dismutase,[343] catalase, and glutathione peroxidase,[344] and by ubiquitous radical scavenging agents, such as glutathione. However, despite these protective systems, it is clear that a number of antibiotics derive their biological effects, at least in part, through production of superoxide radical.

### 7.14.2.16 Diazo and Diazonium Compounds

#### 7.14.2.16.1 Kinamycin

A reinvestigation of the kinamycin antibiotics (155) revealed that these natural products, originally thought to be cyanocarbazoles, actually contain diazofluorene moieties.[345,346] The biological target(s) of the kinamycins are not known; however, the DNA-damaging potential of these agents has been investigated.[347] Initially, three possible mechanisms by which the kinamycins might damage DNA were considered (Scheme 38): (i) protonation of the diazo group would yield a diazonium ion (157) that could act as an alkylating agent; (ii) reduction of the diazo group would lead to loss of nitrogen gas and formation of a reactive radical anion (**158**); or (iii) oxidation of the diazo group could lead to loss of dinitrogen with formation of a resonance-stabilized, carbon-centered radical (**159**).

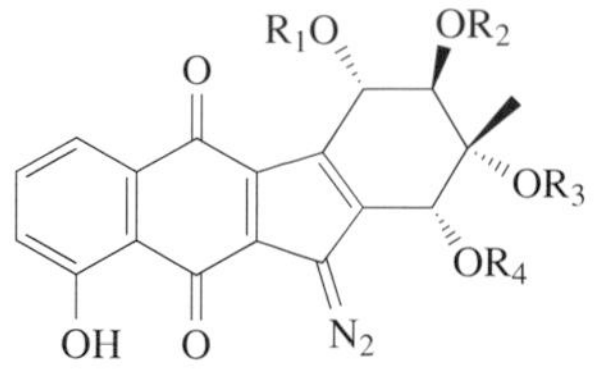

(**155**)

Kinamycin A: $R_1$=H, $R_2$=$R_3$=$R_4$=Ac
Kinamycin B: $R_1$=$R_2$=$R_4$=H, $R_3$=Ac
Kinamycin C: $R_3$=H, $R_1$=$R_2$=$R_4$=Ac
Kinamycin D: $R_1$=$R_3$=H, $R_2$=$R_4$=Ac

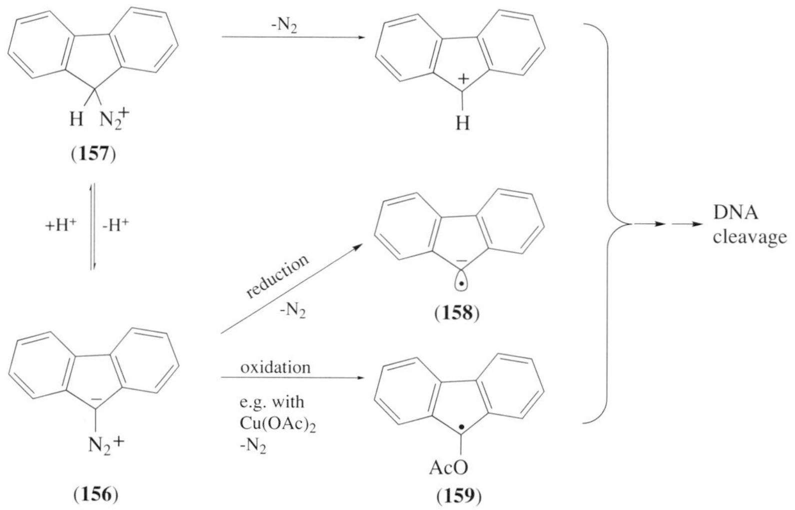

**Scheme 38**

Subjecting 9-aminofluorene to diazotization conditions, followed by addition of the products to a DNA-containing assay mixture, did not lead to DNA cleavage, even in the presence of CuCl, which is known to trigger DNA damage by diazonium ions (see Section 7.14.2.16.3 below). The fate of any diazonium ion (**157**) formed in this experiment remains unclear. Experiments with simple models of the kinamycin diazo functionality demonstrate that DNA cleavage by the diazo compound (**156**) is triggered by copper(II) acetate, and that cleavage is not triggered by CuCl.[347] Thus, it is suggested that DNA cleavage by the kinamycin model compound involves metal-mediated, one-electron oxidation of the diazo functional group, followed by loss of dinitrogen and concomitant formation of a carbon-centered radical (**159**). In earlier synthetic studies, it was found that treatment of (**156**) with copper(II) acetate affords reasonable yields of fluorenone pinacol diacetate, presumably resulting from the dimerization of radical (**159**).[348] It was suggested that DNA cleavage in this system could actually stem from peroxy radicals formed by the reaction of molecular oxygen with the resonance-stabilized carbon radical (**159**). In addition, it was noted that an internal redox reaction between the diazo and quinone functional groups of the kinamycins could trigger DNA damage by the natural products.[347]

### 7.14.2.16.2 *Diazoketones*

A number of α-diazoketones with cytotoxic and anticancer properties have been isolated from natural sources (Figure 13).[349] DNA has not been suggested as a primary target for these agents; however, one of these α-diazoketones, 6-diazo-5-oxo-L-norleucine (**160**), has been found to cleave

DNA, though at relatively high concentrations (30–100 mM), in a plasmid-nicking assay.[350] Involvement of carbon-centered radicals in DNA cleavage by (160) was suggested, although a detailed mechanism for their formation was not proposed.

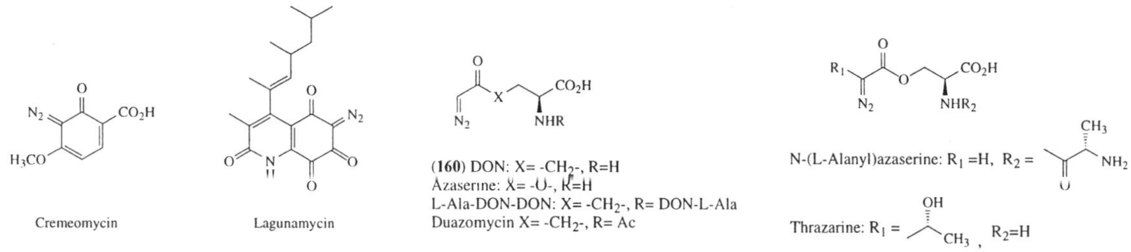

**Figure 13** α-Diazoketone-containing antibiotics.

Reactions similar to those proposed for the diazo-containing antibiotic kinamycin can be imagined for α-diazoketone natural products. In addition, it is known that photochemical processes involving α-diazoketones may yield carbenes and electrophilic ketenes capable of damaging DNA.[351]

### 7.14.2.16.3 *Benzenediazonium ions*

In certain species of mushrooms, enzymatic metabolism of naturally occurring hydrazines is thought to produce aromatic diazonium ions.[352] These diazonium ion metabolites are carcinogenic.[353] Both DNA cleavage[354] and DNA adduct[355] formation have been reported for the mushroom diazonium species 4-(hydroxymethyl)benzenediazonium ion (161). *In vitro* reactions of (161) with nucleosides have yielded adducts (162) and (163), in which the diazonium substituent has been replaced by C-8 of adenine or guanine, respectively.[355]

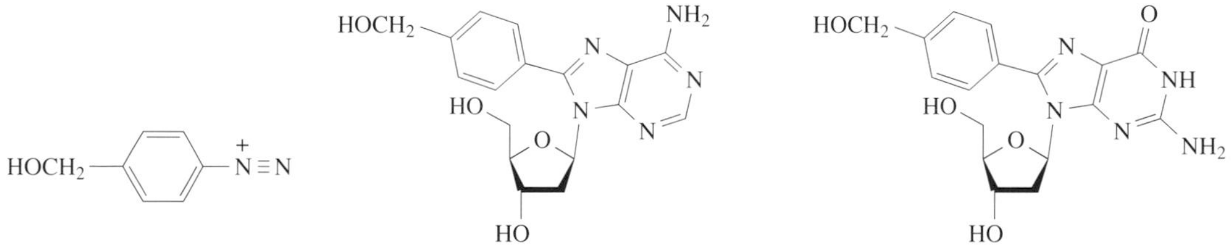

Based upon radical scavenging data, ESR experiments and other mechanistic experiments, it has been suggested that DNA cleavage by (161) results from the formation of carbon-centered phenyl radicals.[354] Previous work with synthetic benzenediazonium ions demonstrated that efficient DNA cleavage by these compounds requires the presence of an agent, such as CuCl, that can serve as a one-electron reducing agent.[356,357] It is believed that one-electron reduction of the diazonium ion followed by loss of dinitrogen affords a phenyl radical that can cleave DNA (Scheme 39). Interestingly, DNA cleavage reported for (161) does not appear to require added reducing agents.[354] The authors suggest that the formation of phenyl radicals from (161) is the result of a Gomberg–Bachman reaction;[358,359] however, the reason for the differences in conditions required for DNA cleavage by (161) and those for other reported diazonium ions is unclear.

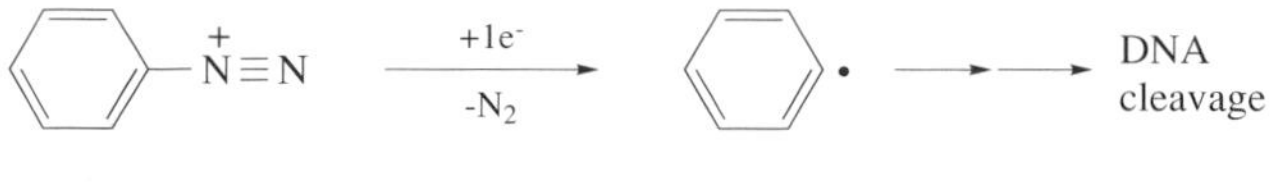

**Scheme 39**

Photolysis of benzenediazonium salts also leads to DNA cleavage; under these conditions it is postulated that DNA damage is due to the formation of arene cations.[360]

### 7.14.2.17  Enediynes

Enediyne-containing antibiotics have captured the attention of many researchers[361–363] because of their potent biological activity, the synthetic challenge that they pose, and also, perhaps, because of the unexpected and intriguing chemical reactions involved in the cleavage of DNA by these natural products. Enediynes undergo cycloaromatization reactions to yield carbon-centered radicals that cause DNA strand cleavage via abstraction of hydrogen atoms from the deoxyribose backbone (Scheme 40). The enediyne antibiotics are discussed in detail in Chapter 7.15 of this volume.

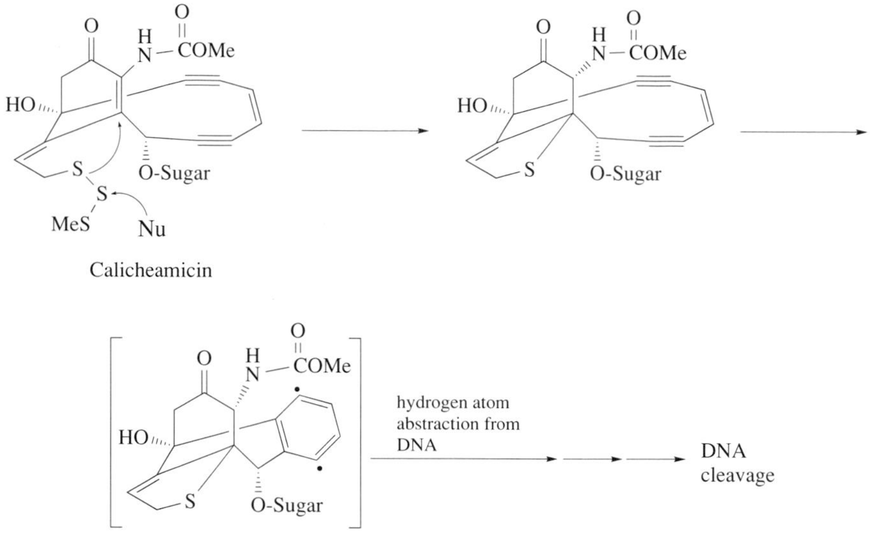

Scheme 40

### 7.14.2.18  Photochemically Activated Agents

#### 7.14.2.18.1  *Light-dependent DNA damage not involving covalent adducts*

There are a number of mechanisms, not involving the formation of stable covalent attachments, by which photoexcited states of organic molecules can damage DNA.[35–37] Energy transfer from a photoexcited state of an organic molecule to ground state molecular oxygen leads to the formation of singlet oxygen which can react with DNA,[38,39,364] selectively causing damage to guanines.[365–367] Some of the damage caused by singlet oxygen leads to spontaneous strand cleavage,[39] although the detailed chemistry of these processes remains to be elucidated. Some photoexcited states of organic molecules are potent oxidizing agents and can damage DNA by direct hydrogen atom abstraction or by electron transfer oxidation of the heterocyclic bases.[35–37] Guanine is the most easily oxidized base and is probably the most common target for oxidations involving electron transfer.[368–370] Finally, the photoexcited states of some organic molecules can reduce molecular oxygen to superoxide radical,[35–37] which ultimately causes DNA damage through the formation of hydroxyl radical (see Section 7.14.2.15).

A variety of natural products have been found to cause photosensitized DNA damage. For example, photolysis of daunomycin (**113a**)[35] and gilvocarcin V (**164**)[371] leads to the production of superoxide radical. Photolysis of simple anthraquinones leads to direct abstraction of hydrogen atoms from DNA and to oxidation of guanines through electron transfer reactions.[372] Tetracyclines[35] and gilvocarcin V[371] act as singlet oxygen photosensitizers. Upon irradiation with 360 nm light, camptothecin can directly abstract hydrogen atoms from DNA or produce superoxide radical, depending upon the reaction conditions.[373] Certain metal complexes of bleomycin (**147**) catalyze light-dependent DNA damage.[310]

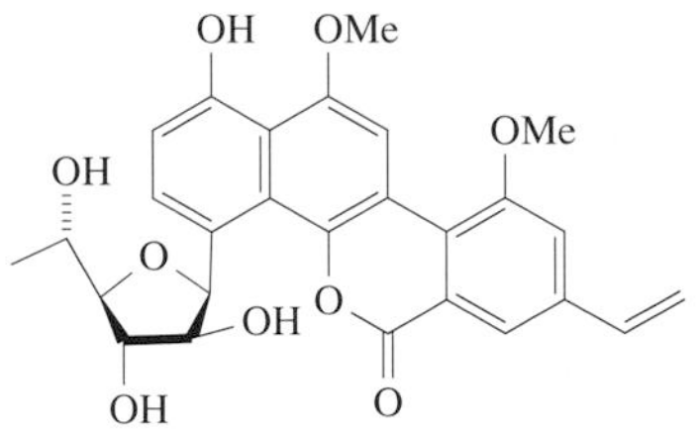

**(164)**

### 7.14.2.18.2 *Formation of covalent photoadducts*

A number of natural products engage in photochemically triggered cycloaddition reactions with DNA that result in the formation of stable covalent adducts. Perhaps the best characterized of these natural products are the psoralens (**165**), a group of furocoumarin-containing plant metabolites.[374–376] Intercalation of psoralens at appropriate sites in double-helical DNA positions the molecule for 2 + 2 cycloaddition reactions with pyrimidine bases. Upon absorption of a 300–400 nm photon, psoralens undergo a cycloaddition reaction involving either the 3,4-double bond of the pyrone moiety or the 4′,5′-double bond of the psoralen's furan ring with the 5,6-double bond of an adjacent pyrimidine.

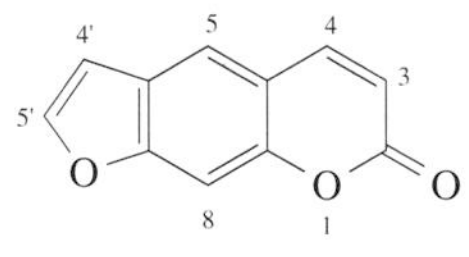

**(165)**

Photoreaction of psoralens with DNA yields monoadducts and also interstrand cross-links that are the result of two sequential photoreactions. Cross-links are possible only when the first photoreaction involves the furan double bond of the psoralen, such that the initial DNA adduct still contains an intact coumarin system that is capable of absorbing light between 300 and 400 nm. In addition, cross-link formation requires the presence of a second pyrimidine on the strand opposite the initial adduct. Thus, monoadducts formed at 5′-PyPu or 5′-PuPy sequences can undergo a second photoreaction with the adjacent pyrimidine on the opposite strand to afford a DNA cross-link. Analysis of the yields of various monoadducts and cross-links shows that thymine is more reactive than cytosine in these photoaddition reactions and that 5′-TA sequences are preferred sites of cross-link formation for the psoralens. Monoadducts and cross-links, e.g., (**166**), of psoralens with DNA have been isolated and fully characterized.[374–376]

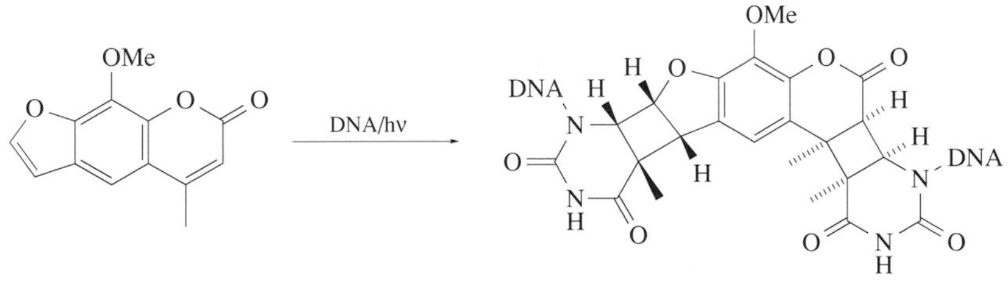

**(166)**

Another example of DNA photoadduction is provided by the naturally occurring C glycoside containing gilvocarcin antibiotics. Structurally similar members of this class of antibiotics include chrysomycin A and ravidomycin.[377] Notably, analogues in this class of antibiotics that contain vinyl side chains are more potent cytotoxins than those that do not. In the case of gilvocarcin V (**164**), intercalation of the benzonaphthopyranone moiety into DNA results in complexes that, upon irradiation with UV light (> 300 nm), yield covalent adducts with DNA. Digestion of this modified DNA with 0.1 N HCl at 100 °C results in release of 5,6-adducted pyrimidine bases. Isolation and characterization of the product (**167**) shows that adduct formation is the result of a 2 + 2 cycloaddition between the vinyl group of gilvocarcin V and the 5,6-double bond of thymine.[378] The acidic workup conditions lead to isomerization of the furanose substituent of gilvocarcin V.

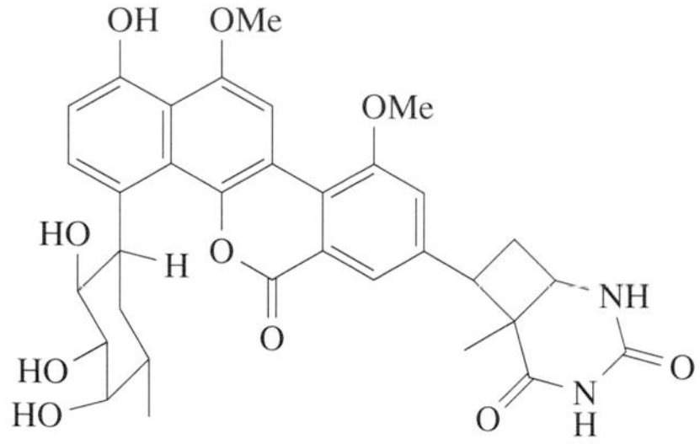

(167)

### 7.14.2.19 Restriction and Methylation Enzymes

Restriction enzymes catalyze the sequence-specific hydrolysis of double-stranded DNA. Restriction enzymes occur in microorganisms as part of restriction–methylation systems consisting of DNA-cleaving enzyme/DNA-methylating enzyme pairs that recognize a common sequence. Sequence-specific methylation of its own DNA protects the host against DNA degradation by its own restriction enzymes. Restriction–modification systems appear to serve as a prokaryotic immune system that protects these organisms against foreign DNA that might enter the cell.[379] The enzymes of restriction–methylation systems in microorganisms formally can be considered secondary metabolite natural products. The chemistry and enzymology of restriction–methylation enzyme systems has been reviewed.[379,380]

### 7.14.3 REFERENCES

1. J. Davies, *Mol. Microbiol.*, 1990, **4**, 1227.
2. D. H. Williams, M. J. Stone, P. R. Hauck, and S. K. Rahman, *J. Nat. Prod.*, 1989, **52**, 1189.
3. R. P. Hertzberg and R. K. Johnson, *Annu. Rep. Med. Chem.*, 1993, **28**, 167.
4. W. A. Remers, in "Textbook of Organic, Medicinal and Pharmaceutical Chemistry," eds. J. W. Delgado and W. A. Remers, Lippincott, Philadelphia, PA, 1991, p. 313.
5. W. O. Foye (ed.), "Cancer Chemotherapeutic Agents," American Chemical Society, Washington, DC, 1995.
6. L. H. Hurley, *J. Med. Chem.*, 1989, **32**, 2027.
7. D. Henderson and L. H. Hurley, *Nature Med.*, 1995, **1**, 525.
8. D. L. Eaton and E. P. Gallagher, *Annu. Rev. Pharmacol. Toxicol.*, 1994, **34**, 135.
9. P. Brookes and P. D. Lawley, *Biochem. J.*, 1961, **80**, 496.
10. J. A. Hartley, in "Molecular Aspects of Anticancer Drug–DNA Interactions," eds. S. Neidle and M. Waring, CRC Press, Boca Raton, FL, 1993, vol. 1, p. 1.
11. A. Pullman and B. Pullman, *Q. Rev. Biophys.*, 1981, **14**, 289.
12. B. Singer, *Nature*, 1976, **264**, 333.
13. P. O. P. T'so, in "Basic Principles in Nucleic Acid Chemistry," Academic Press, New York, 1974, vol. 2, p. 1.
14. P. D. Lawley and P. Brookes, *Biochem. J.*, 1963, **89**, 127.
15. A. M. Maxam and W. Gilbert, *Methods Enzymol.*, 1980, **65**, 499.
16. W. B. Mattes, J. A. Hartley, and K. W. Kohn, *Biochim. Biophys. Acta*, 1986, **868**, 71.
17. R. A. O. Bennett, P. S. Swerdlow, and L. F. Povirk, *Biochemistry*, 1993, **32**, 3188.
18. T. Lindahl and A. Andersson, *Biochemistry*, 1972, **11**, 3618.
19. H. Sugiyama, T. Fujiwara, A. Ura, T. Tashiro, K. Yamamoto, S. Kawanishi, and I. Saito, *Chem. Res. Toxicol.*, 1994, **7**, 673.
20. C.-H. Wong and B. F. L. Li, *Nucleic Acids Res.*, 1994, **22**, 882.
21. P. D. Lawley, *Chem.-Biol. Interact.*, 1973, **7**, 127.
22. B. Singer, M. Kroger, and M. Carrano, *Biochemistry*, 1978, **17**, 1246.
23. A. P. Breen and J. A. Murphy, *Free Radical Biol. Med.*, 1995, **18**, 1033.
24. G. Pratviel, J. Bernadou, and B. Meunier, *Angew. Chem. Int. Ed. Engl.*, 1995, **34**, 746.
25. C. von Sonntag, U. Hagen, A. Schon-Bopp, and D. Schulte-Frohlinde, *Adv. Radiat. Biol.*, 1981, **9**, 109.
26. M. F. Zady and J. L. Wong, *J. Org. Chem.*, 1980, **45**, 2373.
27. H. Sies, *Angew. Chem. Int. Ed. Engl.*, 1986, **25**, 1058.
28. E. Gajewski, G. Rao, Z. Nackerdien, and M. Dizaroglu, *Biochemistry*, 1990, **29**, 7876.
29. M. R. Barvian and M. M. Greenberg, *J. Am. Chem. Soc.*, 1995, **117**, 8291.
30. T. D. Tullius, B. A. Dombroski, M. E. A. Churchill, and L. Kam, *Methods Enzymol.*, 1987, **155**, 537.
31. A. P. Breen and J. A. Murphy, *J. Chem. Soc. Chem. Commun.*, 1993, 191.
32. K. C. Nicolaou and W.-M. Dai, *Angew. Chem. Int. Ed. Engl.*, 1991, **30**, 1387.

33. D. J. Jebaratnam, S. Kugabalasooriar, H. Chen, and D. Arya, *Tetrahedron Lett.*, 1995, **36**, 3123.
34. C. G. Riordan and P. Wei, *J. Am. Chem. Soc.*, 1994, **116**, 2189.
35. I. E. Kochevar and D. A. Dunn, in "Bioorganic Photochemistry: Photochemistry and the Nucleic Acids," ed. H. Morrison, Wiley, New York, 1990, vol. 1, p. 273.
36. B. Epe, in "DNA and Free Radicals," eds. B. Halliwell and O. I. Aruoma, Ellis Horwood, New York, 1993, p. 41.
37. N. Paillous and P. Vicendo, *J. Photochem. Photobiol.*, 1993, **20**, 203.
38. H. Sies, *Mutat. Res.*, 1993, **299**, 183.
39. E. R. Blazek, J. G. Peak, and M. J. Peak, *Photochem. Photobiol.*, 1989, **49**, 607.
40. B. H. Johnston, *Methods Enzymol.*, 1992, **212**, 180.
41. B. N. Ames, *Science*, 1983, **221**, 1256.
42. K. Hemminki, *Carcinogenesis*, 1993, **14**, 2007.
43. T. Lindahl, *Nature*, 1993, **362**, 709.
44. A.-A. Stark, *Annu. Rev. Microbiol.*, 1980, **34**, 235.
45. M. Gniazdowski and C. Cera, *Chem. Rev.*, 1996, **96**, 619.
46. B. N. Ames and M. K. Shigenaga, in "DNA and Free Radicals," eds. B. Halliwell and O. I. Aruoma, Ellis Horwood, New York, 1993, p. 1.
47. J. A. Imlay and S. Linn, *Science*, 1988, **240**, 1302.
48. R. S. Sohal and R. Weindruch, *Science*, 1996, **273**, 59.
49. D. E. Thurston, in "Molecular Aspects of Anticancer Drug–DNA Interactions," eds. S. Neidle and M. Waring, CRC Press, Boca Raton, FL, 1993, vol. 1, p. 55.
50. W. A. Remers and B. S. Iyengar, in "Cancer Chemotherapeutic Agents," ed. W. O. Foye, American Chemical Society, Washington, DC, 1995, p. 577.
51. K. W. Kohn, in "Antibiotics III. Mechanism of Action of Antimicrobial and Antitumor Agents," eds. J. W. Corcoran and F. E. Hahn, Springer, New York, 1975, vol. 3, p. 3.
52. J. A. Mountzouris and L. H. Hurley, in "Bioorganic Chemistry: Nucleic Acids," ed. S. M. Hecht, Oxford University Press, New York, 1996, vol. 1, p. 288.
53. L. H. Hurley, *J. Antibiot.*, 1977, **30**, 349.
54. S. J. Morris, D. E. Thurston, and T. G. Nevell, *J. Antibiot.*, 1990, **43**, 1286.
55. L. H. Hurley, T. Reck, D. E. Thurston, D. R. Langley, K. G. Holden, R. P. Hertzberg, J. R. E. Hoover, G. Gallagher, Jr., L. F. Faucette, S.-M. Mong, and R. K. Johnson, *Chem. Res. Toxicol.*, 1988, **1**, 258.
56. K. W. Kohn, D. Glaubiger, and C. L. Spears, *Biochim. Biophys. Acta*, 1974, **361**, 288.
57. R. L. Petrusek, G. L. Anderson, T. F. Garner, F. L. Quinton, L. Fannin, D. J. Kaplan, S. G. Zimmer, and L. H. Hurley, *Biochemistry*, 1981, **20**, 1111.
58. L. H. Hurley, C. Gairola, and M. Zmijewski, *Biochim. Biophys. Acta*, 1977, **475**, 521.
59. F. L. Boyd, D. Stewart, W. A. Remers, M. D. Barkley, and L. H. Hurley, *Biochemistry*, 1990, **29**, 2387.
60. M. L. Kopka, D. S. Goodsell, I. Baikalov, K. Grzeskowiak, D. Cascio, and R. E. Dickerson, *Biochemistry*, 1994, **33**, 13 593.
61. K. W. Kohn and C. L. Spears, *J. Mol. Biol.*, 1970, **51**, 551.
62. L. H. Hurley, in "DNA Adducts: Identification and Biological Significance," eds. K. Hemminki, A. Dipple, D. E. G. Shuker, F. F. Kadlubar, D. Segerback, and H. Bartsch, IARC Scientific Publications, Lyon, 1994, vol. 125, p. 295.
63. F. L. Boyd, S. F. Cheatham, W. A. Remers, G. C. Hill, and L. H. Hurley, *J. Am. Chem. Soc.*, 1990, **112**, 3279.
64. S. N. Rao and W. A. Remers, *J. Med. Chem.*, 1990, **33**, 1701.
65. R. Kizu, P. H. Draves, and L. H. Hurley, *Biochemistry*, 1993, **32**, 8712.
66. D. E. Thurston and L. H. Hurley, *Correl. Pharmacostruct.*, 1983, **8b**, 957.
67. J.-J. Wang, G. C. Hill, and L. H. Hurley, *J. Med. Chem.*, 1992, **35**, 2995.
68. J. W. Lown, A. V. Joshua, and J. S. Lee, *Biochemistry*, 1982, **21**, 419.
69. K. E. Rao and J. W. Lown, *Chem. Res. Toxicol.*, 1990, **3**, 262.
70. K. E. Rao and J. W. Lown, *Biochemistry*, 1992, **31**, 12 076.
71. K. Ishiguro, K. Takahashi, K. Yazawa, S. Sakiyama, and T. Arai, *J. Biol. Chem.*, 1981, **256**, 2162.
72. G. C. Hill and W. A. Remers, *J. Med. Chem.*, 1991, **34**, 1990.
73. M. J. Zmijewski, Jr., K. Miller-Hatch, and M. Goebel, *Antimicrob. Agents Chemother.*, 1982, **21**, 787.
74. G. C. Hill, T. P. Wunz, N. E. MacKenzie, P. R. Gooley, and W. A. Remers, *J. Med. Chem.*, 1991, **34**, 2079.
75. T. Hayashi and Y. Nawata, *J. Chem. Soc. Perkin Trans. 2*, 1983, 335.
76. J. Zaccardi, M. Alluri, J. Ashcroft, V. Bernan, J. D. Korshalla, G. O. Morton, M. Siegel, R. Tsao, D. R. Williams, W. Maiese, and G. A. Ellestad, *J. Org. Chem.*, 1994, **59**, 4045.
77. F. Tomita, K. Takahashi, and K. Shimizu, *J. Antibiot.*, 1983, **36**, 463.
78. T. Sato, F. Hirayama, F. Saito, K. Takeshi, and H. Kaniwa, *J. Antibiot.*, 1991, **44**, 1367.
79. C. G. Hill, T. P. Wunz, and W. A. Remers, *J. Comput.-Aided Mol. Des.*, 1988, **2**, 91.
80. R. M. Williams, T. Glinka, M. E. Flanagan, R. Gallegos, H. Coffman, and D. Pei, *J. Am. Chem. Soc.*, 1992, **114**, 733.
81. R. M. Williams, M. E. Flanagan, and T. N. Tippie, *Biochemistry*, 1994, **33**, 4086.
82. K. Kimura, T. Morinaga, N. Miyata, and G. Kawashihi, *J. Antibiot.*, 1989, **42**, 1838.
83. Y. Pommier, G. Kohlhagen, C. Bailly, M. Waring, A. Mazumder, and K. W. Kohn, *Biochemistry*, 1996, **35**, 13 303.
84. M. D'Ambrosio, A. Guerriero, G. Chiasera, and F. Pietra, *Tetrahedron*, 1996, **52**, 8899.
85. D. C. Radisky, E. S. Radisky, L. R. Barrows, B. R. Copp, R. A. Kramer, and C. M. Ireland, *J. Am. Chem. Soc.*, 1993, **115**, 1632.
86. B. O. de Lumen, S. J. Kazeniac, and R. H. Forsythe, *J. Food Sci.*, 1978, **43**, 698.
87. O. A.-L. Hsieh, A.-S. Huang, and S. S. Chang, *J. Food Sci.*, 1981, **47**, 16.
88. L. J. Marnett, H. K. Hurd, M. C. Hollstein, D. E. Levin, H. Esterbauer, and B. N. Ames, *Mutat. Res.*, 1985, **148**, 25.
89. B. Reindl and H.-J. Stan, *J. Agric. Food Chem.*, 1982, **30**, 849.
90. E. Eder and C. Hoffman, *Chem. Res. Toxicol.*, 1993, **6**, 486.
91. L. J. Marnett, in "DNA Adducts: Identification and Biological Significance," eds. K. Hemminki, A. Dipple, D. E. G. Shuker, F. F. Kadlubar, D. Segerback, and H. Bartsch, IARC Scientific Publications, Lyon, 1994, vol. 125, p. 151.
92. A. K. Chaudhary, M. Nokubo, G. R. Reddy, S. N. Yeola, J. D. Morrow, I. A. Blair, and L. J. Marnett, *Science*, 1994, **265**, 1580.

93. Y. F. M. Chaw, L. E. Crane, P. Lange, and R. Shapiro, *Biochemistry*, 1980, **19**, 5525.
94. H. Huang, M. S. Solomon, and P. B. Hopkins, *J. Am. Chem. Soc.*, 1992, **114**, 9240.
95. G. D. Manners, G. W. Ivie, and J. T. MacGregor, *Toxicol. Appl. Pharmacol.*, 1978, **45**, 629.
96. C. Aujard, Y. Moule, S. Moreau, and N. Darracq, *Toxicol. Eur. Res.*, 1979, **2**, 273.
97. V. Betina, *Chem.-Biol. Interact.*, 1989, **71**, 105.
98. D. L. Boger and C. M. Baldino, *J. Am. Chem. Soc.*, 1993, **115**, 11 418.
99. J. M. Jay, *Appl. Environ. Microbiol.*, 1982, **44**, 525.
100. H. H. Wasserman and C. M. Baldino, *Bioorg. Med. Chem. Lett.*, 1995, **5**, 3033.
101. K. B. Diamond, G. R. Warren, and J. H. Cardellina, III, *J. Ethnopharmacol.*, 1985, **14**, 99.
102. K.-H. Lee, I. H. Hall, C. O. Starnes, S. A. El Gabaly, T. G. Waddell, R. I. Hadgraft, C. G. Ruffner, and I. Weidner, *Science*, 1977, **196**, 533.
103. S. Danishefsky, *Acc. Chem. Res.*, 1979, **12**, 66.
104. M. Cory, in "Cancer Chemotherapeutic Agents," ed. W. O. Foye, American Chemical Society, Washington, DC, 1995, p. 311.
105. L. H. Hurley and P. H. Draves, in "Molecular Aspects of Anticancer Drug–DNA Interactions," eds. S. Neidle and M. Waring, CRC Press, Boca Raton, FL, 1993, vol. 1, p. 89.
106. L. H. Li, D. H. Swenson, S. L. Schpok, S. L. Kuentzel, B. D. Dayton, and W. C. Krueger, *Cancer Res.*, 1982, **42**, 999.
107. D. H. Swenson, L. H. Li, L. H. Hurley, J. S. Rokem, G. L. Petzold, B. D. Dayton, T. L. Wallace, A. H. Lin, and W. C. Krueger, *Cancer Res.*, 1982, **42**, 2821.
108. L. H. Hurley, V. L. Reynolds, D. H. Swenson, G. L. Petzold, and T. A. Scahill, *Science*, 1984, **226**, 843.
109. A. Asai, S. Nagamura, and H. Saito, *J. Am. Chem. Soc.*, 1994, **116**, 4171.
110. H. Sugiyama, K. Ohmori, K. L. Chan, M. Hosada, A. Asai, H. Saito, and A. Saito, *Tetrahedron Lett.*, 1994, **34**, 2179.
111. D. L. Boger and D. S. Johnson, *Angew. Chem. Int. Ed. Engl.*, 1996, **35**, 1438.
112. L. H. Lin and L. H. Hurley, *Biochemistry*, 1990, **29**, 9503.
113. W. C. Krueger and M. D. Prairie, *Chem.-Biol. Interact.*, 1987, **62**, 281.
114. N. Y. Theriault, W. C. Krueger, and M. D. Prairie, *Chem.-Biol. Interact.*, 1988, **65**, 187.
115. L. H. Hurley, C.-S. Lee, J. P. McGovern, M. A. Warpehoski, M. A. Mitchell, R. C. Kelly, and P. A. Aristoff, *Biochemistry*, 1988, **27**, 3886.
116. L. H. Hurley, M. A. Warpehoski, C.-S. Lee, J. P. McGovern, T. A. Scahill, R. C. Kelly, M. A. Mitchell, N. A. Wicnienski, I. Gebhard, P. D. Johnson, and V. S. Bradford, *J. Am. Chem. Soc.*, 1990, **112**, 4633.
117. M. A. Warpehoski and D. E. Harper, *J. Am. Chem. Soc.*, 1995, **117**, 2951.
118. M. A. Warpehoski and L. H. Hurley, *Chem. Res. Toxicol.*, 1988, **1**, 315.
119. D. L. Boger, R. S. Coleman, B. J. Invergo, S. M. Sakya, T. Ishizaki, S. A. Munk, H. Zarrinmayeh, P. A. Kitos, and S. C. Thompson, *J. Am. Chem. Soc.*, 1990, **112**, 4623.
120. D. L. Boger and R. M. Garbaccio, *Bioorg. Med. Chem.*, 1997, **5**, 263.
121. D. L. Boger and W. Yun, *J. Am. Chem. Soc.*, 1994, **116**, 5523.
122. T. J. Zsido, J. M. Woynarowski, R. M. Baker, L. S. Gawron, and T. A. Beerman, *Biochemistry*, 1991, **30**, 3733.
123. M. Nakagawa, Y.-H. Hsu, A. Hirota, S. Shima, and M. Nakayama, *J. Antibiot.*, 1989, **42**, 218.
124. M. Chu-Moyer and S. J. Danishefsky, *Tetrahedron Lett.*, 1993, **34**, 3025.
125. H. Niwa, M. Ojika, K. Wakamatsu, K. Yamada, I. Hirono, and K. Matsushita, *Tetrahedron Lett.*, 1983, **24**, 4117.
126. T. Kushida, M. Uesugi, Y. Sugiura, H. Kigoshi, H. Tanaka, J. Hirokawa, M. Ojika, and K. Yamada, *J. Am. Chem. Soc.*, 1994, **116**, 479.
127. T. C. McMorris, M. J. Kelner, W. Wang, L. A. Estes, M. A. Montoya, and R. Taetle, *J. Org. Chem.*, 1992, **57**, 6876.
128. T. C. McMorris, M. J. Kelner, W. Wang, S. Moon, and R. Taetle, *Chem. Res. Toxicol.*, 1990, **3**, 574.
129. M. J. Kelner, T. C. McMorris, L. Estes, M. Rutherford, M. Montoya, J. Goldstein, K. Samson, R. Starr, and R. Taetle, *Biochem. Pharmacol.*, 1994, **48**, 403.
130. T. C. McMorris, M. J. Kelner, R. K. Chadha, J. S. Siegel, S.-S. Moon, and M. M. Moya, *Tetrahedron*, 1989, **45**, 5433.
131. K. Maida, T. Takeuchi, K. Nitta, K. Yagishita, R. Utahara, T. Osato, M. Ueda, S. Kondo, Y. Okami, and H. Umezawa, *J. Antibiot. Ser. A*, 1956, **9**, 75.
132. M. R. Hansen and L. H. Hurley, *Acc. Chem. Res.*, 1996, **29**, 249.
133. J. T. Millard and M. M. White, *Biochemistry*, 1993, **32**, 2120.
134. H. L. White and J. R. White, *Biochemistry*, 1969, **8**, 1030.
135. G. N. Bennett, *Nucleic Acids Res.*, 1982, **10**, 4581.
136. D. Sun, M. Hansen, J. J. Clement, and L. H. Hurley, *Biochemistry*, 1993, **32**, 8068.
137. M. Hansen, S. Yun, and L. H. Hurley, *Chem. Biol.*, 1995, **2**, 229.
138. S. Pavlopoulos, W. Bicknell, D. J. Craik, and G. Wickham, *Biochemistry*, 1996, **35**, 9314.
139. D. Sun and L. H. Hurley, *Chem. Biol.*, 1995, **2**, 457.
140. M. Hara, M. Yoshida, and H. Nakano, *Biochemistry*, 1990, **29**, 10 449.
141. K. L. Chan, H. Sugiyama, I. Saito, and M. Hara, *Tetrahedron Lett.*, 1991, **32**, 7719.
142. N. Fujii, T. Katsuyama, E. Kobayashi, M. Hara, and H. Nakano, *J. Antibiot.*, 1995, **48**, 768.
143. A. P. Breen and J. A. Murphy, *J. Chem. Soc. Chem. Commun.*, 1993, 191.
144. M. Hansen, S.-J. Lee, J. M. Cassady, and L. H. Hurley, *J. Am. Chem. Soc.*, 1996, **118**, 5553.
145. I. R. McConnell and R. C. Garner, in "DNA Adducts: Identification and Biological Significance," eds. K. Hemminki, A. Dipple, D. E. G. Shuker, F. F. Kadlubar, D. Segerback, and H. Bartsch, IARC Scientific Publications, Lyon, 1994, vol. 125, p. 49.
146. S. Gopalkrishnan, M. P. Stone, and T. M. Harris, *J. Am. Chem. Soc.*, 1989, **111**, 7232.
147. J. M. Essigmann, R. G. Croy, A. M. Nadzan, W. R. Busby, Jr., V. N. Reinhold, G. Buchi, and G. N. Wogan, *Proc. Natl. Acad. Sci. USA*, 1977, **74**, 1870.
148. R. S. Iyer, M. W. Voehler, and T. M. Harris, *J. Am. Chem. Soc.*, 1994, **116**, 8863.
149. F.-L. Yu, J.-X. Huang, W. Bender, Z. Wu, and J. C. S. Chang, *Carcinogenesis*, 1991, **12**, 997.
150. R. S. Iyer and T. M. Harris, *Chem. Res. Toxicol.*, 1993, **6**, 313.
151. S. W. Baertschi, K. D. Raney, M. P. Stone, and T. M. Harris, *J. Am. Chem. Soc.*, 1988, **110**, 7929.
152. R. S. Iyer, B. F. Coles, K. D. Rancy, R. Thier, F. P. Guengerich, and T. M. Harris, *J. Am. Chem. Soc.*, 1994, **116**, 1603.

153. W. W. Johnson, T. M. Harris, and F. P. Guengerich, *J. Am. Chem. Soc.*, 1996, **118**, 8213.
154. S. Gopalkrishnan, T. M. Harris, and T. P. Stone, *Biochemistry*, 1990, **29**, 10 438.
155. V. M. Raney, T. M. Harris, and M. P. Stone, *Chem. Res. Toxicol.*, 1993, **6**, 64.
156. M. Benasutti, S. Ejadi, M. D. Whitlow, and E. L. Loechler, *Biochemistry*, 1988, **27**, 472.
157. E. A. Bailey, R. S. Iyer, M. P. Stone, T. M. Harris, and J. M. Essigmann, *Proc. Natl. Acad. Sci. USA*, 1996, **93**, 1535.
158. S. Han, C. M. Harris, T. M. Harris, H.-Y. H. Kim, and S. J. Kim, *J. Org. Chem.*, 1996, **61**, 174.
159. M. F. Denissenko, A. Pao, M.-s. Tang, and G. P. Pfeifer, *Science*, 1996, **274**, 430.
160. T. Hata, F. Koga, Y. Sano, K. Kanamori, A. Matsumae, R. Sugawara, T. Shima, S. Ito, and S. Tomizawa, *J. Antibiot. Ser. A*, 1954, **7**, 107.
161. R. W. Armstrong, M. E. Salvati, and M. Nguyen, *J. Am. Chem. Soc.*, 1992, **114**, 3144.
162. A. Terawaki and J. Greenberg, *Nature*, 1966, **209**, 481.
163. T. Tsuchida, R. Sawa, Y. Takahashi, H. Iinuma, T. Sawa, H. Naganawa, and T. Takeuchi, *J. Antibiot.*, 1995, **48**, 1148.
164. W. G. Schulz, R. A. Nieman, and E. B. Skibo, *Proc. Natl. Acad. Sci. USA*, 1995, **92**, 11 854.
165. T. Hata, Y. Sano, R. Sugawara, A. Matsumae, K. Kanamorei, T. Shima, and T. Hoshi, *J. Antibiot. Ser. A*, 1956, **9**, 141.
166. A. C. Sartorelli and S. Rockwell, *Oncol. Res.*, 1994, **6**, 501.
167. M. Tomasz, in "Molecular Aspects of Anticancer Drug–DNA Interactions," eds. S. Neidle and M. Waring, CRC Press, Boca Raton, FL, 1993, vol. 2, p. 312.
168. V. N. Iyer and W. Szybalski, *Proc. Natl. Acad. Sci. USA*, 1963, **50**, 355.
169. V. N. Iyer and W. Szybalski, *Science*, 1964, **145**, 55.
170. H. W. Moore, *Science*, 1977, **197**, 527.
171. M. Tomasz and R. Lipman, *Biochemistry*, 1981, **20**, 5056.
172. M. Tomasz, R. Lipman, D. Chowdary, J. Pawlak, G. L. Verdine, and K. Nakanishi, *Science*, 1987, **235**, 1204.
173. H. Kohn and N. Zein, *J. Am. Chem. Soc.*, 1983, **105**, 4105.
174. S. J. Danishefsky and M. Egbertson, *J. Am. Chem. Soc.*, 1986, **106**, 6424.
175. M. Tomasz, *Chem. Biol.*, 1995, **2**, 575.
176. N. Zein and H. Kohn, *J. Am. Chem. Soc.*, 1986, **108**, 296.
177. S. S. Pan, P. A. Andrews, C. J. Glover, and N. R. Bachur, *J. Biol. Chem.*, 1984, **259**, 959.
178. A. C. Sartorelli, *Cancer Res.*, 1988, **48**, 775.
179. M. Tomasz, D. Chowdary, R. Lipman, S. Shimotakahara, D. Veiro, V. Walker, and G. L. Verdine, *Proc. Natl. Acad. Sci. USA*, 1986, **83**, 6702.
180. R. Bizanek, B. F. McGuinness, K. Nakanishi, and M. Tomasz, *Biochemistry*, 1992, **31**, 3084.
181. R. T. Dorr, G. T. Bowden, D. S. Alberts, and J. D. Liddil, *Cancer Res.*, 1985, **45**, 3510.
182. M. Tomasz, R. Lipman, M. S. Lee, G. L. Verdine, and K. Nakanishi, *Biochemistry*, 1987, **26**, 2010.
183. J. T. Millard, M. F. Weidner, S. Raucher, and P. B. Hopkins, *J. Am. Chem. Soc.*, 1990, **112**, 3637.
184. S. P. Teng, S. A. Woodson, and D. M. Crothers, *Biochemistry*, 1989, **28**, 3901.
185. V. Li and H. Kohn, *J. Am. Chem. Soc.*, 1991, **113**, 275.
186. V.-S. Li, D. Choi, Z. Wang, L. S. Jimenez, M.-s. Tang, and H. Kohn, *J. Am. Chem. Soc.*, 1996, **118**, 2326.
187. M. Sastry, R. Fiala, R. Lipman, M. Tomasz, and D. J. Patel, *J. Mol. Biol.*, 1995, **247**, 275.
188. S. M. Rink, R. Lipman, S. C. Alley, P. B. Hopkins, and M. Tomasz, *Chem. Res. Toxicol.*, 1996, **9**, 382.
189. J. W. Lown, A. Begleiter, D. Johnson, and A. R. Morgan, *Can. J. Biochem.*, 1976, **54**, 110.
190. M. Tomasz, *Chem.-Biol. Interact.*, 1976, **13**, 89.
191. J. Woo, S. T. Sigurdsson, and P. B. Hopkins, *J. Am. Chem. Soc.*, 1993, **115**, 1199.
192. H. Huang, T. K. Pratnum, and P. B. Hopkins, *J. Am. Chem. Soc.*, 1994, **116**, 2703.
193. T. Fukuyama and S. Goto, *Tetrahedron Lett.*, 1989, **30**, 6491.
194. K. F. McClure and S. J. Danishefsky, *J. Org. Chem.*, 1991, **56**, 850.
195. R. M. Williams and S. R. Rajski, *Tetrahedron Lett.*, 1992, **33**, 2929.
196. T. W. Petry, G. T. Bowden, R. J. Huxtable, and I. G Sipes, *Cancer Res.*, 1984, **44**, 1505.
197. M. Wiessler, in "DNA Adducts: Identification and Biological Significance," eds. K. Hemminki, A. Dipple, D. E. G. Shuker, F. F. Kadlubar, D. Segerback, and H. Bartsch, IARC Scientific Publications, Lyon, 1994, vol. 125, p. 165.
198. I. N. H. White and A. R. Mattocks, *Biochem. J.*, 1972, **128**, 291.
199. R. L. Reed, K. G. Ahern, G. D. Pearson, and D. R. Buhler, *Carcinogenesis*, 1988, **9**, 1355.
200. P. P. Wickramanayake, B. L. Arbogast, D. R. Buhler, M. L. Deinzer, and A. L. Burlingame, *J. Am. Chem. Soc.*, 1985, **107**, 2485.
201. J. Woo, S. T. Sigurdsson, and P. B. Hopkins, *J. Am. Chem. Soc.*, 1993, **115**, 3407.
202. M. F. Weidner, S. T. Sigurdsson, and P. B. Hopkins, *Biochemistry*, 1990, **29**, 9225.
203. D. H. Phillips, in "DNA Adducts: Identification and Biological Significance," eds. K. Hemminki, A. Dipple, D. E. G. Shuker, F. F. Kadlubar, D. Segerback, and H. Bartsch, IARC Scientific Publications, Lyon, 1994, vol. 125, p. 131.
204. E. C. Miller, A. B. Swanson, D. H. Phillips, T. L. Fletcher, A. Liem, and J. A. Miller, *Cancer Res.*, 1983, **43**, 1124.
205. E. W. Boberg, A. Liem, E. C. Miller, and J. A. Miller, *Carcinogenesis*, 1987, **8**, 531.
206. R. W. Wiseman, T. R. Fennell, J. A. Miller, and E. C. Miller, *Cancer Res.*, 1985, **45**, 3096.
207. R. R. Herr, H. K. Jahnke, and A. D. Argoudelis, *J. Am. Chem. Soc.*, 1967, **89**, 4808.
208. R. A. Bennett and A. E. Pegg, *Cancer Res.*, 1981, **30**, 112.
209. R. N. Loeppky, *ACS Symp. Ser.*, 1994, **553**, 1.
210. L. K. Keefer, T. Anjo, D. Wade, T. Wang, and C. S. Yang, *Cancer Res.*, 1987, **47**, 447.
211. H. Kakeya, M. Imoto, Y. Takahashi, H. Naganawa, T. Takeuchi, and K. Umezawa, *J. Antibiot.*, 1993, **46**, 1716.
212. Z. Guo, A. McGill, L. Yu, L. Libing, J. Li, J. Ramirez, and P. G. Wang, *Bioorg. Med. Chem. Lett.*, 1996, **6**, 573.
213. S. Nakamura and H. Umezawa, *J. Antibiot. Ser. A*, 1955, **8**, 66.
214. R. A. McClelland, R. Panicucci, and A. M. Rauth, *J. Am. Chem. Soc.*, 1987, **109**, 4308.
215. A. J. Varghese and G. F. Whitmore, *Cancer Res.*, 1983, **43**, 78.
216. A. Zahoor, M. V. M. LaFleur, R. C. Knight, H. Loman, and D. I. Edwards, *Biochem. Pharmacol.*, 1987, **36**, 3299.
217. W. A. Denny and W. R. Wilson, *J. Med. Chem.*, 1986, **29**, 879.
218. M. P. Hay, H. H. Lee, W. R. Wilson, P. B. Roberts, and W. A. Denny, *J. Med. Chem.*, 1995, **38**, 1928.

219. T. H. Broschard, M. Wiessler, C.-W. von der Lieth, and H. H. Schmeiser, *Carcinogenesis*, 1994, **15**, 2331.
220. W. Pfau, H. H. Schmeiser, and M. Wiessler, *Chem. Res. Toxicol.*, 1991, **4**, 581.
221. H. H. Schmeiser, C. A. Bieler, M. Wiessler, C. van Ypersele de Strihou, and J.-P. Cosyns, *Cancer Res.*, 1996, **56**, 2025.
222. W. Pfau, B. L. Pool-Zobel, C.-W. von der Lieth, and M. Wiessler, *Cancer Lett.*, 1990, **55**, 7.
223. F. F. Kadlubar, in "DNA Adducts: Identification and Biological Significance," eds. K. Hemminki, A. Dipple, D. E. G. Shuker, F. F. Kadlubar, D. Segerback, and H. Bartsch, IARC Scientific Publications, Lyon, 1994, vol. 125, p. 199.
224. Z. S. Galiegue, B. Bailleul, and L. M. Loucheux, *Carcinogenesis*, 1983, **4**, 249.
225. M. Hara, K. Asano, I. Kawamoto, T. Takiguchi, S. Katsumata, K. Takahashi, and H. Nakano, *J. Antibiot.*, 1989, **42**, 1768.
226. N. Hirayama and E. S. Matsuzawa, *Chem. Lett.*, 1993, **11**, 1957.
227. Y. Kanda and T. Fukuyama, *J. Am. Chem. Soc.*, 1993, **115**, 8451.
228. M. Hara, Y. Saitoh, and H. Nakano, *Biochemistry*, 1990, **29**, 5676.
229. S. B. Behroozi, W. Kim, and K. S. Gates, *J. Org. Chem.*, 1995, **60**, 3964.
230. A. Asai, M. Hara, S. Kakita, Y. Kanda, M. Yoshida, H. Saito, and Y. Saitoh, *J. Am. Chem. Soc.*, 1996, **118**, 6802.
231. K. Schank, A. Frisch, and B. Zwanenburg, *J. Org. Chem.*, 1983, **48**, 4580.
232. A. Fidder, G. W. H. Moes, A. G. Scheffer, G. P. ver der Schans, R. A. Baan, L. P. A. de Jong, and H. P. Benschop, *Chem. Res. Toxicol.*, 1994, **7**, 199.
233. S. J. Behroozi, W. Kim, J. Dannaldson, and K. S. Gates, *Biochemistry*, 1996, **35**, 1768.
234. K. Mitra, W. Kim, J. S. Daniels, and K. S. Gates, *J. Am. Chem. Soc.*,1997, **119**, 11 691.
235. W. Kim, J. Dannaldson, and K. S. Gates, *Tetrahedron Lett.*, 1996, **37**, 5337.
236. B. Ganley, S. J. Behroozi, L. Breido, R. Bahktiar, and K. S. Gates, 1997, submitted for publication.
237. B. S. Reddy and C. V. Rao, *ACS Symp. Ser.*, 1994, **546**, 164.
238. W. Kim and K. S. Gates, *Chem. Res. Toxicol.*, 1997, **10**, 296.
239. T. Prestera, Y. Zhang, S. R. Spencer, C. A. Wilczak, and P. Talalay, *Adv. Enzyme Regul.*, 1993, **33**, 281.
240. B. S. Davidson, T. F. Molinski, L. R. Barrows, and C. M. Ireland, *J. Am. Chem. Soc.*, 1991, **113**, 4709.
241. P. A. Searle and T. F. Molinski, *J. Org. Chem.*, 1994, **59**, 660.
242. C. Takahashi, A. Numata, Y. Ito, E. Matsumura, H. Araki, H. Iwaki, and K. Kushida, *J. Chem. Soc. Perkin Trans. 1*, 1994, 1859.
243. P. J. Curtis, D. Greatbanks, B. Hesp, A. F. Cameron, and A. A. Freer, *J. Chem. Soc. Perkin Trans. 1*, 1977, 180.
244. T. Chattergi and K. S. Gates, *Bioorg. Med. Chem. Lett.*, 1998, **8**, 535.
245. J. P. Brown, *Mutat. Res.*, 1980, **75**, 243.
246. R. H. Thompson, "Naturally Occurring Quinones," Academic Press, London, 1971.
247. A. H.-J. Wang, in "Molecular Aspects of Anticancer Drug–DNA Interactions," eds. S. Neidle and M. Waring, CRC Press, Boca Raton, FL, 1993, vol. 1, p. 32.
248. R. B. Silverman, "The Organic Chemistry of Drug Design and Drug Action," Academic Press, San Diego, CA, 1992, p. 220.
249. J. B. Chaires, N. Dattagupta, and D. M. Crothers, *Biochemistry*, 1982, **21**, 3933.
250. J. Butler and B. M. Hoey, in "DNA and Free Radicals," eds. B. Halliwell and O. I. Aruoma, Ellis Horwood, New York, 1993, p. 243.
251. G. Gaudiano and T. H. Koch, *Chem. Res. Toxicol.*, 1991, **4**, 2.
252. H. P. Misra and I. Fridovich, *J. Biol. Chem.*, 1972, **247**, 188.
253. L. Weiner, *Methods Enzymol.*, 1994, **233**, 92.
254. J. Fisher, B. R. J. Abdella, and K. E. McLane, *Biochemistry*, 1985, **24**, 3562.
255. G. H. Meguerian, *J. Am. Chem. Soc.*, 1955, **77**, 5019.
256. J. W. Lown, S.-K. Sim, K. C. Majumdar, and R.-Y. Chang, *Biochem. Biophys. Res. Commun.*, 1977, **76**, 705.
257. V. Berlin and W. A. Haseltine, *J. Biol. Chem.*, 1981, **256**, 4747.
258. E. Feinstein, E. Canaani, and L. M. Weiner, *Biochemistry*, 1993, **32**, 13 156.
259. H. Eliot, L. Gianni, and C. Myers, *Biochemistry*, 1984, **23**, 928.
260. J. Hajdu and E. C. Armstrong, *J. Am. Chem. Soc.*, 1981, **103**, 232.
261. G. V. Rumyantseva, L. M. Weiner, E. I. Frolova, and O. S. Fedorova, *FEBS Lett.*, 1989, **242**, 397.
262. R. Cone, S. K. Hasan, J. W. Lown, and A. R. Morgan, *Can. J. Biochem.*, 1976, **54**, 219.
263. M. M. Pater and S. Pak, *Nature*, 1974, **250**, 786.
264. H. Nakazawa, F. E. Chou, P. A. Andrews, and N. R. Bachur, *J. Org. Chem.*, 1981, **46**, 1493.
265. A. Parraga, M. Orozco, and J. Portugal, *Eur. J. Biochem.*, 1992, **208**, 227.
266. S. Akauzawa, H. Yamaguchi, T. Masuda, and Y. Ueno, *Mutat. Res.*, 1992, **266**, 63.
267. C. Auclair and C. Paoletti, *J. Med. Chem.*, 1981, **24**, 289.
268. K. Muller, H.-S. Huang, and W. Wiegrebe, *J. Med. Chem.*, 1996, **39**, 3132.
269. M. Yagi, T. Nishimura, H. Suzuki, and N. Tanaka, *Biochem. Biophys. Res. Commun.*, 1981, **98**, 642.
270. M. Egholm and T. H. Koch, *J. Am. Chem. Soc.*, 1989, **11**, 8291.
271. G. Gaudiano, K. Resing, and T. H. Koch, *J. Am. Chem. Soc.*, 1994, **116**, 6537.
272. B. M. Kacinski and W. D. Rupp, *Cancer Res.*, 1984, **44**, 3489.
273. B. Lambert, P. Laugaa, B. P. Roques, and J. B. le Pecq, *Mutat. Res.*, 1986, **166**, 243.
274. M. Boldt, G. Gaudino, M. J. Haddadin, and T. H. Koch, *J. Am. Chem. Soc.*, 1989, **111**, 2283.
275. G. Gaudiano, M. Frigerio, C. Sangsurasak, P. Bravo, and T. H. Koch, *J. Am. Chem. Soc.*, 1992, **114**, 5546.
276. M. Chatterjee and S. E. Rokita, *J. Am. Chem. Soc.*, 1994, **116**, 1690.
277. A. J. Lin, R. S. Pardini, L. A. Cosby, B. J. Lillis, C. W. Shansky, and A. C. Sartorelli, *J. Med. Chem.*, 1973, **16**, 1268.
278. S. P. Mayalarp, R. H. J. Hargreaves, J. Butler, C. O'Hare, and J. A. Hartley, *J. Med. Chem.*, 1996, **39**, 531.
279. Y.-G. Gao, Y.-C. Liaw, Y.-K. Li, G. A. van der Marel, J. H. van Boom, and W. H.-J. Wang, *Proc. Natl. Acad. Sci. USA*, 1991, **88**, 4845.
280. A. H.-J. Wang, Y.-G. Gao, Y.-C. Liaw, and Y.-k. Li, *Biochemistry*, 1991, **30**, 3812.
281. F. Leng, R. Savkur, I. Fokt, T. Przewloka, W. Priebe, and J. B. Chaires, *J. Am. Chem. Soc.*, 1996, **118**, 4731.
282. D. J. Taatjes, G. Gaudiano, K. Resing, and T. H. Koch, *J. Med. Chem.*, 1996, **39**, 4135.
283. A. Chenna and B. Singer, *Chem. Res. Toxicol.*, 1995, **8**, 865.

284. D. E. Stack, J. Byun, M. L. Gross, E. G. Rogan, and E. L. Cavalieri, *Chem. Res. Toxicol.*, 1996, **9**, 851.
285. Y. J. Abul-Hajj, K. Tabakovic, and I. Tabakovic, *J. Am. Chem. Soc.*, 1995, **117**, 6144.
286. C. Auclair, B. Dugue, B. Meunier, and C. Paoletti, *Biochemistry*, 1986, **25**, 1240.
287. G. Pratviel, J. Bernadou, T. Ha, G. Meunier, S. Cros, B. Meunier, B. Gillet, and E. Guittet, *J. Med. Chem.*, 1986, **29**, 1350.
288. A. Albini and S. Pietra, in "Heterocyclic N-Oxides," CRC Press, Boca Raton, FL, 1991.
289. E. A. Peterson, D. C. Gillespie, and F. D. Cook, *Can. J. Microbiol.*, 1965, **12**, 221.
290. G. R. Clemo and A. F. Daglish, *J. Chem. Soc.*, 1950, 1481.
291. U. Hollstein, and P. L. Butler, *Biochemistry*, 1972, **11**, 1345.
292. R. M. Behki and S. M. Lesley, *J. Bacteriol.*, 1972, **109**, 250.
293. K. Nagai, B. J. Carter, J. Xu, and S. M. Hecht, *J. Am. Chem. Soc.*, 1991, **113**, 5099.
294. K. Nagai and S. M. Hecht, *J. Biol. Chem.*, 1991, **266**, 23 994.
295. T. Nunoshiba and H. Nishioka, *Mutat. Res.*, 1989, **217**, 203.
296. W. Suter, A. Rosselet, and F. Knusel, *Antimicrob. Agents Chemother.*, 1978, **13**, 770.
297. L. Beutin, E. Preller, and B. Kowalski, *Antimicrob. Agents Chemother.*, 1981, **20**, 336.
298. A. Monge, J. A. Palop, A. L. de Cerain, G. Senador, F. J. Martinez-Crespo, Y. Sainz, S. Narro, E. Garcia, C. de Miguel, M. Gonzalez, E. Hamilton, A. J. Barker, E. D. Clarke, and D. T. Greenhow, *J. Med. Chem.*, 1995, **38**, 1786.
299. J. M. Brown, *Br. J. Cancer*, 1993, **67**, 1163.
300. J. S. Daniels and K. S. Gates, *J. Am. Chem. Soc.*, 1996, **118**, 3380.
301. K. Laderoute and A. M. Rauth, *Biochem. Pharmacol.*, 1986, **35**, 3417.
302. J. S. Daniels and K. S. Gates, 1997, submitted for publication.
303. W. Lytollis, R. T. Scannell, H. An, V. S. Murty, K. S. Reddy, J. R. Barr, and S. M. Hecht, *J. Am. Chem. Soc.*, 1995, **117**, 12 683.
304. J. R. Barr, V. S. Murty, Y. Keiichi, S. Singh, D. H. Smith, and S. M. Hecht, *Chem. Res. Toxicol.*, 1988, **1**, 204.
305. R. T. Scannell, J. R. Barr, V. S. Murty, K. S. Reddy, and S. M. Hecht, *J. Am. Chem. Soc.*, 1988, **110**, 3650.
306. U. S. Singh, R. T. Scannell, H. An, B. J. Carter, and S. M. Hecht, *J. Am. Chem. Soc.*, 1995, **117**, 12 691.
307. R. Bhat and S. M. Hadi, *Mutat. Res.*, 1994, **313**, 39.
308. C. Wasser, F. Silva, and E. Rodriquez, *Experientia*, 1990, **46**, 500.
309. S. M. Hecht, in "Cancer Chemotherapeutic Agents," ed. W. O. Foye, American Chemical Society, Washington, DC, 1995, p. 369.
310. S. A. Kane and S. M. Hecht, *Prog. Nucleic Acid Res. Mol. Biol.*, 1994, **49**, 313.
311. L. F. Povirk, *Mutat. Res.*, 1996, **355**, 71.
312. J. Stubbe, J. W. Kozarich, W. Wu, and D. E. Vanderwall, *Acc. Chem. Res.*, 1996, **29**, 322.
313. J. Stubbe and J. W. Kozarich, *Chem. Rev.*, 1987, **87**, 1107.
314. L. F. Povirk, M. Hogan, and N. Dattagupta, *Biochemistry*, 1979, **18**, 96.
315. M. A. Chien, A. P. Grollman, and S. B. Horowitz, *Biochemistry*, 1977, **16**, 3641.
316. H. Sugiyama, R. E. Kilkuskie, S. M. Hecht, G. A. ver der Marel, and J. H. van Boom, *J. Am. Chem. Soc.*, 1985, **107**, 7765.
317. R. P. Hertzberg, M. J. Caranfa, and S. M. Hecht, *Biochemistry*, 1988, **27**, 3164.
318. S. A. Kane, H. Sasaki, and S. M. Hecht, *J. Am. Chem. Soc.*, 1995, **117**, 9107.
319. M. J. Levy and S. M. Hecht, *Biochemistry*, 1988, **27**, 2647.
320. J. Kuwahara and Y. Sugiura, *Proc. Natl. Acad. Sci. USA*, 1988, **85**, 2459.
321. C. Bailly and M. J. Waring, *J. Am. Chem. Soc.*, 1995, **117**, 7311.
322. W. Wu, D. E. Vanderwall, S. M. Lui, X.-J. Tang, C. Turner, J. W. Kozarich, and J. Stubbe, *J. Am. Chem. Soc.*, 1996, **118**, 1281.
323. M. J. Absalon, J. Stubbe, and J. W. Kozarich, *Biochemistry*, 1995, **34**, 2065.
324. R. A. Manderville, J. F. Ellena, and S. M. Hecht, *J. Am. Chem. Soc.*, 1995, **117**, 7891.
325. S. M. Hecht, in "DNA and RNA Cleavers and Chemotherapy of Cancer and Viral Diseases," ed. B. Meunier, Kluwer, Dordrecht, 1996, p. 77.
326. M. V. Keck and S. M. Hecht, *Biochemistry*, 1995, **34**, 12 029.
327. J. Wilshire and D. T. Sawyer, *Acc. Chem. Res.*, 1979, **12**, 105.
328. T. A. Dix, K. M. Hess, M. A. Medina, R. W. Sullivan, S. L. Tilly, and T. L. L. Webb, *Biochemistry*, 1996, **35**, 4578.
329. S. A. Lesko, R. J. Lorentzen, and P. O. P. Ts'o, *Biochemistry*, 1980, **19**, 3023.
330. B. Halliwell and J. M. C. Gutteridge, *Methods Enzymol.*, 1990, **186**, 1.
331. A. U. Khan and M. Kasha, *Proc. Natl. Acad. Sci. USA*, 1994, **91**, 12 365.
332. Y. Luo, Z. Han, S. M. Chin, and S. Linn, *Proc. Natl. Acad. Sci. USA*, 1994, **91**, 12 438.
333. W. K. Pogozelski, T. J. McNeese, and T. D. Tullius, *J. Am. Chem. Soc.*, 1995, **117**, 6428.
334. D. A. Wink, R. W. Nims, J. E. Saavedra, W. E. J. Utermahlen, and P. C. Ford, *Proc. Natl. Acad. Sci. USA*, 1994, **91**, 6604.
335. E. S. Henle, Y. Luo, and S. Linn, *Biochemistry*, 1996, **35**, 12 212.
336. V. C. Bode, *J. Mol. Biol.*, 1967, **26**, 125.
337. H. S. Rosenkrantz and S. Rosenkrantz, *Arch. Biochem. Biophys.*, 1971, **146**, 483.
338. R. A. Morgan, R. L. Cone, and T. M. Elgert, *Nucleic Acids Res.*, 1976, **3**, 1139.
339. K. Hiramoto, R. Aso-o, H. Ni-iyama, S. Hikage, T. Kato, and K. Kikugawa, *Mutat. Res.*, 1996, **359**, 17.
340. H.-J. Rhaese, E. Freese, and M. S. Melzer, *Biochim. Biophys. Acta*, 1968, **155**, 491.
341. K. Brawn and I. Fridovich, *Arch. Biochem. Biophys.*, 1981, **206**, 414.
342. D.-H. Chin and I. H. Goldberg, *Biochemistry*, 1986, **25**, 1009.
343. I. Fridovich, *Acc. Chem. Res.*, 1972, **5**, 321.
344. S. Soboll, S. Grundel, J. Harris, V. Kolb-Bachofen, B. Ketterer, and H. Sies, *Biochem. J.*, 1995, **311**, 889.
345. S. J. Gould, N. Tamayo, C. R. Melville, and M. C. Cone, *J. Am. Chem. Soc.*, 1994, **116**, 2207.
346. S. Mithani, G. Weeratunga, N. J. Taylor, and G. I. Dmitrienko, *J. Am. Chem. Soc.*, 1994, **116**, 2209.
347. D. Arya and D. J. Jebaratnam, *J. Org. Chem.*, 1995, **60**, 3268.
348. T. Shirafuji, Y. Yamamoto, and H. Nozaki, *Tetrahedron*, 1971, **27**, 5353.

349. J. N. McGuire, S. R. Wilson, and K. L. Rinehardt, *J. Antibiot.*, 1995, **48**, 516.
350. K. Hiramoto, T. Fujino, and K. Kikugawa, *Mutat. Res.*, 1996, **360**, 95.
351. K. Nakatani, S. Maekawa, K. Tanabe, and I. Saito, *J. Am. Chem. Soc.*, 1995, **117**, 10 635.
352. A. E. Ross, D. L. Nagel, and B. Toth, *J. Agric. Food Chem.*, 1982, **30**, 521.
353. B. Toth, D. Nagel, and A. Ross, *Br. J. Cancer*, 1982, **46**, 417.
354. K. Hiramoto, M. Kaku, T. Kato, and K. Kikugawa, *Chem.-Biol. Interact.*, 1995, **94**, 21.
355. K. Hiramoto, M. Kaku, A. Sueyoshi, M. Fujise, and K. Kikugawa, *Chem. Res. Toxicol.*, 1995, **8**, 356.
356. J. Griffiths and J. A. Murphy, *J. Chem. Soc. Chem. Commun.*, 1992, 24.
357. D. Arya and D. J. Jebaratnam, *Tetrahedron Lett.*, 1995, **36**, 4369.
358. E. L. Eliel, J. G. Saha, and S. Meyerson, *J. Org. Chem.*, 1965, **30**, 2451.
359. H. Zollinger, *Acc. Chem. Res.*, 1973, **6**, 335.
360. J. P. Behr, *J. Chem. Soc. Chem. Commun.*, 1989, 101.
361. A. L. Smith and K. C. Nicolaou, *J. Med. Chem.*, 1996, **39**, 2103.
362. A. G. Myers, S. B. Cohen, and B.-M. Kwon, *J. Am. Chem. Soc.*, 1994, **116**, 1670.
363. I. H. Goldberg, *Acc. Chem. Res.*, 1991, **24**, 191.
364. S. M. Bishop, M. Malone, D. Phillips, A. W. Parker, and M. C. R. Symons, *J. Chem. Soc. Chem. Commun.*, 1994, 871.
365. C. Sheu and C. S. Foote, *J. Am. Chem. Soc.*, 1993, **115**, 10 446.
366. C. Sheu and C. S. Foote, *J. Am. Chem. Soc.*, 1995, **117**, 474.
367. S. Raoul and J. Cadet, *J. Am. Chem. Soc.*, 1996, **118**, 1892.
368. I. Saito, M. Takayama, H. Sugiyama, and K. Nakatani, *J. Am. Chem. Soc.*, 1995, **117**, 6406.
369. J. Cadet, M. Berger, G. W. Buchko, P. C. Joshi, S. Raoul, and J.-L. Ravanat, *J. Am. Chem. Soc.*, 1994, **116**, 7403.
370. P. M. Cullis, M. E. Malone, and L. A. Merson-Davies, *J. Am. Chem. Soc.*, 1996, **188**, 2775.
371. A. E. Alegria, C. M. Krishna, R. K. Elespuru, and P. Riesz, *Photochem. Photobiol.*, 1989, **49**, 257.
372. D. T. Breslin and G. B. Schuster, *J. Am. Chem. Soc.*, 1996, **118**, 2311.
373. J. W. Lown and H.-H. Chen, *Biochem. Pharmacol.*, 1980, **29**, 905.
374. J. Cadet, in "DNA Adducts: Identification and Biological Significance," eds. K. Hemminki, A. Dipple, D. E. G. Shuker, F. F. Kadlubar, D. Segerback, and H. Bartsch, IARC Scientific Publications, Lyon, 1994, vol. 125, p. 245.
375. F. P. Gasparro, *Photochem. Photobiol.*, 1996, **63**, 553.
376. J. E. Hearst, *Chem. Res. Toxicol.*, 1989, **2**, 69.
377. M. Greenstein, T. Monji, R. Yeung, W. M. Maiese, and R. J. White, *Antimicrob. Agents Chemother.*, 1986, **29**, 861.
378. L. R. McGee and R. Misra, *J. Am. Chem. Soc.*, 1990, **112**, 2386.
379. G. G. Wilson, *Nucleic Acids Res.*, 1991, **19**, 2539.
380. S. P. Bennett and S. E. Halford, *Curr. Top. Cell Regul.*, 1989, **30**, 57.

# 7.15
# DNA-damaging Enediyne Compounds

ZHEN XI and IRVING H. GOLDBERG
*Harvard Medical School, Boston, MA, USA*

| | | |
|---|---|---|
| 7.15.1 | INTRODUCTION | 554 |
| 7.15.2 | BIOSYNTHESIS OF NATURALLY OCCURRING ENEDIYNE ANTIBIOTICS | 554 |
| 7.15.3 | STRUCTURE AND MECHANISM OF ENEDIYNE ANTIBIOTICS | 556 |
| | 7.15.3.1    *Mechanistic Considerations of Activation of Enediynes: Implications for Antibiotic Action* | 556 |
| |    *7.15.3.1.1   Bergman rearrangement* | 556 |
| |    *7.15.3.1.2   Myers rearrangement* | 556 |
| |    *7.15.3.1.3   Mechanistic models and molecular modeling* | 557 |
| | 7.15.3.2    *Neocarzinostatin* | 558 |
| |    *7.15.3.2.1   Structure* | 558 |
| |    *7.15.3.2.2   Activation* | 559 |
| |    *7.15.3.2.3   DNA damage* | 566 |
| | 7.15.3.3    *Calicheamicins* | 568 |
| |    *7.15.3.3.1   Structure* | 568 |
| |    *7.15.3.3.2   Activation* | 569 |
| |    *7.15.3.3.3   Trisulfide chemistry* | 570 |
| |    *7.15.3.3.4   Role of the oligosaccharide amino group* | 570 |
| |    *7.15.3.3.5   DNA damage* | 570 |
| | 7.15.3.4    *Esperamicin* | 571 |
| |    *7.15.3.4.1   Structure* | 571 |
| |    *7.15.3.4.2   Activation* | 572 |
| |    *7.15.3.4.3   DNA damage* | 573 |
| | 7.15.3.5    *Dynemicin* | 574 |
| |    *7.15.3.5.1   Structure* | 574 |
| |    *7.15.3.5.2   Activation* | 575 |
| |    *7.15.3.5.3   DNA damage* | 575 |
| | 7.15.3.6    *Kedarcidin* | 575 |
| | 7.15.3.7    *C-1027* | 578 |
| |    *7.15.3.7.1   Structure* | 578 |
| |    *7.15.3.7.2   Activation* | 578 |
| |    *7.15.3.7.3   Reversibility and reactivity of the 1,4-benzenoid biradical* | 578 |
| |    *7.15.3.7.4   DNA damage* | 579 |
| | 7.15.3.8    *Maduropeptin* | 580 |
| | 7.15.3.9    *Namenamicin* | 581 |
| 7.15.4 | ROLE OF APOPROTEIN | 582 |
| 7.15.5 | DNA/RNA DAMAGE MECHANISMS | 582 |
| | 7.15.5.1    *Chemistry of DNA/RNA Damage* | 582 |
| | 7.15.5.2    *Role of Dioxygen* | 583 |
| | 7.15.5.3    *Adduct Formation* | 585 |
| 7.15.6 | REFERENCES | 586 |

## 7.15.1 INTRODUCTION

In 1965, Ishida *et al.* isolated a new antibiotic, neocarzinostatin, from *Streptomyces carzinostaticus* var. F-41, initially identified as a simple protein (MW = 11 100),[1,2] which was capable of inhibiting DNA synthesis and inducing DNA degradation in cells.[3] Efforts to understand the biochemical behavior of this new agent led to the finding that its actual DNA-damaging activity comes from a previously unidentified nonprotein chromophore (NCS-chrom)[4-8] through a free-radical mechanism involving hydrogen abstraction from the sugar residues.[9-11] Efforts[12-16] to determine the structure of NCS-chrom (1) culminated in a unique structure by Edo *et al.* in 1985,[17] possessing a novel bicyclo[7.3.0]dodecadiynene core unit.

The subsequent disclosure in 1987 of the structures of two other new antibiotics, the calicheamicins and the esperamicins, possessing an enediyne moiety, and that their transformations were closely linked to the well-known Bergman reaction, by scientists from Lederle Laboratories[18,19] and Bristol-Myers,[20,21] respectively, has sparked enormous interest, especially from the synthetic community.[22-30] In 1997, this growing superfamily of naturally occurring enediyne antibiotics contained eight members. The enediyne antibiotics can be conveniently grouped into three classes: (i) calicheamicins (2), esperamicins (3), and namenamicin (8), (ii) dynemicins (4), and (iii) chromoproteins, including neocarzinostatin, kedarcidin, C-1027, and maduropeptin (Figure 1).

The biological activities of these compounds mainly come from their ability to damage DNA, leading to cell death. The term "enediyne," also called "diynene," includes enediyne, enyneallene, enynecumulene, and their closely related analogues, unless specified otherwise. As defined by the scope and the space limitations of this chapter, we shall focus on the chemical aspects of enediyne compounds with demonstrated *in vivo/in vitro* DNA damage activity. As the literature related to enediyne compounds numbered well over 1000 by the end of 1996, selection here is based on historical reasons, leading references, and seminal research papers. Readers interested in aspects of organic synthesis, biology, and clinical applications are advised to seek further reading on these subjects. Some of the excellent recent books and reviews on these topics should be referred to.[25-46] Readers interested in a broader treatment of the chemistry of DNA damage are referred to Chapter 7.11.

## 7.15.2 BIOSYNTHESIS OF NATURALLY OCCURRING ENEDIYNE ANTIBIOTICS

In appreciation of the intriguing molecular architecture of the enediyne antibiotics, it is of interest to review their molecular assembly within the producing organism. Biosynthetic studies may not only provide useful pharmacophores and improve the fermentation production of the known enediyne antibiotics, but also lead to new congeners via a biocombinatorial chemistry approach.[47-49] Several studies have been published dealing with the biosynthetic pathway of the enediyne antibiotics. They include the biosynthesis of NCS-chrom (1),[50] dynemicin A (4),[51] and esperamicin A$_1$ (3).[52]

Biosynthetic studies[50] on NCS-chrom (1) showed that the *N*-methyl of the fucosamine and the *O*-methyl of the naphthoate moieties are derived from methionine via *S*-adenosylmethionine and that the cyclic carbonate carbonyl comes from carbonate. Acetate incorporation studies showed that the C$_{12}$ naphthoic acid ring is derived from hexaketide. The C$_{14}$ cyclic carbonate/bicyclo-[7.3.0]dodecadienediyne ring system appears to be derived from a minimum of eight head-to-tail coupled acetate units (a linear C$_{18}$ polyketide precursor) via the oleate–crepenynate biosynthetic pathway for polyalkynates.

Biosynthetic studies[51] on dynemicin A (4) concluded that the bicyclo[7.3.1]diynene unit and the anthraquinone unit are biosynthesized separately from two heptaketides, which are formed from head-to-tail condensation of seven acetate units, and later assembled. The carboxylic acid linked to C-5 of the A ring is derived from a separate acetic acid condensed at C-5 and the *O*-methyl group from methionine.

Biosynthetic studies[52] on esperamicin A$_1$ (3) revealed that acetate, pyruvate, L-methionine, and D-glucose are the precursors of esperamicin A$_1$, while L-methionine, L-cysteine, and sulfate provide the sulfur source. The enediyne ring of esperamicin A$_1$ is derived from head-to-tail condensation of seven acetate units, and the uncoupled carbon comes from the C-2 of an acetate unit (an octaketide precursor with the loss of the C-1 at the end acetate unit). The two carbons of the yne moiety are derived from separate acetate units. L-[methyl-$^{13}$C]methionine incorporation showed that the *S*-methyl of the trisulfide and the thiosugar and the *O*-methyl of the amino sugar, the aromatic chromophore, and the carbamate moiety are derived from L-methionine via *S*-adenosylmethionine. All four sulfur atoms can be derived from sulfate.

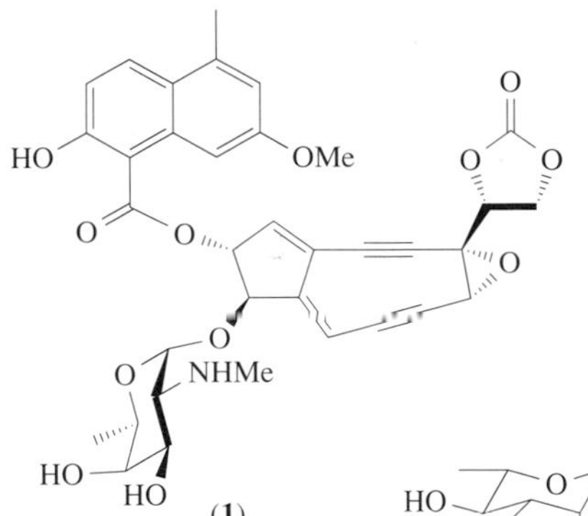

(**1**)
Neocarzinostatin chromophore

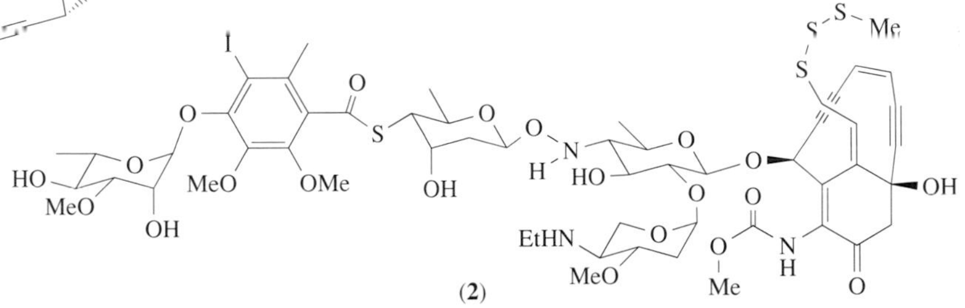

(**2**)
Calicheamicin $\gamma_1^{I}$

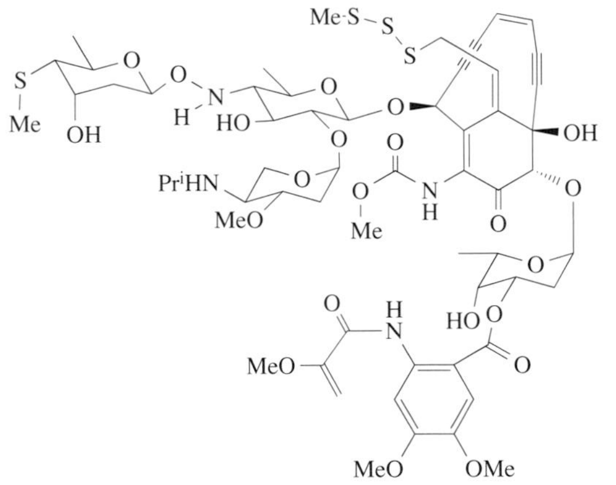

(**3**)
Esperamicin $A_1$

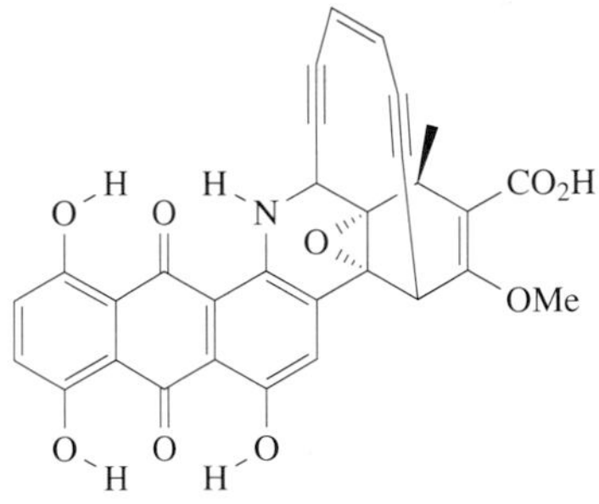

(**4**)
Dynemicin A

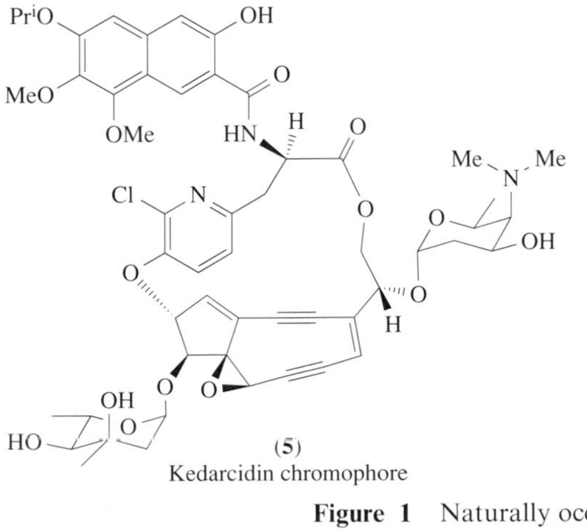

(**5**)
Kedarcidin chromophore

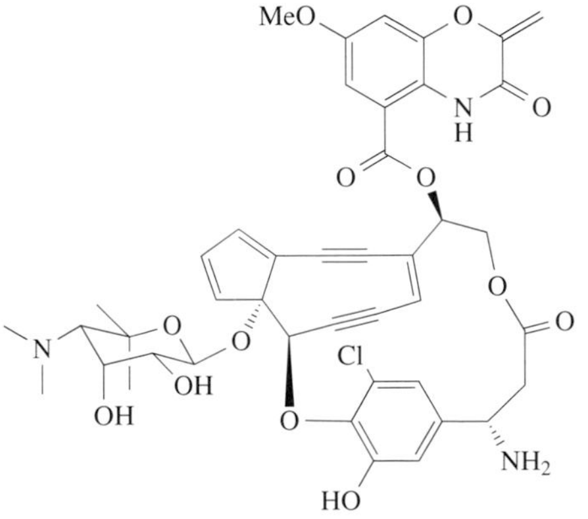

(**6**)
C-1027 chromophore

**Figure 1** Naturally occurring enediyne antibiotics.

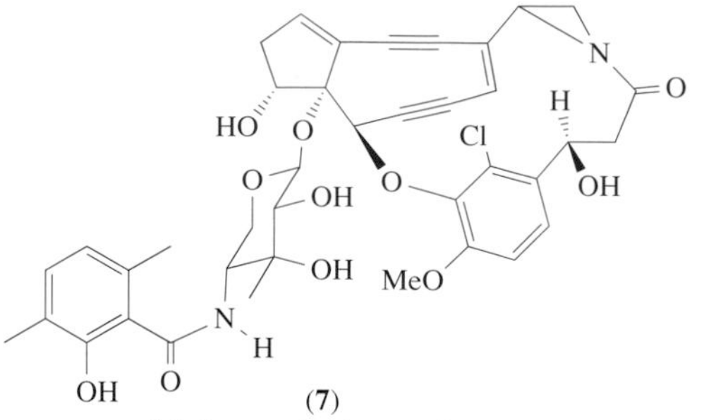

(7)
Maduropeptin chromophore

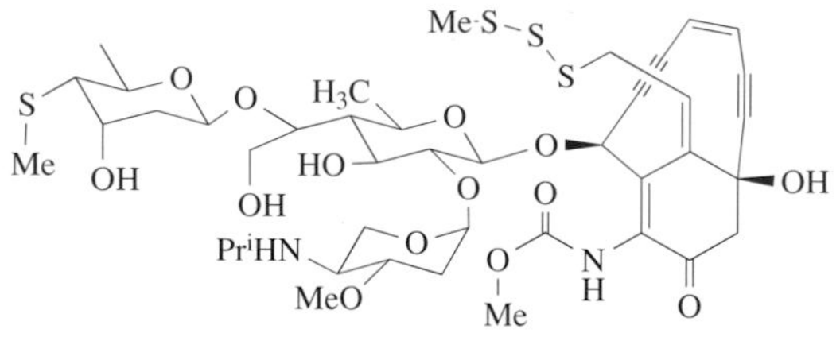

(8)
Namenamicin

**Figure 1** (continued)

It seems that two distinct sets of enediyne bioassembly account for the biosynthetic pathways for the enediyne antibiotics. Biosynthesis of the esperamicins, calicheamicins, and dynemicins may share the same bioarchitecture, whereas NCS-chrom (1) belongs to another class.

## 7.15.3 STRUCTURE AND MECHANISM OF ENEDIYNE ANTIBIOTICS

### 7.15.3.1 Mechanistic Considerations of Activation of Enediynes: Implications for Antibiotic Action

#### 7.15.3.1.1 *Bergman rearrangement*

Long before enediynes were recognized as biomolecules, they were already well appreciated by a few pioneers in the field of physical organic chemistry.[53–55] The seminal discovery of the reactive 1,4-benzenoid biradical intermediacy in the aromatization of synthetic enediyne compounds by Bergman and co-workers in 1972[55–59] laid the foundation for understanding the cycloaromatization of natural enediynes[53,54] and the subsequent mechanistic elucidation of the bioactive enediyne antibiotics (Scheme 1).

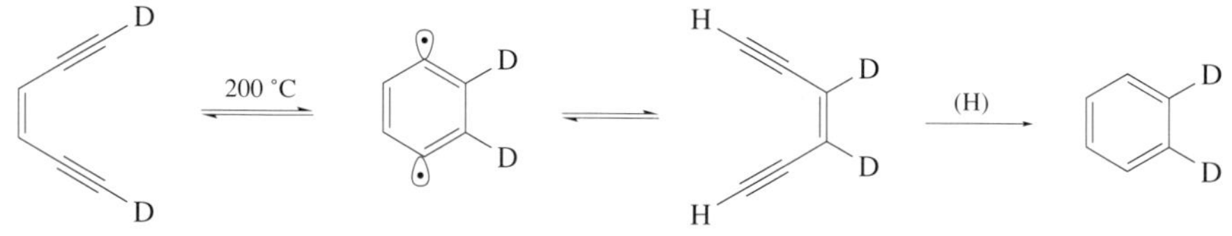

**Scheme 1**

Molecular calculations[60–63] and experimental data[64–67] on the thermodynamics of the typical Bergman rearrangement show that it has the activation parameters $\Delta H^{\neq} = 28.2 \pm 0.5$ kcal mol$^{-1}$ and $\Delta S^{\neq} = -10.2 \pm 1$ cal K$^{-1}$ mol$^{-1}$.[66]

#### 7.15.3.1.2 *Myers rearrangement*

In 1989, shortly after the disclosure of the structures and mechanism of the calicheamicins and esperamicins, Myers and co-workers[68–70] and Saito and co-workers[71] with insight studied the rearrangement of the enyneallene system, which is related to the neocarzinostatin chromophore. (Z)-1,2,4-Heptatriene-6-yne was clearly shown to cycloaromatize under thermal conditions via α,3-dehydrotoluene biradical (another 1,4-benzenoid biradical) to give toluene upon hydrogen atom abstraction (Scheme 2, Myers reaction). This typical Myers reaction yields the activation parameters $\Delta H^{\neq} = 21.8 \pm 0.5$ kcal mol$^{-1}$ and $\Delta S^{\neq} = -11.6 \pm 1.5$ cal K$^{-1}$ mol$^{-1}$. The Myers enyneallene rearrangement is more exothermic with lower activation energy than the Bergman enediyne rearrangement. Molecular calculations by Koga and Morokuma also reached the same conclusion.[60] It should be noted that the Myers reaction is closely related to the Bergman reaction; the biradical in the Myers reaction is a $\sigma,\pi$-biradical, whereas a $\sigma,\sigma$-biradical exists in the Bergman reaction. It is believed that this difference may have profound influence on their mechanistic behaviors.[70,72]

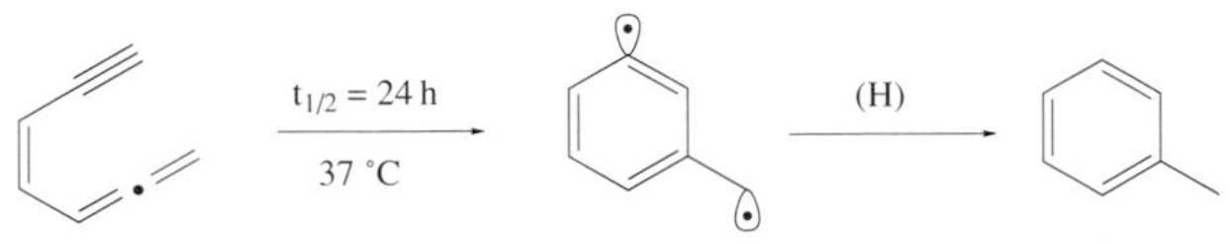

**Scheme 2**

### 7.15.3.1.3 Mechanistic models and molecular modeling

The ease of cycloaromatization of enediyne compounds is certainly an important issue in terms of the usefulness of the enediyne, especially for its consideration as a prodrug aimed at a biotarget. Considering the thermal nature of enediyne rearrangements, the cycloaromatization is expected to occur at an appreciable rate at physiologically relevant temperatures, i.e., around 37 °C, after activation within a biosystem.

The work by Nicolaou and co-workers provides an empirical rule regarding the ease of cycloaromatization of monocyclic enediynes, the c–d distance rule:[73,74] the distance between the alkyne termini (c–d distance between the ends (the cyclization sites) of the 1,5-diyne-3-ene system) is correlated with the ground-state strain energy; increased strain implies increased reactivity and ease of cycloaromatization. Monocyclic enediynes with c–d distances in the range $3.2 \pm 0.1$ Å are predicted to cyclize spontaneously at ambient temperature at an appreciable rate. A 10-membered ring enediyne cyclized smoothly at 37 °C ($t_{1/2} = 18$ h) in the presence of cyclohexadiene via the biradical to give the tetralin. These studies provide an explanation of why naturally occurring enediyne antibiotics contain either nine- or 10-membered ring enediyne cores, as they can be easily cycloaromatized at physiological temperature. Nature usually incorporates a protective braking mechanism on the active nine- or 10-membered ring of enediyne cores, making the naturally occurring enediyne antibiotics prodrugs. It is of interest to note that work by Buchwald and co-workers[75] and others[76,77] in the mid-1990s made use of transition metal-mediated cycloaromatization by adjusting the c–d distance in the enediyne through coordination chemistry, although its pharmaceutical application could not be predicted at that time.

Owing to strong constraint exerted on the ring system, the cycloaromatization can be slowed or inhibited, as in most of the naturally occurring enediyne antibiotics (through bridgehead $sp^2$ hybridization, epoxide insertion, and possibly constrained rotation of certain elements). Snyder,[78,79] Snyder and Tipsword,[80] and Magnus and co-workers,[81–83] based on molecular calculations and experimental data, pointed out the limitation of the c–d distance rule, and suggested that the reactivity of an enediyne is dependent not only on the strain energy of the ground state, but also on that of the transition state. The ease of cycloaromatization of an enediyne is critically determined by the difference in the strain energy between the ground state and the transition state. An enediyne with a smaller strain energy difference between the ground and transition states undergoes easier cycloaromatization.

Some practical methods have been suggested for examining the reactivity of an enediyne.[25] Based on the view that the transition state of enediyne cycloaromatization is product-like, Maier demonstrated that the reactivity of a bicyclic enediyne correlates well with the difference in strain energies between the postactivated product (transition-state mimic) and the enediyne (ground state).[25] The reactivity of an enediyne is influenced not only by steric effects, but also by electronic effects. For example, annelation of aryl systems, such as benzene and naphthalene on the enediyne moiety, has a retarding effect on the cycloaromatization, although the c–d distance may not be very different.[84]

Several important issues await clarification post-1997. Is the Bergman cyclization of enediyne antibiotics reversible under physiological conditions? Does the bifunctional biradical function through a concerted or a sequential process, and how? What is the impact of the results on DNA damage? It is generally assumed that cycloaromatization is very fast, but a systematic quantitative study is still lacking. More detailed discussion can be found in Section 7.15.3.7.3.

Is a biradical the real intermediate in enediyne cycloaromatization? Some *ab initio* calculations have been directed to this issue. Kraka and Cremer showed that the transition state of cycloaromatization of (Z)-hex-1,5-diyne-3-ene does not possess significant biradical character, although its geometry resembles that of the biradical.[61] A concerted process has been suggested,[62] but a biradical mechanism may better explain the data from DNA damage experiments involving isotope selection effects. Radical anion, radical cation, and ionic intermediates have been reported in the cyclization of enediynes.[70,85] Their involvement in DNA damage may arise through hydrogen atom abstraction from the sugar residue of nucleic acids, although the mechanism used by naturally occurring enediyne antibiotics remains to be proven.[86]

### 7.15.3.2  Neocarzinostatin

Neocarzinostatin (NCS), the first DNA-damaging enediyne antibiotic to be described, belongs not only to a superfamily of antibiotics containing enediyne structures but also to a superfamily of macromolecular protein antibiotics.[44,87,88] It was only after it was appreciated that the active component of the macromolecular complex was a small nonprotein chromophore that it was possible to begin to understand its interaction with DNA in molecular terms. These studies, in conjunction with those on the related calicheamicin/esperamicin agents (see Sections 7.15.3.3 and 7.15.3.4), have uncovered novel mechanisms of DNA damage, wherein a biradical of the active drug acts as a sequence-selective, bistrand-reactive agent through hydrogen abstraction from the sugar residues of DNA.

Neocarzinostatin (holo-NCS), a pale-yellow powder, was isolated from the fermentation broth of *Streptomyces carzinostaticus* var. F-41 from a soil sample collected in Sendai, Japan.[1,89] NCS was initially identified as a simple protein (MW = 11 100)[1,2] and later shown to be a chromoprotein, which is composed of an apoprotein (apo-NCS, 113 amino acid residues) and a tightly and specifically bound ($K_d = 10^{-9}$ M), methanol-extractable nonprotein chromophore (NCS-chrom (**1**)).[4-8,90,91] NCS-chrom (**1**), the biologically active component of NCS, is an extremely labile chromophore sensitive to light, heat, pH, and nucleophilic and reducing agents after extraction from NCS ($t_{1/2} = 30$ s at 0 °C, pH 8, aqueous).[4-8,14] The active form of NCS in DNA damage was proposed to be a free radical species long before the structural determination of NCS-chrom (**1**), based on biochemical studies of DNA damage induced by this drug.[9,92] The activation of duplex DNA-damaging NCS and NCS-chrom (**1**) requires cofactors such as a thiol or borohydride.[93]

NCS blocks replicon initiation selectively,[94] causing inhibition of DNA synthesis, and induces DNA degradation in cells.[3,93] NCS has seen limited clinical use, mainly in Japan, because of limited effectiveness and significant toxicity, especially allergic reactions due to the administration of a foreign protein, apo-NCS. It has been used against acute leukemia and certain solid tumors (stomach, colon, kidney, and bladder) with variable results.[95-97] Through its bioconjugation[98-100] to raise the selectivity of its toxicity, the poly(styrene-co-maleic-acid)–NCS conjugate (SMANCS)[101] is attracting significant clinical interest.

#### 7.15.3.2.1  *Structure*

The structural determination of NCS-chrom (**1**) was accomplished through the combined efforts of several research groups, using state-of-the-art instrumentation such as bioassay, UV, c.d., MS, NMR, chemical synthesis, biosynthetic studies, and molecular modeling.[12-17,50,102-107]

The partial structure of NCS-chrom (**1**) was first proposed by Hensens *et al.* in 1983[16] and consisted of three main subunits that are important in its ability to complex with DNA: a 2-hydroxy-5-methyl-7-methoxy-1-naphthoate (initially identified as a 2-hydroxy-7-methyl-5-methoxy-1-naphthoate,[12] later revised by Shikuya *et al.*'s chemical synthesis[13]), a 2,6-dideoxy-2-methyl-aminogalactose[102] (*N*-methylfucosamine), and an interconnecting carbon-rich $C_{12}$-subunit bearing a cyclic ethylene carbonate and an epoxide. No evidence was found for a diacyl peroxide structure, as previously proposed by Edo *et al.*[108] In 1985, Edo *et al.*[17] proposed that the $C_{12}$-subunit is an unprecedented strained bicyclo[7.3.0]dodecadienediyne with two alkynic bonds in a nine-membered ring with *trans* configuration at C-10 and C-11, based primarily on NMR study of a stable chlorohydrin derivative of NCS-chrom (**1**). This unique bicyclo[7.3.0]dodecadienediyne core structure has been supported by a later biosynthetic study[50] and chemical synthesis.[104-106] The elegant synthesis of the full NCS-chrom (**1**) aglycone by Myers *et al.* represents a landmark in the total synthesis of NCS-chrom.[107] Further, they found that the carbohydrate residue, *N*-methylfucosamine, stabilizes NCS-chrom (**1**) from degradation, as the synthesized aglycone degrades much more rapidly than NCS-chrom (**1**) both in solution and in the neat form.

In 1988, Myers *et al.*[103] deduced the absolute stereochemistry of NCS-chrom (**1**), based on structural determination of postactivated NCS-chrom–methyl thioglycolate adducts, thus completing the structural determination of NCS-chrom (**1**) as shown in Figure 1. The absolute configuration in NCS-chrom (**1**) is as follows: 4*S*, 5*R*, 10*R*, 11*R*, 13*R* with an α-linked D-galactose derivative.[103] The total synthesis of NCS-chrom (**1**) continues to be pursued.[106,107]

The apoprotein of NCS dramatically stabilizes NCS-chrom (**1**). Apo-NCS protects NCS-chrom (**1**) from light, bases, and thiols, such as glutathione.[6,7,109,110] It is interesting that Zein discovered that NCS and apo-NCS have proteolytic activity.[111] A possible role for the apoprotein in drug action is discussed in Section 7.15.4.

### 7.15.3.2.2 *Activation*

NCS-chrom (**1**) is extremely labile under physiological conditions. Activation of NCS-chrom (**1**) can be initiated by light,[109] heat,[112] pH,[4,5,8] radicals,[113,114] and nucleophiles such as thiols and NaBH$_4$.[7,9,93,115]

#### *(i) Thiol-dependent activation*

In addition to biochemical support for involvement of a free radical species in DNA damage,[9] evidence for radical involvement upon activation of NCS by thiol was also observed using ESR spectroscopy.[10,11] Thiol activation of NCS-chrom (**1**) involved adduction of the thiol and marked rearrangement of the NCS-chrom (**1**) structure,[16] in association with major spectral changes.[5] The finding[116,117] that thiol-activated NCS-chrom abstracted $^3$H into the drug from C-5′ of deoxyribose of a thymidylate residue in duplex DNA led to the proposal[117] that nucleophilic addition of thiol to NCS-chrom (**1**) converted the drug into a biradical species responsible for the hydrogen abstraction. The bicyclic dienediyne core and a leaving group (epoxide or halogen group) are required for diyl formation.[118] In 1987, Myers,[68] in assigning the structure of thiol-postactivated NCS-chrom (**11**) adduct from the NMR and MS data of Hensens *et al.*,[16] proposed the mechanism for NCS-chrom (**1**) activation by thiol shown in Scheme 3.

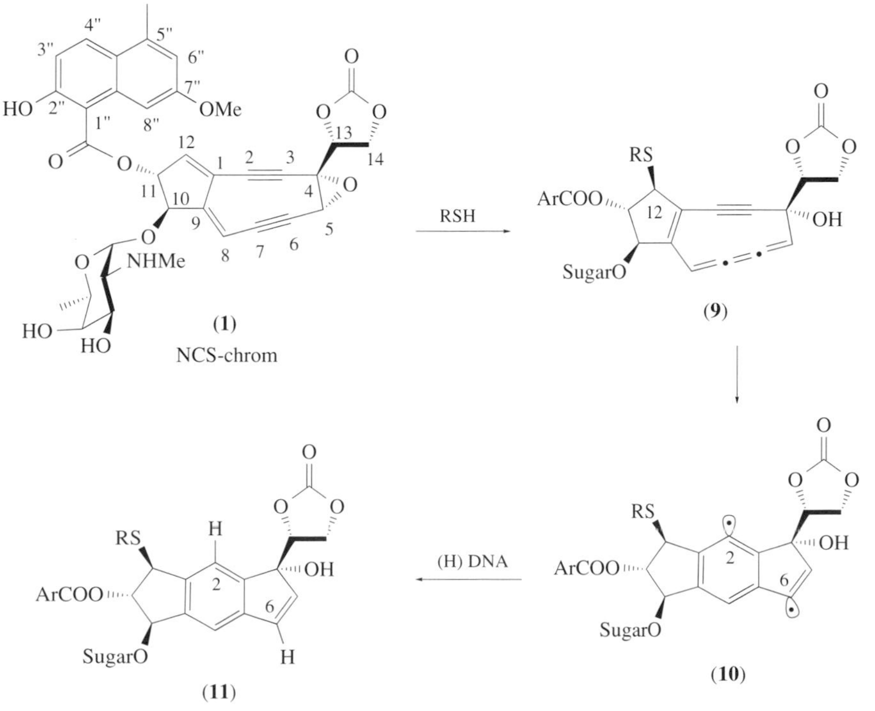

**Scheme 3**

The thiol-dependent activation of NCS-chrom (**1**) involves nucleophilic attack by thiol at C-12 in a *trans* configuration[119] to the naphthoate at C-11 and epoxide opening at C-5 to generate the nine-membered ring enynecumulene (**9**), which cyclizes between C-3 and C-7 to form a 2,6-indacene biradical (**10**).[68,103] The biradical abstracts hydrogen atoms from hydrogen sources such as solvent and/or DNA to give postactivated thiol–NCS-chrom adduct tetrahydroindacene (**11**).

The involvement of nucleophilic attack by an ionic thiolate in triggering the activation was supported by the correlation of DNA damage with pH[120–122] and basicity/nucleophilicity of the thiols.[123] The internal participation of the amino group of *N*-methylfucosamine as a base was proposed from experiments in organic solvents[124,125] and supported by the observation that the amino group is located directly above C-12 at a distance of $\sim 5$ Å, or approximately the van der Waals range of a sulfur atom in the crystal structure of NCS.[126] The relevance of the above results to the situation with DNA at physiological conditions remains to be examined, since the otherwise sound chemical rationale may not provide a profound kinetic advantage under these conditions.[123] The contribution from the charge neutralization offered by the ammonium cation (reasonably assuming that the amino group is protonated under physiological conditions), however, should not be ignored.[127,128]

Spectral evidence for the enynecumulene (**9**) has been obtained.[125,129] The formation of enyne-cumulene (**9**) from (**1**) was pseudo-first-order ($t_{1/2} = 1.5$ h ($-70\,^{\circ}$C, 0.2 M methyl thioglycolate), $k_1 = (1.2 \pm 0.1) \times 10^{-4}$ s$^{-1}$). The half-life of enynecumulene (**9**) was estimated to be $\sim 0.5$ s at 37 °C and $\sim 2$ h at $-38\,^{\circ}$C. The indacene biradical (**10**) was assumed to be short-lived, but its hydrogen abstraction rate has not been determined.

Hydrogen abstraction into C-2 and C-6 of tetrahydroindacene (**11**) was supported by deuterium incorporation experiments. Myers *et al.* observed deuterium incorporation into the C-2 and C-6 positions in organic solvents.[103] Chin *et al.* found that in the absence of DNA, deuterium is incorporated from borodeuteride in $D_2O$ into the C-12, C-6, and C-2 positions, whereas in the presence of DNA significant protium was found at C-6 and C-2, together with deuterium at C-12, showing for the first time that the biradical form of an enediyne abstracts two DNA hydrogens, presumably from the deoxyriboses of closely opposed sites on the complementary DNA strands.[130] These experiments provided support for the role of both drug radical centers in bistranded lesion formation. It was also found that, contrary to prevailing opinion, the sulfhydryl hydrogen was not incorporated into the drug radical centers.[130] Selective abstraction of $^2$H from C-5′ of thymidylate into C-6 and C-1′ of cytidylate into C-2 were observed using specific deuterium-labeled oligo-nucleotides under DNA damage conditions.[131,132] These results established that the bioactive form of thiol-activated NCS-chrom (**1**) is 2,6-dehydroindacene biradical (**10**), and also determined the direction in which the drug molecule lies in the minor groove.

It is interesting that in the experiments by Chin *et al.*[130] the solvent contributes exchangeable hydrogen into C-2/C-6, in addition to the hydrogen from the borohydride, which suggests that a polar process exists or the radical at C-2/C-6 is in equilibrium with a 2,6-dipolar resonance form. Further study is warranted, especially because of its implication for DNA damage.

It has been proposed that the rate-determining step in DNA damage by NCS-chrom (**1**) and thiols, such as glutathione (GSH), cysteine (CySH), and methyl thioglycolate (MTG), under physiological conditions is thiol addition to NCS-chrom (**1**), via a ternary complex of DNA, thiol, and NCS-chrom (**1**).[125] The basicity/nucleophilicity of the thiol is important in NCS-chrom activation; in addition, affinity of the thiol for DNA may contribute to ternary complex formation.[123] The active DNA binding species is enynecumulene (**9**),[125] which determines the sequence specificity of DNA damage and undergoes cycloaromatization while bound to DNA. Myers *et al.* suggested that the 2,6-dehydroindacene biradical (**10**) is a highly reactive and poorly selective intermediate (for related topics, see Section 7.15.3.7.3). The yield-determining step in the formation of postactivated thiol–NCS-chrom adduct (**11**) from NCS-chrom (**1**) and thiol is the quenching of the biradical intermediate by hydrogen atom transfer.[125]

Chin and Goldberg found that the nonexchangeable α-hydrogen (on the carbon α to the sulfur) (and to a lesser extent the β-hydrogen) of γ-L-glutamyl-DL-cysteinylglycine can be incorporated into the C-2 position in the postactivated thiol–NCS-chrom adduct, but not from the D-form of the peptide.[133,134] These results suggest that the steric geometry of the thiol, which donates the hydrogen, is relatively fixed after nucleophilic addition, thus favoring an internal 1,5-hydrogen shift from the attached L-form thiol at C-12, as shown in Scheme 4. The major non-DNA hydrogen source, however, was found to be the carbon-bound hydrogen of solvent methanol.[133] A similar intra-molecular hydrogen atom transfer had been observed by Wender and Tebbe using a monocyclic analogue of NCS-chrom (**1**).[135] Consistent with these studies are experiments by McAfee and Ashley, showing a deuterium isotope effect on DNA damage, in which NCS-chrom (**1**) is activated by

[2-$^2$H$_2$]thioglycolate.[136] It is also possible, although not studied, that a 1,3-hydrogen shift is involved in the standard thiol-dependent activation mechanism and leads to radical quenching at C-2 (see below).

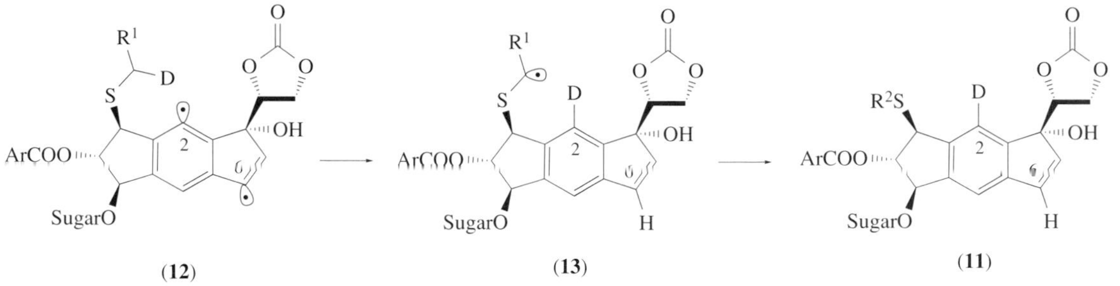

**(12)** **(13)** **(11)**

**Scheme 4**

Sugiyama *et al.* in 1994 reported that the reaction of NCS-chrom (**1**) with 4-hydroxythiophenol (HTP) gave an intramolecular cyclized product as the major product.[137] The reaction involves thiol addition to generate 2,6-dehydroindacene biradical (**14**), followed by intramolecular C-2 radical addition to the phenyl ring of the adducted thiol at C-12, furnishing the intramolecular cyclized product (**15**), as shown in Scheme 5. These results account for the observation that double-stranded lesion formation induced by NCS-chrom (**1**) with HTP is very inefficient compared with that using other nonaromatic thiols such as GSH.[123] Single-stranded lesion formation due to action of the radical at C-6 is unaffected.[123]

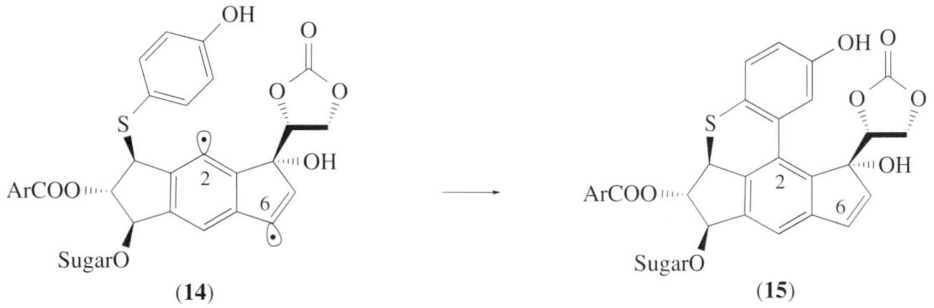

**(14)** **(15)**

**Scheme 5**

Hirama and co-workers observed that the aerobic treatment of NCS-chrom (**1**) with very low concentrations of thiol MTG ($1.5 \times 10^{-5}$ M, 3 equiv. to NCS-chrom (**1**)) in methanolic acetic acid leads to the formation of the indacene-12-one derivative (**18**).[138] A plausible mechanism was proposed, as shown in Scheme 6, which involves hydroperoxy radical (generated by thiol and dioxygen) attack at C-12, followed by intramolecular 1,3-hydrogen shift of the benzylic hydrogen at C-12 to C-2. This is accompanied by hydroperoxide homolysis leading to the indacene-12-one derivative (**18**), as supported by the deuteration experiments. These observations suggest that the cycloaromatization of NCS-chrom (**1**) triggered by hydroperoxy radical might become important under physiological conditions when nucleophilic thiols are insufficient.[138]

The relative ease of intramolecular hydrogen atom transfer to the C-2 radical of 2,6-dehydro-indacene biradical (**10**), as discussed above, could account for the predominance of single-strand over double-strand DNA lesions by NCS.

*(ii) Acid-induced activation*

Although thiol (or borohydride) stimulates DNA damage more than 1000-fold, NCS-chrom (**1**) cleaves duplex DNA weakly in its absence, and this reaction is favored under acidic conditions,[122] suggesting that the initiating event is the acid-induced opening of the epoxide, which facilitates weak nucleophile, such as water, attack at C-12.

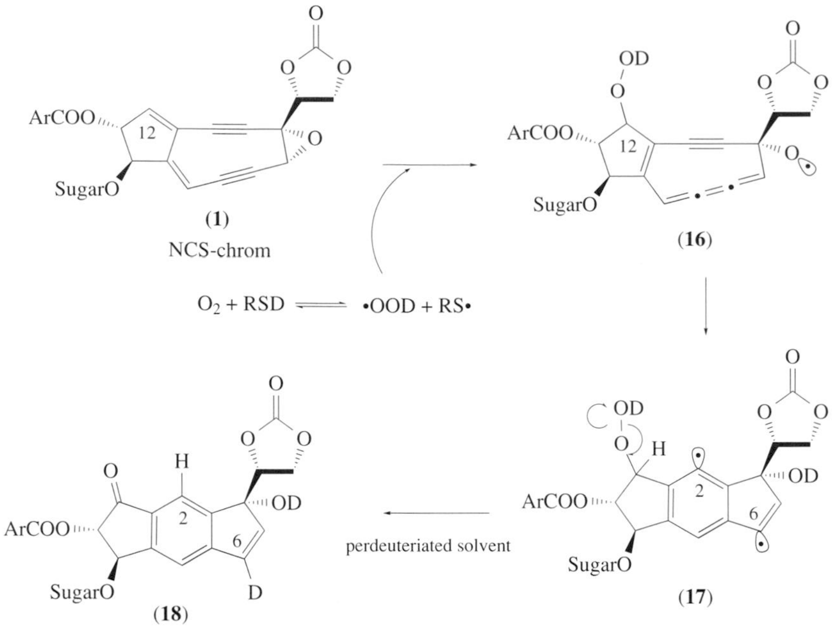

**Scheme 6**

### (iii)  Light-induced activation

NCS-chrom (**1**) is known to be labile even toward ambient light in the absence of cofactors such as thiol.[139] Apo-NCS protects NCS-chrom (**1**) from light. UV irradiation (245 nm) of NCS-chrom (**1**) causes DNA damage.[109]

Gomibuchi and Hirama proposed a photoactivation mechanism, based on the isolation of photo-activated drug products in the absence of DNA.[140] This mechanism, shown in Scheme 7, involves Norrish type II fragmentation and Fuchs-type naphthoate participation[141] through excitation of the naphthoate moiety of NCS-chrom (**1**). The correlation of the proposed mechanism with photo-activated DNA damage by NCS-chrom (**1**) remains to be seen. The existence of the ketene acetal intermediate (**22**) and its binding ability to DNA need to be examined.

### (iv)  Based-catalyzed activation

It has been known for years that in the absence of thiol, duplex DNA is not a target for damage by NCS-chrom, at least not an efficient one (see Section 7.15.3.2.2(ii)), and that the drug decomposes very quickly under basic or neutral conditions when not complexed with DNA or apoprotein.[5,8,44,122]

In 1993, Kappen and Goldberg discovered that bulged DNA can be highly efficient and specifically cleaved by NCS-chrom (**1**) at the bulge site in the absence of thiol at pH ≥ 6.[142,143] The cleavage was restricted to a target nucleotide at the 3′ side of the bulged DNA and was entirely due to 5′ chemistry. Later, it was found that HIV TAR RNA can also be cleaved in a similar fashion, albeit much less efficiently.[144] Based on structural determination of the postactivated NCS-chrom drug product in the presence and absence of bulged DNA, a base-catalyzed mechanism of NCS-chrom (**1**) was proposed, as shown in Scheme 8.[145,146]

In this mechanism, the spirolactone cumulene (**25**) is generated stereospecifically via a general base-catalyzed intramolecular Michael addition at C-12 by the enolate anion (**1b**), which is a resonance form of the naphtholate anion (**1a**) of NCS-chrom (**1**), resulting in the formation of the 2,6-dehydroindacene biradical (**26**). This series of reactions occurs spontaneously, and in the absence of bulged DNA the biradical is quenched by other hydrogen sources, such as methanol in the solvent, to yield spirolactones (**27**). It is very interesting that the cyclic spirolactone (**29**) is only generated in the presence of substrate bulged DNA. Hence, it has been proposed that in the presence of bulged DNA, the spirolactone cumulene (**25**), which has been implicated as the binding species that searches for the favored DNA binding site, is in equilibrium between bound and free forms,

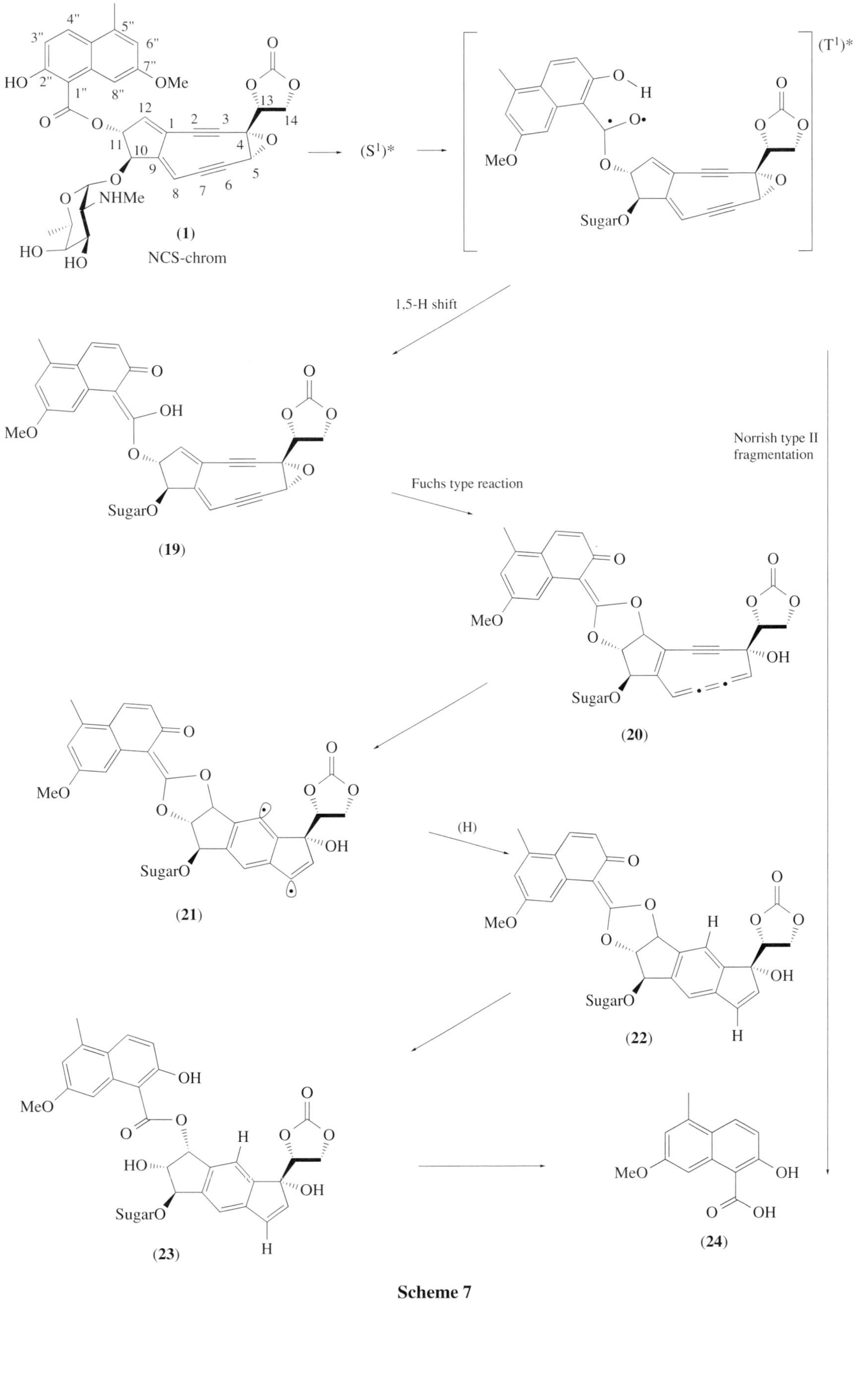
NCS-chrom
(S1)*
(T1)*
1,5-H shift
Fuchs type reaction
Norrish type II fragmentation
(H)
MeO
SugarO
OH
(1)
(19)
(20)
(21)
(22)
(23)
(24)

**Scheme 7**

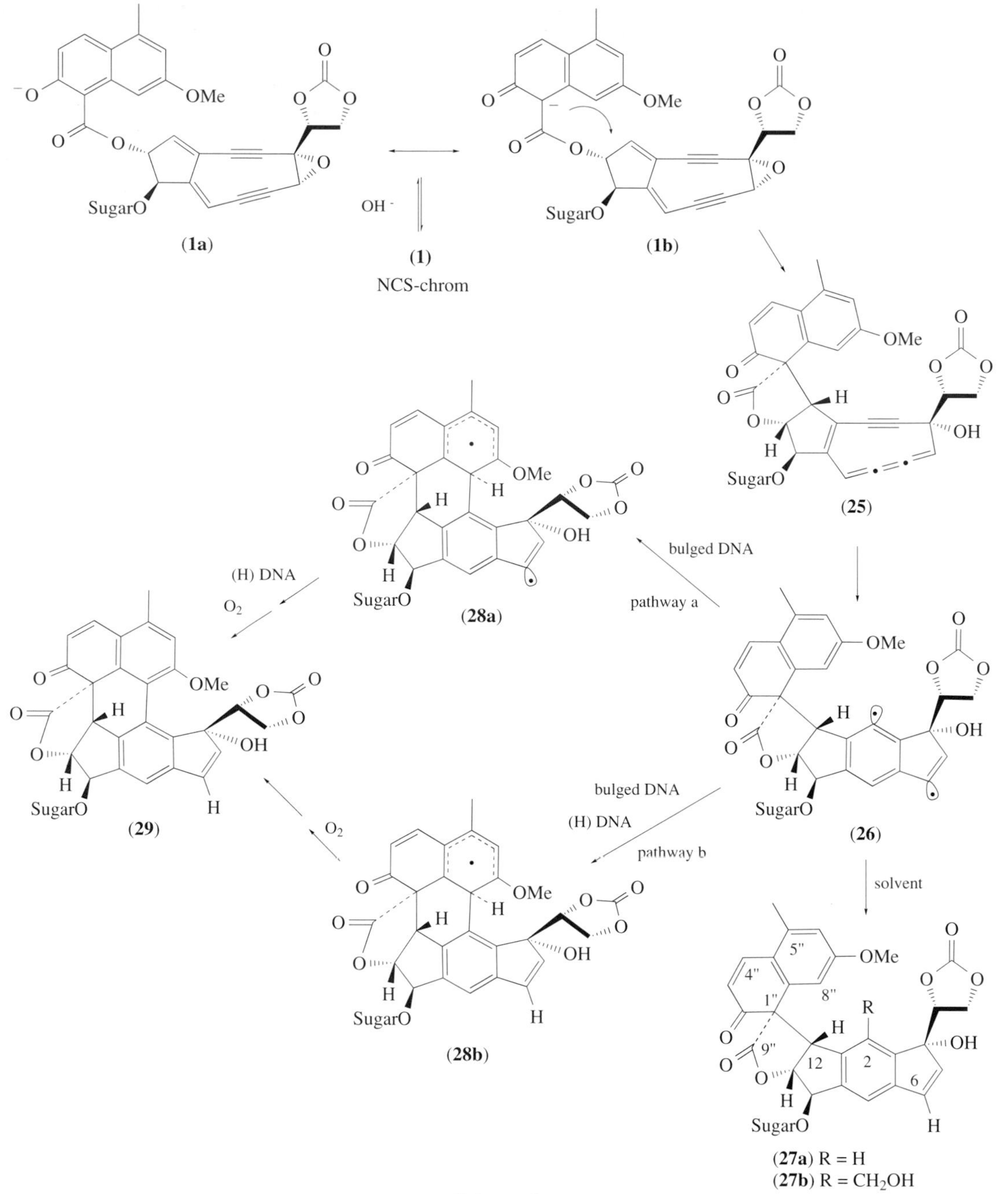

**Scheme 8**

which lead to cyclic spirolactone (**29**) and to spirolactones (**27**), respectively, via spirolactone biradical (**26**). It is not certain whether quenching of the C-2 radical occurs before (pathway a) or after (pathway b) (or concertedly with) abstraction of the hydrogen atom from the DNA. While kinetic factors might favor the former mechanism, the fact that the isostructural analogue of (**26**), (**27a**), binds the bulged DNA ($K_d \approx 0.1\ \mu M$), whereas (**29**), and presumably (**28**), do not,[147] suggest that (**26**) is the DNA-abstracting species. Pathway b is also consistent with a solution structure of the (**27a**)-bulged DNA complex in which the DNA hydrogen abstracting radical at C-6 is only 2.2 Å from the target H-5′.[148]

It is striking that the cyclospirolactone (**29**) is the predominant product (90% yield) in the presence of bulged DNA, whereas it is barely detected (<5% yield) in its absence. This implies that bulged DNA acts as a catalyst to promote cyclospirolactone (**29**) formation, either through the exclusion of hydrogen sources (in particular, solvent) from accessing the C-2 radical in (**26**) or through the induction of a conformation change in (**26**) (thus lowering the transition-state energy favoring formation of cyclospirolactone (**29**)). Recent studies of solvent effects provide support for the involvement of both mechanisms, with the latter predominating.[149]

NMR studies on the complex of the spirolactone (**27a**) and a two-base bulged DNA reveal that the spirolactone (**27a**) functions as a double-decker intercalator (connected by the spirolactone ring) with the tetrahydroindacene moiety stacking with the base pair above and the naphthoate moiety stacking with the base pair below the bulge.[148] The wedge-shaped drug fits tightly into the triangular prism pocket formed by the two looped-out bases and the neighboring base pairs via the major groove. The drug carbohydrate sits at the center of the major groove, limiting further drug penetration. The two drug rings mimic helical DNA bases and complement the bent DNA structure. This structure provides insight into the design of bulge specific DNA/RNA binding molecules. Further, point mutation experiments with RNA/DNA bulged substrates showed that the presence of a 2'-OH group on A (*ribo*-A) of A·T 5' to the bulge significantly inhibited damage at the bulge and (**27a**) binding, probably through a steric clash between the 2'-OH of the A and the 7-methoxy of the naphthoate moiety.[150] These studies suggest that deletion of the 7-methoxy moiety would lead to better binding. The base-catalyzed reaction has also been used to probe for bulge sites in long, single-stranded viral DNAs.[151]

### (v) Apo-NCS directed thiol activation

In 1992, Saito and co-workers reported that when holo-NCS was activated by 2-thioethanol (BME) in the absence of DNA, a new product was formed that is not (**11**) and is formally a 1:1:1 adduct of NCS-chrom (**1**), BME, and water (59% yield).[152] Based on the assignment of structure of the adduct as (**33**), a mechanism was proposed as shown in Scheme 9, which involves an unusual alternative rearrangement of NCS-chrom (**1**), resulting in a 3,7-zwitterionic intermediate or polar pathway.

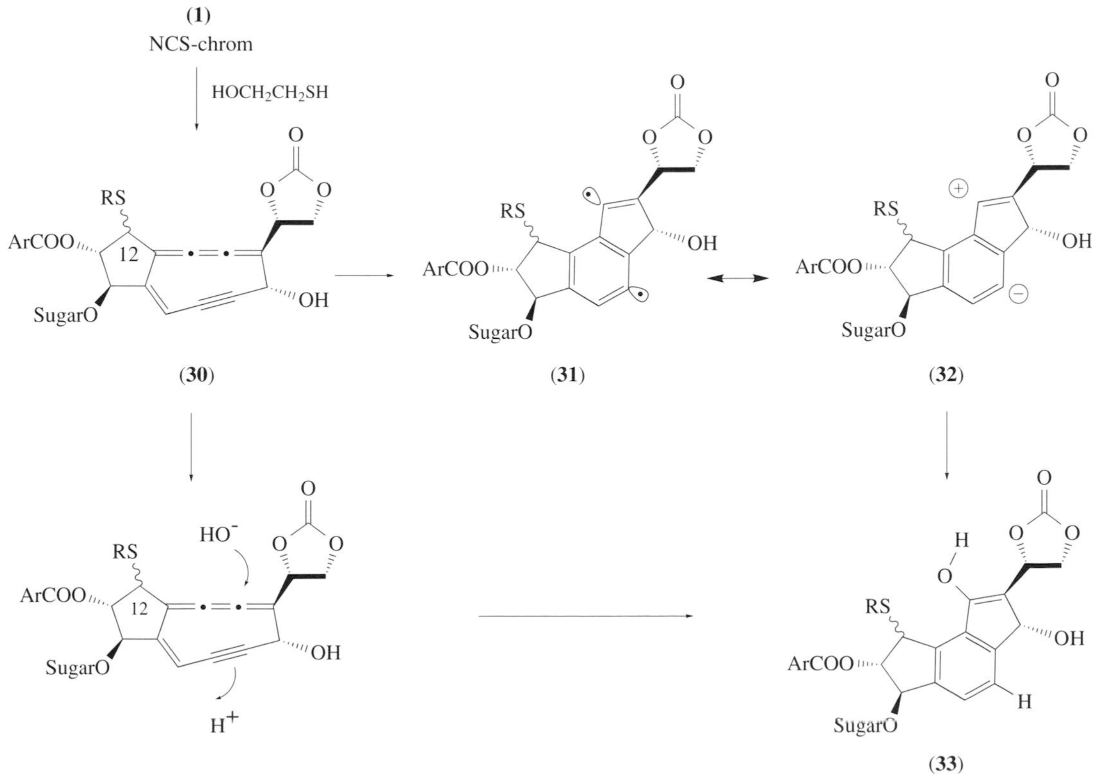

**Scheme 9**

In 1993, Chin and Goldberg pointed out that the above pathway (Scheme 9) does not lead to DNA cleavage, and only small thiols such as BME, dithiothreitol (DTT), and L-cysteine, but not GSH, gave the apo-NCS-directed thiol-activation product.[110] They suggested that the apo-NCS-bound NCS-chrom (**1**) is shielded from GSH and it is the free NCS-chrom (**1**) in equilibrium with the bound form that undergoes conversion to the duplex DNA bioactive form (**10**). They also called into question structure (**33**) in Scheme 9, as the enol proton assigned by Saito and co-workers did not exchange with solvent water in a reasonable time. This was disputed later by Saito and co-workers by invoking of an unusually strong hydrogen bond with the carbonate group.[153]

In 1996, Myers *et al.*, based on studies of model compounds and Saito and co-workers' compound (**33**), in fact, reassigned the structure of compound (**33**) in Scheme 9 as hydroxyisochromene (**37**) in Scheme 10 and proposed an apo-NCS-directed thiol-activation mechanism as shown in Scheme 10.[154] This mechanism involves a transannular nucleophilic addition to the triple bond, followed by protonation of the resultant vinyl carbanion to furnish a substituted α,3-dehydrotoluene biradical (**35**). This is followed by fragmentation of the α-epoxy radical (**35**) yielding another benzylic radical, i.e., α,2-dehydrotoluene biradical (**36**), which reacts with water in a polar process to produce hydroxyisochromene (**37**).

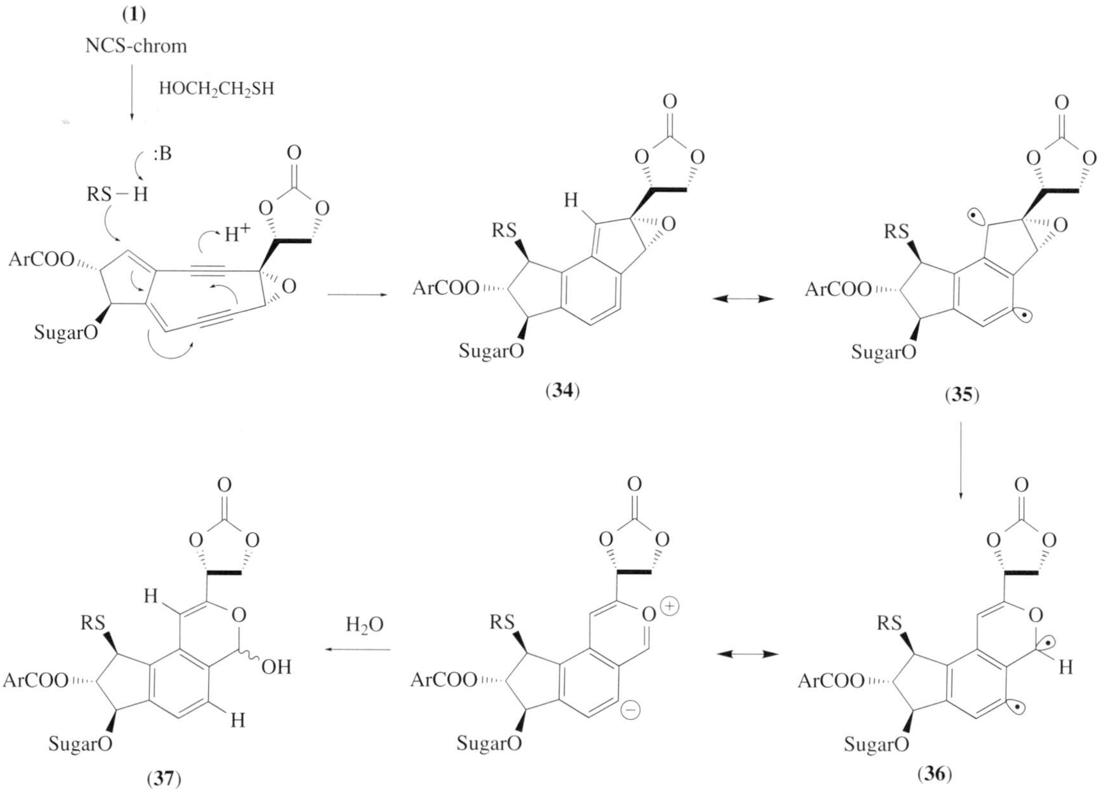

**Scheme 10**

It is interesting that Chin found that hexose thiols produce only tetrahydroindacene product (**11**), instead of the hydroxyisochromene (**37**), from the apoprotein-bound chromophore.[155] The experimental data suggest that this reaction is not due to dissociation of the chromophore from apoprotein and that it occurs while NCS-chrom is still bound in the protein pocket.[155]

As Myers *et al.* pointed out,[154] the protein-directed inactivation of NCS-chrom by small thiols may provide a self-protective mechanism for the organism that produces NCS (*Streptomyces carzinostaticus*).

### 7.15.3.2.3  DNA damage

NCS and its chromophore have been shown to damage duplex DNA, bulged DNA, RNA/DNA hybrids, and RNA.[88,156] Most DNA damage reactions studied need cofactors such as thiols[44] or photoactivation.[109] Bulged DNA and TAR RNA, however, were cleaved by base-catalyzed activation of NCS-chrom (**1**) (discussed in Section 7.15.3.2.2(iv)).[142-144,151]

Early studies on *in vitro* DNA scission revealed that NCS-chrom (**1**) produced primarily single-stranded (SS) breaks with base selectivity, but not sequence selectivity.[157–159] About 75% of breaks were at T residues (T > A ≫ C > G). Whereas SS lesions occur mainly at breaks due to 5' chemistry or to a much lesser degree at abasic sites due to 4' chemistry, double-stranded (DS) lesions (DS break or abasic site with a closely opposed SS break) involve a mixture of chemistry (5' and 1' or 4') at a staggered lesion site (two residues apart). DS lesions are sequence-specific (either AGC·GCT or AGT·ACT) (attack sites underlined).[160–163] DS lesions at AGC·GCT sites consist of an apyrimidinic abasic site due to 2-deoxyribonolactone formation (1' chemistry) at the C residue of AGC and a break due to 5' nucleoside aldehyde formation (5' chemistry) at the T residue two nucleosides to the 3' side on the complementary strand.[160–163] Detailed analysis of the DS lesions induced by GSH-activated NCS-chrom (**1**) at a model AGT·ACT in the AP-1 transcription factor binding site showed that 89% of the DS lesions at the T of AGT were due to C-4' hydrogen abstraction and 11% to C-5' hydrogen abstraction.[164] It was reported that 74% of a 4' chemistry was in the form of the 4'-hydroxylated abasic lesion and the remainder in a 3'-phosphoglycolate-ended fragment. The break at the T of ACT on the complementary strand was more than 90% due to 5' chemistry, generating mainly 5'-nucleoside aldehyde. Deuterium abstraction experiments,[131,132] molecular modeling studies,[165] and NMR-based solution structure of the complex formed between the GSH-postactivated drug adduct and an AGC·GCT site-containing oligonucleotide[127,128] indicate that the DS lesions result from the concerted action of a single drug on both strands. The radical center at C-2 of the drug is involved in attack at the C of AGC (and the T of AGT), while that at C-6 abstracts hydrogen from the T two bases to the 3' side on the complementary strand. The C-6 radical is involved in both SS and DS lesions, whereas that at C-2 is involved only in DS lesions.[163,164] Deuteriation and point mutations in DS lesion sites have been found dramatically to alter the chemistry of DNA damage.[166–168] These findings provide support for the role of minor groove microstructure in determining the chemical mechanism of DNA damage and underscore the usefulness of NCS-chrom (**1**) as a probe of DNA microheterogeneity. That sequence is not the sole determinant of local geometry is shown by the finding of selective strand scission opposite sites of single-base bulges, independent of the specific sequence.[169]

Thiol-activated NCS-chrom (**1**) has also been found to generate staggered (two nucleotides in the 3'-direction) DS lesions in DNA–RNA hybrids, involving 5' chemistry on the DNA strand and 1' chemistry on the RNA strand. The strong deuterium isotope effect on H-1' abstraction at the damaged uridylate confirmed that the abasic site formation was due to 1' chemistry.[170]

NCS-chrom (**1**) has also been found to cleave nucleosome DNA with primary damage in the linker DNA.[171]

NCS-chrom (**1**) binds DNA reversibly, with a dissociation rate of 0.75 s$^{-1}$.[172] Drug binding to duplex DNA is a two-step process. External binding is followed by intercalation of the chromophore between adjacent base pairs.[173] Drug binding causes the DNA helix to unwind by 21° and lengthen by 3.3 Å for each bound NCS-chrom (**1**) molecule.[174] The naphthoate group is the intercalating moiety. Minor-groove binding was inferred, since bulky groups in the major groove failed to interfere with drug binding, whereas minor-groove binders such as netropsin and distamycin did,[172] especially at A/T-rich regions which are the favored binding sites of these minor-groove binders. The microheterogeneity in the B-DNA due to sequence determines the nature of NCS-chrom (**1**) binding.[175] The 2-amino group of G in the minor groove of DNA is important in sequence-specific chromophore binding. NCS-chrom (**1**) binds duplex DNA with $K_d \approx 10^{-6}$ M with an overall preference for T- and A-rich regions of DNA.[5] A drug–DNA binding model was proposed in which the active portion (bicyclo[7.3.0]dodecadienediyne) of the chromophore is positioned in the minor groove of DNA, with which it makes important hydrophobic and other contacts, by intercalation of the naphthoate moiety and electrostatic interaction of the positively charged amino sugar moiety due to the negative potential of the minor groove.[121] The enynecumulene species (**9**) is likely to be the effective binding species.[125] Cycloaromatization of the enynecumulene (**9**) to the diyl (**10**) after binding to the DNA leads to hydrogen atom abstraction from deoxyribose carbons—C-5', C-1', and C-4'—strategically situated in the minor groove near the radical centers on NCS-chrom (**1**). Molecular modeling,[165] binding and cleavage studies using single-site-containing oligonucleotides,[176] and NMR studies[127,128] also support the above proposal.

In the NMR-derived model, the naphthoate moiety intercalates at the 5'-AG/5'-CT step, while the carbohydrate unit and the reduced enediyne ring, forming a right-handed hydrophobic surface, fit into the curvature of the right-handed DNA in the minor groove. The formation of the complex is accompanied by a significant conformational change of the DNA, which unwinds and stretches at the intercalation site. The intermolecular contacts are dominated by van der Waals contacts, while the sequence specificity of the binding has been attributed to two sets of specific interactions:

hydrophobic interactions between the drug and the amino groups of G residues in the floor of the minor groove and drug carbohydrate unit recognition of the T·A base pair as the site for inter-calation of the naphthoate moiety.

### 7.15.3.3　Calicheamicins

The calicheamicins are produced by the fermentation of *Micromonospora echinospora* ssp. *cal-ichensis* from a chalky soil sample collected in Texas. The isolated calicheamicins are named after their relative chromatographic mobility and structural features. Optimization of strain and fermentation conditions yielded calicheamicin $\gamma_1^1$ **(2)**,[177] which is over 1000 times more effective against the murine tumor in mice models than adriamycin, one of the most clinically effective antitumor antibiotics.[178] About 20 calicheamicins were isolated, of which seven have been struc-turally identified (Figure 2). Calicheamicin $\gamma_1^1$ **(2)** will be used here as the representative of the calicheamicin family, since most of the reported work is related to this species.

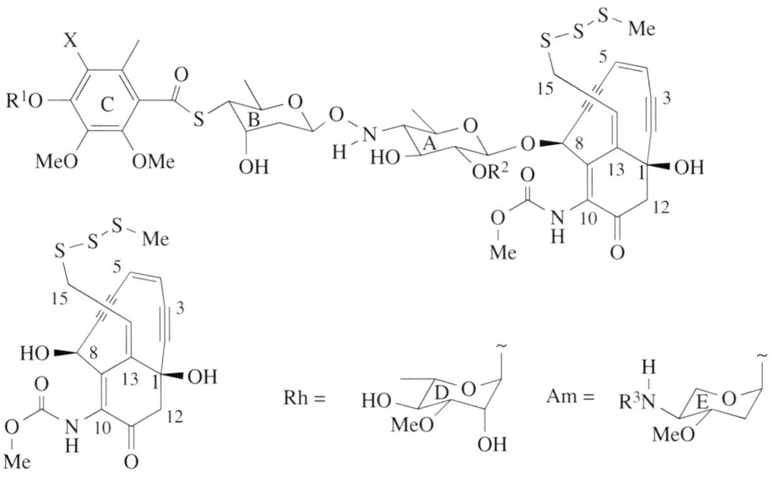

(**38**) Calicheamicinone

| Calicheamicin | X | $R^1$ | $R^2$ | $R^3$ |
|---|---|---|---|---|
| $\beta_1^{Br}$ | Br | Rh | Am | Me$_2$CH- |
| $\gamma_1^{Br}$ | Br | Rh | Am | Et- |
| $\alpha_2^{I}$ | I | H | Am | Et- |
| $\alpha_3^{I}$ | I | Rh | H | |
| $\beta_1^{I}$ | I | Rh | Am | Me$_2$CH- |
| $\gamma_1^{I}$ (**2**) | I | Rh | Am | Et- |
| $\delta_1^{I}$ | I | Rh | Am | Me- |

**Figure 2**　Chemical structures of calicheamicins.

### *7.15.3.3.1　Structure*

In 1987, Lee and co-workers disclosed their discovery of the calicheamicins and the structure of calicheamicin $\gamma_1^1$ **(2)**.[18,19,179] The great interest generated by the discovery of the calicheamicins, together with the esperamicins (see Section 7.15.3.4), comes not only from the structural uniqueness of these natural products, but also from the intriguing relationship between the aromatization of their enediyne moieties to that of the Bergman reaction.[56] Based on the earlier neocarzinostatin work, it followed naturally that DNA damage by a biradical would be expected to account for the biological activity of the calicheamicins.[117]

The structure of calicheamicin $\gamma_1^I$ (**2**) was determined by a combination of degradation studies and spectroscopic methods. The original structure has been revised at C-8 of the aglycone.[179] The stereochemistry shown in calicheamicinone (**38**) in Figure 2 was also supported by the X-ray structure of esperamicin X.[20]

Calicheamicin $\gamma_1^I$ (**2**) is composed of two main units: an aryloligosaccharide segment with a unique N—O glycosidic linkage and an unprecedented bicyclo[7.3.1]tridec-4-ene-2,6-diyne aglycone, namely calicheamicinone (**38**), containing a methyl trisulfide moiety and a bridgehead enone. The aryloligosaccharide segment consists of four glycosidic units and a fully substituted iodo-thiobenzoate (C ring, 2,3-methoxy-4-oxy-5-iodo-6-methylbenzoyl moiety). The four sugar units are 4,6-dideoxy-4-(hydroxyamino)-$\beta$-D-glucopyranoside (A ring), 2,6-dideoxy-4-$S$-thio-$\beta$-D-ribo-hexopyranoside (B ring), 3-$O$-methyl-$\alpha$-L-rhamnopyranoside (D ring), and 2,4-dideoxy-4-($N$-ethylamino)-3-$O$-methyl-$\alpha$-L-xylopyranoside. The unique N—O linkage between ring A and ring B was verified by X-ray crystallography.[179] The structure of calicheamicin $\gamma_1^I$ (**2**) and its absolute stereochemistry were confirmed by the elegant total synthesis by Nicolaou and co-workers in 1992.[22,38,180]

Calicheamicin $\gamma_1^I$ (**2**) is fairly soluble in organic solvents such as methanol, DMSO, methylene chloride, and THF, but insoluble in water. Its aqueous solubility can be dramatically improved in the presence of DNA.

### 7.15.3.3.2 *Activation*

The relationship of the activated species of the enediyne antibiotic calicheamicin to Bergman's 1,4-dehydrobenzene biradical and its DNA-damaging potential provide the basis for its scientific interest. A proposed mechanism is shown in Scheme 11.[45]

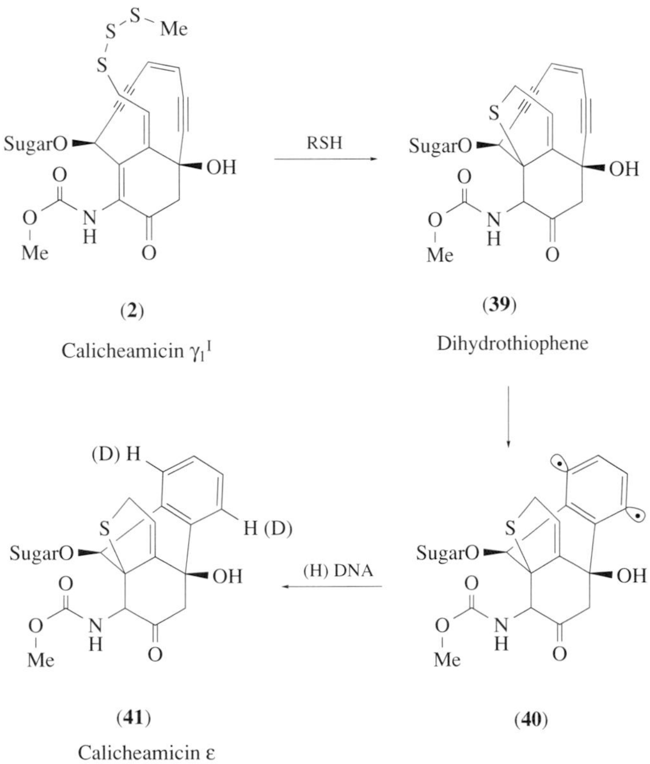

(2)

Calicheamicin $\gamma_1^I$

(39)

Dihydrothiophene

(41)

Calicheamicin $\epsilon$

(40)

**Scheme 11**

Activation involves nucleophilic attack at the allylic trisulfide by thiol or thiophile, resulting in an allylic thiolate or thiol, which triggers an intramolecular Michael addition on the $\alpha,\beta$-unsaturated ketone to give the dihydrothiophene derivative (39). This changes C-9 to $sp^3$ from $sp^2$ and thus drives the 3-ene-1,5-diyne core to the Bergman reaction yielding 1,4-dehydrobenzene biradical (40). The biradical abstracts hydrogen atoms from sugar residues of DNA or solvent to give calicheamicin $\varepsilon$ (41).

The biradical intermediate has been supported by trapping with nonexchangeable deuteriated organic solvent to give a 1,4-dideuteriated benzene moiety in calicheamicin $\varepsilon$ (41).[19] Deuteriation experiments involving DNA damage showed that the bioactive form is a biradical.[181–183] The existence of dihydrothiophene (39) has been observed by low-temperature NMR. The relatively long solution half-life of dihydrothiophene (39) ($t_{1/2} = 4.5 \pm 1.5$ s, 37 °C, MeOH) suggests that the dihydrothiophene (39) is the actual species responsible for sequence-specific DNA binding and subsequent DNA damage.[184] The overall rate-determining step in the activation is the nucleophilic attack by thiol. Breakdown of the allylic trisulfide is complex.[184–187]

### 7.15.3.3.3  Trisulfide chemistry

Ellestad *et al.* observed that in the early stage of calicheamicin activation, the initial nucleophilic attack by thiol upon the unsymmetrical allylic trisulfide moiety yields several sulfur-containing enediyne species, with the relatively stable disulfide-containing species as the major product.[185] Detailed study of the reaction of calicheamicin $\gamma_1^I$ (2) with GSH in the presence of DNA by Myers *et al.* revealed that the calicheamicin-GSH disulfide (42) showed sequence specificity identical to that of calicheamicin $\gamma_1^I$ (2) in DNA damage.[187] Of four possible direct S—S bond exchange products between GSH and calicheamicin, three were identified, while the fourth was assigned as the allylic thiol in the form of dihydrothiophene enediyne (39) due to its transient nature. The calicheamicin-GSH disulfide (42) was found to be the major product with a prolonged half-life, while dihydrothiophene (39), the thiosulfenic acid (43), and the calicheamicin-GSH trisulfide (44) were found as minor products (see Scheme 12).

$$\gamma_1^I\text{-SSSMe} + \text{GSH} \longrightarrow \gamma_1^I\text{-S-SG} + \gamma_1^I\text{-SSH} + \gamma_1^I\text{-SS-SG} + \text{dihydrothiophene}$$
$$\quad\quad (2) \quad\quad\quad\quad\quad\quad\quad (42) \quad\quad (43) \quad\quad (44) \quad\quad\quad (39)$$

**Scheme 12**

The trisulfide chemistry, yielding the fairly stable bioactive disulfide calicheamicin $\gamma_1^I$, has provided a good handle for its further modification to form conjugates with monoclonal antibodies.[188]

### 7.15.3.3.4  Role of the oligosaccharide amino group

Ellestad and co-workers[189] observed that derivatives of calicheamicin $\gamma_1^I$ (2) lacking the 4-ethyl-amino sugar exhibited a DNA cutting pattern identical to that of calicheamicin $\gamma_1^I$ (2), but were 2–3 orders magnitude less efficient, and that acetylation of the 4-ethylamino group in calicheamicin $\gamma_1^I$ (2) also lowered the cleavage efficiency but not the specificity. They concluded that "the basicity of the 4-ethylamino sugar is important in the cleaving efficiency, most likely because of its catalytic role in the activation of the trisulfide group."[189] They also observed that derivatives of calicheamicin $\gamma_1^I$ (2) lacking the amino group are unreactive toward thiols in organic solvents in the absence of added triethylamine, whereas calicheamicin $\gamma_1^I$ (2) itself requires no added amine for thiol activation.[185] However, no solid evidence has been found that the amino group offers significant kinetic advantage in DNA damage under physiological conditions.[187,190]

### 7.15.3.3.5  DNA damage

Calicheamicin $\gamma_1^I$ (2) cleaves DNA sequence specifically with predominantly DS lesions.[189,191,192] The frequency of bases attacked is C $\gg$ T > A $\approx$ G. Dedon *et al.* observed bistranded DNA damage to the virtual exclusion of SS lesions in plasmid DNA.[193] Without expression of drug-induced abasic sites as strand breaks, calicheamicin $\gamma_1^I$ (2) activated by 10 mM GSH produced equal numbers of

DS and SS breaks in plasmid DNA. However, cleavage of drug-induced abasic sites with hydrazine or putrescine resulted in a dramatic increase in the appearance of bistranded damage.[193] Tetranucleotide pyrimidine tracts, such as TCCT and CTCT, were found to be prominent binding/cleavage sites, while other sites such as GCCT, TCCG, TCCC, TCTC, ACCT, and TCCA were also cleaved, with the extent dependent upon the sequence and the flanking sequences (the underlined sites are the damage sites).[189,191,192] Studies with an AGGA/TCCT-containing duplex DNA indicated that the DS lesions are staggered by three nucleotides to the 3′-side of the cleavage site. The C-5′ in the C of TCCT or T of CTCT was attacked to give 3′-phosphate and 5′-nucleoside aldehyde due to 5′ chemistry, and C-4′ at the C of AGGACC to give 3′-phosphoglycolate due to 4′ chemistry.[189] Townsend and co-workers demonstrated by deuterium abstraction into the drug that the 5′-hydrogen was abstracted by the C-6 radical, whereas the 4′-hydrogen was abstracted by the C-3 radical of the 3,6-biradical.[182]

Minor-groove binders inhibit the DNA damage by calicheamicins, suggesting that the drug binds DNA in the minor groove.[189] Further, studies showing 5′(*S*)-hydrogen abstraction also place calicheamicin $\gamma_1^I$ (2) in the minor groove of duplex DNA.[183]

Extensive studies have been carried out to reveal the origins of the sequence-selective DS DNA damage by calicheamicin $\gamma_1^I$ (2) through molecular modeling,[194,195] NMR,[196–199] affinity cleavage,[187,190,200–204] physicochemical measurements such as c.d. and salt effects,[205–208] and organic synthesis.[209,210] The association of calicheamicin with specific DNA sequences is dominated by nonionic, hydrophobic forces rather than electrostatic forces.[207] The aryl-tetrasaccharide domain is believed to be most responsible for the sequence-selective binding of calicheamicin $\gamma_1^I$ (2) to duplex DNA.[211] The unusual N—O bond in calicheamicin, with its distinctive torsional preferences, pre-organizes the two halves of the molecule into a shape that complements the shape of the minor-groove binding sites.[196] The iodo group of the aryl-tetrasaccharide interacts with the 2-amino group of 5′-deoxyguanosine, providing stabilization of approximately 1 kcal mol$^{-1}$.[203,204] It is believed that the calicheamicins induce conformational changes in the binding domain of DNA to accommodate the interactions necessary for selectivity and high affinity in a thermodynamic manner, with hydrophobic forces as the major energetic contribution to binding. The tetrapyrimidine tracts in DNA, possessing conformational flexibility, become the choice of the sequence-selective binding domain of the drug. Studies on the sequence-selective recognition domain of the calicheamicins have identified a new type of DNA sequence-selective minor groove binders—the oligosaccharides[210,212,213]—which have made their way into cell biology.[214,215]

It is natural to think that DNA might function as a catalytic template for the activation of an enediyne, although studies on the interaction of duplex DNA with calicheamicins offer no clear evidence of such a process.[187,190]

### 7.15.3.4 Esperamicin

The esperamicins were isolated from the fermentation broth of *Actinomadura verrucosospora* ATCC 39334 from a soil sample collected in Pto, Esperanza, Misiones, Argentina.[216–220] They are among the most potent antibiotics, as are the calicheamicins, with a broad spectrum of antimicrobial and murine antitumor activities.[221] In this family, seven of them have been structurally characterized, as summarized in Figure 3 (the numbering system here is related to the calicheamicins, see Section 7.15.3.3).

#### 7.15.3.4.1 *Structure*

In 1987, Golik and co-workers[20,21] reported the structures of the esperamicins at the same time as Lee *et al.* disclosed the structures of the calicheamicins. The structure of esperamicin X (post-activated form of esperamicin A₁ (3) lacking the trisaccharide moiety) was determined by X ray crystallography. The absolute stereochemistry has also been determined.[222–228] The structure of esperamicin A₁ (3) was also supported by biosynthetic studies.[229]

The core structure of the esperamicins is very close to that of the calicheamicins. The structure of esperamicin A₁ (3) consists of a bicyclo[7.3.1]tridec-4-ene-2,6-diyne aglycone, which contains a methyl trisulfide moiety and a bridgehead enone. The enediyne core of esperamicin is only different at position C-12, having a hydroxy group compared with calicheamicinone (38). To this enediyne core are attached a trisaccharide via a glycosidic linkage at C-8 and a 2-deoxy-L-fucoseanthranilate

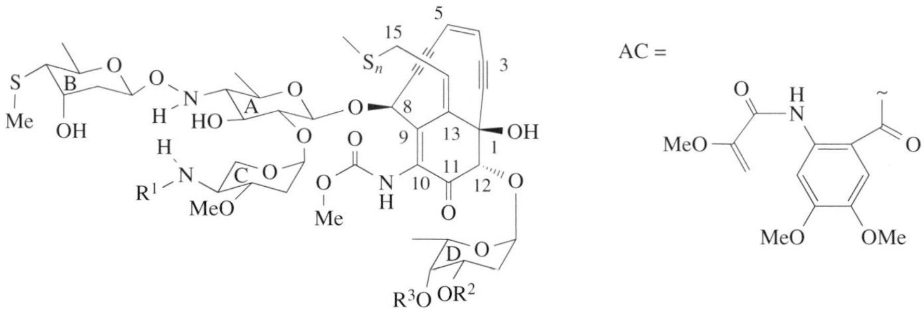

| Esperamicin | $n$ | $R^1$ | $R^2$ | $R^3$ |
|---|---|---|---|---|
| $A_1$ (**3**) | 3 | $(Me)_2CH-$ | AC | H |
| $A_{1b}$ | 3 | Et- | AC | H |
| $A_{1c}$ | 3 | Me | AC | H |
| P | 4 | $(Me)_2CH-$ | AC | H |
| $A_2$ | 3 | $(Me)_2CH-$ | H | AC |
| $A_{2b}$ | 3 | Et- | H | AC |
| $A_{2c}$ | 3 | Me | H | AC |

**Figure 3**   Chemical structures of esperamicins.

moiety via a glycosidic linkage at C-12. The trisaccharide contains an unusual hydroxyamino sugar, 4,6-dideoxy-4-(hydroxyamino)-$\beta$-D-glucopyranoside (A ring), 2,6-dideoxy-4-$S$-methylthio-$\beta$-D-ribohexopyranoside (B ring), and 2,4-dideoxy-4-($N$-isopropylamino)-3-$O$-methyl-$\alpha$-L-xylo-pyranoside (C ring). Ring A connects to ring B via an unusual N—O $\beta$-glycosidic linkage at C-4' and to ring C via an $\alpha$-glycosidic linkage at C-2'. Esperamicin C is the form of esperamicin $A_1$ (**3**) lacking the deoxyfucoseanthranilate moiety at C-12.

### 7.15.3.4.2 *Activation*

Given the structural similarity between the esperamicins and the calicheamicins, they are assumed to have similar activation mechanisms. The proposed activation mechanism of esperamicin, identical to that of calicheamicin (Scheme 11, Section 7.15.3.3.2), is based on experimental results in organic solvents.[21] The bioactive form is assumed to be the 1,4-dehydrobenzene biradical.

It is interesting that esperamicin $A_1$ (**3**) was reported to produce the rearranged trisaccharide (**48**) as the major product, instead of esperamicin Z (**49**), in the presence of calf thymus DNA (Scheme 13).[227,230] The major production of the rearranged trisaccharide (**48**) was suggested to occur by a 1,3-water-mediated hydrogen shift from C-8 to C-6 in the 1,4-dehydrobenzene biradical (Scheme 13, pathway a), and was later suspected to be the degradation product of esperamicin Z (**49**) (Scheme 13, pathway b).[231] It is possible that both pathways contribute to the experimental outcome, as DNA damage experiments with esperamicin $A_1$ (**3**) showed mainly SS lesions (see below). The 1,3-hydrogen shift has been suggested in the activation of NCS-chrom (**1**) by a peroxy radical (see Section 7.15.3.2.2), although a water-mediated hydrogen shift seems unlikely. It is possible that the 1,3-hydrogen shift is highly geometry demanding, i.e., only certain conformers are able to carry it out efficiently. The formation of predominantly SS lesions may be simply the result of the kinetic differentiation of hydrogen quenching by the geometry-defined biradical.

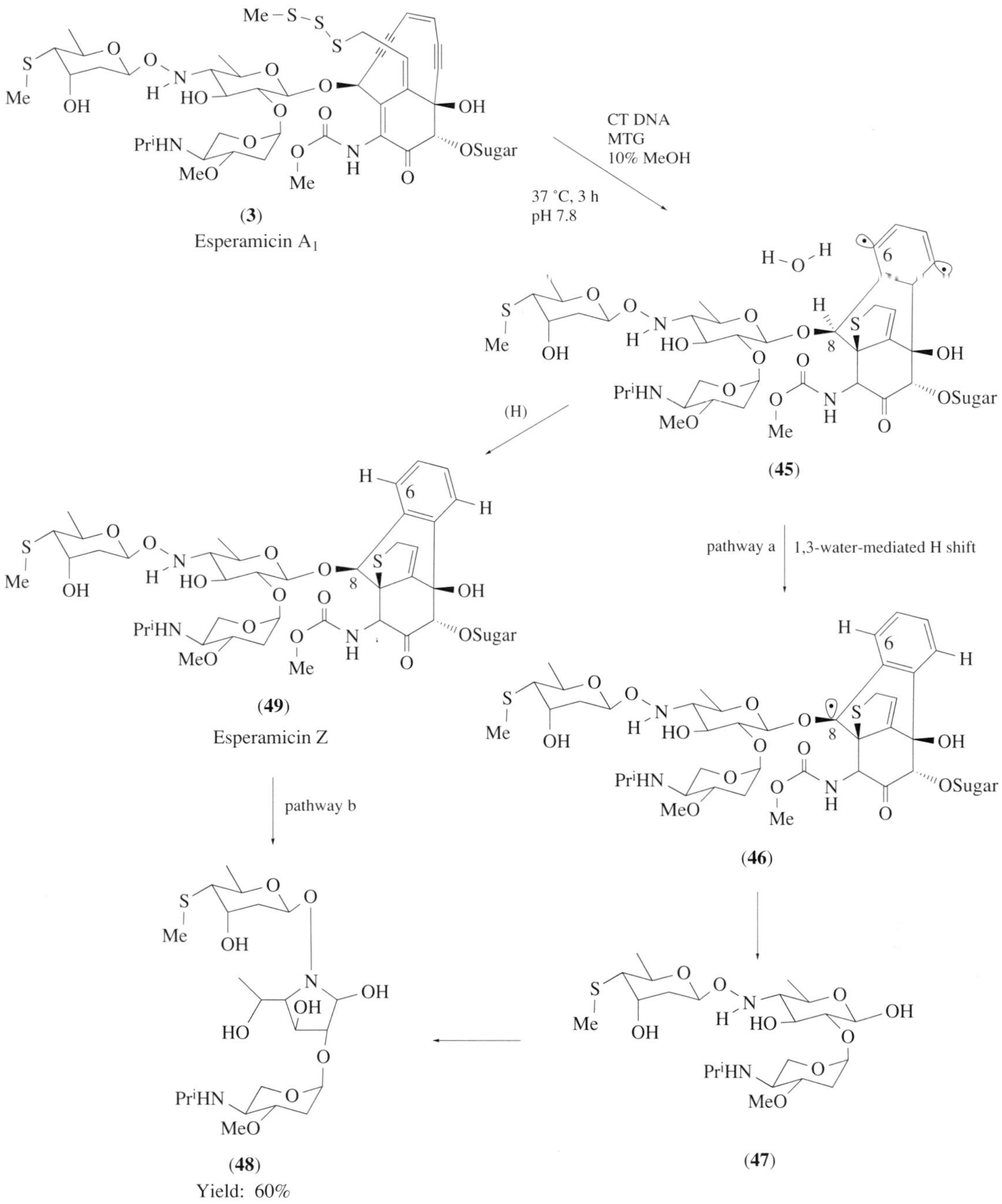

**Scheme 13**

Light- or heat-induced activation of esperamicin, leading to DNA damage, has been observed.[109,232] Several pathways were suggested, including forced cycloaromatization of the aglycone and base-assisted intramolecular participation of the methoxycarbamate group. These proposals await experimental examination.

### 7.15.3.4.3 DNA damage

Earlier studies on DNA damage showed that esperamicin $A_1$ (**3**) produces only SS lesions with poor sequence selectivity (T > C > A > G) via 5′ chemistry.[109,232–237] A cleavage site was often found to be accompanied by another cleavage site staggered three nucleotides to the 3′-side on the complementary DNA strand. This was attributed to bidirectional binding of esperamicin $A_1$ (**3**) to

DNA in the minor groove.[230] Dedon and co-workers, however, recently demonstrated that DS lesions are also produced by esperamicin A₁ (3).[231] DS lesions were shown to be 20–25% of the total damage via 1′ and 5′ chemistry. Esperamicin C, a structurally close analogue of calicheamicin $\gamma^1_1$ (2), lacking the deoxyfucoseanthranilate moiety of esperamicin A₁ (3), was found to produce DS lesions via 4′ and 5′ chemistry, with a similar sequence selectivity to calicheamicin $\gamma^1_1$ (2), albeit with lower efficiency.

Whether the radical at C-3 in the 1,4-dehydrobenzene biradical abstracts the 5′-hydrogen of one of the deoxyribose residues, while the radical at C-6 abstracts either the 1′- or 4′-hydrogen from the complementary strand, remains to be examined experimentally.[235,236] Intramolecular hydrogen quenching such as associated with the 1,3-hydrogen shift or an ill-positioned biradical was suggested to be responsible for SS lesions as the major DNA damage.[230,231,235]

Two major binding models, i.e., a minor–major groove binding model and an intercalative binding model, have been constructed to accommodate the experimental results with esperamicin A₁.[230,238] The minor–major groove binding model involves the deoxyfucoseanthranilate moiety lying in the major groove as a C-clamp, while the rest of esperamicin A₁ (3) binds in the minor groove. This model explains the inhibition of esperamicin A₁ induced DNA damage by minor groove binders, as well as the inhibition of osmium tetraoxide oxidation of the 5,6-double bond in the thymine base.[230] The intercalative binding model involves intercalation of the *N*-(2-methoxyacrylyl)anthranilate group with placement of the enediyne core in the minor groove. This was supported by an NMR-derived binding model of esperamicin A₁ (3) to oligonucleotides[239] and physicochemical evidence of intercalation.[231] The binding constant of esperamicin A₁ (3) for DNA was estimated to be $K_a = 1.2(\pm 0.1) \times 10^5$ M$^{-1}$.[231,237] Hydrophobic binding was considered as the major force in the binding of esperamicin A₁ (3) to DNA.[237] An "induced-fit" model was introduced to explain the binding behavior and the observed sequence selectivity.[237,240]

Esperamicin A₁ (3) was shown to damage the nucleosome linker selectivity.[171] There have been reports showing that esperamicin A₁ (3) may also cleave RNA.[156,241]

### 7.15.3.5  Dynemicin

Dynemicin A (4), a violet-colored antibiotic, was isolated from the fermentation broth of *Micromonospora chersina* sp. nov. No. M965-1 obtained from a soil sample collected in Gujarat State, India, in 1989.[242,243] It exhibits *in vitro* and *in vivo* antibacterial activity and potent antitumor activity against a variety of cancer cell lines with low toxicity, and prolongs significantly the life span of mice implanted with P388 leukemia and B16 melanoma cells.[221,244]

#### 7.15.3.5.1  Structure

The structure of dynemicin A (4) is unique among known naturally occurring enediyne antibiotics, possessing features of both the anthracycline and enediyne antibiotics. The chemical structure of dynemicin A (4), determined by spectroscopic and X-ray crystallographic analysis of its triacetate derivative in 1990,[245] is also supported by biosynthetic studies.[51] The absolute stereochemistry was assigned from molecular modeling studies[246] and confirmed by Myers *et al.*'s total synthesis.[247] Other efforts at total synthesis include the preparation of di- and tri-*O*-methyldynemicin A methyl ester derivatives[248] and racemic dynemicin A.[249]

In 1990, deoxydynemicin (50), a variant of dynemicin A (4), was isolated as the coproduct of dynemicin A (4) from the fermentation broth of *Micromonospora globosa* MG331-hF5 from a soil sample collected on Mt. Minobu, Yamanashi Prefecture, Japan. Deoxydynemicin (50) has a hydrogen atom at C-15 instead of a hydroxyl group as in dynemicin A (4).

Dynemicin A (4) contains a bicyclo[7.3.1]tridec-3-ene-1,5-diyne-8,9-epoxide system fused to a 9-amino-1,4,6-trihydroxyanthraquinone chromophore to form a heptacyclic ring system. The molecule is geometrically rigid. The enediyne ring is perpendicular to the plane of the anthraquinone against the epoxide ring. The absolute stereochemistry of dynemicin A (4) is 2*S*, 3*S*, 4*S*, 7*R*, 8*R*. It is interesting that the L-shaped dynemicin actually has a right-handed twist. This appears to be a general theme regarding the active enediyne core in naturally occurring enediyne antibiotics. The anthraquinone moiety of dynemicin A (**4**) functions as both a minor groove intercalator and the activation site via bioreduction.

Dynemicin A (**4**) is fairly soluble in DMSO, DMF, and dioxane, slightly soluble in chloroform, ethyl acetate, and methanol, but insoluble in water.

### 7.15.3.5.2 Activation

Dynemicin A (**4**) can be activated by light, pH, thiols, and NADPH (reduced nicotinamide adenine dinucleotide phosphate).[250–254] The epoxide moiety is considered as a "chemical lock." The opening of the epoxide ring is believed to be the key step in the activation of dynemicin A (**4**).[80,246,255] Several pathways have been proposed to account for the observed activation under different conditions (Scheme 14).[46,244,246,252] The activation pathways involve the conversion of dynemicin A (**4**) to hydroquinone (**51**) via a two-electron or two-sequential, one-electron transfer by bioreductant or photon, followed by the epoxide ring opening to produce the quinonemethide (**52b**) or its imine tautomer (**52a**). Nucleophilic attack or protonation on the quinonemethide or its imine tautomer gives the diol (**53**) or the alcohol (**56**), unlocking the enediyne for the Bergman reaction, with biradical generation and subsequent hydrogen abstraction to give dynemicin H (**58a**), N (**58b**), or S (**58c**).

Intercalation of the anthraquinone moiety of dynemicin A (**4**), placing the enediyne core into the minor groove, is believed to be the binding motif for its interaction with DNA,[80,246,256–258] as minor groove binders and intercalators interfere with DNA damage by dynemicin A (**4**).[250,254] Myers *et al.* suggested that the E-ring hydroxy groups are beneficial for binding and that the carboxylate group at C-5 of the A ring plays a critical role in the DNA damage process by destabilizing the DNA–drug complex, based on the studies of dynemicin A (**4**) and its synthetic analogues.[259] They proposed that dynemicin A (**4**) must dissociate from the DNA-binding complex before reductive activation can take place, and suggested that dynemicin A (**4**) is highly evolved in nature to balance the reaction rate and cleavage efficiency.

### 7.15.3.5.3 DNA damage

Dynemicin A (**4**) cleaves duplex DNA, producing SS lesions as the major damage and DS lesions as the minor damage, with low sequence selectivity (TA$\underline{N}$ = TG$\underline{N}$ > CG$\underline{N}$ > CA$\underline{N}$ > TT$\underline{N}$ > GG$\underline{N}$ (damage site underlined), where N = G ≈ C > A ≈ T).[250–254] DNA damage by dynemicin A (**4**) is significantly enhanced by thiols, NADPH,[250] or visible light. Sugiura and co-workers demonstrated that dynemicin A (**4**) cleaves conformational flexible regions of DNA, such as the DNA B–Z junction, bulges, and nick sites rather selectively.[260–264] It has been shown that duplex DNA gaps such as 5′-Pu_Pu·3′-PyPuPy were preferentially cleaved at the underlined Py. The observed minor DS lesions were shown to be cooperative lesions three nucleotides away due to two molecules of dynemicin A (**4**). It was found that 1′ chemistry was responsible for the major DNA damage, whereas 4′ chemistry accounted for the minor damage, based on gel analysis.[253]

### 7.15.3.6 Kedarcidin

Kedarcidin, a buff-colored solid, was isolated from the fermentation broth of an *Actinomycete* strain L585-6 from a soil sample collected in Maharashtra State, India.[265] It is an acidic chromoprotein (p$I$ = 3.65) with an apparent molecular weight of 12 400 Da. The chromoprotein consists of a highly unstable, ethyl acetate-extractable chromophore and apoprotein with 114 amino acid residues.[266,267] Several variants have been found; the main variant is discussed here. Kedarcidin shows potent antitumor activity against implanted P388 leukemia and B16 melanoma in mice, and also potent antimicrobial activity against Gram-positive but not Gram-negative bacteria.[221,265]

The structure of the kedarcidin chromophore (**5**) was determined using a combination of NMR, MS, and chemical degradation.[268,269] Kedarcidin chromophore (**5**) consists of an 8,9-epoxybicyclo-[7.3.0]dodecadienediyne core with an 11-membered 2′-chloroazatyrosyl bridge between C-4 and

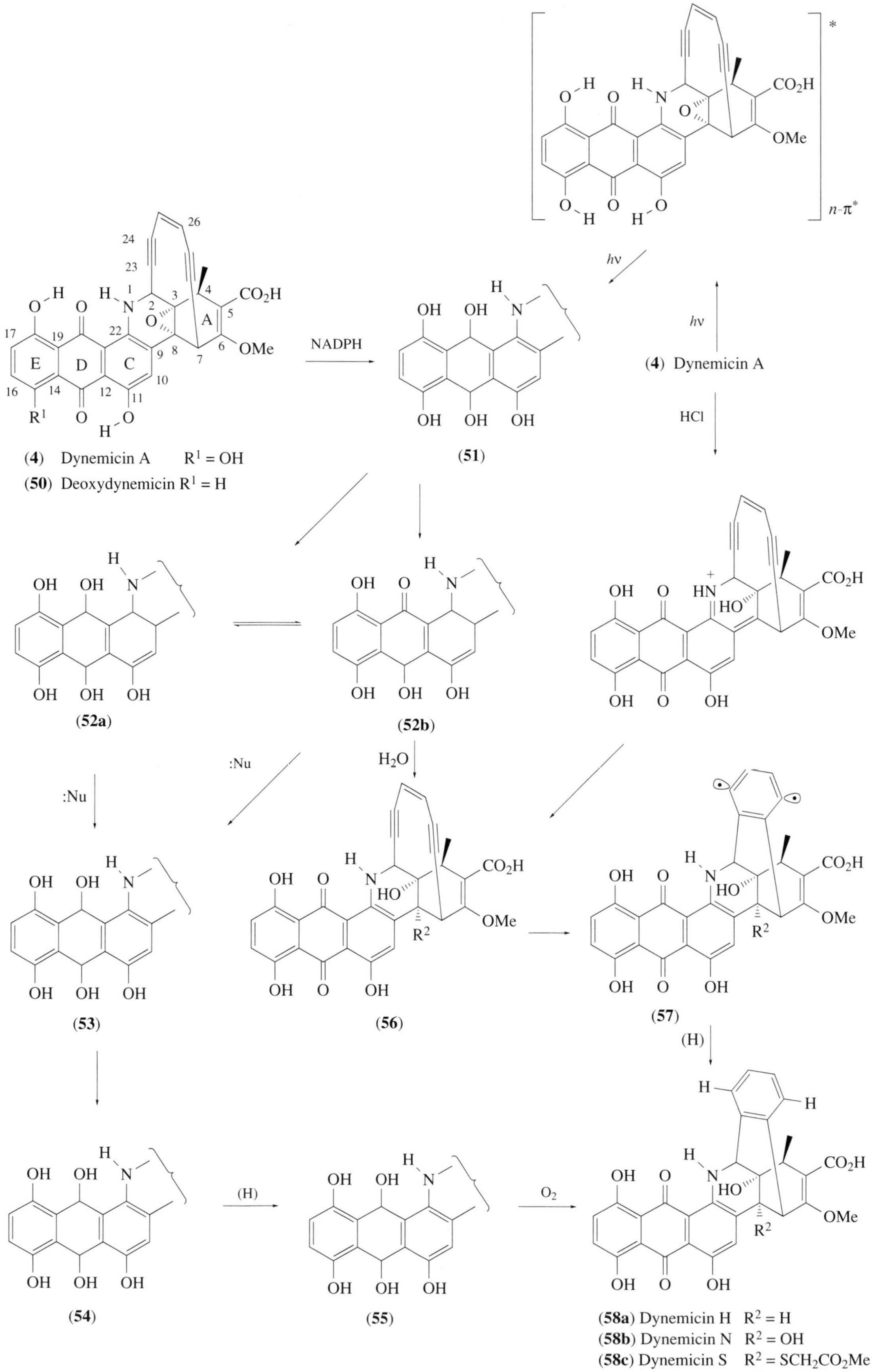
(4) Dynemicin A R1 = OH
(50) Deoxydynemicin R1 = H
NADPH
(51)
hν
hν
(4) Dynemicin A
HCl
n-π*
(52a)
(52b)
:Nu
:Nu
H2O
(53)
(56)
(57)
(H)
(54)
(H)
(55)
O2
(58a) Dynemicin H R2 = H
(58b) Dynemicin N R2 = OH
(58c) Dynemicin S R2 = SCH2CO2Me
(58d) Dynemicin L R2 = Cl

**Scheme 14**

C-11, a new amino sugar 2,4-dideoxy-4-*N*-(dimethylamino)-L-fucopyranoside, namely ked-arosamine with an α-linkage, an α-L-mycaroside, and a 3-hydroxy-6-isopropoxy-7,8-dimethoxy-2-naphthoyl moiety. The absolute stereochemistry in the aglycone was assigned as 8*R*, 9*R*, 10*S*, 11*R*, 13*R*, 8′*R* on the basis of the observed NOE (nuclear Overhauser effect) data and of a model of the NaBH$_4$ reduction product.

The apoprotein of kedarcidin stabilizes the chromophore. Kedarcidin and apokedarcidin have been shown to possess protease-like activity, cleaving histones selectively.[111,270,271]

An activation mechanism was proposed, as shown in Scheme 15, based on NaBH$_4$/NaBD$_4$ reduction.[269] The mechanism involves initial nucleophilic attack on the less hindered side of C-12, followed by double-bond migration and opening of the 8,9-epoxide ring. Subsequent cyclo-aromatization of the enediyne releases the strain to give the 1,4-dehydrobenzene biradical (**60**). The biradical abstracts hydrogen from DNA or solvent. This pathway was supported by deuterium incorporation from solvent into the C-3 and C-6 positions in the reduced form. DNA damage studies showed that the kedarcidin chromophore (**5**) is also activated by thiols such as BME and DTT. Analysis of the postactivated drug product in the DNA damage reaction has not been reported.

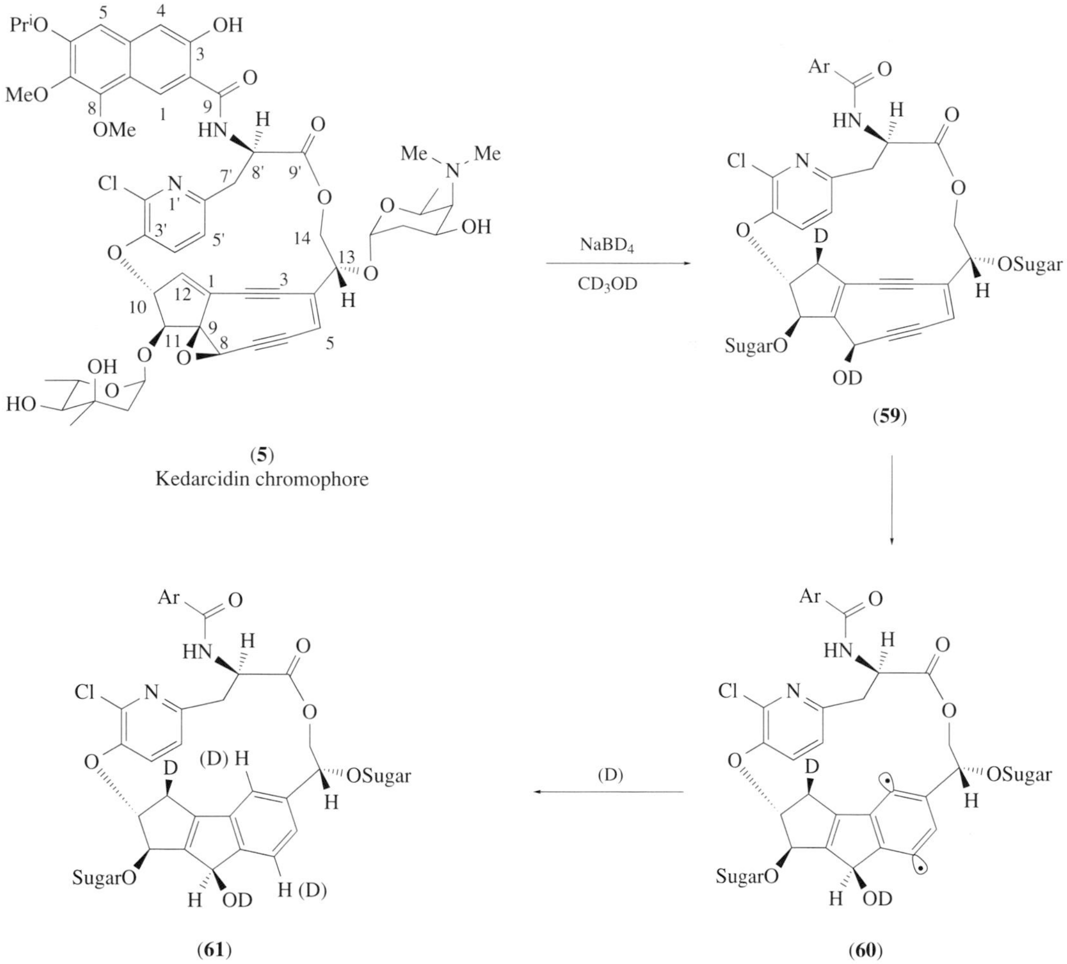

(**5**)
Kedarcidin chromophore

(**59**)

(**61**)

(**60**)

**Scheme 15**

DNA damage studies showed that kedarcidin and its chromophore cleave duplex DNA with the preferred sequence TCCT<u>N</u> (N = T, C, G, or A), producing SS lesions with apparent 4′ chemistry accompanied by minor 5′ chemistry.[111,272] The presence of certain cations such as Ca$^{2+}$ and Mg$^{2+}$ inhibits the DNA damage. A siderophore-like chelation of the naphthoate moiety to Ca$^{2+}$ or Mg$^{2+}$ was suggested to be responsible through interruption of the DNA binding process.

### 7.15.3.7 C-1027

C-1027, a white powder, was isolated from the fermentation broth of *Streptomyces globisporus* C-1027 from a soil sample collected from Qian-jiang County, Hu-bei Province, China.[273] C-1027 is an acidic chromoprotein (p$I$ = 3.5–3.7) with an apparent molecular weight of 15 000 Da.[274] The chromoprotein consists of a labile, methanol-extractable chromophore ($K_d$ = 6.88 × 10$^{-5}$ M), and an apoprotein, with 110 amino acid residues, showing aminopeptidase activity.[275–278] A monoclonal antibody raised against C-1027 apoprotein was reported to recognize not only C-1027 apoprotein but also the holoprotein C-1027.[279] The antibiotic C-1027 shows potent antimicrobial activity against Gram-positive and some strains of Gram-negative bacteria, and also has antitumor activity against implanted leukemia L1210, P388, and sarcoma 180 in mice.[273,280–282] It is believed that the highly potent cytotoxicity of C-1027 (effective in cell culture at 10$^{-16}$–10$^{-14}$ M) comes from its DNA-cleaving ability.[281–285]

#### 7.15.3.7.1 Structure

In 1993, Minami and co-workers determined the structure of the postactivated C-1027 chromophore (**63**) and the structure of C-1027 chromophore (**6**).[275,286,287] The relative stereochemistry of the chromophore (**6**) was deduced from molecular modeling.[288] The absolute stereochemistry of the chromophore (**6**) with 8$S$, 9$R$, 13$S$, 18$S$ was reported to be the mirror enantiomer of the macrocyclic enediyne core from molecular modeling.[289] C-1027 chromophore (**6**) consists of three main units: a macrocyclic enediyne core, an amino sugar residue and a benzoxazolinate residue. The macrocyclic enediyne core contains a bicyclo[7.3.0]dodeca-4,10,12-triene-2,6-diyne with an 11-membered bridge containing a 3′-chloro-5′-hydroxy-$\beta$-tyrosine residue between C-4 and C-8. The amino sugar 4-deoxy-4-*N*-dimethylamino-5,5-dimethyl-$\beta$-D-ribopyranoside was confirmed by synthesis.[290] The benzoxazolinate residue, characterized as the same degradation product of the unidentified chromophore of the DNA-cleaving holoprotein auromomycin,[291–293] is 3,4-dihydro-7-methoxy-2-methylene-3-oxo-2*H*-1,4-benzoxazine-5-carboxylate.

The apoprotein of C-1027 is rich in $\beta$-sheet structure and has 47% homology with apo-NCS.[276,294] It functions as both a carrier to stabilize the labile chromophore and an aminopeptidase or protease, possibly to assist the DNA-damaging activity of the chromophore.[278] It has been suggested that the alkynic bonds in the chromophore are stabilized by (i) hydrophobic interactions with the bottom of the pocket of the apoprotein, (ii) orbital interaction with the Cys$^{36}$–Cys$^{45}$ disulfide bond, (iii) van der Waals contact with Pro$^{76}$, and (iv) $\pi$–$\pi$ stacking with the benzene moiety of the chromophore (**6**).[288]

#### 7.15.3.7.2 Activation

C-1027 chromophore (**6**) is sensitive to light, base, and thiols. It is relatively more stable in acidic aqueous solution than in basic solution. C-1027 chromophore (**6**) does not need any external activator for biradical formation. The proposed activation mechanism is shown in Scheme 16.[287] Bergman cycloaromatization of the bicyclo[7.3.0]dodeca-4,10,12-triene-2,6-diyne core generates 1,4-dehydrobenzene biradical (**62**), which abstracts hydrogen from carbon-bound hydrogen sources to yield benzodihydropentalene (**63**). Deuteration experiments showed that the deuteriums are incorporated into the C-3 and C-6 positions. It seems that the activation of C-1027 chromophore (**6**) is thermally controlled, probably involving the rotation of the tyrosine bridge.

#### 7.15.3.7.3 Reversibility and reactivity of the 1,4-benzenoid biradical

It has been generally assumed that the 1,4-dehydrobenzene biradical is very reactive, similar to a phenyl radical. The rate-determining step in the cycloaromatization cascade studied is usually Bergman cyclization, and thus the subsequent reactions of the biradical are kinetically unobservable. Semmelhack *et al.* showed the dependence of the half-life of 10-membered synthetic enediynes on the concentration of 1,4-cyclohexadiene (hydrogen source) and called attention to the contribution of hydrogen abstraction by the biradical.[295,296] The observation by Hirama and co-workers that

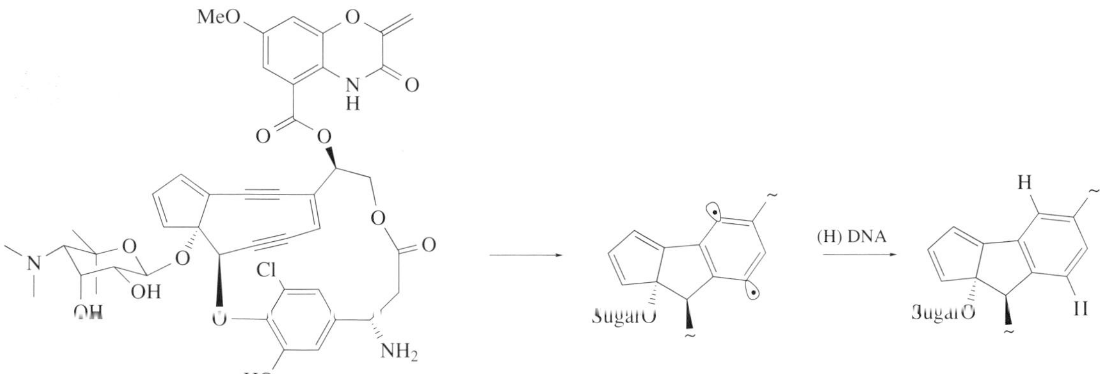

(6)
C-1027 chromophore

(62)

(63)
Benzodihydropentalene

**Scheme 16**

C-1027 chromophore (**6**) cycloaromatizes with a kinetic solvent isotope effect shows that hydrogen abstraction of the biradical is the rate-determining step in the cycloaromatization of C-1027 chromophore (**6**), instead of Bergman cyclization.[297,298] The fact that strong deuterium isotope effects on hydrogen atom abstraction from DNA tend to be observable in nine-membered enediyne antibiotics, such as NCS-chrom[132,166,170,299,300] and C-1027,[301] may be related to the lower reactivity of the biradical. These experiments also provide evidence for a kinetically significant reversibility of Bergman cyclization in the cyclic enediyne system. Chen and co-workers suggested that the reactivity of a biradical is much less than that of a phenyl radical and that the reactivity of a biradical can be tuned by adjusting the singlet–triplet splitting gap of the biradical using through-space and through-bond effects.[302,303] They found that the 9,10-dehydroanthracene biradical—a 1,4-dehydrobenzene-type biradical—abstracts hydrogen unusually slowly.[304] The rates of hydrogen abstraction by the 9,10-dehydroanthracene biradical from acetonitrile and 2-propanol were measured directly by flash photolysis/transient absorption spectroscopy, giving second-order rate constants of $k_{MeCN} = 1.1$ $(\pm 0.2) \times 10^3$ M$^{-1}$ s$^{-1}$ and $k_{2\text{-PrOH}} = 6.5(\pm 0.6) \times 10^3$ M$^{-1}$ s$^{-1}$ at room temperature, which are 100–200 times lower than the corresponding rate constants for the phenyl or 9-anthryl radical. The rate of the retro-Bergman reaction for the 9,10-dehydroanthracene biradical is relatively fast, $k \approx 4 \times 10^5$ s$^{-1}$ at room temperature.[304] The large singlet–triplet splitting gap observed in the 1,4-dehydrobenzene biradical results in more of the less reactive singlet biradical, which may have higher selectivity.[29,302] Assuming sequential hydrogen atom abstraction by the biradical, "the first hydrogen abstraction by a *p*-benzyne-type diradical proceeds rather slowly (and presumably selectively) and yields the two-hundred fold more reactive phenyl radical," thus allowing local flexibility in the two hydrogen abstractions by the biradical. This translates into more DS DNA lesions.[31]

### 7.15.3.7.4 *DNA damage*

C-1027 chromophore (**6**) is responsible for most of the biological activities of the chromoprotein.[282,305] C-1027 cleaves duplex DNA without external activators such as thiols. It generates mainly DS lesions with a two- or one-nucleotide 3′-stagger within pentanucleotide sequences such as CTTTT/AAAAG, ATAAT/ATTAT, CTTTA/TAAAG, CTCTT/AAGAG, and especially GTTAT/ATAAC or GTTAT/ATAAC (damaging sites are underlined), involving 4′ chemistry for the top strand and 5′ or 1′ chemistry for the bottom strand.[301,306] It also generates SS lesions, mainly at A and T residues (A ⩾ T ≫ C > G), involving 4′ chemistry.[306,307]

In their detailed study of DNA damage by C-1027 using the model sequence GTTA$_1$T/ATA$_2$A$_3$C in oligonucleotides, Xu *et al.* found two types of DS lesions by a single molecule of C-1027 chromophore (**6**), one involving A$_1$ via 4′ chemistry and A$_2$ via 1′ chemistry with an unusual one-nucleotide 3′ stagger, and the other involving A$_1$ via 4′ chemistry and A$_3$ via 5′ chemistry with a two-nucleotide 3′-stagger.[301] DS lesions predominate when solvent methanol is excluded, suggesting that one of the two radicals in the biradical is more exposed to solvent and more easily quenched by the methanol. The $A_1$ lesion is the sole lesion in SS damage. Lesions at $A_2$ and $A_3$ occur only in bistranded damage. A deuterium isotope effect at C-1′ of $A_2$ ($k_H/k_D = 7.05$, one of the highest isotope effects reported in the literature[308]) resulted in the substantial shuttling of the damage site from $A_2$ to $A_3$, which proves the 1′ chemistry at $A_2$ and also supports a single binding mode of the drug being responsible for the DS lesions at $A_2$ and $A_3$.[301]

It has been reported that C-1027 cleaves tRNA$^{Phe}$ with selectivity at its anticodon region in the presence of $Mg^{2+}$.[241] Sugiura and co-workers suggest that tRNA may serve as a catalytic template for the cycloaromatization of C-1027 chromophore (**6**), as the chromophore cycloaromatizes three times faster in the presence of tRNA than in its absence, although no experimental data were shown.[241]

C-1027 damages the linker of nucleosomal DNA in nuclei selectively.[309] Beerman and co-workers found that C-1027 cleaved DNA 285 times more efficiently in cells than in a cell-free environment and displayed a preference for intracellular organelle-localized DNA in the order episome > mitochondria $\gg$ genome.[284]

Minor groove binders such as netropsin and distamycin A inhibit DNA damage by C-1027, which suggests that the binding of the drug is in the minor groove.[306,307] The benzoxazolinate group was found to be the intercalating moiety in the postactivated form of C-1027 chromophore (**6**).[310] The presence of the benzoxazolinate group significantly increases the ability to bind to DNA ($K_d = 2.0 \times 10^{-6}$ M compared with $8.3 \times 10^{-4}$ M without the benzoxazolinate group).[310] The two-nucleotide 3′-stagger pattern in the DS lesions also supports the intercalation model.[301] Recent NMR studies also support this model.[311] It appears that the binding of C-1027 chromophore (**6**) to DNA involves both intercalation by the benzoxazolinate group and minor groove binding, which places the enediyne core closes to the sugar residues of the complementary strands, so as to permit DS lesions. The locations of the abstracted hydrogens in postactivated C-1027 chromophore (**6**) have not been determined.

### 7.15.3.8 Maduropeptin

Maduropeptin was isolated from the fermentation broth of *Actinomadura madurae* H710-49 from a soil sample collected in Germany. It was identified as a chromoprotein consisting of a 1:1 complex of an acidic, water-soluble protein (32 kDa, p$I$ = 4.75) and a tightly bound nine-membered ring enediyne chromophore. The chromoprotein exhibited potent inhibitory activity against Gram-positive bacteria and tumor cells and had strong *in vivo* antitumor effects against P388 leukemia and B16 melanoma implanted in mice.[312]

Extraction of holo-maduropeptin gave rise to compounds with structures (**64**), which were determined by low-temperature NMR and MS, as shown in Scheme 17.[313] It is assumed that compounds (**64**) are artifacts of the isolation procedure; different structures have been proposed for the actual parent maduropeptin chromophore, including (**7**) and a 4,5-epoxide version of (**7**).[313,314] If (**7**) is the parent chromophore, the holoantibiotic presumably functions as a prodrug. C-1027-chromophore (**6**) has a similar structure to (**7**). The bioactive artifacts (**64**) are, for convenience, also collectively called maduropeptin chromophore here.

The maduropeptin chromophore (**64**) is composed of a bicyclo[7.3.0]dodecadienediyne core with a 10-membered amide bridge between C-4 and C-8, a new sugar 4-*N*-amino-4-deoxy-3-*C*-methylribopyranoside, namely madurosamine, and a 2-hydroxy-3,6-dimethylbenzoyl moiety with amide linkage to madurasamine. The chlorine group at C-1 of the tetrasubstituted benzene ring of the bridge is found to be large enough to prevent the bridging system from rotating in molecular models. The absolute stereochemistry of the maduropeptin chromophore has not been determined. The stereochemistry shown is the relative stereochemistry with the enantiomer we prefer in view of the structure of C-1027 chromophore (**6**), which is the mirror enantiomer of the reported structure.[313]

The apoprotein of maduropeptin dramatically stabilizes the chromophore. Maduropeptin and its apoprotein have also been shown to have proteolytic activity.[111,271,315]

An activation mechanism was proposed as shown in Scheme 17, based on isolation of the cycloaromatized product. The activation needs no thiol, but is pH sensitive. The cycloaromatized product (**66**) is still bioactive, possibly by alkylation of DNA or a repair enzyme.[313–315]

Maduropeptin and its chromophore (**64a**) damage duplex DNA *in vitro* without a bioreductive activator, producing a mixture of SS and DS lesions.[111,314,315] The DNA lesions involve 4′ chemistry,

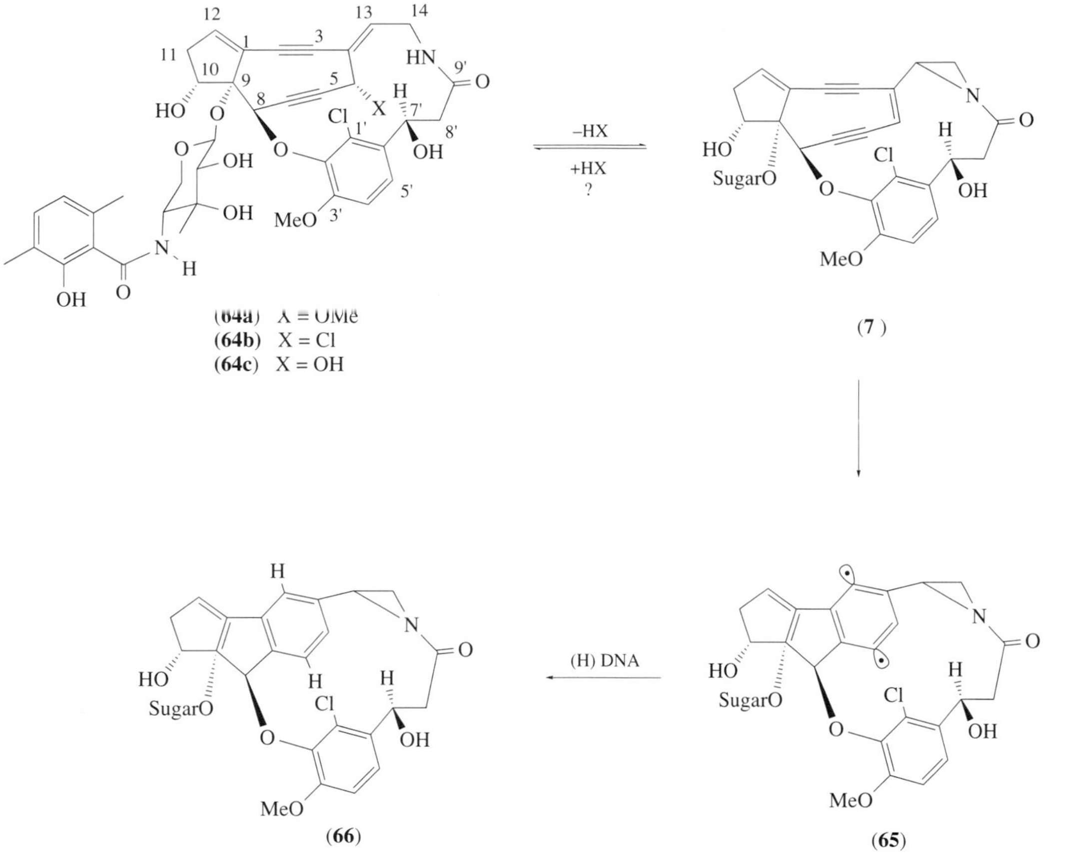

**Scheme 17**

based on gel analysis. The presence of certain cations, such as $Ca^{2+}$ and $Mg^{2+}$, inhibits the DNA damage, probably through siderophore-like chelation of these cations by $\beta$-hydroxyamide and *o*-hydroxybenzamide moieties in the maduropeptin chromophore.

### 7.15.3.9 Namenamicin

Namenamicin (**8**), the newest member of the naturally occurring enediyne antibiotics, was isolated from the thin encrusting orange marine ascidian *Polysyncraton lithostrotum* (Order, Aplou-sobranchia; Family, Didemnidae) collected from Namenalala Island, Fuji Islands.[316] It exhibited potent antimicrobial activity and *in vitro* cytotoxicity with a mean $IC_{50}$ of 3.5 ng ml$^{-1}$ and *in vivo* antitumor activity in a P388 leukemia model in mice.

The structure of namenamicin (**8**) is similar to that of the calicheamicins and esperamicins. It consists of the same aglycone as the calicheamicins—calicheamicinone (**38**)—and a trisaccharide segment through a glycosidic $\beta$-linkage to the C-8 position of the aglycone (see Section 7.15.3.3). The trisaccharide segment, which has a C—O $\beta$-linkage between the A ring and B ring instead of an N—O $\beta$-linkage, is different from esperamicin C only at the C-4 position of the A ring (see Section 7.15.3.4). The absolute stereochemistry at C-7 of the A ring is not yet defined.

It is expected that namenamicin (**8**) has an activation mechanism similar to that of calicheamicin $\gamma_1^1$ (**2**) (see Scheme 11). Namenamicin cleaves DNA in a similar manner to esperamicin C, with primary recognition of the sequence 5′-TTT, but with lower efficiency and a different cleavage pattern than those for calicheamicin $\gamma_1^1$ (**2**). The DNA cleavage chemistry has not been identified.

The fact that all of the naturally occurring enediyne antibiotics prior to namenamicin (**8**) have been the products of actinomycetes (prokaryotes) and given namenamicin's extremely low and variable yield (1 mg from 1 kg of frozen tissue) from the marine ascidian, suggest a microbial origin for namenamicin (**8**). The true biosynthetic origin of namenamicin (**8**) was still under investigation in 1997.

## 7.15.4  ROLE OF APOPROTEIN

It is interesting that the naturally occurring enediyne antibiotics discovered so far can be easily divided into two major classes according to their enediyne core: one with a 10-membered ring, such as the calicheamicins (**2**), the esperamicins (**3**), namenamicin (**8**), and the dynemicins (**4**), and the other with a nine-membered ring, such as the chromophores of neocarzinostatin (**1**), kedarcidin (**5**), C-1027 (**6**), and maduropeptin (**7**). All nine-membered ring-containing enediyne antibiotics are very labile and associated with an apoprotein. All 10-membered ring-containing enediyne antibiotics are relatively stable, and have not been found to be associated with an apoprotein. Whether or not these 10-membered ring-containing enediyne antibiotics are also associated with an apoprotein *in vivo* remains to be seen. It is possible that isolation of the 10-membered ring-containing enediyne antibiotics without apoprotein is due to their relative stability upon extraction.[314]

NMR, molecular modeling, and X-ray crystallography have been used to study the three-dimensional structure of the apoprotein and holoprotein of neocarzinostatin,[126,317–320] C-1027,[276,288] and kedarcidin.[267] Kim *et al.* showed that in the X-ray crystal structure of holo-NCS, the labile NCS-chrom (**1**) is situated inside a hydrophobic pocket, with Phe78 over the top of the dienediyne core to protect the C-12 position from external thiol attack, with Phe52 and the disulfide bond of Cys37 and Cys47 situated at the bottom of the dienediyne core to protect the C-12 position from intramolecular attack by the naphthoate group, and with the 8,9-epoxide situated deep inside the pocket to avoid protonation.[126]

The apoproteins not only appear to function as protective carriers of the chromophores, assisting the labile chromophore to survive intact to the biotarget, but also have proteolytic activity.[270,271,278,315,321,322]

Early studies by Montgomery and co-workers showed that chromoprotein antitumor antibiotics (including macromomycin) possess proteolytic activities.[321,322] Sakata *et al.* found that holoprotein C-1027 has an aminopeptidase activity of one-fifteenth that of porcine kidney enzyme (EC 3.4.11.2) by use of Phe-MCA (L-phenylalanyl-4-methylcoumaryl-7-amide) as the substrate.[278] In a more detailed study of enediyne antibiotics, Zein *et al.* found that the chromoproteins of NCS, kedarcidin, and maduropeptin possess proteolytic activity against histones *in vitro*, with histone H1 as a preferred substrate. Each chromoprotein generates a unique set of cleavage products of histone H1. The kedarcidin apoprotein appears to be less specific than the kedaracidin chromoprotein. It has been proposed that the selective proteolytic activities of the apoproteins are important *in vivo* in the delivery of the enediynes intact to the DNA in nuclear chromatin.[271]

A model of endocytosis, in which the holoprotein enters into the cell intact, has been suggested to account for the action of the chromoprotein on cells.[278,323]

## 7.15.5  DNA/RNA DAMAGE MECHANISMS

Enediyne antibiotics, after release of the chemical lock on the enediyne unit, cycloaromatize through a Bergman-type reaction to generate a reactive biradical. This biradical, if already properly positioned inside the minor groove of duplex DNA, can abstract hydrogen atoms from deoxyribose of one or both strands of duplex DNA and translocate into one or two sugar radicals, which ultimately generate oxidative single- or double-stranded DNA lesions. The presence of dioxygen or its substitute facilitates the expression of the DNA damage. The damaged genetic material induces apoptosis of the cell.[283,324–327] The interactions between drug and DNA consist of mainly hydrophobic forces and/or intercalation, which determine the so-called sequence selectivity or, more accurately, *loci* selectivity (microstructure selectivity). The relevance of *in vitro* sequence-selective DNA damage by these enediyne antibiotics to their *in vivo* activity has been actively pursued and remains under investigation in 1997.[171,309,328–334] It is known that DS lesions are more significant lesions biologically in view of the correlation of DS lesions with the lethality of the drug,[335] the mutagenicity of bistranded DNA lesions,[336–339] and the effect of DS breaks on recombinatorial activity that leads to DNA rearrangements involved in carcinogenesis.[340]

### 7.15.5.1  Chemistry of DNA/RNA Damage

Theoretical studies have shown that all hydrogens of deoxyribose of DNA can be easily abstracted, with the 2′-hydrogen being least likely.[341,342] In B-form duplex DNA, four of the seven sugar hydrogens, i.e., 1′-H, 2′-H(*R*), 4′-H, and 5′-H(*S*), point into the minor groove.[343] An enediyne-generated biradical situated in the minor groove can theoretically abstract any of these sugar hydrogens from each of the complementary DNA strands. The formation of the ensuing sugar radicals is independent of dioxygen. The majority of DNA damage arises from the addition of dioxygen to the biradical-generated sugar radical to form a sugar peroxy radical. The sugar peroxy radical breaks down and results in oxidative DNA damage in the form of strand breaks or abaic sites. Under anaerobic conditions, covalent drug–DNA adducts become the major form of DNA damage.[44]

Three main attack sites and the consequent damage products have been characterized:[39,44,344] (i) 5′ chemistry—abstraction of the 5′-hydrogen of the DNA sugar residue results in a break with a 3′-phosphate and a 5′-nucleoside aldehyde, with oxygen derived from dioxygen addition;[345–347] this reaction also produces single nucleotide gaps yielding 5′ and 3′ phosphate ends, formate (derived from 3′-formylphosphate-ended DNA fragment) and a four-carbon fragment as minor lesions;[92,348,349] (ii) 4′ chemistry—4′ attack leads either to an alkali-labile 4′-hydroxylated abasic site or to a break with a 3′-phosphoglycolate and a 5′-phosphate;[166,299,350] (iii) 1′ chemistry—abstraction of a hydrogen atom from C-1′ results in an alkali-labile abasic site having a 2-deoxyribonolactone residue.[161] The 4′-hydroxylated abasic site is pH sensitive, and gives a significant amount of strand breakage with a 5′-phosphate and a 3′-phosphopentenaldehyde, moving anomalously slower than the Maxam–Gilbert marker, when subjected to electrophoresis on a gel at pH 9.1.[299] The abasic site can be expressed by treatment with base, such as hydrazine, which cleaves mainly 4′-hydroxylated abasic sites to form a 3′-phosphopyridazine,[299,351] and putrescine, which cleaves all abasic sites. These DNA damage pathways are shown in Scheme 18.[44] It should be noted that Giese and co-workers proposed an alternative pathway for the 4′ chemistry, which would produce strand breakage anaerobically,[352,353] an event that has not been observed experimentally in the enediyne-mediated DNA damage reaction.

DNA·RNA hybrid structures have been shown to undergo damage by NCS-chrom (**1**) via 1′ chemistry on the RNA strand and 5′ chemistry on the DNA strand.[170] Abstraction of the 1′-H of uridylate results in an alkali-labile abasic site and/or a break with 3′-phosphate and 5′-phosphate ends. A plausible pathway for RNA damage is shown in Scheme 19. Note that in the absence of metal ions or acidic conditions, a Criegee-type rearrangement is probably not involved in transformation of the sugar peroxy radicals.[354–356] The generation of a direct strand break in RNA is proposed to involve base-catalyzed cleavage of a $\beta$-hydroxyperoxide.[170]

### 7.15.5.2 Role of Dioxygen

Dioxygen is needed for expression of most of the DNA damage as strand breakage and abasic site formation, although it has not been found to be involved in activation of the enediyne antibiotics. It has been shown that the oxygen atom in the 5′-nucleoside aldehyde, generated at the cleavage site, comes from dioxygen in isotope-labeling experiments.[345] In a study of thiol-activated NCS-chrom-mediated DNA damage, kinetic analysis showed that the initial reaction of the chromophore (**1**) with a single molecule of thiol occurs in the absence of dioxygen and that the drug activation is rapidly followed by the uptake of 1 equiv. of dioxygen and the subsequent utilization of at least an additional sulfhydryl group, which occurs only in the presence of dioxygen.[346]

Nitroaromatic sensitizers such as misonidazole can substitute for dioxygen.[357] The effectiveness of various nitroimidazoles in DNA-damage reactions is correlated with their electron affinity, as measured by their one-electron reduction potential, and is inversely related to the concentration of bioreductant such as thiols. In the case of NCS-chrom (**1**)-induced DNA damage, single-nucleotide gaps with phosphate at each end are the main lesions in the presence of misonidazole. 3′-Formylphosphate-ended DNA, an energy-rich formyl donor, has been proposed as an intermediate in this reaction.[348] The reaction pathway involves formation of a misonidazole–sugar radical adduct (nitroxide radical adduct), followed by homolysis of the N—O bond (instead of heterolytic cleavage and one-electron oxidation) of the adduct to form a sugar oxyradical, with the oxygen having come from the misonidazole.[358] This unusual cleavage mechanism has received confirmation in model radiosensitizing systems.[359] The sugar oxyradical undergoes $\beta$-fragmentation to break the DNA strand. This reaction is formally identical to the minor reaction involved in the dioxygen-dependent reaction, where 5′ chemistry gives rise to formate and phosphate at both ends of the break (Scheme 20).[87]

Note that the trapping of the deoxyribose radical by dioxygen or misonidazole is proposed to yield a sugar peroxy radical or oxy radical, respectively. The possibility of the involvement of a

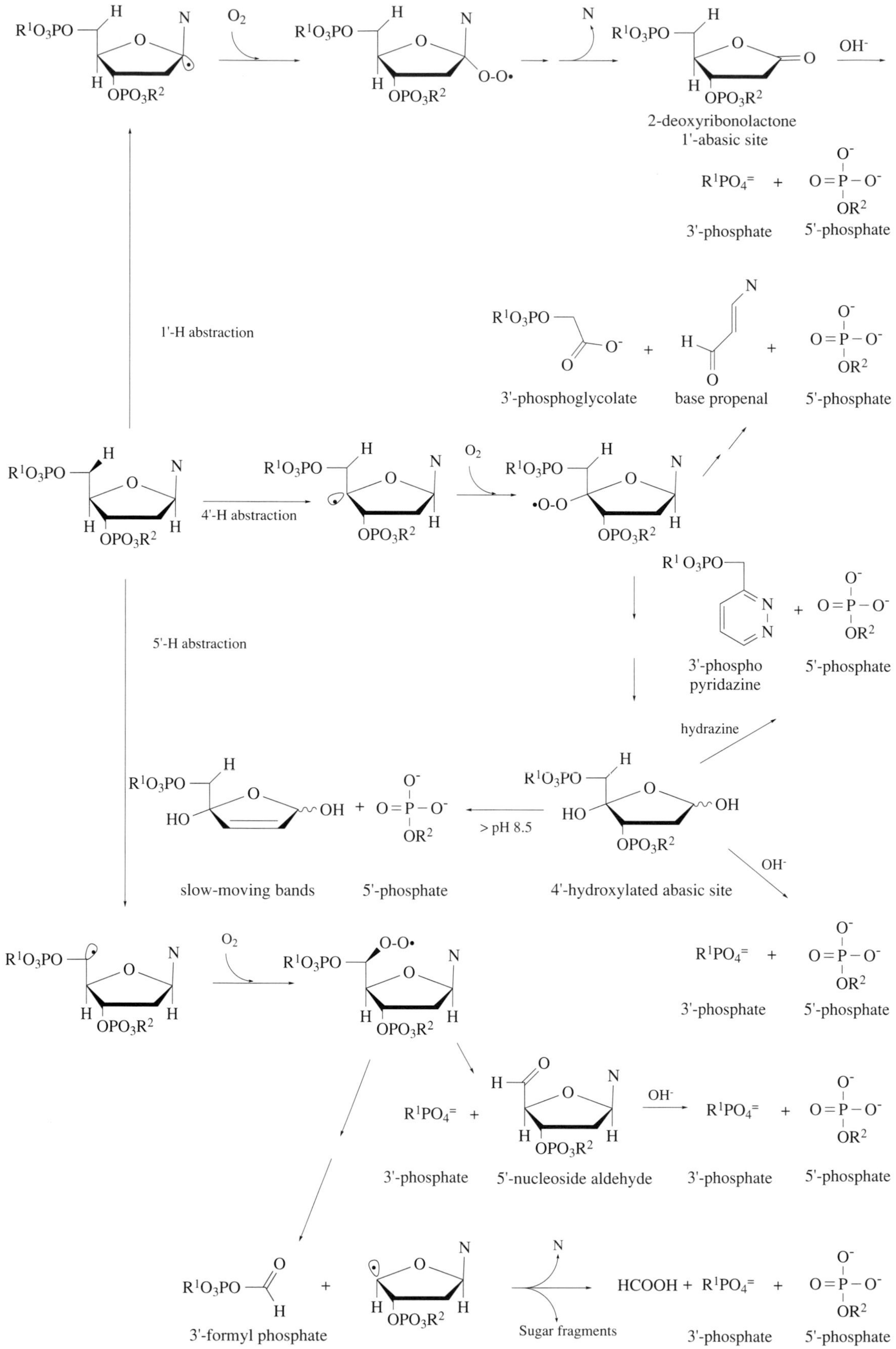

2-deoxyribonolactone
1'-abasic site

$R^1PO_4^=$ + 5'-phosphate

3'-phosphate   5'-phosphate

1'-H abstraction

3'-phosphoglycolate   base propenal   5'-phosphate

4'-H abstraction

5'-H abstraction

3'-phospho
pyridazine     5'-phosphate

hydrazine

> pH 8.5

slow-moving bands   5'-phosphate       4'-hydroxylated abasic site

OH⁻

3'-phosphate   5'-phosphate

3'-phosphate   5'-nucleoside aldehyde   3'-phosphate   5'-phosphate

Sugar fragments

3'-formyl phosphate

HCOOH + $R^1PO_4^=$ +

3'-phosphate   5'-phosphate

**Scheme 18**

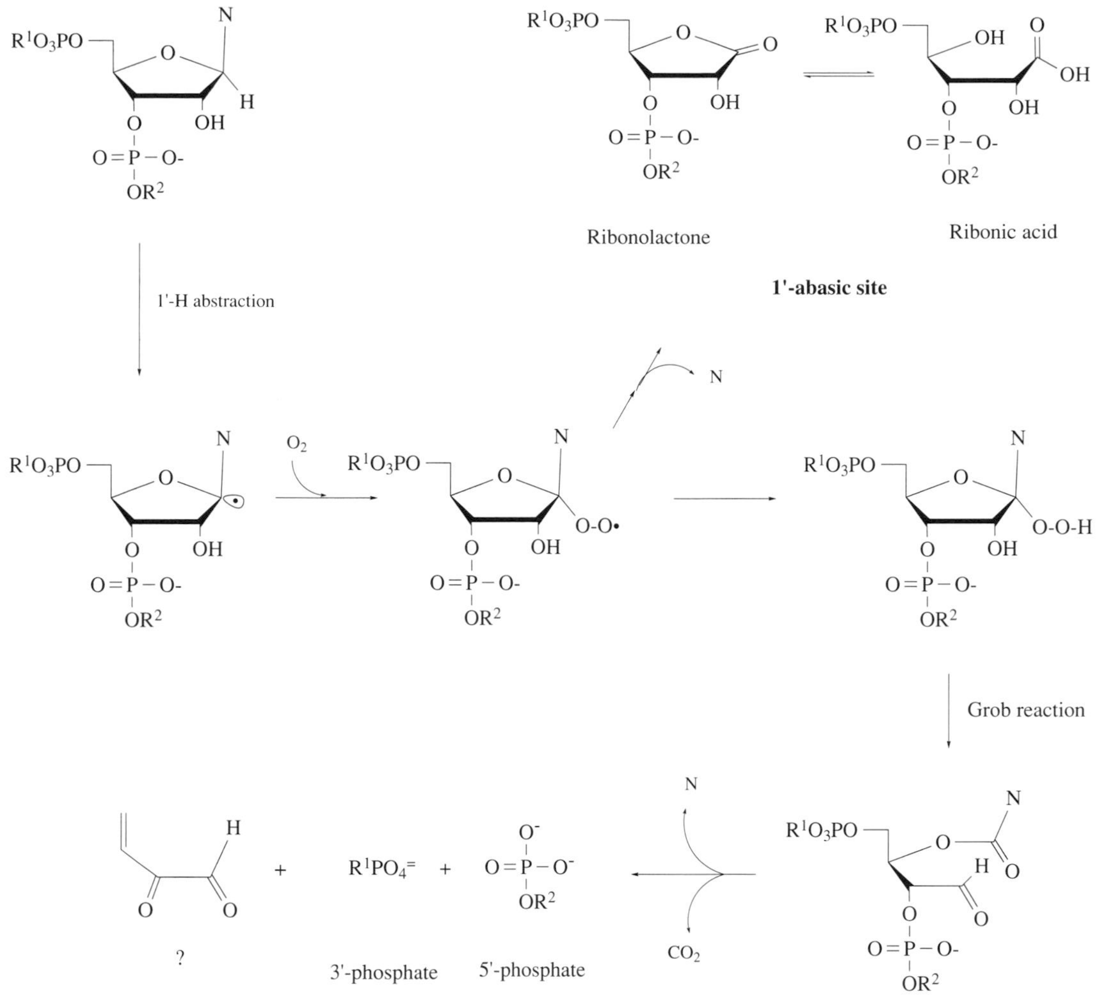

**Scheme 19**

tetraoxide intermediate on the complementary DNA strands in oxy radical formation in the dioxygen-based reaction has been raised,[164] but there is no supporting experimental data for this proposal. Such a mechanism would obviate the involvement of thiol in the reduction of a peroxy radical intermediate, consistent with the oxidative damage produced by those enediynes, such as C-1027, which do not require thiol in producing strand breaks and abasic sites. The outcome of the above paradigm reflects on the different degradation pathways of DNA damage, especially in the case of 4′ chemistry. As predicted from Scheme 20, the production of 3′-phosphoglycolate-ended DNA increases at the expense of the 4′-hydroxylated products if misonidazole is used as a dioxygen substitute under anaerobic conditions.[166]

### 7.15.5.3 Adduct Formation

Since the DNA bound postactivated enediyne core is an unsaturated system, the newly generated sugar radical close by has the ability to add back on to the unsaturated system to form a covalent adduct or even an interstrand cross-linked product. The actual process is one of kinetic differentiation, depending on the relative position of the sugar radical to the unsaturated system and the ease of translocation of the sugar radical by an intramolecular or intermolecular process, such as quenching by dioxygen or other hydrogen sources, such as solvent or thiol. Experimental results, in particular, emphasize the competition of adduct formation with dioxygen quenching. The formation of novel drug–DNA adducts, involving DNA deoxyribose moieties, has been observed, especially

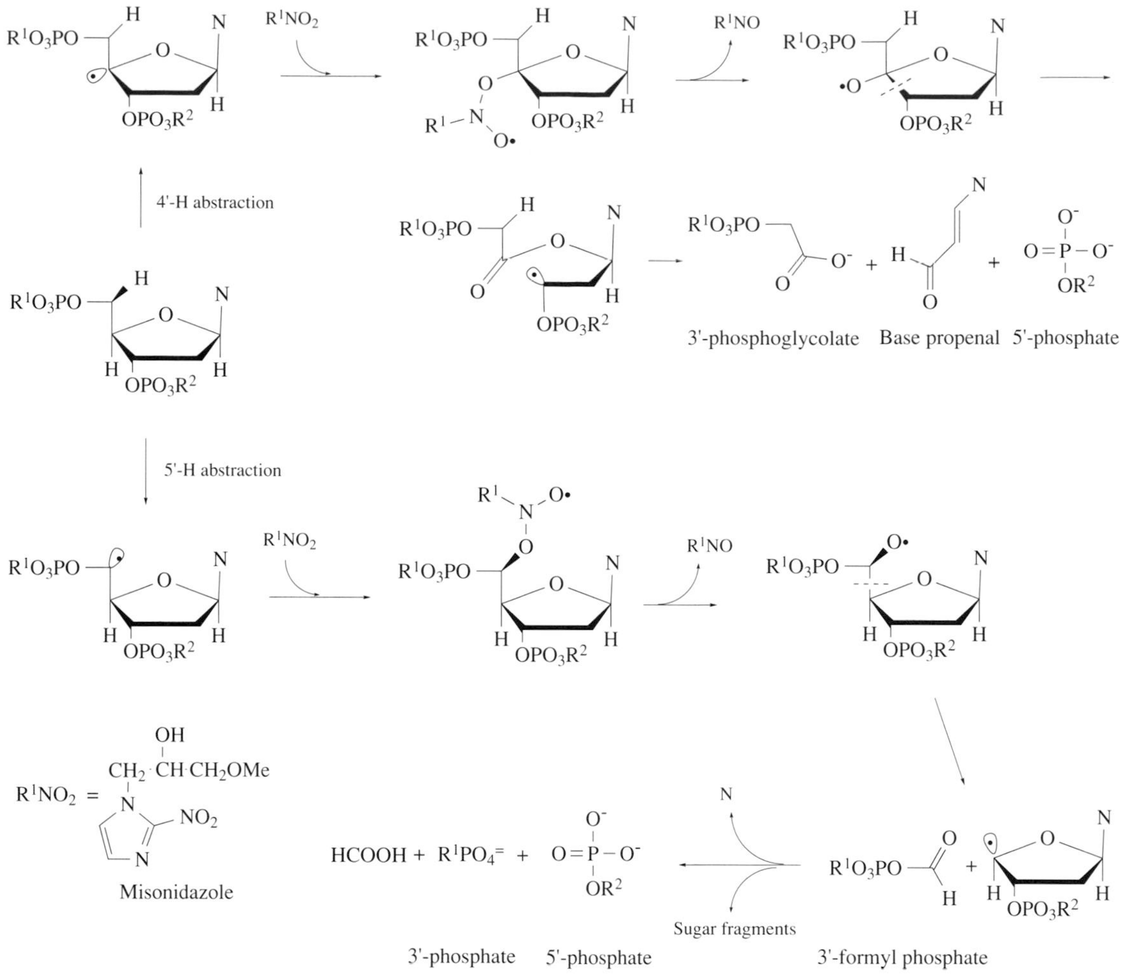

**Scheme 20**

under anaerobic conditions.[87] In the case of NCS, a drug–deoxyribose adduct, involving C-5′ of deoxyribose of DNA, was isolated under both aerobic and anaerobic conditions.[360–363] A cross-linked product of activated C-1027 and double-stranded DNA has also been observed.[364] Further, a covalent monoadduct between NCS chromophore and a specific ribonucleotide on the RNA overhang of an RNA–DNA hybrid has been characterized.[365] The significant increase in drug–adduct or cross-linked complex formation under anaerobic conditions may have important implications for its cytotoxicity in the central regions of large tumors, where oxygen tension is low.

## ACKNOWLEDGMENTS

The authors wish to thank former and current members of the Goldberg Laboratory for their experimental contributions to the field of enediyne antibiotic research and for comments on and criticisms of this chapter. This work was supported by US Public Health Service grant GM 53793 from NIH.

## 7.15.6  REFERENCES

1. N. Ishida, K. Miyazaki, K. Kumagai, and M. Rikimaru, *J. Antibiot.*, 1965, **18**, 68.
2. J. Meienhofer, H. Maeda, C. B. Glaser, J. Czombos, and K. Kuromizu, *Science*, 1972, **178**, 875.
3. Y. Ono, Y. Yatanabe, and N. Ishida, *Biochim. Biophys. Acta*, 1966, **119**, 46.
4. M. A. Napier, B. Holmquist, D. J. Strydom, and I. H. Goldberg, *Biochem. Biophys. Res. Commun.*, 1979, **89**, 635.

5. L. F. Povirk and I. H. Goldberg, *Biochemistry*, 1980, **19**, 4773.
6. L. S. Kappen and I. H. Goldberg, *Biochemistry*, 1980, **19**, 4786.
7. L. S. Kappen, M. A. Napier, and I. H. Goldberg, *Proc. Natl. Acad. Sci. USA*, 1980, **77**, 1970.
8. M. A. Napier, B. Holmquist, D. J. Strydom, and I. H. Goldberg, *Biochemistry*, 1981, **20**, 5602.
9. L. S. Kappen and I. H. Goldberg, *Nucleic Acids Res.*, 1978, **5**, 2959.
10. K. Edo, C. Iseki, N. Ishida, T. Horie, G. Kusano, and S. Nozoe, *J. Antibiot.*, 1980, **33**, 1586.
11. R. P. Sheridan and R. K. Gupta, *Biochem. Biophys. Res. Commun.*, 1981, **99**, 213.
12. K. Edo, S. Katamine, F. Kitame, N. Ishida, Y. Koide, G. Kusano, and S. Nozoe, *J. Antibiot.*, 1980, **33**, 347.
13. M. Shikuya, K. Toyooka, and S. Kubota, *Tetrahedron Lett.*, 1984, **25**, 1171.
14. M. A. Napier, I. H. Goldberg, O. D. Hensens, R. S. Dewey, J. M. Liesch, and G. Albers-Schonberg, *Biochem. Biophys. Res. Commun.*, 1981, **100**, 1703.
15. G. Albers-Schonberg, R. S. Dewey, O. D. Hensens, J. M. Liesch, M. A. Napier, and I. H. Goldberg, *Biochem. Biophys. Res. Commun.*, 1980, **95**, 1351.
16. O. D. Hensens, R. S. Dewey, J. M. Liesch, M. A. Napier, R. A. Reamer, J. L. Smith, G. Albers-Schonberg, and I. H. Goldberg, *Biochem. Biophys. Res. Commun.*, 1983, **113**, 538.
17. K. Edo, M. Mizugaki, Y. Koide, H. Seto, K. Furihata, N. Otake, and N. Ishida, *Tetrahedron Lett.*, 1985, **26**, 331.
18. M. D. Lee, T. S. Dunne, M. M. Siegel, C. C. Chang, G. O. Morton, and D. B. Borders, *J. Am. Chem. Soc.*, 1987, **109**, 3464.
19. M. D. Lee, T. S. Dunne, C. C. Chang, G. A. Ellestad, M. M. Siegel, G. O. Morton, W. J. McGahren, and D. B. Borders, *J. Am. Chem. Soc.*, 1987, **109**, 3466.
20. J. Golik, J. Clardy, G. Dubay, G. Groenewold, H. Kawaguchi, M. Konishi, B. Krishnan, H. Ohkuma, K. Saitoh, and T. W. Doyle, *J. Am. Chem. Soc.*, 1987, **109**, 3461.
21. J. Golik, G. Dubay, G. Groenewold, H. Kawaguchi, M. Konishi, B. Krishnan, H. Ohkuma, K. Saitoh, and T. W. Doyle, *J. Am. Chem. Soc.*, 1987, **109**, 3462.
22. K. C. Nicolaou, C. W. Hummel, M. Nakada, K. Shibayama, E. N. Pitsinos, H. Saimoto, Y. Mizuno, K.-U. Baldenius, and A. L. Smith, *J. Am. Chem. Soc.*, 1993, **115**, 7593.
23. S. A. Hitchcock, S. H. Boyer, M. Y. Chumoyer, S. H. Olson, and S. J. Danishefsky, *Angew. Chem., Int. Ed. Engl.*, 1994, **33**, 858.
24. R. L. Halcomb, in "Enediyne Antibiotics as Antitumor Agents," eds. D. B. Borders and T. W. Doyle, Dekker, New York, 1995, p. 383.
25. M. E. Maier, *Synlett*, 1995, 13.
26. K. K. Wang, *Chem. Rev.*, 1996, **96**, 207.
27. H. Lhermitte and D. S. Grierson, *Contemp. Org. Synth.*, 1996, **3**, 93.
28. H. Lhermitte and D. S. Grierson, *Contemp. Org. Synth.*, 1996, **3**, 41.
29. J. W. Grissom, G. U. Gunawardena, D. Klingberg, and D. Huang, *Tetrahedron*, 1996, **52**, 6453.
30. S. J. Danishefsky and M. D. Shair, *J. Org. Chem.*, 1996, **61**, 16.
31. P. Chen, *Angew. Chem., Int. Ed. Engl.*, 1996, **35**, 1478.
32. A. L. Smith and K. C. Nicolaou, *J. Med. Chem.*, 1996, **39**, 2103.
33. B. Meunier (ed.), "DNA and RNA Cleavers and Chemotherapy of Cancer and Viral Diseases," Kluwer, Dordrecht, 1996.
34. D. B. Borders and T. W. Doyle (eds.), "Enediyne Antibiotics as Antitumor Agents," Dekker, New York, 1995.
35. K. Nicolaou, *Chem. Br.*, 1994, **30**, 33.
36. J. A. Murphy and J. Griffiths, *Nat. Prod. Rep.*, 1993, 551.
37. K. C. Nicolaou, A. L. Smith, and E. W. Yue, *Proc. Natl. Acad. Sci. USA*, 1993, **90**, 5881.
38. K. C. Nicolaou, *Angew. Chem., Int. Ed. Engl.*, 1993, **32**, 1377.
39. P. C. Dedon and I. H. Goldberg, *Chem. Res. Toxicol.*, 1992, **5**, 311.
40. P. C. Dedon and I. H. Goldberg, in "Nucleic Acid Targeted Drug Design," eds. C. L. Propst and T. J. Perun, Dekker, New York, 1992, p. 475.
41. K. C. Nicolaou and A. L. Smith, *Acc. Chem. Res.*, 1992, **25**, 497.
42. K. C. Nicolaou, W. M. Dai, S. C. Tsay, V. A. Estevez, and W. Wrasidlo, *Science*, 1992, **256**, 1172.
43. I. H. Goldberg, *Adv. DNA Sequence Specific Agents*, 1992, **1**, 319.
44. I. H. Goldberg, *Acc. Chem. Res.*, 1991, **24**, 191.
45. M. D. Lee, G. A. Ellestad, and D. B. Borders, *Acc. Chem. Res.*, 1991, **24**, 235.
46. K. C. Nicolaou and W. M. Dai, *Angew. Chem., Int. Ed. Engl.*, 1991, **30**, 1387.
47. J. Rohr, *Angew. Chem., Int. Ed. Engl.*, 1995, **34**, 881.
48. D. M. Rothstein, in "Enediyne Antibiotics as Antitumor Agents," eds. D. B. Borders and T. W. Doyle, Dekker, New York, 1995, p. 107.
49. N. Sakata, S. Ikeno, M. Hori, M. Hamada, and T. Otani, *Biosci. Biotechnol. Biochem.*, 1992, **56**, 1592.
50. O. D. Hensens, J. L. Giner, and I. H. Goldberg, *J. Am. Chem. Soc.*, 1989, **111**, 3295.
51. Y. Tokiwa, M. Miyoshi-Saitoh, H. Kobayashi, R. Sunaga, M. Konishi, T. Oki, and S. Iwasaki, *J. Am. Chem. Soc.*, 1992, **114**, 4107.
52. K. S. Lam, J. A. Veitch, J. Golik, B. Krishnan, S. E. Klohr, K. J. Volk, S. Forenza, and T. W. Doyle, *J. Am. Chem. Soc.*, 1993, **115**, 12340.
53. J. Mayer and F. Sondheimer, *J. Am. Chem. Soc.*, 1966, **88**, 603.
54. N. Darby, C. U. Kim, J. A. Salaun, K. W. Shelton, S. Takada, and S. Masamune, *J. Chem. Soc., Chem. Commun.*, 1971, 1516.
55. R. R. Jones and R. G. Bergman, *J. Am. Chem. Soc.*, 1972, **94**, 660.
56. R. G. Bergman, *Acc. Chem. Res.*, 1973, **6**, 25.
57. T. P. Lockhart, C. B. Mallon, and R. G. Bergman, *J. Am. Chem. Soc.*, 1980, **102**, 5976.
58. T. P. Lockhart and R. G. Bergman, *J. Am. Chem. Soc.*, 1981, **103**, 4091.
59. T. P. Lockhart, P. B. Comita, and R. G. Bergman, *J. Am. Chem. Soc.*, 1981, **103**, 4082.
60. N. Koga and K. Morokuma, *J. Am. Chem. Soc.*, 1991, **113**, 1907.
61. E. Kraka and D. Cremer, *J. Am. Chem. Soc.*, 1994, **116**, 4929.

62. R. Lindh and B. J. Persson, *J. Am. Chem. Soc.*, 1994, **116**, 4963.
63. R. Lindh, T. J. Lee, A. Bernhardsson, B. J. Persson, and G. Karlstrom, *J. Am. Chem. Soc.*, 1995, **117**, 7186.
64. P. G. Wenthold, J. A. Paulino, and R. R. Squires, *J. Am. Chem. Soc.*, 1991, **113**, 7414.
65. P. G. Wenthold and R. R. Squires, *J. Am. Chem. Soc.*, 1994, **116**, 6401.
66. W. R. Roth, H. Hopf, and C. Horn, *Chem. Ber.*, 1994, **127**, 1765.
67. J. W. Grissom, T. L. Calkins, H. A. McMillen, and Y. Jiang, *J. Org. Chem.*, 1994, **59**, 5833.
68. A. G. Myers, *Tetrahedron Lett.*, 1987, **28**, 4493.
69. A. G. Myers, E. Y. Kuo, and N. S. Finney, *J. Am. Chem. Soc.*, 1989, **111**, 8057.
70. A. G. Myers, P. S. Dragovich, and E. Y. Kuo, *J. Am. Chem. Soc.*, 1992, **114**, 9369.
71. R. Nagata, H. Yamanaka, E. Okazaki, and I. Saito, *Tetrahedron Lett.*, 1989, **30**, 4995.
72. P. G. Wenthold, S. G. Wierschke, J. J. Nash, and R. R. Squires, *J. Am. Chem. Soc.*, 1993, **115**, 12 611.
73. K. C. Nicolaou, Y. Ogawa, G. Zuccarello, E. J. Schweiger, and T. Kumazawa, *J. Am. Chem. Soc.*, 1988, **110**, 4866.
74. K. C. Nicolaou, G. Zuccarello, C. Riemer, V. A. Estevez, and W. M. Dai, *J. Am. Chem. Soc.*, 1992, **114**, 7360.
75. B. P. Warner, S. P. Millar, R. D. Broene, and S. L. Buchwald, *Science*, 1995, **269**, 814.
76. Y. Wang and M. G. Finn, *J. Am. Chem. Soc.*, 1995, **117**, 8045.
77. K. Ohe, M. Kojima, K. Yonehara, and S. Uemura, *Angew. Chem., Int. Ed. Engl.*, 1996, **35**, 1823.
78. J. P. Snyder, *J. Am. Chem. Soc.*, 1989, **111**, 7630.
79. J. P. Snyder, *J. Am. Chem. Soc.*, 1990, **112**, 5367.
80. J. P. Snyder and G. E. Tipsword, *J. Am. Chem. Soc.*, 1990, **112**, 4040.
81. P. Magnus, S. Fortt, T. Pitterna, and J. P. Snyder, *J. Am. Chem. Soc.*, 1990, **112**, 4986.
82. P. Magnus, P. Carter, J. Elliott, R. Lewis, J. Harling, T. Pitterna, W. E. Bauta, and S. Fortt, *J. Am. Chem. Soc.*, 1992, **114**, 2544.
83. P. Magnus and R. A. Fairhurst, *J. Chem. Soc., Chem. Commun.*, 1994, 1541.
84. K. C. Nicolaou, W. M. Dai, Y. P. Hong, K. K. Baldridge, J. S. Siegel, and S. C. Tsay, *J. Am. Chem. Soc.*, 1993, **115**, 7944.
85. D. Ramkumar, M. Kalpana, B. Varghese, S. Sankararaman, M. N. Jagadeesh, and J. Chandrasekhar, *J. Org. Chem.*, 1996, **61**, 2247.
86. P. Magnus, S. A. Eisenbeis, W. C. Rose, N. Zein, and W. Solomon, *J. Am. Chem. Soc.*, 1993, **115**, 12 627.
87. I. H. Goldberg and L. S. Kappen, in "Enediyne Antibiotics as Antitumor Agents," eds. D. B. Borders and T. W. Doyle, Dekker, New York, 1995, p. 327.
88. I. H. Goldberg, L. S. Kappen, Y. J. Xu, A. Stassinopoulos, X. Zeng, Z. Xi, and C. F. Yang, in "DNA and RNA Cleavers and Chemotherapy of Cancer and Viral Diseases," ed. B. Meunier, Kluwer, Dordrecht, 1996, p. 1.
89. J. Shoji, *J. Antibiot., Ser. A*, 1961, **14**, 27.
90. Y. Koide, F. Ishii, K. Hasuda, Y. Koyama, K. Edo, S. Kitamine, F. Kitame, and N. Ishida, *J. Antibiot.*, 1980, **33**, 342.
91. Y. Koide, A. Ito, F. Ishii, Y. Koyama, K. Edo, and N. Ishida, *J. Antibiot.*, 1982, **35**, 766.
92. T. Hatayama and I. H. Goldberg, *Biochemistry*, 1980, **19**, 5890.
93. T. A. Beerman and I. H. Goldberg, *Biochem. Biophys. Res. Commun.*, 1974, **59**, 1254.
94. L. F. Povirk and I. H. Goldberg, *Biochemistry*, 1982, **21**, 5857.
95. S. S. Lagha, D. D. von Hoff, M. Rozencweig, D. Abraham, M. Slavik, and M. Muggia, *Oncology*, 1976, **33**, 256.
96. H. Maeda, S. Sakamoto, and J. Ogata, *Antimicrob. Agents Chemother.*, 1977, **11**, 941.
97. I. Satake, K. Tari, M. Yamamoto, and H. Nishimura, *J. Urol.*, 1985, **133**, 87.
98. D. L. Urdal and S. Hakomori, *J. Biol. Chem.*, 1980, **225**, 10 509.
99. U. Gottschalk, A. Maibucher, H. Menke, and W. Kohnlein, *J. Antibiot.*, 1990, **43**, 1051.
100. Y. Kohgo, H. Kondo, J. Kato, K. Sasaki, N. Tsushima, T. Nishisato, M. Hirayama, K. Fujikawa, N. Shintani, Y. Mogi, and Y. Niitsu, *Jpn. J. Cancer Res.*, 1990, **81**, 91.
101. H. Maeda, in "Enediyne Antibiotics as Antitumor Agents," eds. D. B. Borders and T. W. Doyle, Dekker, New York, 1995, p. 363.
102. K. Edo, Y. Akiyama, K. Saito, M. Mizugaki, Y. Koide, and N. Ishida, *J. Antibiot.*, 1986, **39**, 1615.
103. A. G. Myers, P. J. Proteau, and T. M. Handel, *J. Am. Chem. Soc.*, 1988, **110**, 7212.
104. P. A. Wender, J. A. McKinney, and C. Mukai, *J. Am. Chem. Soc.*, 1990, **112**, 5369.
105. A. G. Myers, P. M. Harrington, and E. Y. Kuo, *J. Am. Chem. Soc.*, 1991, **113**, 694.
106. P. Magnus, R. Carter, M. Davies, J. Elliott, and T. Pitterna, *Tetrahedron*, 1996, **52**, 6283.
107. A. G. Myers, M. Hammond, Y. Wu, J.-N. Xiang, P. M. Harrington, and E. Y. Kuo, *J. Am. Chem. Soc.*, 1996, **118**, 10 006.
108. K. Edo, M. Ito, N. Ishida, Y. Koide, A. Ito, M. Haga, T. Takahashi, Y. Suzuki, and G. Kusano, *J. Antibiot.*, 1982, **35**, 106.
109. Y. Uesawa, J. Kuwahara, and Y. Sugiura, *Biochem. Biophys. Res. Commun.*, 1989, **164**, 903.
110. D.-H. Chin and I. H. Goldberg, *J. Am. Chem. Soc.*, 1993, **115**, 9341.
111. N. Zein, in "DNA and RNA Cleavers and Chemotherapy of Cancer and Viral Diseases," ed. B. Meunier, Kluwer, Dordrecht, 1996, p. 53.
112. K. Edo, Y. Akiyama-Murai, K. Saito, M. Mizugaki, Y. Koide, and N. Ishida, *J. Antibiot.*, 1988, **41**, 1272.
113. V. Favaudon, *Biochimie*, 1983, **65**, 593.
114. V. Favaudon, R. L. Charnas, and I. H. Goldberg, *Biochemistry*, 1985, **24**, 250.
115. T. A. Beerman, R. Poon, and I. H. Goldberg, *Biochim. Biophys. Acta*, 1977, **475**, 294.
116. R. L. Charnas and I. H. Goldberg, *Biochem. Biophys. Res. Commun.*, 1984, **122**, 642.
117. L. S. Kappen and I. H. Goldberg, *Nucleic Acids Res.*, 1985, **13**, 1637.
118. S. H. Lee and I. H. Goldberg, *Mol. Pharmacol.*, 1988, **33**, 396.
119. O. D. Hensens and I. H. Goldberg, *J. Antibiot.*, 1989, **42**, 761.
120. R. Poon, T. A. Beerman, and I. H. Goldberg, *Biochemistry*, 1977, **16**, 486.
121. M. A. Napier and I. H. Goldberg, *Mol. Pharmacol.*, 1983, **23**, 500.
122. D. H. Chin and I. H. Goldberg, *Biochemistry*, 1986, **25**, 1009.
123. P. C. Dedon and I. H. Goldberg, *Biochemistry*, 1992, **31**, 1909.
124. A. G. Myers, P. M. Harrington, and B. M. Kwon, *J. Am. Chem. Soc.*, 1992, **114**, 1086.

125. A. G. Myers, S. B. Cohen, and B. M. Kwon, *J. Am. Chem. Soc.*, 1994, **116**, 1670.

126. K. H. Kim, B. M. Kwon, A. G. Myers, and D. C. Rees, *Science*, 1993, **262**, 1042.

127. X. Gao, A. Stassinopoulos, J. Gu, and I. H. Goldberg, *Bioorg. Med. Chem.*, 1995, **3**, 795.

128. X. Gao, A. Stassinopoulos, J. S. Rice, and I. H. Goldberg, *Biochemistry*, 1995, **34**, 40.

129. A. G. Myers and P. J. Proteau, *J. Am. Chem. Soc.*, 1989, **111**, 1146.

130. D. H. Chin, C. H. Zeng, C. E. Costello, and I. H. Goldberg, *Biochemistry*, 1988, **27**, 8106.

131. S. M. Meschwitz and I. H. Goldberg, *Proc. Natl. Acad. Sci. USA*, 1991, **88**, 3047.

132. S. M. Meschwitz, R. G. Schultz, G. W. Ashley, and I. H. Goldberg, *Biochemistry*, 1992, **31**, 9117.

133. D. H. Chin and I. H. Goldberg, *Biochemistry*, 1993, **32**, 3611.

134. D. H. Chin and I. H. Goldberg, *J. Am. Chem. Soc.*, 1992, **114**, 1914.

135. P. A. Wender and M. J. Tebbe, *Tetrahedron Lett.*, 1991, **32**, 4863.

136. S. E. McAfee and G. W. Ashley, *Nucleic Acids Res.*, 1992, **20**, 805.

137. H. Sugiyama, T. Fujiwara, and I. Saito, *Tetrahedron Lett.*, 1994, **35**, 8825.

138. T. Tanaka, K. Fujiwara, and M. Hirama, *Tetrahedron Lett.*, 1990, **31**, 5947.

139. M. A. Napier, L. S. Kappen, and I. H. Goldberg, *Biochemistry*, 1980, **19**, 1767.

140. T. Gomibuchi and M. Hirama, *J. Antibiot.*, 1995, **48**, 738.

141. M. Lamothe and P. L. Fuchs, *J. Am. Chem. Soc.*, 1993, **115**, 4483.

142. L. S. Kappen and I. H. Goldberg, *Science*, 1993, **261**, 1319.

143. L. S. Kappen and I. H. Goldberg, *Biochemistry*, 1993, **32**, 13138.

144. L. S. Kappen and I. H. Goldberg, *Biochemistry*, 1995, **34**, 5997.

145. O. D. Hensens, D. H. Chin, A. Stassinopoulos, D. L. Zink, L. S. Kappen, and I. H. Goldberg, *Proc. Natl. Acad. Sci. USA*, 1994, **91**, 4534.

146. O. D. Hensens, G. L. Helms, D. L. Zink, D.-H. Chin, L. S. Kappen, and I. H. Goldberg, *J. Am. Chem. Soc.*, 1993, **115**, 11030.

147. C. F. Yang, A. Stassinopoulos, and I. H. Goldberg, *Biochemistry*, 1995, **34**, 2267.

148. A. Stassinopoulos, J. Ji, X. Gao, and I. H. Goldberg, *Science*, 1996, **272**, 1943.

149. Z. Xi and I. H. Goldberg, unpublished data.

150. L. K. Kappen, Z. Xi, and I. H. Goldberg, *Bioorg. Med. Chem.*, 1997, **5**, 1221.

151. A. Stassinopoulos and I. H. Goldberg, *Biochemistry*, 1995, **34**, 15359.

152. H. Sugiyama, K. Yamashita, M. Nishi, and I. Saito, *Tetrahedron Lett.*, 1992, **33**, 515.

153. H. Sugiyama, K. Yamashita, T. Fujiwara, and I. Saito, *Tetrahedron*, 1994, **50**, 1311.

154. A. G. Myers, S. P. Arvedson, and R. W. Lee, *J. Am. Chem. Soc.*, 1996, **118**, 4725.

155. D.-H. Chin, personal communication.

156. J. M. Battigello, M. Cui, S. Roshong, and B. J. Carter, *Bioorg. Med. Chem.*, 1995, **3**, 839.

157. T. Hatayama, I. H. Goldberg, M. Takeshita, and A. P. Grollman, *Proc. Natl. Acad. Sci. USA*, 1978, **75**, 3603.

158. M. Takeshita, L. S. Kappen, A. P. Grollman, M. Eisenberg, and I. H. Goldberg, *Biochemistry*, 1981, **20**, 7599.

159. S. H. Lee and I. H. Goldberg, *Biochemistry*, 1989, **28**, 1019.

160. L. S. Kappen, C. Q. Chen, and I. H. Goldberg, *Biochemistry*, 1988, **27**, 4331.

161. L. S. Kappen and I. H. Goldberg, *Biochemistry*, 1989, **28**, 1027.

162. L. F. Povirk and I. H. Goldberg, *Proc. Natl. Acad. Sci. USA*, 1985, **82**, 3182.

163. L. F. Povirk, C. W. Houlgrave, and Y. H. Han, *J. Biol. Chem.*, 1988, **263**, 19263.

164. P. C. Dedon, Z. W. Jiang, and I. H. Goldberg, *Biochemistry*, 1992, **31**, 1917.

165. A. Galat and I. H. Goldberg, *Nucleic Acids Res.*, 1990, **18**, 2093.

166. L. S. Kappen, I. H. Goldberg, B. L. Frank, L. Worth, Jr., D. F. Christner, J. W. Kozarich, and J. Stubbe, *Biochemistry*, 1991, **30**, 2034.

167. L. S. Kappen and I. H. Goldberg, *Biochemistry*, 1992, **31**, 9081.

168. L. S. Kappen and I. H. Goldberg, *Proc. Natl. Acad. Sci. USA*, 1992, **89**, 6706.

169. L. D. Williams and I. H. Goldberg, *Biochemistry*, 1988, **27**, 3004.

170. X. Zeng, Z. Xi, L. S. Kappen, W. Tan, and I. H. Goldberg, *Biochemistry*, 1995, **34**, 12435.

171. L. Yu, I. H. Goldberg, and P. C. Dedon, *J. Biol. Chem.*, 1994, **269**, 4144.

172. D. Dasgupta and I. H. Goldberg, *Biochemistry*, 1985, **24**, 6913.

173. D. Dasgupta, D. S. Auld, and I. H. Goldberg, *Biochemistry*, 1985, **24**, 7049.

174. L. F. Povirk, N. Dattagupta, B. C. Warf, and I. H. Goldberg, *Biochemistry*, 1981, **20**, 4007.

175. D. Dasgupta and I. H. Goldberg, *Nucleic Acids Res.*, 1986, **14**, 1089.

176. A. Stassinopoulos and I. H. Goldberg, *Bioorg. Med. Chem.*, 1995, **3**, 713.

177. A. A. Fantini and R. T. Testa, in "Enediyne Antibiotics as Antitumor Agents," eds. D. B. Borders and T. W. Doyle, Dekker, New York, 1995, p. 29.

178. F. E. Durr, R. E. Wallace, R. T. Testa, and N. A. Kuck, in "Enediyne Antibiotics as Antitumor Agents," eds. D. B. Borders and T. W. Doyle, Dekker, New York, 1995, p. 127.

179. M. D. Lee, T. S. Dunne, C. C. Chang, M. M. Siegel, G. O. Morton, G. A. Ellestad, W. J. McGahren, and D. B. Borders, *J. Am. Chem. Soc.*, 1992, **114**, 985.

180. K. C. Nicolaou, C. W. Hummel, M. Nakada, E. N. Pitsinos, A. L. Smith, K. Shibayama, and H. Saimoto, *J. Am. Chem. Soc.*, 1992, **114**, 10082.

181. N. Zein, W. J. McGahren, G. O. Morton, J. Ashcroft, and G. A. Ellestad, *J. Am. Chem. Soc.*, 1989, **111**, 6888.

182. J. J. De Voss, C. A. Townsend, W.-D. Ding, G. O. Morton, G. A. Ellestad, N. Zein, A. B. Tabor, and S. L. Schreiber, *J. Am. Chem. Soc.*, 1990, **112**, 9669.

183. J. J. Hangeland, J. J. De Voss, J. A. Heath, C. A. Townsend, W.-D. Ding, J. S. Ashcroft, and G. A. Ellestad, *J. Am. Chem. Soc.*, 1992, **114**, 9200.

184. J. J. De Voss, J. J. Hangeland, and C. A. Townsend, *J. Am. Chem. Soc.*, 1990, **112**, 4554.

185. G. A. Ellestad, P. R. Hamann, N. Zein, G. O. Morton, M. M. Siegel, M. Pastel, D. B. Borders, and W. J. McGahren, *Tetrahedron Lett.*, 1989, **30**, 3033.

186. W. J. McGahren, W.-D. Ding, and G. A. Ellestad, in "Enediyne Antibiotics as Antitumor Agents," eds. D. B. Borders and T. W. Doyle, Dekker, New York, 1995, p. 75.

187. A. G. Myers, S. B. Cohen, and B. M. Kwon, *J. Am. Chem. Soc.*, 1994, **116**, 1255.
188. L. M. Hinman, P. R. Hamann, R. Wallace, A. T. Menendez, F. E. Durr, and J. Upeslacis, *Cancer Res.*, 1993, **53**, 3336.
189. N. Zein, M. Poncin, R. Nilakantan, and G. A. Ellestad, *Science*, 1989, **244**, 697.
190. M. Chatterjee, P. J. Smith, and C. A. Townsend, *J. Am. Chem. Soc.*, 1996, **118**, 1938.
191. N. Zein, A. M. Sinha, W. J. McGahren, and G. A. Ellestad, *Science*, 1988, **240**, 1198.
192. G. A. Ellestad, W.-D. Ding, N. Zein, and C. A. Townsend, in "Enediyne Antibiotics as Antitumor Agents," eds. D. B. Borders and T. W. Doyle, Dekker, New York, 1995, p. 137.
193. P. C. Dedon, A. A. Salzberg, and J. Xu, *Biochemistry*, 1993, **32**, 3617.
194. R. C. Hawley, L. L. Kiessling, and S. L. Schreiber, *Proc. Natl. Acad. Sci. USA*, 1989, **86**, 1105.
195. D. R. Langley, J. Golik, B. Krishnan, T. W. Doyle, and D. L. Beveridge, *J. Am. Chem. Soc.*, 1994, **116**, 15.
196. S. Walker, K. G. Valentine, and D. Kahne, *J. Am. Chem. Soc.*, 1990, **112**, 6428.
197. S. Walker, J. Murnick, and D. Kahne, *J. Am. Chem. Soc.*, 1993, **115**, 7954.
198. L. G. Paloma, J. A. Smith, W. J. Chazin, and K. C. Nicolaou, *J. Am. Chem. Soc.*, 1994, **116**, 3697.
199. N. Ikemoto, R. A. Kumar, T. T. Ling, G. A. Ellestad, S. J. Danishefsky, and D. J. Patel, *Proc. Natl. Acad. Sci. USA*, 1995, **92**, 10 506.
200. S. Walker, R. Landovitz, W. D. Ding, G. A. Ellestad, and D. Kahne, *Proc. Natl. Acad. Sci. USA*, 1992, **89**, 4608.
201. S. C. Mah, C. A. Townsend, and T. D. Tullius, *Biochemistry*, 1994, **33**, 614.
202. T. Li, Z. Zeng, V. A. Estevez, K. U. Baldenius, K. C. Nicolaou, and G. F. Joyce, *J. Am. Chem. Soc.*, 1994, **116**, 3709.
203. C. Bailly and M. J. Waring, *J. Am. Chem. Soc.*, 1995, **117**, 7311.
204. M. Chatterjee, S. C. Mah, T. D. Tullius, and C. A. Townsend, *J. Am. Chem. Soc.*, 1995, **117**, 8074.
205. W. Ding and G. A. Ellestad, *J. Am. Chem. Soc.*, 1991, **113**, 6617.
206. G. Krishnamurthy, W. Ding, L. O'Brien, and G. A. Ellestad, *Tetrahedron*, 1994, **50**, 1341.
207. G. Krishnamurthy, M. D. Brenowitz, and G. A. Ellestad, *Biochemistry*, 1995, **34**, 1001.
208. G. Krishnamurthy, W. Ding, and G. A. Ellestad, in "DNA and RNA Cleavers and Chemotherapy of Cancer and Viral Diseases," ed. B. Meunier, Kluwer, Dordrecht, 1996, p. 37.
209. R. L. Halcomb, *Proc. Natl. Acad. Sci. USA*, 1994, **91**, 9197.
210. K. C. Nicolaou, B. M. Smith, K. Ajito, H. Komatsu, L. Gomez-Paloma, and Y. Tor, *J. Am. Chem. Soc.*, 1996, **118**, 2303.
211. J. Drak, N. Iwasawa, S. Danishefsky, and D. M. Crothers, *Proc. Natl. Acad. Sci. USA*, 1991, **88**, 7464.
212. D. Kahne, *Chem. Biol.*, 1995, **2**, 7.
213. C. Zimmer and U. Wahnert, *Prog. Biophys. Mol. Biol.*, 1986, **47**, 31.
214. S. N. Ho, S. H. Boyer, S. L. Schreiber, S. J. Danishefsky, and G. R. Crabtree, *Proc. Natl. Acad. Sci. USA*, 1994, **91**, 9203.
215. C. Liu, B. M. Smith, K. Ajito, H. Komatsu, L. Gomez-Paloma, T. Li, E. A. Theodorakis, K. C. Nicolaou, and P. K. Vogt, *Proc. Natl. Acad. Sci. USA*, 1996, **93**, 940.
216. M. Konishi, H. Ohkuma, K.-I. Saitoh, H. Kawaguchi, J. Golik, G. Dubay, G. Groenewold, B. Krishnan, and T. W. Doyle, *J. Antibiot.*, 1985, **38**, 1605.
217. J. A. Beutler, P. Clark, A. B. Alvarado, and J. Golik, *J. Nat. Prod.*, 1994, **57**, 629.
218. M. Iwami, S. Kiyoto, M. Nishikawa, H. Terano, M. Kohsaka, H. Aoki, and H. Imanaka, *J. Antibiot.*, 1985, **38**, 853.
219. K. S. Lam, J. A. Veitch, J. Golik, W. C. Rose, T. W. Doyle, and S. Forenza, *J. Antibiot.*, 1995, **48**, 1497.
220. K. S. Lam and S. Forenza, in "Enediyne Antibiotics as Antitumor Angents," eds. D. B. Borders and T. W. Doyle, Dekker, New York, 1995, p. 161.
221. A. M. Casazza and S. L. Kelley, in "Enediyne Antibiotics as Antitumor Agents," eds. D. B. Borders and T. W. Doyle, Dekker, New York, 1995, p. 283.
222. J. Golik, in "Enediyne Antibiotics as Antitumor Agents," eds. D. B. Borders and T. W. Doyle, Dekker, New York, 1995, p. 187.
223. J. Golik, H. Wong, D. M. Vyas, and T. W. Doyle, *Tetrahedron Lett.*, 1989, **30**, 2497.
224. M. D. Wittman, R. L. Halcomb, S. J. Danishefsky, J. Golik, and D. Vyas, *J. Org. Chem.*, 1990, **55**, 1979.
225. J. Golik, T. W. Doyle, G. VanDuyne, and J. Clardy, *Tetrahedron Lett.*, 1990, **43**, 6149.
226. R. L. Halcomb, M. D. Wittman, S. H. Olson, S. J. Danishefsky, J. Golik, H. Wong, and D. Vyas, *J. Am. Chem. Soc.*, 1991, **113**, 5080.
227. J. Golik, H. Wong, B. Krishnan, D. M. Vyas, and T. W. Doyle, *Tetrahedron Lett.*, 1991, **32**, 1851.
228. J. Golik, B. Krishnan, T. W. Doyle, G. VanDuyne, and J. Clardy, *Tetrahedron Lett.*, 1992, **33**, 6049.
229. K. S. Lam and J. A. Veitch, in "Enediyne Antibiotics as Antitumor Agents," eds. D. B. Borders and T. W. Doyle, Dekker, New York, 1995, p. 217.
230. D. R. Langley, J. Golik, B. Krishnan, T. W. Doyle, and D. L. Beveridge, *J. Am. Chem. Soc.*, 1994, **116**, 15.
231. L. Yu, J. Golik, R. Harrison, and P. Dedon, *J. Am. Chem. Soc.*, 1994, **116**, 9733.
232. Y. Uesawa and Y. Sugiura, *Biochemistry*, 1991, **30**, 9242.
233. B. H. Long, J. Golik, S. Forenza, B. Ward, R. Rehfuss, J. C. Dabrowiak, J. J. Catino, S. T. Musial, K. W. Brookshire, and T. W. Doyle, *Proc. Natl. Acad. Sci. USA*, 1989, **86**, 2.
234. Y. Sugiura, Y. Uesawa, Y. Takahashi, J. Kuwahara, J. Golik, and T. W. Doyle, *Proc. Natl. Acad. Sci. USA*, 1989, **86**, 7672.
235. H. Kishikawa, Y.-P. Jiang, J. Goodisman, and J. C. Dabrowiak, *J. Am. Chem. Soc.*, 1991, **113**, 5434.
236. D. F. Christner, B. L. Frank, J. W. Kozarich, J. Stubbe, J. Golik, T. W. Doyle, I. E. Rosenberg, and B. Krishnan, *J. Am. Chem. Soc.*, 1992, **114**, 8763.
237. M. Uesugi, T. Kusakabe, and Y. Sugiura, *Biochim. Biophys. Acta*, 1995, **1261**, 99.
238. D. R. Langley, in "Enediyne Antibiotics as Antitumor Agents," eds. D. B. Borders and T. W. Doyle, Dekker, New York, 1995, p. 239.
239. N. Ikemoto, R. A. Kumar, P. C. Dedon, S. J. Danishefsky, and D. J. Patel, *J. Am. Chem. Soc.*, 1994, **116**, 9387.
240. M. Uesugi and Y. Sugiura, *Biochemistry*, 1993, **32**, 4622.
241. R. Totsuka, Y. Aizawa, M. Uesugi, Y. Okuno, T. Matsumoto, and Y. Sugiura, *Biochem. Biophys. Res. Commun.*, 1995, **208**, 168.
242. M. Konishi, H. Ohkuma, K. Matsumoto, T. Tsuno, H. Kamei, T. Miyaki, T. Oki, H. Kawaguchi, G. D. VanDuyne, and J. Clardy, *J. Antibiot.*, 1989, **42**, 1449.

243. M. Miyoshi-Saitoh, N. Morisaki, Y. Tokiwa, S. Iwasaki, M. Konishi, K. Saitoh, and T. Oki, *J. Antibiot.*, 1991, **44**, 1037.
244. M. Konishi and T. Oki, in "Enediyne Antibiotics as Antitumor Agents," eds. D. B. Borders and T. W. Doyle, Dekker, New York, 1995, p. 301.
245. M. Konishi, H. Ohkuma, T. Tsuno, T. Oki, G. D. VanDuyne, and J. Clardy, *J. Am. Chem. Soc.*, 1990, **112**, 3715.
246. D. R. Langley, T. W. Doyle, and D. L. Beveridge, *J. Am. Chem. Soc.*, 1991, **113**, 3495.
247. A. G. Myers, M. E. Fraley, N. J. Tom, S. B. Cohen, and D. J. Madar, *Chem. Biol.*, 1995, **2**, 33.
248. J. Taunton, J. L. Wood, and S. L. Schreiber, *J. Am. Chem. Soc.*, 1993, **115**, 10 378.
249. M. D. Shair, T. Y. Yoon, and S. J. Danishefsky, *Angew. Chem., Int. Ed. Engl.*, 1995, **34**, 1721.
250. Y. Sugiura, T. Shiraki, M. Konishi, and T. Oki, *Proc. Natl. Acad. Sci. USA*, 1990, **87**, 3831.
251. Y. Sugiura, T. Arakawa, M. Uesugi, T. Shiraki, H. Ohkuma, and M. Konishi, *Biochemistry*, 1991, **30**, 2989.
252. T. Shiraki and Y. Sugiura, *Biochemistry*, 1990, **29**, 9795.
253. T. Shiraki, M. Uesugi, and Y. Sugiura, *Biochem. Biophys. Res. Commun.*, 1992, **188**, 584.
254. T. Arakawa, T. Kusakabe, J. Kuwahara, M. Otsuka, and Y. Sugiura, *Biochem. Biophys. Res. Commun.*, 1993, **190**, 362.
255. M. F. Semmelhack, J. Gallagher, and D. Cohen, *Tetrahedron Lett.*, 1990, **31**, 1521.
256. P. A. Wender, R. C. Kelly, S. Beckham, and B. L. Miller, *Proc. Natl. Acad. Sci. USA*, 1991, **88**, 8835.
257. M. G. Cardozo and A. J. Hopfinger, *Biopolymers*, 1993, **33**, 377.
258. M. G. Cardozo and A. J. Hopfinger, *Mol. Pharmacol.*, 1991, **40**, 1023.
259. A. G. Myers, S. B. Cohen, N. J. Tom, D. J. Madar, and M. E. Fraley, *J. Am. Chem. Soc.*, 1995, **117**, 7574.
260. Q. Guo, M. Lu, M. Shahrestanifar, R. D. Sheardy, and N. R. Kallenbach, *Biochemistry*, 1991, **30**, 11 735.
261. A. Ichikawa, T. Kuboya, T. Aoyama, and Y. Sugiura, *Biochemistry*, 1992, **31**, 6784.
262. T. Kusakabe, K. Maekawa, A. Ichikawa, M. Uesugi, and Y. Sugiura, *Biochemistry*, 1993, **32**, 11 669.
263. T. Kusakabe, M. Uesugi, and Y. Sugiura, *Biochemistry*, 1995, **34**, 9944.
264. Y. Sugiura and T. Kusakabe, In "DNA and RNA Cleavers and Chemotherapy of Cancer and Viral Diseases," ed. B. Meunier, Kluwer, Dordrecht, 1996, p. 65.
265. K. S. Lam, G. A. Hesler, D. R. Gustavson, A. R. Crosswell, J. M. Veitch, S. Forenza, and K. Tomita, *J. Antibiot.*, 1991, **44**, 472.
266. S. J. Hofstead, J. A. Matson, A. R. Malacko, and H. Marquardt, *J. Antibiot.*, 1992, **45**, 1250.
267. K. L. Constantine, K. L. Colson, M. Wittekind, M. S. Friedrichs, N. Zein, J. Tuttle, D. R. Langley, J. E. Leet, D. R. Schroeder, K. S. Lam, B. T. Farmer, II, W. J. Metzler, R. E. Bruccoleri, and L. Mueller, *Biochemistry*, 1994, **33**, 11 438.
268. J. E. Leet, D. R. Schroeder, S. J. Hofstead, J. Golik, K. L. Colson, S. Huang, S. E. Khlor, T. W. Doyle, and J. A. Matson, *J. Am. Chem. Soc.*, 1992, **114**, 7946.
269. J. E. Leet, D. R. Schroeder, D. R. Langley, K. L. Colson, S. Huang, S. E. Klohr, M. S. Lee, J. Golik, S. J. Hofstead, T. W. Doyle, and J. A. Matson, *J. Am. Chem. Soc.*, 1993, **115**, 8432.
270. N. Zein, A. M. Casazza, T. W. Doyle, J. E. Leet, D. R. Schroeder, W. Solomon, and S. G. Nadler, *Proc. Natl. Acad. Sci. USA*, 1993, **90**, 8009.
271. N. Zein, P. Reiss, M. Bernatowicz, and M. Bolgar, *Chem. Biol.*, 1995, **2**, 451.
272. N. Zein, K. L. Colson, J. E. Leet, D. R. Schroeder, W. Solomon, T. W. Doyle, and A. M. Casazza, *Proc. Natl. Acad. Sci. USA*, 1993, **90**, 2822.
273. J. L. Hu, Y. C. Xue, M. Y. Xie, R. Zhang, T. Otani, Y. Minami, Y. Yamada, and T. Marunaka, *J. Antibiot.*, 1988, **41**, 1575.
274. T. Otani, Y. Minami, T. Marunaka, R. Zhang, and M. Xie, *J. Antibiot.*, 1988, **41**, 1580.
275. T. Otani, Y. Minami, K. Sakawa, and K. Yoshida, *J. Antibiot.*, 1991, **44**, 564.
276. T. Otani, *J. Antibiot.*, 1993, **46**, 791.
277. T. Otani, T. Yasuhara, Y. Minami, T. Shimazu, R. Zhang, and M. Y. Xie, *Agric. Biol. Chem.*, 1991, **55**, 407.
278. N. Sakata, K. S. Tsuchiya, Y. Moriya, H. Hayashi, M. Hori, T. Otani, M. Nagai, and T. Aoyagi, *J. Antibiot.*, 1992, **45**, 113.
279. K. S-Tsuchiya, M. Arita, M. Hori, and T. Otani, *J. Antibiot.*, 1994, **47**, 787.
280. Y. S. Zhen, X. Y. Ming, B. Yu, T. Otani, H. Saito, and Y. Yamada, *J. Antibiot.*, 1989, **42**, 1294.
281. Y. J. Xu, D. D. Li, and Y. S. Zhen, *Cancer Chemother. Pharmacol.*, 1990, **27**, 41.
282. Y. Sugimoto, T. Otani, S. Oie, K. Wierzba, and Y. Yamada, *J. Antibiot.*, 1990, **43**, 417.
283. B. Jiang, D. D. Li, and Y. S. Zhen, *Biochem. Biophys. Res. Commun.*, 1995, **208**, 238.
284. R. J. Cobuzzi, Jr., S. K. Kotsopoulos, T. Otani, and T. A. Beerman, *Biochemistry*, 1995, **34**, 583.
285. M. M. McHugh, J. M. Woynarowski, L. S. Gawron, T. Otani, and T. A. Beerman, *Biochemistry*, 1995, **34**, 1805.
286. Y. Minami, K. Yoshida, R. Azuma, M. Saeki, and T. Otani, *Tetrahedron Lett.*, 1993, **34**, 2633.
287. K. Yoshida, Y. Minami, R. Azuma, M. Saeki, and T. Otani, *Tetrahedron Lett.*, 1993, **34**, 2637.
288. Y. Okuno, M. Otsuka, and Y. Sugiura, *J. Med. Chem.*, 1994, **37**, 2266.
289. I. Sato, Y. Akahori, K. Iida, and M. Hirama, *Tetrahedron Lett.*, 1996, **37**, 5135.
290. K. Iida, T. Ishii, M. Hirama, T. Otani, Y. Minami, and K. Yoshida, *Tetrahedron Lett.*, 1993, **34**, 4079.
291. L. K. Kappen, I. H. Goldberg, and T. S. A. Samy, *Biochemistry*, 1979, **18**, 5123.
292. Y. Kumada, T. Miwa, N. Naoi, K. Watanabe, H. Naganawa, T. Takita, H. Umezawa, H. Nakamura, and Y. Iitaka, *J. Antibiot.*, 1983, **36**, 200.
293. M. Shibuya, H. Sakurai, T. Maeda, E. Nashiwaki, and M. Saito, *Tetrahedron Lett.*, 1986, **27**, 1351.
294. N. Sakata, S. Minamitani, T. Kanbe, M. Hori, M. Hamada, and K. Edo, *Biol. Pharm. Bull.*, 1993, **16**, 26.
295. M. F. Semmelhack, T. Neu, and F. Foubelo, *Tetrahedron Lett.*, 1992, **33**, 3277.
296. M. F. Semmelhack, T. Neu, and F. Foubelo, *J. Org. Chem.*, 1994, **59**, 5038.
297. K.-I. Yoshida, Y. Minami, T. Otani, Y. Tada, and M. Hirama, *Tetrahedron Lett.*, 1994, **35**, 5253.
298. K. Iida and M. Hirama, *J. Am. Chem. Soc.*, 1995, **117**, 8875.
299. B. L. Frank, L. Worth, Jr., D. F. Christner, J. W. Kozarich, J. Stubbe, L. S. Kappen, and I. H. Goldberg, *J. Am. Chem. Soc.*, 1991, **113**, 2271.
300. L. S. Kappen, I. H. Goldberg, S. H. Wu, J. Stubbe, L. Worth, Jr., and J. W. Kozarich, *J. Am. Chem. Soc.*, 1990, **112**, 2797.

301. Y. J. Xu, Z. Xi, Y. S. Zhen, and I. H. Goldberg, *Biochemistry*, 1995, **34**, 12451.
302. C. F. Logan and P. Chen, *J. Am. Chem. Soc.*, 1996, **118**, 2113.
303. R. Li, R. Smith, and H. I. Kenttamaa, *J. Am. Chem. Soc.*, 1996, **118**, 5056.
304. M. J. Schottelius and P. Chen, *J. Am. Chem. Soc.*, 1996, **118**, 4896.
305. T. Matsumoto, Y. Okuno, and Y. Sugiura, *Biochem. Biophys. Res. Commun.*, 1993, **195**, 659.
306. Y. J. Xu, Y. S. Zhen, and I. H. Goldberg, *Biochemistry*, 1994, **33**, 5947.
307. Y. Sugiura and T. Matsumoto, *Biochemistry*, 1993, **32**, 5548.
308. L. Worth, Jr., B. L. Frank, D. F. Chritner, M. J. Absalon, J. Stubbe, and J. W. Kozarich, *Biochemistry*, 1993, **32**, 2601.
309. Y. J. Xu, D. D. Li, and Y. S. Zhen, *Sci. Sin., Ser. B* (*Engl. Ed.*), 1992, **8**, 814.
310. L. Yu, S. Mah, T. Otani, and P. Dedon, *J. Am. Chem. Soc.*, 1995, **117**, 8877.
311. Y. Okuno, T. Iwashita, T. Otani, and Y. Sugiura, *J. Am. Chem. Soc.*, 1996, **118**, 4729.
312. M. Hanada, H. Ohkuma, T. Yonemoto, K. Tomita, M. Ohbayashi, H. Kamie, T. Miyaki, M. Konishi, H. Kawaguchi, and S. Forenza, *J. Antibiot.*, 1991, **44**, 403.
313. D. R. Schroeder, K. L. Colson, S. E. Klohr, N. Zein, D. R. Langley, M. S. Lee, J. A. Matson, and T. W. Doyle, *J. Am. Chem. Soc.*, 1994, **116**, 9351.
314. T. W. Doyle and D. B. Borders, in "Enediyne Antibiotics as Antitumor Agents," eds. D. B. Borders and T. W. Doyle, Dekker, New York, 1995, p. 1.
315. N. Zein, W. Solomon, K. L. Colson, and D. R. Schroeder, *Biochemistry*, 1995, **34**, 11591.
316. L. A. McDonald, T. L. Capson, G. Krishnamurthy, W. Ding, G. A. Ellestad, V. S. Bernan, W. M. Maiese, P. Lassota, C. Discafani, R. A. Kramer, and C. M. Ireland, *J. Am. Chem. Soc.*, 1996, **118**, 10898.
317. X. Gao, *J. Mol. Biol.*, 1992, **225**, 125.
318. E. Adjadj, E. Quiniou, J. Mispelter, V. Favaudon, and J. M. Lhoste, *Eur. J. Biochem.*, 1992, **203**, 505.
319. A. Teplyakov, G. Obmolova, K. Wilson, and K. Kuromizu, *Eur. J. Biochem.*, 1993, **213**, 737.
320. T. Tanaka, M. Hirama, M. Fujita, S. Imajo, and M. Ishiguro, *J. Chem. Soc., Chem. Commun.*, 1993, 1205.
321. A. Zaheer, S. Zaheer, and R. Montgomery, *J. Biol. Chem.*, 1985, **260**, 11787.
322. A. Zaheer, S. Zaheer, and R. Montgomery, *Biochim. Biophys. Acta*, 1985, **841**, 261.
323. T. Oda, F. Sato, and H. Maeda, *J. Natl. Cancer Inst.*, 1987, **79**, 1205.
324. T. L. Hartsell, L. M. Hinman, P. R. Hamann, and N. F. Schor, *J. Pharmacol. Exp. Ther.*, 1996, **277**, 1158.
325. M. Cortazzo and N. F. Schor, *Cancer Res.*, 1996, **56**, 1199.
326. A. Hiatt, R. Merlock, S. Mauch, and W. Wrasidlo, *Bioorg. Med. Chem.*, 1994, **2**, 315.
327. K. C. Nicolaou, E. N. Pitsinos, E. A. Theodorakis, H. Saimoto, and W. Wrasidlo, *Chem. Biol.*, 1994, **1**, 57.
328. M. T. Kuo and T. S. Samy, *Biochim. Biophys. Acta*, 1978, **518**, 186.
329. T. A. Beerman, G. Mueller, and H. Grimmond, *Mol. Pharmacol.*, 1983, **23**, 493.
330. M. M. McHugh, J. Woynarowski, and T. Beerman, *Biochim. Biophys. Acta*, 1982, **696**, 7.
331. L. S. Kappen, T. E. Ellenberger, and I. H. Goldberg, *Biochemistry*, 1987, **26**, 384.
332. B. L. Smith, G. B. Bauer, and L. F. Povirk, *J. Biol. Chem.*, 1994, **269**, 30587.
333. J. E. Grimwade, E. B. Cullinan, and T. A. Beerman, *Biochim. Biophys. Acta*, 1988, **950**, 102.
334. P. N. Kuduvalli, C. A. Townsend, and T. D. Tullius, *Biochemistry*, 1995, **34**, 3899.
335. T. Hatayama and I. H. Goldberg, *Biochim. Biophys. Acta*, 1979, **563**, 59.
336. E. Eisenstadt, M. Wolf, and I. H. Goldberg, *J. Bacteriol.*, 1980, **144**, 656.
337. L. F. Povirk and R. J. Steighner, *Mutat. Res.*, 1989, **214**, 13.
338. L. F. Povirk and I. H. Goldberg, *Nucleic Acids Res.*, 1986, **14**, 1417.
339. L. F. Povirk and I. H. Goldberg, *Biochimie*, 1987, **69**, 815.
340. D. A. Brenner, A. C. Smigochi, and R. D. Camerini-Otero, *Mol. Cell. Biol.*, 1985, **5**, 684.
341. A. O. Colson and M. D. Sevilla, *J. Phys. Chem.*, 1995, **99**, 3867.
342. K. Miaskiewicz and R. Osman, *J. Am. Chem. Soc.*, 1994, **116**, 232.
343. W. Saenger, "Principles of Nucleic Acid Structure," Springer, New York, 1984.
344. G. Pratviel, J. Bernadou, and B. Meunier, *Angew. Chem., Int. Ed. Engl.*, 1995, **34**, 746.
345. D. H. Chin, S. A. Carr, and I. H. Goldberg, *J. Biol. Chem.*, 1984, **259**, 9975.
346. L. F. Povirk and I. H. Goldberg, *J. Biol. Chem.*, 1983, **258**, 11763.
347. L. S. Kappen and I. H. Goldberg, *Biochemistry*, 1983, **22**, 4872.
348. D. H. Chin, L. S. Kappen, and I. H. Goldberg, *Proc. Natl. Acad. Sci. USA*, 1987, **84**, 7070.
349. H. Kawabata, H. Takeshita, T. Fujiwara, H. Sugiyama, T. Matsuura, and I. Saito, *Tetrahedron Lett.*, 1989, **30**, 4263.
350. I. Saito, H. Kawabata, T. Fujiwara, H. Sugiyama, and T. Matsuura, *J. Am. Chem. Soc.*, 1989, **111**, 8302.
351. H. Sugiyama, H. Kawabata, T. Fujiwara, Y. Dannoue, and I. Saito, *J. Am. Chem. Soc.*, 1990, **112**, 5252.
352. B. Giese, X. Beyrich-Graf, P. Erdmann, L. Giraud, P. Imwinkelried, S. N. Muller, and U. Schwitter, *J. Am. Chem. Soc.*, 1995, **117**, 6146.
353. B. Giese, X. Beyrich-Graf, P. Erdmann, M. Petretta, and U. Schwitter, *Chem. Biol.*, 1995, **2**, 367.
354. J. Stubbe and J. W. Kozarich, *Chem. Rev.*, 1987, **87**, 1107.
355. S. McGall, J. Stubbe, and J. W. Kozarich, *J. Org. Chem.*, 1991, **56**, 48.
356. R. J. Duff, E. de Vroom, A. Geluk, S. M. Hecht, G. A. van der Marel, and J. H. van Boom, *J. Am. Chem. Soc.*, 1993, **115**, 3350.
357. L. S. Kappen and I. H. Goldberg, *Proc. Natl. Acad. Sci. USA*, 1984, **81**, 3312.
358. L. S. Kappen, T. R. Lee, C. C. Yang, and I. H. Goldberg, *Biochemistry*, 1989, **28**, 4540.
359. C. Nese, M. N. Schuchmann, S. Steenken, and C. von Sonntag, *J. Chem. Soc., Perkins Trans. 2*, 1995, 1037.
360. L. F. Povirk and I. H. Goldberg, *Proc. Natl. Acad. Sci. USA*, 1982, **79**, 369.
361. L. F. Povirk and I. H. Goldberg, *Nucleic Acids Res.*, 1982, **10**, 6255.
362. L. F. Povirk and I. H. Goldberg, *Biochemistry*, 1984, **23**, 6304.
363. L. F. Povirk and I. H. Goldberg, *Biochemistry*, 1985, **24**, 4035.
364. Y.-J. Xu, Y. S. Zhen, and I. H. Goldberg, *J. Am. Chem. Soc.*, 1997, **119**, 1133.
365. P. Zheng, C. L. Liu, Z. Xi, R. D. Smith, and I. H. Goldberg, *Biochemistry*, 1998, **37**, 1706.

# 7.16
# DNA Topoisomerase Inhibitors

TIMOTHY L. MacDONALD, MARC A. LABROLI, and
JETZE J. TEPE
*University of Virginia, Charlottesville, VA, USA*

7.16.1   INTRODUCTION                                                                                              594

    *7.16.1.1   Physiological Roles of the DNA Topoisomerases*                                             594
    *7.16.1.2   DNA Topoisomerases in Chemotherapy*                                                          594
    *7.16.1.3   Inhibitors of the DNA Topoisomerases*                                                        595
        *7.16.1.3.1   DNA binding agents: intercalators and minor groove binders*                        595
        *7.16.1.3.2   Direct protein binding agents*                                                     595
        *7.16.1.3.3   "Cleavable complex" mediators*                                                     596
    *7.16.1.4   Cellular Pharmacology of Topoisomerase-directed Cell Death*                                  596

7.16.2   DNA TOPOISOMERASE AND ITS INHIBITORS                                                                       597

    *7.16.2.1   Catalytic Mechanism of DNA Topoisomerase I*                                                  598
        *7.16.2.1.1   Step 1: DNA recognition and binding*                                               598
        *7.16.2.1.2   Step 2: Prestrand passage cleavage–religation equilibrium*                         598
        *7.16.2.1.3   Step 3: Single-strand passage*                                                     599
        *7.16.2.1.4   Step 4: Poststrand passage cleavage–religation equilibrium*                        599
        *7.16.2.1.5   Step 5: Enzyme turnover*                                                           599
    *7.16.2.2   Classes of Cleavable Complex Mediators*                                                      599
        *7.16.2.2.1   Camptothecin and its derivatives*                                                  599
        *7.16.2.2.2   Indolocarbazoles*                                                                  600
        *7.16.2.2.3   Bulgarein*                                                                         600
        *7.16.2.2.4   The naphthacenequinones: UCE6, UCE1022 and saintopin*                              600
        *7.16.2.2.5   The benzo[c]phenanthridines nitidine, chelerythrine, isofagaridine, and fagaridine*   601
        *7.16.2.2.6   Coralyne and its derivatives*                                                      601

7.16.3   DNA TOPOISOMERASE II AND ITS INHIBITORS                                                                    601

    *7.16.3.1   Catalytic Mechanism of DNA Topoisomerase II*                                                 602
        *7.16.3.1.1   Step 1: DNA recognition and binding*                                               602
        *7.16.3.1.2   Step 2: Prestrand passage cleavage–religation equilibrium*                         602
        *7.16.3.1.3   Step 3: Double-strand passage*                                                     603
        *7.16.3.1.4   Step 4: Poststrand passage cleavage–religation equilibrium*                        603
        *7.16.3.1.5   Step 5: ATP hydrolysis and enzyme turnover*                                        603
    *7.16.3.2   Site of Cleavable Complex Formation in the Catalytic Cycle*                                  603
    *7.16.3.3   Classes of Cleavable Complex Mediators*                                                      603
        *7.16.3.3.1   Intercalative agents*                                                              604
        *7.16.3.3.2   "Minimal" intercalative agents*                                                    607
        *7.16.3.3.3   Dual inhibitors*                                                                   607
        *7.16.3.3.4   Nonintercalative agents*                                                           609
    *7.16.3.4   A Composite Pharmacophore for Several Classes of Topoisomerase II Agents*                     610

7.16.4   CONCLUSION                                                                                                 611

7.16.5   REFERENCES                                                                                                 611

## 7.16.1  INTRODUCTION

The quantity of DNA necessary to encode the human genome requires that it be condensed into a conglomerate of turns, twists, tangles, knots, and supercoils. For the efficient reading of cellular genetic information during transcription and replication, this three-dimensional topological maze must be unraveled into a linear array of nucleotides. DNA topoisomerases have evolved as the cellular machinery to effect the various alterations in DNA topology necessary for efficient cellular functioning. Two types of eukaryotic DNA topoisomerases, topoisomerase I and topoisomerase II, have been defined according to their catalytic mechanism.

DNA topoisomerase II was first identified in the mid-1980s as the biochemical target for a diverse group of antitumor agents, and the clinical utility of these drugs and our knowledge of their molecular and cellular mechanisms of action continue to expand at an exceptional pace. Topoisomerase II has been established to be among the most vital and preeminent cancer chemotherapeutic targets with two of the top five antitumor agents, etoposide and adriamycin, targeting this enzyme. Topoisomerase I has additionally been demontrated to be an exciting target for chemotherapy and a number of camptothecin derivatives have reached clinical relevance. These enzymes have been demonstrated to be inhibited by a number of structurally diverse classes of molecules. It is the principal goal of this chapter to review the classes of molecules responsible for inhibition of the DNA topoisomerase, the modes of inhibition by these classes of agents, the physiological and the chemotherapeutic roles of these enzymes, and the cellular consequences of inhibition by these classes of agents.

### 7.16.1.1  Physiological Roles of the DNA Topoisomerases

Topoisomerase I alters the topology of DNA by a single linking number through the creation of a single-stranded nick in a DNA helix, followed by passage of the intact strand through the resulting nick and subsequent resealing.[1-3] Topoisomerase I is involved in the knotting/unknotting and relaxation of supercoiled DNA and although the enzyme plays a large role in replication,[4] recombination,[5] and transcription,[6] it is not essential for survival of the cell.[1]

Topoisomerase II modifies DNA topology by two linking numbers via the double-stranded cleavage of duplex DNA followed by the ATP-dependent passage of an adjacent, intact helix through the transient break. DNA religation and ATP hydrolysis precede enzyme turnover, completing the catalytic cycle.[1-3] Topoisomerase II, similarly to topoisomerase I, catalyzes the knotting/unknotting and relaxation of supercoiled DNA, but also demonstrates the ability to remove and introduce DNA catenanes. Topoisomerase II is essential for the survival of cells[7] and has been shown to be involved in DNA replication,[8,9] transcription,[10] and recombination.[11] The enzyme also plays a critical role during cell division in the segregation of daughter chromosomes,[7] separation of sister chromatids,[7] and chromosomal condensation/decondensation.[12,13] Topoisomerase II has also been demonstrated to be a major component of the nuclear matrix and scaffolding, although the actual function of the enzyme within this role has yet to be elucidated.[14]

### 7.16.1.2  DNA Topoisomerases in Chemotherapy

The realization of the indispensable nature of the topoisomerases in the maintenance of DNA and in its efficient processing has predictably led to the emergence of these enzymes as leading targets for chemotherapy. As with most enzymes, "targeting" the topoisomerases can be effected through "classical" modes of inhibition involving direct binding of the drug to the enzyme. However, the inhibition of the topoisomerases can also occur through several alternative pathways, including processes that involve binding to DNA through either intercalation or minor groove association or involve the formation of a ternary complex of drug–DNA–enzyme, in which the enzyme has cleaved DNA and formed a concomitant covalent association with the broken strand(s) of duplex DNA. Only this last mode of topoisomerase inhibition, termed the induction of a "cleavable complex," is the "hallmark" of and generally presumed essential criterion for topoisomerase-mediated cell death.

Hence this chapter will focus on agents, often termed "topoisomerase poisons" or "cleavable complex mediators" in the literature, that initiate the formation of a cleavable complex.

### 7.16.1.3 Inhibitors of the DNA Topoisomerases

Inhibition of the topoisomerases can proceed through any of several distinct mechanisms, which cannot be conveyed by the "traditional" kinetic inhibitor analyses. As mentioned in the previous section, inhibition of catalytic activity may occur through binding to the substrate, DNA, or to the enzyme, or through the induction of an inhibitor–DNA–enzyme cleavable complex. The potential sites of drug binding for inhibition of the DNA topoisomerases are illustrated in Scheme 1 and are discussed individually below. Agents may bind to the enzymes, to DNA, or to the enzyme–DNA complex. The ternary complex of enzyme–DNA–drug, termed the "cleavable complex," is thought to represent the critical cytotoxic lesion.

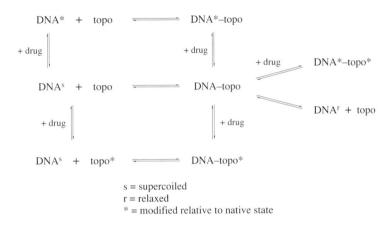

s = supercoiled
r = relaxed
* = modified relative to native state

**Scheme 1**

#### 7.16.1.3.1 *DNA binding agents: intercalators and minor groove binders*

The binding of small molecules to DNA can occur through two well characterized modes, generally termed intercalation and minor groove binding. Such DNA association may serve to alter the conformation of DNA to inhibit enzyme recognition or possibly to promote the formation of an inhibitor–DNA–enzyme ternary complex, to block access of the enzyme to the critical substrate reactive sites, or to "rigidify" DNA to restrict its profound topological alteration during catalysis. Virtually all intercalative agents and most groove binding agents will inhibit the topoisomerases if their stability constants are sufficiently large.[15] At low concentrations, several groove binding agents, including Hoechst 33258 and 33342, are highly efficient in generating drug-induced cleavable complexes of topoisomerase I; however, at higher concentrations, these agents inhibit cleavable complex formation.[16-18] Analogously to groove binding agents, intercalators may exhibit multiple effects on cleavable complex formation with one or occasionally both of the topoisomerases. Thus, Capranico and co-workers[19] characterized inhibitors of topoisomerase II activity according to three classes: type 1, drugs which exhibit the dual induction/suppression phenomenon of induction of cleavable complex formation at low concentrations and suppression at high concentrations; type 2, agents which exhibit only induction of cleavable complex formation, without appreciable inhibition of ternary complex formation; and type 3, agents which only suppress cleavable complex induction. This chapter will focus on the type 1 and 2 classes of agents which promote cleavable complex formation.

#### 7.16.1.3.2 *Direct protein binding agents*

A number of agents bind directly to the topoisomerases either prior to DNA association or subsequent to the formation of a catalytic enzyme–DNA complex. In general, this group of agents inhibits catalysis without the formation of cleavable complexes.[20] For topoisomerase I, a number of compounds, including $\beta$-lapachone,[21] chebulagic acid, and corilagin,[22] inhibit cleavable complex formation either by binding directly to the protein or to the noncovalent enzyme–DNA complex. For topoisomerase II, the coumarin antibiotics novobiocin and coumermycin compete with ATP by binding to the cofactor site and inhibit catalytic turnover without the formation of cleavable complexes.[20] Other agents, including the dioxopiperazines[23] and merbarone,[24] may bind directly to topoisomerase II in the DNA–enzyme complex, perhaps covalently,[23] to form a ternary complex in which DNA has not undergone strand cleavage. Nonetheless, several in this class are undergoing intensive investigation for their potential chemotherapeutic utility.[25]

### 7.16.1.3.3   *"Cleavable complex" mediators*

Cleavable complex formation remains the "hallmark" of topoisomerase-directed antitumor agents. These ternary complexes of drug, DNA, and enzyme appear to be closely related to intermediates formed during the course of the catalytic cycle of the enzyme. Although there are enormous variations in the structures of complex mediators for each enzyme, there may exist some commonalities in the binding sites for these agents in the cleavable complex. Although the drug could theoretically bind to the protein, to DNA, or to a site common to both (see Scheme 1), it has been hypothesized that the binding site for the drug in the topoisomerase II ternary complex resides at the interface between the enzyme and DNA and exhibits an intercalative or intercalation-like domain. This drug–DNA interaction is proposed to induce an altered DNA conformation that mimics an intermediate along the reaction coordinate and thereby results in a tightly bound enzyme–DNA complex; in terms of describing the mode of enzyme inhibition, the drug-induced DNA deformation engendered in the cleavable complex could be viewed as a kind of "transition state analogue." Similar proposals could be advanced for the topoisomerase I cleavable complexes.

Studies on the catalytic mechanism of topoisomerase II and its derangement by drugs are consistent with the outlines of this proposal,[1,26,27] but these studies require that the drug-induced cleavable complex does not represent a single entity. In all likelihood, the pre- and poststrand passed cleavable complexes have distinct DNA conformations and the single- and double-stranded DNA break states of each of these intermediates may possess different conformations. Indeed, a range of altered DNA conformations induced by drugs may lead to cleavable complexes, which would be consistent with the extraordinary structural alteration in DNA catalyzed by these enzymes. Thus, although the DNA duplex may be the principal binding site for topoisomerase II agents in the cleavable complex, there may be no singular definition for the requisite DNA–drug conformation responsible for cleavable complex formation. Presumably, the differences in DNA cleavage pattern, the "cleavage fingerprint," observed for each drug class reflect subtle DNA conformational differences induced by the drug in the local DNA region proximate to the cleavage site.[1,3,26–30] The exceptional structural diversity of topoisomerase-directed agents, possibly unique in enzyme "inhibition," may reflect the hypothetical diversity in DNA–drug conformations found in the cleavable complex and the complementarity between drug structure and DNA "deformability" necessary to induce a cleavable complex.

### 7.16.1.4   Cellular Pharmacology of Topoisomerase-directed Cell Death

The relationship between cleavable complex formation and cytotoxicity is not explicit. For a single agent assayed in cell culture or in *ex vivo* tissue samples, the level of cleavable complex formation correlates with cytotoxic potency (e.g., a "classical" pharmacologic dose–response relationship exists, barring extensive intercalation).[1,26,29,31–34] However, for different agents assayed in a single cell line, the absolute levels of cleavable complex formation do not correlate with cytotoxic potency (e.g., an inter-agent 1 : 1 correlation of cleavable complex concentration with cytotoxicity does not exist).[26,28,32,35–37] Thus, "all cleavable complexes are not created equal" in translating the cleavable complex into a cytotoxic response and the level of apoptotic "signal" originating from the cleavable complex is dependent on ill-defined characteristics of an individual drug-induced complex. A number of hypotheses regarding the relationship between cleavable complex and cytotoxicity have been advanced to rationalize this behavior. Although most of the existing studies have been directed at topoisomerase II, many hypotheses could hold for both topoisomerases.

As noted above, strongly intercalative topoisomerase II agents inhibit cleavable complex formation *in vitro* (e.g., auto-inhibition), although the significance of this observation for *in vivo* cytotoxicity is not clear.[3,26–29] However, cleavable complex formation concomitant with inhibition of some other DNA-processing enzymes is known to depress the cytotoxic response; for example, the inhibition of DNA polymerase α by aphidicolin concurrent with topoisomerase I or II cleavable complex induction dramatically diminishes the cytotoxic response of both the polymerase- and topoisomerase-targeting agents.[38–41] Moreover, inhibition of a range of DNA-processing enzymes is a routine (although not universal) consequence of DNA intercalation, since such drug–DNA association can affect either substrate recognition (the presumed mode of topoisomerase II auto-inhibition) or subsequent enzymic catalysis. Clearly, the sensitivity to intercalation-induced inhibition of DNA-processing enzymes is a complex phenomenon, responding to such factors as stability constant, association/dissociation rates, and drug-induced DNA distortion, and cannot be predicted *a priori*.

Other proposed factors in the "translation" of a cleavable complex into a cytotoxic response include the pattern of DNA cleavage induced by each class of agents. The cleavage pattern is a "fingerprint" of each drug and may relate to its ability to produce cleavable complexes in genes that are actively expressed in the cell and consequently actively engaged with the topoisomerases. Such cleavable complex "fingerprint"-dependent, gene-selective damage has been postulated to contribute to cell selective cytotoxicity.[28,29,35,42–48] In addition, the kinetics of dissociation of the cleavable complex may have additional or alternative importance in "revealing" or "resealing" the drug-induced DNA damage recognized by the "DNA surveillance mechanism" (described below) and may thereby relate to the ability of an agent to initiate a cytotoxic signal.[48] The rates for release of drug-induced cleavable complex upon "wash out" or challenge by other DNA-processing operations are not widely available for all classes of agents, but preliminary data demonstrate that the rates of cleavable complex dissociation are highly variable.[26]

For DNA topoisomerase II-cleavable complex mediators, the relative amounts of cleavable complexes with DNA single- and double-stranded breaks vary significantly between different classes of agents. Double-stranded breaks have been proposed to exhibit greater intrinsic cytotoxicity.[48–52] In addition, the intrinsic signalling by the cleavable complexes from the two topoisomerase II isozymes could be different.[32] Finally, the precise reaction in the catalytic cycle affected during cleavable complex formation has been postulated to relate differentially to the expression of cytotoxicity.[26,53–55]

A finding of fundamental importance to the development of topoisomerase-directed agents has been the recognition that cell death induced by these agents occurs through apoptosis.[41,52–54] The apoptotic signal appears to involve recognition by p53 of topoisomerase agent-induced DNA damage and to proceed through bcl-2.[58–63] Thus, mutation of p53 or hyperexpression of bcl-2 impairs the ability of the cell to undergo drug-induced cell death by agents directed at the topoisomerases. The mechanism of cell death has important implications for the increasing use of topoisomerase agents in combinational chemotherapeutic protocols. Although combinational chemotherapy is thought to be directed principally at attacking the heterogeneity of tumors, the interactions that can occur at the cellular level, between agents that exhibit cytotoxicity through independent mechanisms or between antitumor agents and biological response modifiers, are complex and not predictable. Many single agent combinations result in synergism of their cytotoxic effects, for example tumor necrosis factor and etoposide.[64,65] However, it is increasingly recognized that in cultured cells, the combined use of single agents can antagonize their individual cytotoxic effects.[66] The mechanisms through which the cytotoxicity of the cleavable complex is expressed and is modulated by other signal transduction pathways are not fully elucidated. This understanding has critical implications for the future discovery and development of topoisomerase-directed agents.

## 7.16.2 DNA TOPOISOMERASE AND ITS INHIBITORS

Topoisomerase I was first isolated from *Escherichia coli* in 1971[67] and has since been identified in most cells as a monomeric enzyme with a mass ranging from 90 to 135 kDa.[1] In contrast to the prokaryotic topoisomerase I (and III) enzymes, which can only relax negatively supercoiled DNA,[68] eukaryotic topoisomerase I can catalyze the relaxation of positively and negatively supercoiled DNA. This topoisomerase I-mediated DNA relaxation will decrease the linkage number per catalytic cycle by one, where the linkage number is defined by the number of times two strands cross one another in a covalently close circular DNA.[69] This section will describe the catalytic mechanism and the cleavable complex mediators of topoisomerase I.

### 7.16.2.1   Catalytic Mechanism of DNA Topoisomerase I

The mechanism of the catalytic relaxation of supercoiled DNA mediated by topoisomerase I occurs through a sequence of events which can be separated into five discrete steps (Figure 1).[1]

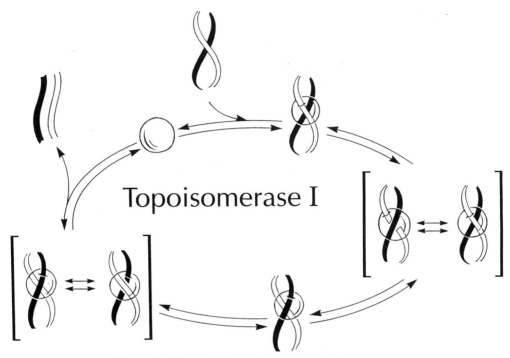

**Figure 1**   The catalytic cycle of DNA topoisomerase I.

#### *7.16.2.1.1   Step 1: DNA recognition and binding*

The interaction of topoisomerase I with DNA is primarily determined by the topological state of the DNA. Topoisomerase I interacts preferentially with bent[70] or supercoiled DNA[71] and DNA crossovers,[72] as illustrated by the 5–10-fold more efficient interaction of the enzyme with supercoiled DNA than with relaxed or nicked molecules.[73] An additional, but less important, determinant of DNA recognition by topoisomerase I is the primary DNA sequence. The enzyme binds to DNA through a noncovalent interaction in the absence of a divalent cation,[1] although the interaction is dependent on the ionic strength of the media and decreases with an increase in ionic strength.[74]

#### *7.16.2.1.2   Step 2: Prestrand passage cleavage–religation equilibrium*

Although the mapping of DNA cleavage sites has shown that 90% of the enzyme is covalently linked to thymidine residues,[75] many DNA cleavage sites do not conform to a specific sequence. In addition, some strong sites for noncovalent DNA binding have been shown to promote little cleavage.[76,77] Even though there does not appear to be a single consensus DNA sequence for recognition by the enzyme, topoisomerase I exhibits a bias for a cleavage site with four nucleotides at the 5′ end of the break ($-4$ to $-1$) and a single nucleotide at the 3′ ($+1$) site. This indicates that the enzyme has a significant interaction with the 5′ site and to a much lesser extent with the 3′ end of the DNA backbone. Thus, upon binding to a supercoiled double-stranded DNA helix, the enzyme will cleave only the strand which contains the recognized sequence.[75] The minimum requirement for topoisomerase I-mediated DNA single-strand cleavage is an 18 base pair (bp) long oligomer.[78]

Following the binding of the enzyme to an active cleavage site, the enzyme will nick a single strand of the supercoiled DNA helix and form a single strand cleavage–religation equilibrium. The enzyme initiates a covalent linkage to the DNA backbone through a ribosyl 3′-*O*-phosphotyrosine ester bond, leaving a free hydroxyl group at the 5′ end of the DNA backbone.[79] This equilibrium between cleaved and religated DNA strongly favors the religated form over the cleaved intermediate. The presence of a divalent cation accelerates acquisition of this equilibrium, but is not required, and an energy source is not required for cleavage.

### 7.16.2.1.3 Step 3: Single-strand passage

The strand passage event takes place after establishment of the cleavage–religation equilibrium. The cleaved intermediate will permit the passage of another single DNA strand and, thereby, relax supercoiled DNA and unknot knotted DNA. In addition to passing a second single strand through the cleaved intermediate, rotation around the single strand of double-stranded DNA will also relax the supercoiled helix.[80] Owing to these modes of single-strand cleavage and strand passage, the topoisomerase I is not able to decatenate DNA, unless the DNA contains a preexisting nick.

### 7.16.2.1.4 Step 4: Poststrand passage cleavage–religation equilibrium

Following the strand passage a second cleavage–religation equilibrium is established with the religated form of the DNA backbone again the favored intermediate.

### 7.16.2.1.5 Step 5: Enzyme turnover

After the religation of the DNA backbone, topoisomerase I will process along the DNA and unwind the DNA superhelical twists.[81] Once the DNA is completely relaxed, the enzyme will dissociate from DNA and be able to initiate another round of catalysis. Unlike topoisomerase II, topoisomerase I does not require the hydrolysis of ATP to exert its catalytic activity.[82]

## 7.16.2.2 Classes of Cleavable Complex Mediators

The topoisomerase I–DNA binary complex is the central intermediate in the catalytic cycle and consequently any event which alters its rate of formation and/or stability will disrupt the catalytic cycle. Camptothecin and its derivatives were the earliest examples of antitumor drugs whose mechanism of action was intimately related to the inhibition of topoisomerase I activity. Since the discovery of their ability to stabilize the binary complex, numerous other topoisomerase I inhibitors have been shown to act via the stabilization of the binary complex. The known topoisomerase I cleavable complex mediators are shown as structures (1)–(17).

### 7.16.2.2.1 Camptothecin and its derivatives

Camptothecin (CPT, **1**) is a plant alkaloid from the *Camptotheca acuminata* tree.[83] In *in vitro* assay, camptothecin exhibited excellent topoisomerase I inhibitory activity but poor water solubility ($2.5 \times 10^{-3}$ mg ml$^{-1}$) and unpredictable toxicity. In an attempt to improve both the toxicological profile and the water solubility, a variety of derivatives have been synthesized. Two camptothecin derivatives, topotecan and irenotecan, have successfully completed Phase I and II testing.

Topotecan (SKF 104864, **2**) is a semisynthetic analogue which incorporates a basic side chain at the 9-position of the A-ring of 10-hydroxycamptothecin. The basic amine-containing side chain renders the drug amenable to formation of ammonium ion salts, thereby improving the aqueous solubility at physiological pH relative to camptothecin.[84]

Irenotecan (CPT-11, **3**) is another semisynthetic analogue of camptothecin with improved water solubility, owing to its basic side chain. Irenotecan serves as a prodrug and is hydrolyzed *in vivo* to its active metabolite, 7-ethyl-10-hydroxycamptothecin (SN-38, **4**).[83]

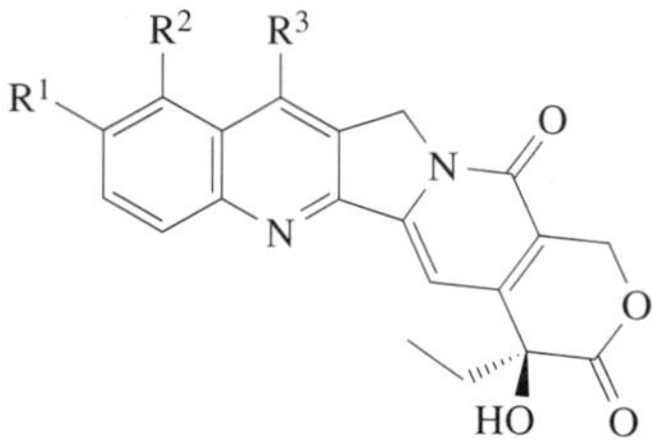

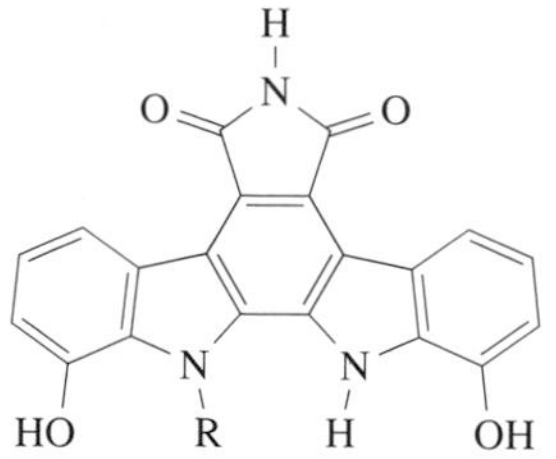

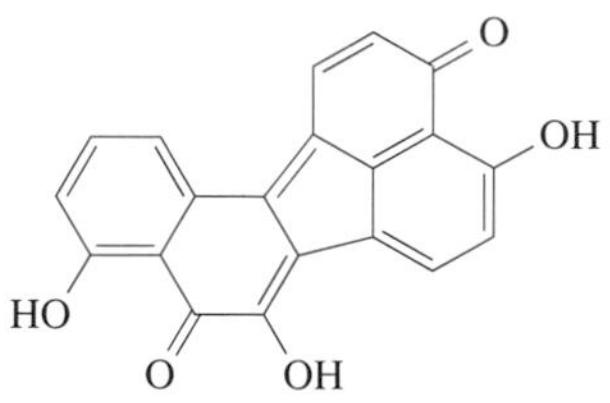

(1) $R^1, R^2, R^3 = H$
(2) $R^1 = OH, R^2 = Me_2NHCH_2–, R^3 = H$

(3) $R^1 = $ [structure], $R^2 = H, R^3 = Et$

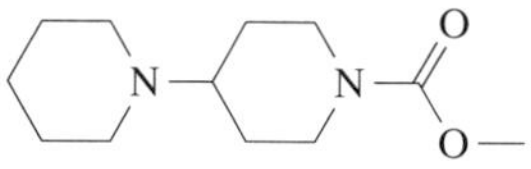

(4) $R^1 = OH, R^2 = H, R^3 = Et$
(5) $R^1, R^3 = H, R^2 = NH_2$

(6) $R = H$
(7) $R = $ glycosyl

(8)

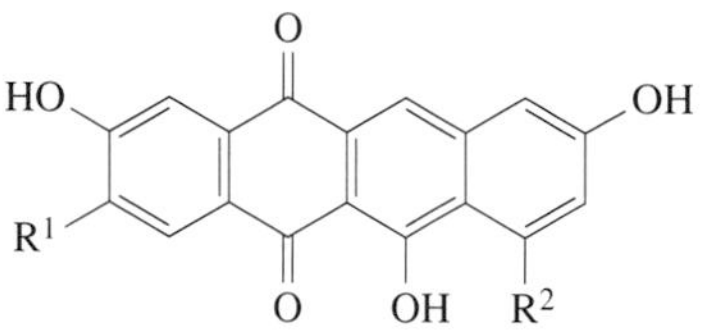

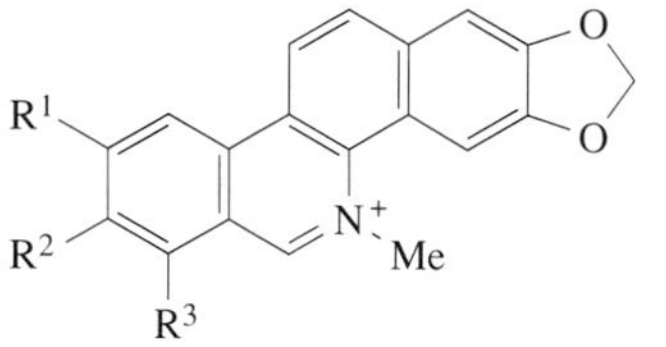

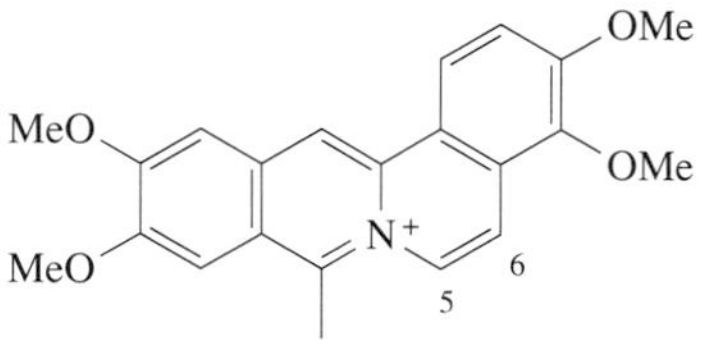

(9) $R^1 = Me, R^2 = –CH_2C(O)CH_2CHOHMe$
(10) $R^1 = H, R^2 = –OSO_3H$
(11) $R^1 = H, R^2 = OH$

(12) $R^1, R^2 = OMe, R^3 = H$
(13) $R^1 = H, R^2, R^3 = Me$
(14) $R^1, R^2 = H, R^3 = Me$
(15) $R^1, R^3 = H, R^2 = Me$

(16)
(17) C-5, C-6 = dihydro

9-Aminocamptothecin (**5**) is another camptothecin derivative undergoing clinical trials. Problems associated with the water solubility of 9-aminocamptothecin have been overcome by employing an infusion formulation containing polyethylene glycol and phosphoric acid.[85]

### 7.16.2.2.2   Indolocarbazoles

BE-13793C (**6**) is a novel indolocarbazole antibiotic isolated from an *Actinomycete* culture following screening for topoisomerase mediated antitumor activity.[86] The low aqueous solubility of BE-13793C prompted the synthesis of the glycosylated derivative ED-110 (**7**). This compound demonstrated both improved water solubility and increased cytotoxicity relative to the aglycone. In addition to its ability to stabilize the topoisomerase I–DNA binary complex as significantly as CPT, ED-110 was also shown to intercalate into DNA.[86]

### 7.16.2.2.3   Bulgarein

Bulgarein (**8**) is a blue pigment isolated from a fungus (*Heteroconium* sp.).[87] It is unique in that it possesses DNA winding, not unwinding, activity; bulgarein is the only molecule known to stabilize the topoisomerase I–DNA binary complex and wind DNA. Additionally, it was observed that at high concentrations bulgarein inhibited topoisomerase I-mediated DNA cleavage, a behavior common for strongly intercalative agents. This result is believed to arise from the intercalation of bulgarein with DNA, which promotes positively supercoiled DNA and renders it a poor substrate for topoisomerase I.[87]

### 7.16.2.2.4   The naphthacenequinones: UCE6, UCE1022 and saintopin

The polyhydroxylated naphthacenequinone antibiotic UCE6 (**9**) was isolated from *Actinomycetes* culture.[88] The structurally related compounds UCE1022 (**10**)[89] and saintopin (**11**)[90] were isolated by culturing soil samples containing *Paeliomyces* species. Both UCE6 and UCE1022 are highly selective topoisomerase I inhibitors. An advantage of UCE1022 is that the sulfate group is highly ionized at physiological pH, which confers significant water solubility relative to both UCE6 and saintopin. Saintopin, while being an effective inhibitor of topoisomerase I, also displays significant topoisomerase II activity (see Section 7.16.3.3.3).

### 7.16.2.2.5  The benzo[c]phenanthridines nitidine, chelerythrine, isofagaridine, and fagaridine

The first three of these benzo[c]phenanthridine alkaloids, nitidine (**12**), chelerythrine (**13**), and isofagaridine (**14**), were isolated from a plant extract of the roots of *Zanthoxylum nitidum*,[91] while fagaridine (**15**) was isolated from *Fargara xanthoxyloides*.[92] All four were found to be comparable at inhibiting topoisomerase I-mediated relaxation.[91] However, only nitidine and fagaridine were found to stabilize effectively the topoisomerase I–DNA binary complex. In addition, fagaridine was observed to intercalate into DNA. The mechanism by which chelerythrine and isofagaridine inhibit topoisomerase I has not been established.

### 7.16.2.2.6  Coralyne and its derivatives

A wide array of substituted dibenzo[a,g]quinolizinium compounds have been investigated on the basis of their structural similarity to the naturally occurring protoberberine alkaloids. Coralyne (**16**) and selected derivatives (e.g., 5,6-dihydrocoralyne (**17**)) were found to stabilize the covalent topoisomerase I–DNA binary complex at relatively low concentrations.[93,94] As the concentrations were increased, certain derivatives were found to inhibit topoisomerase I-mediated DNA relaxation by inhibiting the formation of the binary complex via direct interaction with the DNA, leading to altered DNA topology and diminished topoisomerase I affinity.[93,94]

## 7.16.3  DNA TOPOISOMERASE II AND ITS INHIBITORS

Topoisomerase II is a homodimeric enzyme that modulates the topological state of both positively and negatively supercoiled DNA by generating a transient double strand break, followed by passage of another double-stranded DNA segment through the break. Analogously to topoisomerase I, the type II enzyme catalyzes the knotting/unknotting and relaxation of supercoiled DNA. However, unlike topoisomerase I, topoisomerase II is also capable of catalyzing DNA catenation/decatenation and is, therefore, essential for the viability of the eukaryotic cell. Another difference between the two topoisomerases is that topoisomerase II requires the hydrolysis of a high-energy cofactor, ATP, to exert catalytic activity.

In eukaryotic cells topoisomerase II consists of two isozymes,[95] topoisomerase II$\alpha$ and topoisomerase II$\beta$, with molecular masses of 170 kDa and 180 kDa, respectively. The two isozymes exhibit different characteristics throughout the cell life cycle that have important implications for the cytotoxicities of topoisomerase II inhibitors. The main differences between the isozymes are as follows:

(i) The distribution of the two isoforms throughout the cell cycle. Topoisomerase II$\alpha$ shows an increase in activity through the cell cycle with its peak at the G2/M phase, whereas the activity of topoisomerase II$\beta$ remains relatively stable throughout the cell cycle.[96]

(ii) The cellular responsibilities of the two isozymes. Topoisomerase II$\alpha$ appears to be essential for cell proliferation and DNA replication, whereas topoisomerase II$\beta$ appears to be primarily involved in DNA transcription.[95]

(iii) The sensitivity to chemotherapeutic agents. The isozymes exhibit different sensitivities to chemotherapeutic agents; for example, topoisomerase II$\alpha$ has been shown to be more sensitive to teniposide.[32]

In view of the differences between the two isoforms, topoisomerase II$\alpha$ is thought to be the target for most chemotherapeutic drugs. Although the existence of topoisomerase II$\beta$ should not be disregarded, the following discussion will be focused on the $\alpha$-isozyme.

### 7.16.3.1  Catalytic Mechanism of DNA Topoisomerase II

Five discrete steps in the catalytic sequence of topoisomerase II have been identified by Osheroff.[1] These steps are depicted in Figure 2 and are individually discussed below.

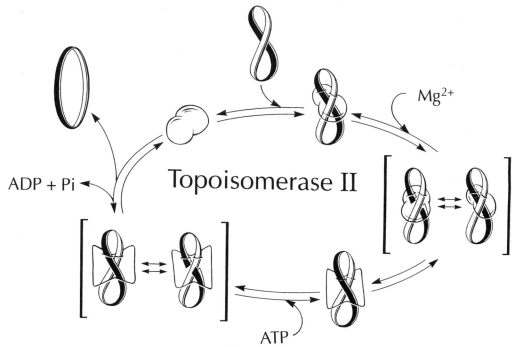

**Figure 2**   The catalytic cycle of DNA topoisomerase II.

#### 7.16.3.1.1   *Step 1: DNA recognition and binding*

Recognition of a binding site by topoisomerase II is determined by the topological state and the nucleotide sequence of the DNA helix. The topological state of the DNA is the most important factor in the recognition of a binding site by the enzyme, indicated by a 3–10-fold increase in enzyme–DNA binding with supercoiled DNA compared with linear DNA molecules.[97] In addition, 90% of the enzyme is located at helix–helix juxtapositions.[71,98] Studies have revealed that DNA binding is additionally directed by a primary nucleotide sequence consisting of 16 nucleotides which has a guanine residue at the +4 position and a thymine at the +2 position.[99] During the recognition and binding step, topoisomerase II interacts with DNA noncovalently[100] and does not require cofactors, such as ATP[101] or magnesium.[102]

#### 7.16.3.1.2   *Step 2: Prestrand passage cleavage–religation equilibrium*

Following the formation of the noncovalently bound DNA–topoisomerase II complex, the enzyme will nick the noncoding DNA strand.[103] The enzyme then forms another symmetrically displaced single-stranded break, in which the enzyme dimers are covalently bound to the 5′ ends of the DNA strands through a phosphotyrosine ester bond.[103,104] Each DNA strand is cleaved by one subunit of the dimeric enzyme. In the presence of a divalent cation, the DNA–topoisomerase II complex establishes a double-stranded prestrand passage cleavage–religation equilibrium, which favors religation by ~99:1. Although $Ca^{2+}$ or $Mn^{2+}$ will promote DNA cleavage, only $Mg^{2+}$ can catalyze the cleavage and subsequent strand passage.[105]

### 7.16.3.1.3  Step 3: Double-strand passage

Following the establishment of the prestrand passage cleavage–religation equilibrium, topoisomerase II passes an intact DNA helix through the double-stranded cleaved intermediate. As a consequence of the 20 Å diameter of a DNA double helix, topoisomerase II must undergo a considerable conformational change during this process. Topoisomerase II-mediated strand passage requires the binding of Mg-ATP, which is thought to allow access to the requisite conformational change or to stabilize the altered enzyme conformation.

### 7.16.3.1.4  Step 4: Poststrand passage cleavage–religation equilibrium

After DNA double-strand passage, a second cleavage–religation equilibrium is established which serves to reseal the double-stranded DNA break and to maintain the integrity of the DNA double helix. Similarly to the prestrand passage cleavage–religation equilibrium, religation is favored over cleavage by $\sim 95:5$.[106]

### 7.16.3.1.5  Step 5: ATP hydrolysis and enzyme turnover

Hydrolysis of the enzyme-bound ATP to ADP and inorganic phosphate is a requirement for the enzyme to turnover for the next catalytic cycle.[106] Upon hydrolysis of ATP, topoisomerase II undergoes a conformational change which is necessary to produce the enzyme state capable of release from the site of catalysis and reinitiation of another round of catalysis. Nonhydrolyzable cofactors, such as APP(NH)P, will promote topoisomerase II-induced DNA strand passage, but will inhibit the regeneration of topoisomerase II for the next catalytic cycle.

### 7.16.3.2  Site of Cleavable Complex Formation in the Catalytic Cycle

Inhibition of topoisomerase II-mediated DNA relaxation can occur at several points in the catalytic cycle. Inhibition may be a consequence of the disruption of DNA–topoisomerase II binding by deformation of the DNA by intercalators or groove binding agents, of a stabilization of the DNA cleavage intermediate, of the inhibition of DNA double-helix strand passage, of the inhibition of religation, or of the inhibition of ATPase activity of the enzyme. Stabilization of the intermediates which possess either single- or double-stranded DNA–enzyme-bound breaks produces the ternary cleavable complexes of drug, DNA, and enzyme. Hence both intercalative and nonintercalative cleavable complex mediators are thought to stabilize the pre- and poststrand passage of DNA cleavage intermediates in the catalytic cycle through their association with these species.

Studies by Osheroff and co-workers[107] have shown that the quinolone derivatives CP-115,953, CP-67,804, and Ro 15-0216 enhance the stability of the pre- and poststrand passage DNA-cleaved intermediates without affecting the rates of religation. The formation of drug-induced cleavable complexes through inhibition of strand passage has also been shown to occur by several mechanistically (intercalative and nonintercalative) and structurally disparate compounds, including amsacrine, genistein, and CP-115,953. The stabilization of the drug-induced cleavable complex can also be enhanced by inhibition of the rate of religation. Intercalators and nonintercalators, such as amsacrine, etoposide, teniposide, adriamycin, ellipticine, and mitoxanthrone, inhibit topoisomerase II-mediated religation of the cleaved DNA intermediate during the pre- and poststrand passage equilibrium.[108–110] Inhibition of the rate of religation and the increase in the formation of the cleaved intermediate both result in a retardation of the rate of the cleavage–religation equilibrium. The promotion of the cleavable complex induced by these agents illustrates the diversity of both inhibitor structure and catalytic step inhibited in the formation of these species.

### 7.16.3.3  Classes of Cleavable Complex Mediators

The topoisomerase II inhibitors may be classified according to their specific point of intervention in the catalytic cycle. The most common inhibitory mechanism appears to be related to the formation and/or stability of the covalent binary topoisomerase II–DNA complex. Those inhibitors whose mechanism of action involves the formation of a ternary complex have been shown to involve both intercalative and nonintercalative inhibitor binding modes. The known intercalators may be further subdivided into categories on the basis of their induction and/or suppression of the formation of the topoisomerase II–DNA binary complex. As noted above, the first class of inhibitors, type 1 intercalators, demonstrate seemingly dichotomous behavior in that these agents can function both to induce and to suppress the formation of the cleavable complex in a concentration-dependent manner. The second and third classes, types 2 and 3, have been assigned on the basis of their affecting only induction or suppression regardless of their concentration.[95]

### 7.16.3.3.1 Intercalative agents

Examples of each class of intercalative agents discussed below are shown as structures (18)–(37).

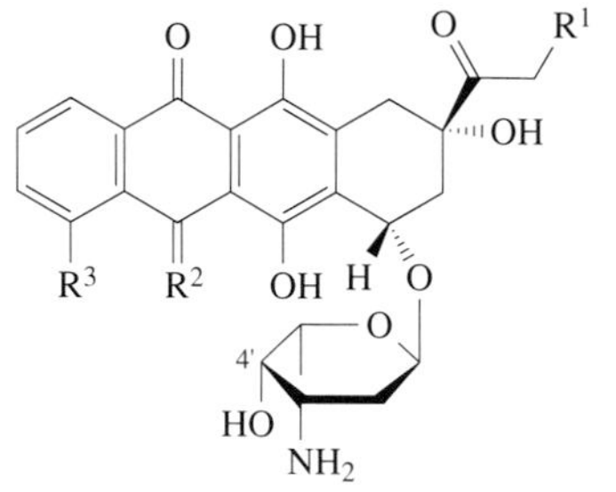

(18) R = H
(19) R = OH

(20)

(21)  R$^1$ = H, R$^2$ = O, R$^3$ = OMe
(22)  R$^1$ = OH, R$^2$ = O, R$^3$ = OMe
(23)  4'-epimer of (22)
(24)  R$^1$ = H, R$^2$ = O, R$^3$ = H
(25)  R$^1$ = H, R$^2$ = NH, R$^3$ = OMe

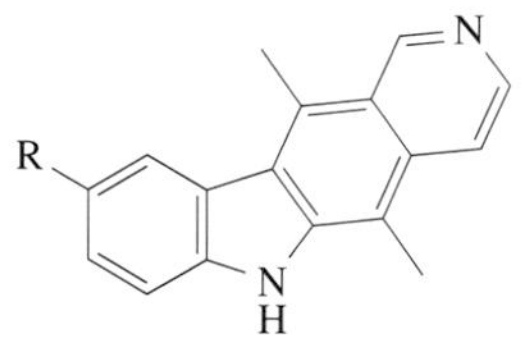

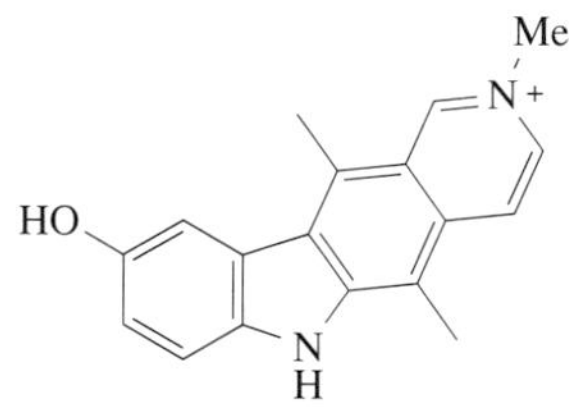

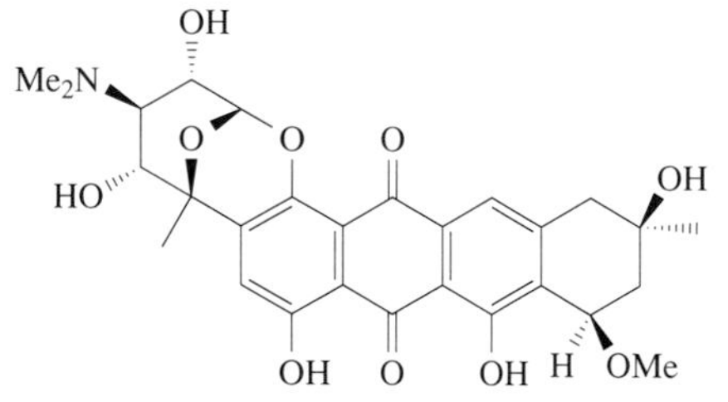

(26)

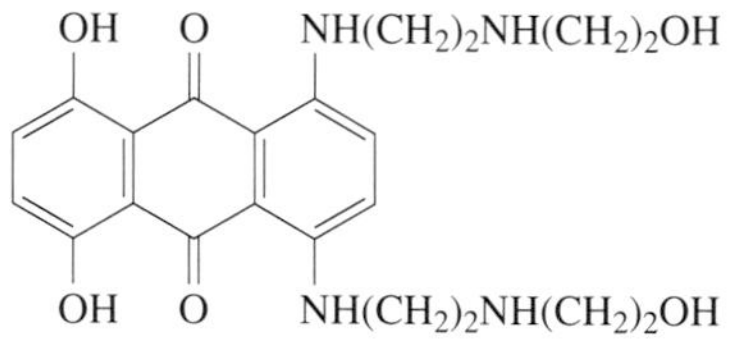

(27)

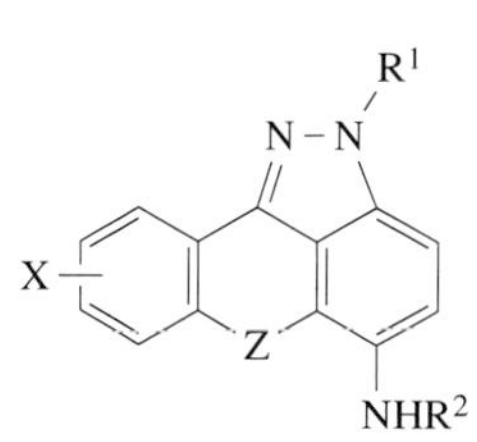

(28) Z = C=O
(29) Z = S
X = H, OH
R$^1$, R$^2$ = H, alkyl, substituted aminoalkyl

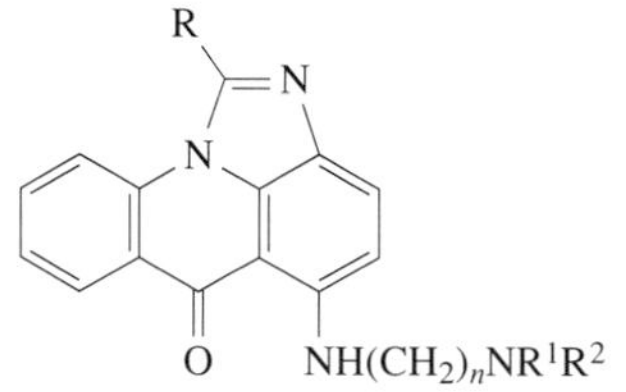

(30) R = H, Me, R$^1$, R$^2$ = alkyl

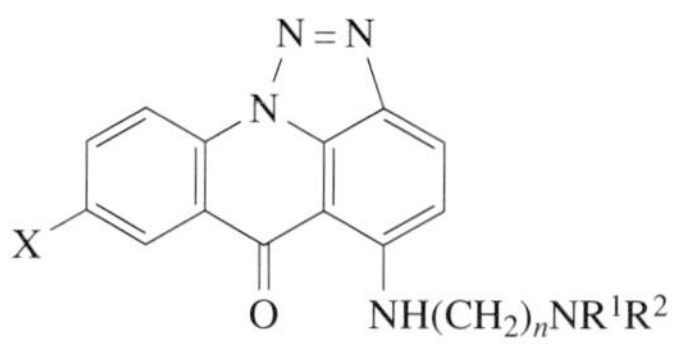

(31) X = H, OH, R$^1$ = H, Me,
R$^2$ = alkyl, substituted alkyl

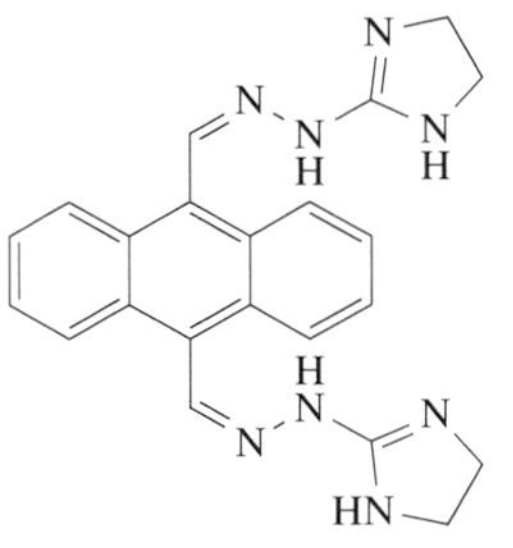

(32)

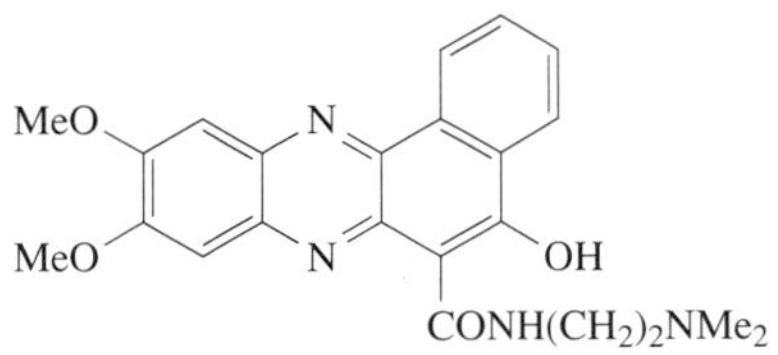

(33)

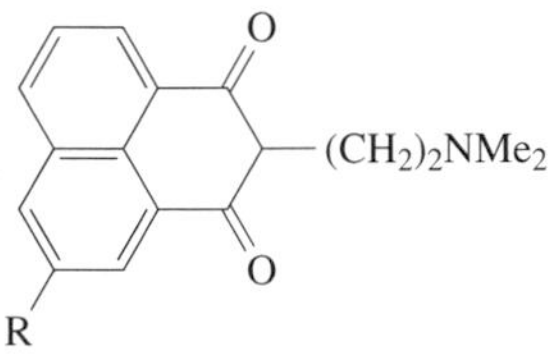

(34) R = NH$_2$
(35) R = NO$_2$

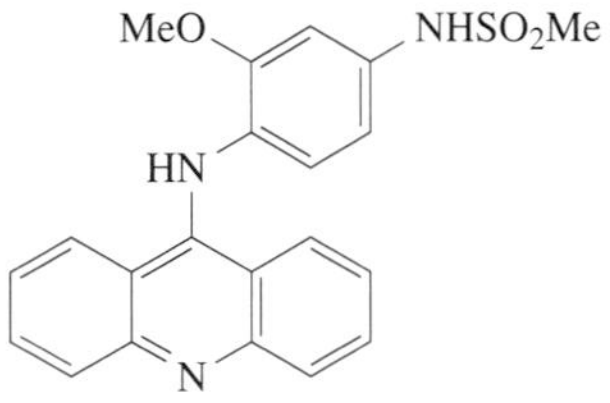

(36)

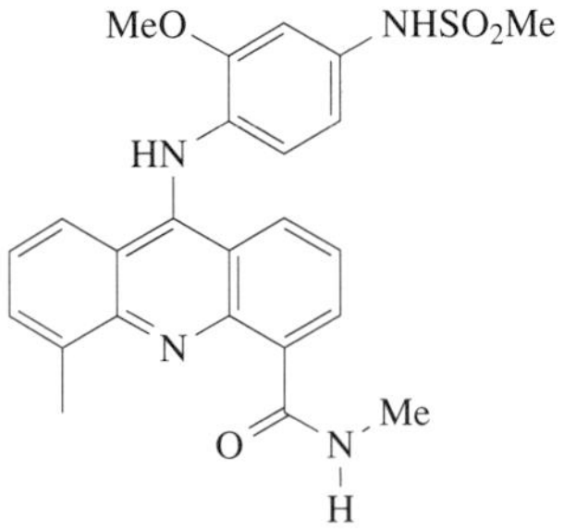

(37)

## (i) Ellipticine and its derivatives

Ellipticine (18) is a pyridoindole alkaloid isolated from the leaves of several *Ochrosia* species.[111] The toxic side effects associated with ellipticine prompted efforts to improve its pharmacological profile. Hydroxylation of ellipticine *in vivo* leads to 9-hydroxyellipticine (19) as the major metabolite.[112] This hydroxylated derivative is more active and less toxic than the parent compound. Furthermore, quaternization of the pyrido nitrogen of 9-hydroxyellipticine, providing the 9-hydroxy-2-methylellipticinium ion (20), resulted in improved water solubility without a significant loss of activity.[112] Ellipticine-induced DNA cleavage appears to result from enhancement of the forward rate of cleavage in the ternary complex versus inhibition of the religation process. It has been observed that, in addition to its intercalative interaction with DNA, ellipticine binds directly to topoisomerase II in the absence of DNA.[113]

## (ii) Anthracyclines

Daunorubicin (daunomycin, 21) and doxorubicin (adriamycin, 22) were isolated from cultures of *Streptomyces* and represent the prototypical anthracycline antibiotics. The cytotoxicity of these agents has been associated with several modes of action, in addition to the now recognized inhibition of topoisomerase II.[114] The anthraquinone system can undergo reduction to the semiquinone radical, which has been shown to initiate iron-catalyzed oxy-radical formation leading to DNA damage. Furthermore, it was demonstrated that adriamycin linked to agarose beads, and thus precluded from entering the cell, displayed cytotoxicity, presumably via radical-induced disruption of the cell membrane.[115] The ability of the anthracyclines to initiate free radical damage is believed to be the mechanistic basis of their observed cardiotoxicity. The cardiac effects which attend the use of the natural anthracyclines has proved a significant impetus for the synthesis of derivatives with improved toxicological profiles.

For example, epirubicin (4′-*epi*-doxorubicin, 23), which differs from doxorubicin by the epimerization of the 4′-hydroxy group on the amino sugar portion, has been shown to be less cardiotoxic.[116] In addition, idarubicin (4-demethoxydaunomycin, 24), in which the methoxy group in the aglycone has been substituted with a hydrogen, is less cardiotoxic and shows significant activity in several cell lines.[117] 5-Iminodaunomycin (25), which incorporates a quasi-iminoquinone moiety, also displays little or no cardiotoxicity, while maintaining its ability to function as a topoisomerase II inhibitor.[118]

Although structurally different from both daunomycin and adriamycin, menogaril (26) has been shown to induce cleavable complex formation in the presence of the binary enzyme–DNA complex.[119] Menogaril, a weak intercalator, is taking part in phase II trials in 1997.

### (iii)  Anthracenediones

Mitoxanthrone (27) is a synthetic compound designed to maintain the DNA intercalative binding of the anthracyclines by retaining the quinone–dihydroquinone array of the anthraquinone moiety. The DNA-binding ability is further enhanced by the addition of the alkylamino side chains which interact with the minor groove of the DNA. However, this derivative does not undergo reduction as readily to the semiquinone radical and thus displays diminished cardiotoxicity.[120]

### (iv)  Modified anthracenediones

This broad class of compounds, which includes the anthrapyrazoles (28),[121] benzothiopyranoindazoles (29),[122] imidazoacridinones (30),[123] triazoloacridinones (31),[124] and acridine-4-carboxamides (see Section 7.16.3.3.2), is characterized by two distinct structural features: (i) a tri- or tetracyclic chromophore and (ii) one or two side chains containing an (aminoalkyl)amino residue. These classes of compounds have been shown not to induce free-radical formation and to display considerable cytotoxicity both *in vitro* and *in vivo*.

### (v)  Anthracenes

Numerous anthracene derivatives have been developed as potential antineoplastic agents. Of these derivatives, bisanthrene (32) is the most active compound against a wide range of tumors, including breast cancer and acute leukemia.[125]

### (vi)  Benzophenazines

The benzophenazine derivative NC-190 (33) has been shown to be a type II intercalative inhibitor effecting only stabilization of the cleavable complex without suppression.[126]

### (vii)  Benzoisoquinolinediones

The benzoisoquinolinediones amonafide (34) and mitonafide (35) have been demonstrated to stabilize topoisomerase II-cleavable complexes, although these agents may inhibit a broader range of DNA-processing enzymes.[127]

### (viii)  Anilinoacridines

The examination of 9-anilinoacridine for topoisomerase II-mediated activity was prompted by the antineoplastic activity of this family of agents. Extensive studies of the structure–activity relationships of both the acridine and pendant ring resulted in the development of 4′-(9-acridinylamino)methanesulfon-m-anisidine (m-AMSA, amsacrine, 36). The anilinoacridines are moderate DNA intercalators whose mode of action appears to be inhibition of the religation step in the ternary complex. The hepatotoxicity associated with amsacrine has limited its clinical utility and provided the impetus to find less toxic analogues. To this end, CI-921 (37) was developed; the N-methylcarboxamide group provides improved water solubility relative to amsacrine.[128]

### 7.16.3.3.2 "Minimal" intercalative agents

The antineoplastic activity of the acridine derivatives resulted in extensive efforts to refine their pharmacological profile further. While initial attempts focused on increasing the efficiency of DNA intercalation, it became apparent that there was a problem associated with correlating *in vitro* binding affinities with *in vivo* cytotoxicity for most analogues of this family. The problem was believed to result from the failure of the drug to be transported efficiently to the nucleus, the site of action of the drug. It had been demonstrated that some molecules with large binding constants were unable to be distributed to sites which were remote from the site of administration. Problems with drug distribution may be overcome in some cases by improving water solubility (e.g., amsacrine · HCl vs. CI-921 · HCl, 0.12 mg ml$^{-1}$ vs. 0.72 mg ml$^{-1}$)[129] However, increased hydrophilicity must be balanced against the hydrophobicity required to diffuse successfully through the lipophilic membrane structure. Membrane penetration generally occurs by the neutral species when ionizable functionalities are present; thus, the difference in p$K_a$ between amsacrine (p$K_a$ = 7.43) vs. CI-921 (p$K_a$ = 6.40) results in a 10-fold concentration advantage for the uncharged CI-921 species at physiological pH.[129] Additionally, because the distribution into tissue will take place by passive diffusion, the magnitude of the diffusion constant for molecules will critically affect their ability to reach an intracellular target. Molecules with large diffusion constants will have difficulty reaching their nuclear target. These problems have led to efforts to develop a class of antineoplastic agents, the "minimal" intercalators, which attempt to balance favorable distribution properties with moderate binding affinities. Although these compounds bear a structural resemblance to the broad class of anthracenedione derivatives, they have been treated separately to illustrate rational approaches to overcoming the distribution problems associated with hydrophobic intercalative agents. A variety of structural classes were synthesized to alter both the DNA-binding affinity and the drug hydrophobicity relative to the parent compound 9-aminoacridine-4-carboxamide (38). The acridine-4-carboxamides (39),[130] and also the angular analogue (40)[131] and the "2–1" analogue (41),[132] had both lower binding constants and lower p$K_a$ values relative to the parent compound (38), while retaining *in vivo* cytoxicity. Additionally, compounds with less aromatic chromophores, including dibenzo[1,4]dioxin-1-carboxamide (42)[133] and the phenylbenzimides (43),[134] had even lower DNA binding constants and p$K_a$ values, while retaining activity. It should be noted, however, that the mechanism of cytotoxicity of the "logical conclusion" of this series, the phenylbenzimides (43), does *not* appear to involve the stabilization of a topoisomerase II-cleavable complex.

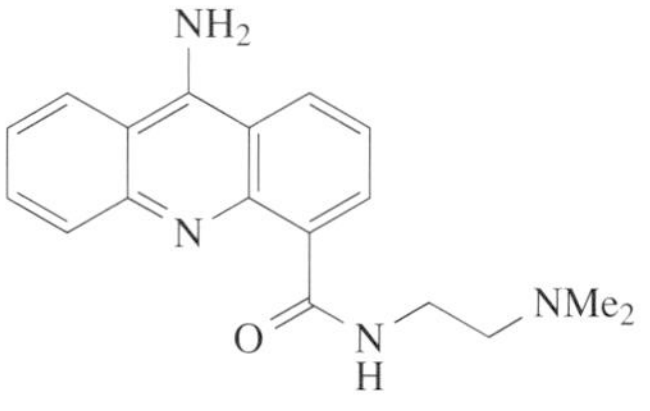

(38) 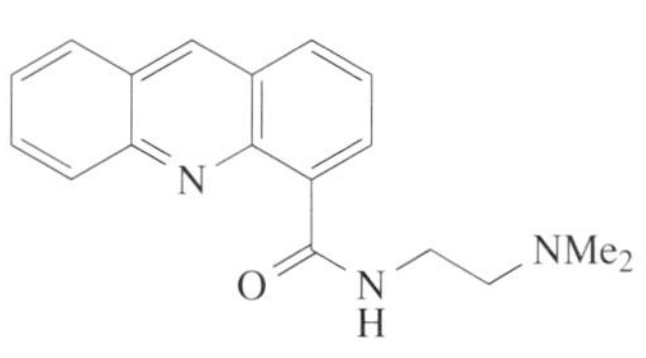          (39) 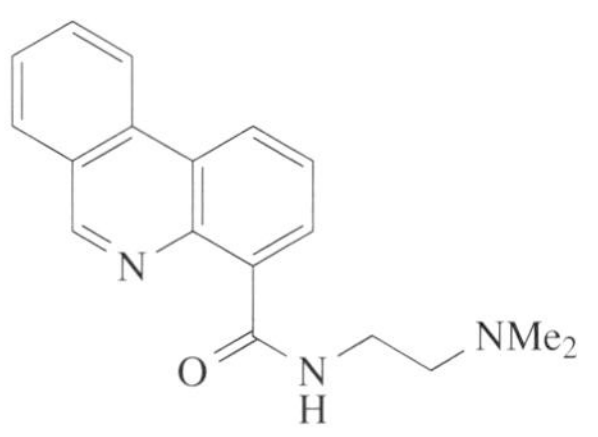          (40)

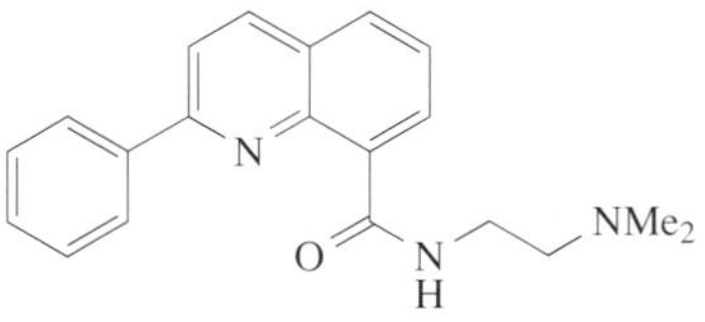

(41) 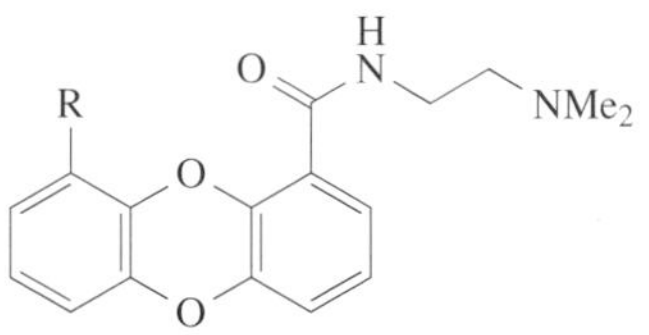          (42) R = H, Cl 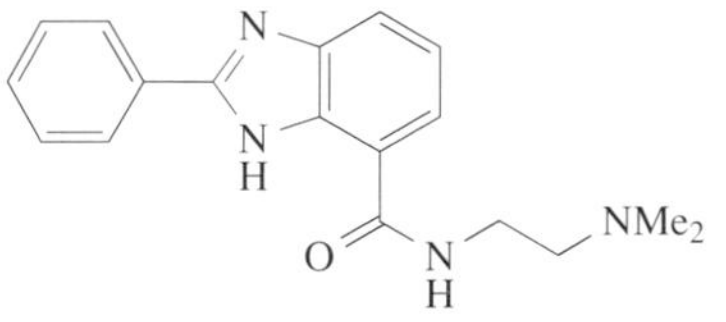          (43)

### 7.16.3.3.3 Dual inhibitors

A growing class of antineoplastic agents, including saintopin (44) (see Section 7.16.2.2.4), actinomycin D (45), and intoplicine (46), have been shown to inhibit both topoisomerases I and II through the formation of drug-induced cleavable complexes. The stabilization of the binary complex for both topoisomerases I and II suggests a common underlying mechanism of cleavable complex induction. Importantly, these agents may be able to circumvent drug resistance caused by the alteration of the enzyme levels or structure of a single topoisomerase target; it has been shown that human colon cancers, which demonstrate a low sensitivity to topoisomerase II-directed chemotherapeutic agents, are highly sensitive to camptothecin derivatives.[135]

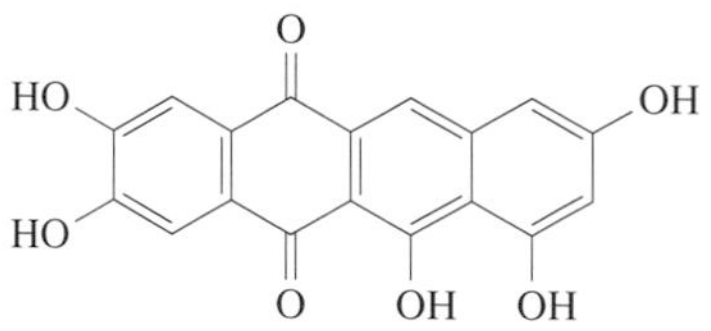

(44) 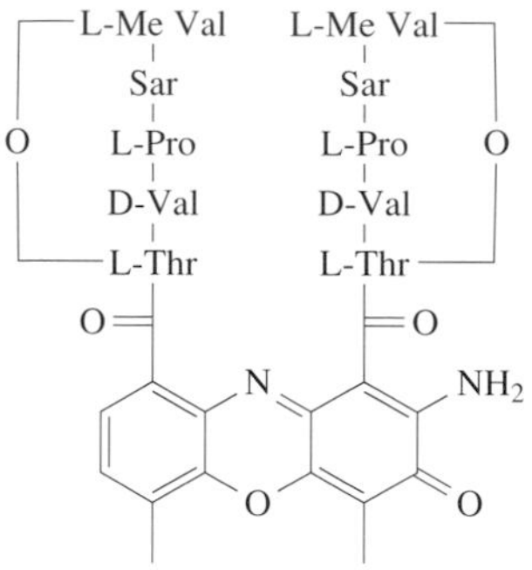              (45) 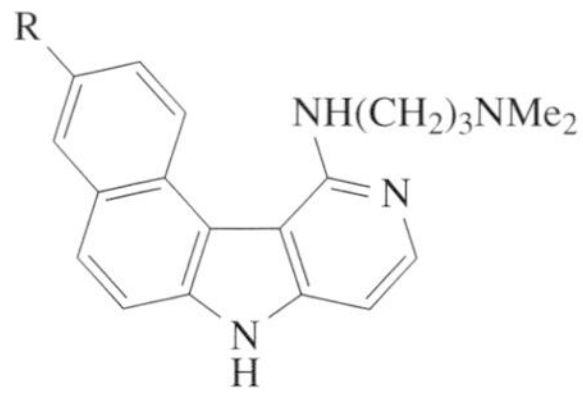              (46) R = OH
                                    (47) R = H

### (i) Actinomycin D

Actinomycin D (**45**) was isolated from a soil sample of *Actinomyces*. It is composed of a heterotricyclic chromophore, 2-amino-4,6-dimethylphenoxazin-3-one-1,9-dicarboxylic acid, appended to two pentapeptide side chains. It was first demonstrated to form a ternary complex of topoisomerase II;[136] since then it has also been shown to stimulate topoisomerase I-mediated DNA cleavage at specific sites.[137]

### (ii) Saintopin

Saintopin (**44**) was isolated from a *Pacilomyces* species. Saintopin has been shown to induce topoisomerase I-mediated DNA cleavage comparable to camptothecin, and also topoisomerase II-mediated DNA cleavage comparable to both amsacrine and etoposide.[138]

### (iii) Intoplicine and its derivatives

Intoplicine (**46**) and its derivatives may be viewed conceptually as the synthetic deconstruction and reconstruction of the tetracyclic ellipticine skeleton. Removal of the C-ring of ellipticine results in the contraction to a tricyclic pyridoindole, which is then re-expanded to the tetracyclic system via benzannulation on either the [*e*] or [*g*] face of the indole. The addition of the alkylamine side chain has been found to enhance binding affinity to DNA, presumably via minor groove interaction. These derivatives are important because they display dual inhibition of both topoisomerases I and II activities. Intoplicine has been found to bind to topoisomerase II alone, but to topoisomerase I only in the presence of DNA.[139] Results obtained using surface-enhanced Raman scattering has suggested that there are two binding modes for these derivatives to DNA. The 7*H*-benzo[*e*]pyrido[4,3-*b*]indole derivative (**47**) was found to be the most potent topoisomerase I inhibitor, while the hydroxy-11*H*-benzo[*g*]pyrido[4,3-*b*]indole derivative (**48**) was discovered to inhibit topoisomerase II activity.[140] For the topoisomerase I ternary complex, the drug appears to be fully intercalated into the DNA and undergoes little change in either the presence or absence of enzyme. However, during the formation of the topoisomerase II ternary complex, a distinct change in the ellipticine environment was observed, suggesting that a portion of the drug is now accessible for interaction with the enzyme.

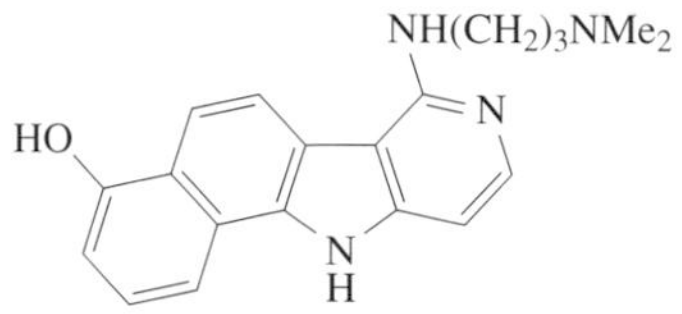

(48)

### 7.16.3.3.4 *Nonintercalative agents*

Examples of each class of nonintercalative topoisomerase II-cleavable complex mediators are shown as structures (49)–(54).

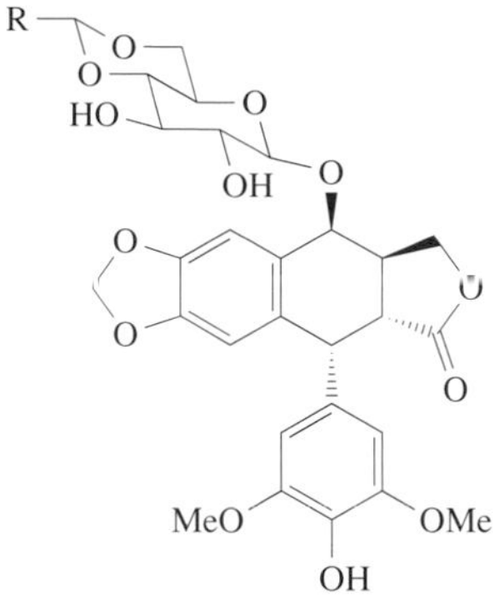

(49) R = Me

(50) R =

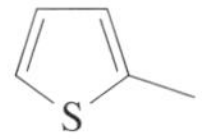

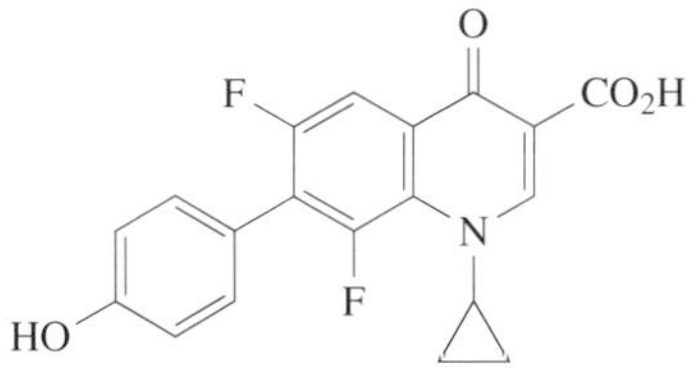

(51)

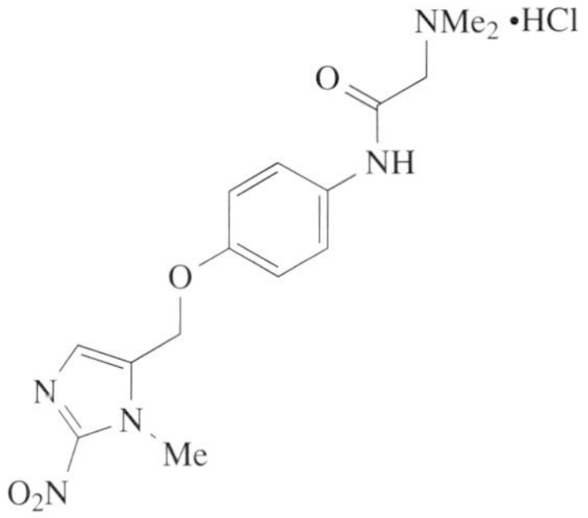

(52)

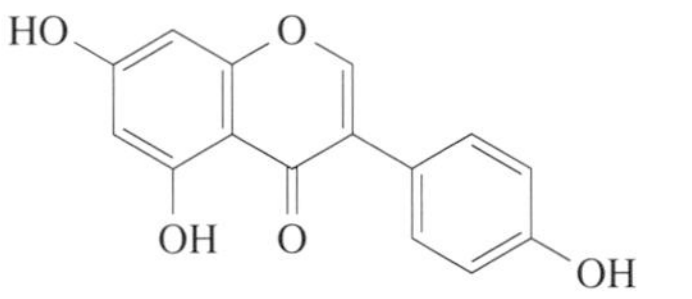

(53)

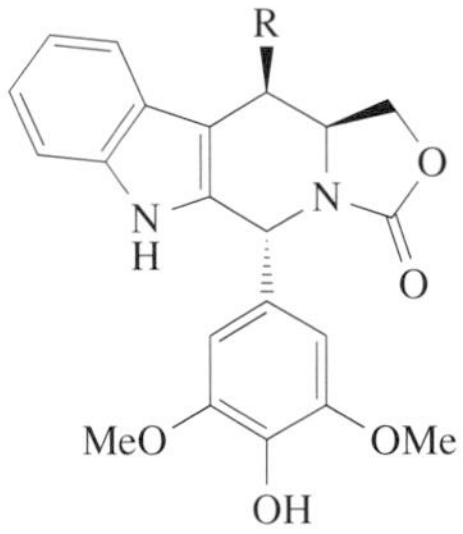

(54) R = substituted anilino

### (i) *Epipodophyllotoxins*

The epipodophyllotoxins etoposide and teniposide are semisynthetic derivatives of podophyllotoxin. In the 1950s, podophyllotoxin entered clinical trials for the treatment of cancer, but was discontinued owing to its high toxicity. The toxicity was successfully reduced by epimerizing the C-1 position and appending an acetal-containing glycosyl side chain. The C-4,6-acetal was found to be essential for the activity of these glycosylated derivatives, presumably to balance the hydrophilicity of the glycosyl moiety with the hydrophobicity required for membrane permeability. Podophyllotoxin does not interact with DNA and etoposide (49) and teniposide (50) react only weakly, implicating a nonintercalative mechanism of action. Additionally, derivatives wherein R = aniline and benzylamine derivatives have demonstrated potent topoisomerase II activity.[141]

### (ii) *Quinolines*

The quinoline derivative CP 115,953 (51)[142] is structurally related to the potent antibiotic ciprofloxacin, a known inhibitor of DNA gyrase, the prokaryotic topoisomerase II. The mechanism of action appears to involve the stimulation of the pre- and poststrand passage DNA cleavage; no inhibition of the rate of religation was observed.

### (iii) *Nitroimidazoles*

The nitroimidazole derivative Ro 15-0216 (52) displays remarkable sequence specificity, inducing only a single cleavage site on both a pBR322 DNA molecule and a *T. thermophilia* rRNA gene, which has a sequence homology for the five base pairs immediately flanking the cleavage site. Additionally, the Ro 15-0216 cleavage point (position 101) was distinct from the single major cleavage on pBR-322 DNA site induced by amonafide (position 1830).[143] The ternary Ro 15-0216–topoisomerase II–DNA complex was readily reversible and was determined to be ~70 times less stable than the complex generated in the presence of amsacrine. Ro 15-0216 did not inhibit topoisomerase II-mediated DNA relaxation. The mode of action appears to be stimulation of the rate of enzyme-mediated DNA cleavage.

### (iv) Isoflavones

The mechanism of action of the isoflavone derivative genistein is associated with an increase in the rate of the enzyme-mediated DNA cleavage.[144] Additionally, genistein (**53**) has been observed to inhibit the strand passage event and hydrolysis of the ATP cofactor. Furthermore, etoposide was observed to reverse the inhibition of strand passage induced by genistein, suggesting that genistein and etoposide share a common interaction domain in the ternary complex.

### (v) Azatoxin

In attempts to discover novel topoisomerase II inhibitors with increased efficacy, a series of established topoisomerase II inhibitors, including both intercalative and nonintercalative, were examined by molecular modeling and a composite model pharmacophore was proposed (Figure 3). Through the use of this model, a hybrid of etoposide aglycone and ellipticine, azatoxin (**54**), was developed. Azatoxin has been shown to possess activity in the *in vitro* formation of topoisomerase II-mediated DNA strand breaks comparable to etoposide.[145] In addition, azatoxin was demonstrated to be a dual inhibitor of topoisomerase II and of tubulin polymerization and each macromolecular target can be exclusively targeted with structural modifications of the azatoxin core.[146] Because the azatoxin skeleton is synthetically readily accessible, a large number of derivatives have been prepared. Several of these compounds have exhibited increased activity relative to azatoxin and are in preclinical evaluation.[147]

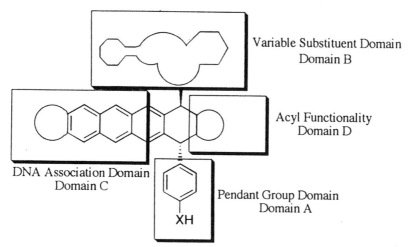

**Figure 3**   A composite pharmacophore for the expression of DNA topoisomerase II-mediated DNA cleavage activity.

### 7.16.3.4   A Composite Pharmacophore for Several Classes of Topoisomerase II Agents

Although not universally adopted as the mechanism underlying the formation of drug-induced topoisomerase II-cleavable complexes, most data support the hypothesis that topoisomerase II-active agents bind to DNA in the cleavable complex and induce a particular DNA deformation that stabilizes an enzyme-mediated DNA cleavage intermediate. This mechanism is supported regardless of whether the drug binds initially to DNA or to a DNA–enzyme complex. Although a diversity of drug-induced DNA conformations may promote the formation of a cleavable complex, the nature of the binding site and the range of potential induced DNA deformations are such that a composite pharmacophore, which encompasses much of the structural diversity observed among topoisemerase II inhibitors, could be defined.[148] This model was developed by superimposing the structures of several classes of topoisomerase II-directed agents, including the anthracyclines, epipodophyllotoxins, ellipticines, and anthracenediones,[149] and "carving out" the common structural features among these drug classes.

Our original model, illustrated in Figure 3, described three domains: (i) a planar polyaromatic array responsible for intercalation or intercalation-like DNA association (Figure 3: domain C); (ii) a pendant group with a hydrogen bond donor ~5.5 Å below the plane of the intercalation domain, responsible for binding in the minor groove of DNA (domain A), and (iii) a variable substituent domain also responsible for minor groove binding, but more "accommodating" with regard to acceptable structures (domain B). To this original model, a fourth structural subunit, the acyl functionality (domain D), can be assigned, which resides in the minor groove and is responsible for the conformational flexibility and disposition of the groove binding domains. Although admittedly unrefined, this model suggested the development of hybrid molecules of known topoisomerase II agents, one of which, an ellipticine–epipodophyllotoxin hybrid, was azatoxin.[150–152]

It is obvious that the domains of this model are intimately associated and that a target molecule must be analyzed as a "whole," not simply as the "sum of its parts." The change in topoisomerase II activity engendered by a structural change in one part of an agent can be amplified or compensated by changes at other molecular sites. In particular, the intercalation domain (domain C) and the acyl functionality (domain D) interact closely in defining the three-dimensional relationships of topoisomerase II-cleavable complex mediators. Nonetheless, this reductionist approach has proven insightful in analyzing the structure–activity relationships of novel topoisomerase II agents and in guiding the synthesis of new target molecules.[153,154]

## 7.16.4  CONCLUSION

The DNA topoisomerases are among the most vital and preeminent cancer chemotherapeutic targets. Topoisomerase II was first identified as the biochemical target for a diverse group of antitumor agents in the mid-1980s and the clinical utility of these drugs and our knowledge of their molecular and cellular mechanisms of action continue to expand at an exceptional pace. Topoisomerase I has entered the clinical arena and promises to contribute importantly to our battery in the treatment of cancer. The number of new topoisomerase-directed agents and our knowledge of the structure–activity relationships of known agents has been greatly enlarged These discoveries have been due largely to the enhanced sensitivity of researchers to the pivotal roles of the topoisomerases as chemotherapeutic targets and to the development of facile *in vitro* and *in vivo* protocols for the assay of topoisomerase involvement. As the search continues for antitumor agents that will improve current protocols and expand the range of treatable diseases, the discovery of new topoisomerase-directed agents and understanding the fundamental mechanisms underlying the clinical profiles of these agents may provide the insights necessary to unlock the promise that these agents hold for future cancer treatment.

## 7.16.5  REFERENCES

1. N. Osheroff, *Pharmacol. Ther.*, 1989, **41**, 223.
2. J. C. Wang, *Annu. Rev. Biochem.*, 1985, **54**, 665.
3. L. F. Liu, *Annu. Rev. Biochem.*, 1989, **58**, 351.
4. L. Yang, M. S. Wold, J. J. Li, T. J. Kelly, and L. F. Liu, *Proc. Natl. Acad. Sci. USA*, 1987, **84**, 950.
5. B. D. Halligan, J. L. Davis, K. A. Edwards, and L. F. Liu, *J. Biol. Chem.*, 1982, **257**, 3995.
6. G. Fleischmann, G. Pflugfelder, E. K. Steiner, K. Javaherian, G. C. Howard, J. C. Wang, and C. S. R. Elgin, *Proc. Natl. Acad. Sci. USA*, 1984, **81**, 6958.
7. S. DiNardo, K. Voelkel, and R. Sternglanz, *Proc. Natl. Acad. Sci. USA*, 1984, **81**, 2616.
8. S. J. Brill, S. DiNardo, K. Voelkel-Meiman, and R. Sternglanz, *Nature (London)*, 1987, **326**, 414.
9. Y. Ishimi, K. Sugasawa, F. Hanoaka, T. Eki, and J. Hurwitz, *J. Biol. Chem.*, 1992, **267**, 462.
10. G. C. Glikin and D. Blangy, *EMBO J.*, 1986, **5**, 151.
11. Y.-S. Bae, I. Kawasaki, H. Ikeda, and L. F. Liu, *Proc. Natl. Acad. Sci. USA*, 1988, **85**, 2076.
12. W. C. Earnshaw, B. Halligan, C. A. Cooke, M. M. S. Heck, and L. F. Liu, *J. Cell Biol.*, 1985, **100**, 1706.
13. W. C. Earnshaw and M. M. S. Heck, *J. Cell Biol.*, 1985, **100**, 1716.
14. M. Berrios, N. Osheroff, and P. A. Fisher, *Proc. Natl. Acad. Sci. USA*, 1985, **82**, 4142.

15. T. A. Beerman, J. M. Woynarowski, and M. M. McHugh, in "DNA Topoisomerases in Cancer," ed. M. E. Potmesil and W. E. Ross, Oxford University Press, London, 1991, p. 172.
16. L. F. Liu, A. Y. Chen, C. Yu, and B. Gatto, *Proc. Natl. Acad. Sci. USA*, 1993, **90**, 8131.
17. L. F. Liu, A. Y. Chen, C. Yu, A. Bodley, and L. F. Peng, *Cancer Res.*, 1993, **53**, 1332.
18. T. A. Beerman, M. M. McHugh, R. Sigmund, J. W. Lown, K. E. Rao, and Y. Bathini, *Biochim. Biophys. Acta*, 1992, **1151**, 53.
19. G. Capranico, F. Zunino, K. W. Kohn, and Y. Pommier, *Biochemistry*, 1990, **29**, 562.
20. C. K. Mirabelli, F. H. Drake, K. B. Tan, S. R. Per, T. D. Y. Chung, R. D. Woessner, R. K. Johnson, S. T. Crooke, and M. R. Mattern, in "DNA Topoisomerases in Cancer," ed. M. E. Potmesil and W. E. Ross, Oxford University Press, London, 1991, p. 133.
21. C. J. Li, L. Averboukh, and A. B. Pardee, *J. Biol. Chem.*, 1993, **268**, 22 463.
22. S. M. Hecht, D. E. Berry, L. MacKenzie, E. A. Shultis, and J. A. Chan, *J. Org. Chem.*, 1992, **57**, 420.
23. J. Roca, R. Ishida, J. M. Berger, T. Andoh, and J. C. Wang, *Proc. Natl. Acad. Sci. USA*, 1994, **91**, 1781.
24. W. T. Beck, R. Kim, and M. Chen, *Cancer Chemother. Pharmacol.*, 1994, **34**, 14.
25. B. Holm, P. B. Jensen, and M. Sehested, *Cancer Chemother. Pharmacol.*, 1996, **38**, 203.
26. A. H. Corbett and N. Osheroff, *Chem. Res. Toxcol.*, 1993, **6**, 585.
27. N. Osheroff, E. L. Zechiedrich, and K. C. Gale, *BioEssays*, 1991, **13**, 269.
28. Y. Pommier, *Cancer Chemother. Pharmacol.*, 1993, **32**, 103.
29. J. Cummings and J. F. Smyth, *Ann. Oncol.*, 1993, **4**, 533.
30. C. H. Freudenreich and K. N. Kreuzer, *Proc. Natl. Acad. Sci. USA*, 1994, **91**, 11 007.
31. S. H. Elsea, P. R. McGuirk, T. D. Gootz, M. Moynihan, and N. Osheroff, *Antimicrob. Agents Chemother.*, 1993, **37**, 2179.
32. D. Hochhauser and A. L. Harris, *Cancer Treat. Rev.*, 1993, **19**, 181.
33. A. G. J. Vanderzee, S. Dejong, W. N. Keith, H. Holleme, H. Boonstra, and E. G. E. Devires, *Cancer Res.*, 1994, **54**, 749.
34. E. Noviello, M. G. Aluigi, G. Cimoli, E. Rovini, A. Mazzoni, S. Parodi, F. Desessa, and P. Russo, *Mutat. Res.*, 1994, **311**, 21.
35. Y. Pommier, K. W. Kohn, G. Capranico, and C. Jaxel, in "Molecular Biology of DNA Topoisomerase and its Application to Chemotherapy," ed. T. Andoh, H. Ikeda, and M. Oguro, CRC Press, Boca Raton, FL, 1993, p. 215.
36. G. E. Francis, M. C. Tejedor, J. J. Berney, C. M. Chresta, C. Delgado, and P. Patel, *Leukemia*, 1994, **8**, 121.
37. D. P. Figgitt, W. A. Denny, S. A. Gamage, and R. K. Ralph, *Anti-Cancer Drug Des.*, 1994, **9**, 199.
38. F. Cortes, J. Pinero, and T. Ortiz, *Mutat. Res.*, 1993, **303**, 71.
39. A. Haldane, G. J. Finlay, and B. C. Baguley, *Oncol. Res.*, 1993, **5**, 133.
40. C. Holm, J. Covey, D. Kerrigan, and Y. Pommier, *Cancer Res.*, 1989, **49**, 6365.
41. R. Bertrand, D. Kerrigan, M. Sarang, and Y. Pommier, *Biochem. Pharmacol.*, 1991, **42**, 77.
42. G. Capranico, P. D. Isabella, S. Tinelli, M. Bigioni, and F. Zunino, *Biochemistry*, 1993, **32**, 3028.
43. P. D. Isabella, G. Capranico, M. Palumbo, C. Sissi, A. J. Krapcho, and F. Zunino, *Mol. Pharmacol.*, 1993, **43**, 715.
44. J. F. Riou, D. Lefervre, and G. Riou, *Biochemistry*, 1989, **28**, 9104.
45. Y. Pommier, G. Capranico, A. Orr, and K. W. Kohn, *J. Mol. Biol.*, 1991, **222**, 909.
46. Y. Pommier, P. N. Cockerill, K. W. Kohn, and W. T. Garrad, *J. Virol.*, 1990, **64**, 419.
47. D. A. Gewirtz, M. S. Orr, F. A. Fornari, J. K. Randolph, J. C. Yalowich, M. K. Ritke, L. F. Povrik, and R. T. Bunch, *Cancer Res.*, 1993, **53**, 3547.
48. L. P. G. Wakelin, G. J. Atwell, G. W. Rewcastle, and W. A. Denny, *J. Med. Chem.*, 1987, **30**, 855.
49. Y. Pommier, D. Kerrigan, J. M. Covey, C. S. Kao-Shan, and J. Whang-Peng, *Cancer Res.*, 1988, **48**, 512.
50. R. T. Bunch, L. F. Povirk, M. S. Orr, J. K. Randolph, F. A. Fornari, and D. A. Gewirtz, *Biochem. Pharmacol.*, 1994, **47**, 317.
51. M. Delvaeye, V. Verovski, W. De Neve, and G. Storme, *Anticancer Res.*, 1993, **13**, 1533.
52. V. Pierson, A. Pierre, Y. Pommier, and P. Gros, *Cancer Res.*, 1988, **48**, 1404.
53. M. J. Robinson, A. H. Corbett, and N. Osheroff, *Biochemistry*, 1993, **32**, 3638.
54. A. H. Corbett, D. Hong, and N. Osheroff, *J. Biol. Chem.*, 1993, **268**, 14 394.
55. M. J. Robinson and N. Osheroff, *Biochemistry*, 1991, **30**, 1807.
56. M. A. Barry, J. E. Reynolds, and A. Eastmen, *Cancer Res.*, 1993, **53**, 2349.
57. M. K. L. Collins and A. L. Rivas, *Trends Biochem. Sci.*, 1993, **18**, 307.
58. B. W. Stewart, *J. Natl. Cancer Inst.*, 1994, **86**, 1286.
59. D. P. Lane, *Nature (London)*, 1993, **362**, 786.
60. C. C. Harris, *Science*, 1993, **262**, 1980.
61. W. G. Nelson and M. B. Kastan, *Mol. Cell Biol.*, 1994, **14**, 1815.
62. S. Kamesaki, H. Kamesaki, T. J. Jorgensen, A. Tanizawa, Y. Pommier, and J. Cossman, *Cancer Res.*, 1993, **53**, 4251.
63. R. Ascaso, J. Marvel, M. K. L. Collins, and A. L. Rivas, *Eur. J. Immunol.*, 1994, **24**, 537.
64. G. Cimoli, M. Valenti, S. Parodi, A. Mazzoni, F. Desessa, P. Conte, and P. Russo, *Oncol. Res.*, 1993, **5**, 311.
65. J. M. Norgaard, S. T. Langkjer, J. Ellegaard, T. Palshof, N. Clausen, and P. Hokland, *Leukemia Res.*, 1993, **17**, 689.
66. H. Yamazaki, A. Dilworth, C. E. Myers, and K. B. Sinha, *Prostate*, 1993, **23**, 25.
67. J. C. Wang, *J. Mol. Chem.*, 1971, **55**, 523.
68. K. S. Srivenugopal, D. Lockshon, and R. D. Morris, *Biochemistry*, 1984, **23**, 1899.
69. J. C. Wang, *Enzymes*, 1981, **14**, 331.
70. S. Krogh, U. H. Mortensen, O. Westergaard, and B. Bonven, *J. Nucleic Acids Res.*, 1991, **19**, 1235.
71. M. Caserta, A. Amadei, G. Camilloni, and E. Di Mauro, *Biochemistry*, 1990, **29**, 8152.
72. E. L. Zechiedrich and N. Osheroff, *EMBO J.*, 1990, **9**, 4555.
73. M. T. Muller, W. P. Pfund, V. B. Mehta, and D. K. Trask, *EMBO J.*, 1985, **4**, 1237.
74. B. Schmitt, U. Buhre, and H.-P. Vosberg, *Eur. J. Biochem.*, 1984, **144**, 127.
75. C. Jaxel, G. Capranico, D. Kerrigan, K. W. Kohn, and Y. Pommier, *J. Biol. Chem.*, 1991, **266**, 20 418.
76. B. J. Newman and N. D. F. Grindley, *Cell*, **38**, 463.
77. M. Gellert, *Annu. Rev. Biochem.*, 1981, **50**, 879.

78. M. D. Been, R. R. Burgess, J. J. Champoux, *Nucleic Acids Res.*, 1984, **12**, 3097.

79. J. J. Champoux, *J. Biochem.*, 1981, **256**, 4805.

80. J. J. Champoux, in "DNA Topology and its Biological Effects," ed. N. R. Cozzarelli and J. C. Wang, Cold Spring Harbor Laboratory Press, Cold Spring Harbor, N.Y., 1990, p. 217.

81. D. E. Pulleyblank and M. J. Ellison, *Biochemistry*, 1982, **21**, 1155.

82. L. Peller, *Biochemistry*, 1976, **15**, 141.

83. G. J. Creemerst, B. Lund, and J. Verweij, *Cancer Treat. Rev.*, 1994, **20**, 73.

84. W. D. Kinksbury, J. C. Boehm, D. R. Jakas, K. G. Holden, S. M. Hecht, G. Gallagher, M. J. Caranfa, F. L. McCabe, L. F. Faucette, R. K. Johnson, and R. P. Hertzberg, *J. Med. Chem.*, 1991, **34**, 98.

85. E. Rubin, V. Wood, A. Bharti, D. Trites, C. Lynch, S. Hurwitz, S. Bartel, S. Levy, A. Rosowsky, D. Toppmeyer, and D. Kufe, *Clin. Cancer Res.*, 1995, **1**, 269.

86. T. Yoshinari, A. Yamada, D. Uemura, K. Nomura, H. Arakawa, K. Kojiri, E. Yoshida, H. Suda, and A. Okura, *Cancer Res.*, 1993, **53**, 490.

87. N. Fujii, Y. Yamashita, Y. Saitoh, and H. Nakano, *J. Biol. Chem.*, 1993, **268**, 13 160.

88. N. Fujii, Y. Yamashita, S. Chiba, Y. Uosaki, Y. Saitoh, Y. Tuji, and H. Nakano, *J. Antibiot.*, 1993, **46**, 1173.

89. N. Fujii, Y. Yamashita, K. Ando, T. Agatsuma, Y. Saitoh, K. Gomi, Y. Nishiie, and H. Nakano, *J. Antibiot.*, 1994, **47**, 949.

90. Y. Yamashita, Y. Saitoh, K. Takahashi, H. Ohno, and H. Nakano, *J. Antibiot.*, 1990, **43**, 1344.

91. S.-D. Fang, L.-K. Wang, and S. M. Hecht, *J. Org. Chem.*, 1993, **58**, 5025.

92. W. M. Messner, M. Tin-Wa, H. H. S. Fong, C. Bevelle, N. R. Farnsworth, D. J. Abraham, and J. Trojanck, *J. Pharm. Sci.*, 1972, **61**, 1858.

93. L.-K. Wang, B. D. Rogers, and S. M. Hecht, *Chem. Res. Toxicol.*, 1996, **9**, 75.

94. D. Makhey, B. Gatto, C. Yu, A. Liu, L. F. Liu, and E. J. LaVoie, *Biol. Med. Chem.*, 1996, **4**, 781.

95. E. Prosperi, E. Sala, C. Negri, C. Oliani, R. Supino, G. Astraldi Ricotti, and G. Bottiroli, *Anticancer Res.*, 1992, **12**, 2093.

96. R. Woessner, M. Mattern, C. Mirabelli, R. Johnson, and F. Drake, *Cell Growth Diff.*, 1991, **2**, 209.

97. N. Osheroff, E. R. Shelton, and D. L. Brutlag, *J. Biol. Chem.*, 1983, **258**, 9536.

98. N. Osheroff and D. L. Brutlag, in "Mechanism of DNA Replication and Recombination," ed. N. R. Cozzarelli, Alan R. Liss, New York, 1983, p. 55.

99. M. Sander and T.-S. Hsieh, *Nucleic Acids Res.*, 1988, **13**, 1057.

100. M. Sander, T.-S. Hsieh, A. Udvardi, and P. Schedl, *J. Biol. Chem.*, 1987, **194**, 219.

101. N. Osheroff, *J. Biol. Chem.*, 1986, **261**, 9944.

102. N. Osheroff, *Biochemistry*, 1987, **26**, 6402.

103. A. H. Andersen, K. Christiansen, E. L. Zechiedrich, P. S. Jensen, N. Osheroff, and O. Westergaard, *Biochemistry*, 1989, **28**, 6237.

104. W. Nelson and M. Kastan, *Mol. Cell. Biol.* 1994, **14**, 1815.

105. S. Kamesaki, H. Kamisaki, T. Jorgensen, A. Tanizawa, Y. Pommier, and J. Crossman, *J. Cancer Res.*, 1993, **53**, 4251.

106. R. Bertand, D. Kerrigan, M. Sarang, and Y. Pommier, *Biochem. Pharmacol.*, 1991, **42**, 77.

107. M. Robinson, A. Martin, T. Gootz, P. McGuirk, M. Moynihan, J. Sutcliffe, and N. Osheroff, *J. Biol. Chem.*, 1991, **266**, 14 185.

108. M. Robinson and N. Osheroff, *Biochemistry*, 1991, **30**, 1807.

109. M. Robinson and N. Osheroff, *Biochemistry*, 1990, **29**, 2511.

110. N. Osheroff, *Biochemistry*, 1989, **28**, 6157.

111. L. K. Dalton, S. Demerac, B. C. Elmes, J. W. Loder, J. M. Swan, and T. Teitei, *Aust. J. Chem.*, 1967, **20**, 2715.

112. K. M. Tewey, G. L. Chen, E. M. Nelson, and L. F. Liu, *J. Biol. Chem.*, 1984, **259**, 9182.

113. S. J. Froelich-Ammon, M. W. Patchan, N. Osheroff, and R. B. Thompson, *J. Biol. Chem.*, 1995, **270**, 14 998.

114. N. R. Bachur, F. Yu, R. Johnson, R. Hickey, Y. Wu, and L. Malkas, *Mol. Pharmacol.*, 1992, **41**, 993.

115. K. M. Tewey, T. C. Rowe, L. Yang, B. D. Halligan, and L. F. Liu, *Science*, 1984, **266**, 466.

116. R. J. Cersosimo and W. K. Hong, *J. Clin. Oncol.*, 1986, **4**, 425.

117. I. Ganzina, M. A. Pacciarini, and N. D. Pietro, *Invest. New Drugs*, 1986, **4**, 85.

118. C. E. Myers, J. R. Muindi, and J. Zweier, *J. Biol. Chem.*, 1988, **262**, 11 571.

119. C. B. Hendricks, A. Y. Chen, A. Y. Yu, A. Boddley, and L. F. Liu, *Proc. Am. Assoc. Cancer Res.*, 1992, **33**, 433.

120. C. E. Morreal, R. J. Bernacki, M. Hillman, A. Atwood, and D. Cartonia, *J. Med. Chem.*, 1990, **33**, 490.

121. H. D. Hollis Showalter, J. L. Johnson, J. M. Hoftiezer, W. R. Turner, L. M. Werbel, W. R. Leopold, J. L. Shillis, R. C. Jackson, and E. F. Elslager, *J. Med. Chem.*, 1987, **30**, 121.

122. H. D. Hollis Showalter, M. M. Angelo, E. M. Berman, G. D. Kanter, D. F. Ortwine, S. G. Ross-Keten, A. D. Sercel, W. R. Turner, L. M. Werbel, D. F. Worth, E. F. Elslager, W. R. Leopold, and J. L. Shillis, *J. Med. Chem.*, 1988, **31**, 1527.

123. W. M. Cholody, S. Martelli, J. Paradziej-Lukowicz, and J. Konopa, *J. Med. Chem.*, 1990, **33**, 49.

124. W. M. Cholody, S. Martelli, and J. Konopa, *J. Med. Chem.*, 1990, **33**, 2852.

125. T. P. Wunz, R. T. Dorr, D. S. Alberts, C. L. Tunget, J. Einspahr, S. Milton, and W. A. Remers, *J. Med. Chem.*, 1987, **30**, 1313.

126. T. Yamagishi, S. Nakaike, T. Ikeda, H. Ikeya, and S. Otomo, *Cancer Chemother. Pharmacol.*, 1996, **38**, 29.

127. K. E. Miller, J. M. Grace, and T. L. MacDonald, *Biol. Med. Chem. Lett.*, 1994, **4**, 1643.

128. W. R. Leopold, T. H. Corbett, D. P. Griswold, J. Plowman, and B. C. Baguley, *J. Natl. Cancer Inst.*, 1987, **79**, 343.

129. B. C. Baguley, W. A. Denny, G. J. Atwell, G. J. Finlay, G. W. Rewcastle, S. J. Twigden, and W. R. Wilson, *Cancer Res.*, 1984, **44**, 3245.

130. G. W. Rewcastle, W. A. Denny, and B. C. Baguley, *J. Med. Chem.*, 1987, **30**, 843.

131. G. J. Atwell, C. D. Bos, B. C. Baguley, and W. A. Denny, *J. Med. Chem.*, **31**, 1048.

132. G. J. Atwell, B. C. Baguley, and W. A. Denny, *J. Med. Chem.*, 1989, **32**, 396.

133. H. H. Lee, B. D. Palmer, M. Boyd, B. C. Baguley, and W. A. Denny, *J. Med. Chem.*, 1992, **35**, 258.

134. W. A. Denny, G. W. Rewcastle, and B. C. Baguley, *J. Med. Chem.*, 1990, **33**, 814.

135. B. C. Giovanella, J. S. Stehlin, M. E. Wall, M. C. Wani, A. W. Nicholas, L. F. Liu, R. Silber, and M. Potmesil, *Science*, 1989, **246**, 1046.

136. W. E. Ross, D. Glaubiger, and K. W. Kohn, *Biochim. Biophys. Acta*, 1979, **562**, 41.
137. K. Wasserman, J. Markovits, C. Jaxel, G. Capranico, K. W. Kohn, and Y. Pommier, *Mol. Pharmacol.*, 1990, **38**, 38.
138. F. Leteurtre, A. Fujimori, A. Tanizawa, A. Chabra, A. Mazumder, G. Kohlhagen, H. Nakano, and Y. Pommier, *J. Biol. Chem.*, 1994, **269**, 28 702.
139. B. Poddevin, J.-F. Riou, F. Lavelle, and Y. Pommier, *Mol. Pharmacol.*, 1993, **44**, 767.
140. C. H. Nguyen, J.-M. Lhoste, F. Lavelle, M.-C. Bissery, and E. Bisagni, *J. Med. Chem.*, 1990, **33**, 1519.
141. S. J. Cho, Y. Kashiwada, K. F. Bastow, Y.-C. Cheng, and K.-H. Lee, *J. Med. Chem.*, 1996, **39**, 1396.
142. M. J. Robinson, B. A. Martin, T. D. Gootz, P. R. McGuirk, M. Moynihan, J. A. Sutcliffe, and N. Osheroff, *J. Biol. Chem.*, 1991, **266**, 14 585.
143. B. S. Sorensen, P. S. Jensen, A. H. Andersen, K. Christiansen, K. Alsner, B. Thomsen, and O. Westergaard, *Biochemistry*, 1990, **29**, 9507.
144. J. Markovits, C. Linassier, P. Fosse, J. Couprie, J. Pierre, A. Jacquemin-Sablon, J.-M. Saucier, J.-B. Le Pecq, and A. K. Larsen, *Cancer Res.*, 1989, **49**, 5111.
145. J. S. Madalengoitia and T. L. Macdonald, *Tetrahedron Lett.*, 1993, **34**, 6237.
146. F. Leteurtre, J. S. Madalengoitia, A. Orr, T. J. Guzi, E. Lehnert, T. L. Macdonald, and Y. Pommier, *Cancer Res.*, 1992, **52**, 4478.
147. J. J. Tepe, J. S. Madalengoitia, K. M. Slunt, K. W. Werbovetz, P. G. Spoors, and T. L. Macdonald, *J. Med. Chem.*, 1996, **39**, 2188.
148. T. L. Macdonald, E. Lehnert, J. Loper, K. Chow, and W. E. Ross, in "DNA Topoisomerases in Cancer," ed. M. E. Potmesil and W. E. Ross, Oxford University Press, London, 1991, p. 199.
149. C. Harris, *Science*, 1993, **262**, 1980.
150. F. Leteurtre, J. S. Madalengoitia, A. Orr, T. Guzi, E. K. Lehnert, T. L. Macdonald, and Y. Pommier, *Cancer Res.*, 1992, **52**, 4478.
151. F. Leteurtre, D. L. Sackett, J. S. Madalengoitia, G. Kohlhagen, T. L. Macdonald, E. Hamel, K. D. Paull, and Y. Pommier, *Biochem. Pharmacol.*, 1995, **19**, 1283.
152. E. K. Lehnert, K. E. Miller, J. S. Madalengoitia, T. J. Guzi, and T. L. Macdonald, *Bioorg. Med. Chem. Lett.*, 1994, **4**, 1643.
153. S. J. Cho, A. Tropsha, M. Suffness, Y.-C. Cheng, and K.-H. Lee, *J. Med. Chem.*, 1996, **39**, 1383.
154. S. J. Cho, Y. Kashiwada, K. F. Bastow, Y.-C. Chen, and K.-H. Lee, *J. Med. Chem.*, 1996, **39**, 1396.

# 7.17
# DNA Selection and Amplification

## DIPANKAR SEN
*Simon Fraser University, Burnaby, BC, Canada*

| | | |
|---|---|---|
| 7.17.1 | INTRODUCTION | 615 |
| 7.17.2 | GOALS OF DNA SELEX | 619 |
| | 7.17.2.1 *Binders* | 619 |
| | 7.17.2.2 *Catalysts* | 619 |
| | 7.17.2.3 *Addressing Problems of Molecular Biology* | 619 |
| | 7.17.2.4 *Enabling Other Kinds of Combinatorial Searches* | 619 |
| 7.17.3 | TECHNIQUES ASSOCIATED WITH SELEX | 619 |
| | 7.17.3.1 *Synthetic Oligonucleotides* | 619 |
| | 7.17.3.2 *The Polymerase Chain Reaction* | 620 |
| | 7.17.3.3 *Generation of Single-stranded DNA from Double Helices* | 620 |
| | 7.17.3.4 *Cloning and Sequencing of DNA* | 620 |
| | 7.17.3.5 *Mutagenesis and DNA shuffling* | 621 |
| 7.17.4 | STRATEGIES FOR SELEX | 621 |
| | 7.17.4.1 *Classes of Libraries* | 621 |
| | 7.17.4.1.1 *"Blanket" libraries* | 622 |
| | 7.17.4.1.2 *"Shotgun" libraries* | 622 |
| | 7.17.4.1.3 *"Focused" libraries* | 623 |
| | 7.17.4.2 *Theoretical Considerations for the Optimization of SELEX* | 623 |
| 7.17.5 | THE UTILITY OF DNA SELEX | 624 |
| | 7.17.5.1 *Selection of Specific Binders* | 624 |
| | 7.17.5.1.1 *Small molecules* | 624 |
| | 7.17.5.1.2 *Protein-binding aptamers* | 626 |
| | 7.17.5.1.3 *Nucleic acid-binding aptamers* | 629 |
| | 7.17.5.2 *Selection of Novel DNA Catalysts* | 630 |
| | 7.17.5.2.1 *Nuclease: the catalysis of RNA cleavage* | 632 |
| | 7.17.5.2.2 *Ligase* | 633 |
| | 7.17.5.2.3 *Chelatase: porphyrin metallation* | 634 |
| | 7.17.5.3 *"Blended" SELEX* | 635 |
| | 7.17.5.4 *SURF* | 636 |
| | 7.17.5.5 *Encoded Libraries* | 636 |
| 7.17.6 | DISCUSSION | 637 |
| | 7.17.6.1 *The Scope and Limitations of SELEX* | 637 |
| | 7.17.6.2 *DNA vs. RNA* | 638 |
| | 7.17.6.3 *Nucleic Acid Aptamers vs. Protein Aptamers* | 638 |
| | 7.17.6.4 *Future Directions* | 639 |
| 7.17.7 | REFERENCES | 640 |

## 7.17.1 INTRODUCTION

In modern organisms the two cellular nucleic acids, DNA and RNA, play roles that are, on the whole, significantly different from those played by the variety of cellular proteins. Gold *et al.*[1]

have discussed the historical perception of DNA and RNA as essentially one-dimensional objects, the linear repositories of genetic information. Proteins, on the other hand, although linear polymers themselves, have been appreciated for their complexity of three-dimensional folding, and the diversity of their function has been understood as stemming precisely from the diversity of their three-dimensional folding. In other words, DNA and RNA have been viewed as linear "tapes," whereas proteins, while coded for by DNA and RNA, have been regarded as complex "shapes." The primacy of proteins—in addition to their astonishing versatility—in catalyzing the thousands of chemical reactions that constitute the metabolism of even the simplest organism has rightly been correlated with their three-dimensional complexity of shape. Catalytic proteins, or enzymes, typically contain certain key structural features such as substrate-binding clefts and active sites with "keyhole"-like structural precision (binding only *specific* substrates and stabilizing only *specific* transition states of reactions). By contrast, the potential of the "tape"-like polymers DNA and RNA to fold and attain anything like the structural precision and complexity of proteins has been viewed as necessarily limited.

The above view of the linearity of nucleic acids (and the structural complexity of proteins) has perhaps persisted because double-helical nucleic acids (such as DNA, as it is found in all extant free-living organisms) may genuinely be regarded as linear "tapes," whose information content is indeed transcribed in a linear fashion to produce messenger RNA, which in turn is translated linearly by ribosomes for protein synthesis. The case for the other, naturally occurring RNAs, however, is different. Thus, both transfer RNAs (tRNAs) and ribosomal RNAs (rRNAs) exercise their crucial roles in protein synthesis by virtue of forming complex three-dimensional shapes. The details of the folded structures of a number of tRNAs (which work as "adapters," or translating devices, in protein synthesis, translating the four-letter, nucleotide, alphabet of messenger RNAs to the 20 letter, amino-acid, alphabet of the coded-for proteins) have been known since the 1970s, when a number of crystal structures of tRNAs were solved. The crystal structures revealed that even the minuscule tRNA molecules ($\sim 60$ nucleotides long) folded to give highly complex and precise three-dimensional shapes, which appeared, furthermore, to incorporate a number of "protein-like" features such as clefts and defined binding sites for metal ions and polyamines.

That the ability of single-stranded RNA molecules to fold to complex shapes might endow some of them with enzyme-like abilities was confirmed in the early 1980s with the discovery of the catalytic RNA molecules[2,3] RNase P (a phosphodiester hydrolase)[2] and the self-splicing ribosomal RNA from the protozoan *Tetrahymena* (which catalyzes two distinct *trans*-phosphorylation reactions).[3] Intensive studies on these catalytic RNAs, or "ribozymes," have revealed them to be both effective and specific catalysts. Although neither one was as catalytically potent as, say, ribonuclease proteins, they appeared nevertheless to share numerous "enzymatic" properties with the equivalent protein enzymes (reviewed by Cech and Bass[4]). These included substrate specificity (with the implied presence of highly discriminating "active sites" within ribozymes), the requirement for the *folded* structures (as opposed to the denatured, tape-like structures) of the RNAs for catalytic activity, and similarities in the overall strategies for catalysis. Although the notion that RNA could be catalytic had been aired in the literature,[5–8] it was the discovery of ribozymes that led to a general and radical reappraisal of the "tape"-like conception of nucleic acids, and of RNA in particular.

With the catalytic ability of RNA clearly demonstrated, discussion focused on the intrinsic ("chemical") differences between proteins and nucleic acids and how these differences might affect the catalytic possibilities for these two classes of biopolymers. Whereas proteins, with their diversity of chemical functionalities, seemed well suited to catalyze many different kinds of reactions, the dearth of useful functionalities in RNA and DNA seemed to impose natural limits on the reactions that might be catalyzed by them. In proteins, amino acid side chains consist of hydrophobic residues (both aliphatic and aromatic), uncharged hydrophilic residues, and positively and negatively charged hydrophilic residues. The heterocyclic bases round in RNA and DNA, by contrast, resemble one another closely in terms of their structure, hydrophobicity/hydrophilicity, and acid–base properties. At neutral pH, they are poor acids and poor bases (the $pK_a$ values closest to the neutral for unmodified DNA and RNA bases are 10.0 and 4.5, respectively), whereas the side chain functionalities of the amino acids histidine and cysteine, within proteins, typically have $pK$ values of 8 and 6 (which makes them particularly good acid–base catalysts). Major differences are also found in the ways that nucleic acid and protein secondary and tertiary structural motifs are organized. Protein secondary structures, such as $\alpha$-helices and $\beta$-sheets, are held together primarily by hydrogen bonds formed between their peptide backbone functionalities. This arrangement allows the diverse side chain functionalities of the amino acids to remain unencumbered and free to participate in both binding and catalytic functions. By contrast, it is precisely the varying units within DNA and RNA (i.e., the bases) that must hydrogen bond to one another to generate nucleic acid secondary and tertiary structures. As a consequence, in many simply folded RNA and DNA structures, the hydrogen-bonded and $\pi$-stacked bases lie in the "interior," whereas the chemically unvarying, and negatively charged, sugar–phosphate backbones are presented on the "outside."

In light of these profound structural differences, Cech and Bass[4] considered that although the bases of RNA and DNA were clearly capable of contributing to the formation of complex folded structures, they did not seem promising in terms of being able to catalyze a broad range of chemistries. Thus, although the naturally occurring RNAses P and the *Tetrahymena* ribozyme were moderately effective enzymes, RNA in general might not be able to catalyze a *broad* range of reactions quite so effectively.

In addition to sparking interest in the purely "polymer" properties of nucleic acids, the discovery of naturally catalytic RNAs engendered the hypothesis of the "RNA World".[9][10] The RNA World was a postulated stage in the evolution of life, prior to the evolution of a diversity of proteins, in which RNA molecules capable of self-replication and catalyzing other kinds of metabolic reactions constituted a primitive "life," with the RNA serving simultaneously as genetic material and catalyst ("genotype" and "phenotype"). The RNA World hypothesis sparked new interest in determining the range of substrate-binding and catalytic possibilities for RNA. Could RNA fold, for example, to form a variety of binding pockets? If so, how might such binding pockets compare with those formed by proteins? Also, what kinds of chemistries might be catalyzed by RNA, given the intrinsic chemical limitations of the nucleic acid bases and nucleotides? The structure of RNA did incorporate the 2'-hydroxy groups on the sugars, and these might have catalytic utility. However, DNA lacked even this functionality. Did this mean that DNA would be capable of catalyzing even fewer kinds of reactions than RNA?

The structural and functional possibilities of nucleic acids have become amenable to experimental investigation owing to the development of techniques for the *in vitro* selection and amplification of interesting DNA and RNA sequences out of large "libraries" of such sequences. This technology is known as SELEX (Systematic Evolution of Ligands by Exponential enrichment),[11] and has its intellectual antecedents in experiments carried out in the 1970s into aspects of molecular evolution,[12,13] which utilized error-prone RNA polymerases such as Q$\beta$ replicase to generate RNA sequence diversity, and to "evolve" certain RNA sequences. Such procedures, however, were limited in terms of what they could do. The modern, versatile SELEX techniques have arisen, by contrast, from the simultaneous utilization of a number of "new" and powerful technologies. These include: the efficient and automated synthesis of DNA and RNA oligonucleotides; amplification techniques such as the polymerase chain reaction (PCR); various kinds of chromatographic, electrophoretic, and other techniques for the separation of molecules and molecular complexes; efficient *in vitro* transcription, involving T7 and other RNA polymerases; and also a number of the established technologies of molecular biology, including reverse transcription, DNA cloning, and DNA sequencing.

What, then, is SELEX? It is an iterative technique for enriching, and eventually selecting, out of a large starting pool (a "library" of $\sim 10^{15}$ individual DNA or RNA molecules, each with a distinct sequence) molecules with certain desirable properties. These properties might include the ability to bind some target molecule of choice or to catalyze a reaction. In order to carry out SELEX, the major requirement is that some scheme for *selection* should exist, which can be used for separating out, in progressively enriching cycles, the "desirable" nucleic acid molecules out of the larger pool. Selection schemes may exploit differences in binding affinities of the members of a nucleic acid library for target molecules, differences in gel mobilities, or in filter-binding characteristics. Selection schemes may also seek to inactivate selectively DNA or RNA molecules that are not "desirable" within a pool. Generically, each cycle of SELEX consists of the following steps (shown in Figure 1): (i) the generation of a starting pool of diverse DNA or RNA molecules; (ii) a selection step, that enriches for the desirable molecules within the pool; and (iii) an amplification step, to generate multiple copies of the molecules selected. This cycle of selection and amplification is carried out until an appropriate level of enrichment of the desirable DNA or RNA molecules (perhaps up from a few individuals out of the starting pool of $10^{15}$ to $\sim 50\%$ of the final pool) has been achieved. Following this, the selected DNA or RNA molecules can be "cloned" to sequester individual sequences from one another. DNA from any individual clone can now be isolated autonomously, amplified, and its individual properties studied. SELEX-enriched, "useful" DNA molecules are often referred to as "aptamers",[14] in particular those DNA molecules that have been selected for binding to target molecules.

To date, SELEX experiments have been carried out with both RNA and DNA (and, in the case of DNA, with both single- and double-stranded DNA, depending on the objective of the selection). Since DNA is the primary focus of this volume, this chapter deals chiefly with DNA SELEX (RNA

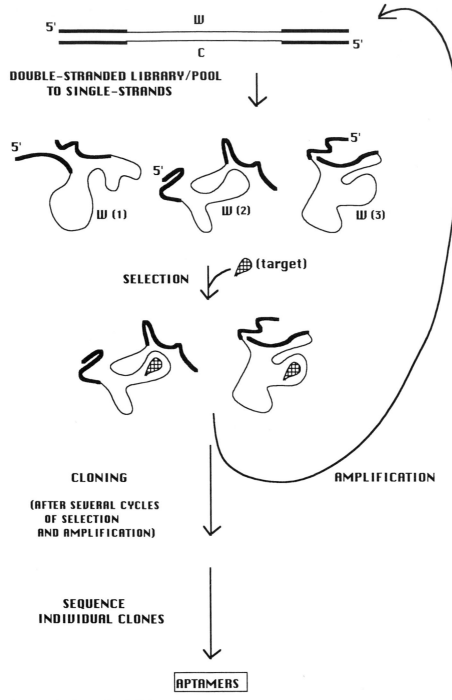

**Figure 1** A general scheme for SELEX, showing a selection and amplification scheme on the basis of the binding of single-stranded DNA molecules to a specified target molecule. Other variants of SELEX use double-stranded DNA molecules or utilize chemical and/or structural changes in individual DNA molecules as bases for selection.

is the subject of Volume 6 in this series). Discussion will therefore be focused primarily on DNA selection experiments, although there are often close methodological similarities between SELEX experiments carried out with RNA and with DNA. In instances where both RNA and DNA SELEX experiments have been carried out towards a particular objective, and where a comparison of the two selections is illuminating, the relevant RNA experiment is discussed. This chapter also does not aspire to be a comprehensive review of all DNA SELEX experiments carried out to date; consequently, a few selected experimental reports have been chosen to be discussed in detail.

## 7.17.2 GOALS OF DNA SELEX

### 7.17.2.1 Binders

A major goal is to find DNA molecules that bind strongly and specifically to different target molecules, including proteins, peptides, nucleic acids, a variety of small organic molecules, and other targets. The rationale for undertaking these kinds of searches includes understanding the nature of binding sites in nucleic acids, the derivation of target- or "antigen"-binding DNA "antibodies," and finding specific inhibitors of enzymes and other proteins.

### 7.17.2.2 Catalysts

The goal is to find DNA catalysts ("deoxyribozymes" or "DNAzymes"). The rationale here is to explore and define the range of chemical reactions catalyzable by DNA (and by nucleic acids in general). A "practical" rationale for many of these selections is the derivation of catalytic DNA molecules of potential biomedical utility.

### 7.17.2.3 Addressing Problems of Molecular Biology

SELEX has been used to address a variety of important molecular biological problems, in addition to structural and functional questions about DNA itself. Specific goals include finding optimal binding sites for DNA-binding proteins, the determination of consensus sequences for the formation of DNA triple strands, and the study of such intrinsic properties of DNA as sequence-dependent bending.

### 7.17.2.4 Enabling Other Kinds of Combinatorial Searches

The power of DNA SELEX has been harnessed to carry out combinatorial searches involving nonpeptide and non-nucleotide libraries ("encoded combinatorial chemistry").

## 7.17.3 TECHNIQUES ASSOCIATED WITH SELEX

### 7.17.3.1 Synthetic Oligonucleotides

A number of methods for the chemical synthesis of DNA have been developed, and DNA oligomers of up to 150 deoxynucleotides can be made efficiently and rapidly in automated DNA synthesizers. These machines are commonly capable of DNA synthesis at up to the 10 μmol scale; however, for the generation of random-sequence oligonucleotides for SELEX, synthesis is typically carried out at the 0.2 μmol scale. The flexibility and the mild conditions of the synthetic chemistry used, together with the commercial availability of a wide variety of activated nucleotides suitable for use in these machines, have made possible the synthesis of an astonishing variety of both standard and chemically modified oligonucleotides. Synthetic DNA oligonucleotides may incorporate any or all of the four *ribo*nucleotides and a variety of non-natural or modified bases and sugars; they may have methylphosphonate or phosphorothioate backbones; there is also the ability to incorporate complex mixtures of deoxynucleotides (say, equimolar mixtures of dA, dG, dC, and dT) at one or more specified sites upon the DNA.

Excellent reviews by Gait[13] and in *Current Protocols of Molecular Biology*,[16] Sections 2.11 and 2.12, describe the chemistry used for automated DNA synthesis and the utility of synthetic oligonucleotides.

### 7.17.3.2   The Polymerase Chain Reaction

The polymerase chain reaction (PCR) is a relatively recent technology that has had a wide impact on numerous aspects of molecular biology. PCR is, broadly, a means for the amplification of DNA, and its development is central to the subsequent development of SELEX. There are numerous excellent reviews (as well as an ever-increasing number of specialist books) on the methods and uses of PCR, including recent reviews by Arnheim and Erlich[17] and Mullis and Falona.[18] the technical aspects of PCR are covered thoroughly in Section 15 of *Current Protocols of Molecular Biology*.[16] PCR, as used in SELEX, not only provides a means for the amplification of DNA, but is also useful for introducing mutations into DNA pools (PCR mutagenesis is discussed in Section 7.17.3.5).

### 7.17.3.3   Generation of Single-stranded DNA from Double Helices

A significant number of DNA SELEX experiments have utilized single-stranded, as opposed to double-stranded, random sequence DNA molecules. However, PCR amplifications of DNA necessarily yield double-stranded DNA. For those experiments which must utilize single-stranded DNA, one of the two strands (i.e., either the "Watson" or the "Crick" strand) must be separated from its complementary strand and recovered as quantitatively as possible. Two distinct methods have been used by SELEX investigators to obtain single-stranded DNA: (i) by the asymmetric PCR amplification of a specified strand and (ii) by using a biotin–avidin affinity system for the selective immobilization of a specified strand of the double-stranded DNA on a matrix, followed by the recovery of the other strand. Details of these two techniques can be found in references 19 and 20, respectively.

In brief, the first technique involves PCR amplifications carried out with nonequal concentrations of the two required oligonucleotide primers. PCR carried out in this way leads to the accumulation of that DNA single strand which was primed by the oligonucleotide primer present in excess.

The second technique, which has been used preferentially in more recent studies involving DNA SELEX, takes advantage of the extraordinarily high binding affinity between biotin and the protein avidin ($K_{assoc} \sim 10^{16}$ M$^{-1}$). PCR amplification of DNA is carried out with *one* of the two required primers carrying a biotin moiety covalently attached to its 5′ end (biotinylated nucleotide phosphoramidites, suitable for use in automated DNA synthesizers, are available commercially). The resulting biotinylated, double-stranded DNA is immobilized (via the biotin) onto a matrix containing avidin, and the nonbiotinylated DNA strand is dissociated from its immobilized complementary strand by washing with 0.1 M sodium hydroxide. Overall, this biotin-mediated method appears to be both more reproducible and quantitative than the asymmetric PCR method.

### 7.17.3.4   Cloning and Sequencing of DNA

These two molecular biological techniques are described in detail in Sections 7 and 3.16 of *Current Protocols in Molecular Biology*.[16] The term "cloning" as applied to DNA in the context of SELEX refers to the sequestration, identification, and amplification of *individual* DNA sequences out of heterogeneous or complex DNA populations (such as libraries containing $\sim 10^{15}$ distinct DNA molecules). DNA aptamers enriched in SELEX experiments are typically ligated into a DNA cloning vector (usually a bacterial plasmid) and introduced into recipient *E. coli* bacteria at low DNA concentrations, such that no bacterium receives more than a single molecule of the recombinant plasmid. The transformed bacteria are grown on agar plates at low density, such that the progeny of individual cells form colonies. Bacteria from any given colony can be grown up and harvested for plasmids containing one *individual* aptamer sequence from the final SELEX pool.

The sequencing of DNA, like DNA synthesis, has been automated and can be carried out in automated DNA sequencers. There are, in general, two methodologies for sequencing DNA: the enzymatic method of Sanger *et al.*[21] ("dideoxy" sequencing) and the chemical method of Maxam and Gilbert.[22] Although each method has its unique advantages, it is the Sanger method that is the more widely used, and this is also the methodology compatible with automated sequencers. The preparation of sequence "ladders" from a given piece of DNA (which can be analyzed manually, using denaturing electrophoresis, or in automated DNA sequencers) is easily accomplished using sequencing kits available from a number of suppliers.

### 7.17.3.5 Mutagenesis and DNA shuffling

In the context of SELEX, mutagenesis (the introduction of changes in the sequence of either a specific DNA molecule, or in a pool of DNA molecules) can be achieved both during the chemical synthesis of oligonucleotides and during the enzymatic replication and amplification of the oligonucleotides. Certain DNA polymerases, such as *Taq* polymerase (commonly used for PCR), are relatively error prone, and introduce mutations into the replicated sequence at a rate of $2 \times 10^{-4}$ per nucleotide for every PCR cycle.[23] A number of investigators have modified the reaction conditions for polymerization to make *Taq* polymerase even more error prone.[24,25] This is accomplished by adding $Mn^{2+}$ ions to the replication cocktail and by the use of higher than usual concentrations of both dNTPs and magnesium. Typically, these changes introduce errors of up to $7 \times 10^{-3}$ per nucleotide for every pass of the polymerase.[25]

A different strategy, also PCR connected, has been devised[26] in which the misincorporation of nucleotides into replicated DNA is stimulated by the inclusion of deoxyinosine triphosphate (dITP) in the replication cocktail. Inosine has the ability to base pair with all of the natural DNA nucleotides; it is, moreover, acceptable as a substrate by *Taq* polymerase. Therefore, the inclusion of dITP in the replication solution leads to the introduction of substantial numbers of mutations (up to $3.7 \times 10^{-3}$ per base pair).

Mutations during the chemical synthesis of DNA are easy to achieve. Generalized mutations at specific sites in a DNA molecule can be introduced by allowing the synthesizer to utilize different nucleotide phosphoramidite cocktails (containing, say, 30% G and 70% A, instead of 100% A) during synthesis. Equimolar incorporation of two or more nucleotides at a site is easily carried out; more complicated ratios of nucleotides (such as above) often require a manual preparation of the relevant phosphoramidite cocktails. The "focused" strategy for SELEX (see below), in particular, utilizes libraries of sequence variants off a single, "core", sequence; in such cases, one might incorporate 10% each of G, C, and T, and 70% A, at a site where there is an A in the core sequence. In addition to this method, certain perturbations in the synthetic cycles used by synthesizers can be used to introduce random deletions and other changes into oligonucleotides.[27] Also, a "universal base phosphoramidite" is commercially available (Clontech), which, when incorporated synthetically into DNA, leads to incorporation of all four DNA nucleotides at complementary sites during subsequent PCR amplifications.

Whereas the above mutagenic techniques are powerful in their ability to both introduce and maintain sequence diversity in evolving pools of DNA, an even more powerful technique for generating sequence diversity has been described by Stemmer,[28] that of "*in vitro* recombination." Briefly, a partial fragmentation of the members of a pool of DNA is initially carried out by using the enzyme deoxyribonuclease I (DNase I). The resulting DNA fragments are then reassembled, or recombined, at random, via a PCR-like primer-extension step. The effect is to obtain "recombinant" DNA molecules, where an individual "recombined" molecule may contain elements that were originally present in entirely separate DNA molecules. This technology, being combinatorial (unlike the mutagenic technologies described above), is capable of creating significantly greater sequence diversity than the purely mutagenic techniques.

### 7.17.4 STRATEGIES FOR SELEX

#### 7.17.4.1 Classes of Libraries

Joyce[29,30] has categorized three different SELEX approaches that it is possible to take in the selection (or evolution) of nucleic acid molecules with particular desirable properties.

The generation of initial pools ("libraries") of random sequences, containing perhaps $10^{13}$–$10^{16}$ individual sequences, is the starting point of all SELEX procedures. The practical upper limit for library diversity is typically $10^{16}$. This is a consequence of the amounts of DNA that are synthesizable in a typical DNA synthesizer. For oligomers 100–150 nucleotides long, synthesis is best carried out at the 0.2 μmol scale. Of the DNA recovered, usually only $\sim 10\%$ is actually utilizable for SELEX (the remainder have chemical lesions of one sort or another, which prevent their amplification by PCR techniques).

However, whatever the degree of library diversity, it is possible to start with at least three different *kinds* of "randomness" in the initial libraries, depending on the nature of the experimental goal. The three categories are outlined below.

### 7.17.4.1.1  "Blanket" libraries

These are libraries in which all of the theoretically possible variants of sequence for a given length of random-sequence DNA are represented. In other words, such a library contains a sufficient number of individuals such that every single possible sequence variant is encompassed. Given that any "random" nucleotide position can, by definition, contain any of the four nucleotides, this sets a natural limit (for a library with $\sim 10^{15}$ individual sequences) on the maximum length of fully representative random sequences that such a library can accommodate. If every possible sequence is to be represented at least once in a library of $10^{15}$ individuals, then the random sequence can be no longer than 25 nucleotides ($4^{25} \approx 10^{15}$). If one assumes that a *useful* library ought to contain at least 100 copies of each individual sequence, this further limits the maximum length of the random sequence to 21–22 nucleotides.

Although this relatively modest length for the random DNA might appear limiting for most SELEX goals, such fully representative libraries are able to find use in, for instance, the determination of optimal recognition sequences for DNA-binding proteins or in the formation of DNA triple helices. In addition, such libraries can be made use of in the techniques of "encoded combinatorial chemistry"[31] (see below).

### 7.17.4.1.2  "Shotgun" libraries

For most SELEX applications, comprehensive libraries containing every conceivable sequence variant within a random region are abandoned in favor of "shotgun" libraries, which, although containing large numbers of different sequences, fall far short of all theoretically possible sequences. Thus, a library of $10^{15}$ individuals is typically designed to include 100 random nucleotides in every DNA molecule (to represent every single possible sequence variant within a 100-nucleotide random stretch would necessitate the synthesis of $10^{60}$ distinct DNA molecules!). An assumption implicit in the design of shotgun libraries is that the $10^{15}$ individuals of such a library are nevertheless representative of the comprehensive (and impossible to achieve) $10^{60}$-strong library. An interesting aspect of shotgun libraries is that since no individual $10^{15}$-member library can sample more than a minuscule portion of the theoretical sequence space, it is likely that no two such libraries would share any single individual sequence. Hence it is conceivable that the same selection carried out in parallel with two separate shotgun libraries might yield distinct and mutually unrelated solutions.

For all their limitations, it is interesting to note that shotgun libraries have, to date, yielded successful DNA (and RNA) "solutions" to a variety of problems. This suggests that shotgun libraries do represent sufficiently large populations of distinct DNA molecules, from which "solutions" to many different "problems" can still be isolated.

In constructing a shotgun library for SELEX there is, of course, no requirement that there be a single, continuous stretch of random sequence. Although a number of experiments have utilized such "continuous stretch" libraries, others have distributed their random sequences around elements of rationally designed sequence. The search for novel catalytic nucleic acids, in particular, has often incorporated elements of rational design (of the three catalytic DNA molecules discussed below, two, the lead-dependent phosphodiesterase and the prophyrin metallase, were derived from unstructured libraries; the DNA ligase, by contrast, was selected from a random library that incorporated significant elements of rational design). The rationally designed elements are typically stable secondary/tertiary structural motifs of DNA or RNA (such as hairpin loops, pseudoknots, and guanine quadruplexes), or else they are the known binding sites of certain small molecules. An important instance of this sort was the embedding of an ATP-binding sequence motif into an otherwise random region of RNA as part of a SELEX search for novel ribozymes with kinase activity.[32,33] Other important instances have built-in "active sites" adjacent to the random RNA and DNA elements, in the context of searching for novel RNA and DNA catalysts for self-ligation[34,35] and RNA cleavage.[36] Further, Hamm[37] has derived structurally distinct RNA aptamers for binding to the antigen binding site of a monoclonal antibody using random sequences that were constrained to form either a stem-loop structure or a guanine quadruplex.

### 7.17.4.1.3 *"Focused" libraries*

A third kind of random library is a so-called "focused" library, in which a closely related group of partially randomized sequences are derived from some known "core" sequence of interest. To generate such a library, the core sequence is resynthesized, but in such a way as to introduce a certain probability of mutation at specified nucleotide sites within it. The result of such a synthesis is a cluster of related sequences surrounding the dominant core or "wild-type" sequence. Focused libraries have been used with particular success for *in vitro* evolution experiments with RNA (such as for evolving calcium-dependent and DNA-cleaving variants of the naturally occurring Class I ribozyme from *Tetrahymena*[38,39]).

However, the major utility of focused libraries in DNA SELEX has been to define important sequence features of successful DNA molecules isolated using a "shotgun" strategy. In these instances, "successful" sequences of DNA, typically high-affinity target binders or novel catalysts, have been individually resynthesized to generate secondary, focused libraries. After a number of rounds of selection with these focused libraries, the identities of the truly indispensable nucleotides in the core sequence have been revealed, because they remain invariant through the secondary selection, or else co-vary as pairs of nucleotides (particularly in instances where base pairing between two defined sites is important for the "active" secondary or tertiary structure of the folded aptamer).

Breaker and Joyce[40] have provided an interesting statistical analysis of the extents to which individual bases in a random sequence of defined length would need to be mutagenized to give the optimal diversity to members of a focused library.

### 7.17.4.2 Theoretical Considerations for the Optimization of SELEX

In carrying out a DNA SELEX experiment, whether for the purpose of isolating a novel binder or a novel catalyst, a number of parameters have to be evaluated regarding the optimal way to attempt SELEX. Several variables need to be considered, for example, whether to carry out SELEX with RNA or DNA; the type of random library to be used (see above); the length of random sequence within the molecules in the library; the number of individual variants within the library; and whether to attempt a "selection" experiment (chiefly a process of isolating a small number of "useful" DNA molecules out of a large, pre-existing pool, without enlarging the diversity of pools by introducing mutations into them), or an "evolution" experiment (selection coupled to a significant degree of mutagenesis during each amplification step).[30] Regarding the use of RNA vs. DNA, the evidence seems to be emerging that both polymers are comparably effective; other than that, few definitive answers have emerged regarding many of the above questions. For instance, regarding the question of the length of random sequence optimal for use in different types of SELEX experiments, the answer remains resolutely obscure, other than a consensus of opinion that a minimum of $\sim 20$ random nucleotides are required.

An attempt to systematize SELEX experiments and to provide a theoretical framework for optimal SELEX was made by Irvine *et al.*,[41] who considered an experiment to find DNA (or RNA) binders for a protein target. Irvine *et al.* made an assumption that during each step of binding or selection, the target molecule and the DNA molecules in the library reached a binding equilibrium. They further postulated that a starting library did not contain simply two classes of DNA molecules, the "binders" and the "nonbinders," but rather a spectrum of binders of progressively varying affinities for the target. Given this, in the early rounds of selection, the very few DNA molecules with the highest affinity for the target would have to compete for binding to the target molecules with a significantly larger pool of medium to poor binders. For this reason, it was recommended that in the early rounds of SELEX, relatively high concentrations of the target molecule to be used for selection, so as not to lose the few high-affinity DNA molecules within the starting pool. In the later rounds of SELEX (with considerably enriched populations of the high binders present in the pools), however, more and more stringent selection procedures, as well as lower target molecule concentrations, would have to be used to concentrate the high-affinity ligands away from the medium-affinity ligands.

Irvine *et al.*[41] suggested that the presence of complex populations of binders within the starting DNA library made the process of enrichment of the highest-affinity ligands a complex one, requiring a larger number of selection-and-amplification cycles than might simply be predicted on the basis of the initial abundance of the high binders as a fraction of the starting library.

## 7.17.5 THE UTILITY OF DNA SELEX

### 7.17.5.1 Selection of Specific Binders

#### 7.17.5.1.1 *Small molecules*

*(i) Organic dyes*

The first SELEX experiments to find nucleic acid aptamers that recognized and bound small, organic, target molecules were carried out by Ellington and Szostak, first with RNA[14] and then with DNA.[42] In both cases, aptamers were selected for binding to a series of heterocyclic dyes (such as Cibacron Blue 3Ga, Reactive Red 120, and Reactive Blue 4 (Sigma Chemicals)) that were commercially available covalently linked to cross-linked agarose beads. In both sets of SELEX experiments, the strategies used were comparable. The later, DNA, selection was carried out as follows: a library of 157-nucleotide long DNA oligomers, containing 120 random nucleotides flanked by fixed-sequence, primer-binding sites (for PCR) on either side, was used. This starting library had approximately $2 \times 10^{13}$–$3 \times 10^{13}$ distinct sequences within it. Asymmetric PCR techniques were used to generate single-stranded DNA molecules from the double-stranded library; the folded, single-stranded oligomers were then applied to the individual dye affinity columns in a relatively high-salt buffer (0.5 M LiCl, 20 mM Tris-Cl (pH 7.6), 1 mM $MgCl_2$) to minimize charge–repulsive interactions between the negatively charged dyes and the DNA. Unbound DNA molecules in each case were washed off with four column volumes of the binding buffer, and the bound DNA molecules were eluted by washing the column with three volumes of water (with the expectation that the salt-folded DNA molecules would unfold in water). Although in the first three to four rounds of selection progressively higher fractions of the total DNA had bound to the columns (starting with less than 1% bound in the first round), much of this could be accounted for by adventitious or artifactual binding of certain DNA molecules to the column matrix itself (this was shown by a lack of selective binding by these early DNA pools to their appropriate dye columns). To correct for this phenomenon, counterselection procedures were employed, involving first passaging the DNA pools through columns containing a noncognate dye and then applying the flow-through, or unbound, DNA from those columns to columns containing the cognate, target dye. The effect of this kind of prior "negative" selection was to improve significantly the specificity of binding in the later rounds of selection.

Dye-binding DNA molecules enriched in this way were ultimately cloned and the DNA from individual clones were sequenced. Analysis of the sequences revealed that while there was no significant similarity between 17 clones that bound Cibacron Blue (CB), the eight clones that bound Reactive Green 19 (RG) included five that shared an 18-nucleotide motif. This motif, synthesized as an autonomous oligomer, did not bind well to the RG column; however, a small degree of further stabilization of the folded motif (shown in Figure 2) provided an excellent and specific binder of RG.

Comparison of these DNA data with the earlier SELEX search for RNA aptamers for these same dyes[14] provided some interesting insights: first, the RNA versions of successful DNA dye-binding sequences proved to be nonbinders, as did the DNA versions of RNA aptamer sequences. In fact, there appeared to be no discernible similarity between the sequences of the DNA aptamers and the RNA aptamers. Since the chemical differences between DNA and RNA consist of (i) the 2′-hydroxy group and (ii) the 5-methyl group in thymine (but not in uracil), the relative influences of these two differences between DNA and RNA were investigated separately. A specific DNA aptamer was resynthesized but with deoxyuracil (dU) replacing its thymine residues. This modification appeared not to affect its binding ability. It was therefore concluded that it was the 2′-hydroxy group of ribose that significantly impacted on the differently folded structures for a given sequence of RNA and DNA.

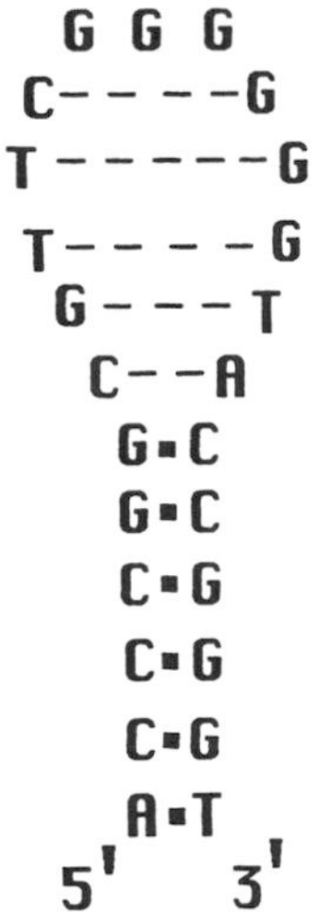

```
        G G G
      C----G
      T-----G
      T----G
       G---T
       C--A
       G=C
       G=C
       C=G
       C=G
       C=G
       A=T
     5'    3'
```

**Figure 2** The structure of a small DNA molecule specific for dye binding.

### (ii) ATP

This particular SELEX experiment could be taken as a paradigm for the procedures and steps taken to enrich for, and then optimize, DNA binders for small-molecule targets. A search for DNA binders for ATP was made by Huizenga and Szostak.[43] ATP is an ideal small-molecule target, being a ubiquitous biochemical and a molecule which serves as the major energy source for most biochemical reactions. For this reason, it is a common cofactor, as well as substrate, for many cellular enzymes and other proteins. For this SELEX experiment, DNA oligomers containing 72 random nucleotide positions (out of 112 total nucleotides) were utilized. The starting library contained $\sim 2 \times 10^{14}$ individual sequences; selections were carried out on affinity columns that had ATP covalently attached to agarose beads via an 11-atom linker. By the seventh round of selection, $\sim 18\%$ of the DNA pool was binding to the ATP column; DNA from the eighth round was therefore cloned in order to isolate individual "binding" sequences from the pool. Seventeen clones were sequenced and were found to have substantially different sequences from one another. To characterize in detail the essential sequence (and, therefore, structural) elements of an individual, strongly binding clone, its DNA was resynthesized as small, overlapping fragments of the original sequence, which were then individually tested for their ability to bind ATP. This analysis led to the identification of a 42-nucleotide fragment with the "random" region of the original aptamer as the ATP-binding element.

Whereas this early part of the selection process had utilized a "shotgun" strategy for selecting out ATP-binding DNA molecules from a vast starting pool, the 42-nucleotide sequence (5'-GTGCT TGGGG GAGTA TTGCG GAGGA AAGCG GCCCT GCTGA AG) identified as the ATP-binding sequence was used as the "core" sequence for the resynthesis of a "focused" library (see above). This resynthesis incorporated the same primer-binding sequences that had flanked the original central sequence (above), but the central sequence itself was synthesized so that a given nucleotide within it had a 70% chance of remaining the same and a 10% chance of being mutated to each of the other three possible nucleotides (e.g., a G to an A, C, or T). This resynthesized, focused library was then subjected to four more rounds of ATP-binding selection and, from the final pool, DNA molecules were again cloned, and of these clones, 45 were sequenced. Analysis of the sequences of these focused clones readily revealed important details of the ATP-binding motif, which was found to contain two conserved (and, therefore, required for ATP-binding) stretches of sequence separated by an unconserved region.

From the above analysis, it was possible to design a 25-nucleotide oligomer, which contained the ATP-binding region in its entirety. This "minimal" DNA aptamer was synthesized and its association with ATP measured. Dissociation constants of ATP from this aptamer gave values of $6 \pm 3$ μM. To determine which specific features or functionalities of ATP were actually important for aptamer binding, various constituent parts of ATP, such as adenosine (and 2'-deoxyadenosine) were examined for their aptamer-binding ability. It was found that structural elements of both the base and sugar moieties of ATP were required for recognition by the aptamer, whereas the triphosphate group was not. Further exploration of the aptamer-ATP recognition elements with adenosine relatives and analogues such as 1-methyladenosine, inosine, and 7-deaza-adenosine, revealed that, specifically, the N-7, N-6, and N-1 positions of adenosine were required for binding. By contrast, substitutions at the C-8 position did not interfere with binding.

To determine the sequence elements of the *aptamer* necessary for ATP binding, the sequences of individual clones derived from the focused selection were compared. From the analysis of invariant and covarying nucleotides, a model for the folded structure of the ATP-binding aptamer was proposed, consisting of two guanine quartets (Figure 3) flanked by two short Watson–Crick base-paired double-helical stems. Confirmation for this model was sought by resynthesizing the aptamer with nucleotide analogues (such as 7-deaza-2'-deoxyadenosine and O-6-methyl-2'-deoxyguanosine) substituting for individual conserved deoxyadenosine and deoxyguanosine residues within the binding motif. These studies revealed that N-7, O-6, and N-2 residues of the conserved guanosines were important—either for direct interactions with the bound ATP or for holding together the folded structure of the aptamer necessary for ATP binding.

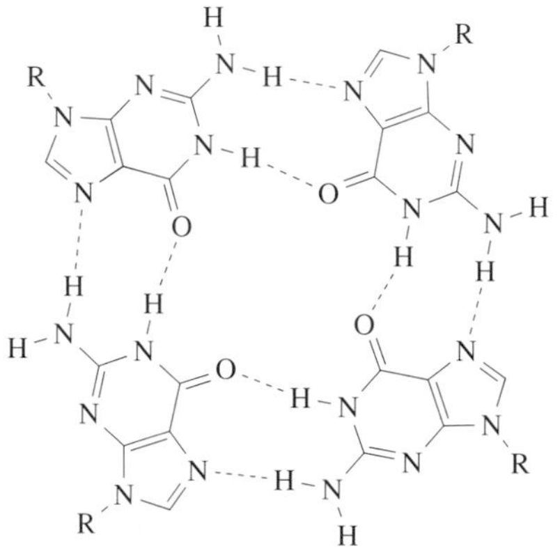

**Figure 3**    A guanine quartet, with four guanine bases hydrogen bonded to each other in a circular fashion. The dotted lines indicate hydrogen bonds; R represents the deoxyribose sugar moieties.

An interesting comparison of this work can be made with a prior and similar selection carried out to find RNA ligands for ATP.[44] That search yielded a binding motif composed of an asymmetric internal loop within a Watson–Crick RNA double helix, a very different secondary structure from that proposed for the DNA aptamers. Overall, a comparison of the DNA and RNA ATP-binding motifs yielded the following insights: although the RNA and DNA aptamers had very different base sequences, the minimal functional structures were similar in size (32 nucleotides for RNA and 25 for DNA); also, both RNA and DNA aptamers bound ATP, AMP, and adenosine (all with comparable affinities), and in neither case did the charged triphosphate play a role in aptamer binding. Both RNA and DNA aptamers required functionalities on the base and the sugar moieties of ATP for binding, although the 3'-hydroxy group of ATP was more important for the DNA aptamer and the 2'-hydroxy group for the RNA aptamer. The numbers of individual "successful" sequences for ATP binding found in the DNA and the RNA selections were comparable, recapitulating the earlier RNA and DNA studies of Ellington and Szostak.[14,42] In other words, it appeared that both RNA and DNA were comparably capable of folding to form complex binding pockets, although different sequences were used by the two polymers. It was concluded, overall, that the 2'-hydroxy groups of RNA did not impact strongly on the relative ability of RNA (with respect to DNA) to fold to form complex and functionally equivalent pockets.

### 7.17.5.1.2   *Protein-binding aptamers*

#### (i)   Thrombin

DNA aptamers specific for binding to a number of proteins, including human immunodeficiency virus (HIV-1) reverse transcriptase,[45] the human enzyme neutrophil elastase,[46] and others, have been described. An interesting conclusion from these experiments is that there appears to exist a sufficient repertoire of folded DNA structures to be able to bind specifically to even those proteins that do not interact with either DNA or RNA *in vivo*. The first, and most illuminating, data on this subject were derived from a DNA SELEX experiment against the human protein thrombin.[47] Thrombin is a glycoprotein that plays an important role in the blood-clotting cascade. SELEX was carried out with a library of 96-nucleotide DNA molecules (containing 60 random nucleotides) and, out of aptamers cloned after only five rounds of selection, a tightly conserved, 15-nucleotide (nt), thrombin-binding consensus sequence (dGGTTG GTGTG GTTGG) was derived. When this 15-nt sequence was synthesized as an autonomous DNA oligomer, it was sufficient for thrombin binding (with a low dissociation constant of ~25–200 nM).

Certain technical innovations in the selection procedure were thought to contribute to the rapid derivation of this highly specific and tightly binding aptamer. For the selection, thrombin was immobilized (via noncovalent binding) on a column of the lectin, concanavalin A (con A), for which it has a known binding affinity. The random DNA pools, first cleared of purely con A- and matrix-binding DNA molecules, were added to the thrombin–con A columns. DNA molecules that bound specifically to thrombin were then eluted from the column by competitive elution with a different con A ligand, α-methylmannoside (rather than by dissociation with water or with EDTA-containing buffers).

The 15-nucleotide aptamer for thrombin has been found to have promising anticoagulant properties[48] and has been undergoing tests for possible use as an anticlotting pharmaceutical. The folded structure of this very small aptamer has been the subject of a number of high-resolution structural studies:[49–53] X-ray crystallography and NMR spectroscopy. The folded structure has been found to consist not of a Watson–Crick double helix, but of a guanine quadruplex,[54–58] consisting of two stacked guanine quartets. This structure is shown schematically in Figure 4. X-ray crystallography has further revealed that thrombin binding does not induce a significant conformational change in the aptamer's folded structure.[51] The aptamer binding site upon thrombin is at the latter's "cationic exosite", normally the binding site of its natural substrate, fibrinogen.

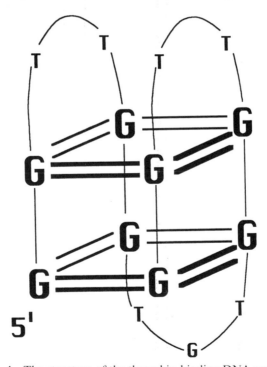

**Figure 4** The structure of the thrombin-binding DNA aptamer.

Guanine quadruplex structure, such as formed by the thrombin-binding aptamer, have emerged as a particularly common structural motif for DNA aptamers (although not so common for RNA aptamers). Other DNA aptamers thought to adopt this structure include those for ATP (see above) and those for anionic prophyrins (see below). A DNA oligomer catalytic for prophyrin metallation, derived from the porphyrin-binding aptamers, has also been proposed to contain guanine quartets (see below).

As a matter of interesting comparison, an RNA SELEX experiment was also carried out for the binding of thrombin, using similar protocols.[59] In common with every other SELEX experiment carried out with both DNA and RNA, the thrombin-binding RNA aptamers appeared to have little sequence similarity with the DNA aptamers; also, they were predicted to form folded structures that were not guanine quadruplexes. Functionally, too, the RNA and DNA aptamers appeared to bind to different regions of thrombin (DNA at the so-called fibrinogen exosite and RNA at the heparin binding site). Although the RNA aptamer in this instance also had anticoagulant activity, the DNA and RNA aptamers—perhaps owing to their different binding sites on thrombin—did not compete with one another in a thrombin-binding assay.

### (ii) The determination of DNA consensus sequences for protein binding

Whereas the search for DNA aptamers for proteins such as thrombin and HIV-1 reverse transcriptase were inspired in part by the wish to find inhibitors for those proteins, another category of protein-binding SELEX experiments have been attempted to try and address certain difficult questions in molecular biology. A major effort has been dedicated, for instance, to the determination of consensus *double-stranded* DNA sequences for the binding of certain DNA-binding proteins. Similar methodologies of SELEX have been used to determine the binding site consensus sequences for two *E. coli* proteins, Lrp (leucine-responsive regulatory protein)[60] and MetJ (methionine repressor),[61] both of which are important in the control of bacterial gene expression.

Lrp is a leucine-binding regulatory protein which is thought to participate in the expression of numerous operons in *E. coli*. Prior to SELEX, a consensus sequence for the binding sites of Lrp was obtained by comparing Lrp binding sites. The subsequent SELEX experiment was therefore carried out with the goal of not only confirming, without prior bias, the preferred binding sequence(s) of the protein, but also to determine whether in the presence of the effector, leucine, the sequence specificity of the protein changed. Similar questions were also addressed with the MetJ protein (which binds its preferred double-stranded binding site with different affinities in the presence and absence of its corepressor, methionine).

The selection steps for these two studies took advantage of retarded electrophoretic mobilities (in nondenaturing gels) of DNA molecules bound to proteins such as Lrp and MetJ compared with free DNA molecules with no bound protein. Thus, for each selection step, the pools of double-stranded DNA molecules were incubated with the appropriate protein and the mixtures were run in nondenaturing gels. Sections of each gel with detectable (in later rounds) DNA–protein complexes, or gel sections that corresponded to the expected positions of the protein–DNA complexes (in the early rounds), were selectively excised and extracted for DNA. The eluted DNA was then purified and amplified for the next round by standard methods.

The results of both sets of SELEX experiments revealed that the preferred double-stranded binding sites for the two proteins corresponded well with the consensus sequences established previously. However, SELEX did provide additional insights that would have been hard to obtain by more conventional methods. Thus, it was found that Lrp, in the presence of bound leucine, did not prefer to bind to some other sequence of DNA. Likewise, it was shown that the consensus binding sequence for MetJ, in the presence of bound methionine, did vary subtly from the sequence preferred by MetJ alone. Moreover, the importance for MetJ binding of certain base pairs not directly contacted by the bound protein was established.

### (iii) Investigations into the editing function of a tRNA synthetase enzyme

The ability of SELEX to clarify certain kinds of molecular biological problems has been highlighted by a particularly elegant experiment carried out by Hale and Schimmel.[62] In a key step of the cellular mechanism for protein synthesis, tRNAs are primed at their 3′ termini with amino acids. These priming reactions are catalyzed by a class of enzymes called aminoacyl-tRNA synthetases. In a given organism, for each amino acid, there is a designated tRNA as well as a designated tRNA synthetase. The alanyl-tRNA synthetase, for example, must correctly couple together the *alanyl*-tRNA with *alanine*, by way of formation (by the enzyme) of an activated form of alanine, alanyl adenylate (analyl-AMP). Accuracy in this coupling of amino acid to its designated tRNA is obviously vital for the integrity of protein synthesis. It is thought that certain tRNA synthetases have additional "editing" functions which serve to remove (via hydrolysis) any wrongly activated aminoacyal adenylates or inappropriately formed aminoacyl-tRNA linkages between noncognate amino acids and tRNAs.

The mechanisms of the editing steps of different tRNA synthetases are not fully understood. Isoleucyl-tRNA synthetase has the property that, in addition to activating its cognate amino acid, isoleucine, it also activates to some extent ($\sim 5\%$) the structurally similar amino acid valine. This unacceptably large error rate is then corrected by the enzyme by means of a selective hydrolysis of valyl-AMP. The enzyme's editing function is thought to be triggered by the binding to the enzyme of isoleucyl tRNA (which is, of course, the ultimate acceptor of the aminoacyl residue). An early hypothesis[63] had postulated that the acceptor function of the tRNA (involving initially the 2′-hydroxy group of its terminal adenosine nucleotide) was critical for the editing reaction. Hale and Schimmel[62] decided to test this hypothesis by isolating, using SELEX, DNA aptamers that bound specifically to the complex between the isoleucyl-tRNA synthetase and valyl-AMP. Such aptamers, once obtained, could be used to test whether the absence of the acceptor 2′-hydroxy group of the tRNA (for which the DNA aptamer, substituting for the tRNA, could offer no comparable functionality) might nevertheless stimulate the editing function of the enzyme.

SELEX experiments were carried out with a library of single-stranded DNA molecules, each containing a 25-nucleotide random stretch. DNA aptamers selected for specific binding to immobilized isoleucyl-tRNA synthetase (complexed to a stable analogue of valyl-AMP) were found to trigger the hydrolysis of valyl-AMP with an efficiency comparable to that of the natural trigger, isoleucyl-tRNA. It was therefore concluded that the editing function of the enzyme was functionally separate from its amino acid binding function, and also that editing was triggered by a conformational change in the enzyme induced by tRNA (or by its mimic, the DNA aptamer), a conformational change that was able to discriminate the activated form of the correct amino acid from those of incorrect ones.

### 7.17.5.1.3 *Nucleic acid-binding aptamers*

#### (i) *Triplexes*

Certain double-helical DNA sequences (generally those which have one purine-rich strand and one pyrimidine-rich strand) are able to interact specifically with a third, substantially purine- or substantially pyrimidine-containing single-stranded DNA or RNA, to form triple-helical structures (triplexes) (reviewed in reference 64). So-called "purine motif" triplexes are formed when a purine-rich third strand binds in the major groove of the above kind of double-helical DNA, the third strand hydrogen bonding to, and lying *antiparallel* to, the purine strand within the duplex. Conversely, polypyrimidine-containing third strands can form triplexes by hydrogen bonding to, and running *parallel* to, the duplex's purine strand. In the former type of triplex, the canonical base triplets that are formed are G*G–C and A*A–T (with the Hoogsteen hydrogen-bonded interaction between the third strand and the duplex indicated by asterisks); in the latter, the canonical pairs are C+*G–C and T*A–T. However, aside from these, various kinds of noncanonical triplets, such as G*T–A, and others, can also form, particularly when these are constrained to form sandwiched between canonical base triplets. The noncanonical base triplets are generally significantly less stable than the canonical ones.

A number of investigators have used SELEX to try to define comprehensively the various kinds of noncanonical base triplets that may contribute (however marginally) to the stability of triplexes. The first to report such an experiment were Pei *et al.*,[65] who used an immobilized double-helical DNA as target (the purine-rich strand of the "target" duplex contained the sequence 5′ AAGAG AGAGA GAGG GA). A random-sequence RNA library was tested for third-strand forming ability with this duplex, and of the RNA sequences cloned at the end (as complementary DNA or cDNA), most were found to be extremely rich in C and U (although also containing additional bases, which appeared to take part in forming small internal loops, hairpin loops, and noncanonical triplets). Another study utilizing RNA SELEX on an immobilized, duplex, DNA target has also been reported.[66]

An all-DNA SELEX experiment (featuring a single-stranded DNA or fixed sequence as target and a randomized pool of DNA duplexes) was reported by Hardenbol and Van Dyke.[67] They used a mild, noncompetitive, selection procedure called "REPSA," in which the selection step was carried out in solution, following incubation of the fixed-sequence third strand (5′ TGGGT GGGGT GGGGT GGGT) with members of the random-sequence duplex library. The experiment was designed such that any duplex that participated in triplex formation with the above third strand was protected from cleavage by the restriction endonuclease BsgI (which was chosen because it would cut within the random region of the duplex, regardless of the sequence). Hence, only those duplexes that survived cleavage (and, therefore, destruction) by forming triple helices in a given round were allowed to proceed to the next round of selection. Ultimately, the enriched triplex-forming DNA molecules were cloned and sequenced; examination of the sequences revealed, in addition to the expected, evidence that a novel noncanonical triplet, G*A–T, might be viable within an all-DNA triplex.

The advantage of the REPSA technique of selection used in this study (which could also be applicable to the determination of protein-binding consensus DNA sequences) was that it did not aim to separate physically triplex-forming sequences from those that did not. Rather, a simple inactivation by hydrolysis, carried out under mild conditions, was used to eliminate the unwanted duplexes. A technique of this sort could, in theory, be used to carry out SELEX experiments with ligands whose physical and chemical properties were not well known. In addition to this, the mildness and noncompetitive aspect of the "selection" step would permit the survival of even relatively weak interactions, which might not survive elimination in experiments utilizing more robust "separation" techniques.

### (ii) Novel sequences that form naturally bent DNA duplexes

Short double helices ($< 150$ base pairs) of DNA behave, hydrodynamically, like straight and rigid cylinders. These hydrodynamic properties of DNA, as a function of DNA size, have been studied extensively, using viscosity, electrophoretic mobility, and rotational relaxation measurements. While the rod-like behavior of short DNA duplexes generally holds true, and is independent of the sequence of the DNA, in the early 1980s evidence was found that certain sequences gave rise to systematic "bends" within double helices. Anomalous gel mobilities, as well as anomalous rotational relaxation times, were observed for certain short restriction fragments derived from the kinetoplast DNA minicircles of a number of protozoans, including *Leishmania tarentolae*.[68,69] These initial discoveries led to an intensive study of this novel phenomenon of sequence-dependent DNA bending (reviewed by Crothers *et al.*[70] and Hagerman[71]). Not only is systematic bending an interesting physicochemical property of DNA, but it has also been implicated in the *in vivo* regulation of the expression of certain genes, and also in special cases of genetic recombination.

A few sequence motifs within DNA have been proposed to contribute to systematic bending of the duplex. These include runs of homopolymeric dA·dT base pairs, when they occur at intervals of 10.5 base pairs (in phase with complete turns of the double helix), the motif $A_3T_3$, and, to lesser extents, a variety of nontract motifs (which contribute correspondingly smaller bending effects). Certain "junction bases," such as C and T in the tract 5′ CAAAAT, have also been considered to enhance the extent of bending.

In order to explore comprehensively the range of sequence motifs, known and unknown, that might contribute to DNA bending, Beutel and Gold[72] carried out SELEX experiments with a library of double-stranded, 104 base-pair DNA molecules (each duplex containing a central core of 30 random base pairs). The selection criterion used was that of reduced electrophoretic mobility, similar to those used for the determination of preferred DNA binding sites for the proteins Lrp and MetJ (see above). Beutel and Gold enriched electrophoretically for those 104 bp DNA molecules that migrated anomalously in a gel, as if they were 114 bp long. This selection scheme is summarized in Figure 5. After several rounds of enrichment, the altered-mobility DNA pool was cloned and 30 clones from this pool were isolated and sequenced. It was found that all 30 clones had distinct sequences—not a particularly surprising finding, since the selection method used had been a "non-competitive" one (like REPSA, but unlike those used in target-binding SELEX experiments, where aptamers compete for binding to a limited number of target molecules). A number of the cloned sequences from this study conformed to motifs already implicated in sequence-dependent DNA bending; however, other results appeared to be in partial disagreement with motifs defined previously. Most importantly, a completely unheralded dinucleotide base step, CA/TG, was found to be important in DNA bending.

### 7.17.5.2 Selection of Novel DNA Catalysts

The discovery of catalytic RNA molecules in the 1980s[2,3] paved the way for an "RNA World" hypothesis for the origin of life.[8–10] This hypothesis postulated that prior to the evolution of protein enzymes, catalytic RNAs catalyzed a variety of chemical reactions that constituted the metabolism of primitive, RNA-based "organisms." Since RNA has both genetic (or "coding") capabilities and

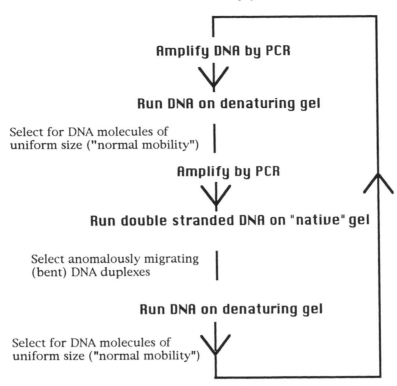

**Figure 5** Schematic diagram of the gel electrophoresis-based selection scheme utilized to derive naturally bent double-helical DNA molecules.

some catalytic capabilities, it could have functioned in the RNA World both as genetic material and as catalyst, thereby embodying, simultaneously, the genetic concepts of "genotype" and "phenotype." However, an examination of the chemistries catalyzed by the ribozymes found in modern organisms (presumably such ribozymes are vestiges of a much larger group of primordial RNA catalysts, whose functions have been taken over by proteins) indicates a vary narrow range—essentially hydrolytic and transesterfication modifications of phosphodiesters. The case for a metabolically diverse RNA World could therefore only be strengthened by demonstrations that RNA, or DNA, could catalyze a wider range of metabolic reactions.

Inspired by this need to demonstrate RNA's catalytic versatility, a number of research groups have undertaken to define the range of reactions that can be catalyzed by RNA and DNA; this is an effort made practical by the development of the technologies of SELEX.

Two distinct approaches have been taken towards the isolation of novel RNA and DNA catalysts (reviewed in references 1, 29, 30, 40, and 73–76). The first of these utilizes a so-called "direct" approach, in which RNA and DNA molecules capable of catalyzing a given reaction are made to modify themselves in the process of exercising their catalysis, and are then selected out of the larger pool of noncatalysts by virtue of their self-modification. Once selected out of an initial library, catalytic RNA or DNA molecules can be amplified according to the standard protocols of SELEX, and subjected to further rounds of catalysis (with concomitant self-modification) and amplification. The end products of such direct selection procedures are RNA or DNA molecules that are effective self-modifiers. Strictly speaking, their activity at this point is stoichiometric, not catalytic. They can, however, be converted into true catalysts (where each catalytic molecule is capable of "turning over" substrate, i.e., processing multiple substrate molecules) by a process of physically unlinking a "catalytic" portion of the self modifying DNA or RNA from a "substrate" component. This "direct" selection approach towards the isolation of novel ribozymes (and DNAzymes) has in fact been very effective, suffering only from the limitation that the substrates for such catalysts necessarily contain significant nucleic acid components. Nevertheless, certain innovations, such as the incorporation of non-nucleotide functionalities (such as an amide group, specifically targeted for hydrolysis by a modified version of a pre-existing ribozyme[77]), and the incorporation of binding sites for molecules such as biotin and ATP into the folded structures of RNA and DNA molecules, have permitted the isolation of a more diverse class of RNA catalysts.

The second approach towards finding novel nucleic acid catalysts has focused on the isolation of RNA and DNA molecules capable of binding to transition-state analogues (TSAs) for particular reactions. Pauling,[78] Jencks,[79] and others have proposed that enzymes catalyze reactions on the basis of their preferential binding to, and stabilization of, the transition states of reactions. This idea has been tested and confirmed in numerous ways, perhaps most elegantly by the derivation of catalytic antibodies (reviewed by Schultz and Lerner[80]). In this technology, TSAs have been used as antigens to elicit immune responses in mice. Monoclonal antibodies derived from the immunized mice have then been shown to catalyze the relevant reaction. In an analogous fashion, it has been considered that it should be possible to use SELEX to obtain DNA and RNA aptamers for binding TSAs; these aptamers should then behave as catalysts.

Much of the recent effort dedicated to finding novel catalytic RNA and DNA molecules (whether using the "direct" or the TSA methods) has focused on RNA rather than on DNA, not least owing to the need to test the RNA World hypothesis, and also to the fact that the only catalytic nucleic acids discovered in nature are, indeed, RNA molecules. A number of novel and often remarkably efficient and specific RNA enzymes have been derived, most of them using "direct" selection techniques. Some of the activities derived include: RNA cleavage (self-cleaving RNAs);[81–85] RNAs that catalyze the formation of phosphoanhydrides[86] and phosphodiesters (in the process of ligating two molecules of RNA);[34,87–89] RNA phophorylation;[32,33] the utilization of a biotin-based alkylating agent (*N*-biotinyl-*N'*-iodoacetylethylenediamine) to obtain self-alkylating RNAs;[90] a transacetylation;[91] an amide bond formation;[92] and an amide bond hydrolysis.[77] The list of new catalytic RNAs obtained using transition-state analogues is much smaller: an isomerase of a bridged biphenyl[93] and an RNA that catalyzes the metallation of porphyrins.[94] The last two are most notable for their utilization of substrates that are not derived from nucleotides or nucleic acids.

Recently, however, a number of DNA catalysts have also been obtained using techniques analogous to those used for new ribozymes. The rationale for searching for new DNA catalysts ("deoxyribozymes" or "DNAzymes") has been twofold: for investigating the inherent ability of DNA to fold to form appropriate active sites and to catalyze reactions (particularly in comparison with RNA counterparts); and also to investigate the importance of the 2'-hydroxy group of RNA for catalytic purposes (a number of naturally occurring ribozymes make mechanistic use of the 2'-hydroxy groups; one would like to know whether DNA, although lacking this functionality, would nevertheless be able to catalyze the same reactions). There is also a purely practical rationale for deriving novel DNA catalysts for reactions already known to be catalyzable by RNA. This relates to the much greater chemical stability of DNA compared with RNA. A number of ribozymes are being considered for biomedical uses; from the perspective of chemical robustness (especially *in vivo*), however, DNA might be a preferable polymer to RNA.

Three distinct kinds of catalytic DNAs have been derived (interestingly, for all three reactions, RNA catalysts have also been derived utilizing SELEX). These three very different DNA enzymes, compared among themselves, and also with their RNA (and, where appropriate, protein) counterparts, have been particularly informative about the range and possibilities of nucleic acid-mediated catalysis.

The three types of DNAzymes isolated are: (i) nucleases, which are DNA molecules that cleave an RNA phosphodiester, utilizing lead and a number of other divalent cations as cofactors; (ii) a ligase, which is a DNA molecule that catalyzes DNA ligation; and (iii) a chelatase, which is a DNA molecule that catalyzes the insertion of zinc and copper ions into porphyrins. The first two of these DNAzymes were obtained using "direct" selection techniques, whereas the third was obtained via the use of a stable TSA.

### 7.17.5.2.1 *Nuclease: the catalysis of RNA cleavage*

The derivation of these earliest reported DNAzymes was carried out by Breaker and Joyce.[95] The original strategy was to utilize lead ions complexed directly to random DNA sequences to cleave a single ribonucleotide phosphodiester present within the DNA. Lead ions were chosen as potential cofactors because the first $pK_a$ of water molecules coordinated to $Pb^{2+}$ ions is ~7.7. This bound water molecule, substantially deprotonated at physiological pH, is particularly suitable as a general base catalyst for RNA cleavage (prior to the derivation of this DNAzyme, both naturally occurring tRNA molecules, and also artificial "leadzyme" ribozymes,[82–84] had been shown to self-cleave efficiently in the presence of lead).

To obtain lead-dependent, hydrolytic DNAzymes, Breaker and Joyce[76] utilized an ingenious variation on the "direct" selection approach, which was called "catalytic elution." This is shown schematically in Figure 6. Random-sequence, single-stranded DNA molecules were generated which all contained (i) a biotin residue at the 5'-terminus and (ii) a single ribonucleotide residue at a fixed position within the 5' constant-sequence region. The biotin residue was used to immobilize each DNA molecule to an avidin-containing matrix. The $\sim 10^{14}$ distinct molecules (containing 50 random nucleotides each) were first allowed to fold up in high salt. 1 mM $Pb^{2+}$ was then added and any self-cleaving DNA molecules within the pool (which thereby detached themselves from the matrix) were collected after a period of incubation. After five rounds of selection and amplification, excellent lead-dependent DNAzymes were isolated and cloned. These had $k_{observed}$ values of $\sim 1$ min$^{-1}$, which represented $10^5$-fold rate enhancements over the same hydrolysis in the presence of lead, but uncatalyzed by the DNAzyme.

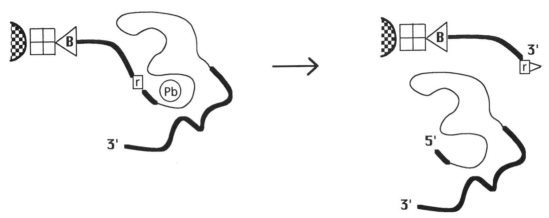

**Figure 6** Schematic diagram of the derivation of DNA molecules catalytic for the lead-dependent hydrolysis of an internal ribonucleotide phosphodiester (r). The individual DNA molecules of the library are immobilized as single-strands on to a matrix (via a biotin–avidin linkage). DNA molecules capable of self-cleavage are released from the matrix, and can then be collected and amplified for subsequent rounds of SELEX and, eventually, cloning.

Subsequent to this initial report, Breaker and Joyce[36] used related strategies to generate other phosphodiester-hydrolyzing DNAzymes, which in turn specifically utilized $Zn^{2+}$, $Mn^{2+}$, and $Mg^{2+}$ ions as cofactors. A pool of the $Mg^{2+}$-dependent DNAzymes were improved by a secondary, "focused" selection process; an individual aptamer from this improved pool had a $k_{observed}$ value of 0.02 min$^{-1}$, which represented a $10^5$-fold enhancement over the magnesium-dependent, but DNA-uncatalyzed, rate of self-cleavage of the aptamer.

### 7.17.5.2.2 *Ligase*

An elegant demonstration of the ability of RNA to catalyze its own ligation had been provided by Bartel and Szostak,[34] who had incorporated elements of rational design into their SELEX experiment by positioning the two reacting functionalities (a 3'-ribose hydroxy and a 5'-triphosphate on two separate pieces of RNA) facing one another, and stabilized by base pairing to a "guide" or "template" segment of the larger of the two RNA molecules. Cuenoud and Szostak[35] attempted to derive a catalytically self-ligating DNA by following a comparable strategy. The reactive groups, a 5'-deoxyribose hydroxy and a 3'-phosphoroimidazole, were again present on distinct DNA molecules (the 5'-hydroxy-containing DNA had a region of 116 random nucleotides downstream from the 5' end; the 5' end of the other DNA molecule had a biotin moiety). In the early rounds, the two reactive functionalities were allowed to orient next to each other with the help of an external DNA "guide" molecule, which base paired to and aligned the two reacting DNA molecules close together (in later rounds, the external guide was no longer necessary; the two reacting DNAs were found to bind one another). The coupling reaction was allowed to proceed in the presence of zinc ions (in addition to the magnesium present to fold the DNA molecules). Selection was begun with $10^{14}$ different sequences; after nine rounds of selection, zinc-dependent self-ligases were obtained; these were found to also function in the presence of copper, but did not require magnesium for catalysis.

Subsequent to the isolation of these catalytic aptamers, one aptamer was refashioned to give a catalytic DNA fragment and two substrate DNA fragments. The sequence and predicted folded structure of one of these assemblies is shown in Figure 7. The catalytic fragment behaved like a true enzyme, ligating together the two substrate DNA molecules with numerous turnovers. This modified incarnation of the DNAzyme had a $k_{cat}$ value of 0.07 min$^{-1}$ (the RNA catalyst initially derived by Bartel and Szostak[34] had, by comparison, a $k_{cat}$ of 0.06 min$^{-1}$, which represented a 10$^6$-fold rate enhancement over that obtained by simply positioning the two reactive groups upon the "guide" RNA. The latest versions of the RNA ligases,[76,87] however, show $k_{obs}$ values of ~100 min$^{-1}$).

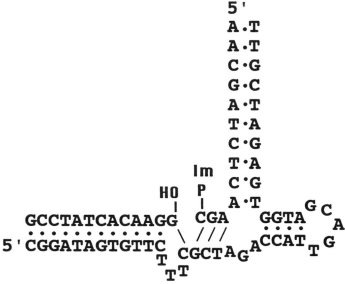

**Figure 7** An assemblage of three DNA molecules that comprise a catalytic system for DNA ligation. Of the two small substrate DNA molecules held in place by the larger, catalytic subunit, one has a 5'-hydroxy group (the nucleophile) positioned adjacent to a phosphoimidazolide (Im-P) group attached to the 3'-hydroxy group of the other substrate DNA molecule.

### 7.17.5.2.3 Chelatase: porphyrin metallation

In 1990, Cochran and Schultz[96] had reported the isolation of a catalytic monoclonal antibody that catalyzed the insertion of copper and zinc ions into mesoporphyrin IX (MPIX) (**1**) to give the metallated form (**2**). This antibody had been obtained in response to immunization by *N*-methylmesoporphyrin IX (NMM) (**3**), a distorted porphyrin, which behaves like a TSA for the metallation of prophyrins by enzymes called ferrochelatases (from the heme-biosynthetic pathways of different organisms). This reaction, and the TSA NMM are shown in Scheme 1.

To test whether DNA SELEX methods could be used in an analogous way to derive a catalytic DNA for porphyrin metallation, Li *et al.*[97] isolated DNA aptamers that specifically bound NMM with submicromolar affinity. All of the derived aptamers, when sequenced, revealed one or more 15–30-nucleotide stretches of a highly guanine-rich sequence; these were the only sites of homology within otherwise unrelated aptamers. Footprinting experiments carried out on selected aptamers in the presence of NMM confirmed these G-rich stretches to be the binding sites for NMM and for other prophyrins (which bound with lower affinities).

Two of these aptamers, PS2 and PS5, and also their prophyrin binding sites, synthesized as autonomous DNA oligomers, were tested for catalytic activity. A very guanine- and thymine-rich 33-nucleotide oligomer, PS5.ST1, was found to catalyze the insertion of copper(II) and zinc ions into mesoporphyrin IX.[98] The catalyzed copper insertion reaction followed Michaelis–Menten kinetics at fixed copper concentrations, with a $k_{cat}$ value of 13.7 h$^{-1}$ a Michaelis constant ($K_M$) of 2.9 mM, and a net catalyzed rate over background ($k_{cat}/k_{uncat}$) of ~1400. Interestingly, this $k_{cat}/k_{uncat}$ value was comparable to that of the catalytic antibody derived using the same TSA by Cochran and Schultz.[96] The $K_M$ value of the DNAzyme, however, was ~60-fold poorer than that of the catalytic antibody, and also comparably poorer than those of a number of naturally occurring ferrochelatase enzymes.

The DNAzyme PS5.ST1 showed a number of novel features, including inhibition of its activity by magnesium and enhancement by potassium. The latter was consistent with the existence of

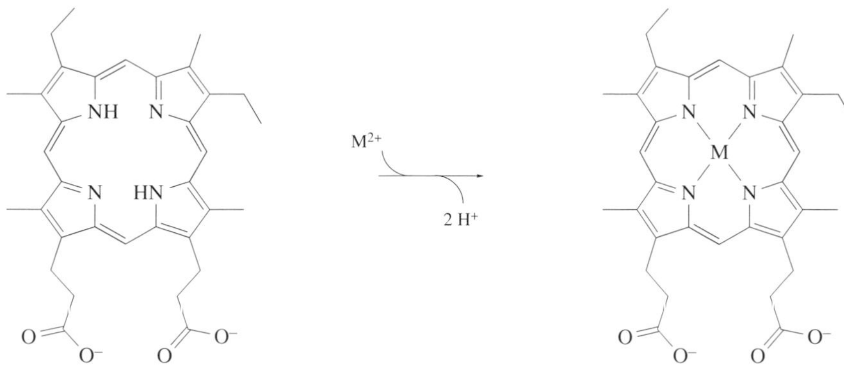

**Scheme 1**

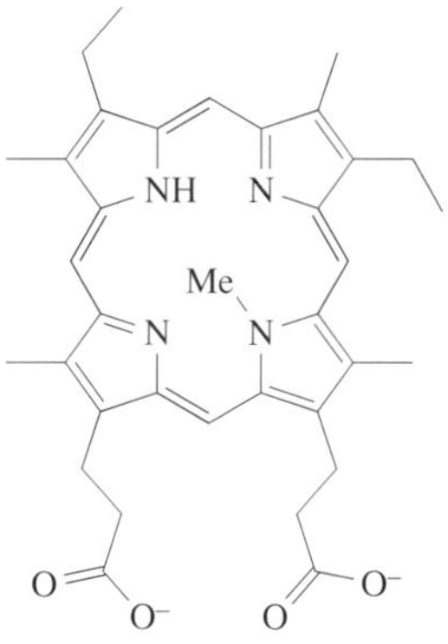

guanine quartets[54-58] (Figure 3) within the enzyme's folded structure. Catalysis was found to be inhibited by NMM, the TSA for the reaction, but not significantly by the reaction products.

Subsequently, Li and Sen[99] carried out a detailed investigation of the properties of the above DNAzyme. Following various steps of optimization, the DNAzyme showed a moderate (about six-fold) increase in $k_{cat}$, but a dramatic 70-fold decrease in $K_M$ compared with the earlier data, derived under unoptimized assay conditions. The DNAzyme's catalytic efficiency, $k_{cat}/K_M$, was now 32 500 $M^{-1}$ $min^{-1}$; the major change in $K_M$ implied that the optimization process had worked significantly to improve the DNAzyme's affinity for its substrate, MPIX.

Concurrent with the reporting of this DNAzyme in the literature, an RNA catalyst for the same reaction was also reported by Conn *et al.*[94] The 35-nucleotide ribozyme was derived using a protocol directly analogous to that used to obtain the catalytic DNA. The RNA catalyst had excellent enzymatic parameters, with a $k_{cat}/K_M$ value of 126 000 $M^{-1}$ $min^{-1}$ and a $k_{cat}/k_{uncat}$ value of $\sim$460. The 35-nucleotide RNA, very guanine rich (like the catalytic DNA), was predicted to fold into a stem-loop structure, with a high proportion of its guanine residues present in the loop.

### 7.17.5.3 "Blended" SELEX

The earliest experiments on DNA SELEX used only the four nucleotides found naturally in DNA for the creation of their "random" libraries. However, as discussed earlier, these four varying units provide far less functional and structural diversity to DNA than, say, do the 20, and diverse, amino acids to proteins. Amino acids supply to proteins widely varying physical and chemical characteristics, such as diverse hydrophobicities, anionic or cationic character, and the ability to form or not form hydrogen bonds, whereas the natural DNA bases are all similar to one another in terms of the above characteristics. A number of workers with SELEX have therefore attempted to supply additional elements of structural and functional diversity to DNA, via non-natural functionalities added to DNA either (i) prior to the selection process itself or (ii) subsequent to the completion of selection, but with rational consideration given to possible effects of any modification on the continuing effectiveness of the selected aptamer. A number of postmodifications, in particular, have been attempted in an effort to protect potentially therapeutic aptamers from degradative enzymes encountered *in vivo*.

With regard to premodifications, only certain categories of nucleotide analogues are acceptable as substrates to the enzymes used in SELEX. An interesting instance of SELEX involving pre-modified DNA was carried out with a DNA library containing the non-natural nucleotide 5-(1-pentynyl)-2′-deoxyuridine in place of thymidine by Latham *et al.*[100] The target molecule for this selection was the glycoprotein thrombin (see Section 7.17.5.1.2). The methodology for this selection was very similar to those used to derive RNA and DNA aptamers for binding thrombin (see above); analysis of cloned aptamers from this selection, however, yielded sequences that were unrelated in sequences to both the unmodified DNA and RNA aptamers derived against thrombin. When standard thymidine was substituted for pentynyl dU residues in an aptamer from this premodified selection, thrombin binding was abolished; it was therefore clear that the modified nucleotide played an important role in the folding of the pentynyl dU-containing aptamer, and possibly also in the specific interactions between that folded aptamer and thrombin.

### 7.17.5.4  SURF

Despite the power of a conventional SELEX experiment, the range of macromolecules that may be investigated for novel target-binding or for novel catalytic activities is restricted to the polymers RNA and DNA. This is because SELEX depends methodologically on a number of steps involving enzymes (such as during the amplification step and in the cloning of aptamers). Therefore, only those biopolymers can be subjected to SELEX that are recognized and treated as substrates by replicative enzymes, both *in vitro* and within the bacteria used for cloning. To define an alternative mode of iterative selection, such as SELEX, but one not dependent on enzymes, Ecker *et al.*[101] devised a technique called synthetic unrandomization of randomized fragments (SURF). The method itself was inspired by similar techniques that had been used for peptides;[102-104] the immediate advantage that SURF offered was that it would tolerate the incorporation into DNA of a variety of nucleotide analogues, as well as other organic functionalities. SURF starts with four pools of randomized, short oligonucleotides, in all of which one fixed position has a known nucleotide in it, whereas the other positions contain all of the nucleotides (or analogues) with equal probability. Thus, hexanucleotide starting pools might have sequences of the form of XXGXXX, XXCXXX, XXAXXX, and XXTXXX (where the sites marked X have an equal probability of having any of the possible nucleotides). These original pools are screened against the target, and the most "active" starting pool of the four (say, XXCXXX) is resynthesized, except that now it has one other position with a known nucleotide in it (such as XXCXGX, XXCXCX, XXCXAX, and XXCXTX), and so on, with each step of repetitive synthesis generating increasingly less complex sets of oligonucleotide pools, until, at the end, a unique sequence is arrived at.

It can be seen that while it is possible to utilize an enormous array of nucleotide analogues and other, nonbiological functionalities with this technique, it is nevertheless limited in that only rather short oligomers may be systematically evolved in this way (larger oligonucleotides would present a formidable synthetic task). However, even with this limitation on oligonucleotide size, SURF has been used to identify useful compounds which have antiviral properties. Thus Wyatt *et al.*,[105] working with phosphorothioate oligonucleotides, identified the sequence $T_2G_4T_2$, which inhibited HIV infection *in vitro*. Interestingly, $T_2G_4T_2$ with standard phosphodiester backbones did not show the anti-HIV activity; therefore, the phosphorothioate backbone was an absolute requirement. Also, this oligomer was active not as a single-stranded oligomer, but rather as a parallel-stranded guanine quadruplex.

### 7.17.5.5  Encoded Libraries

Brenner and Lerner, in an important 1992 paper,[31] introduced new ideas for harnessing the powerful techniques of molecular biology (that make SELEX of DNA and RNA possible) to carry out other kinds of combinatorial searches—searches involving functionalities and molecules not amenable to facile amplification or cloning.

"Genetic" methods of selection and amplification, such as used for SELEX, are strictly limited to DNA and RNA (as mentioned earlier). Recently, it has become possible to carry out SELEX-like experiments with random peptide sequences, using the ingenious technology of "phage display."[106,107] Here, peptide sequences that are either random, or subject to certain structural constraints, are displayed upon the surfaces of filamentous bacteriophages (with the peptides fused to one or more of the phages' coat proteins). The phage carrying a given peptide sequence of interest can be identified readily by cloning the phage and looking at the DNA sequence of the phage's genome. In other words, the identity of the displayed peptide is encoded in the DNA sequence of the phage's genome, and the latter can be accessed at will with the standard techniques of molecular biology.

Brenner and Lerner[31] proposed that a comparable "encoding" process could be applied generally to "tag" with DNA compounds not composed of either peptides or nucleotides. In this way, random libraries of carbohydrates, for instance, might be tagged—with a unique DNA sequence covalently attached to (and synthesized in parallel with) a given carbohydrate sequence. The facile reading of the DNA sequence attached to a given carbohydrate of interest, but of unknown sequence, could be utilized to decode the latter.

Subsequent to Brenner and Lerner's conceptual paper, Fenniri *et al.*[108] described the creation of model encoded "reaction cassettes" for the sensitive detection of both the making and breaking of chemical bonds (shown conceptually in Figure 8). In a prototype experiment, Fenniri *et al.* synthesized a tagged molecule, consisting of a peptide (L-Ala)$_2$–L-Tyr–(L-Ala)$_2$ (a known substrate for α-chymotrypsin) attached to a specific DNA sequence. This chimeric molecule was immobilized on an inert matrix and subjected, separately, to treatment with a variety of protease enzymes, some of which would be expected to cleave it and others not. Following treatment with the individual enzymes, only the α-chymotrypsin and the proteinase K-treated samples yielded free DNA molecules (i.e., released from the matrix), which were then detectable by PCR amplification. The success of this prototype experiment indicated that such encoded selections might in fact be carried out, and thereby enlarge enormously the scope of DNA SELEX.

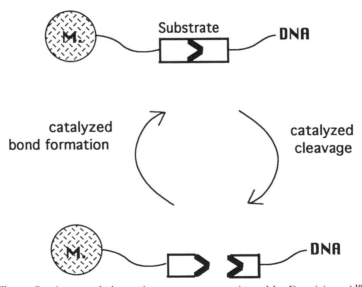

**Figure 8** An encoded reaction cassette, as envisaged by Fenniri *et al.*[108]

## 7.17.6 DISCUSSION

### 7.17.6.1 The Scope and Limitations of SELEX

A number of workers have written perceptively on the strengths and shortcomings of SELEX.[1,29,30,73,74,76] A central issue in these discussions has been the nature and extent of the complete universe of folded DNA and RNA shapes and, also, the extent to which it is possible to learn about this universe from SELEX experiments carried out with limited, "shotgun" libraries. If a given SELEX experiment, using a 100-nucleotide DNA library, yields a single solution to a particular binding problem, does it necessarily imply that that was the only conceivable solution from the universe of DNA sequences 100 nucleotides long? Of course, a number of SELEX experiments have yielded not one but a number of solutions to a given problem. Comparison of the sequences of these successful aptamers has indicated that whereas some are variants on a discernible "consensus" sequence ("local" optima in sequence space), others represent unrelated sequences—in other words, *independent* folding solutions ("global" optima) to the challenge of binding a given shape.[74] For catalytic activity, it has been estimated[73,87] that possibly thousands of independent solutions exist for the design of self-ligating RNA molecules. However, it is true that a few SELEX experiments have yielded only a single sequence solution out of $5 \times 10^{14}$ (such as for RNA binders of cyano-cobalamin[109]); even so, such a result does not necessarily imply that there was only that single solution to the problem, even within the particular starting library used. Rather, it is possible that the selected sequence was merely the simplest or the most common solution to the problem.[73] Since many of the selection steps carried out in SELEX require aptamers to compete for binding to a limited number of target molecules, there is frequently the elimination of modest-affinity binders during the course of SELEX. In addition, any smaller DNA molecules accidentally generated in the pool during selection have a replicative advantage during the amplification steps (see, e.g., reference 97). Conversely, aptamers with complex or very stable folded structures may not be replicated as well as others whose folded structures are more easily disrupted during PCR. Ellington[74] described this selective advantage of smaller and simpler aptamers as the "tyranny of small sequences," whereby it is often the smallest "satisfactory" sequences rather than the best such sequences that are obtained at the end of a SELEX procedure.

A general observation made from the characterization of many different target-binding aptamers (both RNA and DNA) is that successful aptamers are generally of modest size (15–25 nucleotides long). Ellington[74] made three general observations about aptamers and their targets: (i) that the tighter-binding aptamers are also generally the most discriminating aptamers; this is conceptually reasonable, since both a tight-binding association, and a more discriminating association, may result from the presence of extensive contacts between target and aptamer; (ii) that larger target molecules generally bind better; and (iii) that the presence of heterocyclic rings in the target molecules make them superior targets for SELEX. This last finding, again, is not hard to rationalize, given that the DNA and RNA bases are themselves heterocyclic rings. The SELEX experiments with ATP as target (see above), in particular, have demonstrated that the contacts formed between ATP and the binding aptamers are thoroughgoing, involving multiple functionalities of the ATP molecule, and suggesting a "snug" fit between target and aptamer.

### 7.17.6.2  DNA vs. RNA

One of the more intriguing questions surrounding SELEX has been whether it might reveal RNA (relative to DNA or, indeed, chemically modified versions of RNA and DNA) to be the superior polymer in terms of its ability to bind different targets and catalyze different reactions. Although the data set is limited, it appears that RNA and DNA are comparably competent in both these respects. It has been a consistent finding, however, that DNA and RNA find independent sequence solutions to a particular SELEX problem. Examination of an RNA version of a successful DNA aptamer, and vice versa, have invariably revealed that the versions do not function as aptamers. In a particularly interesting case discussed above, separate SELEX experiments carried out with RNA and with DNA against thrombin (a reasonably large and complex target) have yielded comparably effective but utterly different RNA and DNA inhibitors, which actually exercised their inhibitory effects by binding to separate sites on the target protein.

In terms of the relative catalytic potentials of RNA and DNA, there are now three clear cases for comparison: DNA and RNA catalysts for (i) the cleavage of an RNA phosphodiester, (ii) DNA and RNA ligation, and (iii) the metallation of porphyrins. RNA–DNA comparisons in all of these cases are likely to be instructive.

### 7.17.6.3  Nucleic Acid Aptamers vs. Protein Aptamers

The active sites of protein enzymes are frequently lined with hydrophobic residues, which contribute to the sequestration of substrates within active sites, away from the polar aqueous solvent. This importance of hydrophobic linings of binding pockets in proteins appears, from first considerations, not to be replicable by folded RNA and DNA structures. Cech and Bass[4] speculated that stacking interactions between nucleic acid bases and aromatic substrates might provide a functional equivalent of the kinds of hydrophobic interactions found in proteins. Interestingly, however, SELEX experiments with RNA have demonstrated the existence of pockets for the specific and tight binding of small hydrophobic molecules, including hydrophobic amino acids such as valine and tryptophan.[110,111]

Gold *et al.*[1] have provided an excellent discussion of the manifest success of small oligonucleotide aptamers (both DNA and RNA) as good binders. As discussed earlier, pre-SELEX thinking on the ability of DNA and RNA to bind specifically to a variety of shapes and molecules had postulated that the generality of hydrogen bonding of the DNA and RNA bases to each other would lead to their unavailability to interact specifically with a variety of target molecules. SELEX experiments have shown this not to be a problem in practice. Conversely, the fact that extensive base pairing can exist in even short oligonucleotides (such as in ~25 nucleotide aptamers) leads to their folding to form stable and often rigid folded structures, with defined three-dimensional shapes. Oligopeptides containing ~25 amino acid residues, by contrast, tend not to form such stable, defined structures. Phage-display selections of short peptides specific for binding target molecules (of the kind investigated using DNA and RNA SELEX) have generally not yielded specific and tight binders such as one finds with DNA and RNA aptamers.[1]

### 7.17.6.4 Future Directions

In the future we may expect many refinements in the overall technology of SELEX. Improvements in the design of random libraries will undoubtedly occur—possibly incorporating more and more elements of rational design into the "random" elements of sequence.[73] A much larger set of nucleotide analogues, as well as other, unrelated, organic functionalities, will likely be utilized, and it may be possible to find clever ways to carry out conventional, enzyme-based SELEX experiments with larger and larger numbers of nucleotide analogues. The selection protocols, too, will undoubtedly see much refinement; it may be possible to alleviate the selection pressures or interaptamer competition inherent in many selection protocols. Under less competitive selection pressures, some of the suboptimal sequences for a particular SELEX goal, that are generally not retrievable in standard experiments, may be recoverable. The study of such suboptimal aptamers may contribute to a deeper understanding of binding and catalytic mechanisms, and also to improved estimates of the frequencies with which binding "solutions" to specific targets may occur in a library—and may, indeed, exist in the universe of binding shapes possible to DNA and RNA.

The thinking about DNA as a catalyst is in its early days—only three major kinds of catalyst have been described (above). There are two rationales for investigating the catalytic properties of DNA. First, from the perspective of the RNA World hypothesis, it is important to have comparisons of DNA and RNA catalysts for the same reaction. These comparisons should help clarify whether, as a polymer, RNA has distinct chemical or structural advantages (involving, for instance, the 2′-hydroxy group) that make it capable of catalyzing reactions that DNA would be incapable of catalyzing. For those reactions which are already known to be catalyzed by RNA (and also those yet to be discovered in nature) for which the 2′-hydroxy group *is* important, can there be compensating functionalities, cofactors, or mechanisms that DNA may utilize to catalyze these same reactions?

A broader question arising from these above considerations is the extent to which *any* polymer, that contains approximately four variable monomeric units, can fold to give highly variable and specific structures and binding pockets complementary to a variety of shapes. It is entirely likely that prior to the evolution of any RNA World there were "worlds" of other, simpler, self-replicating polymers that were not necessarily nucleotide based or even remotely related to RNA and DNA. It would be interesting to have a sense of the kinds of chemistries that such primordial polymers might have been capable to catalyzing. DNA, although chemically similar to RNA, is poorer than RNA in terms of functional diversity; therefore, it offers us a window for exploring, in a general way, the possibilities of polymers folding to give enzyme-like active sites.

In this respect, the accumulating evidence that DNA is able to fold to different and complex shapes, and to catalyze reactions, is of profound importance, for it allows us to make general statements about the kinds of activities that are the special dominion of RNA, and about others that might be attainable by polymers containing comparable levels of internal diversity as RNA. The fact that DNA *is* a polymer that can be subjected to the enormous range of experiments made possible by SELEX makes it a polymeric system unlike any other for addressing the above questions.

To date, SELEX experiments have indicated that DNA and RNA invariably find different but comparably effective solutions to many binding and catalytic problems. However, the comparisons need to be deeper.

Another important aspect to the discovery of novel DNA catalysts (and, indeed, of novel DNA binding aptamers) is the possibility of their use in biomedical applications. Aptamers such as those which bind thrombin (and thereby interfere with the blood-clotting cascade) have been noted for their potential pharmaceutical value. Undoubtedly, many more such DNA aptamers, and also aptamers containing modified DNA and RNA, will be tested for their pharmaceutical possibilities,[1] Existing and novel catalytic DNAs are also important in this respect. The naturally occurring RNA "Hammerhead" ribozyme, and variants of it, have been shown to destroy specifically viral messenger RNAs within infected cells (reviewed in reference 112). DNAzymes that are able to carry out comparable functions would also undoubtedly be of much interest to the biomedical community.

## 7.17.7 REFERENCES

1. L. Gold, B. Polisky, O. Uhlenbeck, and M. Yarus, *Annu. Rev. Biochem.*, 1995, **64**, 763.
2. K. Guerrier-Takada, T. Gardiner, N. Marsh, N. Pace, and S. Altman, *Cell*, 1983, **35**, 849.
3. K. Kruger, P. J. Grabowski, A. J. Zaug, J. Sands, D. E. Gottschling, and T. R. Cech, *Cell*, 1982, **31**, 147.
4. T. R. Cech and B. L. Bass, *Annu. Rev. Biochem.*, 1986, **55**, 599.
5. F. H. C. Crick, *J. Mol. Biol.*, 1968, **38**, 367.
6. L. E. Orgel, *J. Mol. Biol.*, 1968, **38**, 381.
7. C. R. Woese, in "Exobiology," ed. C. Ponneperuma, Elsevier, Amsterdam, 1968, p. 301.
8. H. B. White, III, *J. Mol. Evol.*, 1976, **7**, 101.
9. W. Gilbert, *Nature (London)*, 1986, **319**, 618.
10. S. A. Benner, A. D. Ellington, and A. Traver, *Proc. Natl. Acad. Sci. USA*, 1989, **86**, 7054.
11. C. Tuerk and L. Gold, *Science*, 1990, **249**, 505.
12. F. R. Kramer, D. R. Mills, P. E. Cole, T. Nishihara, and S. Spiegelman, *J. Mol. Biol.*, 1974, **89**, 719.
13. M. T. Gait (ed.), "Oligonucleotide Synthesis: a Practical Approach," IRL Press, Oxford, 1986.
14. A. D. Ellington and J. W. Szostak, *Nature (London)*, 1990, **346**, 818.
15. L. E. Orgel, *Proc. R. Soc. London (Biol.)*, 1979, **205**, 435.
16. F. Ausubel, R. Brent, R. E. Kingston, D. D. Moore, J. G. Seidman, J. A. Smith, and K. Struhl (eds), "Current Protocols in Molecular Biology," Wiley-Interscience, New York, 1987.
17. N. Arnheim and H. Erlich, *Annu. Rev. Biochem.*, 1992, **61**, 131.
18. K. B. Mullis and F. A. Falona, *Methods Enzymol*, 1987, **155**, 335.
19. U. B. Gyllenstein and H. A. Erlich, *Proc. Natl. Acad. Sci. USA*, 1988, **85**, 7652.
20. T. Hultman, S. Stahl, T. Moks, and M. Uhlen, *Nucleosides Nucleotides*, 1988, **7**, 629.
21. F. Sanger, S. Nicklen, and A. R. Coulson, *Proc. Natl. Acad. Sci. USA*, 1977, **74**, 5463.
22. A. M. Maxam and W. Gilbert, *Proc. Natl. Acad. Sci. USA*, 1977, **74**, 560.
23. R. K. Saiki, T. A. Bergawan, G. T. Horn, K. B. Mullis, and H. A. Erlich, *Nature (London)*, 1986, **324**, 163.
24. D. W. Leung, E. Chen, and D. V. Goeddel, *Technique*, 1989, **1**, 11.
25. R. C. Cadwell and G. F. Joyce, *PCR Methods Appl.*, 1992, **2**, 28.
26. J. H. Spee, W. M. de Vos, and O. P. Kuipers, *Nucleic Acids Res.*, 1993, **21**, 777.
27. D. K. Treiber and J. R. Williamson, *Nucleic Acids Res.*, 1995, **23**, 3603.
28. W. P. C. Stemmer, *Nature (London)*, 1994, **370**, 389.
29. G. F. Joyce, *Sci. Am.*, 1992, **267**, 90.
30. G. F. Joyce, *Curr. Opin. Struct. Biol.*, 1994, **4**, 331.
31. S. Brenner and R. A. Lerner, *Proc. Natl. Acad. Sci. USA*, 1992, **89**, 5381.
32. J. R. Lorsch and J. W. Szostak, *Nature (London)*, 1994, **371**, 31.
33. J. R. Lorsch and J. W. Szostak, *Biochemistry*, 1995, **34**, 15315.
34. D. P. Bartel and J. W. Szostak, *Science*, 1993, **261**, 1411.
35. B. Cuenoud and J. W. Szostak, *Nature (London)*, 1995, **375**, 611.
36. R. R. Breaker and G. F. Joyce, *Chem. Biol.*, 1995, **2**, 655.
37. J. Hamm, *Nucleic Acids Res.*, 1996, **24**, 2220.
38. A. A. Beaudry and G. F. Joyce, *Science*, 1992, **257**, 635.
39. N. Lehman and G. F. Joyce, *Nature (London)*, 1993, **361**, 182.
40. R. R. Breaker and G. F. Joyce, *Trends Biotechnol*, 1994, **12**, 268.
41. D. Irvine, C. Tuerk, and L. Gold, *J. Mol. Biol.*, 1991, **222**, 739.
42. A. D. Ellington and J. W. Szostak, *Nature (London)*, 1992, **355**, 850.
43. D. E. Huizenga and J. W. Szostak, *Biochemistry*, 1995, **34**, 656.
44. M. Sassanfar and J. W. Szostak, *Nature (London)*, 1993, **364**, 550.
45. D. J. Schneider, J. Feigon, Z. Hotomsky, and L. Gold, *Biochemistry*, 1995, **34**, 9599.
46. Y. Lin, Q. Qiu, S. C. Gill, and S. D. Jayasena, *Nucleic Acids Res.*, 1994, **22**, 5229.
47. L. C. Bock, L. C. Griffin, J. A. Latham, E. H. Vermaas, and J. J. Toole, *Nature (London)*, 1992, **355**, 564.
48. L. R. Paborsky, S. N. McCurdy, L. C. Griffin, J. J. Toole, and L. L. K. Leung, *J. Biol. Chem.*, 1993, **268**, 20808.
49. R. F. Macaya, P. Schultz, F. W. Smith, J. A. Roe, and J. Feigon, *Proc. Natl. Acad. Sci. USA*, 1993, **90**, 3745.
50. K. Y. Wang, S. MaCardy, R. G. Shea, S. Swaminathan, and P. H. Bolton, *Biochemistry*, 1993, **32**, 1899.
51. K. Padmanabhan, K. P. Padmanabhan, J. D. Ferrara, J. E. Sadler, and A. Tulinsky, *J. Biol. Chem.*, 1993, **268**, 17651.
52. K. Y. Wang, S. H. Krawczyk, N. Bischofberger, S. Swaminathan, and P. H. Bolton, *Biochemistry*, 1993, **32**, 11285.

53. P. Schultze, R. F. Macaya, and J. Feigon, *J. Mol. Biol.*, 1994, **235**, 1532.
54. D. Sen and W. Gilbert, *Nature (London)*, 1988, **334**, 364.
55. W. I. Sundquist and A. Klug, *Nature (London)*, 1989, **342**, 825.
56. J. R. Williamson, M. K. Raghuraman, and T. R. Cech, *Cell*, 1989, **59**, 871.
57. D. Sen and W. Gilbert, *Nature (London)*, 1990, **344**, 410.
58. J. R. Williamson, *Annu. Rev. Biophys. Biomol. Struct.*, 1994, **23**, 703.
59. M. F. Kubik, A. W. Stephens, D. A. Schneider, R. Marlar, and D. Tassett, *Nucleic Acids Res.*, 1994, **22**, 2619.
60. Y. Cui, Q. Wang, G. D. Stormo, and J. M. Calvo, *J. Bacteriol.*, 1995, **177**, 4872.
61. Y.-Y. He, P. G. Stockley, and L. Gold., *J. Mol. Biol.*, 1996, **255**, 55.
62. S. P. Hale and P. Schimmel, *Proc. Natl. Acad. Sci. USA*, 1996, **93**, 2755.
63. A. N. Baldwin and P. Berg., *J. Biol. Chem.*, 1966, **241**, 839.
64. M. D. Frank-Kamenitskii and S. M. Mirkin, *Annu. Rev. Biochem.*, 1995, **64**, 65.
65. D. Pei, H. D. Ulrich, and P. G. Schultz *Science* 1991, **253**, 1408.
66. G. A. Soukoup, A. E. Ellington, and L. J. Maher, *J. Mol. Biol.*, 1996, **259**, 216.
67. P. Hardenbol and M. W. Van Dyke, *Proc. Natl. Acad. Sci. USA*, 1996, **93**, 2811.
68. J. C. Marini, S. D. Levene, D. M. Crothers, and P. T. Englund, *Proc. Natl. Acad. Sci. USA*, 1982, **79**, 7664; erratum: 1983, **80**, 7678.
69. H.-M. Wu and D. M. Crothers, *Nature (London)*, 1984, **308**, 509.
70. D. M. Crothers, T. E. Haran, and J. G. Nadeau, *J. Biol. Chem.*, 1990, **265**, 7093.
71. P. J. Hagerman, *Annu. Rev. Biochem.*, 1990, **59**, 755.
72. B. A. Beutel and L. Gold, *J. Mol. Biol.*, 1992, **228**, 803.
73. J. R. Lorsch and J. W. Szostak, *Acc. Chem. Res.*, 1996, **29**, 103.
74. A. D. Ellington, *Ber. Bunsenges. Phys. Chem.*, 1994, **98**, 1115.
75. I. Hirao and A. D. Ellington, *Curr. Biol.*, 1995, **5**, 1017.
76. R. R. Breaker and G. F. Joyce, *Chem. Rev.*, 1997, **95**, 371.
77. X. Dai, A. D. De Mesmaeker, and G. F. Joyce, *Science*, 1995, **267**, 237.
78. L. Pauling, *Sci. Am.*, 1948, **36**, 51.
79. W. P. Jencks, "Catalysis in Chemistry and Enzymology," McGraw-Hill, New York, 1969.
80. P. G. Schultz and R. A. Lerner, *Science*, 1995, **269**, 1835.
81. K. P. Williams, S. Ciafre, and G. P. Tocchini-Valentini, *EMBO J.*, 1995, **14**, 4551.
82. B. Dichtl, T. Pan, A. B. DiRenzo, and O. C. Uhlenbeck, *Nucleic Acids Res.*, 1993, **21**, 531.
83. T. Pan and O. C. Uhlenbeck, *Biochemistry*, 1992, **31**, 3887.
84. T. Pan and O. C. Uhlenbeck, *Nature (London)*, 1992, **358**, 560.
85. T. Pan, B. Dichtl, and O. C. Uhlenbeck, *Biochemistry*, 1994, **33**, 9561.
86. K. B. Chapman and J. W. Szostak, *Chem. Biol.*, 1995, **2**, 655.
87. E. H. Ekland, J. W. Szostak, and D. P. Bartel, *Science*, 1995, **269**, 364.
88. E. H. Ekland and D. P. Bartel, *Nucleic Acids Res.*, 1995, **23**, 3231.
89. E. H. Ekland and D. P. Bartel, *Nature (London)*, 1996, **382**, 373.
90. C. Wilson and J. W. Szostak, *Nature (London)*, 1995, **374**, 777.
91. M. Ilangasekare, G. Sanchez, T. Nickles, and M. Yarus, *Science*, 1995, **267**, 643.
92. P. A. Lohse and J. W. Szostak, *Nature (London)*, 1996, **381**, 442.
93. J. R. Prudent, T. Uno, and P. G. Schultz, *Science*, 1994, **264**, 1924.
94. M. M. Conn, J. R. Prudent, and P. G. Schultz, *J. Am. Chem. Soc.*, 1996, **118**, 7012.
95. R. R. Breaker and G. F. Joyce, *Chem. Biol.*, 1994, **1**, 23.
96. A. G. Cochran and P. G. Schultz, *Science*, 1990, **249**, 781.
97. Y. Li, C. R. Geyer, and D. Sen, *Biochemistry*, 1996, **35**, 6911.
98. Y. Li and D. Sen, *Nature Struct. Biol.*, 1996, **3**, 743.
99. Y. Li and D. Sen, *Biochemistry*, 1997, **36**, 5589.
100. J. A. Latham, R. Johnson, and J. J. Toole, *Nucleic Acids Res.*, 1994, **22**, 2817.
101. D. J. Ecker, T. A. Vickers, R. Hanacek, V. Driver, and K. Anderson, *Nucleic Acids Res.*, 1993, **21**, 1853.
102. H. M. Geysen, S. J. Rodda, and T. J. Mason, *Mol. Immunol.*, 1986, **23**, 709.
103. R. A. Houghten, C. Pinilla, S. E. Blondelle, J. R. Appel, C. T. Dooley, and J. H. Cuervo, *Nature (London)*, 1991, **354**, 84.
104. R. A. Owens, P. D. Gesellchen, B. J. Houchins, and R. D. DiMarchi, *Biochem. Biophys. Res. Commun.*, 1991, **181**, 402.
105. J. R. Wyatt, T. A. Vickers, J. L. Roberson, R. W. Buckheit, T. Klimkait, D. DeBaets, P. W. Davis, B. Rayner, J. L. Imbach, and D. J. Ecker, *Proc. Natl. Acad. Sci. USA*, 1994, **91**, 1356.
106. J. K. Scott and G. P. Smith, *Science*, 1990, **249**, 386.
107. S. E. Cwirla, E. A. Peters, R. W. Barrett, and W. J. Dower, *Proc. Natl. Acad. Sci. USA*, 1990, **87**, 6378.
108. H. Fenniri, K. Janda, and R. A. Lerner, *Proc. Natl. Acad. Sci. USA*, 1995, **92**, 2278.
109. J. R. Lorsch and J. W. Szostak, *Biochemistry*, 1994, **33**, 973.
110. I. Majerfeld and M. Yarus, *Nature Struct. Biol.*, 1994, **1**, 287.
111. M. Famulok and J. W. Szostak, *J. Am. Chem. Soc.*, 1992, **114**, 3990.
112. J. Ohkawa, T. Koguma, T. Kohda, and K. Taira, *J. Biochem. (Tokyo)*, 1995, **118**, 251.

# 7.18
# Cloning as a Tool for Organic Chemists

## JOHN D. PICKERT and BENJAMIN L. MILLER
*University of Rochester, NY, USA*

| | | |
|---|---|---|
| 7.18.1 | INTRODUCTION | 643 |
| 7.18.2 | SELECTION OF A TARGET DNA SEQUENCE | 645 |
| 7.18.3 | SELECTION OF AN EXPRESSION SYSTEM | 646 |
| | *7.18.3.1 Common Vectors for Expression in* E. coli | 648 |
| 7.18.4 | OBTAINING OLIGONUCLEOTIDES IN QUANTITY USING THE POLYMERASE CHAIN REACTION | 652 |
| | *7.18.4.1 Choosing a Source of Template for PCR* | 654 |
| | *7.18.4.2 Choosing Primers for PCR* | 655 |
| | *7.18.4.3 The Polymerase* | 658 |
| | *7.18.4.4 Was the PCR Successful?* | 658 |
| | *7.18.4.5 Troubleshooting the PCR* | 659 |
| | *7.18.4.6 Purification of PCR Products* | 660 |
| 7.18.5 | RESTRICTION DIGESTS | 660 |
| 7.18.6 | DEPHOSPHORYLATION | 661 |
| 7.18.7 | LIGATION | 661 |
| 7.18.8 | TRANSFORMATION OF DNA INTO *E. COLI* | 662 |
| 7.18.9 | SCREENING FOR INSERT | 664 |
| 7.18.10 | VERIFICATION OF INSERT IDENTITY | 664 |
| 7.18.11 | LARGE SCALE PRODUCTION OF PROTEIN IN *E. COLI* CULTURE | 665 |
| | *7.18.11.1 SDS–PAGE Analysis of Protein Expression Levels* | 665 |
| | *7.18.11.2 Isolation and Purification of the Fusion Protein from a Large Scale Culture* | 666 |
| | *7.18.11.3 Cleavage of the Fusion Protein* | 667 |
| | *7.18.11.4 Purification of Desired Polypeptide Segment* | 667 |
| 7.18.12 | CASE STUDIES IN CLONING | 667 |
| | *7.18.12.1 Case Study: Cloning and Derivatization of SH3 Domains for Ligand Binding Analysis and NMR Structural Studies* | 668 |
| | *7.18.12.2 Case Study: Production of Zinc Finger Domains for Analysis of Protein–DNA Interactions* | 669 |
| 7.18.13 | CONCLUSIONS | 671 |
| 7.18.14 | REFERENCES | 671 |

## 7.18.1 INTRODUCTION

The advent of methods for the effective manipulation of DNA, allowing for the rapid cloning of nucleic acid sequences of interest from a variety of sources,[1,2] triggered an explosion in many biological fields. While cloning techniques have primarily been the stock in trade of molecular biologists, chemists have recently begun to use these techniques as well, as an integral part of programs in natural product synthesis[3] (e.g., the identification of FK506 binding protein)[4] ligand design (e.g., cathepsin K inhibitors),[5] combinatorial chemistry,[6] and analysis of protein–nucleic acid[7,8] and protein–protein interactions (e.g., in the analysis of the binding of herpes simplex virus VP16 to its target protein).[9] This chapter seeks to survey some of the most general methods for cloning a gene (or portion of a gene) and expressing its encoded protein (or subdomain of a protein). It should be stressed at the outset that this review is by no means exhaustive; the enormous variety of expression systems, amplification protocols, and purification systems which are either commercially available or described in the primary literature preclude such a treatment in the space allotted.[10–12] Numerous review volumes and collections of laboratory protocols are available, perhaps the most comprehensive manual is by Sambrook.[13] Rather, a few general methods for typical cloning tasks, as well as uses to which these gene products have been put are described. This can then serve as a starting point from which the reader may derive a successful cloning protocol. The enormous number of choices available at each stage of the process is, unfortunately, only half of the problem (in terms of information) facing a chemist interested in cloning. Much as an organic chemist is unlikely to include detailed experimental procedures for running flash chromatography or operating a rotary evaporator in every paper, many of the molecular biology techniques associated with cloning experiments are so routine (for biologists) that they are neither discussed, nor referenced. This chapter attempts to correct this omission: liberal referencing to detailed experimental procedures is made throughout. Many monographs are available which describe cloning techniques. One of the best and most accessible is by Watson *et al.*[14]

A chemist casually perusing this chapter might question the advisability of pursuing a cloning project vs. leaving the task to a molecular biologist, and based on our training as chemists, that is a reasonable question. From a historical perspective, most chemistry was done completely in the "small molecule" realm; synthetic organic chemists typically spent their time synthesizing natural products, or compounds based on natural product leads. For those in academia, biological testing was to be done only in collaboration with a research group in a department of biology or allied field, while those in the industrial sector generally focused strictly on synthetic goals, shipping the final compounds off for analysis to another department of the company.

Developments in both chemistry and biology have caused this model to be radically altered. The polymerase chain reaction (PCR), as well as generalized expression systems for producing a protein in *Escherichia coli* or more complex cell types, has considerably simplified the task of cloning a protein or polypeptide. Methods of rapid and precise structural analysis,[15,16] (e.g., multidimensional NMR spectroscopy)[17] developed by both chemists and biologists, have shown chemists that the structure of biological molecules is both more interesting and more accessible than the relatively ill-defined "cartoons" typically shown in introductory biology texts. From the chemist's perspective, the availability of this structural information coupled with advances in molecular modeling (i.e., *de novo* ligand design)[18] and combinatorial chemistry have allowed chemists to generate structural hypotheses rapidly and synthesize molecules intended to test those hypotheses.

The cloning process typically proceeds through several planning and experimental stages, which are in many respects loosely analogous to the various stages involved in the synthesis of a small molecule (Figure 1). (i) A gene or subgene of interest is identified. (ii) An expression system (a bacterial plasmid or other vector designed to introduce the DNA sequence of interest into a host organism) is chosen. (iii) A strategy for obtaining the DNA sequence of interest in sufficient quantities for cloning is developed and implemented. (iv) The target DNA sequence is incorporated into the expression vector. (v) The expression vector is introduced into the host organism, and a series of experiments is run to verify that the vector placed into the host organism carries the correct sequence. (vi) Large quantities of the host organism are grown, and induced to express (or overexpress) the desired polypeptide sequence. (vii) The overexpressed polypeptide is isolated from the host organism, purified, and characterized. Each of these stages will be examined in turn.

A significant and welcome difference between the production of a synthetic compound and a cloned protein or polypeptide is that cloning provides an infinitely renewable source of the final product. Once an expression system has been constructed and implemented in a particular cell type, it is only necessary to "grow up", or culture, additional amounts of the expressing cell type in order to obtain more of the protein or polypeptide. Of course, this is also true of many natural products, although isolation procedures are typically more arduous than those for cloned proteins. The concept of co-opting biology for the production of "unnatural" products in beginning to find numerous applications, as exemplified by the efforts of Khosla and co-workers[19] in the generation of designed polyketide synthases.

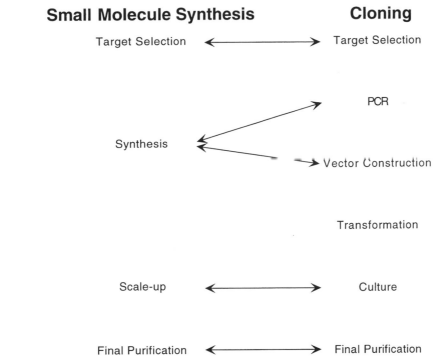

**Figure 1** Comparison of the steps involved in small molecule synthesis vs. a cloning experiment.

## 7.18.2 SELECTION OF A TARGET DNA SEQUENCE

Two factors must be kept in mind when selecting a nucleotide sequence for incorporation into a cloning strategy, in addition to the scientific goals of the cloning project. These are the length and composition of the amino acid sequence, and the relationship of these factors to the probability of obtaining a properly folded, active polypeptide or protein. In all cases, it is the polypeptide, and not the oligonucleotide, which is the primary deciding factor, since several strategies are available for cloning long stretches of DNA (although these can be more complex and error prone than typically observed for the cloning of a short stretch of DNA). This results in two general types of polypeptide that are the target of a cloning experiment: full length protein and shorter subdomains.

First, an entire protein is cloned and expressed. In some respects, this is the simplest case, since the sequence needed for cloning is governed by the published full length sequence of the gene and not subject to the researcher's decision. However, unless the full length protein is of relatively low molecular weight, problems can occur with respect to expression, necessitating the use of expression systems in insect or mammalian cells (see below). Furthermore, the probability of obtaining mis-folded material increases as the size and complexity of an expressed polypeptide increases.

A compelling, but potentially misleading, assertion is that the entire tertiary structure of any arbitrary protein is completely encoded by its amino acid sequence. Although this maxim, originating with early protein denaturation and refolding experiments carried out on ribonuclease,[20] is correct in a general sense, it ignores the complexity of protein folding pathways. Although the active form of a protein is in general the global minimum energy conformer, many proteins must pass through several low energy (local minimum) conformers on their way to this final folded state.[21] It is not uncommon for an expressed polypeptide to become "trapped" in one of these local minima, unable to undergo transition to the final (and presumably active) folded structure. Furthermore, many proteins require the assistance of "foldases" or chaperonins in order to fold properly.[22-24] Finally, many proteins require posttranslational modification in order to exist in their active form.

Despite these caveats, numerous full length proteins have been expressed at high levels and demonstrated to have native structure and/or full activity. Typical examples might include, in the smaller range of proteins, basic pancreatic trypsin inhibitor (BPTI),[25] ubiquitin,[26] and metallo-thionein.[27] Larger proteins have also been expressed, although these frequently require eukaryotic expression systems. For example, several isoforms of protein kinase C have been cloned and expressed in insect cells.[28]

A second possible strategy for the selection of a target cloning sequence is to clone a sub-sequence of the full protein. In general, sequences targeted in these types of cloning experiment are regions (or "domains") of the protein that are known (or expected) to have a well-defined tertiary structure when excised from the full protein. Such domains may be identified by homology mapping[29] to known sequences. For example, a known sequence may be compared to databases of nucleic acids, such as those maintained at the US National Center for Biological Information (NCBI), Bethesda, MD, using the BLAST searching algorithm;[30] subsequent homology modeling by simple sequence analysis or through more complex computational methods[31] designed to suggest potential tertiary structures may then be performed. Alternatively, these domains may be identified by an iteractive process of cloning and deletion analysis of a full length protein (e.g., protein kinase C[32]).[33–36] In another example, deletion analysis of syntaxin-1a was used to determine the minimal domain responsible for binding to synaptotagmin.[37] In addition to providing generally higher levels of protein expression, the targeting of a protein subdomain is potentially preferable for ligand binding studies, since the probability of finding a compound capable of binding to the specific region of interest is higher.

Finally, the specific use to which the cloned product will be put must also be considered in the selection of a target. For example, one convenient method of selectively labeling the product polypeptide at the amino terminus with biotin[38–41] or a fluorescent tag[42] is via periodate oxidation of an amino-terminal serine or threonine residue followed by reaction with a hydrazide.[43] Therefore, if this kind of modification is desired, it is obviously necessary to select a sequence (or mutagenize the natural sequence) such that an amino-terminal serine or threonine is obtained. On the other hand, the precise, native amino acid sequence may be the desired target if NMR or X-ray crystallographic studies are planned.

### 7.18.3   SELECTION OF AN EXPRESSION SYSTEM

After a target DNA sequence is identified, the next step in a cloning project is the selection of an expression system. An expression system is comprised of two parts: the host organism for the DNA to be expressed, and a carrier (or "vector") for this DNA that can be incorporated into the host organism. The selection of a host organism is primarily dependent on the size, expected solubility, and need for posttranslational modification of the target protein or polypeptide. The majority of "small" (<200 or so amino acids) proteins and polypeptides can be expressed most conveniently in *E. coli*, which is the simplest organism to manipulate in the laboratory, since it grows rapidly and can be induced to provide the largest quantities of protein.[44] Since this also corresponds to the primary experience of the authors, the majority of our discussion will focus on *E. coli*-based cloning and protein expression methods. Numerous *E. coli* cell lines are available, and their genotypes and phenotypes have been mapped extensively, allowing for a case by case evaluation of the suitability of a particular cell type to overproduce large quantities of protein.

A few of the most common *E. coli* cell lines available from commercial sources and used in cloning experiments include BL21, JM109, DH5α, and XL1-Blue. BL21 cells serve as hosts for a λ prophage carrying the T7 RNA polymerase,[45] and as a B-strain of *E. coli*, BL21 is deficient in both the *lon* and *ompT* proteases.[46] This can improve stability of the overproduced protein.

Initial transformation of fusion constructs carries its own set of problems; for example, DNA in wild-type *E. coli* strains is methylated at some positions. DNA that is not methylated (e.g., a plasmid constructed from PCR products) is recognized as foreign by a set of restriction enzymes, and degraded. In order to get around this problem, JM109,[47] DH5α,[48] and XL1-Blue[49] cell types all carry an *hsdR* mutation that prevents this restriction of nonmethylated sequences. However, methylation itself is not prevented, so DNA may be isolated from these cells following transformation and retransformed into another cell type that does restrict methylated sequences. Two other mutations common to these cell lines are *recA1* and *endA1*; *recA1* eliminates homologous recombination (improving the stability of large plasmids), while *endA1* eliminates nonspecific endonuclease activity.

XL1-Blue cells are characterized by a particularly high transformation efficiency (up to $5 \times 10^8$ colonies $\mu g^{-1}$ of DNA),[50] which can be particularly useful in the early stages of a cloning effort. XL1-Blue also carries the β-galactosidase (*lacZ*) gene, allowing for cells to be selected via a blue/white screen on Xgal (5-bromo-4-chloro-3-indolyl-β-D-galactoside, (1)) plates.[47]

DH5α and JM109 strains both carry the *lacZ*DM15 mutation; this provides α-complementation of the β-galactosidase gene and concomitant blue/white screening. JM109 (as well as several other cell types) also carries *lacI*q, which provides for overproduction of the *lac* repressor. This can assist in limiting the amount of a potentially toxic polypeptide or fusion protein which is produced.

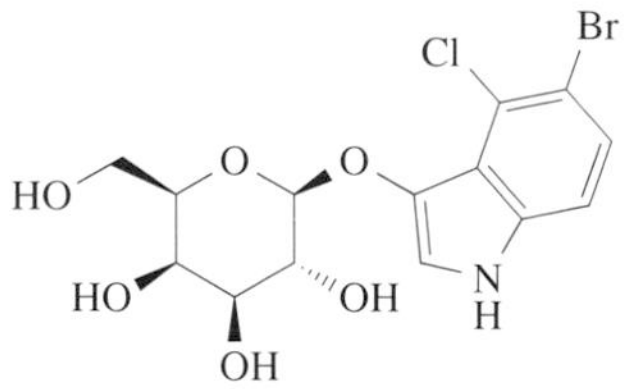

(1)

In many cases, selection of a particular host strain requires weighing transformation efficiency (particularly useful during initial propagation and examination of products from PCR-based cloning protocols) vs. overexpression ability (i.e., the ability of any particular cell type to support large scale protein overexpression). It is not uncommon initially to transform into one cell type for initial analysis, then retransform into another for bulk production. Extensive tables of *E. coli* genotypes and phenotypes may be found in various manufacturers' catalogues (i.e., New England Biolabs) as well as on the World Wide Web.

Despite its general advantages, there are some disadvantages to cloning in *E. coli*. The organism does not have the cellular machinery to provide many common posttranslational modifications of proteins (such as glycosylation, phosphorylation, acylation, or, in some cases, disulfide bond formation), which can lead to products that are misfolded or inactive for other reasons.[51] If initial experiments suggest that *E. coli* cannot serve as a suitable host, baculovirus-infected SF9 cells provide an alternative expression host.[52,53] Although these are somewhat more demanding in their requirements, they are less so than mammalian cells.

After selecting an appropriate host, a vector must be selected to introduce the target DNA into the host cell. The vector, typically a circular DNA plasmid or viral phage, contains genes controlling a selection mechanism, a promoter region, and coupled repressor which can be deactivated by addition of a cofactor to the growth medium, and a multicloning site (MCS) (Figure 2).

Frequently, the vector will also incorporate a sequence encoding a protein for fusion to the target DNA which, when expressed, allows the target protein to be purified away from other cellular components readily. The selection mechanisms incorporated into all commonly used expression vectors are a way of ensuring that only those cells carrying the plasmid grow; furthermore, additional selection schemes are available to distinguish cells carrying plasmid only from cells carrying plasmid plus insert. For example, selection via antibiotic resistance may be coupled with the interruption of a β-galactosidase gene, which allows for the identification of those cells carrying full length plasmid by a blue/white selection scheme.[54]

Expression vectors typically incorporate a strong promoter region under the control of a coexpressed repressor; for example, plasmid vectors frequently include the *trc*,[55] *tac*,[56,57] or *lacUV5*[47] promoter, which is regulated by the *lac* repressor. If one of these promoters and the *lac* repressor are incorporated into the same plasmid, the presence of the repressor will typically prevent expression of the protein (i.e., the repressor binds to the promoter region). Expression of the protein can be induced by the addition of isopropyl-β-thio-D-galactopyranoside (IPTG, (**2**)) to the growth medium, which binds to the *lac* gene product, causing it to dissociate from the promoter region of the plasmid. This type of careful regulation of expression is essential, since the overexpression of proteins places considerable stress on the resources of a cell. As a general result, a culture of bacteria is typically not induced to overexpress with IPTG until a substantial cell density has been reached, since essentially all of the energy of the cells will go into producing protein following induction. Other mechanisms for inducing transcription, such as the thermal inactivation of a repressor,[58] have also been used.

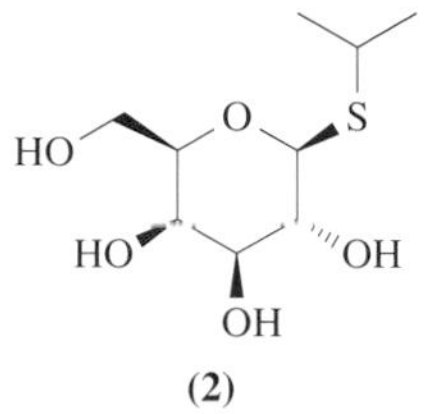

(2)

Vectors may be constructed so as to provide the product with either the precise amino- and carboxy-terminal residues derived from the natural sequence (or modified as desired), or fused to another protein. These "fusion proteins" can be advantageous in that they usually provide a method

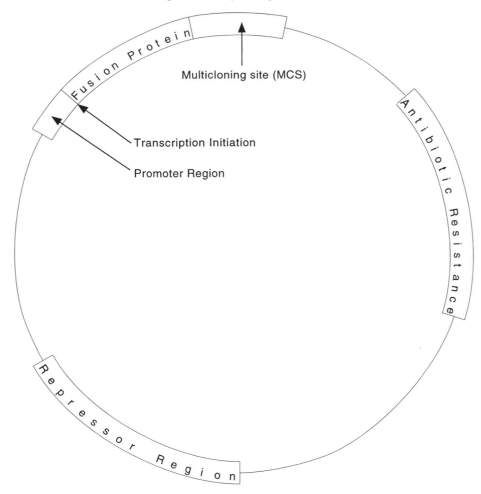

**Figure 2**  General features of a vector for cloning in *E. coli*.

of affinity purification, and they may also help to solubilize the product. Fusion proteins which are available include glutathione-*S*-transferase (GST); maltose binding protein (MBP); His-tag vectors, which incorporate a series of histidine residues for metal ion binding, and others (see below). In all cases in which a fusion protein is produced, the purification of the desired protein involves passing the cellular mixture over an affinity column (for example, glutathione–agarose for GST fusions, maltose–agarose for MBP fusions, or nickel–agarose for His-tag fusions) followed by elution of the fusion protein using excess ligand (glutathione, maltose, or a nickel salt). After isolation of the fusion protein, proteolytic cleavage of the fusion protein allows for subsequent purification of the desired sequence. Typical enzymes for proteolytic cleavage include thrombin, factor $X_a$, and enterokinase; variants of all commonly used vectors are available which allow for any of these to be used. Careful planning is required to determine which enzyme will be used, since all have advantages and disadvantages. For example, while thrombin is the least expensive of these enzymes to use, it is also frequently less selective than factor $X_a$ or enterokinase.

### 7.18.3.1  Common Vectors for Expression in *E. coli*

As discussed above, a wide variety of vectors are available for cloning a gene in *E. coli*. Some of the general features of vectors have already been described, and several vectors will now be considered in more detail. Vector systems are generally divided into three types: phage vectors, phagemid vectors, and plasmid vectors. Since they are the most commonly employed, the focus will be on plasmid vectors, and phage and phagemid vectors will be mentioned briefly. Each of these vectors has specific applications, advantages, and disadvantages.

Bacteriophage λ is a phage virus that inserts itself into a host bacterium, undergoes many rounds of replication, and is subsequently released as many new infectious virus particles.[59] This great infective and replicative ability of bacteriophage λ was the driving force behind the initial experiments into the use of this phage for transporting small segments of foreign DNA into *E. coli*. Some of the advantages of a bacteriophage λ vector are perhaps best illustrated by use of the λ-ZAP vector as a specific example (Figure 3). Like all phage vectors, λ-ZAP is a linear, double stranded DNA; packaging of λ-ZAP as viral particles allows it to be introduced into the host, where it can circularize. The vector can then go on to insert itself into the *E. coli* genome, or replicate as a new phage. Additionally, λ-ZAP carries the phagemid (a phage–plasmid hybrid) pBluescript (Figure 4), which, *in vivo*, is excised from its host vector. Like all vectors, λ-ZAP presents a variety of restriction sites for cloning. Perhaps most importantly, λ-ZAP can accommodate foreign DNA inserts up to 10 kb in length. This is a general feature of all bacteriophage λ vectors and a result of their linear nature.[54]

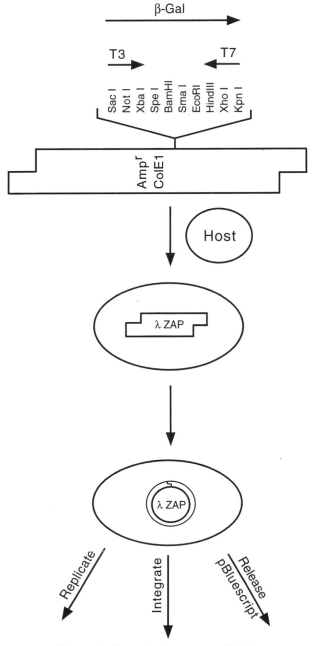

**Figure 3**   Bacteriophage vector λZAP.

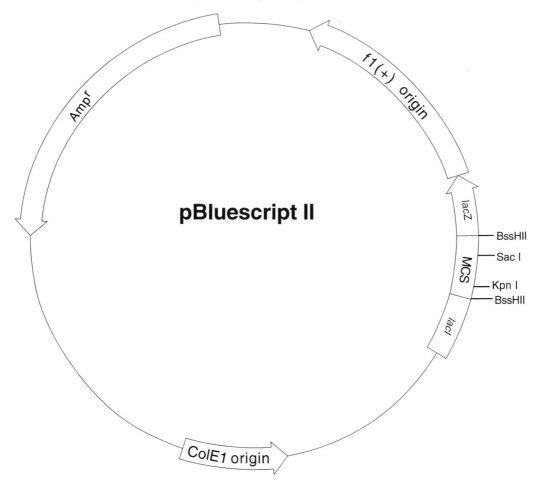

**Figure 4**   Map of pBluescript II, a phagemid vector.

The pBluescript phagemid vector contains the *lacZ* gene, which codes for the protein β-gal-actosidase (β-gal). β-Galactosidase generates a blue color when presented with a medium containing Xgal (see above). Therefore, the pBluescript phagemid offers the advantage of color selection in the determination of bacterial colonies containing the full construct (known as "recombinant plaques"). When foreign DNA is inserted into the vector, the *lacZ* gene is disrupted (since the relevant restriction sites are in the interior of the *lacZ* sequence), thereby disrupting the production of β-gal. As a result, recombinant plaques with foreign DNA inserted will appear white.

The simplest and most commonly used vector systems are the plasmid vectors, a subset of which is listed in Table 1. In general, plasmid vectors are characterized by several factors: their copy number (i.e., the number of copies of a vector that a cell can support; this is generally a self regulating function of the plasmid), the resistance or selection mechanism, whether a fusion protein is encoded and what kind, and the specific sites for restriction enzymes present in the multicloning site. As examples, four families of plasmid vectors will be discussed: pGEX, His-Patch ThioFusion (pThioHis), pET, and pMal. All of these vectors are designed for the production of fusion proteins; two examples of vectors lacking the capability of fusion protein production will be described in Section 7.18.12.

The pGEX family of plasmid vectors (Figure 5) is based on producing target polypeptides as GST fusions,[60] and is comprised of many variations. In general, pGEX plasmid vectors tend to have high levels of expression, exhibit efficient site-specific cleavage of GST fusion proteins (i.e., cleavage of the fusion protein to allow separation of the desired polypeptide sequence from GST occurs with high specificity), and allow for rapid affinity-based purification.[61]

Three representative members of the pGEX family are pGEX-KG, pGEX-KT, and pGEX-3X. The pGEX-KT plasmid vector was developed as a modification of pGEX-KG, which itself was

**Table 1**  Selected plasmid vectors for transformation in *E. coli.*

| Plasmid name | Supplier | Fusion | Comments | Ref. |
|---|---|---|---|---|
| pGEX-KG | Pharmacia Biotech | GST | Fusion protein cleaved by thrombin | 62 |
| pGEX-KT | Pharmacia Biotech | GST | Fusion protein cleaved by thrombin | 63 |
| pGEX-3X | Pharmacia Biotech | GST | Fusion protein cleaved by factor $X_a$ | 61 |
| pThioHis | Invitrogen | Thioredoxin (trxA) | pThioHis A, B, and C are available; each differ with respect to reading frames. Fusion protein removed by washing through a metal-chelating resin or an affinity chromatography resin. | 64 |
| pET | Novagen | Oligo–histidine sequence | Wide variety of pET vectors available. Various forms of each pET vector differ with respect to reading frames. Common proteases used are thrombin, factor $X_a$, and enterokinase; these proteases vary depending on the specific pET vector. | 65 |
| pMAL | New England Biolabs | MBP | Fusion protein cleaved by factor $X_a$ | 66,67 |

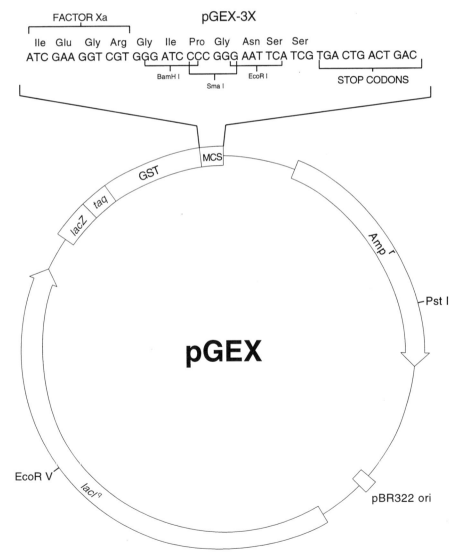

**Figure 5**  Map of the pGEX family of plasmid vectors.

originally prepared in the hope of improving thrombin cleavage of the GST fusion protein. Although cleavage of GST is significantly improved in pGEX-KG, the resulting polypeptide retains several foreign amino acids at the *N*-terminus. With pGEX-KT the presence of nonnative amino acids in the polypeptide is reduced, while still maintaining efficient thrombin cleavage.[63]

Another commonly used pGEX plasmid vector is pGEX-3X, which encodes a factor $X_a$ consensus cleavage sequence adjacent to the GST fusion protein. Thus, factor $X_a$, rather than thrombin, is the protease that cleaves the fusion protein from the desired polypeptide. Efficient cleavage is often observed with pGEX-3X, but in general the cleavage seems to be a bit slower and less efficient than when thrombin is used as the protease, as in some of the other pGEX plasmid vectors (e.g., pGEX-KG, pGEX-KT, and pGEX-KN).[61-63] One disadvantage of factor $X_a$ is that it is a significantly more expensive reagent than thrombin.

Another commonly used family of plasmid vectors is the His-Patch ThioFusion (pThioHis) vectors.[64] Three variations of this vector are available (pThioHis A, B, and C), which differ by the reading frame encoded (Figure 6). As with the pGEX plasmid vectors, pThioHis vectors also contain a sequence encoding a fusion protein, called thioredoxin (trxA). trxA incorporates a metal-binding domain, which allows for purification of the fusion construct on metal-chelating resins.[68] However, the primary use of trxA is its ability to increase the solubility of the fusion protein, particularly in those cases where a large number of disulfide bonds are formed.[69-71] This makes pThioHis (or another vector which provides trxA fusions) an attractive choice if initial experiments suggest that solubility is a problem.

The pET vectors are particularly versatile, in that they are available with a huge variety of amino- and carboxy-terminal fusion peptides to monitor expression levels, improvement of solubility, and affinity purification (Figure 7). In all cases, the T7 promoter is used to provide high expression levels.[72,73] Expression level monitoring may be done with an 11 amino acid epitope at the *N*-terminus which is recognized by a T7 antibody. *C*-Terminal or *N*-terminal fusion to an oligohistidine sequence allows for purification over a $Ni^{2+}$ affinity column.[74-78] These vectors also incorporate an S-tag, allowing for another method of affinity purification.[79] As with pThioHis, several pET vectors also permit expression of fusion proteins incorporating trxA.

MBP has also found use in expression systems as part of a strategy of fusion protein overexpression and affinity purification. The pMAL family of vectors[66,67] (Figure 8) produce such MBP-target fusion proteins, which may be purified on a maltose-derivatized affinity column. These vectors provide a factor $X_a$ site for proteolytic cleavage of the fusion protein following purification.[80]

## 7.18.4 OBTAINING OLIGONUCLEOTIDES IN QUANTITY USING THE POLYMERASE CHAIN REACTION

Once a sequence of interest has been identified and an expression system chosen, the next step in a cloning procedure is to obtain large quantities of this sequence (termed the "insert") in a form suitable for introduction into the expression vector. With very few exceptions, this is done by amplifying the insert from some template source using PCR.[81-84] Several textbooks describing PCR in detail, including specific protocols are available including one by Newton.[85] Prior to the discovery of PCR, the analysis and cloning of genes was a very tedious and time-consuming process.[86] With the advent of PCR in the middle 1980s, however, the analysis of genes became significantly easier experimentally, as well as requiring much less time. In short, PCR allows millions of copies of a specified segment of DNA to be generated from a small quantity of template DNA. While PCR has been used to produce oligonucleotides for applications as diverse as genomic sequencing,[87-89] and forensic analysis,[90] the authors confine this discussion to methods that provide products suitable for introduction into a protein expression system.

PCR is a relatively simple reaction to run, requiring eight major reactants: the DNA template containing the gene of interest, two short oligonucleotide "primers" that bracket the gene of interest as well as providing specific recognition sites for restriction endonucleases, a polymerase enzyme with its corresponding buffer, and a mixture of all four deoxynucleotides. In addition, an apparatus capable of repetitive cycles of very rapid, accurate heating and cooling (a "thermocycler") is necessary. The reaction proceeds through three stages (Figure 9).

First, the reaction mixture is heated to separate the double-stranded DNA template (melting; typically this is done at 94 °C for 30 s). Second, a cooling (annealing) stage (at 35–55 °C) allows the oligonucleotide primers to bind to their complementary sequences on the two single-stranded halves

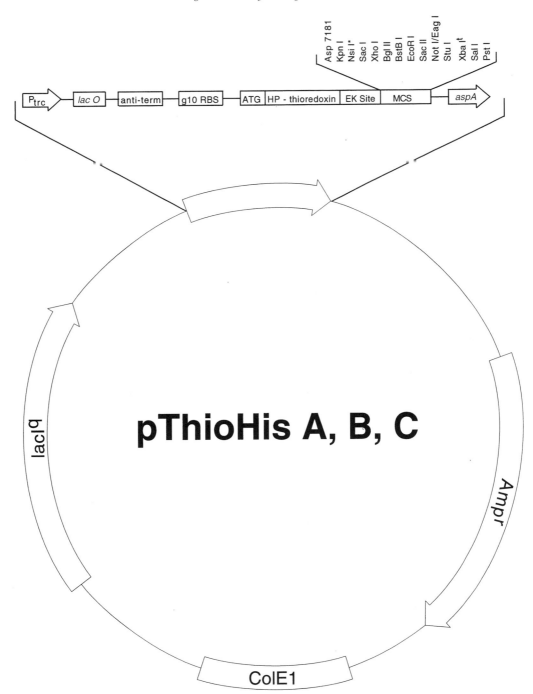

**Figure 6** Map of the pThioHis family of plasmid vectors.

of the template. Third, the temperature of the reaction mixture is raised to 72°C, allowing the polymerase to copy each of the template strands in the 5′ to 3′ direction, starting from the 3′ end of each primer (extension). After this extension period the temperature is raised to the melting temperature again and the cycle is repeated 25 to 30 times. The great advantage of PCR is that each cycle causes a doubling of the number of "templates" available; thus, the amount of oligonucleotide corresponding to the sequence bracketed by the primers increases exponentially with the number of cycles (Figure 10). Of course, the actual product obtained is dependent both on the template that is used and on the design of oligonucleotide primers that can selectively bind to the desired segment of the template.

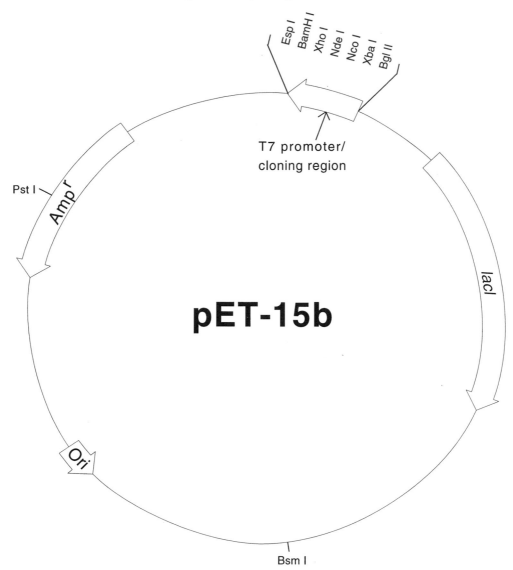

Esp I
BamH I
Xho I
Nde I
Nco I
Xba I
Bgl II

T7 promoter/
cloning region

Pst I

Amp$^r$

pET-15b

lacI

Ori

Bsm I

**Figure 7**   Map of the plasmid vector pET-15b.

### 7.18.4.1  Choosing a Source of Template for PCR

Although the PCR can be used to amplify and detect DNA from miniscule samples (hence its popularity in forensic analysis), template DNA is preferably obtained from a higher purity sample than the typical bloodstain. This is primarily due to the fact that higher purity starting samples increase the probability that the correct sequence will be amplified (or, to put it another way, the fewer types of DNA sequence that are present, the fewer opportunities exist for amplification of undesired sequences due to mispriming). High quality DNA templates for PCR are available from several sources; most commonly, sequences are amplified from a cDNA library, or subcloned from another plasmid.

A cDNA (or "cloned DNA") library is generated for a particular cell type by first isolating all of the cell's messenger RNA (mRNA), and preparing DNA clones of it.[91,92] Therefore, in principle a cDNA library should contain DNA corresponding to all of the different proteins that are expressed in a cell. Such libraries are available from a variety of manufacturers. Subcloning DNA from another plasmid (i.e., cloning a smaller portion of a previously prepared plasmid) is perhaps the simplest method of obtaining template DNA. In all cases, circular DNA plasmids should be cleaved prior to their use in PCR, since the supercoiling of plasmid DNA can reduce the efficiency of primer annealing and extension.

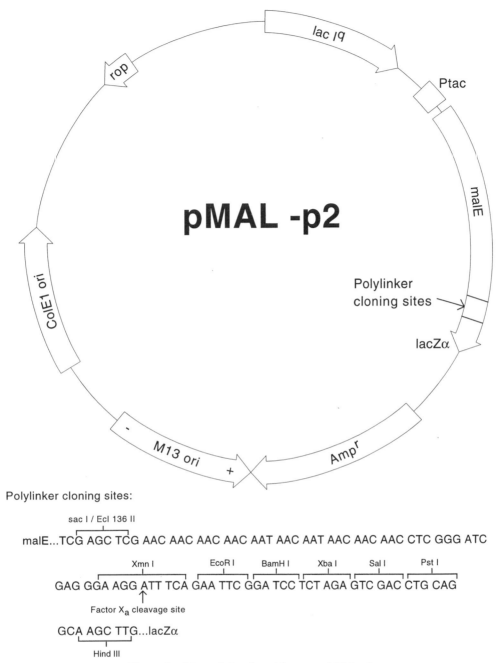

**Figure 8** Map of the plasmid vector pMAL-p2.

### 7.18.4.2 Choosing Primers for PCR

The careful design of oligonucleotide primers is equally critical for successful PCR as the choice of template (Gu[93]; while primarily discussion of *in situ* PCR methods, general descriptions of primer design are included). Two general strategies may be considered (Figure 11). One is the use of a single set of primers to target the segment of interest directly. This is obviously the simplest method, and is recommended when a relatively large amount of template is available. If considerably less template is available, or if that template is part of a diverse mixture of oligonucleotides such as a cDNA library, a second strategy may be considered if significant amounts of product are not obtained from the simple, single-stage PCR. In this multistage (or "nested PCR") strategy,[94] an initial set of primers is designed to amplify a segment of the template larger than the desired segment. The product of this PCR is then used as the template for a second PCR, which incorporates a second set of primers designed to amplify the segment of interest from this larger piece. In such

Double-stranded DNA template

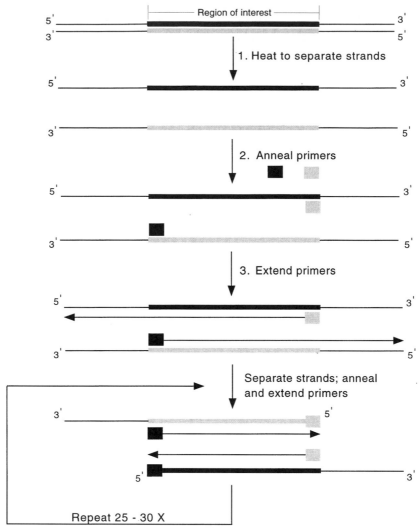

**Figure 9**   The polymerase chain reaction (PCR).

cases amplification of a segment that is slightly larger than the segment of interest, followed by amplification of the segment of interest from the larger piece, should provide sufficient quantities of the desired DNA for cloning.

Regardless of the chosen strategy, the first thing that must be addressed in the design of primers is that they complement the region of the template just outside the region that is going to be amplified. Although the specific sequence length for primers is highly variable from experiment to experiment, typically 18 to 40 bases are sufficient to ensure selectivity and stability. Second, the primers must be designed so as to provide a PCR product with 5′ and 3′ ends suitable for ligation into the vector. Frequently, this means incorporating several nucleotides corresponding to the restriction sites that are to be used in subsequent steps of the cloning process. Third, the downstream primer (corresponding to the nucleotides which will eventually encode the carboxyl terminus of the polypeptide) should contain three nucleotides corresponding to a "stop" codon prior to the restriction site, in order to prevent the transcription machinery of the eventual host of the construct from continuing to synthesize protein past the desired carboxyl terminus. Fourth, primers must be designed such that the gene incorporated into the final construct is in the correct "reading frame"; that is, the three nucleotides which encode the protein of interest must immediately follow (or be a multiple of three nucleotides away from) the final three nucleotides encoding the protein to which the material of interest will be fused. Finally, primers should be designed with an overhang following the restriction site in order to give the restriction enzyme a few bases to "grab" onto. Typically, a

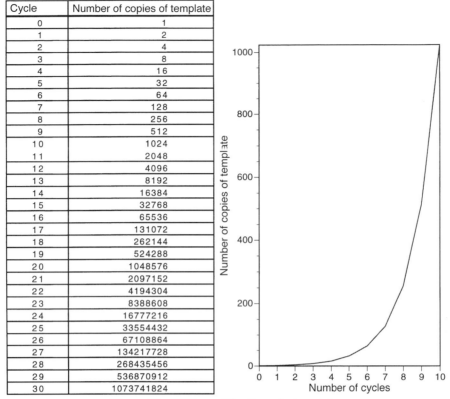

| Cycle | Number of copies of template |
|-------|------------------------------|
| 0 | 1 |
| 1 | 2 |
| 2 | 4 |
| 3 | 8 |
| 4 | 16 |
| 5 | 32 |
| 6 | 64 |
| 7 | 128 |
| 8 | 256 |
| 9 | 512 |
| 10 | 1024 |
| 11 | 2048 |
| 12 | 4096 |
| 13 | 8192 |
| 14 | 16384 |
| 15 | 32768 |
| 16 | 65536 |
| 17 | 131072 |
| 18 | 262144 |
| 19 | 524288 |
| 20 | 1048576 |
| 21 | 2097152 |
| 22 | 4194304 |
| 23 | 8388608 |
| 24 | 16777216 |
| 25 | 33554432 |
| 26 | 67108864 |
| 27 | 134217728 |
| 28 | 268435456 |
| 29 | 536870912 |
| 30 | 1073741824 |

**Figure 10** PCR amplification of a template.

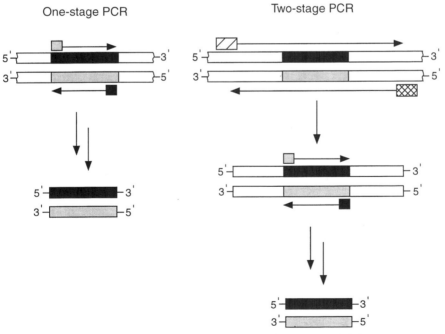

**Figure 11** Single (one-stage) vs. "nested" (two-stage) PCR.

stretch of two to six T nucleotides will suffice. One should be aware, however, that too long an overhang could result in the primer being pulled off the template. Three additional factors should be evaluated as primers are designed: the "primability" of the primer–template duplex the stability of the primer–template duplex, and the probability that the primers will self-bind (known as "primer dimer" formation). Several computer programs are available that can assist in this process. One that the authors have used extensively is Amplify 1.0.[95]

A third option, which completely obviates the need to obtain a full length template, is to use the "total gene synthesis" method. In this method, the gene is synthesized as a series of overlapping fragments (Figure 12).[96] These fragments then serve as both template and primer, and are brought together in a single PCR.

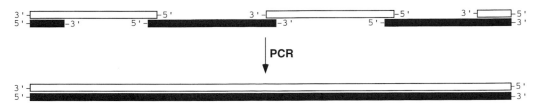

**Figure 12**   Total gene synthesis. Overlapping oligonucleotides (top) are subjected to PCR, providing a full length gene (bottom).

Primers are available by contract synthesis from numerous commercial vendors (in addition to in-house oligonucleotide synthesis facilities, which are becoming quite common). These vendors can generally provide a sample of desalted primer within a few days.

### 7.18.4.3   The Polymerase

Another key component of the PCR reaction is the polymerase enzyme. There are many polymerases to choose from, each with its own advantages and disadvantages, so making a choice of polymerase is frequently a process of trial and error. Two of the most commonly used polymerases are Taq DNA polymerase and its variants, from the bacterium *Thermus aquaticus*,[97,98] and Pfu DNA polymerase from the marine archaebacterium *Pyrococcus furiosus*. As a consequence of their origins in thermophilic bacteria, both of these enzymes are stable at the high temperatures that are necessary for the denaturing step of the PCR reaction. An advantage of Taq DNA polymerase over Pfu is that Taq has a more than fivefold greater nucleotide incorporation rate than Pfu. On the other hand, the error rate for Taq polymerase is on the order of $10^{-5}$ (i.e., one misincorporated nucleotide per 10 000 bases),[99,100] while Pfu DNA polymerase has the lowest rate of misincorporation of nucleotides of all thermostable DNA polymerases, due to its 3′ to 5′ proof-reading ability.[101–105] Therefore, it is often advantageous to carry out the PCR in parallel with both enzymes. In addition, there are a number of variations of Taq DNA polymerase, identified here by their trade names. A few of these include AmplitaqGold,[106] TaqStart antibody, which blocks the polymerase activity during PCR setup so as to eliminate primer dimer formation and production of nonspecific products,[107] and LA Taq, used for long range amplification.

Other, nonthermostable DNA polymerase enzymes, such as *E. coli* DNA polymerase and Klenow fragment DNA polymerase, are not appropriate for the PCR, since thermal cycling causes them to lose activity relatively quickly. Both of these enzymes were used successfully for PCR prior to the discovery of the more stable thermophilic enzymes.[108] However, they can be useful in other contexts. A mixture of Taq and Pfu is also available from at least one manufacturer.[106] In principle, this combines the speed of Taq-based PCR reactions with the proof-reading ability of Pfu. This should be particularly useful for cloning large stretches of DNA (i.e., the manufacturer suggests that the Taq–Pfu mixture may be used for amplification of up to 35 kb).[109]

### 7.18.4.4   Was the PCR Successful?

Determination of the success of the PCR is accomplished by running agarose gel electrophoresis on the product mixture (the specific details are not discussed here; for a general protocol see Southern).[110] Following completion of the desired number of cycles of PCR, a few microliters of the reaction mixture are loaded onto a 1–2% agarose gel alongside a DNA marker. Ethidium bromide (**3**) is added to the gel during casting and to the running buffer. Since ethidium bromide is an efficient DNA intercalator, it inserts itself between the base pairs of the DNA during electrophoresis. This provides a bright fluorescence when the gel is viewed under UV light, indicating the position and number of bands on the gel. While the position of a band on the gel indicates whether a product of roughly the correct number of nucleotides has been obtained, the intensity of the band should give some indication of the extent of amplification: the more intense the band, the more amplification. An important precaution to note is that shining short wavelength UV light (usually 254 nm) on DNA for extended periods of time can cause interstrand cross-linking of the DNA, thereby rendering it useless for further steps in the cloning process. As a result, all agarose gels should be viewed using a longer wavelength light source, or exposure to the 254 nm light should be kept to a minimum. A successful PCR will look somewhat like that shown in Figure 13: a single, clearly defined band (the PCR product) is evident midway down the gel, along with more diffuse bands at the bottom, corresponding to unreacted primers.

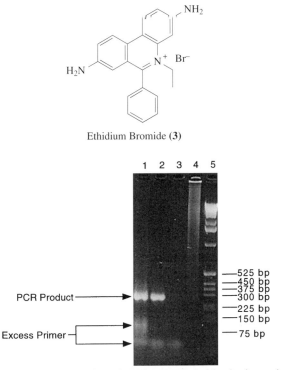

Ethidium Bromide (**3**)

**Figure 13** 1.5% Agarose gel in a PCR reaction. A 75 bp DNA ladder is shown in the lane to the far right, lane 5. Lane 1 shows a fairly sharp, intense band for the desired PCR product around 300 bp, along with diffuse bands representing excess primer. Lane 2 shows an equally intense but sharper band representing the desired PCR product, with less diffusion of the primer band. Lane 3 shows diffuse primer band but no band representing PCR product, and lane 4 represents a failed reaction.

### 7.18.4.5 Troubleshooting the PCR

In an operational sense, PCR is not a particularly complicated experiment. However, there are many variables involved, and occasionally agarose gel electrophoresis of the product mixture will reveal that little or no amplification of the desired fragment was obtained. In such instances it becomes necessary to troubleshoot the experiment.

The first factors that should be given consideration are the temperatures at which each segment of the PCR protocol is carried out and the amount of time allowed for each period of the PCR. For example, possible reasons for insufficient amplification may be that the extension time is too short, the temperature at which extension is carried out is too high, the annealing temperature is too high, or similar problems. It can be expected that different templates and different primers may require slight variations in temperature and time of reaction. If changing the reaction conditions does not seem to help with the overall yield of the PCR it may be necessary to try different primers. For reasons that are not obvious, the primers may not be binding well with template, thereby shutting down the reaction. The length of sequence complemented by the primer may be increased; alternatively, if the precise starting and ending points for the polypeptide are not critical, new primers can be designed to target a slightly different gene segment. Finally, another possibility is to try using a different polymerase.

### 7.18.4.6  Purification of PCR Products

Following completion of the PCR, purification of the product segment is carried out by agarose gel electrophoresis as described above (purification of the DNA fragment and determination of the level of success of the reaction can be carried out simultaneously). Following electrophoresis, it is necessary to separate the DNA from the agarose itself, as well as remove any ethidium bromide used to visualize the DNA, since both of these can interfere with subsequent steps in the cloning process. This may be done in one of several ways. First, several manufacturers supply kits, which adsorb the DNA to a solid support, allowing dissolved agarose and other contaminants to be washed through. These kits have a distinct advantage in terms of simplicity; however, less expensive methods are available using common laboratory materials that are equally efficient. A relatively simple method of isolating and purifying DNA fragments, which does not require the purchase of a kit, is to spin the agarose through silanized glass wool. Silanized glass wool is prepared by soaking glass wool in a solution of 5% dichlorodimethylsilane in hexane for 24 h. After disposing of the silanizing solution, the glass wool should be allowed to stand uncovered in a fume hood for 24 h, then dried thoroughly in a laboratory oven. Alternatively, a gel slice may be squeezed between sheets of plastic film. Regardless of the method of retrieving the DNA fragment from the agarose gel, further purification by phenol/chloroform extraction, followed by ethanol precipitation, must be carried out.

### 7.18.5  RESTRICTION DIGESTS

After isolating the PCR product and obtaining significant quantities of the appropriate expression vector, the next step is to incorporate the PCR product into the circular DNA plasmid. Prior to doing this, a space must be inserted into the vector (i.e., it must be converted from a circular plasmid to a linear form), and the ends of the insert (the PCR product) must be converted to a form that is complementary to the ends of the cleaved vector. To do this, both vector and plasmid are cleaved with restriction endonucleases. As described above, the choice of which restriction enzymes to use is actually made early in the process of designing the cloning protocol; PCR primers are chosen to match equivalent restriction sequences within the multicloning site of the vector. Restriction digests provide the complementary sequences of DNA ("sticky ends" or "blunt ends") which allow proper matching between insert and vector during ligation. Restriction digests can be carried out as either double restriction digests, in which both enzymes simultaneously cut the DNA, or a series of single digests in which the DNA is cut at one site by the first enzyme, and then at the other site by a second enzyme. Regardless of how the restriction is performed, the vector must contain the same restriction sites that were incorporated into the primer sequence so that the ends of the restricted DNA are compatible and can subsequently be ligated together.

Common restriction enzymes can be obtained from a variety of manufacturers and are typically supplied with solutions of their required buffer and cofactors (if any). The amount of enzyme needed for a restriction, as well as the duration of the reaction, varies significantly depending both on the amount of DNA, the type of enzyme being used, and whether one is attempting a single- or double-restriction. Therefore, it is prudent to refer to instructions packaged with the particular enzyme from a manufacturer with regard to setup and execution of the restriction digest. There are various charts and tables in the literature[111,112] as well as those available directly from the manufacturer that can provide such information. However, there are several experimental parameters which most restriction digests have in common. First, digests are typically carried out at 37 °C. Reaction times can be anywhere from 1 h to overnight, depending on the level of activity of the specific enzyme used. The reaction can either be allowed to continue until the enzyme loses activity, or alternatively, the enzyme may be inactivated by heating the reaction mixture at 60 °C for 20 min. The completeness and specificity of a restriction digest can be determined by agarose gel electrophoresis, using unrestricted DNA as a control sample. Gels should be examined for sharp, well-defined bands; the presence of diffuse bands can be an indication of incomplete restriction digest, while multiple bands can be an indication of nonspecific cleavage of the oligonucleotide, known as "star activity".[113] If the former is observed, it is possible to continue the digest directly within the agarose. (Although, to the author's knowledge, a procedure for agarose embedded restriction digest has not been published in a peer-reviewed journal, a protocol is described in the current New England Biolabs catalogue)[114] Star activity may be avoided in some cases by reducing the reaction time, or by altering the buffer composition. As with the PCR, gel electrophoresis may be used both to examine the completeness of the reaction and as a first step in purification of the oligonucleotide away from the restriction enzyme(s).

### 7.18.6 DEPHOSPHORYLATION

Restriction digest creates free phosphate groups on the 5′ ends of the DNA. Regardless of whether a single or double restriction digest is done, the 5′ phosphate groups of the vector must be removed, if restriction digest provides identical termini, in order to prevent self-ligation (i.e., recyclization) of the vector from taking place during the subsequent ligation step. If the ends of the restricted DNA are not identical, dephosphorylation is not necessary, since the noncomplementary ends of the vector should not be able to ligate one another. Two enzymes that are commonly used for dephosphorylation reactions are calf intestinal alkaline phosphatase (CIP)[115] or bacterial alkaline phosphatase (BAP);[116,117] in both cases, the reaction is typically allowed to proceed for 30 min at 37 °C. Following dephosphorylation, complete inactivation of the enzyme is essential if the subsequent ligation reaction is to work efficiently. BAP is more active than CIP and as a result is much more resistant to heat and detergent inactivation. CIP can either be inactivated by proteinase K or by heating to 65 °C for 1 h or 75 °C for 10 min. Gel purification followed by DNA precipitation is once again required following dephosphorylation.

### 7.18.7 LIGATION

In order for a piece of DNA to be inserted into an expression system such as *E. coli*, it must be in circular form; linear DNA cannot be inserted into a cell readily, unless packaged as a phage virus, and in any case will not replicate. This means that the amplified DNA segment must be chemically inserted, or ligated into the vector system. Following ligation, the insert/vector construct can then be used to "transfect" the cell.

A prerequisite to ligation is that the DNA sequence that is to be cloned as well as the vector DNA should be cut with the same enzymes. Two types of ligation are then possible: blunt-end ligation (Figure 14) and cohesive-end ligation (Figure 15). A cohesive-end ligation is one in which the ends of the DNA that are to be ligated have been restricted such that the ends remain staggered. A blunt-end ligation is one in which the ends of DNA that are to be ligated have been restricted such that the ends remain flush. Cohesive-end ligations tend to be more facile than blunt-end ligations, which are often inefficient reactions requiring very high concentrations of ligase and blunt-ended DNA.[118]

A ligation reaction is in essence a bimolecular reaction of insert and plasmid (or a termolecular reaction, if the ligase is counted), followed by a macrocyclization. As is the case with small molecule bimolecular and macrocyclization reactions, specific concentrations of reagents (insert and vector) will improve the odds of success. Commonly used vector:insert ratios that seem to work well are 1:1 or 1:3.[119] In order to achieve such ratios, the amount of vector and insert must be quantitated just prior to ligation.

Two commonly used methods for the quantitation of DNA are UV spectroscopy and agarose gel electrophoresis utilizing ethidium bromide fluorescence. With UV spectroscopy the concentration of DNA in the sample can be obtained by measuring the absorbance at 260 nm ($OD_{260}$); an $OD_{260}$ of 1.0 represents about 50 $\mu$g ml$^{-1}$ of double-stranded DNA.[120] For example, a 500 $\mu$l sample with an $OD_{260}$ of 0.043 contains ~1.075 $\mu$g of DNA. An important experimental caveat is that quantitation of DNA by UV absorbance will only result in an accurate measurement if the sample is pure. Both ethidium bromide (left over from agarose gel electrophoresis) and salts from buffers can compromise an absorbance measurement (Figure 16), so purification of the DNA by phenol:chloroform extraction followed by ethanol precipitation is crucial.

A second method that is commonly used for the quantitation of DNA involves ethidium bromide fluorescence, as observed in agarose gel electrophoresis. Since the intensity of fluorescence is proportional to the amount of DNA present on the gel, running the DNA sample on an agarose gel alongside a series of standards allows comparison of fluorescence intensities and estimation of the quantity of DNA in the sample. This method of quantitation is more useful when there are very small amounts of DNA present or when the DNA cannot be purified away from substances that also absorb UV light.[120]

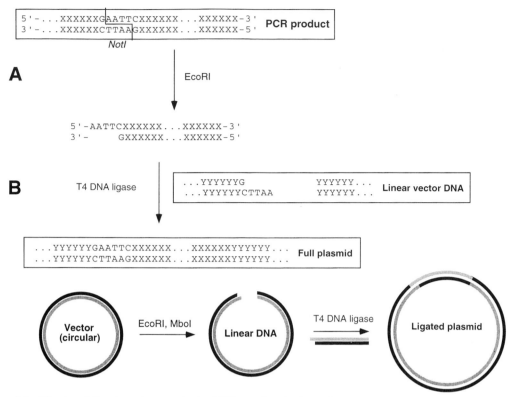

**Figure 14**   Blunt-end ligation. At step A, the PCR product is cleaved with one restriction enzyme, producing one "sticky" end. At step B, this processed PCR product is added with T4 DNA ligase to an appropriately restricted vector; this produces a ligated, recircularized plasmid.

A common enzyme used to perform ligation is T4 DNA ligase. Ligation reaction conditions vary depending on the type of ligation being performed; in general, blunt-end ligations are run for 4–18 h at 15–20 °C, while cohesive-end ligations are allowed to proceed for 3 h at ambient temperature, or overnight at 4–8 °C.[119] Evaluation of the level of success of a ligation experiment is only realistically possible by transformation (introducing the DNA into cells). The amounts of material involved are generally too small to be readily observable on agarose gel electrophoresis. Ligation reactions are typically carried directly to transformation experiments, without purification.

### 7.18.8   TRANSFORMATION OF DNA INTO *E. COLI*

Transformation is the process of inserting recombinant DNA into an expression system such as *E. coli*. Once inserted into a cell, replication of the cell provides an infinitely renewable source of the DNA construct. Alternatively, the DNA can be induced to undergo transcription to give the corresponding RNA, followed by translation to provide the protein.

A requirement of transformation is that the cells be "competent", which means that the cells are in a state such that the uptake of DNA is possible. Although the precise mechanism by which DNA enters a cell is at this point unknown, competency involves changes in the porosity and electrostatic characteristics of the cell membrane that allow for DNA to enter the cell. There are different levels of competency and the more competent the cells, the better the transformation efficiency. Competent cells may be prepared using a variety of experimental protocols, or purchased from a commercial vendor. Competent cells purchased from commercial vendors are of high quality and often provide transformation efficiencies of greater than $10^8$ colonies $\mu g^{-1}$ of supercoiled plasmid DNA. A downfall of purchasing competent cells, however, is that they are very expensive. Protocols for preparing competent cells in the laboratory fall into two general categories: those used to produce cells capable of undergoing chemical transformation ("chemically competent" cells), and those producing cells suitable for electroporation ("electrocompetent" cells). Both preparations involve washing a culture of cells with buffer containing rubidium chloride or calcium chloride. Self-

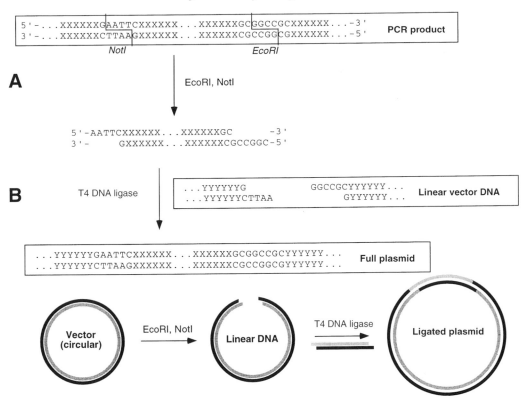

**Figure 15** Sticky-end ligation. The PCR product is digested with two restriction endonucleases at step A, producing two noncomplementary "sticky" ends. As in Figure 14, this product is then mixed in step B with T4 DNA ligase and an appropriately restricted vector, to provide the ligated plasmid.

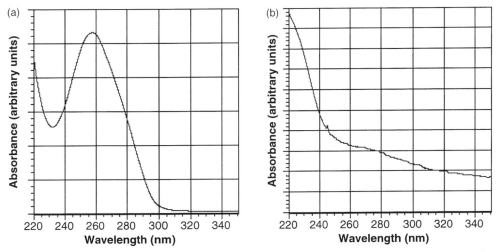

**Figure 16** UV traces of plasmid DNA. Sample (a) is a pure sample, showing the expected overall shape and absorbance maximum at 260 nm: Sample (b) is contaminated with ethidium bromide, making it difficult to determine the actual concentration of DNA present.

preparation of competent cells can often give transformation efficiencies of $10^7$ colonies $\mu g^{-1}$ of supercoiled plasmid DNA. The competency of cells can be tested by transforming with a known amount of (unrestricted) plasmid DNA.

Once the competent cells have been obtained, transformation can take place. Transformations can be carried out either by chemical methods or by electroporation. Chemical methods involve subjecting the cell/DNA mixture to several cycles of icing and heating. Electroporation involves the application of a burst of a strong electrical current to the cells in the presence of the DNA plasmid to be transformed, inducing a rapid transitory change in the voltage of the cell membrane that allows for an influx of DNA. This is a significantly more efficient (10- to 20-fold) method of transformation than chemical methods.[121]

Regardless of the method of transformation, the next step in the cloning process is to grow the cells overnight on an agar media containing a selection agent that will allow for the growth of only those cells containing the complete vector–insert construct. As described previously, all vectors for bacterial transformation encode some sort of antibiotic resistance gene; for ampicillin resistance, this is a $\beta$-lactamase, while for chloramphenicol the gene is chloramphenicol acetyl transferase (CAT). Those cells on the agar plate that do not carry the construct will therefore have no resistance to the antibiotic, and will die. Of course, a very small fraction of *E. coli* will be naturally antibiotic resistant; furthermore, the antibiotic itself is subject to thermal degradation. Therefore, it is important to note that cells appearing on plates following more than 24 h of incubation at 37 °C are less likely to carry the plasmid of interest.

### 7.18.9   SCREENING FOR INSERT

After 18–24 h of incubation, several bacterial colonies should be visible on the agar plate. These colonies could comprise cells containing the full length recombinant DNA (the desired result), cells carrying vector (lacking insert) that was not restricted completely in the original digest and religated, or antibiotic-resistant bacteria. Differentiating the desired recombinants from the set of "background" colonies constitutes the next task in a cloning project. So the question now is which colonies represent cells containing the recombinant DNA, the DNA of interest.

The screening of colonies for insert is accomplished first by "picking" each colony with a sterile pipette tip, and culturing the cells overnight at 37 °C in 1–3 ml Luria–Bertani (LB) media containing the selection agent. After pelleting the resultant cells in a microcentrifuge, a purification step is performed to isolate the DNA from other cellular components. Several DNA purification kits are available from commercial vendors (generally described as plasmid miniprep kits), and are to be recommended due to their simplicity and reliability. Following purification of the plasmid DNA, agarose gel electrophoresis of the putative recombinant plasmids next to a control lane of the initial vector should indicate whether an insert is present.[122] Alternatively, double restriction digest of the DNA using the same two enzymes initially used in the cloning, followed by agarose gel electrophoresis, should reveal bands for both insert and vector.

Several control experiments can be run to indicate how many background colonies should be expected. Frequently, the presence of a large number of background colonies indicates that restriction digest of the plasmid prior to ligation was inefficient; in the authors' experience, this is particularly likely to be problematic in double restriction digests involving removal of a short segment of the multicloning site from the plasmid. In order to test this, a control ligation of doubly restricted plasmid (without insert) and subsequent transformation will provide an indication of the efficiency of the restriction digest and the ability of the plasmid to religate. Alternatively, a restriction digest may be performed following ligation at a site which should have been removed in the initial double digest. Provided the insert does not contain this restriction site, those vectors that took up the insert will not be cleaved in this reaction.

### 7.18.10   VERIFICATION OF INSERT IDENTITY

After it has been determined that a particular bacterial colony indeed carries the full recombinant plasmid, DNA sequencing should be carried out to verify the identity of the insert. The most common method used is Sanger sequencing (Figure 17).[123,124] Since the sequencing reaction is analogous to a PCR reaction (differing primarily in the fact that only a single primer is used, combined with low concentrations of dideoxynucleotides (4)–(7) in order to provide for random termination of the polymerization reaction), it can be readily performed assuming that there is access to a PCR thermal cycler (and the necessary radioisotope licenses). Alternatively, a number of service facilities (commercial or academic) are in existence. For sequencing of relatively short oligonucleotides, an alternative method known as Maxam–Gilbert sequencing is preferred.[125,126]

Unlike Sanger sequencing, which involves DNA *synthesis*, Maxam–Gilbert sequencing relies on base-specific *degradation* reactions of the original DNA.

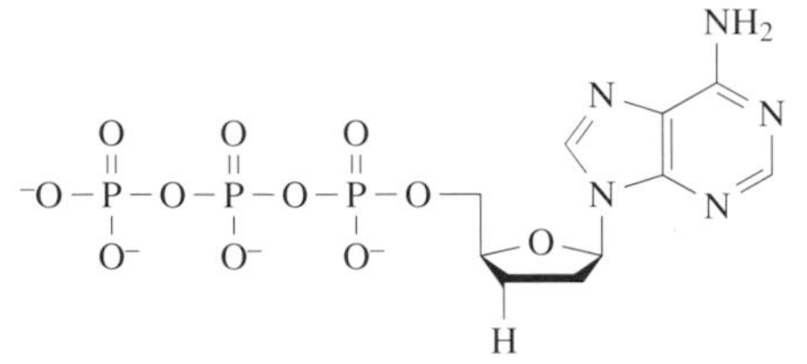

ddATP (4)

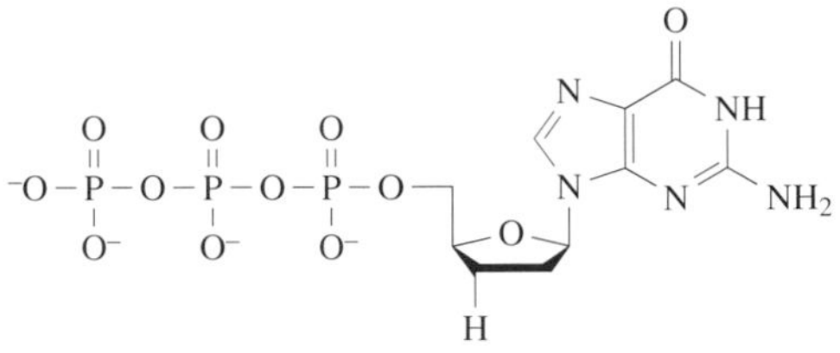

ddGTP (5)

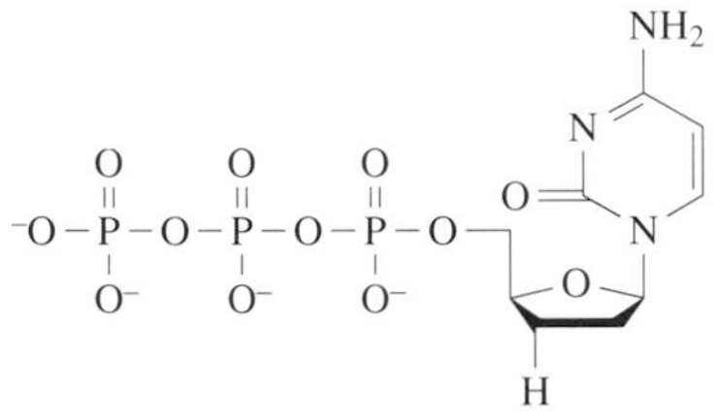

ddCTP (6)

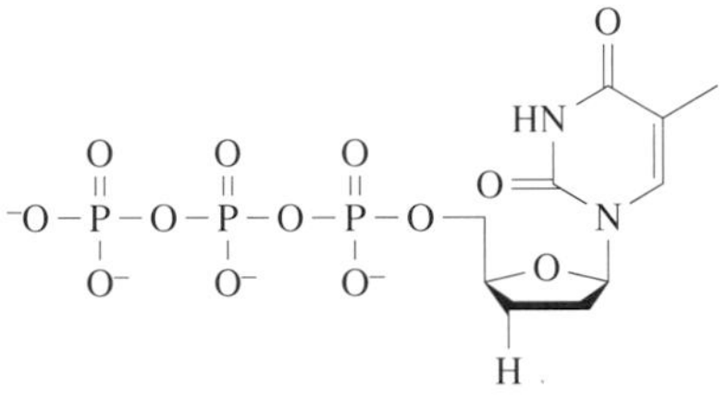

ddTTP (7)

## 7.18.11  LARGE SCALE PRODUCTION OF PROTEIN IN *E. COLI* CULTURE

Once a particular transformant has been selected and the sequence of the insert verified, full scale production of protein may begin. A typical "production scale" protocol for *E. coli* culture is as follows. First, a colony from an agarose plate of the transformant is picked and grown up overnight at 37°C with agitation (12 h) in 10 ml LB-ampicillin (or LB-chloramphenicol). This is then added to a 1 L sterile solution of LB-ampicillin and agitated at 37°C. Cells are grown to an $OD_{600}$ of 0.6 and then transfused (i.e., induced to produce the fusion protein) by the addition of IPTG. IPTG activates the lacZ promoter, which causes essentially all of the cellular machinery of the bacterium to be devoted to production of the fusion protein encoded by the plasmid. Other additives such as diethyl maleate or sodium arsenate may also occasionally be added to increase expression levels.[127] Following addition of IPTG, cells are typically allowed to continue to agitate at 37°C for 6–10 h. As with virtually all of the techniques discussed thus far, the specific amount of time allowed for large scale culture growth is highly variable; the desire to express as much protein as possible from a culture must be balanced against the potential for protein degradation and rise (and predominance) of antibiotic-resistant, nonoverexpressing strains of bacteria. Cultures grown in minimal media (see below) may take substantially longer to grow. In either case, production of protein is typically monitored by removing 1 ml aliquots of cells from the culture at various time points (1–2 h intervals); these are then lysed and subjected to denaturing sodium dodecylsulfate polyacrylamide gel electrophoresis (SDS–PAGE).

### 7.18.11.1  SDS–PAGE Analysis of Protein Expression Levels

A rapid assessment of the level of success for any protein expression experiment may be made by growing small (1–2 ml) cultures of cells. These cultures should be started by diluting 50–100 $\mu$l of an overnight culture; induction and incubation of the test culture should proceed as with a larger culture. It is particularly useful to test several different concentrations of IPTG (or other inducing agent) in order to determine the optimal concentration. After the culture period is complete, samples are centrifuged in microcentrifuge tubes and the supernatant poured off.

To analyze the contents of these pelleted cells, SDS–PAGE is then performed.[128] Conceptually similar to the agarose gel electrophoresis performed on DNA, SDS–PAGE uses a thin (1–2 mm) gel composed of polymerized acrylamide. Typically 30% acrylamide/1% polyacrylamide solutions in a Tris buffer containing the detergent SDS are polymerized using ammonium persulfate. Since the mobility of a protein during electrophoresis is dependent both on its molecular weight and its conformation, proteins must be denatured prior to SDS–PAGE analysis so that their migration is

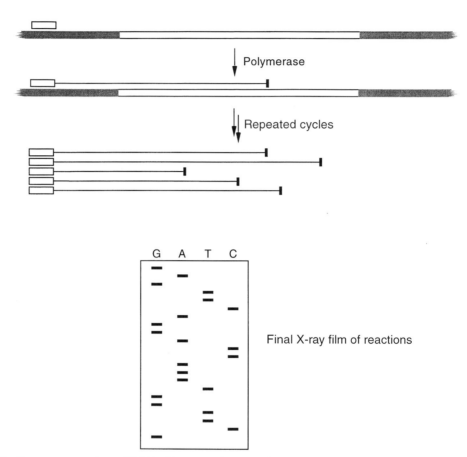

**Figure 17**   Sanger sequencing of DNA. Four reaction vessels containing the indicated components are subjected to cycles of heating and cooling in a protocol analogous to the PCR. Since each vessel contains a small proportion of a dideoxyoligonucleotide, this results in chain termination at random positions along the sequence of interest. Running the resultant reaction mixtures out on a sequencing (polyacrylamide) gel and subsequent exposure of the gel to X-ray film allows the sequence to be visualized.

dependent solely on the molecular weight. This is done by resuspending the pelleted cells in a "loading buffer" containing SDS and dithiothreitol (to reduce disulfide bonds), and a marker dye, and then boiling the solution for 60 s. The samples are then loaded on the gel and electrophoresed in parallel with a standard solution. Visualization of bands on the gel corresponding to various proteins is accomplished by staining with Coomassie blue.[129,130] Other more sensitive methods of staining (such as silver staining[131]) are available, but the need for such methods is generally indicative of a low level of expression. Clear differences in the amount of expressed protein should be observable between cultures induced with IPTG and those without induction.

### 7.18.11.2   Isolation and Purification of the Fusion Protein from a Large Scale Culture

A comprehensive source of information and experimental protocols for protein purification is presented by Deutscher.[132] After cooling the culture solution to room temperature, the cells should be centrifuged; typically, 30 min at 5000–16 000 $\times$ *g* is sufficient. The supernatant (growth medium) is then poured off (caution: this still may contain live *E. coli*, and should be treated as a biohazard; sterilize with bleach prior to disposal). The cellular mass remaining is then resuspended in a lysis buffer (frequently phosphate-buffered saline (PBS), pH 7.4). The precise selection of lysis buffer will vary somewhat depending on the specific nature of the protein; for example, detergent may be added to the buffer solution if solubility of the fusion protein is a concern. Regardless of the composition of the buffer, it is generally desirable that cells be resuspended in as little of the liquid as possible (for example, 10–15 ml of resuspension buffer per liter of culture is usually sufficient). The reason for this is primarily one of purification; since this resuspension volume will typically be loaded on some sort of column for purification, a requirement will be to minimize the volume, much as in loading a flash (silica) column.

Several methods are available for lysing the resuspended cells. One of the simplest is to add a few milligrams of lysozyme to the cellular suspension, then freeze the mixture at $-20\,^{\circ}$C.[133] On thawing, lysis of the cells should be complete. Alternatively, cells may be put through a french press,[134] or lysed using sonication or glass beads plus vortexing. Once cell lysis is complete, the mixture is ready for purification.

Isolation and purification of the expressed protein from the rest of the cellular constituents is dependent on the particular expression system used. If, as is generally the case, a fusion protein was produced which allows for affinity purification, the particular affinity protocol is followed for that protein. For example, purification of GST fusions involves loading the lysed material onto a glutathione–agarose affinity column, washing the column with several volumes of buffer to remove those materials which do not bind to the affinity matrix, then eluting the fusion protein with buffer containing glutathione. Proteins or polypeptides expressed without fused affinity domains must, obviously, be purified by other methods.

### 7.18.11.3 Cleavage of the Fusion Protein

If the purified fusion protein is not suitable "as is" for further experimentation, it must be cleaved and purified further. As previously discussed, the specific protease to be used for cleavage is determined both by the nature of the construct and by the design of the fusion. Much like restriction digests, the specific reagents used and composition of the reaction mixture for proteolytic cleavage will be determined by the specific enzyme used. A trial run is generally helpful in determining reaction time. Aliquots can be removed from the reaction mixture at specific time points, and run on SDS–PAGE to determine the extent of cleavage and the presence of any products of nonspecific cleavage or further degradation of the cleaved material.

### 7.18.11.4 Purification of Desired Polypeptide Segment

Purification of the final polypeptide segment away from the previously fused domain used in affinity purification is, in most cases, a repetition of the procedure for purification of the fusion protein. However, at this point the fraction which is saved is that which does not bind to the affinity column, rather than the one that does. Typically, these affinity procedures provide polypeptide of reasonable (but not ideal) purity. Frequently, it is also necessary to subject the partially pure polypeptide to further chromatographic purification, particularly if the material is to be used for NMR analysis. One often-used method is to purify the polypeptide by ion exchange chromatography, or size exclusion chromatography (examples are given in Section 7.18.12). The purity of the polypeptide can be assessed at each stage by SDS–PAGE analysis.

### 7.18.12 CASE STUDIES IN CLONING

Thus far, the various steps in cloning have been discussed as somewhat isolated, disconnected events; this has been useful as a way to describe the various options that are available during cloning. In an effort to clear up any confusion that may have resulted from this method of presentation and to show how cloning projects may be developed from beginning to end, two sets of cloning experiments have been described, carried out by chemists alone or in collaboration with molecular biologists. In addition to the intrinsically interesting nature of these studies (in one case, cloning as a way to examine DNA binding specificity; in the other, cloning as a prerequisite of structural analysis, ligand design, and combinatorial library screening), these cases serve as proof that cloning is, indeed, possible for chemists.

### 7.18.12.1   Case Study: Cloning and Derivatization of SH3 Domains for Ligand Binding Analysis and NMR Structural Studies

A profoundly active area of research for both chemists and biologists has been the study of the various small molecule–protein and protein–protein interactions that are involved in the transmission of, and reactions to, extracellular stimuli. The collective set of these interactions, termed signal transduction pathways,[135] are frequently extraordinarily complex, often interact with one another, and involve individual proteins that may have molecular weights of 200 000 or more. Crucial to the understanding of signal transduction pathways was the discovery that many (if not most) of the proteins involved are made up of component domains of well-defined structure that have individual functions that contribute to the overall function of the protein. These domains were identified in part by the analysis of sequence homologies and by the cloning of various subdomains of prototypical signal transduction proteins.

One of the most common of these protein subdomains is the *Src* (pronounced sark) *Homology 3* (SH3) domain.[136–139] The SH3 domain was initially identified after comparisons were made between the amino acid sequence of Src (a tyrosine kinase) and the proteins myosin IB, phospholipase C-$\gamma$, Crk, $\alpha$-spectrin, and several yeast proteins.[140–142] At the outset, the function of the SH3 domain in Src itself or of any of the homologous SH3 domains in other proteins was unclear.[143] As a means of discovering the function of the SH3 domain at a structural level, several groups began to clone these subdomains.

Discussion here will focus primarily on the efforts of Schreiber and co-workers, and their work with the SH3 domains of Src and the p85 subunit of phosphotidylinositol-3-kinase (PI3K). These examples are particularly illustrative, because cloning of this domain was carried out in order to provide material for several purposes, including NMR structural analysis and combinatorial library evaluation. Results obtained with this domain were subsequently extended to the SH3 domains of Src and Hck, again for use in structural studies[144,145] and ligand generation via combinatorial library synthesis.[146–148]

The cloning of the SH3 domain from the p85 subunit of PI3K incorporated a slightly different strategy from those discussed previously, in that no fusion protein was employed. Instead, Schreiber and co-workers utilized the "expression cassette" PCR strategy (ECPCR, Figure 18), in which the insert incorporates a ribosome binding site and translational spacer prior to the start codon for the desired sequence.[149,150] A multistage cloning strategy was employed (Figure 19).[151–153]

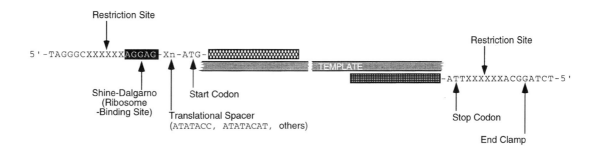

Following PCR:

**Figure 18**   Expression cassette PCR.

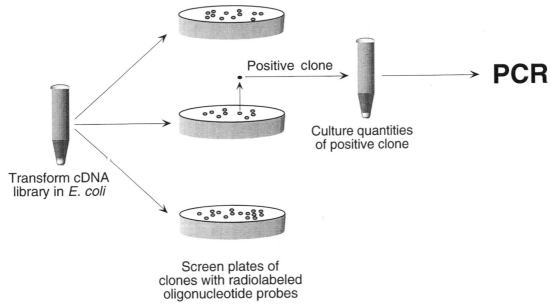

**Figure 19** Strategy used by Yu and co-workers[151-153] in the cloning of the SH3 domain from PI3K.

First, *E. coli* cells were transformed with a cDNA library in the λgt10 vector constructed from human endothelial cells and plated out on culture dishes containing standard growth medium. This allowed for replication of the entire cDNA library to be carried out and separation of each individual gene to occur, since individual bacterial plaques carry only one vector or foreign gene sequence. Direct PCR amplification of sequences from cDNA libraries is also possible.[154] After bacterial plaques had been allowed to grow, nitrocellulose filters were laid on each dish, to create an image of the colonies that had grown there. Each nitrocellulose filter was then hybridized[155] with radioactive ($^{32}$P-labeled) oligonucleotides corresponding to a portion of the PI3K sequence. Subsequent exposure of the nitrocellulose filters to X-ray film allowed for the identification of those colonies carrying the PI3K gene. Larger quantities of those bacterial colonies producing a positive response to the hybridization screen were then grown and used as the source of DNA template for PCR. The PCR reaction employed Pfu polymerase and the oligonucleotide primers 5'-CGC GCG GAA TTC GAA GGA GAT ATA CAT ATG AGT GCT GAG GGG-3' and 5'-CGC GCG CTG CAT GGA TCC TTA GGG AGG CGA GAT TTT TTT CCT-3'. These primers were designed to incorporate the ribosome binding site, translational spacer, and *Eco*RI and *Bam*H1 restriction sites in addition to sequences flanking the SH3 domain from PI3K (Figure 20). Following PCR, double restriction digest of the PCR product, and purification of the restricted material, the insert was ligated into the pLM1 plasmid vector (appropriately digested with *Eco*R1 and *Bam*H1) using T4 DNA ligase. The ligation mixture was used to transform the BL21 (DE3) strain of *E. coli*. Clones containing insert were sequenced using Sanger sequencing (chain termination).

One potential disadvantage of this cloning strategy was that it did not provide the SH3 domain as a fusion protein, so purification of the material from large scale culture (induced at an $OD_{600}$ of 0.6 with 1 mM IPTG, then allowed to incubate at 37 °C for 10 h) was not possible by affinity chromatographic methods. However, the authors reported that purification by DEAE-Sephacel ion exchange followed by Sephacryl S-100 gel filtration chromatography provided material that gave a single band on an SDS–PAGE gel. Yields of purified polypeptide were reported to be on the order of 50 mg L$^{-1}$ of culture.

### 7.18.12.2 Case Study: Production of Zinc Finger Domains for Analysis of Protein–DNA Interactions

The second example of cloning discussed here is the expression of zinc finger domains. The designation "zinc finger" describes a fairly small structural domain (35 to 40 amino acids) found in a wide variety of transcription factors. Given the overall topic of this volume, cloning of zinc finger proteins is particularly appropriate; a significant proportion of the current understanding of factors that govern sequence-specific protein–DNA recognition has derived from the study of zinc fingers.

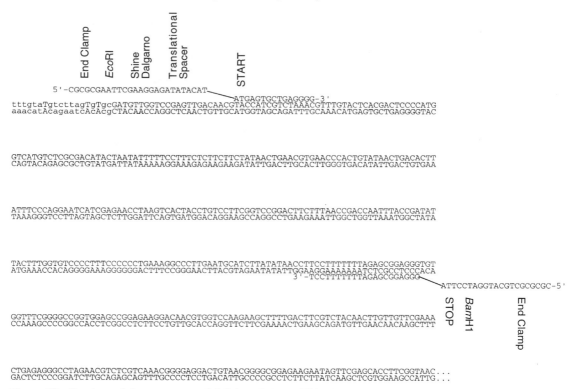

**Figure 20**   Portion of the sequence of full-length PI3K[156] indicating the sequence and positioning of PCR primers utilized by Yu *et al.*

Zinc finger domains, as their name suggests, are structurally organized by the coordination of four cysteine and/or histidine residues to a single $Zn^{2+}$ ion.[157-160] The domain is composed of a single $\alpha$ helix, joined by a loop to a two- or three-stranded $\beta$ sheet (Figure 21).[161] Recognition of DNA base pairs occurs via interactions between residues in the helical portion of the finger and the DNA major groove, allowing for each finger specifically to address a three base-pair region. The full interaction between an individual domain and its DNA binding site is on the order of five to six base pairs; however, only three of these involve specific recognition.[162] Recognition of larger DNA sequences is accomplished by linking several zinc finger domains together.

The discussion will focus on the erythroid-specific transcription factor GATA-1. Cloning of the full length protein was originally reported by two groups. Tsai *et al.* in 1989[163] reported using a strategy that employed transient high level expression of a mouse erythroleukemia cDNA library transfected into monkey kidney COS-1 cells, coupled with hybridization to bacterial colonies. Evans and Felsenfeld[164] simultaneously reported cloning of the chicken homologue. The human variant was reported in 1990;[165] this work employed the previously identified chicken cDNA clone to screen a human bone marrow cDNA library. Subsequently, a large number of related proteins have been discovered in a wide range of organisms.[166] These proteins all bind a consensus DNA sequence 5'-GATA-3'. To examine the ways in which GATA-1 recognizes its binding site, Clore and co-workers[167] cloned and overexpressed the DNA binding domain of GATA-1 and determined its structure bound to a synthetic double-stranded DNA 16-mer by solution NMR.

The DNA-binding domain examined by Clore and co-workers consisted of amino acids 158 to 223 of chicken GATA-1. The requisite DNA sequence was amplified from a cDNA library using primers designed to incorporate *Nde*I and *Bam*HI restriction sites. Following restriction digest of the purified PCR product and subsequent purification, this DNA insert was ligated into a pET11A plasmid vector that had been analogously digested with *Nde*I and *Bam*HI. Transformation of this construct was carried out with the BL21 (DE3) strain of *E. coli*. After identifying and sequencing a bacterial colony carrying the full plasmid, large scale preparation of the polypeptide was carried out. Isotopic labeling ($^{15}N$ and/or $^{13}C$), a requirement for the NMR analysis of a molecule this size,[168] was accomplished by culturing cells in "minimal media", which contain only labeled glucose and ammonium chloride as sources of these elements. Markley and Kainosho review NMR of isotopically labeled proteins and procedures for carrying out isotopic labeling.[168]

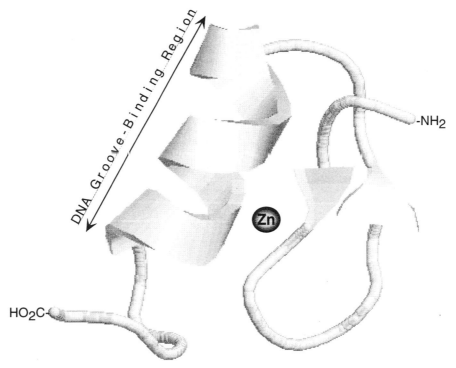

**Figure 21** Secondary structure representation of a zinc finger domain. The structure depicted here corresponds to that described by Clore and co-workers.[161]

Following isolation and lysis of cultured cells, the polypeptide was purified by a three-stage procedure. Since the plasmid was constructed in a manner such that a fusion protein was not produced, affinity purification was not performed, although presumably an affinity column could have been constructed which incorporated DNA carrying the consensus GATA-1 binding site. Instead, crude cellular lysate was chromatographed on a weak anion exchange resin (DEAE-Sepharose Fast Flow), using a pH 8.0 Tris buffer containing 5 mM dithiothreitol and 5 mM EDTA for elution. After concentrating material isolated from this column, a second purification step was carried out using a cation exchange column (S-Sepharose Fast Flow; elution with a 0–1 M NaCl gradient in the Tris buffer described previously). A final purification step involving reverse-phase HPLC (C-8 column, gradient elution in acetonitrile/water/0.1% trifluoroacetic acid) provided material sufficiently pure for NMR analysis.

Zinc fingers have continued to serve as a fertile research area. Significant effort has gone into the determination of the mechanisms whereby individual zinc fingers recognize their target sequences (including other NMR and X-ray crystallographic structural studies, as well as combinatorial libraries/phage display). As these rules have been determined, designer proteins have been generated from linked zinc fingers, many of which display exceptionally high affinity for DNA.[169]

## 7.18.13 CONCLUSIONS

The authors have attempted in this chapter to give an overview of some of the most common methods for the cloning and overexpression of proteins. As stated at the outset, the range of cloning techniques and methods available is immense; however, most cloning experiments may be accomplished with a small subset of these. Most importantly, these are all techniques that are readily accessible to chemists.

## 7.18.14 REFERENCES

1. D. Jackson, R. Symons, and P. Berg, *Proc. Natl. Acad. Sci. USA*, 1972, **69**, 2904.
2. S. Cohen, A. Chang, H. Boyer, and R. Helling, *Proc. Natl. Acad. Sci. USA*, 1973, **70**, 3240.
3. J. J. Siekierka, M. J. Staruch, S. H. Y. Hung, and N. H. Sigal, *J. Immunol.*, 1989, **143**, 1580.

4. M. W. Harding, A. Galat, D. E. Uehling, and S. L. Schreiber, *Nature*, 1989, **341**, 758.
5. S. K. Thompson, S. M. Halbert, M. J. Bossard, et al., *Proc. Natl. Acad. Sci. USA*, 1997, **94**, 14 249.
6. K. S. Lam, S. E. Salmon, E. M. Hersh, V. J. Hruby, W. M. Kazmierski, and R. J. Knapp, *Nature*, 1991, **354**, 82.
7. L. J. Beamer and C. O. Pabo, *J. Mol. Biol.*, 1992, **227**, 177.
8. L. Fairall, J. W. R. Schwabe, L. Chapman, J. T. Finch, and D. Rhodes, *Nature*, 1993, **366**, 483.
9. M. Uesugi, O. Nyanguile, H. Lu, A. J. Levine, and G. L. Verdine, *Science*, 1997, **277**, 1310.
10. D. V. Goeddel (ed.), "Gene Expression Technology," *Methods Enzymol.*, 1991, 185.
11. A. J. Harwood (ed.), "Basic DNA and RNA Protocols," Humana Press, Totowa, NJ, 1996.
12. D. M. Glover and B. D. Hames (eds.), "DNA Cloning: A Practical Approach," 2nd edn., IRL Press, Oxford, 1995.
13. J. Sambrook, E. F. Fritsch, and T. Maniatis, "Molecular Cloning: A Laboratory Manual," Cold Spring Harbor Press, Plainview, NY, 1989.
14. J. D. Watson, M. Gilman, J. Witkowski, and M. Zoller, "Recombinant DNA," 2nd edn., Scientific American Books, New York, 1992.
15. G. C. K. Roberts (ed.), "NMR of Macromolecules: A Practical Approach," IRL Press, Oxford, 1993.
16. G. M. Clore and A. M. Gronenborn, *Trends Biotech.*, 1998, **16**, 22.
17. K. Wüthrich, "NMR of Proteins and Nucleic Acids," John Wiley and Sons, New York, 1984.
18. H. J. Bohm and G. Klebe, *Angew. Chem. Int. Ed. Engl.*, 1996, **35**, 2588.
19. C. Khosla, *Chem. Rev.*, 1997, **97**, 2577.
20. C. B. Anfinsen, E. Haber, M. Sela, and F. H. White, *Proc. Natl. Acad. Sci. USA*, 1961, **47**, 1309.
21. M. E. Goldberg, *J. Mol. Biol.*, 1969, **46**, 441.
22. B. Bukau and A. L. Horwich, *Cell*, 1998, **92**, 351.
23. W. A. Fenton and A. L. Horwich, *Protein Sci.*, 1997, **6**, 743.
24. J. Martin and F. U. Hartl, *Curr. Opin. Struct. Biol.*, 1997, **7**, 41.
25. K. Wüthrich and G. Wagner, *Trends Biochem. Sci.*, 1978, **3**, 227.
26. K. D. Wilkinson and T. K. Audhya, *J. Biol. Chem.*, 1981, **256**, 235.
27. G. F. Nordberg, M. Nordberg, M. Piscator, and O. Vesterberg, *Biochem. J.*, 1972, **126**, 491.
28. C. H. Cabell, G. M. Verghese, N. B. Rankl, D. J. Burns, and P. J. Blackshear, *Proc. Assoc. Am. Physicians*, 1996, **108**, 37.
29. T. E. Meyer, G. Tollin, and M. A. Cusanovich, *Biochemie*, 1994, **76**, 480.
30. T. L. Madden, R. L. Tatusou, and J. Zhang, *Methods Enzymol.*, 1996, **266**, 131.
31. J. Greer, *Proteins*, 1990, **7**, 317.
32. Y. Ono, T. Fujii, K. Igarashi, T. Kuno, C. Tanaka, U. Kikkawa, and Y. Nishizuka, *Proc. Natl. Acad. Sci. USA*, 1989, **86**, 4868.
33. D. J. Burns and R. M. Bell, *J. Biol. Chem.*, 1991, **266**, 18330.
34. A. F. G. Quest, E. S. G. Bardes, and R. M. Bell, *J. Biol. Chem.*, 1994, **269**, 2953.
35. A. F. G. Quest, E. S. G. Bardes, and R. M. Bell, *J. Biol. Chem.*, 1994, **269**, 2961.
36. J.-H. Luo, S. Kahn, K. O'Driscoll, and I. B. Weinstein, *J. Biol. Chem.*, 1993, **268**, 3715.
37. Y. Kee and R. H. Scheller, *J. Neurosci.*, 1996, **16**, 1975.
38. E. A. Bayer and M. Wilchek, *Methods Enzymol.*, 1990, **184**, 138.
39. S. Laveille, G. Chassaing, J. C. Beaujouan, Y. Torrens, and A. Marquet, *Int. J. Peptide Protein Res.*, 1984, **24**, 480.
40. N. M. Green, L. Konirvxny, E. J. Toms, and R. C. Valentine, *Biochem. J.*, 1971, **125**, 781.
41. B. Ghebrehiwet, S. Bossone, A. Erdei, and K. B. M. Reid, *J. Immunol. Methods*, 1988, **110**, 251.
42. K. R. Gee, W.-C. Sun, D. H. Klaubert, R. P. Haugland, R. H. Upson, T. H. Steinberg, and M. Poot, *Tetrahedron Lett.*, 1996, **37**, 7905.
43. K. F. Geoghegan and J. G. Stroh, *Bioconj. Chem.*, 1992, **3**, 138.
44. L. Gold, *Methods Enzymol.*, 1990, **185**, 11.
45. F. W. Studier, A. H. Rosenberg, J. J. Dunn, and J. W. Dubendorff, *Methods Enzymol.*, 1990, **185**, 60.
46. J. Grodberg and J. J. Dunn, *J. Bacteriol.*, 1988, **170**, 1245.
47. C. Yanisch-Perron, J. Viera, and J. Messing, *Gene*, 1985, **33**, 103.
48. D. M. Woodcock, P. J. Crowther, S. Jefferson, and W. P. Diver, *Nucleic Acids Res.*, 1989, **17**, 3469.
49. W. O. Bullock, J. M. Fernandez, and J. M. Short, *BioTechniques*, 1987, **5**, 376.
50. R. Bolivar, R. Rodriguez, P. J. Greene, M. Betlach, H. L. Heyneker, H. W. Boyer, J. Crosa, and S. Falkow, *Gene*, 1977, **2**, 95.
51. P. Balbas and F. Bolivar, *Methods Enzymol.*, 1990, **185**, 14.
52. D. R. O'Reilly, L. Miller, and V. A. Luckow, "Baculovirus Expression Vectors: A Laboratory Manual," W. H. Freeman, New York, 1992.
53. C. D. Richardson (ed.), "Baculovirus Expression Protocols," Humana Press, Totowa, NJ, 1995.
54. J. M. Short, J. M. Fernandez, J. A. Sorge, and W. D. Huse, *Nucleic Acids Res.*, 1988, **16**, 7583.
55. E. Amman and J. Brosius, *Gene*, 1985, **40**, 183.
56. E. Amman, J. Brosius, and M. Ptashne, *Gene*, 1983, **25**, 167.
57. H. De Boer, L. J. Comstock, and M. Vasser, *Proc. Natl. Acad. Sci. USA*, 1983, **80**, 21.
58. C. Götting, G. Thierbach, A. Pühler, and J. Kalinowski, *Biotechniques*, 1998, **24**, 362.
59. J. Parnell, H. Lodish, and D. Baltimore, "Molecular Cell Biology," 2nd edn., Scientific American Books, New York, 1990, p. 254.
60. II. A. Shuman, T. J. Silhavy, and J. R. Beckwith, *J. Biol. Chem.*, 1980, **255**, 168.
61. D. B. Smith and K. S. Johnson, *Gene*, 1988, **67**, 31.
62. K. Guan and J. E. Dixon, *Anal. Biochem.*, 1991, **192**, 262.
63. D. J. Hakes and J. E. Dixon, *Anal. Biochem.*, 1992, **202**, 293.
64. E. A. DiBlasio, S. Kovacic, K. L. Grant, P. F. Schendel, and J. M. McCoy, *Biol. Technol.*, 1993, **11**, 187.
65. A. H. Rosenberg, B. N. Lade, D. Chui, S. Lin, J. J. Dunn and F. W. Studier, *Gene*, 1987, **56**, 125.
66. C. Guan, P. Li, P. D. Riggs, and H. Inouye, *Gene*, 1988, **67**, 21.
67. C. V. Maina, P. D. Riggs, Λ. G. Granden, *et al.*, *Gene*, 1988, **74**, 365.
68. M. E. Bayer, *J. Gen. Microbiol.*, 1968, **53**, 395.

69. A. Holmgren, *Ann. Rev. Biochem.*, 1985, **54**, 237.
70. J. M. McCoy, *Bio/Technology*, 1993, **11**, 187.
71. A. I. Derman, D. Belin, and J. Beckwith, *Science*, 1993, **262**, 1744.
72. B. A. Moffatt and F. W. Studier, *J. Mol. Biol.*, 1986, **189**, 113.
73. D. Chui, S. Lin, J. J. Dunn, and F. W. Studier, *Gene*, 1987, **56**, 125.
74. F. H. Arnold, *Bio/Technology*, 1991, **9**, 151.
75. N. Dekker, M. Cox, R. Boelens, C. P. Verrijzer, P. C. Van der Vliet, and R. Kaptein, *Nature*, 1993, **362**, 852.
76. D. E. Fisher, L. A. Paernt, and P. A. Sharp, *Cell*, 1993, **72**, 467.
77. A. Hoffmann and R. G. Roeder, *Nucleic Acids Res.*, 1991, **19**, 6337.
78. S. M. Keyse and E. A. Emslie, *Nature*, 1992, **359**, 644.
79. J. S. Kim and R. T. Raines, *Protein Sci.*, 1993, **2**, 348.
80. K. Nagai and H. C. Thøgersen, *Methods Enzymol.*, 1987, **153**, 461.
81. R. K. Saiki, S. J. Scharf, F. Faloona, K. B. Mullis, G. T. Horn, H. A. Erlich, and N. Arnheim, *Science*, 1985, **230**, 1350.
82. K. Mullis and F. Faloona, *Methods Enzymol.*, 1987, **55**, 335.
83. B. A. White (ed.), "PCR Protocols: Current Methods and Applications," Methods in Molecular Biology, 1993, 15.
84. M. A. Innis, D. H. Gelfand, J. J. Sninsky, and T. J. White (eds.), "PCR Protocols: A Guide to Methods and Applications," Academic Press, San Diego, 1990.
85. C. R. Newton (ed.), "PCR: Essential Data Series," John Wiley and Sons, New York, 1995.
86. M. Gilman, J. D. Watson, J. Witkowski, and M. Zoller, "Recombinant DNA," Scientific American Books, New York, 1992, p. 79.
87. J. C. Venter, H. O. Smith, and L. Hood, *Nature*, 1996, **381**, 364.
88. X. Huang, M. D. Adams, H. Zhou, and A. R. Kerlavage, *Genomics*, 1997, **46**, 37.
89. R. A. Clayton, O. White, K. A. Ketchum, and J. C. Venter, *Nature*, 1997, **387**, 459.
90. S. A. Westwood and D. J. Werrett, *Forensic Sci. Int.*, 1990, **45**, 201.
91. U. Gubler and B. J. Hoffman, *Gene*, 1983, **25**, 263.
92. T. L. Pauls and M. W. Berchtold, *Methods Enzymol.*, 1993, **217**, 102.
93. J. Gu (ed.), "*In situ* PCR and Related Technology," Birkhäuser, Boston, 1995.
94. W. Bloch, *Biochemistry*, 1991, **30**, 2735.
95. Amplify 1.0: W. Engels, Genetics Department, University of Wisconsin.
96. D. R. Casimiro, P. E. Wright, and H. J. Dyson, *Structure*, 1997, **5**, 1407.
97. A. Chien, D. B. Edgar, and J. M. Trela, *J. Bacteriol.*, 1976, **127**, 1550.
98. R. K. Saiki, D. H. Gelfand, S. Stoffel, S. J. Scharf, R. Higuchi, G. T. Horn, K. B. Mullis, and H. A. Erlich, *Science*, 1988, **239**, 487.
99. K. A. Eckert and T. A. Kunkel, *Nucleic Acids Res.*, 1990, **18**, 3739.
100. H. A. Erlich, D. Gelfand, and J. Sninsky, *J. Sci.*, 1991, **252**, 1643.
101. J. Cline, J. C. Braman, and H. H. Hogrefe, *Nucleic Acids Res.*, 1996, **24**, 3546.
102. P. Keohavong and W. G. Thilly, *Proc. Nucleic Acids Res. USA*, 1989, **86**, 9253.
103. R. S. Cha and W. G. Tilly, in "PCR Primer," eds. C. W. Dieffenbach and G. S. Dveksler, Cold Spring Harbor Laboratory Press, Cold Spring Harbor, NY, 1995.
104. K. S. Lundberg, D. D. Shoemaker, M. W. W. Adams, J. M. Short, J. A. Sorge, and E. J. Mathur, *Gene*, 1991, **108**, 1.
105. J. M. Flaman, T. Frebourg, V. Moreau, F. Charbonnier, C. Martin, C. Ishioka, S. H. Friend, and R. Iggo, *Nucleic Acids Res.*, 1994, **22**, 3259.
106. C. Kebelmann-Betzing, K. Seeger, S. Dragon, G. Schmitt, A. Möricke, T. A. Schild, G. Henze, and B. Beyermann, *Biotechniques*, 1998, **24**, 154.
107. D. E. Kellogg, I. Rybalkin, S. Chen, N. Mukhamedova, T. Vlasik, P. D. Siebert, and A. Chenchik, *Biotechniques*, 1994, **16**, 1134.
108. S. J. Scharf, G. T. Horn, and H. A. Erlich, *Science*, 1986, **233**, 1076.
109. S. Wilton and L. Lim, *Trends Genet.*, 1996, **12**, 458.
110. E. M. Southern, *Methods Enzymol.*, 1979, **68**, 152.
111. R. J. Roberts and D. Macelis, *Nucleic Acids Res.*, 1991, **19**, 2077.
112. R. J. Roberts and D. Macelis, *Nucleic Acids Res.*, 1994, **22**, 3628.
113. M. Nasri and D. Thomas, *Nucleic Acids Res.*, 1986, **14**, 811.
114. New England Biolabs catalogue, New England Biolabs, Inc., Beverly, MA, 1998.
115. D. A. Baunoch, M. Xu, A. Tewari, J. K. Christman, and M. A. Lane, *Oncogene*, 1992, **7**, 2351.
116. L. B. Nguyen, S. Shefer, G. Salen, J. Y. Chiang, and M. Patel, *Hepatology*, 1996, **24**, 1468.
117. H. Donis-Keller, A. M. Maxam, and W. Gilbert, *Nucleic Acids Res.*, 1977, **4**, 2527.
118. J. Sambrook, E. F. Fritsch, and T. Maniatis, "Molecular Cloning: A Laboratory Manual," 2nd edn., Cold Spring Harbor Laboratory Press, Cold Spring Harbor, NY, 1989, vol. 1, p. 5.
119. K. Doyle (ed.), "Promega Protocols and Applications Guide," Promega Corporation, 1996, p. 45.
120. J. Sambrook, E. F. Fritsch, and T. Maniatis, "Molecular Cloning: A Laboratory Manual," 2nd edn., Cold Spring Harbor Laboratory Press, Cold Spring Harbor, NY, 1989, vol. 3, p. E. 5., 1.74.
121. W. J. Dower, J. F. Miller, and C. W. Ragsdale, *Nucleic Acids Res.*, 1988, **16**, 6127.
122. B. Li, K. Y. Pilcher, T. E. Wyman, and C. A. Machida, *Biotechniques*, 1997, **23**, 603.
123. F. Sanger, S. Nicklen, and A. R. Coulson, *Proc. Natl. Acad. Sci. USA*, 1977, **74**, 5463.
124. F. Sanger and A. R. Coulson, *FEBS Lett.*, 1978, **87**, 107.
125. A. M. Maxam, W. Gilbert, *Proc. Natl. Acad. Sci. USA*, 1977, **74**, 560.
126. A. M. Maxam, W. Gilbert, *Methods Enzymol.*, 1980, **65**, 499.
127. M. T. Prosperi, D. Ferbus, D. Rouillard, and G. Goubin, *FEBS Lett.*, 1998, **423**, 39.
128. B. S. Dunbar, "Two-Dimensional Electrophoresis and Immunological Techniques" Plenum Press, New York, 1987.
129. C. M. Wilson, *Anal. Biochem.*, 1979, **9**, 263.
130. C. M. Wilson, *Methods Enzymol.*, 1983, **91**, 236.
131. C. R. Merril and D. Goldman, in "Two-Dimensional Gel Electrophoresis of Proteins: Methods and Applications," eds. J. E. Celis and R. Bravo, Academic Press, New York, 1984, p. 111.

132. M. P. Deutscher, *Methods Enzymol.*, 1990, **182**.
133. W. Wickner and A. Kornberg, *J. Biol. Chem.*, 1974, **249**, 6244.
134. E. Bjurstron, *Chem. Eng.*, 1985, **92**, 126.
135. K. Hinterding, D. Alonso-Diaz, and H. Waldmann, *Angew. Chem. Int. Ed. Engl.*, 1998, **37**, 688.
136. A. Musacchio, M. Wilmanns, and M. Saraste, *Prog. Biophys. Mol. Biol.*, 1994, **61**, 283.
137. F. Sicheri and J. Kuriyan, *Curr. Opin. Struct. Biol.*, 1997, **7**, 777.
138. P. Bork, J. Schultz, and C. P. Ponting, *Trends Biochem. Sci.*, 1997, **22**, 296.
139. S. Zhou and L. C. Cantley, *Trends Biochem. Sci.*, 1995, **20**, 470.
140. V.-P. Lehto, V.-M. Wasenius, P. Salven, and M. Saraste, *Nature*, 1988, **334**, 388.
141. M. L. Stahl, C. R. Ferenz, K. L. O. Kelleher, R. W. Kriz, and J. L. Knopf, *Nature*, 1988, **332**, 269.
142. B. J. Mayer, M. Hamaguchi, and H. Hanafusa, *Nature*, 1988, **332**, 272.
143. A. Musacchio, T. Gibson, V.-P. Lehto, and M. Saraste, *FEBS Lett.*, 1992, **307**, 55.
144. S. Feng, J. K. Chen, H. Yu, J. A. Simon, and S. L. Schreiber, *Science*, 1994, **266**, 1241.
145. J. K. Chen and S. L. Schreiber, *Angew. Chem. Int. Ed. Engl.*, 1995, **34**, 487.
146. T. M. Kapoor, A. H. Andreotti, and S. L. Schreiber, *J. Am. Chem. Soc.*, 1998, **120**, 23.
147. J. P. Morken, T. M. Kapoor, S. Feng, F. Shirai, and S. L. Schreiber, *J. Am. Chem. Soc.*, 1998, **120**, 30.
148. S. Feng, T. M. Kapoor, F. Shirai, A. P. Combs, and S. L. Schreiber, *Chem. Biol.*, 1996, **3**, 661.
149. K. D. MacFerrin, M. P. Terranova, S. L. Schreiber, and G. L. Verdine, *Proc. Natl. Acad. Sci. USA*, 1990, **87**, 1937.
150. K. D. MacFerrin, L. Chen, M. P. Terranova, S. L. Schreiber, and G. L. Verdine, *Methods Enzymol.*, 1993, **217**, 79.
151. Hongtao Yu, Ph.D. Thesis, Harvard University, 1994.
152. H. Yu, S. Koyama, D. C. Dalgarno, T. B. Shin, L. D. Zydowsky, and S. L. Schreiber, *FEBS Lett.*, 1993, **324**, 93.
153. H. Yu, J. K. Chen, S. Feng, D. C. Dalgarno, A. W. Brauer, and S. L. Schreiber, *Cell*, 1994, **76**, 933.
154. T. H. You and R. L. Scholl, *Biotechniques*, 1998, **24**, 574.
155. W. D. Benton and R. W. Davis, *Science*, 1977, **196**, 180.
156. E. Y. Skolnik, B. Margolis, M. Mohammadi, E. Lowenstein, R. Fischer, A. Drepps, A. Ullrich, and J. Schlessinger, *Cell*, 1991, **65**, 83.
157. J. M. Berg, *Ann. Rev. Biophys. Biophys. Chem.*, 1990, **19**, 405.
158. A. Klug and D. Rhodes, *Trends Biochm. Sci.*, 1987, **12**, 464.
159. G. H. Jacobs, *EMBO J.*, 1992, **11**, 4507.
160. J. M. Berg, *Acc. Chem. Res.*, 1995, **28**, 14.
161. J. G. Omichinski, G. M. Clore, E. Appella, K. Sakaguchi, and A. M. Gronenborn, *Biochemistry*, 1990, **29**, 9324.
162. J. Miller, A. D. McLachlan, and A. Klug, *EMBO J.*, 1985, **4**, 1609.
163. S.-F. Tsai, D. I. K. Martin, L. I. Zon, A. D. D'Andrea, G. G. Wong, and S. H. Orkin, *Nature*, 1989, **339**, 446.
164. T. Evans and G. Felsenfeld, *Cell*, 1989, **58**, 577.
165. C. D. Trainor, T. Evans, G. Felsenfled, and M. S. Boguski, *Nature*, 1990, **343**, 92.
166. G. R. Teakle and P. M. Gilmartin, *Trends Biol. Sci.* 1998, **23**, 100.
167. J. G. Omichinski, G. M. Clore, O. Schaad, G. Felsenfeld, C. Trainor, E. Appella, S. J. Stahl, and A. M. Gronenborn, *Science*, 1993, **261**, 438.
168. J. L. Markley and M. Kainosho, in "NMR of Macromolecules: A Practical Approach," ed. G. C. K. Roberts, IRL Press, Oxford, 1993, p. 101.
169. J.-S. Kim and C. O. Pabo, *Proc. Natl. Acad. Sci. USA*, 1998, **95**, 2812.

# Author Index

This Author Index comprises an alphabetical listing of the names of the authors cited in the text and the references listed at the end of each chapter in this volume.

Each entry consists of the author's name, followed by a list of numbers, for example

Templeton, J. L., 366, 385$^{233}$ (350, 366), 387$^{370}$ (363)

For each name, the page numbers for the citation in the reference list are given, followed by the reference number in superscript and the page number(s) in parentheses of where that reference is cited in the text. Where a name is referred to in text only, the page number of the citation appears with no superscript number. References cited in both the text and in the tables are included.

Although much effort has gone into eliminating inaccuracies resulting from the use of different combinations of initials by the same author, the use by some journals of only one initial, and different spellings of the same name as a result of the transliteration processes, the accuracy of some entries may have been affected by these factors.

Abdel-Meguid, S. S., 488$^{13}$ (478)
Abelson, J., 367$^{12}$ (345)
Abildgaard, F., 79$^{17}$ (59)
Aboul-ela, F., 338$^{136}$ (325)
Abraham, R. J., 474$^{97}$ (441, 443)
Abramova, T. V., 242$^{86}$ (164), 244$^{174}$ (182)
Absalon, M. J., 422$^{61}$ (380), 422$^{62}$ (380), 425$^{252}$ (420), 425$^{253}$ (420)
Abul-ela, F., 79$^{4}$ (56, 59)
Acevedo, O. L., 330, 337$^{57}$ (319, 322), 338$^{178}$ (330), 338$^{179}$ (330)
Ackerman, T., 50$^{45}$ (29)
Adam, S., 337$^{87}$ (323)
Adam, W., 424$^{207}$ (411)
Adamiak, R. W., 147$^{119}$ (117), 147$^{120}$ (117)
Adams, A. D., 147$^{96}$ (114)
Adams, S. P., 112, 146$^{55}$ (112)
Adawadkar, P., 475$^{170}$ (463, 464, 465, 466), 475$^{171}$ (463, 464, 466)
Addess, K. J., 368$^{40}$ (352), 474$^{113}$ (446, 448)
Adhya, S., 281$^{12}$ (253)
Adinarayana, M., 422$^{68}$ (381, 403, 406, 410)
Adleman, L. M., 13$^{63}$ (11, 12)
Admas, G. E., 425$^{239}$ (418)
Advani, S., 243$^{111}$ (168), 243$^{114}$ (168)
Affolter, M., 489$^{18}$ (479), 489$^{19}$ (479)
Afonina, I. A., 13$^{50}$ (10)
Agarwal, K., 489$^{79}$ (486)
Aggarwal, A. K., 488$^{8}$ (478), 489$^{24}$ (479)

Agrawal, K. C., 473$^{35}$ (432)
Agrawal, S., 149$^{237}$ (121, 124), 149$^{238}$, 150$^{307}$ (125), 150$^{332}$ (127), 151$^{333}$ (127), 151$^{384}$ (133), 152$^{401}$ (135), 152$^{429}$ (139), 237, 238, 241$^{16}$ (169, 170), 241$^{18}$ (155), 242$^{40}$ (156, 159), 245$^{269}$ (190), 248$^{461}$ (237), 248$^{466}$ (238), 308$^{4}$ (285), 309$^{49}$ (292, 295), 309$^{52}$ (292, 295), 310$^{103}$ (299, 301)
Aguilera, J. A., 425$^{244}$ (418)
Ahle, D., 246$^{341}$ (204), 246$^{342}$ (204)
Ahlert, K., 244$^{158}$ (177, 190)
Aiken, C. R., 283$^{119}$ (266)
Akagi, M., 245$^{245}$ (190)
Akasaka, K., 52$^{144}$ (43)
Akerman, B., 101$^{84}$ (88)
Akhtar, S., 241$^{11}$ (155), 241$^{12}$ (155)
Akiyama, T., 148$^{177}$
Akman, S. A., 102$^{159}$ (92)
Al-Kazwini, A. T., 425$^{239}$ (418)
Alakhov, V. Y., 244$^{202}$ (187, 190)
Alber, T., 489$^{41}$ (482)
Albergo, D. D., 52$^{149}$ (43)
Albericio, F., 147$^{80}$ (114), 147$^{98}$ (114)
Albericio, R., 336$^{12}$ (315)
Alberts, D. S., 549$^{181}$ (516)
Alderfer, J. L., 474$^{91}$ (441)
Alexandrescu, A., 101$^{92}$ (88)
Alexandrova, L. A., 246$^{297}$ (194)
Al'fonsov, V. A., 152$^{410}$ (138)
Alkema, D., 51$^{61}$ (33)
Allain, F. H.-T., 79$^{4}$ (56, 59)
Allard, P., 147$^{113}$ (117)

Allawi, H. T., 51$^{71}$ (33)
Allen, D. J., 246$^{299}$ (194, 195), 246$^{300}$ (195)
Allen, F. S., 100$^{64}$ (86)
Alluri, M., 547$^{76}$ (500)
Almer, H., 152$^{421}$ (139), 152$^{437}$ (139)
Altenbach, C., 489$^{38}$ (481)
Altman, S., 52$^{166}$
Altmann, K.-H., 248$^{450}$ (235), 308$^{16}$ (288), 308$^{25}$ (288)
Altona, C., 367$^{24}$ (347, 353)
Alul, R. H., 114, 147$^{75}$ (114)
Alvarado-Urbina, G., 146$^{46}$ (112)
Alves, A. M., 243$^{140}$ (171), 309$^{35}$ (289)
Amabilino, D. B., 368$^{85}$ (361)
Amann, R. I., 241$^{32}$ (154, 158)
Ames, B. N., 422$^{57}$ (378), 547$^{41}$ (494), 547$^{46}$ (494)
Amin, S., 476$^{218}$ (470), 476$^{219}$ (470)
Amirhaeri, S., 100$^{18}$ (82), 100$^{39}$ (85, 93), 102$^{182}$ (93)
Anand, N. N., 337$^{66}$ (321)
Anderson, C. F., 473$^{19}$ (428, 432)
Anderson, G. L., 547$^{57}$ (495)
Anderson, J. R., 101$^{91}$ (88)
Anderson, P., 284$^{174}$ (274, 277)
Anderson, R. G. W., 242$^{52}$ (159)
Andersson, A., 422$^{63}$ (380), 546$^{18}$ (493)
Andrade, L. K., 242$^{56}$ (159, 160)
Andrade, M., 12$^{12}$ (4)
Andrews, P. A., 549$^{177}$ (516)
Andrews, P. C., 339$^{218}$ (333), 339$^{219}$ (333)

Andrews, R. S., 338[178] (330)
Andrus, A., 146[58] (112, 113, 122), 146[59] (113), 147[131] (119), 147[132] (119, 130, 131, 132), 149[222] (122)
Angus, K., 147[92] (114), 147[93] (114)
Anjaneyulu, P. S. R., 246[338] (203)
Anslyn, E., 247[364] (211)
Ansorge, W., 243[95] (167), 243[105] (167)
Anthony-Cahill, S. J., 489[38] (481)
Antoine, E., 12[26] (5)
Antoshechkin, I., 338[160] (328), 424[192] (410)
Antsypovich, S. I., 242[39] (157)
Aota, R., 248[426] (229)
Apostol, B., 367[12] (345)
Appleby, D. W., 50[56] (32)
Applequist, J., 50[41] (24, 28)
Arai, H., 241[18] (155)
Arai, T., 547[71] (497)
Arcamone, F., 473[32] (429, 432)
Ares, M., 367[14] (345, 367)
Aristoff, P. A., 548[115] (505)
Arkin, M. R., 101[103] (89)
Armitage, B., 424[185] (410), 424[187] (410)
Armitage, I. M., 52[145] (43)
Armstrong, R. W., 549[161] (513, 514)
Arnaud, M., 489[37] (481)
Arndt, K., 80[81] (75)
Arnold, E., 490[91] (488)
Arnold, L., 146[54] (112)
Arnold, Jr., L. J., 242[41] (161), 308[13] (287, 292), 309[83] (295)
Arnott, S., 79[14] (58)
Aronovitch, J., 425[240] (418)
Aruoma, O. I., 423[124] (397, 410), 423[125] (397, 410), 547[36] (494, 544), 547[46] (494)
Arvidson, D. N., 12[29] (6)
Arya, D., 547[33] (494)
Asada, K., 476[185] (466)
Asai, A., 548[109] (505, 506), 548[110] (505)
Ashcroft, J., 547[76] (500)
Ashley, G. W., 369[99] (362, 366), 423[101] (387, 388, 393, 394), 425[255] (421), 560
Aso, Y., 150[305] (123, 136)
Asseline, U., 243[116] (168), 243[119] (168, 190), 244[168] (181), 244[169] (181), 245[226] (190, 236), 245[230] (190), 248[452] (237)
Astashkina, T. G., 100[59] (86)
Asteriadis, G. T., 147[72] (113)
Atcher, R. W., 425[220] (411)
Atherton, F. R., 146[4] (106)
Atkinson, T. C., 146[37] (111)
Atkinson, T., 147[107] (117)
Atwell, G. J., 474[100] (444, 449), 475[125] (451), 475[126] (451)
Aubertin, A.-M., 309[73] (294)
Audic, A., 246[331] (202)
Audrup, H., 284[194] (278)
Aujard, C., 548[96] (503)
Aurelian, L., 309[79] (294), 309[80] (295)
Aurup, H., 248[429] (230), 282[46] (257, 278)
Avino, A., 147[98] (114)

Avino, A. M., 151[373] (133)
Aviñó, A. M., 243[100] (167)
Ayala, E., 148[159]
Azhayev, A., 242[45] (157, 190), 243[136] (171, 190), 243[137] (171, 228), 245[219] (190), 245[220] (190), 245[240] (190), 248[419] (227), 248[420] (227)
Azhayeva, E., 243[138] (171, 190), 248[419] (227), 248[420] (227)
Azzawi, A., 244[161] (178)

Bachur, N. R., 549[177] (516)
Baer, E., 145[2] (106)
Baertschi, S. W., 548[151] (512)
Baguley, B. C., 475[126] (451)
Bahl, C. P., 146[25] (108)
Baikalov, I., 547[60] (497)
Bailey, E. A., 549[157] (513)
Bailleul, B., 367[6] (344, 367)
Bailly, C., 459, 473[13] (428, 436), 474[80] (437), 474[111] (446, 448), 475[144] (453, 459), 475[157] (459), 476[198] (468), 547[83] (501)
Bain, G., 489[33] (480)
Bain, J. D., 337[99] (324)
Baker, B. F., 246[323] (200)
Baker, D. C., 308[11] (287, 298)
Baker, R. M., 548[122] (506)
Baker, T. A., 367[20] (346), 368[31] (350)
Balasubramaniam, T. N., 309[46] (292, 294, 301)
Balatskaya, S. V., 102[167] (92, 95)
Balbi, A., 244[174] (182)
Bald, R., 244[209] (190)
Baldino, C. M., 503, 548[98] (503), 548[100] (503)
Baldwin, R. L., 369[107] (365)
Balgobin, N., 148[138] (120)
Balis, F. M., 339[227] (334)
Balland, A., 146[48] (112)
Baltimore, D., 489[35] (481)
Bancroft, C., 13[64] (12)
Bandaru, R., 310[114] (299, 304)
Banerjee, A. R., 53[208] (49)
Bannwarth, W., 151[378] (133), 241[21] (185, 234), 244[188] (185), 247[391] (219)
Bannworth, W., 310[141] (306)
Bansal, M., 79[16] (59)
Banville, D. L., 474[93] (441, 442, 443), 474[94] (441, 443)
Baranowski, T. C., 476[179] (467)
Barany, G., 152[447] (143)
Barawkar, D. A., 247[357] (209), 247[358] (209)
Barbier, C., 243[120] (168, 169)
Barcelo, F., 52[187] (47)
Bardella, F., 338[141] (326)
Barkley, M. D., 547[59] (495, 497)
Barkley, R. M., 424[147] (400, 406)
Barnekow, F., 150[324] (126), 150[325] (126)
Barnett, R. W., 146[57] (112)
Barone, A. D., 100[5] (82)
Barrio, J. R., 52[143] (43)
Barron, M. D., 103[208] (95)
Barrows, L. R., 547[85] (501, 529)

Barstad, P. A., 243[104] (168)
Bartel, D. P., 52[174] (46), 633, 634
Barthet, C., 247[378] (216)
Bartholomew, J. C., 103[223] (96)
Barton, H. J., 475[135] (452, 453), 475[136] (452, 453), 475[149] (457)
Barton, J. K., 88, 100[30] (85, 88), 100[31] (85, 88), 101[85] (88), 101[86] (88), 383
Bartsch, H., 547[62] (497, 504), 547[91] (502), 548[145] (511, 512)
Bartunek, P., 102[191] (94)
Barvian, M. R., 422[34] (373, 402, 403), 422[35] (373, 402, 403, 404), 424[146] (400), 424[147] (400, 406), 546[29] (494)
Bash, P., 368[37] (351, 352)
Bashkin, J. K., 13[61] (11)
Bashkin, J., 101[120] (90), 101[122] (90)
Basile, L. A., 101[92] (88)
Bass, B. L., 279, 284[180] (275, 277, 279), 284[182] (276), 616, 617, 639
Bass, W. J., 103[232] (96)
Basu, S., 241[12] (155)
Bates, A. D., 12[25] (5)
Bates, P. J., 244[183] (184)
Batey, R. T., 79[2] (56)
Battigello, J.-M., 103[245] (98)
Battiste, J. L., 473[62] (434)
Batyeva, E. S., 152[410] (138)
Bauer, J., 151[376] (133)
Baumeister, K., 247[362] (211)
Bax, A., 79[19] (59), 79[40] (64), 80[72] (72, 73), 80[74] (72)
Baxter, A. D., 297, 310[88] (296, 297), 310[89] (296, 297, 298)
Bayer, E., 146[60] (113)
Bayever, E., 12[9] (4, 5, 9)
Bayley, C. R., 281[14] (253)
Bazin, H., 148[184] (121)
Beabealashvili, R., 103[200] (95)
Beachy, P. A., 489[25] (479)
Beal, P. A., 247[376] (215)
Beardsley, G. P., 336[28] (316), 424[169] (407), 424[174] (407)
Beaton, G., 309[40] (292, 293), 310[90] (296, 297)
Beattie, K. L., 51[73] (33)
Beaucage, S. L., 12[4] (2, 3), 100[5] (82), 108, 119, 130, 146[30] (108, 112, 130, 131, 132, 137), 147[108] (117, 123, 138, 143, 144), 147[132] (119, 130, 131, 132)
Bec, C. L., 309[78] (294)
Beck, S., 12[16] (4), 161
Beck, T. A., 242[41] (161)
Becker, A., 247[413] (225)
Bedford, G. R., 474[97] (441, 443)
Been, M. D., 52[173] (46)
Beer, M., 281[15] (253)
Beerman, T. A., 548[122] (506), 580
Beggs, J. D., 242[42] (156)
Beher, D., 421[10] (372, 417)
Behmoaras, T., 422[64] (380)
Behr, J.-P., 338[140] (326)
Behrens, C., 242[46] (158, 190)
Behrens, G., 422[65] (381, 389), 422[66] (381, 389), 423[106] (389), 423[130] (398, 400, 405), 424[155] (402)

Beigelman, L., 284[188] (277, 279)
Beijer, B., 149[219] (121)
Beiter, A. H., 151[372] (133)
Beiter, A., 148[160] (120)
Belagaje, R., 149[220] (126)
Bellido, D., 242[59] (160)
Belloc, F., 242[47] (158)
Bellon, L., 245[277] (190), 309[36] (289, 293, 299, 302), 310[109] (299, 302)
Belotserkovskii, B. P., 102[193] (94, 95), 102[197] (95)
Delousov, E. S., 13[50] (10)
Benassan, R. V., 424[211] (411)
Benasutti, M., 549[156] (513)
Bender, R., 146[46] (112)
Bender, W., 548[149] (512)
Benfield, P. A., 489[38] (481)
Bengström, M., 241[6] (154, 155), 241[7] (155)
Benight, A. S., 368[96] (362, 366), 369[98] (362)
Benkovic, S. J., 194, 246[298] (194), 246[299] (194, 195)
Benner, S. A., 247[367] (211), 282[38] (256, 271), 282[59] (258), 304, 310[113] (299, 304), 310[132] (304), 324, 327
Bennett, C. F., 242[56] (159, 160), 242[57] (159), 308[1] (285, 287), 309[72] (294)
Bennett, G. N., 548[135] (508)
Bennett, J. E., 422[51] (378)
Bennett, M. R., 308[2] (285)
Bennett, R. A. O., 546[17] (493)
Bennett, W. P., 12[15] (4)
Bennua, B. C. B., 282[84] (260)
Benseler, F., 149[247] (123), 149[248] (123), 245[215] (190), 248[429] (230), 282[46] (257, 278), 282[73] (259), 283[95] (264), 283[120] (267), 284[194] (278)
Benson, S. W., 424[159] (404)
Berendsen, R. G., 52[157] (45)
Berens, C., 151[345] (129)
Berezovskii, M. V., 242[65] (161)
Berg, H., 339[191] (330)
Berg, R., 12[6] (4)
Berger, J. M., 12[24] (5), 485
Berger, M., 424[171] (407), 424[203] (411)
Bergmann, F., 150[329] (127), 151[378] (133), 310[141] (306)
Bergot, B. J., 152[450] (144)
Bergstrom, D. E., 13[62] (11)
Berlin, Y. A., 244[164] (178, 179)
Berman, H., 51[89] (34)
Berman, H. M., 473[10] (428, 436), 473[22] (430, 435)
Bernadou, J., 101[113] (89)
Bernan, V., 547[76] (500)
Berner, S., 152[408] (137), 152[409] (137, 138), 248[451] (236, 237)
Bernier, J.-L., 101[109] (89)
Berressem, R., 283[146] (272), 322
Berzal, H. A., 284[165] (274, 277, 279, 280), 284[166] (274, 277)
Berzal-Herranz, A., 369[117] (367)
Betina, V., 548[97] (503)
Beveridge, D. L., 473[24] (431)

Bévierre, M. O., 248[450] (235)
Bevilacqua, P. C., 248[448] (234), 248[449] (234)
Beyreuther, K., 421[10] (372, 417)
Beyrich-Graf, X., 422[72] (382, 393, 403), 423[110] (390, 393), 423[115] (391, 393, 397, 403)
Bhan, P., 338[126] (325)
Bhanot, O. S., 315, 336[1] (314), 336[9] (315)
Bhat, B., 310[126] (302), 310[128] (303)
Bhat, V., 150[275] (120)
Bhatia, D., 147[85] (114), 147[106] (116)
Bhattacharya, B. K., 332, 339[209] (332), 339[212] (332)
Bhattacharyya, A., 102[132] (92, 93), 102[133] (92)
Bhattacharyya, S. N., 425[232] (417)
Bhongle, N. N., 152[420] (139)
Bi, G., 244[185] (184)
Biala, E., 147[119] (117), 147[120] (117)
Bianchi, A., 102[145] (92, 94)
Biancotto, G., 245[217] (190)
Bickelhaupt, F., 152[411] (138)
Bicknell, W., 368[38] (351), 548[138] (509)
Bidaine, A., 151[345] (129)
Bidan, G., 247[378] (216)
Bienveau, C., 424[162] (406)
Biernat, J., 146[51] (112)
Bigey, P., 242[76] (162)
Bigger, C. A. H., 247[402] (222, 225)
Bilimoria, M. H., 424[167] (407)
Billeter, M., 489[18] (479), 489[21] (479)
Billwitz, H., 339[190] (330)
Bindig, U., 338[137] (325), 339[188] (330), 339[192] (330, 331, 332)
Birch-Hirschfeld, E., 146[63] (113), 146[64] (113), 146[65] (113), 152[452] (144)
Birg, F., 243[128] (169)
Birikh, K. R., 244[164] (178, 179)
Bisagni, E., 102[183] (93)
Bischofberger, N., 244[201] (187)
Bischoff, R., 241[4] (154), 241[5] (168)
Bishop, P., 488[16] (478, 479), 489[36] (481)
Bizanek, R., 549[180] (516)
Bizdona, E., 152[425] (139)
Bjelland, S., 424[166] (407)
Blackburn, E. H., 367[18] (346), 476[202] (469)
Blackburn, G. M., 472[1] (428, 432, 436, 444, 452, 456), 476[217] (470), 476[217] (470)
Blaho, J. A., 100[18] (82)
Blair, I. A., 547[92] (502)
Blake, R. D., 51[74] (33), 52[182] (46), 52[183] (46)
Blakely, W. F., 423[121] (397, 410)
Blanchard, S., 473[60] (434)
Blankemeyer, J. T., 100[47] (84)
Blanks, R., 243[97] (168), 243[108] (168)
Blattner, F. R., 368[57] (354)
Blazek, E. R., 547[39] (494, 544)
Blencowe, B. J., 337[85] (323)
Blöcker, H., 150[289] (122, 126), 150[312] (126)
Blommers, M. J. J., 310[120] (300)

Blonski, C., 151[335] (127)
Bloomfield, V. A., 50[28] (20), 50[29] (20)
Blsagni, E., 476[180] (467, 468), 476[197] (468)
Blumenfeld, K. S., 50[57] (32)
Blumenfeld, M., 242[39] (157)
Boado, R. J., 245[257] (190), 246[317], 246[318] (199)
Boal, J. H., 149[260] (121, 124)
Bobst., A. M., 247[392] (219, 220), 247[393] (219, 220)
Bobst, E. V., 247[393] (219, 220)
Bock, L. C., 13[59] (11)
Boczkowska, M., 309[56] (292, 306)
Bode, J., 102[180] (93)
Bodenhausen, G., 79[38] (64), 80[71] (72, 73)
Bodepudi, V., 338[152] (327), 338[159] (328), 339[221] (334), 424[192] (410), 424[201] (410)
Boelens, R., 489[66] (484)
Bogacki, R., 282[40] (256, 261, 272)
Boger, D. L., 282[20] (253), 503, 505, 506
Böhm, E., 424[155] (402)
Boiteux, S., 424[172] (407), 424[191] (410)
Boiziau, C., 309[45] (292, 293)
Bollen, A., 242[37] (155), 242[68] (161, 166)
Bolli, M., 128[8] (4)
Bollmark, M., 309[44] (292, 293)
Bolo, N., 52[159] (45), 80[84] (75)
Bolton, P. H., 422[61] (380), 422[62] (380)
Bonfils, B., 147[86] (114)
Bonfils, E., 242[70] (161, 190), 243[118] (168, 181, 190), 243[119] (168, 190), 245[263] (190)
Bongartz, J.-P., 309[73] (294)
Bonora, G. M., 146[66] (113), 146[67] (113), 292
Booher, M. A., 368[74] (359)
Boom, J. H., 79[45] (67)
Boone, S. J., 242[43] (156, 173, 188, 190)
Boosalis, M. S., 282[76] (260)
Borah, B., 337[89] (323)
Bordella, F., 146[38] (111)
Borgias, B. A., 79[34] (64)
Borowy-Borowski, H., 336[10] (315)
Borst, P., 367[3] (343, 348)
Bos, E. S., 243[131] (170)
Bosch, L., 367[15] (345), 367[16] (345)
Bose, N. K., 147[82] (114)
Bossmann, S. H., 242[91] (165)
Botchway, S. W., 423[135] (398), 424[179] (409, 410)
Botfiled, M. C., 283[134] (270)
Bothe, E., 422[52] (378), 422[68] (381, 403, 406, 410), 423[133] (398, 405, 406), 423[140] (399), 424[154] (402), 424[155] (402), 425[241] (418)
Bottka, S., 244[178] (183)
Bottomley, L. A., 101[91] (88)
Botuyan, M. V., 80[94] (76)
Boublikova, P., 101[129] (92), 102[137] (92), 102[156] (92)
Boulard, Y., 51[86] (34)
Bourdouxhe-Housiaux, C., 474[80] (437)

Bourgeois, W., 282[21] (254)

Bourson, J., 242[47] (158)

Boutorine, A. S., 242[47] (158), 243[107] (167, 168), 244[204] (188)

Bowden, G. T., 549[181] (516)

Bowness, K. M., 309[35] (289)

Bowser, C. A., 247[382] (217)

Box, H. C., 425[247] (419), 425[248] (419)

Boyd, F. L., 547[59] (495, 497), 547[63] (497)

Boyer, H. W., 283[114] (266, 267, 270)

Boyer, P. D., 51[111] (38), 52[135] (41)

Boykin, D. W., 457, 474[90] (441), 475[140] (453, 457, 458), 475[141] (453, 457, 458)

Boykin, D., 473[17] (428), 473[56] (433, 435, 452, 460), 475[149] (457), 475[151] (457)

Bracco, L. P., 150[285] (122)

Bradford, V. S., 548[116] (505)

Bradkova, E., 146[54] (112)

Bradková, E., 150[283] (122), 150[284] (122)

Bradley, D. H., 202, 246[334] (202, 203)

Bradley, J.-C., 151[341] (129)

Brakel, C. L., 246[309] (197)

Branch, A. D., 367[19] (346)

Brandenburg, G., 368[65] (355)

Brando, R., 310[142] (306)

Brandts, J. F., 30, 50[49] (30), 50[50] (31)

Brankamp, R. G., 103[215] (96)

Brassett, C., 474[84] (439)

Bratu, D. P., 13[49] (9)

Braunschweiler, L., 79[39] (64)

Breaker, R. B., 147[72] (113)

Breen, A. P., 421[15] (372), 546[23] (493, 494), 546[31] (494, 511), 548[143] (510, 511)

Breiner, K. M., 423[83] (385, 395)

Breipohl, G., 311[143] (306)

Brennan, C. A., 283[121] (267)

Brennan, R. G., 283[127] (268)

Brenner, D. J., 425[245] (418)

Brenner, S. L., 489[38] (481)

Breslauer, K. J., 12[36] (8), 33, 50[9] (16, 21, 29), 50[19] (18, 31), 51[59] (33), 51[62] (33), 459

Breslin, D. T., 424[186] (410)

Bresloff, J. L., 473[58] (433)

Breslow, R., 247[364] (211)

Brey, W., 475[170] (463, 464, 465, 466)

Bridgham, J., 241[31] (154, 158)

Bridson, P. K., 147[124] (118)

Brill, W., 244[154] (175)

Brill, W. K.-D., 152[435] (139)

Broca, C., 79[15] (59)

Broder, S., 309[45] (292, 293)

Brody, E., 367[12] (345)

Broeders, N. L. H. L., 149[251]

Brookes, P., 103[234] (96)

Broom, A. D., 248[423] (228)

Broseta, D., 52[160] (45), 80[79] (74)

Brousseau, R., 147[95] (114)

Brown, B. M., 489[75] (486)

Brown, D. D., 368[82] (360)

Brown, D. J. S., 80[101] (76)

Brown, D. M., 101[130] (92), 114, 117, 147[73] (114, 117), 321, 334

Brown, E. L., 149[220] (126)

Brown, F. K., 310[137] (305)

Brown, J. M., 147[137] (120, 127), 148[146] (127)

Brown, K. D., 309[81] (295)

Brown, R. K., 421[7] (372)

Brown, T., 242[42] (156), 243[151] (175), 243[153] (175), 244[195] (187), 244[198] (187, 189, 190), 247[361] (210)

Browne, K. A., 310[101] (299, 300, 301)

Broyde, S., 476[216] (470), 476[218] (470)

Broyles, S. S., 103[226] (96)

Bruening, G., 284[163] (274), 284[164] (274)

Brugghe, H. F., 336[11] (315)

Bruice, T. C., 300, 310[101] (299, 300, 301)

Bruice, T. W., 309[26] (288)

Bruist, M. F., 488[11] (478)

Brumbaugh, J. A., 246[294] (193)

Brunar, H., 338[171] (329), 338[172] (329)

Brunden, M. J., 147[68] (113)

Brunel, C., 12[26] (5)

Brünger, A. T., 80[86] (76), 80[89] (76)

Brusov, R. V., 489[78] (486)

Bruzik, K. S., 148[150] (127), 149[225]

Brysch, W., 152[401] (135)

Buc, H., 101[68] (86), 103[202] (95), 103[203] (95)

Buchanan, J. G., 283[87] (263)

Buchardt, O., 12[6] (4)

Buchi, G., 548[147] (512)

Büchi, H., 149[221] (122)

Buchko, G. W., 424[203] (411), 425[237] (418)

Buciak, J. L., 246[319] (199)

Buck, F., 79[68] (72)

Buck, G. A., 152[440] (142)

Buck, H. M., 149[251] (123)

Buckle, M., 103[203] (95)

Buczak, H., 475[140] (453, 457, 458), 475[141] (453, 457, 458)

Budzinski, E. E., 425[247] (419), 425[248] (419)

Buettner, G. R., 425[233] (417)

Bull, H. G., 150[281] (122)

Bulychev, N. V., 242[47] (158), 338[165] (329)

Burcham, P. C., 421[4] (372)

Burg, J., 247[355] (209)

Burger, J., 423[108] (389), 423[110] (390, 393)

Burger, R. M., 423[102] (388, 394)

Burgers, P. M. J., 151[350] (130)

Burgess, S., 244[195] (187)

Bürglin, T., 489[19] (479)

Burik, A., 149[206] (121)

Burke, J. M., 284[165] (274, 277, 279, 280), 284[166] (274, 277)

Burkhoff, A. M., 101[124] (90), 101[125] (90)

Burley, S. K., 283[117] (266, 268)

Burmeister, J., 244[161] (178)

Burrows, C. J., 89, 100[32] (85, 89), 100[33] (85, 89), 101[108] (89), 101[110] (89), 103[210] (96)

Burzynski, J., 152[451] (144)

Busby, S., 101[68] (86)

Busby, Jr., W. R., 548[147] (512)

Busson, R., 310[106] (299, 300)

Butcher, S. E., 284[176] (274)

Butler, W. O., 309[30] (289, 295)

Butterfield, K., 305, 310[117] (299, 305)

Buxton, G. V., 423[127] (398)

Buzayan, J. M., 284[163] (274), 284[164] (274)

Byrd, S., 476[221] (470)

Byrne, B. C., 13[52] (10)

Cacciapuoti, J. O., 103[224] (96)

Caddle, M. S., 102[145] (92, 94)

Cadet, J., 422[43] (375), 422[57] (378), 423[128] (398, 407), 423[129] (398, 407), 424[162] (406), 424[171] (407)

Cal, S., 338[154] (327)

Calabretta, B., 310[91] (296)

Caldwell, J. C., 80[87] (76)

Calendar, R., 367[28] (348)

Caligiuri, M. A., 12[21] (4)

Calladine, C. R., 79[16] (59), 100[9] (82)

Calnan, B. J., 80[75] (73)

Campbell, J. L., 12[31] (6)

Campbell, R. D., 102[134] (92, 93)

Candeias, L. P., 423[132] (398)

Cantor, C. R., 12[19] (4), 51[112] (38, 41, 42, 43)

Cao, X., 310[116] (299, 305)

Capaldi, D. C., 12[12] (4)

Capel, B., 367[5] (344, 367)

Capobianco, M. L., 369[112] (366)

Carbonaux, C., 80[100] (76)

Carcuro, A., 369[112] (366)

Cardellina, III, J. H., 548[101] (503)

Cardullo, R. A., 241[19] (155)

Carey, J., 281[11] (253)

Carey, M., 489[68] (484)

Carlson, D. V., 244[175] (182)

Carlson, J. O., 103[216] (96)

Carlstedt-Duke, J., 489[66] (484)

Carmalt, D., 281[15] (253)

Carmical, J. R., 424[193] (410)

Carpenter, M. L., 100[55] (85)

Carr, S. A., 423[118] (395)

Carrano, M., 546[22] (493)

Carter, B. J., 103[245] (98)

Cartwright, I. L., 101[71] (86), 101[79] (87)

Caruthers, M. H., 12[4] (2, 3), 50[3] (16), 51[69] (33), 100[5] (82), 108, 126, 130, 131, 146[28] (108, 112, 126, 130), 146[29] (108, 112, 126, 130), 293, 295

Casadevall, A., 100[63] (86)

Casale, R., 149[259] (121, 124), 222, 247[404] (222), 327

Casas, C., 242[75] (162), 242[77] (162)

Cascio, D., 475[131] (452, 456), 547[60] (497)

Case, D. A., 80[87] (76)

Case-Green, S. C., 51[63] (33)

Cassady, J. M., 548[144] (511)

Cassidy, S. A., 468, 476[195] (468)

Castro, M. M., 338[136] (325)

Cate, J. H., 50, 53[209] (50), 53[210] (50)

Catlin, J. C., 108, 146[15] (108)

Catteau, J.-P., 101[109] (89)
Catterall, H., 424[160] (405)
Caulfield, T. J., 310[98] (298), 310[104] (299, 301)
Cavanaugh, P., 310[110] (299, 303)
Cazenave, C., 243[117] (168), 309[45] (292, 293)
Cech, D., 242[55] (159), 244[158] (177, 190), 244[208] (190)
Cech, T. R., 12[42] (8), 49, 52[138] (41, 45, 49, 50), 52[139] (41, 45, 49, 50), 53[202] (49), 53[209] (50), 274, 279, 280, 616, 617, 639
Cech, T., 103[221] (96)
Cedergren, R. J., 152[455] (144)
Cedergren, R., 284[190] (278), 284[191] (278)
Cera, C., 547[45] (494)
Cerny, R. L., 245[259] (190)
Chaddha, M., 150[273] (124)
Chadha, R. K., 548[130] (508)
Chai, M., 147[92] (114), 147[94] (114)
Chaires, J. B., 100[51] (84), 101[90] (88), 451
Chaix, C., 149[216] (120, 121), 149[217]
Chakerian, V., 241[31] (154, 158)
Chakhmakhcheva, O. G., 152[427] (139), 244[200] (187, 190)
Chalikian, T. V., 50[19] (18, 31), 51[91] (34), 51[96] (35)
Chamberlin, A. R., 337[99] (324)
Chambers, R. W., 283[90] (263), 336[10] (315)
Champoux, J. J., 80[115] (77)
Chan, C., 246[312] (197)
Chan, H., 309[72] (294)
Chan, K. L., 548[110] (505), 548[141] (510)
Chandrasekaran, S., 310[92] (296)
Chang, C., 204, 246[341] (204)
Chang, C.-A., 246[342] (204)
Chang, D.-H., 423[120] (397)
Chang, J. C. S., 548[149] (512)
Chang, S. S., 547[87] (502)
Chang, W., 337[76] (322)
Changchien, L., 52[131] (41)
Chao, J., 247[353] (209), 247[354] (209)
Chapman, L., 489[63] (484, 485), 489[67] (484)
Charetier, E., 80[105] (76)
Chari, M. V., 339[226] (334)
Charlier, M., 425[231] (417)
Charnecki, S. E., 103[238] (96)
Charretier, E., 474[65] (434)
Chartrand, P., 284[193] (278)
Charubala, R., 148[150] (127), 148[153] (120)
Chassignol, M., 168, 242[83] (164), 242[85] (164), 243[115] (168), 243[119] (168, 190), 244[165] (181), 244[171] (181), 245[243] (190)
Chastain, M., 473[49] (433, 469)
Chatgilialoglu, C., 422[47] (376), 422[48] (376)
Chatterjee, M., 241[27] (159), 241[28] (159), 242[49] (159)
Chattopadhyaya, J. B., 147[129] (118), 147[133] (120), 148[138] (120), 148[140] (120), 149[208] (126), 149[209] (126),

149[246] (126), 149[261] (123), 150[267] (126), 151[338] (127)
Chaturredi, S., 309[46] (292, 294, 301)
Chaudhary, A. K., 547[92] (502)
Chaudhuri, N. C., 282[33] (255)
Chaw, Y. F. M., 548[93] (502)
Cheatham, S. F., 547[63] (497)
Chehab, F. F., 241[10] (158)
Chen, C. B., 100[27] (85, 86)
Chen, C.-H. B., 101[81] (87), 101[83] (87)
Chen, H., 547[33] (494)
Chen, J., 247[370] (215)
Chen, J. H., 13[55] (10)
Chen, J.-H., 490[94] (488)
Chen, J. K., 12[5] (4)
Chen, J.-K., 244[175] (182), 244[176] (183), 310[87] (296), 310[91] (296)
Chen, L., 241[26] (158)
Chen, P., 86, 425[254] (421), 579
Chen, Q., 100[54] (85)
Chen, S.-l., 474[103] (444, 446)
Chen, X., 101[108] (89)
Chen, Y., 101[107] (89)
Chen, Y.-O., 309[68] (293)
Cheng, C. C., 422[28] (373, 374, 376, 380), 422[29] (373, 374, 380)
Cheng, J. C. Y., 283[92] (263)
Cheng, J.-W., 80[98] (76)
Cheng, Y.-K., 51[100] (36)
Chenier, J. H. B., 422[54] (378)
Chernov, B. K., 80[112] (77)
Cheruvallath, Z. S., 12[12] (4)
Chestney, D. L., 423[99] (387)
Cheung, S., 80[81] (75)
Chevion, M., 425[234] (417)
Chevrie, K., 244[212] (190)
Chevrier, M., 243[117] (168)
Chiang, M. Y., 309[72] (294)
Chiasera, G., 547[84] (501)
Chin, D.-H., 423[118] (395)
Chiu, C.-Y., 310[91] (296)
Chládek, S., 148[144] (120), 148[145] (120)
Chmielewski, J., 488[16] (478, 479), 489[36] (481)
Cho, B. P., 338[124] (325)
Choi, D., 549[186] (517)
Chollet, A., 148[159] (120), 170, 323
Chollet-Damerius, A., 337[81] (323)
Choo, Y., 489[71] (485), 489[72] (486)
Chou, S. H., 79[42] (65), 80[82] (75), 80[98] (76)
Chou, T.-C., 475[137] (452, 453)
Chow, C. S., 88, 100[31] (85, 88), 101[104] (89)
Chow, F., 146[47] (112)
Chow, T. Y.-K., 245[266] (190)
Chowdary, D., 549[172] (515, 516), 549[179] (516)
Chowdhary, B. C., 368[71] (357)
Chowrira, B. M., 284[165] (274, 277, 279, 280), 284[176] (274)
Christner, D. F., 422[81] (385), 422[82] (385), 423[85] (385), 423[89] (385, 394, 395, 421)
Christodoulou, C., 147[137] (120, 127), 148[146] (127)
Christopherson, M. S., 338[148] (327), 339[211] (332)

Christy, M., 103[208] (95)
Chu, B. C., 368[87] (362)
Chu, B. C. F., 170, 243[135] (170)
Chu, C. K., 283[88] (263), 283[89] (263), 308[11] (287, 298)
Chu, D. T. W., 475[164] (462)
Chu-Moyer, M., 506, 548[124] (507)
Churchill, M. E. A., 546[30] (494)
Chypre, C., 246[343] (205)
Ciszewski, L. A., 339[210] (332)
Claes, P., 310[106] (299, 300)
Claesen, C. A. A., 148[165] (120), 148[166] (120)
Claesen, C., 151[369] (133)
Clardy, J., 490[91] (488)
Clark, D. F., 150[278] (122), 150[279] (122)
Clark, J. M., 424[174] (407)
Clark, T., 424[182] (410)
Clark, V. M., 146[5] (106)
Clegg, R. M., 241[9] (158)
Cleland, W. W., 52[135] (41), 282[65] (258)
Clement, J. J., 475[164] (462), 476[224] (471), 548[136] (508, 509)
Clementi, E., 475[147] (456)
Climie, S. C., 245[214] (190)
Clivio, P., 246[331] (202), 246[332] (202), 327
Clore, G. M., 475[172] (465), 475[174] (465), 489[60] (484), 489[61] (484), 670
Clusel, C., 368[88] (362)
Clyne, J., 246[287] (192), 246[288] (192)
Cocuzza, A. J., 241[35] (155), 243[139] (171), 247[362] (211)
Coenen, A. J. J. M., 149[251]
Coffman, H., 547[80] (500, 541)
Cohen, D., 51[92] (35)
Cohen, G., 473[47] (433)
Cohen, J. S., 243[103] (167), 244[166] (181), 244[177] (183), 248[467] (238), 308[12] (287, 292), 309[45] (292, 293)
Cohen, S., 146[52] (112)
Cohn, M., 282[64] (258)
Cole, D. L., 12[12] (4)
Cole, P. E., 52[156] (45, 48), 53[199] (48, 49), 53[201] (48)
Cole-Strauss, A., 13[52] (10)
Coleman, R. S., 246[328] (202), 246[329] (202)
Coles, B. F., 548[152] (512, 513)
Coll, M., 337[88] (323)
Collier, D. A., 100[18] (82), 100[19] (82)
Collingwood, S. P., 297, 310[89] (296, 297, 298)
Collins, M., 243[132] (170)
Collins, M. L., 337[109] (324)
Colocci, N., 317, 336[46] (317)
Colson, A.-O., 422[25] (373, 380, 383, 385, 400, 403, 404, 405), 423[139] (398, 410, 413)
Colson, P., 473[13] (428, 436), 474[80] (437), 475[144] (453, 459), 475[157] (459)
Colston, J. E., 247[381] (216, 217)
Conconi, A., 103[225] (96)
Condon, T. P., 242[56] (159, 160)
Connell, C. R., 241[31] (154, 158), 248[442] (233)

Conner, B. N., 99[3] (82)

Connolly, B. A., 241[20] (154), 242[94], 246[330] (202), 246[333] (202), 282[67] (259), 282[77] (260), 283[102] (265, 270), 336[4] (314, 315), 336[5] (314, 315)

Conti, M., 337[87] (323)

Conway, N. E., 248[478] (240), 249[482] (240), 249[483] (240)

Cook, A. F., 246[309] (197), 309[47] (292, 294)

Cook, G. P., 422[35] (373, 402, 403, 404), 425[229] (411, 413, 419)

Cook, P. D., 242[56] (159, 160), 242[57] (159), 247[368] (214), 248[432] (231), 301, 305, 308[9] (285, 287, 300), 308[11] (287, 298), 310[105] (299, 300), 310[109] (299, 302)

Cook, P. R., 100[55] (85)

Cook, R. M., 241[23] (155)

Cooper, B. C., 369[104] (363)

Cooper, J. P., 367[21] (347, 353)

Copp, B. R., 547[85] (501, 529)

Coppola, G., 309[47] (292, 294)

Corby, N. S., 146[6] (106)

Corcoran, J. W., 547[51] (495)

Cordes, E. H., 150[281] (122)

Corey, D., 147[83] (114)

Cormier, J. F., 303, 310[129] (303, 304)

Corongiu, G., 475[147] (456)

Cory, M., 474[119] (449, 450), 474[120] (449, 450), 548[104] (504)

Cosman, M., 476[218] (470), 476[219] (470)

Cosstick, R., 248[477] (240), 282[61] (258, 265), 282[67] (259), 283[104] (265), 297, 310[88] (296, 297), 330, 336[5] (314, 315)

Costello, A. J. R., 79[65] (72)

Cotton, R. G. H., 102[134] (92, 93)

Coull, J. M., 241[4] (154), 241[5] (168), 242[69] (161), 242[93] (166)

Coulmé, J. J., 309[45] (292, 293)

Coulson, A. R., 100[24] (82)

Courey, A. J., 244[181] (184)

Courtney, S. H., 476[215] (469)

Coury, J. E., 101[91] (88)

Cousineau, B., 284[190] (278)

Coutts, S. M., 243[104] (168)

Coviello, G. M., 476[203] (469)

Cowart, M., 246[299] (194, 195)

Cowburn, D., 80[73] (72, 73)

Cowsert, L. M., 242[57] (159)

Cox, G. B., 146[44] (112)

Cox, T., 245[241] (190)

Coyne, M. Y., 241[15] (158)

Cozzarelli, N. R., 368[29] (348), 368[30] (348)

Craig, M. E., 43, 52[147] (43, 44)

Craig, M. L., 101[127] (90)

Craik, D. J., 548[138] (509)

Cramer, F., 108, 146[15] (108)

Crane, L. E., 548[93] (502)

Cravador, A., 242[68] (161, 166)

Crawford, I. P., 283[125] (268)

Crawford, J. L., 80[117] (77)

Crea, R., 146[39] (111)

Crich, D., 422[69] (382), 422[70] (382)

Crick, F. H. C., 81, 87, 95, 106, 477, 488[1] (477)

Crooke, S. T., 152[444] (142), 308[1] (285, 287), 308[3] (285), 309[27] (288), 309[28] (289), 311[146] (307), 311[147] (288)

Crosby, N. T., 247[403] (222)

Crothers, D. M., 36, 45, 48, 49, 357, 368[55] (354, 365), 368[69] (357), 436, 473[12] (428, 436), 473[58] (433), 474[106] (444, 446), 475[153] (459), 475[163] (460), 476[210] (469), 549[184] (517), 630

Crow, S., 475[138] (452, 453)

Croy, R. G., 548[147] (512)

Cruickshank, K. A., 194, 245[278] (191, 197), 245[279] (191, 197), 246[295] (194), 246[312] (197)

Cruz, P., 284[173] (274)

Cubie, H., 247[361] (210)

Cucco, C., 310[91] (296)

Cuenoud, B., 489[50] (482), 633

Cui, M., 103[245] (98)

Cullinan, D., 79[46] (67, 73, 76), 80[99] (76)

Cullis, P. M., 423[136] (398, 401, 402), 424[150] (401, 402), 424[181] (409, 410)

Cummings, J., 12[23] (5)

Cummins, L., 309[40] (292, 293), 310[109] (299, 302)

Cummins, L. L., 247[368] (214), 308[15] (287), 310[127] (302)

Cundiffe, E., 473[7] (428, 429, 432, 433, 436, 444, 452, 453)

Cunningham, R. P., 422[61] (380)

Cushing, T. D., 282[43] (256)

Cushman, C., 309[80] (295)

Cushman, C. D., 283[105] (265), 320

Cushman, M., 244[175] (182), 244[176] (183)

Czapski, G., 425[234] (417)

Czombos, J., 244[178] (183)

D'Ambrosio, M., 547[84] (501)

D'Aurora, V., 101[66] (86)

D'Cunha, J., 102[166] (92)

D'Souza, D. J., 368[77] (359, 360), 368[78] (359)

Dabkowski, W., 293, 309[43] (292, 293)

Dabrow, M., 473[22] (430, 435)

Dabrowiak, J. C., 101[90] (88), 101[112] (89)

Dahl, B. H., 152[414] (138), 152[415] (138)

Dahl, O., 138, 147[128] (118), 147[129] (118), 149[206], 149[207], 293

Dahlberg, J. E., 102[150] (92, 93)

Dahlén, P., 246[327] (201)

Dahlman, K., 489[66] (484)

Dahm, S. C., 284[178] (275)

Dai, J., 103[208] (95)

Dai, W.-M., 546[32] (494)

Dam, R. J., 247[362] (211)

Damha, M. J., 147[126] (118), 147[127] (118)

Damle, V., 50[41] (24, 28)

Dan, A., 248[417] (226), 248[418] (226)

Dandley, J.-P., 79[58] (72)

Daniher, A. T., 248[421] (228, 229)

Danilevskaya, O. N., 102[167] (92, 95), 102[197] (95)

Danishefsky, A. T., 101[85] (88), 101[92] (88)

Danishefsky, S. J., 506, 548[103] (504), 548[124] (507), 549[174] (515, 516)

Dannenberg, J. J., 101[88] (88)

Dannoue, Y., 423[95] (386)

Darke, P. L., 246[300] (195)

Darracq, N., 548[96] (503)

Das, B. P., 475[140] (453, 457, 458)

Daskalov, H. P., 147[122] (117), 147[123] (117)

Daub, G. W., 151[353] (130, 131)

Daubendiek, S. L., 309[62] (292), 363

Daugherty, M. A., 423[83] (385, 395)

Dautant, A., 283[136] (271)

Daval-Valentin, G., 476[187] (466, 467)

Davanloo, P., 52[145] (43)

Daves, Jr., G. D., 263, 283[91] (263), 283[92] (263)

David, S. S., 339[215] (333)

Davies, D. R., 476[181] (466)

Davies, J., 546[1] (492)

Davies, M. J., 424[160] (405)

Davis, D., 79[40] (64)

Davis, D. R., 310[125] (302)

Davis, G. R., 243[109] (168)

Davis, J. T., 337[96] (323)

Davis, M. A., 425[221] (411)

Davis, P., 149[259] (121, 124)

Davison, A., 247[361] (210)

Davison, E. C., 338[128] (325)

Dawar, H., 243[111] (168)

Dawidzik, J. D., 425[249] (419)

Dawson, M. I., 241[24] (158)

Day, L. A., 100[63] (86)

Dayn, A., 102[184] (93, 94, 95)

Dayton, B. D., 548[106] (504, 505), 548[107] (504, 505)

DeBear, J. S., 147[69] (113)

DeDionisio, L. A., 12[5] (4)

DeGrado, W. F., 489[38] (481), 489[47] (482)

DeGraff, W. G., 425[220] (411)

DeLong, R. K., 244[185] (184)

DeNapoli, L., 369[113] (366)

DePillis, G. D., 337[75] (322)

DeVoe, H., 51[72] (33)

Dean, N. M., 308[16] (288), 308[25] (288)

Dean, N., 310[109] (299, 302)

Decarroz, C., 422[43] (375), 423[128] (398, 407), 423[129] (398, 407)

Decout, J.-L., 243[124] (168), 243[129] (169, 184)

Dedon, P. C., 425[238] (418), 570, 574

Deeble, D. J., 424[156] (403)

Dejean, E., 247[378] (216)

Delaglio, F., 79[19] (59)

Delancy, W., 309[47] (292, 294)

Delbarre, A., 474[104] (444, 446), 475[127] (451)

Delcourt, S. G., 51[74] (33)

Delecki, D., 310[110] (299, 303)

Delepierre, M., 474[104] (444, 446), 475[128] (451)

Delgado, J. W., 546[4] (492)

Demeny, T., 473[24] (431)

Demeunynck, M., 101[106] (89)

Dempcy, R. O., 310[101] (299, 300, 301)

Denissenko, M. F., 12[13] (4)

Dennis, J., 283[129] (268)

Denny, W. A., 368[38] (351), 368[39] (351), 449, 474[76] (437, 440), 474[86] (440), 475[125] (451), 475[126] (451)

Depart, F., 282[53] (257)

Depierreux, C., 242[70] (161, 190)

Derrick, W. B., 284[178] (275)

Dervan, P. B., 46, 86, 215, 246[306] (196), 246[321] (199), 247[374] (215), 247[375] (215), 281[2] (252, 253), 281[7] (253), 282[19] (253), 283[138] (271), 283[139] (271), 317, 319, 324

Deshmukh, H. M., 248[423] (228)

Desjarlais, J. R., 489[70] (485)

Deutsch, J., 146[52] (112)

Devadoss, C., 424[187] (410)

Devasagayam, T. P. A., 424[212] (411)

Devivar, R. V., 339[209] (332)

Devlin, T., 149[237] (121, 124), 149[238], 151[384] (133), 309[49] (292, 295), 309[52] (292, 295)

Devreese, B., 248[428] (230)

Dewanjee, M. K., 241[13] (159)

Dewey, T. G., 42, 52[140] (42)

DiMichele, L. J., 243[109] (168)

Diamond, K. B., 548[101] (503)

Dickerhof, W. A., 245[258] (190)

Dickerson, R. E., 79[15] (59), 79[16] (59), 80[118] (77), 82, 99[3] (82), 100[6] (82), 100[7] (82), 456

Dickmann, S., 339[199] (330)

Diekmann, S., 241[9] (158), 283[122] (267), 283[123] (267)

Diener, T. O., 367[8] (344, 346)

Dietz, T. M., 425[219] (411)

Dikshit, A., 150[273] (124)

Dimock, S., 310[126] (302), 310[128] (303)

Ding, D., 100[64] (86)

Ding, Z.-M., 103[249] (98)

Dipple, A., 247[402] (222, 225)

Dissinger, S., 247[381] (216, 217)

Divakar, K. J., 147[114] (117)

Dix, T. A., 423[117] (395)

Dixon, D. A., 367[25] (347)

Dixon, D. W., 474[90] (441)

Dixon, W. J., 100[34] (85, 90)

Dizaroglu, M., 546[28] (494)

Dizdaroglu, M., 102[159] (92)

Dobrikov, M. I., 242[65] (161), 247[385] (218)

Dobrynin, V. N., 102[167] (92, 95), 102[197] (95)

Docherty, K., 102[186] (93)

Dodd, C., 248[442] (233)

Dodds, D. R., 100[5] (82)

Doetsch, P. W., 424[190] (410)

Dohtsa, K., 151[340] (127)

Doktycz, M. J., 51[73] (33)

Dolinnaya, N. G., 369[110] (366)

Dolinnaya, N., 246[342] (204)

Dombroski, B. A., 100[34] (85, 90), 101[116] (89, 90)

Dombrowski, B. A., 281[9] (253)

Domdey, H., 367[12] (345)

Domenico, P., 150[275] (120)

Dominiak, G., 247[351] (208)

Donahue, J. M., 336[1] (314), 336[9] (315)

Doolittle, R. F., 488[10] (478)

Dormady, S. J., 51[73] (33)

Doronin, S. V., 247[385] (218)

Doroshow, J. H., 102[159] (92)

Dorr, R. T., 549[181] (516)

Dosanjh, H. S., 53[194] (47)

Doty, P., 52[147] (43, 44)

Doudna, J. A., 284[204] (280, 281)

Dougherty, J. P., 243[132] (170)

Douglas, K. T., 100[60] (86)

Douglas, M. E., 248[433] (231)

Doulci, T., 422[57] (378)

Dousset, P., 151[335] (127)

Downing, K., 368[66] (356)

Doyle, T. W., 423[89] (385, 394, 395, 421)

Drach, J. C., 282[57] (257)

Draganescu, A., 101[117] (89)

Draper, D. E., 100[48] (84, 87), 469

Draper, K., 308[14] (287)

Draves, P. H., 103[248] (98)

Dreef, C. E., 151[369] (133), 151[374] (133)

Dreef-Tromp, C. M., 150[270] (120), 150[271] (120)

Drescher, C. W., 12[27] (5)

Drew, H., 79[15] (59), 80[118] (77)

Drew, H. R., 79[62] (72), 99[3] (82)

Dreyer, G. B., 101[80] (87), 215

Driller, H., 149[214] (122)

Driver, G. B., 308[14] (287)

Driver, V. B., 309[26] (288)

Drobny, G., 79[42] (65)

Drobny, G. P., 51[119] (39)

Droge, P., 368[30] (348)

Du, S. M., 368[45] (353), 368[46] (353)

Dubendorff, J. W., 150[285] (122)

Duchange, N., 246[286] (192)

Ducharme, Y., 305, 310[138] (305)

Duchesne, J., 52[151] (43)

Duck, P. D., 146[46] (112)

Duckett, D. R., 102[143] (92)

Duckworth, G., 247[361] (210)

Duh, J.-L., 247[392] (219, 220)

Duker, N. J., 424[170] (407)

Duncan, L., 147[91] (114), 147[92] (114)

Dunn, D. A., 547[35] (494, 544)

Duplaa, A.-M., 149[217], 149[218] (121), 338[120] (325), 338[123] (325)

Dupraz, B., 475[128] (451)

Dupret, D., 245[263] (190)

Dupuy, C. G., 246[302] (195)

Durand, M., 244[212] (190)

Durland, R. H., 336[35] (316), 338[150] (327), 338[152] (327), 339[217] (333), 339[224] (334)

Durland, R. S., 339[221] (334)

Durrant, I., 243[144] (172, 174), 243[145] (172, 174)

Dussy, A., 423[114] (391)

Duval-Valentin, G., 102[183] (93)

Dvorak, M., 102[191] (94)

Dykstra, C. C., 475[142] (453, 457, 458)

Dzau, V. J., 368[90] (362)

Eadie, J. S., 147[131] (119)

Earnshaw, D. J., 310[88] (296, 297)

Earnshaw, W. C., 12[32] (8)

Eaton, D. L., 546[8] (492, 513)

Ebel, S., 242[42] (156)

Ebinger, K., 336[45] (317)

Ebright, R. H., 489[55] (483)

Ecker, D. J., 242[57] (159), 308[14] (287), 310[127] (302), 636

Eckstein, F., 151[347] (130), 151[348] (130), 241[14] (155), 243[108] (168), 243[116] (168), 245[216] (190), 248[429] (230), 248[430] (230), 249[486] (240), 249[487] (240), 282[46] (257, 278), 282[60] (258), 283[95] (264), 283[102] (265, 270), 284[194] (278), 284[202] (280), 291, 308[8] (285), 309[54] (291), 324

Eckstein, F. A. L. D., 284[162] (274)

Ede, N. J., 243[99] (167)

Eder, E., 547[90] (502)

Edgasr, A. R., 283[87] (263)

Edge, M. D., 146[37] (111)

Edison, A. S., 79[17] (59)

Edwards, W. D., 475[133] (452, 453)

Efcavitch, J. W., 147[131] (119), 148[204] (126)

Efimov, V. A., 152[427] (139), 244[200] (187, 190)

Efstratiadis, A., 102[194] (94)

Eftedal, I., 424[166] (407)

Egan, W., 152[446] (143), 248[468] (238), 248[469] (238), 283[108] (265)

Egbertson, M., 549[174] (515, 516)

Eggleston, A. K., 367[25] (347)

Egholm, M., 12[6] (4)

Eglander, S. W., 80[83] (75)

Egli, M., 304, 310[132] (304)

Ehrlich, S. D., 369[105] (364)

Eich, G., 79[38] (64)

Eichhorn, G. L., 100[62] (86)

Eickhoff, H., 146[63] (113)

Eida, K., 151[382] (133)

Eide, L., 424[166] (407)

Eigen, M., 38, 39, 51[110] (38), 51[115] (39, 43, 44, 46)

Eisenbeiss, F., 146[56] (112)

Eisenberg, D., 51[122] (40)

Eisenberg, H., 473[47] (433)

Eisenberg, M., 338[162] (328)

Eisenberg, W., 248[474] (239)

Ejadi, S., 549[156] (513)

Ekker, S. C., 489[25] (479)

Ekland, E. H., 52[174] (46)

Elcock, A. H., 51[93] (35)

Eldrup, A. B., 309[65] (293)

Elghanian, R., 13[53] (10)

Elgin, S. C. R., 101[71] (86), 101[79] (87)

Elie, C., 423[114] (391)

Elison, K. E., 368[90] (362)

Ellenberger, T., 488[5] (477, 478, 480, 482, 484), 489[37] (481)

Ellestad, G. A., 547[76] (500), 570

Ellington, A. D., 13[58] (10, 624, 626, 638)

Elliot, J., 474[77] (437, 440)

Elove, G. A., 52[124] (40)

Emson, P. C., 241[18] (155)

Endo, M., 242[38] (156), 242[87] (164), 310[85] (295)
Engberts, J. B. F. N., 244[207] (190)
Engel, I., 489[33] (480)
Engels, J. W., 146[26] (108), 148[164] (120), 148[167] (120), 297, 322
Engels, J., 150[310] (126)
Englander, S. W., 474[64] (434)
Englisch, U., 244[179] (184)
Enjolras, N., 368[88] (362)
Epe, B., 424[191] (410), 547[36] (494, 544)
Ercolani, L., 336[27] (316)
Erdmann, P., 422[72] (382, 393, 403), 423[109] (389), 423[114] (391)
Erdmann, V. A., 244[208] (190), 244[209] (190)
Erfle, H., 146[57] (112)
Erickson, R. P., 337[59] (319)
Erie, D. A., 369[97] (362)
Eritja, E., 147[80] (114)
Eritja, R., 151[373] (133), 152[399] (140, 141), 242[59] (160), 243[98], 243[100] (167), 245[231] (190), 282[76] (260), 326, 336[12] (315), 336[14] (315)
Erkelens, C., 79[57] (72)
Erlich, H. A., 241[29] (155)
Ernst, R. R., 79[25] (62), 79[37] (64)
Escarceller, M., 243[98] (167), 243[100] (167)
Eschenmoser, A., 324, 337[104] (324)
Escudé, C., 283[149] (273), 476[180] (467, 468)
Esposito, F., 103[215] (96)
Essigmann, J. M., 424[198] (410), 548[147] (512), 549[157] (513)
Esterbauer, H., 547[88] (502, 503)
Estes, L. A., 548[127] (508)
Estes, L., 548[129] (508)
Esteve, A., 424[198] (410)
Evans, D. A., 424[161] (405)
Evans, F. E., 338[124] (325)
Evans, M. R., 243[144] (172, 174), 243[145] (172, 174)
Evans, M. S., 425[247] (419)
Evans, T., 102[194] (94), 670
Evertsz, E. M., 50[40] (24, 28, 36)
Ezaz-Nikpay, K., 329, 337[76] (322), 338[170] (329)

Fabbro, D., 308[16] (288), 308[25] (288), 309[63] (292, 306)
Fabris, D., 337[96] (323)
Faeber, P., 282[78] (260)
Fägerstam, L., 52[125] (40)
Fagerstam, L. G., 52[126] (40)
Fahey, R. C., 425[230] (417), 425[244] (418)
Fahy, E., 243[109] (168)
Fairall, L., 489[62] (484), 489[63] (484, 485)
Fairley, T. A., 474[119] (449, 450)
Fairman, R., 489[38] (481)
Fajur, S. R., 284[184] (276, 277)
Fallon, J., 151[342] (129)
Fan, J.-Y., 475[165] (462, 463)
Fang, Z.-Q., 152[419] (139)
Fannin, L., 547[57] (495)

Fantone, J. C., 421[8] (372)
Farkas, S., 148[156] (120), 148[162] (120)
Farooqui, F., 146[62] (113)
Farr, R. N., 283[92] (263), 283[94] (263)
Fasold, K. I., 473[28] (428)
Fass, D., 489[37] (481)
Fathi, R., 294, 309[47] (292, 294)
Faucette, L. F., 547[55] (495)
Faucon, B., 476[187] (466, 467)
Favre, A., 242[84] (164), 246[331] (202), 246[332] (202), 284[196] (279)
Fazakerley, G. V., 336[29] (316), 336[30] (316), 336[33] (316), 337[112] (324), 337[113] (324), 338[123] (325)
Fedin, E. I., 151[388] (133)
Fedor, M. J., 284[168] (274)
Fedoroff, O. Y., 80[114] (77)
Feigon, J., 283[151] (273)
Felder, E., 245[233] (190)
Felding, J., 309[65] (293)
Feldstein, P. A., 284[167] (274), 284[202] (280)
Felsenfeld, G. D., 476[181] (466), 476[182] (466), 489[60] (484), 489[61] (484), 670
Feng, J. A., 488[12] (478, 479)
Fensholdt, J., 248[436] (232), 248[437] (232)
Fenton, H. J. H., 100[57] (85)
Fera, B., 79[68] (72)
Ferentz, A. E., 282[41] (256)
Fermandjian, S., 80[109] (76)
Fernandez-Forner, D., 147[80] (114)
Fernandez-Saiz, M., 475[176] (466)
Ferrari, M. E., 50[13] (17)
Ferré-D'Amaré, A. R., 489[40] (482)
Ferruti, P., 146[67] (113)
Fersht, A., 52[137] (41)
Fersht, A. R., 283[109] (266), 283[110] (266)
Festy, B., 473[48] (433)
Fiala, R., 476[218] (470), 476[219] (470), 549[187] (517)
Fidanza, J. A., 248[470] (238), 248[471] (238), 249[479] (240), 249[481] (240), 249[483] (240), 283[103] (265)
Fiel, R. J., 474[91] (441)
Fielden, E. M., 425[239] (418)
Figueroa, N., 52[159] (45), 80[84] (75)
Fikus, M., 337[94] (323)
Filippov, S. A., 102[167] (92, 95)
Filzen, G. F., 422[70] (382)
Finch, J. T., 284[157] (274, 276)
Fino, J. R., 244[155] (175)
Fire, A., 363, 368[91] (362, 363, 366)
Fischer, M. A., 50[23] (20)
Fischer, R. W., 295, 309[50] (292, 295)
Fishelevich, R., 309[80] (295)
Fisher, C. L., 475[174] (465)
Fisher, E. F., 100[5] (82), 126
Fisher, H. F., 50[10] (16, 31)
Fisher, R. J., 475[174] (465)
Flaherty, K. M., 284[156] (274, 276)
Flanagan, M. E., 547[80] (500, 541), 547[81] (500, 541)
Flecher, M. C., 474[114] (446, 448)
Fleischmann, R. D., 12[17] (4)
Flick, M. B., 425[242] (418)

Fliss, M., 13[64] (12)
Flores, C., 241[19] (155)
Fok, V., 244[185] (184)
Foldes-Papp, Z., 146[63] (113), 146[64] (113)
Földesi, A., 148[190] (121)
Folsom, V., 246[313] (198)
Folta-Stogniew, E., 474[66] (434)
Fontain, E., 151[376] (133)
Fontanel, M.-L., 243[147] (173), 243[148] (173), 244[156] (176, 177), 244[157] (177, 205)
Foote, C. S., 424[205] (411), 424[208] (411)
Ford, E., 367[14] (345, 367)
Forster, A. C., 284[170] (274)
Forsythe, R. H., 547[86] (502)
Fountain, M., 52[177] (45, 49)
Fouqué, B., 246[343] (205)
Fourrey, J.-L., 132, 150[301] (123, 136), 151[335] (127), 151[364] (132), 151[365] (132), 246[331] (202), 246[332] (202), 284[196] (279)
Fowler, J. F., 425[222] (411)
Fox, J. J., 150[274] (120), 282[86] (262), 283[88] (263), 283[89] (263)
Fox, K. R., 100[12] (82), 102[152] (92), 102[162] (92), 468
Fox, K., 475[150] (457, 458)
Fox, M., 309[26] (288)
Foye, W. O., 546[5] (492), 547[50] (495, 497, 499, 505, 520), 548[104] (504)
Fraga, D., 245[242] (190)
François, J.-C., 243[120] (168, 169), 243[121] (168), 244[169] (181), 244[170] (181)
Francois, P., 147[125] (118)
Frank, B. L., 422[81] (385), 422[82] (385), 423[85] (385), 423[89] (385, 394, 395, 421)
Frank-Kamenetskii, M. D., 102[151] (92), 102[167] (92, 95), 102[193] (94, 95)
Frank, R., 150[289] (122, 126)
Frankel, A. D., 473[60] (434), 473[62] (434)
Franklin, R. E., 81, 82, 99[1] (81)
Fraser, C. M., 12[17] (4)
Fraser, W., 147[99] (114)
Fratini, A. V., 99[3] (82)
Frederick, C. A., 283[114] (266, 267, 270)
Freedman, L. P., 489[66] (484)
Freier, S. M., 242[57] (159), 247[368] (214), 308[14] (287), 308[25] (288), 309[27] (288), 310[109] (299, 302), 310[127] (302)
Frenkel, K., 424[168] (407)
Fresco, J. R., 46, 52[182] (46), 52[183] (46)
Freudenberg, S., 473[28] (428)
Freund, H. G., 425[247] (419), 425[248] (419)
Frey, P. A., 283[131] (270)
Friedman, A., 368[76] (359)
Friedman, A. E., 53[195] (47)
Friedman, R. A., 50[11] (17)
Friedman, R. A. G., 474[107] (444, 446)
Friedmann, T., 101[130] (92)
Friere, F., 50[9] (16, 21, 29)
Friess, S. L., 51[110] (38)

Fritsch, E. F., 100[26] (83)

Fritsch, V., 308[19] (288, 299, 300, 301, 305), 310[105] (299, 300), 310[107] (299, 301)

Fritz, H.-J., 150[289] (122, 126)

Fritzsche, H., 337[112] (324)

Froehler, B. C., 150[290] (122), 152[423] (139), 152[428] (139), 282[50] (257, 265, 273), 282[56] (257, 273), 283[142] (271, 273), 283[148] (272), 317, 318

Fröhlich, T., 337[107] (324)

Frolova, E. I., 242[86] (164), 247[372] (215)

Fronza, G., 103[213] (96)

Frost, A. A., 51[120] (40)

Frostell-Karlsson, A., 52[126] (40)

Frye, R. A., 147[89] (114)

Fu, D. J., 245[215] (190), 282[51] (257, 278), 284[183] (276, 277, 278), 284[184] (276, 277)

Fu, J. M., 79[52] (71), 79[59] (72)

Fuciarelli, A. F., 423[121] (397, 410)

Fujii, M., 148[175] (120), 148[186] (121)

Fujii, N., 548[142] (510)

Fujimoto, J., 424[165] (407)

Fujimoto, K., 242[62] (161), 248[465] (237), 336[8] (314, 315)

Fujiwara, T., 423[86] (385, 386), 423[95] (386), 546[19] (493)

Fukazawa, T., 148[204] (126)

Fukui, K., 242[44] (156), 244[173] (182)

Fukui, T., 338[174] (329)

Fulcrand, G., 336[54] (319)

Fuller, W., 432, 473[44] (432)

Fung, S., 241[31] (154, 158), 248[443] (233)

Furano, A. V., 103[204] (95)

Furimsky, E., 422[54] (378)

Furlong, J. C., 102[141] (92), 102[190] (94)

Furnari, B. A., 489[33] (480)

Fürste, J. P., 244[208] (190), 244[209] (190)

Furukubo-Tokunaga, K., 489[18] (479)

Furuta, M., 151[397] (140)

Furuta, T., 152[402] (136)


Gabbay, E. J., 463, 466, 474[89] (441), 475[170] (463, 464, 465, 466), 475[171] (463, 464, 466)

Gabler, B., 337[110] (324)

Gabrielides, C. N., 336[1] (314), 336[9] (315)

Gaffney, B. L., 146[61] (113), 148[152], 149[213] (120), 259

Gairola, C., 547[58] (495)

Gait, M. J., 149[266], 150[289] (122, 126), 150[332] (127), 151[333] (127), 241[18] (155), 243[144] (172, 174), 243[145] (172, 174), 283[101] (265, 280), 284[187] (277, 279), 284[188] (277, 279)

Gait, M., 241[16] (169, 170)

Gajewski, E., 423[124] (397, 410), 423[126] (397, 410), 546[28] (494)

Galas, D. J., 281[3] (252)

Galazka, G., 102[135] (92), 102[136] (92)

Gale, E. F., 473[7] (428, 429, 432, 433, 436, 444, 452, 453)

Gall, A. A., 245[241] (190), 246[308] (197)

Gallagher, E. P., 546[8] (492, 513)

Gallagher, Jr., G., 547[55] (495)

Gallardo, H. F., 369[100] (363)

Gallegos, R., 547[80] (500, 541)

Gallo, F. J., 369[98] (362)

Gallois, B., 283[136] (271)

Gallouzi, I., 12[26] (5)

Galluppi, G. R., 146[55] (112)

Gamblin, S. J., 367[27] (348)

Gamper, H., 100[44] (85, 96)

Gamper, H. B., 13[50] (10)

Gamper, Jr., H. B., 336[7] (314, 315), 337[90] (323)

Ganesh, K. N., 247[357] (209), 247[358] (209)

Ganguly, T., 424[170] (407)

Gannett, P., 103[207] (95)

Gao, H., 146[61] (113)

Gao, Q., 368[37] (351, 352), 474[103] (444, 446), 474[105] (444, 446)

Gao, X., 474[112] (446, 448)

Gao, X. L., 12[46] (9)

Gao, Y.-G., 474[79] (437), 474[82] (437, 440)

Garbaccio, R. M., 548[120] (506)

Garbay-Jaureguiberry, C., 474[115] (446)

Garbesi, A., 369[112] (366)

Garcia, R. G., 147[98] (114)

Gard, J. K., 13[61] (11)

Garegg, P. J., 146[31] (109, 139, 144)

Garestier, T., 12[37] (8), 51[102] (36), 52[187] (47)

Garg, B. S., 147[105] (116)

Garner, G. A., 309[35] (289)

Garner, R. C., 548[145] (511, 512)

Garner, T. F., 547[57] (495)

Garnier, F., 425[231] (417)

Gasche, J., 246[331] (202)

Gasparutto, D., 149[217] (122)

Gassen, H. G., 338[173] (329)

Gaur, R. K., 168, 241[22] (154, 168), 243[110] (168)

Gawron, L. S., 548[122] (506)

Gdaniec, Z., 147[119] (117), 147[120] (117)

Geacintov, N. E., 476[218] (470), 476[219] (470)

Gebeyehu, G., 246[346] (206), 246[347] (206)

Gebhard, I., 548[116] (505)

Gee, J. E., 338[153] (327)

Gee, Y. G., 147[130] (119)

Gehring, W. J., 489[18] (479), 489[19] (479)

Geierstanger, B. H., 473[8] (428, 435, 436, 452, 456), 490[85] (487)

Geiger, J. H., 283[118] (266, 268)

Geiger, T., 308[16] (288), 308[25] (288), 309[63] (292, 306)

Gelbin, A., 473[24] (431)

Gelinas, R., 102[179] (93)

Gentle, D., 474[111] (446, 448)

George, D. L., 102[184] (93, 94, 95)

Gerlach, W. L., 284[164] (274)

Gerlt, J. A., 422[61] (380), 422[62] (380)

Gerrard, S. R., 283[148] (272)

Gerry, N. P., 247[373] (215)

Gerstnev, J. A., 152[450] (144)

Geselowitz, D. A., 242[54] (159), 242[60] (160)

Gessner, R. V., 368[51] (354)

Ghafouripour, A. K., 241[13] (159)

Ghatlia, N. D., 242[91] (165)

Ghirlando, R., 489[60] (484)

Ghosh, I., 489[36] (481)

Ghosh, S. S., 243[109] (168)

Gibbons, G. H., 368[90] (362)

Gibbs, E. J., 89, 101[114] (89)

Gibson, K. J., 194, 246[298] (194), 246[299] (194, 195)

Gibson, N. J., 424[200] (410)

Giese, B., 389, 391, 394, 422[72] (382, 393, 403), 423[108] (389), 423[109] (389), 583

Giessner-Prettre, C., 79[21] (60), 79[58] (72)

Gilbert, B. C., 423[107] (389, 391), 424[160] (405)

Gilbert, D. E., 50[4] (16), 99[4] (82)

Gilbert, W., 52[130] (41), 82, 93, 95, 100[23] (82), 102[195] (95), 620

Gilboa, E., 369[100] (363)

Gildea, B., 282[74] (259)

Gildea, B. D., 166, 242[93] (166)

Gilham, P. T., 146[9] (106, 107, 119), 147[68] (113)

Gill, S. J., 50[52] (31)

Gillam, S., 147[107] (117)

Gillen, M. F., 146[46] (112)

Gillespie, P., 339[222] (334)

Gillies, N. E., 425[230] (417)

Gilmore, J. L., 147[103] (115)

Gimisis, T., 422[47] (376), 422[48] (376)

Ginsberg, M. D., 421[9] (372)

Giovannangéli, C., 243[130] (169, 184)

Giovannangeli, C., 244[169] (181), 244[171] (181), 245[235] (190)

Giraldo, R., 12[39] (8)

Giralt, E., 146[38] (111), 147[80] (114)

Girard, L., 423[109] (389)

Giraud, L., 422[72] (382, 393, 403)

Girgis, N. S., 146[62] (113)

Gish, G., 249[486] (240), 249[487] (240), 309[54] (291)

Glatstein, E. I., 425[223] (411)

Glaubiger, D., 547[56] (495)

Glazer, A. N., 474[101] (444, 446, 448)

Glazer, P. M., 13[51] (10)

Glick, G. D., 246[337] (203), 282[42] (256), 363

Glinka, T., 547[80] (500, 541)

Glonek, T., 79[65] (72)

Glover, C. J., 549[177] (516)

Glover, J. N. M., 489[45] (482, 483)

Gluseppin, M., 473[28] (428)

Glusker, J. P., 247[405] (223)

Gmeiner, W. H., 245[264] (190), 247[398] (221)

Gniazdowski, M., 547[45] (494)

Göbel, T., 423[109] (389)

Goddard, III, W. A., 12[31] (6), 483

Godinger, D., 425[240] (418)

Godovikova, T. S., 242[66] (161), 244[192] (186), 244[193] (186)

Goebel, M., 547[73] (499)

Gokhovskii, S. L., 489[78] (486)

Gokota, T., 248[422] (228), 248[425] (228)

Golas, T., 337[94] (323)

Gold, B., 103[207] (95), 103[208] (95)

Gold, L., 13[56] (10), 615, 630, 639

Goldberg, I. H., 422[26] (373, 385, 394), 422[36] (373), 423[84] (385, 394, 421), 423[85] (385), 425[238] (418), 425[255] (421), 476[209] (469), 560, 562, 565

Goldberg, J. M., 101[85] (88), 101[87] (88)

Golden, B. L., 53[209] (50), 53[210] (50)

Goldman, M. E., 244[175] (182)

Goldstein, J., 548[129] (508)

Goldstein, R. F., 369[98] (362)

Golik, J., 423[89] (385, 394, 395, 421), 571

Goll, J. G., 422[29] (373, 374, 380), 422[41] (374)

Gonzalez, C., 247[368] (214), 308[14] (287)

Goodchild, J., 282[48] (257, 278)

Goodfellow, P., 367[5] (344, 367)

Gooding, A. R., 53[209] (50), 53[210] (50)

Goodisman, J., 101[115] (89)

Goodman, B. K., 422[35] (373, 402, 403, 404), 422[46] (376)

Goodman, M. F., 282[76] (260)

Goodsell, D., 475[147] (456)

Goodsell, D. S., 475[131] (452, 456), 547[60] (497)

Goodwin, J. T., 246[337] (203), 282[42] (256)

Goody, R. S., 150[276] (120)

Gooley, P. R., 547[74] (499)

Goosen, N., 102[164] (92)

Gopalkrishnan, S., 476[221] (470), 476[222] (470), 548[146] (512), 549[154] (513)

Gordon, D., 490[84] (487)

Gorenstein, D. G., 79[52] (71), 79[53] (71), 368[44] (353), 441, 474[98] (441)

Gorin, M. B., 246[322] (200)

Gorn, V. V., 336[7] (314, 315)

Görner, H., 423[133] (398, 405, 406)

Gorski, J., 102[166] (92), 102[185] (93)

Gortz, H.-H., 151[336] (127), 151[337] (127)

Gosling, R. G., 81, 99[1] (81)

Gosselin, G., 422[33] (373, 379, 394, 395)

Goswami, B., 338[157] (328)

Goto, K., 149[231] (121, 135)

Gottikh, M., 245[230] (190)

Gottschling, D. E., 284[158] (274, 275)

Götz, J., 151[376] (133)

Goudgaon, N. M., 247[400] (221, 240)

Gough, A. N., 473[41] (432)

Gough, G. R., 113, 147[68] (113), 147[69] (113)

Gough, G. W., 102[141] (92), 103[206] (95)

Goutte, C., 489[28] (480), 489[29] (480)

Gouyette, A., 424[194] (410)

Gouzaev, A. P., 245[240] (190)

Gouzaev, A., 248[419] (227)

Goyne, T., 422[38] (374)

Goyne, T. E., 422[37] (373)

Grabale, J., 283[114] (266, 267, 270)

Grable, J. C., 283[115] (266, 267, 270)

Grabowski, P. J., 284[158] (274, 275)

Graeser, E., 149[247] (123), 283[120] (267)

Graff, D., 309[40] (292, 293)

Graham, D., 244[195] (187)

Graham, D. R., 101[66] (86)

Grahe, G., 150[278] (122)

Grajkowski, A., 149[235] (122), 283[132] (270), 309[56] (292, 306)

Gralla, J., 50[43] (28, 45, 49), 52[155] (45)

Gralla, J. D., 102[161] (92)

Grandi, T., 244[174] (182)

Grant, D., 336[48] (317), 336[49] (317)

Grant, D. M., 474[99] (441, 451, 457)

Graves, B. J., 489[43] (482)

Graves, D. E., 100[51] (84)

Gray, D. M., 50[8] (16), 103[208] (95)

Grdina, D. J., 425[220] (411)

Green, A. P., 152[451] (144)

Green, J. B., 473[51] (433, 469)

Green, M., 475[162] (460)

Green, M. R., 473[16] (428)

Green, P., 283[114] (266, 267, 270)

Green, P. J., 283[115] (266, 267, 270)

Green, R., 52[134] (41)

Greenberg, J., 549[162] (513)

Greenberg, M. M., 422[34] (373, 402, 403), 422[35] (373, 402, 403, 404), 424[146] (400), 424[147] (400, 406), 425[229] (411, 413, 419), 546[29] (494)

Greenberg, W. A., 337[72] (322)

Greenblatt, M. S., 12[15] (4)

Greene, A. R., 146[37] (111)

Greenstock, C. L., 423[127] (398)

Greider, C. W., 12[41] (8)

Grein, T., 338[177] (330)

Greulich, K. O., 146[63] (113)

Grevatt, P. C., 336[1] (314), 336[9] (315)

Grewal, R. S., 244[176] (183)

Grieger, R. A., 50[36] (22, 27)

Griessinger, C., 79[47] (67)

Griffey, R. H., 247[368] (214), 310[109] (299, 302), 310[125] (302), 311[147] (288)

Griffin, L. C., 13[59] (11)

Grigg, G. W., 102[162] (92)

Griller, D., 422[49] (377)

Grimm, K., 150[275] (120)

Grimm, S., 309[72] (294)

Grindley, N. D. F., 488[13] (478)

Gritzen, E. F., 12[12] (4)

Gröger, G., 246[293] (193)

Grollman, A. P., 338[160] (328), 338[162] (328), 424[192] (410), 424[195] (410)

Grone, D. L., 246[294] (193)

Gronenborn, A. M., 475[172] (465), 475[174] (465), 489[60] (484), 489[61] (484)

Gronostajski, R. M., 281[12] (253)

Groody, E. P., 151[358] (130), 151[359] (130)

Grosjean, H., 52[179] (46)

Gross, M. L., 246[325] (201)

Grosshans, C. A., 53[202] (49)

Grotjahn, L., 283[102] (265, 270)

Grover, N., 422[39] (374, 376), 422[40] (374, 376)

Groves, J. T., 244[189] (185)

Grueneberg, D. A., 103[235] (96)

Gruff, E. S., 243[126] (169)

Grunberg, S. M., 423[91] (385)

Grunstein, M., 12[34] (8)

Gryan, G. P., 336[27] (316)

Gryaznov, S. M., 136, 150[304] (123, 136), 245[228] (190), 245[232] (190), 246[342] (204), 248[464] (237), 296, 310[87] (296), 310[91] (296), 311[145] (307)

Gryn, J., 13[52] (10)

Grzekowiak, K., 150[303] (123, 136), 151[349] (130)

Grzesiek, S., 79[19] (59)

Grzybowski, J., 243[153] (175), 247[361] (210)

Guarnieri, F., 13[64] (12)

Guckian, K. M., 282[34] (255)

Guendouz, A., 243[129] (169, 184)

Guengerich, F. P., 548[152] (512, 513), 549[153] (512)

Guéron, M., 51[116] (39, 45), 52[159] (45), 52[160] (45), 80[79] (74), 80[84] (75), 80[106] (76), 474[65] (434)

Guerrier-Takada, C., 52[166]

Guerriero, A., 547[84] (501)

Guest, C. R., 246[302] (195)

Guga, P., 309[56] (292, 306)

Guggenheimer, R. A., 281[12] (253)

Guhrs, K.-H., 146[63] (113), 146[64] (113)

Guinosso, C. G., 338[179] (330)

Guinosso, C. J., 248[432] (231), 310[127] (302)

Gulevich, Y., 476[179] (467)

Gumport, R. I., 283[119] (266), 283[121] (267)

Guo, Q., 476[204] (469), 476[214] (469)

Gupta, K. C., 150[327] (127, 137), 152[406] (137), 152[407] (137), 242[78] (163), 243[101] (167), 243[110] (168), 245[218] (190), 245[222] (190)

Gupta, N., 422[39] (374, 376), 422[40] (374, 376)

Gupta, S. V., 424[167] (407)

Gurskii, G. V., 489[78] (486)

Guschlbauer, W., 337[112] (324), 337[113] (324), 338[123] (325)

Gustafsson, J.-Ä., 489[66] (484)

Gutell, R. R., 368[54] (354)

Gutierrez, A. J., 317, 336[44] (317, 318), 336[47] (317)

Guy, A., 337[112] (324), 337[113] (324), 338[120] (325), 338[123] (325)

Guzaev, A., 147[81] (114)


Ha, J.-H., 50[23] (20)

Haasnoot, C. A. G., 52[152] (44), 52[157] (45), 79[57] (72)

Habener, J. F., 336[27] (316)

Habhoub, N., 243[124] (168)

Habus, I., 149[237] (121, 124), 149[238], 245[269] (190), 309[49] (292, 295), 309[82] (295), 310[103] (299, 301)

Haces, A., 246[313] (198)

Hachmann, J. P., 243[104] (168)

Hacker, A., 367[5] (344, 367)

Hadgraft, R. I., 548[102] (503)

Haeberli, P., 310[97] (298)

Hagen, M. D., 148[145] (120), 148[149] (120)

Hagen, U., 546[25] (493, 494)

Hagenberg, L., 338[173] (329)

Hagerhorst, J., 474[65] (434)

Hagerman, P. J., 367[21] (347, 353), 475[154] (459), 630

Haginoya, N., 246[316] (198)

Hahn, F. E., 547[51] (495)

Hahn, S., 283[118] (266, 268)

Hakimelahi, G. H., 149[236] (122), 150[114] (126)

Hakoshima, T., 368[41] (352), 490[87] (488)

Halder, T. C., 246[315] (198)

Hall, D. B., 242[89] (165)

Hall, I. H., 548[102] (503)

Hall, J., 242[80] (163), 242[81] (163, 235)

Hall, J. E., 475[142] (453, 457, 458)

Hall, J. G., 368[39] (351), 474[86] (440)

Hall, K. B., 51[80] (34)

Hall, R. H., 146[8] (106, 109, 139)

Hallick, L. M., 103[217] (96), 103[223] (96)

Halliwell, B., 421[8] (372), 421[9] (372), 423[124] (397, 410), 423[125] (397, 410), 547[36] (494, 544), 547[46] (494)

Hamamoto, S., 151[393] (139), 151[396] (139), 152[400]

Hammer, R. P., 152[447] (143)

Hammes, G. G., 51[111] (38), 51[114] (38, 40, 41)

Hammond-Kosack, M. C. U., 102[186] (93)

Hamoir, G., 147[125] (118)

Hampel, A., 284[163] (274), 284[173] (274)

Hampel, K., 283[155] (274)

Hampel, K. J., 476[192] (467)

Hamy, F., 474[111] (446, 448), 475[157] (459)

Han, H., 53[190] (47)

Han, S., 549[158] (513)

Han, Y.-H., 425[250] (420)

Hanawalt, P. C., 421[3] (372)

Hanecak, R., 309[26] (288)

Häner, R., 242[80] (163), 242[81] (163, 235), 248[431] (230), 308[5] (285), 308[25] (288)

Hangeland, J. J., 423[90] (385, 394, 421)

Hanna, M. M., 202, 246[334] (202, 203), 246[336] (203), 247[380] (216), 247[381] (216, 217)

Hanna, N. B., 150[277] (120, 121), 283[99] (265)

Hanold, D., 367[9] (344, 346)

Hansen, L. F., 474[116] (448), 474[117] (448)

Hansen, M., 476[224] (471), 476[225] (471), 548[136] (508, 509), 548[137] (509)

Hansen, M. R., 548[132] (508, 509)

Hanson, C. V., 103[220] (96), 103[224] (96)

Hansske, F., 282[80] (260)

Hanvey, J. C., 100[18] (82), 100[19] (82), 102[147] (92, 95), 102[148] (92)

Happ, E., 127, 148[144], 148[145]

Hara, M., 548[140] (510), 548[141] (510)

Haralambidis, J., 147[91] (114), 147[92] (114), 194

Härd, T., 489[66] (484)

Harden, D. B., 475[133] (452, 453), 475[135] (452, 453), 475[138] (452, 453)

Harding, J. D., 246[313] (198)

Hardy, S. F., 367[13] (345)

Hare, D. R., 79[27] (64, 67), 79[42] (65), 80[82] (75), 80[90] (76)

Harel, P., 338[120] (325)

Harindranath, N., 149[260] (121, 124)

Harley, C. D., 476[203] (469)

Harnden, D. G., 473[26] (428)

Harper, D. E., 548[117] (505)

Harper, J. W., 51[118] (39), 80[76] (73)

Harris, C. C., 12[14] (4), 12[15] (4)

Harris, C. M., 225, 248[414] (225), 248[415] (225)

Harris, R. K., 474[99] (441, 451, 457)

Harris, T. M., 248[414] (225), 248[415] (225)

Harrison, K. A., 310[138] (305)

Harrison, S. C., 367[27] (348), 488[8] (478), 489[37] (481), 489[45] (482, 483)

Harrison, S. D., 489[62] (484)

Harshey, R. M., 102[165] (92), 103[249] (98)

Harshman, K. D., 281[2] (252, 253)

Hartley, J. A., 474[77] (437, 440), 546[10] (493, 513), 546[16] (493)

Hartman, M., 152[452] (144)

Hartmann, B., 80[109] (76)

Hartwig, A., 423[123] (397, 410)

Harvey, R. G., 247[412] (224)

Harvey, S. C., 100[6] (82), 100[18] (82), 103[212] (96)

Hasan, A., 338[169] (329)

Hasegawa, Y., 148[199] (121)

Haseltine, W. A., 423[91] (385), 423[92] (385)

Hashimoto, H., 309[74] (294)

Hassnoot, C. A. G., 79[41] (64)

Hata, T., 146[49] (112, 123), 147[122] (117), 147[123] (117), 148[172] (120), 148[185] (121), 149[205] (121), 149[210] (121)

Hatahet, Z., 424[173] (407)

Hauck, P. R., 546[2] (492)

Haugland, R. P., 474[101] (444, 446, 448)

Hawkins, B. L., 80[74] (72)

Hawkins, M. E., 339[227] (334)

Haworth, I. S., 103[208] (95)

Hayag, M. S., 243[104] (168)

Hayakawa, M., 309[51] (292, 295)

Hayakawa, T., 336[32] (316)

Hayakawa, Y., 135, 148[168] (121, 135), 148[169] (120)

Hayashi, T., 547[75] (499)

Hayashibara, K. C., 281[17] (253, 256), 281[18] (253, 256), 282[44] (256)

Hayes, J. A., 147[69] (113)

Hayes, J. J., 82, 84, 100[16] (82), 100[21] (82, 84, 85), 101[120] (90), 101[121] (90)

Hearst, J., 284[174] (274, 277)

Hearst, J. E., 96, 100[43] (85, 96), 100[44] (85, 96), 103[220] (96), 103[222] (96)

Heath, J. A., 423[90] (385, 394, 421)

Heathcliffe, G. R., 146[37] (111)

Hebert, E., 103[212] (96)

Hebert, N., 248[472] (238)

Hecht, S. M., 241[1] (154), 244[190] (186), 248[453] (237), 248[459] (237), 281[1] (252, 253, 266), 282[70] (259), 309[77] (294), 532, 533, 534, 535, 539

Heeb, N. V., 247[367] (211)

Heerema, N. A., 12[20] (4)

Hegg, L. A., 284[168] (274)

Heikonen, W., 146[40] (111), 149[215] (120, 121)

Heikkilä, J., 148[184] (121), 148[203] (121), 149[209] (126)

Heiner, C., 241[31] (154, 158), 248[442] (233)

Heintz, N. H., 102[145] (92, 94)

Heisterberg-Moutsis, G., 146[40] (111)

Heitman, J., 283[124] (267)

Hélène, C., 12[37] (8), 47, 51[102] (36), 52[185] (46), 52[187] (47), 80[107] (76), 102[183] (93), 152[445] (142), 242[47] (158), 242[83] (164), 243[107] (167, 168), 243[117] (168), 244[167] (181), 244[168] (181), 245[235] (190), 245[236] (190), 283[137] (271), 283[149] (273), 422[64] (380), 467, 468, 488[17] (478, 479)

Helman, W. P., 423[127] (398)

Helveston, N. M., 152[451] (144)

Hemminki, K., 547[42] (494), 547[62] (497, 504), 548[145] (511, 512)

Henderson, D., 546[7] (492, 494, 506)

Henderson, E. E., 244[205] (189)

Henderson, R. E. L., 281[10] (253)

Hendrix, C., 12[7] (4), 230

Henichart, J.-P., 473[13] (428, 436)

Henle, E. S., 425[236] (417)

Henner, W. D., 423[91] (385), 423[92] (385)

Hennighausen, L., 281[6] (252)

Henrichson, C., 146[31] (109, 139, 144)

Herdering, W., 336[41] (317), 337[116] (325), 339[188] (330)

Herdewijn, P., 12[7] (4), 146[53] (112), 148[157], 148[158], 300

Hering, G., 151[375] (133), 151[376] (133)

Herman, B., 241[11] (155)

Hermes, J. D., 282[65] (258)

Hernandez, L. I., 476[215] (469)

Herold, M., 368[56] (354)

Heron, E., 241[31] (154, 158)

Herr, W., 102[187] (94)

Herrlein, M. K., 369[116] (366)

Herschlag, D., 46, 49, 284[199] (279)

Hertel, K. J., 284[179] (275)

Hertzberg, R. P., 101[74] (86), 101[75] (86)

Hess, K. M., 423[117] (395)

Hesse, L., 421[10] (372, 417)

Heuer, C., 102[196] (95)

Heus, H. A., 339[198] (330)

Hewitt, J. M., 310[131] (303)

Hicke, B. J., 425[217] (411), 425[218] (411)

Hicks, M., 284[173] (274)

Higson, A. P., 310[88] (296, 297)

Hilbers, C. W., 52[152] (44), 52[156] (45, 48), 79[31] (64)

Hildenbrand, K., 423[130] (398, 400, 405), 423[142] (400, 405, 407)

Hill, C. G., 547[79] (500)

Hill, F., 339[220] (333)

Hill, G. C., 248[476] (239)

Hill, T. S., 339[201] (331, 333)

Hillen, W., 102[196] (95)

Hillenkamp, F., 244[209] (190), 244[210] (190)

Hills, M. T., 368[44] (353)

Himmelsbach, F., 148[155] (120), 148[160] (120)

Hingerty, B. E., 476[216] (470), 476[218] (470)

Hinrichsen, R. D., 245[242] (190)

Hinz, H. J., 282[52] (257)

Hinz, M., 247[412] (224)

Hiort, C., 101[89] (88)

Hirayama, F., 547[78] (500)

Hirokawa, J., 548[126] (508)

Hirono, I., 548[125] (507, 508)

Hirose, M., 148[168] (121, 135), 148[169] (121)

Hirota, A., 548[123] (506)

Hirsch, J. A., 489[24] (479)

Hissung, A., 422[71] (382)

Histand, G., 152[430] (139), 248[460] (237), 309[75] (294)

Ho, N.-H., 149[237] (121, 124), 151[384] (133), 151[385] (133), 309[52] (292, 295)

Ho, P. L. C., 476[203] (469)

Hobbs, C., 310[98] (298)

Hobbs, F. W., 247[362] (211)

Hobbs, P. D., 241[24] (158)

Hochstrasser, R. A., 246[302] (195)

Hodge, P., 146[22] (108)

Hodge, R., 149[259] (121, 124)

Hodges, R. R., 249[482] (240)

Hodges, R. S., 489[57] (483)

Hoekstra, D., 244[207] (190)

Hoepfner, R. W., 103[218] (96), 103[229] (96)

Hoess, R. H., 489[47] (482)

Hoff, H., 244[207] (190)

Hoffer, M., 282[85] (261)

Hoffman, C., 547[90] (502)

Hofsteenge, J., 244[154] (175)

Hogan, M. E., 245[256] (190)

Hogrefe, R. I., 242[41] (161), 308[13] (287, 292)

Hohlhenrich, K. H., 281[15] (253)

Hoke, G. D., 308[14] (287)

Holden, K. G., 547[55] (495)

Holder, S. B., 146[55] (112)

Holland, D., 243[140] (171)

Hollenberg, P. F., 423[99] (387)

Hollingsworth, M. A., 103[199] (95)

Holloman, W. K., 13[52] (10)

Hollstein, M. C., 547[88] (502, 503)

Hollstein, M., 12[14] (4), 12[15] (4)

Holmberg, L., 309[33] (289)

Holmlin, R. E., 242[89] (165)

Höltke, H.-J., 247[355] (209)

Holy, A., 148[178] (120)

Hommel, U., 368[56] (354)

Honda, S., 148[193] (120)

Honig, B., 473[21] (428, 432)

Honnens, J., 151[392] (139)

Hood, L. E., 241[17] (158), 248[441] (233), 248[442] (233)

Hoogerhout, P., 150[270] (120)

Hoover, J. R. E., 547[55] (495)

Hope, I. A., 489[42] (482)

Hopkins, N. H., 367[1] (343, 346)

Hopkins, P. B., 101[123] (90, 96), 103[236] (96), 103[239] (96), 332, 502, 518

Hoppe, N., 146[20] (108)

Hopper, M. J., 146[45] (112)

Horinouchi, Y., 148[186], 148[187]

Horiuchi, M., 368[90] (362)

Horn, T., 146[39] (111), 147[130] (119), 204

Horne, D. A., 247[375] (215)

Horner, J. H., 423[111] (391)

Hornung, C., 241[21] (185, 234)

Horowitz, D. M., 282[76] (260)

Horton, P., 248[416] (225)

Horvath, S. J., 488[11] (478)

Horwitz, S. B., 423[102] (388, 394)

Hosada, M., 548[110] (505)

Hoshi, T., 549[165] (514)

Hoshiko, T., 310[109] (299, 302)

Hostomsky, Z., 146[54] (112)

Hosur, R. V., 79[49] (68), 80[91] (76)

Houard, S., 242[68] (161, 166)

Houssier, C., 473[13] (428, 436), 474[80] (437), 475[144] (453, 459), 475[157] (459)

Hovinen, J., 147[81] (114)

Howard, F. B., 323, 337[84] (323), 337[89] (323)

Howard, J. A., 422[54] (378)

Hoyng, C. F., 149[254] (123)

Hrebabecky, H., 148[178]

Hsieh, H. P., 282[30] (255, 263)

Hsieh, O. A.-L., 547[87] (502)

Hsieh, S.-H., 473[24] (431)

Hsieh, W.-T., 100[18] (82)

Hsiung, H. M., 150[309] (126)

Hsu, L.-Y., 282[57] (257)

Hsu, Y.-H., 548[123] (506)

Htun, H., 83, 100[25] (83), 102[150] (92, 93)

Hu, S.-L., 243[133] (170)

Huang, A.-S., 547[87] (502)

Huang, C.-H., 490[92] (488)

Huang, C.-Y., 337[63] (320), 337[64] (321)

Huang, D.-L., 247[364] (211)

Huang, F., 309[42] (292, 293)

Huang, H., 548[94] (502)

Huang, J., 310[114] (299, 304), 310[133] (304)

Huang, J.-X., 548[149] (512)

Huang, Q., 309[47] (292, 294)

Huang, T., 336[48] (317), 336[49] (317)

Hubbell, W. L., 489[38] (481)

Huber, E., 246[311] (197, 209)

Huckriede, B. D., 79[51] (68)

Hudson, R. H. E., 368[63] (355, 356), 368[64] (355)

Hughes, P., 248[442] (233)

Huie, E. M., 310[135] (304)

Hunkapiller, M. W., 241[31] (154, 158), 248[443] (233)

Hunkapiller, T. J., 248[443] (233)

Hunkeler, M. J., 246[313] (198)

Hunter, J. C., 309[80] (295)

Hunter, W. G., 50[36] (22, 27)

Hunter, W. N., 338[125] (325), 424[200] (410)

Hunziker, J., 338[171] (329)

Hurd, H. K., 547[88] (502, 503)

Hurley, L. H., 422[20] (372), 422[21] (372), 471, 472, 473[2] (428, 435, 436, 456, 471), 473[6] (428, 435, 436, 456), 475[165] (462, 463), 475[166] (462, 463), 476[224] (471), 476[225] (471), 497, 505, 506, 508, 546[6] (492, 494), 546[7] (492, 494, 506), 547[52] (495, 497), 547[53] (495), 548[105] (504, 505, 506), 548[107] (504, 505)

Hurskainen, P., 246[326] (201), 246[327] (201)

Hurst, G. D., 246[303] (195)

Hurwitz, J., 281[12] (253)

Hüsken, D., 242[80] (163), 242[81] (163, 235), 308[25] (288)

Huskens, J., 149[252], 150[272] (120)

Hustedt, E. J., 247[389] (219), 247[390] (219)

Hutchinson, F., 413, 424[213] (411, 413), 424[214] (411, 413)

Hutter, M., 424[182] (410)

Huynh-Dinh, T., 242[61] (161)

Hwu, J. R., 150[314] (126)

Hyde, J. E., 103[222] (96)

Hyrup, B., 310[132] (304)

Iadarola, P. I., 283[113] (266)

Iaiza, P., 151[378] (133)

Ialongo, G., 422[48] (376)

Ichikawa, T., 338[121] (325), 338[122] (325, 328)

Ide, H., 242[73] (162), 248[457] (237)

Iden, C. R., 338[159] (328), 424[204] (411)

Igolen, J., 147[113] (117)

Ihara, T., 242[92] (166)

Iida, A., 424[164] (407)

Iimura, S., 148[202] (121)

Iinuma, H., 549[163] (514)

Iitiä, A., 246[327] (201)

Ikeda, H., 248[465] (237)

Ikeda, T., 248[422] (228)

Ikehara, M., 146[14] (108), 330

Ikeuchi, T., 248[455] (237)

Ikura, M., 80[72] (72, 73)

Ikuta, S., 336[15] (315, 330)

Imai, J., 148[192] (121)

Imai, K., 148[174], 148[175]

Imanishi, H., 490[89] (488)

Imbach, J.-L., 248[476] (239), 282[53] (257), 309[48] (292, 295), 309[84] (295)

Imlay, J. A., 547[47] (494)

Imura, A., 339[223] (334)

Imwinkelreid, P., 422[72] (382, 393, 403)

Inagawa, T., 244[199] (187)

Inokawa, T., 242[71] (162), 242[72] (162)

Inoue, H., 246[305] (196), 279, 283[150] (273), 334

Inoue, T., 284[198] (279)
Invergo, B. J., 548[119] (506)
Ippel, J. H., 79[51] (68)
Ireland, C. M., 547[85] (501, 529)
Iribarren, A. M., 149[257] (121, 122), 246[282] (191), 283[97] (264), 283[98] (264)
Isaacs, S. T., 100[43] (85, 96)
Ishido, Y., 147[79] (114), 148[199]
Ishiguro, K., 547[71] (497)
Ishikawa, H., 338[163] (328), 424[202] (410)
Ishikawa, T., 336[8] (314, 315)
Ishizaki, T., 548[119] (506)
Isildar, M., 422[71] (382)
Iso, Y., 248[465] (237)
Isobe, H., 243[107] (167, 168)
Itakura, K., 108, 150[302] (123, 136), 151[355] (130), 282[76] (260)
Ito, S., 549[160] (513)
Ito, T., 148[174] (120), 148[179] (121)
Ivanov, N. L., 152[405] (137)
Ivanov, V. I., 80[112] (77)
Ivanova, E. M., 242[86] (164)
Ivanovskaya, M. G., 369[110] (366)
Ivarsson, B., 52[125] (40)
Iversen, P., 12[9] (4, 5, 9)
Iverson, B. L., 242[79] (163, 164)
Iverson, G. M., 243[104] (168)
Ivie, G. W., 548[95] (503)
Iwai, S., 338[133] (325)
Iwamura, M., 152[402] (136)
Iwane, K., 244[173] (182)
Iwase, R., 150[294] (123)
Iyengar, B. S., 547[50] (495, 497, 499, 505, 520)
Iyer, R. P., 149[237] (121, 124), 149[238] (122), 151[384] (133), 151[385] (133), 152[401] (135), 152[442] (142), 241[2] (154), 241[3] (154), 248[468] (238), 248[469] (238), 283[108] (265), 308[4] (285), 309[49] (292, 295), 309[52] (292, 295)
Iyer, R. S., 548[148] (512), 548[150] (512), 549[157] (513)
Iyer, V. N., 295, 515, 518, 549[168] (515), 549[169] (515, 518)
Izant, J. G., 337[59] (319)

Jablonski, J.-A., 242[67] (161, 162, 165)
Jacobsen, J. P., 474[116] (448), 474[117] (448)
Jacobson, K., 149[222] (122)
Jacobson, K. B., 51[73] (33)
Jaeger, J. A., 51[69] (33)
Jaffe, E. K., 282[64] (258)
Jäger, A., 237, 248[453] (237)
Jain, M. L., 150[314] (126)
Jain, S. C., 473[18] (428), 474[71] (435)
Jajoo, K. K., 248[416] (225)
Jakob, P., 151[376] (133)
James, J. L., 79[67] (72, 73)
James, T. L., 56, 79[8] (56), 79[28] (64), 80[92] (76), 80[93] (76)
Jankowska, J., 152[418] (139)
Jansen, K., 475[152] (458, 459), 475[156] (459)

Janssen, G., 148[158] (120)
Jardetzky, C. D., 79[11] (57)
Jardetzky, O., 79[30] (64)
Jäschke, A., 244[208] (190), 244[209] (190)
Jaworski, A., 100[18] (82)
Jay, J. M., 548[99] (503)
Jayaram, B., 50[24] (20)
Jayaraman, K., 130, 147[97] (114)
Jebaratnam, D. J., 547[33] (494)
Jelen, F., 101[131] (92)
Jen-Jacobson, L., 283[111] (266, 267), 283[132] (270)
Jencks, W. P., 52[136] (41), 632
Jenkins, B. G., 474[91] (441)
Jenkins, T. C., 244[183] (184)
Jenkins, Y., 242[90] (165), 242[91] (165)
Jensen, K. F., 103[209] (96)
Jensen, L. J., 474[117] (448)
Jensen, M. A., 247[362] (211)
Jenssen, J. R., 337[96] (323)
Jeon, C. J., 489[79] (486)
Jeppesen, C., 102[160] (92), 102[163] (92, 94)
Jerina, D. M., 247[406] (223), 247[407] (223)
Jetter, M. C., 338[129] (325)
Jhurani, P., 149[254] (123)
Jiang, H., 100[39] (85, 93)
Jiang, M.-Y., 152[455] (144)
Jiang, Y.-P., 425[257] (421)
Jiang, Z., 149[237] (121, 124), 152[401] (135)
Jiang, Z.-W., 425[238] (418)
Jiao, X.-Y., 422[69] (382)
Jie, L., 148[176] (120)
Jimenez, L. S., 549[186] (517)
Jin, R., 338[157] (328)
Jina, A. N., 241[24] (158)
Jiricny, J., 339[194] (330)
Joachimiak, A., 283[116] (266, 268)
Joglekar, S. P., 248[423] (228)
Johansson, C. S., 249[485] (240)
John, D. C. A., 100[60] (86)
Johnson, A. D., 489[27] (480), 489[28] (480)
Johnson, D. S., 548[111] (505, 506)
Johnson, F., 329, 338[159] (328), 338[160] (328), 424[192] (410), 424[201] (410)
Johnson, K. A., 248[449] (234)
Johnson, L. K., 242[57] (159)
Johnson, P. D., 548[116] (505)
Johnson, P. F., 488[2] (477, 478, 480, 482, 484)
Johnson, R. C., 488[12] (478, 479)
Johnson, R. K., 546[3] (492), 547[55] (495)
Johnson, R., 13[60] (11)
Johnson, S., 149[237] (121, 124), 149[238], 309[49] (292, 295)
Johnson, W. W., 549[153] (512)
Johnsson, B., 52[125] (40)
Johnsson, K., 338[128] (325)
Johnston, B. H., 80[119] (78), 83, 100[25] (83), 100[40] (85, 93), 102[149] (92)
Johnston, J. F., 308[15] (287), 309[63] (292, 306)
Johnston, L. J., 422[74] (383, 398, 407)
Johnston, S. S., 476[223] (470)

Jolles, B., 102[171] (93)
Jones, A. S., 281[14] (253)
Jones, B., 241[31] (154, 158)
Jones, C., 489[36] (481)
Jones, C. R., 79[52] (71), 79[53] (71)
Jones, D. S., 243[104] (168)
Jones, G. D. D., 423[132] (398), 423[143] (400), 424[157] (403)
Jones, M. N., 51[70] (33)
Jones, R., 475[149] (457)
Jones, R. A., 150[293] (122, 123), 152[439] (139), 282[71] (259), 282[77] (259)
Jones, R. J., 282[56] (257, 273)
Jones, R. L., 473[11] (428, 433, 436, 444), 473[41] (432), 474[74] (436), 475[151] (457), 475[169] (463)
Jones, S. K., 475[142] (453, 457, 458)
Jones, S. S., 147[137] (120, 127), 148[139]
Jönsson, U., 52[125] (40)
Joos-Guba, G., 247[405] (223)
Jørgensen, P. N., 248[434] (232), 248[435] (232)
Jorgensen, T. J. D., 147[70] (113)
Jortner, J., 474[79] (437), 474[91] (441), 475[140] (453, 457, 458), 476[221] (470)
Joseph, S., 369[117] (367)
Josephson, S., 149[246] (126), 284[166] (274, 277), 284[176] (274)
Joshi, B. V., 310[94] (297)
Joshi, P. C., 424[203] (411)
Joshua, A. V., 547[68] (497, 498, 499, 529)
Jou, J., 247[380] (216)
Jovin, T. M., 50[40] (24, 28, 36), 51[98] (36), 80[116] (77), 102[158] (92)
Joyce, G. F., 13[57] (10), 621, 623, 632, 633
Juby, C., 245[266] (190)
Juby, C. D., 147[95] (114)
Juliano, R. L., 241[11] (155), 241[12] (155)
Jungell-Nortamo, A., 241[6] (154, 155)
Juodawlkis, A., 336[54] (319)
Jurkiewicz, B. A., 425[233] (417)
Just, G., 304, 310[111] (299, 304), 310[112] (299, 304)

Kaas, R. L., 146[43] (112)
Kabakov, A. E., 102[157] (92)
Kabanov, A. V., 187, 244[196] (187), 244[197] (187), 245[261] (190), 309[76] (294)
Kadlubar, F. F., 547[62] (497, 504), 547[91] (502), 548[145] (511, 512)
Kadrmas, J. L., 367[23] (347, 353)
Kady, I. O., 244[189] (185)
Kadyko, M. Y., 151[388] (133)
Kagiya, T., 425[235] (417)
Kaiser, E. T., 489[22] (479)
Kaiser, K., 245[244] (190)
Kaiser, R. J., 241[17] (158), 248[441] (233), 248[442] (233)
Kajino, H., 148[191] (121)
Kalbfleisch, S., 241[25] (158)
Kalinkina, A. L., 152[427] (139), 244[200] (187, 190)

Kallenbach, N. R., 50[56] (32), 51[64] (33), 51[106] (36), 100[49] (84), 100[52] (85, 86), 101[70] (86), 101[77] (86), 103[243] (98), 103[244] (98), 367[22] (347, 353), 469

Kam, L., 546[30] (494)

Kamaike, K., 147[79] (114), 148[199]

Kamen, B. A., 242[52] (159)

Kamimura, T., 148[193], 148[194], 149[205], 149[210]

Kamiya, H., 424[202] (410)

Kamiya, M., 51[66] (33)

Kan, L.-S., 282[36] (256, 262, 271), 282[37] (256, 262, 271)

Kan, Y. W., 241[10] (158)

Kanamori, K., 549[160] (513), 549[165] (514)

Kandimalla, E. R., 242[40] (159)

Kaneda, Y., 368[90] (362), 424[185] (410)

Kanehara, H., 309[58] (292)

Kaneto, M., 425[235] (417)

Kang, D. S., 102[172] (93)

Kang, T. W., 423[108] (389)

Kaniwa, H., 547[78] (500)

Kanne, D., 100[43] (85, 96)

Kanou, M., 325, 337[118] (325), 337[119] (325)

Kansal, V. K., 242[61] (161)

Kao, J. L.-F., 248[421] (228, 229)

Kaplan, B. E., 282[76] (260)

Kaplan, D. J., 547[57] (495)

Kappen, L. S., 100[45] (85)

Kaptein, R., 489[66] (484)

Karam, L. R., 424[148] (401)

Kardo, J. L., 146[43] (112)

Karim, A. S., 249[485] (240)

Karl, R., 151[376] (133)

Karlovsky, P., 101[129] (92), 101[131] (92)

Karlsson, R., 52[125] (40), 52[126] (40), 148[138]

Karpeisky, A., 284[188] (277, 279)

Karplus, M., 80[86] (76)

Karslake, C., 80[94] (76)

Karwowski, B., 309[56] (292, 306)

Kasai, H., 424[164] (407)

Kaspersen, F. M., 243[131] (170)

Katagiri, N., 151[355] (130)

Kataoka, M., 152[416] (138)

Kato, H., 148[191] (121), 148[192] (121)

Katoh, M., 51[84] (34)

Katsuyama, T., 548[142] (510)

Kattan, G. F., 336[53] (319)

Katzhendler, J., 146[52] (112)

Kaumaya, P. T. P., 489[57] (483)

Kavka, K. S., 146[55] (112)

Kawabata, H., 423[86] (385, 386), 423[95] (386)

Kawaguchi, H., 490[89] (488), 490[90] (488)

Kawai, G., 337[118] (325), 338[121] (325)

Kawai, S. H., 304, 310[111] (299, 304)

Kawanishi, S., 424[163] (407), 546[19] (493)

Kawasaki, A. M., 310[127] (302)

Kawasaki, E. S., 241[15] (158)

Kawase, Y., 338[133] (325), 338[163] (328)

Kawashihi, G., 547[82] (500)

Kawashima, E. H., 148[159], 170

Kawashima, M., 151[397]

Kawata, H., 244[162] (178)

Kawczynski, W., 337[105] (324)

Kay, L. E., 80[72] (72, 73)

Kay, P. S., 279, 284[198] (279)

Kazakov, S. A., 100[59] (86)

Kazeniac, S. J., 547[86] (502)

Kazimierczuk, Z., 337[94] (323), 337[95] (323), 338[137] (325)

Kean, J. M., 100[48] (84, 87), 103[233] (96)

Kearns, D. R., 52[158] (45), 52[163] (45), 79[23] (60), 79[64] (72)

Keel, R. A., 473[46] (433), 475[169] (463)

Keepers, D. B., 79[28] (64)

Kehler, J., 309[65] (293), 309[66] (293)

Kehne, A., 337[116] (325), 339[183] (330), 339[184] (330)

Kell, B., 283[152] (274)

Kellenbach, E., 489[66] (484)

Keller, C. F., 284[165] (274, 277, 279, 280), 284[195] (279)

Keller, T. H., 248[431] (230)

Kelley, F. J., 421[7] (372)

Kellogg, G. W., 79[56] (71)

Kelly, R. C., 548[115] (505), 548[116] (505)

Kelman, Z., 368[35] (350), 368[36] (350)

Kelner, M. J., 548[127] (508), 548[128] (508)

Kemp, D., 282[67] (259)

Kempe, T., 146[47] (112), 170

Kemper, B., 102[143] (92)

Kennard, O., 283[136] (271)

Kenner, G. W., 146[6] (106)

Kent, M. A., 242[43] (156, 173, 188, 190), 243[146] (172, 190)

Kent, S. B. H., 248[442] (233)

Kercher, M. A., 12[28] (6)

Kerremans, L., 311[144] (307)

Kers, A., 152[422] (139)

Kers, I., 152[422] (139)

Kerwin, S. M., 462, 463, 475[165] (462, 463), 475[166] (462, 463)

Kesicki, E. A., 246[329] (202)

Kesselheim, C., 423[108] (389), 423[110] (390, 393)

Kessler, C., 246[310] (197, 209), 246[311] (197, 209), 247[355] (209), 247[356] (209)

Kessler, D. J., 245[256] (190)

Keyes, R. S., 247[394] (220)

Khalimskaya, L. M., 246[297] (194)

Khorana, H. G., 106, 146[9] (106, 107, 119), 146[10] (106, 119), 149[212] (120, 126), 149[220] (126)

Kiber-Herzog, L., 283[152] (274), 283[153] (274)

Kido, K., 196, 246[305] (196)

Kiely, J. S., 148[170] (120), 246[323] (200)

Kierzek, P., 150[303] (123, 136)

Kierzek, R., 150[286] (122), 151[349] (130), 151[395], 245[280] (191), 247[349] (207), 248[448] (234), 248[449] (234), 283[135] (270)

Kiessling, L. L., 339[222] (334)

Kigoshi, H., 548[126] (508)

Kilpatrick, J. E., 79[12] (57)

Kilpatrick, M. W., 102[186] (93)

Kim, H.-K., 476[193] (467)

Kim, H.-Y., 248[416] (225)

Kim, H.-Y. H., 549[158] (513)

Kim, J. L., 283[117] (266, 268)

Kim, J.-M., 476[193] (467)

Kim, K., 422[60] (380)

Kim, N. W., 476[203] (469)

Kim, P. S., 489[48] (482), 489[49] (482)

Kim, S.-G., 151[382] (133), 244[199] (187), 283[154] (274)

Kim, S. J., 248[415] (225), 248[416] (225)

Kim, S. K., 476[193] (467)

Kim, Y., 283[115] (266, 267, 270), 283[118] (266, 268)

Kimura, K., 547[82] (500)

Kimura, Y., 422[45] (376)

Kinas, R., 151[388] (133)

King, A., 102[164] (92)

King, G., 51[118] (90)

King, G. C., 80[76] (73)

Kinkel, J. N., 146[56] (112)

Kinsella, T. J., 425[222] (411), 425[223] (411)

Kinzuka, Y., 244[199] (187)

Kipp, S. A., 244[185] (184), 309[79] (294)

Kirchner, J. J., 247[389] (219), 247[390] (219)

Kirkegaard, K., 103[202] (95)

Kirolos, M. A., 475[145] (453, 459)

Kirschner, K., 368[56] (354)

Kirshenbaum, M. R., 101[93] (88)

Kiselev, V. I., 244[196] (187), 244[197] (187), 283[140] (271)

Kiselyov, A. S., 476[178] (467, 468), 476[179] (467)

Kiselyova, V., 474[90] (441)

Kishikawa, H., 425[257] (421)

Kisner, D., 12[11] (4)

Kissinger, C. R., 489[23] (479, 480)

Kistenmacher, T. J., 100[62] (86)

Kisters-Wolke, B. V., 283[126] (268)

Kitamura, M., 248[424] (228)

Kitos, P. A., 548[119] (506)

Kiyokawa, M., 246[316] (198)

Kizu, R., 547[65] (497)

Kladde, M. P., 102[166] (92), 102[185] (93)

Klappe, K., 244[207] (190)

Klein, R. S., 282[86] (262)

Kleinschmidt, A. K., 152[452] (144)

Klevan, L., 246[346] (206), 246[347] (206)

Klig, L. S., 283[125] (268)

Klotz, L. C., 52[183] (46)

Klug, A., 284[157] (274, 276)

Klump, H. H., 51[70] (33)

Klyne, W., 56, 79[10] (56)

Klysik, J., 100[18] (82), 102[135] (92), 102[136] (92)

Kmiec, E. B., 13[52] (10)

Knaf, A., 242[55] (159)

Kneale, G. G., 101[128] (90)

Knitt, D. E., 369[102] (363)

Knorr, R., 241[21] (185, 234)

Knorre, V. D., 242[66] (161)

Knowland, J., 422[59] (380)

Kobayashi, E., 548[142] (510)

Kobayashi, S., 103[246] (98)

Koch, T. H., 425[217] (411), 425[218] (411), 474[122] (450, 451), 529

Kochel, T. J., 103²²⁷ (96), 103²²⁸ (96), 103²²⁹ (96)
Kochevar, I. E., 547³⁵ (494, 544)
Kochi, J. K., 424¹⁴⁹ (401, 402)
Kochoyan, M., 474⁶⁵ (434)
Kodra, J. T., 309⁶⁶ (293)
Koga, F., 549¹⁶⁰ (513)
Kohlhagen, G., 547⁸³ (501)
Kohli, V., 146⁴⁸ (112), 149²¹⁵ (120, 121)
Kohn, H., 549¹⁷³ (515, 516), 549¹⁷⁶ (516)
Kohn, K. W., 546¹⁶ (493), 547⁵¹ (495), 547⁵⁶ (495)
Köhnlein, W., 424²¹⁴ (411, 413)
Kohno, K., 149²⁶⁴, 150²⁹⁷ (123)
Kohwi-Shigematsu, T., 100⁴¹ (85, 93, 94), 102¹⁷³ (93, 94), 102¹⁷⁵ (93)
Kohwi, Y., 100⁴¹ (85, 93, 94), 102¹⁷³ (93, 94), 102¹⁷⁵ (93)
Kois, P., 339²¹⁰ (332)
Koizumi, M., 284¹⁶⁹ (274)
Koktycz, M. J., 369⁹⁸ (362)
Koleck, M. P., 147⁶⁹ (113)
Koller, T., 103²²⁵ (96)
Kollman, P. A., 80⁸⁷ (76)
Kolpak, F. J., 80¹¹⁷ (77)
Koltzenburg, G., 422⁶⁵ (381, 389), 422⁶⁶ (381, 389)
Komano, T., 103²⁴⁶ (98)
Komatsu, H., 338¹²¹ (325), 338¹²² (325, 328)
Komatsu, Y., 284¹⁶⁹ (274)
Komiyama, M., 242³⁸ (156), 242⁷¹ (162), 310⁸⁵ (295)
Konarska, M. M., 367¹³ (345)
Kondo, H., 474¹²¹ (450, 451)
Kondo, S., 548¹³¹ (508)
Kondoleon, S. K., 103²¹⁷ (96)
Konevetz, D. A., 242⁵⁸ (160)
Kong, S. B., 475¹³⁶ (452, 453)
Kong, X. P., 368³² (350, 361)
Koning, T. M. G., 79³² (64)
Konishi, A., 283¹⁵⁰ (273)
Konishi, K., 490⁹³ (488)
Konishi, M., 103²⁴² (98)
Konishi, Y., 242⁷³ (162), 248⁴⁵⁷ (237)
Kool, E. T., 244²¹³ (190), 255, 282³² (255, 263), 282³³ (255), 292, 309⁶² (292), 332, 357, 361, 363
Koole, L. H., 149²⁵¹ (123), 149²⁵² (123)
Koopman, P., 367⁵ (344, 367)
Kopka, M. L., 475¹⁴⁷ (456), 547⁶⁰ (497)
Kornberg, A., 12¹ (2, 3), 12² (2)
Kornberg, T. B., 489²³ (479, 480)
Korobka, A., 80⁹⁹ (76)
Korshalla, J. D., 547⁷⁶ (500)
Korshun, V. A., 244¹⁶⁴ (178, 179)
Kortes, R. A., 101¹⁰⁷ (89)
Koshiyama, H., 490⁹⁰ (488)
Koshkin, A. A., 242⁴⁷ (158)
Kosiyama, H., 490⁸⁹ (488)
Kosora, N., 150²⁷⁵ (120)
Kössel, H., 108, 146²¹ (108)
Köster, H., 108, 146¹⁹ (107, 108, 112), 146²⁰ (108), 149²¹⁵ (120, 121)

Kostina, E. V., 244²⁰⁴ (188)
Kosylanska, A., 309⁵⁶ (292, 306)
Kotschi, U., 146⁵⁶ (112)
Kouchakdjian, M., 338¹⁶² (328)
Koura, K., 148¹⁹⁷ (123), 148¹⁹⁸ (121), 149²¹⁰ (121)
Kowalczykowski, S. C., 367²⁵ (347)
Kozarich, J. W., 244¹⁹¹ (186)
Koziolkiewicz, M., 283¹³² (270), 283¹³³ (270), 309⁵⁶ (292, 306), 310⁹¹ (296)
Kozlov, A. G., 50¹³ (17)
Kozlowski, S. A., 80⁸⁰ (75), 80¹²¹ (78)
Kozlowski, S., 50⁵⁷ (32)
Kramer, F. R., 9, 12⁴⁷ (9, 10), 12⁴⁸ (9), 13⁴⁹ (9)
Krappen, L. S., 425²⁵⁶ (421)
Kraszewski, A., 150³⁰³ (123, 136), 151³⁴⁹ (130), 152⁴¹⁸ (139), 152⁴²² (139)
Krauch, T., 337⁹⁸ (324)
Krawczyk, S. H., 338¹³⁰ (325), 338¹³¹ (325)
Krecmerová, M., 148¹⁷⁸ (120)
Kreig, A. M., 309⁴⁷ (292, 294)
Kremer, A. B., 336²⁸ (316)
Kremer, F. J. B., 148¹⁶⁶ (120)
Kremer, W., 490⁸⁵ (487)
Kremsky, J., 149²⁵⁹ (121, 124)
Kremsky, J. N., 170, 243¹³² (170)
Krepinsky, J. J., 245²⁷¹ (190)
Kretzmer, G., 473²⁸ (428)
Krisch, R. E., 425²⁴² (418)
Krishnamoorthy, C. R., 474⁶⁸ (434, 440), 474⁶⁹ (434, 440)
Krishnamurthy, G., 100⁵⁰ (84)
Krishnan, B., 423⁸⁹ (385, 394, 395, 421)
Krist, B., 248⁴⁵¹ (236, 237)
Krivonos, A. V., 244¹⁹⁶ (187)
Kroger, M., 546²² (493)
Krokan, H. E., 12⁴⁴ (8)
Krolikiewicz, K., 282⁸⁴ (260)
Kropelin, K. H., 146²⁰ (108)
Krotz, A. H., 12¹² (4), 151³⁸¹ (133)
Krueger, W. C., 548¹⁰⁶ (504, 505), 548¹⁰⁷ (504, 505)
Krug, R. R., 50³⁶ (22, 27)
Kruger, K., 284¹⁵⁸ (274, 275)
Krugh, T. R., 473³ (428, 435, 436), 476²¹⁶ (470)
Krumholz, L., 241³² (154, 158)
Kruse, L. I., 310¹⁰⁴ (299, 301), 310¹¹⁰ (299, 303)
Kryak, D. D., 247³⁹³ (219, 220)
Krynetskaya, N. F., 150²⁹¹ (122)
Krzymanska-Olejnik, E., 243¹⁵⁰ (174), 246²⁸³ (191), 247³⁵¹ (208)
Ku, L., 147¹³⁰ (119)
Kubista, M., 475¹³⁶ (459)
Kucera, L. S., 282⁵⁷ (257)
Kueng, E., 151³⁷⁸ (133)
Kuentzel, S. L., 548¹⁰⁶ (504, 505)
Kugabalasooriar, S., 547³³ (494)
Kuhn, H., 339²¹⁵ (333)
Kuijpers, W. H. A., 149²⁵¹ (123), 149²⁵² (123), 150²⁷² (120), 243¹⁰² (120), 243¹³¹ (170)

Kuimelis, R. G., 282⁶³ (258, 259, 265), 282⁶⁶ (259, 280), 284¹⁶² (274), 336⁶ (314)
Kuklenyik, Z., 101⁶⁵ (86)
Kulikowski, K., 146²⁰ (108), 149²¹⁵ (120, 121)
Kulka, M., 309⁷⁹ (294), 309⁸⁰ (295)
Kumar, A., 243¹¹¹ (168), 243¹¹² (168), 245²²¹ (190), 245²²³ (190)
Kumar, C. V., 101⁸⁷ (88)
Kumar, P., 147⁸² (114), 147⁸⁴ (114), 172
Kumar, R. A., 473⁶¹ (434)
Kumar, S., 53¹⁹⁴ (47)
Kumar, V. A., 247³⁶⁰ (210), 337⁷¹ (322)
Kume, A., 146⁴⁹ (112, 123)
Kumegawa, M., 148¹⁷⁷ (120, 121)
Kumkins, V., 152⁴²⁵ (139)
Kundrot, C. E., 53²⁰⁹ (50), 53²¹⁰ (50)
Kung, P.-P., 152⁴³⁹ (139)
Kuntz, I. D., 100⁵⁴ (85)
Kuntz, J. D., 476²⁰⁶ (469)
Kuperman, R., 245²¹⁴ (190)
Kupferschmitt, G., 79⁶⁸ (72)
Kurfürst, R., 243¹¹⁹ (168, 190), 244¹⁶⁹ (181)
Kuriyan, J., 368³² (350, 361), 368³⁴ (350)
Kurkinen, N. A., 103²¹⁷ (96)
Kuroda, R., 103²⁴⁷ (98)
Kurpiewski, M. R., 283¹¹¹ (266, 267), 283¹³² (270)
Kurucsev, T., 473⁴² (432)
Kushida, T., 548¹²⁶ (508)
Kushlam, D. M., 369⁹⁹ (362, 366)
Kusukawa, N., 476¹⁸⁵ (466)
Kutney, R., 310¹¹⁰ (299, 303)
Kutterer, K. M. K., 310¹²⁴ (301)
Kutyavin, I. V., 242⁶⁴ (161), 243¹²⁷ (169), 245²⁷⁰ (190), 246³⁰⁸ (197), 314, 336⁷ (314, 315)
Kuwabara, M., 422³⁸ (374)
Kuyl-Yeheskiely, E., 310¹³⁶ (305), 310¹⁴² (306)
Kuziemko, G. M., 424¹⁹⁸ (410)
Kwiatkowski, M., 149²⁰⁸ (126), 149²⁰⁹ (126)
Kwinkowski, M., 102¹³⁵ (92)
Kwok, D. I., 283⁹² (263)
Kypr, J., 337⁷³ (322)

Labourier, E., 12²⁶ (5)
Labreton, J., 310¹⁰⁵ (299, 300)
Labuda, D., 284¹⁹¹ (278), 284¹⁹² (278)
Lacey, C. J., 242⁷⁵ (162), 242⁷⁷ (162)
Lacey, S. W., 242⁵² (159)
Ladbury, J. E., 50⁵¹ (31), 51⁶⁵ (33)
Lahm, A., 281⁵ (252)
Lai, M. M., 367¹⁰ (344)
Laigle, A., 102¹⁷¹ (93)
Lakshman, M. K., 247⁴⁰⁶ (223), 247⁴⁰⁷ (223)
Lamberson, C. R., 474¹¹⁹ (449, 450), 474¹²⁰ (449, 450)
Lamm, G. M., 246³⁴⁵ (205, 206), 323
Lamm, M. E., 473⁴³ (432)

Lamminmäki, U., 247[363] (211)

Lamond, A. I., 12[32] (8), 149[219] (121)

Lampe, S., 282[28] (254), 330

Lan, T., 282[31] (255, 263), 283[129] (268)

Lancelot, G., 488[17] (478, 479)

Landegren, U., 247[363] (211)

Lander, N., 151[358] (130)

Landgraf, B., 151[376] (133)

Landgraf, R., 100[28] (85, 86)

Lane, A. N., 51[117] (39), 79[7] (56), 79[30] (64), 80[101] (76)

Lane, W. S., 337[76] (322)

Lange, P., 548[93] (502)

Langley, D. R., 547[55] (495)

Langlois, B., 337[87] (323)

Langman, S., 424[150] (401, 402)

Langner, D., 311[143] (306)

Langowski, J., 369[107] (365)

Lanier, A. C., 475[169] (463)

Lanier, A. L., 473[46] (433)

Lankhorst, P. P., 79[57] (72)

Lanza, A., 103[213] (96)

Lapidot, A., 79[67] (72, 73)

Larsen, T. A., 475[131] (452, 456)

Larson, C. J., 281[1] (252, 253, 266)

Larson, C. S., 147[70] (113)

Larson, J. E., 100[18] (82), 102[174] (93)

Laryea, A., 476[218] (470)

Latham, J. A., 13[59] (11), 13[60] (11), 636

Latimer, L. J. B., 476[192] (467)

Latimer, L. J. P., 283[143] (271, 272), 283[155] (274)

Lauder, S. D., 367[25] (347)

Laugaa, P., 475[127] (451), 475[128] (451)

Laughton, C., 475[142] (453, 457, 458)

Laughton, C. A., 244[183] (184)

Lauhon, C. T., 476[205] (469)

Laval, F., 424[196] (410)

Laval, J., 424[172] (407), 424[194] (410)

Lavery, R., 79[16] (59), 80[110] (77)

Lavrick, O. I., 247[385] (218)

Lawley, P. D., 103[234] (96)

Lawlor, J. M., 245[268] (190), 245[274] (190)

Lawrence, T. S., 425[221] (411)

Lawson, C. L., 283[116] (266, 268)

Le, D. B., 336[27] (316)

LeCuyer, K. A., 52[128] (41)

Lebedev, A. V., 242[47] (158), 244[174] (182), 294, 309[57] (292, 294, 306)

Lebleu, B., 152[444] (142), 308[6] (285), 308[12] (287, 292), 309[27] (288), 309[28] (289), 311[146] (307)

Lebowitz, J., 102[144] (92)

Lebreton, J., 305, 308[19] (288, 299, 300, 301, 305), 310[107] (299, 301), 310[108] (299, 301, 302)

Lecocq, J. P., 146[48] (112)

Lederer, F., 424[194] (410)

Lee, C.-S., 548[115] (505), 548[116] (505)

Lee, H., 247[412] (224)

Lee, H.-J., 151[391] (133)

Lee, J. S., 283[141] (271), 283[143] (271, 272)

Lee, K., 310[119] (299, 306)

Lee, K.-H., 548[102] (503)

Lee, M. S., 549[182] (516)

Lee, P. C. C., 424[209] (411)

Lee, S.-J., 548[144] (511)

Lee, Y. R., 151[344] (129)

Lefebvre, A., 80[109] (76)

Lefèvre, J.-F., 79[30] (64)

Lehman, J. M., 103[216] (96)

Lehn, J.-M., 474[123] (449)

Leikauf, E., 150[324] (126), 150[325] (126)

Lemaire, D. G. E., 423[140] (399), 423[141] (400), 424[189] (410)

Lemmen, P., 148[138] (120)

Lemonidis, K. M., 311[147] (288)

Lemster, T., 12[22] (5)

Lenardo, M., 489[35] (481)

Lenders, J.-P., 242[68] (161, 166)

Leng, F., 475[124] (451)

Leng, M., 102[170] (93, 94), 102[177] (93), 103[211] (96), 103[212] (96)

Lenhart, W. C., 310[131] (303)

Leon, R., 242[36] (155)

Leonard, G. A., 338[125] (325), 424[200] (410)

Leonard, N. J., 150[315] (126), 281[10] (253), 324

Leonard, P., 282[22] (254)

Leone, G., 12[48] (9)

Leonetti, J.-P., 309[28] (289)

Leong, T., 282[73] (259)

Leontis, N. B., 45, 367[23] (347, 353), 368[44] (353)

Lerch, B., 149[212] (120, 126)

Lerman, L. S., 432, 472, 473[37] (432, 433), 473[38] (432, 433)

Lerner, D. B., 79[64] (72)

Lerner, L., 79[48] (67)

Leroy, J., 52[162] (45)

Leroy, J. L., 474[65] (434)

Leroy, J.-L., 51[116] (39, 45), 80[79] (74), 80[84] (75)

Leserman, L. D., 309[28] (289)

Lesiak, K. B., 422[44] (376)

Lesnik, E. A., 247[368] (214), 310[127] (302)

Lesnik, E., 311[147] (288)

Lesnikowski, Z. J., 247[401] (221, 222), 249[490] (240, 241)

Lesser, D. R., 283[111] (266, 267), 283[132] (270)

Lester, C. C., 476[216] (470)

Letsinger, R. L., 12[3] (2), 13[53] (10), 13[54] (10), 107, 108, 110, 130, 136, 146[12] (107, 119), 146[17] (108, 110, 111), 147[75] (114), 147[90] (114), 149[227] (122), 237, 294, 366

Leumann, C., 12[8] (4)

Leupin, W., 474[100] (444, 449), 475[125] (451), 489[20] (479)

Levenson, C. H., 241[15] (158), 241[29] (155)

Lever, D. C., 151[342] (129)

Levin, D. E., 547[88] (502, 503)

Levin, J. R., 100[34] (85, 90)

Levina, A. S., 242[65] (161), 246[297] (194)

Levine, I. N., 42, 52[123] (40, 42)

Levis, J. T., 283[105] (265), 309[30] (289, 295)

Levis, J., 309[80] (295)

Levy, M. J., 248[453] (237)

Lewis, A. F., 339[201] (331, 333), 339[217] (333)

Lewis, E. S., 51[110] (38)

Lewis, J. G., 336[47] (317), 336[49] (317)

Lewis, L., 147[97] (114)

Lewis, M., 12[28] (6)

Lgolen, J., 474[104] (444, 446)

Lhomme, J., 101[106] (89)

Li, B. F. L., 102[154] (92), 147[112] (117)

Li, H., 309[42] (292, 293)

Li, L. H., 548[106] (504, 505), 548[107] (504, 505)

Li, V., 549[185] (517)

Li, V.-S., 549[186] (517)

Li, X., 338[175] (330)

Li, Y., 248[448] (234), 634, 635

Lian, Y., 284[175] (274)

Liang, G., 103[207] (95)

Liang, W., 422[39] (374, 376)

Liao, P. M.-L., 103[232] (96)

Liaw, Y.-C., 474[79] (437), 474[82] (437, 440)

Liddil, J. D., 549[181] (516)

Liese, T., 149[215] (120, 121)

Light-Wahl, K. J., 246[325] (201)

Lilley, D. M. J., 51[107] (36), 52[177] (45, 49), 82, 100[17] (82), 102[132] (92, 93), 102[133] (92), 103[205] (95), 103[206] (95)

Lima, W. F., 309[26] (288)

Liman, U., 338[143] (326)

Limoli, C. L., 425[224] (411, 419), 425[246] (419)

Lin, A. H., 548[107] (504, 505)

Lin, F.-T., 101[107] (89)

Lin, K.-Y., 339[229] (334)

Lin, L. H., 548[112] (505)

Lin, L.-N., 50[49] (30), 50[50] (31)

Lin, P. K. T., 30, 334, 337[65] (321, 334), 337[67] (321, 334), 339[228] (334)

Lin, R. J., 367[12] (345)

Lin, S.-Y., 476[179] (467)

Lin, W. C., 245[214] (190)

Lincoln, P., 101[84] (88), 101[105] (89)

Lindahl, T., 421[1] (372), 422[63] (380), 489[32] (480), 546[18] (493), 547[43] (494)

Linde, S., 339[220] (333)

Lindh, I., 146[31] (109, 139, 144)

Lingner, J., 12[42] (8)

Linn, S., 425[236] (417), 547[47] (494)

Lipman, R., 549[171] (515, 516), 549[172] (515, 516)

Lippard, S., 79[43] (63, 65, 67, 69, 70, 71, 74, 76, 77)

Lipscomb, L. A., 474[96] (441, 443, 444), 474[115] (446)

Liquier, J., 50[6] (16)

Liu, B., 489[23] (479, 480), 489[27] (480)

Liu, D., 309[62] (292)

Liu, E., 147[89] (114)

Liu, J., 298, 325, 337[117] (325)

Liu, L. F., 367[28] (348)

Liu, W.-C., 146[46] (112)

Liu, W.-Y., 245[276] (190)

Liu, X., 283[96] (264), 309[60] (292), 310[96] (297, 298)

Liu, Y., 50[17] (17), 52[177] (45, 49), 53[204] (49), 53[206] (49), 634, 635
Liukkonen, L., 246[327] (201)
Livache, T., 246[343] (205), 247[378] (216)
Live, D. H., 80[73] (72, 73)
Livingstone, J. R., 50[15] (17, 29)
Livshits, M. A., 369[108] (365)
Lloyd, D., 310[91] (296)
Lloyd, D. H., 12[5] (4)
Lloyd, Jr., R. M., 336[54] (319)
Lloyd, R. S., 424[193] (410)
Loakes, D., 339[220] (333)
Loechler, E. L., 103[235] (96)
Lohman, T. M., 17, 473[19] (428, 432)
Lohse, P. A., 52[175] (46)
Loke, S. L., 244[166] (181)
Lokhov, S. G., 242[64] (161), 244[174] (182)
Lombardy, R. L., 475[142] (453, 457, 458)
Long, E. C., 101[86] (88), 101[100] (89)
Lönnberg, H., 150[280] (122), 242[45] (157, 190), 243[136] (171, 190), 243[137] (171, 228), 245[219] (190), 245[220] (190), 248[419] (227), 248[420] (227)
Lonnberg, H., 368[79] (359)
Loontiens, F. G., 475[155] (459)
Lorsch, J. R., 52[171] (46)
Lotys, C., 146[58] (112, 113, 122)
Love, R., 283[115] (266, 267, 270)
Lovell-Badge, R., 367[5] (344, 367)
Lövgren, T., 246[327] (201)
Lowe, G., 100[55] (85)
Lown, J. W., 317, 336[40] (317), 473[33] (429, 432, 452), 474[69] (434, 440), 474[77] (437, 440), 497, 547[68] (497, 498, 499, 529), 547[69] (497)
Lu, M., 476[204] (469), 476[214] (469)
Lu, P., 12[28] (6), 80[81] (75)
Lubon, H., 281[6] (252)
Lucas, M., 149[236] (122)
Lucchesi, B. R., 421[8] (372)
Ludwig, J., 245[215] (190), 248[429] (230)
Luebke, K. J., 369[109] (365)
Luetke, K., 245[272] (190)
Luger, K., 12[33] (8)
Luisi, B. F., 283[116] (266, 268)
Lukhtanov, E. A., 243[127] (169), 245[270] (190), 336[7] (314, 315)
Lüking, S., 282[62] (258, 259, 265, 272)
Lumelsky, N., 52[166] (46)
Lundbäck, T., 51[95] (35)
Lundh, K., 52[125] (40)
Lunec, J., 421[7] (372)
Lunsford, W. B., 108, 130, 146[27] (108, 130)
Luo, W., 245[264] (190)
Luo, Y., 425[236] (417)
Lurquin, P. F., 100[46] (84)
Luttinger, A., 367[26] (348)
Ly, D., 424[185] (410)
Lyamichev, V. I., 102[167] (92, 95), 102[193] (94, 95)
Lyamichev, V., 102[151] (92)
Lysov, Y. P., 489[78] (486)
Lyubchenko, Y. L., 369[108] (365)

Ma, D. D. F., 242[84] (164)

Ma, M. Y.-X., 245[214] (190)
Ma, R.-I., 247[397] (221)
Ma, Y., 147[125] (118)
Maag, H., 232, 248[439] (232)
Maass, G., 52[148] (43, 45, 48)
Maass, K., 102[180] (93)
MacGregor, J. T., 548[95] (503)
MacKellar, C., 244[195] (187)
MacKellar, S. L., 241[17] (158)
MacKenzie, N. E., 547[74] (499)
MacMillan, A. M., 148[142], 148[143], 320
Macaulay, V. M., 244[183] (184)
Macaya, R. F., 283[151] (273)
Maccubbin, A. E., 425[247] (419), 425[248] (419)
Macgregor, Jr., R. B., 50[53] (31)
Macura, S., 79[25] (62)
Maddry, J. A., 303, 304, 310[130] (303)
Maffini, M., 245[217] (190)
Mag, M., 148[157] (120), 148[164] (120), 297
Magda, D., 242[79] (163, 164), 242[82] (164)
Magrath, D. I., 152[453] (144)
Mahadevan, V., 12[3] (2), 108, 110, 146[17] (108, 110, 111), 146[18] (108, 110, 111)
Maher, L. J., 368[59] (354)
Maher, III, L. J., 103[199] (95)
Maher, L. J. I., 52[186] (46)
Maida, K., 548[131] (508)
Maier, T., 247[350] (208)
Maiese, W., 547[76] (500)
Major, E. O., 102[182] (93)
Majors, R. E., 146[45] (112)
Majumdar, A., 79[49] (68)
Majumdarm, A., 80[91] (76)
Makaturova, E., 101[131] (92)
Makaturova-Rasovska, E., 102[153] (92)
Makino, K., 150[282] (122), 242[73] (162), 248[456] (237), 248[457] (237), 309[58] (292)
Maksakova, G. A., 242[66] (161)
Male, D., 473[4] (428, 434, 435, 436, 452)
Maler, B. A., 489[66] (484)
Malfoy, B., 102[170] (93, 94)
Malhotra, A., 368[44] (353)
Malhotra, S., 243[112] (168)
Malmgren, H., 368[71] (357)
Malmqvist, M., 52[125] (40), 52[127] (40)
Malone, M. E., 424[181] (409, 410)
Mamaev, S. V., 100[59] (86)
Mandal, P. C., 425[232] (417)
Manes, T., 102[173] (93, 94)
Maniatis, T., 100[26] (83)
Mann, J. S., 244[163] (178, 180)
Mann, M., 147[98] (114)
Mann, N., 309[35] (289)
Manners, G. D., 548[95] (503)
Manning, A. N., 242[40] (159)
Manning, G. S., 473[20] (428, 432), 474[107] (444, 446)
Manoharan, M., 231, 242[56] (159, 160), 242[57] (159), 248[432] (231), 309[29] (289)
Manzini, G., 337[70] (321)

Marasco, Jr., C. J., 337[74] (322)
Marchand, C., 476[198] (468), 476[199] (468)
Marden, J. J., 103[238] (96)
Marino, J. P., 369[102] (363)
Marion, D., 79[36] (64, 65)
Mark, F., 423[131] (398)
Markham, A. F., 146[37] (111)
Markiewicz, W. T., 147[78] (114), 147[124] (118)
Markley, J. L., 79[17] (59), 670
Markovits, J., 475[127] (451)
Marky, L. A., 282[12] (259)
Marky, L. M., 50[39] (23, 26, 28, 29, 30), 51[94] (35)
Marmorstein, R., 489[68] (484)
Marmorstein, R. Q., 283[116] (266, 268)
Marnett, L. J., 339[231] (334), 421[4] (372), 502, 547[88] (502, 503), 547[91] (502)
Marrot, L., 102[177] (93), 102[178] (93), 103[212] (96)
Marshall, W. S., 309[40] (292, 293)
Marsters, J. C., 244[201] (187)
Martin, A. M., 424[190] (410)
Martin-Bianco, E., 489[23] (479, 480)
Martin, D., 339[194] (330)
Martin, F. H., 325, 338[136] (325)
Martin, J. C., 308[23] (288)
Martin, J. F., 337[57] (319, 322)
Martin, M.-T., 476[197] (468)
Martin, P., 308[5] (285), 308[25] (288)
Martin, R., 146[56] (112)
Martinez, C. S., 490[96] (488)
Martinez, F. N., 423[111] (391)
Maruenda, H., 242[50] (159)
Marugg, J. E., 147[128] (118), 149[206] (121), 149[207] (121)
Maruo, Y., 242[92] (166)
Marzilli, L. G., 80[77] (73), 86, 100[62] (86), 101[65] (86)
Mascarenas, J. L., 282[44] (256)
Masegi, T., 148[193] (121), 148[194] (121), 149[210] (121)
Massari, M. E., 489[33] (480)
Masters, C. L., 421[10] (372, 417)
Mastruzzo, L., 242[84] (164)
Masuda, N., 150[299] (123)
Mathews, D., 53[206] (49)
Mathies, R. A., 474[101] (444, 446, 448)
Matray, T. J., 422[35] (373, 402, 403, 404), 424[152] (402, 408), 424[177] (408)
Matsmura, A., 51[84] (34)
Matsubara, K., 338[132] (325)
Matsuda, A., 246[316] (198), 248[417] (226), 248[418] (226)
Matsukara, M., 309[45] (292, 293)
Matsuki, S., 338[132] (325)
Matsumae, A., 549[160] (513), 549[165] (514)
Matsumoto, Y., 242[88] (165)
Matsumura, K., 242[87] (164)
Matsura, T., 423[86] (385, 386), 423[119] (397)
Matsushita, K., 548[125] (507, 508)
Matsuura, T., 423[103] (388)
Matsuzaki, J., 148[185] (121), 149[210] (121)

Mattaj, I. W., 473[63] (434)

Mattes, W. B., 546[16] (493)

Matteucci, M. D., 100[5] (82), 130, 149[253], 150[290] (122), 150[311] (126, 130), 152[423] (139), 152[438] (139), 282[56] (257, 273), 283[145] (271), 305, 308[18] (288, 299, 305), 308[23] (288), 310[116] (299, 305), 325, 334

Matthaei, H., 338[173] (329)

Matthes, H. W. D., 146[42] (112)

Matthews, B. W., 283[127] (268)

Matthews, J. R., 489[33] (480)

Mattingly, P. G., 244[155] (175)

Matulicadamic, J., 282[69] (259), 298, 310[97] (298)

Maurizot, J. C., 244[212] (190)

Max, E. E., 149[260] (121, 124)

Maxam, A. M., 82, 93, 95, 100[23] (82), 620

Maxwell, A., 12[25] (5)

Mayer, R., 488[17] (478, 479)

Mazumder, A., 339[227] (334), 422[27] (373), 422[61] (380), 547[83] (501)

Mazzarelli, J. M., 283[113] (266)

Mazzei, M., 244[174] (182)

McAuley-Hecht, K. E., 424[200] (410)

McBride, L. J., 100[5] (82), 119, 131, 147[131] (119)

McBridge, L. J., 150[286] (122)

McBurney, A., 421[7] (372)

McCaffey, A., 242[41] (161)

McCallum, K., 369[101] (363)

McCammon, J. A., 51[93] (35)

McCarthy, J. G., 100[38] (85, 92, 94)

McClarin, J. A., 283[114] (266, 267, 270)

McClaugherty, H., 130, 151[362] (130)

McClean, J., 247[361] (210)

McClellan, J. A., 102[138] (92), 102[198] (95)

McClymont, J. D., 423[136] (398, 401, 402)

McCollum, C., 146[58] (112, 113, 122), 146[59] (113), 149[222] (122)

McConnaughie, A. W., 473[14] (428, 436), 473[17] (428), 475[161] (460, 461)

McConnell, I. R., 548[145] (511, 512)

McDonald, C. D., 103[199] (95)

McDonald, J. J., 281[10] (253)

McDonough, K. A., 51[106] (36)

McElroy, E. B., 310[114] (299, 304), 310[133] (304)

McFail-Isom, L., 101[91] (88)

McGall, G. H., 423[97] (386, 387), 423[100] (387)

McGhee, J. D., 474[108] (446)

McGovern, J. P., 548[115] (505), 548[116] (505)

McGuinness, B. F., 549[180] (516)

McHugh, P. J., 422[59] (380)

McKay, D. B., 284[156] (274, 276)

McKee, T. L., 368[80] (359)

McKnight, C. J., 489[48] (482), 489[49] (482)

McKnight, S. L., 488[2] (477, 478, 480, 482, 484)

McLaughlin, L. W., 149[247] (123), 149[248] (123), 151[361] (130), 222, 243[97] (168), 243[108] (168), 245[215]

(190), 247[404] (222), 248[470] (238), 248[471] (238), 249[479] (240), 249[480] (240), 282[29] (255, 260), 282[30] (255, 263), 283[103] (265), 283[107] (265), 284[162] (274), 284[183] (276, 277, 278), 322, 327, 336[3] (314)

McLean, M. J., 100[18] (82), 102[155] (92), 102[174] (93)

McManus, J., 146[51] (112)

McMinn, D. L., 147[101] (115)

McMorris, T. C., 548[127] (508), 548[128] (508)

McNeillie, D., 474[97] (441, 443)

McSwiggen, J. A., 280, 284[181] (276, 280)

Meadows, R. P., 79[33] (64)

Medina, M. A., 423[117] (395)

Meehan, T., 244[163] (178, 180)

Meeuwenoord, N. J., 310[142] (306)

Mei, H.-Y., 101[97] (88), 101[98] (88)

Melik-Nubarov, N. S., 244[196] (187), 244[197] (187)

Mellema, J.-R., 52[157] (45)

Melnikova, A. F., 103[200] (95)

Melvin, T., 423[135] (398), 424[179] (409, 410)

Menchen, S., 241[31] (154, 158)

Meng, B., 310[112] (299, 304)

Menkhoff, S., 339[181] (330)

Mentzel, H., 339[196] (330)

Merchant, A., 246[338] (203)

Merenkova, I. M., 369[110] (366)

Mergny, J.-L., 102[183] (93), 476[187] (466, 467)

Mernagh, D. R., 101[128] (90)

Merrifield, R. B., 108, 146[16] (108)

Merril, C. R., 241[34] (155)

Merriman, M. C., 244[186] (185)

Mersmann, K., 284[187] (277, 279)

Merson-Davies, L. A., 424[181] (409, 410)

Mertens, R., 337[105] (324), 337[106] (324)

Meschwitz, S. M., 425[255] (421)

Messere, A., 369[113] (366)

Mestre, B., 242[74] (162)

Metelev, V. G., 150[291] (122)

Metz, J. T., 79[59] (72)

Meunier, B., 101[113] (89), 379

Meyer, K. L., 246[336] (203)

Meyer, R. B., 13[50] (10), 195, 319

Meyer, Jr., R. B., 147[96] (114)

Meyer, T. J., 422[42] (374, 377)

Meyers, R. A., 308[22] (288)

Meyers, R. E., 243[132] (170)

Miaskiewicz, K., 422[24] (373, 380, 383, 385, 400, 404, 405), 424[175] (407), 424[176] (407)

Michael, B. D., 425[230] (417)

Michael, M. A., 146[62] (113)

Michaux, C., 473[13] (428, 436)

Michel, B., 283[130] (268)

Michels, W. J., 52[172] (46)

Michelson, A. M., 146[7] (106, 109)

Michnicka, M. J., 80[76] (73)

Middendorf, L. R., 246[294] (193)

Middleton, P. J., 248[440] (233), 310[99] (298)

Midoux, P., 242[70] (161, 190)

Mielewczyk, S., 147[119] (117), 147[120] (117), 208

Mihara, H., 489[22] (479)

Mikita, T., 336[28] (316)

Miles, H. T., 323, 337[84] (323), 337[89] (323), 476[182] (466)

Milhand, P. G., 309[73] (294)

Millar, D. P., 246[302] (195)

Millard, J. T., 82, 84, 100[22] (82, 84, 85), 101[123] (90, 96), 103[232] (96), 103[236] (96)

Miller, E. M., 425[222] (411)

Miller-Hatch, K., 547[73] (499)

Miller, J., 338[160] (328), 424[175] (407), 424[176] (407)

Miller, P. S., 244[185] (184), 245[254] (190), 282[70] (259), 283[105] (265), 294, 308[13] (287, 292), 309[77] (294), 309[79] (294), 320, 321

Miller, R. A., 242[79] (163, 164), 242[82] (164)

Milligan, J. F., 152[438] (139), 282[56] (257, 273), 283[145] (271), 308[23] (288)

Milligan, J. R., 338[131] (325), 425[244] (418)

Min, C., 282[43] (256)

Minakawa, N., 246[316] (198)

Minchenkova, L. E., 80[112] (77)

Minisci, F., 423[142] (400, 405, 407)

Mir, K. U., 51[63] (33)

Mirabelli, C. K., 337[57] (319, 322)

Mirau, P. A., 52[163] (45)

Mirau, P., 12[46] (9)

Mirkin, C. A., 13[53] (10), 13[54] (10)

Mirkin, S. M., 102[151] (92), 102[167] (92, 95), 102[184] (93, 94, 95)

Mirzabekov, A. D., 103[200] (95)

Mishra, R. K., 150[268] (123), 150[269] (123)

Misiura, K., 243[144] (172, 174), 243[145] (172, 174), 309[56] (292, 306), 309[70] (293)

Misra, K., 149[256], 149[258] (121), 150[268] (123), 150[269] (123)

Misra, R., 247[352] (208, 209)

Misra, V., 80[106] (76)

Mitas, M., 103[208] (95)

Mitchell, J. B., 425[220] (411), 425[223] (411)

Mitchell, J. E., 102[198] (95)

Mitchell, L. M., 241[34] (155)

Mitchell, M. A., 548[115] (505), 548[116] (505)

Mitra, C. K., 79[22] (60), 80[120] (78)

Mitsuhira, Y., 149[231] (121, 135)

Miura, H., 424[202] (410)

Miura, K., 148[197] (123), 149[210] (210)

Miyaki, T., 490[89] (488)

Miyashiro, H., 79[54] (71)

Miyata, N., 547[82] (500)

Miyoshi, K., 146[24] (108, 112)

Mizan, S., 283[152] (274), 283[153] (274)

Mizuguchi, M., 309[58] (292)

Moazed, D., 473[55] (433)

Modak, A. S., 13[61] (11), 147[137] (120, 127), 148[146] (127), 185

Model, P., 283[124] (267)

Modrich, P., 421[2] (372)

Mohan, V., 242[56] (159, 160), 310[125] (302)

Mokrosz, J. L., 475[133] (452, 453), 475[134] (452, 453), 475[138] (452, 453), 475[139] (452, 453)

Mokrosz, M., 475[133] (452, 453), 475[138] (452, 453)

Mol, C. D., 12[45] (8)

Molko, D., 149[216] (120, 121), 149[217] (122)

Möller, U., 242[55] (159)

Monforte, J. A., 282[47] (257, 278), 284[174] (274, 277)

Mong, S., 490[92] (488)

Mong, S.-M., 547[55] (495)

Monia, B. P., 247[368] (214), 292, 308[15] (287), 308[16] (288), 309[63] (292, 306), 310[109] (299, 302), 311[147] (288)

Monsigny, M., 242[70] (161, 190)

Montenay-Garestier, T., 244[169] (181), 244[170] (181)

Montesarchio, D., 369[113] (366)

Montgomery, J. A., 310[130] (303)

Montoya, M. A., 548[127] (508), 548[129] (508)

Mooberry, E. S., 79[17] (59)

Moody, H. M., 149[251] (123)

Moolenaar, G. F., 102[164] (92)

Moon, S., 548[128] (508)

Moon, S.-H., 151[391] (133)

Moon, S.-S., 548[130] (508)

Moore, H. W., 45, 515, 549[170] (515)

Moore, M., 241[24] (158)

Moore, P. B., 52[161] (45)

Moore, R. N., 310[131] (303)

Moore, W. J., 51[121] (40)

Moraillon, A., 474[65] (434)

Mordan, W., 241[31] (154, 158)

Moreau, S., 548[96] (503)

Morgan, A. R., 283[141] (271), 283[143] (271, 272)

Mori, K., 167, 183, 243[103] (167), 244[166] (181), 244[177] (183), 309[45] (292, 293)

Mori, T., 150[319] (126)

Morii, T., 101[99] (89), 101[101] (89)

Morimoto, M., 242[44] (156)

Morimoto, S., 489[39] (482)

Morin, C., 149[265] (123)

Morinaga, T., 547[82] (500)

Morishita, R., 368[90] (362)

Morningstar, M. L., 424[198] (410)

Moroney, S. E., 337[97] (324), 337[98] (324)

Morris, J., 152[450] (144)

Morris, M. D., 51[73] (33)

Morris, S. J., 547[54] (495)

Morrison, H., 547[35] (494, 544)

Morrison, L. E., 245[278] (191, 197), 246[312] (197), 246[315] (198)

Morrow, J. D., 547[92] (502)

Morton, G. O., 547[76] (500)

Morvan, F., 309[36] (289, 293, 299, 302), 310[109] (299, 302)

Moschel, R. C., 247[402] (222, 225)

Moser, H. E., 242[80] (163), 244[154] (175), 248[450] (235), 283[138] (271),

308[5] (285), 308[25] (288), 310[121] (300)

Mossing, M. C., 488[15] (478)

Moulds, C., 336[47] (317), 336[49] (317), 338[131] (325)

Moule, Y., 548[96] (503)

Mountzouris, J. A., 473[2] (428, 435, 436, 456, 471), 547[52] (495, 497)

Moya, M. M., 548[130] (508)

Mucic, R. C., 13[53] (10), 13[54] (10)

Mueller, J. E., 368[93] (353)

Mühlegger, K., 152[408] (137), 152[409] (137, 138), 246[311] (197, 209), 247[333] (209)

Mujumdar, R., 247[354] (209)

Mujumdar, S., 247[354] (209)

Mukae, M., 242[73] (162), 248[457] (237)

Mukkala, V.-M., 246[327] (201), 247[363] (211)

Mullah, B., 150[293] (122, 123)

Muller, B. C., 101[96] (88)

Müller, F., 241[21] (185, 234)

Muller-Hill, B., 283[126] (268)

Muller, J. G., 101[110] (89), 101[111] (89)

Müller, L., 80[70] (72, 73)

Muller, M., 308[16] (288), 308[25] (288), 309[63] (292, 306)

Muller, S. N., 422[72] (382, 393, 403)

Muller, S. R., 473[28] (428)

Muller, W., 436, 473[12] (428, 436)

Mullis, K. B., 241[33] (155), 620

Multhaup, G., 421[10] (372, 417), 421[12] (372, 417)

Munk, S. A., 282[20] (253)

Munson, B. R., 475[141] (453, 457, 458)

Murakami, A., 242[73] (162), 248[456] (237), 248[457] (237)

Murata-Kamiya, N., 424[202] (410)

Murchie, A. I. H., 102[141] (92), 102[143] (92)

Murphy, C. J., 242[91] (165)

Murphy, F. L., 284[201] (280)

Murphy, J. A., 421[15] (372), 546[23] (493, 494), 546[31] (494, 511), 548[143] (510, 511)

Murray, K. K., 12[18] (4)

Murre, C., 489[33] (480)

Murugesan, N., 423[94] (386), 423[98] (386)

Musier-Forsyth, K., 152[447] (143)

Muthini, S., 242[43] (156, 173, 188, 190), 243[146] (172, 190)

Myrick, M. A., 336[35] (316), 338[152] (327)

Nackerdien, Z., 546[28] (494)

Nadzan, A. M., 548[147] (512)

Nagahara, S., 242[73] (162), 248[456] (237), 248[457] (237)

Nagai, K., 248[459] (237)

Nagai, M., 148[182] (121)

Nagaich, A. K., 149[256] (123)

Nagamura, S., 548[109] (505, 506)

Naganawa, H., 549[163] (514)

Nagata, K., 281[12] (253)

Naghibi, H., 50[48] (30)

Naito, T., 490[90] (488)

Nakagawa, M., 548[123] (506)

Nakai, M., 338[122] (325, 328)

Nakama, M., 368[90] (362)

Nakamaye, K. L., 249[488] (240)

Nakamura, E., 243[107] (167, 168)

Nakamura, T. M., 12[42] (8)

Nakamuta, H., 51[84] (34)

Nakanishi, K., 549[172] (515, 516), 549[180] (516)

Nakanishi, T., 148[201] (121), 148[202] (121)

Nakano, H., 245[275] (190, 235), 248[422] (228), 248[421] (228)

Nakano, S., 51[84] (34)

Nakaseko, Y., 489[65] (484)

Nakashima, H., 244[199] (187)

Nakatani, K., 424[188] (410)

Nakatsuji, Y., 248[458] (237), 248[462] (237)

Nakaura, M., 248[462] (237)

Nakayama, K., 148[188] (121)

Nakayama, M., 548[123] (506)

Namane, A., 246[286] (192)

Nambiar, K. P., 336[6] (314)

Namgoong, S.-Y., 102[165] (92)

Nara, H., 319, 336[52] (319)

Narang, Ch. K., 146[51] (112)

Narang, S. A., 108, 146[25] (108), 147[117] (117, 120)

Narekian, N. D., 248[478] (240)

Narui, S., 150[318] (126)

Natt, F., 308[25] (288)

Nawata, Y., 547[75] (499)

Naylor, L. H., 102[140] (92)

Neckers, L. M., 242[60] (160), 244[166] (181)

Nedderman, A. N. R., 337[67] (321, 334)

Neenhold, H., 246[324] (201)

Neidle, S., 422[22] (372), 457, 473[4] (428, 434, 435, 436, 452), 473[9] (428, 436), 474[67] (434, 435, 440, 441), 474[72] (435), 475[134] (452, 453), 475[142] (453, 457, 458), 476[196] (468), 546[10] (493, 513), 547[49] (495, 497), 548[105] (504, 505, 506), 549[167] (514, 515, 516, 517)

Neilson, L., 311[147] (288)

Neilson, T., 51[61] (33), 51[69] (33)

Nejedly, K., 102[135] (92)

Nelson, H. C. M., 488[7] (478)

Nelson, J. A., 102[181] (93)

Nelson, J. S., 147[96] (114)

Nelson, J. W., 473[57] (433), 476[208] (469)

Nelson, M. G., 309[74] (294)

Nelson, P. S., 147[89] (114)

Nemer, M. J., 149[236], 151[346] (130)

Nerdal, W., 80[90] (76)

Neş, I. F., 103[209] (96)

Nesbitt, S., 284[174] (274, 277)

Ness, J. V., 339[204] (332)

Neta, P., 422[55] (378, 402)

Netzel, T. L., 245[278] (191, 197), 245[279] (191, 197), 246[312] (197)

Neubauer, G., 147[98] (114)

Neuhaus, D., 489[65] (484)

Neuman, J.-M., 337[87] (323)

Neumann, E., 50[45] (29)
Neumann, R. D., 242[54] (159)
Neumann, U., 244[154] (175)
Neuner, P., 243[96] (234), 243[149] (173, 174), 245[281] (191), 246[282] (191), 247[379] (216), 248[446] (234), 248[447] (234)
Nevell, T. G., 547[54] (495)
Neville, D. M., 473[43] (432)
Newbury, S. F., 102[198] (95)
Newcomb, M., 423[99] (387), 423[111] (391)
Newman, A., 367[12] (345)
Newman, E. M., 337[77] (322)
Newman, P. C., 336[4] (314, 315), 336[5] (314, 315)
Newton, C. R., 146[37] (111)
Newton, G. L., 425[230] (417)
Neyhart, G. A., 422[28] (373, 374, 376, 380), 422[29] (373, 374, 380)
Ng, J. Y.-Y., 425[244] (418)
Ng, M. M. P., 337[102] (324), 337[103] (324)
Ng, P. G., 146[32] (109, 139, 144)
Nguyen, C. H., 476[180] (467, 468), 476[187] (466, 467)
Nguyen, M., 549[161] (513, 514)
Nibedita, R., 80[91] (76)
Nichols, R., 339[218] (333), 339[219] (333)
Nicieza, R. G., 338[154] (327)
Nicklen, S., 100[24] (82)
Nicklin, P., 308[16] (288), 308[25] (288)
Nicolaou, K. C., 546[32] (494), 557, 569
Nicolis, S., 367[5] (344, 367)
Nieborowska-Skorska, M., 310[91] (296)
Niece, R. L., 152[440] (142)
Niedballa, U., 282[82] (260), 282[83] (260)
Nielsen, A. H., 152[453] (144)
Nielsen, J., 147[128] (118), 147[129] (118), 149[206] (121), 149[207] (121)
Nielsen, P. E., 12[6] (4), 82, 100[20] (82), 102[160] (92), 102[163] (92, 94)
Nieman, R. A., 549[164] (514)
Nifant'ev, E. E., 152[405] (137)
Nikiforov, T. T., 246[330] (202), 246[333] (202)
Nikolov, D. B., 283[117] (266, 268)
Nikonowicz, E. P., 79[61] (72)
Nilsson, M., 368[71] (357)
Nir, S., 244[207] (190)
Nishijima, Y., 248[422] (228)
Nishimoto, H., 148[177] (120)
Nishimoto, S., 425[235] (417)
Nishimura, S., 338[163] (328), 424[164] (407)
Nishio, H., 337[114] (324), 337[115] (324, 325)
Nishira, S., 474[121] (450, 451)
Nishiyama, S., 149[205] (121)
Nishiyama, Y., 490[89] (488), 490[93] (488)
Nissen, M. S., 490[83] (486)
Nitta, K., 548[131] (508)
Niwa, H., 548[125] (507, 508)
Nobori, T., 148[192] (121)
Nokubo, M., 547[92] (502)

Nolan, C., 473[27] (428)
Noller, H. F., 473[55] (433)
Norden, B., 88, 101[84] (88), 101[89] (88), 451, 458
Nordheim, A., 102[188] (94), 102[189] (94)
Nordhein, S. A., 80[121] (78)
Nordhoff, E., 244[209] (190), 244[210] (190)
Norman, R. O. C., 423[107] (389, 391)
Normolle, D. P., 425[221] (411)
Novick, S. L., 244[207] (190)
Noyori, R., 148[168] (121, 135), 148[169] (120), 295
Nunn, C. M., 475[142] (453, 457, 458), 475[143] (453, 457, 458)
Nunota, K., 245[275] (190, 235), 248[424] (228)
Nwosu, V. U., 336[5] (314, 315)
Nye, J. A., 489[43] (482)
Nyilas, A., 147[135] (120), 148[190] (121)

O'Brien, J. A., 244[175] (182)
O'Connor, T. R., 424[194] (410)
O'Donnell, M., 368[32] (350, 361), 368[33] (350)
O'Donnell, M. J., 248[472] (238), 248[473] (238)
O'Donnell-Maloney, M. J., 12[19] (4)
O'Keeffe, T., 242[69] (161)
O'Neal, K. T., 489[47] (482)
O'Neill, P., 423[132] (398), 423[135] (398), 424[157] (403), 424[179] (409, 410), 425[239] (418)
Oas, T. G., 423[83] (385, 395)
Obendorf, M. S. W., 424[212] (411)
Oberhauser, B., 245[260] (190)
Oda, Y., 338[163] (328)
Oefner, C., 281[4] (252), 281[5] (252)
Oesch, F., 247[413] (225)
Ogata, R., 424[215] (411)
Ogawa, T., 147[79] (114)
Ogihara, T., 368[90] (362)
Ogilvie, K. K., 144, 149[236] (122), 151[346] (130), 151[351] (130), 152[403] (136), 152[404] (136), 284[190] (278), 303, 310[129] (303, 304)
Ohandley, S. F., 476[216] (470)
Ohashi, Y., 248[424] (228)
Ohbayashi, M., 490[89] (488), 490[93] (488)
Ohdate, M., 476[185] (466)
Ohkuma, H., 490[89] (488), 490[90] (488)
Ohkusa, N., 490[93] (488)
Ohmichi, T., 51[84] (34)
Ohmori, K., 423[93] (386), 548[110] (505)
Ohomori, H., 337[118] (325), 337[119] (325)
Ohrui, H., 282[86] (262)
Ohsumi, S., 150[282] (122)
Ohtsuka, E., 108, 146[14] (108)
Ohtsuua, E., 284[169] (274)
Oivanen, M., 150[280] (122)
Ojika, M., 548[125] (507, 508), 548[126] (508)
Ojwang, J. O., 103[235] (96)
Okami, Y., 548[131] (508)
Oki, T., 103[242] (98)
Okruszek, A., 151[387] (133), 151[388] (133), 309[56] (292, 306)

Okupniak, J., 150[303] (123, 136)
Olejnik, J., 243[150] (174)
Olivas, W. M., 338[145] (326, 327)
Olmsted, M. C., 50[30] (20), 50[31] (20)
Olsen, D. B., 282[46] (257, 278), 284[194] (278)
Olsen, G. J., 52[165] (120)
Olson, W. K., 369[97] (362), 473[24] (431)
Omichinski, J. G., 489[60] (484), 489[61] (484)
Önal, A. M., 422[75] (383), 424[189] (410)
Ono, A., 324, 336[32] (316), 336[52] (319), 337[69] (321, 324), 337[114] (324), 338[176] (330), 339[206] (332), 339[207] (332)
Onrust, R., 368[32] (350, 361)
Oosting, R. S., 310[142] (306)
Openshaw, H. T., 146[4] (106)
Opitz, J., 423[133] (398, 405, 406)
Oretskaya, T. S., 242[39] (157)
Orgel, L. E., 170, 243[126] (169), 243[135] (170)
Ornstein, R., 424[176] (407)
Orrell, K. G., 79[6] (56)
Osaki, Y., 149[205] (121)
Osato, T., 548[131] (508)
Osawa, T., 152[402] (136)
Osborne, M. R., 247[403] (222)
Osborne, S. E., 369[102] (363)
Oser, A., 242[48] (158)
Osgood, S. A., 243[104] (168)
Osman, R., 422[24] (373, 380, 383, 385, 400, 404, 405), 424[175] (407), 424[176] (407)
Osterman, D. G., 337[75] (322)
Östlin, H., 52[125] (40)
Otsuka, M., 103[247] (98)
Ott, J., 283[122] (267)
Otting, G., 489[18] (479), 489[21] (479)
Otvos, L., 336[45] (317), 337[73] (322)
Otwinowski, Z., 283[116] (266, 268)
Ouporov, I. V., 368[44] (353)
Ouryupin, A. B., 151[388] (133)
Ovcharenko, A. V., 244[196] (187), 244[197] (187)
Overman, L. B., 50[13] (17)
Ozaki, H., 248[422] (228), 248[425] (228), 249[479] (240), 249[480] (240), 283[103] (265)
Ozaki, S., 148[177] (120)

Pabo, C. A., 488[4] (477, 478, 480, 482, 484)
Pabo, C. O., 12[30] (6)
Pace, B., 52[165] (120)
Pace, N. R., 52[165] (120)
Paces, V., 146[54] (112)
Padgett, R. A., 367[13] (345)
Padmapriya, A. A., 150[307] (125)
Paillous, N., 547[37] (494, 544)
Palecek, E., 82, 92, 100[8] (82), 100[35] (85, 92), 101[129] (92), 101[131] (92), 102[135] (92), 102[136] (92)
Palm, G., 146[47] (112)
Palom, Y., 336[14] (315), 336[15] (315, 330)
Pan, G., 245[272] (190)
Pan, S. S., 549[177] (516)

Pan, T., 368[54] (354)
Pan, Y., 310[92] (296)
Paner, T. M., 369[98] (362)
Panigrahi, G. B., 245[271] (190)
Pankiewicz, K. W., 339[210] (332)
Pannecouque, C., 310[106] (299, 300)
Pao, A., 12[13] (4)
Paolella, G., 282[49] (257, 278)
Paranawithana, S. R., 101[102] (89)
Pardi, A., 339[198] (330)
Pardue, M. L., 103[221] (96)
Pares, F., 490[96] (488)
Pari, G. S., 248[474] (239)
Parikh, S. S., 12[45] (8)
Paris, P. L., 282[34] (255)
Park, C., 12[31] (6)
Parker, A. N., 476[179] (467)
Parker, A. W., 423[135] (398), 424[179] (409, 410)
Parker, J. R., 309[35] (289)
P'arraga, A., 103[231] (96)
Parsons, J. M., 309[35] (289)
Partridge, W. M., 245[257] (190), 246[317] (199), 246[318] (199)
Pasternack, R. F., 89, 101[114] (89)
Patek, D., 247[354] (209)
Patel, D. J., 338[162] (328), 440, 473[61] (434), 474[85] (439, 440), 474[112] (446, 448), 476[218] (470), 476[219] (470), 549[187] (517)
Patel, D. S., 474[87] (440)
Patel, R., 369[114] (366)
Pathak, T., 151[375] (133), 151[376] (133)
Patil, S. V., 248[463] (237)
Patrzyc, H. B., 425[247] (419)
Patterson, S. E., 475[158] (460)
Paul, C. H., 150[306] (125)
Pavlopoulos, S., 548[138] (509)
Pavlova, M. N., 102[197] (95)
Pawlak, J., 549[172] (515, 516)
Peak, J. G., 547[39] (494, 544)
Peak, M. J., 547[39] (494, 544)
Pearlman, D. A., 80[87] (76)
Pearson, C. E., 100[11] (82)
Pearson, L., 100[28] (85, 86)
Pearson, R. G., 51[120] (40)
Peattie, D. A., 52[130] (41)
Pecenka, V., 102[191] (94)
Pecinka, P., 101[131] (92)
Pecoraro, V. L., 282[65] (258)
Pedersen, J. B., 474[116] (448)
Pedone, P. V., 489[60] (484)
Pedrini, A. M., 103[213] (96)
Pedroso, E., 146[38] (111), 147[80] (114)
Pedroso, P., 152[450] (144)
Peek, M. E., 474[96] (441, 443, 444), 474[115] (446)
Pei, D., 547[80] (500, 541), 629
Peisach, J., 423[102] (388, 394)
Pelham, H. R. B., 368[82] (360)
Pellegrini, M., 489[55] (483)
Peloquin, R., 475[160] (460, 461)
Penney, C. L., 152[403] (136)
Péoc'h, D., 248[476] (239), 310[128] (303)
Perbost, M., 309[36] (289, 293, 299, 302), 310[126] (302)
Percival-Smith, A., 489[20] (479), 489[26] (479)

Pérez, J., 149[259] (121, 124)
Periasamy, A., 241[11] (155)
Périgaud, C., 422[33] (373, 379, 394, 395)
Perkocha, L., 367[28] (348)
Permana, P. A., 475[164] (462)
Perreault, J. P., 284[190] (278), 284[191] (278)
Perrin, C., 246[331] (202)
Perrin, D. M., 100[28] (85, 86)
Perrin, K. A., 310[134] (304)
Perrotta, A. T., 52[173] (46)
Perrouault, L., 242[83] (164), 242[85] (164), 245[235] (190), 245[236] (190)
Perry, L., 338[178] (330)
Persson, B., 52[125] (40), 52[126] (40)
Pestov, D. G., 102[184] (93, 94, 95)
Pestov, N. B., 244[164] (178, 179)
Peters, III, W. A., 12[27] (5)
Petersen, G. V., 368[65] (355)
Petersen, K. H., 242[46] (158, 190)
Petersheim, M., 50[42] (28)
Peterson, G. V., 310[123] (300)
Peterson, M. A., 310[122] (300)
Peticolas, W. L., 80[111] (77)
Petretta, M., 423[109] (389), 423[115] (391, 393, 397, 403)
Petrie, C. R., 241[25] (158), 245[253] (190), 246[348] (206, 207)
Petrova, M., 152[425] (139)
Petrovskii, P. V., 151[388] (133)
Petrullo, L. A., 424[173] (407)
Petrusek, R. L., 547[57] (495)
Pettijohn, D. E., 103[216] (96), 103[226] (96)
Pettitt, B. M., 51[100] (36)
Petzold, G. L., 548[107] (504, 505), 548[108] (505)
Peukert, S., 423[113] (391, 393)
Peyman, A., 152[441] (142), 308[21] (288)
Peyrottes, S., 309[48] (292, 295), 309[84] (295)
Pfeiderer, W., 245[233] (190)
Pfeifer, G. P., 12[13] (4)
Pfenniger, O., 103[216] (96)
Pfister, M., 148[150] (127), 148[162] (120)
Pflaum, M., 424[191] (410)
Pfleiderer, W., 127, 149[239] (123), 149[240] (123), 150[328] (127), 150[329] (127), 151[357] (130), 151[370] (133), 244[205] (189), 244[206] (189), 247[350] (208)
Pgnonec, P. P., 489[40] (482)
Philippsen, P., 52[148] (43, 45, 48)
Phillips, J., 308[25] (288)
Phillips, L. R., 248[468] (238), 283[108] (265)
Phillips, S. E. V., 489[34] (480), 489[74] (486)
Piantini, U., 79[37] (64)
Piatyszek, M. K., 476[203] (469)
Piccialli, G., 369[113] (366)
Piccirilli, J. A., 246[338] (203), 284[205] (281)
Pieken, W. A., 282[46] (257, 278), 284[194] (278), 337[102] (324)
Piel, N., 149[247], 151[361] (130), 282[73] (259), 283[120] (267)
Pieles, U., 216, 242[80] (163), 244[154]

(175), 244[179] (184), 246[345] (205, 206), 247[379] (216), 310[120] (300)
Pieles, V., 308[25] (288)
Pierce, J. W., 489[35] (481)
Pietra, F., 547[84] (501)
Pietrasiak, D., 309[70] (293)
Piette, J., 100[44] (85, 96)
Pikkemaat, J. A., 367[24] (347, 353)
Pilch, D. S., 12[36] (8), 50[47] (30, 34, 36), 51[99] (36), 51[101] (36)
Pindur, U., 12[22] (5)
Pingoud, A., 283[102] (265, 270)
Piotto, M., 368[44] (353)
Pirrung, M. C., 151[341] (129), 151[342] (129)
Pistorius, A. M. A., 148[165] (120), 148[166] (120)
Pitié, M., 242[77] (162)
Pitzer, K. S., 79[12] (57)
Piulats, J., 242[59] (160)
Pjura, P., 475[147] (456)
Plaskon, R. R., 474[96] (441, 443, 444)
Plateau, P., 52[159] (45), 80[84] (75)
Plattner, J. J., 475[164] (462)
Pleij, C. W., 367[15] (345), 367[16] (345)
Pley, H. W., 284[156] (274, 276)
Pliasunova, O. A., 242[58] (160)
Plon, S. E., 244[181] (184)
Plum, G. E., 12[36] (8), 50[35] (21), 50[38] (23, 28, 29, 30, 34), 51[87] (34), 51[91] (34)
Podell, E., 53[209] (50), 53[210] (50)
Podomore, I. D., 424[150] (401, 402)
Podyminogin, M. A., 13[50] (10)
Pohl, F. M., 80[116] (77)
Pokrovsky, A. G., 242[58] (160)
Polach, K. J., 103[241] (96)
Polte, T., 100[50] (84)
Polucci, P., 476[196] (468)
Polushin, N. N., 248[427] (229)
Pomerantz, J. L., 12[30] (6)
Pommier, Y., 547[83] (501)
Pommier, Y. G., 339[227] (334)
Pon, R. T., 146[41] (112), 147[76] (114), 147[77] (114), 171
Pons, A., 243[98] (167)
Pope, B. H., 309[35] (289)
Pope, L. E., 101[73] (86)
Popelis, J., 152[425] (139)
Popov, S. G., 146[23] (108)
Pörschke, D., 39, 50[55] (32), 51[115] (39, 43, 44, 46), 52[141] (42), 52[142] (42)
Porter, K., 309[42] (292, 293)
Porter, K. W., 338[166] (329), 338[167] (329)
Portogal, J., 475[146] (456)
Post, C. B., 79[33] (64)
Potaman, V. N., 308[10] (285)
Potapov, V. K., 248[464] (237)
Potekhin, S. A., 50[44] (28, 29, 30)
Potier, P., 247[361] (210)
Potter, B. V. L., 283[102] (265, 270)
Poulter, C. D., 80[74] (72)
Poverenny, A. M., 102[157] (92)
Povirk, L. F., 425[250] (420), 425[251] (420), 546[17] (493)
Povsic, T. J., 196, 246[306] (196), 283[144] (271, 272), 319

Power, M. J., 283[87] (263)
Powers, R., 79[61] (72)
Powers, T., 52[131] (41)
Pownall, S., 147[92] (114), 147[94] (114)
Prado, F. R., 79[58] (72)
Prairie, M. D., 548[113] (505), 548[114] (505)
Prakash, G., 368[67] (356, 357, 359), 368[73] (359, 360, 366)
Prasad, A. K., 150[300] (123)
Prasad, C. V. C., 310[104] (299, 301)
Praseuth, D., 242[83] (164), 243[124] (168), 243[128] (169)
Pratviel, G., 101[113] (89)
Precigoux, G., 283[136] (271)
Prelog, V., 79[10] (56)
Presnell, S. R., 474[96] (441, 443, 444)
Price, G. B., 100[11] (82)
Price, M. A., 90, 101[119] (90)
Prichodko, T. A., 246[297] (194)
Priebe, W., 473[34] (429, 432), 475[124] (451)
Priestley, E. S., 338[171] (329)
Prior, G. M., 152[451] (144)
Prise, K. M., 425[230] (417)
Pritchard, C. E., 147[74] (114, 127)
Prive, G. G., 475[129] (452)
Proba, Z. A., 149[236]
Prober, J. M., 247[362] (211)
Profenno, L., 52[177] (45, 49)
Projan, S. J., 423[102] (388, 394)
Prouty, C. P., 310[104] (299, 301)
Prowse, K. R., 476[203] (469)
Pruss, D., 82, 100[16] (82)
Ptashne, M., 489[68] (484)
Pucher, H.-J., 102[180] (93)
Pudlo, J. S., 152[438] (139), 283[145] (271)
Pudovik, A. N., 152[410] (138)
Puga, E., 244[172] (181)
Puglisi, J. D., 473[60] (434)
Pullman, A., 475[148] (456), 546[11] (493)
Pullman, B., 474[79] (437), 474[91] (441), 475[140] (453, 457, 458), 475[148] (456), 476[221] (470), 546[11] (493)
Purcell, M., 474[111] (446, 448)
Purmal, A. A., 150[291] (122)
Puskás, L. G., 244[178] (183)
Pyle, A. M., 52[172] (46), 52[176] (45, 49), 88, 100[30] (85, 88), 101[99] (89), 101[100] (89)

Qian, X., 489[79] (486)
Qian, Y. Q., 489[18] (479), 489[21] (479)
Qu, F.-C., 309[68] (293)
Quadrifoglio, F., 337[70] (321)
Quaedflieg, P. J. L. M., 149[251] (123), 310[136] (305)
Quesada, M. A., 474[101] (444, 446, 448)
Quigley, G. J., 368[41] (352), 368[51] (354), 474[103] (444, 446), 490[86] (488), 490[87] (488)
Quignard, E., 79[45] (67)
Quinton, F. L., 547[57] (495)
Quong, M. W., 489[33] (480)

Rabinovich, D., 474[103] (444, 446)

Rabow, L. E., 422[62] (380), 423[96] (386), 423[97] (386, 387)
Radisky, D. C., 547[85] (501, 529)
Radisky, E. S., 547[85] (501, 529)
Radler, E., 337[96] (323)
Raff, M., 241[31] (154, 158)
Raha, M., 13[51] (10)
Rahamim, E., 146[52] (112)
Rahman, S. K., 546[2] (492)
Rahn, R. O., 425[216] (411)
Raillard, S. A., 422[18] (372)
Rajagopal, J., 284[204] (280, 281)
Rajagopal, P., 80[96] (76, 77)
Rajeev, K. G., 337[71] (322)
Rajur, S. B., 248[473] (238), 282[29] (255, 260), 282[51] (257, 278), 283[113] (266), 283[128] (268), 336[3] (314)
Ramachandran, K. L., 242[67] (161, 162, 165)
Ramage, R., 151[339] (127)
Ramasamy, K. S., 246[323] (200), 247[368] (214), 283[99] (265), 327
Rammler, D. H., 150[308] (126)
Ramzaeva, N., 282[26] (254)
Rana, T. M., 246[324] (201)
Rana, V. S., 336[38] (316)
Randall, R. E., 243[153] (175)
Randles, J. W., 367[9] (344, 346)
Raner, G., 89, 101[115] (89)
Raney, K. D., 476[221] (470), 548[151] (512), 548[152] (512, 513)
Raney, V. M., 476[221] (470), 549[155] (513)
Rao, G., 546[28] (494)
Rao, K. E., 547[69] (497), 547[70] (497, 499)
Rao, M. V., 247[361] (210)
Rao, P. Y., 246[347] (206)
Rao, S. N., 547[64] (497)
Rao, T. S., 148[147] (120), 331
Raoul, S., 424[203] (411), 424[206] (411)
Raphael, A. L., 101[88] (88), 101[94] (88)
Rapoport, H., 100[43] (85, 96), 148[170] (120), 148[171] (120)
Rappaport, H. P., 338[147] (327)
Rashid, R., 423[131] (398)
Rasovska, E., 102[137] (92)
Rastogi, R. C., 147[85] (114)
Ratliff, R. L., 50[8] (16)
Ratmeyer, L., 12[5] (4), 51[82] (34), 475[159] (460), 475[160] (460, 461), 476[211] (469)
Raucher, S., 103[236] (96)
Rauman, B. E., 489[75] (486), 489[76] (486, 487)
Ravanat, J.-L., 424[203] (411)
Ravikumar, V. T., 12[12] (4)
Ravin, A. J., 367[23] (347, 353)
Ray, K. A., 244[155] (175)
Rayford, J., 336[35] (316)
Rayner, B., 248[476] (239)
Rayner, G., 282[53] (257)
Raynor, B., 309[48] (292, 295), 309[84] (295)
Reardon, B. J., 490[84] (487)
Reck, T., 547[55] (495)
Recknor, M., 241[31] (154, 158)

Record, Jr., M. T., 50[15] (17, 29), 50[16] (17), 101[127] (90)
Record, M. T., 473[19] (428, 432)
Reddy, G. R., 339[231] (334), 547[92] (502)
Reddy, M. P., 146[62] (113)
Reed, M. W., 13[50] (10), 147[96] (114)
Reese, C. B., 107, 108, 144, 149[245] (126), 149[261] (123), 150[330] (127), 150[331] (127), 151[356] (130), 152[454] (144), 246[284] (192), 283[96] (264), 292, 297, 298, 309[60] (292), 310[94] (297), 310[96] (297, 298)
Reeves, R., 490[83] (486)
Refregiers, M., 102[171] (93)
Regan, J. B., 152[446] (143), 248[468] (238), 248[469] (238), 283[108] (265)
Regberg, T., 146[31] (109, 139, 144)
Regner, M., 248[437] (232)
Regnier, F. E., 241[5] (168)
Rehrauer, W. M., 367[25] (347)
Reich, C., 52[165]
Reid, B. R., 79[26] (64), 79[42] (65), 80[82] (75), 80[90] (76)
Reid, L. S., 245[214] (190)
Reidling, J. C., 247[380] (216)
Reindl, B., 547[89] (502)
Reiner, T., 148[150] (127)
Reinhold, V. N., 548[147] (512)
Remaud, G., 147[129] (118)
Remers, W. A., 546[4] (492), 547[50] (495, 497, 499, 505, 520), 547[59] (495, 497)
Ren, W.-Y., 339[210] (332)
Ren, X.-F., 282[34] (255), 282[35] (255)
Renhofa, R., 152[425] (139)
Repkova, M. N., 242[65] (161)
Resendex-Perez, D., 489[18] (479)
Reszka, A. P., 244[183] (184)
Reszka, K., 474[77] (437, 440)
Revankar, G. R., 283[99] (265), 332, 334
Revet, B. M. J., 473[45] (433)
Rey, I., 245[243] (190)
Reynolds, M. A., 242[41] (161), 308[13] (287, 292)
Reynolds, P. E., 473[7] (428, 429, 432, 433, 436, 444, 452, 453)
Reynolds, R. C., 310[130] (303)
Reynolds, V. L., 548[108] (505)
Rhee, S., 241[24] (158)
Rhee, Y., 338[157] (328)
Rhinehart, R. L., 336[7] (314, 315)
Rhodes, D., 12[39] (8)
Rhodes, L. M., 52[146] (43)
Ricca, D. J., 282[50] (257, 265, 273), 283[142] (271, 273)
Rice, M. C., 13[52] (10)
Rich, A., 368[37] (351, 352), 368[41] (352), 474[81] (437, 439, 440), 474[103] (444, 446), 476[181] (466), 490[86] (488), 490[87] (488)
Richards, K. H., 147[121] (117)
Richardson, C. D., 147[95] (114)
Richardson, P. L., 246[325] (201)
Richert, C., 310[113] (299, 304), 310[132] (304)

Richmond, M. H., 473⁷ (428, 429, 432, 433, 436, 444, 452, 453)
Richmond, T. J., 12³³ (8)
Richterich, P., 249⁴⁸⁹ (240)
Rickwood, D., 473⁴ (428, 434, 435, 436, 452)
Rideout, D., 463, 475¹³⁷ (452, 453), 475¹⁶⁷ (463)
Rider, P., 149²¹⁹ (121)
Riesner, D., 52¹⁴⁸ (43, 45, 48), 52¹⁵¹ (43)
Klesz, P., 423¹⁷⁹ (398, 407), 423¹⁷⁹ (398, 407)
Rietveld, K., 367¹⁵ (345)
Rigler, R., 52¹⁵³ (45)
Rijk, E. A. V., 148¹⁶⁶ (120)
Rill, R. L., 101⁶⁷ (86)
Rimsky, S., 101⁶⁸ (86)
Ringel, I., 146⁵² (112)
Rink, S. M., 103²³⁹ (96)
Rinkel, L. J., 79⁵⁰ (68)
Rinnehart, K. L., 473⁵⁴ (433), 473⁵⁴ (433)
Rio, P., 103²¹¹ (96)
Riordan, C. G., 547³⁴ (494)
Rippe, K., 50⁴⁰ (24, 28, 36), 51⁹⁸ (36)
Ritter, A., 422⁶⁵ (381, 389)
Rivalle, C., 245²³⁵ (190), 245²³⁶ (190)
Rivera, R. R., 489³³ (480)
Rizk, I., 151³⁴⁷ (130), 151³⁴⁸ (130)
Rizzarelli, E., 242⁸⁴ (164)
Robert-Nicoud, M., 102¹⁵⁸ (92)
Roberts, C., 337¹⁰⁸ (324)
Roberts, E. S., 423⁹⁹ (387)
Roberts, J. W., 367¹ (343, 346)
Roberts, R. J., 475¹⁷⁵ (466)
Roberts, R. W., 36, 48, 368⁶⁹ (357)
Robertson, C. W., 247³⁶² (211)
Robertson, D. L., 13⁵⁷ (10)
Robertson, H. D., 367¹⁹ (346)
Robertson, M., 152⁴⁴⁰ (142)
Robins, M. J., 282⁸⁰ (260), 300, 310¹²² (300)
Robins, R. K., 283⁹⁹ (265)
Robinson, B. H., 247³⁸⁶ (219), 247³⁸⁷ (219)
Robinson, C. R., 489⁷⁷ (486)
Robinson, H., 337¹¹¹ (324), 474⁷⁹ (437), 474⁸² (437, 440)
Robledo-Luiggi, C., 490⁹⁶ (488)
Robles, J., 147⁸⁰ (114)
Roby, C. D., 339²¹¹ (332)
Roche, A. C., 242⁷⁰ (161, 190)
Rockwell, S., 549¹⁶⁶ (514, 516)
Roder, H., 52¹²⁴ (40)
Roderick, S. L., 488¹⁴ (478)
Rodger, A., 101⁸⁹ (88)
Rodgers, M. A. J., 424²⁰⁹ (411)
Rodolfo, C., 103³¹³ (96)
Rodrigues, N. R., 102¹³⁴ (92, 93)
Rodriguez, L. O., 423⁹² (385)
Roduit, J.-P., 198, 246³¹⁴ (198)
Roeder, R. G., 489⁴⁰ (482)
Roelen, H. C. P. F., 336¹¹ (315)
Roget, A., 205, 244¹⁵⁷ (177, 205), 246³³⁹ (204), 246³⁴³ (205), 247³⁷⁸ (216)

Roig, V., 243¹¹⁹ (168, 190), 244¹⁷⁰ (181)
Rokem, J. S., 548¹⁰⁷ (504, 505)
Rokita, S. E., 89, 100³² (85, 89), 100³³ (85, 89), 101¹⁰⁸ (89), 101¹¹⁰ (89)
Romano, L. J., 247⁴¹⁰ (223), 247⁴¹¹ (223, 224)
Romanova, E. A., 242³⁹ (157)
Romer, R., 52¹⁵¹ (43)
Romero, R., 103²⁰⁸ (95), 150²⁸¹ (122)
Rong, F.-G., 247³⁹⁹ (221)
Rönnberg, B., 52¹²⁶ (40)
Rönnberg, I., 52¹²⁵ (40), 52¹²⁶ (40)
Rooney, T., 100⁵⁰ (84)
Roongta, V. R., 79⁵⁹ (72), 79⁶¹ (72)
Roos, H., 52¹²⁵ (40)
Roques, B. P., 474¹⁰⁴ (444, 446), 474¹¹⁵ (446), 475¹²⁷ (451), 475¹²⁸ (451)
Rösch, R., 152⁴⁵² (144), 242⁷⁸ (163), 245²²⁵ (190), 246²⁹³ (193)
Rose, S. J., 248⁴³⁹ (232)
Rosemeyer, H., 282²¹ (254), 282²⁷ (254), 283¹⁰⁰ (265)
Rosen, M., 79⁴⁴ (67), 80¹⁰³ (76)
Rosenberg, I., 339²¹⁰ (332), 423⁸⁹ (385, 394, 395, 421)
Rosenberg, J. M., 283¹¹² (266), 283¹¹⁴ (266, 267, 270)
Rosenbohm, C., 147⁷⁰ (113)
Rosenstein, S. P., 52¹⁷³
Rosenthal, A., 243⁹⁵ (167), 244¹⁵⁸ (177, 190)
Ross, A. B., 423¹²⁷ (398)
Ross, B. M., 103²³⁰ (96), 103²³¹ (96)
Ross, K. C., 147⁹⁹ (114)
Ross, S. A., 103²¹⁰ (96)
Rossi, F., 12²⁶ (5)
Rossman, T. G., 424¹⁶⁸ (407)
Rothberg, K. G., 242⁵² (159)
Rothschild, K. J., 243¹⁵⁰ (174)
Rougée, M., 242⁴⁷ (158), 244¹⁷¹ (181), 283¹⁴⁹ (273), 476¹⁸⁷ (466, 467)
Roughton, A. L., 310¹¹³ (299, 304)
Rould, M. A., 489⁷⁶ (486, 487)
Roush, W., 12¹⁰ (4, 5, 9)
Rousseau, N., 102¹⁷⁰ (93, 94)
Routier, S., 101¹⁰⁹ (89)
Routledge, A., 147⁹⁹ (114)
Roux, P., 247³⁸⁵ (218)
Rouzina, I., 50²⁸ (20), 50²⁹ (20)
Royappa, A. T., 150³⁰⁶ (125)
Royer, C. A., 50²⁰ (20, 31)
Royo, M., 243¹⁰⁰ (167)
Rozenski, J., 248⁴²⁸ (230), 310¹⁰⁶ (299, 300)
Rozners, E., 152⁴²⁵ (139)
Ruan, K.-C., 245²⁷⁶ (190)
Ruben, D. J., 80⁷¹ (72, 73)
Rubin, C. M., 102¹⁶⁹ (93)
Rubin, E., 368⁸⁰ (359), 368⁸¹ (360), 369¹¹¹ (366)
Rück, A., 242⁷⁸ (163)
Ruf, K., 149²⁴⁴ (123)
Ruff, E., 336⁴⁵ (317)
Ruffner, C. G., 548¹⁰² (503)
Ruffner, D. E., 284¹⁷¹ (274)
Rüger, R., 246³¹¹ (197, 209)

Ruiz-Perez, C., 338¹⁴¹ (326)
Rumney, IV, S., 244²¹³ (190)
Runkel, L., 102¹⁸⁸ (94)
Running, J. A., 246²⁸⁷ (192), 246²⁸⁸ (192)
Rupert, L. A. M., 244²⁰⁷ (190)
Ruppert, T., 421¹⁰ (372, 417)
Russell, G. A., 424¹⁴⁹ (401, 402)
Russo, A., 425²²⁰ (411), 425²²³ (411)
Russu, I., 474⁶⁶ (434)
Rusting, R. L., 421⁵ (372)
Ruterjans, H., 79⁶⁸ (72)
Ruth, J. L., 246²⁹⁴ (193)
Rutherford, M., 548¹²⁹ (508)
Rutkowski, R., 489⁴⁹ (482)
Ryan, K., 309⁶² (292), 361
Ryan, M. J., 149²²⁰ (126)
Ryder, U., 245²⁸¹ (191), 246²⁸² (191)
Rye, R. S., 474¹⁰¹ (444, 446, 448)
Ryte, A., 311¹⁴³ (306)

Saavedra, R. A., 241¹⁷ (158)
Sabattier, R., 425²³¹ (417)
Sadana, K. L., 152⁴⁰⁴ (136)
Sadowski, P. D., 245²⁷¹ (190), 245²⁷² (190)
Saenger, W., 282⁴⁵ (256, 257)
Saga, Y., 242³⁸ (156)
Sagi, G., 336⁴⁵ (317)
Sagi, J., 336⁴⁵ (317), 337⁷³ (322)
Saha, A. K., 298, 303, 310⁹⁸ (298), 310¹⁰⁴ (299, 301)
Saha-Möller, C. R., 424²⁰⁷ (411)
Sahasrabudhe, P. V., 336³¹ (316)
Saiki, R. K., 241²⁹ (155)
Saint-Ruf, G., 103²¹² (96)
Saison-Behmoaras, T., 243¹²⁰ (168, 169), 243¹²¹ (168), 244¹⁶⁹ (181), 244¹⁷⁰ (181), 245²⁴³ (190)
Saito, A., 548¹¹⁰ (505)
Saito, F., 547⁷⁸ (500)
Saito, H., 548¹⁰⁹ (505, 506), 548¹¹⁰ (505)
Saito, I., 382, 412, 422⁷³ (382), 423⁸⁶ (385, 386), 423⁹³ (386), 424¹⁶³ (407), 424¹⁸³ (410), 425²²⁵ (411, 413, 419), 425²²⁶ (411, 413, 419), 489³⁹ (482), 546¹⁹ (493), 548¹⁴¹ (510, 556, 565, 566)
Sakaguchi, K., 489⁶⁴ (484)
Sakai, F., 490⁸⁹ (488), 490⁹⁰ (488)
Sakano, K., 425²³⁵ (417)
Sakata, T., 242⁶² (161)
Sakiyama, S., 547⁷¹ (497)
Sakya, S. M., 548¹¹⁹ (506)
Salazar, M., 80⁹⁸ (76), 80¹¹⁴ (77)
Salisbury, S. A., 147⁷⁴ (114, 127)
Salo, H., 242⁴⁵ (157, 190)
Salunkhe, M., 152⁴³⁰ (139), 248⁴⁶⁰ (237), 248⁴⁶³ (237), 309⁷⁵ (294)
Salvati, M. E., 549¹⁶¹ (513, 514)
Samaha, R. R., 52¹³⁴ (41)
Sambrook, J., 100²⁶ (83), 644
Samiotaki, M., 247³⁶³ (211)
Sampath, U., 247³⁷² (215), 248⁴²¹ (228, 229)
Sams, K. P., 310¹³⁴ (304)

Samson, K., 548[129] (508)

Samstag, W., 283[106] (265)

Samuni, A., 425[234] (417), 425[240] (418)

Sánchez-García, I., 489[73] (486)

Sanchez, J., 338[154] (327)

Sanders, J. Z., 241[17] (158), 248[441] (233), 248[442] (233)

Sands, J., 284[158] (274, 275)

Sandström, A., 148[140] (120), 149[208] (126)

Saneyoshi, M., 247[383] (217), 247[384] (217)

Sanford, D. G., 476[216] (470)

Sangen, O., 245[275] (190, 235), 248[422] (228), 248[424] (228)

Sanger, F., 82, 100[24] (82), 620

Sanghvi, Y. S., 308[9] (285, 287, 300), 308[11] (287, 298), 309[36] (289, 293, 299, 302), 310[105] (299, 300), 310[108] (299, 301, 302), 311[146] (307)

Sano, Y., 549[160] (513), 549[165] (514)

Sansom, P. I., 242[82] (164)

SantaLucia, Jr., J., 339[197] (330)

SantaLucia, J. J., 283[135] (270)

Santi, D. V., 337[75] (322)

Santiago, D., 474[90] (441), 490[96] (488)

Santini, R., 79[52] (71)

Santocroce, C., 369[113] (366)

Sarai, A., 51[66] (33), 53[192] (47)

Sardaro, M., 310[110] (299, 303)

Sarin, P. S., 248[454] (237), 248[455] (237)

Saris, C. P., 338[161] (328)

Sarker, S., 310[122] (300)

Sarma, M. H., 474[65] (434), 475[141] (453, 457, 458), 475[147] (456)

Sarma, R. H., 474[65] (434), 475[141] (453, 457, 458), 475[147] (456)

Sartorelli, A. C., 549[166] (514, 516), 549[178] (516)

Sartorius, J., 475[176] (466)

Sarvazyan, A. P., 50[18] (18, 31), 50[19] (18, 31), 51[91] (34), 51[96] (35)

Sasaki, M., 51[84] (34)

Sasisekharan, V., 242[40] (159)

Sasmor, H., 308[15] (287), 308[25] (288), 310[127] (302)

Sassa-Dwight, S., 102[161] (92)

Sastry, M., 549[187] (517)

Sastry, S. S., 103[230] (96), 103[231] (96)

Sathe, G. M., 146[46] (112)

Sathyanarayana, S., 245[218] (190), 245[222] (190)

Sathynarayana, S., 147[84] (114)

Sato, T., 547[78] (500)

Sato, Y., 150[297] (123)

Satoh, H., 103[247] (98)

Satoh, M., 148[185] (121)

Satyanarayana, S., 101[90] (88)

Saucier, J. M., 473[48] (433)

Sauer, R. T., 488[4] (477, 478, 480, 482, 484), 488[10] (478), 489[75] (486), 489[76] (486, 487)

Sauerwald, R., 146[48] (112)

Sauvaigo, S., 246[343] (205)

Sawa, R., 549[163] (514)

Sawa, T., 549[163] (514)

Sawai, H., 244[162] (178)

Sayeed, V. A., 150[275] (120)

Sayer, J. M., 247[406] (223), 247[407] (223)

Sayre, L. M., 421[11] (372, 417)

Scahill, T. A., 548[108] (505), 548[116] (505)

Scalfi-Happ, C., 127, 148[144] (120), 148[145] (120)

Scanlon, D., 146[37] (111)

Scaria, P. V., 476[191] (467)

Schäfer, T., 423[109] (389), 423[110] (390, 393)

Schairer, W. C., 310[104] (299, 301)

Schaller, H., 149[212] (120, 126)

Schanze, K. S., 245[279] (191, 197)

Schärer, O. D., 339[232] (334)

Scheek, R. M., 80[88] (76)

Scheit, K. H., 282[78] (260)

Schepartz, A., 246[325] (201)

Scherrer, J., 245[277] (190)

Scheublin, R. A., 147[111] (117)

Scheuer-Larsen, C., 248[435] (232)

Schevitz, R. W., 283[116] (266, 268)

Schiavon, O., 146[67] (113)

Schier, A. F., 489[18] (479)

Schifman, A. L., 152[403] (136)

Schill, G., 367[2] (343)

Schimmel, P. R., 51[111] (38), 51[112] (38, 41, 42, 43), 52[146] (43)

Schinazi, R. F., 247[400] (221, 240), 247[401] (221, 222), 249[490] (240, 241), 319

Schirmeister, H., 148[150] (127), 148[151] (127)

Schlepegrell, R., 423[123] (397, 410)

Schlicksupp, A., 421[10] (372, 417)

Schlingensiepen, K. H., 152[401] (135)

Schlingensiepen, R., 152[401] (135)

Schmid, C. W., 102[169] (93)

Schmid, N., 338[140] (326)

Schmidt, B., 248[439] (232)

Schmidt, D., 241[21] (185, 234), 244[188] (185), 247[391] (219)

Schmidt, G., 151[334] (127)

Schmidt, J., 79[68] (72)

Schmidt, S., 284[188] (277, 279)

Schmidt, T., 79[68] (72)

Schmir, M., 473[45] (433)

Schmitt, P., 339[233] (335)

Schmitz, A., 281[3] (252)

Schmitz, G. G., 247[356] (209)

Schmitz, U., 80[92] (76), 80[93] (76)

Schneider, B., 473[24] (431)

Schneider, H.-J., 463, 475[168] (462, 463), 475[176] (466)

Schneider, K., 282[59] (258)

Schneiderwind, R. G. K., 149[228] (122), 149[229] (122)

Schoen, C. D., 12[48] (9)

Scholler, J. K., 245[241] (190)

Scholten, P. M., 102[189] (94)

Scholtissek, S., 283[120] (267)

Schon-Bopp, A., 546[25] (493, 494)

Schönberger, A., 424[207] (411)

Schönwälder, K.-H., 247[405] (223)

Schoppe, A., 282[52] (257)

Schottelius, M. J., 425[254] (421)

Schpok, S. L., 548[106] (504, 505)

Schroeder, S. A., 79[52] (71), 79[53] (71)

Schubert, F., 177, 242[55] (159), 244[158] (177, 190)

Schuchmann, H.-P., 422[50] (378), 422[53] (378), 423[131] (398)

Schuchmann, M. N., 422[53] (378)

Schugerl, K., 473[28] (428)

Schulhof, J. C., 149[223] (122), 149[224] (122)

Schulman, R. G., 52[158] (45)

Schulte-Frohlinde, D., 394, 422[52] (378), 422[65] (381, 389), 423[105] (389), 423[106] (389), 424[154] (402), 424[155] (402), 424[189] (410), 546[25] (493, 494)

Schulte-Herbrüggen, T., 310[132] (304)

Schultz, P. G., 489[46] (482), 489[80] (486), 632, 634

Schultz, R. G., 12[5] (4)

Schultz, W. A., 424[212] (411)

Schultze, P., 283[151] (273)

Schulz, B. S., 148[154] (120), 148[155] (120), 148[160]

Schulz, W. G., 549[164] (514)

Schuster, G. B., 424[185] (410), 424[186] (410)

Schwabe, J. W. R., 489[59] (484), 489[63] (484, 485)

Schwager, C., 243[95] (167), 243[105] (167)

Schwartz, A., 102[178] (93), 103[212] (96)

Schwartz, M. E., 147[72] (113)

Schwarz, M. W., 148[151] (127), 151[370] (133), 151[371] (133)

Schwarz, R., 151[376] (133)

Schweitzer, B. A., 282[32] (255, 263), 282[34] (255), 332

Schweitzer, B. I., 79[56] (71)

Schweitzer, M., 283[106] (265)

Schweizer, B., 12[8] (4)

Schwitter, U., 422[72] (382, 393, 403), 423[114] (391), 423[115] (391, 393, 397, 403)

Schwyzer, R., 245[233] (190)

Schyolkina, A. K., 80[112] (77)

Scott, G. K., 310[88] (296, 297)

Scott, S., 309[35] (289)

Scott, W. G., 284[157] (274, 276)

Scozzari, A. N., 12[12] (4)

Scremin, C. L., 152[449] (144), 245[217] (190)

Searle, C. E., 247[402] (222, 225)

Searle, M. S., 368[38] (351), 368[39] (351), 474[86] (440)

Secrist, III, J. A., 310[130] (303)

Seeberg, E., 424[166] (407)

Seeberger, P. H., 152[434] (139), 309[39] (292, 293)

Seela, F., 254, 317, 324, 330, 331, 332, 336[41] (317), 337[105] (324), 337[106] (324), 338[134] (325, 330, 331, 332), 338[137] (325), 339[181] (330), 339[182] (330, 331)

Seeman, N. C., 12[40] (8), 13[55] (10), 51[64] (33), 51[106] (36), 100[52] (85, 86), 100[53] (85), 101[77] (86), 220, 353, 354

Segawa, H., 242[44] (156)

Segerback, D., 547[62] (497, 504), 547[91] (502), 548[145] (511, 512)
Seibel, G. L., 80[87] (76)
Seibl, R., 247[355] (209), 247[356] (209)
Seidel, A., 247[413] (225)
Seidman, M. M., 13[51] (10)
Seifert, J.-M., 151[346] (130)
Seifert, W. E., 248[438] (232), 310[99] (298), 310[100] (298)
Sekiguchi, A., 284[169] (274)
Sekine, M., 146[49] (112, 123), 147[122] (117), 147[123] (117), 148[172] (120), 148[183] (121), 149[205], 149[210]
Seliger, H., 108, 127, 137, 150[327] (127, 137), 151[334] (127), 151[336] (127), 152[406] (137), 152[407] (137), 236, 242[78] (163), 245[225] (190), 246[293] (193), 248[451] (236, 237)
Seligy, V. L., 100[46] (84)
Seneviratne, P. A., 51[71] (33)
Senior, M. M., 51[59] (33), 51[62] (33)
Senn, M., 423[110] (390, 393)
Sepiol, J., 337[95] (323)
Sera, T., 489[46] (482)
Serafini, A. N., 241[13] (159)
Serafinowska, H. T., 148[146] (127), 148[147] (120)
Sergeev, D. S., 242[64] (161)
Sergeyev, D. S., 244[192] (186), 244[193] (186)
Serra, M. J., 473[52] (433, 469)
Sessler, J. L., 242[79] (163, 164), 242[82] (164)
Seth, D. M., 336[35] (316), 338[152] (327), 339[221] (334)
Sethuram, B., 424[155] (402)
Seto, Y., 244[162] (178)
Severin, E. S., 244[196] (187), 244[197] (187)
Sevilla, M. D., 422[25] (373, 380, 383, 385, 400, 403, 404, 405), 423[139] (398, 410, 413)
Seville, M. D., 423[138] (398)
Sfakianakis, G. N., 241[13] (159)
Sgaramella, V., 369[105] (364)
Shabarova, Z. A., 150[291] (122), 242[39] (157), 246[342] (204)
Shafer, R. H., 476[191] (467), 476[206] (469)
Shah, K., 246[324] (201)
Shahrestanifar, M., 103[244] (98)
Shakked, Z., 79[16] (59)
Shapiro, L., 79[27] (64, 67)
Shapiro, R., 283[90] (263)
Shapiro, T. A., 367[4] (343)
Sharma, A. K., 147[105] (116), 147[106] (116)
Sharma, P., 147[84] (114), 147[105] (116)
Sharp, K. A., 473[21] (428, 432)
Sharp, P. A., 12[70] (6)
Sharpf, C., 146[56] (112)
Sharpless, T. W., 150[278] (122)
Shaw, B. R., 293, 309[42] (292, 293), 329
Shaw, J., 246[314] (198)
Shaw, R., 424[159] (404)
Shay, J. W., 476[203] (469)
Shea, R., 474[77] (437, 440)

Shea, R. G., 244[201] (187)
Sheardy, R. D., 103[244] (98), 220
Sheils, C. J., 282[34] (255), 282[35] (255)
Shen, C.-K. J., 103[220] (96)
Shen, L. L., 475[164] (462)
Shen, N.-Z., 152[419] (139)
Shepherd, R. E., 101[107] (89)
Sheppard, R. C., 146[36] (111)
Sherman-Gold, R., 242[36] (155)
Sherrington, D. C., 146[22] (108)
Sheu, C., 424[205] (411), 424[208] (411)
Shibata, Y., 244[163] (178, 180)
Shibutani, S., 338[160] (328), 338[162] (328), 424[192] (410), 424[197] (410)
Shieh, H. S., 473[22] (430, 435)
Shield, L. S., 473[54] (433)
Shields, T. P., 422[78] (383)
Shigenaga, M. K., 547[46] (494)
Shiiba, T., 242[88] (165)
Shiloach, J., 475[174] (465)
Shima, S., 548[123] (506)
Shima, T., 549[160] (513), 549[165] (514)
Shimada, Y., 283[150] (273)
Shimbara, N., 425[235] (417)
Shimidzu, T., 150[296] (123), 150[298] (123), 242[44] (156), 244[173] (182), 248[422] (228), 248[424]
Shimizu, K., 547[77] (500)
Shimizu, M., 100[19] (82), 102[147] (92, 95)
Shimotakahara, S., 549[179] (516)
Shindo, H., 51[66] (33), 53[192] (47)
Shinomiya, M., 103[247] (98)
Shinozaki, K. Y., 148[197] (123), 149[210] (121)
Shinozuka, K., 244[162] (178), 244[166] (181), 248[467] (238), 283[152] (274)
Shiraki, T., 103[242] (98)
Shire, D., 151[335] (127)
Shirnamé-Moré, L., 424[168] (407)
Shishido, Y., 151[397]
Shishkin, G. V., 242[64] (161), 246[297] (194)
Shoji, Y., 241[11] (155), 244[199] (187)
Shore, D., 369[107] (365)
Shuey, S. W., 151[342] (129), 151[343] (129)
Shugar, D., 337[94] (323), 337[95] (323)
Shuker, D. E. G., 547[62] (497, 504), 547[91] (502), 548[145] (511, 512)
Shulman, R. G., 52[156] (45, 48)
Sibanda, S., 147[137] (120, 127), 148[139] (120)
Sidransky, D., 12[14] (4)
Siebenlist, U., 102[195] (95), 103[201] (95)
Siedenberg, D., 473[28] (428)
Siedlecki, J. M., 246[328] (202)
Siegel, J. S., 548[130] (508)
Siegel, M., 547[76] (500)
Sierzchala, A., 151[387] (133)
Sies, H., 424[212] (411), 546[27] (494), 547[38] (494, 540, 544)
Siftan, S. A., 422[41] (374)
Sigel, A., 100[28] (85, 86), 100[33] (85, 89), 101[84] (88), 101[113] (89), 103[245] (98)
Sigel, H., 100[28] (85, 86), 100[33] (85, 89),

101[84] (88), 101[113] (89), 103[245] (98)
Sigler, P. B., 283[116] (266, 268), 283[118] (266, 268), 283[129] (268)
Sigman, D. C., 244[187] (185, 215)
Sigman, D. S., 86, 100[27] (85, 86), 100[28] (85, 86), 101[66] (86), 101[69] (86)
Sigurdsson, S. T., 248[430] (230)
Siitari, H., 246[327] (201)
Sik, V., 79[6] (56)
Sil'nikov, V. N., 242[64] (161), 242[66] (161)
Simic, M. G., 424[148] (401)
Simms, D. A., 246[347] (206)
Simomura, M., 489[39] (482)
Simon, M. I., 424[210] (411), 488[11] (478)
Simpson, P. J., 421[8] (372)
Sinclair, A., 51[61] (33)
Sinden, R. R., 103[215] (96), 103[216] (96)
Sindona, G., 150[330] (127)
Singer, B., 546[12] (493), 546[22] (493)
Singh, D., 247[360] (210)
Singh, M., 146[23] (108), 146[42] (112)
Singh, M. P., 248[476] (239)
Singh, N., 50[10] (16, 31)
Singh, P., 147[97] (114)
Singh, R. K., 149[258] (121), 150[273] (124)
Singh, U. C., 80[87] (76)
Singleton, S. F., 12[36] (8), 51[101] (36)
Singman, C. N., 147[75] (114)
Sinha, N., 241[14] (155)
Sinha, N. D., 132, 146[51] (112)
Sinha, N. K., 369[97] (362)
Sinsheimer, J. S., 474[113] (446, 448)
Sitlani, A., 422[77] (383)
Siwkowski, A. M., 284[175] (274)
Siyanova, E. Y., 102[184] (93, 94, 95)
Sjölander, S., 52[125] (40)
Skalski, B., 147[119] (117), 147[120] (117)
Skibo, E. B., 549[164] (514)
Sklenar, H., 80[110] (77)
Sklenář, V., 50[4] (16), 79[54] (71)
Sklenar, V., 80[123] (78), 99[4] (82)
Skolik, J. J., 148[200] (121), 149[209] (126)
Skone, P. A., 148[141] (120)
Skorobogaty, A., 101[112] (89)
Skorski, T., 310[91] (296)
Slama-Schwok, A., 474[123] (449)
Sligar, S. G., 50[21] (20)
Slim, G., 280, 283[101] (265, 280)
Sluka, P., 151[376] (133)
Slupphaug, G., 12[44] (8)
Smirnov, V., 152[399] (140)
Smith, A. J., 152[440] (142)
Smith, C. A., 243[153] (175)
Smith, C. C., 309[80] (295)
Smith, C. L., 12[19] (4)
Smith, D. J. H., 151[354] (130)
Smith, D. P., 339[215] (333)
Smith, J. A., 248[474] (239)
Smith, J. C., 474[68] (434, 440), 474[69] (434, 440)
Smith, L. M., 233, 248[441] (233), 248[442] (233)
Smith, L., 12[9] (4, 5, 9)
Smith, M. R., 12[27] (5)

Smith, M., 147[107] (117)
Smith, R. D., 246[325] (201)
Smith, S. S., 337[77] (322)
Smith, S., 283[107] (265), 283[128] (268)
Smith, T. H., 242[43] (156, 173, 188, 190)
Smith, T. M., 246[303] (195)
Smrt, J., 146[54] (112), 148[161] (122)
Smyth, A. P., 248[474] (239)
Smyth, J. F., 12[23] (5)
Snapka, R. M., 475[164] (462)
Sobell, H. M., 474[71] (435)
Sobkowski, M., 152[418] (139), 152[422] (139)
Sobol, Jr., R., 337[118] (325), 337[119] (325)
Söderlund, H., 241[7] (155)
Sogo, J. M., 103[225] (96)
Sohal, R. S., 421[6] (372), 547[48] (494)
Solans, X., 338[141] (326)
Soler, J. F., 339[210] (332)
Soll, D. G., 52[179] (46)
Söll, D., 146[14] (108)
Solomon, J. J., 336[1] (314), 336[9] (315)
Solomon, M. S., 332, 339[213] (332, 333), 548[94] (502)
Soloway, A. H., 247[399] (221)
Song, P.-S., 96, 103[214] (96)
Sonveaux, E., 147[125] (118)
SooChan, P., 246[347] (206)
Sood, A., 338[166] (329), 338[167] (329)
Sørensen, O. W., 79[37] (64), 79[47] (67)
Sorensen, U. S., 284[188] (277, 279)
Sorm, F., 148[161] (122)
Sottofattori, E., 244[174] (182)
Soukchareun, S., 245[273] (190)
Soukup, G. A., 245[259] (190)
Soussou, W., 282[39] (256, 261, 272)
Southern, E. M., 51[63] (33), 658
Sowers, L. C., 316, 336[29] (316), 336[30] (316), 337[77] (322), 424[169] (407)
Soyfer, V. N., 308[10] (285)
Spaltenstein, A., 247[386] (219), 247[387] (219)
Spaney, T., 150[327] (127, 137)
Spassky, A., 101[68] (86), 101[69] (86), 103[202] (95)
Spassova, M., 339[210] (332)
Spears, C. L., 547[56] (495), 547[61] (497)
Spencer, R. D., 52[143] (43)
Spencer, R. J., 103[240] (96)
Spicer, E., 152[440] (142)
Spielmann, H. P., 474[118] (448, 449)
Spielvogel, B. F., 338[167] (329)
Spiess, E., 149[222] (122)
Spinolo, J., 12[9] (4, 5, 9)
Spiro, T. G., 100[62] (86)
Spitzer, R., 79[12] (57)
Spolar, R. S., 50[15] (17, 29), 50[16] (17)
Sponer, J., 102[191] (94)
Spotheim-Maurizot, M., 425[231] (417)
Springer, J., 241[31] (154, 158)
Springer, R. H., 338[178] (330), 338[179] (330)
Sproat, B. S., 117, 149[257] (121, 122), 150[289] (122, 126), 243[95] (167), 243[96] (234), 243[105] (167), 243[106] (167), 245[281] (191), 246[282] (191),

246[345] (205, 206), 247[379] (216), 248[433] (231), 248[445] (234), 248[446] (234), 282[49] (257, 278), 283[97] (264), 283[98] (264)
Spychala, J., 473[17] (428), 475[142] (453, 457, 458), 475[160] (460, 461)
Srinivasachar, K., 152[449] (144)
Srinivasan, A. R., 473[24] (431)
Srivasta, S., 309[35] (289)
Srivastava, S. C., 247[352] (208, 209)
Staacke, D., 283[126] (268)
Stabinsky, Z., 100[5] (82)
Stafford, III, W. F., 489[38] (481)
Stafford, R. S., 425[216] (411)
Stahl, D. A., 241[32] (154, 158)
Stallard, R. L., 241[21] (185, 234)
Stamm, M. R., 246[304] (195), 246[348] (206, 207)
Stan, H.-J., 547[89] (502)
Standal, R., 12[44] (8)
Stark, A.-A., 547[44] (494, 511)
Stark, M. R., 489[32] (480)
Starnes, C. O., 548[102] (503)
Starr, R., 548[129] (508)
Staub, A., 146[48] (112)
Staubli, A. B., 339[225] (334)
Stawinski, J., 146[31] (109, 139, 144), 293
Stec, W. J., 133, 148[150] (127), 149[225] (122), 283[133] (270), 293
Steenken, S., 423[106] (389), 423[122] (397), 424[178] (408, 409, 410), 424[180] (409, 410)
Stegemann, J., 243[95] (167), 243[105] (167)
Steigerwald, M. L., 424[161] (405)
Steighner, R. J., 425[250] (420), 425[251] (420)
Stein, A., 53[200] (48, 49)
Stein, C. A., 48, 49, 243[103] (167), 244[166] (181), 248[467] (238), 293, 309[45] (292, 293)
Stein, P. C., 248[434] (232)
Stein, R., 150[281] (122)
Stein, S., 242[51] (159), 242[53] (159), 245[258] (190)
Steinbrecher, T., 247[413] (225)
Steiner, R., 242[78] (163)
Steitz, J. A., 367[1] (343, 346)
Steitz, T. A., 488[3] (477, 478, 480, 482, 484), 488[11] (478)
Steker, H., 339[188] (330)
Stelzner, A., 146[63] (113)
Stemp, E. D. A., 101[103] (89)
Stempien, M., 246[287] (192)
Stenberg, E., 52[125] (40)
Stengele, K.-P., 148[163]
Stephenson, D., 79[6] (56)
Stephenson, M. L., 289, 309[37] (289)
Sterk, P., 12[16] (4)
Stern, A. M., 101[66] (86)
Stern, M. K., 244[186] (185)
Stern, S., 281[15] (253)
Stevens, S. Y., 369[103] (363)
Stewart, D., 547[59] (495, 497)
Stewart, G. M., 474[100] (444, 449)
Stezowski, J. J., 247[405] (223), 247[412] (224)

Stöcklein-Schneiderwind, R., 151[375] (133), 151[376] (133)
Stockwell, D. L., 194, 246[295] (194)
Stoddardt, J. F., 368[85] (361)
Stols, L. M., 246[315] (198)
Stone, M. J., 337[67] (321, 334), 546[2] (492)
Stone, M. P., 248[415] (225)
Stone, T. P., 549[154] (513)
Storhoff, J. J., 13[53] (10), 13[54] (10)
Stormo, G. D., 284[171] (274)
Stote, R., 424[166] (407)
Strand, E. A., 248[414] (225)
Straub, A., 247[405] (223)
Straub, K., 100[43] (85, 96)
Straume, M., 50[9] (16, 21, 29)
Strauss, J. K., 368[59] (354)
Strauss, U. P., 473[42] (432)
Strekowska, A., 475[134] (452, 453), 475[135] (452, 453)
Strekowski, A., 475[139] (452, 453)
Strekowski, L., 452, 473[56] (433, 435, 452, 460), 475[132] (452, 453), 475[133] (452, 453), 476[178] (467, 468), 476[179] (467)
Strickland, J. A., 474[94] (441, 443), 474[95] (441, 443)
Striepeke, S., 241[14] (155)
Stringer, B. K., 100[47] (84)
Strobel, O. K., 247[393] (219, 220), 247[394] (220)
Strobel, S. A., 246[306] (196), 283[139] (271), 284[189] (278)
Strömberg, R., 146[31] (109, 139, 144)
Strout, M. P., 12[21] (4)
Struhl, K., 12[35] (8)
Stubbe, J., 244[191] (186)
Stuiver, M. H., 489[33] (480)
Stukenberg, P. T., 368[33] (350)
Stumpe, A., 146[50] (112), 146[51] (112)
Sturtevant, J. M., 17, 50[14] (17), 50[17] (17), 51[65] (33)
Suami, T., 473[54] (433)
Subasinghe, C., 243[103] (167), 244[166] (181), 244[177] (183), 248[467] (238), 309[45] (292, 293)
Suck, D., 281[4] (252), 281[5] (252)
Sufrin, J. R., 337[74] (322)
Sugawara, K., 490[93] (488)
Sugawara, R., 549[160] (513), 549[165] (514)
Suggs, J. W., 101[72] (86), 103[232] (96)
Sugimoto, N., 51[69] (33), 51[77] (34)
Sugiura, Y., 103[242] (98), 575, 580
Sugiyama, H., 412, 422[73] (382), 423[86] (385, 386), 423[93] (386), 424[183] (410), 424[188] (410), 425[225] (411, 413, 419), 425[227] (411, 413, 419), 546[19] (493), 548[110] (505), 548[141] (510), 561
Suh, D., 476[213] (469)
Suh, W.-C., 101[127] (90)
Suhadolnik, R. J., 244[205] (189)
Sullenger, B. A., 369[100] (363)
Sullivan, J. K., 102[144] (92)
Sullivan, K. M., 102[141] (92), 103[206] (95)
Sullivan, R. W., 423[117] (395)

Sulston, I., 149[219] (121)
Summers, M. F., 80[77] (73)
Summerton, J., 301, 310[102] (299, 301)
Sumner, A. T., 473[25] (428)
Sumner-Smith, M., 245[214] (190)
Sun, D., 103[250] (98)
Sun, D. K., 248[455] (237)
Sun, J. S., 12[37] (8), 244[169] (181), 244[170] (181)
Sun, S., 246[338] (203)
Sund, C., 246[326] (201)
Sundaralingam, M., 473[18] (428)
Sundquist, W. I., 243[133] (170)
Sung, W. L., 147[115] (117), 147[116] (117)
Surovaya, A. N., 486, 489[78] (486)
Suzdaltseva, Y. G., 244[202] (187, 190), 245[261] (190), 309[76] (294)
Suzuki, M., 490[95] (488)
Suzuki, T., 150[282] (122)
Svendsen, M. L., 248[435] (232)
Svinarchuk, F. P., 242[58] (160)
Swain, A., 367[5] (344, 367)
Swaminathan, S., 282[56] (257, 273)
Swann, P. F., 147[112] (117)
Swanson, B. J., 425[219] (411)
Swanson, P. C., 369[103] (363), 369[104] (363)
Swayze, E. E., 310[109] (299, 302), 310[126] (302), 310[128] (303)
Swenson, D. H., 548[106] (504, 505), 548[107] (504, 505)
Swerdlow, P. S., 546[17] (493)
Swern, D., 424[159] (404)
Switzer, C. Y., 309[74] (294), 318, 337[100] (324)
Symons, M. C. R., 423[134] (398), 423[136] (398, 401, 402), 424[150] (401, 402)
Symons, R. H., 284[160] (274), 284[170] (274)
Syvänen, A.-C., 241[6] (154, 155), 241[7] (155)
Szabo, T., 152[421] (139), 152[422] (139)
Szabolcs, A., 336[45] (317)
Szemzo, A., 336[45] (317), 337[73] (322)
Szewczak, A. A., 53[210] (50)
Szostak, J. W., 13[58] (10), 52[171] (46), 52[174] (46), 624, 625, 626, 633, 634
Szybalski, W., 515, 518, 549[168] (515), 549[169] (515, 518)

Taagaard, M., 147[128] (118), 149[207] (123)
Tabatadse, D. R., 246[297] (194)
Tabdjoun, A., 338[158] (328)
Tabone, J. C., 241[25] (158), 246[303] (195), 246[304] (195)
Taboury, J. A., 337[87] (323)
Taetle, R., 548[127] (508), 548[128] (508)
Tahara, K., 474[121] (450, 451)
Tahara, S., 149[231] (121, 135), 149[264] (123)
Tahmassebi, D. C., 282[34] (255)
Taillandier, E., 337[87] (323), 474[123] (449)
Tainer, J. A., 12[45] (8)

Taira, K., 282[69] (259)
Takabatake, T., 476[185] (466)
Takagi, M., 151[340] (127), 242[92] (166)
Takahashi, K., 547[71] (497), 547[77] (500)
Takahashi, Y., 242[88] (165)
Takai, K., 244[199] (187)
Takaki, M., 151[397] (140)
Takaku, H., 148[147], 148[173], 328
Takano, T., 79[15] (59), 80[118] (77)
Takasugi, M., 102[183] (93)
Takayama, M., 424[163] (407), 424[188] (410)
Takeda, N., 242[88] (165)
Takeda, Y., 488[14] (478)
Takenaka, S., 151[340] (127), 242[92] (166), 451
Takeshi, K., 547[78] (500)
Takeshita, H., 423[119] (397)
Takeshita, M., 424[197] (410)
Takesue, Y., 148[177]
Takeuchi, T., 248[458] (237)
Taktakishvili, M. O., 338[158] (328)
Talanian, R. V., 489[48] (482), 489[49] (482)
Tallman, K. A., 422[58] (378, 379)
Talwar, G. P., 243[111] (168)
Tamatsukuri, S., 151[366] (132), 242[63] (161), 306, 310[141] (306)
Tamura, A., 50[48] (30)
Tan, N.-W., 102[154] (92)
Tan, R., 473[62] (434)
Tan, W., 151[385] (133), 152[401] (135)
Tanaka, H., 548[126] (508)
Tanaka, K., 242[44] (156), 244[173] (182), 248[465] (237), 336[8] (314, 315)
Tanaka, S., 80[118] (77)
Tanaka, T., 132, 151[358] (130), 151[359] (130), 242[62] (161), 242[63] (161)
Tang, J. X., 237, 248[475] (239), 295, 310[86] (295)
Tang, J.-Y., 150[307] (125), 151[390] (133), 151[395], 152[420] (139), 152[429] (139), 248[461] (237), 248[474] (239), 248[475] (239), 310[86] (295)
Tang, M.-s., 12[13] (4), 549[159] (513), 549[186] (517)
Tang, X.-Q., 246[338] (203)
Tanimura, H., 148[172] (120), 148[204] (126)
Tanious, F. A., 474[67] (434, 435, 440, 441), 474[75] (437, 441), 475[133] (452, 453), 475[134] (452, 453), 476[178] (467, 468), 476[179] (467)
Tanious, F., 247[401] (221, 222)
Tannious, F. A., 475[140] (453, 457, 458), 475[141] (453, 457, 458)
Tanooka, H., 424[164] (407)
Tapley, Jr., K. J., 96, 103[214] (96)
Tarkoy, M., 12[8] (4)
Tarköy, M., 282[55] (257, 273)
Tarrasón, G., 242[59] (160)
Tashiro, T., 546[19] (493)
Tashlitsky, V. N., 242[39] (157)
Taylor, J. W., 483, 489[57] (483)
Taylor, R. A., 310[88] (296, 297)
Tazi, J., 12[26] (5)

Tchou, J., 338[160] (328), 424[192] (410), 424[195] (410)
Teare, J., 243[113] (168)
Teasdale, R., 309[47] (292, 294)
Teebor, G. W., 424[168] (407)
Teigelkamp, S., 242[42] (156)
Teitelbaum, H., 80[83] (75)
Telser, J., 245[278] (191, 197), 245[279] (191, 197), 246[312] (197)
Templeton, N. S., 488[13] (478)
Temsamani, J., 309[64] (293), 310[103] (299, 301)
Tener, G. M., 151[352] (130)
Teng, K., 305, 310[115] (299, 305)
Teng, S. P., 549[184] (517)
Tenhunen, J., 241[7] (155)
Téoule, R., 149[223] (122), 149[224] (122), 243[147] (173), 243[148] (173), 244[156] (176, 177), 244[157] (177, 205), 246[339] (204), 246[343] (205), 247[378] (216), 337[112] (324), 337[113] (324), 338[120] (325), 338[123] (325)
Terada, K., 150[297] (123)
Terawaki, A., 549[162] (513)
Terhorst, T. J., 283[148] (272)
Tesser, G. I., 148[165] (120), 148[166] (120)
Teulade-Flchou, M.-P., 474[123] (449)
Thaden, J., 245[254] (190)
Theaker, P. D., 283[87] (263)
Thederahn, T., 422[38] (374)
Theisen, P., 149[222] (122), 178
Thelin, M., 152[422] (139), 152[426] (139, 144), 293, 309[71] (293)
Theriault, N. Y., 151[346] (130)
Theriault, N., 152[404] (136)
Thiebe, R., 52[148] (43, 45, 48)
Thier, R., 548[152] (512, 513)
Thomas, E. J., 305, 310[117] (299, 305)
Thomas, G. A., 80[111] (77)
Thomas, H., 282[24] (254)
Thomas, M. J., 247[380] (216)
Thompson, J. A., 152[451] (144)
Thompson, M. S., 422[42] (374, 377)
Thompson, S. C., 548[119] (506)
Thomson, J. B., 245[216] (190)
Thorp, H. H., 422[28] (373, 374, 376, 380), 422[29] (373, 374, 380), 423[83] (385, 395)
Thrane, H., 248[436] (232), 248[437] (232)
Thuong, N. T., 152[445] (142), 168, 236, 242[47] (158), 242[70] (161, 190), 243[115] (168), 243[116] (168), 244[165] (181), 244[167] (181), 245[226] (190, 236), 245[230] (190), 248[452] (237), 283[137] (271)
Thurston, D. E., 547[49] (495, 497), 547[54] (495)
Ti, G. S., 149[213] (120)
Tibanyenda, N., 52[152] (44)
Tidwell, R. R., 475[142] (453, 457, 458)
Tilly, S. L., 423[117] (395)
Time, R. W., 424[166] (407)
Timmers, C. M., 310[136] (305)
Timmons, S. E., 103[229] (96)
Tinoco, I. J., 52[150] (43), 52[180] (46)
Tinoco, Jr., I., 338[136] (325), 476[207] (469), 476[208] (469)

Tinoco, Jr., J., 473[49] (433, 469), 473[57] (433)

Tinsley, J. H., 245[256] (190)

Tippie, T. N., 547[81] (500, 541)

Tirumala, S., 337[96] (323)

Titmas, R. C., 146[23] (108), 146[42] (112)

Tivel, K. L., 242[56] (159, 160)

Tjian, R., 489[44] (482)

Tocík, Z., 146[54] (112), 150[283] (122), 150[284] (122)

Tocqué, B., 245[243] (190)

Toda, S., 490[93] (488)

Todd, A. R., 106, 109, 139, 146[3] (106, 130), 146[4] (106)

Tojanasakul, Y., 309[31] (289)

Tokuyama, H., 243[107] (167, 168)

Tollin, P., 283[147] (272)

Tolman, G. L., 52[143] (43)

Toma, P. H., 339[219] (333)

Tomasz, J., 338[166] (329), 338[167] (329)

Tomasz, M., 242[50] (159)

Tomita, F., 547[77] (500)

Tomita, Y., 148[173]

Tomizawa, S., 549[160] (513)

Tondelli, L., 369[112] (366)

Tong, G., 245[268] (190), 245[274] (190)

Toole, J. J., 13[59] (11), 13[60] (11)

Tor, Y., 246[321] (199)

Torchia, D. A., 80[72] (72, 73)

Torigai, H., 152[402] (136)

Torigoe, H., 51[66] (33), 53[192] (47)

Torkelson, S., 241[24] (158)

Tornaletti, S., 103[213] (96)

Tosquellas, G., 282[53] (257)

Toulmé, J.-J., 422[64] (380)

Townsend, C. A., 423[90] (385, 394, 421), 571

Townsend, L. B., 282[57] (257), 282[81] (260, 262, 263)

Trainor, C., 489[61] (484)

Trainor, G. L., 247[362] (211), 304, 308[18] (288, 299, 305), 309[57] (292, 294, 306), 310[135] (304), 310[140] (306)

Tran-Cong, Q., 248[462] (237)

Tran-Dinh, S., 337[87] (323)

Tran-Thi, Q.-H., 339[196] (330)

Trapane, T. L., 338[126] (325), 339[211] (332)

Travers, A. A., 489[62] (484)

Travnicek, M., 102[191] (94)

Tregear, G. W., 147[91] (114), 147[92] (114), 147[93] (114)

Tribolet, R., 101[93] (88)

Trichtinger, T., 148[153] (120), 148[155] (120)

Trinh, L., 475[174] (465)

Tritz, R., 284[173] (274), 284[174] (274, 277)

Troll, W., 424[168] (407)

Tromp, M., 149[254] (123), 149[255] (135), 151[361] (130)

Tronche, C., 422[35] (373, 402, 403, 404), 422[58] (378, 379), 423[111] (391)

Trumbore, C. N., 425[242] (418)

Tsang, J., 421[17] (372)

Tsao, R., 547[76] (500)

Tsay, S.-C., 150[314] (126)

Tschudin, R., 79[19] (59), 80[72] (72, 73)

Tseng, B. Y., 309[30] (289, 295), 309[81] (295)

Ts'o, P. O. P., 50[54] (32), 282[36] (256, 262, 271), 282[37] (256, 262, 271), 308[13] (287, 292), 309[30] (289, 295), 309[80] (295), 339[206] (332), 339[207] (332)

Tsuchida, T., 549[163] (514)

Tsuchiya, M., 148[197] (123), 148[198], 149[210]

Tsukahara, S., 283[154] (274)

Tsuno, T., 490[90] (488)

Tsutsumi, Y., 425[225] (411, 413, 419), 425[228] (411, 413)

Tuerk, C., 13[56] (10)

Tuite, E., 101[84] (88)

Tuli, D. K., 338[175] (330)

Tullius, T. D., 85, 89, 90, 100[27] (85, 86), 100[34] (85, 90), 101[115] (89), 101[116] (89, 90)

Tullius, T., 101[117] (89)

Tung, C.-H., 242[53] (159), 245[258] (190)

Turner, A. F., 149[263] (123)

Turner, D. H., 38, 42, 49, 50[12] (17, 34), 50[42] (28), 51[61] (33), 51[69] (33), 52[140] (42), 52[149] (43), 53[204] (49), 53[205] (49)

Turner, G., 147[74] (114, 127)

Turner, R., 489[44] (482)

Turro, N. J., 101[87] (88)

Tuschl, T., 245[216] (190), 248[429] (230)

Twigden, S. J., 475[125] (451)

Tworowska, I., 293, 309[43] (292, 293)

Tyagarajan, K., 282[47] (257, 278)

Tyagi, S., 12[47] (9, 10), 13[49] (9)

Ubasawa, A., 147[109] (117), 147[110] (117), 148[139] (117), 148[139]

Uchida, K., 101[99] (89)

Uchimura, Y., 476[185] (466)

Uchiyama, M., 148[191] (121)

Uddin, A. H., 368[64] (355)

Ueda, K., 103[246] (98)

Ueda, M., 548[131] (508)

Ueda, S., 148[173] (120), 148[179] (121)

Ueda, T., 324, 336[32] (316), 337[69] (321, 324), 337[114] (324), 338[176] (330)

Ueno, M., 482, 489[51] (482)

Uesugi, M., 548[126] (508)

Uesugi, S., 338[163] (328), 338[174] (329)

Ugarkar, B. G., 150[275] (120)

Ugarte, E., 368[88] (362)

Ughetto, G., 368[41] (352), 490[86] (488), 490[87] (488)

Ugi, I., 148[138] (120), 148[196] (121), 149[228] (122), 149[229] (122)

Uhlenbeck, O. C., 284[161] (274, 275), 284[171] (274)

Uhlmann, E., 146[26] (108), 306

Ulyanov, N., 80[92] (76), 80[95] (76)

Umezawa, H., 548[131] (508)

Unger, K. K., 146[56] (112)

Ungers, G. E., 369[100] (363)

Upadhya, K., 244[159] (178, 190)

Upson, D. A., 310[98] (298), 310[104] (299, 301)

Ura, A., 546[19] (493)

Urakami, K., 148[193] (121), 148[197] (123), 149[210]

Urata, H., 245[245] (190)

Urbaniczky, C., 52[125] (40)

Urdea, M. S., 119, 147[130] (119), 192, 204

Usdin, K., 103[204] (95)

Usman, N., 149[259] (121, 124), 152[455] (144), 245[277] (190), 282[69] (259), 284[188] (277, 279), 284[191] (278), 310[97] (298)

Ussery, D. W., 103[218] (96), 103[219] (96)

Utahara, R., 548[131] (508)

Uznanski, B., 149[235] (122)

Vaghefi, M. M., 242[41] (161)

Valet, G., 242[48] (158)

Valeur, B., 242[47] (158)

Vandendriessche, F., 310[106] (299, 300)

Vanderhaeghe, A., 146[53] (112)

Vanderhaeghe, H., 148[157] (120), 148[158] (120)

Vanderwall, D. E., 422[81] (385)

Varani, G., 51[58] (33), 56, 79[4] (56, 59)

Varaprasad, C. V., 338[165] (329)

Varenne, J., 132, 150[301] (123, 136), 151[335] (127), 151[364] (132)

Varma, R. S., 308[20] (288), 308[22] (288)

Vasquez, K. M., 245[256] (190)

Vasseur, J.-J., 295, 309[36] (289, 293, 299, 302), 309[48] (292, 295)

Vasseur, M., 368[88] (362), 369[110] (366)

Vaughn, M. R., 50[8] (16)

Veal, J. M., 101[67] (86)

Veeneman, G. H., 151[361] (130), 243[131] (170)

Veiro, D., 549[179] (516)

Venjaminova, A. G., 242[65] (161)

Venkataraman, G., 242[40] (159)

Venkatesan, H., 147[100] (115)

Venkatramanan, M. K., 475[140] (453, 457, 458), 475[141] (453, 457, 458)

Vera, M., 490[96] (488)

Verderame, M., 473[35] (432)

Verdine, G. L., 246[289] (192, 193), 246[290] (192, 193), 256, 281[1] (252, 253, 266), 281[17] (253, 256), 282[41] (256), 282[43] (256), 320, 325, 329

Verheggen, I., 12[7] (4)

Vermaas, E. H., 13[59] (11)

Vermeersch, H., 147[125] (118)

Vermeulen, N. M. J., 241[25] (158)

Veronese, F., 146[67] (113)

Vershon, A. K., 489[27] (480)

Veselkov, A. G., 102[197] (95)

Vesnaver, G., 33, 51[67] (33)

Vicendo, P., 547[37] (494, 544)

Vickers, T., 337[59] (319)

Vickers, T. A., 242[57] (159)

Vieira, A. J. S. C., 423[122] (397), 424[180] (409, 410)

Vigneron, J.-P., 474[123] (449)

Vilaró, S., 242[59] (160)

Villafranca, J. J., 474[92] (441)
Vinayak, R. S., 241[17] (158), 243[152] (175)
Vincze, A., 281[10] (253)
Vinograd, J., 473[45] (433)
Vinogradov, S. V., 244[196] (187), 244[197] (187), 245[261] (190), 309[76] (294)
Virosco, J. S., 245[241] (190)
Visse, R., 102[164] (92)
Viswanadham, G., 310[123] (300)
Vizsolyi, J. P., 149[263] (123)
Vlach, J., 102[191] (94)
Vlassov, V. V., 100[59] (86)
Vlieghe, D., 283[136] (271)
Vo, C. D., 336[27] (316)
Voehler, M. W., 548[148] (512)
Vogelstein, B., 12[14] (4)
Vogt, N., 102[170] (93, 94)
Vojtiskova, M., 102[146] (92), 102[151] (92)
Voldne, G., 424[166] (407)
Volkman, B. F., 490[85] (487)
Volkov, E. M., 242[39] (157)
Voloshin, O. N., 102[151] (92), 102[167] (92, 95), 102[193] (94, 95)
Vorbruggen, H., 282[82] (260), 282[83] (260)
Vorlickova, M., 337[73] (322)
Vosberg, H.-P., 249[488] (240)
Voss, E. W., 369[103] (363)
Voss, H., 243[95] (167)
Vu, H., 146[58] (112, 113, 122), 147[97] (114), 149[222] (122)
Vuister, G. W., 79[19] (59)
Vuocolo, E., 246[309] (197)
Vyle, J. S., 282[67] (259), 284[205] (281)

Wachter, L., 161, 162, 165, 242[67] (161, 162, 165)
Wada, T., 151[377] (133), 309[58] (292)
Waddell, T. G., 548[102] (503)
Wade, T., 150[319] (126)
Wadhwani, S., 152[438] (139)
Wadwani, S., 282[56] (257, 273), 283[145] (271), 283[148] (272)
Waggoner, A. S., 247[353] (209)
Wagner, E., 245[260] (190)
Wagner, G., 80[78] (73)
Wagner, J. R., 422[43] (375), 422[74] (383, 398, 407), 423[128] (398, 407), 423[129] (398, 407), 424[162] (406), 424[171] (407)
Wagner, R. W., 336[47] (317), 336[48] (317)
Wagonner, A. S., 247[354] (209)
Wahl, F. O., 151[339] (127)
Wainwright, M., 473[30] (428, 429, 432)
Wakabayashi, S., 149[232] (121, 135), 149[234] (121, 135)
Wakamatsu, K., 548[125] (507, 508)
Wakelin, L. P. G., 368[38] (351), 368[39] (351), 474[76] (437, 440), 474[86] (440)
Waldner, A., 308[19] (288, 299, 300, 301, 305), 310[105] (299, 300), 310[107] (299, 301)

Walker, P. A., 282[76] (260)
Walker, R. T., 282[77] (260)
Walker, V., 549[179] (516)
Walker, W. T., 150[276] (120)
Wallace, J. C., 425[247] (419), 425[248] (419)
Wallace, S. S., 424[173] (407)
Wallace, T. L., 548[107] (504, 505)
Wallis, M. P., 147[99] (114)
Walsh, P. S., 241[29] (155)
Walter, B., 283[126] (268)
Walter, M., 367[5] (344, 367)
Walter, T., 247[356] (209)
Wang, A., 473[5] (428, 435, 436)
Wang, A. C., 79[19] (59)
Wang, A. H., 368[41] (352)
Wang, A. H.-J., 336[27] (316), 337[88] (323), 337[111] (324), 474[79] (437), 474[82] (437, 440), 490[86] (488), 490[87] (488)
Wang, B. C., 283[114] (266, 267, 270)
Wang, D., 310[112] (299, 304)
Wang, E., 50[4] (16), 99[4] (82)
Wang, G., 13[51] (10), 13[62] (11)
Wang, H., 368[46] (353)
Wang, J. C., 103[202] (95)
Wang, J.-J., 547[67] (497)
Wang, S., 298, 357, 368[68] (356, 357, 359), 368[70] (357)
Wang, W., 548[127] (508), 548[128] (508)
Wang, Y.-H., 474[68] (434, 440), 475[132] (452, 453)
Wang, Y., 80[111] (77), 100[61] (86)
Wang, Y. L., 368[43] (353)
Wang, Z., 102[165] (92)
Ward, B., 101[112] (89)
Ward, J. F., 418, 425[224] (411, 419), 425[243] (418)
Waring, M., 103[248] (98)
Waring, M. J., 432, 473[7] (428, 429, 432, 433, 436, 444, 452, 453), 473[7] (428, 429, 432, 433, 436, 444, 452, 453), 474[67] (434, 435, 440, 441), 474[70] (434, 440), 475[146] (456), 476[198] (468)
Waring, R. B., 284[206] (281)
Warner, B. D., 147[130] (119)
Warpehoski, M. A., 548[115] (505), 548[116] (505)
Warren, G. R., 548[101] (503)
Wasner, M., 244[205] (189), 244[206] (189)
Wasserman, H. H., 548[100] (503)
Wasserman, S. A., 368[29] (348)
Wasserman, Z. R., 489[38] (481)
Watanabe, K. A., 150[274] (120), 282[81] (260, 262, 263), 283[88] (263), 283[89] (263)
Watanabe, S., 148[180]
Watanabe, S. M., 246[348] (206, 207)
Watanabe, T., 103[247] (98)
Waters, T. R., 282[79] (260)
Watkins, B. E., 148[170] (120), 148[171] (120)
Watson, B. D., 421[9] (372)
Watson, J. D., 81, 87, 95, 106, 367[1] (343, 346), 477, 488[1] (477), 644
Watson, R. A., 475[133] (452, 453), 475[134] (452, 453), 475[135] (452, 453)

Watson, W. P., 424[200] (410)
Way, W., 423[138] (398)
Waychunas, C., 310[98] (298)
Waychunes, C., 310[110] (299, 303)
Wayner, D. D. M., 422[49] (377)
Webb, R. F., 146[8] (106, 109, 139)
Webb, T. L. L., 423[117] (395)
Webb, T. R., 149[253] (123)
Weber, G., 244[197] (187)
Webster, G., 475[134] (452, 453)
Weeks, K. M., 53[191] (47)
Wegher, B. J., 423[121] (397, 410)
Wei, P., 547[34] (494)
Weidner, H., 52[129] (41)
Weidner, I., 548[102] (503)
Weidner, M. F., 100[34] (85, 90), 101[123] (90, 96)
Weiler, J., 150[328] (127)
Weimann, G., 149[212] (120, 126)
Weindruch, R., 421[6] (372), 547[48] (494)
Weiner, A. M., 367[1] (343, 346)
Weiner, P., 80[87] (76)
Weinfeld, M., 425[237] (418)
Weinrich, S. L., 476[203] (469)
Weintraub, H., 102[179] (93)
Weis, A. L., 310[131] (303)
Weiss, M. A., 283[134] (270)
Weissberger, A., 51[110] (38)
Weith, H. L., 241[4] (154), 244[175] (182), 244[176] (183)
Welch, C. J., 147[133] (120), 147[134] (120), 148[184] (121), 149[209] (126), 149[211] (121)
Welch, T. W., 422[29] (373, 374, 380), 422[41] (374)
Weller, D., 338[178] (330)
Weller, D. D., 310[102] (299, 301)
Wells, R. D., 82, 100[6] (82), 100[18] (82), 102[136] (92), 102[145] (92, 94), 103[209] (96), 103[212] (96)
Weltman, J. K., 249[485] (240)
Wemer, M. H., 475[172] (465), 475[174] (465)
Wemmer, D. E., 12[43] (8), 79[42] (65), 80[82] (75)
Wempen, I., 283[89] (263)
Wendeborn, S., 310[118] (299, 306)
Wengel, J., 147[70] (113), 232, 300
Wensel, T. G., 53[197] (48)
Wenzel, T., 282[21] (254), 282[23] (254)
Werner, R. K., 241[13] (159)
West, M. D., 476[203] (469)
Westbrook, J., 473[24] (431)
Westerink, H., 52[157] (45)
Westhof, E., 50[34] (21), 51[88] (34)
Westler, W. M., 79[17] (59)
Westra, J. G., 338[161] (328)
Wetmur, J. G., 51[75] (34)
Wheeler, K. T., 422[44] (376)
Wheeler, P., 310[126] (302)
Whelan, A., 425[230] (417)
White, H. L., 548[134] (508)
White, J. D., 338[178] (330)
White, J. R., 548[134] (508)
White, M. M., 548[133] (508, 509)
White, M. W., 103[237] (96)
White, S. A., 100[48] (84, 87), 469

Whitlow, M. D., 549[156] (513)
Whittler, G., 283[153] (274)
Wickham, G., 548[138] (509)
Wickstrom, E., 241[12], 294, 309[57] (292, 294, 306), 309[78] (294)
Wicnienski, N. A., 548[116] (505)
Widlanski, T. S., 304, 310[114] (299, 304), 310[133] (304)
Widmer, R. W., 103[225] (96)
Widom, J., 103[241] (96)
Wiemer, D. F., 306, 310[119] (299, 306)
Wiesehahn, G. P., 103[222] (96)
Wiesler, W. T., 151[383] (133), 309[41] (292, 293)
Wiesler, W., 50[3] (16)
Wiewiorowski, M., 150[303] (123, 136), 151[349] (130)
Wightman, R. H., 150[302] (123, 136), 151[355] (130)
Wilcken-Bergmann, B., 283[126] (268)
Wilk, A., 149[235] (122), 149[260] (121, 124), 151[386] (133), 283[133] (270), 309[56] (292, 306)
Will, D. W., 175, 242[42] (156), 243[151] (175), 243[153] (175), 244[195] (187), 244[198] (187, 189, 190), 311[143] (306)
Will, S. G., 241[15] (158)
Williams, B. D., 247[381] (216, 217)
Williams, D. H., 337[67] (321, 334), 546[2] (492)
Williams, D. M., 283[95] (264), 336[5] (314, 315)
Williams, D. R., 547[76] (500)
Williams, J. C., 51[63] (33)
Williams, L. D., 100[38] (85, 92, 94), 101[78] (86, 98), 101[91] (88), 443, 444, 446
Williams, P. S., 423[107] (389, 391)
Williams, R. M., 547[80] (500, 541), 547[81] (500, 541)
Williamson, D., 79[55] (71)
Williamson, J. R., 49, 473[62] (434), 476[201] (469)
Willie, G., 147[111] (117)
Willis, M. C., 425[217] (411), 425[218] (411)
Williston, S., 50[50] (31)
Wilm, M., 147[98] (114)
Wilmanska, D., 475[126] (451)
Wilson, E. O., 473[31] (428, 432)
Wilson, J. H., 53[197] (48)
Wilson, J. S., 282[80] (260)
Wilson, W. D., 48, 428, 467, 472[1] (428, 432, 436, 444, 452, 456), 473[11] (428, 432, 436, 444), 473[17] (428), 474[67] (434, 435, 440, 441), 474[68] (434, 440), 475[132] (452, 453), 475[133] (452, 453), 476[178] (467, 468), 476[179] (467), 490[96] (488)
Wincott, F., 245[277] (190), 282[69] (259)
Wing, R., 79[15] (59)
Wing, R. M., 99[3] (82)
Winkeler, H.-D., 339[181] (330)
Winkeler, K. A., 244[186] (185)
Winter, E., 490[84] (487)
Wintermeyer, W., 52[153] (45)
Winters, R. S., 490[84] (487)
Wintzerith, M., 146[48] (112)

Wirth, M., 474[122] (450, 451)
Wise, D. S., 282[57] (257)
Wiseman, T., 50[50] (31)
Withka, J., 422[61] (380)
Wittig, B., 368[53] (354)
Wittmann, R., 293, 309[69] (293)
Wittmer, T., 423[108] (389)
Witzel, H., 422[71] (382)
Wogan, G. N., 548[147] (512)
Wohlrab, F., 99, 100[18] (82), 100[19] (82), 102[174] (93), 102[176] (93), 103[251] (99)
Woisard, A., 242[84] (164), 246[331] (202), 246[332] (202)
Wolberger, C., 489[27] (480), 489[32] (480)
Wold, B. J., 52[186] (46)
Wold, B., 476[186] (466)
Wolf, D. E., 241[19] (155)
Wolf, P., 423[132] (398)
Wolf, R. M., 308[19] (288, 299, 300, 301, 305), 310[105] (299, 300), 310[107] (299, 301)
Wolffe, A. P., 82, 100[15] (82), 100[16] (82), 101[120] (90), 101[121] (90)
Wollenzien, P., 243[113] (168)
Wolter, A., 146[50] (112), 146[51] (112)
Wolters, M., 368[53] (354)
Wong, C.-H., 546[20] (493)
Wong, C.-W., 102[154] (92)
Wong, J. L., 546[26] (494)
Woo, J., 244[180] (184)
Woo, R. J., 474[96] (441, 443, 444)
Woo, S., 241[31] (154, 158)
Wood, M. L., 424[198] (410)
Wood, P., 309[35] (289)
Wood, S., 339[194] (330)
Woodbury, C. P., 50[33] (21)
Woodson, S. A., 101[110] (89)
Woodsworth, M. L., 283[143] (271, 272)
Woodworth, M. L., 337[68] (321), 424[198] (410)
Wooters, J. L., 243[132] (170)
Workman, C., 245[277] (190)
Worner, K., 282[27] (254)
Worth, Jr., L., 422[81] (385), 422[82] (385), 423[85] (385), 423[116] (394, 395, 421), 425[256] (421)
Woynarowski, J. M., 548[122] (506)
Wreesmann, C. T. J., 149[266]
Wright, B., 474[97] (441, 443)
Wright, M., 242[82] (164)
Wright, W. E., 476[203] (469)
Wu, B. Y., 489[57] (483)
Wu, C. C. L., 425[244] (418)
Wu, J. C., 337[75] (322)
Wu, J. Q., 50[53] (31)
Wu, J. W., 50[3] (16)
Wu, R., 146[25] (108), 149[220] (126)
Wu, S. H., 423[101] (387, 388, 393, 394), 425[256] (421)
Wu, T. F., 284[190] (278)
Wu, T.-P., 245[276] (190)
Wu, W., 425[253] (420), 539
Wu, Z., 548[149] (512)
Wunner, W. H., 152[451] (144)
Wunz, T. P., 547[74] (499), 547[79] (500)

Wüthrich, K., 79[20] (59, 65), 79[36] (64, 65), 80[78] (73), 489[18] (479), 489[21] (479), 539
Wydra, R. L., 475[138] (452, 453), 475[150] (457, 458), 475[151] (457)
Wydro, R. M., 13[50] (10)
Wykes, E. J., 146[55] (112)
Wyman, J., 50[52] (31)
Wyrzykiewicz, T. K., 147[78] (114)

Xi, Z., 423[84] (385, 394, 421)
Xiang, G., 282[39] (256, 261, 272), 282[40] (256, 261, 272)
Xiang, Y., 13[52] (10)
Xie, J., 149[237] (121, 124), 248[421] (228, 229)
Xie, M., 310[122] (300)
Xodo, L. E., 337[70] (321)
Xu, C., 423[94] (386), 423[98] (386)
Xu, Q. H., 152[447] (143)
Xu, R., 476[216] (470)
Xu, S.-Q., 368[91] (362, 363, 366)
Xu, Y., 368[52] (354), 423[84] (385, 394, 421)
Xu, Y.-Z., 246[292] (193), 336[2] (314), 336[13] (315)
Xu, Z.-S., 152[419] (139)

Yagi, H., 247[409] (223)
Yagishita, K., 548[131] (508)
Yamada, K., 548[125] (507, 508), 548[126] (508)
Yamaguchi, T., 247[383] (217), 247[384] (217)
Yamaizumi, Z., 424[164] (407)
Yamakage, S., 148[186] (121), 148[189] (121)
Yamammoto, H., 490[93] (488)
Yamamoto, K., 546[19] (493)
Yamamoto, K. R., 489[66] (484)
Yamamoto, N., 244[199] (187)
Yamana, K., 245[275] (190, 235), 248[422] (228), 248[424]
Yamaoka, T., 422[45] (376)
Yanagi, K., 475[129] (452)
Yang, J. H., 284[191] (278), 284[192] (278)
Yang, J.-Y., 100[5] (82)
Yang, M., 369[114] (366)
Yang, S. K., 53[199] (48, 49)
Yang, Z.-W., 152[419] (139)
Yannopoulos, C. G., 310[112] (299, 304)
Yanofsky, C., 283[125] (268)
Yao, Q., 422[70] (382)
Yao, S., 247[401] (221, 222), 310[92] (296)
Yartseva, I. V., 338[158] (328)
Yashiro, M., 242[88] (165)
Yau, L., 151[356] (130)
Yawman, A. M., 310[98] (298), 310[104] (299, 301), 310[110] (299, 303)
Yazawa, K., 547[71] (497)
Ye, X., 473[61] (434)
Yen, E., 309[39] (292, 293)
Yen, S.-F., 474[69] (434, 440), 474[75] (437, 441)
Yeola, S. N., 547[92] (502)

Ylikoski, J., 246[326] (201), 246[327] (201)
Yocum, R. R., 488[10] (478)
Yokota, H. A., 103[223] (96)
Yokoyama, S., 283[154] (274)
Yoneda, F., 248[465] (237), 336[8] (314, 315)
Yonei, S., 424[165] (407)
Yoneyama, M., 51[84] (34)
Yoo, D. J., 422[58] (378, 379)
Yoon, C., 422[38] (374), 475[147] (456)
Yoon, H. S., 489[79] (486)
Yoon, K., 13[52] (10), 52[180] (46)
Yoshida, K., 475[168] (462, 463)
Yoshida, M., 548[140] (510)
Yoshimura, Y., 248[418] (226)
Yoshinari, K., 242[72] (162)
Youderian, P., 12[29] (6)
Young, K. E., 489[25] (479)
Young, P. R., 473[10] (428, 436)
Young, S. L., 338[130] (325)
Yu, A., 103[208] (95)
Yu, C., 424[187] (410)
Yu, D., 149[237] (121, 124), 149[238], 151[384] (133), 151[385] (133), 152[401] (135), 309[49] (292, 295), 309[52] (292, 295)
Yu, F.-L., 548[149] (512)
Yu, H., 247[353] (209), 247[354] (209)
Yu, S., 147[76] (114), 147[77] (114)
Yue, S., 474[101] (444, 446, 448)
Yun, S., 548[137] (509)
Yun, W., 548[121] (506)
Yunes, M. J., 103[238] (96)

Zaccardi, J., 547[76] (500)
Zacharias, W., 99, 100[18] (82), 100[39] (85, 93), 103[252] (99)
Zachau, H. G., 52[148] (43, 45, 48)
Zady, M. F., 546[26] (494)
Zagursky, R. J., 241[35] (155), 247[362] (211)
Zahn, K., 368[57] (354)
Zahr, S., 149[230] (122)
Zain, R., 152[422] (139), 152[426] (139, 144), 309[44] (292, 293)
Zajc, B., 247[408] (223)
Zakin, M. M., 246[286] (192)
Zakrzewska, K., 80[108] (76)

Zalewski, R. I., 148[200] (121), 149[209] (126)
Zamaletdinova, G. U., 152[410] (138)
Zamboni, M., 422[48] (376)
Zamecnik, P. C., 238, 241[19] (155), 248[466] (238), 289, 309[37] (289)
Zannis-Hadjopoulos, M., 100[11] (82)
Zapp, M. L., 473[16] (428), 475[162] (460)
Zappia, G., 148[147]
Zaramella, S., 292, 309[61] (292)
Zarling, D. A., 241[24] (158)
Zarrinkar, P. P., 49, 52[133] (41, 49), 53[203] (49)
Zarrinmayeh, H., 282[20] (253)
Zarytova, V. F., 242[64] (161), 242[65] (161), 242[86] (164), 244[192] (186), 244[193] (186)
Zastawny, T. H., 424[190] (410), 424[193] (410)
Zatyrova, V. P., 246[297] (194)
Zaug, A. J., 284[158] (274, 275)
Zebrowska, A., 246[293] (193)
Zeh, D., 150[327] (127, 137)
Zein, N., 549[173] (515, 516), 549[176] (516), 559, 582
Zelenko, O., 244[154] (175)
Zemlicka, J., 150[288] (122)
Zendegui, J. G., 147[97] (114)
Zenke, N., 243[105] (167)
Zerial, A., 243[128] (169)
Zhang, G., 147[75] (114)
Zhang, H. C., 283[93] (263)
Zhang, L., 368[90] (362)
Zhang, P., 339[218] (333), 339[219] (333)
Zhang, Q.-M., 424[165] (407)
Zhang, R.-G., 283[116] (266, 268)
Zhang, S., 101[107] (89)
Zhang, W., 310[122] (300)
Zhang, X., 102[165] (92), 147[71] (113)
Zhang, Y., 158, 241[15] (158), 247[380] (216)
Zhang, Y.-B., 309[68] (293)
Zhang, Z., 248[474] (239), 248[475] (239), 310[86] (295)
Zhao, B. P., 245[271] (190)
Zhao, H., 150[293] (122, 123)

Zhao, M., 473[17] (428), 473[56] (433, 435, 452, 460), 475[142] (453, 457, 458), 475[158] (460)
Zhao, Q., 245[269] (190)
Zhen, Y., 423[84] (385, 394, 421)
Zheng, G., 103[229] (96)
Zheng, P., 101[111] (89)
Zheng, Q., 246[292] (193), 336[16] (315)
Zhong, M., 476[215] (469)
Zhong, Y. Y., 51[82] (34)
Zhou, B.-W., 244[172] (181)
Zhou, D. M., 282[69] (259)
Zhou, F. X., 474[96] (441, 443, 444)
Zhou, K., 53[209] (50), 53[210] (50)
Zhou, L., 152[449] (144), 248[414] (225), 248[416] (225)
Zhou, W., 149[237] (121, 124)
Zhou, X.-X., 147[133] (120), 147[134] (120), 148[140] (120), 148[196] (121)
Zhou, Y., 247[410] (223), 247[411] (223, 224)
Zhu, G., 79[19] (59)
Zhu, T., 242[51] (159), 245[258] (190)
Zhu, Z., 247[353] (209)
Zhuze, A. L., 489[78] (486)
Ziehler-Martin, J. P., 282[76] (260)
Zillman, M. A., 309[62] (292)
Zillmann, M. A., 368[92] (362, 363)
Zimmer, C., 489[82] (486)
Zimmer, D. P., 79[3] (56, 73)
Zimmer, S. G., 547[57] (495)
Zimmerman, S. C., 339[233] (335), 449, 474[119] (449, 450), 474[120] (449, 450)
Zinder, N. D., 283[130] (268)
Zinkel, S. S., 368[55] (354, 365)
Zipse, H., 423[112] (391)
Zmijewski, M., 547[58] (495)
Zmijewski, Jr., M. J., 547[73] (499)
Zoltewicz, J. A., 150[278] (122), 150[279] (122)
Zon, G., 12[9] (4, 5, 9), 51[82] (34), 79[54] (71), 80[97] (76), 148[204] (126)
Zorbas, H., 100[11] (82)
Zounes, M., 247[368] (214), 310[127] (302)
Zounes, M. C., 308[14] (287)
Zsido, T. J., 548[122] (506)
Zuckermann, R., 147[83] (114)
Zynchlinski, H. V., 149[230]

# Subject Index

**PHILIP AND LESLEY ASLETT**
*Marlborough, Wiltshire, UK*

---

Every effort has been made to index as comprehensively as possible, and to standardize the terms used in the index in line with the IUPAC Recommendations. In view of the diverse nature of the terminology employed by the different authors, the reader is advised to search for related entries under the appropriate headings.

The index entries are presented in letter-by-letter alphabetical sequence. Compounds are normally indexed under the parent compound name, with the substituent component separated by a comma of inversion. An entry with a prefix/locant is filed after the same entry without any attachments, and in alphanumerical sequence. For example, 'diazepines', '1,4-diazepines', and '2,3-dihydro-1,4-diazepines' will be filed as:-

   diazepines
   1,4-diazepines
   1,4-diazepines, 2,3-dihydro-

The Index is arranged in set-out style, with a maximum of three levels of heading. Location references refer to volume number (in bold) and page number (separated by a comma); major coverage of a subject is indicated by bold, elided page numbers; for example;

   triterpene cyclases, **299–320**
      amino acids, 315

*See* cross-references direct the user to the preferred term; for example,

   olefins *see* alkenes

*See also* cross-references provide the user with guideposts to terms of related interest, from the broader term to the narrower term, and appear at the end of the main heading to which they refer, for example,

   thiones
      *see also* thioketones

abasic site, formation, 493
acetal, ribo-, 257
acetaldehyde, bromo- (BAA), applications, DNA
    structure probes, 93
acetaldehyde, chloro- (CAA), applications, DNA
    structure probes, 93
acetaldehydes, halo-, applications, DNA structure
    probes, 93
acetate, as esperamicin A₁ precursor, 554
acetate, cholesterylamino-, oligonucleotide attachment,
    160
acetate, hexadecylamino-, oligonucleotide attachment,
    160
acetate, N-methylanilinium trifluoro-, as coupling
    reaction activator, 138
acetic acid, dichloro-, effects, on purine bases, 121
2-(acetoxymethyl)benzoyl, as protecting group, 121
acridine, 9-anilino-, antineoplastic agent, 606
acridine, 6-chloro-9-(p-chlorophenoxy)-2-methoxy-,
    synthesis, 237
acridine, 6-chloro-9-phenoxy-2-methoxy-, coupling, 211
acridine-4-carboxamide, 9-amino-, 607
acridine-4-carboxamides
    2-1 analogues, cytotoxicity, 607
    angular analogues, cytotoxicity, 607
    cytotoxicity, 607
    structural features, 606
acridines, multi-
    bisintercalation, 451
    synthesis, 451
acridinones, imidazo-, structural features, 606
acridinones, triazolo-, structural features, 606
4′-(9-acridinylamino)methanesulfon-m-anisidine *see*
    amsacrine
acridinyl phosphoramidites, synthesis, 181
acrylate, methyl, reactions, with 5-chloromercury-2′-
    deoxyuridine, 197
*Actinomadura madurae*, maduropeptin, 580
*Actinomadura verrucosospora*, esperamicin, 571
*Actinomyces* spp., actinomycin D, 608
Actinomycetales
    BE-13793C, 600
    kedarcidin, 575
    UCE6, 600
actinomycin D
    applications, 429
        antineoplastic agents, 607
    DNA damage, redox-dependent, 529
    occurrence, 608
actinomycins, antineoplastic activity, 472
acyclic linkers, in nucleoside analogues, 273
adamantane carbonyl chloride, as coupling reaction
    activators, 139
1-adamantanol, use as chemical switch, 463
adamantylethyl phosphoramidites, synthesis, 187
adenine, 2-amino-, in oligodeoxynucleotides, 323
adenine, 8-bromo-, in oligodeoxynucleotides, 325
adenine, N²-imidazolylpropyl-2-amino-, insertion, into
    oligonucleotides, 214
adenine, N⁶-methoxyl-, in oligodeoxynucleotides, 324
adenine, N⁶-methyl-, 324
adenine, 2-oxo- (iso-G), hydrogen bonding pattern, 323
adenosine
    aptamer-binding ability, 625
    depurination, 122
    as starting material, for ribonucleoside
        phosphoramidites, 231
adenosine, 1-methyl-, aptamer-binding ability, 625
adenosine phosphoramidites, insertion, into
    oligonucleotides, 229
adenosine 5′-triphosphatase (ATPase), activity, 603
adenosine 5′-triphosphate (ATP)
    and DNA topoisomerase II, 601

in SELEX studies, 625
S-adenosylmethionine, in neocarzinostatin chromophore,
    554
Adriamycin *see* doxorubicin
aflatoxin B₁
    metabolic activation, 512
    sequence specificity, 513
    synthesis, 512
aflatoxins
    DNA damage induction, 511
    from, *Aspergillus* spp., 470
    studies, 492
aging, and DNA damage, 372
aglycones, synthesis, 558
alanyl adenylate (alanyl-AMP), synthesis, 628
alanyl-AMP *see* alanyl adenylate (alanyl-AMP)
aldehydes
    DNA adducts of, 502
    DNA damage induction, 502
aldehydes, α,β-unsaturated, DNA adduct formation by,
    502
alkoxyl radicals
    DNA damage induction, 510
    from vinyl epoxides, 510
alkylating agents
    biotin-based, 632
    DNA damage induction, 372
allopurinol, as synthon, 332
allylic trisulfide, nucleophilic attack by thiophile, 570
altromycin B
    DNA damage induction, 508
    intercalation, 471
    NMR studies, 471
amidates, synthesis, 109
amides, antisense oligonucleotide linkages, 300
amines, antisense oligonucleotide linkages, 301
amino sugar, 554
amonafide, 609
    inhibition, 606
m-AMSA *see* amsacrine
amsacrine, 609
    as DNA topoisomerase II inhibitor, 603
    hepatotoxicity, 606
amyloid precursor protein (APP), and Alzheimer's
    disease, 372
aniline, 2,4-dinitro-, as intramolecular fluorescence
    quencher, 208
aniline hydrochloride, N,N-dimethyl-, as activator, 130
anilinium trichloroacetate, N-methyl-, as coupling
    reaction activator, 138
anilinoacridines, as DNA topoisomerase II inhibitors, 606
p-anisyldiphenylmethyl groups, in oligonucleotide
    synthesis, 107
o-anisyl-1-naphthylmethyl, as trityl protecting groups,
    126
p-anisyl-1-naphthylphenylmethyl, as trityl protecting
    groups, 126
anthracenediones, as DNA topoisomerase II inhibitors,
    606
anthracenes, as DNA topoisomerase II inhibitors, 606
anthracenyl deoxyguanosine phosphoramidites,
    synthesis, 222
anthracyclines
    applications, 428
    as DNA topoisomerase II inhibitors, 605
    formaldehyde-mediated covalent attachment to DNA,
        530
    intercalation, 446
    use in pyrimidine triplex stabilization, 467
    x-ray crystal structure, 436
anthramycin, cardiotoxicity, 497
anthranilate, N-(2-methoxyacrylyl), intercalation, 574
anthrapyrazoles, structural features, 606

anthraquinone, 9-amino-1,4,6-trihydroxy-, in dynemicin A, 574
anthraquinone phosphoramidites, synthesis, 183
anthraquinones
  biosynthesis, 554
  DNA binding studies, 468
  photochemical DNA damage, 544
  threading intercalating agents, 451
anthraquinonylmethyl deoxyribonucleoside phosphoramidites, insertion, into oligodeoxyribonucleotides, 228
antibiotics
  actinomycin D, 433, 529
  altromycin B, 471
  anthramycin, 497
  applications, DNA structure probes, 98
  arugamycin, 351
  azinomycin B, 513
  barminomycin I, 500
  bioxalomycins, 499
  C-1027, 554
  calicheamicins, 554
  carzinophilin, 513
  CC-1065, 504
  chromoproteins, 554
  chrysomycin A, 545
  clecarmycins, 510
  cleomycin, 539
  cyanocyclines, 499
  cyclopropane-containing, 504, 506
  daunomycin, 527
  discorhabdin A, 501
  doxorubicin, 433, 527, 528
  duocarmycin A, 505
  duocarmycin SA, 505
  dynemicins, 554, 574
  ecteinascidin, 501, 743
  elsamicin, 529
  epinardin D, 501
  esperamicins, 554, 571
  gilvocarcin V, 544, 545
  hedamycin, 509
  oxazolidine, DNA adduct formation, 499
  phleomycin, 539
  platomycins, 539
  pluramycins, 471, 508
  protein, 558
  quinocarcin, 500
  ravidomycin, 545
  renieramycins, 497
  safracins, 497
  saframycin A, 529
  saframycins, 497
  sibiromycin, 497
  sirodesmins, 527
  streptomycin, mechanisms, 433
  streptonigrin, 528
  tallysomycin, 539
  tetrazomine, 500
  varacin, 527
  victomycin, 539
  YA-56, 539
  zorbamycin, 539
  zorbonamycins, 539
  *see also* antitumor antibiotics; enediyne antibiotics; quinoxaline antibiotics
antibody–DNA interactions, disulfide hairpin, 363
antigene strategy, 466
antiinfective agents
  azicemicins, 514
  azomycin, 521
  myxin, 532
  norfloxacin, 462

pluramycins, 508
  quinolones, 462
antimalarial agents, quinacrine, 428
antineoplastic agents
  actinomycin D, 429, 607
  9-anilinoacridine, 606
  bisanthracycline, 451
  bisantrene, 440, 606
  bleomycin, 536
  camptothecin, 594
  daunomycin, 527
  ditercalinium, 444
  doxorubicin, 527, 594
  dual inhibitors, 607
  etoposide, 594
  intoplicine, 607
  mitomycin C, 514
  myrocin C, 506
  pluramycins, 508
  quinobenzoxazines, 462
  quinocarcin, 500
  saintopin, 607
  tetrazomine, 500
  varacin, 527
antisense oligonucleotides, 142
  with altered DNA backbones, **285–311**
    design issues, 306
    early studies, 289
    future research, 307
  amide linkages, 300
  amine linkages, 301
  applications, 286
  binding, with mRNA, 286
  binding affinity, assessment, 288
  carbamate linkages, 301
  carbon linkages, 305
  cellular uptake, 289
  chimeric, 287, 294
  classification, 289
  definitions, 286
  dephosphono linkages, 298
  ether linkages, 305
  formacetal linkages, 305
  guanidine linkages, 300
  hydroxylamine linkages, 302
  melting temperature, 288
  MMI linkages, 302
  nitrogen-containing linkages, 299
  non-phosphate backbone linkages, properties, 299
  non-phosphate sugar linkages, 306
  oxygen-containing linkages, 304
  phosphate-modified linkages, 290
    anionic, 291
    cationic, 294
    neutral, 294
  phosphodiester linkages, 287
  silicon-containing linkages, 303
  sulfide linkages, 304
  sulfonamide linkages, 304
  sulfonate linkages, 304
  sulfone linkages, 304
  sulfur-containing linkages, 304
  synthesis
    automated, 285
    economic factors, 289
    large-scale, 289
  $T_m$ values, 288
  urea linkages, 301
  use of term, 286
antisense strand *see* noncoding strand
antitrypanosomal agents, berenil, 452, 459
antitumor antibiotics, DNA strand breakage induction, 420

aphidicolin, DNA polymerase α inhibition, 597
aponeocarzinostatin, proteolytic activity, 559
apoprotein, 558
 proteolytic activity, 582
 roles, in enediyne antibiotics, 582
apoptosis
 bcl-2 and, 597
 and enediyne antibiotics, 582
 p53 and, 597
 topoisomerase-directed agents and, 597
APP *see* amyloid precursor protein (APP)
aptamers
 applications, pharmaceutical, 640
 characterization, 638
 footprinting experiments, 634
 nucleic acid-binding, 629
 nucleic acid vs. protein, 638
 protein-binding, 626
 SELEX studies, 624
 structure, 627
 structures, guanine–quadruplex, 627
 synthesis, 625
 use of term, 617
*Aristolochia* spp., aristolochic acid, 522
aristolochic acid
 DNA adducts with, 522
 occurrence, 522
 reductive activation, 522
aromatic compounds, nitro-, DNA damage induction,
  521
aromatic compounds, planar, in water, dimerization, 432
aromatic compounds, unfused
 binding to calf thymus DNA, 452
 dichroism, 455
 NMR spectra, 455
 synthesis, 452
 viscosity, 455
aromatic sensitizers, nitro-, misonidazole, 583
aromatic systems
 small, partial intercalation, 463
 unfused, base-pair propeller twist, 452
Arrhenius activation energy, nucleic acids, 41
arugamycin, structure, 351
ascorbate, superoxide production by, 540
A,T-hook peptides, 486
ATP *see* adenosine 5′-triphosphate (ATP)
ATPase *see* adenosine 5′-triphosphatase (ATPase)
austocystins, DNA damage induction, 511
avidin
 binding affinity, with biotin, 620
 interactions, with biotin, 154
6-azacytosine, in oligodeoxynucleotides, 322
6-azathymine, in oligodeoxynucleotides, 319
azatoxin(s)
 as DNA topoisomerase II inhibitors, 610
 as dual inhibitors, 610
 as tubulin polymerization inhibitors, 610
azatyrosyl, chloro-, in kedarcidin chromophore, 575
azicemicins, possible DNA reactions of, 514
azinomycin B
 DNA reactions of, 513
 occurrence, 513
 sequence specificity, 513
aziridines, DNA alkylation, 513
azomycin
 DNA reactions with, 521
 glyoxal formation from, 521
 oxygen radical production by, 521
 reductive activation of, 521

BAA *see* acetaldehyde, bromo- (BAA)
bacterial alkaline phosphatase (BAP), applications, in
 dephosphorylation, 661

bacteriophages, filamentous, 637
bacteriophage λ, as expression vector, 649
BAP *see* bacterial alkaline phosphatase (BAP)
barminomycin I, 531
 DNA adduct formation, 500
basic-helix-loop-helix peptides, 480
 IEB E47, 480
 MyoD, expression, 481
 USF, 480
 *see also* helix-turn-helix peptides
basic-leucine zipper proteins *see* bZip
basic zipper proteins *see* bZip
BE-13793C, occurrence, 600
benzaldoxime, *syn*-4-nitro-, in oligonucleotide synthesis,
  117
benzamide, *o*-hydroxy-, 580
benzenediazonium ion, 4-(hydroxymethyl)-, DNA
 reactions of, 543
benzenediazonium ions, occurrence, 543
benzenes, alkenyl-
 DNA adducts with, 519
 metabolic activation of, 519
benzenes, trihydroxy-, DNA cleavage, mechanisms, 535
benzenes, 1,2,4-trihydroxy-, DNA damage induction, 535
1,4-benzenoid biradical
 discovery, 556
 reactivity, 578
benzimidazole trifluoromethane sulfonate, as coupling
 reaction activator, 138
benzimides, phenyl-, cytotoxicity, 607
benzodihydropentalene, formation, 578
4*H*-1,3,2-benzodioxaphosphorin-4-one, 2-chloro-, in *H*-
 phosphonate synthesis, 139
3*H*-1,2-benzodithiol-3-one 1,1-dioxide, in oligonucleotide
 synthesis, 142
3*H*-1,2-benzodithiol-3-one 1-oxide
 DNA cleavage by, 524
 oxygen radical production, 524
 reactions
  with nitrogen nucleophiles, 526
  with thiol, 523
benzoic acid esters, *p*-azidotetrafluoro-, reactions, with
 oligodeoxyribonucleotides, 161
benzoisoquinolinediones, as DNA topoisomerase II
 inhibitors, 606
benzophenanthridineer deprotection, 119
benzothiopyranoindazoles, structural features, 606
1,2,4-benzotriazine 1,4-dioxide, 3-amino, 534
 DNA damage induction, 534
 hydroxyl radical production by, 534
benzotriazole, 1-hydroxy-, as coupling reaction activator,
  138
2*H*-1,4-benzoxazine 5-carboxylate, 3,4-dihydro-7-
 methoxy-2-methylene-3-oxo-, in C-1027
 chromophore, 578
benzoyl, 2-(2,4-dinitrobenzene sulfenyloxymethyl)-
 (DNBSB), as 5′-hydroxyl protecting group, 127
benzoyl, 2-hydroxy-3,6-dimethyl-, in maduropeptin
 chromophore, 580
benzoyl, 4-methoxy-, use in phosphodiester chemistry,
  120
benzoyl, 3-methoxy-4-phenoxy-, deoxyadenosine
 protection, 123
benzoyl, 2-(methylthiomethoxymethyl)- (MTMT), as 5′-
 hydroxyl protecting group, 127
benzyl, as phosphate protecting groups, 130
berenil
 calorimetric studies, 459
 hydrodynamic studies, 459
 structure, 452
Bergman reaction, 554
 σ,σ-biradical, 556
 and Myers reaction, compared, 556

Bergman rearrangement
  activation parameters, 556
  molecular calculations, 556
  thermodynamics, 556
Biacore system, nucleic acid kinetic studies, 40
bicyclo-deoxynucleosides, effects, of DNA triplex
    formation, 273
bicyclodeoxynucleosides, synthesis, 257
bicyclo[7.3.1]diynene, biosynthesis, 554
bicyclo[7.3.0]dodecadienediyne, in neocarzinostatin
    chromophore, 554
bicyclo[7.3.0]dodecadiynene, synthesis, 554
bicyclo[7.3.0]dodeca-4,10,12-trien-2,6-diyne, in C-1027
    chromophore, 578
bicyclo[7.3.1]tridec-4-ene-2,6-diyne *see* calicheamicinone
bicyclo[7.3.1]tridec-3-ene-1,5-diyne-8,9-epoxide, in
    dynemicin A, 574
binding sites
  metal, probing, 258, 259
  SELEX studies, 619
biochemical assays, interpretation, 265
biochemical processes
  effect of nucleoside analogues on, 265
  nucleoside analogue probes, **251–284**
biotin
  binding affinity, with avidin, 620
  incorporation, into oligonucleotides, 171
  interactions
    with avidin, 154
    with streptavidin, 154
biotin phosphoramidites, applications, 173
biotin–streptavidin complexes, applications, 174
bioxalomycins
  NMR spectra, 500
  occurrence, 499
  relationship, with naphthyridinomycin, 500
bipyridine deoxyribonucleoside phosphoramidites,
    synthesis, 214
bipyridinyl phosphoramidites, synthesis, 185
$\sigma,\pi$-biradical, in Myers reaction, 556
$\sigma,\sigma$-biradical, in Bergman reaction, 556
bisacylphosphite, synthesis, 139
bisanthracycline, high antineoplastic activity, 451
bisantrene
  antineoplastic activity, 440, 606
  use in the synthesis of threading intercalating agents,
    440
bisintercalators
  affinity, increase in, 444
  echinomycin, 448
  Flexi-Di, 446
  and neighbor exclusion, 444
  rigid, neighbor exclusion, 449
  specificity in binding, concerted interactions, 448
  threading, 449
  TOTO, 448
  triostin, 448
  YOYO, 448
1,1-bis(4-methoxyphenyl)-1-pyrenyl methyl, as 5′-
    hydroxyl protecting group, 127
bisphosphates, in DNA damage studies, 390
bistriazolylphosphites, in phosphoramidite synthesis, 132
bistricarbonyl, DNA cross-linking by, 503
bleomycin
  antitumor activity, 186
  applications, DNA structure probes, 98
  DNA binding, 536, 538
  DNA cleavage by, 536
  DNA damage induction, 384, 394
  DNA strand breakage induction, 420
  metal binding, 537
  sequence specificity, 539
blood clotting, cascade, 626, 640

BME *see* ethanol, 2-mercapto- (BME)
BNCT *see* boron neutron capture therapy (BNCT)
boranophosphates, synthesis, 293
borohydride, as cofactor, 558
boron neutron capture therapy (BNCT)
  applications, 221
    cancer treatment, 293
bracken fern, carcinogenicity, 507
BsgI, DNA cleavage protection, 629
bulgarein
  as DNA topoisomerase I inhibitor, 600
  DNA winding activity, 600
  occurrence, 600
2,3-butanedione, possible DNA damage induction, 503
2′-*t*-butyldimethylsilyl group (TBDMS), for 2′-hydroxyl
    group protection, 144
*t*-butylphenoxyacetyl, as protecting group, 120
butyryl, 4-(methylthiomethoxy)- (MTMB), as 5′-
    hydroxyl protecting group, 127
bZip
  DNA binding motif, 482
  motif, truncation, 482
  peptides, 482
    isolation by *in vitro* selection, 482

C-1027
  activation, 578
  applications, 578
  DNA damage, 579
    inhibition, 580
  kinetic solvent isotope effect, 578
  occurrence, 578
  structure, 554, 578
  tRNA cleavage, anticodon region, 580
CAA *see* acetaldehyde, chloro- (CAA)
calicheamicin $\gamma_1^1$
  affinity cleavage, 571
  DNA damage, DS lesions, 570
  efficacy, 568
  solubility, 569
  structure, 568
  tetrasaccharide domain, 571
  total synthesis, 569
calicheamicin–glutathione disulfide, formation, 570
calicheamicin–glutathione trisulfide, formation, 570
calicheamicinone
  in calicheamicin $\gamma_1^1$, 569
  stereochemistry, 569
calicheamicin(s)
  activation, 569
    trisulfide chemistry, 570
  DNA damage, 570
    induction, 385
  DNA strand breakage induction, 421
  enediyne moiety, 554
  occurrence, 568
  role of oligosaccharide amino group, 570
  structure, 568
calicheamicin $\varepsilon$, formation, 570
calorimetry
  differential scanning, 28
  isothermal titration, 31
  in nucleic acid hybridization reactions, 21
*Camptotheca acuminata*, camptothecin, 599
camptothecin, 9-amino-, water solubility, 600
camptothecin(s)
  antitumor agents, 594
  as DNA topoisomerase I inhibitors, 599
  irenotecan, 599
  occurrence, 599

photochemical DNA damage, 544
topotecan, 599
cancer
　boron neutron capture therapy, 293
　and DNA damage, 372
caproic acid, *N*-biotinyl-6-amino-, conjugation, 197
caproic acid, digoxigenin-*O*-succinyl-ε-amino-,
　conjugation, 197
carbamate, methoxy-, in esperamicin, 573
carbamates, antisense oligonucleotide linkages, 301
carbocyclic analogues, 256
carbocyclic phosphoramidites, synthesis, 235
carbodiimide, use in formation of dumb-bell DNAs, 366
carbodiimide hydrochloride, 1-(3-dimethylaminopropyl)-
　3-ethyl (EDC), in oligonucleotide coupling, 158
carbonyl, 2,4-dinitrophenylethoxy- (DNPEoc), as 5′-
　hydroxyl protecting group, 127
carbonyl, fluorenylmethoxy (FMOC), deoxyadenosine
　protection, 123
carbonyl, *p*-phenylazophenyloxy- (PAPoc), as 5′-
　hydroxyl protecting group, 127
carboranyluracil, localization, 319
carcinogenesis, and DNA damage, 582
carcinogens
　aflatoxins, 470
　2-aminofluorene, 470
　covalent, 470
　nitrosamines, 521
　polycyclic aromatic hydrocarbons, 470
carzinophilin
　DNA reactions of, 513
　occurrence, 513
catalytic elution, use of term, 632
catalytic RNA *see* ribozymes
catenane(s)
　kinetoplastid, 343
　structure, 361
　synthesis, 361
CC-1065
　analogues
optosis
cell lysis *see* cytolysis
cells, competent, in transformation, 662
cellulose, solid supports derived from, 110
ceric ammonium nitrate, trityl group removal, 126
chain-termination method
　applications, in cloning, 664
　for DNA sequencing, 620
chaperonins, roles, in protein folding, 645
chebulagic acid, as direct protein binding agent, 595
chelatases, isolation, 632, 634
chelerythrine, occurrence, 601
chemical biology, use of term, 6
chemical cleavage method
　applications, in cloning, 664
　DNA sequencing, 620
　modified, in DNA structure probing, 82
chemistry
　encoded combinatorial, SELEX studies, 619
　and molecular biology, relevance, 6
chemotherapy
　combinational, topoisomerase agents in, 597
　DNA topoisomerases in, 594
chlorambucil–oligonucleotide conjugates, synthesis, 197
cholesterol–oligonucleotide conjugates, carcinoma cell
　studies, 188
cholesteryl nucleoside phosphoramidites, synthesis, 189
cholesteryl phosphoramidites, synthesis, 187
cholic acid, esters, 159
chromatin, 582
　structure, 8
chromatography, ion-exchange, in oligonucleotide
　purification, 144

chromomycin, DNA binding, 9
chromoproteins
　endocytosis model, 582
　structure, 554
α-chymotrypsin, substrates, 637
CI-921, water solubility, 606
Cibacron Blue 3GA, in SELEX processes, 624
cinnamoyl, α-phenyl-, deoxyadenosine protection, 123
CIP *see* calf intestine alkaline phosphatase (CIP)
circular DNA
　double-stranded
　　enzymatic closure, 364
　　nonenzymatic closure, 365
　　occurrence, 343
　　polymerase recognition, rolling circles, 363
　single-stranded
　　closure, 366
　　enzymatic closure, 366
　　ligation, 366
　　nonenzymatic closure, 366
circular RNA
　occurrence, 344
　products, synthesis, 367
cleavable complex
　and cytotoxic potency, correlation, 596
　topoisomerase inhibition, 594
cleavable complex mediators, 596
　classes, 599, 603
　intercalating agents, 604
　　minimal, 607
　topoisomerase poisons, 594
clecarmycins, possible DNA damage induction, 510
cleomycin, structure, 539
cloning
　background colonies, 664
　case studies, 667
　cDNA libraries, 654
　for chemists, **643–674**
　dephosphorylation in, 661
　full length proteins, 645
　historical background, 643
　homology mapping in, 646
　insert identity
　　chain-termination method, 664
　　chemical cleavage method, 664
　　verification, 664
　insert screening, 664
　ligation in, 661
　overview, 644
　processes, 644
　　bulk oligonucleotide production, 652
　　expression system selection, 646
　　target DNA sequence selection, 645
　　vector selection, 647
　strategies, developments, 644
　subdomains, 645
　*see also* molecular cloning
cobalamin, cyano-, RNA binders, 638
cobalt(III)-bleomycin complexes, structure, 539
cofactors, non-hydrolyzable, 603
complementary DNA (cDNA) *see* cDNA
con A *see* concanavalin A (con A)
concanavalin A (con A), thrombin immobilization, 627
conformation, switching, 359
conjugate groups, attachment to DNA, **153–249**
controlled-pore glass (CPG), in automated
　oligonucleotide synthesis, 112, 156
copper phenanthroline, applications, DNA structure
　probes, 86
copper TMPyP(4), Watson–Crick base pair disruption,
　443
coralyne, 5,6-dihydro-, inhibition, 601
coralyne(s)

as DNA topoisomerase I inhibitors, 601
  inhibition, 601
  use in pyrimidine triplex stabilization, 467
cordycepin, 5′-cholesteryl-, as HIV-1 inhibitor, 189
corilagin, as direct protein binding agent, 595
coumarin phosphoramidites, synthesis, 182
coumermycin, as direct protein binding agent, 596
counterion condensation effect, nucleic acids, 20
coupling reaction
  activators, 138
  internucleotidic, mechanisms, 137
CP-67,804, DNA studies, 603
CP-115,953
  DNA studies, 603
  as DNA topoisomerase II inhibitor, 603
  structure, 609
CPG see controlled-pore glass (CPG)
Crick, Francis Harry Compton (1916– ), DNA structure
    studies, 81
Criegée rearrangement, in DNA damage, 387, 394
Cro protein, helix-turn-helix motifs, 478
cross-links, involving psoralens, 545
Ctmp see piperidin-4-yl, 1-[(2-chloro-4-methyl)phenyl]-
    4′-methoxy- (Ctmp)
CV-1 cells, SV40 DNA replication inhibition, 169
cyanine phosphoramidites, synthesis, 177
cyanogen bromide, use in the formation of dumb-bell
    DNAs, 366
cyclines, cyano-, occurrence, 499
cycloaromatization
  of enediyne compounds, 557
  transition-metal mediated, 557
cyclodextrins, anthryl(alkylamino)-, 463
cyclohexadiene, 557
cyclophanes, cationic, partial intercalation, 466
cyclopropanes, DNA damage induction, 504
cyclospirolactone, formation, 564
cysteine
  DNA damage, 560
  functionalities, 616
  pK, 616
L-cysteine, as esperamicin $A_1$ sulfur source, 554
cytidine, $N^6$-amino-2′-deoxy-, synthesis, 271
cytidine, 2′-deoxy-5-methyl-, 259
cytidine-5′-triphosphate, 5-[(4-azidophenacyl)thio]-,
    synthesis, 216
cytochromes P450, and aflatoxin epoxide, 470
cytolysis, mechanisms, 667
cytosine
  bromination, 96
  exocyclic amino group, protection groups, 120
  iodination, 96
cytosine, $N^4$-alkyl-, in oligodeoxynucleotides, 320
cytosine, $N^4$-methoxy-, in oligodeoxynucleotides, 321
cytosine, $C^5$-propynyl-, in oligodeoxynucleotides, 322
cytotoxic potency, and cleavable complex, correlation,
    596


2-dansylethoxycarbonyl (Dnseoc), as 5′-hydroxyl
    protecting group, 127
dansyl phosphoramidites, synthesis, 178
DAPI see indole, 4′,6-diamino-2-phenyl- (DAPI)
daunomycin
  formaldehyde-mediated covalent attachment to DNA,
    530
  occurrence, 605
  photochemical DNA damage, 544
  quinone methide, UV-vis spectra, 529
daunomycin, 5-imino-, inhibitor, 605
daunorubicin see daunomycin
DCC see dicyclohexylcarbodiimide (DCC)
7-deazaadenine, in oligodeoxynucleotides, 330

7-deazaadenosine, aptamer-binding ability, 625
7-deaza-8-azapurines, in oligodeoxynucleotides, 331
7-deaza-2′-deoxyadenosine, in aptamer synthesis, 626
1-deazaguanine, 254
3-deazaguanine, in oligodeoxynucleotides, 330
7-deazaguanine, 513
  $pK_a$, 331
3-deazapurines, 253
7-deazapurines, 253
4-decyloxytrityl, as 5′-hydroxyl protecting group, 127
9,10-dehydroanthracene biradical, 579
  retro-Bergman reaction, 579
1,4-dehydrobenzene biradicals
  structure, 572
2,6-dehydroindacene biradical
  conformation change, 564
  formation, 562
α,2-dehydrotoluene biradical, formation, 566
α,3-dehydrotoluene biradical, 556
demethylation, in oligonucleotide synthesis, 130
deoxyadenosine, 8-bromo-, reactions, with
    aminoalkylthiolate salts, 198
2′-deoxyadenosine, aptamer-binding ability, 625
11-deoxydaunomycin, quinone methide, NMR spectra,
    529
2′-deoxy-3-deazaadenosine, in oligodeoxynucleotides,
    329
4-deoxy-4-$N$-dimethylamino-5,5-dimethyl-$\beta$-D-
    ribopyranoside, in C-1027 chromophore,
    578
deoxydynemicin, occurrence, 574
2′-deoxyformycin, 333
2-deoxy-L-fucose, in esperamicins, 571
deoxyfucose-anthranilate, 572
deoxyguanosine, depurination, 122
deoxyguanosine, 8-oxo-
  mutagenicity, 410
  oxidative stress, 411
2′-deoxyguanosine, $O^6$-methyl-, in aptamer synthesis,
    626
5′-deoxyguanosine, 571
deoxyinosine, covalent DNA adduct structure
    determination, 495
deoxyinosine triphosphate (dITP), in DNA mutation,
    621
2′-deoxyisoinosine, synthesis, 325
deoxyluminarosine phosphoramidites, synthesis, 208
2-deoxyribonolactone, formation, 567
2′-deoxyribonolactone
  in DNA damage, 373
  lesion formation, 376
  mechanisms, 377
deoxyribonuclease I (DNase I)
  in DNA fragmentation, 621
  footprinting, 252
  disadvantages, 252
deoxyribonucleoside phosphoramidites
  development, 130
  EDTA-functionalized, 215
  from
    adenosine, 231
    5-iododeoxyuridine, 194
  synthesis, 166, 192
deoxyribonucleoside phosphoramidites, 5′-aminated,
    synthesis, 234
deoxyribonucleoside phosphoramidites, aminoalkylated
  from, 2-mercapto-2′-deoxyadenosine, 196
  synthesis, 195
deoxyribonucleoside phosphoramidites, biotinylated,
    synthesis, 205
deoxyribonucleoside phosphoramidites, dansylated,
    incorporation, into oligodeoxyribonucleotides,
    209

deoxyribonucleoside phosphoramidites, 2,4-dinitrophenyl-, insertion, into oligonucleotides, 210

deoxyribonucleoside phosphoramidites, mercaptoalkylated, synthesis, 203

2′,5′-deoxyribonucleoside-3′-O-phosphoramidites, 5′-mercapto-, synthesis, 234

4′-deoxyribonucleosides, synthesis, 264

2′-deoxyribose, conformation, 67

deoxyribosyl radicals, oxidation, 415

deoxyribozymes
  catalytic efficiency, 635
  inhibition, 634
  search for, 632
  SELEX studies, 619

2′-deoxyribozymes, 278

5′-deoxythymine, 5′-amino-, 233

deoxyuridine, 6-amino-, incorporation, into oligodeoxyribonucleotides, 199

deoxyuridine, 5-bromo-
  DNA-containing, sensitization, 416
  in DNA damage, 411

deoxyuridine, 5-formyl-, synthesis, 407

deoxyuridine, 5-hydroxymethyl-, synthesis, 407

deoxyuridine, 5-iodo-
  in DNA damage, 411
  as starting material, for deoxyribonucleoside phosphoramidites, 194

deoxyuridine phosphoramidites, 5-iodo-, synthesis, 220

2′-deoxyuridine, 2′-iodo-, photolysis, 382

2′-deoxyuridine, 5-(1-pentenyl)-, SELEX studies, 636

2′-deoxyuridine phosphoramidites, 5-[S-(2,4-dinitrophenyl)thio]-, synthesis, 203

2′-deoxyuridine, 5-(1-propynyl)-, DNA triplex stabilization, 272

2′-deoxyuridine, 5-thiocyanato-, synthesis, 202

2′-deoxyuridin-2′-yl, synthesis, 382

2′-deoxyuridin-5-yl, roles, in DNA strand breakage, 411

DEPC *see* pyrocarbonate, diethyl (DEPC)

dephosphorylation, in cloning, 661

dephostatin
  occurrence, 521
  possible DNA reactions with, 521

depsipeptides, 352
  structure, 352

depurination, in oligonucleotide chain, 122

deuteration, and DNA damage, 567

diacetyl, possible DNA damage induction, 503

diamine, *N*-biotinyl-*N*′-iodoacetylethylene-, applications, 632

1,8-diazabicyclo[5,4,0]undec-7-ene (DBU), in oligonucleotide synthesis, 114

diazo compounds, DNA damage induction, 541

diazoic acid, methyl-, as intermediate, 520

diazonium compounds, DNA damage induction, 541

diazonium ions, 543

dibenzo[1,4]dioxin-1-carboxamide, cytotoxicity, 607

dicarbonyl, DNA damage induction, 503

dichroism
  flow, 432, 455
  linear, in furamidine DNA binding studies, 458

1,3-dicyclohexylcarbodiimide, 111

dicyclohexylcarbodiimide (DCC), in oligonucleotide synthesis, 107

2,4-dideoxy-4-*N*-(dimethylamino)-L-fucopyranoside, in kedarcidin chromophore, 575

2,4-dideoxy-4-(*N*-ethylamino)-3-*O*-methyl-α-L-xylopyranoside, in calicheamicin γ₁¹, 569

2,4-dideoxy-4-(*N*-isopropylamino)-3-*O*-methyl-α-L-xylopyranoside, in esperamicin, 572

2,6-dideoxy-L-*lyxo*-hexose *see* 2-deoxy-L-fucose

2,6-dideoxy-2-methylaminogalactose (*N*-methylfucosamine), 558

2,6-dideoxy-4-*S*-methylthio-β-D-ribohexopyranoside, in esperamicin, 572

2,6-dideoxy-4-*S*-thio-β-D-ribohexopyranoside, in calicheamicin γ₁¹, 569

4,6-dideoxy-4-(hydroxyamino)-β-D-glucopyranoside
  in calicheamicin γ₁¹, 569
  in esperamicin, 572

dienediyne, 582

diesters, synthesis, 109

digoxigenins
  oligonucleotide conjugation, 160
  oligonucleotide labeling, 209

diimidazole, 1,1′-carbonyl-, reactions, with oligodeoxyribonucleotides, 162

dimerization domain, DNA binding, 482

2,4-dinitrophenyl (DNP) groups, addition, to oligonucleotides, 174

dinucleotide binding sites, intercalation, 436

dinucleotide phosphoramidites, in oligonucleotide synthesis, 135

dinucleotides, crystallization, 435

diol epoxide derivatives, and polycyclic aromatic hydrocarbons, 470

dioxygen, roles, in DNA damage, 583

direct protein binding agents
  chebulagic acid, 595
  corilagin, 595
  coumermycin, 596
  β-lapachone, 595
  novobiocin, 596

discorhabdin A
  occurrence, 501
  possible DNA reactions of, 501

distamycin
  DNA binding, 567
  minor groove, 486

c-d distance rule, in monocyclic enediynes, 557

disulfide
  in bicyclic oligonucleotide synthesis, 360
  hairpin, antibody–DNA interactions, 363

ditercalinium
  intercalation, 446
  x-ray crystal structure, 444

1,2-dithiolan-3-one 1-oxides, DNA damage induction, 522

1,2-dithiole-3-thiones
  DNA damage induction, 526
  oxygen radicals from, 526

dithiole-3-thiones/thiol mixtures, superoxide production by, 540

2,4-dithiopyrimidines, synthesis, 260

dithiothreitol (DTT), 565

dithymidylyl phosphorothioate, reactions, with *N*-(1-pyrene)maleimide, 240

dITP *see* deoxyinosine triphosphate (dITP)

1,5-diyne-3-ene system, reactivity, 557

diynene *see* enediyne

DMS *see* sulfate, dimethyl (DMS)

DMTr group, as 5′-hydroxyl protecting group, 126

DNA
  adenine derivatives, 323
  with altered bases, **313–339**
    future research, 335
  5′-aminoalkylation, 165
  applications, 2, 7
    catalysts, 11
    computation, 11
    diagnostic tools, 9
    therapeutic agents, 9
  azoles, 333
  B-form double helix, 432
  branched, 356
    synthesis, 204

bulged, as catalyst, 564
C1′-position, and DNA damage, 373
C2′-position, and DNA damage, 380
C3′-position, and DNA damage, 383
C4′-position, and DNA damage, 384
C5′-position, and DNA damage, 394
catalysts
    selection, 630
    SELEX studies, 619
chemiluminescent detection, 161
chemistry, 7
    future research, 8
    and medicine, 4
classical intercalation model, features, 432
C-linked nucleosides, 332
closed circular superhelical, and intercalation
        unwinding angle, 433
conjugate group attachment, **153–249**
cyclization, 366
cytosine derivatives, 320
disulfide bond formation, 366
folding, complex, 617
3′-formylphosphate-ended, 583
functions, 342
groove-binding agents, 435, 595
groove shape, RNA compared, 452
guanine derivatives, 327
heterocyclic bases, 616
historical perceptions, 615
hydrodynamic properties, 630
imine formation, 366
interactions, with natural products, 5, 9
ionizing radiation, targeting, 375
kinetics
    determination, 37
    relaxation, 37
knotted, 347
as ligand, 10
linear, viscosity, 433
linker, damage, 567
loops
    hairpin, 629
    internal, 629
major groove, α-helices in, 477
maxicircles, 343
methylation, 646
microheterogeneity, 567
minicircles, 343, 630
minor groove binders
    distamycin, 567
    netropsin, 567
and molecular biology, overview, **1–13**
as molecular scaffold, 10
mutagenesis, 621
nucleosome, cleavage, 567
as organic natural product, 1
physical properties, 7
plasmid, 570
    in cloning, 647
preorganization, 360
protons
    exchangeable, 60
    non-exchangeable, 60
purine analogues, 329
purine modifications, 323
pyrimidine modifications, 314
    polycyclic heterocycles, 334
γ-radiolysis, 375, 398, 417
reactions, with intercalating agents, 470
relaxation, DNA topoisomerase I-mediated, 597
reporter group attachment, **153–249**
scission, *in vitro* studies, 567
selection, **615–641**

*in vitro*, 617
SELEX studies, 617
shuffling, 621
single-stranded, from double helices, 620
supercoiled, catalytic relaxation mechanism, 598
T4
    binding of pyrrolo[1,4]benzodiazepines to, 495
    CC-1065 binding to, 505
as tapes, 615
telomeres, 346
thymine derivatives, 314
topologically modified
    future research, 367
    synthesis, 363
topological modification, **341–369**
transformation, into *Escherichia coli*, 662
viral, probe for bulge sites, 565
*see also* cDNA; circular DNA; dsDNA; ssDNA
DNA adducts
    with aflatoxins, 511
    with aldehydes, 502
    with alkenylbenzenes, 519
    with barminomycin I, 500
    with bioxalomycins, 499
    with CC-1065, 504
    with cyanocyclines, 499
    with duocarmycins, 504
    with epoxides, 508
    with leinamycin, 522
    with mitomycin C, 514
    with naphthyridinomycin, 499
    with quinocarcin, 499
    reactions, with pyrrolo[1,4]benzodiazepines, 495
    with renieramycins, 497
    with safracins, 497
    with saframycins, 497
    with tetrazomine, 499
DNA alkylation
    by pluramycins, sequence specificity, 508
    by quinones, 531
DNA amplification *see* gene amplification
DNA backbone
    altered, 287
    in antisense applications, **285–311**
    religation, 599
DNA bending
    by CC-1065, 506
    circular permutation, 354
    sequence motifs, 630
DNA binding
    dimerization domain, 482
    helix-type-dependent, 460
    intercalation, 7, 428
    minor-groove, 428
        by CC-1065, 505
    molecular, 7
    negative cooperation, neighbor exclusion, 446
    SELEX studies, 619
DNA binding modes
    sequence-dependent, 456
    unfused aromatic compounds, 455
DNA binding peptides, 352, **477–490**
    based on protein motifs, 477
DNA binding proteins, sequences, 622
DNA cleavage
    affinity, 253
    by 5-alkyl resorcinols, 536
    by bleomycin, 536
    by CC-1065, 504
    by diazo and diazonium compounds, 541
    by diazoketones, 542
    by 6-diazo-5-oxo-L-norleucine, DNA damage
        induction, 542

by duocarmycins, 504
by epoxides, 508
by kinamycin, 541
by leinamycin, 522
by resorcinols, 535
pattern, 596
with $\alpha,\beta$-unsaturated aldehydes, 502
DNA cloning *see* molecular cloning
DNA conformation
  determination, 75
    computational methods, 75
  molecular dynamics simulations, 76
  structural parameters, 76
DNA cross-linking
  agents, pyrrole-derived, 514
  by azinomycin B, 513
  by formaldehyde, 502
  by FR 66979, 518
  by FR 900482, 518
  by isochrysohermidine, 503
  by mitomycin C, 514
  by pyrrolizidine alkaloids, 518
  and DNA structure probes, 96
DNA damage
  adduct formation, 585
  and aging, 372
  alkaline labile lesion formation, mechanisms, 387
  amplification, 401, 419
  attack sites, 583
  biological consequences, 494
  C1′-position reactions, 373
  C2′-position reactions, 380
  C3′-position reactions, 383
  C4′-position reactions, 384
  C5′-position reactions, 394
  and cancer, 372
  and carcinogenesis, 582
  chemistry, **371–425**, 554, 582
  covalent drug–DNA adducts, 492, 582
  and deuteration, 567
  deuterium isotope effect, 560
  dioxygen in, 583
  effects, 34
  enediyne compound-induced, **553–592**
  from UV radiation, 561
  interstrand transfer, 419
  mechanisms, 558, 582
  modes, 372
  and mutation, 372
  natural product-induced, 495
  natural products classification, 492
  and nitroimidazoles, 583
  oxidative, 582
  oxidizing agent-induced, 384
  and 3′-phosphoglycolate, 583
  and 3′-phosphopentenaldehyde, 583
  and 3′-phosphopyridazine, 583
  prevalence, 372
  protection agents, 417
  of pyrrolizidine alkaloids, by oxidative activation, 518
  radiosensitization, 583
  rate-determining step, 560
  sensitization, 417
  sequence-specificity, 560
  and sugar peroxyradical, 582
  via oxidation, 376
DNA–DNA duplexes, $T_m$ values, 288
DNA drug complexes, structure, 351
DNA duplexes
  $^{13}$C HMQC spectra, 73
  $^{15}$N HMQC spectra, 72
  base pairs
    identification, 66

lifetimes, 76
  twist, 76
  bent, 630
  B-form, non-exchangeable proton distances, 66
  formation, 43
  glycosidic angles, determination, 76
  HETCOR studies, 71
  hydration, 34
  $^1$H NMR spectra, 60
  $^{31}$P NMR spectra
    limitations, 72
    proton-decoupled, 70
  NOESY spectra, 65
    fingerprint region, 65
DNA footprinting
  interference, 252, 253
  protection, 252
DNA function
  effects of purines on, 410
  effects of pyrimidines on, 407
  overview, 2
DNA intercalators, **427–476**
  applications, DNA structure probes, 83
  structure, 433
  topological changes, 446
DNA ligase (ATP), applications, 364
DNA modification
  covalent
    by natural products, **491–552**
    mechanisms, 493
DNA polymerase $\alpha$-, inhibition by aphidicolin, 597
DNA polymerase III
  mechanisms, 349
  sliding clamp, 350
DNA polymerases
  activity, 194
  mutagenesis, 621
DNA processing enzymes, cytotoxic response depression, 597
DNA–proflavine complex, intercalation sites, 435
DNA–protein binding *see* protein–DNA binding
DNA recognition, by topoisomerase I, 598
DNA recombination
  *in vitro*, 621
  junctions, 346
DNA repair enzyme *see* DNA ligase (ATP)
DNA replication
  problems, 346
  rolling circles, 346
  topology and, 346
DNA–RNA damage, chemistry, 582
DNA–RNA duplexes, $T_m$ values, 288
DNA–RNA hybrids, 567
  damage, 566
  duplexes, 34
DNase I *see* deoxyribonuclease I (DNase I)
DNA sequences
  conformation switching, 359
  consensus double-stranded, 628
  diversity, 621
  nucleoside analogue incorporation, 259
  random, 622
  sequestration, 620
  viral, detection, 205
DNA sequencing
  automated, 620
  chemical cleavage method, 620
    modified, 82
  techniques, 620
DNA strand breakage
  antitumor antibiotic-induced, 420
  2′-deoxyuridin-5-yl in, 411
  direct, 380

mechanisms, 387
dsDNA, 418
    mechanisms, 418
DNA structure, 342
    A-form, 76, 82
    B-form, 76, 82
    branched, applications, 352
    circular, 354
    cuboid, 354
    determination
        via NMR spectroscopy, **55–80**
        via x-ray crystallography, 55
    early studies, 81
    effects of purines on, 410
    effects of pyrimidines on, 407
    elements, 56
    eukaryotic, 82
    knotted, research, 353
    man-made
        applications, 352
        non-linear topology, 352
    modifications, 3
    molecular probes of, **81–103**
    nomenclature, 56
    non-B, 82
    overview, 1
    parameters, helical, 58
    probes
        DNA intercalators, 83
        metal complexes, 85
        selection, 83
        transition metals, 85
    Watson–Crick base pair alignments, 58, 66
    Z-form, 77, 82
        assays, 96
        $^1$H NMR spectra, 78
DNA surveillance, mechanisms, 597
DNA synthesis
    automated, 617, 619
    chemical, 619
    importance of, 2
    inhibition, 554
    large-scale, 142
    mechanisms, 2
    phosphoramidite methodology, 130, 292
    ultrafast, 120
DNA topoisomerase I
    as cancer chemotherapeutic target, 594
    cleavage/religation equilibrium, 599
    DNA binding, 598
    DNA cleavage sites, 598
    DNA recognition, 598
    enzyme turnover, 599
    mechanisms, 594
        catalytic, 598
    post-strand passage cleavage/religation equilibrium,
        599
    pre-strand passage cleavage/religation equilibrium, 598
    strand passage, 599
DNA topoisomerase I inhibitors, 597
    benzo[c]phenanthridines, 601
    bulgarein, 600
    coralynes, 601
    indolocarbazoles, 600
    naphthacenequinones, 600
DNA topoisomerase II
    adenosine 5′-triphosphatase activity, inhibition, 603
    and adenosine 5′-triphosphate, 601
    ATP hydrolysis and enzyme turnover, 603
    as cancer chemotherapeutic target, 594
    cleavable complex mediators, non-intercalating agents,
        609
    composite pharmacophore, 610

double strand passage, 603
    inhibition, 462
    mechanisms, 594
        catalytic, 602
    post strand passage cleavage/religation equilibrium, 603
    pre-strand passage/religation equilibrium, 602
    recognition of binding site, 602
DNA topoisomerase II inhibitors, 601
DNA topoisomerase IIα
    molecular mass, 601
    and topoisomerase IIβ compared, 601
DNA topoisomerase IIβ, molecular mass, 601
DNA topoisomerase inhibitors, **593–614**
    direct protein binding agents, 595
    intercalation, 595
    mechanisms, 595
    minor groove binding, 595
    overview, 594
DNA topoisomerase(s)
    in chemotherapy, 594
    cytotoxicity, 596
    future research, 611
    inhibition
        cleavable complex, 594
        modes, 595
    overview, 347
    physiological roles, 594
    poisons, cleavable complex mediators, 594
DNA triplexes, 270
    applications, 142
    canonical base triplets, 629
    formation, 366
    5-halogenated pyrimidines, 316
    pH-independent, 271
    'purine motif', 629
    pyrimidine–purine–pyrimidine, 270
        stability, 271
    SELEX studies, 629
    stability, 47
        studies, 273
    structure, 77
DNAzymes *see* deoxyribozymes
DNBSB *see* benzoyl, 2-(2,4-dinitrobenzene
        sulfenyloxymethyl)- (DNBSB)
DNP groups *see* 2,4-dinitrophenyl (DNP) groups
DODC *see* oxadicarbocyanine, 3,3′-diethyl- (DODC)
dodecanucleotides, synthesis, 317
double-stranded DNA *see* dsDNA
doxorubicin, 528, 536
    applications, antitumor agents, 594
    as DNA topoisomerase II inhibitor, 603
    occurrence, 605
    oxidative generation of formaldehyde by, 530
drugs, intercalating agents, discovery, 428
drug transport, problems, 607
dsDNA
    breakage, 418
    detection, 166
dumb-bell DNA
    as decoys, 362
    use of carbodiimide in, 366
    use of cyanogen bromide in, 366
dumb-bell RNA, as decoys, 362
duocarmycin
    analogues
        with altered reactivity, 506
        sequence selectivity, 505
duocarmycin A, properties, 505
duocarmycin B$_1$, properties, 505
duocarmycin B$_2$, properties, 505
duocarmycin C$_1$, properties, 505
duocarmycin C$_2$, properties, 505
duocarmycins

cytotoxicity, mechanisms, 506
DNA reactions with, 504
occurrence, 505
duocarmycin SA, properties, 505
dyes
aromatic, DNA binding, 83
heterocyclic, in SELEX processes, 624
intercalating agents, discovery, 428
organic, in SELEX processes, 624
dynemicin A
activation, 575
anthracycline features, 574
binding motif, 575
DNA B-Z junction cleavage, 575
DNA damage, 575
occurrence, 574
reduced nicotinamide adenine dinucleotide phosphate
activation, 575
structure, 574
dynemicin A methyl ester, tri-*O*-methyl-, synthesis, 574
dynemicin H, synthesis, 575
dynemicin N, synthesis, 575
dynemicin(s)
applications, DNA structure probes, 98
structure, 554
dynemicin S, synthesis, 575

EB *see* ethidium bromide (EB)
EC 6.5.1.1 *see* DNA ligase (ATP)
EC 3.4.21.9 *see* enteropeptidase
EC 4.1.2.28 *see* 2-dehydro-3-deoxy-D-pentonate aldolase
echinomycin
bisintercalation, 488
with neighbor exclusion, 446
specificity, 448
ecteinascidin 743
DNA adducts of, 501
occurrence, 501
ED-110, synthesis, 600
EDC *see* carbodiimide hydrochloride, 1-(3-
dimethylaminopropyl)-3-ethyl (EDC)
electron affinic agents, DNA damage sensitization, 417
electron transfer, involving photoexcited states, 544
electrophiles, reactions, with DNA, 493
electroporation, mechanisms, 663
ellipticine
as DNA topoisomerase II inhibitor, 603
occurrence, 605
ellipticine, 9-hydroxy-, toxicity, 605
ellipticines, as DNA topoisomerase II inhibitors, 605
ellipticinium ion, 9-hydroxy-2-methyl-, water solubility,
605
elsamicin, DNA damage, redox-dependent, 529
endocytosis, chromoprotein model, 582
enediyne antibiotics
with apoprotein, 582
and apoptosis, 582
biosynthesis, 554
classes, 554, 582
classification, 582
DNA/RNA damage mechanisms, 582
DNA damage induction, 544
mechanism, 556
molecular modeling, 557
structure, 556
enediyne(s)
aromatization, 556
cycloaromatization, 557
*ab initio* calculations, 557
definition, 554
DNA damage induction, 385, 394, **553–592**
ionic intermediates, 557
nomenclature, 554

radical anion, 557
radical cation, 557
reactivity, 557
reviews, 554
in vivo/in vitro DNA damage activity, 554
enediyne(s), monocyclic, c-d distance rule, 557
enterokinase *see* enteropeptidase
enteropeptidase, applications, proteolytic cleavage, 648
envelope (conformation), definition, 56
enyne-allene, nomenclature, 554
enyne-cumulene
half-life, 560
nomenclature, 554
enzyme–nitroaromatic systems, superoxide production
by, 540
enzymes
methylation, DNA modification by, 546
restriction, DNA modification by, 546
structural features, 616
epinardin D
occurrence, 501
possible DNA adduct reactions of, 501
epipodophyllotoxins
as DNA topoisomerase II inhibitors, 609
etoposide, 609
teniposide, 609
epirubicin, cardiotoxicity, 605
episulfonium, from, leinamycin, 524
epoxide–DNA, and intercalation, 470
epoxides, DNA damage induction, 508
8,9-epoxybicyclo[7.3.0]dodecadienediyne, in kedarcidin
chromophore, 575
Erhlich, Paul
dye research, 428
magic bullet hypothesis, 428
ERYF1 *see* GATA-1
*Escherichia coli*
cell lines, 646
availability, 646
mutations, 646
in cloning techniques, 646
disadvantages, 647
DNA transformation into, 662
in expression systems, 646
expression vectors, 646
proteins, bulk production, 665
esperamicin, DNA damage induction, 385
esperamicin A$_1$
biosynthetic studies, 554
carbamate moiety, 554
intercalative binding model, 574
$K_a$, 574
minor-major groove binding model, 574
nucleosome linker damage, 574
esperamicin C, DNA damage, 573
esperamicins
activation, 572
2-deoxy-L-fucose, 571
DNA damage, 571, 573
DNA strand breakage induction, 421
enediyne moiety, 554
occurrence, 571
structure, 571
esperamicin X, x-ray crystal structure, 569, 571
esperamicin Z, 572
estragole
carcinogenic action, 519
DNA damage induction, 519
estrogens, metabolic activation of DNA adduct
formation, 532
ethanol, $\beta$-cyano-, in oligonucleotide synthesis, 127
ethanol, 2-mercapto- (BME), 565
ethers, antisense oligonucleotide linkages, 305

ethidium
    DNA binding, 457, 460
    intercalation, 446
        unwinding angle, 432
    RNA binding, bulged bases, 469
ethidium bromide (EB)
    applications
        DNA quantitation, 661
        DNA structure probes, 84
        in polymerase chain reaction evaluation, 658
    use with electrophoresis gels, 428
ethidium–RNA complex
    major and minor grooves, 435
    phosphate groups, 435
    steric clash, 435
ethyl, β-cyano-, as phosphate protecting groups, 130
ethyl, p-nitrophenyl-, as phosphate protecting groups, 130
ethyl, 2,2,2-trichloro-, as phosphate protecting groups,
    130
etoposide
    applications, antitumor agents, 594
    as DNA topoisomerase II inhibitor, 603, 609
europium(III)–texaphyrin–oligodeoxyribonucleotide
    conjugate, synthesis, 163
evolution
    molecular, early studies, 617
    'RNA World' hypothesis, 617, 630
expression system
    components, 646
    selection, for cloning, 646
expression vectors
    for *Escherichia coli*, 648
    promoter regions, 647
    types of, 648
Eyring transition state enthalpy, nucleic acids, 41

factor $X_a$, applications, proteolytic cleavage, 648
*Fagara xanthoxyloides see Zanthoxylum senegalense*
fagaridine, occurrence, 601
fagaronine chloride, reverse transcriptase activity, 182
Fe(II)EDTA *see* ferrous ethylenediamine tetraacetate
    (Fe(II)EDTA)
Fenton reaction, 540
    metal ions in, 417
ferrocene–oligonucleotide conjugates, synthesis, 166
ferrochelatase(s), in porphyrin metallation, 634
ferrous ethylenediamine tetraacetate (Fe(II)EDTA),
    applications, DNA structure probes, 89
*Fibrobacter* spp., classification, 158
Flexi-Di, x-ray crystal structure, 446
fluorene, 2-*N,N*-acetoxyacetylamino- (N-AcO-AAF),
    applications, DNA structure probes, 96
fluorene, 2-amino-, 2D-NMR, 470
fluorene, *N*-(2′-deoxyguanosin-8-yl)-2-(acetylamino)-,
    roles, in mutagenesis, 223
fluorene, diazo-, occurrence, 541
fluorescein isothiocyanate, reactions, with
    aminoalkylated oligonucleotides, 193
fluoresceinyl phosphoramidites, synthesis, 177
fluorophores, groove-binding, 238
FMOC *see* carbonyl, fluorenylmethoxy (FMOC)
formacetals, antisense oligonucleotide linkages, 305
formaldehyde, DNA cross-linking by, 502
formamidine, in oligonucleotide synthesis, 123
Fpg protein, in DNA repair, 410
FR 66979
    DNA damage induction, 518
    reduction-dependent DNA reaction, 518
FR 900482
    DNA damage induction, 518
    reduction-dependent DNA reaction, 518
Fractogel, in oligonucleotide synthesis, 113
Franklin, Rosalind Elsie (1920–58), DNA structure
    studies, 81

fucosamine, *N*-methyl- *see* 2,6-dideoxy-2-
    methylaminogalactose
fullerene–oligonucleotide conjugates
    applications, 167
    synthesis, 167
fullerenes, reactions, with 5′-mercaptoalkylated
    oligonucleotides, 167
furamidine
    DNA binding, 457, 460
    DNA complexes, molecular modeling comparisons,
        458
    structure, 452
furamine
    DNA complexes, molecular modeling comparisons,
        458
    structure, 452
furan, 2 hydroperoxytetrahydro (THF OOH),
    applications, DNA structure probes, 95
furanone, lesion formation, mechanisms, 377
3-(2*H*)-furanones, 4-hydroxy-5-methyl-, superoxide
    production by, 540
furans, diphenyl-
    DNA minor-groove vs. intercalation features, 457
    structure, 457
fusion proteins
    applications, 647
    bulk production
        isolation, 666
        purification, 666
    cleavage, 667
    isolation, 648

β-D-galactoside, 5-bromo-4-chloro-3-indolyl-, structure,
    646
GATA-1, cloning, 670
gene amplification, **615–641**
gene expression, inhibition, 286
genes, cloning, 643–674
genistein, as DNA topoisomerase II inhibitor, 603, 610
GF-1 *see* GATA-1
gilvocarcin V, photochemical DNA damage, 544
D-glucose, as esperamicin $A_1$ precursor, 554
glutathione (GSH), 559
    DNA damage, 560
glycine, γ-L-glutamyl-DL-cysteinyl-, 560
glycosylation
    acid-catalyzed
        for the synthesis of purine nucleosides, 260
        for the synthesis of pyrimidine nucleosides, 260
glyoxal
    applications, DNA structure probes, 95
    formation, 521
Gosling, Raymond G., DNA structure studies, 81
Gram-negative bacteria, 575
Gram-positive bacteria, kedarcidin activity against, 575
Grob fragmentation, in DNA damage, 394
guanidines, antisense oligonucleotide linkages, 300
guanine, $N^2$-(anthracen-9-ylmethyl)-, in
    oligodeoxynucleotides, 327
guanine, $N^7$-cyanoborane-, in oligodeoxynucleotides, 329
guanine, 7-hydro-8-oxo-, in oligodeoxynucleotides, 328
guanine, $N^2$-imidazolylpropyl-, insertion, into
    oligonucleotides, 214
guanine, $C^6$-O-methyl-, in oligodeoxynucleotides, 328
guanine, $N^7$-methyl-, dodecamer, synthesis, 329
guanine, $C^8$-oxo-7,8-dihydro-6-O-methyl-, in
    oligodeoxynucleotides, 329
guanosine, depurination, 122
guanosine, 3′-hydroxyl-, analogues, 279
guanosine, $O^6$-methyl-, use in hairpin ribozyme studies,
    277
guanosine nucleoside, reaction, with methanesulfonyl
    chloride, 118

hapten phosphoramidites, synthesis, 175
hedamycin, 509
helenalin, 503
α-helices
  in the DNA major groove, 477
    helix-turn-helix peptides, 477
helix-turn-helix motifs
  Cro protein, 478
  Hin recombinase, 478
  homeodomain, 479
  peptide truncation, 480
  reviews, 478
helix-turn-helix peptides
  in α-helical DNA binding proteins, 478
  *see also* basic-helix-loop-helix peptides
hepatitis delta virusoid, topology, 344
heptaketides, biosynthesis, 554
(*Z*)-1,2,4-heptatriene-6-yne, cycloaromatization via
    α,3-dehydrotoluene biradical, 556
HETCOR *see* heteronuclear correlation experiment
    (HETCOR)
*Heteroconium spp.*, bulgarein, 600
heterocyclic *N*-oxides, DNA damage induction, 532
heteronuclear correlation experiment (HETCOR),
    improvements, 71
4-hexadecyloxytrityl, as 5′-hydroxyl protecting group,
    127
hexaketide, as intermediate, 554
2-hexenal, DNA adduct formation by, 502
Hin recombinase, helix-turn-helix motifs, 478
histidine(s)
  functionalities, 616
  p*K*, 616
histone H1, 582
HIV *see* human immunodeficiency virus (HIV)
HIV-1 reverse transcriptase (HIV-1 RT)
  binding, 626
  photoaffinity labeling studies, 217
Hoechst 33258, applications, groove binding agents, 595
Hoechst 33342, applications, groove binding agents, 595
holoprotein auromomycin, 578
homeodomain
  engrailed, 479
  helix-turn-helix motifs, 479
homeodomain DNA binding, heterodimeric, 479
homology mapping, in cloning, 646
human immunodeficiency virus (HIV)
  dumb-bell RNAs and, 362
  HIV-1
    inhibitors, 187, 189
    sequences, 363
  TAR RNA
    cleavage, 562
    intercalation, 469
hydrazines, 570
  superoxide production by, 540
hydrogen atoms, abstraction, from DNA backbone,
    493
hydrogen peroxide, DNA damage induction, 540
hydrolases, isolation, 632
hydroperoxyl radical, 561
hydroperoxyl radical, DNA cleavage by, 540
hydrophobic forces, in intercalation, 432
hydroquinone, 575
hydroquinone-*O,O*′-diacetic acid (QDA), 114
β-hydroxyamide, 580
hydroxylamine(s)
  applications, DNA structure probes, 93
  superoxide production by, 540
hydroxyl radicals
  formation via Fenton reaction, 540
  from hydrogen peroxide, 540
  reactivity, 398

β-hydroxyperoxide, base-catalyzed cleavage, 583
(hydroxystyryl)diisopropylsilyl (HSDIS), as
    photochemically removable protecting groups,
    129
(hydroxystyryl)dimethylsilyl (HSDMS), as
    photochemically removable protecting groups, 129
hymenovin, mutagenesis, 503
hypochromism, induction, 452
hypoxanthines, in oligodeoxynucleotides, 325

idarubicin, cardiotoxicity, 605
IEB E47, peptide, conformation, 481
illudins
  carcinogenicity, 508
  possible DNA damage induction, 507
  toxicity, 508
imidazole, in oligodeoxynucleotides, 334
imidazole, 4,5-dichloro-, in oligonucleotide synthesis, 133
imidazole, *N*-methyl-, in oligonucleotide synthesis, 108
imidazole hydrochloride, *N*-methyl-, as coupling reaction
    activator, 138
*N²*-3-imidazolepropionic acid, in oligodeoxynucleotides,
    327
imidazoles, nitro-
  in DNA damage, 583
  as DNA topoisomerase II inhibitors, 609
  one-electron reduction potential, 583
  Ro 15-0216, 609
imidazoles, 2-nitro-, reductive activation of, 521
imidazolium trifluoromethane sulfonate, *N*-methyl-, as
    coupling reaction activator, 138
imines, DNA damage induction, 495
imino protons
  exchange rates, 74
  exchange times, measurement, 75
indacene, tetrahydro-
  deuterium incorporation, 560
  formation, 560
2,6-indacene biradical, formation, 560
indacene-12-ones, synthesis, mechanisms, 561
indole, 4′,6-diamino-2-phenyl- (DAPI)
  fluorescence quantum yield, 459
  minor-groove binding, 457
  sequence-dependent binding mode, 456
  structure, 452
indolocarbazoles, as DNA topoisomerase I inhibitors,
    600
inosine
  aptamer-binding ability, 625
  use in hammerhead ribozyme studies, 276
intercalating agents
  acridine orange, applications, 428
  anilinoacridines, 606
  anthracenediones, 606
    modified, 606
  anthracenes, 606
  anthracyclines, 605
    applications, 428
  applications, 428
  association rate constants, 434
  benzoisoquinolinediones, 606
  benzophenazines, 606
  chemically switched, 461
  coupling, 226
  covalent, 470
  design of, major and minor grooves, 440
  discovery, 428
  DNA binding, 7
  as DNA topoisomerase II inhibitors, 604
  double-decker, 565
  ellipticines, 605
  ethidium bromide, applications, 428
  furamine, 458

future research, 472
incorporation, into oligonucleotides, 236
minimal
  acridine-4-carboxamides, 607
  9-aminoacridine-4-carboxamide, 607
  as DNA topoisomerase II inhibitors, 607
peptides, 487
porphyrins, 441
proflavine, 435
pyrimidines S1- S3, 458
quinacrine, applications, 428
quinacrine mustard, 470
  applications, 428
reactions, with DNA, 470
self-assembly, 461
simple systems, binding sites, 436
threading, 437
  anthraquinone, 451
  nogalamycin, 437
  synthetic, 440
triplex, naphthylquinoline, 468
variations in, 435
*see also* bisintercalators; DNA intercalators; RNA
  intercalators; trisintercalators
intercalating agents, multi-, 451, 461
intercalating agents, unfused aromatic, 452
intercalation
  barrier to, steric clash, 441
  dinucleotide binding sites, 436
  with distorted nucleic acid structures, 469
  and epoxide–DNA, 470
  inhibition, in multi-aromatic compounds, 451
  into distorted duplexes, 466
  into multistranded helices, 466
  models, Lerman's classical, 432, 433
  neighbor exclusion, 444
  NMR signal downfield shift, 441
  nonclassical, furamidine model, 459
  parallel, 437
  partial
    model systems, 463
    and protein–DNA recognition, 463
    of protein side chains, 465
    small aromatic systems, 463
  perpendicular, 437
  RNA and DNA duplexes compared, 433
  specificity, 436
  tetraplex, 469
  threading
    intercalation geometry, 441
    substituent position, 441
  triplex, 466
intercalation binding mode, 428
intercalation sites
  backbone torsional angles, 435
  essential features, 439
  formation, 434
  for nogalamycin, 438
  structural details, 435
  sugar conformation, 435
  variations in, 435
intercalation unwinding angle, 432
  and closed circular superhelical DNA, 433
intercalative binding model, in esperamicin $A_1$, 574
International Union of Biochemistry and Molecular
    Biology (IUBMB), DNA structure nomenclature, 56
International Union of Pure and Applied Chemistry
    (IUPAC), DNA structure nomenclature, 56
interpretation, of biochemical results, 265
intoplicine
  as antineoplastic agent, 607
  structure, 608
introns

folding, kinetics, 49
  pre-mRNA, 345
iodonin, DNA damage induction, 532
ionizing radiation, DNA damage induction, 397
irenotecan
  characteristics, 599
  structure, 599
iron–bleomycin complexes, DNA cleavage, mechanisms,
    537
irradiation, use in deprotection, 129
3-isoadenine, in DNA hexamer, 324
isochromene, hydroxy-, formation, 566
isochrysohermidine, DNA cross-linking by, 503
isofagaridine, occurrence, 601
isoflavones
  as DNA topoisomerase II inhibitors, 610
  genistein, 610
iso-G *see* adenine, 2-oxo- (iso-G)
isolated spin-pair approximation (ISPA)
  definition, 63
  limitations, 64
isoleucyl tRNA synthases, properties, 628
isophthalic acid, 5-nitro-, in phosphoramidite synthesis,
    158
2-(isopropylthiomethoxymethyl)benzoyl (PTMT), as
    5'-hydroxyl protecting group, 127
isosteres, nonpolar, in synthetic oligodeoxynucleotides,
    332
ISPA *see* isolated spin-pair approximation (ISPA)
IUBMB *see* International Union of Biochemistry and
    Molecular Biology (IUBMB)
IUPAC *see* International Union of Pure and Applied
    Chemistry (IUPAC)

kapurimycin A3
  DNA damage induction, 510
  possible alkoxyl radicals from, 510
Karplus equation
  in DNA structure determination, 64
  nucleic acids, 72
kedarcidin
  acidity, 575
  activation, 577
  activity against Gram-positive bacteria, 575
  DNA damage, 577
  histone cleavage, 577
  occurrence, 575
  siderophore chelation, 577
  structure, 554, 575
kedarosamine, structure, 575
ketene acetal, DNA binding, 562
2-keto-3-deoxy-D-xylonate aldolase *see* 2-dehydro-3-
    deoxy-D-pentonate aldolase
ketones, *t*-butyl, in DNA damage studies, 393
ketones, diazo-, DNA damage induction, 542
kinamycins, possible DNA damage induction, 541
Klyne–Prelog notation, nucleic acid structure, 56

*lac* genes
  in cloning, 647
  *lacZ*
    in cloning, 646
    encoding, 650
lactones, α,β-unsaturated, possible DNA damage
    induction, 503
β-lapachone, as direct protein binding agent, 595
LCAA *see* long chain alkylamine (LCAA)
lead ions, as cofactors, 632
leinamycin
  enzyme inactivation by, 526
  occurrence, 522
  reactions, with thiols, 523
  thiol-activated DNA damage induction, 522

*Leishmania tarentolae*, DNA minicircles, 630
leptosins
    possible DNA damage induction, 527
    reactions, with thiol, 527
Lerman, classical intercalation model, 432
leucine-responsive regulatory protein (Lrp) *see* Lrp
Lewis acids, in trityl group removal, 126
life, origin of, 'RNA World' hypothesis, 630
ligand–DNA binding, affinity cleavage, 253
ligands, unwinding angle, 433
ligases, isolation, 632, 633
ligation
    blunt-end, 661
    in cloning, 661
    cohesive-end, 661
    enzymatic method, 366
    mechanisms, 661
    nonenzymatic methods, 366
lipid–phosphorothioate oligonucleotides, HIV-1
        replication inhibition, 187
lissoclinotoxin A
    possible DNA damage induction, 527
    reactions, with thiol, 527
long chain alkylamine (LCAA), in controlled-pore glass,
        112
Lrp, SELEX studies, 628
lung cancer, research, 4
lutetium(III)–texaphyrin complexes, conjugation, 164

macrocycles
    as probes of nucleic acid dynamics, 449
    synthesis, 449
macrocyclic compounds, bisintercalation, 449
maduropeptin
    activation, 580
    applications, 580
    DNA damage, 580
    occurrence, 580
    structure, 554, 580
madurosamine, structure, 580
magic bullets, Erhlich's hypothesis, 428
makaluvamine F
    DNA reactions, 501
    occurrence, 501
maleimide, *N*-(1-pyrene)-, reactions, with dithymidylyl
        phosphorothioate, 240
malonate, diethyloxo-, trityl group removal, 126
malondialdehyde
    DNA adducts of, 502
    in oligodeoxynucleotides, 334
maltose binding protein (MBP), applications, in
        expression systems, 652
manganese porphyrins, DNA damage induction, 373
mannoside, α-methyl-, in thrombin studies, 627
marine sponges
    discorhabdin A, 501
    epinardin D, 501
    makaluvamine F, 501
    renieramycins, 497
markers, isotropic, nucleobase conjugation, 219
Maxam–Gilbert marker, 583
Maxam–Gilbert method (DNA sequencing) *see* chemical
        cleavage method
MBP *see* maltose binding protein (MBP)
MBP-1 peptide, interaction, 484
medicine, and DNA chemistry, 4
menadione *see* quinone, 2-methyl-1,4-naphtho-
menogaril, 606
    quinone methide, UV-vis spectra, 529
merbarone, binding, 596
mercury-2′-deoxyuridine, 5-chloro-, reactions, with
        methyl acrylate, 197
mesitylene sulfonyl chloride (MS-Cl), in oligonucleotide
        synthesis, 107

mesoporphyrin IX, *N*-methyl- (NMM), in porphyrin
        metallation, 634
mesoporphyrin IX (MPIX), in porphyrin metallation, 634
messenger RNA *see* mRNA
metal complexes, applications, DNA structure probes, 85
metal ions
    aqueous, in DNA structure probing, 86
    and DNA damage, 417
metalloporphyrins
    applications, DNA structure probes, 89
    in DNA damage, 379
methanesulfonyl chloride, reaction, with guanosine
        nucleoside, 118
methanol, as source of hydrogen, 560
methidiumpropylethylenediamine tetraacetate–iron(II)
        complex (MPE-EDTA/Fe(II)), applications,
        DNA structure probes, 86
L-methionine, as esperamicin A$_1$ precursor, 554
methionine repressor *see* MetJ
methionine(s), as starting material(s), for enediyne
        antibiotics, 554
methoxylamine, applications, DNA structure probes, 93
methyl, as phosphate protecting groups, 130
MetJ, SELEX studies, 628
mice, immune responses, 632
*Micrococcus luteus*, DNA polymerases, 220
*Micromonospora chersina*, dynemicin, 574
*Micromonospora echinospora* ssp. *calichensis*,
        calicheamicin, 568
*Micromonospora globosa*, deoxydynemicin, 574
minor-major groove binding model, in esperamicin A$_1$,
        574
misonidazole
    as dioxygen substitute, 583
    DNA damage sensitization, 417
mitomycin C
    acidic activation, 516
    conjugation, to oligonucleotides, 159
    DNA damage, redox-dependent, 529
    mechanism of DNA reactions, 515
    monofunctional vs. bifunctional activation, 516
    oxygen radical production by, 517
    reductive activation of, 515
    sequence specificity of cross-linking, 516
mitomycins, DNA damage induction, 514
mitonafide, inhibition, 606
mitoxantrone
    DNA binding, 606
    as DNA topoisomerase II inhibitor, 603
mixed-sequence probes, in oligonucleotide synthesis, 126
MMTr group, as 5′-hydroxyl protecting group, 126
MNU *see* urea, *N*-methyl-*N*-nitroso- (MNU)
molecular biology
    addressing problems, 619
    and chemistry, relevance, 6
    DNA, overview, **1–13**
    techniques, 6, 8
molecular cloning
    techniques, 620
    vectors, 620
molecular dynamics simulations, DNA conformation, 76
molecularity
    in nucleic acid transition curves, 23
    pseudomonomolecular behavior, 32
molecular modeling, enediyne antibiotics, 557
molecular probes, of DNA structure, 81–103
molecular recognition, applications, 355
monoclonal antibodies, 570
monocrotaline, properties, 518
Moxyl *see* xanthen-9-yl, 9-*p*-methoxyphenyl- (Moxyl)
MPE-EDTA/Fe(II) *see*
        methidiumpropylethylenediamine tetraacetate–
        iron(II) complex (MPE-EDTA/Fe(II))

MPIX *see* mesoporphyrin IX (MPIX)
mRNA
 binding, with antisense oligonucleotides, 286
 *see also* pre-mRNA
MSNT *see* 1,2,4-triazole, 1-(mesitylenesulfonyl)-3-nitro-
  (MSNT)
mutation, and DNA damage, 372
α-L-mycaroside, structure, 575
*Mycobacterium tuberculosis*, genome, targeting, 232
mycotoxins, metabolically-activated, 511
Myers reaction
 activation parameters, 556
 and Bergman reaction, compared, 556
 σ,π-biradical, 556
MyoD, peptide, expression, 481
myrocin C
 antitumor properties, 506
 possible DNA reactions of, 506
myxin, DNA damage induction, 532

N-AcO-AAF *see* fluorene, 2-*N*,*N*-acetoxyacetylamino-
  (N-AcO-AAF)
NADPH *see* reduced nicotinamide adenine dinucleotide
  phosphate (NADPH)
namenamicin
 activation, 581
 applications, 581
 cytotoxicity, 581
 occurrence, 581
 structure, 554, 581
naphthacenequinones, as DNA topoisomerase I
  inhibitors, 600
naphthalene diimides
 DNA dynamics, 441
 kinetics of association, 441
 use in the synthesis of threading intercalating agents,
  440
1-naphthoate, 2-hydroxy-5-methyl-7-methoxy-, 558
naphthoyl, deoxyadenosine protection, 123
2-naphthoyl, 3-hydroxy-6-isopropoxy-7,8-dimethoxy-,
  structure, 575
naphthyridinomycin
 covalent DNA adducts with, 499
 occurrence, 499
 reduction-dependent DNA reaction, 499
National Center for Biological Information (NCBI),
  databases, 646
natural products
 carbonyl-containing 502
 covalent DNA modification, **491–552**
 DNA damage induction, 492, 495
  classification, 492
 interactions, with DNA, 5, 9
 oxazolidine-containing, 499
 structure, developments, 644
NC-190, inhibition, 606
NCBI *see* National Center for Biological Information
  (NCBI)
NCS *see* neocarzinostatin (NCS)
NCS-chrom *see* neocarzinostatin chromophore
  (NCS-chrom)
neighbor exclusion principle, 448
 in intercalation, 444
 negative cooperation, 446
neocarzinostatin, holo-, activation by 2-mercaptoethanol,
  565
neocarzinostatin chromophore (NCS-chrom)
 absolute stereochemistry, 559
 activation
  acid-induced, 561
  apo-neocarzinostatin-directed, 565
  base-catalyzed, 562
  by nucleophiles, 559

 by radicals, 559
 by sodium borohydride, 559
 by [2-²H₂]-thioglycolate, 560
 light-induced, 561
 mechanisms, 559
 thiol-induced, 559, 560
 aerobic treatment, 561
 biosynthesis studies, 554
 and bulged DNA cleavage, 562
 cofactors
  borohydride, 558
  thiol, 558
 DNA binding, 567
 DNA damage, 560
  activity, 554
 as probe of DNA microheterogeneity, 567
 reactions, with 4-hydroxy thiophenol, 561
 structure, 558
neocarzinostatin (NCS)
 applications, DNA structure probes, 98
 bioconjugation, 558
 biosynthesis, 558
 cofactors
  borohydride, 558
  thiol, 558
 crystal structure, 560
 DNA damage, 566
 DNA/RNA hybrid damage, 566
  induction, 373, 380, 385, 394
 DNA strand breakage induction, 421
 occurrence, 554
 oxygen radical production by, 540
 proteolytic activity, 559
 toxicity, 558
neomycin, nonintercalation, 433
netropsin
 minor groove DNA binding, 456, 460, 486, 567
 molecular comparisons, 457
 torsional flexibility, 452
neutrophil elastase, SELEX studies, 626
NF-E1 *see* GATA-1
nickel complexes, applications, DNA structure probes, 89
nitidine, occurrence, 601
nitrogen mustards, applications, DNA structure probes,
  96
nitrosamines
 DNA damage induction, 521
 metabolic activation, 521
*N*-nitroso compounds, DNA damage induction, 520
nitroxide, nucleobase conjugation, 219
nitroxides, DNA damage sensitization, 417
NMM *see* mesoporphyrin IX, *N*-methyl- (NMM)
NMR *see* nuclear magnetic resonance (NMR)
nogalamycin
 as intercalating agent, 437
 intercalation sites, 438
 structure, 351
nomenclature, DNA structure, 56
noncalorimetric methods, 21
noncoding strand, 466
norfloxacin, 462
novobiocin, as direct protein binding agent, 596
nuclear magnetic resonance (NMR)
 applications, 72
 DNA structure determination, 55–80
  advantages, 55
 linewidths, 75
nucleic acid duplexes, nonradioactive probes for, 448
nucleic acid probes, design, 161
nucleic acids
 Arrhenius activation energy, 41
 association, mononucleotide, 42
 backbone, torsion angles, 56, 434

**nucleic acids**        *Subject Index*        724

backbone atoms, thermal motions, 434
base pairs
  dynamics, 74
  opening, 45
bases, thermal motions, 434
catalysis, 616
conformation
  backbone, 72
  *endo*, 57
  envelope, 56
  *exo*, 57
  twist, 56
counterion condensation effect, 20
coupling constants, 64
  $J_{P-H3'}$, 72
  vicinal, 64, 68
cross-relaxation rates, 62
dihedral angles, 72
distorted, intercalation, 469
double helices
  dissociation, 44
  formation, 43
duplexes, 33
  unusual, 36
equilibrium melting profiles, analysis, 21
exchange formalism, 74
extrathermodynamic parameters, 18
  cooperative unit size, 19
  equilibrium constants, 20
  free energy, 20
  thermal melting temperature, 18
Eyring transition state enthalpy, 41
folding
  kinetics, 36
  thermodynamics, 16
hairpin helices, formation, 44
helical parameters, pictorial definition, 59
hybridization, 16
  calorimetry, 21
  probe, 35
hydrogen bonding, 270
ID spectra, 66
imino proton exchange rates, 74
interactions, with proteins, 266
intercalation sites, formation, 434
Karplus equation, 72
kinetics, **15–53**
  Biacore system, 40
  determination, 37
  interpretation, 41
  manual mixing, 41
  mixing, 39
  NMR, 39
  rate-limiting step, 41
  steady-state, 41
  stopped flow, 40
  temperature jump, 38
loop–loop complexes, association rates, 46
monomeric units, 56
nearest neighbor thermodynamic data, 34
$^{13}$C NMR spectra, 73
$^{1}$H NMR spectra, 59
$^{15}$N NMR spectra, 72
$^{31}$P NMR spectra, 69
NOESY spectra, 62
polymer properties, 617
process rates, 42
proton exchange, 45
protons
  exchangeable, 66
  nonexchangeable, 64
  stereospecific assignment, 66
pseudorotation angles, 68

reaction rate
  laws, 41
  temperature dependence, 41
reactions, diffusion-controlled, 42
secondary structures, dynamic motions, 434
stacking, rates, 42
structure
  inter-proton distances, 62
  Klyne–Prelog notation, 56
  pseudorotation angles, 57
sugar moieties, 287
synthesis, historical background, 106
tertiary structures, dynamic motions, 434
thermodynamic parameters, 16
  adiabatic compressibility, 18
  common reference states, 18
  enthalpy change, 17, 29
  entropy change, 17
  experimental determination, 21
  free energy change, 16
  from transition curves, 22
  heat capacity, 28
  heat capacity change, 17, 29
  interpretation, 32
  standard states, 18
  van't Hoff enthalpy, 21, 30
  via calorimetry, 30
  via differential scanning calorimetry, 28
  via isothermal titration calorimetry, 31
  via UV spectra, 22
thermodynamics, **15–53**
  importance of, 16
torsion angles, 64, 72
  glycosidic, 57
triplexes, 36
  intermolecular, 46
  kinetics, 46
  stability, 47
volumetric properties, 18
  apparent molar adiabatic compressibility, 32
  apparent molar volume, 32
  densitometry, 31
  measurement, 31
  molar sound velocity increment, 32
  ultrasonic velocimetry, 31
nucleic acids, single-stranded, 32
  structures, 33
nucleobases
  boronated, 221
  conjugation, 219
  functionalization, 191
  pyrrole-containing, copolymerization, 216
  reactivity, 397
nucleophiles, neocarzinostatin chromophore activation, 559
5'-nucleoside aldehyde, formation, 567
nucleoside analogues
  acyclic carbohydrate linkers, 257
  with altered base residues, 271
  with altered carbohydrate residues, 273
  with altered phosphate linkages, 273
  base-modified, 276
    in protein–DNA probing, 266
    related group, 255
  carbohydrate structure and functionality, alterations, 256
  containing acyclic linkers, 273
  C-protonated, 273
  design, 253
  glycolysis, 260
  hydrogen bonding alteration, 253
  hydrophobic functionality alteration, 253
  impact, on pyrimidine–purine–pyrimidine DNA triplex stability, 271

incorporation into DNA sequence, synthetic
    procedures, 259
phosphate linkage alteration, 258
phosphate-modified, in protein–nucleic acid probing,
    268
as probes, for biochemical processes, **251–284**
in RNA catalysis, 274
site-specific placing
    in DNA sequences, 252
    in RNA sequences, 252
studies
    Eco RI restriction endonuclease, 266
    TBP-TATA box complex, 266
    *trp* repressor-operator, 266
sugar-modified, 144
    in protein–nucleic acid probing, 268
use in conjunction with affinity cleavage, 253
use in conjunction with footprinting, 252
nucleoside analogues, 2'-*O*-allyl, 256
nucleoside analogues, 2'-deoxy-2'-fluoro, 256
nucleoside analogues, 2'-*O*-methyl, 256
nucleoside *H*-phosphonates, for the synthesis of
    polynucleotides, 139
C2'-nucleoside radicals
    reactivity
        under aerobic conditions, 382
        under anaerobic conditions, 381
nucleosides
    modifications, 259, 264
        carbohydrate portions, 226
    protonation, 122
    synthesis, 259
nucleosides, arabinosyl, 257
nucleosides, 7-deazapurine, use in hammerhead ribozyme
    studies, 276
nucleosides, α-(ribo)-, 257
C-nucleosides, synthesis, 262
nucleotides
    activated, 619
    misincorporation, 621
    selection, for cloning, 645
    structure, 56
nucleotides, 2'-*O*-methyl-, use in hairpin ribozyme studies,
    279

*Ochrosia* spp., ellipticine, 605
octaketide, 554
ODNs *see* oligodeoxynucleotides (ODNs)
oligodeoxynucleotide duplexes
    $^{31}$P NMR spectra, 69
    structure, determination, 56
oligodeoxynucleotides (ODNs)
    $N^4$-alkylcytosine in, 320
    2-aminoadenine in, 323
    $N^2$-(anthracen-9-ylmethyl)guanine in, 327
    6-azacytosine in, 322
    6-azathymine in, 319
    8-bromoadenine in, 325
    5-bromouracil in, 316
    $N^7$-cyanoboraneguanine in, 329
    7-deazaadenine in, 330
    7-deaza-8-azapurines in, 331
    3-deazaguanine in, 330
    2'-deoxy-3-deazaadenosine in, 329
    $N^3$-ethylthymine in, 315
    $O^2$-ethylthymine in, 314
    5-fluorouracil in, 316
    7-hydro-8-oxoguanine in, 328
    hypoxanthines in, 325
    imidazole in, 334
    $N^2$-3-imidazolepropionic acid in, 327
    5-iodouracil in, 316

malondialdehyde in, 334
$N^4$-methoxycytosine in, 321
$N^6$-methoxyladenine in, 324
$C^6$-*O*-methylguanine in, 328
2'-*O*-methylpseudoisocytidine in, 332
$C^4$-*O*-methylthymine in, 315
$O^2$-methylthymine in, 314
3-nitropyrrole in, 333
$C^8$-oxo-7,8-dihydro-6-*O*-methylguanine in, 329
phenothiazine in, 334
phenoxazine in, 334
$C^5$-propynylcytosine in, 322
$C^5$-propynyl pyrimidines in, 317
$C^5$-propynyluracil in, 317
spermine in, 322
$C^5$-substituted pyrimidines, hybridization properties,
    315
synthetic, nonpolar isosteres, 332
thiazoleuracil in, 318
$C^5$-thiazoleuracil in, 318
$C^6$-thiopurine, synthesis, 327
2-thiothymine in, 314
4-thiothymine in, 315
2-thiouracil in, 314
4-thiouracil in, 315
zwitterionic oligomers, 318
oligodeoxyribonucleoside methylphosphonates,
    synthesis, 160
oligodeoxyribonucleotide-5'-phosphorothioate,
    nucleophilicity, 169
oligodeoxyribonucleotides
    aminoalkylation, 154
    biotinylation, 199
    reactions
        with *p*-azidotetrafluorobenzoic acid esters, 161
        with 1,1'-carbonyldiimidazole, 162
oligodeoxyribonucleotides, 5'-aminoalkylated
    applications, 159
    conjugation, 159
oligodeoxyribonucleotides, 5'-thiolated, 234
oligomers
    biotinylation, 170
    fluorescent, as probes, 234
    radioiodination, 170
    zwitterionic, 318
oligonucleoside phosphorothioates, antihuman
    cytomegalovirus activity, 239
oligonucleotide, digoxigenin labeling, 209
oligonucleotide–peptide conjugates, synthesis, 114
oligonucleotides
    aminoalkylation, phosphoramidites for, 155
    applications, 144
        therapeutic, 142
    biotinylation, 166, 171
        by nick translation, 206
    5'-biotinylation, 170
    bulk production, via polymerase chain reaction,
        652
    carboxylation, 157
    chemical synthesis, 154
    cross-linking efficiency, 159
    deprotection, 192
    depurination, 122
    derivatization, 233
    deuterium-labeled, 560
    functionalization
        labeling, 154
        solid supports, 114
        at 3'-terminus, 190
        at 5'-terminus, 154
    labeling, 185
    mercaptoalkylation, phosphoramidites for, 167
    with modified internucleotide linkages, 265

with nonbridging phosphate–oxygen atom, properties, 291
purification, 144
  by reversed-phase HPLC, 127
radiolabeling, 135
random-sequence, 619
retrosynthesis, 106
self-complementary, synthesis, 32
synthesis, **105–152**
  automated, 112, 144
  capping, 143
  coupling, 143
  detritylation, 143
  dichlorophenylphosphate in, 106
  diphenylchlorophosphate in, 106
  exocyclic amino function protection, 119
  historical background, 106
  5′-hydroxy group protection, 126
  imide and lactam functions protection, 117
  mixed-sequence probes, 126
  no protection option, 123, 136
  oxidation, 143
  phosphate protection, 130
  phosphodiester approach, 107
  *H*-phosphonate approach, 109, 144, 237
  phosphoramidite approach, 108, 142
  phosphotriester approach, 107
  protecting groups in, 117
  purine bases, protection, 121
  solid-phase, 142, 195, 237
  solid supports for, 110
  'spacer arm', 112
  using acyl protecting groups, 124
  washing, 143
triplex-forming, 159
*see also* antisense oligonucleotides
oligonucleotides, aminoalkylated
  reactions, with fluorescein isothiocyanate, 193
  synthesis, 229
oligonucleotides, 5′-aminoalkylated
  condensation, 159
  reactions
    with Denny–Jaffe reagents, 160
    with dithiobis(*N*-succinimidyl)propionate, 168
oligonucleotides, bicyclic, 360
  affinity, 360
  disulfide, 360
  specificity, 360
oligonucleotides, branched, triplexes, 355
oligonucleotides, circular, 356
  DNA-binding properties, studies, 357
  sequence specificity, 359
oligonucleotides, dansylated, as fluorescent labeled compounds, 316
oligonucleotides, europium(III)-labeled, synthesis, 211
oligonucleotides, lariat-shaped, triplexes, 359
oligonucleotides, 5′-mercaptoalkylated
  reactions, with fullerenes, 167
  synthesis, 167
oligonucleotides, stem-loop, 356, 359
oligonucleotides, 5′-thiolated, synthesis, 167
oligonucleotides, topologically-modified, molecular recognition, 355
oligoribonucleotides, hydrolysis, 164
oligoribonucleotides, aminoalkylated, reactions, with ethylene diaminetetraacetic anhydride, 201
oligosaccharide amino group, in calicheamicins, 570
oligosaccharides, 571
  in calicheamicin $\gamma_1^1$, 569
oncoprotein, human, ETS1, DNA major groove recognition, 465
osmium tetroxide, applications, DNA structure probes, 92

oxadicarbocyanine, 3,3′-diethyl- (DODC), applications, DNA structure probes, 85
oxathiaphospholanes, as reagents, 133
oxidizing agents, DNA damage induction, 384
oxygen radicals
  from
    anthramycin, 497
    3*H*-1,2-benzodithiol-3-one 1-oxide, 524
    leinamycin, 524
    mitomycin C, 517
    phenazine *N*-oxides, 533
    polysulfides, 524
    quinones, 528
    renieramycins, 499
    resorcinols, 535
    safracins, 499
    saframycins, 499
    sibiromycin, 497
  production via reduction of molecular oxygen, 540

*Paecilomyces*
  saintopin, 600, 608
  UCE1022, 600
PAGE *see* polyacrylamide gel electrophoresis (PAGE)
PAs *see* pyrrolizidine alkaloids (PAs)
patulin, possible DNA damage, 503
PBDs *see* pyrrolo[1,4]benzodiazepines (PBDs)
pBluescript phagemid, as expression vector, 650
PCR *see* polymerase chain reaction (PCR)
PEG *see* polyethylene glycol (PEG)
*N*-pent-4-enoyl (PNT), as protecting group, 121
peptide minor groove binders, A,T-hook, 486
peptide nucleic acid (PNA)
  antisense oligonucleotide linkages, 306
  hybridization properties, 306
  synthesis, 306
peptides
  disulfide-linked, synthesis, 483
  DNA binding, 352, **477–490**
  as intercalating agents, 487
  minor groove binders, 486
    distamycin-based, 486
    netropsin-based, 486
  monomeric, synthesis, 483
  random sequences, studies, 637
  zinc finger motif, 484
peroxyl radicals
  reactivity, 402
  reduction, 395
  synthesis, 402
*Pfu* DNA polymerase, applications, in polymerase chain reaction, 658
phage display, in random peptide studies, 637
phage display library, 485
phages
  in cloning, 647
  *see also* bacteriophages
1,10-phenanthroline–cuprous ion complexes, applications, DNA structure probes, 86
phenazathionium chloride, 3-*N*-(4-carboxybutyl)methylamino-7-dimethylamino-, esters, 159
phenazine-di-*N*-oxide-oligonucleotide, DNA cleavage, 533
phenazine *N*-oxides
  DNA damage induction, 532
  oxygen radicals from, 533
phenothiazine, in oligodeoxynucleotides, 334
phenoxazine, in oligodeoxynucleotides, 334
phenoxyacetyl, in oligonucleotide synthesis, 123
*t*-phenoxyacetyl, deoxyadenosine protection, 123
phenyl, *p*-chloro-, as phosphate protecting groups, 130
phenyl radicals, carbon-centered, 543

phleomycin, 539
3′-phosphate, formation, 571
phosphate diesters, elimination, rate constants, 390
phosphate hydrolysis, enzymatic, 372
phosphate mapping, studies, 270
phosphate protecting groups, in oligonucleotide
　　synthesis, 130
phosphates, phenyl chloro-, as coupling reaction
　　activator, 139
phosphinates, synthesis, 297, 298
phosphine, tributyl-, use in phosphate protection, 130
phosphite, tristriazolyl-, in *H*-phosphonate synthesis, 139
phosphitylation reagents
　　2-cyano-1,1-dimethylethyl phosphorodichloridite, 130
　　2,2,2-trichloro-1,1-dimethylethyl
　　　　phosphorodichloridite, 130
phosphodiesterases, lead-dependent, 622
phosphodiester bonds, conformation, 76
phosphodiester hydrolases, discovery, 616
phosphodiester linkages, in antisense oligonucleotides,
　　287
phosphodiesters, internucleotide, in protein–DNA
　　　　complexes, 268
phosphodiesters, transesterification, 630
phosphoglycaldehydes, structure, 384
3′-phosphoglycolate, and DNA damage, 403, 583
phosphonamidites, methyl-, synthesis, 161
phosphonates, aminoalkyl-, synthesis, 294
phosphonates, methyl-
　　linkage, 259
　　in oligonucleotides, 619
　　synthesis, 240, 287, 294
*H*-phosphonates
　　in oligonucleotide synthesis, 109
　　overactivation, 139
　　synthesis, 139
3′-phosphopentenaldehyde, and DNA damage, 583
3′-phosphopyridazine, and DNA damage, 583
phosphoramidates, synthesis, 294–297
phosphoramidates, *N,N*-diisopropyl-, 131
phosphoramidimidates, synthesis, 295
phosphoramidites
　　*O*-allyl-protected, 135
　　applications, 154
　　*O*-aryl-protected, 135
　　availability, 620
　　biotinylation, 171
　　*N,N*-diisopropyl-, structure, 131
　　mercaptoalkylation, 167
　　for oligonucleotide aminoalkylation, 155
　　for oligonucleotide mercaptoalkylation, 167
　　reporter and conjugate group attachment, 154
　　support-bound, 137
　　synthesis, 156
　　universal base, 621
phosphoramidites, aminoalkylated, synthesis, 191
phosphoramidites, *β*-cyanoethyl, in oligonucleotide
　　synthesis, 132
phosphoramidites, dichloro-, in phosphoramidite
　　synthesis, 132
phosphoramidites, methoxy, in oligonucleotide synthesis,
　　132
phosphoramidites, polyalkyl, synthesis, 187
phosphorodichloridate, 2,2,2-trichloro-1,1-dimethylethyl
　　(TCDME), development, 130
phosphorodichloridite, 2-cyano-1,1-dimethylethyl, as
　　phosphitylation reagent, 130
phosphorodithioates, synthesis, 293
phosphorofluoridates, synthesis, 293
3′-phosphoroimidazoles, in SELEX studies, 633
phosphoroselenoates, synthesis, 293
phosphorotetrazolides, synthesis, 137
phosphorothioate(s)

analogues, synthesis, 265
antisense mechanisms, 292
applications, 280
　　in ribozyme cleavage, 280
chirality, 306
clinical trials, 290
in oligonucleotides, 619
studies, 258
synthesis, 287, 291, 296, 298
phosphorothioate(s), 3′-bridging, 281
phosphorothioic acid(s) *see* phosphorothioate(s)
phosphotidylinositol-3-kinase (PI3K), p85 subunits, 668
phosphotriesters, synthesis, 107, 295
phosphotyrosinyl oligonucleotides, detection, 174
phosphotyrosinyl phosphoramidites, synthesis, 174
photoactivation, of neocarzinostatin chromophore, 562
photoadducts, covalent, formation, 545
photochemically-activated agents
　　DNA adduct formation, 544
　　DNA damage, 544
photolabile supports, in oligonucleotide synthesis, 115
phototrityl group, 129
PI3K *see* phosphotidylinositol-3-kinase (PI3K)
piperazines, dioxo-, binding, 596
piperidin-4-yl, 1-[(2-chloro-4-methyl)phenyl]-4′-
　　methoxy- (Ctmp), 2′-hydroxyl group
　　protection, 144
pivaloyl chloride, as coupling reaction activators, 139
Pixyl *see* xanthen-9-yl, 9-phenyl- (Pixyl)
5′-pixyl groups, as protecting groups, 126
5′-*S*-pixyl groups, as protecting groups, 126
plasmid vectors
　　applications, in cloning, 650
　　characteristics, 650
　　families, 650
　　　　His-Patch ThioFusion, 652
　　　　pET, 652
　　　　pGEX, 650
　　　　pMAL, 652
platomycins, 539
pluramycin A, possible alkoxyl radicals from, 510
pluramycinones
　　DNA adducts of, 508
　　DNA binding by, 508
pluramycins
　　DNA adducts of, 508
　　DNA binding by, 508
　　occurrence, 508
PNA *see* peptide nucleic acid (PNA)
podophyllotoxin, as topoisomerase II inhibitor, 609
polyacrylamide gel electrophoresis (PAGE)
　　in DNA damage analysis, 385
　　in oligonucleotide purification, 144
polyamides, solid supports derived from, 110
polycations, RNA major-groove binding, 428
polycyclic aromatic hydrocarbons (PAH)
　　carcinogenicity, 222
　　cell transformation, mechanisms, 223
　　and diol epoxide derivatives, 470
　　reactions, with DNA, 470
polyd(A-T)$_2$, comparative binding studies, 455
polyd(G-C)$_2$, comparative binding studies, 455
poly(deoxyribonucleotide):poly(deoxyribonucleotide)
　　ligase (AMP-forming) *see* DNA ligase (ATP)
polydeoxyribonucleotide synthase (ATP) *see* DNA ligase
　　(ATP)
polyethylene glycol–oligonucleotide conjugates,
　　synthesis, 190
polyethylene glycol (PEG), roles, in cellular processes,
　　190
polyethylene glycol phosphoramidites, synthesis, 190
polyethylene glycol–polystyrene (PEG–PS), in
　　oligonucleotide synthesis, 113

polymerase chain reaction (PCR), 346
  advantages, 653
  applications, in oligonucleotide bulk production, 652
  asymmetric, 620
  development, 617
  evaluation, 658
  expression cassette strategy, 668
  historical background, 652
  mechanisms, 620
  mutagenesis, 620
  polymerases, choice of, 658
  primer-binding sites, 624
  primers, 154
    availability, 658
    choice of, 655
  processes, 652
  products, purification, 660
  templates, choice of, 654
  total gene synthesis method, 658
  troubleshooting, 659
polymerases, choice of, for polymerase chain reaction, 658
polymers, primordial, 639
polynucleotide ligase *see* DNA ligase (ATP)
polynucleotides, nucleoside *H*-phosphonates, 139
polypeptides
  in cloning, 645
  purification, 667
polypurines, cleavage, 410
polystyrene, solid supports derived from, 110
poly(styrene-co-maleic acid)–neocarzinostatin conjugate (SMANCS), 558
polysulfide/thiol mixtures, superoxide production by, 540
polysulfides
  DNA damage induction, 524
  formed by leinamycin, 526
  oxygen radicals from, 524
*Polysyncraton lithostrotum*, namenamicin, 581
'popcorn copolymer', in oligonucleotide synthesis, 110
porphyrin, tetraphenyl-, ring current–chemical shift map, 443
porphyrin metallases, derivation, 622
porphyrins
  destabilization, 444
  as intercalating agents, 441
  metallation, 627, 634
  stabilization, 444
porphyrins, cationic, interactions with DNA, 441
porphyrins, metallotris(methylpyridinium)-, activated, 162
potassium permanganate, applications, DNA structure probes, 92
pre-mRNA, introns, 345
'primitive life', theories, 617
probe hybridization, nucleic acids, 35
probes
  molecular design, 451
  nonradioactive, for nucleic acid duplexes, 448
  nucleoside analogues, **251–284**
  polybiotinylated, 204
  for specific distorted conformations, in DNA and RNA, 444
  synthetic methods, 451
prodrug, enediyne as, 557
proflavine, 457
propanoate, 2-oxo- *see* pyruvate
2-propanol, 1,1,1,3,3,3-hexafluoro-, trityl group removal, 126
propidium iodide, applications, DNA structure probes, 85
propionate, dithiobis(*N*-succinimidyl)-, reactions, with 5′-aminoalkylated oligonucleotides, 168
Protein Data Bank, DNA structures, 55

protein-DNA
  interactions, analysis, 669
  recognition
    and partial intercalation, 463
    topological variations, 349
protein–DNA binding, sequence determination, 628
protein–DNA complexes, probing, 266
protein–nucleic acid complexes, probing, 268
proteins
  alpha helices, 616
  amino acid side chains, 616
  amino-terminal labeling, 646
  beta sheets, 616
  binding, 616
  binding pockets, 638
  bulk production, in *Escherichia coli*, 665
  catalysis, 616
  expression, SDS-polyacrylamide gel electrophoresis analysis, 665
  folding studies, 645
  full length, cloning, 645
  historical perceptions, 615
  interactions, with nucleic acids, 266
  structure, inter-proton distances, 62
  tertiary structure, encoding, 645
  *see also* fusion proteins
proteolytic cleavage, enzymes for, 648
protonation, of nucleosides, 122
protozoans, DNA minicircles, 630
pseudoisocytidine, 2′-*O*-methyl-
  from, pseudouridine, 262
  in oligodeoxynucleotides, 332
pseudoisocytosine, 255
pseudorotaxanes, structure, 361
pseudouridine, as starting material, for 2′-*O*-methyl pseudoisocytidine, 262
psicothymidines, protection, 227
psoralen, 4′-[(3-carboxyproprionamido)methyl]-4,5,8-trimethyl-, synthesis, 161
psoralen, 4,5′,8-trimethyl-, applications, DNA structure probes, 96
psoralen–oligonucleotide conjugates, synthesis, 169
psoralen phosphoramidites, synthesis, 184
psoralens
  applications, DNA structure probes, 96
  light-dependent DNA reactions, 545
psorospermin, DNA damage induction, 511
ptaquiloside, DNA adducts of, 507
purine, 2-amino-, 254
purine bases, protection, 121
purine-2′-deoxyriboside, 2-amino-, 259
purine-2′-deoxyriboside, 2,6-diamino-, 259
purine nucleotides, hydroxy radical adducts, 409
purine–pyrimidine–purine triplexes, kinetics, 47
purine radicals, reactivity, 408
purine(s)
  alkene cation radicals, 410
  aminoalkylation, 191
  applications, hammerhead ribozyme studies, 276
  effects, on DNA, 410
purines, 8-oxo-
  in DNA damage, 410
  synthesis, 409
putrescine, in DNA damage, 570
pyranoside, 3-*O*-methyl-α-L-rhamno-, in calicheamicin $\gamma_1^1$, 569
pyrazine C-nucleoside, 255
pyrenyl phosphoramidites, synthesis, 178, 224
pyridine *N*-oxides, in oligonucleotide synthesis, 108
pyridone–2-aminopurine duplex, CD spectra, 332
2-pyridone C-nucleoside, 254
pyridoxine–peptide–oligonucleotide conjugates, synthesis, 159

pyrimidine alkene cation radicals
  reactivity, 405
  synthesis, 405
pyrimidine hydroxyl radicals, intranucleotidyl hydrogen
      atom abstraction, 400
pyrimidine–purine–pyrimidine triplexes, kinetics, 46, 48
pyrimidine radicals, synthesis, 398
pyrimidines
  aminoalkylation, 191
  effects, on DNA, 407
pyrimidines, formamido-, in DNA damage, 410
pyrimidines, 5-halo-
  applications, radiosensitizers, 411
  DNA damage induction, 382
  DNA damage sensitization, 417
  triplexes, 316
pyrimidines, $C^5$-propynyl, in oligodeoxynucleotides, 317
4-pyrimidinone, 254
2,6-[1*H*,3*H*]-pyrimidinone, 4-amino-1-(*β*-D-
      ribofuranosyl)-, x-ray crystal analysis, 272
2-pyrimidinone-2'-deoxyribosides, synthesis, 259
pyrindamycin A, occurrence, 505
pyrindamycin B, occurrence, 505
pyrocarbonate, diethyl (DEPC), applications, DNA
      structure probes, 94
*Pyrococcus furiosus*, polymerases, 658
pyrrole, 3-nitro-, in oligodeoxynucleotides, 333
pyrrolizidine alkaloids (PAs)
  DNA damage by, 518
  oxidative activation, 518
pyrrolo[1,4]benzodiazepines (PBDs)
  covalent adduct with DNA, 495
  DNA damage induction, 495
  oxygen radical production by, 497
  sequence selectivity, 497
  stability of DNA adducts, 497
  structure–function analysis, 495
pyrroloiminoquinones, possible DNA adduct formation
      by, 501
pyruvate, as esperamicin A₁ precursor, 554

quinacrine
  as antimalarial agent, 428
  binding, 457
  use in cytogenetic map preparation, 428
  use in pyrimidine triplex stabilization, 467
quinacrine mustard
  in DNA binding, 470
  use in cytogenetic map preparation, 428
quinobenzoxazines
  antineoplastic activity, 462
  self-assembly, 463
quinocarcin
  disproportionation of, 541
  DNA damage, 500
  occurrence, 499
  superoxide production by, 541
quinoid compounds, redox cycling, 528
quinoline, naphthyl-, triplex intercalator, 468
quinoline, 2-phenyl-, 468
quinolines
  CP-115,953, 609
  as DNA topoisomerase II inhibitors, 609
quinolones
  antibacterial activity, 462
  CP-67,804, 603
  CP-115,953, 603
  norfloxacin, 462
  Ro 15-0216, 603
quinone, 2-methyl-1,4-naphtho-, irradiation, 375
quinone methides
  DNA alkylation by, 529
  tautomers, 575

quinones
  chemical reduction of, 528
  DNA alkylation by, 531
  DNA damage induction, 527
  enzymatic reduction of, 528
  formation of quinone methides from, 529
  redox cycling, 528
quinoxaline antibiotics
  echinomycin, 444
  intercalation, 487
  triostin, 444
quinoxaline di-*N*-oxide, 2-carboxy, possible DNA
      damage induction, 533
quinoxaline *N*-oxides
  DNA damage induction, 533
  hydroxyl radical from, 533
  oxygen radicals from, 533

radicals
  neocarzinostatin chromophore activation, 559
  reactions, with DNA, 493
radiolabeling, of oligonucleotides, 135
radiosensitization, DNA damage studies, 583
radiosensitizers, 5-halopyrimidines, 411
Reactive Blue 4, in SELEX processes, 624
Reactive Red 120, in SELEX processes, 624
reagents, base-specific, as DNA structure probes, 91
recombinant DNA technology, advantages, 621
redox cycling
  with nitro compounds, 521
  of quinones and quinoid compounds, 528
reduced nicotinamide adenine dinucleotide phosphate
      (NADPH), in dynemicin activation, 575
reductant/*N*-oxide systems, superoxide production by,
      540
reductive activation, of 2-nitroimidazoles, 521
renieramycins
  DNA adduct formation by, 497
  DNA damage, redox-activated, 497
  imines formed from, 497
  occurrence, 497
replicon, blockage, 558
reporter groups, attachment to DNA, **153–249**
resorcinol–copper mixtures, superoxide production by,
      540
resorcinols
  DNA binding by, 536
  DNA damage induction, 535
  oxygen radicals from, 535
resorcinols, 5-alkenyl, DNA cleavage, 535
resorcinols, alkyl-, DNA binding by, 536
resorcinols, 5-alkyl, DNA cleavage, 535
restriction digests, applications, in cloning, 660
restriction endonucleases
  applications, in cloning, 660
  DNA cleavage protection, 629
  Eco RI
    nucleoside analogue probing, 266
    studies, 270
    X-ray crystal structure, 266
retrorsine, properties, 518
reversed-phase HPLC, in oligonucleotide purification,
      127, 144
rhodium complexes, DNA damage induction, 383
ribonuclease P
  catalysis, 617
  discovery, 616
ribonuclease(s)
  RNAse H, activity, 287, 292
  studies, 645
ribonucleoside, 2-aminopurine, in hairpin ribozyme
      studies, 277
ribonucleosides, arabinosyl-, synthesis, 264

ribonucleosides, carbohydrate analogues, synthesis, 264
ribonucleosides, xylo-, synthesis, 264
ribonucleotides
   hydrolysis, 632
   incorporation, into DNA, 619
ribopyranoside, 4-*N*-amino-4-deoxy-3-*C*-methyl-, in
   maduropeptin, 580
3'-ribose hydroxyls, in SELEX studies, 633
ribosomal RNA *see* rRNA
ribozymes
   applications
      biomedical, 632
      therapeutic, 142
   association, with substrates, 45
   catalysis, 190, 616, 617
   for the catalysis of ligations, 367
   cleavage, 280
   discovery, 616
   'leadzymes', 632
   mechanisms, 274
   sequences, 363
   substrate specificity, 616
   in *Tetrahymena* spp., 616, 617
      catalysis, 279
      kinetic studies, 277
ribozymes, 2'-fluoro-substituted, 278
ribozymes, hairpin, 274
ribozymes, hammerhead, 274
   structure, 274
RNA
   A-form duplex, 460
   binding pockets, 617
   chemical limitations, 617
   double helices, formation, 43
   folding
      complex, 617
      kinetics, 48
   groove shape, DNA compared, 452
   heterocyclic bases, 616
   historical perceptions, 615
   hydrolysis, 163, 211
   intercalation structures, 433
   kinetics, relaxation, 37
   lariats, 345
   pseudoknots, 345
   selection, *in vitro*, 617
   SELEX studies, 617
   self-alkylating, 632
   self-cleaving, 632
   synthesis, automated, 617
   as tapes, 615
   topological variants, 344
   triplexes, stability, 47
   *see also* circular RNA; dumb-bell RNA; tRNA
RNA catalysis, use of nucleoside analogues in, 274
RNA catalysts, primordial, 630
RNA damage
   chemistry, 582
   mechanisms, 582
RNA–DNA hybrids, cleavage, bleomycin-mediated, 539
RNA–ethidium complex, intercalation sites, 435
RNA grooves, groove-binding compounds, 435
RNA intercalators, **427–476**
RNA ligations, ribozymes as catalyst, 367
RNA probes, biotinylation, 197
RNA replication, rolling circles, 346
RNA–RNA duplexes, Tm values, 288
RNAse P *see* ribonuclease P
RNA synthesis, 144
'RNA World' hypothesis, 617, 630
Ro 15-0216, as DNA topoisomerase II inhibitor, 609
rRNA
   conformational complexity, 616

discovery, 616
rubiflavin A, possible alkoxyl radicals from, 510
ruthenium(II), tris(4,7-diphenylphenanthroline)-,
   applications, DNA structure probes, 88
ruthenium(IV) complexes, oxo-, DNA damage induction,
   374
ruthenium complexes, luminescence, 165
ruthenium–oligonucleotide conjugates, in ssDNA
   detection, 166

safracins
   DNA adduct formation by, 497
   DNA damage, redox-activated, 497
   imines formed from, 497
   occurrence, 497
saframycin A, DNA damage, redox-dependent, 529
saframycins
   DNA adduct formation by, 497
   DNA damage, redox-activated, 497
   imines formed from, 497
   occurrence, 497
   oxygen radical protection by, 499
   redox cycling by, 499
safrole
   carcinogenic action, 519
   DNA damage induction, 519
saintopin
   applications, antineoplastic agents, 607
   occurrence, 600, 608
Sanger's method (DNA sequencing) *see* chain-
   termination method
Schulte–Frohlinde fragmentation, in DNA damage, 394
SDS-PAGE *see* SDS-polyacrylamide gel electrophoresis
SDS-polyacrylamide gel electrophoresis
   processes, 665
   protein expression analysis, 665
sealase *see* DNA ligase (ATP)
Seeman laboratory, DNA research, 353
selectivity
   *loci*-, 582
   microstructure, 582
SELEX
   binder selection, 624
   blended, 635
   definitions, 617
   development, 617
   DNA vs. RNA, 638
   future research, 639
   goals, 619
   libraries
      'blanket', 621
      classes, 621
      encoded, 636
      'focused', 623
      randomness, 622
      'shotgun', 622
   limitations, 637
   optimization, 623
   REPSA procedures, 629
   scope, 637
   strategies, 621
      focused, 621
   techniques, 617, 619
   utility, 624
self-assembly, of quinobenzoxazines, 463
Sepharose, covalent attachment, 168
SH3 domain
   cloning, 668
   identification, 668
$\beta$-sheet peptides, in DNA major groove, 486
sibiromycin, cardiotoxicity, 497
signal transduction, pathways, studies, 668
silane, 3-aminopropyl triethoxy-, use in solid supports,
   111

single-stranded DNA *see* ssDNA
singlet oxygen, reaction, with DNA, 544
sirodesmins
    possible DNA damage induction, 527
    reactions, with thiol, 527
SLE *see* systemic lupus erythematosus (SLE)
sliding clamp, DNA polymerase III, 350
sodium borohydride, neocarzinostatin chromophore
    activation, 559
solid supports
    for functionalized oligonucleotide synthesis, 114
    for oligonucleotide synthesis, overview, 110
    tentacle support, 113
    universal, 113
spermine, in oligodeoxynucleotides, 322
spin labels, 238
spirolactone
    cyclic, formation, 562
    as double-decker intercalator, 565
spirolactone biradical, 562
spirolactone cumulene, formation, via Michael addition,
    562
spliceosome, functions, 345
Src homology 3 domain *see* SH3 domain
ssDNA
    detection, 165
    labeling, 198
    selective cleavage, 164
Stains-All *see* thiacarbocyanine bromide, 4,5:4′,5′-
    dibenzo-3,3′-diethyl-9-methyl-
steric clash, barrier to intercalation, 441
sterigmatocystin, DNA damage induction, 511
streptavidin, interactions, with biotin, 154
*Streptomyces* spp.
    azinomycin B, 513
    bioxalomycins, 499
    carzinophilin, 513
    CC-1065, 504
    cyanocyclines, 499
    daunomycin, 605
    dephostatin, 521
    doxorubicin, 605
    leinamycin, 522
    naphthyridinomycin, 499
    pluramycins, 508
    quinocarcin, 499
    safracins, 497
    saframycins, 497
    tetrazomine, 499
*Streptomyces carzinostaticus*, neocarzinostatin, 554, 558
*Streptomyces globisporus*, C-1027, 578
streptomycin, nonintercalation, 433
streptonigrin, 528, 536
streptozocin, DNA damage induction, 520
structural biology
    developments, 644
    scope, 6
substrates, association, with ribozymes, 45
succinyl linkers, in oligonucleotide synthesis, 142
sugar oxyradicals, formation, 583
sugar peroxyradicals
    DNA damage, 582
    formation, 583
sugar puckers
    in A-form DNA structure, 76
    in B-form DNA structure, 76
    use of term, 56
    in Z-form DNA structure, 77
sugar radicals, 582
sugars
    conformation, 64, 67
    coupling constants, 68
    proton distances, 68

pseudorotation angles, 68
sulfate, dimethyl (DMS), applications, DNA structure
    probes, 95
sulfides, antisense oligonucleotide linkages, 304
sulfonamides, antisense oligonucleotide linkages, 304
sulfonates, antisense oligonucleotide linkages, 304
sulfones, antisense oligonucleotide linkages, 304
sulfonyl, *o*-nitrophenyl-, as protecting groups, 123
superoxide
    cellular protection from, 541
    DNA damage due to, 540
    hydroxyl radical formation from, 540
    production, 378
SURF, techniques, 636
SW15 peptide, binding, 484
synthetic urandomization of randomized fragments
    (SURF) *see* SURF
Systematic Evolution of Ligands by Exponential
    Enrichment (SELEX) *see* SELEX
systemic lupus erythematosus (SLE), treatment, 168

tallysomycin, 539
tannic acid, DNA damage, 536
*Taq* DNA polymerase(s)
    applications, in polymerase chain reaction, 658
    mutagenesis, 621
TATA-binding protein (TBP), structural studies, 465
TBP *see* TATA-binding protein (TBP)
TBP eukaryotic transcription factor, studies, 268
telomerase, reviews, 8
telomeres, 469
    development, 346
teniposide, as DNA topoisomerase II inhibitor, 603, 609
ternary complex
    drug-DNA-enzyme, 594
    formation, 603
terpyridine deoxyribonucleoside phosphoramidites,
    synthesis, 215
tetraacetic anhydride, ethylene diamine-, reactions, with
    aminoalkylated oligoribonucleotides, 201
4-17-tetrabenzo[*a,c,g,i*]fluorenylmethyl, 127
tetra-*n*-butylammonium fluoride (TBAF), in
    oligonucleotide synthesis, 114
tetracyclines, photochemical DNA damage, 544
*Tetrahymena* spp., ribozymes, 277, 279, 616, 617
tetralin, synthesis, 557
tetraplexes
    conformation, 469
    intercalation, 469
tetrazole, nitrophenyl-, as coupling reaction activator, 138
tetrazole, thioethyl-, as coupling reaction activator, 138
1*H*-tetrazole, as coupling reaction activator, 138
1*H*-tetrazole, 5-trifluoromethyl-, as coupling reaction
    activator, 138
tetrazolides, diisopropyl-, in oligonucleotide synthesis,
    133
tetrazomine
    disproportionation of, 541
    DNA damage, 500
    occurrence, 499
    superoxide production by, 541
therapeutics, oligonucleotide-based, future research, 335
*Thermus aquaticus*, polymerases, 658
THF-OOH *see* furan, 2-hydroperoxytetrahydro- (THF-
    OOH)
thiacarbocyanine bromide, 4,5:4′,5′-dibenzo-3,3′-diethyl-
    9-methyl-, applications, DNA structure probes,
    85
thioates, synthesis, 109
thiobenzoate, iodo-, in calicheamicin $\gamma_1{}^1$, 569
thiobenzoate, 2,3-methoxy-4-oxy-5-iodo-6-
    methylbenzoyl iodo-, in calicheamicin $\gamma_1{}^1$, 569
4-thiodeoxyribonucleosides, aminoalkylation, 205

4-thio-2'-deoxyuridine phosphoramidite, incorporation, into oligonucleotides, 202

β-thio-D-galactopyranoside, isopropyl- (IPTG), applications, in protein expression, 647

thioglycolate, methyl (MTG), DNA damage, 560

[2-$^2$H$_2$]thioglycolate, neocarzinostatin chromophore activation, 560

C$^6$-thioguanine, phosphoramidite monomer, 327

thiol
    as cofactor, 558
    1,5-hydrogen shift, 560

thiolate salts, aminoalkyl-, reactions, with 8-bromodeoxyadenosine, 198

thiol–neocarzinostatin chromophore adduct, formation, yield-determining step, 560

thiols
    basicity, 560
    hexose, 566
    nucleophilicity, 560

thiophene, dihydro-, formation, 570

thiophenol, 4-hydroxy- (HTP), reactions, with neocarzinostatin chromophore, 561

thiophile, activation of allylic trisulfide, 570

C$^6$-thiopurine, in oligodeoxynucleotides, 327

6-thiopurines, synthesis, 260

2-thiopyrimidines, synthesis, 260

4-thiopyrimidines, synthesis, 260

thioredoxin (trxA), encoding, 652

4'-thioribonucleosides, synthesis, 264

thiosugar, 554

thiosulfenic acid, formation, 570

2-thiothymine, in oligodeoxynucleotides, 314

4-thiothymine, in oligodeoxynucleotides, 315

2-thiouracil, in oligodeoxynucleotides, 314

4-thiouracil, in oligodeoxynucleotides, 315

thrombin
    applications, proteolytic cleavage, 648
    aptamers
        anticoagulant activity, 627
        structure, 627
    in SELEX studies, 626

thymidine, 6-amino-, synthesis, 261

thymidine, 5,6-dihydro-, synthesis, 407

thymidine, 5,6-dihydro-5,6-dihydroxy-, synthesis, 407

thymidine, 5,6-dihydro-5-hydroxy-, synthesis, 407

thymidine glycol, polymerase blocking activity, 407

thymidine phosphoramidites, synthesis, 231

thymidine phosphoramidites, trifluoro-, incorporation, into oligonucleotides, 221

thymid-5-yl, 5,6-dihydro-
    DNA strand breakage induction, 401
    reactivity, 403
    synthesis, 401

thymine, $N^3$-ethyl-, in oligodeoxynucleotides, 315

thymine, $O^2$-ethyl-, in oligodeoxynucleotides, 314

thymine, $C^4$-$O$-methyl-, in oligodeoxynucleotides, 315

thymine, $O^2$-methyl-, in oligodeoxynucleotides, 314

TMPyP(4)
    imino proton NMR signals, 441
    intercalation, 441
        selectivity, 443

DL-α-tocopherol phosphoramidites, synthesis, 189

*p*-tolylmethyl, as trityl protecting groups, 126

topoisomerase I *see* DNA topoisomerase I

topoisomerase II *see* DNA topoisomerase II

topoisomerase-directed agents, future development, 597

topotecan
    characteristics, 599
    structure, 599

total gene synthesis, method, 658

TOTO
    DNA binding
        fluorescence enhancement, 448

sequence selectivity, 448
    from thiazole orange, binding, 448
    specificity, 448

Tramtrack, cocrystal structure, 484

transfer RNA *see* tRNA

transfer RNA synthetase *see* tRNA synthetases

transformation
    mechanisms, 662
    requirements, 662

transition metals, applications, DNA structure probes, 85

transition state analogues (TSAs), binding, 632

triazine di-*N*-oxides
    DNA damage induction, 534
    as hypoxia-selective redox-activated cytotoxin, 534

1,2,4-triazole, 1-(mesitylenesulfonyl)-3-nitro- (MSNT), in oligonucleotide synthesis, 117

2,2,2-tribromoethyl, as phosphate protecting groups, 130

2,2,2-trichloroethyl phosphorodichloridite, in oligonucleotide synthesis, 108

triesters, synthesis, 109

2,4,6-triisopropylbenzene sulfonyl chloride (TPS-CL), in oligonucleotide synthesis, 107

triostin
    bisintercalation, with neighbor exclusion, 446
    specificity, 448
    X-ray crystal structure, 444

triostin A
    bisintercalation, 488
    structure, 352

triphenylmethyl, as protecting groups, 126

5'-triphosphates, in SELEX studies, 633

trisaccharide, 571

trisbenzoyloxy trityl, as protecting groups, 123

4,4',4''-trisbenzoyloxy trityl, as 5'-hydroxyl protecting groups, 126

4,4',4''-tris-(4,5-dichlorophthalimido) trityl, as 5'-hydroxyl protecting groups, 126

trisintercalators
    and anthraquinone threading intercalating agents, 451
    threading, 452

4,4',4''-tris-levulinyloxy trityl, as 5'-hydroxyl protecting groups, 126

tris(phenanthroline)–metal complexes, applications, DNA structure probes, 88

tris(1,10-phenanthroline)ruthenium(II), applications, DNA structure probes, 88

tris(1,10-phenanthroline)zinc(II), applications, DNA structure probes, 88

tRNA
    as 'adapter', 616
    association, with anticodon loops, 46
    conformational complexity, 616
    folding, kinetics, 48
    structures, three-dimensional, 616

tRNA synthetases, SELEX studies, 628

tRNA synthetases, aminoacyl-, catalysis, 628

*trp* repressors
    binding affinity, 268
    X-ray crystal structure, 268

trxA *see* thioredoxin (trxA)

tryptophan, binding, 638

TSAs *see* transition state analogues (TSAs)

tumors, boron neutron capture therapy, 221

twist (conformation), definition, 56

tyrosine, 578

β-tyrosine, 3'-chloro-5'-hydroxy-, in C-1027 chromophore, 578

tyrosine esters, ribosyl 3' $O$-phospho-, 599

UCE6, occurrence, 600

UCE1022, occurrence, 600

ultrafast, DNA synthesis, 120

uracil, 5-bromo-, in oligodeoxynucleotides, 316

uracil, (*E*)-5-(2-bromovinyl)-, in dodecanucleotides, 317
uracil, 5-fluoro-, in oligodeoxynucleotides, 316
uracil, 5-iodo-, in oligodeoxynucleotides, 316
uracil, $C^5$-propynyl-, in oligodeoxynucleotides, 317
uracil, thiazole-, in oligodeoxynucleotides, 318
uracil, $C^5$-thiazole-, in oligodeoxynucleotides, 318
uracil-5-yl, DNA damage induction, 413
uracil-5-yl radicals, synthesis, 413
urea, *N*-methyl-*N*-nitroso- (MNU)
   decomposition, 520
   DNA methylation by, 520
ureas, antisense oligonucleotide linkages, 301
uridine phosphoramidites
   insertion, into oligomers, 229
   synthesis, 202
uridine phosphoramidites, 5-(thioalkyl)-, synthesis,
   203
uridine-5′-triphosphate, 5-[(4-azidophenacyl)thio]-,
   synthesis, 216
urushiol, DNA damage, 536
USF, peptide, conformation, 482
UV radiation, DNA damage, 561


valine, binding, 638
valyl-AMP, hydrolysis, 628
van't Hoff enthalpy
   determination, 21
      from transition curves, 23
van der Waals, 560
van der Waals interactions, in intercalation, 432
varacin
   possible DNA damage induction, 527
   reactions, with thiol, 527
versicolorins, DNA damage induction, 511
victomycin, 539
viroids, topology, 344
viscometric titration, 433
vitamin B$_{12}$ *see* cobalamin, cyano-

Watson, James Dewey (1928– ), DNA structure studies,
   81
Watson–Crick base pair alignments, DNA structure, 58,
   66
Watson–Crick duplexes, canonical, formation, 20

xanthenyl, as protecting groups, 126
xanthen-9-yl, 9-*p*-methoxyphenyl- (Moxyl), as
   5′-hydroxyl protecting groups, 126
xanthen-9-yl, 9(4-octadecyloxyphenyl)-, as 5′-hydroxyl
   protecting group, 127
xanthen-9-yl, 9-phenyl- (Pixyl), as 5′-hydroxyl protecting
   groups, 126
xanthine, as universal base, 326
xanthosine, synthesis, 260
X-ray crystallography, DNA structure determination, 55


YA-56, 539
YOYO
   DNA binding, fluorescence enhancement, 448
   from thiazole orange, binding, 448


*Zanthoxylum nitidum*
   chelerythrine, 601
   isofagaridine, 601
   nitidine, 601
*Zanthoxylum senegalense*
   fagaridine, 601
   fagaronine chloride, 182
*λ*-ZAP, as expression vector, 649
zinc finger(s), 484
   applications, 669
   double, 484
   research, 671
   single, 484
   triple, 485
zorbamycin, 539
zorbonamycins, 539

WITHDRAWAL